Tables of Integrals (constants of integrations have been omitted to save space)

Elementary forms

1 $\displaystyle\int u^n \, du = \frac{u^{n+1}}{n+1} \quad (n \neq -1)$

2 $\displaystyle\int \frac{du}{u} = \ln |u|$

3 $\displaystyle\int \sin u \, du = -\cos u$

4 $\displaystyle\int \cos u \, du = \sin u$

5 $\displaystyle\int \sec^2 u \, du = \tan u$

6 $\displaystyle\int \csc^2 u \, du = -\cot u$

7 $\displaystyle\int \sec u \tan u \, du = \sec u$

8 $\displaystyle\int \csc u \cot u \, du = -\csc u$

9 $\displaystyle\int \tan u \, du = -\ln |\cos u|$

10 $\displaystyle\int \cot u \, du = \ln |\sin u|$

11 $\displaystyle\int \sec u \, du = \ln |\sec u + \tan u|$

12 $\displaystyle\int \csc u \, du = \ln |\csc u - \cot u|$

13 $\displaystyle\int \sin^2 u \, du = \tfrac{1}{2}u - \tfrac{1}{4}\sin 2u$

14 $\displaystyle\int \cos^2 u \, du = \tfrac{1}{2}u + \tfrac{1}{4}\sin 2u$

15 $\displaystyle\int \frac{du}{\sqrt{a^2 - u^2}} = \sin^{-1}\frac{u}{a}$

16 $\displaystyle\int \frac{du}{a^2 + u^2} = \frac{1}{a}\tan^{-1}\frac{u}{a}$

17 $\displaystyle\int \frac{du}{u\sqrt{u^2 - a^2}} = \frac{1}{a}\sec^{-1}\left|\frac{u}{a}\right|$

18 $\displaystyle\int \sinh u \, du = \cosh u$

19 $\displaystyle\int \cosh u \, du = \sinh u$

20 $\displaystyle\int \operatorname{sech}^2 u \, du = \tanh u$

21 $\displaystyle\int \operatorname{csch}^2 u \, du = -\coth u$

22 $\displaystyle\int \operatorname{sech} u \tanh u \, du = -\operatorname{sech} u$

23 $\displaystyle\int \operatorname{csch} u \coth u \, du = -\operatorname{csch} u$

24 $\displaystyle\int u \, dv = uv - \int v \, du$

25 $\displaystyle\int e^u \, du = e^u$

26 $\displaystyle\int a^u \, du = \frac{a^u}{\ln a}$

Trigonometric Forms

27 $\displaystyle\int \tan^2 u \, du = \tan u - u$

28 $\displaystyle\int \cot^2 u \, du = -\cot u - u$

29 $\displaystyle\int \sin^n u \, du = -\frac{\sin^{n-1} u \cos u}{n} + \frac{n-1}{n}\int \sin^{n-2} u \, du$

30 $\displaystyle\int \cos^n u \, du = \frac{\cos^{n-1} u \sin u}{n} + \frac{n-1}{n}\int \cos^{n-2} u \, du$

31 $\displaystyle\int \tan^n u \, du = \frac{1}{n-1}\tan^{n-1} u - \int \tan^{n-2} u \, du$

32 $\displaystyle\int \cot^n u \, du = -\frac{1}{n-1}\cot^{n-1} u - \int \cot^{n-2} u \, du$

33 $\displaystyle\int \sec^n u \, du = \frac{1}{n-1}\sec^{n-2} u \tan u + \frac{n-2}{n-1}\int \sec^{n-2} u \, du$

34 $\displaystyle\int \csc^n u \, du = -\frac{1}{n-1}\csc^{n-2} u \cot u + \frac{n-2}{n-1}\int \csc^{n-2} u \, du$

35 $\displaystyle\int u^n \sin u \, du = -u^n \cos u + n \int u^{n-1} \cos u \, du$

36 $\displaystyle\int u^n \cos u \, du = u^n \sin u - n \int u^{n-1} \sin u \, du$

Inverse Trigonometric Forms

37 $\displaystyle\int \sin^{-1} u \, du = u \sin^{-1} u + \sqrt{1 - u^2}$

38 $\displaystyle\int \cos^{-1} u \, du = u \cos^{-1} u - \sqrt{1 - u^2}$

39 $\displaystyle\int \tan^{-1} u \, du = u \tan^{-1} u - \frac{1}{2}\ln(1 + u^2)$

40 $\displaystyle\int u \sin^{-1} u \, du = \frac{2u^2 - 1}{4}\sin^{-1} u + \frac{u\sqrt{1 - u^2}}{4}$

41 $\displaystyle\int u \cos^{-1} u \, du = \frac{2u^2 - 1}{4}\cos^{-1} u - \frac{u\sqrt{1 - u^2}}{4}$

42 $\displaystyle\int u \tan^{-1} u \, du = \frac{1}{2}(u^2 + 1)\tan^{-1} u - \frac{u}{2}$

SECOND EDITION

CALCULUS

with Analytic Geometry

SECOND EDITION

CALCULUS

with Analytic Geometry

M. A. Munem   Macomb Community College

D. J. Foulis   University of Massachusetts

Worth Publishers, Inc.

Calculus with Analytic Geometry, *second edition*

Printed in the United States of America

Library of Congress Catalog No. 83-50583

ISBN: 0-87901-236-6

First printing, January 1984

EDITOR: Anne Vinnicombe

PRODUCTION: George Touloumes and Pat Lawson

DESIGNER: Malcolm Grear Designers

ILLUSTRATOR: Mel Erikson Art Services

TYPOGRAPHER: York Graphic Services, Inc.

PRINTER AND BINDER: Kingsport Press

COVER: *Sinusoidal surface*. Computer graphics by Thomas Banchoff, David Laidlaw, and David Margolis.

Worth Publishers, Inc.
444 Park Avenue South
New York, New York 10016

CONTENTS

PREFACE

PURPOSE

The second edition of *Calculus* has been extensively rewritten and improved. Its basic purpose, however, remains the same—to provide the necessary background in calculus and analytic geometry for students of mathematics, engineering, physics, chemistry, economics, or the life sciences.

PREREQUISITES

Students who use this textbook should know the basic principles of algebra, geometry, and trigonometry covered in the usual precalculus mathematics courses. No previous study of analytic geometry is required.

OBJECTIVES

This book was written and revised with two main objectives in mind:

First, to provide the reader with a competent working knowledge of calculus. We have taken pains to insure adequate coverage of the traditional topics and techniques, especially those that will be needed in courses such as engineering, physics, chemistry, and other sciences. Furthermore, we have tried to attune the textbook to the two outstanding mathematical trends of our time—the computer revolution and the burgeoning use of mathematical models in the life sciences, the social sciences, economics, and business.

Second, to complement and enhance effective classroom teaching and to allay students' ''math anxiety.'' Our own classroom experience has shown that successful teaching is fostered by a textbook that provides clear and complete explanations, cogent illustrations, step-by-step procedures, pertinent examples and problems, and a multitude of topical, real-world applications. Such a textbook should build student confidence by beginning with familiar material, explaining new concepts in terms of ideas already well understood, offering worked-out examples that show in detail how problems are to be solved, and providing opportunities for students to reinforce and test their understanding via problem sets in which the level of difficulty increases gradually.

SPECIAL FEATURES

Aspects of the book that have worked well have been carried over from the first edition and a number of new features, designed to enhance readability and stimulate student interest, have been introduced.

1 *Examples and Figures*   In the numerous examples, we offer guidance to students in organizing their work and in deciding when a shortcut may be reasonable. The examples are worked in detail with all substitutions shown. Figures and graphs are used to enhance verbal explanations whenever possible.

2 *Problem Sets*   Problems at the end of each section progress from simple drill-type exercises, amply illustrated by worked-out examples, to more demanding conceptual questions. To solve the odd-numbered problems requires a level of understanding sufficient for most purposes, whereas the even-numbered problems toward the end of each problem set are more challenging and probe for deeper understanding. To generate student interest, the problems include applications to a wide variety of fields—not only to engineering, geometry, and the physical sciences but also to biology, business, the earth sciences, ecology, economics, medicine, navigation, and the social sciences. Answers to odd-numbered problems are given in the back of the book and solutions to all problems are provided in a *Solutions Manual* for instructors.

In each problem set, a group of problems that provide a good representation of the main ideas of the section has been identified with numerals *printed in color*. This feature is helpful for students preparing for quizzes and exams, and it can be used by instructors in selecting problems for assignment.

3 *Review Problem Sets*   The review problems at the end of each chapter can be used in a variety of ways: Instructors may wish to use them for supplementary or extra-credit assignments or for quizzes and exams; students may wish to scan them to pinpoint areas where further study is needed. In some places, these problems purposely are not arranged by section so that students can gain experience in recognizing types of problems as well as in solving them.

4 *Use of Calculators*   In keeping with the recommendations of the National Council of Teachers of Mathematics (NCTM) and the Mathematical Association of America (MAA), we have de-emphasized the use of tables in favor of the use of scientific calculators. We feel obliged to prepare students to function outside the classroom, where virtually everyone using mathematics in a practical way—from the actuary to the zoologist—routinely employs a calculator. In this book, we have tried to assign the calculator its rightful place—a tool, useful at times, unnecessary at other times. Problems and examples that call for the use of a calculator are marked with the special symbol Ⓒ.

5 *Computer Graphics*   The use of computers to generate graphs is fascinating and illuminating to students and instructors alike. We have included a number of computer-generated graphs in Chapters 3, 13, and 14.

6 *SI (Metric) Units*   In the majority of physics and engineering problems we have employed the SI (Systeme International) system of units, which is rapidly replacing the customary British and U.S. systems in all applied work.

7 *Historical Notes and Portraits*   Brief historical notes and portraits of prominent mathematicians serve to remind the reader that mathematics is a *human* endeavor.

8 *Design*   In this edition, a more open design is used for ease of reading and better comprehension. Newly defined terms are indicated in boldface type; important results and formulas are highlighted in color and set off by ruled boxes; and the symbol ∎ is used to signal the end of a proof or the end of the solution to an example.

9 *Formulas*   Pertinent formulas from geometry, trigonometry, and calculus are listed in the appendix and inside the front and back covers of the book.

ARRANGEMENT OF TOPICS

Although the table of contents clearly identifies the topics covered and their order of presentation, the following supplementary comments may be helpful.

1 *Trigonometry* Because deficiencies in trigonometry plague many calculus students, a substantial review of trigonometry precedes coverage of limits and continuity of the trigonometric functions in Chapter 1. At the request of many of our first-edition users, we have introduced *early differentiation of the trigonometric functions* in Chapter 2. This provides for more meaningful applications of the chain rule and enlarges the scope of the applications of the derivative in Chapter 3.

2 *Algebraic Signs and Zeros of Functions* An expanded section emphasizing the use of *test values* to determine algebraic signs of functions and presenting both the *bisection method* and *Newton's method* for approximating zeros is placed in Chapter 2. This material is particularly well-adapted to the use of computers and calculators.

3 *Concepts Used in Physics and Engineering* The tools and techniques that are needed by students of engineering and science are developed as early as possible in the text. The important idea of setting up and solving simple *differential equations* is given suitable emphasis.

4 *Geometry* Throughout the book, the reader is continually encouraged to visualize analytic relationships in geometric form. This is particularly true for the treatment of *vectors*—all concepts involving vectors are first introduced geometrically.

5 *New Material* To reflect the current needs and interests of students taking calculus, new material has been added on the following topics:

    (a) Harmonic oscillators (Chapter 4)
    (b) Linear differential equations (Chapters 4 and 7)
    (c) Logistic growth model (Chapter 7)
    (d) Method of least squares (Chapter 14)
    (e) Jacobians (Chapter 15)
    (f) Conservative vector fields and potentials (Chapter 15)

6 *Optional Material* Because a topic that one person regards as optional could be deemed essential by others, we have refrained from marking any of the material in this book as "optional." However, material that could be considered optional has been placed at the ends of sections or chapters, where it can be omitted without loss of continuity.

PACE

With adequately prepared students, the entire book can be covered in three semesters or in five quarters. In general, chapters 1 through 5 include enough material for a first-semester course, chapters 6 through 11 are suitable for the second semester, and the remaining chapters can be covered in the third semester. The book is written for maximum flexibility; there are many ways to arrange the material coherently to conform to a wide variety of teaching situations.

SUPPLEMENTARY MATERIALS

Study Guide An accompanying *Study Guide* is available for students who require more drill or more assistance in any topic. The *Study Guide* conforms with the arrangement of topics in the book and contains many carefully graded fill-in statements and problems broken down into simple units. Study objectives and tests are also included for each chapter. Answers are given for all problems in the *Study Guide* in order to encourage the building of skills and confidence.

COMPUCALC—Computer Calculus Supplement   A microcomputer diskette and workbook by Robert J. Weaver of *Mt. Holyoke College*, designed to be used in conjunction with the textbook, are available for students with access to a microcomputer. COMPUCALC consists of 19 ''user friendly'' computer programs, many of them employing high-resolution graphics, and related workbook projects that guide the student to make full use of the computer's ability to vividly demonstrate the conceptual and computational power of calculus.

Instructor's Resource Manual   This manual provides a comprehensive testing program, closely coordinated with the textbook. It includes an examination for each chapter and three comprehensive examinations. Suggestions are offered on how each topic might be presented and alternative sequences of topics are suggested.

Solutions Manual   The step-by-step solution to each problem in the textbook is available in this manual for instructors. By glancing through the worked-out solutions, an instructor can select those problems that will provide the kind of practice students need for each section of the book.

ACKNOWLEDGMENTS

By the time a book reaches its second edition, a great many people have contributed to its development. We wish to thank the many individuals who have provided remarks and helpful suggestions on the first edition. In particular, we would like to express our gratitude to the following people: Nancy Angle, *University of Colorado, Denver*; Frank Anger, *University of Puerto Rico*; Olga Beaver, *Williams College*; Murray Eisenberg, *University of Massachusetts, Amherst*; Robert C. Gebhardt, *County College of Morris*; David Hayes, *College of Notre Dame*; Herbert Kamowitz, *University of Massachusetts, Boston*; Eleanor Killam, *University of Massachusetts, Amherst*; William McKinley, *Eastern Montana College*; Bruce McQuarrie, *Worcester Polytechnic Institute*; Robert Piziak, *Baylor University*; Karl Rehmer, *Blackburn College*; Franklin Schroeck, Jr., *Florida Atlantic University*; Ronald Smit, *University of Portland*; Charles Stone, *DeKalb Community College*; Thomas Tredon, *Lord Fairfax Community College*; James Wahab, *University of South Carolina, Columbia*; Harry Whitcomb, *The Philadelphia College of Pharmacy and Science*; Robert White, *North Carolina State University*.

In preparing this second edition, we were fortunate to have the advice and guidance of many reviewers; the knowledge and skills they shared with us have greatly enhanced this book. For their contribution, we thank:

Robert B. Burckel,
Kansas State University

Ray E. Collings,
Tri-County Technical College

Daniel G. Dewey,
College of the Holy Cross

Roger T. Douglass,
Alfred University

Bruce Edwards,
University of Florida

August J. Garver,
University of Missouri, Rolla

Louis M. Herman,
Kansas State University

Frank E. Higginbotham,
University of Puerto Rico, Rió Piedras

Jerry Johnson,
Oklahoma State University

Roy Kelly,
Jackson Community College

Gary Lippman,
*California State University,
Hayward*

J. J. Malone,
Worcester Polytechnic Institute

James L. Moseley,
West Virginia University

Laurence Small,
Los Angeles Pierce College

Howard E. Taylor,
West Georgia College

John von Zellen,
Macomb Community College

Elaine Kirley Whittlesy,
Sierra College

Special thanks are due to Hyla Gold Foulis for reviewing each successive stage of manuscript, proofreading, and solving problems; to Steve Fasbinder, *Oakland University*, for reviewing the manuscript, reading page proof, and solving problems; and to Jarema Chypchar, Barbara Bloom, and Lela Grant for reading page proofs. We also extend our appreciation to the problem checkers: Roger Douglass, *Alfred University*; Kathy Franklin, Larry Small, and Ann Watkins, *Los Angeles Pierce College*, John Spellman and Ricardo Torrejon, *Southwest Texas State University*, and Elaine Kirley Whittlesy, *Sierra College*. Finally, we wish to express our gratitude to the staff at Worth Publishers, especially Rory Baruth, for their constant help and encouragement.

M. A. Munem

January, 1984

D. J. Foulis

INTRODUCTION

Mathematics has been developed, cultivated, and refined not only for its practical applications to science, commerce, and industry, but also because of its aesthetic attractions. The ancient Egyptians developed some of the fundamental ideas of trigonometry for the very practical purpose of relocating property lines after the periodic Nile floods; the early Greeks, however, studied the conic sections largely because they were intrigued with the beautiful geometry of these graceful curves. Nearly two thousand years later, the Renaissance astronomer Johannes Kepler (1571–1630) discovered that the paths of the planets as they orbit the sun are conic sections (ellipses). Thus, mathematical ideas, originally developed because their elegance appealed to the human mind, later found practical application in astronomy and celestial mechanics.

The Italian mathematician, astronomer, and physicist Galileo Galilei (1564–1642), a contemporary of Kepler, showed that objects thrown into the air also

Johannes Kepler

Galileo Galilei

follow conic sections (parabolas). Galileo's experiments and meditation on the principles of mechanics moved him to declare that the "Book of Nature is written in mathematical characters." Within fifty years of this astonishing proclamation, Sir Isaac Newton (1642–1727) in England and Gottfried Leibniz (1646–1716) in Germany independently created *calculus*—the mathematical language with which many chapters of Galileo's "Book of Nature" have been and continue to be written. Fired with enthusiasm by this major intellectual triumph, generations of mathematicians and scientists over the next two centuries developed and perfected the subject of classical mechanics—the foundation of most branches of engineering.

Two more chapters, relativity and quantum mechanics, were added to the "Book of Nature" in the twentieth century. As Galileo had foreseen, these chapters are also "written in mathematical characters," among which the *derivatives* and *integrals* of Newton and Leibniz' calculus are especially prominent.

What accounts for the conspicuous role of calculus in Galileo's "Book of Nature"? The fundamental reason is that the world is full of change and calculus is the mathematics of change. For the same reason, calculus is profoundly involved in the burgeoning use of *mathematical models* in the life and social sciences, economics, and business. These mathematical models usually take the form of *differential equations* that express relationships among the rates of change of variable quantities.

In this book, we stress the idea of calculus as the mathematics of change and we emphasize the use of differential equations as mathematical models for real-world phenomena. We hope that our readers will be able to recapture some of the excitement felt by the architects of calculus when they first began to realize the immense power and beauty of their creation. Some of these readers, we like to think, may themselves help to write the next chapter in Galileo's "Book of Nature."

Sir Isaac Newton

Gottfried Leibniz

1 FUNCTIONS AND LIMITS

We begin this chapter with a brief review of some precalculus mathematics, after which we introduce the two critical ideas upon which calculus is based—*function* and *limit*. Using the idea of a limit of a function, we formulate and study the important concept of *continuity*.

1.1 Real Numbers, Inequalities, and Absolute Value

The familiar idea of a **number scale, number line,** or **coordinate axis** makes it possible to visualize real numbers as points along an infinite straight line (Figure 1).

Figure 1

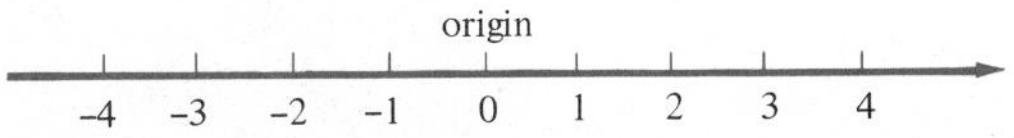

The real number corresponding to a point on the line is called the **coordinate** of the point. An arrowhead may be used to indicate the direction (to the right in Figure 1) in which the coordinates are increasing. The point with coordinate 0 is called the **origin,** and the point with coordinate 1 is called the **unit point.** The distance between the origin and the unit point is called the **unit distance** on the number scale. On a horizontal number scale, points to the right of the origin have positive coordinates, and points to the left of the origin have negative coordinates. The set of all coordinates of points on the line is called the set of **real numbers** and is denoted by the special symbol $\mathbb{R}$.

Two real numbers x and y in $\mathbb{R}$ can be combined by the usual arithmetic operations to yield new real numbers $x + y$, $x - y$, xy, and (provided $y \neq 0$) x/y. If y is positive, we can also raise y to the power x and obtain a definite real number y^x; however, if y is negative, y^x is not always defined as a real number. For instance, $(-1)^{1/2} = \sqrt{-1}$ is not defined as a real number.

1

The real numbers 1, 2, 3, 4, 5, and so on are called the **natural numbers,** or the **positive integers** (Figure 2). The **integers** consist of all the natural numbers, the negatives of the natural numbers, and zero (Figure 3). Real numbers that can be

Figure 2

Figure 3

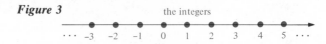

written in the form a/b, where a and b are integers and $b \neq 0$, are called **rational numbers.** The **irrational numbers** are those real numbers that are not rational. A real number is irrational if and only if its decimal representation is **nonterminating** and **nonrepeating.** Examples of irrational numbers are

$$\sqrt{2} = 1.4142135 \cdots \qquad \text{and} \qquad \pi = 3.1415926 \cdots$$

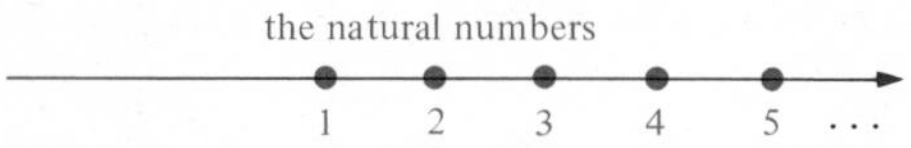

Inequalities

Figure 4

Two different real numbers x and y can always be compared to determine which is greater. If the point with coordinate x lies to the left of the point with coordinate y on the number line (Figure 4), we say that y is **greater than** x (or, equivalently, that x is **less than** y), and we write

$$y > x \qquad \text{or} \qquad x < y$$

Note that

$x > 0$ means that x is positive.
$x < 0$ means that x is negative.

The integer 0 is neither positive nor negative.

A statement of the form $x < y$ (or $y > x$) is called an **inequality.** You may already be familiar with the following rules for handling inequalities.

Let a, b, and c denote real numbers.

   1 *Trichotomy* One and only one of the following is true:
$$a < b \qquad \text{or} \qquad b < a \qquad \text{or} \qquad a = b$$

   2 *Transitivity* If $a < b$ and $b < c$, then $a < c$.

   3 *Addition* If $a < b$, then $a + c < b + c$.

   4 *Subtraction* If $a < b$, then $a - c < b - c$.

   5 *Multiplication* If $a < b$ and $c > 0$, then $ac < bc$.

   6 *Division* If $a < b$ and $c > 0$, then $\dfrac{a}{c} < \dfrac{b}{c}$.

According to the addition and subtraction rules, you can add or subtract the same quantity on both sides of an inequality. The multiplication and division rules permit you to multiply or divide both sides of an inequality by a *positive* number. According to the following rule, if you multiply or divide both sides of an inequality by a *negative* number, you must *reverse the inequality.*

> **7** *Inequality reversing*   If $a < b$ and $c < 0$, then $ac > bc$ and $\dfrac{a}{c} > \dfrac{b}{c}$.

For instance, if you multiply both sides of the inequality $2 < 5$ by -3, you must reverse the inequality and write $-6 > -15$; that is, $-15 < -6$. Indeed, on a number line, -15 lies to the left of -6.

Sometimes we know only that a certain inequality does *not* hold. If $x < y$ does not hold, then by the trichotomy rule either $x > y$ or $x = y$. In this case, we say that x is **greater than or equal to** y, and we write

$$x \geq y \qquad (\text{or, equivalently, } y \leq x)$$

Statements of the form $y < x$ (or $x > y$) are called **strict** inequalities; those of the form $y \leq x$ (or $x \geq y$) are called **nonstrict.**

If we write $x < y < z$, we mean that $x < y$ *and* $y < z$. Likewise, $x \geq y > z$ means that $x \geq y$ *and* $y > z$. This notation for combined inequalities is used only when the inequalities run in the *same direction*.

Statements analogous to Rules 2 through 7 can be made for nonstrict inequalities and for combined inequalities. For instance, if you know that

$$-1 \leq 3 - \frac{x}{2} < 5$$

you can multiply all members by -2, provided that you reverse the inequalities to obtain

$$2 \geq x - 6 > -10$$

or

$$-10 < x - 6 \leq 2$$

If you are asked to "solve" an equation, it's usually clear what is wanted: namely, the value (or values) of the variable (or variables) that makes the equation true. Similarly, you solve an inequality by finding all values of the variable (or variables) that make the inequality true. The set of all such values is called the **solution set** of the inequality.

__EXAMPLE 1__   Solve the inequality $x + 3 < 5x - 1$.

SOLUTION

$x + 3 < 5x - 1$	(Given)
$3 < 4x - 1$	(We subtracted x from both sides.)
$4 < 4x$	(We added 1 to both sides.)
$1 < x$	(We divided both sides by 4.)

Thus, the solution set consists of all real numbers that are greater than 1 (Figure 5). ■

Figure 5

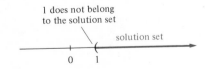

In Figure 5, the solution set consists of one connected piece. Such sets, called **intervals,** often arise as the solution sets of inequalities.

DEFINITION 1   **Bounded Intervals**

Let a and b be fixed real numbers with $a < b$.

(i) The **open interval** (a, b) with *endpoints* a and b is the set of all real numbers x such that $a < x < b$.

(ii) The **closed interval** $[a, b]$ with *endpoints* a and b is the set of all real numbers x such that $a \leq x \leq b$.

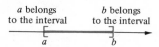

(iii) The **half-open interval** $[a, b)$ with *endpoints* a and b is the set of all real numbers x such that $a \leq x < b$.

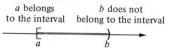

(iv) The **half-open interval** $(a, b]$ with *endpoints* a and b is the set of all real numbers x such that $a < x \leq b$.

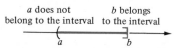

Notice that a closed interval contains its endpoints, but an open interval does not. A half-open interval (also called a **half-closed** interval) contains one of its endpoints, but not the other.

   Unbounded intervals, which extend indefinitely to the right or left, are written with the aid of the special symbols $+\infty$ and $-\infty$, called **positive infinity** and **negative infinity,** respectively.

DEFINITION 2   **Unbounded Intervals**

Let a be a fixed real number.

(i) $(a, +\infty)$ is the set of all real numbers x such that $a < x$.

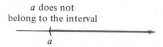

(ii) $(-\infty, a)$ is the set of all real numbers x such that $x < a$.

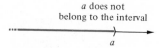

(iii) $[a, +\infty)$ is the set of all real numbers x such that $a \leq x$.

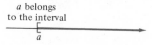

(iv) $(-\infty, a]$ is the set of all real numbers x such that $x \leq a$.

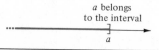

It must be emphasized that $+\infty$ and $-\infty$ are just convenient symbols—*they are not real numbers* and should not be treated as if they were. In the notation for unbounded intervals, we usually write ∞ rather than $+\infty$. For instance, $(5, \infty)$ denotes the set of real numbers greater than 5. The notation $(-\infty, \infty)$ is often used to denote the set $\mathbb{R}$ of all real numbers.

In Examples 2 and 3, solve the inequality and show the solution set on a number line.

<u>EXAMPLE 2</u> $-11 \leq 2x - 3 < 7$

SOLUTION   We begin by adding 3 to all members to help isolate x in the middle:

$$-8 \leq 2x < 10$$

Then we divide all members by 2, to obtain

$$-4 \leq x < 5$$

Figure 6

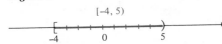
[-4, 5)

Therefore, the solution set is the half-open interval $[-4, 5)$ (Figure 6).   ■

<u>EXAMPLE 3</u> $x^2 + 3x + 2 \geq 0$

SOLUTION   We begin by factoring $x^2 + 3x + 2 = (x + 1)(x + 2)$ and rewriting the given inequality as

$$(x + 1)(x + 2) \geq 0$$

Here, equality holds exactly when $x = -1$ or $x = -2$. Also, the strict inequality

$$(x + 1)(x + 2) > 0$$

holds if and only if $x + 1$ and $x + 2$ have the same algebraic sign. First, let's consider the case in which both $x + 1$ and $x + 2$ are positive. Then

$$x + 1 > 0 \quad \text{and} \quad x + 2 > 0$$

that is, $x > -1 \quad \text{and} \quad x > -2$

Note that if $x > -1$, then $x > -2$ automatically holds; hence, both $x + 1$ and $x + 2$ will be positive precisely when

$$x > -1$$

Next, we consider the case in which both $x + 1$ and $x + 2$ are negative. Then

$$x + 1 < 0 \quad \text{and} \quad x + 2 < 0$$

that is, $x < -1 \quad \text{and} \quad x < -2$

Note that if $x < -2$, then $x < -1$ automatically holds; hence, both $x + 1$ and $x + 2$ will be negative precisely when

$$x < -2$$

It follows that

$$x^2 + 3x + 2 \geq 0$$

holds if and only if

$$x \leq -2 \quad \text{or} \quad x \geq -1$$

Figure 7

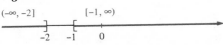
$(-\infty, -2]$ $[-1, \infty)$

In other words, the solution set consists of the two intervals $(-\infty, -2]$ and $[-1, \infty)$ (Figure 7).   ■

Absolute Value

The idea of absolute value plays an important role in analytic geometry and calculus, especially in expressions that involve the distance between two points on a number line.

DEFINITION 3   **Absolute Value**

If x is a real number, then the **absolute value** of x, denoted $|x|$, is defined as follows:

$$|x| = \begin{cases} x & \text{if } x \geq 0 \\ -x & \text{if } x < 0 \end{cases}$$

For example, $|7| = 7$ because $7 \geq 0$, $|0| = 0$ because $0 \geq 0$, and $|-3| = -(-3) = 3$ because $-3 < 0$. Notice that *the absolute value of a real number is always nonnegative*.

Geometrically, the absolute value of the real number x is the distance between the point P whose coordinate is x and the origin O, regardless of whether P is to the right or left of O (Figure 8). More generally,

$|x - y|$ is the number of units of distance between the point with coordinate x and the point with coordinate y.

This statement holds no matter which point is to the left of the other (Figure 9a and b), and, of course, it also holds when $x = y$. For simplicity, we refer to the distance $|x - y|$ as the **distance between the numbers x and y.**

Figure 8

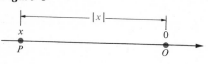

Figure 9

(a)   (b)

Notice that

$$|-x| = |x| \qquad \text{and} \qquad |x| = \sqrt{x^2}$$

The absolute value of a product of two numbers is equal to the product of their absolute values; that is,

$$|xy| = |x||y|$$

The last rule can be derived as follows:

$$|xy| = \sqrt{(xy)^2} = \sqrt{x^2y^2} = \sqrt{x^2}\sqrt{y^2} = |x||y|$$

A similar rule applies to quotients; if $y \neq 0$,

$$\left|\frac{x}{y}\right| = \frac{|x|}{|y|}$$

Many of the properties of absolute value can be established by considering all possible cases in which the quantities involved are positive, negative, or zero.

EXAMPLE 4   Show that $-|x| \le x \le |x|$ holds for every real number x.

SOLUTION   If $x \ge 0$, then $|x| = x$, so $x \le |x|$ holds. Also, if $x \ge 0$, then $-|x| = -x \le 0 \le x$, so $-|x| \le x$ holds. Thus, $-|x| \le x \le |x|$ holds when $x \ge 0$. On the other hand, if $x < 0$, then $|x| = -x$, so $-|x| = x < 0 < -x = |x|$; hence, $-|x| \le x \le |x|$ also holds when $x < 0$. ∎

One of the most important properties of absolute value is the **triangle inequality** given in the following theorem.

THEOREM 1

Triangle Inequality

If a and b are real numbers, then $|a + b| \le |a| + |b|$.

PROOF   By the preceding example, $-|a| \le a \le |a|$ and $-|b| \le b \le |b|$. Adding these inequalities, we obtain

$$-(|a| + |b|) \le a + b \le |a| + |b|$$

If $a + b \ge 0$, then $|a + b| = a + b \le |a| + |b|$. On the other hand, if $a + b < 0$, then $|a + b| = -(a + b) \le |a| + |b|$. In either case, we have $|a + b| \le |a| + |b|$. ∎

In solving equations and inequalities involving absolute values, you will find the following rules useful.

Let u and v be real numbers, and suppose a is a positive real number. Then

(i) $|u| = a$ if and only if $u = a$ or $u = -a$.
(ii) $|u| = |v|$ if and only if $u = v$ or $u = -v$.
(iii) $|u| < a$ if and only if $-a < u < a$.
(iv) $|u| > a$ if and only if $u < -a$ or $u > a$.

Rules (i) and (ii) are obvious from the definition of absolute value. Rules (iii) and (iv) are easy to understand and remember if you keep in mind that $|u|$ is the distance between u and 0.

In Examples 5 to 7, solve each equation or inequality.

EXAMPLE 5 $|1 - 2x| = 7$

SOLUTION   By Rule (i), $|1 - 2x| = 7$ is true if and only if

$$1 - 2x = 7 \quad \text{or} \quad 1 - 2x = -7$$

that is,

$$-2x = 6 \quad \text{or} \quad -2x = -8$$

Therefore, the equation has two solutions, $x = -3$ and $x = 4$. In other words, the solution set consists of the numbers -3 and 4. ∎

EXAMPLE 6 $|3x - 2| < 4$

SOLUTION   By Rule (iii), the inequality is true if and only if

$$-4 < 3x - 2 < 4$$

that is,
$$-2 < 3x < 6$$

or
$$-\tfrac{2}{3} < x < 2$$

Therefore, the solution set is the interval $(-\tfrac{2}{3}, 2)$.

EXAMPLE 7 $|3y + 2| > 5$

SOLUTION   By Rule (iv), the inequality is true if and only if

$$3y + 2 < -5 \qquad \text{or} \qquad 3y + 2 > 5$$

that is,
$$3y < -7 \qquad \text{or} \qquad 3y > 3$$

Hence, the inequality is true if and only if

$$y < -\tfrac{7}{3} \qquad \text{or} \qquad y > 1$$

In other words, the solution set consists of the two intervals $(-\infty, -\tfrac{7}{3})$ and $(1, \infty)$.

Problem Set 1.1

1 Answer true or false. Use suitable rules for inequalities to justify the true statements, and give examples to show that the false statements really are false.

(a) If x is a positive number, then $5x$ is a positive number.

(b) If $x < 3$ and $y > 3$, then $x < y$.

(c) If $x \le y$, then $-5x \le -5y$.

(d) If $x^2 \le 9$, then $x \le 3$.

(e) If $x \ge 2$ and $y > x$, then $y > 0$.

2 Show that if $x \ne 0$, then $x^2 > 0$. (*Hint*: By the trichotomy rule, if $x \ne 0$, then either $x > 0$ or $x < 0$. Consider the two cases $x > 0$ and $x < 0$ separately.)

3 Use the rules for inequalities to show that $\tfrac{3}{28} < \tfrac{25}{233}$. [*Hint*: Begin by showing that $3(233) < 28(25)$. Then use the division rule for inequalities.]

4 Show that if $0 < x < y$, then $1/x > 1/y$. (*Hint*: Use the division rule for inequalities twice.)

5 (a) Under what conditions is $-x > 0$? Explain. (b) Under what conditions is $-x < 0$? Explain. (c) Under what conditions is $-x = 0$? Explain.

6 (a) If $0 < x < y$, show that $x^2 < y^2$. (b) If $0 < x < y$, show that $\sqrt{x} < \sqrt{y}$.

7 If $x^2 \ge 9$, is it necessarily true that $x \ge 3$? Explain.

8 If $x > 0$, is it necessarily true that $1/x > 0$? Why?

9 The following are solution sets for certain inequalities. Illustrate each solution set on a number line.

(a) $[-2, 3)$

(b) All numbers x such that $-3 < x \le 4$ and also $-6 \le x < 2$

(c) All numbers belonging to $[-2, 0]$ or to $[-\tfrac{1}{2}, 1]$ or to both of these intervals

(d) All numbers belonging to $(0, \infty)$ or to $(-\infty, 0)$

(e) All numbers x that belong to both intervals $(0, \infty)$ and $(3, \infty)$

(f) All numbers x that belong to both intervals $(-\infty, -2]$ and $(-\infty, -5)$

10 Use interval notation to represent each shaded set below.

(a)
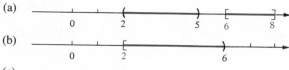

(b)

(c)
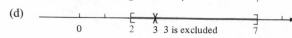

(d)

(e) $\cdots$

In Problems 11 to 24, solve each inequality. Express the solution in interval notation, and illustrate it on a number line.

11 $10x < 18 + 4x$

12 $\frac{9}{4} < \frac{5}{2} + \frac{2}{3}x$

13 $2 \le 5 - 3x < 11$

14 $3 < 5x \le 2x + 11$

15 $3 > -4 - 4x \ge -8$

16 $\frac{2}{x} - 4 < \frac{3}{x} - 8$

17 $x^2 > 9$

18 $\frac{3}{1-x} \le 1$

19 $x^2 - x - 2 < 0$

20 $\frac{5}{3-x} \ge 2$

21 $x^2 \le 4$

22 $\frac{3+x}{3-x} \le 1$

23 $2x^2 + 5x - 12 < 0$

24 $0 < \frac{x-1}{2x-1} < 2$

In Problems 25 to 30, solve each equation.

25 $|x - 3| = 2$

26 $|x - 5| = |3x - 1|$

27 $|3y + 2| = 5$

28 $|t - 2| = |3 - 5t|$

29 $|5x| = 3 - x$

30 $|y^2 + y - 6| = 0$

In Problems 31 to 39, solve each inequality. Express the solution in interval notation, and illustrate it on a number line.

31 $|2x - 5| < 1$

32 $|4y - 6| \le 3$

33 $|3t - 5| > 2$

34 $|3 - 5s| \ge 5$

35 $|x - 2| < 0.1$

36 $|3u + 1| \ge 0.02$

37 $|x + 3| > 0.5$

38 $|9 - 2z| \ge |7z|$

39 $|x + 2| < 0.2$

40 If x and y are real numbers, show that
(a) $|x| - |y| \le |x - y|$ (b) $||x| - |y|| \le |x - y|$

41 A rectangular solar collector is to have a height of 1.5 meters, but its length is still to be determined. What is the range of values for this length if the collector provides 400 watts per square meter, and if it must provide a total of between 2000 and 3500 watts, inclusive.

42 Suppose that a and b are real numbers with $a < b$. Show that every real number x that belongs to the interval $[a, b]$ can be expressed in the form $x = ta + (1 - t)b$, where t is a real number that belongs to the interval $[0, 1]$.

43 An operator-assisted station-to-station phone call from Boston, MA, to Tucson, AZ, costs $2.25 plus $0.38 for each additional minute after the first 3 minutes. Several such calls were made, each costing between $6.05 and $8.71, inclusive. What is the range in minutes for the lengths of these calls?

44 Suppose that families with an income, after deductions, of over $19,200 but not over $23,200 must pay an income tax of $3260 plus 28 percent of the amount over $19,200. What is the range of possible values for the income, after deductions, of families whose income tax liability is between $3484 and $4044?

45 A car can travel 220 miles on a full tank of gas. How many *full* tanks of gas would it need to travel at least 1314 miles?

46 In each case, use the triangle inequality to show that the inequality holds under the given hypotheses.
(a) If $|x - 2| < \frac{1}{2}$ and $|y - 2| < \frac{1}{3}$, then $|x - y| < \frac{5}{6}$.
(b) If $|x + 2| < \frac{1}{2}$ and $|y + 2| < \frac{1}{3}$, then $|x - y| < \frac{5}{6}$.
(c) If $|x - y| < \frac{1}{2}$ and $|x + 2| < \frac{1}{3}$, then $|y + 2| < \frac{5}{6}$.

47 A technician determines that an electronic circuit fails to operate because the resistance between points A and B, 1200 ohms, exceeds the specifications, which call for a resistance of no less than 400 ohms and no greater than 900 ohms. The circuit can be made to satisfy the specifications by adding a shunt resistor R ohms, $R > 0$, as in Figure 10. After adding the shunt resistor, the resistance between points A and B will be $1200R/(1200 + R)$ ohms. What are the possible values of R?

Figure 10

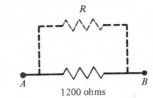

R

A B

1200 ohms

1.2   The Cartesian Coordinate System

René Descartes

In Section 1.1, we saw that a point P on a number line can be specified by a real number x called its coordinate. Similarly, by using a Cartesian coordinate system, named in honor of the French philosopher and mathematician René Descartes (1596–1650), we can specify a point P in the plane with *two* real numbers, also called *coordinates*.

A **Cartesian coordinate system** consists of two perpendicular number lines, called **coordinate axes,** which meet at a common origin O (Figure 1). Ordinarily, one of the number lines, called the **x axis,** is horizontal, and the other, called the **y axis,** is vertical. Numerical coordinates increase to the right along the x axis and upward along the y axis. We usually use the same scale (that is, the same unit distance) on the two axes, although in some figures, space considerations make it convenient to use different scales.

Figure 1

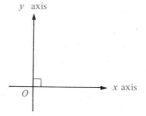

If P is a point in the plane, the **coordinates** of P are the coordinates x and y of the points where perpendiculars from P meet the two axes (Figure 2). The x coordinate is called the **abscissa** of P, and the y coordinate is called the **ordinate** of P. The coordinates of P are traditionally written as an ordered pair (x, y) enclosed in parentheses, with the abscissa first and the ordinate second. (Unfortunately, this is the same symbolism used for an open interval; however, it is always clear from the context what is intended.)

To **plot** the point P with coordinates (x, y) means to draw Cartesian coordinate axes and to place a dot representing P at the point with abscissa x and ordinate y. You can think of the ordered pair (x, y) as the numerical "address" of P. The correspondence between P and (x, y) seems so natural that in practice we identify the point P with its address (x, y) by writing $P = (x, y)$. With this identification in mind, we call an ordered pair of real numbers (x, y) a **point,** and we refer to the set of all such ordered pairs as the **Cartesian plane,** or the **xy plane.**

The x and y axes divide the plane into four regions called **quadrants,** denoted Q_{I}, Q_{II}, Q_{III}, and Q_{IV} and called the **first, second, third,** and **fourth** quadrants (Figure 3). Notice that a point on a coordinate axis belongs to no quadrant.

Figure 2

Figure 3

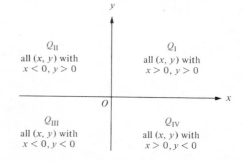

The Distance Formula

One of the attractive features of the Cartesian coordinate system is that there is a simple formula that gives the distance between two points in terms of their coordinates. If P_1 and P_2 are two points in the Cartesian plane, we denote the distance between P_1 and P_2 by $|\overline{P_1P_2}|$.

THEOREM 1

The Distance Formula

If $P_1 = (x_1, y_1)$ and $P_2 = (x_2, y_2)$ are two points in the Cartesian plane, then the distance between P_1 and P_2 is given by

$$|\overline{P_1P_2}| = \sqrt{(x_2 - x_1)^2 + (y_2 - y_1)^2}$$

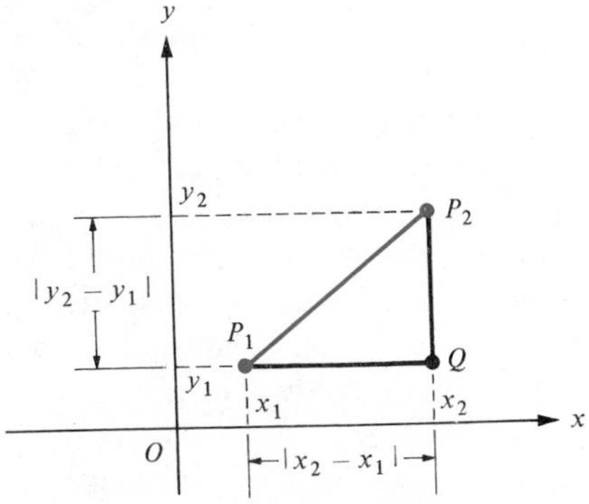

The distance formula is a simple consequence of the Pythagorean theorem, as can be seen in Figure 4. Notice that $\overline{P_1P_2}$ is the hypotenuse of the right triangle P_1QP_2, where $Q = (x_2, y_1)$. We have

$$|\overline{P_1Q}| = |x_2 - x_1| \qquad \text{and} \qquad |\overline{QP_2}| = |y_2 - y_1|$$

Hence, by the Pythagorean theorem,

$$|\overline{P_1P_2}|^2 = |x_2 - x_1|^2 + |y_2 - y_1|^2$$
$$= (x_2 - x_1)^2 + (y_2 - y_1)^2$$

Taking the square root on both sides of the last equation, we obtain the distance formula in Theorem 1.

Figure 4

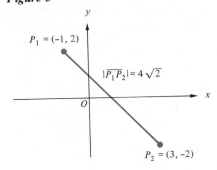

EXAMPLE 1   Find the distance $|\overline{P_1P_2}|$ if $P_1 = (-1, 2)$ and $P_2 = (3, -2)$. Plot P_1 and P_2.

SOLUTION
$$|\overline{P_1P_2}| = \sqrt{[3 - (-1)]^2 + (-2 - 2)^2}$$
$$= \sqrt{4^2 + (-4)^2}$$
$$= \sqrt{32}$$
$$= 4\sqrt{2}$$

Thus, in Figure 5 the distance between P_1 and P_2 is $4\sqrt{2}$ units.

Figure 5

$P_1 = (-1, 2)$

$|\overline{P_1P_2}| = 4\sqrt{2}$

$P_2 = (3, -2)$

Sometimes a calculator may be useful for finding the distance between two points in the plane. However, in using a calculator, you must keep in mind that your answer may be only an approximation of the true value. For instance, using an eight-digit calculator, we could write the distance $|\overline{P_1P_2}|$ in Example 1 as

$$|\overline{P_1P_2}| = 4\sqrt{2} \approx 5.6568542 \text{ units}$$

Note that we use a wiggly equals sign to mean *approximately equal to*. In this book, problems or examples for which the use of a calculator is recommended are marked with the symbol ⓒ.

ⓒ **EXAMPLE 2**   Let $P = (31.42, -17.04)$ and $Q = (13.75, 11.36)$. Find $\overline{PQ}$ by using a calculator. Round off your answer to two decimal places.

SOLUTION
$$|\overline{PQ}| = \sqrt{(13.75 - 31.42)^2 + [11.36 - (-17.04)]^2}$$
$$\approx 33.45 \text{ units}$$

Graphs in the Cartesian Plane

Recall (from precalculus mathematics) the following definition of the graph of an equation.

DEFINITION 1   **Graph of an Equation**

> The **graph** of an equation involving the variables x and y is the set of all points (x, y) in the Cartesian plane, and only those points, whose coordinates x and y satisfy the equation.

For instance, consider the equation

$$x^2 + y^2 = 9$$

We can rewrite this equation as

$$\sqrt{x^2 + y^2} = \sqrt{9}$$

or as

$$\sqrt{(x - 0)^2 + (y - 0)^2} = 3$$

By the distance formula (Theorem 1), the last equation holds if and only if the point $P = (x, y)$ is 3 units from the origin $O = (0, 0)$. Therefore, the graph of $x^2 + y^2 = 9$ is a circle of radius 3 units with its center at the origin (Figure 6).

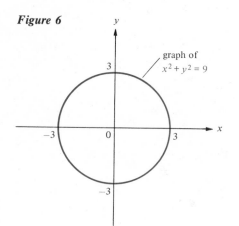

Figure 6

graph of $x^2 + y^2 = 9$

If we are given a curve in the Cartesian plane, we can ask whether there is an equation for which it is the graph. Such an equation is called an **equation for the curve,** or an **equation of the curve.** For instance, $x^2 + y^2 = 9$ is an equation for the circle in Figure 6. Two equations in x and y are said to be **equivalent** if they have the same graph. For example, $x^2 + y^2 = 9$ is equivalent to $\sqrt{x^2 + y^2} = 3$. We often use an equation for a curve to designate the curve; for instance, if we speak of "the circle $x^2 + y^2 = 9$," we mean "the circle for which $x^2 + y^2 = 9$ is an equation."

If $r > 0$, then the circle of radius r with center (h, k) consists of all points (x, y) such that the distance between (x, y) and (h, k) is r units (Figure 7). Using the distance formula (Theorem 1), we can write an equation for this circle as

$$\sqrt{(x - h)^2 + (y - k)^2} = r$$

or, equivalently,

> $$(x - h)^2 + (y - k)^2 = r^2$$

This last equation is called the **standard form** for an equation of a circle of radius r with center (h, k).

Figure 7

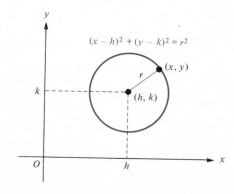

$(x - h)^2 + (y - k)^2 = r^2$

EXAMPLE 3   Find an equation for a circle of radius 5 with center $(-2, 3)$.

SOLUTION   Here $r = 5$ and $(h, k) = (-2, 3)$, so, in standard form, the equation of the circle is

$$[x - (-2)]^2 + (y - 3)^2 = 5^2$$

or

$$(x + 2)^2 + (y - 3)^2 = 25$$

If desired, we can expand the squares,

$$x^2 + 4x + 4 + y^2 - 6y + 9 = 25$$

and combine the constant terms to obtain an equivalent equation in the form

$$x^2 + y^2 + 4x - 6y - 12 = 0$$

The last equation can be restored to standard form by "completing the squares." The work is arranged as follows:

$$x^2 + y^2 + 4x - 6y - 12 = 0$$
$$x^2 + 4x \quad\quad + y^2 - 6y \quad\quad = 12$$
$$x^2 + 4x + 4 + y^2 - 6y + 9 = 12 + 4 + 9$$
$$(x + 2)^2 + (y - 3)^2 = 25$$

Here, we added 4 to both sides of the equation to change $x^2 + 4x$ to the perfect square $x^2 + 4x + 4$, and we added 9 to both sides to change $y^2 - 6y$ to the perfect square $y^2 - 6y + 9$. Notice that, in general:

An expression of the form $x^2 + kx$ becomes a perfect square if you add $\left(\dfrac{k}{2}\right)^2$

to obtain

$$x^2 + kx + \left(\frac{k}{2}\right)^2 = \left(x + \frac{k}{2}\right)^2$$

The process of creating a perfect square from an expression of the form $x^2 + kx$ by adding the square of half the coefficient of x is called **completing the square.** Be careful, this particular procedure can be used *only when the coefficient of x^2 is* 1.

EXAMPLE 4   Sketch the graph of $x^2 + y^2 + 2x + 8y - 8 = 0$.

SOLUTION   Completing the squares, we have

$$x^2 + 2x \quad\quad + y^2 + 8y \quad\quad = 8$$
$$x^2 + 2x + \left(\frac{2}{2}\right)^2 + y^2 + 8y + \left(\frac{8}{2}\right)^2 = 8 + \left(\frac{2}{2}\right)^2 + \left(\frac{8}{2}\right)^2$$
$$x^2 + 2x + \quad 1 \quad + y^2 + 8y + \quad 16 \quad = 8 + 1 + 16$$
$$(x + 1)^2 + (y + 4)^2 = 25$$

Therefore, the graph is a circle of radius $\sqrt{25} = 5$ units with center $(-1, -4)$ (Figure 8).

Figure 8

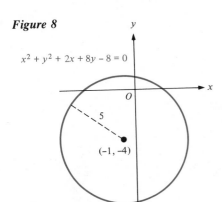

$x^2 + y^2 + 2x + 8y - 8 = 0$

EXAMPLE 5   Find an equation in the form $x^2 + y^2 + Ax + By + C = 0$ and an equation in the standard form of the circle containing the three points $(-2, 5)$, $(1, 4)$, and $(-3, 6)$.

SOLUTION   We substitute the x and y coordinates of the three points into the equation $x^2 + y^2 + Ax + By + C = 0$ and thus obtain the three simultaneous equations

$$\begin{cases} -2A + 5B + C = -29 \\ A + 4B + C = -17 \\ -3A + 6B + C = -45 \end{cases}$$

Solving these simultaneous linear equations in the usual way, we get $A = -2$, $B = -18$, and $C = 57$. Thus, an equation of the circle is

$$x^2 + y^2 - 2x - 18y + 57 = 0$$

Completing the squares, we have

$$x^2 - 2x + 1 + y^2 - 18y + 81 = -57 + 1 + 81$$

$$(x - 1)^2 + (y - 9)^2 = 25$$

Thus, the circle has radius 5 units and center $(1, 9)$.   ∎

Problem Set 1.2

1 Plot each point and indicate which quadrant or coordinate axis contains it.

(a) $(1, 6)$

(b) $(-2, 3)$

(c) $(4, 1)$

(d) $(4, -2)$

(e) $(-1, -4)$

(f) $(0, 2)$

(g) $(-3, 0)$

(h) $(0, -4)$

© **2** Plot the four points $A = (\pi, \sqrt{2})$, $B = (-\sqrt{3}, \sqrt{2})$, $C = (\sqrt{5}, -\sqrt{2})$, and $D = (\frac{3}{4}, -\frac{27}{5})$.

In Problems 3 to 6, plot the point P and determine the coordinates of points Q, R, and S such that (a) the line segment $\overline{PQ}$ is perpendicular to the x axis and is bisected by it; (b) the line segment $\overline{PR}$ is perpendicular to the y axis and is bisected by it; (c) the line segment $\overline{PS}$ passes through the origin and is bisected by it.

3 $P = (3, 2)$

4 $P = (-4, -3)$

5 $P = (-1, 3)$

6 $P = \left(\frac{\sqrt{3}}{2}, -\frac{1}{2}\right)$

In Problems 7 to 18, find the distance between the two points.

7 $(7, 10)$ and $(1, 2)$

8 $(-1, 7)$ and $(2, 11)$

9 $(7, -1)$ and $(7, 3)$

10 $(-4, 7)$ and $(0, -8)$

11 $(-6, 3)$ and $(3, -5)$

12 $(0, 4)$ and $(-4, 0)$

13 $(0, 0)$ and $(-8, -6)$

14 $(t, 4)$ and $(t, 8)$

15 $(-3, -5)$ and $(-7, -8)$

16 $(-\frac{1}{2}, -\frac{3}{2})$ and $(-3, -\frac{5}{2})$

17 $(2, -t)$ and $(5, t)$

18 $(a, b + 1)$ and $(a + 1, b)$

© In Problems 19 and 20, use a calculator to find the distance between the two points. Round off your answer to three decimal places.

19 $(-2.714, 7.111)$ and $(3.135, 4.982)$

20 $(\pi, \frac{53}{4})$ and $(-\sqrt{17}, \frac{211}{5})$

In Problems 21 to 24, (a) use the distance formula and the converse of the Pythagorean theorem to show that triangle ABC is a right triangle, and (b) find the area of triangle ABC.

21 $A = (1, 1)$, $B = (5, 1)$, $C = (5, 7)$

22 $A = (-1, -2)$, $B = (3, -2)$, $C = (-1, -7)$

23 $A = (0, 0)$, $B = (-3, 3)$, $C = (2, 2)$

24 $A = (-2, -5)$, $B = (9, \frac{1}{2})$, $C = (4, \frac{21}{2})$

25 Show that the points $A = (-2, -3)$, $B = (3, -1)$, $C = (1, 4)$, and $D = (-4, 2)$ are the vertices of a square.

26 Show that the distance between the points (x_1, y_1) and (x_2, y_2) is the same as the distance between the point $(x_1 - x_2, y_1 - y_2)$ and the origin.

27 If $A = (-5, 1)$, $B = (-6, 5)$, and $C = (-2, 4)$, determine whether triangle ABC is isosceles.

28 Check the derivation of the distance formula (Theorem 1) for the case in which one point is in the second quadrant and the other point is in the third quadrant.

29 Find all values of t so that the distance between the points $(-2, 3)$ and (t, t) is 5 units.

30 If P_1, P_2, and P_3 are points in the plane, then P_2 lies on the line segment $\overline{P_1P_3}$ if and only if $|\overline{P_1P_3}| = |\overline{P_1P_2}| + |\overline{P_2P_3}|$. Illustrate this geometric fact with diagrams.

In Problems 31 to 33, determine whether P_2 lies on the line segment $\overline{P_1P_3}$ by checking whether $|\overline{P_1P_3}| = |\overline{P_1P_2}| + |\overline{P_2P_3}|$ (see Problem 30).

31 $P_1 = (1, 2)$, $P_2 = (0, \frac{5}{2})$, $P_3 = (-1, 3)$

32 $P_1 = (-\frac{7}{2}, 0)$, $P_2 = (-1, 5)$, $P_3 = (2, 11)$

33 $P_1 = (2, 3)$, $P_2 = (3, -3)$, $P_3 = (-1, -1)$

34 On a Cartesian coordinate grid, an aircraft carrier is detected by radar at point $A = (52, 71)$ and a submarine is detected by sonar at point $S = (47, 83)$. If distances are measured in nautical miles, how far is the carrier from a point on the surface of the water directly over the submarine?

In Problems 35 and 36, find the equations in standard form for the circles in the Cartesian plane that satisfy the given conditions.

35 (a) Radius 3 and center $(0, 2)$

    (b) Radius 2 and center $(-1, 4)$

    (c) Radius 5 and center $(3, 4)$

36 (a) Center at $(1, 6)$ and containing the point $(-2, 2)$

    (b) Radius 4 and containing the points $(-3, 0)$ and $(5, 0)$

    (c) Points $(3, 7)$ and $(-3, -1)$ are the endpoints of a diameter

In Problems 37 to 44, find the radius r and the coordinates (h, k) of the center of the circle for each equation, and sketch a graph of the circle.

37 $(x + 1)^2 + (y - 2)^2 = 9$

38 $(x + 3)^2 + (y - 10)^2 = 100$

39 $x^2 + y^2 + 2x + 4y + 4 = 0$

40 $x^2 + y^2 - x - y - 1 = 0$

41 $4x^2 + 4y^2 + 8x - 4y + 1 = 0$ (*Hint:* Begin by dividing by 4.)

42 $3x^2 + 3y^2 - 6x + 9y = 27$

43 $4x^2 + 4y^2 + 4x - 4y + 1 = 0$

44 $4x^2 + 4y^2 + 12x + 20y + 25 = 0$

In Problems 45 to 47, find the equation in the standard form of the circle or circles in the xy plane that satisfy the conditions given.

45 Containing the points $(-3, 1)$, $(7, 1)$, and $(-7, 5)$

46 Containing the points $(1, 7)$, $(8, 6)$, and $(7, -1)$

47 Radius $\sqrt{17}$, center on the x axis, and containing the point $(0, 1)$. (There are *two* circles in this case.)

48 A point $P = (x, y)$ moves so that it is always twice as far from the point $(6, 0)$ as it is from the point $(0, 3)$. (a) Find the equation of the curve traced out by the point P. (b) Sketch a graph of this curve.

49 Salesperson A is allowed to sell encyclopedias anywhere within the circle $x^2 + y^2 = 100$, while salesperson B is allowed to sell them within the circle $(x - 20)^2 + y^2 = 144$. Do their sales territories overlap? Explain.

50 Find conditions on the constants a, b, c, d, r, and R so that the circle $(x - a)^2 + (y - b)^2 = r^2$ will intersect the circle $(x - c)^2 + (y - d)^2 = R^2$.

51 Show that the equation of the circle in the standard form $(x - h)^2 + (y - k)^2 = r^2$ can always be rewritten in the alternative form $x^2 + y^2 + Ax + By + C = 0$, where $A = -2h$, $B = -2k$, and $C = h^2 + k^2 - r^2$. (*Hint:* Expand the squares and collect the terms.)

52 Show that the equation $x^2 + y^2 + Ax + By + C = 0$ represents a circle with center at the point $(h, k) = (-A/2, -B/2)$ and radius $r = \frac{1}{2}\sqrt{A^2 + B^2 - 4C}$, provided that $A^2 + B^2 > 4C$.

1.3   Straight Lines and Their Slopes

Perhaps the simplest curves in the plane are straight lines (called simply *lines*) and circles. In Section 1.2, we derived equations for circles by using the distance formula. In this section, we derive equations for lines by using the idea of *slope*.

    In ordinary language, the word "slope" refers to a steepness, an incline, or a deviation from the horizontal. For instance, we speak of a ski slope or the slope of a roof. In mathematics, the word "slope" has a precise meaning.

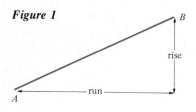

Figure 1

Consider the inclined line segment $\overline{AB}$ in Figure 1. The horizontal distance between A and B is called the **run,** and the vertical distance between A and B is called the **rise.** The ratio of rise to run is called the **slope** of the line segment and is traditionally denoted by the symbol m. Thus, by definition,

$$\textbf{slope of } \overline{AB} = m = \frac{\text{rise}}{\text{run}}$$

If the line segment $\overline{AB}$ is turned so that it becomes more nearly vertical, then the rise increases, the run decreases, and the slope $m =$ rise/run becomes very large. When the line segment becomes vertical, the slope $m =$ rise/run becomes undefined since the denominator is zero. In this case, we sometimes say that the slope is *infinite*.

If the line segment $\overline{AB}$ is horizontal, its rise is zero, so its slope $m =$ rise/run is zero. If $\overline{AB}$ slants downward to the right as in Figure 2, its rise is considered negative; hence, its slope $m =$ rise/run is negative. (The run is always regarded as being nonnegative.)

Figure 2

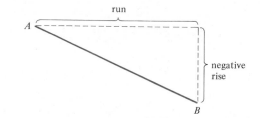

Now, set up a Cartesian coordinate system and consider the line segment $\overline{AB}$, where $A = (x_1, y_1)$ and $B = (x_2, y_2)$ (Figure 3). Here, the rise is $y_2 - y_1$, the run is $x_2 - x_1$, and so the slope m is given by

$$m = \frac{y_2 - y_1}{x_2 - x_1}$$

Of course, Figure 3 represents a special situation in which B lies above and to the right of A; however, you can check the other possible cases and see that the slope m of $\overline{AB}$ is always given by the preceding formula. Therefore, we have the following theorem.

Figure 3

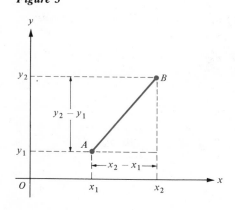

THEOREM 1   **The Slope Formula**

Let $A = (x_1, y_1)$ and $B = (x_2, y_2)$ be any two points in the Cartesian plane. Then, provided that $x_1 \neq x_2$, the slope m of the line segment $\overline{AB}$ is given by

$$m = \frac{y_2 - y_1}{x_2 - x_1}$$

EXAMPLE 1   If $A = (8, -2)$ and $B = (3, 7)$, find the slope m of $\overline{AB}$.

SOLUTION $m = \dfrac{7 - (-2)}{3 - 8} = \dfrac{9}{-5} = -\dfrac{9}{5}$

Consideration of the similar triangles in Figure 4 shows that two parallel line segments $\overline{AB}$ and $\overline{CD}$ have the same slope. Similarly, if two line segments $\overline{AB}$ and $\overline{CD}$ lie on the same infinite line, as in Figure 5, they have the same slope. The common slope of *all* the segments of a line L is called the **slope of L.**

From the fact that two parallel line *segments* have the same slope, it follows that two parallel lines have the same slope. Conversely, it is easy to see that two distinct lines having the same slope must be parallel, and we have the following theorem.

Figure 4

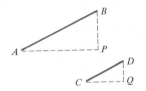

Figure 5

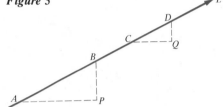

THEOREM 2

Figure 6

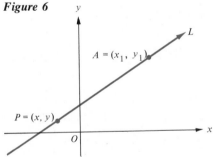

Parallelism Condition

> Two distinct nonvertical lines are parallel if and only if they have the same slope.

Now, consider a nonvertical line L with slope m and containing a point $A = (x_1, y_1)$ (Figure 6). If $P = (x, y)$ is any other point on L, then, by Theorem 1, $m = (y - y_1)/(x - x_1)$; hence,

$$y - y_1 = m(x - x_1)$$

Note that this equation holds even if $P = A$, when it simply reduces to $0 = 0$. In fact, we claim that it is an *equation of the line L* in the sense that not only do all points (x, y) on L satisfy this equation, but, conversely, any point (x, y) that satisfies this equation lies on line L. (The converse follows from Theorem 2.) The equation

$$y - y_1 = m(x - x_1)$$

is called the **point-slope form** for an equation of L.

Figure 7

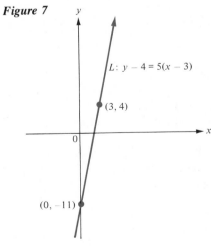

EXAMPLE 2   Let L be the line of slope 5 containing the point $(3, 4)$. Write an equation of L in point-slope form, determine where L intersects the y axis, draw a diagram showing L and the coordinate axes, and decide whether the point $(4, 9)$ belongs to L.

SOLUTION   The point-slope form for an equation of the line L is $y - 4 = 5(x - 3)$. This equation can be rewritten as $y = 5x - 11$. If L intersects the y axis at the point $(0, b)$, then $b = 5(0) - 11 = -11$. Since both $(0, -11)$ and $(3, 4)$ belong to L, it is easy to draw L by drawing the line through these two points (Figure 7). If we put $x = 4$, $y = 9$ in the equation for L, we obtain $9 = 5(4) - 11$, which is true. Therefore, $(4, 9)$ does belong to L.   ■

Figure 8

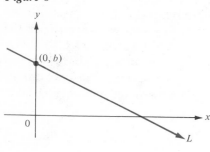

Now suppose that L is any nonvertical line with slope m. Since L is not parallel to the y axis, it must intersect it at some point $(0, b)$ (Figure 8). The ordinate b of this intersection point is called the **y intercept** of L. Since $(0, b)$ belongs to L, we can write the point-slope equation $y - b = m(x - 0)$ for L. This equation simplifies to

$$y = mx + b$$

which is called the **slope-intercept form** for an equation of L.

EXAMPLE 3 In 1982 the Solar Electric Company showed a profit of $3.17 per share, and it expects this figure to increase by $0.24 per share per year. Counting the years so that 1982 corresponds to $x = 0$ and successive years correspond to $x = 1, 2, 3$, and so forth, find the equation $y = mx + b$ of the line that will enable the company to predict its profit y per share during future years. Draw a graph showing this line, and find the predicted profit per share in 1990.

SOLUTION   When $x = 0$, $y = 3.17$; hence, $3.17 = m(0) + b$, and so $b = 3.17$. Thus, $y = mx + 3.17$. When x increases by 1, y increases by 0.24; hence, $m = 0.24$. The equation, therefore, is $y = 0.24x + 3.17$. In 1990, $x = 8$ and $y = (0.24)(8) + 3.17 = 5.09$. The predicted profit per share in 1990 is $5.09 (Figure 9). ∎

Figure 9

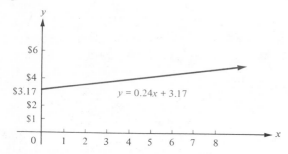

Figure 10

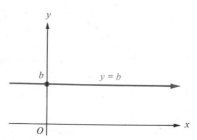

A horizontal line has slope zero; hence, such a line has the equation $y = 0(x) + b$, or simply

$$y = b$$

in slope-intercept form (Figure 10). The equation $y = b$ places no restriction whatsoever on the abscissa x of a point (x, y) on the horizontal line, but requires that all the ordinates y have the same value b.

Of course, a vertical line has an undefined slope, so its equation cannot be written in slope-intercept form at all. However, since all points on a vertical line must have the same abscissa, an equation of such a line can be written as

$$x = a$$

where a is the common value of all these abscissas (Figure 11).

Figure 11

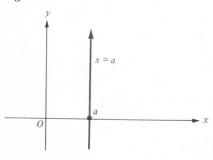

The equation of any line can be put into the form

$$Ax + By + C = 0$$

where A, B, and C are constants and not both A and B are zero. This is called the **general form** of an equation of a line. If $B \neq 0$, the equation $Ax + By + C = 0$ can be rewritten

$$y = -\frac{A}{B}x + \frac{-C}{B}$$

and therefore represents a line with slope $m = -A/B$ and y intercept $b = -C/B$. On the other hand, if $B = 0$, then $A \neq 0$ and the equation can be rewritten in the form $x = -C/A$, which represents a vertical line.

Perpendicular Lines

In Theorem 2 we have seen that two nonvertical lines are parallel if and only if they have the same slope. Theorem 3 gives a condition for two nonvertical lines to be perpendicular.

THEOREM 3

Perpendicularity Condition

Two nonvertical lines are perpendicular if and only if the slope of one of the lines is the negative of the reciprocal of the slope of the other line.

PROOF

Let the two lines be L_1 and L_2, and suppose that their slopes are m_1 and m_2, respectively. The condition that the slope of either one of the lines be the negative of the reciprocal of the slope of the other can be written as

$$m_1 m_2 = -1$$

Neither the angle between the lines nor their slopes are affected if we place the origin O at the point where L_1 and L_2 intersect (Figure 12). Starting at O on L_1, we move 1 unit to the right and $|m_1|$ units vertically to arrive at the point $A = (1, m_1)$ on L_1. Likewise, the point $B = (1, m_2)$ is on line L_2. By the Pythagorean theorem and its converse, triangle AOB is a right triangle if and only if

$$|\overline{AB}|^2 = |\overline{OA}|^2 + |\overline{OB}|^2$$

Using the distance formula (Theorem 1, Section 2), we find that

$$|\overline{AB}|^2 = (1-1)^2 + (m_1 - m_2)^2 = m_1^2 - 2m_1m_2 + m_2^2$$
$$|\overline{OA}|^2 = (1-0)^2 + (m_1 - 0)^2 = 1 + m_1^2$$

and

$$|\overline{OB}|^2 = (1-0)^2 + (m_2 - 0)^2 = 1 + m_2^2$$

Therefore, the condition $|\overline{AB}|^2 = |\overline{OA}|^2 + |\overline{OB}|^2$ is equivalent to

$$m_1^2 - 2m_1m_2 + m_2^2 = 1 + m_1^2 + 1 + m_2^2$$

The last equation simplifies to $m_1 m_2 = -1$, and the proof is complete. ■

Figure 12

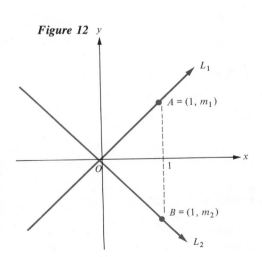

EXAMPLE 4   If $A = (-1, 2)$ and $B = (4, -5)$, find the slope m_1 of a line that is perpendicular to the line segment $\overline{AB}$.

SOLUTION   The slope m_2 of the line containing the points A and B is given by

$$m_2 = \frac{y_2 - y_1}{x_2 - x_1} = \frac{-5 - 2}{4 - (-1)} = -\frac{7}{5}$$

Therefore, by the perpendicularity condition (Theorem 3),

$$m_1 = -\frac{1}{m_2} = -\frac{1}{-\frac{7}{5}} = \frac{5}{7}$$

EXAMPLE 5   Let L be the line $3x - y - 1 = 0$. Find **(a)** an equation of the line L_1 that contains the point $(-1, 2)$ and is parallel to L; **(b)** an equation of the line L_2 that contains the point $(-1, 2)$ and is perpendicular to L. Sketch the graphs of these lines.

SOLUTION

(a) In slope-intercept form, the equation of L is $y = 3x - 1$; hence, L has slope $m = 3$. Since L_1 is to be parallel to L, the slope of L_1 must be $m_1 = m = 3$. Because L_1 contains $(-1, 2)$, its equation is $y - 2 = 3[x - (-1)]$ in point-slope form, or $y = 3x + 5$ in slope-intercept form.

(b) Since L_2 is to be perpendicular to L, it follows that the slope of L_2 must be $m_2 = -1/m = -\frac{1}{3}$. Therefore, because L_2 contains $(-1, 2)$, its equation in point-slope form is $y - 2 = -\frac{1}{3}[x - (-1)]$, or, in slope-intercept form, $y = -\frac{1}{3}x + \frac{5}{3}$ (Figure 13).

Figure 13

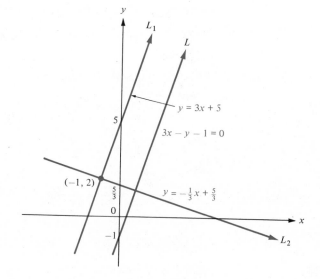

If two different lines in the plane are not parallel, they will intersect in a single point. For instance, in Figure 13, the lines L_1 and L_2 meet in the point $(-1, 2)$. In order to find the point at which two nonparallel lines intersect, it is only necessary to solve the equations of the two lines simultaneously.

Problem Set 1.3

In Problems 1 to 6, find the slope of the line that contains the two points.

1 $(6, 2)$ and $(3, 7)$

2 $(3, -2)$ and $(5, -6)$

3 $(14, 7)$ and $(2, 1)$

4 $(2, 2)$ and $(-4, -1)$

5 $(-5, 3)$ and $(6, 8)$

6 $(1, 3)$ and $(-1, -1)$

In Problems 7 to 18, find an equation in point-slope form of the line L.

7 L contains the point $(5, 4)$ and has slope $m = 2$.

8 L contains the point $(6, 1)$ and has slope $m = -4$.

9 L contains the point $(3, 2)$ and has slope $m = \frac{1}{4}$.

10 L contains the point $(-5, -1)$ and has slope $m = 0$.

11 L contains the point $(7, -2)$ and has slope $m = -3$.

12 L contains the point $(0, 2)$ and has slope $m = -\frac{2}{3}$.

13 L contains the point $(\frac{1}{2}, \frac{2}{3})$ and has slope $m = 0$.

14 L contains the points $(7, 11)$ and $(-1, 1)$.

15 L contains the points $(3, 2)$ and $(4, 8)$.

16 L contains the point $(7, 2)$ and is parallel to the line segment $\overline{AB}$, where $A = (\frac{1}{3}, 1)$ and $B = (-\frac{2}{3}, \frac{3}{5})$.

17 L contains the points $(-3, 4)$ and $(-4, 4)$.

18 L contains the point $(-1, 2)$ and is perpendicular to the line segment $\overline{AB}$, where $A = (\frac{3}{5}, \frac{2}{3})$ and $B = (-\frac{2}{5}, \frac{1}{3})$.

19 Write an equation for each of the following lines. (a) The line containing the point $(-2, 3)$ and perpendicular to the y axis. (b) The line containing the point $(-2, 3)$ and perpendicular to the x axis.

20 Show that if $x_1 \neq x_2$, then an equation of the line containing the two points (x_1, y_1) and (x_2, y_2) is

$$y = \frac{y_2 - y_1}{x_2 - x_1} x + \frac{x_2 y_1 - x_1 y_2}{x_2 - x_1}$$

In Problems 21 to 24, rewrite each equation in slope-intercept form, find the slope m and the y intercept b, and sketch the graph.

21 $3x - 2y = 6$

22 $5x - 2y - 10 = 0$

23 $y + 1 = 0$

24 $x = -\frac{3}{5}y + \frac{7}{5}$

In Problems 25 to 32, find an equation of the line L in (a) point-slope form, (b) slope-intercept form, and (c) general form.

25 L has slope $m = -3$ and y intercept $b = 5$.

26 L has slope $m = \frac{4}{5}$ and intersects the x axis at $(-3, 0)$.

27 L intersects the x and y axes at $(3, 0)$ and $(0, 5)$.

28 L contains the points $(\frac{7}{2}, \frac{5}{3})$ and $(\frac{2}{5}, -6)$.

29 L contains the point $(4, -4)$ and is parallel to the line $2x - 5y + 3 = 0$.

30 L contains the point $(-3, \frac{2}{3})$ and is perpendicular to the y axis.

31 L contains the point $(-3, \frac{2}{3})$ and is perpendicular to the line $5x + 3y - 1 = 0$.

32 L is the perpendicular bisector of the line segment $\overline{AB}$, where $A = (3, -2)$ and $B = (7, 6)$.

33 Find a real number B so that the graph of $3x + By - 5 = 0$ has y intercept $b = -4$.

34 Suppose that the line L intersects the axes at $(a, 0)$ and $(0, b)$. If $a \neq 0$ and $b \neq 0$, show that the equation of L can be written in **intercept form**

$$\frac{x}{a} + \frac{y}{b} = 1$$

In Problems 35 to 38, suppose that m_1 and m_2 are the slopes of the distinct lines L_1 and L_2, respectively. Indicate whether the lines are (a) parallel, (b) perpendicular, or (c) neither parallel nor perpendicular. Then, supposing that L_1 contains the point $(3, 2)$ and L_2 contains the point $(-2, 5)$, draw L_1 and L_2 on the same diagram.

35 $m_1 = \frac{2}{3}$ and $m_2 = \frac{4}{6}$

36 $m_1 = \frac{2}{3}$ and $m_2 = \frac{3}{2}$

37 $m_1 = \frac{3}{2}$ and $m_2 = -\frac{2}{3}$

38 $m_1 = -1$ and $m_2 = 1$

39 If $A = (-5, -2)$, $B = (1, -1)$, $C = (4, 4)$, and $D = (-2, 3)$, show that the quadrilateral $ABCD$ is a parallelogram. (*Hint:* Show that opposite sides have the same slope.)

40 There are two circles of radius $\sqrt{10}$ that are tangent to the line $3x + y = 6$ at the point $(3, -3)$. Find the equations of these circles in standard form. (*Hint:* The tangent line to a circle is perpendicular to the radius at the point of tangency.)

41 (a) Determine d so that the line containing $A = (d, 3)$ and $B = (-2, 1)$ is perpendicular to the line containing $C = (5, -2)$ and $D = (1, 4)$. (b) Determine k so that the line containing $E = (k, 3)$ and $B = (-2, 1)$ is parallel to the line containing $C = (5, -2)$ and $D = (1, 4)$.

42 Prove that

$$\left(\frac{x_1 + x_2}{2}, \frac{y_1 + y_2}{2} \right)$$

is the midpoint of the line segment between (x_1, y_1) and (x_2, y_2).

43 Use the midpoint formula in Problem 42 to find the midpoint of the line segment between each pair of points.

(a) (8, 1) and (7, 3)   (b) (9, 3) and (−5, 7)

(c) (−1, 1) and (5, 3)   (d) (1, −3) and (5, 8)

44 If $m_1 \neq m_2$, show that the line $y = m_1 x + b_1$ intersects the line $y = m_2 x + b_2$ at the point

$$\left(\frac{b_2 - b_1}{m_1 - m_2}, \ \frac{m_1 b_2 - m_2 b_1}{m_1 - m_2} \right)$$

45 A car rental company leases automobiles for $22 per day plus $0.20 per mile. Write an equation for the cost y dollars in terms of the distance x miles driven if the car is leased for N days. If $N = 3$, sketch a graph of the equation.

46 Find the distance (measured perpendicularly) between the point $(-4, 3)$ and the line $y = 3x - 5$ by carrying out the following steps: (a) Find the equation of the line through $(-4, 3)$ which is perpendicular to the line $y = 3x - 5$. (b) Find the point (x_1, y_1) at which the line obtained in part (a) meets the line $y = 3x - 5$. (c) Use the distance formula to find the distance between $(-4, 3)$ and (x_1, y_1).

47 If a piece of property is *depreciated linearly* over a period of n years, then its value y dollars at the end of x years is given by $y = c[1 - (x/n)]$, where c dollars is the original value of the property. An apartment building built in 1975 and originally worth $400,000 is being depreciated linearly over a period of 40 years. Sketch a graph showing the value y dollars of the apartment building x years after it was built, and determine its value in the year 1995.

48 Show that the line $Ax + By + C = 0$ is perpendicular to the line $-Bx + Ay + D = 0$.

49 In 1980, tests showed that water in a lake was polluted with 7 milligrams of mercury compounds per 1000 liters of water. Cleaning up the lake became an immediate priority, and environmentalists determined that the pollution level would drop at the rate of 0.75 milligram of mercury compounds per 1000 liters of water per year if all their recommendations were followed. If 1980 corresponds to $x = 0$ and successive years correspond to $x = 1, 2, 3$, and so on, find the equation $y = mx + b$ of the line that allows the environmentalists to predict the pollution level y in future years if their recommendations are followed. Sketch the graph of the equation and determine when the lake will be free of mercury pollution according to this graph.

1.4   Functions

Advances in our scientific understanding of the world often result from the discovery that things depend on one another in definite ways. For instance, the gravitational attraction between two material bodies depends on the distance between them, and the pitch of a guitar string depends on its tension. The idea that a quantity y depends upon another quantity x is nicely symbolized by the **mapping notation**

$$x \longmapsto y$$

This notation indicates that to each value of x there corresponds a uniquely determined value of y; or, as mathematicians say, each value of x is "mapped onto" a corresponding value of y. Another name for a mapping is a *function*.

DEFINITION 1   **Function**

A **function** is a rule or correspondence

$$x \longmapsto y$$

that assigns to each real number x in a certain set X one and only one real number y.

In Definition 1, the set X is called the **domain** of the function, and y is called the **dependent** variable since its value depends on the value of x. Because x is permitted to have any value in the domain X, we refer to x as the **independent variable.** The set of values assumed by y as x runs through all values in X is called the **range** of the function. If $x \longmapsto y$, we say that y is the **image** of x under the function.

Scientific calculators have special keys for some of the more important functions. By entering a number x and touching, for instance, the $\sqrt{x}$ key, you obtain a vivid demonstration of the mapping $x \longmapsto \sqrt{x}$ as the display changes from x to its image $\sqrt{x}$ under the square-root function. For instance,

$$4 \longmapsto 2$$

$$25 \longmapsto 5$$

$$2 \longmapsto \sqrt{2} \approx 1.414$$

Programmable calculators and microcomputers have "user-definable" keys that can be programmed for whatever function $x \longmapsto y$ may be required. The program for the required function is the actual rule whereby y is to be calculated from x. Each user-definable key is marked with a letter of the alphabet or other symbol, so that, after the key has been programmed for a particular function, the letter or symbol can be used as the "name" of the function.

The use of letters of the alphabet to designate functions is not restricted exclusively to calculating machines. Although any letters of the alphabet can be used to designate functions, the letters f, g, and h as well as F, G, and H are most common. (Letters of the Greek alphabet are also used.) For instance, if we wish to designate the square-root function $x \longmapsto \sqrt{x}$ by the letter f, we write

$$x \xrightarrow{\;f\;} \sqrt{x} \qquad \text{or} \qquad f : x \longmapsto \sqrt{x}$$

If $f : x \longmapsto y$ is a function, it is customary to write the value of y that corresponds to x as $f(x)$, read "f of x." In other words, $f(x)$ is the image of x under the function f. For instance, if $f : x \longmapsto \sqrt{x}$ is the square-root function, then

$$f(4) = \sqrt{4} = 2$$

$$f(25) = \sqrt{25} = 5$$

$$f(2) = \sqrt{2} \approx 1.414$$

and, in general, for any nonnegative value of x,

$$f(x) = \sqrt{x}$$

If $f : x \longmapsto y$, then, for every value of x in the domain of f, we have

$$y = f(x)$$

an equation relating the dependent variable y to the independent variable x. Conversely, when an equation of the form

$$y = \text{an expression involving } x$$

determines a function $f : x \longmapsto y$, we say that the function f is *defined by*, or *given by*, the equation. For instance, the equation

$$y = 3x^2 - 1$$

defines a function $f : x \longmapsto y$, so that

$$y = f(x) = 3x^2 - 1$$

or simply

$$f(x) = 3x^2 - 1$$

When a function f is defined by an equation, you can determine, by substitution, the image $f(a)$ corresponding to a particular value $x = a$.

EXAMPLE 1   Let f be the function defined by $f(x) = 5x^2 + 3x$. Find the indicated values.

(a) $f(2)$ (b) $f(-2)$ (c) $f(t^3)$

(d) $[f(-1)]^2$ (e) $f(t + h)$ (f) $f(-x)$

SOLUTION

(a) $f(2) = 5(2)^2 + 3(2) = 20 + 6 = 26$

(b) $f(-2) = 5(-2)^2 + 3(-2) = 20 - 6 = 14$

(c) $f(t^3) = 5(t^3)^2 + 3(t^3) = 5t^6 + 3t^3$

(d) $[f(-1)]^2 = [5(-1)^2 + 3(-1)]^2 = (5 - 3)^2 = 2^2 = 4$

(e) $f(t + h) = 5(t + h)^2 + 3(t + h) = 5t^2 + 10th + 5h^2 + 3t + 3h$

(f) $f(-x) = 5(-x)^2 + 3(-x) = 5x^2 - 3x$ ∎

If a function $f: x \longmapsto y$ is defined by an equation, you may assume (unless otherwise stated) that its domain consists of all values of x for which the equation makes sense and determines a unique corresponding real number y. Then the range of the function is automatically determined, since it consists of the set of all values of y that correspond, by the equation that defines the function, to values of x in the domain.

EXAMPLE 2   Find the domain of the function defined by

(a) $f(x) = \dfrac{1}{x - 1}$ (b) $g(x) = \sqrt{4 - x}$ (c) $h(x) = 3x - 5$

SOLUTION

(a) The domain of f is the set of all real numbers except 1. In other words, it consists of the interval $(-\infty, 1)$ together with the interval $(1, \infty)$.

(b) The expression $\sqrt{4 - x}$ represents a real number if and only if $4 - x \geq 0$, that is, $x \leq 4$. Therefore, the domain of g is the interval $(-\infty, 4]$.

(c) Since the expression $3x - 5$ represents a real number for all real values of x, the domain of h is the set $\mathbb{R}$ of all real numbers.   ∎

Functions that arise in applied mathematics may have restrictions imposed on their domains by physical or geometric circumstances. For instance, the function $x \longmapsto \pi x^2$ that expresses the correspondence between the radius x and the area πx^2 of a circle would have its domain restricted to the interval $(0, \infty)$, since a circle must have a positive radius.

The particular letters used to denote the dependent and independent variables are of no importance in themselves—the important thing is the rule by which a definite value of the dependent variable is assigned to each value of the independent variable. In applied work, variables other than x and y are often used because physical and geometric quantities are designated by conventional symbols. For instance, the radius of a circle is often designated by r and its area by A. Thus, the function $f: r \longmapsto A$ that assigns to each positive value of r the corresponding value of A is given by

$$A = f(r) = \pi r^2$$

In dealing with a function f, it is important to distinguish among the *function itself*

$$f : x \longmapsto y$$

which is a rule, the *image*

$$y \quad \text{OF} \quad f(x)$$

which is a number depending on x, and the *equation*

$$y = f(x)$$

which relates the dependent variable y to the independent variable x. Nevertheless, people tend to take shortcuts and speak, incorrectly, of "the function $f(x)$" or "the function $y = f(x)$." Similarly, in applied mathematics, people often say that "y is a function of x," for instance, "current is a function of voltage." Although we avoid these practices when absolute precision is required, we indulge in them whenever it seems convenient and harmless.

The Graph of a Function

The **graph** of a function f is defined to be the graph of the corresponding equation $y = f(x)$. In other words, the graph of f is the set of all points (x, y) in the Cartesian plane such that x is in the domain of f and $y = f(x)$.

For instance, if m and b are constants, then the graph of the function

$$f(x) = mx + b$$

is the same as the graph of the equation

$$y = mx + b$$

a line with slope m and y intercept b. For this reason, a function of the form $f(x) = mx + b$ is called a **linear function.**

EXAMPLE 3   Sketch the graph of the function $f(x) = \frac{3}{4}x + 2$.

SOLUTION   The graph of f is the same as the graph of the equation

$$y = \tfrac{3}{4}x + 2$$

a line with slope $m = \frac{3}{4}$ and y intercept $b = 2$ (Figure 1). ■

Graphs of functions that are not linear are often (but not always) smooth curves in the Cartesian plane. For instance, the graph of $f(x) = 1 + \sqrt{x - 1}$ is shown in Figure 2. Sketching such a graph can be considerably more challenging than sketching a straight line, although the use of a calculator to determine points on the graph will often give a good indication of its general shape. Incidentally, most of the microcomputers now on the market have graph-plotting capabilities.

In scientific work, a graph showing the relationship between two variable quantities is often obtained by actual measurement. For instance, Figure 3 shows the blood pressure p (in millimeters of mercury, abbreviated "mm of Hg") in an artery of a healthy person plotted against time t (in seconds). Such a curve may be regarded as the graph of a function $p = f(t)$, even though it may not be clear how to write a "mathematical formula" giving p in terms of t.

Figure 1

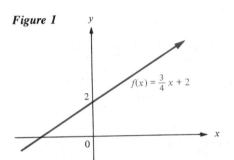

Figure 2

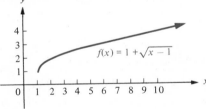

Figure 3

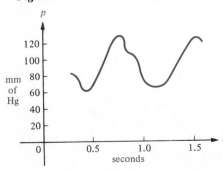

Figure 4

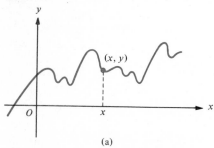

(a)

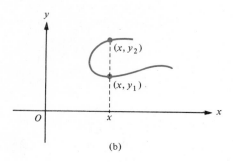

(b)

It is important to realize that *not every curve in the Cartesian plane is the graph of a function*. Indeed, the definition of a function (Definition 1) requires that there be one and *only one* value of y corresponding to each value of x in the domain. Thus, on the graph of a function, *we cannot have two points (x, y_1) and (x, y_2) with the same abscissa x and different ordinates y_1 and y_2*. Hence, we have the following test.

Vertical-Line Test

> A set of points in the Cartesian plane is the graph of a function if and only if no vertical straight line intersects the set more than once.

EXAMPLE 4   Which of the curves in Figure 4 is the graph of a function?

SOLUTION   By the vertical-line test, the curve in Figure 4a is the graph of a function, but the curve in Figure 4b is not. ∎

Graph Sketching

A basic graph-sketching procedure is the following.

The Point-Plotting Method

> To sketch the graph of $y = f(x)$, select several values of x in the domain of f, calculate the corresponding values of $f(x)$, plot the resulting points, and connect the points with a smooth curve. The more points you plot, the more accurate your sketch will be.

After you have learned some calculus, you'll find that you can sketch accurate graphs by plotting relatively few points. This is a case of substituting knowledge for tedious labor.

EXAMPLE 5   Use the point-plotting method to sketch the graph of $f(x) = x^2$, where the domain of f is restricted by the condition that $x > 0$.

SOLUTION   The domain of f is the interval $(0, \infty)$, so we begin by selecting several values of x in this interval and calculating the corresponding values of $f(x) = x^2$, as in the table in Figure 5. We then plot the points $(x, f(x))$ from the table and connect them with a smooth curve (Figure 5). Because the domain of f consists only of positive numbers, the point $(0, 0)$ is excluded from the graph. This excluded point is indicated by a small open circle. Notice that we have used different scales on the x and y axes to obtain a figure with reasonable dimensions. ∎

Figure 5

x	$f(x) = x^2$
1	1
2	4
3	9
4	16
5	25
6	36

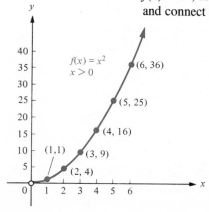

The point-plotting method requires us to *guess* about the shape of the graph between or beyond known points, and so it must be used with caution (see Problems 45 and 46). If the function is fairly simple, the point-plotting method usually works pretty well; however, more complicated functions may require more advanced methods that we study later.

Figure 6

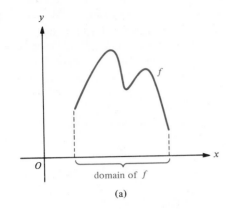

(a)

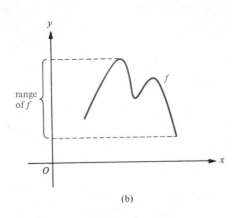

(b)

The domain and range of a function are easily found from its graph. Indeed, as Figure 6 illustrates, *the domain of a function is the set of all abscissas of points on its graph* (Figure 6a) and *the range of a function is the set of all ordinates of points on its graph* (Figure 6b).

For each function in Examples 6 to 10, determine the domain, sketch the graph, and determine the range.

EXAMPLE 6 $f(x) = |x|$

SOLUTION   The independent variable x can take on any value, so the domain is the set $\mathbb{R}$ of all real numbers. For $x \geq 0$, we have $f(x) = x$; hence, the portion of the graph corresponding to $x \geq 0$ is part of a line of slope 1 starting at the origin and extending upward into quadrant I. For $x < 0$, we have $f(x) = -x$, so the corresponding portion of the graph is part of a line of slope -1. The graph is sketched in Figure 7. Evidently, $f(x)$ cannot be negative but can take on any nonnegative value. Thus, the range of f is the interval $[0, \infty)$. We call $f(x) = |x|$ the **absolute-value function.**

Figure 7

EXAMPLE 7 $g(x) = \dfrac{x^2 - 4}{x - 2}$

SOLUTION   The fraction

$$\frac{x^2 - 4}{x - 2}$$

is defined for all values of x *except* $x = 2$ (which makes the denominator zero). Therefore, the domain of g consists of the two intervals $(-\infty, 2)$ and $(2, \infty)$. Note that

$$x^2 - 4 = (x + 2)(x - 2)$$

hence, for $x \neq 2$,

$$\frac{x^2 - 4}{x - 2} = \frac{(x + 2)(x - 2)}{x - 2} = x + 2$$

Therefore, provided that $x \neq 2$,

$$g(x) = x + 2$$

Figure 8

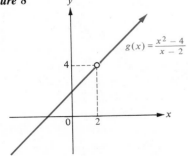

It follows that the graph of g consists of all points on the line $y = x + 2$ *except for the point* $(2, 4)$, which is excluded (Figure 8). Evidently, the range of g is all real numbers except 4; that is, the range consists of the two intervals $(-\infty, 4)$ and $(4, \infty)$.

Figure 9

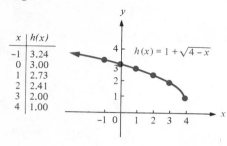

x	h(x)
−1	3.24
0	3.00
1	2.73
2	2.41
3	2.00
4	1.00

Figure 10

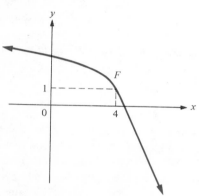

© **EXAMPLE 8** $h(x) = 1 + \sqrt{4 - x}$

SOLUTION   Since $\sqrt{4 - x}$ is defined only when $4 - x \geq 0$, that is, when $x \leq 4$, the domain of h is the interval $(-\infty, 4]$. We select several values of x in this interval and calculate the corresponding values of

$$h(x) = 1 + \sqrt{4 - x}$$

as in the table in Figure 9. Here, we have used a calculator and rounded off values of $h(x)$ to two decimal places. (If a calculator isn't available, the table of square roots in Appendix D may be used.) Using the information in this table and the point-plotting method, we sketch the graph as in Figure 9. Evidently, the range of h is the interval $[1, \infty)$. ∎

EXAMPLE 9 $F(x) = \begin{cases} 1 + \sqrt{4 - x} & \text{if } x < 4 \\ -2x + 9 & \text{if } x \geq 4 \end{cases}$

SOLUTION   A function such as F that is defined by using different equations in different intervals is called a **piecewise-defined** function. Here x can take on any value, so the domain of F is $\mathbb{R}$. In sketching the graph, we must consider separately the portions to the right and to the left of the vertical line $x = 4$. To the right of $x = 4$, the graph is a portion of the line of slope -2, starting at the point $(4, 1)$ and extending downward to the right. To the left of $x = 4$, the graph coincides with the curve sketched in Figure 9. These parts, *taken together*, form the graph of F (Figure 10). Evidently, the range of F is $\mathbb{R}$. ∎

© **EXAMPLE 10** $G(x) = \dfrac{1}{1 - x}$

SOLUTION   The domain of G consists of the two intervals $(-\infty, 1)$ and $(1, \infty)$. If x is near 1 but a little smaller than 1, then $1 - x$ is small and positive, so $G(x) = 1/(1 - x)$ is very large and positive. Similarly, if x is near 1 but a little larger than 1, then $1 - x$ is negative and small in absolute value, so $G(x)$ is negative and very large in absolute value. Also, if x is very large in absolute value, then so is $1 - x$, and it follows that $G(x)$ is very small in absolute value. Plotting a few points and keeping these facts in mind, we can sketch the graph of G (Figure 11). From this graph, we can see that the range of G consists of all real numbers except for 0; that is, it consists of the two intervals $(-\infty, 0)$ and $(0, \infty)$. ∎

Figure 11

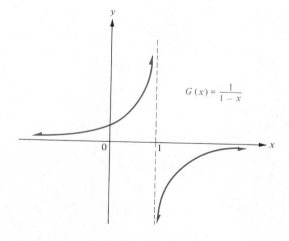

Problem Set 1.4

In Problems 1 to 14, let

$$f(x) = 2x + 1 \qquad g(x) = x^2 - 3x - 4 \qquad h(x) = \sqrt{3x + 5}$$
$$F(x) = \frac{x - 2}{3x + 7} \qquad G(x) = \sqrt[3]{x^3 - 4} \qquad H(x) = |2 - 5x|$$

Find the indicated values.

1 $f(-3)$ **2** $F(\frac{7}{3})$ **3** $h(-\frac{1}{3})$

4 $H(-4)$ **5** $G(\sqrt[3]{31})$ **6** $[h(-1)]^2$

7 $g(0)$ **8** $f(1/a)$ **9** $H(c + 2)$

10 $f[(x - 1)/2]$ **11** $F(a/3)$ **12** $G(\sqrt{b})$

©**13** $g(4.718)$ ©**14** $h(2.003)$

In Problems 15 and 16, find the domain of each function.

15 (a) $f(x) = 1 - 4x^2$ (b) $g(x) = (x + 2)^{-1}$
     (c) $h(x) = \sqrt{x}$ (d) $F(x) = \sqrt{5 - 3x}$
     (e) $G(x) = 7/(5 - 6x)$ (f) $K(x) = (4 - 5x)^{-1/2}$

16 (a) $g(x) = (9 + x^4)^{3/4}$ (b) $h(x) = \dfrac{1}{x + |x|}$

     (c) $F(x) = \dfrac{x^3 - 8}{x^2 - 4}$ (d) $H(x) = \sqrt{\dfrac{x - 2}{x - 4}}$

17 Use the vertical-line test to determine which of the curves in Figure 12 are graphs of functions.

Figure 12

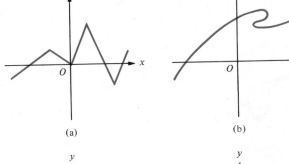

(a)   (b)

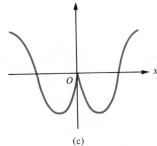   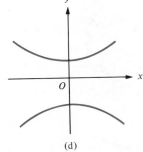

(c)   (d)

18 Is the graph of the equation $x^2 + y^2 = 9$ the graph of a function? Why or why not?

In Problems 19 to 32, determine the domain, sketch the graph, and determine the range of the given function.

19 $f(x) = 4x - 1$ **20** $g(x) = 5$

21 $h(x) = |3x|$ **22** $F(x) = |2x - 3|$

23 $G(x) = \dfrac{9x^2 - 4}{3x - 2}$ **24** $H(x) = \sqrt{9 - x^2}$

©**25** $p(x) = \sqrt{x - 1}$ **26** $q(x) = \dfrac{x}{|x|}$

27 $r(x) = \begin{cases} 5 + x & \text{if } x \le 3 \\ 9 - \dfrac{x}{3} & \text{if } x > 3 \end{cases}$

28 $P(x) = \begin{cases} -3 & \text{if } x < -1 \\ -1 & \text{if } -1 \le x \le 1 \\ 2 & \text{if } x > 1 \end{cases}$

29 $Q(x) = \begin{cases} 1 - x & \text{if } x < 1 \\ \sqrt{x - 1} & \text{if } x \ge 1 \end{cases}$

©**30** $R(x) = x^{-2}$

©**31** $S(x) = \dfrac{1}{x + 2}$

©**32** $T(x) = x + \dfrac{1}{x}$

In Problems 33 and 34, for each function f, find and simplify the expression

$$\frac{f(x + h) - f(x)}{h} \qquad \text{for } h \ne 0$$

33 (a) $f(x) = 4x - 1$ (b) $f(x) = 5$ (c) $f(x) = x^2 + 3$

34 (a) $f(x) = x^2 + x$ (b) $f(x) = 1/\sqrt{x}$ (c) $f(x) = 1/x$

35 The function $f : C \longmapsto F$ given by the equation $F = \frac{9}{5}C + 32$ converts the temperature C in degrees Celsius to the corresponding temperature F in degrees Fahrenheit. Find $f(0)$, $f(15)$, $f(-10)$, and $f(55)$; write the results, using both function and mapping notation.

36 For the function f given in Problem 35, solve the equation $f(x) = x$ for x and interpret the result.

37 An ecologist investigating the effect of air pollution on plant life finds that the percentage, $p(x)$ percent, of diseased trees and shrubs at a distance x kilometers from an industrial city is given by $p(x) = 32 - (3x/50)$ for $50 \le x \le 500$. Sketch a graph of the function p, and find $p(50)$, $p(100)$, $p(200)$, $p(400)$, and $p(500)$.

38 A closed box with a square base x centimeters by x centimeters has a volume of 100 cubic centimeters. Express the total surface area A of the exterior of this box as a function of x. (Assume that the thickness of the walls is negligible.)

39 An airline chart shows that the temperature T at an altitude of $h = 15,000$ feet is $T = 5°$ F. At an altitude of $h = 20,000$ feet, $T = -15°$ F. Supposing that T is a linear function of h, obtain an equation that defines this function, sketch its graph, and find the temperature at an altitude $h = 30,000$ feet.

40 The speed v of a car is measured during a certain time interval and found to vary according to the equation

$$v = \begin{cases} 60t & \text{for } 0 \le t < 5 \\ 300 & \text{for } t \ge 5 \end{cases}$$

where v is measured in meters per minute and the elapsed time t is measured in seconds. If V is the speed measured in kilometers per hour and T is the elapsed time measured in minutes, find V as a function of T.

©**41** In physics, the (absolute) pressure P in newtons* per square meter at a point h meters below the surface of a body of water is

shown to be a linear function of h. When $h = 0$, $P = 1.013 \times 10^5$ newtons per square meter. When $h = 1$ meter, $P = 2.003 \times 10^5$ newtons per square meter. Obtain the equation that defines P as a function of h, and use it to find the pressure P at a depth of $h = 100$ meters.

©**42** The *period $T(l)$* seconds of a simple pendulum of length l meters swinging along a small arc is given approximately by $T(l) = 2\pi\sqrt{l/9.807}$. Using a calculator and rounding off your answer to three decimal places, find $T(0.1)$, $T(1)$, $T(1.5)$, and $T(0.2484)$.

43 A baseball diamond is a square 90 feet on each side. A player is running from home plate to first base at a rate of 30 feet per second. Express the runner's distance s from second base as a function of the time t since he left home plate for $0 \le t \le 3$.

44 In Problem 43, suppose the runner continues around the bases at a constant speed of 30 feet per second. Express his distance s from second base as a function of t for $0 \le t \le 12$.

©**45** A person sketches the graph of $f(x) = 4x^4 - 14x^3 + 20x^2 - 5x$ for $x \ge 0$ by plotting five points and connecting them with a smooth curve, as shown in Figure 13. However, the graph is not correct as shown. Find the error.

Figure 13

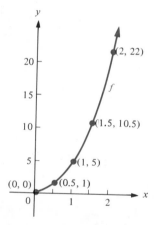

x	$f(x)$
0	0
0.5	1
1	5
1.5	10.5
2	22

©**46** A person sketches the graph of $f(x) = x^3 + 3x^2$ by plotting seven points and connecting them with a smooth curve, as shown in Figure 14. Criticize this person's work.

Figure 14

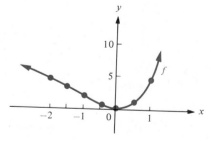

x	$f(x)$
-2	4
-1.5	3.38
-1	2
-0.5	0.63
0	0
0.5	0.88
1	4

*See page 388 for a discussion of SI units.

1.5   Types of Functions

Grouping, or *classifying,* is a familiar technique in the natural sciences for dealing with the immense diversity of things in the real world. For instance, in biology, plants and animals are divided into various phyla, classes, orders, families, genera, and species. In much the same way, functions can be grouped, or classified, by singling out important features possessed by some functions but not by others. In this section we describe certain types, or classes, of functions that are considered in calculus. Among these are *even* functions, *odd* functions, *polynomial* functions, *rational* functions, *algebraic* functions, and *transcendental* functions.

Even and Odd Functions

Consider the graphs in Figure 1. The graph of f (Figure 1a) is **symmetric about the y axis;** that is, the portion of the graph to the right of the y axis is the mirror image of the portion to the left of it. Specifically, if the point (x, y) belongs to the graph of f, then so does the point $(-x, y)$. In other words, $f(-x) = f(x)$. Similarly, the graph of g (Figure 1b) is **symmetric about the origin** because if the point (x, y) belongs to the graph, then so does the point $(-x, -y)$; that is, $g(-x) = -g(x)$.

Figure 1

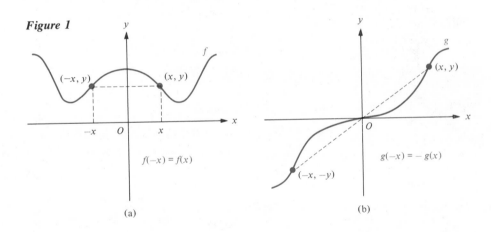

(a)   (b)

A function whose graph is symmetric about the y axis is called an *even function*; a function whose graph is symmetric about the origin is called an *odd function*. This is stated more formally in the following definition.

DEFINITION 1   **Even and Odd Functions**

> **(i)** A function f is said to be **even** if, for every number x in the domain of f, $-x$ is also in the domain of f and
> $$f(-x) = f(x)$$
> **(ii)** A function f is said to be **odd** if, for every number x in the domain of f, $-x$ is also in the domain of f and
> $$f(-x) = -f(x)$$

Figure 2

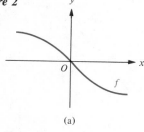

(a)

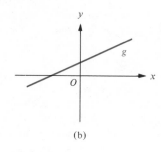

(b)

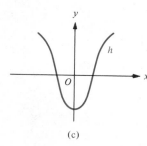

(c)

EXAMPLE 1   Determine whether each of the functions whose graphs are shown in Figure 2 is even, odd, or neither.

SOLUTION   In Figure 2a, the graph of f is symmetric about the origin; thus, $f(-x) = -f(x)$ and f is an odd function. In Figure 2b, the graph of g is symmetric neither about the y axis nor about the origin, so g is neither even nor odd. In Figure 2c, the graph of h is symmetric about the y axis; thus, $h(-x) = h(x)$ and h is an even function.   ∎

EXAMPLE 2   Determine whether each function is even, odd, or neither.

      (a) $f(x) = x^4$ **(b)** $g(x) = x - 1$
      (c) $h(x) = 2x^2 - 3|x|$ **(d)** $F(x) = x^5$

SOLUTION

      (a) $f(-x) = (-x)^4 = x^4 = f(x)$, so f is an even function.

      (b) $g(-x) = -x - 1$, while $g(x) = x - 1$ and $-g(x) = -x + 1$. Since we have neither $g(-x) = g(x)$ nor $g(-x) = -g(x)$, g is neither even nor odd.

      (c) $h(-x) = 2(-x)^2 - 3|-x| = 2x^2 - 3|x| = h(x)$, so h is an even function.

      (d) $F(-x) = (-x)^5 = -x^5 = -F(x)$, so F is an odd function.   ∎

There are many functions that are neither even nor odd; however, if a function is found to be either even or odd, the job of sketching its graph becomes easier because of the symmetry involved.

EXAMPLE 3   Sketch the graph of $F(x) = x^2$.

SOLUTION   The domain of F is the set $\mathbb{R}$. Because

$$F(-x) = (-x)^2 = x^2 = F(x)$$

the function F is even and its graph is symmetric about the y axis. We already sketched the portion of this graph for $x > 0$ in Section 1.4 (Figure 5, page 26). The full graph includes the mirror image of this portion on the other side of the y axis and the point (0, 0) (Figure 3).   ∎

Figure 3

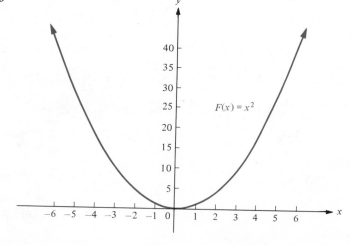

Polynomial Functions

A function f of the form

$$f(x) = a_n x^n + a_{n-1} x^{n-1} + \cdots + a_2 x^2 + a_1 x + a_0$$

where n is a nonnegative integer and the **coefficients**

$$a_n, a_{n-1}, \ldots, a_2, a_1, a_0$$

are constant real numbers, is called a **polynomial function.** If $a_n \neq 0$, we say that this polynomial function has **degree** n. For instance,

$$f(x) = 8x^3 - 3x^2 + 5x + 7$$

is a polynomial function of degree 3 with coefficients

$$a_3 = 8 \qquad a_2 = -3 \qquad a_1 = 5 \qquad a_0 = 7$$

A polynomial function of the form

$$f(x) = b$$

is called a **constant function;** its graph is a line of slope zero with y intercept b (Figure 4). If $b \neq 0$, the polynomial function $f(x) = b$ has degree zero. The constant polynomial function $f(x) = 0$, all of whose values are zero, is not assigned any degree.

As we mentioned in Section 1.4, a polynomial function of the form

$$f(x) = mx + b$$

is called a **linear function** because its graph is a line with slope m and y intercept b. Therefore, nonconstant linear functions are the same as polynomial functions of degree 1. A particularly important linear function is the **identity function**

$$f(x) = x$$

The graph of the identity function is a line of slope 1 passing through the origin (Figure 5). Notice that the graph consists of all points for which the abscissa equals the ordinate.

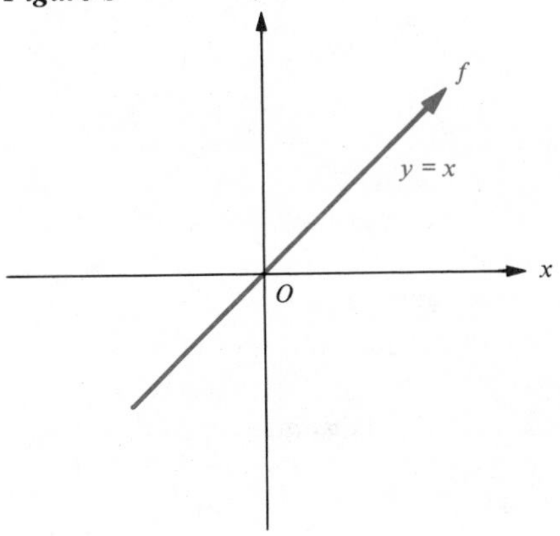

A polynomial function of degree 2 is called a **quadratic function.** Thus, a quadratic function has the form

$$f(x) = ax^2 + bx + c$$

where a, b, and c are constants and $a \neq 0$. By completing the square, such a quadratic function can be rewritten in the alternative form shown in the following theorem.

Figure 4

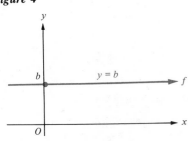

Figure 5

THEOREM 1   **Quadratic Function Theorem**

The quadratic function

$$f(x) = ax^2 + bx + c$$

can be rewritten in the form

$$f(x) = a(x - h)^2 + k$$

where $\qquad h = -\dfrac{b}{2a} \qquad$ and $\qquad k = f(h)$

We leave the proof of this theorem as an exercise (Problem 42). Example 4 shows how the theorem works.

<u>EXAMPLE 4</u> Rewrite the quadratic function $f(x) = 3x^2 + 12x + 1$ in the form $f(x) = 3(x - h)^2 + k$ by completing the square.

SOLUTION   To prepare for completing the square, we factor the coefficient 3 of x^2 out of the first two terms:

$$f(x) = 3(x^2 + 4x) + 1$$

By adding

$$\left(\frac{4}{2}\right)^2 = 2^2 = 4$$

to $x^2 + 4x$, we obtain the perfect square

$$x^2 + 4x + 4 = (x + 2)^2$$

Therefore we have

$$f(x) = 3(x^2 + 4x + 4) + 1 - 3(4)$$

where we have subtracted $3(4)$ to compensate for the addition of 4 to $x^2 + 4x$. [Notice that the 4 we added will be multiplied by 3, which is why we subtracted $3(4)$.] It follows that

$$f(x) = 3(x + 2)^2 - 11$$

Thus, we have rewritten $f(x)$ in the form

$$f(x) = 3(x - h)^2 + k$$

with $h = -2$ and $k = -11$.   ■

Once a quadratic function is written in the form

$$f(x) = a(x - h)^2 + k$$

it's fairly easy to sketch the graph. Indeed, the graph will have the same general shape as the graph of $F(x) = x^2$ in Figure 3, although it may be "shifted," it may be "stretched" or "flattened" vertically, and it may be turned upside down (Figure 6). The graph of $f(x) = a(x - h)^2 + k$, a type of curve called a **parabola**, always has either a "low point" (Figure 6a) or a "high point" (Figure 6b) called its **vertex.** The graph "opens upward" (Figure 6a) if $a > 0$, and it "opens downward" (Figure 6b) if $a < 0$. In either case, the vertex is (h, k).

<u>EXAMPLE 5</u> Sketch the graph of

$$f(x) = -2(x - 3)^2 + 5$$

SOLUTION   The function has the form

$$f(x) = a(x - h)^2 + k$$

with $a = -2$, $h = 3$, and $k = 5$; hence, its graph is a parabola, opening downward, with vertex ("highest point") $(3, 5)$. This information indicates the general appearance of the graph. By plotting a few points, we can obtain a reasonably accurate sketch of the parabola (Figure 7).   ■

We study parabolas and other related curves in more detail in Chapter 9.

Figure 6

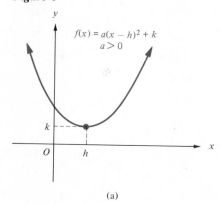

(a)

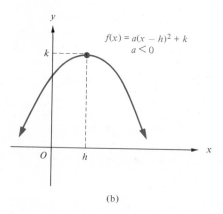

(b)

Figure 7

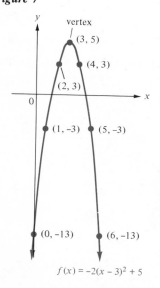

Illuminated fountain, St. Louis, Missouri, showing parabolic jets of water

Algebra of Functions

Sometimes functions are classified by the way in which they are formed from simpler functions. For instance, new functions can be formed from old ones by *addition, subtraction, multiplication,* or *division.* Thus, if $f(x) = x^2 - 2$ and $g(x) = -\frac{1}{2}x + 1$, we can form the new function $h(x) = f(x) + g(x) = x^2 - \frac{1}{2}x - 1$ simply by adding $f(x)$ and $g(x)$. Naturally, we refer to the function h as the *sum* of the functions f and g and write $h = f + g$ (Figure 8). Notice that the graph of h is obtained from the graphs of f and g by adding corresponding ordinates; for instance, $h(-2) = f(-2) + g(-2)$.

It should be clear that any two functions with intersecting domains can be added as shown.* In a similar way, subtraction, multiplication, and division of functions can be accomplished. The following definition shows exactly how this is done.

Figure 8

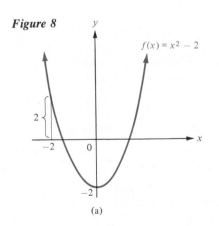

(a)

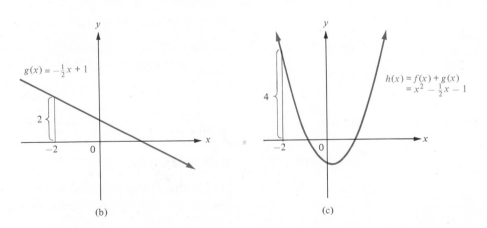

(b)

(c)

*This idea is crucial in applied mathematics, since the functions that describe natural phenomena (for example, light waves or sound waves) often add when the phenomena are combined.

DEFINITION 2   **Sum, Difference, Product, and Quotient of Functions**

> Let f and g be two functions with domains that intersect. We define functions $f + g, f - g, f \cdot g$, and f/g by the following equations:
>
> $$(f + g)(x) = f(x) + g(x)$$
>
> $$(f - g)(x) = f(x) - g(x)$$
>
> $$(f \cdot g)(x) = f(x) \cdot g(x)$$
>
> $$\left(\frac{f}{g}\right)(x) = \frac{f(x)}{g(x)}$$

In each case, the domain of the defined function consists of all values of x common to the domains of f and g, except that in the fourth case the values of x for which $g(x) = 0$ are excluded.

Geometrically, the graph of the sum, difference, product, or quotient of f and g has at each point an ordinate that is the sum, difference, product, or quotient, respectively, of the ordinates of the graphs of f and g at the corresponding points. For sums or differences, this is usually relatively easy to visualize.

EXAMPLE 6   Let $f(x) = x^2 + 3$ and $g(x) = 2x - 1$. Evaluate:

(a) $(f + g)(x)$ **(b)** $(f - g)(x)$ **(c)** $(f \cdot g)(x)$ **(d)** $\left(\dfrac{f}{g}\right)(x)$

SOLUTION

(a) $(f + g)(x) = f(x) + g(x) = (x^2 + 3) + (2x - 1) = x^2 + 2x + 2$

(b) $(f - g)(x) = f(x) - g(x) = (x^2 + 3) - (2x - 1) = x^2 - 2x + 4$

(c) $(f \cdot g)(x) = f(x) \cdot g(x) = (x^2 + 3)(2x - 1) = 2x^3 - x^2 + 6x - 3$

(d) $\left(\dfrac{f}{g}\right)(x) = \dfrac{f(x)}{g(x)} = \dfrac{x^2 + 3}{2x - 1}$

Rational and Algebraic Functions

The sum, difference, or product of two polynomials is again a polynomial, but the quotient of two polynomials is generally not a polynomial. For instance,

$$\frac{3x^2 - x + 1}{4x^5 - x^3 + 1}$$

is not a polynomial. This observation suggests the following definition.

DEFINITION 3   **Rational Function**

> A function f defined by an equation of the form $f(x) = p(x)/q(x)$, where p and q are polynomial functions and q is not the constant zero function, is called a **rational function.**

The domain of the rational function defined by $f(x) = p(x)/q(x)$ consists of all values of x for which $q(x) \neq 0$. The graphs of rational functions, which can take on a variety of geometric shapes, are discussed later. Note that *polynomial functions themselves are special kinds of rational functions*—just let the denominator $q(x)$ in Definition 3 be the constant function $q(x) = 1$. Some examples of rational functions are

$$f(x) = \frac{x}{x + 1} \qquad g(x) = 2/x \qquad h(x) = x^2 + 2x + 1 \qquad F(x) = x \qquad H(x) = \frac{x^2 - 1}{x - 1}$$

It is easy to see that:

> The sum, product, difference, or quotient of rational functions is again a rational function.

The class of rational functions is not large enough to include many of the functions that we encounter in calculus; for instance, it is possible to obtain functions that are not rational simply by extracting roots of rational functions. A case in point is the absolute-value function.

$$h(x) = |x| = \sqrt{x^2}$$

which is *not* a rational function. Thus, we are led to the following definition.

DEFINITION 4   **Elementary Algebraic Function**

> An **elementary algebraic function** is a function that can be formed by a finite number of algebraic operations (these operations being addition, subtraction, multiplication, division, and extraction of positive integer roots), starting with the identity function and constant functions.

Some examples of elementary algebraic functions are

$$f(x) = \sqrt{x^2} \qquad g(x) = \frac{x}{\sqrt{x^2 + 5}} \qquad F(x) = \frac{\sqrt[3]{x + 1} + 1}{\sqrt[5]{\sqrt{x^2 - 2} + 2}}$$

Note that any rational function is automatically an elementary algebraic function.

Using the methods already discussed, you can sketch the graphs of some of the simpler algebraic functions.

EXAMPLE 7   Sketch the graph of the **square-root function** $f(x) = \sqrt{x}$.

SOLUTION   Because the square root of a negative number is undefined (as a real number), the domain of f is the interval $[0, \infty)$. Using the point-plotting method, we obtain the graph in Figure 9. ■

Figure 9

x	$f(x) = \sqrt{x}$
0	0
1	1
4	2
9	3
16	4
25	5

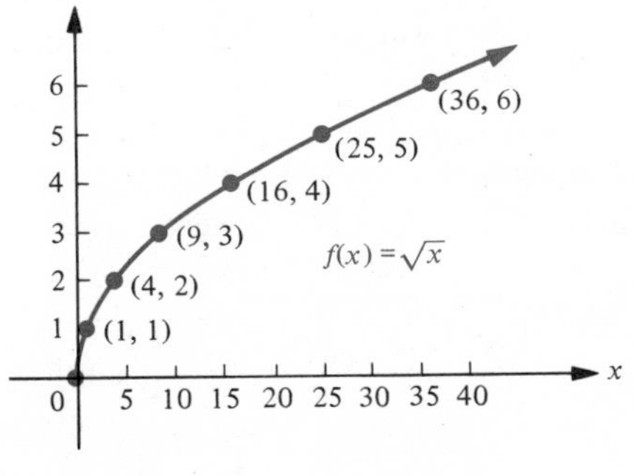

In advanced courses, a more inclusive class of functions, called **algebraic functions** (without the adjective "elementary"), is defined. Broadly speaking, these are the functions that are accessible by algebraic means. The remaining functions, those that are not algebraic, are called **transcendental functions,** since they transcend purely algebraic methods. For instance, the trigonometric functions, which we review in Section 1.6, are transcendental functions; so are the exponential, logarithmic, and hyperbolic functions, which we study in Chapter 7.

Discontinuous Functions

Many functions considered in elementary calculus have graphs that are "connected" in the sense that they consist of one continuous piece. Such functions, which are said to be *continuous,* are discussed in detail in Section 1.9. In order to fully understand and appreciate the nature of continuous functions, it is sometimes useful to examine specific functions that are not continuous. One of the more interesting discontinuous functions is the *greatest-integer function,* which, like the absolute-value function, has its own special symbol.

DEFINITION 5   **Greatest-Integer Function**

> If x is a real number, the symbol $[[x]]$ denotes the greatest integer not exceeding x; that is, $[[x]]$ is the integer that is nearest to x but is less than or equal to x. The **greatest-integer function** is the function f defined by $f(x) = [[x]]$.

Note that $[[x]]$ is the unique integer satisfying the condition

$$[[x]] \leq x < [[x]] + 1$$

For instance, $[[3.7]] = 3$, $[[\frac{9}{10}]] = 0$, $[[3.234334]] = 3$, $[[-2.7]] = -3$, $[[-2.34334]] = -3$, $[[-\frac{1}{2}]] = -1$, $[[\sqrt{3}]] = 1$, $[[2]] = 2$, and $[[-2]] = -2$. A table of values of $[[x]]$ for $-3 \leq x < 4$ follows, and the corresponding graph of $f(x) = [[x]]$ is shown in Figure 10:

$$[[x]] = \begin{cases} -3 & \text{for } -3 \leq x < -2 \\ -2 & \text{for } -2 \leq x < -1 \\ -1 & \text{for } -1 \leq x < 0 \\ 0 & \text{for } 0 \leq x < 1 \\ 1 & \text{for } 1 \leq x < 2 \\ 2 & \text{for } 2 \leq x < 3 \\ 3 & \text{for } 3 \leq x < 4 \end{cases}$$

In Figure 10, we use small dots to emphasize that the left-hand endpoints of the horizontal line segments belong to the graph, and we use small open circles to indicate that the right-hand endpoints do not belong to the graph. The discontinuous nature of the greatest-integer function is apparent from its graph.

Figure 10

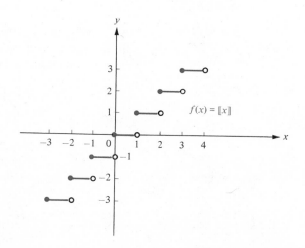

$f(x) = [[x]]$

Problem Set 1.5

1 For each function in Figure 11, find the domain and range and
determine whether the function is even, odd, or neither.

Figure 11

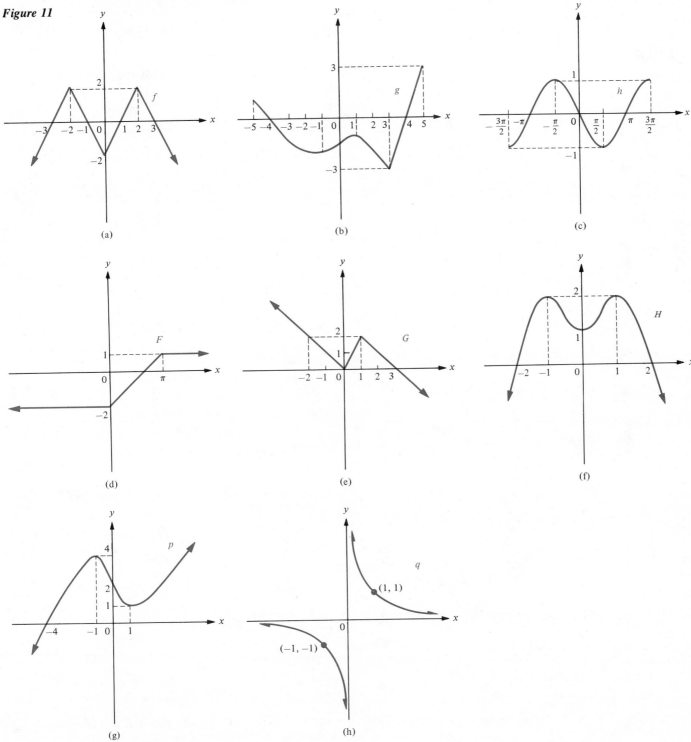

(a)

(b)

(c)

(d)

(e)

(f)

(g)

(h)

2 Is there any function that is *both* even and odd? If not, why not? If so, give an example.

In Problems 3 to 11, decide whether the function is even, odd, or neither.

3 $f(x) = x^4 + 3$

4 $g(x) = -x^4 + 2x^2 + 1$

5 $f(x) = x^4 + x$

6 $g(t) = t^2 + |t|$

7 $F(x) = 5x^3 + 7x$

8 $f(t) = -t^3 + 7t$

9 $h(x) = \sqrt{8x^3 + x}$

10 $f(y) = \dfrac{\sqrt{y^2 + 1}}{|y|}$

11 $f(x) = \dfrac{x + 1}{x^2 + 1}$

12 (a) Show that the function $F(x) = \sqrt{|x|}$ is even. (b) Using the symmetry implied by the fact that F is even, and Figure 9, sketch the graph of F.

In Problems 13 to 20, decide whether the function is a polynomial function. If it is a polynomial function, indicate the degree (if any) and identify the coefficients.

13 $f(x) = 6x^2 - 3x - 8$

14 $f(x) = x^{-3} + 2x$

15 $g(x) = (x - 3)(x - 2) - x^3$

16 $f(x) = 2^{-1}$

17 $F(x) = \sqrt{2}x^4 - 5^{-1}x^3 + 20$

18 $f(x) = 210x^{117} - 11x - 40$

19 $g(x) = 0$

20 $h(x) = \sqrt[3]{x^3 - 6x^2 + 12x - 8}$

21 Explain in your own words the distinction between the constant function f defined by the equation $f(x) = 2$ and the real number 2.

22 Is the function $f(x) = x^{-1} + [(x - 1)/x]$ a constant function?

In Problems 23 to 26, determine a linear function f that satisfies the condition.

23 $f(2) = 5$ and $f(-3) = 7$

24 $f(2x + 3) = 2f(x) + 3$

25 $f(5x) = 5f(x)$

26 $f(x + 7) = f(x) + f(7)$

27 A number r is called a *zero* of a function f if $f(r) = 0$. Prove that every nonconstant linear function has a zero.

28 Let f be a function with domain $\mathbb{R}$. Prove that

$$f(tc + (1 - t)d) = tf(c) + (1 - t)f(d)$$

holds for every three numbers c, d, and t if and only if f is linear.

29 Sketch the graph of $f(x) = ax^2$ for (a) $a = 3$, (b) $a = -3$, (c) $a = \frac{1}{3}$, and (d) $a = -\frac{1}{3}$.

30 Sketch the graph of $f(x) = 2x^2 + k$ for (a) $k = 0$, (b) $k = 1$, and (c) $k = -1$.

In Problems 31 to 36, rewrite each function f in the form $f(x) = a(x - h)^2 + k$ by completing the square.

31 $f(x) = x^2 + 2x - 4$

32 $f(x) = x^2 - 12x + 5$

33 $f(x) = 3x^2 - 10x - 2$

34 $f(x) = 2x^2 + 3x - 1$

35 $f(x) = -2x^2 + 6x + 3$

36 $f(x) = \frac{3}{2}x^2 - 6x - 7$

In Problems 37 to 40, sketch the graph of each function.

37 $f(x) = 2(x - 1)^2 + 3$

38 $h(x) = -\frac{1}{2}(x + 1)^2 - 4$

39 $g(x) = -2(x + 2)^2 - 3$

40 $p(x) = \frac{1}{2}(x + 1)^2 - \frac{1}{2}$

41 Let $f(x) = -3x^2 - 12x - 1$. (a) By completing the square, rewrite f in the form $f(x) = a(x - h)^2 + k$. (b) Sketch the graph of f.

42 Prove Theorem 1. [*Hint:* Begin by rewriting

$$f(x) = ax^2 + bx + c$$

as

$$f(x) = a\left(x^2 + \frac{b}{a}x \quad\right) + c$$

Then complete the square.]

43 In each case, evaluate $(f + g)(x)$, $(f - g)(x)$, $(f \cdot g)(x)$, and $(f/g)(x)$.

(a) $f(x) = 2x - 5$ and $g(x) = x^2 + 1$

(b) $f(x) = \sqrt{x}$ and $g(x) = x^2 + 4$

(c) $f(x) = 3x + 5$ and $g(x) = 7 - 4x$

(d) $f(x) = \sqrt{x + 3}$ and $g(x) = 1/x$

(e) $f(x) = |x|$ and $g(x) = |x - 2|$

(f) $f(x) = ax + b$ and $g(x) = cx + d$

44 Suppose that f and g are even functions. Show that $f + g, f - g$, $f \cdot g$, and f/g are also even functions.

45 Specify whether the algebraic function is a rational function.

(a) $f(x) = \dfrac{3x}{x - 1}$

(b) $g(x) = \dfrac{x + 1}{\sqrt[3]{2x^2 + 5}}$

(c) $f(x) = x^2 + 2x + 1$

(d) $f(t) = \dfrac{t^2}{2t^3 + 5}$

(e) $f(t) = \dfrac{6t^2}{\sqrt[5]{t + 1}}$

46 Show that

$$f(x) = \frac{x}{1 - x} - \frac{1}{1 + x}$$

is a rational function by rewriting it as a ratio of polynomial functions. What is the domain of f?

In Problems 47 to 53, sketch the graph of the function and specify its domain and its range.

47 $H(x) = |x + 1| - |x|$

48 $h(x) = -3|x| + x$

49 $f(x) = [\![3x]\!]$

50 $h(x) = [\![x]\!] + x$

51 $f(x) = [\![\frac{1}{3}x]\!]$

52 $G(x) = [\![|x|]\!]$

53 $g(x) = |[\![x]\!]|$

54 Let f be any function with domain $\mathbb{R}$. (a) Define a function g by the equation $g(x) = [f(x) + f(-x)]/2$. Prove that g is even. (b) Define a function h by the equation $h(x) = [f(x) - f(-x)]/2$. Prove that h is odd. (c) Prove that $f(x) = g(x) + h(x)$ holds for all x. Thus, conclude that *any function with domain $\mathbb{R}$ is the sum of an even function and an odd function.* (d) Suppose that G is an even function with domain $\mathbb{R}$, that H is an odd function with domain $\mathbb{R}$, and that $f(x) = G(x) + H(x)$ holds for all x. Prove that $G(x) = g(x)$ and that $H(x) = h(x)$ for all values of x. (e) Show that f is even if and only if $f(x) = g(x)$ holds for all x. (f) Show that f is odd if and only if $f(x) = h(x)$ holds for all x.

55 The **signum function** (abbreviated sgn) is defined by

$$\text{sgn } x = \begin{cases} \dfrac{|x|}{x} & \text{if } x \neq 0 \\ 0 & \text{if } x = 0 \end{cases}$$

(a) Find sgn (-2), sgn (-3), sgn 0, sgn 2, sgn 3, and sgn 151. (b) Prove that $|x| = x \text{ sgn } x$ is true for all values of x. (c) Prove that sgn $(ab) = (\text{sgn } a)(\text{sgn } b)$ is true for all values of a and b. (d) Sketch the graph of the signum function. (e) Find the domain and the range of the signum function. (f) Sketch the graph of the function f defined by $f(x) = \text{sgn } (x - 1)$. (g) Explain why the sgn function is discontinuous.

© **56** In economics, the *profit function P* from the sale of goods is related to the *revenue function R* from the sale and the *cost C* of producing the goods by the equation $P = R - C$. Suppose that the revenue function R (in dollars) and the cost function C (in dollars) for a product are given by $R(x) = 25x + (x^2/250)$ and $C(x) = 100 + 3x + (x^2/30)$, where x is the number of units of the product manufactured and sold. Find the equation that describes the profit function P; also find $P(350)$, $P(375)$, and $P(400)$. Sketch the graph of P, and find the smallest and largest values of the production level x for which the profit is positive. Also find the production level for which the profit is maximum.

© **57** The sum of all expenses, such as insurance, rent, and salaries, incurred by a manufacturer even when no items are produced is known as the *fixed cost F*. The part of the total cost of production C that varies with the level of production x is called the *variable cost V*. (a) Write an equation for C in terms of F and V. (b) If $F = 500$ dollars and V (in dollars) is given by $V(x) = x^2 + 4x$, find a formula for $C(x)$ and sketch the graph of C.

1.6 Trigonometric Functions

This section is intended as a concise review of the basic trigonometry that you will need in the rest of the book. To facilitate your review, much of the material is presented in somewhat condensed, "outline" form.

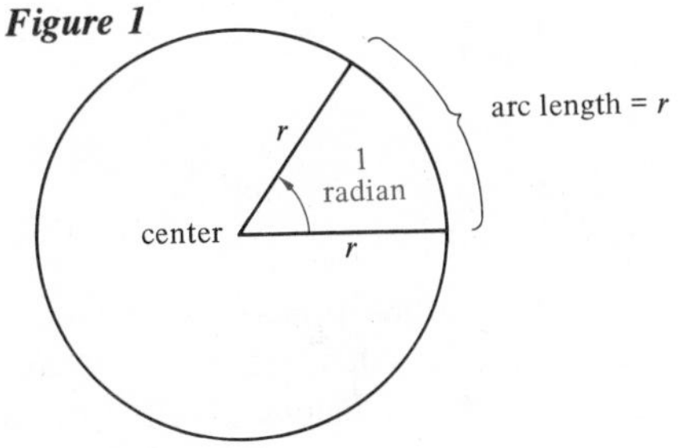

Radian Measure of Angles

Although the degree measure of angles is used in most elementary applications of trigonometry, more advanced applications (especially those that involve calculus) require radian measure. **One radian** is the measure of an angle that has its vertex at the center of a circle (that is, a **central angle**) and intercepts an arc on the circle equal in length to the radius r (Figure 1).

A central angle of 2 radians in a circle of radius r intercepts an arc of length $2r$ on the circle, a central angle of $\frac{3}{4}$ radian intercepts an arc of length $\frac{3}{4}r$, and so forth. More generally, if a central angle AOB of θ radians (θ is the Greek letter *theta*) intercepts an arc $\overset{\frown}{AB}$ of length s on a circle of radius r (Figure 2), then we have

$$s = r\theta$$

Figure 1

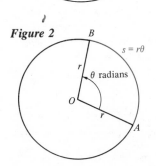

arc length $= r$

center

r

1 radian

r

Figure 2

B

$s = r\theta$

θ radians

O

r

A

It follows that the radian measure θ of angle AOB is given by the formula

$$\theta = \frac{s}{r}$$

Figure 3

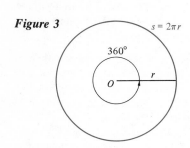

For instance, if a central angle in a circle of radius $r = 27$ meters intercepts an arc of length $s = 9$ meters, then the measure of the angle in radians is given by $\theta = s/r = \frac{9}{27} = \frac{1}{3}$ radian.

A central angle of 360° corresponds to 1 revolution; hence it intercepts an arc $s = 2\pi r$ equal to the entire circumference of the circle (Figure 3). Therefore, if θ is the radian measure of the 360° angle,

$$\theta = \frac{s}{r} = \frac{2\pi r}{r} = 2\pi \text{ radians}$$

that is, $360° = 2\pi$ radians, or

$$180° = \pi \text{ radians}$$

You can use this relationship to convert degrees to radians and vice versa. In particular,

$$1° = \frac{\pi}{180} \text{ radian} \qquad \text{and} \qquad 1 \text{ radian} = \left(\frac{180}{\pi}\right)°$$

so we have the following rule:

Multiply degrees by $\pi/180°$ to convert to radians, and multiply radians by $180°/\pi$ to convert to degrees.

For instance,

$$60° = \frac{\pi}{180°} (60°) \text{ radians} = \frac{\pi}{3} \text{ radians}$$

and

$$\frac{13\pi}{10} \text{ radians} = \frac{180°}{\pi} \left(\frac{13\pi}{10}\right) = 234°$$

Note that when radian measures are expressed as rational multiples of π, we may leave them in that form rather than write them as decimals. Also, when angles are measured in radians, the word "radian" is often omitted. For instance, rather than writing $60° = \pi/3$ radians, we write $60° = \pi/3$. Therefore, *when no unit of angular measure is indicated, it is always understood that radian measure is intended.*

Table 1 gives the corresponding degree and radian measures of certain special angles.

Table 1

Degree measure	30°	45°	60°	90°	120°	135°	150°	180°	270°	360°
Radian measure	$\frac{\pi}{6}$	$\frac{\pi}{4}$	$\frac{\pi}{3}$	$\frac{\pi}{2}$	$\frac{2\pi}{3}$	$\frac{3\pi}{4}$	$\frac{5\pi}{6}$	π	$\frac{3\pi}{2}$	2π

Figure 4

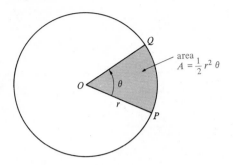

A **sector** OPQ of a circle is the region inside the circle bounded by the arc $\overset{\frown}{PQ}$ and the radial segments $\overline{OP}$ and $\overline{OQ}$ (Figure 4). If the central angle POQ has a measure of θ radians and r is the radius of the circle, then the area A of the sector is given by the formula

$$A = \frac{\theta r^2}{2}$$

Indeed, the area of the circle is πr^2, and the sector occupies a fraction $\theta/(2\pi)$ of this area, so

$$A = \frac{\theta}{2\pi} \cdot \pi r^2 = \frac{\theta r^2}{2}$$

The Trigonometric Functions

Figure 5

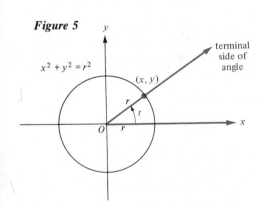

In order to define the six trigonometric functions, we begin with any real number t and measure out an angle of $|t|$ radians, starting at the positive x axis and turning counterclockwise about the origin if $t \geq 0$ and clockwise if $t < 0$ (Figure 5). Then we choose any positive number r and draw the circle $x^2 + y^2 = r^2$ of radius r with center at the origin. Let (x, y) be the point at which the terminal side of the angle meets the circle. Now the values at t of the six trigonometric functions are defined by

$$\sin t = \frac{y}{r} \qquad \csc t = \frac{r}{y}$$

$$\cos t = \frac{x}{r} \qquad \sec t = \frac{r}{x}$$

$$\tan t = \frac{y}{x} \qquad \cot t = \frac{x}{y}$$

The domain of both the sine and cosine functions is the set $\mathbb{R}$ of all real numbers. For each of the remaining four functions, the domain is the set of values of t for which the denominator of the defining fraction is nonzero. By using similar triangles, you can see that the values of the six trigonometric functions depend only on the real number t, and not on the choice of the radius r of the circle $x^2 + y^2 = r^2$. If you look at the ratios that define the six trigonometric functions, you will also see that

$$\tan t = \frac{\sin t}{\cos t} \qquad \cot t = \frac{\cos t}{\sin t}$$

$$\sec t = \frac{1}{\cos t} \qquad \csc t = \frac{1}{\sin t}$$

Therefore, if you know the values of $\sin t$ and $\cos t$, you can easily calculate the values of $\tan t$, $\cot t$, $\sec t$, and $\csc t$.

By using a little geometry, you can find the exact values of the trigonometric functions at the special values of t shown in Table 2. (Dashes in the table indicate that the function is undefined at the corresponding value of t.)

Table 2

Value of t	0	$\pi/6$	$\pi/4$	$\pi/3$	$\pi/2$	π
Corresponding degrees	$0°$	$30°$	$45°$	$60°$	$90°$	$180°$
$\sin t$	0	$1/2$	$\sqrt{2}/2$	$\sqrt{3}/2$	1	0
$\cos t$	1	$\sqrt{3}/2$	$\sqrt{2}/2$	$1/2$	0	-1
$\tan t$	0	$\sqrt{3}/3$	1	$\sqrt{3}$	—	0
$\sec t$	1	$2\sqrt{3}/3$	$\sqrt{2}$	2	—	-1
$\csc t$	—	2	$\sqrt{2}$	$2\sqrt{3}/3$	1	—
$\cot t$	—	$\sqrt{3}$	1	$\sqrt{3}/3$	0	—

A scientific calculator can be used to find approximate values of the trigonometric functions for values of the independent variable other than those shown in Table 2. (If a calculator isn't available, Table 1 in Appendix D can be used instead.)

ⓒ **EXAMPLE 1** Use a calculator to evaluate the following:

(a) $\sin (2\pi/7)$ **(b)** $\sec 1.098$

SOLUTION

(a) We set the calculator in radian mode, press the π key (or enter 3.141592654), multiply by 2, divide by 7, and press the SIN key to obtain

$$\sin \frac{2\pi}{7} \approx 0.781831483$$

(For most practical purposes, this value would be rounded off to three or four decimal places.)

(b) Most calculators have trigonometric keys only for SIN, COS, and TAN. To find $\sec 1.098$, we use the fact that $1/\cos 1.098 = \sec 1.098$. Thus, we set the calculator in radian mode, enter 1.098, press the COS key, and then press the $1/x$ key to obtain

$$\sec 1.098 \approx 2.195979642$$ ∎

On most calculators, the number in the display register is denoted by x. Pressing the $1/x$ key gives you the reciprocal of x, pressing the x^2 key gives you the square of x, and, likewise, pressing the SIN, COS, or TAN key gives you $\sin x$, $\cos x$, or $\tan x$. This notation is consistent with the custom of denoting the independent variable of a function by x (unless there are good reasons for doing otherwise). But be careful—note that x, used in this way as the independent variable for a trigonometric function, must not be confused with the abscissa of the point (x, y) in Figure 5.

Figure 6 shows the graphs of the six trigonometric functions. Notice the "wavelike" periodic appearance of these graphs. When the graphs are sketched over larger intervals, the geometric shapes shown in Figure 6 repeat themselves

Figure 6

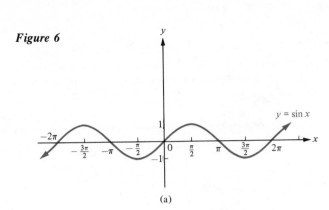

(a)

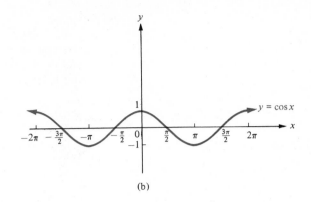

(b)

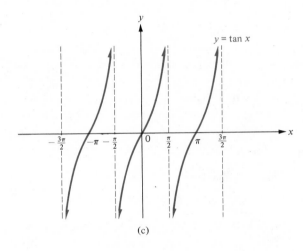

(c)

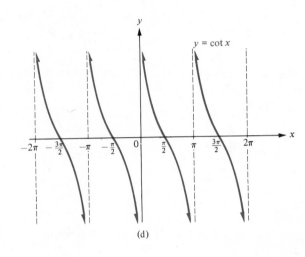

(d)

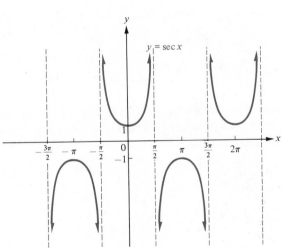

(e)

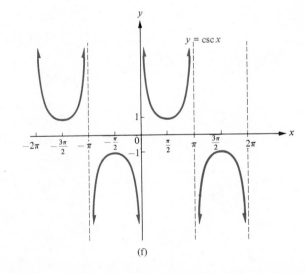

(f)

indefinitely. It is precisely the wavelike nature of the sine and cosine functions that makes them so useful in applied mathematics. Indeed, many natural phenomena, from electromagnetic waves to the ebb and flow of the tides, are periodic, and so these functions are indispensable in the construction of mathematical descriptions or models for such phenomena.

Trigonometric Identities

The six trigonometric functions satisfy certain identities that are routinely derived in precalculus mathematics courses. For your convenience, these standard identities are listed on the inside front cover of this textbook.

In Examples 2 and 3, use the standard trigonometric identities to simplify the given expression.

EXAMPLE 2 $(\sin x + \cos x)^2 - \sin 2x$

SOLUTION   We use the Pythagorean identity

$$\sin^2 x + \cos^2 x = 1$$

and the double-angle identity

$$\sin 2x = 2 \sin x \cos x$$

Thus,

$$
\begin{aligned}
(\sin x + \cos x)^2 - \sin 2x &= \sin^2 x + 2 \sin x \cos x + \cos^2 x - \sin 2x \\
&= \sin^2 x + \cos^2 x + 2 \sin x \cos x - \sin 2x \\
&= 1 + \sin 2x - \sin 2x \\
&= 1
\end{aligned}
$$

EXAMPLE 3 $\sin (x + \pi)$

SOLUTION   We use the addition formula

$$\sin (x + y) = \sin x \cos y + \sin y \cos x$$

with $y = \pi$. Recall that

$$\cos \pi = -1 \qquad \text{and} \qquad \sin \pi = 0$$

Thus,
$$
\begin{aligned}
\sin (x + \pi) &= \sin x \cos \pi + \sin \pi \cos x \\
&= (\sin x)(-1) + (0)(\cos x) \\
&= -\sin x
\end{aligned}
$$

Right-Triangle Trigonometry

If we denote one of the two acute angles in a right triangle by θ, we can abbreviate the lengths of the side *opposite* θ, the *hypotenuse,* and the side *adjacent* to θ as opp, hyp, and adj, respectively (Figure 7). Then

Figure 8

$\sin \theta = \dfrac{\text{opp}}{\text{hyp}}$	$\csc \theta = \dfrac{\text{hyp}}{\text{opp}}$
$\cos \theta = \dfrac{\text{adj}}{\text{hyp}}$	$\sec \theta = \dfrac{\text{hyp}}{\text{adj}}$
$\tan \theta = \dfrac{\text{opp}}{\text{adj}}$	$\cot \theta = \dfrac{\text{adj}}{\text{opp}}$

Figure 7

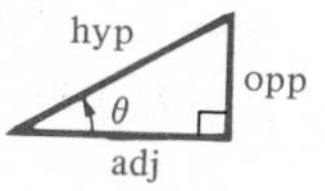

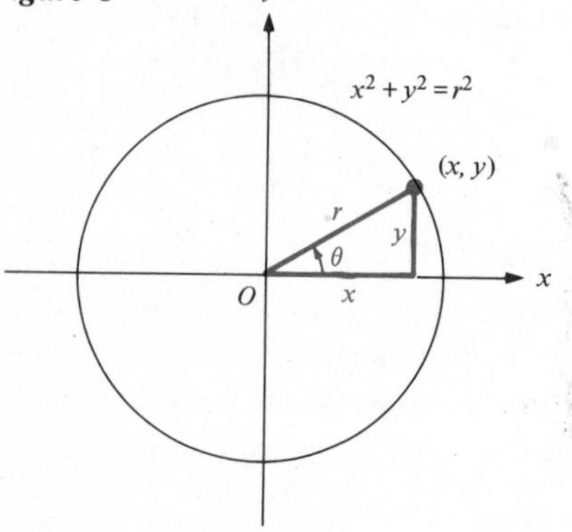

To see that these relationships hold, just let

$$\text{opp} = y \qquad \text{hyp} = r \qquad \text{adj} = x$$

and place the right triangle in the Cartesian plane as in Figure 8.

Figure 9

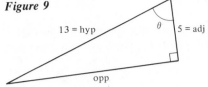

Figure 10

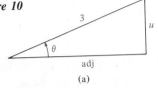

(a)

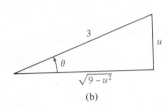

(b)

Figure 11

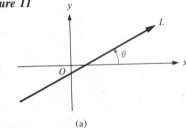

(a)

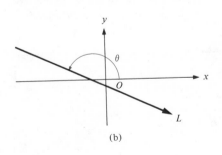

(b)

Figure 12

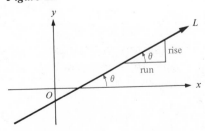

EXAMPLE 4 Find the values of the six trigonometric functions of the acute angle θ in Figure 9.

SOLUTION   For angle θ, we have

$$\text{adj} = 5 \qquad \text{and} \qquad \text{hyp} = 13$$

but opp isn't given. However, by the Pythagorean theorem,

$$\text{adj}^2 + \text{opp}^2 = \text{hyp}^2$$

so that

$$\text{opp}^2 = \text{hyp}^2 - \text{adj}^2 = 13^2 - 5^2 = 144$$

Therefore,

$$\text{opp} = \sqrt{144} = 12$$

It follows that

$$\sin \theta = \frac{\text{opp}}{\text{hyp}} = \frac{12}{13} \qquad \csc \theta = \frac{\text{hyp}}{\text{opp}} = \frac{13}{12}$$

$$\cos \theta = \frac{\text{adj}}{\text{hyp}} = \frac{5}{13} \qquad \sec \theta = \frac{\text{hyp}}{\text{adj}} = \frac{13}{5}$$

$$\tan \theta = \frac{\text{opp}}{\text{adj}} = \frac{12}{5} \qquad \cot \theta = \frac{\text{adj}}{\text{opp}} = \frac{5}{12}$$

EXAMPLE 5 If $0 < \theta < \pi/2$ and $u = 3 \sin \theta$, find $\tan \theta$ in terms of u.

SOLUTION   Let θ be the angle shown in Figure 10a, so that

$$\sin \theta = \frac{u}{3} \qquad \text{and} \qquad u = 3 \sin \theta$$

Using the Pythagorean theorem, we have

$$\text{adj} = \sqrt{\text{hyp}^2 - \text{opp}^2} = \sqrt{9 - u^2}$$

(Figure 10b). Therefore,

$$\tan \theta = \frac{\text{opp}}{\text{adj}} = \frac{u}{\sqrt{9 - u^2}}$$

If a line L in the Cartesian plane is not parallel to the x axis, then the smallest counterclockwise angle θ from the positive x axis to L is called the **inclination** of L. Figure 11a shows a case in which $0 < \theta < \pi/2$, and Figure 11b shows a case in which $\pi/2 < \theta < \pi$. If L is parallel to the x axis, we define $\theta = 0$. If L is not perpendicular to the x axis, the slope m of L is given by

$$m = \tan \theta$$

Indeed, if $0 < \theta < \pi/2$, the right triangle in Figure 12 shows that

$$m = \frac{\text{rise}}{\text{run}} = \frac{\text{opp}}{\text{adj}} = \tan \theta$$

We leave it as an exercise for you to show that $m = \tan \theta$ also holds in the remaining cases, where $\theta = 0$ and where $\pi/2 < \theta < \pi$ (Problem 43).

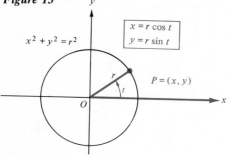

Figure 13

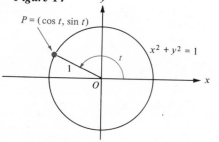

Figure 14

Trigonometric Inequalities

Consider Figure 13, in which t is the radian measure of the angle from the positive x axis to the line segment $\overline{OP}$ from the center of the circle $x^2 + y^2 = r^2$ to the point $P = (x, y)$. Because

$$\cos t = \frac{x}{r} \qquad \text{and} \qquad \sin t = \frac{y}{r}$$

it follows that

$$\boxed{x = r \cos t \qquad \text{and} \qquad y = r \sin t}$$

An important special case is obtained by using the **unit circle,** that is, the circle of radius $r = 1$ with center at the origin (Figure 14). In this case we obtain

$$\boxed{x = \cos t \qquad \text{and} \qquad y = \sin t}$$

so that $P = (\cos t, \sin t)$. This result can be used to help confirm the inequalities in the following theorem.

THEOREM 1   **Comparison of sin t and t**

> If $0 < t < \dfrac{\pi}{2}$, then $0 < \sin t < t$.

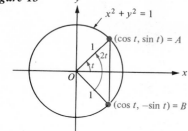

Figure 15

Let's make a geometric argument for the plausibility of Theorem 1. Consider the point A whose coordinates are $(\cos t, \sin t)$ in Figure 15. Point A lies in quadrant I because $0 < t < \pi/2$. If $B = (\cos t, -\sin t)$, then the length of the chord $\overline{AB}$ is given by $|\overline{AB}| = 2 \sin t$. Since t is measured in radians and since the circle in Figure 15 has radius 1, the arc between A and B has length $2t$. The geometrically obvious fact that the chord $\overline{AB}$ has positive length and is shorter than the circular arc between A and B implies that $0 < 2 \sin t < 2t$, or $0 < \sin t < t$, as claimed in Theorem 1. Because $\sin(-t) = -\sin t$, Theorem 1 implies that

$$\boxed{0 < |\sin t| < |t| \qquad \text{holds whenever} \qquad 0 < |t| < \frac{\pi}{2}}$$

In Section 1.8, we need the following two theorems.

THEOREM 2   **Fundamental Trigonometric Inequalities**

> If $0 < t < \dfrac{\pi}{2}$, then $0 < \cos t < \dfrac{\sin t}{t} < 1$.

Again, we give a geometric argument to support this theorem. Since $0 < t < \pi/2$, the point $P = (\cos t, \sin t)$ lies in quadrant I (Figure 16); hence, $\cos t > 0$ and $\sin t > 0$. The area A_C of the circular sector POQ is given by

$$A_C = \frac{tr^2}{2} = \frac{t \cdot 1^2}{2} = \frac{1}{2}t$$

Figure 16

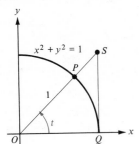

(see Figure 4). In right triangle OQS, we have

$$\tan t = \frac{\text{opp}}{\text{adj}} = \frac{|\overline{QS}|}{|\overline{OQ}|} = \frac{|\overline{QS}|}{1}$$

so that

$$|\overline{QS}| = \tan t$$

Therefore, the area A_T of right triangle OQS (one-half height times base) is given by

$$A_T = \tfrac{1}{2}|\overline{QS}|\,|\overline{OQ}| = \tfrac{1}{2}(\tan t)(1) = \tfrac{1}{2}\tan t$$

Evidently, the area A_C of the circular sector POQ is smaller than the area A_T of triangle OQS; that is,

$$A_C < A_T$$

or

$$\tfrac{1}{2}t < \tfrac{1}{2}\tan t$$

Therefore,

$$t < \tan t$$

It follows that

$$t < \frac{\sin t}{\cos t}$$

or

$$t\cos t < \sin t$$

that is,

$$\cos t < \frac{\sin t}{t}$$

The inequality

$$\frac{\sin t}{t} < 1$$

follows from Theorem 1. Combining the inequalities we have obtained, we have

$$0 < \cos t < \frac{\sin t}{t} < 1$$

as claimed in Theorem 2.

THEOREM 3

Comparison of $1 - \cos t$ and $t^2/2$

If $0 < |t| < \pi$, then $|1 - \cos t| < \dfrac{t^2}{2}$.

PROOF

From the half-angle formula

$$\sin^2\frac{t}{2} = \frac{1}{2}(1 - \cos t)$$

(see inside front cover), we have

$$|1 - \cos t| = \left|2\sin^2\frac{t}{2}\right| = 2\left|\sin^2\frac{t}{2}\right|$$

If $0 < |t| < \pi$, then $0 < |t/2| < \pi/2$; hence, by the remark following Theorem 1,

$$\left|\sin\frac{t}{2}\right| < \left|\frac{t}{2}\right|$$

Therefore,

$$\left|\sin^2\frac{t}{2}\right| = \left|\sin\frac{t}{2}\right|^2 < \left|\frac{t}{2}\right|^2 = \frac{t^2}{4}$$

and it follows that

$$|1 - \cos t| = 2\left|\sin^2\frac{t}{2}\right| < 2\left(\frac{t^2}{4}\right) = \frac{t^2}{2}$$

$\blacksquare$

Problem Set 1.6

In Problems 1 to 6, s denotes the length of the arc intercepted on a circle of radius r by a central angle of θ radians. Find the missing quantity.

1 $r = 2$ meters, $\theta = 1.65$ radians, $s = ?$

2 $r = 1.8$ centimeters, $\theta = 8$ radians, $s = ?$

3 $r = 9$ feet, $s = 12$ feet, $\theta = ?$

4 $s = 4\pi$ kilometers, $\theta = \pi/2$ radians, $r = ?$

5 $r = 12$ inches, $\theta = 5\pi/18$ radian, $s = ?$

6 $r = 5$ meters, $s = 13\pi$ meters, $\theta = ?$

7 Convert each degree measure to radians. Do *not* use a calculator. Write your answers as rational multiples of π.

(a) $30°$ (b) $45°$ (c) $90°$ (d) $120°$

(e) $-150°$ (f) $520°$ (g) $72°$ (h) $67.5°$

(i) $-330°$ (j) $450°$ (k) $21°$ (l) $-360°$

[C] **8** Use a calculator to convert each degree measure to an approximate radian measure expressed as a decimal. Round off all answers to four decimal places.

(a) $7°$ (b) $33.333°$ (c) $-11.227°$

(d) $571°$ (e) $1229°$ (f) $0.0425°$

9 Convert each radian measure to degrees. Do *not* use a calculator.

(a) $\dfrac{\pi}{2}$ (b) $\dfrac{\pi}{3}$ (c) $\dfrac{\pi}{4}$ (d) $\dfrac{\pi}{6}$

(e) $\dfrac{2\pi}{3}$ (f) $-\pi$ (g) $\dfrac{3\pi}{5}$ (h) $-\dfrac{5\pi}{2}$

(i) $\dfrac{9\pi}{4}$ (j) $-\dfrac{3\pi}{8}$ (k) 7π (l) $-\dfrac{\pi}{14}$

[C] **10** Use a calculator to convert each radian measure to an approximate degree measure. Round off all answers to four decimal places.

(a) $\frac{2}{3}$ (b) -2

(c) 200 (d) $7\pi/12$

(e) 2.7333 (f) 1.5708

11 Indicate both the degree measure and the radian measure of the angle formed by (a) $\frac{3}{8}$ of a clockwise revolution, (b) $\frac{25}{6}$ counterclockwise revolutions, and (c) the hands of a clock at 4:00.

[C] **12** A **nautical mile** may be defined as the arc length intercepted on the surface of the earth by a central angle of measure 1 minute ($\frac{1}{60}$ of a degree). The radius of the earth is 2.09×10^7 feet. How many feet are there in a nautical mile?

13 Use the formula $A = \theta r^2/2$ to find the area A of a circular sector of radius r with central angle θ radians if (a) $r = 7$ centimeters and $\theta = 3\pi/14$, (b) $r = 9$ inches and $\theta = 13\pi/9$.

14 A radar beam has an effective range of 70 kilometers and sweeps through an angle of $135°$. What is the effective area (in square kilometers) swept by the radar beam?

[C] In Problems 15 to 22, use a calculator to find the approximate values of the six trigonometric functions. Be sure that the calculator is set in degree or radian mode, as the case may be.

15 $2\pi/7$ **16** $5\pi/21$

17 $-17\pi/3$ **18** $7°$

19 1.7764 **20** $-231.4°$

21 $48°$ **22** $16.19°$

23 How could you check to make sure that a calculator is set in degree mode? (*Hint:* We know that $\sin 30° = \frac{1}{2}$.)

[C] **24** Using a calculator, verify the entries in Table 2.

25 Use the standard identities (inside front cover) to simplify each expression.

(a) $(1 - \cos t)(1 + \cos t)$ (b) $2 \sin t \cos t \csc t$

(c) $\sec^2 t \, (\csc^2 t - 1)(\sin t + 1) - \csc t$

(d) $\dfrac{1 + \cot^2 t}{\sec^2 t}$ (e) $\dfrac{\cos t - 1}{\sec t - 1}$

26 Show that

(a) $\cos\left(\dfrac{\pi}{2} - t\right) = \sin t$ (b) $\sin\left(\dfrac{\pi}{2} - t\right) = \cos t$

(c) $\tan(t + \pi) = \tan t$ (d) $\tan\left(t + \dfrac{\pi}{2}\right) = -\cot t$

27 (a) Use the fact that $45° + 30° = 75°$ to find the *exact* values of $\sin 75°$ and $\cos 75°$. (b) Find the *exact* values of $\tan 75°$, $\cot 75°$, $\sec 75°$, and $\csc 75°$.

28 Using the standard trigonometric identities, prove that $\cos 3t = 4\cos^3 t - 3\cos t$. Then use this to show that $\cos(\pi/9)$ is a solution of the equation $8x^3 - 6x - 1 = 0$.

29 Simplify.

(a) $\dfrac{\sin^2 2t}{(1 + \cos 2t)^2} + 1$ (b) $\dfrac{\cos^4 t - \sin^4 t}{\sin 2t}$

(c) $\cos^2 2t - \sin^2 t$ (d) $\tan t - \csc t \, (1 - 2\cos^2 t) \sec t$

(e) $\cos(s - t)\cos t - \sin(s - t)\sin t$

30 Suppose that $\sin t = \frac{12}{13}$ and $\cos s = -\frac{4}{5}$, where $\pi/2 < t < \pi$ and $\pi/2 < s < \pi$. Find the *exact* values of

(a) $\sin(s - t)$ (b) $\cos(s + t)$ (c) $\cot(s - t)$

In Problems 31 to 38, find the values of the six trigonometric functions of the acute angle θ.

31
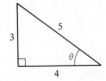

32
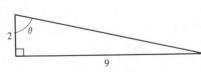

33   **34**

35   **36**
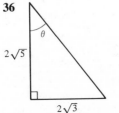

37
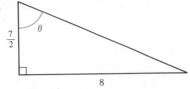

38

39 If $0 < \theta < \pi/2$ and $x = 2 \sin \theta$, find $\sec \theta$ in terms of x.

40 If $0 < \theta < \pi/2$ and $x = 3 \tan \theta$, find $\csc \theta$ in terms of x.

41 If $0 < \theta < \pi/2$ and $z = 5 \sec \theta$, find $\cos \theta$ in terms of z.

42 If $0 < \theta < \pi/2$ and $3u = 2 \tan \theta$, find $\sin \theta$ in terms of u.

43 If L is a line with inclination θ and slope m, show that $m = \tan \theta$ in the cases where $\theta = 0$ and where $\pi/2 < \theta < \pi$.

44 An irrigation ditch has a cross section in the shape of an isosceles trapezoid that is wider at the top than at the bottom (Figure 17). The bottom and equal sides of the trapezoid are each 2 meters long. If θ is the acute angle between the horizontal and the side of the ditch, show that the cross-sectional area A of the ditch is given by $A = 4 \sin \theta \, (1 + \cos \theta)$.

Figure 17
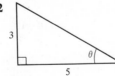

© **45** From a point on level ground 75 meters from the base of a television transmitting tower, the angle of elevation of the top of the tower is $68.17°$. Find the height h of the tower (Figure 18).

Figure 18
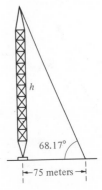

© **46** A high-altitude military reconnaissance jet photographs a missile silo under construction near a small town. The jet is at an altitude of 57,000 feet, and the angles of depression of the town and silo are 60° and 30°, respectively (Figure 19). Assuming that the jet, the silo, and the town lie in the same vertical plane, find the distance between the town and the silo.

Figure 19

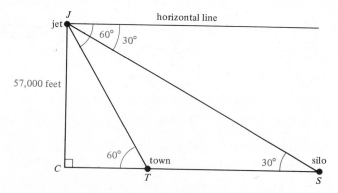

© **47** A rectangular panel to collect solar energy rests on flat ground and is tilted toward the sun. The edge resting on the ground is 3.217 meters long, and the upper edge is 1.574 meters above the ground. The panel is located near Chicago, and its latitude is 41.8°. Solar engineers recommend that the angle between the

panel and the ground be equal to the latitude of its location. Assuming that this recommendation has been followed, find the surface area of the panel. (Round off your answer to three decimal places.)

© **48** Biologists studying the migration of birds are following a migrating flock in a light plane. The birds are flying at a constant altitude of 1200 feet, and the plane is following at a constant altitude of 1700 feet. The biologists must maintain a distance of at least 600 feet between the plane and the flock to avoid disturbing the birds; therefore, they must monitor the angle of depression of the flock from the plane. Find the maximum allowable angle of depression, rounded off to the nearest degree.

49 (a) Use the half-angle formulas

$$\sin^2 \frac{x}{2} = \frac{1}{2}(1 - \cos x) \quad \text{and} \quad \cos^2 \frac{x}{2} = \frac{1}{2}(1 + \cos x)$$

to help show that

$$\sin^2 \frac{x}{2} < \cos^2 \frac{x}{2} \quad \text{for} \quad 0 < x < \frac{\pi}{2}$$

(b) Using the result of part (a), show that $0 < \sin t < \cos t$ holds for $0 < t < \pi/4$. (*Hint:* Let $x = 2t$.) (c) Use the result of part (b) to show that $0 < \tan t < 1$ holds for $0 < t < \pi/4$.

50 For small values of t, $\sin t \approx \tan t$. Prove this by establishing the inequality $|\sin t - \tan t| < t^2/2$ for $0 < |t| < \pi/4$.

1.7   Limits of Functions

The basic concept upon which calculus depends is the **limit** of a function. The idea of a limit is easy to grasp intuitively. For instance, imagine a square metal plate that is expanding uniformly because it is being heated. If x is the length of an edge, the area of the plate is given by $A = x^2$. Evidently, if x comes closer and closer to 3 centimeters, the area A comes closer and closer to 9 square centimeters. We express this by saying that

as x approaches 3, x^2 approaches 9 as a *limit*.

In symbols, we write

$$\lim_{x \to 3} x^2 = 9$$

where the notation "$x \to 3$" indicates that x comes closer and closer to 3 and "lim" stands for "the limit of."

Figure 1

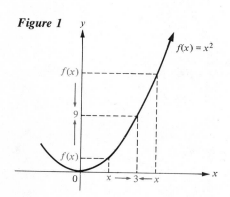

EXAMPLE 1   If $f(x) = x^2$, show graphically that $\lim_{x \to 3} f(x) = 9$.

SOLUTION   From Figure 1 we can see clearly that as x approaches 3, the function values $f(x)$ approach 9 as a limit.   ■

More generally, if f is a function and a is a number, we understand the notation

$$\lim_{x \to a} f(x) = L$$

which is read "the limit of $f(x)$ as x approaches a is L," to mean that $f(x)$ comes closer and closer to the number L as x comes closer and closer to the number a. Although we give a more formal definition at the end of this section, you can acquire a working understanding of limits by considering further examples and geometric illustrations.

EXAMPLE 2 Determine $\lim_{x \to 4} (5x + 7)$.

SOLUTION   As x comes closer and closer to 4, $5x$ comes closer and closer to 20, and $5x + 7$ comes closer and closer to 27. Therefore, $\lim_{x \to 4} (5x + 7) = 27$. ∎

Unfortunately, it isn't always possible to determine the limit of a function by simple arithmetic considerations as in the preceding example. In the first place, the function values may jump around so erratically that they never settle down and approach a limit, in which case we say that the *limit does not exist*. (Later we give some examples in which limits do not exist; however, in the present section, our examples are chosen so that all required limits exist.) In the second place, the function may be so complicated that the limit, even though it exists, is not evident by superficial inspection.

For instance, let

$$f(x) = \frac{3x^2 - 4x - 4}{x - 2}$$

and consider the problem of determining $\lim_{x \to 2} f(x)$. Here, it is not immediately clear just how $f(x)$ behaves as x approaches 2; however, we can gain some insight into this behavior by using a calculator to find some values of $f(x)$ as x gets closer and closer to 2 but stays less than 2. These values are shown in the following table:

x	1	1.25	1.50	1.75	1.90	1.99	1.999
$f(x) = \dfrac{3x^2 - 4x - 4}{x - 2}$	5	5.75	6.50	7.25	7.70	7.97	7.997

Similarly, the following table shows some values of $f(x)$ as x gets closer and closer to 2 but stays greater than 2:

x	3	2.75	2.50	2.25	2.10	2.01	2.001
$f(x) = \dfrac{3x^2 - 4x - 4}{x - 2}$	11	10.25	9.50	8.75	8.30	8.03	8.003

Evidently, as x gets closer and closer to 2, $f(x)$ gets closer and closer to 8. We are tempted to guess that

$$\lim_{x \to 2} \frac{3x^2 - 4x - 4}{x - 2} = 8$$

This guess can be verified by some elementary algebra, as in the following example.

EXAMPLE 3   Determine $\lim\limits_{x \to 2} \dfrac{3x^2 - 4x - 4}{x - 2}$.

SOLUTION   Let

$$f(x) = \frac{3x^2 - 4x - 4}{x - 2}$$

Then the domain of f consists of all real numbers x except for $x = 2$ (which makes the denominator zero). We are only concerned with the values of $f(x)$ as x *approaches* 2—what happens when x *reaches* 2 is not in question here. For $x \neq 2$,

$$f(x) = \frac{3x^2 - 4x - 4}{x - 2} = \frac{(3x + 2)(x - 2)}{x - 2} = 3x + 2$$

Therefore, as x comes closer and closer to 2, $f(x)$ comes closer and closer to 8; that is,

$$\lim_{x \to 2} f(x) = \lim_{x \to 2} \frac{3x^2 - 4x - 4}{x - 2} = \lim_{x \to 2} (3x + 2) = 8$$

The limit found in the preceding example can be illustrated geometrically by sketching a graph of the function

$$f(x) = \frac{3x^2 - 4x - 4}{x - 2}$$

Figure 2

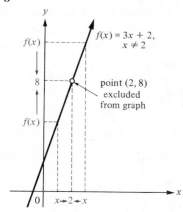

(Figure 2). Since $f(x) = 3x + 2$ holds for $x \neq 2$, this graph is a line with the point (2, 8) excluded. Figure 2 makes it clear that the value of $f(x)$ can be made to come as close to 8 as we please simply by choosing x to be sufficiently close to 2 (but not equal to 2).

We see from Figure 2 that:

> In finding the limit of $f(x)$ as x approaches a, it does not matter how f is defined *at a* (or even whether it is defined there at all). The only thing that does matter is how f is defined for values of x *near a.*

Sometimes, you can use special algebraic tricks to simplify an expression so that its limit becomes apparent.

© **EXAMPLE 4**   Find $\lim\limits_{x \to 1} \dfrac{\sqrt{x} - 1}{x - 1}$ and check your result numerically with a calculator.

SOLUTION   The desired limit, if it exists at all, is certainly not clear as things stand. But, by multiplying both the numerator and the denominator of the fraction by $\sqrt{x} + 1$, we "rationalize the numerator" and obtain

$$\frac{\sqrt{x} - 1}{x - 1} = \frac{(\sqrt{x} - 1)(\sqrt{x} + 1)}{(x - 1)(\sqrt{x} + 1)} = \frac{x - 1}{(x - 1)(\sqrt{x} + 1)} = \frac{1}{\sqrt{x} + 1}$$

Now, as x comes closer and closer to 1, $\sqrt{x}$ comes closer and closer to 1, $\sqrt{x} + 1$ comes closer and closer to 2, and $1/(\sqrt{x} + 1)$ comes closer and closer to $\frac{1}{2}$. Therefore,

$$\lim_{x \to 1} \frac{\sqrt{x} - 1}{x - 1} = \lim_{x \to 1} \frac{1}{\sqrt{x} + 1} = \frac{1}{2}$$

Using a calculator, we find several values of $(\sqrt{x} - 1)/(x - 1)$ for values of x less than 1, but approaching 1. Thus,

x	0.9	0.99	0.999	0.9999
$\dfrac{\sqrt{x} - 1}{x - 1}$	0.51317	0.50126	0.50013	0.50001

Likewise, for values of x larger than 1 but approaching 1, we obtain the following:

x	1.1	1.01	1.001	1.0001
$\dfrac{\sqrt{x} - 1}{x - 1}$	0.48809	0.49876	0.49988	0.49999

Here we have clear numerical evidence for our result that

$$\lim_{x \to 1} \frac{\sqrt{x} - 1}{x - 1} = \frac{1}{2}$$

After you have decided, on the basis of one argument or another, that

$$\lim_{x \to a} f(x) = L$$

you can always use a calculator to check your result. The calculator is also a marvelous tool for *guessing* limits that you may not be able to obtain by other means. However, no finite amount of numerical information—no matter how persuasive it may seem—can conclusively establish the exact value of a limit.

For instance, suppose that a friend, who has stored a "mystery function" f in a programmable calculator, challenges you to use the calculator and find $\lim_{x \to 0} f(x)$. Let's say the results of your calculations are as follows:

x	± 1	± 0.1	± 0.01	± 0.001	± 0.0001
$f(x)$	1102	13	2.11	2.0011	2.00001

So, you say, "There's really no mystery at all—it's clear that $\lim_{x \to 0} f(x) = 2$."

"Ah," says your friend, "but the function I stored in the calculator was

$$f(x) = 1100x^2 + 1.999999$$

and its limit as $x \to 0$ isn't 2 as you guessed; it's really 1.999999."

In calculus, there's an infamous "mystery function" that has puzzled generations of students: It's the function

$$f(x) = \frac{\sin x}{x}$$

Consider, for instance, the problem of finding

$$\lim_{x \to 0} \frac{\sin x}{x}$$

It's difficult to think of any algebraic trick that will simplify (sin x)/x so that its limit as $x \to 0$ will become apparent. Using a calculator (in *radian* mode), you can verify the entries in the following table:

x	± 1	± 0.1	± 0.01	± 0.001
$\dfrac{\sin x}{x}$	0.84147	0.99833	0.99998	0.999999833

You may be tempted to conclude from this table that

$$\lim_{x \to 0} \frac{\sin x}{x} = 1$$

but, remember, this is only a *guess*. In this case, however, it turns out that the guess is correct! As we prove in Section 1.8, it really is true that the limit is 1.

Until now, we have treated the idea of limit intuitively; however, in order to make conclusive arguments, a more formal definition of limit is required. To say, "$f(x)$ comes closer and closer to L as x comes closer and closer to a" simply lacks precision. How close is $f(x)$ to L? How close is x to a?

In careful arguments about limits, mathematicians customarily use the Greek letters ϵ and δ (called *epsilon* and *delta*) to denote positive real numbers that indicate how close $f(x)$ is to L and how close x is to a. Evidently, to say that $f(x)$ is close to L is equivalent to saying that $|f(x) - L|$ is small. Similarly, x is close to a when $|x - a|$ is small. Thus, to claim that $\lim_{x \to a} f(x) = L$ is to claim that if we take any positive number ϵ, *no matter how small*, we can always find a sufficiently small positive number δ such that $|f(x) - L| < \epsilon$ holds whenever $0 < |x - a| < \delta$. In most cases, the value of δ will depend on the value of ϵ, and the smaller the ϵ we take, the smaller the δ that will be required.

In the above discussion, the condition $0 < |x - a|$ means that $x \neq a$ and reflects our statement that in finding the limit of $f(x)$ as x approaches a, the value of $f(x)$ when $x = a$ does not matter. As Figure 3 shows, the condition $0 < |x - a| < \delta$ means that x lies in the open interval $(a - \delta, a + \delta)$, but $x \neq a$. Similarly, the condition $|f(x) - L| < \epsilon$ means that $f(x)$ lies in the open interval $(L - \epsilon, L + \epsilon)$.

Figure 3

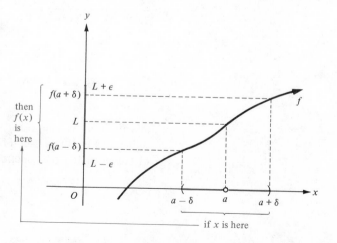

Geometrically, $\lim_{x \to a} f(x) = L$ means that, for $x \neq a$, we can *guarantee* that $f(x)$ is in *any* given small open interval around L if we make sure that x is in a *suitably small* open interval around a.

Now, we summarize these considerations in a formal definition of limit.

DEFINITION 1   **Limit**

> Let f be a function defined on some open interval containing a, except possibly at the number a itself. The statement $\lim_{x \to a} f(x) = L$ means that, for each positive number ϵ, there exists a positive number δ such that $|f(x) - L| < \epsilon$ holds whenever $0 < |x - a| < \delta$.

EXAMPLE 5   Given $\epsilon = 0.03$, determine a positive δ so that $|(3x + 7) - 1| < \epsilon$ whenever $0 < |x - (-2)| < \delta$.

SOLUTION   We have

$$|(3x + 7) - 1| = |3x + 6| = |3(x + 2)| = 3|x + 2| \quad \text{and} \quad |x - (-2)| = |x + 2|$$

hence, we must find a positive δ such that

$$3|x + 2| < 0.03 \quad \text{holds whenever} \quad 0 < |x + 2| < \delta$$

The condition $3|x + 2| < 0.03$ is equivalent to $|x + 2| < 0.03/3 = 0.01$; hence, we must determine a positive δ such that

$$|x + 2| < 0.01 \quad \text{holds whenever} \quad 0 < |x + 2| < \delta$$

Obviously, $\delta = 0.01$ works. So does any *smaller* positive value of δ.   ∎

EXAMPLE 6   Use Definition 1 to prove that $\lim_{x \to -2} (3x + 7) = 1$.

SOLUTION   Let $\epsilon > 0$. We must find $\delta > 0$ such that $|(3x + 7) - 1| < \epsilon$ holds whenever $0 < |x - (-2)| < \delta$, that is, whenever $0 < |x + 2| < \delta$. Just as in Example 5, we have

$$|(3x + 7) - 1| = 3|x + 2|$$

Therefore, the condition $|(3x + 7) - 1| < \epsilon$ is equivalent to $3|x + 2| < \epsilon$, that is, $|x + 2| < \epsilon/3$. Hence, we must determine a positive δ such that

$$|x + 2| < \frac{\epsilon}{3} \quad \text{holds whenever} \quad 0 < |x + 2| < \delta$$

Obviously, $\delta = \epsilon/3$ works. So does any *smaller* positive value of δ. This proves that

$$\lim_{x \to -2} (3x + 7) = 1$$ ∎

In the seventeenth and eighteenth centuries, the creators of calculus had to manage with something like the intuitive idea of limit that we first introduced—and they had no electronic calculators with which they could quickly check their results. In finding limits, they had to depend on geometric and numerical insights that were developed and sharpened by laborious calculations. It wasn't until 1821 that the great French mathematician Augustin Louis Cauchy (1789–1857) proposed a more precise definition of limit. Definition 1 is a somewhat refined and perfected version of Cauchy's idea of the meaning of limit.

Augustin Louis Cauchy

Problem Set 1.7

In Problems 1 to 6, determine the limit and sketch a graph of each function to illustrate the limit involved.

1 $\lim\limits_{x \to 4} 3x$

2 $\lim\limits_{x \to 1} (3x - 6)$

3 $\lim\limits_{x \to -2} (2 - 3x)$

4 $\lim\limits_{x \to 5} \dfrac{2}{x}$

5 $\lim\limits_{x \to 3} \dfrac{x^2 - 9}{x - 3}$

6 $\lim\limits_{x \to 1/2} |1 - 2x|$

In Problems 7 to 14, use elementary algebra to simplify each expression so that its limit becomes apparent. Then C check your result numerically by using a calculator to fill in the blanks in the accompanying table.

7 $\lim\limits_{x \to 2} \dfrac{x^2 - 5x + 6}{x - 2}$

x	1	1.9	1.99	1.999	1.9999
$\dfrac{x^2 - 5x + 6}{x - 2}$					

x	3	2.1	2.01	2.001	2.0001
$\dfrac{x^2 - 5x + 6}{x - 2}$					

8 $\lim\limits_{x \to 1} \dfrac{x^3 - 1}{x^2 - 1}$

x	0	0.9	0.99	0.999	0.9999
$\dfrac{x^3 - 1}{x^2 - 1}$					

x	2	1.1	1.01	1.001	1.0001
$\dfrac{x^3 - 1}{x^2 - 1}$					

9 $\lim\limits_{x \to -1} \dfrac{x^2 - 2x - 3}{x + 1}$

x	-2	-1.1	-1.01	-1.001	-1.0001
$\dfrac{x^2 - 2x - 3}{x + 1}$					

x	0	-0.9	-0.99	-0.999	-0.9999
$\dfrac{x^2 - 2x - 3}{x + 1}$					

10 $\lim\limits_{x \to -2} \dfrac{x + 2}{x^2 + 3x + 2}$

x	-3	-2.1	-2.01	-2.001	-2.0001
$\dfrac{x + 2}{x^2 + 3x + 2}$					

x	-1	-1.9	-1.99	-1.999	-1.9999
$\dfrac{x + 2}{x^2 + 3x + 2}$					

11 $\lim\limits_{t \to 4} \dfrac{\sqrt{t} - 2}{t - 4}$

t	3	3.9	3.99	3.999	3.9999
$\dfrac{\sqrt{t} - 2}{t - 4}$					

t	5	4.1	4.01	4.001	4.0001
$\dfrac{\sqrt{t} - 2}{t - 4}$					

12 $\lim\limits_{x \to 0} \dfrac{|x|}{3 - \sqrt{9 + |x|}}$

x	± 1	± 0.1	± 0.01	± 0.001	± 0.0001
$\dfrac{|x|}{3 - \sqrt{9 + |x|}}$					

13 $\lim\limits_{h \to 1} \dfrac{1 - h}{1 - (1/h)}$

h	0	0.9	0.99	0.999	0.9999
$\dfrac{1 - h}{1 - (1/h)}$					

h	2	1.1	1.01	1.001	1.0001
$\dfrac{1 - h}{1 - (1/h)}$					

14 $\lim\limits_{x\to1}\dfrac{\sqrt[3]{x}-1}{x-1}$ [Hint: Multiply numerator and denominator by $(\sqrt[3]{x})^2+\sqrt[3]{x}+1$.]

x	0	0.9	0.99	0.999	0.9999
$\dfrac{\sqrt[3]{x}-1}{x-1}$					

x	2	1.1	1.01	1.001	1.0001
$\dfrac{\sqrt[3]{x}-1}{x-1}$					

15 Suppose that a is a positive constant. By rationalizing the numerator, make an informal argument to show that

$$\lim_{x\to0}\frac{\sqrt{a^2+x}-a}{x}=\frac{1}{2a}.$$

C **16** In Problem 15, take $a=1000$. Using a calculator, find values of $(\sqrt{a^2+x}-a)/x$ for $x=1, 0.1, 0.01,$ and 0.001. How do you explain your result for $x=0.001$ in view of the fact that the limit as $x\to0$ should be $1/(2a)=0.0005$?

C In Problems 17 to 20, use a calculator to make a *guess* about the value of the indicated limit. (Remember, there's no guarantee that your guess has any validity.)

17 $\lim\limits_{x\to0}\dfrac{1-\cos x}{x}$

18 $\lim\limits_{x\to0}\dfrac{\sin 2x}{\sin 3x}$

19 $\lim\limits_{t\to0}\dfrac{\tan t}{t}$

20 $\lim\limits_{\theta\to0}\dfrac{1-\cos\theta}{\theta^2}$

21 If f is a function and $\lim\limits_{x\to a}f(x)=L$, then exactly one of these cases holds: (i) f is defined at a and $f(a)=L$, (ii) f is defined at a and $f(a)\ne L$, or (iii) f is not defined at a. Illustrate these three possibilities for the specific functions given below by determining $\lim\limits_{x\to2}f(x)$ and sketching a graph for each case.

(a) $f(x)=x+2$

(b) $f(x)=\begin{cases}x+2 & \text{if }x\ne2\\6 & \text{if }x=2\end{cases}$

(c) $f(x)=\dfrac{x^2-4}{x-2}$

22 In medicine, it is often assumed that the *reaction* $R(x)$ to a dose of size x of a drug is given by $R(x)=Ax^2(B-x)$, where A and B are certain positive constants depending on the specific drug being administered and the particular patient being treated. The *sensitivity* $S(x)$ of the patient to a dose of size x is defined by

$$S(x)=\lim_{h\to0}\frac{R(x+h)-R(x)}{h}$$

where, in taking the limit as $h\to0$, the dose size x is to be held constant. Use elementary algebra to simplify the fraction so that its limit becomes apparent, and thus obtain a formula for $S(x)$.

In Problems 23 to 28, for the given ϵ, determine a positive δ so that $|f(x)-L|<\epsilon$ holds whenever $0<|x-a|<\delta$.

23 $f(x)=4x-1$, $L=11$, $a=3$, $\epsilon=0.01$, $\lim\limits_{x\to3}(4x-1)=11$

24 $f(x)=3-4x$, $L=7$, $a=-1$, $\epsilon=0.02$, $\lim\limits_{x\to-1}(3-4x)=7$

25 $f(x)=\dfrac{x^2-25}{x-5}$, $L=10$, $a=5$, $\epsilon=0.01$, $\lim\limits_{x\to5}\dfrac{x^2-25}{x-5}=10$

26 $f(x)=x+1$, $L=0$, $a=-1$, $\epsilon=0.1$, $\lim\limits_{x\to-1}(x+1)=0$

27 $f(x)=\dfrac{x+1}{2}$, $L=3$, $a=5$, $\epsilon=0.1$, $\lim\limits_{x\to5}\dfrac{x+1}{2}=3$

28 $f(x)=x^2$, $L=4$, $a=2$, $\epsilon=0.1$, $\lim\limits_{x\to2}x^2=4$

In Problems 29 to 34, establish that each limit is correct by direct use of Definition 1. That is, for $\epsilon>0$, find $\delta>0$ so that $|f(x)-L|<\epsilon$ holds whenever $0<|x-a|<\delta$.

29 $\lim\limits_{x\to4}(2x-5)=3$

30 $\lim\limits_{x\to0}(2-5x)=2$

31 $\lim\limits_{x\to3}(4x-1)=11$

32 $\lim\limits_{x\to4}\dfrac{x^2-16}{x-4}=8$

33 $\lim\limits_{x\to3}a=a$, where a is constant

34 $\lim\limits_{x\to2}|x-2|=0$

35 Use Definition 1 to establish the following properties of limits: (a) If f is a constant function given by $f(x)=c$, then $\lim\limits_{x\to a}f(x)=c$; that is, $\lim\limits_{x\to a}c=c$. (b) If f is the identity function given by $f(x)=x$, then $\lim\limits_{x\to a}f(x)=a$; that is, $\lim\limits_{x\to a}x=a$.

36 When we speak of *the* limit as x approaches a of $f(x)$ and use the notation $\lim\limits_{x\to a}f(x)=L$, we tacitly assume that there can be at most one number L such that, for each $\epsilon>0$, there exists $\delta>0$ such that $|f(x)-L|<\epsilon$ holds whenever $0<|x-a|<\delta$. Prove that this tacit assumption is justified. (*Hint:* Suppose that there were two different values of L, say L_1 and L_2, satisfying the given condition, and let $\epsilon=\frac{1}{2}|L_1-L_2|$.)

1.8   Properties of Limits of Functions

Although graphs, "hunches," and numerical evidence generated by a calculator can provide means for guessing the value of a limit, such a value must be viewed with suspicion until it is confirmed. The formal ϵ, δ definition (Definition 1, Section 1.7) is a reliable means for confirming the true value of a limit, but it provides few clues for *finding* such a value in the first place. Furthermore, direct use of the definition can be difficult—it may not be easy to find a δ that works for a given ϵ.

In practical work, limits are often determined by using certain *properties* that can be shown to follow deductively from the ϵ, δ definition. If you have managed a good intuitive grasp of the idea of a limit, most of these properties should seem quite reasonable, if not obvious. We recommend that you become familiar with these properties and their use before studying their proofs (in Appendix B). At first, in order to show you how the limit properties work, we have deliberately chosen some very simple examples in which the limits themselves are more or less apparent. Later in this section we use these properties to help prove some important theorems about limits of trigonometric functions.

The Constant and Identity Properties

$$\mathbf{1} \ \lim_{x \to a} c = c \qquad (c \text{ is any constant})$$

$$\mathbf{2} \ \lim_{x \to a} x = a$$

EXAMPLE 1   Find:

(a) $\displaystyle\lim_{x \to -7} 5$ (b) $\displaystyle\lim_{x \to -7} x$

SOLUTION

(a) $\displaystyle\lim_{x \to -7} 5 = 5$ (Property 1)

(b) $\displaystyle\lim_{x \to -7} x = -7$ (Property 2)

The Constant-Multiplier Property

Suppose that $\displaystyle\lim_{x \to a} f(x) = L$, and let c be a constant. Then

$$\mathbf{3} \ \lim_{x \to a} [cf(x)] = c[\lim_{x \to a} f(x)] = cL$$

EXAMPLE 2   Find $\displaystyle\lim_{x \to -7} 5x$.

SOLUTION $\displaystyle\lim_{x \to -7} 5x = 5(\lim_{x \to -7} x)$ (Property 3)

$= 5(-7)$ (Property 2)

$= -35$

The Addition, Subtraction, Multiplication, and Division Properties

Suppose that $\lim\limits_{x \to a} f(x) = L$ and $\lim\limits_{x \to a} g(x) = M$. Then

4 $\lim\limits_{x \to a} [f(x) + g(x)] = \lim\limits_{x \to a} f(x) + \lim\limits_{x \to a} g(x) = L + M$

5 $\lim\limits_{x \to a} [f(x) - g(x)] = \lim\limits_{x \to a} f(x) - \lim\limits_{x \to a} g(x) = L - M$

6 $\lim\limits_{x \to a} [f(x) \cdot g(x)] = [\lim\limits_{x \to a} f(x)][\lim\limits_{x \to a} g(x)] = LM$

7 $\lim\limits_{x \to a} \dfrac{f(x)}{g(x)} = \dfrac{\lim\limits_{x \to a} f(x)}{\lim\limits_{x \to a} g(x)} = \dfrac{L}{M}$ provided that $M \neq 0$

EXAMPLE 3 Find:

(a) $\lim\limits_{x \to 3} (x + 1)$ **(b)** $\lim\limits_{x \to 3} (x - 1)$ **(c)** $\lim\limits_{x \to 3} (x + 1)(x - 1)$ **(d)** $\lim\limits_{x \to 3} \dfrac{x + 1}{x - 1}$

SOLUTION

(a) $\lim\limits_{x \to 3} (x + 1) = \lim\limits_{x \to 3} x + \lim\limits_{x \to 3} 1$ (Property 4)

$= 3 + 1$ (Properties 2 and 1)

$= 4$

(b) $\lim\limits_{x \to 3} (x - 1) = \lim\limits_{x \to 3} x - \lim\limits_{x \to 3} 1$ (Property 5)

$= 3 - 1$ (Properties 2 and 1)

$= 2$

(c) $\lim\limits_{x \to 3} (x + 1)(x - 1) = [\lim\limits_{x \to 3} (x + 1)][\lim\limits_{x \to 3} (x - 1)]$ (Property 6)

$= (4)(2)$

$= 8$

(d) $\lim\limits_{x \to 3} \dfrac{x + 1}{x - 1} = \dfrac{\lim\limits_{x \to 3} (x + 1)}{\lim\limits_{x \to 3} (x - 1)}$ (Property 7)

$= \dfrac{4}{2}$

$= 2$

The Power, Root, and Absolute-Value Properties

Suppose that $\lim\limits_{x \to a} f(x) = L$, and let n be a positive integer. Then

8 $\lim\limits_{x \to a} [f(x)]^n = [\lim\limits_{x \to a} f(x)]^n = L^n$

9 $\lim\limits_{x \to a} \sqrt[n]{f(x)} = \sqrt[n]{\lim\limits_{x \to a} f(x)} = \sqrt[n]{L}$ if $L > 0$ and n is a positive integer or if $L \leq 0$ and n is an odd positive integer

10 $\lim\limits_{x \to a} |f(x)| = |\lim\limits_{x \to a} f(x)| = |L|$

EXAMPLE 4   Find:

$$\textbf{(a)} \lim_{x \to 4} (3 - x)^2 \qquad \textbf{(b)} \lim_{x \to 4} \sqrt[3]{3 - x} \qquad \textbf{(c)} \lim_{x \to 4} |3 - x|$$

SOLUTION

$$\textbf{(a)} \lim_{x \to 4} (3 - x)^2 = [\lim_{x \to 4} (3 - x)]^2 \qquad \text{(Property 8)}$$

$$= (3 - 4)^2 \qquad \text{(Properties 5, 2, and 1)}$$
$$= 1$$

$$\textbf{(b)} \lim_{x \to 4} \sqrt[3]{3 - x} = \sqrt[3]{\lim_{x \to 4} (3 - x)} \qquad \text{(Property 9)}$$

$$= \sqrt[3]{-1}$$
$$= -1$$

$$\textbf{(c)} \lim_{x \to 4} |3 - x| = |\lim_{x \to 4} (3 - x)| \qquad \text{(Property 10)}$$

$$= |-1|$$
$$= 1 \qquad\qquad\qquad\qquad\qquad \blacksquare$$

The addition and multiplication properties (Properties 4 and 6) can be extended to three or more functions. By using Properties 1 through 10 in combination, you can evaluate limits of a wide variety of algebraic functions.

EXAMPLE 5   Find $\lim_{t \to 2} (4t^2 + 5t - 7)$.

SOLUTION

$$\lim_{t \to 2} (4t^2 + 5t - 7) = \lim_{t \to 2} 4t^2 + \lim_{t \to 2} 5t + \lim_{t \to 2} (-7) \qquad \text{(Property 4)}$$

$$= 4 \lim_{t \to 2} t^2 + 5 \lim_{t \to 2} t + \lim_{t \to 2} (-7) \qquad \text{(Property 3)}$$

$$= 4 \lim_{t \to 2} t^2 + 5(2) + \lim_{t \to 2} (-7) \qquad \text{(Property 2)}$$

$$= 4 \lim_{t \to 2} t^2 + 10 + (-7) \qquad \text{(Property 1)}$$

$$= 4(\lim_{t \to 2} t)^2 + 3 \qquad \text{(Property 8)}$$

$$= 4(2)^2 + 3 = 19 \qquad \text{(Property 2)} \qquad \blacksquare$$

EXAMPLE 6   Find $\lim_{y \to 3} \sqrt[3]{\dfrac{y^2 + 5y + 3}{y^2 - 1}}$.

SOLUTION   Proceeding as in Example 5, we have

$$\lim_{y \to 3} (y^2 + 5y + 3) = 3^2 + 5(3) + 3 = 27$$

and

$$\lim_{y \to 3} (y^2 - 1) = 3^2 - 1 = 8$$

Hence, by Property 7,

$$\lim_{y \to 3} \frac{y^2 + 5y + 3}{y^2 - 1} = \frac{27}{8}$$

Therefore, using Property 9, we obtain

$$\lim_{y \to 3} \sqrt[3]{\frac{y^2 + 5y + 3}{y^2 - 1}} = \sqrt[3]{\lim_{y \to 3} \frac{y^2 + 5y + 3}{y^2 - 1}}$$
$$= \sqrt[3]{\frac{27}{8}} = \frac{3}{2}$$

As you may have noticed, for all examples given so far in this section, $\lim_{x \to a} f(x)$ turns out to be the same as $f(a)$. Although this is true for a large class of "well-behaved" functions, it isn't always true (see Problem 21b, Section 1.7). In general, the value of $\lim_{x \to a} f(x)$ is *independent* of the value of $f(a)$; in fact, $f(a)$ need not even be defined. A more precise statement of this independence is provided by the following property.

The Independence Property

> Suppose that $f(x) = g(x)$ holds for all values of x in an open interval containing a, except possibly for $x = a$. Then, if $\lim_{x \to a} g(x) = L$, it follows that
>
> **11** $\lim_{x \to a} f(x) = \lim_{x \to a} g(x) = L$

EXAMPLE 7   Find $\lim_{x \to 7} \dfrac{x^2 - 49}{x - 7}$.

SOLUTION   In Section 1.7, we found limits such as this *informally* by simplifying the fraction so that the limit became apparent. The independence property *justifies* such an argument. Let's see how. First, note that we cannot use the division property (Property 7) because the limit of the denominator is zero. However,

$$\frac{x^2 - 49}{x - 7} = x + 7$$

holds for all values of x except $x = 7$. Therefore, by using Property 11 with $f(x) = (x^2 - 49)/(x - 7)$ and $g(x) = x + 7$, we have

$$\lim_{x \to 7} \frac{x^2 - 49}{x - 7} = \lim_{x \to 7} (x + 7) = 14$$

where we used Properties 4, 1, and 2 to evaluate the last limit.

The next property involves the situation shown in Figure 1 in which the graph of a function f is "squeezed," or "pinched," between the graphs of functions g and h.

Figure 1

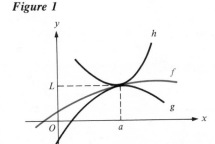

The Squeezing Property

> Suppose that $g(x) \leq f(x) \leq h(x)$ holds for all values of x in an open interval containing a, except possibly for $x = a$. Then
>
> **12** If $\lim_{x \to a} g(x) = L$ and $\lim_{x \to a} h(x) = L$, then it follows that $\lim_{x \to a} f(x) = L$.

EXAMPLE 8   Find $\displaystyle\lim_{x\to 0} x \cos \frac{1}{x}$.

SOLUTION   Because values of the cosine always lie between -1 and 1 inclusive, we have

$$\left| x \cos \frac{1}{x} \right| = |x| \left| \cos \frac{1}{x} \right| \le |x| \cdot 1 = |x|$$

for all values of x other than 0. Therefore,

$$-|x| \le x \cos \frac{1}{x} \le |x|$$

holds for all values of x other than 0. By Properties 10, 3, and 2,

$$\lim_{x\to 0} |x| = 0 \qquad \text{and} \qquad \lim_{x\to 0} (-|x|) = 0$$

Hence, by Property 12,

$$\lim_{x\to 0} x \cos \frac{1}{x} = 0$$

The following properties can be used to rewrite limits in equivalent alternative forms.

The Alternative-Form Properties

13 $\displaystyle\lim_{x\to a} f(x) = L$	if and only if	$\displaystyle\lim_{x\to a} |f(x) - L| = 0$
14 $\displaystyle\lim_{x\to a} f(x) = L$	if and only if	$\displaystyle\lim_{h\to 0} f(a + h) = L$

EXAMPLE 9   Find $\displaystyle\lim_{x\to 0} \sin x$.

SOLUTION   By the remark following Theorem 1 in Section 1.6,

$$0 < |\sin x| < |x|$$

holds for every value of x in the open interval $(-\pi/2, \pi/2)$, except for $x = 0$. By Properties 1, 10, and 2,

$$\lim_{x\to 0} 0 = 0 \qquad \text{and} \qquad \lim_{x\to 0} |x| = \left| \lim_{x\to 0} x \right| = |0| = 0$$

Therefore, by the squeezing property (Property 12),

$$\lim_{x\to 0} |\sin x| = 0$$

Now, we can use Property 13 (with $L = 0$) to conclude that

$$\lim_{x\to 0} \sin x = 0$$

EXAMPLE 10   Find $\displaystyle\lim_{x\to 0} \cos x$.

SOLUTION   By Theorem 3 in Section 1.6,

$$0 \le |\cos x - 1| < \tfrac{1}{2}x^2$$

holds for every value of x in the open interval $(-\pi, \pi)$ except for $x = 0$. By Properties 1, 3, 8, and 2, we have both

$$\lim_{x \to 0} 0 = 0$$

and

$$\lim_{x \to 0} \tfrac{1}{2}x^2 = \tfrac{1}{2} \lim_{x \to 0} x^2 = \tfrac{1}{2} \cdot 0^2 = 0$$

Therefore, by the squeezing property (Property 12),

$$\lim_{x \to 0} |\cos x - 1| = 0$$

Hence, by Property 13, it follows that

$$\lim_{x \to 0} \cos x = 1 \qquad \blacksquare$$

THEOREM 1   **Limits of sin x and cos x**

$$\lim_{x \to a} \sin x = \sin a \qquad \text{and} \qquad \lim_{x \to a} \cos x = \cos a$$

PROOF   We prove the first statement and leave the second as an exercise (Problem 71). By Property 14, we can find $\lim\limits_{x \to a} \sin x$ by finding $\lim\limits_{h \to 0} \sin (a + h)$. Using the addition formula for the sine, Properties 4 and 3, and the results of Examples 9 and 10, we have

$$\lim_{h \to 0} \sin (a + h) = \lim_{h \to 0} (\sin a \cos h + \cos a \sin h)$$

$$= \lim_{h \to 0} (\sin a \cos h) + \lim_{h \to 0} (\cos a \sin h)$$

$$= (\sin a) \lim_{h \to 0} \cos h + (\cos a) \lim_{h \to 0} \sin h$$

$$= (\sin a) \cdot 1 + (\cos a) \cdot 0$$

$$= \sin a$$

Therefore, by Property 14,

$$\lim_{x \to a} \sin x = \sin a \qquad \blacksquare$$

EXAMPLE 11   Find $\lim\limits_{x \to \pi} \tan x$.

SOLUTION $\displaystyle \lim_{x \to \pi} \tan x = \lim_{x \to \pi} \frac{\sin x}{\cos x} = \frac{\lim\limits_{x \to \pi} \sin x}{\lim\limits_{x \to \pi} \cos x}$ (Property 7)

$$= \frac{\sin \pi}{\cos \pi}$$

$$= \frac{0}{-1}$$

$$= 0 \qquad \blacksquare$$

In Section 1.7, our calculator work suggested that

$$\lim_{x \to 0} \frac{\sin x}{x} = 1$$

Now we can prove this important fact.

THEOREM 2   **Limit of (sin x)/x**

$$\lim_{x \to 0} \frac{\sin x}{x} = 1$$

PROOF   By Theorem 2 in Section 1.6,

$$\cos x < \frac{\sin x}{x} < 1$$

holds for $0 < x < \pi/2$. If $-\pi/2 < x < 0$, then $0 < -x < \pi/2$, and we have

$$\cos (-x) < \frac{\sin (-x)}{-x} < 1$$

Because $\cos (-x) = \cos x$ and $\sin (-x) = -\sin x$, it follows that

$$\cos x < \frac{\sin x}{x} < 1$$

also holds for $-\pi/2 < x < 0$. Therefore, it holds for every value of x in the open interval $(-\pi/2, \pi/2)$, except for $x = 0$. Hence, by the squeezing property (Property 12) together with the facts that

$$\lim_{x \to 0} \cos x = 1 \qquad \text{and} \qquad \lim_{x \to 0} 1 = 1$$

we can conclude that

$$\lim_{x \to 0} \frac{\sin x}{x} = 1$$

  ■

EXAMPLE 12   Evaluate $\lim\limits_{x \to 0} \dfrac{\sin 5x}{x}$.

SOLUTION   Let $y = 5x$, and note that $y \to 0$ when $x \to 0$. Because $x = y/5$, we have

$$\lim_{x \to 0} \frac{\sin 5x}{x} = \lim_{y \to 0} \frac{\sin y}{y/5} = \lim_{y \to 0} \left(5\frac{\sin y}{y} \right)$$

$$= 5 \lim_{y \to 0} \frac{\sin y}{y} \qquad \text{(Property 3)}$$

$$= 5(1) \qquad \text{(Theorem 2)}$$

$$= 5$$

  ■

The method used in Example 12—substituting y for $5x$—is generalized and made more precise by the following property.

The Substitution Property*

> Suppose that $\lim\limits_{y \to c} f(y)$ exists and that $\lim\limits_{x \to a} g(x) = c$. Then, provided that $f(g(x))$
> is defined and $g(x) \neq c$ for all values of x in an open interval containing a, except
> possibly for $x = a$, it follows that
>
> **15** $\lim\limits_{x \to a} f(g(x)) = \lim\limits_{y \to c} f(y)$

EXAMPLE 13 Find $\lim\limits_{x \to 0} \cos \left(x^2 - \dfrac{\pi}{3} \right)$.

SOLUTION   Here, the idea is to make the substitution $y = x^2 - (\pi/3)$, so that
$\cos [x^2 - (\pi/3)] = \cos y$. Informally, we would proceed by noting that as $x \to 0$,
$y \to -\pi/3$; hence,

$$\lim\limits_{x \to 0} \cos \left(x^2 - \dfrac{\pi}{3} \right) = \lim\limits_{y \to -\pi/3} \cos y = \cos \left(-\dfrac{\pi}{3} \right) \qquad \text{(Theorem 1)}$$

$$= \dfrac{1}{2}$$

This procedure is justified by taking $f(y) = \cos y$, $g(x) = x^2 - (\pi/3)$, $a = 0$, and
$c = -(\pi/3)$ in Property 15. ∎

THEOREM 3   **Limit of $(1 - \cos x)/x$**

> $$\lim\limits_{x \to 0} \dfrac{1 - \cos x}{x} = 0$$

PROOF   By the half-angle formula,

$$\sin^2 \dfrac{x}{2} = \dfrac{1 - \cos x}{2}$$

Therefore,

$$\dfrac{1 - \cos x}{x} = \dfrac{2 \sin^2 (x/2)}{x} = \dfrac{\sin^2 (x/2)}{x/2}$$

Now we use the substitution property (Property 15) with $y = x/2$. Note that $y \to 0$
when $x \to 0$. Therefore,

$$\lim\limits_{x \to 0} \dfrac{1 - \cos x}{x} = \lim\limits_{x \to 0} \dfrac{\sin^2 (x/2)}{x/2} = \lim\limits_{y \to 0} \dfrac{\sin^2 y}{y} \qquad \text{(Property 15)}$$

$$= \lim\limits_{y \to 0} \left(\sin y \dfrac{\sin y}{y} \right) \qquad \text{(Property 6)}$$

$$= (\lim\limits_{y \to 0} \sin y)\left(\lim\limits_{y \to 0} \dfrac{\sin y}{y} \right)$$

$$= (0)(1) = 0 \qquad \text{(Theorems 1 and 2)} \qquad \blacksquare$$

*The function $h(x) = f(g(x))$ obtained by evaluating f at $g(x)$ is called the *composition* of f
and g. The composition of functions is studied in detail in Section 2.7.

Problem Set 1.8

In Problems 1 to 66, find the limit by using the properties and theorems of this section.

1 $\lim\limits_{x \to 4} 5$

2 $\lim\limits_{x \to 5} 5$

3 $\lim\limits_{x \to -4} \pi$

4 $\lim\limits_{x \to 0} \pi x$

5 $\lim\limits_{x \to \pi} x$

6 $\lim\limits_{x \to \pi} \cos \pi$

7 $\lim\limits_{x \to 2} 7x$

8 $\lim\limits_{x \to -2} [x + \cos (\pi/3)]$

9 $\lim\limits_{x \to 4} (x - 3)$

10 $\lim\limits_{y \to 6} (y - 6)$

11 $\lim\limits_{t \to -3} (2t + 1)$

12 $\lim\limits_{x \to 1} (x + 1)(x - 1)$

13 $\lim\limits_{x \to 3} x(2x - 1)$

14 $\lim\limits_{s \to 0} s(s - 1)(s + 1)$

15 $\lim\limits_{t \to -2} t(2t + 1)(t - 1)$

16 $\lim\limits_{q \to 2} \dfrac{4q}{3q + 2}$

17 $\lim\limits_{x \to 3} \dfrac{3x + 2}{2x + 5}$

18 $\lim\limits_{y \to 1} \dfrac{y - 1}{y + 1}$

19 $\lim\limits_{y \to 5} (y^2 - 2y + 1)$

20 $\lim\limits_{t \to -1} (3t^7 - 2t^5 + 4)$

21 $\lim\limits_{s \to -2} (5 - 3s - s^2)$

22 $\lim\limits_{x \to -1} |3x^3 - 2x^2 + 5x - 1|$

23 $\lim\limits_{u \to 2} \dfrac{u^2 + u + 1}{u^2 + 2u}$

24 $\lim\limits_{t \to -2} \dfrac{t^3 - 5t}{t + 3}$

25 $\lim\limits_{y \to 3} \sqrt{y + 1} \, (2y - 3)$

26 $\lim\limits_{x \to 1} \dfrac{\sqrt{4 - x^2}}{2 + x}$

27 $\lim\limits_{z \to 3} \sqrt{\left| \dfrac{2z - 15}{z + 1} \right|}$

28 $\lim\limits_{x \to 8/3} \sqrt{\dfrac{9x^2 - 64}{3x - 8}}$

29 $\lim\limits_{z \to -3} \sqrt[3]{\dfrac{z - 4}{6z^2 + 2}}$

30 $\lim\limits_{y \to 1} \sqrt[3]{\dfrac{27y^3 + 4y - 4}{y^{10} + 4y^2 + 3y}}$

31 $\lim\limits_{t \to 1/2} \dfrac{t^2 + 1}{1 + \sqrt{2t + 8}}$

32 $\lim\limits_{z \to 2} \dfrac{2z^2 - 5z + 2}{z - 2}$

33 $\lim\limits_{t \to 3} \dfrac{t^2 + t - 12}{t - 3}$

34 $\lim\limits_{w \to -5} \dfrac{w^2 - 25}{w + 5}$

35 $\lim\limits_{x \to 5/2} \dfrac{4x^2 - 25}{2x - 5}$

36 $\lim\limits_{x \to -3} \left| \dfrac{x^2 + 4x + 3}{x + 3} \right|$

37 $\lim\limits_{h \to 0} \dfrac{(3 + h)^2 - 9}{h}$

38 $\lim\limits_{x \to 0} \dfrac{\sqrt{x + 2} - \sqrt{2}}{x}$

39 $\lim\limits_{y \to 0} \dfrac{2 - \sqrt{4 - y}}{y}$

40 $\lim\limits_{v \to 4} \dfrac{v^{5/2} - 16v^{1/2}}{v - 4}$

41 $\lim\limits_{x \to 1} \dfrac{(1/\sqrt{x}) - 1}{1 - x}$

42 $\lim\limits_{z \to -1} \dfrac{z^2 + 4z + 3}{z^2 - 1}$

43 $\lim\limits_{t \to 4} \dfrac{(1/\sqrt{t}) - \frac{1}{2}}{t - 4}$

44 $\lim\limits_{t \to 0} \left(\sqrt{1 + \dfrac{1}{|t|}} - \sqrt{\dfrac{1}{|t|}} \right)$

45 $\lim\limits_{x \to 0} x \sin \dfrac{1}{x}$

46 $\lim\limits_{y \to 2} \dfrac{\sqrt{y^3} - \sqrt{8}}{y^2 - 4}$

47 $\lim\limits_{x \to \pi} \cos x$

48 $\lim\limits_{t \to 0} \sin t \cos \dfrac{1}{t}$

49 $\lim\limits_{x \to \pi/6} \dfrac{\cos x}{\sin x}$

50 $\lim\limits_{\theta \to \pi} \left| \dfrac{1}{\cos \theta} \right|$

51 $\lim\limits_{t \to \pi} \dfrac{\sin t}{t}$

52 $\lim\limits_{x \to \pi/4} \sec x$

53 $\lim\limits_{h \to 0} \sin [(\pi/3) + h^3]$

54 $\lim\limits_{w \to 0} w \csc w$

55 $\lim\limits_{x \to 0} \dfrac{\sin 6x}{x}$

56 $\lim\limits_{x \to 0} \dfrac{x}{\sin 3x}$

57 $\lim\limits_{x \to 0} \dfrac{\sin 2x}{\sin 5x}$

58 $\lim\limits_{t \to 0} \dfrac{1 - \cos 2t}{\sin t}$

59 $\lim\limits_{\theta \to 0} \dfrac{\sin^2 \theta}{\theta^2}$

60 $\lim\limits_{x \to 0} \dfrac{\sin x - \cos x \sin x}{x^2}$

61 $\lim\limits_{u \to 0} \dfrac{1 - \cos^2 u}{u^2}$

62 $\lim\limits_{x \to 3} \dfrac{x - 3}{\sin (x - 3)}$

63 $\lim\limits_{x \to \pi} \dfrac{\cos x/2}{(x/2) - (\pi/2)}$

64 $\lim\limits_{t \to 0} \dfrac{\tan 4t}{2t}$

65 $\lim\limits_{\theta \to 0} \dfrac{\tan 2\theta}{\sin \theta}$

66 $\lim\limits_{v \to \pi} \dfrac{1 + \cos v}{(\pi - v)^2}$

In Problems 67 to 70, find

$$\lim_{h \to 0} \frac{f(x + h) - f(x)}{h}$$

for the given function f. (In calculating the limit as $h \to 0$, treat x as a constant.)

67 $f(x) = x^2 + 1$

68 $f(x) = \dfrac{1}{\sqrt{x}}$

69 $f(x) = \sqrt{x}$

70 $f(x) = \dfrac{1}{x}$

71 Complete the proof of Theorem 1 by showing that $\lim\limits_{x \to a} \cos x = \cos a$.

72 Use mathematical induction to prove that Property 8 follows from Property 6.

73 Prove that Property 3 follows from Properties 6 and 1.

74 Prove that Property 11 follows from Property 12.

75 Prove that Property 5 follows from Properties 4 and 3.

76 Suppose $\lim_{x \to a} f(x) = 0$ and that there exists a constant B such that $|g(x)| \leq B$ for all values of x in some open interval containing a, except possibly for $x = a$. Use the squeezing property (Property 12) to prove that $\lim_{x \to a} [f(x)g(x)] = 0$.

77 In economics, the *profit function* P gives the profit $P(x)$ dollars to the manufacturer if x units of a commodity are produced and sold. The *marginal profit* at production level x is defined by

$$P'(x) = \lim_{h \to 0} \frac{P(x + h) - P(x)}{h}$$

where, in taking the limit as $h \to 0$, x is to be held constant. Suppose that the profit in dollars from the manufacture and sale of x quartz travel alarm clocks is

$$P(x) = 20x - 50,000 - \frac{x^2}{1000}$$

Find the marginal profit $P'(x)$.

1.9 Continuous Functions

As we mentioned, there are many functions f for which $\lim_{x \to a} f(x)$ is the same as $f(a)$. For such a well-behaved function, there's no problem in finding $\lim_{x \to a} f(x)$—all you have to do is evaluate $f(a)$. A function that has this useful property is said to be *continuous at the number a*.

DEFINITION 1

Continuous Function at a Number

A function f is **continuous at the number** a if the following conditions hold:

(i) $f(a)$ is defined

(ii) $\lim_{x \to a} f(x)$ exists

(iii) $\lim_{x \to a} f(x) = f(a)$

Figure 1

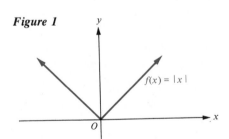
$f(x) = |x|$

EXAMPLE 1 Is the absolute-value function $f(x) = |x|$ (Figure 1) continuous at 0?

SOLUTION

(i) Here, $f(0) = |0| = 0$, so $f(0)$ is defined.

(ii) By Property 10 in Section 1.8,

$$\lim_{x \to 0} f(x) = \lim_{x \to 0} |x| = |0| = 0$$

so $\lim_{x \to 0} f(x)$ exists.

(iii) By (i) and (ii) above,

$$\lim_{x \to 0} f(x) = 0 = f(0)$$

so f is continuous at 0.

EXAMPLE 2 Show that the absolute-value function $f(x) = |x|$ is actually continuous at every number a.

SOLUTION   Here, $f(a) = |a|$ is defined. Also, just as in Example 1,

$$\lim_{x \to a} f(x) = \lim_{x \to a} |x| = |a| = f(a)$$

so f is continuous at every number a.

If any of conditions (i), (ii), or (iii) of Definition 1 fail, we say that f is **discontinuous** at the number a. Condition (iii),

$$\lim_{x \to a} f(x) = f(a)$$

means that $f(x)$ comes closer and closer to $f(a)$ as x comes closer and closer to a. This condition will fail if the function value takes a sudden "jump" when $x = a$.

EXAMPLE 3   Is the function

$$f(x) = \begin{cases} \dfrac{2x^2 + 3x + 1}{x + 1} & \text{if } x \neq -1 \\[2mm] 3 & \text{if } x = -1 \end{cases}$$

continuous at -1?

Figure 2

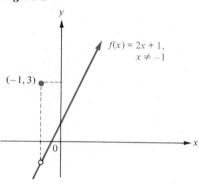

SOLUTION   Here, $f(-1) = 3$, so $f(-1)$ is defined. For $x \neq -1$,

$$f(x) = \frac{2x^2 + 3x + 1}{x + 1} = \frac{(2x + 1)(x + 1)}{x + 1} = 2x + 1$$

Hence, by Property 11 of Section 1.8,

$$\lim_{x \to -1} f(x) = \lim_{x \to -1} (2x + 1) = -1$$

so $\lim_{x \to -1} f(x)$ exists. However, because $\lim_{x \to -1} f(x) = -1$ and $f(-1) = 3$, we have

$$\lim_{x \to -1} f(x) \neq f(-1)$$

and it follows that f is discontinuous at -1. The graph of f (Figure 2) is the line $y = 2x + 1$ everywhere except at $x = -1$, where a "hole" appears. The defined value of f at -1 produces a point $(-1, 3)$ on the graph of f but above this hole. ∎

Figure 3

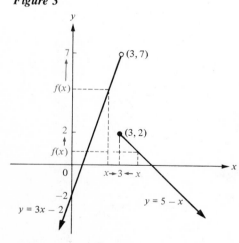

Another example of a discontinuous function is provided by

$$f(x) = \begin{cases} 3x - 2 & \text{if } x < 3 \\ 5 - x & \text{if } x \geq 3 \end{cases}$$

The graph of f (Figure 3) visually indicates a discontinuity at the number $x = 3$. In order to study this apparent discontinuity, it is convenient to introduce the idea of a **one-sided limit,** that is, a limit of $f(x)$ as x approaches 3 through values entirely on one side of 3. Figure 3 clearly shows that the numerical value $f(x)$ approaches 7 as x approaches 3 through values that are always less than 3. We write this fact symbolically as

$$\lim_{x \to 3^-} f(x) = 7$$

the condition that x approaches 3 **from the left** being denoted by $x \to 3^-$.

Likewise, the condition that x approaches 3 through values that are always greater than 3, so that x approaches 3 **from the right,** is denoted by $x \to 3^+$. From Figure 3 we see that

$$\lim_{x \to 3^+} f(x) = 2$$

Because the two one-sided limits of $f(x)$ as $x \to 3^-$ and as $x \to 3^+$ are not the same, it follows that $\lim_{x \to 3} f(x)$ cannot exist. Therefore (as we have already guessed from our inspection of Figure 3), f is discontinuous at the number 3.

A formal definition of one-sided limits can be patterned after Definition 1 in Section 1.7 (Problem 44). The basic idea is simple. Indeed, $\lim_{x \to a^-} f(x) = L$ just means that we can make $|f(x) - L|$ as small as we please by taking x close enough to a but *less than a.* Likewise, $\lim_{x \to a^+} f(x) = R$ means that we can make $|f(x) - R|$ as small as we please by taking x close enough to a but *greater than a.* If $L \neq R$ (as in Figure 3), then $\lim_{x \to a} f(x)$ *cannot exist.* More generally, we have the following theorem.

THEOREM 1

Limits and One-Sided Limits

> The limit $\lim_{x \to a} f(x)$ exists and equals L if and only if both of the one-sided limits
>
> $\lim_{x \to a^-} f(x)$ and $\lim_{x \to a^+} f(x)$ exist and have the common value L.

We leave the proof of this theorem as an exercise (Problems 47 to 50).

One-sided limits satisfy appropriate analogs of the properties in Section 1.8. For instance, if $\lim_{x \to a^+} f(x) = L$ and $\lim_{x \to a^+} g(x) = M$, then

$$\lim_{x \to a^+} [f(x) + g(x)] = L + M$$

$$\lim_{x \to a^+} [f(x) \cdot g(x)] = LM$$

and so forth.

*In Examples 4 and 5, (**a**) sketch the graph of the function, (**b**) find the one-sided limits of the function as $x \to a^-$ and as $x \to a^+$, (**c**) determine the limit of the function as $x \to a$ (if this limit exists), and (**d**) decide whether the function is continuous at the number a.*

EXAMPLE 4 $g(x) = \begin{cases} 3 - x^2 & \text{if } x \leq 1 \\ 1 + x^2 & \text{if } x > 1 \end{cases} ; \, a = 1$

SOLUTION

(**a**) A sketch of the graph is shown in Figure 4.

(**b**) $\lim_{x \to 1^-} g(x) = \lim_{x \to 1^-} (3 - x^2) = 2$

$\lim_{x \to 1^+} g(x) = \lim_{x \to 1^+} (1 + x^2) = 2$

(**c**) Because both of the one-sided limits exist and have the same value, 2, it follows that $\lim_{x \to 1} g(x) = 2$.

Figure 4

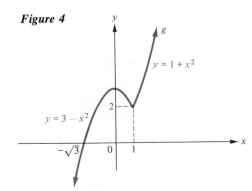

$y = 1 + x^2$

$y = 3 - x^2$

(d) Here, $g(1) = 3 - 1^2 = 2$, so g is defined at 1. Also, $\lim\limits_{x \to 1} g(x)$ exists and

$$\lim_{x \to 1} g(x) = 2 = g(1)$$

hence, g is continuous at 1.   ▪

$$\underline{\text{EXAMPLE 5}} \quad f(x) = \begin{cases} |x - 2| & \text{if } x \neq 2 \\ 1 & \text{if } x = 2 \end{cases}; a = 2$$

SOLUTION

(a) A sketch of the graph of f is shown in Figure 5.

Figure 5

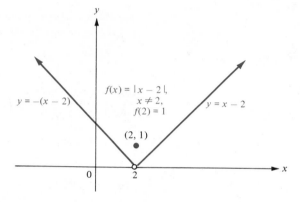

(b) By Property 11 of Section 1.8,

$$\lim_{x \to 2} f(x) = \lim_{x \to 2} |x - 2|$$

hence, by Property 10 of Section 1.8,

$$\lim_{x \to 2} f(x) = \left| \lim_{x \to 2} (x - 2) \right| = |0| = 0$$

It follows from Theorem 1 that

$$\lim_{x \to 2^-} f(x) = \lim_{x \to 2^+} f(x) = \lim_{x \to 2} f(x) = 0$$

(c) We have already seen in part (b) that $\lim\limits_{x \to 2} f(x) = 0$.

(d) $f(2) = 1$, $\lim\limits_{x \to 2} f(x) = 0 \neq 1$; hence, f is discontinuous at 2.   ▪

Properties of Continuous Functions

Suppose that f and g are two functions that are continuous at the number a. Then $f(a)$ and $g(a)$ are both defined, so that $(f + g)(a) = f(a) + g(a)$ is defined. Moreover, by Property 4 of Section 1.8,

$$\lim_{x \to a} (f + g)(x) = \lim_{x \to a} [f(x) + g(x)] = \lim_{x \to a} f(x) + \lim_{x \to a} g(x)$$

$$= f(a) + g(a) = (f + g)(a)$$

We conclude that $f + g$ is continuous at a. Similar reasoning yields analogous results for the difference, product, and quotient (see Appendix B for some of the details), and we have the following.

Basic Algebraic Properties of Continuous Functions

> **1** If f and g are continuous at a, then so are $f + g$, $f - g$, and $f \cdot g$.
>
> **2** If f and g are continuous at a and $g(a) \neq 0$, then f/g is continuous at a.
>
> **3** A polynomial function is continuous at every number.
>
> **4** A rational function is continuous at every number for which it is defined.

In Examples 6 and 7, use the properties of continuous functions to determine the numbers at which the given function is continuous.

EXAMPLE 6 $f(x) = |x| + x$

SOLUTION   Notice that $f = g + h$, where $g(x) = |x|$ and $h(x) = x$. In Example 2, we saw that the absolute-value function $g(x) = |x|$ is continuous at every number. By Property 3, the polynomial function $h(x) = x$ is also continuous at every number. Hence, by Property 1, $f = g + h$ is continuous at every number. The graph of $f(x) = |x| + x$ shows no holes or sudden jumps (Figure 6). ∎

EXAMPLE 7 $f(x) = \dfrac{2}{x - 1}$

SOLUTION   Here, f is a rational function that is defined at every number x except $x = 1$. By Property 4, f is continuous at every number x, except for $x = 1$. The graph of f (Figure 7) clearly shows the discontinuity at $x = 1$. ∎

The following property of continuous functions is related to the substitution property for limits (Property 15 in Section 1.8).

The Substitution Property for Continuous Functions*

> **5** If g is continuous at a and f is continuous at $g(a)$, then the function h defined by $h(x) = f(g(x))$ is continuous at a.

EXAMPLE 8   Show that the function $h(x) = |x|^3$ is continuous at every number a.

SOLUTION   If we let $g(x) = |x|$ and $f(x) = x^3$, we see that

$$h(x) = |x|^3 = f(|x|) = f(g(x))$$

We have already shown that the absolute-value function $g(x) = |x|$ is continuous at every number a. Because $f(x) = x^3$ is a polynomial function, it is continuous at the number $g(a)$ by Property 3. Hence, h is continuous at every number a by the substitution property (Property 5). The graph of $h(x) = |x|^3$ shows no holes or sudden jumps (Figure 8). ∎

*See page 123 for a restatement of this property in terms of the idea of function composition.

Figure 6
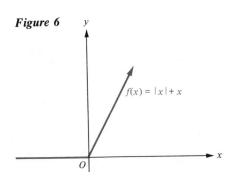

Figure 7
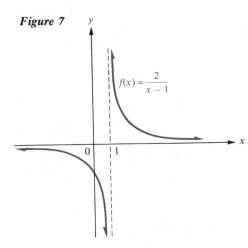

Figure 8
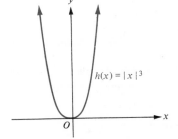

The following continuity properties of the trigonometric functions follow from Theorem 1 in Section 1.8 (Problem 46).

Continuity Properties of the Trigonometric Functions

> **6** The sine and cosine functions are continuous at every number.
>
> **7** The tangent, cotangent, secant, and cosecant functions are continuous at every number for which they are defined.

In Examples 9 and 10, use the properties of continuous functions to determine the numbers at which the given function is continuous.

<u>**EXAMPLE 9**</u> $f(x) = \tan x$

SOLUTION   Because

$$f(x) = \tan x = \frac{\sin x}{\cos x}$$

$f(x)$ is undefined when $\cos x = 0$; that is, when

$$x = \pm\frac{\pi}{2},\ \pm 3 \cdot \frac{\pi}{2},\ \pm 5 \cdot \frac{\pi}{2}$$

and so on. Therefore, by Property 7, f is continuous at all real numbers except for odd integer multiples of $\pi/2$. The graph in Figure 9 clearly shows the periodic discontinuities of the tangent function.   ∎

Figure 9

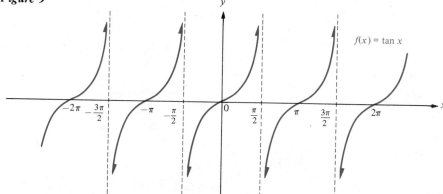

<u>**EXAMPLE 10**</u> $F(t) = 2 \cos\left(\pi t - \frac{\pi}{6}\right)$

SOLUTION   The polynomial function

$$g(t) = \pi t - \frac{\pi}{6}$$

is continuous at every number by Property 3, and the cosine function is continuous at every number by Property 6; hence, by the substitution property (Property 5), the function

$$h(t) = \cos\left(\pi t - \frac{\pi}{6}\right) = \cos g(t)$$

is also continuous at every number. The constant function

$$f(t) = 2$$

is continuous at every number by Property 3; hence, by Property 1,

$$F(t) = 2 \cos\left(\pi t - \frac{\pi}{6}\right) = f(t) \cdot h(t)$$

is continuous at every number. (See Figure 10 for the graph of F.)

Figure 10

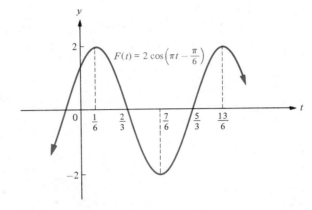

Continuity on an Interval

To say that a function f is **continuous on an open interval** means, by definition, that f is continuous at every number in the interval. For instance, the function

$$f(x) = \sqrt{9 - x^2}$$

is continuous on the open interval $(-3, 3)$ (Figure 11).

Similarly, to say that a function f is **continuous on a closed interval $[a, b]$** means, by definition, that f is continuous on the open interval (a, b) and that f satisfies the following "one-sided" continuity conditions at the endpoints a and b:

$$\lim_{x \to a^+} f(x) = f(a) \qquad \text{and} \qquad \lim_{x \to b^-} f(x) = f(b)$$

For instance, the function $f(x) = \sqrt{9 - x^2}$ is actually continuous on the closed interval $[-3, 3]$ (Figure 11).

Figure 11

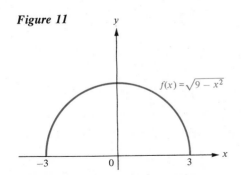

<u>EXAMPLE 11</u>   Determine whether the rational function

$$f(x) = \frac{x + 7}{x^2 - 36}$$

is continuous on the following intervals: $(-\infty, -6)$, $[-6, 6]$, $(-6, 6)$, and $(6, \infty)$.

SOLUTION   Here, $f(x)$ is defined for all real numbers x except for the values $x = -6$ and $x = 6$ (which make the denominator zero). Thus, the rational function f is continuous on $(-\infty, -6)$, $(-6, 6)$, and $(6, \infty)$ by Property 4. Since -6 and 6 do not belong to the domain of f, it follows that f is not continuous on the closed interval $[-6, 6]$.

Problem Set 1.9

In Problems 1 to 14, (a) sketch a graph of the function; (b) find the one-sided limits of the function as x approaches a from the right and from the left; (c) determine the limit of the function as x approaches a (if this limit exists); and (d) using the definition of continuity, decide whether the function is continuous at the number a.

1 $f(x) = \begin{cases} 5 + x & \text{if } x \leq 3 \\ 9 - x & \text{if } x > 3 \end{cases}; a = 3$

2 $F(x) = \begin{cases} -1 & \text{if } x < 0 \\ 0 & \text{if } x = 0 \\ 1 & \text{if } x > 0 \end{cases}; a = 0$

3 $g(x) = \begin{cases} 3 + x & \text{if } x \leq 1 \\ 3 - x & \text{if } x > 1 \end{cases}; a = 1$

4 $G(x) = \begin{cases} 2x - 1 & \text{if } x < 1 \\ x^2 & \text{if } x \geq 1 \end{cases}; a = 1$

5 $H(x) = \begin{cases} |x - 5| & \text{if } x \neq 5 \\ 2 & \text{if } x = 5 \end{cases}; a = 5$

6 $H(x) = \begin{cases} \dfrac{x - 2}{|x - 2|} & \text{if } x \neq 2 \\ 1 & \text{if } x = 2 \end{cases}; a = 2$

7 $f(x) = \begin{cases} 2 - x & \text{if } x > 1 \\ x^2 & \text{if } x \leq 1 \end{cases}; a = 1$

8 $Q(x) = \begin{cases} \dfrac{1}{x - 2} & \text{if } x \neq 2 \\ 0 & \text{if } x = 2 \end{cases}; a = 2$

9 $F(x) = \begin{cases} \dfrac{x^2 - 9}{x - 3} & \text{if } x \neq 3 \\ 2 & \text{if } x = 3 \end{cases}; a = 3$

10 $R(x) = \begin{cases} 3 + x^2 & \text{if } x < -2 \\ 0 & \text{if } x = -2 \\ 11 - x^2 & \text{if } x > -2 \end{cases}; a = -2$

11 $S(x) = 5 + |6x - 3|, a = \frac{1}{2}$

12 $g(x) = [\![x]\!] + [\![5 - x]\!], a = 4$

13 $f(x) = \dfrac{x^2 - 2x - 3}{x + 1}, a = -1$

14 $T(x) = [\![1 - x]\!] + [\![x - 1]\!], a = 1$

In Problems 15 to 34, use the properties of continuous functions to determine the numbers at which the function is continuous.

15 $f(x) = 2|x|$

16 $g(x) = |1 - x|$

17 $h(x) = x - 2|x|$

18 $F(x) = \dfrac{x + 1}{x - 1}$

19 $G(x) = \dfrac{1}{x}$

20 $f(x) = \left| \dfrac{1}{x} \right|$

21 $g(x) = \dfrac{x^2 - 4x + 3}{x - 1}$

22 $h(x) = \begin{cases} x^3 & \text{if } x \leq 0 \\ x^2 & \text{if } x > 0 \end{cases}$

23 $F(x) = \dfrac{x^2 - 4x + 3}{x - 1} - \dfrac{x^2 - 2x - 3}{x + 1} + 1$

24 $G(x) = \dfrac{1}{|x| + 1}$

25 $H(x) = \left| \dfrac{3}{x - 2} + 4 \right|$

26 $T(x) = \sqrt{\dfrac{1}{x^2 + 1}}$

27 $f(x) = \cot x$

28 $f(x) = \sec \dfrac{x}{2}$

29 $g(x) = \csc x$

30 $h(t) = \tan |t|$

31 $f(t) = \dfrac{1 - \sin t}{\cos t}$

32 $f(x) = \begin{cases} \dfrac{\sin 2x}{x} & \text{if } x \neq 0 \\ 1 & \text{if } x = 0 \end{cases}$

33 $h(x) = \begin{cases} \tan x & \text{if } x \leq \pi/4 \\ \sqrt{2} \sin x & \text{if } x > \pi/4 \end{cases}$

34 $f(x) = \begin{cases} \sin \dfrac{1}{x} & \text{if } x \neq 0 \\ 0 & \text{if } x = 0 \end{cases}$

In Problems 35 to 40, determine whether each function is continuous or discontinuous on each interval.

35 $f(x) = \sqrt{4 - x^2}$ on $[-2, 2]$, $[2, 3]$, $(-2, 2)$, and $(-1, 5)$

36 $g(x) = \dfrac{3}{x + 1}$ on $(-\infty, 1)$, $(-3, -1)$, $(-\infty, -1)$, $(-1, \infty)$, $[-1, \infty)$, and $[-2, 2]$

37 $F(x) = \dfrac{x + 6}{x^2 - 36}$ on $(-\infty, 6]$, $(-\infty, -4]$, $(-6, \infty)$, $[-6, 9]$, and $[-7, \infty)$

38 $f(x) = \dfrac{|x - 5|}{x - 5}$ on $[-1, 1]$, $(-1, 1)$, $(-5, \infty)$, $(-\infty, 5]$, and $[-8, 6]$

39 $G(x) = \dfrac{4x - 3}{16x^2 - 9}$ on $[-\frac{3}{4}, 0]$, $[-\frac{1}{2}, 0]$, $(-\frac{3}{4}, \infty)$, $[-2, \infty)$, and $(-1, -\frac{3}{4})$

40 $f(x) = \cot x$ on $\left(-\dfrac{\pi}{2}, \dfrac{\pi}{2}\right), \left[-\dfrac{\pi}{2}, \dfrac{\pi}{2}\right], (0, \pi)$, and $[0, \pi]$

41 The weight of an object is given by

$$w(x) = \begin{cases} ax & \text{if } x \le R \\ \dfrac{b}{x^2} & \text{if } x > R \end{cases}$$

where x is the distance of the object from the center of the earth, R is the radius of the earth, and a and b are constants. What relation must exist between these constants if w is to be a continuous function? Assuming that w is continuous, sketch its graph.

42 If a hollow sphere of radius a is charged with one unit of static electricity, the field intensity E at a point P depends as follows on the distance x from the center of the sphere to P:

$$E(x) = \begin{cases} 0 & \text{if } 0 \le x < a \\ \dfrac{1}{2a^2} & \text{if } x = a \\ \dfrac{1}{x^2} & \text{if } x > a \end{cases}$$

(a) Sketch a graph of E. (b) Discuss the continuity of E.

43 If g is defined by the equation

$$g(x) = \begin{cases} \dfrac{x^3 - 1}{x - 1} & \text{if } x \ne 1 \\ a & \text{if } x = 1 \end{cases}$$

determine the value of a so that g will be continuous at 1.

44 Give a formal definition of (a) $\lim\limits_{x \to a^-} f(x) = L$ and (b) $\lim\limits_{x \to a^+} f(x) = R$ patterned after Definition 1 in Section 1.7. (*Hint:* Note, for instance, that $0 < |x - a| < \delta$ and $x < a$ if and only if $-\delta < x - a < 0$.)

45 Find $\lim\limits_{x \to 2^+} \sqrt{x - 2}$ and explain why $\lim\limits_{x \to 2} \sqrt{x - 2}$ does not exist.

46 Use Theorem 1 in Section 1.8 and Properties 2 and 3 of continuous functions to prove the continuity Properties 6 and 7 of the trigonometric functions.

47 Suppose that $\lim\limits_{x \to a} f(x) = L$. Give an informal argument to explain why it should follow that $\lim\limits_{x \to a^+} f(x) = L$.

48 Suppose that $\lim\limits_{x \to a} f(x) = L$. Use the formal definition in Problem 44 to prove that $\lim\limits_{x \to a^+} f(x) = L$.

49 Suppose that $\lim\limits_{x \to a^+} f(x) = \lim\limits_{x \to a^-} f(x) = L$. Give an informal argument to explain why it should follow that $\lim\limits_{x \to a} f(x) = L$.

50 Suppose that $\lim\limits_{x \to a^+} f(x) = \lim\limits_{x \to a^-} f(x) = L$. Use the formal definitions to prove that $\lim\limits_{x \to a} f(x) = L$.

51 State the analog of Property 11 of Section 1.8 for limits from the right.

52 State the analog of Property 12 of Section 1.8 for limits from the left.

Review Problem Set, Chapter 1

1 Assume that a and b are real numbers such that $a < b$. Which of the following are necessarily true and which could be false?
(a) $a < 5b$ (b) $-3a < -3b$
(c) $5a < -(-5b)$ (d) $a + 5 < b + 5$
(e) $1/a > 1/b$ (f) $a - 7 > b - 7$

In Problems 2 to 13, solve the inequality and illustrate the solution on the number line.

2 $x - 5 \le 7$

3 $3x + 2 > 8$

4 $3x - 2 \ge 1 + 2x$

5 $5 \ge 8x - 3 \ge -6$

6 $\dfrac{x - 1}{3} \ge 2 + \dfrac{x}{2}$

7 $x^2 + x - 20 < 0$

8 $x^2 - 6x - 7 \le 0$

9 $2x^2 + 9 < 5$

10 $\dfrac{x - 2}{x + 3} > 0$

11 $\dfrac{2x - 1}{x - 6} < 0$

12 $\dfrac{5x - 1}{x - 2} \le 1$

13 $\dfrac{x - 4}{x + 2} \ge 3$

14 State the conditions under which each inequality is true.
(a) $\dfrac{1}{x} < \dfrac{1}{10}$ (b) $\dfrac{1}{x} < \dfrac{1}{100}$ (c) $\dfrac{1}{x^2 - 1} < \dfrac{1}{1000}$
(d) $-x < x$ (e) $-x^2 < x^2$

15 Give two different examples of real numbers a, b, c, and d with $a < b$ and $c < d$ but $ac > bd$.

16 For positive real numbers a and b, show that

$$\dfrac{2ab}{a + b} \le \sqrt{ab} \le \dfrac{a + b}{2}$$

17 A student on academic probation must earn a C in his calculus course to avoid being expelled from college. Since he is lazy and has no interest in earning a grade any higher than necessary, he wants his final numerical grade for calculus to lie in the interval [70, 80). This grade will be determined by taking three-fifths of his classroom average, which is 68, plus two-fifths of his score on the final exam. Since he has always had trouble with inequalities, he can't figure out the range in which his score on the final exam should fall to allow him to stay in school. Can you help him?

C **18** A runner leaves a starting point and runs along a straight road at a steady speed of 8.8 miles per hour. After a while, she turns around and begins to run back to the starting point. How fast must she run on the way back so that her average speed for the entire run will be greater than 8 miles per hour but no greater than 8.5 miles per hour? (*Caution:* The average speed for her entire run *cannot* be found by adding 8.8 miles per hour to her return speed and dividing by 2.)

19 Find the distance between the numbers in each pair and illustrate your answer on a number line.

    (a) -3 and 4

    (b) 0 and -5

    C (c) -2.735 and $-\pi$

    (d) -3.2 and 4.1

    (e) $-\frac{2}{3}$ and $\frac{5}{2}$

    C (f) 1.42 and $\sqrt{2}$

20 Use the absolute-value notation to write an expression for the distance between the two given numbers. Simplify your answer.

    (a) -6 and 5
    (b) $\dfrac{1}{x}$ and $\dfrac{1}{x + 1}$

    (c) $\dfrac{3}{13}$ and $\dfrac{4}{17}$
    (d) $\dfrac{x}{x + 1}$ and $\dfrac{x - 1}{x}$

In Problems 21 to 26, solve the absolute-value equation.

21 $|x + 1| = 3$
    22 $|2x - 3| = 5$

23 $|2y + 1| = 5$
    24 $|2t + 3| = |t + 2|$

25 $|2u^2 - u - 2| = 0$
    26 $|5 - 3z| = 2z$

In Problems 27 to 34, solve the absolute-value inequality and show your solution on a number line.

27 $|2x + 5| \le 6$
    28 $|3x + 4| \le 2$

29 $|1 - x| < 0.01$
    30 $|1 - 4x| \le x$

31 $|7x - 6| > x$
    32 $\dfrac{1}{|x - 1|} \ge 3$

33 $\dfrac{1}{|2x + 3|} \le \dfrac{1}{4}$
    34 $\dfrac{1}{|3x - 1|} \ge 5$

35 Determine which equations or inequalities are true for all values of the variables.

    (a) $|x^2 - 4| = |x - 2||x + 2|$
    (b) $|x^2 - 4| = x^2 + 4$

    (c) $|-x|^2 = x^2$
    (d) $|x - y| = |y - x|$

    (e) $|x^2 + 3x| \le x^2 + |3x|$
    (f) $|x - y| \le |x| + |y|$

36 It is predicted that the number N of cars to be produced by an automobile industry in the coming year will satisfy the inequality $|N - 4,000,000| < 500,000$. Describe the anticipated production as an interval of real numbers.

In Problems 37 to 42, plot each pair of points and find the distance between them by using the distance formula.

37 $(1, 1)$ and $(4, 5)$

38 $(-1, 2)$ and $(5, -7)$

39 $(-3, 2)$ and $(2, 14)$

C **40** $(4.71, -3.22)$ and $(0, \pi)$

41 $(-2, -5)$ and $(-2, 3)$

C **42** $(\sqrt{\pi}, \pi)$ and $\left(\dfrac{1 + \sqrt{2}}{2}, \dfrac{31}{7}\right)$

C **43** What is the perimeter of the triangle with vertices $(-1, 5)$, $(8, -7)$, and $(4, 1)$? Round off your answer to two decimal places.

44 Are the points $(-5, 4)$, $(7, -11)$, $(12, -11)$, and $(0, 4)$ the vertices of a parallelogram? Justify your answer.

In Problems 45 to 48, find the radius r and the coordinates (h, k) of the center of each circle, and sketch a graph of the circle.

45 $(x - 2)^2 + (y + 3)^2 = 25$

46 $x^2 + 2x + y^2 + 2y + 1 = 0$

47 $x^2 + y^2 - 3x + 4y + 4 = 0$

48 $4x^2 + 4y^2 + 4x - 4y + 1 = 0$

In Problems 49 to 52, find the slope of the line segment $\overline{AB}$ and an equation in point-slope form of the line containing A and B.

49 $A = (3, -5)$, $B = (2, 2)$

50 $A = (0, 7)$, $B = (5, 0)$

51 $A = (1, 2)$, $B = (-3, -4)$

52 $A = (\frac{3}{2}, \frac{2}{3})$, $B = (\frac{1}{6}, -\frac{5}{6})$

In Problems 53 and 54, sketch the line L that contains the point P and has slope m, and find an equation in point-slope form for L.

53 $P = (5, 2)$, $m = -\frac{3}{5}$

54 $P = (-\frac{2}{3}, \frac{1}{2})$, $m = \frac{3}{2}$

In Problems 55 and 56, (a) find an equation of the line that contains the point P and is parallel to the line segment $\overline{AB}$, and (b) find an equation of the line that contains the point P and is perpendicular to line segment $\overline{AB}$.

55 $P = (7, -5)$, $A = (1, 8)$, $B = (-3, 2)$

56 $P = (\frac{2}{5}, \frac{1}{3})$, $A = (\frac{7}{3}, -\frac{3}{5})$, $B = (1, \frac{2}{5})$

In Problems 57 and 58, rewrite each equation in slope-intercept form, find the slope m and the y intercept b of the graph, and sketch the graph.

57 $4x - 3y + 2 = 0$

58 $\frac{2}{3}x - \frac{1}{5}y + 3 = 0$

In Problems 59 to 62, find an equation of the line L in (a) point-slope form, (b) slope-intercept form, and (c) general form.

59 L contains the point $(-7, 1)$ and has slope $m = 3$.

60 L contains the points $(2, 5)$ and $(1, -3)$.

61 L contains the point $(1, -2)$ and is parallel to the line $7x - 3y + 2 = 0$.

62 L contains the point $(3, -4)$ and is perpendicular to the line $2x - 5y + 4 = 0$.

63 If (a, b) is a point on the circle $(x - h)^2 + (y - k)^2 = r^2$, find an equation of the line that is tangent to the circle at (a, b). (*Hint:* The tangent line to a circle at a point is perpendicular to the radius drawn from the center to the point.)

64 An advertising agency claims that a furniture store's revenue will increase by $20 per month for each additional dollar spent on advertising. The current average monthly sales revenue is $140,000 with an expenditure of $100 per month for advertising. Find the equation that relates the store's expected average monthly sales revenue y to the total expenditure x for advertising. Find the value of y if $x = 400.

In Problems 65 to 76, let $f(x) = 3x^2 - 4$, $g(x) = 6 - 5x$, and $h(x) = 1/x$. Find the indicated values.

65 $f(-3)$

66 $h(\frac{1}{2})$

67 $g(\frac{6}{5})$

68 $h[h(x)]$

69 $f(x) - f(2)$

70 $f(x + k) - f(x)$

71 $f[g(x)]$

72 $g\left(\dfrac{1}{4 + k}\right)$

73 $g(x) + g(-x)$

74 $\sqrt{f(-|x|)}$

75 $\dfrac{h(x + k) - h(x)}{k}$

76 $\dfrac{1}{h(4 + k)}$

In Problems 77 to 82, find the domain of each function.

77 $f(x) = \dfrac{1}{x - 1}$

78 $g(x) = \dfrac{1}{\sqrt{4 - x^2}}$

79 $h(x) = \sqrt{1 + x}$

80 $F(x) = \dfrac{|x|}{1 - x}$

81 $G(x) = (x^2 - 1)^{3/2}$

82 $H(x) = \dfrac{3}{|x| - x}$

83 Which of the graphs in Figure 1 are graphs of functions?

Figure 1

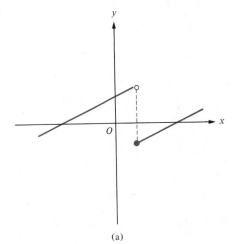

(a)

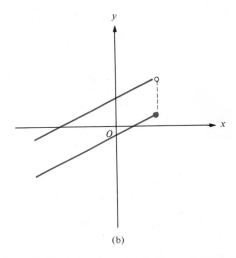

(b)

84 Which of the graphs in Figure 2 are graphs of functions?

Figure 2

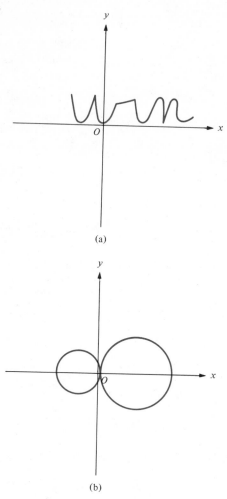

(a)

(b)

In Problems 85 to 90, find the domain of the function, sketch its graph, and find the range of the function. ©️ If you wish, use a calculator to help sketch more accurate graphs.

85 $f(x) = 5x - 3$

86 $g(x) = 3 - \dfrac{x}{5}$

87 $F(x) = 2\sqrt{x - 2}$

88 $G(x) = x^{1/3}$

89 $h(x) = \begin{cases} x^2 & \text{if } x > 0 \\ -x^2 & \text{if } x \le 0 \end{cases}$

90 $H(x) = \begin{cases} x & \text{if } x < 0 \\ 2x & \text{if } 0 \le x \le 1 \\ 3x^3 - 1 & \text{if } x > 1 \end{cases}$

©️ **91** The power delivered by a wind-powered generator is given by

$$P(x) = kx^3 \text{ horsepower}$$

where x is the speed of the wind in miles per hour and k is a constant depending on the size of the generator and its effi-

ciency. For a certain wind-powered generator, the constant $k = 3.38 \times 10^{-4}$. Sketch the graph of the function P for this generator, and determine how many horsepower are generated when the wind speed is 35 miles per hour.

92 In economics, it is often assumed that the demand for a commodity is a function of selling price, that is, $q = f(p)$, where p dollars is the selling price per unit of the commodity and q is the number of units that will sell at that price. (a) If p increases, would you expect q to increase or decrease? What does this mean about the graph of f? (b) Write an equation for the total amount of money $F(p)$ in dollars spent by consumers for the commodity if the selling price per unit is p dollars. (c) What would it mean to say that there is a value p_0 dollars for which $f(p_0) = 0$?

93 In economics, it is often assumed that the number of units s of a commodity that producers will supply to the marketplace is a function of the selling price p dollars per unit, that is, $s = g(p)$. (a) If p increases, would you expect s to increase or decrease? What does this mean about the graph of g? (b) Write an equation for the total amount of money (in dollars) $G(p)$ spent by consumers for the commodity if the selling price per unit is p dollars and all supplied units are purchased. (c) What would it mean to say that there is a value p_1 dollars for which $g(p_1) = 0$?

94 A manufacturer finds that 100,000 electronic calculators are sold per month at a price of \$50 each, but that only 60,000 are sold per month if the price is \$75 each. Suppose that the graph of the demand function f for these calculators is a straight line in the pq plane (see Problem 92). (a) Find a formula for $f(p)$, where p dollars is the selling price per calculator and $q = f(p)$ is the number per month that will sell at that price. (b) Find the selling price per calculator if the monthly demand is 80,000 calculators. (c) What price would be so high that no calculators would be sold?

In Problems 95 to 100, determine without drawing a graph whether the function is even, odd, or neither, and discuss any symmetry of the graph.

95 $f(x) = 5x^5 + 3x^3 + x$

96 $g(x) = (x^4 + x^2 + 1)^{-1}$

97 $h(x) = (x + 1)x^{-1}$

98 $F(x) = -x^3|x|$

99 $G(x) = x^{80} - 5x^6 + 9$

100 $H(x) = \dfrac{\sqrt{x}}{1 + x}$

In Problems 101 to 110, find

(a) $(f + g)(x)$ (b) $(f - g)(x)$ (c) $(f \cdot g)(x)$ (d) $(f/g)(x)$

101 $f(x) = x + 2,\ g(x) = 3x - 4$

102 $f(x) = x^2 + 2x,\ g(x) = x^2 - 2x$

103 $f(x) = \dfrac{1}{x - 1},\ g(x) = \dfrac{1}{x + 1}$

104 $f(x) = \dfrac{x+3}{x-2}$, $g(x) = \dfrac{x}{x-2}$

105 $f(x) = x^4$, $g(x) = \sqrt{x+1}$

106 $f(x) = x$, $g(x) = |x-2| - x$

107 $f(x) = |x|$, $g(x) = -x$

108 $f(x) = \sqrt{1+x^2}$, $g(x) = \pi|x|$

109 $f(x) = x^{2/3} + 1$, $g(x) = \sqrt{x}$

110 $f(x) = \dfrac{|x|}{x}$, $g(x) = \dfrac{-x}{|x|}$

In Problems 111 to 122, find the vertex of the graph of the quadratic function, determine whether the graph opens upward or downward, sketch the graph, and find the domain and range of the function.

111 $f(x) = 4x^2$

112 $g(x) = \frac{1}{4}x^2$

113 $h(x) = -\frac{1}{4}x^2$

114 $F(x) = 3x^2 + 2$

115 $G(x) = 3(x-2)^2 + 1$

116 $H(x) = \frac{1}{3}[(x+3)^2 + 2]$

117 $f(x) = x^2 - 3x + 2$

118 $g(x) = 6x^2 + 13x - 5$

119 $h(x) = -6x^2 - 7x + 20$

120 $G(x) = -2x^2 + x + 10$

121 $F(x) = 10x - 25 - x^2$

122 $H(x) = 7x + 2x^2 - 39$

©123 Suppose a manufacturer of sports trophies knows that the total cost C (in dollars) of making x thousand trophies is given by $C = 600 + 60x$, and the corresponding sales revenue R (in dollars) is given by $R = 300x - 4x^2$. Let $P(x)$ be the manufacturer's profit if x thousand trophies are manufactured and sold. (a) Find a formula for $P(x)$. (b) Find $P(10)$ and $P(35)$. (c) Find the largest and smallest values of the production level x, rounded off to the nearest trophy, for which the profit is positive.

124 Suppose that f is a quadratic function and that $f(x_1) = f(x_2) = 0$. Show that the x coordinate of the vertex of the graph of f is $\frac{1}{2}(x_1 + x_2)$.

In Problems 125 to 130, s denotes the length of the arc intercepted on a circle of radius r by a central angle of θ radians. Find the missing quantity.

125 $r = 5$ meters, $\theta = 0.57$ radian, $s = ?$

126 $r = 40$ centimeters, $s = 4$ centimeters, $\theta = ?$

127 $s = 3\pi$ feet, $\theta = \pi$ radians, $r = ?$

128 $r = 13$ kilometers, $\theta = \dfrac{3\pi}{7}$ radians, $s = ?$

129 $r = 2$ meters, $s = \pi$ meters, $\theta = ?$

130 $s = 17\pi$ micrometers, $\theta = \dfrac{5\pi}{6}$ radians, $r = ?$

131 Convert each degree measure to radian measure. Do not use a calculator. Write your answer as a rational multiple of π.

(a) $80°$ (b) $570°$ (c) $-355°$

(d) $-810°$ (e) $-310°$ (f) $765°$

©132 Use a calculator to convert each degree measure to an approximate radian measure, rounded off to four decimal places.

(a) $5°$ (b) $27.7533°$ (c) $-17.173°$ (d) $35.2819°$

133 Convert each radian measure to degree measure. Do not use a calculator.

(a) $\dfrac{2\pi}{5}$ (b) $-\dfrac{13\pi}{4}$ (c) $-\dfrac{7\pi}{8}$

(d) $\dfrac{35\pi}{3}$ (e) $\dfrac{51\pi}{4}$ (f) $\dfrac{18\pi}{5}$

©134 Use a calculator to convert each radian measure to an approximate degree measure, rounded off to four decimal places.

(a) 5 (b) 3.9 (c) -7.63 (d) -21.403

In Problems 135 and 136, find the area $A = \frac{1}{2}r^2\theta$ of a sector of a circle of radius r with central angle θ.

135 $r = 25$ centimeters, $\theta = \pi/6$ **136** $r = 3.5$ meters, $\theta = 60°$

137 The minute hand on a tower clock is 0.6 meter long. How far does the tip of the hand travel in 4 minutes?

138 In Problem 137, what is the area of the sector swept out by the minute hand in 4 minutes?

©139 Find the approximate diameter of the moon if its disk subtends an angle of $0.5°$ at a point on the earth 240,000 miles away (Figure 3). (*Hint:* Approximate the diameter $|\overline{DE}|$ by the length of the arc $\overset{\frown}{BC}$.)

Figure 3

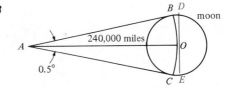

©140 A satellite in a circular orbit above the earth is known to have a speed of 9.92 kilometers per second. In 10 seconds it moves along an arc that subtends an angle of $0.75°$ at the center of the earth. If the radius of the earth is 6371 kilometers, how high is the satellite above the surface of the earth?

©In Problems 141 to 158, use a calculator to find the approximate values of the trigonometric functions.

141 $\sin 27.33°$

142 $\sin 421.25°$

143 $\cos 53.47°$

144 $\cos(-113.81°)$

145 $\tan(-117.2583)$

146 $\tan(-281.5236)$

147 $\sec 16.43°$

148 $\sec(-248.2°)$

149 $\csc \dfrac{4\pi}{5}$

150 $\csc 5.132$

151 $\cot(-3.18)$

152 $\cot(-7.167)$

153 $\cos(-19.213)$

154 $\csc 18.113$

155 $\sin 5.015$

156 $\cot \sqrt{3}$

157 $\cos \dfrac{\sqrt{2}\pi}{4}$

158 $\sin[\tan(-71.32)]$

In Problems 159 to 178, simplify each expression.

159 $\dfrac{\sin(-\theta)}{\cos(-\theta)}$

160 $\dfrac{-\sin(-\alpha)}{-\cos(-\alpha)}$

161 $\csc x - \cos x \cot x$

162 $\sec \theta - \sin \theta \tan \theta$

163 $\csc^2 t \tan^2 t - 1$

164 $(\cot x + 1)^2 - \csc^2 x$

165 $\dfrac{\sec^2 u + 2 \tan u}{1 + \tan u}$

166 $\dfrac{\sec \beta}{\cot \beta + \tan \beta}$

167 $\dfrac{\sin^2\theta + 2\cos^2\theta}{\sin \theta \cos \theta} - 2 \cot \theta$

168 $\dfrac{1}{\csc y - \cot y} - \dfrac{1}{\csc y + \cot y}$

169 $\cos(360° - \theta)$

170 $\tan(2\pi - \beta)$

171 $\sin(270° + \alpha)$

172 $\cos(270° - \phi)$

173 $\sin(2\pi + t)$

174 $\cot\left(\dfrac{3\pi}{2} + x\right)$

175 $\sin 37° \cos 23° + \cos 37° \sin 23°$

176 $\dfrac{\tan(\pi/5) + \tan(\pi/20)}{1 - \tan(\pi/5)\tan(\pi/20)}$

177 $\sin x \cos y - \sin\left(x + \dfrac{\pi}{2}\right)\sin(-y)$

178 $\cos(\pi - t) - \tan t \cos\left(\dfrac{\pi}{2} - t\right)$

179 Use the fact that $7\pi/12 = (\pi/4) + (\pi/3)$ to find the exact numerical value of

(a) $\sin \dfrac{7\pi}{12}$

(b) $\cos \dfrac{7\pi}{12}$

(c) $\tan \dfrac{7\pi}{12}$

180 Assume that α is in quadrant IV, $\cos \alpha = \frac{3}{5}$, β is in quadrant I, $\sin \beta = \frac{8}{17}$, γ is in quadrant II, $\cos \gamma = -\frac{24}{25}$, θ is in quadrant II, and $\sin \theta = \frac{5}{13}$. Find the exact numerical value of each expression.

(a) $\sin(\alpha + \beta)$ (b) $\cos(\gamma + \theta)$ (c) $\sin(\beta + \theta)$

(d) $\sin(\alpha - \gamma)$ (e) $\cos(\beta - \gamma)$ (f) $\sin(\beta - \gamma)$

(g) $\tan(\beta - \gamma)$ (h) $\sec(\beta - \gamma)$ (i) $\sin(\theta - \gamma)$

(j) $\cos(\beta - \theta)$

In Problems 181 to 188, simplify each expression.

181 $\cos^2 2x - \sin^2 2x$

182 $1 - 2\sin^2 \dfrac{t}{2}$

183 $2\sin \dfrac{t}{2} \cos \dfrac{t}{2}$

184 $\cos^4 2\theta - \sin^4 2\theta$

185 $2\sin^2 \dfrac{\theta}{2} + \cos \theta$

186 $\dfrac{\sin 4\pi t}{4 \sin \pi t \cos \pi t}$

187 $\dfrac{\tan \omega t}{1 - \tan^2 \omega t}$

188 $\dfrac{\cos^2(\nu/2) - \cos \nu}{\sin^2(\nu/2)}$

189 If $0 < \theta < \pi/2$ and $x = \sqrt{7}\sin \theta$, find $\cot \theta$ in terms of x.

190 If $0 < t < \pi/2$ and $x = \sqrt{11}\tan t$, find $\cos t$ in terms of x.

Ⓒ **191** A broadcasting antenna tower 200 meters high is to be held vertical by three cables running from a point 10 meters below the top of the tower to concrete anchors sunk in the ground (Figure 4). If the cables are to make angles of 60° with the horizontal, how many meters of cable will be required?

Figure 4

Ⓒ **192** An electronic echo locator on a commercial fishing boat indicates a school of fish at a slant distance of 575 meters from the boat with an angle of depression of 33.60°. What is the depth of the school of fish?

Ⓒ In Problems 193 and 194, use elementary algebra to simplify each expression so that its limit becomes apparent. Then check your results numerically by using a calculator to fill in the blanks in the accompanying table.

193 $\lim\limits_{x \to 2} \dfrac{x^3 - 8}{1 - (2/x)}$

x	1.9	1.99	1.999	1.9999
$\dfrac{x^3 - 8}{1 - (2/x)}$				

x	2.1	2.01	2.001	2.0001
$\dfrac{x^3 - 8}{1 - (2/x)}$				

194 $\lim\limits_{x \to 1} \dfrac{x - 1}{\sqrt{x + 24} - 5}$

x	0.9	0.99	0.999	0.9999
$\dfrac{x - 1}{\sqrt{x + 24} - 5}$				

x	1.1	1.01	1.001	1.0001
$\dfrac{x - 1}{\sqrt{x + 24} - 5}$				

In Problems 195 to 200, determine a positive number δ for the given ϵ so that $|f(x) - L| < \epsilon$ holds whenever $0 < |x - a| < \delta$.

195 $f(x) = 2x - 7$, $a = -1$, $L = -9$, $\epsilon = 0.01$

196 $f(x) = 1 - 5x$, $a = 3$, $L = -14$, $\epsilon = 0.02$

197 $f(x) = 5x + 1$, $a = -2$, $L = -9$, $\epsilon = 0.002$

198 $f(x) = \dfrac{4x^2 - 9}{2x - 3}$, $a = \dfrac{3}{2}$, $L = 6$, $\epsilon = 0.001$

199 $f(x) = \dfrac{25x^2 - 1}{5x + 1}$, $a = -\dfrac{1}{5}$, $L = -2$, $\epsilon = 0.01$

200 $f(x) = 2x - 7$, $a = -1$, $L = -9$, ϵ an arbitrarily small positive number

In Problems 201 to 222, use the properties of limits to evaluate each limit.

201 $\lim\limits_{t \to 5} (6t^2 + t - 4)$

202 $\lim\limits_{y \to 2} \dfrac{3y + 5}{4y^2 + 5y - 4}$

203 $\lim\limits_{t \to 1} \dfrac{1 - t^3}{1 - t^2}$

204 $\lim\limits_{z \to 5/2} \dfrac{4z^2 - 25}{2z - 5}$

205 $\lim\limits_{h \to 0} \dfrac{1}{h} \left(\dfrac{6 + h}{3 + 2h} - 2 \right)$

206 $\lim\limits_{x \to 0} \dfrac{1}{x} \left[1 - \dfrac{1}{(x + 1)^2} \right]$

207 $\lim\limits_{t \to 1} \dfrac{\sqrt{4 - t^2}}{2 + t}$

208 $\lim\limits_{h \to -1} \dfrac{3 - \sqrt{h^2 + h + 9}}{h^3 + 1}$

209 $\lim\limits_{x \to 1} \dfrac{1 - x}{2 - \sqrt{x^2 + 3}}$

210 $\lim\limits_{t \to 0} \dfrac{\sqrt{6 + t} - \sqrt{6}}{t}$

211 $\lim\limits_{x \to 9} \dfrac{\sqrt{x} - 3}{x - 9}$

212 $\lim\limits_{t \to 0} \dfrac{\sqrt[3]{5 + t} - \sqrt[3]{5}}{t}$

213 $\lim\limits_{t \to \pi/2} t \sin t \cos t$

214 $\lim\limits_{y \to \pi/4} y \sin^4 y$

215 $\lim\limits_{x \to \pi/6} \sin^3 x \cos^2 x$

216 $\lim\limits_{w \to 0} (w^2 - \cos \pi w)$

217 $\lim\limits_{t \to 0} \dfrac{\sin 13t}{t}$

218 $\lim\limits_{x \to 0} \dfrac{x}{\sin 47x}$

219 $\lim\limits_{u \to 0} \dfrac{\sin 19u}{\sin 7u}$

220 $\lim\limits_{y \to 0} \dfrac{\sin \sqrt[3]{y}}{\sqrt[3]{y}}$

221 $\lim\limits_{x \to \pi/2} \dfrac{\tan x}{\sec x + 1}$

222 $\lim\limits_{x \to 0} \dfrac{1 - \cos(\sin x)}{\sin x}$

223 Let $f(x) = 3x - 1$. We know that $\lim\limits_{x \to a} (3x - 1) = 3a - 1$.

(a) How close to a must you choose x so that $f(x)$ is within $\epsilon > 0$ of $3a - 1$? (b) If $\epsilon = 0.01$, how close to a must you choose x? Would x in the interval $a - 0.1 < x < a + 0.1$ be close enough? Explain.

224 Show that if

$$\lim\limits_{x \to 0} \frac{f(x)}{x} = L \qquad \text{and} \qquad b \neq 0$$

then

$$\lim\limits_{x \to 0} \frac{f(bx)}{x} = bL$$

In Problems 225 to 228, evaluate each one-sided limit.

225 $\lim\limits_{t \to 3^-} \dfrac{t - 3}{t^2 - 9}$

226 $\lim\limits_{y \to 2^+} \dfrac{|2 - y|}{y^2 - 4}$

227 $\lim\limits_{x \to 0^+} \dfrac{3}{2 + \sqrt{x}}$

228 $\lim\limits_{x \to 2^-} (3 + [\![2x - 4]\!])$

In Problems 229 to 232, sketch the graph of the given function, and find the indicated limit if it exists. If the limit does not exist, give the reason.

229 $f(x) = \begin{cases} 2x - 3 & \text{if } x \geq \frac{3}{2} \\ 6 - 4x & \text{if } x < \frac{3}{2} \end{cases}$; $\lim\limits_{x \to (3/2)^-} f(x)$, $\lim\limits_{x \to (3/2)^+} f(x)$, $\lim\limits_{x \to 3/2} f(x)$

230 $h(x) = \begin{cases} x^2 + 2 & \text{if } x < 1 \\ 4 - x & \text{if } x \geq 1 \end{cases}$; $\lim\limits_{x \to 1^-} h(x)$, $\lim\limits_{x \to 1^+} h(x)$, $\lim\limits_{x \to 1} h(x)$

231 $g(x) = \begin{cases} \dfrac{x^2 - 4}{x - 2} & \text{if } x \neq 2 \\ 1 & \text{if } x = 2 \end{cases}$; $\displaystyle\lim_{x \to 2^-} g(x),\ \lim_{x \to 2^+} g(x),\ \lim_{x \to 2} g(x)$

232 $f(x) = \begin{cases} \dfrac{5x + 5}{|x + 1|} & \text{if } x \neq -1 \\ 0 & \text{if } x = -1 \end{cases}$; $\displaystyle\lim_{x \to -1^-} f(x),\ \lim_{x \to -1^+} f(x),\ \lim_{x \to -1} f(x)$

In Problems 233 to 236, sketch the graph of the given function and indicate whether the function is continuous at $x = a$.

233 $f(x) = \begin{cases} \dfrac{x^2 - 9}{x - 3} & \text{if } x \neq 3 \\ 6 & \text{if } x = 3 \end{cases}$; $a = 3$

234 $g(x) = \begin{cases} \dfrac{x^2 - 1}{x - 1} & \text{if } x \neq 1 \\ \frac{1}{2} & \text{if } x = 1 \end{cases}$; $a = 1$

235 $f(x) = \begin{cases} \sqrt{\dfrac{x - 1}{x^2 - 1}} & \text{if } x \neq 1 \\ \frac{1}{2}\sqrt{2} & \text{if } x = 1 \end{cases}$; $a = 1$

236 $h(x) = \begin{cases} \dfrac{2 - x}{2 - |x|} & \text{if } x \neq 2 \\ 1 & \text{if } x = 2 \end{cases}$; $a = 2$

237 Taxi fare is 90 cents plus 50 cents for each quarter-mile or portion thereof. If we let $f(x)$ denote the fare for a ride of x miles, sketch the graph of f and indicate where it is discontinuous.

238 Assume that it takes 0.5 calorie of heat to raise the temperature of 1 gram of ice 1 degree Celsius, that it takes 80 calories to melt the ice at 0°C, and that it takes 1 calorie to raise the temperature of 1 gram of water 1 degree Celsius. Suppose that $-40 \leq x \leq 20$, and let $Q(x)$ be the number of calories of heat required to raise 1 gram of water from -40°C to x°C. Sketch the graph of Q, and indicate where Q is discontinuous.

239 The air freight charge for shipping merchandise between two cities is given by the equation

$$C(x) = \begin{cases} 0.5x & \text{if } 0 < x < 10 \\ 0.4x & \text{if } 10 \leq x < 30 \\ 0.3x & \text{if } 30 \leq x \end{cases}$$

where x is the number of pounds being shipped and $C(x)$ is the charge in dollars. Sketch the graph of C, and find those values of x for which C is discontinuous.

240 A function f is defined by the equation

$$f(x) = \begin{cases} -x & \text{if } x \leq 0 \\ x & \text{if } 0 < x \leq 1 \\ 2 - x & \text{if } 1 < x < 2 \\ 0 & \text{if } x \geq 2 \end{cases}$$

(a) Sketch the graph of f, and discuss the continuity of f at 0, 1, and 2. (b) Determine the constants A, B, C, D, and E so that

$$f(x) = Ax + B + C|x| + D|x - 1| + E|x - 2|$$

241 Determine the values of the constants A and B so that the function f is continuous at every real number and sketch the graph of the resulting function.

$$f(x) = \begin{cases} 3x & \text{if } x \leq 2 \\ Ax + B & \text{if } 2 < x < 5 \\ -6x & \text{if } x \geq 5 \end{cases}$$

242 Let $[\![x]\!]$ denote the greatest integer not exceeding x, and consider the function $f(x) = [\![1/x]\!]$ for $x > 0$. Sketch a graph of f, and indicate where f is discontinuous.

243 Determine whether each function is continuous or discontinuous on each of the indicated intervals.

(a) $f(x) = \dfrac{3}{2x - 1}$; $[-1, 1]$, $[-\frac{1}{2}, \frac{1}{2}]$, $(-1, \frac{1}{2})$, $[\frac{1}{2}, \infty)$

(b) $g(x) = \begin{cases} 3x - 2 & \text{if } x < 1 \\ 2 - x & \text{if } 1 \leq x \leq 2 \end{cases}$; $(-\infty, 1)$, $(1, 2)$, $[1, 2]$

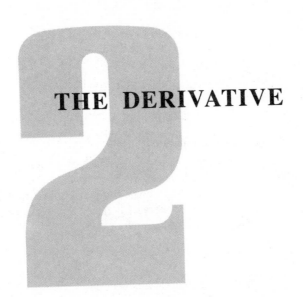

THE DERIVATIVE

The limit concept, introduced in Chapter 1, is used in this chapter to define a mathematical procedure called *differentiation*. A variety of problems that cannot be handled by strictly algebraic techniques—including problems involving the rate of change of a variable quantity—can be solved by this procedure. From a geometric point of view, such problems can be interpreted as questions involving a tangent line to the graph of a function.

2.1   Rates of Change and Slopes of Tangent Lines

In this section we use the concept of limit to solve two apparently unrelated problems; later, we see that they are really the same problem. The first problem is to find the *rate of change of a variable quantity,* for instance, the time rate of change of distance (speed). The second problem is to find the *slope of the tangent line to the graph of a function at a given point.*

Speed of an Automobile

Suppose you are driving along a straight road from city A to city B, perhaps at a variable *rate of speed r*. The *distance d* of your automobile from city A depends on the elapsed *time t* since the start of the journey (Figure 1). Suppose that the functions f and g give the distance d and the speed r at time t, so that

$$d = f(t) \qquad \text{and} \qquad r = g(t)$$

For instance, if the rate of speed r is constant, say $r = 55$ miles per hour, we have the familiar formula

$$\text{distance} = \text{rate} \times \text{time}$$

or

$$d = rt = 55t$$

Figure 1

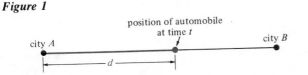

Hence, in this case, f and g are given by

$$f(t) = 55t \qquad \text{and} \qquad g(t) = 55$$

In the more general case in which your speed is variable, the functions f and g are more complicated.

We now find a relationship between the distance function f and the speed function g. Let's choose and (temporarily) fix a value t of the time variable. Thus, at time t, your automobile is $d = f(t)$ miles from city A, and the speedometer reads $r = g(t)$ miles per hour. Suppose a short additional interval of time h passes. At time $t + h$, your automobile is at a distance $f(t + h)$ miles from city A (Figure 2), and its speed is $g(t + h)$ miles per hour. Evidently, the automobile has gone $f(t + h) - f(t)$ miles during the time interval h; hence its **average speed** (distance divided by time) during the time interval h is

$$\text{average speed} = \frac{f(t + h) - f(t)}{h} \text{ miles per hour}$$

Figure 2

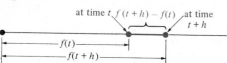

If the time interval h is very short, the speedometer reading $g(t)$ at time t will not differ much from the speedometer reading $g(t + h)$ at the slightly later time $t + h$. Furthermore, during this short interval of time, the speedometer readings should be approximately the same as the average rate of speed,

$$g(t) \approx \frac{f(t + h) - f(t)}{h}$$

As the time interval h becomes shorter and shorter, this approximation will become more and more accurate. In other words, the speedometer reading at the instant t is given by

$$g(t) = \lim_{h \to 0} \frac{f(t + h) - f(t)}{h}$$

This equation expresses the promised relationship between the distance function f and the speed function g. In fact, it shows that you can calculate or derive the instantaneous speed $g(t)$ from the distance function f by evaluating a suitable limit.

Instantaneous Rate of Change in General

The above considerations concerning the rate of change of distance with respect to time can be generalized to any variable quantities whatsoever. Indeed, let x and y denote variable quantities, and suppose that y depends on x, so that $y = f(x)$, where f is a suitable function.

To find the rate of change of y per unit change in x, we naturally begin by considering a change in x, say from the value x_1 to the value x_2. Now, let

$$y_1 = f(x_1) \qquad \text{and} \qquad y_2 = f(x_2)$$

so that as x changes from x_1 to x_2, y undergoes a corresponding change from y_1 to y_2.

It is traditional to denote the change in x by the symbol Δx (read "delta x"),* so that

$$\Delta x = x_2 - x_1$$

*The expression Δx must be regarded as a *single symbol* representing a change in the variable x; it must *not* be thought of as a product of Δ and x.

Similarly, the resulting change in y is denoted by the symbol Δy (read "delta y"), so that

$$\Delta y = y_2 - y_1 = f(x_2) - f(x_1)$$

The ratio of the change in y to the change in x that produced it is called the *average rate of change of y per unit change in x* (or *with respect to x*). More formally, we have the following definition.

DEFINITION 1   **Average Rate of Change**

If $y = f(x)$, then the ratio

$$\frac{\Delta y}{\Delta x} = \frac{y_2 - y_1}{x_2 - x_1} = \frac{f(x_2) - f(x_1)}{x_2 - x_1}$$

is called the **average rate of change** *of y with respect to x as x changes from x_1 to x_2.*

Since $\Delta x = x_2 - x_1$, then $x_2 = x_1 + \Delta x$, and we can also write

$$\frac{\Delta y}{\Delta x} = \frac{f(x_1 + \Delta x) - f(x_1)}{\Delta x}$$

If the average rate of change of y with respect to x approaches a limiting value as Δx approaches 0, it seems reasonable to refer to this limiting value as the *instantaneous rate of change of y with respect to x*; hence, we make the following definition.

DEFINITION 2   **Instantaneous Rate of Change**

If $y = f(x)$, we define the **instantaneous rate of change** *of y with respect to x at the instant when $x = x_1$* to be

$$\lim_{\Delta x \to 0} \frac{\Delta y}{\Delta x} = \lim_{\Delta x \to 0} \frac{f(x_1 + \Delta x) - f(x_1)}{\Delta x}$$

© **EXAMPLE 1**   A metal cube with an edge length x is expanding uniformly as a consequence of being heated. Find **(a)** the average rate of change of its volume with respect to edge length as x increases from 2 to 2.01 centimeters; **(b)** the instantaneous rate of change of its volume with respect to edge length at the instant when $x = 2$ centimeters.

SOLUTION   Let y denote the volume of the cube, so that $y = x^3$ cubic centimeters.

(a) When $x = 2$ centimeters, $y = 2^3 = 8$ cubic centimeters. If x increases by $\Delta x = 0.01$ centimeter to 2.01 centimeters, then y increases to $(2.01)^3$ cubic centimeters. Thus, a change in x of $\Delta x = 0.01$ centimeter produces a corresponding change in y of

$$\Delta y = (2.01)^3 - 2^3 = 0.120601 \text{ cubic centimeter}$$

Hence, the average rate of change of y with respect to x over the interval Δx is given by

$$\frac{\Delta y}{\Delta x} = \frac{0.120601}{0.01} = 12.0601 \text{ cubic centimeters per centimeter of edge length}$$

(b) More generally, if x changes by an amount Δx from 2 to $2 + \Delta x$ centimeters, then y changes by a corresponding amount

$$\Delta y = (2 + \Delta x)^3 - 2^3 = 8 + 12\,\Delta x + 6(\Delta x)^2 + (\Delta x)^3 - 8$$
$$= 12\,\Delta x + 6(\Delta x)^2 + (\Delta x)^3 \qquad \text{cubic centimeters}$$

Therefore, the required instantaneous rate of change of y with respect to x is given by

$$\lim_{\Delta x \to 0} \frac{\Delta y}{\Delta x} = \lim_{\Delta x \to 0} \frac{12\,\Delta x + 6(\Delta x)^2 + (\Delta x)^3}{\Delta x} = \lim_{\Delta x \to 0} [12 + 6\,\Delta x + (\Delta x)^2]$$
$$= 12 \text{ cubic centimeters per centimeter of edge length} \qquad ▪$$

Figure 3

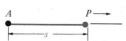

In calculating the rate of change of one variable with respect to another, it is not necessary to use the symbols y and x. For instance, suppose that a particle P is moving away from a starting point A along a straight line (Figure 3). It is traditional to denote the distance between A and P at time t by the letter s. If $s = f(t)$, the *speed** of the particle at the instant when $t = t_1$ is given by

$$\lim_{\Delta t \to 0} \frac{\Delta s}{\Delta t} = \lim_{\Delta t \to 0} \frac{f(t_1 + \Delta t) - f(t_1)}{\Delta t}$$

EXAMPLE 2   A particle is moving along a straight line in such a way that at the end of t seconds, its distance s in meters from the starting point is given by $s = 2t^2 + t$. Find **(a)** the average speed $\Delta s/\Delta t$ of the particle during the interval of time from $t = 3$ to $t = 5$ seconds and **(b)** the instantaneous speed of the particle when $t = 3$ seconds.

SOLUTION   Here $s = f(t) = 2t^2 + t$.

(a) When $t = 3$ seconds, $s = 2(3)^2 + 3 = 21$ meters. As t increases by $\Delta t = 2$ to $3 + 2 = 5$ seconds, s increases to $s + \Delta s = 2(5)^2 + 5 = 55$ meters. Thus, during these 2 seconds, the particle moves through a distance

$$\Delta s = 55 - 21 = 34 \text{ meters}$$

Its average speed during the interval of time from $t = 3$ to $t = 5$ seconds is therefore

$$\frac{\Delta s}{\Delta t} = \frac{34}{2} = 17 \text{ meters per second}$$

(b) The change in s as the value of t changes (by an amount Δt) from 3 to $3 + \Delta t$ seconds is given by

$$\Delta s = f(3 + \Delta t) - f(3) = [2(3 + \Delta t)^2 + (3 + \Delta t)] - [2(3)^2 + 3] \qquad \text{meters}$$

---

*Actually, there is a difference between *speed* and *velocity*; however, for the simple cases considered in this section, it isn't necessary to distinguish between the two. (See page 143.)

The average speed during the interval of time Δt between 3 and $3 + \Delta t$ seconds is

$$\frac{\Delta s}{\Delta t} = \frac{f(3 + \Delta t) - f(3)}{\Delta t} \qquad \text{meters per second}$$

Hence, the instantaneous speed of the particle when $t = 3$ seconds is, by Definition 2,

$$\lim_{\Delta t \to 0} \frac{\Delta s}{\Delta t} = \lim_{\Delta t \to 0} \frac{f(3 + \Delta t) - f(3)}{\Delta t}$$

$$= \lim_{\Delta t \to 0} \frac{[2(3 + \Delta t)^2 + (3 + \Delta t)] - [2(3)^2 + 3]}{\Delta t}$$

$$= \lim_{\Delta t \to 0} \frac{[18 + 12\,\Delta t + 2(\Delta t)^2 + 3 + \Delta t] - 21}{\Delta t}$$

$$= \lim_{\Delta t \to 0} \frac{13\,\Delta t + 2(\Delta t)^2}{\Delta t}$$

$$= \lim_{\Delta t \to 0} (13 + 2\,\Delta t) = 13 \text{ meters per second} \qquad \blacksquare$$

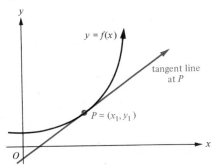

Figure 4

Slope of a Tangent Line to a Graph

Suppose you want to draw the tangent line to the graph of a function f at a point P (Figure 4). Since the tangent line is the line that contains P and "best approximates" the graph of f near P, it's easy to sketch it roughly "by eye." However, suppose you need to draw the tangent line accurately. Since a straight line in the plane is completely determined once you know its slope and one point P on it, you only need to find the slope m of the tangent line.

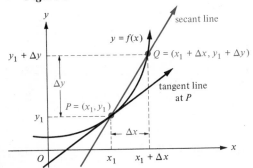

Figure 5

Figure 5 shows a point $Q = (x_1 + \Delta x, y_1 + \Delta y)$ on the graph of f near the point $P = (x_1, y_1)$. Thus, Δx and Δy are the differences between the coordinates of Q and the corresponding coordinates of P. A line segment such as $\overline{PQ}$ joining two points of a curve is called a **secant,** and the line containing P and Q is called a **secant line.** Because both P and Q lie on the graph of f,

$$y_1 = f(x_1) \qquad \text{and} \qquad y_1 + \Delta y = f(x_1 + \Delta x)$$

Therefore,

$$\Delta y = f(x_1 + \Delta x) - f(x_1)$$

By the slope formula, the slope of the secant line through $P = (x_1, f(x_1))$ and $Q = (x_1 + \Delta x, f(x_1 + \Delta x))$ is given by

$$\frac{f(x_1 + \Delta x) - f(x_1)}{x_1 + \Delta x - x_1} = \frac{\Delta y}{\Delta x}$$

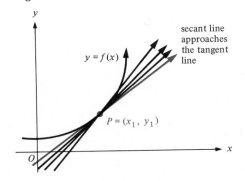

Figure 6

Now, if we let Δx approach 0, the point Q will move along the curve $y = f(x)$ and approach the point P; furthermore, the secant line will pivot about the point P and approach the tangent line (Figure 6). Thus as Δx approaches 0, the slope $\Delta y/\Delta x$ of the secant line approaches the slope m of the tangent line; that is,

$$m = \lim_{\Delta x \to 0} \frac{\Delta y}{\Delta x} = \lim_{\Delta x \to 0} \frac{f(x_1 + \Delta x) - f(x_1)}{\Delta x}$$

The preceding considerations lead us to the following formal definition.

DEFINITION 3

Tangent Line to a Graph

Let f be a function defined at least in some open interval containing the number x_1, and let $y_1 = f(x_1)$. If the limit

$$m = \lim_{\Delta x \to 0} \frac{f(x_1 + \Delta x) - f(x_1)}{\Delta x}$$

exists, we say that the line in the xy plane containing the point (x_1, y_1) and having slope m is the **tangent line** *to the graph of f at* (x_1, y_1).

In Examples 3 and 4, find the slope m of the tangent line to the graph of the given function f at the indicated point P. Sketch a graph of f showing the tangent line at P.

Figure 7

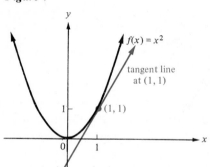

EXAMPLE 3 $f(x) = x^2$, $P = (1, 1)$

SOLUTION $$m = \lim_{\Delta x \to 0} \frac{f(1 + \Delta x) - f(1)}{\Delta x}$$

$$= \lim_{\Delta x \to 0} \frac{(1 + \Delta x)^2 - 1^2}{\Delta x}$$

$$= \lim_{\Delta x \to 0} \frac{1 + 2\,\Delta x + (\Delta x)^2 - 1}{\Delta x}$$

$$= \lim_{\Delta x \to 0} (2 + \Delta x) = 2 \qquad \text{(Figure 7)}$$

EXAMPLE 4 $f(x) = \dfrac{1}{3x - 2}$, $P = (2, \tfrac{1}{4})$

SOLUTION $$m = \lim_{\Delta x \to 0} \frac{f(2 + \Delta x) - f(2)}{\Delta x}$$

Figure 8

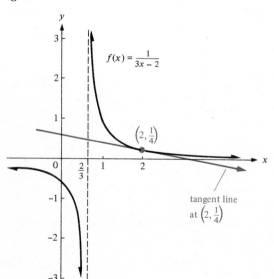

$$= \lim_{\Delta x \to 0} \frac{\dfrac{1}{3(2 + \Delta x) - 2} - \dfrac{1}{3(2) - 2}}{\Delta x}$$

$$= \lim_{\Delta x \to 0} \frac{\dfrac{1}{4 + 3\,\Delta x} - \dfrac{1}{4}}{\Delta x}$$

$$= \lim_{\Delta x \to 0} \frac{\dfrac{4 - (4 + 3\,\Delta x)}{(4 + 3\,\Delta x)(4)}}{\Delta x}$$

$$= \lim_{\Delta x \to 0} \frac{-3\,\Delta x}{16 + 12\,\Delta x} \cdot \frac{1}{\Delta x}$$

$$= \lim_{\Delta x \to 0} \frac{-3}{16 + 12\,\Delta x}$$

$$= -\frac{3}{16} \qquad \text{(Figure 8)}$$

Problem Set 2.1

1 At a certain instant the speedometer of an automobile reads r miles per hour. During the next $\frac{1}{4}$ second the automobile travels 20 feet. Estimate r from this information.

2 Explain why the answer in Problem 1 is only an estimate for r and need not be exactly the same as r.

In Problems 3 and 4, assume that $y = f(x)$ as given. (a) Find the average rate of change of y with respect to x as x changes from x_1 to x_2. (b) Find the instantaneous rate of change of y with respect to x at the instant when $x = x_1$.

[C] **3** $y = f(x) = x^2 + x + 1$, $x_1 = 3$, $x_2 = 3.5$

4 $y = f(x) = \dfrac{4}{x}$, $x_1 = 5$, $x_2 = 6$

In Problems 5 to 8, a particle is moving along a straight line according to the equation given, where s is the distance in meters of the particle from its starting point at the end of t seconds. Find (a) the average speed $\Delta s / \Delta t$ of the particle during the interval of time from $t = t_1$ to $t = t_2$ and (b) the instantaneous speed of the particle when $t = t_1$.

5 $s = 6t^2$, $t_1 = 2$, $t_2 = 3$ **6** $s = 7t^3$, $t_1 = 1$, $t_2 = 2$

7 $s = t^2 + t$, $t_1 = 3$, $t_2 = 4$ [C] **8** $s = \dfrac{2}{5 - t}$, $t_1 = 1$, $t_2 = 1.1$

In Problems 9 to 18, find the slope m of the tangent line to the graph of each function at the indicated point, sketch the graph, and show the tangent line at that point.

9 $f(x) = 2x - x^2$ at $(1, 1)$ **10** $f(x) = (x - 2)^2$ at $(-2, 16)$

11 $f(x) = x^2 - 4x$ at $(3, -3)$

12 $f(x) = x^3$ at $(-1, -1)$

13 $f(x) = 3 + 2x - x^2$ at $(0, 3)$

14 $f(x) = \dfrac{1}{x}$ at $\left(\dfrac{1}{2}, 2\right)$

15 $f(x) = \dfrac{3}{x + 2}$ at $(1, 1)$

16 $f(x) = \sqrt{x - 3}$ at $(7, 2)$

17 $f(x) = \sqrt{x + 1}$ at $(3, 2)$

18 $f(x) = \sqrt{9 - 4x}$ at $(-4, 5)$

19 An object falls from rest according to the equation $s = 16t^2$, where s is the number of feet through which it falls during the first t seconds after being released. Find (a) the average speed during the first 5 seconds of fall and (b) the instantaneous speed at the end of this 5-second interval.

20 A projectile is fired vertically upward and is s feet above the ground t seconds after being fired, where $s = 256t - 16t^2$. Find (a) the speed of the projectile 4 seconds after being fired, (b) the time in seconds required for the projectile to reach its maximum height (at which point its speed is 0 feet per second), and (c) the maximum height to which the projectile ascends.

21 An equilateral triangle made of sheet metal is expanding because it is being heated. Its area A is given by $A = (\sqrt{3}/4)x^2$ square centimeters, where x is the length of one side in centimeters. Find the instantaneous rate of change of A with respect to x at the instant when $x = 10$ centimeters.

22 A spherical balloon of radius R meters has volume $V = \frac{4}{3}\pi R^3$ cubic meters. Find the instantaneous rate of change of V with respect to R at the moment when $R = 5$ meters.

23 The pressure P of a gas depends on its volume V according to Boyle's law, $P = C/V$, where C is a constant. Suppose that $C = 2000$, that P is measured in pounds per square inch, and that V is measured in cubic inches. Find (a) the average rate of change of P with respect to V as V increases from 100 to 125 cubic inches and (b) the instantaneous rate of change of P with respect to V at the instant when $V = 100$ cubic inches.

24 Ecologists studying *acid rain* determine that the concentration C of sulfur dioxide in parts per million at a distance x kilometers from a coal-burning power plant is given by $C = 800/x^2$. Find the instantaneous rate of change of sulfur dioxide concentration in parts per million per kilometer at $x = 10$ kilometers.

25 In respiratory physiology it has been found that a person's respiration rate R in breaths per minute is related to the partial pressure P in newtons per square meter of carbon dioxide in the lungs by an equation of the form $R = AP - B$, where A and B are constants depending on the particular individual involved. If $A = 0.0044$ and $B = 10.4$ for a certain person, what is the instantaneous rate of change of R with respect to P when $P = 8000$ newtons per square meter?

26 Respiratory physiologists sometimes use the simplified mathematical model $p = a - b \sin ct$ for the air pressure p (in newtons per square meter) in the lungs of a person t seconds after inspiration begins. Here a denotes ambient atmospheric pressure, 1.013×10^5 newtons per square meter at sea level, and b and c are constants depending on the particular individual involved. If $b = 100$ and $c = 1.26$, find the instantaneous rate of change of p with respect to time t at the instant when $t = 0$ (the beginning of inspiration). [*Hint:* Use the fact that $\lim_{x \to 0} (\sin x)/x = 1$.]

27 The current I in amperes in a certain electric circuit is given by the formula $I = 100/R$, where R is the resistance in ohms. Find the instantaneous rate of change of I with respect to R when $R = 10$ ohms.

2.2   The Derivative of a Function

In Section 2.1 we found that if x and y are two variables related by an equation $y = f(x)$, then the instantaneous rate of change of y with respect to x when x has the value x_1 is given by

$$\lim_{\Delta x \to 0} \frac{\Delta y}{\Delta x} = \lim_{\Delta x \to 0} \frac{f(x_1 + \Delta x) - f(x_1)}{\Delta x}$$

We also found that the slope m of the tangent line to the graph of f at the point $(x_1, f(x_1))$ is given by the same limit $m = \lim_{\Delta x \to 0} (\Delta y / \Delta x)$. Therefore, *the problem of finding the rate of change of one variable with respect to another and the problem of finding the slope of the tangent line to a graph are both solved by calculating the same limit.*

Limits of the form $\lim_{\Delta x \to 0} \Delta y / \Delta x$ arise so often in calculus that it is useful to introduce some special notation and terminology for them. If $y = f(x)$, then a quotient of the form

$$\frac{\Delta y}{\Delta x} = \frac{f(x + \Delta x) - f(x)}{\Delta x}$$

is called a **difference quotient.** The limit of such a difference quotient as Δx approaches 0 defines a new function f', read "f prime," by the equation

$$f'(x) = \lim_{\Delta x \to 0} \frac{f(x + \Delta x) - f(x)}{\Delta x}$$

Since the function f' is derived from the original function f, it is called the *derivative* of f. Thus, we have the following definition.

DEFINITION 1

The Derivative

Given a function f, the function f' defined by

$$f'(x) = \lim_{\Delta x \to 0} \frac{f(x + \Delta x) - f(x)}{\Delta x} = \lim_{\Delta x \to 0} \frac{\Delta y}{\Delta x}$$

is called the **derivative** of f.

In the definition it is understood that the *domain of the derivative function f' is the set of all numbers x in the domain of f for which the limit of the difference quotient exists.* In calculating this limit, you must be careful to treat x as a constant while letting Δx approach zero.

In Examples 1 and 2, find $f'(x)$ for the given function by direct use of Definition 1.

EXAMPLE 1 $f(x) = x^3$

SOLUTION $\displaystyle f'(x) = \lim_{\Delta x \to 0} \frac{f(x + \Delta x) - f(x)}{\Delta x} = \lim_{\Delta x \to 0} \frac{(x + \Delta x)^3 - x^3}{\Delta x}$

$\displaystyle \qquad = \lim_{\Delta x \to 0} \frac{x^3 + 3x^2\,\Delta x + 3x(\Delta x)^2 + (\Delta x)^3 - x^3}{\Delta x}$

$\displaystyle \qquad = \lim_{\Delta x \to 0} [3x^2 + 3x\,\Delta x + (\Delta x)^2] = 3x^2 \qquad \blacksquare$

EXAMPLE 2 $f(x) = \sqrt{x}$

SOLUTION   For $x > 0$, we have

$$f'(x) = \lim_{\Delta x \to 0} \frac{f(x + \Delta x) - f(x)}{\Delta x} = \lim_{\Delta x \to 0} \frac{\sqrt{x + \Delta x} - \sqrt{x}}{\Delta x}$$

$$= \lim_{\Delta x \to 0} \frac{(\sqrt{x + \Delta x} - \sqrt{x})(\sqrt{x + \Delta x} + \sqrt{x})}{\Delta x(\sqrt{x + \Delta x} + \sqrt{x})} = \lim_{\Delta x \to 0} \frac{(x + \Delta x) - x}{\Delta x(\sqrt{x + \Delta x} + \sqrt{x})}$$

$$= \lim_{\Delta x \to 0} \frac{\Delta x}{\Delta x(\sqrt{x + \Delta x} + \sqrt{x})} = \lim_{\Delta x \to 0} \frac{1}{\sqrt{x + \Delta x} + \sqrt{x}}$$

$$= \frac{1}{\sqrt{x} + \sqrt{x}} = \frac{1}{2\sqrt{x}}$$

The Derivative Notations

The derivative was invented independently by Isaac Newton and Gottfried Leibniz in the seventeenth century. Newton used the notation $\dot{s}$ to denote the time rate of change $\lim_{\Delta t \to 0} (\Delta s / \Delta t)$ of a variable quantity s, where $s = f(t)$. Thus, Newton wrote $\dot{s}$ for what we write as $f'(t)$, the value of the derivative f' at the time t. Newton's notation is still used in many physics textbooks.

Leibniz, on the other hand, realizing that the numerical value of a derivative is the limit of $\Delta y / \Delta x$, wrote this limit as dy/dx; that is,

$$\frac{dy}{dx} = \lim_{\Delta x \to 0} \frac{\Delta y}{\Delta x} = f'(x)$$

Henceforth, we make extensive use of the **Leibniz notation;** however, until the "differentials" dy and dx are given separate meanings (in Section 4.1), we do not regard dy/dx as a fraction, only as a convenient symbol for the value of a derivative.

In the eighteenth century, the French mathematician Joseph Louis Lagrange (1736–1813) introduced the notation f' for the derivative of the function f. Lagrange is perhaps best known for his book *Mécanique analytique,* which not only summarized all previous work in mechanics since the time of Newton, but also contained many of Lagrange's own highly original and important contributions to the field. The Lagrange notation f' is the preferred notation for the derivative whenever precision and absolute clarity are demanded. Indeed, using this notation, you can easily distinguish between the derivative f' (which is a *function*) and the *numerical value* $f'(x)$ of the derivative function at the number x.

The operation of finding the derivative f' of a function f or of finding the value $f'(x)$ is called **differentiation.** Thus, the incomplete symbol

$$\frac{d}{dx}$$

Joseph Louis Lagrange

can be regarded as an instruction to differentiate whatever follows. For instance, the result of Example 1 can be written in the Leibniz notation as

$$\frac{d}{dx}x^3 = 3x^2$$

Popular alternative notation for the symbol d/dx is the simpler symbol D_x (or sometimes just D if the independent variable is understood), which is called the **differentiation operator.** Thus, if $y = x^3$, then

$$D_x y = D_x x^3 = 3x^2$$

In Example 2, we showed that if $f(x) = \sqrt{x}$, then $f'(x) = 1/(2\sqrt{x})$. This useful result can be rewritten in Leibniz and operator notation as follows:

$$\frac{d}{dx}\sqrt{x} = D_x\sqrt{x} = \frac{1}{2\sqrt{x}}$$

Preference for one notation over another is often just a matter of taste and convenience. In the remainder of this book, we use whichever symbolism seems appropriate to the problem at hand. Also, as the following example shows, we can use symbols other than y and x to denote the dependent and independent variables.

EXAMPLE 3   A body dropped from rest will fall s meters in t seconds according to the equation $s = 4.9t^2$. Find the speed ds/dt of the falling body at the end of t seconds.

SOLUTION

$$\frac{ds}{dt} = \lim_{\Delta t \to 0} \frac{\Delta s}{\Delta t}$$

$$= \lim_{\Delta t \to 0} \frac{4.9(t + \Delta t)^2 - 4.9t^2}{\Delta t}$$

$$= \lim_{\Delta t \to 0} \frac{4.9[t^2 + 2t\,\Delta t + (\Delta t)^2] - 4.9t^2}{\Delta t}$$

$$= \lim_{\Delta t \to 0} \frac{9.8t\,\Delta t + 4.9(\Delta t)^2}{\Delta t}$$

$$= \lim_{\Delta t \to 0} (9.8t + 4.9\,\Delta t)$$

$$= 9.8t$$

Up to this point, we have taken care to distinguish between the derivative f' of a function f and the value $f'(x)$ of this derivative function at the number x. In practice, however, it is customary to use the word "derivative" to refer to both the derived function f' and the value $f'(x)$ of this function at x.

In summary, if

$$y = f(x)$$

then the *instantaneous rate of change of y with respect to x,* or, what is the same thing, the *slope of the tangent line to the graph of f at the point* (x, y), is given by

$$\frac{dy}{dx} = D_x y = D_x f(x) = f'(x) = \lim_{\Delta x \to 0} \frac{f(x + \Delta x) - f(x)}{\Delta x} = \lim_{\Delta x \to 0} \frac{\Delta y}{\Delta x}$$

Differentiability and Continuity

Consider the function f defined by the equation

$$f(x) = \begin{cases} 5 - 2x & \text{if } x < 3 \\ 4x - 13 & \text{if } x \geq 3 \end{cases}$$

Figure 1

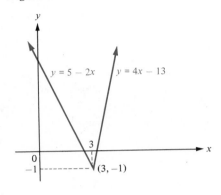

(See Figure 1.) Since $\lim\limits_{x \to 3} f(x) = -1 = f(3)$, it follows that f is continuous at the number 3. However, if we form the difference quotient

$$\frac{f(3 + \Delta x) - f(3)}{\Delta x} = \frac{f(3 + \Delta x) + 1}{\Delta x}$$

and calculate its limits as Δx approaches zero both from the right and from the left,* we obtain

$$\lim_{\Delta x \to 0^+} \frac{f(3 + \Delta x) - f(3)}{\Delta x} = \lim_{\Delta x \to 0^+} \frac{[4(3 + \Delta x) - 13] + 1}{\Delta x} = \lim_{\Delta x \to 0^+} \frac{4 \Delta x}{\Delta x} = 4$$

whereas

$$\lim_{\Delta x \to 0^-} \frac{f(3 + \Delta x) - f(3)}{\Delta x} = \lim_{\Delta x \to 0^-} \frac{[5 - 2(3 + \Delta x)] + 1}{\Delta x} = \lim_{\Delta x \to 0^-} \frac{-2 \Delta x}{\Delta x} = -2$$

Since the right and left limits of the difference quotient are not equal, the limit of the difference quotient cannot exist; that is, the derivative $f'(3)$ cannot exist. The non-existence of the derivative of f at 3 might have been anticipated from the graph in Figure 1, since this graph *has no tangent line* at $(3, -1)$.

In general, we define the **derivative from the right** of a function f by

$$f'_+(x) = \lim_{\Delta x \to 0^+} \frac{f(x + \Delta x) - f(x)}{\Delta x}$$

Similarly, the **derivative from the left** of f is defined by

$$f'_-(x) = \lim_{\Delta x \to 0^-} \frac{f(x + \Delta x) - f(x)}{\Delta x}$$

Figure 2

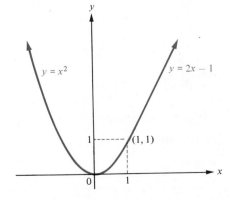

Thus, for the function graphed in Figure 1, $f'_+(3) = 4$ and $f'_-(3) = -2$; hence, $f'(3)$ cannot exist. More generally, *the derivative $f'(x)$ exists and has the value A if and only if both of the one-sided derivatives $f'_+(x)$ and $f'_-(x)$ exist and have the common value A.*

<u>EXAMPLE 4</u>   Let the function f be defined by

$$f(x) = \begin{cases} x^2 & \text{if } x < 1 \\ 2x - 1 & \text{if } x \geq 1 \end{cases}$$

(Figure 2). Find the one-sided derivatives $f'_+(1)$ and $f'_-(1)$, and determine $f'(1)$, if it exists.

*Note that the definition of $f(3 + \Delta x)$ depends on whether Δx is positive or negative.

SOLUTION   Here,

$$f'_+(1) = \lim_{\Delta x \to 0^+} \frac{f(1 + \Delta x) - f(1)}{\Delta x}$$

$$= \lim_{\Delta x \to 0^+} \frac{[2(1 + \Delta x) - 1] - 1}{\Delta x} = \lim_{\Delta x \to 0^+} \frac{2\,\Delta x}{\Delta x} = 2$$

Also,

$$f'_-(1) = \lim_{\Delta x \to 0^-} \frac{f(1 + \Delta x) - f(1)}{\Delta x}$$

$$= \lim_{\Delta x \to 0^-} \frac{(1 + \Delta x)^2 - 1}{\Delta x} = \lim_{\Delta x \to 0^-} (2 + \Delta x) = 2$$

Since $f'_+(1) = f'_-(1) = 2$, we conclude that $f'(1)$ exists and equals 2. This example shows that a function defined "piecewise" can have a derivative at the boundary number between the "pieces." ∎

DEFINITION 2

Differentiable Function

> A function f is said to be **differentiable** at the number x if f is defined at least on some open interval containing x and $f'(x)$ exists.

Evidently, f is differentiable at x if and only if both of the one-sided derivatives $f'_+(x)$ and $f'_-(x)$ exist and are equal. A function f is said to be **differentiable on the open interval (a, b)** if it is differentiable at each number in this interval. If a function is differentiable at each number in its domain, it is called a **differentiable function.**

Geometrically, to say that a function f is differentiable at a number x is to say that the graph of f has a tangent line with slope $f'(x)$ at the point $(x, f(x))$. Obviously, if a graph has a tangent line at a point, it cannot have a discontinuity at that point. Theorem 1 confirms this analytically.

THEOREM 1

Continuity of a Differentiable Function

> If a function f is differentiable at the number x, then it is continuous at x.

PROOF   Assume that f is differentiable at the number x. By Property 14 on page 64, f will be continuous at x if $\lim_{\Delta x \to 0} f(x + \Delta x) = f(x)$. Since the limit of a product is the product of the limits, we have

$$\lim_{\Delta x \to 0} [f(x + \Delta x) - f(x)] = \lim_{\Delta x \to 0} \left[\frac{f(x + \Delta x) - f(x)}{\Delta x} \Delta x \right]$$

$$= \left[\lim_{\Delta x \to 0} \frac{f(x + \Delta x) - f(x)}{\Delta x} \right] \left(\lim_{\Delta x \to 0} \Delta x \right)$$

$$= f'(x) \cdot 0 = 0$$

Therefore, since the limit of a sum is the sum of the limits, we have

$$\lim_{\Delta x \to 0} f(x + \Delta x) = \lim_{\Delta x \to 0} [f(x + \Delta x) - f(x) + f(x)]$$

$$= \lim_{\Delta x \to 0} [f(x + \Delta x) - f(x)] + \lim_{\Delta x \to 0} f(x) = 0 + f(x) = f(x) \quad ∎$$

Figure 3

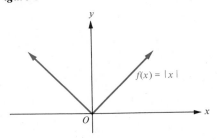

Although, as Theorem 1 shows, a differentiable function is automatically continuous, *there are continuous functions that are not differentiable*. The simplest example is the function f defined by $f(x) = |x|$ (Figure 3). Note that f is continuous at the number 0, but it is not differentiable at 0 since $f'_+(0) = 1$ and $f'_-(0) = -1$. A similar example is provided by the function graphed in Figure 1.

Approximating Derivatives with a Calculator

A calculator or computer may be used to find the approximate value of a derivative by using the fact that

$$f'(x) \approx \frac{f(x + \Delta x) - f(x)}{\Delta x}$$

holds for small values of Δx. Up to a point, the approximation becomes more and more accurate as Δx is chosen to be smaller and smaller. However, if Δx is chosen to be too small, the calculator may not be able to resolve the difference between $f(x + \Delta x)$ and $f(x)$, and the result of the calculation is useless. For most practical purposes, the value $\Delta x = x/10^4$ will provide a good approximation.

© **EXAMPLE 5**   Use a calculator to find the approximate value of $f'(\pi/3)$ if $f(x) = \sin x$. Use $\Delta x = (\pi/3)/10^4$.

SOLUTION   With the calculator in radian mode, we find that

$$f'\left(\frac{\pi}{3}\right) \approx \frac{\sin\,[(\pi/3) + \Delta x] - \sin\,(\pi/3)}{\Delta x} \approx 0.499956$$

[The exact value of $f'(\pi/3)$ happens to be 0.5.]   ■

Problem Set 2.2

In Problems 1 to 10, find $f'(x)$ by direct use of Definition 1.

1 $f(x) = 4x + 7$

2 $f(x) = 13 - 7x$

3 $f(x) = \frac{5}{11}$

4 $f(x) = 3 + \sqrt{x}$

5 $f(x) = x^2 + 4x$

6 $f(x) = 2x^3 - 1$

7 $f(x) = 2x^3 - 4x$

8 $f(x) = \frac{x^3}{2} + \frac{3}{2}x$

9 $f(x) = \frac{2}{x}$

10 $f(x) = \frac{-7}{x - 3}$

In Problems 11 to 16, find the required derivative by direct use of Definition 1.

11 $s = \frac{3}{t - 1}$, $\dfrac{ds}{dt} = ?$

12 $s = \frac{t}{t + 1}$, $D_t s = ?$

13 $f(v) = \sqrt{v - 1}$, $f'(v) = ?$

14 $\dfrac{d}{du}\,(\sqrt{1 - 9u^2}) = ?$

15 $y = \frac{2}{x + 1}$, $D_x y = ?$

16 $h(t) = \frac{1}{\sqrt{t + 1}}$, $h'(t) = ?$

In Problems 17 to 20, find $f'(x_1)$ for the given value of x_1 by direct calculation of

$$\lim_{\Delta x \to 0} \frac{f(x_1 + \Delta x) - f(x_1)}{\Delta x}$$

17 $f(x) = 1 - 2x^2$; $x_1 = -1$

18 $f(x) = \sin x$; $x_1 = 0$
(Use Theorem 2, page 66.)

19 $f(x) = \frac{7}{2x - 1}$; $x_1 = 3$

20 $f(x) = \cos x$; $x_1 = 0$
(Use Theorem 3, page 67.)

In Problems 21 to 23, find the value of the derivative for each function at the number indicated.

21 $f'(4)$ if $f(x) = \dfrac{1}{x - 1}$

22 $D_t s$ at $t = 3$ if $s = \sqrt{2t + 3}$

23 $\dfrac{dy}{dx}$ at $x = 2$ if $y = \dfrac{2}{2x + 1}$

24 Find D_IP if $P = I^2R$ and R is a constant.

25 Rewrite your answers to odd-numbered Problems 1 to 15 in both Leibniz and operator notation.

26 Given that $s = f(t)$, write the value of the derivative $f'(t)$ in as many different ways as you can.

In Problems 27 and 28, find the indicated derivative.

27 $D_t s$ if $s = 16t^2 + 30t + 10$

28 $\dfrac{du}{dv}$ if $u = 16v^2 + 30v + 10$

In Problems 29 to 36, (a) sketch the graph of f, (b) determine whether f is continuous at the number x_1, and (c) determine whether f is differentiable at x_1 by finding $f'_+(x_1)$ and $f'_-(x_1)$.

29 $f(x) = x^2 - 2x$; $x_1 = 3$

30 $f(x) = \begin{cases} 2x + 9 & \text{if } x \le -1 \\ 5 - 2x & \text{if } x > -1 \end{cases}$; $x_1 = -1$

31 $f(x) = \begin{cases} 3x - 2 & \text{if } x \le 3 \\ 10 - x & \text{if } x > 3 \end{cases}$; $x_1 = 3$

32 $f(x) = \begin{cases} \sqrt{4 - x} & \text{if } x < 4 \\ (4 - x)^2 & \text{if } x \ge 4 \end{cases}$; $x_1 = 4$

33 $f(x) = \begin{cases} x^2 & \text{if } x \le 2 \\ 6 - x & \text{if } x > 2 \end{cases}$; $x_1 = 2$

34 $f(x) = \begin{cases} |x + 2| & \text{if } x \ge -2 \\ 0 & \text{if } x < -2 \end{cases}$; $x_1 = -2$

35 $f(x) = \begin{cases} (x - 1)^3 & \text{if } x > 0 \\ \frac{3}{2}x^2 + 3x - 1 & \text{if } x \le 0 \end{cases}$; $x_1 = 0$

36 $f(x) = 1 - |x - 3|$; $x_1 = 3$

37 (a) Explain, in your own words, why it is geometrically reasonable that a differentiable function must be continuous. (b) Is it geometrically reasonable to believe that every continuous function is differentiable? Why or why not?

38 Suppose that

$$f(x) = \begin{cases} x^2 & \text{if } x < -1 \\ ax + b & \text{if } x \ge -1 \end{cases}$$

Find values of the constants a and b so that $f'(-1)$ exists.

In Problems 39 and 40, sketch the graph of the function and indicate where the function is not differentiable.

39 $f(x) = |3x - 1|$ **40** $f(x) = |x^3 - 1|$

© In Problems 41 to 44, use a calculator to find the approximate value of the indicated derivative.

41 $f'(\pi/3)$ for $f(x) = \cos x$ **42** $f'(\pi/4)$ for $f(x) = \tan x$

43 $f'(10)$ for $f(x) = \sqrt{x - 1}$ **44** $f'(\pi/6)$ for $f(x) = \sqrt{\sin x}$

2.3   Basic Algebraic Rules for Differentiation

In Section 2.2 we differentiated functions by direct use of the definition of the derivative as a limit of a difference quotient. Direct calculation of derivatives in this way can be tedious, even for the relatively simple functions considered up to now. Relief from this tedium is forthcoming—there are general rules for differentiation which permit straightforward calculation of derivatives. In this section, we present rules for differentiating sums, products, powers, quotients, and square roots. At first, we simply state these rules informally and illustrate their applications in examples. Later in the section, we state them precisely and give rigorous proofs.

RULE 1   **Constant Rule**

> The derivative of a constant function is the zero function. In symbols, if c is a constant, then
>
> $$D_x c = 0 \qquad \text{or} \qquad \frac{dc}{dx} = 0$$

In using the constant rule, it is often convenient to deliberately confuse a constant number such as 7 and the constant function f which it determines according to the equation $f(x) = 7$. Thus, we often write "$D_x(7) = 0$" or "$d7/dx = 0$" instead of writing "f' is the function defined by the equation $f'(x) = 0$." Since *you can differentiate only functions—never numbers*—an expression of the form $D_x(7) = 0$ could only be interpreted sensibly as above.

EXAMPLE 1   Let f be the constant function defined by the equation $f(x) = 5 + \pi$. Find f'.

SOLUTION   By the constant rule, f' is the constant function defined by the equation $f'(x) = 0$.   ■

EXAMPLE 2   Find $D_t(5 + \sqrt{3})$.

SOLUTION   By the constant rule, $D_t(5 + \sqrt{3}) = 0$.   ■

RULE 2   **Identity Rule**

> The derivative of the identity function is the constant function 1. In symbols,
>
> $$D_x x = 1 \qquad \text{or} \qquad \frac{dx}{dx} = 1$$

EXAMPLE 3   If f is the function defined by $f(x) = x$, find f'.

SOLUTION   By the identity rule, f' is the constant function defined by $f'(x) = 1$.   ■

RULE 3   **Power Rule**

> The derivative of a positive integer power of x is the exponent of x times x raised to the next lower power. In symbols, if n is a fixed positive integer, then
>
> $$D_x x^n = nx^{n-1} \qquad \text{or} \qquad \frac{d}{dx} x^n = nx^{n-1}$$

EXAMPLE 4   Differentiate the function $f(x) = x^7$.

SOLUTION   By the power rule, $f'(x) = 7x^{7-1} = 7x^6$.   ■

EXAMPLE 5   If $u = t^{13}$, find $\dfrac{du}{dt}$.

SOLUTION   By the power rule, $\dfrac{du}{dt} = \dfrac{d}{dt} t^{13} = 13t^{12}$.   ■

RULE 4   **Homogeneous Rule**

> The derivative of a constant times a function is the constant times the derivative of the function. In symbols, if c is a constant and u is a differentiable function of x, then
>
> $$D_x(cu) = cD_x u \qquad \text{or} \qquad \frac{d}{dx}(cu) = c\frac{du}{dx}$$

EXAMPLE 6 Differentiate the function $f(x) = 5x^4$.

SOLUTION Using both the homogeneous rule and the power rule, we have

$$f'(x) = D_x(5x^4) = 5D_x x^4 = 5(4x^3) = 20x^3$$

■

EXAMPLE 7 Find $D_x\left(\dfrac{2x^7}{3}\right)$.

SOLUTION By the homogeneous and power rules,

$$D_x\left(\frac{2x^7}{3}\right) = D_x\left(\frac{2}{3}x^7\right) = \frac{2}{3}D_x x^7 = \frac{2}{3}(7x^6) = \frac{14}{3}x^6 = \frac{14x^6}{3}$$

■

EXAMPLE 8 If c is a constant and n is a positive integer, find $\dfrac{d}{dx}(cx^n)$.

SOLUTION $\dfrac{d}{dx}(cx^n) = c\dfrac{d}{dx}x^n = c(nx^{n-1}) = ncx^{n-1}$

■

One important consequence of the homogeneous rule is that

$$D_x(-u) = -D_x u$$

This follows from putting $c = -1$ in Rule 4.

Many functions encountered in practice are (or can be rewritten as) sums of simpler functions. For instance, the polynomial function $f(x) = 2x^2 + 5x - 1$ is a sum of $2x^2$, $5x$, and -1. Thus, one of the most useful differentiation rules is the following.

RULE 5 **Sum Rule, or Addition Rule**

> The derivative of a sum is the sum of the derivatives. In symbols, if u and v are differentiable functions of x, then
>
> $$D_x(u + v) = D_x u + D_x v \qquad \text{or} \qquad \frac{d}{dx}(u + v) = \frac{du}{dx} + \frac{dv}{dx}$$

EXAMPLE 9 Find $D_x(3x^5 + 11x^8)$.

SOLUTION We have

$$\begin{aligned}
D_x(3x^5 + 11x^8) &= D_x(3x^5) + D_x(11x^8) &&\text{(sum rule)} \\
&= 3D_x x^5 + 11D_x x^8 &&\text{(homogeneous rule)} \\
&= 3(5x^4) + 11(8x^7) &&\text{(power rule)} \\
&= 15x^4 + 88x^7
\end{aligned}$$

■

If u, v, and w are three differentiable functions of x, then, by the sum rule,

$$\begin{aligned}
D_x(u + v + w) &= D_x[u + (v + w)] \\
&= D_x u + D_x(v + w) \\
&= D_x u + D_x v + D_x w
\end{aligned}$$

More generally, *the derivative of the sum of three or more differentiable functions is the sum of their derivatives.* This rule is also called the **sum rule.**

Using the sum, homogeneous, power, identity, and constant rules, *you can differentiate any polynomial function term by term*. This is illustrated by the following examples.

EXAMPLE 10   Given $f(x) = 3x^{100} - 24x^3 + 7x^2 - x - 2$, find $f'(x)$.

SOLUTION

$$\begin{aligned}
f'(x) &= D_x(3x^{100} - 24x^3 + 7x^2 - x - 2) \\
&= D_x(3x^{100}) + D_x(-24x^3) + D_x(7x^2) + D_x(-x) + D_x(-2) \\
&= 300x^{99} - 72x^2 + 14x - 1
\end{aligned}$$

EXAMPLE 11   Find $\dfrac{dy}{du}$ if $y = \sqrt{2}u^5 - \dfrac{u^4}{4} + 5u^3 + u + \pi^2$.

SOLUTION

$$\begin{aligned}
\frac{dy}{du} &= 5\sqrt{2}u^4 - \frac{1}{4}(4u^3) + 15u^2 + 1 + 0 \\
&= 5\sqrt{2}u^4 - u^3 + 15u^2 + 1
\end{aligned}$$

The rule for differentiating the product of two functions is more complicated than the rule for differentiating their sum. Early in the development of calculus, Leibniz found that, in general, *the derivative of a product is not the product of the derivatives*. He did manage to find the correct rule, which is the following.

RULE 6

Product Rule, or Multiplication Rule

The derivative of the product of two functions is the first function times the derivative of the second function plus the derivative of the first function times the second function. In symbols, if u and v are differentiable functions of x, then

$$D_x(uv) = u(D_x v) + (D_x u)v \qquad \text{or} \qquad \frac{d}{dx}(uv) = u\frac{dv}{dx} + \frac{du}{dx}v$$

EXAMPLE 12   Find $D_x[(3x^2 + 1)(7x^3 + x)]$ by using the multiplication rule.

SOLUTION

$$\begin{aligned}
D_x[(3x^2 + 1)(7x^3 + x)] &= (3x^2 + 1)[D_x(7x^3 + x)] + [D_x(3x^2 + 1)](7x^3 + x) \\
&= (3x^2 + 1)(21x^2 + 1) + (6x)(7x^3 + x) \\
&= (63x^4 + 24x^2 + 1) + (42x^4 + 6x^2) \\
&= 105x^4 + 30x^2 + 1
\end{aligned}$$

EXAMPLE 13   Suppose that f and g are differentiable functions at the number 2 and that $f(2) = 1$, $g(2) = 10$, $f'(2) = \frac{1}{2}$, and $g'(2) = 3$. If $h = f \cdot g$, find $h'(2)$.

SOLUTION   By the multiplication rule,

$$h'(x) = f(x) \cdot g'(x) + f'(x) \cdot g(x)$$

so that

$$h'(2) = f(2) \cdot g'(2) + f'(2) \cdot g(2)$$

or

$$h'(2) = (1)(3) + (\tfrac{1}{2})(10) = 8$$

RULE 7   **Reciprocal Rule**

> The derivative of the reciprocal of a function is the negative of the derivative of the function divided by the square of the function. In symbols, if v is a differentiable function of x, then
>
> $$D_x\left(\frac{1}{v}\right) = -\frac{D_x v}{v^2} \qquad \text{or} \qquad \frac{d}{dx}\left(\frac{1}{v}\right) = -\frac{dv/dx}{v^2}$$

EXAMPLE 14   Find $D_x\left(\dfrac{1}{x}\right)$.

SOLUTION   By the reciprocal rule,

$$D_x\left(\frac{1}{x}\right) = -\frac{D_x x}{x^2} = -\frac{1}{x^2}$$

RULE 8   **Quotient Rule**

> The derivative of a quotient of two functions is the denominator times the derivative of the numerator minus the numerator times the derivative of the denominator, all divided by the square of the denominator. In symbols, if u and v are differentiable functions of x, then
>
> $$D_x\left(\frac{u}{v}\right) = \frac{v D_x u - u D_x v}{v^2} \qquad \text{or} \qquad \frac{d}{dx}\left(\frac{u}{v}\right) = \frac{v\dfrac{du}{dx} - u\dfrac{dv}{dx}}{v^2}$$

EXAMPLE 15   Find $D_x\left(\dfrac{x^2}{x^3 + 7}\right)$.

SOLUTION   By the quotient rule,

$$D_x\left(\frac{x^2}{x^3 + 7}\right) = \frac{(x^3 + 7)D_x x^2 - x^2 D_x(x^3 + 7)}{(x^3 + 7)^2}$$

$$= \frac{(x^3 + 7)(2x) - x^2(3x^2)}{(x^3 + 7)^2}$$

$$= \frac{14x - x^4}{(x^3 + 7)^2}$$

EXAMPLE 16   Suppose that f and g are differentiable functions at the number 3 and that $f(3) = -2$, $g(3) = -5$, $f'(3) = 3$, and $g'(3) = 1$. If $h = f/g$, find the slope of the tangent line to the graph of h at the point $(3, \frac{2}{5})$.

SOLUTION   By the quotient rule,

$$h'(x) = \frac{g(x)f'(x) - f(x)g'(x)}{[g(x)]^2}$$

so the required slope is given by

$$h'(3) = \frac{g(3)f'(3) - f(3)g'(3)}{[g(3)]^2} = \frac{(-5)(3) - (-2)(1)}{(-5)^2} = -\frac{13}{25}$$

Since a rational function is a quotient of polynomial functions and you can now differentiate any polynomial function, *you can use the quotient rule to differentiate any rational function.* Example 15 above illustrates the technique.

In Example 2 in Section 2.2 (page 93), we obtained the following result.

RULE 9   **Square-Root Rule**

> The derivative of the square root of x is the reciprocal of twice the square root of x. In symbols,
>
> $$D_x\sqrt{x} = \frac{1}{2\sqrt{x}} \qquad \text{or} \qquad \frac{d}{dx}\sqrt{x} = \frac{1}{2\sqrt{x}}$$

EXAMPLE 17   Find $D_x\dfrac{1}{\sqrt{x}}$.

SOLUTION   By the reciprocal and square-root rules,

$$D_x\frac{1}{\sqrt{x}} = -\frac{D_x\sqrt{x}}{(\sqrt{x})^2} = -\frac{1/(2\sqrt{x})}{x} = \frac{-1}{2x\sqrt{x}}$$

It's interesting to note that because

$$\sqrt{x} = x^{1/2} \qquad \text{and} \qquad \frac{1}{2}x^{(1/2)-1} = \frac{1}{2\sqrt{x}}$$

the square-root rule is just an extension of the power rule (Rule 3) to the case $n = \frac{1}{2}$. The power rule can also be generalized to arbitrary integer powers as follows.

RULE 10   **Power Rule for Integer Exponents**

> If n is any integer, then
>
> $$D_x x^n = nx^{n-1} \qquad \text{or} \qquad \frac{d}{dx}x^n = nx^{n-1}$$

For $n = 1$, this formula gives $dx/dx = 1 \cdot x^0 = 1$, in conformity with the identity rule, except when $x = 0$, since 0^0 is undefined. It is traditional to overlook this slight difficulty and simply interpret 0^0 to be 1 *for purposes of this rule only.* Rule 10 permits you to use the same formula as in Rule 3, even when n is a negative integer or zero.

EXAMPLE 18   Given $f(x) = -2/x^3$, find $f'(x)$.

SOLUTION   Using the homogeneous rule and the power rule for integer exponents, we have

$$f'(x) = D_x\left(\frac{-2}{x^3}\right) = D_x(-2x^{-3}) = -2D_x(x^{-3}) = (-2)(-3x^{-3-1}) = 6x^{-4}$$

EXAMPLE 19   Find $\dfrac{d}{dx}\left(x^2 + \dfrac{\sqrt{5}}{x^2} - \dfrac{\pi}{x^5}\right)$.

SOLUTION
$$\frac{d}{dx}\left(x^2 + \frac{\sqrt{5}}{x^2} - \frac{\pi}{x^5}\right) = \frac{d}{dx}(x^2 + \sqrt{5}x^{-2} - \pi x^{-5})$$
$$= 2x - 2\sqrt{5}x^{-3} + 5\pi x^{-6}$$

∎

Later we show that the power rule $D_x x^n = nx^{n-1}$ actually works for *any* constant real number n.

Proofs of the Basic Differentiation Rules

We now give precise statements and rigorous proofs of the basic differentiation rules. It is convenient to prove the theorems in an order other than that in which the rules were stated.

THEOREM 1   **Constant Rule**

> If c is a constant and f is the constant function defined by $f(x) = c$, then f is differentiable at every number x and f' is the function defined by $f'(x) = 0$.

PROOF
$$f'(x) = \lim_{\Delta x \to 0} \frac{f(x + \Delta x) - f(x)}{\Delta x} = \lim_{\Delta x \to 0} \frac{c - c}{\Delta x} = \lim_{\Delta x \to 0} 0 = 0$$

∎

THEOREM 2   **Identity Rule**

> If f is the function defined by $f(x) = x$, then f is differentiable at every number x and f' is the constant function defined by $f'(x) = 1$.

PROOF
$$f'(x) = \lim_{\Delta x \to 0} \frac{f(x + \Delta x) - f(x)}{\Delta x} = \lim_{\Delta x \to 0} \frac{x + \Delta x - x}{\Delta x} = \lim_{\Delta x \to 0} 1 = 1$$

∎

THEOREM 3   **Sum Rule**

> Let f and g be functions both of which are differentiable at the number x_1, and let $h = f + g$. Then h is also differentiable at x_1 and
> $$h'(x_1) = f'(x_1) + g'(x_1)$$

PROOF
$$h'(x_1) = \lim_{\Delta x \to 0} \frac{h(x_1 + \Delta x) - h(x_1)}{\Delta x}$$
$$= \lim_{\Delta x \to 0} \frac{[f(x_1 + \Delta x) + g(x_1 + \Delta x)] - [f(x_1) + g(x_1)]}{\Delta x}$$
$$= \lim_{\Delta x \to 0} \frac{f(x_1 + \Delta x) - f(x_1) + g(x_1 + \Delta x) - g(x_1)}{\Delta x}$$
$$= \lim_{\Delta x \to 0} \left[\frac{f(x_1 + \Delta x) - f(x_1)}{\Delta x} + \frac{g(x_1 + \Delta x) - g(x_1)}{\Delta x}\right]$$
$$= \lim_{\Delta x \to 0} \frac{f(x_1 + \Delta x) - f(x_1)}{\Delta x} + \lim_{\Delta x \to 0} \frac{g(x_1 + \Delta x) - g(x_1)}{\Delta x}$$
$$= f'(x_1) + g'(x_1)$$

∎

THEOREM 4   **Product Rule**

> Let f and g be functions both of which are differentiable at the number x_1, and let $h = f \cdot g$. Then h is also differentiable at x_1 and
> $$h'(x_1) = f(x_1) \cdot g'(x_1) + f'(x_1) \cdot g(x_1)$$

PROOF

$$h'(x_1) = \lim_{\Delta x \to 0} \frac{h(x_1 + \Delta x) - h(x_1)}{\Delta x}$$

$$= \lim_{\Delta x \to 0} \frac{f(x_1 + \Delta x) \cdot g(x_1 + \Delta x) - f(x_1) \cdot g(x_1)}{\Delta x}$$

We now use a curious but effective algebraic trick—the expression $f(x_1 + \Delta x) \cdot g(x_1)$ is subtracted from the numerator and then added back again (which, of course, leaves the value of the numerator unchanged). The result is

$$h'(x_1) = \lim_{\Delta x \to 0} \frac{f(x_1 + \Delta x) \cdot g(x_1 + \Delta x) - f(x_1 + \Delta x) \cdot g(x_1) + f(x_1 + \Delta x) \cdot g(x_1) - f(x_1) \cdot g(x_1)}{\Delta x}$$

$$= \lim_{\Delta x \to 0} \left[f(x_1 + \Delta x) \frac{g(x_1 + \Delta x) - g(x_1)}{\Delta x} + \frac{f(x_1 + \Delta x) - f(x_1)}{\Delta x} g(x_1) \right]$$

$$= \left[\lim_{\Delta x \to 0} f(x_1 + \Delta x) \right] \cdot \left[\lim_{\Delta x \to 0} \frac{g(x_1 + \Delta x) - g(x_1)}{\Delta x} \right]$$

$$+ \left[\lim_{\Delta x \to 0} \frac{f(x_1 + \Delta x) - f(x_1)}{\Delta x} \right] \cdot \left[\lim_{\Delta x \to 0} g(x_1) \right]$$

$$= \left[\lim_{\Delta x \to 0} f(x_1 + \Delta x) \right] \cdot g'(x_1) + f'(x_1) \left[\lim_{\Delta x \to 0} g(x_1) \right]$$

Since f is differentiable at x_1, it is continuous at x_1 (Theorem 1, Section 2.2); hence,

$$\lim_{\Delta x \to 0} f(x_1 + \Delta x) = \lim_{x \to x_1} f(x) = f(x_1)$$

Also, since $g(x_1)$ is a constant,

$$\lim_{\Delta x \to 0} g(x_1) = g(x_1)$$

It follows that

$$h'(x_1) = f(x_1) \cdot g'(x_1) + f'(x_1) \cdot g(x_1)$$
∎

THEOREM 5   **Homogeneous Rule**

> Let g be a function that is differentiable at the number x_1, and let c be a constant. Let the function h be defined by $h(x) = cg(x)$. Then h is differentiable at x_1 and
> $$h'(x_1) = cg'(x_1)$$

PROOF   Let f be the constant function defined by $f(x) = c$. By Theorem 1, $f'(x) = 0$. Evidently,

$$h(x) = cg(x) = f(x) \cdot g(x)$$

Therefore, by Theorem 4,

$$h'(x_1) = f(x_1) \cdot g'(x_1) + f'(x_1) \cdot g(x_1) = cg'(x_1) + 0 \cdot g(x_1) = cg'(x_1)$$
∎

THEOREM 6

Power Rule

> Let n be an integer greater than 1, and let f be the function defined by $f(x) = x^n$. Then f is differentiable at every number x, and f' is the function defined by
>
> $$f'(x) = nx^{n-1}$$

PROOF

The proof proceeds by mathematical induction,* starting with $n = 2$. For $n = 2$ we have, by the product and identity rules,

$$f'(x) = D_x x^2 = D_x(x \cdot x) = x(D_x x) + (D_x x)x$$
$$= x + x = 2x = 2x^{2-1}$$

Hence, the theorem holds when $n = 2$. Now, if we assume that n is greater than 2 and that the theorem holds for exponents less than n, Theorems 4 and 2 imply that

$$f'(x) = D_x x^n = D_x(x^{n-1} \cdot x) = x^{n-1}(D_x x) + (D_x x^{n-1})x$$
$$= x^{n-1} + [(n-1)x^{n-2}]x = x^{n-1} + (n-1)x^{n-1}$$
$$= nx^{n-1}$$

THEOREM 7

Reciprocal Rule

> Let g be a function that is differentiable at x_1, and suppose that $g(x_1) \neq 0$. Let h be the function defined by $h(x) = 1/g(x)$. Then h is differentiable at x_1 and
>
> $$h'(x_1) = -\frac{g'(x_1)}{[g(x_1)]^2}$$

PROOF

Since g is differentiable at x_1, it is defined in some open interval about x_1 and it is continuous at x_1. Therefore, for values of x close to x_1, the numerical values of $g(x)$ come close to $g(x_1)$. Since $g(x_1) \neq 0$, the numerical values $g(x)$ must differ from zero for values of x sufficiently close to x_1. This shows that $h(x) = 1/g(x)$ is defined at least in a small open interval around x_1. We have

$$h'(x_1) = \lim_{\Delta x \to 0} \frac{h(x_1 + \Delta x) - h(x_1)}{\Delta x} = \lim_{\Delta x \to 0} \frac{\dfrac{1}{g(x_1 + \Delta x)} - \dfrac{1}{g(x_1)}}{\Delta x}$$

$$= \lim_{\Delta x \to 0} \frac{1}{\Delta x}\left[\frac{g(x_1)}{g(x_1) \cdot g(x_1 + \Delta x)} - \frac{g(x_1 + \Delta x)}{g(x_1) \cdot g(x_1 + \Delta x)} \right]$$

$$= \lim_{\Delta x \to 0} \left[\frac{1}{\Delta x} \cdot \frac{g(x_1) - g(x_1 + \Delta x)}{g(x_1) \cdot g(x_1 + \Delta x)} \right]$$

$$= \lim_{\Delta x \to 0} (-1)\left[\frac{g(x_1 + \Delta x) - g(x_1)}{\Delta x} \cdot \frac{1}{g(x_1) \cdot g(x_1 + \Delta x)} \right]$$

$$= (-1)\left[\lim_{\Delta x \to 0} \frac{g(x_1 + \Delta x) - g(x_1)}{\Delta x} \right]\left[\lim_{\Delta x \to 0} \frac{1}{g(x_1) \cdot g(x_1 + \Delta x)} \right]$$

$$= (-1)g'(x_1) \cdot \frac{1}{g(x_1) \lim_{\Delta x \to 0} g(x_1 + \Delta x)}$$

*See Appendix C for a review of mathematical induction.

Since g is continuous at x_1,

$$\lim_{\Delta x \to 0} g(x_1 + \Delta x) = \lim_{x \to x_1} g(x) = g(x_1)$$

Thus,

$$h'(x_1) = (-1)g'(x_1) \cdot \frac{1}{g(x_1) \cdot g(x_1)} = -\frac{g'(x_1)}{[g(x_1)]^2} \qquad \blacksquare$$

THEOREM 8

Quotient Rule

Let f and g be functions both of which are differentiable at the number x_1, and suppose that $g(x_1) \neq 0$. Then, if $h = f/g$, it follows that h is differentiable at x_1 and

$$h'(x_1) = \frac{g(x_1) \cdot f'(x_1) - f(x_1) \cdot g'(x_1)}{[g(x_1)]^2}$$

PROOF

Note that $h = f \cdot (1/g)$; hence, by the product and reciprocal rules,

$$h'(x_1) = f(x_1) \cdot \frac{-g'(x_1)}{[g(x_1)]^2} + f'(x_1) \cdot \frac{1}{g(x_1)}$$

$$= \frac{-f(x_1) \cdot g'(x_1)}{[g(x_1)]^2} + \frac{g(x_1) \cdot f'(x_1)}{[g(x_1)]^2}$$

$$= \frac{g(x_1) \cdot f'(x_1) - f(x_1) \cdot g'(x_1)}{[g(x_1)]^2} \qquad \blacksquare$$

THEOREM 9

Power Rule for Integer Exponents

If the function f is defined by $f(x) = x^n$, where n is any fixed integer, then f is differentiable and

$$f'(x) = nx^{n-1}$$

Here we understand that

(i) If $n \leq 0$, then x can be any number except 0.

(ii) If $n = 1$, we interpret 0^0 as being the number 1 (for purposes of this theorem only).

PROOF

Theorem 6 takes care of the case $n \geq 2$, while Theorem 2 takes care of the case $n = 1$. For $n = 0$, $x^n = x^0 = 1$ (except for $x = 0$); hence, for $x \neq 0$ and $n = 0$, $f'(x) = D_x x^n = D_x 1 = 0 = 0 \cdot x^{-1} = 0 \cdot x^{0-1} = nx^{n-1}$, as desired. Finally, suppose that $n < 0$ and that $x \neq 0$. Note that $-n$ is a positive integer; hence, by what was already proved for positive exponents and by Theorem 7,

$$f'(x) = D_x x^n = D_x\left(\frac{1}{x^{-n}}\right) = -\frac{D_x x^{-n}}{(x^{-n})^2}$$

$$= -\frac{-nx^{-n-1}}{x^{-2n}}$$

$$= nx^{-n-1+2n} = nx^{n-1} \qquad \blacksquare$$

Problem Set 2.3

In Problems 1 to 40, differentiate each function by applying the basic rules, Rules 1 through 10.

1 $f(x) = 3x^2$

2 $g(x) = \frac{3}{7}x^7$

3 $h(x) = -5x^4$

4 $G(t) = \frac{8}{11}t^{11}$

5 $F(y) = -\frac{4y^3}{3}$

6 $H(v) = -\frac{v^6}{30}$

7 $H(t) = 5t - 7$

8 $G(w) = \sqrt{3}(7 - 8w)$

9 $f(x) = x^5 - 3x^3 + 1$

10 $f(x) = \frac{5}{6}x^6 - 9x^4$

11 $f(x) = \frac{x^{10}}{2} + \frac{x^5}{5} + 6$

12 $F(x) = \frac{x^4}{4} - \frac{x^3}{3} + 1$

13 $f(t) = t^8 - 2t^7 + 3t + 1$

14 $f(t) = 3t^2 + 7t + 17$

15 $F(x) = \frac{3}{x^2} + \frac{4}{3x}$

16 $f(t) = \frac{1}{3t^3} - \frac{1}{2t^2} + 1$

17 $f(y) = \frac{5}{y^5} - \frac{25}{y}$

18 $f(u) = \frac{1}{u} - \frac{3}{u^3} + \sqrt{u}$

19 $g(x) = 3x^{-2} - 7x^{-1} + 6\sqrt{x}$

20 $G(x) = \frac{1}{3}x^{-3} - \frac{1}{2}x^{-2} + \frac{11}{\sqrt{x}}$

21 $f(x) = \frac{2}{5x} - \frac{\sqrt{2}}{3x^2} + \frac{1}{\sqrt{x}}$

22 $f(x) = \sqrt{x}(x^3 - x)$

23 $F(x) = x^2(3x^3 - 1)$

24 $f(x) = (x^2 + 1)(2x^3 + 5)$

25 $G(x) = (x^2 + 3x)(x^3 - 9x)$

26 $g(x) = (3x - x^2)(3x^3 - 4)$

27 $f(y) = \sqrt{y}(4y^2 + 7)$

28 $f(t) = (6t^2 + 7)^2$

29 $f(x) = (x^3 - 8)\left(\frac{2}{x} - 1\right)$

30 $f(x) = \left(\frac{1}{x} + 3\right)\left(\frac{2}{x} + 7\right)$

31 $g(x) = \left(\frac{1}{x^2} + 3\right)\left(\frac{2}{x^3} + x\right)$

32 $g(u) = \left(u^2 + \frac{1}{u}\right)\left(u - \frac{1}{u^3}\right)$

33 $f(x) = \frac{2x + 7}{3x - 1}$

34 $f(x) = \frac{3x^2}{x - 2}$

35 $g(x) = \frac{2x^2 + x + 1}{x^2 - 3x + 2}$

36 $G(t) = \frac{t^3}{2t^4 + 5}$

37 $F(t) = \frac{3t^2 + 7}{t^2 - 1}$

38 $f(x) = \frac{x^2 - 19}{x^2 + 19}$

39 $f(x) = \left(\frac{3x + 1}{x + 2}\right)(x + 7)$

40 $p(x) = \frac{x + 1}{\sqrt{x}}$

41 Find $f'(2)$ in each case.

(a) $f(x) = \frac{1}{3}x^3 - 1$

(b) $f(x) = (1/x^3) - 1$

(c) $f(x) = (x^2 + 1)(1 - x)$

(d) $f(x) = \left(\frac{1}{x} + 2\right)\left(\frac{3}{x} - 1\right)$

(e) $f(x) = \frac{x}{x^2 + 2}$

(f) $f(x) = \frac{2x^2}{x + 7}$

42 Suppose that f, g, and h are differentiable functions. Let k be a function defined by $k(x) = f(x) \cdot g(x) \cdot h(x)$. Use the product rule to show that

$$k'(x) = f(x) \cdot g(x) \cdot h'(x) + f(x) \cdot g'(x) \cdot h(x) + f'(x) \cdot g(x) \cdot h(x)$$

43 Use the result of Problem 42 to differentiate the following functions.

(a) $f(x) = (2x - 5)(x + 2)(x^2 - 1)$

(b) $f(x) = (1 - 3x)^2(2x + 5)$

(c) $f(x) = \left(\frac{1}{x^2} + 1\right)(3x - 1)(x^2 - 3x)$

(d) $f(x) = (2x^2 + 7)^3$

44 Using the sum and the homogeneous rules, show that the derivative of a difference is the difference of the derivatives; that is, $D_x(u - v) = D_x u - D_x v$.

45 Let f and g be differentiable functions at the number 1, and let $f(1) = 1$, $f'(1) = 2$, $g(1) = \frac{1}{2}$, and $g'(1) = -3$. Use the differentiation rules to find

(a) $(f + g)'(1)$

(b) $(f - g)'(1)$

(c) $(2f + 3g)'(1)$

(d) $(fg)'(1)$

(e) $\left(\frac{f}{g}\right)'(1)$

(f) $\left(\frac{g}{f}\right)'(1)$

46 Suppose that f, g, and h are differentiable functions at the number 2, and let $f(2) = -2$, $f'(2) = 3$, $g(2) = -5$, $g'(2) = 1$, $h(2) = 2$, and $h'(2) = 4$. Use the differentiation rules to find

(a) $(f + g + h)'(2)$

(b) $(2f - g + 3h)'(2)$

(c) $(fgh)'(2)$

(d) $\left(\frac{fg}{h}\right)'(2)$

47 Find the slope of the tangent line to the graph of the function f at the point whose x coordinate is 4.

(a) $f(x) = x^3 - 4x^2 - 1$

(b) $f(x) = \frac{3}{4x - 2}$

48 Determine the rate of change of volume with respect to the radius (a) of a sphere and (b) of a right circular cylinder with fixed height h.

49 Find the slope of the tangent line to the graph of $f(x) = x/(x^3 - 2)$ at the point $(1, -1)$.

50 For a thin lens of constant focal length p, the object distance x and the image distance y are related by $(1/x) + (1/y) = 1/p$ (Figure 1). (a) Solve for y in terms of x and p. (b) Find the rate of change of y with respect to x.

Figure 1

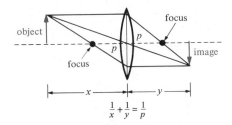

$$\frac{1}{x} + \frac{1}{y} = \frac{1}{p}$$

51 Criticize the following erroneous argument: We wish to compute the value of the derivative of $f(x) = 2x^2 + 3x - 1$ at $x = 2$. To this end, we put $x = 2$ and we have $f(2) = 2(2)^2 + 3(2) - 1 = 13$. But $D_x13 = 0$, so $f'(2) = 0$.

52 Show that the reciprocal rule is a special case of the quotient rule when the numerator is the constant function $f(x) = 1$.

53 An object is moving along a straight line in such a way that at the end of t seconds its distance s in feet from the starting point is given by $s = 8t + (2/t)$, with $t > 0$. Find the speed of the object at the instant when $t = 2$ seconds.

C **54** In a certain electric circuit, the electric power P (in watts) is related to the load resistance R (in ohms) by the equation $P = 100R(0.5 + R)^{-2}$ for $R \geq 0$. Find the rate of change dP/dR of the power P with respect to the resistance R when $R = 10$ ohms.

55 Wildlife biologists predict that the population N of a certain endangered species after t years will be given by the equation $N = (3t + 150)(50 - t)$ for $0 \leq t \leq 50$ years. If this prediction is correct, find the rate of change dN/dt of the population 20 years from now. (*Note:* Here N really isn't a continuous variable, but we treat it as if it were, in order to obtain an approximation to the rate at which the population will be declining in 20 years.)

56 The cost C (in dollars) to a refinery for refining x million gallons of gasoline in 1 month is given by the equation $C = 120,000 + 200,000x - 16,000x^2$. Find the rate of change dC/dx of C with respect to x when $x = 4.5$ million gallons.

C **57** Environmentalists determine that if untreated sewage is discharged into a lake, it will decrease the amount A of dissolved oxygen in moles per cubic meter of water in such a way that

$$A = 14.4 \left(\frac{t + 8}{t^2 + 16t + 65} \right)$$

where t is the number of days the sewage has been flowing into the lake. Find the rate of change dA/dt of dissolved oxygen per day when $t = 4$ days.

2.4   Tangent and Normal Lines

Suppose that the function f is differentiable at x_1, so that $f'(x_1)$ is the slope of the tangent line to the graph of f at the point $(x_1, f(x_1))$. If $y_1 = f(x_1)$, an equation of this tangent line in point-slope form is

$$y - y_1 = f'(x_1)(x - x_1)$$

Figure 1

EXAMPLE 1   Find an equation of the tangent line to the graph of $f(x) = 4 - x^2$ at the point $(1, 3)$. Sketch the graph.

SOLUTION   Here, $f'(x) = -2x$, so that $f'(1) = -2$. An equation of the tangent line is

$$y - 3 = -2(x - 1) \qquad \text{or} \qquad y = -2x + 5$$

(See Figure 1.)

The **normal line** to the graph of f at the point (x_1, y_1) is defined to be the line through (x_1, y_1) that is perpendicular to the tangent line at (x_1, y_1) (Figure 2). Since $f'(x_1)$ is the slope of the tangent line to the graph of f at (x_1, y_1) it follows from Theorem 3 in Section 1.3 that $-1/f'(x_1)$ is the slope of the normal line at (x_1, y_1). Consequently, an equation of the normal line in point-slope form is

$$y - y_1 = \frac{-1}{f'(x_1)}(x - x_1)$$

Figure 2

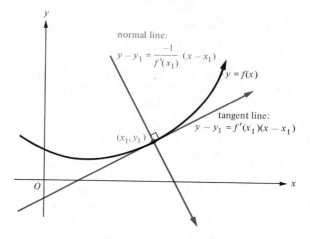

EXAMPLE 2   Find the equations of the tangent and normal lines to the graph of $f(x) = 1/x$ at the point $(\frac{1}{2}, 2)$. Illustrate graphically.

SOLUTION   Here, $f'(x) = -1/x^2$, so that $f'(\frac{1}{2}) = -1/(\frac{1}{2})^2 = -4$. Therefore, an equation of the tangent line is

$$y - 2 = -4(x - \tfrac{1}{2}) \qquad \text{or} \qquad y = -4x + 4$$

Since the tangent line at $(\frac{1}{2}, 2)$ has slope -4, the normal line at this point has slope $-1/(-4) = \frac{1}{4}$. Hence, an equation of the normal line is

$$y - 2 = \frac{1}{4}\left(x - \frac{1}{2}\right) \qquad \text{or} \qquad y = \frac{x}{4} + \frac{15}{8}$$

(See Figure 3.)

Figure 3

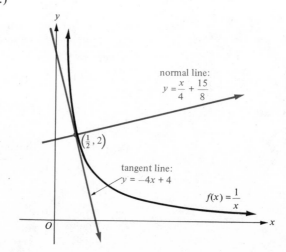

It is sometimes useful to be able to find the point (or points) $(x_1, f(x_1))$ on the graph of a differentiable function f at which the tangent line has a prescribed direction. If the slope of a line in this direction is m, it is only necessary to solve the equation $f'(x_1) = m$ for x_1 in order to find the desired value (or values) of x_1. This technique is illustrated by the following example.

EXAMPLE 3   If $f(x) = 2x^2 - x$, find the point on the graph of f where the tangent line is parallel to the line $3x - y - 4 = 0$, find an equation of the tangent line at this point, and sketch the graph.

SOLUTION   Here, $f'(x) = 4x - 1$. The line $3x - y - 4 = 0$ has slope 3; hence, the x coordinate x_1 of the desired point must satisfy the equation $f'(x_1) = 3$; that is, $4x_1 - 1 = 3$. It follows that $x_1 = 1$; hence, the desired point on the graph of f is given by $(x_1, f(x_1)) = (1, f(1)) = (1, 1)$. An equation of the tangent line at $(1, 1)$ is

$$y - 1 = 3(x - 1) \qquad \text{or} \qquad y = 3x - 2$$

(See Figure 4.)

Figure 4

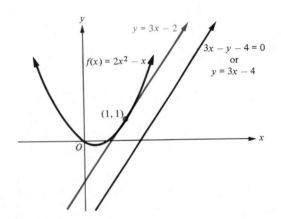

Horizontal Tangents

A tangent line with slope zero is called a **horizontal tangent.**

EXAMPLE 4   Find all points on the graph of $f(x) = x^3 - 3x^2 + 5$ at which the tangent line is horizontal.

SOLUTION   The slope of the tangent line at $(x, f(x))$ is given by

$$f'(x) = 3x^2 - 6x$$

The tangent line at the point $(x, f(x))$ is horizontal if and only if $f'(x) = 0$; that is, if

$$3x^2 - 6x = 0 \qquad \text{or} \qquad 3x(x - 2) = 0$$

The solutions of the last equation are $x = 0$ and $x = 2$. Since

$$f(0) = 0^3 - 3(0)^2 + 5 = 5$$

and $$f(2) = 2^3 - 3(2)^2 + 5 = 1$$

the graph of f has horizontal tangents at the two points $(0, 5)$ and $(2, 1)$.

In Figure 5, we have sketched the graph of $f(x) = x^3 - 3x^2 + 5$ and drawn the horizontal tangents at $(0, 5)$ and $(2, 1)$. Notice that the point $(0, 5)$ is at the "crest of a hill," and the point $(2, 1)$ is at the "bottom of a valley" on this graph. Thus, the point $(0, 5)$ is higher than all its immediate neighboring points on the graph, although the graph eventually climbs even higher. A point such as $(0, 5)$ is called a **relative maximum point** of the graph of f. Similarly, a point such as $(2, 1)$ that is lower than all its immediate neighboring points on the graph is called a **relative minimum point** of the graph of f (Figure 6). The following definitions apply to the x coordinates (abscissas) of such points.

Figure 5

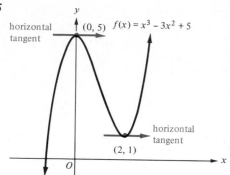

Figure 6

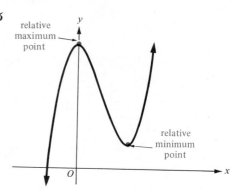

DEFINITION 1

Relative Maximum

A function f is said to have a **relative maximum** (or a **local maximum**) at a number c if there is an open interval I containing c such that f is defined on I and $f(c) \geq f(x)$ holds for every number x in I.

Figure 7

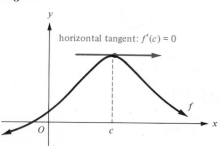

DEFINITION 2

Relative Minimum

A function f is said to have a **relative minimum** (or a **local minimum**) at a number c if there is an open interval I containing c such that f is defined on I and $f(c) \leq f(x)$ holds for every number x in I.

If a function f has either a relative maximum or a relative minimum at a number c, we say that f has a **relative extremum** at c. It is geometrically clear that if a function f has a relative extremum at a number c and if the graph of f has a nonvertical tangent line at $(c, f(c))$, then this tangent line must be horizontal; that is, $f'(c) = 0$ (Figure 7). This important result is proved in the following theorem.

THEOREM 1

Necessary Condition for Relative Extrema

If the function f has a relative extremum at the number c and if f is differentiable at c, then $f'(c) = 0$.

PROOF

We consider the case in which f has a relative maximum at c. The case of a relative minimum can be handled similarly (Problem 36). Because f is differentiable at c,

$$f'(c) = \lim_{\Delta x \to 0} \frac{f(c + \Delta x) - f(c)}{\Delta x}$$

Using Property 14 on page 64, we can rewrite this limit in the alternative form

$$f'(c) = \lim_{x \to c} \frac{f(x) - f(c)}{x - c}$$

We must prove $f'(c) = 0$. If we prove that $f'(c)$ can be neither positive nor negative, then it must be zero, and our argument will be complete. We prove that $f'(c)$ is not negative by showing that if it were, then, contrary to hypothesis, f could not have a relative maximum at c. [That $f'(c)$ is not positive can be shown similarly.] Thus, suppose $f'(c)$ is negative. Since we can make $[f(x) - f(c)]/(x - c)$ as close as we please to the negative number $f'(c)$ by taking x sufficiently close to c (but not equal to c), there is a small open interval I containing c such that $[f(x) - f(c)]/(x - c)$ is negative if x is different from c and belongs to I.

If x belongs to I and $x < c$, then $x - c < 0$ and

$$\frac{f(x) - f(c)}{x - c} < 0$$

But, if a fraction and its denominator are both negative, then the numerator must be positive; hence,

$$\text{if } x \text{ belongs to } I \text{ and } x < c \quad \text{then} \quad f(x) > f(c)$$

The last statement contradicts the hypothesis that f has a relative maximum at c, since it says that the function values $f(x)$ are larger than $f(c)$ for values of x slightly to the left of c. This is the promised contradiction, and the proof is complete. ∎

We study relative extrema in more detail in Sections 3.2 and 3.3. As you can see, horizontal tangents will have an important role to play in this connection.

Problem Set 2.4

In Problems 1 to 14, find the equations of the tangent and normal lines to the graph of the function at the indicated point. Illustrate graphically in Problems 1 to 5.

1 $f(x) = 2x^2 - 7$ at $(2, 1)$

2 $F(x) = 5 + 2x - x^2$ at $(0, 5)$

3 $g(x) = x^2 + x + 1$ at $(1, 3)$

4 $G(x) = (x - 1)^2$ at $(1, 0)$

5 $h(x) = \dfrac{2}{x - 1}$ at $(3, 1)$ **6** $H(x) = x\sqrt{x}$ at $(9, 27)$

7 $p(x) = x^3 - 8x^2 + 9x + 20$ at $(4, -8)$

8 $P(x) = x + \dfrac{2}{\sqrt{x}}$ at $(4, 5)$

9 $q(x) = 4x^4 - 4x^3 - 25x^2 + x + 6$ at $(3, 0)$

10 $Q(x) = \dfrac{1}{x^2 + 1}$ at $(0, 1)$

11 $r(x) = \dfrac{x + 1}{x - 1}$ at $(2, 3)$

12 $R(x) = \dfrac{1 + x + x^2}{3 - x + x^2}$ at $(1, 1)$

13 $s(x) = \dfrac{x}{\sqrt{x} - 1}$ at $(4, 4)$

14 $S(x) = ax^2 + bx + c$ at $(0, c)$

15 Find the point where the tangent line to the graph of $f(x) = 2\sqrt{x}$ at $(1, 2)$ crosses (a) the x axis and (b) the y axis.

16 Find the point where the normal line to the graph of $f(x) = 2/x$ at $(1, 2)$ crosses (a) the x axis and (b) the y axis.

17 At what point on the curve $y = x^2 + 8$ is the slope of the tangent line 16? Write an equation for this tangent line.

18 Suppose that the function f is differentiable at a and that $f'(a) \neq 0$. Show that the tangent line to the graph of f at the point $(a, f(a))$ intersects the x axis at the point with x coordinate $a - [f(a)/f'(a)]$.

In Problems 19 to 24, find a point on the graph of the given function where the tangent or normal line satisfies the indicated condition, and then write an equation for this line.

19 The tangent line to $f(x) = x - x^2$ is parallel to the line $x + y - 2 = 0$.

20 The tangent line to $f(x) = 2x^3 - x^2$ is parallel to the line $4x - y + 3 = 0$.

21 The normal line to $f(x) = \sqrt{x}$ is parallel to the line $4x + y - 4 = 0$.

22 The normal line to $f(x) = x - (1/x)$ is parallel to the line $x + 2y - 3 = 0$.

23 The tangent line to $f(x) = 5 + x^2$ intersects the x axis at the point $(2, 0)$.

24 The normal line to $f(x) = 3x^2 + 2x + 1$ contains the point $(9, 5)$.

In Problems 25 to 34, find all points on the graph of the function at which the tangent line is horizontal.

25 $f(x) = (x - 3)(x - 2)$ **26** $g(x) = x + x^{-1}$

27 $F(x) = 3x^2 + 5x + 6$ **28** $G(x) = x^3 - 6x^2 + 9x + 4$

29 $h(x) = \dfrac{x^3}{3} + 2x^2 - 5x - 1$ **30** $H(x) = x^{-1} + x^{-2}$

31 $q(x) = 2\sqrt{x} - x + 1$

32 $Q(x) = \dfrac{x^2 + 2x - 1}{x - 3}$

33 $r(x) = \dfrac{x^2}{x + 1}$

34 $R(x) = ax^3 + bx^2 + cx + d$

35 Determine a value of the constant b so that the graph of $y = x^2 + bx + 17$ has a horizontal tangent at $(2, 21 + 2b)$.

36 Complete the proof of Theorem 1 by considering the case in which f has a relative minimum at c.

37 Let $f(x) = x^3$. (a) Show that the graph of f has a horizontal tangent at $(0, 0)$. (b) Show that f has neither a relative maximum nor a relative minimum at 0. (c) Explain why the results in (a) and (b) do not contradict Theorem 1.

38 In Section 1.5, we showed that the vertex of the parabola $f(x) = ax^2 + bx + c$ is

$$\left(-\frac{b}{2a}, f\left(-\frac{b}{2a}\right)\right)$$

Derive this same result by using Theorem 1.

39 If the graph of a function f has a horizontal tangent at the point $(c, f(c))$, what is the equation of the normal line at this point?

2.5   Rules for Differentiating Trigonometric Functions

In Section 1.8, we showed that

$$\lim_{x \to 0} \frac{\sin x}{x} = 1 \quad \text{and} \quad \lim_{x \to 0} \frac{1 - \cos x}{x} = 0$$

(See Theorem 2, page 66, and Theorem 3, page 67.) Using these two special limits, we can prove the following theorem, which provides a rule for the derivative of the sine function.

THEOREM 1   **Derivative of the Sine Function**

> The sine function is differentiable at each real number, and the derivative of the sine function is the cosine function. In symbols,
>
> $$D_x \sin x = \cos x \quad \text{or} \quad \frac{d}{dx} \sin x = \cos x$$

PROOF

$$D_x \sin x = \lim_{\Delta x \to 0} \frac{\sin (x + \Delta x) - \sin x}{\Delta x}$$

We have $\sin (x + \Delta x) = \sin x \cos \Delta x + \sin \Delta x \cos x$ by the addition formula for sine; hence,

$$D_x \sin x = \lim_{\Delta x \to 0} \frac{\sin x \cos \Delta x + \sin \Delta x \cos x - \sin x}{\Delta x}$$

$$= \lim_{\Delta x \to 0} \left[\frac{\sin x(\cos \Delta x - 1)}{\Delta x} + \frac{\sin \Delta x \cos x}{\Delta x} \right]$$

$$= \lim_{\Delta x \to 0} \left[(-\sin x)\frac{1 - \cos \Delta x}{\Delta x} + \cos x \frac{\sin \Delta x}{\Delta x} \right]$$

$$= (-\sin x) \lim_{\Delta x \to 0} \frac{1 - \cos \Delta x}{\Delta x} + \cos x \lim_{\Delta x \to 0} \frac{\sin \Delta x}{\Delta x}$$

$$= (-\sin x)(0) + (\cos x)(1) = \cos x$$

EXAMPLE 1   Differentiate $f(x) = x^2 \sin x$.

SOLUTION   By the product rule, the power rule, and Theorem 1,

$$f'(x) = D_x(x^2 \sin x) = (D_x x^2)(\sin x) + x^2(D_x \sin x)$$
$$= 2x \sin x + x^2 \cos x$$

Figure 1 shows the graph of the sine function and its derivative, the cosine function, on the same coordinate system. Because

$$\cos x = \sin \left(x + \frac{\pi}{2} \right)$$

the graph of $y = \cos x$ "lags behind" the graph of $y = \sin x$ by $\pi/2$ units, although both graphs have exactly the same shape. The fact that the cosine function is the derivative (rate of change) of the sine function is nicely illustrated by Figure 1. For instance, notice that when the sine function is increasing most rapidly—namely at $x = 0$, $\pm 2\pi$, $\pm 4\pi$, and so on—its derivative, the cosine function, takes on its maximum value 1. Likewise, when the sine function is decreasing most rapidly—namely at $x = \pm \pi$, $\pm 3\pi$, $\pm 5\pi$, and so on—its derivative, the cosine function, takes on its minimum value -1.

Figure 1

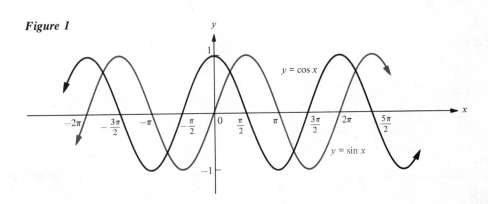

According to Theorem 1 in Section 2.4, when the sine function takes on maximum or minimum values, its derivative, the cosine function, should have the value 0. This is also confirmed by the graphs in Figure 1. Indeed, when $x = \pm\pi/2$, $\pm 3\pi/2$, $\pm 5\pi/2$, and so on, $\sin x$ takes on maximum or minimum values ± 1, and $\cos x = 0$.

The rule for the derivative of the cosine is established in Theorem 2, whose proof is similar to the proof of Theorem 1.

THEOREM 2

Derivative of the Cosine Function

> The cosine function is differentiable at each real number, and the derivative of the cosine function is the negative of the sine function. In symbols,
>
> $$D_x \cos x = -\sin x \qquad \text{or} \qquad \frac{d}{dx} \cos x = -\sin x$$

PROOF

$$D_x \cos x = \lim_{\Delta x \to 0} \frac{\cos (x + \Delta x) - \cos x}{\Delta x}$$

We have $\cos (x + \Delta x) = \cos x \cos \Delta x - \sin x \sin \Delta x$ by the addition formula for cosine; hence,

$$D_x \cos x = \lim_{\Delta x \to 0} \frac{\cos x \cos \Delta x - \sin x \sin \Delta x - \cos x}{\Delta x}$$

$$= \lim_{\Delta x \to 0} \left[\frac{\cos x (\cos \Delta x - 1)}{\Delta x} - \frac{\sin x \sin \Delta x}{\Delta x} \right]$$

$$= \lim_{\Delta x \to 0} \left[(-\cos x) \frac{1 - \cos \Delta x}{\Delta x} - \sin x \frac{\sin \Delta x}{\Delta x} \right]$$

$$= (-\cos x) \lim_{\Delta x \to 0} \frac{1 - \cos \Delta x}{\Delta x} - \sin x \lim_{\Delta x \to 0} \frac{\sin \Delta x}{\Delta x}$$

$$= (-\cos x)(0) - (\sin x)(1)$$

$$= -\sin x$$

EXAMPLE 2   Differentiate $g(t) = \dfrac{\cos t}{t^2 + 4}$.

SOLUTION   By the quotient rule and Theorem 2,

$$g'(t) = D_t \frac{\cos t}{t^2 + 4} = \frac{(t^2 + 4)(D_t \cos t) - (\cos t)[D_t (t^2 + 4)]}{(t^2 + 4)^2}$$

$$= \frac{(t^2 + 4)(-\sin t) - (\cos t)(2t)}{(t^2 + 4)^2}$$

$$= \frac{-(t^2 + 4)(\sin t) - 2t \cos t}{(t^2 + 4)^2}$$

Now the derivatives of the remaining trigonometric functions are easily obtained by using the derivatives of the sine and cosine functions together with the differentiation rules for quotients and reciprocals.

THEOREM 3   **Derivatives of Tan, Cot, Sec, and Csc Functions**

(i) $D_x \tan x = \sec^2 x$ **(ii)** $D_x \cot x = -\csc^2 x$

(iii) $D_x \sec x = \sec x \tan x$ **(iv)** $D_x \csc x = -\csc x \cot x$

hold for all values of x in the domains of the respective functions.

PROOF   We prove (i) and (iii), leaving (ii) and (iv) as exercises (Problem 45).

(i) $D_x \tan x = D_x \dfrac{\sin x}{\cos x} = \dfrac{\cos x\, D_x \sin x - \sin x\, D_x \cos x}{\cos^2 x}$

$$= \frac{\cos x \cos x - \sin x(-\sin x)}{\cos^2 x} = \frac{\cos^2 x + \sin^2 x}{\cos^2 x}$$

$$= \frac{1}{\cos^2 x} = \sec^2 x$$

(iii) $D_x \sec x = D_x \dfrac{1}{\cos x} = \dfrac{-D_x \cos x}{\cos^2 x} = \dfrac{-(-\sin x)}{\cos^2 x}$

$$= \frac{1}{\cos x} \cdot \frac{\sin x}{\cos x} = \sec x \tan x$$

EXAMPLE 3   Find $D_x(2x \tan x)$.

SOLUTION $D_x(2x \tan x) = (D_x\, 2x)(\tan x) + 2x(D_x \tan x)$
$$= 2 \tan x + 2x \sec^2 x$$

EXAMPLE 4   Find $\dfrac{dy}{dt}$ if $y = \dfrac{1}{\sec t + 1}$.

SOLUTION $\dfrac{dy}{dt} = \dfrac{d}{dt}\left(\dfrac{1}{\sec t + 1}\right) = \dfrac{-\dfrac{d}{dt}(\sec t + 1)}{(\sec t + 1)^2} = \dfrac{-\sec t \tan t}{(\sec t + 1)^2}$

EXAMPLE 5   Find equations for the tangent and normal lines to the graph of $f(x) = 2 \cot x$ at the point $(\pi/6, 2\sqrt{3})$.

SOLUTION   Here,

$$f'(x) = \frac{d}{dx}(2 \cot x) = 2 \frac{d}{dx} \cot x = 2(-\csc^2 x) = -2 \csc^2 x$$

so that

$$f'\left(\frac{\pi}{6}\right) = -2 \csc^2 \frac{\pi}{6} = -2(2)^2 = -8$$

Thus, an equation of the tangent line at $(\pi/6, 2\sqrt{3})$ is

$$y - 2\sqrt{3} = -8\left(x - \frac{\pi}{6}\right) \qquad \text{or} \qquad y = -8x + \frac{4\pi}{3} + 2\sqrt{3}$$

and an equation of the normal line at the same point is

$$y - 2\sqrt{3} = \frac{1}{8}\left(x - \frac{\pi}{6}\right) \qquad \text{or} \qquad y = \frac{x}{8} + 2\sqrt{3} - \frac{\pi}{48}$$

EXAMPLE 6   Find all values of x with $-\pi < x \le \pi$ that are x coordinates of points where the graph of $y = \sin x - \cos x$ has a horizontal tangent.

SOLUTION   Here,

$$\frac{dy}{dx} = \frac{d}{dx}(\sin x - \cos x) = \frac{d}{dx}\sin x - \frac{d}{dx}\cos x = \cos x - (-\sin x)$$

$$= \cos x + \sin x$$

The condition for a horizontal tangent is $dy/dx = 0$; that is,

$$\cos x + \sin x = 0 \qquad \text{or} \qquad \sin x = -\cos x$$

Squaring both sides of the last equation, we obtain

$$\sin^2 x = \cos^2 x \qquad \text{or} \qquad \sin^2 x = 1 - \sin^2 x$$

that is,

$$2\sin^2 x = 1 \qquad \text{or} \qquad \sin^2 x = \tfrac{1}{2}$$

Therefore,

$$\sin x = \pm\frac{1}{\sqrt{2}} = \pm\frac{\sqrt{2}}{2}$$

The values of x between $-\pi$ and π that satisfy the last equation are

$$x = -\frac{3\pi}{4},\ -\frac{\pi}{4},\ \frac{\pi}{4},\ \frac{3\pi}{4}$$

Because we squared both sides of an equation, we may have introduced extraneous roots, so we must check each of these possible values of x in the original equation $\cos x + \sin x = 0$. Doing so, we find that the only solutions are

$$x = -\frac{\pi}{4} \qquad \text{and} \qquad x = \frac{3\pi}{4}$$

■

The functions $\sin x$ and $\cos x$ are said to be **cofunctions** of each other. The same terminology is applied to $\tan x$ and $\cot x$ as well as to $\sec x$ and $\csc x$. As you can see from Theorems 1, 2, and 3, the rule for differentiating a cofunction is obtained from the rule for differentiating the corresponding function by changing function to cofunction and introducing a negative sign. Thus, if you learn only the three rules

$$\begin{array}{c} D_x \sin x = \cos x \\ D_x \tan x = \sec^2 x \\ D_x \sec x = \sec x \tan x \end{array}$$

you can easily recall the corresponding cofunction rules

$$\begin{array}{c} D_x \cos x = -\sin x \\ D_x \cot x = -\csc^2 x \\ D_x \csc x = -\csc x \cot x \end{array}$$

whenever you need them.

Problem Set 2.5

In Problems 1 to 30, differentiate the function.

1 $f(x) = 7 \sin x$

2 $g(x) = -x \cos x$

3 $h(t) = 4 \sin t - t \cos t$

4 $F(r) = \sqrt{r} \cos r$

5 $g(x) = 3 \tan x + \sec x$

6 $G(t) = t^7 - 5 \cot t$

7 $H(y) = 8 \sec y - \frac{1}{3}y^6$

8 $f(z) = 4 \csc z - 3 \sec z$

9 $g(r) = r^4 \sin r + 4 \csc r$

10 $H(u) = (\sqrt{u} + 5)(\cos u)$

11 $f(z) = \cot z + \sqrt{z} \tan z$

12 $f(t) = t\sqrt{t} \sin t - \cot t + \pi$

13 $p(x) = \sin x \cos x$

14 $F(v) = 2 \sin v(1 - \tan v)$

15 $g(y) = -7 \cot y \csc y$

16 $f(\theta) = \cos^2 \theta - \sin^2 \theta$

17 $H(x) = 3 \sec x(1 - \tan x)$

18 $h(x) = \sin x \cos x \tan x$

19 $f(\theta) = \dfrac{2 \sin \theta}{\theta + 5}$

20 $G(y) = \dfrac{7 \cos y}{\sqrt{y}}$

21 $p(x) = \dfrac{x^2 + 5}{\cos x - 1}$

22 $P(x) = \dfrac{\cos x}{5 + 3 \sin x}$

23 $q(t) = \dfrac{3 \tan t}{\sec t + 4}$

24 $Q(u) = \dfrac{1 - \sin u}{2 + \sin u}$

25 $r(y) = \dfrac{3 + \cos y}{3 - \cos y}$

26 $R(v) = \dfrac{1 - \cot v}{1 + \cot v}$

27 $f(z) = \dfrac{\csc z}{1 + \tan z}$

28 $S(w) = \dfrac{3}{\sqrt{w} \cos w}$

29 $F(\theta) = \dfrac{3 \sec \theta}{2 \cos \theta - \sin \theta}$

30 $q(t) = \dfrac{\sqrt{t}}{\sin t - 2 \cos t}$

In Problems 31 to 36, find equations for the tangent and normal lines to the graph of each function at the given point.

31 $f(x) = 2 \sin x$ at $\left(\dfrac{\pi}{6}, 1\right)$

32 $g(x) = 4 \cos x$ at $\left(\dfrac{\pi}{3}, 2\right)$

33 $h(x) = 3 \tan x$ at $\left(\dfrac{\pi}{4}, 3\right)$

34 $F(x) = -3 \cot x$ at $\left(-\dfrac{\pi}{6}, 3\sqrt{3}\right)$

35 $g(x) = 2x - 5 \sin x$ at $(\pi, 2\pi)$

36 $H(x) = x - \tan x$ at (π, π)

In Problems 37 and 38, find all values of x with $-\pi < x \le \pi$ that are x coordinates of points where the tangent to the graph of the given equation is horizontal.

37 $y = \sin x + \cos x$

38 $y = \cos^2 x + 2 \sin x$

39 If air resistance is neglected, it can be shown that a projectile shot into the air at an angle θ with the horizontal will reach a maximum height h given by

$$h = \frac{v^2 \sin^2 \theta}{2g}$$

where v is the muzzle velocity of the projectile and g is the acceleration of gravity (Figure 2). Assuming that v and g are constant, find a formula for the rate of change $dh/d\theta$ of maximum height with respect to the angle θ. (*Hint:* Because $\sin^2 \theta = \sin \theta \sin \theta$, you can use the multiplication rule to help find the derivative of $\sin^2 \theta$.)

Figure 2

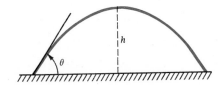

40 A 2-meter fence stands 3 meters from a high wall (Figure 3). The length L of the shortest ladder that can reach the wall from outside the fence and that makes an angle θ with the horizontal is given by $L = 3 \sec \theta + 2 \csc \theta$. Find the rate of change of L with respect to θ at $\theta = \pi/6$.

Figure 3

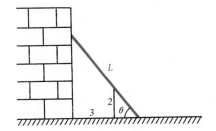

41 An object is oscillating about the origin on the x axis in such a way that its x coordinate at time t seconds is given by $x = 2 \sin t$. Find the rate of change dx/dt of x with respect to t at the instant when the object, moving to the right, passes through the origin.

42 Suppose that a is a constant. (a) Using the substitution property for limits, show that

$$\lim_{\Delta x \to 0} \frac{\sin (a\, \Delta x)}{\Delta x} = a \quad \text{and} \quad \lim_{\Delta x \to 0} \frac{1 - \cos (a\, \Delta x)}{\Delta x} = 0$$

(b) Using the results of part (a) and arguments similar to those in Theorems 1 and 2, show that

$$D_x \sin ax = a \cos ax \quad \text{and} \quad D_x \cos ax = -a \sin ax$$

43 A tractor is pulling a heavy sledge weighing W newtons with a cable making an angle θ with the horizontal (Figure 4). If μ is the coefficient of sliding friction between the sledge and the ground, then the tension T newtons in the cable is given by

$$T = \frac{\mu W}{\cos \theta + \mu \sin \theta}$$

Assuming that μ and W are constants, find a formula for the rate of change $dT/d\theta$ of the tension T with respect to the angle θ.

Figure 4

[C]**44** Because of seasonal variations in demand, the revenue R to a swimsuit manufacturer on the xth day of the year, starting with $x = 1$ on January 1, is given by

$$R = 3800 - 1800 \cos \frac{2\pi}{365}x \quad \text{dollars}$$

Find the rate of change dR/dx of revenue per day on June 1 when $x = 152$. (*Hint:* Use the result in Problem 42b.)

45 Complete the proof of Theorem 3 by proving parts (ii) and (iv).

2.6 Function Composition

We now introduce the idea of *function composition;* in Section 2.7 we use this idea to develop an important differentiation rule called the *chain rule*.

Suppose we have three variable quantities y, u, and x. If y depends on u and if u, in turn, depends on x, then clearly y depends on x. In other words, if y is a function of u and u is a function of x, then y is a function of x. For instance, suppose that

$$y = u^2 \quad \text{and} \quad u = 3x - 1$$

Then, substituting the value of u from the second equation into the first equation, we find that

$$y = (3x - 1)^2$$

More generally, if

$$y = f(u) \quad \text{and} \quad u = g(x)$$

then, substituting u from the second equation into the first equation, we find that

$$y = f(g(x))$$

To avoid a pileup of parentheses, we often replace the outside parentheses in the last equation by square brackets and write

$$y = f[g(x)]$$

If f and g are functions, the equation $y = f[g(x)]$ defines a new function h:

$$h(x) = f[g(x)]$$

The function h obtained by "chaining" f and g together in this way is called the *composition* of f and g and is written $h = f \circ g$. The idea of function composition is made precise in the following definition.

DEFINITION 1   **Composition of Functions**

Let f and g be two functions satisfying the condition that at least one number in the range of g belongs to the domain of f. Then the **composition** of f and g, in symbols $f \circ g$, is the function defined by the equation

$$(f \circ g)(x) = f[g(x)]$$

Evidently, the domain of the composite function $f \circ g$ is the set of all values of x in the domain of g such that $g(x)$ belongs to the domain of f. The range of $f \circ g$ is just the set of all numbers of the form $f[g(x)]$ as x runs through the domain of $f \circ g$.

EXAMPLE 1   Let $f(x) = 3x - 1$ and $g(x) = x^3$. Find

(a) $(f \circ g)(2)$ (b) $(g \circ f)(2)$
(c) $(f \circ g)(x)$ (d) $(g \circ f)(x)$
(e) $(f \circ f)(x)$ ©(f) $(f \circ g)(3.007)$

SOLUTION

(a) $(f \circ g)(2) = f[g(2)] = f(2^3) = f(8) = 3(8) - 1 = 23$
(b) $(g \circ f)(2) = g[f(2)] = g[3(2) - 1] = g(5) = 5^3 = 125$
(c) $(f \circ g)(x) = f[g(x)] = f(x^3) = 3x^3 - 1$
(d) $(g \circ f)(x) = g[f(x)] = g(3x - 1) = (3x - 1)^3$
(e) $(f \circ f)(x) = f[f(x)] = f(3x - 1) = 3(3x - 1) - 1 = 9x - 4$
(f) $(f \circ g)(3.007) = f[g(3.007)] = 3(3.007)^3 - 1 \approx 80.57$

EXAMPLE 2   Let $f(x) = 3x - 1$, $g(x) = x^3$, and $h(x) = \sqrt{x}$. Find $[f \circ (g \circ h)](x)$.

SOLUTION

$$[f \circ (g \circ h)](x) = f[(g \circ h)(x)] = f\{g[h(x)]\} = f[g(\sqrt{x})] = f[(\sqrt{x})^3]$$
$$= f(x^{3/2}) = 3x^{3/2} - 1$$

EXAMPLE 3   Let $f(x) = 3x - 1$, $g(x) = x^3$, and $p(x) = \frac{1}{3}(x + 1)$. Show that $f \circ (g + p)$ is not the same function as $(f \circ g) + (f \circ p)$.

SOLUTION   On the one hand,

$$[f \circ (g + p)](x) = f[(g + p)(x)] = f[g(x) + p(x)] = f[x^3 + \frac{1}{3}(x + 1)]$$
$$= 3[x^3 + \frac{1}{3}(x + 1)] - 1 = 3x^3 + x$$

On the other hand,

$$[(f \circ g) + (f \circ p)](x) = (f \circ g)(x) + (f \circ p)(x) = f[g(x)] + f[p(x)]$$
$$= f(x^3) + f[\frac{1}{3}(x + 1)]$$
$$= 3x^3 - 1 + 3[\frac{1}{3}(x + 1)] - 1$$
$$= 3x^3 + x - 1$$

Therefore, the functions $f \circ (g + p)$ and $(f \circ g) + (f \circ p)$ are not the same.

Although the symbolism $f \circ g$ for the composition of f and g looks vaguely like some kind of a "product," you must not confuse it with the actual product $f \cdot g$ of f and g. Whereas $f \cdot g = g \cdot f$, note (in Example 1) that $f \circ g \neq g \circ f$.

Whereas $f \cdot (g + p) = (f \cdot g) + (f \cdot p)$, note (in Example 3) that $f \circ (g + p) \neq (f \circ g) + (f \circ p)$. It does turn out, however, that function composition is **associative;** that is,

$$f \circ (g \circ h) = (f \circ g) \circ h$$

(Problem 12). Because of the associative property of composition, parentheses aren't really necessary when three or more functions are to be composed. Thus we simply write $f \circ g \circ h, f \circ g \circ h \circ p$, and so forth. Composing a function f with itself is called **iteration**. The successive iterates of f are $f \circ f, f \circ f \circ f, f \circ f \circ f \circ f$, and so forth. For an interesting account of the use of function iteration in building mathematical models for phenomena in physics, engineering, economics, and the life sciences, see Douglas R. Hofstadter, "Strange attractors: Mathematical patterns delicately poised between order and chaos," *Scientific American,* Vol. 245, No. 5, pp. 22–43, November 1981.

Using a programmable calculator, you can see a vivid demonstration of function composition. Suppose that the f and g keys are programmed with functions of your choice. If you enter a number x and touch the g key, you see the mapping

$$x \longmapsto g(x)$$

take place, and the number $g(x)$ appears in the display. Now, if you touch the f key, the mapping

$$g(x) \longmapsto f[g(x)]$$

will take place. Therefore, after you enter the number x, you can perform the composite mapping

$$x \longmapsto (f \circ g)(x)$$

Figure 1

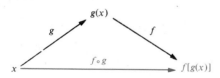

by *touching first the g key, then the f key*. This fact is represented by the diagram in Figure 1.

In using a programmable calculator or a computer, it is important to be able to tell when a complicated function can be obtained as a composition of simpler functions. The same skill is essential in calculus. If

$$h(x) = (f \circ g)(x) = f[g(x)]$$

let's agree to call g the *inside function* and f the *outside function* because of the positions they occupy in the expression $f[g(x)]$. In order to see that h can be obtained as a composition $h = f \circ g$, you must be able to recognize the inside function g and the outside function f in the equation that defines h.

EXAMPLE 4   Express the function $h(x) = (3x + 2)^2$ as a composition $h = f \circ g$ of two functions f and g.

SOLUTION   Here we can take the inside function to be $g(x) = 3x + 2$ and the outside function to be $f(x) = x^2$, so that

$$(f \circ g)(x) = f[g(x)] = f(3x + 2) = (3x + 2)^2 = h(x)$$

The substitution property for limits (page 67) and the substitution property for continuous functions (page 73) can be restated in terms of function composition.

For instance, the *substitution property for continuous functions* can be restated as follows:

> If f and g are functions, g is continuous at a, and f is continuous at $g(a)$, then $f \circ g$ is continuous at a.

We leave it as an exercise for you to restate the *substitution property for limits* in terms of function composition (Problem 44).

Problem Set 2.6

In Problems 1 to 10, let $f(x) = x - 3$ and $g(x) = x^2 + 4$. Find the indicated value.

1 $(f \circ g)(4)$ **2** $(f \circ g)(\sqrt{2})$

C **3** $(g \circ f)(4.73)$ C **4** $(g \circ f)(-2.08)$

5 $(f \circ f)(3)$ **6** $(g \circ g)(-3)$

7 $[f \circ (g \circ f)](2)$ **8** $[(f \circ g) \circ f](2)$

9 $(f \circ g)(x)$ **10** $(g \circ f)(x)$

11 Let $f(x) = \sin x$, $g(x) = x^2$, and $h(x) = \cos x$. Find a formula for the given function.

  (a) $f \circ g$ (b) $g \circ f$ (c) $g \circ g$

  (d) $g \circ (f + h)$ (e) $g \circ (f/h)$ (f) $(f/h) \circ (h/f)$

  (g) $f \circ (g \circ h)$ (h) $(f \circ g) \circ h$

12 Show that function composition is associative; that is, show that $f \circ (g \circ h) = (f \circ g) \circ h$ for any three functions f, g, and h (provided, of course, that either side of the equation is defined).

In Problems 13 to 24, find (a) $(f \circ g)(x)$, (b) $(g \circ f)(x)$, and (c) $(f \circ f)(x)$.

13 $f(x) = x^2$, $g(x) = \sqrt{x}$

14 $f(x) = x^3 + 1$, $g(x) = \sqrt[3]{x - 1}$

15 $f(x) = \tan x$, $g(x) = \sqrt{x}$

16 $f(x) = x + x^{-1}$, $g(x) = \sqrt{x - 1}$

17 $f(x) = |x|$, $g(x) = \csc x$

18 $f(x) = 1/x$, $g(x) = 1/x$

19 $f(x) = x^2 - 1$, $g(x) = 1 + \cos x$

20 $f(x) = ax + b$, $g(x) = cx + d$

21 $f(x) = \dfrac{1}{2x - 3}$, $g(x) = 2x - 3$

22 $f(x) = \dfrac{Ax + B}{Cx + D}$, $g(x) = \dfrac{ax + b}{cx + d}$

23 $f(x) = \dfrac{3x - 1}{x}$, $g(x) = \dfrac{1}{3 - x}$

24 $f(x) = 2$, $g(x) = 7$

In Problems 25 to 32, let $f(x) = 4x$, $g(x) = x^2 - 3$, and $h(x) = \sqrt{x}$. Express each function as a composition of functions chosen from f, g, and h.

25 $F(x) = \sqrt{x^2 - 3}$ **26** $G(x) = (\sqrt{x})^2 - 3$

27 $H(x) = 2\sqrt{x}$ **28** $K(x) = 4x^2 - 12$

29 $Q(x) = 4\sqrt{x}$ **30** $q(x) = 16x^2 - 3$

31 $r(x) = \sqrt[4]{x}$ **32** $s(x) = x^4 - 6x^2 + 6$

In Problems 33 to 40, express each function h as a composition $h = f \circ g$ of two simpler functions f and g.

33 $h(x) = \cos^7 x$ **34** $h(x) = \sin(x^7 + 1)$

35 $h(x) = 1 - \tan^2 x$ **36** $h(x) = 5 \csc |x|$

37 $h(x) = \sqrt{\dfrac{x + 1}{x - 1}}$ **38** $h(x) = \dfrac{1}{(4x + 5)^5}$

39 $h(x) = \dfrac{|x + 1|}{x + 1}$ **40** $h(x) = \sqrt{1 - \sqrt{x - 1}}$

41 Suppose that user-definable keys f and g on a programmable calculator are programmed so that $f : x \longmapsto 3x^2 - 2$ and $g : x \longmapsto 5x - 3$. Draw a mapping diagram (see Figure 1) to show the effect of entering a number x and touching (a) first the g key, then the f key; (b) first the f key, then the g key.

42 Give examples of (a) a function f for which $f \circ f$ is not the same as $f \cdot f$ and (b) a function g for which $g \circ g$ is the same as $g \cdot g$.

43 Explain why the function $f(x) = \sin |x|$ is continuous at $x = 0$.

44 Restate the substitution property for limits (page 67) in terms of function composition.

45 Explain why the function $g(x) = \sin(\sin x)$ is continuous on $\mathbb{R}$.

46 A baseball diamond is a square 90 feet on each side. Suppose a ball is hit directly toward third base at the rate of 50 feet per second. Let y denote the distance in feet of the ball from first base, let x denote its distance from home plate, and let t denote the elapsed time in seconds since the ball was hit. Here y is a function of x, say $y = f(x)$, and x is a function of t, say $x = g(t)$. (a) Find formulas for $f(x)$ and $g(t)$. (b) Find a formula for $(f \circ g)(t)$. (c) Explain why $y = (f \circ g)(t)$.

47 Show that $(f \circ g)(x)$ can be obtained graphically as follows: Start at the point $(x, 0)$ on the x axis. Move vertically to the graph of g, then horizontally to the graph of $y = x$, then vertically to the graph of f, and finally horizontally to the point $(0, y)$ on the y axis (Figure 2). Conclude that $y = (f \circ g)(x)$.

Figure 2

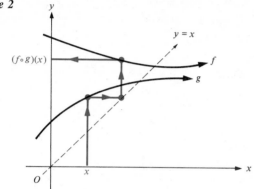

48 Show how to modify the procedure in Problem 47 to find the values $f(x)$, $(f \circ f)(x)$, $(f \circ f \circ f)(x)$, $(f \circ f \circ f \circ f)(x)$, and so on of the successive iterates of f at x. The set consisting of x and these values is called the *orbit* of x under the iterates of f. (See Douglas R. Hofstadter's article, referred to on page 122.)

[C] **49** In the life sciences, a **simple recursive model** for population growth is a function f that predicts the population $f(p)$ of an organism 1 unit of time later if the current population is p. (For instance, the unit of time might be the gestation period of the organism.) Thus, according to this model, if p is the current population, then $f(p)$ is the population 1 unit of time later, $f[f(p)] = (f \circ f)(p)$ is the population 2 units of time later, $f[(f \circ f)(p)] = (f \circ f \circ f)(p)$ is the population 3 units of time later, and so forth. One of the most popular simple recursive growth models is the **discrete logistic model** $f(p) = Bp - Ap^2$, where A and B are nonnegative constants depending on the birth and death rates of the organism. Suppose that $B = 2.6$ and $A = 0.08$ for a certain organism in a particular habitat. Starting with $p = 4$ organisms, use the discrete logistic model to predict the population 1, 2, 3, 4, 5, and 6 units of time later according to the discrete logistic model.

[C] **50** Draw an accurate graph of the function $f(p) = 2.6p - 0.08p^2$ for $0 \le p \le 32.5$. (Use a calculator to help plot points.) (a) Use the graphical procedure obtained in Problem 48 to work Problem 49. (b) Use the same graphical procedure, starting with several different values of the initial population p. Can you draw any (tentative) conclusions?

2.7   The Chain Rule

Suppose that $y = (x^2 + 5x)^3$ and we wish to find dy/dx. One approach is to expand $(x^2 + 5x)^3$ and then differentiate the resulting polynomial. Thus,

$$y = (x^2 + 5x)^3 = x^6 + 15x^5 + 75x^4 + 125x^3$$

and so

$$\frac{dy}{dx} = 6x^5 + 75x^4 + 300x^3 + 375x^2$$

Another approach is to let

$$u = x^2 + 5x$$

so that

$$y = u^3$$

Then, $\dfrac{dy}{du} = 3u^2$ and $\dfrac{du}{dx} = \dfrac{d}{dx}(x^2 + 5x) = 2x + 5$

from which we might conclude that

$$\frac{dy}{dx} = \frac{dy}{du}\frac{du}{dx} = 3u^2(2x + 5) = 3(x^2 + 5x)^2(2x + 5)$$

$$= 6x^5 + 75x^4 + 300x^3 + 375x^2$$

The second approach produced the correct result, but there's a catch to it! The expressions dy/du and du/dx are just symbols for derivatives in which the "numerators" and "denominators" have not yet been given any separate meanings, so we weren't really justified in supposing that

$$\frac{dy}{dx} = \frac{dy}{du}\frac{du}{dx}$$

In fact, the legitimacy of this calculation is guaranteed by one of the most important differentiation rules in calculus—the **chain rule.** Although we give a precise statement and proof of the chain rule later (see Theorem 1), we begin with the following informal version.

The Chain Rule

> If y is a differentiable function of u and u is a differentiable function of x, then y is a differentiable function of x and
>
> $$\frac{dy}{dx} = \frac{dy}{du}\frac{du}{dx}$$

EXAMPLE 1   If $y = u^4$ and $u = 2x^2 + 3x - 1$, find dy/dx.

SOLUTION $$\frac{dy}{dx} = \frac{dy}{du}\frac{du}{dx} = 4u^3(4x + 3) = 4(2x^2 + 3x - 1)^3(4x + 3)$$ ∎

Perhaps the best way to understand the meaning of the chain rule is to think of it in terms of rates of change. The equation

$$\frac{dy}{dx} = \frac{dy}{du}\frac{du}{dx}$$

just says that the rate of change of y per unit change in x is equal to the rate of change of y per unit change in u times the rate of change of u per unit change in x. For instance, if y is increasing *twice* as fast as u and u is increasing *three times* as fast as x, then y is increasing *six times* as fast as x.

Of course, the chain rule can be written in operator notation as

$$D_x y = (D_u y)(D_x u)$$

If we let $y = f(u)$, then $D_u y = f'(u)$, and the chain rule takes the form

> $$D_x f(u) = f'(u)D_x u$$

This is probably the most practical form of the chain rule for routine calculation of derivatives.

EXAMPLE 2   Use the chain rule to find $D_x\sqrt{x^2 + 1}$.

SOLUTION   If we let $f(u) = \sqrt{u}$ and $u = x^2 + 1$, then $f(u) = \sqrt{x^2 + 1}$, $f'(u) = 1/(2\sqrt{u})$, and $D_x u = D_x(x^2 + 1) = 2x$. Therefore,

$$D_x\sqrt{x^2 + 1} = D_x f(u) = f'(u)D_x u = \frac{1}{2\sqrt{u}}(2x) = \frac{x}{\sqrt{x^2 + 1}}$$ ∎

The following crude but effective memory device shows the basic pattern of the chain rule:

$$D\, f(\text{whatever}) = f'(\text{whatever}) \cdot D(\text{whatever})$$

For instance, using this device with f as the square-root function, we would have

$$D\sqrt{\text{whatever}} = \frac{1}{2\sqrt{\text{whatever}}} \cdot D(\text{whatever})$$

Thus, the calculation in Example 2 could have been abbreviated as follows:

$$D_x\sqrt{x^2 + 1} = \frac{1}{2\sqrt{x^2 + 1}} \cdot D_x(x^2 + 1)$$

$$= \frac{1}{2\sqrt{x^2 + 1}}(2x)$$

$$= \frac{x}{\sqrt{x^2 + 1}}$$

The chain rule is often used to calculate derivatives of the form $D_x u^n$, where u is a differentiable function of x and n is an integer. Thus, letting $f(u) = u^n$, so that $f'(u) = nu^{n-1}$, we obtain the important formula

$$\boxed{D_x u^n = nu^{n-1}D_x u}$$

In other words, we have the pattern

$$D(\text{whatever})^n = n(\text{whatever})^{n-1}\, D(\text{whatever})$$

In Examples 3 to 6, find the derivative of each function with the aid of the chain rule.

EXAMPLE 3 $f(x) = (x^2 + 5x)^{100}$

SOLUTION $f'(x) = D_x(x^2 + 5x)^{100}$
$= 100(x^2 + 5x)^{99}\, D_x(x^2 + 5x)$
$= 100(x^2 + 5x)^{99}(2x + 5)$

EXAMPLE 4 $F(x) = \dfrac{1}{(3x - 1)^4}$

SOLUTION $F'(x) = D_x\dfrac{1}{(3x - 1)^4} = D_x(3x - 1)^{-4}$

$= -4(3x - 1)^{-4-1}\, D_x(3x - 1)$
$= -4(3x - 1)^{-5}(3)$
$= -12(3x - 1)^{-5}$

EXAMPLE 5 $g(x) = \sin^3 x$

SOLUTION $g'(x) = D_x(\sin x)^3 = 3(\sin x)^2(D_x \sin x)$
$= 3\sin^2 x \cos x$

EXAMPLE 6 $G(t) = (t^2 + 6t)^{10}(1 - 3t)^4$

SOLUTION

$$
\begin{aligned}
G'(t) &= D_t[(t^2 + 6t)^{10}(1 - 3t)^4] \\
&= [D_t(t^2 + 6t)^{10}](1 - 3t)^4 + (t^2 + 6t)^{10}[D_t(1 - 3t)^4] \\
&= 10(t^2 + 6t)^9[D_t(t^2 + 6t)](1 - 3t)^4 + (t^2 + 6t)^{10}[4(1 - 3t)^3 \, D_t(1 - 3t)] \\
&= 10(t^2 + 6t)^9(2t + 6)(1 - 3t)^4 + (t^2 + 6t)^{10}[4(1 - 3t)^3(-3)] \\
&= 10(t^2 + 6t)^9(2t + 6)(1 - 3t)^4 - 12(t^2 + 6t)^{10}(1 - 3t)^3 \\
&= 2(t^2 + 6t)^9(1 - 3t)^3[5(2t + 6)(1 - 3t) - 6(t^2 + 6t)] \\
&= 2(t^2 + 6t)^9(1 - 3t)^3(-36t^2 - 116t + 30) \\
&= -4(t^2 + 6t)^9(1 - 3t)^3(18t^2 + 58t - 15)
\end{aligned}
$$

Sometimes in calculating derivatives, you have to use the chain rule repeatedly.

EXAMPLE 7   Find $D_x(\sqrt{x^3 - 1})^5$.

SOLUTION   A first use of the chain rule gives

$$
D_x(\sqrt{x^3 - 1})^5 = 5(\sqrt{x^3 - 1})^4 D_x \sqrt{x^3 - 1}
$$

Now, to work out $D_x\sqrt{x^3 - 1}$, we use the chain rule again. Thus,

$$
\begin{aligned}
D_x(\sqrt{x^3 - 1})^5 &= 5(\sqrt{x^3 - 1})^4 D_x\sqrt{x^3 - 1} \\
&= 5(\sqrt{x^3 - 1})^4 \frac{1}{2\sqrt{x^3 - 1}} D_x(x^3 - 1) \\
&= \tfrac{5}{2}(\sqrt{x^3 - 1})^3(3x^2) \\
&= \tfrac{15}{2} x^2(\sqrt{x^3 - 1})^3
\end{aligned}
$$

If u is a differentiable function of x, we can combine the chain rule with the rules for differentiating the sine, tangent, and secant functions to obtain the following:

$$
\begin{aligned}
D_x \sin u &= \cos u \, D_x u \\
D_x \tan u &= \sec^2 u \, D_x u \\
D_x \sec u &= \sec u \tan u \, D_x u
\end{aligned}
$$

Similar rules hold for the corresponding cofunctions:

$$
\begin{aligned}
D_x \cos u &= -\sin u \, D_x u \\
D_x \cot u &= -\csc^2 u \, D_x u \\
D_x \csc u &= -\csc u \cot u \, D_x u
\end{aligned}
$$

In Examples 8 to 10, find the derivative of each function with the aid of the chain rule.

EXAMPLE 8 $f(x) = \sin (5x^2)$

SOLUTION $\begin{aligned}[t] f'(x) &= D_x \sin (5x^2) = \cos (5x^2) \, D_x 5x^2 \\ &= [\cos (5x^2)](10x) \\ &= 10x \cos (5x^2) \end{aligned}$

EXAMPLE 9 $F(t) = \csc \sqrt{4t^2 + 3}$

SOLUTION

$$F'(t) = D_t \csc \sqrt{4t^2 + 3} = -\csc \sqrt{4t^2 + 3} \, \cot \sqrt{4t^2 + 3} \, D_t \sqrt{4t^2 + 3}$$

$$= -\csc \sqrt{4t^2 + 3} \, \cot \sqrt{4t^2 + 3} \, \frac{1}{2\sqrt{4t^2 + 3}} D_t(4t^2 + 3)$$

$$= -\csc \sqrt{4t^2 + 3} \, \cot \sqrt{4t^2 + 3} \, \frac{1}{2\sqrt{4t^2 + 3}} (8t)$$

$$= \frac{-4t \csc \sqrt{4t^2 + 3} \, \cot \sqrt{4t^2 + 3}}{\sqrt{4t^2 + 3}}$$

EXAMPLE 10 $g(\theta) = [3 + \tan (2\theta - \pi)]^4$

SOLUTION

$$g'(\theta) = D_\theta[3 + \tan (2\theta - \pi)]^4 = 4[3 + \tan (2\theta - \pi)]^3 D_\theta[3 + \tan (2\theta - \pi)]$$
$$= 4[3 + \tan (2\theta - \pi)]^3 D_\theta \tan (2\theta - \pi)$$
$$= 4[3 + \tan (2\theta - \pi)]^3 \sec^2(2\theta - \pi) \, D_\theta(2\theta - \pi)$$
$$= 4[3 + \tan (2\theta - \pi)]^3 \sec^2(2\theta - \pi)(2)$$
$$= 8[3 + \tan (2\theta - \pi)]^3 \sec^2(2\theta - \pi)$$

The chain rule is actually a rule for differentiating the composition $f \circ g$ of two functions. To see this, let

$$y = f(u) \qquad \text{and} \qquad u = g(x)$$

so that

$$y = f(u) = f[g(x)] = (f \circ g)(x)$$

Thus,

$$(f \circ g)'(x) = \frac{dy}{dx} = \frac{dy}{du} \frac{du}{dx} = f'(u)g'(x) = f'[g(x)]g'(x)$$

and we have the chain rule expressed as

$$\boxed{(f \circ g)'(x) = f'[g(x)]g'(x)}$$

This statement of the chain rule can be recalled when needed by using the following memory device:

> The derivative of the composite of two functions is the derivative of the outside function taken at the value of the inside function times the derivative of the inside function.

EXAMPLE 11   If f and g are functions such that $f'(4) = 6$, $g(3) = 4$, and $g'(3) = 5$, find $(f \circ g)'(3)$.

SOLUTION   By the chain rule,

$$(f \circ g)'(3) = f'[g(3)]g'(3) = f'(4) \cdot 5 = 6 \cdot 5 = 30$$

Now, using function composition, we can give a formal statement and proof of the chain rule.

THEOREM 1   **The Chain Rule**

> If g is a differentiable function at the number x_1 and f is a differentiable function at the number $g(x_1)$, then $f \circ g$ is a differentiable function at the number x_1 and
>
> $$(f \circ g)'(x_1) = f'[g(x_1)]g'(x_1)$$

PROOF   Using the substitution property for limits (page 67) with $x = x_1 + \Delta x$, we have

$$g'(x_1) = \lim_{\Delta x \to 0} \frac{g(x_1 + \Delta x) - g(x_1)}{\Delta x} = \lim_{x \to x_1} \frac{g(x) - g(x_1)}{x - x_1}$$

Let $u_1 = g(x_1)$. Then we also have

$$f'[g(x_1)] = f'(u_1) = \lim_{u \to u_1} \frac{f(u) - f(u_1)}{u - u_1}$$

Likewise,

$$(f \circ g)'(x_1) = \lim_{x \to x_1} \frac{(f \circ g)(x) - (f \circ g)(x_1)}{x - x_1} = \lim_{x \to x_1} \frac{f[g(x)] - f[g(x_1)]}{x - x_1}$$

provided that the limit exists. Now define a function Q by

$$Q(u) = \begin{cases} \dfrac{f(u) - f(u_1)}{u - u_1} & \text{if } u \neq u_1 \\[2mm] f'(u_1) & \text{if } u = u_1 \end{cases}$$

Then

$$\lim_{u \to u_1} Q(u) = \lim_{u \to u_1} \frac{f(u) - f(u_1)}{u - u_1} = f'(u_1) = f'[g(x_1)]$$

Also, for $u \neq u_1$,

$$f(u) - f(u_1) = Q(u)(u - u_1)$$

Note that the last equation holds even if $u = u_1$, when it simply reduces to $0 = 0$. Now, we substitute $u = g(x)$ in the last equation and divide both sides by $x - x_1$ to obtain

$$\frac{f[g(x)] - f[g(x_1)]}{x - x_1} = Q[g(x)] \frac{g(x) - g(x_1)}{x - x_1}$$

for $x \neq x_1$. Therefore, by the multiplication property for limits,

$$\lim_{x \to x_1} \frac{f[g(x)] - f[g(x_1)]}{x - x_1} = \lim_{x \to x_1} Q[g(x)] \lim_{x \to x_1} \frac{g(x) - g(x_1)}{x - x_1}$$

or

$$(f \circ g)'(x_1) = \lim_{x \to x_1} Q[g(x)] \cdot g'(x_1)$$

provided that $\lim_{x \to x_1} Q[g(x)]$ exists. The proof will be complete if we can show that this limit is equal to $f'[g(x_1)]$. Because g is differentiable at x_1, it follows from Theorem 1 on page 96 that g is continuous at x_1. Therefore,

$$\lim_{x \to x_1} g(x) = g(x_1) = u_1$$

and it follows from the substitution property for limits that

$$\lim_{x \to x_1} Q[g(x)] = \lim_{u \to u_1} Q(u) = f'[g(x_1)]$$ ∎

Problem Set 2.7

In Problems 1 to 4, find the required derivative by using the chain rule.

1 $y = \sqrt{u}$, $u = x^2 + x + 1$, find dy/dx.

2 $y = u^3 - 2u^{1/2}$, $u = x^2 + 2x$, find dy/dx.

3 $y = u^{-5}$, $u = x^4 + 1$, find dy/dx.

4 $y = u$, $u = (7 - x^2)(7 + x^2)^{-1}$, find $D_x y$.

In Problems 5 to 62, find the derivative of each function with the aid of the chain rule.

5 $f(x) = (5 - 2x)^{10}$

6 $f(x) = (2x - 3)^8$

7 $f(y) = \dfrac{1}{(4y + 1)^5}$

8 $F(t) = (2t^4 - t + 1)^{-4}$

9 $g(x) = (3x^2 + 7)^2(5 - 3x)^3$

10 $G(t) = (5t^2 + 1)^2(3t^4 + 2)^4$

11 $f(x) = \left(3x + \dfrac{1}{x}\right)^2(6x - 1)^5$

12 $f(t) = (3t - 1)^{-1}(2t + 5)^{-3}$

13 $g(y) = (7y + 3)^{-2}(2y - 1)^4$

14 $f(u) = \left(6u + \dfrac{1}{u}\right)^{-5}(2u - 2)^7$

15 $f(x) = \left(\dfrac{x^2 + x}{1 - 2x}\right)^4$

16 $f(t) = \left(\dfrac{1 + t^2}{1 - t^2}\right)^5$

17 $F(x) = \left(\dfrac{3x + 1}{x^2}\right)^3$

18 $f(x) = \left(\dfrac{16x}{x^2 - 7}\right)^{-3}$

19 $f(x) = \dfrac{1}{\sqrt{x}} = (\sqrt{x})^{-1}$

20 $F(x) = \dfrac{1}{\sqrt{x^2 + 1}}$

21 $g(x) = \sqrt{x^2 + 2x - 1}$

22 $f(x) = \sqrt{\sqrt{x}} = x^{1/4}$

23 $f(t) = \sqrt{t^4 - t^2 + \sqrt{3}}$

24 $g(y) = \sqrt{y^3 - y + \sqrt{y}}$

25 $F(x) = (x - \sqrt{x})^4$

26 $Q(s) = \dfrac{\sqrt{1 + s^3}}{s}$

27 $f(x) = 5 \sin 7x$

28 $f(x) = 8 \cos (3x + 5)$

29 $g(x) = 4 \sin 6x^2$

30 $g(t) = 3 \sin (5t^2 + t)$

31 $h(x) = \sin \sqrt{x}$

32 $H(s) = s^2 \sin s^3$

33 $g(t) = \sin^4 3t$

34 $g(x) = \cos^2 5x - \sin^2 5x$

35 $H(x) = \cos (\sin x)$

36 $f(t) = (1 - 2 \sin 3t)^5$

37 $f(x) = \sqrt{\cos 5x}$

38 $G(x) = \dfrac{4 - \cos 3x}{x^2}$

39 $H(x) = \dfrac{\sin x}{1 + \cos 5x}$

40 $g(x) = \dfrac{\sin x - x \cos x}{\sqrt{\cos x}}$

41 $H(t) = \dfrac{27}{\sin 2t} + \dfrac{35}{\cos 2t}$

42 $g(r) = \tan 5r^4$

43 $g(t) = \cot (3t^5)$

44 $h(r) = \sec (\sqrt{r} - r)$

45 $F(u) = \csc \sqrt{u^2 + 1}$

46 $g(s) = \cot \dfrac{7}{s}$

47 $h(x) = \sqrt{1 + \sec 5x}$

48 $g(t) = \tan \dfrac{t}{t + 2}$

49 $h(t) = \sec^2 7t - \tan^2 7t$

50 $g(x) = \csc^2 15x - \cot^2 15x$

51 $H(s) = \sec^4 13s - \tan^4 13s$

52 $g(x) = (\tan x + \sec x)^3$

53 $g(x) = x^3 \tan^5 2x$

54 $f(t) = \dfrac{\cot 3t}{t^2 + 1}$

55 $H(x) = \dfrac{2x}{1 + \sec 5x}$

56 $g(t) = \tan 3t \sec 3t$

57 $f(x) = \frac{1}{3}x^2 - \cot^3 2x$

58 $G(r) = \frac{3}{2}r^2 \csc^5 3r$

59 $g(t) = \dfrac{\sec^2 3t}{t^3}$

60 $f(\theta) = \left(\dfrac{\theta}{\tan \theta}\right)^3$

61 $f(x) = \sin(\tan 5x^2)$

62 $g(x) = \sec(\csc^2 7x)$

In Problems 63 to 70, use the information in the table and the chain rule (Theorem 1) to evaluate the given quantity.

x	$f(x)$	$f'(x)$	$g(x)$	$g'(x)$
2	2	4	5	-4
5	5	-3	2	6

63 $(f \circ g)'(2)$

64 $(f \circ g)'(5)$

65 $(g \circ f)'(5)$

66 $(g \circ g)'(5)$

67 $h'(5)$ where $h(x) = [f(x)]^2$

68 $H'(2)$ where $H(x) = (f \circ g)(6x - 7)$

69 $F'(2)$ where $F(x) = [g(x)]^4 f(x)$

70 $G'(5)$ where $G(x) = g(x)\sqrt{f(x)}$

71 Show that the chain rule can be expressed by the formula $(f \circ g)' = (f' \circ g) \cdot g'$.

72 Explain the distinction between $D_x[f(7x + 3)]$ and $f'(7x + 3)$.

73 Let f and g be functions such that $f(5) = -3$, $f'(5) = 10$, $f'(7) = 20$, $g(5) = 7$, $g'(5) = \frac{1}{4}$, and $g'(7) = \frac{2}{3}$. Find $(f \circ g)'(5)$. (Be careful—there is more information here than you need.)

74 A 50-kilohertz radio frequency oscillation with the equation $y_1 = 10 \sin(10^5 \pi t)$ is amplitude-modulated by an audio frequency tone with the equation $y_2 = \sin(2\pi 10^4 t)$ to produce a signal with the equation $y = y_2 y_1$. Find a formula for the rate of change dy/dt of y with respect to time t.

⃝C 75 An alternating-current generator provides a current I amperes according to the equation $I = 30 \sin 120\pi t$, where t is the time in seconds. Find the rate of change dI/dt of the current with respect to time at $t = 0.97$ second. Round off your answer to the nearest whole number of amperes per second.

⃝C 76 A space shuttle in orbit is a distance y kilometers from the plane of the equator t minutes after orbital insertion in accordance with the equation $y = 8000 \cos[(\pi t/40) - (2\pi/9)]$. Find the rate of change dy/dt of y with respect to t at the instant when $t = 10$ minutes. Round off your answer to the nearest tenth of a kilometer per minute.

⃝C 77 Because of predation by wolves, the moose population N in a certain region fluctuates in accordance with the equation $N = 5000 + 1000 \sin 0.25t$, where t is the time in years since wildlife biologists began making observations. Find the rate of change dN/dt of the number of moose with respect to time when $t = 10$ years. Round off your answer to the nearest integer.

⃝C 78 An industrial plant is dumping A thousand gallons of waste per day into a nearby stream according to the equation $A = 24 + 18 \sin(\pi t/4)$, where t is the number of months that the plant has been in operation. Find the rate of change of A with respect to t when $t = 9$ months. Round off your answer to the nearest integer.

79 The labor force $F(x)$ persons required by a certain industry to manufacture x units of a product is given by $F(x) = 6\sqrt{x}$. At present, there is a demand for 40,000 units of the product, and this demand is increasing at a constant rate of 10,000 units per year. (a) Write an equation for the function $u(t)$ that gives the number of units that will be demanded t years from now. (b) Write a formula for $(F \circ u)(t)$ and interpret the resulting expression. (c) Find $(F \circ u)'(t)$ and interpret the resulting expression. (d) Find $(F \circ u)'(5)$.

2.8 Implicit Functions and Implicit Differentiation

An equation such as

$$y = 3x^2 - 5x + 12$$

that is already solved for y in terms of x is said to give y **explicitly** as a function of x. On the other hand, an equation such as

$$xy + 1 = 2x - y$$

which can be solved for y in terms of x but is not solved for y as it stands, is said to give y **implicitly** as a function of x. (Shortly, we shall give a more precise definition of an implicit function.) In this section we introduce a technique called *implicit differentiation* which allows us to calculate the derivative dy/dx of an implicitly given function without bothering to solve explicitly for y in terms of x.

Implicit Differentiation

The equation $2x + 3y = 1$ can be solved for y in terms of x to give $y = \frac{1}{3} - \frac{2}{3}x$, and so we have $dy/dx = -\frac{2}{3}$. The same result can be obtained directly from the original equation $2x + 3y = 1$ simply by differentiating both sides, term by term, to get $2 + 3(dy/dx) = 0$ and then solving for $dy/dx = -\frac{2}{3}$. The latter technique is called **implicit differentiation.** More generally, we have the following.

Procedure for Implicit Differentiation

Given an equation that determines y implicitly as a differentiable function of x, calculate dy/dx as follows:

Step 1 Differentiate both sides of the equation with respect to x; that is, apply d/dx to both sides of the equation, term by term. In so doing, keep in mind that y is regarded as a function of x, and use the *chain rule* when necessary to differentiate expressions involving y.

Step 2 The result of step 1 will be an equation involving not only x and y but also dy/dx. Solve this equation for the desired derivative dy/dx.

When the procedure for implicit differentiation is executed, the result is often an equation that gives dy/dx in terms of *both* x and y. In this case, in order to calculate the numerical value of dy/dx, it is necessary to know not only the numerical value of x but also the numerical value of y.

The procedure for implicit differentiation can be used legitimately only if it is known that the equation in question really does determine y implicitly as a differentiable function of x. However, in what follows we routinely apply the procedure and simply assume that this requirement is fulfilled.

In Examples 1 to 5, use implicit differentiation to solve each problem.

<u>EXAMPLE 1</u>   If $x^3 - 3x^2y^4 + 4y^3 = 6x + 1$, find dy/dx.

SOLUTION   We begin by differentiating both sides of the given equation, term by term, with respect to x:

$$\frac{d}{dx}x^3 - \frac{d}{dx}3x^2y^4 + \frac{d}{dx}4y^3 = \frac{d}{dx}6x + \frac{d}{dx}1$$

$$3x^2 - 3(2x)y^4 - 3x^2\left(4y^3\,\frac{dy}{dx}\right) + 12y^2\,\frac{dy}{dx} = 6 + 0$$

$$3x^2 - 6xy^4 - 12x^2y^3\,\frac{dy}{dx} + 12y^2\,\frac{dy}{dx} = 6$$

This completes step 1 of the procedure for implicit differentiation. To carry out step 2, we must solve the last equation for dy/dx in terms of x and y. Subtracting $3x^2 - 6xy^4$ from both sides of the equation, we obtain

$$-12x^2y^3\,\frac{dy}{dx} + 12y^2\,\frac{dy}{dx} = -3x^2 + 6xy^4 + 6$$

Now, factoring dy/dx out of the left side and multiplying both sides of the equation by -1, we have

$$(12x^2y^3 - 12y^2)\,\frac{dy}{dx} = 3x^2 - 6xy^4 - 6$$

Hence
$$\frac{dy}{dx} = \frac{3x^2 - 6xy^4 - 6}{12x^2y^3 - 12y^2} = \frac{x^2 - 2xy^4 - 2}{4y^2(x^2y - 1)}$$

■

EXAMPLE 2   If $(2x + y)^3 - (3x - y)^3 = y^5$, find dy/dx.

SOLUTION   Differentiating the equation term by term with respect to x, we obtain

$$3(2x + y)^2\,\frac{d}{dx}(2x + y) - 3(3x - y)^2\,\frac{d}{dx}(3x - y) = 5y^4\,\frac{dy}{dx}$$

or
$$3(2x + y)^2\left(2 + \frac{dy}{dx}\right) - 3(3x - y)^2\left(3 - \frac{dy}{dx}\right) = 5y^4\,\frac{dy}{dx}$$

Thus,

$$6(2x + y)^2 + 3(2x + y)^2\,\frac{dy}{dx} - 9(3x - y)^2 + 3(3x - y)^2\,\frac{dy}{dx} = 5y^4\,\frac{dy}{dx}$$

Rewriting the last equation so that the terms involving dy/dx are on the left, we have

$$3(2x + y)^2\,\frac{dy}{dx} + 3(3x - y)^2\,\frac{dy}{dx} - 5y^4\,\frac{dy}{dx} = 9(3x - y)^2 - 6(2x + y)^2$$

or
$$[3(2x + y)^2 + 3(3x - y)^2 - 5y^4]\,\frac{dy}{dx} = 9(3x - y)^2 - 6(2x + y)^2$$

Therefore,

$$\frac{dy}{dx} = \frac{9(3x - y)^2 - 6(2x + y)^2}{3(2x + y)^2 + 3(3x - y)^2 - 5y^4}$$

■

EXAMPLE 3 If $x \cos y + y \sin x = 5$, find dy/dx.

SOLUTION

$$\frac{d}{dx}(x \cos y) + \frac{d}{dx}(y \sin x) = \frac{d}{dx}5$$

$$\cos y + x \frac{d}{dx}\cos y + \frac{dy}{dx}\sin x + y \frac{d}{dx}\sin x = 0$$

$$\cos y - x \sin y \frac{dy}{dx} + \frac{dy}{dx}\sin x + y \cos x = 0$$

$$(\sin x - x \sin y)\frac{dy}{dx} = -y \cos x - \cos y$$

Multiplying both sides of the last equation by -1 and solving for dy/dx, we obtain

$$\frac{dy}{dx} = \frac{y \cos x + \cos y}{x \sin y - \sin x}$$

EXAMPLE 4 Find the equations of the tangent and normal lines to the graph of $\dfrac{x}{y} + \dfrac{x^2}{y^2} = 6$ at the point $(2, 1)$.

SOLUTION   Differentiating the given equation term by term with respect to x, we obtain

$$\frac{d}{dx}\left(\frac{x}{y}\right) + \frac{d}{dx}\left(\frac{x^2}{y^2}\right) = \frac{d}{dx}6$$

$$\frac{y - x\dfrac{dy}{dx}}{y^2} + \frac{y^2 \dfrac{d}{dx}x^2 - x^2 \dfrac{d}{dx}y^2}{y^4} = 0$$

$$\frac{y - x\dfrac{dy}{dx}}{y^2} + \frac{y^2(2x) - x^2\left(2y \dfrac{dy}{dx}\right)}{y^4} = 0$$

$$\frac{y - x\dfrac{dy}{dx}}{y^2} + \frac{2xy - 2x^2 \dfrac{dy}{dx}}{y^3} = 0$$

To clear fractions, we multiply both sides of the last equation by y^3:

$$y^2 - xy \frac{dy}{dx} + 2xy - 2x^2 \frac{dy}{dx} = 0$$

Hence,

$$(2x^2 + xy)\frac{dy}{dx} = y^2 + 2xy$$

and so

$$\frac{dy}{dx} = \frac{y^2 + 2xy}{2x^2 + xy}$$

Therefore, when $x = 2$ and $y = 1$, we have

$$\frac{dy}{dx} = \frac{1^2 + 2(2)(1)}{2(2)^2 + 2(1)} = \frac{5}{10} = \frac{1}{2}$$

Consequently, the slope of the tangent line at (2, 1) is $\frac{1}{2}$, the slope of the normal line is $-1/\frac{1}{2} = -2$, and the corresponding equations are

Tangent line: $y - 1 = \frac{1}{2}(x - 2)$ or $y = \frac{1}{2}x$

Normal line: $y - 1 = -2(x - 2)$ or $y = -2x + 5$ ∎

EXAMPLE 5   Given the equation $x^3 + xy + y^3 = 2$, consider x to be a differentiable function of y and find dx/dy.

SOLUTION   Applying d/dy to both sides of $x^3 + xy + y^3 = 2$, we obtain

$$3x^2 \frac{dx}{dy} + x + \frac{dx}{dy}y + 3y^2 = 0$$

Solving this equation for dx/dy, we have

$$\frac{dx}{dy} = \frac{-x - 3y^2}{3x^2 + y}$$ ∎

The idea of an implicit function is made more precise by the following definition.

DEFINITION 1   **Implicit Function**

A continuous function f, defined at least on an open interval, is said to be **implicit** in an equation involving the variables x and y provided that when y is replaced by $f(x)$ in this equation, the resulting equation is true for all values of x in the domain of f.

Figure 1

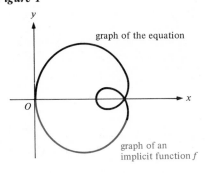

graph of the equation

graph of an implicit function f

Thus, a continuous function f is implicit in an equation if and only if the graph of f is contained in the graph of the equation (Figure 1).

In Examples 6 to 8, if possible, find the functions implicit in the given equation by solving for y in terms of x.

EXAMPLE 6 $7 + x = y^2 - 3y$

SOLUTION   We have $y^2 - 3y + (-7 - x) = 0$. Using the quadratic formula, we obtain

$$y = \frac{3 \pm \sqrt{(-3)^2 - 4(1)(-7 - x)}}{2(1)}$$

$$= \frac{3 \pm \sqrt{4x + 37}}{2}$$

The last equation does not give y as an explicit function of x because of the ambiguous $\pm$ sign. However, the function f defined by

$$f(x) = \frac{3 + \sqrt{4x + 37}}{2}$$

is implicit in the given equation. But so is the function h defined by

$$h(x) = \frac{3 - \sqrt{4x + 37}}{2}$$

Figure 2

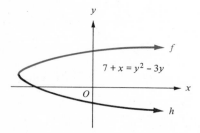

Figure 2 shows us geometrically just what's happening here. Indeed, the graph of $7 + x = y^2 - 3y$ is a parabola with a horizontal axis of symmetry. By the vertical-line test (page 26), this parabola is not the graph of a function. However, it can be broken up into two pieces, each of which is the graph of a function. The graph of the implicit function f is the top half of the parabola, and the graph of h is the bottom half. ∎

EXAMPLE 7 $x^2 + y^2 + 1 = 0$

SOLUTION   This equation has no (real) solution, so it cannot define y implicitly as a function of x. ∎

EXAMPLE 8 $\dfrac{2y - 3}{4y + 7} = 1 - x$

SOLUTION   Clearing the fraction, we have

$$2y - 3 = (4y + 7)(1 - x) = 4y - 4xy + 7 - 7x$$

Bringing the terms involving y to the left, we have

$$2y - 4xy = 7x - 10 \qquad \text{or} \qquad (2 - 4x)y = 7x - 10$$

It follows that

$$y = \frac{7x - 10}{2 - 4x}$$

so the function

$$f(x) = \frac{7x - 10}{2 - 4x}$$

Figure 3

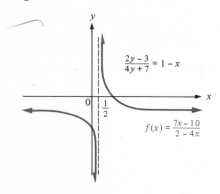

is implicit in the given equation. Here, the graph of the original equation (Figure 3) satisfies the vertical-line test; hence, it is the graph of a function, namely, the implicit function f. Notice that f is essentially the only implicit function for the given equation; other implicit functions can be obtained only by restricting the domain of f. ∎

As these examples show, given an equation involving x and y, any one of the following can happen:

1 There are two or more functions implicit in the equation (Example 6).

2 There is no function implicit in the equation (Example 7).

3 There is (essentially) only one function implicit in the equation (Example 8).

Furthermore, even if there is a function f implicit in the equation, it may or may not be possible to find a practical mathematical formula for $f(x)$. For instance, the graph of the equation

$$x = \frac{y^5}{40} + \frac{13}{600}y^3$$

Figure 4

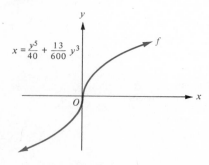

(Figure 4) satisfies the vertical-line test, and so it is the graph of a function f. Although the implicit function f is a perfectly definite function, whose graph appears in Figure 4, it isn't possible to find an elementary algebraic formula for $f(x)$.

Problem Set 2.8

In Problems 1 to 26, find dy/dx by using implicit differentiation.

1 $9x^2 + 4y^2 = 36$

2 $4xy^2 + 3x^2y = 2$

3 $x^2y - xy^2 + x^2 = 7$

4 $xy^2 + x^3 + y^3 = 5$

5 $x^2 - 3xy + y^2 = 3x - 4$

6 $xy^3 + 2y^3 = x^2 - 4y^2$

7 $x^{-1} + y^{-1} = 8y$

8 $x^2 - \sqrt{xy} - y = 0$

9 $x^4y + \sqrt{xy} = 5$

10 $y\sqrt{x} + \sqrt{y} = 9$

11 $y(x^2 - y^2)^3 = x$

12 $(4x - 1)^3 = 5y^3 + 2$

13 $\dfrac{x^2}{y} - y = \dfrac{x}{2} - \dfrac{4}{y^2}$

14 $(5x^2y + 4)^7 = x^3$

15 $\sqrt{x + y} + \sqrt{x - y} = 6$

16 $\dfrac{x}{x - y} + \dfrac{y}{x} = 4$

17 $y = \sin(2x + y)$

18 $x \cos y = (x + y)^2$

19 $\tan xy + xy = 2$

20 $\tan^2 x + \tan^2 y = 4$

21 $\sin^2 x + \cos^2 y = 1$

22 $\sec(x + y) + \csc(x + y) = 5$

23 $\cos xy + y^2 = 2$

24 $\sin x \tan y = y^3$

25 $\cot(3x + y) = 5xy$

26 $\sec(x^2 + y^2) = 5x^3$

In Problems 27 to 32, find the equations of the tangent and normal lines to the graph of the implicit function determined by each of the following equations at the given point.

27 $x^2 + xy + 2y^2 = 28$ at $(2, 3)$

28 $x^3 - 3xy^2 + y^3 = 1$ at $(2, -1)$

29 $\sqrt{2x} + \sqrt{3y} = 5$ at $(2, 3)$

30 $x^2 - 2\sqrt{xy} - y^2 = 52$ at $(8, 2)$

31 $\sin xy = y$ at $(\pi/2, 1)$

32 $(5 + \tan xy)^2 = 36$ at $(\pi/12, 3)$

In Problems 33 to 36, consider x to be a function of y and find dx/dy by implicit differentiation.

33 $3x^2 + 5xy = 2$

34 $x^2y^2 = x^2 + y^2$

35 $x^2 = y^2 - y$

36 $\sqrt{xy} + xy^4 = 5$

In Problems 37 to 46, find all functions implicit in each equation by solving for y in terms of x.

37 $5x - 4y = 6$

38 $5x^2 - 4y^2 = 6$

39 $xy + y^2 = x$

40 $3xy^2 + 4y + x = 0$

41 $x = \dfrac{2y - 1}{3y + 1}$

42 $x^2 - x^2y + y^3 = y^2$

43 $x = y^4$

44 $y^3(y^2 + 4) = x$

45 $y^3 - 3y^2 + 3y = 3(x + 1)$

46 $\dfrac{x}{y} + \dfrac{y}{x} = 2$

47 Show that the function $f(x) = \frac{3}{4}\sqrt{16 - x^2}$ is implicit in the equation $(x^2/16) + (y^2/9) = 1$. What does this imply about the relationship between the graph of f and the graph of the equation $(x^2/16) + (y^2/9) = 1$?

48 Show that the function $g(x) = -\frac{3}{4}\sqrt{16 - x^2}$ is implicit in the equation $(x^2/16) + (y^2/9) = 1$. What does this imply about the relationship between the graph of g and the graph of the equation $(x^2/16) + (y^2/9) = 1$? How is the graph of g related to the graph of the function f of Problem 47?

49 Both of the functions f and g defined by $f(x) = \frac{3}{4}\sqrt{16 - x^2}$ and $g(x) = -\frac{3}{4}\sqrt{16 - x^2}$ are implicit in the equation $(x^2/16) + (y^2/9) = 1$. (a) Calculate $f'(x)$ directly from $f(x) = \frac{3}{4}\sqrt{16 - x^2}$. (b) Calculate $g'(x)$ directly from $g(x) = -\frac{3}{4}\sqrt{16 - x^2}$. (c) Calculate dy/dx from the equation $(x^2/16) + (y^2/9) = 1$ by implicit differentiation. (d) Show that the answer to part (c) is compatible with the answer to part (a). (e) Show that the answer to part (c) is also compatible with the answer to part (b).

50 Interpret the answers to parts (a), (b), and (c) of Problem 49 in terms of the slopes of tangent lines to the graphs of the function f, the function g, and the equation $(x^2/16) + (y^2/9) = 1$, respectively. Thus, explain the "compatibility" found in the answers to parts (d) and (e) of Problem 49.

51 Find the slope of the tangent line to the graph of the implicit function determined by the equation $(x^2/30) - (y^2/20) = 1$ at the point $(6, -2)$ (a) by implicit differentiation and (b) by solving the given equation to obtain y explicitly as a function f of x and then finding the value of $f'(x)$ when $x = 6$.

52 In a certain tropical habitat, the predator-prey relationship between the population x of boa constrictors (the predator) and the population y of wild pigs (the prey) is given by the equation $70x^2 + y^2 - 2000x - 150y + 18,600 = 0$. Find the rate of change dy/dx of pigs with respect to boas when $x = 10$ and $y = 80$.

53 A civil engineer is designing a curved access road leading from a point O on a state highway to a point P on an interstate highway. She sets up a Cartesian coordinate system with O as the origin and the state highway along the x axis (Figure 5). With distances in meters, $P = (150, 100)$. The access road is to be tangent to the state highway at O and tangent to the interstate highway at P. To provide for acceleration up to interstate highway speeds, the curvature of the access road is to decrease uniformly from O to P. To meet these conditions, the access road is designed to follow the graph of the equation $4x^2 + 3(y - 200)^2 = 120{,}000$. Find the x coordinate of the point I where the interstate highway intersects the state highway.

Figure 5

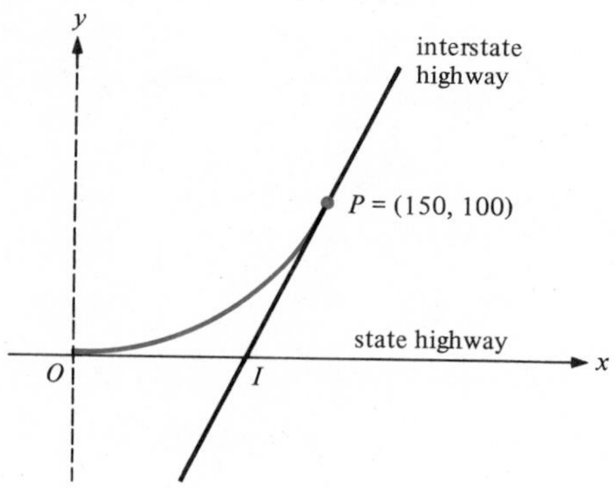

[Figure 5: graph showing the interstate highway line passing through $P = (150, 100)$, the state highway along the x axis, point O at origin, point I on the x axis, and the curved access road.]

2.9 The Rational-Power Rule

In Theorem 9 of Section 2.3, we established the power rule

$$\frac{d}{dx}x^n = nx^{n-1}$$

for integer exponents. In this section, we show that the same rule works for rational exponents. We begin by considering the case in which the exponent is the reciprocal of a positive integer.

Recall that if n is a positive integer, then

$$x^{1/n} = \sqrt[n]{x}$$

where $\sqrt[n]{x}$ denotes the principal nth root of x. Thus, if n is even and x is nonnegative, $\sqrt[n]{x}$ denotes the nonnegative number whose nth power is x. Similarly, if n is odd, $\sqrt[n]{x}$ denotes the real number whose nth power is x, but there is no restriction on the algebraic sign of x.

If the positive integer n is even, then the graph of $y = \sqrt[n]{x}$ has the general appearance shown in Figure 1a; but if n is odd, the graph looks like the curve in Figure 1b. In either case, it is visually apparent that the graph of $y = \sqrt[n]{x}$ has a definite tangent line at every point, except perhaps at $(0, 0)$.* If $x \neq 0$ and there is a tangent line at the point $(x, \sqrt[n]{x})$, then its slope must be given by the derivative $d\sqrt[n]{x}/dx$. Thus, Figure 1 provides graphical evidence that $d\sqrt[n]{x}/dx$ exists, provided that $\sqrt[n]{x}$ is defined and $x \neq 0$. In what follows, we shall assume this. A more formal argument can be given by using Theorem 1 on page 406.

Now, suppose that n is a positive integer, and let

$$y = \sqrt[n]{x}$$

Then,

$$y^n = x$$

Figure 1

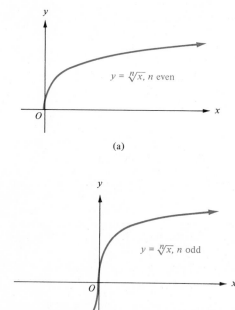

(a)

$y = \sqrt[n]{x}$, n even

(b)

$y = \sqrt[n]{x}$, n odd

*In fact, for n odd, the graph does have a tangent line at $(0, 0)$, namely, the y axis (see page 201).

Applying the implicit differentiation procedure to the last equation and assuming that $x \neq 0$, we have

$$\frac{d}{dx} y^n = \frac{d}{dx} x$$

or

$$ny^{n-1} \frac{dy}{dx} = 1$$

Therefore,

$$\frac{dy}{dx} = \frac{1}{ny^{n-1}} = \frac{y}{ny^n}$$

Because $y = \sqrt[n]{x}$ and $y^n = x$, the last equation can be rewritten as

$$\frac{d}{dx} \sqrt[n]{x} = \frac{\sqrt[n]{x}}{nx}$$

Using the alternative notation $x^{1/n}$ for $\sqrt[n]{x}$, we can rewrite this equation as

$$\frac{d}{dx} x^{1/n} = \frac{x^{1/n}}{nx} = \frac{1}{n} x^{1/n} x^{-1}$$

or

$$\frac{d}{dx} x^{1/n} = \frac{1}{n} x^{(1/n)-1}$$

provided $x \neq 0$.

EXAMPLE 1   If $f(x) = x^{1/9}$, find $f'(x)$.

SOLUTION   For $x \neq 0$,

$$f'(x) = \frac{d}{dx} x^{1/9} = \frac{1}{9} x^{(1/9)-1} = \frac{1}{9} x^{-8/9}$$

If desired, the same result can be written as

$$\frac{d}{dx} \sqrt[9]{x} = \frac{\sqrt[9]{x}}{9x}$$

It's interesting to notice that if $n = 2$, the differentiation rule

$$\frac{d}{dx} x^{1/n} = \frac{1}{n} x^{(1/n)-1}$$

gives

$$\frac{d}{dx} \sqrt{x} = \frac{d}{dx} x^{1/2} = \frac{1}{2} x^{-1/2} = \frac{1}{2\sqrt{x}}$$

which is in conformity with the square-root rule obtained by other means in Section 2.3.

Now we can prove the following theorem, which shows that the rule for differentiating rational powers of x follows the same pattern as the rule for differentiating integer powers of x.

THEOREM 1

Power Rule for Rational Exponents

Let $r = m/n$ be a rational number, reduced to lowest terms so that n is a positive integer and the integers m and n have no common factors. Then

$$\frac{d}{dx} x^r = rx^{r-1}$$

holds for all values of x for which $x^r = (x^{1/n})^m$ is defined, except possibly for $x = 0$. It also holds for $x = 0$, provided that n is odd and $m > n$.

PROOF

Suppose that x^r is defined and that $x \neq 0$. Then, using the chain rule, the rule for differentiating integer powers, and the rule obtained above for differentiating $x^{1/n}$, we have

$$\frac{d}{dx} x^r = \frac{d}{dx} (x^{1/n})^m = m(x^{1/n})^{m-1} \frac{d}{dx} x^{1/n} = mx^{(m-1)/n} \left(\frac{1}{n} x^{(1/n)-1} \right)$$

$$= \frac{m}{n} x^{(m-1)/n+(1/n)-1} = \frac{m}{n} x^{(m/n)-1} = rx^{r-1}$$

To complete the proof, assume that $m > n$ and n is odd. Let $f(x) = x^r$, noting that $r > 1$. Thus, when $x = 0$, $rx^{r-1} = 0$, and so we must prove that $f'(0) = 0$. We have

$$f'(0) = \lim_{\Delta x \to 0} \frac{f(0 + \Delta x) - f(0)}{\Delta x} = \lim_{\Delta x \to 0} \frac{[(\Delta x)^{1/n}]^m - 0}{\Delta x}$$

$$= \lim_{\Delta x \to 0} \frac{[(\Delta x)^{1/n}]^m}{[(\Delta x)^{1/n}]^n} = \lim_{\Delta x \to 0} [(\Delta x)^{1/n}]^{m-n} = 0 \quad \blacksquare$$

If we combine the power rule for rational exponents in Theorem 1 with the chain rule, we obtain the following important result:

If r is a rational number and u is a differentiable function of x, then

$$\frac{d}{dx} u^r = ru^{r-1} \frac{du}{dx} \qquad \text{or} \qquad D_x u^r = ru^{r-1} D_x u$$

Of course, if $u = 0$, this rule may not apply, unless $r = m/n$, where m and n have no common factors, $m > n > 0$, and n is odd.

In Examples 2 to 7, find the derivative.

EXAMPLE 2 $f(x) = x^{3/2}$

SOLUTION For $x > 0$,

$$f'(x) = \frac{d}{dx} x^{3/2} = \frac{3}{2} x^{(3/2)-1} = \frac{3}{2} x^{1/2} = \frac{3}{2} \sqrt{x} \quad \blacksquare$$

EXAMPLE 3 $g(t) = t^{-7/3}$

SOLUTION   For $t \neq 0$,

$$g'(t) = \frac{d}{dt} t^{-7/3} = -\frac{7}{3} t^{(-7/3)-1} = -\frac{7}{3} t^{-10/3}$$

EXAMPLE 4 $h(x) = \sqrt[3]{2x^2 + 3}$

SOLUTION $h'(x) = \dfrac{d}{dx} \sqrt[3]{2x^2 + 3} = \dfrac{d}{dx}(2x^2 + 3)^{1/3}$

$$= \frac{1}{3}(2x^2 + 3)^{(1/3)-1} \frac{d}{dx}(2x^2 + 3)$$

$$= \tfrac{1}{3}(2x^2 + 3)^{-2/3}(4x) = \tfrac{4}{3}x(2x^2 + 3)^{-2/3}$$

EXAMPLE 5 $F(x) = (1 - x)^{4/5}(1 + x^2)^{-2/3}$

SOLUTION

$$F'(x) = \frac{d}{dx}(1 - x)^{4/5}(1 + x^2)^{-2/3}$$

$$= \left[\frac{d}{dx}(1 - x)^{4/5}\right](1 + x^2)^{-2/3} + (1 - x)^{4/5}\left[\frac{d}{dx}(1 + x^2)^{-2/3}\right]$$

$$= [\tfrac{4}{5}(1 - x)^{-1/5}(-1)](1 + x^2)^{-2/3} + (1 - x)^{4/5}[-\tfrac{2}{3}(1 + x^2)^{-5/3}(2x)]$$

$$= (1 - x)^{-1/5}(1 + x^2)^{-5/3}[-\tfrac{4}{5}(1 + x^2) - \tfrac{2}{3}(1 - x)(2x)]$$

$$= \tfrac{4}{15}(1 - x)^{-1/5}(1 + x^2)^{-5/3}(2x^2 - 5x - 3)$$

This calculation illustrates a useful rule: To factor an algebraic sum in which each term contains a rational power of the same expression, take as a common factor the *smallest* rational power to which the expression is raised. For instance, in the third line of the calculation, the expressions $(1 - x)^{-1/5}$ and $(1 - x)^{4/5}$ occur. Since $-\frac{1}{5} < \frac{4}{5}$, we factor out $(1 - x)^{-1/5}$. Likewise, since $-\frac{5}{3} < -\frac{2}{3}$, we factor out $(1 + x^2)^{-5/3}$.

EXAMPLE 6 $G(x) = \sin \sqrt[3]{x^4 + 5}$

SOLUTION $G'(x) = \dfrac{d}{dx} \sin \sqrt[3]{x^4 + 5} = \cos \sqrt[3]{x^4 + 5} \, \dfrac{d}{dx} \sqrt[3]{x^4 + 5}$

$$= \cos \sqrt[3]{x^4 + 5} \, \frac{d}{dx}(x^4 + 5)^{1/3}$$

$$= \cos \sqrt[3]{x^4 + 5} \left[\frac{1}{3}(x^4 + 5)^{-2/3} \frac{d}{dx}(x^4 + 5)\right]$$

$$= \cos \sqrt[3]{x^4 + 5} \, [\tfrac{1}{3}(x^4 + 5)^{-2/3}(4x^3)]$$

$$= \frac{4x^3 \cos \sqrt[3]{x^4 + 5}}{3(x^4 + 5)^{2/3}}$$

$$= \frac{4x^3 \sqrt[3]{x^4 + 5} \cos \sqrt[3]{x^4 + 5}}{3(x^4 + 5)}$$

In the last step, we rationalized the denominator by multiplying numerator and denominator of the fraction by $(x^4 + 5)^{1/3} = \sqrt[3]{x^4 + 5}$.

EXAMPLE 7 $H(u) = (1 + \tan u^2)^{-5/8}$

SOLUTION   If $\tan u^2 \neq -1$,

$$H'(u) = -\tfrac{5}{8}(1 + \tan u^2)^{(-5/8)-1}[(\sec^2 u^2)(2u)]$$
$$= -\tfrac{5}{4}u(1 + \tan u^2)^{-13/8} \sec^2 u^2$$

■

Problem Set 2.9

In Problems 1 to 30, find the derivative of each function.

1 $f(x) = \sqrt[5]{x}$

2 $f(x) = \sqrt[7]{x^2}$

3 $g(x) = 36x^{-4/9}$

4 $f(x) = 21x^{5/7}$

5 $h(t) = (1 - t)^{-2/3}$

6 $f(x) = \dfrac{1}{\sqrt[5]{x^4}}$

7 $f(u) = \left(1 + \dfrac{2}{u}\right)^{3/4}$

8 $g(s) = \sqrt[3]{\dfrac{9 - s^2}{9 + s^2}}$

9 $g(x) = x^{-1/2} + x^{-1/3} + x^{-1/4}$

10 $f(x) = \sqrt{x} + \sqrt[3]{x} + \sqrt[4]{x}$

11 $f(t) = \sqrt[5]{t^3} - \sqrt[4]{t}$

12 $g(y) = \sqrt{y^4 - y + \sqrt[3]{y}}$

13 $g(x) = \sqrt[10]{\dfrac{x}{x + 1}}$

14 $f(x) = (x + \sqrt[5]{x})(x - 2\sqrt[3]{x})$

15 $h(x) = (1 + x)^{-3/4}(2x + 1)^{1/2}$

16 $f(t) = \dfrac{t}{\sqrt[4]{36 - t^2}}$

17 $f(t) = \sqrt[4]{t + 2}\sqrt[5]{t + 5}$

18 $g(x) = \sqrt[3]{x}(1 + 2\sqrt{x})$

19 $f(t) = \sqrt[5]{\sin t}$

20 $g(x) = \sqrt[7]{\cos 3x}$

21 $g(x) = \cos^{3/4} x$

22 $h(t) = \sin^{5/7}(4t - 1)$

23 $F(t) = \sec^{3/2}(4t^2 + 1)$

24 $G(s) = 3 \cot \sqrt[3]{4s^2}$

25 $H(y) = \tan \sqrt[3]{y} \csc \sqrt[4]{y}$

26 $R(\theta) = \left(\dfrac{\sin \theta - 1}{\cos \theta}\right)^{1/3}$

27 $P(z) = z^{2/3} \cos z^{-2/3}$

28 $Q(\phi) = \left(\dfrac{\cos \phi + 1}{1 - \cos \phi}\right)^{4/3}$

29 $f(\theta) = \sqrt[3]{3 + \cos^4 3\theta}$

30 $g(x) = (7 - \sec^2 x)^{-4/5}$

In Problems 31 to 36, find equations of the tangent and normal lines to the graph of each equation at the indicated point.

31 $y = \sqrt[3]{4x^2 + 23}$ at $(1, 3)$

32 $y = \left(\dfrac{1 - x}{1 + x}\right)^{1/5}$ at $(1, 0)$

33 $y = \left(\dfrac{3x^3 + 1}{x^3 - 1}\right)^{1/7}$ at $(0, -1)$

34 $y = \left(\dfrac{x}{5x^2 + 3}\right)^{3/2}$ at $(3, \tfrac{1}{64})$

35 $2x^{1/3} - y^{2/3} = x$ at $(1, 1)$

36 $y + \sqrt[4]{15 + 2 \sin xy} = \tfrac{5}{2}$ at $(\pi/3, \tfrac{1}{2})$

37 Find $D_x(\sqrt[4]{x})$ (a) by writing $\sqrt[4]{x} = \sqrt{\sqrt{x}}$ and using the chain rule and (b) by writing $\sqrt[4]{x} = x^{1/4}$ and using the power rule for rational exponents.

38 In the last part of the proof of Theorem 1, (a) where did we use the assumption that n is odd, and (b) where did we use the assumption that $m > n$?

39 During a flu epidemic, public health officials develop the mathematical model $N = 1200t^{3/2}$ to predict the number of people in a certain region who will have contracted the flu t days after the beginning of the epidemic, $0 \leq t \leq 30$. Find the rate dN/dt at which the flu will be spreading 25 days after the beginning of the epidemic.

40 Suppose that the population P of a certain culture of bacteria is given by the equation $P = 1000\sqrt[4]{t^5 + 10t^2 + 9}$, where t is the time in hours since the culture was started. Find the rate dP/dt of bacterial population growth when $t = 2$ hours.

2.10   Higher-Order Derivatives

As we have seen, a derivative can be interpreted as a rate of change. For instance, the derivative dP/dt of the consumer price index P with respect to time t can be taken as a measure of monetary inflation. When government economists say that the rate of inflation is decreasing, they are talking about a rate of change of a rate of change—a derivative of a derivative.

The idea of a "second derivative" also arises in the study of the motion of a particle P along a linear scale (Figure 1). Call the scale the s axis, and denote the variable coordinate of P by s, so that

Figure 1

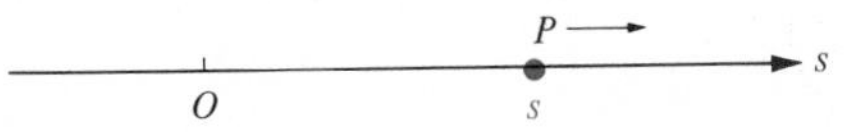

$$s = f(t)$$

where f is a function determining the location of P at time t. The equation $s = f(t)$ is called the **law of motion,** or the **equation of motion,** of the particle.

The **velocity** v of the particle P is defined to be the instantaneous rate of change of its position coordinate s with respect to time:

$$v = \frac{ds}{dt}$$

The absolute value of the velocity is called the **speed** of the particle:

$$\text{speed} = |v|$$

In physics, the instantaneous rate of change of velocity with respect to time is called the **acceleration** a of P:

$$a = \frac{dv}{dt}$$

Therefore, $$a = \frac{d}{dt}v = \frac{d}{dt}\left(\frac{ds}{dt}\right)$$

so the acceleration is the derivative of the derivative, or, as we say, the **second derivative** of the position coordinate s with respect to time. If distances along the s axis are given in meters and time t is measured in seconds, then the velocity v is given in meters per second (m/s). Consequently, the acceleration a is given in meters per second per second read "meters per second squared" (m/s^2).

EXAMPLE 1   Suppose that a particle is moving along the s axis according to the law of motion $s = 100t^2 + 10t + 5$, with s in meters and t in seconds. Find the values of v and a when $t = 0.5$ second.

SOLUTION   Here,

$$v = \frac{ds}{dt} = \frac{d}{dt}(100t^2 + 10t + 5) = 200t + 10$$

and $$a = \frac{dv}{dt} = \frac{d}{dt}(200t + 10) = 200$$

Hence, when $t = 0.5$ second,

$$v = 110 \text{ m/s} \quad \text{and} \quad a = 200 \text{ m/s}^2 \quad \blacksquare$$

EXAMPLE 2   Suppose that a particle is oscillating about the origin on the s axis according to the law of motion $s = 10 \cos [(\pi/12)t - (\pi/3)]$, with s in centimeters and t in seconds. Find the values of v and a when $t = 6$ seconds.

SOLUTION   Here,

$$v = \frac{ds}{dt} = \frac{d}{dt} 10 \cos \left(\frac{\pi}{12}t - \frac{\pi}{3} \right) = -\frac{10\pi}{12} \sin \left(\frac{\pi}{12}t - \frac{\pi}{3} \right)$$

$$= -\frac{5\pi}{6} \sin \left(\frac{\pi}{12}t - \frac{\pi}{3} \right)$$

and

$$a = \frac{dv}{dt} = \frac{d}{dt} \left[-\frac{5\pi}{6} \sin \left(\frac{\pi}{12}t - \frac{\pi}{3} \right) \right] = \left(-\frac{5\pi}{6} \right)\left(\frac{\pi}{12} \right) \cos \left(\frac{\pi}{12}t - \frac{\pi}{3} \right)$$

$$= -\frac{5\pi^2}{72} \cos \left(\frac{\pi}{12}t - \frac{\pi}{3} \right)$$

Hence, when $t = 6$ seconds,

$$v = -\frac{5\pi}{6} \sin \left(\frac{\pi}{2} - \frac{\pi}{3} \right) = -\frac{5\pi}{6} \sin \frac{\pi}{6} = -\frac{5\pi}{6}\left(\frac{1}{2} \right) = -\frac{5\pi}{12} \text{ cm/s}$$

and

$$a = -\frac{5\pi^2}{72} \cos \left(\frac{\pi}{2} - \frac{\pi}{3} \right) = -\frac{5\pi^2}{72} \cos \frac{\pi}{6}$$

$$= -\frac{5\pi^2}{72}\left(\frac{\sqrt{3}}{2} \right) = -\frac{5\pi^2\sqrt{3}}{144} \text{ cm/s}^2 \quad \blacksquare$$

Derivatives of Order n

More generally, if f is any function that is differentiable on some open interval, then the derivative f' is again a function defined on this open interval, and we can ask whether f' is differentiable on the interval. If it is, then its derivative $(f')'$ is written for simplicity as

> f'' (read "f double prime")

We call f'' the **second-order derivative,** or simply the **second derivative,** of the function f. For instance, if a particle moves along a linear scale according to the law of motion $s = f(t)$, then

$$v = f'(t) \quad \text{and} \quad a = f''(t)$$

There is nothing to prevent us from successively taking derivatives of a function as many times as we please, provided that the derived functions remain differentiable at each stage. Thus, if f is a function and if $f, f',$ and f'' are differentiable on an open interval, we can form the **third-order derivative,** or **third derivative,**

> $f''' = (f'')'$

If f can be successively differentiated n times in this way, we say that f is **n times differentiable,** and we write its **nth-order derivative,** or **nth derivative,** as

$$\overbrace{f'' \cdots ''}^{n} \quad \text{or} \quad f^{(n)}$$

The parentheses around the n are to prevent its being confused with an exponent. For instance,

$$f'' = f^{(2)} \quad \text{and} \quad f''' = f^{(3)}$$

EXAMPLE 3 Find all the higher-order derivatives of the polynomial function $f(x) = 15x^4 - 8x^3 + 3x^2 - 2x + 4$.

SOLUTION
$$f'(x) = f^{(1)}(x) = 60x^3 - 24x^2 + 6x - 2$$
$$f''(x) = f^{(2)}(x) = 180x^2 - 48x + 6$$
$$f'''(x) = f^{(3)}(x) = 360x - 48$$
$$f''''(x) = f^{(4)}(x) = 360$$
$$f'''''(x) = f^{(5)}(x) = 0$$

Since $f^{(4)}$ is a constant function, all subsequent derivatives are zero; that is,

$$f^{(n)}(x) = 0 \quad \text{for} \quad n \geq 5$$

Just as with first derivatives, we often deliberately ignore the distinction between the nth-order derived function $f^{(n)}$ and the value $f^{(n)}(x)$ of this function at the number x; both are referred to as "the *nth derivative.*"

Operator notation for higher-order derivatives is self-explanatory:

$$D_x^n f(x) = f^{(n)}(x)$$

The corresponding Leibniz notation is developed as follows: If

$$y = f(x)$$

so that

$$\frac{dy}{dx} = f'(x)$$

then the second derivative is given by

$$\frac{d\left(\dfrac{dy}{dx}\right)}{dx} = D_x^2 f(x) = f''(x)$$

The symbolism

$$\frac{d\left(\dfrac{dy}{dx}\right)}{dx}$$

for the second derivative is cumbersome. Formal algebraic manipulation, as if actual fractions were involved, converts this to

$$\frac{d^2 y}{(dx)^2}$$

In practice, the parentheses in the "denominator" are usually dropped, and the second derivative is written as d^2y/dx^2. Similar notation is used for higher-order derivatives (assuming these exist), as shown in Table 1.

Table 1

$y = f(x)$	Prime	Operator	Leibniz
1 First derivative	$y' = f'(x)$	$D_x y = D_x f(x)$	$\dfrac{dy}{dx} = \dfrac{d}{dx} f(x)$
2 Second derivative	$(y')' = y'' = f''(x)$	$D_x(D_x y) = D_x^2 y = D_x^2 f(x)$	$\dfrac{d^2 y}{dx^2} = \dfrac{d^2}{dx^2} f(x)$
3 Third derivative	$(y'')' = y''' = f'''(x)$	$D_x(D_x^2 y) = D_x^3 y = D_x^3 f(x)$	$\dfrac{d^3 y}{dx^3} = \dfrac{d^3}{dx^3} f(x)$
$\vdots$	$\vdots$	$\vdots$	$\vdots$
n nth derivative	$(y^{(n-1)})' = y^{(n)} = f^{(n)}(x)$	$D_x(D_x^{n-1} y) = D_x^n y = D_x^n f(x)$	$\dfrac{d^n y}{dx^n} = \dfrac{d^n}{dx^n} f(x)$

EXAMPLE 4   If $y = 2x^2 + \dfrac{1}{x^2}$, find

(a) $D_x y$ (b) $D_x^2 y$ (c) $D_x^3 y$

SOLUTION

(a) $D_x y = D_x\left(2x^2 + \dfrac{1}{x^2}\right) = 4x - \dfrac{2}{x^3}$

(b) $D_x^2 y = D_x\left(4x - \dfrac{2}{x^3}\right) = 4 + \dfrac{6}{x^4}$

(c) $D_x^3 y = D_x\left(4 + \dfrac{6}{x^4}\right) = -\dfrac{24}{x^5}$

EXAMPLE 5   Let $y = \sqrt{x}$. Find $d^n y/dx^n$ for all values of n.

SOLUTION   Here, $y = \sqrt{x} = x^{1/2}$. Using direct calculation, we have

$$\frac{dy}{dx} = \frac{1}{2} x^{-1/2}$$

$$\frac{d^2 y}{dx^2} = \frac{d}{dx}\left(\frac{1}{2} x^{-1/2}\right) = \frac{1}{2}\left(-\frac{1}{2}\right) x^{-3/2}$$

$$\frac{d^3 y}{dx^3} = \frac{d}{dx}\left[\frac{1}{2}\left(-\frac{1}{2}\right) x^{-3/2}\right] = \frac{1}{2}\left(-\frac{1}{2}\right)\left(-\frac{3}{2}\right) x^{-5/2}$$

$$\frac{d^4 y}{dx^4} = \frac{d}{dx}\left[\frac{1}{2}\left(-\frac{1}{2}\right)\left(-\frac{3}{2}\right) x^{-5/2}\right] = \frac{1}{2}\left(-\frac{1}{2}\right)\left(-\frac{3}{2}\right)\left(-\frac{5}{2}\right) x^{-7/2}$$

The pattern is now emerging; in fact,

$$\frac{d^n y}{dx^n} = (-1)^{n+1} \frac{1 \cdot 3 \cdot 5 \cdot 7 \cdots (2n-3)}{2^n} x^{-(2n-1)/2}$$

apparently holds for $n \geq 2$. (A formal proof of this fact can be made by using mathematical induction.)

EXAMPLE 6 Let $f(x) = x \cos 2x$. Find

(a) $f'(x)$ (b) $f''(x)$ (c) $f'''(x)$

(d) $f'(0)$ (e) $f''(\pi/6)$ (f) $f'''(-\pi/4)$

SOLUTION

(a) $f'(x) = \cos 2x - 2x \sin 2x$

(b) $f''(x) = -2 \sin 2x - 2 \sin 2x - 4x \cos 2x$
$$= -4 \sin 2x - 4x \cos 2x$$

(c) $f'''(x) = -8 \cos 2x - 4 \cos 2x + 8x \sin 2x$
$$= -12 \cos 2x + 8x \sin 2x$$

(d) By (a), $f'(0) = \cos [2(0)] - 2(0) \sin [2(0)]$
$$= \cos 0 - 0 = 1$$

(e) By (b), $f''\left(\dfrac{\pi}{6}\right) = -4 \sin \left[2\left(\dfrac{\pi}{6}\right)\right] - 4\left(\dfrac{\pi}{6}\right) \cos \left[2\left(\dfrac{\pi}{6}\right)\right]$

$$= -4 \sin \dfrac{\pi}{3} - \dfrac{2\pi}{3} \cos \dfrac{\pi}{3}$$

$$= -4\left(\dfrac{\sqrt{3}}{2}\right) - \dfrac{2\pi}{3}\left(\dfrac{1}{2}\right)$$

$$= -2\sqrt{3} - \dfrac{\pi}{3}$$

(f) By (c), $f'''\left(-\dfrac{\pi}{4}\right) = -12 \cos \left[2\left(-\dfrac{\pi}{4}\right)\right] + 8\left(-\dfrac{\pi}{4}\right) \sin \left[2\left(-\dfrac{\pi}{4}\right)\right]$

$$= -12 \cos \left(-\dfrac{\pi}{2}\right) - 2\pi \sin \left(-\dfrac{\pi}{2}\right)$$

$$= -12(0) - 2\pi(-1)$$

$$= 2\pi$$

Many of the differentiation rules developed earlier have generalizations for higher-order derivatives. (Most of these are proved by mathematical induction on n, the order of the derivative involved.) For instance, the addition rule and the homogeneous rule work for the nth-order derivative; that is,

$$D_x^n[f(x) + g(x)] = D_x^n f(x) + D_x^n g(x)$$

and $$D_x^n[cf(x)] = cD_x^n f(x) \qquad c \text{ constant}$$

The product (or Leibniz) rule for higher-order derivatives $(f \cdot g)^{(n)}$ is more complicated. For example, if $n = 2$, we have

$$(f \cdot g)'' = (f \cdot g' + f' \cdot g)' = (f \cdot g')' + (f' \cdot g)'$$
$$= f \cdot g'' + f' \cdot g' + f' \cdot g' + f'' \cdot g$$

Hence,

$$(f \cdot g)'' = f \cdot g'' + 2f' \cdot g' + f'' \cdot g$$

As Example 7 illustrates, higher-order derivatives can be found by implicit differentiation.

EXAMPLE 7   If $x^2 - 2y^2 = 4$, find $D_x y$ and $D_x^2 y$ by implicit differentiation.

SOLUTION   Differentiating both sides of $x^2 - 2y^2 = 4$ with respect to x, we have

$$2x - 4yD_x y = 0$$

hence,

$$D_x y = \frac{2x}{4y} = \frac{x}{2y}$$

Differentiating both sides of the last equation with respect to x and using the quotient rule, we obtain

$$D_x^2 y = \frac{2yD_x x - xD_x(2y)}{(2y)^2} = \frac{2y - 2xD_x y}{4y^2} = \frac{y - xD_x y}{2y^2}$$

Substituting $D_x y = x/(2y)$ into the last equation and simplifying, we find that

$$D_x^2 y = \frac{y - x[x/(2y)]}{2y^2} = \frac{2y^2 - x^2}{4y^3}$$

Finally, since $x^2 - 2y^2 = 4$, we can rewrite this result as

$$D_x^2 y = \frac{-4}{4y^3}$$

or

$$D_x^2 y = -\frac{1}{y^3}$$ ∎

Problem Set 2.10

In Problems 1 to 6, a particle is moving along the s axis according to the given law of motion $s = f(t)$. Find $v = ds/dt$ and $a = dv/dt$.

1 $s = t^3 + 2t^2 + 3t - 5$, s in meters, t in seconds

2 $s = (t^2 + 1)^{-1}$, s in centimeters, t in seconds

3 $s = 3 \sin \pi t - 2 \cos 2\pi t$, s in meters, t in seconds

4 $s = t\sqrt{t^2 + 4}$, s in miles, t in hours

5 $s = \frac{5}{2}t^{5/2} + \frac{2}{3}t^{3/2}$, s in kilometers, t in hours

6 $s = \frac{1}{2}gt^2 + v_0 t + s_0$, where g, v_0, and s_0 are constants

7 In Problems 1, 3, and 5, find v and a at the instant when $t = 1$ unit.

8 If air resistance is neglected, a dropped object will fall through $s = 4.9t^2$ meters during a time interval of t seconds. (a) Find the acceleration of the falling object. (b) Find a formula for the velocity of the object after it has dropped through s meters.

In Problems 9 to 36, find the first and second derivatives of the function.

9 $f(x) = 5x^3 + 4x + 2$

10 $g(x) = x^2(x^2 + 7)$

11 $f(t) = 7t^5 - 23t^2 + t + 9$

12 $F(x) = x^3(x + 2)^2$

13 $G(x) = (x^2 - 3)(x^4 + 3x^2 + 9)$

14 $f(u) = (u^2 + 1)^3$

15 $g(t) = t^3\sqrt{t} - 5t$

16 $f(x) = x - \dfrac{3}{x}$

17 $f(x) = x^2 - \dfrac{1}{x^3}$

18 $g(x) = \left(x + \dfrac{1}{x}\right)^2$

19 $f(u) = \dfrac{2u}{2 - u}$

20 $F(v) = \sqrt{v} + \dfrac{1}{\sqrt{v}}$

21 $f(t) = \sqrt{t^2 + 1}$

22 $g(y) = \sqrt{3y + 1}$

23 $F(r) = \left(1 - \sqrt{r}\right)^2$

24 $h(x) = \dfrac{x}{\sqrt{x^2 + 1}}$

25 $f(x) = 7 \cos 11x$

26 $f(t) = -6 \sin (5 - 2t)$

27 $F(\theta) = \sin 2\theta + \cos 3\theta$

28 $h(x) = (x + \sin x)^4$

29 $H(t) = 2 \csc 7t$

30 $p(y) = 5 \tan^3 4y$

31 $Q(\theta) = \theta \cot 3\theta$

32 $S(x) = \sqrt{1 + \sin 5x}$

33 $G(x) = \sin \dfrac{x}{x + 1}$

34 $y = \sqrt{1 + \sec x}$

35 $y = -4 \sec 5x$

36 $y = x^2 \sin \dfrac{1}{x}$

37 In Problem 17, find $f'(-1)$ and $f''(-1)$.

38 In Problem 24, find $h'(\sqrt{2})$ and $h''(\sqrt{2})$.

39 In Problem 27, find $F'(\pi/6)$ and $F''(\pi/6)$.

40 In Problem 34, find the value of dy/dx and of d^2y/dx^2 when $x = \pi/4$.

41 Find all higher-order derivatives of the function $f(x) = 7x^4 - 5x^3 + 8x^2 - 3x + 81$.

42 If $f(t) = (\sqrt{t+1})^{-1}$, find $f^{(10)}(t)$.

43 If $y = \sqrt{x^2 - 1}$, find dy/dx, d^2y/dx^2, and d^3y/dx^3.

44 Develop a formula for $D_x^n(\sqrt[3]{x})$, $x \neq 0$.

45 Develop a formula for $D_x^n(1/x)$, $x \neq 0$.

46 Find $D_x^{70} \sin x$.

47 If $y = 3x^2 + 2x$, evaluate and simplify the expression $x^2(d^2y/dx^2) - 2x(dy/dx) + 2y$.

48 Assume the Leibniz rule

$$(f \cdot g)'' = f \cdot g'' + 2f' \cdot g' + f'' \cdot g$$

for second derivatives. (a) Derive the Leibniz rule

$$(f \cdot g)''' = f \cdot g''' + 3f' \cdot g'' + 3f'' \cdot g' + f''' \cdot g$$

for third derivatives. (b) Assuming that $f(1) = -3$, $g(1) = \frac{1}{2}$, $f'(1) = -1$, $g'(1) = 4$, $f''(1) = 16$, and $g''(1) = \frac{2}{3}$, find $(f \cdot g)''(1)$.

49 Let g be a twice-differentiable function such that $g(2) = 3$, $g'(2) = \frac{1}{3}$, and $g''(2) = 5$. Define the function f by the equation $f(x) = x^4 g(x)$. Find the numerical value of $f''(2)$.

50 Find formulas for $f'(x)$ and $f''(x)$ if

(a) $f(x) = \begin{cases} x^2 & \text{if } x \leq 1 \\ 2x - 1 & \text{if } x > 1 \end{cases}$

(b) $f(x) = \begin{cases} \dfrac{x^2}{2} & \text{if } x \geq 0 \\ -\dfrac{x^2}{2} & \text{if } x < 0 \end{cases}$

51 A particle is moving along a linear scale according to the given equation of motion, where s is the distance in feet from the origin at the end of t seconds. Find the time when the instantaneous acceleration is zero if

(a) $s = t^3 - 6t^2 + 12t + 1$, $t \geq 0$ (b) $s = \sqrt{1 + t}$, $t \geq 0$

(c) $s = 5t + \dfrac{2}{t+1}$, $t \geq 0$

52 Let f be a twice-differentiable function. A particle Q moves along the graph of f in such a way that the x coordinate of Q at time t is $x = g(t)$, where g is a twice-differentiable function. A particle P moves along the y axis in such a way that the y coordi-

nate of P is always the same as the y coordinate of Q. Find a formula for the acceleration of P at time t.

53 Explain why the nth derivative of a rational function is again a rational function.

54 Let f and g be twice-differentiable functions. Suppose that $g(-1) = 27$, $g'(-1) = -1$, $g''(-1) = \frac{1}{4}$, $f(27) = 2$, $f'(27) = -4$, and $f''(27) = 1$. Find the numerical value of $(f \circ g)''(-1)$.

In Problems 55 to 58, find d^2y/dx^2 in terms of x and/or y.

55 $x^4 + y^4 = 64$ **56** $x^3 + y^3 = 16$

57 $x^2 \sin 3y = 4$ **58** $x \cot y = y$

59 During the early years of the U.S. space program, experiments were conducted on human volunteers riding rocket-driven sleds to determine the physiological effects of large accelerations. One such sled, moving in accordance with an equation of the form $s = kt^2$, where k is a constant, was able to attain a velocity of 500 meters per second in 2 seconds. What distance (in meters) did the sled travel in these 2 seconds?

60 If a and v are the acceleration and velocity of a particle moving on the s axis, show that $a = v(dv/ds)$.

61 If N is the number of people who have contracted a disease t days after the beginning of an epidemic, then the rate R at which the disease is spreading is given by $R = dN/dt$. Public health officials often feel that the epidemic is "under control" when $dR/dt = 0$. Suppose that epidemiologists use the mathematical model $N = 500(20t + 18t^2 - t^3)$ to predict the number N of people who will have contracted a certain virulent new form of flu t days after the beginning of an epidemic. If the model is correct, after how many days will the epidemic be brought under control?

2.11   Algebraic Signs and Zeros of Functions

If a function f satisfies the condition $f(x) > 0$ for all values of x in an interval, we say that f is **positive** on the interval. Likewise, we say that f is **negative** on an interval if $f(x) < 0$ for all values of x in the interval. To say that f is positive (or negative) on an interval means that, over this interval, the graph of f lies entirely above (or entirely below) the x axis. The x coordinates of points where the graph of f touches or crosses the x axis are called the x **intercepts** of the graph. In other words, the x intercepts of the graph of f are the solutions, or roots, of the equation

$$f(x) = 0$$

For this reason, an x intercept of the graph of f is also called a **zero** of f.

For instance, from the graph shown in Figure 1 of the function

$$f(x) = 4x^4 - 4x^3 - 25x^2 + x + 6$$

you can see that $f(x)$ is positive on the intervals $(-\infty, -2)$, $(-\frac{1}{2}, \frac{1}{2})$, and $(3, \infty)$ and negative on the intervals $(-2, -\frac{1}{2})$ and $(\frac{1}{2}, 3)$. Evidently, the x intercepts of the graph of f are -2, $-\frac{1}{2}$, $\frac{1}{2}$, and 3. Notice that the intervals on which f is positive are separated by zeros of f from the intervals on which f is negative.

Figure 1

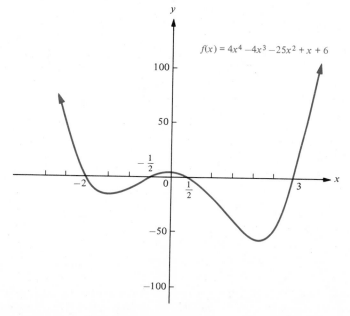

$$f(x) = 4x^4 - 4x^3 - 25x^2 + x + 6$$

In this section, we use the ideas of continuity and the derivative to study the algebraic signs and zeros of a function. One of our basic tools is the following important theorem about continuous functions.

THEOREM 1   **The Intermediate-Value Theorem**

Suppose that the function f is continuous on the closed interval $[a, b]$. Then if k is any number between $f(a)$ and $f(b)$, there is at least one number c in the interval $[a, b]$ such that $f(c) = k$.

Although the formal proof of this theorem is best left to more advanced courses, Figure 2 should make it seem plausible. Indeed, since f is continuous, its graph consists of one connected piece with no holes and no sudden jumps. The points $(a, f(a))$ and $(b, f(b))$ lie on this graph, one below and the other above the horizontal line $y = k$. It seems clear that the graph of f must intersect the line $y = k$, say, at the point (c, k). Therefore, $f(c) = k$. Note that the intermediate-value theorem just guarantees the *existence* of such a number c, but it doesn't tell us how to actually *find* such a number.

Figure 2

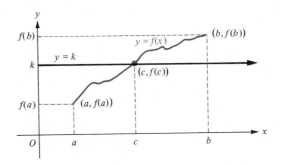

EXAMPLE 1   Show that there is a number c between 1 and 2 such that $c^5 - 2c^3 = 9$.

SOLUTION   The polynomial function

$$f(x) = x^5 - 2x^3$$

is continuous on the interval [1, 2]. Also, $f(1) = -1$ and $f(2) = 16$. Because 9 is between -1 and 16, Theorem 1 guarantees the existence of at least one number c between 1 and 2 such that $f(c) = 9$; that is,

$$c^5 - 2c^3 = 9$$

A function f is said to **change sign** on an interval I if there are numbers a and b in I such that $f(a)$ and $f(b)$ have opposite algebraic signs. As a consequence of Theorem 1, we can state the following useful property of continuous functions.

Change-of-Sign Property

> A continuous function cannot change sign on an interval unless it has a zero in that interval.

To see how the change-of-sign property follows from Theorem 1, let f be a continuous function on an interval I, and suppose that a and b are numbers in I for which $f(a)$ and $f(b)$ have opposite algebraic signs. It follows that 0 is between $f(a)$ and $f(b)$; hence, by Theorem 1, there is at least one number c between a and b such that $f(c) = 0$.

EXAMPLE 2   Show that the function $f(x) = x^3 + x - 1$ has a zero in the interval [0, 1].

SOLUTION   The polynomial function f is continuous on the interval [0, 1], and because $f(0) = -1$ and $f(1) = 1$, it changes its sign on this interval. Therefore, by the change-of-sign property, f has a zero on the interval [0, 1].

The change-of-sign property justifies the following simple procedure.

Procedure for Finding the Intervals on Which a Function f is Positive or Negative

Step 1 Find all the numbers at which f is undefined or discontinuous or takes the value 0, and arrange these numbers in increasing order: $x_1, x_2, x_3, \ldots,$ x_n. These numbers will divide the x axis into $n + 1$ open intervals: $(-\infty, x_1)$, (x_1, x_2), (x_2, x_3), $\ldots$, (x_{n-1}, x_n), and (x_n, ∞).

Step 2 By the change-of-sign property, you may conclude that f cannot change sign on any of the intervals (x_1, x_2), (x_2, x_3), $\ldots$, (x_{n-1}, x_n). Therefore, f is either positive or negative on each of these intervals. The same is true for the unbounded intervals $(-\infty, x_1)$ and (x_n, ∞), provided that they are contained in the domain of f.

Step 3 Select any convenient "test number" from each interval considered in step 2, and evaluate the function f at each test number. The algebraic sign of f at each test number determines the algebraic sign of f on the entire interval containing that test number.

In Examples 3 and 4, determine the intervals on which the function is positive or negative.

<u>EXAMPLE 3</u> $f(x) = 4x^2 + 8x - 5$

SOLUTION   We follow the steps in the procedure above.

Step 1 The polynomial function f is defined and continuous at every real number x. The zeros of f are the roots of the quadratic equation

$$4x^2 + 8x - 5 = 0$$

Factoring the left side of this equation, we have

$$(2x + 5)(2x - 1) = 0$$

so the roots are

$$x = -\tfrac{5}{2} \qquad \text{and} \qquad x = \tfrac{1}{2}$$

These two numbers divide the x axis into the three open intervals

$$(-\infty, -\tfrac{5}{2}) \qquad (-\tfrac{5}{2}, \tfrac{1}{2}) \qquad (\tfrac{1}{2}, \infty)$$

Step 2 The function f is either positive or negative on each of the intervals obtained in step 1.

Step 3 From the interval $(-\infty, -\tfrac{5}{2})$ we select, say, the test number -3. Because

$$f(-3) = 4(-3)^2 + 8(-3) - 5 = 7 > 0$$

we conclude that f is positive on the entire interval $(-\infty, -\tfrac{5}{2})$. From the interval $(-\tfrac{5}{2}, \tfrac{1}{2})$ we select, say, the test number 0. Because

$$f(0) = 4(0)^2 + 8(0) - 5 = -5 < 0$$

we conclude that f is negative on the entire interval $(-\tfrac{5}{2}, \tfrac{1}{2})$. Finally, from the interval $(\tfrac{1}{2}, \infty)$ we select, say, the test number 1. Because

$$f(1) = 4(1)^2 + 8(1) - 5 = 7 > 0$$

we conclude that f is positive on the entire interval $(\frac{1}{2}, \infty)$. Figure 3 summarizes our findings. ∎

Figure 3

algebraic sign of $f(x) = 4x^2 + 8x - 5$

EXAMPLE 4 $f(x) = \dfrac{(x + 3)(x - 2)}{(x + 1)(x - 1)}$

SOLUTION   Here, $f(x)$ is undefined when the denominator of the fraction

$$\frac{(x + 3)(x - 2)}{(x + 1)(x - 1)}$$

is zero, that is, when $x = -1$ and $x = 1$. Also, $f(x) = 0$ when the numerator is zero, that is, when $x = -3$ and when $x = 2$. We arrange these four numbers in the order $-3, -1, 1, 2$ to determine the five open intervals

$$(-\infty, -3) \quad (-3, -1) \quad (-1, 1) \quad (1, 2) \quad (2, \infty)$$

on which the function f is either positive or negative. Selecting a convenient test number from each interval, say $-4, -2, 0, \frac{3}{2}$, and 3, and evaluating $f(x)$ at each test number, we obtain

$$f(-4) = \tfrac{2}{5} \quad f(-2) = -\tfrac{4}{3} \quad f(0) = 6 \quad f(\tfrac{3}{2}) = -\tfrac{9}{5} \quad f(3) = \tfrac{3}{4}$$

Figure 4   algebraic sign of $f(x) = \frac{(x + 3)(x - 2)}{(x + 1)(x - 1)}$

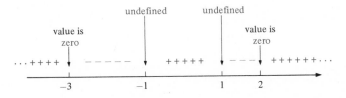

The algebraic signs of $f(x)$ at these test numbers determine the algebraic signs of f on the corresponding intervals, as shown in Figure 4. ∎

The change-of-sign property can also be used to help locate the zeros of a continuous function f by the following **bisection method:** By sketching a rough graph, or by trial and error, find a closed interval $[a, b]$ for which $f(a)$ and $f(b)$ have opposite algebraic signs. By the change-of-sign property, f must have a zero in this interval. Using the midpoint $c = \frac{1}{2}(a + b)$, divide the interval $[a, b]$ into two subintervals $[a, c]$ and $[c, b]$. If $f(a)$ and $f(c)$ have opposite algebraic signs, then f must have a zero in the interval $[a, c]$. If $f(c) = 0$, then c itself is a zero of f. The only other possibility is that $f(a)$ and $f(c)$ have the same algebraic sign, in which case $f(c)$ and $f(b)$ must have opposite algebraic signs (why?) and f must have a zero in the interval $[c, b]$. Continuing in this way, you can locate a zero of f as accurately as you wish by confining it in successively shorter and shorter intervals.

Ⓒ **EXAMPLE 5**   In Example 2, we used the change-of-sign property to show that the continuous function $f(x) = x^3 + x - 1$

has a zero between 0 and 1. Starting with the interval $[0, 1]$, use the bisection method twice in succession to locate a zero of f with more accuracy.

SOLUTION   The midpoint of the interval $[0, 1]$ is $\frac{1}{2}(0 + 1) = 0.5$. A simple calculation reveals that

$$f(0) = -1 \quad \text{and} \quad f(0.5) = -0.375$$

so $f(0)$ and $f(0.5)$ have the same algebraic sign. It follows that $f(0.5)$ and $f(1)$ must have opposite algebraic signs. Indeed, they do:

$$f(0.5) = -0.375 \quad \text{and} \quad f(1) = 1$$

so f has a zero between 0.5 and 1. Now, for our second application of the bisection method, we start with the interval $[0.5, 1]$. The midpoint of this interval is $\frac{1}{2}(0.5 + 1) = 0.75$. Here we have

$$f(0.5) = -0.375 \quad \text{and} \quad f(0.75) = 0.171875$$

so $f(0.5)$ and $f(0.75)$ have opposite algebraic signs. It follows that f has a zero between 0.5 and 0.75. ■

Newton's Method

In the seventeenth century, Isaac Newton discovered an elegant method, using derivatives, for estimating the zeros of functions. This method, which is usually much more efficient than the bisection method, is especially well suited to modern electronic calculators and computers. The basic idea of **Newton's method** is illustrated in Figure 5. Suppose that z is a zero of the differentiable function f and that a is an approximation to z. If we start at the point $(a, f(a))$ and follow the graph of f to the point where it intersects the x axis, we obtain the exact value of z. If, instead, we start at the point $(a, f(a))$ and follow the *tangent line* to the point $(b, 0)$ where it intersects the x axis, we obtain an approximation b to the zero z. In many cases, b *is a better approximation to z than a was.*

In Figure 5, an equation of the tangent line to the graph of f at $(a, f(a))$ is

$$y - f(a) = f'(a)(x - a)$$

Because the point $(b, 0)$ is on this line, we must have

$$0 - f(a) = f'(a)(b - a)$$

If $f'(a) \neq 0$, we can rewrite the last equation as

$$-\frac{f(a)}{f'(a)} = b - a$$

or

$$\boxed{b = a - \frac{f(a)}{f'(a)}}$$

Thus, the basic idea of Newton's method is to replace an approximation a to a zero of f by what we presume is a better approximation b given by the equation above. But then we can take this "better" approximation b and substitute it back for a in the equation to obtain, we presume, an even better approximation:

$$c = b - \frac{f(b)}{f'(b)}$$

Figure 5

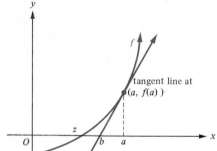

This process can be repeated as often as we wish to obtain (we hope!) better and better approximations to a zero of f.

The discussion above is summarized as follows.

Newton's Method

Let x_1 be a first approximation to a zero of the differentiable function f. Then successive approximations

$$x_2, x_3, \ldots, x_n, x_{n+1}, \ldots$$

are given by repeated use of the equation

$$x_{n+1} = x_n - \frac{f(x_n)}{f'(x_n)}$$

which gives the $(n + 1)$st approximation in terms of the nth approximation.

In using Newton's method, the first approximation x_1 can be obtained by an "educated guess," by sketching a rough graph of f, or by using the bisection method.

In more advanced courses in numerical analysis, conditions are developed to guarantee that Newton's method "works" in the sense that the successive approximations $x_1, x_2, x_3, \ldots$ really do come closer and closer to a zero of f. Roughly, these conditions are that $|f'(x)|$ is not too small and $|f''(x)|$ is not too large for values of x close to x_1. However, rather than bothering to check these conditions, it's usually easier just to use a calculator and see what happens. When the method works, you'll soon see the successive approximations "settling down" and coming closer and closer to the desired number. When the number in the display of the calculator stops changing as you repeat the calculation, you have found an approximation to a zero of the function within the limits of accuracy of the calculator. When the method fails, you'll soon see the successive "approximations" behaving erratically and not "settling down" (see Problems 53 to 56).

© **EXAMPLE 6**   In Example 5, we used the bisection method to show that the function

$$f(x) = x^3 + x - 1$$

has a zero between 0.5 and 0.75. Using a calculator and Newton's method, find this zero to as many decimal places as possible.

SOLUTION   Here,

$$f'(x) = 3x^2 + 1$$

and the formula

$$x_{n+1} = x_n - \frac{f(x_n)}{f'(x_n)}$$

becomes

$$x_{n+1} = x_n - \frac{(x_n)^3 + x_n - 1}{3(x_n)^2 + 1}$$

We're looking for a zero on the interval [0.5, 0.75], so let's take the midpoint

$$x_1 = \tfrac{1}{2}(0.5 + 0.75) = 0.625$$

as our first approximation. Then, using a 10-digit calculator, we find that

$$x_2 = 0.625 - \frac{(0.625)^3 + 0.625 - 1}{3(0.625)^2 + 1} = 0.685251799$$

Repeating the process with the value of x_2 just obtained, we have

$$x_3 = x_2 - \frac{(x_2)^3 + x_2 - 1}{3(x_2)^2 + 1} = 0.682335090$$

Note that in the passage from x_2 to x_3 the first two decimal places haven't changed. This is an indication that the successive approximations are settling down as desired. Let's continue the process:

$$x_4 = x_3 - \frac{(x_3)^3 + x_3 - 1}{3(x_3)^2 + 1} = 0.682327804$$

$$x_5 = x_4 - \frac{(x_4)^3 + x_4 - 1}{3(x_4)^2 + 1} = 0.682327804$$

Because $x_4 = x_5$, we have found the desired zero z to within the limits of accuracy of the 10-digit calculator:

$$z \approx 0.682327804$$ ∎

You can use Newton's method to find approximate solutions to equations that you cannot solve by the usual algebraic techniques.

Figure 6

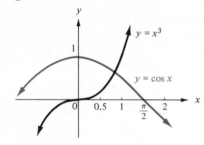

© **EXAMPLE 7** Find an (approximate) solution of the equation $x^3 = \cos x$.

SOLUTION   A rough sketch of the graphs of $y = x^3$ and $y = \cos x$ on the same coordinate system (Figure 6) shows that they intersect at a point whose abscissa is approximately 0.8. In order to use Newton's method, we rewrite the given equation in the form

$$x^3 - \cos x = 0$$

Thus, we are looking for a zero of the function

$$f(x) = x^3 - \cos x$$

and we know that $x_1 = 0.8$ is a reasonable first approximation. Here,

$$f'(x) = 3x^2 + \sin x$$

and we have

$$x_{n+1} = x_n - \frac{(x_n)^3 - \cos x_n}{3(x_n)^2 + \sin x_n}$$

Using the last formula and a 10-digit calculator, we obtain the following successive approximations to the desired zero:

$$\begin{array}{ll} x_1 = 0.8 & x_4 = 0.865474034 \\ x_2 = 0.870034801 & x_5 = 0.865474033 \\ x_3 = 0.865494102 & x_6 = 0.865474033 \end{array}$$

Thus, to within the limits of accuracy of the 10-digit calculator, the solution of the equation $x^3 = \cos x$ is $x \approx 0.865474033$. ∎

Problem Set 2.11

In Problems 1 to 12, use the intermediate-value theorem (Theorem 1) to show that there is a number c between the two given numbers that satisfies the equation.

1 c between 0 and 1, $4c^3 - c^2 = 2$

2 c between 1 and 2, $2c^4 - 4c^3 + 8c = 7.07$

3 c between 2 and 3, $c^3 + 3c^2 - 9c = 10$

4 c between 2 and 3, $\dfrac{\sqrt{8c - 15}}{c} = \dfrac{2}{3}$

5 c between 0 and 1, $\dfrac{c^3 + 5}{\sqrt{c + 1}} = 4$

© **6** c between 2.5 and 2.6, $c^4 - 8c^2 + c = -6$

7 c between -2 and -1, $2c^3 - 3c^2 - 12c = 1$

© **8** c between 1.4 and 1.5, $\dfrac{1}{c^4 - 4c^3 + 4c^2} = \sqrt{3}$

9 c between $3\pi/4$ and π, $\sin c + 2 \cos 2c = 1$

© **10** c between $\pi/6$ and $\pi/3$, $2 \csc c + \cot c = 4$

© **11** c between 0 and $\pi/6$, $c + \sin c = 1$

12 c between $\pi/6$ and $\pi/2$, $\dfrac{\sin c}{2 + \cos c} = \dfrac{1}{4}$

In Problems 13 to 18, show that the function has a zero in the interval by using the change-of-sign property. (You need not try to find the value of the zero.)

13 $f(x) = x^5 - 2x^3 - 1$ in $[1, 2]$

© **14** $g(x) = x^4 + 6x^3 - 18x^2$ in $[2.1, 2.2]$

© **15** $f(x) = x^5 - 2x^3 - 1$ in $[1.5, 1.6]$

© **16** $g(x) = x^4 + 6x^3 - 18x^2$ in $[-8.2, -8.1]$

17 $h(x) = \sin x + 2 \cos 2x$ in $[\pi/4, \pi/3]$

© **18** $F(x) = \sin^2 x + 2 \cos x$ in $[1.9, 2]$

In Problems 19 to 34, determine the intervals on which each function is positive or negative.

19 $f(x) = x^2 - 6x + 8$

20 $g(x) = 25x^2 - 20x + 4$

21 $F(x) = 2x(x - 3)(x + 5)$

22 $G(x) = x^4 - 7x^2 + 12$

23 $h(x) = (2x + 1)(3x - 1)(x - 2)$

24 $H(x) = \sqrt{x - 1}(x - 2)|x - 3|$

25 $q(x) = \dfrac{2x - 1}{3x - 2}$

26 $Q(x) = \dfrac{(3x + 4)(2x - 1)}{(x + 2)(x - 1)}$

27 $r(x) = \dfrac{(x + 1)(x - 2)}{(x + 3)(x - 4)}$

28 $R(x) = \dfrac{x^2 - 2x - 15}{x^2 - 2x + 10}$

29 $s(x) = \dfrac{(x - 2)^2(x + 1)(2x - 1)}{(x + 3)(x - 1)^2(x + 4)}$

30 $S(x) = \dfrac{1 + x}{(1 - x^2)^2}$

31 $f(x) = 3x^{2/3} - x^{5/3}$

32 $F(x) = (x - 1)^{1/3}x^{-2/3}$

33 $g(x) = \begin{cases} x^2 - 1 & \text{if } x \geq 1 \\ 1 - x^2 & \text{if } x < 1 \end{cases}$

34 $G(x) = \begin{cases} \sqrt{25 - x^2} & \text{if } x \leq 4 \\ 7 - x & \text{if } x > 4 \end{cases}$

© **35** In Problem 15, use the bisection method twice in succession, starting with the interval $[1.5, 1.6]$, to locate a zero of $f(x) = x^5 - 2x^3 - 1$ with greater accuracy.

© **36** In Problem 16, use the bisection method twice in succession, starting with the interval $[-8.2, -8.1]$, to locate a zero of $g(x) = x^4 + 6x^3 - 18x^2$ with greater accuracy.

© **37** In Problem 17, use the bisection method twice in succession, starting with the interval $[\pi/4, \pi/3]$, to locate a zero of $h(x) = \sin x + 2 \cos 2x$ with greater accuracy.

© **38** In Problem 18, use the bisection method four times in succession, starting with the interval $[1.9, 2]$, to locate a zero of $F(x) = \sin^2 x + 2 \cos x$ with greater accuracy.

© In Problems 39 to 44, use a calculator and Newton's method to find a zero of the function to as many decimal places as possible. Start with the value of x_1 as a first approximation.

39 The function $f(x) = x^5 - 2x^3 - 1$ of Problems 15 and 35; $x_1 =$ midpoint of the interval $[1.5, 1.6]$

40 The function $g(x) = x^4 + 6x^3 - 18x^2$ of Problems 16 and 36; $x_1 =$ midpoint of the interval $[-8.2, -8.1]$

41 The function $h(x) = \sin x + 2 \cos 2x$ of Problems 17 and 37; x_1 = midpoint of the interval $[\pi/4, \pi/3]$

42 The function $F(x) = \sin^2 x + 2 \cos x$ of Problems 18 and 38; x_1 = midpoint of the interval $[1.9, 2]$

43 $G(x) = x^3 - 7x + 7$; $x_1 = -3.5$

44 $H(x) = x^3 - 4x^2 - 2x + 4$; $x_1 = 4.5$

Ⓒ In Problems 45 to 48, use Newton's method to solve the equation.

45 $2x^3 - 4x^2 + 5x = 7$

46 $15x^5 + 13x^3 = 1$

47 $x = \dfrac{\sin x}{x}$

48 $x^{3/2} + x = 1 - 2x^{1/2}$

Ⓒ **49** The ancient Babylonians discovered the following rule for approximating the square root of a positive number k: Begin by making a reasonable estimate a of $\sqrt{k}$. Then $b = \frac{1}{2}[a + (k/a)]$ will be an even better estimate. Repeat the process to get better and better estimates. (a) Use this method to obtain $\sqrt{2}$ to as many decimal places as you can, starting with 1 as a first approximation. (b) Show that this method of estimating square roots is just a special case of Newton's method.

Ⓒ **50** The function $G(x) = x^3 - 7x + 7$ of Problem 43 has *two* zeros on the interval $[1.3, 1.7]$. Find these two zeros to as many decimal places as you can.

51 Using Newton's method, derive a rule for obtaining successively better and better approximations for the nth root of a positive number k.

Ⓒ **52** In order to work Problem 56 on page 194 it will be necessary to solve the equation $(t - a)\sqrt{t + a} - b = 0$ for t, where $a = 324$ and $b = 4.32 \times 10^5$. (a) Use the change-of-sign property to show that there is a solution of this equation on the interval from $t = 5000$ to $t = 6000$. (b) Use Newton's method to find this solution to as many decimal places as you can.

53 For the function f whose graph appears in Figure 7, x_1 appears to be a reasonable first approximation to the zero z. However, if you use Newton's method, the second approximation

$$x_2 = x_1 - \frac{f(x_1)}{f'(x_1)}$$

Figure 7

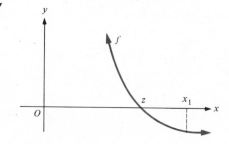

will turn out to be a rather poor estimate for z. Explain why. [*Hint:* Sketch the tangent line to the graph at $(x_1, f(x_1))$.]

Ⓒ **54** Let $f(x) = (1 - 4x)/(1 + 4x)$. Obviously, the only zero z of f is $z = \frac{1}{4}$. Try to use Newton's method with $x_1 = 1$ to obtain an approximation to the zero z. What goes wrong? (*Hint:* See Problem 53.)

55 Let $f(x) = 2 + 3x^2 - x^4$. If z is a zero of f, then $2 + 3z^2 - z^4 = 0$, so $(z^2)^2 - 3(z^2) - 2 = 0$. Using the quadratic formula, we find that $z^2 = \frac{1}{2}(3 + \sqrt{17})$; hence, $z = \pm\sqrt{\frac{1}{2}(3 + \sqrt{17})}$. Try to use Newton's method with $x_1 = 1$ to obtain an approximation to a zero of f. What happens?

Ⓒ **56** With the aid of a calculator, use the point-plotting method to sketch an accurate graph of $f(x) = 2 + 3x^2 - x^4$ for $-2 \le x \le 2$. Using this graph, explain geometrically what goes wrong in the attempt to use Newton's method in Problem 55.

Ⓒ **57** The equation $8x^3 - 6x - 1 = 0$ is used in the proof that it is impossible to trisect a 60° angle with straightedge and compass. (a) Use the change-of-sign property to show that this equation has a root on each of the following intervals: $[-1, -0.5]$, $[-0.5, 0]$, and $[0.5, 1]$. (b) Use Newton's method to find the three roots of $8x^3 - 6x - 1 = 0$ as accurately as you can. (c) The root in the interval $[0.5, 1]$ is actually $\cos 20°$. Use a calculator to check this.

Ⓒ **58** The depth x to which a floating sphere of radius r sinks in fresh water is a root of the equation $x^3 - 3rx^2 + 4r^3s = 0$, where s is the specific gravity of the sphere. The specific gravity of Spanish mahogany is $s = 0.774$. Find the depth to which a sphere of radius $r = 0.4$ meter made of Spanish mahogany will sink in fresh water.

59 Suppose k is a positive constant. By applying Newton's method to the function $f(x) = x^{-1} - k$, develop the formula $b = 2a - ka^2$ for replacing an approximation a to k^{-1} by a "better" approximation b.

60 In Problem 59, show that b really is a better approximation to k^{-1} than a was, provided that $0 < a < 2k^{-1}$.

Ⓒ **61** In thermodynamics, the volume v in cubic meters of 1 mole of carbon dioxide at 0°C and a pressure of 10^5 newtons per square meter is a root of the equation $v^3 - av^2 + bv - c = 0$, where $a = 2.28 \times 10^{-2}$, $b = 3.60 \times 10^{-6}$, and $c = 1.51 \times 10^{-10}$. Use Newton's method to find v. (Round off your answer to three significant digits.)

62 Let f be a differentiable function, suppose that x_1 is a first approximation to a zero of f, and define $F(x) = x - [f(x)/f'(x)]$. Show that the sequence of successive approximations to a zero of f provided by Newton's method, starting with x_1 as a first approximation, is the orbit of x_1 under the iterates of F. (See Problem 48 on page 124.)

Review Problem Set, Chapter 2

1 An object is thrown straight upward with such a speed that at any time its distance s in feet from the surface of the earth is given by the equation $s = 200t - 16t^2$, where t is measured in seconds. Find (a) the average speed during the third second of motion, (b) the instantaneous speed when $t = 2$ and when $t = 3$, and (c) the highest point reached by the object.

2 The motion of a particle along the x axis is given by the equation $x = 2t - (t^2/2)$, where x is the coordinate of the particle at time t. Distance is in meters and time in seconds. (a) Determine whether the particle is moving in the positive or negative direction when $t = 0$. (b) Find the instantaneous speed when $t = 1$. (c) When does the particle change its direction of motion?

3 The side of a square metal plate is 20 inches when its temperature is 50°. If its temperature is raised to 75° in 10 minutes, each side expands 0.2 inch. Find (a) the average rate of change of the area of the plate per inch change in the length of a side and (b) the average rate of change of the area per degree change in temperature.

4 In optics, the *magnification* at x of a lens that projects the point x of one number line onto the point $y = f(x)$ of another number line is defined to be the instantaneous rate of change of y with respect to x. Suppose that a lens projects x to $y = x^3$. What is its magnification at $x = 2$?

In Problems 5 to 8, find the equations of the tangent and normal lines to the graph of the function at the indicated point.

5 $f(x) = x^2 - 4x + 2$ at $(4, 2)$

6 $g(x) = \dfrac{4}{x + 1}$ at $(2, \frac{4}{3})$

7 $f(x) = \frac{1}{3}x^4$ at $(\frac{3}{2}, \frac{27}{16})$

8 $f(x) = 96x - \frac{1}{2}x^3$ at $(0, 0)$

9 Indicate which of the functions in Figure 1 is (i) continuous on (a, b), (ii) differentiable on (a, b), and (iii) continuous and differentiable on (a, b)

Figure 1

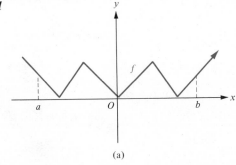

(a)

Figure 1 (*Continued*)

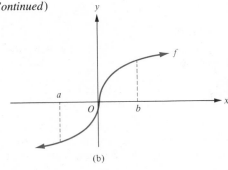

(b)

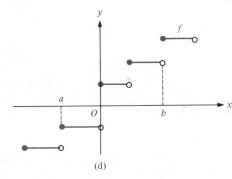

(c)

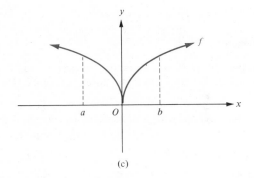

(d)

10 Suppose that f is a differentiable function at $x = a$. Show that

$$f'(a) = \lim_{h \to 0} \frac{f(a + h) - f(a - h)}{2h}$$

In Problems 11 to 18, (a) find

$$\frac{f(x + \Delta x) - f(x)}{\Delta x}$$

and simplify; (b) using the result of (a), find

$$f'(x) = \lim_{\Delta x \to 0} \frac{f(x + \Delta x) - f(x)}{\Delta x}$$

by actually calculating the limit.

11 $f(x) = 3x - 2$

12 $f(x) = x - x^2$

13 $f(x) = x^2 + x + 1$

14 $f(x) = 2x^3 - 1$

15 $f(x) = \frac{1}{2}(x^2 - 4x + 3)$

16 $f(x) = \dfrac{1}{x - 1}$

17 $f(x) = 1 + \dfrac{2}{x}$

18 $f(x) = -\pi^2$

19 Verify your answers to part (b) of Problems 11, 13, 15, and 17 by using the basic algebraic rules for differentiation.

20 We know that the derivative of a product is not, in general, the same as the product of the derivatives. Are there any integers n and m for which $D_x(x^n x^m)$ turns out to be the same as $(D_x x^n)(D_x x^m)$?

21 Let f be a function defined by $f(x) = |x| + |x + 1|$. (a) Sketch the graph of f. (b) Determine the points at which f is not differentiable.

22 Suppose that the function g is continuous at 0, and f is a function defined by $f(x) = xg(x)$. Is f differentiable at 0? If so, find $f'(0)$ in terms of g.

23 Suppose that f and g are differentiable functions at 7 and that $f(7) = 10$, $f'(7) = 3$, $g(7) = 5$, and $g'(7) = -\frac{1}{30}$. Find

(a) $(f + g)'(7)$ (b) $(f - g)'(7)$ (c) $(fg)'(7)$

(d) $\left(\dfrac{f}{g}\right)'(7)$ (e) $\left(\dfrac{f + 3g}{f}\right)'(7)$ (f) $(f + 2g)'(7)$

(g) $\left(\dfrac{f}{f + g}\right)'(7)$

24 Suppose that f is a differentiable function on $\mathbb{R}$ such that $f(x + y) = f(x) + f(y)$ holds for all numbers x and y in $\mathbb{R}$. (a) Prove that $f(0) = 0$. (b) Prove that $f'(0) = f'(x)$ holds for all numbers x in $\mathbb{R}$.

In Problems 25 to 28: (a) Indicate whether the function f is continuous at $x = 2$. (b) Find $f'_-(a)$ and $f'_+(a)$. (c) Is f differentiable at $x = a$? (d) Sketch the graph.

25 $f(x) = \begin{cases} 4 - 3x & \text{if } x \le 2 \\ x^2 - 6 & \text{if } x > 2 \end{cases}$; $a = 2$

26 $f(x) = 2 + |x - 1|$; $a = 1$

27 $f(x) = \begin{cases} 7 - x & \text{if } x \le -3 \\ 10 & \text{if } x > -3 \end{cases}$; $a = -3$

28 $f(x) = [[x - 2]]$; $a = 2$

ⓒ In Problems 29 and 30, use a calculator to find the approximate value of the indicated derivative. Use $\Delta x = a/10^4$.

29 $f'(a)$ where $f(x) = \sqrt{x^2 + 1}$ and $a = 3$

30 $g'(a)$ where $g(x) = \cot \sqrt{x}$ and $a = 0.6169$

In Problems 31 to 80, use the differentiation rules to find the derivative.

31 $f(x) = 7x^5$

32 $g(x) = -4x^{-2}$

33 $h(x) = \frac{1}{2}x^{-3}$

34 $F(x) = \pi x^{3/2}$

35 $G(x) = \sqrt{2}x^{2/3}$

36 $H(x) = \dfrac{-2}{x^{3/4}}$

37 $f(x) = 5x^2 - 7x + 11$

38 $g(x) = \pi(1 - \pi)$

39 $h(x) = -9x^3 + 3x^2 - x + 22^7$

40 $F(x) = 7x^{2/3} - \frac{2}{3}x^{-7} + \sqrt{43}$

41 $G(t) = 5t^{21} + 15t^4 - 5t^{-2} + 8$

42 $H(u) = 3.1u^{2.4} + 1.2u^{-4.2}$

43 $f(x) = (x^2 + 2x + 1)(2x^3 + 5)$

44 $g(t) = (\sqrt{t} + 1)(2t + \sqrt{t} - 2)$

45 $h(s) = (6s^4 + 5s^2 - s^{-1})^{27}$

46 $F(v) = (2\sqrt{v} - 3v^{-2})^{99}$

47 $G(x) = \dfrac{x^2 + 3x - 1}{x + 2}$

48 $H(t) = \dfrac{\sqrt{t} + 3}{t - 1}$

49 $f(x) = \sqrt{x}(x^2 + 2x + 1)$

50 $g(t) = 2\sqrt{t + 1}\,(2t + 1)$

51 $h(x) = \left(1 - \dfrac{1}{x}\right)\left(1 + \dfrac{3}{x^2}\right)$

52 $F(y) = \dfrac{y^2 - 5y + 4}{\sqrt{y + 1}}$

53 $G(x) = \sqrt{x^2 + 12}$

54 $H(z) = \sqrt{z^2 + \sqrt{z^2 + 1}}$

55 $f(x) = 8\sqrt[3]{x} - \dfrac{x}{\sqrt{x} - 1}$

56 $g(x) = \dfrac{x}{x - \sqrt{x}}$

57 $h(x) = (x^2 + 7)^3 \sqrt{x - 7}$

58 $f(y) = \dfrac{1 + (1/y)}{1 - (1/y)}$

59 $g(t) = \left(\dfrac{t^2 - 3t + 2}{t^2 + 2t + 5}\right)^{3/2}$

60 $h(q) = \sqrt{q + \sqrt{q + \sqrt{q}}}$

61 $f(x) = \dfrac{ax + b}{cx + d}$, where a, b, c, and d are constants

62 $g(x) = \left(\dfrac{ax + b}{cx + d}\right)^r$, where a, b, c, d, and r are constants

63 $f(x) = \dfrac{1}{2}\sin 2x$

64 $g(t) = t \sin \dfrac{1}{t}$

65 $h(t) = -\frac{3}{2}\cos 4t$

66 $F(x) = \sqrt{x + 1}\,\tan(x + 1)$

67 $G(x) = \frac{5}{3}\sin(3x - 1)$

68 $H(\theta) = \cot(\sqrt{\theta} + \theta)$

69 $f(t) = \dfrac{\cos t}{1 + \csc t}$

70 $g(y) = \sqrt[3]{\cos 5y}$

71 $h(\theta) = 3 \sin \theta \cos 2\theta$

72 $F(x) = x \tan \sqrt{3x}$

73 $G(s) = \frac{2}{3} \tan \frac{3}{2}s - \frac{3}{4} \cot \frac{4}{3}s$

74 $H(x) = 2 \sin (\csc x)$

75 $f(x) = \dfrac{2 \csc x - 3 \cot x}{x + 1}$

76 $g(x) = \cot^{3/2} 5x$

77 $h(\theta) = \dfrac{\theta}{\sec \theta + \tan \theta}$

78 $q(y) = \sqrt{\sin \sqrt{y}}$

79 $f(t) = a \cos (\omega t - \phi) + k$, where a, ω, ϕ, and k are constants

80 $g(x) = \begin{cases} 3 \cos x & \text{if } x \geq 0 \\ 4x^2 + 3 & \text{if } x < 0 \end{cases}$

C **81** Some people believe that we are all subject to periodic variations, called *biorhythms*, in our physical stamina, emotional well-being, and intellectual ability. The physical cycle, for instance, is supposed to have a period of 23 days. Assume that, on a scale from 0 to 10, a certain person's physical stamina p is given by $p = 5 + 3 \cos (6.5564t - 5.804)$ on day number t, starting with $t = 1$ on January 1. Find the rate of change of p with respect to t on January 21.

82 In the study of FM (frequency-modulated) radio transmission, the function $f(t) = A \cos (\omega_c t + bt \cos \omega_m t)$ represents the strength $f(t)$ of an FM signal at time t seconds. Here A is the amplitude of the carrier signal, $\omega_c/(2\pi)$ is the frequency of the carrier signal, b is the amplitude of the modulating signal, and $\omega_m/(2\pi)$ is the frequency of the modulating signal. Find $f'(t)$ in terms of A, ω_c, b, ω_m, and t.

83 If k is a rational number and $k \geq 1$, show that the tangent line at the point $(1, 1)$ to the graph of $f(x) = x^k$ intersects the y axis at a point that is $k - 1$ units below the origin.

84 Show that if n is an odd positive integer, the lines tangent to the graph of $f(x) = x^n$ at the points $(1, 1)$ and $(-1, -1)$ are parallel.

85 Let $y = (1 - u)/(1 + u)$ and $u = (3 - x)/(2 + x)$. Use the chain rule to find dy/dx.

86 The electric field intensity E on the axis of a uniformly charged ring at a point x units from the center of the ring is given by the formula $E = Qx/(a^2 + x^2)^{3/2}$, where a and Q are constants. Find a formula for the rate of change of field intensity E with respect to distance x along the axis.

In Problems 87 to 90, find a point on the graph of the function where the tangent or normal line satisfies the indicated condition. Then write an equation for this line.

87 The tangent line to $f(x) = 3x^2 - 2x + 1$ is perpendicular to the line $x + 4y - 1 = 0$.

88 The tangent line to $g(x) = x/(x - 1)$, $x > 1$, intersects the y axis at $(0, 4)$.

89 The normal line to $h(x) = \cos x$, $0 \leq x \leq \pi/2$, is parallel to the line $2x - y + 4 = 0$.

90 The normal line to $F(x) = x/\sqrt{x^2 + 1}$ is parallel to the line $x + y = 1$.

In Problems 91 to 96, find all points on the graph of the function at which the tangent line is horizontal.

91 $f(x) = x^3 - 3x - 2$

92 $g(x) = 3x^{4/3} - 12x^{1/3} + 2$

93 $h(x) = \dfrac{x}{x^2 + 4}$

94 $F(x) = x\sqrt{3 - x}$

95 $G(x) = \sin^2 x + \cos x$, $-\pi \leq x \leq \pi$

96 $H(x) = x + \cot x$, $0 \leq x \leq 2\pi$

97 Suppose that the interval $(1, 3)$ is contained in the domain of the function f, that f is differentiable at 2, and that $f(x) \leq f(2)$ for all values of x in the interval $(1, 3)$. By quoting an appropriate theorem, show that $f'(2) = 0$.

98 Let $g(x) = x - \sin x$. (a) Show that the graph of g has a horizontal tangent at $(0, 0)$. (b) Does g have a relative extremum at $x = 0$? Why or why not?

In Problems 99 to 102, suppose that $h = f \circ g$. Use the given information to find $h'(a)$. (*Caution:* There may be more information than you need.)

99 $a = 2$, $f(2) = 8$, $g(2) = 0$, $f'(0) = 12$, $g'(2) = -1$

100 $a = \pi$, $f(\pi) = 3$, $g(\pi) = \pi$, $f'(\pi) = 1/\pi$, $g'(\pi) = \pi$

101 $a = \sqrt{2}$, $f'[g(\sqrt{2})] = \sqrt{2}$, $g'(\sqrt{2}) = \sqrt{2}$

102 $a = 1.732$, $g(1.732) = 7.007$, $f'(7.007) = 0.2$, $g'(1.732) = 5$

103 Suppose that f gives the area of a square as a function of the length of one of its diagonals. Express f as a composition of two other functions.

104 In economics, a **linear supply-demand-price model**

$$s = Ap - B \qquad \text{and} \qquad q = b - ap$$

relates the number of units s of a commodity supplied by manufacturers, the number of units q demanded by consumers, and the price p per unit of the commodity. Here A, B, a, and b are positive constants. Of course, if the demand exceeds the supply, the manufacturers will be inclined to increase the price. A **recursive pricing model** is a formula

$$\text{new price} = f(\text{old price})$$

that gives the new price per unit of the commodity as a function f of the old price 1 unit of time earlier. Consider the recursive pricing model given by

$$f(p) = p + k(q - s)$$

where k is a positive constant. (a) If p is the current price per unit of the commodity, what is the economic interpretation of the orbit of p,

$$p, f(p), (f \circ f)(p), (f \circ f \circ f)(p), \ldots$$

under the iterates of f? (b) When $p_0 = f(p_0)$, we refer to p_0 as the **equilibrium price** of the commodity. Give the economic interpretation of the equilibrium price, and find a formula for p_0 in terms of A, B, a, and b.

105 Let $f(x) = \alpha + \beta x$, where α and β are constants and $\beta \neq 1$. Let x_1 be any real number and consider the successive images

$$x_2 = f(x_1), \quad x_3 = (f \circ f)(x_1), \quad x_4 = (f \circ f \circ f)(x_1), \ldots$$

of x_1 under the iterates of f. (a) Show that there exists exactly one number x_0 such that $x_0 = f(x_0)$. (b) Show that $|f(x) - x_0| = |\beta| |x - x_0|$ holds for every real number x. (c) Show that $|x_{n+1} - x_0| = |\beta|^n |x_1 - x_0|$ holds for every positive integer n. [*Hint:* Use part (b) and mathematical induction.] (d) If $|\beta| < 1$, use the result of part (c) to show that the successive images of x_1 under the iterates of f come closer and closer to x_0.

106 In Problem 105, show that if $|\beta| > 1$ and $x_1 \neq x_0$, then the successive images of x_1 under the iterates of f have absolute values that become larger and larger without bound. What happens if $\beta = -1$?

C **107** In Problem 105, let $\alpha = 2.7$ and take $x_1 = 5$. Use a calculator to find $x_2, x_3, \ldots, x_{10}$ for (a) $\beta = 0.9$ and (b) $\beta = 1.1$.

108 The recursive pricing model in Problem 104 is said to be *stable* if, as time goes on, the successive prices given by the model approach the equilibrium price p_0. Using the results of Problems 105 and 106, find necessary and sufficient conditions in terms of the constants A, B, a, b, and k for the recursive pricing model to be stable.

In Problems 109 to 116, use implicit differentiation to find $D_x y$.

109 $3x^3 + 2y^3 = 1$

110 $x^4 + 4x^2y^2 = 25$

111 $5x^2 + 8y^3 = 16\sqrt{x + 1}$

112 $4x^3 - 5xy^2 + y^3 = 18$

113 $x - \sin y = 0$

114 $\sin y = \cos x$

115 $1 + y = x \cos y$

116 $y \tan x^2 - xy^4 = 15$

117 In Problem 109, find $D_x^2 y$ in terms of x and y.

118 In Problem 110, find $D_x^2 y$ in terms of x and y.

In Problems 119 to 124, find the equations of the tangent and normal lines to the graph of the implicit function determined by the equation at the point indicated.

119 $y^2 + 2xy = 16$ at $(3, 2)$

120 $x^2 + 4xy + y^2 + 3 = 0$ at $(2, -1)$

121 $x^3 - xy^2 + y^3 = 8$ at $(2, 2)$

122 $y\sqrt{2x + 1} = y + 1$ at $(4, \frac{1}{2})$

123 $x - \cos y = 0$ at $(\frac{1}{2}, \pi/3)$

124 $\pi \tan (x - y) = 2x$ at $(\pi/2, \pi/4)$

125 Let (a, b) be a point on the circle $x^2 + y^2 = r^2$. Using implicit differentiation, confirm the fact (already known from elementary geometry) that the tangent line to the circle at (a, b) is perpendicular to the radius at (a, b).

126 The curve $(x^2 + y^2)^2 = x^2 - y^2$, which was first studied by the Swiss mathematician Jakob Bernoulli in 1694, is called a **lemniscate** (Figure 2). Use implicit differentiation to find the coordinates of the four points on the lemniscate at which the tangent line is horizontal.

Figure 2

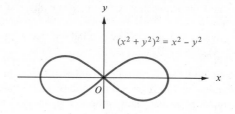

$(x^2 + y^2)^2 = x^2 - y^2$

127 The curve $x^3 + y^3 = 3xy$, which was first studied by Descartes in 1638, is called the **folium of Descartes** (Figure 3). Use implicit differentiation to find the coordinates of the point P in quadrant I where the folium has a horizontal tangent.

Figure 3

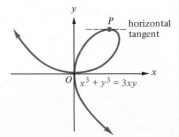

P — horizontal tangent

$x^3 + y^3 = 3xy$

128 The curve $x^2 + xy + y^2 = 4$ (an *ellipse*) intersects the y axis in two different points. Show that the tangent lines to the ellipse at these two points are parallel.

In Problems 129 to 136, find the second derivative of each function.

129 $f(x) = 4x^5 + 3x^2 - x + 1$

130 $g(x) = x - (1/x)$

131 $h(t) = \dfrac{t}{t + 1}$

132 $F(s) = \sqrt{2s + 1}$

133 $G(u) = (u^{1/4} + 3)^7$

134 $H(w) = \sqrt[3]{w} - 4\sqrt[3]{1 + w}$

135 $f(\theta) = \theta^2 \cos 3\theta$

136 $g(x) = (x^2 + 7) \sin (x/7)$

In Problems 137 to 142, find the indicated higher-order derivative.

137 $f^{(3)}(x)$ if $f(x) = 8x^{3/2} - 9x^{4/3}$

138 d^4y/dx^4 if $y = 3x^4 - 2x^3 + 7x^2 - 5x + 7$

139 y''' if $y = (x - 1)/(x + 1)$

140 $f^{(n)}(x)$ if $f(x) = a_n x^n + a_{n-1} x^{n-1} + \cdots + a_1 x + a_0$, where a_n, $a_{n-1}, \ldots , a_1$, and a_0 are constants

141 $d^6 (\cos ax)/dx^6$, where a is a constant

142 $g^{(4n+1)}(\theta)$ if $g(\theta) = a \sin b\theta$, where a and b are constants

143 Find all higher-order derivatives of the function $f(x) = 6x^5 + 7x^4 - 8x^3 - 9x^2 + 10x + 11$.

144 If $f(x) = |x|^3$, find formulas for $f'(x)$ and $f''(x)$.

145 Suppose that f, g, and h are functions which are differentiable at the number -2 and that $f(-2) = 1$, $f'(-2) = -3$, $f''(-2) = -4$, $g(-2) = 4$, $g'(-2) = -\frac{1}{2}$, $g''(-2) = -3$, $h(-2) = 6$, $h'(-2) = -8$, and $h''(-2) = 7$. Find

(a) $(fg)''(-2)$

(b) $(fh)''(-2)$

(c) $(f + g)''(-2)$

(d) $(g - h)''(-2)$

(e) $(fgh)''(-2)$

(f) $\left(\dfrac{f}{g}\right)''(-2)$

146 The **curvature** κ of the graph of the function f at the point $(x, f(x))$ is defined to be

$$\kappa = \frac{f''(x)}{\{1 + [f'(x)]^2\}^{3/2}}$$

The equation of a semicircle of radius r with center at $(0, 0)$ is $y = \sqrt{r^2 - x^2}$. Find the curvature of this semicircle at the point $(x, \sqrt{r^2 - x^2})$, where $-r < x < r$.

147 (a) If $(x - a)^2 + y^2 = b^2$, where a and b are constants, show that

$$yD_x^2 y + 1 + (D_x y)^2 = 0$$

(b) If $x^2 + y^2 = a^2$, where a is a constant, show that $[1 + (y')^2]^3 = a^2(y'')^2$.

148 Suppose that g is a function defined by $g(t) = \sqrt{1 - f(t)}$, where $f(-2) = -3$, $f'(-2) = 3$, and $f''(-2) = 5$. Find $g''(-2)$.

In Problems 149 to 152, a particle is moving along a horizontal s axis according to the given equation of motion. Find the velocity v and the acceleration a of the particle.

149 $s = 6t^2 - 2t^3$

150 $s = 64t^2 - 16t^4$

151 $s = 3 \sin 2\pi t$

152 $s = \cos^2 t$

153 Find a formula for $f''(x)$ if

$$f(x) = \begin{cases} x^3 & \text{if } x \leq 1 \\ 3x - 2 & \text{if } x > 1 \end{cases}$$

154 A particle moving along an s axis according to an equation of motion of the form $s = A \cos (2\pi v t - \phi) + k$, where A, v, ϕ, and k are constants, is said to be undergoing *simple harmonic motion* with *amplitude A*, *frequency v*, *phase angle ϕ*, and *equilibrium point k*. For simple harmonic motion, show that

$$\frac{d^2 s}{dt^2} + 4\pi^2 v^2 s = 4\pi^2 v^2 k$$

In Problems 155 and 156, use the intermediate-value theorem to show that there is a number c between the two given numbers that satisfies the equation.

155 c between 1 and 2, $4c^3 - 7c^2 + 2c = \sqrt{5}$

156 c between 0 and $\frac{1}{4}$, $5 \sin 2\pi c - 4 \tan \pi c = 0.707$

In Problems 157 to 162, determine the intervals on which the function is positive or negative.

157 $f(x) = 3x(2x - 1)(3x - 2)$

158 $g(x) = x^3 + 8x^2 + 11x - 20$

159 $h(x) = \dfrac{8x^2 + 22x + 15}{6x^2 + 13x - 5}$

160 $F(x) = 6x^{11/5} - 13x^{6/5} + 6x^{1/5}$

161 $G(x) = \dfrac{1 - \sin x}{1 + \tan x}$ for $-\pi \leq x \leq \pi$

162 $H(x) = (x - 1)^{2/3} \tan x$ for $-\pi \leq x \leq \pi$

In Problems 163 to 166, show that the function has a zero in the interval by using the change-of-sign property.

163 $f(x) = x^3 + x^2 + x + 5$ in $[-2, -1]$

Ⓒ **164** $g(x) = 1 + 3x^{5/3} - 15x^{2/3}$ in $[4, 5]$

Ⓒ **165** $h(x) = \cos x - \sqrt{x}$ in $[0, 1]$

Ⓒ **166** $F(x) = 3 \sin x - 2 \sin (x/2) - 1$ in $[0.5, 0.6]$

Ⓒ **167** In Problem 163, use the bisection method twice in succession, starting with the interval $[-2, -1]$, to locate a zero of f with greater accuracy.

Ⓒ **168** In Problem 166, use the bisection method four times in succession, starting with the interval $[0.5, 0.6]$, to locate a zero of F with greater accuracy.

In Problems 169 to 172, use a calculator and Newton's method to find a zero of the function to as many decimal places as possible. Start with the value of x_1 as a first approximation.

169 The function f of Problem 163; $x_1 = -1.5$

170 The function g of Problem 164; $x_1 = 4.5$

171 The function h of Problem 165; $x_1 = 0.5$

172 The function F of Problem 166; $x_1 = 0.55$

In Problems 173 and 174, use Newton's method to solve the equation.

173 $x^5 + x = 17$

174 $x^3 + x = 3 - \sin x$

175 The graph of $y = x^4 + 2x^2 - 4x + 1$ has a horizontal tangent at exactly one point. Find the x coordinate of this point to as many decimal places as you can.

176 In optics, it is necessary to solve the equation

$$\frac{1}{2} = \left(\frac{\sin x}{x} \right)^2$$

in order to calculate the half-width of the central maximum for Fraunhofer diffraction by a single slit. Solve this equation to as many decimal places as you can.

177 Let

$$f(x) = \frac{x + 1}{x^2 - 4}$$

Note that $f(0) = -\frac{1}{4}$ and $f(3) = \frac{4}{5}$ have opposite algebraic signs, yet there is no number c between 0 and 3 such that $f(c) = 0$. Explain why this does not contradict the change-of-sign property.

178 Suppose that f is a continuous function on an interval I and that f takes on only integer values. Use the intermediate-value theorem to conclude that f must be a constant function on the interval I.

3

APPLICATIONS OF THE DERIVATIVE

In Chapter 2 we saw that the derivative of a function can be interpreted as the slope of the tangent line to its graph. In the present chapter we exploit this fact and develop techniques for using derivatives to determine important geometric features of the graph of a function. In Sections 3.2 through 3.5, computer-generated graphs are used to illustrate these features. The chapter also includes graph sketching and applications of derivatives to problems in such diverse fields as geometry, engineering, physics, and economics.

3.1 The Mean-Value Theorem

The driving distance between Kansas City and St. Louis is 250 miles. If you drive this distance in 5 hours, then your *average* speed is 250/5 = 50 miles per hour. Of course, your instantaneous speed—your speedometer reading—probably will not remain constant during your 5-hour drive; sometimes it will be more than 50 miles per hour, sometimes less. But it seems clear that at some point along the way, your instantaneous speed must be exactly 50 miles per hour. *You can't average m miles per hour on a journey without attaining an instantaneous speed of m miles per hour somewhere along the way.* This rather obvious fact corresponds to a mathematical theorem, called the *mean-value theorem,* which we study in this section. (In mathematics, the word "mean" is often used as a synonym for "average.")

The mean-value theorem (or *law of the mean,* as it is sometimes called) applies to variable quantities in general, not just to distance, rate, and time for an automobile trip. Indeed, let x and y be variable quantities, and suppose that y is a function of x, say,

$$y = f(x)$$

Consider a closed interval $[a, b]$ contained in the domain of f. As x changes from $x = a$ to $x = b$, the variable y will change from $y = f(a)$ to $y = f(b)$, and the *aver-*

age rate of change of y with respect to x over the interval [a, b] is given by the change in y divided by the change in x, namely,

$$\frac{f(b) - f(a)}{b - a}$$

If c is any number between a and b, then the *instantaneous rate of change of y with respect to x when x = c* is given by

$$f'(c)$$

provided, of course, that f is differentiable at c. The condition that there be a value of c between a and b at which the instantaneous rate of change equals the average rate of change over the interval [a, b] is expressed by the equation

$$f'(c) = \frac{f(b) - f(a)}{b - a}$$

The equation above has an interesting geometric interpretation (Figure 1). Consider the points $A = (a, f(a))$ and $B = (b, f(b))$ on the graph of f. By the slope formula, the slope of the secant $\overline{AB}$ is given by

$$\frac{f(b) - f(a)}{b - a}$$

However, if $P = (c, f(c))$, then

$$f'(c) = \text{slope of the tangent line at } P$$

Because lines with the same slope are parallel, the condition

$$f'(c) = \frac{f(b) - f(a)}{b - a}$$

means that the tangent line to the graph of f at P is parallel to the secant $\overline{AB}$. Thus, the mean-value theorem can be stated *informally* either in terms of rate of change:

> The average rate of change of a smoothly varying quantity over an interval is equal to the instantaneous rate of change of the quantity at some point on the interval.

Or geometrically:

> If A and B are two points on a continuous curve, and if the curve has a tangent line at each point between A and B, then there is at least one point P between A and B at which the tangent line to the curve is parallel to the secant $\overline{AB}$.

EXAMPLE 1 Figure 2 shows a computer-generated graph of $f(x) = \sqrt[3]{x}$. Find a point $P = (c, f(c))$ on the graph of f between $A = (0, f(0)) = (0, 0)$ and $B = (8, f(8)) = (8, 2)$ at which the tangent line is parallel to the secant $\overline{AB}$.

SOLUTION   By the slope formula, the slope of the secant $\overline{AB}$ is given by

$$\frac{2 - 0}{8 - 0} = \frac{1}{4}$$

Figure 1

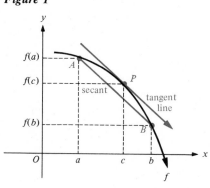

Figure 2

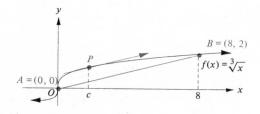

By the geometric version of the mean-value theorem, there is a point $P = (c, f(c)) = (c, \sqrt[3]{c})$, with c between 0 and 8, at which the slope of the tangent is equal to $\frac{1}{4}$; that is,

$$f'(c) = \tfrac{1}{4}$$

Now,

$$f'(x) = D_x\sqrt[3]{x} = D_x x^{1/3} = \tfrac{1}{3}x^{-2/3}$$

so we want

$$\tfrac{1}{3}c^{-2/3} = \tfrac{1}{4}$$

that is, $c^{-2/3} = \tfrac{3}{4}$ or $c^{2/3} = \tfrac{4}{3}$

If we cube both sides of the last equation, we get

$$c^2 = \frac{64}{27}$$

so that $c = \pm\sqrt{\dfrac{64}{27}} = \pm\dfrac{8}{3\sqrt{3}} = \pm\dfrac{8\sqrt{3}}{9}$

Since we require that c belong to the interval between 0 and 8, we must reject the negative solution and conclude that

$$c = \frac{8\sqrt{3}}{9}$$

Therefore, $P = (c, f(c)) = \left(\dfrac{8\sqrt{3}}{9},\ \sqrt[3]{\dfrac{8\sqrt{3}}{9}}\right)$ ∎

The mean-value theorem is one of the most important theorems in all of calculus. We use it in the present chapter to help develop techniques of graphing and in Chapter 5 in connection with the idea of an integral. For these purposes, we require a more formal treatment of the theorem. In spite of the strong intuitive appeal of the informal versions already given, it turns out that the proof of the mean-value theorem is a bit tricky. The usual approach is to begin with the following theorem, which is of interest in its own right.

THEOREM 1   **Extreme-Value Theorem**

> If a function f is defined and continuous on a closed interval $[a, b]$, then f takes on a maximum (largest) value at some number in $[a, b]$ and a minimum (smallest) value at some number in $[a, b]$.

Geometrically, the property of continuous functions expressed by Theorem 1 is easy to accept since it just says that a continuous curve drawn from point A to point B, as in Figure 3, has a highest point C and a lowest point D. Surprisingly, a formal proof of this obvious fact requires a rather sophisticated study of the structure of the real number system $\mathbb{R}$ and so is best left to more advanced courses in analysis.

Using Theorem 1, we can now prove a special case of the mean-value theorem which was discovered by the French mathematician Michel Rolle (1652–1719) in 1690.

Figure 3

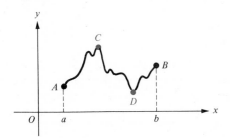

THEOREM 2

Rolle's Theorem

> Let f be a continuous function on the closed interval $[a, b]$ such that f is differentiable on the open interval (a, b) and $f(a) = f(b) = 0$. Then there is at least one number c in the open interval (a, b) such that $f'(c) = 0$.

PROOF

If f is a constant function on $[a, b]$, then we have $f'(c) = 0$ for *any* choice of c in (a, b). If f is not constant on $[a, b]$, then f must take on either a positive or a negative value somewhere on (a, b). We consider the case in which f takes on a positive value somewhere on (a, b); the other case is handled similarly (Problem 35). By the extreme-value theorem, f takes on a maximum value at some number, say c, in $[a, b]$. For the case under consideration, we must have $f(c) > 0$, so c is in (a, b). By the necessary condition for relative extrema (Theorem 1, Section 2.4), we have $f'(c) = 0$. ∎

Figure 4

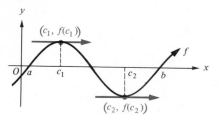

From a geometric point of view, Rolle's theorem can be interpreted as follows: *The graph of a differentiable function must have at least one horizontal tangent between any two x intercepts* (Figure 4).

EXAMPLE 2   Show that the hypotheses of Rolle's theorem are satisfied for the function $f(x) = x^{4/3} - 3x^{1/3}$ on the interval $[0, 3]$, and find a value of c in the open interval $(0, 3)$ such that $f'(c) = 0$.

SOLUTION   For $x \neq 0$, we have

$$f'(x) = \tfrac{4}{3}x^{1/3} - x^{-2/3} = \tfrac{1}{3}x^{-2/3}(4x - 3)$$

Thus, although f is continuous on all of $\mathbb{R}$, and in particular on $[0, 3]$, it is not differentiable at 0. However, it is differentiable on the open interval $(0, 3)$, and this is all that is required. Because

$$f(x) = x^{4/3} - 3x^{1/3} = x^{1/3}(x - 3)$$

the requirements that

$$f(0) = f(3) = 0$$

Figure 5

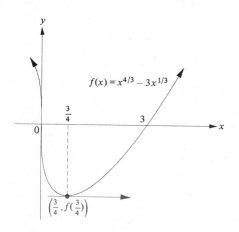

$$f(x) = x^{4/3} - 3x^{1/3}$$

$\left(\tfrac{3}{4}, f(\tfrac{3}{4})\right)$

are also satisfied. Therefore, Rolle's theorem guarantees that there is a number c in the interval $(0, 3)$ such that

$$f'(c) = 0$$

that is,

$$\tfrac{1}{3}c^{-2/3}(4c - 3) = 0$$

Solving the last equation for c, we find that $c = \tfrac{3}{4}$ and, indeed, $\tfrac{3}{4}$ is in the open interval $(0, 3)$. Figure 5 shows a computer-generated graph of f with the horizontal tangent at $(c, f(c))$, $c = \tfrac{3}{4}$. ∎

In Example 2, it turned out that the equation $f'(c) = 0$ had only one solution c, which, of course, belonged to the interval $(a, b) = (0, 3)$. In general, however, the equation $f'(c) = 0$ may have many solutions, some that fall into the interval (a, b) and others that do not. Furthermore, you may not be able to solve the equation $f'(c) = 0$ by the usual algebraic techniques—notice that Rolle's theorem doesn't tell you *how to find* c; it just guarantees that c *exists*. Of course, you can always use the bisection method or Newton's method to find an approximate value of c to any desired degree of accuracy.

Now we are in a position to give a formal statement and proof of the mean-value theorem.

<u>THEOREM 3</u>   **The Mean-Value Theorem**

> If the function f is defined and continuous on the closed interval $[a, b]$ and is differentiable on the open interval (a, b), then there exists at least one number c in (a, b) such that
>
> $$f'(c) = \frac{f(b) - f(a)}{b - a}$$

PROOF

The slope of the secant line through the two points $(a, f(a))$ and $(b, f(b))$ is

$$\frac{f(b) - f(a)}{b - a}$$

Therefore, if

$$g(x) = f(a) + \frac{f(b) - f(a)}{b - a}(x - a)$$

then

$$y = g(x)$$

is an equation of the secant line (Problem 37). Notice that

$$g(a) = f(a) \qquad \text{and} \qquad g(b) = f(b)$$

Consider the function h defined by

$$h(x) = f(x) - g(x)$$

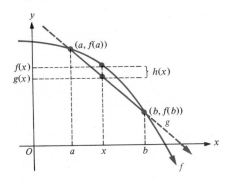

Figure 6

for $a \le x \le b$ (Figure 6). We claim that the function h satisfies the hypotheses of Rolle's theorem (Theorem 2). Indeed, f is continuous on $[a, b]$ and g is continuous on $\mathbb{R}$, so $h = f - g$ is continuous on $[a, b]$. Likewise, f is differentiable on (a, b) and g is differentiable on $\mathbb{R}$, so $h = f - g$ is differentiable on (a, b). Also,

$$h(a) = f(a) - g(a) = 0 \qquad \text{and} \qquad h(b) = f(b) - g(b) = 0$$

Therefore, applying Rolle's theorem to the function h, we can conclude that there exists a number c in the interval (a, b) such that

$$h'(c) = 0$$

Now, $$h'(x) = f'(x) - g'(x) = f'(x) - \frac{f(b) - f(a)}{b - a}$$

and so $$f'(c) - \frac{f(b) - f(a)}{b - a} = 0$$

that is, $$f'(c) = \frac{f(b) - f(a)}{b - a}$$

∎

The conclusion of the mean-value theorem is often written in the equivalent form

$$f(b) - f(a) = f'(c)(b - a) \qquad a < c < b$$

EXAMPLE 3   Let $f(x) = x^3 + 2x^2 + 1$. Verify the hypotheses of the mean-value theorem for f on the interval $[0, 3]$, and find a numerical value of c in the interval $(0, 3)$ such that $f(3) - f(0) = f'(c)(3 - 0)$.

SOLUTION   The hypotheses of the mean-value theorem are satisfied because the polynomial function f is continuous and differentiable at every number. Here we have

$$f(3) = 3^3 + 2(3)^2 + 1 = 46$$

$$f(0) = 0^3 + 2(0)^2 + 1 = 1$$

and

$$f'(x) = 3x^2 + 4x$$

so the condition $f(3) - f(0) = f'(c)(3 - 0)$ is equivalent to

$$46 - 1 = (3c^2 + 4c)(3) \qquad \text{or} \qquad 9c^2 + 12c - 45 = 0$$

The last equation can be rewritten as

$$3c^2 + 4c - 15 = 0 \qquad \text{or} \qquad (3c - 5)(c + 3) = 0$$

Figure 7

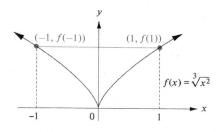

Thus, $c = \frac{5}{3}$ or $c = -3$. Since we require that c belong to the interval $(0, 3)$, we must reject the solution $c = -3$. Therefore, the desired number is $c = \frac{5}{3}$.   ∎

In using the mean-value theorem, you must make certain that all the hypotheses are satisfied. For instance, consider the function $f(x) = \sqrt[3]{x^2} = x^{2/3}$ on the interval $[-1, 1]$. Figure 7 shows a computer-generated graph of f. Notice that there is no point on the graph of f where the tangent line is parallel to the horizontal secant line containing the points $(-1, f(-1))$ and $(1, f(1))$. This does not contradict the mean-value theorem, since the hypothesis that f be differentiable on the open interval $(-1, 1)$ fails. Indeed, f is not differentiable at the number 0 in this interval.

Problem Set 3.1

1 There are *two* points, P and Q, between A and B on the continuous curve in Figure 8 at which the tangent line is parallel to the secant $\overline{AB}$. Does this contradict the mean-value theorem?

2 Give an explicit example of a function f and an interval $[a, b]$ such that f is continuous on $[a, b]$, f is differentiable on (a, b), and there are *infinitely many* points on the graph of f between $A = (a, f(a))$ and $B = (b, f(b))$ at which the tangent is parallel to the secant $\overline{AB}$. (*Hint:* Although there are some complicated examples, there is at least one *very simple* example.)

In Problems 3 to 8, find a numerical value of c such that $a < c < b$ and the tangent line to the graph of each function f at $(c, f(c))$ is parallel to the secant between the points $(a, f(a))$ and $(b, f(b))$. Sketch the graph of f, and show the tangent line and the secant.

3 $f(x) = x^2$, $a = 2$, $b = 4$

4 $f(x) = \sqrt{x}$, $a = 4$, $b = 9$

5 $f(x) = x^3$, $a = 1$, $b = 3$

6 $f(x) = \dfrac{1}{x - 1}$, $a = 1.5$, $b = 1.6$

7 $f(x) = \sin x$, $a = 0$, $b = \pi$

8 $f(x) = x^2 + 3x - 1$, $a = -1$, $b = 1$

Figure 8

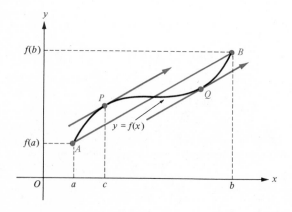

In Problems 9 to 16, verify the hypotheses of Rolle's theorem for each function f on the indicated closed interval $[a, b]$. Then find all numbers c in the open interval (a, b) for which $f'(c) = 0$.

9 $f(x) = x^2 - 3x$, $[a, b] = [0, 3]$

10 $f(x) = x^2 - 5x + 6$, $[a, b] = [2, 3]$

11 $f(x) = x^3 - 3x^2 - x + 3$, $[a, b] = [-1, 3]$

12 $f(x) = \sqrt{x}(x^3 - 1)$, $[a, b] = [0, 1]$

13 $f(x) = x^{3/4} - 2x^{1/4}$, $[a, b] = [0, 4]$

14 $f(x) = x^3 - 3x$, $[a, b] = [-\sqrt{3}, \sqrt{3}]$

15 $f(x) = \sin x$, $[a, b] = [0, 4\pi]$

16 $f(x) = \sqrt{1 - \cos x}$, $[a, b] = \left[-\dfrac{\pi}{2}, \dfrac{\pi}{2} \right]$

In Problems 17 to 26, verify the hypotheses of the mean-value theorem for each function f on the indicated closed interval $[a, b]$ and then find all numbers c in the open interval (a, b) for which $f(b) - f(a) = f'(c)(b - a)$.

17 $f(x) = \frac{1}{6}x^2$, $[a, b] = [2, 6]$

18 $f(x) = x^3 + x - 1$, $[a, b] = [0, 2]$

19 $f(x) = 2x^3$, $[a, b] = [0, 2]$

20 $f(x) = \sqrt{x}$, $[a, b] = [1, 4]$

21 $f(x) = \dfrac{x - 1}{x + 1}$, $[a, b] = [0, 3]$

22 $f(x) = \sqrt{x + 1}$, $[a, b] = [3, 8]$

23 $f(x) = \sqrt{25 - x^2}$, $[a, b] = [-3, 4]$

24 $f(x) = \dfrac{x^2 - 2x - 3}{x + 4}$, $[a, b] = [-1, 3]$

25 $f(x) = x - \cos x$, $[a, b] = \left[\dfrac{\pi}{2}, \dfrac{5\pi}{2} \right]$

26 $f(x) = 2x + 5 \sin^2 2x$, $[a, b] = [0, \pi]$

In Problems 27 to 32, the conclusion of the mean-value theorem fails for each function on the interval indicated. Sketch the graph of the function, and determine which hypothesis of the mean-value theorem fails to hold.

27 $f(x) = \sqrt{|x|}$, $[-1, 1]$ **28** $g(x) = \dfrac{3}{x - 2}$, $[1, 3]$

29 $f(x) = \begin{cases} x^2 + 1 & \text{if } x < 1 \\ 3 - x & \text{if } x \geq 1 \end{cases}$; $[0, 3]$

30 $G(x) = x - [[x]]$, $[-1, 1]$

31 $f(x) = \tan x$, $[0, \pi]$ **32** $f(x) = |\sin x|$, $\left[-\dfrac{\pi}{6}, \dfrac{\pi}{6} \right]$

33 The functions whose graphs are shown in Figure 9 fail to satisfy the hypotheses of the mean-value theorem on the interval from a to b. In each case, determine which hypothesis fails.

Figure 9

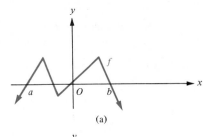

(a)

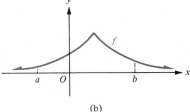

(b)

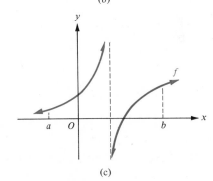

(c)

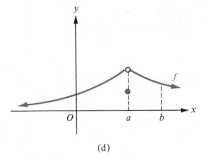

(d)

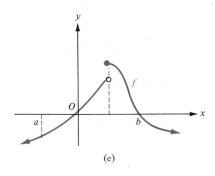

(e)

34 Let f be a differentiable function (and therefore continuous) at each number x in some open interval, and suppose that $|f'(x)| \leq 1$ holds for each such x. Use the mean-value theorem to show that $|f(b) - f(a)| \leq |b - a|$ for any two numbers a and b in the open interval.

35 Complete the proof of Rolle's theorem by taking care of the case in which f takes on a negative value somewhere on (a, b).

36 Show that $|\sin x - \sin y| \leq |x - y|$ holds for all numbers x and y by using the result in Problem 34.

37 In connection with the proof of the mean-value theorem, show that

$$y = f(a) + \frac{f(b) - f(a)}{b - a}(x - a)$$

is an equation of the secant line containing the two points $(a, f(a))$ and $(b, f(b))$. (*Hint:* Use the point-slope form of the equation of a line.)

38 For all values of x and y in the interval $(-\pi/2, \pi/2)$, show that $|\tan x - \tan y| \geq |x - y|$.

C In Problems 39 and 40, use a calculator and Newton's method to find an approximation to a value of c on the open interval (a, b) for which $f(b) - f(a) = f'(c)(b - a)$.

39 $f(x) = \sin x$, $[a, b] = \left[\dfrac{\pi}{6}, \dfrac{\pi}{3}\right]$

40 $f(x) = 2x^5 + x^4 - 3x^3 + x^2 - x - 1$, $[a, b] = [-1, 1]$

41 Let $f(x) = x^5 + 2x^3 - 5x - 10$. (a) Use the intermediate-value theorem to show that f has a zero between 1 and 2. (b) Use Rolle's theorem to show that f cannot have two different zeros between 1 and 2. [*Hint:* If $1 < a < b < 2$ and $f(a) = f(b) = 0$, then f' would have a zero between 1 and 2.]

42 Suppose that the function f is differentiable on the open interval (a, b) and that the one-sided limits $\lim_{x \to a^+} f(x)$ and $\lim_{x \to b^-} f(x)$ exist and are finite. Prove that there exists a number c in (a, b) such that $\lim_{x \to b^-} f(x) - \lim_{x \to a^+} f(x) = (b - a)f'(c)$.

43 Let f be a second-degree polynomial function. Given any two numbers a and b, find a formula for c strictly between a and b such that $f(b) - f(a) = (b - a)f'(c)$. [*Hint:* There are constants A, B, and C such that $f(x) = Ax^2 + Bx + C$.]

44 Let F and G be two functions, both of which are continuous on $[a, b]$ and differentiable on (a, b). Define a function f on $[a, b]$ by the equation

$$f(x) = [G(a) - G(b)]F(x) - [F(a) - F(b)]G(x)$$

for $a \leq x \leq b$. Verify that f satisfies the hypotheses of the mean-value theorem, and draw the conclusion that for some value of c strictly between a and b,

$$\frac{F'(c)}{G'(c)} = \frac{F(b) - F(a)}{G(b) - G(a)}$$

(provided that the denominators do not vanish).

45 Give an alternative proof of the mean-value theorem using the function $\psi(x) = (x - a)f(b) - (x - b)f(a) - (b - a)f(x)$ rather than the function h in the proof of Theorem 3.

46 Suppose that the function f is differentiable on an open interval containing both x and $x + \Delta x$. Show that there exists a number c between x and $x + \Delta x$ such that $f(x + \Delta x) = f(x) + f'(c) \, \Delta x$.

Figure 1

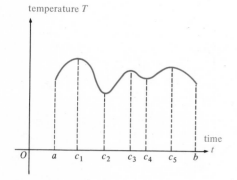

temperature T

time t

3.2 Monotonicity and the First-Derivative Test

Many applications of mathematics involve finding where the graph of a function is *rising* or *falling* and locating *high* and *low* points on the graph. For instance, suppose the graph in Figure 1 represents the recorded temperature of a patient in an intensive-care unit. In monitoring the patient's condition, the hospital staff is concerned with whether the temperature is rising or falling and with the high and low temperatures that occur over a given interval of time. On the graph, the patient's temperature T is plotted as a function of the time t. Notice that T is increasing in the time intervals $[a, c_1]$, $[c_2, c_3]$, and $[c_4, c_5]$ and decreasing in the time intervals $[c_1, c_2]$, $[c_3, c_4]$, and $[c_5, b]$. High temperatures (relative, or local, maxima) occur at times c_1, c_3, and c_5; and low temperatures (relative, or local, minima) occur at times c_2 and c_4.

The idea of an *increasing* or *decreasing* function is made precise by the following definitions.

DEFINITION 1   **Increasing and Decreasing Functions**

> Let the interval I be contained in the domain of the function f.
>
> **(i)** f is **increasing** on I if for every two numbers a and b in I with $a < b$ we have $f(a) < f(b)$.
>
> **(ii)** f is **decreasing** on I if for every two numbers a and b in I with $a < b$ we have $f(a) > f(b)$.

DEFINITION 2   **Monotone Function**

> A function f is said to be **monotone** on an interval I if it is either increasing or decreasing on I.

Figure 2

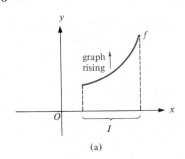

(a)

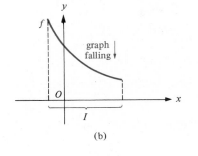

(b)

To say that a function f is *increasing* on an interval I means that the graph of f *rises* as you move to the right along the interval (Figure 2a). Similarly, to say that f is *decreasing* on I means that the graph of f *falls* as you move to the right along the interval (Figure 2b).

Now, suppose that f is a function with a *positive* derivative $f'(x)$ at each number x in some interval I. Then, for each value of x in I, the tangent line to the graph of f at the point $(x, f(x))$ is *rising* to the right since its slope $f'(x)$ is positive (Figure 3). Each such tangent line is a good approximation to the graph of f near the point of tangency, so it stands to reason that this graph also must be *rising* to the right. For similar reasons, a function with a *negative* derivative on an interval should be *decreasing* on this interval. In fact, we have the following theorem.

Figure 3

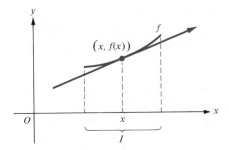

THEOREM 1   **Test for Increasing and Decreasing Functions**

> Assume that the function f is continuous on the interval I and that f is differentiable at every number in I, except possibly for the endpoints of I.
>
> **(i)** If $f'(x) > 0$ for every number x in I, except possibly for the endpoints of I, then f is increasing on I.
>
> **(ii)** If $f'(x) < 0$ for every number x in I, except possibly for the endpoints of I, then f is decreasing on I.

PROOF

(i) Suppose that $f'(x) > 0$ for every number x in I, except possibly for the endpoints of I, and let a and b be numbers in I with $a < b$. By the mean-value theorem (Section 3.1), there exists a number c with $a < c < b$ such that

$$f(b) - f(a) = f'(c)(b - a)$$

By hypothesis, $f'(c) > 0$. Also, since $a < b$, we have $b - a > 0$, and it follows that $f(b) - f(a) > 0$; that is, $f(a) < f(b)$.

(ii) The proof of (ii) is similar to the proof of (i), except that $f'(c) < 0$, so $f(b) - f(a) < 0$; that is, $f(a) > f(b)$. ∎

By applying the procedure for finding the intervals on which a function is positive or negative (page 152) to the *derivative f'* of a function f, you can use Theorem 1 to find the intervals where the function f is monotone (that is, either increasing or decreasing).

In Examples 1 and 2, find the intervals where each function is monotone.

EXAMPLE 1 $f(x) = x^3 - 3x + 1$

SOLUTION   Here,

$$f'(x) = 3x^2 - 3 = 3(x^2 - 1) = 3(x - 1)(x + 1)$$

and the continuous function f' cannot change sign on any of the intervals $(-\infty, -1)$, $(-1, 1)$, or $(1, \infty)$. We select convenient test numbers, say -2, 0, and 2, from these intervals and evaluate $f'(-2) = 9$, $f'(0) = -3$, and $f'(2) = 9$. Thus, we see that

f' is positive on $(-\infty, -1)$

f' is negative on $(-1, 1)$

and f' is positive on $(1, \infty)$

Figure 4a summarizes these findings. Therefore, by Theorem 1, we can conclude that

f is increasing on $(-\infty, -1]$

f is decreasing on $[-1, 1]$

and f is increasing on $[1, \infty)$ ∎

Figure 4

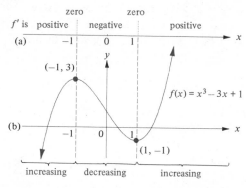

In using Theorem 1, note that we can conclude that f is monotone on an interval *including its endpoints* if f' is either positive or negative on the interval *excluding its endpoints*. Thus, for instance, the number 1 belongs both to the interval $[-1, 1]$ on which f is decreasing and to the interval $[1, \infty)$ on which f is increasing. This is perfectly consistent with Definition 1. Figure 4b shows a computer-generated graph of $f(x) = x^3 - 3x + 1$. The intervals on which f is increasing or decreasing are indicated on the graph.

EXAMPLE 2 $g(x) = x^{2/3}$

SOLUTION   Here,

$$g'(x) = \frac{2}{3}x^{-1/3} = \frac{2}{3x^{1/3}}$$

and although $g'(x)$ is never equal to zero, it is undefined for $x = 0$. Thus, g' has a

Figure 5

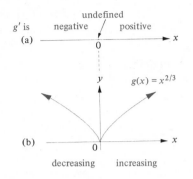

Figure 6

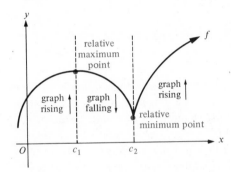

constant algebraic sign on the intervals $(-\infty, 0)$ and $(0, \infty)$. Using, say, test numbers -1 and 1 in these intervals, we have $g'(-1) = -\frac{2}{3}$ and $g'(1) = \frac{2}{3}$, which shows that

$$g' \text{ is negative on } (-\infty, 0)$$

and

$$g' \text{ is positive on } (0, \infty)$$

Figure 5a displays this information. Therefore,

$$g \text{ is decreasing on } (-\infty, 0]$$

and

$$g \text{ is increasing on } [0, \infty)$$

Figure 5b shows a computer-generated graph of $g(x) = x^{2/3}$. The intervals on which g is increasing or decreasing are indicated on the graph.

Critical Numbers

After you have found the intervals on which the graph of a continuous function f is rising or falling, you will see that a point on the graph separating a rising portion from a falling portion is either the "crest of a hill" or the "bottom of a valley" (Figure 6). Of course, the x coordinates of such local high or low points are just the numbers at which f has a relative maximum or minimum (Definitions 1 and 2, page 112). Locating these numbers is made easier by the following definition and theorem.

DEFINITION 3

Critical Number

> A number c is called a **critical number** for a function f if f is defined at c and either f is not differentiable at c or $f'(c) = 0$.

THEOREM 2

Relative Extrema and Critical Numbers

> If a function f has a relative extremum at the number c, then c is a critical number for f.

PROOF

Suppose that f has a relative extremum at c. On one hand, if f is not differentiable at c, then by Definition 3, c is a critical number for f. On the other hand, if f is differentiable at c, then by the necessary condition for relative extrema (Theorem 1, page 112), it follows that $f'(c) = 0$, and again (by Definition 3) c is a critical number for f.

According to Theorem 2, if you wish to locate all the relative extrema of a function, you can begin by finding all its critical numbers. Be careful, though—Theorem 2 does *not* say that if c is a critical number for f, then f necessarily has a relative extremum at c.*

EXAMPLE 3   Find all critical numbers for each function:

(a) $f(x) = 2x^3 - 3x^2 - 12x + 1$ (b) $g(x) = |x|$ (c) $h(x) = \sin x$

*See Example 7 on page 178.

SOLUTION

(a) Here, $f'(x) = 6x^2 - 6x - 12 = 6(x^2 - x - 2) = 6(x + 1)(x - 2)$ is defined for every real number x, and $f'(x) = 0$ for $x = -1$ and $x = 2$. Hence, -1 and 2 are the critical numbers for f.

(b) Here $g'(0)$ is undefined, so 0 is a critical number for g. Because $g'(x) = -1$ for $x < 0$ and $g'(x) = 1$ for $x > 0$, there are no numbers x for which $g'(x) = 0$. Hence, 0 is the only critical number for g.

(c) We have $h'(x) = \cos x$ for every real number x. Now, $\cos x = 0$ for $x = \pm\pi/2, \pm3\pi/2, \pm5\pi/2$, and so on. Hence, the critical numbers for h are all the odd integer multiples of $\pi/2$. ■

After you have found all the critical numbers for a function, you must *test* each such number to see whether it corresponds to a relative extremum. The simplest tests for relative extrema make use of first or second derivatives.

Figure 7

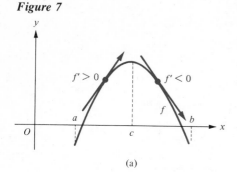

(a)

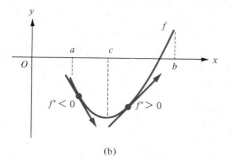

(b)

The First-Derivative Test

Consider Figure 7a, in which the continuous function f has a critical number at c, a positive first derivative on the open interval (a, c) just to the left of c, and a negative first derivative on the open interval (c, b) just to the right of c. By Theorem 1, f is increasing on $[a, c]$ and decreasing on $[c, b]$; hence, f has a *relative maximum at* c. Similarly, in Figure 7b, the continuous function f has a critical number at c, a negative first derivative on the open interval (a, c) just to the left of c, and a positive first derivative on the open interval (c, b) just to the right of c. Again by Theorem 1, f is decreasing on $[a, c]$ and increasing on $[c, b]$; hence, f has a *relative minimum at* c. These observations are summarized in the following theorem.

THEOREM 3 | **First-Derivative Test for Relative Extrema**

> Let the function f be defined and continuous on the open interval (a, b), assume that the number c belongs to (a, b), and suppose that f is differentiable at every number in (a, b) except possibly at c.
>
> **(i)** If the derivative f' is positive on (a, c) and negative on (c, b), then f has a relative maximum at c.
>
> **(ii)** If the derivative f' is negative on (a, c) and positive on (c, b), then f has a relative minimum at c.

In case (i) of Theorem 3, not only is there a relative maximum at c, but also $f(c) \geq f(x)$ holds for *every* x in the interval (a, b) (Figure 7a). Similarly, in case (ii) of Theorem 3 (Figure 7b), $f(c) \leq f(x)$ holds for *every* x in the interval (a, b).

In Examples 4 to 7, use the first-derivative test to find all numbers at which the function has a relative maximum or minimum.

EXAMPLE 4 $f(x) = x^3 - 2x^2 + x + 1$

SOLUTION   Here, $f'(x) = 3x^2 - 4x + 1 = (3x - 1)(x - 1)$, so the only critical numbers for f are the roots $x = \frac{1}{3}$ and $x = 1$ of the equation $f'(x) = 0$. To decide whether f has a relative maximum or minimum at either of these numbers, we use the first-derivative test. The procedure for determining the intervals on which f' is positive or negative shows that

$$f' \text{ is positive on } (-\infty, \tfrac{1}{3})$$

$$f' \text{ is negative on } (\tfrac{1}{3}, 1)$$

and

$$f' \text{ is positive on } (1, \infty)$$

By the first-derivative test (Theorem 3), we conclude that f has a relative maximum at $\tfrac{1}{3}$ and a relative minimum at 1. Notice that $f(\tfrac{1}{3}) = (\tfrac{1}{3})^3 - 2(\tfrac{1}{3})^2 + \tfrac{1}{3} + 1 = \tfrac{31}{27}$ and $f(1) = 1^3 - 2(1)^2 + 1 + 1 = 1$. All this information is summarized in the following table:

Figure 8

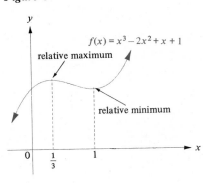

$f(x) = x^3 - 2x^2 + x + 1$

relative maximum

relative minimum

x	$f(x)$	$f'(x)$	Conclusion
$x < \tfrac{1}{3}$		positive	f is increasing
$x = \tfrac{1}{3}$	$\tfrac{31}{27}$	zero	f has a relative maximum
$\tfrac{1}{3} < x < 1$		negative	f is decreasing
$x = 1$	1	zero	f has a relative minimum
$x > 1$		positive	f is increasing

Figure 8 shows a computer-generated graph of f. The relative maximum and relative minimum points are indicated on the graph.

EXAMPLE 5 $g(x) = \begin{cases} 2x + 10 & \text{if } x \le -2 \\ x^2 + 2 & \text{if } x > -2 \end{cases}$

SOLUTION   If $x < -2$, then $g'(x) = 2$; hence, there are no critical numbers for g that are less than -2. If $x > -2$, then $g'(x) = 2x$; hence, since $g'(x) = 0$ when $x = 0$, the only critical number for g that is greater than -2 is 0. It remains to be seen whether -2 itself is a critical number for g. To find out, we calculate the one-sided derivatives of g:

$$g'_-(-2) = 2 \quad \text{and} \quad g'_+(-2) = -4$$

Since these are different, $g'(-2)$ does not exist, so -2 is a critical number for g. Now, proceeding as in Example 4, we obtain the following table:

Figure 9

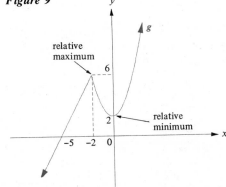

relative maximum

relative minimum

x	$g(x)$	$g'(x)$	Conclusion
$x < -2$		positive	g is increasing
$x = -2$	6	undefined	g has a relative maximum
$-2 < x < 0$		negative	g is decreasing
$x = 0$	2	zero	g has a relative minimum
$x > 0$		positive	g is increasing

Figure 9 shows a computer-generated graph of g. The relative extrema are indicated on the graph.

EXAMPLE 6   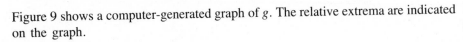 $h(x) = x + \dfrac{1}{x}$

SOLUTION   Here, $h'(x) = 1 - (1/x^2)$ for $x \neq 0$. To find the critical numbers, we set $h'(x) = 0$, so that

$$1 - \frac{1}{x^2} = 0 \qquad \text{or} \qquad x^2 = 1$$

Thus, $x = -1$ and $x = 1$ are the roots of $h'(x) = 0$, and it follows that -1 and 1 are the only critical numbers for h. Now we have the following:

Figure 10

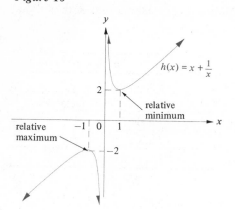

x	$h(x)$	$h'(x)$	Conclusion
$x < -1$		positive	h is increasing
$x = -1$	-2	zero	h has a relative maximum
$-1 < x < 0$		negative	h is decreasing
$x = 0$	undefined	undefined	h is undefined
$0 < x < 1$		negative	h is decreasing
$x = 1$	2	zero	h has a relative minimum
$x > 1$		positive	h is increasing

Figure 10 shows a computer-generated graph of h. The relative extrema are indicated on the graph. ∎

EXAMPLE 7 $F(x) = x^3 - 3x^2 + 3x$

SOLUTION   Here,

$$F'(x) = 3x^2 - 6x + 3 = 3(x^2 - 2x + 1)$$
$$= 3(x - 1)^2$$

so the only critical number for F is 1. We have the following:

x	$F(x)$	$F'(x)$	Conclusion
$x < 1$		positive	F is increasing
$x = 1$	1	zero	F has neither a local maximum nor a local minimum
$x > 1$		positive	F is increasing

Thus F is increasing on both $(-\infty, 1]$ and $[1, \infty)$; that is, it is increasing on all $\mathbb{R}$. Consequently, although F has a critical number, it has no relative extremum. Figure 11 shows a computer-generated graph of F. ∎

Figure 11

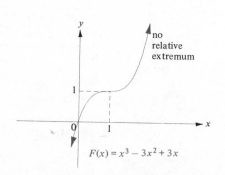

Problem Set 3.2

1 For each graph in Figure 12, indicate the intervals on which the function is increasing, decreasing, or constant.

Figure 12

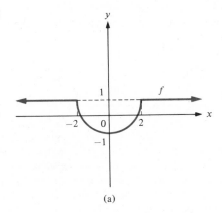

(a)

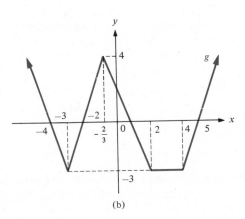

(b)

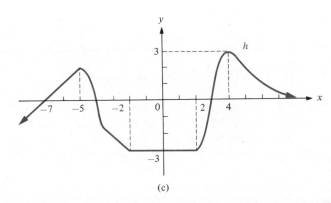

(c)

2 Suppose a friend, who is just beginning to study calculus, says to you, "I can't understand how a number c can belong to an interval on which a function f is increasing and at the same time belong to an interval on which f is decreasing." Can you clear this up for your friend?

In Problems 3 to 32, find the intervals on which the function is increasing or decreasing. Sketch the graphs for Problems 3 to 8 only.

3 $f(x) = x^2 - 6x - 2$ **4** $g(x) = 4x^2 - 8x + 3$

5 $h(x) = -3x^2 + 6x + 2$ **6** $F(x) = -2x^2 - 8x + 5$

7 $G(x) = -x^3 + 12x$

8 $q(x) = -x^3 - 6x^2 - 9x + 2$

9 $f(x) = x^3 - 12x + 11$ **10** $g(x) = x^3 + x^2 - 5x$

11 $h(x) = -4x^3 - 18x^2 - 27x$

12 $F(x) = x^3 - 3x^2 + 3x + 7$

13 $G(x) = x^3 - 6x^2 + 9x + 1$

14 $H(x) = 3x^4 + 8x^3 - 18x^2 + 12$

15 $f(x) = x^4 - 8x^2 + 1$

16 $g(x) = x^4 + 4x^3 + 6x^2 + 4x - 1$

17 $h(x) = 3x^5 - 5x^3$

18 $F(x) = 4x^5 - 25x^4 + 40x^3$

19 $G(x) = 1 - x^{1/3}$ **20** $H(x) = x^{2/3}(x^2 - 16)$

21 $p(x) = x - 4x^{-2}$ **22** $q(x) = \sqrt{x} - 9x^{-1}$

23 $r(x) = \sqrt{x} + 4x^{-1}$ **24** $P(x) = x^2 - 3x^{-2}$

25 $Q(x) = x^{4/3} + 4x^{1/3}$ **26** $R(x) = x^{7/5} - 8x^{3/5}$

27 $U(x) = \dfrac{x}{x^2 - 1}$ **28** $V(x) = \dfrac{x + 1}{x^2 + 2x + 2}$

29 $S(x) = \sin x$ **30** $s(x) = 3 \cos 2x$

31 $T(x) = \tan x$ **32** $t(x) = \sin x + \cos x$

In Problems 33 to 52, find all critical numbers of the function. Then use the first-derivative test to find all numbers at which the function has a relative maximum or minimum. Sketch the graph in Problems 33 to 38 only.

33 $f(x) = 7 + 12x - 2x^2$ **34** $g(x) = 6 - 5x - 6x^2$

35 $h(x) = x^3 - 27x$ **36** $F(x) = 2x^3 - 3x^2 - 4$

37 $G(x) = x^3 - x^2 - x - 1$ **38** $H(x) = -x^3 + x^2 - x$

39 $f(x) = 2x^3 + 3x^2 - 12x$

40 $g(x) = 2x^3 + x^2 - 20x + 1$

41 $h(x) = x^4 - 4x$ **42** $F(x) = (x - 1)^2(x - 2)^2$

43 $G(x) = x^4 - 4x^3 + 4x^2 + 6$

44 $H(x) = 3x^4 + 4x^3 - 12x^2 + 2$

45 $f(x) = 2\sqrt{x} - x$

46 $g(x) = x - 3\sqrt[3]{x}$

47 $h(x) = \dfrac{3}{x - 2}$

48 $r(x) = \dfrac{x - 1}{x^2 - 2x}$

49 $f(x) = \begin{cases} x^2 - 5 & \text{if } x < 4 \\ 10 - 3x & \text{if } x \geq 4 \end{cases}$

50 $g(x) = \begin{cases} \sqrt{25 - x^2} & \text{if } x \leq 4 \\ 7 - x & \text{if } x > 4 \end{cases}$

51 $h(x) = \begin{cases} x^2 + 4 & \text{if } x \geq 1 \\ 8 - 3x & \text{if } x < 1 \end{cases}$

52 $q(x) = \begin{cases} x^2 + 10x + 21 & \text{if } x < -4 \\ x^2 + 2x - 11 & \text{if } x \geq -4 \end{cases}$

In Problems 53 and 54, sketch a graph of a function having the indicated properties.

53 $f(1) = 2, f(5) = 4, f'(x) < 0$ for $x < 1, f'(x) > 0$ for $1 < x < 5$, and $f'(x) < 0$ for $x > 5$

54 $f(1) = 2, f(2) = 1, f'(x) > 0$ for $x < 1, f'(x) < 0$ for $1 < x < 2$, and $f'(x) > 0$ for $x > 2$

55 Suppose that f is a continuous function on the closed interval $[a, b]$ and that $f(a) = f(b)$. Show that f has at least one critical number on the open interval (a, b). (*Hint:* Use the mean-value theorem.)

56 Prove the following "converse" of part (i) of Theorem 1: If the function f has a continuous first derivative f' on an open interval I and if f is increasing on I, then $f'(x) \geq 0$ holds for all numbers x in I.

57 Let a, b, c, and d be constants; suppose $a > 0$; and let $f(x) =$ $ax^3 + bx^2 + cx + d$. Show that f is an increasing function on $\mathbb{R}$ if and only if $b^2 \leq 3ac$.

58 Suppose that f is a continuous function on the closed interval $[a, b]$ and that f is monotone on the open interval (a, b). Prove that f is monotone on the closed interval $[a, b]$.

59 Let $f(x) = \cos 2x + 2 \sin x$ for $-2\pi \leq x \leq 2\pi$. (a) Find all the critical numbers for f. (b) Find the intervals on which f is increasing or decreasing. (c) Find all numbers at which f has a relative maximum or minimum.

ⓒ **60** In thermodynamics, the *van der Waals gas law*

$$p = \frac{RT}{v - b} - \frac{a}{v^2}$$

relates the pressure p in atmospheres of a gas, the temperature T of the gas in kelvins, and the volume v in liters of 1 mole of the gas. Here $R = 8.206 \times 10^{-2}$ is the universal gas constant, and a and b are constants depending on the gas in question. For carbon dioxide, $a = 3.59$ and $b = 0.0427$. For carbon dioxide at a temperature $T = 260$ kelvins, consider p to be a function of v for $v > b$. (a) Find all critical numbers for this function. (*Hint:* Use Newton's method.) (b) Find the intervals on which p is increasing or decreasing. (c) Find all numbers at which f has a relative maximum or minimum. (*Caution:* Don't forget that $v > b$, so critical numbers less than b have no physical meaning.)

61 Suppose that the concentration C in milligrams per liter of a certain drug in the bloodstream t hours after it is administered orally is given by

$$C(t) = \frac{7t}{t^2 + 3t + 4} \qquad t \geq 0$$

(a) Find the interval in which C is increasing and the interval in which it is decreasing. (b) At what time t is the concentration C maximum? (c) What is the maximum concentration C?

3.3 Concavity and the Second-Derivative Test

We saw in Section 3.2 that the algebraic sign of the *first derivative* of a function determines whether the graph is rising or falling. Here we see that the algebraic sign of the *second derivative* determines whether the graph is bending upward ("cup-shaped") or bending downward ("cap-shaped"). We also show that the algebraic sign of the second derivative can be used to test whether a critical number for a function yields a relative maximum or minimum.

Figure 1a shows a cup-shaped graph. Notice that as the point P on this graph moves to the right, the tangent line at P turns counterclockwise and its *slope increases*. We say that such a graph is *concave upward*. Similarly, in Figure 1b, the graph is cap-shaped, and as the point P moves to the right, the tangent line at P turns clockwise and its *slope decreases*. We say that such a graph is *concave downward*. These simple geometric considerations lead us to the following formal definition.

Figure 1

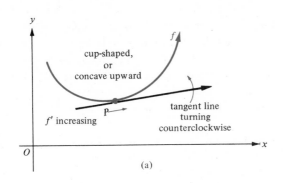

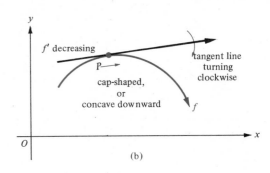

(a)

(b)

<u>DEFINITION 1</u>   **Concavity of a Graph**

Let f be a differentiable function on an open interval I.

(i) The graph of f is said to be **concave upward** on I if f' is an increasing function on I (Figure 1a).

(ii) The graph of f is said to be **concave downward** on I if f' is a decreasing function on I (Figure 1b).

Using the test for increasing and decreasing functions (Theorem 1, page 173), we see that if $(f')'$ is positive on an open interval, then f' is increasing on that interval, so the graph of f is concave upward on the interval by Definition 1. Similarly, if $(f')'$ is negative on an open interval, then f' is decreasing on the interval and the graph of f is concave downward on the interval. Thus, we have the following theorem.

<u>THEOREM 1</u>   **Test for Concavity of a Graph**

Let the function f be twice differentiable on the open interval I.

(i) If f'' is positive on I, then the graph of f is concave upward on I.

(ii) If f'' is negative on I, then the graph of f is concave downward on I.

<u>EXAMPLE 1</u>   Find the intervals where the graph of $f(x) = x^3 - 9x^2 + 24x - 20$ is concave upward or downward.

SOLUTION   Here,

$$f'(x) = 3x^2 - 18x + 24$$

and

$$f''(x) = 6x - 18 = 6(x - 3)$$

Therefore

$$f''(x) < 0 \qquad \text{for } x < 3$$

$$f''(x) > 0 \qquad \text{for } x > 3$$

Figure 2a shows this information. By Theorem 1, we may conclude that the graph of f is concave downward (cap-shaped) on the interval $(-\infty, 3)$ and concave upward (cup-shaped) on the interval $(3, \infty)$. Figure 2b shows a computer-generated graph of $f(x) = x^3 - 9x^2 + 24x - 20$. The intervals where the graph is concave downward or upward are indicated.

Figure 2

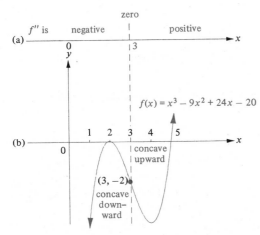

A point on a smooth graph that separates a portion that is concave upward from a portion that is concave downward [for instance, the point $(3, -2)$ in Figure 2b] is called a *point of inflection*. At a point of inflection, a graph may "cut across its own tangent line" (Figure 3). The notion of a point of inflection is made more precise by the following definition.

DEFINITION 2

Figure 3

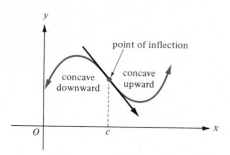

Point of Inflection

A point $(c, f(c))$ is called a **point of inflection** of the graph of a function f if the graph has a tangent line at this point and if there is an open interval I containing the number c such that for every pair of numbers a and b in I with $a < c < b$, $f''(a)$ and $f''(b)$ exist and have opposite algebraic signs.

Intuitively, to say that $(c, f(c))$ is a point of inflection of f means that the second derivative f'' "changes its algebraic sign" at c. This can happen only if either $f''(c) = 0$ or $f''(c)$ is not defined. For convenience, we record this useful fact in the following theorem.

THEOREM 2

Necessary Condition for a Point of Inflection

If $(c, f(c))$ is a point of inflection of the graph of the function f, then either $f''(c) = 0$ or $f''(c)$ is not defined.

Be careful! Theorem 2 does *not* say that the condition $f''(c) = 0$ guarantees a point of inflection at $(c, f(c))$—for instance, see Example 3 below. The theorem does say that if you are looking for a point of inflection, you need only check points $(c, f(c))$ for which either $f''(c) = 0$ or $f''(c)$ is not defined.

In Examples 2 and 3, find all points of inflection of the graph of the function.

EXAMPLE 2 $f(x) = x^3 - 6x^2 + 3x + 13$

SOLUTION   Here,

$$f'(x) = 3x^2 - 12x + 3$$

and

$$f''(x) = 6x - 12 = 6(x - 2)$$

Therefore, $f''(x)$ is defined for all numbers x in $\mathbb{R}$, and the equation $f''(x) = 0$ has only one solution, namely, $x = 2$. Therefore, by Theorem 2, the only possible point of inflection of the graph of f is $(2, f(2))$. Because

$$f''(x) = 6(x - 2) < 0 \qquad \text{for } x < 2$$

and

$$f''(x) = 6(x - 2) > 0 \qquad \text{for } x > 2$$

it follows that f'' really does change its algebraic sign at 2; hence, $(2, f(2)) = (2, 3)$ is the only point of inflection of the graph of f. Figure 4 shows a computer-generated graph of $f(x) = x^3 - 6x^2 + 3x + 13$. The point of inflection is indicated on the graph. ■

Figure 4

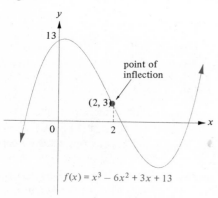

$f(x) = x^3 - 6x^2 + 3x + 13$

Figure 5

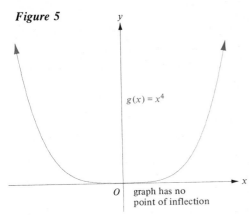

$g(x) = x^4$

O graph has no point of inflection

EXAMPLE 3 $g(x) = x^4$

SOLUTION   We have

$$g'(x) = 4x^3 \qquad \text{and} \qquad g''(x) = 12x^2$$

Therefore, $g''(x)$ is defined for all numbers x in $\mathbb{R}$, and the equation $g''(x) = 0$ has only one solution, namely, $x = 0$. Therefore, by Theorem 2, the only possible point of inflection of the graph of g is $(0, g(0)) = (0, 0)$. However, because

$$g''(x) = 12x^2 > 0 \qquad \text{for } x < 0$$

and

$$g''(x) = 12x^2 > 0 \qquad \text{for } x > 0$$

it follows that $(0, 0)$ is *not* a point of inflection. Figure 5 shows a computer-generated graph of $g(x) = x^4$. The graph of g has no point of inflection. Because g' is an increasing function on $\mathbb{R}$, the graph of g is concave upward on $\mathbb{R}$. ∎

In economics, psychology, exercise science, and many other fields, a **law of diminishing returns** is said to be operating if a certain "input" produces a corresponding "output" in such a way that:

1 An increase in input causes an increase in output.

2 As input is increased, a point is reached beyond which additional increases in input result in progressively smaller increases in output.

In economics, the input might be labor or capital and the output production or profit. In psychology, the input might be effort and the output achievement; in exercise science, the input might be training time and the output athletic accomplishment; and so forth.

Using calculus, we can analyze the law of diminishing returns as follows: Suppose that the input is represented quantitatively by the variable x and the output by the variable y. We assume that $y = f(x)$, where f is a continuous function possessing at least two derivatives. Condition 1 simply means that f is an increasing function, in other words, that f' is positive. Condition 2 may be interpreted to mean that there is a point $(c, f(c))$ on the graph of f (Figure 6) such that f' is increasing to the left of c and decreasing to the right of c. In other words, the **point of diminishing returns** $(c, f(c))$ is just the point of inflection of the graph of f. By Theorem 2, we have $f''(c) = 0$ for the point of diminishing returns.

Figure 6

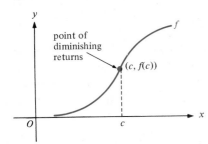

point of diminishing returns → $(c, f(c))$

The Second-Derivative Test

Sometimes, when the first derivative test for relative extrema is inconvenient, it is possible to test critical numbers by using the second derivative. The geometric basis for the **second-derivative test** is easily seen and remembered as follows: Figure 7a shows the graph of a function f and a critical number c such that $f''(c) > 0$. The condition $f''(c) > 0$ indicates that the graph of f is concave upward near the point $(c, f(c))$; hence, f has a relative minimum at c. Similarly, Figure 7b shows a critical number c for f such that $f''(c) < 0$, so the graph of f is concave downward near $(c, f(c))$ and f has a relative maximum at c.

The second-derivative test is stated and proved formally in the following theorem.

Figure 7

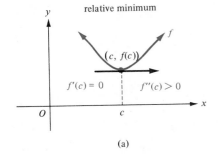

relative minimum

$(c, f(c))$

$f'(c) = 0$ $f''(c) > 0$

(a)

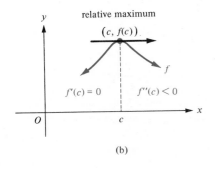

relative maximum

$(c, f(c))$

$f'(c) = 0$ $f''(c) < 0$

(b)

THEOREM 3   **Second-Derivative Test for Relative Extrema**

> Let the function f be differentiable on the open interval I, and suppose that c is a number in I such that $f'(c) = 0$ and $f''(c)$ exists.
>
> (i)   If $f''(c) > 0$, then f has a relative minimum at c.
>
> (ii) If $f''(c) < 0$, then f has a relative maximum at c.
>
> (iii) If $f''(c) = 0$, then the test is inconclusive.

PROOF

We prove (i) only, since the proof of (ii) is similar. [See Problem 48 for part (iii).] Thus, assume $f'(c) = 0$ and $f''(c) > 0$. By definition of $f''(c) = (f')'(c)$ and the fact that $f'(c) = 0$, we have

$$(f')'(c) = \lim_{x \to c} \frac{f'(x) - f'(c)}{x - c} = \lim_{x \to c} \frac{f'(x)}{x - c}$$

Hence, we can make the ratio $f'(x)/(x - c)$ as close as we please to the positive number $f''(c)$ simply by taking x sufficiently close to c (but not equal to c). In particular, if x is close enough to c (but different from c), then $f'(x)/(x - c)$ will have to be positive. Therefore, there must be an open interval (a, b) containing c such that if x is different from c and belongs to (a, b), then $f'(x)/(x - c) > 0$. It follows that if $a < x < c$, then $x - c < 0$ and $f'(x)/(x - c) > 0$, so that $f'(x) < 0$. Similarly, if $c < x < b$, we have $x - c > 0$ and $f'(x)/(x - c) > 0$, from which $f'(x) > 0$ follows. Thus, slightly to the left of c the derivative $f'(x)$ is negative, while slightly to the right of c it is positive. By the first-derivative test (Theorem 3, page 176) we conclude that f has a relative minimum at c.   ∎

In Examples 4 and 5, use the second-derivative test to find all numbers at which the function has a relative maximum or minimum.

EXAMPLE 4 $f(x) = x^3 - 6x^2 + 9x$

SOLUTION   Here,

$$f'(x) = 3x^2 - 12x + 9 = 3(x - 1)(x - 3)$$

and

$$f''(x) = 6x - 12 = 6(x - 2)$$

To find the critical numbers for f, we set $f'(x) = 0$:

$$3(x - 1)(x - 3) = 0$$

Therefore the critical numbers are 1 and 3. Using Theorem 3, we can determine whether these critical numbers correspond to relative maxima or minima by finding the algebraic signs of the second derivative at 1 and at 3. We have

$$f''(1) = 6(1 - 2) = -6 < 0$$

and

$$f''(3) = 6(3 - 2) = 6 > 0$$

Hence, f has a relative maximum at 1 and a relative minimum at 3. Figure 8 shows a computer-generated graph of $f(x) = x^3 - 6x^2 + 9x$. The relative extrema are indicated on the graph.   ∎

Figure 8

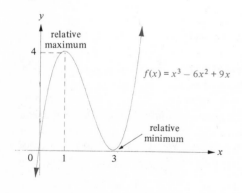

EXAMPLE 5 $g(x) = \sin x$

SOLUTION   Here,

$$g'(x) = \cos x$$

and $$g''(x) = -\sin x$$

The critical numbers for g are the solutions of the equation $g'(x) = 0$, that is, of

$$\cos x = 0$$

Now, $\cos x = 0$ if and only if x is an odd integer multiple of $\pi/2$; that is,

$$x = (2n + 1)\frac{\pi}{2} = n\pi + \frac{\pi}{2}$$

for $n = 0, \pm1, \pm2, \pm3$, and so on. We have

$$g''\left(n\pi + \frac{\pi}{2}\right) = -\sin\left(n\pi + \frac{\pi}{2}\right)$$

But by the addition formula for the sine (inside front cover),

$$\sin\left(n\pi + \frac{\pi}{2}\right) = \sin n\pi \cos\frac{\pi}{2} + \sin\frac{\pi}{2}\cos n\pi = 0 \cdot 0 + 1 \cdot \cos n\pi$$

$$= \cos n\pi$$

Therefore,

$$g''\left(n\pi + \frac{\pi}{2}\right) = -\cos n\pi$$

Now, $$\cos n\pi = \begin{cases} 1 & \text{if } n \text{ is an } even \text{ integer} \\ -1 & \text{if } n \text{ is an } odd \text{ integer} \end{cases}$$

Hence, $$g''\left(n\pi + \frac{\pi}{2}\right) = \begin{cases} -1 & \text{if } n \text{ is even} \\ 1 & \text{if } n \text{ is odd} \end{cases}$$

Therefore, by the second-derivative test, $g(x) = \sin x$ takes on relative maxima when

$$x = n\pi + \frac{\pi}{2} \qquad \text{for } n = 0, \pm2, \pm4, \pm6, \ldots$$

that is, when x is $\pi/2$ plus an integer multiple of 2π. Similarly, $g(x) = \sin x$ takes on relative minima when

$$x = n\pi + \frac{\pi}{2} = n\pi + \pi - \frac{\pi}{2}$$

$$= (n + 1)\pi - \frac{\pi}{2} \qquad \text{for } n = \pm1, \pm3, \pm5, \ldots$$

that is, when x is $-\pi/2$ plus an integer multiple of 2π. Figure 9 shows a computer-generated graph of $g(x) = \sin x$. ∎

Figure 9

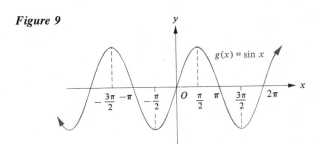

Of course, we really didn't need to use calculus to obtain the results in the last example—we already knew from elementary trigonometry when the sine function takes on its maximum value of 1 and its minimum value of -1. However, the technique illustrated in Example 5 can be applied to more complicated functions for which the result isn't known in advance.

Problem Set 3.3

In Problems 1 to 34, find the intervals on which the graph of the function is concave upward or downward. Also, find all points of inflection. Sketch the graph for Problems 1 to 6 only.

1 $f(x) = x^2 - x - 6$

2 $g(x) = -x^2 + 2x + 8$

3 $h(x) = (3x + 2)(1 - x)$

4 $F(x) = x^3 + 1$

5 $G(x) = (x + 1)^3$

6 $H(x) = 5 + 12x - x^3$

7 $f(x) = x^3 + 6x^2 - 14$

8 $g(x) = \frac{1}{3}x^3 + \frac{1}{2}x^2 - 2x - \frac{1}{4}$

9 $h(x) = 2x^3 - \frac{1}{2}x^2 - 7x + 2$

10 $P(x) = 4x^3 - 18x^2 + 15x + 5$

11 $Q(x) = x^3 - 6x^2 + 9x + 1$

12 $R(x) = x^3 + 3x^2 + 5$

13 $p(x) = x^4 + 4x^2 - 16$

14 $q(x) = 8x - 2x^2 - x^4$

15 $r(x) = -x^4$

16 $u(x) = x^4 + 2x^3 - 12x^2$

17 $v(x) = x^4 - 4x^3 - 18x^2$

18 $w(x) = \frac{1}{3}x^4 - \frac{1}{2}x^2$

19 $f(x) = 2x^5 - 20x^3$

20 $g(x) = 3x^5 - 5x^3$

21 $h(x) = 2x + \dfrac{2}{x}$

22 $F(x) = x + \dfrac{1}{\sqrt{x}}$

23 $G(x) = x^2 + 5x^{-2}$

24 $H(x) = (x - 1)^{8/3} + (x - 1)^2$

25 $p(x) = x^{5/3}$

26 $q(x) = (x + 2)^{1/3}x^{-2/3}$

27 $r(x) = \dfrac{1}{x^2 + x}$

28 $s(x) = 1 + (x - 2)^{2/3}$

29 $P(x) = \dfrac{5x}{x^2 + 4}$

30 $Q(x) = 1 + \left(\dfrac{x}{x - 1}\right)^{1/3}$

31 $R(x) = x - \sqrt{x}$

32 $f(x) = |x^2 - 1|$

33 $g(x) = \cos x$

34 $h(x) = \sin 2x$

In Problems 35 to 46, use the second-derivative test to find all numbers at which the function has a relative maximum or minimum. Sketch the graph for Problems 35 to 40 only.

35 $f(x) = x^2 - 5x + 4$

36 $g(x) = x^2(2x + 7)$

37 $h(x) = x(x - 2)^2$

38 $F(x) = 2x^3 + x^2 + x - 3$

39 $G(x) = x^2 - x^3$

40 $H(x) = x^3 - x^2 + 2x - 2$

41 $f(x) = x^4 - 6x^2 + 4$

42 $g(x) = 3x^4 + 8x^3 - 18x^2 + 12$

43 $h(x) = \dfrac{x^2}{x - 1}$

44 $F(x) = \dfrac{x^2 + x + 1}{x^2}$

45 $G(x) = \sin x - \cos x$

46 $H(x) = x - \sin x$

47 In Figure 10, consider the graphs of the given functions and the displayed interval $[a, e]$. In each case, the interval $[a, e]$ is broken up into four subintervals $[a, b]$, $[b, c]$, $[c, d]$, and $[d, e]$. Assume that the given functions are twice-differentiable on the interior of each subinterval. Determine on which of these subintervals the given function (i) is increasing, (ii) is decreasing, (iii) has a graph that is concave upward, and (iv) has a graph that is concave downward. (v) Find all inflection points.

Figure 10

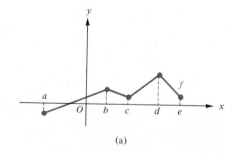

(a)

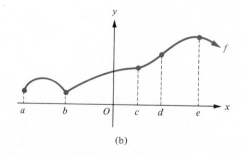

(b)

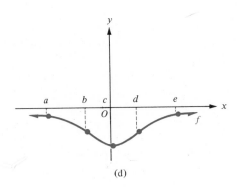

(c)

(d)

48 Show that if $f'(c) = 0$ and $f''(c) = 0$, as in part (iii) of Theorem 3, then the test is inconclusive. Do this by considering the three functions $f(x) = x^4$, $g(x) = x^3$, and $h(x) = -x^4$, and take $c = 0$.

In Problems 49 to 64, use whatever method you wish to find all numbers at which the function has a relative maximum or minimum.

49 $f(x) = 2x^3 - 3x^2 - 4$

50 $g(x) = \frac{5}{3}x^3 - \frac{7}{2}x^2 - 6x + 4$

51 $h(x) = x^4 - 4x$

52 $F(x) = x^4 + 4x^3 + 6x^2 + 4x - 1$

53 $G(x) = 3x^4 - 4x^3 - 12x^2$

54 $H(x) = \dfrac{x}{(x + 1)^2}$

55 $p(x) = \dfrac{2x}{x + 2}$

56 $q(x) = x^2 - 3x^{-2}$

57 $r(x) = x^{4/5}$

58 $P(x) = 3x^{2/3} - x^{5/3}$

59 $f(x) = \begin{cases} 10 - 3x & \text{if } x \geq 2 \\ x^2 & \text{if } x < 2 \end{cases}$

60 $g(x) = \begin{cases} x^2 - 1 & \text{if } x \leq 2 \\ x^2 - 8x + 15 & \text{if } x > 2 \end{cases}$

61 $h(x) = \begin{cases} x^3 & \text{if } x \leq 1 \\ (x - 2)^2 & \text{if } x > 1 \end{cases}$

62 $F(x) = \begin{cases} 1 + x^{-2} & \text{if } x \neq 0 \\ 0 & \text{if } x = 0 \end{cases}$

63 $S(x) = \sin x + \csc x$

64 $T(x) = x - \tan x$

65 If the function f is differentiable on the open interval I, some textbooks define the graph of f to be concave upward if f satisfies the following condition: For every pair of distinct numbers a and b in I, the point $(b, f(b))$ on the graph of f lies strictly above the tangent line to the graph of f at $(a, f(a))$. Show that this condition holds if and only if, for every pair of distinct numbers a and b in I, $f(b) > f(a) + f'(a)(b - a)$.

66 Assume that the function f is differentiable on the open interval I and that the graph of f is concave upward on I in the sense of Definition 1. If a and b are distinct numbers in I, prove that $f(b) > f(a) + f'(a)(b - a)$.

67 Some textbooks define the graph of the function f to be concave upward on the interval I provided that, for any two distinct numbers a and b in I, the portion of the graph between $(a, f(a))$ and $(b, f(b))$ lies below the secant between $(a, f(a))$ and $(b, f(b))$. Draw a diagram illustrating this condition.

68 Suppose that the function f is differentiable on the open interval I and that the graph of f is concave upward on I. Prove that if a and b are two numbers in I, then $0 < t < 1$ implies that $f[ta + (1 - t)b] < tf(a) + (1 - t)f(b)$.

69 In educational psychology, **learning curves** are often used to indicate a person's achievement or performance (as measured by a standardized test) as a function of the total time spent studying the subject in question. These curves often exhibit a point of diminishing returns (see page 183). Explain why study time is being spent most efficiently at the point of diminishing returns.

70 As a result of a time-and-motion study, a manufacturer determines that between starting time (8 A.M.) and lunchtime (12 noon), the average worker has produced p units of the product after working t hours, where $p = 22t + 9t^2 - 3t^3$. (a) How many units of the product does the average worker produce between 8 A.M. and 12 noon? (b) At what time in the morning does the point of diminishing returns occur for the average worker? (c) At what time in the morning is the average worker performing at maximum efficiency? (d) When the average worker is performing at maximum efficiency, how many units of the product per hour is he or she producing? ⓒ(e) Sketch the graph of p as a function of t for $0 \leq t \leq 4$. If you wish, a calculator may be used to help determine the coordinates of points on the graph and thus improve the accuracy of the sketch.

3.4   Absolute Extrema

People often find it necessary to determine the maximum (largest) or minimum (smallest) value of a variable quantity. Thus, a wage earner is interested in maximizing after-tax income, an environmentalist wishes to minimize air or water pollution, a business person is concerned with maximum profit, and a conservationist wishes to minimize energy consumption. Often problems such as these can be translated into the mathematical question of finding the extreme values of functions. In Sections 3.2 and 3.3, we used the first and second derivatives of a function to locate its *relative* extrema. Notice, however, that such a relative maximum or minimum need not be the actual largest or smallest value of the function. To emphasize this fact, we often refer to the largest and smallest values of a function (if they exist) as its *absolute* maximum and its *absolute* minimum.

Figure 1 illustrates the distinction between absolute and relative extrema. Notice that the function f has relative maximum values at p and r; however, the function value $f(r)$ is larger than the function value $f(p)$—in fact, $f(r)$ is the largest of all the function values $f(x)$ for $a \le x \le b$. Thus, we say that f *takes on an absolute maximum value $f(r)$ at r.* Similarly, f has a relative minimum value at q; however, the function value $f(b)$ is smaller than the function value $f(q)$—in fact, $f(b)$ is the smallest of all the function values $f(x)$ for $a \le x \le b$. Thus, we say that f *takes on an absolute minimum value $f(b)$ at b.*

This discussion leads us to the following general definition.

Figure 1

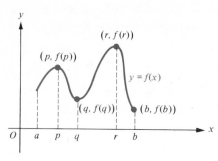

DEFINITION 1

Absolute Maximum and Minimum

Suppose that the interval I is contained in the domain of the function f, and let c and d be numbers in I.

(i) If $f(c) \ge f(x)$ holds for all numbers x in I, we say that *on the interval I the function f takes on its* **absolute maximum** *value $f(c)$ at the number c.*

(ii) If $f(d) \le f(x)$ holds for all numbers x in I, we say that *on the interval I the function f takes on its* **absolute minimum** *value $f(d)$ at the number d.*

In dealing with either absolute or relative extrema, it is essential to keep in mind the distinctions among the *number c* at which the function f takes on the extreme value; the *extreme value itself,* which is $f(c)$; and the corresponding *extreme point* $(c, f(c))$ on the graph of f.

Now, suppose you are given a continuous function f on an interval I and asked to find the maximum and minimum values of f and where they occur on I. To begin with, if the adjective "relative" is not used, you are usually safe in assuming that the *absolute* extrema are wanted. If I is a *closed* interval $[a, b]$, Theorem 1 on page 167 guarantees that these absolute extrema exist. An absolute extremum of f may occur at an endpoint of the interval I, in which case it is called an **endpoint extremum.** According to Definitions 1 and 2 on page 112, an absolute extremum of f that occurs at a number other than an endpoint of I is automatically a relative extremum of f. Therefore, by Theorem 2 on page 175, if f has an absolute extremum at a number c, other than an endpoint of I, then c is a critical number for f. These considerations provide the basis for the following.

Procedure for Finding the Absolute Extrema of a Continuous Function on a Closed Interval

To find the absolute extrema of a continuous function f on a closed interval $[a, b]$, carry out the following steps:

Step 1 Find all critical numbers c for the function f on the open interval (a, b).

Step 2 Calculate the function values $f(c)$ for each of the numbers c obtained in step 1.

Step 3 Calculate the function values $f(a)$ and $f(b)$ at the endpoints a and b of the interval.

Step 4 Conclude that the largest of all the numbers calculated in steps 2 and 3 is the absolute maximum of f on $[a, b]$ and the smallest of these numbers is the absolute minimum of f on $[a, b]$.

If the function f is not continuous on the interval I, or if I is not a closed interval $[a, b]$, then perhaps the most effective method for finding the absolute extrema of f on I (when they exist) is to sketch the graph of f.

In Examples 1 to 6, find the absolute maximum and minimum values of the function on the given interval, and indicate where they occur.

<u>EXAMPLE 1</u> $f(x) = 2x^2$ on $[-3, 1]$

SOLUTION   Here, f is continuous on $[-3, 1]$, so we can follow the procedure given above.

Step 1 Since $f'(x) = 4x$, the only critical number for f in the open interval $(-3, 1)$ is 0.

Step 2 $f(0) = 2(0)^2 = 0$

Step 3 $f(-3) = 2(-3)^2 = 18$ and $f(1) = 2(1)^2 = 2$

Step 4 The largest of the numbers $f(-3) = 18$, $f(0) = 0$, and $f(1) = 2$ is 18; the smallest of these numbers is 0. We conclude that on the interval $[-3, 1]$ the function f takes on an absolute maximum value of 18 at -3 and an absolute minimum value of 0 at 0. Figure 2 shows a computer-generated graph of f. ∎

<u>EXAMPLE 2</u> $f(x) = \sqrt{9 - x^2}$ on $[-3, 3]$

SOLUTION   Again, we follow the procedure:

Step 1 Since

$$f'(x) = \frac{-2x}{2\sqrt{9 - x^2}} = \frac{-x}{\sqrt{9 - x^2}}$$

the only critical number for f in the open interval $(-3, 3)$ is 0.

Step 2 $f(0) = \sqrt{9 - 0^2} = 3$

Step 3 $f(-3) = \sqrt{9 - (-3)^2} = 0$ and $f(3) = \sqrt{9 - 3^2} = 0$

Step 4 The largest of the numbers $f(-3) = 0$, $f(0) = 3$, and $f(3) = 0$ is 3; the smallest of these numbers is 0. We conclude that on the interval $[-3, 3]$ the function f takes on an absolute maximum value of 3 at 0; it takes on an absolute minimum value of 0 at -3 and again at 3. Figure 3 shows a computer-generated graph of f. ∎

Figure 2

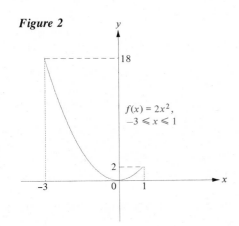

$f(x) = 2x^2,$
$-3 \leqslant x \leqslant 1$

Figure 3

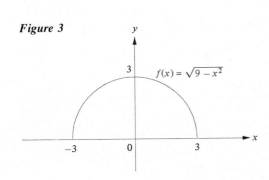

$f(x) = \sqrt{9 - x^2}$

EXAMPLE 3 $g(x) = \begin{cases} x^2 - 2x + 2 & \text{if } x \geq 0 \\ x^2 + 2x + 2 & \text{if } x < 0 \end{cases}$ on $[-\frac{1}{2}, \frac{3}{2}]$

SOLUTION   To determine whether g is continuous on $[-\frac{1}{2}, \frac{3}{2}]$, we examine the one-sided limits at 0, the only number in question:

$$\lim_{x \to 0^-} g(x) = \lim_{x \to 0^-} (x^2 + 2x + 2) = 2$$

and

$$\lim_{x \to 0^+} g(x) = \lim_{x \to 0^+} (x^2 - 2x + 2) = 2$$

Because these one-sided limits both have the same value, 2, we have

$$\lim_{x \to 0} g(x) = 2 = g(0)$$

and so g is continuous at 0. Thus, g is continuous on $[-\frac{1}{2}, \frac{3}{2}]$, and we can use our procedure:

Step 1 We have

$$g'(x) = \begin{cases} 2x - 2 & \text{if } 0 < x < \frac{3}{2} \\ 2x + 2 & \text{if } -\frac{1}{2} < x < 0 \end{cases}$$

To determine whether g is differentiable at 0, we calculate the derivatives from the left and right at 0:

$$g'_-(0) = 2 \quad \text{and} \quad g'_+(0) = -2$$

Since these are unequal, $g'(0)$ does not exist and 0 is a critical number for g on the interval $(-\frac{1}{2}, \frac{3}{2})$. Evidently, $g'(x) = 0$ for x in the interval $(-\frac{1}{2}, \frac{3}{2})$ only when $x = 1$. Thus, in the open interval $(-\frac{1}{2}, \frac{3}{2})$, the critical numbers for g are 0 and 1.

Step 2 $g(0) = 2 \quad$ and $\quad g(1) = 1$

Step 3 $g(-\frac{1}{2}) = (-\frac{1}{2})^2 + 2(-\frac{1}{2}) + 2 = \frac{5}{4} \quad$ and $\quad g(\frac{3}{2}) = (\frac{3}{2})^2 - 2(\frac{3}{2}) + 2 = \frac{5}{4}$

Step 4 The largest of the numbers $g(-\frac{1}{2}) = \frac{5}{4}$, $g(0) = 2$, $g(1) = 1$, and $g(\frac{3}{2}) = \frac{5}{4}$ is 2; the smallest is 1. We conclude that on the interval $[-\frac{1}{2}, \frac{3}{2}]$ the function g takes on an absolute maximum value of 2 at 0 and an absolute minimum value of 1 at 1. Figure 4 shows a computer-generated graph of g. ∎

Figure 4

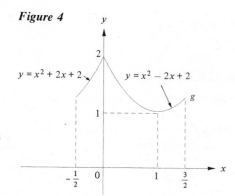

EXAMPLE 4 $h(x) = \dfrac{2 + x - x^2}{2 - x + x^2}$ on $[-2, 1]$

SOLUTION   Because the quadratic equation $2 - x + x^2 = 0$ has no real roots, the denominator $2 - x + x^2$ is nonzero for all real values of x; hence, the rational function h is continuous on $\mathbb{R}$. In particular, it is continuous on $[-2, 1]$, so we can apply our procedure:

Step 1 Using the quotient rule and simplifying, we have

$$h'(x) = \frac{(2 - x + x^2)(1 - 2x) - (2 + x - x^2)(-1 + 2x)}{(2 - x + x^2)^2} = \frac{4(1 - 2x)}{(2 - x + x^2)^2}$$

Because the denominator $(2 - x + x^2)^2$ is nonzero for all real values of x, the only critical number for h is $\frac{1}{2}$. Thus, $\frac{1}{2}$ is the only critical number for h in the open interval $(-2, 1)$.

Figure 5

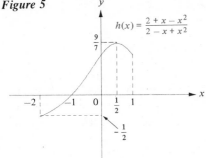

$$h(x) = \frac{2 + x - x^2}{2 - x + x^2}$$

Figure 6

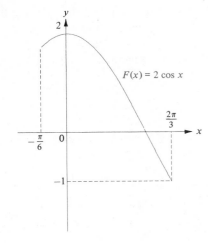

$$F(x) = 2 \cos x$$

Figure 7

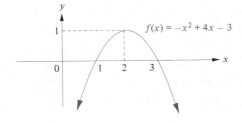

$$f(x) = -x^2 + 4x - 3$$

Step 2 $h(\frac{1}{2}) = \frac{9}{7}$

Step 3 $h(-2) = -\frac{1}{2}$ and $h(1) = 1$

Step 4 The largest of the numbers $h(-2) = -\frac{1}{2}$, $h(\frac{1}{2}) = \frac{9}{7}$, and $h(1) = 1$ is $\frac{9}{7}$; the smallest is $-\frac{1}{2}$. We conclude that on the interval $[-2, 1]$ the function h takes on an absolute maximum value of $\frac{9}{7}$ at $\frac{1}{2}$ and an absolute minimum value of $-\frac{1}{2}$ at -2. Figure 5 shows a computer-generated graph of h. ∎

EXAMPLE 5 $F(x) = 2 \cos x$ on $\left[-\dfrac{\pi}{6}, \dfrac{2\pi}{3}\right]$

SOLUTION   The function F is continuous on the closed interval, so we can apply our procedure:

Step 1 We know that $F'(x) = -2 \sin x$. Now, $\sin x = 0$ if and only if x is an integer multiple of π, and the only integer multiple of π between $-\pi/6$ and $2\pi/3$ is 0; hence, 0 is the only critical number of F on $(-\pi/6, 2\pi/3)$.

Step 2 $F(0) = 2 \cos 0 = 2$

Step 3 $F\left(-\dfrac{\pi}{6}\right) = 2 \cos\left(-\dfrac{\pi}{6}\right) = 2\left(\dfrac{\sqrt{3}}{2}\right) = \sqrt{3}$

and $F\left(\dfrac{2\pi}{3}\right) = 2 \cos\dfrac{2\pi}{3} = 2\left(-\dfrac{1}{2}\right) = -1$

Step 4 The largest of the numbers $F(-\pi/6) = \sqrt{3}$, $F(0) = 2$, and $F(2\pi/3) = -1$ is 2; the smallest is -1. We conclude that on the interval $[-\pi/6, 2\pi/3]$ the function F takes on an absolute maximum value of 2 at 0 and an absolute minimum value of -1 at $2\pi/3$. Figure 6 shows a computer-generated graph of F. ∎

EXAMPLE 6 $f(x) = -x^2 + 4x - 3$ on $(-\infty, \infty)$

SOLUTION   Since the interval is not of the form $[a, b]$, the procedure we have been using is not applicable. However, we can still argue that an absolute extremum of f on the open interval $(-\infty, \infty)$ would automatically be a relative extremum; hence, it could only occur at a critical number for f. Now,

$$f'(x) = -2x + 4$$

so the only such critical number is 2. Since $f'(x) > 0$ for $x < 2$ and since $f'(x) < 0$ for $x > 2$, it follows that f is increasing on the interval $(-\infty, 2]$ and decreasing on the interval $[2, \infty)$. Therefore, f takes on an absolute maximum value of $f(2) = 1$ at 2, but it has no absolute minimum value on $(-\infty, \infty)$. Figure 7 shows a computer-generated graph of f. The graph of f is a parabola opening downward. ∎

In Example 6, the interval $(-\infty, \infty)$ is the domain of the function f, and the absolute maximum of f on $(-\infty, \infty)$ is the largest of *all* function values of f. More generally, we say that a function f *takes on the* **absolute maximum value** $f(c)$ *at the number* c if c is a number in the domain of f and if $f(c) \geq f(x)$ holds for every number x in the domain of f. Note that there is no reference here to any interval, it being understood that the entire domain of f is under consideration. Of course, there is a similar understanding when we speak of the **absolute minimum value** of a function f. Often, we even drop the word "absolute" and speak simply of the **maximum value** or the **minimum value** of the function f.

In Examples 7 and 8, find the maximum and minimum values (if they exist) of the function, and indicate where they occur.

EXAMPLE 7 $f(x) = \dfrac{x^2 - 1}{x^2 + 1}$

Figure 8

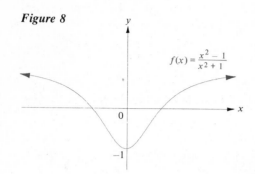

SOLUTION   Because the denominator $x^2 + 1$ is nonzero for all real numbers, the rational function f is defined and continuous on $\mathbb{R}$. Hence, an absolute extremum of f would automatically be a relative extremum and could only occur at a critical number for f. Using the quotient rule and simplifying, we have

$$f'(x) = \frac{4x}{(x^2 + 1)^2}$$

so 0 is the only critical number for f. Since $f'(x) < 0$ for $x < 0$ and since $f'(x) > 0$ for $x > 0$, it follows that f is decreasing on $(-\infty, 0]$ and increasing on $[0, \infty)$. Therefore, f takes on an absolute minimum value of $f(0) = -1$ at 0, but it has no absolute maximum value. Figure 8 shows a computer-generated graph of f.   ∎

©**EXAMPLE 8** $G(x) = x + \dfrac{1}{\sqrt{x - 1}}$

Figure 9

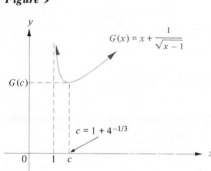

SOLUTION   The domain of G is the interval $(1, \infty)$. For any number x in this interval, we have

$$G'(x) = 1 - \frac{1}{2\sqrt{(x - 1)^3}}$$

Solving the equation $G'(c) = 0$, we find that the only critical number for G is

$$c = 1 + 4^{-1/3} \approx 1.63$$

For $1 < x < c$ we have $G'(x) < 0$, and for $x > c$ we have $G'(x) > 0$. Therefore, G is decreasing on $(1, c]$ and increasing on $[c, \infty)$. It follows that G has an absolute minimum value of $G(c) \approx 2.89$ at c. Here G has no absolute maximum. Figure 9 shows a computer-generated graph of G.   ∎

The methods introduced above are especially useful for solving practical problems involving maxima and minima.

EXAMPLE 9   Suppose that the distance d in kilometers that a certain car can travel on one tank of gasoline at a speed of v kilometers per hour is given by the equation $d = 8v - (v/4)^2$. What speed maximizes the distance d (and thus minimizes the fuel consumption)?

SOLUTION   We must find the value of v for which the function

$$f(v) = 8v - \left(\frac{v}{4}\right)^2 \qquad \text{for } 0 < v$$

takes on an absolute maximum. Here,

$$f'(v) = 8 - \frac{v}{8}$$

so the only critical number for f is $v = 64$. Because $f'(v) > 0$ for v in the interval $(0, 64)$ and $f'(v) < 0$ for v in the interval $(64, \infty)$, it follows that f is increasing on

(0, 64] and decreasing on [64, ∞). Therefore, on the interval (0, ∞), f takes on an absolute maximum at 64. Consequently, a speed of $v = 64$ kilometers per hour will yield a maximum distance of $f(64) = 256$ kilometers. ∎

In Sections 3.7 and 3.9, we undertake a more detailed and systematic study of applied problems involving maxima and minima.

Problem Set 3.4

In Problems 1 to 32, find the absolute maximum and minimum values of the function on the given interval, and indicate where they occur. Sketch the graph in Problems 1 to 10 only.

1 $f(x) = -x^2$ on $[-2, 1]$

2 $g(x) = 4x + 3$ on $[0, 2]$

3 $h(x) = -2x$ on $[-1, 2]$

4 $F(x) = \dfrac{1}{x}$ on $[1, 4]$

5 $G(x) = \sqrt{4 - x^2}$ on $[-2, 2]$

6 $H(x) = -x^2 + 5x - 4$ on $[0, 5]$

7 $f(x) = (x + 1)^2$ on $[-2, 1]$

8 $g(x) = |x - 2| + 3$ on $[1, 5]$

9 $h(x) = \begin{cases} x + 2 & \text{if } x < 1 \\ x^2 - 3x + 5 & \text{if } x \geq 1 \end{cases}$ on $[-6, 5]$

10 $f(x) = \begin{cases} 2x - 1 & \text{if } x \leq 2 \\ 2x^2 - 5 & \text{if } x > 2 \end{cases}$ on $[-3, 4]$

11 $g(x) = \begin{cases} \dfrac{1}{x + 1} & \text{if } x \neq -1 \\ 1 & \text{if } x = -1 \end{cases}$ on $[-2, 3]$

12 $H(x) = \begin{cases} |x - 2| & \text{if } x \neq 2 \\ 4 & \text{if } x = 2 \end{cases}$ on $[1, 4]$

13 $p(x) = |x - 1|$ on $[0, 3]$

14 $q(x) = \sqrt{8 - 2x - x^2}$ on $[-3, 2]$

15 $r(x) = \sqrt{1 + x}$ on $[-1, 8]$

16 $P(x) = \dfrac{5}{(x - 3)^2}$ on $[1, 5]$

17 $Q(x) = x^3 + 3x^2 - 9x$ on $[-5, 4]$

18 $R(x) = x^3 + 5x - 4$ on $[-4, 0]$

19 $F(x) = \dfrac{2x + 1}{2x - 3}$ on $[-1, 1]$

20 $G(x) = \dfrac{x + 4}{x - 2}$ on $[-4, 1]$

21 $H(x) = \dfrac{1}{x^2 + 1}$ on $[-2, 1]$

22 $f(x) = \dfrac{3}{x - 1}$ on $[0, \frac{2}{3}]$

23 $g(x) = \dfrac{x}{x^2 + 2}$ on $[-1, 4]$

24 $h(x) = \dfrac{3x}{\sqrt{4x^2 + 1}}$ on $[-1, 1]$

25 $f(x) = (x + 2)^{2/3}$ on $[-4, 3]$

26 $g(x) = 1 - (x - 2)^{2/3}$ on $[-5, 5]$

27 $s(x) = -2 \sin x$ on $\left[0, \dfrac{3\pi}{4}\right]$

28 $S(x) = \sin x - \cos x$ on $[0, \pi]$

29 $t(x) = x - \tan x$ on $\left[-\dfrac{\pi}{4}, \dfrac{\pi}{4}\right]$

30 $T(x) = 3 \cos 2x$ on $\left[\dfrac{\pi}{6}, \dfrac{3\pi}{4}\right]$

31 $f(x) = 4 - x^2$ on $(-\infty, \infty)$

32 $g(x) = x^2 - 2x - 8$ on $(-\infty, \infty)$

In Problems 33 to 50, find the maximum and minimum values (if they exist) of the function and indicate where they occur.

33 $f(x) = 3x^2 - 6x + 2$

34 $g(x) = 2x^3 - 3x^2 + 3$

35 $h(x) = x^3 - 12x + 5$

36 $F(x) = -x^3 - 6x^2 + 15x$

37 $G(x) = x^4 - 8x^2 + 8$

38 $H(x) = 3x^4 - 4x^3 - 12x^2$

39 $p(x) = x^5 - 5x^4$

40 $q(x) = 2x^5 - 10x^4$

41 $r(x) = \dfrac{x + 3}{x^2 + 16}$

42 $P(x) = \dfrac{x - 5}{x^2 - 16}$

43 $Q(x) = \dfrac{x^2}{1 + x^2}$

44 $R(x) = \dfrac{1 + x^2}{1 - x^2}$

45 $f(x) = -(x + 1)^{2/3}$

46 $g(x) = 1 - (1 + x)^{3/2}$

47 $h(x) = \sqrt{\dfrac{x}{1 - x}}$

48 $F(x) = \dfrac{x^3}{\sqrt{x^3 - 1}}$

49 $s(x) = \sqrt{1 + \sin^2 x}$

50 $S(x) = 1 + 2 \sin x$

51 One hundred animals belonging to an endangered species are placed in a protected game preserve. Suppose that the mathematical model

$$p = 100 \, \frac{t^2 + 5t + 25}{t^2 + 25}$$

is used to predict the population p of these animals in the preserve after t years. According to this model, after how many years is the population p a maximum?

52 Show that the absolute maximum value of the function

$$f(x) = \frac{1}{1 + |x|} + \frac{1}{1 + |x - 4|}$$

is $\frac{2}{3}$ and occurs when $x = 2$.

53 The range R in meters of a projectile is given by the formula $R = (v_0^2 \sin 2\theta)/g$, where v_0 is the muzzle velocity in meters per second, g is the acceleration of gravity in meters per second per second, and θ is the angle of elevation at which the projectile is fired. Find the angle of elevation that gives the maximum range.

54 The deflection of a beam 40 feet long supported at the ends and loaded at a point 30 feet from the left end is expressed by the equation

$$y = \frac{P}{3EI}\left(200x - \frac{x^3}{8}\right)$$

where P, E, and I are constants; x is the distance in feet from the left end; and $0 \le x \le 40$. Find the maximum deflection and where it occurs.

55 In medicine, it is often assumed that the **reaction** R to a dose of size x of a drug is given by an equation of the form $R = Ax^2(B - x)$, where A and B are certain positive constants. The **sensitivity** of the body to a dose of size x is defined to be the derivative dR/dx of the reaction with respect to the dose. (a) For what value of x is the reaction R maximum? (b) What is the maximum value of R? (c) For what value of x is the sensitivity dR/dx maximum?

©**56** A 4-ton weight is to be suspended from two identical cables fixed to points A and B (Figure 10). The distance between A and B is 36 feet, the perpendicular distance from the weight to the line $\overline{AB}$ is x feet, and the cable weighs 3 pounds per running foot. The resulting tension in the cable is given by

$$T = \frac{4000\sqrt{324 + x^2}}{x} + \frac{972}{x} + 3x \qquad \text{pounds}$$

Find the value of x that minimizes the tension. (*Hint:* You will have to use Newton's method to find an approximation to the critical number.)

Figure 10

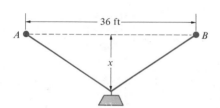

3.5   Asymptotes and Limits Involving Infinity

Figure 1

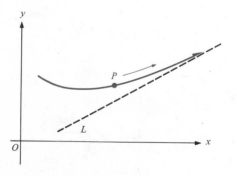

In Sections 3.2 to 3.4, we discussed monotonicity, concavity, points of inflection, and extreme points of a graph. Information about these features can be used to help sketch the graph. Here, we introduce another useful concept—the idea of an asymptote. This idea is illustrated in Figure 1, which shows a graph "approaching" a straight line L in the sense that, as the point P moves along the graph, the distance between P and L approaches zero. When this is so, we say that L is an **asymptote** of the graph. Although an asymptote isn't really part of the graph, it helps us to visualize how the graph behaves in distant regions of the xy plane.

As the point P in Figure 1 moves along the graph and approaches the line L, the distance $|\overline{OP}|$ between P and the origin O becomes larger and larger without bound, or, as we say, $|\overline{OP}|$ *approaches infinity*. Thus, in order to study asymptotes, we must introduce a new kind of limit—a *limit involving infinity*.

Figure 2

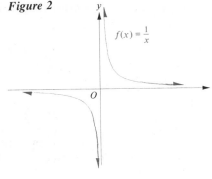

Limits at Infinity and Horizontal Asymptotes

The computer-generated graph of $f(x) = 1/x$ in Figure 2 shows that as x becomes larger and larger without bound, the function values $f(x)$ approach 0. We express this fact symbolically by writing

$$\lim_{x \to +\infty} \frac{1}{x} = 0$$

Also, as x decreases through negative values in such a way that $|x|$ increases without bound, the function values $f(x)$ again approach 0; symbolically,

$$\lim_{x \to -\infty} \frac{1}{x} = 0$$

More generally, if f is a function defined at least on an unbounded open interval of the form (a, ∞), the notation

$$\lim_{x \to +\infty} f(x) = B$$

means that, for each positive number ϵ, there exists a positive number N in the interval (a, ∞) such that

$$|f(x) - B| < \epsilon \qquad \text{holds whenever} \qquad x > N$$

In other words, $|f(x) - B|$ *can be made as small as we please by taking x to be sufficiently large*. Similarly, if f is a function defined at least on an unbounded open interval of the form $(-\infty, a)$, the notation

$$\lim_{x \to -\infty} f(x) = B$$

means that $|f(x) - B|$ *can be made as small as we please by taking x to be negative with $|x|$ sufficiently large* (see Problem 60). In other words,

$$\lim_{x \to -\infty} f(x) = B \qquad \text{means that} \qquad \lim_{t \to +\infty} f(-t) = B$$

Limits as $x \to +\infty$ or as $x \to -\infty$ are called **limits at infinity.**

Geometrically, the condition

$$\lim_{x \to +\infty} f(x) = B$$

means that as the point $P = (x, f(x))$ moves farther and farther, without bound, to the right along the graph of f, its y coordinate approaches B; in other words, *the line $y = B$ is a horizontal asymptote of the graph of f* (Figure 3). Likewise, the condition

$$\lim_{x \to -\infty} f(x) = b$$

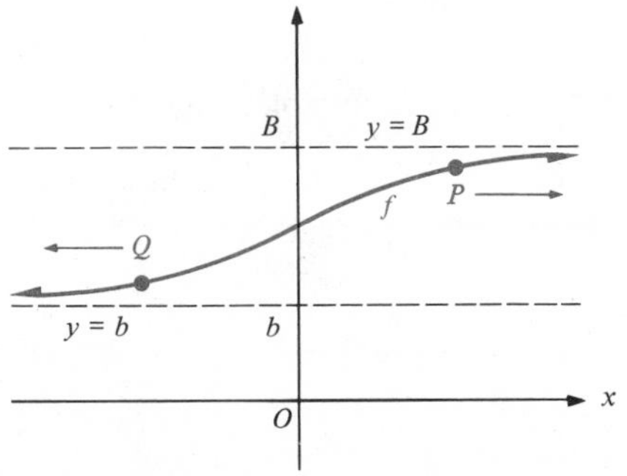

means that as the point $Q = (x, f(x))$ moves farther and farther, without bound, to the left along the graph of f, its y coordinate approaches b; in other words, *the line $y = b$ is a horizontal asymptote of the graph of f* (Figure 3). Evidently, the graph of a function can have *at most* two horizontal asymptotes—one corresponding to a limit as $x \to +\infty$ and one corresponding to a limit as $x \to -\infty$. As a matter of fact, a *rational* function can have at most *one* horizontal asymptote (Problem 62).

Figure 3

Figure 4

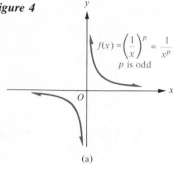

(a)

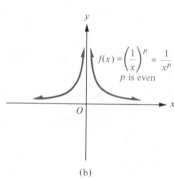

(b)

In calculating limits at infinity, it is often useful to keep in mind that for any *positive integer p*,

$$\lim_{x \to +\infty} \left(\frac{1}{x}\right)^P = \lim_{x \to +\infty} \frac{1}{x^p} = 0 \quad \text{and} \quad \lim_{x \to -\infty} \left(\frac{1}{x}\right)^P = \lim_{x \to -\infty} \frac{1}{x^p} = 0$$

These facts are illustrated for odd values of p in Figure 4a and for even values of p in Figure 4b.

In dealing with limits at infinity of rational functions, it is often helpful to divide numerator and denominator by the highest power of the independent variable appearing in the fraction. The following examples illustrate the techniques involved.

In Examples 1 to 3, find the horizontal asymptotes (if any) of the graph of the function.

__EXAMPLE 1__ $f(x) = \dfrac{5x^2}{2x^2 + 3}$

SOLUTION   We must calculate

$$\lim_{x \to +\infty} \frac{5x^2}{2x^2 + 3} \quad \text{and} \quad \lim_{x \to -\infty} \frac{5x^2}{2x^2 + 3}$$

As x gets large, both the numerator and the denominator of the fraction get large, making it difficult to say offhand just what happens to the fraction. However, dividing both the numerator and the denominator by x^2, we have

$$\frac{5x^2}{2x^2 + 3} = \frac{\dfrac{5x^2}{x^2}}{\dfrac{2x^2}{x^2} + \dfrac{3}{x^2}} = \frac{5}{2 + \dfrac{3}{x^2}}$$

As $x \to +\infty$, $1/x^2$ approaches 0 (Figure 4b), so $3/x^2$ approaches 0. Therefore,

$$\lim_{x \to +\infty} \frac{5x^2}{2x^2 + 3} = \lim_{x \to +\infty} \frac{5}{2 + \dfrac{3}{x^2}} = \frac{5}{2 + 0} = \frac{5}{2}$$

Similarly, as $x \to -\infty$, $1/x^2$ approaches 0 (Figure 4b), so $3/x^2$ approaches 0, and we have

$$\lim_{x \to -\infty} \frac{5x^2}{2x^2 + 3} = \lim_{x \to -\infty} \frac{5}{2 + \dfrac{3}{x^2}} = \frac{5}{2 + 0} = \frac{5}{2}$$

It follows that the line $y = \frac{5}{2}$ is a horizontal asymptote, and the only horizontal asymptote of the graph of

$$f(x) = \frac{5x^2}{2x^2 + 3}$$

Figure 5 shows a computer-generated graph of f. ∎

Figure 5

$$f(x) = \frac{5x^2}{2x^3 + 3}$$

__EXAMPLE 2__ $g(x) = \dfrac{2x}{\sqrt{x^2 + 4}}$

SOLUTION   Here the function is not rational; nevertheless, we try our trick of dividing numerator and denominator by some power of x. We divide by x itself, so that the numerator simplifies to 2. Thus, we have

$$\lim_{x \to +\infty} g(x) = \lim_{x \to +\infty} \frac{2}{\frac{1}{x}\sqrt{x^2 + 4}} = \lim_{x \to +\infty} \frac{2}{\sqrt{\frac{x^2 + 4}{x^2}}} = \lim_{x \to +\infty} \frac{2}{\sqrt{1 + \frac{4}{x^2}}} = 2$$

Therefore, $y = 2$ is a horizontal asymptote. For negative values of x, $\sqrt{1/x^2} = 1/(-x)$, and so, dividing numerator and denominator by $-x$, we have

$$\lim_{x \to -\infty} g(x) = \lim_{x \to -\infty} \frac{-2}{\frac{1}{-x}\sqrt{x^2 + 4}} = \lim_{x \to -\infty} \frac{-2}{\sqrt{\frac{x^2 + 4}{x^2}}}$$

$$= \lim_{x \to -\infty} \frac{-2}{\sqrt{1 + \frac{4}{x^2}}} = -2$$

Hence, $y = -2$ is another horizontal asymptote. Figure 6 shows a computer-generated graph of g. ■

EXAMPLE 3 $h(x) = 7x^2 + 3x^3$

SOLUTION   Factoring, we have

$$h(x) = x^2(7 + 3x)$$

As x grows larger and larger without bound, so does x^2, so does $7 + 3x$, and so does $h(x) = x^2(7 + 3x)$. We express this fact by writing

$$\lim_{x \to +\infty} h(x) = +\infty$$

Of course, $+\infty$ *is not a real number*, so the graph of h has no horizontal asymptote corresponding to $x \to +\infty$. If x is negative and $|x|$ grows larger and larger without bound, then x^2 grows larger and larger without bound. However, $7 + 3x$ becomes negative as soon as $x < -\frac{7}{3}$, so $h(x) = x^2(7 + 3x)$ becomes negative and its absolute value grows larger and larger without bound. We express this fact by writing

$$\lim_{x \to -\infty} h(x) = -\infty$$

Again, $-\infty$ *is not a real number*, so the graph of h has no horizontal asymptote at all. Figure 7 shows a computer-generated graph of h. ■

Infinite Limits and Vertical Asymptotes

Consider the rational function $f(x) = 4/(2x - 3)$. Note that f is defined everywhere except at $x = \frac{3}{2}$. Figure 8 shows a computer-generated graph of f. We see from the graph that as x approaches $\frac{3}{2}$ from the left, $f(x)$ decreases without bound. In fact, $f(x)$ can be made smaller than any preassigned number simply by taking x close enough to $\frac{3}{2}$ but less than $\frac{3}{2}$. This is expressed symbolically by writing

$$\lim_{x \to (3/2)^-} \frac{4}{2x - 3} = -\infty$$

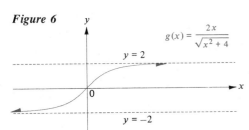

Figure 6

$$g(x) = \frac{2x}{\sqrt{x^2 + 4}}$$

$y = 2$

$y = -2$

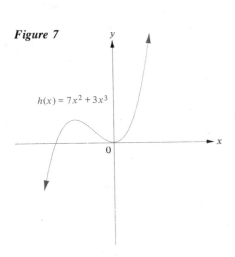

Figure 7

$$h(x) = 7x^2 + 3x^3$$

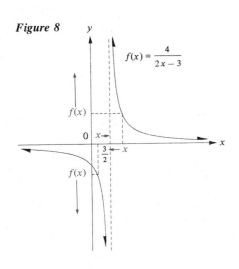

Figure 8

$$f(x) = \frac{4}{2x - 3}$$

Figure 8 also shows that as x approaches $\frac{3}{2}$ from the right, $f(x)$ increases without bound. In fact, $f(x)$ can be made larger than any preassigned number simply by taking x close enough to $\frac{3}{2}$ but greater than $\frac{3}{2}$. This is expressed symbolically by writing

$$\lim_{x \to (3/2)^+} \frac{4}{2x - 3} = +\infty$$

In general, there are four possibilities for such **one-sided infinite limits:**

1 The limit of $f(x)$ as x approaches a from the right is positive infinity; that is, $\lim_{x \to a^+} f(x) = +\infty$ (Figure 9a).

2 The limit of $f(x)$ as x approaches a from the left is positive infinity; that is, $\lim_{x \to a^-} f(x) = +\infty$ (Figure 9b).

3 The limit of $f(x)$ as x approaches a from the right is negative infinity; that is, $\lim_{x \to a^+} f(x) = -\infty$ (Figure 9c).

4 The limit of $f(x)$ as x approaches a from the left is negative infinity; that is, $\lim_{x \to a^-} f(x) = -\infty$ (Figure 9d).

In all four cases, *the vertical line $x = a$ is an asymptote of the graph of the function f.* If

$$\lim_{x \to a^+} f(x) = +\infty \qquad \text{and} \qquad \lim_{x \to a^-} f(x) = +\infty$$

we simply write

$$\lim_{x \to a} f(x) = +\infty$$

Likewise, the notation

$$\lim_{x \to a} f(x) = -\infty$$

is understood to mean that

$$\lim_{x \to a^+} f(x) = -\infty \qquad \text{and} \qquad \lim_{x \to a^-} f(x) = -\infty$$

In dealing with limits of functions of the form $f(x) = p(x)/q(x)$, keep in mind that if the denominator of a fraction is close to zero while the numerator is close to any number other than zero, the fraction will tend to have a large absolute value. More precisely:

If $\lim_{x \to a} p(x) = L \neq 0$ and $\lim_{x \to a} q(x) = 0$ then $\lim_{x \to a} \left| \dfrac{p(x)}{q(x)} \right| = +\infty$

Naturally, the same thing holds for limits from the right or for limits from the left.

In Examples 4 and 5, for the given function and the indicated value of a, find
(a) $\lim_{x \to a^+} f(x)$, **(b)** $\lim_{x \to a^-} f(x)$, *and* **(c)** $\lim_{x \to a} f(x)$ *if these limits are defined.*
(d) *Determine whether the graph of f has a vertical asymptote $x = a$.*

EXAMPLE 4 $f(x) = \tan x, \ a = \pi/2$

Figure 9

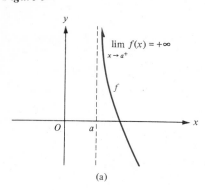

$\lim_{x \to a^+} f(x) = +\infty$

(a)

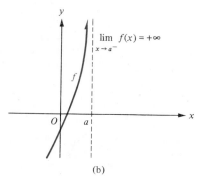

$\lim_{x \to a^-} f(x) = +\infty$

(b)

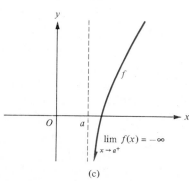

$\lim_{x \to a^+} f(x) = -\infty$

(c)

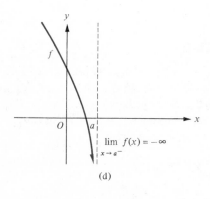

$\lim_{x \to a^-} f(x) = -\infty$

(d)

SOLUTION   We have $\tan x = \sin x / \cos x$. Note that $\lim\limits_{x \to \pi/2} \sin x = 1$ and $\lim\limits_{x \to \pi/2} \cos x = 0$; hence,

$$\lim_{x \to \pi/2} \left| \frac{\sin x}{\cos x} \right| = +\infty$$

(a) For values of x slightly larger than $\pi/2$, the numerator, $\sin x$, is positive and the denominator, $\cos x$, is negative; hence, the fraction $\sin x / \cos x$ is negative with a large absolute value. Therefore,

$$\lim_{x \to (\pi/2)^+} \tan x = \lim_{x \to (\pi/2)^+} \frac{\sin x}{\cos x} = -\infty$$

(b) For values of x slightly smaller than $\pi/2$, both the numerator and the denominator are positive; hence, the fraction $\sin x / \cos x$ is positive and large. Therefore,

$$\lim_{x \to (\pi/2)^-} \tan x = \lim_{x \to (\pi/2)^-} \frac{\sin x}{\cos x} = +\infty$$

(c) From (a) and (b), we have

$$\lim_{x \to (\pi/2)^+} \tan x = -\infty \qquad \text{whereas} \qquad \lim_{x \to (\pi/2)^-} \tan x = +\infty$$

Hence, $\tan x$ does not have a limit, finite or infinite, as $x \to \pi/2$.

(d) From the result of either (a) or (b), we can conclude that $x = \pi/2$ is a vertical asymptote of the graph of the tangent function (Figure 10). As a matter of fact, since $\tan x$ repeats itself periodically whenever x is increased or decreased by π, it follows that the graph of the tangent function has periodic asymptotes, each separated from the next by π units. ∎

Figure 10

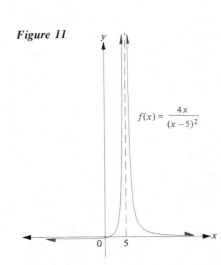

$y = \tan x$

EXAMPLE 5 $f(x) = \dfrac{4x}{(x - 5)^2}$, $a = 5$

Figure 11

$f(x) = \dfrac{4x}{(x - 5)^2}$

SOLUTION   The limit of the numerator is $\lim\limits_{x \to 5} 4x = 20$. As x approaches 5, the denominator, $(x - 5)^2$, approaches 0 through positive values; hence, for values of x near 5, the fraction is positive and very large. Therefore,

(a) $\lim\limits_{x \to 5^+} f(x) = +\infty$

(b) $\lim\limits_{x \to 5^-} f(x) = +\infty$

(c) $\lim\limits_{x \to 5} f(x) = +\infty$

(d) The graph of f has a vertical asymptote $x = 5$. As a matter of fact, the graph of f also has the x axis as a horizontal asymptote (Problem 25). Figure 11, shows a computer-generated graph of f. ∎

Locating Asymptotes

As we have seen, horizontal asymptotes of the graph of a function f, if they exist, can be located by calculating $\lim_{x \to +\infty} f(x)$ and $\lim_{x \to -\infty} f(x)$. To spot potential vertical asymptotes $x = a$ of a function of the form p/q, just look for values of a for which $q(a) = 0$. You must be careful, though—there may be such values of a that do not give vertical asymptotes, nor does this procedure always yield all possible vertical asymptotes (see Problem 63). However, if p and q are polynomial functions with no common factors of degree 1 or more, so that the fraction p/q is in reduced form, then the graph of $f = p/q$ has a vertical asymptote $x = a$ if and only if a is a zero of q.

EXAMPLE 6   Find all horizontal and vertical asymptotes of the graph of

$$f(x) = \frac{2x^2 + 1}{2x^2 - 3x}$$

SOLUTION   To find any horizontal asymptotes, we must calculate the limits of $f(x)$ as $x \to -\infty$ and as $x \to +\infty$. For this purpose, we divide numerator and denominator by x^2, so that

$$f(x) = \frac{2 + (1/x^2)}{2 - (3/x)}$$

This makes it clear that

$$\lim_{x \to -\infty} f(x) = \frac{2}{2} = 1 \quad \text{and} \quad \lim_{x \to +\infty} f(x) = \frac{2}{2} = 1$$

Hence, $y = 1$ is a horizontal asymptote of the graph of f.

To find vertical asymptotes, we must find the zeros of the denominator. Factoring the denominator,

$$2x^2 - 3x = x(2x - 3),$$

we find that the zeros are 0 and $\frac{3}{2}$. Now,

$$f(x) = \frac{2x^2 + 1}{x(2x - 3)}$$

and for values of x slightly smaller than 0, $2x^2 + 1$ is positive, x is negative, and $2x - 3$ is negative; hence, $f(x)$ is positive. It follows that

$$\lim_{x \to 0^-} f(x) = +\infty$$

and this is enough to show that $x = 0$ (the y axis) is indeed a vertical asymptote of the graph of f. Although it isn't needed for our present purposes, we notice that $f(x)$ is negative for values of x slightly larger than 0; hence,

$$\lim_{x \to 0^+} f(x) = -\infty$$

For values of x slightly smaller than $\frac{3}{2}$, $2x^2 + 1$ is positive, x is positive, and $2x - 3$ is negative; hence $f(x)$ is negative and

$$\lim_{x \to (3/2)^-} f(x) = -\infty$$

Figure 12

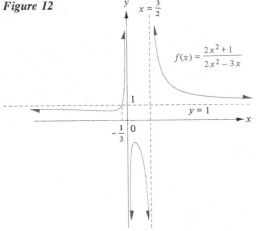

$f(x) = \dfrac{2x^2 + 1}{2x^2 - 3x}$

Therefore, the vertical line $x = \frac{3}{2}$ is also an asymptote of the graph of f. We note in passing that

$$\lim_{x \to (3/2)^+} f(x) = +\infty$$

Figure 12 shows a computer-generated graph* of f.   ■

If a function f is given or can be rewritten in the form

$$f(x) = mx + b + g(x)$$

where m and b are constants, $m \neq 0$, and

$$\lim_{x \to -\infty} g(x) = 0 \qquad \text{or} \qquad \lim_{x \to +\infty} g(x) = 0$$

then the line $y = mx + b$ is an asymptote of the graph of f. Such an asymptote, which is neither horizontal nor vertical, is called an **oblique asymptote.**

Figure 13

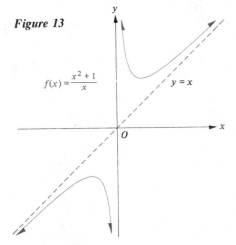

$f(x) = \dfrac{x^2 + 1}{x}$

EXAMPLE 7   Find the oblique asymptote of the graph of

$$f(x) = \frac{x^2 + 1}{x}$$

SOLUTION   Here,

$$f(x) = x + \frac{1}{x} \qquad \text{and} \qquad \lim_{x \to +\infty} \frac{1}{x} = 0$$

This shows that $y = x$ is an oblique asymptote of the graph of f. We note in passing that

$$\lim_{x \to -\infty} \frac{1}{x} = 0$$

also holds. Figure 13 shows a computer-generated graph of f.   ■

Vertical Tangent Lines

Limits involving infinity have uses other than locating asymptotes of graphs. For instance, the following definition uses the idea of an infinite limit.

DEFINITION 1   **Vertical Tangent Line**

If the function f is continuous at the number a and

$$\lim_{\Delta x \to 0} \left| \frac{f(a + \Delta x) - f(a)}{\Delta x} \right| = +\infty$$

we say that the vertical line $x = a$ is the **tangent line** to the graph of f at the point $(a, f(a))$.

*Notice that the graph of f crosses its own asymptote at the point $(-\frac{1}{3}, 1)$.

Figure 14

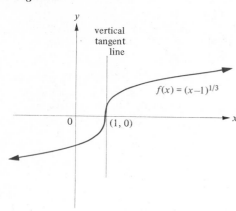

$f(x) = (x-1)^{1/3}$

EXAMPLE 8   Show that the graph of $f(x) = (x - 1)^{1/3}$ has a vertical tangent line at $(1, 0)$.

SOLUTION   Here, f is continuous at 1 and

$$\frac{f(1 + \Delta x) - f(1)}{\Delta x} = \frac{[(1 + \Delta x) - 1]^{1/3} - 0}{\Delta x}$$

$$= \frac{(\Delta x)^{1/3}}{\Delta x}$$

$$= \frac{1}{(\Delta x)^{2/3}}$$

Therefore,

$$\lim_{\Delta x \to 0} \left| \frac{f(1 + \Delta x) - f(1)}{\Delta x} \right| = \lim_{\Delta x \to 0} \left| \frac{1}{(\Delta x)^{2/3}} \right| = +\infty$$

and it follows that the graph of f has a vertical tangent line $x = 1$ at $(1, 0)$. Figure 14 shows a computer-generated graph of f. ∎

Problem Set 3.5

In Problems 1 to 24, evaluate the limit.

1 $\lim_{x \to +\infty} \dfrac{1 + 6x}{-2 + x}$

2 $\lim_{x \to -\infty} \dfrac{2x^2 + x + 1}{-4x^2 + 5x + 10}$

3 $\lim_{x \to +\infty} \dfrac{5x^2 - 7x + 3}{8x^2 + 5x + 1}$

4 $\lim_{x \to -\infty} \dfrac{7x^3 + 3x + 1}{x^3 - 2x + 3}$

5 $\lim_{x \to -\infty} \dfrac{x^{100} + x^{99}}{x^{101} - x^{100}}$

6 $\lim_{x \to +\infty} \dfrac{x^{99} + x^{98}}{x^{100} - x^{99}}$

7 $\lim_{t \to +\infty} \dfrac{8t}{\sqrt[4]{3t^4 + 5}}$

8 $\lim_{x \to -\infty} \dfrac{6x^2}{\sqrt[3]{5x^6 - 1}}$

9 $\lim_{x \to +\infty} (5x^2 - 3x)$

10 $\lim_{x \to -\infty} \dfrac{x^3 - 5x^2}{3x}$

11 $\lim_{x \to 1^+} \dfrac{2x}{x - 1}$

12 $\lim_{x \to 2^-} \dfrac{x^2}{x - 2}$

13 $\lim_{x \to 0^+} \dfrac{\sqrt{4 + 3x^2}}{5x}$

14 $\lim_{x \to 3^+} \dfrac{x^2 + 5x + 1}{x^2 - 2x - 3}$

15 $\lim_{x \to 4^-} \dfrac{2x^2 + 3x - 2}{x^2 - 3x - 4}$

16 $\lim_{t \to 5^-} \dfrac{\sqrt{25 - t^2}}{t - 5}$

17 $\lim_{x \to 1^-} \dfrac{x^2 - 1}{|x^2 - 1|}$

18 $\lim_{x \to 2^-} \dfrac{[\![2 - x]\!]}{2 - x}$

19 $\lim_{x \to 2^-} \dfrac{x^2 + 1}{x - 2}$

20 $\lim_{z \to 2^+} \dfrac{z^2 + 1}{z - 2}$

21 $\lim_{t \to -1^+} \left(\dfrac{3}{t + 1} - \dfrac{5}{t^2 - 1} \right)$

22 $\lim_{x \to -\infty} \dfrac{1 + \sqrt[5]{x}}{1 - \sqrt[5]{x}}$

23 $\lim_{x \to (\pi/2)^+} \sec x$

24 $\lim_{\theta \to +\infty} \dfrac{\sin \theta}{\theta}$

25 (a) Show that the x axis is a horizontal asymptote of the graph of $f(x) = 4x/(x - 5)^2$. (b) Show that the graph of f crosses its own *asymptote* at $(0, 0)$.

26 Find $\lim_{x \to +\infty} x \sin (1/x)$. (*Hint:* Let $t = 1/x$ and rewrite the limit in terms of a limit as $t \to 0^+$.)

In Problems 27 to 46, find all horizontal and vertical asymptotes of the graph of the function. Sketch the graphs in Problems 27 to 32 only.

27 $f(x) = \dfrac{7x}{2x - 5}$

28 $g(x) = \dfrac{5x}{3x + 1}$

29 $h(x) = \dfrac{1 - 2x}{3 + 5x}$

30 $F(x) = \dfrac{5x + 1}{3x - 2}$

31 $G(x) = \dfrac{-2}{(x - 1)^2}$

32 $H(x) = \dfrac{3}{(x + 1)^2}$

33 $p(x) = \dfrac{x^2}{x^2 + 4}$

34 $q(x) = \dfrac{x^2 + 1}{x^2 + 9}$

35 $r(x) = \dfrac{3x}{\sqrt{2x^2 + 1}}$

36 $P(x) = \dfrac{\sqrt[3]{25 - x^2}}{5 - x}$

37 $Q(x) = \sqrt{\dfrac{x}{x-2}}$

38 $R(x) = \dfrac{x+2}{\sqrt{1-x}}$

39 $u(x) = \dfrac{x^2 - 1}{x}$

40 $v(x) = \dfrac{4x^2 + 1}{x^3}$

41 $U(x) = \dfrac{3x^2 + 1}{2x^2 - 7x}$

42 $V(x) = \dfrac{4x^2}{\sqrt{x^2 + 5x + 4}}$

43 $f(x) = \cot x$

44 $g(x) = \sec x$

45 $h(x) = \csc x$

46 $F(x) = x \sin (1/x)$

In Problems 47 to 52, find the oblique asymptote of the graph of the function. Sketch the graphs in Problems 47 and 48 only.

47 $f(x) = \dfrac{x^2 - 1}{x}$

48 $g(x) = \dfrac{2x^2 - x + 1}{x}$

49 $h(x) = 3x - 2 + \dfrac{1}{x^2 + 1}$

50 $F(x) = \dfrac{2x^3 + x^2 + 5x + 1}{x^2 + 2}$ (*Hint:* Begin by using long division of polynomials to divide the numerator by the denominator, thus obtaining a quotient polynomial $Q(x)$ and a remainder polynomial $R(x)$. Then rewrite $F(x)$ as $Q(x) + [R(x)/(x^2 + 2)]$.)

51 $G(x) = \dfrac{x^2 + 2x}{x + 1}$

52 $H(x) = 2x + 2 - x \sin (1/x)$

In Problems 53 to 58, show that the graph of the function has a vertical tangent line at the indicated point.

53 $f(x) = 1 + \sqrt[3]{x}$ at $(0, 1)$

54 $g(x) = x + \sqrt[3]{x}$ at $(0, 0)$

55 $h(x) = 1 + (x - 1)^{2/3}$ at $(1, 1)$

56 $F(x) = -3 - (x - 2)^{2/3}$ at $(2, -3)$

57 $G(x) = -2 - \sqrt[5]{x - 1}$ at $(1, -2)$

58 $H(x) = (x + 1)^{1/3} x^{2/3}$ at $(-1, 0)$

©59 Use a calculator to check your answers to Problems 1 and 3 by evaluating the given expression for successively larger and larger values of x.

60 Write a formal definition involving positive numbers ϵ and N for the statement $\lim_{x \to -\infty} f(x) = B$.

©61 Use a calculator to check your answers to Problems 21 and 23.

62 Using long division of polynomials, you can rewrite any rational function in the form $f(x) = p(x) + [r(x)/q(x)]$, where $p(x)$, $r(x)$, and $q(x)$ are polynomials and the degree of $r(x)$ is less than the degree of $q(x)$. Use this fact to show that a rational function can have at most one horizontal asymptote.

63 (a) Show that even though $x = 0$ makes the denominator x of $(\sqrt{1 + x} - 1)/x$ zero, $x = 0$ is not a vertical asymptote of the graph of the function

$$f(x) = \frac{\sqrt{1 + x} - 1}{x}$$

(b) Show that even though $x = 0$ does not make the denominator $1 + x$ of

$$\frac{1 + \csc x}{1 + x}$$

zero, nevertheless, $x = 0$ is a vertical asymptote of the graph of the function

$$F(x) = \frac{1 + \csc x}{1 + x}$$

64 The amplitude $A(x)$ of a particle forced to oscillate in a resisting medium is given by the equation

$$A(x) = \frac{1}{\sqrt{(1 - x^2)^2 + kx^2}}$$

where x is the ratio of the forcing frequency to the natural frequency of oscillation and k is a constant that measures the damping caused by the resisting medium. (a) Sketch the graph of A for the values $k = 0$, $k = \frac{1}{2}$, and $k = 1$. (b) Show that the graph of A has a vertical asymptote when $k = 0$, but that for $0 < k < 4$ the graph of A has no vertical asymptote.

65 In physiology, **Wiess' law** relates the electric current I (in milliamperes) required to excite a living tissue and the duration t (in milliseconds) of the electrical stimulus by the equation

$$I = \frac{a + bt}{t}$$

where a and b are positive constants. (a) Find the horizontal asymptote of the graph of I as a function of t. (b) Explain why the constant b is called the **threshold strength** of the electrical stimulus.

66 Suppose that animal scientists use the mathematical model

$$W = 0.012 + \frac{3t^2 + 6t}{4t^2 + 7t + 8}$$

to predict the weight W (in kilograms) of a laboratory guinea pig t months after it is born. According to this model, the weight of the growing guinea pig asymptotically approaches the weight of an adult. What is the approximate weight of an adult guinea pig according to this model?

3.6   Graph Sketching

Nearly all the graphs shown in Sections 3.2 to 3.5 were computer-generated. In order to generate such graphs you need to be knowledgeable about graph-sketching techniques. Ordinarily, it is not possible to generate the *entire* graph of a function, so a program has to be developed to produce a suitable portion illustrating significant features of the graph. For this purpose, methods of calculus may be used to make a rough sketch of the desired graph. The techniques introduced in Sections 3.2 to 3.5 provide the basis for the following systematic procedure for sketching the graph of a function f:

Step 1 Determine the domain of f, the numbers at which f is discontinuous, and the zeros of f. Use this information and the procedure on page 152 to find the intervals on which f is positive or negative.

Step 2 Check whether f is even, odd, or neither.

Step 3 Find all asymptotes of the graph of f.

Step 4 Calculate the derivative f', find all critical numbers of f, and apply the procedure on page 152 to f' to find the intervals on which f' is positive or negative. Thus, find the intervals on which f is increasing or decreasing.

Step 5 Calculate the second derivative f'', and apply the procedure on page 152 to find the intervals on which f'' is positive or negative. Thus find the intervals on which the graph of f is concave upward or concave downward. Find all points of inflection of the graph.

Step 6 Locate all points on the graph of f at which a relative or absolute extremum occurs.

Step 7 Use the information gained in steps 1 through 6 together with the point-plotting method to sketch the graph of f.

The steps in this procedure are more or less independent of one another, so if you have difficulty with the details of any one step, it may be best to move on to the next. For instance, in step 1, if it proves difficult to find the zeros of f, it may be best to carry out the rest of the procedure and thus obtain a preliminary rough sketch of the graph. This rough sketch may enable you to find reasonable first approximations to the zeros of f. Then you can go back to step 1 and use, say, Newton's method to locate the zeros with greater precision. Finally, using this information, you can improve the accuracy of your preliminary rough sketch.

The graph-sketching procedure given above does not exhaust *all* the techniques for sketching graphs. As you practice, you may discover some tricks of your own. For instance, in carrying out step 1, it's usually a good idea, whenever possible, to select 0 as one of the test numbers for determining the algebraic signs of f. The number $f(0)$, which is often easy to calculate, is called the **y intercept** of the graph because it is the y coordinate of the point $(0, f(0))$ where the graph crosses the y axis. Some additional graph-sketching methods are presented in Chapters 7 and 9.

© *In Examples 1 to 5, sketch the graph of the function.*

__EXAMPLE 1__ $f(x) = (x - 1)^2(x - 4)$

SOLUTION   We carry out the seven steps of the graph-sketching procedure.

Step 1 Expanding the product $(x - 1)^2(x - 4)$, we see that

$$f(x) = x^3 - 6x^2 + 9x - 4$$

is a polynomial function; hence, it is defined and continuous at every real number. The advantage of the factored form is that it quickly provides the zeros of f, namely 1 and 4. Thus, f maintains constant algebraic signs on the three intervals $(-\infty, 1)$, $(1, 4)$, and $(4, \infty)$. We select, say, the three test numbers 0, 2, and 5 from these intervals and calculate: $f(0) = -4$, $f(2) = -2$, and $f(5) = 16$. This information will be useful in step 7, when we are ready to plot points. For now, it gives us the algebraic signs of f on the three intervals (Figure 1).

Step 2 Evidently, $f(x) = x^3 - 6x^2 + 9x - 4$ is neither even nor odd, so its graph is symmetric about neither the y axis nor the origin.

Step 3 Since f is a polynomial function, its graph has no asymptotes.

Step 4 Here,

$$\begin{aligned} f'(x) &= 3x^2 - 12x + 9 \\ &= 3(x^2 - 4x + 3) \\ &= 3(x - 1)(x - 3) \end{aligned}$$

Hence, the critical numbers for f are 1 and 3. Again using test numbers, we determine the algebraic signs of f' on the intervals $(-\infty, 1)$, $(1, 3)$, and $(3, \infty)$, as shown in Figure 2. Thus, f is increasing on $(-\infty, 1]$, decreasing on $[1, 3]$, and increasing again on $[3, \infty)$.

Step 5 Since $f'(x) = 3x^2 - 12x + 9$, we have

$$f''(x) = 6x - 12 = 6(x - 2)$$

Hence, f'' is defined at every real number, and $x = 2$ is the only solution of the equation $f''(x) = 0$. Evidently, f'' is negative on the interval $(-\infty, 2)$ and positive on the interval $(2, \infty)$ (Figure 3). It follows that the graph of f is concave downward on $(-\infty, 2)$, is concave upward on $(2, \infty)$, and has a point of inflection at $(2, f(2)) = (2, -2)$.

Step 6 Applying either the first- or the second-derivative test to the critical numbers obtained in step 4, we find that f takes on a relative maximum value of $f(1) = 0$ at 1 and a relative minimum value of $f(3) = -4$ at 3.

Step 7 Using the information in steps 1 through 6, we can already sketch a rough graph of f. By plotting a few additional points, we can improve the precision of our sketch until it becomes sufficiently accurate to suit our purposes. Figure 4 shows our final result.   ∎

Figure 1

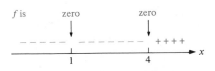

Figure 2

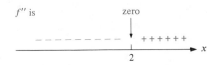

Figure 3

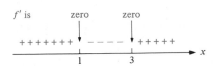

Figure 4

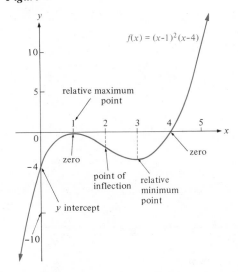

__EXAMPLE 2__ $g(x) = x^2 - \dfrac{1}{x}$

SOLUTION

Step 1 Here g is defined and continuous at every real number except 0. To find the zeros of g, we must solve the equation

$$x^2 - \frac{1}{x} = 0$$

that is,

$$x^2 = \frac{1}{x} \qquad \text{or} \qquad x^3 = 1$$

Thus, 1 is the only zero of g. Using test numbers, say, -2, $\frac{1}{2}$, and 2, we find that $g(-2) = \frac{9}{2}$, $g(\frac{1}{2}) = -\frac{7}{4}$, and $g(2) = \frac{7}{2}$. Hence, g is positive on $(-\infty, 0)$, negative on $(0, 1)$, and positive on $(1, \infty)$ (Figure 5).

Step 2 The function $g(x) = x^2 - (1/x)$ is neither even nor odd.

Step 3 We have

$$\lim_{x \to 0^-} \left(x^2 - \frac{1}{x} \right) = +\infty \qquad \text{and} \qquad \lim_{x \to 0^+} \left(x^2 - \frac{1}{x} \right) = -\infty$$

so the y axis is a vertical asymptote of the graph of g. Because

$$\lim_{x \to \pm\infty} \left(x^2 - \frac{1}{x} \right) = +\infty$$

the graph has no horizontal asymptote.

Step 4 Here $g'(x) = 2x + (1/x^2)$. To find the critical numbers of g, we must solve the equation

$$2x + \frac{1}{x^2} = 0$$

that is,

$$2x = \frac{-1}{x^2} \qquad \text{or} \qquad x^3 = -\frac{1}{2}$$

Thus, the only critical number of g is

$$-\sqrt[3]{\frac{1}{2}} = \frac{-1}{\sqrt[3]{2}} = -\frac{(\sqrt[3]{2})^2}{2}$$
$$= -\tfrac{1}{2}\sqrt[3]{4} \approx -0.79$$

Using test numbers, we obtain the information about the algebraic signs of g' shown in Figure 6. Thus, g is decreasing on $(-\infty, -\frac{1}{2}\sqrt[3]{4}]$, increasing on $[-\frac{1}{2}\sqrt[3]{4}, 0)$, and increasing on $(0, \infty)$.

Step 5 We have

$$g''(x) = 2 - \frac{2}{x^3}$$
$$= 2\left(1 - \frac{1}{x^3} \right)$$

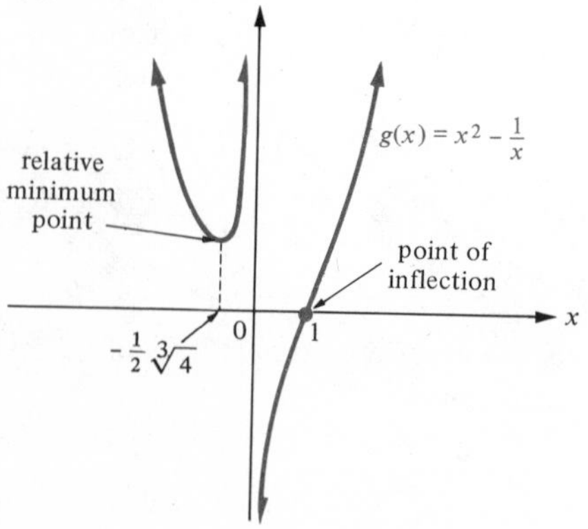

so that g'' is undefined at 0 and its only zero is 1. Using test numbers, we obtain the information shown in Figure 7. Thus, the graph of g is concave upward on $(-\infty, 0)$, is concave downward on $(0, 1)$, is concave upward on $(1, \infty)$, and has a point of inflection at $(1, g(1)) = (1, 0)$.

Step 6 By either the first- or the second-derivative test, g has a relative minimum value of $g(-\frac{1}{2}\sqrt[3]{4}) \approx 1.89$ at $-\frac{1}{2}\sqrt[3]{4}$.

Step 7 The graph of $g(x) = x^2 - (1/x)$ is sketched in Figure 8.

Figure 5

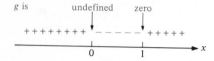

Figure 6

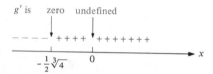

Figure 7

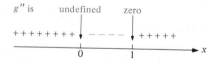

Figure 8

EXAMPLE 3 $h(x) = \dfrac{4x}{1 + x^2}$

SOLUTION

Step 1 Because the denominator $1 + x^2$ is never zero, the rational function h is defined and continuous at every real number. Evidently, the only zero of h is 0. Figure 9 shows the algebraic signs of h on the intervals $(-\infty, 0)$ and $(0, \infty)$.

Step 2 Evidently, $h(-x) = -h(x)$, so h is odd and its graph is symmetric about the origin.

Step 3 We have

$$\lim_{x \to \pm\infty} \frac{4x}{1 + x^2} = \lim_{x \to \pm\infty} \frac{4/x}{(1/x^2) + 1} = \frac{0}{0 + 1} = 0$$

so the x axis is a horizontal asymptote of the graph of h.

Step 4 Using the quotient rule and simplifying, we find that

$$h'(x) = \frac{4(1 - x^2)}{(1 + x^2)^2}$$

so the critical numbers of h are -1 and 1. The algebraic signs of h on the intervals $(-\infty, -1)$, $(-1, 1)$, and $(1, \infty)$ are shown in Figure 10. Thus, h is decreasing on $(-\infty, -1]$, increasing on $[-1, 1]$, and decreasing on $[1, \infty)$.

Step 5 Again using the quotient rule and simplifying, we find that

$$h''(x) = \frac{8x(x^2 - 3)}{(1 + x^2)^3}$$

so that $h''(x) = 0$ for $x = -\sqrt{3}$, $x = 0$, and $x = \sqrt{3}$. Figure 11 shows the algebraic signs of h'' on the intervals $(-\infty, -\sqrt{3})$, $(-\sqrt{3}, 0)$, $(0, \sqrt{3})$, and $(\sqrt{3}, \infty)$. Thus, the graph of h is concave upward on $(-\sqrt{3}, 0)$ and $(\sqrt{3}, \infty)$, whereas it is concave downward on $(-\infty, -\sqrt{3})$ and $(0, \sqrt{3})$. Consequently, the graph of h has points of inflection at $(-\sqrt{3}, h(-\sqrt{3})) = (-\sqrt{3}, -\sqrt{3})$, $(0, h(0)) = (0, 0)$, and $(\sqrt{3}, h(\sqrt{3})) = (\sqrt{3}, \sqrt{3})$.

Step 6 By either the first- or the second-derivative test, h has a relative minimum value of $h(-1) = -2$ at -1 and a relative maximum value of $h(1) = 2$ at 1.

Step 7 The graph is sketched in Figure 12.   ■

Figure 9

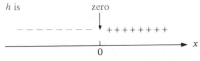

Figure 10

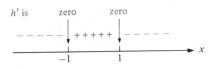

Figure 11

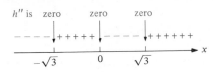

Figure 12

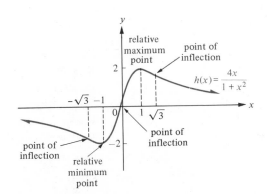

EXAMPLE 4 $F(x) = x^{1/3}(x-1)^{2/3}$

SOLUTION

Step 1 The function F is defined and continuous at every real number, and its zeros are 0 and 1. Figure 13 shows the algebraic signs of F on the intervals $(-\infty, 0)$, $(0, 1)$, and $(1, \infty)$.

Step 2 The function F is neither even nor odd.

Step 3 The graph of F has no horizontal or vertical asymptotes. As a matter of fact, it can be shown that $y = x - \frac{2}{3}$ is an oblique asymptote of the graph of F; however, because the proof is a bit tricky (Problem 38), we neglect this fact in working the rest of the example.

Step 4 Calculating the derivative F' and simplifying, we obtain

$$F'(x) = \frac{3x - 1}{3x^{2/3}(x-1)^{1/3}}$$

Thus, F' is undefined at 0 and 1. Also, $F'(x) = 0$ only for $x = \frac{1}{3}$, so 0, $\frac{1}{3}$, and 1 are the critical numbers for F. Figure 14 shows the intervals on which F' is positive or negative. Thus, F is increasing on $(-\infty, 0]$, $[0, \frac{1}{3}]$, and $[1, \infty)$, whereas it is decreasing on $[\frac{1}{3}, 1]$. Because F is increasing on the adjacent intervals $(-\infty, 0]$ and $[0, \frac{1}{3}]$, it is actually increasing on the entire interval $(-\infty, \frac{1}{3}]$.

Step 5 Calculating the second derivative F'' and simplifying, we obtain

$$F''(x) = \frac{-2}{9x^{5/3}(x-1)^{4/3}}$$

Thus, F'' is undefined at 0 and 1 and has no zeros. Figure 15 shows the intervals on which F'' is positive or negative. Thus, the graph of F is concave upward on $(-\infty, 0)$ and concave downward on $(0, 1)$ and $(1, \infty)$. Because the graph is concave upward on $(-\infty, 0)$ and concave downward on $(0, 1)$, there is a possible point of inflection at the point $(0, F(0)) = (0, 0)$. But, by Definition 2 on page 182, $(0, 0)$ can be a point of inflection only if the graph of F has a tangent line at $(0, 0)$. Since F' is not defined at 0, the only possibility is a *vertical* tangent line at $(0, 0)$. Now F is continuous at 0, and we have

$$\lim_{\Delta x \to 0}\left|\frac{F(0 + \Delta x) - F(0)}{\Delta x}\right| = \lim_{\Delta x \to 0}\left|\frac{(\Delta x)^{1/3}(\Delta x - 1)^{2/3}}{\Delta x}\right|$$

$$= \lim_{\Delta x \to 0}\left|\frac{\Delta x - 1}{\Delta x}\right|^{2/3}$$

$$= +\infty$$

Hence, by Definition 1 on page 201, the graph of F does have a vertical tangent line at $(0, 0)$. Therefore, $(0, 0)$ is indeed a point of inflection of the graph of F.

Step 6 By, say, the first-derivative test, the function F takes on a relative maximum value of $F(\frac{1}{3}) = \frac{1}{3}\sqrt[3]{4} \approx 0.53$ at $\frac{1}{3}$ and a relative minimum value of $F(1) = 0$ at 1.

Step 7 The graph is sketched in Figure 16.

Figure 13

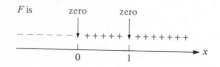

Figure 14

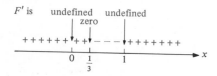

Figure 15

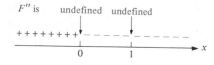

Figure 16

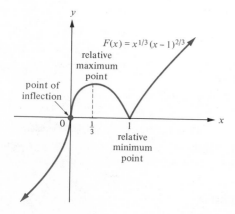

EXAMPLE 5 $G(x) = \sin^2 x + 2 \cos x$ for $-\pi \le x \le \pi$

SOLUTION

Step 1 The function G is continuous on $[-\pi, \pi]$. To find the zeros of G, we must solve the equation

$$\sin^2 x + 2 \cos x = 0$$

for x between $-\pi$ and π. Using, say, Newton's method, we find that there are two solutions, namely

$$x_1 \approx -2.00 \quad \text{and} \quad x_2 \approx 2.00$$

The algebraic signs of G on the intervals $[-\pi, x_1)$, (x_1, x_2), and $(x_2, \pi]$ are shown in Figure 17.

Step 2 Because $G(-x) = \sin^2(-x) + 2 \cos(-x) = \sin^2 x + 2 \cos x = G(x)$, it follows that G is an even function and its graph is symmetric about the y axis.

Step 3 The graph of G has no asymptotes.

Step 4 We have

$$G'(x) = 2 \sin x \cos x - 2 \sin x$$
$$= 2 \sin x (\cos x - 1)$$

Hence, the only critical number for G on the open interval $(-\pi, \pi)$ is 0. If we consider the one-sided derivatives at the endpoints of the domain of G, we also have $G'_+(-\pi) = 0$ and $G'_-(\pi) = 0$. Evidently (Figure 18), G is increasing on $[-\pi, 0]$ and decreasing on $[0, \pi]$.

Step 5 Using the fact that $2 \sin x \cos x = \sin 2x$ (see inside front cover), we can rewrite G' as $G'(x) = \sin 2x - 2 \sin x$. Thus,

$$G''(x) = 2 \cos 2x - 2 \cos x$$
$$= 2(\cos 2x - \cos x)$$

Because $\cos 2x = 2 \cos^2 x - 1$ (see inside front cover), we can rewrite G'' as

$$G''(x) = 2(2 \cos^2 x - 1 - \cos x)$$
$$= 2(2 \cos^2 x - \cos x - 1)$$
$$= 2(2 \cos x + 1)(\cos x - 1)$$

Therefore, $G''(x) = 0$ when $\cos x = -\frac{1}{2}$ and when $\cos x = 1$, that is (for $-\pi < x < \pi$), when $x = \pm 2\pi/3$ and when $x = 0$. The algebraic signs of G'' are shown in Figure 19. Thus, the graph of G is concave upward on $(-\pi, -2\pi/3)$ and $(2\pi/3, \pi)$, whereas it is concave downward on $(-2\pi/3, 0)$ and $(0, 2\pi/3)$. Consequently, the graph of G has points of inflection at $(-2\pi/3, G(-2\pi/3)) = (-2\pi/3, -\frac{1}{4})$ and at $(2\pi/3, G(2\pi/3)) = (2\pi/3, -\frac{1}{4})$. Notice that even though $G''(0) = 0$, the graph of G does *not* have a point of inflection at $(0, G(0)) = (0, 2)$.

Step 6 Checking the values of G at its critical number 0 and at the two endpoints of its domain, we find that $G(-\pi) = -2$, $G(0) = 2$, and $G(\pi) = -2$. Hence, G takes on an absolute maximum value of 2 at 0 and an absolute minimum value of -2 at $-\pi$ and π.

Step 7 The graph of G is sketched in Figure 20.

Figure 17

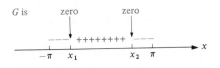

Figure 18

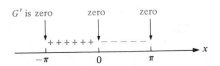

Figure 19

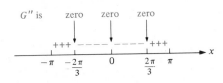

Figure 20

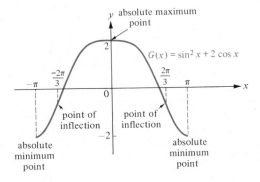

Problem Set 3.6

In Problems 1 to 36, sketch the graph of each function.

1 $f(x) = (x + 1)^2(x - 2)$

2 $g(x) = x^3 - 6x^2 + 9x - 4$

3 $h(x) = x^3 - 3x + 2$

4 $F(x) = 10 + 12x - 3x^2 - 2x^3$

5 $G(x) = 4x^2 - x^4$

6 $H(x) = x^4 - x^2 + 2$

7 $P(x) = (3 + x)^2(1 - x)^2$

8 $Q(x) = \frac{1}{3}x^3 + \frac{1}{2}x^2 - 2x + 1$

9 $f(x) = \frac{1}{12}(x^4 - 6x^3 - 18x^2)$

10 $g(x) = x^5 - 3x^4$

11 $h(x) = x + \dfrac{9}{x}$

12 $F(x) = x^2 + \dfrac{8}{x}$

13 $G(x) = x + \dfrac{3}{x^2}$

14 $H(x) = \dfrac{x + 2}{x - 2}$

15 $f(x) = \dfrac{3x}{x^2 + 9}$

16 $g(x) = \dfrac{x^2 + 4}{x^2 + 2}$

17 $h(x) = \dfrac{x^2 + 3x}{x + 4}$

18 $p(x) = \dfrac{x + 1}{x^2 + 4x + 5}$

19 $q(x) = \dfrac{(1 - x)^3}{2 - 3x}$

20 $r(x) = \dfrac{x^3}{x^2 + 2x + 4}$

21 $f(x) = 1 - (x - 2)^{2/3}$

22 $g(x) = (3 + x)^{1/3}(1 - x)^{2/3}$

23 $h(x) = (x + 1)^2 x^{1/3}$

24 $u(x) = \dfrac{x^{2/3}}{x + 1}$

25 $v(x) = \sqrt{x} - \dfrac{1}{\sqrt{x}}$

26 $w(x) = \dfrac{x}{\sqrt{x^2 + 1}}$

27 $U(x) = \dfrac{x^2 + 1}{\sqrt{x^2 + 4}}$

28 $V(x) = \sqrt{\dfrac{9 - x}{9 + x}}$

29 $f(x) = \sin 2x, \ 0 \le x \le 2\pi$

30 $g(x) = \sin x + \cos x, \ 0 \le x \le 2\pi$

31 $f(x) = \frac{1}{3} + \frac{2}{3} \cos 2x, \ 0 \le x \le 2\pi$

32 $h(x) = \sin 2x + 2 \cos x, \ -\pi \le x \le \pi$

33 $g(x) = x + 2 \cos (x/2), \ 0 \le x \le 2\pi$

34 $f(x) = x - \tan x, \ -\pi \le x \le \pi$

35 $f(x) = \sec x + \cos x, \ 0 \le x \le 2\pi$

36 $h(x) = 10 \csc x - 5 \cot x, \ -\pi < x < \pi$

37 In an engine, the distance y from the center of the wrist pin on a piston to the center of the crankshaft is a function of the angle x through which the crankshaft has turned from top dead center. If a denotes the stroke of the piston and b denotes the length of the connecting rod, we have $y = \frac{1}{2}(a \cos x + \sqrt{4b^2 - a^2 \sin^2 x})$. If $a = 8$ centimeters and $b = 16$ centimeters, sketch a graph of y as a function of x for $-\pi \le x \le \pi$.

38 Let $F(x) = x^{1/3}(x - 1)^{2/3}$ (Figure 16). (a) Show that for $x \ne 0$, $F(x) = x[1 - (1/x)]^{2/3}$. (b) For $x \ne 0$, let $\Delta t = 1/x$ and show that

$$F(x) - x = \frac{(1 - \Delta t)^{2/3} - 1}{\Delta t}$$

Conclude that

$$\lim_{x \to \pm\infty} [F(x) - x] = \lim_{\Delta t \to 0} \frac{(1 - \Delta t)^{2/3} - 1}{\Delta t}$$

(c) Let $G(t) = t^{2/3}$. Using the fact that $G'(1) = \frac{2}{3}$, show that

$$\lim_{\Delta t \to 0} \frac{(1 - \Delta t)^{2/3} - 1}{\Delta t} = -\frac{2}{3}$$

(d) Combine the results of parts (b) and (c) to show that $y = x - \frac{2}{3}$ is an oblique asymptote of the graph of F.

39 Use a calculator to help confirm that $y = x - \frac{2}{3}$ is an oblique asymptote of the graph of $F(x) = x^{1/3}(x - 1)^{2/3}$ (Figure 16, page 208) by calculating $F(x) - x$ for (a) $x = 100$, (b) $x = 1000$, and (c) $x = 1,000,000$. (d) Try the calculation for $x = 10^{10}$, and explain the curious result.

3.7   Applied Maximum and Minimum Problems

In Section 3.4, we applied the methods of calculus to solve some practical problems that required finding maximum or minimum values of certain quantities; however, in all these problems, the function to be maximized or minimized was given in advance. In this section, we continue our study of applied maximum and minimum problems, but now part of the problem will be to *write an appropriate function* on the basis of the information given in the problem. For working these problems, we recommend the following systematic procedure:

Step 1 Attack the problem with determination! Just "have at it" and you will often find that it is not as formidable as it appears. Begin by reading the problem carefully (several times if necessary). Make certain that you understand just which quantity is to be maximized or minimized.

Step 2 Select a suitable symbol for the quantity to be maximized or minimized; for purposes of this discussion, call it Q. Determine the remaining quantities or variables upon which Q depends, and select symbols for these variables. If it is feasible, draw a diagram and label the various parts with the corresponding symbols.

Step 3 Express the quantity Q whose extreme value is desired in terms of a formula involving the variables upon which Q depends. If the formula involves more than one of these variables, use the conditions given in the statement of the problem to find relationships among the variables that can be used to eliminate all but one independent variable from the formula.

Step 4 You now have $Q = f(x)$, where (for purposes of this discussion) x denotes the single variable upon which Q was found to depend in step 3 and f is the function determined by this dependence. If there are constraints on the quantity x imposed by the physical nature of the problem or by other practical considerations, specify these constraints explicitly. Apply the methods of Section 3.4 to find the desired absolute extremum of $f(x)$ subject to the imposed constraints on x.

Step 5 Reread the problem and make sure you have answered the specific question or questions asked.

In practice, the elimination of all but one of the variables upon which Q depends in step 3 is often the trickiest part of the procedure. Sometimes it cannot be carried out at all because not enough relationships among these variables are given by the statement of the problem. In this case, the procedure given above fails, and more sophisticated methods must be used. (See Sections 14.8 and 14.9.)

EXAMPLE 1   A recreation department plans to build a rectangular playground enclosed by 1000 meters of fence. What are the dimensions of the rectangular playground of maximum area?

SOLUTION   We carry out the suggested procedure:

Step 1 The quantity to be maximized is the area of the playground.

Step 2 Let A denote the area of the playground. Of course, A will depend on the length l and the width w of the playground. Here the diagram is so simple that it may seem ridiculous to bother drawing it. We do so anyway (Figure 1)

Figure 1

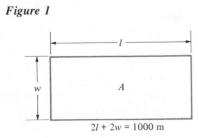

$2l + 2w = 1000$ m

because it fixes ideas, displays the notation, and gives us a start on the problem.

Step 3 Evidently, $A = lw$. Now, provided that we can find a suitable relationship between l and w, we may be able to eliminate either l or w from the formula $A = lw$. But only 1000 meters of fence are available, and the perimeter of the playground is $2l + 2w$ meters; so

$$2l + 2w = 1000 \qquad \text{or} \qquad l + w = 500$$

Solving the last equation for (say) l, we find that

$$l = 500 - w$$

Substituting $l = 500 - w$ into $A = lw$, we obtain

$$A = (500 - w)w = 500w - w^2$$

Step 4 We now have $A = f(w)$, where

$$f(w) = 500w - w^2$$

Since the dimensions w and l of the playground cannot be negative, we have $w \geq 0$ and $l = 500 - w \geq 0$; that is,

$$0 \leq w \leq 500$$

(Actually, w and l must be positive, but with the *closed* interval [0, 500] we can use the procedure on page 189, and no real harm comes from including the endpoints.) We must find the value of w that gives the absolute maximum value of $f(w) = 500w - w^2$ on the closed interval [0, 500]. Here,

$$f'(w) = 500 - 2w$$

so $w = 250$ gives the only critical number of f on the open interval (0, 500). Calculating the values of f at this critical number and at the endpoints 0 and 500, we find that $f(0) = 0$, $f(250) = 62,500$, and $f(500) = 0$. Hence, f takes on an absolute maximum value of 62,500 square meters when $w = 250$ meters and $l = 500 - w = 250$ meters.

Step 5 The specific question was: "What are the dimensions of the rectangular playground of maximum area?" The answer is: "The playground should be a square with length and width both equal to 250 meters."

After some experience in solving maximum-minimum problems, you may develop an intuition that suggests a somewhat informal treatment of the details of the suggested procedure. In fact, people often just "set the first derivative equal to zero" for a solution, perhaps "checking with the second derivative" to see whether a maximum or minimum is obtained. Of course, informal solutions must be viewed with skepticism since the desired extremum (if one exists at all) might occur at an *endpoint* and the second derivative test (when it works) indicates only a *relative* extremum. Thus, although we proceed more informally in the remainder of this section, we advise you to fill in all details whenever you have the slightest doubt.

© **EXAMPLE 2** Equal squares are cut off at each corner of a rectangular piece of cardboard 8 inches wide by 15 inches long, and an open-topped box is formed by turning up the sides (Figure 2). Find the length x of the sides of the squares that must be cut off to produce a box of maximum volume.

Figure 2

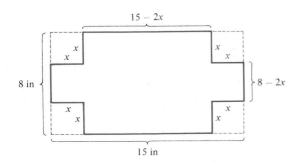

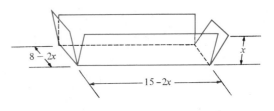

SOLUTION   The quantity to be maximized is the volume V of the box. From Figure 2, we see that the height of the box is x inches, the width is $8 - 2x$ inches, and the length is $15 - 2x$ inches. Thus,

$$V = x(8 - 2x)(15 - 2x) = 4x^3 - 46x^2 + 120x \qquad \text{for} \qquad 0 \le x \le 4$$

Here, $$\frac{dV}{dx} = 12x^2 - 92x + 120 = 4(x - 6)(3x - 5)$$

Setting dV/dx equal to zero and solving for x, we find that $x = \frac{5}{3}$ or $x = 6$. Because 6 does not belong to the interval $(0, 4)$, the only critical number in this interval is $\frac{5}{3}$. Evidently, V takes on a maximum value of $4(\frac{5}{3})^3 - 46(\frac{5}{3})^2 + 120(\frac{5}{3}) \approx 90.74$ cubic inches when $x = \frac{5}{3}$ inches. ■

EXAMPLE 3   A cylindrical stainless-steel boiler with an open top has a copper bottom. If the price of the copper per square unit is 5 times the price of the stainless steel per square unit, find the dimensions of the most economical boiler of fixed volume V cubic units.

Figure 3

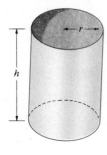

SOLUTION   Let C be the total cost of the materials for the boiler. Our problem is to find the radius r and the height h of the boiler of volume V for which C is minimum (Figure 3). Suppose the stainless steel costs k dollars per square unit. Then, the copper costs $5k$ dollars per square unit. The cost of the stainless steel for the boiler is k times the lateral surface area:

$$\text{cost of stainless steel} = k \cdot 2\pi rh = 2k\pi rh \qquad \text{dollars}$$

Likewise, the cost of the copper for the base is $5k$ times the area of the base:

$$\text{cost of copper} = 5k \cdot \pi r^2 = 5k\pi r^2 \qquad \text{dollars}$$

Thus, the total cost is given by

$$C = 2k\pi rh + 5k\pi r^2 \qquad \text{dollars}$$

Now, we must eliminate one of the two variables r or h from the last equation. This is accomplished by using the fact that the volume V of a cylinder is the area of the base times its height:

$$V = \pi r^2 h$$

Therefore,

$$h = \frac{V}{\pi r^2}$$

Substituting this into the equation for the total cost C, we find that

$$C = 2k\pi r\left(\frac{V}{\pi r^2}\right) + 5k\pi r^2 = \frac{2kV}{r} + 5k\pi r^2$$

Now, keeping in mind that V is constant, we have

$$\frac{dC}{dr} = -\frac{2kV}{r^2} + 10k\pi r$$

Setting dC/dr equal to zero, we obtain the equation

$$10k\pi r = \frac{2kV}{r^2}$$

or

$$10k\pi r^3 = 2kV$$

Hence, the critical value of r is

$$r = \sqrt[3]{\frac{2kV}{10k\pi}} = \sqrt[3]{\frac{V}{5\pi}}$$

Substituting this value of r into the equation $h = V/(\pi r^2)$ and simplifying, we find that

$$h = \sqrt[3]{\frac{25V}{\pi}}$$

It seems reasonable to assume that there *is* a most economical design for the boiler. Since we found only *one* critical number r, it is tempting to suppose that it produces the desired minimum cost, so that the most economical boiler has dimensions

$$r = \sqrt[3]{\frac{V}{5\pi}} \quad \text{and} \quad h = \sqrt[3]{\frac{25V}{\pi}}$$

It is important to realize that this conclusion is only an educated guess. If great importance were attached to the problem—for instance, if we were going into the boiler business—we would certainly complete the formal analysis to show that these dimensions really do produce an absolute minimum cost (Problem 36). ∎

Figure 4

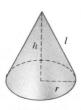

© **EXAMPLE 4**   A conical tent with no floor (Figure 4) is to have a capacity of 1000 cubic meters. Find the dimensions that minimize the amount of canvas required.

SOLUTION   The lateral surface area of the cone is given by

$$S = \pi r l = \pi r\sqrt{r^2 + h^2}$$

whereas the volume is given by

$$1000 = \tfrac{1}{3}\pi r^2 h$$

From the last equation,

$$h = \frac{3000}{\pi r^2}$$

Hence, $$S = \pi r\sqrt{r^2 + \left(\frac{3000}{\pi r^2}\right)^2} = \sqrt{\pi^2 r^4 + \left(\frac{3000}{r}\right)^2}$$

In order to minimize S, it is sufficient to minimize the quantity

$$Q = \pi^2 r^4 + \left(\frac{3000}{r}\right)^2$$

under the radical in the equation for S. Here, $r > 0$ and

$$\frac{dQ}{dr} = 4\pi^2 r^3 - \frac{2(3000)^2}{r^3}$$

Setting dQ/dr equal to zero, we find only one positive critical number, namely,

$$r = \sqrt[6]{\frac{3000^2}{2\pi^2}} \approx 8.77 \text{ m}$$

The corresponding value of h is given by

$$h = \frac{3000}{\pi r^2} \approx 12.41 \text{ m}$$

Again, if desired, it isn't difficult to supply a conclusive argument that these dimensions correspond to an absolute minimum amount of canvas (Problem 36). ■

Figure 5

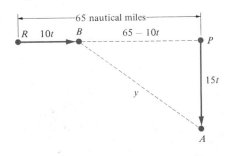

⌐65 nautical miles⌐

R $10t$ B $65 - 10t$ P

$15t$

y

A

© **EXAMPLE 5** Ship A is 65 nautical miles due east of ship B and is sailing south at 15 knots (nautical miles per hour), while ship B is sailing east at 10 knots. If the ships continue on their respective courses, find the minimum distance between them and when it occurs.

SOLUTION   In Figure 5, P shows the original position of ship A, and R shows the original position of ship B. After t hours have elapsed, B will have moved $10t$ nautical miles while A will have moved $15t$ nautical miles. By the Pythagorean theorem, the distance y between A and B at time t is given by

$$y = \sqrt{(15t)^2 + (65 - 10t)^2} = \sqrt{325t^2 - 1300t + 4225}$$

nautical miles. Clearly, y is minimum when the quantity

$$325t^2 - 1300t + 4225 = 325(t^2 - 4t + 13)$$

is minimum, that is, when the quantity

$$Q = t^2 - 4t + 13$$

is minimum. Here,

$$\frac{dQ}{dt} = 2t - 4$$

and the only critical number is $t = 2$. Since $dQ/dt < 0$ for $0 < t < 2$ and $dQ/dt > 0$ for $t > 2$, it follows that Q is decreasing for $0 \le t \le 2$ and increasing for $t \ge 2$. Therefore, Q, hence also y, takes on its absolute minimum value when $t = 2$ hours. The absolute minimum distance between the ships is therefore given by

$$y = \sqrt{30^2 + (65 - 20)^2} = \sqrt{2925} \approx 54.08 \text{ nautical miles}$$ ■

In engineering and physics, many laws state—or can be reformulated so as to state—that physical motions or transformations take place in such a way that certain quantities are maximized or minimized. For instance, in optics, the **principle of Fermat** states that light follows a path that minimizes the time of transit. Pierre de Fermat (1601–1665), a French jurist for whom mathematics was a hobby, discovered this "principle of least time" in 1657. He also did important pioneering work in analytic geometry, number theory, and the theory of probability. The following example shows how Fermat's principle can be used to solve a problem in optics.

Pierre de Fermat

Figure 6

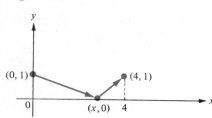

© **EXAMPLE 6**   In Figure 6, a light beam starts at the point $(0, 1)$ on the y axis, strikes a horizontal mirror along the x axis at the point $(x, 0)$, and is reflected back up to the point $(4, 1)$. Use Fermat's principle to find the value of x. (Assume that light travels with a constant velocity c.)

SOLUTION   The distance from $(0, 1)$ to $(x, 0)$ is given by

$$\sqrt{(x - 0)^2 + (0 - 1)^2} = \sqrt{x^2 + 1}$$

so the time required for the light beam to go from $(0, 1)$ to $(x, 0)$ is $\sqrt{x^2 + 1}/c$. Similarly, the time required for the light beam to go from $(x, 0)$ to $(4, 1)$ is given by

$$\frac{\sqrt{(4 - x)^2 + (1 - 0)^2}}{c} = \frac{\sqrt{x^2 - 8x + 17}}{c}$$

Therefore, the total time of transit from point $(0, 1)$ to point $(4, 1)$ is given by

$$T = \frac{1}{c}(\sqrt{x^2 + 1} + \sqrt{x^2 - 8x + 17})$$

Thus,

$$\frac{dT}{dx} = \frac{x}{c\sqrt{x^2 + 1}} + \frac{x - 4}{c\sqrt{x^2 - 8x + 17}}$$

Setting dT/dx equal to zero and solving for x, we obtain

$$x\sqrt{x^2 - 8x + 17} = -(x - 4)\sqrt{x^2 + 1}$$

$$x^2(x^2 - 8x + 17) = (x - 4)^2(x^2 + 1)$$

$$x^4 - 8x^3 + 17x^2 = x^4 - 8x^3 + 17x^2 - 8x + 16$$

so $8x = 16$ and $x = 2$. Thus, 2 is the only critical number for the continuous function expressing T in terms of x. Clearly $0 \le x \le 4$. When $x = 0$, $T = (1/c)(1 + \sqrt{17}) \approx 5.12/c$; when $x = 2$, $T = (1/c)(\sqrt{5} + \sqrt{5}) \approx 4.47/c$; and when $x = 4$, $T = (1/c)(\sqrt{17} + 1) \approx 5.12/c$. Therefore, T assumes its absolute minimum value when $x = 2$. ∎

The following example is conceptually similar to Example 6 but involves a different physical situation.

Figure 7

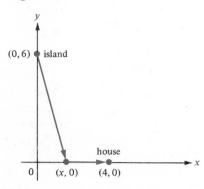

© **EXAMPLE 7**   Juan lives on an island 6 kilometers away from a straight beach, and his friend Rita lives 4 kilometers up the beach. Juan can row his boat 3 kilometers per hour, and he can walk 5 kilometers per hour on the beach. Find the minimum time required for Juan to reach Rita's house from the island.

SOLUTION   Establish a coordinate system with the straight beach running along the x axis and with the island at the point $(0, 6)$ on the y axis (Figure 7). Then Rita's house is located at the point $(4, 0)$ along the x axis. Suppose that Juan rows his boat from his island to the point $(x, 0)$ on the beach and then walks from $(x, 0)$ to Rita's house. Reasoning just as in Example 6, but taking into account the different rates of speed for rowing and walking, we see that the time of transit is given by

$$T = \frac{\sqrt{(0 - 6)^2 + (x - 0)^2}}{3} + \frac{\sqrt{(4 - x)^2 + (0 - 0)^2}}{5}$$

that is, $\quad T = \dfrac{\sqrt{x^2 + 36}}{3} + \dfrac{4 - x}{5} \qquad$ for $0 \le x \le 4$

Thus,

$$\frac{dT}{dx} = \frac{x}{3\sqrt{x^2 + 36}} - \frac{1}{5}$$

Setting dT/dx equal to zero and solving for x, we find that

$$5x = 3\sqrt{x^2 + 36}$$

$$25x^2 = 9(x^2 + 36)$$

$$16x^2 = 324$$

$$4x^2 = 81$$

and so $x \doteq \pm\frac{9}{2}$. Here, $-\frac{9}{2}$ turns out to be an extraneous root introduced by squaring, and $\frac{9}{2}$ does not belong to the interval $(0, 4)$. Therefore, there are no critical numbers in the interval $(0, 4)$, and so we must have an *endpoint extremum*. When $x = 0$, we have

$$T = \frac{\sqrt{0^2 + 36}}{3} + \frac{4 - 0}{5} = 2 + \frac{4}{5} = 2.8 \text{ hours}$$

whereas for $x = 4$, we have

$$T = \frac{\sqrt{4^2 + 36}}{3} + \frac{4 - 4}{5} = \frac{2\sqrt{13}}{3} \approx 2.4 \text{ hours}$$

Thus, for least time of transit, approximately 2.4 hours, Juan should row directly to Rita's house.

Problem Set 3.7

1 A rectangular observation deck is to be built overlooking a scenic lake. Find the dimensions of the deck if its area is to be 100 square meters and its perimeter is to be a minimum.

2 A rancher with 10,000 meters of available fencing intends to enclose a rectangular field adjacent to a straight river. If the side along the river requires no fencing, find the dimensions of the field with largest area.

3 Find two nonnegative real numbers whose sum is 20 and whose product is (a) maximum and (b) minimum.

4 A real estate developer wishes to enclose a rectangular plot by a fence and then divide it into two plots by another fence parallel to one of the sides. What are the dimensions of the largest area that can be enclosed by using a total of 1800 meters of fencing?

5 Advertising fliers are to be made from rectangular sheets of paper that contain 400 square centimeters of printed message. If the margins at the top and bottom are each 5 centimeters and the margins at the sides are each 2 centimeters, what should be the dimensions of the fliers if the total page area is to be a minimum?

6 In New Mattoon, two straight roads and a highway mark off a triangular piece of land (Figure 8). The side of the triangle lying along the highway is 100 meters long, and the perpendicular distance from the intersection of the two roads to the highway is 80 meters. Within this triangle, the city intends to zone a rectangular plot of land, fronting along the highway, for commercial use. The park department will plant shrubs in the remaining portion of the triangle. What is the maximum possible area of the rectangular plot of commercially zoned land?

Figure 8

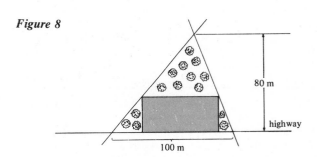

7 A small college is planning to build a jogging track and a field house. The jogging track is to be 1 kilometer in circumference and will consist of two parallel, straight segments connected by two semicircles (Figure 9). The field house will be built on the rectangular plot of land enclosed by the two parallel segments of track and the diameters of the two semicircles. Find the area of the largest possible rectangular plot of land for the field house.

Figure 9

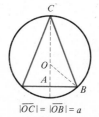

field house

8 A wire 24 centimeters long is to be cut into two pieces. A circle will be formed from one piece and a square from the other. How much wire should be used to form the square if the total area enclosed by the square and the circle is to be (a) a minimum and (b) a maximum?

9 A child's sandbox is to be made by cutting equal squares from the corners of a square sheet of galvanized iron and turning up the sides. If each side of the sheet of galvanized iron is 2 meters long, what size squares should be cut from the corners to maximize the volume of the sandbox?

10 A right circular cone is to be inscribed in a sphere of fixed radius *a* units. Figure 10 shows a vertical cross section through the center O of the sphere. Find the height $h = |\overline{AC}|$ and the radius $r = |\overline{AB}|$ of the base of the cone with maximum volume V. (*Hint:* Of course, your answer will depend on the radius *a* of the sphere.)

Figure 10

$$|\overline{OC}| = |\overline{OB}| = a$$

C **11** A standard "55-gallon" oil drum is a cylinder, with closed top and bottom, of radius 11 inches and height 34 inches. Do these dimensions minimize the amount of metal in the drum for a given fixed volume $V = 4114\pi$ cubic inches? If not, what dimensions would minimize the amount of metal in the drum? (Neglect the metal used in the crimped seams of the drum.)

12 The Norman style of architecture is characterized by massive construction, carved decorations, and rectangular windows and doors surmounted by semicircular arches. Find the height *h* and the width *w* of a Norman door of maximum area if its perimeter is 7 meters (Figure 11).

Figure 11

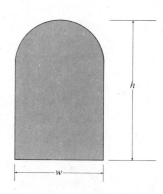

13 A rectangular box with a square base is to be constructed so as to have a volume of 1 cubic meter. Find the height *h* and the length *x* of an edge of the base of such a box if the amount of material used for its construction is to be minimum and (a) the box has an open top, (b) the box has a closed top. (Neglect the thickness of the material and extra material needed for making seams.)

14 In Problem 13, suppose that the material for the bottom and (if there is one) the top of the box costs *c* times as much per square meter as the material for the sides. Find the dimensions of the most economical box.

15 A rod-shaped bacterium has the form of a circular cylinder of length *l* and radius *r*, closed at the ends by two hemispheres (Figure 12). Find the dimensions *l* and *r* of a rod-shaped bacterium of fixed volume *V* if its total surface area is to be a minimum. (*Hint:* The volume of a sphere of radius *r* is $\frac{4}{3}\pi r^3$ cubic units, and its surface area is $4\pi r^2$ square units.)

Figure 12

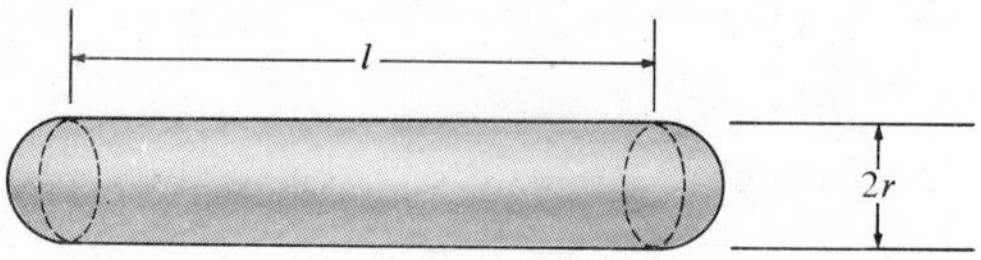

16 The logo for a world's fair is to consist of a transparent right circular cone circumscribing a sphere (Figure 13). For a sphere of fixed radius *a*, find the height *h* and the base radius *r* of the circumscribing cone of minimum volume.

Figure 13

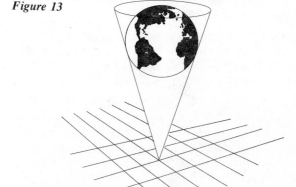

17 A cable television company has its master antenna located at point A on the bank of a straight river 1 kilometer wide (Figure 14). It is going to run a cable from A to a point P on the opposite bank of the river and then straight along the bank to a town T situated 3 kilometers downstream from A. It costs \$15 per meter to run the cable under the water and \$9 per meter to run the cable along the bank. What should be the distance from P to T in order to minimize the total cost of the cable?

Figure 14

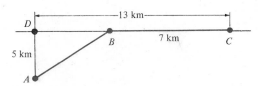

18 In Problem 17, suppose there is an increase in the cost of laying underwater cable, but the cost of laying cable along the bank remains \$9 per meter. Is there a cost for laying underwater cable for which it is most economical to run the cable straight across the river from point A to point Q?

©**19** An offshore oil well is located at point A, which is 13 kilometers from the nearest point Q on a straight shoreline. The oil is to be piped from A to a terminal at a point T on the shoreline by piping it straight under water to a point P on the shoreline between Q and T and then to T by a pipe along the shoreline. Suppose that the distance $|\overline{QT}|$ is 10 kilometers, that it costs \$90,000 per kilometer to lay underwater pipe, and that it costs \$60,000 per kilometer to lay the pipe along the shoreline. What should be the distance from P to T in order to minimize the cost of laying the pipe?

©**20** Ship A is sailing north at 20 knots, while ship B is sailing west at 30 knots. At the start, ship B is 50 nautical miles east of ship A. If they continue their respective courses, find the minimum distance between the ships.

21 If a resistor of R ohms is connected across a battery of E volts with an internal resistance of r ohms, a current of I amperes will flow and generate P watts of power. Here P is given by $P = I^2R$, where $I = E/(R + r)$. Find the resistance R if the power P is to be maximum.

©**22** Ornithologists have determined that some species of birds prefer to fly over land, when possible, rather than over water. It is believed that this may have to do with unfavorable air currents that are more prevalent over water. Suppose that a certain species of bird, released on an island A which is 5 kilometers from the nearest point D along a straight shoreline, flies straight to the point B on the shoreline and then flies along the shoreline the remaining distance $|\overline{BC}| = 7$ kilometers to its nesting area C, which is 13 kilometers from point D (Figure 15). Assume that the bird instinctively chooses a flight path that minimizes its total energy expenditure. If the bird uses w joules of energy per kilometer to fly over water and l joules of energy per kilometer to fly over land, find the ratio w/l.

Figure 15

23 A farmer is planning to construct a silo in the form of a circular cylinder capped by a hemisphere. Material for the hemisphere will cost twice as much per square unit as material for the cylinder. Find the ratio of the total height of the silo to its radius that provides the most economical structure for a given fixed volume. Neglect the thickness of the silo and the waste in construction.

24 Suppose that the concentration, in parts per million, of particulate matter in the atmosphere caused by an industrial plant is proportional to the rate at which the plant discharges the particulate matter and inversely proportional to the square of the distance from the plant. If two industrial plants A and B are located 12 kilometers apart, and if plant A emits particulate matter 8 times as fast as plant B, at what distance x from A along the line segment $\overline{AB}$ will the atmospheric concentration of particulate matter be minimum?

25 In sociology, a simple model for the spread of a rumor predicts that the rate R at which the rumor is spreading in a population of size N is proportional to the product of the number of people x in the population who have already heard the rumor and the number of people $N - x$ who have not yet heard it. According to this model, what fraction of the population will have already heard the rumor when the rumor is being spread the fastest? (Although x represents a whole number here, you may treat it as a continuous variable.)

26 The strength of a rectangular beam is proportional to the product of its width and the square of its depth. Find the dimensions of the strongest beam that can be cut from a circular cylindrical log (with bark removed) of radius 10 centimeters.

27 Light travels with speed c in air and with (slower) speed v in water. Figure 16 shows the path by which a ray of light travels from a point A in the air to a point B on the surface of the water and then to a point W in the water. Here, α, the *angle of incidence*, is the angle between $\overline{AB}$ and the normal to the surface of the water, while β, the *angle of refraction*, is the angle between $\overline{BW}$ and this normal.

(a) Show that the total time required for the light to go from A to W is

$$T = \frac{\sqrt{a^2 + x^2}}{c} + \frac{\sqrt{b^2 + (k - x)^2}}{v}$$

where $a = |\overline{AD}|$, $x = |\overline{DB}|$, $b = |\overline{EW}|$, and $k = |\overline{DE}|$.

(b) Show that when x has the value that makes T minimum, then $(\sin \alpha)/\sin \beta = c/v$ (**Snell's law of refraction**).

Figure 16

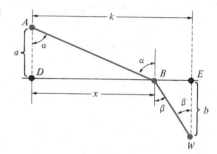

28 A plastic cup in the shape of a right circular cone is to have a slant height of 8 centimeters (Figure 17). Find the vertex angle θ for which the volume of the cone will be a maximum.

Figure 17   **Figure 18**

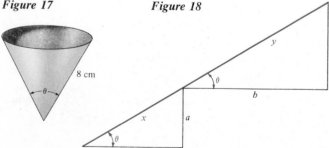

29 In Figure 18, assume that a and b are positive constants and that θ, x, and y are variables.

(a) Express x and y in terms of θ, a, and b.

(b) Show that $x + y$ is minimum when $\tan \theta = \sqrt[3]{a/b}$.

(c) Using the result of part (b), show that the minimum value of $x + y$ is $(a^{2/3} + b^{2/3})^{3/2}$.

[C] **30** An iron pipe of length l is to be carried horizontally around the corner from a corridor of width 1 meter into a corridor of width 1.5 meters (Figure 19). Find the largest value of l for which this can be accomplished without bending the pipe. Neglect the thickness of the pipe. (*Hint:* Use the results of Problem 29.)

Figure 19

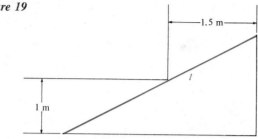

[C] **31** A steel cable to guy an electric pole is to pass over a 3-meter wall (Figure 20). If the wall is 8 meters from the pole, what is the length l of the shortest cable that will serve this purpose? (*Hint:* Use the results of Problem 29.)

Figure 20

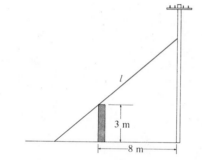

[C] **32** A spherical ball of radius r settles slowly into a full conical glass of water, causing the water to overflow. The line segments connecting the vertex of the cone to the points on the upper rim each make an angle of θ radians with the central axis of the glass. If the conical glass has height a, find the value of r for which the amount of overflow is a maximum.

33 A rain gutter with a cross section in the shape of an isosceles trapezoid (Figure 21) is to have sides of length 8 centimeters and a horizontal bottom of length 8 centimeters. If the sides make equal angles θ with the horizontal, find the value of θ that produces a trapezoid of maximum area.

Figure 21

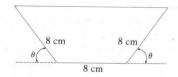

34 Figure 22 shows a first circle of radius a and part of a second circle of radius r whose center is on the first circle. Let l be the length of the arc of the second circle that lies inside the first circle. Show that l is a maximum when the angle θ satisfies the equation $\cot \theta = \theta$. (Here a is a constant, while θ and r can vary.)

Figure 22

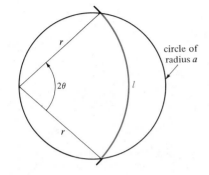

circle of radius a

35 Find the dimensions of the right triangle ABC with hypotenuse $|\overline{AC}| = c$, a given constant, if the area of the triangle is a maximum.

36 In Examples 3 and 4, pages 213 to 215, complete the formal analyses to show that the given solutions are correct.

37 A sodium vapor lamp L is to be placed on top of a pole of height x meters to furnish illumination at a busy traffic intersection T (Figure 23). The foot P of the pole must be located 30 meters from T. If $r = |\overline{LT}|$ is the distance from the lamp to the point T and α is the angle PTL, then the intensity of illumination I at T will be proportional to the sine of α and inversely proportional to r^2; thus, $I = c (\sin \alpha)/r^2$, where c is a constant. Find the value of x that maximizes I.

Figure 23

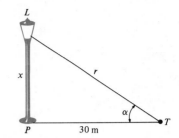

38 In Figure 24, a light beam starts at the point $(p, 0)$ on the x axis, strikes the circular mirror $x^2 + y^2 = 1$ at the point R, and is reflected to the point $(0, q)$ on the y axis. Use Fermat's principle to show that the angle of incidence α is equal to the angle of reflection β.

Figure 24

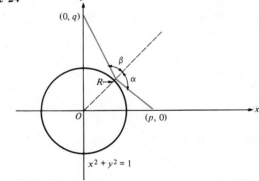

3.8   Related Rates

In this section we use derivatives to solve problems involving the rates of change of quantities that vary in time. In a typical *related-rates* problem, the variable quantities are related by an equation that either is given or can be derived from information supplied in the statement of the problem. The rates of change with respect to time of all but one of these quantities are also given, and the problem is to find the rate of change with respect to time of the remaining quantity. This is accomplished by differentiating the equation implicitly (see Section 2.8) with respect to time, substituting the known values of the variables and their time derivatives into the resulting equation, and then solving for the value of the unknown time derivative. The method is illustrated in the following examples.

Ⓒ **EXAMPLE 1**   Petroleum from a leaking offshore well forms a circular oil slick on the surface of the water. If the area of the oil slick is increasing at the rate of 5 square kilometers per day, at what rate is the radius of the oil slick increasing at the instant when the radius is 3 kilometers?

Figure 1

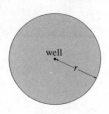

SOLUTION   Let A be the area of the oil slick, and let r be its radius (Figure 1). Then

$$A = \pi r^2$$

Here the area A and the radius r are functions of time t in days since the beginning of the leak. We want to find dr/dt when $r = 3$. Differentiating both sides of the equation $A = \pi r^2$ with respect to t, we find that

$$\frac{dA}{dt} = 2\pi r \frac{dr}{dt}$$

We are given that

$$\frac{dA}{dt} = 5 \text{ km}^2/\text{day}$$

Hence, at the instant when $r = 3$ kilometers, we have

$$5 = 2\pi(3) \frac{dr}{dt}$$

so that

$$\frac{dr}{dt} = \frac{5}{6\pi} \approx 0.265 \text{ km/day}$$

EXAMPLE 2   A guard 6 feet tall is walking away from a lamppost that is 20 feet high. If he is walking at the rate of 4.4 feet per second, at what rate is the length of his shadow increasing when he is 12 feet away from the base of the lamppost?

SOLUTION   In Figure 2, let x denote the length in feet of the guard's shadow at t seconds, and let y denote the distance in feet of the guard from the lamppost at t seconds. We want to find dx/dt when $y = 12$. In Figure 2, triangle ACE is similar to triangle BCD; hence,

Figure 2

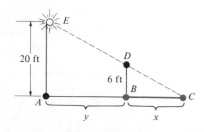

$$\frac{y + x}{20} = \frac{x}{6}$$

Therefore,

$$6y + 6x = 20x \quad \text{or} \quad 7x = 3y$$

Differentiating both sides of the last equation with respect to t, we find that

$$7 \frac{dx}{dt} = 3 \frac{dy}{dt}$$

We are given that

$$\frac{dy}{dt} = 4.4 \text{ ft/s}$$

Hence,

$$7 \frac{dx}{dt} = 3(4.4)$$

and it follows that

$$\frac{dx}{dt} = \frac{3(4.4)}{7} \approx 1.89 \text{ ft/s}$$

Note that we did not need to use the information that the guard is 12 feet away from the lamppost. In other words, the rate dx/dt at which the guard's shadow is lengthening is a constant, independent of his distance from the lamppost.

Figure 3

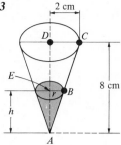

© **EXAMPLE 3**   Water is running out of a conical funnel at the constant rate of 3 cubic centimeters per second. The funnel has a radius of 2 centimeters and a height of 8 centimeters (Figure 3). How fast is the water level dropping when it is 3 centimeters from the top of the funnel?

SOLUTION   Let t denote the time in seconds that has elapsed since the water started to run out of the funnel, and let V denote the volume in cubic centimeters of the water in the funnel at t seconds. In Figure 3, let $h = |\overline{AE}|$ and $r = |\overline{EB}|$. We must find dh/dt at the instant when $|\overline{DE}| = 3$, that is, when $h = 8 - |\overline{DE}| = 5$. Since triangle ADC is similar to triangle AEB, we have

$$\frac{|\overline{AD}|}{|\overline{AE}|} = \frac{|\overline{DC}|}{|\overline{EB}|}$$

so that

$$r = |\overline{EB}| = |\overline{DC}|\, \frac{|\overline{AE}|}{|\overline{AD}|} = 2\left(\frac{h}{8}\right) = \frac{h}{4}$$

At any time t, the volume V of water in the funnel can be expressed as the volume of a cone (Figure 3):

$$V = \frac{1}{3}\pi r^2 h = \frac{1}{3}\pi\left(\frac{h}{4}\right)^2 h$$

Hence,

$$V = \frac{\pi}{48}h^3$$

Differentiating both sides of the last equation with respect to t, we find that

$$\frac{dV}{dt} = \frac{3\pi}{48}h^2\,\frac{dh}{dt}$$

Since V is *decreasing* at the rate of 3 cubic centimeters per second, we have

$$\frac{dV}{dt} = -3 \text{ cm}^3/\text{s}$$

Thus, when $h = 5$,

$$-3 = \frac{3\pi}{48}(5)^2\,\frac{dh}{dt}$$

and so

$$\frac{dh}{dt} = \frac{48(-3)}{3\pi(5)^2} = -\frac{48}{25\pi} \approx -0.61 \text{ cm/s}$$

The negative sign shows that the water level is *dropping*.

© **EXAMPLE 4**   Two highways intersect at right angles. Car A on one highway is 0.5 kilometer from the intersection and is moving toward it at 96 kilometers per hour, while car B on the other highway is 1 kilometer from the intersection and is moving away from it at 88 kilometers per hour. At what rate is the distance between the two cars changing at this instant?

SOLUTION   In Figure 4, we have represented the two highways by the x and y axes of a Cartesian coordinate system. Let y denote the distance in kilometers of car A from the origin, and let x denote the distance of car B from the origin at time t hours. Then the distance s in kilometers between the two cars at time t hours is given by

$$s = \sqrt{x^2 + y^2}$$

Figure 4

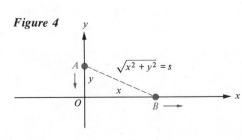

Differentiating both sides of this equation with respect to t, we find that

$$\frac{ds}{dt} = \frac{1}{2\sqrt{x^2 + y^2}}\frac{d}{dt}(x^2 + y^2) = \frac{1}{2s}\left(2x\frac{dx}{dt} + 2y\frac{dy}{dt}\right) = \frac{1}{s}\left(x\frac{dx}{dt} + y\frac{dy}{dt}\right)$$

Since car A is moving toward O at 96 kilometers per hour, we have

$$\frac{dy}{dt} = -96 \text{ km/h}$$

Since car B is moving away from O at 88 kilometers per hour, we also have

$$\frac{dx}{dt} = 88 \text{ km/h}$$

Thus, at the instant when $y = 0.5$ and $x = 1$ kilometer,

$$s = \sqrt{1^2 + (0.5)^2} = \sqrt{1.25}$$

and $\quad\dfrac{ds}{dt} = \dfrac{1}{s}\left(x\dfrac{dx}{dt} + y\dfrac{dy}{dt}\right) = \dfrac{1}{\sqrt{1.25}}[1(88) + 0.5\,(-96)] = \dfrac{40}{\sqrt{1.25}}$

$$\approx 35.78 \text{ km/h}$$

EXAMPLE 5   The pressure P and the volume V of a sample of air that is undergoing adiabatic expansion are related by the equation $PV^{1.4} = C$, where C is a constant. At a certain instant the volume of such a sample is 4 cubic inches, the pressure is 4000 pounds per square inch, and the volume is increasing at the instantaneous rate of 2 cubic inches per second. At what rate is the pressure changing at this instant?

SOLUTION   Differentiation of both sides of $PV^{1.4} = C$ with respect to t gives

$$P(1.4)V^{0.4}\frac{dV}{dt} + \frac{dP}{dt}V^{1.4} = 0$$

so that

$$\frac{dP}{dt} = -\frac{1.4P}{V}\frac{dV}{dt}$$

When $V = 4$, $P = 4000$, and $dV/dt = 2$, we have

$$\frac{dP}{dt} = -\frac{1.4(4000)}{4}(2) = -2800 \text{ pounds per square inch per second}$$

EXAMPLE 6   An airplane flies at a height of 9 kilometers in the direction of an observer on the ground at a speed of 800 kilometers per hour. Find the rate of change of the angle of elevation of the plane from the observer at the instant when this angle is $\pi/3$ radians.

SOLUTION   Denote the angle of elevation by θ, and let x be the horizontal distance between the plane and the observer (Figure 5). Then

$$\cot \theta = x/9 \quad\text{and}\quad \frac{dx}{dt} = -800 \text{ km/h}$$

Differentiating both sides of $\cot \theta = x/9$ with respect to t, we obtain

$$-\csc^2 \theta \frac{d\theta}{dt} = \frac{1}{9}\frac{dx}{dt} = -\frac{800}{9}$$

Figure 5

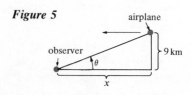

At the instant when $\theta = \pi/3$, we have $\csc \theta = 2/\sqrt{3}$, so

$$-\frac{4}{3}\frac{d\theta}{dt} = -\frac{800}{9} \quad \text{or} \quad \frac{d\theta}{dt} = \frac{200}{3} \text{ rad/h}$$

(about 63.66° per minute). ■

©Problem Set 3.8

1 A stone thrown into a still pond causes a circular ripple to spread. If the radius of the circle is increasing at the rate of 0.85 meter per second, how fast is the area enclosed by the ripple increasing when the radius is 5 meters?

2 A culture of bacteria occupies a circular region in a petri dish. An experimental bactericide, placed in the petri dish, causes the area of the circular region occupied by the bacteria to decrease at the rate of 3 square centimeters per hour. At what rate is the radius r of the circular region changing at the instant when $r = 2$ centimeters?

3 A circular metal plate expands when heated so that its radius increases at the constant rate of 0.02 centimeter per second. At what rate is the surface area of one side of the plate increasing when the radius is 4 centimeters?

4 The length of each side of a square is increasing at the constant rate of 2 meters per second. Find the rate of increase of the area and of the perimeter of the square at the instant when its side is 3 meters long.

5 Each side of a closed cubical box is decreasing at the rate of 10 centimeters per minute. (a) How fast is the volume of the box decreasing at the instant when its side length is 20 centimeters? (b) How fast is its total surface area (four sides plus top and bottom) decreasing at the instant when its side length is 20 centimeters?

6 Show that if each side of a cube is increasing at the constant rate of 0.1 unit per second, then, at the instant when the side length of the cube is 10 units, the volume will be increasing at the rate of 30 cubic units per second. Does this mean that during the next second exactly 30 cubic units will be added to the volume? Explain.

7 If the base of a triangle is increasing at the constant rate of 6 meters per minute and the altitude is increasing at the constant rate of 2 meters per minute, what is the rate of change of the area of the triangle at the instant when its base is 8 meters long and its height is 10 meters?

8 The area A of a rectangle is decreasing at the constant rate of 9 square centimeters per second. At any instant, the length l of the rectangle is decreasing twice as fast as the width w. At a certain instant the rectangle is a 1-centimeter by 1-centimeter square. At this instant, how rapidly is the width decreasing?

9 A streetlight is on top of a pole 4 meters high. A person 1.8 meters tall is 6 meters from the base of the pole and is walking directly toward the pole at the constant rate of 2 meters per second. (a) At what rate is the person's shadow shortening? (b) At what rate is the tip of the person's shadow moving?

10 A person 6 feet tall is walking toward a wall at the rate of 4 feet per second. Directly behind the person and 40 feet from the wall is a spotlight 3 feet above ground level. How fast is the length of the person's shadow on the wall changing at the moment when the person is exactly halfway between the spotlight and the wall? Is the shadow lengthening or shortening?

11 At 2:00 P.M. a ship, which is steaming north at 20 knots, is 4 nautical miles west of a lighthouse. At the same time, a motor launch, traveling at 10 knots, is proceeding south from a point 4 nautical miles east of the lighthouse. Find the rate of change of the distance between the ship and the launch at 3:30 P.M., assuming that they maintain their respective speeds and courses.

12 Suppose that a particle is moving in the xy plane along the line $y = mx + b$. Show that the speed of the particle at time t is given by $\sqrt{m^2 + 1}\ |dx/dt|$.

13 A television camera is located on the ground 16 kilometers from the point where a rocket is launched vertically. At what rate is the distance from the camera to the rocket increasing at the instant when the rocket is 14 kilometers high and rising at the rate of 3 kilometers per second?

14 An airplane is being pulled into a hangar by a rope from a pulley to the nose of the plane. If the pulley is 3 meters higher than the nose of the plane and the rope is being pulled in at a constant rate of 1 meter per second, how fast is the plane coming into the hangar when the length of the rope from the pulley to the nose of the plane is 12 meters?

15 A girl on a bike rides north 8 kilometers and then turns east. If she rides at a steady rate of 16 kilometers per hour, at what rate is the distance, measured directly between the girl and her starting point, changing 2 hours after she leaves that point?

16 In a baseball game (which is played on a square-shaped diamond 90 feet on each side) between the Brewers and the Cubs, Joe (who is playing left field for the Brewers) hits the ball while at bat and runs toward first base at the rate of 20 feet per second. Just as Joe starts toward first base, his teammate Tony, who had taken a 10-foot lead off second base, starts running toward third base at the same rate of speed. How fast is the distance between Joe and Tony changing at the instant when Tony reaches third base?

17 A boat is moored to a dock by a rope. A person on the dock begins pulling in the rope at the rate of 72 feet per minute. The person's hands (holding the rope) are 5 feet above the level of the bow of the boat (where the rope is fastened). How fast is the boat moving toward the dock at the instant when 13 feet of rope is out?

18 A spherical snowball is melting so that its volume is decreasing at the rate of 0.17 cubic meter per minute. Find the rate at which its radius is decreasing at the instant when the volume of the snowball is 0.4 cubic meter.

19 A ladder 4 meters long is leaning against the vertical wall of a house. If the base of the ladder is pulled horizontally away from the house at the rate of 0.7 meter per second, how fast is the top of the ladder sliding down the wall at the instant when it is 2 meters from the ground?

20 A spherical weather balloon is being inflated with helium so that its volume is increasing at the rate of 0.75 cubic meter per minute. At the moment when the radius of the balloon is 0.25 meter, (a) how fast is the radius of the balloon increasing and (b) how fast is the surface area of the balloon increasing?

21 A spherical tank of radius R filled with liquid to a height h contains $V = \frac{1}{3}\pi h^2(3R - h)$ cubic units of liquid (Figure 6). A spherical gasoline tank has a radius of $R = 20$ feet. Gasoline is being pumped out of this tank at the rate of 200 gallons per minute. (One gallon is approximately 0.134 cubic foot.) At what rate is the level of gasoline in the tank dropping at the instant when $h = 5$ feet?

Figure 6

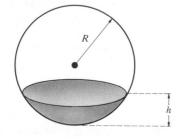

22 Water is being pumped at the rate of 1.5 cubic meters per minute into a swimming pool 20 meters long and 10 meters wide. The depth of the pool decreases uniformly from 7 meters at one end to 1 meter at the other end. How fast is the surface level of the water rising at the instant when its depth at the deeper end is 6 meters?

23 A horizontal eaves trough is 20 feet long and has a cross section in the shape of an isosceles triangle 8 inches across at the top and 10 inches deep. Because of a heavy rainstorm, the water in the trough is rising at the rate of 0.5 inch per minute at the instant when it is 5 inches deep. How fast is the volume of water in the trough increasing at this instant?

24 Water is stored in a reservoir in the shape of a right circular cone with an open base facing upward. The radius of the base is 20

meters, and the depth of the reservoir from top to vertex is 5 meters. The water in the reservoir evaporates at the rate of 0.00005 cubic meter per hour for each square meter of its surface exposed to the air. Show that, because of evaporation, the water level will drop at a uniform rate (independent of the depth of the water), and find this rate.

25 Sand is falling at the rate of 2 cubic meters per minute upon the tip of a sandpile, which maintains the form of a right circular cone whose height is the same as the radius of the base. Find the rate at which the height of the pile is increasing at the instant when the pile is 6 meters high.

26 A cylindrical tank with vertical axis has a radius of 10 inches. The tank has a circular hole in its base with a 1-inch radius through which water pours out with a velocity of $v = 8\sqrt{h}$ feet per second, where h is the height in feet of the surface of the water above the base. How rapidly is the water level dropping at the instant when $h = 5$ feet?

27 A water tank has the shape of an inverted right circular cone with a radius of 5 meters at the top and a height of 12 meters. At the instant when the water in the tank is 6 meters deep, more water is being poured in at the rate of 10 cubic meters per minute. Find the rate at which the surface level of the water is rising at this instant.

28 A water tank has the shape of an inverted right circular cone with a radius of R units at the top and a height (from vertex to top) of H units. A small hole with cross-sectional area k square units allows water to leak out of the tank at the vertex with a velocity of $\sqrt{2gh}$ units per second, where the depth of the water in the tank is h units and g is the acceleration of gravity in units per second squared. If water is being pumped into the tank at a uniform rate of c cubic units per second, find a formula for the rate at which the water level is changing.

29 A certain microorganism has a spherical shape and a density of 1 gram per cubic centimeter. It grows by absorbing nutrients through its surface at the rate of $A(10^{-4})$ gram per hour, where A is its surface area. If its radius is now 10^{-3} centimeter, what will be its radius after 2 hours?

30 If the volume of 1 unit of mass of a substance at the temperature Θ is V, then $dV/d\Theta$ is called the *coefficient of cubical expansion* of the substance. The volume of 1 gram of water is given by $V = 1 + (8.38)(10^{-6})(\Theta - 4)^2$ cubic centimeters, where Θ is the temperature of the water in degrees Celsius. Find the coefficient of cubical expansion of water at 10°C and the rate of change of the volume of 1 gram of water at 10°C if its temperature is decreasing at the rate of 1.5°C per minute.

31 **Boyle's law** for the expansion of a gas held at constant temperature states that the pressure P and the volume V are related by the equation $PV = C$, where C is a constant. Suppose that a sample of 1000 cubic inches of gas is under a pressure of 150 pounds per square inch, but that the pressure is decreasing at the instantaneous rate of 5 pounds per square inch per second. Find the instantaneous rate of increase of the volume.

32 Wildlife biologists studying the predator-prey relationship between owls (the predator) and field mice (the prey) in a certain habitat have determined that the owl population y and the field mouse population x obey the predator-prey model

$$(x - 800)^2 + 400(y - 50)^2 = 100,000$$

When the owl population is 55, the field mouse population is 1100, and the field mouse population is increasing by 40 field mice per month, how fast is the owl population changing according to this model?

In Problems 33 to 37, consider the triangle ABC (Figure 7). Assume that θ is decreasing at the rate of $\frac{1}{30}$ radian per second. Find the indicated rate of change by using the given information.

Figure 7

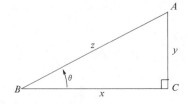

33 dy/dt when $\theta = \pi/3$ given that x is a constant equal to 12 centimeters

34 dz/dt when $\theta = \pi/4$ given that y is a constant equal to $10\sqrt{2}$ centimeters

35 dx/dt when $y = 20$ centimeters given that z is a constant equal to 40 centimeters

36 dz/dt when $y = x$ if x remains 1.6 kilometers at all times

37 dz/dt when $x = 1$ meter and $z = 2$ meters if x and y are both changing and y is increasing at the rate of $\frac{2}{15}$ meter per second

38 A parachutist is descending at a constant (but unknown) velocity from an unknown height directly toward the center of a horizontal circular target of unknown radius (Figure 8). She carries a small optical device that measures the angle θ (in radians) subtended by the circular target and calculates the quantity T given by $T = (\sin \theta)/(d\theta/dt)$. Show that she will hit the target in T seconds.

Figure 8

parachutist

circular target

39 A ladder 3 meters long leans against a house. The upper end slips down the wall at the rate of 1.5 meters per second. How fast is the ladder turning when it makes an angle of $\pi/6$ radian with the ground?

40 A weight is lifted by a rope that passes over a pulley and down to a truck that is moving at 5 feet per second. If the pulley is 100 feet above the level of the truck, how fast is the weight rising when the rope from the pulley to the truck makes an angle of $\pi/4$ radian with the ground?

41 In an isosceles triangle, whose equal sides are each 3 centimeters long, the vertex angle increases at the rate of $\pi/90$ radian per second. How fast is the area of the triangle increasing when the vertex angle is $\pi/3$ radians?

42 A man is walking along a straight sidewalk at the rate of 2 meters per second. A searchlight on the ground 12 meters from the sidewalk is kept trained on him. At what rate is the searchlight revolving when the man is 7 meters away from the point on the sidewalk nearest the light?

43 An airplane is flying horizontally at a speed of 400 miles per hour at an elevation of 10,000 feet toward an observer on the ground. Find the rate of change of the angle of elevation of the airplane from the observer when the angle is $\pi/4$.

44 The hands of a tower clock are 4.5 and 6 feet long. How fast are the tips of the hands approaching each other at four o'clock?

45 The angle of elevation of the sun is decreasing at the rate of 14 degrees per hour. What is the rate of increase of the length of the shadow cast by a tree 4 meters tall when the angle of elevation of the sun is $\pi/6$?

46 The crankshaft of an engine is turning at a constant rate of N revolutions per second (Figure 9a). Suppose that the arm $\overline{OA}$ is a units long and the connecting rod $\overline{AP}$ is b units long. Let θ be the angle through which the crankshaft has turned from the top-dead-center position of the piston. (a) By considering Figure 9b, show that the distance $x = |\overline{OP}|$ units between the center of the crankshaft O and the wrist pin P is given by the equation $x = a \cos \theta + \sqrt{b^2 - a^2 \sin^2 \theta}$. (b) Using the result of part (a), obtain a formula for dx/dt in terms of a, b, θ, and N. (*Hint:* $d\theta/dt = 2\pi N$ radians per second.)

Figure 9

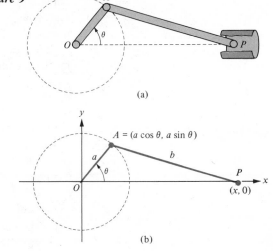

(a)

(b)

47 A television crew is televising a 100-meter dash from a position 10 meters from the track and in line with the finish line (Figure 10). At the instant that the winner of the race is 10 meters from the finish line, the camera, focused on her, is turning 0.5 radian per second. How fast is she running?

Figure 10

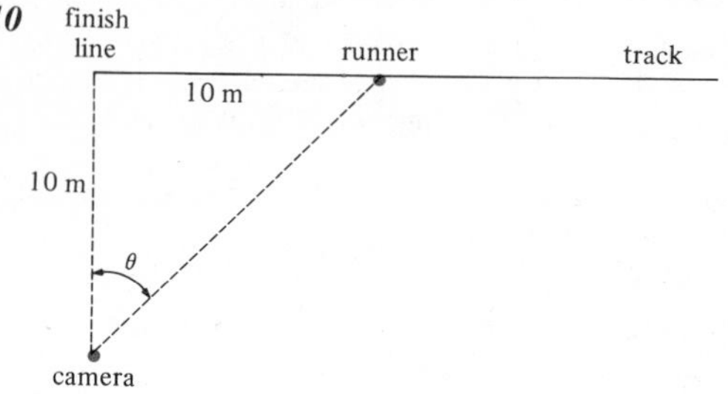

finish line

10 m   runner   track

10 m

θ

camera

48 A radar station R can determine the *range* $r = |\overline{RT}|$ of a target T and the *bearing angle* θ of the target from due north (Figure 11).

If the target is moving, a *Doppler radar* can also determine the rate of change dr/dt of the range by measuring the frequency change of the reflected signal caused by the Doppler effect. Assuming that the target T is moving along a straight line (called its *track*), show that by using two Doppler radars located in known positions, it is possible to determine the speed of the target and its track.

Figure 11

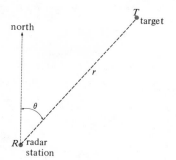

north

T target

r

θ

R radar station

3.9 Applications to Economics and Business

Although calculus was originally developed by people interested in astronomy, mechanics, physics, and engineering, it is currently finding more and more applications in the worlds of economics, manufacturing, advertising, and marketing. Economists and business managers routinely use the methods of calculus to study relationships among such variables as advertising costs, demand for goods and services, inflation rates, interest rates, inventories, prices, production costs, production levels, profits, revenues, supplies of goods, supplies of raw materials, taxes, unemployment, and wages. Here we can give only a brief survey of one aspect of the application of calculus to economics and business.

We consider a **producer** supplying a particular **commodity** (goods or services) to **consumers.** We understand that the producer could range from a small manufacturing plant to an entire industry. As our basic independent variable, we choose the number x of units of the commodity produced in a given period of time by the producer in question. We assume that the variable x, which is called the **production level,** is under the control of the producer and that

$$0 \leq x \leq M$$

where M is the maximum possible production level. In order to be able to apply the methods of calculus, we treat x as if it were a continuous variable.

We denote by $C(x)$ the cost in dollars to the producer for producing x units of the commodity in the given period of time. The portion of the total cost of production $C(x)$ that is independent of the production level x is called the **fixed cost.** This fixed cost may include capital expenses for new equipment, depreciation on existing equipment, insurance premiums, salaries of permanent employees, taxes, advertising costs, and so forth, as well as overhead expenses for rent, lighting, and heating of the plant. The remaining part of the total cost of production—the part that depends on the production level—is called the **variable cost.** This variable cost may include the cost of raw materials, the cost of fuel and electricity actually used in the manufacturing process, costs of packaging, wages for part-time workers, overtime wages for permanent workers, and so forth. Thus,

$$C(x) = \text{the fixed cost} + \text{the variable cost}$$

If we assume that the variable cost is zero when the production level is zero, then we have

$$C(0) = \text{the fixed cost}$$

and therefore,

$$C(x) - C(0) = \text{the variable cost}$$

To specify a realistic cost function C, it is necessary to determine both the fixed and the variable costs. Usually, the fixed cost can be found by consulting suitable accounting records. The variable cost may be more difficult to determine—past production records must be studied; changing costs of raw materials, energy, and labor at various production levels must be considered; and differences in the efficiency of plant operation at various production levels must be taken into account. Often, an exact formula for variable cost cannot be found, in which case a reasonable approximation, based on experience and insight, must be used.

In economics, the derivative of a function is usually referred to as a **marginal.** Thus, the derivative C' of the cost function is called the **marginal-cost function,** and $C'(x)$ is called the **marginal cost** at production level x. Hence, the marginal cost is the rate at which the production cost changes per unit change in production. In practical situations, the production level x is usually a rather large number—so large that, in comparison to x, the number 1 can be considered to be very small. Then we have

$$C'(x) = \lim_{\Delta x \to 0} \frac{C(x + \Delta x) - C(x)}{\Delta x} \approx \frac{C(x + 1) - C(x)}{1} = C(x + 1) - C(x)$$

In other words, *when the production level x is rather large, the marginal cost $C'(x)$ can be regarded as a good approximation to the cost $C(x + 1) - C(x)$ of producing one more unit.* Under ordinary circumstances, an increase in production level will result in an increase in cost; that is, C is an increasing function, and the marginal cost $C'(x)$ is always positive.

Often, as the production level increases, the manufacturing process becomes more efficient and, at higher production levels, it costs less to produce one more unit of the commodity than at lower production levels. When the *most efficient production level* is reached, the marginal cost will have a minimum value. Above this most efficient production level, the advantages of mass production begin to be offset by such factors as stress put on workers and production machinery, scarcity of raw materials in large quantities, the necessity of paying overtime wages, and so forth.

ⓒ **EXAMPLE 1** Decibel Records is planning to produce a new album featuring recording star Jason Dean. The accounting department estimates the fixed cost (including the recording session) on a daily pro rata basis over the projected production lifetime of the album to be $500 per day. On the basis of experience with previous albums, management projects that the variable cost for the production of x albums per day (including royalty payments) is given by

$$ax^3 - bx^2 + cx \qquad \text{dollars} \qquad \text{for } 0 \le x \le 8000$$

where $a = 1.25 \times 10^{-7}$, $b = 1.5 \times 10^{-3}$, and $c = 7$.

(a) Find the cost function $C(x)$.

(b) Find the marginal-cost function $C'(x)$.

(c) Find the exact cost of manufacturing the 1000th album.

(d) Use the marginal cost to find the approximate cost of manufacturing the 1000th album.

(e) Find the most efficient production level.

SOLUTION

(a) $C(x) = 500 + ax^3 - bx^2 + cx$
$\qquad = (1.25 \times 10^{-7})x^3 - (1.5 \times 10^{-3})x^2 + 7x + 500$

$$\text{for } 0 \le x \le 8000$$

(b) $C'(x) = (3.75 \times 10^{-7})x^2 - (3 \times 10^{-3})x + 7 \qquad \text{for } 0 < x < 8000$

(c) The exact cost of manufacturing the 1000th album is given by $C(1000) - C(999) = 4.376125$ dollars.

(d) $C'(999) \approx 4.37725$ dollars

(e) To find the minimum value of $C'(x)$, we find its derivative: $C''(x) = (7.5 \times 10^{-7})x - (3 \times 10^{-3})$. Thus, the only critical number for C' on the interval $(0, 8000)$ is

$$\frac{3 \times 10^{-3}}{7.5 \times 10^{-7}} = 4000$$

For $0 < x < 4000$, $C''(x) < 0$, whereas for $4000 < x < 8000$, $C''(x) > 0$. Therefore, C' is decreasing on the interval $(0, 4000)$ and increasing on the interval $(4000, 8000)$. It follows that $C'(x)$ has an absolute minimum value when the production level is 4000 albums per day. ∎

The **revenue** $R(q)$ is the total income to the producer from the sale of q units of the commodity. For instance, if each unit of the commodity sells for p dollars, then the revenue from the sale of q units is given by

$$R(q) = pq \qquad \text{dollars}$$

Often q, the number of units sold, will depend on the price p per unit. Ordinarily, the higher the price p, the smaller the number q of units sold, and vice versa. A **demand equation** relates the price p per unit of the commodity and the number of units q that will be demanded at that price by consumers in a given period of time. If we assume that this demand is met by the producer, then we can take $q = x$, the production level. For the remainder of this section, we make this reasonable assumption.

By means of suitable *market research*, it may be possible to determine a demand equation relating the price p per unit and the demand x units of the commodity. In this equation, we may regard p as a (perhaps implicity defined) function of x. The revenue function R is then determined by the formula

$$\boxed{R(x) = px}$$

EXAMPLE 2 Market research conducted by Decibel Records indicates that 3000 albums per day will be sold at a price (to dealers) of $8 per album. (Dealers will add their own markup to arrive at the retail price.) For each $0.50 increase in the wholesale price, it is projected that the resulting increase in the retail price will result in 500 fewer album sales per day.

(a) Find the demand equation for the albums.

(b) Find the daily revenue $R(x)$ dollars to Decibel Records at production level x.

SOLUTION

(a) Let n denote the number of $0.50 increases in the price per album above the price of $8 per album. Then the price p per album is given by

$$p = 8.00 + 0.50n \qquad \text{dollars}$$

and the daily demand x for albums is given by

$$x = 3000 - 500n$$

From the last equation, we have

$$n = \frac{3000 - x}{500}$$

Substituting this value of n into the equation for p, we obtain the demand equation

$$p = 8.00 + 0.50\left(\frac{3000 - x}{500}\right)$$

$$= 11 - \frac{x}{1000}$$

(b) Since $R(x) = px$, we have

$$R(x) = 11x - \frac{x^2}{1000} \qquad \blacksquare$$

The **marginal revenue** $R'(x)$ is the rate at which the revenue changes per unit change in x. Again, for large values of x, the marginal revenue $R'(x)$ is a good approximation to the additional revenue $R(x + 1) - R(x)$ generated by the sale of one more unit of the commodity.

ⓒ **EXAMPLE 3** Let $R(x) = 11x - (x^2/1000)$ dollars be the daily revenue to Decibel Records if x albums per day are produced and sold, $0 \leq x \leq 8000$.

(a) Find the marginal revenue $R'(x)$.

(b) Approximately how much additional revenue per day would be generated by increasing the daily production level from 3000 to 3001 albums?

(c) For what daily production level x will the revenue be maximum?

SOLUTION

(a) $R'(x) = 11 - \dfrac{x}{500} \qquad$ for $0 < x < 8000$

(b) $R'(3000) = 11 - \dfrac{3000}{500} = \$5$

(c) Setting $R'(x) = 0$, we find that the only critical number for R on the interval $(0, 8000)$ is 5500. Checking the values of $R(x)$ at this critical number and at the two endpoints of the interval $[0, 8000]$, we find that $R(x)$ takes on a maximum value of \$30,250 per day when 5500 albums per day are produced and sold to retailers. ∎

Actually, it's usually not the revenue that a producer wants to maximize—it's the **profit,** that is, the revenue minus the cost of production. If $C(x)$ is the cost of producing x units of a commodity and $R(x)$ is the revenue from the sale of these x units, then the profit $P(x)$ from the sale is given by

$$P(x) = R(x) - C(x)$$

Again, if the production level x is large, the **marginal profit** $P'(x)$ can be regarded as the approximate additional profit $P(x + 1) - P(x)$ generated by the manufacture and sale of one more unit of the commodity.

© **EXAMPLE 4**   For Decibel Records, suppose that the daily cost and revenue functions for Jason Dean albums at a production level of x albums per day are given by

$$C(x) = (1.25 \times 10^{-7})x^3 - (1.5 \times 10^{-3})x^2 + 7x + 500 \qquad \text{dollars}$$

and $R(x) = 11x - (10^{-3})x^2 \qquad \text{dollars} \qquad \text{for } 0 \le x \le 8000$

(a) Find the daily profit $P(x)$ for the manufacture and sale of x albums.

(b) Find the marginal profit $P'(x)$.

(c) For what daily production level x will the profit be maximum?

(d) What price per album should be charged by Decibel Records when the production level is set to yield maximum profit?

SOLUTION

(a) $P(x) = R(x) - C(x)$
$$= -(1.25 \times 10^{-7})x^3 + (5 \times 10^{-4})x^2 + 4x - 500 \qquad \text{dollars}$$
$$\text{for } 0 \le x \le 8000$$

(b) $P'(x) = -(3.75 \times 10^{-7})x^2 + (10^{-3})x + 4 \qquad \text{for } 0 < x < 8000$

(c) Using the quadratic formula and a calculator, we find that the roots of the quadratic equation $P'(x) = 0$ are (approximately) -2194 and 4861. Rejecting -2194 because it does not belong to the interval $(0, 8000)$, we find that 4861 is the only critical number for P in the interval $(0, 8000)$. Checking the values of $P(x)$ at this critical number and at the endpoints of the interval $[0, 8000]$, we find that a maximum profit of \$16,400.89 per day is achieved by manufacturing and selling 4861 albums per day.

(d) Since $R(x) = px$, we have

$$p = \frac{R(x)}{x} = 11 - (10^{-3})x \qquad \text{dollars}$$

Hence, when $x = 4861$, the price per album is

$$11 - (10^{-3})(4861) = 6.139 \qquad \text{dollars} \qquad ∎$$

A negative profit is called a *loss.* At a production level for which the profit is zero, the producer is said to *break even.*

© **EXAMPLE 5**   Using the profit function for Decibel Records obtained in Example 4a, find the production levels for which the producer breaks even.

SOLUTION   Using Newton's method to find the (approximate) zeros of the profit function P in the interval $[0, 8000]$, we find that production levels of 123 and 7958 albums per day cause the producer to break even. Production levels of fewer than 123 albums per day or more than 7958 albums per day will result in a net loss for the producer.   ∎

A study of cost, revenue, and profit (illustrated for Decibel Records in the previous examples) is called a **marginal analysis** because of its reliance on marginals (that is, derivatives) to investigate changes in cost, revenue, and profit at various levels of production. In conducting a marginal analysis, you should consider the following questions:

1 What period of time is involved when we speak of production level x? Is it units per day, per week, per production run, or what? In what units is the commodity measured? What is the maximum possible production level?

2 What is the fixed cost (prorated over the period of time involved in the specification of the production level)? What is the variable cost at production level x? What is the cost $C(x)$ dollars? What is the marginal cost $C'(x)$? What is the most efficient production level?

3 What is the demand equation for the commodity? What is the revenue $R(x)$ dollars? What is the marginal revenue $R'(x)$? For what production level will the revenue be maximum?

4 What is the profit $P(x)$ dollars at production level x? What is the marginal profit $P'(x)$? For what production level will the profit be maximum? According to the demand equation, what price per unit should be charged by the producer when the production level is adjusted to yield maximum profit? At what production level or levels does the producer break even?

Problem Set 3.9

© In Problems 1 to 4, assume that $C(x)$ is the cost in dollars of producing x units of a certain commodity during a particular interval of time. (a) Find the fixed cost. (b) Find the variable cost. (c) Find the marginal cost $C'(x)$. (d) Find the exact cost of producing the 901st unit of the commodity. (e) Use the marginal cost to estimate the cost of producing the 901st unit of the commodity. (f) Find the most efficient production level.

1 $C(x) = -x^2 + 2500x + 10{,}000, \ 0 \le x \le 1000$

2 $C(x) = 150 + 8\sqrt{x}, \ 0 \le x \le 1500$

3 $C(x) = 0.00003x^3 - 0.18x^2 + 500x + 5000, \ 0 \le x \le 2500$

4 $C(x) = \dfrac{26{,}000x + 150{,}000}{3x + 1000}, \ 0 \le x \le 1500$

© **5** If C is a cost function, the corresponding **average-cost function** $\overline{C}$ is defined by $\overline{C}(x) = C(x)/x$. Thus, $\overline{C}(x)$ is the average cost per unit if x units are produced. In Problems 1 and 3, find $\overline{C}(900)$.

6 Under normal economic circumstances, it often works out that the production cost per unit of a commodity is minimized by setting the production level so that the production cost per unit is equal to the marginal cost. Explain why.

7 Under normal economic circumstances, it often works out that the most efficient production level is the abscissa of a point of inflection of the graph of the cost function. Explain why.

8 Under normal economic circumstances, it often works out that the production level for which minimum production cost per unit is obtained is somewhat larger than the most efficient production level. Explain why.

© **9** A production-cost analysis conducted by the Mattoon Steel Company, Inc., shows that the fixed cost for a production run of a new alloy is $100,000 and the variable cost for a production run of x tons of the alloy is

$$\frac{x^3}{900} - 3x^2 + 4000x \qquad \text{dollars} \qquad \text{for } 0 \le x \le 1000$$

(a) Find the cost function $C(x)$. (b) Find the marginal cost $C'(x)$. (c) Find the exact cost of producing the 500th ton of the new alloy during a production run. (d) Use the marginal cost to find the approximate cost of producing the 500th ton of the new alloy during a production run. (e) Find the most efficient production level for the new alloy.

© 10 The Gainesburg Tire and Rubber Company, Inc., manufactures tires for dirt bikes. The fixed cost for production of these tires per week is $15,000 and the variable cost for production of x tires per week is

$$(2 \times 10^{-7})x^3 - (3 \times 10^{-3})x^2 + 20x \qquad \text{dollars}$$
$$\text{for } 0 \le x \le 10,000$$

(a) Find the cost function $C(x)$. (b) Find the marginal cost $C'(x)$. (c) Find the exact cost of manufacturing the 1001st tire. (d) Use the marginal cost to find the approximate cost of manufacturing the 1001st tire. (e) Find the most efficient production level for the tires.

© 11 Solar Electronics manufactures personal walkaround stereo cassette players. The fixed cost per day for producing these walkarounds is $4000 and the variable cost for the production of x walkarounds per day is

$$(5 \times 10^{-6})x^3 - (1.5 \times 10^{-2})x^2 + 40x \qquad \text{dollars}$$
$$\text{for } 0 \le x \le 3000$$

(a) Find the cost function $C(x)$. (b) Find the marginal cost $C'(x)$. (c) Find the exact cost of manufacturing the 201st walkaround. (d) Use the marginal cost to find the approximate cost of manufacturing the 201st walkaround. (e) Find the most efficient production level for the walkarounds.

© 12 An analysis of production costs conducted by the Molar Brush Company shows that the fixed cost per day for manufacturing toothbrushes is $500 and the variable daily cost for manufacturing x toothbrushes is $30\sqrt{x}$ dollars for $0 \le x \le 10,000$. (a) Find the cost function $C(x)$. (b) Find the marginal cost $C'(x)$. (c) Find the exact cost of manufacturing the 5001st toothbrush. (d) Use the marginal cost to find the approximate cost of manufacturing the 5001st toothbrush. (e) Find the most efficient production level for the toothbrushes?

© 13 Market research conducted by Mattoon Steel indicates that for each production run of its new alloy, it will sell 600 tons at $4000 per ton; however, for each $500 increase in the price per ton, 100 fewer tons will be sold. (a) Find the demand equation for the alloy. (b) Find the revenue $R(x)$ in dollars for the alloy at production level x. (c) Find the marginal revenue $R'(x)$. (d) Approximately how much additional revenue would be generated by increasing the production level from 500 to 501 tons during a production run? (e) For what production level will the revenue be maximum?

© 14 Market research conducted by Gainesburg Tire and Rubber indicates that at a price of $24 per tire it will sell 3000 dirt-bike tires per week; however, each $1 increase in the price per tire will result in the sale of 500 fewer tires per week. (a) Find the demand equation for the tires. (b) Find the revenue $R(x)$ in dollars if x tires per week are manufactured and sold. (c) Find the marginal revenue $R'(x)$. (d) Approximately how much additional weekly revenue would be generated by increasing the production level from 1000 to 1001 tires per week? (e) For what production level will the revenue be maximum?

© 15 Market research conducted by Solar Electronics indicates that at a price of $60 per unit, it will sell 2000 personal walkaround stereo cassette players per day; however, for each $5 increase in price per unit, 250 fewer units will be sold per day. (a) Find the demand equation for the walkarounds. (b) Find the daily revenue $R(x)$ for the manufacture and sale of x walkarounds per day. (c) Find the marginal revenue $R'(x)$. (d) Approximately how much additional daily revenue would be generated by increasing the production level from 200 to 201 walkarounds per day? (e) For what production level will the revenue be maximum?

© 16 Market research conducted by the Molar Brush Company indicates that 6000 toothbrushes per day will be sold at $0.90 per toothbrush, but that each $0.25 increase in the price per brush will result in 1000 fewer sales per day. (a) Find the demand equation for the toothbrushes. (b) Find the daily revenue $R(x)$ if x toothbrushes per day are manufactured and sold. (c) Find the marginal revenue $R'(x)$. (d) Approximately how much additional daily revenue would be generated by increasing the production level from 5000 to 5001 toothbrushes per day? (e) For what production level will the revenue be maximum?

© In Problems 17 to 20, use the indicated information to find (a) the profit $P(x)$ at production level x, (b) the marginal profit $P'(x)$, (c) the production level for which the profit is maximum, (d) the price per unit that the producer should charge when the production level is adjusted to yield maximum profit, and (e) the production level(s) for which the producer breaks even.

17 The information in Problems 9 and 13 concerning the production of the new alloy by Mattoon Steel.

18 The information in Problems 10 and 14 concerning the production of dirt-bike tires by Gainesburg Tire and Rubber.

19 The information in Problems 11 and 15 concerning the production of personal walkaround stereo cassette players by Solar Electronics.

20 The information in Problems 12 and 16 concerning the production of toothbrushes by Molar Brush.

© In Problems 21 and 22, use the given information to conduct a marginal analysis.

21 Rolar Bicycle Company is planning to bring out a new, lightweight, and relatively inexpensive 10-speed bike. Management

estimates a daily fixed cost of $1700 and a variable cost of

$$\frac{23x^3}{10,000} - \frac{69x^2}{100} + 159x \qquad \text{dollars per day}$$

for a daily production run of x bikes. Market research shows that at $150 per bike, 75 bikes will be purchased per day, but each $10 increase in price per bike will result in 15 fewer bikes per day being purchased. Because of limited production facilities, no more than 125 bikes per day can be manufactured.

22 Decibel Records is planning to produce an album featuring the trumpet virtuoso Adrian Scott. The recording will be made by using a state-of-the-art digital process that is considerably more expensive than ordinary recording methods. Management estimates a daily fixed cost of $800 and a variable cost of

$$(6.5 \times 10^{-7})x^3 - (3.9 \times 10^{-3})x^2 + 9.8x \qquad \text{dollars per day}$$

for a daily production run of x albums. Market research shows that 500 albums per day will be sold to distributors at $10 per album, but each $1 increase in price per album will result in 50 fewer albums being purchased. Because of limited production facilities for these high-quality records, no more than 1000 records per day can be produced.

23 A cable television company plans to begin operations in a small town. It foresees that about 600 people will subscribe to the service if the price per subscriber is $10 per month; but for each $0.10 increase in the price per month, 4 of the original 600 people will decide not to subscribe. The fixed cost to the company is $2000 per month, and each subscription costs the company $3 per month for maintenance. (a) What price per month per subscription will bring in the greatest revenue to the company? (b) What price will bring in the greatest profit to the company?

24 An automobile leasing agency rents cars to members of a teachers' credit union and discounts its total bill to these members by 2 percent for each rented car in excess of 12. For how many cars rented to the members would the receipts to the agency be maximum?

25 For a certain commodity, suppose that at a price of p_1 dollars per unit there is a demand for q_1 units, but for each k-dollar increase in the price per unit, r fewer units will be sold. (a) Show that the demand equation for the commodity has the form $\alpha q + p = \beta$, where $\alpha = k/r$ and $\beta = \alpha q_1 + p_1$. (b) Show that unless the price p per unit of the commodity is less than β dollars, there will be no demand for the commodity. (c) Explain why the demand q units for the commodity must satisfy $0 \le q \le \beta/\alpha$. (d) Assuming that the demand is met, so that $q = x = $ the production level, write a formula for the revenue $R(x)$ at production level x. (e) Assuming that $q = x$, find the production level that maximizes the revenue.

26 Suppose that q units of a certain commodity are demanded when the producer charges a price p per unit. The quantity E

defined by

$$E = -\frac{p}{q} \frac{dq}{dp}$$

is called the **elasticity of demand** with respect to price. (a) Explain why under normal economic conditions you might expect that $E \ge 0$. (b) Assuming that the demand is met by the producer, so that $q = x = $ the level of production, show that the rate of change of revenue R with respect to price p is given by

$$\frac{dR}{dp} = x(1 - E)$$

(c) Economists say that the demand is **elastic** if $E > 1$. If the demand is elastic, explain why a slight increase in price p will result in a loss of revenue.

27 A *linear cost function* has the form $C(x) = C_0 + Ax$ for $0 \le x \le M$, where $C_0 \ge 0$ and $A > 0$. Here, C_0 is the fixed cost, M is the maximum possible production level, and each additional unit of the commodity costs the same amount, A dollars, to produce. Such a linear cost function applies to situations in which production efficiency is the same at all possible levels of production, so there is no advantage in mass production. For a commodity with a linear cost function, suppose the demand equation has the form $\alpha q + p = \beta$ with $\alpha > 0$ and $\beta > 0$ (see Problem 25). Assume that the demand is met by production, so that $q = x$. (a) Show that the profit function P has the form $P(x) = -\alpha x^2 + (\beta - A)x - C_0$ for $0 \le x \le M_1$, where M_1 is the smaller of the two numbers M and β/α (or their common value if they are equal). (b) If $0 < \beta - A \le 2\alpha M$, show that maximum profit is obtained at the production level $x = (\beta - A)/(2\alpha)$.

28 In Problem 26, show that the elasticity of demand E is unaffected by a change in the units in which the commodity is measured (for instance, kilograms rather than pounds) and that it is unaffected by a change in monetary units (for instance, cents rather than dollars per unit of the commodity).

29 A *quadratic cost function* has the form $C(x) = C_0 + bx - ax^2$ for $0 \le x \le M$, where $C_0 \ge 0$ and $a \ne 0$. Of course, it is required that $C(x) \ge 0$ for $0 \le x \le M$. (a) Explain why the condition $a > 0$ corresponds to the situation in which production becomes more efficient as the production level increases. (b) Under normal economic conditions, the marginal $C'(x)$ will be positive for $0 < x < M$. Show that this is so if and only if $b > 0$ and $a < b/(2M)$. (c) Show that the average production cost per unit of the commodity at the maximum production level M is given by $(C_0/M) + b - aM$. (d) If $a > 0$, what is the most efficient production level? (e) If $0 < a < b/(2M)$, show that the condition $C(x) \ge 0$ for $0 \le x \le M$ holds automatically.

30 For the quadratic cost function of Problem 29, assume that $0 < a < b/(2M)$ and that the demand equation has the form $\alpha q + p = \beta$ with $\alpha > 0$ and $\beta > 0$ (see Problem 25). Suppose that the demand is met by production, so that $q = x$. (a) If P is the profit function, write a formula for $P(x)$. (b) What production level will generate maximum profit?

Review Problem Set, Chapter 3

In Problems 1 to 6, verify the hypotheses of Rolle's theorem for each function f on the indicated closed interval $[a, b]$, and then find all numbers c in the open interval (a, b) for which $f'(c) = 0$.

1 $f(x) = x^2 + 6x - 7$, $[a, b] = [-7, 1]$

2 $f(x) = x^3 - x$, $[a, b] = [0, 1]$

3 $f(x) = 4x^3 - 21x^2 + 25$, $[a, b] = [-1, 5]$

4 $f(x) = 2x^3 - 27x^2 + 25x$, $[a, b] = [0, 1]$

5 $f(x) = \cos x$, $[a, b] = \left[-\dfrac{\pi}{2}, \dfrac{3\pi}{2}\right]$

6 $f(x) = \sin^2 x$, $[a, b] = [0, 2\pi]$

7 Which of the following functions, for one reason or another, fail to satisfy the hypotheses of Rolle's theorem for the closed interval $[a, b]$? In each case where a failure occurs, indicate the hypothesis that fails.

(a) $f(x) = x^2 - 25$, $[a, b] = [-5, 5]$

(b) $f(x) = 1 - |x|$, $[a, b] = [-1, 1]$

(c) $f(x) = 1 - (x - 2)^{2/3}$, $[a, b] = [1, 3]$

(d) $f(x) = \begin{cases} 4 - x^2 & \text{if } x < 1 \\ 5 - 2x & \text{if } x \geq 1 \end{cases}$, $[a, b] = [-2, \frac{5}{2}]$

8 Use Rolle's theorem to prove that if $f'(x) > 0$ for $a < x < b$, then there is at most one x for which $a < x < b$ and $f(x) = 0$.

9 Which of the functions whose graphs are shown in Figure 1 fail to satisfy the hypotheses of the mean-value theorem on the interval $[a, b]$? In each case where a failure occurs, indicate the hypothesis that fails.

Figure 1

(a)

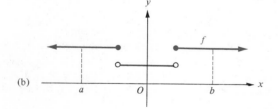

(b)

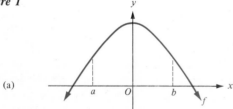

(c)

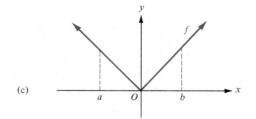

(d)

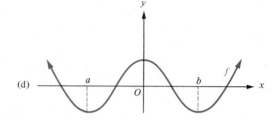

(e)

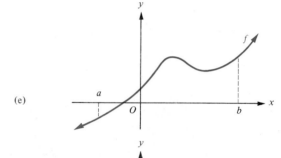

(f)

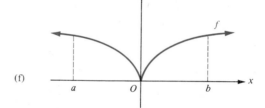

(g)

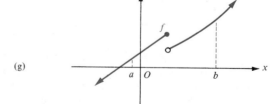

(h)

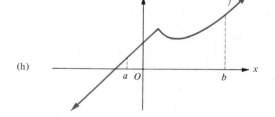

10 Find all values of c such that $-1 < c < 1$ and the tangent line to the graph of $f(x) = 3x^4 + 2x^3 - x^2 + x - 1$ at $(c, f(c))$ is parallel to the secant between the points $(-1, f(-1))$ and $(1, f(1))$.

In Problems 11 to 16, verify the hypotheses of the mean-value theorem for each function f on the indicated closed interval $[a, b]$. Then find all numbers c in the open interval (a, b) for which $f(b) - f(a) = f'(c)(b - a)$.

11 $f(x) = \sqrt{x}$, $[a, b] = [1, 4]$

12 $f(x) = \dfrac{x - 4}{x + 4}$, $[a, b] = [0, 4]$

13 $f(x) = x^3 - 2x^2 + 3x - 2$, $[a, b] = [0, 2]$

14 $f(x) = \begin{cases} \dfrac{3 - x^2}{2} & \text{if } x \le 1 \\ \dfrac{1}{x} & \text{if } x > 1 \end{cases}$, $[a, b] = [0, 2]$

15 $f(x) = x + \sin^2 x$, $[a, b] = [0, 2\pi]$

16 $f(x) = x + \sqrt{1 - \cos x}$, $[a, b] = \left[-\dfrac{\pi}{2}, \dfrac{\pi}{2}\right]$

ⓒ In Problems 17 and 18, use a calculator and Newton's method to find an approximation to a value of c in the open interval (a, b) for which $f(b) - f(a) = f'(c)(b - a)$.

17 $f(x) = \cos x$, $[a, b] = \left[0, \dfrac{\pi}{3}\right]$

18 $f(x) = x^4 + x^3 + 2x^2 - x + 3$, $[a, b] = [-1, 1]$

19 Use Rolle's theorem to prove that if f'' exists on the open interval (a, b) and if f'' maintains a constant algebraic sign on this interval, then f has at most one critical number in the interval (a, b).

20 Use the mean-value theorem to prove that $\sqrt{a} < (a + 1)/2$ for $0 < a < 1$. [*Hint:* Let $f(x) = \sqrt{x}$ and apply the mean-value theorem on the interval $[a, 1]$. Use the fact that if $0 < c < 1$, then $0 < \sqrt{c} < 1$.]

21 For each graph in Figure 2, indicate the intervals on which the function is increasing or decreasing?

Figure 2

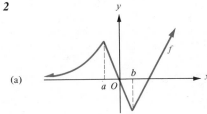

(a)

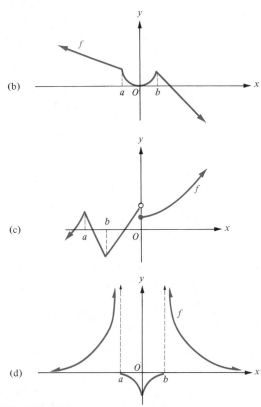

(b)

(c)

(d)

22 Suppose that f and g are increasing functions on the interval I. (a) Is $3f$ increasing on I? Why? (b) Is $-3f$ increasing on I? Why? (c) Is $f + g$ increasing on I? Why? (d) Is $f \cdot g$ increasing on I? Why?

In Problems 23 to 30, indicate the intervals on which each function is increasing or decreasing.

23 $f(x) = x^3 + 3x^2 - 2$

24 $g(x) = x^3 + 6x^2 + 9x + 3$

25 $g(x) = \sqrt[3]{x}(x - 4)^2$

26 $h(x) = x^3 + \dfrac{4}{x}$

27 $f(x) = \begin{cases} x + 1 & \text{if } x \le 0 \\ x^2 & \text{if } x > 0 \end{cases}$

28 $g(x) = \begin{cases} 1/x & \text{if } x < 0 \\ (x - 1)^2 & \text{if } 0 \le x \le 2 \\ 1/x^3 & \text{if } x > 2 \end{cases}$

29 $F(x) = \cos^2 x$, $-2\pi \le x \le 2\pi$

30 $G(x) = \dfrac{x}{2} - \sin x$, $0 \le x \le 4\pi$

In Problems 31 to 52, find all the critical numbers of the function. Then use either the first- or the second-derivative test to find all numbers at which the function has a relative maximum or minimum.

31 $f(x) = x^3 - x^2 + 1$

32 $g(x) = 2x^3 + 3x^2 - 12x - 2$

33 $h(x) = 2x^3 - 9x^2 + 12x + 1$ **34** $F(x) = \frac{1}{3}x^3 + \frac{1}{2}x^2 - 2x$

35 $G(x) = \frac{1}{3}x^3 + x^2 - 8x + 1$ **36** $H(x) = x^4 - 6x^2 + 8x$

37 $f(x) = (x^2 - 9)^2$ **38** $g(x) = (x - 3)^2(x + 1)^2$

39 $h(x) = x^2 - x^{-2}$ **40** $F(x) = x(x + 1)^{-1}$

41 $G(x) = x^2 + 2x^{-1}$ **42** $H(x) = (x^2 - 16)^{-1}$

43 $p(x) = \dfrac{16}{x^2 + 4}$ **44** $q(x) = x\sqrt{x - 1}$

45 $r(x) = \sqrt{2x^2 + 9}$ **46** $P(x) = 10(x + 2)^{-1/2}$

47 $Q(x) = x + \sin x, \ -2\pi \leq x \leq 2\pi$

48 $R(x) = x + \cos 2x, \ 0 \leq x \leq 2\pi$

49 $f(x) = \cos^2 x + 2 \cos x, \ -2\pi \leq x \leq 2\pi$

50 $g(x) = \sin 2x + 2 \cos x, \ 0 \leq x \leq 2\pi$

51 $h(x) = \begin{cases} 2x + 1 & \text{if } x \geq 1 \\ 4 - x^2 & \text{if } x < 1 \end{cases}$

52 $F(x) = \begin{cases} \sqrt{x + 1} & \text{if } x < 0 \\ 1 + \sqrt{x} & \text{if } x \geq 0 \end{cases}$

In Problems 53 to 66, indicate the intervals on which the graph of each function is concave upward or downward. Also find all points of inflection of each graph.

53 $f(x) = x^3 - 8x$ **54** $g(x) = x^3 - 6x^2 + 9x + 1$

55 $h(x) = -2x^3 + 4x^2 + 5$ **56** $F(x) = x^2(x^2 - 6)$

57 $G(x) = 2x^3 + 4x^2 + 2x + 1$

58 $H(x) = x^4 - 8x^3 + 64x + 8$

59 $p(x) = 2x^4 + 4x^3 - 24x^2 + x - 3$

60 $q(x) = \dfrac{2x}{(x + 3)^2}$

61 $r(x) = \dfrac{3x}{x^2 - 9}$ **62** $s(x) = \dfrac{x + 1}{x^2 + 1}$

63 $P(x) = 2 \cos 2x, \ 0 \leq x \leq 2\pi$

64 $Q(x) = 8 \sin x + \sin 2x, \ 0 \leq x \leq 2\pi$

65 $R(x) = x + \tan x$

66 $f(x) = \begin{cases} 1 + \sin 2x & \text{if } -\pi \leq x < 0 \\ \cos 2x & \text{if } 0 \leq x \leq \pi \end{cases}$

67 As a result of a time-and-motion study, management at a factory determines that a crew of x persons can unload y trucks per day, where

$$y = \frac{x^2}{25}\left(3 - \frac{x}{6}\right) \qquad \text{for } 0 \leq x \leq 16$$

If too many people are put on the work crew, they get in each other's way and the efficiency of the crew decreases. (a) For what value of x does the point of diminishing returns occur? (b) For what value of x will the maximum number of trucks per day be unloaded? (c) Sketch the graph of y as a function of x. (Although x can assume only integer values, you may treat it as a continuous variable for this problem.)

68 Suppose that f and g are continuous functions, both of which have relative minima at x_1. Given that $f(x_1) > 0$ and $g(x_1) > 0$, show that the product function fg has a relative minimum at x_1.

69 Let a, b, c, and d be constant numbers such that $ad \neq bc$ and $c \neq 0$. Show that the graph of the function f defined by the equation $f(x) = (ax + b)/(cx + d)$ has no points of inflection.

70 Suppose that the function f is defined by the equation $f(x) = ax^3 + bx^2 + cx + d$, where a, b, c, and d are constants with $a > 0$. Show that the graph of f has exactly one point of inflection, the point

$$\left(-\frac{b}{3a}, f\left(-\frac{b}{3a}\right)\right)$$

Also show that f' is increasing when $x > -b/(3a)$.

In Problems 71 to 78, find the absolute maximum and minimum values of the function on the given interval and indicate where they occur. ⒸSketch the graph of the function.

71 $f(x) = x^3 - 6x^2 + 9x + 1, \ [0, 4]$

72 $g(x) = x^3 - 2x + 1, \ [-1, 1]$

73 $h(x) = (x - 1)^3, \ [-1, 2]$

74 $F(x) = x(x^2 + 2)^{-3/2}, \ [0, \infty)$

75 $G(x) = \dfrac{x + 1}{x - 1}, \ [-3, 3]$

76 $H(x) = (x^2 + x)^{2/3}, \ [-2, 3]$

77 $f(x) = \sin^2 x + \cos x, \ [-\pi, \pi]$

78 $g(x) = \sqrt{3} \sec x - \tan x, \ \left[0, \dfrac{\pi}{4}\right]$

In Problems 79 to 88, find the maximum and minimum values (if they exist) of the function and indicate where they occur.

79 $f(x) = 1 + x - x^2$ **80** $g(x) = x^3 + x - 1$

81 $h(x) = x^4 - 2x^3 + 1$ **82** $F(x) = x^4 + 4x - 3$

83 $G(x) = \dfrac{x + 1}{x^2 + 1}$ **84** $H(x) = \dfrac{x^2}{x^2 + 4}$

85 $p(x) = x + \dfrac{1}{\sqrt{x}}$ **86** $q(x) = x\sqrt{x + 1}$

87 $r(x) = 1 - 2 \sin^2 x$ **88** $R(x) = \sqrt{\sec x}$

In Problems 89 to 102, evaluate the limit.

89 $\lim\limits_{x\to+\infty} \dfrac{8x^2 + x - 3}{4x^2 + 71}$

90 $\lim\limits_{x\to-\infty} \dfrac{x^2 + 1}{5x + 3}$

91 $\lim\limits_{t\to-\infty} \dfrac{5t}{t^2 + 1}$

92 $\lim\limits_{t\to+\infty} \dfrac{3t^{-2} + 7t^{-3}}{7t^{-2} + 5t^{-3}}$

93 $\lim\limits_{h\to+\infty} \dfrac{h^2 - 3h}{\sqrt{5h^4 + 7h^2 + 3}}$

94 $\lim\limits_{x\to-\infty} \dfrac{\sqrt{7x^6 + 5x^4 + 7}}{x^4 + 2}$

95 $\lim\limits_{y\to-\infty} (4y^2 - 7y)$

96 $\lim\limits_{\theta\to+\infty} \left(\theta - \theta\cos\dfrac{1}{\theta}\right)$

97 $\lim\limits_{t\to3^-} \dfrac{t}{t^2 - 9}$

98 $\lim\limits_{y\to2^+} \dfrac{\sqrt{y} - 2}{y^2 - 4}$

99 $\lim\limits_{x\to0^+} \dfrac{x^2 - 3}{x^2 - x}$

100 $\lim\limits_{x\to0^-} \csc x$

101 $\lim\limits_{x\to\pi} \cot x$

102 $\lim\limits_{x\to2^+} (3 + [\![2x - 4]\!])$

In Problems 103 to 110, find all horizontal and vertical asymptotes of the graph of the function. ⒸSketch the graph.

103 $f(x) = \dfrac{x + 3}{2x + 1}$

104 $g(x) = \dfrac{1}{x(x + 1)} - \dfrac{1}{x}$

105 $h(x) = \dfrac{1}{x^2 + 1}$

106 $F(x) = \dfrac{x + 2}{x - 2}$

107 $G(x) = \dfrac{x}{\sqrt{1 + x}}$

108 $H(x) = \dfrac{5x}{\sqrt{4x^2 + 1}}$

109 $f(x) = \tan 2x$

110 $g(x) = \tan (x/2)$

In Problems 111 to 114, find the oblique asymptote of the graph of the function.

111 $f(x) = \dfrac{2x^2 + 3x + 1}{x}$

112 $g(x) = \dfrac{3x^3 - 2x^2 + 5x - 2}{x^2 + 1}$

113 $h(x) = 1 - x - \dfrac{x - 2}{x^2 - 1}$ **114** $F(x) = 2x - x\cos\dfrac{1}{x}$

In Problems 115 and 116, show that the graph of the function has a vertical tangent line at the indicated point.

115 $f(x) = \sqrt[3]{x + 1}$ at $(-1, 0)$ **116** $g(x) = x^{1/5} - 1$ at $(0, -1)$

Ⓒ In Problems 117 to 130, sketch the graph of the function.

117 $f(x) = x^3 - 8x$

118 $g(x) = x^3 - 4x^2$

119 $h(x) = x^4 - 2x^2 + 1$

120 $F(x) = 2x^4 - 8x^3$

121 $G(x) = x^{2/3}(x - 1)$

122 $H(x) = x^{2/3}(x^2 - 1)$

123 $f(x) = \dfrac{2x - 1}{3x + 2}$

124 $g(x) = \dfrac{x^3}{x^2 + x - 6}$

125 $h(x) = x + \dfrac{1}{\sqrt{x}}$

126 $F(x) = \sqrt{\dfrac{1 - x}{1 + x}}$

127 $G(x) = \dfrac{x}{\sqrt{x^2 + 1}}$

128 $H(x) = 5\sin\left(\pi x - \dfrac{\pi}{3}\right)$

129 $f(x) = \cos 2x + 2\cos x$

130 $g(x) = x + \cot x$

Ⓒ**131** Solar Electronics is planning to hire and train several new workers for its calculator division. As a result of previous experience, management determines that the mathematical model

$$n = 1 + \dfrac{99t^2}{100 + t^2}$$

accurately predicts the number n of calculators per hour that can be assembled by a worker after t hours of on-the-job training. (a) Treating n as a continuous variable, plot a graph of n as a function of t for $t \ge 0$. (b) After how many hours of on-the-job training is the worker learning most rapidly? (c) What is the meaning of the horizontal asymptote of the graph of n as a function of t?

Ⓒ**132** As a result of delicate experiments conducted by the zoologist T. Weis-Fogh, a number of mathematical models have been constructed for the motion of insect wings in flight. One such model for locust wings is

Forewing: $\theta = 0.6\cos(107t - 2.10)$

Hindwing: $\theta = \cos(107t - 1.57)$

where θ is the angle between the wing and the horizontal at time t seconds. Sketch graphs on the same $t\theta$ coordinate system for the motions of the forewing and the hindwing of a locust according to this model. Notice that the forewing and the hindwing are slightly "out of phase." How many complete wing beats per second does the locust make?

133 Find the value of x, $0 \le x \le 5$ (Figure 3), that minimizes the given expression.

(a) $\overline{|AP|} + \overline{|PB|}$ (b) $\overline{|AP|}^2 + \overline{|PB|}^2$

(c) $\overline{|AP|}^2 - \overline{|PB|}^2$ (d) Area ACP + area BDP

Figure 3

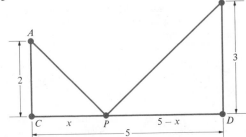

134 A long rectangular sheet of tin is 8 centimeters wide. Find the depth of the V-shaped trough of maximum cross-sectional area that can be made by bending the plate along its central longitudinal axis.

135 Find the dimensions of the rectangle of greatest area that can be inscribed in (a) a circle of radius 5 meters, (b) a semicircle of radius 5 meters, (c) an isosceles triangle of base 10 meters and altitude 10 meters, and (d) an isosceles trapezoid with bases of 10 and 6 meters and an altitude of 8 meters.

136 A long sheet of paper is 8 inches wide. One corner of the paper is folded over (Figure 4). Find the value of x that gives the right triangle ABC the least possible area.

Figure 4

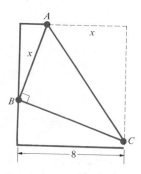

137 The range of a projectile is given by the formula $R = (v_0^2 \sin 2\theta)/g$, where v_0 is the muzzle velocity, g is the gravitational acceleration, and θ is the angle of elevation. Find the angle of elevation that gives the maximum range.

138 A cylindrical can without a top is to be formed from sheet metal of uniform thickness and is to weigh 100 grams. What is the relationship between its height h and its base radius r if the volume V of the can is to be maximum?

139 A newly independent country designs a flag that consists of a red rectangular region divided by a green stripe. The perimeter of the flag is to be 8 meters, and the red part is to have an area of 2 square meters. What are the dimensions of the green stripe of largest possible area?

140 In the theory of probability, it is shown that the function L defined by

$$L(p) = \frac{n!}{k!(n-k)!}p^k(1-p)^{n-k} \qquad \text{for } 0 \le p \le 1$$

has a value equal to the probability of exactly k successes in n independent trials when the probability of success in each trial is p. Here, of course, n and k are integers and $0 \le k \le n$. An important problem in statistics is to estimate the value of p on the basis of experimentally determined values of n and k. The **method of maximum likelihood** uses, for such an estimate, the value of p that maximizes $L(p)$. Find this value of p in terms of the constants n and k.

141 Suppose that n identical nickel-cadmium cells are to be arranged in series-parallel to furnish current to the motor of an experimental electric car. The motor has a resistance of R ohms, while each cell has an internal resistance of r ohms and an electromotive force of E volts. In the arrangement, x cells are to be connected in series so that the battery will have a net electromotive force of xE volts and an internal resistance of x^2r/n ohms. The current delivered to the motor is given by

$$I = \frac{xE}{R + (x^2r/n)}$$

Solve for the value of x that will maximize the current I. (Although x represents a whole number here, you may treat it as a continuous variable.)

142 If air resistance is neglected, it can be shown that the stream of water projected from a fire hose satisfies the equation

$$y = mx - \frac{g}{2}(1 + m^2)\left(\frac{x}{v}\right)^2$$

where m is the slope of the nozzle, v is the velocity of the stream at the nozzle, y is the height of the stream x units from the nozzle, and g is the acceleration of gravity (Figure 5). Assume that v and g are positive constants. (a) For a fixed value of m, find the value of x for which the height y of the stream is maximum. (b) For a fixed value of m, find the distance d from the nozzle at which the stream hits the ground. (c) Find the value of m for which the stream hits the ground at the greatest distance from the nozzle. (d) Find the value of m for which the water reaches the greatest height on a vertical wall x units from the nozzle.

Figure 5

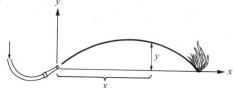

143 Ship A is anchored 3 kilometers directly from a point B off the shore of a lake (Figure 6). Opposite a point D, 5 kilometers farther along the shore, another ship E is anchored 9 kilometers directly from point D. A boat is to take some passengers from ship A to a point C on shore and proceed to ship E. Find the shortest course of the boat.

Figure 6

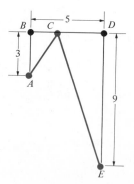

144 A weight of 1000 kilograms hanging 2 meters from a point A at one end of a lever is to be lifted by an upward force F from the other end B. Suppose that the lever is to weigh 10 kilograms per meter (Figure 7). Find the length of the lever if the force at end B is to be a minimum.

Figure 7

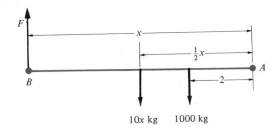

10x kg   1000 kg

145 A sector ABC of central angle θ is cut out from a circular sheet of metal of radius a (Figure 8), and the remainder is used to form a conical shape. Find the maximum volume of such a cone.

Figure 8

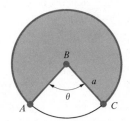

146 A right circular cone whose generators make an angle θ with its central axis is inscribed in a sphere of fixed radius a. For what value of θ will the lateral area of the cone be greatest?

147 A projectile fired from the foot of a 30° slope will strike the slope at a horizontal distance x given by the equation

$$x = \frac{v_0^2}{g}\left[\sin 2\theta - \frac{1}{\sqrt{3}}(1 + \cos 2\theta)\right]$$

where v_0 is the muzzle velocity, θ is the angle of elevation of the gun, g is the acceleration of gravity, and air resistance is neglected (Figure 9). For what value of θ will the projectile reach the farthest distance up the slope?

Figure 9

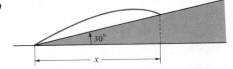

148 A ladder 8 meters long leans against a building. At what angle should the ladder be inclined to give maximum headroom under it at a point 2 meters from the building?

149 A gutter is to be made out of a long sheet of metal 30 centimeters wide by turning up strips 10 centimeters long on each side so that they make equal angles θ with the vertical (Figure 10).

For what angle θ will the carrying capacity of the gutter be maximum?

Figure 10

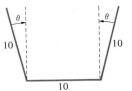

150 The **arithmetic mean** $\overline{x}$ of the numbers $x_1, x_2, x_3, \ldots, x_n$ is defined by

$$\overline{x} = \frac{x_1 + x_2 + x_3 + \cdots + x_n}{n}$$

Show that $\overline{x}$ is the value of x that minimizes the function

$$f(x) = (x - x_1)^2 + (x - x_2)^2 + (x - x_3)^2 + \cdots + (x - x_n)^2$$

151 A heavy wooden crate is being dragged along a horizontal surface by a chain making an angle θ with the horizontal. The force F exerted on the chain is given by the equation

$$F = \frac{400}{0.4 \sin \theta + \cos \theta} \qquad \text{newtons}$$

Find the angle θ for which the force F is minimum.

152 A real estate company manages an apartment building containing 80 units. When the rent for each unit is $250 per month, all apartments are occupied. However, for each $10 increase in monthly rent per unit, one of the units becomes vacant. Each vacant unit costs the management $15 per month for taxes and upkeep, and each occupied unit costs the management $65 per month for taxes, service, upkeep, and water. What rent should be charged for a maximum profit?

153 A concession at a large city park rents bicycles on a daily basis. All 100 of its bikes are rented at a rate of $10 per bike per day, but for each $1 increase in the daily rental rate per bike, 10 fewer bikes will be rented per day. What daily rental rate per bike will yield the maximum revenue to the concession?

154 A manufacturer of sports trophies knows that the daily cost of manufacturing x trophies is $60 + 6x$ dollars and that the corresponding sales revenue is $30x - \frac{2}{5}x^2$ dollars. Find the daily production level for trophies that will maximize the profit.

155 The manufacturer of a certain microcomputer finds that in order to sell x microcomputers each week, he must price them at $\sqrt{5,000,000 - 2x^2}$ dollars each. How many microcomputers per week will bring the largest total revenue?

156 A garage owner finds that she can sell x tires per week at p dollars per tire, where $p = \frac{1}{3}(375 - 5x)$ and the cost C in dollars of obtaining x tires per week to sell is expressed by $C = 500 + 15x + (x^2/5)$. Find the number of tires she must sell per week and the price she should charge per tire to maximize the profit.

157 The labor force y required by a firm to manufacture x units of a certain product is given by the equation $y = \frac{1}{2}\sqrt{x}$. Find the instantaneous rate at which the labor force should be increasing if, at present, there is a demand for 40,000 units of the product, but the demand is increasing at the constant rate of 10,000 units per year.

158 The demand q in thousands of boxes per week for a detergent is given by $q = (1000/p) - 30$, where p is the price in dollars per box of detergent. The current price is $p = \$0.83$ per box, but inflation is increasing the price at the rate of $\$0.01$ per month. Find the current instantaneous rate of change of the demand.

159 The price of apples in a certain marketing area is given by the equation $p = 2 + 60(30 + x)^{-1}$, where x is the supply in thousands of bushels and p is the price in dollars per bushel. At what rate will the price per bushel be changing if the current supply of 10,000 bushels is decreasing at the rate of 200 bushels per day?

☐ 160 A boat sails parallel to a straight beach at a constant speed of 19.2 kilometers per hour, staying 6.4 kilometers offshore. How fast is it approaching a lighthouse on the shoreline at the instant it is exactly 8 kilometers from the lighthouse?

161 A 10-meter ladder leans against a vertical wall. The lower end is pulled horizontally away from the wall at a constant rate of 2 meters per second, and the top slides down the wall. Find the rate at which the top moves (a) when the lower end is 3 meters from the wall and (b) when the upper end is 8 meters above ground level.

162 Water is flowing through a hole in the bottom of a hemispherical bowl of radius 10 centimeters. At the instant the water is 6 centimeters deep, the rate of flow is 5 cubic centimeters per minute. Find the rate at which the surface of the water is falling at this instant.

163 A highway crosses a railroad track at right angles. A car traveling at the constant rate of 40 kilometers per hour goes through the intersection 2 minutes before the engine of a train traveling at the constant rate of 36 kilometers per hour goes through the intersection. At what rate are the car and engine separating 10 minutes after the train goes through the intersection?

164 The height of a right circular cylinder is increasing at the rate of 3 centimeters per minute, and the radius of the base is decreasing at the rate of 2 centimeters per minute. Find the rate at which the volume of the cylinder is changing when the height is 10 centimeters and the radius of the base is 4 centimeters.

165 A boy is going to lift a weight W by means of a rope mounted on a pulley (Figure 11) that is 12 meters above the ground. At the start the weight is resting on the ground, and the boy, stand-

ing directly under the pulley, grasps the rope 2 meters above the ground. If the boy, holding fast to the rope and keeping his

Figure 11

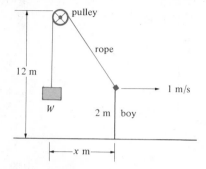

hand 2 meters above the ground, walks away at a constant rate of 1 meter per second, how fast is the weight rising 2 seconds after he starts to walk?

166 From a point 0.5 kilometer from a straight road, a searchlight is kept trained upon a car which travels along the road at a constant speed of 80 kilometers per hour. At what rate is the beam of light turning when the car is at the nearest point of the road?

167 A balloon rises from the ground 1000 feet from an observer and ascends vertically at the rate of 20 feet per second. How fast is the angle of elevation changing at the observer's position at the instant when the balloon is 2000 feet from the ground?

168 A liquid nitrogen tank has the shape of a horizontal cylinder with a radius of 2 meters and a length of 10 meters. The tank is vented to allow the evaporating liquid nitrogen to escape. At the instant when the tank is half full, the surface level of the liquid nitrogen is dropping at the rate of 2 centimeters per hour. Find the rate of evaporation, in cubic meters per hour, at this instant.

☐ 169 Conduct a marginal analysis using the following information: Polar Oil Company is looking into the possibility of setting up a plant to manufacture synthetic fuel (synfuel) from coal. It projects that by using a new process, up to 1400 barrels per day could be produced at a fixed cost of $1500 per day and a variable cost of

$$(3.5 \times 10^{-6})x^3 - (1.05 \times 10^{-2})x^2 + 40.5x \qquad \text{dollars}$$

where x is the number of barrels produced per day. Market research indicates that at a price of $40 per barrel, 700 barrels per day can be sold; but for each $1 increase in the price per barrel, 100 fewer barrels per day will be sold.

170 In Problem 169, calculate the elasticity of demand with respect to price when the price is $40 per barrel. (See Problem 26, page 235.)

4 ANTIDIFFERENTIATION AND DIFFERENTIAL EQUATIONS

In the previous chapters we have seen that differentiation has applications to problems in areas ranging from geometry and physics to business and economics. In this chapter we introduce *antidifferentiation*, which reverses the procedure of differentiation and allows us to find functions that have a given function as their derivative. We show how antidifferentiation permits us to solve simple *differential equations*, and we relate antidifferentiation to a number of practical problems. However, before going into these matters, we introduce the idea of *differentials*, since *differential notation* is useful in connection with antidifferentiation.

4.1   Differentials and Linear Approximation

In Section 2.2 we introduced the Leibniz notation dy/dx for the derivative; however, we were careful to point out that dy/dx should not be regarded as being a fraction until the "numerator" dy and the "denominator" dx are given separate meanings. Shortly, we provide such meanings for the so-called "differentials" dx and dy.

Leibniz himself regarded dx and dy as being "infinitesimals," that is, quantities that, although they are nonzero, are smaller in magnitude than any finite quantity. He imagined that in the limit, Δx and Δy somehow become "infinitesimal quantities" dx and dy, respectively, so that the difference quotient $\Delta y/\Delta x$ becomes the derivative dy/dx. Leibniz's point of view has persisted, and even today some mathematicians and most engineers and scientists prefer to think of dx and dy as infinitesimals.

Part of Leibniz's concept can be salvaged by regarding dy/dx as being an honest ratio, if not of infinitesimals, then of differentials dy and dx and by rewriting the equation $dy/dx = f'(x)$ as $dy = f'(x)\,dx$. The last equation will serve as a *definition* of the differential dy, provided that an appropriate meaning can be given to the differential dx. This is accomplished as follows.

<u>DEFINITION 1</u>   **Differentials**

Let f be a function, and let x and y be variables so related that $y = f(x)$. Then the **differential** dx is a quantity that can take on (or be assigned) any value in $\mathbb{R}$. If x is any number in the domain of f for which $f'(x)$ exists, then the **differential** dy is defined by

$$dy = f'(x)\ dx$$

<u>EXAMPLE 1</u>   If $y = f(x) = 3x^2 - 2x + 1$, find dy.

SOLUTION   Here $f'(x) = 6x - 2$, so $dy = (6x - 2)\ dx$.

Notice the distinction between the differential dx of the independent variable x and the differential dy of the dependent variable y. Whereas dx can be given any value whatsoever, dy depends for its value on x and dx (not to mention f). For instance, in the example above when $x = 1$ and $dx = 0.002$, we have

$$dy = (6 - 2)(0.002) = 0.008$$

Because of Definition 1, from now on we can regard the derivative dy/dx as an actual fraction with numerator dy and denominator dx, and we can manipulate such fractions according to the usual rules of algebra.* In particular, an equation involving derivatives expressed in Leibniz notation may be converted to an equation involving differentials by "clearing fractions" as usual. For instance, if

$$y = u + v$$

then the equation

$$\frac{dy}{dx} = \frac{du}{dx} + \frac{dv}{dx}$$

can be rewritten as

$$dy = du + dv$$

simply by multiplying both sides by dx and canceling in the usual way.

In Table 1, some of the standard formulas for derivatives have been converted into formulas for differentials. In these formulas u and v are variables that are assumed to be differentiable functions of x, c is a constant, and n is a constant rational exponent. Formulas for the differentials of cotangent, secant, and cosecant functions can be obtained in a similar manner (Problems 21, 22, and 23).

<u>EXAMPLE 2</u>   Let $y = 47x^3 - 21x^2 + 3x^{-1}$. Find dy.

SOLUTION

$$dy = 141x^2\ dx - 42x\ dx - 3x^{-2}\ dx$$

or

$$dy = (141x^2 - 42x - 3x^{-2})\ dx$$

Table 1

Derivatives	Differentials
1 $\dfrac{dc}{dx} = 0$	I $dc = 0$
2 $\dfrac{d(cu)}{dx} = c\dfrac{du}{dx}$	II $d(cu) = c\ du$
3 $\dfrac{d(u + v)}{dx} = \dfrac{du}{dx} + \dfrac{dv}{dx}$	III $d(u + v) = du + dv$
4 $\dfrac{d(uv)}{dx} = u\dfrac{dv}{dx} + v\dfrac{du}{dx}$	IV $d(uv) = u\ dv + v\ du$
5 $\dfrac{d\left(\dfrac{u}{v}\right)}{dx} = \dfrac{v\dfrac{du}{dx} - u\dfrac{dv}{dx}}{v^2}$	V $d\left(\dfrac{u}{v}\right) = \dfrac{v\ du - u\ dv}{v^2}$
6 $\dfrac{d(cu^n)}{dx} = ncu^{n-1}\dfrac{du}{dx}$	VI $d(cu^n) = ncu^{n-1}\ du$
7 $\dfrac{d}{dx}\sin u = \cos u\dfrac{du}{dx}$	VII $d(\sin u) = \cos u\ du$
8 $\dfrac{d}{dx}\cos u = -\sin u\dfrac{du}{dx}$	VIII $d(\cos u) = -\sin u\ du$
9 $\dfrac{d}{dx}\tan u = \sec^2 u\dfrac{du}{dx}$	IX $d(\tan u) = \sec^2 u\ du$

*Here, of course, we are assuming that $dx \neq 0$.

EXAMPLE 3 Let $y = \sqrt{3 - x^5}$. Find dy.

SOLUTION $\quad dy = \dfrac{d(3 - x^5)}{2\sqrt{3 - x^5}} = \dfrac{-5x^4}{2\sqrt{3 - x^5}} \, dx$

EXAMPLE 4 Let $z = \tan \dfrac{x}{2}$. Find dz.

SOLUTION $\quad dz = \sec^2 \dfrac{x}{2} \, d\left(\dfrac{x}{2}\right) = \sec^2 \dfrac{x}{2} \dfrac{dx}{2} = \dfrac{1}{2} \sec^2 \dfrac{x}{2} \, dx$

EXAMPLE 5 Let x and y be differentiable functions of a third variable t, and suppose that $x^3 + 4x^2y + y^3 = 2$. Find the relationship between dx and dy.

SOLUTION "Taking the differential" on both sides of the given equation, we obtain

$$d(x^3 + 4x^2y + y^3) = d(2) \quad \text{or} \quad d(x^3) + 4d(x^2y) + d(y^3) = 0$$

Therefore,

$$3x^2 \, dx + 4[x^2 \, dy + y(2x \, dx)] + 3y^2 \, dy = 0$$

or

$$3x^2 \, dx + 4x^2 \, dy + 8xy \, dx + 3y^2 \, dy = 0$$

By collecting terms involving like differentials, we may rewrite the last equation in the equivalent form

$$(3x^2 + 8xy) \, dx + (4x^2 + 3y^2) \, dy = 0$$

Beginning calculus students often carelessly confuse derivatives and differentials. Derivatives and differentials are closely related—but *they are not the same*. The *differential* of a variable u is always written with a *small d* as du; the *derivative* of u with respect to x is a *ratio* $\dfrac{du}{dx}$ of differentials. Of course, the *derivative* $\dfrac{du}{dx}$ can also be written in operator notation with a *capital D* as $D_x u$. Beware of equations with a differential on one side and a derivative on the other side—an equation such as $du = D_x u$ simply makes no sense! A *correct* equation would be

$$\dfrac{du}{dx} = D_x u \quad \text{or} \quad du = (D_x u) \, dx$$

Suppose that u is a function of v and, in turn, v is a function of x. According to the chain rule, we have

$$\dfrac{du}{dx} = \dfrac{du}{dv} \dfrac{dv}{dx}$$

provided that the derivatives du/dv and dv/dx exist. We pointed out in Section 2.7 that the Leibniz notation makes the chain rule—a rather deep and important fact—*look* obvious. Now that du/dx, du/dv, and dv/dx are actual fractions, can't we regard the chain rule as an algebraic triviality? The answer is still *no*, since the differential dv in du/dv is the differential of v *regarded as an independent variable* (upon which u depends), while the differential dv in dv/dx is the differential of v

regarded as a dependent variable (depending, in fact, on x). Because of this distinction between the dv in du/dv and the dv in dv/dx, we cannot conclude that

$$\frac{du}{dx} = \frac{du}{dv} \cdot \frac{dv}{dx}$$

on purely algebraic grounds. It is the chain rule that justifies the algebraic manipulation rather than vice versa.

In casual calculation with differentials, one does not bother to distinguish between differentials of dependent and independent variables. Miraculously, such carelessness rarely causes any difficulty. The "miracle," of course, is really the chain rule!

A geometric interpretation of differentials can be obtained as follows: Let $y = f(x)$, and assume that f is differentiable at x_1. If we calculate the differential dy when $x = x_1$, we have

$$dy = f'(x_1)\, dx$$

Recall that dx can take on or be assigned any value in $\mathbb{R}$. Thus, let

$$dx = \Delta x$$

denote an increase in the value of x from x_1 to $x_1 + \Delta x$ (Figure 1). Then

$$\Delta y = f(x_1 + \Delta x) - f(x_1)$$

is the corresponding increase in the value of y from the value $f(x_1)$ to the value $f(x_1 + \Delta x)$ as determined by moving up *along the graph of f*. However, since $f'(x_1)$ is the slope of the tangent line to the graph of f at $(x_1, f(x_1))$, it follows that

$$dy = f'(x_1)\, dx$$

gives the corresponding increase in the value of y as determined by moving up *along the tangent line*.

Figure 1

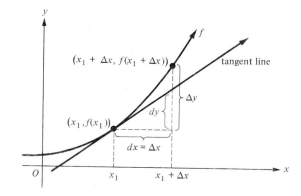

Because the tangent line to the graph of f is a good approximation to the graph itself near the point of tangency, it shouldn't make very much difference whether we move up along the graph of f or up along the tangent line, provided that we don't move very far from the point of tangency. In other words, dy should be a good approximation to Δy provided that $dx = \Delta x$ and Δx is sufficiently small. Of course,

Figure 1 shows a rather special situation in which Δx is positive, the graph of f is concave upward, and f is an increasing function. However, by sketching graphs showing other possibilities,* you can quickly convince yourself that

$$\Delta y \approx dy \qquad \text{if } dx = \Delta x \quad \text{and} \quad \Delta x \text{ is ``small''}$$

© **EXAMPLE 6**   Let $y = f(x) = 3x^2 - 2x + 4$, and put $x_1 = 1$ and $dx = \Delta x = 0.02$.

(a) Calculate $\Delta y = f(x_1 + \Delta x) - f(x_1)$ exactly.

(b) Find an approximation for Δy, using $dy = f'(x_1)\, dx$.

(c) Determine the error $\Delta y - dy$ involved in the approximation $\Delta y \approx dy$.

SOLUTION

(a) $\Delta y = f(x_1 + \Delta x) - f(x_1) = f(1.02) - f(1) = 5.0812 - 5 = 0.0812$

(b) Since $f'(x) = 6x - 2$, it follows that

$$\Delta y \approx dy = f'(x_1)\, dx = [6(1) - 2](0.02) = 0.08$$

(c) The error is given by $\Delta y - dy = 0.0812 - 0.08 = 0.0012$.

© **EXAMPLE 7**   Use differentials to estimate $\sqrt{35}$.

SOLUTION   We know that $\sqrt{36} = 6$, so $\sqrt{35}$ will be a little smaller than 6. To estimate the numerical value of $\sqrt{35}$, let

$$y = \sqrt{x} \qquad x_1 = 36 \qquad dx = \Delta x = -1$$

Then

$$\Delta y = \sqrt{x_1 + \Delta x} - \sqrt{x_1} = \sqrt{35} - \sqrt{36} = \sqrt{35} - 6$$

hence,

$$\sqrt{35} = 6 + \Delta y$$

If we knew the exact numerical value of Δy, we could determine the exact numerical value of $\sqrt{35}$ by using the last equation. Now the idea is to use the fact that $\Delta y \approx dy$, so that

$$\sqrt{35} \approx 6 + dy$$

Here

$$dy = d\sqrt{x} = \frac{1}{2\sqrt{x}}\, dx = \frac{dx}{2\sqrt{x}}$$

Thus, putting $x = x_1 = 36$ and $dx = \Delta x = -1$, we find that

$$dy = \frac{-1}{2\sqrt{36}} = -\frac{1}{12}$$

It follows that

$$\sqrt{35} \approx 6 + dy = 6 - \tfrac{1}{12} \approx 5.9167$$

(Incidentally, the correct value of $\sqrt{35}$ rounded off to four decimal places is 5.9161, so our estimate using differentials is fairly accurate.)

*See the proof of Theorem 1 on page 249 for a more conclusive argument.

Figure 2

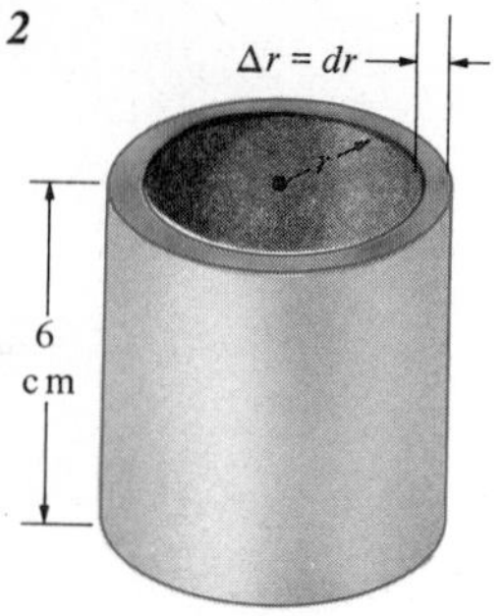

$\Delta r = dr$

6 cm

© **EXAMPLE 8**   Use differentials to find the approximate volume of a right circular cylindrical shell (Figure 2) 6 centimeters high whose inner radius is 2 centimeters and whose thickness is 0.1 centimeter.

SOLUTION   The volume of a right circular cylinder is its height times the area of its base. If V denotes the volume of a (solid) cylinder of height 6 centimeters and radius r centimeters, then $V = 6\pi r^2$ cubic centimeters. Figure 2 shows a cylinder of height 6 centimeters and radius $r = 2$ centimeters inside a larger concentric cylinder of height 6 centimeters and radius $r + \Delta r = 2 + 0.1 = 2.1$ centimeters. The difference ΔV in the volumes of the two cylinders is the required volume of the cylindrical shell. We put $dr = \Delta r = 0.1$ centimeter and use the approximation

$$\Delta V \approx dV = d(6\pi r^2) = 12\pi r \, dr$$

Hence, for $r = 2$ centimeters and $dr = 0.1$ centimeter, we have

$$\Delta V \approx 12\pi(2)(0.1) = 2.4\pi \approx 7.5 \text{ cubic centimeters} \qquad \blacksquare$$

© **EXAMPLE 9**   The radius of a spherical steel ball bearing is measured to be 1.5 centimeters, and it is known that the error involved in this measurement does not exceed 0.1 centimeter. The volume V of the ball bearing is calculated from its measured radius by using the standard formula $V = \frac{4}{3}\pi r^3$. Estimate the possible error in the calculated volume.

SOLUTION   The true value of the radius is $1.5 + \Delta r$, where Δr is the measurement error. We are given that $|\Delta r| \le 0.1$. The true value of the volume is $\frac{4}{3}\pi(1.5 + \Delta r)^3$, while the value of the volume calculated from the measurement of the radius is $\frac{4}{3}\pi(1.5)^3$. The difference $\Delta V = \frac{4}{3}\pi(1.5 + \Delta r)^3 - \frac{4}{3}\pi(1.5)^3$ represents the error in the calculated volume. We put $dr = \Delta r$ and estimate ΔV by dV as follows:

$$\Delta V \approx dV = d(\tfrac{4}{3}\pi r^3) = 4\pi r^2 \, dr = 4\pi(1.5)^2 \, \Delta r = 9\pi \, \Delta r$$

Therefore,

$$|\Delta V| \approx |9\pi \, \Delta r| = 9\pi|\Delta r| \le 9\pi(0.1) = 0.9\pi$$

so the possible error is bounded in absolute value by about $0.9\pi \approx 2.8$ cubic centimeters $\qquad \blacksquare$

The Linear-Approximation Theorem

The approximation $\Delta y \approx dy$ is called a *linear* approximation because it is based on the geometric idea that the tangent line to a curve is a good approximation to the curve near the point of tangency (Figure 1). In scientific work, an approximation of any kind is of limited use unless there is some way to specify bounds on the *error* that might be involved. The **error** E in the linear approximation $\Delta y \approx dy$ is given by

$$\begin{aligned} E &= \text{true value} - \text{approximate value} \\ &= \Delta y - dy \\ &= [f(x_1 + \Delta x) - f(x_1)] - [f'(x_1) \, dx] \\ &= f(x_1 + \Delta x) - f(x_1) - f'(x_1) \, \Delta x \end{aligned}$$

If we solve the last equation for $f(x_1 + \Delta x)$, we can rewrite the linear approximation $\Delta y \approx dy$ in the equivalent form

$$f(x_1 + \Delta x) = f(x_1) + f'(x_1)\, \Delta x + E$$

In other words,

$$f(x_1 + \Delta x) \approx f(x_1) + f'(x_1)\, \Delta x \qquad \text{with error } E$$

Here we do not discuss the question of putting bounds on the numerical value of the error E; such matters are dealt with in Section 10.5. Instead, we prove the following theorem, which shows formally that for "small" values of Δx, the linear approximation is indeed very accurate.

THEOREM 1   **Linear-Approximation Theorem**

If the function f is differentiable at the number x_1, then for all values of Δx for which $x_1 + \Delta x$ is in the domain of f,

$$f(x_1 + \Delta x) = f(x_1) + f'(x_1)\, \Delta x + (\Delta x)\epsilon$$

where the quantity ϵ satisfies the condition that

$$\lim_{\Delta x \to 0} \epsilon = 0$$

PROOF   For each value of Δx such that $x_1 + \Delta x$ is in the domain of f, we define

$$\epsilon = \begin{cases} \dfrac{f(x_1 + \Delta x) - f(x_1)}{\Delta x} - f'(x_1) & \text{if } \Delta x \neq 0 \\ 0 & \text{if } \Delta x = 0 \end{cases}$$

Then for $\Delta x \neq 0$ we have

$$f(x_1 + \Delta x) = f(x_1) + f'(x_1)\, \Delta x + (\Delta x)\epsilon$$

If $\Delta x = 0$, the last equation still holds, since both sides reduce to $f(x_1)$. Therefore, the equation holds for all values of Δx for which $x_1 + \Delta x$ is in the domain of f. We also have

$$\begin{aligned}
\lim_{\Delta x \to 0} \epsilon &= \lim_{\Delta x \to 0} \left[\frac{f(x_1 + \Delta x) - f(x_1)}{\Delta x} - f'(x_1) \right] \\
&= \lim_{\Delta x \to 0} \frac{f(x_1 + \Delta x) - f(x_1)}{\Delta x} - f'(x_1) \\
&= f'(x_1) - f'(x_1) = 0
\end{aligned}$$

and the proof is complete.

In Theorem 1, note that

$$E = (\Delta x)\epsilon$$

This formulation of the error E as a product emphasizes the fact that E approaches zero very rapidly as Δx approaches zero, since *both* factors approach zero as Δx approaches zero.

Problem Set 4.1

In Problems 1 to 20, find dy in terms of x and dx.

1 $y = 3x^2$

2 $y = 1 - x - x^4$

3 $y = x^3 - 4x + 5$

4 $y = \dfrac{3}{2x} - 5x$

5 $y = \dfrac{7x - 2}{3x + 1}$

6 $y = \dfrac{x^2 - 3x + 2}{2x^2 + x + 1}$

7 $y = (3x^2 + 2)^3$

8 $y = (2x^4 - x)^{-3}$

9 $y = \left(\dfrac{1 + x^2}{1 + x}\right)^4$

10 $y = \dfrac{3 - x^2}{\sqrt{x}}$

11 $y = \sqrt{9 - 3x^2}$

12 $y = x^4\sqrt[5]{x^2 + 2}$

13 $y = \sqrt{\dfrac{x - 3}{x + 3}}$

14 $y = \dfrac{x}{\sqrt{x^2 + 5}}$

15 $y = \dfrac{\sqrt{3x + 1}}{x^2 + 7}$

16 $y = \dfrac{x^3}{\sqrt[3]{x + 1}}$

17 $y = 3 \sin 2x$

18 $y = \cos \sqrt{x}$

19 $y = 2 \tan^2 x$

20 $y = \sqrt{\sin \pi x}$

21 Find a formula for $d(\cot u)$.

22 Find a formula for $d(\sec u)$.

23 Find a formula for $d(\csc u)$.   **24** Find a formula for $d\sqrt{u}$.

In Problems 25 to 36, assume that x and y are differentiable functions of t which are related as shown. Find the relationship between dx and dy by "taking the differential" on both sides of each equation.

25 $x^2 + y^2 = 36$

26 $9x^2 = 36 - 4y^2$

27 $9x^2 - 16y^2 = 144$

28 $x^{2/3} + y^{2/3} = 4$

29 $\sqrt[3]{x} = 2 - \sqrt[3]{y}$

30 $x^2 + xy + 3y^2 = 51$

31 $2x^3 + 5y^3 = 13 + 4xy^2 - xy$

32 $\sqrt{1 - x} + \sqrt{1 - y} = 9$

33 $x^3 + y^3 = \sqrt[3]{x + y}$

34 $\sin x + \cos y = 0$

35 $\tan^2 x + \tan^2 y = 1$

36 $1 + \cot xy + xy = 0$

© In Problems 37 to 44, set dx equal to the given value of Δx and use the indicated value of x_1. (a) Use a calculator to find the numerical value of $\Delta y = f(x_1 + \Delta x) - f(x_1)$ to at least four decimal places. (b) Find an approximation for Δy, using $dy = f'(x_1)\,dx$. (c) Determine the error $\Delta y - dy$ involved in the approximation $\Delta y \approx dy$.

37 f is defined by $y = 3x^2 + 1$, $x_1 = 1$, and $\Delta x = 0.1$.

38 f is defined by $y = -5x^2 + x$, $x_1 = 2$, and $\Delta x = 0.02$.

39 f is defined by $y = -2x^2 + 4x + 1$, $x_1 = 2$, and $\Delta x = 0.4$.

40 f is defined by $y = 2x^3 + 5$, $x_1 = -1$, and $\Delta x = 0.05$.

41 f is defined by $y = 9/\sqrt{x}$, $x_1 = 9$, and $\Delta x = -1$.

42 f is defined by $y = 3/(x + 4)$, $x_1 = 3$, and $\Delta x = -2$.

43 f is defined by $y = 4 \cos x$, $x_1 = \pi/3$, and $\Delta x = 0.01$.

44 f is defined by $y = -\sec x$, $x_1 = 0$, and $\Delta x = 0.07$.

© In Problems 45 to 56, use differentials to approximate each expression.

45 $\sqrt{9.06}$

46 $\sqrt{48.8}$

47 $(3.07)^3$

48 $\dfrac{1}{1.98}$

49 $x^2 + 2x - 3$ at $x = 1.07$

50 $\dfrac{1}{\sqrt[5]{31}}$

51 $\sqrt[4]{15}$

52 $\sqrt[3]{0.000063}$

53 $(10)^{-1/2}$

54 $\sin(\pi/6 + 0.01)$

55 $\cos 61°$ (*Hint:* First change 61° to radians.)

56 $\tan 44°$

© **57** Use differentials to find the approximate volume of a spherical shell whose inner radius is 3 inches and whose thickness is $\frac{3}{32}$ inch.

58 The volume of a sphere of radius r is $V = \frac{4}{3}\pi r^3$. Note that $dV/dr = 4\pi r^2$. The surface area of a sphere of radius r is $A = 4\pi r^2$. Is it just an accident that the same expression, $4\pi r^2$, shows up as dV/dr and as A?

59 The edge of a cube is measured to be 10 centimeters with a possible error of 0.02 centimeter. Use differentials to find an approximate upper bound for the error involved in calculating its volume to be $10^3 = 1000$ cubic centimeters.

© **60** The region between two concentric circles in the plane is called an **annulus.** Find (a) the exact area of an annulus with inner radius 5 meters and outer radius 5.03 meters, (b) an approximation to the exact area found in part (a) by using differentials, and (c) the error involved in the approximation.

© **61** The altitude of a certain right circular cone is a constant equal to 2 meters. If the radius of its base is increased from 100 centimeters to 105 centimeters, use differentials to find the approximate increase in the volume of the cone.

62 A particle moves along a straight line in accordance with the equation $s = \frac{1}{3}t^3 - 2t + 3$, where t is the elapsed time in seconds and s is the directed distance, measured in meters, from the origin to the particle. Use differentials to find the approximate distance covered by the particle in the interval from $t = 2$ to $t = 2.1$ seconds.

[C] **63** Use differentials to find approximately how many cubic centimeters of chromium plate must be applied to coat the lateral surface of a cylindrical rod of radius 2.34 centimeters to a thickness of 0.01 centimeter if the rod is 30 centimeters long.

[C] **64** The period of oscillation of a pendulum of length L units is given by $T = 2\pi\sqrt{L/g}$, where g is the acceleration of gravity in units of length per second squared and T is in seconds. Use differentials to find the approximate percent that the pendulum in a grandfather clock should be lengthened if the clock gains 3 minutes in 24 hours.

65 The attractive force between unlike electrically charged particles is expressed by $F = k/x^2$, where x is the distance between the particles and k is a certain constant. If x is increased by 2 percent, use differentials to find the approximate percent decrease in F.

66 The total cost in dollars of producing x toys is

$$C = \frac{x^3}{15,000} - \frac{3x^2}{100} + 11x + 75$$

and each toy is sold at \$10. (a) Find the total profit P as a function of x. (b) Find dP in terms of x and dx. (c) When the production level changes from $x = 350$ to $x = 355$, what is the approximate change in P?

67 The law for adiabatic expansion of a certain gas is $PV^{1.7} = C$, where V is the volume of the gas, P is the pressure of the gas, and C is a constant. Derive the following equation: $(dP/P) + (1.7 \, dV/V) = 0$.

68 If f is a differentiable function at the number x_1, show that there is a linear function g such that

$$\lim_{x \to x_1} \frac{f(x) - g(x)}{x - x_1} = 0$$

(*Hint:* Let $\Delta x = x - x_1$ in Theorem 1.)

[C] **69** An angle is measured to be 31.4° with an error that is known not to exceed 0.05°. The sine of the angle is then found by using a calculator to be $\sin 31.4° = 0.521009632$. Use differentials to estimate the maximum possible error in the calculated value of the sine of the angle. (Neglect any error caused by inaccuracies inherent in the calculator.)

4.2   Antiderivatives

In applied mathematics it often happens that we know the derivative of a function and would like to find the function itself. For instance, we might know the velocity ds/dt of a particle and wish to find its equation of motion $s = f(t)$, or we might wish to find the revenue function for a certain product when we know the marginal revenue. The solutions of such problems require us to "undo" the operation of differentiation; that is, we are required to *anti*differentiate.

Suppose that g and f are functions and that the derivative of g is f. Then we say that g is an *antiderivative* of f. For instance, since

$$D_x(x^2) = 2x$$

we say that $g(x) = x^2$ is an antiderivative of $f(x) = 2x$. If C is any constant, then we also have

$$D_x(x^2 + C) = 2x$$

and it follows that $x^2 + C$ is another antiderivative of $2x$. This is geometrically clear (Figure 1), since the graph of $y = x^2 + C$ is obtained by shifting the graph of $y = x^2$ vertically by C units, and this does not change the slope of the tangent line for a given value of the abscissa x.

In general, we define antiderivatives as follows.

Figure 1

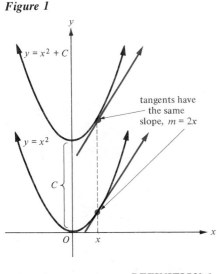

tangents have the same slope, $m = 2x$

DEFINITION 1   **Antiderivative**

> A function g is called an **antiderivative** of a function f on a set of numbers I if $g'(x) = f(x)$ holds for every value of x in I. The procedure of finding antiderivatives is called **antidifferentiation**.

If we say that g is an antiderivative of f without explicitly mentioning the set I in Definition 1, then it is to be understood that I is the entire domain of f, so that $g'(x) = f(x)$ holds for all values of x in the domain of f.

EXAMPLE 1 Show that $g(x) = \dfrac{x+1}{x-1}$ is an antiderivative of $f(x) = \dfrac{-2}{(x-1)^2}$.

SOLUTION   By the quotient rule,

$$g'(x) = \frac{(x-1) - (x+1)}{(x-1)^2} = \frac{-2}{(x-1)^2}$$ ∎

Since the derivative of a constant function is the zero function, it follows that *any constant function is an antiderivative of the zero function*. The following important theorem shows, conversely, that *the constant functions are the only antiderivatives of the zero function*.

THEOREM 1   **Antidifferentiation of the Zero Function**

> Let g be a function such that $g'(x) = 0$ holds for all values of x in some open interval I. Then g has a constant value on I.

PROOF   It is enough to prove that the value of g at a number a in I is the same as the value of g at every other number b in I. By the mean-value theorem (Section 3.1), there exists a number c between a and b such that

$$g(b) - g(a) = g'(c)(b-a) = 0(b-a) = 0$$

Thus, $g(a) = g(b)$, and the theorem is proved.   ∎

The following theorem, which is a direct consequence of Theorem 1, tells us how to find *all* the antiderivatives of a function on an open interval, provided that we know one such antiderivative.

THEOREM 2   **Antidifferentiation on an Open Interval**

> Suppose that g is an antiderivative of the function f on the open interval I. Then a function h with domain I is an antiderivative of f on I if and only if $h = g + C$ for some constant C.

PROOF   If $h = g + C$, then $h' = g' = f$, so that h is an antiderivative of f on I. Conversely, suppose that h is an antiderivative of f on I. Then, the function $h - g$ satisfies $(h-g)' = h' - g' = f - f = 0$ on the open interval I. It follows from Theorem 1 that there exists a constant C such that $h - g = C$; that is, $h = g + C$.   ∎

EXAMPLE 2   Given that the function $g(x) = \frac{1}{2}x^2$ is an antiderivative of the function $f(x) = x$, find all antiderivatives of f.

SOLUTION   Here, the interval I is $\mathbb{R}$. By Theorem 2, the antiderivatives of f are all functions h of the form $h(x) = \frac{1}{2}x^2 + C$, where C is a constant.   ∎

Notation for Antiderivatives

Antiderivatives are traditionally written by using a special symbolism which has some of the same advantages as the Leibniz notation for derivatives and which, in fact, was used by Leibniz himself. This symbolism can be understood by thinking of the differential dy as an "infinitesimal bit of y" and imagining that y is the "sum" of all these infinitesimals. Leibniz used a stylized letter s, written $\int$, for such "summations," so that

$$y = \int dy$$

would symbolize the idea that "y is the sum of all its own differentials." Johann Bernoulli, a contemporary of Leibniz, suggested that the process of assembling infinitesimals so as to get a whole, or complete, quantity should properly be called **integration** rather than summation. People accepted Bernoulli's suggestion; hence, the symbol $\int$ is referred to as the **integral sign.**

Now suppose that g is an antiderivative of f, so that

$$g' = f$$

If we let

$$y = g(x)$$

then

$$dy = g'(x)\,dx = f(x)\,dx$$

and we have

$$y = \int dy = \int f(x)\,dx$$

that is,

$$g(x) = \int f(x)\,dx$$

For this reason, the notation

$$\int f(x)\,dx$$

is used to denote an antiderivative of the function f. If C is any constant, then $g(x) + C$ is also an antiderivative of f; therefore, we make the following definition.

DEFINITION 2   **Integral Notation for Antiderivatives**

> The notation $\int f(x)\,dx = g(x) + C$, where C denotes an arbitrary constant, means that the function g is an antiderivative of the function f, so that $g'(x) = f(x)$ holds for all values of x in the domain of f.

In Definition 2, the constant C is called the **constant of integration,** and the function f [or the expression $f(x)$] is called the **integrand*** of the expression $\int f(x)\,dx$. We often say that the integrand f [or $f(x)$] is *under the integral sign.*

*Sometimes the entire expression $f(x)\,dx$, including the differential dx, is referred to as the *integrand.*

In order to verify a statement of the form
$$\int f(x)\ dx = g(x) + C$$
it is only necessary to check that $D_x g(x) = f(x)$.

In Examples 3 and 4, verify the given equation.

EXAMPLE 3 $\displaystyle\int x^2\ dx = \tfrac{1}{3}x^3 + C$

SOLUTION $D_x(\tfrac{1}{3}x^3) = x^2$

EXAMPLE 4 $\displaystyle\int dx = x + C$

SOLUTION $\displaystyle\int dx = \int 1\ dx = x + C$, since $D_x(x) = 1$

The procedure of **evaluating**

$$\int f(x)\ dx$$

that is, the procedure of finding an antiderivative g of f and writing

$$\int f(x)\ dx = g(x) + C$$

is called **indefinite integration.** The adjective "indefinite" is presumably used because the constant C can have any value and therefore is not definitely determined by the function f. Because of the arbitrary nature of the constant C,

$$\int f(x)\ dx$$

which is called the **indefinite integral** of the function f, does not represent a particular quantity or function; hence, care should be taken in manipulating this expression. Since the problem of evaluating an indefinite integral $\int f(x)\ dx$ is essentially the same as that of finding an antiderivative g of the integrand f, we use the terms "*indefinite integration*" and "*antidifferentiation*" interchangeably from now on. The fact that differentiation and antidifferentiation "undo" each other is expressed by the following formulas:

$$D_x \int f(x)\ dx = f(x) \qquad \text{and} \qquad \int f'(x)\ dx = f(x) + C$$

Basic Algebraic Rules for Antidifferentiation

Since antidifferentiation (or indefinite integration) "undoes" differentiation, each rule or formula for differentiation will yield a corresponding rule or formula for antidifferentiation when "read backward." Some of these rules are as follows:

Let $f, g, f_1, f_2, \ldots, f_m$ be functions that have antiderivatives; let $a, a_1, a_2, \ldots,$ a_m be constants; and let n be a rational number with $n \neq -1$. Then:

1 $\displaystyle\int x^n \, dx = \frac{x^{n+1}}{n+1} + C$

2 $\displaystyle\int af(x) \, dx = a \int f(x) \, dx$

3 $\displaystyle\int [f(x) + g(x)] \, dx = \int f(x) \, dx + \int g(x) \, dx$

4 $\displaystyle\int [a_1 f_1(x) + a_2 f_2(x) + \cdots + a_m f_m(x)] \, dx =$

$$a_1 \int f_1(x) \, dx + a_2 \int f_2(x) \, dx + \cdots + a_m \int f_m(x) \, dx$$

Each of these four basic algebraic rules may be verified simply by differentiating the quantity on the right side of the equation and showing that the result is the expression under the integral sign on the left side (Problem 41).

<u>EXAMPLE 5</u>   Evaluate $\displaystyle\int (5x^2 - 3x + 7) \, dx$.

SOLUTION

$$\int (5x^2 - 3x + 7) \, dx = 5 \int x^2 \, dx + (-3) \int x \, dx + 7 \int dx \qquad \text{(Rule 4)}$$

$$= 5\left(\frac{x^{2+1}}{2+1} + C_1\right) - 3\left(\frac{x^{1+1}}{1+1} + C_2\right) + 7(x + C_3) \, \text{(Rule 1)}$$

$$= \tfrac{5}{3}x^3 - \tfrac{3}{2}x^2 + 7x + 5C_1 - 3C_2 + 7C_3$$

$$= \tfrac{5}{3}x^3 - \tfrac{3}{2}x^2 + 7x + C$$

where $C = 5C_1 - 3C_2 + 7C_3$.   ∎

In practice, when the basic rules are used to antidifferentiate (that is, to evaluate indefinite integrals), the individual constants of integration that arise are immediately combined into one constant. Thus, the solution given above would ordinarily be condensed as follows:

$$\int (5x^2 - 3x + 7) \, dx = 5 \int x^2 \, dx - 3 \int x \, dx + 7 \int dx = \tfrac{5}{3}x^3 - \tfrac{3}{2}x^2 + 7x + C$$

Just as the derivative of a product is not the product of the derivatives, neither is the antiderivative of a product the product of the antiderivatives. In fact, there is no simple general formula for an antiderivative of a product, nor is there such a formula for an antiderivative of a quotient. However, by actually *performing the indicated multiplication or division*, you may be able to rewrite a product or quotient in such a form that it can be antidifferentiated by using the basic rules given above. This is illustrated in the following examples.

In Examples 6 to 9, evaluate the antiderivative.

<u>EXAMPLE 6</u> $\displaystyle\int (x + 2)(3x - 1) \, dx$

SOLUTION $\displaystyle\int (x + 2)(3x - 1)\,dx = \int (3x^2 + 5x - 2)\,dx$

$$= 3\int x^2\,dx + 5\int x\,dx - 2\int dx$$

$$= 3\left(\frac{x^3}{3}\right) + 5\left(\frac{x^2}{2}\right) - 2x + C$$

$$= x^3 + \tfrac{5}{2}x^2 - 2x + C \qquad\blacksquare$$

EXAMPLE 7 $\displaystyle\int \frac{x^4 + 3x^2 + 5}{x^2}\,dx$

SOLUTION $\displaystyle\int \frac{x^4 + 3x^2 + 5}{x^2}\,dx = \int \left(\frac{x^4}{x^2} + 3\frac{x^2}{x^2} + \frac{5}{x^2}\right)\,dx$

$$= \int (x^2 + 3 + 5x^{-2})\,dx$$

$$= \int x^2\,dx + 3\int dx + 5\int x^{-2}\,dx$$

$$= \frac{x^3}{3} + 3x + 5\frac{x^{-1}}{-1} + C$$

$$= \frac{x^3}{3} + 3x - \frac{5}{x} + C \qquad\blacksquare$$

EXAMPLE 8 $\displaystyle\int (y\sqrt[3]{y} + 1)^2\,dy$

SOLUTION $\displaystyle\int (y\sqrt[3]{y} + 1)^2\,dy = \int (y^{4/3} + 1)^2\,dy = \int (y^{8/3} + 2y^{4/3} + 1)\,dy$

$$= \int y^{8/3}\,dy + 2\int y^{4/3}\,dy + \int dy$$

$$= \frac{y^{11/3}}{11/3} + \frac{2y^{7/3}}{7/3} + y + C$$

$$= \frac{3y^{11/3}}{11} + \frac{6y^{7/3}}{7} + y + C \qquad\blacksquare$$

EXAMPLE 9 $\displaystyle\int \left(\frac{2}{u^3} - \frac{3}{\sqrt[4]{u}}\right)\,du$

SOLUTION $\displaystyle\int \left(\frac{2}{u^3} - \frac{3}{\sqrt[4]{u}}\right)\,du = \int (2u^{-3} - 3u^{-1/4})\,du$

$$= 2\int u^{-3}\,du - 3\int u^{-1/4}\,du$$

$$= 2\frac{u^{-2}}{-2} - 3\frac{u^{3/4}}{3/4} + C = -u^{-2} - 4u^{3/4} + C \qquad\blacksquare$$

Antidifferentiation of Trigonometric Functions

The formulas obtained in Section 2.5 for the derivatives of the trigonometric functions can be reversed to obtain the following formulas for antiderivatives:

$$1 \int \sin u \, du = -\cos u + C \qquad 2 \int \cos u \, du = \sin u + C$$

$$3 \int \sec^2 u \, du = \tan u + C \qquad 4 \int \csc^2 u \, du = -\cot u + C$$

$$5 \int \sec u \tan u \, du = \sec u + C \qquad 6 \int \csc u \cot u \, du = -\csc u + C$$

Each of these formulas is easily verified by differentiating the right side and observing that the result is the quantity under the integral sign on the left side (Problem 43).

In Examples 10 to 12, evaluate the antiderivative.

EXAMPLE 10 $\quad \int (3 \sin u - 2 \cos u) \, du$

SOLUTION
$$\int (3 \sin u - 2 \cos u) \, du = 3 \int \sin u \, du - 2 \int \cos u \, du$$
$$= 3(-\cos u) - 2 \sin u + C$$
$$= -3 \cos u - 2 \sin u + C \qquad \blacksquare$$

EXAMPLE 11 $\quad \int \sec x (\tan x + \sec x) \, dx$

SOLUTION
$$\int \sec x (\tan x + \sec x) \, dx = \int (\sec x \tan x + \sec^2 x) \, dx$$
$$= \int \sec x \tan x \, dx + \int \sec^2 x \, dx$$
$$= \sec x + \tan x + C \qquad \blacksquare$$

EXAMPLE 12 $\quad \int (\sqrt{t} + \csc^2 t) \, dt$

SOLUTION
$$\int (\sqrt{t} + \csc^2 t) \, dt = \int t^{1/2} \, dt + \int \csc^2 t \, dt$$
$$= \frac{t^{3/2}}{3/2} - \cot t + C$$
$$= \tfrac{2}{3} t^{3/2} - \cot t + C \qquad \blacksquare$$

Problem Set 4.2

In Problems 1 to 4, verify that the function g is an antiderivative of the function f.

1 $f(x) = 12x^2 - 6x + 1$, $g(x) = 4x^3 - 3x^2 + x - 1$

2 $f(x) = \dfrac{4x}{x^4 - 2x^2 + 1}$, $g(x) = \dfrac{1 + x^2}{1 - x^2}$

3 $f(u) = u \sin u^2$, $g(u) = -\tfrac{1}{2} \cos u^2$

4 $f(t) = (t - 1)^3$, $g(t) = \tfrac{1}{4}t^4 - t^3 + \tfrac{3}{2}t^2 - t + 753$

In Problems 5 to 40, evaluate each antiderivative (indefinite integral).

5 $\int (3x^2 - 4x - 5) \, dx$ 　　　　 **6** $\int (x^3 - 3x^2 + 2x - 4) \, dx$

7 $\int (2t^3 - 4t^2 - 5t + 6)\, dt$

8 $\int (2 + 3y^2 - 8y^3)\, dy$

35 $\int (2\sec x + \csc x)(2\sec x - \csc x)\, dx$

9 $\int (3u^4 - 2u^3 - u - 1)\, du$

10 $\int (\frac{2}{3}z^5 - \frac{3}{5}z^2 + \frac{4}{7}z - \frac{1}{9})\, dz$

36 $\int \tan^2 x\, dx$ (*Hint:* $1 + \tan^2 x = \sec^2 x$)

11 $\int (4x^{-2} - 3x^{-4} + 1)\, dx$

12 $\int \left(\frac{2}{w^2} + 5w - 3\right) dw$

37 $\int (x^{2/3} - 4\sin x + 5\cos x)\, dx$

13 $\int \frac{2t^6 + 5t^2 + 7}{t^2}\, dt$

14 $\int (2x^3 - 1)(x^2 + 5)\, dx$

38 $\int 4\cot^2 u\, du$

15 $\int \left(t^2 + 3t + \frac{1}{t^2}\right) dt$

16 $\int \left(\sqrt{x} - \frac{1}{x^4}\right) dx$

39 $\int (2\sqrt{x} + 3\sec^2 x - 4\csc x \cot x)\, dx$

17 $\int (2x^3 - 1)\left(\frac{1}{x^2} + 5\right) dx$

18 $\int \frac{x^3 - 1}{x - 1}\, dx$

40 $\int \cos^2 \frac{x}{2}\, dx \left[Hint: \cos^2 \frac{x}{2} = \frac{1}{2}(1 + \cos x) \right]$

19 $\int (4t^2 + 3)^2\, dt$

20 $\int \left(\frac{7}{y^3} - 4\right)^2 dy$

41 Verify the basic algebraic rules for antidifferentiation on page 255.

21 $\int \sqrt{w}(w - 2)\, dw$

22 $\int \sqrt[3]{u}(3u + 11)\, du$

42 Explain why an antiderivative of a polynomial function is again a polynomial function.

23 $\int \frac{25x^3 - 1}{\sqrt{x}}\, dx$

43 Verify the rules for antidifferentiation of trigonometric functions on page 257.

24 $\int \left(\sqrt{2x} + 2x\sqrt{x} + \frac{1}{\sqrt{x}}\right) dx$

44 Let f, g, and h be defined by the equations $f(x) = -2/x^3$, $g(x) = 1/x^2$, and

25 $\int \frac{(\sqrt{x} - 1)^2}{\sqrt{x}}\, dx$

26 $\int \frac{t^3 + 2t^2 - 3}{\sqrt[3]{t}}\, dt$

$$h(x) = \begin{cases} \dfrac{1 - 2x^2}{x^2} & \text{if } x < 0 \\ \dfrac{1}{x^2} & \text{if } x > 0 \end{cases}$$

27 $\int (2\cos u + 4\sin u)\, du$

28 $\int (3\sec^2 u - 4\csc^2 u)\, du$

respectively. Show that both g and h are antiderivatives of f but that there is no constant such that $h = g + C$. Does this contradict Theorem 2? Explain.

29 $\int \sec x\,(3\tan x - 2\sec x)\, dx$

45 (a) Give an example to show that $\int f(x)\, dx \neq f(x) \int dx$.

30 $\int (2\sin t - 3\cos t + 11)\, dt$

   (b) Give an example to show that $\int f(x)g(x)\, dx \neq [\int f(x)\, dx][\int g(x)\, dx]$.

31 $\int \csc y\,(2\cot y - 7\csc y)\, dy$

32 $\int \frac{dx}{\cos^2 x}$

   (c) Give an example to show that $\int \dfrac{f(x)}{g(x)}\, dx \neq \dfrac{\int f(x)\, dx}{\int g(x)\, dx}$.

33 $\int (5\csc^2 t + 7\sec^2 t + 4)\, dt$

34 $\int \frac{5\cos u + 4}{\sin^2 u}\, du$

46 Prove that *on an open interval, two antiderivatives of the same function must differ by a constant.*

4.3 The Method of Substitution, or Change of Variable

In order to evaluate $\int x(x^2 + 5)^{100}\, dx$ using only the basic algebraic rules, it would be necessary to expand $(x^2 + 5)^{100}$ by the binomial theorem, multiply through by x, and then antidifferentiate the resulting expression term by term. The required calculation could be quite tedious, to say the least. Fortunately, there is a simpler way to proceed, namely, by "changing the variable" from x to a new variable $u = x^2 + 5$. Notice that if $u = x^2 + 5$, then we have $du = 2x\, dx$, or $x\, dx = \frac{1}{2}\, du$, and so

$$\int x(x^2 + 5)^{100}\, dx = \int (x^2 + 5)^{100} x\, dx$$

$$= \int u^{100}\, \frac{1}{2}\, du = \frac{1}{2} \int u^{100}\, du = \frac{1}{2}\, \frac{u^{101}}{101} + C$$

Substitution of $u = x^2 + 5$ into the last expression gives

$$\int x(x^2 + 5)^{100}\, dx = \frac{(x^2 + 5)^{101}}{202} + C$$

The method illustrated here is called **change of variable,** or **substitution.**

There is more going on than meets the eye in the calculation above, since the integral notation has compressed what could otherwise be a complicated argument (using the chain rule) into a few simple calculations. In the same way that the basic rules for antidifferentiation are obtained by reading the basic rules for differentiating backward, the method of substitution (or change of variable) is just the chain rule "read backward." However, we defer the formal justification of this method to the end of the section and concentrate here on seeing just how it works.

Procedure for Evaluating $\int f(x)\, dx$ by Change of Variable (Substitution)

Step 1 Find a portion of the integrand $f(x)$ that is especially "prominent," in the sense that if it were replaced by a single new variable, say u, then the integrand would be noticeably simpler. Set u equal to this portion. The resulting equation will have the form

$$u = g(x)$$

Step 2 Using the equation $u = g(x)$ obtained in step 1, find the differential

$$du = g'(x)\, dx$$

Step 3 Using the two equations $u = g(x)$ and $du = g'(x)\, dx$ obtained in steps 1 and 2, rewrite the entire integrand *including* dx, in terms of u and du only.

Step 4 Evaluate the resulting indefinite integral in terms of u.

Step 5 Using the equation $u = g(x)$ of step 1, rewrite the answer obtained in step 4 in terms of the original variable x.

There is never a guarantee of success when the method of substitution is used—one can only try and see what happens. After the algebraic manipulations required in step 3 are carried out, it may happen that the resulting integral (involving u) is more complicated than the original integral (involving x). However, if the procedure fails for one choice of u, it may still work for another—just try again.

In Examples 1 to 5, use the procedure for change of variable (substitution) and the basic rules to evaluate the antiderivative (indefinite integral).

<u>EXAMPLE 1</u> $\displaystyle\int \sqrt{7x + 2}\, dx$

SOLUTION   We carry out the steps in the procedure.

Step 1 Let's try $u = 7x + 2$.

Step 2 $du = 7\, dx$

Step 3 Since $du = 7 \, dx$, it follows that $dx = \frac{1}{7} \, du$; hence,

$$\int \sqrt{7x + 2} \, dx = \int \sqrt{u} \cdot \frac{1}{7} \, du = \int \frac{1}{7}\sqrt{u} \, du$$

Step 4 $\displaystyle\int \sqrt{7x + 2} \, dx = \int \frac{1}{7}\sqrt{u} \, du = \frac{1}{7} \int u^{1/2} \, du = \frac{1}{7} \frac{u^{3/2}}{3/2} + C$

$$= \frac{2}{21} u^{3/2} + C$$

Step 5 Since $u = 7x + 2$, we have

$$\int \sqrt{7x + 2} \, dx = \tfrac{2}{21}(7x + 2)^{3/2} + C \qquad \blacksquare$$

<u>EXAMPLE 2</u> $\displaystyle\int \frac{x^2 \, dx}{(x^3 + 4)^5}$

SOLUTION

Step 1 Let's try $u = x^3 + 4$.

Step 2 $du = 3x^2 \, dx$

Step 3 Notice the $x^2 \, dx$ in the numerator of the integrand and the same expression multiplied by 3 in step 2. Solving the equation in step 2 for $x^2 \, dx$, we find that

$$x^2 \, dx = \tfrac{1}{3} \, du$$

Thus, $\displaystyle\int \frac{x^2 \, dx}{(x^3 + 4)^5} = \int \frac{\frac{1}{3} \, du}{u^5} = \int \frac{1}{3} u^{-5} \, du$

Step 4 $\displaystyle\int \frac{x^2 \, dx}{(x^3 + 4)^5} = \int \frac{1}{3} u^{-5} \, du = \frac{1}{3} \int u^{-5} \, du$

$$= \frac{1}{3} \frac{u^{-4}}{-4} + C = -\frac{1}{12} u^{-4} + C$$

Step 5 $\displaystyle\int \frac{x^2 \, dx}{(x^3 + 4)^5} = -\frac{1}{12}(x^3 + 4)^{-4} + C \qquad \blacksquare$

<u>EXAMPLE 3</u> $\displaystyle\int x^2 \sqrt{3 - 2x} \, dx$

SOLUTION

Step 1 Let's try $u = 3 - 2x$.

Step 2 $du = -2 \, dx$

Step 3 The integrand contains three factors: x^2, $\sqrt{3 - 2x}$, and dx. The substitution $u = 3 - 2x$ will change the factor $\sqrt{3 - 2x}$ into $\sqrt{u}$. We can solve the equation in step 2 for the differential dx, so that $dx = -\frac{1}{2} \, du$. It only remains to rewrite the factor x^2 in terms of u. To this end, we solve the equation $u = 3 - 2x$ in step 1 for x, so that

$$x = \tfrac{1}{2}(3 - u)$$

and

$$x^2 = \tfrac{1}{4}(3 - u)^2 = \tfrac{1}{4}(9 - 6u + u^2) = \tfrac{9}{4} - \tfrac{3}{2}u + \tfrac{1}{4}u^2$$

Therefore, we have

$$\int x^2\sqrt{3-2x}\ dx = \int (\tfrac{9}{4} - \tfrac{3}{2}u + \tfrac{1}{4}u^2)\sqrt{u}(-\tfrac{1}{2}\ du)$$

$$= \int (-\tfrac{9}{8}\sqrt{u} + \tfrac{3}{4}u\sqrt{u} - \tfrac{1}{8}u^2\sqrt{u})\ du$$

$$= \int (-\tfrac{9}{8}u^{1/2} + \tfrac{3}{4}u^{3/2} - \tfrac{1}{8}u^{5/2})\ du$$

Step 4 $\displaystyle\int x^2\sqrt{3-2x}\ dx = \int (-\tfrac{9}{8}u^{1/2} + \tfrac{3}{4}u^{3/2} - \tfrac{1}{8}u^{5/2})\ du$

$$= -\tfrac{9}{8}\int u^{1/2}\ du + \tfrac{3}{4}\int u^{3/2}\ du - \tfrac{1}{8}\int u^{5/2}\ du$$

$$= -\frac{9}{8}\frac{u^{3/2}}{3/2} + \frac{3}{4}\frac{u^{5/2}}{5/2} - \frac{1}{8}\frac{u^{7/2}}{7/2} + C$$

$$= -\tfrac{3}{4}u^{3/2} + \tfrac{3}{10}u^{5/2} - \tfrac{1}{28}u^{7/2} + C$$

Step 5 $\displaystyle\int x^2\sqrt{3-2x}\ dx$

$$= -\tfrac{3}{4}(3-2x)^{3/2} + \tfrac{3}{10}(3-2x)^{5/2} - \tfrac{1}{28}(3-2x)^{7/2} + C \quad\blacksquare$$

EXAMPLE 4 $\displaystyle\int \sin 9x\ dx$

SOLUTION

Step 1 Let $u = 9x$.

Step 2 $du = 9\ dx$

Step 3 From step 2, we have $dx = \tfrac{1}{9}\ du$; hence,

$$\int \sin 9x\ dx = \int \sin u\ (\tfrac{1}{9}\ du) = \int \tfrac{1}{9}\sin u\ du$$

Step 4 $\displaystyle\int \sin 9x\ dx = \int \tfrac{1}{9}\sin u\ du = \tfrac{1}{9}\int \sin u\ du = \tfrac{1}{9}(-\cos u) + C$

$$= -\tfrac{1}{9}\cos u + C$$

Step 5 $\displaystyle\int \sin 9x\ dx = -\tfrac{1}{9}\cos 9x + C$

EXAMPLE 5 $\displaystyle\int \frac{\sin 7x\ dx}{(1 + \cos 7x)^4}$

SOLUTION

Step 1 Let $u = 1 + \cos 7x$.

Step 2 $du = -7\sin 7x\ dx$

Step 3 $\displaystyle\int \frac{\sin 7x\ dx}{(1 + \cos 7x)^4} = \int -\frac{1}{7}\frac{du}{u^4}$

Step 4 $\displaystyle\int \frac{\sin 7x\ dx}{(1 + \cos 7x)^4} = \int -\frac{1}{7}\frac{du}{u^4} = -\frac{1}{7}\int u^{-4}\ du$

$$= -\frac{1}{7}\frac{u^{-3}}{-3} + C = \frac{1}{21}u^{-3} + C$$

Step 5 $\int \dfrac{\sin 7x \, dx}{(1 + \cos 7x)^4} = \dfrac{1}{21}(1 + \cos 7x)^{-3} + C$

After you have become familiar with the method of substitution, you may wish to abbreviate the step-by-step procedure suggested above. The following example illustrates how this can be done.

<u>EXAMPLE 6</u> Evaluate $\int \dfrac{\sec \sqrt{t} \tan \sqrt{t} \, dt}{\sqrt{t}}$.

SOLUTION   Let $u = \sqrt{t}$, so that $du = 1/(2\sqrt{t}) \, dt$ and $dt/\sqrt{t} = 2 \, du$. Then

$$\int \frac{\sec \sqrt{t} \tan \sqrt{t} \, dt}{\sqrt{t}} = \int \sec \sqrt{t} \tan \sqrt{t} \frac{dt}{\sqrt{t}} = \int \sec u \tan u \, (2 \, du)$$

$$= 2 \int \sec u \tan u \, du = 2 \sec u + C$$

$$= 2 \sec \sqrt{t} + C$$

Justification of the Method of Substitution

The following theorem can be used to justify the method of substitution, or change of variable.

THEOREM 1   **Substitution**

> Suppose that
> $$\int h(u) \, du = H(u) + C$$
> Then, if g is a differentiable function,
> $$\int h[g(x)]g'(x) \, dx = H[g(x)] + C$$

PROOF   Because

$$\int h(u) \, du = H(u) + C$$

it follows that $\qquad H'(u) = h(u)$

We need to show that the derivative of $H[g(x)]$ with respect to x is the integrand $h[g(x)]g'(x)$. By the chain rule,

$$\frac{d}{dx}H[g(x)] = H'[g(x)]g'(x) = h[g(x)]g'(x)$$

and the proof is complete.

Let's see how Theorem 1 justifies the method of substitution. Suppose we want to evaluate

$$\int f(x) \, dx$$

by the method of substitution. In step 1 of our procedure, we choose a suitable portion of the integrand $f(x)$ and call it $u = g(x)$. Step 2 produces the differential

$$du = g'(x) \, dx$$

In step 3, we use the last equation to rewrite $f(x)$ and dx in terms of u and du. Suppose the result is

$$f(x)\ dx = h(u)\ du$$

where h is a suitable function. What the last equation really means is that

$$f(x) = h(u)\ \frac{du}{dx}$$

or

$$f(x) = h[g(x)]g'(x)$$

Now, step 4 produces an equation of the form

$$\int h(u)\ du = H(u) + C$$

and step 5 yields the result

$$\int f(x)\ dx = \int h[g(x)]g'(x)\ dx = H[g(x)] + C$$

in conformity with Theorem 1.

Problem Set 4.3

In Problems 1 to 50, use substitution to evaluate each antiderivative (indefinite integral). (In some cases a suitable substitution is suggested.)

1 $\int (4x + 3)^4\ dx,\ u = 4x + 3$

2 $\int t(4t^2 + 7)^9\ dt,\ u = 4t^2 + 7$

3 $\int x\sqrt{4x^2 + 15}\ dx,\ u = 4x^2 + 15$

4 $\int \dfrac{3x\ dx}{(4 - 3x^2)^8}$

5 $\int \dfrac{s\ ds}{\sqrt[3]{5s^2 + 16}},\ u = 5s^2 + 16$ **6** $\int \dfrac{(8t + 2)\ dt}{(4t^2 + 2t + 6)^{17}}$

7 $\int (1 - x^{3/2})^{5/3}\sqrt{x}\ dx,\ u = 1 - x^{3/2}$

8 $\int (x^2 - 6x + 9)^{11/3}\ dx,\ u = x - 3$

9 $\int \dfrac{x^2\ dx}{(4x^3 + 1)^7}$ **10** $\int \dfrac{x^2 + 1}{\sqrt{x^3 + 3x}}\ dx$

11 $\int (5t^2 + 1)\sqrt[4]{5t^3 + 3t - 2}\ dt$ **12** $\int \dfrac{\sqrt[3]{1 + 1/(2t)}}{t^2}\ dt$

13 $\int \dfrac{2x^2 - 1}{(6x^3 - 9x + 1)^{3/2}}\ dx$ **14** $\int \dfrac{\sqrt{1 + \sqrt{x}}}{\sqrt{x}}\ dx$

15 $\int \left(x + \dfrac{5}{x}\right)^{21}\left(\dfrac{x^2 - 5}{x^2}\right)\ dx$ **16** $\int (49x^2 - 42x + 9)^{6/7}\ dx$

17 $\int x\sqrt{5 - x}\ dx$ **18** $\int x^2\sqrt{1 + x}\ dx$

19 $\int \dfrac{t\ dt}{\sqrt{t + 1}}$ **20** $\int \dfrac{y + 2}{\sqrt[3]{2 - y}}\ dy$

21 $\int \dfrac{2x\ dx}{(2 - x)^{2/3}}$ **22** $\int (x + 2)^2\sqrt{1 + x}\ dx$

23 $\int \sqrt[3]{3x^2 + 5}\ x^3\ dx$ **24** $\int \sqrt[4]{x^3 + 1}\ x^5\ dx$

25 $\int \dfrac{t^2\ dt}{\sqrt{t + 4}}$ **26** $\int \dfrac{y\ dy}{\sqrt{3 - y}}$

27 $\int 2 \sin 35x\ dx$

28 $\int (7 \sin 5x + 3 \cos 7x)\ dx$

29 $\int 8 \cos (16x - 1)\ dx$

30 $\int 5 \cos (8 - 3x)\ dx$

31 $\int \sec^2 11x \, dx$

32 $\int -\csc^2 5x \, dx$

33 $\int \dfrac{dt}{\sin^2 3t}$

34 $\int \dfrac{dy}{\cos^2 5y}$

35 $\int \sec(2y+1)\tan(2y+1)\,dy$

36 $\int \tan^2(3t+7)\,dt$

37 $\int -\sec \dfrac{t}{5} \tan \dfrac{t}{5}\,dt$

38 $\int \csc 10z \cot 10z \, dz$

39 $\int \cos x \cos(\sin x)\,dx, \ u = \sin x$

40 $\int x^3 \sec 10x^4 \tan 10x^4 \, dx$

41 $\int x^3 \csc^2 7x^4 \, dx, \ u = 7x^4$

42 $\int \dfrac{\sin \sqrt{x+1}\,dx}{\sqrt{x+1}}$

43 $\int \dfrac{\sin x \, dx}{(2 + \cos x)^2}$

44 $\int \dfrac{\sec^2 x \, dx}{(3 + 2\tan x)^3}$

45 $\int \cos 2y \sqrt{5 + \sin 2y}\,dy$

46 $\int \tan 3t \sec 3t \sqrt{4 + \sec 3t}\,dt$

47 $\int \dfrac{\cot \sqrt{x} \csc \sqrt{x}\,dx}{\sqrt{x}}$

48 $\int (\sin 2x)^{-1/3} \cos 2x \, dx$

49 $\int \dfrac{\sec 3\theta \tan 3\theta \, d\theta}{\sqrt{1 - 5 \sec 3\theta}}$

50 $\int \dfrac{\cot(\sqrt{x}/2) \csc^2 (\sqrt{x}/2)\,dx}{\sqrt{x}}$

51 (a) Evaluate $\int \sin x \cos x \, dx$ using the substitution $u = \sin x$. (b) Evaluate $\int \sin x \cos x \, dx$ using the identity $\sin x \cos x = \frac{1}{2}\sin 2x$. (c) Show that the answers to parts (a) and (b) are consistent with each other. [*Hint:* Use the identity $\sin^2 x = \frac{1}{2}(1 - \cos 2x)$.]

52 Evaluate $\int \sin mx \cos nx \, dx$, where m and n are constants, by using the identity $\sin a \cos b = \frac{1}{2}\sin(a+b) + \frac{1}{2}\sin(a-b)$.

53 Evaluate

$$\int \frac{x \, dx}{\sqrt{x+1}}$$

by two different methods, and show that the answers are consistent with each other. (a) Use the substitution $u = x + 1$. (b) Use the substitution $u = \sqrt{x + 1}$.

54 Given that f is a function with domain $(-1, \infty)$ such that $f'(x) = 2(1+x)^{-2}$ and $f(0) = 0$, find f.

55 Evaluate $\int x^2 \sqrt{5x - 1}\,dx$ by two different methods, and show that the answers are consistent with each other. (a) Use the substitution $u = 5x - 1$. (b) Use the substitution $u = \sqrt{5x - 1}$.

56 Suppose that $g'(x) = (1 + x)^{-2}$ holds for all values of x except for $x = -1$. Given that $\lim\limits_{x \to +\infty} g(x) = \lim\limits_{x \to -\infty} g(x) = 0$, find g.

4.4 Differential Equations

Mathematical models for real-world phenomena often take the form of equations relating various quantities and their rates of change. Since the rate of change of a quantity is represented mathematically by a derivative, such equations often involve derivatives or differentials. For instance, in the biochemistry of digestion, the enzyme trypsin, which helps break proteins into amino acids, is produced in the small intestine from an inactive proenzyme called trypsinogen. If y is the amount of trypsin in the small intestine at time t, it is known that the rate dy/dt at which trypsin is produced is proportional to the product of the amount of trypsinogen present and the amount of trypsin present. Thus,

$$\frac{dy}{dt} = k(A - y)(B + y)$$

where k is the constant of proportionality, A is the amount of trypsinogen present when $t = 0$, and B is the amount of trypsin present when $t = 0$.

An equation such as

$$\frac{dy}{dt} = k(A - y)(B + y)$$

that involves derivatives or differentials of an unknown function is called a **differential equation.** To *solve* a differential equation is to find the unknown function or functions that satisfy the equation.*

The simplest type of differential equation has the form

$$\frac{dy}{dx} = f(x) \qquad \text{or} \qquad dy = f(x) \, dx$$

Here f is a given function, and y is an unknown function of x. Evidently,

$$y = g(x)$$

is a solution of this differential equation if and only if g is an antiderivative of f. Any one solution $y = g(x)$ is called a **particular solution** of the differential equation. Given any such particular solution $y = g(x)$, we can write

$$y = g(x) + C$$

where C is an arbitrary constant, to obtain other particular solutions.

Conversely, by Theorem 2 of Section 4.2, any particular solution of the differential equation

$$\frac{dy}{dx} = f(x)$$

on an open interval I has the form

$$y = g(x) + C$$

where g is an antiderivative of f and C is a suitable constant. In this sense,

$$y = g(x) + C$$

where C is an arbitrary constant and g is an antiderivative of f, represents the **complete solution** of the differential equation. Since

$$\int f(x) \, dx = g(x) + C$$

the complete solution of the differential equation $dy/dx = f(x)$ can be written as

$$y = \int f(x) \, dx$$

Thus:

The complete solution of the differential equation

$$dy = f(x) \, dx$$

is given by

$$y = \int f(x) \, dx$$

---

*For the solution of the differential equation for the formation of trypsin, see Problem 49 on page 516.

<u>EXAMPLE 1</u>   Find the complete solution of the differential equation $dy = 6x^2\, dx$.

SOLUTION $y = \displaystyle\int 6x^2\, dx = 6 \int x^2\, dx = 6\dfrac{x^3}{3} + C = 2x^3 + C$ ∎

The circumstances that give rise to differential equations often entail additional conditions called **side conditions, initial conditions, boundary conditions,** or **constraints.** For instance, in the differential equation for the formation of trypsin, we have the initial condition that $y = A$ when $t = 0$. Such additional conditions can be used to single out the relevant particular solution from the complete solution.

<u>EXAMPLE 2</u>   Find the complete solution of the differential equation $dy/dx = \cos 2x$. Then find the particular solution satisfying the side condition that $y = 1$ when $x = \pi/12$.

SOLUTION   The complete solution is given by

$$y = \int \cos 2x\, dx = \tfrac{1}{2}\sin 2x + C$$

where we evaluated the antiderivative by using the substitution $u = 2x$. Putting $y = 1$ and $x = \pi/12$ into the complete solution, we obtain

$$1 = \frac{1}{2}\sin 2\!\left(\frac{\pi}{12}\right) + C = \frac{1}{2}\sin\frac{\pi}{6} + C = \frac{1}{4} + C$$

from which it follows that

$$C = 1 - \tfrac{1}{4} = \tfrac{3}{4}$$

Substituting this value of C into the complete solution, we find that the particular solution satisfying the side condition is

$$y = \tfrac{1}{2}\sin 2x + \tfrac{3}{4}$$ ∎

The differential equation

$$\frac{dy}{dx} = f(x)$$

is said to be **separable** since it can be rewritten in the form

$$dy = f(x)\, dx$$

in which the variables x and y are "separated" in the sense that all expressions involving x are on one side of the equation and all expressions involving y are on the other side. More generally:

A differential equation that can be rewritten in the form

$$G(y)\, dy = F(x)\, dx$$

where F and G are functions, is called **separable.** The **general solution** of such a differential equation, obtained by separating the variables as shown and then antidifferentiating both sides of the resulting equation, is given by

$$\int G(y)\, dy = \int F(x)\, dx$$

In solving a separable differential equation by the indicated method, you can combine the two constants of integration corresponding to $\int G(y)\,dy$ and $\int F(x)\,dx$ into a single constant C. If a side condition is given, it can be used to determine the value of C.

EXAMPLE 3   Find the general solution of the differential equation $y' = 4x^2y^2$. Then find the particular solution satisfying the side condition that $x = 1$ when $y = -1$.

SOLUTION   Here y' is used as an abbreviation for dy/dx, so the given differential equation has the form

$$\frac{dy}{dx} = 4x^2y^2 \qquad \text{or} \qquad dy = 4x^2y^2\,dx$$

We can separate the variables if we divide both sides of the last equation by y^2 to obtain

$$\frac{dy}{y^2} = 4x^2\,dx \qquad (\text{provided } y \neq 0)$$

Thus,

$$\int y^{-2}\,dy = \int 4x^2\,dx$$

so

$$\frac{y^{-1}}{-1} + C_1 = 4\frac{x^3}{3} + C_2$$

or

$$\frac{1}{y} = C_0 - \frac{4x^3}{3}$$

where $C_0 = C_1 - C_2$. The last equation may be rewritten as

$$y = \frac{3}{3C_0 - 4x^3}$$

Therefore, letting $C = 3C_0$, we have the general solution

$$y = \frac{3}{C - 4x^3}$$

Putting $x = 1$ and $y = -1$ in the general solution, we obtain

$$-1 = \frac{3}{C - 4} \qquad \text{or} \qquad C = 1$$

Hence, the desired particular solution is

$$y = \frac{3}{1 - 4x^3} \qquad\qquad\blacksquare$$

When you solve a separable differential equation by separating the variables and antidifferentiating, you can lose certain solutions. For instance, in the preceding example, when we divided by y^2 to separate the variables, we assumed that $y \neq 0$. However, as you can check, the constant function given by $y = 0$ *is* a solution of the original differential equation $y' = 4x^2y^2$. Notice that the solution $y = 0$ cannot be obtained from the general solution

$$y = \frac{3}{C - 4x^3}$$

by assigning a value to the constant C. This shows that *the general solution of a separable differential equation might not be the complete solution.* A solution (such as $y = 0$ in the example above) that cannot be obtained directly from the general solution by assigning a value to the constant of integration is called a **singular solution.**

The general solution of a separable differential equation involving x and y might be an equation that only implicitly determines y as a function of x. This is illustrated by the following example.

EXAMPLE 4   Find the general solution of the differential equation $x \, dx + y \, dy = 0$.

SOLUTION   Separating the variables and antidifferentiating, we have

$$x \, dx = -y \, dy \qquad \text{so} \qquad \int x \, dx = \int (-y) \, dy$$

hence, $\dfrac{x^2}{2} + C_1 = -\dfrac{y^2}{2} + C_2 \qquad \text{or} \qquad \dfrac{x^2}{2} + \dfrac{y^2}{2} = C_2 - C_1$

Therefore, $x^2 + y^2 = 2(C_2 - C_1)$, that is, $x^2 + y^2 = C$, where we have put $C = 2(C_2 - C_1)$. Thus, the general solution of $x \, dx + y \, dy = 0$ is

$$x^2 + y^2 = C \qquad\qquad ■$$

Second-Order Differential Equations

Up to now, we have considered only "first-order" differential equations, that is, equations involving only first derivatives. By definition, the highest order of all the derivatives involved in a differential equation is called the **order** of the equation. For instance,

$$\frac{d^2y}{dx^2} + 2\frac{dy}{dx} = x$$

is a second-order differential equation, and

$$y^{(n)} = x^3 + 7$$

is an nth-order differential equation.

Sometimes, the general solution of a second-order differential equation can be obtained by two *successive* antidifferentiations. The resulting solution will involve *two* arbitrary constants that cannot be combined into one constant. *Two* side conditions are required to determine these constants.

EXAMPLE 5   Find the general solution of the second-order differential equation $y'' = -2x + 1$, and then find the particular solution that satisfies the following two initial-value conditions: $y = -1$ when $x = 0$ and $y' = 1$ when $x = 0$.

SOLUTION   Since

$$y'' = \frac{d}{dx}y'$$

we can write the differential equation as

$$\frac{d}{dx}y' = -2x + 1 \qquad \text{or} \qquad d(y') = (-2x + 1) \, dx$$

Antidifferentiating both sides of the last equation, we obtain

$$y' = \int (-2x + 1) \, dx = -x^2 + x + C_1$$

Therefore, we have

$$\frac{dy}{dx} = -x^2 + x + C_1 \qquad \text{or} \qquad dy = (-x^2 + x + C_1) \, dx$$

Again, we antidifferentiate both sides of the last equation and obtain

$$y = \int (-x^2 + x + C_1) \, dx = -\frac{x^3}{3} + \frac{x^2}{2} + C_1 x + C_2$$

Thus, the general solution is

$$y = -\frac{x^3}{3} + \frac{x^2}{2} + C_1 x + C_2$$

Notice, here, that there is no way to combine the two constants C_1 and C_2 into a single constant since C_1 is a multiplier of the variable x.

To find the particular solution corresponding to the given initial conditions, we begin by substituting $y = -1$ and $x = 0$ into the general solution to obtain

$$-1 = C_2$$

Then we substitute $y' = 1$ and $x = 0$ into the previously obtained equation

$$y' = -x^2 + x + C_1$$

and find that

$$1 = C_1$$

Substituting these values of C_1 and C_2 into the general solution, we obtain the desired particular solution,

$$y = -\frac{x^3}{3} + \frac{x^2}{2} + x - 1$$

Linear Motion

Differential equations are important in the study of the motion of physical objects (cars, projectiles, balls, planets, electrons, and so forth). Often we disregard the size, shape, and orientation of such objects and think of them as particles. Here we consider the motion of a particle P along a linear scale, which we call the s axis (Figure 1). The equation of motion of P,

$$s = f(t)$$

Figure 1

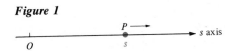

gives the coordinate s of P in terms of the elapsed time t since some arbitrary (but fixed) initial instant. The instantaneous velocity v and the instantaneous acceleration a of P are given by the equations

$$v = \frac{ds}{dt} = f'(t) \qquad \text{and} \qquad a = \frac{dv}{dt} = \frac{d^2s}{dt^2} = f''(t)$$

(See Section 2.10.)

If v or a is a known function of t, subject to suitable initial conditions, it may be possible to solve the resulting differential equations for the law of motion $s = f(t)$. This is illustrated in the following examples.

In Examples 6 and 7, find the law of motion $s = f(t)$ from the information given.

EXAMPLE 6 $a = 2t - t^2$; $v = 0$ when $t = 0$ and $s = 0$ when $t = 0$.

SOLUTION   We have $dv/dt = a = 2t - t^2$, so $dv = (2t - t^2)\, dt$. A first antidifferentiation gives

$$v = \int dv = \int (2t - t^2)\, dt = t^2 - \frac{t^3}{3} + C_1$$

Since $v = 0$ when $t = 0$, it follows that $C_1 = 0$. Also,

$$\frac{ds}{dt} = v = t^2 - \frac{t^3}{3} \qquad \text{so} \qquad ds = \left(t^2 - \frac{t^3}{3} \right) dt$$

and a second antidifferentiation gives

$$s = \int ds = \int \left(t^2 - \frac{t^3}{3} \right) dt = \frac{t^3}{3} - \frac{t^4}{12} + C_2$$

Since $s = 0$ when $t = 0$, it follows that $C_2 = 0$ and the equation of motion is

$$s = \frac{t^3}{3} - \frac{t^4}{12}$$

■

EXAMPLE 7   A car is braked to a stop with constant deceleration. The car stops 8 seconds after the brakes are applied and travels 200 feet during this time. Find the law of motion of the car during this 8-second interval. Also, find the acceleration and determine the speed of the car at the instant the brakes are first applied.

SOLUTION   Represent the car by a particle on the s axis moving, say, to the right. Start reckoning time t from the instant when the brakes are first applied, and place the origin at the position of the car when $t = 0$. Here the acceleration a is constant, and we have

$$s = 0 \qquad \text{when } t = 0$$
$$s = 200 \qquad \text{when } t = 8$$
$$v = 0 \qquad \text{when } t = 8$$

Since $dv/dt = a$ and a is constant, then $dv = a\, dt$ and

$$v = \int dv = \int a\, dt = a \int dt = at + C_1$$

Therefore, $ds/dt = v = at + C_1$, so $ds = (at + C_1)\, dt$ and

$$s = \int ds = \int (at + C_1)\, dt = \tfrac{1}{2}at^2 + C_1 t + C_2$$

Since $s = 0$ when $t = 0$, it follows that $C_2 = 0$, and we have

$$s = \tfrac{1}{2}at^2 + C_1 t$$

Substituting $v = 0$ and $t = 8$ into the equation $v = at + C_1$, we find that

$$0 = 8a + C_1 \qquad \text{or} \qquad C_1 = -8a$$

Now we substitute $C_1 = -8a$ into the equation $s = \frac{1}{2}at^2 + C_1 t$ to obtain

$$s = \frac{1}{2}at^2 - 8at$$

Finally, we substitute $s = 200$ and $t = 8$ into the last equation and find that

$$200 = \frac{1}{2}a(8)^2 - 8a(8) \quad \text{or} \quad 200 = 32a - 64a = -32a$$

from which it follows that during the entire 8-second interval,

$$a = -\frac{200}{32} = -\frac{25}{4} \text{ ft/s}^2$$

Here the negative sign indicates *deceleration*. By substituting this value of a into the equation $s = \frac{1}{2}at^2 - 8at$, we obtain the law of motion

$$s = \frac{1}{2}\left(\frac{-25}{4}\right)t^2 - 8\left(\frac{-25}{4}\right)t \quad \text{or} \quad s = 50t - \frac{25t^2}{8}$$

The speed of the car at the instant when $t = 0$ is obtained by putting $t = 0$ in the equation $v = at + C_1$ to obtain

$$v = C_1 = -8a = -8(-\tfrac{25}{4}) = 50 \text{ ft/s} \qquad \blacksquare$$

An object falling near the surface of the earth experiences some force because of air resistance; however, in many situations this force is negligible, especially when the object has a high density, is more or less "streamlined," and has not attained a very high velocity. Such an object falls with a constant acceleration g, called the **acceleration of gravity.** The value of g is approximately 32 feet per second squared or 980 centimeters per second squared, or 9.8 meters per second squared.

If such an object P is projected straight upward, it is usually convenient to take the s axis pointing straight upward with the origin at the surface of the earth (Figure 2). Then the acceleration a of P is negative (it will slow down, stop, then begin to fall back down); hence,

$$a = \frac{d^2s}{dt^2} = \frac{dv}{dt} = -g$$

On the other hand, if the object P is dropped or thrown straight downward from an initial height h, it is usually convenient to take the s axis pointing downward with the origin h units above the surface of the earth (Figure 3). Then the acceleration a of P is positive; hence,

$$a = \frac{d^2s}{dt^2} = \frac{dv}{dt} = g$$

Figure 2

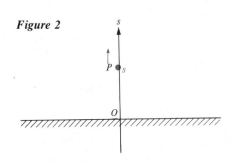

Figure 3

© **EXAMPLE 8** An iron ball is thrown vertically upward, starting 2 meters above the ground, with an initial velocity of 8 meters per second. How many seconds will elapse before the ball strikes the ground?

SOLUTION We set up the coordinate axis as in Figure 2. Here, $dv/dt = -g$, so $dv = -g\,dt$ and

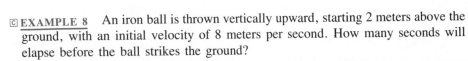

When $t = 0$, $v = 8$ meters per second, so

$$8 = -g(0) + C_1 \qquad \text{or} \qquad C_1 = 8$$

Therefore

$$v = -gt + 8$$

The last equation can be rewritten as

$$\frac{ds}{dt} = -gt + 8 \qquad \text{or} \qquad ds = (-gt + 8)\, dt$$

Antidifferentiating, we find that

$$s = \int (-gt + 8)\, dt = -\tfrac{1}{2}gt^2 + 8t + C_2$$

When $t = 0$, $s = 2$ meters; hence, substituting these values into the last equation yields

$$2 = -\tfrac{1}{2}g(0)^2 + 8(0) + C_2 \qquad \text{or} \qquad C_2 = 2$$

Therefore, the equation of motion of the ball is

$$s = -\tfrac{1}{2}gt^2 + 8t + 2$$

where $g = 9.8$ meters per second squared. Thus,

$$s = -4.9t^2 + 8t + 2$$

When the ball strikes the surface of the earth, $s = 0$, so that

$$0 = -4.9t^2 + 8t + 2 \qquad \text{or} \qquad 4.9t^2 - 8t - 2 = 0$$

Solving the last equation by the quadratic formula, we find that

$$t = \frac{8 \pm \sqrt{64 - 4(4.9)(-2)}}{2(4.9)} = \frac{8 \pm \sqrt{103.2}}{9.8}$$

Because t is positive when the ball lands, we reject the negative solution and conclude that

$$t = \frac{8 + \sqrt{103.2}}{9.8} \approx 1.85 \text{ seconds}$$

Problem Set 4.4

In Problems 1 to 8, find the complete solution of each differential equation.

1 $\dfrac{dy}{dx} = 5x^4 + 3x^2 + 1$

2 $\dfrac{dy}{dx} = 20x^3 - 6x^2 + 17$

3 $\dfrac{dy}{dx} = \dfrac{6}{x^2} + 15x^2 + 10$

4 $y' = \dfrac{(x^2 - 4)^2}{2x^2}$

5 $\dfrac{dy}{dx} = \sqrt{7x^3}$

6 $dy = (5t + 12)^3\, dt$

7 $\dfrac{ds}{dt} = t^{-2} + \sin t$

8 $D_\theta u = \sqrt[3]{\theta} - \csc^2 5\theta$

In Problems 9 to 14, find the particular solution of the given differential equation that satisfies the indicated side condition.

9 $\dfrac{dy}{dx} = 5 - 3x$; $y = 4$ when $x = 0$

10 $\dfrac{dy}{dx} = 3x^2 + x$; $y = -2$ when $x = 1$

11 $\dfrac{dy}{dt} = t^3 + \dfrac{1}{t^2}$; $y = 1$ when $t = -2$

12 $y' = \sqrt{x} + 2$; $y = 5$ when $x = 4$

13 $y' = 3 \sin \dfrac{x}{2}$; $y = 1$ when $x = \dfrac{\pi}{3}$

14 $\dfrac{ds}{dt} = \sec^2 6t$; $s = -1$ when $t = 0$

In Problems 15 to 26, find the general solution of each separable differential equation.

15 $y' = x(x^2 - 3)^4$

16 $(3x^2 + 2x + 1)^5 \, dy = (6x + 2) \, dx$

17 $\sqrt{x^3 + 7} \, dy = x^2 \, dx$

18 $\dfrac{ds}{dt} = (t + 1)^2 t^3$

19 $\sqrt{2x + 1} \, dy = y^2 \, dx$

20 $(y^2 - \sqrt{y}) \, dy = (x^2 + \sqrt{x}) \, dx$

21 $\dfrac{dy}{dx} = \dfrac{x\sqrt[3]{y^4 + 7}}{5y^3}$

22 $y\dfrac{dy}{dx} = x^3 \sqrt{10y^2 + 1}$

23 $\dfrac{dy}{dx} = \dfrac{\sin 2x}{\cos 3y}$

24 $\csc x \cos y \, dx + \tan x \tan y \, dy = 0$

25 $y^2 \, dx - \csc x \, dy = 0$

26 $\cos^2 3t \sin 4s \, ds - \cos^2 4s \, dt = 0$

In Problems 27 to 32, find the particular solution of the given separable differential equation that satisfies the indicated side condition.

27 $\dfrac{dx}{y} = \dfrac{dy}{2 - x^{3/2}}$; $y = 2$ when $x = 9$

28 $\dfrac{ds}{dt} = \dfrac{t^2}{\sqrt{t^3 + 1}}$; $s = \frac{1}{2}$ when $t = 2$

29 $t^{2/3} \, dW = (1 - t^{1/3})^3 \, dt$; $W = -1$ when $t = 8$

30 $\dfrac{dy}{dx} = \sqrt{\dfrac{y}{x}}$; $y = 4$ when $x = 1$

31 $\sec 3t \, ds + \csc 2s \, dt = 0$; $s = \dfrac{\pi}{3}$ when $t = \dfrac{\pi}{2}$

32 $\csc y \, dx + \cos^2 x \, dy = 0$; $y = \dfrac{\pi}{2}$ when $x = \dfrac{\pi}{4}$

In Problems 33 to 42, find the general solution of each second-order differential equation.

33 $\dfrac{d^2y}{dx^2} = 3x^2 + 2x + 1$

34 $y'' = (5x + 1)^4$

35 $y'' = \sqrt[3]{4x + 5}$

36 $S'' = \dfrac{5}{(t + 7)^3}$

37 $\dfrac{d^2s}{dt^2} = 2t^4 + 3$

38 $y'' = (x + 1)^2$

39 $\dfrac{d^2y}{dx^2} = 0$

40 $D_x^2 y = 1$

41 $\dfrac{d^2y}{dx^2} = \cos 2x$

42 $y'' = \sin 3x$

In Problems 43 to 50, find the particular solution of the given differential equation that satisfies the indicated side conditions.

43 $\dfrac{d^2y}{dx^2} = 6x + 1$; $y = 2$ and $y' = 3$ when $x = 0$

44 $\dfrac{d^2y}{dx^2} = \sqrt{x}$; $y = 3$ and $y' = 2$ when $x = 9$

45 $\dfrac{d^2s}{dt^2} = 2$; $s = 0$ when $t = 1$ and $s = 0$ when $t = -3$

46 $\dfrac{d^2y}{dx^2} = 3x^2$; $y = -1$ when $x = 0$ and $y = 9$ when $x = 2$

47 $y'' = 3(2 + 5x)^2$; $y = 2$ and $y' = -1$ when $x = 1$

48 $\dfrac{d^2s}{dt^2} = \sqrt[4]{5t - 4}$; $s = 2$ and $s' = -3$ when $t = 4$

49 $y'' = \sin\dfrac{x}{2}$; $y' = 0$ when $x = \pi$ and $y = 2$ when $x = 0$

50 $\dfrac{d^2s}{dt^2} = 2 \sec^2 t \tan t$; $s = 0$ and $\dfrac{ds}{dt} = 0$ when $t = 0$

51 A particle moving along a straight line has the equation of motion $s = f(t)$, where t is in seconds and s is in meters. Its velocity v satisfies the equation $v = t^2 - 8t + 15$. If $s = 1$ when $t = 0$, find s when $t = 3$.

52 A particle, starting with an initial velocity of 25 meters per second, moves in a straight line through a resisting medium which decreases the velocity of the particle at a constant rate of 10 meters per second each second. How far will the particle travel before coming to rest?

©**53** The brakes on a certain car can stop the car in 200 feet from a speed of 55 miles per hour. Assume that when the brakes are applied, the car has a constant negative acceleration. (a) How much time in seconds is required to bring the car to a stop from 55 miles per hour? (b) If the car is brought to a stop from 55 miles per hour, how far will it have moved by the time its speed is reduced to 25 miles per hour? (Recall that 1 mile is 5280 feet.)

©**54** From the top edge of a building 20 meters high, a stone is thrown vertically upward with an initial velocity of 30 meters per second. (a) In how many seconds will the stone strike the ground? (b) How high will the stone rise? (c) How fast will the stone be falling when it hits the ground?

☐ 55 A balloon is rising at the constant rate of 10 feet per second and is 100 feet from the ground at the instant when the aeronaut drops her binoculars. (a) How long will it take the binoculars to strike the ground? (b) With what speed will the binoculars strike the ground?

56 A projectile is fired vertically upward by a cannon with an initial velocity of v_0 meters per second. At what speed will the projectile be moving when it returns and strikes the hapless cannoneer? (Neglect air resistance.)

☐ 57 A balloon is rising vertically at the constant rate of 1 meter per second and has reached an altitude of 8 meters at the instant when an assistant on the ground directly under the balloon attempts to toss the aeronaut's binoculars up to her. What is the minimum velocity with which the binoculars should be thrown

straight upward if they are released from the assistant's hand at an altitude of 2 meters?

58 Suppose that a particle P moves along the s axis with a constant acceleration a. Let v_0 be the velocity of the particle when $t = 0$, and let s_0 be its coordinate when $t = 0$. Show that (a) $v = at + v_0$ and (b) $s = \frac{1}{2}at^2 + v_0t + s_0$.

59 A stone is dropped from a height of h meters with zero initial velocity, and it hits the ground T seconds later. Show that $h = 4.9T^2$.

60 Consider the differential equation $dy/dt = k(A - y)(B + y)$ for the formation of trypsin in the small intestine. Assuming that $A > B$, determine the time t at which trypsin is being formed most rapidly.

4.5 Applications of Differential Equations

In Section 4.4, we saw an application of differential equations to linear-motion problems. Further simple applications to geometry, physics, and economics are given in this section.

A first-order differential equation of the form

$$\frac{dy}{dx} = g(x)$$

where g is a given function, can be interpreted geometrically as a condition on the slope dy/dx of the tangent line to the graph of $y = f(x)$, where f is an unknown function. The complete solution of the differential equation gives all functions f whose graphs satisfy this condition. A side condition amounts to an additional requirement that the graph of f contain a specified point.

<u>EXAMPLE 1</u>

(a) Find all curves in the plane satisfying the condition that the slope of the tangent at each point is 3 times the abscissa of that point.

(b) Sketch graphs of several such curves on the same xy coordinate system.

(c) Find the equation of that particular curve satisfying the condition in part (a) which contains the point (1, 2).

SOLUTION

(a) The differential equation expressing the condition on the slope of the tangent is

$$\frac{dy}{dx} = 3x$$

The complete solution of this differential equation is

$$y = \int 3x\, dx = 3 \int x\, dx = 3\,\frac{x^2}{2} + C$$

Figure 1

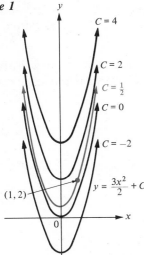

Thus, an equation of a curve satisfying the given condition is of the form

$$y = \frac{3x^2}{2} + C$$

where C is a constant. There is a different curve for each different value of C.

(b) The curves corresponding to $C = -2$, $C = 0$, $C = \frac{1}{2}$, $C = 2$, and $C = 4$ are sketched in Figure 1.

(c) Here we have to impose the side condition that $y = 2$ when $x = 1$. Setting $y = 2$ and $x = 1$ in the equation $y = \frac{3}{2}x^2 + C$ and solving for C, we find that $C = 2 - \frac{3}{2} = \frac{1}{2}$. The desired equation is, therefore,

$$y = \tfrac{3}{2}x^2 + \tfrac{1}{2}$$

Work Done by a Variable Force

Suppose that a *constant* force F, having a direction parallel to the s axis, acts on a particle P which moves along this axis from an original position with coordinate s_0 to a final position with coordinate s_1 (Figure 2). Then, by definition, the force does an amount of work W on the particle given by

$$W = F \cdot (s_1 - s_0)$$

A positive (respectively, a negative) force is understood to act in the positive (respectively, the negative) direction along the s axis.

We now take up the problem of calculating the work done when the applied force is *not necessarily constant* but still acts in a direction parallel to the s axis. We suppose that P starts at an initial position with coordinate s_0, and we denote by W the net work done by the variable force F in moving P from its original position to the position s (Figure 3). Since F may not be constant, we cannot calculate W simply by multiplying F by $s - s_0$.

Figure 2

original position of P · · · final position of P

In this case, suppose that P moves from s to $s + \Delta s$, causing the net work done by the force to change from W to $W + \Delta W$ (Figure 4). Here, ΔW is the work done by F in moving P from s to $s + \Delta s$. Although the force F might change as P is so moved, it should not change much if Δs is very small. Thus, if F is the force acting on P at position s, we should have

$$\Delta W \approx F \, \Delta s \qquad \text{or} \qquad \frac{\Delta W}{\Delta s} \approx F$$

Figure 3

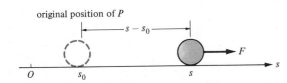

original position of P

As Δs approaches 0, this approximation should become better and better; hence, we should have

$$\frac{dW}{ds} = \lim_{\Delta s \to 0} \frac{\Delta W}{\Delta s} = F$$

Therefore, the net work W done by the variable force F satisfies the differential equation

$$\boxed{dW = F \, ds}$$

Figure 4

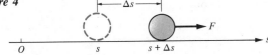

with the initial condition that $W = 0$ when $s = s_0$.

EXAMPLE 2   The force F acting on a particle P is given by $F = 1/s^2$, where s is the coordinate of P. How much work is done by this force in moving P from $s = 1$ to $s = 9$?

SOLUTION   Let W denote the net work done by the force in moving P from the point with coordinate 1 to the point with coordinate s. We want to find the value of W when $s = 9$. Since

$$dW = F\,ds = \frac{1}{s^2}\,ds$$

it follows that

$$W = \int \frac{1}{s^2}\,ds = \frac{-1}{s} + C$$

When $s = 1$, we have $W = 0$, so that

$$0 = \frac{-1}{1} + C$$

hence, $C = 1$ and

$$W = \frac{-1}{s} + 1$$

When $s = 9$, we have

$$W = -\tfrac{1}{9} + 1 = \tfrac{8}{9} \qquad \text{unit of work}$$

EXAMPLE 3   Consider the apparatus shown in Figure 5, in which a particle P that is attached to a stanchion by a perfectly elastic spring can slide without friction along the horizontal s axis. The particle starts at $s = 0$ and is pulled by a force F to a final position $s = b$. At the start, when $s = 0$, assume that the spring is relaxed and $F = 0$. Find the work done by F in pulling P from $s = 0$ to $s = b$.

SOLUTION   Because the spring is perfectly elastic, **Hooke's law** requires that the force F be proportional to the displacement s; that is,

$$F = ks$$

where k is a constant called the **spring constant.** (The stiffer the spring, the larger the value of k.) Let W be the net work done by F in pulling P from the origin to the position s, so that $dW = F\,ds = ks\,ds$. Antidifferentiation gives

$$W = \int ks\,ds = k \int s\,ds = k\frac{s^2}{2} + C$$

When $s = 0$, we have $W = 0$, so that $0 = k(0^2/2) + C$, and it follows that $C = 0$ and $W = ks^2/2$. Thus, when $s = b$, we have

$$W = \frac{kb^2}{2}$$

Figure 5

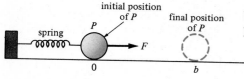

initial position of P

final position of P

spring

P

F

0

b

s

EXAMPLE 4   A perfectly elastic spring is stretched from its relaxed position through 2 meters. When it is extended these 2 meters, the stretching force on the spring is 3 newtons. How much work is done?

SOLUTION   Reasoning as in Example 3, we have $F = ks$, so $k = F/s$. Because $F = 3$ newtons when $s = 2$ meters, it follows that

$$k = \frac{F}{s} = \frac{3}{2} \text{ newtons per meter}$$

Using the result of Example 3, we have

$$W = k\frac{b^2}{2} = \frac{3}{2} \cdot \frac{2^2}{2} = 3 \text{ newton-meters}$$

In the international system (SI) of units, one newton-meter (1 N · m) is defined to be one **joule** (1 J). (One pound of force is approximately 4.45 newtons.) Therefore, the work done in stretching the spring is 3 joules. ∎

Remarks on Setting Up Differential Equations

As we mentioned, many engineers and scientists like to think of dx as an "infinitesimal bit of x," and they prefer to regard $\int dx$ as a "summation" of all the infinitesimal bits of x to give the quantity x. That is, apart from an additive constant of integration, $x = \int dx$. They persist in this point of view, in spite of its possible lack of "mathematical rigor," because it allows them to set up differential equations quickly and easily. Thus, for instance, in the textbook *Physics* by Paul Tipler, Worth Publishers, Inc., New York, 1976, on page 175, we find, "This definition is equivalent to saying that the work dW done over a very short interval dx is $dW = F_x \, dx$."

In 1960, the mathematical logician Abraham Robinson (1918–1974) discovered a way to make Leibniz's infinitesimals mathematically rigorous and thus to justify the kind of reasoning that engineers and scientists have indulged in for more than three hundred years. In the remainder of this book, we use arguments involving infinitesimals whenever it is convenient to do so. In every case, rigorous arguments can be made to justify the result in question.

Newton's Law of Universal Gravitation

The following example illustrates a typical physical application of the infinitesimal point of view.

EXAMPLE 5   According to **Newton's law of universal gravitation,** two particles with masses m_1 and m_2 kilograms which are separated by a distance of s meters attract each other with a force of

$$F = G\frac{m_1 m_2}{s^2} \qquad \text{newtons}$$

where G is a constant given by

$$G = 6.672 \times 10^{-11} \text{ N} \cdot \text{m}^2/\text{kg}^2$$

If 1 kilogram of lead is distributed uniformly along the x axis between the origin and the point with x coordinate 2 meters, how much gravitational force does it exert on a 1-kilogram particle P situated on the x axis 1 meter to the left of the origin?

SOLUTION   Let F denote the total gravitational force of attraction on P of the portion of the mass lying on the interval $[0, x]$ (Figure 6). The linear density of lead on the interval $[0, 2]$ is

$$\tfrac{1}{2} = 0.5 \text{ kg/m}$$

hence, the infinitesimal mass dm of the portion of lead on the subinterval $[x, x + dx]$ is given by

$$dm = 0.5 \, dx \qquad \text{kg}$$

Figure 6

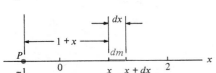

Since the distance between P and dm is $1 + x$ meters (Figure 6), this infinitesimal mass exerts an infinitesimal gravitational force dF on the unit mass P given by

$$dF = G\frac{1\,dm}{(1+x)^2} = 0.5G\frac{dx}{(1+x)^2} \qquad \text{newtons}$$

Therefore, $\quad F = \int dF = \int 0.5G\frac{dx}{(1+x)^2} = 0.5G\int\frac{dx}{(1+x)^2}$

Making the change of variable $u = 1 + x$, so that $du = dx$, we find that

$$F = 0.5G\int\frac{dx}{(1+x)^2} = 0.5G\int u^{-2}\,du = 0.5G\frac{u^{-1}}{-1} + C$$

$$= C - \frac{0.5G}{u} = C - \frac{0.5G}{1+x}$$

Here we have the side condition that $F = 0$ when $x = 0$, so that

$$0 = C - \frac{0.5G}{1} \qquad \text{and} \qquad C = 0.5G$$

It follows that

$$F = 0.5G - \frac{0.5G}{1+x} = 0.5G\left(1 - \frac{1}{1+x}\right) = 0.5G\frac{x}{1+x}$$

Putting $x = 2$, we find that

$$F = 0.5G\left(\frac{2}{3}\right) = \frac{G}{3} = \frac{6.672 \times 10^{-11}}{3} = 2.224 \times 10^{-11} \text{ newtons}$$

Applications to Economics

Here we give two very simple examples to illustrate how the total-cost or total-revenue function can be found if the marginal-cost or the marginal-revenue function is known. We use K for the constant of integration since C is used to represent the cost function. We denote the revenue function by R.

EXAMPLE 6   The marginal cost for producing x tennis rackets is given by $dC/dx = 200/\sqrt{x}$ dollars per racket. Find the cost C of manufacturing x rackets if $C = \$4100$ when $x = 100$.

SOLUTION   The cost function satisfies the differential equation

$$dC = 200x^{-1/2}\,dx$$

Therefore, $\quad C = 200\int x^{-1/2}\,dx = 200\frac{x^{1/2}}{1/2} + K = 400\sqrt{x} + K$

The side condition that $C = 4100$ when $x = 100$ implies that

$$4100 = 400\sqrt{100} + K = 4000 + K$$

Hence, $K = 100$, and so

$$C = 400\sqrt{x} + 100 \qquad \text{dollars}$$

EXAMPLE 7   The marginal revenue for a digital watch is expressed by

$$\frac{dR}{dx} = 60,000 - 40,000(1 + x)^{-2} \qquad \text{dollars per 1000 watches}$$

where x represents the demand in thousands of watches. Express the total sales revenue in terms of x, given that for $x = 1$ (thousand watches), the total sales revenue is \$38,000. If the demand increases to $x = 4$ (thousand watches), what is the total sales revenue?

SOLUTION

$$R = \int [60,000 - 40,000(1 + x)^{-2}]\, dx = 60,000 \int dx - 40,000 \int (1 + x)^{-2}\, dx$$

The change of variable $u = 1 + x$ in the second integral gives

$$\int (1 + x)^{-2}\, dx = \int u^{-2}\, du = -u^{-1} + K_1 = -\frac{1}{1 + x} + K_1$$

Hence,

$$R = 60,000x + \frac{40,000}{1 + x} + K \qquad \text{where } K = -40,000K_1$$

Putting $x = 1$ and $R = 38,000$ in the last equation and solving for K, we find that $K = -42,000$; hence,

$$R = 60,000x + \frac{40,000}{1 + x} - 42,000$$

When $x = 4$, we have $R = 240,000 + 8000 - 42,000 = \$206,000$.   ■

Problem Set 4.5

In Problems 1 to 6, the slope dy/dx of the tangent line to the graph of a function $y = f(x)$ is given by a differential equation. Find the function f if its graph contains the indicated point (a, b).

1 $\dfrac{dy}{dx} = 1 - 3x$, $(a, b) = (-1, 4)$

2 $\dfrac{dy}{dx} = x^2 + 1$, $(a, b) = (-3, 5)$

3 $\dfrac{dy}{dx} = \left(\dfrac{y}{x}\right)^2$, $(a, b) = (2, 1)$

4 $\dfrac{dy}{dx} = 2xy^2$, $(a, b) = (0, 1)$

5 $\dfrac{dy}{dx} = -3 \cos 3x$, $(a, b) = \left(\dfrac{\pi}{3}, 2\right)$

6 $\dfrac{dy}{dx} = 2 \tan^2 x \sec x$, $(a, b) = (0, 1)$

7 A curve in the plane contains the point $(0, 1)$ and satisfies the condition that at each point on the curve, the product of the ordinate and the slope of the tangent is the abscissa. Find an equation of the curve.

8 The tangent to a certain curve at $(0, 3)$ makes an angle of $\pi/4$ with the positive x axis, and at each point P on the curve, the normal to the curve at P intersects the x axis at a point Q in such a way that the vertical projection of the segment $\overline{PQ}$ on the x axis has constant length (independent of P). Find an equation of such a curve.

In Problems 9 to 14, the force F acting on a particle P moving along the s axis is given in terms of the coordinate s of P. Find the work done by F in moving P from $s = s_0$ to $s = s_1$. In each case, assume F is in newtons and s is in meters.

9 $F = 2s$, $s_0 = 1$, $s_1 = 5$

10 $F = 400s\sqrt{1 + s^2}$, $s_0 = 0$, $s_1 = 3$

11 $F = \sqrt{s - 1}$, $s_0 = 1$, $s_1 = 10$

12 $F = (1 + s)^{2/3}$, $s_0 = -7$, $s_1 = 7$

13 $F = \sin\dfrac{s}{2}$, $s_0 = 0$, $s_1 = \pi$

14 $F = \sin s \cos s$, $s_0 = 0$, $s_1 = \dfrac{\pi}{2}$

15 A perfectly elastic spring is stretched from its relaxed position through 0.5 meter. When it is stretched 0.5 meter, the stretching force on the spring is 40 newtons. How much work is done?

16 Consider the apparatus shown in Figure 5, but suppose that the spring is relaxed when the particle is at the position $x = c$. Assume that P is pulled to the position $s = a$ with $a > c$ and then to the position $s = b$ with $b > a$. Let F_a be the force on P when $s = a$, and let F_b be the force on P when $s = b$. (a) Show that the spring constant is given by

$$k = \frac{1}{b - a}(F_b - F_a)$$

(b) Show that $c(F_b - F_a) = aF_b - bF_a$. (c) Show that the work $W_{a,b}$ done in stretching the spring from $s = a$ to $s = b$ is given by $W_{a,b} = \frac{1}{2}(b - a)(F_a + F_b)$.

17 A perfectly elastic spring with spring constant $k = 150$ pounds per inch is stretched through 6 inches. At the start of the stretching, the spring is not relaxed, and the force on the spring is 300 pounds. How much work in foot-pounds is done in stretching the spring through the 6 inches?

C 18 A bucket containing sand is lifted, starting at ground level, at a constant speed of 2 meters per second. The bucket itself weighs 12 newtons and at the start is filled with 30 kilograms of sand. As the bucket is lifted, sand runs out a hole in the bottom at a constant rate of 1 kilogram per second. How much work (in joules) is done in lifting the bucket up to the height at which the last of the sand runs out? (*Hint:* One kilogram weighs approximately 9.8 newtons.)

19 A homogeneous sphere of mass M and radius r exerts a gravitational attraction on a particle P of mass m that is the same as if all the mass of the sphere were concentrated at its center—provided that the distance from P to the center of the sphere is not less than r. If G is Newton's universal constant of gravitation, show that the work $W(R)$ done against the force of gravity by carrying P from the surface of the sphere to a point R units from the center of the sphere is given by

$$W(R) = GMm\left(\frac{1}{r} - \frac{1}{R}\right)$$

C 20 (a) In Problem 19, show that the work required to move the particle P infinitely far from the sphere, starting from the surface of the sphere, is given by GMm/r. (b) If the particle P is shot away from the surface of the sphere with an initial velocity v, then its initial kinetic energy is given by $\frac{1}{2}mv^2$. If all this energy is used in work against the force of gravity, show that the small-

est velocity v_{escape} for which the particle will continue to move forever away from the sphere is given by $v_{escape} = \sqrt{2GM/r}$. This is called the **escape velocity.** (c) The radius of the earth is approximately 6.371×10^6 meters, and its mass is approximately 5.983×10^{24} kilograms. Neglecting air resistance, find the escape velocity for the earth. (d) The radius of the moon is approximately 1.738×10^6 meters and its mass is approximately 7.347×10^{22} kilograms. Find the escape velocity for the moon.

C 21 Using the result of Problem 19, determine how many joules of work are required to transport 1 kilogram of mass from the surface of the earth to the orbit of the moon. Neglect the gravitational attraction of the moon, take the radius of the earth to be 6.37×10^6 meters, take the mass of the earth to be 5.98×10^{24} kilograms, and suppose that the orbit of the moon is approximately a circle of radius 3.80×10^8 meters with center at the center of the earth.

22 A homogeneous sphere so dense that the escape velocity from its surface exceeds the speed of light c is called a **black hole.** Using the result in Problem 20b, show that a homogeneous sphere of mass M will form a black hole if its radius is less than the **Schwarzschild radius** $2GM/c^2$.

23 If a mass of M kilograms is distributed uniformly along the x axis between the origin and the point $x = a$ meters, $a > 0$, show that the gravitational force of attraction F of this mass on a 1-kilogram particle P situated at the point $x = -b$, $b > 0$, is given by $F = MG/[b(a + b)]$.

24 Two thin homogeneous rods, each of mass M and each 1 meter long, are placed end to end along the x axis so that they meet at the origin. Let $F(w)$ denote the gravitational force of attraction of the right-hand rod on the portion of the left-hand rod that lies on the interval $[-1, w - 1]$ for $0 \le w \le 1$. Show that F satisfies the differential equation $(1 - w)(2 - w)\, dF = M^2G\, dw$ with the initial-value condition $F(0) = 0$.

25 The marginal cost C' of manufacturing x nylon backpacks is given by $C'(x) = 5 + (8/\sqrt{x})$ dollars per backpack. (a) Find the cost C of manufacturing x backpacks if it costs \$1200 to manufacture 100 backpacks. (b) If each backpack sells for \$21, find the manufacturer's profit P if x backpacks are manufactured and sold.

26 For the production of a certain commodity, suppose that the marginal cost C' is a quadratic function of the production level x, C_0 is the fixed cost, I is the most efficient production level, A is the average production cost per unit at production level I, and $B = C'(I)$ is the marginal cost at production level I. (a) Find the cost function C in terms of x, C_0, I, A, and B. (b) What is the economic significance of the condition $A > B + (C_0/I)$?

27 The marginal cost dC/dx of producing x thousand cans of dog food is given by $dC/dx = 30x^{-2/3}$ dollars per 1000 cans. Given that 8000 cans can be produced for \$2600: (a) What is the fixed cost? (b) How much will it cost to produce 125,000 cans?

C 28 Suppose that the marginal cost C' of producing x electric razors per week is a quadratic function of x and that the fixed cost is $5000 per week. Assume that the most efficient production level is 20,000 razors per week and that at this production level the average production cost per razor is $15. Finally, suppose that when operating at the most efficient production level, it costs $10.75 to manufacture one more razor. Thus, taking the marginal cost $C'(20,000)$ to be $10.75, find a formula for the cost $C(x)$ of manufacturing x electric razors per week.

29 The cost C of manufacturing x thousand cassette tapes is $C(x) = 600 + 60x$ dollars. The corresponding marginal revenue is $R'(x) = 400 - 8x$ dollars per 1000 tapes. Find the number of tapes (in thousands) that will maximize the manufacturer's profit.

C 30 In Problem 28, suppose that the marginal weekly revenue is given by $R'(x) = 40 - (x/500)$ dollars per razor. Find (a) the production levels for which the manufacturer breaks even and (b) the production level for which the manufacturer's weekly profit is maximum. [Assume that $R(0) = 0$.]

31 The cost of production of a newspaper in a small town is $C = 100 + 3.5x$ dollars per month, where x is the number of subscriptions to the paper. The marginal revenue is $R'(x) = 13 - (x/40)$ dollars per subscriber per month, and $R(0) = 0$. (a) Find R as a function of x. (b) Find the monthly profit P as a function of x. (c) What value of x will maximize the profit? (d) If the value of x is that which maximizes the profit, what price does each subscriber pay per month for a subscription?

C 32 In Problem 28, find the production level for which the production cost per razor is minimum.

33 The Frostbite Frozen Foods Company can manufacture x thousand frozen macaroni and cheese dinners for $700x + 5000$ dollars. The marginal revenue is $R'(x) = 1030 - x$ dollars per 1000 dinners. How many macaroni and cheese dinners should be manufactured to maximize (a) the revenue and (b) the profit? (c) When the profit is maximized, what price does Frostbite charge for a single macaroni dinner? [Assume that $R(0) = 0$.]

4.6 The Harmonic-Oscillator Equation

There are a number of real-world quantities that oscillate or vibrate in a uniform manner, repeating themselves periodically in definite intervals of time. Examples include alternating electric currents; sound waves; light waves, radio waves, and other electromagnetic waves; pendulums, mass-spring systems, and other mechanical oscillators; tides, geysers, seismic waves, the seasons, climatic cycles, and other periodic phenomena of interest in the earth sciences; and biological phenomena ranging from a human heartbeat to periodic variation of the population of a plant or animal species. The simplest mathematical model for such quantities is

$$y = A \cos(\omega t - \phi)$$

where y is the oscillating quantity

t is time

A is a positive constant called the **amplitude** of the oscillation

ω (the Greek letter omega) is a positive constant called the **angular frequency** of the oscillation

ϕ (the Greek letter phi) is a constant called the **phase angle**

The equation $y = A \cos(\omega t - \phi)$ gives the value of y at time t, where t is measured from an arbitrary but fixed initial moment when $t = 0$. As we shall see, the constants A, ω, and ϕ determine certain specific features of the oscillation. We define the positive constant ν (the Greek letter nu) by

$$\nu = \frac{\omega}{2\pi}$$

and we refer to ν as the **frequency** of the oscillation. If t is measured in seconds,

Heinrich Hertz

then ν represents the number of complete oscillations, or **cycles,** per second. One cycle per second is called a **hertz** (abbreviated Hz) in honor of the German physicist Heinrich Hertz (1857–1894), who discovered radio waves in the late 1880s. The quantity T defined by

$$T = \frac{1}{\nu}$$

is called the **period** of the oscillation.

The graph of $y = A \cos(\omega t - \phi)$ (Figure 1) resembles the graph of the cosine function, except that it is stretched vertically by a factor of A (the amplitude) and makes one complete oscillation (cycle) whenever t increases by $T = 1/\nu$ units (Problem 8). A quantity y oscillating in accordance with the equation $y = A \cos(\omega t - \phi)$ is said to be undergoing **simple harmonic oscillation.** Any physical device capable of producing a quantity undergoing simple harmonic oscillation is called a **harmonic oscillator.**

In order to derive a differential equation for a harmonic oscillator, suppose that

$$y = A \cos(\omega t - \phi)$$

Figure 1

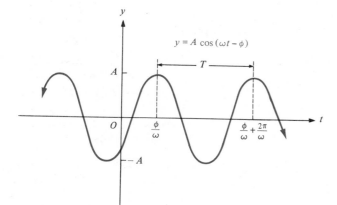

Then
$$\frac{dy}{dt} = -\omega A \sin(\omega t - \phi)$$

and
$$\frac{d^2 y}{dt^2} = -\omega^2 A \cos(\omega t - \phi)$$

Therefore, $\quad \dfrac{d^2 y}{dt^2} = -\omega^2 y \quad$ or $\quad \dfrac{d^2 y}{dt^2} + \omega^2 y = 0$

The second-order differential equation

$$\frac{d^2 y}{dt^2} + \omega^2 y = 0$$

is called the **harmonic-oscillator equation.** The calculation above proves the following theorem.

THEOREM 1   **The Harmonic-Oscillator Equation**

If A, ω, and ϕ are any three constants, then

$$y = A \cos(\omega t - \phi)$$

is a solution of the harmonic-oscillator equation

$$\frac{d^2 y}{dt^2} + \omega^2 y = 0$$

When two simple harmonic oscillations (for instance, musical tones) combine by addition, the resulting oscillation is called a **superposition** of the original oscillations. Thus, a superposition of the oscillations represented by

$$y_1 = A_1 \cos(\omega_1 t - \phi_1) \quad \text{and} \quad y_2 = A_2 \cos(\omega_2 t - \phi_2)$$

is represented by
$$y = y_1 + y_2$$

THEOREM 2   **The Superposition Theorem**

> **(i)** The superposition of any two solutions of a harmonic-oscillator equation is again a solution of that equation.
>
> **(ii)** Any constant multiple of a solution of a harmonic-oscillator equation is again a solution of that equation.

PROOF   We prove (i) here and leave the proof of (ii) as an exercise (Problem 23). Thus, suppose that y_1 and y_2 are solutions of the harmonic-oscillator equation

$$\frac{d^2y}{dt^2} + \omega^2 y = 0$$

so that $$\frac{d^2y_1}{dt^2} + \omega^2 y_1 = 0 \quad \text{and} \quad \frac{d^2y_2}{dt^2} + \omega^2 y_2 = 0$$

Adding corresponding sides of the last two equations, we obtain

$$\frac{d^2y_1}{dt^2} + \frac{d^2y_2}{dt^2} + \omega^2 y_1 + \omega^2 y_2 = 0$$

or $$\frac{d^2(y_1 + y_2)}{dt^2} + \omega^2(y_1 + y_2) = 0$$

Therefore, the superposition $y_1 + y_2$ satisfies the original differential equation. ■

By combining parts (i) and (ii) of Theorem 2, we obtain the following result: If k_1 and k_2 are constants and if y_1 and y_2 are solutions of the harmonic-oscillator equation

$$\frac{d^2y}{dt^2} + \omega^2 y = 0$$

then $$y = k_1 y_1 + k_2 y_2$$

is also a solution of the equation. In particular, taking $k_1 = 1$ and $k_2 = -1$, we see that

$$y = y_1 - y_2$$

is a solution of the equation. Thus, *the difference of any two solutions of a harmonic-oscillator equation is again a solution of that equation.*

If y is a solution of the harmonic-oscillator equation

$$\frac{d^2y}{dt^2} + \omega^2 y = 0$$

then $$\mathcal{E} = \left(\frac{dy}{dt}\right)^2 + \omega^2 y^2$$

can be considered to be a measure of the total *energy* stored in the oscillator. In various special cases, $\mathcal{E}$ may have to be multiplied by a suitable constant K to obtain the actual energy

$$E = K\mathcal{E}$$

in appropriate units (Problems 17 and 21). Thus, Theorem 3 can be regarded as a *conservation-of-energy* theorem for harmonic oscillators.

THEOREM 3

Conservation of Energy for a Harmonic Oscillator

Let y be a solution of the harmonic-oscillator equation

$$\frac{d^2y}{dt^2} + \omega^2 y = 0$$

and let

$$\mathscr{E} = \left(\frac{dy}{dt}\right)^2 + \omega^2 y^2$$

Then $\mathscr{E}$ is a constant.

PROOF

$$\frac{d\mathscr{E}}{dt} = 2\frac{dy}{dt}\frac{d^2y}{dt^2} + 2\omega^2 y\frac{dy}{dt} = 2\frac{dy}{dt}\left(\frac{d^2y}{dt^2} + \omega^2 y\right) = 2\frac{dy}{dt}(0) = 0$$

Therefore, by Theorem 1 in Section 4.2, $\mathscr{E}$ is a constant. ∎

Now, suppose that $y = g(t)$ is a solution of the harmonic-oscillator equation

$$\frac{d^2y}{dt^2} + \omega^2 y = 0$$

and let

$$\mathscr{E} = \left(\frac{dy}{dt}\right)^2 + \omega^2 y^2 = [g'(t)]^2 + \omega^2 [g(t)]^2$$

By Theorem 3, $\mathscr{E}$ is a constant, independent of the value of t. In particular, we can find $\mathscr{E}$ by setting $t = 0$:

$$\mathscr{E} = [g'(0)]^2 + \omega^2 [g(0)]^2$$

Therefore, if both $g(0) = 0$ and $g'(0) = 0$, the energy of the oscillation is 0. Intuitively, this would mean that there is no oscillation. Theorem 4 confirms our intuition.

THEOREM 4

The Case of Energy Zero

Suppose that $y = g(t)$ is a solution of the harmonic-oscillator equation

$$\frac{d^2y}{dt^2} + \omega^2 y = 0$$

Then, if $g(0) = 0$ and $g'(0) = 0$, it follows that $g(t) = 0$ for all values of t.

PROOF

Since both $g(0) = 0$ and $g'(0) = 0$, we have

$$\mathscr{E} = [g'(0)]^2 + \omega^2 [g(0)]^2 = 0$$

Because $\mathscr{E}$ is a constant (Theorem 3), it follows that

$$0 = [g'(t)]^2 + \omega^2 [g(t)]^2$$

holds for all values of t. Because neither $[g'(t)]^2$ nor $\omega^2[g(t)]^2$ can be negative, both must be 0 for all values of t. In particular,

$$g'(t) = 0 \qquad \text{for all values of } t$$

Hence, by Theorem 1 in Section 4.2, g is a constant function. Therefore,

$$g(t) = g(0) = 0 \qquad \text{for all values of } t \qquad \blacksquare$$

According to Theorem 1, $y = A \cos(\omega t - \phi)$ is a solution of the harmonic-oscillator equation

$$\frac{d^2 y}{dt^2} + \omega^2 y = 0$$

In Theorem 5 below, we show that $y = A \cos(\omega t - \phi)$ is, in fact, the *complete* solution of this equation; that is, all possible solutions can be obtained by assigning suitable values to the constants A and ϕ. To begin with, suppose that f is a differentiable function and that $\omega \neq 0$. Consider the point

$$P = \left(f(0), \frac{f'(0)}{\omega} \right)$$

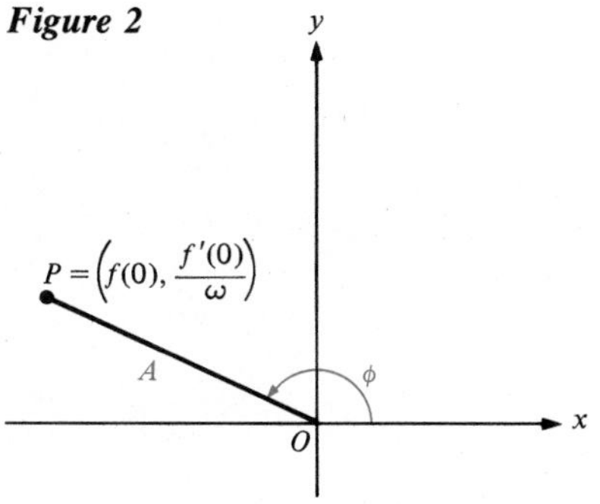

in the xy plane (Figure 2), let

$$A = |\overline{OP}| = \sqrt{[f(0)]^2 + \left[\frac{f'(0)}{\omega} \right]^2}$$

and take ϕ to be the radian measure of the smallest positive angle from the positive x axis to the line segment $\overline{OP}$. [If $P = (0, 0)$, just take $\phi = 0$.] Then, if $A \neq 0$,

$$\cos \phi = \frac{f(0)}{A} \qquad \text{and} \qquad \sin \phi = \frac{f'(0)/\omega}{A}$$

so that even if $A = 0$, we have

$$A \cos \phi = f(0) \qquad \text{and} \qquad \omega A \sin \phi = f'(0)$$

Figure 2

THEOREM 5

Complete Solution of the Harmonic-Oscillator Equation

Let ω be a nonzero constant, and suppose that $y = f(t)$ is a solution of the second-order differential equation

$$\frac{d^2 y}{dt^2} + \omega^2 y = 0$$

Choose numbers A and ϕ with $A \geq 0$ and $0 \leq \phi < 2\pi$ such that

$$A \cos \phi = f(0) \qquad \text{and} \qquad \omega A \sin \phi = f'(0)$$

(Figure 2). Then

$$f(t) = A \cos(\omega t - \phi)$$

PROOF

Let $$g(t) = f(t) - A \cos(\omega t - \phi)$$

By hypothesis, $f(t)$ is a solution of the differential equation

$$\frac{d^2 y}{dt^2} + \omega^2 y = 0$$

and by Theorem 1, so is $A \cos(\omega t - \phi)$. Hence, by Theorem 2, $g(t)$ is also a solution of this differential equation.

Now

$$g(0) = f(0) - A \cos (-\phi) = f(0) - A \cos \phi = 0$$

and

$$g'(t) = f'(t) + \omega A \sin (\omega t - \phi)$$

so

$$g'(0) = f'(0) + \omega A \sin (-\phi) = f'(0) - \omega A \sin \phi = 0$$

Therefore, by Theorem 4,

$$g(t) = 0 \qquad \text{for all values of } t$$

Consequently,

$$f(t) = A \cos (\omega t - \phi) \qquad \text{for all values of } t \qquad ∎$$

EXAMPLE 1   Solve the differential equation

$$\frac{d^2y}{dt^2} + 4y = 0$$

with the initial conditions that $y = 3$ and $dy/dt = 6$ when $t = 0$.

SOLUTION   Here $\omega^2 = 4$, so $\omega = 2$. In Theorem 5,* we seek a solution $y = f(t)$ for which

$$f(0) = 3 \qquad \text{and} \qquad f'(0) = 6$$

Thus, in Figure 2 we take

$$P = \left(f(0), \frac{f'(0)}{\omega} \right) = (3, \tfrac{6}{2}) = (3, 3)$$

The resulting diagram (Figure 3) shows that

$$A = \sqrt{3^2 + 3^2} = 3\sqrt{2} \qquad \text{and} \qquad \phi = \frac{\pi}{4}$$

Therefore, the desired solution $y = f(t)$ is given by

$$y = 3\sqrt{2} \cos \left(2t - \frac{\pi}{4} \right) \qquad ∎$$

Figure 3

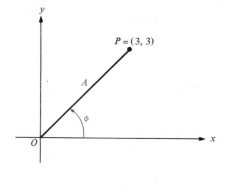

Figure 4

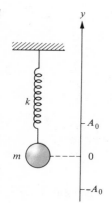

Mass-Spring Systems

The simplest mechanical harmonic oscillators are **mass-spring systems.** Such a system consists of a mass m suspended by a perfectly elastic spring with spring constant k (Figure 4). The mass of the spring itself and air friction are to be neglected. A vertical y axis is set up so that when the mass and spring are hanging in equilibrium, the y coordinate of m is zero. The mass is lifted to the position with coordinate $y = A_0$ and is released at time $t = 0$ with an initial velocity $v_0 = 0$.

*In using Theorem 5, you must be careful not to confuse $y = f(t)$, which is the solution of the differential equation, and the y in Figure 2, which is just the label on the vertical axis.

Naturally, the mass bobs up and down between A_0 and $-A_0$. At any time t, the velocity of m is given by

$$v = \frac{dy}{dt}$$

and its acceleration is

$$a = \frac{dv}{dt} = \frac{d^2y}{dt^2}$$

By Hooke's law (see page 276), the unbalanced force F exerted on the mass m by the spring is given by

$$F = -ky$$

(The minus sign is present because when y is *negative*, the spring is pulling *up* on the mass.) According to **Newton's (second) law of motion**—force equals mass times acceleration—we have

$$F = ma$$

It follows that

$$ma = -ky \qquad \text{or} \qquad a + \frac{k}{m}y = 0$$

In other words, the vibrating mass in Figure 4 satisfies the differential equation

$$\frac{d^2y}{dt^2} + \frac{k}{m}y = 0$$

This is just a special case of the harmonic-oscillator equation

$$\frac{d^2y}{dt^2} + \omega^2 y = 0$$

with

$$\omega^2 = \frac{k}{m} \qquad \text{or} \qquad \omega = \sqrt{\frac{k}{m}}$$

EXAMPLE 2   For the mass-spring system in Figure 4, find **(a)** the equation of motion and **(b)** the frequency of oscillation.

SOLUTION   In Theorem 5 we take $\omega = \sqrt{k/m}$ and seek a solution $y = f(t)$ satisfying the initial conditions

$$f(0) = A_0 \qquad \text{and} \qquad f'(0) = v_0 = 0$$

Thus, in Figure 2 we have

$$P = \left(f(0), \frac{f'(0)}{\omega} \right) = (A_0,\ 0)$$

and it follows that

$$A = \sqrt{(A_0)^2 + 0^2} = A_0 \qquad \text{and} \qquad \phi = 0$$

Therefore, $f(t) = A \cos(\omega t - \phi) = A_0 \cos \sqrt{\dfrac{k}{m}}\, t$

Hence:

(a) The equation of motion is

$$y = A_0 \cos \sqrt{\frac{k}{m}}\, t$$

(b) The frequency of oscillation is given by

$$\nu = \frac{\omega}{2\pi} = \frac{1}{2\pi}\sqrt{\frac{k}{m}}$$

Problem Set 4.6

In Problems 1 to 6, solve the differential equation with the indicated initial-value conditions.

1 $\dfrac{d^2y}{dt^2} + 9y = 0$; $y = 0$ and $\dfrac{dy}{dt} = 3$ when $t = 0$

2 $\dfrac{d^2y}{dt^2} + 2y = 0$; $y = 2\sqrt{3}$ and $\dfrac{dy}{dt} = 2\sqrt{2}$ when $t = 0$

3 $f''(t) + 3f(t) = 0$; $f(0) = -1$ and $f'(0) = 0$

4 $\dfrac{d^2y}{dx^2} + 5y = 0$; $y = -4$ and $\dfrac{dy}{dx} = -4\sqrt{5}$ when $x = 0$

5 $D_t^2 y + y = 0$; $y = 0$ and $D_t y = 1$ when $t = 0$

© **6** $y'' + 4y = 0$; $y = 4$ and $y' = 6$ when $x = 0$

7 In Problems 1 and 3, find (a) the frequency ν of the oscillation and (b) the period T of the oscillation.

8 Account for the general features of the graph in Figure 1.

© In Problems 9 to 12, sketch the graph of a quantity y oscillating in a simple harmonic manner and satisfying the conditions given.

9 $A = 1$, $\nu = 50$ hertz, $\phi = \pi/2$

10 $A = 5$, $\nu = \dfrac{1}{28}$ cycle/day, $\phi = -3\pi/14$

11 $A = 220$, $\nu = 60$ hertz, $\phi = 0$

12 $A = 5$, $\nu = 88.5$ megahertz, $\phi = 0.8$

© In Problems 13 to 16, find (a) the equation of motion of the mass-spring system and (b) its frequency by using the information given.

13 $k = 19.6$ newtons per meter, $m = 0.1$ kilogram

14 A force of 40 newtons stretches the spring 0.5 meter, and $m = 2$ kilograms.

15 $T = 1$ second and $A_0 = 0.8$ meter

16 $m = 0.5$ kilogram; when m is suspended from the spring, it stretches 0.2 meter. (*Hint:* A mass of m kilograms weighs mg newtons, where g is the acceleration of gravity.)

17 For a mass-spring system (Figure 4), the *potential energy* stored in the spring is given by $\frac{1}{2}ky^2$, and the *kinetic energy* of the mass m is given by $\frac{1}{2}m(dy/dt)^2$. The *total energy* E is the sum of the potential and kinetic energies. If $\mathcal{E}$ is the quantity defined in Theorem 3, find a constant K such that $E = K\mathcal{E}$.

18 Consider a simple pendulum of length l (Figure 5). When the bob of mass m is displaced θ radians from its equilibrium position O, the restoring force F is the component of its weight mg in the direction tangent to the circular arc s; hence,

$$m\frac{d^2s}{dt^2} = -mg \sin \theta$$

or

$$\frac{d^2s}{dt^2} + g \sin \theta = 0$$

Figure 5

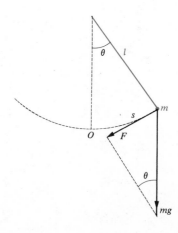

If the pendulum is swinging through a small arc, then

$$\frac{\sin \theta}{\theta} \approx 1 \qquad \text{so} \qquad \sin \theta \approx \theta$$

and we can write (approximately)

$$\frac{d^2 s}{dt^2} + g\theta = 0$$

(a) Using the fact that $s = l\theta$, show that the last differential equation can be rewritten in the form of the harmonic-oscillator equation. (b) Assuming that the pendulum is swinging through a small arc, find a formula for the (approximate) period of its oscillation.

19 An electric circuit consisting of a coil with inductance L henries and a capacitor with a capacitance of C farads is called an **LC-circuit** (Figure 6). If the capacitor is charged with Q_0 coulombs and the switch is closed, a current will begin to flow in the circuit and the capacitor will begin to discharge. If Q is the charge in coulombs on the capacitor at time t, then the voltage drop across the capacitor at this time is given by Q/C. The current I in amperes flowing in the circuit at time t is given by $I = dQ/dt$, and the voltage drop across the inductor is given by $L(dI/dt)$. From **Kirchhoff's (first) law,** we have

$$L\frac{dI}{dt} + \frac{Q}{C} = 0$$

Figure 6

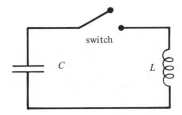

(a) Show that the current in the LC-circuit satisfies the harmonic-oscillator equation

$$\frac{d^2 I}{dt^2} + \omega^2 I = 0 \qquad \text{with } \omega = \frac{1}{\sqrt{LC}}$$

(b) Show that the charge Q on the capacitor satisfies the harmonic-oscillator equation with the same ω as in part (a). (c) Assuming that $Q = Q_0$ and $I = 0$ when $t = 0$, write an equation for Q at time t. (d) Using the initial conditions in part (c), write an equation for I at time t. (e) Find a formula for the frequency ν with which the LC-circuit oscillates.

20 An astronaut on the surface of the moon finds that a certain pendulum has a period 2.45 times as long as it does on the surface of the earth. From this information, she is able to calculate the acceleration of gravity on the moon. How does she do this, and what is her answer? (See Problem 18.)

21 In Problem 19, the *electrostatic energy* stored in the capacitor at any given time is given by $Q^2/(2C)$, and the *magnetic energy* stored in the inductor is given by $LI^2/2$. The *total energy* E stored in the LC-circuit is the sum of these two energies. If $\mathcal{E}$ is the quantity defined in Theorem 3, find a constant K such that $E = K\mathcal{E}$.

22 By Theorem 1, if B and C are any two constants, then $B \cos t$ and $C \sin t$ are solutions of the harmonic-oscillator equation $(d^2y/dt^2) + \omega^2 y = 0$. By the superposition theorem (Theorem 2), $B \cos t + C \sin t$ is also a solution of this equation. Therefore, by Theorem 5, there must exist constants A and ϕ such that $B \cos t + C \sin t = A \cos (t - \phi)$ holds for all values of t. Find A and ϕ in terms of B and C.

23 Prove part (ii) of Theorem 2.

24 If $F(t)$ is a given function, then the differential equation $(d^2y/dt^2) + \omega^2 y = F(t)$ is called the **inhomogeneous** harmonic-oscillator equation with **driving function** $F(t)$. Suppose that $y_p = f_p(t)$ is any particular solution of this inhomogeneous equation. Show that $y = A \cos (\omega t - \phi) + f_p(t)$ is also a solution of the inhomogeneous equation for any value of the constants A and ϕ.

25 If B is a constant, find a solution of the second-order differential equation $(d^2y/dt^2) + \omega^2 y = B$.

26 In Problem 24, show that $y = A \cos (\omega t - \phi) + f_p(t)$ is the complete solution of the inhomogeneous harmonic-oscillator equation.

27 In Figure 7, suppose that the point P is moving counterclockwise around a circle of radius A with a constant angular speed of ω radians per second. Let y be the coordinate of the projection of P on a vertical y axis with the origin at the level of the center of the circle. Show that y undergoes simple harmonic motion with angular frequency ω.

Figure 7

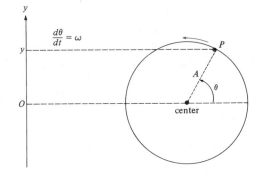

28 If B and ω are constants, find the complete solution of the inhomogeneous harmonic-oscillator equation $(d^2y/dt^2) + \omega^2 y = B$.

29 In Problem 27, explain the geometric meaning of the phase angle ϕ for the simple harmonic oscillation of y. Assume the initial condition $\theta = \theta_0$ when $t = 0$ (Figure 7).

4.7   Areas by the Method of Slicing

Figure 1

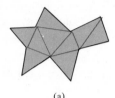

(a)

(b)

One of the most important applications of antidifferentiation is finding the areas of regions in the plane. If such a region has a boundary consisting of a finite number of line segments, it can be subdivided into a finite number of nonoverlapping triangles (Figure 1a), and its area can be found simply by summing up the areas of these triangles. However, if the region is bounded by curves (Figure 1b), it may not be immediately clear how to calculate its area. One clever method is to cut out the region, weigh it with a chemist's analytical balance, and divide its weight by the weight of 1 square unit of the same paper.

In this section, we present a method for calculating areas that is simple, can give exact results, and does not require access to an analytical balance. Here we develop this method, called the **method of slicing,** on the basis of Leibniz's idea that differentials represent infinitesimals. This will set the stage for the more rigorous and formal treatment of areas given in Chapter 5.

To find the area of a region in the plane by the method of slicing, begin by choosing or setting up a convenient *reference* axis. This could be the x axis, the y axis, or another number scale of your own choosing. For the sake of generality, let's call it the *s axis* (Figure 2a). At each point along this axis, construct a perpendicular line intersecting the region in a line segment of length l. Notice that l is a function of s. Suppose that the entire region lies between the perpendiculars at $s = a$ and at $s = b$, where $a < b$.

Figure 2

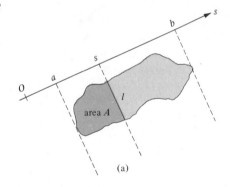

(a)

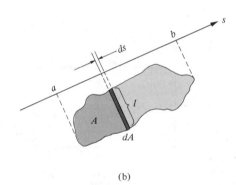

(b)

Now let A denote the area of that portion of the region between the perpendiculars at a and at s (Figure 2a). Notice that A is a function of s and $A = 0$ when $s = a$. If s is increased by an infinitesimal amount ds, then A increases by a corresponding infinitesimal amount dA (Figure 2b). Apparently, dA is virtually the area of a small rectangle of length l and width ds; that is,

$$dA = l\ ds$$

Therefore, A can be obtained as a function of s by solving this differential equation with the initial condition

$$A = 0 \qquad \text{when } s = a$$

The value of A when s = b is the desired area of the region.

Figure 3

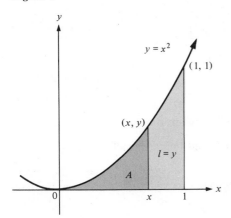

In Examples 1 to 4, find the area of the region by the method of slicing.

EXAMPLE 1   The region in the first quadrant of the *xy* plane bounded above by the parabola $y = x^2$, below by the *x* axis, and on the right by the vertical line $x = 1$.

SOLUTION   Figure 3 shows the region in question. Let's take the *x* axis as the axis of reference. Then the differential equation for the area *A* is

$$dA = l\, dx$$

where, by Figure 3,

$$l = y = x^2$$

Thus, $$A = \int l\, dx = \int x^2\, dx = \frac{x^3}{3} + C$$

Since $A = 0$ when $x = 0$, we have $C = 0$ and

$$A = \frac{x^3}{3}$$

Hence, when $x = 1$,

$$A = \frac{1^3}{3} = \frac{1}{3} \text{ square unit}$$

Figure 4

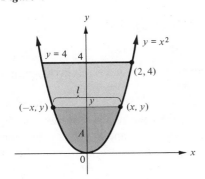

EXAMPLE 2   The region in the *xy* plane bounded below by the parabola $y = x^2$ and above by the horizontal line $y = 4$

SOLUTION   In Figure 4, we take the *y* axis as the reference axis for the method of slicing, so that

$$dA = l\, dy$$

From Figure 4,

$$l = 2x$$

where $x \geq 0$ and (x, y) lies on the graph of $y = x^2$. Therefore,

$$x = \sqrt{y} \quad \text{and} \quad l = 2x = 2\sqrt{y}$$

Hence,

$$A = \int l\, dy = \int 2\sqrt{y}\, dy = 2 \int y^{1/2}\, dy = 2\frac{y^{3/2}}{3/2} + C = \frac{4}{3}y^{3/2} + C$$

Since $A = 0$ when $y = 0$, it follows that $C = 0$ and

$$A = \tfrac{4}{3}y^{3/2}$$

Thus, when $y = 4$,

$$A = \tfrac{4}{3}(4^{3/2}) = \tfrac{32}{3} \text{ square units}$$

EXAMPLE 3   The region in the *xy* plane between the curve $y = \sqrt{x}$ and the curve $y = x^3$

SOLUTION   The region in question is shown in Figure 5. If we take the *x* axis as the axis of reference for the method of slicing, then $dA = l\, dx$. Notice that

$$l = \sqrt{x} - x^3$$

Figure 5

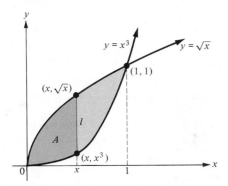

and therefore

$$A = \int l \, dx = \int (\sqrt{x} - x^3) \, dx = \int x^{1/2} \, dx - \int x^3 \, dx$$

$$= \frac{2}{3} x^{3/2} - \frac{x^4}{4} + C$$

When $x = 0$, $A = 0$, and so $C = 0$. When $x = 1$,

$$A = \tfrac{2}{3}(1)^{3/2} - \tfrac{1}{4} = \tfrac{5}{12} \text{ square unit}$$

■

Figure 6

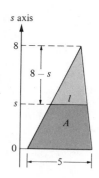

s axis

EXAMPLE 4   A triangle with base 5 meters and height 8 meters

SOLUTION   In Figure 6, we take the reference s axis perpendicular to the base of the triangle with the origin at the level of the base, so that $dA = l \, ds$. From Figure 6 and similar triangles,

$$\frac{l}{8 - s} = \frac{5}{8} \qquad \text{hence} \qquad l = \frac{5}{8}(8 - s)$$

The differential equation $dA = l \, ds$ can now be solved to obtain

$$A = \int l \, ds = \int \frac{5}{8}(8 - s) \, ds = 5 \int ds - \frac{5}{8} \int s \, ds = 5s - \frac{5}{16} s^2 + C$$

Since $A = 0$ when $s = 0$, we have $C = 0$; hence,

$$A = 5s - \tfrac{5}{16} s^2$$

The area of the entire triangle is obtained by putting $s = 8$ and calculating

$$A = (5)(8) - (\tfrac{5}{16})(8)^2 = 20 \text{ square meters}$$

(This, of course, corresponds to the result obtained by the usual formula of one-half the height times the base.)

■

If f is a function and $f(x) \geq 0$ for $a \leq x \leq b$, then the region bounded by the graph of $y = f(x)$, the x axis, and the vertical lines $x = a$ and $x = b$ is called the region **under the graph of f between $x = a$ and $x = b$** (Figure 7a). You can find the area of such a region by using the method of slicing with the x axis as the reference axis and with $l = y = f(x)$ (Figure 7b). Thus, in Figure 7b, we have

$$dA = l \, dx = f(x) \, dx$$

and so

$$A = \int f(x) \, dx$$

Figure 7

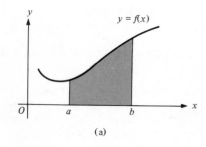

(a)

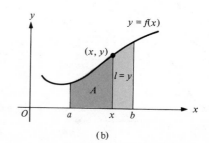

(b)

where the constant of integration is determined by the condition

$$A = 0 \qquad \text{when } x = a$$

Then, *the area of the region is the value of A when x = b.*

EXAMPLE 5 Find the area of the region under the graph of $y = x^2 - 2x + 8$ between $x = -2$ and $x = 4$ (Figure 8).

SOLUTION   Since the graph lies above the x axis between $x = -2$ and $x = 4$, we can use the method suggested above. We have

$$A = \int (x^2 - 2x + 8) \, dx = \frac{x^3}{3} - x^2 + 8x + C$$

Since $A = 0$ when $x = -2$, we have

$$0 = \frac{(-2)^3}{3} - (-2)^2 + 8(-2) + C = -\frac{68}{3} + C$$

Therefore, $C = \frac{68}{3}$ and

$$A = \frac{x^3}{3} - x^2 + 8x + \frac{68}{3}$$

Thus, when $x = 4$, we have

$$A = \frac{4^3}{3} - 4^2 + 8(4) + \frac{68}{3} = 60 \text{ square units}$$

Figure 8

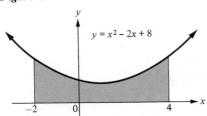

Remarks on the Method of Slicing and the Idea of Area

Although our "derivation" of the differential equation

$$dA = l \, ds$$

for area (Figure 2) might make it seem as if an approximation is involved, nevertheless, it is possible to prove that the differential equation holds *exactly*, and, with the appropriate choice of the constant of integration,

$$A = \int l \, ds$$

gives the area *exactly*. Also, it can be shown that the reference axis can be chosen arbitrarily—the answer will always be the same. Of course, skillful selection of the reference axis may simplify the actual details of the calculation.

The proof of these facts requires a formal definition of **area.** In spite of the fact that the intuitive idea of area seems reasonably clear, it is surprisingly difficult to define it formally. One has to begin by making clear just what is meant by a region in the plane and then define exactly what is meant by the area of such a region. To do this in general by modern standards of rigor is beyond the scope of most calculus textbooks, including this one. However, in Chapter 5 we make a start in this direction by studying the special case of the area under the graph of a function by using the important idea of a *definite integral*. In Section 5.6 we show how to use the definite integral to find more general areas by the method of slicing.

Problem Set 4.7

In Problems 1 to 8, sketch the indicated region and use the method of slicing to find its area.

1 The region in the first quadrant of the xy plane bounded above by the parabola $y = x^2$, below by the x axis, and on the right by the vertical line $x = 2$

2 The region bounded above by the graph of $y = -x^2 + 3x - 2$ and below by the x axis

3 The region bounded below by the parabola $y = 1 + 2x^2$ and above by the horizontal line $y = 9$

4 The region between the curve $y = 4 - x^2$ and the x axis

5 The region in the first quadrant of the xy plane bounded above by $y = 2x^3$, below by the x axis, and on the right by the vertical line $x = 2$

6 The region bounded above by one arch of the sine curve $y = \sin x$ and bounded below by the x axis

7 The region bounded below by one branch of the curve $y = \sec^2 x$ and bounded above by the horizontal line $y = 2$

8 The region bounded by $y = x^3$, the x axis, and the vertical lines $x = -2$ and $x = 2$

In Problems 9 to 13, find the area of the shaded region in the figure by the method of slicing, using the indicated axis as the reference axis. In each case, compare your answer with the area obtained by using elementary geometry.

9

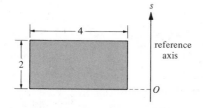

10

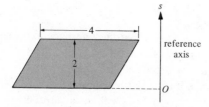

11

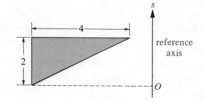

12

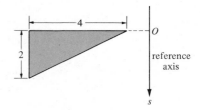

13

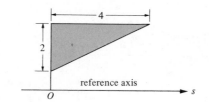

14 Calculate the area of the shaded region in Figure 9 by the method of slicing in two ways: (a) using the x axis as the reference axis; (b) using the y axis as the reference axis.

Figure 9

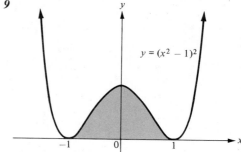

$y = (x^2 - 1)^2$

In Problems 15 to 28, calculate the area of the given region by the method of slicing using the given axis as the reference axis. Sketch the region in the xy plane.

15 The region bounded above by $y = \sqrt{x - 2}$, on the left by $x = 2$, on the right by $x = 6$, and below by $y = 0$. Take the x axis as the reference axis.

16 Same as Problem 15, but take the y axis as the reference axis.

17 The region between $y^2 = x$ and $y - x + 2 = 0$. Take the y axis as the reference axis.

18 The region between $y = x^2 - 6x + 8$ and $y = -x^2 + 4x - 3$. Take the x axis as the reference axis.

19 The triangular region bounded by the lines $y + 2x - 2 = 0$, $y - x - 5 = 0$, and $y = 7$. Take the y axis as the reference axis.

20 The region between $x = y^2 - 4$ and $x = 2 - y^2$. Take the y axis as the reference axis.

21 The region between $y^2 = 1 - x$ and $y = x + 5$. Take the y axis as the reference axis.

22 The region under the curve $y = x\sqrt{25 - x^2}$ between $x = -5$ and $x = 0$. Take the x axis as the reference axis.

23 The region between $y = 4 - x^2$ and $y = x^2 - 2$. Take the x axis as the reference axis.

24 The region between $y = x^2$ and $y = 2x$. Take the x axis as the reference axis.

25 The region between $y = x^2$ and $x = y^2$. Take the x axis as the reference axis.

26 The region between $y = x$, $y = 2 - x$, and $y = 0$. Take the y axis as the reference axis.

27 The region between $y = 4 - x^2$ and $y = -2$. Take the x axis as the reference axis.

28 The region between $y = x^4 + 1$ and $y = 17$. Take the y axis as the reference axis.

Review Problem Set, Chapter 4

In Problems 1 to 6, find dy in terms of x and dx.

1 $y = x^3 - x$

2 $y = x \cos 2x$

3 $y = \dfrac{x^2 + 5}{2x + 1}$

4 $y = \sqrt{4 - x^2}$

5 $y = 3 \cot^2 x$

6 $y = \sec \sqrt{x}$

In Problems 7 to 12, assume that x and y are differentiable functions of a third variable and are related as shown. Find the relationship between dx and dy.

7 $x^3 + y^3 - 6xy = 2$

8 $x^4 + y^4 - 4xy = 13$

9 $\sec^2 x + \csc^2 y = 4$

10 $\tan xy + xy = 3$

11 $\sin \pi xy = 12\dfrac{y}{x}$

12 $\tan \sqrt{xy} = x^2 - y^2$

© In Problems 13 to 18, let $dx = \Delta x$ and use the indicated value of x_1. (a) Use a calculator to find the numerical value of $\Delta y = f(x_1 + \Delta x) - f(x_1)$ to at least four decimal places. (b) Find an approximation for Δy, using $dy = f'(x_1)\, dx$. (c) Determine the **percent of error** $|(\Delta y - dy)/\Delta y| \times 100$ percent involved in the approximation $\Delta y \approx dy$.

13 $y = f(x) = x^2 + 1$, $x_1 = 2$, $\Delta x = 0.01$

14 $y = f(x) = \sqrt{x}$, $x_1 = 1$, $\Delta x = 0.23$

15 $y = f(x) = x^3$, $x_1 = 2$, $\Delta x = 0.02$

16 $y = f(x) = \dfrac{1}{x}$, $x_1 = 2$, $\Delta x = -0.5$

17 $y = f(x) = 2 \sin x$, $x_1 = \dfrac{\pi}{6}$, $\Delta x = 0.01$

18 $y = f(x) = -\csc x$, $x_1 = \dfrac{\pi}{2}$, $\Delta x = 0.06$

© In Problems 19 to 24, use differentials to approximate each expression.

19 $\sqrt{36.1}$

20 $x^2 + 2x - 3$ at $x = -3.02$

21 $\cos\left(\dfrac{\pi}{3} + 0.1\right)$

22 $\cot 44°$

23 $\dfrac{1}{1.02}$

24 $\sqrt[4]{0.0015}$

25 Use differentials to find the approximate area of a walk 1 meter wide around a city square which is 100 meters on each side, not counting the walk.

26 A large spherical tank of inside radius r has a thin metal wall of thickness Δr. Use differentials to write a formula for the approximate volume of metal in the wall.

27 Derive an approximate formula for the volume of metal in a can having the shape of a cube of edge x if the thickness Δx of the metal is small.

© **28** Sketch the graph of the function defined by $y = 1/x$, show a typical value x_0 in the domain of this function, indicate a small change $\Delta x = dx$ in the value of x from x_0 to $x_0 + \Delta x$, and show the corresponding values of Δy and dy on the graph. Explain why Δy is approximately equal to $-\Delta x/x^2$. Use this fact to find a decimal approximation of $\frac{1}{102}$. Indicate how much error is involved in this approximation.

© **29** The diameter of a bar in the shape of a right circular cylinder is measured as 4.2 centimeters with a possible error of 0.05 centimeter. Give a reasonable estimate (using differentials) for the possible error in the cross-sectional area computed from the measured diameter.

Ⓒ **30** Suppose that the mathematical model

$$p = (1.013 \times 10^5) - 115 \sin (1.24t)$$

is used to represent the air pressure p in newtons per square meter t seconds after inspiration begins in the lungs of a human volunteer during a study of the respiratory effects of air pollution. Using differentials, find a formula for the approximate change dp in lung air pressure during the first Δt seconds after inspiration begins, assuming that Δt is small.

In Problems 31 to 60, evaluate each antiderivative.

31 $\int (3x^4 + 4x^2 + 11)\, dx$

32 $\int (4x^3 + 3x^2 - x + 91)\, dx$

33 $\int 3t\sqrt[3]{t}\, dt$

34 $\int (1 + 2t)^5\, dt$

35 $\int \sqrt[7]{3t + 9}\, dt$

36 $\int x^2(x^3 - 1)^{40}\, dx$

37 $\int x^2(x^3 + 8)^{17}\, dx$

38 $\int x(x^2 + 4)^{-1/3}\, dx$

39 $\int \dfrac{x^7\, dx}{\sqrt[5]{x^8 + 13}}$

40 $\int \dfrac{(\sqrt{x} - 3)^{44}\, dx}{\sqrt{x}}$

41 $\int x\sqrt{7 + x}\, dx$

42 $\int \dfrac{3t\, dt}{\sqrt{t + 5}}$

43 $\int x^5\sqrt{x^3 + 1}\, dx$

44 $\int \dfrac{x^8\, dx}{\sqrt{x^3 + 1}}$

45 $\int (3 \cos x - 2 \sin x)\, dx$

46 $\int \sec 2x(2 \tan 2x + \sec 2x)\, dx$

47 $\int 2 \sin 3x\, dx$

48 $\int u \sin 3u^2\, du$

49 $\int (2 - 3 \cos 4t)\, dt$

50 $\int x \cos (4x^2 - 1)\, dx$

51 $\int x \sec x^2 \tan x^2\, dx$

52 $\int \sqrt{\tan \theta} \sec^2 \theta\, d\theta$

53 $\int (2 + 3 \cot \beta)^{3/2} \csc^2 \beta\, d\beta$

54 $\int \dfrac{\cos x}{\sin^2 x}\, dx$

55 $\int \dfrac{\cos v\, dv}{(1 - \sin v)^4}$

56 $\int \csc^2 (\sin x) \cos x\, dx$

57 $\int \dfrac{\sec 3x \tan 3x}{(\sec 3x + 8)^{10}}\, dx$

58 $\int \dfrac{\csc^2 u\, du}{\sqrt{5 + \cot u}}$

59 $\int \dfrac{a + b \cos x}{(ax + b \sin x)^4}\, dx$

60 $\int \dfrac{dt}{\sin t\,(\sin t + \cos t)}$

In Problems 61 to 76, find the general solution of each differential equation.

61 $\dfrac{dy}{dx} = 2x + 1$

62 $\dfrac{dy}{dx} = (x - 4)(3x - 2)$

63 $\dfrac{ds}{dt} = \dfrac{1}{(3 - t)^2}$

64 $\dfrac{dy}{dx} = \dfrac{1 + x}{\sqrt{x}}$

65 $\dfrac{dy}{dx} = (1 - x^{3/2})^{15}\sqrt{x}$

66 $\dfrac{dy}{dx} = \dfrac{\sqrt{1 + \sqrt{x}}}{\sqrt{x}}$

67 $\dfrac{d^2y}{dx^2} = 3 - 2x + 6x^2$

68 $\dfrac{d^2y}{dx^2} = \dfrac{1}{(1 - x)^4}$

69 $\sqrt{y}\, dx + \sqrt{x}\, dy = 0$

70 $\sqrt{x^2 + 1}\, y\, dy + x\, dx = 0$

71 $\dfrac{dy}{dx} = x - 3 \cos x$

72 $\dfrac{d^2y}{dx^2} = 4x^2 - \sin 2x$

73 $y' = \sqrt{x} - \sec^2 x$

74 $ds = (\cos 3t - \csc^2 3t)\, dt$

75 $\sin x \cos^2 y\, dx + \cos^2 x\, dy = 0$

76 $\sec t\, ds + (1 + s)^2\, dt = 0$

In Problems 77 to 84, solve the differential equation subject to the indicated side conditions.

77 $\dfrac{dy}{dx} = 2x^3 + 2x + 1;\ y = 0$ when $x = 0$

78 $\dfrac{dy}{dx} = x^{-1/3};\ y = 0$ when $x = 1$

79 $\dfrac{dy}{dx} = \dfrac{x}{\sqrt{1 - x^2}};\ y = -1$ when $x = 0$

80 $\dfrac{dy}{dx} = x^2(1 + x^3)^{10};\ y = 2$ when $x = 0$

81 $\dfrac{d^2y}{dx^2} = x^3 + 1;\ y = 0$ and $y' = 1$ when $x = 0$

82 $\dfrac{d^2y}{dx^2} = \dfrac{1}{x^3};\ y = 2$ and $y' = 1$ when $x = 1$

83 $\dfrac{ds}{dt} = t - 3 \sin t;\ s = 0$ when $t = 0$

84 $\sec y\, dx + \csc x\, dy = 0;\ y = \dfrac{\pi}{2}$ when $x = 0$

85 The slope of the tangent line to the graph of $y = f(x)$ at the point with abscissa x is given by $dy/dx = x\sqrt{x^2 + 5}$. Find the function f if its graph contains the point $(2, 6)$.

86 The slope of the tangent line to the graph of the function f at the point $(x, f(x))$ is $x \csc^2 x^2$. Find the function f if its graph contains the point $(\sqrt{\pi/2}, 3)$.

87 (a) Show that there is no function f satisfying the differential equation $f'(x) = 3x^2 + 1$ such that $f(0) = 0$ and $f(1) = 3$. (b) Show that there is a function f satisfying the differential equation $f''(x) = 3x^2 + 1$ such that $f(0) = 0$ and $f(1) = 3$. Explain, contrasting part (a) with part (b).

88 Solve the differential equation $dy/dx = |x| + |x - 1| + |x - 2|$ subject to the initial-value condition that $y = 1$ when $x = 0$. (*Hint:* $\int |x| \, dx = x|x|/2 + C$.)

In Problems 89 to 92, the force F in newtons acting on a particle P moving along the s axis is given in terms of the coordinate s of P. Find the work (in joules) done by F in moving P from $s = s_0$ to $s = s_1$ meters.

89 $F = 3s - 1$, $s_0 = \frac{1}{3}$, $s_1 = 6$

90 $F = \cos 2s$, $s_0 = 0$, $s_1 = \pi/4$

91 $F = \sec^2 \dfrac{s}{2}$, $s_0 = 0$, $s_1 = 2\pi/3$

92 $F = \sqrt{1 + \sqrt{s}}$, $s_0 = 0$, $s_1 = 1$

93 A perfectly elastic spring is compressed 6 inches from its natural length because a weight of 2 tons is placed on it. How much work (in foot-pounds) is done?

94 What constant negative acceleration is required to bring a train to rest in 500 meters if it is initially going at 44 meters per second?

© **95** An athlete running the 100-meter dash maintains a constant acceleration for the first 15 meters and thereafter has zero acceleration. What must the acceleration be if the athlete is to run the race in 10 seconds?

© **96** A certain type of motorcycle can be brought to a stop from a speed of 72 kilometers per hour in 4 seconds. How long will it take to bring it to a stop from a speed of 96 kilometers per hour? Assume the same constant deceleration in both cases.

97 A stone is thrown straight down with an initial velocity of 96 feet per second from a bridge 256 feet above a river. (a) How many seconds elapse before the stone hits the water? (b) What will be the velocity of the stone when it strikes the water? (Take $g = 32$ feet per second squared.)

98 Two particles P and P^* are moving on the s axis with constant accelerations a and a^*, respectively. When $t = 0$, particle P^* is k units to the right of particle P, where $k > 0$; particle P is moving with a velocity v_0; and particle P^* is moving with a velocity v_0^*. The question is, "Will the two particles collide?"

(a) If $a^* < a$, show that the two particles will eventually collide. (b) If $a^* = a$, show that the two particles will eventually collide if and only if $v_0^* < v_0$. (c) If $a^* > a$, show that the two particles will eventually collide if and only if $v_0^* < v_0$ and $2(a^* - a)k < (v_0^* - v_0)^2$. (*Hint:* Let s and s^* denote the position coordinates of P and P^*, respectively, at time t. Put $y = s^* - s$, and show that y satisfies the differential equation $d^2y/dt^2 = a^* - a$ with the initial conditions that $y = k$ and $dy/dt = v_0^* - v_0$ when $t = 0$. Collision occurs if $y = 0$ for some positive value of t.)

99 A 2000-newton weight is suspended 50 meters below a windlass by a cable weighing 10 newtons per meter. Neglecting friction, determine how many joules of work will be required to lift the weight through the 50 meters.

100 Let W be the work done in charging a capacitor having (constant) capacitance C with Q coulombs. If E is the voltage drop across the capacitor and I is the current in amperes flowing into the capacitor, then $dW = E \, dQ$, $Q = CE$, and $dQ = I \, dt$, where t denotes time in seconds. (a) Show that the instantaneous *power* dW/dt required to charge the capacitor is given by $dW/dt = EI$. (b) Assuming that $W = 0$ when $E = 0$, prove that $W = \frac{1}{2}CE^2$.

101 Suppose that m kilograms of mass are distributed uniformly along the s axis between $s = a$ and $s = b$ meters, where $0 < a < b$. An M-kilogram particle P is placed on the s axis at the origin. Find the net gravitational force exerted on P by the distributed mass.

© **102** In the study of the transport of oxygen (O_2) by the blood from the lungs to the tissues of the body, it has been found that the *conversion of hemoglobin to oxyhemoglobin* is governed by a differential equation of the form $(1 + kx^n)^2 \, dy - nkx^{n-1} \, dx = 0$, where x is the partial pressure in newtons per square meter of O_2 dissolved in the plasma and y is the ratio of the amount of O_2 combined with hemoglobin to the maximum amount of O_2 that could combine with the hemoglobin. For normal human blood, the constants n and k are given approximately by $n = 2.7$ and $k = 2.4 \times 10^{-10}$. (a) Using the initial-value condition that $y = 0$ when $x = 0$, solve the differential equation for arbitrary positive values of the constants n and k. (b) Sketch a graph of y as a function of x for normal human blood, $0 \le x \le 1.5 \times 10^4$ newtons per square meter. (c) Find the coordinates of the point of inflection of the graph in part (b).

103 A company manufacturing jogging shorts expects a marginal revenue of $dR/dx = 10 - (x/12,500)$ dollars per pair manufactured. Assume that $R = 0$ when $x = 0$. The marginal cost is $dC/dx = 6$ dollars per pair manufactured, and $C = \$400$ when $x = 0$. (a) Find the total revenue expected if x pairs of jogging shorts are manufactured. (b) Find the total cost of manufacturing x pairs of jogging shorts. (c) How many pairs should be manufactured to maximize the total revenue? (d) How many pairs should be manufactured to maximize the profit? (e) How much does the company charge per pair when the profit is maximized?

104 The expected revenue for the sale of x sweaters is given by

$$R = x\left(27 - \frac{x}{1000}\right) \qquad \text{dollars}$$

while the marginal cost is given by $dC/dx = 700/\sqrt{x}$ dollars per sweater. When $x = 0$, then $C = \$500$. (a) Find the cost as a function of x. (b) Show that the maximum profit is achieved by selling 10,000 sweaters. (c) What would be the price per sweater if 10,000 sweaters were sold?

In Problems 105 to 108, solve the differential equation with the indicated initial-value conditions.

105 $\dfrac{d^2y}{dt^2} + 25y = 0$; $y = 0$ and $\dfrac{dy}{dt} = 5$ when $t = 0$

106 $\dfrac{d^2y}{dx^2} + 7y = 0$; $y = 0$ and $\dfrac{dy}{dx} = \sqrt{7}$ when $x = 0$

107 $y'' + 36y = 0$; $y = 2\sqrt{3}$ and $y' = 12$ when $x = 0$

C **108** $D_t^2 s + 9s = 0$; $s = 9$ and $D_t s = 11$ when $t = 0$

C **109** A car with faulty shock absorbers is approximately a mass-spring system, and so it vibrates more or less in a simple harmonic manner. Suppose that the equation for the displacement y meters of the body of such a car is $(d^2y/dt^2) + 100y = 0$, where $y = 0.02$ meter and $dy/dt = 0.24$ meter per second when $t = 0$ seconds. Find (a) a formula for the displacement y meters of the car body at time t seconds, (b) the frequency of the vibration, and (c) the period of the vibration.

C **110** A mass of 0.5 kilogram is attached to a spring with constant $k = 4.5$ newtons per meter. A vertical y axis is set up so that when the mass and spring are hanging in equilibrium, the y coordinate of the mass is zero. The mass is lifted to the position

with coordinate $y = 0.2$ meter and is released at time $t = 0$ with initial velocity $dy/dt = 0$. Find (a) the equation of motion of the mass-spring system and (b) its frequency.

In Problems 111 to 120, sketch the region and use the method of slicing to find its area.

111 The region in the first quadrant of the xy plane bounded above by the parabola $y = x^2 + 1$, below by the x axis, on the left by the y axis, and on the right by the line $x = 3$. Use the x axis as the reference axis.

112 The region in the first quadrant of the xy plane bounded above by the graph of $y = \sqrt{x}$ and to the right by the line $x = 4$. Use the x axis as the reference axis.

113 The region bounded above by $y = 1$, on the left by $y = x^2$, and on the right by $y = x$. Use the y axis as the reference axis.

114 The region under the graph of $y = 1 - |x|$ between $x = -1$ and $x = 1$. Use the y axis as the reference axis.

115 The region bounded above by $y = 4x - x^2$ and below by the x axis.

116 The region bounded below by one branch of $y = \csc^2 x$ and above by $y = 2$.

117 The region between the x axis and one complete cycle of $y = \sin x$.

118 The region between $y = x^4$ and $y = x^2$.

119 The region in the second quadrant under the graph of $y = \sqrt{x + 1}$. Use the x axis as the reference axis.

120 Same region as in Problem 119, but use the y axis as the reference axis.

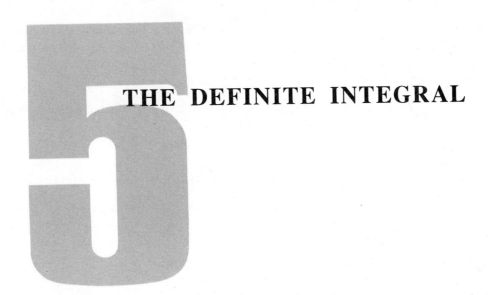

THE DEFINITE INTEGRAL

In Chapter 4, we introduced the antiderivative, or indefinite integral, and indicated how it could be used to help solve differential equations. In particular, we showed how to use the indefinite integral to calculate areas by the method of slicing. There our study of area was preliminary and exploratory because it was based on the intuitive idea that a differential can be considered to be an infinitesimal.

In the present chapter, we introduce, by means of a formal definition, the important idea of a *definite integral,* and we show how to use the definite integral to calculate area. A profound connection between definite and indefinite integrals—the *fundamental theorem of calculus*—is proved in Section 5.4. Finally, in Section 5.6, we indicate how the definite integral and the fundamental theorem of calculus can be used to provide a more formal justification for the method of slicing.

5.1 The Sigma Notation for Sums

The formal definition of the definite integral involves the sum of many terms and thus calls for some special notation. In mathematical symbolism, the capital Greek letter **sigma,** which is written

$$\sum$$

and corresponds to the letter S, stands for the words

"the sum of all terms of the form . . ."

For instance, rather than writing $1 + 2 + 3 + 4 + 5 + 6$, we can write $\sum k$, and $1^2 + 2^2 + 3^2 + 4^2 + 5^2 + 6^2$ can be written as $\sum k^2$, provided that we make the convention that k is to run through all integer values from 1 to 6. If we wish to include the range of values of k as part of the sigma notation, we write, for instance,

$$\sum_{1 \le k \le 6} \qquad \text{or} \qquad \sum_{k=1}^{k=6} \qquad \text{or} \qquad \sum_{k=1}^{6}$$

Thus, $\sum_{k=1}^{6} k^2$ means "the sum of all terms of the form k^2 as k runs through the integers from 1 to 6."

In Examples 1 to 3, write out the given sum explicitly and then find its numerical value.

__EXAMPLE 1__ $\displaystyle\sum_{k=1}^{7} k^2$

SOLUTION $\displaystyle\sum_{k=1}^{7} k^2 = 1^2 + 2^2 + 3^2 + 4^2 + 5^2 + 6^2 + 7^2$

$$= 1 + 4 + 9 + 16 + 25 + 36 + 49 = 140$$ ∎

__EXAMPLE 2__ $\displaystyle\sum_{k=1}^{4} 2^k$

SOLUTION $\displaystyle\sum_{k=1}^{4} 2^k = 2^1 + 2^2 + 2^3 + 2^4 = 2 + 4 + 8 + 16 = 30$ ∎

__EXAMPLE 3__ $\displaystyle\sum_{k=1}^{n} (3^k - 3^{k-1})$

SOLUTION

$$\sum_{k=1}^{n} (3^k - 3^{k-1}) = (3^1 - 3^0) + (3^2 - 3^1) + (3^3 - 3^2) + \cdots$$

$$+ (3^{n-1} - 3^{n-2}) + (3^n - 3^{n-1})$$

If we rewrite this sum in the form

$$\sum_{k=1}^{n} (3^k - 3^{k-1}) = -3^0 + (3^1 - 3^1) + (3^2 - 3^2) + \cdots + (3^{n-1} - 3^{n-1}) + 3^n$$

then the terms within the parentheses cancel, and we have

$$\sum_{k=1}^{n} (3^k - 3^{k-1}) = -3^0 + 3^n = 3^n - 1$$ ∎

In Examples 1 to 3 above, the variable k, which runs from 1 to 7 in $\sum_{k=1}^{7} k^2$, from 1 to 4 in $\sum_{k=1}^{4} 2^k$, and from 1 to n in $\sum_{k=1}^{n} (3^k - 3^{k-1})$, is called the **summation**

index. There is no particular reason to use k for the summation index—any letter of the alphabet will do. For instance,

$$\sum_{k=1}^{4} 2^k = \sum_{i=1}^{4} 2^i = \sum_{j=1}^{4} 2^j = 2^1 + 2^2 + 2^3 + 2^4$$

Note that if f is a function and if the integers from 1 to n belong to the domain of f, then

$$\sum_{k=1}^{n} f(k) = f(1) + f(2) + f(3) + \cdots + f(n)$$

For instance, if f is a constant function, say $f(x) = C$ for all values of x, then

$$\sum_{k=1}^{n} f(k) = f(1) + f(2) + f(3) + \cdots + f(n) = \underbrace{C + C + C + \cdots + C}_{n \text{ terms}} = nC$$

The fact expressed by this equation is often written simply as

$$\sum_{k=1}^{n} C = nC$$

Thus,

$$\sum_{k=1}^{7} 5 = (7)(5) = 35 \qquad \sum_{k=1}^{n} (-2) = -2n \qquad \sum_{i=1}^{100} 1 = 100$$

and so forth.

The sigma notation is especially handy for indicating the sum of the terms of a sequence of numbers. A *sequence of numbers* consisting of a first number a_1, a second number a_2, a third number a_3, and so on, is written as $a_1, a_2, a_3, \ldots$. Thus, the kth term in this sequence is a_k, and the sum of, say, the first six terms can be written

$$\sum_{k=1}^{6} a_k = a_1 + a_2 + a_3 + a_4 + a_5 + a_6$$

Similarly, the sum of the second, third, and fourth terms of the sequence can be written

$$\sum_{k=2}^{4} a_k = a_2 + a_3 + a_4$$

Sometimes, in dealing with sequences, it is convenient to start with a *zeroth term* a_0. For such a sequence, the sum of the terms a_k, with k running from 0 to n, can be expressed as

$$\sum_{k=0}^{n} a_k = a_0 + a_1 + a_2 + \cdots + a_n$$

Some basic properties of summation, which are easy to establish either by inspection or by using the principle of mathematical induction, are as follows:

Basic Properties of Summation

Let $a_0, a_1, a_2, \ldots, a_n$ and $b_0, b_1, b_2, \ldots, b_n$ denote sequences of numbers, and let A, B, and C be constant numbers. Then:

1 *Constant property* $\displaystyle\sum_{k=1}^{n} C = nC$

2 *Homogeneous property* $\displaystyle\sum_{k=1}^{n} Ca_k = C \sum_{k=1}^{n} a_k$

3 *Additive property* $\displaystyle\sum_{k=1}^{n} (a_k + b_k) = \sum_{k=1}^{n} a_k + \sum_{k=1}^{n} b_k$

4 *Linear property* $\displaystyle\sum_{k=1}^{n} (Aa_k + Bb_k) = A \sum_{k=1}^{n} a_k + B \sum_{k=1}^{n} b_k$

EXAMPLE 4   Given that $\displaystyle\sum_{k=1}^{9} a_k = 121$ and $\displaystyle\sum_{k=1}^{9} b_k = 77$, find $\displaystyle\sum_{k=1}^{9} (3a_k - 2b_k)$.

SOLUTION   By Property 4,

$$\sum_{k=1}^{9} (3a_k - 2b_k) = \sum_{k=1}^{9} [3a_k + (-2)b_k] = 3 \sum_{k=1}^{9} a_k + (-2) \sum_{k=1}^{9} b_k$$

$$= 3(121) - 2(77) = 209$$

THEOREM 1   **Special Sums**

(i) *Sum of successive integers* $\displaystyle\sum_{k=1}^{n} k = \frac{n(n+1)}{2}$

(ii) *Sum of successive squares* $\displaystyle\sum_{k=1}^{n} k^2 = \frac{n(n+1)(2n+1)}{6}$

(iii) *Sum of successive cubes* $\displaystyle\sum_{k=1}^{n} k^3 = \frac{n^2(n+1)^2}{4}$

(iv) *Sum of successive powers* $\displaystyle\sum_{k=0}^{n} C^k = \frac{1 - C^{n+1}}{1 - C}$ if $C \neq 0, 1$

geometric series

PROOF   Although formulas (i) to (iv) can be derived in various ways, perhaps the quickest way to verify them is by using the principle of mathematical induction.* Here we prove (i) by mathematical induction. We leave the proofs of (ii), (iii), and (iv) as exercises (Problems 24, 26, and 28). If $n = 1$, the formula in (i) becomes

$$1 = \frac{1(1+1)}{2}$$

*For a review of the principle of mathematical induction, see Appendix C.

which is obviously true. Now, we assume that (i) holds for a particular value of n, and try to show that it holds for the next value, $n + 1$. Thus, we assume that

$$\sum_{k=1}^{n} k = \frac{n(n+1)}{2}$$

Adding $n + 1$ to both sides of the last equation we find that

$$\sum_{k=1}^{n} k + (n+1) = \frac{n(n+1)}{2} + (n+1)$$

that is,

$$\sum_{k=1}^{n+1} k = \frac{n(n+1)}{2} + \frac{2(n+1)}{2} = \frac{n(n+1) + 2(n+1)}{2}$$

Therefore, we have

$$\sum_{k=1}^{n+1} k = \frac{(n+1)[(n+1)+1]}{2}$$

and our proof by induction of formula (i) is complete.   ■

In Examples 5 to 9, use the basic properties of summation and Theorem 1 to evaluate each sum.

EXAMPLE 5 $\displaystyle\sum_{k=1}^{20} (2k^2 - 3k + 1)$

SOLUTION $\displaystyle\sum_{k=1}^{20} (2k^2 - 3k + 1) = \sum_{k=1}^{20} 2k^2 + \sum_{k=1}^{20} (-3k) + \sum_{k=1}^{20} 1$ (Property 3)

$$= 2 \sum_{k=1}^{20} k^2 - 3 \sum_{k=1}^{20} k + 20 \text{ (Properties 2 and 1)}$$

$$= 2\frac{(20)(21)(41)}{6} - 3\frac{(20)(21)}{2} + 20$$

$$= 5130 \qquad \text{[Theorem 1 (ii) and (i)]} ■$$

EXAMPLE 6 $\displaystyle\sum_{k=1}^{n} k^2(5k + 1)$

SOLUTION $\displaystyle\sum_{k=1}^{n} k^2(5k+1) = \sum_{k=1}^{n} (5k^3 + k^2) = 5 \sum_{k=1}^{n} k^3 + \sum_{k=1}^{n} k^2$

$$= \frac{5}{4}n^2(n+1)^2 + \frac{n(n+1)(2n+1)}{6}$$

$$= n(n+1)\left[\frac{5n(n+1)}{4} + \frac{2n+1}{6}\right]$$

$$= n(n+1)\frac{15n(n+1) + 2(2n+1)}{12}$$

$$= \frac{n(n+1)(15n^2 + 19n + 2)}{12} ■$$

EXAMPLE 7 $1 + \dfrac{1}{2} + \dfrac{1}{4} + \dfrac{1}{8} + \cdots + \dfrac{1}{2^n}$

SOLUTION

$$1 + \frac{1}{2} + \frac{1}{4} + \frac{1}{8} + \cdots + \frac{1}{2^n} = \sum_{k=0}^{n} \frac{1}{2^k} = \sum_{k=0}^{n} \left(\frac{1}{2}\right)^k$$

$$= \frac{1 - (\frac{1}{2})^{n+1}}{1 - \frac{1}{2}} = 2 - \frac{1}{2^n} \qquad \text{[Theorem 1 (iv)]} \ \blacksquare$$

EXAMPLE 8 The sum of the first 100 odd positive integers

SOLUTION   As k runs through the integers from 1 to 100, $2k - 1$ runs through the first 100 odd positive integers. Thus,

$$\sum_{k=1}^{100} (2k - 1) = 2\left(\sum_{k=1}^{100} k\right) - \sum_{k=1}^{100} 1 = 2\left[\frac{100(101)}{2}\right] - 100 = 10{,}000 \qquad \blacksquare$$

EXAMPLE 9 $\displaystyle\sum_{k=1}^{n} C^k \qquad C \neq 0, 1$

SOLUTION   Formula (iv) in Theorem 1 cannot be used directly here because k starts with the value 1 rather than 0. However, by "splitting off the zeroth term" in part (iv) of Theorem 1, we see that

$$C^0 + \sum_{k=1}^{n} C^k = \sum_{k=0}^{n} C^k = \frac{1 - C^{n+1}}{1 - C}$$

Therefore,

$$\sum_{k=1}^{n} C^k = \frac{1 - C^{n+1}}{1 - C} - C^0 = \frac{1 - C^{n+1}}{1 - C} - 1 = \frac{1 - C^{n+1}}{1 - C} - \frac{1 - C}{1 - C}$$

$$= \frac{1 - C^{n+1} - (1 - C)}{1 - C} = \frac{C - C^{n+1}}{1 - C} = C\frac{1 - C^n}{1 - C} \qquad \blacksquare$$

The Area Under a Parabola

In Example 1, page 291, we used the method of slicing to calculate the area A under the parabola $y = x^2$ between $x = 0$ and $x = 1$, and we found that $A = \frac{1}{3}$ square unit. However, we never really proved that the method of slicing works. At best, our argument was merely plausible since it was based on Leibniz's idea that a differential can be thought of as an infinitesimal. Now, using the sigma notation, we can give a more conclusive argument that the area in question really is $\frac{1}{3}$ square unit.

In Figure 1a, the interval [0, 1] is subdivided into n subintervals

$$\left[0, \frac{1}{n}\right], \left[\frac{1}{n}, \frac{2}{n}\right], \left[\frac{2}{n}, \frac{3}{n}\right], \left[\frac{3}{n}, \frac{4}{n}\right], \ldots, \left[\frac{n-1}{n}, \frac{n}{n}\right]$$

of equal length $1/n$. Above each subinterval we have formed a corresponding circumscribed rectangle. Evidently, the height of the kth circumscribed rectangle is

Figure 1

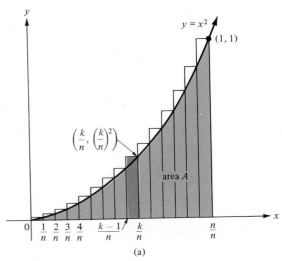

(a)

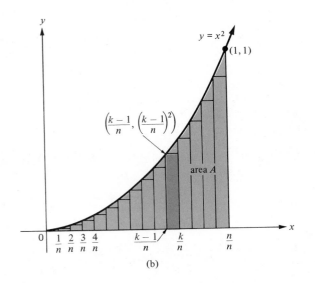

(b)

$(k/n)^2$, and its area is $(1/n)(k/n)^2$. An estimate of the area A can now be obtained by adding the areas of the n circumscribed rectangles:

$$A \approx \sum_{k=1}^{n} \frac{1}{n}\left(\frac{k}{n}\right)^2$$

Now n is a fixed number (a constant) in the summation, so we can use Property 2, page 302, to write

$$\sum_{k=1}^{n} \frac{1}{n}\left(\frac{k}{n}\right)^2 = \sum_{k=1}^{n} \frac{1}{n^3}k^2 = \frac{1}{n^3}\sum_{k=1}^{n} k^2$$

So, by part (ii) of Theorem 1, we have

$$A \approx \frac{1}{n^3}\sum_{k=1}^{n} k^2 = \frac{1}{n^3}\left[\frac{n(n+1)(2n+1)}{6}\right]$$

$$= \frac{(n+1)(2n+1)}{6n^2} = \frac{2n^2+3n+1}{6n^2}$$

or

$$A \approx \frac{1}{3} + \frac{1}{2n} + \frac{1}{6n^2}$$

Moreover, since the approximating rectangles are circumscribed,

$$A \leq \frac{1}{3} + \frac{1}{2n} + \frac{1}{6n^2}$$

In Figure 1b, we again subdivided the interval $[0, 1]$ into n subintervals of equal length, but we formed the inscribed rather than the circumscribed rectangles. Evidently, the height of the kth inscribed rectangle is $[(k-1)/n]^2$, and its area is $(1/n)[(k-1)/n]^2$. Again, we obtain an estimate of the area A by adding the areas of the n inscribed rectangles:

$$A \approx \sum_{k=1}^{n} \frac{1}{n}\left(\frac{k-1}{n}\right)^2$$

Now,

$$\sum_{k=1}^{n} \frac{1}{n}\left(\frac{k-1}{n}\right)^2 = \sum_{k=1}^{n} \left(\frac{1}{n}\right)^3 (k-1)^2 = \left(\frac{1}{n}\right)^3 \sum_{k=1}^{n} (k^2 - 2k + 1)$$

$$= \frac{1}{n^3}\left[\frac{n(n+1)(2n+1)}{6} - \frac{2n(n+1)}{2} + n\right]$$

$$= \frac{2n^2 - 3n + 1}{6n^2}$$

so

$$A \approx \frac{2n^2 - 3n + 1}{6n^2} = \frac{1}{3} - \frac{1}{2n} + \frac{1}{6n^2}$$

Moreover, since the approximating rectangles are inscribed,

$$\frac{1}{3} - \frac{1}{2n} + \frac{1}{6n^2} \le A$$

The above considerations show that for every positive integer n,

$$\frac{1}{3} - \frac{1}{2n} + \frac{1}{6n^2} \le A \le \frac{1}{3} + \frac{1}{2n} + \frac{1}{6n^2}$$

As n becomes larger and larger, both

$$\frac{1}{3} - \frac{1}{2n} + \frac{1}{6n^2} \quad \text{and} \quad \frac{1}{3} + \frac{1}{2n} + \frac{1}{6n^2}$$

approach $\frac{1}{3}$ as a limit. Since the constant number A is "trapped" between two quantities, both of which can be made to come as close to $\frac{1}{3}$ as we please, it follows from the squeezing property for limits* (page 63) that $A = \frac{1}{3}$.

---

*Actually, the squeezing property applies to limits as a *continuous* variable x approaches a real number a—and here we have limits as the *integer-valued* variable n becomes larger and larger without bound. However, the basic principle is the same.

Problem Set 5.1

In Problems 1 to 8, write out each sum explicitly and then find its numerical value.

1 $\displaystyle\sum_{k=1}^{6} (2k + 1)$

2 $\displaystyle\sum_{k=1}^{5} 7k^2$

3 $\displaystyle\sum_{j=1}^{4} \sin \frac{\pi}{j}$

4 $\displaystyle\sum_{j=3}^{7} \frac{j}{j - 2}$

5 $\displaystyle\sum_{k=2}^{6} \frac{1}{k(k - 1)}$

6 $\displaystyle\sum_{i=-1}^{3} 3^i$

7 $\displaystyle\sum_{j=0}^{3} \frac{1}{j^2 + 3}$

8 $\displaystyle\sum_{k=-1}^{3} \frac{k}{k + 2}$

In Problems 9 and 10, assume that $\sum_{k=1}^{19} a_k = 23$, $\sum_{k=1}^{19} b_k = 99$, and $\sum_{k=1}^{19} c_k = -14$. Find the numerical value of the indicated sum.

9 $\displaystyle\sum_{k=1}^{19} (5a_k + 3b_k)$

10 $\displaystyle\sum_{k=1}^{19} (2a_k - 3b_k + 4c_k - 5)$

In Problems 11 to 20, use the basic properties of summation and Theorem 1 to evaluate the sum.

11 $\displaystyle\sum_{k=1}^{50} (2k + 3)$

12 $\displaystyle\sum_{i=1}^{30} i(i - 1)$

13 $\displaystyle\sum_{k=1}^{100} 5^k$

14 $\displaystyle\sum_{k=0}^{100} (5^{k+1} - 5^k)$

15 $\displaystyle\sum_{k=1}^{n} k(k + 1)$

16 $\displaystyle\sum_{k=1}^{100} \left(\frac{1}{k} - \frac{1}{k + 1}\right)$

17 $\displaystyle\sum_{k=1}^{n-1} k^2$

18 $\displaystyle\sum_{k=1}^{n} (a_k - a_{k-1})$

19 $\displaystyle\sum_{k=1}^{n} (k - 1)^2$

20 $\displaystyle\sum_{j=1}^{100} \frac{1}{10^j}$

21 Using the method of circumscribed and inscribed rectangles, find the area under the parabola $y = x^2$ between $x = 1$ and $x = 2$.

22 Using the method of circumscribed and inscribed rectangles, find the area under the graph of $y = x^3$ between $x = 0$ and $x = 1$.

23 Give an alternative proof of part (i) of Theorem 1 by completing the following argument: Let $\sum_{k=1}^{n} k = S$. Then

$$S = 1 + 2 + 3 + \cdots + (n - 1) + n$$

Writing the same sum in the opposite order, we have

$$S = n + (n - 1) + \cdots + 3 + 2 + 1$$

Adding the last two equations and combining corresponding terms, we obtain

$$2S = (n + 1) + (n + 1) + \cdots + (n + 1)$$

from which the desired formula follows.

24 Use the principle of mathematical induction to prove part (ii) of Theorem 1.

25 Prove the **telescoping series property:**

$$\sum_{k=1}^{n} (b_k - b_{k-1}) = b_n - b_0$$

(*Hint:* Notice that part of each term cancels with part of the next term.)

26 Use the principle of mathematical induction to prove part (iii) of Theorem 1.

27 Use the result in Problem 25 to show that *the sum of the first n odd positive integers is n^2*. [*Hint:* Let $b_k = k^2$.]

28 Use the principle of mathematical induction to prove part (iv) of Theorem 1.

29 Use the result in Problem 27 to give an alternative proof of part (i) of Theorem 1.

30 Use the triangle inequality (Theorem 1, page 7) and the principle of mathematical induction to prove the generalized triangle inequality:

$$\left| \sum_{k=1}^{n} a_k \right| \leq \sum_{k=1}^{n} |a_k|$$

31 Give an alternative proof of part (iv) of Theorem 1 by completing the following argument: Let $C \neq 0, 1$ and put $\sum_{k=0}^{n} C^k = S$. Then $SC = \sum_{k=0}^{n} C^{k+1}$, so $S - SC = C^0 - C^{n+1}$, and the desired formula follows from the last equation.

5.2   The Definite Integral

Figure 1

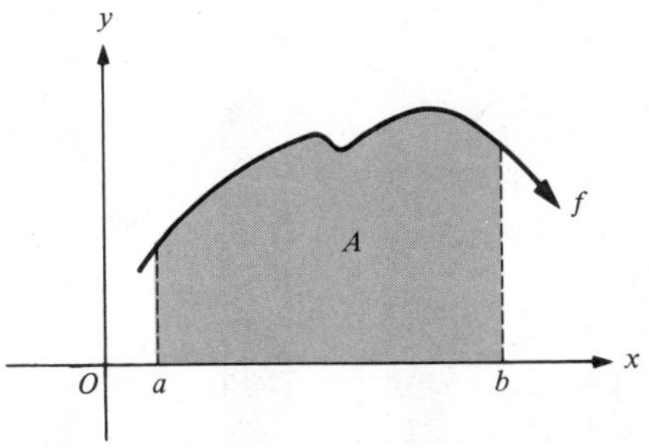

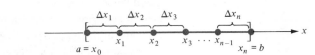

The method of circumscribed and inscribed rectangles, used in Section 5.1 to find the area under the parabola $y = x^2$ between $x = 0$ and $x = 1$, is easily generalized. Indeed, suppose that f is a continuous function such that $f(x) \geq 0$ for $a \leq x \leq b$, and we wish to find the area A under the graph of f between $x = a$ and $x = b$ (Figure 1).

By a **partition** of the interval $[a, b]$, we mean a sequence of n subintervals

$$[x_0, x_1], [x_1, x_2], [x_2, x_3], \ldots, [x_{n-1}, x_n]$$

where $x_0 = a$ and $x_n = b$ (Figure 2). The subinterval $[x_{k-1}, x_k]$ is called the **kth subinterval** in the partition, and its length is denoted by

$$\Delta x_k = x_k - x_{k-1}$$

Thus, the length of the first subinterval is $\Delta x_1 = x_1 - x_0$, the length of the second subinterval is $\Delta x_2 = x_2 - x_1$, and so on. There is *no* requirement that all subintervals in the partition have the same length. Denote the partition

$$[x_0, x_1], [x_1, x_2], [x_2, x_3], \ldots, [x_{n-1}, x_n]$$

by the symbol $\mathscr{P}$.

Figure 2

By the **norm** of the partition $\mathscr{P}$, in symbols

$$\|\mathscr{P}\|$$

we mean the largest of the numbers $\Delta x_1, \Delta x_2, \Delta x_3, \ldots, \Delta x_n$ representing the lengths of the n subintervals in $\mathscr{P}$. Note that two or more of the numbers $\Delta x_1, \Delta x_2, \Delta x_3, \ldots, \Delta x_n$ may have the same largest value $\|\mathscr{P}\|$. In particular, if all the subintervals in the partition $\mathscr{P}$ happen to have the same length, then

$$\Delta x_1 = \Delta x_2 = \Delta x_3 = \cdots = \Delta x_n = \frac{b-a}{n}$$

so that

$$\|\mathscr{P}\| = \frac{b-a}{n}$$

in this special case.

We now choose a number from each subinterval in the partition $\mathscr{P}$. Let c_1 denote the number chosen from the first subinterval $[x_0, x_1]$, let c_2 denote the number chosen from the second subinterval $[x_1, x_2]$, and so forth. Thus, c_k denotes the number chosen from the kth subinterval (Figure 3). The partition $\mathscr{P}$, together with the chosen numbers $c_1, c_2, \ldots, c_n$, is called an **augmented partition** and is denoted by $\mathscr{P}^*$.

On each subinterval of the partition $\mathscr{P}$, we now construct a rectangle as in Figure 4. Notice that the kth rectangle has the kth subinterval $[x_{k-1}, x_k]$ of length Δx_k as its base and extends to the point $(c_k, f(c_k))$ on the graph of f. Since the height of the kth rectangle is $f(c_k)$, its area ΔA_k is given by

$$\Delta A_k = f(c_k)\, \Delta x_k$$

Figure 3

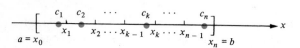

Figure 4

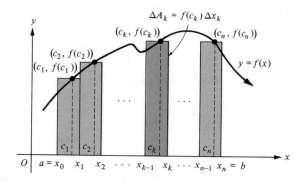

The sum

$$\sum_{k=1}^{n} \Delta A_k = \sum_{k=1}^{n} f(c_k)\, \Delta x_k$$

of the areas ΔA_k of the rectangles determined by the augmented partition $\mathscr{P}^*$ is called the **Riemann sum** corresponding to $\mathscr{P}^*$ for the function f. This terminology is used in honor of the German mathematician Bernhard Riemann (1826–1866), who did some of the early definitive work on the problem of giving a precise mathematical formalization of the integral of Newton and Leibniz.

Bernhard Riemann

Evidently, the Riemann sum provides an approximation for the area A under the graph of f between $x = a$ and $x = b$, that is,

$$A \approx \sum_{k=1}^{n} \Delta A_k = \sum_{k=1}^{n} f(c_k)\, \Delta x_k$$

If an exceedingly large number of very skinny rectangles are involved in the Riemann sum, this approximation will be quite accurate.

<u>EXAMPLE 1</u>   Find the Riemann sum for the function $f(x) = 1 + x^3$ on the interval $[-1, 2]$, using the augmented partition $\mathscr{P}^*$ consisting of the six subintervals

$$[-1, -\tfrac{1}{2}],\ [-\tfrac{1}{2}, 0],\ [0, \tfrac{1}{2}],\ [\tfrac{1}{2}, 1],\ [1, \tfrac{3}{2}],\ [\tfrac{3}{2}, 2]$$

with $c_1 = -\tfrac{1}{2}$ $c_2 = 0$ $c_3 = \tfrac{1}{2}$ $c_4 = 1$ $c_5 = \tfrac{3}{2}$ $c_6 = 2$

Sketch a graph showing the six rectangles corresponding to this Riemann sum.

SOLUTION   For the given partition, each of the six subintervals has length $\tfrac{1}{2}$ unit, that is,

$$\Delta x_1 = \Delta x_2 = \Delta x_3 = \Delta x_4 = \Delta x_5 = \Delta x_6 = \tfrac{1}{2}$$

Also,

$$f(c_1) = 1 + (-\tfrac{1}{2})^3 = \tfrac{7}{8} \qquad f(c_4) = 1 + 1^3 = 2$$

$$f(c_2) = 1 + 0^3 = 1 \qquad\qquad f(c_5) = 1 + (\tfrac{3}{2})^3 = \tfrac{35}{8}$$

$$f(c_3) = 1 + (\tfrac{1}{2})^3 = \tfrac{9}{8} \qquad\quad f(c_6) = 1 + 2^3 = 9$$

Therefore, the required Riemann sum is given by

$$\sum_{k=1}^{6} f(c_k)\, \Delta x_k$$

$$= f(c_1)\, \Delta x_1 + f(c_2)\, \Delta x_2 + f(c_3)\, \Delta x_3 + f(c_4)\, \Delta x_4 + f(c_5)\, \Delta x_5 + f(c_6)\, \Delta x_6$$

$$= (\tfrac{7}{8})(\tfrac{1}{2}) + (1)(\tfrac{1}{2}) + (\tfrac{9}{8})(\tfrac{1}{2}) + (2)(\tfrac{1}{2}) + (\tfrac{35}{8})(\tfrac{1}{2}) + (9)(\tfrac{1}{2})$$

$$= \tfrac{147}{16}$$

Figure 5 shows the six rectangles whose areas are the terms in the Riemann sum. ∎

As you can see from Figure 5, the Riemann sum

$$\sum_{k=1}^{6} f(c_k)\, \Delta x_k = \tfrac{147}{16}$$

provides an approximation to the area under the graph of $f(x) = 1 + x^3$ between $x = -1$ and $x = 2$. For a better approximation, we could take an augmented partition with a smaller norm (and hence with more rectangles).

In general, if f is a continuous function and $f(x) \geq 0$ for $a \leq x \leq b$, the approximation

$$A \approx \sum_{k=1}^{n} f(c_k)\, \Delta x_k$$

for the area A under the graph of f between $x = a$ and $x = b$ becomes better and

Figure 5

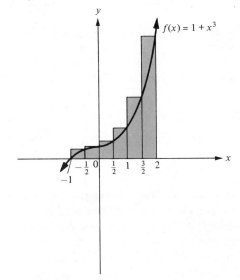

better as the number of rectangles increases and their widths decrease. Since this can be accomplished by taking the norm $\|\mathscr{P}\|$ of the partition smaller and smaller, we might expect that

$$\lim_{\|\mathscr{P}\| \to 0} \sum_{k=1}^{n} f(c_k) \, \Delta x_k$$

should exist and give the *exact* area A.

As the discussion above indicates, the area under the graph of a continuous function may be found by evaluating the limit of a Riemann sum. However, nothing in the construction of the Riemann sum

$$\sum_{k=1}^{n} f(c_k) \, \Delta x_k$$

corresponding to an augmented partition for a function f requires that the function be continuous, that its values be nonnegative, or that there be any particular association with area—*all that is required is that $f(x)$ be defined for every x in the interval $[a, b]$.* As you will see in Chapter 6, there are many quantities other than areas that can be obtained as limits of Riemann sums. Thus, we make the following definition.

DEFINITION 1   **The Definite (Riemann) Integral**

> If the limit $\displaystyle\lim_{\|\mathscr{P}\| \to 0} \sum_{k=1}^{n} f(c_k) \, \Delta x_k$ exists, then the function f is said to be **integrable**
>
> *on $[a, b]$ in the sense of Riemann.* If f is integrable, then the **definite (Riemann) integral** of f on the interval $[a, b]$ is defined by
>
> $$\int_a^b f(x) \, dx = \lim_{\|\mathscr{P}\| \to 0} \sum_{k=1}^{n} f(c_k) \, \Delta x_k$$

The limit indicated in Definition 1 is to be understood in the following sense: To say that a number L is the limit of the Riemann sum $\sum_{k=1}^{n} f(c_k) \, \Delta x_k$ as the norm $\|\mathscr{P}\|$ approaches zero means that for each positive number ϵ (no matter how small), there exists a positive number δ (depending on ϵ) such that

$$\left| \sum_{k=1}^{n} f(c_k) \, \Delta x_k - L \right| < \epsilon$$

holds for every augmented partition $\mathscr{P}^*$ with $\|\mathscr{P}\| < \delta$.

The notation $\int_a^b f(x) \, dx$ for the definite integral is deliberately chosen to be similar to the notation $\int f(x) \, dx$ for the indefinite integral (antiderivative). The advantage of this is seen in Section 5.4. Because of the similarity of notation, you must be careful not to confuse the *definite* integral $\int_a^b f(x) \, dx$ with the *indefinite* integral $\int f(x) \, dx$. Note that the definite integral is a *number*—a limit of Riemann sums—whereas the indefinite integral represents all *functions* that are antiderivatives of the integrand.

We call $\int_a^b f(x)\,dx$ the **definite integral from a to b of the function f.** The interval $[a, b]$ is called the **interval of integration,** and the numbers a and b are known as the **lower** and the **upper limits* of integration.** The function f, or the expression $f(x)$, is called the **integrand,** just as it is for the indefinite integral:

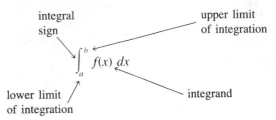

If the function f is integrable on the closed interval $[a, b]$ in the sense of Riemann, we simply say that f is **integrable on** $[a, b]$ or that $\int_a^b f(x)\,dx$ **exists.** If $\int_a^b f(x)\,dx$ exists, it can be shown that it is a uniquely determined real number (Problem 38).

Existence of the Definite Integral

Virtually every function encountered in practical scientific work is integrable on any closed interval contained in its domain. This includes not only all continuous functions and all monotone (increasing or decreasing) functions, but also all functions that are bounded and "piecewise continuous" or "piecewise monotone" as well. The proofs of the integrability of such functions can be found in more advanced textbooks on analysis; here, we content ourselves with precise statements of some of the existence theorems for definite integrals.

THEOREM 1

Existence of the Definite Integral of a Continuous Function

If f is a continuous function on the closed interval $[a, b]$, then f is integrable on $[a, b]$.

Figure 6

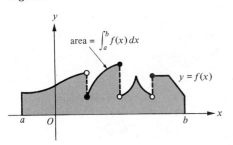

For instance, a polynomial function is integrable on any closed interval since polynomial functions are continuous.

The function whose graph is shown in Figure 6 is not continuous, but it is **piecewise continuous** on the interval $[a, b]$ in the sense that $[a, b]$ can be partitioned into a finite number of subintervals such that f is continuous on each subinterval. The function f is also **bounded** on the interval $[a, b]$ in the sense that there are fixed numbers A and B such that $A \leq f(x) \leq B$ holds for all values of x in $[a, b]$. The integrability of any such function is ensured by the following theorem.

THEOREM 2

Existence of the Definite Integral of a Piecewise Continuous, Bounded Function

If f is a bounded and piecewise continuous function on the closed interval $[a, b]$, then f is integrable on $[a, b]$.

*This use of the word "limit" must not be confused with the limit of a function or with the limit of Riemann sums in Definition 1.

Figure 7

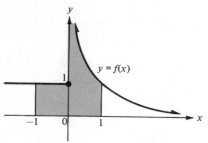

The function defined by

$$f(x) = \begin{cases} \dfrac{1}{x} & \text{for } x > 0 \\[2mm] 1 & \text{for } x \le 0 \end{cases}$$

is piecewise continuous but not bounded on the interval $[-1, 1]$ (Figure 7). A glance at Figure 7 suggests that great care should be taken in talking about the "area" of the region under the graph of an unbounded function. Theorem 3 gives further evidence for the necessity of such care.

THEOREM 3

Boundedness of Integrable Functions

> If f is defined and integrable on $[a, b]$, then f is bounded on $[a, b]$.

For instance, the function f whose graph appears in Figure 7 is *not* integrable on $[-1, 1]$, since it is unbounded on this interval.

The following theorem can be useful in deciding whether certain functions are integrable.

THEOREM 4

Change of a Function at Finitely Many Points

> If f is defined and integrable on $[a, b]$ and if h is also defined on $[a, b]$ and satisfies $h(x) = f(x)$ for all but a finite number of values of x in $[a, b]$, then h is also integrable on $[a, b]$ and
>
> $$\int_a^b h(x)\, dx = \int_a^b f(x)\, dx$$

Calculation of Definite Integrals by Direct Use of the Definition

The definite integral $\int_a^b f(x)\, dx$ of any integrable function f can be calculated by selecting any sequence of augmented partitions

$$\mathscr{P}_1^*, \mathscr{P}_2^*, \mathscr{P}_3^*, \ldots, \mathscr{P}_n^*, \ldots$$

such that $\lim_{n \to +\infty} \|\mathscr{P}_n\| = 0$, calculating the Riemann sums corresponding to each augmented partition, and finding the limit as $n \to +\infty$ of the resulting sequence of Riemann sums. In doing this, it is often convenient to consider only partitions of $[a, b]$ consisting of n subintervals of the *same length*

$$\Delta x = \frac{b - a}{n}$$

Furthermore, in augmenting these partitions, it is often helpful to select the numbers $c_1, c_2, c_3, \ldots, c_n$ at the *left endpoints* (or at the *right endpoints*) of the subintervals.

In Examples 2 and 3, evaluate the definite integral directly by calculating a limit of Riemann sums. Use partitions consisting of subintervals of equal lengths, and augment the partitions by using the left or right endpoints of the subintervals as indicated. If possible, interpret the result as an area.

EXAMPLE 2 $\displaystyle\int_0^2 x^3\, dx$ (left endpoints)

SOLUTION   Let $\mathscr{P}_n$ denote the partition of the interval $[0, 2]$ into n subintervals of equal lengths

$$\Delta x = \frac{2 - 0}{n} = \frac{2}{n}$$

Figure 8

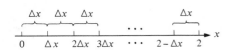

(Figure 8). Thus, the augmented partition $\mathscr{P}_n^*$ consists of the n subintervals

$$[0, \Delta x], \ [\Delta x, 2\,\Delta x], \ [2\,\Delta x, 3\,\Delta x], \ \ldots, \ [2 - \Delta x, 2]$$

with $c_1 = 0, \ c_2 = \Delta x, \ c_3 = 2\,\Delta x, \ \ldots, \ c_n = 2 - \Delta x$

Evidently, for $k = 1, 2, 3, \ldots, n$, we have

$$c_k = (k - 1)\,\Delta x = (k - 1)\frac{2}{n} = \frac{2(k - 1)}{n}$$

Consequently, the Riemann sum corresponding to $\mathscr{P}_n^*$ for the function $f(x) = x^3$ is given by

$$\sum_{k=1}^n f(c_k)\,\Delta x_k = \sum_{k=1}^n (c_k)^3\,\Delta x = \sum_{k=1}^n \left[\frac{2(k-1)}{n}\right]^3 \frac{2}{n} = \sum_{k=1}^n \frac{16}{n^4}(k - 1)^3$$

Using the properties of summation in Section 5.1, we have

$$\sum_{k=1}^n f(c_k)\,\Delta x_k = \sum_{k=1}^n \frac{16}{n^4}(k - 1)^3 = \frac{16}{n^4}\sum_{k=1}^n (k - 1)^3$$

$$= \frac{16}{n^4}\sum_{k=1}^n (k^3 - 3k^2 + 3k - 1)$$

$$= \frac{16}{n^4}\left(\sum_{k=1}^n k^3 - 3\sum_{k=1}^n k^2 + 3\sum_{k=1}^n k - \sum_{k=1}^n 1\right)$$

$$= \frac{16}{n^4}\left[\frac{n^2(n + 1)^2}{4} - 3\frac{n(n + 1)(2n + 1)}{6} + 3\frac{n(n + 1)}{2} - n\right]$$

$$= \frac{16}{n^4}\left[\frac{n^4 - 2n^3 + n^2}{4}\right] = 4\left(1 - \frac{2}{n} + \frac{1}{n^2}\right)$$

Figure 9

Therefore,

$$\int_0^2 x^3\, dx = \lim_{n \to +\infty}\sum_{k=1}^n f(c_k)\,\Delta x_k = \lim_{n \to +\infty} 4\left(1 - \frac{2}{n} + \frac{1}{n^2}\right) = 4$$

Since $x^3 \geq 0$ for $0 \leq x \leq 2$, we may interpret the result

$$\int_0^2 x^3\, dx = 4$$

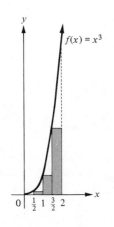

as stating that the area under the graph of $f(x) = x^3$ between $x = 0$ and $x = 2$ is 4 square units. Because $f(x) = x^3$ is an *increasing* function on the interval $[0, 2]$, *inscribed* rectangles are obtained by forming augmented partitions $\mathscr{P}_n^*$ for which $c_1, c_2, c_3, \ldots, c_n$ are the *left* endpoints of the subintervals; for instance, Figure 9 shows the case where $n = 4$.

Figure 10

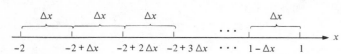

EXAMPLE 3 $\displaystyle\int_{-2}^{1} x \, dx$ (right endpoints)

SOLUTION   Let $\mathscr{P}_n$ denote the partition of the interval $[-2, 1]$ into n subintervals of equal lengths

$$\Delta x = \frac{1 - (-2)}{n} = \frac{3}{n}$$

(Figure 10). Thus, the augmented partition $\mathscr{P}_n^*$ consists of the n subintervals

$$[-2, -2 + \Delta x], \ [-2 + \Delta x, -2 + 2 \Delta x], \ [-2 + 2 \Delta x, -2 + 3 \Delta x],$$
$$\ldots, [1 - \Delta x, 1]$$

with $c_1 = -2 + \Delta x, \ c_2 = -2 + 2 \Delta x, \ c_3 = -2 + 3 \Delta x, \ \ldots, c_n = 1$

Evidently,

$$c_k = -2 + k \, \Delta x = -2 + k\left(\frac{3}{n}\right) = \frac{3k}{n} - 2$$

Consequently, the Riemann sum corresponding to $\mathscr{P}_n^*$ for the function $f(x) = x$ is given by

$$\sum_{k=1}^{n} f(c_k) \, \Delta x_k = \sum_{k=1}^{n} c_k \, \Delta x = \sum_{k=1}^{n} \left(\frac{3k}{n} - 2\right)\left(\frac{3}{n}\right) = \sum_{k=1}^{n} \left(\frac{9k}{n^2} - \frac{6}{n}\right)$$

$$= \sum_{k=1}^{n} \frac{9k}{n^2} - \sum_{k=1}^{n} \frac{6}{n} = \frac{9}{n^2} \sum_{k=1}^{n} k - \frac{6}{n} n$$

$$= \frac{9}{n^2}\left[\frac{n(n + 1)}{2}\right] - 6 = \frac{9}{2}\left(\frac{n + 1}{n}\right) - 6$$

$$= \frac{9}{2}\left(1 + \frac{1}{n}\right) - 6$$

Therefore,

$$\int_{-2}^{1} x \, dx = \lim_{n \to +\infty} \sum_{k=1}^{n} f(c_k) \, \Delta x_k = \lim_{n \to +\infty} \left[\frac{9}{2}\left(1 + \frac{1}{n}\right) - 6\right] = \frac{9}{2} - 6 = -\frac{3}{2}$$

Since the function $f(x) = x$ has negative values on the interval $[-2, 1]$, we *cannot* interpret the integral

$$\int_{-2}^{1} x \, dx = -\frac{3}{2}$$

as an area.   ∎

The Definite Integral $\displaystyle\int_{a}^{b} f(x) \, dx$ for $a \geq b$

In the definition of the definite integral $\int_a^b f(x) \, dx$, it was assumed that $a < b$, so that $[a, b]$ is a closed interval. For certain purposes, it is convenient to be able to deal with expressions such as $\int_a^b f(x) \, dx$ without worrying about whether $a < b$. This is accomplished by the following definition.

DEFINITION 2   **The Definite Integral** $\displaystyle\int_a^b f(x)\, dx$ **for** $a \geq b$

(i) If f is any function and a is a number in the domain of f, then we define

$$\int_a^a f(x)\, dx = 0$$

(ii) If $a > b$ and f is integrable on $[b, a]$, then we define

$$\int_a^b f(x)\, dx = -\int_b^a f(x)\, dx$$

EXAMPLE 4   Evaluate

(a) $\displaystyle\int_{-2}^{-2} x\, dx$ (b) $\displaystyle\int_{1}^{-2} x\, dx$

SOLUTION

(a) By part (i) of Definition 2, $\displaystyle\int_{-2}^{-2} x\, dx = 0$.

(b) By Example 3 above, $\displaystyle\int_{-2}^{1} x\, dx = -\frac{3}{2}$. Therefore, by part (ii) of Definition 2,

$$\int_1^{-2} x\, dx = -\int_{-2}^{1} x\, dx = -(-\tfrac{3}{2}) = \tfrac{3}{2}$$

∎

Problem Set 5.2

In Problems 1 to 4, find the Riemann sum for each function on the prescribed interval, using the indicated augmented partition. Also, sketch the graph of the function on the interval, showing the rectangles corresponding to the Riemann sum.

1 $f(x) = 3x + 1$ on $[0, 3]$. $\mathscr{P}^*$ consists of $[0, \frac{1}{2}]$, $[\frac{1}{2}, 1]$, $[1, \frac{3}{2}]$, $[\frac{3}{2}, 2]$, $[2, \frac{5}{2}]$, and $[\frac{5}{2}, 3]$ with $c_1 = \frac{1}{2}$, $c_2 = 1$, $c_3 = \frac{3}{2}$, $c_4 = 2$, $c_5 = \frac{5}{2}$, and $c_6 = 3$.

2 $f(x) = 2x^2$ on $[0, 3]$. $\mathscr{P}^*$ consists of $[0, \frac{1}{2}]$, $[\frac{1}{2}, 1]$, $[1, \frac{3}{2}]$, $[\frac{3}{2}, 2]$, $[2, \frac{5}{2}]$, and $[\frac{5}{2}, 3]$ with $c_1 = \frac{1}{4}$, $c_2 = \frac{3}{4}$, $c_3 = \frac{5}{4}$, $c_4 = \frac{7}{4}$, $c_5 = \frac{9}{4}$, and $c_6 = \frac{11}{4}$.

3 $f(x) = 1/x$ on $[1, 3]$. $\mathscr{P}^*$ consists of $[1, \frac{3}{2}]$, $[\frac{3}{2}, 2]$, $[2, \frac{5}{2}]$, and $[\frac{5}{2}, 3]$ with $c_1 = \frac{5}{4}$, $c_2 = \frac{7}{4}$, $c_3 = \frac{9}{4}$, and $c_4 = \frac{11}{4}$.

4 $f(x) = 1/(2 + x)$ on $[-1, 2]$. $\mathscr{P}^*$ consists of $[-1, -\frac{1}{2}]$, $[-\frac{1}{2}, 0]$, $[0, \frac{1}{2}]$, $[\frac{1}{2}, 1]$, $[1, \frac{3}{2}]$, and $[\frac{3}{2}, 2]$ with $c_1 = -1$, $c_2 = -\frac{1}{2}$, $c_3 = 0$, $c_4 = \frac{1}{2}$, $c_5 = 1$, and $c_6 = \frac{3}{2}$.

In Problems 5 to 14, evaluate each definite integral directly by calculating a limit of Riemann sums. Use partitions consisting of sub-intervals of equal lengths, and augment the partitions by using the left or right endpoints of the subintervals as indicated.

5 $\displaystyle\int_0^3 2x\, dx$ (right endpoints)   **6** $\displaystyle\int_0^3 2x\, dx$ (left endpoints)

7 $\displaystyle\int_4^7 (2x - 6)\, dx$ (left endpoints)

8 $\displaystyle\int_1^3 (9 - x^2)\, dx$ (right endpoints)

9 $\displaystyle\int_{-2}^{-1} (x^2 - x - 2)\, dx$ (right endpoints)

10 $\displaystyle\int_0^2 (x^3 + 2)\, dx$ (right endpoints)

11 $\displaystyle\int_0^2 (x^3 + 2)\, dx$ (left endpoints)

12 $\displaystyle\int_{-2}^{-1} (4 - x^2)\, dx$ (right endpoints)

13 $\int_{-3}^{0} (1 - 2x^2) \, dx$ (left endpoints)

14 (a) $\int_{1}^{2} (x^2 - 4x + 2) \, dx$ (left endpoints)

(b) $\int_{1}^{2} (x^2 - 4x + 2) \, dx$ (right endpoints)

In Problems 15 to 28, indicate whether each definite integral exists and give a reason for your answer.

15 $\int_{1}^{1000} \frac{1}{x} \, dx$

16 $\int_{0}^{1} \frac{1}{x} \, dx$

17 $\int_{-1}^{1} |x| \, dx$

18 $\int_{-1}^{1} \frac{x + 1}{\sqrt{x}} \, dx$

19 $\int_{1}^{0} x^4 \, dx$

20 $\int_{1}^{100} [[x]] \, dx$

21 $\int_{0}^{3} f(x) \, dx$, where $f(x) = \begin{cases} x & \text{for } 0 \le x < 1 \\ x - 1 & \text{for } 1 \le x < 2 \\ 1 - x^2 & \text{for } 2 \le x \le 3 \end{cases}$

22 $\int_{0}^{2} f(x) \, dx$, where $f(x) = \begin{cases} 7 & \text{for } x \neq 1 \\ 2 & \text{for } x = 1 \end{cases}$

23 $\int_{0}^{\pi} \tan x \, dx$

24 $\int_{-\pi/4}^{\pi/4} \sec x \, dx$

25 $\int_{0}^{\pi} \cos 2x \, dx$

26 $\int_{-\pi/2}^{\pi/2} \sin |x| \, dx$

27 $\int_{\pi/2}^{\pi/2} \tan x \, dx$

28 $\int_{\pi}^{-\pi} \sin^2 x \, dx$

In Problems 29 to 34, evaluate the definite integral.

29 $\int_{3}^{0} 2x \, dx$ (see Problem 5)

30 $\int_{3}^{1} (9 - x^2) \, dx$ (see Problem 8)

31 $\int_{-1}^{-2} (x^2 - x - 2) \, dx$ (see Problem 9)

32 $\int_{-1}^{-2} (4 - x^2) \, dx$ (see Problem 12)

33 $\int_{3}^{3} (x^4 - 2x^3 + x^2 - x + 5) \, dx$

34 $\int_{\pi}^{\pi} \tan x \, dx$

35 Use Definition 1 to evaluate $\int_{1}^{2} 7 \, dx$ as a limit of Riemann sums. Interpret the result geometrically.

© **36** It can be shown that $\int_{1}^{2} (1/x) \, dx \approx 0.693$ (correct to three decimal places). Estimate $\int_{1}^{2} (1/x) \, dx$, using a Riemann sum involving 10 circumscribed rectangles with equal bases.

37 Use Definition 1 to prove that if $a < b$, then $\int_{a}^{b} 1 \, dx = b - a$.

38 Prove that *if a definite integral exists, it is unique;* that is, prove that there cannot be two *different* numbers L_1 and L_2 that satisfy the conditions on L given immediately after Definition 1. (*Hint:* Suppose $L_1 \neq L_2$ and let $\epsilon = \frac{1}{2}|L_1 - L_2|$.)

5.3 Basic Properties of the Definite Integral

The basic properties of the definite integral can be proved deductively from the formal definition given in Section 5.2 (Definition 1). For the most part, we do not give such proofs; however, we state these properties as theorems and interpret them, when possible, in terms of our intuitive idea of area.

Figure 1

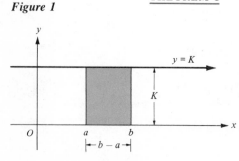

THEOREM 1   **Integral of a Constant Function**

If K is a constant, then $\displaystyle\int_{a}^{b} K \, dx = K(b - a)$.

If $a < b$ and $K > 0$, then Theorem 1 can be interpreted geometrically as stating that a rectangle with width $b - a$ and height K has an area of $K(b - a)$ square units (Figure 1). If $b < a$, then by Definition 2 in Section 5.2,

$$\int_{a}^{b} K \, dx = -\int_{b}^{a} K \, dx$$

and the same geometric interpretation applies to a rectangle of height K units and width $a - b$. Finally, if $a = b$, the equation

$$\int_a^a K\, dx = K(a - a)$$

just says that $0 = 0$.

Two special cases of Theorem 1 are of interest:

$$\int_a^b dx = \int_a^b 1\, dx = b - a \qquad \text{and} \qquad \int_a^b 0\, dx = 0(b - a) = 0$$

EXAMPLE 1   Find $\int_{-2}^{33} (-7)\, dx$.

SOLUTION   By Theorem 1,

$$\int_{-2}^{33} (-7)\, dx = (-7)[33 - (-2)] = -245$$

THEOREM 2   **Homogeneous Property**

If f is a function, K is a constant, and $\int_a^b f(x)\, dx$ exists, then $\int_a^b Kf(x)\, dx$ exists and

$$\int_a^b Kf(x)\, dx = K \int_a^b f(x)\, dx$$

Figure 2 shows the underlying geometric reason for the homogeneous property; namely, when f is multiplied by K, the rectangles whose areas are involved in the Riemann sum have their heights multiplied by K, and so their areas are multiplied by K.

Figure 2

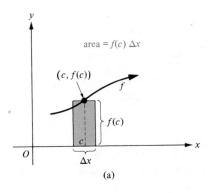
(a)

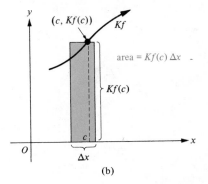
(b)

EXAMPLE 2   Given that $\int_1^{2.7} f(x)\, dx = -13$, find $\int_1^{2.7} 52f(x)\, dx$.

SOLUTION   By Theorem 2,

$$\int_1^{2.7} 52f(x)\, dx = 52 \int_1^{2.7} f(x)\, dx = 52(-13) = -676$$

Figure 3

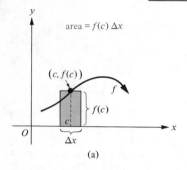

(a)

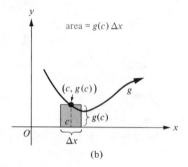

(b)

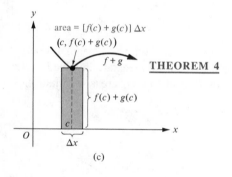

(c)

THEOREM 3

Additive Property

Let f and g be functions, and suppose that $\int_a^b f(x)\,dx$ and $\int_a^b g(x)\,dx$ exist. Then $\int_a^b [f(x) + g(x)]\,dx$ exists and

$$\int_a^b [f(x) + g(x)]\,dx = \int_a^b f(x)\,dx + \int_a^b g(x)\,dx$$

The underlying geometric reason for the additive property of the definite integral can be seen in Figure 3, which shows corresponding rectangles involved in Riemann sums for f, g, and $f + g$. The area of such a rectangle for $f + g$ is $[f(c) + g(c)]\,\Delta x = f(c)\,\Delta x + g(c)\,\Delta x$; hence, its area is the sum of the areas of the corresponding rectangles for f and for g.

EXAMPLE 3 Given that $\int_{-7}^{13} f(x)\,dx = 2.77$ and $\int_{-7}^{13} g(x)\,dx = -1.32$, find $\int_{-7}^{13} [f(x) + g(x)]\,dx$.

SOLUTION By Theorem 3,

$$\int_{-7}^{13} [f(x) + g(x)]\,dx = \int_{-7}^{13} f(x)\,dx + \int_{-7}^{13} g(x)\,dx$$
$$= 2.77 + (-1.32) = 1.45 \qquad ∎$$

The homogeneous and additive properties of the definite integral can be combined to yield the following property.

THEOREM 4

Linear Property

Let A and B be constants, and suppose that f and g are functions for which $\int_a^b f(x)\,dx$ and $\int_a^b g(x)\,dx$ exist. Then $\int_a^b [Af(x) + Bg(x)]\,dx$ exists and

$$\int_a^b [Af(x) + Bg(x)]\,dx = A \int_a^b f(x)\,dx + B \int_a^b g(x)\,dx$$

EXAMPLE 4 Given that $\int_2^3 x^4\,dx = \frac{211}{5}$ and $\int_2^3 x\,dx = \frac{5}{2}$, find $\int_2^3 (10x^4 + 16x)\,dx$.

SOLUTION By Theorem 4,

$$\int_2^3 (10x^4 + 16x)\,dx = 10 \int_2^3 x^4\,dx + 16 \int_2^3 x\,dx = 10(\tfrac{211}{5}) + 16(\tfrac{5}{2}) = 462 \qquad ∎$$

By using the principle of mathematical induction, the linear property can be extended to more than two functions. Thus, for example, we have

$$\int_0^2 (4x^3 - 3x^2 + 7x - 8)\,dx$$

$$= 4 \int_0^2 x^3\,dx + (-3) \int_0^2 x^2\,dx + 7 \int_0^2 x\,dx + (-8) \int_0^2 dx$$

$$= 4 \int_0^2 x^3\,dx - 3 \int_0^2 x^2\,dx + 7 \int_0^2 x\,dx - 8 \int_0^2 dx$$

THEOREM 5   **Additivity with Respect to the Interval of Integration**

> Suppose that the function f is integrable on a closed bounded interval I and that a, b, and c are three numbers in I. Then $\int_a^c f(x)\,dx$, $\int_a^b f(x)\,dx$, and $\int_b^c f(x)\,dx$ exist and
>
> $$\int_a^c f(x)\,dx = \int_a^b f(x)\,dx + \int_b^c f(x)\,dx$$

Figure 4

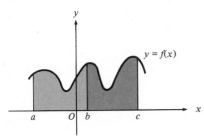

For the special case in which $a < b < c$ and $f(x) \geq 0$ for $a \leq x \leq c$, Theorem 5 has the following geometric interpretation: The area under the graph of $y = f(x)$ from $x = a$ to $x = c$ is the sum of the area from $x = a$ to $x = b$ and the area from $x = b$ to $x = c$ (Figure 4). From this special case, the result for other orderings of a, b, and c can be derived by using Definition 2 of Section 5.2. For instance, suppose that $a < c < b$. Then

$$\int_a^b f(x)\,dx = \int_a^c f(x)\,dx + \int_c^b f(x)\,dx$$

Solving the last equation for $\int_a^c f(x)\,dx$, we obtain

$$\int_a^c f(x)\,dx = \int_a^b f(x)\,dx - \int_c^b f(x)\,dx = \int_a^b f(x)\,dx + \int_b^c f(x)\,dx$$

We leave it as an exercise for you to check the remaining cases (Problem 59).

EXAMPLE 5   Given that $\int_{-1}^2 f(x)\,dx = 7$ and that $\int_2^3 f(x)\,dx = -5$, find $\int_{-1}^3 f(x)\,dx$.

SOLUTION   By Theorem 5,

$$\int_{-1}^3 f(x)\,dx = \int_{-1}^2 f(x)\,dx + \int_2^3 f(x)\,dx = 7 + (-5) = 2$$

■

Although we illustrated Theorems 1 through 5 geometrically for only those cases in which the definite integrals can be interpreted in terms of area, we must emphasize that these theorems hold in *all* cases, even if no such interpretation is possible. In the next section, we present a general method for evaluating definite integrals. Meanwhile, the following theorem gives some special integrals to aid in the illustration of the basic properties.

THEOREM 6   **Two Special Integrals**

> (i) $\displaystyle\int_a^b x\,dx = \frac{1}{2}(b^2 - a^2)$ (ii) $\displaystyle\int_a^b x^2\,dx = \frac{1}{3}(b^3 - a^3)$

PROOF   We prove (i) and leave the proof of (ii) as an exercise (Problem 60). To begin with, suppose that $a < b$. Let $\mathscr{P}_n$ denote the partition of the interval $[a, b]$ into n subintervals of equal length

$$\Delta x = \frac{b - a}{n}$$

Thus, $\mathcal{P}_n$ consists of the n subintervals

$$[a, a + \Delta x], [a + \Delta x, a + 2\,\Delta x], \ldots, [a + (n - 1)\,\Delta x, b]$$

and the kth subinterval is

$$[a + (k - 1)\,\Delta x, a + k\,\Delta x]$$

for $k = 1, 2, 3, \ldots, n$. To obtain an augmented partition $\mathcal{P}_n^*$, we choose the numbers

$$c_1, c_2, \ldots, c_n$$

to be the right-hand endpoints of the corresponding subintervals, so that

$$c_k = a + k\,\Delta x$$

for $k = 1, 2, 3, \ldots, n$. Thus, the Riemann sum for $f(x) = x$ corresponding to $\mathcal{P}_n^*$ is

$$\sum_{k=1}^{n} f(c_k)\,\Delta x_k = \sum_{k=1}^{n} c_k\,\Delta x = \sum_{k=1}^{n} (a + k\,\Delta x)\,\Delta x$$

$$= \sum_{k=1}^{n} \left(a + k\,\frac{b - a}{n}\right)\left(\frac{b - a}{n}\right)$$

$$= \sum_{k=1}^{n} \left[\frac{a(b - a)}{n} + k\left(\frac{b - a}{n}\right)^2\right]$$

$$= \sum_{k=1}^{n} \frac{a(b - a)}{n} + \sum_{k=1}^{n} k\left(\frac{b - a}{n}\right)^2$$

$$= \frac{a(b - a)}{n} \sum_{k=1}^{n} 1 + \left(\frac{b - a}{n}\right)^2 \sum_{k=1}^{n} k$$

$$= \frac{a(b - a)}{n}\,n + \left(\frac{b - a}{n}\right)^2 \frac{n(n + 1)}{2}$$

$$= a(b - a) + (b - a)^2\,\frac{n + 1}{2n}$$

$$= (b - a)\left[a + (b - a)\left(\frac{1}{2} + \frac{1}{2n}\right)\right]$$

Therefore,

$$\int_a^b x\,dx = \lim_{n \to +\infty} \sum_{k=1}^{n} f(c_k)\,\Delta x_k = \lim_{n \to +\infty} (b - a)\left[a + (b - a)\left(\frac{1}{2} + \frac{1}{2n}\right)\right]$$

$$= (b - a)[a + (b - a)(\tfrac{1}{2})] = \tfrac{1}{2}(b - a)(b + a) = \tfrac{1}{2}(b^2 - a^2)$$

This proves (i) for $a < b$. If $a = b$, then (i) is obviously true since both sides of the equation are zero. Finally, suppose $b < a$. From what has already been proved, we have

$$\int_b^a x\,dx = \tfrac{1}{2}(a^2 - b^2)$$

and it follows from Definition 2 in Section 5.2 that

$$\int_a^b x\, dx = -\int_b^a x\, dx = -\tfrac{1}{2}(a^2 - b^2) = \tfrac{1}{2}(b^2 - a^2) \qquad ■$$

In Examples 6 to 8, use the basic properties of definite integrals to evaluate each expression.

EXAMPLE 6 $\displaystyle\int_2^3 (x - 1)^2\, dx$

SOLUTION   By the linear property (Theorem 4), we have

$$\int_2^3 (x - 1)^2\, dx = \int_2^3 (x^2 - 2x + 1)\, dx = \int_2^3 x^2\, dx - 2\int_2^3 x\, dx + \int_2^3 dx$$

Therefore, by Theorem 6 and Theorem 1,

$$\int_2^3 (x - 1)^2\, dx = \tfrac{1}{3}(3^3 - 2^3) - 2[\tfrac{1}{2}(3^2 - 2^2)] + (3 - 2) = \tfrac{7}{3} \qquad ■$$

EXAMPLE 7 $\displaystyle\int_{-3}^2 |x|\, dx$

SOLUTION   For $-3 \le x \le 0$ we have $|x| = -x$, whereas for $0 \le x \le 2$ we have $|x| = x$. This suggests that we use Theorem 5 to break the interval of integration $[-3, 2]$ into the two intervals $[-3, 0]$ and $[0, 2]$. Thus, we have

$$\int_{-3}^2 |x|\, dx = \int_{-3}^0 |x|\, dx + \int_0^2 |x|\, dx = \int_{-3}^0 (-x)\, dx + \int_0^2 x\, dx$$

$$= -\int_{-3}^0 x\, dx + \int_0^2 x\, dx = -\{\tfrac{1}{2}[0^2 - (-3)^2]\} + \tfrac{1}{2}(2^2 - 0^2)$$

$$= \tfrac{9}{2} + 2 = \tfrac{13}{2} \qquad ■$$

EXAMPLE 8 $\displaystyle\int_{-2}^4 f(x)\, dx$, where $f(x) = \begin{cases} x^2 - 1 & \text{for } x < 0 \\ x - 1 & \text{for } 0 \le x < 1 \quad \text{(Figure 5)} \\ 3 & \text{for } x \ge 1 \end{cases}$

SOLUTION

Figure 5

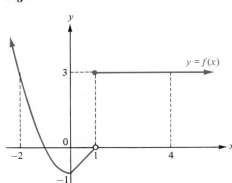

$$\int_{-2}^4 f(x)\, dx = \int_{-2}^0 f(x)\, dx + \int_0^1 f(x)\, dx + \int_1^4 f(x)\, dx$$

$$= \int_{-2}^0 (x^2 - 1)\, dx + \int_0^1 (x - 1)\, dx + \int_1^4 3\, dx$$

$$= \int_{-2}^0 x^2\, dx - \int_{-2}^0 dx + \int_0^1 x\, dx - \int_0^1 dx + 3\int_1^4 dx$$

$$= \tfrac{1}{3}[0^3 - (-2)^3] - [0 - (-2)] + \tfrac{1}{2}(1^2 - 0^2) - (1 - 0) + 3(4 - 1)$$

$$= \tfrac{8}{3} - 2 + \tfrac{1}{2} - 1 + 9 = \tfrac{55}{6} \qquad ■$$

Note, in the above calculation, that we have

$$\int_0^1 f(x)\, dx = \int_0^1 (x - 1)\, dx$$

in spite of the fact that

$$f(1) \neq 1 - 1$$

(Figure 5). This is justified by Theorem 4 of Section 5.2, since

$$f(x) = x - 1$$

for all values of x in [0, 1] except $x = 1$.

THEOREM 7   **Nonnegative Property**

> If f is an integrable function on the interval $[a, b]$ and if $f(x) \geq 0$ for all values of x in $[a, b]$, then $\displaystyle\int_a^b f(x)\, dx \geq 0$.

Figure 6

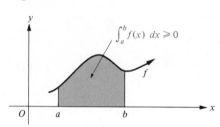

Theorem 7 is geometrically evident since if $f(x) \geq 0$ for $a \leq x \leq b$, then

$$\int_a^b f(x)\, dx$$

is supposed to represent the area under the graph of $y = f(x)$ between $x = a$ and $x = b$ (Figure 6).

EXAMPLE 9   Show that $\int_0^1 x^3\, dx \leq \int_0^1 x\, dx$.

SOLUTION   For $0 \leq x \leq 1$, $x^3 \leq x$ holds; hence, $x - x^3 \geq 0$. By Theorem 7,

$$\int_0^1 (x - x^3)\, dx \geq 0$$

that is,

$$\int_0^1 x\, dx - \int_0^1 x^3\, dx \geq 0$$

From the last inequality,

$$\int_0^1 x\, dx \geq \int_0^1 x^3\, dx$$

that is,

$$\int_0^1 x^3\, dx \leq \int_0^1 x\, dx$$ ■

The argument given in the last example is quite general. Indeed, suppose that f and g are integrable functions on the interval $[a, b]$ such that $f(x) \leq g(x)$. Then, $g(x) - f(x) \geq 0$, so that by Theorem 7

$$\int_a^b [g(x) - f(x)]\, dx \geq 0$$

that is,

$$\int_a^b g(x)\, dx - \int_a^b f(x)\, dx \geq 0 \qquad \text{or} \qquad \int_a^b f(x)\, dx \leq \int_a^b g(x)\, dx$$

Therefore, we have the following theorem.

THEOREM 8   **Comparison**

> If f and g are integrable functions on the interval $[a, b]$ and if $f(x) \le g(x)$ holds for all values of x in $[a, b]$, then $\displaystyle\int_a^b f(x)\,dx \le \int_a^b g(x)\,dx$.

Theorem 8 provides the basis for the following theorem.

THEOREM 9   **Absolute-Value Property**

> If f is integrable on the interval $[a, b]$, then so is $|f|$ and
> $$\left| \int_a^b f(x)\,dx \right| \le \int_a^b |f(x)|\,dx$$

A rigorous proof of Theorem 9 would require an argument to show that $|f|$ is integrable, but this is beyond the scope of this book. However, assuming that f and $|f|$ are integrable functions on the interval $[a, b]$, we can derive the inequality in Theorem 9 as follows: Applying Theorem 8 to the inequality

$$-|f(x)| \le f(x) \le |f(x)|$$

we obtain

$$\int_a^b -|f(x)|\,dx \le \int_a^b f(x)\,dx \le \int_a^b |f(x)|\,dx$$

that is,

$$-\int_a^b |f(x)|\,dx \le \int_a^b f(x)\,dx \le \int_a^b |f(x)|\,dx$$

It follows from Rule (iii) on page 7 that

$$\left| \int_a^b f(x)\,dx \right| \le \int_a^b |f(x)|\,dx$$

The following property of the definite integral plays an important role in our forthcoming proof of the fundamental theorem of calculus.

THEOREM 10   **Mean-Value Theorem for Integrals**

> Suppose that f is a continuous function on the interval $[a, b]$. Then there exists a number c in $[a, b]$ such that
> $$f(c) \cdot (b - a) = \int_a^b f(x)\,dx$$

PROOF   By Theorem 1, page 167, the continuous function f on the closed interval $[a, b]$ takes on a maximum value, say, B, and a minimum value, say, A. Thus, $A \le f(x) \le B$ holds for all values of x in $[a, b]$. By Theorem 8, then,

$$\int_a^b A\,dx \le \int_a^b f(x)\,dx \le \int_a^b B\,dx$$

By Theorem 1,

$$\int_a^b A\ dx = A(b - a) \qquad \text{and} \qquad \int_a^b B\ dx = B(b - a)$$

Hence,

$$A(b - a) \le \int_a^b f(x)\ dx \le B(b - a)$$

But $b - a > 0$, so the latter inequality can be rewritten as

$$A \le \frac{1}{b - a} \int_a^b f(x)\ dx \le B$$

Now the intermediate-value theorem for continuous functions (Theorem 1, page 150) implies that f takes on every value between any two of its values. Thus, since A and B are two such values of f and since $[1/(b - a)] \int_a^b f(x)\ dx$ lies between these two values, there must exist a number c in $[a, b]$ such that

$$f(c) = \frac{1}{b - a} \int_a^b f(x)\ dx$$

that is,

$$f(c) \cdot (b - a) = \int_a^b f(x)\ dx$$ ∎

If $f(x) \ge 0$ for $a \le x \le b$, the condition $f(c)(b - a) = \int_a^b f(x)\ dx$ (Figure 7) means that the area under the curve $y = f(x)$ between $x = a$ and $x = b$ is the same as the area of a rectangle whose base is the interval $[a, b]$ and whose height is $f(c)$.

Figure 7

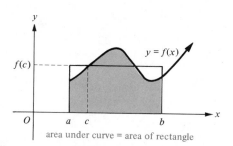

area under curve = area of rectangle

Thus, if the curve $y = f(x)$ were "flattened out" between $x = a$ and $x = b$ so as to have constant height $f(c)$, then the area under the curve would stay the same. In this sense, the number

$$f(c) = \frac{1}{b - a} \int_a^b f(x)\ dx$$

represents an "average value," or a "mean value," of the function f between $x = a$ and $x = b$. We make this idea official with the following definition.

DEFINITION 1   **Mean Value of a Function on an Interval**

Let f be an integrable function on the interval $[a, b]$. Then the **mean value** of f on $[a, b]$ is given by

$$\frac{1}{b - a} \int_a^b f(x)\ dx$$

The mean-value theorem for integrals just says that *a continuous function f on an interval [a, b] takes on its own mean value at some number c on this interval*.

EXAMPLE 10   Find the mean value of the function $f(x) = x^2$ on the interval $[1, 4]$, and find a value of c on this interval such that $f(c)$ gives this mean value.

SOLUTION   The desired mean value is given by

$$\frac{1}{4 - 1} \int_1^4 x^2 \, dx = \frac{1}{3}\left[\frac{1}{3}(4^3 - 1^3)\right] = 7$$

We need to find a value of c with $1 \le c \le 4$ such that $f(c) = c^2 = 7$. Evidently, $c = \sqrt{7}$. ∎

In a definite integral

$$\int_a^b f(x) \, dx$$

there's no particular reason to use x as the variable of integration—any other letter would do. In fact,

$$\int_a^b f(x) \, dx$$

is a *number* depending on a, b, and the function f; it really has nothing at all to do with x. In other words, *the variable of integration in a definite integral is a "dummy variable,"* so that, for instance,

$$\int_0^1 x^2 \, dx = \int_0^1 t^2 \, dt = \int_0^1 s^2 \, ds = \int_0^1 y^2 \, dy = \tfrac{1}{3}(1^3 - 0^3) = \tfrac{1}{3}$$

More generally,

$$\boxed{\int_a^b f(x) \, dx = \int_a^b f(t) \, dt}$$

Often it is necessary to consider a definite integral in which one or both of the limits of integration are variable quantities. For example, if the upper limit of integration is a variable quantity and the lower limit of integration is fixed, then the value of the integral will be a function of the upper limit. Since it is customary to use the symbol x for the independent variable whenever possible, we might be tempted to write such a function as

$$g(x) = \int_a^x f(x) \, dx$$

however, the x in the expression $f(x) \, dx$ is the (dummy) variable of integration and must not be confused with the independent variable x in the expression $g(x)$ and in the upper limit of integration. Therefore, in such a situation we use another symbol for the variable of integration and write, for instance,

$$\boxed{g(x) = \int_a^x f(t) \, dt}$$

Problem Set 5.3

In Problems 1 to 30, use the basic properties of the definite integral to evaluate the expression.

1 $\int_3^4 2\,dx$

2 $\int_{-5}^4 (7 + \pi)\,dx$

3 $\int_{-2}^7 (-dx)$

4 $\int_{0.5}^{0.75} (1 + \sqrt{2} - \sqrt{3})\,dx$

5 $\int_2^1 dx$

6 $\int_{-2}^{-4} (-4)\,dx$

7 $\int_{-\pi}^{\pi} 2\,dx$

8 $\int_{\pi}^{\pi} 2\,dx$

9 $\int_1^3 5x\,dx$

10 $\int_3^{-2} (-3x)\,dx$

11 $\int_5^1 (-2x)\,dx$

12 $\int_1^{-1} (4 - 3x)\,dx$

13 $\int_{-2}^3 (2x + 1)\,dx$

14 $\int_{2/3}^{3/4} \left(\frac{x}{2} - 1\right) dx$

15 $\int_1^2 (x + x^2)\,dx$

16 $\int_2^1 (x^2 - 1)\,dx$

17 $\int_{-2}^3 (3x^2 - 2x + 1)\,dx$

18 $\int_{-2}^{-3} (-2x^2 + 4x + 5)\,dx$

19 $\int_3^{-2} (3x - 1)(2x + 3)\,dx$

20 $\int_1^{-1} (2x + 1)^2\,dx$

21 $\int_3^{-2} 4x(2x - 7)\,dx$

22 $\int_1^0 \frac{x^2 - 25}{x - 5}\,dx$

23 $\int_{-1}^0 x\,dx + \int_0^1 x\,dx$

24 $\int_{-1}^a x\,dx + \int_a^1 x\,dx$

25 $\int_0^{\pi} (2x - 1)\,dx + \int_{\pi}^4 (2x - 1)\,dx$

26 $\int_{-1}^a x^2\,dx - \int_1^a x^2\,dx$

27 $\int_{-2}^1 |x|\,dx$

28 $\int_{-1}^2 |x - 1|\,dx$

29 $\int_{-1}^2 |x^3|\,dx$

30 $\int_3^{-2} [\![x]\!]\,dx$

In Problems 31 to 34, evaluate the given expression. You may assume that $\int_0^{\pi/2} \cos x\,dx = 1$ and $\int_0^{\pi/2} \sin^2 x\,dx = \pi/4$.

31 $\int_0^{\pi/2} (3 \cos x - 2 \sin^2 x)\,dx$

32 $\int_0^{\pi/2} \cos^2 x\,dx$ (*Hint:* $\sin^2 x + \cos^2 x = 1$.)

33 $\int_{\pi/2}^0 3 \cos x\,dx$

34 $\int_0^{\pi/2} \cos 2x\,dx$ (*Hint:* $\cos 2x = 1 - 2 \sin^2 x$.)

In Problems 35 to 38, evaluate the definite integral.

35 $\int_0^4 f(x)\,dx$, where $f(x) = \begin{cases} 2x^2 & \text{for } 0 \le x \le 2 \\ 4x & \text{for } 2 < x \le 4 \end{cases}$

36 $\int_{-1}^1 f(x)\,dx$, where $f(x) = \begin{cases} 1 & \text{for } x \ne 0 \\ 0 & \text{for } x = 0 \end{cases}$

37 $\int_{-2}^3 f(x)\,dx$, where $f(x) = \begin{cases} 1 - x & \text{for } -2 \le x \le 0 \\ 1 + x & \text{for } 0 < x \le 3 \end{cases}$

38 $\int_0^1 f(x)\,dx$, where $f(x) = \begin{cases} 0 & \text{for } x = 0 \\ x + 1 & \text{for } 0 < x < 1 \\ 0 & \text{for } x = 1 \end{cases}$

39 Use the nonnegative property or the comparison property to decide whether each inequality holds. Do not evaluate the definite integrals involved.

(a) $\int_0^1 x\,dx \le \int_0^1 dx$

(b) $\int_1^2 x^2\,dx < \int_1^2 x\,dx$

(c) $0 \le \int_0^{\pi} \sin x\,dx$

(d) $0 \le \int_0^1 \frac{dx}{1 + x^2}$

(e) $\int_0^1 x^5\,dx \le \int_0^1 x^6\,dx$

(f) $\int_0^{\pi/2} \sin x\,dx \le \int_0^{\pi/2} x\,dx$

40 If $0 < K \le f(x)$ holds for all values of x in $[a, b]$, prove that $0 < \int_a^b f(x)\,dx$. [*Hint:* Use the comparison theorem and the fact that $\int_a^b K\,dx = K(b - a)$.]

In Problems 41 to 48, find the mean value of the function f on the interval $[a, b]$ and find a value of c in $[a, b]$ such that $f(c)$ is equal to this mean value.

41 $f(x) = x + 5$ on $[1, 3]$

42 $f(x) = x^2$ on $[-3, -1]$

43 $f(x) = x^2 - 2x + 3$ on $[-1, 5]$

44 $f(x) = (x - 2)(x + 3)$ on $[-3, 2]$

45 $f(x) = |x|$ on $[-2, 5]$

46 $f(x) = x$ on $[a, b]$

47 $f(x) = Ax + B$ on $[a, b]$

48 $f(x) = x^2$ on $[-a, a]$

In Problems 49 to 58, use the properties of the definite integral to justify each statement.

49 $0 \le \int_{\pi/4}^{3\pi/4} (1 + \cos x) \, dx$

50 $\int_0^1 x^4 \, dx \le \int_0^1 x \, dx$

51 $\int_0^{0.8} \tan x \, dx = \int_0^{0.8} \tan t \, dt$

52 $\int_0^{1/2} x^6 \, dx \le \int_0^{1/2} t^2 \, dt$

53 $\int_\pi^{2\pi} \sin x \, dx = \int_0^{2\pi} \sin x \, dx - \int_0^\pi \sin x \, dx$

54 $\int_{-1000}^{1000} (x^{1776} + \sqrt{|x|^{1699}}) \, dx \ge 0$

55 $\int_5^1 \sqrt{x^2 + 1} \, dx = \int_0^1 \sqrt{x^2 + 1} \, dx - \int_0^5 \sqrt{x^2 + 1} \, dx$

56 $\int_4^{-1} \sqrt[3]{5x^2 + 3} \, dx - \int_4^2 \sqrt[3]{5x^2 + 3} \, dx + \int_{-1}^2 \sqrt[3]{5x^2 + 3} \, dx = 0$

57 $\int_3^4 \frac{dx}{1 + x^2} - \int_5^5 \frac{dx}{1 + x^2} = \int_6^4 \frac{dt}{1 + t^2} + \int_3^6 \frac{dy}{1 + y^2}$

58 $\int_a^b \frac{dx}{\sqrt{1 + x^2}} + \int_b^c \frac{dy}{\sqrt{1 + y^2}} + \int_c^a \frac{dz}{\sqrt{1 + z^2}} = 0$

59 Assuming that Theorem 5 is correct for the special case in which $a < b < c$, and considering that the case $a < c < b$ has already been discussed, check the result for the remaining cases: $b < a < c$, $b < c < a$, $c < a < b$, and $c < b < a$. Also, check the result for the cases in which two of the numbers a, b, and c are equal.

60 Prove part (ii) of Theorem 6.

61 The force required to stretch a spring from its relaxed position through s units is given by $F = ks$, where k is the spring con-

stant. Find the mean value of F if the spring is stretched from $s = a$ to $s = b$. Also find a value of c between a and b such that the value of F when $s = c$ is the same as this mean value.

62 Suppose that f is a continuous function such that $\int_a^b f(x) \, dx = 0$ for *every* closed interval $[a, b]$. Prove that f is the zero function.

63 Is it true that when a perfectly elastic spring is stretched, the work done is equal to the product of the mean force (see Problem 61) and the distance through which the spring is stretched? Justify your answer.

64 Suppose that f and g are integrable functions on $[a, b]$ such that $|f(x) - g(x)| \le K$ for every number x in $[a, b]$, where K is a positive constant. Prove that

$$\left| \int_a^b f(x) \, dx - \int_a^b g(x) \, dx \right| \le K(b - a)$$

(*Hint:* Use the absolute-value property.)

65 Prove that Theorem 10 continues to hold even if $a \ge b$; that is, show that if f is continuous on the closed interval between a and b, then there exists a number c on this interval such that $f(c) \cdot (b - a) = \int_a^b f(x) \, dx$.

66 If f and g are integrable functions on $[a, b]$, prove that

$$\left| \int_a^b [f(x) + g(x)] \, dx \right| \le \int_a^b |f(x)| \, dx + \int_a^b |g(x)| \, dx$$

67 Find formulas for (a) $\int_a^x dt$, (b) $\int_a^x t \, dt$, and (c) $\int_a^x t^2 \, dt$.

68 If $f(x) = Ax^2 + Bx + C$, find a formula for

$$\frac{d}{dx} \int_a^x f(t) \, dt$$

(*Hint:* Use the results of Problem 67.)

5.4   The Fundamental Theorem of Calculus

In this section, we state and prove a theorem that establishes a profound connection between antidifferentiation and definite integration—a theorem so useful and so important that it is known as the *fundamental theorem of calculus*. We begin by giving an informal argument that leads to this theorem; then we illustrate the usefulness of the theorem in several examples; and finally we give a formal proof of the theorem.

You might already suspect that a connection exists between the indefinite and the definite integrals. After all, in Section 4.7, we used indefinite integration to find areas by the method of slicing, whereas our definition of the definite integral in Section 5.2 was suggested by the idea of the area under a curve. Here is an *informal* argument to make this connection more explicit:

Figure 1

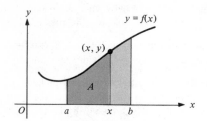

Suppose that f is a continuous function on the interval $[a, b]$ and that $f(x) \geq 0$ for $a \leq x \leq b$. Let A denote the area under the graph of f between a and x (Figure 1). By the method of slicing (see page 292), we have

$$dA = f(x)\, dx$$

Let g be any antiderivative of f, so that

$$\int f(x)\, dx = g(x) + C$$

where C is the constant of integration. Because $dA = f(x)\, dx$, it follows that

$$A = \int dA = \int f(x)\, dx = g(x) + C$$

When $x = a$, we have $A = 0$ (Figure 1); hence,

$$0 = g(a) + C \qquad \text{or} \qquad C = -g(a)$$

Substituting $C = -g(a)$ into the equation $A = g(x) + C$, we find that

$$A = g(x) - g(a)$$

The total area under the graph of f between a and b is obtained by letting $x = b$ in the last equation. But the total area under the graph of f between a and b is also given by the definite integral

$$\int_a^b f(x)\, dx$$

and so it follows that

$$\int_a^b f(x)\, dx = g(b) - g(a)$$

Although the argument given above requires that $f(x) \geq 0$ for $a \leq x \leq b$, the final result holds even if this requirement is not satisfied. Thus, we have the following *informal version* of the **fundamental theorem of calculus:**

> If $\int f(x)\, dx = g(x) + C$, then $\int_a^b f(x)\, dx = g(b) - g(a)$.

A formal statement and proof of the theorem are given shortly (Theorem 2, page 335).

EXAMPLE 1   Use the fundamental theorem of calculus to evaluate $\int_1^2 4x^3\, dx$.

SOLUTION   Because

$$\int 4x^3\, dx = x^4 + C$$

we can take $g(x) = x^4$ and conclude that

$$\int_1^2 4x^3\, dx = g(2) - g(1) = 2^4 - 1^4 = 15 \qquad\blacksquare$$

The simple notational device introduced in the following definition makes the fundamental theorem of calculus easier to state and to use.

DEFINITION 1   **Special Notation**

If g is any function and if $g(a)$ and $g(b)$ are defined, then

$$g(x) \Big|_a^b = g(b) - g(a)$$

The notation $g(x) \Big|_a^b$ is read "$g(x)$ *evaluated between* $x = a$ *and* $x = b$." For instance, in the previous example with $g(x) = x^4$, we can write

$$\int_1^2 4x^3 \, dx = x^4 \Big|_1^2 = 2^4 - 1^4 = 15$$

Using this special notation, we can rewrite the fundamental theorem of calculus in the form

$$\int_a^b f(x) \, dx = \left[\int f(x) \, dx \right] \Big|_a^b$$

Indeed, if

$$\int f(x) \, dx = g(x) + C$$

then

$$\left[\int f(x) \, dx \right] \Big|_a^b = [g(x) + C] \Big|_a^b = [g(b) + C] - [g(a) + C] = g(b) - g(a)$$

Notice how the constant of integration C cancels out in the calculation above. Therefore, *in using the fundamental theorem of calculus to evaluate a definite integral, the constant of integration in the corresponding indefinite integral can safely be neglected.*

In Examples 2 and 3, use the fundamental theorem of calculus to evaluate the definite integral.

EXAMPLE 2 $\displaystyle\int_{-1}^2 (x^3 - 3x^2 + 2x + 3) \, dx$

SOLUTION

$$\int_{-1}^2 (x^3 - 3x^2 + 2x + 3) \, dx = \left[\int (x^3 - 3x^2 + 2x + 3) \, dx \right] \Big|_{-1}^2$$

$$= \left(\frac{x^4}{4} - x^3 + x^2 + 3x \right) \Big|_{-1}^2$$

$$= \left[\frac{2^4}{4} - 2^3 + 2^2 + 3(2) \right]$$

$$\qquad - \left[\frac{(-1)^4}{4} - (-1)^3 + (-1)^2 + 3(-1) \right]$$

$$= \frac{27}{4} \qquad\qquad \blacksquare$$

EXAMPLE 3 $\displaystyle\int_0^{\pi/3} \sin x \, dx$

SOLUTION $\displaystyle\int_0^{\pi/3} \sin x \, dx = \left[\int \sin x \, dx \right] \Big|_0^{\pi/3} = (-\cos x) \Big|_0^{\pi/3}$

$$= \left(-\cos \frac{\pi}{3}\right) - (-\cos 0) = \left(-\frac{1}{2}\right) - (-1) = \frac{1}{2} \quad \blacksquare$$

Note that the fundamental theorem of calculus,

$$\int_a^b f(x) \, dx = \left[\int f(x) \, dx \right] \Big|_a^b$$

works even if $a \geq b$ (Problem 69).

EXAMPLE 4   Evaluate $\displaystyle\int_3^1 \frac{t^3 + 9}{t^2} \, dt$ by using the fundamental theorem of calculus.

SOLUTION

$$\int_3^1 \frac{t^3 + 9}{t^2} \, dt = \int_3^1 (t + 9t^{-2}) \, dt$$

$$= \left[\int (t + 9t^{-2}) \, dt \right] \Big|_3^1 = \left(\frac{t^2}{2} + \frac{9t^{-1}}{-1} \right) \Big|_3^1$$

$$= \left(\frac{t^2}{2} - \frac{9}{t} \right) \Big|_3^1 = \left(\frac{1^2}{2} - \frac{9}{1} \right) - \left(\frac{3^2}{2} - \frac{9}{3} \right) = -10 \quad \blacksquare$$

Sometimes it is useful to split the interval of integration into two (or more) subintervals before using the fundamental theorem of calculus.

EXAMPLE 5   Use the fundamental theorem of calculus to evaluate $\int_0^2 |1 - x| \, dx$.

SOLUTION   For $x \leq 1$, we have $1 - x \geq 0$; hence, $|1 - x| = 1 - x$. On the other hand, if $x > 1$, we have $1 - x < 0$; hence, $|1 - x| = -(1 - x) = x - 1$. This suggests that we split the interval of integration $[0, 2]$ into the two subintervals $[0, 1]$ and $[1, 2]$. Thus, by Theorem 5 in Section 5.3,

$$\int_0^2 |1 - x| \, dx = \int_0^1 |1 - x| \, dx + \int_1^2 |1 - x| \, dx = \int_0^1 (1 - x) \, dx + \int_1^2 (x - 1) \, dx$$

$$= \left(x - \frac{x^2}{2} \right) \Big|_0^1 + \left(\frac{x^2}{2} - x \right) \Big|_1^2 = \left(\frac{1}{2} - 0 \right) + \left[0 - \left(-\frac{1}{2} \right) \right] = 1 \quad \blacksquare$$

Change of Variable in a Definite Integral

Suppose we want to evaluate the *definite* integral

$$\int_1^2 \frac{x^2 \, dx}{(x^3 + 2)^2}$$

Since we propose to use the fundamental theorem of calculus, we begin by using the change of variable $u = x^3 + 2$ to evaluate the corresponding *indefinite* integral. Thus,

$$du = 3x^2 \, dx \qquad \text{so} \qquad x^2 \, dx = \tfrac{1}{3} \, du$$

and we have

$$\int \frac{x^2 \, dx}{(x^3 + 2)^2} = \int \frac{\tfrac{1}{3} \, du}{u^2} = \frac{1}{3} \int u^{-2} \, du = \frac{1}{3} \frac{u^{-1}}{-1} + C$$

$$= -\frac{1}{3u} + C = \frac{-1}{3(x^3 + 2)} + C$$

Therefore,

$$\int_1^2 \frac{x^2 \, dx}{(x^3 + 2)^2} = \frac{-1}{3(x^3 + 2)} \, \Big|_1^2 = \frac{-1}{3(8 + 2)} - \frac{-1}{3(1 + 2)} = \frac{7}{90}$$

The calculation above can be shortened by *changing the limits of integration in accordance with the change of variable*. Indeed, the limits of integration in the original definite integral refer to the variable x, a fact that is sometimes emphasized by writing

$$\int_{x=1}^{x=2} \frac{x^2 \, dx}{(x^3 + 2)^2} \qquad \text{rather than} \qquad \int_1^2 \frac{x^2 \, dx}{(x^3 + 2)^2}$$

To make the change of variable $u = x^3 + 2$, we notice that

$$\text{when } x = 1 \qquad \text{we have } u = 1^3 + 2 = 3$$

and

$$\text{when } x = 2 \qquad \text{we have } u = 2^3 + 2 = 10$$

Thus, we can abbreviate our previous calculation as follows:

$$\int_{x=1}^{x=2} \frac{x^2 \, dx}{(x^3 + 2)^2} = \int_{u=3}^{u=10} \frac{\tfrac{1}{3} \, du}{u^2} = \frac{1}{3} \int_3^{10} u^{-2} \, du = \frac{1}{3} \left(\frac{-1}{u} \right) \Big|_3^{10}$$

$$= \frac{1}{3} \left(\frac{-1}{10} \right) - \frac{1}{3} \left(\frac{-1}{3} \right) = \frac{7}{90}$$

More generally, if you change the variable from, say, x to, say, $u = g(x)$ in a *definite* integral, not only must you change the integrand just as you do for indefinite integrals, but also you must *change the limits of integration*. Thus

$$\boxed{\int_a^b f[g(x)] g'(x) \, dx = \int_{g(a)}^{g(b)} f(u) \, du}$$

In Examples 6 and 7, use a change of variable to evaluate the integral.

<u>EXAMPLE 6</u> $\displaystyle \int_0^1 x \sqrt{9 - 5x^2} \, dx$

SOLUTION   We make the change of variable $u = 9 - 5x^2$, noting that $du = -10x \, dx$, or $x \, dx = -\tfrac{1}{10} \, du$. Also, $u = 9$ when $x = 0$, and $u = 4$ when $x = 1$.

Therefore,

$$\int_0^1 x\sqrt{9-5x^2}\,dx = \int_{x=0}^{x=1} \sqrt{9-5x^2}\,x\,dx$$

$$= \int_{u=9}^{u=4} \sqrt{u}\left(\frac{-1}{10}\right)du = \frac{-1}{10}\int_9^4 \sqrt{u}\,du$$

$$= \frac{1}{10}\int_4^9 u^{1/2}\,du = \frac{1}{10}\left.\frac{u^{3/2}}{\frac{3}{2}}\right|_4^9$$

$$= \frac{1}{15}9^{3/2} - \frac{1}{15}4^{3/2} = \frac{19}{15}$$

EXAMPLE 7 $\displaystyle\int_0^{\pi/3} \cos^2 t \sin t\,dt$

SOLUTION   We make the change of variable $u = \cos t$, noting that $du = -\sin t\,dt$, or $\sin t\,dt = -du$. Also, $u = 1$ when $t = 0$, and $u = \frac{1}{2}$ when $t = \pi/3$. Therefore,

$$\int_0^{\pi/3} \cos^2 t \sin t\,dt = \int_1^{1/2} u^2(-du) = -\int_1^{1/2} u^2\,du = \int_{1/2}^1 u^2\,du$$

$$= \left.\frac{u^3}{3}\right|_{1/2}^1 = \frac{1}{3} - \frac{1}{24} = \frac{7}{24}$$

The Derivative of an Integral

Suppose that

$$\int f(x)\,dx = g(x) + C$$

so that

$$\frac{d}{dx}g(x) = f(x)$$

Now, by the fundamental theorem of calculus,

$$\int_a^x f(t)\,dt = g(x) - g(a)$$

where we have written the variable of integration as t to avoid confusing it with the upper limit of integration x. Since $g(a)$ is a constant, we have

$$\frac{d}{dx}\int_a^x f(t)\,dt = \frac{d}{dx}[g(x) - g(a)] = \frac{d}{dx}g(x) - \frac{d}{dx}g(a) = f(x) - 0 = f(x)$$

Thus we have

$$\boxed{\frac{d}{dx}\int_a^x f(t)\,dt = f(x)}$$

a result that can be regarded as an **alternative version of the fundamental theorem of calculus** because it is so closely related to the fundamental theorem. A formal statement and proof of this result are given shortly (Theorem 1).

In Examples 8 to 10, find the indicated derivative.

EXAMPLE 8 $\dfrac{d}{dx}\displaystyle\int_0^x \dfrac{1}{5+t^2}\,dt$

SOLUTION $\dfrac{d}{dx}\displaystyle\int_0^x \dfrac{1}{5+t^2}\,dt = \dfrac{1}{5+x^2}$ ■

EXAMPLE 9 $\dfrac{d}{dv}\displaystyle\int_{\pi/2}^v \csc x\,dx$

SOLUTION $\dfrac{d}{dv}\displaystyle\int_{\pi/2}^v \csc x\,dx = \csc v$ ■

EXAMPLE 10 $\dfrac{d}{dx}\displaystyle\int_x^1 \sqrt{4+t}\,dt$

SOLUTION $\dfrac{d}{dx}\displaystyle\int_x^1 \sqrt{4+t}\,dt = \dfrac{d}{dx}\left(-\displaystyle\int_1^x \sqrt{4+t}\,dt\right)$

$$= -\dfrac{d}{dx}\int_1^x \sqrt{4+t}\,dt$$

$$= -\sqrt{4+x}$$ ■

The derivative with respect to x of a definite integral whose limits of integration are functions of x can be found by combining the formula given above and the chain rule. The technique is illustrated in the following examples.

In Examples 11 and 12, find the indicated derivative.

EXAMPLE 11 $\dfrac{dy}{dx}$ if $y = \displaystyle\int_3^{x^2} (5t+7)^{25}\,dt$

SOLUTION We put $u = x^2$, so that $y = \displaystyle\int_3^u (5t+7)^{25}\,dt$ and

$$\dfrac{dy}{du} = \dfrac{d}{du}\int_3^u (5t+7)^{25}\,dt = (5u+7)^{25} = (5x^2+7)^{25}$$

Therefore, by the chain rule,

$$\dfrac{dy}{dx} = \dfrac{dy}{du}\,\dfrac{du}{dx} = (5x^2+7)^{25}(2x)$$ ■

EXAMPLE 12 $D_x y$ if $y = \displaystyle\int_{-x}^{3x+2} \sqrt{1+t^2}\,dt$

SOLUTION $y = \displaystyle\int_{-x}^0 \sqrt{1+t^2}\,dt + \int_0^{3x+2} \sqrt{1+t^2}\,dt$

$$= -\int_0^{-x} \sqrt{1+t^2}\,dt + \int_0^{3x+2} \sqrt{1+t^2}\,dt$$

hence, $D_x y = -\sqrt{1+(-x)^2}\,(-1) + \sqrt{1+(3x+2)^2}\,(3)$

$$= \sqrt{1+x^2} + 3\sqrt{1+(3x+2)^2}$$ ■

Proof of the Fundamental Theorem of Calculus

We now give a formal proof of the fundamental theorem of calculus. Here we begin by giving a formal statement and proof of the alternative version involving the derivative of an integral. Using this as a basis, we can then prove the fundamental theorem itself.

THEOREM 1

The Derivative of an Integral

Let f be a continuous function on the closed interval $[b, c]$, and suppose that a is a fixed number in this interval. Define the function g with domain $[b, c]$ by

$$g(x) = \int_a^x f(t)\, dt$$

for x in $[b, c]$. Then g is differentiable on the open interval (b, c), and

$$g'(x) = f(x)$$

holds for all x in (b, c). Furthermore, g is continuous on $[b, c]$, and at the endpoints b and c, we have

$$g'_+(b) = f(b) \qquad \text{and} \qquad g'_-(c) = f(c)$$

PROOF

Suppose that x belongs to the open interval (b, c) and that Δx is small enough that $x + \Delta x$ also belongs to (b, c). Then

$$g(x) = \int_a^x f(t)\, dt \qquad \text{and} \qquad g(x + \Delta x) = \int_a^{x+\Delta x} f(t)\, dt$$

Therefore, by Definition 2 in Section 5.2 and Theorem 5 in Section 5.3,

$$g(x + \Delta x) - g(x) = \int_a^{x+\Delta x} f(t)\, dt - \int_a^x f(t)\, dt = \int_a^{x+\Delta x} f(t)\, dt + \int_x^a f(t)\, dt$$

$$= \int_x^a f(t)\, dt + \int_a^{x+\Delta x} f(t)\, dt = \int_x^{x+\Delta x} f(t)\, dt$$

Since f is continuous on the interval $[b, c]$, it is also continuous on the closed subinterval between x and $x + \Delta x$. By Theorem 10 in Section 5.3, there exists a number x^* on the closed interval between x and $x + \Delta x$ such that

$$\int_x^{x+\Delta x} f(t)\, dt = f(x^*)[(x + \Delta x) - x] = f(x^*)\, \Delta x$$

Consequently,

$$g(x + \Delta x) - g(x) = f(x^*)\, \Delta x \qquad \text{or} \qquad \frac{g(x + \Delta x) - g(x)}{\Delta x} = f(x^*)$$

Since x^* lies between x and $x + \Delta x$, it follows that x^* approaches x as Δx approaches zero. Therefore,

$$g'(x) = \lim_{\Delta x \to 0} \frac{g(x + \Delta x) - g(x)}{\Delta x} = \lim_{x^* \to x} f(x^*) = f(x)$$

where we have used the continuity of f in the last equation. This establishes the desired result for values of x on the open interval (b, c). The proof that one-sided derivatives of g at the endpoints b and c give the values of f at these endpoints is similar and is left as an exercise (Problem 68a). Because g is differentiable on the open interval (b, c), it follows from Theorem 1, page 96, that g is continuous on (b, c). Using the fact that the one-sided derivatives of g exist at the endpoints b and c, and arguing as in Theorem 1, page 96, we can easily show that g is continuous on the closed interval $[b, c]$ (Problem 68b).   ∎

In Theorem 1, notice that the function g is an antiderivative of the function f on the open interval (b, c). This fact is often phrased informally as follows: *A continuous function f always has an antiderivative g.*

THEOREM 2

The Fundamental Theorem of Calculus

Let f and g be continuous functions on the closed interval $[a, b]$, and suppose that

$$g'(x) = f(x)$$

holds for all values of x in the open interval (a, b). Then

$$\int_a^b f(x) \, dx = g(b) - g(a)$$

PROOF

Define a function G with domain $[a, b]$ by

$$G(x) = \int_a^x f(t) \, dt$$

for x in $[a, b]$. By Theorem 1, G is continuous on $[a, b]$, and

$$G'(x) = f(x)$$

holds for all x in (a, b). On the open interval (a, b), we have $G' = f = g'$; hence,

$$(G - g)' = 0$$

By Theorem 1, page 252, there exists a constant C such that

$$G(x) - g(x) = C$$

for all values of x in (a, b).

Since G and g are continuous from the right at a, we have

$$C = \lim_{x \to a^+} C = \lim_{x \to a^+} [G(x) - g(x)]$$

$$= \lim_{x \to a^+} G(x) - \lim_{x \to a^+} g(x) = G(a) - g(a)$$

But

$$G(a) = \int_a^a f(t) \, dt = 0$$

Hence,

$$C = 0 - g(a) = -g(a)$$

and so

$$G(x) - g(x) = -g(a)$$

holds for all values of x in (a, b). Using the last equation and the fact that G and g are continuous from the left at b, we have

$$-g(a) = \lim_{x \to b^-} [-g(a)] = \lim_{x \to b^-} [G(x) - g(x)]$$

$$= \lim_{x \to b^-} G(x) - \lim_{x \to b^-} g(x) = G(b) - g(b)$$

Therefore,

$$G(b) = g(b) - g(a)$$

that is,

$$\int_a^b f(t)\, dt = g(b) - g(a)$$

Consequently,

$$\int_a^b f(x)\, dx = \int_a^b f(t)\, dt = g(b) - g(a)$$

and the theorem is proved.

Problem Set 5.4

In Problems 1 to 45, use the fundamental theorem of calculus to evaluate each definite integral.

1 $\int_0^2 3x\, dx$

2 $\int_1^{14} 2\, dx$

3 $\int_{-1}^4 (-t)\, dt$

4 $\int_5^0 (-4u)\, du$

5 $\int_1^{-3} 5x^4\, dx$

6 $\int_0^{16} z^{5/4}\, dz$

7 $\int_2^3 (3x + 4)\, dx$

8 $\int_{-3}^{-1} (4 - 8x + 3x^2)\, dx$

9 $\int_1^5 (x^3 - 3x^2 + 1)\, dx$

10 $\int_1^3 (x - 1)(x^2 + x + 1)\, dx$

11 $\int_0^1 (x^2 + 2)^2\, dx$

12 $\int_1^5 \dfrac{x^4 - 16}{x^2 + 4}\, dx$

13 $\int_0^8 (2 - \sqrt[3]{t})^2\, dt$

14 $\int_1^{32} \dfrac{1 + \sqrt[5]{t^2}}{\sqrt[3]{t}}\, dt$

15 $\int_0^{\pi/6} \cos x\, dx$

16 $\int_{-\pi/4}^{\pi/3} \sin t\, dt$

17 $\int_0^{\pi/4} \sec t \tan t\, dt$

18 $\int_{\pi/4}^{\pi/6} \csc^2 y\, dy$

19 $\int_0^{\pi/3} \sec^2 u\, du$

20 $\int_{\pi/6}^{\pi/2} \csc z \cot z\, dz$

21 $\int_0^1 \dfrac{y^2\, dy}{(y^3 + 1)^5}$

22 $\int_0^1 (2x + 3)^{10}\, dx$

23 $\int_{-1}^1 \sqrt{1 - x}\, dx$

24 $\int_0^1 \sqrt{4 - 3x}\, dx$

25 $\int_{1/4}^3 \dfrac{dx}{\sqrt{1 + x}}$

26 $\int_0^2 \dfrac{x\, dx}{(4 + x^2)^{3/2}}$

27 $\int_0^2 x^2\sqrt[3]{x^3 + 1}\, dx$

28 $\int_0^1 \dfrac{x + 3}{\sqrt{x^2 + 6x + 2}}\, dx$

29 $\int_7^{10} \dfrac{x\, dx}{\sqrt{x - 6}}$

30 $\int_3^0 (x + 2)\sqrt{x + 1}\, dx$

31 $\int_0^\pi 2 \sin 3x\, dx$

32 $\int_0^{\pi/3} (2 + \cos 3t)\, dt$

33 $\int_1^0 \sec^2 \dfrac{\pi x}{4}\, dx$

34 $\int_0^1 \sec \dfrac{\pi t}{4} \tan \dfrac{\pi t}{4}\, dt$

35 $\int_{1/2}^1 \csc \dfrac{\pi t}{3} \cot \dfrac{\pi t}{3}\, dt$

36 $\int_{\pi/2}^{\pi/4} \csc^2 \left(\dfrac{x}{2} - \dfrac{\pi}{2}\right) dx$

37 $\int_0^{\pi/2} \sin^2 x \cos x\, dx$

38 $\int_{7\pi/12}^{11\pi/12} \dfrac{\cos 2x}{\sin^2 2x}\, dx$

39 $\int_{\pi/4}^0 \cos^3 \theta \sin \theta\, d\theta$

40 $\int_{-1}^1 \sqrt{|t| + t}\, dt$

41 $\int_0^3 |3 - x^2|\, dx$

42 $\int_{-1}^3 \sqrt[3]{2(|x| - x)}\, dx$

43 $\int_0^3 y|2 - y|\, dy$

44 $\int_{-1}^3 [\![x]\!] x\, dx$

45 $\int_{-3}^5 f(x)\, dx$, where $f(x) = \begin{cases} (1 - x)^{3/2} & \text{for } x \le 0 \\ (x + 4)^{1/2} & \text{for } x > 0 \end{cases}$

46 Does the following computation violate the nonnegative property of definite integrals?

$$\int_{-1}^1 \dfrac{1}{x^2}\, dx = \left[-\dfrac{1}{x}\right]\Bigg|_{-1}^1 = -1 - 1 = -2$$

In Problems 47 to 58, use the alternative version of the fundamental theorem of calculus to find the derivative.

47 $\dfrac{d}{dx} \displaystyle\int_0^x (t^2 + 1)\, dt$

48 $\dfrac{d}{dx} \displaystyle\int_1^x (w^3 - 2w + 1)\, dw$

49 $\dfrac{d}{dx} \displaystyle\int_{-1}^x \dfrac{ds}{1 + s^2}$

50 $\dfrac{d}{dx} \left(\displaystyle\int_0^x \dfrac{ds}{1 + s} + \displaystyle\int_2^x \dfrac{ds}{1 + s} \right)$

51 $\dfrac{d}{dx} \displaystyle\int_0^x \sin t^4\, dt$

52 $\dfrac{d}{dx} \displaystyle\int_{-\pi}^x \sec^3 t\, dt$

53 $D_x \displaystyle\int_{-1}^x \sqrt{t^2 + 4}\, dt$

54 $D_x \displaystyle\int_x^1 (t^3 - 3t + 1)^{10}\, dt$

55 $D_x \displaystyle\int_x^1 (w^{10} + 3)^{25}\, dw$

56 $D_x \displaystyle\int_x^4 \sqrt[3]{4s^2 + 7}\, ds$

57 $\dfrac{d}{dx} \left(\displaystyle\int_x^0 \sqrt[3]{t^2 + 1}\, dt + \displaystyle\int_0^x \sqrt[3]{t^2 + 1}\, dt \right)$

58 $D_x^2 \displaystyle\int_1^x \dfrac{1}{1 + t^2}\, dt$

In Problems 59 to 66, use the alternative version of the fundamental theorem of calculus together with the chain rule to find dy/dx.

59 $y = \displaystyle\int_1^{3x} (5t^3 + 1)^7\, dt$

60 $y = \displaystyle\int_1^{5x+1} \dfrac{dt}{9 + t^2}$

61 $y = \displaystyle\int_1^{8x+2} (w - 3)^{15}\, dw$

62 $y = \displaystyle\int_1^{x-1} \sqrt{s^2 - 1}\, ds$

63 $y = \displaystyle\int_{-x}^0 \sqrt{t + 2}\, dt$

64 $y = \displaystyle\int_{x^2+1}^2 \sqrt[3]{u - 1}\, du$

65 $y = \displaystyle\int_x^{3x^2+2} \sqrt[4]{t^4 + 17}\, dt$

66 $y = \displaystyle\int_{x^3}^{x-x^2} \sqrt{t^3 + 1}\, dt$

67 Given that u and v are differentiable functions of x and that f is a continuous function, justify the equation

$$\dfrac{d}{dx} \int_u^v f(t)\, dt = f(v)\, \dfrac{dv}{dx} - f(u)\, \dfrac{du}{dx}$$

68 Complete the proof of Theorem 1 by attending to the following details: (a) Prove that $g'_+(b) = f(b)$ and that $g'_-(c) = f(c)$. (b) Prove that g is continuous on $[b, c]$.

69 Show that the fundamental theorem of calculus works even if $a \geq b$; that is, show that if f and g are continuous functions on the closed interval $[b, a]$ and if $g'(x) = f(x)$ for x in the open interval (b, a), then $\int_a^b f(x)\, dx = g(b) - g(a)$.

70 The operator of a trucking company wants to determine the optimal period of time T (in months) between overhauls of a truck. Let the rate of depreciation of the truck be given by $f(t)$, where t is the time in months since the last overhaul. If K represents the fixed cost of an overhaul, explain why T is the value of t that minimizes $g(t) = t^{-1} \left[K + \displaystyle\int_0^t f(x)\, dx \right]$.

71 The marginal cost of processing a certain grade of tuna fish is given by $200 - 30\sqrt{x}$, where x is the number of cans in thousands and the cost is in dollars. What would be the total increase in cost if production were increased from 4000 to 25,000 cans?

72 Find the maximum value of the function

$$f(x) = \int_0^x (|t| - |t - 1|)\, dt$$

on the interval $[-1, 2]$.

73 (a) Show that $\int_0^x |t|\, dt = x|x|/2$ by considering the separate cases $x \leq 0$ and $x > 0$. (b) Use the result in part (a) and the alternative version of the fundamental theorem of calculus to prove that

$$\dfrac{d}{dx} \left(\dfrac{x|x|}{2} \right) = |x|$$

74 Use the method suggested by Problem 73 to find an antiderivative of the function $f(x) = |x|^n$.

75 Find the mean value of each function on the interval given.
(a) $f(x) = x^2 + 1$ on $[1, 4]$ (b) $f(x) = x^3 - 1$ on $[1, 3]$
(c) $f(x) = \sqrt{x}$ on $[1, 9]$ (d) $f(x) = |x| + 1$ on $[a, b]$

76 Show that $\int_0^{-x} f(t)\, dt + \int_0^x f(-t)\, dt$ is a constant by taking the derivative with respect to x. (Assume that f is continuous.) What is the value of this constant?

In Problems 77 to 80, sketch graphs of f and g.

77 $f(t) = \begin{cases} 1 - t^2 & \text{for } t \leq 0 \\ 1 + t^2 & \text{for } t > 0 \end{cases}$, $g(x) = \displaystyle\int_{-1}^x f(t)\, dt$

78 $f(t) = \begin{cases} 1 - t^2 & \text{for } t \leq 0 \\ t^2 & \text{for } t > 0 \end{cases}$, $g(x) = \displaystyle\int_{-1}^x f(t)\, dt$

79 $f(t) = [\![t]\!]$ for $-3 \leq t \leq 3$, $g(x) = \displaystyle\int_0^x f(t)\, dt$ for $-3 \leq x \leq 3$

80 $f(t) = \begin{cases} -2 & \text{for } t \leq 0 \\ t & \text{for } 0 < t \leq 5 \\ 2t & \text{for } t > 5 \end{cases}$, $g(x) = \displaystyle\int_{-1}^x f(t)\, dt$

81 Assume that f is a differentiable function and that f' is continuous. Explain why $D_x \displaystyle\int_a^x f(t)\, dt = \displaystyle\int_a^x D_t f(t)\, dt + f(a)$.

5.5   Approximation of Definite Integrals—The Trapezoidal Rule and Simpson's Rule

The fundamental theorem of calculus enables us to determine the numerical value of a definite integral $\int_a^b f(x)\, dx$ provided that we can find an antiderivative g of the function f. However, in practical applications of calculus, it is sometimes difficult or impossible to find an antiderivative g of f such that $g(b)$ and $g(a)$ can be calculated explicitly. In these cases, *numerical methods of approximation* can be used to estimate the value of the definite integral to within acceptable error bounds.

In this section, we present some methods for estimating the value of a definite integral $\int_a^b f(x)\, dx$ by formulas that use the values of $f(x)$ at only a finite number of points on the interval $[a, b]$. These methods involve only simple computations and thus lend themselves well to the use of calculators or computers.

Perhaps the most obvious way to approximate the definite integral

$$\int_a^b f(x)\, dx$$

is to use its very definition as a limit of Riemann sums. Thus, we can select a partition $\mathscr{P}$ with a small norm, augment $\mathscr{P}$ to obtain $\mathscr{P}^*$, evaluate the corresponding Riemann sum, and observe that

$$\sum_{k=1}^n f(c_k)\, \Delta x_k \approx \int_a^b f(x)\, dx$$

Figure 1 illustrates how this procedure, with $n = 5$, is used to estimate

$$\int_0^1 \frac{1}{1 + x^2}\, dx$$

The areas of the five rectangles are shown in Figure 1, and the sum of these areas, which is approximately 0.834, provides an estimate for the integral above. The true value of this integral, rounded off to four decimal places, is 0.7854, so the estimate

$$\int_0^1 \frac{1}{1 + x^2}\, dx \approx 0.834$$

is not particularly good. To obtain a reasonably accurate estimate by this method, we would have to partition the interval $[0, 1]$ into a very large number of subintervals.

Figure 1

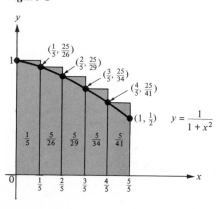

The Trapezoidal Rule

Figure 1 makes it clear why the above estimation of

$$\int_0^1 \frac{1}{1 + x^2}\, dx$$

is so crude—the five rectangles overhang the region under the graph of

$$y = \frac{1}{1 + x^2}$$

so that their total area is considerably larger than the desired area.

Evidently, a much more accurate estimate is obtained by adding the areas of the five trapezoids shown in Figure 2. Each of these trapezoids has the same width, namely, $\frac{1}{5}$ unit. Since the area of a trapezoid is given by one-half the sum of the lengths of its two parallel edges times the distance between these edges, the total area of the five trapezoids in Figure 2 is

$$\left(\frac{1 + \frac{25}{26}}{2}\right)\frac{1}{5} + \left(\frac{\frac{25}{26} + \frac{25}{29}}{2}\right)\frac{1}{5} + \left(\frac{\frac{25}{29} + \frac{25}{34}}{2}\right)\frac{1}{5} + \left(\frac{\frac{25}{34} + \frac{25}{41}}{2}\right)\frac{1}{5} + \left(\frac{\frac{25}{41} + \frac{1}{2}}{2}\right)\frac{1}{5}$$

$$= \left(\frac{1}{2} + \frac{25}{26} + \frac{25}{29} + \frac{25}{34} + \frac{25}{41} + \frac{1}{4}\right)\frac{1}{5}$$

The numerical value of this total area, rounded off to four decimal places, is 0.7837. Therefore, by this "trapezoidal method,"

$$\int_0^1 \frac{1}{1 + x^2}\, dx \approx 0.7837$$

The **trapezoidal rule** for estimating definite integrals in general is given by the following theorem.

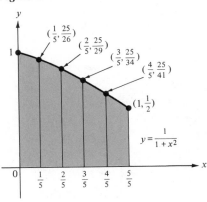

Figure 2

$y = \frac{1}{1 + x^2}$

points: $\left(\frac{1}{5}, \frac{25}{26}\right)$, $\left(\frac{2}{5}, \frac{25}{29}\right)$, $\left(\frac{3}{5}, \frac{25}{34}\right)$, $\left(\frac{4}{5}, \frac{25}{41}\right)$, $\left(1, \frac{1}{2}\right)$

THEOREM 1   **Trapezoidal Rule**

Let the function f be defined and integrable on the closed interval $[a, b]$. For each positive integer n, define

$$T_n = \left(\frac{y_0}{2} + y_1 + y_2 + \cdots + y_{n-1} + \frac{y_n}{2}\right)\Delta x$$

where

$$\Delta x = \frac{b - a}{n} \qquad \text{and} \qquad y_k = f(a + k\,\Delta x) \qquad \text{for } k = 0, 1, 2, \ldots, n$$

Then

$$T_n \approx \int_a^b f(x)\, dx$$

The approximation becomes better and better as n increases in the sense that

$$\lim_{n \to +\infty} T_n = \int_a^b f(x)\, dx$$

PROOF   Let $\mathscr{P}_n$ be the partition of $[a, b]$ consisting of n subintervals of equal length

$$\Delta x = \frac{b - a}{n}$$

Thus, the subintervals in $\mathscr{P}_n$ are

$$[x_0, x_1],\ [x_1, x_2],\ [x_2, x_3],\ \ldots,\ [x_{n-1}, x_n]$$

where

$$x_0 = a \qquad x_n = b \qquad \text{and} \qquad x_k - x_{k-1} = \Delta x$$

for $k = 1, 2, \ldots, n$ (Figure 3). Evidently,

$$x_1 = a + \Delta x, \quad x_2 = a + 2\,\Delta x, \quad x_3 = a + 3\,\Delta x, \quad \ldots, \quad x_k = a + k\,\Delta x, \quad \ldots,$$
$$\text{and} \qquad x_n = a + n\,\Delta x = b$$

Figure 3

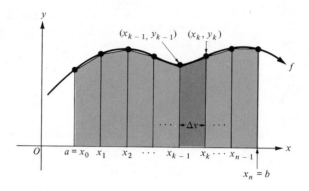

Now, augment the partition $\mathscr{P}_n$ by choosing the points $c_1, c_2, c_3, \ldots, c_n$, where

$$c_1 = x_1, \quad c_2 = x_2, \quad c_3 = x_3, \quad \ldots, \quad c_k = x_k, \quad \ldots \qquad \text{and} \qquad c_n = x_n$$

Then

$$f(c_k) = f(x_k) = f(a + k\,\Delta x) = y_k$$

and the corresponding Riemann sum is given by

$$\sum_{k=1}^{n} f(c_k)\,\Delta x_k = \sum_{k=1}^{n} y_k\,\Delta x = \left(\sum_{k=1}^{n} y_k \right) \Delta x = (y_1 + y_2 + \cdots + y_n)\,\Delta x$$

$$= \left(\frac{y_0}{2} + y_1 + y_2 + \cdots + y_{n-1} + \frac{y_n}{2} \right) \Delta x - \frac{y_0}{2}\,\Delta x + \frac{y_n}{2}\,\Delta x$$

$$= T_n + \frac{y_n - y_0}{2}\,\Delta x = T_n + \frac{f(b) - f(a)}{2}\,\Delta x$$

$$= T_n + \frac{[f(b) - f(a)](b - a)}{2n}$$

Therefore,

$$T_n = \sum_{k=1}^{n} f(c_k)\,\Delta x_k - \frac{[f(b) - f(a)](b - a)}{2n}$$

so that

$$\lim_{n \to +\infty} T_n = \lim_{n \to +\infty} \sum_{k=1}^{n} f(c_k)\,\Delta x_k - \lim_{n \to +\infty} \frac{[f(b) - f(a)](b - a)}{2n}$$

$$= \int_a^b f(x)\,dx - 0 = \int_a^b f(x)\,dx \qquad \blacksquare$$

If $f(x) \geq 0$ for $a \leq x \leq b$, then Theorem 1 admits a simple geometric explanation since the quantity T_n represents the total area of all the trapezoids shown in Figure 3 (Problem 16).

© *In Examples 1 and 2, use the trapezoidal rule, $T_n \approx \int_a^b f(x)\, dx$, with the indicated value of n to estimate the given definite integral.*

EXAMPLE 1 $\displaystyle\int_0^1 \sqrt{1 + x^3}\, dx;\ n = 4$

SOLUTION   Here $[a, b] = [0, 1]$ and $n = 4$, so

$$\Delta x = \frac{b - a}{n} = \frac{1 - 0}{4} = \frac{1}{4}$$

Also,

$$y_k = f(0 + k\,\Delta x) = \sqrt{1 + (k/4)^3}$$

for $k = 0, 1, 2, 3, 4$; hence,

$$y_0 = \sqrt{1 + 0^3} = 1 \qquad y_1 = \sqrt{1 + (\tfrac{1}{4})^3} = \frac{\sqrt{65}}{8}$$

$$y_2 = \sqrt{1 + (\tfrac{2}{4})^3} = \sqrt{\tfrac{9}{8}} \qquad y_3 = \sqrt{1 + (\tfrac{3}{4})^3} = \frac{\sqrt{91}}{8} \qquad y_4 = \sqrt{1 + 1^3} = \sqrt{2}$$

Therefore,

$$T_4 = \left(\frac{y_0}{2} + y_1 + y_2 + y_3 + \frac{y_4}{2}\right)\Delta x$$

$$= \left(\frac{1}{2} + \frac{\sqrt{65}}{8} + \sqrt{\frac{9}{8}} + \frac{\sqrt{91}}{8} + \frac{\sqrt{2}}{2}\right)\left(\frac{1}{4}\right)$$

$$\approx (0.50 + 1.01 + 1.06 + 1.19 + 0.71)(0.25) = (4.47)(0.25) \approx 1.12$$

Thus,

$$\int_0^1 \sqrt{1 + x^3}\, dx \approx 1.12$$

EXAMPLE 2 $\displaystyle\int_1^2 \frac{dx}{1 + x^2};\ n = 5$

SOLUTION   Here

$$\Delta x = \frac{2 - 1}{5} = \frac{1}{5} \qquad \text{and} \qquad y_k = \frac{1}{1 + (1 + k/5)^2} = \frac{25}{25 + (5 + k)^2}$$

for $k = 0, 1, 2, 3, 4, 5$. Therefore,

$$T_5 = \left(\tfrac{1}{4} + \tfrac{25}{61} + \tfrac{25}{74} + \tfrac{25}{89} + \tfrac{25}{106} + \tfrac{25}{250}\right)\left(\tfrac{1}{5}\right)$$

$$\approx (0.2500 + 0.4098 + 0.3378 + 0.2809 + 0.2358 + 0.1000)(0.2) \approx 0.323$$

hence,

$$\int_1^2 \frac{dx}{1 + x^2} \approx 0.323$$

The following theorem, whose proof can be found in more advanced textbooks, provides an upper bound for the error involved in using the trapezoidal rule.

THEOREM 2

Error Bound for Trapezoidal Rule

Suppose that f'' is defined and continuous on $[a, b]$ and that M is the maximum value of $|f''(x)|$ for x in $[a, b]$. Then, if T_n is the approximation to $\int_a^b f(x)\, dx$ given by the trapezoidal rule,

$$\left| T_n - \int_a^b f(x)\, dx \right| \le M\, \frac{(b - a)^3}{12n^2}$$

Ⓒ **EXAMPLE 3**   Use the trapezoidal rule to estimate $\int_1^2 dx/x$ with $n = 6$, and use Theorem 2 to find a bound for the error of the estimate.

SOLUTION   We have

$$\Delta x = \frac{2 - 1}{6} = \frac{1}{6} \quad \text{and} \quad y_k = \frac{1}{1 + k/6} = \frac{6}{6 + k}$$

for $k = 0, 1, 2, 3, 4, 5, 6$; hence,

$$y_0 = \tfrac{6}{6} = 1 \quad y_1 = \tfrac{6}{7} \quad y_2 = \tfrac{6}{8} \quad y_3 = \tfrac{6}{9} \quad y_4 = \tfrac{6}{10} \quad y_5 = \tfrac{6}{11} \quad y_6 = \tfrac{6}{12}$$

Therefore,

$$\begin{aligned} T_6 &= (\tfrac{1}{2} + \tfrac{6}{7} + \tfrac{6}{8} + \tfrac{6}{9} + \tfrac{6}{10} + \tfrac{6}{11} + \tfrac{1}{4})(\tfrac{1}{6}) \\ &\approx (0.5 + 0.8571 + 0.75 + 0.6667 + 0.6 + 0.5455 + 0.25)(0.1667) \\ &\approx 0.6950 \end{aligned}$$

so that

$$\int_1^2 \frac{dx}{x} \approx 0.695$$

Here,

$$f(x) = \frac{1}{x} \qquad f'(x) = -\frac{1}{x^2} \quad \text{and} \quad f''(x) = \frac{2}{x^3}$$

Since f'' is a decreasing function on $[1, 2]$, its maximum value on this interval is taken on at the left endpoint 1. Hence, in Theorem 2,

$$M = f''(1) = \frac{2}{1^3} = 2$$

In Theorem 2 we also put $b = 2$, $a = 1$, and $n = 6$, to conclude that the error in the estimate above does not exceed

$$M\, \frac{(b - a)^3}{12n^2} = (2)\, \frac{1^3}{432} = \frac{1}{216}$$

Since $\frac{1}{216} < 0.005$, the estimation $\int_1^2 dx/x \approx 0.695$ is correct to at least two decimal places.

Simpson's Rule

A third method for approximating the value of a definite integral is known as **Simpson's rule,** or the **parabolic rule,** and is usually more efficient than either the direct use of the definition or the trapezoidal rule. The method is based on the use of a number of adjacent regions having the shape shown in Figure 4 to approximate the area under the graph of f. (See Problem 33.) The result is the following theorem, whose proof is analogous to the proof of Theorem 1 and is therefore omitted.

THEOREM 3

Figure 4

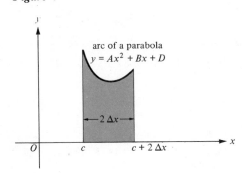

arc of a parabola
$y = Ax^2 + Bx + D$

Simpson's Parabolic Rule

Let the function f be defined and integrable on the closed interval $[a, b]$. For each positive integer n, define

$$S_{2n} = \frac{\Delta x}{3}(y_0 + 4y_1 + 2y_2 + 4y_3 + 2y_4 + \cdots + 2y_{2n-2} + 4y_{2n-1} + y_{2n})$$

where

$$\Delta x = \frac{b - a}{2n} \quad \text{and} \quad y_k = f(a + k\,\Delta x) \quad \text{for } k = 0, 1, 2, \ldots, 2n$$

Then

$$S_{2n} \approx \int_a^b f(x)\,dx$$

with the approximation becoming better and better as n increases in the sense that

$$\lim_{n \to +\infty} S_{2n} = \int_a^b f(x)\,dx$$

© *In Examples 4 and 5, use Simpson's parabolic rule, $S_{2n} \approx \int_a^b f(x)\,dx$, to estimate the definite integral, using the indicated value of n.*

EXAMPLE 4 $\displaystyle\int_0^1 (1 + x)^{-1}\,dx; \ n = 4$

SOLUTION   Here, the interval $[0, 1]$ must be subdivided into $2n = 8$ parts, each of length $\Delta x = (1 - 0)/8 = \frac{1}{8}$. Thus,

$$y_k = f(a + k\,\Delta x) = f\left(0 + \frac{k}{8}\right) = \left(1 + \frac{k}{8}\right)^{-1}$$

so

$$y_0 = (1 + 0)^{-1} = 1 \qquad y_1 = (1 + \tfrac{1}{8})^{-1} = \tfrac{8}{9} \qquad y_2 = (1 + \tfrac{2}{8})^{-1} = \tfrac{8}{10}$$
$$y_3 = (1 + \tfrac{3}{8})^{-1} = \tfrac{8}{11} \qquad y_4 = (1 + \tfrac{4}{8})^{-1} = \tfrac{8}{12} \qquad y_5 = (1 + \tfrac{5}{8})^{-1} = \tfrac{8}{13}$$
$$y_6 = (1 + \tfrac{6}{8})^{-1} = \tfrac{8}{14} \qquad y_7 = (1 + \tfrac{7}{8})^{-1} = \tfrac{8}{15} \qquad y_8 = (1 + \tfrac{8}{8})^{-1} = \tfrac{1}{2}$$

Therefore,

$$S_{2n} = S_8 = \frac{\Delta x}{3}(y_0 + 4y_1 + 2y_2 + 4y_3 + 2y_4 + 4y_5 + 2y_6 + 4y_7 + y_8)$$

$$= \tfrac{1}{24}(1 + \tfrac{32}{9} + \tfrac{16}{10} + \tfrac{32}{11} + \tfrac{16}{12} + \tfrac{32}{13} + \tfrac{16}{14} + \tfrac{32}{15} + \tfrac{1}{2})$$

$$\approx \tfrac{1}{24}(1 + 3.5556 + 1.6000 + 2.9091 + 1.3333 + 2.4615$$
$$+ \ 1.1429 + 2.1333 + 0.5000)$$

$$= \tfrac{1}{24}(16.6357) \approx 0.6932$$

Consequently, $\int_0^1 (1 + x)^{-1}\, dx \approx 0.6932$. Incidentally, the correct value, rounded off to five decimal places, is $\int_0^1 (1 + x)^{-1}\, dx \approx 0.69315$. ∎

EXAMPLE 5 $\displaystyle\int_0^1 \sqrt{1 - x^4}\, dx;\; n = 3$

SOLUTION   Here

$$\Delta x = \frac{1 - 0}{6} = \frac{1}{6}$$

$$y_k = \sqrt{1 - \left(\frac{k}{6}\right)^4} = \sqrt{\frac{6^4 - k^4}{6^4}} = \sqrt{\frac{1296 - k^4}{1296}}$$

for $k = 0, 1, 2, 3, 4, 5, 6$. Therefore,

$$S_{2n} = S_6 = \tfrac{1}{18}\left(1 + 4\sqrt{\tfrac{1295}{1296}} + 2\sqrt{\tfrac{1280}{1296}} + 4\sqrt{\tfrac{1215}{1296}} + 2\sqrt{\tfrac{1040}{1296}} + 4\sqrt{\tfrac{671}{1296}} + 0\right)$$

$$\approx \tfrac{1}{18}(1 + 3.9985 + 1.9876 + 3.8730 + 1.7916 + 2.8782)$$

$$= \tfrac{1}{18}(15.5289) \approx 0.8627 \approx \int_0^1 \sqrt{1 - x^4}\, dx$$ ∎

The following theorem, whose proof can be found in more advanced textbooks, provides an upper bound for the error involved in using Simpson's parabolic rule.

THEOREM 4   **Error Bound for Simpson's Parabolic Rule**

Suppose that the fourth derivative $f^{(4)}(x)$ is defined and continuous on $[a, b]$ and that N is the maximum value of $|f^{(4)}(x)|$ for x in $[a, b]$. If S_{2n} is the approximation to $\int_a^b f(x)\, dx$ given by Simpson's parabolic rule, then

$$\left[S_{2n} - \int_a^b f(x)\, dx \right] \le N\frac{(b - a)^5}{2880 n^4}$$

©**EXAMPLE 6**   Use Simpson's parabolic rule to estimate $\int_1^2 dx/x$, making certain that the error of estimation does not exceed 0.001.

SOLUTION   Here,

$$f(x) = \frac{1}{x} \qquad f'(x) = -\frac{1}{x^2} \qquad f''(x) = \frac{2}{x^3} \qquad f'''(x) = -\frac{6}{x^4} \qquad f^{(4)}(x) = \frac{24}{x^5}$$

Since $f^{(4)}$ is a decreasing function on $[1, 2]$, its maximum value on this interval is taken on at the left-hand endpoint 1. Hence, in Theorem 4,

$$N = |f^{(4)}(1)| = \frac{24}{1^5} = 24 \qquad \text{and} \qquad b - a = 2 - 1 = 1$$

so the error of estimation cannot exceed

$$N\frac{(b - a)^5}{2880 n^4} = 24\frac{1}{2880 n^4} = \frac{1}{120 n^4}$$

Therefore, we require that $1/(120 n^4) \le 0.001$, that is, $25/3 \le n^4$. The smallest value of n satisfying the latter inequality is $n = 2$. Thus, we estimate $\int_1^2 dx/x$, using $S_{2n} = S_4$. We have

$$\Delta x = \frac{2-1}{4} = \frac{1}{4}$$

so

$$y_0 = 1 \qquad y_1 = \tfrac{4}{5} \qquad y_2 = \tfrac{4}{6} \qquad y_3 = \tfrac{4}{7} \qquad y_4 = \tfrac{1}{2}$$

Hence,

$$S_{2n} = \tfrac{1}{12}(1 + \tfrac{16}{5} + \tfrac{8}{6} + \tfrac{16}{7} + \tfrac{1}{2}) = \tfrac{1747}{2520}$$

It follows that $\int_1^2 dx/x \approx \tfrac{1747}{2520}$ with an error not exceeding $\tfrac{1}{1000}$. In fact, if we round off to five decimal places, $\tfrac{1747}{2520} \approx 0.69325$, while the correct value of $\int_1^2 dx/x$, rounded off to five decimal places, is 0.69315. ∎

Theorem 4 has an interesting consequence: *Simpson's parabolic rule gives the exact value of $\int_a^b f(x)\, dx$ if f is a polynomial function of degree not exceeding 3.* The reason for this is just that for such a polynomial function, $f^{(4)}$ is the zero function; hence, in Theorem 4, $N = 0$. Thus, taking $n = 1$ in Theorem 3, we obtain the following theorem.

THEOREM 5   **Prismoidal Formula**

> Let f be any polynomial function whose degree does not exceed 3. Then
>
> $$\int_a^b f(x)\, dx = \frac{b-a}{6}\left[f(a) + 4f\left(\frac{a+b}{2}\right) + f(b)\right]$$

EXAMPLE 7   Use the prismoidal formula to evaluate $\int_0^2 (x^3 + 1)\, dx$.

SOLUTION $\displaystyle\int_0^2 (x^3 + 1)\, dx = \tfrac{2}{6}[(0^3 + 1) + 4(1^3 + 1) + (2^3 + 1)] = 6$ ∎

Problem Set 5.5

C In Problems 1 to 12, use the trapezoidal rule, $T_n \approx \int_a^b f(x)\, dx$, with the indicated value of n to estimate the integral.

1 $\displaystyle\int_0^1 \frac{dx}{1+x^2}$; $n = 4$

2 $\displaystyle\int_1^3 \frac{dx}{x}$; $n = 3$

3 $\displaystyle\int_2^8 \frac{dx}{1+x}$; $n = 6$

4 $\displaystyle\int_0^3 \sqrt{9 - x^2}\, dx$; $n = 6$

5 $\displaystyle\int_0^1 \frac{dx}{\sqrt{1+x^4}}$; $n = 5$

6 $\displaystyle\int_0^1 \frac{dx}{1+x^3}$; $n = 4$

7 $\displaystyle\int_2^8 (4 + x^2)^{-1/3}\, dx$; $n = 6$

8 $\displaystyle\int_2^3 \sqrt{1 + x^2}\, dx$; $n = 7$

9 $\displaystyle\int_1^2 \frac{dx}{x\sqrt{1+x}}$; $n = 5$

10 $\displaystyle\int_{\pi/2}^{\pi} \frac{\sin x}{x}\, dx$; $n = 4$

11 $\displaystyle\int_0^{\pi/4} \tan x\, dx$; $n = 4$

12 $\displaystyle\int_0^{\pi} \frac{\sin x}{1+x}\, dx$; $n = 6$

C **13** Use Theorem 2 to find an error bound for the estimate in Problem 1.

C 14 Use Theorem 2 to find an error bound for the estimate in Problem 2.

15 Suppose that $f(x) \geq 0$ for all x in $[a, b]$ and that the graph of f is concave downward on $[a, b]$. Explain geometrically why $T_n \leq \int_a^b f(x)\, dx$ in this case.

16 If $f(x) \geq 0$ for $a \leq x \leq b$, show that the quantity T_n in Theorem 1 represents the total area of all the trapezoids in Figure 3.

C In Problems 17 to 26, use Simpson's parabolic rule, $S_{2n} \approx \int_a^b f(x)\, dx$, with the indicated value of n to estimate the integral.

17 $\displaystyle\int_{-1}^{1} \frac{dx}{1 + x^2}$; $n = 2$

18 $\displaystyle\int_{0}^{4} x^2\sqrt{x + 1}\, dx$; $n = 4$

19 $\displaystyle\int_{0}^{8} \frac{dx}{x^3 + x + 1}$; $n = 4$

20 $\displaystyle\int_{2}^{10} \frac{dx}{1 + x^3}$; $n = 4$

21 $\displaystyle\int_{0}^{2} x\sqrt{9 - x^3}\, dx$; $n = 3$

22 $\displaystyle\int_{0}^{2} \frac{dx}{\sqrt{1 + x^2}}$; $n = 2$

23 $\displaystyle\int_{0}^{2} \sqrt{1 + x^4}\, dx$; $n = 4$

24 $\displaystyle\int_{0}^{2} \sqrt[3]{1 - x^2}\, dx$; $n = 4$

25 $\displaystyle\int_{0}^{\pi/2} \sqrt{\cos x}\, dx$; $n = 3$

26 $\displaystyle\int_{\pi/6}^{\pi/3} \csc x\, dx$; $n = 4$

C 27 Find the smallest value of n for which the error involved in the estimation $S_{2n} \approx \int_1^2 dx/x$ does not exceed 0.0001. (Use Theorem 4.)

C 28 Use Simpson's parabolic rule, $S_{2n} \approx \int_a^b f(x)\, dx$, with $n = 1$ to estimate $\int_{2.5}^{2.7} dx/x$. Give an upper bound for the error of estimation.

C 29 On geometric grounds, $4 \int_0^1 \sqrt{1 - x^2}\, dx = \pi$. Why? Using Simpson's parabolic rule, $S_{2n} \approx \int_0^1 \sqrt{1 - x^2}\, dx$, to estimate $\int_0^1 \sqrt{1 - x^2}\, dx$, give an estimate for π. Use $n = 2$.

C 30 Use the procedure of Problem 29 to estimate π taking $n = 5$. Compare the result with the correct value of π, which is 3.14159. . . .

31 (a) Use the prismoidal formula to prove that
$$\int_{-a}^{a} (Ax^3 + Bx^2 + Cx + D)\, dx = \frac{a}{3}(2Ba^2 + 6D)$$
(b) Prove the formula in part (a) directly using the fundamental theorem of calculus.

32 Prove the prismoidal formula directly using the fundamental theorem of calculus.

33 Prove that by a suitable choice of the three coefficients A, B, and D, the graph of $y = Ax^2 + Bx + D$ can be made to pass through any three points of the form (c, p), $(c + \Delta x, q)$, and $(c + 2\Delta x, r)$ (Figure 4).

C 34 A loaded freighter is anchored in still water. At water level, the boat is 200 feet long and for each $k = 0, 1, 2, \ldots, 20$ has breadth y_k at a distance $10k$ feet from the bow. Assume that $y_0 = 0$ and $y_{20} = 0$. (a) Use Simpson's parabolic rule to write a formula giving the approximate area of the water-level section of the boat. (b) Recalling an exploit of Archimedes, write a formula for the approximate number of tons of freight that should be removed to raise the level of the boat by 1 foot. (Assume that the water weighs 64 pounds per cubic foot.)

Figure 1

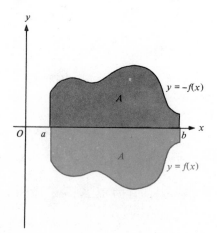

5.6 Areas of Regions in the Plane

We already used the definite integral to calculate areas of regions under the graphs of continuous functions. In this section, we develop methods for using the definite integral to find areas of more general regions in the plane.

Suppose that f is a continuous function on $[a, b]$ and that $f(x) \leq 0$ for $a \leq x \leq b$. Thus, the graph of $y = f(x)$ lies *under* the x axis between $x = a$ and $x = b$ (Figure 1). Let A denote the area of the region under the x axis and above the graph of $y = f(x)$ between $x = a$ and $x = b$. Notice that the graph of $y = -f(x)$ is the mirror image *above* the x axis of the graph of $y = f(x)$; hence, A is also the area under the graph of $y = -f(x)$ between $x = a$ and $x = b$. Therefore,

$$A = \int_a^b -f(x)\, dx = -\int_a^b f(x)\, dx$$

Figure 2

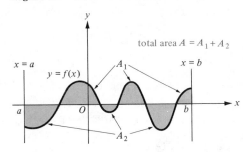

More generally, suppose that f is a continuous function on $[a, b]$ and that $f(x)$ takes on *both* positive and negative values for $a \le x \le b$ (Figure 2). If you wish to calculate the area A of the region *between the graph of f and the x axis* from $x = a$ to $x = b$, you can begin by dividing the interval $[a, b]$ into subintervals over which f is either positive or negative. If $[c, d]$ is a subinterval of $[a, b]$ and f is *positive* over (c, d), then

$$\int_c^d f(x) \, dx$$

gives the area *under* the graph of f between c and d; however, if f is *negative* over (c, d), then the area *above* the graph of f between c and d is given by

$$-\int_c^d f(x) \, dx$$

The total area A is obtained by adding the areas corresponding to all such subintervals.

Figure 3

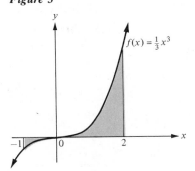

__EXAMPLE 1__   Find the area A between the graph of $f(x) = \frac{1}{3}x^3$ and the x axis from $x = -1$ to $x = 2$.

SOLUTION   A sketch of the graph of f (Figure 3) shows that it falls below the x axis on the interval $[-1, 0]$. Therefore, we cannot find A simply by calculating

$$\int_{-1}^2 \tfrac{1}{3}x^3 \, dx$$

since the area below the x axis provides a negative contribution to this integral. However, by splitting the interval $[-1, 2]$ into two subintervals, we obtain the area A as follows:

$$A = -\int_{-1}^0 \tfrac{1}{3}x^3 \, dx + \int_0^2 \tfrac{1}{3}x^3 \, dx = -\left[\tfrac{1}{12}x^4\right]\Big|_{-1}^0 + \left[\tfrac{1}{12}x^4\right]\Big|_0^2$$

$$= \tfrac{1}{12} + \tfrac{16}{12} = \tfrac{17}{12} \text{ square units}$$

Areas by Slicing Using the Definite Integral

The method of slicing, introduced in Section 4.7, enables us to find areas of regions by solving a differential equation of the form $dA = l \, ds$. We now show how such areas can be calculated by using the definite integral.

In this section we consider only regions R in the plane that satisfy the following two conditions:

1 The boundary of R consists of a finite number of straight line segments or smooth arcs which can meet in a finite number of "corners," or "vertices."

2 The region R is **bounded** in the sense that there is an upper bound to the distances between the points of R.

Figure 4

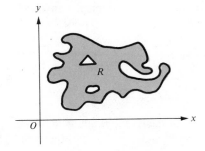

A region R satisfying conditions 1 and 2 will be called an **admissible** region. The region R shown in Figure 4 is an example of an admissible region. Notice that such a region is permitted to have a finite number of "holes," provided that the boundaries of these holes satisfy condition 1. The boundaries of the holes, if any, are regarded as part of the boundary of R.

We assume that any admissible region R in the plane has an area A square units associated with it. The following theorem shows how to calculate A in terms of a definite integral.

<u>THEOREM 1</u>   **Areas by Slicing Using the Definite Integral**

Let R be an admissible region, and choose a convenient reference axis called, say, the s axis (Figure 5a). At each point along the s axis, erect a perpendicular line, and suppose that the region R is entirely contained between the two perpendiculars at $s = a$ and $s = b$, $a < b$. Let the perpendicular at s intersect R in one or more line segments of total length $l(s)$. Then the area A of the region R is given by

$$A = \int_a^b l(s)\, ds$$

Figure 5

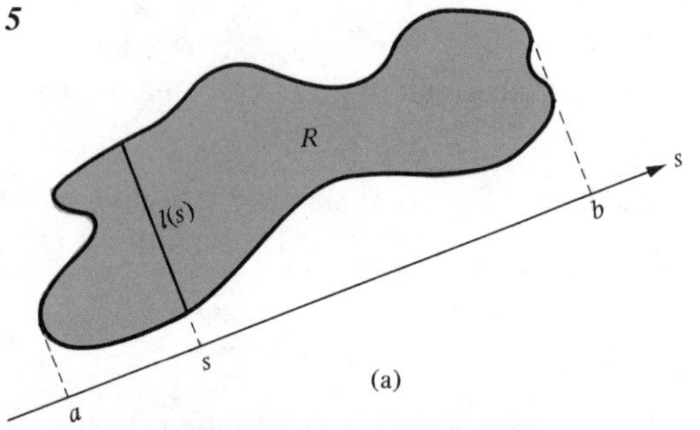

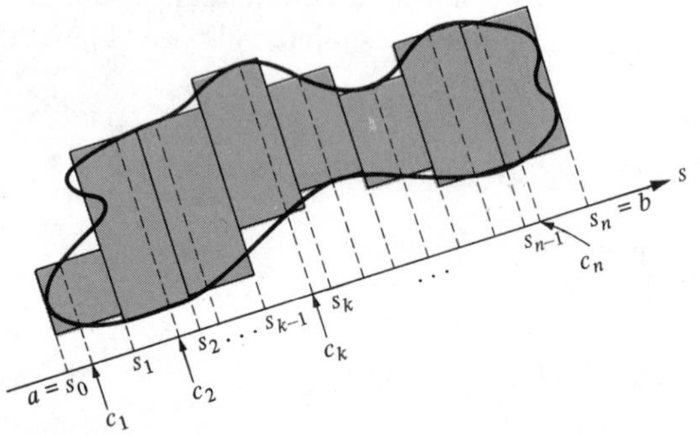

(a)

(b)

Figure 6

Although a rigorous proof of Theorem 1 is beyond the scope of this textbook, we give an informal argument to indicate its plausibility. Figure 5b shows a partition of the reference axis consisting of n subintervals

$$[s_0, s_1],\ [s_1, s_2],\ \ldots,\ [s_{k-1}, s_k],\ \ldots,\ [s_{n-1}, s_n]$$

where $a = s_0$ and $b = s_n$. We have augmented this partition by choosing numbers $c_1, c_2, c_3, \ldots, c_n$ from the successive subintervals. Above each subinterval $[s_{k-1}, s_k]$ we have constructed a rectangle of width

$$\Delta s_k = s_k - s_{k-1}$$

and of length $l(c_k)$. For $k = 0, 1, 2, 3, \ldots, n$, the area of the kth rectangle is

$$l(c_k)\, \Delta s_k$$

Evidently, the desired area A is approximated by the sum of the areas of all these rectangles:

$$A \approx \sum_{k=1}^n l(c_k)\, \Delta s_k$$

As the norm of the partition approaches zero and the rectangles become narrower and more numerous, the approximation obviously should improve. The sum on the right is a Riemann sum, and its limit as the norm of the partition approaches zero is, by definition, the definite integral in Theorem 1. Therefore, the result

$$A = \int_a^b l(s)\, ds$$

seems geometrically reasonable.

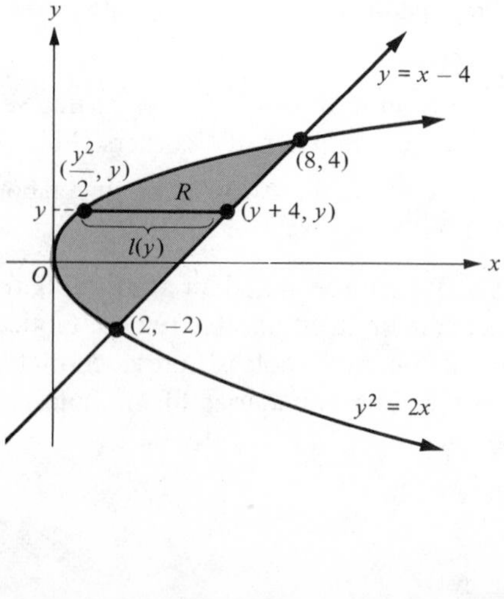

<u>EXAMPLE 2</u>   Find the area A of the region R bounded by the graphs of the equations $y^2 = 2x$ and $y = x - 4$ (Figure 6).

SOLUTION   We use Theorem 1, taking the y axis as the reference axis. To determine the points of intersection of the two graphs, we solve the two equations $y^2 = 2x$ and $y = x - 4$ simultaneously. Substituting $y = x - 4$ into $y^2 = 2x$, we have

$$(x - 4)^2 = 2x$$

$$x^2 - 10x + 16 = 0$$

Factoring to solve the latter equation, we obtain

$$(x - 2)(x - 8) = 0 \qquad \text{so} \qquad x = 2 \qquad \text{or} \qquad x = 8$$

When $x = 2$, then

$$y = x - 4 = 2 - 4 = -2$$

When $x = 8$, then

$$y = x - 4 = 8 - 4 = 4$$

Therefore, the two graphs meet at $(2, -2)$ and at $(8, 4)$.

The point on the graph of $y^2 = 2x$ with ordinate y has abscissa $x = y^2/2$, and the point on the graph of $y = x - 4$ with ordinate y has abscissa $x = y + 4$ (Figure 6). Therefore,

$$l(y) = (y + 4) - \frac{y^2}{2} \qquad \text{for } -2 \le y \le 4$$

By Theorem 1,

$$A = \int_{-2}^{4} l(y) \, dy = \int_{-2}^{4} \left(y + 4 - \frac{y^2}{2} \right) dy$$

$$= \left(\frac{y^2}{2} + 4y - \frac{y^3}{6} \right) \Big|_{-2}^{4}$$

$$= \tfrac{40}{3} - (-\tfrac{14}{3}) = 18 \text{ square units} \qquad \blacksquare$$

The Area Between the Graphs of Two Functions

Figure 7

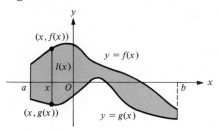

Let's consider the problem of finding the area A of an admissible region bounded above by the graph of $y = f(x)$, below by the graph of $y = g(x)$, on the left by the vertical line $x = a$, and on the right by the vertical line $x = b$ (Figure 7). If we take the x axis as the reference axis, then $l(x)$ is the distance between the points $(x, f(x))$ and $(x, g(x))$; hence,

$$l(x) = f(x) - g(x)$$

Therefore, by Theorem 1,

$$A = \int_{a}^{b} [f(x) - g(x)] \, dx$$

In Examples 3 to 5, find the area A of the indicated region.

<u>EXAMPLE 3</u>   The region between the graphs of $y = x + 2$ and $y = x^2$.

Figure 8

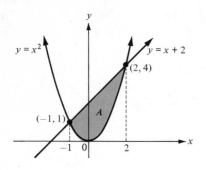

SOLUTION   To determine the points of intersection of the two graphs, we solve the two equations $y = x + 2$ and $y = x^2$ simultaneously and find that the intersection points are

$$(-1, 1) \quad \text{and} \quad (2, 4)$$

(Figure 8). Notice that the graph of $y = x + 2$ lies above the graph of $y = x^2$ between $x = -1$ and $x = 2$. Therefore, using the formula above with $f(x) = x + 2$ and $g(x) = x^2$, we find that

$$A = \int_{-1}^{2} [(x + 2) - x^2]\, dx = \left(\frac{x^2}{2} + 2x - \frac{x^3}{3} \right) \Big|_{-1}^{2}$$

$$= \frac{10}{3} - \left(-\frac{7}{6} \right) = \frac{9}{2} \text{ square units} \qquad \blacksquare$$

Figure 9

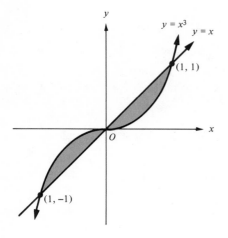

EXAMPLE 4   The region bounded by $y = x$ and $y = x^3$.

SOLUTION   The two graphs are sketched in Figure 9. By symmetry, the portion of the region in quadrant III has the same area as the portion in quadrant I. Therefore, the area A of the entire region is twice the area of the portion in quadrant I. Consequently,

$$A = 2 \int_0^1 (x - x^3)\, dx = 2 \left(\frac{x^2}{2} - \frac{x^4}{4} \right) \Big|_0^1 = 2 \left(\frac{1}{4} - 0 \right) = \frac{1}{2} \text{ square unit} \qquad \blacksquare$$

EXAMPLE 5   The region bounded above by $y = \cos x$, below by $y = \sin x$, on the left by $x = 0$, and on the right by $x = \pi/4$ (Figure 10).

SOLUTION

$$A = \int_0^{\pi/4} (\cos x - \sin x)\, dx = (\sin x + \cos x) \Big|_0^{\pi/4}$$

$$= \left(\frac{\sqrt{2}}{2} + \frac{\sqrt{2}}{2} \right) - (0 + 1) = \sqrt{2} - 1 \text{ square unit} \qquad \blacksquare$$

Figure 10

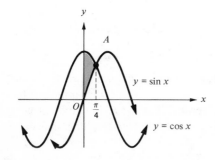

Areas Bounded by Graphs of "Piecewise-Defined" Functions

The following example illustrates the method for finding the area between the graph of a "piecewise-defined" function and the x axis.

EXAMPLE 6

(a) Find the area between the graph of $y = f(x)$ and the x axis from $x = -2$ to $x = 5$, where f is the function defined by

$$f(x) = \begin{cases} 2 + \dfrac{x^3}{4} & \text{if } x < 0 \\ x^2 - x - 2 & \text{if } 0 \le x < 3 \\ 16 - 4x & \text{if } 3 \le x \end{cases}$$

(b) Find $\displaystyle\int_{-2}^{5} f(x)\, dx$.

Figure 11

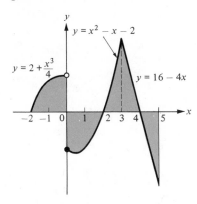

SOLUTION

(a) Figure 11 shows a sketch of the graph of f with the desired area A shaded. From the figure, we see that

$$A = \int_{-2}^{0} \left(2 + \frac{x^3}{4}\right) dx - \int_{0}^{2} (x^2 - x - 2)\, dx + \int_{2}^{3} (x^2 - x - 2)\, dx$$

$$+ \int_{3}^{4} (16 - 4x)\, dx - \int_{4}^{5} (16 - 4x)\, dx$$

$$= \left(2x + \frac{x^4}{16}\right)\Big|_{-2}^{0} - \left(\frac{x^3}{3} - \frac{x^2}{2} - 2x\right)\Big|_{0}^{2} + \left(\frac{x^3}{3} - \frac{x^2}{2} - 2x\right)\Big|_{2}^{3}$$

$$+ (16x - 2x^2)\Big|_{3}^{4} - (16x - 2x^2)\Big|_{4}^{5}$$

$$= 3 + \frac{10}{3} + \frac{11}{6} + 2 + 2 = \frac{73}{6} \text{ square units}$$

(b) $\displaystyle \int_{-2}^{5} f(x)\, dx = \int_{-2}^{0} \left(2 + \frac{x^3}{4}\right) dx + \int_{0}^{3} (x^2 - x - 2)\, dx$

$$+ \int_{3}^{5} (16 - 4x)\, dx$$

$$= \left(2x + \frac{x^4}{16}\right)\Big|_{-2}^{0} + \left(\frac{x^3}{3} - \frac{x^2}{2} - 2x\right)\Big|_{0}^{3} + (16x - 2x^2)\Big|_{3}^{5}$$

$$= 3 - \frac{3}{2} + 0 = \frac{3}{2}$$ ∎

Again, in the example above, notice that

$$\int_{-2}^{5} f(x)\, dx = \frac{3}{2}$$

does *not* give the area between the x axis and the graph of $y = f(x)$ from $x = -2$ to $x = 5$ because the area below the x axis provides a negative contribution to the integral.

Problem Set 5.6

In Problems 1 to 14, sketch the graph of each function, and find the area between the graph and the x axis from $x = a$ to $x = b$.

1 $f(x) = 1 - x^2$; $a = -1$, $b = 1$

2 $g(x) = x^2 - 2$; $a = 0$, $b = 1$

3 $h(x) = x^3 - x$; $a = -1$, $b = 1$

4 $F(x) = x^2 - 9$; $a = -3$, $b = 3$

5 $G(x) = x^3$; $a = -2$, $b = 2$

6 $H(x) = x^2 - 6x + 5$; $a = 1$, $b = 3$

7 $f(x) = x^3 - 4x^2 + 3x$; $a = 0$, $b = 2$

8 $g(x) = x^3 - 6x^2 + 8x$; $a = 0$, $b = 4$

9 $f(x) = \frac{1}{3}(x - x^3)$; $a = -1$, $b = 2$

10 $f(x) = x^n$; $a = 0$, $b = 1$, where $n \geq 1$

11 $f(x) = \sin 2x$; $a = 0$, $b = \dfrac{\pi}{2}$

12 $g(x) = \cos \dfrac{x}{3}$; $a = 0$, $b = \pi$

13 $h(x) = \sec^2 x$; $a = -\dfrac{\pi}{4}$, $b = \dfrac{\pi}{4}$

14 $F(x) = \sin x \cos x$; $a = -\pi$, $b = \pi$

In Problems 15 and 16, find the area A of the shaded region in the indicated figure in two ways: (a) by using the x axis as reference axis and (b) by using the y axis as reference axis.

15

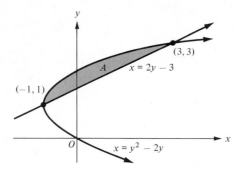

16

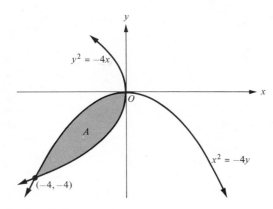

In Problems 17 to 40, (a) sketch the graphs of the two equations, (b) find the points of intersection of the two graphs, and (c) find the area of the region bounded by the two graphs.

17 $f(x) = x^2$ and $g(x) = 2x + \frac{5}{4}$

18 $y = \dfrac{x^2}{4}$ and $7x - 2y = 20$

19 $f(x) = -x^2 - 4$ and $g(x) = -8$

20 $y = \sqrt{x}$ and $y = x$

21 $y = x^2 - x$ and $y = x$

22 $f(x) = x^3$ and $g(x) = x$

23 $f(x) = -x^2 + 4x$ and $g(x) = x^2$

24 $f(x) = x^2 + 1$ and $g(x) = 4x + 1$

25 $f(x) = 10x - x^2$ and $g(x) = 3x - 8$

26 $f(x) = x^2 - 8x$ and $g(x) = -x^2$

27 $x = (y - 2)^2$ and $x = y$

28 $y^2 = 3x$ and $y = x$

29 $x = 6y^2 - 3$ and $x + 3y = 0$

30 $x = 4y^2 - 1$ and $8x - 6y + 3 = 0$

31 $f(x) = x^3$ and $g(x) = \sqrt[3]{x}$

32 $y = 2x^2 - x^3$ and $y = 2x - x^2$

33 $y = x^2$ and $y = x^4$

34 $y = |x|$ and $y = x^4$

35 $x = y^2 - 2$ and $x = 6 - y^2$

36 $x = y^2 - y$ and $x = y - y^2$

37 $f(x) = x|x|$ and $g(x) = x^3$

38 $f(x) = -x$ and $g(x) = 2x - 3x^2$

39 $f(x) = \cos x$ and $g(x) = 1 - \cos x$ for $-\dfrac{\pi}{3} \le x \le \dfrac{\pi}{3}$

40 $f(x) = 3 - |x|$ and $g(x) = \sec^2 \dfrac{\pi x}{4}$

In Problems 41 to 44, find the area of the region bounded by the graphs of the given equations.

41 $y = x + 6$, $y = \frac{1}{2}x^2$, $x = 1$, $x = 4$

42 $y = x^3$, $y = 12 - x^2$, $x = 0$

43 $y = 2 - x$, $y = x^2$, and above $y = \sqrt[3]{x}$

44 $y = x^3$, $y = 0$, and $y = 3x - 2$ (first quadrant)

In Problems 45 to 50, sketch a graph of the function f, find the area between the graph of f and the x axis from $x = a$ to $x = b$, and find $\int_a^b f(x)\, dx$.

45 $f(x) = \begin{cases} x^3 & \text{for } -2 \le x \le 1 \\ \sqrt{x} & \text{for } 1 < x \le 4 \\ 10 - 2x & \text{for } 4 < x \le 7 \\ 2x - 18 & \text{for } 7 < x \le 12 \end{cases}$; $a = -2$, $b = 12$

46 $f(x) = \begin{cases} -x - 3 & \text{for } -5 \le x < -2 \\ x^2 + 2x - 1 & \text{for } -2 \le x \le 1 \\ 2 & \text{for } 1 < x \le 4 \end{cases}$; $a = -5$, $b = 4$

47 $f(x) = \begin{cases} x^2 + 6x - 7 & \text{for } -7 \le x \le -6 \\ -x^2 - 4x + 5 & \text{for } -6 < x \le 0 \\ |x - 5| & \text{for } 0 < x \le 8 \end{cases}$; $a = -7$, $b = 8$

48 $f(x) = \begin{cases} x\sqrt{x^2 - 4} & \text{for } -3 \le x \le -2 \\ -x^2 & \text{for } -2 < x \le 0 \\ 3 - x & \text{for } 0 < x \le 4 \\ \sqrt{2x + 1} & \text{for } 4 < x \le 6 \end{cases}$; $a = -3$, $b = 6$

49 $f(x) = \begin{cases} x^3 - 1 & \text{for } -1 \le x < 0 \\ x^2 & \text{for } 0 \le x < 2 \\ 2[\![x]\!] & \text{for } 2 \le x < 4 \\ -\sqrt{x - 4} & \text{for } 4 \le x \le 8 \end{cases}$; $a = -1$, $b = 8$

50 $f(x) = \begin{cases} \dfrac{x}{(x^2 + 1)^2} & \text{for } -3 \leq x < 0 \\ \sqrt[3]{2x} & \text{for } 0 \leq x < 4 \\ x^2 - 12x + 32 & \text{for } 4 \leq x \leq 10 \\ 5 & \text{for } 10 < x \leq 12 \end{cases}$; $a = -3$, $b = 12$

51 Use the method of slicing to find the area of the triangle whose vertices are the points (1, 1), (2, 4), and (4, 3).

52 Use the method of slicing to find the area of the trapezoid whose vertices are the points (1, 1), (6, 1), (2, 4), and (5, 4).

Review Problem Set, Chapter 5

In Problems 1 to 6, write out each sum explicitly and then find its numerical value.

1 $\displaystyle\sum_{k=1}^{5} (5k + 3)$

2 $\displaystyle\sum_{i=1}^{3} 5(i + 1)^2$

3 $\displaystyle\sum_{k=0}^{4} \frac{k}{k + 1}$

4 $\displaystyle\sum_{i=0}^{3} \frac{1}{i^2 + 1}$

5 $\displaystyle\sum_{k=1}^{4} \cos \frac{\pi}{k}$

6 $\displaystyle\sum_{k=0}^{6} \tan \frac{k\pi}{3}$

In Problems 7 to 10, evaluate the sum by using the basic properties of summation.

7 $\displaystyle\sum_{k=1}^{n} k(2k - 1)$

8 $\displaystyle\sum_{j=1}^{n} (6^{j+1} - 6^j)$

9 $\displaystyle\sum_{j=0}^{n} (3^j + 3^{j+1})$

10 $\displaystyle\sum_{k=0}^{n} (k + 1)^3$

11 Find the sum of the first 1000 even integers:

$$2 + 4 + 6 + 8 + \cdots + 2000$$

12 Use the principle of mathematical induction to show that $\sum_{k=1}^{n} (2k - 1) = n^2$.

13 For doing a certain job, you are offered $0.01 the first day, $0.03 the second day, $0.09 the third day, $0.27 the fourth day, and so on, so that your daily earnings are tripled each day. (a) Using the Σ notation, write an expression for your total earnings after 14 days of work. (b) Use the basic properties of summation to evaluate the sum in part (a), and write the answer in dollars.

14 Evaluate $\displaystyle\sum_{k=1}^{n} \frac{1}{k^2 + k}$. $\left(\textit{Hint: } \dfrac{1}{k^2 + k} = \dfrac{1}{k} - \dfrac{1}{k + 1}.\right)$

15 Evaluate the Riemann sum $\sum_{k=1}^{n} f(c_k) \Delta x_k$ corresponding to the augmented partition $[0, \frac{1}{4}], [\frac{1}{4}, \frac{1}{2}], [\frac{1}{2}, \frac{3}{4}], [\frac{3}{4}, 1]$ with $c_1 = \frac{1}{8}$, $c_2 = \frac{3}{8}$, $c_3 = \frac{5}{8}$, and $c_4 = \frac{7}{8}$ if f is given by $f(x) = x^2$. Interpret this Riemann sum as an approximation to a certain area.

16 (a) Suppose that f is an integrable function on the interval [0, 1]. Use the definition of the definite integral to express

$$\lim_{n \to +\infty} \frac{1}{n} \sum_{k=1}^{n} f\left(\frac{k}{n}\right)$$

as a certain integral.

(b) Write $\displaystyle\lim_{n \to +\infty} \sum_{k=1}^{n} \frac{k^5}{n^6}$ as a definite integral.

In Problems 17 to 20, find the Riemann sum corresponding to each augmented partition for the indicated function, sketch the graph of the function, and show the Riemann sum as a sum of areas of rectangles.

17 $f(x) = x^3 - x$, $\mathcal{P}_4^*$ is the partition of [1, 3] into four equal subintervals with c_1, c_2, c_3, and c_4 chosen so that all approximating rectangles are inscribed

18 $f(x) = x^2 + 1$, $\mathcal{P}_4^*$ is the partition of [1, 4] into four equal subintervals with c_1, c_2, c_3, and c_4 chosen at the midpoints of these subintervals

19 Same as Problem 17, except that all approximating rectangles are to be circumscribed.

20 $f(x) = 16 - x^2$, $\mathcal{P}_4^*$ is the partition of [0, 4] into four equal subintervals with c_1, c_2, c_3, and c_4 chosen so that all approximating rectangles are circumscribed

In Problems 21 to 24, evaluate each definite integral directly by calculating a limit of Riemann sums. Use partitions consisting of subintervals of equal lengths, and augment these partitions by using left or right endpoints of the subintervals as indicated.

21 $\displaystyle\int_{1}^{4} (x^2 + 1)\, dx$ (left endpoints)

22 $\displaystyle\int_{1}^{3} (x^3 + x)\, dx$ (left endpoints)

23 $\displaystyle\int_{1}^{4} (x^2 + 1)\, dx$ (right endpoints)

24 $\displaystyle\int_{-1}^{0} (x^2 - 2x + 3)\, dx$ (right endpoints)

In Problems 25 to 32, indicate whether each definite integral exists and give a reason for your answer.

25 $\int_1^{100} \dfrac{[[x]]}{x}\, dx$

26 $\int_1^2 \dfrac{dx}{[[x]]}$

27 $\int_1^2 \sqrt{1 - x^2}\, dx$

28 $\int_0^1 \dfrac{dx}{1 + x^2}$

29 $\int_0^{\pi/2} \dfrac{dx}{\cos x}$

30 $\int_{-1}^1 f(x)\, dx$, where $f(x) = \begin{cases} x^2 & \text{for } x < 0 \\ x^4 & \text{for } x \geq 0 \end{cases}$

31 $\int_0^{\pi} \cot x\, dx$

32 $\int_0^{2\pi} \csc \dfrac{x}{2}\, dx$

In Problems 33 and 34, use the indicated previous result to evaluate the definite integral.

33 $\int_4^1 (x^2 + 1)\, dx$ (see Problem 21)

34 $\int_3^1 (x^3 + x)\, dx$ (see Problem 22)

35 Evaluate $\int_{\pi/2}^{\pi/2} \cot x\, dx$.

36 Is $\int_{\pi/2}^{\pi/2} \tan x\, dx$ defined? If so, what is its value?

37 Is $f(x) = [[x]]$ continuous on $[0, 3]$? Does $\int_0^3 f(x)\, dx$ exist? Why or why not?

38 By using the definition of the definite integral, prove that if K is a constant, then $\int_a^b K\, dx = K(b - a)$.

In Problems 39 to 48, assume that $\int_1^3 f(x)\, dx = 6$, $\int_3^5 f(x)\, dx = 7$, $\int_3^5 g(x)\, dx = 8$, and $\int_0^{\pi/2} \cos^2 x\, dx = \pi/4$. Use the properties of definite integrals to evaluate each expression.

39 $\int_1^3 (-5)f(x)\, dx$

40 $\int_3^1 7f(x)\, dx$

41 $\int_3^5 [f(x) + 3g(x)]\, dx$

42 $\int_5^3 [4g(x) - 2f(x)]\, dx$

43 $\int_1^5 f(x)\, dx$

44 $\int_{\pi/2}^{\pi/2} \cos^2 x\, dx$

45 $\int_1^5 h(x)\, dx$, where $h(x) = \begin{cases} 4f(x) & \text{for } 1 \leq x < 3 \\ -2g(x) & \text{for } 3 \leq x \leq 5 \end{cases}$

46 $\int_1^3 F(x)\, dx$, where $F(x) = \begin{cases} 0 & \text{for } x = 1 \\ f(x) & \text{for } 1 < x < 3 \\ 52 & \text{for } x = 3 \end{cases}$

47 $\int_0^{\pi/2} (4 - 3\cos^2 x)\, dx$

48 $\int_5^1 H(x)\, dx$, where $H(x) = \begin{cases} 1 + f(x) & \text{for } 1 \leq x \leq 3 \\ g(x) - 1 & \text{for } 3 < x \leq 5 \end{cases}$

49 True or false?

(a) $\int_1^{10} f(x)\, dx - \int_6^{10} f(x)\, dx \stackrel{?}{=} \int_1^6 f(x)\, dx$

(b) $\int_{-2}^3 g(x)\, dx - \int_4^3 g(x)\, dx \stackrel{?}{=} \int_{-2}^4 g(x)\, dx$

50 Suppose that f is an integrable function on $[a, b]$ and that $|f(x)| \leq K$ holds for all values of x in $[a, b]$, where K is a constant. Prove that $|\int_a^b f(x)\, dx| \leq K|b - a|$.

In Problems 51 to 55, without evaluating the integrals involved, show that the inequality holds.

51 $\int_1^3 x\, dx \leq \int_1^3 x^3\, dx$

52 $\int_0^1 \dfrac{dx}{1 + x^2} \geq \int_0^1 \dfrac{dx}{1 + x}$

53 $\int_0^1 x^5\, dx \leq \int_0^1 x\, dx$

54 $\int_{\pi/6}^{\pi/3} \cos x\, dx \leq \int_{\pi/6}^{\pi/3} \dfrac{\sin x}{x}\, dx$

55 $\int_0^{\pi/4} \sin x\, dx \leq \int_0^{\pi/4} \cos x\, dx$

56 Assume that f and g are continuous functions on $[a, b]$ and that K is a constant. Show that

$$2K \int_a^b f(x)g(x)\, dx \leq \int_a^b [f(x)]^2\, dx + K^2 \int_a^b [g(x)]^2\, dx$$

(*Hint:* $0 \leq [f(x) - Kg(x)]^2$.)

57 Verify the inequality in Problem 51 by using the fundamental theorem of calculus to evaluate the integrals.

58 Use Problem 56 to derive the **Cauchy-Bunyakovski-Schwarz inequality:** If f and g are continuous functions on $[a, b]$, then

$$\left[\int_a^b f(x)g(x)\, dx \right]^2 \leq \int_a^b [f(x)]^2\, dx \int_a^b [g(x)]^2\, dx$$

[*Hint:* First take care of the case in which $\int_a^b [g(x)]^2\, dx = 0$. Then, assuming that $\int_a^b [g(x)]^2\, dx \neq 0$, let K in Problem 56 be the ratio of $\int_a^b f(x)g(x)\, dx$ to $\int_a^b [g(x)]^2\, dx$.]

In Problems 59 to 64, find the mean value of the function f on the indicated interval, and find a value of c in the interval for which $f(c)$ is equal to this mean value.

59 $f(x) = 3x + 1$ on $[0, 4]$

60 $f(x) = x^2 + 2$ on $[0, 3]$

61 $f(x) = 4 - x^2$ on $[0, 2]$

62 $f(x) = 4 + 3x^2$ on $[-1, 1]$

63 $f(x) = |x|$ on $[-1, 3]$

64 $f(x) = x|x|$ on $[-\frac{1}{2}, 1]$

In Problems 65 to 88, use the fundamental theorem of calculus and the properties of definite integrals to evaluate the integral.

65 $\int_0^3 (4x + 3) \, dx$

66 $\int_{-4}^0 3y^2 \, dy$

67 $\int_0^4 6\sqrt{x} \, dx$

68 $\int_8^{27} 9\sqrt[3]{t} \, dt$

69 $\int_0^1 (2t + 3t^2) \, dt$

70 $\int_0^3 (3u - 1)(u^2 + 1) \, du$

71 $\int_{-1}^1 (z^2 + 2)^2 \, dz$

72 $\int_0^1 5(x - \sqrt{x})^2 \, dx$

73 $\int_{-1}^3 (x + |x|) \, dx$

74 $\int_{-1}^3 |x + 1| \, dx$

75 $\int_{-1}^1 \sqrt{t + 1} \, dt$

76 $\int_{-1}^4 x^2|x| \, dx$

77 $\int_{-1}^3 \frac{2x \, dx}{(1 + x^2)^2}$

78 $\int_0^1 x^3\sqrt{x^4 + 1} \, dx$

79 $\int_0^4 (|x - 1| + |x - 2|) \, dx$

80 $\int_0^2 \frac{x \, dx}{\sqrt{3x + 10}}$

81 $\int_0^{\pi/8} \sin 4x \, dx$

82 $\int_0^{\pi/4} \tan u \sec^2 u \, du$

83 $\int_{\pi/6}^\pi \sin y \cos y \, dy$

84 $\int_{\pi/12}^{\pi/6} \frac{\sec^2 2\theta \, d\theta}{\tan^3 2\theta}$

85 $\int_0^{2\pi} |\sin x| \, dx$

86 $\int_0^{2\pi} \cos |x| \, dx$

87 $\int_0^3 f(x) \, dx$, where $f(x) = \begin{cases} 1 - x & \text{for } 0 \le x < 1 \\ x^2 - 1 & \text{for } 1 \le x < 2 \\ x + 1 & \text{for } 2 \le x \le 3 \end{cases}$

88 $\int_{-1}^2 g(x) \, dx$, where $g(x) = \begin{cases} -\sqrt{|x|} & \text{for } -1 \le x < 0 \\ \sqrt{x + 1} & \text{for } 0 \le x < 1 \\ x\sqrt{1 + x^2} & \text{for } 1 \le x < 2 \end{cases}$

In Problems 89 to 99, use the alternative version of the fundamental theorem of calculus to find the derivative.

89 $D_x \int_3^x (4t + 1)^{300} \, dt$

90 $\frac{d}{dx} \int_2^x (3w^2 - 7)^{15} \, dw$

91 $g'(x)$, where $g(x) = \int_1^x (8t^{17} + 5t^2 - 13)^{40} \, dt$

92 $h''(t)$, where $h(t) = \int_0^t \sqrt{1 + x^{16}} \, dx$

93 $D_x \int_x^{1000} \frac{t^2 \, dt}{\sqrt{t^4 + 8}}$

94 $\frac{d}{dx} \int_x^0 |w| \, dw$

95 $g''(t)$, where $g(t) = \int_t^0 \sqrt{1 + x^2} \, dx + \int_0^t \sqrt{1 + w^2} \, dw$

96 $h'(t)$, where $h(t) = \int_t^0 \frac{dx}{1 + x^2} + \int_1^t \frac{dx}{1 + x^2}$

97 $D_x \int_1^{x^2} \frac{t^2 \, dt}{1 + t^2}$

98 $\frac{d}{dx} \int_{3x+1}^{x^2} \frac{t + \sqrt{t}}{t^3 + 5} \, dt$

99 $D_t \int_{4t+3}^{5t^2+t} \cos(w^5 + 1) \, dw$

100 Assuming that all required derivatives and integrals exist:

(a) Find $\dfrac{d}{dx} \left[\int_a^x f[g(t)]g'(t) \, dt - \int_{g(a)}^{g(x)} f(u) \, du \right]$.

(b) Conclude that $\displaystyle\int_a^b f[g(t)]g'(t) \, dt = \int_{g(a)}^{g(b)} f(u) \, du$.

101 Find the maximum value of each function g on the interval indicated.

(a) $g(x) = \displaystyle\int_0^x (t^2 - 4t + 4) \, dt$ on $[0, 4]$

(b) $g(x) = \displaystyle\int_0^x (\sqrt{t} - t) \, dt$ on $[0, 1]$

102 Suppose that the function f is piecewise continuous and bounded on the interval $[a, b]$, and define g by $g(x) = \int_a^x f(t) \, dt$. Prove that g is continuous on $[a, b]$.

In Problems 103 to 110, indicate whether each statement is true or false. You may assume that all required derivatives and integrals exist, that functions appearing in denominators are nonzero, and so forth. If the statement is false, give a specific example to show it is false. (Such an example is called a **counterexample.**)

103 $\displaystyle\int_a^b f(x)g(x) \, dx \overset{?}{=} \int_a^b f(x) \, dx \int_a^b g(x) \, dx$

104 $\displaystyle\int_a^b f(x)g(x) \, dx \overset{?}{=} f(x) \int_a^b g(x) \, dx + g(x) \int_a^b f(x) \, dx$

105 $\displaystyle\int_a^b \frac{f(x)}{g(x)} \, dx \overset{?}{=} \frac{\displaystyle\int_a^b f(x) \, dx}{\displaystyle\int_a^b g(x) \, dx}$

106 $\int_a^b f(x)g(x)\ dx \overset{?}{=} [f(x)h(x)]\Big|_a^b - \int_a^b h(x)f'(x)\ dx$, where $h'(x) = g(x)$

107 $\int_{-a}^a f(x)\ dx \overset{?}{=} 0$, if f is an odd function

108 $\int_a^x f(kt)\ dt \overset{?}{=} \frac{1}{k}\int_{ka}^{kx} f(t)\ dt$, where k is a constant

109 $\left|\int_a^b g(x)\ dx\right| \overset{?}{=} \int_a^b |g(x)|\ dx$

110 If f is a decreasing function, then the function g defined by $g(x) = \int_0^x f(t)\ dt$ is also decreasing.

© In Problems 111 to 116, use the trapezoidal rule, $T_n \approx \int_a^b f(x)\ dx$, with the indicated value of n to estimate the integral.

111 $\int_0^2 x\sqrt{16 - x^3}\ dx; n = 4$

112 $\int_1^2 \sqrt{4 + x^3}\ dx; n = 4$

113 $\int_0^{10} \sqrt[3]{125 + x^3}\ dx; n = 5$

114 $\int_4^8 \sqrt{64 - x^2}\ dx; n = 8$

115 $\int_0^{\pi/3} \sec x\ dx; n = 6$

116 $\int_0^{\pi/2} \sin^3 x\ dx; n = 7$

© In Problems 117 to 122, use Simpson's parabolic rule, $S_{2n} \approx \int_a^b f(x)\ dx$, with the indicated value of n to estimate the integral.

117 $\int_0^8 \frac{3x}{1 + x^3}\ dx; n = 4$

118 $\int_0^4 \sqrt{16 - x^2}\ dx; n = 2$

119 $\int_2^8 \frac{x\ dx}{\sqrt[3]{3 + x^3}}; n = 3$

120 $\int_0^5 \frac{x^3\ dx}{\sqrt{1 + x^3}}; n = 3$

121 $\int_0^{\pi/2} \frac{dx}{2 + \sin x}; n = 3$

122 $\int_0^1 \cos \frac{\pi\sqrt{x}}{2}\ dx; n = 4$

In Problems 123 to 134, (a) sketch the graphs of the equations, (b) find the points of intersection of these graphs, and (c) find the area of the region bounded by these graphs.

123 $y = \frac{1}{4}x^3$ and $y = x, x \geq 0$

124 $y = x^3$, the y axis, and $y = -27$

125 $y = 9 - x^2$ and $y = x^2$

126 $2y^2 + 9x = 36$ and $3x + 2y = 0$

127 $y = 2x^2$ and $y = x^2 + 2x + 3$

128 $y^2 = -4(x - 1)$ and $y^2 = -2(x - 2)$

129 $2x + 3y + 1 = 0$ and $x + 3 = (y - 1)^2$

130 $x^2y = x^2 - 1, y = 1, x = 1$, and $x = 4$

131 $y = 4 - x^2$ and $y = 4 - 4x$

132 $x = \frac{1}{4}y^2 - 1$ and $y = 4x - 16$

133 $y^3 = 9x, y^2 = -3(x - 6)$, and $y = -3$

134 $y^2 = -16(x - 1)$ and $y^2 = \frac{16}{3}(x + 3)$

135 Find the area of the region bounded by the curve $y = -\frac{2}{27}x^3$ and the tangent to this curve at $(3, -2)$.

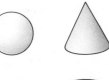

6 APPLICATIONS OF THE DEFINITE INTEGRAL

In Chapter 5 we used definite integrals to find the area of plane regions. Here we apply definite integrals to the problem of finding the volume of three-dimensional regions. In addition, we consider applications of definite integrals to the calculation of arc length, surface area, consumer's surplus, blood flow, work, and energy.

6.1 Volumes of Solids of Revolution

Figure 1

The techniques used in Section 5.6 can be modified to express the volume of a three-dimensional region S as an integral. We assume that any three-dimensional region S which has a "reasonable shape" has a definite volume $V(S)$ cubic units associated with it. We refer to such a region as an **admissible three-dimensional region,** or simply as a **solid.** In particular, any three-dimensional region having the following two properties is a solid:

1 The boundary of S consists of a finite number of smooth surfaces that can intersect in a finite number of edges. These edges, in turn, can intersect in a finite number of vertices.

2 S is **bounded** in the sense that there is an upper bound to the distances between the points of S.

For instance, a solid spherical ball, a solid right circular cone, a solid cube, or the solid region between two coaxial right circular cylinders would satisfy the above conditions (Figure 1).

A solid consisting of all points lying between an admissible plane region B_1 and a second admissible plane region B_2 obtained by parallel translation of B_1 is called a **solid cylinder with bases B_1 and B_2** (Figure 2). All the line segments joining points on the base B_1 to corresponding points on the base B_2 are parallel to one another. If all these line segments are perpendicular to the bases, then the solid cylinder is

Figure 2

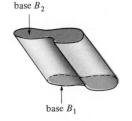

base B_2

base B_1

357

Figure 3

solid right cylinder

Figure 4

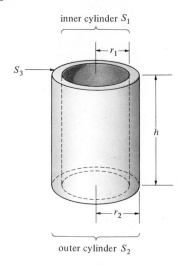

inner cylinder S_1

S_3

r_1

h

r_2

outer cylinder S_2

called a **solid right cylinder** (Figure 3). The distance, measured perpendicularly, between the two bases of a solid cylinder is called its **height.** Henceforth, we assume that *the volume of a solid right cylinder is its height times the area of one of its bases.*

If one solid S_1 is contained in a similar but slightly larger solid S_2, then the three-dimensional region S_3 consisting of all points in S_2 that are not in S_1 is sometimes called a **shell** (Figure 4). Notice that

$$V(S_3) = V(S_2) - V(S_1)$$

that is, *the volume of the shell is the difference between the volumes of the larger and the smaller solids.* For instance, in Figure 4, the volume of the right cylindrical shell is given by

$$V(S_3) = V(S_2) - V(S_1) = \pi r_2^2 h - \pi r_1^2 h$$

The Method of Circular Disks

In this section and the next we develop methods for finding the volumes of solids called *solids of revolution*. Solids of revolution are formed as follows: Let R be an admissible plane region, and let l be a line lying in the same plane as R but touching R, if at all, only at boundary points of R (Figure 5a). The solid S "swept out," or generated, when R is revolved about the line l as an axis is called a **solid of revolution** (Figure 5b).

Consider the special case in which R is the region under the graph of a continuous nonnegative function f between $x = a$ and $x = b$ (Figure 6a). Denote by S the solid of revolution generated by revolving R about the x axis (Figure 6b). Figure 6c shows an "infinitesimal" portion dV of the volume V of S consisting of a circular disk of infinitesimal thickness dx, perpendicular to the axis of revolution and intersecting it at the point with coordinate x. Evidently, the radius r of this disk is given by $r = f(x)$.

Figure 5

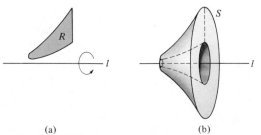

(a)   (b)

The disk in Figure 6c can be thought of as a solid right cylinder whose base is a circle of radius r and whose height is dx. The area of its base is πr^2; hence, its volume dV is given by

$$dV = \pi r^2\, dx = \pi [f(x)]^2\, dx$$

The total volume V of the solid S should be obtained by "summing"—that is to say, integrating—all the infinitesimal volumes dV of such disks as x runs from a to b.

Figure 6

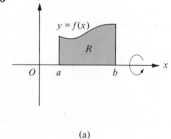

(a)

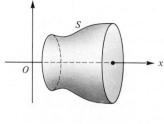

(b)

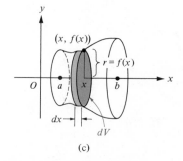

(c)

The shape of a solid of revolution is found in these elevators

Therefore, we should have

$$V = \int_{x=a}^{x=b} dV = \int_{a}^{b} \pi[f(x)]^2 \, dx = \pi \int_{a}^{b} [f(x)]^2 \, dx$$

The calculation of volumes of solids of revolution about the x axis by the formula

$$V = \pi \int_{a}^{b} [f(x)]^2 \, dx$$

is called the **method of circular disks.** Although our derivation of the formula using Leibnizian infinitesimals may not be mathematically rigorous, the formula is correct. In Section 6.3, we outline a more conclusive argument.

In Examples 1 and 2, use the method of circular disks to find the volume V of the solid S generated by revolving the region R under the graph of the given function f on the indicated interval [a, b] about the x axis. Sketch the graph of f and the solid S.

Figure 7

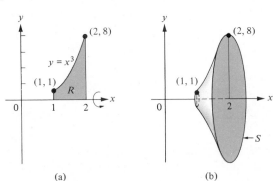

(a)　　　　(b)

Figure 8

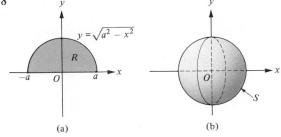

(a)　　　　(b)

EXAMPLE 1　$f(x) = x^3$ on $[1, 2]$

SOLUTION　　Here,

$$V = \pi \int_{1}^{2} [f(x)]^2 \, dx$$

$$= \pi \int_{1}^{2} (x^3)^2 \, dx$$

$$= \pi \int_{1}^{2} x^6 \, dx = \pi \left(\frac{x^7}{7} \right) \Big|_{1}^{2}$$

$$= \pi \left(\frac{128}{7} - \frac{1}{7} \right) = \frac{127}{7} \pi \text{ cubic units (Figure 7)} \quad \blacksquare$$

EXAMPLE 2　$f(x) = \sqrt{a^2 - x^2}$ on $[-a, a]$

SOLUTION　　Here,

$$V = \pi \int_{-a}^{a} [f(x)]^2 \, dx = \pi \int_{-a}^{a} (\sqrt{a^2 - x^2})^2 \, dx$$

$$= \pi \int_{-a}^{a} (a^2 - x^2) \, dx = \pi \left(a^2 x - \frac{x^3}{3} \right) \Big|_{-a}^{a}$$

$$= \pi \left[\left(a^3 - \frac{a^3}{3} \right) - \left(-a^3 + \frac{a^3}{3} \right) \right] = \pi \frac{4a^3}{3} = \frac{4}{3} \pi a^3 \quad \blacksquare$$

The graph of $f(x) = \sqrt{a^2 - x^2}$ on $[-a, a]$ is a semicircle (Figure 8a), and the corresponding solid of revolution is a sphere of radius a (Figure 8b). The formula $V = \frac{4}{3} \pi a^3$, obtained here by the method of circular disks, is the familiar formula for the volume of a sphere of radius a.

Of course, a plane region can be revolved about the y axis rather than the x axis, and again a solid of revolution is generated. For instance, suppose that R is a plane region bounded by the y axis, the horizontal lines $y = a$ and $y = b$, where $a < b$, and the graph of $x = g(y)$, where the function g is continuous and $g(y) \geq 0$ for

Figure 9

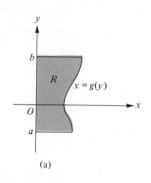

(a)

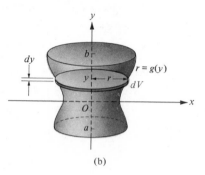

(b)

$a \leq y \leq b$ (Figure 9a). Figure 9b shows the solid of revolution S generated by revolving R about the y axis. In Figure 9b,

$$dV = \pi r^2 \, dy = \pi [g(y)]^2 \, dy$$

hence, $$V = \int_{y=a}^{y=b} dV = \int_a^b \pi [g(y)]^2 \, dy = \pi \int_a^b [g(y)]^2 \, dy$$

Use of the formula

$$V = \pi \int_a^b [g(y)]^2 \, dy$$

to calculate the volume of a solid of revolution about the y axis is also called the **method of circular disks.**

EXAMPLE 3   Find the volume of the solid S generated by revolving the region R bounded by the y axis, the line $y = 4$, and the graph of $y = x^2$ with $x \geq 0$ about the y axis. Use the method of circular disks, and sketch both R and S.

SOLUTION   Solving the equation $y = x^2$ for x in terms of y and using the fact that $x \geq 0$, we have $x = \sqrt{y}$ (Figure 10). By the method of circular disks,

$$V = \pi \int_0^4 (\sqrt{y})^2 \, dy = \pi \int_0^4 y \, dy = \pi \left(\frac{y^2}{2} \right) \Big|_0^4 = 8\pi \text{ cubic units} \qquad \blacksquare$$

Figure 10

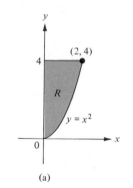

(a)

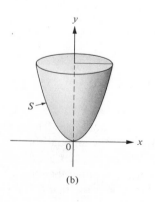

(b)

The Method of Circular Rings

Volumes of solids of revolution more general than those considered above can be found by using the *method of circular rings*. This method works as follows: Suppose that f and g are nonnegative continuous functions on the interval $[a, b]$ such that $f(x) \geq g(x)$ holds for all values of x in $[a, b]$, and let R be the plane region bounded by the graphs of f and g between $x = a$ and $x = b$ (Figure 11a). Let S be the solid generated by revolving R about the x axis (Figure 11b). Here we consider an infinitesimal portion dV of the volume V of S consisting of a circular ring, or washer, of infinitesimal thickness dx, perpendicular to the axis of revolution and centered at the point with coordinate x. The base of this circular ring is the region

Figure 11

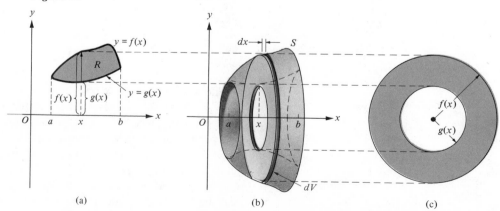

(a)   (b)   (c)

between two concentric circles of radii $f(x)$ and $g(x)$ (Figure 11c); hence, the area of this base is $\pi[f(x)]^2 - \pi[g(x)]^2$ square units. It follows that

$$dV = \{\pi[f(x)]^2 - \pi[g(x)]^2\}\, dx$$

Therefore,

$$V = \int_{x=a}^{x=b} dV = \int_a^b \{\pi[f(x)]^2 - \pi[g(x)]^2\}\, dx = \pi \int_a^b \{[f(x)]^2 - [g(x)]^2\}\, dx$$

Thus, we have the formula

$$V = \pi \int_a^b \{[f(x)]^2 - [g(x)]^2\}\, dx$$

for finding the volume of a solid of revolution about the x axis (Figure 11) by the **method of circular rings.**

EXAMPLE 4   Using the method of circular rings, find the volume V of the solid S generated by revolving the region R about the x axis, where R is bounded by the curves $y = x^2$ and $y = x + 2$.

SOLUTION   The points of intersection of the two curves are $(2, 4)$ and $(-1, 1)$ (Figure 12a). By the method of circular rings (Figure 12b), we have

$$V = \pi \int_{-1}^2 [(x + 2)^2 - (x^2)^2]\, dx = \pi \int_{-1}^2 (x^2 + 4x + 4 - x^4)\, dx$$

$$= \pi \left(\frac{x^3}{3} + 2x^2 + 4x - \frac{x^5}{5} \right) \Big|_{-1}^2 = \pi \left[\frac{184}{15} - \left(-\frac{32}{15} \right) \right]$$

$$= \frac{72\pi}{5} \text{ cubic units} \qquad \blacksquare$$

Naturally, the method of circular rings works for solids S generated by revolving plane regions R about the y axis rather than the x axis. Thus, in Figure 13a, the plane region R is bounded on the right by the graph of $x = F(y)$, on the left by the graph of $x = G(y)$, above by $y = b$, and below by $y = a$, where $a < b$. If R is revolved about the y axis, then a solid of revolution S is generated (Figure 13b). The circular

Figure 12

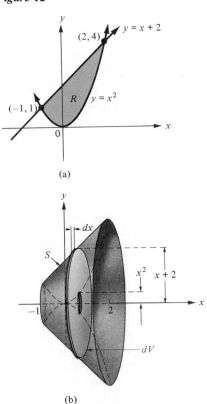

(a)

(b)

Figure 13

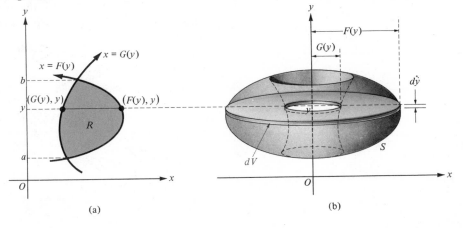

(a)

(b)

ring perpendicular to the axis of revolution and centered at the point $(0, y)$ has infinitesimal volume dV given by

$$dV = \pi\{[F(y)]^2 - [G(y)]^2\}\, dy$$

hence,

$$V = \int_{y=a}^{y=b} dV = \int_a^b \pi\{[F(y)]^2 - [G(y)]^2\}\, dy$$

Therefore, to find the volume of a solid of revolution about the y axis (Figure 13) by the **method of circular rings,** we use the formula

$$V = \pi \int_a^b \{[F(y)]^2 - [G(y)]^2\}\, dy$$

Figure 14

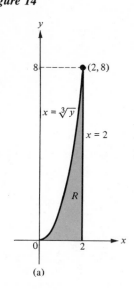

(a)

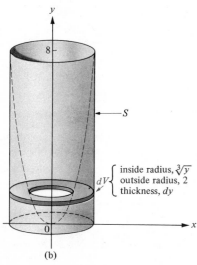

$dV \begin{cases} \text{inside radius, } \sqrt[3]{y} \\ \text{outside radius, } 2 \\ \text{thickness, } dy \end{cases}$

(b)

EXAMPLE 5   Use the method of circular rings to find the volume V of the solid of revolution S generated by revolving the region R about the y axis, where R is the plane region bounded on the right by the graph of $x = 2$, on the left by the graph of $y = x^3$, and below by the x axis. Sketch R and S.

SOLUTION   The region R and the solid S are shown in Figure 14a and b, respectively. Let $F(y) = 2$ and $G(y) = \sqrt[3]{y}$. By the method of circular rings,

$$V = \pi \int_0^8 \{[F(y)]^2 - [G(y)]^2\}\, dy$$

so

$$V = \pi \int_0^8 [4 - (\sqrt[3]{y})^2]\, dy = \pi\left(4y - \frac{3}{5} y^{5/3}\right)\Bigg|_0^8$$

$$= \pi\left(32 - \frac{96}{5}\right) = \frac{64\pi}{5} \text{ cubic units}$$

The method of circular rings is also effective for solids generated by revolving plane regions about axes *other than the x or the y axis.* This is illustrated by the following examples.

EXAMPLE 6   Use the method of circular rings to find the volume of the solid S obtained by revolving the region R about the line $x = 6$, where R is bounded by the graphs of $y^2 = 4x$ and $x = 4$.

SOLUTION   From Figure 15, the inside radius of the circular ring at level y is 2 units, and its outside radius is $6 - (y^2/4)$ units; hence,

$$dV = \pi\left[\left(6 - \frac{y^2}{4}\right)^2 - 2^2\right] dy$$

Integrating, we obtain

$$V = \int_{y=-4}^{y=4} dV = \int_{-4}^4 \pi\left[\left(6 - \frac{y^2}{4}\right)^2 - 2^2\right] dy$$

$$= \pi \int_{-4}^4 \left(\frac{y^4}{16} - 3y^2 + 32\right) dy = \pi\left(\frac{y^5}{80} - y^3 + 32y\right)\Bigg|_{-4}^4$$

$$= \pi\left[\frac{384}{5} - \left(-\frac{384}{5}\right)\right] = \frac{768\pi}{5} \text{ cubic units}$$

Figure 15

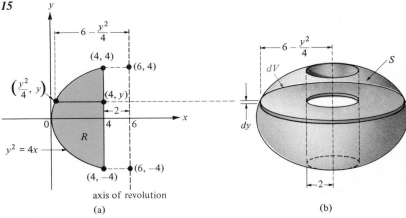

(a)

(b)

Figure 16

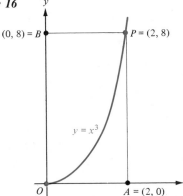

EXAMPLE 7   In Figure 16, the curve OP has the equation $y = x^3$. Find the volume of the solid of revolution generated by revolving the region

(a) OBP about the line $y = 8$

(b) OAP about the line $x = 2$

(c) OAP about the line $y = 8$

SOLUTION

(a) When the region OBP is revolved about the axis $y = 8$, the infinitesimal rectangle of height $8 - x^3$ and width dx shown in Figure 17 sweeps out a circular *disk* of thickness dx and radius $8 - x^3$. Its volume is given by

$$dV = \pi(8 - x^3)^2\, dx = \pi(64 - 16x^3 + x^6)\, dx$$

Therefore,

$$V = \int_{x=0}^{x=2} dV = \int_0^2 \pi(64 - 16x^3 + x^6)\, dx$$
$$= \pi\left(64x - 4x^4 + \frac{x^7}{7}\right)\Bigg|_0^2$$
$$= \frac{576\pi}{7} \text{ cubic units}$$

Figure 17

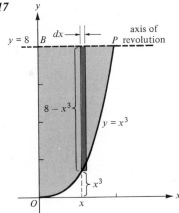

(b) When the region OAP is revolved about the axis $x = 2$, the infinitesimal rectangle of height dy and length $2 - \sqrt[3]{y}$ shown in Figure 18 sweeps out a circular *disk* of thickness dy and radius $2 - \sqrt[3]{y}$. Its volume is given by

$$dV = \pi(2 - \sqrt[3]{y})^2\, dy = \pi(4 - 4y^{1/3} + y^{2/3})\, dy$$

Therefore,

$$V = \int_{y=0}^{y=8} dV = \int_0^8 \pi(4 - 4y^{1/3} + y^{2/3})\, dy$$
$$= \pi\left(4y - 3y^{4/3} + \frac{3}{5}y^{5/3}\right)\Bigg|_0^8$$
$$= \frac{16\pi}{5} \text{ cubic units}$$

Figure 18

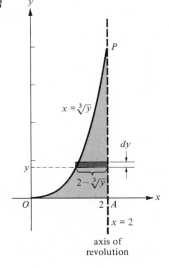

Figure 19

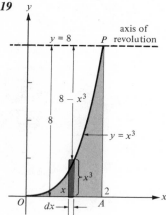

(c) When the region *OAP* is revolved about the axis $y = 8$, the infinitesimal rectangle of height x^3 and width dx shown in Figure 19 sweeps out a circular *ring* of thickness dx with inside radius $8 - x^3$, outside radius 8, and infinitesimal volume

$$dV = \pi[8^2 - (8 - x^3)^2]\, dx = \pi(16x^3 - x^6)\, dx$$

Therefore,

$$V = \int_{x=0}^{x=2} dV = \int_0^2 \pi(16x^3 - x^6)\, dx$$

$$= \pi\left(4x^4 - \frac{x^7}{7}\right)\Bigg|_0^2 = \frac{320\pi}{7} \text{ cubic units}$$

■

Problem Set 6.1

In Problems 1 to 8, find the volume of the solid generated by revolving the region under the graph of each function over the indicated interval about the *x* axis.

1 $f(x) = 3x^2$; $[-1, 3]$ **2** $g(x) = 3\sqrt{x}$; $[1, 4]$

3 $h(x) = \sqrt{9 - x^2}$; $[-1, 3]$ **4** $G(x) = |x|$; $[-2, 1]$

5 $F(x) = \sqrt{2 + x^2}$; $[0, 2]$ **6** $f(x) = |x| - x$; $[-3, 2]$

7 $g(x) = \sec x$; $\left[0, \dfrac{\pi}{4}\right]$ **8** $f(x) = \tan x$; $\left[0, \dfrac{\pi}{3}\right]$

In Problems 9 to 16, find the volume of the solid generated by revolving the region bounded by the graphs of the given equations about the *y* axis.

9 $y = x^3$, $y = 8$, and $x = 0$ **10** $y^2 = x$, $y = 4$, and $x = 0$

11 $y^2 = 4x$, $y = 4$, and $x = 0$

12 $y = x^2 + 2$, $y = 4$, and $x = 0$ (first quadrant)

13 $y^2 = x^3$, $y = 8$, and $x = 0$ **14** $y = 2x^3$, $y = 2$, and $x = 0$

15 $x = \sqrt{\cos(\pi y/4)}$, $y = 0$, and $y = 1$

16 $x = \csc(\pi y/6)$, $y = 1$, and $y = 2$

In Problems 17 to 30, find the volume of the solid generated by rotating the region bounded by the given curves about the indicated axis. Use the method of circular disks or the method of circular rings.

17 $y = x^2$ and $y = 2x$ about the *x* axis

18 $y = x^3$ and $y^2 = x$ about the *x* axis

19 $y = x^2$ and $y = x$ about the *y* axis

20 $y = x^2 + 4$ and $y = 2x^2$ about the *y* axis

21 $y = x^3$, $x = 2$, and the *x* axis about the *y* axis

22 $y = 2x$, $y = x$, and $x + y = 6$ about the *x* axis

23 $y^2 = 4x + 16$, and the *y* axis about the *y* axis

24 $y = 3x$, $y = x$, and $x + y = 8$ about the *y* axis

25 $y = x^2$ and $y^2 = x$ about the line $x = -1$

26 $y = x^3$, $x = 0$, and $y = 8$ about the line $y = 8$

27 $y = 4x - x^2$ and $y = x$ about the line $x = 3$

28 $y = x^2 - x$ and $y = 3 - x^2$ about the line $y = 4$

29 $y = \cos x$, $y = \sin x$, $x = 0$, and $x = \pi/4$ about the *x* axis [*Hint:* $\cos^2 x = \frac{1}{2}(1 + \cos 2x)$ and $\sin^2 x = \frac{1}{2}(1 - \cos 2x)$.]

30 $x = \cos y + \sin y$, $x = \cos y - \sin y$, $y = 0$, and $y = \pi/2$ about the *y* axis

In Problems 31 to 38, find the volume of the solid generated when the given region in Figure 20 is rotated about the axis indicated. In Figure 20, the curve *OP* has the equation $y = 3x^2$ for $0 \le x \le 2$.

Figure 20

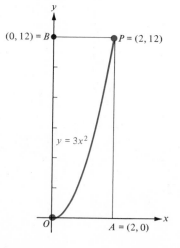

31 OAP about the x axis

32 OBP about the x axis

33 OBP about the y axis

34 OAP about the y axis

35 OAP about the line $\overline{AP}$

36 OBP about the line $\overline{AP}$

37 OBP about the line $\overline{BP}$

38 OAP about the line $\overline{BP}$

39 Find a general formula for the volume of the solid generated when the region OAP in Figure 20 is revolved about the axis $x = a$, for $a \geq 2$.

40 A **torus,** or **anchor ring,** is a doughnut-shaped solid (Figure 21) generated by revolving a circular region R about an axis in its plane that does not cut the region R. Find the volume V of such a torus if the radius of R is a and the distance from the center of R to the axis of revolution is b. (*Hint:* In working this problem by the method of circular rings, you will encounter the integral $\int_{-a}^{a} \sqrt{a^2 - x^2}\, dx$. Notice that this integral represents the area of a semicircle of radius a; hence its value is $\frac{1}{2}\pi a^2$.)

Figure 21

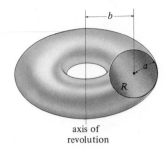

axis of
revolution

6.2 The Method of Cylindrical Shells

In this section we present an alternative method for finding the volume of a solid of revolution based on cylindrical shells rather than on circular disks or rings. We begin by considering the volume V of a solid right circular cylinder of fixed height h, but with variable radius x (Figure 1a). Here

$$V = \pi x^2 h$$

If the radius x is increased by a small amount $\Delta x = dx$, then the volume V is increased by a corresponding amount ΔV (Figure 1b). Evidently, ΔV is the volume of a thin cylindrical shell of height h with inner radius x and thickness $\Delta x = dx$. Using the differential to approximate ΔV, we have

$$\Delta V \approx dV = d(\pi x^2 h) = 2\pi x h\, dx$$

As Δx approaches zero, this approximation becomes better and better. Thus, *a cylindrical shell of height h, inside radius x, and infinitesimal thickness dx has an infinitesimal volume*

$$dV = 2\pi x h\, dx$$

You can remember the formula $dV = 2\pi x h\, dx$ by imagining that the cylindrical shell (Figure 1b) is cut vertically and "unwrapped" to form a rectangular slab of height h and thickness dx (Figure 1c). The length of this slab is approximately the inner circumference of the shell, $2\pi x$ units; hence, its volume is approximately

$$(2\pi x)(h)(dx) = 2\pi x h\, dx \qquad \text{cubic units}$$

Now, let R be a region in the xy plane bounded above by the curve $y = f(x)$, below by the curve $y = g(x)$, on the left by $x = a$, and on the right by $x = b$, where $0 < a < b$ and $f(x) \geq g(x)$ for $a \leq x \leq b$ (Figure 2a). Let S be the solid

Figure 1

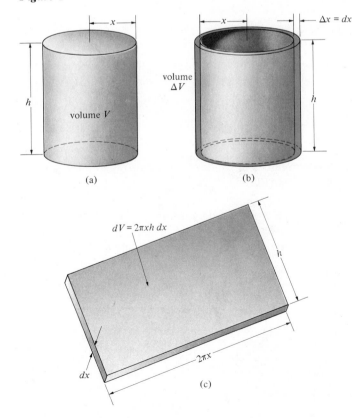

Figure 2

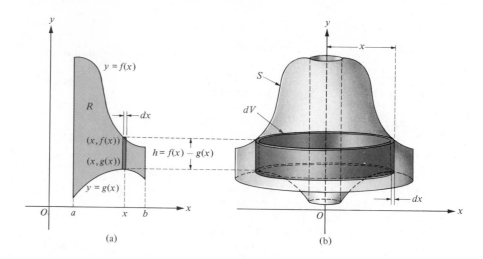

(a)   (b)

of revolution generated by revolving R about the y axis (Figure 2b). Consider the rectangle with infinitesimal width dx and height $h = f(x) - g(x)$ situated above the point $(x, 0)$, as in Figure 2a. As R revolves about the y axis to generate S, this rectangle sweeps out an infinitesimal portion dV of the volume of the solid S having the shape of a cylindrical shell of height $h = f(x) - g(x)$, inside radius x, and infinitesimal thickness dx. Hence, by the argument in the preceding paragraph,

$$dV = 2\pi xh \, dx = 2\pi x[f(x) - g(x)] \, dx$$

Figure 3

Integrating the differential equation above, and noting that x runs from a to b, we obtain

$$V = \int_{x=a}^{x=b} dV = \int_{a}^{b} 2\pi x[f(x) - g(x)] \, dx = 2\pi \int_{a}^{b} x[f(x) - g(x)] \, dx$$

Therefore, the volume V of the solid generated by revolving the region bounded by $y = f(x)$, $y = g(x)$, $x = a$, and $x = b$ (Figure 2a) about the y axis is given by

$$\boxed{V = 2\pi \int_{a}^{b} x[f(x) - g(x)] \, dx}$$

Use of this formula to calculate volumes of solids of revolution about the y axis is called the **method of cylindrical shells.**

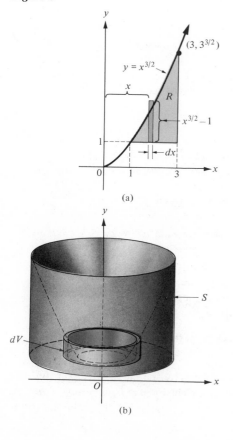

EXAMPLE 1   Let R be the plane region bounded by the graphs of $y = x^{3/2}$, $y = 1$, $x = 1$, and $x = 3$ (Figure 3a), and let S be the solid generated by revolving R about the y axis (Figure 3b). Use the method of cylindrical shells to find the volume V of S.

SOLUTION

$$V = 2\pi \int_{1}^{3} x(x^{3/2} - 1) \, dx = 2\pi \int_{1}^{3} (x^{5/2} - x) \, dx$$

$$= 2\pi \left(\frac{2}{7} x^{7/2} - \frac{x^2}{2} \right) \Big|_{1}^{3} = 2\pi \left\{ \left[\frac{2}{7}(3)^{7/2} - \frac{3^2}{2} \right] - \left[\frac{2}{7}(1)^{7/2} - \frac{1^2}{2} \right] \right\}$$

$$= 2\pi \left(\frac{54\sqrt{3} - 30}{7} \right) \text{ cubic units} \quad \blacksquare$$

Figure 4

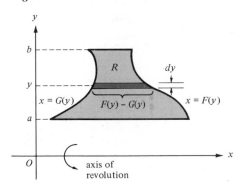

With obvious modifications, the method of cylindrical shells works for solids generated by revolving plane regions about the *x* axis—or, indeed, about any axis lying in the plane of the region. For instance, consider a region *R* in the *xy* plane bounded on the right by the graph of $x = F(y)$, on the left by the graph of $x = G(y)$, below by the horizontal line $y = a$, and above by the horizontal line $y = b$ (Figure 4). Then the volume *V* of the solid generated by revolving *R* about the *x* axis is given by

$$V = 2\pi \int_a^b y[F(y) - G(y)]\, dy$$

EXAMPLE 2 Find the volume of the solid generated by revolving the region *R* bounded by the curves $y = \frac{1}{2}x$ and $y = \sqrt{x}$ about the *x* axis.

SOLUTION   On the right boundary of *R* we have $y = \frac{1}{2}x$, so that $x = F(y) = 2y$; on the left boundary of *R*, we have $y = \sqrt{x}$, so that $x = G(y) = y^2$ (Figure 5a). The two boundary curves intersect at (0, 0) and (4, 2). Thus, by the method of cylindrical shells,

$$V = 2\pi \int_0^2 y(2y - y^2)\, dy = 2\pi \int_0^2 (2y^2 - y^3)\, dy = 2\pi \left(\frac{2y^3}{3} - \frac{y^4}{4} \right)\Bigg|_0^2$$

$$= 2\pi \left(\frac{16}{3} - \frac{16}{4} \right) = \frac{8\pi}{3} \text{ cubic units}$$

Figure 5

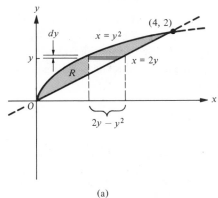

(a)

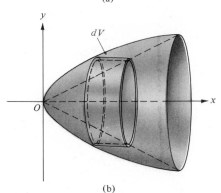

(b)

Comparison of the Ring and Shell Methods

In Figure 6, suppose that the plane region *R* is revolved about the indicated axis of revolution to generate a solid of revolution *S*. The volume *V* of *S* can be calculated either by the method of circular rings or by the method of cylindrical shells; the result will be the same. Of course, the computations involved may be easier if one method is used rather than the other.

Figure 6 contrasts the method of circular rings (Figure 6a) with the method of cylindrical shells (Figure 6b). Notice that the circular rings are generated by infinitesimal rectangles *perpendicular* to the axis of revolution, whereas the cylindrical shells are generated by infinitesimal rectangles *parallel* to the axis of revolution. For the method of circular rings (Figure 6a),

$$V = \pi \int_a^b (r_2^2 - r_1^2)\, ds$$

Figure 6

axis of revolution

method of circular rings:
$$V = \pi \int (r_2^2 - r_1^2)\, ds$$

(a)

axis of revolution

method of cylindrical shells:
$$V = 2\pi \int rh\, ds$$

(b)

where r_1 is the inner radius, r_2 is the outer radius, and ds is the thickness of the circular ring. For the method of cylindrical shells (Figure 6b),

$$V = 2\pi \int_a^b rh \, ds$$

where r is the inner radius, h is the height, and ds is the thickness of the cylindrical shell. In either case, the limits of integration are chosen so that the rectangles in Figure 6a or 6b sweep out the region R as the variable of integration goes from a to b.

It is neither necessary nor desirable to memorize the formulas for volumes by the ring or shell methods. Indeed, after some practice sketching or visualizing circular rings and cylindrical shells, you can quickly set up the appropriate integrals.

Figure 7

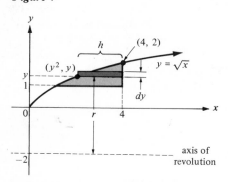

EXAMPLE 3   Let R be the plane region bounded by the graphs of $y = \sqrt{x}$, $y = 1$, and $x = 4$ (Figure 7). Find the volume V of the solid S generated by revolving R about the line $y = -2$.

SOLUTION   Let's try the method of cylindrical shells. (In Problem 27, we ask you to calculate the same volume by using the method of circular rings.) When revolved about the line $y = -2$, the infinitesimal rectangle of length h and width dy shown in Figure 7 generates a cylindrical shell of radius r, thickness dy, height h, and volume $dV = 2\pi rh \, dy$. From Figure 7, we see that

$$r = y + 2 \qquad \text{and} \qquad h = 4 - y^2$$

hence,

$$dV = 2\pi rh \, dy = 2\pi(y + 2)(4 - y^2) \, dy = 2\pi(-y^3 - 2y^2 + 4y + 8) \, dy$$

As y runs from 1 to 2, the infinitesimal rectangle in Figure 7 sweeps out the region R. Hence,

$$V = \int_{y=1}^{y=2} dV = \int_1^2 2\pi(-y^3 - 2y^2 + 4y + 8) \, dy$$

$$= 2\pi \left(-\frac{y^4}{4} - \frac{2y^3}{3} + 2y^2 + 8y \right) \Bigg|_1^2 = 2\pi \left(\frac{44}{3} - \frac{109}{12} \right) = \frac{67\pi}{6} \text{ cubic units} \quad \blacksquare$$

Problem Set 6.2

In Problems 1 to 10, use the method of cylindrical shells to find the volume V of the solid of revolution generated by revolving each region R about the y axis.

1 R is bounded by the graphs of $y = x^2 + 1$, $y = 0$, $x = 0$, and $x = 1$.

2 R is bounded by the graphs of $y = \sqrt{x} + x^2$, $y = 0$, $x = 1$, and $x = 4$.

3 R is bounded by the graphs of $y = \sqrt{x^2 + 1}$, $y = 0$, $x = 0$, and $x = 1$.

4 R is bounded by the graphs of $y = \sqrt{x + 1}$, $y = 0$, $x = 3$, and $x = 8$.

5 R is bounded by the graphs of $y = x^3$, $y = x^2 + 1$, $x = 0$, and $x = 1$.

6 R is bounded by the graphs of $y = \sqrt[3]{x}$, $y = 1$, $x = 1$, and $x = 27$.

7 R is bounded by the graphs of $3x - 2y + 1 = 0$, $y = x$, $x = 1$, and $x = 3$.

8 R is bounded by the graphs of $y = \sqrt{1 - x^2}$, $y = -\sqrt{1 - x^2}$, and $x = 0$ and lies in the first and fourth quadrants.

9 R is bounded by the graphs of $y = \sin x^2$, $y = 0$, $x = \sqrt{\pi/2}$, and $x = \sqrt{\pi}$.

10 R is bounded by the graphs of $y = \cos x^2$, $y = \sin x^2$, $x = 0$, and $x = \sqrt{\pi}/2$.

In Problems 11 to 16, use the method of cylindrical shells to find the volume V of the solid of revolution generated by revolving each region R about the x axis.

11 R is bounded by the graphs of $y = x^3$, $y = 27$, $x = 0$.

12 R is bounded by the graphs of $y = \sqrt{x}$, $y = 2$, and $x = 0$.

13 R is bounded by the graphs of $x = 4\sqrt{y}$, $y = 4$, and $x = 0$.

14 R is bounded by the graphs of $x = y^2\sqrt{1 + y^4}$, $x = 0$, and $y = 1$.

15 R is bounded by the graphs of $x = y^2 + 1$, $x = y\sqrt{1 + y^3}$, $y = 0$, and $y = 1$.

16 R is bounded by the graphs of $y = x^n$, $y = x^m$, $x = 0$, and $x = 1$, where $n < m$.

In Problems 17 to 20, use the method of cylindrical shells to rework the indicated problem.

17 Problem 5 in Problem Set 6.1

18 Problem 22 in Problem Set 6.1

19 Problem 25 in Problem Set 6.1

20 Problem 28 in Problem Set 6.1

In Problems 21 to 26, use either the method of circular rings or the method of cylindrical shells to find the volume V of the solid of revolution generated by revolving each region about the indicated axis.

21 R is bounded by the line $y = 16$ and the parabola $y = x^2$; about the x axis.

22 R is the same region as in Problem 11, but the axis of revolution is $y = 27$.

23 R is the same region as in Problem 21, but the axis of revolution is $x = 20$.

24 R is bounded by the graphs of $y = 2x^3 - 9x^2 + 12x$, $y = 0$, $x = 0$, and $x = 3$; about the x axis.

25 R is the region in the first quadrant above the graph of $y = x^2$ and below the graph of $y = x + 2$; about the line $x = -2$.

26 R is the region in the second quadrant above the graph of $y = -x^3$ and below the graph of $y = 3x^2$; about the line $y = -3$.

27 Use the method of circular rings to rework Example 3 on page 368.

28 Use the method of cylindrical shells to show that the volume of a right circular cone is one-third of its height times the area of its base.

29 Find the volume of the solid generated by revolving the right triangle with vertices at $(a, 0)$, $(b, 0)$, and (a, h) about the y axis. Assume that $0 < a < b$ and $h > 0$.

30 Let V be the volume of the solid of revolution generated by revolving a region R, which lies to the right of the y axis, about the y axis. If b is a positive constant, show that the volume of the solid of revolution generated by revolving R about the axis $x = -b$ is given by $V + 2\pi bA$, where A is the area of R.

31 Let R be the region between $x = -a$ and $x = a$ bounded above by the graph of $y = b\sqrt{1 - (x/a)^2}$ and bounded below by the x axis, where a and b are positive constants. The football-shaped solid S generated by revolving R about the x axis is called an **ellipsoid of revolution.** Find a formula for the volume V of S in terms of a and b.

32 Rework Problem 40 in Problem Set 6.1, using the method of cylindrical shells.

33 A cylindrical hole is bored through the center of a sphere of unknown radius. However, the length of the hole is known to be L units. Show that the volume of the portion of the sphere that remains is equal to the volume of a sphere of diameter L.

6.3 Volumes by the Method of Slicing

Figure 1

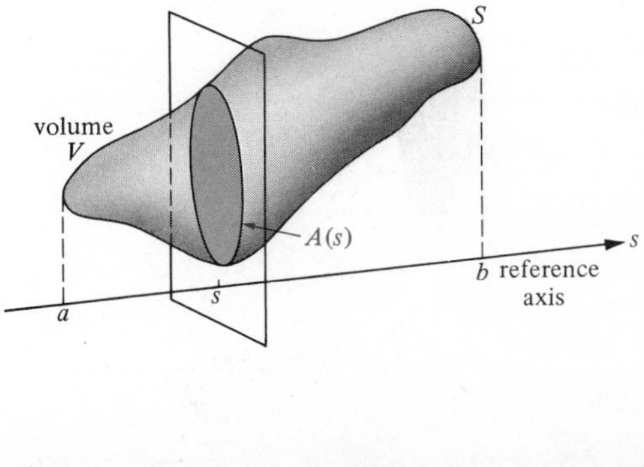

In Sections 4.7 and 5.6, we studied the method of slicing for determining the areas of admissible plane regions. In this section we use an analogous method, also called the **method of slicing,** for finding the volumes of solids. Actually, this method is just a generalization of the method of circular disks or circular rings presented in Section 6.1.

In order to calculate the volume V of a solid S by the method of slicing, select a convenient reference axis, and let $A(s)$ be the area of the cross section of S intercepted by the plane perpendicular to the reference axis at the point with coordinate s (Figure 1). Suppose that the entire solid S is contained between the plane at $s = a$ and the plane at $s = b$, where $a < b$.

Figure 2

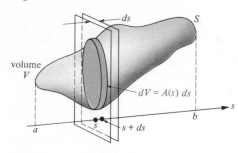

Figure 3

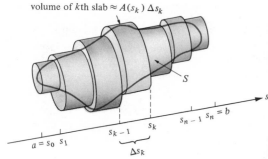

volume of kth slab $\approx A(s_k)\,\Delta s_k$

In Figure 2, let dV denote the infinitesimal volume of the portion of the solid S between the plane perpendicular to the reference axis at the point with coordinate s and the corresponding plane at the point with coordinate $s + ds$. Here ds represents an infinitesimal increase in the reference coordinate. Evidently, dV is the volume of an infinitesimal solid cylinder of height ds with base area $A(s)$; hence,

$$dV = A(s)\,ds$$

The total volume V of the solid S should be obtained by summing—that is, integrating—all the infinitesimal volumes dV as s runs from a to b. Therefore, we should have

$$V = \int_a^b A(s)\,ds$$

This argument in which we use Leibnizian infinitesimals, although perhaps not mathematically rigorous, produces the correct result. A more conclusive argument, based directly on the definition of the definite integral, can be made as follows:

We partition the reference axis into n subintervals

$$[s_0, s_1], \ [s_1, s_2], \ \ldots, \ [s_{k-1}, s_k], \ \ldots, \ [s_{n-1}, s_n]$$

(Figure 3). Above each subinterval $[s_{k-1}, s_k]$ of this partition we have constructed a circumscribed cylindrical slab of thickness Δs_k. The cross-sectional area of such a slab is approximately $A(s_k)$, so its volume is approximately $A(s_k)\,\Delta s_k$. Therefore,

$$V \approx \sum_{k=0}^{n} A(s_k)\,\Delta s_k$$

As the norm of the partition becomes smaller and smaller and the number of slabs increases while the slabs get thinner and thinner, the approximation becomes better and better. In the limit, as the norm of the partition approaches zero,

$$\sum_{k=0}^{n} A(s_k)\,\Delta s_k \quad \text{approaches} \quad \int_a^b A(s)\,ds$$

by definition; therefore, we have

$$\boxed{V = \int_a^b A(s)\,ds}$$

Figure 4

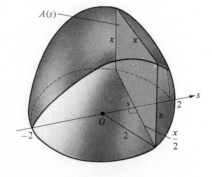

EXAMPLE 1   Find the volume of a solid whose base is a circle of radius 2 units if all cross sections perpendicular to a fixed diameter of the base are squares.

SOLUTION   Figure 4 shows the square cross section of the solid at a distance s units from the center O along the fixed diameter. The reference axis is taken to lie along this diameter. If x is the length of one side of the square, then, from Figure 4 and the Pythagorean theorem,

$$s^2 + \left(\frac{x}{2}\right)^2 = 2^2 \qquad \text{or} \qquad \frac{x^2}{4} = 4 - s^2$$

Therefore, the area of the square is

$$A(s) = x^2 = 4(4 - s^2) = 16 - 4s^2$$

By the method of slicing, the required volume is

$$V = \int_{-2}^{2} A(s)\, ds = \int_{-2}^{2} (16 - 4s^2)\, ds = [16s - \tfrac{4}{3}s^3]\Big|_{-2}^{2}$$

$$= \tfrac{128}{3} \text{ cubic units}$$

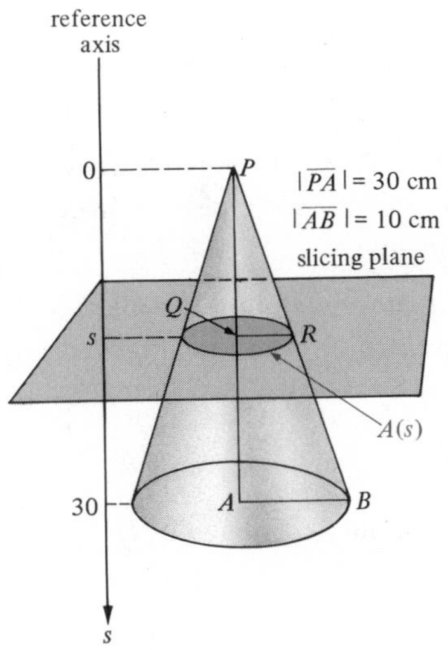

EXAMPLE 2   Find the volume of a solid right circular cone of height 30 centimeters if the base radius is 10 centimeters.

SOLUTION   We choose the reference axis (Figure 5) pointing downward. The cross section of the cone cut by the slicing plane at level s is a circle of radius $|\overline{QR}|$ and area $A(s) = \pi|\overline{QR}|^2$. We find $|\overline{QR}|$ in terms of s by similar triangles as follows:

$$\frac{|\overline{QR}|}{|\overline{PQ}|} = \frac{|\overline{AB}|}{|\overline{PA}|}$$

that is,

$$\frac{|\overline{QR}|}{s} = \frac{10}{30} \quad \text{or} \quad |\overline{QR}| = \frac{1}{3}s$$

Consequently,

$$A(s) = \pi|\overline{QR}|^2 = \pi\left(\frac{1}{3}s\right)^2 = \frac{\pi s^2}{9}$$

By the method of slicing, the volume of the cone is therefore given by

$$V = \int_{0}^{30} A(s)\, ds = \int_{0}^{30} \frac{\pi s^2}{9}\, ds$$

$$= \left(\frac{\pi s^3}{27}\right)\Big|_{0}^{30} = \frac{\pi}{27}(30)^3 = 1000\pi \text{ cubic centimeters}$$

This coincides with the volume as calculated by the familiar formula—one-third the height times the area of the base.

☐ **EXAMPLE 3**   Gasoline is stored in a spherical tank of radius $r = 10$ meters. How many cubic meters of gasoline are in the tank if the surface of the gasoline is 3 meters below the center of the tank? Use a calculator to find the number of cubic meters rounded off to two decimal places.

SOLUTION   The cross section of the tank cut by the slicing plane at level s is a circle of radius $|\overline{QP}|$ and area $A(s) = \pi|\overline{QP}|^2$ (Figure 6). We find $|\overline{QP}|^2$ in terms of s by the Pythagorean theorem as follows:

$$|\overline{QP}|^2 + |\overline{QC}|^2 = |\overline{CP}|^2$$

that is,

$$|\overline{QP}|^2 = |\overline{CP}|^2 - |\overline{QC}|^2$$

$$= r^2 - |s|^2 = r^2 - s^2$$

Consequently,

$$A(s) = \pi(r^2 - s^2) = \pi(100 - s^2)$$

Figure 5

reference axis

$|\overline{PA}| = 30$ cm
$|\overline{AB}| = 10$ cm
slicing plane

$A(s)$

Figure 6

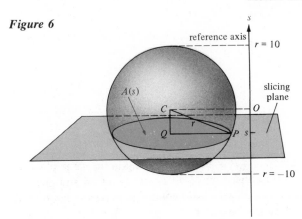

reference axis
$r = 10$

slicing plane

$r = -10$

Notice that the surface of the gasoline is at level $s = -3$. Therefore, by the method of slicing, the volume of the portion of the sphere occupied by the gasoline is given by

$$V = \int_{-10}^{-3} A(s)\,ds = \int_{-10}^{-3} \pi(100 - s^2)\,ds = \pi\left(100s - \frac{s^3}{3}\right)\Bigg|_{-10}^{-3}$$

$$= \pi\left[100(-3) - \frac{(-3)^3}{3}\right] - \pi\left[100(-10) - \frac{(-10)^3}{3}\right]$$

$$= -291\pi + \frac{2000}{3}\pi = \frac{1127}{3}\pi \approx 1180.19 \text{ cubic meters} \qquad \blacksquare$$

Figure 7

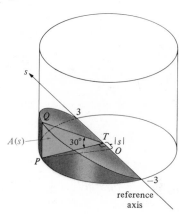

reference axis

EXAMPLE 4   A solid right circular cylinder has a radius of 3 units. A wedge is cut from this cylinder by a plane through a diameter of the base and is inclined to the base at an angle of 30°. Find the volume of the wedge.

SOLUTION   Let the reference axis lie along the intersection of the plane and the base of the cylinder, with the origin O at the center of the base (Figure 7). The cutting plane for the method of slicing intersects the wedge in a triangular cross section TPQ, whose area

$$A(s) = \tfrac{1}{2}|\overline{PQ}| \cdot |\overline{PT}|$$

we must find in terms of the coordinate s of the point T. Triangle OTP is a right triangle, and $|\overline{OT}| = |s|$ while $|\overline{OP}| = 3$; hence,

$$|\overline{OP}|^2 = |\overline{OT}|^2 + |\overline{PT}|^2$$

so

$$|\overline{PT}|^2 = |\overline{OP}|^2 - |\overline{OT}|^2 = 9 - |s|^2 = 9 - s^2$$

Referring to right triangle TPQ, we have

$$|\overline{PQ}| = |\overline{PT}| \tan 30°$$

Therefore,

$$A(s) = \tfrac{1}{2}|\overline{PQ}| \cdot |\overline{PT}| = \tfrac{1}{2}|\overline{PT}| \tan 30° \,|\overline{PT}|$$

$$= \tfrac{1}{2}|\overline{PT}|^2 \tan 30°$$

Since $|\overline{PT}|^2 = 9 - s^2$,

$$A(s) = \frac{1}{2}(9 - s^2)\tan 30° = \frac{1}{2}(9 - s^2)\left(\frac{\sqrt{3}}{3}\right)$$

By the method of slicing,

$$V = \int_{-3}^{3} A(s)\,ds = \int_{-3}^{3} \frac{1}{2}(9 - s^2)\frac{\sqrt{3}}{3}\,ds$$

$$= \frac{\sqrt{3}}{6}\left(9s - \frac{s^3}{3}\right)\Bigg|_{-3}^{3} = \frac{\sqrt{3}}{6}[18 - (-18)]$$

$$= \frac{\sqrt{3}}{6}(36) = 6\sqrt{3} \text{ cubic units} \qquad \blacksquare$$

Figure 8

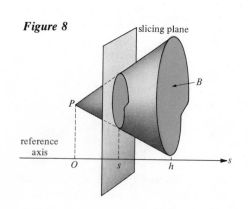

slicing plane

reference axis

If B is an admissible plane region and P is a point not lying in the same plane as B, then the three-dimensional solid consisting of all points lying on straight line segments between P and points of B is called a **solid cone** with **vertex** P and **base** B (Figure 8). The perpendicular distance h between the vertex P and the base B is called the **height** of the cone.

If a reference axis is chosen perpendicular to the base B, then a slicing plane at a distance s from the vertex P will intercept the cone in a cross-sectional region which is similar to the base B. Furthermore, the linear dimensions of this cross section are proportional to its distance s from P. Since the area $A(s)$ of the cross section is proportional to the *square* of its linear dimensions, $A(s)$ is proportional to s^2. Thus,

$$A(s) = Ks^2$$

where K is a constant. When $s = h$, then

$$A(s) = A(h) = \text{area of base } B$$

Therefore,

$$A(h) = Kh^2 \qquad \text{so} \qquad K = \frac{A(h)}{h^2}$$

Figure 9

Hence,

$$A(s) = \frac{A(h)}{h^2}s^2$$

By the method of slicing,

$$V = \int_0^h A(s)\, ds = \int_0^h \frac{A(h)}{h^2}s^2\, ds = \frac{A(h)}{h^2} \int_0^h s^2\, ds$$

$$= \frac{A(h)}{h^2}\left(\frac{s^3}{3}\right)\bigg|_0^h = \frac{A(h)}{h^2} \cdot \frac{h^3}{3} = \frac{h}{3}A(h)$$

Hence, *the volume of a solid cone is given by one-third of its height times the area of the base.*

EXAMPLE 5   The Great Pyramid of Cheops at Al Giza, Egypt, has a square base 230 meters on each side and a height of 147 meters (Figure 9). Find the volume V of the Great Pyramid.

SOLUTION   The pyramid is a solid cone with height 147 meters and base area $(230)^2$ square meters. Hence, its volume is

$$V = \tfrac{1}{3}(147)(230)^2 = 2{,}592{,}100 \text{ cubic meters} \qquad \blacksquare$$

Problem Set 6.3

1 A certain solid has a circular base of radius 3 units. If cross sections perpendicular to one of the diameters of the base are squares, find the volume of the solid.

2 Find the volume of a solid whose base is a circle of radius 5 centimeters if all cross sections perpendicular to a fixed diameter of the base are equilateral triangles.

3 A monument is 30 meters high. A horizontal cross section x meters above the base is an equilateral triangle whose sides are $(30 - x)/15$ meters long. Find the volume of the monument.

4 A tower is 24 meters tall. A horizontal cross section of the tower x meters from its top is a square whose sides are $\frac{1}{13}(x + 1.5)$ meters long. Find the volume of the tower.

5 Find the volume of a spherical shell whose inside diameter is Y_1 units and whose outside diameter is Y_2 units.

6 The base of a solid is the region in the xy plane bounded by the oval-shaped curve $(x/4)^2 + (y/3)^2 = 1$ (an *ellipse*). Every cross section of the solid perpendicular to the x axis is a semicircle. Find the volume of the solid.

7 The base of a solid lies in a plane, and the height of the solid is 5 meters. Find the volume of the solid if the area of a cross section parallel to the base and s meters above the base is given by the equation (a) $A(s) = 3s^2 + 2$, (b) $A(s) = s^2 + s$.

8 Use the method of slicing to find the volume of a solid right circular cylinder of radius r and height h, taking the cross sec-

tions perpendicular to a fixed diameter of the base. You may assume that $\int_{-r}^{r} \sqrt{r^2 - s^2}\, ds = \frac{1}{2}\pi r^2$.

9 The Department of Public Works of East Mattoon intends to cut down a diseased elm tree 4 feet in diameter. They first cut out a wedge bounded below by a horizontal plane and bounded above by a plane that meets the horizontal plane along a diameter of the tree at a 45° angle. What is the volume of the wedge?

10 A solid right circular cylinder has a radius of r units. A wedge is cut from this cylinder by a plane through a diameter of the base circle, and the plane inclines to this base circle at an angle θ. Find a formula for the volume of this wedge.

11 The plans for a wave-guide antenna are shown in Figure 10. All dimensions are in meters. Each cross section perpendicular to the central axis (the x axis) is an oval-shaped region whose area is $\pi y^2 / 2$ square meters, where y is the distance from the center of the oval to its upper edge. The upper and lower boundaries of these cross sections are parabolas $y = \frac{1}{12}x^2 + 1$ and $y = -\frac{1}{12}x^2 - 1$. Find the volume enclosed by the wave guide if its length is 1.5 meters.

Figure 10

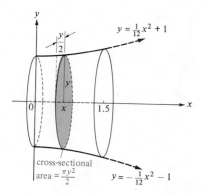

cross-sectional area $= \dfrac{\pi y^2}{2}$

12 The portion of a cone (not necessarily a circular cone) formed by cutting off the top by a plane is called a **frustum.** If the cutting plane is parallel to the base, then the distance between the cutting plane and the base is called the **height** h of the frustum. Show that the volume V of the frustum shown in Figure 11 is given by $V = (h/3)(A + \sqrt{Aa} + a)$ cubic units.

Figure 11

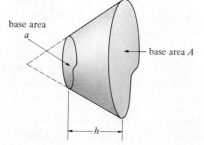

base area a

base area A

h

13 Show that the *volume of a solid cylinder* (not necessarily a circular cylinder and not necessarily a right cylinder) is given by *its height times the area of one of its bases*.

C **14** The reentry vehicles used during Projects Mercury, Gemini, and Apollo were each shaped roughly like a frustum of a right circular cone (Figure 12). Find the volume of such a reentry vehicle if its base is a circle of radius 2.7 meters, its top is a circle of radius 1.4 meters, and its height is 2.8 meters.

Figure 12

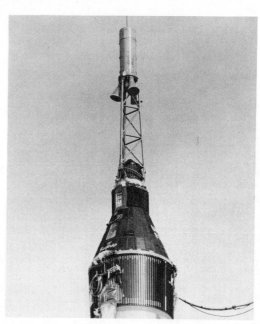

15 Find the volume of the **spherical segment of one base** shown in Figure 13 if the radius of the sphere is r units and the height of the segment is h units.

Figure 13

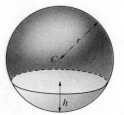

16 Find the volume of a solid whose cross section made by a plane perpendicular to a reference axis at the point with coordinate s has area given by

$$A(s) = \begin{cases} as^2 + bs + c & \text{for } 0 \le s \le h \\ 0 & \text{otherwise} \end{cases}$$

Express this volume in terms of $A_0 = A(0)$, $A_1 = A(h/2)$, and $A_2 = A(h)$. (*Hint:* Use the prismoidal formula, Theorem 5 on page 345.)

17 A tent is made by stretching canvas from a circular base of radius a to a semicircular rib erected at right angles to the base and meeting the base at the ends of a diameter. Find the volume enclosed by the tent.

18 Find the volume of the **spherical segment of two bases** shown in Figure 14 if the sphere has radius r, the height of the segment is h_2, and the lower base of the segment is h_1 units from the bottom of the sphere.

Figure 14

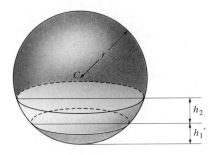

Figure 15

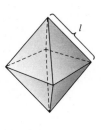

19 A regular **octahedron** is a solid bounded by eight congruent equilateral triangles (Figure 15). Find the volume of an octahedron if each of its eight bounding triangles has sides of length l.

20 Find the volume of the **spherical sector** shown in Figure 16.

Figure 16

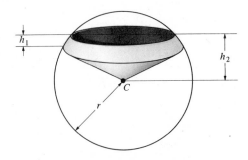

21 A tower is 60 meters high, and every horizontal cross section is a square. The vertices of these squares lie in four congruent parabolas whose planes pass through the central axis of the tower and which open outward away from this central axis. Each of the four parabolas has its vertex in the upper square base of the tower, and each of these parabolas has a horizontal axis. The diagonals of the upper and lower bases of the tower are 2 and 12 meters long, respectively. Find the volume enclosed by the tower.

22 Two right circular cylinders of radius r have central axes meeting at right angles. Find the volume V of the solid common to both cylinders.

6.4 Arc Length and Surface Area

In this section, we give two further geometric applications of the definite integral: the calculation of arc length and the calculation of the area of a surface of revolution.

The Arc Length of a Curve

Figure 1

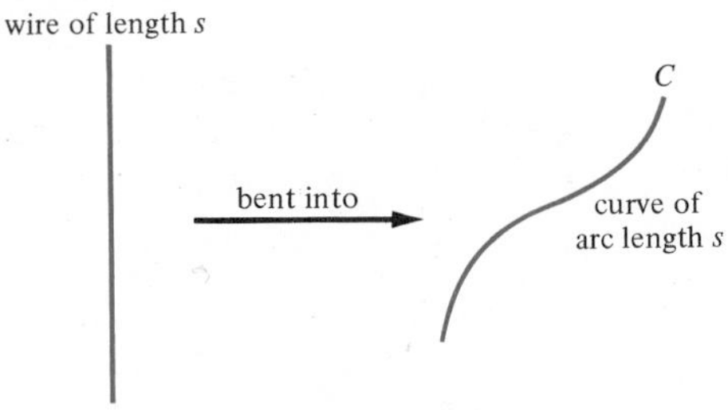

wire of length s

bent into

C

curve of arc length s

If a straight piece of wire of length s is bent into a curve C, we understand that the curve C has **arc length** s (Figure 1). For instance, the arc length of a circle of diameter D is known to be given by $s = \pi D$. Here we do not attempt to give a formal definition* of arc length, but we assume that you have an intuitive understanding of this concept.

Now, let f be a function with a continuous first derivative on some open interval I containing the closed interval $[a, b]$. We give an informal derivation of a formula for the arc length of the portion of the graph of f between the points $(a, f(a))$ and

*See Problems 41 and 42 (pages 382 and 383) for an indication of a formal development.

Figure 2

$(b, f(b))$ (Figure 2). Let s denote the arc length of the portion of the graph of f between the points $(a, f(a))$ and $(x, f(x))$, where $a \le x \le b$. If x is increased by an infinitesimal amount dx, then $y = f(x)$ will change by a corresponding infinitesimal amount dy, and likewise s will increase by an infinitesimal amount ds (Figure 3).

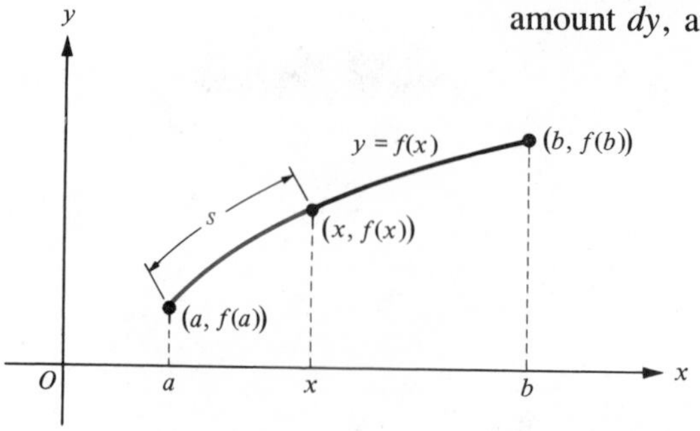

Notice the infinitesimal right triangle with legs $|dx|$ and $|dy|$ and with hypotenuse ds. By the Pythagorean theorem,

$$(ds)^2 = (dx)^2 + (dy)^2$$
$$= \left[1 + \left(\frac{dy}{dx} \right)^2 \right] (dx)^2$$

Hence,

$$ds = \sqrt{1 + \left(\frac{dy}{dx} \right)^2}\, dx = \sqrt{1 + [f'(x)]^2}\, dx$$

Figure 3

Integration of the differential equation above gives the arc length

$$s = \int_{x=a}^{x=b} ds$$

of the graph of f between the point with abscissa a and the point with abscissa b. Thus,

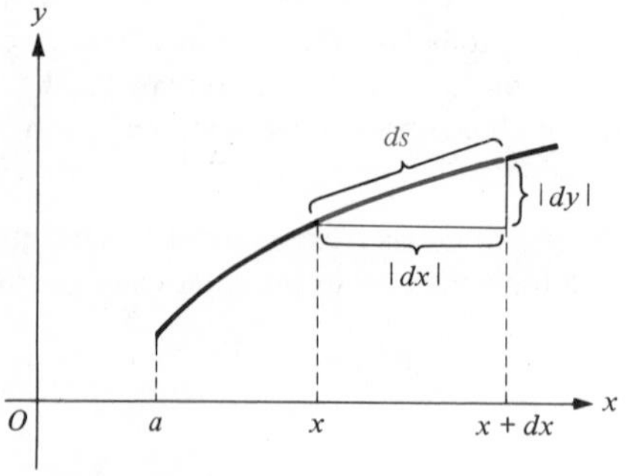

$$s = \int_a^b \sqrt{1 + [f'(x)]^2}\, dx$$

© **EXAMPLE 1**   Find the arc length of the graph of the function $f(x) = x^{2/3} - 1$ between the points $(8, 3)$ and $(27, 8)$.

SOLUTION   Here $f'(x) = \frac{2}{3}x^{-1/3}$, so the desired arc length is given by

$$s = \int_8^{27} \sqrt{1 + [f'(x)]^2}\, dx = \int_8^{27} \sqrt{1 + \tfrac{4}{9}x^{-2/3}}\, dx$$
$$= \int_8^{27} \sqrt{\frac{9x^{2/3} + 4}{9x^{2/3}}}\, dx = \int_8^{27} \sqrt{9x^{2/3} + 4}\, \frac{dx}{3x^{1/3}}$$

Making the change of variable $u = 9x^{2/3} + 4$ and noting that $du = 6x^{-1/3}\, dx$, $u = 40$ when $x = 8$, and $u = 85$ when $x = 27$, we have

$$s = \int_{x=8}^{x=27} \sqrt{9x^{2/3} + 4}\, \frac{dx}{3x^{1/3}} = \int_{u=40}^{u=85} \sqrt{u}\, \frac{du}{18}$$
$$= \frac{u^{3/2}}{27} \bigg|_{40}^{85} = \frac{85^{3/2} - 40^{3/2}}{27} \approx 19.65 \text{ units} \quad\blacksquare$$

© **EXAMPLE 2**   Set up an integral representing the arc length of the portion of the parabola $y = x^2$ between the origin and the point $(2, 4)$. Then use Simpson's parabolic rule, $S_{2n} \approx \int_0^2 \sqrt{1 + [f'(x)]^2}\, dx$, with $n = 2$ to estimate this integral and thus the arc length.

SOLUTION   Here $dy/dx = 2x$, so the desired arc length is given by

$$s = \int_0^2 \sqrt{1 + (2x)^2}\, dx = \int_0^2 \sqrt{1 + 4x^2}\, dx$$

To find the quantities needed for Simpson's parabolic rule with $n = 2$, we subdivide the closed interval $[0, 2]$ into $2n = 4$ equal subintervals by means of the points $x_k = k/2$ for $k = 0, 1, 2, 3, 4$. Now let

$$y_k = \sqrt{1 + 4(x_k)^2} = \sqrt{1 + k^2}$$

so that

$$y_0 = 1 \qquad y_1 = \sqrt{2} \qquad y_2 = \sqrt{5} \qquad y_3 = \sqrt{10} \qquad \text{and} \qquad y_4 = \sqrt{17}$$

Then

$$S_{2n} = S_4 = \frac{1/2}{3}(y_0 + 4y_1 + 2y_2 + 4y_3 + y_4)$$

$$= \frac{1}{6}(1 + 4\sqrt{2} + 2\sqrt{5} + 4\sqrt{10} + \sqrt{17})$$

Evaluating S_4 and rounding off to two decimal places, we have

$$S_4 \approx 4.65$$

hence, the desired arc length is approximately 4.65 units. (Incidentally, the correct value of the desired arc length, rounded off to four decimal places, is 4.6468 units.) ■

If we express the equation of the curve between two points in the form $x = g(y)$, where g' is a continuous function in the closed interval $[c, d]$, then the arc length of the graph of g between $(g(c), c)$ and $(g(d), d)$ is given by the formula

$$s = \int_c^d \sqrt{1 + [g'(y)]^2} \, dy = \int_c^d \sqrt{1 + \left(\frac{dx}{dy}\right)^2} \, dy$$

EXAMPLE 3 Find the arc length of the graph of the equation $8x = y^4 + (2/y^2)$ from $(\frac{3}{8}, 1)$ to $(\frac{33}{16}, 2)$.

SOLUTION   Here $x = \frac{1}{8}y^4 + \frac{1}{4}y^{-2}$, so

$$\frac{dx}{dy} = \frac{1}{2}y^3 - \frac{2}{4}y^{-3} = \frac{y^3}{2} - \frac{1}{2y^3}$$

The desired arc length is given by

$$s = \int_1^2 \sqrt{1 + \left(\frac{dx}{dy}\right)^2} \, dy = \int_1^2 \sqrt{1 + \left(\frac{y^3}{2} - \frac{1}{2y^3}\right)^2} \, dy$$

$$= \int_1^2 \sqrt{1 + \frac{y^6}{4} - \frac{1}{2} + \frac{1}{4y^6}} \, dy = \int_1^2 \sqrt{\frac{y^6}{4} + \frac{1}{2} + \frac{1}{4y^6}} \, dy$$

$$= \int_1^2 \sqrt{\left(\frac{y^3}{2} + \frac{1}{2y^3}\right)^2} \, dy = \int_1^2 \left(\frac{y^3}{2} + \frac{1}{2y^3}\right) dy$$

$$= \left(\frac{y^4}{8} - \frac{1}{4y^2}\right)\Bigg|_1^2 = \frac{33}{16} \text{ units} \qquad\qquad ■$$

Figure 4

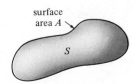
surface area A

S

The Area of a Surface of Revolution

Once again, we do not attempt to give a formal definition of the "surface area" of a solid S (Figure 4). Suffice it to say that if the solid S has surface area A, then the same amount of paint required to apply a uniformly thin coat to the surface of S would be required to apply a uniformly thin coat to a flat surface of area A.

Figure 5

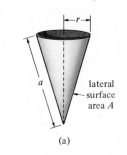

(a)

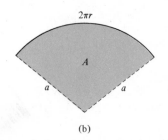

(b)

Figure 6

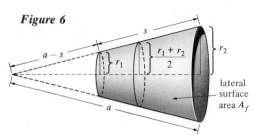

Figure 7

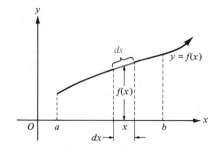

(a)

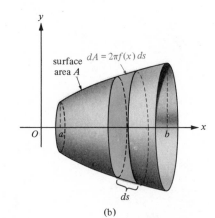

(b)

To find the surface area A of the right circular cone shown in Figure 5a, we cut the cone along the dashed line and flatten it out to form a sector of a circle, as in Figure 5b. The slant height a of the cone is the radius of the sector, and the arc length of the sector is the circumference $2\pi r$ of the base of the cone.

The sector in Figure 5b is part of a circle of area πa^2 with circumference $2\pi a$. The ratio of the area A of the sector to the area πa^2 of the circle is the same as the ratio of the arc length $2\pi r$ of the sector to the total arc length $2\pi a$ of the circle; that is,

$$\frac{A}{\pi a^2} = \frac{2\pi r}{2\pi a} \quad \text{or} \quad A = \pi a r$$

Now, consider the frustum of a right circular cone shown in Figure 6. The surface area A_f of this frustum is evidently the difference between the surface area of the large cone with slant height a and base radius r_2 and the surface area of the small cone with slant height $a - s$ and base radius r_1 (Figure 6). Thus, by the formula previously obtained for the surface area of a cone,

$$A_f = \pi a r_2 - \pi(a - s)r_1 = \pi a(r_2 - r_1) + \pi s r_1$$

By similar triangles, $a/r_2 = (a - s)/r_1$, so

$$ar_1 = ar_2 - sr_2 \quad \text{or} \quad a = \frac{sr_2}{r_2 - r_1}$$

Therefore,

$$A_f = \pi \frac{sr_2}{r_2 - r_1}(r_2 - r_1) + \pi s r_1 = \pi s r_2 + \pi s r_1 = \pi s(r_1 + r_2) = 2\pi \frac{r_1 + r_2}{2}s$$

Notice that $(r_1 + r_2)/2$ is the radius of the cross section of the frustum midway between its two bases and $2\pi(r_1 + r_2)/2$ is the circumference of this midsection. Hence, *the surface area A_f of a frustum of a right circular cone is given by*

$$A_f = 2\pi \frac{r_1 + r_2}{2}s$$

the circumference of its midsection times its slant height.

We now consider the problem of finding the surface area A of the surface of revolution generated by revolving the portion of the graph of the nonnegative continuous function f between the lines $x = a$ and $x = b$ about the x axis (Figure 7). Let *ds* denote the infinitesimal arc length of the portion of the graph of *f* above the interval of infinitesimal length *dx*, as shown in Figure 7a, and let *x* denote the coordinate of the center of this interval. When the infinitesimal arc of length *ds* is revolved about the *x* axis, it generates an infinitesimal frustum of a cone of slant height *ds* whose midsection has radius *f*(*x*) (Figure 7b). The surface area of this infinitesimal frustum is

$$dA = 2\pi f(x)\, ds$$

Assume that the function *f* has a continuous first derivative, so that $ds = \sqrt{1 + [f'(x)]^2}\, dx$; hence,

$$dA = 2\pi f(x)\sqrt{1 + [f'(x)]^2}\, dx$$

The desired surface area *A* can now be obtained by integrating *dA*, so, for revolution about the *x* axis,

$$A = \int_a^b 2\pi f(x)\sqrt{1 + [f'(x)]^2}\, dx$$

that is,

$$A = \int_{x=a}^{x=b} 2\pi f(x)\, ds \qquad \text{where } ds = \sqrt{1 + [f'(x)]^2}\, dx$$

EXAMPLE 4   Let *m* be a positive constant. Find the area of the surface of revolution generated by revolving the graph of *f*(*x*) = *mx* between *x* = 0 and *x* = *b*, where *b* is a positive constant, about the *x* axis. Interpret the result geometrically.

SOLUTION   When the graph of *f* between *x* = 0 and *x* = *b* is revolved about the *x* axis, it generates a right circular cone of height *b* with base radius *mb* (Figure 8). Here *f*'(*x*) = *m*, so the integral formula for surface area gives

$$A = \int_0^b 2\pi mx\sqrt{1 + m^2}\, dx$$

$$= (2\pi\sqrt{1 + m^2})m\int_0^b x\, dx$$

$$= (2\pi\sqrt{1 + m^2})m\frac{b^2}{2} = \pi mb^2\sqrt{1 + m^2}$$

Since the slant height of the cone is the distance from the origin to the point (*b*, *mb*), the slant height is $\sqrt{b^2 + m^2 b^2} = b\sqrt{1 + m^2}$. Therefore, the formula

$$A = \pi mb^2\sqrt{1 + m^2} = \pi(b\sqrt{1 + m^2})mb$$

corresponds to the previously derived formula

$$A = \pi(\text{slant height})(\text{base radius})$$

for the surface area of a right circular cone.   ■

Figure 8

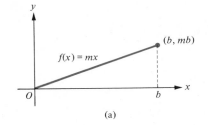

(a)

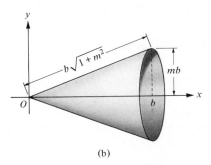

(b)

ⓒ **EXAMPLE 5**   Find the area of the surface obtained by revolving the curve $y = \sqrt{x}$ between *x* = 1 and *x* = 4 about the *x* axis. Sketch the curve and the surface.

Figure 9

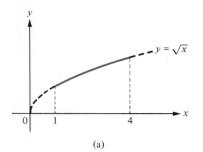

(a)

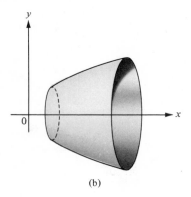

(b)

SOLUTION   The curve, part of a parabola, and the corresponding paraboloid of revolution are shown in Figure 9. Here $dy/dx = 1/(2\sqrt{x})$, and

$$A = 2\pi \int_1^4 \sqrt{x}\sqrt{1 + \left(\frac{dy}{dx}\right)^2}\, dx = 2\pi \int_1^4 \sqrt{x}\sqrt{1 + \frac{1}{4x}}\, dx$$

$$= 2\pi \int_1^4 \sqrt{x + \frac{1}{4}}\, dx$$

We make the change of variable $u = x + \frac{1}{4}$, noting that $du = dx$, so

$$A = 2\pi \int_{x=1}^{x=4} \sqrt{x + \frac{1}{4}}\, dx = 2\pi \int_{u=5/4}^{u=17/4} \sqrt{u}\, du = 2\pi \left(\frac{2}{3}u^{3/2}\right)\Big|_{5/4}^{17/4}$$

$$= \frac{4\pi}{3}\left[\left(\frac{17}{4}\right)^{3/2} - \left(\frac{5}{4}\right)^{3/2}\right] \approx 30.85 \text{ square units} \quad\blacksquare$$

EXAMPLE 6   Find the surface area of a sphere of radius r.

SOLUTION   The sphere of radius r is generated by revolving the semicircle whose equation is

$$y = \sqrt{r^2 - x^2} \qquad -r \le x \le r$$

about the x axis (Figure 10). Here,

$$\frac{dy}{dx} = \frac{-x}{\sqrt{r^2 - x^2}}$$

Figure 10

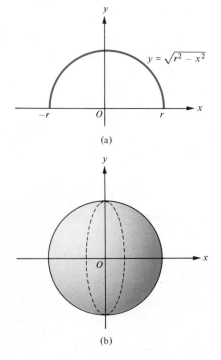

(a)

(b)

is not defined on the closed interval $[-r, r]$. But it is defined on the open interval $(-r, r)$. Thus, we let ϵ be a small positive number, calculate the surface area of the portion of the sphere generated by revolving the part of the semicircle between $x = -r + \epsilon$ and $x = r - \epsilon$ about the x axis, and then take the limit as ϵ approaches zero. The desired area A is, therefore, given by

$$A = \lim_{\epsilon \to 0^+} \int_{-r+\epsilon}^{r-\epsilon} 2\pi\sqrt{r^2 - x^2}\sqrt{1 + \left(\frac{dy}{dx}\right)^2}\, dx$$

$$= \lim_{\epsilon \to 0^+} 2\pi \int_{-r+\epsilon}^{r-\epsilon} \sqrt{r^2 - x^2}\sqrt{1 + \frac{x^2}{r^2 - x^2}}\, dx$$

$$= \lim_{\epsilon \to 0^+} 2\pi \int_{-r+\epsilon}^{r-\epsilon} r\, dx = \lim_{\epsilon \to 0^+} 2\pi r x\Big|_{-r+\epsilon}^{r-\epsilon}$$

$$= \lim_{\epsilon \to 0^+} 2\pi r[(r - \epsilon) - (-r + \epsilon)]$$

$$= \lim_{\epsilon \to 0^+} 2\pi r(2r - 2\epsilon) = 4\pi r^2$$

This confirms the familiar formula $A = 4\pi r^2$ for the surface area of a sphere. $\blacksquare$

If the axis of revolution is the y axis, then the corresponding formula for the surface area of the surface of revolution is given by

$$A = \int_{y=c}^{y=d} 2\pi g(y)\, ds \qquad \text{where } ds = \sqrt{1 + [g'(y)]^2}\, dy$$

Figure 11

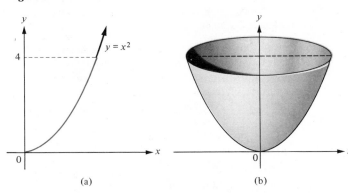

(a)   (b)

$\boxed{\text{c}}$ **EXAMPLE 7** Find the area of the surface obtained by revolving the curve $y = x^2$ between $(0, 0)$ and $(2, 4)$ about the y axis. Sketch the curve and the surface.

SOLUTION   The curve, part of a parabola, and the corresponding paraboloid of revolution are shown in Figure 11. Here, $x = g(y) = \sqrt{y}$, and we have

$$A = 2\pi \int_{y=0}^{y=4} x \, ds = 2\pi \int_0^4 x \sqrt{1 + \left(\frac{dx}{dy}\right)^2} \, dy$$

$$= 2\pi \int_0^4 \sqrt{y} \sqrt{1 + \left(\frac{1}{2\sqrt{y}}\right)^2} \, dy = 2\pi \int_0^4 \sqrt{y + \frac{1}{4}} \, dy$$

$$= 2\pi \left[\frac{2}{3}\left(y + \frac{1}{4}\right)^{3/2}\right]\Big|_0^4 = \frac{4\pi}{3}\left[\left(\frac{17}{4}\right)^{3/2} - \left(\frac{1}{4}\right)^{3/2}\right]$$

$$\approx 36.18 \text{ square units}$$

Problem Set 6.4

In Problems 1 to 14, find the arc length of the graph of each equation between the indicated points.

1 $y = 4x + 3$ from $(0, 3)$ to $(2, 11)$

2 $y = -2x + 1$ from $(-1, 3)$ to $(2, -3)$

3 $y = mx + b$ from $(0, b)$ to $(a, ma + b)$

4 $y = \dfrac{x^3}{6} + \dfrac{1}{2x}$ from $\left(1, \dfrac{2}{3}\right)$ to $\left(3, \dfrac{14}{3}\right)$

$\boxed{\text{c}}$ **5** $y = x^{3/2}$ from $(0, 0)$ to $(4, 8)$

$\boxed{\text{c}}$ **6** $x = \frac{1}{4}y^{2/3}$ from $(0, 0)$ to $(1, 8)$

$\boxed{\text{c}}$ **7** $x = y^{3/2} + 4$ from $(5, 1)$ to $(12, 4)$

8 $y = (1 - x^{2/3})^{3/2}$ from $\left(\dfrac{1}{8}, \dfrac{3\sqrt{3}}{8}\right)$ to $(1, 0)$

$\boxed{\text{c}}$ **9** $x = \frac{2}{3}(y - 5)^{3/2}$ from $(0, 5)$ to $(\frac{2}{3}, 6)$

10 $x = \dfrac{y^4}{8} + \dfrac{1}{4y^2}$ from $\left(\dfrac{3}{8}, 1\right)$ to $\left(\dfrac{33}{16}, 2\right)$

11 $12xy = 4x^4 + 3$ from $(1, \frac{7}{12})$ to $(3, \frac{109}{12})$

12 $x = \dfrac{y^5}{5} + \dfrac{1}{12y^3}$ from $\left(\dfrac{17}{60}, 1\right)$ to $\left(\dfrac{3077}{480}, 2\right)$

13 $x = \dfrac{y^3}{3} + \dfrac{1}{4y}$ from $\left(\dfrac{7}{12}, 1\right)$ to $\left(\dfrac{67}{24}, 2\right)$

$\boxed{\text{c}}$ **14** $y = \displaystyle\int_1^x \sqrt{t^4 + t^2 - 1} \, dt$ from $(1, 0)$ to

$\left(3, \displaystyle\int_1^3 \sqrt{t^4 + t^2 - 1} \, dt\right)$

$\boxed{\text{c}}$ In Problems 15 to 18, set up an integral representing the arc length of each curve, and then use Simpson's parabolic rule with $n = 2$ (that is, with *four* subintervals) to estimate this arc length.

15 $y = \dfrac{1}{x}$ from $(1, 1)$ to $\left(2, \dfrac{1}{2}\right)$

16 $y = x^3$ from $(1, 1)$ to $(2, 8)$

17 $y = \displaystyle\int_1^x \dfrac{dt}{t}$ from $(1, 0)$ to $\left(2, \displaystyle\int_1^2 \dfrac{dt}{t}\right)$

18 $y = \sin x$ from $(0, 0)$ to $(\pi, 0)$

19 Suppose that a particle P moves in the xy plane in such a way that, at time t, its x and y coordinates are given by $x = f(t)$ and $y = g(t)$, where f and g are functions with continuous first derivatives. Write an integral that gives the arc length of the path of P between the instant when $t = a$ and the instant when $t = b$, where $a < b$. [*Hint:* $(ds)^2 = (dx)^2 + (dy)^2$.]

20 Let f be a function with a continuous first derivative, and let Δs denote the arc length of the graph of f between the point $(a, f(a))$ and the point $(a + \Delta x, f(a + \Delta x))$. If Δl denotes the length of the straight line segment joining the point $(a, f(a))$ and the point $(a + \Delta x, f(a + \Delta x))$, prove that $\lim_{\Delta x \to 0} (\Delta l / \Delta s) = 1$.

In Problems 21 to 26, find the area of the surface of revolution obtained by revolving each curve about the x axis.

21 $y = 3x + 2$ between $(0, 2)$ and $(3, 11)$

22 $y^2 = kx$, where k is a positive constant and $y \geq 0$, between $(a, \sqrt{ka})$ and $(b, \sqrt{kb})$, where $0 < a < b$

© 23 $y = x^3$ between (0, 0) and (2, 8)

24 $y = \sqrt{2x - x^2}$ between $\left(\dfrac{1}{2}, \dfrac{\sqrt{3}}{2}\right)$ and $\left(\dfrac{3}{2}, \dfrac{\sqrt{3}}{2}\right)$

© 25 $y = \dfrac{x^4}{8} + \dfrac{1}{4x^2}$ between $\left(1, \dfrac{3}{8}\right)$ and $\left(2, \dfrac{33}{16}\right)$

26 $y = \dfrac{\sqrt{2}}{4} x\sqrt{1 - x^2}$ between (0, 0) and $\left(\dfrac{1}{2}, \dfrac{\sqrt{6}}{16}\right)$

In Problems 27 to 30, find the area of the surface of revolution obtained by revolving each curve about the y axis.

© 27 $y = 4x^2$ between (0, 0) and (3, 36)

© 28 $y^2 = x^3$ between (0, 0) and (4, 8)

29 $y^3 = 9x$ between (0, 0) and $\left(\dfrac{8}{9}, 2\right)$

© 30 $x = \dfrac{y^3}{12} + \dfrac{1}{y}$ between $\left(\dfrac{13}{12}, 1\right)$ and $\left(\dfrac{7}{6}, 2\right)$

© In Problems 31 and 32, use Simpson's parabolic rule with $n = 2$ (that is, with *four* subintervals) to estimate the area of the surface of revolution obtained by revolving each curve about the x axis.

31 $y = \sin x$ between (0, 0) and $(\pi, 0)$

32 $y = \frac{2}{3}\sqrt{9 - x^2}$ between (0, 2) and $(2, \frac{2}{3}\sqrt{5})$

33 Find a formula for the *total* surface area (lateral surface plus base) of a right circular cone of height h with base radius r.

34 (a) If the linear dimensions of a right circular cone are multiplied by a positive constant k, how is its total surface area affected? Why? (b) Complete the following sentence: If the linear dimensions of a solid S are multiplied by a positive constant k, then the surface area of S is _____ .

35 The **astroid**, or **tetracuspid**, $x^{2/3} + y^{2/3} = 1$ (Figure 12) was first studied by Johann Bernoulli in 1691–1692. Find (a) the arc length of the astroid and (b) the area of the surface of revolution obtained by revolving the portion of the astroid in the first and second quadrants about the x axis.

36 Suppose that the function f is nonnegative and has a continuous derivative on $[a, b]$. Let S_0 denote the arc length of the graph of f between $(a, f(a))$ and $(b, f(b))$, and suppose that A_0 is the area of the surface of revolution generated when this graph is revolved about the x axis. Let k be a positive constant, and define a function g by $g(x) = f(x) + k$. Express the area A of the surface generated by revolving the graph of g between $(a, g(a))$ and $(b, g(b))$ about the x axis in terms of S_0 and A_0.

37 One cubic centimeter of a certain substance is formed into n equal spheres. (a) What is the total surface area of all these spheres? (b) What is the limit as n grows larger and larger without bound of the total surface area in part (a)? (c) Assuming that the rate at which this substance dissolves in a certain solvent is proportional to its surface area, explain why it dissolves more rapidly if it is first ground into a fine powder.

38 An organism in the shape of a sphere with a semipermeable surface is suspended in a fluid having the same density, d grams per cubic centimeter, as the organism. Nutrients diffuse through the surface from the fluid into the organism at the rate of k grams per second per square centimeter of surface, and waste products diffuse into the surrounding fluid at the same rate. To sustain itself, the organism requires at least b grams of nutrients per second per gram of its own total weight. Find an upper bound R for the radius r of the organism.

39 A sphere of radius r is inscribed in a right circular cylinder of radius r. Two planes perpendicular to the central axis of the cylinder cut off a **spherical zone** of area A on the surface of the sphere. Show that the same two planes cut off a region on the cylinder with the same surface area A.

© 40 The *Torrid Zone* is the region on the surface of the earth between the Tropic of Cancer and the Tropic of Capricorn (Figure 13). The Tropic of Cancer is the parallel of latitude that is 23.45° north of the equator, and the Tropic of Capricorn is the parallel of latitude that is 23.45° south of the equator. Find the area of the Torrid Zone if the radius of the earth is 6.371×10^3 kilometers. (*Hint:* Use the result of Problem 39.)

Figure 12

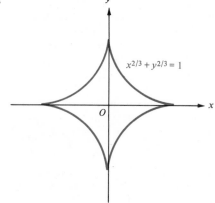

$x^{2/3} + y^{2/3} = 1$

Figure 13

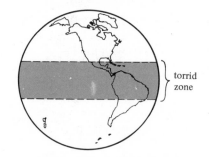

torrid zone

41 Let f be a continuous function on the interval $[a, b]$. A partition $\mathscr{P}$ of $[a, b]$ consisting of the subintervals $[x_0, x_1]$, $[x_1, x_2]$,

$[x_2, x_3], \ldots, [x_{n-1}, x_n]$ determines points $P_0 = (x_0, f(x_0))$, $P_1 = (x_1, f(x_1))$, $P_2 = (x_2, f(x_2)), \ldots, P_n = (x_n, f(x_n))$ on the graph of f above the endpoints of the subintervals in the partition $\mathcal{P}$. (Figure 14 shows the case $n = 4$.) The polygon consisting of the line segments $\overline{P_0P_1}, \overline{P_1P_2}, \overline{P_2P_3}, \ldots, \overline{P_{n-1}P_n}$ is called a **polygon inscribed in the graph of** f. Let

$$L(\mathcal{P}) = \sum_{k=1}^{n} |\overline{P_{k-1}P_k}|$$

denote the total length of the inscribed polygon corresponding to the partition $\mathcal{P}$. If

$$s = \lim_{\|\mathcal{P}\| \to 0} L(\mathcal{P})$$

exists, then the graph of f is called a **rectifiable curve,** and s is called its **arc length.** If $\Delta x_k = x_k - x_{k-1}$, show that

$$L(\mathcal{P}) = \sum_{k=1}^{n} \sqrt{(\Delta x_k)^2 + [f(x_k) - f(x_{k-1})]^2}$$

Figure 14

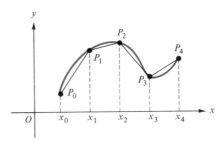

42 In Problem 41, assume that f has a continuous first derivative on $[a, b]$. Using the mean-value theorem, show that for each $k = 1, 2, 3, \ldots, n$, there exists a number c_k in the interval $[x_{k-1}, x_k]$ such that $f(x_k) - f(x_{k-1}) = f'(c_k) \, \Delta x_k$. Using this result, rewrite $L(\mathcal{P})$ as a Riemann sum, and thus obtain the formula $s = \int_a^b \sqrt{1 + [f'(x)]^2} \, dx$ for arc length.

6.5   Work, Force, and Energy

In Chapter 4 we used antidifferentiation to calculate the work done by a variable force. Here we show how the definite integral can be used in calculating work, force, and energy.

Work Done by a Variable Force

In Section 4.5, we showed that if a particle P moves along the s axis under the influence of a possibly variable force F acting parallel to the s axis (Figure 1), then the net work W done by F on P satisfies the differential equation $dW = F \, ds$. If a and b are coordinates of two points on the s axis, we can take the definite integral from $s = a$ to $s = b$ of both sides of the differential equation to obtain

Figure 1

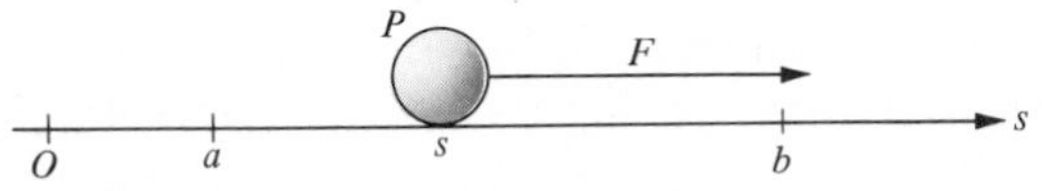

$$\int_{s=a}^{s=b} dW = \int_a^b F \, ds$$

Since

$$\int_{s=a}^{s=b} dW = W \Big|_{s=a}^{s=b}$$

represents the difference between the value of W at $s = b$ and at $s = a$, then

$$\int_a^b F \, ds = W \Big|_{s=a}^{s=b}$$

gives the work done by F in moving P from a to b.

EXAMPLE 1   A force F given by $F = (s^3/3) + 1$ pounds acts on a particle P on the s axis and moves the particle from $s = 2$ to $s = 5$ feet. How much work is done?

SOLUTION   The work done is given by

$$\int_2^5 F\, ds = \int_2^5 \left(\frac{s^3}{3} + 1\right) ds = \left(\frac{s^4}{12} + s\right)\Bigg|_2^5 = \frac{215}{4} = 53.75 \text{ ft} \cdot \text{lb} \qquad \blacksquare$$

EXAMPLE 2   A spring has a natural length of 0.25 meter. If a force of 30 newtons is required to stretch the spring 0.05 meter, how much work is done in stretching the spring 0.15 meter?

SOLUTION   By Hooke's law, the force F on the spring is proportional to its displacement: $F = ks$. When $s = 0.05$ meter, we have $F = 30$ newtons; hence, $k = 30/0.05 = 600$ newtons per meter. Therefore, $F = 600s$, and the work done is given by

$$\int_0^{0.15} F\, ds = \int_0^{0.15} 600s\, ds = 300s^2 \Bigg|_0^{0.15} = 6.75 \text{ joules} \qquad \blacksquare$$

Work Done in Pumping a Liquid

A container C contains a liquid weighing w units of force per cubic unit of volume (Figure 2a). It is desired to pump some of this liquid up above the rim of the container to a certain height and then discharge it. To calculate the work done by the pump, we establish a vertical s axis with its origin at the level up to which the liquid is to be pumped. For simplicity, let the positive direction on the s axis be downward.

Assume that the level of the liquid at the start of the pumping is $s = a$ and that its level at the end is $s = b$. Suppose that the cross-sectional area of the surface of the liquid at level s is $A(s)$ square units. The infinitesimal slab of liquid between level s and level $s + ds$ has a volume of $A(s)\, ds$ cubic units and weighs $wA(s)\, ds$ units of force (Figure 2b). The infinitesimal work done in raising this slab through s units to the level of the origin is given by $dW = swA(s)\, ds$, so the *total work* required to pump the liquid from level $s = a$ to level $s = b$ is given by

$$\boxed{\int_{s=a}^{s=b} dW = \int_a^b swA(s)\, ds = w \int_a^b sA(s)\, ds}$$

© **EXAMPLE 3**   Water, which weighs 9800 newtons per cubic meter, fills a hemispherical reservoir of radius 5 meters. The water is pumped out of the reservoir up to a level 6 meters above the rim until the surface of the remaining water is 4 meters below the rim of the reservoir. How much work is done?

SOLUTION   In Figure 3, we have chosen a vertical reference axis pointing straight down through the center of the reservoir with its origin 6 meters above the rim. When the upper surface of the water is at the position with coordinate s, its radius a satisfies

$$a^2 + (s - 6)^2 = 5^2$$

by the Pythagorean theorem. Hence, the cross-sectional area $A(s)$ is given by

$$A(s) = \pi a^2 = \pi[25 - (s - 6)^2]$$

At the start of the pumping, $s = 6$ meters; at the end of the pumping, $s = 10$ meters.

Figure 2

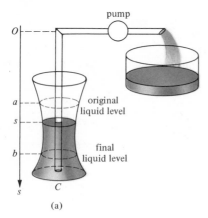
(a)

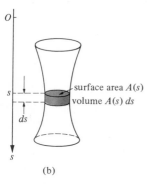
(b)

Figure 3

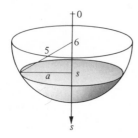

Therefore, the work done is given by

$$w \int_{6}^{10} sA(s)\, ds = 9800 \int_{6}^{10} s\pi[25 - (s-6)^2]\, ds$$

$$= 9800\pi \left(-\frac{s^4}{4} + 4s^3 - \frac{11s^2}{2} \right) \Bigg|_{6}^{10}$$

$$= (5.9584 \times 10^6)\pi \approx 1.8719 \times 10^7 \text{ joules} \qquad \blacksquare$$

Compression or Expansion of a Gas

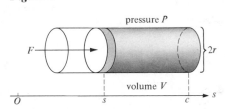

Figure 4

pressure P

$F \longrightarrow$

$2r$

volume V

$O \qquad s \qquad c \qquad s$

It requires work to compress a gas; conversely, when a gas expands, it does work upon its surroundings. Consider, for instance, a quantity of gas in a cylinder of radius r closed by a movable piston (Figure 4). Set up a reference axis parallel to the central axis of the cylinder, denote the coordinate of the piston by s, and let c be the coordinate of the end of the cylinder.

Denote the pressure of the gas by P units of force per unit area, and let V represent the volume of the gas in cubic units. The force F on the piston is given by

$$F = \pi r^2 P$$

the product of its cross-sectional area and the pressure of the gas. The volume V of the gas is the volume of a cylinder of radius r and height $c - s$; hence,

$$V = \pi r^2 (c - s) = \pi r^2 c - \pi r^2 s$$

Therefore,

$$dV = -\pi r^2\, ds \qquad \text{or} \qquad ds = \frac{-dV}{\pi r^2}$$

The work done on the gas by the force F in moving the piston from s to $s + ds$ is given by

$$dW = F\, ds = (\pi r^2 P)\left(\frac{-dV}{\pi r^2} \right) = -P\, dV$$

Consequently, if the gas is compressed from an initial volume V_0 to a final volume V_1, the *total work* done is given by

$$W = \int_{V=V_0}^{V=V_1} dW = \int_{V_0}^{V_1} (-P)\, dV = -\int_{V_0}^{V_1} P\, dV$$

By switching the limits of integration, we can remove the negative sign, so that

$$W = \int_{V_1}^{V_0} P\, dV$$

Notice that if the gas is being compressed, its volume decreases, so that $V_1 < V_0$ and W is positive.

If the gas is expanding, it does just as much work on its surroundings as would be required to compress it back to its original volume. Hence, *if a gas expands from an original volume V_0 to a final volume V_1, it does an amount of work given by*

$$W = \int_{V_0}^{V_1} P\, dV$$

© **EXAMPLE 4**   One cubic foot of air at an initial pressure of 50 pounds per square inch expands adiabatically (that is, without transfer of any heat energy) to a final volume of 3 cubic feet according to the adiabatic gas law $P = kV^{-1.4}$, where k is a constant. Find the work done.

SOLUTION   When $V = 1$, $P = 50$ pounds per square inch $= (50)(144)$ pounds per square foot. Therefore, $(50)(144) = k(1^{-1.4})$, so $k = (50)(144) = 7200$. Hence, $P = 7200V^{-1.4}$, $V_0 = 1$, $V_1 = 3$, and

$$W = \int_{V_0}^{V_1} P \, dV = \int_1^3 7200V^{-1.4} \, dV$$

$$= \left(\frac{7200}{-0.4} V^{-0.4} \right) \Big|_1^3$$

$$= -18,000(3^{-0.4} - 1^{-0.4}) \approx 6400.9 \text{ ft} \cdot \text{lb}$$ ∎

Force Caused by Fluid Pressure

If an admissible plane region R is exposed to a fluid under pressure, there is a resulting **hydrostatic force** F on R. If the pressure of the fluid is constant over the region R, say with a value of P units of force per square unit of area, then $F = PA$, where A is the area of R. However, if P is not constant over R, integration is required to find F.

Here we consider a special case of the problem of calculating F—the situation in which the plane of the region R is vertical and R is submerged in an incompressible liquid of constant density. If such a liquid has w units of weight per cubic unit of its volume, then a column of this liquid h units high and with unit cross-sectional area has a volume of $h \cdot 1$ cubic units and a weight of $w \cdot h \cdot 1 = wh$ units (Figure 5). This weight causes a total force of wh units spread over the 1 square unit at the bottom of the column; hence, the pressure at the bottom of this column is given by $P = wh$. In words, *the pressure at a point in an incompressible liquid of constant density is the product of its weight per unit volume and the distance of the point below the surface of the liquid.*

Now, let R be an admissible plane region placed *vertically* beneath the surface of an incompressible liquid of constant weight per unit volume w (Figure 6). Establish a vertical s axis for reference, pointing downward, with its origin at the surface level of the liquid. Denote the total length of a horizontal cross section of R at level s units below the surface of the liquid by $l(s)$. The pressure P at depth s is given by ws units of force per square unit of area; hence, the infinitesimal force on the strip of height ds at depth s (Figure 6) is given by

$$dF = P \, dA = (ws)[l(s) \, ds] = wsl(s) \, ds$$

Therefore, if R lies entirely between the horizontal lines $s = a$ and $s = b$, the total *hydrostatic force* on R is given by

$$F = \int_{s=a}^{s=b} dF = \int_a^b wsl(s) \, ds = w \int_a^b sl(s) \, ds$$

Figure 5

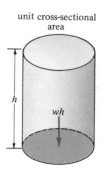

unit cross-sectional area

h

wh

Figure 6

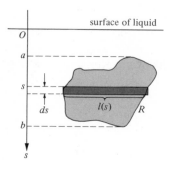

surface of liquid

O

a

s

ds $l(s)$ R

b

s

© **EXAMPLE 5**   Find the force exerted on the semicircular end of a trough if the trough is full of water and the semicircular end has a radius of 3 feet. Assume that the density of water is given by $w = 62.4$ pounds per cubic foot.

Figure 7

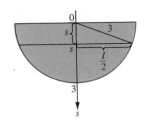

SOLUTION   From Figure 7 and the Pythagorean theorem,

$$\left(\frac{l}{2}\right)^2 + s^2 = 3^2 \qquad \text{or} \qquad l = 2\sqrt{9 - s^2}$$

The desired force is given by

$$F = w \int_0^3 sl \, ds = 62.4 \int_0^3 2s\sqrt{9 - s^2} \, ds = 62.4[-\tfrac{2}{3}(9 - s^2)^{3/2}]\Big|_0^3$$

$$= (62.4)(18) = 1123.2 \text{ lb}$$

∎

Energy

If a particle P of constant mass m moves along a linear scale because of an unopposed (possibly variable) force F (Figure 8), then **Newton's (second) law of motion**—*force equals mass times acceleration*—can be written

Figure 8

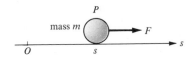

$$F = ma \qquad \text{or} \qquad F = m\frac{dv}{dt} \qquad \text{or} \qquad F = m\frac{d^2s}{dt^2}$$

Thus, Newton's law of motion is really a second-order differential equation whose solution (subject to suitable initial-value conditions) gives the equation of motion of the particle. To obtain this solution, two successive antidifferentiations are required. In order to carry out these antidifferentiations, we need to know F explicitly as a function of time t.

Even if we don't know F explicitly as a function of t, it is possible to carry out the first of the two antidifferentiations, at least formally. This can be done either by multiplying by dt and integrating or by multiplying by ds and integrating.

Multiplying $F = m(dv/dt)$ by dt and integrating, we have

$$\int F \, dt = \int m \, dv = m \int dv = mv + C$$

The quantity mv, mass times velocity, is called the **linear momentum** of the particle P and plays an important role in dynamics. Thus, the idea of momentum arises naturally from Newton's law by a first antidifferentiation *with respect to time*.

Multiplying $F = m(dv/dt)$ by ds and integrating, we have

$$\int F \, ds = \int m\frac{dv}{dt} \, ds = m \int \frac{ds}{dt} \, dv = m \int v \, dv = m\left(\frac{1}{2}v^2\right) + C = \frac{1}{2}mv^2 + C$$

The quantity $\frac{1}{2}mv^2$, one-half mass times the square of velocity, is called the **kinetic energy** of the particle P. Thus, the notion of kinetic energy arises naturally from Newton's law by a first antidifferentiation *with respect to distance*.

As we mentioned earlier (see Figure 1), the work done by the force F in moving the particle from $s = a$ to $s = b$ is given by

$$W\Big|_{s=a}^{s=b} = \int_a^b F \, ds$$

Since

$$\int F \, ds = \tfrac{1}{2}mv^2 + C$$

we have

$$W\Big|_{s=a}^{s=b} = \int_a^b F \, ds = \tfrac{1}{2}mv^2\Big|_{s=a}^{s=b} = \tfrac{1}{2}mv_b^2 - \tfrac{1}{2}mv_a^2$$

where v_b is the velocity of P at $s = b$ and v_a is the velocity of P at $s = a$.

The equation

$$W \Big|_{s=a}^{s=b} = \tfrac{1}{2}mv_b^2 - \tfrac{1}{2}mv_a^2$$

which expresses the fact that the work done by the force F in moving the particle P from $s = a$ to $s = b$ is equal to the resulting change in the kinetic energy of P, is called the **work-energy theorem.**

The work that would be done by the force F in moving the particle P from its present position s to an arbitrary but fixed reference position—say, the point with coordinate b—is called the **potential energy** V of P. Thus,

$$V = \int_s^b F \, ds = - \int_b^s F \, ds$$

By the alternative version of the fundamental theorem of calculus (Theorem 1, page 334) it follows that

$$\frac{dV}{ds} = - \frac{d}{ds} \int_b^s F \, ds = -F$$

that is,

$$F = - \frac{dV}{ds}$$

The last equation can be taken as an equivalent definition of potential energy V.

The kinetic energy $\tfrac{1}{2}mv^2$ of the particle P represents the capacity of P to do work owing to its *motion*. The potential energy V of P also represents the capacity of P to do work, not because it is moving, but because of *where it is located*. For instance, if the particle is connected a spring and the spring is stretched, then work can be done by letting the spring return to its natural length. The **total energy** E of the particle is defined to be the sum of its kinetic energy and its potential energy

$$E = \tfrac{1}{2}mv^2 + V$$

EXAMPLE 6   Prove the **law of conservation of energy** for the particle P discussed above; that is, show that the total energy E remains constant as the particle moves along the s axis.

SOLUTION   It will be enough to show that $dE/dt = 0$. We have

$$\frac{dE}{dt} = \frac{d}{dt}\left(\frac{1}{2}mv^2 + V\right) = mv\frac{dv}{dt} + \frac{dV}{dt} = m\frac{dv}{dt}v + \frac{dV}{ds}\frac{ds}{dt} = Fv - Fv = 0 \quad \blacksquare$$

In the **International System of units** (SI), the basic unit of mass is the **kilogram** (approximately 2.2 pounds), the basic unit of distance is the **meter** (approximately 39.37 inches), and the basic unit of time is the **second.** One **newton** is the force necessary to accelerate one kilogram of mass by one meter per second per second. If mass, distance, time, and force are measured in kilograms, meters, seconds, and newtons, respectively, then energy—kinetic, potential, or total—and work are measured in **joules.** In lifting a 1-pound weight through 1 foot, you do approximately 1.36 joules of work.

Vertical Motion in a Gravitational Field

An object of mass m kilograms near the surface of the earth is acted upon by a constant force of gravity F newtons and if allowed to fall, accelerates with a constant acceleration $g \approx 9.8$ meters per second per second. We refer to the gravitational force F as the **weight** of the object. According to Newton's law,

$$F = mg$$

the weight of an object (near the surface of the earth) is obtained by multiplying its mass by the acceleration of gravity. In particular, *one kilogram weighs approximately 9.8 newtons.*

In the following examples, suppose that a projectile P of mass m is fired vertically upward from the surface of the earth at time $t = 0$ with an initial velocity v_0 (Figure 9).* Take the s axis pointing straight upward with its origin at the surface of the earth. Then the (constant) force F of gravity on the projectile is acting downward; hence F is negative. The acceleration a of P is also negative; hence, $a = -g$ and

$$F = ma = -mg$$

We take $s = 0$, the surface of the earth, as the reference position for the calculation of the potential energy V of P. Thus,

$$V = \int_s^0 F \, ds = \int_s^0 (-mg) \, ds = (-mgs)\Big|_s^0 = 0 - (-mgs) = mgs$$

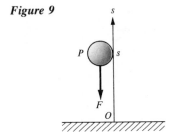

Figure 9

EXAMPLE 7   Find the equation of motion $s = f(t)$ of P.

SOLUTION   We have

$$\frac{dv}{dt} = a = -g \qquad \text{or} \qquad dv = -g \, dt$$

Hence,

$$v = \int (-g) \, dt = -gt + C_1$$

When $t = 0$, we have $v = v_0$, and it follows that $C_1 = v_0$. Therefore,

$$\boxed{v = -gt + v_0}$$

Because $v = ds/dt$, we can rewrite the last equation as

$$\frac{ds}{dt} = -gt + v_0 \qquad \text{or} \qquad ds = (-gt + v_0) \, dt$$

Hence,

$$s = \int (-gt + v_0) \, dt = -\tfrac{1}{2}gt^2 + v_0 t + C_2$$

Since $s = 0$ when $t = 0$, it follows that $C_2 = 0$, and the equation of motion is

$$\boxed{s = -\tfrac{1}{2}gt^2 + v_0 t}$$

∎

*In these examples we assume that the *air resistance* can be neglected and that the *decrease in the force of gravity* as the particle moves away from the center of the earth can also be neglected.

EXAMPLE 8   Find the speed $|v|$ in terms of s.

SOLUTION   From Example 7,

$$v = -gt + v_0 \qquad \text{and} \qquad s = -\tfrac{1}{2}gt^2 + v_0 t$$

Eliminating t from these two equations, we find by simple algebra that

$$v^2 = v_0^2 - 2gs$$

Hence,

$$\boxed{|v| = \sqrt{v_0^2 - 2gs}}$$

EXAMPLE 9   Find the maximum height h to which the projectile climbs in terms of the initial velocity v_0.

SOLUTION   To maximize s, we set

$$\frac{ds}{dt} = v = 0$$

In Example 8, we found that

$$v^2 = v_0^2 - 2gs$$

so, setting $v = 0$ and $s = h$, we obtain

$$0 = v_0^2 - 2gh$$

or

$$\boxed{h = \frac{v_0^2}{2g}}$$

EXAMPLE 10   Find the kinetic energy and the potential energy of P as functions of time t.

SOLUTION   By the result of Example 7, we have

$$\tfrac{1}{2}mv^2 = \tfrac{1}{2}m(-gt + v_0)^2 = \tfrac{1}{2}mg^2t^2 - mgv_0 t + \tfrac{1}{2}mv_0^2$$

Similarly, the potential energy is given by

$$V = mgs = mg(-\tfrac{1}{2}gt^2 + v_0 t) = -\tfrac{1}{2}mg^2t^2 + mgv_0 t$$

EXAMPLE 11   Find the (constant) total energy E of P, and explain what happens to the kinetic energy and to the potential energy of P as it climbs from $s = 0$ to $s = h$.

SOLUTION   By the results of Example 10, we have

$$E = \tfrac{1}{2}mv^2 + V = \tfrac{1}{2}mv_0^2$$

As P climbs from $s = 0$ to $s = h$, there is a trade-off between the kinetic and potential energies. At the start, when $t = 0$ and $s = 0$, all the energy is kinetic, and the potential energy is zero. As P climbs from $s = 0$ to $s = h$, the kinetic energy decreases and the potential energy increases. Finally, when $s = h$, the kinetic energy is zero (because $v = 0$), and all the energy is potential.

Problem Set 6.5

c **1** How much work in foot-pounds is done in stretching a spring of natural length 12 inches from 15 to 18 inches if the final stretching force is 25 pounds?

c **2** A spring has a natural length of 30 centimeters. A force of 200 newtons is required to compress the spring by 3 centimeters. How much work is done in compressing the spring from its natural length to a length of 25 centimeters?

c **3** A reservoir is in the form of a hemisphere of radius 3 meters. If it is filled with salt water weighing 10,110 newtons per cubic meter, how much work (in joules) is required to pump all the salt water out over the rim of the reservoir?

c **4** Water is pumped directly up from the surface of a lake into a water tower. The tank of this water tower is a vertical right circular cylinder of height 20 feet, with radius 5 feet, whose bottom is 60 feet above the surface of the lake. The pump is driven by a 1.5-horsepower motor. Neglecting friction, how long will it take to fill the tank with water? Assume that water weighs 62.4 pounds per cubic foot and that 1 horsepower is 33,000 foot-pounds per minute.

c **5** A conical cistern is 6 meters across the top and 4.5 meters deep and is filled to within 1.5 meters of the top with rainwater weighing 9800 newtons per cubic meter. Find the work in joules done in pumping the water over the top to empty the tank.

c **6** An elevator weighing 10,000 newtons is lifted through 60 meters by winding its cable onto a winch. Neglecting friction, how much work is done if the cable weighs 150 newtons per meter?

c **7** A water trough is 6 feet long and has a cross section consisting of an isosceles trapezoid with altitude 4 feet, upper base of 3 feet, and lower base of 2 feet. If the trough is filled with water weighing 62.4 pounds per cubic foot, how much work will be required to pump all this water up to a level 20 feet above the top of the trough?

c **8** An object immersed in water is buoyed up by a force equal to the weight of the water that the object displaces (**Archimedes' principle**). Find the work required to completely submerge a spherical float of negligible weight if its diameter is 0.3 meter. Assume that the water is in such a large reservoir that its level rises negligibly as we submerge the float and that the water weighs 9800 newtons per cubic meter.

c **9** A cable that is 25 meters long hangs vertically from a winch. If the cable weighs 20 newtons per meter, how much work is done in winding the cable onto the winch?

10 A uniform girder of length L meters and weight w newtons lies horizontally on the ground. How much work in joules will be required to pivot the girder about one of its ends into a vertical position?

c **11** The piston in a cylinder compresses a gas adiabatically from 50 to 25 cubic inches. Assuming that $PV^{1.3} = 100$, where P is the pressure in pounds per square inch and V is the volume in cubic inches, how much work in *foot-pounds* is done on the gas?

12 Gas being compressed in a cylinder satisfies the equation $PV^\gamma = $ constant, where γ is a constant greater than 1, if the compression takes place adiabatically (that is, without heat passing into or out of the cylinder). If the compression takes place isothermally (that is, at constant temperature), then the gas satisfies the equation $PV = $ constant. If a gas is compressed from an initial volume V_0 to a final volume V_1, will more work be required for adiabatic or for isothermal compression?

c **13** Steam expands adiabatically according to the law $PV^{1.4} = $ constant. How much work is done if 0.5 cubic meter of steam at a pressure of 2×10^4 newtons per square meter expands by 60 percent?

c **14** A small compressor is used to supply pressurized air for filling tires. The compressor is driven by a 1500-watt electric motor, and its air tank has a volume of 0.02 cubic meter. Suppose we start from a normal atmospheric pressure of 1.013×10^5 newtons per square meter. How long will be required to bring the pressure in the air tank up to 5×10^5 newtons per square meter? Assume that the compression takes place adiabatically, so that $PV^{1.4} = $ constant, and recall that one **watt** is one joule per second.

c In Problems 15 to 21, use the fact that water weighs 9800 newtons per cubic meter (approximately 62.4 pounds per cubic foot).

15 What force must be withstood by a vertical dam 30 meters long and 6 meters deep if the water level is at the top of the dam?

16 A water main in the shape of a horizontal cylinder 1.9 meters in radius is half filled with water. Find the force on the circular gate that closes the main.

17 A rectangular oil can is filled with oil weighing 9114 newtons per cubic meter. What is the force on one side of the can that is 0.2 meter wide and 0.4 meter high?

18 A tank in the shape of a horizontal right circular cylinder of radius 0.5 meter is half filled with water and half filled with oil. The oil, which weighs 9114 newtons per cubic meter, floats on top of the water. Find the total force caused by these liquids on the circular end of the tank.

19 The face of a vertical dam has the shape of an isosceles trapezoid of altitude 16 feet with an upper base of 42 feet and a lower base of 30 feet. Find the total force exerted by the water on the dam when the water is 12 feet deep.

20 A horizontal tank in the shape of a right circular cylinder of radius 2 meters is sealed, and water is forced in until the tank is exactly half full. Find the total force on one end of the tank caused by the resulting air and water pressure. Take normal atmospheric pressure to be 1.013×10^5 newtons per square meter.

21 A water trough has a cross section in the shape of an isosceles triangle with its apex pointing downward. If the trough is 0.6 meter wide and 0.4 meter deep, find the force on one end when it is full of water.

22 Show that for a region R submerged as shown in Figure 10, the force F caused by a liquid weighing w units of weight per unit volume is given by

$$F = \frac{w}{2} \int_a^b [h(s)]^2 \, ds$$

Figure 10

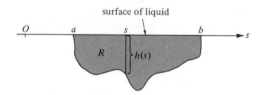

surface of liquid

23 A 5-cent piece has a mass of about 5.2×10^{-3} kilogram. (a) How many newtons does a 5-cent piece weigh? (b) How many joules of work are required to lift a 5-cent piece 1 meter? (c) If 5 cents will buy 1.8×10^6 joules of electric energy, approximately how many 5-cent pieces could be lifted 1 meter for 5 cents' worth of electricity?

24 An elastic spring with a spring constant of k newtons per meter is stretched from its relaxed position through L meters. How much (potential) energy in joules is stored in the stretched spring?

25 (a) Is the work necessary to stretch a spring from 21 to 22 centimeters the same as the work necessary to stretch it from 22 to 23 centimeters? Explain. (b) Is the work necessary to lift a 5-cent piece from 21 to 22 centimeters above ground level the same as the work necessary to lift it from 22 to 23 centimeters? Explain.

26 A weight of 1 metric ton (1000 kilograms) is dropped from a height of h feet, and it hits the ground with the same kinetic energy as that possessed by a car of mass 1.2 metric tons traveling 88 kilometers per hour. Find h.

27 Given that the potential energy of a relaxed spring is zero and that the spring constant is 500 newtons per meter, how far must the spring be stretched so that its potential energy is 200 joules?

28 Two particles P_1 and P_2 with masses m_1 and m_2 are moving on the s axis with variable velocities v_1 and v_2. A variable force of attraction exists between the two particles, so that the force exerted on P_1 by P_2 is F_1 and the force exerted on P_2 by P_1 is F_2.

Given that $-F_1 = F_2$, prove that the sum $m_1v_1 + m_2v_2$ of the linear momentum of P_1 and the linear momentum of P_2 remains constant during the motion.

29 A particle P is moving along the positive s axis under the influence of a force F. The total energy E of P is zero, and its potential energy is given by $V = -1/s$. (a) Find the velocity $v = ds/dt$ of P in terms of s. (b) If $s = 25$ meters and v is negative when $t = 0$, find the equation of motion of P. (c) If $s = 25$ meters and v is positive when $t = 0$, find the equation of motion of P and show that the velocity v of P is positive for all $t > 0$. (d) Given the information in part (c), find $\lim_{t \to +\infty} v$.

30 Consider the mass-spring system in Figure 11 in which the mass m is suspended by a perfectly elastic spring with spring constant k. The mass of the spring itself and air friction are to be neglected. The origin on the vertical s axis is chosen so that when the mass and spring are hanging in equilibrium, the s coordinate of m is zero. The mass m is lifted to the position with coordinate $s = A_0$ and released at time $t = 0$ with an initial velocity $v_0 = 0$.

Figure 11

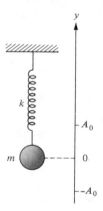

(See the discussion at the end of Section 4.6.) (a) Given that the potential energy V is zero when the mass passes through the origin, find a formula giving V in terms of s. (b) Find the (constant) total energy E of the mass-spring system. (c) Find the velocity of the mass when it passes through the position $s = 0$ the first time after it is released. (d) Discuss the trade-off between the potential and the kinetic energies as the mass oscillates.

31 Two positively charged particles with charges of 1 coulomb apiece repel each other with a force equal to k/r^2 newtons, where r is the distance in meters between the particles and the constant k is given by $k = 8.99 \times 10^9$. Suppose that one of the particles is fixed at the origin and the other particle, call it P, is free to move along the positive s axis. Find the potential energy V of P in terms of s, given that $\lim_{s \to +\infty} V = 0$.

32 Figure 12 shows an s axis with its origin at the center of a homogeneous solid sphere of radius r meters and with constant density of w kilograms per cubic meter. A particle P of mass m kilograms at the point with coordinate s on the axis will experience a gravitational force F given (in newtons) by

$$F = \begin{cases} \dfrac{-4Gm\pi r^3 w}{3s^2} & \text{for } s \geq r \\[2mm] -\dfrac{4}{3}Gm\pi w s & \text{for } 0 \leq s \leq r \end{cases}$$

where $G = 6.672 \times 10^{-11}$ N·m²/kg² is Newton's universal constant of gravitation. (a) Find the potential energy V of P in terms of s if $\lim\limits_{s\to +\infty} V = 0$. (b) Find the *escape velocity* for P starting on the surface of the sphere.

33 Suppose that a projectile P of mass m is fired vertically upward from the surface of the earth at time $t = 0$ with an initial velocity v_0. Neglecting air resistance and neglecting the decrease in the force of gravity as the projectile moves away from the center of the earth, find a formula for the kinetic energy of P in terms of its height s above the surface of the earth.

Figure 12

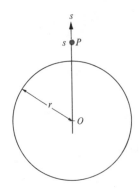

[C]34 Work Problem 33, continuing to neglect air resistance, but taking into account the decrease in the force of gravity as the projectile moves away from the center of the earth. Take the radius of the earth to be 6.371×10^6 meters and the mass of the earth to be 5.983×10^{24} kilograms.

6.6 Applications to Economics and the Life Sciences

We now present a few simple, but typical, applications of the definite integral to economics and the life sciences.

Consumer's Surplus

Suppose that the **demand equation**

$$p = f(q)$$

relates the price p per unit of a commodity and the number of units q of the commodity that will be demanded at that price by consumers in a given period of time. Usually, fewer units of the commodity are demanded as the price per unit increases, and when the price reaches a sufficiently large value, say c dollars per unit, none of the commodity is demanded. Thus, we assume that

$$c = f(0)$$

Now suppose that the actual market price, p_0 dollars per unit, is less than c dollars per unit and that the corresponding demand, q_0 units, is positive. Thus,

$$p_0 = f(q_0)$$

Clearly, those consumers who are willing to pay more than p_0 dollars per unit benefit from the fact that the price is only p_0 dollars per unit. Economists measure this benefit, the **consumer's surplus,** by the integral

$$\text{consumer's surplus} = \int_{p_0}^{c} q \, dp$$

We now derive an alternative formula for the consumer's surplus in terms of the function f. To begin with, notice that the revenue R dollars to the producer for the sale of q units of the product at p dollars per unit is given by

$$R = pq$$

If we take the differential on both sides of the last equation, we obtain

$$dR = p\,dq + q\,dp \qquad \text{or} \qquad q\,dp = -p\,dq + dR$$

Therefore,

$$\text{consumer's surplus} = \int_{P_0}^{c} q\,dp = \int_{p=p_0}^{p=c} (-p)\,dq + \int_{p=p_0}^{p=c} dR$$

$$= -\int_{p=p_0}^{p=c} p\,dq + R\Big|_{p=p_0}^{p=c}$$

Now, since $q = q_0$ when $p = p_0$ and $q = 0$ when $p = c$, we can rewrite the last equation as

$$\text{consumer's surplus} = -\int_{q=q_0}^{q=0} p\,dq + R\Big|_{p=p_0}^{p=c} = \int_{0}^{q_0} p\,dq + pq\Big|_{p=p_0,\ q=q_0}^{p=c,\ q=0}$$

$$= \int_{0}^{q_0} p\,dq + c \cdot 0 - p_0 q_0$$

$$= \int_{0}^{q_0} p\,dq - p_0 q_0$$

Therefore, since $p = f(q)$, we have

$$\boxed{\text{consumer's surplus} = \int_{0}^{q_0} f(q)\,dq - f(q_0)q_0}$$

EXAMPLE 1   The demand equation for coal in a particular marketing area is $p = 200(600 - 10q - q^2)$, where q is the demand in thousands of tons per week and p is the price in dollars per thousand tons. If the current price is \$80,000 per thousand tons, find the consumer's surplus.

SOLUTION   Here, $f(q) = 200(600 - 10q - q^2)$ and $p_0 = \$80,000$. Solving the quadratic equation

$$80,000 = 200(600 - 10q_0 - q_0^2)$$

for q_0, and recalling that q_0 cannot be negative, we find that $q_0 = 10$. Therefore,

$$\text{consumer's surplus} = \int_{0}^{q_0} f(q)\,dq - f(q_0)q_0$$

$$= \int_{0}^{10} 200(600 - 10q - q^2)\,dq - (80,000)(10)$$

$$= 200\left(600q - 5q^2 - \frac{q^3}{3}\right)\Big|_{0}^{10} - 800,000$$

$$= \frac{700,000}{3} \approx \$233,333.33 \qquad \blacksquare$$

Production over a Period of Time

When a new production process is set in motion, the rate of production is often rather slow at first because of "bugs" in the production techniques, unfamiliarity with new methods, and so forth. As the bugs are worked out and the personnel become accustomed to the new methods, the rate of production ordinarily increases and, after a suitable period of time, approaches a steady value.

Thus, let x denote the number of units of a certain commodity produced in the first t units of working time by a new process, so the derivative dx/dt represents the rate of production at time t. Then, according to the fundamental theorem of calculus, the total number of units of the commodity produced during the interval of time from $t = a$ to $t = b$ is given by

$$x\Big|_{t=a}^{t=b} = \int_{t=a}^{t=b} \frac{dx}{dt}\, dt$$

EXAMPLE 2   Solar Electric Company has set up a new production line to manufacture small wind-powered generators to be used as alternative energy sources. The rate of production t weeks after the start is $dx/dt = 300[1 - 400(t + 20)^{-2}]$ generators per week. How many generators are produced during the fifth week of operation?

SOLUTION   From the end of the fourth week, when $t = 4$, to the end of the fifth week, when $t = 5$, the number of generators produced is

$$\int_4^5 300[1 - 400(t + 20)^{-2}]\, dt = 300\left(t + \frac{400}{t + 20}\right)\Big|_4^5 = 6300 - 6200$$

$$= 100 \text{ generators} \qquad \blacksquare$$

Pollution

The rate at which pollutants are introduced into a given ecosystem may vary with time as a consequence of a number of factors. For instance, the rate at which a factory dumps pollutants into a lake may increase as the level of production increases and as antipollution devices in the factory wear out and become less efficient. If we denote by x the number of units of pollutant accumulated in a given ecosystem after t units of time, the rate of pollution of the ecosystem is given by the derivative dx/dt. Consequently, the total number of units of pollutant accumulated in the ecosystem during the interval of time from $t = a$ to $t = b$ is given by

$$x\Big|_{t=a}^{t=b} = \int_{t=a}^{t=b} \frac{dx}{dt}\, dt$$

EXAMPLE 3   Filters at the O.L. Factory, designed to remove sulfur dioxide from vented air, are replaced every 90 days. However, t days after the filters are changed, they allow sulfur dioxide to escape into the atmosphere at the rate of $25\sqrt{t}/10$ pounds per day. How many pounds of sulfur dioxide are introduced into the atmosphere over one 90-day interval?

SOLUTION   Making the change of variable $u = t/10$, we have

$$\int_0^{90} 25 \sqrt{\frac{t}{10}} \, dt = 250 \int_0^9 u^{1/2} \, du = 250 \left(\frac{2}{3} u^{3/2} \right) \Big|_0^9 = 4500 \text{ lb}$$ ∎

Blood Flow in the Circulatory System

Figure 1

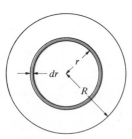

Provided that certain factors (such as pressure and viscosity) are held within pre-scribed limits, blood will flow smoothly through a cylindrical blood vessel in such a way that the velocity v of flow increases continuously from a value close to zero at the wall of the vessel to a maximum value at its center. Figure 1 shows a cross section of the blood vessel perpendicular to its central axis and an infinitesimal circular ring of width dr at a distance r from the center. Assume that the velocity v of blood flow depends only on r; then the infinitesimal volume of blood dV flowing across the circular ring in unit time will be given by

$$dV = v(2\pi r \, dr)$$

the product of the velocity of flow across the circular ring and the area of this ring. If R denotes the radius of the blood vessel, the total volume of blood passing the entire cross section in unit time will be given by

$$V = \int_{r=0}^{r=R} dV = 2\pi \int_{r=0}^{r=R} vr \, dr$$

In 1842, the French physician Jean-Louis-Marie Poiseuille (1799–1869) deter-mined experimentally that for nonturbulent blood flow,

$$v = k(R^2 - r^2)$$

where k is a constant depending on the viscosity of the blood, the length of the blood vessel, and the difference in blood pressure between the two ends of the vessel.

ⓒ **EXAMPLE 4**   Suppose that blood is flowing through an artery of radius $R = 0.25$ centimeter according to **Poiseuille's equation** $v = k(R^2 - r^2)$ with $k = 90$. Find the rate of flow of blood through the artery.

SOLUTION   The velocity v of the blood flow r centimeters from the center of the artery is given by

$$v = k(R^2 - r^2) = 90[(0.25)^2 - r^2] = 5.625 - 90r^2 \qquad \text{cm/s}$$

Hence, the rate of flow, measured by the total volume V of blood passing a cross section of the artery in 1 second, is

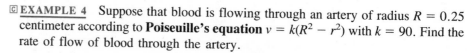

$$V = 2\pi \int_0^R vr \, dr = 2\pi \int_0^{0.25} (5.625 - 90r^2)r \, dr = 2\pi \int_0^{0.25} (5.625r - 90r^3) \, dr$$

$$= 2\pi \left(\frac{5.625r^2}{2} - \frac{90r^4}{4} \right) \Big|_0^{0.25} = 2\pi \left[\frac{5.625(0.25)^2}{2} - \frac{90(0.25)^4}{4} \right]$$

$$\approx 0.55 \text{ cm}^3/\text{s}$$ ∎

Problem Set 6.6

1 Consider the demand equation $q = 100(4 - \sqrt{4p})$. (a) Solve the demand equation for p in terms of q. (b) Calculate the consumer's surplus when the price is $p_0 = \$1$ per unit, using the definition of consumer's surplus as the integral $\int_{p_0}^{c} q \, dp$. (c) Calculate the consumer's surplus, using the alternative formula $\int_{0}^{q_0} f(q) \, dq - f(q_0)q_0$, where $p = f(q)$.

2 Sketch a graph showing an arbitrary demand curve $p = f(q)$ and a point (q_0, p_0) on this curve. Shade in an appropriate region on your figure whose area corresponds to the consumer's surplus when the price is p_0 dollars per unit.

3 On the basis of market surveys, a manufacturer determines that 5000 souvenirs will be sold per week in a certain resort area if the price is $1 per souvenir, but that for every 5-cent increase in price per souvenir, 500 fewer souvenirs per week will be sold. Find the demand equation for the souvenirs, and calculate the consumer's surplus if the souvenirs are priced at $1.25 each.

4 Suppose that the **supply equation** $p = g(q)$ relates the price per unit of a commodity and the number of units q of the commodity that the producers would be willing to supply per unit of time at that price. Supposing that the producers would be unwilling to supply any of the commodity if the price drops as low as c dollars per unit, so that $c = g(0)$, and that the current selling price of the commodity is p_0 dollars per unit, economists define the **producer's surplus** as the value of $\int_{c}^{p_0} q \, dp$. If $p_0 = g(q_0)$, show that the producer's surplus is given by $p_0q_0 - \int_{0}^{q_0} g(q) \, dq$.

5 Heron Motors has set up an assembly line for their new steam-powered automobile and expects to be producing them at the rate of $30\sqrt{t}$ automobiles per week at the end of t weeks. How many automobiles do they expect to produce during the first 36 weeks of production?

6 Suppose that the rate of production of a new product is

$$A\left[1 - \left(\frac{k}{t + k}\right)^{p}\right]$$

units per week at the end of t weeks, where A and k are positive constants and p is a constant greater than 1. Find a formula for the number of units produced during the nth week of production.

7 A factory is dumping pollutants into a lake at the rate of $t^{2/3}/600$ tons per week, where t is the time in weeks since the factory commenced operations. After 10 years of operation, how much pollutant has the factory dumped into the lake?

8 In Problem 7, assume that natural processes can remove up to 0.015 ton of pollutant per week from the lake and that there was no pollution in the lake when the factory commenced operations 10 years ago. How many tons of pollutant have now accumulated in the lake?

9 A cylindrical blood vessel has a radius $R = 0.1$ centimeter, and blood is flowing through this vessel with a velocity $v = 0.30 - 30r^2$ centimeters per second at points r centimeters from the center. Find the rate of flow of the blood.

10 If blood is flowing in a blood vessel according to Poiseuille's equation $v = k(R^2 - r^2)$, show that the volume V of blood that flows across a fixed cross section of the vessel in unit time is given by $V = (\pi k/2)R^4$.

Review Problem Set, Chapter 6

In Problems 1 to 12, use any convenient method to find the volume of the solid generated by revolving the region bounded by the given curves about the indicated axis.

1 $y = \sqrt[3]{x}$, $y = 0$, and $x = 8$ about the x axis

2 $y = x^2$, $y = 0$, and $x = 1$ about the y axis

3 $x^2 + 4 = 4y$, $x = 0$, and $y = x$ about the x axis

4 $y = x^3$, $x = 1$, and $y = 0$ about the line $x = -2$

5 $y = x^2$ and $y = 2x$ about the y axis

6 $x^2 = 4(1 - y)$ and $y = 0$ about the line $y = 3$

7 $y = 2/x$ and $y = 3 - x$ about the y axis

8 $y^2 = 4x$ and $x = 4$ about the line $x = -2$

9 $y = x^{2/3}$, $y = 1$, and $y = 4$ about the y axis

10 The loop of $y^2 = x^4(x + 4)$ about the y axis

11 $y = x$, $x = 4$, and $y = 0$ about the line $y = -2$

12 $y = \sqrt{x}$ and $y = 18 - x^2$, to the right of $x = 1$, about the y axis

13 Find the volume of the solid generated by revolving the region bounded by a circle of radius 3 centimeters about a line in its plane that is 7 centimeters from its center.

14 A solid is generated by revolving the region bounded by the graph of $y = f(x)$, the lines $x = 0$ and $x = a$, and the x axis about the x axis. Its volume for all values of a is given by the equation $V = a^2 + 7a$. Find a formula for $f(x)$.

15 A solid **paraboloid of revolution** is generated by revolving the region bounded by the graph of $y = b\sqrt{x/a}$, the x axis, and the line $x = a$ about the x axis. (a) Sketch the paraboloid of revolution. (b) Find the volume of the solid by the method of circular disks. (c) Find the volume of the solid by the method of cylindrical shells.

16 A vessel has the form generated by revolving the graph of $y = \frac{1}{4}x^2$ about the y axis. If 4 cubic units of liquid leak out of the bottom of the vessel per minute, at what rate is the depth of liquid changing at the instant when the depth is 4 units?

17 Sketch the graph of the portion of the parabola $Ax^2 = B^2y$ lying in the first quadrant, where A and B are positive constants. Find the volume of the solid generated when the region bounded by this parabola, the y axis, and the line $y = A$ is revolved about the y axis. Compare the answer with the volume of a cylinder of height A and base radius B.

18 For each positive integer n, let U_n denote the volume of the solid generated by revolving the region under the graph of $y = x^n$ between $x = 0$ and $x = 1$ about the y axis. Let V_n be the volume of the solid generated by revolving the same region about the x axis. Evaluate

(a) $\lim_{n \to +\infty} U_n$ (b) $\lim_{n \to +\infty} V_n$ (c) $\lim_{n \to +\infty} \dfrac{U_n}{V_n}$

19 Find the volume of a solid whose cross section perpendicular to the x axis is a square x units on a side for $0 \le x \le 2$.

20 Find the volume of a solid whose cross section perpendicular to the x axis is an equilateral triangle $|x|$ units on a side for $-2 \le x \le 3$.

21 A spire is 26 feet high. A horizontal cross section x feet above the base is a rectangle whose long side is $(26 - x)/13$ feet long and whose short side is $(26 - x)/20$ feet long. Find the volume of the spire.

22 A **conoid** is a wedge-shaped solid whose lateral surface is generated by a line segment which moves so as always to be parallel to a fixed plane and which has one endpoint on a fixed circle and the other endpoint on a fixed line, with both the circle and the straight line being perpendicular to the fixed plane (Figure 1). If the distance from the fixed line to the plane of the circle is h units and the circle has radius a units, find the volume enclosed by the conoid.

Figure 1

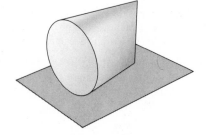

23 A solid is generated by a variable regular hexagon which moves so that its plane is always perpendicular to a given diameter of a fixed circle of radius a, with the center of the hexagon lying in this diameter and two opposite vertices lying on the circle. Find the volume of the solid.

24 Liquid in a reservoir evaporates at a rate (cubic units per unit time) proportional to the surface area of the liquid exposed to the air. Show that no matter what the shape of the reservoir, the surface of the liquid drops at a constant rate.

25 A football-shaped solid is swept out by a variable square that is perpendicular to the x axis and whose center lies on the x axis. In sweeping out the solid, one vertex of the variable square follows the curve $y = \frac{1}{2}\sqrt{49 - x^2}$ in the xy plane between $x = -7$ and $x = 7$. Find the volume of the solid.

26 A horn-shaped solid is generated by a variable circle whose plane turns about a fixed line. The point of the circle nearest the line describes a quadrant $\overset{\frown}{AB}$ of a circle of radius a, and the radius of the variable circle is $c\theta$, where θ denotes the angle between the variable plane and its initial position when it passes through A. Show that the volume of the horn is given by $V = \frac{1}{192}\pi^4 c^2(8a + 3\pi c)$.

In Problems 27 to 38, find the arc length of the graph of each equation between the points indicated.

27 $(y - 8)^2 = x^3$ from $(0, 8)$ to $(1, 9)$

28 $y^3 = 4x^2$ from $(4, 4)$ to $(32, 16)$

29 $y = \frac{2}{3}x^{3/2}$ from $(0, 0)$ to $(4, \frac{16}{3})$

30 $y = \frac{1}{8}[x^4 + (2/x^2)]$ from $(1, \frac{3}{8})$ to $(2, \frac{33}{16})$

31 $y^2 = \frac{4}{9}x^3$ from $(0, 0)$ to $(1, \frac{2}{3})$

32 $x = \dfrac{y^5}{5} + \dfrac{1}{12y^3}$ from $(\frac{323}{480}, \frac{1}{2})$ to $(\frac{17}{60}, 1)$

33 $y = (x + 1)^{3/2} + 2$ from $(3, 10)$ to $(8, 29)$

34 $y = \displaystyle\int_0^x \sqrt{t^2 + 2t}\ dt$ from $(0, 0)$ to $\left(1, \displaystyle\int_0^1 \sqrt{t^2 + 2t}\ dt\right)$

35 $(x + 1)^2 = 4y^3$ from $(-1, 0)$ to $(1, 1)$

36 $x = \frac{2}{3}(y + 1)^{3/2}$ from $(\frac{16}{3}, 3)$ to $(18, 8)$

37 $y = \displaystyle\int_1^x \sqrt{2t^4 + t^7 - 1}\ dt$ from $(1, 0)$ to $\left(2, \displaystyle\int_1^2 \sqrt{2t^4 + t^7 - 1}\ dt\right)$

38 $y = \sqrt{4 - x^2}$ from $(-2, 0)$ to $(2, 0)$

© In Problems 39 to 42, set up an integral representing the arc length of each curve, and use Simpson's parabolic rule with $n = 2$ (that is, with *four* subintervals) to estimate this arc length.

39 $y = 2\sqrt{x}$ from $(1, 2)$ to $(4, 4)$

40 $y = 2(1 + x^2)^{1/2}$ from $(0, 2)$ to $(1, 2\sqrt{2})$

41 $y = x^2$ from $(0, 0)$ to $(1, 1)$

42 $y = \displaystyle\int_0^x \dfrac{dt}{1 + t^2}$ from $(0, 0)$ to $\left(2, \displaystyle\int_0^2 \dfrac{dt}{1 + t^2}\right)$

43 Use the arc length formula to verify that the straight-line distance d between the points $P_1 = (x_1, y_1)$ and $P_2 = (x_2, y_2)$ is given by $d = \sqrt{(x_2 - x_1)^2 + (y_2 - y_1)^2}$.

44 Suppose that the function f has a continuous first derivative on an open interval containing the closed interval $[0, 1]$ and that f is monotonically decreasing on $[0, 1]$. Assume that $f(0) = 1$ and $f(1) = 0$. Write an integral for the amount of time T required for a particle P of mass m to slide frictionlessly down the graph of $y = f(x)$ from $(0, 1)$ to $(1, 0)$ under the influence of gravity.

In Problems 45 to 54, find the area of the surface of revolution obtained by revolving each curve about the axis given.

45 $y = 3\sqrt{x}$ from $(1, 3)$ to $(4, 6)$ about the x axis

46 $y = 3x^2$ from $(0, 0)$ to $(2, 12)$ about the y axis

47 $y = x^3$ from $(1, 1)$ to $(3, 27)$ about the x axis

48 $y = 4x^2$ from $(0, 0)$ to $(1, 4)$ about the y axis

49 $y = \frac{1}{3}x^3$ from $(0, 0)$ to $(3, 9)$ about the x axis

50 $y = \dfrac{x^4}{4} + \dfrac{1}{8x^2}$ from $(1, \frac{3}{8})$ to $(3, \frac{1459}{72})$ about the x axis

51 $y = \frac{1}{4}x^2$ from $(0, 0)$ to $(4, 4)$ about the y axis

52 $y^2 = x^3$ from $(1, 1)$ to $(4, 8)$ about the y axis

53 $y^2 = 9 - x$ from $(0, 3)$ to $(9, 0)$ about the x axis

54 $y = \dfrac{x^5}{5} + \dfrac{1}{12x^3}$ from $(1, \frac{17}{60})$ to $(2, \frac{3077}{480})$ about the x axis

55 Find the work done in pumping all the water out of a full cistern in the shape of a hemisphere of radius r feet surmounted by a right circular cylinder of radius r feet and height h feet.

56 A tank is made in the shape of a right circular cylinder surmounted by a frustum of a cone as shown in Figure 2. How much work is done in pumping the tank full of water *from the bottom*?

Figure 2

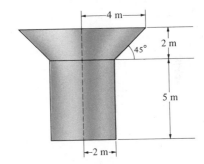

57 A cable that is 30 meters long and weighs 4 newtons per meter hangs vertically from a winch. Find the work done in winding half of the cable onto the winch.

58 A cable that is l units long weighs w units of force per unit of length and hangs vertically from a winch. If a weight W hangs at the end of the cable, how much work is done in winding k units of the cable onto the winch?

59 How much work is done compressing 400 cubic feet of air at 15 pounds per square inch to a volume of 30 cubic feet, provided that the compression is adiabatic ($PV^{1.4} = $ constant)?

60 A vertical masonry dam in the form of an isosceles trapezoid is 70 meters long at the surface of the water, 50 meters long at the bottom, and 20 meters high. Find the total force that it must withstand if the water is level with the top of the dam.

61 A right triangle ABC is submerged in water so that edge $\overline{AB}$ lies on the surface and edge $\overline{BC}$ is vertical. Find the total force on a face of the triangle caused by the pressure of the water if $|\overline{BC}| = 6$ meters and $|\overline{AB}| = 2$ meters.

62 A mass m is falling freely near the earth's surface. At a certain instant, $t = 0$ seconds, the mass is s_0 units above the ground and is falling with a velocity v_0. Take the s axis positively upward with its origin at ground level. (Thus, v_0 is *negative*, and the acceleration of gravity is $-g$.) (a) Find the velocity of the mass in terms of its distance s above the ground. (b) Find the increase in the kinetic energy of the mass between $s = s_0$ and $s = s_1$, where $0 \leq s_1 \leq s_0$. (c) Find the potential energy V of the mass in terms of s if the potential energy is zero when $s = s_0$. (d) Compute the decrease in the potential energy of the mass between $s = s_0$ and $s = s_1$, where $0 \leq s_1 \leq s_0$. (e) Find the kinetic energy K and the potential energy V of the mass in terms of the time t.

63 The velocity of a 100-kilogram sled is $20 - 2t$ meters per second at time t seconds. Find the change in its kinetic energy between $t = 0$ and $t = 10$ seconds. How much work is done in stopping the sled?

64 A rod of length b is spinning n times per second about one of its ends. The material of which the rod is composed has a density of $f(x)$ mass units per unit of length at a distance of x units from the stationary end. Find an integral representing the kinetic energy of the spinning rod.

65 Joe, Jamal, and Gus have set up a lemonade stand. The demand equation for lemonade is $p = -(q^2/100) - (7q/20) + 30$, where p is their asking price in cents per glass and q is the number of glasses demanded. If the asking price is 15 cents per glass, find the consumer's surplus.

66 Rebecca, Joshua, and Miriam have grown 100 pumpkins, which they are planning to sell. They believe that they can sell all their pumpkins at a price of 75 cents each, but that each 25-cent increase in price per pumpkin will result in the sale of 20 fewer pumpkins. What is the consumer's surplus if they sell the pumpkins at the price that brings them the maximum revenue?

67 A former mathematics professor is going into business as an income tax consultant and expects to be receiving $100\sqrt{t}$ dollars per week in consulting fees t weeks after opening a new office. The fixed expenses for office rental, telephone, newspaper advertisements, and so forth will amount to \$200 per week. What is the consultant's net profit after 49 weeks in this new job?

68 Suppose that a new employee at Solar Electric Company can solder $50 - 50/\sqrt{t + 1}$ connections per hour at the end of t working hours. How many connections does the employee solder during the first 8 working hours?

69 Blood is flowing smoothly through a cylindrical artery of radius R in such a way that its velocity r units from the center is given by $v = K(R^2 - r^2)$ units per second. Here K is a constant depending on blood pressure, viscosity, and so forth. A drug is administered which increases the radius of the artery by 5 percent. Calculate the percentage increase in the rate of flow of blood through the artery.

7 TRANSCENDENTAL FUNCTIONS

Algebraic and trigonometric functions, useful as they are, are not sufficient for the applications of mathematics to physics, chemistry, engineering, economics, and the life and earth sciences. In this chapter, we add the inverse trigonometric, logarithmic, exponential, and hyperbolic functions to our repertory. In order to better understand the inverse trigonometric functions and the relationship between logarithmic and exponential functions, we begin by considering the general idea of the inverse of a function.

All the functions that can be built up from algebraic, trigonometric, logarithmic, and exponential functions by addition, subtraction, multiplication, division, composition, and inversion are called **elementary functions.** Although nonelementary functions sometimes have to be used in applied mathematics, the elementary functions are sufficient for our purposes in this textbook.

7.1 Inverse Functions

Many calculators have both a squaring key (marked x^2) and a square-root key (marked $\sqrt{x}$). For nonnegative numbers x, the functions represented by these keys "undo each other" in the sense that

$$x \overset{\text{squaring function}}{\longmapsto\!\!\!\longrightarrow} x^2 \overset{\text{square-root function}}{\longmapsto\!\!\!\longrightarrow} \sqrt{x^2} = x$$

and

$$x \overset{\text{square-root function}}{\longmapsto\!\!\!\longrightarrow} \sqrt{x} \overset{\text{squaring function}}{\longmapsto\!\!\!\longrightarrow} (\sqrt{x})^2 = x$$

Two functions related in such a way that each "undoes" what the other "does" are said to be **inverse** to each other. Not only do most scientific calculators have keys corresponding to a number of important functions, but they also can calculate the values of the inverses of these functions.

In order that two functions f and g be inverses of each other, we must have the following: For every value of x in the domain of f, $f(x)$ is in the domain of g and

$$x \xmapsto{\;f\;} f(x) \xmapsto{\;g\;} g[f(x)] = x$$

Likewise, for every value of x in the domain of g, $g(x)$ is in the domain of f and

$$x \xmapsto{\;g\;} g(x) \xmapsto{\;f\;} f[g(x)] = x$$

In other words, f and g are inverses of each other if and only if

$$(g \circ f)(x) = g[f(x)] = x$$

for every value of x in the domain of f and

$$(f \circ g)(x) = f[g(x)] = x$$

for every value of x in the domain of g. A function f for which such a function g exists is said to be **invertible.**

In Examples 1 and 2, show that the functions f and g are inverses of each other.

<u>EXAMPLE 1</u> $f(x) = 5x - 1$ and $g(x) = \dfrac{x + 1}{5}$

SOLUTION   The domain of f is the set $\mathbb{R}$ of all real numbers. For every real number x, we have

$$(g \circ f)(x) = g[f(x)] = g(5x - 1) = \frac{(5x - 1) + 1}{5} = \frac{5x}{5} = x$$

The domain of g is also the set $\mathbb{R}$ of all real numbers, and for every real number x, we have

$$(f \circ g)(x) = f[g(x)] = f\left(\frac{x + 1}{5}\right) = 5\left(\frac{x + 1}{5}\right) - 1 = x + 1 - 1 = x \qquad \blacksquare$$

<u>EXAMPLE 2</u> $f(x) = \dfrac{2x - 1}{x}$ and $g(x) = \dfrac{1}{2 - x}$

SOLUTION   The domain of f is the set of all nonzero real numbers. If $x \neq 0$, we have

$$(g \circ f)(x) = g[f(x)] = g\left(\frac{2x - 1}{x}\right) = \frac{1}{2 - \left(\frac{2x - 1}{x}\right)} = \frac{x}{\left[2 - \left(\frac{2x - 1}{x}\right)\right]x}$$

$$= \frac{x}{2x - (2x - 1)} = \frac{x}{1} = x$$

The domain of g is the set of all real numbers except 2. If $x \neq 2$, we have

$$(f \circ g)(x) = f[g(x)] = f\left(\frac{1}{2 - x}\right) = \frac{2\left(\frac{1}{2 - x}\right) - 1}{\left(\frac{1}{2 - x}\right)} = \left[2\left(\frac{1}{2 - x}\right) - 1\right](2 - x)$$

$$= 2 - (2 - x) = x \qquad \blacksquare$$

Figure 1

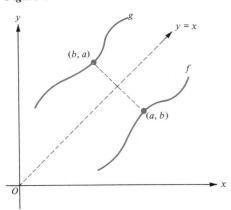

Geometrically, f and g are inverses of each other if and only if the graph of g is the mirror image of the graph of f across the straight line $y = x$ (Figure 1). This fact will be clear if you note first that the mirror image of a point (a, b) across the line $y = x$ is the point (b, a). But if f and g are inverses of each other and if (a, b) belongs to the graph of f, then

$$b = f(a) \qquad \text{so} \qquad g(b) = g[f(a)] = a$$

that is, (b, a) belongs to the graph of g. Similarly, you can show that if (b, a) belongs to the graph of g, then (a, b) belongs to the graph of f (Problem 42).

Not every function is invertible. Indeed, consider the function f whose graph appears in Figure 2. The mirror image of the graph of f across the line $y = x$ isn't the graph of a function, because there is a vertical line l that intersects it more than once. (Recall the vertical-line test, page 26.) Notice that the horizontal line L obtained by reflecting l across the line $y = x$ intersects the graph of f more than once. These considerations provide the basis for the following test (see Problem 43).

Horizontal-Line Test

A function f is invertible if and only if no horizontal straight line intersects its graph more than once.

Figure 2

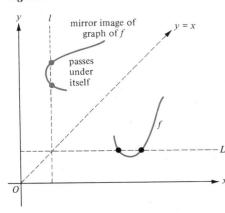

EXAMPLE 3 Use the horizontal-line test to determine whether the functions graphed in Figure 3 are invertible.

Figure 3

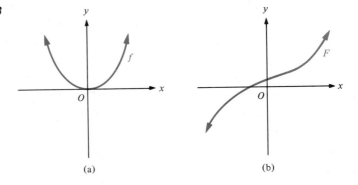

(a)   (b)

SOLUTION

(a) Any horizontal line drawn above the origin will intersect the graph of f *twice*. Therefore, f is not invertible.

(b) No horizontal line intersects the graph of F more than once; hence, F is invertible. ∎

Figure 4

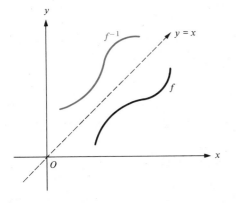

If a function f is invertible, there is *exactly one* function g such that f and g are inverses of each other; indeed, g is the one and only function whose graph is the mirror image of the graph of f across the line $y = x$. We call g the **inverse** of the function f, and we write $g = f^{-1}$. The notation f^{-1} is read "f inverse." If you imagine that the graph of f is drawn with wet ink, then the graph of f^{-1} would be the imprint obtained by folding the paper along the line $y = x$ (Figure 4).

The fact that the graph of f^{-1} is the set of all points (x, y) such that (y, x) belongs to the graph of f provides the basis for the following procedure.

Algebraic Method for Finding f^{-1}

> **Step 1** Write the equation $y = f(x)$ that defines f.
>
> **Step 2** Interchange x and y in the equation obtained in step 1; that is, change y to x and change all x's to y's so that the equation becomes $x = f(y)$.
>
> **Step 3** Solve the equation in step 2 for y in terms of x to get $y = f^{-1}(x)$. This equation defines f^{-1}.

Figure 5

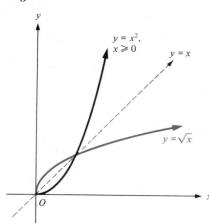

EXAMPLE 4   Use the algebraic method to find the inverse of the function $f(x) = x^2$ for $x \geq 0$, and sketch the graphs of f and f^{-1} on the same coordinate system.

SOLUTION   We carry out the algebraic procedure:

Step 1 $y = x^2$ for $x \geq 0$

Step 2 $x = y^2$ for $y \geq 0$

Step 3 Solving the equation $x = y^2$ for y and using the condition that $y \geq 0$, we obtain
$$y = \sqrt{x}$$

Therefore,
$$f^{-1}(x) = \sqrt{x} \qquad \text{(Figure 5)}$$

EXAMPLE 5   Let $f(x) = 2x + 1$.

(a) Use the algebraic method to find f^{-1}.

(b) Check that $f^{-1}[f(x)] = x$ for all values of x in the domain of f.

(c) Check that $f[f^{-1}(x)] = x$ for all values of x in the domain of f^{-1}.

SOLUTION

(a) We carry out the algebraic procedure.

Step 1 $y = 2x + 1$

Step 2 $x = 2y + 1$

Step 3 Solving the equation $x = 2y + 1$ for y, we find that
$$y = \tfrac{1}{2}(x - 1)$$

Therefore,
$$f^{-1}(x) = \tfrac{1}{2}(x - 1)$$

(b) $f^{-1}[f(x)] = f^{-1}(2x + 1) = \tfrac{1}{2}[(2x + 1) - 1] = x$

(c) $f[f^{-1}(x)] = f[\tfrac{1}{2}(x - 1)] = 2[\tfrac{1}{2}(x - 1)] + 1 = x$

In working with inverses of functions, you must be careful not to confuse $f^{-1}(x)$ and $[f(x)]^{-1}$. Notice that $f^{-1}(x)$ is the value of the function f^{-1} at x, whereas
$$[f(x)]^{-1} = \frac{1}{f(x)}$$

is the reciprocal of the value of the function f at x. For instance, in Example 5 above,
$$f^{-1}(x) = \tfrac{1}{2}(x - 1) \qquad \text{whereas} \qquad [f(x)]^{-1} = \frac{1}{2x + 1}$$

The Inverse-Function Rule

Now we give a rule, called the **inverse-function rule,** that enables us to find the derivative of the inverse of a function. First we give an informal derivation of this rule; then we state the rule more formally as a theorem (Theorem 1).

Suppose that f is an invertible and differentiable function and that

$$y = f(x)$$

Then,

$$x = f^{-1}[f(x)] = f^{-1}(y)$$

hence, using Leibniz notation, we can write the derivative of $f^{-1}(y)$ as

$$\frac{dx}{dy} = \frac{d}{dy} f^{-1}(y)$$

assuming that this derivative exists. However, by the chain rule,

$$\frac{dx}{dy} \frac{dy}{dx} = \frac{dx}{dx} = 1$$

Solving the last equation for dx/dy, we obtain the following:

Inverse-Function Rule Using Leibniz Notation

$$\text{If} \quad \frac{dy}{dx} \neq 0 \quad \text{then} \quad \frac{dx}{dy} = \frac{1}{dy/dx}$$

In words, this rule just says that under appropriate conditions, *the derivative of the inverse of a function is the reciprocal of the derivative of the function.* In Theorem 1 below, we give a more formal statement of this rule.

EXAMPLE 6   Let $y = 3x^2 - 4x + 2$ for $x > \frac{2}{3}$. Find the value of x when $y = 2$, and find the value of dx/dy when $y = 2$.

SOLUTION   When $y = 2$, we have $x > \frac{2}{3}$ and

$$2 = 3x^2 - 4x + 2 \quad \text{or} \quad 3x^2 - 4x = 0$$

that is,

$$(3x - 4)x = 0 \quad \text{and} \quad x > \frac{2}{3}$$

Therefore,

$$x = \frac{4}{3} \quad \text{when} \quad y = 2$$

Now, by the inverse function rule,

$$\frac{dx}{dy} = \frac{1}{dy/dx} = \frac{1}{d(3x^2 - 4x + 2)/dx} = \frac{1}{6x - 4}$$

Therefore, when $y = 2$ and $x = \frac{4}{3}$, we have

$$\frac{dx}{dy} = \frac{1}{6(\frac{4}{3}) - 4} = \frac{1}{4}$$

The Inverse-Function Theorem

Now we give a precise statement of the inverse-function rule in the form of a theorem. In order to understand this theorem, suppose that f is a differentiable and invertible function and

$$y = f(x)$$

so that

$$x = f^{-1}(y)$$

as we saw above. Thus, assuming that $dy/dx \neq 0$ and that f^{-1} is differentiable, we have the inverse-function rule

$$\frac{dx}{dy} = \frac{1}{dy/dx}$$

as before. Notice that since

$$\frac{dx}{dy} = \frac{d}{dy} f^{-1}(y) = (f^{-1})'(y)$$

and

$$\frac{dy}{dx} = \frac{d}{dx} f(x) = f'(x) = f'[f^{-1}(y)]$$

the inverse-function rule can be rewritten in the form

$$(f^{-1})'(y) = \frac{1}{f'[f^{-1}(y)]}$$

In many applications of the inverse-function rule, our main concern is to calculate the derivative $(f^{-1})'$ of the function f^{-1}. Because it is traditional to use x as the independent variable, we often prefer to use the letter x rather than the letter y in the formula derived above. Thus, we have

$$(f^{-1})'(x) = \frac{1}{f'[f^{-1}(x)]}$$

Our motivation is complete, and we can state the theorem.

THEOREM 1

Inverse-Function Rule, or Inverse-Function Theorem

Let f be a function whose domain is an open interval I, assume that f is differentiable on I, and suppose that $f'(c) \neq 0$ for every number c in I. Then f has an inverse f^{-1}, f^{-1} is differentiable, and

$$(f^{-1})'(x) = \frac{1}{f'[f^{-1}(x)]}$$

holds for every number x in the domain of f^{-1}.

Not only does the inverse-function theorem give us a formula for calculating the derivative $(f^{-1})'$ of the inverse of f, but also it guarantees the *existence* of f^{-1} as well as its *differentiability*. For this reason, its proof is too technical to be given in this textbook. However, you can see geometrically why f^{-1} is differentiable by considering the graphs in Figure 6. Recall that the graph of f^{-1} is the mirror image of the graph of f across the line $y = x$. By hypothesis, f is differentiable; hence, the graph of f has a tangent line at each point (u, v) with $v = f(u)$. Evidently, its mirror

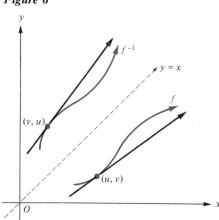

Figure 6

image—the graph of f^{-1}—also has a tangent line at each point (v, u) with $u = f^{-1}(v)$. But to say that the graph of f^{-1} has a tangent line at each point is to say that it is a differentiable function.

Use of the inverse-function rule in the form

$$(f^{-1})'(x) = \frac{1}{f'[f^{-1}(x)]}$$

is illustrated in the following examples.

EXAMPLE 7   Let f be the function defined by $f(x) = x^3 + x + 1$.

(a) Show that f^{-1} exists.

(b) Find $(f^{-1})'(1)$.

SOLUTION

(a) Since $f'(x) = 3x^2 + 1 \neq 0$ for all real numbers x, it follows from Theorem 1 that f^{-1} exists.

(b) Evidently, $f(0) = 1$, so $f^{-1}(1) = f^{-1}[f(0)] = 0$. Hence, by Theorem 1,

$$(f^{-1})'(1) = \frac{1}{f'[f^{-1}(1)]} = \frac{1}{f'(0)} = \frac{1}{3(0)^2 + 1} = 1$$

EXAMPLE 8   Suppose that the hypotheses of the inverse-function theorem are satisfied by the function f, that $f(3) = 7$, and that $f'(3) = 2$. Find $(f^{-1})'(7)$.

SOLUTION   Since $f(3) = 7$, it follows that $f^{-1}(7) = 3$. Hence,

$$(f^{-1})'(7) = \frac{1}{f'[f^{-1}(7)]} = \frac{1}{f'(3)} = \frac{1}{2}$$

Problem Set 7.1

In Problems 1 to 6, show that the functions f and g are inverses of each other.

1 $f(x) = 2x - 3$ and $g(x) = \dfrac{x + 3}{2}$

2 $f(x) = x^3$ and $g(x) = \sqrt[3]{x}$

3 $f(x) = \dfrac{1}{x}$ and $g(x) = \dfrac{1}{x}$

4 $f(x) = \dfrac{2x - 3}{3x - 2}$ and $g(x) = \dfrac{2x - 3}{3x - 2}$

5 $f(x) = \sqrt[3]{x + 8}$ and $g(x) = x^3 - 8$

6 $f(x) = x^2 + 1$ for $x \geq 1$ and $g(x) = \sqrt{x - 1}$

7 Use the horizontal-line test to determine whether the functions graphed in Figure 7 are invertible.

Figure 7

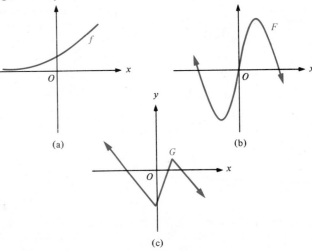

(a)

(b)

(c)

8 Under what conditions is a linear function $f(x) = mx + b$ invertible?

9 The three functions graphed in Figure 8 are invertible. In each case, obtain the graph of the inverse function by reflecting the given graph across the line $y = x$.

Figure 8

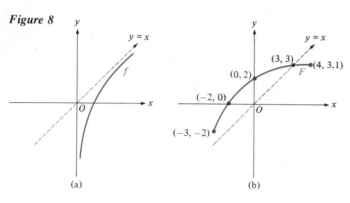

(a)

(b)

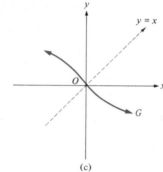

(c)

10 Sketch the graph of $f(x) = |x - 1|$, and determine whether f is invertible.

In Problems 11 to 20, (a) use the algebraic method to find f^{-1}, (b) check that $f^{-1}[f(x)] = x$ for all values of x in the domain of f, (c) check that $f[f^{-1}(x)] = x$ for all values of x in the domain of f^{-1}, and (d) sketch the graphs of f and f^{-1} on the same coordinate system.

11 $f(x) = 2x - 5$

12 $f(x) = \dfrac{3x + 2}{5}$

13 $f(x) = x^3$

14 $f(x) = 1 + \sqrt{x}$

15 $f(x) = 1 - 2x^3$

16 $f(x) = x^5 - 1$

17 $f(x) = -\dfrac{1}{x} - 1$

18 $f(x) = \dfrac{3}{x + 2}$

19 $f(x) = \dfrac{3x - 7}{x + 1}$

20 $f(x) = (x + 3)^5 - 2$

21 If a, b, c, and d are constants such that $ad \neq bc$, use the algebraic method to find the inverse of $f(x) = (ax + b)/(cx + d)$.

22 If $f(x) = \sqrt{4 - x^2}$ for $0 \le x \le 2$, show that f is its own inverse.

In Problems 23 to 29, find the value of x when $y = a$, and use the inverse-function rule $dx/dy = 1/(dy/dx)$ to find the numerical value of dx/dy when $y = a$.

23 $a = 1$, $y = x^5$

24 $a = 64$, $y = x^6$, $x > 0$

25 $a = 4$, $y = x^2 + 2x + 1$, $x > -1$

26 $a = -3$, $y = \dfrac{2x + 3}{x - 1}$

27 $a = -1$, $y = \dfrac{7x - 2}{2x - 7}$

28 $a = \dfrac{1}{2}$, $y = \sin x$, $-\dfrac{\pi}{2} < x < \dfrac{\pi}{2}$

29 $a = \dfrac{2}{3}\sqrt{3}$, $y = \dfrac{x}{\sqrt{x^2 - 1}}$, $x > 1$

30 If f is the function defined by

$$f(x) = \begin{cases} x & \text{if } x < 1 \\ x^2 & \text{if } 1 \le x \le 9 \\ 27\sqrt{x} & \text{if } x > 9 \end{cases}$$

find $(f^{-1})'(x)$ if it exists.

In Problems 31 to 36, use the given information and the inverse-function theorem to find $(f^{-1})'(a)$. (You may assume that the hypotheses of the theorem are satisfied.)

31 $a = 7$, $f(3) = 7$, $f'(3) = 2$

32 $a = 2$, $f(2) = 5$, $f(5) = 2$, $f'(5) = 7$, $f'(2) = 6$

33 $a = -1$, $f(-1) = -2$, $f(4) = -1$, $f'(-1) = -3$, $f'(4) = \frac{1}{7}$

34 $a = \frac{1}{3}$, $f(\frac{1}{3}) = \frac{2}{3}$, $f(1) = \frac{1}{3}$, $f'(1) = \frac{2}{3}$, $f'(\frac{1}{3}) = 1$

35 $a = \dfrac{\sqrt{2}}{2}$, $f\left(\dfrac{\pi}{4}\right) = \dfrac{\sqrt{2}}{2}$, $f'\left(\dfrac{\pi}{4}\right) = \dfrac{\sqrt{2}}{2}$

36 $a = 0$, $f(0) = 0$, $f'(0) = 1$

37 Let f be the function defined by $f(x) = (x + 1)/(2x - 3)$. (a) Show that $f^{-1}(x) = (3x + 1)/(2x - 1)$. (b) Calculate $(f^{-1})'(0)$ by using part (a) and the quotient rule. (c) Calculate $(f^{-1})'(0)$ by using the inverse-function theorem.

38 Let f be the function defined by $f(x) = (2x - 1)/(3x - 2)$. (a) Show that $f^{-1} = f$. (b) Calculate $(f^{-1})'(x)$ by using part (a) and the quotient rule. (c) Calculate $(f^{-1})'(x)$ by using the inverse-function theorem.

39 Let f be the function $f(x) = 2x^2 - x + 1$, $x > \frac{1}{4}$. (a) Use the algebraic method to find f^{-1}. (b) Calculate $(f^{-1})'(2)$ by using part (a). (c) Calculate $(f^{-1})'(2)$ by using the inverse-function theorem.

40 Let f be the function $f(x) = x^3 - x^2 + 1$, $x > \frac{2}{3}$. Use the inverse-function theorem to find $(f^{-1})'(5)$. [Note that $f(2) = 5$, so $f^{-1}(5) = 2$.]

41 Let f be the function $f(x) = (x^3 - 1)/(x^2 + 1)$ for $x > 0$. Since $f(1) = 0$, it follows that $f^{-1}(0) = 1$. Use the inverse-function rule to find $(f^{-1})'(0)$.

42 If g is the inverse of the function f and if (b, a) belongs to the graph of g, show that (a, b) belongs to the graph of f.

43 Complete the argument to justify the horizontal-line test by showing that if no horizontal line intersects the graph of f more than once, then f is invertible.

44 A function f is said to be **one-to-one** if whenever a and b are in the domain of f and $f(a) = f(b)$, it follows that $a = b$. Using the horizontal-line test, show that f is invertible if and only if it is one-to-one.

45 Suppose that the concentration C of an anesthetic in body tissues satisfies an equation of the form $t = f(C)$, where t is the elapsed time since the anesthetic was administered. Assuming that f is invertible, write an equation for C in terms of t.

46 If f is an invertible function, show that *the range of f is the same as the domain of f^{-1}* and that *the domain of f is the same as the range of f^{-1}*.

47 Sketch the graph of any invertible function f; then turn the paper over, rotate it 90° clockwise, and hold it up to the light. Through the paper you will see the graph of f^{-1}. Explain why.

48 If f is an invertible function, show that f^{-1} is also an invertible function and that $(f^{-1})^{-1} = f$.

7.2 Inverse Trigonometric Functions

Figure 1

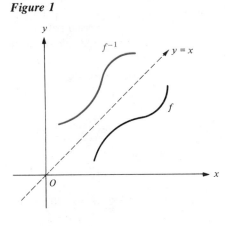

In Section 7.1 we saw that by reflecting the graph of an invertible function f across the line $y = x$ we obtain the graph of the inverse function f^{-1} (Figure 1). However, in order for f to be invertible, no horizontal line can intersect its graph more than once.

Because the sine function is periodic, any horizontal line $y = k$ with $-1 \leq k \leq 1$ intersects the graph of $y = \sin x$ repeatedly (Figure 2), and therefore the sine function is *not* invertible. But consider the portion of the graph of $y = \sin x$ between $x = -\pi/2$ and $x = \pi/2$ (Figure 3a). No horizontal line intersects this curve more than once, so its reflection across the line $y = x$ (Figure 3b) is the graph of a function. This function is often called the *inverse sine* and denoted by $\sin^{-1}$.

Of course, the terminology "inverse sine function" and the notation $\sin^{-1}$ aren't really correct because the sine function (defined on the set of all real numbers) has no inverse. Because of this, some people use the notation Sine (with a *capital S*) for the function in Figure 3a and Sin^{-1} for its inverse in Figure 3b. Another source of difficulty is the possible confusion of $\sin^{-1} x$ with $(\sin x)^{-1} = 1/\sin x = \csc x$. To avoid such confusion, many people use the notation "arcsin x," meaning the arc (or angle) whose sine is x, rather than $\sin^{-1} x$. Because both the notations $\sin^{-1} x$ and arcsin x are in common use in mathematics and its applications, we use them interchangeably. Thus, we make the following definition.

Figure 2

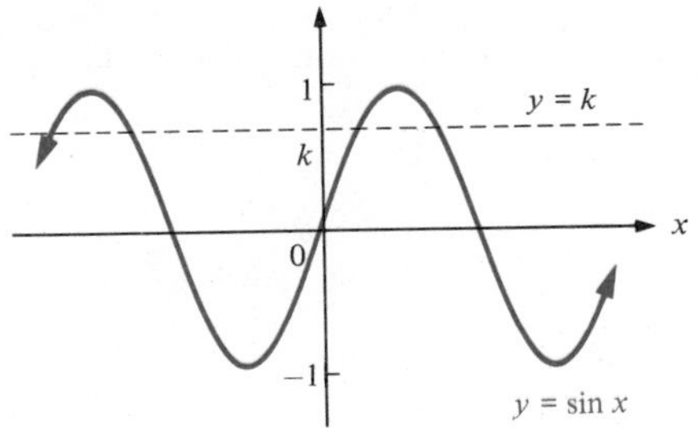

Figure 3

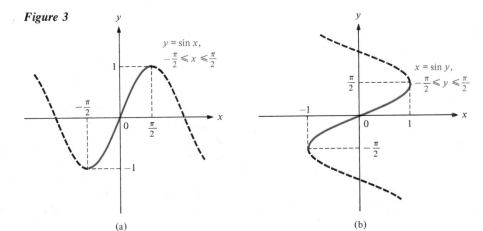

(a)   (b)

DEFINITION 1

The Inverse Sine, or Arcsine

The **inverse sine**, or **arcsine**, function, denoted by $\sin^{-1}$, or arcsin, is defined by

$$\sin^{-1} x = y \qquad \text{if and only if} \qquad x = \sin y \qquad \text{and} \qquad -\frac{\pi}{2} \le y \le \frac{\pi}{2}$$

In words, $\sin^{-1} x$ *is the angle (or number) between* $-\pi/2$ *and* $\pi/2$ *whose sine is* x. The graph of

$$y = \sin^{-1} x \qquad \text{or} \qquad y = \arcsin x$$

Figure 4

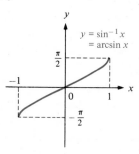

(Figure 4) shows that $\sin^{-1}$ is an increasing function with domain $[-1, 1]$ and range $[-\pi/2, \pi/2]$. The graph appears to be symmetric about the origin, indicating that $\sin^{-1}$ is an odd function; that is,

$$\sin^{-1} (-x) = -\sin^{-1} x$$

A scientific calculator can be used to find approximate values of $\sin^{-1} x$ (or arcsin x). (If a calculator isn't available, you can find these values by reading Appendix D, Table 1 backward.) Some calculators have a SIN^{-1} key; some have a key marked ARCSIN; and some have an INV key which must be pressed before the SIN key to give the inverse sine. Again, you must be careful to put the calculator in radian mode if you want $\sin^{-1} x$ in radians—otherwise, you will get the angle in *degrees* (between $-90°$ and $90°$) whose sine is x. For purposes of calculus, we almost always want the answer in radians.

EXAMPLE 1 Find:

(a) $\sin^{-1} \frac{1}{2}$ (b) $\arcsin (-\sqrt{3}/2)$ ⓒ (c) $\sin^{-1} 0.7321$

SOLUTION

(a) Here a calculator isn't needed because we know that $\pi/6$ radian ($30°$) is an angle whose sine is $\frac{1}{2}$. Thus, because $-\pi/2 \le \pi/6 \le \pi/2$, we have

$$\sin^{-1} \frac{1}{2} = \frac{\pi}{6}$$

(b) Again we don't need a calculator because we know that $-\pi/3$ radians ($-60°$) is an angle whose sine is $-\sqrt{3}/2$. Thus, since $-\pi/2 \le -\pi/3 \le \pi/2$, we have

$$\arcsin \left(\frac{-\sqrt{3}}{2} \right) = -\frac{\pi}{3}$$

(c) Using a 10-digit calculator in radian mode, we obtain

$$\sin^{-1} 0.7321 = 0.821399673$$

(In degree mode, we get $\sin^{-1} 0.7321 = 47.06273457°$.) ∎

The remaining five trigonometric functions—cosine, tangent, cotangent, secant, and cosecant—are also periodic; hence, they are not invertible. However, by restricting each of these functions to suitable intervals on which they satisfy the horizontal-line test, we can define corresponding inverses just as we did for the sine function. Figure 5 shows the intervals usually selected for the cosine and tangent functions. Thus, we make the following definitions.

DEFINITION 2   **The Inverse Cosine, or Arccosine**

> The **inverse cosine,** or **arccosine,** function, denoted by $\cos^{-1}$, or arccos, is defined by
>
> $$\cos^{-1} x = y \qquad \text{if and only if} \qquad x = \cos y \qquad \text{and} \qquad 0 \le y \le \pi$$

DEFINITION 3   **The Inverse Tangent, or Arctangent**

> The **inverse tangent,** or **arctangent,** function, denoted by $\tan^{-1}$, or arctan, is defined by
>
> $$\tan^{-1} x = y \qquad \text{if and only if} \qquad x = \tan y \qquad \text{and} \qquad -\frac{\pi}{2} < y < \frac{\pi}{2}$$

Figure 5

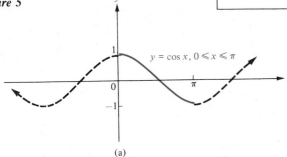

(a)

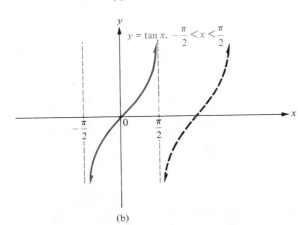

(b)

In words, $\cos^{-1} x$ *is the angle (or number) between* 0 *and* π *whose cosine is* x. Likewise, $\tan^{-1} x$ *is the angle (or number) between* $-\pi/2$ *and* $\pi/2$ *whose tangent is* x.

The graph of $y = \cos^{-1} x$, obtained by reflecting the curve in Figure 5a across the line $y = x$, shows that $\cos^{-1}$ (or arccos) is a decreasing function with domain $[-1, 1]$ and range $[0, \pi]$ (Figure 6a). Notice that the graph of $\cos^{-1}$ is *not* symmetric about the origin or about the y axis. Thus, in spite of the fact that cosine is an even function, $\cos^{-1}$ is not.

The graph of $y = \tan^{-1} x$, obtained by reflecting the curve in Figure 5b across the line $y = x$, shows that $\tan^{-1}$ (or arctan) is an increasing function with domain $\mathbb{R}$ and range $(-\pi/2, \pi/2)$ (Figure 6b). Notice that the lines $y = -\pi/2$ and $y = \pi/2$ are horizontal asymptotes. The graph appears to be symmetric about the origin, indicating that $\tan^{-1}$ is an odd function; that is,

$$\tan^{-1} (-x) = -\tan^{-1} x$$

EXAMPLE 2   Find:

(a) $\cos^{-1} (-\sqrt{2}/2)$ ⓒ **(b)** arccos 0.6675

(c) arctan $\sqrt{3}$ ⓒ **(d)** $\tan^{-1} (-2.498)$

Figure 6

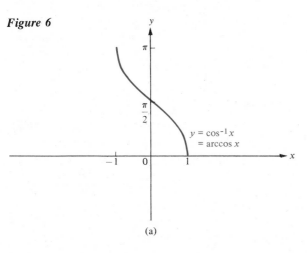

(a)

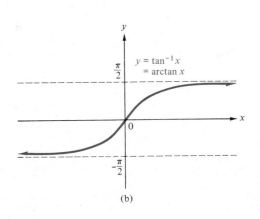

(b)

SOLUTION

(a) If $\cos^{-1}(-\sqrt{2}/2) = y$, then $-\sqrt{2}/2 = \cos y$ and $0 \le y \le \pi$. Therefore, $y = 3\pi/4$.

(b) Using a 10-digit calculator in radian mode, we find that

$$\arccos 0.6675 = 0.839950077$$

(c) If $\arctan \sqrt{3} = y$, then $\sqrt{3} = \tan y$ and $-\pi/2 < y < \pi/2$. Therefore, $y = \pi/3$.

(d) Using a 10-digit calculator in radian mode, we find that

$$\tan^{-1}(-2.498) = -1.190013897$$ ∎

If you enter a number x between -1 and 1 in a calculator, take $\sin^{-1} x$, and then take the sine of the result, you will get x back again. Thus,

$$x \xmapsto{\ \sin^{-1}\ } y \xmapsto{\ \sin\ } x$$

Do you see why? Therefore, we have the identity

$$\sin(\sin^{-1} x) = x \qquad \text{for } -1 \le x \le 1$$

However, if you try the same thing the other way around, first taking the sine and then taking the inverse sine, you might not get your original number back. For instance, starting with $x = 2$, we get (on a 10-digit calculator in radian mode)

$$2 \xmapsto{\ \sin\ } 0.909297427 \xmapsto{\ \sin^{-1}\ } 1.141592654$$

The reason is simply that $\sin^{-1} 0.909297427$ is the number *between $-\pi/2$ and $\pi/2$* whose sine is 0.909297427, and 2 does not lie between $-\pi/2$ and $\pi/2$. However, if you start with a number between $-\pi/2$ and $\pi/2$, take the sine, and then take the inverse sine, you will get your original number back. In other words,

$$\sin^{-1}(\sin x) = x \qquad \text{for } \frac{-\pi}{2} \le x \le \frac{\pi}{2}$$

Similar rules apply to cos and $\cos^{-1}$ and to tan and $\tan^{-1}$ (Problems 41 and 42).

It is sometimes necessary to find the *exact* values of expressions such as $\sin(\tan^{-1}\frac{2}{3})$. (Notice that if you used a calculator, you would obtain only an approximation to the value of this expression.) The following example shows how to find the exact values of such expressions by sketching an appropriate right triangle.

<u>EXAMPLE 3</u>   Find the exact value of $\sin(\tan^{-1}\frac{2}{3})$.

SOLUTION   We begin by sketching an acute angle θ in a right triangle such that $\theta = \tan^{-1}\frac{2}{3}$; that is, $\tan \theta = \frac{2}{3}$. This is accomplished simply by letting opp, the side opposite θ, be 2 units long and adj, the side adjacent to θ, be 3 units long (Figure 7). Then, by the Pythagorean theorem, the length of the hypotenuse is given by

$$\text{hyp} = \sqrt{\text{opp}^2 + \text{adj}^2} = \sqrt{2^2 + 3^2} = \sqrt{13}$$

It follows that

$$\sin\left(\tan^{-1}\frac{2}{3}\right) = \sin\theta = \frac{\text{opp}}{\text{hyp}} = \frac{2}{\sqrt{13}} = \frac{2\sqrt{13}}{13}$$ ∎

Figure 7

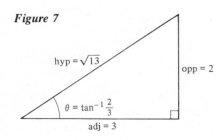
hyp $= \sqrt{13}$
opp $= 2$
$\theta = \tan^{-1}\frac{2}{3}$
adj $= 3$

The method illustrated in Example 3 above works more generally, as in the following example.

EXAMPLE 4   Show that for $-1 \leq x \leq 1$, $\cos (\sin^{-1} x) = \sqrt{1 - x^2}$.

SOLUTION   Again, we begin by letting $\theta = \sin^{-1} x$, so that $\sin \theta = x$. We consider first the case in which θ is an acute angle; that is, $0 < \theta < \pi/2$, or $0 < x < 1$. This allows us to rig up a right triangle with vertex angle θ simply by taking hyp $= 1$ and opp $= x$, so that

$$\sin \theta = \frac{\text{opp}}{\text{hyp}} = \frac{x}{1} = x$$

(Figure 8). By the Pythagorean theorem,

$$\text{hyp}^2 = \text{opp}^2 + \text{adj}^2$$

and it follows that

$$\text{adj} = \sqrt{\text{hyp}^2 - \text{opp}^2} = \sqrt{1^2 - x^2} = \sqrt{1 - x^2}$$

Therefore, for $0 < x < 1$, we have

$$\cos (\sin^{-1} x) = \cos \theta = \frac{\text{adj}}{\text{hyp}} = \frac{\sqrt{1 - x^2}}{1} = \sqrt{1 - x^2}$$

That $\cos (\sin^{-1} x) = \sqrt{1 - x^2}$ when $x = 0$ and when $x = 1$ can be verified by direct calculation. For instance,

$$\cos (\sin^{-1} 0) = \cos 0 = 1 = \sqrt{1 - 0^2}$$

and a similar calculation confirms the equation for $x = 1$. Thus, for $0 \leq x \leq 1$, we have
$$\cos (\sin^{-1} x) = \sqrt{1 - x^2}$$

Now, let's consider the case in which $-1 \leq x < 0$. Then $0 < -x \leq 1$, and by what has already been proved,

$$\cos [\sin^{-1} (-x)] = \sqrt{1 - (-x)^2} = \sqrt{1 - x^2}$$

But $$\cos [\sin^{-1} (-x)] = \cos (-\sin^{-1} x) = \cos (\sin^{-1} x)$$

and it follows that
$$\cos (\sin^{-1} x) = \sqrt{1 - x^2}$$

holds for $-1 \leq x < 0$ as well as for $0 \leq x \leq 1$. Therefore, it holds for $-1 \leq x \leq 1$. ∎

The standard trigonometric identities can be used to help simplify expressions involving inverse trigonometric functions. The method is illustrated in the following example.

EXAMPLE 5   If $-1 \leq x \leq 1$, simplify the expression $\cos (2 \sin^{-1} x)$.

SOLUTION   Let $y = \sin^{-1} x$. By the double-angle formula (see inside front cover),

$$\cos (2 \sin^{-1} x) = \cos 2y = 2 \cos^2 y - 1 = 2[\cos (\sin^{-1} x)]^2 - 1$$

Therefore, using the formula $\cos (\sin^{-1} x) = \sqrt{1 - x^2}$ from the preceding example, we have

$$\cos (2 \sin^{-1} x) = 2(\sqrt{1 - x^2})^2 - 1 = 2(1 - x^2) - 1 = 1 - 2x^2$$ ∎

Figure 8

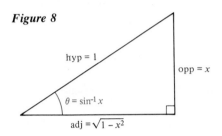
hyp = 1
opp = x
$\theta = \sin^{-1} x$
adj = $\sqrt{1 - x^2}$

The Inverse Cotangent, Secant, and Cosecant

Because the remaining inverse trigonometric functions—$\cot^{-1}$, $\sec^{-1}$, and $\csc^{-1}$ (that is, arccot, arcsec, and arccsc)—are easily calculated by using $\tan^{-1}$, $\cos^{-1}$, and $\sin^{-1}$, you'll rarely find special keys for these functions on scientific calculators.

Figure 9a shows the portion of the graph of the cotangent function that is used, by reflection across the line $y = x$, to form the graph of the inverse cotangent function (Figure 9b). Thus, the inverse cotangent function is defined as follows.

DEFINITION 4   **The Inverse Cotangent, or Arccotangent**

> The **inverse cotangent**, or **arccotangent**, function, denoted by $\cot^{-1}$, or arccot, is defined by
>
> $$\cot^{-1} x = y \qquad \text{if and only if} \qquad x = \cot y \qquad \text{and} \qquad 0 < y < \pi$$

The following identity is useful when you are using a scientific calculator to evaluate $\cot^{-1} x$:

> For all values of x,
> $$\cot^{-1} x = \frac{\pi}{2} - \tan^{-1} x$$

Figure 9

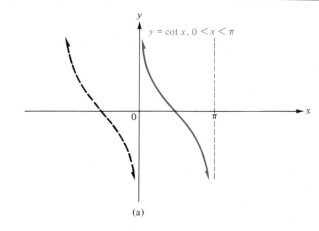

(a)

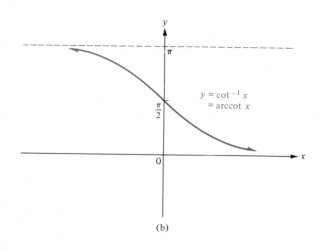

(b)

To confirm this identity, suppose that

$$y = \cot^{-1} x$$

Then $x = \cot y = \tan\left(\dfrac{\pi}{2} - y\right)$ and $0 < y < \pi$

Since $0 < y < \pi$, it follows that

$$-\pi < -y < 0$$

Hence, $\dfrac{\pi}{2} - \pi < \dfrac{\pi}{2} - y < \dfrac{\pi}{2}$

or $-\dfrac{\pi}{2} < \dfrac{\pi}{2} - y < \dfrac{\pi}{2}$

Therefore,

$$x = \tan\left(\frac{\pi}{2} - y\right) \qquad \text{where} \quad -\frac{\pi}{2} < \frac{\pi}{2} - y < \frac{\pi}{2}$$

and so by Definition 3 we have

$$\tan^{-1} x = \frac{\pi}{2} - y$$

that is, $y = \dfrac{\pi}{2} - \tan^{-1} x$

Hence, $\cot^{-1} x = \dfrac{\pi}{2} - \tan^{-1} x$

and the identity is confirmed.

ⓒ **EXAMPLE 6**   Evaluate $\cot^{-1}(-2.334)$.

SOLUTION   Using a 10-digit calculator in radian mode, we obtain

$$\cot^{-1}(-2.334) = \frac{\pi}{2} - \tan^{-1}(-2.334) = 1.570796327 - (-1.166007964)$$

$$= 2.736804291$$

Unfortunately, two *different* definitions for the inverse secant are in wide use; the same is true for the inverse cosecant. One definition makes it relatively easy to evaluate these functions on a scientific calculator, but requires absolute values in certain calculus formulas. The other definition does not require the absolute values, but makes it more difficult to evaluate the functions on a calculator. Because most people these days routinely use a calculator to evaluate trigonometric and inverse trigonometric functions, we prefer the definition that simplifies these calculations. If you use formulas from other sources, it is essential to check which definitions of $\sec^{-1}$ and $\csc^{-1}$ the author chooses to employ.

In this textbook, we choose the portions of the graphs of the secant and cosecant functions shown in Figure 10 for purposes of defining the inverses of these functions. Thus, we make the following definitions.

Figure 10

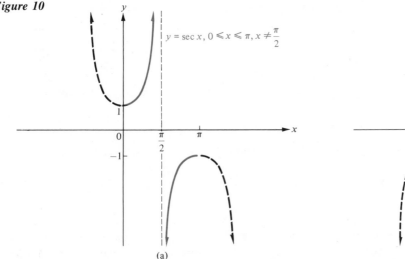
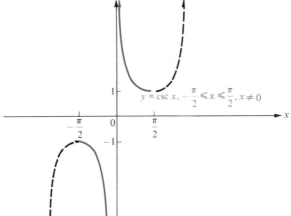

DEFINITION 5   **The Inverse Secant, or Arcsecant**

> The **inverse secant,** or **arcsecant,** function, denoted by $\sec^{-1}$, or arcsec, is defined by
>
> $$\sec^{-1} x = y \quad \text{if and only if} \quad x = \sec y \quad y \neq \frac{\pi}{2} \quad \text{and} \quad 0 \leq y \leq \pi$$

DEFINITION 6   **The Inverse Cosecant, or Arccosecant**

> The **inverse cosecant,** or **arccosecant,** function, denoted by $\csc^{-1}$, or arccsc, is defined by
>
> $$\csc^{-1} x = y \quad \text{if and only if} \quad x = \csc y \quad y \neq 0 \quad \text{and} \quad -\frac{\pi}{2} \leq y \leq \frac{\pi}{2}$$

Figure 11

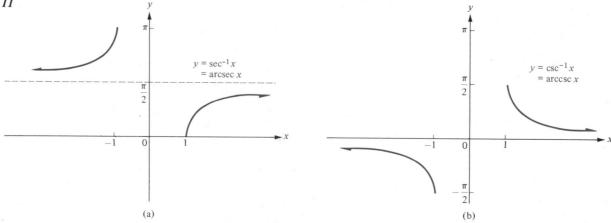

The graphs of the inverse secant and inverse cosecant, obtained by reflecting the graphs in Figure 10 across the line $y = x$, are shown in Figure 11. Notice that the domains of $\sec^{-1}$ and $\csc^{-1}$ both consist of all real numbers x with $|x| \ge 1$.

The following identities, which are consequences of the facts that $\sec y = 1/\cos y$ and $\csc y = 1/\sin y$, are useful for evaluating $\sec^{-1} x$ and $\csc^{-1} x$ with a calculator:

> **1** For $|x| \ge 1$, $\sec^{-1} x = \cos^{-1} \dfrac{1}{x}$ **2** For $|x| \ge 1$, $\csc^{-1} x = \sin^{-1} \dfrac{1}{x}$

We leave the proofs of these identities as exercises (Problems 69 and 70).

ⓒ **EXAMPLE 7**   Evaluate $\csc^{-1}(-3.514)$.

SOLUTION   Using a 10-digit calculator in radian mode, we obtain

$$\csc^{-1}(-3.514) = \sin^{-1}\left(\frac{1}{-3.514}\right) = \sin^{-1}(-0.284575982)$$

$$= -0.288564093$$

[Notice that it isn't necessary to write down the intermediate result $\sin^{-1}(-0.284575982)$—we did so only so you could check your calculator work.] ■

Problem Set 7.2

In Problems 1 to 22, evaluate each expression without using a calculator or tables.

1 $\sin^{-1} 1$

2 $\arcsin \dfrac{\sqrt{3}}{2}$

3 $\arcsin\left(-\dfrac{\sqrt{2}}{2}\right)$

4 $\cos^{-1}\left(-\dfrac{1}{2}\right)$

5 $\arccos 1$

6 $\cos^{-1} \dfrac{\sqrt{3}}{2}$

7 $\sin^{-1} \dfrac{\sqrt{2}}{2}$

8 $\cos^{-1} 0$

9 $\arccos \dfrac{1}{2}$

10 $\arctan 1$

11 $\tan^{-1}(-1)$

12 $\tan^{-1} \dfrac{\sqrt{3}}{3}$

13 $\arctan\left(-\dfrac{\sqrt{3}}{3}\right)$

14 $\tan^{-1} \sqrt{3}$

15 $\cot^{-1}(-1)$

16 $\cot^{-1}\sqrt{3}$

17 $\text{arccot}\left(-\dfrac{\sqrt{3}}{3}\right)$

18 $\sec^{-1}\sqrt{2}$

19 $\text{arcsec}(-2)$

20 $\csc^{-1}\sqrt{2}$

21 $\csc^{-1}2$

22 $\text{arccsc}(-\sqrt{2})$

ⓒ In Problems 23 to 40, use a calculator (or Appendix D, Table 1) to evaluate each expression. Give answers in *radians*.

23 $\arcsin 0.6442$

24 $\arccos 0.6675$

25 $\cos^{-1} 0.9051$

26 $\tan^{-1} 0.2500$

27 $\arctan 2$

28 $\sin^{-1}(-0.5495)$

29 $\tan^{-1}(-3.224)$

30 $\cos^{-1}\left(-\dfrac{1}{8}\right)$

31 $\arcsin(-0.5505)$

32 $\text{arccos}\left(-\dfrac{5}{11}\right)$

33 $\cos^{-1}\dfrac{\sqrt{5}}{4}$

34 $\tan^{-1}\left(-\dfrac{\sqrt{7}}{3}\right)$

35 $\cot^{-1} 3.217$

36 $\text{arcsec } 1.732$

37 $\sec^{-1} 2.718$

38 $\csc^{-1}(-3.709)$

39 $\csc^{-1}(-1.747)$

40 $\text{arccsc}(-5.432)$

41 Show that
  (a) $\cos^{-1}(\cos x) = x$ for $0 \le x \le \pi$
  (b) $\cos(\cos^{-1} x) = x$ for $-1 \le x \le 1$

42 Show that
  (a) $\tan^{-1}(\tan y) = y$ for $-\pi/2 < y < \pi/2$
  (b) $\tan(\tan^{-1} x) = x$ for all values of x
  (c) $\cot^{-1}(\cot y) = y$ for $0 < y < \pi$
  (d) $\cot(\cot^{-1} x) = x$ for all values of x
  (e) $\sec^{-1}(\sec y) = y$ for $y \ne \pi/2$ and $0 \le y \le \pi$
  (f) $\sec(\sec^{-1} x) = x$ for $|x| \ge 1$
  (g) $\csc^{-1}(\csc y) = y$ for $y \ne 0$ and $-\pi/2 \le y \le \pi/2$
  (h) $\csc(\csc^{-1} x) = x$ for $|x| \ge 1$

In Problems 43 to 59, find the exact value of each expression without using a calculator or tables.

43 $\sin\left(\sin^{-1}\dfrac{3}{5}\right)$

44 $\cos^{-1}\left(\cos\dfrac{5\pi}{4}\right)$

45 $\sin^{-1}\left(\sin\dfrac{\pi}{6}\right)$

46 $\tan(\arctan 3)$

47 $\tan^{-1}\left(\tan\dfrac{3\pi}{4}\right)$

48 $\cos^{-1}\left[\cos\left(-\dfrac{\pi}{3}\right)\right]$

49 $\sin\left(\arctan\dfrac{4}{3}\right)$

50 $\cos[\arctan(-2)]$

51 $\cos\left(\sin^{-1}\dfrac{\sqrt{10}}{10}\right)$

52 $\sin(2\arcsin\tfrac{2}{3})$

53 $\tan(\sin^{-1}\tfrac{4}{5})$

54 $\tan[\text{arcsec}(-5)]$

55 $\sec(\cos^{-1}\tfrac{7}{10})$

56 $\csc[\cot^{-1}(-2)]$

57 $\sec(\csc^{-1}\sqrt{2})$

58 $\sec(2\sin^{-1}\tfrac{1}{8})$

59 $\cot(\text{arccsc } 7)$

ⓒ **60** To the nearest thousandth of a radian, find the two vertex angles α and β of a 3-4-5 right triangle (Figure 12).

Figure 12

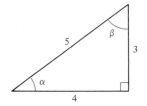

In Problems 61 to 70, show that the given equation is an identity.

61 $\cos^{-1} x = \dfrac{\pi}{2} - \sin^{-1} x$ for $-1 \le x \le 1$

62 $\cos^{-1}(-x) = \pi - \cos^{-1} x$ for $-1 \le x \le 1$

63 $\sin(\cos^{-1} x) = \sqrt{1 - x^2}$ for $-1 \le x \le 1$

64 $\tan(\sin^{-1} x) = \dfrac{x}{\sqrt{1 - x^2}}$ for $-1 < x < 1$

65 $\sin(\arctan x) = \dfrac{x}{\sqrt{1 + x^2}}$ for all values of x

66 $\tan\left(\dfrac{1}{2}\arccos x\right) = \sqrt{\dfrac{1 - x}{1 + x}}$ for $-1 < x < 1$

67 $\tan(\tan^{-1} x + \tan^{-1} y) = \dfrac{x + y}{1 - xy}$ if $xy \ne 1$

68 $\cos(\sin^{-1} x + \sin^{-1} y) = \sqrt{1 - x^2 - y^2 + x^2 y^2} - xy$ for $-1 \le x \le 1$ and $-1 \le y \le 1$

69 $\sec^{-1} x = \cos^{-1}\dfrac{1}{x}$ for $|x| \ge 1$

70 $\csc^{-1} x = \sin^{-1}\dfrac{1}{x}$ for $|x| \ge 1$

In Problems 71 to 74, simplify each expression.

71 $\sin(2\sin^{-1} x)$ for $-1 \le x \le 1$

72 $\sin(\csc^{-1} x)$ for $|x| \ge 1$

73 $\cos(\sin^{-1} x - \cos^{-1} x)$ for $-1 \le x \le 1$

74 $\cos\left(2\,\sec^{-1} x\right)$ for $|x| \ge 1$

75 Solve for x:

$$\tan^{-1}\tfrac{1}{3} + \tan^{-1}\tfrac{1}{2} = \sin^{-1} x$$

76 (a) We know that $\tan^{-1} x$ does not mean $1/\tan x$, but there is a value of x between 0 and 1 for which the equation $\tan^{-1} x = 1/\tan x$ does hold. Use graphs to show this. Ⓒ (b) Use a calculator to find the value of x between 0 and 1 for which $\tan^{-1} x = 1/\tan x$. Round off your answer to four decimal places.

77 A picture a meters high hangs on a wall so that its bottom is b meters above the eye level of an observer. If the observer stands x meters from the wall (Figure 13), show that the angle θ subtended by the picture is given by

$$\theta = \arctan\frac{a + b}{x} - \arctan\frac{b}{x}$$

78 Two parallel lines, each at a distance a from the center of a circle of radius r, cut off the region of area A shown in Figure 14. Show that

$$A = 2a\sqrt{r^2 - a^2} + 2r^2 \arcsin\frac{a}{r}$$

Figure 13

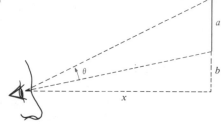

Figure 14

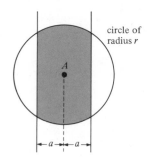

circle of radius r

7.3   Derivatives of and Integrals Yielding Inverse Trigonometric Functions

Rules for differentiation of the inverse trigonometric functions can be obtained informally by implicit differentiation. For instance, suppose that u is a differentiable function of x and that $-1 < u < 1$. Let

$$y = \sin^{-1} u \qquad \text{so that} \qquad \sin y = u$$

Assuming that y is a differentiable function of x, we implicitly differentiate the last equation with respect to x to obtain

$$(\cos y)\,\frac{dy}{dx} = \frac{du}{dx} \qquad \text{or} \qquad \frac{dy}{dx} = \frac{1}{\cos y}\frac{du}{dx}$$

Therefore, using the identity

$$\cos\left(\sin^{-1} u\right) = \sqrt{1 - u^2}$$

(see Example 4 on page 413), we have

$$\frac{d}{dx}\sin^{-1} u = \frac{dy}{dx} = \frac{1}{\cos y}\frac{du}{dx} = \frac{1}{\cos\left(\sin^{-1} u\right)}\frac{du}{dx} = \frac{1}{\sqrt{1 - u^2}}\frac{du}{dx}$$

Thus, we have the differentiation rule

$$\frac{d}{dx}\sin^{-1} u = \frac{1}{\sqrt{1 - u^2}}\frac{du}{dx}$$

(For a formal proof of this rule, see Theorem 2 on page 423.) A similar argument, which again uses implicit differentiation, yields the rule

$$\frac{d}{dx} \cos^{-1} u = \frac{-1}{\sqrt{1 - u^2}} \frac{du}{dx}$$

(Problem 22).

In Examples 1 and 2, differentiate each function.

<u>EXAMPLE 1</u> $f(x) = \sin^{-1} 2x$

SOLUTION $f'(x) = \dfrac{d}{dx} \sin^{-1} 2x = \dfrac{1}{\sqrt{1 - (2x)^2}} \dfrac{d}{dx}(2x) = \dfrac{2}{\sqrt{1 - 4x^2}}$ ■

<u>EXAMPLE 2</u> $g(t) = \cos^{-1} t^4$

SOLUTION $g'(t) = D_t \cos^{-1} t^4 = \dfrac{-D_t t^4}{\sqrt{1 - (t^4)^2}} = \dfrac{-4t^3}{\sqrt{1 - t^8}}$ ■

To obtain the rule for differentiating the inverse tangent, let

$$y = \tan^{-1} u \qquad \text{so that} \qquad \tan y = u$$

Again, assuming that y is a differentiable function of x, we implicitly differentiate the last equation with respect to x and obtain

$$(\sec^2 y) \frac{dy}{dx} = \frac{du}{dx} \qquad \text{or} \qquad \frac{dy}{dx} = \frac{1}{\sec^2 y} \frac{du}{dx}$$

Using the identity $1 + \tan^2 y = \sec^2 y$ (see inside front cover), we have

$$\sec^2 y = 1 + \tan^2 y = 1 + u^2$$

Therefore,

$$\frac{d}{dx} \tan^{-1} u = \frac{1}{1 + u^2} \frac{du}{dx}$$

A similar argument yields the rule

$$\frac{d}{dx} \cot^{-1} u = \frac{-1}{1 + u^2} \frac{du}{dx}$$

(Problem 22).

In Examples 3 and 4, differentiate each function.

<u>EXAMPLE 3</u> $F(x) = \tan^{-1} \dfrac{x}{5}$

SOLUTION $F'(x) = \dfrac{d}{dx} \tan^{-1} \dfrac{x}{5} = \dfrac{1}{1 + (x/5)^2} \dfrac{d}{dx}\left(\dfrac{x}{5}\right)$

$$= \frac{1}{1 + (x^2/25)} \cdot \frac{1}{5} = \frac{5}{25 + x^2}$$ ■

EXAMPLE 4 $G(t) = \cot^{-1} t^2$

SOLUTION $G'(t) = \dfrac{d}{dt} \cot^{-1} t^2 = \dfrac{-d(t^2)/dt}{1 + (t^2)^2} = \dfrac{-2t}{1 + t^4}$ ∎

The rule for differentiating the inverse secant can be obtained by using the identity

$$\sec^{-1} u = \cos^{-1} \frac{1}{u}$$

and the rule for differentiating the inverse cosine. Thus, for $|u| > 1$, we have

$$\frac{d}{dx} \sec^{-1} u = \frac{d}{dx} \cos^{-1} \frac{1}{u} = \frac{-1}{\sqrt{1 - (1/u)^2}} \frac{d}{dx}\left(\frac{1}{u}\right) = \frac{-1}{\sqrt{1 - (1/u^2)}}\left(\frac{-1}{u^2}\right) \frac{du}{dx}$$

$$= \frac{1}{u^2\sqrt{(u^2 - 1)/u^2}} \frac{du}{dx} = \frac{1}{u^2\sqrt{u^2 - 1}/\sqrt{u^2}} \frac{du}{dx}$$

Now, to simplify the denominator in the last expression, recall that $\sqrt{u^2} = |u|$ and $u^2 = |u|^2$; hence,

$$\frac{u^2\sqrt{u^2 - 1}}{\sqrt{u^2}} = \frac{u^2\sqrt{u^2 - 1}}{|u|} = \frac{|u|^2}{|u|}\sqrt{u^2 - 1} = |u|\sqrt{u^2 - 1}$$

Therefore,

$$\boxed{\frac{d}{dx} \sec^{-1} u = \frac{1}{|u|\sqrt{u^2 - 1}} \frac{du}{dx}}$$

A similar argument (Problem 22) yields the rule

$$\boxed{\frac{d}{dx} \csc^{-1} u = \frac{-1}{|u|\sqrt{u^2 - 1}} \frac{du}{dx}}$$

In Examples 5 and 6, differentiate each function.

EXAMPLE 5 $f(x) = \sec^{-1} (5x - 7)$

SOLUTION $f'(x) = \dfrac{d}{dx} \sec^{-1} (5x - 7) = \dfrac{1}{|5x - 7|\sqrt{(5x - 7)^2 - 1}} \dfrac{d}{dx}(5x - 7)$

$$= \frac{5}{|5x - 7|\sqrt{25x^2 - 70x + 48}}$$ ∎

EXAMPLE 6 $h(t) = \csc^{-1} t^3$

SOLUTION $h'(t) = D_t \csc^{-1} t^3 = \dfrac{-D_t(t^3)}{|t^3|\sqrt{(t^3)^2 - 1}} = \dfrac{-3t^2}{|t|^3\sqrt{t^6 - 1}} = \dfrac{-3}{|t|\sqrt{t^6 - 1}}$ ∎

Further calculations involving derivatives of inverse trigonometric functions are illustrated by the following examples.

EXAMPLE 7   Given that $\tan^{-1} x + \tan^{-1} y = \pi/2$, find dy/dx.

SOLUTION   We use implicit differentiation. Differentiation of both sides of the given equation with respect to x yields

$$\frac{1}{1 + x^2} + \frac{dy/dx}{1 + y^2} = 0 \quad \text{or} \quad \frac{dy}{dx} = -\frac{1 + y^2}{1 + x^2}$$

EXAMPLE 8   Let $f(x) = x \sin^{-1} x$ for $-1 \le x \le 1$. Find f' and f'', and sketch the graph of f.

SOLUTION   For $-1 < x < 1$ we have

$$f'(x) = \sin^{-1} x + \frac{x}{\sqrt{1 - x^2}}$$

and

$$f''(x) = \frac{1}{\sqrt{1 - x^2}} + \frac{1}{(\sqrt{1 - x^2})^3} = \frac{2 - x^2}{(\sqrt{1 - x^2})^3}$$

Now

$$f(-x) = -x \sin^{-1} (-x) = -(-x \sin^{-1} x) = f(x)$$

so, f is an even function. Also, $f'(0) = 0$ and $f''(0) = 2 > 0$, so the graph of f shows a relative minimum at $(0, 0)$. Since $f''(x) > 0$ for $-1 < x < 1$, the graph of f is concave upward (Figure 1).

Figure 1

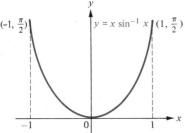

$(-1, \frac{\pi}{2})$ $y = x \sin^{-1} x$ $(1, \frac{\pi}{2})$

© **EXAMPLE 9**   A high tower stands at the end of a level road. A truck driver approaches the tower at the rate of 50 miles per hour. The tower rises 500 feet above the level of the driver's eyes. How fast is the angle θ subtended by the tower increasing at the instant when the distance x between the driver and the base of the tower is 1200 feet (Figure 2)?

SOLUTION   Since $\tan \theta = 500/x$, we have $\theta = \tan^{-1} (500/x)$. Therefore,

$$\frac{d\theta}{dt} = \frac{\dfrac{d}{dt}\left(\dfrac{500}{x}\right)}{1 + \left(\dfrac{500}{x}\right)^2} = \frac{-\left(\dfrac{500}{x^2}\right)}{1 + \dfrac{500^2}{x^2}}\frac{dx}{dt} = -\frac{500}{x^2 + 500^2}\frac{dx}{dt}$$

Figure 2

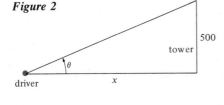

driver x tower   500 θ

Now, $\dfrac{dx}{dt} = -50 \text{ mi/h} = -\dfrac{(50 \times 5280)}{3600} \text{ ft/s} \approx -73.33 \text{ ft/s}$

hence, when $x = 1200$,

$$\frac{d\theta}{dt} \approx -\frac{500}{1200^2 + 500^2} (-73.33) \approx 0.0217 \text{ rad/s}$$

Integrals Yielding Inverse Trigonometric Functions

Integration of the six inverse trigonometric functions involves the use of techniques* that we have not yet introduced. However, we can obtain integration formulas that *yield* the inverse trigonometric functions simply by reversing the differentiation rules obtained above. For instance, we have the following three formulas:

*See Section 8.4.

$$1 \int \frac{du}{\sqrt{1-u^2}} = \sin^{-1} u + C, \text{ for } |u| < 1$$

$$2 \int \frac{du}{1+u^2} = \tan^{-1} u + C, \text{ for } u \text{ in } \mathbb{R}$$

$$3 \int \frac{du}{u\sqrt{u^2-1}} = \sec^{-1} |u| + C, \text{ for } |u| > 1$$

Formulas 1 and 2 follow directly from the rules for the derivatives of $\sin^{-1} u$ and $\tan^{-1} u$. To establish Formula 3, recall that for $u \neq 0$,

$$\frac{d}{du} |u| = \frac{|u|}{u}$$

Hence,

$$\frac{d}{du} \sec^{-1} |u| = \frac{\frac{d}{du} |u|}{|u|\sqrt{|u|^2-1}} = \frac{\left(\frac{|u|}{u}\right)}{|u|\sqrt{u^2-1}} = \frac{1}{u\sqrt{u^2-1}}$$

holds for $|u| > 1$. It follows that

$$\int \frac{du}{u\sqrt{u^2-1}} = \sec^{-1} |u| + C \qquad \text{for } |u| > 1$$

Formulas 1, 2, and 3 can easily be generalized as in the following theorem.

THEOREM 1

Integrals Yielding Inverse Trigonometric Functions

(i) $\displaystyle\int \frac{dx}{\sqrt{a^2-x^2}} = \sin^{-1} \frac{x}{a} + C$ for $a > 0$ and $|x| < a$

(ii) $\displaystyle\int \frac{dx}{a^2+x^2} = \frac{1}{a} \tan^{-1} \frac{x}{a} + C$

(iii) $\displaystyle\int \frac{dx}{x\sqrt{x^2-a^2}} = \frac{1}{a} \sec^{-1} \frac{|x|}{a} + C$ for $|x| > a > 0$

PROOF

We prove (i) and leave the similar proofs of (ii) and (iii) as exercises (Problems 73 and 74). To prove (i), we make the substitution $x = au$, so that $dx = a\, du$, and we use Formula 1 above. Thus, since $a > 0$,

$$\int \frac{dx}{\sqrt{a^2-x^2}} = \int \frac{a\, du}{\sqrt{a^2-a^2u^2}} = \int \frac{a\, du}{\sqrt{a^2}\sqrt{1-u^2}}$$

$$= \int \frac{a\, du}{a\sqrt{1-u^2}} = \int \frac{du}{\sqrt{1-u^2}}$$

$$= \sin^{-1} u + C = \sin^{-1} \frac{x}{a} + C$$ ∎

EXAMPLE 10 Evaluate $\displaystyle\int \frac{dx}{\sqrt{9-x^2}}$.

SOLUTION   By part (i) of Theorem 1, $\int \dfrac{dx}{\sqrt{9 - x^2}} = \sin^{-1} \dfrac{x}{3} + C.$ ∎

EXAMPLE 11   Evaluate $\int \dfrac{dx}{\sqrt{1 - 16x^2}}$.

SOLUTION   Put $u = 4x$, so $u^2 = 16x^2$ and $dx = du/4$. Then

$$\int \frac{dx}{\sqrt{1 - 16x^2}} = \int \frac{du}{4\sqrt{1 - u^2}} = \frac{1}{4} \sin^{-1} u + C = \frac{1}{4} \sin^{-1} 4x + C$$ ∎

EXAMPLE 12   Find $\displaystyle\int_0^{5\sqrt{3}} \dfrac{dx}{25 + x^2}$.

SOLUTION   By part (ii) of Theorem 1,

$$\int_0^{5\sqrt{3}} \frac{dx}{25 + x^2} = \left(\frac{1}{5} \tan^{-1} \frac{x}{5}\right)\bigg|_0^{5\sqrt{3}} = \frac{1}{5} \tan^{-1} \sqrt{3} - \frac{1}{5} \tan^{-1} 0$$

$$= \frac{1}{5}\left(\frac{\pi}{3}\right) - 0 = \frac{\pi}{15}$$ ∎

EXAMPLE 13   Evaluate $\displaystyle\int_0^{\sqrt{3}} \dfrac{\tan^{-1} x}{1 + x^2} \, dx.$

SOLUTION   Put $u = \tan^{-1} x$, and note that $du = dx/(1 + x^2)$, $u = 0$ when $x = 0$, and $u = \pi/3$ when $x = \sqrt{3}$. Therefore,

$$\int_0^{\sqrt{3}} \frac{\tan^{-1} x}{1 + x^2} \, dx = \int_0^{\pi/3} u \, du = \left(\frac{u^2}{2}\right)\bigg|_0^{\pi/3} = \frac{\pi^2}{18}$$ ∎

EXAMPLE 14   Evaluate $\displaystyle\int_{3\sqrt{2}}^{6} \dfrac{dx}{x\sqrt{x^2 - 9}}$.

SOLUTION   By part (iii) of Theorem 1,

$$\int_{3\sqrt{2}}^{6} \frac{dx}{x\sqrt{x^2 - 9}} = \left(\frac{1}{3} \sec^{-1} \frac{|x|}{3}\right)\bigg|_{3\sqrt{2}}^{6} = \frac{1}{3} \sec^{-1} 2 - \frac{1}{3} \sec^{-1} \sqrt{2}$$

$$= \frac{1}{3}\left(\frac{\pi}{3}\right) - \frac{1}{3}\left(\frac{\pi}{4}\right) = \frac{\pi}{36}$$ ∎

A Proof of the Differentiation Rule for Inverse Sine

The formal proofs of the rules for differentiating the inverse trigonometric functions provide nice illustrations of the use of the inverse-function theorem (Theorem 1, page 406). As an example, we give the formal proof of the rule for differentiating the inverse sine. Similar arguments (see Problems 75 and 76) will provide formal proofs of the rules for the remaining inverse trigonometric functions.

THEOREM 2   **The Derivative of Sin^{-1}**

Let $f(x) = \sin x$ for $-\pi/2 < x < \pi/2$. Then f is invertible, f^{-1} is differentiable, and

$$(f^{-1})'(x) = \frac{1}{\sqrt{1 - x^2}}$$

PROOF   We have

$$f'(x) = \cos x \neq 0 \qquad \text{for } x \text{ in the interval } \left(-\frac{\pi}{2}, \frac{\pi}{2}\right)$$

Hence, by the inverse-function theorem, f is invertible, f^{-1} is differentiable, and

$$(f^{-1})'(x) = \frac{1}{f'[f^{-1}(x)]}$$

But, by Definition 1 on page 410,

$$f^{-1}(x) = \sin^{-1} x$$

Therefore, using the identity

$$\cos(\sin^{-1} x) = \sqrt{1 - x^2}$$

we have

$$(f^{-1})'(x) = \frac{1}{f'[f^{-1}(x)]} = \frac{1}{\cos(\sin^{-1} x)} = \frac{1}{\sqrt{1 - x^2}}$$

and the proof is complete.

The differentiation rule

$$\frac{d}{dx} \sin^{-1} u = \frac{1}{\sqrt{1 - u^2}} \frac{du}{dx}$$

as originally stated, is obtained by combining Theorem 2 with the chain rule.

Problem Set 7.3

In Problems 1 to 21, differentiate each function.

1 $f(x) = \sin^{-1} 3x$

2 $g(x) = \cos^{-1} 7x$

3 $h(x) = \tan^{-1} \dfrac{x}{5}$

4 $H(x) = \cot^{-1} \dfrac{2x}{3}$

5 $G(t) = \sec^{-1} t^3$

6 $f(x) = \csc^{-1} x^2$

7 $f(t) = \cot^{-1}(t^2 + 3)$

8 $F(x) = \tan^{-1} \dfrac{2x}{1 - x^2}$

9 $g(x) = \csc^{-1} \dfrac{3}{2x}$

10 $f(r) = \tan^{-1} \dfrac{r + 2}{1 - 2r}$

11 $h(u) = \sec^{-1} \dfrac{1}{\sqrt{1 - u^2}}$

12 $f(x) = x\sqrt{4 - x^2} + 4 \sin^{-1} \dfrac{x}{2}$

13 $f(s) = \sin^{-1} \dfrac{2}{s} + \cot^{-1} \dfrac{s}{2}$

14 $g(t) = t \cos^{-1} 2t - \frac{1}{2}\sqrt{1 - 4t^2}$

15 $G(r) = \sec^{-1} r + \csc^{-1} r$

16 $F(x) = \sec^{-1} \sqrt{x^2 + 9}$

17 $g(x) = x^2 \cos^{-1} 3x$

18 $h(t) = t(\sin^{-1} t)^3 - 3t$

19 $H(x) = \dfrac{1}{x^2} \tan^{-1} \dfrac{5}{x}$

20 $F(x) = \dfrac{\sec^{-1} \sqrt{x}}{x^2 + 1}$

21 $g(x) = \dfrac{\csc^{-1}(x^2 + 1)}{\sqrt{x^2 + 1}}$

22 Use implicit differentiation to derive the rules for differentiating
(a) $\cos^{-1} u$ (b) $\cot^{-1} u$ (c) $\csc^{-1} u$

In Problems 23 to 26, use implicit differentiation to find $D_x y$.

23 $x \sin^{-1} y = x + y$

24 $\cos^{-1} xy = \sin^{-1}(x + y)$

25 $\tan^{-1} x + \cot^{-1} y = \pi/2$

26 $\sec^{-1} x + \csc^{-1} y = \pi/2$

27 If $y = \tan\left(2 \tan^{-1} \dfrac{x}{2}\right)$, show that $\dfrac{dy}{dx} = \dfrac{4(1 + y^2)}{4 + x^2}$.

28 If $y = \displaystyle\int_0^{\tan^{-1} 2x} (5 + u^2)^{20} \, du$, find $D_x y$.

In Problems 29 to 48, evaluate each integral.

29 $\displaystyle\int \frac{dx}{\sqrt{4 - x^2}}$

30 $\displaystyle\int \frac{dt}{\sqrt{16 - 9t^2}}$

31 $\displaystyle\int \frac{dx}{\sqrt{9 - 4x^2}}$

32 $\displaystyle\int \frac{dy}{\sqrt{25 - 11y^2}}$

33 $\displaystyle\int \frac{dt}{\sqrt{1 - 9t^2}}$

34 $\displaystyle\int \frac{dx}{x^2 + 9}$

35 $\displaystyle\int \frac{dy}{4 + 9y^2}$

36 $\displaystyle\int \frac{du}{9u^2 + 1}$

37 $\displaystyle\int \frac{dx}{4x^2 + 9}$

38 $\displaystyle\int \frac{dx}{x\sqrt{x^2 - 4}}$

39 $\displaystyle\int \frac{dt}{t\sqrt{16t^2 - 25}}$

40 $\displaystyle\int \frac{du}{u\sqrt{9u^2 - 100}}$

41 $\displaystyle\int \frac{4\, dx}{x\sqrt{x^2 - 16}}$

42 $\displaystyle\int_0^{1/2} \frac{dt}{\sqrt{1 - t^2}}$

43 $\displaystyle\int_{-3}^{3} \frac{dx}{\sqrt{12 - x^2}}$

44 $\displaystyle\int_0^{2} \frac{2\, du}{\sqrt{8 - u^2}}$

45 $\displaystyle\int_0^{3} \frac{dt}{3 + t^2}$

46 $\displaystyle\int_{-1}^{1} \frac{dx}{4 + x^2}$

47 $\displaystyle\int_{-2}^{-\sqrt{2}} \frac{dt}{t\sqrt{t^2 - 1}}$

48 $\displaystyle\int_{\sqrt{2}/2}^{1} \frac{du}{u\sqrt{4u^2 - 1}}$

In Problems 49 to 58, evaluate each integral by substitution. (In some cases, an appropriate substitution is indicated.)

49 $\displaystyle\int \frac{\cos x}{\sqrt{36 - \sin^2 x}}\, dx,\ u = \sin x$ **50** $\displaystyle\int \frac{\sec^2 t}{1 + \tan^2 t}\, dt$

51 $\displaystyle\int \frac{x}{4 + x^4}\, dx,\ u = x^2$ **52** $\displaystyle\int \frac{dx}{7 + (3x - 1)^2}$

53 $\displaystyle\int \frac{\cos (x/2)}{1 + \sin^2 (x/2)}\, dx,\ u = \sin \frac{x}{2}$

54 $\displaystyle\int \frac{\sec^2 t\, dt}{\sqrt{1 - 9 \tan^2 t}}$

55 $\displaystyle\int_{\pi/6}^{\pi/3} \frac{\csc t \cot t}{1 + 9 \csc^2 t}\, dt,\ u = 3 \csc t$

56 $\displaystyle\int_{\sqrt{2}/2}^{\sqrt{3}/2} \frac{\cos^{-1} x}{\sqrt{1 - x^2}}\, dx$

57 $\displaystyle\int_1^{8} \frac{dt}{t^{2/3}(1 + t^{2/3})}$ **58** $\displaystyle\int_0^{2} \frac{\cot^{-1} (x/2)}{4 + x^2}\, dx$

59 Evaluate $\displaystyle\int \frac{dx}{\sqrt{a^2 - b^2 x^2}},\ b \neq 0.$

60 Evaluate $\displaystyle\int \frac{du}{4u^2 - 4u + 5}.$

61 Sketch the graph of $y = \sin^{-1} (\cos x) + \cos^{-1} (\sin x)$ in the interval $[0, 2\pi]$.

62 Find the acute angle between the tangents to the curves $y = \tan^{-1} x$ and $y = \cot^{-1} x$ at their point of intersection.

©63 A picture 2.13 meters high is hung on a wall in such a way that its lower edge is 2.74 meters above the level of an observer's eye. How far from the wall should the observer stand in order to maximize the angle subtended by the picture?

64 Solve Problem 63 for the case in which the picture is h units high and its lower edge is a units above the observer's eye level.

©65 A ladder 15 meters long is leaning against a vertical wall of an office building. A window washer pulls the bottom of the ladder horizontally away from the building so that the top of the ladder slides down the wall at the rate of 2 meters per minute. How fast is the angle between the ladder and the ground changing when the bottom of the ladder is 6 meters from the wall?

© 66 A missile rises vertically from a point on the ground 10 miles from a radar station. If the missile is rising at the rate of 4000 feet per minute at the instant when it is 2000 feet high, find the rate of change of the angle of elevation of the missile from the radar station at this instant.

©67 A police officer in a patrol car is approaching an intersection at 80 feet per second. When she is 210 feet from the intersection, a car crosses it, traveling at right angles to the police car's path at 60 feet per second. If the officer trains her spotlight on the other car, how fast is the light beam turning 2 seconds later, provided that both vehicles continue at their original rates?

© 68 Romeo needs a ladder to elope with Juliet. However, for protection, the Capulets' palazzo is surrounded by an 8-foot-high wall located 4 feet from the outer wall of the palazzo (Figure 3). What is the shortest ladder that Romeo must bring?

Figure 3

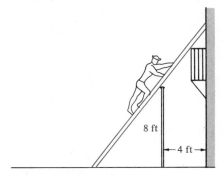

69 If $0 < x < \dfrac{\pi}{2}$, calculate $\displaystyle\int_{\cos x}^{\sin x} \frac{du}{\sqrt{1 - u^2}}.$

70 Find the area of the region under the graph of $y = 3/(9 + x^2)$ between $x = 0$ and $x = \sqrt{3}$.

71 Find the volume of the solid generated when the region bounded by the graphs of $y = 1/\sqrt{1 + x^2}$, $x = 1$, $x = 0$, and $y = 0$ is rotated about the x axis.

72 (a) Show that $\pi = \displaystyle\int_0^{1/2} \frac{6\,dx}{\sqrt{1 - x^2}}$ (b) Use Simpson's parabolic rule with $n = 2$ (four subintervals) to estimate the definite integral in part (a) and thus estimate π.

73 Prove part (ii) of Theorem 1.

74 Prove part (iii) of Theorem 1.

75 Use the inverse-function theorem to give a formal proof of the rule for differentiating $\tan^{-1} x$.

76 Use the inverse-function theorem to give a formal proof of the rule for differentiating $\sec^{-1} x$.

7.4   The Natural Logarithm Function

Recall from precalculus courses that if

$$x = b^y$$

then y is called the **logarithm** of x to the **base** b, and we write

$$y = \log_b x$$

Figure 1

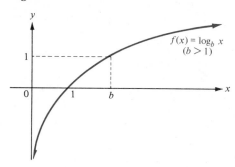

A sketch of the graph of the function $f(x) = \log_b x$ (Figure 1) reveals that f is continuous on the interval $(0, \infty)$. Furthermore, the graph appears to have a tangent line at every point, so f is evidently differentiable.

Assuming that the function $f(x) = \log_b x$ is differentiable, we can calculate its derivative as follows: Suppose that $b > 0$, $b \neq 1$, and $x > 0$. Then, using basic properties of the logarithm familiar from precalculus courses, we have

$$f'(x) = \lim_{\Delta x \to 0} \frac{f(x + \Delta x) - f(x)}{\Delta x} = \lim_{\Delta x \to 0} \frac{\log_b (x + \Delta x) - \log_b x}{\Delta x}$$

$$= \lim_{\Delta x \to 0} \frac{1}{\Delta x} \log_b \frac{x + \Delta x}{x} = \lim_{\Delta x \to 0} \frac{1}{x} \cdot \frac{x}{\Delta x} \log_b \left(1 + \frac{\Delta x}{x}\right)$$

$$= \frac{1}{x} \lim_{\Delta x \to 0} \frac{x}{\Delta x} \log_b \left(1 + \frac{\Delta x}{x}\right) = \frac{1}{x} \lim_{\Delta x \to 0} u \log_b \left(1 + \frac{1}{u}\right)$$

where we have put $u = x/\Delta x$. As Δx approaches zero from the right, u approaches $+\infty$; hence,

$$f'(x) = \frac{1}{x} \lim_{u \to +\infty} u \log_b \left(1 + \frac{1}{u}\right) = \frac{1}{x} \lim_{u \to +\infty} \log_b \left(1 + \frac{1}{u}\right)^u$$

$$= \frac{1}{x} \log_b \left[\lim_{u \to +\infty} \left(1 + \frac{1}{u}\right)^u\right]$$

If we provisionally define

$$e = \lim_{u \to +\infty} \left(1 + \frac{1}{u}\right)^u$$

assuming that the limit exists, then $f'(x) = (1/x) \log_b e$; that is,

$$D_x \log_b x = \frac{1}{x} \log_b e$$

The constant e can be evaluated to as many decimal places as necessary by various methods. A reasonably good estimate can be obtained by using a scientific calculator.

© EXAMPLE 1   Use a calculator to estimate $e = \lim_{u \to +\infty} \left(1 + \dfrac{1}{u}\right)^u$.

SOLUTION   The following table shows values of $[1 + (1/u)]^u$, rounded off to four decimal places, for selected values of u.

u	10	100	1000	10,000	100,000
$\left(1 + \dfrac{1}{u}\right)^u$	2.5937	2.7048	2.7169	2.7181	2.7183

Evidently, $e \approx 2.7183$. It can be shown that rounded off to 10 decimal places,

$$e \approx 2.7182818285$$

∎

Leonhard Euler

The symbol e is used in honor of the great Swiss mathematician Leonhard Euler (1707–1783), who was among the first to recognize the importance of this number in calculus. For one thing, the formula

$$D_x \log_b x = \frac{1}{x} \log_b e$$

becomes especially simple if we choose the base b to be Euler's constant e; indeed, $\log_e e = 1$, so that

$$D_x \log_e x = \frac{1}{x}$$

It follows that

$$\int \frac{1}{x}\, dx = \log_e x + C$$

and consequently

$$\int_1^x \frac{1}{t}\, dt = (\log_e t)\Big|_1^x = \log_e x - \log_e 1 = \log_e x - 0 = \log_e x$$

Unfortunately, the discussion above suffers from a number of logical gaps arising from the fact that, whereas it is clear what is meant by b^y when y is an integer or even a rational number, there could be some question about the meaning of b^y when y is not a rational number. For instance, what does $57^{\sqrt{2}}$ really mean? This difficulty can be avoided by starting with the equation

$$\log_e x = \int_1^x \frac{1}{t}\, dt$$

as the definition of a logarithm and then using logarithms to define b^y. This is exactly what we are going to do; however, to avoid the possibility of circular reasoning, we use the symbol "ln" rather than "$\log_e$" for the function so defined. These considerations lead us to the following definition.

DEFINITION 1   **The Natural Logarithm Function**

The **natural logarithm** function, denoted by ln, is defined by

$$\ln x = \int_1^x \frac{1}{t}\, dt \qquad \text{for } x > 0$$

Figure 2

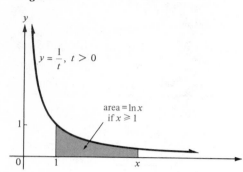

Evidently, the domain of ln is the interval $(0, \infty)$, and for $x > 1$, ln x can be interpreted geometrically as the area under the graph of $y = 1/t$, $t > 0$, between $t = 1$ and $t = x$ (Figure 2). From Definition 1, we have

$$\ln 1 = \int_1^1 \frac{1}{t}\, dt = 0$$

Also, for $0 < x < 1$,

$$\ln x = \int_1^x \frac{1}{t}\, dt = -\int_x^1 \frac{1}{t}\, dt$$

(Figure 3). Therefore,

1 $\ln x < 0$	if $0 < x < 1$	
2 $\ln x = 0$	if $x = 1$	
3 $\ln x > 0$	if $x > 1$	

Figure 3

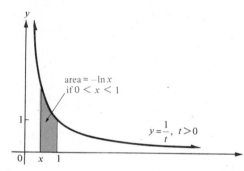

The Derivative of the Natural Logarithm

Since, by definition,

$$\ln x = \int_1^x \frac{1}{t}\, dt \qquad \text{for } x > 0$$

it follows from the alternative version of the fundamental theorem of calculus (Theorem 1, page 334) that for $x > 0$,

$$D_x \ln x = \frac{1}{x}$$

Combining this result with the chain rule, we obtain the following differentiation rule.

THEOREM 1

The Derivative of ln u

If u is a differentiable function of x and $u > 0$, then

$$D_x \ln u = \frac{1}{u} D_x u$$

EXAMPLE 2   Given $y = \ln (5x + 7)$, find $D_x y$.

SOLUTION   Applying Theorem 1 with $u = 5x + 7$, we get

$$D_x y = \frac{1}{5x + 7} D_x(5x + 7) = \frac{5}{5x + 7}$$

EXAMPLE 3   Find $D_x\left(\ln \dfrac{x}{x + 3}\right)$.

SOLUTION $D_x\left(\ln \dfrac{x}{x + 3}\right) = \dfrac{x + 3}{x} D_x\left(\dfrac{x}{x + 3}\right)$

$$= \frac{x + 3}{x} \cdot \frac{3}{(x + 3)^2} = \frac{3}{x(x + 3)}$$

EXAMPLE 4   Given $f(x) = 5x \ln \sqrt{\cos x}$, find $f'(x)$.

SOLUTION   Using the product rule, we have

$$f'(x) = 5 \ln \sqrt{\cos x} + 5x \, \frac{1}{\sqrt{\cos x}} \cdot \frac{-\sin x}{2\sqrt{\cos x}} = 5\left(\ln \sqrt{\cos x} - \frac{x}{2} \tan x\right) \quad \blacksquare$$

EXAMPLE 5   If $\ln (xy) + 3x + y = 5$, use implicit differentiation to find dy/dx.

SOLUTION   Here, we have

$$\frac{d}{dx} (\ln xy) + \frac{d(3x)}{dx} + \frac{dy}{dx} = \frac{d}{dx} 5$$

or

$$\frac{1}{xy}\left(y + x \, \frac{dy}{dx}\right) + 3 + \frac{dy}{dx} = 0$$

Therefore,

$$\frac{1}{x} + \frac{1}{y} \frac{dy}{dx} + 3 + \frac{dy}{dx} = 0 \quad \text{or} \quad \frac{dy}{dx} = -\frac{3 + (1/x)}{1 + (1/y)} = -\frac{3xy + y}{xy + x} \quad \blacksquare$$

Integrals Yielding the Natural Logarithm

For $u \neq 0$, we have

$$D_u|u| = D_u\sqrt{u^2} = \frac{2u}{2\sqrt{u^2}} = \frac{u}{|u|}$$

hence, by Theorem 1,

$$D_u \ln |u| = \frac{1}{|u|} D_u|u| = \frac{1}{|u|} \cdot \frac{u}{|u|} = \frac{u}{|u|^2} = \frac{u}{u^2} = \frac{1}{u}$$

Rewriting the latter equation in terms of antidifferentiation, we obtain the following theorem.

THEOREM 2   **Integration of $1/u$**

> For $u \neq 0$, $\displaystyle\int \frac{1}{u} \, du = \ln |u| + C$.

In Examples 6 to 10, evaluate the given integral.

EXAMPLE 6 $\displaystyle\int \frac{x}{x^2 + 7} \, dx$

SOLUTION   We make the change of variable $u = x^2 + 7$, so that $du = 2x \, dx$ and $x \, dx = \frac{1}{2} \, du$. Then, since $x^2 + 7 > 0$, Theorem 2 gives

$$\int \frac{x}{x^2 + 7} \, dx = \frac{1}{2} \int \frac{1}{u} \, du = \frac{1}{2} \ln |u| + C$$

$$= \frac{1}{2} \ln |x^2 + 7| + C = \frac{1}{2} \ln (x^2 + 7) + C \quad \blacksquare$$

EXAMPLE 7 $\displaystyle\int \frac{(\ln x)^2}{x} \, dx$

SOLUTION   Put $u = \ln x$, so that $du = dx/x$. Thus,

$$\int \frac{(\ln x)^2}{x}\, dx = \int u^2\, du = \frac{u^3}{3} + C = \frac{(\ln x)^3}{3} + C$$ ∎

EXAMPLE 8 $\int \tan x\, dx$

SOLUTION   Noting that $\int \tan x\, dx = \int (\sin x/\cos x)\, dx$, we put $u = \cos x$, so that $du = -\sin x\, dx$. Thus,

$$\int \tan x\, dx = -\int \frac{du}{u} = -\ln |u| + C = -\ln |\cos x| + C$$ ∎

EXAMPLE 9 $\displaystyle\int_{-9}^{-5} \frac{dx}{x + 1}$

SOLUTION   Put $u = x + 1$, so that $du = dx$. When $x = -9$, $u = -8$; and when $x = -5$, $u = -4$. Thus,

$$\int_{-9}^{-5} \frac{dx}{x + 1} = \int_{-8}^{-4} \frac{du}{u} = (\ln |u|)\Big|_{-8}^{-4} = \ln |-4| - \ln |-8| = \ln 4 - \ln 8$$ ∎

EXAMPLE 10 $\displaystyle\int_{0}^{\pi/4} \frac{\sin 2x}{3 + 5\cos 2x}\, dx$

SOLUTION   We put $u = 3 + 5\cos 2x$, so that $du = -10\sin 2x\, dx$, or $\sin 2x\, dx = -\frac{1}{10}\, du$. When $x = 0$, $u = 8$; and when $x = \pi/4$, $u = 3$. Thus,

$$\int_{0}^{\pi/4} \frac{\sin 2x}{3 + 5\cos 2x}\, dx = \int_{8}^{3} \frac{-\frac{1}{10}\, du}{u} = -\frac{1}{10}\int_{8}^{3} \frac{du}{u} = -\frac{1}{10}\ln |u|\Big|_{8}^{3}$$

$$= -\frac{1}{10}(\ln 3 - \ln 8) = \frac{1}{10}(\ln 8 - \ln 3)$$ ∎

Problem Set 7.4

In Problems 1 to 22, differentiate each function.

1 $f(x) = \ln (4x^2 + 1)$

2 $g(x) = \ln (\cos x^2)$

3 $f(x) = \sin (\ln x)$

4 $f(t) = \tan^{-1} (\ln t)$

5 $g(x) = x - \ln (\sin 6x)$

6 $H(x) = \ln (x + \cos x) - \tan^{-1} x$

7 $f(t) = \sin t \ln (t^2 + 7)$

8 $G(x) = \ln (4x + x^2 + 5)$

9 $F(u) = \ln (\ln u)$

10 $f(x) = \ln (\csc x - \cot x)$

11 $h(x) = x \ln \dfrac{\sec x}{5}$

12 $f(r) = \ln \sqrt[5]{1 + 5r^3}$

13 $g(v) = \ln (v^2 \sqrt{v + 1})$

14 $g(t) = \ln (t^3 \ln t^2)$

15 $h(x) = \ln (\cos^2 x)$

16 $g(r) = r^2 \csc (\ln r^2)$

17 $f(x) = \dfrac{1}{6} \ln \dfrac{8x^2}{4x^3 + 1}$

18 $g(x) = \ln \sqrt[3]{\dfrac{x}{x^2 + 1}}$

19 $h(t) = \dfrac{\ln t}{t^3 + 5}$

20 $H(x) = \sqrt{\ln \dfrac{x}{x + 2}}$

21 $g(x) = \dfrac{\ln (\tan^2 x)}{x^2}$

22 $f(x) = \dfrac{\ln [\cot (x/3)]}{x}$

In Problems 23 to 26, find dy/dx by implicit differentiation.

23 $\ln \dfrac{x}{y} + \dfrac{y}{x} = 5$

24 $y = \ln |\sec x + \tan x| - \csc y$

25 $y \ln (\sin x) - xy^2 = 4$

26 $\ln y - \cos (x + y) = 2$

27 (a) Find $D_x \int_{1}^{\ln x} \cos t^2\, dt$. (b) Evaluate dy/dx if $y = \int_{1}^{\cos x} \ln (\tan t^4)\, dt$. (*Hint:* Use the alternative version of the fundamental theorem of calculus, page 332, and the chain rule.)

28 If $a > 0$ and $b > 0$, find

$$D_x\left(\frac{1}{2\sqrt{ab}} \ln \frac{x\sqrt{a} - \sqrt{b}}{x\sqrt{a} + \sqrt{b}}\right)$$

and simplify your answer.

In Problems 29 to 40, evaluate each integral by using a suitable substitution. (In some problems, an appropriate substitution is suggested.)

29 $\displaystyle\int \frac{dx}{7 + 5x}$, $u = 7 + 5x$

30 $\displaystyle\int \frac{\sin x}{9 + \cos x} dx$

31 $\displaystyle\int \cot x \, dx$, $u = \sin x$

32 $\displaystyle\int \frac{dx}{(x + 2) \ln (x + 2)}$

33 $\displaystyle\int \frac{\sec^2 (\ln 4x)}{x} dx$, $u = \ln 4x$

34 $\displaystyle\int \frac{dx}{3\sqrt[3]{x^2} (1 + \sqrt[3]{x})}$

35 $\displaystyle\int \frac{4x \, dx}{x^2 + 7}$

36 $\displaystyle\int \frac{(\ln 5x)^2}{x} dx$

37 $\displaystyle\int \frac{\cos (\ln x)}{x} dx$

38 $\displaystyle\int \frac{\sec^2 x + \sec x \tan x}{\sec x + \tan x} dx$

39 $\displaystyle\int \frac{dx}{x\sqrt{1 - (\ln x)^2}}$

40 $\displaystyle\int \frac{dx}{x[1 + (\ln x)^2]}$

In Problems 41 to 46, evaluate each definite integral.

41 $\displaystyle\int_{1/8}^{1/5} \frac{dx}{x}$

42 $\displaystyle\int_{0.01}^{10} \frac{dt}{t}$

43 $\displaystyle\int_{1}^{\sqrt{6}} \frac{x \, dx}{x^2 + 3}$

44 $\displaystyle\int_{1}^{9} \frac{dx}{\sqrt{x}(1 + \sqrt{x})}$

45 $\displaystyle\int_{1}^{4} \frac{\cos (\ln x) \, dx}{x}$

46 $\displaystyle\int_{0}^{\pi/2} \frac{\sin x}{2 - \cos x} dx$

47 A particle moves along the s axis in such a way that its velocity at time t seconds is $1/(t + 1)$ meters per second. If the particle is at the origin at $t = 0$, find the distance that it travels during the time interval from $t = 0$ to $t = 3$.

48 Find the arc length of the graph of $y = (x^2/4) - (\ln x)/2$ between $(1, \frac{1}{4})$ and $(2, 1 - (\ln 2)/2)$.

49 Find the average (or mean) value M of the function $f(x) = (\ln x^2)/x$ on the interval $[1, 4]$.

50 Find the arc length of the graph of $f(x) = \ln (\cos x)$ between the points $(\pi/4, f(\pi/4))$ and $(\pi/3, f(\pi/3))$.

7.5 Properties of the Natural Logarithm Function

Since the natural logarithm function is defined by an integral (Definition 1, page 427), we can use properties of the integral to derive properties of ln. For instance, we have already shown that ln is a differentiable function on $(0, \infty)$ with $D_x \ln x = 1/x$, from which it follows that ln is a continuous function on $(0, \infty)$ (Theorem 1, page 96). The following theorem establishes one of the most important properties of ln.

THEOREM 1

The Natural Logarithm of a Product

If $a > 0$ and $b > 0$, then $\ln ab = \ln a + \ln b$.

PROOF

Using Theorem 5, page 319, we have

$$\ln ab = \int_{1}^{ab} \frac{dt}{t} = \int_{1}^{a} \frac{dt}{t} + \int_{a}^{ab} \frac{dt}{t} = \ln a + \int_{a}^{ab} \frac{dt}{t}$$

In the last integral, we make the substitution $t = au$, so that $dt = a \, du$. Also $u = 1$ when $t = a$, and $u = b$ when $t = ab$. Thus,

$$\int_{a}^{ab} \frac{dt}{t} = \int_{1}^{b} \frac{a \, du}{au} = \int_{1}^{b} \frac{du}{u} = \ln b$$

It follows that $\ln ab = \ln a + \ln b$.

EXAMPLE 1　Given that $\ln a \approx 0.6931$ and $\ln b \approx 1.0986$, estimate $\ln ab$.

SOLUTION　By Theorem 1, we have

$$\ln ab = \ln a + \ln b \approx 0.6931 + 1.0986 = 1.7917$$　■

THEOREM 2　**The Natural Logarithm of a Quotient**

If $a > 0$ and $b > 0$, then $\ln \dfrac{a}{b} = \ln a - \ln b$.

PROOF　Since $a = (a/b)b$, it follows that $\ln a = \ln [(a/b)b] = \ln (a/b) + \ln b$ by Theorem 1. Solving the last equation for $\ln (a/b)$, we obtain $\ln (a/b) = \ln a - \ln b$.　■

EXAMPLE 2　Given that $\ln a \approx 1.0986$ and $\ln b \approx 1.3863$, estimate $\ln \dfrac{a}{b}$.

SOLUTION　By Theorem 2, we have

$$\ln \frac{a}{b} = \ln a - \ln b \approx 1.0986 - 1.3863 = -0.2877$$　■

If we put $a = 1$ in Theorem 2 and use the fact that $\ln 1 = 0$, we obtain the useful identity

$$\ln \frac{1}{b} = -\ln b \qquad \text{for } b > 0$$

EXAMPLE 3　Given that $\ln b \approx 1.6094$, estimate $\ln \dfrac{1}{b}$.

SOLUTION　$\ln \dfrac{1}{b} = -\ln b \approx -1.6094$　■

Since $\log_b x^k = k \log_b x$, you might expect that $\ln x^k = k \ln x$. However, as we pointed out, there may be some question about the meaning of x^k when k is an irrational number. Thus, the following theorem establishes the desired identity only for the case in which k is rational.

THEOREM 3　**The Natural Logarithm of a Rational Power**

If $x > 0$ and k is a rational number, then $\ln x^k = k \ln x$.

PROOF　We have

$$D_x \ln x^k = \frac{1}{x^k} D_x(x^k) = \frac{1}{x^k} (kx^{k-1}) = \frac{k}{x}$$

and

$$D_x(k \ln x) = k \, D_x \ln x = k \cdot \frac{1}{x} = \frac{k}{x}$$

Therefore,

$$D_x(\ln x^k - k \ln x) = \frac{k}{x} - \frac{k}{x} = 0$$

and it follows from Theorem 1, page 252, that

$$\ln x^k - k \ln x = C$$

for some constant C. Putting $x = 1$ in the last equation, we find that

$$C = \ln 1^k - k \ln 1 = \ln 1 - k \ln 1 = 0 - k \cdot 0 = 0$$

Consequently,

$$\ln x^k - k \ln x = 0 \qquad \text{or} \qquad \ln x^k = k \ln x$$

and our proof is complete. ∎

EXAMPLE 4   Given that $\ln c \approx 0.6931$, estimate $\ln c^{11}$.

SOLUTION   By Theorem 3, we have

$$\ln c^{11} = (11) \ln c \approx (11)(0.6931) \approx 7.624$$ ∎

THEOREM 4   **Monotonicity of the Natural Logarithm**

> **(i)** If $0 < a < b$, then $\ln a < \ln b$.
>
> **(ii)** If $x > 0$, $y > 0$, and $\ln x = \ln y$, it follows that $x = y$.

PROOF   **(i)** Since $D_x \ln x = 1/x > 0$ for $x > 0$, it follows from Theorem 1, page 173, that $\ln$ is an increasing function on $(0, \infty)$.

(ii) Suppose that $x > 0$, $y > 0$, and $\ln x = \ln y$. Then it cannot be that $x < y$, since this would imply that $\ln x < \ln y$, which contradicts $\ln x = \ln y$. Likewise, it cannot be that $y < x$, since this would imply that $\ln y < \ln x$, again contradicting $\ln x = \ln y$. The only remaining possibility is $x = y$. ∎

EXAMPLE 5   Solve the equation $\ln (5 - 3x) + \ln (1 + x) = \ln 4$ for x.

SOLUTION   By Theorem 1, we can rewrite the given equation as

$$\ln[(5 - 3x)(1 + x)] = \ln 4$$

Therefore, by part (ii) of Theorem 4,

$$(5 - 3x)(1 + x) = 4$$

that is, $5 + 2x - 3x^2 = 4 \qquad \text{or} \qquad 3x^2 - 2x - 1 = 0$

Factoring the last equation, we obtain

$$(3x + 1)(x - 1) = 0$$

from which it follows that $x = -\frac{1}{3}$ or $x = 1$.

Notice that $\ln (5 - 3x)$ and $\ln (1 + x)$ are defined only when $5 - 3x > 0$ and $1 + x > 0$, so we must take the trouble to check that these inequalities hold for our alleged solutions $x = -\frac{1}{3}$ and $x = 1$. If $x = -\frac{1}{3}$, then $5 - 3x = 6 > 0$ and $1 + x = \frac{2}{3} > 0$; hence, $x = -\frac{1}{3}$ is indeed a solution. Similarly, if $x = 1$, then $5 - 3x = 2 > 0$ and $1 + x = 2 > 0$; hence, $x = 1$ is also a solution. ∎

The Graph of the Natural Logarithm Function

To evaluate $\ln x$ directly from the definition

$$\ln x = \int_1^x \frac{1}{t}\, dt$$

it is necessary to evaluate the integral on the right. Of course, by using (say) Simpson's parabolic rule, we can obtain an approximate value of the integral and thus an approximate value of $\ln x$. For instance, using the result in Example 6 on page 344, we have

$$\ln 2 = \int_1^2 \frac{1}{t}\, dt \approx 0.69325$$

with an error not exceeding 0.001. Therefore, we can be certain that $\ln 2$ lies between $0.69325 - 0.001$ and $0.69325 + 0.001$; that is,

$$0.69225 \le \ln 2 \le 0.69425$$

Rounding off to two decimal places, we obtain

$$\ln 2 \approx 0.69$$

It follows that

$$\ln 4 = \ln 2^2 = 2 \ln 2 \approx 2(0.69) = 1.38$$

$$\ln 8 = \ln 2^3 = 3 \ln 2 \approx 3(0.69) = 2.07$$

$$\ln \tfrac{1}{2} = -\ln 2 \approx -0.69$$

$$\ln \tfrac{1}{4} = -\ln 4 \approx -1.38$$

$$\ln \tfrac{1}{8} = -\ln 8 \approx -2.07 \qquad \text{and so forth.}$$

From the considerations above, together with the fact that $\ln 1 = 0$, we can compile a rudimentary table of approximate values of the natural logarithm which is useful in plotting a graph of $y = \ln x$ (Figure 1). The table also shows the corresponding values of $dy/dx = 1/x$. By part (i) of Theorem 4, the graph of $y = \ln x$ is always rising as x increases, and since

$$\frac{d^2 y}{dx^2} = \frac{d}{dx}\left(\frac{1}{x}\right) = -\frac{1}{x^2} < 0$$

the graph is concave downward. This information can now be used to sketch a graph of $y = \ln x$ (Figure 1).

Unfortunately, the graph in Figure 1 does not make it clear whether $y = \ln x$ becomes arbitrarily large as x increases or whether the graph has a horizontal asymptote. However, notice that $\ln 4 \approx 1.38$; hence,

$$\ln 4 > 1$$

Therefore, by Theorem 3, if n is any positive integer, we have

$$\ln 4^n = n \ln 4 > n \cdot 1 = n$$

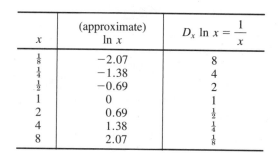

x	(approximate) $\ln x$	$D_x \ln x = \dfrac{1}{x}$
$\frac{1}{8}$	-2.07	8
$\frac{1}{4}$	-1.38	4
$\frac{1}{2}$	-0.69	2
1	0	1
2	0.69	$\frac{1}{2}$
4	1.38	$\frac{1}{4}$
8	2.07	$\frac{1}{8}$

Figure 1

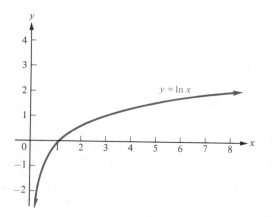

This shows that ln x can be made as large as we please by taking x large enough; for instance, to make ln x larger than 1000, it is only necessary to take x larger than 4^{1000}. It follows that

$$\lim_{x \to +\infty} \ln x = +\infty$$

so the graph of $y = \ln x$ climbs without bound as x grows larger and larger. In particular, the graph of ln x has no horizontal asymptote.

Using the limit just established and the identity $\ln (1/t) = -\ln t$, we find that

$$\lim_{x \to 0^+} \ln x = \lim_{t \to +\infty} \ln \frac{1}{t} = \lim_{t \to +\infty} (-\ln t) = - \lim_{t \to +\infty} \ln t = -\infty$$

In other words, ln x can be made smaller than any given number simply by taking x close enough to 0. The fact that

$$\lim_{x \to 0^+} \ln x = -\infty$$

shows that the graph of $y = \ln x$ has the y axis as a vertical asymptote.

EXAMPLE 6   Sketch the graph of the equation $y = \ln |x|$.

SOLUTION   Since $\ln |x| = \ln |-x|$, the graph of $y = \ln |x|$ is symmetric about the y axis. For positive values of x, the graph coincides with the graph of $y = \ln x$ in Figure 1. By reflecting this portion of the graph across the y axis, we obtain the entire graph of $y = \ln |x|$ (Figure 2).

Figure 2

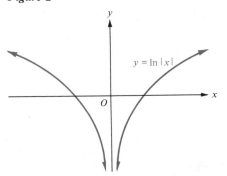

$y = \ln |x|$

THEOREM 5

Range of the Natural Logarithm

The range of ln is the entire real number system $\mathbb{R}$.

PROOF   We must prove that given any real number y, there exists a positive real number x such that $y = \ln x$. Since $\lim_{x \to +\infty} \ln x = +\infty$, there exists a positive real number b so large that $\ln b > y$. Since $\lim_{x \to 0^+} \ln x = -\infty$, there exists a positive real number a so small that $\ln a < y$. Therefore, because $\ln a < y < \ln b$ and ln is a continuous function, the intermediate-value theorem (Theorem 1, page 150) implies that there exists a value of x between a and b such that $\ln x = y$.

Evaluation of Natural Logarithms

Until now, in dealing with the natural logarithm, we have deliberately avoided using a calculator or tables so that you could see clearly how the properties of ln follow directly from its definition as an integral. Approximate values of ln x can be found almost effortlessly by using the LN key on a scientific calculator.*

*If a calculator isn't available, you can use Appendix D, Table 2 to evaluate natural logarithms.

©**EXAMPLE 7** Evaluate

(a) ln 7124   (b) ln 0.05319

SOLUTION   Using a 10-digit calculator, we obtain

(a) ln 7124 = 8.871224644   (b) ln 0.05319 = −2.933884870   ∎

The following examples illustrate the use of the natural logarithm function.

©**EXAMPLE 8** Find the exact volume V of the solid generated by revolving the region bounded by the graphs of $y = x^{-1/2}$, $x = 1$, $x = 4$, and $y = 0$ about the x axis. Also find the numerical value of V, rounded off to two decimal places.

SOLUTION   Using the method of circular disks, we have

$$V = \int_1^4 \pi(x^{-1/2})^2 \, dx = \pi \int_1^4 \frac{1}{x} \, dx$$

Therefore, by Theorem 2 on page 429,

$$V = \pi(\ln |x|)\Big|_1^4 = \pi(\ln |4| - \ln |1|) = \pi \ln 4$$

Using a calculator or Appendix D, Table 2, we find that

$$V = \pi \ln 4 \approx 4.36 \text{ cubic units}$$

∎

©**EXAMPLE 9** To the nearest joule, how many joules of work are required to compress air isothermally* according to the equation $PV = C$ from an initial volume of $V_0 = 10^{-4}$ cubic meter to a final volume of $V_1 = 10^{-5}$ cubic meter if the initial pressure is $P_0 = 10^5$ newtons per square meter?

SOLUTION   Here

$$C = P_0 V_0 = (10^5)(10^{-4}) = 10$$

and the work done during the isothermal compression is given by

$$W = \int_{10^{-5}}^{10^{-4}} P \, dV = \int_{10^{-5}}^{10^{-4}} \frac{C}{V} \, dV = C \int_{10^{-5}}^{10^{-4}} \frac{1}{V} \, dV = 10 \int_{10^{-5}}^{10^{-4}} \frac{1}{V} \, dV$$

(see page 385). Therefore,

$$W = 10(\ln |V|)\Big|_{10^{-5}}^{10^{-4}} = 10(\ln |10^{-4}| - \ln |10^{-5}|) = 10(\ln 10^{-4} - \ln 10^{-5})$$

By Theorem 3, we have

$$\ln 10^{-4} = -4 \ln 10 \quad \text{and} \quad \ln 10^{-5} = -5 \ln 10$$

so that

$$W = 10[-4 \ln 10 - (-5) \ln 10] = 10(5 \ln 10 - 4 \ln 10) = 10 \ln 10$$

Using a calculator (or Appendix D, Table 2), we find that

$$W = 10 \ln 10 \approx 23 \text{ joules}$$

∎

---

*Isothermal compression is compression with the temperature held constant.

Problem Set 7.5

In Problems 1 to 6, assume that $\ln a \approx 0.6931$, $\ln b \approx 1.0986$, and $\ln c \approx 1.6094$. Use the properties of the natural logarithm to estimate each quantity.

1 $\ln ac$

2 $\ln \dfrac{1}{b}$

3 $\ln a^2c^2$

4 $\ln \dfrac{ab}{c}$

5 $\ln \sqrt{b}$

6 $\ln c^{-1/2}$

In Problems 7 to 10, solve each equation for x.

7 $\ln (x - 1) + \ln (x - 2) = \ln 6$

8 $\ln (x^2 - 4) + \ln (x - 2) = \ln 3$

9 $2 \ln (x - 2) = \ln x$

10 $\ln (6 - x - x^2) - \ln (x + 3) = \ln (2 - x)$

In Problems 11 to 17, sketch the graph of each function. Indicate the domain and the range of the function. Also indicate the extreme points and the inflection points of the graph and any horizontal or vertical asymptotes.

11 $g(x) = \ln (2 - x)$

12 $h(x) = \ln (-x)$

13 $F(x) = \ln |x + 1|$

14 $G(x) = \ln \dfrac{4}{x}$

15 $H(x) = x \ln x$

16 $f(x) = \dfrac{x}{\ln x}$

17 $L(x) = x - \ln x$

18 Sketch the graph of the equation $x = \ln y$ for $y > 0$. Use implicit differentiation to find the slope of the graph at the point $(\ln 2, 2)$.

19 Find the equations of the tangent and normal lines to the curve $y = x^2 \ln x$ at the point $(2, 4 \ln 2)$.

20 A particle is moving along a straight line according to the equation of motion $s = \ln [8t/(4t^2 + 5)]$. Find the velocity ds/dt and the acceleration d^2s/dt^2.

In Problems 21 to 24, find the exact area of the region under each curve in terms of logarithms.

21 $y = \dfrac{1}{x}$ between $x = 5$ and $x = 7$

22 $y = \dfrac{4}{x - 1}$ between $x = 2$ and $x = 3$

23 $y = \dfrac{3}{x - 2}$ between $x = 3$ and $x = 4$

24 $y = \dfrac{1}{2x - 1}$ between $x = 2$ and $x = 3$

In Problems 25 to 27, find the exact volume of the solid generated by revolving the region bounded by the given curves about the x axis. Write your answers in terms of logarithms.

25 $(1 + x^2)y^2 = x$, $x = 1$, $x = 4$, $y = 0$

26 $(x - 6)y^2 = x$, $x = 7$, $x = 10$, $y = 0$

27 $(x^2 + 2x)y^2 = x + 1$, $x = 1$, $x = 4$, $y = 0$

28 Given that $\ln 10 \approx 2.3025851$, use the differential to estimate $\ln 10.007$.

© In Problems 29 to 34, evaluate each logarithm.

29 $\ln 4126$

30 $\ln 2.704$

31 $\ln 0.040404$

32 $\ln (7.321 \times 10^8)$

33 $\ln (1.732 \times 10^{-7})$

34 $\ln \pi$

© **35** Find numerical approximations to (a) the area in Problem 21 and (b) the volume in Problem 25. Round off your answers to four decimal places.

36 Show that $\ln (1 + x) \approx x$ if $|x|$ is small and that the approximation becomes more and more accurate as $|x|$ gets smaller and smaller.

37 Determine the coordinates of the point at which the function $f(x) = (\ln x + 1)/x$ takes on an extreme value. Is the extreme value a maximum or a minimum? Is it an absolute or only a relative extremum?

38 Find a general formula for the work done during the isothermal compression of a sample of gas from an initial pressure of P_0 and an initial volume of V_0 to a final pressure of P_1.

39 The speed of the signal in a submarine telegraph cable is proportional to $x^2 \ln (1/x)$, where x is the ratio of the radius of the core to the thickness of the winding. For what value of this ratio will the speed be maximum?

40 A product sells for $25 per unit. The production cost C for x units is expressed by the equation $C = 250 + x(6 + 2 \ln x)$. Find the output level that yields maximum profit, assuming that all manufactured units are sold.

41 A lake polluted by coliform bacteria is treated with bactericidal agents. Environmentalists estimate that t days after the treatment, the number N of viable bacteria per milliliter will be given by

$$N(t) = 100\left(\frac{t}{10} - \ln \frac{t}{10}\right) - 30 \qquad \text{for } 1 \le t \le 12$$

How many days after the treatment does the pollution reach a minimum?

7.6   The Exponential Function

Because the natural logarithm is an increasing function (Theorem 4, page 433), its graph can intersect a horizontal line $y = k$ only once (Figure 1a). Therefore, by the horizontal-line test, ln is an invertible function. Later we prove that ln x is exactly the same as $\log_e x$, from which it will follow that the inverse of ln is the function $f(x) = e^x$. In anticipation of this result, we refer to the inverse of ln as the *exponential function*.

DEFINITION 1   **The Exponential Function**

> The inverse of the natural logarithm function is called the **exponential function,** denoted by **exp.**

Thus, the graph of exp is obtained by reflecting the graph of ln about the line $y = x$ (Figure 1b).

Since the domain of ln is the interval $(0, \infty)$, it follows that the range of exp is $(0, \infty)$. Likewise, since the range of ln is the set $\mathbb{R}$ of all real numbers, it follows that $\mathbb{R}$ is the domain of exp. From the graph of $y = \exp x$ (Figure 1b), we see that exp is an increasing function and that $\exp x > 0$ for all values of x in $\mathbb{R}$. Also,

$$
\begin{aligned}
&& 0 < \exp x < 1 && \text{if } x < 0 \\
&& \exp x = 1 && \text{if } x = 0 \\
\text{and} && \exp x > 1 && \text{if } x > 0
\end{aligned}
$$

The fact that

$$\exp 0 = 1$$

(which later we see corresponds to the fact that $e^0 = 1$) is used freely in what follows. Also, Figure 1b shows that

$$\lim_{x \to -\infty} \exp x = 0 \qquad \text{and} \qquad \lim_{x \to +\infty} \exp x = +\infty$$

Figure 1

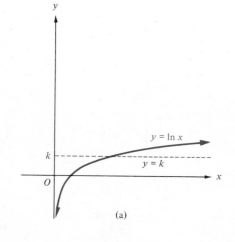

(a)

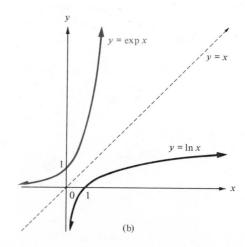

(b)

Properties of the Exponential Function

Because exp is the inverse of ln, we have

	$\exp\,(\ln x) = x$ for $x > 0$
and	$\ln\,(\exp x) = x$ for all x in $\mathbb{R}$

These relations can be used to establish many of the basic properties of the exponential function. For instance, we have the following theorem.

THEOREM 1   **Basic Properties of the Exponential Function**

If x and y are any two real numbers and if k is a rational number, then

    (i) $\exp\,(x + y) = (\exp x)(\exp y)$

    (ii) $\exp\,(kx) = (\exp x)^k$

    (iii) $\exp\,(x - y) = \dfrac{\exp x}{\exp y}$

PROOF   **(i)** Let $A = \exp x$, so that $x = \ln A$; and let $B = \exp y$, so that $y = \ln B$. By Theorem 1 of Section 7.5, $\ln A + \ln B = \ln AB$; hence, $x + y = \ln AB$. It follows that

$$\exp\,(x + y) = \exp\,(\ln AB) = AB = (\exp x)(\exp y)$$

(ii) Again put $A = \exp x$, so that $x = \ln A$. By Theorem 3 of Section 7.5, $kx = k \ln A = \ln A^k$; hence,

$$\exp kx = \exp\,(\ln A^k) = A^k = (\exp x)^k$$

(iii) The proof of part (iii) is left as an exercise (Problem 55).   ∎

Now we replace our provisional definition of e by a formal definition based on the definitions of ln and exp. The clue as to how this should be done comes from the anticipated result $\exp x = e^x$. Putting $x = 1$, we should have $\exp 1 = e$. This leads us to the following formal definition of the constant e.

Figure 2

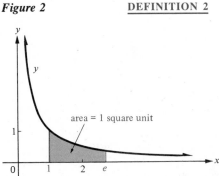

DEFINITION 2   **The Constant e**

By definition, $e = \exp 1$; that is, e is the positive real number for which $\ln e = 1$.

Graphically, e is determined by the condition that the area

$$\int_1^e \frac{1}{x}\,dx$$

between $x = 1$ and $x = e$ under the graph of $y = 1/x$ is exactly 1 square unit (Figure 2). By combining the definition of e with part (ii) of Theorem 1, we obtain the following result.

THEOREM 2   **The Exponential of a Rational Number**

If k is a rational number, then $\exp k = e^k$.

PROOF   Put $x = 1$ in part (ii) of Theorem 1 to obtain $\exp k = (\exp 1)^k$. By Definition 2, $\exp 1 = e$; hence, $\exp k = e^k$. ∎

By Theorem 2, $e^x = \exp x$ holds for all rational numbers x, so it is reasonable to ask whether it also holds when x is irrational. However, as we originally pointed out, if x is irrational, it is perhaps not clear just what is meant by e^x anyway. To clarify matters once and for all, we simply *define* e^x to be equal to $\exp x$ when x is irrational.

DEFINITION 3   **Definition of e^x for Irrational Values of x**

> If x is an irrational number, we define e^x as follows:
>
> $$e^x = \exp x$$

By Theorem 2 and Definition 3, $e^x = \exp x$ holds for all real numbers x, whether rational or irrational. Therefore, $\ln e^x = \ln (\exp x) = x$; that is, e^x is the number whose natural logarithm is x. Similarly, $e^{\ln x} = \exp (\ln x) = x$ holds for $x > 0$. In summary, we have the following:

> **1** $y = e^x$ if and only if $x = \ln y$
>
> **2** $e^{\ln x} = x$ for $x > 0$
>
> **3** $\ln e^x = x$ for all values of x
>
> **4** $e^{x+y} = e^x e^y$ for all values of x and y
>
> **5** $e^{kx} = (e^x)^k$ for all values of x and all rational numbers k
>
> **6** $e^{x-y} = e^x/e^y$ for all values of x and y

Of course, Properties 4, 5, and 6 are just parts (i), (ii), and (iii), respectively, of Theorem 1, rewritten by using Definition 3.

EXAMPLE 1   Simplify each of the following expressions:

(a) $e^{\ln a}$ (b) $e^{-2+2\ln 3}$ (c) $\ln e^{5x}$ (d) $\dfrac{e^{\ln (x^2-9)}}{x - 3}$

SOLUTION

(a) $e^{\ln a} = a$

(b) $e^{-2+2\ln 3} = e^{-2}e^{2\ln 3} = (e^2)^{-1}(e^{\ln 3^2}) = (e^2)^{-1}(3^2) = \dfrac{9}{e^2}$

(c) $\ln e^{5x} = 5x$

(d) $\dfrac{e^{\ln (x^2-9)}}{x - 3} = \dfrac{x^2 - 9}{x - 3} = x + 3$, for $x \neq 3$ ∎

Approximate values of e^x can be found by using a scientific calculator.*

© EXAMPLE 2   Use a calculator with an e^x (or EXP) key to evaluate:

(a) e^1 (b) $e^{\sqrt{2}}$ (c) $e^{-5.0321}$

*If a calculator isn't available, you can use Appendix D, Table 3.

SOLUTION   Using a 10-digit calculator, we find that

(a) $e^1 = e = 2.718281828$

(b) $e^{\sqrt{2}} = e^{1.414213562} = 4.113250377$

(c) $e^{-5.0321} = 6.525093476 \times 10^{-3} = 0.006525093476$ ∎

Differentiation of the Exponential Function

From now on, we can use either the notation e^x or the notation exp x for the inverse of ln evaluated at x. In the following theorem, we obtain a most remarkable result: *The exponential function is its own derivative*.

THEOREM 3   **The Derivative of the Exponential Function**

$$D_x e^x = e^x$$

PROOF*   Let

$$y = e^x$$

so that

$$x = \ln y$$

Differentiating both sides of the last equation with respect to x, we obtain

$$1 = \frac{1}{y} D_x y$$

that is,

$$D_x y = y$$

or

$$D_x e^x = e^x$$ ∎

Combining the result of Theorem 3 with the chain rule, we see that if u is a differentiable function of x, then

$$D_x e^u = e^u D_x u$$

EXAMPLE 3   If $y = e^{x^2 + x}$, find $D_x y$.

SOLUTION $D_x y = D_x e^{x^2 + x} = e^{x^2 + x} D_x(x^2 + x) = e^{x^2 + x}(2x + 1)$ ∎

EXAMPLE 4   Find $f'(x)$ if $f(x) = e^{\cos x}$.

SOLUTION $f'(x) = e^{\cos x} D_x \cos x = e^{\cos x}(-\sin x) = -\sin x \, e^{\cos x}$ ∎

EXAMPLE 5   If $xe^y = \tan^{-1} x$, use implicit differentiation to find dy/dx.

*A more rigorous proof may be given by using the inverse-function theorem (Theorem 1, page 406).

SOLUTION $\quad \dfrac{d}{dx}(xe^y) = \dfrac{d}{dx}\tan^{-1}x \quad$ so $\quad e^y + xe^y\dfrac{dy}{dx} = \dfrac{1}{1+x^2}$

Solving for dy/dx, we obtain

$$\frac{dy}{dx} = \frac{1}{(1+x^2)xe^y} - \frac{e^y}{xe^y} = \frac{1}{(1+x^2)\tan^{-1}x} - \frac{1}{x}$$

EXAMPLE 6 A medical team predicts that the percent P of a city's population that will be infected with a communicable disease t days after the disease is first detected is given by

$$P(t) = 10te^{-t/10} \qquad \text{for } 1 \le t \le 15$$

How many days after detection will the maximum percentage of the population be infected?

SOLUTION $\quad$ We have

$$P'(t) = 10te^{-t/10}(-\tfrac{1}{10}) + 10e^{-t/10} = e^{-t/10}(10-t)$$

Recall that e^x is positive for all values of x; hence,

$$e^{-t/10} > 0$$

for all values of t. It follows that $t = 10$ is a critical value and that

$$P'(t) > 0 \qquad \text{for } 1 < t < 10$$

whereas $\qquad\qquad P'(t) < 0 \qquad \text{for } 10 < t < 15$

Therefore, by the first-derivative test, P takes on a maximum value 10 days after the disease is first detected.

Integration of e^u

Theorem 3 can now be used to obtain the following important integral formula.

THEOREM 4 $\qquad$ **Integration of e^u**

$$\int e^u \, du = e^u + C$$

PROOF $\qquad$ By Theorem 3, $D_u e^u = e^u$; that is, e^u is an antiderivative of e^u.

EXAMPLE 7 Evaluate $\int e^{4x}\,dx$.

SOLUTION $\quad$ Put $u = 4x$, so $du = 4\,dx$ and $dx = \tfrac{1}{4}\,du$. Thus, by Theorem 4,

$$\int e^{4x}\,dx = \int e^u(\tfrac{1}{4}\,du) = \tfrac{1}{4}\int e^u\,du = \tfrac{1}{4}e^u + C = \tfrac{1}{4}e^{4x} + C$$

EXAMPLE 8 Evaluate $\int x^2 e^{x^3}\,dx$.

SOLUTION $\quad$ Put $u = x^3$, so $du = 3x^2\,dx$, or $x^2\,dx = \tfrac{1}{3}\,du$. Thus,

$$\int x^2 e^{x^3}\,dx = \tfrac{1}{3}\int e^u\,du = \tfrac{1}{3}e^u + C = \tfrac{1}{3}e^{x^3} + C$$

EXAMPLE 9 Find $\displaystyle\int \frac{e^x\,dx}{\sqrt{1 - e^{2x}}}$.

SOLUTION   We put $u = e^x$, noting that $du = e^x\,dx$ and that $e^{2x} = (e^x)^2 = u^2$. Therefore,

$$\int \frac{e^x\,dx}{\sqrt{1 - e^{2x}}} = \int \frac{du}{\sqrt{1 - u^2}} = \sin^{-1} u + C = \sin^{-1} e^x + C$$ ∎

EXAMPLE 10   Find the volume V of the solid generated by revolving the region bounded by the curves $y = e^{-4x}$, $x = 0$, $x = 2$, and $y = 0$ about the x axis.

SOLUTION   By the method of circular disks,

$$V = \int_0^2 \pi(e^{-4x})^2\,dx = \pi \int_0^2 e^{-8x}\,dx$$

We put $u = -8x$, so that $du = -8\,dx$ and $dx = -\frac{1}{8}\,du$. When $x = 0$, $u = 0$, and when $x = 2$, $u = -16$; hence,

$$V = \pi \int_0^{-16} e^u \left(-\frac{1}{8}\,du\right) = -\frac{\pi}{8} \int_0^{-16} e^u\,du = -\frac{\pi}{8} e^u \Big|_0^{-16}$$

$$= -\frac{\pi}{8}(e^{-16} - 1) = \frac{\pi}{8}(1 - e^{-16}) \text{ cubic unit}$$ ∎

Problem Set 7.6

1 Simplify each expression.

(a) $e^{\ln 5}$ (b) $e^{-3 \ln 2}$ (c) $e^{3+4 \ln 2}$

(d) $\ln e^{1/x}$ (e) $\ln e^{x-x^2}$ (f) $e^{-\ln (1/x)}$

(g) $\ln e^{x^2-4}$ (h) $e^{(\ln x^2)-4}$ (i) $\dfrac{e^{\ln (x^2-4)}}{x + 2}$

(j) $e^{\ln x - 3 \ln y}$

2 Solve the following equations for x:

(a) $2e^x + 1 = e^{-x}$ (b) $e^x + 20e^{-x} = 21$

Ⓒ **3** Use a calculator with an e^x (or EXP) key to evaluate the given quantity as accurately as you can.

(a) e^{-1} (b) e^{-2} (c) e^3 (d) $\exp (0.5)$

(e) $e^{\sqrt{5}}$ (f) e^e (g) $e^{-3.11}$ (h) e^{-e}

(i) e^{π} (j) $\exp (1 - \sqrt{2})$

Ⓒ **4** Use a calculator with an e^x (or EXP) key to verify each equation for the indicated values of the variables.

(a) $e^x e^y = e^{x+y}$ for $x = \sqrt{2}$, $y = \sqrt{3}$

(b) $e^{x+y} = e^x e^y$ for $x = \sqrt{5}$, $y = -\pi$

(c) $(e^x)^y = e^{xy}$ for $x = \dfrac{\pi}{2}$, $y = 1 - \sqrt{3}$

(d) $e^{x-y} = \dfrac{e^x}{e^y}$ for $x = 3.9$, $y = 2.5$

In Problems 5 to 24, find the derivative of each function.

5 $f(x) = e^{7x}$ **6** $g(t) = (e^{4t})^3$

7 $g(x) = e^{\ln x^3}$ **8** $f(u) = \exp (\sin u)$

9 $f(x) = \cos (\exp x)$ **10** $g(x) = e^{-x} \cos 2x$

11 $f(t) = e^{-2t} \sin t$ **12** $g(r) = \tan^{-1} (\exp r)$

13 $h(x) = e^{x^2+5 \ln x}$ **14** $F(x) = \exp \sqrt{4 - x^2}$

15 $f(t) = e^{t \ln t}$ **16** $g(x) = e^{x^2} \cot 4x$

17 $h(x) = \sec^{-1} (e^{3x})$ **18** $f(x) = e^{\sqrt{x}} \ln \sqrt{x}$

19 $G(s) = (1 - e^{3s})^2$ **20** $f(t) = \dfrac{e^{2t}}{e^t + 1}$

21 $f(x) = \dfrac{1}{\sqrt{2\pi}} \exp \left(-\dfrac{x^2}{2}\right)$ **22** $f(t) = \ln \dfrac{e^t}{t + 1}$

23 $f(x) = (8x^2 - 3x + 1)e^{-x}$ **24** $H(r) = \displaystyle\int_1^{e^r} \sqrt{1 + t^2}\,dt$

In Problems 25 to 28, use implicit differentiation to find dy/dx.

25 $y(1 + e^x) - xy^2 = 7$ **26** $e^y - \sin (x + y) = 2$

27 $x \sin y = e^{x+y}$ **28** $e^{-x} \ln y + e^y \ln x = 4$

29 Show that $y = e^{-3x}$ satisfies the differential equation $y'' + 2y' - 3y = 0$.

30 Show that $y = 5xe^{-4x}$ satisfies the differential equation $(d^2y/dx^2) + 8(dy/dx) + 16y = 0$.

In Problems 31 to 48, evaluate each integral. (In some cases a suitable substitution is suggested.)

31 $\int e^{3x} \, dx$

32 $\int e^{-7x} \, dx$

33 $\int e^{5x+3} \, dx$

34 $\int e^{-4x+5} \, dx$

35 $\int xe^{5x^2} \, dx, \ u = 5x^2$

36 $\int e^{\sin x} \cos x \, dx$

37 $\int \dfrac{e^x \, dx}{1 + e^{2x}}, \ u = e^x$

38 $\int \dfrac{e^{\sqrt[3]{x}}}{\sqrt[3]{x^2}} \, dx$

39 $\int \dfrac{3e^x \, dx}{\sqrt{e^x + 4}}$

40 $\int \dfrac{5e^{-3x}}{(e^{-3x} + 7)^8} \, dx$

41 $\int e^{\cot x} \csc^2 x \, dx, \ u = \cot x$

42 $\int e^{\sec x} \sec x \tan x \, dx$

43 $\int_0^1 e^{2x} \, dx$

44 $\int_0^{\ln 5} e^{-3x} \, dx$

45 $\int_0^1 2x^2(e^{x^3} + 1) \, dx$

46 $\int_1^2 (1 + e^{-x})^2 \, dx$

47 $\int_0^{\pi/2} e^{\sin 2x} \cos 2x \, dx$

48 $\int_0^1 \dfrac{3 + e^{4x}}{e^{4x}} \, dx$

© In Problems 49 to 51, sketch the graph of each function and indicate the intervals where the function is increasing, decreasing, concave upward, and concave downward. Also, locate all extreme points, points of inflection, and asymptotes.

49 $f(x) = e^{-3x}$

50 $f(x) = e^{-x^2}$

51 $f(x) = xe^{-x}$

52 Find the area under the curve $y = e^{2x} - x$ between $x = 1$ and $x = 5$.

53 Find the area bounded by the curves $y = e^{2x}$, $y = e^{3x}$, and $x = 1$.

54 Find the volume of the solid generated by revolving the region under the curve $y = e^{2x}$ between $x = 0$ and $x = 2$ about the x axis.

55 Finish the proof of Theorem 1 by demonstrating that

$$\exp(x - y) = \exp x / \exp y.$$

56 Find the arc length of the graph of $y = (e^x + e^{-x})/2$ from $(0, 1)$ to $(1, (e + e^{-1})/2)$.

57 Let $f(x) = e^x - 1 - x$. Show that $f'(x) \geq 0$ if $x \geq 0$ and that $f'(x) \leq 0$ if $x \leq 0$. Use these facts to prove that $e^x \geq 1 + x$ and that $e^{-x} \geq 1 - x$ for all values of x.

58 If the x intercept of the tangent to the curve $y = e^{-x}$ at the variable point (p, q) is increasing at the constant rate of 5 units per

second, find the rate at which the y intercept is changing when the x intercept is 10 units.

© **59** If the value of a certain piece of property at time t years is given by the equation $V = \$20{,}000 - \$10{,}000e^{-0.1t}$, find the rate of change of V with respect to t when $t = 5$ years.

© **60** The electric current I in amperes flowing in a series circuit having an inductance L henrys, a resistance R ohms, and a constant electromotive force E volts (Figure 3) satisfies the equation

$$I = \frac{E}{R} - \frac{E}{R} \exp\left(-\frac{Rt}{L}\right)$$

where t is the time in seconds after the current begins to flow. If $E = 12$ volts, $R = 5$ ohms, and $L = 0.03$ henry, sketch the graph of I as a function of t.

Figure 3

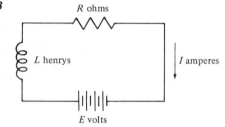

© **61** Atmospheric pressure at altitude h feet above sea level is given by $P = 15e^{-0.0004h}$ pounds per square inch. A jetliner is climbing through 10,000 feet at the rate of 1000 feet per minute. Find the rate of change of external air pressure as measured by a gauge on board the plane.

© **62** Carbon 14 decays radioactively according to the equation $y = y_0 e^{-0.0001212t}$, where y grams is the amount left after t years and y_0 grams is the initial amount. (a) Sketch a graph of y as a function of t for $0 \leq t \leq 10{,}000$ years. (b) Of a 10-gram sample of carbon 14, how much will be left after 10,000 years?

© **63** The total potential audience for an advertising campaign numbers 10,000. The average revenue per response is \$3, and the cost of the campaign is \$300 per day, plus a fixed cost of \$500. To maximize profit, find the number of days that the campaign should continue if the number of responses y is given in terms of the number of campaign days t by $y = (1 - e^{-0.25t})(10{,}000)$.

© **64** The concentration C of a drug in a person's circulatory system decreases as the drug is eliminated by the liver and kidneys or absorbed by other organs. Medical researchers often use the equation $C = C_0 e^{-kt}$ to predict the concentration C at a time t hours after the drug is administered, where C_0 is the initial concentration when $t = 0$ and k is a constant depending on the type of drug. If $C_0 = 3$ milligrams per liter and $k = 0.173$, (a) sketch a graph of C as a function of t for $0 \leq t \leq 4$ hours and (b) find C when $t = 4$ hours.

7.7 Exponential and Logarithmic Functions with Base Other Than e

In this section we use the natural logarithm and the exponential function to define b^x and $\log_b x$ for values of the base b other than e. Of course, b^k is already defined when $b > 0$ and k is a rational number. In fact, we have the following theorem.

THEOREM 1 b^k **for** $b > 0$ **and** k **Rational**

If $b > 0$ and k is a rational number, then

$$b^k = e^{k \ln b}$$

PROOF By part (ii) of Theorem 1 in Section 7.6,

$$[\exp(\ln b)]^k = \exp(k \ln b)$$

that is,

$$(e^{\ln b})^k = e^{k \ln b}$$

Since

$$e^{\ln b} = b$$

it follows that

$$b^k = e^{k \ln b}$$

Theorem 1 gives us the key to the proper definition of b^x when x is irrational.

DEFINITION 1 **Definition of** b^x **for Irrational Values of** x

If $b > 0$ and x is an irrational number, we define

$$b^x = e^{x \ln b}$$

As a consequence of Theorem 1 and Definition 1, if the base b is positive, then

$$b^x = e^{x \ln b}$$

holds for all real numbers x. Notice that

$$b^x > 0$$

for all values of the exponent x.

© **EXAMPLE 1**

(a) Using a calculator with a y^x key, evaluate $\pi^{\sqrt{2}}$.
(b) Evaluate $e^{\sqrt{2} \ln \pi}$, and thus confirm that $\pi^{\sqrt{2}} = e^{\sqrt{2} \ln \pi}$.

SOLUTION

(a) $\pi^{\sqrt{2}} = 5.047497266$
(b) $e^{\sqrt{2} \ln \pi} = 5.047497268$

Notice the discrepancy in the last decimal place caused by inaccuracies inherent in the calculator.

Basic Properties of b^x

Using the equation $b^x = e^{x \ln b}$ and the known properties of the exponential function, we can now derive the basic properties of b^x.

THEOREM 2   **Laws of Exponents**

> Let x and y be real numbers, and suppose that a and b are positive real numbers. Then
>
> **(i)** $b^x b^y = b^{x+y}$ **(v)** $\left(\dfrac{a}{b}\right)^x = \dfrac{a^x}{b^x}$
>
> **(ii)** $\dfrac{b^x}{b^y} = b^{x-y}$ **(vi)** $b^{-x} = \dfrac{1}{b^x}$
>
> **(iii)** $(b^x)^y = b^{xy}$ **(vii)** $\ln b^x = x \ln b$
>
> **(iv)** $(ab)^x = a^x b^x$

PROOF   We prove (i), (iii), (iv), and (v) and leave (ii), (vi), and (vii) as exercises (Problems 53, 54, and 55).

(i) $b^x b^y = e^{x \ln b} e^{y \ln b} = e^{x \ln b + y \ln b} = e^{(x+y) \ln b} = b^{x+y}$

(iii) Since $\ln e^{x \ln b} = x \ln b$, it follows that $\ln b^x = x \ln b$. Therefore,
$$(b^x)^y = e^{y \ln b^x} = e^{yx \ln b} = b^{yx} = b^{xy}$$

(iv) $(ab)^x = e^{x \ln ab} = e^{x(\ln a + \ln b)} = e^{x \ln a + x \ln b} = e^{x \ln a} e^{x \ln b} = a^x b^x$

(v) $\left(\dfrac{a}{b}\right)^x = e^{x \ln (a/b)} = e^{x(\ln a - \ln b)} = e^{x \ln a - x \ln b} = \dfrac{e^{x \ln a}}{e^{x \ln b}} = \dfrac{a^x}{b^x}$ ∎

According to the power rule (Theorem 1, page 140), $D_x x^k = k x^{k-1}$ holds for $x > 0$ and k a constant rational exponent. The following theorem generalizes the power rule to arbitrary (constant) exponents.

THEOREM 3   **General Power Rule**

> Let c be a constant real number, and suppose that u is a differentiable function of x. Then, for $u > 0$,
> $$D_x u^c = c u^{c-1} D_x u$$

PROOF

$$D_x u^c = D_x e^{c \ln u} = e^{c \ln u} D_x(c \ln u) = u^c \left(c \frac{D_x u}{u} \right) = c u^{c-1} D_x u$$ ∎

In Examples 2 and 3, differentiate the given function.

EXAMPLE 2 $f(x) = x^{-e} + e^{-x}$

SOLUTION   Using Theorem 3 to differentiate x^{-e}, we have

$$f'(x) = D_x(x^{-e}) + D_x(e^{-x}) = (-e)x^{-e-1} + e^{-x} D_x(-x) = -e x^{-e-1} - e^{-x}$$ ∎

EXAMPLE 3 $g(x) = (x^3 + 1)^\pi$

SOLUTION

$$g'(x) = D_x(x^3 + 1)^\pi = \pi(x^3 + 1)^{\pi-1} D_x(x^3 + 1) = \pi(x^3 + 1)^{\pi-1}(3x^2)$$ ∎

In applying the general power rule, it is important to realize that the *base is the variable* and the *exponent is constant*. On the other hand, if the *base is constant* and the *exponent is the variable*, you can find the derivative by using the formula

$$b^x = e^{x \ln b}$$

For instance,

$$D_x 2^x = D_x e^{x \ln 2} = e^{x \ln 2} D_x(x \ln 2) = 2^x(\ln 2)$$

More generally, we have the following theorem.

THEOREM 4   **Derivative of b^x**

> If b is a positive constant, then $D_x b^x = b^x \ln b$.

PROOF $$D_x b^x = D_x e^{x \ln b} = e^{x \ln b} D_x(x \ln b) = b^x(\ln b) = b^x \ln b$$ ■

Of course, Theorem 4 can be combined with the chain rule to obtain the formula

> $$D_x b^u = b^u \ln b \, D_x u$$

EXAMPLE 4   If $y = 2^{x^2}$, find dy/dx.

SOLUTION $$\frac{dy}{dx} = 2^{x^2}(\ln 2)(2x) = 2x 2^{x^2} \ln 2 = x 2^{x^2+1} \ln 2$$ ■

EXAMPLE 5   Let $f(x) = 3^{\tan x}$. Find $f'(x)$.

SOLUTION $$f'(x) = 3^{\tan x} \ln 3 \sec^2 x$$ ■

EXAMPLE 6   Find $D_x(5^{x^3} e^{\sin x})$.

SOLUTION $$\begin{aligned} D_x(5^{x^3} e^{\sin x}) &= 5^{x^3} \ln 5(3x^2) e^{\sin x} + 5^{x^3} e^{\sin x} \cos x \\ &= 5^{x^3} e^{\sin x}(3x^2 \ln 5 + \cos x) \end{aligned}$$ ■

The differentiation formula $D_x b^x = b^x \ln b$ yields a corresponding integral formula as follows:

$$\int b^x \, dx = \int \frac{b^x \ln b}{\ln b} \, dx = \frac{1}{\ln b} \int b^x \ln b \, dx = \frac{1}{\ln b} b^x + C$$

Thus, replacing x with u, we have

> $$\int b^u \, du = \frac{b^u}{\ln b} + C \qquad \text{provided that } b > 0 \text{ and } b \neq 1$$

In Examples 7 and 8, evaluate the given integral.

EXAMPLE 7 $$\int_1^2 7^x \, dx$$

SOLUTION $$\int_1^2 7^x \, dx = \frac{7^x}{\ln 7}\Big|_1^2 = \frac{49 - 7}{\ln 7} = \frac{42}{\ln 7}$$ ■

EXAMPLE 8 $\displaystyle\int 5^{\sin 2x} \cos 2x \, dx$

SOLUTION Put $u = \sin 2x$, so that $du = 2 \cos 2x \, dx$. Thus,

$$\int 5^{\sin 2x} \cos 2x \, dx = \frac{1}{2} \int 5^u \, du = \frac{5^u}{2 \ln 5} + C = \frac{5^{\sin 2x}}{2 \ln 5} + C \qquad \blacksquare$$

Logarithmic Differentiation

The properties of the natural logarithm (as developed in Section 7.5) are useful for finding the derivatives of functions involving products, quotients, and powers of other functions. The technique, called **logarithmic differentiation,** works as follows.

Procedure for Logarithmic Differentiation

To find the derivative of a function f, carry out the following steps:

Step 1 Write the equation $y = f(x)$.

Step 2 Take the natural logarithm on both sides of the equation.

Step 3 Differentiate the resulting equation implicitly with respect to x.

In Examples 9 and 10, use logarithmic differentiation to find dy/dx.

EXAMPLE 9 $y = x^x$

SOLUTION Taking the natural logarithm on both sides of the equation $y = x^x$, we obtain

$$\ln y = \ln x^x = x \ln x$$

Differentiating the equation $\ln y = x \ln x$ on both sides with respect to x, we have

$$\frac{1}{y} \frac{dy}{dx} = x \frac{1}{x} + \ln x = 1 + \ln x$$

Therefore,

$$\frac{dy}{dx} = y(1 + \ln x) = x^x(1 + \ln x) \qquad \blacksquare$$

EXAMPLE 10 $y = \dfrac{(x^2 + 5)(5x + 2)^{3/2}}{\sqrt[4]{(3x + 1)(x^3 + 2)}}$

SOLUTION

$$\begin{aligned}
\ln y &= \ln (x^2 + 5) + \ln (5x + 2)^{3/2} - \ln \sqrt[4]{(3x + 1)(x^3 + 2)} \\
&= \ln (x^2 + 5) + \tfrac{3}{2} \ln (5x + 2) - \tfrac{1}{4}[\ln (3x + 1) + \ln (x^3 + 2)]
\end{aligned}$$

so that

$$\frac{1}{y} \frac{dy}{dx} = \frac{1}{x^2 + 5} (2x) + \left(\frac{3}{2}\right) \frac{1}{5x + 2} (5) - \frac{1}{4}\left[\frac{1}{3x + 1} (3) + \frac{1}{x^3 + 2} (3x^2)\right]$$

or

$$\frac{dy}{dx} = y\left[\frac{2x}{x^2 + 5} + \frac{15}{2(5x + 2)} - \frac{3}{4(3x + 1)} - \frac{3x^2}{4(x^3 + 2)}\right]$$

Therefore,

$$\frac{dy}{dx} = \frac{(x^2 + 5)(5x + 2)^{3/2}}{\sqrt[4]{(3x + 1)(x^3 + 2)}} \left[\frac{2x}{x^2 + 5} + \frac{15}{2(5x + 2)} - \frac{3}{4(3x + 1)} - \frac{3x^2}{4(x^3 + 2)} \right] \quad \blacksquare$$

The Function $\log_a$

The base $e = 2.71828 \ldots$ is preferred in all work with logarithms or exponents in calculus because of the relative simplicity of the resulting differentiation and integration formulas. However, other bases are used in some applications. For instance, $\log_{10}$ is used in chemistry for calculating the pH of acids and bases; in computer science, $\log_2$ and $\log_8$ are used in reckoning with "bits" and "bytes" of information.

Recall the definition of a logarithm to the base a: If $a > 0$, $a \neq 1$, and $x > 0$, then we say that *y is the logarithm of x to the base a* and write

$$y = \log_a x$$

to mean that

$$a^y = x$$

In particular, $y = \log_e x$ means that $e^y = x$; that is, it means that $y = \ln x$. In other words,

$$\ln x = \log_e x$$

The following useful theorem can be used to convert logarithms from one base to another.

THEOREM 5 **Base-Changing Formula**

If $a > 0$, $b > 0$, $a \neq 1$, $b \neq 1$, and $x > 0$, then

$$\log_a x = \frac{\log_b x}{\log_b a}$$

PROOF We have

$$b^{\log_a x \log_b a} = (b^{\log_b a})^{\log_a x} = a^{\log_a x} = x$$

Hence, by the definition of logarithm,

$$\log_a x \log_b a = \log_b x$$

and it follows that

$$\log_a x = \frac{\log_b x}{\log_b a} \qquad \blacksquare$$

If we put $b = e$ in Theorem 5 and use the fact that $\log_e x = \ln x$, we obtain the important formula

$$\log_a x = \frac{\ln x}{\ln a}$$

Using this formula and the properties of the natural logarithm, you can verify the properties of $\log_a$, such as the following:

$$\log_a xy = \log_a x + \log_a y \qquad \log_a \frac{x}{y} = \log_a x - \log_a y \qquad \log_a x^y = y \log_a x$$

(Problem 59). Since scientific calculators ordinarily have keys only for $\ln$ and $\log_{10}$, *you must use the base-changing formula if you want to calculate logarithms to other bases*.

Ⓒ **EXAMPLE 11** Use a calculator and the base-changing formula to find $\log_2 3$.

SOLUTION $\qquad \log_2 3 = \dfrac{\ln 3}{\ln 2} = 1.584962501$ ∎

By using the base-changing formula, we can derive the following rule for differentiating logarithmic functions.

THEOREM 6   **Derivative of Logarithmic Functions**

Let u be a differentiable function of x with $u > 0$, and suppose that $a > 0$ with $a \neq 1$. Then

$$D_x \log_a u = \frac{1}{u \ln a} D_x u$$

PROOF $\qquad D_x \log_a u = D_x \dfrac{\ln u}{\ln a} = \dfrac{1}{\ln a} D_x \ln u = \dfrac{1}{\ln a} \cdot \dfrac{D_x u}{u} = \dfrac{1}{u \ln a} D_x u$ ∎

EXAMPLE 12 Find $\dfrac{dy}{dx}$ if $y = \log_{10}(x^2 + 5)$.

SOLUTION   By Theorem 6,

$$\frac{dy}{dx} = \frac{1}{(x^2 + 5) \ln 10} \frac{d}{dx}(x^2 + 5) = \frac{2x}{(x^2 + 5) \ln 10}$$

EXAMPLE 13 Find $D_x(\log_2 \sin x)$.

SOLUTION $\qquad D_x \log_2 \sin x = \dfrac{\cos x}{\sin x \ln 2} = \dfrac{\cot x}{\ln 2}$ ∎

Problem Set 7.7

Ⓒ **1** Use a calculator with a y^x key to find the value of each quantity as accurately as you can.

(a) $2^{\sqrt{2}}$ (b) $2^{-\sqrt{2}}$ (c) 2^{π} (d) $2^{-\pi}$
(e) $\sqrt{2}^{\sqrt{2}}$ (f) π^{π} (g) $(\sqrt{3})^{-\sqrt{5}}$ (h) $3.0157^{2.7566}$

Ⓒ **2** Use a calculator with a y^x key to verify each equation for the indicated values of the variables.

(a) $a^x a^y = a^{x+y}$ for $a = 3.074$, $x = 2.183$, $y = 1.075$

(b) $a^{x+y} = a^x a^y$ for $a = 2.471$, $x = 5.507$, $y = 0.012$

(c) $(a^x)^y = a^{xy}$ for $a = 1.777$, $x = -2.058$, $y = 3.333$

(d) $a^{x-y} = \dfrac{a^x}{a^y}$ for $a = \sqrt{2}$, $x = \sqrt{5}$, $y = \sqrt{3}$

(e) $(ab)^x = a^x b^x$ for $a = \sqrt{7}$, $b = \pi$, $x = \sqrt{\pi}$

(f) $\left(\dfrac{a}{b}\right)^x = \dfrac{a^x}{b^x}$ for $a = 2 + \pi$, $b = \sqrt{2} - 1$, $x = \sqrt{5} - \sqrt{3}$

In Problems 3 to 24, differentiate each function.

3 $f(x) = x^{-3\pi}$

4 $g(t) = t^{\pi - 2}$

5 $h(x) = 6^{-5x}$

6 $f(x) = 2^{7x^2}$

7 $g(x) = 3^{2x+1}$

8 $h(t) = (t^2 + 1)^{3e}$

9 $G(t) = 5^{\sin t}$

10 $H(x) = 3^{\cos x^2}$

11 $h(x) = (x^2 + 5)2^{-7x^2}$

12 $g(t) = (\sin t)3^{5t^2}$

13 $f(x) = \dfrac{2^{x+1}}{x^2 + 5}$

14 $h(x) = 2^{5x} \cot x$

15 $g(x) = \ln (5^x + 5^{-x})$

16 $h(t) = \sqrt[3]{7^t}$

17 $f(x) = \dfrac{3^x - 1}{3^x + 1}$

18 $F(x) = 4^{-3x} \ln (x^2 + 8)$

19 $h(t) = \log_{10} \dfrac{2t}{1 + t}$

20 $f(t) = \log_5 (\ln \cos t)$

21 $F(u) = 3^{\tan u} \log_8 u$

22 $g(t) = \sqrt{\log_5 t}$

23 $f(x) = \dfrac{\log_3 (x^2 + 5)}{x + 2}$

24 $F(x) = \csc x \log_3 (x^4 + 1)$

In Problems 25 and 26, use implicit differentiation to find dy/dx.

25 $x(2^y) + y^2 = 5$

26 $3^{xy} = x^2$

In Problems 27 to 38, use logarithmic differentiation to find dy/dx.

27 $y = x^{\sqrt{x}}$

28 $y = (\cos x)^x$

29 $y = (\sin x^2)^{3x}$

30 $y = (x + 1)^x$

31 $y = (x^2 + 4)^{\ln x}$

32 $y = (\sin x)^{\cos x}$

33 $y = (x^2 + 7)^2(6x^3 + 1)^4$

34 $y = x^2 \sin x^3 \cos (3x + 7)$

35 $y = \dfrac{\sin x \sqrt[3]{1 + \cos x}}{\sqrt{\cos x}}$

36 $y = \dfrac{\tan^2 x}{\sqrt{1 - 4 \sec x}}$

37 $y = \dfrac{x^2 \sqrt[5]{x^2 + 7}}{\sqrt[4]{11x + 8}}$

38 $y = \sqrt{\dfrac{\sec x + \tan x}{\sec x - \tan x}}$

In Problems 39 to 48, evaluate each integral. (In some cases an appropriate change of variable is suggested.)

39 $\displaystyle\int 3^{5x} \, dx, \ u = 5x$

40 $\displaystyle\int \dfrac{5^{\ln x^2}}{x} \, dx$

41 $\displaystyle\int 7^{x^4 + 4x^3}(x^3 + 3x^2) \, dx, \ u = x^4 + 4x^3$

42 $\displaystyle\int 3^{\tan x} \sec^2 x \, dx$

43 $\displaystyle\int \dfrac{2^{\ln (1/x)}}{x} \, dx, \ u = \ln x$

44 $\displaystyle\int 8^{\sec x} \sec x \tan x \, dx$

45 $\displaystyle\int 4^{\cot x} \csc^2 x \, dx$

46 $\displaystyle\int 2^{x \ln x}(1 + \ln x) \, dx$

47 $\displaystyle\int_0^1 5^{-2x} \, dx$

48 $\displaystyle\int_0^{\pi/2} 3^{\sin x} \cos x \, dx$

C **49** Using a calculator and the base-changing formula, evaluate

(a) $\log_2 25$ 　　(b) $\log_3 2$ 　　(c) $\log_8 e$

(d) $\log_\pi 5$ 　　(e) $\log_{\sqrt{2}} 0.07301$

50 Show that for $x > 0$, $\ln x = M \log_{10} x$, where $M = \ln 10$.

C **51** Which number is larger, e^π or π^e?

52 Find the maximum value of $(\ln x)/x$ for $x > 0$, and use the result to answer Problem 51 without evaluating e^π or π^e.

53 Prove that $b^x/b^y = b^{x-y}$ for $b > 0$.

54 Prove that $b^{-x} = 1/b^x$ for $b > 0$.

55 Prove that $\ln b^x = x \ln b$ for $b > 0$.

56 For what value of x is $f(x) = x^{1/x}$, $x > 0$, a maximum?

57 Prove that if $a > 0$ and $b > 0$, then $\log_a b = 1/\log_b a$.

58 Use Problem 57 and Theorem 6 to show that

$$D_x \log_a u = \left(\dfrac{\log_a e}{u}\right) D_x u.$$

59 Prove that if $a > 0$, $a \neq 1$, $x > 0$, and $y > 0$, then

(a) $\log_a xy = \log_a x + \log_a y$, 　(b) $\log_a \dfrac{x}{y} = \log_a x - \log_a y$,

and (c) $\log_a x^y = y \log_a x$.

60 Suppose that $a > 0$ and $a \neq 1$ is a given constant. Find $D_x \log_x a$ for $x > 0$, $x \neq 1$.

61 Find the volume of the solid generated by revolving the region under the graph of $y = 3^x$ between $x = 0$ and $x = 2$ about the x axis.

62 If y is a differentiable function of x, then the (point) **elasticity** of y with respect to x is defined to be the limit

$$\lim_{\Delta x \to 0} \dfrac{\Delta y/y}{\Delta x/x}$$

of the proportional change in y to the proportional change in x. Show that if x and y are positive, then the elasticity is given by dY/dX, where $Y = \log_b y$ and $X = \log_b x$.

7.8 Hyperbolic Functions

Certain combinations of exponential functions, which are related to a curve called a *hyperbola** in much the same way as trigonometric functions are related to a circle, prove to be important in applied mathematics. These functions are called the **hyperbolic functions,** and their similarity to the trigonometric functions is emphasized by calling them **hyperbolic sine, hyperbolic cosine,** and so on. The definitions of the hyperbolic sine and the hyperbolic cosine, abbreviated **sinh** and **cosh,** are as follows:

$$\sinh x = \frac{e^x - e^{-x}}{2} \quad \text{and} \quad \cosh x = \frac{e^x + e^{-x}}{2}$$

People pronounce "sinh" and "cosh" to rhyme with "cinch" and "gosh." These functions satisfy identities that correspond to standard trigonometric identities, except for an occasional switch of plus and minus signs. For instance,

1 $\cosh^2 x - \sinh^2 x = 1$

2 $\sinh(-x) = -\sinh x$

3 $\cosh(-x) = \cosh x$

4 $\sinh(x + y) = \sinh x \cosh y + \sinh y \cosh x$

5 $\cosh(x + y) = \cosh x \cosh y + \sinh x \sinh y$

and so forth.

EXAMPLE 1 Prove the identity $\cosh^2 x - \sinh^2 x = 1$.

SOLUTION

$$\cosh^2 x - \sinh^2 x = \left(\frac{e^x + e^{-x}}{2}\right)^2 - \left(\frac{e^x - e^{-x}}{2}\right)^2$$

$$= \frac{e^{2x} + 2e^x e^{-x} + e^{-2x}}{4} - \frac{e^{2x} - 2e^x e^{-x} + e^{-2x}}{4}$$

$$= \frac{e^{2x} + 2 + e^{-2x} - e^{2x} + 2 - e^{-2x}}{4} = \frac{4}{4} = 1$$

Like the sine function, the hyperbolic sine function is odd, so its graph is symmetric about the origin. Just as $\sin 0 = 0$, so also

$$\sinh 0 = \frac{e^0 - e^{-0}}{2} = \frac{0}{2} = 0$$

For large values of x, e^{-x} is very small, and so the graph of $y = \sinh x$ is very close to but below the graph of $y = e^x/2$ (Figure 1a). Like the cosine function, the hyperbolic cosine function is even, so its graph is symmetric about the y axis. Just as $\cos 0 = 1$, so also

$$\cosh 0 = \frac{e^0 + e^{-0}}{2} = \frac{2}{2} = 1$$

*See Sections 9.4 and 9.7.

Figure 1

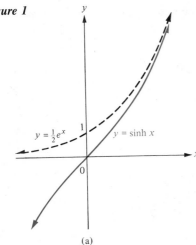

(a)

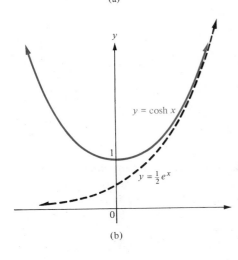

(b)

For large values of x, e^{-x} is very small, and so the graph of $y = \cosh x$ is very close to but above the graph of $y = e^x/2$ (Figure 1b). Unlike the trigonometric sine and cosine, the hyperbolic sine and hyperbolic cosine functions are *not periodic*.

The remaining hyperbolic functions are defined by analogy with the corresponding trigonometric functions as follows:

$$\tanh x = \frac{\sinh x}{\cosh x} = \frac{e^x - e^{-x}}{e^x + e^{-x}} \qquad \text{sech } x = \frac{1}{\cosh x} = \frac{2}{e^x + e^{-x}}$$

$$\coth x = \frac{\cosh x}{\sinh x} = \frac{e^x + e^{-x}}{e^x - e^{-x}} \qquad \text{csch } x = \frac{1}{\sinh x} = \frac{2}{e^x - e^{-x}}$$

We leave it as an exercise for you to sketch the graphs of these functions (Problem 10).

Differentiation of the Hyperbolic Functions

The differentiation rules for the hyperbolic functions closely resemble those for the trigonometric functions except for some algebraic signs. For instance, if u is a differentiable function of x, we have

$$D_x \sinh u = \cosh u\, D_x u \qquad \text{and} \qquad D_x \cosh u = \sinh u\, D_x u$$

To prove the differentiation rule for the hyperbolic sine, we calculate as follows:

$$D_x \sinh u = D_x \frac{e^u - e^{-u}}{2} = \frac{D_x e^u - D_x e^{-u}}{2} = \frac{e^u + e^{-u}}{2} D_x u = \cosh u\, D_x u$$

A similar calculation (Problem 32) proves the differentiation rule for the hyperbolic cosine.

<u>EXAMPLE 2</u>   Find $D_x \sinh (5x + 2)$.

SOLUTION $\begin{aligned} D_x \sinh (5x + 2) &= \cosh (5x + 2) D_x(5x + 2) \\ &= 5 \cosh (5x + 2) \end{aligned}$ ■

<u>EXAMPLE 3</u>   If $y = \cosh (\ln x)$, find dy/dx.

SOLUTION $\dfrac{dy}{dx} = \sinh (\ln x) \dfrac{d}{dx} \ln x = \dfrac{1}{x} \sinh (\ln x)$ ■

<u>EXAMPLE 4</u>   Prove the differentiation rule $D_x \tanh u = \text{sech}^2 u\, D_x u$.

SOLUTION   Assuming that u is a differentiable function of x, we calculate as follows:

$$D_x \tanh u = D_x \frac{\sinh u}{\cosh u} = \frac{\cosh u\, D_x \sinh u - \sinh u\, D_x \cosh u}{\cosh^2 u}$$

$$= \frac{\cosh^2 u - \sinh^2 u}{\cosh^2 u} D_x u = \frac{1}{\cosh^2 u} D_x u = \text{sech}^2 u\, D_x u \qquad ■$$

By inverting the differentiation rules for the hyperbolic functions, we can obtain integration formulas. Thus:

$$\int \sinh u \, du = \cosh u + C \qquad \int \cosh u \, du = \sinh u + C$$

$$\int \operatorname{sech}^2 u \, du = \tanh u + C$$

EXAMPLE 5   Evaluate $\int \sinh 5x \, dx$.

SOLUTION   Put $u = 5x$, so that $dx = \frac{1}{5} \, du$ and

$$\int \sinh 5x \, dx = \int \sinh u \left(\tfrac{1}{5} \, du\right) = \tfrac{1}{5} \int \sinh u \, du = \tfrac{1}{5} \cosh u + C$$

$$= \tfrac{1}{5} \cosh 5x + C$$

The Inverse Hyperbolic Functions

By analogy with the six inverse trigonometric functions, there are six **inverse hyperbolic functions.** Unlike the analogous trigonometric functions, the graphs of sinh, tanh, coth, and csch satisfy the horizontal-line test, so there is no problem about the existence of $\sinh^{-1}$, $\tanh^{-1}$, $\coth^{-1}$, or csch^{-1}. The graphs of sinh, tanh, coth, csch, and their inverses are sketched in Figure 2. In each case, the graph of the inverse function is obtained by reflecting the graph of the function about the line $y = x$.

The hyperbolic cosine and its reciprocal, the hyperbolic secant, are not invertible because their graphs do not satisfy the horizontal-line test. However, the portions of these graphs that lie to the right of the y axis do satisfy the horizontal-line test; hence, their reflections across the line $y = x$ are taken, by definition, to be the graphs of $\cosh^{-1}$ and sech^{-1} (Figure 3).

Since $D_x \sinh x = \cosh x > 0$, it follows that $\sinh^{-1}$ is a differentiable function by the inverse-function theorem (Theorem 1, page 406). Now, suppose that u is a differentiable function of x, and let

$$y = \sinh^{-1} u \qquad \text{so that} \qquad u = \sinh y$$

Implicit differentiation of the last equation gives

$$D_x u = \cosh y \, D_x y$$

Hence,

$$D_x y = \frac{1}{\cosh y} \, D_x u$$

Since $\cosh^2 y - \sinh^2 y = 1$ and $\cosh y > 0$, we have

$$\cosh y = \sqrt{1 + \sinh^2 y} = \sqrt{1 + u^2}$$

and it follows that

$$D_x y = \frac{1}{\sqrt{1 + u^2}} \, D_x u$$

or

$$D_x \sinh^{-1} u = \frac{1}{\sqrt{1 + u^2}} \, D_x u$$

Figure 2

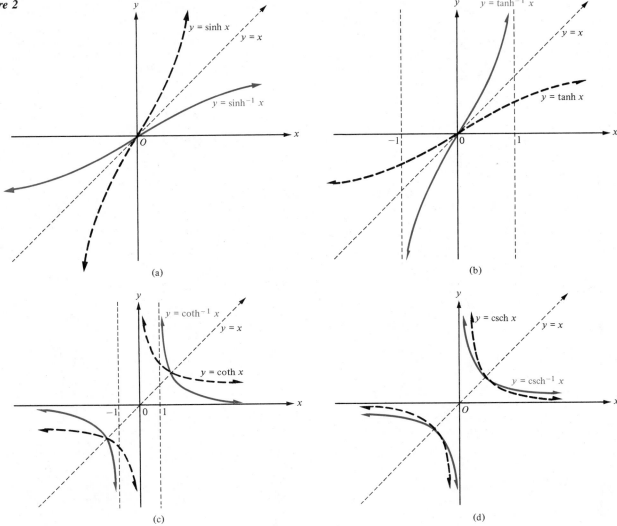

(a)

(b)

(c)

(d)

Figure 3

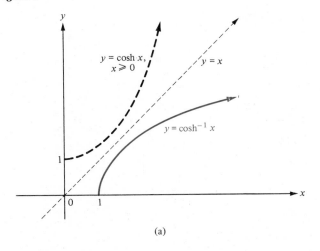

(a)

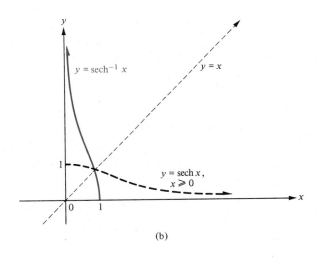

(b)

Similar differentiation rules may be derived for the remaining inverse hyperbolic functions; for instance,

$$D_x \cosh^{-1} u = \frac{1}{\sqrt{u^2 - 1}} D_x u \qquad \text{if } u > 1$$

$$D_x \tanh^{-1} u = \frac{1}{1 - u^2} D_x u \qquad \text{if } |u| < 1$$

(Problem 50).

EXAMPLE 6   If $y = \tanh^{-1} (\sin 2x)$, find $\dfrac{dy}{dx}$.

SOLUTION $\dfrac{dy}{dx} = \dfrac{1}{1 - \sin^2 2x} \dfrac{d}{dx} \sin 2x$

$= \dfrac{2 \cos 2x}{\cos^2 2x} = 2 \sec 2x$

Problem Set 7.8

In Problems 1 to 8, prove each identity.

1 $\sinh (-x) = -\sinh x$ **2** $\cosh (-x) = \cosh x$

3 $\sinh (x + y) = \sinh x \cosh y + \sinh y \cosh x$

4 $\cosh (x + y) = \cosh x \cosh y + \sinh x \sinh y$

5 $1 - \tanh^2 x = \operatorname{sech}^2 x$ **6** $\coth^2 x - 1 = \operatorname{csch}^2 x$

7 $e^x = \sinh x + \cosh x$ **8** $\sinh (\ln x) = \dfrac{x^2 - 1}{2x}$

© **9** Use a calculator with an e^x (or EXP) key to evaluate the following quantities as accurately as you can.

    (a) $\sinh 1.2$ (b) $\cosh (-1.4)$ (c) $\tanh 0.7$

    (d) $\coth 1.3$ (e) $\operatorname{sech} 0.6$ (f) $\operatorname{csch} (-0.9)$

© **10** Sketch the graphs of (a) $y = \tanh x$, (b) $y = \coth x$, (c) $y = \operatorname{sech} x$, and (d) $y = \operatorname{csch} x$.

In Problems 11 to 18, differentiate each function.

11 $f(x) = \sinh (3x^2 + 5)$ **12** $g(x) = \cosh (\ln x)$

13 $f(t) = \ln (\sinh t^3)$ **14** $f(u) = e^{2u} \tanh u$

15 $h(t) = \tanh e^{3t}$

16 $F(r) = \sin^{-1} r \tanh (3r + 5)$

17 $G(s) = \sin^{-1} (\tanh s)$ **18** $f(x) = \displaystyle\int_1^{\tanh x} \frac{dt}{1 + t^2}$

In Problems 19 to 22, use implicit differentiation to find dy/dx.

19 $x^2 = \sinh y$

20 $\sinh x = \cosh y$

21 $\sin x = \sinh y$

22 $\tanh^2 x - 2 \sinh y = \tanh y$

In Problems 23 to 30, evaluate each integral.

23 $\displaystyle\int \cosh 7x \, dx$

24 $\displaystyle\int \sinh \frac{5x}{3} \, dx$

25 $\displaystyle\int \operatorname{sech}^2 3x \, dx$

26 $\displaystyle\int \frac{\operatorname{sech}^2 \sqrt{x}}{\sqrt{x}} \, dx$

27 $\displaystyle\int \frac{\sinh 5x}{\cosh^3 5x} \, dx$

28 $\displaystyle\int \sinh^{10} 3x \cosh 3x \, dx$

29 $\displaystyle\int_0^1 \cosh^3 x \sinh x \, dx$

30 $\displaystyle\int_0^2 \sinh^4 x \cosh x \, dx$

31 Evaluate $\int \cosh (\ln x)\, dx$. (*Hint:* Use the definition of cosh before integrating.)

32 Derive differentiation rules for (a) $\cosh u$, (b) $\coth u$, (c) $\operatorname{sech} u$, and (d) $\operatorname{csch} u$.

33 Show that if A, B, and k are constants, the function $y = A \sinh kx + B \cosh kx$ is a solution of the differential equation $y'' - k^2 y = 0$.

34 Find the area under the curve $y = \sinh x$ between $x = 0$ and $x = 1$.

In Problems 35 to 44, differentiate each function.

35 $f(x) = \sinh^{-1} x^3$

36 $g(x) = \cosh^{-1} (\sec x)$

37 $h(x) = \cosh^{-1} \dfrac{x}{3}$

38 $F(t) = \tanh^{-1} (\sin t)$

39 $G(t) = \tanh^{-1} 5t$

40 $H(t) = \ln \sqrt{t^2 - 1} - t \tanh^{-1} t$

41 $f(x) = x \cosh^{-1} e^x$

42 $g(x) = x \sinh^{-1} x - \sqrt{1 + x^2}$

43 $h(u) = u \tanh^{-1} (\ln u)$

44 $F(w) = \dfrac{\cosh^{-1} w}{w^2 + 4}$

45 Let $y = \sinh^{-1} x$, so that $x = \sinh y = \frac{1}{2}(e^y - e^{-y})$. (a) Show that $(e^y)^2 - 2xe^y - 1 = 0$. (b) Recalling that e^y is always positive, solve the quadratic equation in (a) for e^y and thus show that $e^y = x + \sqrt{x^2 + 1}$. (c) Using the result in (b), show that $\sinh^{-1} x = \ln (x + \sqrt{x^2 + 1})$.

46 Proceeding as in Problem 45, show that

$$\tanh^{-1} x = \frac{1}{2} \ln \frac{1 + x}{1 - x} \qquad \text{for } |x| < 1$$

In Problems 47 to 49, justify the given integral formula by differentiating the right side.

47 $\displaystyle\int \frac{dx}{\sqrt{a^2 + x^2}} = \sinh^{-1} \frac{x}{a} + C$ for $a > 0$

48 $\displaystyle\int \frac{dx}{a^2 - x^2} = \frac{1}{a} \tanh^{-1} \frac{x}{a} + C$ for $|x| < a$

49 $\displaystyle\int \frac{dx}{\sqrt{x^2 - a^2}} = \cosh^{-1} \frac{x}{a} + C$ for $x > a > 0$

50 Derive the differentiation rules for (a) $\cosh^{-1}$ and (b) $\tanh^{-1}$.

51 Use the formula in Problem 47 to evaluate $\displaystyle\int \frac{dx}{\sqrt{9 + x^2}}$.

52 Use the formula in Problem 47 to find the area of the region bounded by the curve $y = 16/\sqrt{16 + x^2}$ and the lines $x = 0$, $y = 0$, and $x = 3\sqrt{2}$.

53 When a flexible cord or chain is suspended from its ends, it hangs in a curve called a **catenary** whose equation has the form $y = a \cosh (x/a)$, where a is a positive constant (Figure 4). Find the length of the catenary between $(-b,\ a \cosh (b/a))$ and $(b,\ a \cosh (b/a))$.

Figure 4

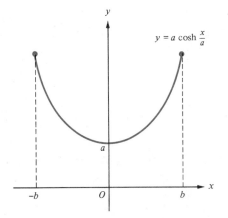

54 The famous Gateway Arch to the West in St. Louis has the shape of an inverted catenary (see Problem 53). The equation of any such arch takes the form $y = h + a[1 - \cosh (x/a)]$, where a is a positive constant, h is the height of the arch, the two bases of the arch lie on the x axis, and the apex of the arch lies on the y axis. Show that the distance between the two bases of such an arch is $2a \cosh^{-1} [1 + (h/a)]$ units.

7.9   Exponential Growth and Decay

One of the most important applications of exponential functions occurs in the study of a quantity that grows or decays at a rate proportional to the quantity itself. For instance, money invested at compound interest increases at a rate that is proportional to the amount of money earning interest. We discuss such growth and decay in this section. We begin by establishing some useful technical results.

The Exponential Function as a Limit

On page 426 we *provisionally* defined e to be the limit as $u \to +\infty$ of $[1 + (1/u)]^u$. However, at that stage of the game, this was not really legitimate since we didn't have Definition 1 on page 445 at our disposal, so the meaning of $[1 + (1/u)]^u$ when u is irrational was in doubt. Now we can prove the following theorem.

THEOREM 1   **e as a Limit**

$$e = \lim_{u \to +\infty} \left(1 + \frac{1}{u}\right)^u = \lim_{u \to -\infty} \left(1 + \frac{1}{u}\right)^u$$

PROOF   In the expression $[1 + (1/u)]^u$, we put $\Delta x = 1/u$, so $u = 1/\Delta x$ and

$$\left(1 + \frac{1}{u}\right)^u = (1 + \Delta x)^{1/\Delta x}$$

Notice that $u \to +\infty$ as $\Delta x \to 0^+$ and that $u \to -\infty$ as $\Delta x \to 0^-$. Hence, the equations to be established can be written

$$e = \lim_{\Delta x \to 0^+} (1 + \Delta x)^{1/\Delta x} = \lim_{\Delta x \to 0^-} (1 + \Delta x)^{1/\Delta x}$$

or simply as the single equation

$$e = \lim_{\Delta x \to 0} (1 + \Delta x)^{1/\Delta x}$$

To prove this, we use Definition 1 on page 445 to write

$$(1 + \Delta x)^{1/\Delta x} = e^{(1/\Delta x)\ln(1+\Delta x)} = \exp\left[\frac{1}{\Delta x}\ln(1 + \Delta x)\right]$$

The proof will be complete once we show that

$$\lim_{\Delta x \to 0}\left[\frac{1}{\Delta x}\ln(1 + \Delta x)\right] = 1$$

for then

$$\lim_{\Delta x \to 0} (1 + \Delta x)^{1/\Delta x} = \exp\left\{\lim_{\Delta x \to 0}\left[\frac{1}{\Delta x}\ln(1 + \Delta x)\right]\right\} = \exp 1 = e$$

will follow from the continuity of the exponential function. Thus, let

$$f(x) = \ln x \qquad \text{for } x > 0$$

so that $\qquad f(1) = \ln 1 = 0 \qquad f'(x) = \dfrac{1}{x} \qquad$ and $\qquad f'(1) = 1$

Now, $\displaystyle\lim_{\Delta x \to 0} \left[\dfrac{1}{\Delta x} \ln (1 + \Delta x) \right] = \lim_{\Delta x \to 0} \dfrac{f(1 + \Delta x)}{\Delta x} = \lim_{\Delta x \to 0} \dfrac{f(1 + \Delta x) - f(1)}{\Delta x}$

$$= f'(1) = 1$$

and our proof is complete.   ■

THEOREM 2   **The Exponential Function as a Limit**

$$e^a = \lim_{h \to +\infty} \left(1 + \dfrac{a}{h} \right)^h = \lim_{h \to -\infty} \left(1 + \dfrac{a}{h} \right)^h$$

PROOF   The equations obviously hold when $a = 0$. Thus, we can assume that $a \neq 0$. First we consider the case in which $a > 0$. Put $u = h/a$, noting that $u \to +\infty$ as $h \to +\infty$. Therefore,

$$\lim_{h \to +\infty} \left(1 + \dfrac{a}{h} \right)^h = \lim_{h \to +\infty} \left(1 + \dfrac{a}{h} \right)^{(h/a)a} = \lim_{u \to +\infty} \left(1 + \dfrac{1}{u} \right)^{ua}$$

$$= \lim_{u \to +\infty} \left[\left(1 + \dfrac{1}{u} \right)^u \right]^a$$

We put $v = [1 + (1/u)]^u$, noting that $\displaystyle\lim_{u \to +\infty} v = e$ by Theorem 1. By Theorem 3 on page 446, v^a is a differentiable function of v; hence, it is a continuous function of v. Therefore,

$$\lim_{h \to +\infty} \left(1 + \dfrac{a}{h} \right)^h = \lim_{u \to +\infty} v^a = (\lim_{u \to +\infty} v)^a = e^a$$

Similar arguments take care of the cases in which $a < 0$ and in which $h \to -\infty$ (Problem 50).   ■

EXAMPLE 1   Evaluate each limit:

(a) $\displaystyle\lim_{x \to 0} (1 + 5x)^{1/x}$ (b) $\displaystyle\lim_{t \to +\infty} \left(1 + \dfrac{3}{t} \right)^{2t}$

SOLUTION

(a) Let $h = 1/x$, so that $x = 1/h$, and note that $h \to +\infty$ as $x \to 0^+$. Thus,

$$\lim_{x \to 0^+} (1 + 5x)^{1/x} = \lim_{h \to +\infty} \left(1 + \dfrac{5}{h} \right)^h = e^5$$

(b) $\displaystyle\lim_{t \to +\infty} \left(1 + \dfrac{3}{t} \right)^{2t} = \lim_{t \to +\infty} \left[\left(1 + \dfrac{3}{t} \right)^t \right]^2$

$$= \left[\lim_{t \to +\infty} \left(1 + \dfrac{3}{t} \right)^t \right]^2 = (e^3)^2 = e^6$$ ■

© EXAMPLE 2   Telephone calls coming into a certain switchboard follow a *Poisson probability distribution*, averaging c calls per minute. The theory of probability gives $P = \displaystyle\lim_{n \to +\infty} c[1 - (c/n)]^{n-1}$ as the probability that exactly one call comes into the switchboard during any given 1-minute period. Find P if $c = 4$ calls per minute.

SOLUTION $P = \lim\limits_{n \to +\infty} \dfrac{c\left(1 - \dfrac{c}{n}\right)^n}{\left(1 - \dfrac{c}{n}\right)} = \dfrac{c \lim\limits_{n \to +\infty}\left(1 + \dfrac{-c}{n}\right)^n}{\lim\limits_{n \to +\infty}\left(1 - \dfrac{c}{n}\right)} = \dfrac{ce^{-c}}{1} = ce^{-c}$

by Theorem 2. For $c = 4$, we have $P = 4e^{-4} \approx 0.073 = \frac{73}{1000}$, and we could expect exactly one incoming call during approximately 73 of 1000 one-minute periods.   ∎

Compound Interest and Present Value

Suppose that a **principal** P dollars is invested for a **term** of t years at a **nominal annual interest rate** r **compounded** n times per year. Thus, the interest is periodically calculated and added to the principal. The time interval between successive conversions of interest into principal, $(1/n)$th of a year, is called the **conversion period.**

   At the end of the first conversion period, an interest of $P(r/n)$ is added to the principal, so that the value of the investment is $P + P(r/n) = P[1 + (r/n)]$ dollars. At the end of the second conversion period, an interest of $P[1 + (r/n)](r/n)$ is added, and the value of the investment is

$$P\left(1 + \frac{r}{n}\right) + P\left(1 + \frac{r}{n}\right)\left(\frac{r}{n}\right) = P\left(1 + \frac{r}{n}\right)\left(1 + \frac{r}{n}\right) = P\left(1 + \frac{r}{n}\right)^2 \qquad \text{dollars}$$

At the end of the third conversion period, an interest of $P[1 + (r/n)]^2(r/n)$ is added, and the value of the investment is

$$P\left(1 + \frac{r}{n}\right)^2 + P\left(1 + \frac{r}{n}\right)^2\left(\frac{r}{n}\right) = P\left(1 + \frac{r}{n}\right)^2\left(1 + \frac{r}{n}\right) = P\left(1 + \frac{r}{n}\right)^3 \qquad \text{dollars}$$

and so on. Evidently, at the end of K conversion periods, the value of the investment is

$$P\left(1 + \frac{r}{n}\right)^K \qquad \text{dollars}$$

Over the term of t years, there will be $K = nt$ conversion periods, so the final value S dollars of the investment will be given by

$$S = P\left(1 + \frac{r}{n}\right)^{nt}$$

© **EXAMPLE 3**   If you invest $P = \$500$ at a nominal annual interest rate of 8 percent (that is, $r = 0.08$), what is the final value S of the investment after a term of $t = 3$ years if the interest is compounded (a) quarterly, (b) monthly, (c) daily?

SOLUTION

    (a) Here $n = 4$, so $S = 500[1 + (0.08/4)]^{4(3)} = 500(1.02)^{12} = \$634.12$.

    (b) Here $n = 12$, so $S = 500[1 + (0.08/12)]^{12(3)} = \$635.12$.

    (c) Here $n = 365$, so $S = 500[1 + (0.08/365)]^{365(3)} = \$635.61$.   ∎

When a bank offers compound interest, it usually specifies not only the nominal annual interest rate r, but also the **effective** simple annual interest rate R, that is, the rate of simple annual interest that would yield the same final value over a 1-year term as the compound interest. Thus,

$$P(1 + R) = P\left(1 + \frac{r}{n}\right)^{n(1)} \quad \text{or} \quad 1 + R = \left(1 + \frac{r}{n}\right)^{n}$$

Therefore,

$$R = \left(1 + \frac{r}{n}\right)^{n} - 1$$

© **EXAMPLE 4**   A bank offers savings accounts at 8 percent nominal annual interest compounded quarterly. Find the corresponding effective simple annual interest rate R.

SOLUTION $$R = \left(1 + \frac{r}{n}\right)^{n} - 1 = \left(1 + \frac{0.08}{4}\right)^{4} - 1 = 0.0824 = 8.24\%$$ ∎

Money that you will receive in the future is worth *less* to you than the same amount of money received now, because you miss out on the interest you could collect by investing the money now. For this reason, economists use the idea of the *present value* of money to be received in the future. If there is an opportunity to invest P dollars at a nominal annual interest rate r compounded n times a year, this principal plus the interest it earns will amount to S dollars after t years, where

$$S = P\left(1 + \frac{r}{n}\right)^{nt}$$

Thus, P dollars in hand *right now* is worth S dollars to be received t years *in the future*. Solving the equation above for P in terms of S, we obtain the equation

$$P = S\left(1 + \frac{r}{n}\right)^{-nt}$$

for the **present value** P dollars of an offer of S dollars to be received t years in the future.

© **EXAMPLE 5**   Find the present value of \$500 to be paid to you 2 years in the future if investments during this period are earning a nominal annual interest of 10 percent compounded monthly.

SOLUTION   Here $S = \$500$, $r = 0.10$, $n = 12$, $t = 2$, and

$$P = S\left(1 + \frac{r}{n}\right)^{-nt} = 500\left(1 + \frac{0.10}{12}\right)^{-12(2)} = \$409.70$$ ∎

Some banks offer savings accounts with interest compounded not quarterly, not weekly, not daily, not hourly, but **continuously.** The formula for continuously compounded interest is obtained by letting $n \to +\infty$ in the formula for compound interest. Thus, using Theorem 2, we have

$$S = \lim_{n \to +\infty} P\left(1 + \frac{r}{n}\right)^{nt} = \lim_{n \to +\infty} P\left[\left(1 + \frac{r}{n}\right)^{n}\right]^{t} = P\left[\lim_{n \to +\infty} \left(1 + \frac{r}{n}\right)^{n}\right]^{t}$$

$$= P(e^{r})^{t} = Pe^{rt}$$

Therefore, if a principal of P dollars is invested for a term of t years at a nominal annual interest rate r, the final value S dollars of the investment is given by

$$S = Pe^{rt}$$

The corresponding effective simple annual interest rate R, obtained by solving the equation $P(1 + R) = Pe^{r(1)}$ for R, is given by

$$R = e^r - 1$$

© **EXAMPLE 6**   The New Mattoon Savings Bank offers a savings account with continuously compounded interest at a nominal annual rate of 8 percent.

  (a) Find the corresponding effective simple annual interest rate R.

  (b) What is the final value S dollars if $P = \$500$ is deposited in such an account for a term of 3 years?

SOLUTION   Here $P = \$500$, $r = 0.08$, and $t = 3$.

  (a) $R = e^r - 1 = e^{0.08} - 1 = 0.0833 = 8.33\%$
  (b) $S = Pe^{rt} = 500e^{0.08(3)} = \$635.62$

Solving the equation

$$S = Pe^{rt}$$

for P in terms of S, r, and t, we obtain the formula

$$P = Se^{-rt}$$

for the **present value** P dollars of an offer of S dollars to be received t years in the future if investments during this period are earning a nominal annual interest rate r compounded continuously.

© **EXAMPLE 7**   Find the present value of $6000 to be paid to you 3 years in the future if investments during this period are earning a nominal annual interest rate of 9 percent compounded continuously.

SOLUTION   Here $S = \$6000$, $r = 0.09$, $t = 3$, and

$$P = Se^{-rt} = 6000e^{-0.09(3)} = \$4580.28$$

Exponential Growth and Decay

Suppose that a positive quantity q is changing in time at a rate dq/dt that is proportional to q itself; that is,

$$\frac{dq}{dt} = kq$$

where k is the constant of proportionality. If $k > 0$, then $dq/dt > 0$, so q is increasing as time goes on; whereas if $k < 0$, then $dq/dt < 0$ and q decreases as time passes. The constant k is called the **growth constant.**

Figure 1

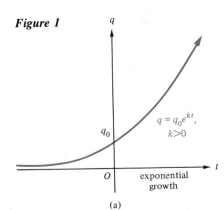

(a)

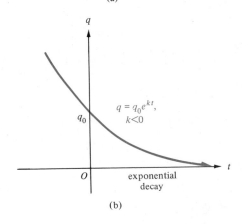

(b)

The differential equation $dq/dt = kq$ is separable, so it can be rewritten as

$$\frac{dq}{q} = k\,dt$$

Integration of both sides of the last equation yields

$$\ln q = kt + C$$

where C is the constant of integration. It follows that

$$e^{\ln q} = e^{kt+C} = e^{kt}e^{C} \qquad \text{or} \qquad q = e^{C}e^{kt}$$

Suppose that q_0 *is the value of* q *when* $t = 0$. Then

$$q_0 = e^{C}e^{k(0)} = e^{C}e^{0} = e^{C}$$

Therefore, the general solution of the differential equation $dq/dt = kq$ is

$$q = q_0 e^{kt}$$

A positive quantity q whose value at time t is given by $q = q_0 e^{kt}$ is said to be **increasing, or growing, exponentially** if $k > 0$, and it is said to be **decreasing, or decaying, exponentially** if $k < 0$. Graphs of q as a function of t for exponential growth and for exponential decay are shown in Figure 1.

<u>EXAMPLE 8</u>   Solve the differential equation $dq/dt = 3q$ with the initial-value condition that $q = 2$ when $t = 0$.

SOLUTION   Separating variables and integrating, we have

$$\frac{dq}{q} = 3\,dt \qquad \text{so that} \qquad \ln q = 3t + C$$

Therefore,

$$e^{\ln q} = e^{3t+C} \qquad \text{or} \qquad q = e^{C}e^{3t}$$

Substituting $q = 2$ and $t = 0$ into the last equation, we find that

$$2 = e^{C}e^{3(0)} = e^{C}e^{0} = e^{C}$$

and it follows that

$$q = 2e^{3t}$$

When one-celled organisms reproduce by simple cell division in a culture containing an unlimited supply of nutrients, the rate at which the population of organisms increases is often proportional to the number of organisms present; hence, the growth is exponential.

Ⓒ **EXAMPLE 9**   Bacteria grown in a certain culture increase at a rate proportional to the number of bacteria present. If there are 2500 bacteria present initially and if the number of bacteria triples in $\frac{1}{2}$ hour, how many bacteria are present after t hours? How many are present after 2 hours? How long will it take before there are 1 million bacteria present?

SOLUTION   Let q be the number of bacteria present after t hours, so that

$$\frac{dq}{dt} = kq$$

where k is the constant of proportionality. Separating variables and solving this differential equation, we find that

$$q = q_0 e^{kt}$$

where $q_0 = 2500$ is the number of bacteria present when $t = 0$. When $t = \frac{1}{2}$, we know that $q = 3q_0 = 7500$; hence,

$$7500 = 2500 e^{k(1/2)} \qquad \text{or} \qquad 3 = e^{k/2}$$

From the last equation, it follows that

$$\frac{k}{2} = \ln 3 \qquad \text{or} \qquad k = 2 \ln 3$$

Therefore,

$$q = q_0 e^{kt} = 2500 e^{(2 \ln 3)t} = 2500 (e^{\ln 3})^{2t} = 2500(3)^{2t} = 2500(9^t)$$

Thus, after t hours, there will be $2500(9^t)$ bacteria in the culture. After 2 hours, the culture will contain

$$2500(9^2) = 202{,}500 \text{ bacteria}$$

The time t required for the number of bacteria to increase to 1 million can be found by solving the equation

$$1{,}000{,}000 = 2500(9^t) \qquad \text{or} \qquad 9^t = \frac{1{,}000{,}000}{2500} = 400$$

Taking the natural logarithm on both sides of the last equation, we find that

$$\ln 9^t = \ln 400 \qquad \text{or} \qquad t \ln 9 = \ln 400$$

Hence,

$$t = \frac{\ln 400}{\ln 9} \approx 2.73 \text{ hours}$$

Radioactive substances decay at a rate proportional to the amount of the substance present; hence, the quantity q of such a substance satisfies a differential equation

$$\frac{dq}{dt} = kq$$

with a *negative* constant k. Separating variables and solving this differential equation, we find that

$$q = q_0 e^{kt}$$

where q_0 is the amount of radioactive material present when $t = 0$. Although the decay constant k gives a measure of how rapidly the substance is decaying, the rate of decay of a radioactive material is usually expressed by its **half-life**—the length of time T during which exactly half of it will decay.

© **EXAMPLE 10** Polonium, a radioactive element discovered by Marie Curie in 1898 and named after her native country, Poland, has a half-life of 140 days. Find the decay constant k, and sketch a graph showing the amount q grams of polonium remaining after t days if the amount when $t = 0$ is $q_0 = 5$ grams.

SOLUTION   The differential equation for q is

$$\frac{dq}{dt} = kq$$

where k is the decay constant. Separating variables and solving this differential equation, we have

$$q = q_0 e^{kt}$$

where $q_0 = 5$ grams. The half-life is $T = 140$ days, so when $t = T$, $q = \frac{1}{2}q_0$; that is,

$$\tfrac{1}{2}q_0 = q_0 e^{kT} \qquad \text{or} \qquad e^{kT} = \tfrac{1}{2}$$

Therefore,

$$kT = \ln\frac{1}{2} = -\ln 2 \qquad \text{so that} \qquad k = -\frac{\ln 2}{T} = -\frac{\ln 2}{140} \approx -0.005$$

The graph of

$$q = 5e^{-0.005t}$$

is sketched in Figure 2.   ∎

Figure 2

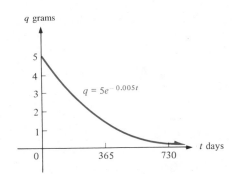

Problem Set 7.9

In Problems 1 to 8, evaluate each limit.

1 $\displaystyle\lim_{h \to +\infty}\left(1 + \frac{6}{h}\right)^{h}$

2 $\displaystyle\lim_{n \to +\infty}\left(1 - \frac{2}{n}\right)^{n}$

3 $\displaystyle\lim_{x \to 0^{+}}(1 - 5x)^{1/x}$

4 $\displaystyle\lim_{n \to -\infty}\left(1 - \frac{3}{n}\right)^{n}$

5 $\displaystyle\lim_{n \to +\infty}\left(1 + \frac{1}{n}\right)^{4n}$

6 $\displaystyle\lim_{h \to +\infty}\left(1 + \frac{3}{h}\right)^{h/5}$

7 $\displaystyle\lim_{u \to 0^{-}}(1 - 4u)^{1/u}$

8 $\displaystyle\lim_{u \to 0}\left(1 + \frac{u}{7}\right)^{1/u}$

© **9** Using a calculator, evaluate $[1 + (6/h)]^{h}$ for successively larger and larger values of h, and thus confirm your answer to Problem 1.

10 Patients arriving at the emergency room of a certain hospital

follow a **Poisson probability distribution,** averaging c patients per hour. According to the theory of probability.

$$P(k) =$$

$$\lim_{n \to +\infty}\frac{n(n-1)(n-2)\cdots(n-k+1)}{k(k-1)(k-2)\cdots 1}\left(\frac{c}{n}\right)^{k}\left(1 - \frac{c}{n}\right)^{n-k}$$

gives the probability that *exactly* k patients will arrive at the emergency room during a given 1-hour period. Evaluate this limit and thus find $P(k)$.

© In Problems 11 to 16, assume that you have invested a principal P dollars at a nominal annual interest rate r compounded n times per year for a term of t years. Calculate (a) the effective simple annual interest rate R and (b) the final value S dollars of your investment.

11 $P = \$1000$, $r = 7$ percent, $n = 1$, $t = 13$ years

12 $P = \$1000$, $r = 7$ percent, $n = 12$, $t = 13$ years

13 $P = \$1000$, $r = 7$ percent, $n = 52$, $t = 13$ years

14 $P = \$1000$, $r = 12$ percent, $n = 12$, $t = 13$ years

15 $P = \$50,000$, $r = 13.5$ percent, $n = 12$, $t = \frac{1}{2}$ year

16 $P = \$25,000$, $r = 15.5$ percent, $n = 52$, $t = \frac{1}{4}$ year

© **17** Suppose that a bank offers to pay a nominal annual interest rate of 8 percent on money left on deposit for a 2-year term. Assume that a principal of $1000 is deposited. Find the final value S dollars after the 2-year term if the interest is compounded (a) annually, (b) semiannually, (c) quarterly, (d) monthly, (e) weekly, (f) daily, (g) hourly, and (e) continuously.

© **18** Find out the nominal annual interest rate r offered by your local savings bank for regular savings accounts and the number of times n per year that the interest is compounded. Using this information, calculate the effective simple annual interest rate R.

© **19** Suppose that someone owes you $100 and offers to pay you this money 6 months from now. What is the present value to you of this offer if you currently have an investment opportunity paying 8 percent nominal annual interest compounded (a) quarterly, (b) monthly, (c) weekly, (d) continuously?

© **20** If a sum of money is invested at a nominal annual interest rate of 7 percent, how long will it take for the investment to double in value if the interest is compounded (a) quarterly, (b) monthly, (c) continuously?

© **21** A savings bank with $28 million in regular savings accounts is paying a nominal annual interest of 5.5 percent compounded quarterly on such accounts. The bank is contemplating offering the same rate of interest, but compounding continuously rather than quarterly. If this plan is implemented, (a) what effective simple annual interest rate R can the bank advertise on its savings accounts and (b) how much more interest will the bank have to pay out per year on these accounts?

© **22** On Jay's 16th birthday, his father promises to give him $50,000 when he turns 21 to help set him up in business. Local banks are offering savings accounts at a nominal annual interest rate of 7 percent compounded continuously. Jay, who has studied the mathematics of finance, says, "Dad, I'll settle for ——— dollars right now!" Fill in the blank appropriately.

© **23** Money deposited in a certain savings account doubles every 9.9 years with the interest compounded continuously. (a) What is the nominal annual interest rate? (b) What is the effective simple annual interest rate?

© **24** The inflation rate in a certain country has been 11 percent per year for the past 5 years. If in that country a loaf of bread now costs $1, how much did it cost 5 years ago?

© **25** How much money should you invest at 8 percent compounded continuously in order to have $40,000 after 15 years?

© **26** Government economists, monitoring the rate of inflation, often report a *monthly* rate r and the corresponding *annual* rate R that

would result if that monthly rate continued for 12 months. The understanding is that, on the average, if an article cost P dollars at the beginning of the month, then it would cost $P + Pr$ dollars at the end of the month; similarly, if an article cost P dollars at the beginning of the year, it would cost $P + PR$ dollars at the end of the year. (a) Show that $R = (1 + r)^{12} - 1$. (b) Find the annual rate of inflation corresponding to a monthly rate of 1 percent.

In Problems 27 to 32, solve each differential equation with the given initial-value condition.

27 $\dfrac{dq}{dt} = 5q$, $q = 2$ when $t = 0$

28 $\dfrac{dy}{dx} = 2y$, $y = 10$ when $x = 0$

29 $\dfrac{dN}{dt} = -4N$, $N = 40$ when $t = 0$

30 $\dfrac{dy}{dx} = -2y$, $y = -10$ when $x = 0$

31 $\dfrac{dq}{dt} = 10 - q$, $q = 3$ when $t = 0$

32 $\dfrac{dx}{dt} = -0.2(80 - x)$, $x = 0$ when $t = 0$

© **33** Bacteria grown in a certain culture increase at a rate proportional to the number of bacteria present. If there are 1000 bacteria initially and if the number of bacteria doubles in 15 minutes, how long will it take before there are 2 million bacteria present?

© **34** Bacteria in a culture have a natural tendency to increase by 25 percent each hour; however, an experimental bactericide present in the culture kills 20 percent of the bacteria each hour. How much time is required for the number of bacteria to double under these conditions?

© **35** Ecologists have determined that the rate at which the population of bears in a protected forest area increases is proportional to the number of bears. If there were 225 bears in 1977 and 239 bears in 1980, how many bears will inhabit the region in 1990?

36 If a quantity q is growing exponentially in time with a growth constant k, show that the percentage of increase of q in 1 unit of time is given by $(e^k - 1) \times 100$ percent.

© **37** The number of bacteria in an unrefrigerated chicken salad triples in 2 hours. In how many *more* hours will the original number of bacteria be multiplied by a factor of 50?

38 Suppose that a quantity q is growing or decaying exponentially in time and that k is the growth or decay constant. Assume that $q = q_1$ when $t = t_1$ and $q = q_2$ when $t = t_2$. If $t_1 \neq t_2$, prove that

$$k = \frac{1}{t_2 - t_1} \ln \frac{q_2}{q_1}$$

C **39** The bacterium *Escherichia coli* is found in the human intestine. When *E. coli* is cultivated under ideal conditions in a biological laboratory, the population grows exponentially and doubles in 20 minutes. Suppose that $q_0 = 10$ *E. coli* cells are placed in a nutrient broth at time $t = 0$ minutes. (a) Write an equation for the number q of *E. coli* bacteria in the colony t minutes later. (b) Find q when $t = 60$ minutes.

40 Prove that the solution $q = q_0 e^{kt}$ of the differential equation $dq/dt = kq$ is actually the *complete* solution. To do this, assume that $q = f(t)$ is a solution and that $q_0 = f(0)$ and then calculate $D_t[f(t)e^{-kt}]$.

C **41** At a certain instant, 100 grams of a radioactive substance is present. After 4 years, 20 grams remains. (a) How much of the substance remains after 8 years? (b) What is the half-life of the substance?

42 If $K \times 100$ percent of a radioactive substance remains after t years, show that the half-life of the substance is T years, where $T = (t \ln 2)/(-\ln K)$.

C **43** If the half-life of polonium is 140 days, how long does it take for 2 grams of polonium to decay to 0.1 gram?

C **44** In the chemical processing of a certain mineral, the rate of change of the amount of mineral remaining is proportional to this amount. If 100 kilograms of mineral has been reduced to 70 kilograms after 8 hours, what quantity of the mineral remains after 24 hours?

C **45** In 1921, President Warren G. Harding presented Marie Curie a gift of 1 gram of radium on behalf of the women of the United States. Using the fact that the half-life of radium is 1656 years, determine how much of the original 1-gram gift was left in 1981.

Marie Curie with President Warren G. Harding, 1921

46 Infusion of a glucose solution into the bloodstream is a standard medical technique. The concentration q of glucose in the bloodstream t minutes after the beginning of the infusion is governed by a differential equation of the form $dq/dt = k(A - q)$, where A and k are positive constants that depend on the rate at which the glucose is converted and removed from the bloodstream. If q_0 is the value of q when $t = 0$, find a formula that expresses q in terms of q_0, A, and k. Assuming that $q_0 < A$, find $\lim_{t \to +\infty} q$ and interpret the result.

C **47** A mobile home initially costs \$28,000. If it depreciates at a rate that is proportional to its value and if it has a value of \$20,000 after 2 years, what is its value after 10 years?

C **48** A cylindrical tank with a vertical central axis is 2 meters high and has a cross-sectional area of 9 square meters. The tank is initially full of benzene, but there is a leak in the bottom and the benzene is running out at a rate proportional to its depth. Find the volume of benzene in the tank at the end of 2 days if the tank is half full at the end of 12 hours.

49 In 1960, the U.S. scientist Willard Libby received the Nobel Prize in physical chemistry for his discovery of the technique of **radiocarbon dating.** When a plant or animal dies, it receives no more of the naturally occurring radioactive carbon ^{14}C from the atmosphere. Libby developed methods for determining the fraction F of the original ^{14}C that is left in a fossil and made an experimental determination of the half-life T of ^{14}C. If F is the fraction of the original ^{14}C left in a fossil, show that the original plant or animal died t years ago, where $t = -T(\ln F/\ln 2)$.

50 Complete the proof of Theorem 2 by considering the cases in which $a < 0$ and in which $h \to -\infty$.

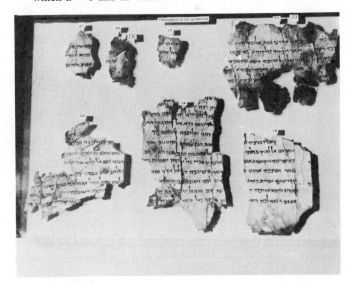

C **51** An ancient scroll is unearthed, and it is determined that it contains only 76 percent of its original ^{14}C. If the half-life of ^{14}C is $T = 5580$ years, use the result of Problem 49 to determine the age of the scroll.

7.10   Mathematical Models for Biological Growth

In this section, we consider some of the mathematical models currently used in the life sciences to describe biological growth. The construction of realistic mathematical models nearly always requires the patient accumulation of experimental data, sometimes over a period of many years. In Figure 1, we plotted points (t, N) showing the population N of the United States in the year $1790 + t$, according to the U.S. Census Bureau. These points seem to lie along a curve that is reminiscent of the graph of exponential growth (Figure 1a in Section 7.9), and they therefore suggest the mathematical model

$$N = N_0 e^{kt}$$

for the growth of the population of the United States.

In 1798, the English economist Thomas Malthus made similar observations about the world population in his *Essay on the Principle of Population*. Because Malthus also proposed a linear model for the expansion of food resources, he forecast that the exponentially growing population would eventually be unable to feed itself. This dire prediction had such an impact on economic thought that the exponential model for population growth came to be known as the **Malthusian model.**

Consider a population growing according to the Malthusian model

$$N = N_0 e^{kt}$$

Of course, N_0 is the population when $t = 0$; but what is the meaning of the growth constant k? To find out, let's consider how the population changes in 1 year. At the beginning of the $(t + 1)$st year, the population is $N_0 e^{kt}$, and at the end of this year it is $N_0 e^{k(t+1)}$. During the year, the population increase is

$$N_0 e^{k(t+1)} - N_0 e^{kt} = N_0(e^{kt+k} - e^{kt}) = N_0(e^{kt}e^k - e^{kt})$$
$$= N_0 e^{kt}(e^k - 1)$$

The percentage of the increase in population during the year is therefore given by

$$\frac{\text{increase during the year}}{\text{population at the beginning of the year}} \times 100\% = \frac{N_0 e^{kt}(e^k - 1)}{N_0 e^{kt}} \times 100\%$$
$$= (e^k - 1) \times 100\%$$

In other words, if the yearly percentage increase in population is expressed as a decimal K, we have

$$K = e^k - 1 \qquad \text{or} \qquad k = \ln(1 + K)$$

© **EXAMPLE 1**   According to the U.S. Census Bureau, the population of the United States in 1980 was $N_0 = 226$ million. Suppose that the population grows according to the Malthusian model at 1.1 percent per year.

(a) Write an equation for the population N million of the United States t years after 1980.

(b) Predict N in the year 2000.

(c) Predict N in the year 2020.

Figure 1

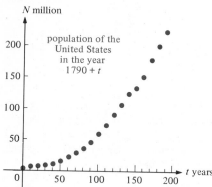

N million

population of the United States in the year $1790 + t$

t years

SOLUTION

(a) Expressed as a decimal, 1.1 percent is $K = 0.011$. Therefore,

$$k = \ln(1 + K) = \ln 1.011 \approx 0.011$$

and we have

$$N = N_0 e^{kt} = 226 e^{0.011t}$$

(b) In the year 2000, $t = 20$ and

$$N = 226 e^{0.011(20)} = 226 e^{0.22} \approx 282 \text{ million}$$

(c) In the year 2020, $t = 40$ and

$$N = 226 e^{0.011(40)} = 226 e^{0.44} \approx 351 \text{ million} \qquad \blacksquare$$

In the example above, the growth constant $k = \ln(1 + K)$, rounded off to three decimal places, is the same as the yearly percentage of population increase expressed as a decimal K. This is no accident. Indeed, suppose that K is small, and let $y = f(x) = \ln x$. By the linear-approximation theorem (Theorem 1, page 249) with $x_1 = 1$ and $\Delta x = K$, we have

$$f(1 + K) \approx f(1) + f'(1)K$$

Here, $f(1) = \ln 1 = 0$, and since $f'(x) = 1/x$, $f'(1) = 1$. Therefore, if K is small,

$$\boxed{k = \ln(1 + K) \approx K}$$

In fact, if $K \le 0.06$ (6 percent per year), then the error in the approximation $k \approx K$ is less than 3 percent.

For any growth model, it is interesting to ask when the growing quantity doubles. If T is the **doubling time** for the Malthusian model $N = N_0 e^{kt}$, then

$$N_0 e^{k(t+T)} = 2N_0 e^{kt} \qquad \text{or} \qquad e^{kt+kT} = 2e^{kt}$$

that is,

$$e^{kt} e^{kT} = 2e^{kt} \qquad \text{or} \qquad e^{kT} = 2$$

Thus,

$$kT = \ln 2$$

and we have the formula

$$\boxed{T = \frac{\ln 2}{k} \approx \frac{\ln 2}{K}}$$

© **EXAMPLE 2**   If the population of the United States grows according to the Malthusian model at 1.1 percent per year, in approximately how many years will it double?

SOLUTION   Here $K = 0.011$, so

$$T \approx \frac{\ln 2}{K} = \frac{\ln 2}{0.011} \approx 63 \text{ years} \qquad \blacksquare$$

Inhibited-Growth Models

When one-celled organisms reproduce by simple cell division in a culture containing an unlimited supply of nutrients, the Malthusian, or exponential, model $N = N_0 e^{kt}$ for the number N of organisms at time t is often quite accurate. In a natural environment, however, growth is often inhibited by various constraints that have a greater and greater effect as time goes on—depletion of the food supply, build-up of toxic wastes, physical crowding, and so on—and the Malthusian model may no longer apply.

If N_0 organisms are introduced into a habitat at time $t = 0$ and the population N of these organisms is plotted as a function of time t, the result is often an **s**-shaped curve (Figure 2). Typically, such a curve shows an increasing rate of growth up to a point of inflection P_I, followed by a declining rate of growth as the population N levels off and approaches the maximum population C that can be supported by the habitat. The horizontal line $N = C$ is an asymptote of the graph, and the constant C is called the **carrying capacity** of the habitat.

There are a number of growth models that produce the characteristic **s**-shaped curve of Figure 2. Many of these models for inhibited growth are obtained by replacing the growth constant k in the differential equation

$$\frac{dN}{dt} = kN$$

for exponential growth by a **variable growth factor** $g(N)$ that decreases as the population N increases. Thus, a **differential equation for inhibited growth** has the form

$$\frac{dN}{dt} = g(N)N$$

where the growth factor $g(N)$ is positive but decreases as N increases.

Perhaps the simplest model for inhibited growth is the **logistic model,** obtained by using a linear growth factor

$$g(N) = k - AN$$

where A and k are positive constants. (Because of the negative sign, $g(N)$ decreases as N increases.) Our calculations will be simpler if we let $C = k/A$, so that $A = k/C$ and

$$g(N) = k - AN = k - \frac{k}{C}N = \frac{C - N}{C}k$$

Hence, the differential equation for logistic growth takes the form

$$\boxed{\frac{dN}{dt} = \frac{C - N}{C}kN}$$

where k and C are positive constants.

To solve this differential equation, we begin by separating variables as follows:

$$\frac{C}{(C - N)N} dN = k\, dt$$

Notice that

$$\frac{C}{(C - N)N} = \frac{1}{C - N} + \frac{1}{N}$$

Figure 2

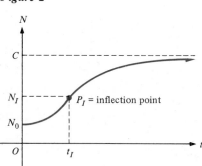

Therefore,

$$\left(\frac{1}{C-N} + \frac{1}{N}\right) dN = k\,dt \qquad \text{or} \qquad \frac{dN}{C-N} + \frac{dN}{N} = k\,dt$$

Integrating both sides of the last equation, we obtain

$$\int \frac{dN}{C-N} + \int \frac{dN}{N} = \int k\,dt \qquad \text{or} \qquad -\ln(C-N) + \ln N = kt + C_1$$

where C_1 is the constant of integration. Using the properties of logarithms, we can rewrite the last equation as

$$\ln \frac{N}{C-N} = kt + C_1 \qquad \text{or} \qquad \frac{N}{C-N} = e^{kt+C_1}$$

Now we take reciprocals of both sides of the last equation to obtain

$$\frac{C-N}{N} = e^{-kt-C_1} = e^{-C_1}e^{-kt} = C_0 e^{-kt}$$

where we have put $C_0 = e^{-C_1}$. Thus,

$$\frac{C}{N} - 1 = C_0 e^{-kt} \qquad \text{or} \qquad \frac{C}{N} = 1 + C_0 e^{-kt}$$

Solving the last equation for N, we obtain the **equation for logistic growth:**

$$N = \frac{C}{1 + C_0 e^{-kt}}$$

If N_0 is the value of N when $t = 0$, we have

$$N_0 = \frac{C}{1 + C_0 e^0} \qquad \text{or} \qquad N_0 = \frac{C}{1 + C_0}$$

Solving the last equation for C_0, we find that

$$C_0 = \frac{C - N_0}{N_0}$$

In what follows, we consider the case in which $0 < N_0 < C$. (The case in which $C \le N_0$ is considered in Problem 8.) Thus, since both $C - N_0$ and N_0 are positive, it follows that C_0 is positive. Therefore, the denominator $1 + C_0 e^{-kt}$ in the equation for logistic growth is greater than 1, and we have $0 < N < C$ for all values of t. Also, since

$$\lim_{t \to +\infty} N = \lim_{t \to +\infty} \frac{C}{1 + C_0 e^{-kt}} = \frac{C}{1 + 0} = C$$

the horizontal line $N = C$ is an asymptote of the graph of the equation. We leave it as an exercise for you to show that the graph has a point of inflection $P_I = (t_I, N_I)$ given by

$$t_I = \frac{1}{k} \ln C_0 \qquad \text{and} \qquad N_I = \frac{C}{2}$$

(Problem 14).

© **EXAMPLE 3**   Suppose that the population N million of the United States grows according to the logistic model

$$N = \frac{C}{1 + C_0e^{-kt}}$$

where t is the time in years since 1780 and where $k = 0.03$. Use the following data: The population in 1780 was 3 million, and the population in 1880 was 50 million.

(a) Find the carrying capacity C.

(b) Find N in the year 1980.

(c) Find N in the year 2000.

(d) Determine when the inflection occurred.

In each case, round off the population to the nearest 10 million.

SOLUTION

(a) From the data given, $N_0 = 3$ million in 1780 when $t = 0$. Thus,

$$C_0 = \frac{C - N_0}{N_0} = \frac{C - 3}{3}$$

and we have

$$N = \frac{C}{1 + [(C-3)/3]e^{-0.03t}} \quad \text{or} \quad N = \frac{3C}{3 + (C-3)e^{-0.03t}}$$

Using the fact that $N = 50$ million in 1880 when $t = 100$, we substitute into the last equation to obtain

$$50 = \frac{3C}{3 + (C-3)e^{-0.03(100)}} \quad \text{or} \quad 50[3 + (C-3)e^{-3}] = 3C$$

Thus,

$$150 + 50Ce^{-3} - 150e^{-3} = 3C \quad \text{or} \quad (3 - 50e^{-3})C = 150(1 - e^{-3})$$

and it follows that the carrying capacity is

$$C = \frac{150(1 - e^{-3})}{3 - 50e^{-3}} \approx 280 \text{ million}$$

(b) Substituting $C = 280$ in the equation obtained in part (a), we have

$$N = \frac{3C}{3 + (C-3)e^{-0.03t}} = \frac{840}{3 + 277e^{-0.03t}}$$

Hence, in 1980, when $t = 200$ years,

$$N = \frac{840}{3 + 277e^{-0.03(200)}} \approx 230 \text{ million}$$

(The correct value according to the 1980 census was 226 million, so our logistic model has predicted the correct value with an error of less than 2 percent.)

(c) In the year 2000, $t = 220$ years and

$$N = \frac{840}{3 + 277e^{-0.03(220)}} \approx 250 \text{ million}$$

[Compare this with the prediction of 282 million according to the Malthusian model (page 469). Notice that the "leveling off" built into the logistic model has apparently taken hold.]

(d) $t_I = \dfrac{1}{k} \ln \dfrac{C - N_0}{N_0} = \dfrac{1}{0.03} \ln \dfrac{280 - 3}{3} = \dfrac{1}{0.03} \ln \dfrac{277}{3} \approx 150$ years

Thus, according to this logistic model, the inflection would have taken place in the year $1780 + 150 = 1930$.   ∎

Problem Set 7.10

© **1** The population of a small country was 10 million in 1980, and it is growing according to the Malthusian model at 3 percent per year. (a) Write an equation for the population N million of the country t years after 1980. (b) Predict N in the year 2000. (c) Find the doubling time T for the population.

2 Suppose that a population that is growing according to the Malthusian model increases by $100K$ percent per year. Write an *exact* formula (not an approximation) for the doubling time T in terms of K.

© **3** The population of a certain city is growing according to the Malthusian model, and it is expected to double in 35 years. Approximately what percent of growth will occur in this population over 1 year?

4 Explain why, for small values of t, logistic growth is approximately the same as Malthusian growth.

© **5** The fruit fly *Drosophila melanogaster* is often used by biologists for genetic experiments because it breeds rapidly and has a short life cycle. Suppose a colony of *D. melanogaster* in a laboratory is observed to double in $T = 2$ days. Assuming a Malthusian model for growth of the colony, determine the approximate daily percentage increase in the size of the colony.

6 Suppose that a population is growing according to the Malthusian model with a doubling time T. If N_0 is the original size of the population when $t = 0$, show that the population t units of time later is given by $N = N_0 2^{t/T}$.

© **7** Suppose that a herd of 300 deer, just introduced into a game preserve, grows according to the logistic model with $k = 0.1$. Assume that the herd has grown to 387 deer after 5 years. (a) Find the carrying capacity C of the habitat for deer. (b) Find the population N of the herd after 7 years. (c) When does the inflection occur in the population of the herd?

8 Consider the solution of the logistic differential equation for the case in which $C \leq N_0$. (a) Show that $C_0 \leq 0$. (b) If $C = N_0$, show that N is constant. (c) If $C < N_0$, show that N decreases as t increases. (d) If $C < N_0$, show that the graph of N is concave upward for $t > 0$ and has the line $N = C$ as a horizontal asymptote.

© **9** On the same (N, t) coordinate system, sketch accurate graphs of (a) the population N of the deer herd in Problem 7 and (b) the population N that would result if the herd grew exponentially with growth constant $k = 0.1$.

10 In the equation for logistic growth, solve for t in terms of N to obtain $t = (1/k) \ln [NC_0/(C - N)]$.

© **11** In 1800, the population of Sweden was 2.35 million. Suppose that the population N has been growing since then according to a logistic model with $k = 0.023$. In 1900 the population of Sweden was 5.14 million. (a) Find the population carrying capacity C of Sweden according to the logistic model. (b) Use the logistic model to estimate the population of Sweden in 1970. (*Note:* In 1970, the population of Sweden was actually 8.08 million.)

12 If a certain population N is thought to be growing according to the logistic model, the value of the constant k is often found experimentally as follows: First, an estimate is made for the carrying capacity C of the habitat. Then values of N are measured corresponding to several different values of t, and

$$y = \ln \frac{N}{C - N}$$

is calculated for each observed value of N. The points (t, y) are plotted on a graph (Figure 3), and a straight line L that "best fits" these points is drawn. The constant k is taken to be the slope of the line L. Justify this procedure. (*Hint:* Use the result of Problem 10 to show that $kt = y - b$, where b is the value of y when $t = 0$.)

Figure 3

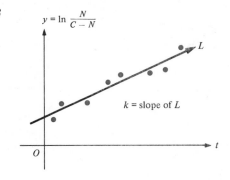

13 Consider the inhibited-growth model given by the differential equation $dN/dt = g(N)N$, where the variable growth factor has the form $g(N) = [(C/N) - 1]k$. Because of its extensive use by psychologists to study the learning process, this is often called the **learning model.** (a) Show that the solution to the differential equation above is $N = C(1 - C_0 e^{-kt})$, where $C_0 = 1 - (N_0/C)$. (b) Show that $N = C$ is a horizontal asymptote of the graph of N as a function of t.

14 Verify the formulas on page 471 for the coordinates (t_I, N_I) of the point of inflection for the logistic model.

©**15** The learning model in Problem 13 often provides a realistic model for population growth. Suppose the deer herd in Problem 7 grows according to this growth model with $C = 700$, $k = 0.05$, and $N_0 = 300$. Sketch the graph of N as a function of t according to this model, and compare it with the graphs sketched in Problem 9.

16 Outline a procedure for determining the constant k for the learning model in Problem 13. The procedure should be similar to that in Problem 12, but with $y = \ln [C/(C - N)]$.

17 Suppose that $0 < N_0 < C$ for the learning model in Problem 13, and show that the graph of N as a function of t has no point of inflection.

18 A **seasonal-growth model** is described by a differential equation of the form $dN/dt = [g(N) \cos (\omega t - \phi)]N$ in which the growth factor $g(N)$ is multiplied by a periodic function of t to account for seasonal variations in the rate of growth (for instance, variations caused by seasonal changes in the availability of food). If $g(N) = k$, for all values of N, where k is a positive constant, solve this differential equation with the initial condition that $N = N_0$ when $t = 0$.

©**19** Suppose that a particular person learning to type is able to type N words per minute after t hours of practice. Assume that at the beginning, when $t = 0$, this person can type 10 words per minute by the "two-finger" method. If N increases according to the learning model in Problem 13 with $C = 100$ and $k = 0.025$, (a) how many words per minute can this person type after $t = 30$ hours of practice and (b) what is the rate of improvement dN/dt in words per minute per hour of practice when $t = 30$ hours?

©**20** An advertising campaign is aimed at a projected 100,000 potential purchasers of a new high-performance luxury sports car. The rate at which more of these potential purchasers become aware of the new sports car because of the advertising is proportional to the number of people, out of the original 100,000, who have not yet become aware of the car. If no one has heard about the car at the start of the campaign and after 10 days 30 percent of the potential purchasers have heard about the car, how many of the potential purchasers have heard about the car after 30 days?

21 The **Gompertz model** is the inhibited-growth model given by the differential equation $dN/dt = g(N)N$, where the variable growth factor has the form $g(N) = (\ln C - \ln N)k$. (a) If C_0 is any constant, show by differentiation that $N = C \exp (-C_0 e^{-kt})$ is a solution of the differential equation. (b) For the solution in part (a), show that $C_0 = \ln (C/N_0)$, where $N = N_0$ when $t = 0$. (c) For the solution in part (a), show that $N = C$ is a horizontal asymptote of the graph.

22 Does the graph of the Gompertz growth equation in part (a) of Problem 21 have a point of inflection? If so, what are its coordinates?

©**23** Suppose that the deer herd in Problem 7 grows according to the Gompertz model in Problem 21 with $N_0 = 300$, $C = 700$, and $k = 0.07$. Sketch the resulting graph of N as a function of t.

7.11   First-Order Linear Differential Equations

If P and Q are continuous functions, then a first-order differential equation of the form

$$\frac{dy}{dx} + P(x)y = Q(x)$$

is said to be **linear.** The growth models considered in Section 7.10 were all obtained by solving first-order linear differential equations.

Another important example of a linear differential equation is obtained from **Newton's law of cooling,** which states that the rate of change of the temperature of a cooling object is proportional to the difference between the temperature of the object and the temperature of the surrounding medium. Thus, if y denotes the temperature of the cooling object at time t and a is the temperature of the surrounding medium, we have

$$\frac{dy}{dt} = -k(y - a)$$

where k is the positive constant of proportionality and the negative sign in front of k is to account for the fact that y decreases as t increases.

© **EXAMPLE 1**   An object with an initial temperature of 150°C is allowed to cool in a room whose temperature is 25°C. If after 10 minutes the temperature of the object has dropped to 120°C, what is its temperature after 30 minutes?

SOLUTION   Let y be the temperature of the object after t minutes, so that by Newton's law of cooling,

$$\frac{dy}{dt} = -k(y - 25)$$

Separating variables and integrating, we have

$$\frac{dy}{y - 25} = -k \, dt \quad \text{so that} \quad \int \frac{dy}{y - 25} = \int (-k) \, dt$$

Therefore,

$$\ln (y - 25) = -kt + C_1 \quad \text{or} \quad y - 25 = e^{-kt + C_1}$$

where C_1 is the constant of integration. It follows that

$$y = 25 + e^{C_1 - kt} = 25 + e^{C_1}e^{-kt} \quad \text{or} \quad y = 25 + C_0 e^{-kt}$$

where we have put $C_0 = e^{C_1}$. Substituting the initial conditions $y = 150$ when $t = 0$ into the equation $y = 25 + C_0 e^{-kt}$, we find that

$$150 = 25 + C_0 e^0 = 25 + C_0 \quad \text{so that} \quad C_0 = 150 - 25 = 125$$

Therefore,

$$y = 25 + 125e^{-kt}$$

Now, when $t = 10$, we have $y = 120$, so that

$$120 = 25 + 125e^{-k(10)} \quad \text{or} \quad e^{-10k} = \frac{120 - 25}{125} = \frac{19}{25}$$

It follows that

$$-10k = \ln \tfrac{19}{25} \quad \text{or} \quad k = -\tfrac{1}{10} \ln \tfrac{19}{25} \approx 0.027$$

Thus, when $t = 30$, we have

$$y = 25 + 125e^{-k(30)} = 25 + 125e^{-(0.027)(30)} \approx 81°C \quad \blacksquare$$

Problems that involve the uniform mixing of various substances often give rise to linear differential equations.

© **EXAMPLE 2**   A tank initially holds 100 gallons of salt water at a concentration of $\frac{3}{10}$ pound of salt per gallon of water. Water containing $\frac{1}{10}$ pound of salt per gallon runs into the tank at the rate of 2 gallons per minute, and the solution, kept uniform by stirring, runs out at the same rate.

(a) Find an equation for the number of pounds q of salt in the tank at the end of t minutes, and sketch the graph of this equation.

(b) Find the concentration of salt in the tank at the end of 25 minutes.

SOLUTION   The concentration of salt in the tank at time t is $q/100$ pounds per gallon. In the infinitesimal time interval dt, 2 dt gallons of salt water carrying $\frac{1}{10}(2\ dt)$ pounds of salt runs into the tank, while 2 dt gallons of salt water carrying $(q/100)(2\ dt)$ pounds of salt flows out of the tank. Thus, during the time interval dt, the amount of salt in the tank changes by the infinitesimal amount

$$dq = \frac{1}{10}(2\ dt) - \frac{q}{100}(2\ dt) = (q - 10)\left(-\frac{dt}{50}\right) \qquad \text{lb}$$

Therefore, q satisfies the differential equation

$$\frac{dq}{q - 10} = -\frac{dt}{50}$$

with the initial condition that

$$q = (\tfrac{3}{10})(100) = 30 \text{ lb} \qquad \text{when } t = 0$$

(a) Integrating both sides of the differential equation above, we have

$$\ln(q - 10) = -\frac{t}{50} + C \qquad \text{or} \qquad q - 10 = e^{C-(t/50)} = e^{C}e^{-t/50}$$

Therefore,

$$q = 10 + C_0 e^{-t/50}$$

where we have put $C_0 = e^C$. Since $q = 30$ when $t = 0$, it follows that

$$30 = 10 + C_0 e^0 = 10 + C_0$$

Hence, $C_0 = 20$ and

$$q = 10 + 20e^{-t/50} \text{ (Figure 1)}$$

(b) At the end of 25 minutes, there are

$$q = 10 + 20e^{-25/50} = 10 + 20e^{-1/2} \qquad \text{pounds of salt}$$

in the tank, so the salt concentration is

$$\frac{10 + 20e^{-1/2}}{100} \approx 0.22 \text{ pound per gallon}$$

■

Figure 1

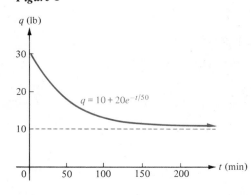

q (lb)

$q = 10 + 20e^{-t/50}$

t (min)

Integrating Factors

Until now, we have been able to solve only first-order linear differential equations that are *separable*; however, in general, a differential equation of the form

$$\frac{dy}{dx} + P(x)y = Q(x)$$

will not be separable. If such a differential equation is not separable, it can still be solved by using what is called an *integrating factor*.

To develop the idea of an integrating factor, we begin by considering a special case of the equation above in which $Q(x)$ is (temporarily) replaced by zero, so that

$$\frac{dy}{dx} + P(x)y = 0$$

Now, this differential equation is separable, and we can rewrite it as

$$\frac{dy}{y} = -P(x)\,dx$$

Integrating both sides of the last equation, we obtain

$$\ln|y| + C_1 = -\int P(x)\,dx \qquad \text{or} \qquad \ln|y| = -C_1 - \int P(x)\,dx$$

where C_1 is the constant of integration. Therefore,

$$|y| = e^{-C_1 - \int P(x)\,dx} = e^{-C_1}e^{-\int P(x)\,dx}$$

Hence, $\qquad\qquad y = \pm e^{-C_1}e^{-\int P(x)\,dx} \qquad \text{or} \qquad y = Ce^{-\int P(x)\,dx}$

where we have put $C = \pm e^{-C_1}$. It follows that

$$ye^{\int P(x)\,dx} = C$$

If we let

$$\phi(x) = e^{\int P(x)\,dx}$$

we can write our solution as

$$y\phi(x) = C$$

The expression $y\phi(x)$ on the left of the last equation deserves further study. First, notice that

$$\frac{d}{dx}\phi(x) = \frac{d}{dx}e^{\int P(x)\,dx} = e^{\int P(x)\,dx}\frac{d}{dx}\int P(x)\,dx = \phi(x)P(x)$$

Therefore,

$$\frac{d}{dx}[y\phi(x)] = \frac{dy}{dx}\phi(x) + y\frac{d}{dx}\phi(x) = \frac{dy}{dx}\phi(x) + y\phi(x)P(x)$$

$$= \phi(x)\left[\frac{dy}{dx} + P(x)y\right]$$

In other words, *if you multiply*

$$\frac{dy}{dx} + P(x)y$$

by the factor

$$\phi(x) = e^{\int P(x)\,dx}$$

the resulting expression is exactly the derivative of

$$y\phi(x)$$

It follows that the linear differential equation

$$\frac{dy}{dx} + P(x)y = Q(x)$$

can be solved by multiplying both sides by the **integrating factor** $\phi(x)$. The technique is illustrated in the following example.

EXAMPLE 3   Solve the linear differential equation $\dfrac{dy}{dx} - \dfrac{2}{x}y = x^2e^x$.

SOLUTION   The given differential equation has the linear form

$$\frac{dy}{dx} + P(x)y = Q(x)$$

with

$$P(x) = -\frac{2}{x} \quad \text{and} \quad Q(x) = x^2e^x$$

Here, the integrating factor is given by

$$\phi(x) = e^{\int P(x)\,dx} = e^{\int (-2/x)\,dx} = e^{-2\ln|x|} = |x|^{-2} = \frac{1}{x^2}$$

Multiplying both sides of the given differential equation by the integrating factor, we obtain

$$\frac{1}{x^2}\left(\frac{dy}{dx} - \frac{2}{x}y\right) = \frac{1}{x^2}(x^2e^x)$$

or

$$\frac{1}{x^2}\frac{dy}{dx} - \frac{2}{x^3}y = e^x$$

Now we know that the left side of the last equation must be the derivative of

$$y\phi(x) = y\frac{1}{x^2} = \frac{y}{x^2}$$

Hence, we can rewrite the equation as

$$\frac{d}{dx}\left(\frac{y}{x^2}\right) = e^x$$

Therefore,

$$\frac{y}{x^2} = \int e^x\,dx = e^x + C$$

and it follows that

$$y = x^2e^x + Cx^2$$

Problem Set 7.11

In Problems 1 to 4, solve each linear differential equation by separating variables.

1 $\dfrac{dy}{dx} + xy = 0$

2 $\dfrac{ds}{dt} + 3s = 3$

3 $\dfrac{dN}{dt} - 0.1N = -100$

4 $dy - 2y\,dx = dx$

© **5** An iron ball at a temperature of 90°C is placed in a water bath at a constant temperature of 2°C. After 10 minutes, the ball has cooled to 25°C. How many *more* minutes is required for the ball to cool to 10°C?

© **6** A body, initially at the temperature 160°F, is allowed to cool in air. The body cools to 100°F after 50 minutes. After 100 minutes, it has cooled to 80°F. What is the air temperature?

© **7** A baked potato at a temperature of 200°F is placed on Melissa's dinner plate. She does not start eating the potato until 7 minutes later. After 4 minutes, the temperature of the potato has dropped to 175°F. How hot is the potato when Melissa starts to eat it if the air temperature is 70°F?

© **8** If a roast at a room temperature of 70°F is placed into a refrigerator with a constant temperature of 35°F, and if the temperature of the roast after 2 hours is 45°F, what is its temperature after 4 hours?

© **9** A tank initially contains 50 gallons of water in which 10 pounds of salt is dissolved. Pure water runs into the tank at the rate of 3 gallons per minute and is uniformly stirred into the solution. Meanwhile, the mixture runs out of the tank at the constant rate of 2 gallons per minute. After how long is only 2 pounds of dissolved salt left in the tank?

© **10** Solve Problem 9 if the water running into the tank is not pure but contains $\frac{1}{30}$ pound of salt per gallon and this water runs into the tank at the rate of 2 gallons per minute.

© **11** Air in a chemical laboratory contains 1 percent hydrogen sulfide. An exhaust fan is turned on which removes air from the room at the rate of 500 cubic feet per minute. Meanwhile, fresh air is drawn into the room through cracks under doors and so forth. If the volume of the room is 10,000 cubic feet, what is the concentration of hydrogen sulfide after 5 minutes?

12 Water containing A kilograms of pollutant per cubic meter flows into a reservoir of constant volume V cubic meters at a rate of R cubic meters per second. Polluted reservoir water is drawn off at the same rate R, so as to maintain constant reservoir volume V. Let y denote the concentration of pollutant in kilograms per cubic meter at time t. (a) Show that y satisfies the differential equation $dy/dt = (R/V)(A - y)$. (b) Find the general solution of this differential equation. (c) Find the particular solution satisfying the condition that $y = y_0$ when $t = 0$.

© **13** A flu-like virus is spreading through a city of population 260,000 at a rate proportional to the product of the number of people already infected and the number of people still uninfected. If 600 people were infected initially and 30,000 people were infected after 10 days, how many people will be infected after 30 days?

14 In Problem 12, suppose that polluted reservoir water is drawn off at a rate of r cubic meters per second, where $r \neq R$, so that the volume V of water in the reservoir does not remain constant. Again, let y denote the reservoir concentration of pollutant in kilograms per cubic meter at time t. (a) If V_0 is the volume of water in the reservoir when $t = 0$, show that $V = V_0 + (R - r)t$. (b) Show that y continues to satisfy the differential equation

$$\frac{dy}{dt} = \frac{R}{V}(A - y)$$

(c) Find the general solution of the differential equation in (b).

(d) Assuming that $y = y_0$ when $t = 0$, show that

$$y = A + (y_0 - A)\left(\frac{V_0}{V}\right)^{R/(R-r)}$$

In Problems 15 to 20, use an integrating factor to solve each linear differential equation.

15 $\dfrac{dy}{dx} - 4y = e^{4x}$ **16** $\dfrac{dy}{dx} - \dfrac{3}{x}y = x^4$

17 $\dfrac{dy}{dt} + 2ty = 2te^{-t^2}$

18 $dy = xy\, dx + e^{x^2/2} \cos x\, dx$

19 $\dfrac{dq}{dt} + q \cos t = e^{-\sin t}$ **20** $D_x y - \dfrac{2y}{x} = x$

21 A circuit consisting of an inductor of L henrys in series with a resistor of R ohms is called an **RL-circuit** (Figure 2). If I is the current in amperes flowing in an RL-circuit because of a driving electromotive force of E volts, then I as a function of time t satisfies the first-order linear differential equation

$$\frac{dI}{dt} + \frac{R}{L}I = \frac{E}{L}$$

(a) If E is constant and if $I = 0$ when $t = 0$, solve this differential equation for I and find $\lim\limits_{t \to +\infty} I$. (b) Assuming that E is a variable, solve the differential equation.

Figure 2

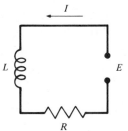

22 Show that the general solution to the linear differential equation $dy/dx + P(x)y = Q(x)$ is given by

$$y = e^{-\int P(x)\, dx} \int Q(x)e^{\int P(x)\, dx}\, dx$$

© **23** Solar Industries has a current inventory of 5000 wind-powered generators, 10 percent of which are defective. Generators are chosen at random from the warehouse and shipped out to customers at the rate of 200 per week. Newly manufactured generators are delivered to the warehouse at the rate of 175 per week. Because of improved quality control, only 5 percent of the newly manufactured generators are defective. What percent of the generators in the warehouse will be defective 1 year (52 weeks) from now? (*Hint:* Let N be the number of defective generators in the warehouse t weeks from now. Treat N as if it were a continuous variable, write a linear differential equation for N, and solve this equation.)

24 The labor force in a certain industry currently consists of F_0 individuals of whom y_0 are skilled workers and $F_0 - y_0$ are unskilled. Workers are quitting randomly at a constant rate of c workers per unit time. Replacements are being hired at constant rates of A skilled workers and B unskilled workers per unit time. Denote by y the number of skilled workers employed by the industry at time t, and treat y as though it were a continuous

variable. (a) Show that the labor force F at time t consists of $F_0 + kt$ individuals, where $k = A + B - c$. (b) Show that y satisfies the first-order linear differential equation

$$\frac{dy}{dt} + \frac{c}{F}y = A$$

(c) Assuming that $k \neq 0$, show that the particular solution of the differential equation in part (b) that satisfies the initial conditions $F = F_0$ and $y = y_0$ when $t = 0$ is

$$y = \frac{A}{c + k}(F_0 + kt) + \left(y_0 - \frac{AF_0}{c + k}\right)\left(\frac{F_0}{F_0 + kt}\right)^{c/k}$$

Review Problem Set, Chapter 7

In Problems 1 to 6, show that the functions f and g are inverses of each other.

1 $f(x) = x^4$ for $x \geq 0$ and $g(x) = \sqrt[4]{x}$

2 $f(x) = 3 + x^3$ and $g(x) = \sqrt[3]{x - 3}$

3 $f(x) = \dfrac{1}{1 - x}$ and $g(x) = \dfrac{x - 1}{x}$

4 $f(x) = x^2 - 3x + 2$ for $x \leq \frac{3}{2}$ and $g(x) = \dfrac{3 - \sqrt{1 + 4x}}{2}$ for $x \geq -\frac{1}{4}$

5 $f(x) = \dfrac{1}{1 + \sin x}$ for $-\dfrac{\pi}{2} < x \leq \dfrac{\pi}{2}$ and

$g(x) = \sin^{-1}\left(\dfrac{1}{x} - 1\right)$ for $x \geq \frac{1}{2}$

6 $f(x) = e^x - e^{-x}$ and $g(x) = \ln(x + \sqrt{x^2 + 4}) - \ln 2$

In Problems 7 to 10, sketch the graph of each function, and then use the horizontal-line test to determine whether the function is invertible.

7 $f(x) = 3x + 5$

8 $g(x) = \dfrac{1}{x}$

9 $h(x) = 3x^2 + 5$

10 $F(x) = \sin x$ for $-\dfrac{\pi}{2} \leq x \leq \dfrac{3\pi}{2}$

11 Sketch the graph of the inverses of f, g, and h in Figure 1.

12 Are there values of the constants a, b, c, and d for which the function $f(x) = \dfrac{ax + b}{cx + d}$ is its own inverse?

Figure 1

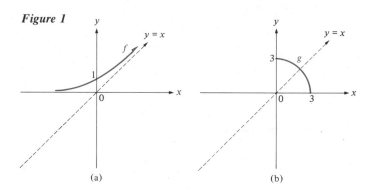

(a)   (b)

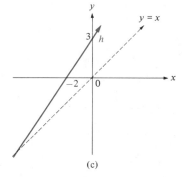

(c)

In Problems 13 to 16, (a) use the algebraic method to find f^{-1}, (b) check that $f^{-1}[f(x)] = x$ for all values of x in the domain of f, (c) check that $f[f^{-1}(x)] = x$ for all values of x in the domain of f^{-1}, and (d) sketch the graphs of f and f^{-1} on the same coordinate system.

13 $f(x) = 7x - 9$

14 $f(x) = 1 - \ln x$

15 $f(x) = \dfrac{4}{x + 1}$

16 $f(x) = e^x + e^{-x}$ for $x \geq 0$

17 If $f(x) = x^5 + 3x^3 + 7$, show that f is invertible.

18 Suppose that A, B, and C are constants with $A > 0$. Let f be the function defined by $f(x) = Ax^2 + Bx + C$ for $x \geq -B/(2A)$. Show that the inverse function f^{-1} is given by

$$f^{-1}(x) = \frac{-B + \sqrt{B^2 - 4AC + 4Ax}}{2A} \quad \text{for } x \geq \frac{4AC - B^2}{4A}$$

19 Suppose that f is an invertible function. If f is increasing, show that f^{-1} is increasing.

20 Beginning students often confuse $f^{-1}(x)$ and $[f(x)]^{-1}$. Is there any continuous function f defined on $\mathbb{R}$ such that f is invertible and $f^{-1}(x) = [f(x)]^{-1}$ for all values of x in $\mathbb{R}$?

In Problems 21 to 26, find the value of x when $y = a$, and use the inverse-function rule to find the numerical value of dx/dy when $y = a$.

21 $a = 1$, $y = \frac{3}{7}x^5$

22 $a = -8$, $y = -\frac{4}{3}x^3$

23 $a = -1$, $y = \dfrac{5x}{x + 2}$

24 $a = \dfrac{\sqrt{3}}{2}$, $y = \cos x$ for $0 < x < \pi$

25 $a = 1$, $y = x + \ln x$

26 $a = 4$, $y = x^5 + 2x^3 + 1$

In Problems 27 to 29, use the given information and the inverse-function theorem to find $(f^{-1})'(a)$. (You may assume that the hypotheses of the theorem are satisfied.)

27 $a = 3$, $f(-5) = 3$, $f'(-5) = 7$

28 $a = 0.1$, $f(0.1) = 1$, $f(1) = 0.1$, $f'(1) = 3$, $f'(0.1) = 2$

29 $a = \pi$, $f(\pi) = e$, $f(e) = \sqrt{2}$, $f(\sqrt{2}) = \pi$, $f'(e) = e$, $f'(\sqrt{2}) = \pi$

30 Let $f(x) = -2x^2 + 8x - 5$ for $x > 2$. (a) Use the algebraic method to find f^{-1}. (b) Calculate $(f^{-1})'(1)$ by using part (a). (c) Calculate $(f^{-1})'(1)$ by using the inverse-function theorem.

🄲 **31** Use a calculator (or Appendix D, Table 1) to evaluate each expression. Give answers in radians.

(a) $\sin^{-1} 0.3750$ (b) $\arccos (-0.3901)$

(c) $\cos^{-1} 0.9273$ (d) $\tan^{-1} 57.29$

(e) $\arctan 1.425$ (f) $\cos^{-1} \frac{1}{8}$

(g) $\arcsin (-\frac{3}{4})$ (h) $\tan^{-1} 8$

(i) $\cos^{-1} (-\frac{10}{11})$ (j) $\cot^{-1} 2.731$

(k) $\sec^{-1} \sqrt{5}$ (l) $\csc^{-1} (-4)$

32 Evaluate each expression without using a calculator or tables.

(a) $\sin^{-1} (-\frac{1}{2})$ (b) $\arccos (-\sqrt{3}/2)$

(c) $\arctan \sqrt{3}$ (d) $\arcsec \sqrt{2}$

In Problems 33 to 38, find the *exact* value of each expression.

33 $\cos (\tan^{-1} \frac{4}{3})$

34 $\sin [\arctan (-\frac{5}{12})]$

35 $\sin [\sin^{-1} (-\frac{12}{13})]$

36 $\tan [\arccos (-\frac{3}{5})]$

37 $\arcsin [\sin (19\pi/14)]$

38 $\sin (\sin^{-1} \frac{2}{3} + \sin^{-1} \frac{3}{4})$

In Problems 39 and 40, show that the given equation is an identity.

39 $\tan (\arccos x) = \dfrac{\sqrt{1 - x^2}}{x}$

40 $\sin \left(\dfrac{1}{2} \cos^{-1} x \right) = \sqrt{\dfrac{1 - x}{2}}$

41 Solve the trigonometric equation $6 \sin^2 t + \sin t - 2 = 0$ for t if $-\pi/2 \leq t \leq \pi/2$.

42 A ray of light passing through a plate of material with parallel faces is displaced, but not deviated (Figure 2); that is, the emerging ray is parallel to the ingoing ray. If d is the amount of displacement, t the thickness of the plate, α the angle of incidence, and β the angle of refraction, show that

$$\beta = \arctan \left(\tan \alpha - \frac{d}{t} \sec \alpha \right)$$

Figure 2

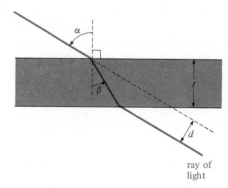

ray of light

In Problems 43 to 56, differentiate each function.

43 $f(x) = 2 \sin^{-1} \dfrac{x}{3}$

44 $g(x) = 4 \tan^{-1} x^2$

45 $h(x) = \sec^{-1} \sqrt{x - 1}$

46 $p(x) = \arccsc \dfrac{1 + x}{1 - x}$

47 $F(t) = \cos^{-1} \sqrt{3t}$

48 $G(y) = \sqrt[5]{\arccos (y^3 + 1)}$

49 $H(u) = u^2 \tan^{-1} u^4$

50 $f(t) = (\cot^{-1} t^2)^4$

51 $g(u) = (\csc^{-1} u)^2$

52 $h(x) = \dfrac{\sec^{-1} (x^2 + 1)}{x}$

53 $q(x) = x^3 \cot^{-1} 5x$

54 $\phi(x) = \dfrac{\sin^{-1} x^2}{\sqrt{x^2 - 1}}$

55 $g(u) = \int_1^{\sin^{-1} u} (17 + x^2)^{34} \, dx$

56 $h(x) = \int_1^{\tan^{-1} x} \left(\dfrac{1 - t^2}{1 + t^2} \right)^{14} dt$

In Problems 57 to 60, find dy/dx by implicit differentiation.

57 $x \cos^{-1} (x + y) - \pi = y^2$

58 $y \tan x^2 - xy^4 = 15$

59 $3x \sin (x - y) + 107 = x^2 y$

60 $y \tan^{-1} x - x \tan^{-1} y = \pi/2$

In Problems 61 to 74, evaluate each integral.

61 $\displaystyle\int \dfrac{dx}{\sqrt{9 - x^2}}$

62 $\displaystyle\int (2 - u^2)^{-1/2} \, du$

63 $\displaystyle\int \dfrac{dx}{x^2 + 36}$

64 $\displaystyle\int \dfrac{2x \, dx}{1 + x^4}$

65 $\displaystyle\int \dfrac{x \, dx}{\sqrt{9 - 2x^4}}$

66 $\displaystyle\int \dfrac{\sec u \tan u \, du}{\sqrt{16 - \sec^2 u}}$

67 $\displaystyle\int \dfrac{dx}{x\sqrt{x^2 - 49}}$

68 $\displaystyle\int \dfrac{dt}{t\sqrt{t^4 - 1}}$

69 $\displaystyle\int \dfrac{\sin x \, dx}{4 + \cos^2 x}$

70 $\displaystyle\int \dfrac{x \sin^{-1} x^2 \, dx}{\sqrt{1 - x^4}}$

71 $\displaystyle\int \dfrac{\cot^{-1} v \, dv}{1 + v^2}$

72 $\displaystyle\int \dfrac{dx}{\sin^2 x \sqrt{1 - \cot^2 x}}$

73 $\displaystyle\int_0^5 \dfrac{dx}{25 + x^2}$

74 $\displaystyle\int_0^{\sqrt{3}} \dfrac{x \, dx}{9 + x^4}$

75 If $y = \sin^{-1} (a \sin t) + \sin^{-1} [a \sin (2 - t)]$, where a is a constant, find the critical value(s) $\pm$.

76 A thin rod (whose weight can be ignored) is 40 centimeters long and passes through the center of a small, heavy ball that is fixed on the rod at a distance of 10 centimeters from one end. The rod is placed in a smooth, hollow hemisphere of radius 40 centimeters, and the slightly rounded ends of the rod slide frictionlessly on the walls of the hemisphere. Find the angle that the rod makes with the horizontal when it finally comes to rest at equilibrium. (At equilibrium the center of mass of the rod and ball is as low as possible.)

77 An isosceles triangle whose two equal sides are each 6 centimeters long has an included angle θ which is increasing at the constant rate of $1°$ per minute. How fast is the area of the triangle changing?

ⓒ **78** Angle BAC of a right triangle is computed from measurements of the opposite side $y = |\overline{BC}|$ and the adjacent side $x = |\overline{AC}|$. (Angle ACB is the right angle.) Use differentials to approxi-

mate the possible error in the calculated value of angle BAC if the measurements of x and y are subject to at most 1 percent error.

In Problems 79 to 126, differentiate each function.

79 $f(x) = \ln (x^2 + 7)$

80 $f(t) = \ln (\cos 3t)$

81 $g(r) = \ln (r\sqrt{r + 2})$

82 $G(u) = \ln (u - 1)^3$

83 $g(x) = x^2 \tan^{-1} (\ln x)$

84 $h(x) = \sin (\ln x^2)$

85 $F(u) = \dfrac{(\ln u)^2}{u}$

86 $G(x) = \sqrt[7]{\ln x}$

87 $f(x) = \ln \dfrac{\sin x}{x}$

88 $H(x) = \ln (x + \sqrt{x^2 + 1})$

89 $g(x) = e^{-4x^3}$

90 $H(x) = -xe^{-x}$

91 $F(x) = \sin^{-1} (e^{-2x})$

92 $g(u) = e^u \cot e^u$

93 $f(t) = \ln \dfrac{e^t + 2}{e^t - 2}$

94 $H(x) = \ln (e^{\sin x} + 5)$

95 $g(x) = \cos^{-1} x^2 - xe^{x^3}$

96 $g(x) = e^x \ln (\sin x)$

97 $f(x) = \tan e^x$

98 $f(x) = e^x(x^2 - 2x + 5)$

99 $f(x) = (3 - e^{4x})^4$

100 $g(x) = \dfrac{e^x - e^{-x}}{e^x + e^{-x}}$

101 $f(x) = \ln \dfrac{e^x + 2}{e^{-x} + 2}$

102 $h(z) = e^{3z} \cos^{-1} e^{2z}$

103 $F(x) = x^{4e}$

104 $g(x) = x^{-17\pi}$

105 $f(x) = 5^{\cos x}$

106 $g(t) = 3^{t^2 + 2}$

107 $f(x) = 7^{\sin x^2}$

108 $g(x) = (x^2 + 7)2^{-5x}$

109 $g(x) = 3^{5x} \cdot 2^{4x^2}$

110 $H(x) = e^{\cos x} \cdot 2^{4x}$

111 $f(t) = \dfrac{\log_7 t}{t}$

112 $g(u) = \log_3 \dfrac{u}{u + 7}$

113 $g(x) = \sqrt[4]{\log_{10} x}$

114 $f(x) = \sqrt[5]{\log_{10} \dfrac{1 + x}{1 - x}}$

115 $g(x) = \cosh e^{4x}$

116 $H(s) = \sinh (\sin^{-1} s)$

117 $g(t) = \operatorname{csch} (e^{-t})$

118 $F(x) = \dfrac{\tanh (\sin x)}{x}$

119 $f(x) = e^{\operatorname{sech} x^2}$

120 $g(u) = \sinh u^2 \tanh 3u$

121 $g(t) = \ln (\tanh t + \operatorname{sech} t)$

122 $f(x) = \tan^{-1} (\sinh x^3)$

123 $g(x) = \sinh^{-1} (3x + 1)$

124 $g(u) = \coth^{-1} e^{5u}$

125 $f(x) = \tanh^{-1} e^x$

126 $F(x) = \coth^{-1} e^{-x^2}$

127 If $(\ln x)/x = (\ln 2)/2$, does it necessarily follow that $x = 2$? Justify your answer. (*Hint:* Sketch a graph of $y = (\ln x)/x$.)

128 If $(\ln x)/x = (\ln \frac{1}{2})/\frac{1}{2}$, does it necessarily follow that $x = \frac{1}{2}$? Justify your answer.

In Problems 129 to 138, use logarithmic differentiation to find dy/dx.

129 $y = (3x)^x$

130 $y = x^{x^3}$

131 $y = (\sin x)^{x^2}$

132 $y = (\cosh x)^{2x}$

133 $y = (\tanh^{-1} x)^{x^3}$

134 $y = x^{\cos^{-1} x}$

135 $y = \dfrac{\cos x \sqrt[3]{1 + \sin^2 x}}{\sin^5 x}$

136 $y = \dfrac{x^2 \sin x}{\sqrt{1 - 3 \tan x \sec 2x}}$

137 $y = \dfrac{x^2(x + 5)^3 \sin 2x}{\sec 3x}$

138 $y = \dfrac{x \cot x}{(x + 1)(x + 3)^2(x + 7)^4}$

139 Let a be a positive constant, and define the function f by $f(x) = a^x$. Use the fact that $f'(0) = a^0 \ln a$ to prove that $\lim\limits_{h \to 0} [(a^h - 1)/h] = \ln a$.

140 Let $0 < a < b$, where a and b are constants, and define the function p by $p(t) = \int_a^b x^t \, dx$, with the understanding that t is to be held fixed while you are calculating $\int_a^b x^t \, dx$. Is the function p continuous at the number -1? (*Hint:* Use the result in Problem 139.)

In Problems 141 to 144, use implicit differentiation to find dy/dx.

141 $xe^{4y} + x \cos y = 2$

142 $y2^x + xe^y = 3$

143 $\cosh (x - y) + \sinh (x + y) = 1$

144 $\log_{10} (x + y) + \log_{10} (x - y) = 2$

In Problems 145 and 146, use the fundamental theorem of calculus and the chain rule to calculate $D_x y$.

145 $y = \displaystyle\int_1^{\ln x} \dfrac{dt}{5 + t^3}$

146 $y = \displaystyle\int_1^{\cosh x} \dfrac{dt}{3 + t^2}$

In Problems 147 to 166, evaluate each integral.

147 $\displaystyle\int \dfrac{dx}{8 + 3x}$

148 $\displaystyle\int \dfrac{\sin (\ln x)}{x} \, dx$

149 $\displaystyle\int \dfrac{\ln x}{x} \, dx$

150 $\displaystyle\int xe^{x^2 - 4} \, dx$

151 $\displaystyle\int e^{\sqrt{x}} \dfrac{dx}{\sqrt{x}}$

152 $\displaystyle\int \dfrac{e^x \, dx}{\sqrt{1 - e^{2x}}}$

153 $\displaystyle\int \dfrac{e^x \, dx}{\cos^2 e^x}$

154 $\displaystyle\int \dfrac{\pi^{1/x}}{x^2} \, dx$

155 $\displaystyle\int 2^x \cdot 5^x \, dx$

156 $\displaystyle\int 3^{2x} \cos 3^{2x} \, dx$

157 $\displaystyle\int_1^4 \dfrac{1/x^2}{1 + (1/x)} \, dx$

158 $\displaystyle\int_0^{1/5} e^{\cosh 5x} \sinh 5x \, dx$

159 $\displaystyle\int \operatorname{csch}^2 x \coth x \, dx$

160 $\displaystyle\int \dfrac{(\sinh^{-1} x)^5}{\sqrt{1 + x^2}} \, dx$

161 $\displaystyle\int \dfrac{e^{\cosh^{-1} x}}{\sqrt{x^2 - 1}} \, dx$

162 $\displaystyle\int \dfrac{dx}{\sqrt{16 + x^2}}$

163 $\displaystyle\int \dfrac{dx}{\sqrt{x^2 - 1}}$

164 $\displaystyle\int \dfrac{dx}{\sqrt{16 - 9x^2}}$

165 $\displaystyle\int \dfrac{dx}{x\sqrt{16 - 4x^2}}$

166 $\displaystyle\int \dfrac{dx}{x\sqrt{16 + 4x^2}}$

167 Using the definition of the natural logarithm, prove that $t > \ln t$ holds for all positive values of t.

168 Using the result of Problem 167, show that $e^t > t$, hence that $e^{nt} > t^n$ holds for all positive integers n if $t > 0$. Conclude that $e^x/x^n > (1/n)^n$ holds for all positive integers n if $x > 0$. (*Hint:* Put $t = x/n$.)

169 Using the result of Problem 168, prove that:

(a) If $x > 0$, then $e^x/x > x/4$.   (b) $\lim\limits_{x \to +\infty} e^x/x = +\infty$.

(c) If $x > 0$, then $e^x/x^2 > x/27$.   (d) $\lim\limits_{x \to +\infty} e^x/x^2 = +\infty$.

170 Prove that if n is a positive integer, then (a) $\lim\limits_{x \to +\infty} e^x/x^n = +\infty$ and therefore (b) $\lim\limits_{x \to +\infty} x^n e^{-x} = 0$.

171 Apply the mean-value theorem to $\ln (1 + x) - \ln 1$, and thus conclude that $x/(1 + x) < \ln (1 + x) < x$ holds for $x > 0$.

172 Show that $\lim\limits_{x \to 0} (\sinh x)/x = 1$. (*Hint:* Use the result of Problem 139.)

In Problems 173 to 176, sketch a graph of the given function, indicating all significant features such as extreme points, inflection points, and asymptotes.

173 $f(x) = x^2 e^{2x}$

174 $g(x) = x - \cosh^{-1} x$

175 $h(x) = x^2 e^{-x}$

176 $F(x) = \begin{cases} \dfrac{\sinh x}{x} & \text{if } x \neq 0 \\ 1 & \text{if } x = 0 \end{cases}$

177 Find the volume of the solid generated by revolving the region bounded by $xy^2 = 1$, $y = 0$, $x = 1$, and $x = e^3$ about the x axis.

178 Find the area under the graph of $y = 4/\sqrt{4x^2 + 9}$ between $x = 0$ and $x = 2\sqrt{2}$.

179 The **normal probability density function** f is defined in probability theory by the equation

$$f(x) = \dfrac{1}{\sigma\sqrt{2\pi}} \exp \left[\dfrac{-(x - \mu)^2}{2\sigma^2} \right]$$

where σ and μ are certain constants called the **standard deviation** and the **mean**, respectively, and $\sigma > 0$. When students ask to be graded "on the curve," the curve in question is the graph of f. For what value of x is $f(x)$ a maximum and what is the maximum value of $f(x)$? $\boxed{c}$ Sketch the graph of f if $\sigma = 10$ and $\mu = 75$.

$\boxed{c}$ **180** In public health, it is sometimes necessary to screen large numbers of people for certain diseases. When this can be done by means of a blood test, the technique of **batch testing** is often more efficient than performing individual blood tests. If N people are to be tested, they are divided into groups of k people, and the blood samples from each person in such a group are mixed into a single batch. Each of the N/k resulting batches is tested for the disease. If such a test result is negative, it is concluded that none of the people in the corresponding group has the disease; if the test result is positive, the people in the group are retested individually. By means of a pilot study, it is determined in advance that a fraction $1 - p$ of the population has the disease. The expected number of tests required is then given by $T = (N/k) + N(1 - p^k)$. Find the value of k that minimizes T for (a) $p = 0.95$, (b) $p = 0.8$, and (c) $p = 0.7$.

$\boxed{c}$ **181** You have just won first prize in the state lottery, and you have your choice of the following: (a) \$30,000 will be placed in a savings account in your name, compounded continuously at the rate of 10 percent per year; (b) \$0.01 will be placed in a fund in your name and the amount will be doubled every 6 months over the next 12 years. Which plan should you choose?

182 Carry out the following steps to show that the function f defined by $f(x) = [1 + (a/x)]^x$ is increasing if a is a positive constant and $x > 0$. (a) Use the mean-value theorem to show that there is a number t with

$$\ln\left(1 + \frac{x}{a}\right) - \ln\frac{x}{a} = \frac{1}{t} \quad \text{and} \quad \frac{x}{a} \le t \le 1 + \frac{x}{a}$$

(b) From part (a), show that $\ln[1 + (a/x)] \ge a/(a + x)$. (c) Use part (b) to prove that $D_x[1 + (a/x)]^x \ge 0$.

183 Interpret the result that $f(x) = [1 + (a/x)]^x$ is increasing (Problem 182) in terms of compound interest.

184 Suppose a bank offers savings accounts at a nominal annual rate r compounded n times per year. By regularly depositing a fixed sum of p dollars in such an account every $(1/n)$th of a year, money can be accumulated for future needs. Such a plan is called a **sinking fund.** If it is required that such a sinking fund yield an amount S dollars after a term of t years, show that the periodic payment p dollars must be

$$p = \frac{Sr}{n\left[\left(1 + \dfrac{r}{n}\right)^{nt} - 1\right]}$$

$\boxed{c}$ **185** A radioactive substance has a half-life of 2 hours. How long does it take it to decay to $\frac{1}{10}$ of its original mass?

$\boxed{c}$ **186** At 2:00 P.M. a dish of cream that has been standing in a warm room contains 150,000 bacteria of a certain kind. At 4:00 P.M. there are 900,000 of these bacteria present. How many were present at 3:00 P.M.?

$\boxed{c}$ **187** Black coffee is poured into a cup at an initial temperature of 200°F. One minute later, the coffee has cooled to 160°F. If the room temperature is 80°F, find how many *additional* minutes will be required for the coffee to cool to 110°F.

188 Radioactive substance A with half-life $(\ln 2)/a$ decomposes into radioactive substance B with half-life $(\ln 2)/b$, where $a \ne b$; that is, when an atom of substance A decomposes, it is transmuted into an atom of substance B. If the number of atoms of substance A at time t is given by $q = q_0 e^{-at}$, and if y denotes the number of atoms of substance B at this same instant, show that y satisfies the differential equation $(dy/dt) + by = aq_0 e^{-at}$.

189 In paleontology, **radioactive dating** is sometimes used to determine the age of plant and animal fossils. The age t is calculated in terms of the ratio y/q of radioactive "daughter" substance B to "mother" substance A in the fossil. In Problem 188 assume that $y = 0$ when $t = 0$. (a) Derive the equation

$$y = \frac{aq_0}{a - b}(e^{-bt} - e^{-at})$$

for the number of atoms of substance B present at time t. (b) Show that

$$t = \frac{1}{a - b}\ln\left(1 + \frac{a - b}{a}\cdot\frac{y}{q}\right)$$

$\boxed{c}$ **190** When a capacitor with capacitance C farads discharges through a resistor with resistance R ohms, the current I in amperes flowing through the resistor at time t satisfies the differential equation $dI/dt + [I/(RC)] = 0$. (a) If $I = I_0$ when $t = 0$, find a formula giving I in terms of t. (b) If $RC = \frac{1}{200}$ and $I_0 = 40$ milliamperes, plot a graph of I as a function of t.

$\boxed{c}$ **191** A new species of plant is introduced on a small island. Assume that 500 plants were introduced initially and that the number N of plants increases according to the logistic model with $k = 0.62$. Suppose that there are 1700 plants on the island 2 years after they were introduced. (a) Find the maximum possible number of plants the island will support. (b) Find N after 10 years. (c) When does the inflection occur? (d) Sketch a graph of N as a function of time t in years.

$\boxed{c}$ **192** An industrial plant dumps 10 pounds of pollutant into a lake every minute. Fresh water from streams runs into the lake at the rate of 100 cubic feet per minute and mixes uniformly with the polluted water, and meanwhile the polluted water runs out of the lake into a river at the same rate. The plant commenced operations 10 years ago, at which time the lake was unpolluted. The lake now tests at $\frac{1}{20}$ pound of pollutant per cubic foot. How many cubic feet of water are there in the lake? (Take 10 years to be 5.2596×10^6 minutes.)

8 TECHNIQUES OF INTEGRATION

Our purpose in this chapter is to consolidate the methods of integration presented previously, introduce some new methods, and give a unified development of the most important basic techniques of integration. There are really only three general procedures for evaluating indefinite integrals:

1 *Substitution,* or *change of variable*

2 *Manipulation of the integrand* using algebraic or other identities so as to convert it into a more manageable form

3 *Integration by parts*

Although we have made extensive use of the first two methods in the preceding chapters, we have yet to introduce a few special tricks of the trade such as *trigonometric substitution* and *partial fractions*. The third method, *integration by parts,* is also introduced in this chapter.

The three general procedures are often used in combination; however, none of these procedures—singly or in combination—is guaranteed to be effective under all circumstances. Frequently, the choice of the appropriate technique is a matter of trial and error guided by experience.

8.1   Integrals of Powers of Sines and Cosines

In this section we use the integration formulas on page 257, in conjunction with appropriate trigonometric identities and changes of variable, to evaluate integrals involving products of powers of sines and cosines.

Integrals of the Form $\int \sin^m x \cos^n x \, dx$

In order to evaluate $\int \sin^m x \cos^n x \, dx$, where m and n are constant exponents, we consider separately the case in which at least one of the exponents is a positive odd integer and the case in which both of the exponents are nonnegative even integers.

Case 1: At Least One of the Exponents m, n Is a Positive Odd Integer

In this case we use the identity $\sin^2 x + \cos^2 x = 1$ to rewrite the integrand either in the form $F(\cos x) \sin x$ or in the form $F(\sin x) \cos x$. In the former case, the substitution $u = \cos x$ is effective, while in the latter case the substitution $u = \sin x$ works.

In Examples 1 to 3, evaluate the given integral.

EXAMPLE 1 $\displaystyle\int \sin^3 x \, dx$

SOLUTION $\displaystyle\int \sin^3 x \, dx = \int \sin^2 x \sin x \, dx = \int (1 - \cos^2 x) \sin x \, dx$

Making the substitution $u = \cos x$, so that $du = -\sin x \, dx$, we have

$$\int (1 - \cos^2 x) \sin x \, dx = \int (1 - u^2)(-du) = -\int (1 - u^2) \, du$$

$$= -\left(u - \frac{u^3}{3}\right) + C = -\cos x + \frac{1}{3} \cos^3 x + C \qquad \blacksquare$$

EXAMPLE 2 $\displaystyle\int \sin^2 x \cos^5 x \, dx$

SOLUTION $\displaystyle\int \sin^2 x \cos^5 x \, dx = \int \sin^2 x \cos^4 x \cos x \, dx$

$$= \int \sin^2 x \, (\cos^2 x)^2 \cos x \, dx$$

$$= \int \sin^2 x \, (1 - \sin^2 x)^2 \cos x \, dx$$

Making the substitution $u = \sin x$, so that $du = \cos x \, dx$, we have

$$\int \sin^2 x \, (1 - \sin^2 x)^2 \cos x \, dx = \int u^2 (1 - u^2)^2 \, du$$

$$= \int u^2 (1 - 2u^2 + u^4) \, du$$

$$= \int (u^2 - 2u^4 + u^6) \, du = \tfrac{1}{3} u^3 - \tfrac{2}{5} u^5 + \tfrac{1}{7} u^7 + C$$

$$= \tfrac{1}{3} \sin^3 x - \tfrac{2}{5} \sin^5 x + \tfrac{1}{7} \sin^7 x + C \qquad \blacksquare$$

EXAMPLE 3 $\displaystyle\int \frac{\sin^3 x}{\sqrt{\cos x}} \, dx$

SOLUTION $\displaystyle\int \frac{\sin^3 x}{\sqrt{\cos x}} \, dx = \int \frac{\sin^2 x}{\sqrt{\cos x}} \sin x \, dx = \int \frac{1 - \cos^2 x}{\sqrt{\cos x}} \sin x \, dx$

Put $u = \cos x$, so that $du = -\sin x \, dx$ and

$$\int \frac{1 - \cos^2 x}{\sqrt{\cos x}} \sin x \, dx = -\int \frac{1 - u^2}{\sqrt{u}} \, du = \int (u^{3/2} - u^{-1/2}) \, du$$

$$= \tfrac{2}{5} u^{5/2} - 2u^{1/2} + C = \tfrac{2}{5} (\cos x)^{5/2} - 2\sqrt{\cos x} + C \qquad \blacksquare$$

In the last example, notice that the technique works even when one of the powers of the sine or cosine is not an integer—just so the other power is an odd positive integer.

Case 2: Both of the Exponents *m*, *n* Are Nonnegative Even Integers

In this case the integral $\int \sin^m x \cos^n x \, dx$ can be evaluated by using the trigonometric identities

$$\sin^2 x = \tfrac{1}{2}(1 - \cos 2x) \qquad \text{and} \qquad \cos^2 x = \tfrac{1}{2}(1 + \cos 2x)$$

which are direct consequences of the double-angle formulas

$$\cos 2x = 1 - 2 \sin^2 x = 2 \cos^2 x - 1$$

In Examples 4 to 6, evaluate the given integral.

EXAMPLE 4 $\int \cos^2 kx \, dx$, where k is a constant, $k \neq 0$

SOLUTION $\displaystyle \int \cos^2 kx \, dx = \int \tfrac{1}{2}(1 + \cos 2kx) \, dx = \tfrac{1}{2} \int dx + \tfrac{1}{2} \int \cos 2kx \, dx$

$$= \frac{x}{2} + \frac{1}{2} \int \cos 2kx \, dx$$

Put $u = 2kx$, so that $du = 2k \, dx$ and $dx = du/(2k)$. Thus,

$$\int \cos 2kx \, dx = \int \cos u \left(\frac{1}{2k} \, du \right) = \frac{1}{2k} \int \cos u \, du = \frac{\sin u}{2k} + C = \frac{\sin 2kx}{2k} + C$$

It follows that

$$\int \cos^2 kx \, dx = \frac{x}{2} + \frac{1}{2} \left(\frac{\sin 2kx}{2k} \right) + C = \frac{x}{2} + \frac{\sin 2kx}{4k} + C \qquad \blacksquare$$

EXAMPLE 5 $\int \sin^4 x \, dx$

SOLUTION $\displaystyle \int \sin^4 x \, dx = \int (\sin^2 x)^2 \, dx = \int [\tfrac{1}{2}(1 - \cos 2x)]^2 \, dx$

$$= \tfrac{1}{4} \int (1 - 2 \cos 2x + \cos^2 2x) \, dx$$

$$= \tfrac{1}{4} \int dx - \tfrac{1}{2} \int \cos 2x \, dx + \tfrac{1}{4} \int \cos^2 2x \, dx$$

$$= \frac{x}{4} - \frac{1}{2} \left(\frac{\sin 2x}{2} \right) + \frac{1}{4} \left(\frac{x}{2} + \frac{\sin 4x}{8} \right) + C$$

where we used the substitution $u = 2x$ to evaluate the middle integral and we used the result of Example 4 with $k = 2$ to evaluate $\int \cos^2 2x \, dx$. Combining terms and simplifying, we have

$$\int \sin^4 x \, dx = \frac{3x}{8} - \frac{\sin 2x}{4} + \frac{\sin 4x}{32} + C \qquad \blacksquare$$

EXAMPLE 6 $\displaystyle\int \sin^4 2x \cos^2 2x \, dx$

SOLUTION

$$\int \sin^4 2x \cos^2 2x \, dx = \int [\tfrac{1}{2}(1 - \cos 4x)]^2 [\tfrac{1}{2}(1 + \cos 4x)] \, dx$$

$$= \tfrac{1}{8} \int (1 - 2 \cos 4x + \cos^2 4x)(1 + \cos 4x) \, dx$$

$$= \tfrac{1}{8} \int (1 - \cos 4x - \cos^2 4x + \cos^3 4x) \, dx$$

$$= \tfrac{1}{8} \int dx - \tfrac{1}{8} \int \cos 4x \, dx - \tfrac{1}{8} \int \cos^2 4x \, dx + \tfrac{1}{8} \int \cos^3 4x \, dx$$

$$= \frac{x}{8} - \frac{\sin 4x}{32} - \frac{1}{8}\left(\frac{x}{2} + \frac{\sin 8x}{16}\right) + \frac{1}{8} \int \cos^2 4x \cos 4x \, dx$$

$$= \frac{x}{16} - \frac{\sin 4x}{32} - \frac{\sin 8x}{128} + \frac{1}{8} \int (1 - \sin^2 4x) \cos 4x \, dx$$

To evaluate the remaining integral, put $u = \sin 4x$, so that $du = 4 \cos 4x \, dx$ and

$$\frac{1}{8} \int (1 - \sin^2 4x) \cos 4x \, dx = \frac{1}{8} \int (1 - u^2)\frac{du}{4}$$

$$= \frac{1}{32}\left(u - \frac{u^3}{3}\right) + C$$

$$= \frac{\sin 4x}{32} - \frac{\sin^3 4x}{96} + C$$

Therefore,

$$\int \sin^4 2x \cos^2 2x \, dx = \frac{x}{16} - \frac{\sin 4x}{32} - \frac{\sin 8x}{128} + \frac{\sin 4x}{32} - \frac{\sin^3 4x}{96} + C$$

$$= \frac{x}{16} - \frac{\sin 8x}{128} - \frac{\sin^3 4x}{96} + C \qquad \blacksquare$$

The following chart summarizes the technique for evaluating integrals of the form

$$\int \sin^m x \cos^n x \, dx$$

where m and n are nonnegative integers:

Case	Procedure
m is odd	Use $\sin^2 x = 1 - \cos^2 x$, and let $u = \cos x$.
n is odd	Use $\cos^2 x = 1 - \sin^2 x$, and let $u = \sin x$.
both m and n are even	Use $\sin^2 x = \tfrac{1}{2}(1 - \cos 2x)$ and $\cos^2 x = \tfrac{1}{2}(1 + \cos 2x)$.

If *both* m and n are odd, apply the technique to the *smaller* exponent.

Integrals of Products of Sines and Cosines

Integrals of the form

$$\int \sin mx \cos nx \, dx \qquad \int \sin mx \sin nx \, dx \qquad \text{and} \qquad \int \cos mx \cos nx \, dx$$

play an important role in the mathematical analysis of periodic phenomena. These integrals can be evaluated by using the following trigonometric identities. (See Problem 38.)

---

1 $\sin s \cos t = \frac{1}{2} \sin (s + t) + \frac{1}{2} \sin (s - t)$

2 $\sin s \sin t = \frac{1}{2} \cos (s - t) - \frac{1}{2} \cos (s + t)$

3 $\cos s \cos t = \frac{1}{2} \cos (s - t) + \frac{1}{2} \cos (s + t)$

EXAMPLE 7   Evaluate $\int \sin 3x \cos 4x \, dx$.

SOLUTION   By Identity 1,

$$\int \sin 3x \cos 4x \, dx = \int [\tfrac{1}{2} \sin 7x + \tfrac{1}{2} \sin (-x)] \, dx = \tfrac{1}{2} \int (\sin 7x - \sin x) \, dx$$

$$= - \frac{\cos 7x}{14} + \frac{\cos x}{2} + C \qquad\blacksquare$$

Problem Set 8.1

In Problems 1 to 8, use the identity $\cos^2 x + \sin^2 x = 1$ and appropriate substitutions to evaluate each integral.

1 $\int \cos^3 x \, dx$

2 $\int \sin^3 4x \, dx$

3 $\int \sin^5 2t \, dt$

4 $\int \cos^5 3v \, dv$

5 $\int \sin^7 2x \cos^3 2x \, dx$

6 $\int \cos^3 x \sin^3 x \, dx$

7 $\int \sin^2 x \cos^3 x \, dx$

8 $\int \sin^3 4x \cos^2 4x \, dx$

In Problems 9 to 14, use the identities $\cos^2 x = \frac{1}{2}(1 + \cos 2x)$ and $\sin^2 x = \frac{1}{2}(1 - \cos 2x)$ and appropriate substitutions to evaluate each integral.

9 $\int \sin^2 3x \, dx$

10 $\int \cos^2 \frac{x}{2} \, dx$

11 $\int \sin^2 \frac{t}{2} \, dt$

12 $\int \cos^4 2x \, dx$

13 $\int \sin^6 u \, du$

14 $\int \sin^2 \pi t \cos^2 \pi t \, dt$

In Problems 15 to 20, evaluate each integral by using Identities 1, 2, or 3 above.

15 $\int \sin 5x \cos 2x \, dx$

16 $\int \sin 4x \cos 2x \, dx$

17 $\int \cos 4x \cos 3x \, dx$

18 $\int \sin 3t \cos 5t \, dt$

19 $\int \sin 7u \sin 3u \, du$

20 $\int \cos 8v \cos 4v \, dv$

In Problems 21 to 34, evaluate each integral.

21 $\int \sin^5 x \cos^2 x \, dx$

22 $\int \sin^4 2x \cos^5 2x \, dx$

23 $\int \frac{\sin^3 x}{\cos^4 x} \, dx$

24 $\int \sin^2 3x \cos^4 3x \, dx$

25 $\int \cos^6 4x \, dx$

26 $\int \sqrt[3]{\sin^2 3x} \cos^5 3x \, dx$

27 $\int_{\pi/4}^{\pi/2} \frac{\cos^3 x}{\sqrt{\sin x}} \, dx$

28 $\int_0^{\pi/3} \sin^2 3x \cos^5 3x \, dx$

29 $\int_0^{1/2} \sqrt[4]{\sin \pi t} \cos^3 \pi t \, dt$

30 $\int_{1/4}^{1/2} \frac{\cos^5 \pi u}{\sin^2 \pi u} \, du$

31 $\displaystyle\int_0^{\pi/8} \sin^4 2x \cos^2 2x \, dx$ **32** $\displaystyle\int_0^{\pi} \sin^8 x \, dx$

33 $\displaystyle\int_0^1 \sin 2\pi x \cos 3\pi x \, dx$ **34** $\displaystyle\int_0^5 \cos \frac{2\pi x}{5} \cos \frac{7\pi x}{5} \, dx$

In Problems 35 to 37, let m and n denote positive integers and verify each formula.

35 $\displaystyle\int_{-\pi}^{\pi} \cos mx \cos nx \, dx = \begin{cases} 0 & \text{if } m \neq n \\ \pi & \text{if } m = n \end{cases}$

36 $\displaystyle\int_{-\pi}^{\pi} \sin mx \sin nx \, dx = \begin{cases} 0 & \text{if } m \neq n \\ \pi & \text{if } m = n \end{cases}$

37 $\displaystyle\int_{-\pi}^{\pi} \cos mx \sin nx \, dx = 0$

38 Verify Identities 1, 2, and 3 on page 489 by using the addition formulas to expand the right sides.

39 Find the volume of the solid generated by revolving one arch of the sine curve about the x axis.

40 Find the volume of the solid generated if the region under one arch of the sine curve is revolved about the line $y = -2$.

41 Find the area under the curve $y = \cos^2 x$ between $x = 0$ and $x = 2\pi$.

42 If n is an odd positive integer, show that $\int_0^{\pi} \cos^n x \, dx = 0$.

43 Suppose that a particle is moving on the s axis according to the law of motion $s = f(t)$, where $v = ds/dt = \sin^2 \pi t$. If $s = 0$ when $t = 0$, find a formula for $f(t)$, and locate the particle when $t = 8$ seconds.

44 Suppose that $f(x) = \sum_{n=1}^{N} a_n \sin nx$, where N is a positive integer and $a_1, a_2, \ldots, a_N$ are constants. Show that

$$a_m = \frac{1}{\pi} \int_{-\pi}^{\pi} f(x) \sin mx \, dx$$

holds for $m = 1, 2, \ldots, N$. (*Hint:* Use Problem 36.)

8.2 Integrals of Powers of Other Trigonometric Functions

In this section we present some techniques for dealing with integrals of the form

$$\int \tan^m u \sec^n u \, du \qquad \text{or} \qquad \int \cot^m u \csc^n u \, du$$

where m and n are nonnegative integers. The simplest integrals of this form are

$$\int \tan u \, du \qquad \int \cot u \, du \qquad (m = 1 \text{ and } n = 0)$$

and

$$\int \sec u \, du \qquad \int \csc u \, du \qquad (m = 0 \text{ and } n = 1)$$

To evaluate the integral of the tangent, we write $\tan u$ as $\sin u/\cos u$ and make the change of variable $v = \cos u$. Thus, since $dv = -\sin u \, du$, we have

$$\int \tan u \, du = \int \frac{\sin u}{\cos u} \, du = \int \frac{-dv}{v} = -\ln |v| + C$$

$$= -\ln |\cos u| + C = \ln |\cos u|^{-1} + C$$

$$= \ln |\sec u| + C$$

A similar maneuver handles the integral of the cotangent (Problem 40).

The integration of the secant and cosecant functions requires a clever trick. To find $\int \sec u \, du$, we write

$$\int \sec u \, du = \int \frac{\sec u(\sec u + \tan u)}{\sec u + \tan u} \, du = \int \frac{\sec^2 u + \sec u \tan u}{\sec u + \tan u} \, du$$

and then we notice that the numerator of the integrand is the derivative of the denominator. Thus, putting $v = \sec u + \tan u$, we have

$$dv = (\sec u \tan u + \sec^2 u)\, du$$

and so $\displaystyle\int \sec u\, du = \int \frac{dv}{v} = \ln |v| + C = \ln |\sec u + \tan u| + C$

A similar trick—writing $\csc u$ as $\csc u(\csc u - \cot u)/(\csc u - \cot u)$—effects the integration of the cosecant (Problem 40). Thus, we have the following integration formulas:

$$\mathbf{1}\ \int \tan u\, du = \ln |\sec u| + C$$

$$\mathbf{2}\ \int \cot u\, du = \ln |\sin u| + C$$

$$\mathbf{3}\ \int \sec u\, du = \ln |\sec u + \tan u| + C$$

$$\mathbf{4}\ \int \csc u\, du = \ln |\csc u - \cot u| + C$$

In Examples 1 and 2, evaluate the given integral.

EXAMPLE 1 $\displaystyle\int \tan 4x\, dx$

SOLUTION   Put $u = 4x$, so that $du = 4\, dx$. Using Formula 1, we have

$$\int \tan 4x\, dx = \int \tan u\, \frac{du}{4} = \frac{1}{4} \int \tan u\, du$$

$$= \tfrac{1}{4} \ln |\sec u| + C = \tfrac{1}{4} \ln |\sec 4x| + C$$

EXAMPLE 2 $\displaystyle\int \frac{dx}{\cos 5x}$

SOLUTION   Put $u = 5x$, so that $du = 5\, dx$. Using Formula 3, we have

$$\int \frac{dx}{\cos 5x} = \int \sec 5x\, dx = \int \sec u\, \frac{du}{5} = \frac{1}{5} \int \sec u\, du$$

$$= \tfrac{1}{5} \ln |\sec u + \tan u| + C$$

$$= \tfrac{1}{5} \ln |\sec 5x + \tan 5x| + C$$

Powers of Tangent, Cotangent, Secant, and Cosecant

In integrating positive integer powers or products of positive integer powers of tan, cot, sec, and csc, the identities

$$1 + \tan^2 u = \sec^2 u \qquad \text{and} \qquad 1 + \cot^2 u = \csc^2 u$$

can often be used to advantage. The following examples illustrate the techniques.

In Examples 3 and 4, evaluate the given integral.

<u>EXAMPLE 3</u> $\displaystyle\int \cot^2 x \, dx$

SOLUTION

$$\int \cot^2 x \, dx = \int (\csc^2 x - 1) \, dx = \int \csc^2 x \, dx - \int dx = -\cot x - x + C \quad \blacksquare$$

<u>EXAMPLE 4</u> $\displaystyle\int \tan^3 2x \, dx$

SOLUTION   We write $\tan^3 2x$ as $\tan 2x \tan^2 2x$ and then use the identity $\tan^2 2x = \sec^2 2x - 1$ to obtain

$$\int \tan^3 2x \, dx = \int \tan 2x \tan^2 2x \, dx = \int \tan 2x \, (\sec^2 2x - 1) \, dx$$

$$= \int (\tan 2x \sec^2 2x - \tan 2x) \, dx$$

$$= \int \tan 2x \sec^2 2x \, dx - \int \tan 2x \, dx$$

To evaluate the integral $\int \tan 2x \sec^2 2x \, dx$, we make the substitution $u = \tan 2x$, so that $du = 2 \sec^2 2x \, dx$ and

$$\int \tan 2x \sec^2 2x \, dx = \int u \frac{du}{2} = \frac{1}{2} \int u \, du = \frac{u^2}{4} + C = \frac{\tan^2 2x}{4} + C$$

The integral $\int \tan 2x \, dx$ is easily evaluated by the method already illustrated in Example 1, page 491. Therefore,

$$\int \tan^3 2x \, dx = \frac{\tan^2 2x}{4} - \frac{1}{2} \ln |\sec 2x| + C \quad \blacksquare$$

To evaluate an integral of the form

$$\int \tan^m x \sec^n x \, dx$$

where m and n are nonnegative integers and either n is even or m is odd, proceed as in the following chart:

Case	Procedure
n is even	Factor $\sec^2 x$ out of the integrand, use the identity $\sec^2 x = 1 + \tan^2 x$ to rewrite what remains in terms of $\tan x$, and let $u = \tan x$.
m is odd	Factor $\sec x \tan x$ out of the integrand, use the identity $\tan^2 x = \sec^2 x - 1$ to rewrite what remains in terms of $\sec x$, and let $u = \sec x$.

Likewise, an integral of the form

$$\int \cot^m x \csc^n x \, dx$$

can be evaluated by following the chart above, but with secant replaced by cosecant and tangent replaced by cotangent.

In Examples 5 and 6, evaluate the given integral.

<u>EXAMPLE 5</u> $\int \tan^6 x \sec^4 x \, dx$

SOLUTION   We follow the procedure for the case in which n is even:

$$\int \tan^6 x \sec^4 x \, dx = \int \tan^6 x \sec^2 x \sec^2 x \, dx$$

$$= \int \tan^6 x (1 + \tan^2 x) \sec^2 x \, dx$$

$$= \int (\tan^6 x + \tan^8 x) \sec^2 x \, dx$$

Now, we put $u = \tan x$ so that $du = \sec^2 x \, dx$ and

$$\int \tan^6 x \sec^4 x \, dx = \int (u^6 + u^8) \, du = \frac{u^7}{7} + \frac{u^9}{9} + C$$

$$= \frac{1}{7} \tan^7 x + \frac{1}{9} \tan^9 x + C$$

<u>EXAMPLE 6</u> $\int \cot^5 x \csc^3 x \, dx$

SOLUTION   Since the integrand involves an odd power of $\cot x$, we follow the procedure for the case in which m is odd, but with functions replaced by cofunctions. Thus,

$$\int \cot^5 x \csc^3 x \, dx = \int \cot^4 x \csc^2 x (\csc x \cot x) \, dx$$

$$= \int (\cot^2 x)^2 \csc^2 x (\csc x \cot x) \, dx$$

$$= \int (\csc^2 x - 1)^2 \csc^2 x (\csc x \cot x) \, dx$$

Putting $u = \csc x$, so that $du = -\csc x \cot x \, dx$, we have

$$\int \cot^5 x \csc^3 x \, dx = \int (u^2 - 1)^2 u^2 (-du) = -\int (u^4 - 2u^2 + 1) u^2 \, du$$

$$= -\int (u^6 - 2u^4 + u^2) \, du$$

$$= -\tfrac{1}{7} u^7 + \tfrac{2}{5} u^5 - \tfrac{1}{3} u^3 + C$$

$$= -\tfrac{1}{7} \csc^7 x + \tfrac{2}{5} \csc^5 x - \tfrac{1}{3} \csc^3 x + C$$

The techniques illustrated above are not particularly helpful for evaluating

$$\int \tan^m x \sec^n x \, dx$$

when m is even and n is odd. Using the identity $\tan^2 x = \sec^2 x - 1$, you can always rewrite such an integral as a sum of constant multiples of integrals of odd powers of $\sec x$ (Problem 42). Methods for evaluating

$$\int \sec^n x \, dx$$

when n is odd and greater than 1 are discussed later (see pages 504 and 529).

Problem Set 8.2

In Problems 1 to 36, evaluate each integral.

1 $\int \cot 4x \, dx$

2 $\int \tan \frac{x}{2} \, dx$

3 $\int \frac{dx}{\cos 3x}$

4 $\int \csc \frac{x}{5} \, dx$

5 $\int \tan^2 \frac{2x}{3} \, dx$

6 $\int \cot^3 5x \, dx$

7 $\int \cot^4 4x \, dx$

8 $\int \tan^3 \frac{\pi t}{2} \, dt$

9 $\int \csc^4 3t \, dt$

10 $\int \sec^6 2x \, dx$

11 $\int \tan^4 2t \sec^2 2t \, dt$

12 $\int \cot^4 3x \csc^4 3x \, dx$

13 $\int \tan^3 5x \sec^5 5x \, dx$

14 $\int \cot^3 \frac{\pi x}{2} \csc^3 \frac{\pi x}{2} \, dx$

15 $\int (\tan 2x + \cot 2x)^2 \, dx$

16 $\int (\sec 3x + \tan 3x)^2 \, dx$

17 $\int \frac{\sec^4 t \, dt}{\sqrt{\tan t}}$

18 $\int \frac{\tan^3 3x \, dx}{\sqrt{\sec 3x}}$

19 $\int \tan^3 7x \sec^4 7x \, dx$

20 $\int \left(\frac{\tan x}{\cos x} \right)^4 dx$

21 $\int \cot 3x \csc^3 3x \, dx$

22 $\int \cot^{7/2} 2x \csc^4 2x \, dx$

23 $\int \tan^3 5x \sec 5x \, dx$

24 $\int \cot^3 \frac{x}{2} \csc^3 \frac{x}{2} \, dx$

25 $\int \tan^3 2x \sqrt{\sec 2x} \, dx$

26 $\int \sqrt{\tan 7x} \sec^4 7x \, dx$

27 $\int \tan^5 x \sec^7 x \, dx$

28 $\int \frac{\csc^4 2\pi x}{\cot^2 2\pi x} \, dx$

29 $\int \frac{\csc^2 8x}{\cot^4 8x} \, dx$

30 $\int \sec^3 2x \tan^5 2x \, dx$

31 $\int x \cot^3 x^2 \csc^3 x^2 \, dx$

32 $\int x^3 \tan^5 x^4 \sec^7 x^4 \, dx$

33 $\int_{\pi/6}^{\pi/9} \cot 3x \, dx$

34 $\int_{\pi/8}^{\pi/6} 5 \sec 2x \, dx$

35 $\int_{\pi/4}^{\pi/2} \cot^4 x \csc^4 x \, dx$

36 $\int_{0}^{\pi/4} \tan^5 x \, dx$

37 Find the area of the region under the curve $y = 5 \tan^2 x$ from $x = -\pi/4$ to $x = \pi/4$.

38 Find the area of the region under the curve $y = \sec x$ from $x = -\pi/3$ to $x = \pi/3$.

39 Find the volume of the solid generated when the region under the curve $y = \sec^2 x$ between $x = 0$ and $x = \pi/3$ is rotated about the x axis.

40 Derive the integration formulas for the cotangent and for the cosecant.

41 Find the arc length of the curve $y = \ln (\sin x)$ from $x = \pi/4$ to $x = \pi/2$.

42 Let m and n be nonnegative integers. If m is even and n is odd, show that $\int \tan^m x \sec^n x \, dx$ can be rewritten as a sum of integrals of the form $\int a \sec^k x \, dx$, where a is a constant and k is an odd positive integer.

43 Show that $\int \csc u \, du = \ln |\tan (u/2)| + C$, and reconcile this with Formula 4 on page 491.

8.3   Integration by Trigonometric Substitution

Algebraic expressions not originally containing trigonometric functions can often be simplified by substituting trigonometric expressions for the variable. This technique, called **trigonometric substitution,** is particularly useful for evaluating integrals involving roots or reciprocals of expressions of the form

$$a^2 - u^2 \qquad a^2 + u^2 \qquad \text{or} \qquad u^2 - a^2$$

where a is a positive constant. For instance, to simplify an integrand involving, say,

$$\sqrt{a^2 - u^2}$$

we might use the substitution

$$u = a \sin \theta$$

to obtain

$$\sqrt{a^2 - u^2} = \sqrt{a^2 - a^2 \sin^2 \theta} = \sqrt{a^2(1 - \sin^2 \theta)} = \sqrt{a^2 \cos^2 \theta} = a \cos \theta$$

Here we are assuming that $-\pi/2 < \theta < \pi/2$, so that $\cos \theta > 0$.

The trigonometric substitution

$$u = a \sin \theta \quad \text{or} \quad \sin \theta = \frac{u}{a}$$

is illustrated geometrically by the right triangle in Figure 1a. Likewise, trigonometric substitutions corresponding to the expressions $a^2 + u^2$ and $u^2 - a^2$ are suggested by the right triangles in Figure 1b and 1c. The three trigonometric substitutions illustrated in Figure 1 are summarized in the following chart.

Figure 1

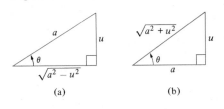

(a)   (b)

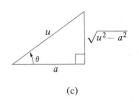

(c)

For the expression	Use the substitution	To obtain	Figure
$a^2 - u^2$	$u = a \sin \theta$	$a^2 - u^2 = a^2 \cos^2 \theta$	1a
$a^2 + u^2$	$u = a \tan \theta$	$a^2 + u^2 = a^2 \sec^2 \theta$	1b
$u^2 - a^2$	$u = a \sec \theta$	$u^2 - a^2 = a^2 \tan^2 \theta$	1c

After using a trigonometric substitution to simplify an integrand and carrying out the integration with respect to the variable θ, you can rewrite the answer in terms of the original variable by referring to the appropriate right triangle in Figure 1.

EXAMPLE 1 Evaluate $\displaystyle\int \frac{x^2 \, dx}{(4 - x^2)^{3/2}}$.

SOLUTION   Make the substitution $x = 2 \sin \theta$ (Figure 2), so that

$$dx = 2 \cos \theta \, d\theta \quad \text{and} \quad 4 - x^2 = 4 - 4 \sin^2 \theta = 4 \cos^2 \theta$$

Figure 2

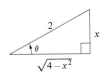

Thus, $\displaystyle\int \frac{x^2 \, dx}{(4 - x^2)^{3/2}} = \int \frac{(2 \sin \theta)^2 (2 \cos \theta \, d\theta)}{(4 \cos^2 \theta)^{3/2}} = \int \frac{\sin^2 \theta}{\cos^2 \theta} \, d\theta$

$$= \int \tan^2 \theta \, d\theta = \int (\sec^2 \theta - 1) \, d\theta$$

$$= \tan \theta - \theta + C$$

By Figure 2, $\tan \theta = x/\sqrt{4 - x^2}$ and $\theta = \sin^{-1}(x/2)$; hence,

$$\int \frac{x^2 \, dx}{(4 - x^2)^{3/2}} = \frac{x}{\sqrt{4 - x^2}} - \sin^{-1} \frac{x}{2} + C$$

EXAMPLE 2 Evaluate $\displaystyle\int \frac{dx}{x^2 \sqrt{x^2 + 9}}$.

SOLUTION   Make the substitution $x = 3 \tan \theta$ (Figure 3), so that

$$dx = 3 \sec^2 \theta \, d\theta \quad \text{and} \quad x^2 + 9 = 9 \tan^2 \theta + 9 = 9 \sec^2 \theta$$

Figure 3

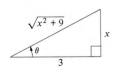

Thus, $\displaystyle\int \frac{dx}{x^2 \sqrt{x^2 + 9}} = \int \frac{3 \sec^2 \theta \, d\theta}{(9 \tan^2 \theta) \sqrt{9 \sec^2 \theta}} = \frac{1}{9} \int \frac{\sec \theta \, d\theta}{\tan^2 \theta}$

$$= \frac{1}{9} \int \frac{1/\cos \theta}{\sin^2 \theta / \cos^2 \theta} \, d\theta = \frac{1}{9} \int \frac{\cos \theta \, d\theta}{\sin^2 \theta}$$

In order to evaluate the last integral, we put $v = \sin\theta$, so that $dv = \cos\theta\, d\theta$ and

$$\frac{1}{9}\int \frac{\cos\theta\, d\theta}{\sin^2\theta} = \frac{1}{9}\int \frac{dv}{v^2} = -\frac{1}{9v} + C = -\frac{1}{9\sin\theta} + C = -\frac{1}{9}\csc\theta + C$$

From Figure 3, $\csc\theta = \sqrt{x^2+9}/x$; hence,

$$\int \frac{dx}{x^2\sqrt{x^2+9}} = -\frac{\sqrt{x^2+9}}{9x} + C$$ ■

EXAMPLE 3 Evaluate $\displaystyle\int \frac{dt}{t^3\sqrt{t^2-25}}$.

Figure 4

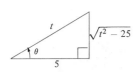

SOLUTION Make the substitution $t = 5\sec\theta$ (Figure 4), so that

$$dt = 5\sec\theta\tan\theta\, d\theta \qquad \text{and} \qquad t^2 - 25 = 25\sec^2\theta - 25 = 25\tan^2\theta$$

Thus,

$$\begin{aligned}
\int \frac{dt}{t^3\sqrt{t^2-25}} &= \int \frac{5\sec\theta\tan\theta\, d\theta}{(125\sec^3\theta)\sqrt{25\tan^2\theta}} \\
&= \frac{1}{125}\int \frac{d\theta}{\sec^2\theta} = \frac{1}{125}\int \cos^2\theta\, d\theta \\
&= \frac{1}{125}\int \frac{1}{2}(1+\cos 2\theta)\, d\theta = \frac{1}{250}\left(\theta + \frac{\sin 2\theta}{2}\right) + C \\
&= \frac{1}{250}\left(\theta + \frac{2\sin\theta\cos\theta}{2}\right) + C = \frac{1}{250}(\theta + \sin\theta\cos\theta) + C
\end{aligned}$$

By Figure 4, $\theta = \sec^{-1}(t/5)$, $\sin\theta = \sqrt{t^2-25}/t$, and $\cos\theta = 5/t$; hence

$$\int \frac{dt}{t^3\sqrt{t^2-25}} = \frac{1}{250}\left(\sec^{-1}\frac{t}{5} + \frac{5\sqrt{t^2-25}}{t^2}\right) + C$$ ■

EXAMPLE 4 Evaluate $\displaystyle\int_{2/3}^{2\sqrt{3}/3} \frac{du}{u\sqrt{9u^2+4}}$.

SOLUTION Make the substitution $3u = 2\tan\theta$, or $u = \frac{2}{3}\tan\theta$, so that

$$du = \tfrac{2}{3}\sec^2\theta\, d\theta \qquad \text{and} \qquad \sqrt{9u^2+4} = \sqrt{4\tan^2\theta + 4} = 2\sec\theta$$

Here, $\theta = \tan^{-1}(3u/2)$, so that $\theta = \pi/4$ when $u = \frac{2}{3}$ and $\theta = \pi/3$ when $u = 2\sqrt{3}/3$. Therefore,

$$\begin{aligned}
\int_{2/3}^{2\sqrt{3}/3} \frac{du}{u\sqrt{9u^2+4}} &= \int_{\pi/4}^{\pi/3} \frac{\frac{2}{3}\sec^2\theta\, d\theta}{\frac{2}{3}\tan\theta \cdot 2\sec\theta} = \frac{1}{2}\int_{\pi/4}^{\pi/3} \frac{\sec\theta\, d\theta}{\tan\theta} \\
&= \frac{1}{2}\int_{\pi/4}^{\pi/3} \frac{1/\cos\theta}{\sin\theta/\cos\theta}\, d\theta = \frac{1}{2}\int_{\pi/4}^{\pi/3} \frac{1}{\sin\theta}\, d\theta \\
&= \frac{1}{2}\int_{\pi/4}^{\pi/3} \csc\theta\, d\theta = \frac{1}{2}\ln|\csc\theta - \cot\theta|\,\Big|_{\pi/4}^{\pi/3} \\
&= \frac{1}{2}\left[\ln\left(\frac{2}{\sqrt{3}} - \frac{1}{\sqrt{3}}\right) - \ln(\sqrt{2}-1)\right] = \frac{1}{2}\ln\frac{1/\sqrt{3}}{\sqrt{2}-1} \\
&= \frac{1}{2}\ln\frac{1}{\sqrt{6}-\sqrt{3}}
\end{aligned}$$ ■

Integrals Involving $ax^2 + bx + c$

If a quadratic expression $ax^2 + bx + c$ occurs in an integrand, it may be useful to rewrite the expression in the form $a(x - h)^2 + k$ by completing the square* and then to make the substitution $u = x - h$.

In Examples 5 and 6, evaluate the given integral.

EXAMPLE 5 $\int \dfrac{dx}{(5 - 4x - x^2)^{3/2}}$

SOLUTION   Completing the square, we have

$$5 - 4x - x^2 = 5 - (x^2 + 4x) = 5 + 4 - (x^2 + 4x + 4)$$

$$= 9 - (x + 2)^2 = 9 - u^2$$

where we have put $u = x + 2$. Thus, $du = dx$, and we have

$$\int \frac{dx}{(5 - 4x - x^2)^{3/2}} = \int \frac{du}{(9 - u^2)^{3/2}}$$

Now, we make the trigonometric substitution $u = 3 \sin \theta$, so that $du = 3 \cos \theta \, d\theta$ and $9 - u^2 = 9 - 9 \sin^2 \theta = 9 \cos^2 \theta$. Thus,

$$\int \frac{du}{(9 - u^2)^{3/2}} = \int \frac{3 \cos \theta \, d\theta}{(9 \cos^2 \theta)^{3/2}} = \int \frac{3 \cos \theta \, d\theta}{27 \cos^3 \theta} = \frac{1}{9} \int \frac{d\theta}{\cos^2 \theta}$$

$$= \frac{1}{9} \int \sec^2 \theta \, d\theta = \frac{1}{9} \tan \theta + C$$

Figure 5

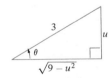

Referring to Figure 5, we find that $\tan \theta = u/\sqrt{9 - u^2}$; hence,

$$\int \frac{dx}{(5 - 4x - x^2)^{3/2}} = \frac{1}{9} \tan \theta + C = \frac{u}{9\sqrt{9 - u^2}} + C = \frac{x + 2}{9\sqrt{5 - 4x - x^2}} + C \quad \blacksquare$$

EXAMPLE 6 $\int \dfrac{x \, dx}{\sqrt{4x^2 + 8x + 5}}$

SOLUTION   Completing the square, we have

$$4x^2 + 8x + 5 = 4(x^2 + 2x) + 5 = 4(x^2 + 2x + 1) + 5 - 4$$

$$= 4(x + 1)^2 + 1 = 4u^2 + 1$$

where we have put $u = x + 1$. Here, $x = u - 1$ and $dx = du$, so that

$$\int \frac{x \, dx}{\sqrt{4x^2 + 8x + 5}} = \int \frac{(u - 1) \, du}{\sqrt{4u^2 + 1}}$$

Now, we make the trigonometric substitution $2u = \tan \theta$, so that $u = \frac{1}{2} \tan \theta$, $du = \frac{1}{2} \sec^2 \theta \, d\theta$, and $4u^2 + 1 = \tan^2 \theta + 1 = \sec^2 \theta$. Thus,

$$\int \frac{(u - 1) \, du}{\sqrt{4u^2 + 1}} = \int \frac{(\frac{1}{2} \tan \theta - 1)(\frac{1}{2} \sec^2 \theta) \, d\theta}{\sqrt{\sec^2 \theta}} = \frac{1}{4} \int (\tan \theta - 2) \sec \theta \, d\theta$$

$$= \frac{1}{4} \int (\sec \theta \tan \theta - 2 \sec \theta) \, d\theta$$

$$= \frac{1}{4} \sec \theta - \frac{1}{2} \ln |\sec \theta + \tan \theta| + C$$

*The technique for completing the square is illustrated in connection with Theorem 1, page 33.

Here we have

$$\sec\theta = \sqrt{\tan^2\theta + 1} = \sqrt{4u^2 + 1} = \sqrt{4x^2 + 8x + 5}$$

and

$$\tan\theta = 2u = 2(x + 1) = 2x + 2$$

Hence,

$$\int \frac{x\,dx}{\sqrt{4x^2 + 8x + 5}} = \frac{1}{4}\sqrt{4x^2 + 8x + 5} - \frac{1}{2}\ln\left|\sqrt{4x^2 + 8x + 5} + 2x + 2\right| + C \quad\blacksquare$$

In the examples above, we have not taken care to impose restrictions on the variables so that quantities under square-root signs are nonnegative, inverse trigonometric functions are properly defined, and so forth. For instance, in order to write $\sec\theta = \sqrt{\tan^2\theta + 1}$ in the last example, we should have restricted θ to values for which $\sec\theta$ is positive. Furthermore, Figures 1 through 5 apply only to quadrant I angles θ. Nevertheless, integral formulas obtained by trigonometric substitution are quite often correct in spite of such negligence—for an indication of why, see Problems 52 through 54. Of course, an integral formula—once obtained by any method, no matter how questionable—*can and should be checked by differentiation* (see Problem 23).

Problem Set 8.3

In Problems 1 to 22, use an appropriate trigonometric substitution to evaluate each integral.

1 $\displaystyle\int \frac{dx}{x^2\sqrt{16 - x^2}}$

2 $\displaystyle\int \frac{\sqrt{9 - x^2}}{x^2}\,dx$

3 $\displaystyle\int \frac{dt}{t^4\sqrt{4 - t^2}}$

4 $\displaystyle\int \frac{y^3\,dy}{\sqrt{4 - y^2}}$

5 $\displaystyle\int \frac{x^2\,dx}{\sqrt{4 - 9x^2}}$

6 $\displaystyle\int \sqrt{9 - 2u^2}\,du$

7 $\displaystyle\int \frac{\sqrt{7 - 4t^2}}{t^4}\,dt$

8 $\displaystyle\int \frac{dx}{x^2(a^2 + x^2)^{3/2}}, \ a > 0$

9 $\displaystyle\int \frac{t\,dt}{\sqrt{t^2 - a^2}}$

10 $\displaystyle\int \frac{x^3\,dx}{\sqrt{x^2 + 4}}$

11 $\displaystyle\int \frac{dx}{x^2\sqrt{1 + x^2}}$

12 $\displaystyle\int \frac{dx}{x^4\sqrt{4 + x^2}}$

13 $\displaystyle\int \frac{dt}{t\sqrt{t^2 + 5}}$

14 $\displaystyle\int \frac{dx}{(x^2 + 9)^2}$

15 $\displaystyle\int \frac{7x^3\,dx}{(4x^2 + 9)^{3/2}}$

16 $\displaystyle\int \frac{dt}{\sqrt{16t^2 + 9}}$

17 $\displaystyle\int \frac{dx}{x^2\sqrt{x^2 - 4}}$

18 $\displaystyle\int \frac{dt}{t^4\sqrt{t^2 - 4}}$

19 $\displaystyle\int \frac{dt}{\sqrt{9t^2 - 4}}$

20 $\displaystyle\int \frac{\sqrt{y^2 - 9}}{y}\,dy$

21 $\displaystyle\int_{3/8}^{2/3} \frac{dx}{\sqrt{4x^2 + 1}}$

22 $\displaystyle\int_3^4 \sqrt{25 - t^2}\,dt$

23 Check your answers to Problems 1, 13, and 19 by differentiation, carefully noting any required restrictions on the values of the variables involved.

24 Evaluate the integral $\int (x\,dx)/\sqrt{1 - x^2}$ in two ways, and reconcile the answers. (a) Use the substitution $u = 1 - x^2$. (b) Use trigonometric substitution. Which method is easier?

In Problems 25 to 30, complete the square in the quadratic expression and then use trigonometric substitution to evaluate the integral.

25 $\displaystyle\int \frac{dt}{(5 - 4t - t^2)^{3/2}}$

26 $\displaystyle\int \frac{x\,dx}{\sqrt{2 - x - x^2}}$

27 $\displaystyle\int \frac{2t}{(t^2 + 3t + 4)^2}\,dt$

28 $\displaystyle\int \frac{dt}{\sqrt{2t^2 - 6t + 5}}$

29 $\displaystyle\int \frac{x\,dx}{\sqrt{1 - x + 3x^2}}$

30 $\displaystyle\int_0^1 \frac{x\,dx}{\sqrt{6 - x - x^2}}$

In Problems 31 to 48, evaluate each integral. (Some of the integrals may be easier to evaluate by using methods *other than* trigonometric substitution or completing the square.)

31 $\displaystyle\int \frac{dx}{(4x^2 - 9)^{3/2}}$

32 $\displaystyle\int x^3\sqrt{x^2 - 1}\,dx$

33 $\displaystyle\int \frac{t\,dt}{t^2 - 4t + 8}$

34 $\displaystyle\int \frac{2x\,dx}{\sqrt{9 + x^2}}$

35 $\displaystyle\int \frac{dx}{\sqrt{-3 + 8x - 4x^2}}$

36 $\displaystyle\int \frac{t - 1}{(t^2 - 2t + 1)^2}\,dt$

37 $\displaystyle\int \frac{x\,dx}{\sqrt{4 - x^2}}$

38 $\displaystyle\int \frac{3}{(x^2 + 6x + 1)^2}\,dx$

39 $\displaystyle\int \sqrt{2 - 3x + x^2}\,dx$

40 $\displaystyle\int \frac{e^{-x}\,dx}{\sqrt{4 + 9e^{-2x}}}$

41 $\displaystyle\int \frac{2x + 2}{\sqrt{x^2 + 2x + 2}}\,dx$

42 $\displaystyle\int \frac{\sec^2 t\,dt}{(\tan^2 t + 9)^{3/2}}$

43 $\displaystyle\int \frac{\sin v\,dv}{(25 - \cos^2 v)^{3/2}}$

44 $\displaystyle\int \frac{dt}{t[(\ln t)^2 - 4]^{3/2}}$

45 $\displaystyle\int_0^1 \frac{t^2\,dt}{(25 - 9t^2)^{3/2}}$

46 $\displaystyle\int_{\ln 2}^{\ln 3} \frac{dz}{e^z - e^{-z}}$

47 $\displaystyle\int_3^6 \frac{\sqrt{t^2 - 9}}{t}\,dt$

48 $\displaystyle\int_0^1 (3y + 2)\sqrt{9y^2 + 12y + 3}\,dy$

49 Find the area bounded by the curves $y = 45/\sqrt{16x^2 - 175}$, $y = 0$, $x = 4$, and $x = 5$.

50 Evaluate $\int (y^2\,dy)/(4 + y^2)^{5/2}$ by making the hyperbolic substitution $y = 2 \sinh t$.

51 Find the volume of the solid generated when the region bounded

by the curves $y = x/(x^2 + 16)^{3/2}$, $y = 0$, and $x = 4$ is revolved about the x axis.

52 Let $a > 0$, and suppose that it is required to integrate an expression involving $\sqrt{a^2 - u^2}$, where $|u| < a$. Since $|u/a| < 1$, we can make the substitution $\theta = \sin^{-1}(u/a)$, where $-\pi/2 < \theta < \pi/2$. Show that $u = a \sin \theta$ and that

(a) $\sqrt{a^2 - u^2} = a \cos \theta$ (b) $du = a \cos \theta\,d\theta$

(c) $\csc \theta = \dfrac{a}{u}$ if $u \neq 0$ (d) $\sec \theta = \dfrac{a}{\sqrt{a^2 - u^2}}$

(e) $\tan \theta = \dfrac{u}{\sqrt{a^2 - u^2}}$ (f) $\cot \theta = \dfrac{\sqrt{a^2 - u^2}}{u^2}$ if $u \neq 0$

53 Using the results in Problem 52, explain why it is not really necessary to assume that u is positive in using the trigonometric substitution $u = a \sin \theta$.

54 Discuss the trigonometric substitution $u = a \tan \theta$ for the case in which u is not necessarily positive.

55 Find the arc length of the curve $y = \sin^{-1} e^x$ from $x = \ln \frac{1}{2}$ to $x = \ln \frac{3}{5}$.

56 A vertical wire of uniform density with total mass M and length l stands with its lower end on the x axis at the point $(a, 0)$. If G denotes the constant of gravitation, calculate the horizontal component of the force of gravity exerted by the wire on a particle of mass m situated at the origin.

57 Solve the separable differential equation

$$x^2\,dy - \sqrt{x^2 - 9}\,dx = 0$$

58 Find the surface area generated by revolving the arc of the curve $x = e^y$ between $(1, 0)$ and $(e^2, 2)$ about the y axis.

8.4   Integration by Parts

Nearly every rule or technique for differentiation can be inverted to yield a corresponding rule or technique for integration. For instance, the technique of integration by change of variable is essentially the inversion of the chain rule for differentiation. In this section we study the integration technique, called *integration by parts*, that results from an inversion of the product rule for differentiation.

According to the product rule,

$$(f \cdot g)' = f' \cdot g + f \cdot g'$$

the function $f \cdot g$ is an antiderivative of $f' \cdot g + f \cdot g'$; that is,

$$\int [f'(x)g(x) + f(x)g'(x)]\,dx = f(x)g(x) + C$$

or

$$\int f'(x)g(x)\,dx + \int f(x)g'(x)\,dx = f(x)g(x) + C$$

The last equation can be rewritten as

$$\int f(x)g'(x)\,dx = f(x)g(x) - \int g(x)f'(x)\,dx$$

provided it is understood that the constant C is absorbed into the constant of integration corresponding to $\int g(x)f'(x)\,dx$.

The equation obtained above is easier to remember if we let

$$u = f(x) \qquad \text{and} \qquad v = g(x)$$

so that

$$du = f'(x)\,dx \qquad \text{and} \qquad dv = g'(x)\,dx$$

and we have

$$\boxed{\int u\,dv = uv - \int v\,du}$$

This is the formula for **integration by parts.**

<u>EXAMPLE 1</u> Evaluate $\int x \sin x\,dx$.

SOLUTION   In order to use integration by parts, we put

$$u = x \qquad \text{and} \qquad dv = \sin x\,dx$$

so that

$$du = dx \qquad \text{and} \qquad v = \int \sin x\,dx = -\cos x + C_0$$

Thus,

$$\int \underset{u}{\underbrace{x}}\,\underset{dv}{\underbrace{\sin x\,dx}} = \int u\,dv = uv - \int v\,du = x(-\cos x + C_0) - \int (-\cos x + C_0)\,dx$$

$$= -x \cos x + xC_0 + \int \cos x\,dx - \int C_0\,dx$$

$$= -x \cos x + xC_0 + \sin x + C - C_0 x$$

$$= -x \cos x + \sin x + C$$

In the preceding example, the constant of integration C_0 arising from the preliminary integration $v = \int \sin x\,dx$ cancels out in the end. In carrying out integration by parts, this always happens (Problem 52); hence, the constant arising from the preliminary integration $v = \int dv$ need not be written into the calculation. Thus, we would write

$$\int x \sin x\,dx = \int u\,dv = uv - \int v\,du = x(-\cos x) - \int (-\cos x)\,dx$$

$$= -x \cos x + \sin x + C$$

In order to use the formula

$$\int u\,dv = uv - \int v\,du$$

to evaluate an integral, you must begin by choosing u and dv. Often the choice can be made in several different ways, and a poor choice of u and dv may result in an integral

$$\int v\,du$$

that is more complicated than the one you started with. Try to choose u and dv so that $\int v\, du$ is as simple as possible.

In Examples 2 to 4, evaluate the given integral using integration by parts.

EXAMPLE 2 $\displaystyle\int x \sec^2 x\, dx$

SOLUTION   Here, there seem to be two reasonable choices:

Choice 1 $u = x \qquad dv = \sec^2 x\, dx$

so that $du = dx \qquad v = \displaystyle\int \sec^2 x\, dx = \tan x$

Choice 2 $u = \sec^2 x \qquad dv = x\, dx$

so that $du = 2 \sec^2 x \tan x\, dx \qquad v = \displaystyle\int x\, dx = \dfrac{x^2}{2}$

If we use choice 1,

$$\int v\, du = \int \tan x\, dx$$

whereas if we use choice 2

$$\int v\, du = \int \left(\frac{x^2}{2}\right) 2 \sec^2 x \tan x\, dx$$

Evidently, $\int v\, du$ is simpler if we use choice 1. Thus, we let

$$u = x \qquad dv = \sec^2 x\, dx$$
$$du = dx \qquad v = \tan x$$

so that

$$\int \underbrace{x}_{u}\ \underbrace{\sec^2 x\, dx}_{dv} = \int u\, dv = uv - \int v\, du = x \tan x - \int \tan x\, dx$$

$$= x \tan x - (-\ln|\cos x|) + C = x \tan x + \ln|\cos x| + C \qquad \blacksquare$$

EXAMPLE 3 $\displaystyle\int \ln x\, dx$

SOLUTION   Here, we don't seem to have much choice, so we put

$$u = \ln x \qquad dv = dx$$
$$du = \frac{1}{x}\, dx \qquad v = x$$

Thus, $\displaystyle\int \underbrace{\ln x}_{u}\ \underbrace{dx}_{dv} = \int u\, dv = uv - \int v\, du = (\ln x)x - \int x\left(\frac{1}{x}\, dx\right)$

$$= x \ln x - \int dx$$
$$= x \ln x - x + C \qquad \blacksquare$$

EXAMPLE 4 $\displaystyle\int x^3 \cos x^2 \, dx$

SOLUTION   Before using integration by parts, we make the substitution

$$t = x^2 \qquad \text{so that} \qquad dt = 2x \, dx$$

and

$$\int x^3 \cos x^2 \, dx = \tfrac{1}{2} \int x^2 \cos x^2 \, (2x \, dx) = \tfrac{1}{2} \int t \cos t \, dt$$

To evaluate the last integral, we use integration by parts with

$$u = t \qquad dv = \cos t \, dt$$

$$du = dt \qquad v = \int \cos t \, dt = \sin t$$

Thus,

$$\int x^3 \cos x^2 \, dx = \tfrac{1}{2} \int \underbrace{t}_{u} \underbrace{\cos t \, dt}_{dv} = \tfrac{1}{2}\left(\int u \, dv \right) = \tfrac{1}{2}\left(uv - \int v \, du \right)$$

$$= \tfrac{1}{2}\left(t \sin t - \int \sin t \, dt \right) = \tfrac{1}{2}(t \sin t + \cos t) + C$$

$$= \tfrac{1}{2}(x^2 \sin x^2 + \cos x^2) + C \qquad \blacksquare$$

In the formula for integration by parts, suppose that u and v are functions of (say) the variable x. Then uv, as well as the indefinite integrals $\int u \, dv$ and $\int v \, du$, depend on x. Consider the definite integral $\int_{x=a}^{x=b} u \, dv$, where we have written the limits of integration as $x = a$ to $x = b$ to emphasize that x (rather than u or v) is the variable in question. We have

$$\int_{x=a}^{x=b} u \, dv = \left(uv - \int v \, du \right) \Bigg|_{x=a}^{x=b} = (uv) \Bigg|_{x=a}^{x=b} - \int_{x=a}^{x=b} v \, du$$

the formula for **integration by parts of definite integrals.**

EXAMPLE 5   Use integration by parts to find $\displaystyle\int_1^e x^2 \ln x \, dx$.

SOLUTION   We put

$$u = \ln x \qquad dv = x^2 \, dx$$

$$du = \frac{1}{x} \, dx \qquad v = \int x^2 \, dx = \frac{x^3}{3}$$

Thus,

$$\int_1^e x^2 \ln x \, dx = \int_{x=1}^{x=e} u \, dv = (uv) \Bigg|_{x=1}^{x=e} - \int_{x=1}^{x=e} v \, du$$

$$= (\ln x)\left(\frac{x^3}{3} \right) \Bigg|_{x=1}^{x=e} - \int_{x=1}^{x=e} \frac{x^3}{3} \frac{dx}{x} = \frac{x^3 \ln x}{3} \Bigg|_{x=1}^{x=e} - \frac{x^3}{9} \Bigg|_{x=1}^{x=e}$$

$$= \frac{e^3}{3} - \frac{e^3}{9} + \frac{1}{9} = \frac{2e^3 + 1}{9} \qquad \blacksquare$$

Repeated Integration by Parts

The formula $\int u\,dv = uv - \int v\,du$ for integration by parts converts the problem of evaluating $\int u\,dv$ to the problem of evaluating $\int v\,du$. By suitable choice of u and dv, we can often arrange it so that $\int v\,du$ is simpler than $\int u\,dv$; however, we may not be able to evaluate $\int v\,du$ directly. A second integration by parts might be necessary to evaluate $\int v\,du$. In fact, several successive integrations by parts could be required.

EXAMPLE 6 Evaluate $\int x^2 e^{2x}\,dx$.

SOLUTION   Put $u = x^2$ and $dv = e^{2x}\,dx$, so $du = 2x\,dx$ and $v = \int e^{2x}\,dx = \frac{1}{2}e^{2x}$. Thus,

$$\int x^2 e^{2x}\,dx = uv - \int v\,du = \frac{1}{2}x^2 e^{2x} - \int x e^{2x}\,dx$$

Although $\int x e^{2x}\,dx$ is simpler than the original $\int x^2 e^{2x}\,dx$, a second integration by parts is required to evaluate it. Thus, put $u_1 = x$ and $dv_1 = e^{2x}\,dx$, so that $du_1 = dx$ and $v_1 = \int e^{2x}\,dx = \frac{1}{2}e^{2x}$. Now,

$$\int x e^{2x}\,dx = u_1 v_1 - \int v_1\,du_1 = \frac{1}{2}x e^{2x} - \int \frac{1}{2}e^{2x}\,dx = \frac{1}{2}x e^{2x} - \frac{1}{4}e^{2x} + C_1$$

Hence,

$$\begin{aligned}
\int x^2 e^{2x}\,dx &= \frac{1}{2}x^2 e^{2x} - \int x e^{2x}\,dx \\
&= \frac{1}{2}x^2 e^{2x} - (\frac{1}{2}x e^{2x} - \frac{1}{4}e^{2x} + C_1) \\
&= \frac{1}{2}x^2 e^{2x} - \frac{1}{2}x e^{2x} + \frac{1}{4}e^{2x} + C
\end{aligned}$$

where $C = -C_1$.

Repeated integration by parts of integrands of the form

$$e^{ax}\sin bx \qquad e^{ax}\cos bx \qquad \sec^n x \qquad \text{or} \qquad \csc^n x$$

where a and b are constants and n is an odd positive integer, leads back to an integrand similar to the original one. When this happens, the resulting equation can often be solved for the unknown integral. This is illustrated by the following examples.

EXAMPLE 7 Evaluate $\int e^x \cos x\,dx$.

SOLUTION   An initial integration by parts with $u = e^x$, $dv = \cos x\,dx$, $du = e^x\,dx$, and $v = \sin x$ gives

$$\int e^x \cos x\,dx = e^x \sin x - \int e^x \sin x\,dx + C_1 \tag{1}$$

A second integration by parts with $u_1 = e^x$, $dv_1 = \sin x\,dx$, $du_1 = e^x\,dx$, and $v_1 = -\cos x$ applied to $\int e^x \sin x\,dx$ gives

$$\int e^x \sin x\,dx = -e^x \cos x + \int e^x \cos x\,dx + C_2 \tag{2}$$

Substitution of (2) into (1) yields

$$\int e^x \cos x \, dx = e^x \sin x - \left(-e^x \cos x + \int e^x \cos x \, dx + C_2\right) + C_1 \qquad (3)$$

or $\qquad \int e^x \cos x \, dx = e^x \sin x + e^x \cos x - \int e^x \cos x \, dx + C_3 \qquad (4)$

where we have put $C_3 = -C_2 + C_1$. Solving (4) for $\int e^x \cos x \, dx$, we obtain

$$2 \int e^x \cos x \, dx = e^x \sin x + e^x \cos x + C_3$$

so that $\qquad \int e^x \cos x \, dx = \dfrac{e^x \sin x + e^x \cos x}{2} + C$

where we have put $C = C_3/2$. ∎

<u>EXAMPLE 8</u>   Evaluate $\displaystyle\int \sec^3 x \, dx$.

SOLUTION   We already noticed in Section 8.2 that the methods used there are not effective in evaluating the integral of $\sec^3 x$; however, we now use integration by parts to find such an integral. Factor $\sec^3 x$ as $\sec x \sec^2 x$ and put

$$u = \sec x \qquad\qquad dv = \sec^2 x \, dx$$
$$du = \sec x \tan x \, dx \qquad v = \int \sec^2 x \, dx = \tan x$$

Thus,

$$\int \sec^3 x \, dx = \int \sec x \sec^2 x \, dx = \sec x \tan x - \int \sec x \tan^2 x \, dx + C_0$$

Since

$$\int \sec x \tan^2 x \, dx = \int \sec x (\sec^2 x - 1) \, dx = \int \sec^3 x \, dx - \int \sec x \, dx$$

it follows that

$$\int \sec^3 x \, dx = \sec x \tan x - \left(\int \sec^3 x \, dx - \int \sec x \, dx\right) + C_0$$
$$= \sec x \tan x - \int \sec^3 x \, dx + \int \sec x \, dx + C_0$$
$$= \sec x \tan x - \int \sec^3 x \, dx + \ln |\sec x + \tan x| + C_0$$

From the last equation,

$$2 \int \sec^3 x \, dx = \sec x \tan x + \ln |\sec x + \tan x| + C_0$$

or

$$\boxed{\int \sec^3 x \, dx = \frac{1}{2} \sec x \tan x + \frac{1}{2} \ln |\sec x + \tan x| + C}$$

where we have put $C = \frac{1}{2}C_0$. ∎

The Tabular Method

Repeated integration by parts can be accomplished by a shortcut called the tabular method. To see how this method works, suppose that u and v are two functions, and consider the table

$$
\begin{array}{lll}
& u & v \\
\text{derivative of } u \rightarrow u_1 & v_1 & \leftarrow \text{antiderivative of } v \\
\text{derivative of } u_1 \rightarrow u_2 & v_2 & \leftarrow \text{antiderivative of } v_1 \\
\quad\vdots \qquad\qquad \vdots & \vdots & \qquad \vdots \\
\text{derivative of } u_{n-1} \rightarrow u_n & v_n & \leftarrow \text{antiderivative of } v_{n-1} \\
\text{derivative of } u_n \rightarrow u_{n+1} & v_{n+1} & \leftarrow \text{antiderivative of } v_n
\end{array}
$$

Multiply each function—except the last—in the first column by the corresponding function in the second column, change the sign of every other product, and add the resulting terms to obtain a sum S:

$$
\begin{array}{lcl}
u \xrightarrow{\text{times}} v & \xrightarrow{(+)} & +uv \\
u_1 \xrightarrow{\text{times}} v_1 & \xrightarrow{(-)} & -u_1 v_1 \\
u_2 \xrightarrow{\text{times}} v_2 & \xrightarrow{(+)} & +u_2 v_2 \\
\vdots \qquad\quad \vdots & & \vdots \\
u_n \xrightarrow{\text{times}} v_n & \xrightarrow{(\mp)} & \mp u_n v_n \\
u_{n+1} \qquad v_{n+1} & & S \quad \leftarrow \text{sum}
\end{array}
$$

Then, as can be proved by using the principle of mathematical induction (Problem 62),

$$
\int u \, dv = S \pm \int u_{n+1} \, dv_{n+1}
$$

where the plus sign is used if n is odd and the minus sign if n is even.

If u is a polynomial of degree n, then the $(n+1)$st derivative of u is zero; hence, $u_{n+1} = 0$, and the tabular method yields the simple formula

$$
\int u \, dv = S + C
$$

where C is the constant of integration.

<u>EXAMPLE 9</u>   Use the tabular method to evaluate $\int (2x^4 - 8x^3)e^{-3x} \, dx$.

SOLUTION   Let

$$
u = 2x^4 - 8x^3 \qquad \text{and} \qquad dv = e^{-3x} \, dx
$$

so that

$$
\int (2x^4 - 8x^3)e^{-3x} \, dx = \int u \, dv
$$

Here

$$
v = \int dv = \int e^{-3x} \, dx = -\tfrac{1}{3}e^{-3x}
$$

(Again, notice that when you are using the tabular method, constants of integration can be neglected until the very end.) Thus, our table takes the form

$$2x^4 - 8x^3 \xrightarrow{\text{times}} -\tfrac{1}{3}e^{-3x} \xrightarrow{(+)} +(2x^4 - 8x^3)(-\tfrac{1}{3}e^{-3x})$$
$$8x^3 - 24x^2 \xrightarrow{\text{times}} \tfrac{1}{9}e^{-3x} \xrightarrow{(-)} -(8x^3 - 24x^2)(\tfrac{1}{9}e^{-3x})$$
$$24x^2 - 48x \xrightarrow{\text{times}} -\tfrac{1}{27}e^{-3x} \xrightarrow{(+)} +(24x^2 - 48x)(-\tfrac{1}{27}e^{-3x})$$
$$48x - 48 \xrightarrow{\text{times}} \tfrac{1}{81}e^{-3x} \xrightarrow{(-)} -(48x - 48)(\tfrac{1}{81}e^{-3x})$$
$$48 \xrightarrow{\text{times}} -\tfrac{1}{243}e^{-3x} \xrightarrow{(+)} +48(-\tfrac{1}{243}e^{-3x})$$
$$0 \qquad -\tfrac{1}{729}e^{-3x} \qquad\qquad S$$

Adding the terms in the last column to obtain S and simplifying, we have

$$\int (2x^4 - 8x^3)e^{-3x}\, dx = S + C$$
$$= (-\tfrac{2}{3}x^4 + \tfrac{16}{9}x^3 + \tfrac{16}{9}x^2 + \tfrac{32}{27}x + \tfrac{32}{81})e^{-3x} + C \qquad \blacksquare$$

Problem Set 8.4

In Problems 1 to 12, use integration by parts to evaluate each integral.

1 $\int x \cos 2x\, dx$

2 $\int x \sin kx\, dx$

3 $\int xe^{3x}\, dx$

4 $\int xe^{-4x}\, dx$

5 $\int \ln 5x\, dx$

6 $\int x \ln 2x\, dx$

7 $\int \cos^{-1} x\, dx$

8 $\int x^3 \ln (x^2)\, dx$

9 $\int \sec^{-1} x\, dx$

10 $\int \sin^{-1} 3x\, dx$

11 $\int t \sec t \tan t\, dt$

12 $\int \tan^{-1} x\, dx$

In Problems 13 to 22, use repeated integration by parts to evaluate each integral.

13 $\int x^2 \sin 3x\, dx$

14 $\int x^2 \sin^2 x\, dx$

15 $\int (3x^2 - 2x + 1) \cos x\, dx$

16 $\int (x^2 - 3x + 2)e^{-x}\, dx$

17 $\int \left(\dfrac{x^2}{2} + x\right)e^{2x}\, dx$

18 $\int x^2 \sec^2 x \tan x\, dx$

19 $\int e^{-x} \cos 2x\, dx$

20 $\int e^{2x} \sin x\, dx$

21 $\int \csc^3 x\, dx$

22 $\int e^{ax} \sin bx\, dx$

In Problems 23 to 26, use a suitable substitution to express the integral in a form so that integration by parts is applicable. Then evaluate the integral.

23 $\int x^3 e^{x^2}\, dx$

24 $\int x^3 \sin 2x^2\, dx$

25 $\int \sqrt{1 + x^2}\, dx$

26 $\int \cos x \tan^{-1} (\sin x)\, dx$

In Problems 27 to 46, evaluate each integral.

27 $\int (2x - 1)e^{-x}\, dx$

28 $\int x \sinh x\, dx$

29 $\int xe^x \sin x\, dx$

30 $\int \ln (1 + t^2)\, dt$

31 $\int (\ln x)^2\, dx$

32 $\int \sin \sqrt{x}\, dx$

33 $\int x \csc^2 x\, dx$

34 $\int \cosh^{-1} x\, dx$

35 $\int \dfrac{x^3\, dx}{\sqrt{1 - x^2}}$

36 $\int \dfrac{x^2}{\sqrt{x^2 - 1}}$

37 $\int x^{11} \cos x^4\, dx$

38 $\int x^{3/2} \cos \sqrt{x}\, dx$

39 $\int_0^{\pi/9} 4x^2 \sin 3x\, dx$

40 $\int_0^1 \dfrac{xe^x}{(1 + x)^2}\, dx$

41 $\int_2^3 \sec^{-1} x\, dx$

42 $\int_{-1}^1 \cos^{-1} x\, dx$

43 $\int_0^{\pi/4} (5x^2 - 3x + 1) \sin x\, dx$

44 $\int_1^e \sin (\ln x)\, dx$

45 $\displaystyle\int_0^{\pi/2} x \sin^2 x \, dx$ **46** $\displaystyle\int_0^3 x \tan^{-1} \sqrt{x} \, dx$

In Problems 47 to 50, use the tabular method to evaluate each integral.

47 $\displaystyle\int x^4 \cos 2x \, dx$

48 $\displaystyle\int (x^3 - 2x^2 + x)e^x \, dx$

49 $\displaystyle\int t^4 e^{-t} \, dt$

50 $\displaystyle\int (x^5 - x^3 + x)e^{-x} \, dx$

51 Use integration by parts to show that

$$\int_0^a x^2 f'''(x) \, dx = a^2 f''(a) - 2af'(a) + 2f(a) - 2f(0)$$

52 It is desired to evaluate $\int F_1(x)F_2(x) \, dx$ by parts. To this end, we put $u = F_1(x)$ and $dv = F_2(x) \, dx$. Then $du = F_1'(x) \, dx$ and $v = G(x) + C_0$, where G is an antiderivative of F_2 and C_0 is the constant of integration. Show that C_0 always cancels in the calculation of $\int F_1(x)F_2(x) \, dx$ by parts.

53 Use integration by parts to prove that

$$\int f(x) \, dx = xf(x) - \int xf'(x) \, dx$$

54 Show that the volume of the solid generated by revolving the region under the graph of $y = \sin x$ between $x = 0$ and $x = \pi$ about the y axis is four times the volume of the solid generated when the same region is revolved about the x axis. (*Hint:* When revolving about the y axis, use cylindrical shells.)

55 Derive a formula for $\int \sec^5 x \, dx$.

56 If a is a positive constant, derive a formula for $\int \sqrt{a^2 + x^2} \, dx$.

57 Suppose that $g'' = f$. Evaluate $\int x^2 f(x) \, dx$.

58 Evaluate $\int \sqrt{2 - 3x + x^2} \, dx$. (*Hint:* Complete the square, change the variable, and then use the result of Problem 56.)

59 Let $f(x) = xe^{-x}$ for $x \geq 0$. Suppose that $f(x)$ takes on its maximum value when $x = a$. Find the area of the region under the graph of f between $x = 0$ and $x = a$.

60 Find the volume of the solid generated if the region described in Problem 59 is rotated about the x axis.

61 If a is a constant, derive a formula for $\int x^a \ln x \, dx$.

62 Using the principle of mathematical induction, justify the tabular method of integration.

8.5   Integration by Partial Fractions—Linear Case

Recall that a *rational function h* has the form

$$h(x) = \frac{P(x)}{Q(x)}$$

where $P(x)$ and $Q(x)$ are polynomials and $Q(x)$ is not the zero polynomial. As we shall see in this section and the next, it is possible to find the integral of a rational function by expanding it into a sum of simple rational functions called **partial fractions** and then finding the integrals of these partial fractions.

A rational function is said to be *proper* if the degree of its denominator is greater than the degree of its numerator. If a rational function is not proper, you can perform a long division

$$\text{denominator} \,\overline{)\,\text{numerator}}^{\text{quotient}} \quad \text{with remainder}$$

and write

$$\frac{\text{numerator}}{\text{denominator}} = \text{quotient} + \frac{\text{remainder}}{\text{denominator}}$$

where the rational function

$$\frac{\text{remainder}}{\text{denominator}}$$

is proper. For instance, the rational function

$$\frac{6x^4 + x^3 + 4x + 4}{2x^2 + x - 1}$$

is not proper; however, by long division we find that

$$\begin{array}{r} 3x^2 - x + 2 \\ 2x^2 + x - 1 \overline{) 6x^4 + x^3 + 4x + 4} \end{array} \qquad \text{with remainder } x + 6$$

so that $\qquad \dfrac{6x^4 + x^3 + 4x + 4}{2x^2 + x - 1} = 3x^2 - x + 2 + \dfrac{x + 6}{2x^2 + x - 1}$

Thus, by means of long division, *a rational function that is not proper can always be rewritten as the sum of a polynomial function and a proper rational function.*

Using the procedure explained above, you can rewrite the integral of a rational function that isn't proper as the sum of the integral of a polynomial function and the integral of a proper rational function. Since there's no difficulty in integrating the polynomial function, the problem is thereby reduced to integrating the proper rational function. Thus, in what follows, we can concentrate on the problem of integrating *proper* rational functions. Of course, we assume that all common factors in numerators and denominators have been canceled, so that the proper rational functions are in *reduced form.*

The Case in Which the Denominator Factors into Distinct Linear Factors

Notice that the two fractions $\dfrac{2}{x - 2}$ and $\dfrac{3}{x + 1}$ can be added to obtain

$$\frac{2}{x - 2} + \frac{3}{x + 1} = \frac{2(x + 1) + 3(x - 2)}{(x - 2)(x + 1)} = \frac{5x - 4}{(x - 2)(x + 1)}$$

It follows that

$$\int \frac{(5x - 4)\,dx}{(x - 2)(x + 1)} = \int \frac{2\,dx}{x - 2} + \int \frac{3\,dx}{x + 1} = 2 \ln |x - 2| + 3 \ln |x + 1| + C$$

Therefore, it is a simple matter to integrate $\dfrac{5x - 4}{(x - 2)(x + 1)}$ *provided that we know it decomposes into the sum of the partial fractions*

$$\frac{2}{x - 2} \qquad \text{and} \qquad \frac{3}{x + 1}$$

Decomposition of a rational function into partial fractions is easiest *when the denominator factors into distinct* (that is, different) *linear factors.* In this case, it is only necessary to *provide a* **partial fraction** *of the form*

$$\frac{\text{constant}}{ax + b}$$

for each of the linear factors ax + b in the denominator. The constant numerators of the partial fractions can be denoted by A, B, C, and so forth.

For instance, the fraction $\dfrac{5x - 4}{(x - 2)(x + 1)}$ is decomposed as follows:

$$\frac{5x - 4}{(x - 2)(x + 1)} = \frac{A}{x - 2} + \frac{B}{x + 1}$$

Another example is provided by

$$\frac{27x^3 - 4x + 13}{(3x - 1)(x + 2)[(x/2) - 7]x} = \frac{C}{3x - 1} + \frac{D}{x + 2} + \frac{E}{(x/2) - 7} + \frac{F}{x}$$

In the case being considered, there are two methods for finding the values of the numerators of the partial fractions.

1 Equating Coefficients

We have

$$\frac{P(x)}{Q(x)} = \text{a sum of partial fractions}$$

where the numerators of the partial fractions are denoted by A, B, C, and so forth. Multiply both sides of this equation by the denominator $Q(x)$ to obtain

$$P(x) = Q(x) \text{ (sum of partial fractions)}$$

and then collect like terms on the right side. Equate the coefficients of powers of x on the left side to the coefficients of corresponding powers of x on the right side, and thus obtain a system of equations involving the unknowns A, B, C, and so forth. Solve this system for the unknowns.

EXAMPLE 1 Find the partial-fraction decomposition of $\dfrac{5x - 4}{(x - 2)(x + 1)}$ by equating coefficients.

SOLUTION   We have

$$\frac{5x - 4}{(x - 2)(x + 1)} = \frac{A}{x - 2} + \frac{B}{x + 1}$$

where the constants A and B must be determined. Multiplying both sides of this equation by the denominator $(x - 2)(x + 1)$ yields

$$5x - 4 = A(x + 1) + B(x - 2) \quad \text{or} \quad 5x - 4 = (A + B)x + (A - 2B)$$

Equating coefficients of like powers of x on both sides of the last equation gives

$$\begin{cases} 5 = A + B \\ -4 = A - 2B \end{cases}$$

Solving (say, by elimination) the system of linear equations above, we find that $A = 2$ and $B = 3$. Therefore,

$$\frac{5x - 4}{(x - 2)(x + 1)} = \frac{2}{x - 2} + \frac{3}{x + 1}$$

2   Substitution

Suppose that

$$\frac{A}{ax + b}$$

is any one of the partial fractions in the decomposition of $P(x)/Q(x)$. Then $ax + b$ is one of the factors of $Q(x)$, so we can write

$$Q(x) = (ax + b)Q_1(x)$$

where $Q_1(x)$ is the remaining part of the denominator. Thus,

$$\frac{P(x)}{(ax + b)Q_1(x)} = \frac{A}{ax + b} + \text{other partial fractions}$$

Multiply both sides of this equation by $ax + b$ to obtain

$$\frac{P(x)}{Q_1(x)} = A + (ax + b)(\text{other partial fractions})$$

If we now put $x = -b/a$, then $ax + b = 0$ and the equation becomes

$$\frac{P(-b/a)}{Q_1(-b/a)} = A$$

Thus, we obtain the numerical value of A. Repeat this procedure for each of the partial fractions.

<u>EXAMPLE 2</u>   Find the partial-fraction decomposition of $\dfrac{5x - 4}{(x - 2)(x + 1)}$ by substitution.

SOLUTION   Again

$$\frac{5x - 4}{(x - 2)(x + 1)} = \frac{A}{x - 2} + \frac{B}{x + 1}$$

where the constants A and B must be determined. Multiply both sides of this equation by $x - 2$, and then substitute $x = 2$ to obtain

$$\frac{5x - 4}{x + 1} = A + (x - 2)\frac{B}{x + 1}$$

$$\frac{10 - 4}{2 + 1} = A + (0)\frac{B}{2 + 1}$$

so that

$$\frac{6}{3} = A \qquad \text{or} \qquad A = 2$$

Similarly, to find B, multiply both sides of the original equation by $x + 1$, and then substitute $x = -1$ to obtain

$$\frac{-5 - 4}{-1 - 2} = B \qquad \text{or} \qquad B = 3$$

Therefore,
$$\frac{5x - 4}{(x - 2)(x + 1)} = \frac{2}{x - 2} + \frac{3}{x + 1}$$ ∎

The method of substitution can be shortened simply by *temporarily "covering up," or "disregarding," appropriate portions of the equation and then substituting*. For instance, to find the constant A in

$$\frac{5x - 4}{(x - 2)(x + 1)} = \frac{A}{x - 2} + \frac{B}{x + 1}$$

we cover up, or disregard, everything on the right side of the equation except A, and we cover up, or disregard, the factor in the denominator on the left side that corresponds to A. This produces the equation

$$\frac{5x - 4}{(x - 2)(x + 1)} = \frac{A}{x - 2} + \frac{B}{x + 1}$$

into which we substitute the value $x = 2$ to obtain

$$\frac{10 - 4}{2 + 1} = A \qquad \text{or} \qquad A = 2$$

Similarly, to find B, we substitute $x = -1$ into

$$\frac{5x - 4}{(x - 2)(x + 1)} = \frac{A}{x - 2} + \frac{B}{x + 1}$$

to obtain

$$\frac{-5 - 4}{-1 - 2} = B \qquad \text{or} \qquad B = 3$$

This method is called the **short method of substitution.**

In Examples 3 and 4, evaluate the given integral.

EXAMPLE 3 $\displaystyle\int \frac{3x - 5}{x^2 - x - 2}\, dx$

SOLUTION $x^2 - x - 2 = (x - 2)(x + 1)$

Hence, $\displaystyle\frac{3x - 5}{x^2 - x - 2} = \frac{3x - 5}{(x - 2)(x + 1)} = \frac{A}{x - 2} + \frac{B}{x + 1}$

By the short method of substitution

$$\frac{6 - 5}{2 + 1} = A \qquad \text{so that} \qquad A = \frac{1}{3}$$

and

$$\frac{-3 - 5}{-1 - 2} = B \qquad \text{so that} \qquad B = \frac{8}{3}$$

Therefore,

$$\int \frac{3x - 5}{x^2 - x - 2}\, dx = \int \frac{\frac{1}{3}}{x - 2}\, dx + \int \frac{\frac{8}{3}}{x + 1}\, dx$$

$$= \tfrac{1}{3} \ln |x - 2| + \tfrac{8}{3} \ln |x + 1| + C$$

$$= \ln \left[|x - 2|^{1/3} |x + 1|^{8/3} \right] + C$$

EXAMPLE 4 $\displaystyle\int \frac{15x^4 - 17x^3 + 42x^2 - 17x - 8}{15x^3 - 17x^2 - 4x}\, dx$

SOLUTION   The integrand is an improper fraction. By long division,

$$
\begin{array}{r}
x \\
15x^3 - 17x^2 - 4x \overline{\smash{\big)}\; 15x^4 - 17x^3 + 42x^2 - 17x - 8} \\
\underline{15x^4 - 17x^3 - 4x^2 } \\
46x^2 - 17x - 8
\end{array}
$$

so that

$$\frac{15x^4 - 17x^3 + 42x^2 - 17x - 8}{15x^3 - 17x^2 - 4x} = x + \frac{46x^2 - 17x - 8}{15x^3 - 17x^2 - 4x}$$

Factoring the denominator, we have

$$15x^3 - 17x^2 - 4x = x(15x^2 - 17x - 4) = x(5x + 1)(3x - 4)$$

Hence,

$$\frac{46x^2 - 17x - 8}{x(5x + 1)(3x - 4)} = \frac{A}{x} + \frac{B}{5x + 1} + \frac{C}{3x - 4}$$

Using (say) the short method of substitution, we find that

$$A = 2 \qquad B = -3 \qquad \text{and} \qquad C = 5$$

Therefore,

$$\int \frac{15x^4 - 17x^3 + 42x^2 - 17x - 8}{15x^3 - 17x^2 - 4x} \, dx$$

$$= \int \left(x + \frac{2}{x} - \frac{3}{5x + 1} + \frac{5}{3x - 4} \right) dx$$

$$= \int x \, dx + 2 \int \frac{dx}{x} - 3 \int \frac{dx}{5x + 1} + 5 \int \frac{dx}{3x - 4}$$

$$= \frac{x^2}{2} + 2 \ln |x| - \frac{3}{5} \ln |5x + 1| + \frac{5}{3} \ln |3x - 4| + K$$

where we have written the constant of integration as K to avoid confusing it with the numerator of the partial fraction $C/(3x - 4)$. Using the properties of logarithms, we can rewrite the solution as

$$\int \frac{15x^4 - 17x^3 + 42x^2 - 17x - 8}{15x^3 - 17x^2 - 4x} \, dx = \frac{x^2}{2} + \ln \frac{x^2 |3x - 4|^{5/3}}{|5x + 1|^{3/5}} + K$$

■

The Case in Which the Denominator Has Repeated Linear Factors

We now consider the case of a rational function $P(x)/Q(x)$ whose denominator $Q(x)$ factors completely into linear factors, *not all of which are distinct*. An example would be

$$Q(x) = (3x - 2)(x - 1)(3x - 2)(x - 1)(x + 1)(x - 1)$$

in which the factor $(3x - 2)$ appears twice and the factor $(x - 1)$ appears three times. The first step in handling such a case is to gather like factors and write them with the aid of exponents. In the example at hand, we have

$$Q(x) = (x - 1)^3 (3x - 2)^2 (x + 1)$$

If the denominator contains a factor of the form $(ax + b)^k$ *where* $k > 1$, *it is necessary to provide k corresponding* **partial fractions** *of the form*

$$\frac{A_1}{ax + b} + \frac{A_2}{(ax + b)^2} + \frac{A_3}{(ax + b)^3} + \cdots + \frac{A_k}{(ax + b)^k}$$

where $A_1, A_2, A_3, \ldots, A_k$ *are constants that need to be determined.* This must be done for *each* factor in the denominator. Nonrepeated factors are handled just as before.

For instance, the partial fraction decomposition for

$$\frac{3x^2 + 4x + 2}{x(x + 1)^2}$$

has the form

$$\frac{3x^2 + 4x + 2}{x(x + 1)^2} = \frac{A}{x} + \frac{B_1}{x + 1} + \frac{B_2}{(x + 1)^2}$$

Although the short method of substitution can no longer be used to determine *all* the unknown constants, it can still be used to find constants, such as A, that correspond to nonrepeated factors. Indeed,

$$\frac{(3)(0)^2 + (4)(0) + 2}{(0 + 1)^2} = A$$

so $A = 2$.

The short method of substitution is also effective for determining the constant numerators of the partial fractions involving the *highest powers* of the repeated factors. (Why?) Here, for instance,

$$\frac{(3)(-1)^2 + (4)(-1) + 2}{(-1)} = B_2$$

so $B_2 = -1$. Unfortunately, the short method of substitution does not work for the constant numerators of the partial fractions involving remaining (lower) powers of the repeated factors, such as B_1 in the present example. These remaining constants can be found by the method of equating coefficients.

For instance, we now have

$$\frac{3x^2 + 4x + 2}{x(x + 1)^2} = \frac{2}{x} + \frac{B_1}{x + 1} - \frac{1}{(x + 1)^2}$$

In order to find B_1, we multiply both sides of the equation by $x(x + 1)^2$ to obtain

$$3x^2 + 4x + 2 = 2(x + 1)^2 + B_1 x(x + 1) - x$$

that is, $\qquad 3x^2 + 4x + 2 = (2 + B_1)x^2 + (3 + B_1)x + 2$

Equating coefficients of like powers, we have $3 = 2 + B_1$ and $4 = 3 + B_1$. Therefore, $B_1 = 1$, and so

$$\frac{3x^2 + 4x + 2}{x(x + 1)^2} = \frac{2}{x} + \frac{1}{x + 1} - \frac{1}{(x + 1)^2}$$

Rational functions for which the denominator factors completely into linear factors can always be expressed in terms of partial fractions of the form

$$\frac{A}{(ax + b)^n}$$

Since

$$\int \frac{A\,dx}{(ax+b)^n}$$

can easily be evaluated using the change of variable $u = ax + b$, we see that the integral of such a rational function can always be evaluated. This is illustrated by the following examples.

In Examples 5 and 6, evaluate the given integral.

EXAMPLE 5 $\displaystyle \int \frac{3x^2 + 4x + 2}{x(x+1)^2}\,dx$

SOLUTION   Using the decomposition above, we have

$$\int \frac{3x^2 + 4x + 2}{x(x+1)^2}\,dx = \int \left[\frac{2}{x} + \frac{1}{x+1} - \frac{1}{(x+1)^2} \right] dx$$

$$= \int \frac{2\,dx}{x} + \int \frac{dx}{x+1} - \int \frac{dx}{(x+1)^2}$$

$$= 2\ln|x| + \ln|x+1| + \frac{1}{x+1} + C$$

$$= \ln|x|^2 + \ln|x+1| + \frac{1}{x+1} + C$$

$$= \ln|x^2(x+1)| + \frac{1}{x+1} + C \qquad \blacksquare$$

EXAMPLE 6 $\displaystyle \int_2^3 \frac{3x-2}{x^3 - x^2}\,dx$

SOLUTION $\displaystyle \frac{3x-2}{x^3 - x^2} = \frac{3x-2}{x^2(x-1)} = \frac{A_1}{x} + \frac{A_2}{x^2} + \frac{B}{x-1}$

By the short method of substitution,

$$\frac{3-2}{1^2} = B \qquad \text{so that} \qquad B = 1$$

and

$$\frac{0-2}{0-1} = A_2 \qquad \text{so that} \qquad A_2 = 2$$

Thus,

$$\frac{3x-2}{x^2(x-1)} = \frac{A_1}{x} + \frac{2}{x^2} + \frac{1}{x-1}$$

Using the method of equating coefficients, we have

$$3x - 2 = A_1 x(x-1) + 2(x-1) + x^2 = (A_1 + 1)x^2 + (-A_1 + 2)x - 2$$

Hence, $A_1 + 1 = 0$ and $-A_1 + 2 = 3$, so that $A_1 = -1$. Therefore,

$$\int_2^3 \frac{3x-2}{x^3 - x^2}\,dx = \int_2^3 \left(\frac{-1}{x} + \frac{2}{x^2} + \frac{1}{x-1} \right) dx$$

$$= \left(-\ln|x| - \frac{2}{x} + \ln|x-1| \right) \Big|_2^3 = \left(\ln \left| \frac{x-1}{x} \right| - \frac{2}{x} \right) \Big|_2^3$$

$$= (\ln \tfrac{2}{3} - \tfrac{2}{3}) - (\ln \tfrac{1}{2} - 1) = \ln \tfrac{4}{3} + \tfrac{1}{3} \qquad \blacksquare$$

Problem Set 8.5

In Problems 1 to 30, evaluate each integral. (*Caution:* Some integrands may require a preliminary long division.)

1 $\int \dfrac{x+1}{x(x-2)}\, dx$

2 $\int \dfrac{x+3}{x^2-x-2}\, dx$

3 $\int \dfrac{31y-9}{6y^2-y-2}\, dy$

4 $\int \dfrac{11t+17}{2t^2+7t-4}\, dt$

5 $\int \dfrac{4t^2-3t-4}{t^3-t^2-2t}\, dt$

6 $\int \dfrac{8x+7}{2x^2+3x+1}\, dx$

7 $\int \dfrac{2x+1}{x^3+x^2-2x}\, dx$

8 $\int \dfrac{3z+1}{z(z^2-4)}\, dz$

9 $\int \dfrac{dx}{x^3-x}$

10 $\int \dfrac{(t+7)\,dt}{(t+1)(t^2-4t+3)}$

11 $\int \dfrac{x^2\,dx}{x^2-x-6}$

12 $\int \dfrac{x^3+2x^2-3x+1}{x^3+3x^2+2x}\, dx$

13 $\int \dfrac{x^3+x^2-9x-3}{x^2+x-12}\, dx$

14 $\int \dfrac{x\,dx}{(x-1)(x+1)(x+2)}$

15 $\int \dfrac{x^3+5x^2-4x-20}{x^2+3x-10}\, dx$

16 $\int \dfrac{x^4+2x^3+1}{x^3-x^2-2x}\, dx$

17 $\int \dfrac{5x^2-7x+8}{x^3+3x^2-4x}\, dx$

18 $\int \dfrac{x^3+5x^2-x-22}{x^2+3x-10}\, dx$

19 $\int \dfrac{x^2\,dx}{x^2+x-6}$

20 $\int \dfrac{5x^3-6x^2-68x-16}{x^3-2x^2-8x}\, dx$

21 $\int \dfrac{y^3-4y-1}{y(y-1)^3}\, dy$

22 $\int \dfrac{2x\,dx}{(x+2)(x^2-1)}$

23 $\int \dfrac{2z+3}{z^2(4z+1)}\, dz$

24 $\int \dfrac{x^2+1}{(x+3)(x^2+4x+4)}\, dx$

25 $\int \dfrac{x+3}{(x+1)^2(x+7)}\, dx$

26 $\int \dfrac{x+4}{(x^2+2x+1)(x-1)^2}\, dx$

27 $\int \dfrac{4x^2-7x+10}{(x+2)(3x-2)^2}\, dx$

28 $\int \dfrac{x^3-3x^2+5x-12}{(x-1)^2(x^2-3x-4)}\, dx$

29 $\int \dfrac{4z^2\,dz}{(z-1)^2(z^2-4z+3)}$

30 $\int \dfrac{(t+2)\,dt}{(t^2-1)(t+3)^2}$

In Problems 31 to 36, evaluate each definite integral.

31 $\displaystyle\int_2^4 \dfrac{x\,dx}{(x+1)(x+2)}$

32 $\displaystyle\int_1^2 \dfrac{5t^2-3t+18}{t(9-t^2)}\, dt$

33 $\displaystyle\int_2^3 \dfrac{4t^5-3t^4-6t^3+4t^2+6t-1}{(t-1)(t^2-1)}\, dt$

34 $\displaystyle\int_1^2 \dfrac{x^5+3x^4-4x^3-x^2+11x+12}{x^2(x^2+4x+5)}\, dx$

35 $\displaystyle\int_3^5 \dfrac{x^2-2}{(x-2)^2}\, dx$

36 $\displaystyle\int_1^2 \dfrac{2z^3+1}{z(z+1)^2}\, dz$

37 Evaluate $\int [(ax+b)/(cx+d)]\, dx$ if $c \neq 0$. $\left(\textit{Hint:}\right.$
$$\dfrac{ax+b}{cx+d} = \dfrac{a}{c} + \dfrac{bc-ad}{c(cx+d)}.\Bigg)$$

38 (a) Evaluate $\displaystyle\int \dfrac{dx}{(x-a)(x-b)}$ if $a \neq b$.

(b) Evaluate $\displaystyle\int \dfrac{dx}{(x-a)(x-b)}$ if $a = b$.

(c) Is it true that $\displaystyle\lim_{a \to b} \int \dfrac{dx}{(x-a)(x-b)} = \int \dfrac{dx}{(x-b)^2}$?

39 Evaluate $\int [(x+1)/(x^2-x-6)]\, dx$ in two ways: (a) By completing the square in the denominator and substituting $u = x - \frac{1}{2}$. (b) By factoring the denominator and using partial fractions.

40 Evaluate $\int [(x+c)/(x-a)^2]\, dx$.

41 Find the area of the region in the first quadrant bounded by the curve $(x+2)^2 y = 4 - x$.

42 Find the volume of the solid generated by revolving the region in Problem 41 about the x axis.

43 In connection with the study of the *production of ions* by radiation, it is necessary to deal with the integral $\int dx/(q-ax^2)$, where q and a are positive constants. Evaluate this integral.

44 The equation $dy/dt = ak[1-(1+b)y][1-(1-b)y]$ applies to the *velocity of the reaction between ethyl alcohol and chloroacetic acid*. Here, a, k, and b represent positive constants. Solve this differential equation.

45 The differential equation $dx/dt = k(a-x)^4$ applies to the *velocity of the reaction between hydrobromic and bromic acids*. Here a and k are positive constants. Solve this differential equation.

46 Find the arc length of the curve $y = \ln(1-x^2)$ from $x = 0$ to $x = \frac{1}{3}$.

47 The marginal-cost function of a certain product is given by
$$C'(x) = \dfrac{400x^2+1300x-900}{x(x-1)(x+3)}$$
where x is the number of units produced. If it costs \$47 to produce 2 units, find a formula for C in terms of x. (Assume that $x \geq 2$.)

48 Explain why every rational function whose denominator factors completely into linear factors has an integral that can be expressed in terms of rational functions and logarithms (of absolute values). Assume—as is, in fact, the case—that such a rational function can always be decomposed into partial fractions.

49 As we mentioned on page 264, the differential equation $dy/dt = k(A - y)(B + y)$ applies to the *formation of trypsin* in the biochemistry of digestion. Assuming that A, B, and k are positive constants, solve this differential equation.

50 Ecologists sometimes use the following simplified **growth model** *for the population q of a sexually reproducing species:* The rate of change of q is the birthrate minus the death rate. The birthrate depends on the frequency of contacts between males and females; hence, it is proportional to q^2. The death rate is proportional to q. Hence, $dq/dt = Bq^2 - Aq$, where B and A are constants of proportionality. Solve this differential equation.

51 Sociologists sometimes use the following simplified model for the *spread of a rumor* in a population: The rate at which the rumor spreads is proportional to the number of contacts between those who have heard it and those who have not. Thus, if p is the proportion of the population who have heard the rumor at time t, so that $1 - p$ is the proportion who have not yet heard it, then $dp/dt = kp(1 - p)$, where k is the constant of proportionality. Solve this differential equation.

8.6   Integration by Partial Fractions— Quadratic Case

We now consider the problem of integrating a rational function $P(x)/Q(x)$ whose denominator $Q(x)$ cannot be factored completely into linear factors. It can be proved that any polynomial $Q(x)$ with real numbers as coefficients can be factored completely into a finite number of polynomials, each of which is either linear or quadratic. Furthermore, the quadratic factors (if any) can be assumed to be **irreducible**—that is, incapable of further factorization into linear polynomials. For instance,

$$x^5 + x^4 - x^3 - 3x^2 + 2 = (x - 1)(x^2 + 2x + 2)(x + 1)(x - 1)$$

as can be confirmed by direct multiplication of the factors on the right. The quadratic polynomial $x^2 + 2x + 2$ cannot be factored into linear factors with real coefficients; hence, it is irreducible.

It is easy to check whether a quadratic polynomial

$$ax^2 + bx + c$$

is irreducible—*it is irreducible if and only if its* **discriminant**

$$b^2 - 4ac$$

is negative. For instance, $x^2 + 2x + 2$ is irreducible because its discriminant $2^2 - (4)(1)(2) = -4$ is negative.

As in Section 8.5, we can assume without loss of generality that $P(x)/Q(x)$ is a proper fraction and that all common factors have been canceled from its numerator and denominator.

The Case in Which the Denominator Involves Distinct Irreducible Quadratic Factors

Assume that the denominator $Q(x)$ of $P(x)/Q(x)$ factors into linear and irreducible quadratic polynomials, but that *none of the quadratic factors are repeated*. Again, we seek to decompose $P(x)/Q(x)$ into suitable partial fractions. This is accomplished as follows:

For each nonrepeated irreducible quadratic factor $ax^2 + bx + c$ in the denominator of $P(x)/Q(x)$, provide a corresponding **partial fraction** *of the form*

$$\frac{Ax + B}{ax^2 + bx + c}$$

Again, it is necessary to determine the numerical values of A and B, say by equating coefficients. Partial fractions corresponding to linear factors of the denominator must be introduced, just as in Section 8.5.

In Examples 1 and 2, decompose the given rational fraction into partial fractions.

EXAMPLE 1 $\dfrac{8x^2 + 3x + 20}{x^3 + x^2 + 4x + 4}$

SOLUTION   Factoring the denominator, we obtain

$$x^3 + x^2 + 4x + 4 = x^2(x + 1) + 4(x + 1) = (x + 1)(x^2 + 4)$$

The quadratic factor $x^2 + 4$ is irreducible; hence we must provide a corresponding partial fraction of the form

$$\frac{Ax + B}{x^2 + 4}$$

We must also introduce a partial fraction corresponding to the linear factor $x + 1$. Consequently,

$$\frac{8x^2 + 3x + 20}{x^3 + x^2 + 4x + 4} = \frac{8x^2 + 3x + 20}{(x + 1)(x^2 + 4)} = \frac{Ax + B}{x^2 + 4} + \frac{C}{x + 1}$$

We can evaluate the constant C by the short method of substitution, just as in Section 8.5; thus,

$$\frac{(8)(-1)^2 + (3)(-1) + 20}{(-1)^2 + 4} = C \qquad \text{so} \qquad C = 5$$

Therefore, $\qquad \dfrac{8x^2 + 3x + 20}{(x + 1)(x^2 + 4)} = \dfrac{Ax + B}{x^2 + 4} + \dfrac{5}{x + 1}$

Multiplying both sides of the last equation by $(x + 1)(x^2 + 4)$ and collecting terms on the right, we have

$$8x^2 + 3x + 20 = (Ax + B)(x + 1) + 5(x^2 + 4)$$

or $\qquad 8x^2 + 3x + 20 = (A + 5)x^2 + (A + B)x + (B + 20)$

Equating coefficients of like powers in the latter equation, we have

$$8 = A + 5 \qquad 3 = A + B \qquad \text{and} \qquad 20 = B + 20$$

Therefore, $B = 0$, $A = 3$, and

$$\frac{8x^2 + 3x + 20}{x^3 + x^2 + 4x + 4} = \frac{8x^2 + 3x + 20}{(x + 1)(x^2 + 4)} = \frac{3x}{x^2 + 4} + \frac{5}{x + 1} \qquad ∎$$

EXAMPLE 2 $\dfrac{3x^3 + 11x - 16}{(x^2 + 1)(x^2 + 4x + 13)}$

SOLUTION $$\frac{3x^3 + 11x - 16}{(x^2 + 1)(x^2 + 4x + 13)} = \frac{Ax + B}{x^2 + 1} + \frac{Cx + D}{x^2 + 4x + 13}$$

Multiplying by $(x^2 + 1)(x^2 + 4x + 13)$, we have

$$3x^3 + 11x - 16 = (Ax + B)(x^2 + 4x + 13) + (Cx + D)(x^2 + 1)$$

or $$3x^3 + 11x - 16 = (A + C)x^3 + (4A + B + D)x^2 \\ + (13A + 4B + C)x + (13B + D)$$

Equating the coefficients, we have

$$\begin{cases} 3 = A + C \\ 0 = 4A + B + D \\ 11 = 13A + 4B + C \\ -16 = 13B + D \end{cases}$$

Solving these simultaneous equations for A, B, C, and D (say, by elimination), we obtain $A = 1$, $B = -1$, $C = 2$, and $D = -3$. Therefore,

$$\frac{3x^3 + 11x - 16}{(x^2 + 1)(x^2 + 4x + 13)} = \frac{x - 1}{x^2 + 1} + \frac{2x - 3}{x^2 + 4x + 13}$$ ∎

As soon as a rational function has been decomposed into a sum of partial fractions, it can be integrated by integrating each of these partial fractions. We have already seen in Section 8.5 that there is no problem in integrating the partial fractions that correspond to the linear factors (repeated or not) in the denominator. The partial fractions corresponding to nonrepeated irreducible quadratic factors in the denominator have the form

$$\frac{Ax + B}{ax^2 + bx + c}$$

The integral

$$\int \frac{Ax + B}{ax^2 + bx + c}\, dx$$

can be handled by completing the square in the denominator (if necessary), as in Section 8.3.

In Examples 3 to 5, evaluate the given integral.

EXAMPLE 3 $$\int \frac{8x^2 + 3x + 20}{(x + 1)(x^2 + 4)}\, dx$$

SOLUTION   In Example 1 (page 517), we saw that the partial-fraction decomposition of the integrand is

$$\frac{8x^2 + 3x + 20}{(x + 1)(x^2 + 4)} = \frac{3x}{x^2 + 4} + \frac{5}{x + 1}$$

Hence, $$\int \frac{8x^2 + 3x + 20}{(x + 1)(x^2 + 4)}\, dx = 3 \int \frac{x\, dx}{x^2 + 4} + 5 \int \frac{dx}{x + 1}$$

$$= \tfrac{3}{2} \ln (x^2 + 4) + 5 \ln |x + 1| + C$$ ∎

EXAMPLE 4 $\displaystyle\int \frac{3x^3 + 11x - 16}{(x^2 + 1)(x^2 + 4x + 13)}\, dx$

SOLUTION   Using the partial-fraction decomposition of the integrand already found in Example 2 (page 517), we have

$$\int \frac{3x^3 + 11x - 16}{(x^2 + 1)(x^2 + 4x + 13)}\, dx$$

$$= \int \frac{x - 1}{x^2 + 1}\, dx + \int \frac{2x - 3}{x^2 + 4x + 13}\, dx$$

$$= \int \frac{x\, dx}{x^2 + 1} - \int \frac{dx}{x^2 + 1} + \int \frac{2x - 3}{x^2 + 4x + 13}\, dx$$

$$= \int \frac{x\, dx}{x^2 + 1} - \int \frac{dx}{x^2 + 1} + \int \frac{2x + 4}{x^2 + 4x + 13}\, dx - 7 \int \frac{dx}{(x + 2)^2 + 9}$$

$$= \frac{1}{2} \ln (x^2 + 1) - \tan^{-1} x + \ln (x^2 + 4x + 13) - \frac{7}{3} \tan^{-1} \left(\frac{x + 2}{3} \right) + C$$

where

$$\int \frac{2x - 3}{x^2 + 4x + 13}\, dx$$

has been found by adding and subtracting 7 in the numerator and splitting the integral into two integrals such that in the first integral the numerator is the derivative of the denominator. The second integral is evaluated by completing the square in the denominator and then substituting $u = x + 2$.   ■

EXAMPLE 5 $\displaystyle\int \frac{3x^3 + 2x - 2}{x^2(x^2 + 2)}\, dx$

SOLUTION   Here we have

$$\frac{3x^3 + 2x - 2}{x^2(x^2 + 2)} = \frac{A}{x} + \frac{B}{x^2} + \frac{Cx + D}{x^2 + 2}$$

By the short method of substitution, $B = -2/(0 + 2) = -1$. Putting $B = -1$ and clearing fractions, we have

$$3x^3 + 2x - 2 = Ax(x^2 + 2) - (x^2 + 2) + (Cx + D)x^2$$

$$= (A + C)x^3 + (D - 1)x^2 + 2Ax - 2$$

Equating coefficients, we obtain $3 = A + C$, $0 = D - 1$, and $2 = 2A$; hence, $A = 1$, $C = 2$, and $D = 1$. Consequently,

$$\int \frac{3x^3 + 2x - 2}{x^2(x^2 + 2)}\, dx = \int \frac{dx}{x} - \int \frac{dx}{x^2} + \int \frac{2x + 1}{x^2 + 2}\, dx$$

$$= \ln |x| + \frac{1}{x} + \int \frac{2x\, dx}{x^2 + 2} + \int \frac{dx}{x^2 + 2}$$

$$= \ln |x| + \frac{1}{x} + \ln (x^2 + 2) + \frac{\sqrt{2}}{2} \tan^{-1} \frac{\sqrt{2}}{2} x + C   ■$$

The Case in Which the Denominator Involves Repeated Irreducible Quadratic Factors

We now consider the case in which the denominator, after being completely factored, involves *repeated* irreducible quadratic factors. After such repeated quadratic factors as well as all linear factors are gathered with the aid of exponents, the denominator is a product of factors having the form $(ax + b)^k$ or $(ax^2 + bx + c)^k$, where $k \geq 1$. The factors of the form $(ax + b)^k$ are handled just as in Section 8.5. *The k* **partial fractions** *corresponding to* $(ax^2 + bx + c)^k$ *are those appearing in the expression*

$$\frac{A_1 x + B_1}{ax^2 + bx + c} + \frac{A_2 x + B_2}{(ax^2 + bx + c)^2} + \cdots + \frac{A_k x + B_k}{(ax^2 + bx + c)^k}$$

Again, A_1, B_1, A_2, B_2, . . . , A_k, B_k must be determined, say, by the method of equating coefficients.

In Examples 6 and 7, decompose the given rational fraction into partial fractions.

EXAMPLE 6 $\dfrac{x^3 + x + 2}{x(x^2 + 1)^2}$

SOLUTION $\dfrac{x^3 + x + 2}{x(x^2 + 1)^2} = \dfrac{A}{x} + \dfrac{Bx + C}{x^2 + 1} + \dfrac{Dx + E}{(x^2 + 1)^2}$

Here, the constant A can be found by the short method of substitution as

$$\frac{0^3 + 0 + 2}{(0^2 + 1)^2} = A \qquad \text{or} \qquad A = 2$$

In order to find B, C, D, and E, we put $A = 2$ and multiply both sides of the equation by $x(x^2 + 1)^2$ to clear fractions, and we obtain

$$x^3 + x + 2 = 2(x^2 + 1)^2 + (Bx + C)x(x^2 + 1) + (Dx + E)x$$

$$= (2x^4 + 4x^2 + 2) + (Bx^4 + Cx^3 + Bx^2 + Cx) + (Dx^2 + Ex)$$

$$= (2 + B)x^4 + Cx^3 + (4 + B + D)x^2 + (C + E)x + 2$$

Equating coefficients of like powers of x on both sides of the equation above, we obtain

$$\begin{cases} 0 = 2 + B \\ 1 = C \\ 0 = 4 + B + D \\ 1 = C + E \end{cases}$$

Solving these equations simultaneously, we have $B = -2$, $C = 1$, $D = -2$, and $E = 0$. Hence,

$$\frac{x^3 + x + 2}{x(x^2 + 1)^2} = \frac{2}{x} - \frac{2x - 1}{x^2 + 1} - \frac{2x}{(x^2 + 1)^2}$$

EXAMPLE 7 $\dfrac{x^5 - 2x^4 + 2x^3 + x - 2}{x^2(x^2 + 1)^2}$

SOLUTION $\dfrac{x^5 - 2x^4 + 2x^3 + x - 2}{x^2(x^2 + 1)^2} = \dfrac{A}{x} + \dfrac{B}{x^2} + \dfrac{Cx + D}{x^2 + 1} + \dfrac{Ex + G}{(x^2 + 1)^2}$

Here we find that $B = -2$ by the short method of substitution. Thus,

$$\frac{x^5 - 2x^4 + 2x^3 + x - 2}{x^2(x^2 + 1)^2} = \frac{A}{x} - \frac{2}{x^2} + \frac{Cx + D}{x^2 + 1} + \frac{Ex + G}{(x^2 + 1)^2}$$

Clearing fractions, we have

$$x^5 - 2x^4 + 2x^3 + x - 2 = Ax(x^2 + 1)^2 - 2(x^2 + 1)^2$$
$$+ (Cx + D)x^2(x^2 + 1) + (Ex + G)x^2$$

$$= (Ax^5 + 2Ax^3 + Ax) - (2x^4 + 4x^2 + 2)$$
$$+ (Cx^5 + Dx^4 + Cx^3 + Dx^2) + (Ex^3 + Gx^2)$$

$$= (A + C)x^5 + (D - 2)x^4 + (2A + C + E)x^3$$
$$+ (D + G - 4)x^2 + Ax - 2$$

Equating coefficients, we have

$$\begin{cases} 1 = A + C \\ -2 = D - 2 \\ 2 = 2A + C + E \\ 0 = D + G - 4 \\ 1 = A \end{cases}$$

Simultaneous solution of these equations yields $A = 1$, $C = 0$, $D = 0$, $E = 0$, and $G = 4$. Hence,

$$\frac{x^5 - 2x^4 + 2x^3 + x - 2}{x^2(x^2 + 1)^2} = \frac{1}{x} - \frac{2}{x^2} + \frac{4}{(x^2 + 1)^2}$$

   In order to integrate rational functions containing repeated irreducible quadratic factors $ax^2 + bx + c$ in the denominator, it is necessary to be able to integrate partial fractions of the form

$$\frac{Ax + B}{(ax^2 + bx + c)^k}$$

After we complete the square, if necessary, in the expression $ax^2 + bx + c$ and make the usual change of variable, the required integral can be brought into the form

$$\int \frac{Cu + D}{(au^2 + q)^k} \, du \qquad \text{or} \qquad C \int \frac{u \, du}{(au^2 + q)^k} + D \int \frac{du}{(au^2 + q)^k}$$

   The integral

$$\int \frac{u \, du}{(au^2 + q)^k}$$

can be handled with ease by the substitution $t = au^2 + q$ (Problem 40). However, the integral

$$\int \frac{du}{(au^2 + q)^k}$$

can present more of a challenge. Since the original polynomial $ax^2 + bx + c$ was irreducible, it can be shown (Problem 41) that $q/a > 0$. Thus, we can make the substitution $u = \sqrt{q/a} \, w$ and obtain

$$\int \frac{du}{(au^2 + q)^k} = \int \frac{\sqrt{q/a} \, dw}{(qw^2 + q)^k} = \frac{\sqrt{q/a}}{q^k} \int \frac{dw}{(w^2 + 1)^k}$$

The required integral can be found provided that we can evaluate an integral of the form

$$\int \frac{dw}{(w^2 + 1)^k}$$

The trigonometric substitution $w = \tan \theta$ converts the last integral into the form $\int \cos^n \theta \, d\theta$, where $n = 2(k - 1)$ (Problem 42); by this or other means (Problems 43 and 44), it can always be evaluated.

EXAMPLE 8   Evaluate $\displaystyle\int \frac{x^3 + x + 2}{x(x^2 + 1)^2} \, dx$.

SOLUTION   Here we must make a decomposition of the integrand into partial fractions. By Example 6,

$$\frac{x^3 + x + 2}{x(x^2 + 1)^2} = \frac{2}{x} - \frac{2x - 1}{x^2 + 1} - \frac{2x}{(x^2 + 1)^2}$$

Hence,

$$\int \frac{x^3 + x + 2}{x(x^2 + 1)^2} \, dx = 2 \int \frac{dx}{x} - \int \frac{2x - 1}{x^2 + 1} \, dx - \int \frac{2x \, dx}{(x^2 + 1)^2}$$

$$= 2 \int \frac{dx}{x} - \int \frac{2x \, dx}{x^2 + 1} + \int \frac{dx}{x^2 + 1} - \int \frac{2x \, dx}{(x^2 + 1)^2}$$

$$= 2 \ln |x| - \ln (x^2 + 1) + \tan^{-1} x + \frac{1}{x^2 + 1} + C$$

where the second and fourth integrals have been evaluated by using the substitution $t = x^2 + 1$.   ∎

EXAMPLE 9   Evaluate $\displaystyle\int \frac{x^5 - 2x^4 + 2x^3 + x - 2}{x^2(x^2 + 1)^2} \, dx$.

SOLUTION   By Example 7,

$$\frac{x^5 - 2x^4 + 2x^3 + x - 2}{x^2(x^2 + 1)^2} = \frac{1}{x} - \frac{2}{x^2} + \frac{4}{(x^2 + 1)^2}$$

Hence,

$$\int \frac{x^5 - 2x^4 + 2x^3 + x - 2}{x^2(x^2 + 1)^2} \, dx = \int \frac{dx}{x} - 2 \int \frac{dx}{x^2} + 4 \int \frac{dx}{(x^2 + 1)^2}$$

$$= \ln |x| + \frac{2}{x} + 4 \left[\frac{\tan^{-1} x}{2} + \frac{x}{2(x^2 + 1)} \right] + C$$

where we have used the trigonometric substitution $x = \tan \theta$ to evaluate $\int dx/(x^2 + 1)^2$. Thus, $dx = \sec^2 \theta \, d\theta$ and

$$\int \frac{dx}{(x^2 + 1)^2} = \int \frac{\sec^2 \theta \, d\theta}{(\tan^2 \theta + 1)^2} = \int \frac{\sec^2 \theta \, d\theta}{(\sec^2 \theta)^2} = \int \frac{d\theta}{\sec^2 \theta}$$

$$= \int \cos^2 \theta \, d\theta = \frac{\theta}{2} + \frac{\sin 2\theta}{4} = \frac{\theta}{2} + \frac{\sin \theta \cos \theta}{2} + C$$

Using an appropriate right triangle, we have $\theta = \tan^{-1} x$, $\sin \theta = x/\sqrt{x^2 + 1}$, and $\cos \theta = 1/\sqrt{x^2 + 1}$, so that

$$\int \frac{dx}{(x^2 + 1)^2} = \frac{\tan^{-1} x}{2} + \frac{1}{2} \frac{x}{\sqrt{x^2 + 1}} \cdot \frac{1}{\sqrt{x^2 + 1}} + C$$

$$= \frac{\tan^{-1} x}{2} + \frac{x}{2(x^2 + 1)} + C$$ ■

Problem Set 8.6

In Problems 1 to 26, evaluate each integral. (*Caution:* Some integrands may require a preliminary long division, and others may be partial fractions to begin with.)

1 $\displaystyle\int \frac{5 \, dx}{(x - 1)(x^2 + 4)}$

2 $\displaystyle\int \frac{x^5 + 9x^3 + 1}{x^3 + 9x} \, dx$

3 $\displaystyle\int \frac{(x + 3) \, dx}{x(x^2 + 1)}$

4 $\displaystyle\int \frac{dy}{y^4 - 16}$

5 $\displaystyle\int \frac{3x^2 + x - 2}{(x - 1)(x^2 + 1)} \, dx$

6 $\displaystyle\int \frac{7x^2 + 6x + 5}{x(x^2 + x + 1)} \, dx$

7 $\displaystyle\int \frac{4x \, dx}{x^4 - 1}$

8 $\displaystyle\int \frac{x - 3}{2x^2 - 12x + 19} \, dx$

9 $\displaystyle\int \frac{2t^2 - t + 1}{t(t^2 + 25)} \, dt$

10 $\displaystyle\int \frac{u^2 - u - 21}{2u^3 - u^2 + 8u - 4} \, du$

11 $\displaystyle\int \frac{16 \, dx}{x(x^2 + 4)^2}$

12 $\displaystyle\int \frac{2x^3 + 9}{x^4 + x^3 + 12x^2} \, dx$

13 $\displaystyle\int \frac{15y^2 - 4y + 12}{3y^3 - y^2 + 12y - 4} \, dy$

14 $\displaystyle\int \frac{x^3 + 2x^2 + 7x + 2}{x^2 + 2x + 5} \, dx$

15 $\displaystyle\int \frac{6x^2 - 8x - 1}{(x - 2)(2x^2 - 3x + 5)} \, dx$

16 $\displaystyle\int \frac{17 \, dy}{(y - 2)(y^2 + 4y + 5)}$

17 $\displaystyle\int \frac{2x^4 - 7x^3 + 31x^2 - 45x + 46}{2x^3 - 7x^2 + 11x - 10} \, dx$

18 $\displaystyle\int \frac{3x^2 + 8x + 6}{x^3 + 4x^2 + 6x + 4} \, dx$

19 $\displaystyle\int \frac{5t^3 - 3t^2 + 2t - 1}{t^4 + 9t^2} \, dt$

20 $\displaystyle\int \frac{dx}{(x^2 + 1)^3}$

21 $\displaystyle\int \frac{x^3 + 4}{x^2(x^2 + 1)^2} \, dx$

22 $\displaystyle\int \frac{2y^2}{y^4 + y^3 + 12y^2} \, dy$

23 $\displaystyle\int \frac{x^5 + 4x^3 + 3x^2 - x + 2}{x^5 + 4x^3 + 4x} \, dx$

24 $\displaystyle\int \frac{4x^2 \, dx}{(x - 1)^2(x^2 - x + 1)}$

25 $\displaystyle\int \frac{4(t + 1) \, dt}{t(t^2 + 2t + 2)^2}$

26 $\displaystyle\int \frac{dx}{x^3 + 3x^2 + 7x + 5}$

In Problems 27 to 32, evaluate each definite integral.

27 $\displaystyle\int_0^3 \frac{t + 10}{(t + 1)(t^2 + 1)} \, dt$

28 $\displaystyle\int_0^1 \frac{dx}{8x^3 + 27}$

29 $\displaystyle\int_0^{1/2} \frac{8x \, dx}{(2x + 1)(4x^2 + 1)}$

30 $\displaystyle\int_1^2 \frac{4 \, dx}{x^3 + 4x}$

31 $\displaystyle\int_1^2 \frac{1 - x^2}{x(x^2 + 1)} \, dx$

32 $\displaystyle\int_2^5 \frac{x^4 - x^3 + 2x^2 - x + 2}{(x - 1)(x^2 + 2)} \, dx$

In Problems 33 to 36, find a change of variable that reduces each integrand to a rational function.

33 $\displaystyle\int \frac{\cos x \, dx}{\sin^3 x + \sin^2 x + 9 \sin x + 9}$

34 $\displaystyle\int \frac{dx}{x\sqrt{x} + x + 1}$

35 $\displaystyle\int \frac{3e^{2x} + 2e^x - 2}{e^{3x} - 1} \, dx$

36 $\displaystyle\int \frac{3e^{3x} + e^x + 3}{(e^{2x} + 1)^2} \, dx$

37 Find the partial-fraction decomposition of

$$\frac{ax^3 + bx^2 + cx + d}{(x^2 + 1)^2}$$

38 Find a formula for $\displaystyle\int \frac{ax^3 + bx^2 + cx + d}{(x^2 + 1)^2} \, dx$.

39 Integrate

$$\int \frac{5x^4 + 6x^2 + 1}{x^5 + 2x^3 + x} \, dx$$

in two ways: (a) by using the substitution $u = x^5 + 2x^3 + x$ and (b) by using partial fractions.

40 Find a formula for $\displaystyle\int \frac{u \, du}{(au^2 + q)^k}$.

41 Suppose that $ax^2 + bx + c$ is irreducible, so that $b^2 - 4ac < 0$. Complete the square, and make the appropriate change of variable so that $ax^2 + bx + c = au^2 + q$. Then prove that $q/a > 0$.

42 Prove that the trigonometric substitution $w = \tan \theta$ converts the integral

$$\int \frac{dw}{(w^2 + 1)^k}$$

to the form $\int \cos^n \theta \, d\theta$, where $n = 2(k - 1)$.

43 By differentiation of the right side, verify that, for $k \geq 2$,

$$\int \frac{dw}{(w^2 + 1)^k} = \frac{1}{2k - 2} \cdot \frac{w}{(w^2 + 1)^{k-1}} + \frac{2k - 3}{2k - 2} \int \frac{dw}{(w^2 + 1)^{k-1}}$$

44 Show that

$$\int \frac{dx}{(x^2 + 1)^k}$$

where k is a positive integer, can always be expressed in terms of rational functions and the inverse tangent function. (*Hint:* Use Problem 43.)

45 Use the formula in Problem 43 to evaluate

(a) $\displaystyle\int \frac{dw}{(w^2 + 1)^2}$ (b) $\displaystyle\int \frac{dw}{(w^2 + 1)^3}$

46 It is an algebraic fact that every rational function can be decomposed into partial fractions. Show, therefore, that the integral of any rational function can be expressed in terms of rational functions, inverse tangents, and logarithms (of absolute values).

47 An integral of the form

$$\int \frac{dx}{(a - bx)^{2/3}(c - x)}$$

where a, b, and c are positive constants, has to be evaluated to determine the time required for a homogeneous sphere of iron to dissolve in an acid bath. Make the change of variable $\sqrt[3]{a - bx} = z$, and show that the integrand then becomes a rational function of z.

48 Find a formula for the integral in Problem 47.

8.7   Integration by Special Substitutions

In this section we examine some special substitutions that can be effective when the integrand contains nth roots of the variable of integration or when it involves rational functions of sines and cosines.

Integration of Functions Containing nth Roots

If the integrand involves an expression of the form $\sqrt[n]{x}$, *then the substitution* $z = \sqrt[n]{x}$ *may be helpful. If the integrand involves both* $\sqrt[n]{x}$ *and* $\sqrt[m]{x}$, *then the substitution* $u = \sqrt[p]{x}$, *where* $p = nm$, *may prove effective.* In the first case, we have $x = z^n$, so that $dx = nz^{n-1} \, dz$; in the second case, we have $x = u^p$, so that $dx = pu^{p-1} \, du$.

In Examples 1 to 3, evaluate the given integral.

__EXAMPLE 1__ $\displaystyle\int \frac{dx}{1 + \sqrt{x}}$

SOLUTION   Put $z = \sqrt{x}$, so that $x = z^2$, $dx = 2z \, dz$, and

$$\int \frac{dx}{1 + \sqrt{x}} = \int \frac{2z \, dz}{1 + z}$$

Since $2z/(1 + z)$ is an improper fraction, we divide numerator by denominator to obtain a quotient of 2 and a remainder of -2. Thus,

$$\frac{2z}{1 + z} = 2 - \frac{2}{1 + z}$$

It follows that

$$\int \frac{dx}{1 + \sqrt{x}} = \int \frac{2z\,dz}{1 + z} = \int \left(2 - \frac{2}{1 + z}\right) dz = 2\int dz - 2\int \frac{dz}{1 + z}$$

$$= 2z - 2\ln|1 + z| + C = 2\sqrt{x} - 2\ln|1 + \sqrt{x}| + C \qquad \blacksquare$$

<u>EXAMPLE 2</u> $\displaystyle\int \frac{dx}{\sqrt{1 + \sqrt[3]{x}}}$

SOLUTION   We could put $z = \sqrt[3]{x}$, but, with some foresight, it seems better to try $z = \sqrt{1 + \sqrt[3]{x}}$. Then

$$z^2 = 1 + \sqrt[3]{x} \qquad z^2 - 1 = \sqrt[3]{x} \qquad \text{and so} \qquad x = (z^2 - 1)^3$$

In particular, we have

$$dx = 3(z^2 - 1)^2(2z\,dz) = 6z(z^2 - 1)^2\,dz$$

so $\displaystyle\int \frac{dx}{\sqrt{1 + \sqrt[3]{x}}} = \int \frac{6z(z^2 - 1)^2\,dz}{z} = 6\int (z^2 - 1)^2\,dz$

$$= 6\int (z^4 - 2z^2 + 1)\,dz = 6\left(\frac{z^5}{5} - \frac{2z^3}{3} + z\right) + C$$

$$= 6\left[\frac{(1 + \sqrt[3]{x})^{5/2}}{5} - \frac{2(1 + \sqrt[3]{x})^{3/2}}{3} + (1 + \sqrt[3]{x})^{1/2}\right] + C \qquad \blacksquare$$

<u>EXAMPLE 3</u> $\displaystyle\int \frac{dt}{\sqrt{t} - \sqrt[3]{t}}$

SOLUTION   Put $u = \sqrt[6]{t}$, so that $t = u^6$ and $dt = 6u^5\,du$. Thus,

$$\int \frac{dt}{\sqrt{t} - \sqrt[3]{t}} = \int \frac{6u^5\,du}{\sqrt{u^6} - \sqrt[3]{u^6}} = \int \frac{6u^5\,du}{u^3 - u^2} = 6\int \frac{u^3\,du}{u - 1}$$

Since $u^3/(u - 1)$ is an improper fraction, we divide numerator by denominator to obtain the quotient $u^2 + u + 1$ and the remainder 1. Hence,

$$\frac{u^3}{u - 1} = u^2 + u + 1 + \frac{1}{u - 1}$$

so that $\displaystyle\int \frac{dt}{\sqrt{t} - \sqrt[3]{t}} = 6\int \left(u^2 + u + 1 + \frac{1}{u - 1}\right) du$

$$= 6\left(\frac{u^3}{3} + \frac{u^2}{2} + u + \ln|u - 1|\right) + C$$

$$= 2\sqrt{t} + 3\sqrt[3]{t} + 6\sqrt[6]{t} + 6\ln|\sqrt[6]{t} - 1| + C \qquad \blacksquare$$

Naturally, if the integrand involves an expression of the form $\sqrt[n]{u}$, where u is a function of x, then the substitution $z = \sqrt[n]{u}$ suggests itself. Such a substitution was effective in Example 2 above. Further examples are given below.

In Examples 4 and 5, evaluate the given integral.

<u>EXAMPLE 4</u> $\displaystyle\int_4^{12} x\sqrt{x - 3}\,dx$

SOLUTION   Put $z = \sqrt{x - 3}$, so that $z^2 = x - 3$, $x = z^2 + 3$, $dx = 2z\,dz$, $z = 1$ when $x = 4$, and $z = 3$ when $x = 12$. Thus,

$$\int_4^{12} x\sqrt{x - 3}\,dx = \int_1^3 (z^2 + 3)z(2z\,dz) = 2\int_1^3 (z^4 + 3z^2)\,dz$$

$$= \left(\tfrac{2}{5} z^5 + 2z^3\right)\Big|_1^3 = \tfrac{756}{5} - \tfrac{12}{5} = \tfrac{744}{5}$$

EXAMPLE 5 $\displaystyle\int \frac{1 + x^2}{(3 + x)^{1/3}}\,dx$

SOLUTION   Put $z = (3 + x)^{1/3}$, so that $z^3 = 3 + x$, $x = z^3 - 3$, $dx = 3z^2\,dz$, and

$$\int \frac{1 + x^2}{(3 + x)^{1/3}}\,dx = \int \frac{1 + (z^3 - 3)^2}{z}(3z^2\,dz) = 3\int (z^7 - 6z^4 + 10z)\,dz$$

$$= \frac{3z^8}{8} - \frac{18z^5}{5} + 15z^2 + C$$

$$= \tfrac{3}{8}(3 + x)^{8/3} - \tfrac{18}{5}(3 + x)^{5/3} + 15(3 + x)^{2/3} + C$$

Integration of Rational Functions of Sine and Cosine

It turns out that *the substitution $z = \tan(x/2)$ reduces any integrand that is a rational function of $\sin x$ and $\cos x$ to a rational function of z.* The appropriate formulas are contained in the following theorem.

THEOREM 1   **Tangent-Half-Angle Substitution**

Suppose that $z = \tan(x/2)$. Then

(i) $\cos x = \dfrac{1 - z^2}{1 + z^2}$ **(ii)** $\sin x = \dfrac{2z}{1 + z^2}$ **(iii)** $dx = \dfrac{2\,dz}{1 + z^2}$

PROOF   Let $z = \tan(x/2)$. Then, using the double-angle formula, we have

$$\cos x = 2\cos^2 \frac{x}{2} - 1 = \frac{2}{\sec^2(x/2)} - 1 = \frac{2}{1 + \tan^2(x/2)} - 1$$

$$= \frac{2}{1 + z^2} - 1 = \frac{1 - z^2}{1 + z^2}$$

so that (i) holds. Also,

$$\sin x = 2\sin\frac{x}{2}\cos\frac{x}{2} = 2\frac{\sin(x/2)}{\cos(x/2)}\cos^2\frac{x}{2} = \tan\frac{x}{2}\left(2\cos^2\frac{x}{2}\right)$$

$$= \tan\frac{x}{2}(\cos x + 1) = z\left(\frac{1 - z^2}{1 + z^2} + 1\right) = \frac{2z}{1 + z^2}$$

so that (ii) holds. To prove (iii), note that $z = \tan(x/2)$ implies that

$$dz = \left(\sec^2\frac{x}{2}\right)\frac{dx}{2} = \frac{1}{2}\left(1 + \tan^2\frac{x}{2}\right)dx = \frac{1 + z^2}{2}\,dx$$

Hence, $dx = 2\,dz/(1 + z^2)$, as desired.

The following examples illustrate the use of the tangent-half-angle substitution.

In Examples 6 and 7, use the tangent-half-angle substitution to evaluate the given integral.

EXAMPLE 6 $\displaystyle\int \frac{dx}{1 - \cos x}$

SOLUTION   Put $z = \tan(x/2)$, so that

$$\cos x = \frac{1 - z^2}{1 + z^2} \qquad dx = \frac{2\,dz}{1 + z^2}$$

and

$$\int \frac{dx}{1 - \cos x} = \int \frac{\dfrac{2\,dz}{1 + z^2}}{1 - \left(\dfrac{1 - z^2}{1 + z^2}\right)} = \int \frac{dz}{z^2}$$

$$= -\frac{1}{z} + C = \frac{-1}{\tan \dfrac{x}{2}} + C$$

$$= -\cot \frac{x}{2} + C \qquad\qquad \blacksquare$$

EXAMPLE 7 $\displaystyle\int \frac{dx}{\sin x - \cos x + 1}$

SOLUTION   Put $z = \tan(x/2)$, so that

$$\sin x = \frac{2z}{1 + z^2} \qquad \cos x = \frac{1 - z^2}{1 + z^2} \qquad dx = \frac{2\,dz}{1 + z^2}$$

and

$$\int \frac{dx}{\sin x - \cos x + 1} = \int \frac{\dfrac{2\,dz}{1 + z^2}}{\dfrac{2z}{1 + z^2} - \dfrac{1 - z^2}{1 + z^2} + 1} = \int \frac{dz}{z^2 + z}$$

$$= \int \left(\frac{1}{z} - \frac{1}{z + 1}\right) dz = \ln |z| - \ln |z + 1| + C$$

$$= \ln \left|\frac{z}{z + 1}\right| + C = \ln \left|\frac{\tan (x/2)}{\tan (x/2) + 1}\right| + C \qquad \blacksquare$$

Problem Set 8.7

In Problems 1 to 6, use an appropriate substitution of the form $z = \sqrt[n]{x}$ and evaluate each integral.

In Problems 7 to 12, use an appropriate substitution of the form $z = \sqrt[n]{u}$, where u is a function of x, and evaluate each integral.

1 $\displaystyle\int \frac{dx}{1 - \sqrt{x}}$

2 $\displaystyle\int \frac{dx}{4 + \sqrt{x}}$

3 $\displaystyle\int \frac{dx}{1 + \sqrt[3]{x}}$

4 $\displaystyle\int \frac{x\,dx}{1 - \sqrt[3]{x}}$

5 $\displaystyle\int \frac{x\,dx}{2 + \sqrt{x}}$

6 $\displaystyle\int \frac{2\sqrt{x}\,dx}{1 + \sqrt[3]{x}}$

7 $\displaystyle\int x^3 \sqrt{2x^2 - 1}\,dx$

8 $\displaystyle\int x^5 \sqrt{5 - 2x^2}\,dx$

9 $\displaystyle\int x \sqrt[3]{3x + 1}\,dx$

10 $\displaystyle\int x^9 \sqrt{1 + 2x^5}\,dx$

11 $\displaystyle\int x^2 (4x + 1)^{3/2}\,dx$

12 $\displaystyle\int x(1 + x)^{2/3}\,dx$

In Problems 13 to 18, use the tangent-half-angle substitution to evaluate each integral.

13 $\displaystyle\int \frac{dx}{3 + 5 \sin x}$

14 $\displaystyle\int \frac{\sin t}{1 + \cos t} dt$

15 $\displaystyle\int \frac{\cos x \, dx}{\sin x \cos x + \sin x}$

16 $\displaystyle\int \frac{dx}{\sin x + \sqrt{3} \cos x}$

17 $\displaystyle\int \frac{d\theta}{\tan \theta - \sin \theta}$

18 $\displaystyle\int \frac{du}{(1 - \cos u)^2}$

In Problems 19 and 20, use the substitution $x = 1/t$, $dx = -dt/t^2$ to simplify each integrand; then evaluate the integral.

19 $\displaystyle\int \frac{dx}{x\sqrt{1 + x^2}}$

20 $\displaystyle\int \frac{dx}{x^2\sqrt{x^2 + 2x}}$

In Problems 21 to 48, evaluate each integral.

21 $\displaystyle\int \frac{1 - \sqrt{t}}{1 + \sqrt[4]{t}} dt$

22 $\displaystyle\int \frac{dy}{y - y^{3/5}}$

23 $\displaystyle\int \frac{dx}{\sqrt[4]{x} + \sqrt{x}}$

24 $\displaystyle\int \frac{dx}{1 + \sqrt{x + 1}}$

25 $\displaystyle\int \frac{x \, dx}{\sqrt[4]{1 - x}}$

26 $\displaystyle\int \frac{x^3 \, dx}{(2 - 3x^2)^{3/4}}$

27 $\displaystyle\int \frac{dx}{x^{1/2} - x^{3/4}}$

28 $\displaystyle\int e^x\sqrt{1 - e^x} \, dx$

29 $\displaystyle\int e^{2x}\sqrt{1 + e^x} \, dx$

30 $\displaystyle\int \sin x \cos x\sqrt{1 + \sin x} \, dx$

31 $\displaystyle\int \sqrt{\frac{1 - x}{x}} \, dx$

32 $\displaystyle\int \frac{1}{(1 + x)^2}\sqrt[3]{\frac{1 - x}{1 + x}} \, dx$

33 $\displaystyle\int \frac{w \, dw}{\sqrt[5]{w + 32}}$

34 $\displaystyle\int \frac{x}{(3x + 1)^2}\sqrt{\frac{1}{3x + 1}} \, dx$

35 $\displaystyle\int \frac{dx}{1 + \sin x + \cos x}$

36 $\displaystyle\int \frac{dt}{\sin t + \cos t}$

37 $\displaystyle\int \frac{\sec t \, dt}{1 + \sin t}$

38 $\displaystyle\int \frac{du}{2 \csc u - \sin u}$

39 $\displaystyle\int_1^4 x\sqrt{x - 1} \, dx$

40 $\displaystyle\int_3^{11} x\sqrt{2x + 3} \, dx$

41 $\displaystyle\int_1^4 \frac{4 - \sqrt{x}}{1 + x} dx$

42 $\displaystyle\int_4^9 \frac{1 - \sqrt{x}}{1 + \sqrt{x}} dx$

43 $\displaystyle\int_{-3}^{-1} \frac{x^2 \, dx}{\sqrt{1 - x}}$

44 $\displaystyle\int_1^{7/3} \frac{1 - \sqrt{3x + 2}}{1 + \sqrt{3x + 2}} dx$

45 $\displaystyle\int_0^{\pi/2} \frac{\cos x}{2 + \sin x} dx$

46 $\displaystyle\int_1^2 \frac{\sqrt{t^4 + 1}}{t} dt$

47 $\displaystyle\int_{\pi/3}^{\pi/2} \frac{dx}{\csc x - \cot x}$

48 $\displaystyle\int_2^3 \frac{dx}{x\sqrt{3x^2 - 2x - 1}}$

49 If $z = \tanh(x/2)$, show that

(a) $\cosh x = \dfrac{1 + z^2}{1 - z^2}$ (b) $\sinh x = \dfrac{2z}{1 - z^2}$

(c) $dx = \dfrac{2 \, dz}{1 - z^2}$

In Problems 50 to 52, use the substitution suggested by Problem 49 to evaluate each integral.

50 $\displaystyle\int \frac{dx}{1 - \sinh x}$

51 $\displaystyle\int \frac{dx}{\cosh x - \sinh x}$

52 $\displaystyle\int \frac{\tanh x}{1 + \cosh x} dx$

53 Find the area of the region bounded by the curves $y = 5x/(1 + \sqrt{x})$, $y = 0$, and $x = 9$.

54 Find the volume generated by revolving the region bounded by the curves given by $y = x + \sqrt{x + 1}$ and $y = 0$ between $x = 0$ and $x = 8$ about the x axis.

8.8   Tables of Integrals and Reduction Formulas

When complicated integrals arise in practical scientific work, people often refer to tables of integrals such as the one inside the covers of this book. The use of computers with symbol-manipulating capabilities to evaluate indefinite integrals is also becoming more widespread. Before you use either a table or a computer to evaluate an indefinite integral, it may be necessary or desirable to simplify the integrand by means of the techniques of integration presented in Sections 8.1 through 8.7.

The following example illustrates the use of a table of integrals.

<u>EXAMPLE 1</u>   Use the table of integrals inside the covers of this book to evaluate

$$\int \frac{\sqrt{7 + 3x^2}}{x^2} \, dx$$

SOLUTION   The expression $\sqrt{7 + 3x^2}$, which appears in the integrand, suggests that we try the part of the table labeled "Forms Involving $\sqrt{a^2 + u^2}$." The left side of Formula 66,

$$\int \frac{\sqrt{a^2 + u^2}}{u^2} \, du = -\frac{\sqrt{a^2 + u^2}}{u} + \ln \left| u + \sqrt{a^2 + u^2} \right|$$

is very similar to the given integral. (Notice that to save space, constants of integration are omitted in the table of integrals.) In order to bring the given integral into a form that matches Formula 66, we let

$$a = \sqrt{7} \qquad \text{and} \qquad u = \sqrt{3} \, x$$

so that

$$u^2 = 3x^2 \qquad \sqrt{a^2 + u^2} = \sqrt{7 + 3x^2} \qquad du = \sqrt{3} \, dx$$

and

$$\int \frac{\sqrt{7 + 3x^2}}{x^2} \, dx = \int \frac{\sqrt{a^2 + u^2}}{u^2/3} \frac{du}{\sqrt{3}} = \frac{3}{\sqrt{3}} \int \frac{\sqrt{a^2 + u^2}}{u^2} \, du$$

$$= \sqrt{3} \left(-\frac{\sqrt{a^2 + u^2}}{u} + \ln \left| u + \sqrt{a^2 + u^2} \right| \right) + C$$

$$= \sqrt{3} \left(-\frac{\sqrt{7 + 3x^2}}{\sqrt{3}x} + \ln \left| \sqrt{3}x + \sqrt{7 + 3x^2} \right| \right) + C$$

$$= -\frac{\sqrt{7 + 3x^2}}{x} + \sqrt{3} \ln \left| \sqrt{3}x + \sqrt{7 + 3x^2} \right| + C$$

Notice that we have restored the constant of integration that was omitted from the table.   ∎

Reduction Formulas

Several formulas in the table of integrals, for instance,

$$\int \sec^n u \, du = \frac{1}{n - 1} \sec^{n-2} u \tan u + \frac{n - 2}{n - 1} \int \sec^{n-2} u \, du$$

reduce the problem of integrating an expression involving a power to the problem of integrating a similar expression involving a smaller power. Such a **reduction formula** may have to be used repeatedly to obtain the final answer. Most reduction formulas are derived by using integration by parts (see Problems 31, 32, and 34).

<u>EXAMPLE 2</u>   Evaluate $\int \sec^5 2x \, dx$.

SOLUTION   We begin by making the substitution $u = 2x$, so that $du = 2 \, dx$, $dx = \frac{1}{2} \, du$, and

$$\int \sec^5 2x \, dx = \int (\sec^5 u)(\tfrac{1}{2} \, du) = \tfrac{1}{2} \int \sec^5 u \, du$$

Now we can apply the reduction formula for powers of the secant [Formula (33)] with $n = 5$ to obtain

$$\int \sec^5 2x \, dx = \frac{1}{2} \int \sec^5 u \, du$$

$$= \frac{1}{2} \left(\frac{1}{5-1} \sec^{5-2} u \tan u + \frac{5-2}{5-1} \int \sec^{5-2} u \, du \right) + C$$

$$= \tfrac{1}{8} \sec^3 u \tan u + \tfrac{3}{8} \int \sec^3 u \, du + C$$

Again, we can apply the same reduction formula, this time with $n = 3$, to obtain

$$\int \sec^5 2x \, dx$$

$$= \frac{1}{8} \sec^3 u \tan u + \frac{3}{8} \left(\frac{1}{3-1} \sec^{3-2} u \tan u + \frac{3-2}{3-1} \int \sec^{3-2} u \, du \right) + C$$

$$= \tfrac{1}{8} \sec^3 u \tan u + \tfrac{3}{16} \sec u \tan u + \tfrac{3}{16} \int \sec u \, du + C$$

Finally, using the standard formula for the integral of the secant, we have

$$\int \sec^5 2x \, dx = \tfrac{1}{8} \sec^3 u \tan u + \tfrac{3}{16} \sec u \tan u + \tfrac{3}{16} \ln |\sec u + \tan u| + C$$

$$= \tfrac{1}{8} \sec^3 2x \tan 2x + \tfrac{3}{16} \sec 2x \tan 2x + \tfrac{3}{16} \ln |\sec 2x + \tan 2x| + C \quad \blacksquare$$

Nonelementary Integrals

There are continuous functions whose indefinite integrals *cannot be expressed* in terms of a finite number of the elementary functions considered in this textbook.* Examples are

$$\int \frac{e^x}{x} \, dx \qquad \int e^{-x^2} \, dx \qquad \int \frac{\sin x}{x} \, dx \qquad \int \sin x^2 \, dx \qquad \text{and} \qquad \int \sqrt{1 + x^4} \, dx$$

Such nonelementary integrals arise in many branches of applied mathematics; for instance, they are employed in the study of the path of a ray of light through the atmosphere, the cooling of the earth, the behavior of a pendulum with a large amplitude, the kinetic theory of gases, and the theory of probability. In Chapter 11, we give a brief indication of how nonelementary integrals are handled (see, for instance, Example 8 on page 698).

*For more details, see D. G. Mead, "Integration," *American Mathematical Monthly*, Vol. 68, No. 2, pp. 152–156, February 1961.

Problem Set 8.8

In Problems 1 to 22, use the table of integrals to evaluate each integral.

1 $\int \dfrac{u \, du}{(3 + 5u)^2}$

2 $\int \dfrac{5x \, dx}{(2 - 3x)^3}$

3 $\int \dfrac{dx}{x^2 \sqrt{5 + x^2}}$

4 $\int (7 + 3x^2)^{-3/2} \, dx$

5 $\int \dfrac{3 \, dy}{\sqrt{11 + 5y^2}}$

6 $\int \dfrac{\sqrt{2 + 7x^2}}{x} \, dx$

7 $\int t^2 \sqrt{13 + 8t^2} \, dt$

8 $\int \dfrac{dw}{w^2 \sqrt{5 - 2w^2}}$

9 $\int \dfrac{\sqrt{3y^2 - 5}}{y^2} \, dy$

10 $\int \dfrac{\sqrt{3 + 5z}}{z^2} \, dz$

11 $\displaystyle\int t^2\sqrt{t^2-5}\ dt$

12 $\displaystyle\int \frac{\sqrt{2-3x^2}}{x^2}\ dx$

13 $\displaystyle\int x^2\sqrt{5-7x^2}\ dx$

14 $\displaystyle\int \frac{\sqrt{2-3u^4}}{u^3}\ du$

15 $\displaystyle\int \frac{dt}{2+3t+4t^2}$

16 $\displaystyle\int \frac{dx}{\sqrt{2+3x+4x^2}}$

17 $\displaystyle\int \frac{dx}{x\sqrt{5-4x+2x^2}}$

18 $\displaystyle\int \frac{\sqrt{16t^2-5t+7}}{t}\ dt$

19 $\displaystyle\int \sin^{-1}(3y+2)\ dy$

20 $\displaystyle\int \tan^{-1}(2t+1)\ dt$

21 $\displaystyle\int x\cos^{-1}5x\ dx$

22 $\displaystyle\int w\sin^{-1}(3w-1)\ dw$

In Problems 23 to 30, use a reduction formula to evaluate each integral.

23 $\displaystyle\int \frac{dx}{(x^2+1)^2}$

24 $\displaystyle\int \frac{dx}{(x^2+1)^3}$

25 $\displaystyle\int \frac{4v^4\ dv}{\sqrt{3+2v}}$

26 $\displaystyle\int \frac{dy}{(y^2+2y+2)^4}$

27 $\displaystyle\int \csc^5 3x\ dx$

28 $\displaystyle\int \cot^5(2x-1)\ dx$

29 $\displaystyle\int \tan^5 7x\ dx$

30 $\displaystyle\int \sec^7(t/2)\ dt$

31 Using integration by parts with $u=\sin^{n-1}ax$ and $dv=\sin ax\ dx$, derive the reduction formula
$$\int \sin^n ax\ dx = -\frac{\sin^{n-1}ax\cos ax}{an}+\frac{n-1}{n}\int \sin^{n-2}ax\ dx$$

32 Derive the reduction formula for powers of the tangent function [Formula (31) in the table inside the front cover].

33 Use the reduction formula of Problem 31 to evaluate (a) $\int \sin^2 ax\ dx$, (b) $\int \sin^3 ax\ dx$, (c) $\int \sin^4 ax\ dx$.

34 Derive the reduction formula for powers of the secant function [Formula (33) in the table inside the front cover].

In Problems 35 to 40, suppose that $I_k=\int_0^{\pi/2}\sin^k x\ dx$ for $k=1, 2, 3, 4, \ldots$, and let n denote a positive integer.

35 (a) Show that $I_1=1$, $I_2=\pi/4$, $I_3=\frac{2}{3}$, and $I_4=3\pi/16$. (b) Use Problem 31 to show that if $n\ge 2$, then $I_n=[(n-1)/n]\ I_{n-2}$.

36 If $k\ge 1$, show that

(a) $I_{2k}=\dfrac{1\cdot 3\cdot 5\cdot 7\cdots(2k-1)}{2\cdot 4\cdot 6\cdot 8\cdots(2k)}\cdot\dfrac{\pi}{2}$

(b) $I_{2k+1}=\dfrac{2\cdot 4\cdot 6\cdot 8\cdots(2k)}{3\cdot 5\cdot 7\cdot 9\cdots(2k+1)}$

37 Use Problem 36 to show that $\pi/2=(2k+1)I_{2k+1}\cdot I_{2k}$.

38 Use Problem 36 to show that $\pi/2=2kI_{2k-1}\cdot I_{2k}$.

39 Show that if $1\le k\le n$, then $I_n\le I_k$. [Hint: For $0\le x\le\pi/2$, $0\le\sin x\le 1$, so that $(\sin x)^n\le(\sin x)^k$.]

40 Use Problems 37, 38, and 39 to show that
$$\frac{1}{2k+1}\cdot\frac{\pi}{2I_{2k}}\le I_{2k}\le\frac{1}{2k}\cdot\frac{\pi}{2I_{2k}}$$

41 Using Problem 40 and part (a) of Problem 36, show that
$$\frac{2}{\pi(2k+1)}\le\left[\frac{1\cdot 3\cdot 5\cdot 7\cdots(2k-1)}{2\cdot 4\cdot 6\cdot 8\cdots(2k)}\right]^2\le\frac{1}{\pi k}$$

42 Use Problem 41 to prove **Wallis's formula** for $\pi/4$, namely,
$$\frac{\pi}{4}=\lim_{k\to+\infty}\frac{2\cdot 4\cdot 4\cdot 6\cdot 6\cdot 8\cdot 8\cdots(2k)\cdot(2k)}{3\cdot 3\cdot 5\cdot 5\cdot 7\cdot 7\cdot 9\cdots(2k-1)(2k+1)}$$

Review Problem Set, Chapter 8

In Problems 1 to 90, evaluate each integral.

1 $\displaystyle\int \cos^3 2x\ dx$

2 $\displaystyle\int \sin^3 4x\cos^2 4x\ dx$

3 $\displaystyle\int \sin^3 3x\cos^3 3x\ dx$

4 $\displaystyle\int \sqrt{\cos x}\ \sin^5 x\ dx$

5 $\displaystyle\int \sin^3(1-2x)\ dx$

6 $\displaystyle\int \sin^3\frac{x}{2}\cos^{3/2}\frac{x}{2}\ dx$

7 $\displaystyle\int \sin^{-2/3}5x\cos^3 5x\ dx$

8 $\displaystyle\int \sin^4\frac{2x}{5}\cos^3\frac{2x}{5}\ dx$

9 $\displaystyle\int \sin^2(2-3x)\ dx$

10 $\displaystyle\int (4+\cos x)(3-\cos x)\ dx$

11 $\displaystyle\int (\sin x-\cos x)^2\ dx$

12 $\displaystyle\int \sin^2(1-2x)\cos^2(1-2x)\ dx$

13 $\displaystyle\int \sin^2 6x\cos^2 6x\ dx$

14 $\displaystyle\int \sin^4 4x\cos^2 4x\ dx$

15 $\displaystyle\int \frac{\cos^3(3t/2)}{\sqrt[3]{\sin(3t/2)}}\ dt$

16 $\displaystyle\int \frac{\cos x}{\sin^4 x}\ dx$

17 $\displaystyle\int \sin 8x\sin 3x\ dx$

18 $\displaystyle\int \cos 13x\cos 2x\ dx$

19 $\int \sin x \sin 2x \sin 3x \, dx$

20 $\int \cos 3x \cos 5x \cos 9x \, dx$

21 $\int \tan^4 (2x - 1) \, dx$

22 $\int \cot^4 (2 - 3x) \, dx$

23 $\int x \tan^3 5x^2 \, dx$

24 $\int x^2 \cot^3 (5 - x^3) \, dx$

25 $\int (\sec t - \tan t)^2 \, dt$

26 $\int \frac{\cos (\tan x)}{\cos^2 x} \, dx$

27 $\int \frac{dx}{(1 - \sin x)^2}$

28 $\int \sqrt{1 + \cos x} \, dx$

29 $\int \sec^4 (1 + 2x) \, dx$

30 $\int \csc^4 (3 - 2x) \, dx$

31 $\int \tan^3 (2 + 3x) \sec^4 (2 + 3x) \, dx$

32 $\int \cot^3 (1 - x) \csc^4 (1 - x) \, dx$

33 $\int \frac{dx}{\sqrt{x^2 + 64}}$

34 $\int \frac{dx}{x^2 \sqrt{81 - x^2}}$

35 $\int \frac{dx}{(1 - x^2)^{5/2}}$

36 $\int \frac{4 \, dx}{x \sqrt{x^6 - 16}}$

37 $\int x \sqrt{x^2 - 4} \, dx$

38 $\int \frac{dt}{(t - 4)\sqrt{t^2 - 8t + 41}}$

39 $\int \frac{dx}{\sqrt{2x - x^2}}$

40 $\int \frac{dt}{\sqrt{1 + 2t - 2t^2}}$

41 $\int \frac{dx}{\sqrt{x^2 + 6x + 13}}$

42 $\int \frac{dx}{\sqrt{8 + 4x - 4x^2}}$

43 $\int x^2 e^{-7x} \, dx$

44 $\int \sqrt{x} \ln 2x \, dx$

45 $\int t^2 \sin^{-1} 2t \, dt$

46 $\int \ln (x^2 + 16) \, dx$

47 $\int (x + 2)e^{3x} \, dx$

48 $\int \ln (x + \sqrt{x^2 + 4}) \, dx$

49 $\int t^3 \cos 3t \, dt$

50 $\int \sin (\ln x) \, dx$

51 $\int \tan^{-1} \sqrt{x} \, dx$

52 $\int e^{\sin t} \left(\frac{t \cos^3 t - \sin t}{\cos^2 t} \right) dt$

53 $\int \frac{\tan^{-1} x}{x^2} \, dx$

54 $\int e^{2x} \sin^2 x \, dx$

55 $\int e^{3x} \cos^2 x \, dx$

56 $\int \sec^3 5x \, dx$

57 $\int x^{11} e^{-x^4} \, dx$

58 $\int x^5 \sin x^2 \, dx$

59 $\int x^3 \cos (-3x^2) \, dx$

60 $\int x^{17} \cos x^6 \, dx$

61 $\int \frac{3y^2 - y + 1}{(y^2 - y)(y + 1)} \, dy$

62 $\int \frac{2x + 1}{x(x + 1)(x + 2)} \, dx$

63 $\int \frac{3x^2 - x + 1}{x^3 - x^2} \, dx$

64 $\int \frac{dx}{x^3(1 + x)}$

65 $\int \frac{t^2 + 6t + 4}{t^4 + 5t^2 + 4} \, dt$

66 $\int \frac{x^2 - 4x - 4}{(x - 2)(x^2 + 9)} \, dx$

67 $\int \frac{t^4 + 4t^3 + 6t^2 + 4t - 3}{t^4 - 1} \, dt$

68 $\int \frac{\sin t \, dt}{\cos^3 t + \cos t}$

69 $\int \frac{x \, dx}{\sqrt[4]{1 + 2x}}$

70 $\int \frac{dt}{\sqrt[4]{t} + 3}$

71 $\int \frac{\sqrt[5]{x^3} + \sqrt[6]{x}}{\sqrt{x}} \, dx$

72 $\int \frac{dy}{\sqrt{y} + y^{3/4}}$

73 $\int \frac{dt}{\sqrt{e^t + 1}}$

74 $\int \frac{\sqrt{x} + 1}{\sqrt{x} - 1} \, dx$

75 $\int \frac{dx}{\sqrt{4 + \sqrt{x + 1}}}$

76 $\int \frac{dy}{y \ln y \, (\ln y + 5)}$

77 $\int \ln \sqrt{x^2 + 3} \, dx$

78 $\int y \ln \sqrt[3]{5y + 2} \, dy$

79 $\int x \sqrt{1 - x^{2/3}} \, dx$

80 $\int \sqrt{\frac{1 + \sqrt{x}}{x}} \, dx$

81 $\int \cos \sqrt[3]{x} \, dx$

82 $\int \frac{e^{2y}}{\sqrt[4]{e^y + 3}} \, dy$

83 $\int \frac{dx}{10 + 11 \cos x}$

84 $\int \frac{\sin x}{8 + \cos x} \, dx$

85 $\int \frac{dy}{3 + 2 \sin y + \cos y}$

86 $\int \frac{\cot x}{\cot x + \csc x} \, dx$

87 $\int \frac{\sec x}{1 + \sin x} \, dx$

88 $\int \frac{dx}{3 - \cos x + 2 \sin x}$

89 $\int \frac{e^{4x}}{\sqrt[4]{e^{2x} + 1}} \, dx$

90 $\int \frac{dx}{a^2 \cos x + b^2 \sin x}$

In Problems 91 to 122, evaluate each definite integral.

91 $\int_0^{\pi/4} \cos x \cos 5x \, dx$

92 $\int_0^{\pi/4} \sin^3 2t \cos^3 2t \, dt$

93 $\int_{\pi/12}^{\pi/8} \tan^3 2x \, dx$

94 $\int_0^1 x \tan^{-1} x \, dx$

95 $\int_1^2 (\ln t)^2 \, dt$

96 $\int_0^{\pi/4} x^2 \sin 2x \, dx$

97 $\int_1^2 t^3 \ln t \, dt$

98 $\int_0^\pi \sin^3 x \, dx$

99 $\displaystyle\int_{-\pi/8}^{\pi/8} |\tan^3 2x|\, dx$

100 $\displaystyle\int_{0}^{1} \cosh^4 x\, dx$

101 $\displaystyle\int_{3}^{3/2} \frac{(9-x^2)^{3/2}}{x^2}\, dx$

102 $\displaystyle\int_{0}^{1/3} \frac{t\, dt}{\sqrt{1-9t^4}}$

103 $\displaystyle\int_{5}^{10} \frac{\sqrt{t^2-25}}{t}\, dt$

104 $\displaystyle\int_{0}^{a} x^2\sqrt{a^2-x^2}\, dx$

105 $\displaystyle\int_{0}^{\pi} \sqrt{1+\cos\frac{x}{3}}\, dx$

106 $\displaystyle\int_{\pi/4}^{\pi/2} \frac{\cot x\, dx}{1-\cos x}$

107 $\displaystyle\int_{3}^{5} \frac{t^2-1}{(t-2)^2}\, dt$

108 $\displaystyle\int_{1}^{2} \frac{5x^2-3x+18}{x(9-x^2)}\, dx$

109 $\displaystyle\int_{0}^{1} \frac{x^2+3x+1}{x^4+2x^2+1}\, dx$

110 $\displaystyle\int_{0}^{1} \frac{t^5\, dt}{(t^2+1)^2}$

111 $\displaystyle\int_{1/2}^{2} \frac{dx}{x\sqrt{5x^2+4x-1}}$

112 $\displaystyle\int_{0}^{1/2} (2x-x^2)^{3/2}\, dx$

113 $\displaystyle\int_{1}^{8} \frac{dx}{x+\sqrt[3]{x}}$

114 $\displaystyle\int_{1}^{4} \frac{\sqrt{x}+1}{\sqrt{x}(x+1)}\, dx$

115 $\displaystyle\int_{2}^{5} \frac{t\, dt}{(t-1)^{3/2}}$

116 $\displaystyle\int_{-1}^{8} \frac{dx}{\sqrt{1+\sqrt{1+x}}}$

117 $\displaystyle\int_{1/4}^{5/4} \frac{dt}{\sqrt{t+1}-\sqrt{t}}$

118 $\displaystyle\int_{16}^{25} \frac{dy}{y-2\sqrt{y}-3}$

119 $\displaystyle\int_{0}^{\ln 4} \frac{dx}{\sqrt{e^{-2x}+2e^{-x}}}$

120 $\displaystyle\int_{0}^{\pi/4} \frac{dx}{1-\sin x+2\cos x}$

121 $\displaystyle\int_{\pi/4}^{\pi/8} \frac{dx}{\sin x+\tan x}$

122 $\displaystyle\int_{0}^{\pi/2} \frac{dt}{4\sin t+3\cos t}$

123 Determine constants A and B so that

$$\frac{c\sin\theta+d\cos\theta}{e\sin\theta+f\cos\theta}=A+B\frac{e\cos\theta-f\sin\theta}{e\sin\theta+f\cos\theta}$$

Then obtain a formula for the integral

$$\int \frac{c\sin\theta+d\cos\theta}{e\sin\theta+f\cos\theta}\, d\theta$$

124 Show that

$$\frac{1}{\sqrt{1-t^3}}<\frac{1}{\sqrt{1-t^2}} \qquad \text{for } 0<t<1$$

Then show that

$$\int_{0}^{x} \frac{dt}{\sqrt{1-t^3}}<\sin^{-1} x \qquad \text{for } 0<x<1$$

125 Evaluate the integral

$$\int \frac{dx}{x\sqrt{x^2-2x+5}}$$

by using the substitution $\sqrt{x^2-2x+5}+x=z$.

126 If

$$f(x)=\int_{1}^{x} \frac{dt}{t+\sqrt{t^2-1}} \qquad \text{for } x\geq 1$$

show that $\frac{1}{2}\ln x\leq f(x)\leq \ln x$.

127 Show that

$$\int_{0}^{x} e^{-y}y^2\, dy=2e^{-x}\left[e^x-1-x-\left(\frac{x^2}{2}\right)\right]$$

128 If

$$g(x)=\int_{1}^{x} f\left(t+\frac{1}{t}\right)\left(\frac{dt}{t}\right)$$

show that $g(1/x)=-g(x)$.

129 Derive the reduction formula

$$\int x^m(\ln x)^n\, dx$$
$$=\frac{x^{m+1}(\ln x)^n}{m+1}-\frac{n}{m+1}\int x^m(\ln x)^{n-1}\, dx \qquad m\neq -1$$

130 Use graphical considerations to prove that

$$\int_{0}^{2\pi} \sin^{2n} t\, dt=\int_{0}^{2\pi} \cos^{2n} t\, dt$$

131 Find the area under the curve $y=\frac{1}{4}x^2-\frac{1}{2}\ln x$ from $x=1$ to $x=4$.

132 Find the area under the curve $y=x^2e^{-x}$ between $x=0$ and $x=1$.

133 Find the area bounded by one arch of $y=\sin^3 x$ from $x=0$ to $x=\pi$.

134 Find the volume of the solid generated by revolving the region bounded by the curves $y=x\ln x$, $y=0$, and $x=4$ about the x axis.

135 Find the volume of the solid generated by revolving the region bounded by one arch of $y=\sin x$ and the line $y=0$ about the line $y=-2$.

136 If the velocity v in meters per second of a particle that is moving along a straight line is expressed by the formula $v=(t+3)/(t^3+t)$, find the distance s in meters that the particle traveled from $t=1$ second to $t=3$ seconds.

137 Evaluate

$$\int_{0}^{\pi/3} \sqrt{1+\left(\frac{dy}{dx}\right)^2}\, dx \text{ if } y=\ln(\cos x).$$

138 Find the arc length of the curve $y=\ln x$ from $x=1$ to $x=\sqrt{3}$.

139 Find the arc length of the curve $y=\ln(\csc x)$ from $x=\pi/6$ to $x=\pi/2$.

140 Find the surface area generated by revolving the arc of the curve $y=e^x$ from $x=0$ to $x=1$ about the x axis.

9
POLAR COORDINATES AND ANALYTIC GEOMETRY

Until now we have specified the position of points in the plane by using Cartesian coordinates; however, in some situations it is more natural to use a different coordinate system. In this chapter, we study the *polar coordinate system,* the conversion from Cartesian to polar coordinates and vice versa, the graphs of polar equations, and area and arc length in polar coordinates. We also study the *conics*—the *ellipse,* the *parabola,* and the *hyperbola*—in both Cartesian and polar coordinates, and we use the idea of rotation and translation of coordinates to simplify their equations.

9.1 Polar Coordinates

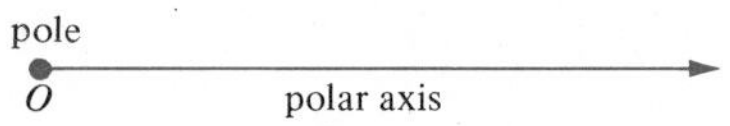

In order to establish a **polar coordinate system** in the plane, we choose a fixed point O called the **pole** and a fixed ray (or half-line) with endpoint O called the **polar axis** (Figure 1). An angle in the **standard position** is understood to have its vertex at the pole O and to have the polar axis as its initial side.

Now let P be any point in the plane and denote by r the distance from the pole O to the point P, so that $r = |\overline{OP}|$ (Figure 2). If $P \neq O$, then P lies on a uniquely determined ray with endpoint O, and this ray forms the terminal side of an angle in the standard position. We denote this angle, measured in degrees or radians (whichever is preferred), by θ, and we refer to the ordered pair (r, θ) as the **polar coordinates** of the point P. (As usual, positive angles are measured counterclockwise.) Thus, in polar coordinates

$$P = (r, \theta)$$

The polar coordinates (r, θ) locate the point P with respect to a "grid" formed by concentric circles with center O and rays emanating from O (Figure 3). The value of r locates P on a circle of radius r, the value of θ locates P on a ray which is the terminal side of the angle θ in standard position, and P itself lies at the intersection of the circle and the ray. For instance, the point with polar coordinates $(r, \theta) = (4, 240°)$ is shown in Figure 3.

Figure 1

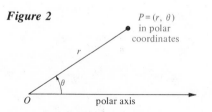

pole
O polar axis

Figure 2

$P = (r, \theta)$
in polar coordinates

r

θ

O polar axis

534

Figure 3

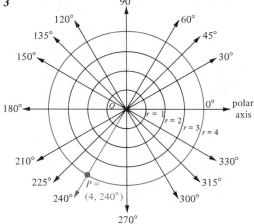

If $r = 0$ in the polar coordinate system, we understand that the point (r, θ) is at the pole O no matter what the angle θ may be. Also, it is convenient to allow the first polar coordinate to be negative, by making the convention that the point $(-r, \theta°)$ is located $|r|$ units from the pole, but on the ray opposite to the $\theta°$ ray, that is, on the ray $\theta° + 180°$ (Figure 4). Thus,

$$(-r, \theta°) = (r, \theta° + 180°)$$

or, in radians,

$$(-r, \theta) = (r, \theta + \pi)$$

For instance, the point $P = (4, 240°)$ in Figure 3 can also be written as $P = (-4, 60°)$.

Unlike the Cartesian coordinate system, a point P has many different representations in the polar coordinate system. Not only do we have $(r, \theta°) = (-r, \theta° + 180°)$ as above, but also we have

$$(r, \theta°) = (r, \theta° + 360°) = (r, \theta° - 360°)$$

since $\pm 360°$ corresponds to a full revolution about the pole. Indeed, if n is any integer, we have

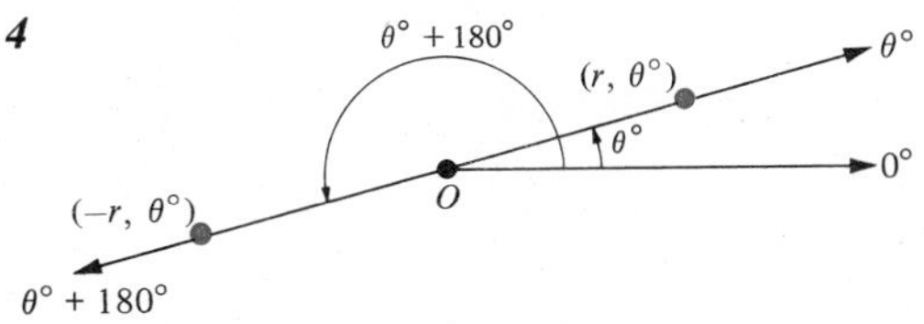

$$(r, \theta°) = (r, \theta° + 360° \cdot n)$$

or, in radians,

$$(r, \theta) = (r, \theta + 2n\pi)$$

Figure 4

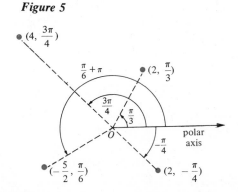

In summary, if $P = (r, \theta°)$ in polar coordinates, then this same point P can be represented by

$$P = (r, \theta° + 360° \cdot n) \qquad \text{or} \qquad P = (-r, \theta° + 180° \cdot (2n + 1))$$

where n is any integer. Likewise, if angles are measured in radians and $P = (r, \theta)$, then P is also represented by

$$P = (r, \theta + 2n\pi) \qquad \text{or} \qquad P = (-r, \theta + (2n + 1)\pi)$$

where n is any integer.

Figure 5

To "*plot* the polar point (r, θ)," or to "*plot* the point (r, θ) in the polar coordinate system," means to draw a diagram showing the pole, the polar axis, and the point P whose polar coordinates are (r, θ).

<u>EXAMPLE 1</u>   Plot the points $(2, \pi/3)$, $(4, 3\pi/4)$, $(-\frac{5}{2}, \pi/6)$, and $(2, -\pi/4)$ in the polar coordinate system.

SOLUTION   To plot the polar point $(2, \pi/3)$, we construct an angle of $\pi/3$ radians (that is, $60°$) in the standard position and then locate the point 2 units from the pole on the terminal side of this angle (Figure 5). The remaining polar points are plotted similarly.

EXAMPLE 2

(a) Plot the point $P = (3, \pi/6)$ in the polar coordinate system.

Give three other polar representations of the point P for which

(b) $r < 0$ and $0 \le \theta < 2\pi$,   **(c)** $r > 0$ and $-2\pi < \theta \le 0$,

(d) $r < 0$ and $-2\pi < \theta \le 0$.

SOLUTION

(a) The point $(3, \pi/6)$ is 3 units from the pole and lies on the ray that is the terminal side of the angle $30° = \pi/6$ radian in the standard position (Figure 6a).

The same point P can also be represented by

(b) $\left(-3, \dfrac{\pi}{6} + \pi\right) = \left(-3, \dfrac{7\pi}{6}\right)$ (Figure 6b)

(c) $\left(3, \dfrac{\pi}{6} - 2\pi\right) = \left(3, -\dfrac{11\pi}{6}\right)$ (Figure 6c)

(d) $\left(-3, \dfrac{7\pi}{6} - 2\pi\right) = \left(-3, -\dfrac{5\pi}{6}\right)$ (Figure 6d)

Figure 6

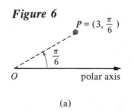
(a)

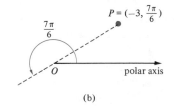
(b)

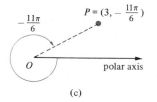
(c)

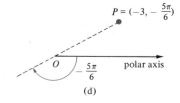
(d)

Conversion of Coordinates

At times it may be advantageous to convert from a Cartesian representation to a polar representation, or vice versa. When we make such a conversion, it is important to realize that the geometric points in the plane do not change—only the method by which they are assigned numerical "addresses" changes.

The usual arrangement is to take the pole for the polar coordinate system at the origin of the Cartesian coordinate system and the polar axis along the positive x axis, so that the positive y axis is the polar ray $\theta = \pi/2$. If we consider the point P whose polar coordinates are (r, θ) with $r \ge 0$, it is clear that the Cartesian coordinates (x, y) of P are given by $x = r \cos \theta$ and $y = r \sin \theta$ (Figure 7). This is certainly true if $r = 0$, while if $r > 0$, it follows from $\cos \theta = x/r$ and $\sin \theta = y/r$.

Now, suppose that a point P has polar coordinates (r, θ) with $r < 0$ and we desire to find the Cartesian coordinates (x, y) of P. Since $(r, \theta) = (-r, \theta + \pi)$ and $-r > 0$, it follows from the equations developed above that

$x = -r \cos (\theta + \pi) = -r(\cos \theta \cos \pi - \sin \theta \sin \pi) = -r(-\cos \theta) = r \cos \theta$

$y = -r \sin (\theta + \pi) = -r(\sin \theta \cos \pi + \cos \theta \sin \pi) = -r(-\sin \theta) = r \sin \theta$

Therefore, the equations

$$x = r \cos \theta \qquad \text{and} \qquad y = r \sin \theta$$

Figure 7

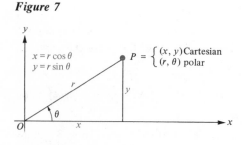

work in all possible cases to convert from the polar coordinates (r, θ) of a point P to the Cartesian coordinates (x, y) of P.

From the last equations, we have

$$x^2 + y^2 = r^2 \cos^2 \theta + r^2 \sin^2 \theta = r^2(\cos^2 \theta + \sin^2 \theta) = r^2$$

so that

$$r = \pm\sqrt{x^2 + y^2}$$

Also, if $x \neq 0$, we have

$$\frac{y}{x} = \frac{r \sin \theta}{r \cos \theta} = \frac{\sin \theta}{\cos \theta} = \tan \theta$$

so that

$$\tan \theta = \frac{y}{x} \qquad \text{for } x \neq 0$$

The equations above do not determine r and θ uniquely because the point P whose Cartesian coordinates are (x, y) has an unlimited number of different representations in the polar coordinate system. In finding polar coordinates of P, you must pay attention to the *quadrant* in which P lies, since this will help to determine θ.

EXAMPLE 3   Convert the given polar coordinates to Cartesian coordinates:

    (a) $(4, 30°)$ **(b)** $(-2, 5\pi/6)$ **(c)** $(-4, -2\pi/3)$

SOLUTION

(a) $(x, y) = (4 \cos 30°, 4 \sin 30°) = \left(4 \cdot \sqrt{3}/2, 4 \cdot \frac{1}{2}\right) = (2\sqrt{3}, 2)$

(b) $(x, y) = \left(-2 \cos \dfrac{5\pi}{6}, -2 \sin \dfrac{5\pi}{6}\right)$

$\qquad\qquad = \left(-2 \cdot \left(-\dfrac{\sqrt{3}}{2}\right), -2 \cdot \dfrac{1}{2}\right) = (\sqrt{3}, -1)$

(c) $(x, y) = \left(-4 \cos \left(-\dfrac{2\pi}{3}\right), -4 \sin \left(-\dfrac{2\pi}{3}\right)\right)$

$\qquad\qquad = \left(-4 \left(-\dfrac{1}{2}\right), -4 \left(-\dfrac{\sqrt{3}}{2}\right)\right) = (2, 2\sqrt{3})$

EXAMPLE 4   Convert the given Cartesian coordinates to polar coordinates with $r \geq 0$ and $-\pi < \theta \leq \pi$:

    (a) $(2, 2)$ **(b)** $(5, -5/\sqrt{3})$ **(c)** $(0, -7)$ **(d)** $(-3, 3)$

SOLUTION

(a) $r = \sqrt{2^2 + 2^2} = \sqrt{8} = 2\sqrt{2}$, and $\tan \theta = \frac{2}{2} = 1$. Since the point lies in the first quadrant, it follows that $0 < \theta < \pi/2$; hence, $\theta = \pi/4$. The polar coordinates are $(2\sqrt{2}, \pi/4)$.

(b) $r = \sqrt{25 + \dfrac{25}{3}} = \dfrac{10}{\sqrt{3}}$ and $\tan \theta = \dfrac{(-5/\sqrt{3})}{5} = \dfrac{-1}{\sqrt{3}}$. Here the point lies in the fourth quadrant, so that $-\pi/2 < \theta < 0$; hence, $\theta = -\pi/6$. The polar coordinates are $(10/\sqrt{3}, -\pi/6)$.

(c) $r = \sqrt{0 + 49} = 7$. Since $x = 0$ and $y < 0$, the point lies on the negative y axis; hence, $\theta = -\pi/2$. The polar coordinates are $(7, -\pi/2)$.

(d) $r = \sqrt{9 + 9} = 3\sqrt{2}$, and $\tan \theta = 3/(-3) = -1$. Since the point lies in the second quadrant, it follows that $\pi/2 < \theta < \pi$; hence, $\theta = 3\pi/4$. The polar coordinates are $(3\sqrt{2}, 3\pi/4)$. ∎

Graphs of Polar Equations

A **polar equation** is an equation relating the polar coordinates r and θ, such as $r = \theta^2$ or $r^2 = 9 \cos 2\theta$. Because a single point P in the plane has a multitude of different polar representations, it is necessary to define the graph of a polar equation with some care.

DEFINITION 1   **Graph of a Polar Equation**

> The **graph** of a polar equation consists of all the points P in the plane that have at least one pair of polar coordinates (r, θ) satisfying the equation.

Thus, if none of the pairs of polar coordinates that represent P satisfy the polar equation, then P does not belong to the graph of this equation. However, in order for P to belong to this graph, it is not necessary for all its polar representations to satisfy the equation—any one will do.

EXAMPLE 5   Sketch the graph of the polar equations:

$$\textbf{(a) } r = 4 \qquad \textbf{(b) } r^2 = 16 \qquad \textbf{(c) } \theta = \frac{\pi}{6} \qquad \textbf{(d) } \theta = -\frac{5\pi}{6}$$

SOLUTION

(a) The graph of $r = 4$ is a circle of radius 4 with center at the pole (Figure 8). Notice, for instance, that the point $P = (4, -\pi)$ belongs to the graph in spite of the fact that not all its representations, such as $(-4, 0)$ or $(-4, 2\pi)$, satisfy the equation $r = 4$.

(b) The equation $r^2 = 16$ is equivalent to $|r| = 4$, and its graph is the same as the graph of $r = 4$ (Figure 8).

(c) The graph of $\theta = \pi/6$ consists of the *entire line* through O making an angle of $\pi/6$ radian (30°) with the polar axis—not just the *ray*, as one might think at first (Figure 9). A point $P = (r, \pi/6 + \pi)$ belongs to the graph of $\theta = \pi/6$ because it can be rewritten as $P = (-r, \pi/6)$.

(d) $\theta = -5\pi/6$ has the same graph as $\theta = \pi/6$ (Figure 9). ∎

Figure 8

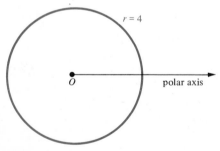

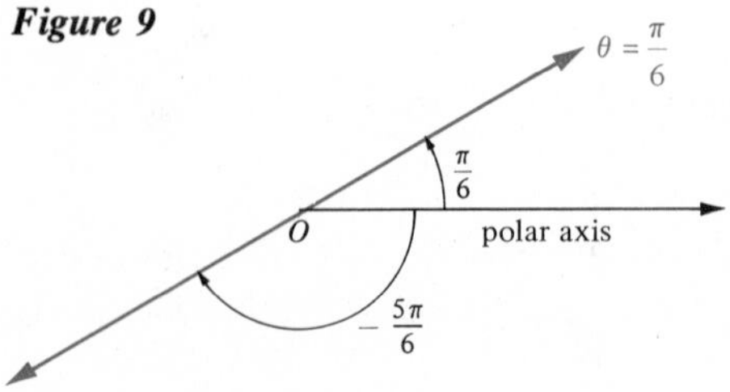

Sometimes it is easy to find the shape of a polar graph by converting the polar equation to a Cartesian equation by means of the equations $x = r \cos \theta$ and $y = r \sin \theta$ and then by sketching the Cartesian graph as usual. Conversely, any Cartesian equation can be converted to a corresponding polar equation simply by substituting $r \cos \theta$ for x and $r \sin \theta$ for y.

EXAMPLE 6   Find a Cartesian equation corresponding to each polar equation and sketch the graph.

$$\textbf{(a) } r = 4 \cos \theta \qquad \textbf{(b) } r \sin \theta = 2 \qquad \textbf{(c) } r = 4 \tan \theta \sec \theta$$

Figure 9

SOLUTION

(a) We multiply both sides of the equation $r = 4 \cos \theta$ by r to obtain $r^2 = 4r \cos \theta$, or $x^2 + y^2 = 4x$; that is, $x^2 - 4x + y^2 = 0$. By adding 4 to both sides of the last equation, we complete the square for the expression $x^2 - 4x$ and obtain

$$x^2 - 4x + 4 + y^2 = 4 \qquad \text{or} \qquad (x - 2)^2 + y^2 = 4$$

Thus, the graph is a circle of radius 2 with center at the point with Cartesian coordinates $(2, 0)$ (Figure 10a).

(b) The equation $r \sin \theta = 2$ is equivalent to $y = 2$, and its graph is therefore a horizontal line (Figure 10b).

Figure 10

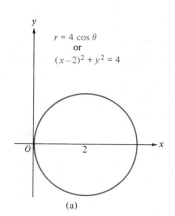

(a)

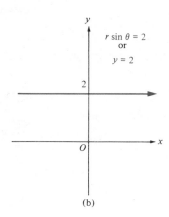

(b)

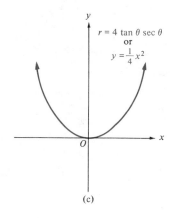

(c)

(c) We have

$$r = 4 \tan \theta \sec \theta = 4 \, \frac{\sin \theta}{\cos \theta} \cdot \frac{1}{\cos \theta}$$

so that
$$r \cos^2 \theta = 4 \sin \theta$$

Multiplication of both sides of the last equation by r gives

$$r^2 \cos^2 \theta = 4r \sin \theta \qquad \text{or} \qquad (r \cos \theta)^2 = 4r \sin \theta$$

that is,
$$x^2 = 4y \qquad \text{or} \qquad y = \tfrac{1}{4}x^2$$

Therefore, the graph is a parabola (Figure 10c). ■

__EXAMPLE 7__ Find a polar equation corresponding to the Cartesian equation $x^2 + (y + 4)^2 = 16$, and sketch the graph.

SOLUTION   The graph is a circle of radius 4 with its center at the point with Cartesian coordinates $(0, -4)$ (Figure 11). Rewriting the equation as

$$x^2 + y^2 + 8y + 16 = 16 \qquad \text{or} \qquad x^2 + y^2 + 8y = 0$$

and substituting $x = r \cos \theta$ and $y = r \sin \theta$, so that $x^2 + y^2 = r^2$, we obtain

$$r^2 + 8r \sin \theta = 0 \qquad \text{or} \qquad r + 8 \sin \theta = 0$$ ■

Figure 11

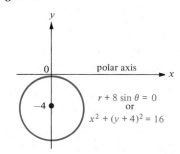

When we multiply a polar equation by r, as we did in Example 6a and 6c, we *may* introduce an extraneous solution $r = 0$ that does not really belong; whereas, when we divide a polar equation by r, as we did in Example 7, we *may* lose a solution

$r = 0$ that really does belong. Each case should be checked! (See Problems 49 to 51.)

The methods used in Examples 6a and 7 can be generalized to show that if a is a positive constant, the graph of a polar equation of either of the two forms

$$r = \pm 2a \cos \theta$$

is a circle of radius a with center at the Cartesian point $(\pm a, 0)$, respectively; and the graph of either of the two equations

$$r = \pm 2a \sin \theta$$

is a circle of radius a with center at the Cartesian point $(0, \pm a)$, respectively, (Problem 53).

Problem Set 9.1

In Problems 1 to 6, plot each point in the polar coordinate system, and then give three other polar representations of the same point for which (a) $r < 0$ and $0 \le \theta < 2\pi$, (b) $r > 0$ and $-2\pi < \theta \le 0$, and (c) $r < 0$ and $-2\pi < \theta \le 0$.

1 $(3, \pi/4)$ **2** $(6, 2\pi/3)$

3 $(-2, \pi/6)$ **4** $(3, 150°)$

5 $(4, 180°)$ **6** $(4, 5\pi/4)$

In Problems 7 to 12, convert the given polar coordinates to Cartesian coordinates.

7 $(7, \pi/3)$ **8** $(0, \pi/3)$

9 $(-2, \pi/4)$ **10** $(6, 13\pi/6)$

11 $(1, -\pi/3)$ **12** $(-5, 150°)$

In Problems 13 to 18, convert the given Cartesian coordinates to polar coordinates (r, θ) with $r \ge 0$ and $-\pi < \theta \le \pi$.

13 $(7, 7)$ **14** $(1, -\sqrt{3})$

15 $(-3, -3\sqrt{3})$ **16** $(-5, 5)$

17 $(0, 7)$ **18** $(-2, 0)$

In Problems 19 and 20, rewrite the answers to Problems 13 to 18 subject to the conditions given.

19 $r \ge 0$, $0 \le \theta < 2\pi$ **20** $r \le 0$, $0 \le \theta < 2\pi$

In Problems 21 and 22, plot the given point P in the polar coordinate system. Then give five other polar representations for the same point.

21 $P = (-3, \frac{187}{6}\pi)$ **22** $P = (4, -\frac{6002}{3}\pi)$

In Problems 23 to 30, sketch the graph of each polar equation.

23 $r = 1$ **24** $r^2 = 9$

25 $\theta = \dfrac{\pi}{2}$ **26** $\theta^2 = \dfrac{25\pi^2}{36}$

27 $\theta = -\dfrac{\pi}{2}$ **28** $\theta = -\dfrac{5\pi}{4}$

29 $|\theta| = \pi$ **30** $\theta^2 - \dfrac{4\pi}{3}\theta + \dfrac{7\pi^2}{36} = 0$

In Problems 31 to 40, convert each polar equation to a Cartesian equation. Sketch the graphs in Problems 31 to 37 only.

31 $r = -3 \cos \theta$ **32** $r = 2 \cos \theta + 2 \sin \theta$

33 $r \cos^2 \theta = \sin \theta$ **34** $r = \sec \theta$

35 $r \cos \theta = -2$ **36** $r = \dfrac{1}{2 \cos \theta - 3 \sin \theta}$

37 $\theta = -\pi/3$ **38** $r^2 \sin 2\theta = 2$

39 $r^2 = \cos 2\theta$ **40** $r = 5\theta$

In Problems 41 to 48, convert each Cartesian equation to a polar equation. Sketch the graphs in Problems 41 to 45 only.

41 $x^2 + y^2 = 25$ **42** $y = -2$

43 $2x + 3y = 6$ **44** $x^2 + y^2 = 4x$

45 $y = -6x^2$ **46** $xy = 1$

47 $(x^2/4) + y^2 = 1$ **48** $x^4 + 2x^2y^2 + y^4 = 4xy$

49 Explain why the graph of $r = 4 \cos \theta$ is the same as the graph of $r^2 = 4r \cos \theta$.

50 Under what conditions will the graph of a first polar equation be the same as the graph of a second polar equation obtained from the first by multiplying both sides by r^n, where n is a positive integer?

51 Explain why the graph of $r^2 + 8r \sin \theta = 0$ is the same as the graph of $r + 8 \sin \theta = 0$.

52 Give a proof—without reference to diagrams—to establish rules governing the conditions under which $(r_1, \theta_1) = (r_2, \theta_2)$ in polar coordinates. Use the fact that $(r_1, \theta_1) = (r_2, \theta_2)$ if and only if $r_1 \cos \theta_1 = r_2 \cos \theta_2$ and $r_1 \sin \theta_1 = r_2 \sin \theta_2$. (*Hint:* Use suitable trigonometric identities.)

53 Suppose that a is a positive constant. (a) Show that the graph of $r = \pm 2a \cos \theta$ is a circle with center at the Cartesian point $(\pm a, 0)$ and radius a. (b) Show that the graph of $r = \pm 2a \sin \theta$ is a circle with center at the Cartesian point $(0, \pm a)$ and radius a.

54 (a) Show that the distance between the point (r_1, θ_1) and the point (r_2, θ_2) in the polar coordinate system is given by $\sqrt{r_1^2 - 2r_1r_2 \cos (\theta_1 - \theta_2) + r_2^2}$. (b) Use the result of part (a) to show that a polar equation of the circle with center (r_0, θ_0) and radius a is $r^2 - 2rr_0 \cos (\theta - \theta_0) + r_0^2 = a^2$.

9.2   Sketching Polar Graphs

Although we managed to sketch certain polar graphs in Section 9.1 by converting the polar equation to Cartesian form and then sketching the graph as usual, there are polar equations that are quite difficult to express in Cartesian form. Thus, in this section we consider the problem of sketching the graph of a polar equation *directly*, without converting it to Cartesian form.

Polar graph sketching can be expedited by the same techniques that are effective for sketching Cartesian graphs. If the polar equation of the graph involves trigonometric functions, then the periodicity of these functions should be taken into consideration. Symmetries of the graph can also be especially helpful.

To plot the graph of a polar equation, it is often useful to start with a fixed value of θ (say $\theta = 0$) and then investigate the corresponding value (or values) of r as θ increases or decreases. It may be helpful to make a table of values of r corresponding to selected values of θ and to plot polar points (r, θ) corresponding to pairs of values in the table. A calculator may be used to help prepare such a table. If r depends continuously on θ, then a sketch of the graph is obtained simply by connecting these points with a continuous curve. With patience, virtually any continuous polar graph can be sketched in this way.

Spiral used as decoration

EXAMPLE 1 Sketch the graph of $r = 1 + \dfrac{6}{\pi} \theta$ for $0 \le \theta \le 2\pi$.

SOLUTION   The table in Figure 1 shows some selected values of θ between 0 and 2π and the corresponding values of r. Plotting the polar points (r, θ) shown in the table and connecting them with a continuous curve, we obtain the desired graph (Figure 1).

If f is either an increasing or a decreasing function, then the graph of the polar equation

$$r = f(\theta)$$

will have a *spiral* shape similar to the curve in Figure 1. For instance, if a is a positive constant, the graph of

$$\boxed{r = a\theta \qquad \text{for} \qquad \theta \ge 0}$$

Figure 1

θ	r
0	1
$\pi/6$	2
$\pi/3$	3
$\pi/2$	4
$2\pi/3$	5
$5\pi/6$	6
π	7
$7\pi/6$	8
$4\pi/3$	9
$3\pi/2$	10
$5\pi/3$	11
$11\pi/6$	12
2π	13

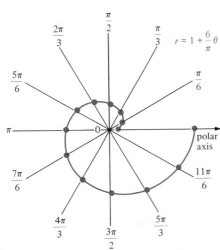

is called an **Archimedean spiral** (Figure 2a), and the graph of

$$r = e^{a\theta}$$

is called a **logarithmic spiral** (Figure 2b).

Figure 2

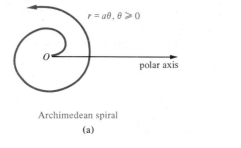

$r = a\theta, \theta \geqslant 0$

polar axis

Archimedean spiral
(a)

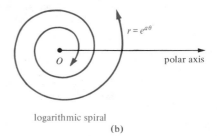

$r = e^{a\theta}$

polar axis

logarithmic spiral
(b)

Symmetries of Polar Graphs

Symmetries of the graph of a polar equation can often be detected by making suitable replacements in the equation and checking whether the new equation is equivalent to the original one. The following table shows some replacements which, when they produce equations equivalent to the original one, imply the type of symmetry indicated:

Replace	An equivalent equation implies
θ by $-\theta$ or θ by $\pi - \theta$ and r by $-r$	Symmetry about the straight line obtained by extending the polar axis (Figure 3a)
θ by $\pi - \theta$ or θ by $-\theta$ and r by $-r$	Symmetry about the straight line through the pole perpendicular to the polar axis (that is, about the line $\theta = \pi/2$) (Figure 3b)
θ by $\theta + \pi$ or r by $-r$	Symmetry about the pole (Figure 3c)

Figure 3

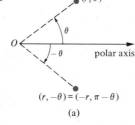

(r, θ)

θ

O $-\theta$ polar axis

$(r, -\theta) = (-r, \pi - \theta)$
(a)

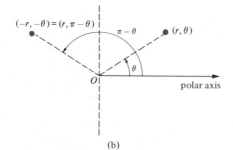

$(-r, -\theta) = (r, \pi - \theta)$ $\pi - \theta$ (r, θ)

θ

O polar axis

(b)

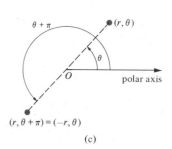

$\theta + \pi$ (r, θ)

θ

O polar axis

$(r, \theta + \pi) = (-r, \theta)$
(c)

Because of the nonuniqueness of representations in polar coordinates, the conditions given above *may* fail even though the graph displays the indicated symmetry. (See Problem 42.) Nevertheless, the given conditions are the ones most likely to be encountered in practice.

In Examples 2 to 5, sketch the graph of the given polar equation and discuss the symmetry.

ⓒ **EXAMPLE 2** $r = 2(1 - \cos \theta)$

SOLUTION   Testing all the rules for symmetry, we observe that the graph is symmetric about the polar axis; indeed, replacing θ by $-\theta$ yields

$$r = 2[1 - \cos (-\theta)] = 2(1 - \cos \theta)$$

which is the same as the original equation. We construct a table giving the polar coordinates of some points on the graph corresponding to selected values of θ. Plotting these points, we sketch the upper half of the graph. The lower half of the graph is then drawn by using the symmetry about the polar axis (Figure 4). This curve is called a *cardioid* because it is "heart-shaped." ■

Figure 4

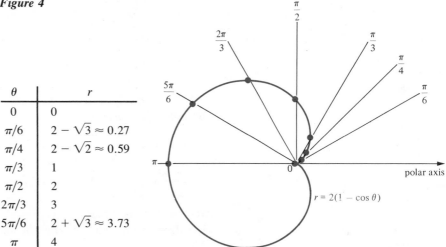

θ	r
0	0
$\pi/6$	$2 - \sqrt{3} \approx 0.27$
$\pi/4$	$2 - \sqrt{2} \approx 0.59$
$\pi/3$	1
$\pi/2$	2
$2\pi/3$	3
$5\pi/6$	$2 + \sqrt{3} \approx 3.73$
π	4

More generally, if a is any positive constant, the polar graph of any one of the four equations

$$r = a(1 \pm \cos \theta) \qquad r = a(1 \pm \sin \theta)$$

is a **cardioid** similar to the curve in Figure 4; but for the cases $r = a(1 + \cos \theta)$ and $r = a(1 \pm \sin \theta)$, it is rotated about the pole through an angle of 180°, or $\mp 90°$. For instance, the graph of $r = 2(1 + \sin \theta)$ is shown in Figure 5. Microphones are often designed with a sensitivity pattern in the shape of a cardioid in order to provide uniform response in front and minimum response in back.

Figure 5

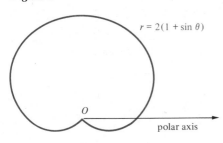

$r = 2(1 + \sin \theta)$

ⓒ **EXAMPLE 3** $r = 1 + 2 \cos \theta$

SOLUTION   Testing all the rules for symmetry, we observe that the graph is symmetric about the polar axis; indeed, replacing θ by $-\theta$ yields

$$r = 1 + 2 \cos (-\theta) = 1 + 2 \cos \theta$$

which is the same as the original equation. We construct a table giving the polar coordinates of some points on the graph corresponding to selected values of θ for $0 \le \theta \le \pi$. Plotting these points, we sketch the portion of the curve shown in Figure 6a. The remaining portion of the curve is then drawn by using the symmetry about the polar axis (Figure 6b). Notice that the graph has *two* loops—an inner loop for $2\pi/3 \le \theta \le 4\pi/3$ and an outer loop for $0 \le \theta \le 2\pi/3$ and $4\pi/3 \le \theta \le 2\pi$. This curve is called a *limaçon,* from the Latin word "limax" for snail. (Notice that the portion of the curve in Figure 6a is similar to the shell of a snail.) ∎

Figure 6

θ	r
0	3
$\pi/6$	$1 + \sqrt{3} \approx 2.73$
$\pi/4$	$1 + \sqrt{2} \approx 2.41$
$\pi/3$	2
$\pi/2$	1
$2\pi/3$	0
$3\pi/4$	$1 - \sqrt{2} \approx -0.41$
$5\pi/6$	$1 - \sqrt{3} \approx -0.73$
π	-1

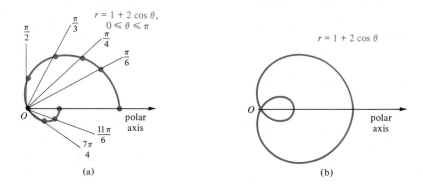

More generally, if a and b are positive constants, the polar graph of any one of the four equations

$$r = a \pm b \cos \theta \qquad r = a \pm b \sin \theta$$

is called a **limaçon.** If $0 < a < b$, the limaçon has an inner loop (Figures 6b and 7a); but if $0 < a/2 < b < a$, it merely has an indentation (Figure 7b). Note that when $a = b$, the limaçon becomes a cardioid. For $0 < b \le a/2$, even the indentation disappears, and the limaçon takes the shape of a slightly distorted circle (Figure 7c).

Figure 7

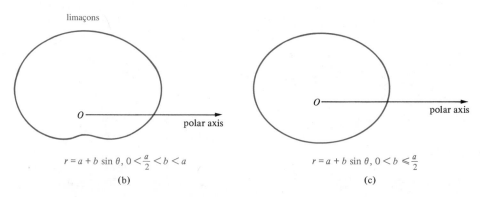

limaçons

$r = a + b \sin \theta, 0 < a < b$
(a)

$r = a + b \sin \theta, 0 < \frac{a}{2} < b < a$
(b)

$r = a + b \sin \theta, 0 < b \le \frac{a}{2}$
(c)

© **EXAMPLE 4** $r^2 = 4 \sin 2\theta$

SOLUTION   Testing all the rules for symmetry, we find that the graph is symmetric about the pole; indeed, since $(-r)^2 = r^2$, the equation remains unchanged if r is replaced by $-r$. Because $r^2 \ge 0$, the equation $r^2 = 4 \sin 2\theta$ can hold only for values of θ for which $\sin 2\theta \ge 0$. For $0 \le \theta \le \pi/2$, we have $\sin 2\theta \ge 0$ and $r = \pm 2\sqrt{\sin 2\theta}$. We construct a table giving the polar coordinates of some points

Figure 8

θ	$r = 2\sqrt{\sin 2\theta}$
0	0
$\pi/12$	$\sqrt{2} \approx 1.41$
$\pi/8$	$\sqrt[4]{8} \approx 1.68$
$\pi/6$	$\sqrt[4]{12} \approx 1.86$
$\pi/4$	2
$\pi/3$	$\sqrt[4]{12} \approx 1.86$
$3\pi/8$	$\sqrt[4]{8} \approx 1.68$
$5\pi/12$	$\sqrt{2} \approx 1.41$
$\pi/2$	0

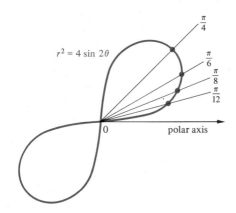

on the graph corresponding to selected values of θ between 0 and $\pi/2$. Plotting these points, we sketch a portion of the graph. The remainder of the graph is then drawn by using the symmetry about the pole (Figure 8). This curve is called a *lemniscate*.

Figure 9

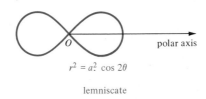

lemniscate

More generally, if a is any positive constant, the polar graph of

$$r^2 = a^2 \sin 2\theta \qquad \text{or} \qquad r^2 = a^2 \cos 2\theta$$

is called a **lemniscate.** The graph of $r^2 = a^2 \cos 2\theta$ is sketched in Figure 9.

Ⓒ **EXAMPLE 5** $r = 3 \sin 3\theta$

SOLUTION   Testing all the rules for symmetry, we find that the graph is symmetric about the line $\theta = \pi/2$; indeed, replacing θ by $-\theta$ and r by $-r$ yields the equation

$$-r = 3 \sin 3(-\theta) = -3 \sin 3\theta$$

which is equivalent to the original equation. We construct a table giving the coordinates of some points on the graph. Plotting these points, we sketch part of the graph. The other part is obtained by using the symmetry about the vertical line $\theta = \pi/2$ (Figure 10).

Figure 10

θ	r
$-\pi/2$	3
$-\pi/3$	0
$-\pi/4$	$(-3\sqrt{2})/2 \approx -2.12$
$-\pi/6$	-3
0	0
$\pi/6$	3
$\pi/4$	$(3\sqrt{2})/2 \approx 2.12$
$\pi/3$	0
$\pi/2$	-3

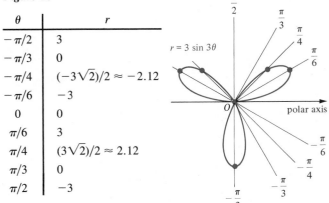

The graph in Figure 10 is called a *three-leaved rose*. More generally, either one of the equations

$$r = a \sin k\theta \qquad \text{or} \qquad r = a \cos k\theta$$

gives an *N*-leaved rose, where

$$N = \begin{cases} k & \text{if } k \text{ is an odd integer} \\ 2k & \text{if } k \text{ is an even integer} \end{cases}$$

Figure 11

$r = a \sin 2\theta$

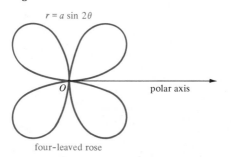

four-leaved rose

For instance, the graph of $r = a \sin 2\theta$, a *four-leaved rose,* is shown in Figure 11.

Direction of a Polar Graph

Now we derive formulas for the direction of a polar graph at a point, that is, for the inclination angle of the tangent line to the polar graph at the point. Thus, consider a polar curve $r = f(\theta)$, where f is a differentiable function of θ, and let l denote the tangent line to the curve at the polar point $P = (r, \theta)$ (Figure 12). Let α be the angle measured from the polar axis to the tangent line l. Thus, since the polar axis coincides with the x axis, $\tan \alpha$ is the slope of the tangent line l in the Cartesian xy coordinate system.

If $P = (x, y)$ in the Cartesian coordinate system, then we have $dy/dx = \tan \alpha$ as usual. Using the formulas for conversion from polar to Cartesian coordinates obtained in Section 9.1, we have

$$x = r \cos \theta = f(\theta) \cos \theta \quad \text{and} \quad y = r \sin \theta = f(\theta) \sin \theta$$

Hence, x and y can be regarded as functions of θ. By the chain rule, $dy/d\theta = (dy/dx)(dx/d\theta)$. Therefore, if $dx/d\theta \neq 0$, we have

$$\tan \alpha = \frac{dy}{dx} = \frac{\dfrac{dy}{d\theta}}{\dfrac{dx}{d\theta}} = \frac{\dfrac{d}{d\theta}(r \sin \theta)}{\dfrac{d}{d\theta}(r \cos \theta)}$$

Consequently, since

$$\frac{d}{d\theta}(r \sin \theta) = \frac{dr}{d\theta} \sin \theta + r \cos \theta \quad \text{and} \quad \frac{d}{d\theta}(r \cos \theta) = \frac{dr}{d\theta} \cos \theta - r \sin \theta$$

it follows that

$$\tan \alpha = \frac{\dfrac{dr}{d\theta} \sin \theta + r \cos \theta}{\dfrac{dr}{d\theta} \cos \theta - r \sin \theta}$$

Figure 12

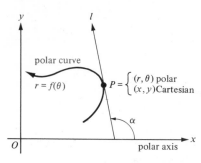

EXAMPLE 6   Find the slope of the tangent line to the three-leaved rose $r = 3 \sin 3\theta$ at the point $(3, \pi/6)$ (Figure 10).

SOLUTION   Here $dr/d\theta = 9 \cos 3\theta = 0$ when $\theta = \pi/6$; hence,

$$\tan \alpha = \frac{\dfrac{dr}{d\theta} \sin \theta + r \cos \theta}{\dfrac{dr}{d\theta} \cos \theta - r \sin \theta} = \frac{0 + 3 \cos \dfrac{\pi}{6}}{0 - 3 \sin \dfrac{\pi}{6}} = -\sqrt{3}$$

The case in which $r = 0$ in the formula for $\tan \alpha$ is of particular interest, since the formula then gives the slope of the tangent line at the pole. Specifically, we have

$$\tan \alpha = \frac{\dfrac{dr}{d\theta} \sin \theta}{\dfrac{dr}{d\theta} \cos \theta} = \frac{\sin \theta}{\cos \theta} = \tan \theta \quad \text{when } r = 0$$

Therefore:

Figure 13

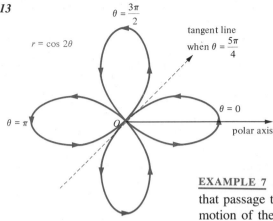

Figure 14

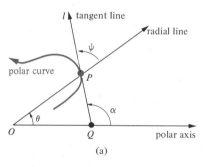

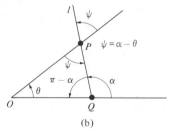

Figure 15

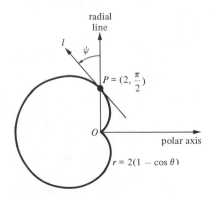

If a polar curve passes through the pole for a particular value of θ, the tangent line to the curve at the pole has a slope equal to tan θ.

A polar graph may pass repeatedly through the pole, and it may have different directions on each passage. (For instance, consider the three-leaved rose in Figure 10.) To find the direction of the curve on any particular passage, simply use the appropriate value of θ corresponding to that passage.

EXAMPLE 7 Find the slope of the tangent to the four-leaved rose $r = \cos 2\theta$ on that passage through the origin for which $\theta = 5\pi/4$ (Figure 13). Also discuss the motion of the point (r, θ) along the curve as θ increases from 0 to 2π.

SOLUTION   When $\theta = 0$, $r = 1$, and we start at the indicated point on the polar axis (Figure 13). As θ increases from 0 to $\pi/4$, $r = \cos 2\theta$ decreases from 1 to 0, and the curve makes its first passage through the pole. As θ continues to increase, the point (r, θ) moves along the four-leaved rose according to the arrows, passing through the pole for the second time when $\theta = 3\pi/4$, for the third time when $\theta = 5\pi/4$, and for the fourth time when $\theta = 7\pi/4$. When θ reaches 2π, the point (r, θ) returns to the starting point. On its third passage through the pole, when $\theta = 5\pi/4$, the slope of the tangent line is equal to tan $(5\pi/4) = 1$. ∎

In dealing with polar curves, it is often convenient to specify the direction of the tangent line l at a polar point $P = (r, \theta)$ by giving the angle ψ measured from the radial line through P to the tangent line l (Figure 14a). (The symbol ψ is the Greek letter *psi*.) Since the three vertex angles of triangle OQP must add up to $180° = \pi$ radians, it follows that $\theta + (\pi - \alpha) + \psi = \pi$; that is, $\psi = \alpha - \theta$ (Figure 14b). Using the trigonometric identity for the tangent of the difference between two angles, we have

$$\tan \psi = \tan (\alpha - \theta) = \frac{\tan \alpha - \tan \theta}{1 + \tan \alpha \tan \theta}$$

Substituting the value of tan α previously derived into the last equation and simplifying, we find that

$$\tan \psi = \frac{r}{dr/d\theta}$$

(Problem 32).

EXAMPLE 8 Find tan ψ at the polar point $(2, \pi/2)$ on the cardioid $r = 2(1 - \cos \theta)$ (Figure 15).

SOLUTION   Here we have $dr/d\theta = 2 \sin \theta$, so that

$$\tan \psi = \frac{r}{dr/d\theta} = \frac{2(1 - \cos \theta)}{2 \sin \theta} = \csc \theta - \cot \theta$$

Therefore, when $\theta = \pi/2$, we have

$$\tan \psi = \csc \frac{\pi}{2} - \cot \frac{\pi}{2} = 1$$

∎

Problem Set 9.2

In Problems 1 to 8, each of the curves is the graph of exactly one of the equations (a) to (h). Match the equation with its corresponding graph.

(a) $r = \sin \theta$ (b) $r = 1 - \sin \theta$ (c) $r = \cos \theta$

(d) $r = 1 + 2 \sin \theta$ (e) $r = 2 + \sin \theta$ (f) $r = 2 \sin 2\theta$

(g) $r = \theta, \theta > 0$ (h) $r^2 = 2 \cos 2\theta$

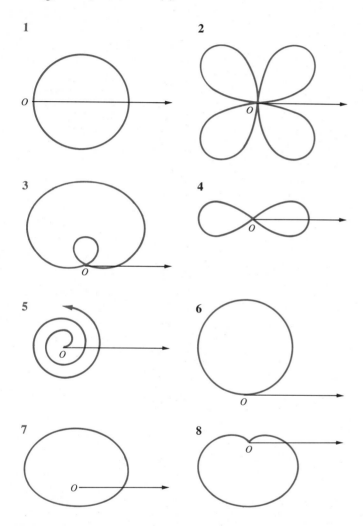

1

2

3

4

5

6

7

8

In Problems 9 to 26, test for symmetry with respect to the polar axis, the line $\theta = \pi/2$, and the pole. Sketch the graph of the equation.

9 $r = 4 \cos \theta$

10 $r \sin \theta = 5$

11 $r \cos \theta = 5$

12 $r = 2$

13 $r = 2 \sin \theta$

14 $\theta = 3$

15 $r = 2 \sin 3\theta$ (three-leaved rose)

16 $r = 2 \cos 2\theta$ (four-leaved rose)

17 $r = 4 \sin 2\theta$ (four-leaved rose)

18 $r = 2 \sin 4\theta$ (eight-leaved rose)

19 $r = 4(1 + \cos \theta)$ (cardioid)

20 $r = 2(1 - \sin \theta)$ (cardioid)

21 $r = 3 - 2 \cos \theta$ (limaçon)

22 $r = 3 + 4 \sin \theta$ (limaçon)

23 $r = 1 - 2 \sin \theta$ (limaçon)

24 $r = 1/\theta, \theta > 0$ (reciprocal spiral)

25 $r^2 = 8 \cos 2\theta$ (lemniscate)

26 $9\theta = \ln r$ (logarithmic spiral)

In Problems 27 to 30, find the slope of the tangent line to the graph of each polar equation at the given point.

27 $r = 3(1 + \cos \theta)$ at $(3, \pi/2)$

28 $r = 2(1 - \sin \theta)$ at $(1, \pi/6)$

29 $r = 8 \sin 2\theta$ at $(0, 0)$

30 $r = \sec^2 \theta$ at $(4, \pi/3)$

31 Suppose that f is a differentiable function, and consider the polar curve $r = f(\theta)$. (a) If $(dr/d\theta) \sin \theta + r \cos \theta = 0$ and $(dr/d\theta) \cos \theta - r \sin \theta \neq 0$, show that the tangent line to the polar curve is horizontal at the polar point (r, θ). (b) Derive a condition for the tangent line to the polar curve at the polar point (r, θ) to be vertical.

32 Fill in all details in the derivation of the formula for $\tan \psi$ (page 547).

In Problems 33 to 37, find $\tan \psi$ at the indicated point (r, θ) on the graph of each polar curve.

33 $r = \sin \theta$ at $(\frac{1}{2}, \pi/6)$ **34** $r = \cos 2\theta$ at $(0, \pi/4)$

35 $r = e^\theta$ at $(e^2, 2)$ **36** $r^2 = \csc 2\theta$ at $(1, \pi/4)$

37 $r = 4 \sec \theta$ at $(4, 0)$

38 If $0 < a < b$, find the slope of *both* tangents to the limaçon $r = a + b \sin \theta$ at the pole.

39 Find all points where the tangent to the limaçon $r = 4 + 3 \sin \theta$ is either horizontal or vertical. (Use the results of Problem 31.)

40 Find the polar coordinates of the tips of the petals of the N-leaved rose $r = 2 \cos k\theta$ by finding the values of θ where r takes on its maximum values. (*Note:* Here it is not necessary to use calculus.)

41 If $0 < b < a$, find the minimum value of r for the limaçon $r = a + b \sin \theta$.

42 Show that the polar graph of $r = \sin (\theta/3)$ is symmetric about the line $\theta = \pm \pi/2$ even though the tests for such symmetry given on page 542 fail. Sketch the graph.

43 Show that the polar graph of $r = f(\theta)$ is (a) symmetric about the line $\theta = 0$ if f is an even function and (b) symmetric about the line $\theta = \pm \pi/2$ if f is an odd function.

44 In view of Problem 42, the symmetry tests given on page 542 may fail to detect actual symmetries. Develop a set of tests that cannot fail.

45 Find and sketch the graph of a polar curve such that at each point of the curve, $\tan \psi = K$, where K is a nonzero constant.

46 (a) If $0 < a/2 < b < a$, show that the limaçon $r = a + b \sin \theta$ has an indentation. (b) If $0 < b \leq a/2$, show that the limaçon has no indentation.

9.3   Area and Arc Length in Polar Coordinates

Geometrically, the two fundamental problems of calculus are finding the slope of the tangent to a curve and determining the area of a region bounded by a curve. In Section 9.2 we attended to the first problem for polar curves; in this section we see how to handle the second. We also develop a formula for arc length of polar curves.

Area of a Region in Polar Coordinates

Consider the curve whose polar equation is $r = f(\theta)$, where f is a continuous function (Figure 1). As θ increases from $\theta = \alpha$ to $\theta = \beta$, the point $P = (f(\theta), \theta)$ moves along the polar curve from $(f(\alpha), \alpha)$ to $(f(\beta), \beta)$ and the radial line segment $\overline{OP}$ sweeps out a plane region. We refer to this region as *the region enclosed by the polar curve between $\theta = \alpha$ and $\theta = \beta$.* Below, we develop a formula for its area.

The simplest polar region is perhaps the circular sector enclosed by the circle of radius r between $\theta = \alpha$ and $\theta = \beta$ (Figure 2). Since the area of the circle is πr^2 and the sector occupies a fraction $(\beta - \alpha)/2\pi$ of the whole circle, the area A of the sector is given by

$$A = \pi r^2 \frac{\beta - \alpha}{2\pi} = \frac{1}{2} r^2(\beta - \alpha)$$

More generally, even if the polar curve is not a circle, when the angle increases from θ to $\theta + d\theta$, the radial segment $\overline{OP}$ in Figure 1 will sweep out a small region which is virtually a sector of a circle of radius $|r| = |f(\theta)|$ (Figure 3). If $d\theta$ is regarded as an infinitesimal, then the infinitesimal area of this sector is given by

$$dA = \tfrac{1}{2}|r|^2 \, d\theta = \tfrac{1}{2}r^2 \, d\theta = \tfrac{1}{2}[f(\theta)]^2 \, d\theta$$

If we "sum," that is, integrate, the areas of all such infinitesimal sectors, running θ from α to β, then we obtain the required area A. Thus,

$$A = \int_\alpha^\beta dA = \int_\alpha^\beta \tfrac{1}{2}[f(\theta)]^2 \, d\theta = \tfrac{1}{2} \int_\alpha^\beta [f(\theta)]^2 \, d\theta$$

This formula is also written as

$$A = \tfrac{1}{2} \int_\alpha^\beta r^2 \, d\theta$$

Figure 1

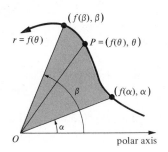

Figure 2

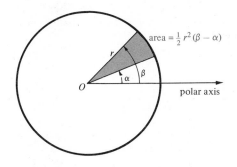

Figure 3

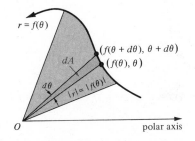

Figure 4

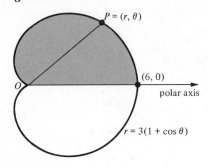

EXAMPLE 1   Find the area of the "top half" of the region inside the cardioid $r = 3(1 + \cos \theta)$ (Figure 4).

SOLUTION   As θ goes from 0 to π, the radial segment $\overline{OP}$ sweeps out the top half of the region inside the cardioid. Therefore, the area A of the region is given by

$$A = \frac{1}{2} \int_0^\pi r^2 \, d\theta = \frac{1}{2} \int_0^\pi [3(1 + \cos \theta)]^2 \, d\theta$$

$$= \frac{9}{2} \int_0^\pi (1 + 2 \cos \theta + \cos^2 \theta) \, d\theta$$

$$= \frac{9}{2} \left(\theta + 2 \sin \theta + \frac{\theta}{2} + \frac{\sin 2\theta}{4} \right) \Bigg|_0^\pi = \frac{27\pi}{4} \text{ square units} \quad \blacksquare$$

Figure 5

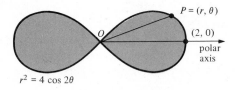

EXAMPLE 2   Find the area enclosed by the lemniscate $r^2 = 4 \cos 2\theta$ (Figure 5).

SOLUTION   Consider the portion of the lemniscate for which $r = 2\sqrt{\cos 2\theta}$. When $\theta = 0$, $r = 2$, and as θ increases, the point $P = (r, \theta)$ moves to the left along the top of the lemniscate until it arrives at the pole O when $\theta = \pi/4$. Thus, as θ goes from 0 to $\pi/4$, the line segment $\overline{OP}$ sweeps out one-fourth of the desired area. Therefore, the area A of the entire region is given by

$$A = 4 \left(\frac{1}{2} \int_0^{\pi/4} r^2 \, d\theta \right) = 2 \int_0^{\pi/4} 4 \cos 2\theta \, d\theta = 4 \sin 2\theta \Bigg|_0^{\pi/4} = 4 \text{ square units} \quad \blacksquare$$

Figure 6

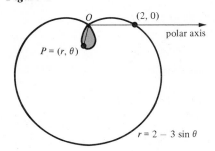

EXAMPLE 3   Find the area of the region enclosed by the inner loop of the limaçon $r = 2 - 3 \sin \theta$ (Figure 6).

SOLUTION   When $\theta = 0$, $r = 2$ and the point $(r, \theta) = (2, 0)$ is on the polar axis. Now as θ begins to increase, $r = 2 - 3 \sin \theta$ begins to decrease, and it reaches 0 when $\theta = \sin^{-1} \frac{2}{3}$. At this point, the radial line segment $\overline{OP}$ begins to sweep out the desired region. When θ reaches the value $\pi/2$, then $r = -1$ and the radial line segment $\overline{OP}$, which points straight downward, has swept out exactly half of the desired area. Therefore, the area A of the region is given by

$$A = 2 \left[\frac{1}{2} \int_{\sin^{-1}(2/3)}^{\pi/2} (2 - 3 \sin \theta)^2 \, d\theta \right]$$

$$= \int_{\sin^{-1}(2/3)}^{\pi/2} (4 - 12 \sin \theta + 9 \sin^2 \theta) \, d\theta$$

$$= (4\theta + 12 \cos \theta + \frac{9}{2} \theta - \frac{9}{4} \sin 2\theta) \Bigg|_{\sin^{-1}(2/3)}^{\pi/2}$$

$$= \frac{17\pi}{4} - \left[\frac{17}{2} \sin^{-1} \frac{2}{3} + 12 \cos \left(\sin^{-1} \frac{2}{3} \right) - \frac{9}{4} \sin \left(2 \sin^{-1} \frac{2}{3} \right) \right]$$

Since
$$\cos \left(\sin^{-1} \frac{2}{3} \right) = \frac{\sqrt{5}}{3}$$

and
$$\sin \left(2 \sin^{-1} \frac{2}{3} \right) = 2 \sin \left(\sin^{-1} \frac{2}{3} \right) \cos \left(\sin^{-1} \frac{2}{3} \right) = \frac{4\sqrt{5}}{9}$$

we have
$$A = \frac{17\pi}{4} - \frac{17}{2} \sin^{-1} \frac{2}{3} - 3\sqrt{5} \text{ square unit} \quad \blacksquare$$

In using the formula $A = \frac{1}{2} \int_\alpha^\beta r^2 \, d\theta$ to find the area of the region enclosed by a polar curve $r = f(\theta)$ between $\theta = \alpha$ and $\theta = \beta$, you must be certain that $\alpha \leq \beta$ and that the radial line segment $\overline{OP}$, where $P = (f(\theta), \theta)$, sweeps *just once* across each point in the interior of the region. For instance, if you wish to find the *total* area inside the limaçon $r = 2 - 3 \sin \theta$ (Figure 6), it would be incorrect to integrate from 0 to 2π. The reason is that as θ goes from 0 to 2π, the radial segment $\overline{OP}$ sweeps *twice* across all the points within the inner loop, so the area of the inner loop gets counted twice.

Area Between Two Polar Curves

To determine the area of a plane region enclosed by two polar curves, we need to know where the curves intersect. Although the points where two Cartesian curves intersect can be found by solving the equations of the curves simultaneously, this procedure does not necessarily yield *all* intersection points of two polar curves. The reason is that an intersection point may have two *different* polar representations, one satisfying the equation of the first curve and the other satisfying the equation of the second curve.

All the intersection points P of the polar graphs of

$$r = f(\theta) \quad \text{and} \quad r = g(\theta)$$

can be found by carrying out the following procedure:

Step 1 Check each polar graph separately to see whether it passes through the pole O. If both do, then O is an intersection point of the graphs.

Step 2 Solve, if possible, the simultaneous equations

$$\begin{cases} r = f(\theta) \\ r = g(\theta + 2n\pi) \end{cases}$$

for r, θ, and n, where $r \neq 0$ and n must be an *integer*. For each such solution, the point $P = (r, \theta)$ is an intersection point of the two graphs.

Step 3 Solve, if possible, the simultaneous equations

$$\begin{cases} r = f(\theta) \\ r = -g(\theta + (2n + 1)\pi) \end{cases}$$

for r, θ, and n, where $r \neq 0$ and n must be an *integer*. For each such solution, the point $P = (r, \theta)$ is an intersection point of the two graphs.

EXAMPLE 4   Find all points of intersection of the circle $r = -6 \cos \theta$ and the limaçon $r = 2 - 4 \cos \theta$.

SOLUTION

Step 1 Since $O = (0, \pi/2)$ is on the circle and $O = (0, \pi/3)$ is on the limaçon, it follows that O is a point of intersection.

Step 2 We solve the simultaneous equations

$$\begin{cases} r = -6 \cos \theta \\ r = 2 - 4 \cos (\theta + 2n\pi) \end{cases} \quad \text{for } r \neq 0, \ n \text{ an integer}$$

Noting that $\cos (\theta + 2n\pi) = \cos \theta$, we rewrite the equations above as

$$r = -6 \cos \theta = 2 - 4 \cos \theta$$

Figure 7

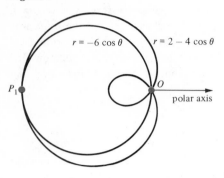

Figure 8

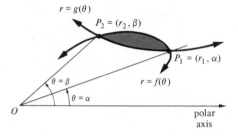

Figure 9

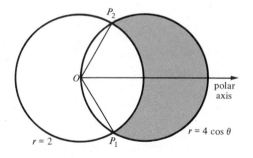

Figure 10

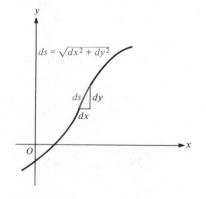

Thus, $-2 \cos \theta = 2$, so that $\cos \theta = -1$, $r = 6$, and $\theta = \pi$. Thus, step 2 yields the intersection point $P_1 = (6, \pi)$.

Step 3 We solve the simultaneous equations

$$\begin{cases} r = -6 \cos \theta \\ r = -[2 - 4 \cos (\theta + (2n + 1)\pi)] & \text{for } r \neq 0, \ n \text{ an integer} \end{cases}$$

Since $\cos (\theta + 2n\pi + \pi) = \cos (\theta + \pi) = -\cos \theta$, we rewrite the equations above as

$$r = -6 \cos \theta = -(2 + 4 \cos \theta)$$

Thus, $2 \cos \theta = 2$, so that $\cos \theta = 1$, $r = -6$, and $\theta = 0$. Therefore, step 3 yields the same intersection point $P_1 = (-6, 0) = (6, \pi)$ as step 2. Hence, the circle and the limaçon intersect in the two points O and P_1 (Figure 7).   ■

Now, suppose we wish to find the area A of the region enclosed by the graphs of the two polar equations

$$r = f(\theta) \qquad \text{and} \qquad r = g(\theta)$$

between two points of intersection

$$P_1 = (r_1, \alpha) \qquad \text{and} \qquad P_2 = (r_2, \beta)$$

(Figure 8). If the region enclosed by the curve $r = g(\theta)$ between $\theta = \alpha$ and $\theta = \beta$ is contained in the region enclosed by the curve $r = f(\theta)$ between $\theta = \alpha$ and $\theta = \beta$, then the desired area A is just the difference of the areas of the two regions. Hence,

$$A = \tfrac{1}{2} \int_\alpha^\beta [f(\theta)]^2 \, d\theta - \tfrac{1}{2} \int_\alpha^\beta [g(\theta)]^2 \, d\theta = \tfrac{1}{2} \int_\alpha^\beta [[f(\theta)]^2 - [g(\theta)]^2] \, d\theta$$

EXAMPLE 5   Find the area of the region inside the circle $r = 4 \cos \theta$ but outside the circle $r = 2$ (Figure 9).

SOLUTION   The two circles intersect at $P_1 = (2, -\pi/3)$ and $P_2 = (2, \pi/3)$. The area A of the region outside the circle $r = 2$ and inside the circle $r = 4 \cos \theta$ between $\theta = -\pi/3$ and $\theta = \pi/3$ is given by

$$\begin{aligned} A &= \tfrac{1}{2} \int_{-\pi/3}^{\pi/3} [(4 \cos \theta)^2 - 2^2] \, d\theta \\ &= \tfrac{1}{2} \int_{-\pi/3}^{\pi/3} (16 \cos^2 \theta - 4) \, d\theta = \tfrac{1}{2} \int_{-\pi/3}^{\pi/3} [\tfrac{16}{2} (1 + \cos 2\theta) - 4] \, d\theta \\ &= \tfrac{1}{2} \int_{-\pi/3}^{\pi/3} (4 + 8 \cos 2\theta) \, d\theta = 2(\theta + \sin 2\theta) \Big|_{-\pi/3}^{\pi/3} \\ &= 2 \left[\left(\frac{\pi}{3} + \frac{\sqrt{3}}{2} \right) - \left(-\frac{\pi}{3} - \frac{\sqrt{3}}{2} \right) \right] = \frac{4\pi}{3} + 2\sqrt{3} \ \text{square units} \end{aligned}$$

   ■

Arc Length of a Polar Curve

For a curve in Cartesian coordinates, $ds = \sqrt{(dx)^2 + (dy)^2}$ gives the differential of arc length (Figure 10). If we convert to polar coordinates, we have

$$dx = d(r \cos \theta) = dr \cos \theta - r \sin \theta \, d\theta$$

and

$$dy = d(r \sin \theta) = dr \sin \theta + r \cos \theta \, d\theta$$

Thus,

$$
\begin{aligned}
(dx)^2 + (dy)^2 &= (dr)^2 \cos^2 \theta - 2r \cos \theta \sin \theta \, (dr)(d\theta) + r^2 \sin^2 \theta \, (d\theta)^2 \\
&\quad + (dr)^2 \sin^2 \theta + 2r \cos \theta \sin \theta \, (dr)(d\theta) + r^2 \cos^2 \theta \, (d\theta)^2 \\
&= (dr)^2(\cos^2 \theta + \sin^2 \theta) + r^2(d\theta)^2(\sin^2 \theta + \cos^2 \theta) \\
&= (dr)^2 + r^2(d\theta)^2
\end{aligned}
$$

so that

$$
\begin{aligned}
ds &= \sqrt{(dx)^2 + (dy)^2} = \sqrt{(dr)^2 + r^2(d\theta)^2} \\
&= \sqrt{\left[\left(\frac{dr}{d\theta}\right)^2 + r^2\right](d\theta)^2}
\end{aligned}
$$

Therefore, in polar coordinates,

$$ds = \sqrt{\left(\frac{dr}{d\theta}\right)^2 + r^2} \, d\theta$$

Figure 11

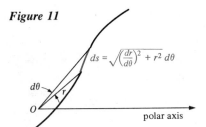

(Figure 11).

The arc length of the portion of a polar curve $r = f(\theta)$ between $\theta = \alpha$ and $\theta = \beta$ is accordingly given by

$$s = \int_\alpha^\beta ds = \int_\alpha^\beta \sqrt{\left(\frac{dr}{d\theta}\right)^2 + r^2} \, d\theta = \int_\alpha^\beta \sqrt{[f'(\theta)]^2 + [f(\theta)]^2} \, d\theta$$

provided that the derivative f' exists and is continuous on the interval $[\alpha, \beta]$.

EXAMPLE 6   Find the total arc length of the cardioid $r = 2(1 - \cos \theta)$ (Figure 12).

Figure 12

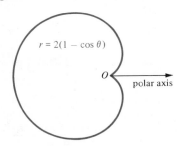

SOLUTION   Because the cardioid is symmetric about the polar axis and its extension, we have

$$
\begin{aligned}
s &= 2 \int_0^\pi \sqrt{\left(\frac{dr}{d\theta}\right)^2 + r^2} \, d\theta \\
&= 2 \int_0^\pi \sqrt{(2 \sin \theta)^2 + 4(1 - 2 \cos \theta + \cos^2 \theta)} \, d\theta \\
&= 4 \int_0^\pi \sqrt{\sin^2 \theta + \cos^2 \theta + 1 - 2 \cos \theta} \, d\theta \\
&= 4 \int_0^\pi \sqrt{2 - 2 \cos \theta} \, d\theta = 4\sqrt{2} \int_0^\pi \sqrt{1 - \cos \theta} \, d\theta
\end{aligned}
$$

Since $2 \sin^2 (\theta/2) = 1 - \cos \theta$, it follows that

$$
\begin{aligned}
s &= 4\sqrt{2} \int_0^\pi \sqrt{2 \sin^2 \frac{\theta}{2}} \, d\theta = 4\sqrt{2} \int_0^\pi \sqrt{2} \sin \frac{\theta}{2} \, d\theta \\
&= 8\left(-2 \cos \frac{\theta}{2}\right) \Big|_0^\pi = 16 \text{ units} \quad \blacksquare
\end{aligned}
$$

Problem Set 9.3

In Problems 1 to 6, find the area enclosed by each polar curve between the indicated values of θ. Sketch the curve.

1 $r = 4\theta$ from $\theta = \pi/4$ to $\theta = 5\pi/4$

2 $r = \sin^2 (\theta/6)$ from $\theta = 0$ to $\theta = \pi$

3 $r = 4/\theta$ from $\theta = \pi/4$ to $\theta = 5\pi/4$

4 $r = 3 \sin 3\theta$ from $\theta = 0$ to $\theta = \pi/3$

5 $r = 3 \csc \theta$ from $\theta = \pi/2$ to $\theta = 5\pi/6$

6 $r = 2 \sec^2 \theta$ from $\theta = 0$ to $\theta = \pi/3$

In Problems 7 to 14, find the area enclosed by each polar curve. In each case sketch the curve, determine the appropriate limits of integration, and make certain that none of the area is being counted twice.

7 $r = 4 \sin \theta$

8 $r = 2(1 + \cos \theta)$

9 $r = 2 \sin 3\theta$

10 $r = 2\sqrt{|\cos \theta|}$

11 $r = 2 + \cos \theta$

12 $r = 1 + 2 \cos \theta$

13 $r = 2 - 2 \sin \theta$

14 $r = 3 \cos \theta + 4 \sin \theta$

In Problems 15 to 22, find all points of intersection of each pair of polar curves. Sketch the two curves, showing these points.

15 $r = -3 \sin \theta$ and $r = 2 + \sin \theta$

16 $r = \cos \theta$ and $r = \sin \theta$

17 $r = 1$ and $r = 2 \cos 3\theta$

18 $r = \theta$ for $\theta \geq 0$ and $r = -\theta$ for $\theta \geq 0$

19 $r = \sin \theta$ and $r = \sin 2\theta$

20 $r = 1 + \cos \theta$ and $r = (1 + \sqrt{2}) \cos \theta$

21 $r = 2 \sin 3\theta$ and $r = -2/\sin \theta$ **22** $r = \theta$ and $r = 2\theta$

In Problems 23 to 27, find all points of intersection of the two curves, sketch the two curves, and find the area of each region described.

23 Inside $r = 6 \sin \theta$ and outside $r = 3$.

24 Inside both $r = 1$ and $r = 1 + \cos \theta$.

25 Inside both $r = 3 \cos \theta$ and $r = 1 + \cos \theta$.

26 Inside $r^2 = 8 \cos 2\theta$ and outside $r = 2$.

27 Inside $r = 6 \cos \theta$ and outside $r = 6(1 - \cos \theta)$.

28 Assuming that $b > a > 0$, find a formula for the area enclosed by the inner loop of the limaçon $r = a + b \sin \theta$.

29 Find the area of the shaded region enclosed by the graph of $r = \theta$ in Figure 13.

Figure 13

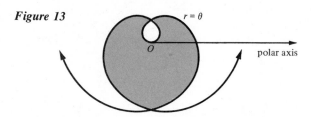

30 Find the area of the region inside the limaçon $r = 2 - 4 \cos \theta$ and outside the circle $r = -6 \cos \theta$ (Figure 7).

In Problems 31 to 38, find the arc length of each polar curve between the indicated values of θ.

31 $r = 6 \sin \theta$ from $\theta = 0$ to $\theta = \pi$

32 $r = -2$ from $\theta = 0$ to $\theta = 2\pi$

33 $r = 4\theta^2$ from $\theta = 0$ to $\theta = \frac{3}{2}$

34 $r = 2(1 + \cos \theta)$ from $\theta = 0$ to $\theta = 2\pi$

35 $r = e^\theta$ from $\theta = 0$ to $\theta = 4\pi$

36 $r = 2 \sin^3 (\theta/3)$ from $\theta = 0$ to $\theta = 3\pi$

37 $r = 3 \cos \theta + 4 \sin \theta$ from $\theta = 0$ to $\theta = \pi/2$

38 $r = -7 \csc \theta$ from $\theta = \pi/4$ to $\theta = 3\pi/4$

39 Find the error in the following argument: Since the graph of the polar equation $r = 4 \cos \theta$ is a circle of radius 2, the circumference of this circle should be given by

$$s = \int_0^{2\pi} \sqrt{\left(\frac{dr}{d\theta}\right)^2 + r^2} \, d\theta$$

$$= \int_0^{2\pi} \sqrt{16 \sin^2 \theta + 16 \cos^2 \theta} \, d\theta = \int_0^{2\pi} 4 \, d\theta = 8\pi$$

(But the circumference of a circle of radius 2 should be only 4π.)

40 Criticize the following naive use of differentials: In Figure 14, ds is virtually the arc length of the portion of the circumference of a circle of radius $|r|$ cut off by the angle $d\theta$ radians; therefore, ds should be given by $|r| \, d\theta$ rather than by $\sqrt{(dr/d\theta)^2 + r^2} \, d\theta$.

Figure 14

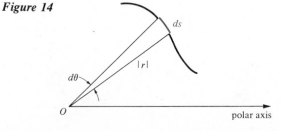

41 Show that, with suitable restrictions, the area of the surface generated by revolving the portion of the polar curve $r = f(\theta)$ between $\theta = \alpha$ and $\theta = \beta$ about the polar axis is

$$A = 2\pi \int_{\alpha}^{\beta} r \sin \theta \sqrt{\left(\frac{dr}{d\theta}\right)^2 + r^2} \, d\theta$$

42 Show that, with suitable restrictions, the area of the surface generated by revolving the portion of the polar curve $r = f(\theta)$ between $\theta = \alpha$ and $\theta = \beta$ about the axis $\theta = \pi/2$ is

$$A = 2\pi \int_{\alpha}^{\beta} r \cos \theta \sqrt{\left(\frac{dr}{d\theta}\right)^2 + r^2} \, d\theta$$

In Problems 43 to 46, use the formulas given in Problems 41 and 42 to find the area of the surface generated by revolving each polar curve about the given axis.

43 $r = 2$ about the polar axis

44 $r = 4$ about the axis $\theta = \pi/2$

45 $r = 5 \cos \theta$ about the polar axis

46 $r^2 = 4 \cos 2\theta$ about the axis $\theta = \pi/2$

9.4 The Ellipse

For the remainder of this chapter we study the **conic sections** (or **conics,** for short). These graceful curves were well known to the ancients; however, their study is immensely enhanced by the use of analytic geometry and calculus. They are obtained by sectioning, or cutting, a right circular cone of two nappes with a suitable plane (Figure 1).

Figure 1

| Circle | Ellipse | Parabola | Hyperbola |

By shining a flashlight onto a white wall, you can see examples of the conic sections. If the axis of the flashlight is perpendicular to the wall, then the illuminated region is *circular* (Figure 2a); if the flashlight is tilted slightly upward, the illuminated region elongates and its boundary becomes an *ellipse* (Figure 2b). As the flashlight is tilted further, the ellipse becomes more and more elongated until it changes into a *parabola* (Figure 2c). Finally, if the flashlight is tilted still further, the edges of the parabola become straighter and it changes into a portion of a *hyperbola* (Figure 2d).

We have already obtained equations for circles, in both Cartesian and polar form. In this section, we derive Cartesian equations for ellipses, and we see how such

Figure 2

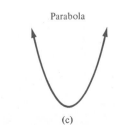 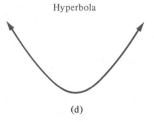

| Circle | Ellipse | Parabola | Hyperbola |

(a)   (b)   (c)   (d)

equations can be simplified by translation of the coordinate axes. Parabolas and hyperbolas in Cartesian form are studied in the next two sections.*

Ellipses are of practical importance in fields ranging from art to astronomy. For instance, a circular object viewed in perspective forms an ellipse (Figure 3), and an orbiting satellite (natural or artificial) moves in an elliptical path. The geometric definition of an ellipse is as follows.

DEFINITION 1   **Ellipse**

Figure 3

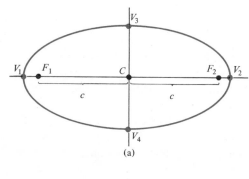

> An **ellipse** is the set of all points P in the plane such that the sum of the distances from P to two fixed points F_1 and F_2 is constant. Here F_1 and F_2 are called the **focal points,** or the **foci,** of the ellipse. The midpoint C of the line segment $\overline{F_1F_2}$ is called the **center** of the ellipse.

Figure 4a shows two fixed pins F_1 and F_2 and a loop of string of length l stretched tightly about them to the point P. Here we have

$$|\overline{PF_1}| + |\overline{PF_2}| + |\overline{F_1F_2}| = l \qquad \text{or} \qquad |\overline{PF_1}| + |\overline{PF_2}| = l - |\overline{F_1F_2}|$$

Figure 4

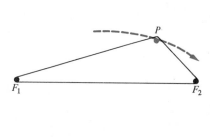 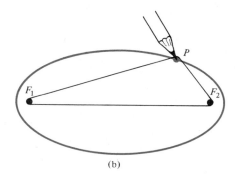

(a)   (b)

Figure 5

(a)

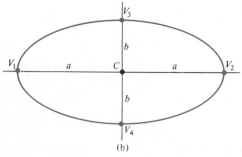

(b)

Hence, as P is moved about, $|\overline{PF_1}| + |\overline{PF_2}|$ always has the constant value $l - |\overline{F_1F_2}|$. Thus, if a pencil point P is inserted into the loop of string and moved so as to keep the string tight, it traces out an ellipse (Figure 4b).

Consider the ellipse in Figure 5a with foci F_1 and F_2 and center C. We denote by c the distance between the center C and either focus F_1 or F_2. Notice that the ellipse is symmetric about the line through F_1 and F_2. Let V_1 and V_2 be the points where this line intersects the ellipse. The center C bisects the line segment $\overline{V_1V_2}$, and the ellipse is symmetric about the line through C perpendicular to $\overline{V_1V_2}$. Let V_3 and V_4 be the points where this perpendicular intersects the ellipse. The four points V_1, V_2, V_3, and V_4 are called the **vertices** of the ellipse. The line segment $\overline{V_1V_2}$ is called the **major axis,** and the line segment $\overline{V_3V_4}$ is called the **minor axis** of the ellipse. Let $2a$ denote the length of the major axis, and let $2b$ denote the length of the minor axis (Figure 5b). The numbers a and b are called the **semimajor axis** and the **semiminor axis** of the ellipse.

If a point P moves along the ellipse in Figure 5, then, by definition, the sum

$$|\overline{PF_1}| + |\overline{PF_2}|$$

does not change. Therefore, its value when P reaches V_1 is the same as its value when P reaches V_3; that is,

$$|\overline{V_1F_1}| + |\overline{V_1F_2}| = |\overline{V_3F_1}| + |\overline{V_3F_2}|$$

*See Section 9.8 for polar equations of the conics.

By symmetry

$$|\overline{V_3F_1}| = |\overline{V_3F_2}|$$

so the equation above can be rewritten

$$|\overline{V_1F_1}| + |\overline{V_1F_2}| = 2|\overline{V_3F_2}|$$

But, by symmetry again,

$$|\overline{V_1F_2}| = |\overline{V_2F_1}|$$

Hence,

$$2|\overline{V_3F_2}| = |\overline{V_1F_1}| + |\overline{V_1F_2}| = |\overline{V_1F_1}| + |\overline{V_2F_1}| = |\overline{V_1V_2}| = 2a$$

from which it follows that

$$|\overline{V_3F_2}| = a$$

Therefore, applying the Pythagorean theorem to the right triangle V_3CF_2 (Figure 6), we find that

$$a^2 = b^2 + c^2$$

Figure 6

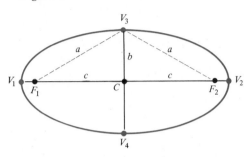

If we place the ellipse of Figure 6 in the xy plane so that its center C is at the origin O and the foci F_1 and F_2 lie on the negative and positive portions of the x axis, respectively, then we can derive an equation of the ellipse as follows.

THEOREM 1

Ellipse Equation in Cartesian Form

An equation of the ellipse with foci at $F_1 = (-c, 0)$ and $F_2 = (c, 0)$ is

$$\frac{x^2}{a^2} + \frac{y^2}{b^2} = 1$$

where a is the semimajor axis, b is the semiminor axis, and $a^2 = b^2 + c^2$.

PROOF

Let $P = (x, y)$ be any point on the ellipse (Figure 7). As Figure 6 shows, when $P = (0, b)$, we have

$$|\overline{PF_1}| + |\overline{PF_2}| = 2a$$

Therefore, by Definition 1, the equation

$$|\overline{PF_1}| + |\overline{PF_2}| = 2a$$

holds for every point $P = (x, y)$ on the ellipse. Using the distance formula, we can rewrite this as

Figure 7

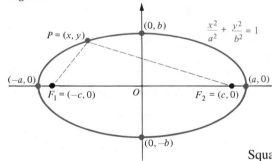

$$\sqrt{(x + c)^2 + y^2} + \sqrt{(x - c)^2 + y^2} = 2a$$

or

$$\sqrt{(x + c)^2 + y^2} = 2a - \sqrt{(x - c)^2 + y^2}$$

Squaring both sides of the last equation, we have

$$x^2 + 2cx + c^2 + y^2 = 4a^2 - 4a\sqrt{(x - c)^2 + y^2} + x^2 - 2cx + c^2 + y^2$$

so that

$$4cx - 4a^2 = -4a\sqrt{(x - c)^2 + y^2} \qquad \text{or} \qquad cx - a^2 = -a\sqrt{(x - c)^2 + y^2}$$

Squaring both sides of the last equation, we obtain

$$c^2x^2 - 2a^2cx + a^4 = a^2(x^2 - 2cx + c^2 + y^2)$$

so that

$$a^4 - a^2c^2 = (a^2 - c^2)x^2 + a^2y^2 \qquad \text{or} \qquad a^2(a^2 - c^2) = (a^2 - c^2)x^2 + a^2y^2$$

Since $a^2 = b^2 + c^2$, we have $a^2 - c^2 = b^2$, and the equation above can be rewritten as

$$a^2b^2 = b^2x^2 + a^2y^2$$

If both sides of the last equation are divided by a^2b^2, the result is

$$1 = \frac{x^2}{a^2} + \frac{y^2}{b^2}$$

as desired. By reversing the argument above, it can be shown that, conversely, if the equation $(x^2/a^2) + (y^2/b^2) = 1$ holds, then the point $P = (x, y)$ is on the ellipse (Problem 48). ∎

Figure 8

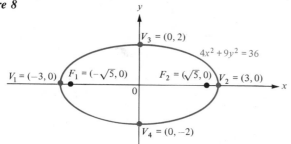

If a and b are positive constants and $a > b$, the Cartesian equation

$$\boxed{\frac{x^2}{a^2} + \frac{y^2}{b^2} = 1}$$

is called the **standard form** for the equation of an ellipse with center at the origin O and with a horizontal major axis.

EXAMPLE 1 Find the coordinates of the four vertices and the two foci of the ellipse $4x^2 + 9y^2 = 36$, and sketch the graph.

SOLUTION   Divide both sides of the equation by 36 to obtain

$$\frac{x^2}{9} + \frac{y^2}{4} = 1$$

that is,

$$\frac{x^2}{a^2} + \frac{y^2}{b^2} = 1$$

with $a = 3$ and $b = 2$. By Theorem 1, this is the equation of an ellipse with vertices $(-3, 0)$, $(3, 0)$, $(0, 2)$, and $(0, -2)$ (Figure 8). Also, the foci are $(-c, 0)$ and $(c, 0)$, where

$$c^2 = a^2 - b^2 = 9 - 4 = 5$$

that is, $c = \sqrt{5}$. Thus, $F_1 = (-\sqrt{5}, 0)$ and $F_2 = (\sqrt{5}, 0)$. ∎

Figure 9

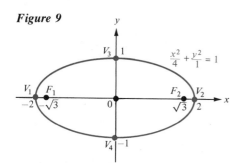

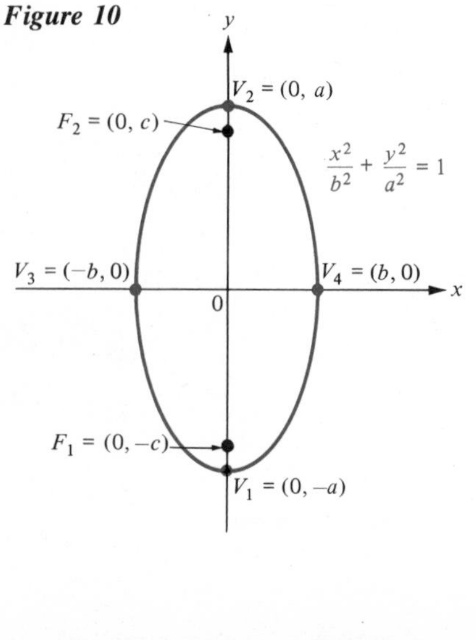

EXAMPLE 2 Find the equation in the standard form of the ellipse with foci $F_1 = (-\sqrt{3}, 0)$ and $F_2 = (\sqrt{3}, 0)$ and vertices $V_1 = (-2, 0)$ and $V_2 = (2, 0)$. Also, find the coordinates of the remaining two vertices, V_3 and V_4, and sketch the graph.

SOLUTION   Here $c = \sqrt{3}$, $a = 2$; hence, $b = \sqrt{a^2 - c^2} = \sqrt{4 - 3} = 1$, and the equation is

$$\frac{x^2}{4} + \frac{y^2}{1} = 1$$

Also, $V_3 = (0, 1)$ and $V_4 = (0, -1)$. The graph appears in Figure 9. ∎

Figure 10

It is not difficult to derive an equation of an ellipse with center at the origin and with a vertical major axis (Figure 10). In this case, the ellipse has foci $F_1 = (0, -c)$

and $F_2 = (0, c)$ on the y axis and vertices $V_1 = (0, -a)$, $V_2 = (0, a)$, $V_3 = (-b, 0)$, and $V_4 = (b, 0)$. The semimajor axis is a, and the semiminor axis is b. The equation can be derived as in Theorem 1, the argument being word for word the same except that the variables x and y interchange their roles. Therefore, the equation is

$$\frac{x^2}{b^2} + \frac{y^2}{a^2} = 1 \qquad \text{where } a > b$$

Figure 11

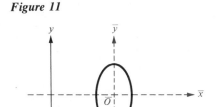

This equation is also called the **standard form** for the equation of an ellipse.

Translation of Axes

The equation of a curve can often be simplified by changing to a different coordinate system. In practice, this is often accomplished by choosing one or both of the new coordinate axes to coincide with an axis of symmetry of the curve. For instance, the equation of the ellipse in Figure 11 would probably be simplified by switching from the xy coordinate system to the $\overline{x}\overline{y}$ coordinate system as shown.

If two Cartesian coordinate systems have corresponding axes that are parallel and have the same positive directions, then we say that these systems are obtained from one another by **translation.**

Figure 12

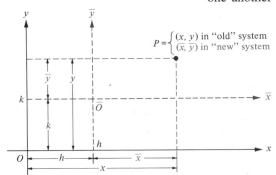

$$P = \begin{cases} (x, y) \text{ in "old" system} \\ (\overline{x}, \overline{y}) \text{ in "new" system} \end{cases}$$

Figure 12 shows a translation of the "old" xy coordinate system to a "new" $\overline{x}\overline{y}$ system whose origin $\overline{O}$ has the "old" coordinates (h, k). Consider the point P in Figure 12 having old coordinates (x, y), but having new coordinates $(\overline{x}, \overline{y})$. Evidently, we have the following *equations for translation of Cartesian coordinates:*

$$\begin{cases} x = h + \overline{x} \\ y = k + \overline{y} \end{cases} \quad \text{or} \quad \begin{cases} \overline{x} = x - h \\ \overline{y} = y - k \end{cases}$$

EXAMPLE 3   Let the $\overline{x}\overline{y}$ axes be obtained from the xy axes by translation in such a way that the origin $\overline{O}$ of the new coordinate system has coordinates $(h, k) = (-3, 4)$ in the old coordinate system. Let P be the point whose old coordinates are $(x, y) = (2, 1)$. Find the coordinates $(\overline{x}, \overline{y})$ of P in the new coordinate system (Figure 13).

Figure 13

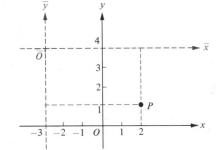

SOLUTION   According to the translation equations,

$$\overline{x} = x - h = 2 - (-3) = 5$$

and

$$\overline{y} = y - k = 1 - 4 = -3$$

Hence, the point P has coordinates $(\overline{x}, \overline{y}) = (5, -3)$ in the new coordinate system. ∎

Translation of axes can be a useful tool for graph sketching. For instance, suppose we want to sketch the graph of the equation

$$4x^2 + 9y^2 - 32x - 36y + 64 = 0$$

Offhand, it is perhaps not clear what the general shape of this graph might be. Of course, we could plot points, but this could become quite tedious. However, as we shall show, we can greatly simplify this equation—and identify its graph—by using the translation equations to switch to an $\overline{x}\overline{y}$ coordinate system.

We begin by rewriting the equation in the form

$$4x^2 - 32x + 9y^2 - 36y = -64$$

or

$$4(x^2 - 8x \quad) + 9(y^2 - 4y \quad) = -64$$

where we have left spaces for completing the squares. Thus,

$$4(x^2 - 8x + 16) + 9(y^2 - 4y + 4) = -64 + 4(16) + 9(4)$$

or

$$4(x - 4)^2 + 9(y - 2)^2 = 36$$

Dividing both sides of the last equation by 36, we obtain

$$\frac{(x - 4)^2}{9} + \frac{(y - 2)^2}{4} = 1$$

which is of the form

$$\frac{\bar{x}^2}{9} + \frac{\bar{y}^2}{4} = 1$$

with

$$\bar{x} = x - 4 \qquad \text{and} \qquad \bar{y} = y - 2$$

The last two equations represent a translation of the coordinate axes, and the new $\bar{x}\bar{y}$ coordinate system has an origin $\bar{O}$ whose old xy coordinates are (4, 2) (Figure 14). With respect to the $\bar{x}\bar{y}$ coordinate system, the graph of

$$\frac{\bar{x}^2}{9} + \frac{\bar{y}^2}{4} = 1$$

is an ellipse with center $\bar{O}$, semimajor axis $\sqrt{9} = 3$, and semiminor axis $\sqrt{4} = 2$, which we have sketched in Figure 14. But this is the same as the graph of the equation

$$\frac{(x - 4)^2}{9} + \frac{(y - 2)^2}{4} = 1$$

which, in turn, is the same as the graph of the original equation

$$4x^2 + 9y^2 - 32x - 36y + 64 = 0$$

Thus, by means of a translation of coordinates, we have obtained an accurate graph with a minimum of effort. (If desired, the new $\bar{x}\bar{y}$ coordinate system can be drawn very lightly and then erased after it has done its job.)

EXAMPLE 4   Show that the graph of $25x^2 + 9y^2 - 100x - 54y = 44$ is an ellipse. Find the coordinates of the center, the vertices, and the foci, and sketch the graph.

SOLUTION   Here we have

$$25(x^2 - 4x \quad) + 9(y^2 - 6y \quad) = 44$$

Completing the squares in the last equation, we obtain

$$25(x^2 - 4x + 4) + 9(y^2 - 6y + 9) = 44 + 25(4) + 9(9)$$

or

$$25(x - 2)^2 + 9(y - 3)^2 = 225$$

Dividing the last equation by 225, we have

$$\frac{(x - 2)^2}{9} + \frac{(y - 3)^2}{25} = 1$$

Figure 14

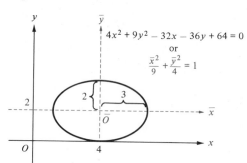

which is of the form

$$\frac{\bar{x}^2}{9} + \frac{\bar{y}^2}{25} = 1$$

with $\qquad\qquad \bar{x} = x - 2 \qquad$ and $\qquad \bar{y} = y - 3$

The last two equations represent a translation of the coordinate axes, and the new $\overline{xy}$ coordinate system has an origin $\bar{O}$ whose old xy coordinates are $(2, 3)$. The graph of

$$\frac{\bar{x}^2}{9} + \frac{\bar{y}^2}{25} = 1$$

is an ellipse with center $\bar{O}$, *vertical* major axis, semimajor axis $a = \sqrt{25} = 5$, and semiminor axis $b = \sqrt{9} = 3$ (Figure 15). The distance c between the center and the foci is given by

$$c = \sqrt{a^2 - b^2} = \sqrt{25 - 9} = 4 \text{ units}$$

Thus, the foci F_1 and F_2 have xy coordinates

$$F_1 = (2, 3 - 4) = (2, -1) \quad \text{and} \quad F_2 = (2, 3 + 4) = (2, 7)$$

and the xy coordinates of the four vertices are

$$(2, -2) \qquad (2, 8) \qquad (-1, 3) \qquad \text{and} \qquad (5, 3)$$

Figure 15

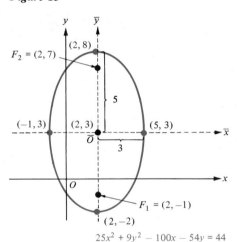

$$25x^2 + 9y^2 - 100x - 54y = 44$$

Notice that a translation of the coordinate axes does not change the *position* or the *shape* of a curve in the plane—it changes only the *equation* of this curve. Thus, by using the translation equations $\bar{x} = x - h$ and $\bar{y} = y - k$, we can see that an equation of the form

$$\frac{(x - h)^2}{a^2} + \frac{(y - k)^2}{b^2} = 1 \quad \text{or} \quad \frac{(x - h)^2}{b^2} + \frac{(y - k)^2}{a^2} = 1$$

where $a > b$, represents an ellipse with center (h, k), semimajor axis a, and semiminor axis b. In the first case, the major axis is horizontal; in the second case, it is vertical.

EXAMPLE 5   Write the equation of the ellipse whose vertices are the points $(-5, 1)$, $(1, 1)$, $(-2, 3)$, and $(-2, -1)$; find the foci, and sketch the graph.

SOLUTION   The horizontal axis is the line segment from $(-5, 1)$ to $(1, 1)$, and its length is $|(-5) - 1| = 6$ units. The vertical axis is the line segment from $(-2, -1)$ to $(-2, 3)$, and its length is $|(-1) - 3| = 4$ units. Therefore, the ellipse has a horizontal major axis,

$$a = \tfrac{6}{2} = 3 \qquad b = \tfrac{4}{2} = 2 \qquad \text{and} \qquad c = \sqrt{a^2 - b^2} = \sqrt{5}$$

Here, the center is at $(-2, 1)$, so $h = -2$ and $k = 1$. Consequently, the equation of the ellipse is

$$\frac{(x + 2)^2}{9} + \frac{(y - 1)^2}{4} = 1$$

The foci are $c = \sqrt{5}$ units on either side of the center; hence, $F_1 = (-\sqrt{5} - 2, 1)$ and $F_2 = (\sqrt{5} - 2, 1)$ (Figure 16). ∎

Figure 16

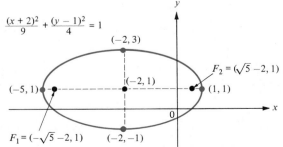

Problem Set 9.4

In Problems 1 to 8, find the coordinates of the vertices and foci of each ellipse, and sketch its graph.

1 $\dfrac{x^2}{16} + \dfrac{y^2}{4} = 1$

2 $\dfrac{x^2}{9} + y^2 = 1$

3 $4x^2 + y^2 = 16$

4 $36x^2 + 9y^2 = 144$

5 $x^2 + 16y^2 = 16$

6 $16x^2 + 25y^2 = 400$

7 $9x^2 + 36y^2 = 4$

8 $x^2 + 4y^2 = 1$

In Problems 9 to 12, find the equation in standard form of the ellipse that satisfies the conditions given.

9 Foci $F_1 = (-4, 0)$ and $F_2 = (4, 0)$; vertices $V_3 = (0, 3)$ and $V_4 = (0, -3)$

10 Vertices $V_1 = (-5, 0)$ and $V_2 = (5, 0)$; horizontal major axis; $c = 3$ units

11 Foci $F_1 = (0, -12)$ and $F_2 = (0, 12)$; vertices $V_1 = (0, -13)$ and $V_2 = (0, 13)$

12 Foci $F_1 = (0, -8)$ and $F_2 = (0, 8)$; semiminor axis $b = 6$ units

13 A new $\overline{x}\overline{y}$ coordinate system is obtained by translating the old xy coordinate system so that the origin $\overline{O}$ of the new system has old xy coordinates $(-1, 2)$. Find the new $\overline{x}\overline{y}$ coordinates of the points whose old xy coordinates are:

(a) $(0, 0)$ (b) $(-2, 1)$ (c) $(3, -3)$

(d) $(-3, -2)$ (e) $(5, 5)$ (f) $(6, 0)$

14 Describe a translation of coordinates that will reduce the equation

$$x^2 + y^2 + 4x - 2y + 1 = 0$$

of a circle in the old xy coordinate system to the simpler form $\overline{x}^2 + \overline{y}^2 = r^2$ in the new $\overline{x}\overline{y}$ system. Find the radius r, and draw a graph showing the circle and the two coordinate systems.

15 A new $\overline{x}\overline{y}$ coordinate system is obtained by translating the old xy coordinate system so that the origin O of the old xy system has coordinates $(-3, 2)$ in the new $\overline{x}\overline{y}$ system. Find the old xy coordinates of the points whose new $\overline{x}\overline{y}$ coordinates are:

(a) $(0, 0)$ (b) $(3, 2)$ (c) $(-3, 4)$

(d) $(\sqrt{2}, -2)$ (e) $(0, -\pi)$ (f) $(-3, 2)$

16 Let b, c, and d be constants. Show that the translation $\overline{x} = x + (b/3)$, $\overline{y} = y + (cb/3) - d - (2b^3/27)$ reduces the cubic equation $\overline{y} = x^3 + bx^2 + cx + d$ to the form $\overline{y} = \overline{x}^3 + p\overline{x}$, where $p = c - (b^2/3)$.

In Problems 17 to 22, use a suitable translation of axes $\overline{x} = x - h$ and $\overline{y} = y - k$ to bring each equation into the standard form for an ellipse. Find the xy coordinates of the center, the vertices, and the foci, and sketch the graph showing both the old xy and the new $\overline{x}\overline{y}$ coordinate systems.

17 $x^2 + 4y^2 + 2x - 8y + 1 = 0$

18 $9x^2 + y^2 - 18x + 2y + 9 = 0$

19 $6x^2 + 9y^2 - 24x - 54y + 51 = 0$

20 $9x^2 + 4y^2 - 18x + 16y - 11 = 0$

21 $16x^2 + 9y^2 - 192x + 36y + 468 = 0$

22 $3x^2 + 4y^2 - 12x + 8y + 4 = 0$

In Problems 23 to 30, find the coordinates of the center, the vertices, and the foci of each ellipse, and sketch its graph.

23 $\dfrac{(x - 1)^2}{9} + \dfrac{(y + 2)^2}{4} = 1$

24 $\dfrac{(x + 2)^2}{16} + \dfrac{(y - 1)^2}{4} = 1$

25 $4(x + 3)^2 + y^2 = 36$

26 $25(x + 1)^2 + 16(y - 2)^2 = 400$

27 $x^2 + 2y^2 + 6x + 7 = 0$

28 $4x^2 + y^2 - 8x + 4y - 8 = 0$

29 $2x^2 + 5y^2 + 20x - 30y + 75 = 0$

30 $9x^2 + 4y^2 + 18x - 16y - 11 = 0$

In Problems 31 to 42, find an equation for the ellipse satisfying the conditions given.

31 Foci $(-4, 1)$ and $(4, 1)$; vertices $(-5, 1)$ and $(5, 1)$

32 Foci $(1, -2)$ and $(1, 2)$; vertices $(1, -4)$ and $(1, 4)$

33 Vertices $(0, -8)$ and $(0, 8)$; containing the point $(6, 0)$

34 Vertices $(0, -3)$ and $(0, 3)$; containing the point $(\frac{2}{3}, 2\sqrt{2})$

35 Center at the origin; containing the points $(4, 0)$ and $(3, 2)$

36 Vertices $(-2\sqrt{3}, 0)$, $(2\sqrt{3}, 0)$, $(0, -4)$, and $(0, 4)$

37 Vertices $(-2, -3)$, $(-2, 5)$, $(-7, 1)$, and $(3, 1)$

38 Foci $(1, 3)$ and $(5, 3)$; major axis 10 units long

39 Center $(1, -2)$; major axis parallel to the y axis, major axis 6 units long; minor axis 4 units long

40 Endpoints of major axis $(-3, 2)$, $(5, 2)$; length of minor axis is 4 units

41 Vertices $(2, -3)$ and $(2, 5)$; foci $(2, -2)$ and $(2, 4)$

42 Foci $(-4, 5)$ and $(2, 5)$; semiminor axis $b = 3$ units

In Problems 43 to 46, use implicit differentiation to find the slopes and the equations of the tangent and normal lines to each ellipse at the point indicated.

43 $x^2 + 9y^2 = 225$ at $(9, 4)$ **44** $4x^2 + 9y^2 = 45$ at $(3, 1)$

45 $x^2 + 4y^2 - 2x + 8y = 35$ at $(3, 2)$

46 $9x^2 + 25y^2 - 50y - 200 = 0$ at $(5, 1)$

47 The segment cut by an ellipse from a line containing a focus and perpendicular to the major axis is called a **focal chord** of the ellipse. (a) Show that $2b^2/a$ is the length of a focal chord of the ellipse whose equation is $b^2x^2 + a^2y^2 = a^2b^2$. (b) Find the length of a focal chord of the ellipse $9x^2 + 16y^2 = 144$.

48 Suppose that a and b are constants such that $a > b > 0$ and let $c = \sqrt{a^2 - b^2}$. Assume that the numbers x and y satisfy the equation $(x^2/a^2) + (y^2/b^2) = 1$, and let $P = (x, y)$, $F_1 = (-c, 0)$, and $F_2 = (c, 0)$. Prove the following without reference to geometric diagrams:

(a) $c|x| < a^2$ (b) $\sqrt{(x - c)^2 + y^2} < 2a$

(c) $|\overline{PF_1}| + |\overline{PF_2}| = 2a$

49 Find the maximum area of a rectangle that can be inscribed in the ellipse $(x^2/a^2) + (y^2/b^2) = 1$ if the sides of the rectangle are parallel to the coordinate axes.

50 A mathematician has accepted a position at a new university situated 6 kilometers from the straight shoreline of a large lake (Figure 17). The professor wishes to build a home that is half as far from the university as it is from the shore of the lake. The possible homesites satisfying this condition lie along a curve. Describe this curve, and find its equation with respect to a coordinate system having the shoreline as the x axis and the university at the point $(0, 6)$ on the y axis.

Figure 17

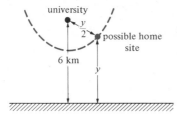

51 How long a loop of string should be used to lay out an elliptical flower bed 20 feet wide and 60 feet long? How far apart should the two stakes (foci) be?

52 Except for minor perturbations, a satellite orbiting the earth moves in an ellipse with the center of the earth at one focus. Suppose that a satellite at perigee (nearest point to center of earth) is 400 kilometers from the surface of the earth and at apogee (farthest point to center of earth) is 600 kilometers from the surface of the earth. Assume that the earth is a sphere of radius 6371 kilometers. Find the semiminor axis b of the elliptical orbit.

53 An arch in the shape of the upper half of an ellipse with a horizontal major axis is to support a bridge over a river 100 meters wide. The center of the arch is to be 25 meters above the surface of the river. Find the equation in standard form for the ellipse.

54 In Figure 4b, show that the semiminor axis b of the ellipse is given by $b = \frac{1}{2}\sqrt{l^2 - 4lc}$, where $c = \frac{1}{2}|\overline{F_1F_2}|$.

55 Let A be the area enclosed by an ellipse with semimajor axis a and semiminor axis b. (a) Show that

$$A = 4 \int_0^a \frac{b}{a}\sqrt{a^2 - x^2}\, dx$$

(b) Evaluate the integral in part (a) and thus obtain the formula $A = \pi ab$ for the *area of an ellipse*.

56 Prove that the normal line to the ellipse in Figure 18 at the point $P = (x, y)$ bisects the angle F_1PF_2. [*Hint:* The equation of the ellipse is $(x^2/a^2) + (y^2/b^2) = 1$. By implicit differentiation, the slope of the normal line at P is $m = a^2y/(b^2x)$. The slopes of $\overline{F_1P}$ and $\overline{F_2P}$, respectively, are

$$m_1 = \frac{y}{x + c} \quad \text{and} \quad m_2 = \frac{y}{x - c}$$

where $F_1 = (-c, 0)$, $F_2 = (c, 0)$, and $c = \sqrt{a^2 - b^2}$. Verify that $(m_1 + m_2)m^2 + 2(1 - m_1m_2)m = m_1 + m_2$. Then show that the last equation implies the desired result.]

Figure 18

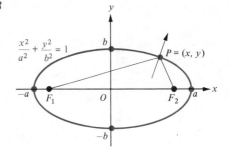

57 A ray of light will be reflected from a curved mirror (Figure 19) in such a way that the angle α between the incident ray and the normal is equal to the angle α between the normal and the reflected ray. Use the result of Problem 56 to show that a ray of light emanating from one focus of an elliptical mirror will be reflected through the other focus. (This is called the **reflecting property** of the ellipse.)

Figure 19

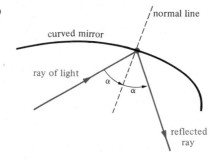

9.5 The Parabola

In Chapter 1 (page 34) we mentioned that the graph of a quadratic function is "a type of curve called a parabola." In this section we give a proof of this fact based on the following geometric definition.

DEFINITION 1 **Parabola**

> A **parabola** is the set of all points P in the plane such that the distance from P to a fixed point F (the **focus**) is equal to the distance from P to a fixed line D (the **directrix**).

Parabolas appear often in the real world. A ball thrown up at an angle travels along a parabolic arc (Figure 1), a main cable in a suspension bridge forms an arc of a parabola (Figure 2), and the familiar "dish antennas" have parabolic cross sections (Figure 3).

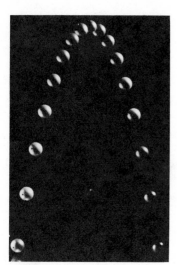

Figure 1   Multiflash photo of a ball thrown into the air

Figure 2   Bay Bridge, San Francisco, showing parabolic cables

Figure 3   Parabolic dish antenna

If the focus F of a parabola is placed on the positive y axis at the point $(0, p)$ and if the directrix D is placed parallel to the x axis and p units below it, the resulting parabola appears as in Figure 4. Its Cartesian equation is derived in the following theorem.

THEOREM 1 **Parabola Equation in Cartesian Form**

> An equation of the parabola with focus $F = (0, p)$ and with directrix $D: y = -p$ is $x^2 = 4py$, or $y = [1/(4p)]x^2$.

PROOF   Let $P = (x, y)$ be any point and let $Q = (x, -p)$ be the point at the foot of the perpendicular from P to the directrix D (Figure 4). The requirement for P to be on the parabola is $|\overline{FP}| = |\overline{PQ}|$; that is,

$$\sqrt{x^2 + (y - p)^2} = \sqrt{(y + p)^2}$$

Figure 4

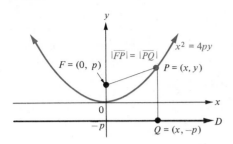

The last equation is equivalent to

$$x^2 + (y - p)^2 = (y + p)^2$$

that is,

$$x^2 + y^2 - 2py + p^2 = y^2 + 2py + p^2 \qquad \text{or} \qquad x^2 = 4py$$ ∎

Obvious modifications of the argument in Theorem 1 provide Cartesian equations for parabolas opening to the *right* (Figure 5b), *downward* (Figure 5c), and to the *left* (Figure 5d). Any one of the equations

$$x^2 = \pm 4py \qquad \text{or} \qquad y^2 = \pm 4px$$

is called the **standard form** for the equation of a parabola.

Figure 5

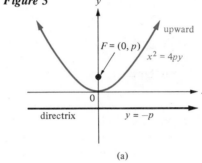

(a)

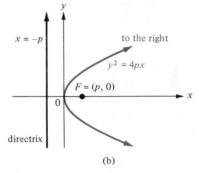

(b)

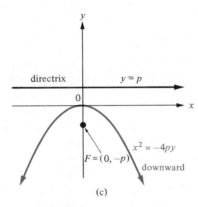

(c)

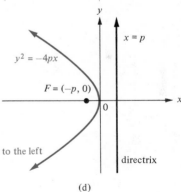

(d)

Figure 6

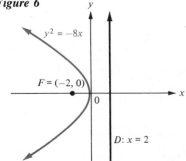

In Examples 1 and 2, find the coordinates of the focus and an equation of the directrix of the given parabola, determine its direction of opening, and sketch the graph.

<u>EXAMPLE 1</u> $y^2 = -8x$

SOLUTION   The equation has the form $y^2 = -4px$ with $p = 2$; hence, it corresponds to Figure 5d. Therefore, the graph is a parabola opening to the left with focus given by

$$F = (-p, 0) = (-2, 0)$$

and directrix D: $x = p$; that is, D: $x = 2$ (Figure 6). ∎

Figure 7

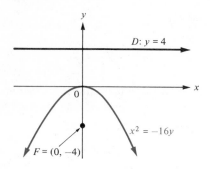

Figure 8

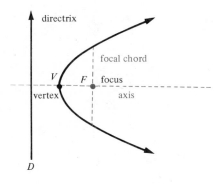

Figure 9

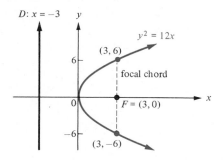

Figure 10

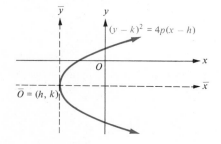

EXAMPLE 2 $x^2 = -16y$

SOLUTION   The equation has the form $x^2 = -4py$ with $p = 4$; hence, it corresponds to Figure 5c. Therefore, the graph is a parabola opening downward with focus given by

$$F = (0, -p) = (0, -4)$$

and directrix $D: y = p$; that is, $D: y = 4$ (Figure 7). ∎

A parabola with focus F and directrix D is evidently symmetric about the line through F perpendicular to D. This line is called the **axis** of the parabola (Figure 8). The axis intersects the parabola at the **vertex** V, which is located midway between the focus and the directrix. The segment cut by the parabola on the line through the focus and perpendicular to its axis is called the **focal chord,** or the **latus rectum,** of the parabola (Figure 8). The following example shows how to find the length of the focal chord.

EXAMPLE 3   A parabola opens to the right, has its vertex at the origin, and contains the point (3, 6). Find its equation, sketch the parabola, and find the length of its focal chord.

SOLUTION   The equation must have the form $y^2 = 4px$. Since the point (3, 6) belongs to the graph, we put $x = 3$ and $y = 6$ in the equation to obtain $36 = 12p$, and we conclude that $p = 3$. Thus, an equation of the parabola is $y^2 = 12x$ (Figure 9). The focus is given by $F = (p, 0) = (3, 0)$. The focal chord lies along the line $x = 3$. Putting $x = 3$ in the equation $y^2 = 12x$ and solving for y, we obtain $y^2 = 36$, $y = \pm 6$. Thus, the points (3, 6) and (3, −6) are the endpoints of the focal chord, and therefore its length is 12 units. ∎

A parabola with its vertex at the origin which opens upward, to the right, downward, or to the left is said to be **in standard position.** If a parabola is in the standard position, then its axis of symmetry is either horizontal or vertical. Conversely, if a parabola has an axis of symmetry that is either horizontal or vertical, then a translation of the coordinate axes to the vertex of the parabola will put it into standard position with respect to the "new" coordinate system.

For instance, if the parabola in Figure 10 has its vertex at the point (h, k) with respect to the old xy coordinate system and opens to the right as shown, then its equation with respect to the new $\bar{x}\bar{y}$ coordinate system is $\bar{y}^2 = 4p\bar{x}$. Since $\bar{x} = x - h$ and $\bar{y} = y - k$, its equation with respect to the old xy coordinate system is $(y - k)^2 = 4p(x - h)$. Similar arguments can be made for parabolas opening upward, downward, or to the left. The following table summarizes the results:

Vertex	Opening	Equation
(h, k)	Upward	$(x - h)^2 = 4p(y - k)$
(h, k)	To the right	$(y - k)^2 = 4p(x - h)$
(h, k)	Downward	$(x - h)^2 = -4p(y - k)$
(h, k)	To the left	$(y - k)^2 = -4p(x - h)$

In particular, the equation of a parabola with a *vertical axis* may be written in the form

$$y = a(x - h)^2 + k$$

where $a = 1/4p$ if the parabola opens upward and $a = -1/4p$ if it opens downward. (See Theorem 1 on page 33.)

In Examples 4 and 5, find the coordinates of the vertex V and the focus F of the given parabola, determine the direction in which it opens, find the equation of its directrix, determine the length of its focal chord, and sketch the graph.

Figure 11

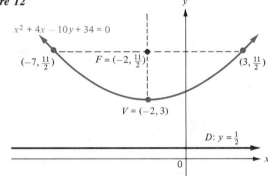

$(y + 1)^2 = -12(x - 2)$

D: x = 5

(-1, 5)

focal chord

F = (-1, -1)

V = (2, -1)

(-1, -7)

<u>EXAMPLE 4</u> $(y + 1)^2 = -12(x - 2)$

SOLUTION   The equation can be written as

$$(y - k)^2 = -4p(x - h)$$

with $p = 3$, $h = 2$, and $k = -1$. Hence, the parabola opens to the left, $V = (2, -1)$, $F = (2 - 3, -1) = (-1, -1)$, and the directrix is $D: x = 5$. Since the x coordinate of the focus is -1, the focal chord lies along the vertical line $x = -1$. Putting $x = -1$ in the equation of the parabola, we obtain $(y + 1)^2 = 36$, so that $y + 1 = \pm 6$; that is, $y = 5$ or $y = -7$. Therefore, the endpoints of the focal chord are $(-1, 5)$ and $(-1, -7)$, and its length is 12 units (Figure 11). ∎

Figure 12

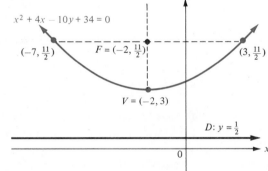

$x^2 + 4x - 10y + 34 = 0$

$(-7, \frac{11}{2})$ $F = (-2, \frac{11}{2})$ $(3, \frac{11}{2})$

$V = (-2, 3)$

$D: y = \frac{1}{2}$

<u>EXAMPLE 5</u> $x^2 + 4x - 10y + 34 = 0$

SOLUTION   Completing the square, we obtain

$$x^2 + 4x + 4 - 10y + 34 = 4$$

or $(x + 2)^2 = 10y - 30$

that is, $(x + 2)^2 = 10(y - 3)$. Thus, $p = \frac{10}{4} = \frac{5}{2}$, and the graph is a parabola opening upward with vertex $V = (-2, 3)$, focus $F = (-2, \frac{11}{2})$, and directrix $D: y = \frac{1}{2}$ (Figure 12). Here, the focal chord lies along the horizontal line $y = \frac{11}{2}$. Putting $y = \frac{11}{2}$ in the equation of the parabola, we obtain $(x + 2)^2 = 10(\frac{11}{2}) - 30 = 25$, so $x = 3$ or $x = -7$. Therefore, the endpoints of the focal chord are $(-7, \frac{11}{2})$ and $(3, \frac{11}{2})$, and its length is 10 units. ∎

Of course, you can find the slope dy/dx of the tangent line (and therefore the slope $-1/(dy/dx)$ of the normal line) to a parabola at a given point by differentiation as usual. (Implicit differentiation may be useful here.) This can be used for locating the vertex of a parabola. For instance, the parabola whose equation has the form $Ax^2 + Bx + C = Ky$, where A, B, C, and K are constants, has a vertical axis (Figure 13). Its tangent line is horizontal only at its vertex. Thus, the vertex can be found by calculating dy/dx and setting it equal to zero, as illustrated in the following example.

Figure 13

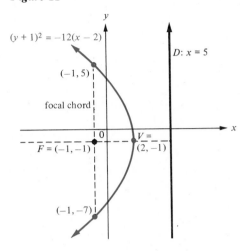

axis

$Ax^2 + Bx + C = Ky$

horizontal tangent at vertex

$-\frac{B}{2A}$

<u>EXAMPLE 6</u>   Find the vertex of the parabola $3y = 2x^2 + 4x + 5$.

SOLUTION   This parabola has a vertical axis of symmetry, so we set dy/dx equal to zero: $dy/dx = \frac{1}{3}(4x + 4) = 0$ for $x = -1$. When $x = -1$, $y = \frac{1}{3}[2(-1)^2 + 4(-1) + 5] = 1$. Therefore, the vertex is $(-1, 1)$. ∎

Figure 14

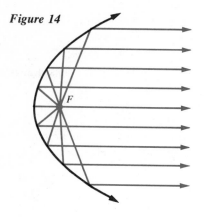

One of the most important properties of a parabola is its **reflecting property:** A ray of light emanating from the focus of a parabolic mirror is always reflected parallel to the axis (Figure 14). In order to demonstrate the reflecting property analytically, we recall that a ray of light striking a curved mirror is reflected so that the angle between the incident ray and the normal to the mirror is the same as the angle between the normal and the reflected ray.

Thus, in Figure 15 we wish to show that angle α is the same as angle β. By implicit differentiation of the equation $4px = y^2$ of the parabola,

$$4p = 2y\left(\frac{dy}{dx}\right) \qquad \text{or} \qquad \frac{dy}{dx} = \frac{2p}{y}$$

so the slope of the normal is $m = -y/(2p)$. The slope of the segment $\overline{FP}$ is $m_1 = y/(x - p)$. In Problem 37, we ask you to show that

$$\tan \alpha = \frac{m - m_1}{1 + mm_1} \qquad \text{and} \qquad \tan \beta = -m$$

Figure 15

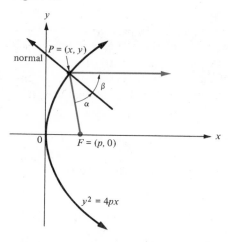

Thus, the reflecting property follows if it can be shown that

$$\frac{m - m_1}{1 + mm_1} = -m$$

The last equation is easy to verify by using the relations

$$m = -\frac{y}{2p} \qquad m_1 = \frac{y}{x - p} \qquad \text{and} \qquad 4px = y^2$$

(Problem 38).

If an intense source of light such as a carbon arc or an incandescent filament is placed at the focus of a parabolic mirror, the light is reflected and projected in a parallel beam. The same principle is used in reverse in a reflecting telescope—parallel rays of light from a distant object are brought together at the focus of a parabolic mirror.

In practice, it is very difficult to manufacture large parabolic mirrors, so it is often necessary to make do with mirrors whose cross section is a portion of a circle approximating the appropriate parabola (Figure 16). It can be shown (Problem 36) that the circle that "best approximates" the parabola near its vertex V has its center C located on the axis of the parabola twice as far from the vertex V as the focus F of the parabola and so has radius $r = 2|\overline{VF}| = 2p$.

Figure 16

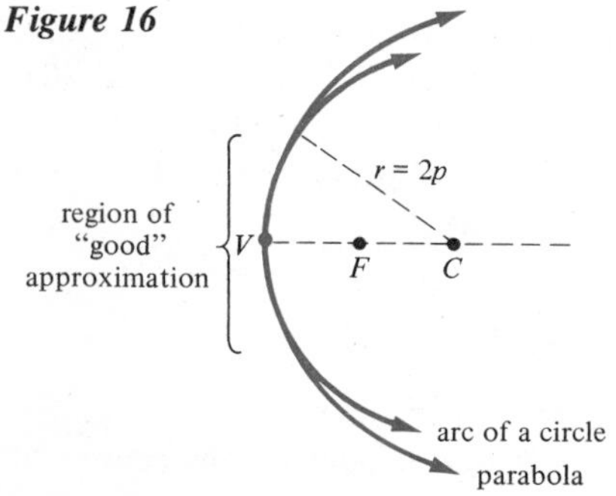

region of "good" approximation

$r = 2p$

arc of a circle

parabola

Problem Set 9.5

In Problems 1 to 6, find the coordinates of the vertex and the focus of the parabola. Also find an equation of the directrix and the length of the focal chord. Sketch the graph.

1 $y^2 = 4x$

2 $y^2 = -9x$

3 $x^2 - y = 0$

4 $x^2 - 4y = 0$

5 $x^2 + 9y = 0$

6 $3x^2 - 4y = 0$

7 Find an equation of the parabola whose focus is the point $(0, 3)$ and whose directrix is the line $y = -3$.

8 Find the vertex of the parabola $y = Ax^2 + Bx + C$, where A, B, and C are constants and $A \neq 0$.

In Problems 9 to 16, find the coordinates of the vertex and the focus of the parabola. Also find an equation of the directrix and the length of the focal chord. Sketch the graph.

9 $(y - 2)^2 = 8(x + 3)$

10 $(y + 1)^2 = -4(x - 1)$

11 $(x - 4)^2 = 12(y + 7)$

12 $(x + 1)^2 = -8y$

13 $y^2 - 8y - 6x - 2 = 0$

14 $2x^2 + 8x - 3y + 4 = 0$

15 $x^2 - 6x - 8y + 1 = 0$

16 $y^2 + 10y - x + 21 = 0$

In Problems 17 to 22, find an equation of the parabola that satisfies the conditions given.

17 Focus at (4, 2) and directrix $x = 6$

18 Focus at (3, −1) and directrix $y = 5$

19 Vertex at (−6, −5) and focus at (2, −5)

20 Vertex at (2, −3) and directrix $x = -8$

21 Axis is parallel to the x axis, vertex $(-\frac{1}{2}, -1)$, and contains the point $(\frac{5}{8}, 2)$.

22 Axis coincides with the y axis and parabola contains the points (2, 3) and (−1, −2).

In Problems 23 to 25, reduce each equation to "simpler" form by a suitable translation of axes. Also, sketch the graph in the old as well as the new coordinate system.

23 $y^2 + 2y - 8x - 3 = 0$

24 $x^2 + 2x + 4y - 7 = 0$

25 $5y = x^2 + 4x + 19$

26 Let A, B, and C be constants with $A > 0$. Show that $y = Ax^2 + Bx + C$ is an equation of a parabola with a vertical axis of symmetry, opening upward. Find the coordinates of the vertex V and the focus F. Find p and the length of the focal chord. Find conditions for the graph to intersect the x axis.

27 Find equations of the tangent and normal lines to each parabola at the point indicated.

(a) $y^2 = 8x$ at (2, −4)

(b) $2y^2 = 9x$ at (2, −3)

(c) $x^2 = -12y$ at (−6, −3)

(d) $x^2 + 8y + 4x - 20 = 0$ at $(1, \frac{15}{8})$

(e) $y^2 - 2y + 10x - 44 = 0$ at $(\frac{9}{2}, 1)$

28 How can you tell immediately (without computation) that $3y - 2 = 4x^2 - 27x + 11$ is the equation of a parabola opening upward?

29 Find the coordinates of the vertex of each parabola by differentiating its equation on both sides and then solving for a horizontal (or vertical) tangent.

(a) $y = x^2 - 2x + 6$

(b) $4x^2 + 24x + 39 - 3y = 0$

(c) $y^2 - 10y = 4x - 21$

(d) $3x = 14y - y^2 - 43$

30 Show that the arc length s of the parabola $4py = x^2$ between its vertex (0, 0) and the point $(a, a^2/(4p))$ is given by

$$s = \frac{a\sqrt{4p^2 + a^2}}{4p} + p \ln \frac{\sqrt{4p^2 + a^2} + a}{2p}$$

31 Find the dimensions of the rectangle of largest area whose base is on the x axis and whose two upper vertices are on the parabola whose equation is $y = 12 - x^2$ (Figure 17).

Figure 17

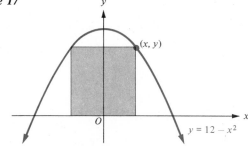

32 Show that, if p is the distance between the vertex and the focus of a parabola, then the focal chord of the parabola has length $4p$.

33 A roadway 400 meters long is held up by a parabolic main cable. The main cable is 100 meters above the roadway at the ends and 4 meters above the roadway at the center. Vertical supporting cables run at 50-meter intervals along the roadway. Find the lengths of these vertical cables. (*Hint:* Set up an xy coordinate system with vertical y axis and having the vertex of the parabola 4 units above the origin.)

34 The surface of a roadway over a stone bridge follows a parabolic curve with the vertex in the middle of the bridge. The span of the bridge is 60 meters, and the road surface is 1 meter higher in the middle than at the ends. How much higher than the ends is a point on the roadway 15 meters from an end?

35 For $a > 0$, let $(0, f(a))$ be the center of the circle tangent to the parabola whose equation is $4py = x^2$ at the points whose x coordinates are $-a$ and a (Figure 18). Find (a) a formula for $f(a)$ in terms of a and p and (b) $\lim_{a \to 0} f(a)$.

Figure 18

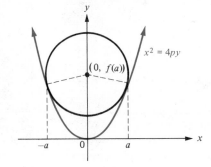

36 Find the circle that "best fits" the parabola $4py = x^2$ at and near its vertex by letting a approach 0 in Figure 18.

37 In Figure 15, show that $\tan \alpha = (m - m_1)/(1 + mm_1)$ and that $\tan \beta = -m$, where m is the slope of the normal line and m_1 is the slope of the segment $\overline{FP}$. [*Hint:* Use the formula

$$\tan (\theta - \theta_1) = \frac{\tan \theta - \tan \theta_1}{1 + \tan \theta \tan \theta_1}$$

from trigonometry.]

38 In Figure 15, verify that $\alpha = \beta$. [Use Problem 37 and the relations $m = -y/2p$, $m_1 = y/(x - p)$, and $4px = y^2$.]

39 Figure 19 shows a cross section of a parabolic dish antenna. Show that the focus F of the antenna is p units above the vertex V, where $p = a^2/(16b)$.

Figure 19

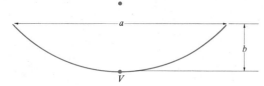

40 A parabola may be thought of as an enormous ellipse with one vertex infinitely far away. To see this, consider the ellipse in Figure 20. If we hold the lower vertex fixed at O and hold the lower focus fixed at $(0, p)$, but allow the upper vertex to approach $+\infty$ along the y axis, then the ellipse approaches the parabola $y = (1/4p)x^2$ as a limiting curve. Prove this.

Figure 20

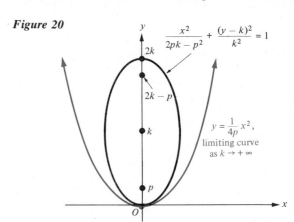

41 A calculus student says, "If you've seen one parabola, you've seen them all." (a) Explain why this is essentially true. (b) Explain why a similar statement cannot be made for ellipses.

9.6 The Hyperbola

Hyperbolas are of practical importance in fields ranging from engineering to navigation. The natural-draft evaporative cooling towers used at large electric power stations have hyperbolic cross sections (Figure 1); a comet or other object moving with more than enough kinetic energy to escape the sun's gravitational pull traces out one branch of a hyperbola (Figure 2); and the long-range radio navigation system known as LORAN locates a ship or plane at the intersection of two hyperbolas (Figure 3).

Figure 1 *Hyperbolic cooling towers for an electric power station*

Figure 2 *Comet photographed through a telescope*

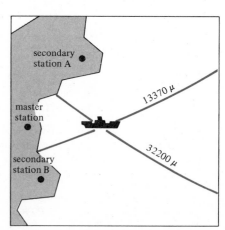

Figure 3 *Synchronized pulses transmitted from three stations locate a ship at the intersection of two hyperbolas plotted on a LORAN chart.*

The geometric definition of a hyperbola is as follows.

Figure 4

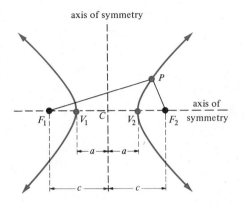

axis of symmetry

Hyperbola

A **hyperbola** is the set of all points P in the plane such that the absolute value of the difference of the distances from P to two fixed points F_1 and F_2 is a constant positive number. Here F_1 and F_2 are called the **focal points,** or the **foci,** of the hyperbola. The midpoint C of the line segment $\overline{F_1F_2}$ is called the **center** of the hyperbola.

Figure 4 shows a hyperbola with foci F_1 and F_2. Notice that the line through the two foci is an axis of symmetry for the hyperbola and so is the perpendicular bisector of the line segment $\overline{F_1F_2}$. The two points V_1 and V_2 where the two branches of the hyperbola intersect the line through F_1 and F_2 are called the **vertices,** and the line segment $\overline{V_1V_2}$ is called the **transverse axis** of the hyperbola. The distance from the center C to either focus is denoted by c, and the distance from the center C to either vertex is denoted by a. Thus, the length of the transverse axis is $2a$, and the distance between the two foci is $2c$.

As the point P in Figure 4 moves along the right-hand branch toward V_2, the difference

$$\left|\overline{PF_1}\right| - \left|\overline{PF_2}\right|$$

Figure 5

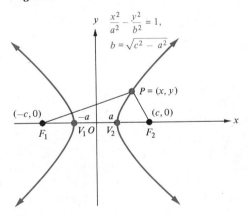

maintains a constant value (Definition 1). When P reaches V_2,

$$\left|\overline{PF_1}\right| - \left|\overline{PF_2}\right| = \left|\overline{V_2F_1}\right| - \left|\overline{V_2F_2}\right| = (c + a) - (c - a) = 2a$$

Therefore, for any point P on the hyperbola,

$$\left|\left|\overline{PF_1}\right| - \left|\overline{PF_2}\right|\right| = 2a$$

by Definition 1.

If we place the hyperbola in Figure 4 in the xy plane so that its center C is at the origin O and the foci F_1 and F_2 lie on the x axis (Figure 5), we can derive its Cartesian equation as in the following theorem.

Hyperbola Equation in Cartesian Form

An equation of the hyperbola with foci $F_1 = (-c, 0)$, $F_2 = (c, 0)$ and vertices $V_1 = (-a, 0)$, $V_2 = (a, 0)$ is

$$\frac{x^2}{a^2} - \frac{y^2}{b^2} = 1 \qquad \text{where } b = \sqrt{c^2 - a^2}.$$

The proof of Theorem 1 is quite similar to the proof of Theorem 1 in Section 9.4; hence, it is left as an exercise (Problem 42). The equation

$$\frac{x^2}{a^2} - \frac{y^2}{b^2} = 1$$

is called the **standard form** for the equation of a hyperbola (Figure 5). We solve this equation for y in terms of x as follows:

$$\frac{y^2}{b^2} = \frac{x^2}{a^2} - 1$$

so

$$y^2 = b^2\left(\frac{x^2}{a^2} - 1\right) = \left(\frac{b^2 x^2}{a^2}\right)\left(1 - \frac{a^2}{x^2}\right)$$

Hence,

$$y = \pm\left(\frac{bx}{a}\right)\sqrt{1 - \frac{a^2}{x^2}}$$

provided that $x \neq 0$. Since

$$\lim_{x \to \pm\infty} \sqrt{1 - \frac{a^2}{x^2}} = 1$$

we see that when x is large in absolute value,

$$y \approx \pm\frac{bx}{a}$$

The two lines whose equations are

$$\boxed{y = \frac{b}{a}x \qquad \text{and} \qquad y = -\frac{b}{a}x}$$

are called the **asymptotes** of the hyperbola. They are good approximations to the hyperbola itself at suitably large distances from the origin.

Although the asymptotes of the hyperbola are *not* part of the hyperbola itself, they are useful in sketching it. For instance, if we wish to sketch the graph of the equation $(x^2/a^2) - (y^2/b^2) = 1$, we begin by sketching the rectangle with height $2b$ and horizontal base $2a$ whose center is at the origin (Figure 6). The asymptotes are then drawn through the two diagonals of this rectangle. If we keep in mind that the vertices of the hyperbola are located at the midpoints of the left and right sides of the rectangle, and that the hyperbola approaches the asymptotes as it moves out away from the vertices, then it becomes an easy matter to sketch the graph (Figure 6).

EXAMPLE 1   Find the coordinates of the foci and the vertices and find equations of the asymptotes of the hyperbola $(x^2/4) - (y^2/1) = 1$. Also sketch the graph.

SOLUTION   The equation has the form

$$\frac{x^2}{a^2} - \frac{y^2}{b^2} = 1$$

with $a = 2$, $b = 1$. Hence, $c = \sqrt{a^2 + b^2} = \sqrt{5}$. Thus, the foci are $F_1 = (-\sqrt{5}, 0)$, $F_2 = (\sqrt{5}, 0)$, and the vertices are $V_1 = (-2, 0)$, $V_2 = (2, 0)$. The asymptotes are given by $y = \frac{1}{2}x$ and $y = -\frac{1}{2}x$ (Figure 7). ∎

Suppose that we wish to find an equation of the hyperbola in Figure 8, which has a *vertical* transverse axis, center at the origin, vertices $V_1 = (0, -b)$ and $V_2 =$

Figure 6

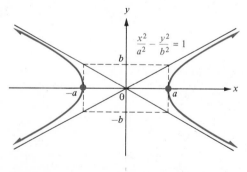

Figure 7

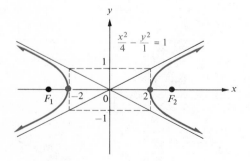

Figure 8

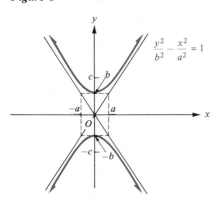

$$\frac{y^2}{b^2} - \frac{x^2}{a^2} = 1$$

$(0, b)$, and foci $F_1 = (0, -c)$ and $F_2 = (0, c)$. Using Theorem 1 but interchanging x with y and interchanging a with b, we obtain the equation

$$\frac{y^2}{b^2} - \frac{x^2}{a^2} = 1 \qquad \text{where } a = \sqrt{c^2 - b^2}$$

This equation is also called the **standard form** for the equation of a hyperbola. The asymptotes are still given by

$$y = \frac{b}{a}x \qquad \text{and} \qquad y = -\frac{b}{a}x$$

(Why?)

Unlike the ellipse equations $(x^2/a^2) + (y^2/b^2) = 1$ and $(x^2/b^2) + (y^2/a^2) = 1$, *there is no requirement that $a > b$ in the hyperbola equation $(x^2/a^2) - (y^2/b^2) = 1$ or in the hyperbola equation $(y^2/b^2) - (x^2/a^2) = 1$.*

<u>EXAMPLE 2</u>   Which of the following hyperbolas have a horizontal transverse axis and which have a vertical transverse axis?

(a) $\dfrac{y^2}{16} - \dfrac{x^2}{9} = 1$ (b) $\dfrac{y^2}{9} - \dfrac{x^2}{4} = 1$ (c) $\dfrac{y^2}{4} - \dfrac{x^2}{4} = 1$

(d) $\dfrac{x^2}{9} - \dfrac{y^2}{4} = 1$ (e) $\dfrac{x^2}{4} - \dfrac{y^2}{9} = 1$ (f) $\dfrac{x^2}{4} - \dfrac{y^2}{4} = 1$

SOLUTION   Whether the hyperbola has a horizontal or a vertical transverse axis does not depend on the relative sizes of the denominators but depends on which term is subtracted from the other. Consequently, **(a)**, **(b)**, and **(c)** have vertical transverse axes, whereas **(d)**, **(e)**, and **(f)** have horizontal transverse axes. ∎

Suppose that a hyperbola has a horizontal or vertical transverse axis and its center is *not* at the origin. By translating the coordinate axes so that the new origin $\overline{O}$ is at the center, we obtain an equation in standard form for the hyperbola with respect to the $\overline{x}\overline{y}$ coordinate system. Therefore, with respect to the original xy coordinate system, an equation of the hyperbola with center $C = (h, k)$ has the form

$$\frac{(x-h)^2}{a^2} - \frac{(y-k)^2}{b^2} = 1 \qquad \text{or} \qquad \frac{(y-k)^2}{b^2} - \frac{(x-h)^2}{a^2} = 1$$

depending on whether the transverse axis is horizontal or vertical, respectively. In either case, the asymptotes have the equations

$$y - k = \frac{b}{a}(x - h) \qquad \text{and} \qquad y - k = -\frac{b}{a}(x - h)$$

and the distance from the center (h, k) to either focus is given by

$$c = \sqrt{a^2 + b^2}$$

EXAMPLE 3   Find the coordinates of the center, the foci, and the vertices of the hyperbola $y^2 - 4x^2 - 8x - 4y - 4 = 0$. Also find equations of its asymptotes, and sketch its graph.

SOLUTION   Completing the squares, we have

$$y^2 - 4y + 4 - 4(x^2 + 2x + 1) = 4$$

or

$$(y - 2)^2 - 4(x + 1)^2 = 4$$

Dividing by 4, we obtain

$$\frac{(y - 2)^2}{4} - \frac{(x + 1)^2}{1} = 1$$

Figure 9

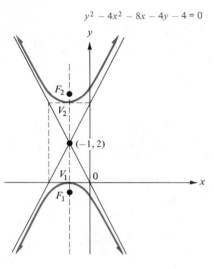

$$y^2 - 4x^2 - 8x - 4y - 4 = 0$$

the equation of a hyperbola with center $(-1, 2)$ and with a vertical transverse axis. Since $a = 1$ and $b = 2$, the equations of the asymptotes are

$$y - 2 = \frac{2}{1}(x + 1) \qquad \text{and} \qquad y - 2 = \frac{-2}{1}(x + 1)$$

that is,

$$y = 2x + 4 \qquad \text{and} \qquad y = -2x$$

Also, $c = \sqrt{a^2 + b^2} = \sqrt{5}$, so $F_1 = (-1, 2 - \sqrt{5})$, $F_2 = (-1, 2 + \sqrt{5})$, $V_1 = (-1, 0)$, and $V_2 = (-1, 4)$ (Figure 9).   ∎

EXAMPLE 4   Simplify the equation $4x^2 - 9y^2 - 24x - 90y - 225 = 0$ by using a suitable translation $x = \bar{x} + h$, $y = \bar{y} + k$. Sketch the graph, showing both the old xy and the new $\bar{x}\bar{y}$ coordinate axes.

SOLUTION   Completing the squares, we have

$$4(x^2 - 6x + 9) - 9(y^2 + 10y + 25) = 225 + 36 - 225$$

or

$$4(x - 3)^2 - 9(y + 5)^2 = 36$$

Dividing by 36, we obtain

$$\frac{(x - 3)^2}{9} - \frac{(y + 5)^2}{4} = 1$$

Thus, if we make the translation

$$\bar{x} = x - 3 \qquad \bar{y} = y + 5$$

Figure 10

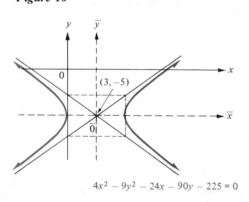

$$4x^2 - 9y^2 - 24x - 90y - 225 = 0$$

the equation simplifies to $\dfrac{\bar{x}^2}{9} - \dfrac{\bar{y}^2}{4} = 1$ (Figure 10).   ∎

EXAMPLE 5   The segment cut off by a hyperbola on a line perpendicular to the transverse axis and passing through a focus is called a **focal chord.** Find a formula for the length of a focal chord of the hyperbola $(y^2/b^2) - (x^2/a^2) = 1$.

SOLUTION   The hyperbola has a vertical transverse axis, and the upper focus is at $(0, c)$, $c = \sqrt{a^2 + b^2}$. Putting $y = c$ in the equation of the hyperbola and solving for x, we obtain $x = \pm a^2/b$. The upper focal chord is therefore the line segment from $(-a^2/b, c)$ to $(a^2/b, c)$; its length is $2a^2/b$.   ∎

Figure 11

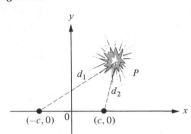

EXAMPLE 6   Two microphones are located at the points $(-c, 0)$ and $(c, 0)$ on the x axis (Figure 11). An explosion occurs at an unknown point P to the right of the y axis. The sound of the explosion is detected by the microphone at $(c, 0)$ exactly T seconds before it is detected by the microphone at $(-c, 0)$. Assuming that sound travels in air at the constant speed of v feet per second, show that the point P must have been located on the right-hand branch of the hyperbola whose equation is $(x^2/a^2) - (y^2/b^2) = 1$, where

$$a = \frac{vT}{2} \quad \text{and} \quad b = \frac{\sqrt{4c^2 - v^2T^2}}{2}$$

SOLUTION   Let d_1 and d_2 denote the distances from P to $(-c, 0)$ and $(c, 0)$, respectively. The sound of the explosion reaches $(-c, 0)$ in d_1/v seconds, and it reaches $(c, 0)$ in d_2/v seconds; hence, $d_1/v - d_2/v = T$, so $d_1 - d_2 = vT$. Putting $a = vT/2$, we notice that the condition $d_1 - d_2 = 2a$ requires that P belong to a hyperbola with foci $F_1 = (-c, 0)$ and $F_2 = (c, 0)$. By Theorem 1, an equation of this hyperbola is $(x^2/a^2) - (y^2/b^2) = 1$, where

$$a = \frac{vT}{2} \quad \text{and} \quad b = \sqrt{c^2 - a^2} = \sqrt{c^2 - \left(\frac{vT}{2}\right)^2} = \frac{\sqrt{4c^2 - v^2T^2}}{2} \qquad ■$$

Problem Set 9.6

In Problems 1 to 8, find the coordinates of the vertices and the foci of each hyperbola. Also find equations of the asymptotes, and sketch the graph.

1 $\dfrac{x^2}{9} - \dfrac{y^2}{4} = 1$

2 $\dfrac{x^2}{1} - \dfrac{y^2}{9} = 1$

3 $\dfrac{y^2}{16} - \dfrac{x^2}{4} = 1$

4 $\dfrac{y^2}{4} - \dfrac{x^2}{1} = 1$

5 $4x^2 - 16y^2 = 64$

6 $49x^2 - 16y^2 = 196$

7 $36y^2 - 10x^2 = 360$

8 $y^2 - 4x^2 = 1$

In Problems 9 to 11, find an equation of the hyperbola that satisfies the conditions given.

9 Vertices at $(-4, 0)$ and $(4, 0)$, foci at $(-6, 0)$ and $(6, 0)$

10 Vertices at $(0, -\frac{1}{2})$ and $(0, \frac{1}{2})$, foci at $(0, -1)$ and $(0, 1)$

11 Vertices at $(-4, 0)$ and $(4, 0)$, the equations of the asymptotes are $y = -\frac{5}{4}x$ and $y = \frac{5}{4}x$.

12 Determine the values of a^2 and b^2 so that the graph of the equation $b^2x^2 - a^2y^2 = a^2b^2$ contains the pair of points (a) $(2, 5)$ and $(3, -10)$ and (b) $(4, 3)$ and $(-7, 6)$.

In Problems 13 to 20, find the coordinates of the center, the vertices, and the foci of each hyperbola. Also find equations of the asymptotes, and sketch the graph.

13 $\dfrac{(x - 1)^2}{9} - \dfrac{(y + 2)^2}{4} = 1$

14 $\dfrac{(x + 3)^2}{1} - \dfrac{(y - 1)^2}{9} = 1$

15 $\dfrac{(y + 1)^2}{16} - \dfrac{(x + 2)^2}{25} = 1$

16 $4x^2 - y^2 - 8x + 2y + 7 = 0$

17 $x^2 - 4y^2 - 4x - 8y - 4 = 0$

18 $16x^2 - 9y^2 + 180y = 612$

19 $9x^2 - 25y^2 + 72x - 100y + 269 = 0$

20 $9x^2 - 16y^2 - 90x - 256y = 223$

21 Find an equation of the hyperbola that satisfies the conditions given. (a) Foci at $(1, -1)$ and $(7, -1)$, length of transverse axis is 2. (b) Vertices at $(-4, 3)$ and $(0, 3)$, foci at $(-\frac{9}{2}, 3)$ and $(\frac{1}{2}, 3)$. (c) Center at $(2, 3)$, one vertex at $(2, 8)$, and one focus at $(2, -3)$.

22 Show that the length of a focal chord of the hyperbola $(x^2/a^2) - (y^2/b^2) = 1$ is $2b^2/a$.

23 Find the length of a focal chord of the hyperbola $x^2 - 8y^2 = 16$.

24 Find equations of the tangent and normal lines to each hyperbola at the point indicated.
    (a) $x^2 - y^2 = 9$ at $(-5, 4)$ (b) $4y^2 - x^2 = 7$ at $(3, -2)$
    (c) $x^2 - 4x - y^2 - 2y = 0$ at $(0, 0)$
    (d) $9(x + 1)^2 - 16(y - 2)^2 = 144$ at $(3, 2)$

25 Find an equation of the hyperbola whose asymptotes are the given lines and that contains the indicated point.
    (a) $y = -2x$ and $y = 2x$, $(1, 1)$
    (b) $y = -2x + 3$ and $y = 2x + 1$, $(1, 4)$
    (c) $y = x + 5$ and $y = -x + 3$, $(2, 4)$

26 Find the volume V of the solid of revolution obtained by revolving the region inside the right-hand branch of the hyperbola $b^2x^2 - a^2y^2 = a^2b^2$ between the vertex and the focal chord about the y axis.

In Problems 27 to 31, use a suitable translation of axes to bring each equation into the standard form for a hyperbola. In each case, sketch the graph and show both the xy and the $\overline{x}\overline{y}$ axes.

27 $3x^2 - y^2 + 12x + 8y = 7$

28 $4x^2 - 25y^2 + 24x + 50y + 22 = 0$

29 $5y^2 - 9x^2 + 10y + 54x - 112 = 0$

30 $x^2 - 4y^2 - 4x - 8y - 4 = 0$

31 $4x^2 - 9y^2 + 16x + 54y = 29$

32 A point moves so that it is equidistant from the point $(2, 0)$ and the circle of radius 3 units with center at $(-2, 0)$. Describe the path of the point.

33 A point moves on the hyperbola $4x^2 - 9y^2 = 27$ with its abscissa increasing at the constant rate of 6 units per second. How fast is the ordinate changing at the point $(3, 1)$?

34 A point moves so that the product of the slopes of the line segments that join it to two given points is 9. Describe the path of the point.

35 Find the shortest (minimum) distance from the point $(3, 0)$ to the hyperbola $y^2 - x^2 = 18$.

36 A region in the plane is bounded by the line $x = 8$ and the hyperbola $x^2 - y^2 = 16$. Find the dimensions of the rectangle of maximum area that can be inscribed in this region.

37 Let m be the slope of the tangent line to the hyperbola $(x^2/a^2) - (y^2/b^2) = 1$ at the point (x_0, y_0), where $x_0 > a$ and $y_0 > 0$. Find $\lim\limits_{x_0 \to +\infty} m$, and relate your answer to the asymptote $y = (b/a)x$.

38 It can be shown that the graph of the equation $xy = 1$ is a hyperbola with center at the origin and with the x and y axes as asymptotes (Figure 12). Find the coordinates of the foci of this hyperbola. (*Hint:* The transverse axis makes a 45° angle with the x axis.)

Figure 12

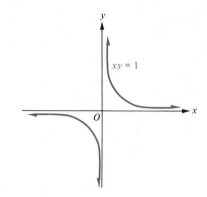

39 A hyperbola is said to be **equilateral** if its two asymptotes are perpendicular. Find an equation of an equilateral hyperbola with horizontal transverse axis and center at the origin. (Denote the distance from the center to a vertex by a.)

40 Sketch the graph of the hyperbola

$$\frac{(y + b)^2}{b^2} - \frac{x^2}{2bp + p^2} = 1$$

Show that as $b \to +\infty$, the upper branch of this hyperbola approaches the parabola $y = (1/4p)x^2$.

41 In Figure 13, hold the center C and the asymptotes fixed, but allow the foci F_1 and F_2 to move in toward the center. What happens to the hyperbola?

Figure 13

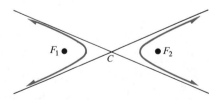

42 Give a proof of Theorem 1.

43 Sound travels with speed s in air, and a bullet travels with speed b from a gun at $(-h, 0)$ to a target at $(h, 0)$ in the xy plane. At what points (x, y) can the boom of the gun and the ping of the bullet hitting the target be heard simultaneously?

9.7   Rotation of Axes

We have studied conics whose axes of symmetry are parallel to the coordinate axes and found that their equations can be "simplified" and brought into standard form by a suitable translation of the Cartesian coordinate axes. Notice that the translation equations

$$\begin{cases} \bar{x} = x - h \\ \bar{y} = y - k \end{cases}$$

are especially simple in the Cartesian coordinate system, whereas the conversion of these equations to polar coordinates would introduce some complication. The Cartesian coordinate system is naturally adapted to translation—the polar coordinate system is not.

On the other hand, the polar coordinate system is naturally adapted to *rotation* about the pole. Figure 1 shows an "old" polar axis, a "new" polar axis obtained by rotating the old polar axis about the pole through the angle ϕ, and a point P in the plane. Evidently, if $P = (r, \theta)$ in the old polar coordinate system, then we have $P = (r, \theta - \phi) = (\bar{r}, \bar{\theta})$ in the new polar coordinate system. Thus, for *polar coordinates*, we have the very simple *rotation equations*

$$\begin{cases} \bar{r} = r \\ \bar{\theta} = \theta - \phi \end{cases}$$

which give the new polar coordinates $(\bar{r}, \bar{\theta})$ of the point P whose old polar coordinates are (r, θ).

Figure 1

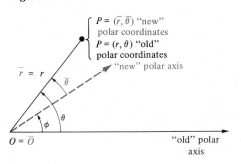

EXAMPLE 1   An equation of the circle of radius $a > 0$ with center at the point $(a, 0)$ on the old horizontal polar axis is $r = 2a \cos \theta$ (Figure 2). Find an equation of the same circle with respect to a new polar axis making the angle $\phi = \pi/2$ with the old polar axis.

SOLUTION   Using the polar rotation equations

$$\begin{cases} \bar{r} = r \\ \bar{\theta} = \theta - \phi \end{cases} \quad \text{or} \quad \begin{cases} r = \bar{r} \\ \theta = \bar{\theta} + \phi \end{cases}$$

we rewrite $r = 2a \cos \theta$ as $\bar{r} = 2a \cos (\bar{\theta} + \phi)$. Putting $\phi = \pi/2$ and noticing that

$$\cos (\bar{\theta} + \phi) = \cos \left(\bar{\theta} + \frac{\pi}{2}\right) = -\sin \bar{\theta}$$

we see that an equation in the new polar coordinate system is $\bar{r} = -2a \sin \bar{\theta}$.   ∎

Figure 2

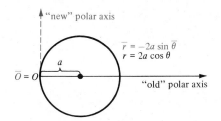

Rotation of Cartesian Axes

The rotation equations for *Cartesian* coordinates are not quite so simple as those for polar coordinates. Nevertheless, it is often necessary to rotate Cartesian coordinates, and we give the appropriate equations for such a rotation in the following theorem.

<u>THEOREM 1</u>

Figure 3

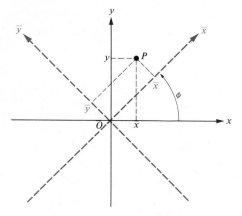

Cartesian Rotation Equations

Suppose that the new $\overline{xy}$ coordinate system has the same origin as the old xy coordinate system, but that the new $\overline{x}$ axis is obtained by rotating the old x axis about the origin through an angle ϕ (Figure 3). Let the point P have old Cartesian coordinates (x, y) and new Cartesian coordinates $(\overline{x}, \overline{y})$. Then

$$\begin{cases} x = \overline{x} \cos \phi - \overline{y} \sin \phi \\ y = \overline{x} \sin \phi + \overline{y} \cos \phi \end{cases}$$

The proof of Theorem 1 is accomplished by converting from Cartesian coordinates to polar coordinates, rotating the polar coordinate system through the angle ϕ, and then converting back to Cartesian coordinates (Problem 28a).

The Cartesian rotation equations in Theorem 1 can be solved for $\overline{x}$ and $\overline{y}$ in terms of x and y to obtain

$$\begin{cases} \overline{x} = x \cos \phi + y \sin \phi \\ \overline{y} = -x \sin \phi + y \cos \phi \end{cases}$$

(see Problem 28b). The last two equations enable us to find the new $\overline{xy}$ coordinates in terms of the old xy coordinates and the rotation angle ϕ.

<u>EXAMPLE 2</u>   The new $\overline{xy}$ coordinate system is obtained by rotating the old xy coordinate system through $-30°$. Find the new $\overline{xy}$ coordinates of the point whose old xy coordinates are $(\sqrt{3}, 2)$.

SOLUTION   Here we have

$$\overline{x} = x \cos (-30°) + y \sin (-30°) = x \frac{\sqrt{3}}{2} + y \left(-\frac{1}{2}\right) = \frac{\sqrt{3}\,x - y}{2}$$

and

$$\overline{y} = -x \sin (-30°) + y \cos (-30°) = -x \left(-\frac{1}{2}\right) + y \left(\frac{\sqrt{3}}{2}\right) = \frac{x + \sqrt{3}\,y}{2}$$

Thus,

$$\overline{x} = \frac{\sqrt{3}\sqrt{3} - 2}{2} = \frac{1}{2} \quad \text{and} \quad \overline{y} = \frac{\sqrt{3} + \sqrt{3}(2)}{2} = \frac{3\sqrt{3}}{2}$$

so

$$(\overline{x}, \overline{y}) = \left(\frac{1}{2}, \frac{3\sqrt{3}}{2}\right)$$

The General Second-Degree Equation

If A, B, C, D, E, and F are constants and at least one of the constants A, B, or C is different from zero, the equation

$$Ax^2 + Bxy + Cy^2 + Dx + Ey + F = 0$$

is called the **general second-degree equation** in x and y. Notice that by taking $A = 1/a^2$, $B = 0$, $C = 1/b^2$, $D = 0$, $E = 0$, and $F = -1$ in the general second-degree equation, we obtain

$$\frac{x^2}{a^2} + \frac{y^2}{b^2} = 1$$

the standard form for the equation of an ellipse. Also, if we set $A = 1$, $B = 0$, $C = 0$, $D = 0$, $E = -4p$, and $F = 0$, we obtain

$$x^2 = 4py$$

the standard form for the equation of a parabola. Finally, by putting $A = 1/a^2$, $B = 0$, $C = -1/b^2$, $D = 0$, $E = 0$, and $F = -1$, we obtain

$$\frac{x^2}{a^2} - \frac{y^2}{b^2} = 1$$

the standard form for the equation of a hyperbola.

Every conic section considered until now has had a Cartesian equation which was a special case of the general second-degree equation

$$Ax^2 + Bxy + Cy^2 + Dx + Ey + F = 0$$

However, the coefficient B has always been zero, so the "mixed" term Bxy did not show up. As we shall soon see, *the mixed term can always be removed from the general second-degree equation by a suitable rotation of the coordinate axes*.

Suppose that the old xy coordinate system is rotated about the origin through the angle ϕ to obtain a new $\overline{xy}$ coordinate system. If we substitute

$$x = \overline{x} \cos \phi - \overline{y} \sin \phi$$

and

$$y = \overline{x} \sin \phi + \overline{y} \cos \phi$$

from the rotation equations (Theorem 1) into the equation

$$Ax^2 + Bxy + Cy^2 + Dx + Ey + F = 0$$

and collect similar terms, we obtain

$$\overline{A}\overline{x}^2 + \overline{B}\overline{xy} + \overline{C}\overline{y}^2 + \overline{D}\overline{x} + \overline{E}\overline{y} + \overline{F} = 0$$

where the coefficients $\overline{A}$, $\overline{B}$, $\overline{C}$, $\overline{D}$, $\overline{E}$, and $\overline{F}$ are given by

$$\overline{A} = A \cos^2 \phi + B \cos \phi \sin \phi + C \sin^2 \phi$$
$$\overline{B} = 2(C - A) \cos \phi \sin \phi + B(\cos^2 \phi - \sin^2 \phi)$$
$$\overline{C} = A \sin^2 \phi - B \cos \phi \sin \phi + C \cos^2 \phi$$
$$\overline{D} = D \cos \phi + E \sin \phi$$
$$\overline{E} = -D \sin \phi + E \cos \phi$$
$$\overline{F} = F$$

(Problem 39). Since

$$\sin 2\phi = 2 \cos \phi \sin \phi$$

and

$$\cos 2\phi = \cos^2 \phi - \sin^2 \phi$$

it follows that $\overline{B}$ can be written in the alternative form

$$\overline{B} = (C - A) \sin 2\phi + B \cos 2\phi$$

Using the equations obtained above, we can now prove the following important theorem.

THEOREM 2

Removal of the Mixed Term by Rotation

Suppose that $B \neq 0$ in the general second-degree equation

$$Ax^2 + Bxy + Cy^2 + Dx + Ey + F = 0$$

If the coordinate system is rotated about the origin through the angle ϕ determined by the conditions

$$\cot 2\phi = \frac{A - C}{B} \qquad \text{and} \qquad 0 < \phi < \frac{\pi}{2}$$

then the mixed term $\overline{B}\overline{x}\overline{y}$ will not appear in the new equation

$$\overline{A}\overline{x}^2 + \overline{C}\overline{y}^2 + \overline{D}\overline{x} + \overline{E}\overline{y} + \overline{F} = 0$$

PROOF

We have

$$\cot 2\phi = \frac{A - C}{B} \qquad \text{or} \qquad \frac{\cos 2\phi}{\sin 2\phi} = \frac{A - C}{B}$$

Hence,

$$B \cos 2\phi = (A - C) \sin 2\phi$$

and it follows that

$$\overline{B} = (C - A) \sin 2\phi + B \cos 2\phi = -(A - C) \sin 2\phi + B \cos 2\phi = 0 \qquad \blacksquare$$

EXAMPLE 3 Rotate the coordinate axes to remove the mixed term from the equation $x^2 - 4xy + y^2 - 6 = 0$, and sketch the graph, showing both the old xy and the new $\overline{x}\overline{y}$ coordinate system.

SOLUTION   The equation has the form $Ax^2 + Bxy + Cy^2 + Dx + Ey + F = 0$ with $A = 1$, $B = -4$, $C = 1$, $D = 0$, $E = 0$, and $F = -6$. According to Theorem 2, the mixed term can be removed by rotating the coordinate system through the angle ϕ, where $0 < \phi < \pi/2$ and

$$\cot 2\phi = \frac{A - C}{B} = \frac{1 - 1}{-4} = 0$$

Hence, we take $2\phi = \pi/2$, so that $\phi = \pi/4$, $\sin \phi = \sqrt{2}/2$, and $\cos \phi = \sqrt{2}/2$. Then, by the rotation equations,

$$x = \overline{x} \cos \phi - \overline{y} \sin \phi = \frac{\sqrt{2}}{2} (\overline{x} - \overline{y})$$

and

$$y = \overline{x} \sin \phi + \overline{y} \cos \phi = \frac{\sqrt{2}}{2} (\overline{x} + \overline{y})$$

Substituting these expressions for x and y into the equation $x^2 - 4xy + y^2 - 6 = 0$, we have

$$\left[\frac{\sqrt{2}}{2} (\overline{x} - \overline{y}) \right]^2 - 4 \left[\frac{\sqrt{2}}{2} (\overline{x} - \overline{y}) \right] \left[\frac{\sqrt{2}}{2} (\overline{x} + \overline{y}) \right] + \left[\frac{\sqrt{2}}{2} (\overline{x} + \overline{y}) \right]^2 - 6 = 0$$

The last equation simplifies to

$$\tfrac{1}{2}(\overline{x} - \overline{y})^2 - 2(\overline{x} - \overline{y})(\overline{x} + \overline{y}) + \tfrac{1}{2}(\overline{x} + \overline{y})^2 = 6$$

or

$$\tfrac{1}{2}(\overline{x}^2 - 2\overline{x}\overline{y} + \overline{y}^2) - 2(\overline{x}^2 - \overline{y}^2) + \tfrac{1}{2}(\overline{x}^2 + 2\overline{x}\overline{y} + \overline{y}^2) = 6$$

Figure 4

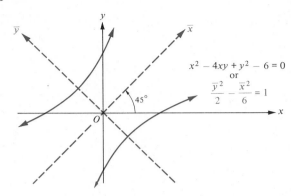

$$x^2 - 4xy + y^2 - 6 = 0$$
or
$$\frac{\bar{y}^2}{2} - \frac{\bar{x}^2}{6} = 1$$

Collecting terms in the equation above, we get

$$-\bar{x}^2 + 3\bar{y}^2 = 6 \qquad \text{or} \qquad \frac{\bar{y}^2}{2} - \frac{\bar{x}^2}{6} = 1$$

a hyperbola (Figure 4). ∎

If $0 < \phi < \pi/2$, then the trigonometric identities

$$\cos 2\phi = \frac{\cot 2\phi}{\sqrt{\cot^2 2\phi + 1}} \qquad \cos \phi = \sqrt{\frac{1 + \cos 2\phi}{2}}$$

and

$$\sin \phi = \sqrt{\frac{1 - \cos 2\phi}{2}}$$

permit us to find $\cos \phi$ and $\sin \phi$ algebraically in terms of the value of $\cot 2\phi$. This is useful in applying Theorem 2, as is shown by the following examples.

In Examples 4 and 5, rotate the coordinate system to remove the mixed term, and  *sketch the graph, showing both the old xy and the new $\bar{x}\bar{y}$ axes.*

EXAMPLE 4 $8x^2 - 4xy + 5y^2 = 144$

SOLUTION   Here $A = 8$, $B = -4$, $C = 5$, $D = 0$, $E = 0$, and $F = -144$; therefore,

$$\cot 2\phi = (A - C)/B = (8 - 5)/-4 = -\tfrac{3}{4}$$

$$\cos 2\phi = \frac{\cot 2\phi}{\sqrt{\cot^2 2\phi + 1}} = \frac{-\tfrac{3}{4}}{\sqrt{(-\tfrac{3}{4})^2 + 1}} = \frac{-\tfrac{3}{4}}{\sqrt{\tfrac{9}{16} + 1}} = \frac{-\tfrac{3}{4}}{\tfrac{5}{4}} = -\frac{3}{5}$$

$$\cos \phi = \sqrt{\frac{1 + \cos 2\phi}{2}} = \sqrt{\frac{1 - \tfrac{3}{5}}{2}} = \sqrt{\frac{2}{10}} = \sqrt{\frac{1}{5}} = \frac{\sqrt{5}}{5}$$

and $\sin \phi = \sqrt{\dfrac{1 - \cos 2\phi}{2}} = \sqrt{\dfrac{1 + \tfrac{3}{5}}{2}} = \sqrt{\dfrac{8}{10}} = \sqrt{\dfrac{4}{5}} = \dfrac{2\sqrt{5}}{5}$

Figure 5

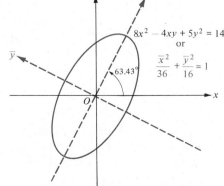

$$8x^2 - 4xy + 5y^2 = 144$$
or
$$\frac{\bar{x}^2}{36} + \frac{\bar{y}^2}{16} = 1$$

Substituting

$$x = \frac{\sqrt{5}}{5}\,\bar{x} - \frac{2\sqrt{5}}{5}\,\bar{y} \qquad \text{and} \qquad y = \frac{2\sqrt{5}}{5}\,\bar{x} + \frac{\sqrt{5}}{5}\,\bar{y}$$

into the given equation, we obtain

$$8\left(\frac{\sqrt{5}}{5}\,\bar{x} - \frac{2\sqrt{5}}{5}\,\bar{y}\right)^2 - 4\left(\frac{\sqrt{5}}{5}\,\bar{x} - \frac{2\sqrt{5}}{5}\,\bar{y}\right)\left(\frac{2\sqrt{5}}{5}\,\bar{x} + \frac{\sqrt{5}}{5}\,\bar{y}\right)$$
$$+ 5\left(\frac{2\sqrt{5}}{5}\,\bar{x} + \frac{\sqrt{5}}{5}\,\bar{y}\right)^2 = 144$$

This equation simplifies to $4\bar{x}^2 + 9\bar{y}^2 = 144$, or $(\bar{x}^2/36) + (\bar{y}^2/16) = 1$, which is an ellipse. Since $\sin \phi = 2\sqrt{5}/5$, it follows that $\phi \approx 63.43°$ (Figure 5). ∎

EXAMPLE 5 $16x^2 - 24xy + 9y^2 - 80x - 190y + 425 = 0$

SOLUTION   Here $A = 16$, $B = -24$, $C = 9$, $D = -80$, $E = -190$, and $F = 425$; therefore,

$$\cot 2\phi = \frac{A - C}{B} = -\frac{7}{24} \qquad \cos 2\phi = \frac{-\frac{7}{24}}{\sqrt{(-\frac{7}{24})^2 + 1}} = -\frac{7}{25}$$

$$\cos \phi = \sqrt{\frac{1 + (-\frac{7}{25})}{2}} = \frac{3}{5} \quad \text{and} \quad \sin \phi = \sqrt{\frac{1 - (-\frac{7}{25})}{2}} = \frac{4}{5}$$

By direct substitution in the equations on page 579, we have

$$\overline{A} = A(\tfrac{3}{5})^2 + B(\tfrac{3}{5})(\tfrac{4}{5}) + C(\tfrac{4}{5})^2 = 0 \qquad \overline{B} = 0$$

$$\overline{C} = A(\tfrac{4}{5})^2 - B(\tfrac{3}{5})(\tfrac{4}{5}) + C(\tfrac{3}{5})^2 = 25 \qquad \overline{D} = D(\tfrac{3}{5}) + E(\tfrac{4}{5}) = -200$$

$$\overline{E} = -D(\tfrac{4}{5}) + E(\tfrac{3}{5}) = -50 \qquad \overline{F} = F = 425$$

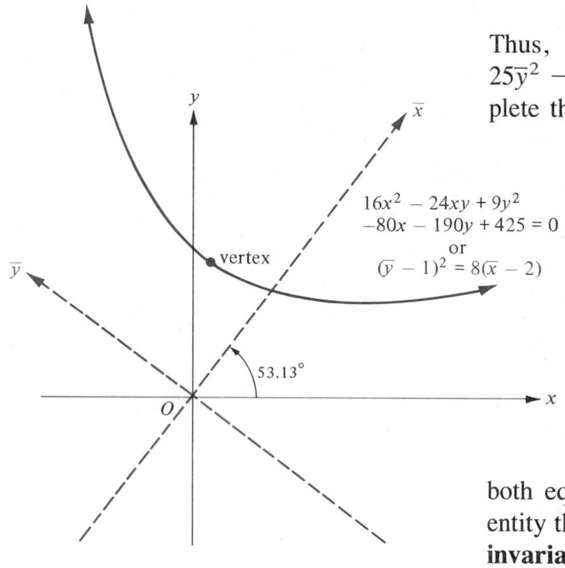

Thus, the equation relative to the new $\overline{x}\overline{y}$ coordinate system is $25\overline{y}^2 - 200\overline{x} - 50\overline{y} + 425 = 0$; that is, $\overline{y}^2 - 8\overline{x} - 2\overline{y} + 17 = 0$. Now, we complete the square to get

$$\overline{y}^2 - 2\overline{y} + 1 = 8\overline{x} - 17 + 1 \qquad \text{or} \qquad (\overline{y} - 1)^2 = 8(\overline{x} - 2)$$

This is a parabola with vertex at $(2, 1)$ in the new coordinate system. Since $\sin \phi = \tfrac{4}{5}$, it follows that $\phi \approx 53.13°$ (Figure 6).

Figure 6

$16x^2 - 24xy + 9y^2$
$-80x - 190y + 425 = 0$
or
$(\overline{y} - 1)^2 = 8(\overline{x} - 2)$

vertex

Although a rotation of the coordinate axes through the angle ϕ changes $Ax^2 + Bxy + Cy^2 + Dx + Ey + F = 0$ into

$$\overline{A}\overline{x}^2 + \overline{B}\overline{x}\overline{y} + \overline{C}\overline{y}^2 + \overline{D}\overline{x} + \overline{E}\overline{y} + \overline{F} = 0$$

both equations have precisely the same graph. This graph, as well as any other entity that remains unchanged as a consequence of the rotation, is called a **rotation invariant.** An **algebraic rotation invariant,** in particular, is any expression involving the coefficients $A, B, C, D, E,$ and F which remains unchanged as the axes are rotated. For instance, $\overline{F} = F$, so F is an algebraic rotation invariant. The following theorem, whose proof is left as an exercise (Problem 44), gives two additional algebraic rotation invariants.

THEOREM 3   **Rotation Invariants**

When the coordinate axes are rotated through an angle ϕ, causing the second-degree equation $Ax^2 + Bxy + Cy^2 + Dx + Ey + F = 0$ to transform into $\overline{A}\overline{x}^2 + \overline{B}\overline{x}\overline{y} + \overline{C}\overline{y}^2 + \overline{D}\overline{x} + \overline{E}\overline{y} + \overline{F} = 0$, then the quantities F, $A + C$, and $B^2 - 4AC$ are invariant in the sense that

    (i) $F = \overline{F}$ **(ii)** $A + C = \overline{A} + \overline{C}$ **(iii)** $B^2 - 4AC = \overline{B}^2 - 4\overline{A}\overline{C}$

The algebraic rotation invariants in Theorem 3 may be used as a check against numerical errors in performing a rotation of axes for a second-degree equation. For instance, to check the numerical calculation in Example 5 above, we have $A = 16$, $B = -24$, $C = 9$, $\overline{A} = 0$, $\overline{B} = 0$, and $\overline{C} = 25$, so that

$$A + C = \overline{A} + \overline{C} = 25 \qquad \text{and} \qquad B^2 - 4AC = \overline{B}^2 - 4\overline{A}\overline{C} = 0$$

We now have the means to sketch the graph of any second-degree equation in x and y. If the equation contains a mixed term, then a suitable rotation of the coordinate axes will remove it. Then, by completing the squares if necessary, we *usually* obtain the standard form for the equation of a circle, an ellipse, a parabola, or a hyperbola. However, in some cases this does not happen (Problem 42). In these exceptional cases, the graph is called a **degenerate conic.** The only possible degenerate conics are (1) the whole xy plane, (2) the empty set, (3) a line, (4) two lines, and (5) a single point.

Assuming that the graph of a given second-degree equation in x and y is a *nondegenerate* conic, it can be identified as follows:

> **1** The conic is a circle or an ellipse if $B^2 - 4AC < 0$.
> **2** The conic is a parabola if $B^2 - 4AC = 0$.
> **3** The conic is a hyperbola if $B^2 - 4AC > 0$.

(See Problem 40.)

EXAMPLE 6   Identify the nondegenerate conic $2x^2 - 12xy - 3y^2 = 84$.

SOLUTION   Here $A = 2$, $B = -12$, and $C = -3$; hence, $B^2 - 4AC = 168 > 0$. Therefore,   the conic is a hyperbola.   ∎

Problem Set 9.7

In Problems 1 to 10, the old horizontal polar axis is rotated through the indicated angle ϕ to obtain a new polar axis. Find the new equation in terms of $\bar{r}$ and $\bar{\theta}$ of the polar curve whose old polar equation is given.

1 $r = 5$; $\phi = \pi/2$

2 $\theta = \pi/3$; $\phi = -\pi/6$

3 $r = 4 \cos \theta$; $\phi = -\dfrac{\pi}{2}$

4 $r = \dfrac{1}{1 - \cos \theta}$; $\phi = \pi$

5 $r = 3 + 5 \sin \theta$; $\phi = \pi$

6 $r = \theta$; $\phi = \pi/3$

7 $r = 3 - 5 \sin \theta$; $\phi = -\dfrac{\pi}{2}$

8 $r = \dfrac{4}{1 + 2 \cos \theta}$; $\phi = \dfrac{\pi}{2}$

9 $r^2 = 25 \cos 2\theta$; $\phi = \dfrac{\pi}{2}$

10 $r = \dfrac{-C}{A \cos \theta + B \sin \theta}$; ϕ arbitrary

In Problems 11 to 19, the old xy axes have been rotated through the angle ϕ to form a new $\bar{x}\bar{y}$ coordinate system. The point P has coordinates (x, y) in the old xy system, and it has coordinates $(\bar{x}, \bar{y})$ in the new $\bar{x}\bar{y}$ system. Find the missing information.

11 $(x, y) = (4, -7)$, $\phi = 90°$, $(\bar{x}, \bar{y}) = ?$

12 $(x, y) = (2, 0)$, $(\bar{x}, \bar{y}) = (1, \sqrt{3})$, $\phi = ?$

13 $(\bar{x}, \bar{y}) = (-3, -3)$, $\phi = \pi/3$ radians, $(x, y) = ?$

14 $(x, y) = (5\sqrt{2}, \sqrt{2})$, $\phi = 45°$, $(\bar{x}, \bar{y}) = ?$

15 $(\bar{x}, \bar{y}) = (-4, -2)$, $\phi = 30°$, $(x, y) = ?$

16 $(\bar{x}, \bar{y}) = (-3\sqrt{2}, \sqrt{2})$, $\phi = 3\pi/4$ radians, $(x, y) = ?$

17 $(x, y) = (-4, 0)$, $\phi = \pi$ radians, $(\bar{x}, \bar{y}) = ?$

18 $(x, y) = (1, -7)$, $\phi = 240°$, $(\bar{x}, \bar{y}) = ?$

19 $(x, y) = (0, 8)$, $\phi = 360°$, $(\bar{x}, \bar{y}) = ?$

20 Find an angle ϕ (if one exists) for which the rotation equations give each of the following:
   (a) $\bar{x} = y$ and $\bar{y} = -x$
   (b) $\bar{x} = -x$ and $\bar{y} = -y$
   (c) $\bar{x} = y$ and $\bar{y} = x$
   (d) $\bar{x} = -y$ and $\bar{y} = -x$

In Problems 21 to 26, the xy axes are rotated $30°$ to form the $\bar{x}\bar{y}$ axes. Express each equation in the other coordinate system.

21 $y^2 = 3x$

22 $\bar{y} = 3\bar{x}$

23 $\bar{x}^2 + \bar{y}^2 = 1$

24 $5x - y = 4$

25 $x^2 + y^2 = 1$

26 $x^2 = 25$

27 By rotating the coordinate axes through $\phi = 45°$ about the origin, show that the graph of $xy = 1$ is a hyperbola with the x and y axes as asymptotes (Figure 7).

Figure 7

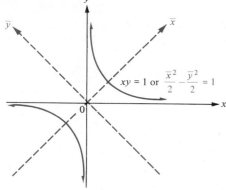

$xy = 1$ or $\dfrac{\bar{x}^2}{2} - \dfrac{\bar{y}^2}{2} = 1$

28 (a) Derive the rotation equations in Theorem 1 by converting to polar coordinates, rotating the polar axis, and reconverting to Cartesian coordinates. (b) Solve the rotation equations in Theorem 1 for $\bar{x}$ and $\bar{y}$ in terms of x and y.

Ⓒ In Problems 29 to 38, determine (a) the angle ϕ of rotation that removes the mixed term $(0 < \phi < \pi/2)$, (b) x and y in terms of $\bar{x}$ and $\bar{y}$, and (c) the new equation in terms of $\bar{x}$ and $\bar{y}$. Then sketch the graph showing both the old xy and the new $\bar{x}\bar{y}$ coordinate systems.

29 $x^2 + 4xy - 2y^2 = 12$

30 $x^2 + 2xy + y^2 + x + y = 0$ **31** $x^2 + 2xy + y^2 = 1$

32 $x^2 + 2xy + y^2 - 4\sqrt{2}x + 4\sqrt{2}y = 0$

33 $9x^2 - 24xy + 16y^2 = 144$

34 $2x^2 + 4\sqrt{3}\,xy - 2y^2 - 4 = 0$ **35** $6x^2 - 6xy + 14y^2 = 45$

36 $17x^2 - 12xy + 8y^2 - 68x + 24y - 12 = 0$

37 $2x^2 + 6xy - 6y^2 + 2\sqrt{10}\,x + 3\sqrt{10}\,y - 16 = 0$

38 $4x^2 - 12xy + 9y^2 - 52x + 26y + 27 = 0$

39 Derive the equations on page 579 for $\bar{A}, \bar{B}, \bar{C}, \bar{D}, \bar{E}$, and $\bar{F}$ in terms of A, B, C, D, E, F, $\sin \phi$, and $\cos \phi$.

40 Suppose that $Ax^2 + Bxy + Cy^2 + Dx + Ey + F = 0$ is an equation of a nondegenerate conic. Show that the conic is:

(a) A circle or ellipse if $B^2 - 4AC < 0$

(b) A parabola if $B^2 - 4AC = 0$

(c) A hyperbola if $B^2 - 4AC > 0$

Also show that in (a) the conic is a circle if $A = C$ and $B = 0$.

41 Assuming the facts given in Problem 40, identify the given conics without bothering to rotate the coordinate system. (Assume nondegeneracy.)

(a) $6x^2 - 4xy + 3y^2 + x + 11y + 10 = 0$

(b) $18x^2 + 12xy + 2y^2 + x - 3y - 4 = 0$

(c) $2x^2 + 2y^2 - 8x + 12y + 1 = 0$

(d) $x^2 - 3xy - 3y^2 + 21 = 0$

42 The following second-degree equations in x and y have graphs that are degenerate conics. In each case, verify that the graph is as described.

(a) $0x^2 + 0xy + 0y^2 + 0x + 0y + 0 = 0$ (the whole xy plane)

(b) $x^2 + y^2 + 1 = 0$ (the empty set)

(c) $2x^2 - 4xy + 2y^2 = 0$ (a line)

(d) $4x^2 - y^2 + 16x + 2y + 15 = 0$ (two intersecting lines)

(e) $x^2 - 2xy + y^2 - 18 = 0$ (two parallel lines)

(f) $x^2 + y^2 - 6x + 4y + 13 = 0$ (a single point)

43 Sketch the graph of $2x^2 + 3xy + 2y^2 = 4$.

44 Prove Theorem 3.

45 In odd Problems 29 to 37, use the algebraic rotation invariants in Theorem 3 to check for a possible error in your calculation.

46 Assume that $AC > 0$. Show that the graph of the equation $Ax^2 + Cy^2 + Dx + Ey + F = 0$ is an ellipse, a circle, a single point, or the empty set.

47 Derive equations of the following conics from the given information. Then rotate the coordinate axes so that the new $\bar{x}$ axis contains the foci of the conic. Find the equation of the conic in the new $\bar{x}\bar{y}$ coordinate system. (a) Foci at $F_1 = (3, 1)$ and $F_2 = (-3, -1)$; distance from the center to either vertex along the major axis is $a = 4$. (b) Foci at $F_1 = (4, 3)$ and $F_2 = (-4, -3)$; distance from the center to either vertex along the transverse axis is $a = 3$.

48 Assume that $AC < 0$. Show that the graph of the equation $Ax^2 + Cy^2 + Dx + Ey + F = 0$ is a hyperbola or a pair of intersecting lines.

49 Find a second-degree equation in x and y whose graph consists of a pair of lines intersecting at the origin and having slopes 3 and -3, respectively.

50 Show that the graph of the equation $Ax^2 + Dx + Ey + F = 0$, where $A \neq 0$, is a parabola, a pair of parallel lines, a single line, or the empty set.

Ⓒ **51** Sketch the graph of $x^2 + 4xy + 4y^2 = 8$.

52 Show that $2A^2 + B^2 + 2C^2$ is a rotation invariant in the sense that $2A^2 + B^2 + 2C^2 = 2\bar{A}^2 + \bar{B}^2 + 2\bar{C}^2$.

53 Sketch the graph of $x^2 - 9xy + y^2 = 12$.

54 In Theorem 2, suppose that $0 < \phi < \pi/2$ and $\cot 2\phi = (A - C)/B$. Prove that

$$\bar{A} = \tfrac{1}{2}[A + C + S\sqrt{(A - C)^2 + B^2}]$$

and

$$\bar{C} = \tfrac{1}{2}[A + C - S\sqrt{(A - C)^2 + B^2}]$$

where $S = B/|B|$.

55 If $B \neq 0$, then show that the graph of the equation $Ax^2 + Bxy + Cy^2 + Dx + Ey + F = 0$ cannot be a circle.

9.8 Eccentricity and Conics in Polar Form

In Sections 9.4 to 9.6, we derived equations in Cartesian form for the ellipse, the parabola, and the hyperbola from three special geometric definitions—one for each type of curve. By using the idea of *eccentricity,* it is possible to give a unified geometric definition for these curves. In this definition, F is a fixed point in the plane, D is a fixed line in the plane, and F does not belong to D.

DEFINITION 1

Figure 1

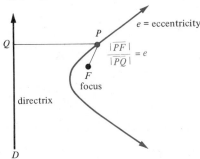

Conics in Terms of Eccentricity, Focus, and Directrix

> Let e be a fixed positive number. A **conic** with **eccentricity** e, **focus** F, and **directrix** D is the set of all points P in the plane such that the distance from P to F, divided by the distance from P to D, is equal to e.

Thus, the point P belongs to the conic if and only if

$$\frac{|\overline{PF}|}{|\overline{PQ}|} = e$$

where Q is the foot of the perpendicular from P to D (Figure 1). Although we use the same symbol e for both the eccentricity of a conic and the base of the natural logarithm, no confusion should result—you can always tell from the context what is intended.

EXAMPLE 1   Find the equation in Cartesian form of the conic with eccentricity $e = 2$ whose focus is at the origin and whose directrix is given by $D: x = -3$.

SOLUTION   From Figure 2, the point $P = (x, y)$ belongs to the given conic if and only if $|\overline{PF}|/|\overline{PQ}| = 2$; that is,

$$\frac{\sqrt{x^2 + y^2}}{3 + x} = 2 \quad\text{or}\quad \sqrt{x^2 + y^2} = 6 + 2x$$

Figure 2

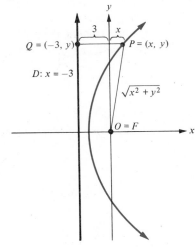

Squaring both sides of the last equation, we have

$$x^2 + y^2 = 36 + 24x + 4x^2$$

or

$$3x^2 + 24x - y^2 = -36$$

Completing the square, we obtain

$$3(x^2 + 8x + 16) - y^2 = 3(16) - 36$$

or

$$3(x + 4)^2 - y^2 = 12$$

that is,

$$\frac{(x + 4)^2}{4} - \frac{y^2}{12} = 1$$

Therefore, the conic is a hyperbola with center at $(-4, 0)$.

By an argument similar to that in the example above, it can be shown that a conic, defined as in Definition 1, is an ellipse, a parabola, or a hyperbola. In fact, we have the following theorem, whose proof is left as an exercise (Problem 56).

THEOREM 1

Conic Theorem

Suppose that a conic with focus F at the origin and directrix D: $x = -d$ has eccentricity e, where e and d are positive. Then exactly one of the following holds:

Case (i) $e < 1$, and the conic is an ellipse with the equation

$$\frac{(x - c)^2}{a^2} + \frac{y^2}{b^2} = 1$$

where $a = \dfrac{ed}{1 - e^2}$, $b = \dfrac{ed}{\sqrt{1 - e^2}}$, and $c = \sqrt{a^2 - b^2} = ae$.

Case (ii) $e = 1$, and the conic is a parabola with the equation

$$4p(x + p) = y^2 \qquad \text{where } p = \frac{d}{2}$$

Case (iii) $e > 1$, and the conic is a hyperbola with the equation

$$\frac{(x + c)^2}{a^2} - \frac{y^2}{b^2} = 1$$

where $a = \dfrac{ed}{e^2 - 1}$, $b = \dfrac{ed}{\sqrt{e^2 - 1}}$, and $c = \sqrt{a^2 + b^2} = ae$.

Conics in Polar Form

Figure 3

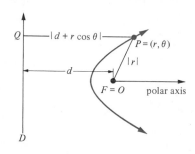

To find a polar equation for a conic of eccentricity e, we place the focus F at the pole O, and we place the directrix D perpendicular to the polar axis and d units to the *left* of the pole, $d > 0$ (Figure 3).

Now consider an arbitrary point $P = (r, \theta)$ in the plane, and let Q be the point at the foot of the perpendicular from P to the directrix D. Switching momentarily to Cartesian coordinates, so that $P = (x, y)$, $Q = (-d, y)$, $x = r \cos \theta$, and $y = r \sin \theta$ (Figure 4), we see that

$$|\overline{PQ}| = |d + x| = |d + r \cos \theta|$$

and $|\overline{PF}| = \sqrt{x^2 + y^2} = |r|$. By definition, P belongs to the conic if and only if

$$\frac{|\overline{PF}|}{|\overline{PQ}|} = e \qquad \text{that is,} \qquad \frac{|r|}{|d + r \cos \theta|} = e$$

Therefore, an equation of the conic in polar form is

$$\left| \frac{r}{d + r \cos \theta} \right| = e \qquad \text{or} \qquad \frac{\pm r}{d + r \cos \theta} = e$$

Figure 4

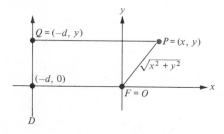

If $P = (r_1, \theta_1)$ satisfies the equation

$$\frac{-r}{d + r \cos \theta} = e$$

then $P = (-r_1, \theta_1 + \pi)$ also satisfies the equation

$$\frac{r}{d + r \cos \theta} = e$$

Hence, we lose no points on the conic by writing its equation as

$$\frac{r}{d + r \cos \theta} = e$$

Solving for r, we obtain a polar equation for the graph in Figure 3:

$$r = \frac{ed}{1 - e \cos \theta}$$

Because $\cos(-\theta) = \cos \theta$, it follows from the test for symmetry (page 542) that the graph is symmetric about the polar axis.

The Ellipse in Polar Form

If $0 < e < 1$, the conic

$$r = \frac{ed}{1 - e \cos \theta}$$

is an *ellipse* (Figure 5). For such an ellipse, r takes on its maximum value $ed/(1 - e)$ when $\theta = 0$, whereas it takes on its minimum value $ed/(1 + e)$ when $\theta = \pi$ (Problem 33). Thus, the points with polar coordinates

$$V_1 = \left(\frac{ed}{1 + e}, \pi \right) \quad \text{and} \quad V_2 = \left(\frac{ed}{1 - e}, 0 \right)$$

are the vertices of the ellipse at the ends of its major axis. It follows that the semimajor axis a of the ellipse is given by

$$a = \frac{1}{2} |\overline{V_1 V_2}| = \frac{1}{2} (|\overline{V_1 F}| + |\overline{F V_2}|) = \frac{1}{2} \left(\frac{ed}{1 + e} + \frac{ed}{1 - e} \right)$$

Simplifying the last expression, we find that

$$a = \frac{ed}{1 - e^2}$$

(Figure 6a). Similarly (Figure 6b), the distance c between the focus F and the center C of the ellipse is given by

$$c = |\overline{FC}| = |\overline{V_1 C}| - |\overline{V_1 F}| = a - \frac{ed}{1 + e} = \frac{ed}{1 - e^2} - \frac{ed}{1 + e}$$

Simplifying the last expression, we find that

$$c = ae$$

Figure 5

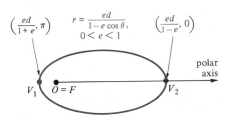

$\left(\frac{ed}{1+e}, \pi \right)$ $r = \frac{ed}{1 - e \cos \theta},$ $\left(\frac{ed}{1-e}, 0 \right)$
$0 < e < 1$

polar axis

V_1 $O = F$ V_2

Figure 6 $a = \frac{ed}{1 - e^2}$

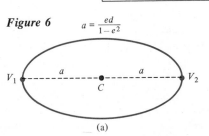

V_1 a a V_2
C

(a)

$c = ae$

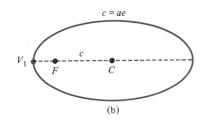

V_1 c
F C

(b)

To find the remaining two vertices of the ellipse

$$r = \frac{ed}{1 - e \cos \theta} \qquad 0 < e < 1$$

consider the angle

$$\theta_a = \cos^{-1} e$$

When $\theta = \theta_a$, we have $\cos \theta = \cos \theta_a = e$, and

$$r = \frac{ed}{1 - e^2} = a$$

Thus, the point with polar coordinates (a, θ_a) belongs to the ellipse. The Cartesian x coordinate of this point is given by

$$a \cos \theta_a = ae = c$$

Hence, the polar point (a, θ_a) is on the ellipse and directly above the center C (Figure 7). Therefore, the polar point (a, θ_a) is the third vertex V_3 of the ellipse, and by symmetry the fourth vertex V_4 has polar coordinates $(a, -\theta_a)$. From right triangle OCV_3 in Figure 7, we have

$$a^2 = b^2 + c^2 \qquad \text{or} \qquad b = \sqrt{a^2 - c^2}$$

Because the ellipse is symmetric about its center C, it has a *second focus* and a *second directrix* located to the right of C (Figure 8). If we denote the original focus and directrix by $F = F_1$ and $D = D_1$, then we can denote the second focus and directrix by F_2 and D_2. Notice that F_2 is c units to the right of the center C, and D_2 is d units to the right of F_2. Thus, D_1 and D_2 are each $c + d$ units from the center C.

© **EXAMPLE 2**   Show that the graph of the polar equation $r = 12/(3 - 2 \cos \theta)$ is an ellipse. Find the eccentricity, the directrices, the center, the foci, and the vertices of the ellipse, and sketch the graph.

SOLUTION   Dividing numerator and denominator of the given fraction by 3 gives

$$r = \frac{4}{1 - \frac{2}{3} \cos \theta} = \frac{\frac{2}{3}(6)}{1 - \frac{2}{3} \cos \theta}$$

which is the polar equation of a conic with focus at the pole, directrix perpendicular to the polar axis and $d = 6$ units to the left of the pole, and eccentricity $e = \frac{2}{3}$. Since $e < 1$, the conic is an ellipse. We have

$$a = \frac{ed}{1 - e^2} = \frac{4}{1 - \frac{4}{9}} = \frac{36}{5} \qquad c = ae = \frac{36}{5} \cdot \frac{2}{3} = \frac{24}{5}$$

and

$$b = \sqrt{a^2 - c^2} = \sqrt{\left(\frac{36}{5}\right)^2 - \left(\frac{24}{5}\right)^2} = \frac{\sqrt{36^2 - 24^2}}{5} = \frac{\sqrt{720}}{5} = \frac{12\sqrt{5}}{5}$$

The pole O is the first focus, the center C is $c = \frac{24}{5}$ units to the right of O, and the second focus is $c = \frac{24}{5}$ units to the right of C. Vertices V_1 and V_2 are $a = \frac{36}{5}$ units to the left and right of C. Vertices V_3 and V_4 are $b = 12\sqrt{5}/5$ units above and below C. The second directrix is perpendicular to the polar axis, and $d = 6$ units to the right of the second focus. We have sketched the ellipse in Figure 9. ∎

Figure 7

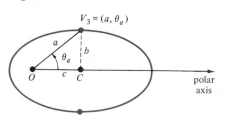

Figure 8

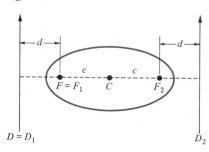

Figure 9

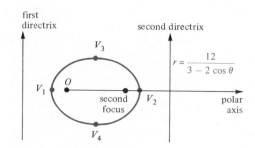

Figure 10

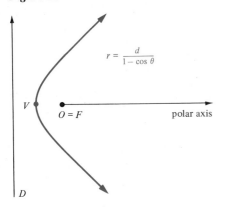

The Parabola in Polar Form

If $e = 1$, the conic

$$r = \frac{ed}{1 - e \cos \theta}$$

is a *parabola* with focus F at the origin and directrix D perpendicular to the polar axis and d units to the left of F (Figure 10). The vertex V of the parabola has polar coordinates $V = (d/2, \pi)$. The parabola has *only one* vertex V, *only one* directrix D, and *only one* focus F; and the vertex is exactly halfway between the focus and the directrix.

The Hyperbola in Polar Form

If $e > 1$, the conic

Figure 11

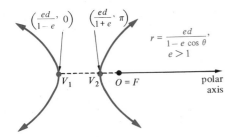

$$r = \frac{ed}{1 - e \cos \theta}$$

is a *hyperbola* with branches opening to the left and right (Figure 11). The two vertices of the hyperbola have polar coordinates

$$V_1 = \left(\frac{ed}{1 - e}, 0\right) \quad \text{and} \quad V_2 = \left(\frac{ed}{1 + e}, \pi\right)$$

Notice that because $e > 1$, the quantity $ed/(1 - e)$ is *negative*, which accounts for the fact that V_1 is to the left of V_2. Here we have

$$|\overline{V_1F}| = -\frac{ed}{1 - e} \qquad |\overline{V_2F}| = \frac{ed}{1 + e}$$

Figure 12

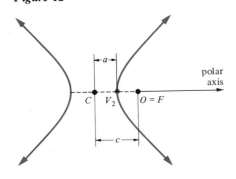

and $\quad |\overline{V_1V_2}| = |\overline{V_1F}| - |\overline{V_2F}| = -\frac{ed}{1 - e} - \frac{ed}{1 + e} = 2\frac{ed}{e^2 - 1}$

It follows that the distance a from the center C of the hyperbola to either vertex is given by $a = \frac{1}{2}|\overline{V_1V_2}|$, or

$$\boxed{a = \frac{ed}{e^2 - 1}}$$

If $c = |\overline{CF}|$ denotes the distance between the center C and the focus F of the hyperbola (Figure 12), then

$$c = |\overline{CF}| = |\overline{CV_2}| + |\overline{V_2F}|$$

$$= a + \frac{ed}{1 + e} = \frac{ed}{e^2 - 1} + \frac{ed}{1 + e} = \left(\frac{ed}{e^2 - 1}\right)e$$

Hence,

$$\boxed{c = ae}$$

Figure 13

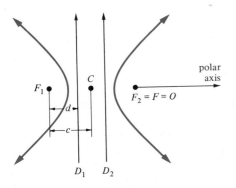

Because the hyperbola is symmetric about its center C, it actually has *two* foci and *two* directrices. If we label them as shown in Figure 13, then focus F_1 is c units to the left of the center C and directrix D_1 is d units to the right of F_1. Therefore, D_1 and D_2 are $c - d$ units to the left and right of C.

Figure 14

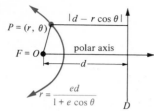

$$r = \frac{ed}{1 + e\cos\theta}$$

Figure 15

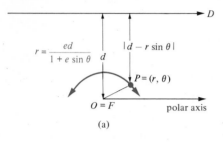

$$r = \frac{ed}{1 + e\sin\theta}$$

(a)

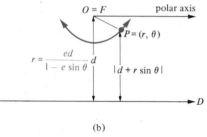

$$r = \frac{ed}{1 - e\sin\theta}$$

(b)

Figure 16

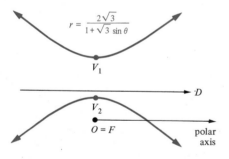

$$r = \frac{2\sqrt{3}}{1 + \sqrt{3}\sin\theta}$$

Figure 17

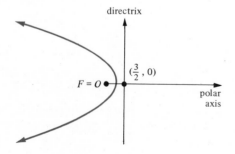

EXAMPLE 3   Find a polar equation of the hyperbola of eccentricity $e = \frac{5}{4}$, one of whose foci F_2 is at the pole with the corresponding directrix D_2 perpendicular to the polar axis and $d = 3$ units to the left of the pole.

SOLUTION $$r = \frac{ed}{1 - e\cos\theta} = \frac{\frac{5}{4}(3)}{1 - \frac{5}{4}\cos\theta} = \frac{15}{4 - 5\cos\theta}$$ ∎

Alternative Polar Forms for Conics

If the focus F of a conic of eccentricity e is placed at the pole O and the directrix D is placed to the *right* of the focus and perpendicular to the polar axis (Figure 14), then a polar equation of the conic is

$$r = \frac{ed}{1 + e\cos\theta}$$

(Problem 49). Likewise, if the directrix D is *parallel* to the polar axis, with the focus F still at the pole O, then a polar equation of the conic is

$$r = \frac{ed}{1 \pm e\sin\theta}$$

where the plus sign is used if the directrix is d units *above* the polar axis (Figure 15a) and the minus sign is used if the directrix is d units *below* the polar axis (Figure 15b). We leave the derivation as an exercise (Problem 50).

©**EXAMPLE 4**   Identify the conic $r = \dfrac{2\sqrt{3}}{1 + \sqrt{3}\sin\theta}$, and sketch its graph.

SOLUTION   The polar equation has the form

$$r = \frac{ed}{1 + e\sin\theta}$$

with $e = \sqrt{3}$ and $d = 2$; hence its graph is a conic with eccentricity $\sqrt{3}$, focus F at the origin, and directrix D parallel to the polar axis and $d = 2$ units above it. Because $e > 1$, the conic is a hyperbola. Vertex V_2 is $ed/(1 + e) = 2\sqrt{3}/(1 + \sqrt{3}) \approx 1.27$ units above O, and vertex V_1 is $-ed/(1 - e) = ed/(e - 1) = 2\sqrt{3}/(\sqrt{3} - 1) \approx 4.73$ units above O (Figure 16).   ∎

EXAMPLE 5   Find a polar equation of the parabola with focus F at the origin and with directrix D perpendicular to the polar axis and $\frac{3}{2}$ units to the right of the origin (Figure 17).

SOLUTION   The equation will have the form

$$r = \frac{ed}{1 + e\cos\theta}$$

with $e = 1$ and $d = \frac{3}{2}$. Thus, the desired equation is

$$r = \frac{\frac{3}{2}}{1 + \cos\theta} \qquad \text{or} \qquad r = \frac{3}{2 + 2\cos\theta}$$ ∎

Problem Set 9.8

In Problems 1 to 6, find a Cartesian equation of the conic whose focus F is at the origin and whose eccentricity and directrix are given. Sketch the graph.

1 Eccentricity $e = \frac{2}{3}$, directrix $x = -\frac{5}{2}$

2 Eccentricity $e = \frac{1}{2}$, directrix $x = -\frac{4}{5}$

3 Eccentricity $e = 1$, directrix $x = -4$

4 Eccentricity $e = 1$, directrix $x = -\frac{1}{3}$

5 Eccentricity $e = 2$, directrix $x = -3$

6 Eccentricity $e = \sqrt{5}$, directrix $x = -1$

ⓒ In Problems 7 to 10, show that the graph of each polar equation is an ellipse; find the eccentricity, the directrices, the center, the foci, and the vertices of the ellipse; and sketch the graph.

7 $r = \dfrac{20}{4 - 3 \cos \theta}$

8 $r = \dfrac{10}{2 - \cos \theta}$

9 $r = \dfrac{20}{4 - \cos \theta}$

10 $r = \dfrac{1}{4 - 2 \cos \theta}$

ⓒ In Problems 11 and 12, show that the graph of each polar equation is a parabola; find the focus, the directrix, and the vertex; and sketch the graph.

11 $r = \dfrac{4}{1 - \cos \theta}$

12 $r = \dfrac{5}{8 - 8 \cos \theta}$

ⓒ In Problems 13 and 14, show that the graph of each polar equation is a hyperbola; find the eccentricity, the directrices, the center, the foci, and the vertices of the hyperbola; and sketch the graph.

13 $r = \dfrac{1}{1 - 2 \cos \theta}$

14 $r = \dfrac{15}{3 - 4 \cos \theta}$

ⓒ In Problems 15 to 24, identify each conic; find the eccentricity, the directrix (directrices), the center (if it exists), the focus (foci), and the vertex (vertices); and sketch the graph.

15 $r = \dfrac{16}{5 - 3 \cos \theta}$

16 $r = \dfrac{16}{5 + 3 \cos \theta}$

17 $r = \dfrac{6}{1 - \cos \theta}$

18 $r = \dfrac{24}{5 - 7 \cos \theta}$

19 $r = \dfrac{10}{1 - \sin \theta}$

20 $r = \dfrac{6}{1 - 2 \sin \theta}$

21 $r = \dfrac{1}{1 + 2 \sin \theta}$

22 $r = \dfrac{6}{10 + 5 \sin \theta}$

23 $r = \dfrac{4}{1 + \cos \theta}$

24 $r = \csc^2 \theta - \csc \theta \cot \theta$

In Problems 25 to 32, find a polar equation for the conic having one focus at the pole and satisfying the conditions given. In each case, the directrix given is the one corresponding to the focus at the pole.

25 Eccentricity is $\frac{1}{3}$, directrix is vertical and 3 units to the left of the pole.

26 Eccentricity is $\frac{1}{4}$, directrix is vertical and 3 units to the right of the pole.

27 Eccentricity is 1, directrix is vertical and 3 units to the right of the pole.

28 Eccentricity is 1, directrix is horizontal and 3 units below the origin.

29 Eccentricity is 3, directrix is vertical and 2 units to the left of the pole.

30 Eccentricity is 2.5, directrix is horizontal and 4 units above the origin.

31 Eccentricity is $\frac{1}{3}$, directrix is horizontal and 5 units above the pole.

32 Directrix is 4 units to the right of the pole; a second directrix is 6 units to the left of the pole.

In Problems 33 to 40, we refer to the ellipse $r = ed/(1 - e \cos \theta)$ with semimajor axis a, semiminor axis b, and distance c from either focus to the center.

33 Show that r takes on its maximum value $ed/(1 - e)$ when $\theta = 0$ and that it takes on its minimum value $ed/(1 + e)$ when $\theta = \pi$.

34 Fill in the details of the argument on page 587 to show that $c = ae$.

35 Show that $ec + ed = a$.

36 Show that $d = b^2/c$.

37 Show that e is the ratio of the distance between the two foci to the length of the major axis.

38 Show that the distance between the two directrices is $2a^2/c$ units.

39 Show that $e = \sqrt{1 - (b/a)^2}$.

40 Derive a formula for e in terms of c and d.

In Problems 41 to 46, we refer to the hyperbola $r = ed/(1 - e \cos \theta)$ with a transverse axis of length $2a$ and distance c from either focus to the center.

41 Fill in the details of the argument on page 589 to show that $c = ae$.

42 Show that $d = c - (a^2/c)$.

43 Show that $ec - ed = a$.

44 Show that the distance between the two directrices is $2a^2/c$ units.

45 Show that e is the ratio of the distance between the two foci to the length of the transverse axis.

46 Derive a formula for e in terms of c and d.

[C]**47** Except for minor perturbations, the orbit of the earth is an ellipse with the sun at one focus. The least and greatest distances (at *perihelion* and *aphelion*) between the earth and the sun have a ratio of approximately $\frac{29}{30}$. Find the approximate eccentricity of the earth's orbit.

48 Suppose that the foci of a particular ellipse lie midway between the center and the vertices. Find the eccentricity.

49 Derive a polar equation of a conic with focus at the pole and directrix perpendicular to the polar axis and d units to the right of the pole.

50 Derive a polar equation of a conic with focus at the pole and directrix parallel to the polar axis.

51 If the two directrices of an ellipse are held fixed and the eccentricity is decreased, what happens to the shape of the ellipse?

52 If the two directrices of a hyperbola are held fixed and the eccentricity is increased, what happens to the shape of the hyperbola?

53 Find all points of intersection of the hyperbola $r = 2/(1 - 2 \cos \theta)$ and the limaçon $r = 2 - 4 \cos \theta$. Sketch the graphs.

54 Let $0 < e < 1$, $0 < d$, and $0 < a < b$. Find the condition or conditions on e and d such that the left vertex of the ellipse $r = ed/(1 - e \cos \theta)$ lies on the inner loop of the limaçon $r = a - b \cos \theta$.

55 Find Cartesian equations of the two directrices of the ellipse $(x^2/a^2) + (y^2/b^2) = 1$, where $a > b > 0$.

56 Let the focus F of a conic with eccentricity $e > 0$ be the origin O of a Cartesian coordinate system, and suppose that the directrix is $D: x = -d$, where $d > 0$. Using Definition 1 directly, prove that $(1 - e^2)x^2 - (2e^2d)x + y^2 = e^2d^2$ is a Cartesian equation of the conic. Using this result, prove Theorem 1.

57 Find Cartesian equations of the two directrices of the hyperbola $(x^2/a^2) - (y^2/b^2) = 1$, where a and b are positive constants.

58 In Problem 57, show that the *eccentricity of the hyperbola* is given by $e = \sqrt{1 + (b/a)^2}$.

Review Problem Set, Chapter 9

In Problems 1 to 6, plot each point in the polar coordinate system, and then give three other polar representations of the same point for which (a) $r < 0$ and $0 \le \theta < 2\pi$, (b) $r > 0$ and $-2\pi < \theta \le 0$, and (c) $r < 0$ and $-2\pi \le \theta < 0$.

1 $(1, \pi/3)$

2 $(2, 5\pi/6)$

3 $(2, 7\pi/4)$

4 $(-1, -11\pi/3)$

5 $(3, \pi)$

6 (π, π)

In Problems 7 to 12, convert the polar coordinates to Cartesian coordinates.

7 $(0, 0)$

8 $(17, 0)$

9 $(17, \pi)$

10 $(-3, \pi/6)$

11 $(11, 3\pi/4)$

12 $(-3, -\pi/3)$

In Problems 13 to 18, convert the Cartesian coordinates to polar coordinates (r, θ) with $r \ge 0$ and $-\pi < \theta \le \pi$.

13 $(17, 0)$

14 $(2, 3)$

15 $(2, -2\sqrt{3})$

16 $(-\sqrt{3}, -1)$

17 $(-17, 17)$

18 $(0, -1)$

In Problems 19 to 22, convert each polar equation into a Cartesian equation.

19 $r^2 \cos 2\theta = 1$

20 $r = \dfrac{1}{3 \cos \theta - 4 \sin \theta}$

21 $r = \sqrt{|\sec \theta|}$

22 $r = \dfrac{ed}{1 - e \cos \theta}$, $0 < e < 1$, $d > 0$

In Problems 23 to 26, convert each Cartesian equation to a polar equation.

23 $y = 2x - 1$

24 $(x - 1)^2 + (y - 3)^2 = 4$

25 $y^2 = 4x$

26 $\dfrac{x^2}{a^2} + \dfrac{y^2}{b^2} = 1$

In Problems 27 to 32, sketch the graph of each polar equation. Discuss whatever symmetry the graph may possess.

27 $r = 2 \cos 3\theta$

28 $r = 2 \cos (3\theta + \pi)$

29 $r = \dfrac{2}{\cos \theta + \sqrt{3} \sin \theta}$

30 $r = \frac{1}{2} + \sin \theta$

31 $r = \dfrac{3}{1 + \sin \theta}$

32 $r = 1 + \frac{1}{2} \sin \theta$

In Problems 33 to 36, find the slope of the tangent line to the polar graph at the point indicated.

33 $r = \dfrac{1}{5 \cos \theta - 3 \sin \theta}$ at $(\frac{1}{5}, 0)$

34 $r = a \sin k\theta$, $a > 0$, k a positive integer at $(0, \pi/k)$

35 $r = 2 - 3 \cos \theta$ at $(2, \pi/2)$

36 $r^2 = -9 \cos 2\theta$ at $(3, \pi/2)$

37 Find all points where the tangent to the limaçon $r = 3 - 4 \cos \theta$ is either horizontal or vertical.

38 In Problems 35 and 36, find $\tan \psi$ at the indicated points.

39 Find $\tan \psi$ in terms of θ for the curve $r = \sin^3 \theta$.

40 Let $P \neq O$ be a point of intersection of the two polar curves $r = f(\theta)$ and $r = g(\theta)$. If ψ_1 is the value of ψ for $r = f(\theta)$ at P and ψ_2 is the value of ψ for $r = g(\theta)$ at P, show that

$$\tan \phi = \frac{\tan \psi_1 - \tan \psi_2}{1 + \tan \psi_1 \tan \psi_2}$$

where ϕ is the angle between the tangent to $r = f(\theta)$ at P and the tangent to $r = g(\theta)$ at P.

41 Find the polar coordinates of all points of intersection of the limaçon $r = 1 - 2 \cos \theta$ and the parabola $r = 2/(1 - \cos \theta)$. Sketch the graphs showing these intersection points.

42 Find the polar coordinates of all points of intersection of the circle $r = 2 \cos \theta$ and the four-leaved rose $r = 2 \cos 2\theta$.

In Problems 43 to 48, find the area enclosed by each polar curve between the indicated values of θ.

43 $r = 1 + \cos \theta$ from $\theta = 0$ to $\theta = \pi/4$

44 $r = -\theta$ from $\theta = 0$ to $\theta = \pi/2$

45 $r = 4 \sin \theta$ from $\theta = 0$ to $\theta = \pi$

46 $r = \dfrac{1}{\sin \theta + \cos \theta}$ from $\theta = 0$ to $\theta = \pi/6$

47 $r = 1 - \cos \theta$ from $\theta = 0$ to $\theta = \pi$

48 $r = e^{2\theta}$ from $\theta = 0$ to $\theta = \pi$

In Problems 49 and 50, find all points of intersection of the two curves, sketch the two curves, and find the area of each region described.

49 Inside the circle $r = 4 \sin \theta$ and below the line $r = \csc \theta$

50 Inside the circle $r = 6 \sin \theta$ and outside the cardioid $r = 2(1 - \sin \theta)$

51 If $b > a > 0$, find the area of the region inside the outer loop but outside the inner loop of the limaçon $r = a + b \sin \theta$.

52 Find the area inside the lemniscate $r^2 = a^2 \cos 2\theta$.

In Problems 53 to 56, find the arc length of the portion of each polar curve between the indicated values of θ.

53 $r = e^{-3\theta}$ for θ between 0 and 2π

54 $r = 5 \sin \theta$ for θ between 0 and π

55 $r = \cos^2 (\theta/2)$ for θ between 0 and π

56 $r = 1 - \cos \theta$ for θ between 0 and 2π

In Problems 57 to 62, identify the type of conic, give its eccentricity, and specify the position of the directrix corresponding to the focus at the pole.

57 $r = \dfrac{17}{1 - \cos \theta}$

58 $r = \dfrac{15}{3 + 5 \sin \theta}$

59 $r = \dfrac{10}{5 - 2 \sin \theta}$

60 $r = \dfrac{3}{2} \csc^2 \dfrac{\theta}{2}$

61 $r = \dfrac{1}{\frac{1}{2} + \sin \theta}$

62 $r = \dfrac{1}{1 + \cos (\theta + \pi/4)}$

In Problems 63 to 68, find a Cartesian equation of the ellipse that satisfies the conditions given.

63 Center $(0, 0)$; one vertex is $(5, 0)$; one focus is $(3, 0)$.

64 Center $(0, 0)$; containing the points $(4, 3)$ and $(6, 2)$; with major axis on the x axis.

65 Major axis 16; minor axis 8; center at the origin; vertices on the y axis.

66 Major axis 20; minor axis 12; center at the origin; vertices on the x axis.

67 Vertices $(0, 0)$ and $(10, 0)$; foci $(1, 0)$ and $(9, 0)$.

68 Foci $(3, -2)$ and $(9, -2)$; minor axis 8.

In Problems 69 to 73, find the coordinates of the center, the vertices, and the foci of each ellipse. Also find the eccentricity and equations of the directrices.

69 $\dfrac{x^2}{8} + \dfrac{y^2}{12} = 1$

70 $144x^2 + 169y^2 = 24{,}336$

71 $9x^2 + 25y^2 + 18x - 50y - 191 = 0$

72 $3x^2 + 4y^2 - 28x - 16y + 48 = 0$

73 $9x^2 + 4y^2 + 72x - 48y + 144 = 0$

74 Find equations of the tangent and normal lines to the ellipse $x^2 + 3y^2 = 21$ at the point $(3, -2)$.

75 At what point(s) on the ellipse $16x^2 + 9y^2 = 400$ does y decrease at the same rate as x increases?

76 The lower base of an isosceles trapezoid is the major axis of an ellipse; the ends of the upper base are points on the ellipse. Show that the length of the upper base of the trapezoid of maximum area is half the length of the lower base.

77 Reduce each of the following equations to an equation of an ellipse in a "simpler" form by using a suitable translation.

(a) $16x^2 + y^2 - 32x + 4y - 44 = 0$

(b) $9x^2 + 4y^2 + 36x - 24y - 252 = 0$

78 A point P moves so that the product of the slopes of the line segments $\overline{PQ}$ and $\overline{PR}$, where $Q = (3, -2)$ and $R = (-2, 1)$, is -6. Find a Cartesian equation of the curve traced out by P, and sketch it.

In Problems 79 to 82, find a Cartesian equation of the parabola satisfying the given conditions. Sketch the graph.

79 Vertex $(0, 0)$, focus on the x axis, and containing the point $(-2, 6)$.

80 Containing the points $(-2, 1)$, $(1, 2)$, and $(-1, 3)$ and whose axis is parallel to the x axis.

81 Vertex $(2, 3)$, containing the point $(4, 5)$, and whose axis is parallel to the y axis.

82 Focus $(6, -2)$ and whose directrix is the line $x - 2 = 0$.

83 Find a Cartesian equation of the circle containing the vertex and the ends of the focal chord of the parabola $y^2 = 8x$.

84 Find equations of the tangent and normal lines to the following parabolas at the point given.

(a) $y^2 + 5x = 0$ at $(-5, 5)$

(b) $4y = 8 + 16x - x^2$ at $(1, \frac{23}{4})$

85 Find the dimensions of the rectangle of maximum area that can be inscribed in the segment of the parabola $x^2 = 24y$ cut off by the line $y = 6$.

86 Show that the line from the focus of the parabola $y^2 = 24x$ to the point where the tangent to the parabola at the point $(24, 24)$ cuts the y axis is perpendicular to the tangent.

87 At what point on the graph of the parabola $y = 4 + 2x - x^2$ is the tangent line parallel to the line $2x + y - 6 = 0$?

88 A parabola whose axis is parallel to the y axis contains the origin and is tangent to the line $x - 2y = 8$ at the point $(20, 6)$. Find an equation of the parabola and sketch the graph.

In Problems 89 to 94, find a Cartesian equation of the hyperbola that satisfies the conditions given.

89 Foci $(-5, 0)$ and $(5, 0)$; length of transverse axis is 4.

90 Vertices $(3, -6)$ and $(3, 6)$; foci $(3, -10)$ and $(3, 10)$.

91 Vertices $(-2, 3)$ and $(6, 3)$; one focus $(7, 3)$.

92 Containing the point $(1, 1)$; with equations of asymptotes $y = -2x$ and $y = 2x$.

93 Center $(0, 0)$; transverse axis on the y axis; length of focal chord 36; distance between its foci 24.

94 Center $(0, 0)$; one focus $(8, 0)$; one vertex $(6, 0)$.

In Problems 95 to 99, find the coordinates of the center, the vertices, and the foci of each hyperbola. Also find the eccentricity and the equations of the asymptotes. Sketch the graph.

95 $x^2 - 9y^2 = 72$ **96** $y^2 - 9x^2 = 54$

97 $x^2 - 4y^2 + 4x + 24y - 48 = 0$ **98** $16x^2 - 9y^2 - 96x = 0$

99 $4y^2 - x^2 - 24y + 2x + 34 = 0$

100 Find equations of the tangent lines to the hyperbola $x^2 - y^2 + 16 = 0$ that are perpendicular to the line $5x + 3y - 15 = 0$.

101 Find equations of the tangent and normal lines to the hyperbola $x^2 - 8y^2 = 1$ at the point $(3, 1)$.

102 Find the points on the hyperbola $x^2 - y^2 - 16 = 0$ that are nearest to the point $(0, 6)$.

103 A point moves on the hyperbola $x^2 - 4y^2 = 20$ in such a way that x is increasing at the rate of 3 units per second. Find the rate at which y is changing when the moving point passes through $(6, -2)$.

104 Two points are 800 meters apart. At one of these points the report of a cannon is heard 1 second later than at the other. Show that the cannon is somewhere on a certain hyperbola, and write an equation for the hyperbola after making a suitable choice of axes. (Consider the velocity of sound to be 332 meters per second.)

Ⓒ**105** The Ellipse in front of the White House in Washington, D.C., has a semimajor axis of 229 meters and a semiminor axis of 195 meters. What is its eccentricity?

106 For a main cable in a suspension bridge, the dimensions H and L shown in Figure 1 are called the *sag* and the *span*, respectively. If the weight of the suspended roadbed (together with that of the cables) is uniformly distributed horizontally, the main cables form parabolas. (a) Taking the origin of an xy coordinate system at the lowest point of a parabolic main cable with sag H and span L, find an equation of the cable. (b) Using the result in part (a), find the length of the cable. Ⓒ (c) For the Ambassador Bridge across the Detroit River, the sag is $H = 63$

meters and the span is $L = 564$ meters. Find the length of one of the main cables in the Ambassador Bridge.

Figure 1

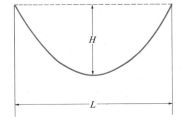

107 The Yale Bowl, completed in 1914, is a football stadium modeled after the Colosseum in Rome. The concrete stands sur-

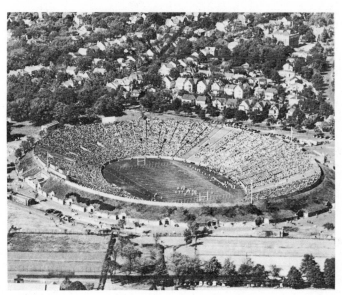

Yale Bowl

round a central region in the shape of an ellipse with a major axis 148 meters long and a minor axis 84 meters long. Find the area of the central region.

108 We know that the eccentricity $e = c/a < 1$ for the ellipse $(x^2/a^2) + (y^2/b^2) = 1$, where $c^2 = a^2 - b^2$. Now, as e comes closer and closer to zero in value, $c/a = e$ approaches zero; hence, c^2/a^2 approaches zero, so that a approaches b in value. (Why?) What is the shape of an ellipse in which a and b are close in value? What if $a = b$? Does this give an indication of how to define a conic with eccentricity $e = 0$? Use sketches and examples to answer these questions.

109 Find the equation into which the polar equation $r^2 \sin 2\theta = 2$ is transformed by rotating the polar axis through $45°$.

110 Find the equation into which the Cartesian equation $\sqrt{x} + \sqrt{y} = 2$ is transformed by rotation of the axes through an angle of $45°$ and elimination of radicals. Is the graph of this equation a parabola?

In Problems 111 to 116, rotate the axes to transform each equation to an equation that has no mixed term. Ⓒ Sketch the graph, showing both the old and the new coordinate axes.

111 $3x^2 + 4\sqrt{3}xy - y^2 = 15$ 112 $4x^2 + 4xy + 4y^2 = 24$

113 $\sqrt{3}x^2 + xy = 11$ 114 $y^2 + xy - 3x = 7$

115 $9x^2 + 16y^2 + 54x - 32y = 48$

116 $x^2 - 6xy + 9y^2 + x - 3y = 4$

In Problems 117 to 121, use the invariant $B^2 - 4AC$ to identify each conic. You may assume that the conic is nondegenerate.

117 $13x^2 + 6\sqrt{3}xy + 7y^2 = 16$ 118 $x^2 + 6xy + y^2 + 8 = 0$

119 $7x^2 + 2\sqrt{3}xy + 5y^2 + 1 = 0$ 120 $x^2 - 3xy + 5y^2 = 16$

121 $81x^2 - 18xy + 5x + y^2 = 41$

INDETERMINATE FORMS, IMPROPER INTEGRALS, AND TAYLOR'S FORMULA

Since the two basic notions of calculus—the derivative and the definite integral—are defined in terms of limits, it should come as no surprise that derivatives and integrals can often be used to help evaluate limits. In this chapter we see how certain limits, whose values are not obvious by inspection, can be found by using derivatives according to special rules called *L'Hôpital's rules*. We also use limits to study the behavior of *improper integrals*. The chapter ends with an investigation of *Taylor's formula,* which can be used to approximate certain functions by polynomials.

10.1 The Indeterminate Form 0/0

From our earlier study of the properties of limits, we know that

$$\lim_{x \to a} \frac{f(x)}{g(x)} = \frac{\lim_{x \to a} f(x)}{\lim_{x \to a} g(x)}$$

provided that $\lim_{x \to a} f(x)$ and $\lim_{x \to a} g(x)$ both exist and that $\lim_{x \to a} g(x)$ is not zero (Property 7, page 61). If $\lim_{x \to a} g(x) = 0$, this rule simply does not apply, and special methods, which we introduce in this section, must be used to find the limit—if it exists.

To begin, note that if the limit of a quotient exists and the limit of the denominator is zero, then *the limit of the numerator must also be zero*. Indeed, if $\lim_{x \to a} g(x) = 0$ and $\lim_{x \to a} [f(x)/g(x)]$ exists, we have

$$\lim_{x \to a} f(x) = \lim_{x \to a} \left[\frac{f(x)}{g(x)} g(x) \right] = \left[\lim_{x \to a} \frac{f(x)}{g(x)} \right] [\lim_{x \to a} g(x)] = \left[\lim_{x \to a} \frac{f(x)}{g(x)} \right] \cdot 0 = 0$$

Therefore, in studying the problem of finding limits of quotients whose denominators approach zero, we need to consider only quotients whose numerators also approach zero. Thus, we are led to the following definition.

DEFINITION 1

Indeterminate Form 0/0

An expression of the form

$$\frac{f(x)}{g(x)}$$

such that both

$$\lim_{x \to a} f(x) = 0 \quad \text{and} \quad \lim_{x \to a} g(x) = 0$$

is said to have the **indeterminate* form** 0/0 at a.

In this definition we are assuming that $g(x) \neq 0$ for values of x near a but different from a, so that the fraction $f(x)/g(x)$ is actually defined for such values of x.

For instance, the fraction $\dfrac{x^2 - x - 2}{x^2 + x - 6}$ has the indeterminate form $\dfrac{0}{0}$ at 2 since

$$\lim_{x \to 2} (x^2 - x - 2) = 0 \quad \text{and} \quad \lim_{x \to 2} (x^2 + x - 6) = 0$$

In Chapter 1 we evaluated limits of such fractions by factoring numerator and denominator; thus,

$$\lim_{x \to 2} \frac{x^2 - x - 2}{x^2 + x - 6} = \lim_{x \to 2} \frac{(x + 1)(x - 2)}{(x + 3)(x - 2)} = \lim_{x \to 2} \frac{x + 1}{x + 3} = \frac{3}{5}$$

Other examples, such as

$$\lim_{x \to 0} \frac{\sin x}{x} \quad \text{and} \quad \lim_{x \to 0} \frac{1 - \cos x}{x}$$

(both of which have the indeterminate form 0/0 at 0) were not so amenable and had to be found by special techniques.

In one of the first textbooks on calculus, the amateur French mathematician L'Hôpital (1661–1704) published a simple and elegant method for handling limits of indeterminate forms. Here, we give an informal statement of **L'Hôpital's rule;** later, in Theorem 2, we give a formal statement and proof.

L'Hôpital's Rule

If $\dfrac{f(x)}{g(x)}$ has the indeterminate form $\dfrac{0}{0}$ at a and $\lim\limits_{x \to a} \dfrac{f'(x)}{g'(x)}$ exists, then

$$\lim_{x \to a} \frac{f(x)}{g(x)} = \lim_{x \to a} \frac{f'(x)}{g'(x)}$$

---

*The word "indeterminate" should not be interpreted as meaning that the limit of such a quotient cannot be determined; it is intended to convey the idea that the limit, if it exists, may not be obvious by inspection.

Notice that $f'(x)/g'(x)$ is *not* the derivative of $f(x)/g(x)$. In using L'Hôpital's rule, you obtain the fraction $f'(x)/g'(x)$ by *separately* differentiating the numerator and denominator of the fraction $f(x)/g(x)$. For instance,*

$$\lim_{x \to 0} \frac{\sin x}{x} = \lim_{x \to 0} \frac{D_x \sin x}{D_x x} = \lim_{x \to 0} \frac{\cos x}{1} = 1$$

In Examples 1 and 2, use L'Hôpital's rule to find the given limit.

EXAMPLE 1 $\displaystyle \lim_{x \to 3} \frac{x^2 - 6x + 9}{x^2 - 7x + 12}$

SOLUTION   Both numerator and denominator approach zero as $x \to 3$, so the fraction has the indeterminate form $0/0$ at 3. By L'Hôpital's rule, we have

$$\lim_{x \to 3} \frac{x^2 - 6x + 9}{x^2 - 7x + 12} = \lim_{x \to 3} \frac{D_x(x^2 - 6x + 9)}{D_x(x^2 - 7x + 12)} = \lim_{x \to 3} \frac{2x - 6}{2x - 7} = \frac{0}{-1} = 0 \qquad \blacksquare$$

EXAMPLE 2 $\displaystyle \lim_{x \to 0} \frac{e^x - e^{-x}}{\ln(x + 1)}$

SOLUTION   The fraction has the indeterminate form $0/0$ at 0, so

$$\lim_{x \to 0} \frac{e^x - e^{-x}}{\ln(x + 1)} = \lim_{x \to 0} \frac{D_x(e^x - e^{-x})}{D_x \ln(x + 1)} = \lim_{x \to 0} \frac{e^x + e^{-x}}{1/(x + 1)}$$

$$= \lim_{x \to 0} (x + 1)(e^x + e^{-x}) = (0 + 1)(e^0 + e^0) = 2 \qquad \blacksquare$$

Sometimes an application of L'Hôpital's rule to an indeterminate form produces a new indeterminate form. When this happens, a second application of L'Hôpital's rule may be necessary; in fact *several* successive applications of the rule might be required to remove the indeterminacy. (However, there are cases in which the indeterminacy stubbornly persists no matter how many times L'Hôpital's rule is applied—see Problem 36 for an example.)

EXAMPLE 3   Evaluate $\displaystyle \lim_{x \to 0} \frac{x^2}{1 - \cos 2x}$.

SOLUTION   The fraction has the indeterminate form $0/0$ at 0. Applying L'Hôpital's rule, we have

$$\lim_{x \to 0} \frac{x^2}{1 - \cos 2x} = \lim_{x \to 0} \frac{D_x x^2}{D_x(1 - \cos 2x)} = \lim_{x \to 0} \frac{2x}{2 \sin 2x} = \lim_{x \to 0} \frac{x}{\sin 2x}$$

---

*A rigorous application of L'Hôpital's rule requires that we first establish the *existence* of

$$\lim_{x \to a} \frac{f'(x)}{g'(x)}$$

before writing

$$\lim_{x \to a} \frac{f(x)}{g(x)} = \lim_{x \to a} \frac{f'(x)}{g'(x)}$$

However, in the interest of brevity, it is customary to proceed more informally.

But the fraction $x/\sin 2x$ still has the indeterminate form 0/0 at 0; hence, we use L'Hôpital's rule a second time to obtain

$$\lim_{x \to 0} \frac{x^2}{1 - \cos 2x} = \lim_{x \to 0} \frac{x}{\sin 2x} = \lim_{x \to 0} \frac{D_x x}{D_x \sin 2x} = \lim_{x \to 0} \frac{1}{2 \cos 2x} = \frac{1}{2}$$ ∎

In using L'Hôpital's rule repeatedly as above, you must make certain that the rule remains applicable at each stage—in particular, you must check that the fraction under consideration really is indeterminate before each application of L'Hôpital's rule. A common error is illustrated by the following *incorrect* calculation:

$$\lim_{x \to 1} \frac{3x^2 - 2x - 1}{x^2 - x} = \lim_{x \to 1} \frac{6x - 2}{2x - 1} = \lim_{x \to 1} \frac{6}{2} = 3?$$

Here, the first step is correct, but the second is not! (Why?) In fact,

$$\lim_{x \to 1} \frac{3x^2 - 2x - 1}{x^2 - x} = \lim_{x \to 1} \frac{6x - 2}{2x - 1} = \frac{6 - 2}{2 - 1} = 4$$

L'Hôpital's rule may be extended in various ways; for instance, it works for one-sided limits as $x \to a^+$ or as $x \to a^-$. As we shall prove (Theorem 3), it also works for limits at infinity; that is, for limits as $x \to +\infty$ or as $x \to -\infty$.

EXAMPLE 4   Evaluate $\displaystyle \lim_{x \to +\infty} \frac{\sin (5/x)}{2/x}$.

SOLUTION   Here, $\displaystyle \lim_{x \to +\infty} \sin (5/x) = \sin [\lim_{x \to +\infty} (5/x)] = \sin 0 = 0$ and $\displaystyle \lim_{x \to +\infty} (2/x) = 0$. Hence, the fraction has the indeterminate form 0/0 at $+\infty$. Applying L'Hôpital's rule, we have

$$\lim_{x \to +\infty} \frac{\sin (5/x)}{2/x} = \lim_{x \to +\infty} \frac{D_x \sin (5/x)}{D_x(2/x)} = \lim_{x \to +\infty} \frac{(-5/x^2) \cos (5/x)}{-2/x^2}$$

$$= \lim_{x \to +\infty} \frac{5}{2} \cos \frac{5}{x} = \frac{5}{2} \cos \left(\lim_{x \to +\infty} \frac{5}{x} \right) = \frac{5}{2} \cos 0 = \frac{5}{2}$$ ∎

Cauchy's Generalized Mean-Value Theorem and a Proof of L'Hôpital's Rule

To prove L'Hôpital's rule, we need the following generalized version of the mean-value theorem (Theorem 3, page 169), which is attributed to the French mathematician Augustin Louis Cauchy (1789–1857).

THEOREM 1   **Cauchy's Generalized Mean-Value Theorem**

Let f and g be functions that are continuous on the closed interval $[a, b]$ and differentiable on the open interval (a, b), and suppose that $g'(x) \neq 0$ for all values of x in (a, b). Then there is a number c in the interval (a, b) such that

$$\frac{f(b) - f(a)}{g(b) - g(a)} = \frac{f'(c)}{g'(c)}$$

PROOF   Notice that if $g(a) = g(b)$, then by Rolle's theorem (Theorem 2, page 168), there is a number x in (a, b) such that $g'(x) = 0$, contrary to the hypothesis of the theorem. Therefore, $g(a) \neq g(b)$, so that $g(b) - g(a) \neq 0$. Define a function K by the equation

$$K(x) = [f(b) - f(a)]g(x) - [g(b) - g(a)]f(x) \qquad \text{for } x \text{ in } [a, b]$$

Evidently, K is continuous on $[a, b]$ since f and g are continuous on $[a, b]$. Likewise, since f and g are differentiable on (a, b), so is K; furthermore,

$$K'(x) = [f(b) - f(a)]g'(x) - [g(b) - g(a)]f'(x) \qquad \text{for } x \text{ in } (a, b)$$

Now,

$$K(b) = [f(b) - f(a)]g(b) - [g(b) - g(a)]f(b) = f(b)g(a) - f(a)g(b)$$

$$K(a) = [f(b) - f(a)]g(a) - [g(b) - g(a)]f(a) = f(b)g(a) - f(a)g(b)$$

and it follows that $K(a) = K(b)$. Therefore, we can apply Rolle's theorem to the function K and conclude that there exists a number c in the open interval (a, b) such that $K'(c) = 0$; that is,

$$0 = [f(b) - f(a)]g'(c) - [g(b) - g(a)]f'(c)$$

We have seen that $g(b) - g(a) \neq 0$; hence, since $g'(c) \neq 0$ by hypothesis, the last equation can be rewritten as

$$\frac{f(b) - f(a)}{g(b) - g(a)} = \frac{f'(c)}{g'(c)}$$

and the theorem is proved.   ∎

Notice that if $g(x) = x$, then $g'(x) = 1$ and the conclusion of Theorem 1 reduces to the conclusion of the original mean-value theorem.

EXAMPLE 5   If $f(x) = x^3 + 12$ and $g(x) = x^2 - 2$, find a number c in the open interval $(0, 2)$ such that

$$\frac{f(2) - f(0)}{g(2) - g(0)} = \frac{f'(c)}{g'(c)}$$

SOLUTION   Theorem 1 shows that such a number c exists but does not tell us how to actually find it. To find c, we calculate as follows: $f(2) = 20$, $f(0) = 12$, $g(2) = 2$, $g(0) = -2$, $f'(c) = 3c^2$, and $g'(c) = 2c$. Therefore, the required condition is

$$\frac{20 - 12}{2 - (-2)} = \frac{3c^2}{2c} \qquad \text{or} \qquad \frac{8}{4} = \frac{3c}{2}$$

It follows that $c = \frac{4}{3}$.   ∎

We now state and prove L'Hôpital's rule for the case in which x approaches a *from the right*. An analogous result holds for the case in which x approaches *a from the left* (Problem 33), and the two results, taken together, provide the justification for our original informal statement of L'Hôpital's rule.

THEOREM 2   **L'Hôpital's Rule for the Indeterminate Form 0/0 as $x \to a^+$**

Let the functions f and g be defined and differentiable on an open interval (a, b), and suppose that $g(x) \neq 0$ for $a < x < b$. Assume that $\lim\limits_{x \to a^+} f(x) = 0$, $\lim\limits_{x \to a^+} g(x) = 0$, and $g'(x) \neq 0$ for $a < x < b$. Then, if $\lim\limits_{x \to a^+} \dfrac{f'(x)}{g'(x)}$ exists, so does $\lim\limits_{x \to a^+} \dfrac{f(x)}{g(x)}$ and
$$\lim_{x \to a^+} \frac{f(x)}{g(x)} = \lim_{x \to a^+} \frac{f'(x)}{g'(x)}$$

PROOF   Define functions F and G as follows:

$$F(x) = \begin{cases} f(x) & \text{if } a < x < b \\ 0 & \text{if } x = a \end{cases} \qquad G(x) = \begin{cases} g(x) & \text{if } a < x < b \\ 0 & \text{if } x = a \end{cases}$$

for all values of x in $[a, b]$. Notice that F coincides with f on the open interval (a, b); hence, F is differentiable on (a, b), and $F'(x) = f'(x)$ for $a < x < b$. It follows that F is continuous on (a, b) and, since $\lim\limits_{x \to a^+} F(x) = \lim\limits_{x \to a^+} f(x) = 0 = F(a)$, F is actually continuous on $[a, b]$. Likewise, G is continuous on $[a, b]$ and differentiable on (a, b) with $G'(x) = g'(x)$ for $a < x < b$.

Now, choose any number x in the open interval (a, b), and notice that F and G are continuous on the closed interval $[a, x]$ and differentiable on the open interval (a, x). Furthermore, for any number t in (a, x), $G'(t) = g'(t) \neq 0$. Applying the generalized mean-value theorem (Theorem 1) to the functions F and G on the interval $[a, x]$, we conclude that there is a number c in the open interval (a, x) such that

$$\frac{F(x) - F(a)}{G(x) - G(a)} = \frac{F'(c)}{G'(c)} \qquad \text{that is,} \qquad \frac{f(x)}{g(x)} = \frac{f'(c)}{g'(c)}$$

Notice that $a < c < x$ and that the value of c might depend on the choice of x. Evidently, as x approaches a from the right, then c—which is "trapped" between a and x—must also approach a from the right. Thus,

$$\lim_{x \to a^+} \frac{f(x)}{g(x)} = \lim_{x \to a^+} \frac{f'(c)}{g'(c)} = \lim_{c \to a^+} \frac{f'(c)}{g'(c)} = \lim_{x \to a^+} \frac{f'(x)}{g'(x)}$$

and the proof is complete.   ∎

The following theorem shows that L'Hôpital's rule is also effective for limits at infinity. We state and prove the theorem only for the case $x \to +\infty$ and leave the statement and proof of the analogous result for the case $x \to -\infty$ as an exercise (Problem 38).

THEOREM 3   **L'Hôpital's Rule at Infinity**

Let the functions f and g be defined and differentiable on an open interval of the form $(k, +\infty)$, where $k > 0$ and $g(x) \neq 0$ for $x > k$. Suppose that $\lim\limits_{x \to +\infty} f(x) = 0$, $\lim\limits_{x \to +\infty} g(x) = 0$, and $g'(x) \neq 0$ for $x > k$. Then, if $\lim\limits_{x \to +\infty} \dfrac{f'(x)}{g'(x)}$ exists, so does $\lim\limits_{x \to +\infty} \dfrac{f(x)}{g(x)}$ and
$$\lim_{x \to +\infty} \frac{f(x)}{g(x)} = \lim_{x \to +\infty} \frac{f'(x)}{g'(x)}$$

PROOF We put $t = 1/x$ for $x > k$, noting that $x = 1/t$, that $0 < t < 1/k$, and that $t \to 0^+$ as $x \to +\infty$. We define functions F and G on the interval $(0, 1/k)$ by the equations

$$F(t) = f\left(\frac{1}{t}\right) \qquad \text{and} \qquad G(t) = g\left(\frac{1}{t}\right) \qquad \text{for } 0 < t < \frac{1}{k}$$

Notice that $\lim_{t\to 0^+} F(t) = \lim_{t\to 0^+} f\left(\frac{1}{t}\right) = \lim_{x\to +\infty} f(x) = 0$. Similarly, $\lim_{t\to 0^+} G(t) = 0$. By the chain rule, F and G are differentiable on $(0, 1/k)$ and we have

$$F'(t) = (-t^{-2})f'\left(\frac{1}{t}\right) \qquad \text{and} \qquad G'(t) = (-t^{-2})g'\left(\frac{1}{t}\right) \qquad \text{for } 0 < t < \frac{1}{k}$$

Applying Theorem 2 to the functions F and G on the interval $(0, 1/k)$, we have

$$\lim_{t\to 0^+} \frac{F(t)}{G(t)} = \lim_{t\to 0^+} \frac{F'(t)}{G'(t)}$$

Hence,

$$\lim_{x\to +\infty} \frac{f(x)}{g(x)} = \lim_{t\to 0^+} \frac{f\left(\frac{1}{t}\right)}{g\left(\frac{1}{t}\right)} = \lim_{t\to 0^+} \frac{F(t)}{G(t)} = \lim_{t\to 0^+} \frac{F'(t)}{G'(t)}$$

$$= \lim_{t\to 0^+} \frac{(-t^{-2})f'\left(\frac{1}{t}\right)}{(-t^{-2})g'\left(\frac{1}{t}\right)} = \lim_{t\to 0^+} \frac{f'\left(\frac{1}{t}\right)}{g'\left(\frac{1}{t}\right)} = \lim_{x\to +\infty} \frac{f'(x)}{g'(x)}$$

and the proof is complete. ∎

Problem Set 10.1

In Problems 1 to 22, use L'Hôpital's rule to evaluate each limit.

1 $\lim_{x\to 0} \dfrac{x + \sin 2x}{x - \sin 2x}$

2 $\lim_{x\to \pi/2} \dfrac{\sin x - 1}{(\pi/2) - x}$

3 $\lim_{x\to -2} \dfrac{2x^2 + 3x - 2}{3x^2 - x - 14}$

4 $\lim_{x\to 1} \dfrac{x^3 - 3x^2 + 5x - 3}{x^2 + x - 2}$

5 $\lim_{x\to 0^+} \dfrac{\sqrt{x}}{\sin \sqrt{x}}$

6 $\lim_{x\to 0} \dfrac{\cos x - \cos 3x}{\sin x^2}$

7 $\lim_{t\to \pi/2} \dfrac{\sin t - 1}{\cos t}$

8 $\lim_{x\to 0} \dfrac{xe^{3x} - x}{1 - \cos 2x}$

9 $\lim_{y\to 0} \dfrac{e^y - 1}{y^3}$

10 $\lim_{t\to 0} \dfrac{t - \sin t}{t^3}$

11 $\lim_{x\to 7} \dfrac{\ln (x/7)}{7 - x}$

12 $\lim_{x\to 0} \dfrac{x - \tan^{-1} x}{x - \sin^{-1} x}$

13 $\lim_{x\to 1} \dfrac{\ln x - \sin (x - 1)}{(x - 1)^2}$

14 $\lim_{t\to 0^+} \dfrac{\ln (e^t + 1) - \ln 2}{t^2}$

15 $\lim_{t\to 0} \dfrac{e^t - e^{-t} - 2 \sin t}{4t^3}$

16 $\lim_{y\to 0} \dfrac{y^2 - y \sin y}{e^y + e^{-y} - y^2 - 2}$

17 $\lim_{x\to \pi/2} \dfrac{\ln (\sin x)}{(\pi - 2x)^2}$

18 $\lim_{x\to 1} \dfrac{\ln x}{x - \sqrt{x}}$

19 $\lim_{x\to +\infty} \dfrac{\sin (7/x)}{5/x}$

20 $\lim_{x\to +\infty} \dfrac{\sin (3/x)}{\tan^{-1} (2/x)}$

21 $\lim_{x\to +\infty} \dfrac{1 - \cos (2/x)}{\tan (3/x)}$

22 $\lim_{x\to +\infty} \dfrac{\sin (1/x)}{\sin (2/x)}$

In Problems 23 and 24, use L'Hôpital's rule and the fundamental theorem of calculus to evaluate each limit.

23 $\lim_{x\to 0} \dfrac{\displaystyle\int_0^x 3 \cos^4 7t \, dt}{\displaystyle\int_0^x e^{5t^2} \, dt}$

24 $\lim\limits_{x \to 0} \dfrac{\displaystyle\int_0^x e^{7t}(4t^3 + t^2 + 11)\, dt}{\displaystyle\int_0^x e^{7t}(-7t^3 + 6t + 8)\, dt}$

In Problems 25 to 30, find a number c satisfying Cauchy's generalized mean-value theorem for each pair of functions on the indicated interval.

25 $f(x) = 2x^3$ and $g(x) = 3x^2 - 1$ on $[0, 2]$

26 $f(x) = \sin x$ and $g(x) = \cos x$ on $[0, \pi/4]$

27 $f(x) = \ln x$ and $g(x) = 1/x$ on $[1, e]$

28 $f(x) = \sin^{-1} x$ and $g(x) = x$ on $[0, 1]$

29 $f(x) = \tan x$ and $g(x) = 4x/\pi$ on $[-\pi/4, \pi/4]$

30 $f(x) = x^2(x^2 - 2)$ and $g(x) = x$ on $[-1, 1]$

31 Find constants a and b such that

$$\lim_{x \to 0} \frac{1}{bx - \sin x} \int_0^x \frac{t^2\, dt}{\sqrt{a + t}} = 1$$

32 Show that

$$\lim_{x \to 1} \frac{nx^{n+1} - (n + 1)x^n + 1}{(x - 1)^2} = \frac{n(n + 1)}{2}$$

33 State and prove the analog of Theorem 2 for the case $x \to a^-$.

34 Explain why the calculation of a derivative directly from the definition always involves the evaluation of a limit of an indeterminate form $0/0$.

35 Explain why the use of L'Hôpital's rule to evaluate $\lim\limits_{x \to 0} \dfrac{\sin x}{x}$ really involves circular reasoning.

36 What happens when you try to find $\lim\limits_{t \to 0^+} (e^{-1/t}/t)$ by repeated applications of L'Hôpital's rule?

37 In Problem 36, make the change of variable $t = 1/x$. Then use the result of Problem 169b in the Review Problem Set for Chapter 7 to evaluate the limit.

38 State and prove the analog of Theorem 3 for the case $x \to -\infty$.

39 The equation $I = (E/R)(1 - e^{-Rt/L})$ gives the current I amperes at time t seconds in an RL-circuit with resistance R ohms, inductance L henrys, and electromotive force E volts. (See Problem 21 in Problem Set 7.11.) If t, E, and L are held fixed, find the limit of I as R approaches zero.

40 Let A be an endpoint of a diameter of a fixed circle with center O and radius r. In Figure 1, $\overline{AQ}$ is tangent to the circle at point A, and $|\overline{AQ}|$ is equal to the length of the arc AP. If B is the point of intersection of the line through Q and P with the line through A and O, find the limiting position of B as P approaches A.

Figure 1

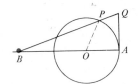

41 A weight hanging from a spring is caused to vibrate by a driving force so that its displacement at time t seconds is given by

$$y = \frac{A}{p^2 - w^2}(\sin wt - \sin pt)$$

where A, p, and w are positive constants with $p \neq w$. Determine the limiting value of y as p approaches w holding A and t fixed.

42 For each positive integer n, let V_n denote the volume of the solid obtained by revolving the region under the graph of $f(x) = x^n$ between $x = 0$ and $x = 1$ about the y axis and let H_n be the volume of the solid obtained by revolving the same region about the x axis. Evaluate

(a) $\lim\limits_{n \to +\infty} V_n$ (b) $\lim\limits_{n \to +\infty} H_n$ (c) $\lim\limits_{n \to +\infty} \dfrac{V_n}{H_n}$

10.2   Other Indeterminate Forms

In Section 10.1 we saw that L'Hôpital's rule can often be used to handle limits of fractions having indeterminate form $0/0$. In this section we discuss additional indeterminate forms, ∞/∞, $\infty \cdot 0$, $\infty - \infty$, 0^0, ∞^0, and 1^∞.

The Indeterminate Form ∞/∞

By methods similar to those used in Section 10.1, it can be proved that L'Hôpital's rule remains effective when *the numerator and denominator become infinite in absolute value*, or, as we say, when the fraction has the **indeterminate form ∞/∞**.

The facts are set forth in the following theorem.

<u>THEOREM 1</u>   **L'Hôpital's Rule for the Indeterminate Form ∞/∞**

Suppose that the functions f and g are defined and differentiable on an open interval I, except possibly at the number a in I. Moreover, assume that $\lim\limits_{x \to a} |f(x)| = +\infty$ and $\lim\limits_{x \to a} |g(x)| = +\infty$. Then, if $g'(x) \neq 0$ for all values of x other than a in I and if $\lim\limits_{x \to a} \dfrac{f'(x)}{g'(x)}$ exists, it follows that $\lim\limits_{x \to a} \dfrac{f(x)}{g(x)}$ also exists and

$$\lim_{x \to a} \frac{f(x)}{g(x)} = \lim_{x \to a} \frac{f'(x)}{g'(x)}$$

This theorem is also correct when $x \to a^+$, $x \to a^-$, $x \to +\infty$, or $x \to -\infty$, provided that obvious changes are made in the hypotheses. We omit the somewhat technical proofs of Theorem 1 and its alternative versions.

In Examples 1 and 2, evaluate the given limit.

<u>EXAMPLE 1</u> $\lim\limits_{x \to +\infty} \dfrac{\ln x}{\sqrt{x}}$

SOLUTION   The fraction has the indeterminate form ∞/∞ at $+\infty$. Thus, by L'Hôpital's rule,

$$\lim_{x \to +\infty} \frac{\ln x}{\sqrt{x}} = \lim_{x \to +\infty} \frac{D_x \ln x}{D_x \sqrt{x}} = \lim_{x \to +\infty} \frac{1/x}{1/(2\sqrt{x})}$$

$$= \lim_{x \to +\infty} \frac{2\sqrt{x}}{x} = \lim_{x \to +\infty} \frac{2}{\sqrt{x}} = 0$$

<u>EXAMPLE 2</u> $\lim\limits_{x \to \pi/2^+} \dfrac{7 \tan x}{5 + \sec x}$

SOLUTION   The fraction has the indeterminate form ∞/∞ at $\pi/2$. By L'Hôpital's rule, we have

$$\lim_{x \to \pi/2^+} \frac{7 \tan x}{5 + \sec x} = \lim_{x \to \pi/2^+} \frac{7 \sec^2 x}{\sec x \tan x} = \lim_{x \to \pi/2^+} \frac{7 \sec x}{\tan x}$$

$$= \lim_{x \to \pi/2^+} \frac{7/\cos x}{\sin x/\cos x} = \lim_{x \to \pi/2^+} \frac{7}{\sin x} = 7$$

Just as was the case for the indeterminate form $0/0$, it may be necessary to apply L'Hôpital's rule several times in order to evaluate a limit of an indeterminate form ∞/∞. Also, it can be shown that *L'Hôpital's rule still holds if the quotient $f'(x)/g'(x)$ approaches $+\infty$ or $-\infty$;* that is,

$$\lim_{x \to a} \frac{f(x)}{g(x)} = \lim_{x \to a} \frac{f'(x)}{g'(x)} = \pm\infty$$

This is illustrated by the following example.

<u>EXAMPLE 3</u>   Evaluate $\lim\limits_{x \to +\infty} \dfrac{e^x}{x^3}$.

SOLUTION   The fraction has the indeterminate form ∞/∞ at $+\infty$. Using L'Hôpital's rule 3 times, we have

$$\lim_{x\to+\infty}\frac{e^x}{x^3} = \lim_{x\to+\infty}\frac{e^x}{3x^2} = \lim_{x\to+\infty}\frac{e^x}{6x} = \lim_{x\to+\infty}\frac{e^x}{6} = +\infty \qquad \blacksquare$$

Other Cases of Indeterminate Forms

We now consider indeterminate forms of the type $\infty\cdot 0$, $\infty - \infty$, 0^0, ∞^0, and 1^∞. These are handled by algebraic manipulation of the indeterminate form so as to reduce it to the type $0/0$ or ∞/∞ so that L'Hôpital's rule can be used. In each case suitable examples are given to illustrate the technique.

The Case $\infty\cdot 0$

If $\lim_{x\to a}|f(x)| = +\infty$ and $\lim_{x\to a}g(x) = 0$, we say that the product $f(x)g(x)$ **has the indeterminate form** $\infty\cdot 0$ **at** a. To find $\lim_{x\to a}f(x)g(x)$ in this case, we can write $f(x)g(x)$ as $g(x)/[1/f(x)]$ or as $f(x)/[1/g(x)]$, thus obtaining the form $0/0$ or the form ∞/∞, respectively, whichever is more convenient.

EXAMPLE 4   Evaluate $\lim_{x\to 0^+} x^2\ln x$.

SOLUTION   We write $x^2\ln x = (\ln x)/x^{-2}$, noting that $(\ln x)/x^{-2}$ has the indeterminate form ∞/∞ at 0. By applying L'Hôpital's rule, we obtain

$$\lim_{x\to 0^+} x^2\ln x = \lim_{x\to 0^+}\frac{\ln x}{x^{-2}} = \lim_{x\to 0^+}\frac{1/x}{-2x^{-3}} = \lim_{x\to 0^+}(-\tfrac{1}{2})x^2 = 0 \qquad \blacksquare$$

The Case $\infty - \infty$

If $\lim_{x\to a}f(x) = +\infty$ and $\lim_{x\to a}g(x) = +\infty$, we say that the difference $f(x) - g(x)$ **has the indeterminate form** $\infty - \infty$ **at** a. By performing the indicated subtraction, the indeterminate form $\infty - \infty$ can usually be converted to the indeterminate form $0/0$.

EXAMPLE 5   Evaluate $\lim_{x\to 0^+}\left(\csc x - \dfrac{1}{x}\right)$.

SOLUTION   The expression has the indeterminate form $\infty - \infty$ at 0; however, if we write $\csc x$ as $1/\sin x$ and subtract, we have

$$\lim_{x\to 0^+}\left(\csc x - \frac{1}{x}\right) = \lim_{x\to 0^+}\left(\frac{1}{\sin x} - \frac{1}{x}\right) = \lim_{x\to 0^+}\frac{x - \sin x}{x\sin x}$$

The fraction has the indeterminate form $0/0$ at 0. Applying L'Hôpital's rule twice, we obtain

$$\lim_{x\to 0^+}\frac{x - \sin x}{x\sin x} = \lim_{x\to 0^+}\frac{1 - \cos x}{\sin x + x\cos x}$$

$$= \lim_{x\to 0^+}\frac{\sin x}{2\cos x - x\sin x} = \frac{0}{2} = 0 \qquad \blacksquare$$

The Case 0^0, ∞^0, or 1^∞

If $f(x) > 0$, $\lim_{x \to a} f(x) = 0$, and $\lim_{x \to a} g(x) = 0$, we say that the expression $f(x)^{g(x)}$ **has the indeterminate form** 0^0 **at** a. Expressions with the indeterminate forms ∞^0 or 1^∞ are defined analogously.* In order to evaluate $\lim_{x \to a} f(x)^{g(x)}$ in such indeterminate cases we carry out the following procedure:

Step 1 Calculate $\lim_{x \to a} [g(x) \ln f(x)] = L$.

Step 2 Conclude that $\lim_{x \to a} f(x)^{g(x)} = e^L$.

This procedure is justified by the observation that

$$\lim_{x \to a} f(x)^{g(x)} = \lim_{x \to a} e^{g(x) \ln f(x)} = e^L$$

Notice that the product $g(x) \ln f(x)$ in step 1 has an indeterminate form at a.

In Examples 6 and 7, evaluate the indicated limit.

<u>EXAMPLE 6</u> $\lim_{x \to 0^+} x^{2x}$

SOLUTION   The indeterminate form is 0^0, so we carry out the procedure above.

Step 1 We must evaluate $\lim_{x \to 0^+} 2x \ln x$. Here the product has the indeterminate form $0 \cdot \infty$ at 0; hence, we rewrite it as

$$2x \ln x = \frac{2 \ln x}{1/x}$$

The resulting fraction has the indeterminate form ∞/∞ at 0. Thus, by L'Hôpital's rule,

$$L = \lim_{x \to 0^+} 2x \ln x = \lim_{x \to 0^+} \frac{2 \ln x}{1/x} = \lim_{x \to 0^+} \frac{D_x(2 \ln x)}{D_x(1/x)}$$

$$= \lim_{x \to 0^+} \frac{2/x}{-1/x^2} = \lim_{x \to 0^+} (-2x) = 0$$

Step 2 $\lim_{x \to 0^+} x^{2x} = e^L = e^0 = 1$

<u>EXAMPLE 7</u> $\lim_{x \to 0^+} (\csc x)^{\sin x}$

SOLUTION   The indeterminate form is ∞^0. Since

$$\lim_{x \to 0^+} (\csc x)^{\sin x} = \lim_{x \to 0^+} \left(\frac{1}{\sin x}\right)^{\sin x}$$

*See J. V. Baxley and E. K. Hayashi, "Indeterminate Forms of Exponential Type," *American Mathematical Monthly*, Vol. 85, No. 6, June–July 1978, pp. 484–486 for an interesting discussion of these cases.

we begin by making the change of variable $u = \sin x$. Notice that $u \to 0^+$ as $x \to 0^+$; hence,

$$\lim_{x \to 0^+} (\csc x)^{\sin x} = \lim_{u \to 0^+} \left(\frac{1}{u}\right)^u$$

Now we use our procedure.

Step 1 $L = \lim_{u \to 0^+} \left[u \ln\left(\frac{1}{u}\right)\right] = \lim_{u \to 0^+} (-u \ln u) = \lim_{u \to 0^+} \frac{-\ln u}{1/u}$

$= \lim_{u \to 0^+} \frac{D_u(-\ln u)}{D_u(1/u)} = \lim_{u \to 0^+} \frac{-1/u}{-1/u^2} = \lim_{u \to 0^+} u = 0$

Step 2 $\lim_{x \to 0^+} (\csc x)^{\sin x} = \lim_{u \to 0^+} \left(\frac{1}{u}\right)^u = e^L = e^0 = 1$

Problem Set 10.2

In Problems 1 to 50, evaluate each limit.

1 $\lim_{x \to \pi/2} \dfrac{1 + \sec x}{\tan x}$

2 $\lim_{x \to 1/2} \dfrac{\sec 3\pi x}{\tan 3\pi x}$

3 $\lim_{x \to +\infty} \dfrac{\ln (17 + x)}{x}$

4 $\lim_{x \to 0^+} \dfrac{1 - \ln x}{e^{1/x}}$

5 $\lim_{x \to +\infty} \dfrac{e^x + 1}{x^4 + x^3}$

6 $\lim_{x \to +\infty} \dfrac{2^x}{x^3}$

7 $\lim_{x \to +\infty} \dfrac{2x^4}{e^{3x}}$

8 $\lim_{x \to +\infty} \dfrac{\ln (x + e^x)}{x}$

9 $\lim_{t \to +\infty} \dfrac{t \ln t}{(t + 2)^2}$

10 $\lim_{x \to +\infty} \dfrac{x + e^{2x}}{\ln x + e^{2x}}$

11 $\lim_{x \to +\infty} x(e^{-x} - 1)$

12 $\lim_{t \to 0} \sin 3t \cot 2t$

13 $\lim_{x \to 0^+} xe^{1/x}$

14 $\lim_{x \to +\infty} x \sin (\pi/x)$

15 $\lim_{x \to 0^+} x(\ln x)^2$

16 $\lim_{x \to \pi/2} \cos 3x \sec 5x$

17 $\lim_{x \to \pi/2} \tan x \tan 2x$

18 $\lim_{x \to 0} \csc x \sin^{-1} x$

19 $\lim_{x \to 1} \left(\dfrac{x}{x - 1} - \dfrac{1}{\ln x}\right)$

20 $\lim_{x \to 1} \left(\dfrac{x}{\ln x} - \dfrac{1}{x \ln x}\right)$

21 $\lim_{x \to 0^+} (\csc x - \csc 2x)$

22 $\lim_{t \to 0} \left(\dfrac{1}{e^t - 1} - \dfrac{1}{t}\right)$

23 $\lim_{x \to +\infty} (x^2 - \sqrt{x^4 + x^2 + 7})$

24 $\lim_{x \to \pi/2} \left(x \tan x - \dfrac{\pi}{2} \sec x\right)$

25 $\lim_{x \to 4} \left(\dfrac{7}{x^2 - x - 12} - \dfrac{1}{x - 4}\right)$

26 $\lim_{x \to 1} \left(\dfrac{n}{x^n - 1} - \dfrac{m}{x^m - 1}\right)$

27 $\lim_{x \to 0^+} x^x$

28 $\lim_{x \to 0^+} (\sinh x)^x$

29 $\lim_{y \to 0^+} (e^y - 1)^y$

30 $\lim_{x \to \pi/2^-} \left(\dfrac{5\pi}{2} - 5x\right)^{\cos x}$

31 $\lim_{x \to \pi/2^-} (\cos x)^{x - (\pi/2)}$

32 $\lim_{x \to \pi/4^-} \left(\dfrac{\pi}{4} - x\right)^{\cos 2x}$

33 $\lim_{x \to 0^+} (\cot x)^{x^2}$

34 $\lim_{x \to 0^+} (\cot x)^{\sin x}$

35 $\lim_{x \to +\infty} \left(\dfrac{x}{x - 2}\right)^x$

36 $\lim_{x \to +\infty} (e^x + x)^{1/x}$

37 $\lim_{n \to +\infty} \sqrt[n]{n}$

38 $\lim_{x \to 0^+} (-\ln x)^x$

39 $\lim_{x \to 0} (1 + \tan x)^{1/x}$

40 $\lim_{x \to +\infty} \left(1 + \dfrac{3}{x}\right)^x$

41 $\lim_{x \to 0} (1 + 2x)^{3/x}$

42 $\lim_{x \to 0^+} (1 + x)^{\ln x}$

43 $\lim_{x \to +\infty} \left(\cos \dfrac{2}{x}\right)^{x^2}$

44 $\lim_{y \to 0} \left(\dfrac{\sin y}{y}\right)^{1/y}$

45 $\lim_{x \to 0} (1 + x)^{\cot x}$

46 $\lim_{x \to 2^-} \left(1 - \dfrac{x}{2}\right)^{\tan \pi x}$

47 $\lim_{x \to 0^-} (1 + x)^{\ln |x|}$

48 $\lim_{x \to 0} (e^{2x} + 2x)^{1/(4x)}$

49 $\lim_{x \to \pi/2} (\sin x)^{\sec x}$

50 $\lim_{x \to 1} x^{1/(1-x)}$

[C]**51** Use a calculator to obtain numerical confirmation of your result in (a) Problem 3, (b) Problem 13, and (c) Problem 27.

52 Let

$$g(x) = \begin{cases} x^2 \sin (1/x) & \text{if } x \neq 0 \\ 0 & \text{if } x = 0 \end{cases}$$

Show that $g'(0) = 0$.

53 Suppose that the function f has the following properties:

$$\lim_{x \to +\infty} f(x) = \lim_{x \to +\infty} f'(x) = \lim_{x \to +\infty} f''(x) = +\infty$$

and

$$\lim_{x \to +\infty} \frac{xf'''(x)}{f''(x)} = 1$$

Evaluate $\lim_{x \to +\infty} \dfrac{xf'(x)}{f(x)}$.

54 Find a value of the constant c such that $\lim_{x \to +\infty} \left(\dfrac{x+c}{x-c} \right)^x = 4$.

55 Evaluate $\lim_{x \to +\infty} \dfrac{1}{x} \int_0^x e^t \, dt$.

10.3   Improper Integrals with Infinite Limits

In Chapter 5, we obtained the areas of admissible regions in the plane by using definite integrals. Recall, however, that an admissible region must be *bounded*. If we wish to find areas of unbounded regions, we have to deal with "improper" integrals.

Consider, for instance, the region R under the graph of $y = 1/x^2$ to the right of $x = 1$ (Figure 1a). Notice that the region R extends indefinitely to the right and therefore is unbounded. Offhand, it is perhaps not clear what is meant by the "area" of such an unbounded region. However, let R_b denote the bounded region under the graph of $y = 1/x^2$ between $x = 1$ and $x = b$ (Figure 1b). The area of R_b is given by

$$\int_1^b \frac{1}{x^2} \, dx = \frac{-1}{x} \bigg|_1^b = 1 - \frac{1}{b}$$

For large values of b, the bounded region R_b might be considered to be a good approximation to the unbounded region R; in fact, it is tempting to write $R = \lim_{b \to +\infty} R_b$. Hence, we might expect that

$$\text{area of } R = \lim_{b \to +\infty} (\text{area of } R_b) = \lim_{b \to +\infty} \left(1 - \frac{1}{b} \right) = 1 \text{ square unit}$$

In general, if f is a function defined on an infinite interval of the form $[a, +\infty)$, and if $f(x) \geq 0$ holds for $x \geq a$, we define *the area of the unbounded region under the graph of f and to the right of $x = a$ to be* $\lim_{b \to +\infty} \int_a^b f(x) \, dx$ (Figure 2). Often we write this area simply as $\int_a^{+\infty} f(x) \, dx$. More generally, we make the following definition.

Figure 1

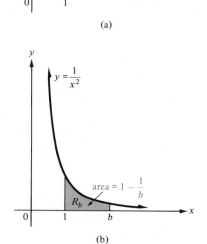

(a)

(b)

Figure 2

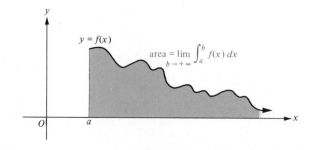

DEFINITION 1   **Improper Integrals with Infinite Upper Limits**

Let f be a function defined at least on the infinite interval $[a, +\infty)$. Suppose that f is integrable on the closed interval $[a, b]$ for every value of b larger than a. Then we define

$$\int_a^{+\infty} f(x)\ dx = \lim_{b \to +\infty} \int_a^b f(x)\ dx$$

provided that the limit exists as a finite number.

The expression $\int_a^{+\infty} f(x)\ dx$, which is often written as $\int_a^{\infty} f(x)\ dx$ for simplicity, is called an **improper integral** with an **infinite upper limit**. If $\lim_{b \to +\infty} \int_a^b f(x)\ dx$ exists as a finite number, then we say that the improper integral $\int_a^{+\infty} f(x)\ dx$ is **convergent**. Otherwise, it is **divergent**. An **improper integral** with an **infinite lower limit** is defined in a similar way by

$$\int_{-\infty}^b f(x)\ dx = \lim_{a \to -\infty} \int_a^b f(x)\ dx$$

provided that the limit exists as a finite number, in which case we say that the improper integral $\int_{-\infty}^b f(x)\ dx$ is **convergent**. Otherwise, it is **divergent**.

In Examples 1 to 4, evaluate the given improper integral (if it is convergent).

EXAMPLE 1 $\displaystyle\int_1^{\infty} \frac{dx}{x^3}$

SOLUTION   By Definition 1, we have

$$\int_1^{\infty} \frac{dx}{x^3} = \lim_{b \to +\infty} \int_1^b \frac{dx}{x^3} = \lim_{b \to +\infty} \left(\frac{-1}{2x^2} \bigg|_1^b \right) = \lim_{b \to +\infty} \left(\frac{-1}{2b^2} + \frac{1}{2} \right) = \frac{1}{2} \quad \blacksquare$$

EXAMPLE 2 $\displaystyle\int_0^{\infty} \frac{dx}{1 + x^2}$

SOLUTION $\displaystyle\int_0^{\infty} \frac{dx}{1 + x^2} = \lim_{b \to +\infty} \int_0^b \frac{dx}{1 + x^2} = \lim_{b \to +\infty} (\tan^{-1} x) \bigg|_0^b$

$$= \lim_{b \to +\infty} (\tan^{-1} b - \tan^{-1} 0) = \lim_{b \to +\infty} \tan^{-1} b = \frac{\pi}{2} \quad \blacksquare$$

EXAMPLE 3 $\displaystyle\int_{-\infty}^3 \frac{dx}{(9 - x)^2}$

SOLUTION $\displaystyle\int_{-\infty}^3 \frac{dx}{(9 - x)^2} = \lim_{a \to -\infty} \int_a^3 \frac{dx}{(9 - x)^2} = \lim_{a \to -\infty} \left(\frac{1}{9 - x} \bigg|_a^3 \right)$

$$= \lim_{a \to -\infty} \left(\frac{1}{9 - 3} - \frac{1}{9 - a} \right) = \frac{1}{6} \quad \blacksquare$$

EXAMPLE 4 $\displaystyle\int_{-\infty}^{0} e^{-x}\,dx$

SOLUTION $\displaystyle\int_{-\infty}^{0} e^{-x}\,dx = \lim_{a \to -\infty} \int_{a}^{0} e^{-x}\,dx = \lim_{a \to -\infty}\left(-e^{-x}\,\Big|_{a}^{0}\right)$

$$= \lim_{a \to -\infty}(e^{-a} - e^{0}) = +\infty$$

Hence, $\displaystyle\int_{-\infty}^{0} e^{-x}\,dx$ is divergent. ■

The following example illustrates the use of improper integrals in finding areas of unbounded regions.

Figure 3

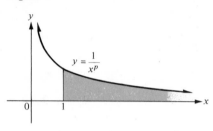

EXAMPLE 5   Suppose that $p > 1$. Find the area A of the region under the curve $y = 1/x^p$ and to the right of the line $x = 1$ (Figure 3).

SOLUTION

$$A = \int_{1}^{\infty} \frac{dx}{x^p} = \lim_{b \to +\infty} \int_{1}^{b} \frac{dx}{x^p} = \lim_{b \to +\infty}\left(\frac{x^{1-p}}{1-p}\,\Big|_{1}^{b}\right)$$

$$= \lim_{b \to +\infty}\left(\frac{b^{1-p}}{1-p} - \frac{1^{1-p}}{1-p}\right) = 0 - \frac{1}{1-p} = \frac{1}{p-1} \qquad \text{square units}$$

since the fact that $p > 1$ implies that $\displaystyle\lim_{b \to +\infty} b^{1-p} = \lim_{b \to +\infty}(1/b^{p-1}) = 0$. ■

Notice that if $p < 1$, then $\displaystyle\lim_{b \to +\infty} b^{1-p} = +\infty$, so that $\int_{1}^{\infty}(dx/x^p)$ diverges and the unbounded region in Figure 3 has an infinite area. If $p = 1$, the integral also diverges (Problem 5); hence,

$$\int_{1}^{\infty} \frac{dx}{x^p} \begin{cases} \text{converges to } \dfrac{1}{p-1} & \text{if } p > 1 \\[2mm] \text{diverges} & \text{if } p \le 1 \end{cases}$$

Improper integrals for which both limits of integration are infinite are defined as follows.

DEFINITION 2   **Improper Integrals with Both Limits Infinite**

Let the function f be defined for all real numbers, and suppose that the improper integrals $\int_{0}^{\infty} f(x)\,dx$ and $\int_{-\infty}^{0} f(x)\,dx$ *both* converge. Then, by definition,

$$\int_{-\infty}^{\infty} f(x)\,dx = \int_{-\infty}^{0} f(x)\,dx + \int_{0}^{\infty} f(x)\,dx$$

and we say that the improper integral $\int_{-\infty}^{\infty} f(x)\,dx$ is **convergent.**

If *either* of the improper integrals $\int_{0}^{\infty} f(x)\,dx$ or $\int_{-\infty}^{0} f(x)\,dx$ is divergent, we say that the **improper integral** $\int_{-\infty}^{\infty} f(x)\,dx$ is **divergent.**

In Examples 6 and 7, evaluate the given improper integral (if it is convergent).

EXAMPLE 6 $\displaystyle\int_{-\infty}^{\infty} \frac{dx}{x^2 - 2x + 2}$

SOLUTION   We have

$$\int_0^\infty \frac{dx}{x^2 - 2x + 2} = \lim_{b \to +\infty} \int_0^b \frac{dx}{x^2 - 2x + 2}$$

$$= \lim_{b \to +\infty} \int_0^b \frac{dx}{(x-1)^2 + 1} = \lim_{b \to +\infty} \left[\tan^{-1}(x-1) \Big|_0^b \right]$$

$$= \lim_{b \to +\infty} [\tan^{-1}(b-1) - \tan^{-1}(-1)] = \frac{\pi}{2} + \frac{\pi}{4} = \frac{3\pi}{4}$$

Likewise,

$$\int_{-\infty}^0 \frac{dx}{x^2 - 2x + 2} = \lim_{a \to -\infty} \int_a^0 \frac{dx}{(x-1)^2 + 1} = \lim_{a \to -\infty} \left[\tan^{-1}(x-1) \Big|_a^0 \right]$$

$$= \lim_{a \to -\infty} [\tan^{-1}(-1) - \tan^{-1}(a-1)] = -\frac{\pi}{4} + \frac{\pi}{2} = \frac{\pi}{4}$$

Therefore, the improper integral is convergent, and

$$\int_{-\infty}^\infty \frac{dx}{x^2 - 2x + 2} = \int_{-\infty}^0 \frac{dx}{x^2 - 2x + 2} + \int_0^\infty \frac{dx}{x^2 - 2x + 2} = \frac{\pi}{4} + \frac{3\pi}{4} = \pi$$

Naturally, we interpret this result geometrically as meaning that the entire unbounded region under the curve

$$y = \frac{1}{x^2 - 2x + 2}$$

has an area of π square units (Figure 4).

Figure 4

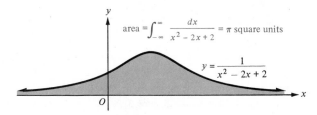

area $= \displaystyle\int_{-\infty}^\infty \frac{dx}{x^2 - 2x + 2} = \pi$ square units

$y = \dfrac{1}{x^2 - 2x + 2}$

EXAMPLE 7 $\displaystyle\int_{-\infty}^\infty x \, dx$

SOLUTION $\displaystyle\int_{-\infty}^\infty x \, dx = \int_{-\infty}^0 x \, dx + \int_0^\infty x \, dx$

provided that the last two improper integrals converge. But they do not converge; in particular,

$$\int_0^\infty x \, dx = \lim_{b \to +\infty} \int_0^b x \, dx = \lim_{b \to +\infty} \frac{b^2}{2} = +\infty$$

Therefore, $\displaystyle\int_{-\infty}^\infty x \, dx$ diverges.

Problem Set 10.3

In Problems 1 to 24, evaluate each improper integral (if it is convergent).

1 $\displaystyle\int_1^\infty \frac{dx}{x\sqrt{x}}$

2 $\displaystyle\int_1^\infty \frac{dx}{(4x+3)^2}$

3 $\displaystyle\int_3^\infty \frac{dx}{x^2+9}$

4 $\displaystyle\int_{-\infty}^0 \frac{dx}{x^2+16}$

5 $\displaystyle\int_1^\infty \frac{dx}{x}$

6 $\displaystyle\int_{-\infty}^2 \frac{dx}{(4-x)^2}$

7 $\displaystyle\int_0^\infty \frac{dx}{(x+1)(x+2)}$

8 $\displaystyle\int_2^\infty \frac{x\,dx}{(x+1)(x+2)}$

9 $\displaystyle\int_0^\infty 4e^{8x}\,dx$

10 $\displaystyle\int_1^\infty \frac{e^{-\sqrt{u}}}{\sqrt{u}}\,du$

11 $\displaystyle\int_1^\infty \frac{x\,dx}{1+x^4}$

12 $\displaystyle\int_e^\infty \frac{dx}{x(\ln x)^2}$

13 $\displaystyle\int_e^\infty \frac{dx}{x\ln x}$

14 $\displaystyle\int_0^\infty e^{-x}\sin x\,dx$

15 $\displaystyle\int_{-\infty}^{-2} \frac{dx}{(x-1)^4}$

16 $\displaystyle\int_{-\infty}^1 \frac{3t\,dt}{(3t^2+1)^3}$

17 $\displaystyle\int_{-\infty}^0 xe^x\,dx$

18 $\displaystyle\int_{-\infty}^\infty (x^2+2x+2)^{-1}\,dx$

19 $\displaystyle\int_{-\infty}^\infty \frac{|x|\,dx}{1+x^4}$

20 $\displaystyle\int_{-\infty}^\infty x^2 e^{-x^3}\,dx$

21 $\displaystyle\int_{-\infty}^\infty \frac{x\,dx}{(x^2+1)^2}$

22 $\displaystyle\int_{-\infty}^\infty \frac{e^x\,dx}{\cosh x}$

23 $\displaystyle\int_{-\infty}^\infty \frac{dx}{a^2+x^2}$

24 $\displaystyle\int_{-\infty}^\infty \mathrm{sech}\,x\,dx$

25 Find the area of the region in the first quadrant under the curve $y = 2^{-x}$ and to the right of the y axis (Figure 5).

Figure 5

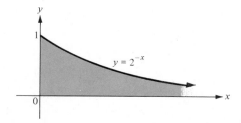

$y = 2^{-x}$

26 Evaluate $\int_0^\infty xe^{-x}\,dx$. (*Hint:* Use integration by parts and L'Hôpital's rule.)

27 For what value of n does the improper integral

$$\int_0^\infty \left(\frac{2}{x+1} - \frac{n}{x+3}\right) dx$$

converge? Evaluate the integral for this value of n.

28 If $\int_{-\infty}^\infty f(x)\,dx$ is convergent, show that $\int_{-\infty}^\infty f(x)\,dx = \lim_{c\to+\infty} \int_{-c}^c f(x)\,dx$.

29 Show that $\lim_{c\to+\infty} \int_{-c}^c \sin x\,dx = 0$.

30 In view of Problems 28 and 29, can you conclude that $\int_{-\infty}^\infty \sin x\,dx = 0$? Explain.

31 Find the area of the unbounded region under the curve $y = \dfrac{1}{e^x + e^{-x}}$.

32 Find the volume of the unbounded solid generated by revolving the region under the curve $y = \sqrt{x}\,e^{-x^2}$, $x \geq 0$, about the x axis.

33 Find the volume of the unbounded solid generated by revolving the region under the curve $y = 1/x$, $x \geq 1$, about the x axis.

34 Show that the surface area of the unbounded surface generated by revolving the curve $y = 1/x$, $x \geq 1$, about the x axis is infinite. [*Hint:* For $x \geq 1$, $(1/x)\sqrt{1 + (-1/x^2)^2} > 1/x$.]

35 Find the area of the unbounded region between the two curves $y = 1/x^2$ and $y = e^{-2x}$ over the interval $[1, \infty)$.

36 Give an example of (a) an unbounded region in the plane with infinite area that, when rotated about the x axis, generates an unbounded solid with finite volume; and (b) an unbounded solid with finite volume that has infinite surface area.

37 A flow of revenue into a business firm may be described by a function f whose value $f(t)$ represents the revenue in dollars per year t years in the future. The idea of *present value* of money to be received in the future applies to such a **revenue stream** (see page 461). If investments earning continuously compounded interest are available at a nominal annual rate r, then the present value $P(r)$ of all future revenue is given by

$$P(r) = \int^\infty e^{-rt} f(t)\,dt$$

If $f(t) = A + Bt$, where A and B are constants, calculate $P(r)$.

38 The formula for present value $P(r)$ given in Problem 37 applies to a stream of revenue from any capital asset, for instance, a *perpetuity*. If you purchase such a "perpetual annuity," it provides you and your heirs an annual income of A dollars for all future time. What is the present value to you of such a perpetuity if $A = \$10{,}000$ per year and the nominal annual interest rate is 10 percent ($r = 0.10$)?

39 The formula in Problem 37 that gives the function P in terms of the function f has important applications outside the world of business and finance. If the improper integral in this formula is convergent, then the function P is called the **Laplace transform** of the function f. If P is the Laplace transform of f, Q is the Laplace transform of g, and a and b are constants, show that $aP + bQ$ is the Laplace transform of $af + bg$.

40 Suppose that P is the Laplace transform of f (see Problem 39) and that $\lim_{t\to+\infty} e^{-rt}f(t) = 0$ for all values of r. If f is differentiable and Q is the Laplace transform of f', show that $Q(r) = rP(r) - f(0)$.

41 The **gamma function,** denoted by Γ (the Greek letter capital gamma), is defined by

$$\Gamma(n) = \int_0^\infty e^{-x}x^{n-1}\, dx$$

for all positive real numbers n. This function, which was discovered by Leonhard Euler, is useful for solving differential equations in applied mathematics. Using integration by parts, show that $\Gamma(n + 1) = n\Gamma(n)$ for $n > 0$. [*Hint:* Evaluate $\Gamma(n + 1)$, using integration by parts with $u = x^n$ and $dv = e^{-x}\, dx$.]

42 Recall that if n is a positive integer, then n **factorial,** denoted by

$n!$, is defined by

$$n! = n(n - 1)(n - 2)\cdots 3 \cdot 2 \cdot 1$$

Also, by special definition, $0! = 1$. Using the result of Problem 41, show that $\Gamma(n + 1) = n!$ holds for all nonnegative integers n. Therefore, the gamma function provides a continuous generalization of the factorial of a nonnegative integer.

43 For the gamma function (defined in Problem 41), show that

$$\Gamma(n + 1) = r^{n+1} \int_0^\infty e^{-rt}t^n\, dt$$

for $n > -1$ and $r > 0$. (*Hint:* Make the change of variable $x = rt$.)

44 Let $f(t) = t^n$, where n is a constant and $n > -1$. If P is the Laplace transform of f (see Problem 39), show that

$$P(r) = \frac{\Gamma(n + 1)}{r^{n+1}}$$

for $r > 0$. (*Hint:* Use the result of Problem 43.)

45 Find the Laplace transform of $f(t) = \sin t$. (See Problem 39.)

46 If n is a nonnegative integer, find the Laplace transform of $f(t) = t^n$. (*Hint:* Combine Problems 44 and 42.)

10.4   Improper Integrals with Unbounded Integrands

The improper integrals considered in Section 10.3 enabled us to calculate the areas of unbounded regions in the xy plane that extend indefinitely to the right or to the left. In this section we consider unbounded regions that extend indefinitely upward or downward.

For instance, consider the unbounded region R under the curve $y = 1/\sqrt{x}$, to the right of the y axis, and to the left of the vertical line $x = 9$ (Figure 1a). A good approximation to this unbounded region is afforded by the bounded region R_ϵ under the curve $y = 1/\sqrt{x}$ between $x = \epsilon$ and $x = 9$, provided that ϵ is a small positive number (Figure 1b). As $\epsilon \to 0^+$, R_ϵ becomes a better and better approximation to R, and it is tempting to write $R = \lim_{\epsilon\to 0^+} R_\epsilon$. Hence, we might expect that

$$\text{area of } R = \lim_{\epsilon\to 0^+} (\text{area of } R_\epsilon) = \lim_{\epsilon\to 0^+} \int_\epsilon^9 \frac{dx}{\sqrt{x}} = \lim_{\epsilon\to 0^+} \left(2\sqrt{x}\, \Big|_\epsilon^9\right)$$

$$= \lim_{\epsilon\to 0^+} (6 - 2\sqrt{\epsilon}) = 6 \text{ square units}$$

In view of this calculation, we write

$$\int_0^9 \frac{dx}{\sqrt{x}} = 6$$

Figure 1

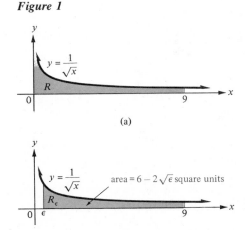

(a)

(b)

However, it is important to realize that, as a definite (Riemann) integral, $\int_0^9 (dx/\sqrt{x})$ *fails to exist* because the integrand $1/\sqrt{x}$ is undefined on $[0, 9]$ and unbounded on $(0, 9]$. Thus,

$$\int_0^9 \frac{dx}{\sqrt{x}}$$

is called an *improper integral*. More generally, we make the following definition.

DEFINITION 1   **Improper Integral at Lower Limit**

> Suppose that the function f is defined on the half-open interval $(a, b]$ and is integrable on every closed interval of the form $[a + \epsilon, b]$ for $0 < \epsilon < b - a$. Then we define the **improper integral** $\int_a^b f(x)\, dx$ by
>
> $$\int_a^b f(x)\, dx = \lim_{\epsilon \to 0^+} \int_{a+\epsilon}^b f(x)\, dx$$
>
> provided that the limit exists as a finite number.

If $\lim_{\epsilon \to 0^+} \int_{a+\epsilon}^b f(x)\, dx$ exists as a finite number, we say that the improper integral $\int_a^b f(x)\, dx$ is **convergent;** otherwise, it is called **divergent.**

Definition 1 takes care of the case in which the definite integral $\int_a^b f(x)\, dx$ may fail to exist because $f(a)$ is undefined. A similar definition,

$$\int_a^b f(x)\, dx = \lim_{\epsilon \to 0^+} \int_a^{b-\epsilon} f(x)\, dx$$

of an **improper integral at the upper limit** covers the case in which $f(b)$ is undefined.

To avoid confusing the improper integrals just defined with ordinary definite integrals, some authors use the notation $\int_{a+}^b f(x)\, dx$ or $\int_a^{b^-} f(x)\, dx$. However, since you can always determine whether an integral is improper by examining the integrand, we do not use this notation in what follows.

In Examples 1 to 3, evaluate the given improper integral (if it is convergent).

EXAMPLE 1 $\displaystyle \int_{3/2}^4 \frac{dx}{(x - \frac{3}{2})^{2/5}}$

SOLUTION   The integral is improper because the integrand is undefined at the lower limit $\frac{3}{2}$. We begin by using the substitution $u = x - \frac{3}{2}$ to evaluate the corresponding indefinite integral. Thus,

$$\int \frac{dx}{(x - \frac{3}{2})^{2/5}} = \int \frac{du}{u^{2/5}} = \frac{5}{3} u^{3/5} + C = \frac{5}{3} \left(x - \frac{3}{2}\right)^{3/5} + C$$

Applying Definition 1, we have

$$\int_{3/2}^4 \frac{dx}{(x - \frac{3}{2})^{2/5}} = \lim_{\epsilon \to 0^+} \int_{(3/2)+\epsilon}^4 \frac{dx}{(x - \frac{3}{2})^{2/5}}$$

$$= \lim_{\epsilon \to 0^+} \left[\frac{5}{3}(x - \frac{3}{2})^{3/5} \Big|_{(3/2)+\epsilon}^4 \right]$$

$$= \lim_{\epsilon \to 0^+} [\frac{5}{3}(\frac{5}{2})^{3/5} - \frac{5}{3}\epsilon^{3/5}] = \frac{5}{3}(\frac{5}{2})^{3/5} \quad \blacksquare$$

EXAMPLE 2 $\displaystyle\int_0^{\pi/2} \sec x \, dx$

SOLUTION   The integral is improper because the integrand is undefined at the upper limit $\pi/2$. Here we have

$$\int_0^{\pi/2} \sec x \, dx = \lim_{\epsilon \to 0^+} \int_0^{(\pi/2)-\epsilon} \sec x \, dx = \lim_{\epsilon \to 0^+} \left[\ln (\sec x + \tan x) \, \Big|_0^{(\pi/2) - \epsilon} \right]$$

$$= \lim_{\epsilon \to 0^+} \left\{ \ln \left[\sec \left(\frac{\pi}{2} - \epsilon \right) + \tan \left(\frac{\pi}{2} - \epsilon \right) \right] - \ln (\sec 0 + \tan 0) \right\}$$

$$= +\infty$$

because $\lim_{\epsilon \to 0^+} \sec [(\pi/2) - \epsilon] = +\infty$ and $\lim_{\epsilon \to 0^+} \tan [(\pi/2) - \epsilon] = +\infty$. Therefore, the improper integral is divergent.   ■

EXAMPLE 3 $\displaystyle\int_0^{\pi/2} \frac{\cos x}{\sqrt{\sin x}} \, dx$

SOLUTION   Here the integral is improper at the lower limit 0. Making the substitution $u = \sin x$, we have

$$\int \frac{\cos x}{\sqrt{\sin x}} \, dx = \int u^{-1/2} \, du = 2u^{1/2} + C = 2\sqrt{\sin x} + C$$

Therefore,

$$\int_0^{\pi/2} \frac{\cos x}{\sqrt{\sin x}} \, dx = \lim_{\epsilon \to 0^+} \int_\epsilon^{\pi/2} \frac{\cos x}{\sqrt{\sin x}} \, dx = \lim_{\epsilon \to 0^+} \left(2\sqrt{\sin x} \, \Big|_\epsilon^{\pi/2} \right)$$

$$= \lim_{\epsilon \to 0^+} (2\sqrt{\sin (\pi/2)} - 2\sqrt{\sin \epsilon}) = 2\sqrt{\sin (\pi/2)} = 2 \quad ■$$

EXAMPLE 4   Let R be the unbounded region under the curve $y = 1/\sqrt{4 - x^2}$ to the right of the y axis and to the left of the vertical line $x = 2$ (Figure 2). Determine whether the region R has a finite area, and if it does, find this area.

SOLUTION   The area in question is given by the improper integral

$$\int_0^2 \frac{dx}{\sqrt{4 - x^2}} = \lim_{\epsilon \to 0^+} \int_0^{2-\epsilon} \frac{dx}{\sqrt{4 - x^2}} = \lim_{\epsilon \to 0^+} \left[\left(\sin^{-1} \frac{x}{2} \right) \Big|_0^{2-\epsilon} \right]$$

$$= \lim_{\epsilon \to 0^+} \left(\sin^{-1} \frac{2 - \epsilon}{2} - \sin^{-1} 0 \right) = \frac{\pi}{2} \text{ square units} \quad ■$$

Figure 2

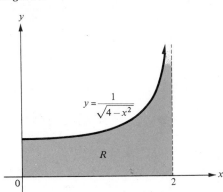

$y = \dfrac{1}{\sqrt{4 - x^2}}$

If a function f is defined at every number on a closed interval $[a, b]$ except for a number c, with $a < c < b$, we also refer to $\int_a^b f(x) \, dx$ as an **improper integral.** Such an improper integral is said to **converge** provided that both improper integrals

$$\int_a^c f(x) \, dx \qquad \text{and} \qquad \int_c^b f(x) \, dx$$

converge. Otherwise, it is said to **diverge.** If the improper integral $\int_a^b f(x) \, dx$ converges, then its value is defined by

$$\int_a^b f(x) \, dx = \int_a^c f(x) \, dx + \int_c^b f(x) \, dx$$

EXAMPLE 5   Evaluate the integral $\int_0^\pi \tan x \, dx$ (if it is convergent).

SOLUTION   Because the integrand is undefined at $\pi/2$, the integral is improper. It is convergent only if both $\int_0^{\pi/2} \tan x \, dx$ and $\int_{\pi/2}^\pi \tan x \, dx$ are convergent. But

$$\int_0^{\pi/2} \tan x \, dx = \lim_{\epsilon \to 0^+} \int_0^{(\pi/2)-\epsilon} \tan x \, dx = \lim_{\epsilon \to 0^+} \left[\ln (\sec x) \, \Big|_0^{(\pi/2)-\epsilon} \right]$$

$$= \lim_{\epsilon \to 0^+} \left[\ln \sec \left(\frac{\pi}{2} - \epsilon \right) - \ln \sec 0 \right] = +\infty$$

Hence, $\int_0^{\pi/2} \tan x \, dx$ diverges. Therefore, $\int_0^\pi \tan x \, dx$ also diverges.   ∎

Figure 3

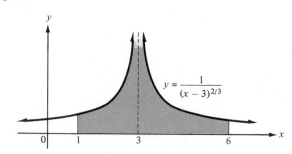

$$y = \frac{1}{(x-3)^{2/3}}$$

EXAMPLE 6   Let R be the unbounded region under the graph of

$$y = \frac{1}{(x-3)^{2/3}}$$

between $x = 1$ and $x = 6$ (Figure 3). Determine whether the region R has a finite area, and if it does, find this area.

SOLUTION   We cannot simply evaluate

$$\int_1^6 \frac{dx}{(x-3)^{2/3}}$$

as if it were an ordinary definite integral because the integrand is not defined at $x = 3$. However, if we split the region R into two portions, one to the left and the other to the right of the vertical line $x = 3$, then the areas of these two portions are given by the improper integrals

$$\int_1^3 \frac{dx}{(x-3)^{2/3}} = \lim_{\epsilon \to 0^+} \int_1^{3-\epsilon} \frac{dx}{(x-3)^{2/3}} = \lim_{\epsilon \to 0^+} \left[3(x-3)^{1/3} \, \Big|_1^{3-\epsilon} \right] = 3(2^{1/3})$$

and

$$\int_3^6 \frac{dx}{(x-3)^{2/3}} = \lim_{\epsilon \to 0^+} \int_{3+\epsilon}^6 \frac{dx}{(x-3)^{2/3}} = \lim_{\epsilon \to 0^+} \left[3(x-3)^{1/3} \, \Big|_{3+\epsilon}^6 \right] = 3(3^{1/3})$$

Therefore, the desired area A is actually given by the convergent *improper* integral

$$A = \int_1^6 \frac{dx}{(x-3)^{2/3}} = \int_1^3 \frac{dx}{(x-3)^{2/3}} + \int_3^6 \frac{dx}{(x-3)^{2/3}}$$

$$= 3(2^{1/3}) + 3(3^{1/3}) \text{ square units}$$ ∎

Problem Set 10.4

In Problems 1 to 24, evaluate each improper integral (if it is convergent).

1 $\displaystyle\int_0^4 \frac{dx}{\sqrt{x}}$

2 $\displaystyle\int_0^9 \frac{dx}{x\sqrt{x}}$

3 $\displaystyle\int_1^{28} \frac{dx}{\sqrt[3]{x-1}}$

4 $\displaystyle\int_0^{\pi/2} \frac{\cos x \, dx}{\sqrt{\sin x}}$

5 $\displaystyle\int_0^1 \frac{\cos \sqrt[3]{x}}{\sqrt[3]{x^2}} \, dx$

6 $\displaystyle\int_0^1 \frac{dx}{(1+x)\sqrt{x}}$

7 $\displaystyle\int_0^{\pi/2} \csc^2 x \, dx$

8 $\displaystyle\int_0^1 \frac{(\ln x)^2}{x} \, dx$

9 $\displaystyle\int_0^4 \frac{dx}{\sqrt{16 - x^2}}$

10 $\displaystyle\int_0^5 \frac{x \, dx}{\sqrt{25 - x^2}}$

11 $\displaystyle\int_0^4 \frac{e^{-\sqrt{x}}}{\sqrt{x}} \, dx$

12 $\displaystyle\int_2^5 \frac{2x - 6}{x^2 - 6x + 5} \, dx$

13 $\displaystyle\int_{1/2}^1 \frac{dt}{t(\ln t)^{2/7}}$

14 $\displaystyle\int_0^1 \frac{1}{x^2} \sin \frac{1}{x} \, dx$

15 $\displaystyle\int_{-1}^1 \frac{dx}{x^3}$

16 $\displaystyle\int_1^3 \frac{x \, dx}{2 - x}$

17 $\displaystyle\int_0^\pi \frac{\sin x}{\sqrt[5]{\cos x}} \, dx$

18 $\displaystyle\int_0^2 \frac{x \, dx}{(x - 1)^{2/3}}$

19 $\displaystyle\int_0^{\pi/2} \sec 2x \, dx$

20 $\displaystyle\int_{-\pi}^\pi \frac{dt}{1 - \cos t}$

21 $\displaystyle\int_0^{\pi/3} \frac{\sec^2 t}{\sqrt{1 - \tan t}} \, dt$

22 $\displaystyle\int_{e/3}^2 \frac{dx}{x(\ln x)^3}$

23 $\displaystyle\int_{-1}^1 \frac{e^x \, dx}{\sqrt[5]{e^x - 1}}$

24 $\displaystyle\int_{-1}^1 \frac{e^{-1/x}}{x^2} \, dx$

25 Show that the integral $\int_0^1 x^n \ln x \, dx$ converges and has the value $-1/(n + 1)^2$ if and only if $n > -1$.

26 (a) Show that $\displaystyle\lim_{\epsilon \to 0^+} \left(\int_{-1}^{-\epsilon} \frac{dx}{x} + \int_\epsilon^1 \frac{dx}{x} \right) = 0$.

(b) From part (a), can you conclude that $\int_{-1}^1 (dx/x) = 0$? Explain.

27 Criticize the following calculation:

$$\int_{-1}^1 \frac{dx}{x} = (\ln |x|) \Big|_{-1}^1 = \ln |1| - \ln |-1| = 0$$

28 Show that the improper integral $\displaystyle\int_0^1 \frac{dx}{1 - x^4}$ is divergent.

29 Find the area of the unbounded region under the curve $y = \dfrac{1}{\sqrt{x(2 - x)}}$ between $x = 0$ and $x = 2$.

30 Find the volume of the unbounded solid generated by revolving the region of Problem 29 about the x axis.

31 Find the volume of the unbounded solid generated by revolving the region bounded by the curve $y = 1/(x - 2)$ and the lines $x = 2$, $x = 4$, and $y = 0$ about the x axis.

32 Find the area of the unbounded region under the curve $y = \ln (1/x)$ between $x = 0$ and $x = 1$.

33 Find the volume of the unbounded solid generated by revolving the region under the curve $y = 1/x^p$ between $x = 0$ and $x = 1$ about the x axis. Assume that $p > 0$.

34 If the function f is defined and continuous on the closed interval $[a, b]$, show that

$$\int_a^b f(x) \, dx = \lim_{\epsilon \to 0^+} \int_{a+\epsilon}^b f(x) \, dx$$

$$\left[\text{Hint: } \int_a^b f(x) \, dx = \int_a^{a+\epsilon} f(x) \, dx + \int_{a+\epsilon}^b f(x) \, dx. \right]$$

35 Find the volume of the unbounded solid generated by revolving the region described in Problem 33 about the y axis.

36 Suppose that f is a continuous nonnegative function defined on the interval $[1, \infty)$ and that R is the unbounded region under the graph of f. Assume that the unbounded solid obtained by revolving the region R about the x axis has a finite volume. Is it necessarily true that the unbounded solid obtained by revolving the region R about the y axis has a finite volume? Explain.

10.5   Taylor Polynomials

In Section 10.1 we introduced Cauchy's generalized mean-value theorem and used it in establishing L'Hôpital's rule. In this section we present another extension of the mean-value theorem which is usually associated with the English mathematician Brook Taylor (1685–1731). This remarkable theorem enables us to approximate various complicated functions by much simpler polynomial functions. The polynomial functions involved in Taylor's theorem are formed according to the following definition.

<u>DEFINITION 1</u>   **Taylor Polynomial**

> Let f be a function having a derivative $f^{(n)}$ of order $n \geq 1$ on an open interval I, and let a be a fixed number in I. Then the **nth-degree Taylor polynomial for the function f at the number a** is the polynomial function P_n defined by
>
> $$P_n(x) = f(a) + \frac{f'(a)}{1!}(x - a) + \frac{f''(a)}{2!}(x - a)^2 + \frac{f'''(a)}{3!}(x - a)^3 + \cdots$$
>
> $$+ \frac{f^{(n)}(a)}{n!}(x - a)^n$$

Recall that if n is a positive integer, then $n!$ is defined by

$$n! = n(n - 1)(n - 2) \cdots 3 \cdot 2 \cdot 1$$

Also, by definition,

$$0! = 1$$

Brook Taylor

<u>EXAMPLE 1</u>   Find the fourth-degree Taylor polynomial P_4 for $f(x) = \sin x$ at the number $a = \pi/6$.

SOLUTION   We arrange the work as follows:

$$f(x) = \sin x \qquad f\left(\frac{\pi}{6}\right) = \frac{1}{2}$$

$$f'(x) = \cos x \qquad f'\left(\frac{\pi}{6}\right) = \frac{\sqrt{3}}{2}$$

$$f''(x) = -\sin x \qquad f''\left(\frac{\pi}{6}\right) = -\frac{1}{2}$$

$$f'''(x) = -\cos x \qquad f'''\left(\frac{\pi}{6}\right) = -\frac{\sqrt{3}}{2}$$

$$f^{(4)}(x) = \sin x \qquad f^{(4)}\left(\frac{\pi}{6}\right) = \frac{1}{2}$$

Hence,

$$P_4(x) = f\left(\frac{\pi}{6}\right) + \frac{f'(\pi/6)}{1!}\left(x - \frac{\pi}{6}\right) + \frac{f''(\pi/6)}{2!}\left(x - \frac{\pi}{6}\right)^2$$

$$+ \frac{f'''(\pi/6)}{3!}\left(x - \frac{\pi}{6}\right)^3 + \frac{f^{(4)}(\pi/6)}{4!}\left(x - \frac{\pi}{6}\right)^4$$

$$= \frac{1}{2} + \frac{\sqrt{3}}{2}\left(x - \frac{\pi}{6}\right) - \frac{1}{4}\left(x - \frac{\pi}{6}\right)^2 - \frac{\sqrt{3}}{12}\left(x - \frac{\pi}{6}\right)^3 + \frac{1}{48}\left(x - \frac{\pi}{6}\right)^4 \quad \blacksquare$$

<u>EXAMPLE 2</u>   Find the Taylor polynomial P_3 of degree 3 for the function f defined by

$$f(x) = \frac{1}{1 + x} \qquad \text{for } x > -1 \text{ at the number } a = 0$$

SOLUTION

$$f(x) = (1 + x)^{-1} \qquad f(0) = 1$$
$$f'(x) = -(1 + x)^{-2} \qquad f'(0) = -1$$
$$f''(x) = 2(1 + x)^{-3} \qquad f''(0) = 2$$
$$f'''(x) = -6(1 + x)^{-4} \qquad f'''(0) = -6$$

Hence,

$$P_3(x) = f(0) + \frac{f'(0)}{1!}(x - 0) + \frac{f''(0)}{2!}(x - 0)^2 + \frac{f'''(0)}{3!}(x - 0)^3$$
$$= 1 - x + x^2 - x^3$$

The Taylor polynomial P_n for the function f at the number a always has the following property, whose proof is left as an exercise (Problem 24):

> The values of the successive derivatives of P_n at the number a—up to and including order n—are equal to the values of the corresponding successive derivatives of f at the number a.

Furthermore, as is clear from Definition 1, the value of f at a is equal to the value of P_n at a. Thus, we have

$$f(a) = P_n(a)$$
$$f'(a) = P'_n(a)$$
$$f''(a) = P''_n(a)$$
$$\vdots$$
$$f^{(n)}(a) = P^{(n)}_n(a)$$

For instance, as you can easily check, in Example 2 above,

$$f(0) = 1 = P_3(0)$$
$$f'(0) = -1 = P'_3(0)$$
$$f''(0) = 2 = P''_3(0)$$

and

$$f'''(0) = -6 = P'''_3(0)$$

Perhaps the most important feature of the Taylor polynomial P_n for the function f at a is that the approximation

$$f(x) \approx P_n(x)$$

often is exceedingly accurate, especially if n is large and x is close to a. For instance, in Example 1 above (rounding off to six decimal places), we have

$$\sin \frac{\pi}{4} = 0.707107 \approx P_4\left(\frac{\pi}{4}\right) = 0.707098$$

Calculators and computers handle transcendental functions by using polynomials whose values approximate the values of these functions. Often, Taylor polynomials are chosen for this purpose because of the accuracy of the approximation $f(x) \approx P_n(x)$. Thus, in applied mathematics—especially in work that involves calculators or computers—it is essential to be able to place bounds on the error that results when the value of a function is approximated by the value of a Taylor polynomial for that function. This error, the difference between the true value $f(x)$ and the estimated value $P_n(x)$, is called the *Taylor remainder* $R_n(x)$. Thus, we make the following definition.

DEFINITION 2

Taylor Remainder

If P_n is the nth-degree Taylor polynomial for the function f at the number a, we define the corresponding **Taylor remainder** to be the function R_n given by

$$R_n(x) = f(x) - P_n(x)$$

Notice that

$$f(x) = P_n(x) + R_n(x)$$

so *the approximation $f(x) \approx P_n(x)$ will be accurate when $|R_n(x)|$ is small*. The extended mean-value theorem, referred to in the introduction to this section, provides an effective means for estimating $|R_n(x)|$.

THEOREM 1

Extended Mean-Value Theorem

Let n be a nonnegative integer, and suppose that f is a function having a derivative $f^{(n+1)}$ of order $n + 1$ on an open interval I. Then if a and b are any two distinct numbers in I, there exists a number c strictly between a and b such that

$$f(b) = f(a) + \frac{f'(a)}{1!}(b - a) + \frac{f''(a)}{2!}(b - a)^2 + \frac{f'''(a)}{3!}(b - a)^3 + \cdots$$
$$+ \frac{f^{(n)}(a)}{n!}(b - a)^n + r_n$$

where $r_n = \dfrac{f^{(n+1)}(c)}{(n + 1)!}(b - a)^{n+1}$.

PROOF

Define the number r_n by

$$r_n = f(b) - f(a) - \frac{f'(a)}{1!}(b - a) - \frac{f''(a)}{2!}(b - a)^2 - \frac{f'''(a)}{3!}(b - a)^3 - \cdots$$
$$- \frac{f^{(n)}(a)}{n!}(b - a)^n$$

Then

$$f(b) = f(a) + \frac{f'(a)}{1!}(b - a) + \frac{f''(a)}{2!}(b - a)^2 + \frac{f'''(a)}{3!}(b - a)^3 + \cdots$$
$$+ \frac{f^{(n)}(a)}{n!}(b - a)^n + r_n$$

Define a new function g with domain I by the equation

$$g(x) = f(x) + \frac{f'(x)}{1!}(b - x) + \frac{f''(x)}{2!}(b - x)^2 + \frac{f'''(x)}{3!}(b - x)^3 + \cdots$$

$$+ \frac{f^{(n)}(x)}{n!}(b - x)^n + r_n\frac{(b - x)^{n+1}}{(b - a)^{n+1}}$$

Then

$$g'(x) = f'(x) + \left[-\frac{f'(x)}{1!} + \frac{f''(x)}{1!}(b - x) \right]$$

$$+ \left[-\frac{2f''(x)}{2!}(b - x) + \frac{f'''(x)}{2!}(b - x)^2 \right]$$

$$+ \left[-\frac{3f'''(x)}{3!}(b - x)^2 + \frac{f^{(4)}(x)}{3!}(b - x)^3 \right] + \cdots$$

$$+ \left[-\frac{nf^{(n)}(x)}{n!}(b - x)^{n-1} + \frac{f^{(n+1)}(x)}{n!}(b - x)^n \right] - \frac{(n + 1)r_n(b - x)^n}{(b - a)^{n+1}}$$

In the expression above, each set of brackets contains a term of the form

$$-\frac{kf^{(k)}(x)}{k!}(b - x)^{k-1}$$

Since

$$\frac{k}{k!} = \frac{k}{k(k - 1)(k - 2)\cdots 1} = \frac{1}{(k - 1)!}$$

this term can be rewritten as

$$-\frac{f^{(k)}(x)}{(k - 1)!}(b - x)^{k-1}$$

and thus cancels with the term

$$\frac{f^{(k)}(x)}{(k - 1)!}(b - x)^{k-1}$$

in the preceding set of brackets.

Performing all such cancellations, we find that all but the last two terms drop out, and we have

$$g'(x) = \frac{f^{(n+1)}(x)}{n!}(b - x)^n - \frac{(n + 1)r_n(b - x)^n}{(b - a)^{n+1}}$$

Going back to our original definition of g and putting $x = a$, we find that

$$g(a) = f(a) + \frac{f'(a)}{1!}(b - a) + \frac{f''(a)}{2!}(b - a)^2 + \frac{f'''(a)}{3!}(b - a)^3 + \cdots$$

$$+ \frac{f^{(n)}(a)}{n!}(b - a)^n + r_n \cdot 1$$

so that $g(a) = f(b)$. Also, putting $x = b$ in the definition of g, we obtain $g(b) = f(b)$; hence, $g(a) = g(b)$. Thus, applying Rolle's theorem to the function g on the closed interval from a to b, we conclude that there is a number c strictly between a and b such that $g'(c) = 0$. Using the formula above for $g'(x)$, we have

$$g'(c) = \frac{f^{(n+1)}(c)}{n!}(b - c)^n - \frac{(n + 1)r_n(b - c)^n}{(b - a)^{n+1}} = 0$$

or

$$\frac{f^{(n+1)}(c)}{n!}(b - c)^n = \frac{(n + 1)r_n(b - c)^n}{(b - a)^{n+1}}$$

Since $b \neq c$, it follows that $b - c \neq 0$, and common factor $(b - c)^n$ on both sides of the last equation can be canceled to yield

$$\frac{f^{(n+1)}(c)}{n!} = \frac{(n + 1)r_n}{(b - a)^{n+1}} \qquad \text{or} \qquad r_n = \frac{f^{(n+1)}(c)}{(n + 1)n!}(b - a)^{n+1}$$

Because $(n + 1)n! = (n + 1)!$, we have

$$r_n = \frac{f^{(n+1)}(c)}{(n + 1)!}(b - a)^{n+1}$$

and the proof is complete.   ■

Notice that if we put $n = 0$ in Theorem 1, we obtain

$$f(b) - f(a) = (b - a)f'(c)$$

where c is strictly between a and b. Therefore, for $n = 0$, the conclusion of Theorem 1 coincides with the conclusion of the mean-value theorem.

As promised, we can now use Theorem 1 to obtain an expression for the Taylor remainder R_n. The appropriate theorem is nothing but a restatement of Theorem 1 as follows.

THEOREM 2   **Taylor's Formula with Lagrange Remainder**

Let f be a function having a derivative $f^{(n+1)}$ of order $n + 1$ on an open interval I, and let a be a fixed number in I. Denote the nth-degree Taylor polynomial for f at a and the corresponding Taylor remainder by P_n and R_n, respectively. Then, for any x in I, we have

$$f(x) = P_n(x) + R_n(x)$$

If $x \neq a$, there is a number c strictly between a and x such that

$$R_n(x) = \frac{f^{(n+1)}(c)}{(n + 1)!}(x - a)^{n+1}$$

PROOF   Put $b = x$ in Theorem 1, noting that

$$f(a) + \frac{f'(a)}{1!}(x - a) + \frac{f''(a)}{2!}(x - a)^2$$

$$+ \frac{f'''(a)}{3!}(x - a)^3 + \cdots + \frac{f^{(n)}(a)}{n!}(x - a)^n = P_n(x)$$

and, therefore, that

$$R_n(x) = r_n = \frac{f^{(n+1)}(c)}{(n + 1)!}(x - a)^{n+1}$$ ■

Theorems 1 and 2, which really make the same statement with different notation, are both referred to as **Taylor's theorem,** or **Taylor's formula.** The expression

$$\frac{f^{(n+1)}(c)}{(n+1)!}(x-a)^{n+1}$$

for $R_n(x)$ given in Theorem 2 is called the **Lagrange form** for the Taylor remainder.

EXAMPLE 3   Find the fourth-degree Taylor polynomial P_4 and the corresponding Taylor remainder R_4 in Lagrange form for the function f defined by

$$f(x) = \frac{1}{x+2} \qquad \text{for } x > -2 \text{ at the number } a = 1$$

SOLUTION   Here we have

$$f'(x) = -(x+2)^{-2} \qquad f''(x) = 2(x+2)^{-3} \qquad f'''(x) = -6(x+2)^{-4}$$

$$f^{(4)}(x) = 24(x+2)^{-5} \qquad \text{and} \qquad f^{(5)}(x) = -120(x+2)^{-6}$$

Hence,

$$f(1) = \tfrac{1}{3} \qquad f'(1) = -\tfrac{1}{9} \qquad f''(1) = \tfrac{2}{27} \qquad f'''(1) = -\tfrac{6}{81} \qquad f^{(4)}(1) = \tfrac{24}{243}$$

and

$$P_4(x) = f(1) + \frac{f'(1)}{1!}(x-1) + \frac{f''(1)}{2!}(x-1)^2 + \frac{f'''(1)}{3!}(x-1)^3 + \frac{f^{(4)}(1)}{4!}(x-1)^4$$

$$= \tfrac{1}{3} - \tfrac{1}{9}(x-1) + \tfrac{1}{27}(x-1)^2 - \tfrac{1}{81}(x-1)^3 + \tfrac{1}{243}(x-1)^4$$

In Lagrange form, the corresponding remainder is given by

$$R_4(x) = \frac{f^{(5)}(c)}{5!}(x-1)^5 = -\frac{120(x-1)^5}{5!(c+2)^6} = -\frac{(x-1)^5}{(c+2)^6}$$

where c is a number strictly between 1 and x.   ∎

©**EXAMPLE 4**   Use the third-degree Taylor polynomial for the function $f(x) = \ln(1+x)$ at the number $a = 0$ to estimate $\ln 1.1$. Then use the Lagrange form of the remainder to place a bound on the error of this estimate.

SOLUTION   Here, we have

$$f'(x) = (1+x)^{-1} \qquad f''(x) = -(1+x)^{-2}$$

$$f'''(x) = 2(1+x)^{-3} \qquad \text{and} \qquad f^{(4)}(x) = -6(1+x)^{-4}$$

Thus, $f(0) = 0 \qquad f'(0) = 1 \qquad f''(0) = -1 \qquad f'''(0) = 2$

so the third-degree Taylor polynomial at $a = 0$ is given by

$$P_3(x) = f(0) + \frac{f'(0)}{1!}(x-0) + \frac{f''(0)}{2!}(x-0)^2 + \frac{f'''(0)}{3!}(x-0)^3$$

$$= 0 + x - \tfrac{1}{2}x^2 + \tfrac{1}{3}x^3$$

Putting $x = 0.1$ in the approximation $f(x) \approx P_3(x)$, we obtain the estimate

$$\ln 1.1 = f(0.1) \approx P_3(0.1) = 0.1 - \tfrac{1}{2}(0.1)^2 + \tfrac{1}{3}(0.1)^3 = \tfrac{143}{1500}$$

or $\ln 1.1 \approx 0.0953333 \ldots$

The error involved in this estimation is given by the Lagrange remainder

$$f(0.1) - P_3(0.1) = R_3(0.1) = \frac{f^{(4)}(c)}{4!}(0.1 - 0)^4$$

$$= \frac{-6(1 + c)^{-4}}{4!}10^{-4} = \frac{-1}{10^4(1 + c)^4(4)}$$

where $0 < c < 0.1$. Since $c > 0$, it follows that

$$|R_3(0.1)| = \frac{1}{10^4(1 + c)^4(4)} < \frac{1}{10^4(4)}$$

Hence,

$$|R_3(0.1)| < \frac{1}{40,000} = 0.000025$$

Notice that $R_3(0.1)$ is negative, so that the estimated value $\ln 1.1 \approx 0.0953333\ldots$ is actually a little larger than the true value of $\ln 1.1$. However, since the absolute value of the error cannot exceed 0.000025, the given estimate is correct to at least four decimal places, and we therefore write $\ln 1.1 \approx 0.0953$. ∎

The following theorem can often be used to determine—in advance—the degree n of the Taylor polynomial required to guarantee that the absolute value of the error involved in the estimation $f(b) \approx P_n(b)$ does not exceed a specified bound.

THEOREM 3   **Error Bound for Taylor Polynomial Approximation**

> Let f be a function having a derivative $f^{(n+1)}$ of order $n + 1$ on an open interval I, and let a and b be two distinct numbers in I. Suppose that M_n is a constant* (depending only on n) and that $|f^{(n+1)}(c)| \leq M_n$ holds for all values of c strictly between a and b. Then if P_n is the nth-degree Taylor polynomial for f at a, the absolute value of the error involved in the estimate $f(b) \approx P_n(b)$ does not exceed
>
> $$M_n\frac{|b - a|^{n+1}}{(n + 1)!}$$

PROOF   By Theorem 2, there is a number c strictly between a and b such that

$$R_n(b) = \frac{f^{(n+1)}(c)}{(n + 1)!}(b - a)^{n+1}$$

Therefore,

$$|f(b) - P_n(b)| = |R_n(b)| = \left|\frac{f^{(n+1)}(c)}{(n + 1)!}(b - a)^{n+1}\right|$$

$$= |f^{(n+1)}(c)|\frac{|b - a|^{n+1}}{(n + 1)!} \leq M_n\frac{|b - a|^{n+1}}{(n + 1)!}$$ ∎

*In practice, it is often possible to *generously* overestimate $|f^{n+1}(c)|$ by M_n and *still* get a very usable error estimate.

Ⓒ **EXAMPLE 5**   Estimate ln 0.99, making certain that the error does not exceed 10^{-7} in absolute value.

SOLUTION   In Theorem 3, we take $f(x) = \ln x$, $a = 1$, and $b = 0.99$. (We have chosen $a = 1$ because it is near 0.99 and we know that ln 1 = 0.) Here,

$$f'(x) = x^{-1} \qquad f''(x) = -x^{-2} \qquad f'''(x) = 2x^{-3} \qquad f^{(4)}(x) = -6x^{-4} \qquad \ldots$$

$$f^{(n)}(x) = (-1)^{n-1}(n-1)! \, x^{-n} \qquad \text{and} \qquad f^{(n+1)}(x) = (-1)^n n! \, x^{-(n+1)}$$

(If desired, this can be established rigorously by induction on n.) Hence, for $0.99 < c < 1$,

$$\left| f^{(n+1)}(c) \right| = \left| (-1)^n n! \, c^{-(n+1)} \right| = \frac{n!}{c^{n+1}} < \frac{n!}{(0.99)^{n+1}}$$

and we can take $M_n = n!/(0.99)^{n+1}$ in Theorem 3. Thus, the absolute value of the error involved in the estimation ln 0.99 = $f(b) \approx P_n(b)$ does not exceed

$$M_n \frac{|0.99 - 1|^{n+1}}{(n+1)!} = \frac{n!}{(0.99)^{n+1}} \frac{(0.01)^{n+1}}{(n+1)!} = \frac{1}{(n+1)(99)^{n+1}}$$

The smallest value of n for which $1/[(n+1)(99)^{n+1}] \leq 10^{-7}$ can be found by trial and error to be $n = 3$; hence, $P_3(0.99)$ approximates ln 0.99 with the desired accuracy. Here,

$$P_3(x) = f(1) + \frac{f'(1)}{1!}(x-1) + \frac{f''(1)}{2!}(x-1)^2 + \frac{f'''(1)}{3!}(x-1)^3$$

$$= 0 + (x-1) - \tfrac{1}{2}(x-1)^2 + \tfrac{1}{3}(x-1)^3$$

so that

$$\ln 0.99 \approx P_3(0.99) = -0.01 - \tfrac{1}{2}(0.01)^2 - \tfrac{1}{3}(0.01)^3 = -0.0100503333 \ldots$$

Since the error involved in this estimation does not exceed $10^{-7} = 0.0000001$ in absolute value, we can be certain that the approximation

$$\ln 0.99 \approx -0.0100503$$

is accurate to six decimal places. (The true value, rounded off to eight decimal places, is ln 0.99 $\approx$ 0.01005034.)   ▪

Ⓒ **EXAMPLE 6**   Estimate sin 40° with an error no more than 10^{-5} in absolute value.

SOLUTION   In Theorem 3 we take

$$f(x) = \sin x \qquad a = \frac{\pi}{4} = 45° \qquad \text{and} \qquad b = \frac{40\pi}{180} = \frac{2\pi}{9} = 40°$$

(We have chosen $a = 45°$ because it is near 40° and its sine is well known.) Here,

$$f'(x) = \cos x \qquad f''(x) = -\sin x \qquad f'''(x) = -\cos x \qquad f^{(4)}(x) = \sin x$$

and so forth. Hence, $f^{(n+1)}(c)$ is one of $\pm\sin c$ or $\pm\cos c$, so that $\left| f^{(n+1)}(c) \right| \leq 1$ and we can take $M_n = 1$ in Theorem 3. Since the absolute value of the error of estimation cannot exceed

$$M_n \frac{|b-a|^{n+1}}{(n+1)!} = (1)\frac{\left| \dfrac{2\pi}{9} - \dfrac{\pi}{4} \right|^{n+1}}{(n+1)!} = \frac{\left(\dfrac{\pi}{36} \right)^{n+1}}{(n+1)!}$$

we must choose n large enough so that $(\pi/36)^{n+1}/(n+1)! \leq 10^{-5}$. By trial and error, the smallest such value is $n = 3$. Since

$$P_3(x) = \sin\frac{\pi}{4} + \frac{\cos\dfrac{\pi}{4}}{1!}\left(x - \frac{\pi}{4}\right) - \frac{\sin\dfrac{\pi}{4}}{2!}\left(x - \frac{\pi}{4}\right)^2 - \frac{\cos\dfrac{\pi}{4}}{3!}\left(x - \frac{\pi}{4}\right)^3$$

is follows that

$$\sin 40° = \sin\frac{2\pi}{9} \approx P_3\left(\frac{2\pi}{9}\right)$$

$$= \frac{\sqrt{2}}{2} + \frac{\sqrt{2}}{2}\left(-\frac{\pi}{36}\right) - \frac{\sqrt{2}}{4}\left(-\frac{\pi}{36}\right)^2 - \frac{\sqrt{2}}{12}\left(-\frac{\pi}{36}\right)^3$$

$$= 0.6427859 \ldots$$

with an error no greater than $10^{-5} = 0.00001$ in absolute value. (The true value, rounded off to seven decimal places, is $\sin 40° \approx 0.6427876$.) ∎

Problem Set 10.5

In Problems 1 to 16, find the Taylor polynomial of degree n at the indicated number a for each function, and write the corresponding Taylor remainder in the Lagrange form.

1 $f(x) = 1/x$, $a = 2$, $n = 6$

2 $g(x) = \sqrt{x}$, $a = 4$, $n = 5$

3 $f(x) = 1/\sqrt{x}$, $a = 100$, $n = 4$

4 $f(x) = \sqrt[3]{x}$, $a = 1000$, $n = 4$

5 $g(x) = (x - 2)^{-2}$, $a = 3$, $n = 5$

6 $f(x) = (1 - x)^{-1/2}$, $a = 0$, $n = 3$

7 $f(x) = \sin x$, $a = 0$, $n = 6$

8 $g(x) = \cos x$, $a = -\pi/3$, $n = 3$

9 $g(x) = \tan x$, $a = \pi/4$, $n = 4$

10 $f(x) = e^{2x}$, $a = 0$, $n = 5$

11 $f(x) = xe^x$, $a = 1$, $n = 3$

12 $f(x) = e^{-x^2}$, $a = 0$, $n = 3$

13 $g(x) = 2^x$, $a = 1$, $n = 3$

14 $f(x) = \ln x$, $a = 1$, $n = 4$

15 $g(x) = \sinh x$, $a = 0$, $n = 4$

16 $f(x) = \ln(\cos x)$, $a = \pi/3$, $n = 3$

© In Problems 17 to 23, use an appropriate Taylor polynomial to approximate each function value with an error of no more than 10^{-5} in

absolute value. In each case, write the answer rounded off to four decimal places.

17 $\sin 1$

18 $\cos 29°$

19 e (*Hint:* $e = e^1$.)

20 $e^{-1.1}$

21 $\ln 0.98$

22 $\ln 17$ [*Hint:* Write $\ln 17 = \ln 16(1 + \frac{1}{16}) = 4\ln 2 + \ln(1 + \frac{1}{16})$.]

23 $\sqrt{9.04}$

24 Use mathematical induction to prove that if P_n is the nth-degree Taylor polynomial at a of the function f, then $f^{(k)}(a) = P_n^{(k)}(a)$ holds for $k = 1, 2, \ldots, n$.

25 Give a bound on the absolute value of the error involved in estimating $\sin 5°$ [that is, $\sin(\pi/36)$] by using the nth-degree Taylor polynomial for $\sin x$ at $a = 0$. (*Hint:* Use the fact that $\pi/36 < \frac{1}{10}$.)

© **26** Give a bound on the absolute value of the error involved in estimating $\sqrt{1 + x}$ by $1 + \frac{1}{2}x$ for $|x| \leq 0.1$.

27 Show that the error of the estimate $\cos x = 1 - (x^2/2)$ does not exceed $x^4/24$ in absolute value. [*Hint:* Notice that $1 - (x^2/2) = P_3(x)$.]

28 (a) Show that the difference between the arc length s of an arc of a circle of fixed radius r and the length of the corresponding chord is given by $s - 2r\sin[s/(2r)]$. (b) Use a third-degree Taylor polynomial to approximate $s - 2r\sin[s/(2r)]$, and give a bound for the absolute value of the error involved in this approximation.

C **29** Use a third-degree Taylor polynomial to approximate the area of the smaller segment cut off by the chord in Problem 28, and give a bound for the absolute value of the error involved in this approximation.

C **30** Use a third-degree Taylor polynomial approximation for sin x at $a = 0$ to find, approximately, a value of $x > 0$ for which $5 \sin x - 4x = 0$.

31 Suppose that p is a nonzero constant and let $P_n(x)$ be the nth-degree Taylor polynomial for $f(x) = (1 + x)^p$ at the number $a = 0$. Find (a) $P_1(x)$, (b) $P_2(x)$, (c) $P_3(x)$, and (d) $P_4(x)$.

32 If a perfectly flexible cable weighs w newtons per meter, has a span of L meters between points of support at the same level, and has a tension of T newtons at its lowest point, then it has a sag of H meters given by

$$H = \frac{T}{w}\left(\cosh \frac{wL}{2T} - 1\right)$$

Using an appropriate second-degree Taylor polynomial and assuming that wL/T is small, obtain an approximate formula for H.

33 The tension T_s in the cable of Problem 32 at either point of support is given by

$$T_s = T \cosh \frac{wL}{2T}$$

Using an appropriate second-degree Taylor polynomial and assuming that wL/T is small, obtain an approximate formula for T_s.

34 Let f be a polynomial function of degree n, and let P_n be the nth-degree Taylor polynomial for f at a. Show that $f(x) = P_n(x)$ for all values of x.

35 (a) Show that, for $x \leq 0$,

$$e^x = 1 + x + \frac{x^2}{2} + \frac{x^3}{6} + \frac{x^4}{24} + R_4(x)$$

where $\dfrac{x^5}{120} \leq R_4(x) \leq 0$.

(b) Replace x by $-t^2$ in part (a) to conclude that

$$e^{-t^2} = 1 - t^2 + \frac{t^4}{2} - \frac{t^6}{6} + \frac{t^8}{24} - r(t)$$

where $0 \leq r(t) \leq \dfrac{t^{10}}{120}$.

(c) If $b > 0$, use part (b) to show that

$$\int_0^b e^{-t^2}\, dt = b - \frac{b^3}{3} + \frac{b^5}{10} - \frac{b^7}{42} + \frac{b^9}{216} - \epsilon$$

where $0 \leq \epsilon \leq \dfrac{b^{11}}{1320}$.

C (d) Evaluate $\int_0^{3/4} e^{-t^2}\, dt$, rounded off to three decimal places.

C **36** Assume that f is a function with a continuous derivative $f^{(n+1)}$ of order $n + 1$ on an open interval I. Let a and b be distinct numbers in I, and let P_n be the nth-degree Taylor polynomial for f at a. Use induction to prove that the value $R_n(b)$ of the corresponding Taylor remainder is given by

$$R_n(b) = \frac{1}{n!} \int_a^b (b - t)^n f^{(n+1)}(t)\, dt$$

This is called the **integral form** of the Taylor remainder.

37 (a) If

$$f(x) = \frac{1}{1 - x}$$

show that the Taylor polynomial P_n for f at $a = 0$ is given by $P_n(x) = 1 + x + x^2 + x^3 + \cdots + x^n$.

(b) Show that the Taylor remainder corresponding to P_n is given by

$$R_n(x) = \frac{x^{n+1}}{1 - x}$$

38 Suppose that f is a function having a derivative of order n on the open interval I and that a is a number in I. Let P be a polynomial function of degree n such that $f(a) = P(a)$ and $f^{(k)}(a) = P^{(k)}(a)$ for $k = 1, 2, \ldots, n$. Prove that P is the nth-degree Taylor polynomial for f at a.

39 Let P_n be the nth-degree Taylor polynomial for f at a, where the function f has a derivative of order n on the open interval I and a belongs to I. Let the function g be defined on I by $g(x) = \int_a^x f(t)\, dt$, and let the function Q be defined on I by $Q(x) = \int_a^x P_n(x)\, dx$. Show that Q is the Taylor polynomial of degree $n + 1$ for g at a. (*Hint:* Use the result of Problem 38.)

40 Suppose that h is a function that has a derivative of order n on the open interval I and that the function g is defined by $g(x) = x^{n+1}h(x)$ for x in I. (a) Prove that $g(0) = 0$ and that $g^{(k)}(0) = 0$ for $k = 1, 2, \ldots, n$. (b) Prove that the nth-degree Taylor polynomial for g at $a = 0$ is the zero polynomial.

41 Assume that f is a function having a derivative of order n on the open interval I and that 0 belongs to I. Suppose that the function h also has a derivative of order n on I and that P is a polynomial of degree n such that $f(x) = P(x) + x^{n+1}h(x)$ holds for all values of x in I. Prove that P is the nth-degree Taylor polynomial for f at 0. [*Hint:* Use part (a) of Problem 40 and Problem 38.]

42 (a) Prove that

$$\frac{1}{1 + x^2} = 1 - x^2 + x^4 - x^6 + \cdots$$

$$+ (-1)^n x^{2n} + \frac{(-1)^{n+1}x^{2n+2}}{1 + x^2}$$

(*Hint:* Use Problem 37.)

(b) If

$$f(x) = \frac{1}{1 + x^2}$$

show that the Taylor polynomial P_{2n} of degree $2n$ for f at $a = 0$ is given by $P_{2n}(x) = 1 - x^2 + x^4 - x^6 + \cdots + (-1)^n x^{2n}$, and show that the corresponding Taylor remainder is given by

$$R_{2n}(x) = \frac{(-1)^{n+1}x^{2n+2}}{1 + x^2}$$

[*Hint:* Use Problem 41 and part (a).]

43 Use part (b) of Problem 42, Problem 39, and the fact that $\tan^{-1} x = \int_0^x dt/(1 + t^2)$ to show that the Taylor polynomial of degree $2n + 1$ for the inverse tangent function at $a = 0$ is given by

$$P_{2n+1}(x) = x - \frac{x^3}{3} + \frac{x^5}{5} - \frac{x^7}{7} + \cdots + (-1)^n \frac{x^{2n+1}}{2n + 1}$$

Review Problem Set, Chapter 10

In Problems 1 to 32, use L'Hôpital's rules to evaluate each limit (if it exists).

1 $\lim\limits_{x \to 0} \dfrac{xe^x}{1 - e^x}$

2 $\lim\limits_{x \to 0} \dfrac{8^x - 2^x}{4x}$

3 $\lim\limits_{x \to 0} \dfrac{\ln(\sec 2x)}{\ln(\sec x)}$

4 $\lim\limits_{x \to 0} \dfrac{\cos 2x - \cos x}{\sin^2 x}$

5 $\lim\limits_{x \to 1^+} \left(\dfrac{x}{\ln x} - \dfrac{1}{1 - x} \right)$

6 $\lim\limits_{x \to 0} \dfrac{e^x - 1}{x^2 - x}$

7 $\lim\limits_{x \to 0^-} \dfrac{2 - 3e^{-x} + e^{-2x}}{2x^2}$

8 $\lim\limits_{x \to 1} \dfrac{2x^3 + 5x^2 - 4x - 3}{x^3 + x^2 - 10x + 8}$

9 $\lim\limits_{x \to 0} \dfrac{\sqrt{1 - x} - \sqrt{1 + x}}{x}$

10 $\lim\limits_{x \to 1^+} \left(\dfrac{1}{x - 1} - \dfrac{1}{\sqrt{x - 1}} \right)$

11 $\lim\limits_{x \to 0^+} x^3(\ln x)^3$

12 $\lim\limits_{x \to 0^+} \dfrac{\ln x}{\cot x}$

13 $\lim\limits_{x \to 1^+} \dfrac{(\ln x)^2}{\sin(x - 1)}$

14 $\lim\limits_{x \to 0} \dfrac{\sqrt{1 + \sin x} - \sqrt{1 - \sin x}}{\tan x}$

15 $\lim\limits_{x \to 1^+} x \sin \dfrac{a}{x}$

16 $\lim\limits_{x \to 0} \csc x \sin(\tan x)$

17 $\lim\limits_{x \to 1} \left(\dfrac{2}{x^2 - 1} - \dfrac{1}{x - 1} \right)$

18 $\lim\limits_{x \to +\infty} \dfrac{\sqrt[3]{1 + x^6}}{1 - x + 2\sqrt{1 + x^2 + x^4}}$

19 $\lim\limits_{x \to 0^+} \dfrac{\sin x}{x} \cdot \dfrac{\sin x}{x - \sin x}$

20 $\lim\limits_{x \to +\infty} (\cosh x - \sinh x)$

21 $\lim\limits_{x \to 1^-} x^{1/(1 - x^2)}$

22 $\lim\limits_{x \to 0^+} \left(\dfrac{\sin x}{x} \right)^{1/x^3}$

23 $\lim\limits_{x \to +\infty} \left(1 + \dfrac{1}{x} \right)^{x^2}$

24 $\lim\limits_{x \to 0} (1 + ax^2)^{a/x}$

25 $\lim\limits_{x \to +\infty} (x^2 + 4)^{1/x}$

26 $\lim\limits_{x \to 4^+} (x - 4)^{x^2 - 16}$

27 $\lim\limits_{x \to 0} (\cos x)^{1/x^2}$

28 $\lim\limits_{x \to 0} (1 + \sin x)^{\cot x}$

29 $\lim\limits_{x \to 0} [\ln(x + 1)]^x$

30 $\lim\limits_{x \to 0^+} (\tan^{-1} x)^{1/\ln x}$

31 $\lim\limits_{x \to 0^+} \left(\ln \dfrac{1}{x} \right)^x$

32 $\lim\limits_{x \to 0} (\sin^{-1} x)^x$

In Problems 33 and 34, find all numbers c satisfying the conclusion of Cauchy's generalized mean-value theorem for each of the functions f and g on the indicated interval $[a, b]$.

33 $f(x) = \sqrt{x + 9}$, $g(x) = \sqrt{x}$, and $[a, b] = [0, 16]$.

34 $f(x) = \sin x$, $g(x) = \cos x$, and $[a, b] = [\pi/6, \pi/3]$.

In Problems 35 and 36, use the fundamental theorem of calculus and L'Hôpital's rule to find each limit.

35 $\displaystyle \lim_{x \to +\infty} \frac{\displaystyle \int_0^x e^t(t^2 - t + 5)\, dt}{\displaystyle \int_0^x e^t(3t^2 + 7t + 1)\, dt}$

36 $\displaystyle \lim_{x \to 0} \frac{\displaystyle \int_0^x (\cos^2 t + 5 \cos t^2)\, dt}{\displaystyle \int_0^x e^{-t^2}\, dt}$

37 Show that if α is a fixed positive number, then $\displaystyle \lim_{x \to 0^+} x^\alpha \ln x = 0$.

38 Find constants a and b so that

$$\lim_{t \to 0} \left(\frac{\sin 3t}{t^3} + \frac{a}{t^2} + b \right) = 0$$

In Problems 39 to 56, evaluate each improper integral (if it is convergent).

39 $\displaystyle \int_1^\infty \frac{dx}{x\sqrt{2x^2 - 1}}$

40 $\displaystyle \int_1^\infty \frac{t\, dt}{(1 + t^2)^2}$

41 $\displaystyle \int_1^\infty \frac{e^{2/t^2}\, dt}{t^3}$

42 $\displaystyle \int_{-\infty}^0 (e^x - e^{2x})\, dx$

43 $\displaystyle \int_1^\infty \frac{x^2 - 1}{x^4}\, dx$

44 $\displaystyle \int_2^\infty \frac{x\, dx}{(x^2 - 1)^{3/2}}$

45 $\displaystyle \int_e^\infty \frac{dx}{x(\ln x)^{7/2}}$

46 $\displaystyle \int_{-\infty}^0 \frac{e^x + 2x}{e^x + x^2}\, dx$

47 $\displaystyle \int_{-\infty}^1 xe^{3x}\, dx$

48 $\displaystyle \int_{-\infty}^\infty x^3 e^{-x}\, dx$

49 $\displaystyle \int_{-3}^1 \frac{dx}{x + 3}$

50 $\displaystyle \int_{-2}^6 \frac{dx}{\sqrt[3]{x + 2}}$

51 $\displaystyle \int_0^1 \frac{e^t\, dt}{\sqrt[3]{e^t - 1}}$

52 $\displaystyle \int_{-2}^2 \frac{dx}{\sqrt[5]{x + 1}}$

53 $\displaystyle \int_0^{3a} \frac{2x\, dx}{(x^2 - a^2)^{2/3}}$

54 $\displaystyle \int_a^{2a} \frac{x^2\, dx}{\sqrt{x^2 - a^2}}$

55 $\displaystyle \int_0^\infty \frac{1}{x^2 + 9}\, dx$

56 $\displaystyle \int_0^\infty \sqrt{x}\, e^{-\sqrt{x}}dx$

57 Find the area of the unbounded region under the graph of

$$y = \frac{1}{x \ln x}$$

and to the right of the line $x = e$.

58 Find the area of the unbounded region under the graph of

$$y = \frac{1}{x(x + 2)^2}$$

and to the right of the line $x = 1$.

59 Let $f(x) = x^2 e^{-ax}$, where a is a positive constant. Find the volume of the unbounded solid generated by revolving the region in the first quadrant under the graph of f: (a) about the x axis, and (b) about the y axis.

60 The gravitational force of attraction F between two particles of masses m_1 and m_2 is given by $F = Gm_1m_2/s^2$, where G is a constant and s is the distance between the particles. Find the work done in moving the particle with mass m_2 along a straight line until it is "infinitely far away" from the other particle if the two particles are initially one unit distance apart.

In Problems 61 to 64, find the Taylor polynomial P_n at a for each function f, and write the corresponding Taylor remainder in the Lagrange form.

61 $f(x) = \sin 2x$, $a = 0$, $n = 3$

62 $f(x) = \dfrac{1}{(1 + x)^2}$, $a = 1$, $n = 3$

63 $f(x) = e^{-x}$, $a = 0$, $n = 7$

64 $f(x) = \cos 3x$, $a = \pi/6$, $n = 6$

© In Problems 65 to 70, use an appropriate Taylor polynomial to estimate each quantity. In each case, round your answer off to five decimal places, and be certain that $|R_n(b)| \le 5/10^6$.

65 $\sin 88°$

66 $\cos \dfrac{59\pi}{180}$

67 $\ln 1.5$

68 $\sqrt[10]{e}$

69 $\sqrt{1.03}$

70 $\displaystyle \int_0^{1/2} \sin t^2\, dt$

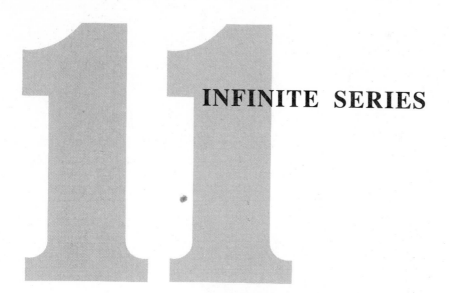

INFINITE SERIES

In Section 10.5 we discussed the use of Taylor polynomials to approximate the value of various functions and observed that the approximation often becomes more and more accurate as the degree of the Taylor polynomial is increased. For instance, the approximation

$$e^x \approx 1 + \frac{x}{1!} + \frac{x^2}{2!} + \frac{x^3}{3!} + \cdots + \frac{x^n}{n!}$$

becomes better and better as the number of terms on the right becomes larger and larger. This suggests that in some appropriate sense, e^x should be given exactly by the "infinite sum" of all terms of the form $x^k/k!$; that is,

$$e^x = 1 + \frac{x}{1!} + \frac{x^2}{2!} + \frac{x^3}{3!} + \cdots + \frac{x^n}{n!} + \frac{x^{n+1}}{(n+1)!} + \cdots$$

Such infinite sums, which are called *infinite series,* are studied in this chapter. Most of the important functions considered in calculus can be represented as a "sum" of an infinite series in which the terms involve powers of the independent variable; such a series is called a *power series*.

In the chapter we consider the following questions: How do we assign meaning to the sum of an infinite series? How can we tell whether a given infinite series has such a sum (that is, whether it is *convergent*)? When and how can we represent a function by a power series? When and how can we differentiate and integrate functions represented by power series?

To answer these questions and to develop a theory of infinite series, we begin by studying the closely related idea of *infinite sequences*.

11.1 Sequences

The word "sequence" is used in ordinary language to mean a succession of things arranged in a definite order. Here, we are concerned with sequences of numbers such as

$$1, 3, 5, 7, 9$$

or such as

$$0, 1, 4, 9, 16, 25, 36, 49, 64, \ldots$$

The individual numbers that appear in a sequence are called the **terms** of the sequence. A sequence having only a finite number of terms (such as the sequence 1, 3, 5, 7, 9) is called a **finite sequence.** Notice that the sequence

$$0, 1, 4, 9, 16, 25, 36, 49, 64, \ldots$$

(whose terms are the "perfect squares" arranged in ascending order) involves an infinite number of terms and is therefore an **infinite sequence.** Since we cannot write down *all* the terms of an infinite sequence, we use the convention of writing the first few terms and then appending three dots to mean "and so forth."

Here, our concern is with infinite sequences only, so from now on we refer to an infinite sequence simply as a **sequence.** Such a sequence is indicated by

$$a_1, a_2, a_3, \ldots, a_n, \ldots$$

where a_1 is the first term, a_2 is the second term, a_3 is the third term, and so forth. The **general term,** or the **nth term,** is here denoted by a_n. *In order to specify the sequence, it is sufficient to provide a rule, or a formula, for the nth term a_n.* For instance, the sequence whose nth term is given by the formula

$$a_n = 3n - 1$$

has first term

$$a_1 = 3(1) - 1 = 2$$

second term

$$a_2 = 3(2) - 1 = 5$$

and so on. The resulting sequence is

$$2, 5, 8, 11, 14, 17, 20, \ldots, 3n - 1, \ldots$$

Any letter with a subscript may be used to denote terms of a sequence—there's nothing special about the letter a. Sometimes a listing of the first few terms of a sequence indicates beyond any reasonable doubt the rule, or formula, for the general term. Examples are

$$1, 2, 3, 4, 5, 6, \ldots \qquad (a_n = n)$$

$$2, 4, 6, 8, 10, 12, \ldots \qquad (b_n = 2n)$$

$$1, \frac{1}{2}, \frac{1}{3}, \frac{1}{4}, \frac{1}{5}, \frac{1}{6}, \ldots \qquad \left(c_n = \frac{1}{n}\right)$$

$$-1, \frac{1}{3}, -\frac{1}{5}, \frac{1}{7}, -\frac{1}{9}, \frac{1}{11}, \ldots \qquad \left(s_n = (-1)^n \frac{1}{2n - 1}\right)$$

However, it can often be very difficult—if not impossible—to determine the intended general rule from an examination of the numerical pattern formed by the first few terms. When there is the slightest doubt, the safe thing is to specify the general term explicitly. (In this connection, see Problem 47.)

It is important to realize that a sequence is more than a mere collection of numbers; indeed, *the numbers in a sequence appear in a definite order,* and *repetitions of these numbers are permitted.* For instance, the following are perfectly legitimate sequences:

$$1, -1, 1, -1, 1, -1, \ldots, (-1)^{n+1}, \ldots$$

and

$$0, 0, 0, 0, 0, 0, 0, \ldots, 0, \ldots$$

In rigorous mathematical treatises, a sequence is defined to be a function f whose domain is the set of positive integers. Then $f(1)$ is called the first term, $f(2)$ the second term, and in general $f(n)$ is called the nth term of the sequence f. From this point of view, the sequence

$$3, \frac{21}{4}, \frac{17}{3}, \frac{93}{16}, \ldots, 6 - \frac{3}{n^2}, \ldots$$

would be identified with the function f whose domain is the positive integers and which is defined by $f(n) = 6 - (3/n^2)$.

The definition of a sequence as a function not only has the advantage of technical precision; it also permits many of the ideas previously developed for functions to be applied directly to sequences. If you are so inclined, we encourage you to regard sequences as being functions; however, in this book, we deal with sequences more informally.

We use the notation $\{a_n\}$ as shorthand for the sequence whose nth term is a_n. For instance, $\{n\}$ denotes the sequence

$$1, 2, 3, 4, 5, 6, \ldots, n, \ldots$$

and

$$\left\{ (-1)^n \frac{1}{2n - 1} \right\}$$

denotes the sequence

$$-1, \frac{1}{3}, -\frac{1}{5}, \frac{1}{7}, -\frac{1}{9}, \frac{1}{11}, \ldots, (-1)^n \frac{1}{2n - 1}, \ldots$$

Consider the sequence $\left\{ \dfrac{n}{3n + 1} \right\}$; that is,

$$\frac{1}{4}, \frac{2}{7}, \frac{3}{10}, \frac{4}{13}, \ldots, \frac{n}{3n + 1}, \ldots$$

Since

$$\frac{n}{3n + 1} = \frac{1}{3 + (1/n)}$$

and $1/n$ grows smaller as n grows larger, it is clear that as we go farther and farther out in this sequence, the terms come closer and closer to the value $\frac{1}{3}$. Thus, we write

$$\lim_{n \to +\infty} \frac{n}{3n + 1} = \frac{1}{3}$$

and we say that the sequence $\left\{ \dfrac{n}{3n + 1} \right\}$ *converges to the limit* $\dfrac{1}{3}$.

More generally, we say that a sequence $\{a_n\}$ *converges to the limit L in case*

$$\lim_{n \to +\infty} a_n = L$$

in the sense that the difference between a_n and L can be made as small as we please in absolute value, provided that n is made sufficiently large. The following definition expresses this idea more formally.

DEFINITION 1 **Convergence of a Sequence**

> We write $\lim\limits_{n \to +\infty} a_n = L$ and say that the sequence $\{a_n\}$ **converges to the limit** L
> provided that for each positive number ϵ, there exists a positive integer N (possi-
> bly depending on ϵ) such that
>
> $$|a_n - L| < \epsilon \qquad \text{whenever } n \geq N$$

A sequence that converges to a limit is called a **convergent** sequence; a sequence
that is not convergent is said to be **divergent.**

In dealing with an unfamiliar sequence, it's a good idea to write out the first few
terms to see whether any "trend" is developing. However, you must be careful
because *the tail end—not the first few terms—determines the convergence or divergence of a sequence.*

*In Examples 1 and 2, determine whether the given sequence converges or diverges.
If it converges, find its limit.*

EXAMPLE 1 $\left\{\dfrac{1}{10^{n-1}}\right\}$

SOLUTION Here the sequence is

$$1, \frac{1}{10}, \frac{1}{100}, \frac{1}{1000}, \frac{1}{10,000}, \ldots, \frac{1}{10^{n-1}}, \ldots$$

and it is clear that the terms are steadily becoming smaller and smaller. By choosing
n large enough, we can make $|(1/10^{n-1}) - 0| = 1/10^{n-1}$ as small as we please;
hence, by Definition 1, the sequence converges to the limit 0. ∎

EXAMPLE 2 $\{(-1)^n\}$

SOLUTION This sequence,

$$-1, 1, -1, 1, -1, \ldots, (-1)^n, \ldots$$

jumps back and forth forever between -1 and 1; hence it cannot approach a limit
and is therefore divergent. ∎

Properties of Limits of Sequences

The calculation of limits of convergent sequences is carried out in much the same
way as the calculation of limits of functions. In fact, the two procedures are closely
related, as is indicated by the following theorem.

THEOREM 1 **Convergence of Sequences and Functions**

> Let the function f be defined on the interval $[1, \infty)$, and define the sequence $\{a_n\}$
> by $a_n = f(n)$ for each positive integer n. Then, if $\lim\limits_{x \to +\infty} f(x) = L$, it follows that
> $\lim\limits_{n \to +\infty} a_n = L$.

The proof of Theorem 1 is left as an exercise (Problem 48).

EXAMPLE 3   Show that the sequence $\left\{\dfrac{\ln n}{n}\right\}$ converges, and find its limit.

SOLUTION   The function $\dfrac{\ln x}{x}$ is indeterminate of the form ∞/∞ at $+\infty$, so we can apply L'Hôpital's rule to obtain

$$\lim_{x \to +\infty} \frac{\ln x}{x} = \lim_{x \to +\infty} \frac{1/x}{1} = 0$$

By Theorem 1, $\lim\limits_{n \to +\infty}\left(\dfrac{\ln n}{n}\right) = 0$; that is, the sequence $\left\{\dfrac{\ln n}{n}\right\}$ converges to the limit 0.   ∎

The following properties of limits of sequences are analogous to the properties of limits of functions (see Section 1.8); therefore, we simply assume them here and illustrate their use by examples.

Properties of Limits of Sequences

Suppose that the sequences $\{a_n\}$ and $\{b_n\}$ converge and that c is a constant. Then:

> **1** $\lim\limits_{n \to +\infty} c = c$
>
> **2** $\lim\limits_{n \to +\infty} ca_n = c \lim\limits_{n \to +\infty} a_n$
>
> **3** $\lim\limits_{n \to +\infty} (a_n \pm b_n) = \left(\lim\limits_{n \to +\infty} a_n\right) \pm \left(\lim\limits_{n \to +\infty} b_n\right)$
>
> **4** $\lim\limits_{n \to +\infty} (a_n b_n) = \left(\lim\limits_{n \to +\infty} a_n\right)\left(\lim\limits_{n \to +\infty} b_n\right)$
>
> **5** If $b_n \neq 0$ for all positive integers n and $\lim\limits_{n \to +\infty} b_n \neq 0$, then
>
> $$\lim_{n \to +\infty} \frac{a_n}{b_n} = \frac{\lim\limits_{n \to +\infty} a_n}{\lim\limits_{n \to +\infty} b_n}$$

The following properties are also useful:

> **6** If k is a positive constant, then
>
> $$\lim_{n \to +\infty} \frac{c}{n^k} = 0$$
>
> **7** If $|c| < 1$, then
>
> $$\lim_{n \to +\infty} c^n = 0$$
>
> **8** If $|c| > 1$, then $\{c^n\}$ diverges.

(See Problems 49 and 50.)

In Examples 4 to 7, use Theorem 1 together with Properties 1 to 8 to determine the limit of the given sequence (provided that it converges).

EXAMPLE 4 $\left\{\dfrac{3n^2 + 7n + 11}{8n^2 - 5n + 3}\right\}$

SOLUTION   Dividing the numerator and the denominator of the given fraction by n^2 and applying Properties 1 through 6, we obtain

$$\lim_{n \to +\infty} \frac{3n^2 + 7n + 11}{8n^2 - 5n + 3} = \lim_{n \to +\infty} \frac{3 + \dfrac{7}{n} + \dfrac{11}{n^2}}{8 - \dfrac{5}{n} + \dfrac{3}{n^2}} = \frac{\lim\limits_{n \to +\infty} \left(3 + \dfrac{7}{n} + \dfrac{11}{n^2}\right)}{\lim\limits_{n \to +\infty} \left(8 - \dfrac{5}{n} + \dfrac{3}{n^2}\right)}$$

$$= \frac{3 + 0 + 0}{8 - 0 + 0} = \frac{3}{8}$$

EXAMPLE 5 $\left\{\dfrac{2^n}{3^{n+1}}\right\}$

SOLUTION $\lim\limits_{n \to +\infty} \dfrac{2^n}{3^{n+1}} = \lim\limits_{n \to +\infty} \dfrac{1}{3}\left(\dfrac{2}{3}\right)^n = \dfrac{1}{3}\lim\limits_{n \to +\infty}\left(\dfrac{2}{3}\right)^n = \dfrac{1}{3}(0) = 0$

Here we used Property 7 to conclude that $\lim\limits_{n \to +\infty} (\tfrac{2}{3})^n = 0$.

EXAMPLE 6 $\left\{n \sin \dfrac{\pi}{2n}\right\}$

SOLUTION $\lim\limits_{x \to +\infty} x \sin \dfrac{\pi}{2x} = \lim\limits_{x \to +\infty} \dfrac{\pi}{2} \cdot \dfrac{\sin (\pi/2x)}{\pi/2x} = \dfrac{\pi}{2} \cdot \lim\limits_{x \to +\infty} \dfrac{\sin (\pi/2x)}{\pi/2x}$

$$= \dfrac{\pi}{2} \lim_{t \to 0^+} \dfrac{\sin t}{t} = \dfrac{\pi}{2}(1) = \dfrac{\pi}{2}$$

where we have put $t = \pi/2x$ and observed that $t \to 0^+$ as $x \to +\infty$. Therefore, by Theorem 1,

$$\lim_{n \to +\infty} n \sin \dfrac{\pi}{2n} = \dfrac{\pi}{2}$$

EXAMPLE 7 $\left\{\dfrac{n^3 + 5n}{7n^2 + 1}\right\}$

SOLUTION   Here

$$\frac{n^3 + 5n}{7n^2 + 1} = \frac{\dfrac{1}{n^2}(n^3 + 5n)}{\dfrac{1}{n^2}(7n^2 + 1)} = \frac{n + \dfrac{5}{n}}{7 + \dfrac{1}{n^2}}$$

As n grows larger and larger, the numerator $n + (5/n)$ grows larger and larger, while the denominator $7 + (1/n^2)$ approaches 7. Therefore, the fraction becomes large without bound as $n \to +\infty$; hence, the sequence diverges. We can express this compactly as

$$\lim_{n \to +\infty} \frac{n^3 + 5n}{7n^2 + 1} = +\infty$$

However, we do *not* say that the sequence *converges* to $+\infty$. The notation $\lim\limits_{n \to +\infty} a_n = +\infty$ is a way of saying that the sequence $\{a_n\}$ *diverges* for the specific reason that its terms become larger and larger without bound.

Monotonic and Bounded Sequences

Consider the sequence $\left\{\dfrac{2n}{5n+3}\right\}$, and notice that its terms

$$\frac{2}{8}, \frac{4}{13}, \frac{6}{18}, \frac{8}{23}, \frac{10}{28}, \ldots, \frac{2n}{5n+3}, \ldots$$

are steadily growing larger and larger. This can be seen algebraically by writing

$$\frac{2n}{5n+3} = \frac{2}{5+(3/n)}$$

and noticing that as n grows larger, $3/n$ gets smaller, so that the fraction grows larger.

More generally, we make the following definition.

DEFINITION 2

Increasing and Decreasing Sequences

A sequence $\{a_n\}$ is said to be **increasing** (respectively, **decreasing**) if $a_n \leq a_{n+1}$ (respectively, $a_n \geq a_{n+1}$) holds for all positive integers n.

A sequence that is either increasing or decreasing is called **monotonic;** otherwise, it is called **nonmonotonic.**

In Examples 8 to 10, determine whether the given sequence is increasing, decreasing, or nonmonotonic.

EXAMPLE 8 $\left\{\dfrac{2n+1}{3n-2}\right\}$

SOLUTION   Here,

$$a_n = \frac{2n+1}{3n-2} \quad \text{and} \quad a_{n+1} = \frac{2(n+1)+1}{3(n+1)-2} = \frac{2n+3}{3n+1}$$

Hence, for any n,

$$a_{n+1} - a_n = \frac{2n+3}{3n+1} - \frac{2n+1}{3n-2} = \frac{(3n-2)(2n+3) - (2n+1)(3n+1)}{(3n+1)(3n-2)}$$

$$= \frac{(6n^2 + 5n - 6) - (6n^2 + 5n + 1)}{(3n+1)(3n-2)} = \frac{-7}{(3n+1)(3n-2)}$$

If n is a positive integer, then $3n+1 > 0$ and $3n-2 > 0$, so it follows that

$$a_{n+1} - a_n < 0 \quad \text{that is,} \quad a_n > a_{n+1}$$

Therefore, the given sequence is decreasing.   ■

EXAMPLE 9 $\left\{\sin\dfrac{n\pi}{2}\right\}$

SOLUTION   The sequence is

$$1, 0, -1, 0, 1, 0, -1, 0, \ldots, \sin\frac{n\pi}{2}, \ldots$$

and it repeats the pattern $1, 0, -1, 0$ again and again; hence, it is nonmonotonic.   ■

EXAMPLE 10 $\left\{\dfrac{n+1}{n+2}\right\}$

SOLUTION   Consider the function

$$f(x) = \frac{x+1}{x+2}$$

Since $f'(x) = \dfrac{(x+2)\cdot 1 - (x+1)\cdot 1}{(x+2)^2} = \dfrac{1}{(x+2)^2} > 0$ for $x \geq 1$

it follows that f is increasing on the interval $[1, \infty)$. In particular,

$$f(n) < f(n+1)$$

holds for all positive integers n; that is, the given sequence is increasing. ∎

Consider the sequence $\left\{\dfrac{n}{n+1}\right\}$, and notice that none of its terms

$$\frac{1}{2}, \frac{2}{3}, \frac{3}{4}, \frac{4}{5}, \ldots, \frac{n}{n+1}, \ldots$$

is larger than 1. Thus, we say that 1 is an *upper bound* for the sequence $\left\{\dfrac{n}{n+1}\right\}$. Similarly, since no term in the sequence $\left\{\dfrac{n}{n+1}\right\}$ is smaller than 0, we say that 0 is a lower bound for the sequence. More generally, we make the following definition.

DEFINITION 3

Bounded Sequences

> A number C (respectively, a number D) is called a **lower bound** (respectively, an **upper bound**) for the sequence $\{a_n\}$ if $C \leq a_n$ (respectively, $a_n \leq D$) holds for every positive integer n.

If the sequence $\{a_n\}$ has a lower bound (respectively, an upper bound), it is said to be **bounded from below** (respectively, **bounded from above**). A sequence is said to be **bounded** if it is bounded from both below and above. For instance, the sequence $\{\sin n\}$ is bounded from above by 1 and from below by -1; the sequence $\{3^n\}$ is bounded from below by 3, but it is not bounded from above; and the sequence $\{(-1)^n 3^n\}$ is bounded neither from above nor from below.

It is easy to show that *a sequence $\{a_n\}$ is bounded if and only if there exists a positive constant M such that*

$$|a_n| \leq M$$

holds for all positive integers n (Problem 56). *If a sequence converges, it must be bounded* (Problem 54); however, *a bounded sequence need not converge* (Problem 55).

EXAMPLE 11   Determine whether the sequence

$$\left\{(-1)^n \frac{2n}{3n+1}\right\}$$

is bounded from above or from below.

SOLUTION   For every positive integer n, we have

$$\left| (-1)^n \frac{2n}{3n+1} \right| = \frac{2n}{3n+1} = \frac{2}{3+(1/n)} < \frac{2}{3}$$

so the given sequence is bounded both from above and from below—from above by $\frac{2}{3}$ and from below by $-\frac{2}{3}$.   ∎

Imagine a sequence $a_1, a_2, a_3, \ldots, a_n, \ldots$ whose terms are growing steadily larger but which has an upper bound D. Thus,

$$a_1 \leq a_2 \leq a_3 \leq \cdots \leq a_n \leq a_{n+1} \leq \cdots \leq D$$

If we think of the terms of this sequence as corresponding to points on the number line (Figure 1), then we see an inevitable "pileup" of these points to the left of the point corresponding to D, and we easily persuade ourselves that the sequence must be converging to a limit somewhere in the interval $[a_1, D]$. In this instance our geometric intuition is confirmed by the following theorem, whose proof is beyond the scope of this book.

Figure 1

THEOREM 2

Convergence of Bounded Monotonic Sequences

> Every increasing sequence that is bounded from above is convergent. Likewise, every decreasing sequence that is bounded from below is convergent.

In Examples 12 and 13, use Theorem 2 to show that the given sequence is convergent.

EXAMPLE 12 $\left\{ \dfrac{n}{e^n} \right\}$

SOLUTION   Consider the function $f(x) = x/e^x$, and notice that

$$f'(x) = \frac{e^x - xe^x}{(e^x)^2} = \frac{1-x}{e^x} < 0 \qquad \text{for } x > 1$$

Hence, f is decreasing on $[1, \infty)$. It follows that $f(n) > f(n+1)$ for all positive integers n; that is, the sequence $\{n/e^n\}$ is decreasing. Since all terms of the sequence are positive, it follows that it is bounded from below by the number 0. Hence, by Theorem 2, the sequence converges.   ∎

ⓒ EXAMPLE 13 $\left\{ \dfrac{5^n}{3^n n!} \right\}$

SOLUTION   Offhand, it isn't clear whether this sequence is increasing, decreasing, or nonmonotonic. In order to gain some insight into its behavior, we calculate a few terms, rounded off to (say) three decimal places:

$$1.667, \ 1.389, \ 0.772, \ 0.322, \ 0.107, \ \ldots$$

Here we have an indication that the sequence *may be* decreasing. To prove that it really is decreasing, denote the general term by a_n, so that

$$a_n = \frac{5^n}{3^n n!}$$

We want to prove that

$$a_n > a_{n+1}$$

holds for all positive integers n. Because both a_n and a_{n+1} are positive, the last inequality is equivalent to

$$\frac{a_n}{a_{n+1}} > 1$$

Now,

$$\frac{a_n}{a_{n+1}} = \frac{5^n/(3^n n!)}{5^{n+1}/[3^{n+1}(n+1)!]} = \frac{5^n}{5^{n+1}} \cdot \frac{3^{n+1}}{3^n} \cdot \frac{(n+1)!}{n!}$$

$$= \frac{1}{5} \cdot 3 \cdot (n+1) = \frac{3}{5}(n+1)$$

Hence, what we want to prove is that

$$\tfrac{3}{5}(n+1) > 1$$

holds for all positive integers n. But this inequality is obviously true, since it is equivalent to the condition that

$$n + 1 > \tfrac{5}{3} \qquad \text{for } n \geq 1$$

Therefore, the given sequence is decreasing, just as we suspected. All terms of the sequence are positive, so 0 is a lower bound. Since the sequence is decreasing and bounded from below, it converges by Theorem 2.   ∎

We bring this section to a close by showing how Definition 1 is used to make formal proofs of theorems on limits of sequences. For this purpose, the following theorem (which is intuitively quite clear) serves as a typical example.

THEOREM 3   **Convergent Monotonic Sequences**

> The limit of a convergent increasing (respectively, decreasing) sequence is an upper bound (respectively, a lower bound) for the sequence.

PROOF   We prove only the part of the theorem pertaining to increasing sequences, since the proof for decreasing sequences is quite similar and only requires reversing some inequalities (Problem 60). Thus, assume that $\{a_n\}$ is monotone increasing and that $\lim_{n \to +\infty} a_n = L$. We must prove that all the terms of the sequence are less than or equal to L. If this were not so, there would be at least one term, say a_q, with $L < a_q$. Thus, let $\epsilon = a_q - L$, so that $\epsilon > 0$. By Definition 1, there exists a positive integer N such that $|a_n - L| < \epsilon$ holds whenever $n \geq N$.

Now, choose the integer n to be larger than both N and q. Since $q < n$, it follows that $a_q \leq a_n$, so that $L < a_q \leq a_n$ and $a_n - L > 0$. Consequently,

$$a_n - L = |a_n - L| < \epsilon = a_q - L$$

from which it follows that $a_n < a_q$, contrary to the fact that $a_q \leq a_n$. Therefore, the supposition that there is a term a_q with $L < a_q$ leads to a contradiction. It follows that no term such as a_q can exist. Thus, L is an upper bound for the sequence, and the theorem is proved.   ∎

Problem Set 11.1

In Problems 1 to 4, write out the first six terms of each sequence. Also write the 100th term.

1 $\{n^2 + 1\}$

2 $\left\{\dfrac{(-1)^{n+1}}{n + 1}\right\}$

3 $\left\{\dfrac{n}{n^2 + 5}\right\}$

4 $\left\{2 + \dfrac{1}{n}\right\}$

In Problems 5 to 8, find an expression for the general term (nth term) of each sequence.

5 $1, \frac{3}{2}, 2, \frac{5}{2}, 3, \frac{7}{2}, \ldots$

6 $1, 0, 1, 0, 1, 0, \ldots$

7 $\frac{1}{2}, \frac{1}{3}, \frac{1}{4}, \frac{1}{5}, \frac{1}{6}, \ldots$

8 $1, 9, 25, 49, 81, 121, \ldots$

In Problems 9 to 26, determine whether each sequence converges or diverges. If it converges, find its limit.

9 $\left\{\dfrac{100}{n}\right\}$

10 $\left\{\dfrac{n^2}{5n^2 + 1}\right\}$

11 $\left\{\dfrac{n^3 - 5n}{7n^3 + 2n}\right\}$

12 $\left\{\dfrac{2n^2 + 1}{9n^2 + 5}\right\}$

13 $\left\{\dfrac{5n^2}{3n + 1}\right\}$

14 $\left\{\dfrac{(-1)^n}{10^n}\right\}$

15 $\left\{\dfrac{2n^2 + n}{n + 1} \sin \dfrac{\pi}{2n}\right\}$

16 $\left\{\dfrac{e^n + e^{-n}}{e^n - e^{-n}}\right\}$

17 $\left\{\dfrac{\ln (n + 1)}{n + 1}\right\}$

18 $\{1 + (\frac{1}{3})^n - (\frac{3}{4})^n\}$

19 $\left\{\dfrac{\ln (1/n)}{\ln (n + 4)}\right\}$

20 $\{\ln (e^n + 2) - \ln (e^n + 1)\}$

21 $\left\{\dfrac{1}{\sqrt{n^2 + 1} - n}\right\}$

22 $\{\ln (e^n + 2) - n\}$

23 $\{n^{1/\sqrt{n}}\}$

24 $\{n^{1/n^2}\}$

25 $\left\{\left(1 + \dfrac{1}{n}\right)^n\right\}$

26 $\left\{\left(1 + \dfrac{5}{n}\right)^n\right\}$

In Problems 27 to 44, determine whether each sequence is increasing, decreasing, or nonmonotonic and whether it is bounded from above or below. Indicate whether the sequence is convergent or divergent.

27 $\left\{\dfrac{2n + 1}{3n + 2}\right\}$

28 $\{\sin n\pi\}$

29 $\{3^n - n\}$

30 $\left\{\dfrac{3^n}{1 + 3^n}\right\}$

31 $\{(-1)^{n^2}\}$

32 $\left\{\dfrac{3n^4}{n + 3^n}\right\}$

33 $\left\{1 + \dfrac{3}{5n - 2}\right\}$

34 $\{\sqrt[n]{2}\}$

35 $\left\{\dfrac{\ln (n + 1)}{n + 2}\right\}$

36 $\left\{\left(\dfrac{-n}{\ln n}\right)^n\right\}$

37 $\left\{\dfrac{n + 5}{n^2 + 6n + 4}\right\}$

38 $\left\{\dfrac{e^{n+1}}{(n + 1)!}\right\}$

39 $\left\{\dfrac{(-1)^n n}{n + 1}\right\}$

40 $\{\sqrt{n + 4} - \sqrt{n + 3}\}$

41 $\left\{1 - \dfrac{2^n}{n}\right\}$

42 $\left\{\dfrac{n^n}{n!}\right\}$

43 $\left\{\dfrac{\sin (n\pi/4)}{n}\right\}$

44 $\left\{\dfrac{1 \cdot 3 \cdot 5 \cdot 7 \cdots (2n - 1)}{n!}\right\}$

45 Give an example to show that the sum $\{a_n + b_n\}$ of two unbounded sequences $\{a_n\}$ and $\{b_n\}$ can be a bounded sequence.

46 Let $a_n = n!/2^n$.

(a) By writing

$$a_n = \left(\frac{1}{2} \cdot \frac{2}{2} \cdot \frac{3}{2}\right)\left(\frac{4}{2} \cdot \frac{5}{2} \cdot \frac{6}{2} \cdots \frac{n}{2}\right)$$

noting that all factors after the first three are greater than or equal to 2 and that there are $n - 3$ such factors, conclude that $a_n \geq \frac{3}{4} \cdot 2^{n-3}$ for $n \geq 3$.

(b) If K is any positive number, use the result of part (a) to show that $a_n > K$ holds provided that

$$n > 3 + \ln \frac{(4K/3)}{\ln 2}$$

(c) Explain why the sequence $\{a_n\}$ has no upper bound.

47 (a) Write out the first six terms of the sequence

$$\{n + (n - 1)(n - 2)(n - 3)(n - 4)(n - 5)(n - 6)\}$$

(b) What is the seventh term of the sequence in part (a)?

(c) What can you conclude about determining the general term of a sequence from an examination of its first few terms?

48 Prove Theorem 1.

49 Prove Property 6 on page 634.

50 Prove Properties 7 and 8 on page 634.

51 If c is a constant, determine whether the sequence $\{c^n\}$ converges or diverges for (a) $c < -1$, (b) $c = -1$, (c) $-1 < c < 1$, (d) $c = 1$, and (e) $c > 1$.

52 Suppose that $\{a_n\}$ and $\{b_n\}$ are two sequences whose corresponding terms agree from some point on; that is, suppose that there is a positive integer k such that $a_n = b_n$ holds for all $n \geq k$. If $\lim_{n \to +\infty} a_n = L$, prove that $\lim_{n \to +\infty} b_n = L$, too.

53 Suppose that the sequence $\{a_n\}$ has the property that $a_{n+1} = \frac{1}{2}(a_n + a_{n-1})$ for $n \geq 2$ and that $a_1 = 1$ while $a_2 = 3$.

(a) Write out the first eight terms of the sequence $\{a_n\}$.

(b) Use mathematical induction to prove that

$$a_n = \frac{7}{3} + \frac{(-1)^n}{3 \cdot 2^{n-3}}$$

(c) Find $\lim_{n \to +\infty} a_n$.

54 Assume that the sequence $\{a_n\}$ is convergent. Show that $\{a_n\}$ is bounded.

55 By means of a suitable example, show that a bounded sequence need not be convergent.

56 Show that the sequence $\{a_n\}$ is bounded if and only if there exists a positive constant M such that $|a_n| \leq M$ holds for all positive integers n.

57 Suppose that the sequence $\{a_n\}$ is convergent and satisfies the condition that $a_{n+1} = A + Ba_n$ for all positive integers n, where A and B are constants and $B \neq 1$. Find $\lim_{n \to +\infty} a_n$. (*Hint:* Take the limit as $n \to +\infty$ on both sides of the equation $a_{n+1} = A + Ba_n$.)

58 In advanced calculus, **Stirling's formula** is proved:

$$\sqrt{2n\pi}\left(\frac{n}{e}\right)^n < n! < \sqrt{2n\pi}\left(\frac{n}{e}\right)^n\left(1 + \frac{1}{12n - 1}\right)$$

Use Stirling's formula to show that the sequence

$$\left\{\frac{n^n}{e^n n!}\right\}$$

has an upper bound.

59 Let $|a| < 1$. Show that (a) $\{na^n\}$ converges to the limit 0 and (b) $\{n^2 a^n\}$ converges to the limit 0.

60 Prove the part of Theorem 3 that pertains to decreasing sequences.

61 Assume that a and b are constants with $a > 1$ and $b > 0$. Show that the sequence $\{n^b/a^n\}$ converges to the limit 0.

11.2 Infinite Series

An indicated sum

$$a_1 + a_2 + a_3 + \cdots + a_n + \cdots$$

of all the terms of an infinite sequence $\{a_n\}$ is called an **infinite series,** or simply a **series.** Using the sigma notation introduced in Section 5.1, we can write this series more compactly as

$$\sum_{k=1}^{\infty} a_k$$

The numbers a_1, a_2, a_3, and so forth are called the **terms** of the series, and a_n is called the **nth term,** or the **general term,** of the series.

Although we cannot literally add an infinite number of terms, it is sometimes useful to assign a numerical value to an infinite series by means of a special definition and to refer to this value as being the "sum" of the series. This is accomplished by using the "partial sums" of the series.

The sum s_n of the first n terms of a series $\sum_{k=1}^{\infty} a_k$ is called the **nth partial sum** of the series; thus,

$$s_n = a_1 + a_2 + a_3 + \cdots + a_n = \sum_{k=1}^{n} a_k$$

The sequence $\{s_n\}$ is referred to as the **sequence of partial sums** of the series. Notice that for each positive integer n,

$$s_{n+1} = s_n + a_{n+1}$$

For instance, the first few partial sums of the infinite series

$$1 + \frac{1}{2} + \frac{1}{4} + \frac{1}{8} + \frac{1}{16} + \cdots + \frac{1}{2^{n-1}} + \cdots$$

are as follows:

First partial sum $= s_1 = 1$

Second partial sum $= s_2 = 1 + \frac{1}{2} = 1.5$

Third partial sum $= s_3 = 1 + \frac{1}{2} + \frac{1}{4} = 1.75$

Fourth partial sum $= s_4 = 1 + \frac{1}{2} + \frac{1}{4} + \frac{1}{8} = 1.875$

Fifth partial sum $= s_5 = 1 + \frac{1}{2} + \frac{1}{4} + \frac{1}{8} + \frac{1}{16} = 1.9375$

We can continue in this way as long as we please. In this case, we find that the sequence s_n of partial sums

$$1, \quad 1.5, \quad 1.75, \quad 1.875, \quad 1.9375, \ldots$$

seems to be approaching 2 as a limit; for example, the 25th partial sum is

$$s_{25} = 1 + \frac{1}{2} + \frac{1}{4} + \frac{1}{8} + \frac{1}{16} + \cdots + \frac{1}{2^{24}} \approx 1.99999998$$

Here it is not difficult to verify that the sequence of partial sums really does converge to 2 as a limit; hence, it seems reasonable to *define* the sum of the infinite series to be 2 and to write

$$2 = 1 + \frac{1}{2} + \frac{1}{4} + \frac{1}{8} + \frac{1}{16} + \cdots + \frac{1}{2^{n-1}} + \cdots$$

More generally, we make the following definition.

DEFINITION 1   **Convergence of an Infinite Series**

> If the sequence $\{s_n\}$ of partial sums of the infinite series $\sum_{k=1}^{\infty} a_k$ converges to a limit $S = \lim\limits_{n \to +\infty} s_n$, we say that the infinite series $\sum_{k=1}^{\infty} a_k$ **converges** and that its **sum** is S.

If the infinite series $\sum_{k=1}^{\infty} a_k$ converges and its sum is S, we write

$$S = \sum_{k=1}^{\infty} a_k$$

Of course, an infinite series that does not converge is said to **diverge.**

In Examples 1 and 2, write out the first five terms of the given series, and then write out the first five terms of the sequence of partial sums of the series. Find a formula for the nth partial sum of the series, determine whether the series converges or diverges, and if it converges, find its sum S.

EXAMPLE 1 $\displaystyle\sum_{k=1}^{\infty} \frac{1}{k(k+1)}$

SOLUTION   Here,

$$\sum_{k=1}^{\infty} \frac{1}{k(k+1)} = \frac{1}{2} + \frac{1}{6} + \frac{1}{12} + \frac{1}{20} + \frac{1}{30} + \cdots$$

The first five partial sums are

$$
\begin{aligned}
s_1 &= a_1 &&= \tfrac{1}{2} \\
s_2 &= s_1 + a_2 = \tfrac{1}{2} + \tfrac{1}{6} &&= \tfrac{2}{3} \\
s_3 &= s_2 + a_3 = \tfrac{2}{3} + \tfrac{1}{12} &&= \tfrac{3}{4} \\
s_4 &= s_3 + a_4 = \tfrac{3}{4} + \tfrac{1}{20} &&= \tfrac{4}{5} \\
s_5 &= s_4 + a_5 = \tfrac{4}{5} + \tfrac{1}{30} &&= \tfrac{5}{6}
\end{aligned}
$$

These first five partial sums *suggest* that the *n*th partial sum might be given by the formula

$$s_n = s_{n-1} + a_n = \frac{n}{n+1}$$

This can be confirmed by mathematical induction on *n*, but we present an alternative argument. By partial fractions,

$$a_k = \frac{1}{k(k+1)} = \frac{1}{k} - \frac{1}{k+1}$$

Hence,

$$
\begin{aligned}
s_n = \sum_{k=1}^{n} a_k &= \sum_{k=1}^{n} \left(\frac{1}{k} - \frac{1}{k+1} \right) \\
&= \left(\frac{1}{1} - \frac{1}{2} \right) + \left(\frac{1}{2} - \frac{1}{3} \right) + \left(\frac{1}{3} - \frac{1}{4} \right) + \cdots + \left(\frac{1}{n} - \frac{1}{n+1} \right) \\
&= 1 - \frac{1}{2} + \frac{1}{2} - \frac{1}{3} + \frac{1}{3} - \frac{1}{4} + \cdots + \frac{1}{n} - \frac{1}{n+1} \\
&= 1 - \frac{1}{n+1} = \frac{n}{n+1}
\end{aligned}
$$

Therefore, we have

$$S = \lim_{n \to +\infty} s_n = \lim_{n \to +\infty} \frac{n}{n+1} = \lim_{n \to +\infty} \frac{1}{1 + (1/n)} = 1$$

Hence, the series converges and $\displaystyle\sum_{k=1}^{\infty} \frac{1}{k(k+1)} = 1$. ∎

EXAMPLE 2 $\displaystyle\sum_{k=1}^{\infty} k(k-1)$

SOLUTION Here,

$$\sum_{k=1}^{\infty} k(k-1) = 0 + 2 + 6 + 12 + 20 + \cdots$$

The first five partial sums are

$$
\begin{aligned}
s_1 &= a_1 & &= 0 \\
s_2 &= s_1 + a_2 = 0 & + 2 &= 2 \\
s_3 &= s_2 + a_3 = 2 & + 6 &= 8 \\
s_4 &= s_3 + a_4 = 8 & + 12 &= 20 \\
s_5 &= s_4 + a_5 = 20 & + 20 &= 40
\end{aligned}
$$

Using the formulas for the sums of successive integers and successive squares from Section 5.1, we find that

$$
\begin{aligned}
s_n &= \sum_{k=1}^{n} k(k-1) = \sum_{k=1}^{n} (k^2 - k) = \sum_{k=1}^{n} k^2 - \sum_{k=1}^{n} k \\
&= \frac{n(n+1)(2n+1)}{6} - \frac{n(n+1)}{2} = \frac{n(n^2-1)}{3}
\end{aligned}
$$

Therefore, we have

$$\lim_{n \to +\infty} s_n = \lim_{n \to +\infty} \frac{n(n^2-1)}{3} = +\infty$$

Thus, the sequence of partial sums—hence also the given series—diverges. ∎

The series

$$\sum_{k=1}^{\infty} \frac{1}{k(k+1)}$$

of Example 1, when rewritten in the form

$$\sum_{k=1}^{\infty} \left(\frac{1}{k} - \frac{1}{k+1} \right)$$

is called a *telescoping series* because of the cancellation that occurs in the calculation of its partial sums. More generally, if $\{b_n\}$ is a sequence, then a series of the form

$$\sum_{k=1}^{\infty} (b_k - b_{k+1})$$

is called a **telescoping series.** The nth partial sum is given by

$$
\begin{aligned}
s_n &= \sum_{k=1}^{n} (b_k - b_{k+1}) = (b_1 - b_2) + (b_2 - b_3) + \cdots + (b_n - b_{n+1}) \\
&= b_1 - \cancel{b_2} + \cancel{b_2} - \cancel{b_3} + \cdots + \cancel{b_n} - b_{n+1} \\
&= b_1 - b_{n+1}
\end{aligned}
$$

Therefore, if $\lim\limits_{n \to +\infty} b_{n+1}$ exists, say, $\lim\limits_{n \to +\infty} b_{n+1} = L$, then we have

$$\sum_{k=1}^{\infty} (b_k - b_{k+1}) = \lim_{n \to +\infty} s_n = \lim_{n \to +\infty} (b_1 - b_{n+1}) = b_1 - L$$

EXAMPLE 3 Show that the series

$$\sum_{k=1}^{\infty} \frac{3}{9k^2 + 3k - 2}$$

converges and find its sum.

SOLUTION By partial fractions,

$$\frac{3}{9k^2 + 3k - 2} = \frac{3}{(3k - 1)(3k + 2)} = \frac{1}{3k - 1} - \frac{1}{3k + 2}$$

$$= \frac{1}{3k - 1} - \frac{1}{3(k + 1) - 1} = b_k - b_{k+1}$$

where $b_k = 1/(3k - 1)$. Therefore,

$$\sum_{k=1}^{\infty} \frac{3}{9k^2 + 3k - 2} = \sum_{k=1}^{\infty} \left[\frac{1}{3k - 1} - \frac{1}{3(k + 1) - 1} \right]$$

$$= \frac{1}{3(1) - 1} - \lim_{n \to +\infty} \frac{1}{3(n + 1) - 1} = \frac{1}{2} - 0 = \frac{1}{2} \quad \blacksquare$$

In the study of infinite series, it is sometimes useful to be able to manufacture a series $\sum_{k=1}^{\infty} a_k$ with a preassigned sequence $\{s_n\}$ of partial sums. In this connection, the equations

$$a_1 = s_1 \quad \text{and} \quad a_n = s_n - s_{n-1} \quad \text{for } n > 1$$

provide the desired series.

EXAMPLE 4 Find an infinite series whose sequence of partial sums is

$$\{s_n\} = \left\{ \frac{3n}{2n + 1} \right\}$$

Determine whether this series converges, and if it converges, find its sum.

SOLUTION Here we have

$$s_n = \frac{3n}{2n + 1} \quad \text{and} \quad s_{n-1} = \frac{3(n - 1)}{2(n - 1) + 1} = \frac{3n - 3}{2n - 1}$$

Hence, for $n > 1$

$$a_n = s_n - s_{n-1} = \frac{3n}{2n + 1} - \frac{3n - 3}{2n - 1} = \frac{(2n - 1)(3n) - (3n - 3)(2n + 1)}{(2n + 1)(2n - 1)}$$

$$= \frac{3}{4n^2 - 1}$$

Here,

$$a_1 = s_1 = \frac{3(1)}{2(1) + 1} = 1$$

which happens to be the same as the value of

$$\frac{3}{4n^2 - 1} \qquad \text{when } n = 1$$

Thus, the desired series is given by

$$\sum_{k=1}^{\infty} a_k = \sum_{k=1}^{\infty} \frac{3}{4k^2 - 1}$$

Since the nth partial sum of the series is s_n, it follows that

$$\sum_{k=1}^{\infty} \frac{3}{4k^2 - 1} = \lim_{n \to +\infty} s_n = \lim_{n \to +\infty} \frac{3n}{2n + 1} = \lim_{n \to +\infty} \frac{3}{2 + (1/n)} = \frac{3}{2}$$

Geometric Series

By definition, a **geometric series** has the form

$$\sum_{k=1}^{\infty} ar^{k-1} = a + ar + ar^2 + ar^3 + \cdots + ar^{n-1} + \cdots$$

in which each term after the first is obtained by multiplying its immediate predecessor by a constant multiplier r. Since r is the ratio between any term (after the first) and its immediate predecessor, we refer to the series as a geometric series with **ratio r.**

Notice that *a geometric series is completely specified by giving its initial term a and its ratio r.* For instance, the geometric series with initial term $a = 1$ and ratio $r = \frac{1}{2}$ is

$$1 + \frac{1}{2} + \frac{1}{4} + \frac{1}{8} + \frac{1}{16} + \cdots + \frac{1}{2^{n-1}} + \cdots$$

A negative ratio r produces alternating algebraic signs; for example, the geometric series

$$\frac{2}{3} - \frac{1}{2} + \frac{3}{8} - \frac{9}{32} + \frac{27}{128} - \frac{81}{512} + \cdots$$

has initial term $a = \frac{2}{3}$ and ratio $r = -\frac{3}{4}$.

By a clever maneuver it is possible to obtain a simple formula for the **nth partial sum s_n of a geometric series** $\sum_{k=1}^{\infty} ar^{k-1}$. In fact, starting with

$$s_n = a + ar + ar^2 + \cdots + ar^{n-1}$$

and multiplying through by r, we obtain

$$s_n r = ar + ar^2 + ar^3 + \cdots + ar^n$$

Subtracting the second equation from the first, we have

$$s_n - s_n r = a - ar^n \qquad \text{or} \qquad s_n(1 - r) = a(1 - r^n)$$

Therefore,

$$s_n = a\left(\frac{1 - r^n}{1 - r}\right) \qquad \text{if } r \neq 1$$

By Property 7 in Section 11.1, if $|r| < 1$, then $\lim\limits_{n \to +\infty} r^n = 0$; hence,

$$\lim_{n \to +\infty} s_n = \lim_{n \to +\infty} a\left(\frac{1 - r^n}{1 - r}\right) = \frac{a}{1 - r} \qquad \text{if } |r| < 1$$

On the other hand, by Property 8 in Section 11.1, if $|r| > 1$, then the sequence $\{r^n\}$ diverges, and it follows that the sequence $\{s_n\}$ also diverges. In the remaining case in which $|r| = 1$, so that $r = 1$ or $r = -1$, it is easy to see that the sequence of partial sums diverges (unless $a = 0$). Thus, we have the following theorem.

THEOREM 1   **Geometric Series**

The geometric series $\sum_{k=1}^{\infty} ar^{k-1}$ with initial term $a \neq 0$ and ratio r converges if $|r| < 1$ and diverges if $|r| \geq 1$. If $|r| < 1$, then

$$\sum_{k=1}^{\infty} ar^{k-1} = \frac{a}{1 - r}$$

In Examples 5 to 7, determine whether the given geometric series converges or diverges, and if it converges, find its sum.

EXAMPLE 5 $\displaystyle\sum_{k=1}^{\infty} \frac{2}{3^{k-1}}$

SOLUTION   Since

$$\sum_{k=1}^{\infty} \frac{2}{3^{k-1}} = \sum_{k=1}^{\infty} 2\left(\frac{1}{3}\right)^{k-1}$$

the given series is indeed geometric with ratio $r = \frac{1}{3}$ and initial term $a = 2$. Because $|r| = \frac{1}{3} < 1$, the series converges and its sum is given by

$$\sum_{k=1}^{\infty} \frac{2}{3^{k-1}} = \frac{a}{1 - r} = \frac{2}{1 - \frac{1}{3}} = 3$$

EXAMPLE 6 $-1 + \frac{2}{3} - \frac{4}{9} + \frac{8}{27} - \frac{16}{81} + \cdots$

SOLUTION   Here $a = -1$ and the ratio is $r = -\frac{2}{3}$. [For instance, the ratio of the fourth term to the third term is $\frac{8}{27} \div (-\frac{4}{9}) = -\frac{2}{3}$.] Since $|r| = \frac{2}{3} < 1$, the series is convergent and

$$-1 + \frac{2}{3} - \frac{4}{9} + \frac{8}{27} - \frac{16}{81} + \cdots = \frac{a}{1 - r} = \frac{-1}{1 - (-\frac{2}{3})} = -\frac{3}{5}$$

EXAMPLE 7 $\displaystyle\sum_{k=1}^{\infty} (\tfrac{3}{2})^k$

SOLUTION   The series is geometric with initial term $a = \tfrac{3}{2}$ and ratio $r = \tfrac{3}{2}$. Since $|r| = \tfrac{3}{2} > 1$, the series is divergent. ∎

Applications of Geometric Series

Geometric series arise quite naturally in many branches of mathematics, as the following examples illustrate.

EXAMPLE 8   The probability of making the point "8" in a game of craps—that is, the probability of rolling an 8 with two dice before rolling a 7—is given by

$$\tfrac{5}{36} + (\tfrac{5}{36})(\tfrac{25}{36}) + (\tfrac{5}{36})(\tfrac{25}{36})^2 + (\tfrac{5}{36})(\tfrac{25}{36})^3 + \cdots$$

Find this probability.

SOLUTION   The displayed series is geometric with initial term $a = \tfrac{5}{36}$ and ratio $r = \tfrac{25}{36}$. Its sum is accordingly given by

$$\frac{a}{1-r} = \frac{\tfrac{5}{36}}{1 - \tfrac{25}{36}} = \frac{5}{11}$$

Therefore, the probability of making the point 8 is $\tfrac{5}{11}$. ∎

EXAMPLE 9   A simple air pump is evacuating a container of volume V. The cylinder of the pump, with the piston at the top, has volume v, and the total mass of air in the container at the outset is M. On the nth stroke of the pump, the mass of air removed from the container is

$$\frac{Mv}{V+v}\left(\frac{V}{V+v}\right)^{n-1}$$

Assuming that the pump operates "forever," what is the total mass of the air removed from the container?

SOLUTION   The total mass removed is given by the sum of the infinite series

$$\frac{Mv}{V+v} + \frac{Mv}{V+v}\left(\frac{V}{V+v}\right) + \frac{Mv}{V+v}\left(\frac{V}{V+v}\right)^2 + \cdots + \frac{Mv}{V+v}\left(\frac{V}{V+v}\right)^{n-1} + \cdots$$

with initial term $a = Mv/(V + v)$ and ratio $r = V/(V + v)$. Its sum is

$$\frac{a}{1-r} = \frac{\dfrac{Mv}{V+v}}{1 - \left(\dfrac{V}{V+v}\right)} = M$$

Thus, *all* the air is removed if the pump operates forever. (Of course, no pump is perfect—valves leak, air leaks around the piston, and so forth—so our answer is only of theoretical interest.) ∎

EXAMPLE 10   Express the infinite repeating decimal 1.267676767 . . . as a ratio of whole numbers.

SOLUTION   1.267676767 . . . = 1.2 + 0.067 + 0.00067 + 0.0000067 + $\cdots$

$$= \tfrac{12}{10} + (\tfrac{67}{1000} + \tfrac{67}{100,000} + \tfrac{67}{10,000,000} + \cdots)$$

The geometric series in the parentheses has initial term $a = \frac{67}{1000}$ and ratio $r = \frac{1}{100}$; hence, it converges and its sum is given by

$$\frac{a}{1-r} = \frac{\frac{67}{1000}}{1 - \frac{1}{100}} = \frac{67}{990}$$

Therefore,

$$1.267676767 \ldots = \tfrac{12}{10} + \tfrac{67}{990} = \tfrac{1255}{990} = \tfrac{251}{198}$$

Problem Set 11.2

In Problems 1 to 6, write out the first five terms of each series, and then write out the first five terms of the sequence $\{s_n\}$ of its partial sums. Find a "simple" formula for the nth partial sum s_n in terms of n, determine whether the series converges or diverges, and if it converges, find its sum $S = \lim\limits_{n \to +\infty} s_n$.

1 $\displaystyle\sum_{k=1}^{\infty} \frac{1}{(k+1)(k+2)}$

2 $\displaystyle\sum_{k=1}^{\infty} \ln\left(1 - \frac{2}{2k+3}\right)$

3 $\displaystyle\sum_{k=1}^{\infty} k(k+1)$

4 $\displaystyle\sum_{k=1}^{\infty} \frac{1}{k^2 + 2k}$

5 $\displaystyle\sum_{k=0}^{\infty} \frac{1}{(2k-1)(2k+1)}$

6 $\displaystyle\sum_{k=1}^{\infty} \frac{2k+1}{k^2(k+1)^2}$

In Problems 7 to 12, find an infinite series with the given sequence of partial sums, determine whether this series converges or diverges, and if it converges, find its sum.

7 $\{s_n\} = \left\{ \dfrac{n}{n+1} \right\}$

8 $\{s_n\} = \left\{ \dfrac{2n}{n+5} \right\}$

9 $\{s_n\} = \left\{ \dfrac{2n^2}{3n+5} \right\}$

10 $\{s_n\} = \{n\}$

11 $\{s_n\} = \{1 - (-1)^n\}$

12 $\{s_n\} = \left\{ 2 - \dfrac{1}{2^{n-1}} \right\}$

In Problems 13 to 29, find the initial term a and the ratio r of each geometric series, determine whether the series converges, and if it converges, find its sum.

13 $\displaystyle\sum_{k=1}^{\infty} \frac{5}{4^{k+1}}$

14 $\displaystyle\sum_{k=1}^{\infty} (\tfrac{3}{5})^{k-1}$

15 $\displaystyle\sum_{k=1}^{\infty} (-\tfrac{2}{3})^{k-1}$

16 $\displaystyle\sum_{k=1}^{\infty} \frac{3}{10^k}$

17 $1 + \tfrac{2}{7} + \tfrac{4}{49} + \tfrac{8}{343} + \cdots$

18 $-\tfrac{5}{8} + \tfrac{25}{64} - \tfrac{125}{512} + \tfrac{625}{4096} - \cdots$

19 $\displaystyle\sum_{k=1}^{\infty} (\tfrac{7}{6})^k$

20 $\displaystyle\sum_{k=0}^{\infty} (-\tfrac{5}{3})^k$

21 $\displaystyle\sum_{k=1}^{\infty} (\tfrac{9}{10})^{k+1}$

22 $\displaystyle\sum_{k=1}^{\infty} \frac{3^{k-1}}{4^{k+1}}$

23 $1 - 1 + 1 - 1 + 1 - 1 + 1 - 1 + \cdots$

24 $0.9 + 0.09 + 0.009 + 0.0009 + \cdots$

25 $\displaystyle\sum_{k=1}^{\infty} 5^{-k}$

26 $\displaystyle\sum_{k=1}^{\infty} e^{1-k}$

27 $\displaystyle\sum_{k=1}^{\infty} \frac{3^{k-1}}{5^k}$

28 $\displaystyle\sum_{k=0}^{\infty} \frac{5^{k+1}}{6^k}$

29 $1 - \tfrac{1}{10} + \tfrac{1}{100} - \tfrac{1}{1000} + \cdots$

30 The series $1 - 1 + 1 - 1 + 1 - 1 + \cdots + (-1)^{n-1} + \cdots$ is geometric with ratio $r = -1$; hence, it diverges. Thus, the calculation $1 - 1 + 1 - 1 + 1 - 1 + \cdots + (-1)^{n-1} + \cdots = (1 - 1) + (1 - 1) + (1 - 1) + \cdots = 0 + 0 + 0 + \cdots = 0$ must be incorrect. What is wrong with this calculation?

In Problems 31 to 34, express each repeating decimal as a ratio of whole numbers by using an appropriate geometric series.

31 $0.33333 \ldots$

32 $1.11111 \ldots$

33 $4.717171 \ldots$

34 $15.712712712 \ldots$

35 Is it true $\displaystyle\lim_{n \to +\infty} \sum_{k=1}^{n} a_k = \sum_{k=1}^{\infty} a_k$? Explain.

36 Find $\displaystyle\lim_{n \to +\infty} \left(1 + \frac{1}{3^2} + \frac{1}{3^4} + \frac{1}{3^6} + \cdots + \frac{1}{3^{2n}}\right)$.

37 In a game of *craps*, the probability that the shooter wins—that is, rolls 7 or 11 on the first throw or rolls a number other than 2, 3, or 12 and then, on a successive roll, repeats this number before rolling a 7—is given by the repeating decimal $0.4929292929 \ldots$ Express this probability as a ratio of whole numbers.

38 A beaker originally contains 10 grams of salt dissolved in 1000 cubic centimeters of water. The following procedure is performed repeatedly: 250 cubic centimeters of salt water is poured out, replaced by 250 cubic centimeters of pure water, and the solution is thoroughly stirred. (a) After the procedure is repeated n times, how many grams of salt have been removed from the beaker? (b) If the procedure is repeated "infinitely often," how much salt remains in the beaker?

39 A rubber ball rebounds to 60 percent of the height from which it was dropped. If it is dropped from a height of 2 meters, how far does it travel before coming to rest?

40 Abner starts walking directly toward a brick wall d meters away at a constant speed of v meters per second. At the same instant, a fly departs from Abner's forehead flying directly toward the brick wall at a constant speed of V meters per second, where $V > v$. Upon arriving at the brick wall, the fly immediately turns about and flies back to Abner's forehead at the same speed, V meters per second. The fly continues to shuttle between Abner's forehead and the wall in this manner until Abner finally reaches the wall.

(a) Show that on the nth round trip from Abner's forehead to the wall and back, the fly covers a distance of

$$\frac{2Vd}{V+v}\left(\frac{V-v}{V+v}\right)^{n-1} \quad \text{meters}$$

(b) Show that the fly requires

$$\frac{2d}{V+v}\left(\frac{V-v}{V+v}\right)^{n-1} \quad \text{seconds}$$

for the nth round trip.

(c) By part (a), the total distance flown by the fly is given by

$$\sum_{n=1}^{\infty} \frac{2Vd}{V+v}\left(\frac{V-v}{V+v}\right)^{n-1} \quad \text{meters}$$

Find this distance by summing the series.

(d) Using part (b), set up and sum a series to determine the total time required for Abner to reach the wall.

(e) Determine the total distance flown by the fly without summing an infinite series.

41 Let $\sum_{k=1}^{\infty} a_k$ be a given infinite series, and let $\{s_n\}$ be its sequence of partial sums. Define the sequence $\{b_n\}$ by

$$b_n = \begin{cases} 0 & \text{if } n = 1 \\ -s_{n-1} & \text{if } n > 1 \end{cases}$$

for each integer $n \geq 1$. Show that the series $\sum_{k=1}^{\infty} a_k$ is term by term exactly the same as the telescoping series $\sum_{k=1}^{\infty} (b_k - b_{k+1})$. Thus, conclude that *any infinite series can be rewritten as a telescoping series*.

42 Show that

$$\sum_{k=1}^{n} (b_k - b_{k+2}) = (b_1 + b_2) - (b_{n+1} + b_{n+2})$$

[*Hint:* $b_k - b_{k+2} = (b_k - b_{k+1}) + (b_{k+1} - b_{k+2})$.]

43 Using Problem 42, show that if $\{b_n\}$ is a convergent sequence with $\lim_{n \to +\infty} b_n = L$, then the series $\sum_{k=1}^{\infty} (b_k - b_{k+2})$ is convergent and $\sum_{k=1}^{\infty} (b_k - b_{k+2}) = b_1 + b_2 - 2L$.

11.3 Properties of Infinite Series

In Section 11.2 we were able to find the sum of certain infinite series by finding "nice" formulas for their partial sums. For instance, the nth partial sum of the telescoping series

$$\sum_{k=1}^{\infty} (b_k - b_{k+1}) \quad \text{is simply} \quad b_1 - b_{n+1}$$

and the nth partial sum of the geometric series

$$\sum_{k=1}^{\infty} ar^{k-1} \quad \text{is just} \quad a\left(\frac{1-r^n}{1-r}\right)$$

Unfortunately, it is not always so easy to find tidy formulas for nth partial sums; hence, it is important to develop alternative methods for determining whether a given series converges or diverges and for dealing with its sum if it converges.

Some of these methods are consequences of the general properties of infinite series which we develop in this section.

The following theorem gives an important property of convergent series.

THEOREM 1   **Necessary Condition for Convergence**

> If the infinite series $\sum_{k=1}^{\infty} a_k$ converges, then $\lim_{n \to +\infty} a_n = 0$.

PROOF   Let $\{s_n\}$ be the sequence of partial sums of the series $\sum_{k=1}^{\infty} a_k$. If $\sum_{k=1}^{\infty} a_k$ converges, then by definition the sequence $\{s_n\}$ converges and $\lim_{n \to +\infty} s_n = S = \sum_{k=1}^{\infty} a_k$. As $n \to +\infty$, then also $n - 1 \to +\infty$, so that $\lim_{n \to +\infty} s_{n-1} = S$. Notice that $a_n = s_n - s_{n-1}$; hence,

$$\lim_{n \to +\infty} a_n = \lim_{n \to +\infty} (s_n - s_{n-1}) = \lim_{n \to +\infty} s_n - \lim_{n \to +\infty} s_{n-1} = S - S = 0 \qquad \blacksquare$$

EXAMPLE 1   Given that $\displaystyle\sum_{k=1}^{\infty} \frac{2^k}{k!}$ converges,* find $\displaystyle\lim_{n \to +\infty} \frac{2^n}{n!}$.

SOLUTION   By Theorem 1, because $\displaystyle\sum_{k=1}^{\infty} \frac{2^k}{k!}$ converges, $\displaystyle\lim_{n \to +\infty} \frac{2^n}{n!} = 0$. $\qquad \blacksquare$

Theorem 1 can be used to show that certain series *diverge:* An immediate consequence of Theorem 1 is that if $\lim_{n \to +\infty} a_n$ is not 0, then $\sum_{k=1}^{\infty} a_k$ cannot converge. We record this fact for future use as follows.

THEOREM 2   **Sufficient Condition for Divergence**

> If $\lim_{n \to +\infty} a_n$ does not exist or if $\lim_{n \to +\infty} a_n$ exists but is different from zero, then the series $\sum_{k=1}^{\infty} a_k$ is divergent.

In Examples 2 and 3, use Theorem 2 to show that the given series diverges.

EXAMPLE 2 $\displaystyle\sum_{k=1}^{\infty} \frac{k+1}{k}$

SOLUTION   Since $\displaystyle\lim_{n \to +\infty} \frac{n+1}{n} = \lim_{n \to +\infty} \left(1 + \frac{1}{n}\right) = 1 \neq 0$, it follows that $\displaystyle\sum_{k=1}^{\infty} \frac{k+1}{k}$ diverges, by Theorem 2. $\qquad \blacksquare$

*As a matter of fact, the series

$$\sum_{k=1}^{\infty} \frac{2^k}{k!}$$

converges to $e^2 - 1$. For the proof, see Example 3 on page 696.

EXAMPLE 3 $\displaystyle\sum_{k=1}^{\infty} (-1)^k$

SOLUTION   Here, $\displaystyle\lim_{n \to +\infty} (-1)^n$ does not exist. By Theorem 2, therefore, $\sum_{k=1}^{\infty} (-1)^k$ diverges.   ∎

Caution: Do not misinterpret Theorem 1. It says that the general term of a convergent series must approach zero, but it does not assert the converse.

> *There are lots of divergent series whose general terms approach zero.*

EXAMPLE 4   Consider the series

$$\sum_{k=1}^{\infty} \ln \frac{k}{k+1}$$

and notice that the general term approaches zero:

$$\lim_{n \to +\infty} \ln \frac{n}{n+1} = \ln \left(\lim_{n \to +\infty} \frac{n}{n+1} \right) = \ln 1 = 0$$

From this, can we conclude that the series $\displaystyle\sum_{k=1}^{\infty} \ln \frac{k}{k+1}$ converges?

SOLUTION   No! *Just because the general term approaches zero is no guarantee that the series converges.* In fact, $\displaystyle\sum_{k=1}^{\infty} \ln \frac{k}{k+1}$ can be rewritten as $\displaystyle\sum_{k=1}^{\infty} [\ln k - \ln (k+1)]$, a telescoping series which diverges since $\displaystyle\lim_{n \to +\infty} \ln (n+1) = +\infty$.   ∎

Many properties of infinite series are analogs of corresponding properties of sequences. For instance, we have the following theorem.

THEOREM 3

Linear Properties of Series

(i) If $\sum_{k=1}^{\infty} a_k$ and $\sum_{k=1}^{\infty} b_k$ are convergent series, then $\sum_{k=1}^{\infty} (a_k + b_k)$ and $\sum_{k=1}^{\infty} (a_k - b_k)$ are also convergent and

$$\sum_{k=1}^{\infty} (a_k \pm b_k) = \sum_{k=1}^{\infty} a_k \pm \sum_{k=1}^{\infty} b_k$$

(ii) If $\sum_{k=1}^{\infty} a_k$ is a convergent series and c is a constant, then $\sum_{k=1}^{\infty} ca_k$ is also convergent and $\sum_{k=1}^{\infty} ca_k = c\sum_{k=1}^{\infty} a_k$. If $\sum_{k=1}^{\infty} a_k$ is a divergent series and c is a nonzero constant, then $\sum_{k=1}^{\infty} ca_k$ is also divergent.

PROOF   We indicate the proof of part (i) and leave the remainder as exercises (Problems 35 and 37). Thus, let $s_n = \sum_{k=1}^{n} a_k$ and $t_n = \sum_{k=1}^{n} b_k$ be the nth partial sums of the two given series. Then

$$s_n + t_n = \sum_{k=1}^{n} a_k + \sum_{k=1}^{n} b_k = \sum_{k=1}^{n} (a_k + b_k)$$

is the nth partial sum of the series $\sum_{k=1}^{\infty} (a_k + b_k)$. Since

$$\lim_{n \to +\infty} (s_n + t_n) = \lim_{n \to +\infty} s_n + \lim_{n \to +\infty} t_n = \sum_{k=1}^{\infty} a_k + \sum_{k=1}^{\infty} b_k$$

$\sum_{k=1}^{\infty} (a_k + b_k)$ converges and its sum is given by

$$\sum_{k=1}^{\infty} (a_k + b_k) = \sum_{k=1}^{\infty} a_k + \sum_{k=1}^{\infty} b_k$$

By a similar argument (Problem 35),

$$\sum_{k=1}^{\infty} (a_k - b_k) = \sum_{k=1}^{\infty} a_k - \sum_{k=1}^{\infty} b_k$$

■

EXAMPLE 5   Find the sum of the series $\sum_{k=1}^{\infty} \left(\dfrac{5}{2^{k-1}} + \dfrac{1}{3^{k-1}} \right)$.

SOLUTION   Notice that $\sum_{k=1}^{\infty} \dfrac{5}{2^{k-1}}$ is a geometric series with initial term $a = 5$ and ratio $= \frac{1}{2}$; hence it converges and

$$\sum_{k=1}^{\infty} \frac{5}{2^{k-1}} = \frac{a}{1-r} = \frac{5}{1 - \frac{1}{2}} = 10$$

Similarly, $\sum_{k=1}^{\infty} \dfrac{1}{3^{k-1}}$ converges and

$$\sum_{k=1}^{\infty} \frac{1}{3^{k-1}} = \frac{1}{1 - \frac{1}{3}} = \frac{3}{2}$$

It follows from part (i) of Theorem 3 that $\sum_{k=1}^{\infty} \left(\dfrac{5}{2^{k-1}} + \dfrac{1}{3^{k-1}} \right)$ converges and

$$\sum_{k=1}^{\infty} \left(\frac{5}{2^{k-1}} + \frac{1}{3^{k-1}} \right) = 10 + \frac{3}{2} = \frac{23}{2}$$

■

The following useful theorem is an immediate consequence of part (i) of Theorem 3.

THEOREM 4   **Divergence of a Series of Sums**

If the series $\sum_{k=1}^{\infty} a_k$ converges and the series $\sum_{k=1}^{\infty} b_k$ diverges, then the series $\sum_{k=1}^{\infty} (a_k + b_k)$ diverges.

PROOF   Assume the contrary; that is, suppose that $\sum_{k=1}^{\infty} (a_k + b_k)$ converges. Since $\sum_{k=1}^{\infty} a_k$ converges, $\sum_{k=1}^{\infty} [(a_k + b_k) - a_k] = \sum_{k=1}^{\infty} b_k$ also converges, by part (i) of Theorem 3, contradicting the hypothesis that $\sum_{k=1}^{\infty} b_k$ diverges. Hence, the assumption that $\sum_{k=1}^{\infty} (a_k + b_k)$ converges must be wrong; that is, $\sum_{k=1}^{\infty} (a_k + b_k)$ must diverge.   ∎

EXAMPLE 6   Determine whether the series

$$\sum_{k=1}^{\infty} \left(\ln \frac{k}{k+1} - \frac{1}{3^k} \right)$$

converges or diverges.

SOLUTION   As we saw in Example 4 on page 652, the series $\sum_{k=1}^{\infty} \ln \frac{k}{k+1}$ diverges. However, the series $\sum_{k=1}^{\infty} \frac{-1}{3^k}$ is a geometric series with initial term $a = -\frac{1}{3}$ and ratio $r = \frac{1}{3}$; hence, it converges. Therefore, by Theorem 4, the series

$$\sum_{k=1}^{\infty} \left(\ln \frac{k}{k+1} - \frac{1}{3^k} \right) = \sum_{k=1}^{\infty} \left(\frac{-1}{3^k} + \ln \frac{k}{k+1} \right)$$

must be divergent.   ∎

Notice that *even if both series $\sum_{k=1}^{\infty} a_k$ and $\sum_{k=1}^{\infty} b_k$ are divergent, the series $\sum_{k=1}^{\infty} (a_k + b_k)$ may be convergent.* For instance, let $a_n = n$ and $b_n = -n$ for all positive integers n.

Calculations with infinite series are often simplified by various manipulations involving the summation index. Such maneuvers are usually obvious and virtually self-explanatory. For instance, it is not necessary to begin a series with $k = 1$. Thus, we can write

$$\sum_{k=0}^{\infty} \frac{1}{2^k} = 1 + \frac{1}{2} + \frac{1}{4} + \frac{1}{8} + \cdots$$

$$\sum_{k=2}^{\infty} \ln \frac{k-1}{k} = \ln \frac{1}{2} + \ln \frac{2}{3} + \ln \frac{3}{4} + \cdots$$

and so forth. Also, there is no particular reason to use the symbol k for the summation index. Indeed it is possible, and often desirable, to change the summation index in an infinite series in much the same way that variables are changed in integrals. For instance, in the series $\sum_{k=1}^{\infty} (1/2^{k-1})$, let us put $j = k - 1$, noting that $j = 0$ when $k = 1$. Then we obtain

$$\sum_{k=1}^{\infty} \frac{1}{2^{k-1}} = \sum_{j=0}^{\infty} \frac{1}{2^j}$$

As the following theorem shows, the first few terms of an infinite series have no effect whatsoever on the convergence or divergence of the series—*it is only the tail end of the series that matters as far as convergence or divergence is concerned.*

<u>THEOREM 5</u>   **Removing the First M Terms from a Series**

If M is a fixed positive integer, then the series $\sum_{k=1}^{\infty} a_k$ converges if and only if the series $\sum_{k=M+1}^{\infty} a_k$ converges. Moreover, if these series converge, then

$$\sum_{k=1}^{\infty} a_k = \sum_{k=1}^{M} a_k + \sum_{k=M+1}^{\infty} a_k$$

PROOF   For $n > M$, we have

$$\sum_{k=1}^{n} a_k = \sum_{k=1}^{M} a_k + \sum_{k=M+1}^{n} a_k$$

Since $\sum_{k=1}^{M} a_k$ is a constant, it follows that $\lim_{n \to +\infty} \sum_{k=1}^{n} a_k$ exists if and only if $\lim_{n \to +\infty} \sum_{k=M+1}^{n} a_k$ exists; that is, $\sum_{k=1}^{\infty} a_k$ converges if and only if $\sum_{k=M+1}^{\infty} a_k$ converges. Assuming that these limits do exist and taking the limit on both sides of the above equation as $n \to +\infty$, we obtain

$$\sum_{k=1}^{\infty} a_k = \sum_{k=1}^{M} a_k + \sum_{k=M+1}^{\infty} a_k$$ ∎

Let us illustrate Theorem 5 using the geometric series $\sum_{k=1}^{\infty} ar^{k-1}$ with $|r| < 1$. Notice that

$$\sum_{k=M+1}^{\infty} ar^{k-1} = ar^{M} + ar^{M+1} + ar^{M+2} + \cdots$$

is also a geometric series with initial term ar^{M}, ratio r, and sum equal to $ar^{M}/(1-r)$. Also, $\sum_{k=1}^{M} ar^{k-1}$ is just the Mth partial sum of $\sum_{k=1}^{\infty} ar^{k-1}$; consequently,

$$\sum_{k=1}^{M} ar^{k-1} = a\left(\frac{1 - r^{M}}{1 - r}\right)$$

Thus, the equation of Theorem 5 becomes

$$\sum_{k=1}^{\infty} ar^{k-1} = \sum_{k=1}^{M} ar^{k-1} + \sum_{k=M+1}^{\infty} ar^{k-1}$$

or

$$\frac{a}{1-r} = a\left(\frac{1 - r^{M}}{1 - r}\right) + \frac{ar^{M}}{1-r}$$

an obvious algebraic identity.

By Theorem 2 in Section 11.1, if the sequence $\{s_n\}$ of partial sums of a series $\sum_{k=1}^{\infty} a_k$ is monotonic and bounded, then the sequence $\{s_n\}$—hence also the series $\sum_{k=1}^{\infty} a_k$—is convergent. In particular, we have the following theorem.

THEOREM 6

Convergence of a Series of Nonnegative Terms Whose Partial Sums Are Bounded

Let $\sum_{k=1}^{\infty} a_k$ be an infinite series whose terms are all nonnegative (that is, $a_k \geq 0$ for all k). If the sequence $\{s_n\}$ of nth partial sums of $\sum_{k=1}^{\infty} a_k$ is bounded above (that is, $s_n \leq M$ for all n, where M is a constant), then the series $\sum_{k=1}^{\infty} a_k$ is convergent.

PROOF

Since $s_{n+1} = s_n + a_{n+1}$ and $a_{n+1} \geq 0$, we have $s_{n+1} \geq s_n$ for all integers $n \geq 1$. Thus, $\{s_n\}$ is an increasing sequence that is bounded above. It follows (Theorem 2, page 638) that $\{s_n\}$ is convergent; hence, the series $\sum_{k=1}^{\infty} a_k$ is convergent. ■

EXAMPLE 7 Use Theorem 6 to show that the series $\sum_{k=1}^{\infty} \dfrac{k-1}{k \cdot 2^k}$ converges.

SOLUTION Clearly, each term of the given series is nonnegative. By Theorem 6, the series converges if the sequence $\{s_n\}$ of partial sums

$$s_n = \sum_{k=1}^{n} \frac{k-1}{k \cdot 2^k}$$

is bounded. Notice that $(k-1)/k < 1$, so that

$$\frac{k-1}{k \cdot 2^k} = \frac{k-1}{k}\left(\frac{1}{2}\right)^k < \left(\frac{1}{2}\right)^k$$

Therefore,

$$s_n = \sum_{k=1}^{n} \frac{k-1}{k \cdot 2^k} < \sum_{k=1}^{n} \left(\frac{1}{2}\right)^k = \frac{1}{2} \cdot \frac{1 - (\frac{1}{2})^n}{1 - \frac{1}{2}} < \frac{1}{2} \cdot \frac{1}{1 - \frac{1}{2}} = 1$$

so $\{s_n\}$ is bounded above by $M = 1$ and $\sum_{k=1}^{\infty} \dfrac{k-1}{k \cdot 2^k}$ converges. ■

Problem Set 11.3

In Problems 1 to 8, show that each series diverges by showing that the general term does not approach zero.

1 $\sum_{k=1}^{\infty} \dfrac{k}{5k+7}$

2 $\sum_{k=1}^{\infty} \ln\left(\dfrac{5k}{12k+5}\right)$

3 $\sum_{k=1}^{\infty} \dfrac{3k^2 + 5k}{7k^2 + 13k + 2}$

4 $\sum_{k=1}^{\infty} \dfrac{e^k}{3e^k + 7}$

5 $\sum_{k=1}^{\infty} \sin \dfrac{\pi k}{4}$

6 $\sum_{k=1}^{\infty} \dfrac{k}{\cos k}$

7 $\sum_{k=1}^{\infty} k \sin \dfrac{1}{k}$

8 $\sum_{k=1}^{\infty} \dfrac{k!}{2^k}$

In Problems 9 to 14, use the linear properties of series to find the sum of each series.

9 $\sum_{k=1}^{\infty} [(\tfrac{1}{3})^k + (\tfrac{1}{4})^k]$

10 $\sum_{k=1}^{\infty} [(\tfrac{1}{2})^{k-1} - (-\tfrac{1}{3})^{k+1}]$

11 $\sum_{k=1}^{\infty} \left[\dfrac{1}{k(k+1)} - \left(\dfrac{3}{4}\right)^{k-1}\right]$

12 $\sum_{k=0}^{\infty} [2(\tfrac{1}{3})^k - 3(-\tfrac{1}{5})^{k+1}]$

13 $\sum_{k=1}^{\infty} \left(\dfrac{2^k + 3^k}{6^k} - \dfrac{1}{7^{k+1}}\right)$

14 $\sum_{k=1}^{\infty} \left(\sin \dfrac{1}{k} + 2^{-k} - \sin \dfrac{1}{k+1}\right)$

15 Does the fact that

$$\lim_{n \to +\infty} \frac{1}{n} = 0$$

guarantee the convergence of the series

$$\sum_{k=1}^{\infty} \frac{1}{k}$$

16 Given that

$$\sum_{k=1}^{\infty} \frac{c^k}{k!}$$

converges for each value of the constant c, find

$$\lim_{n \to +\infty} \frac{c^n}{n!}$$

17 Given that $1 - \frac{1}{2} + \frac{1}{3} - \frac{1}{4} + \frac{1}{5} - \frac{1}{6} + \cdots = \ln 2$, find the sum of the series

$$-2 + 1 - \frac{2}{3} + \frac{2}{4} - \frac{2}{5} + \frac{2}{6} - \frac{2}{7} + \cdots$$

18 Criticize the following calculation: Let $\sum_{k=1}^{\infty} (b_k - b_{k+1})$ be a convergent telescoping series. Then

$$\sum_{k=1}^{\infty} (b_k - b_{k+1}) = \sum_{k=1}^{\infty} b_k - \sum_{k=1}^{\infty} b_{k+1}$$

$$= (b_1 + b_2 + b_3 + \cdots) - (b_2 + b_3 + \cdots) \stackrel{?}{=} b_1$$

19 Show that the series

$$\sum_{k=1}^{\infty} \left[\frac{1}{k(k+1)} - \ln \frac{k}{k+1} \right]$$

diverges.

In Problems 20 to 23, rewrite each series by changing the summation index from k to j as indicated.

20 $\displaystyle\sum_{k=1}^{\infty} ar^{k-1}; \ j = k - 1$

21 $\displaystyle\sum_{k=2}^{\infty} \frac{1}{k(k-1)}; \ j = k - 1$

22 $\displaystyle\sum_{k=M}^{\infty} a_k; \ j = k - M + 1$

23 $\displaystyle\sum_{k=1}^{\infty} a_k; \ j = k + M - 1$

24 Suppose that $\sum_{k=1}^{\infty} (b_k - b_{k+1})$ is a convergent telescoping series. By Theorem 5,

$$\sum_{k=1}^{\infty} (b_k - b_{k+1}) = \sum_{k=1}^{M} (b_k - b_{k+1}) + \sum_{k=M+1}^{\infty} (b_k - b_{k+1})$$

that is,

$$b_1 - \lim_{n \to +\infty} b_n = b_1 - b_{M+1} + \sum_{k=M+1}^{\infty} (b_k - b_{k+1})$$

or $\displaystyle\sum_{k=M+1}^{\infty} (b_k - b_{k+1}) = b_{M+1} - \lim_{n \to +\infty} b_n$

Verify the last equation directly without using Theorem 5.

25 Use the facts that

$$\sum_{k=1}^{\infty} \frac{1}{k(k+1)} = 1 \quad \text{and} \quad \sum_{k=1}^{M} \frac{1}{k(k+1)} = 1 - \frac{1}{M+1}$$

to find the sum of

$$\sum_{k=M+1}^{\infty} \frac{1}{k(k+1)}$$

26 (a) Use Theorem 5 to show that *if two series agree, term by term, except possibly for the first M terms, then either they both converge or else they both diverge.* (b) Show that *changing, deleting, or adding a single term cannot affect the convergence or divergence of a series.*

27 Given that

$$e = \sum_{k=1}^{\infty} \frac{1}{(k-1)!}$$

find the sum of the series

$$1 + \frac{1}{2!} + \frac{1}{3!} + \frac{1}{4!} + \cdots$$

28 If $\sum_{k=1}^{\infty} a_k$ is a convergent series all of whose terms are nonnegative, show that $\sum_{k=1}^{M} a_k \leq \sum_{k=1}^{\infty} a_k$ holds for all positive integers M. (*Hint:* Use Theorem 3 in Section 11.1.)

In Problems 29 to 34, all the series have nonnegative terms. In each case, establish the convergence of the series by proving directly that its sequence of partial sums is bounded above.

29 $\displaystyle\sum_{k=1}^{\infty} \frac{k}{(k+1) \cdot 3^k}$

30 $\displaystyle\sum_{k=1}^{\infty} \frac{(k-1) \ln 3}{4^{k-1}}$

31 $\displaystyle\sum_{k=0}^{\infty} \frac{4^{-k} k}{k^2 + 1}$

32 $\displaystyle\sum_{k=0}^{\infty} \frac{k}{5^k}$

33 $\displaystyle\sum_{k=1}^{\infty} \frac{1}{k^2} \left[Hint: \frac{1}{k^2} \leq \frac{1}{(k-1)k} \text{ for } k \geq 2. \right]$

34 $\displaystyle\sum_{k=1}^{\infty} \frac{1}{k!}$

35 Complete the proof of Theorem 3 by showing that if $\sum_{k=1}^{\infty} a_k$ and $\sum_{k=1}^{\infty} b_k$ are convergent, then so is $\sum_{k=1}^{\infty} (a_k - b_k)$, and

$$\sum_{k=1}^{\infty} (a_k - b_k) = \sum_{k=1}^{\infty} a_k - \sum_{k=1}^{\infty} b_k$$

36 Prove that *if a series of nonnegative terms converges, then its sequence of partial sums must be bounded.*

37 Prove part (ii) of Theorem 3.

11.4   Series of Nonnegative Terms

In Theorem 6 of Section 11.3 we showed that a series of nonnegative terms converges if its sequence of partial sums is bounded. In this section we present the *integral test* and the *comparison tests* for convergence or divergence of series whose terms are nonnegative. We begin with the integral test, which uses the convergence or divergence of an improper integral as a criterion for the convergence or divergence of the series.

The Integral Test

The integral test is based on the comparison of the partial sums of a series of the form $\sum_{k=1}^{\infty} f(k)$ and certain areas under the graph of the function f. Geometrically, the basic idea is quite simple and is illustrated in Figure 1. In Figure 1a, the area under the graph of a continuous, decreasing, nonnegative function f between $x = 1$ and $x = n + 1$ is overestimated by the sum $f(1) + f(2) + f(3) + \cdots + f(n)$ of the areas of the shaded rectangles; that is,

$$\int_{1}^{n+1} f(x)\, dx \le f(1) + f(2) + f(3) + \cdots + f(n)$$

Similarly, in Figure 1b, the area under the graph of the same function f between $x = 1$ and $x = n$ is underestimated by the sum $f(2) + f(3) + f(4) + \cdots + f(n)$ of the shaded rectangles; that is,

$$f(2) + f(3) + f(4) + \cdots + f(n) \le \int_{1}^{n} f(x)\, dx$$

Adding $f(1)$ to both sides of the last inequality, we obtain

$$f(1) + f(2) + f(3) + \cdots + f(n) \le f(1) + \int_{1}^{n} f(x)\, dx$$

In summary, we have the following result. *If f is a continuous, decreasing, nonnegative function defined at least on the closed interval $[1, n + 1]$, where n is a positive integer, then*

$$\int_{1}^{n+1} f(x)\, dx \le f(1) + f(2) + f(3) + \cdots + f(n) \le f(1) + \int_{1}^{n} f(x)\, dx$$

(See Problems 65 and 66 for an analytic derivation of the preceding inequalities.) These inequalities are used to prove the following theorem.

Figure 1

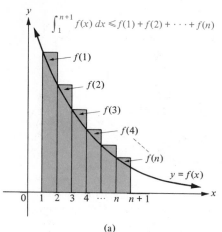

(a)

(b)

THEOREM 1   **The Integral Test**

Suppose that the function f is continuous, decreasing, and nonnegative on the interval $[1, \infty)$.

(i) If the improper integral $\int_{1}^{\infty} f(x)\, dx$ converges, then the infinite series $\sum_{k=1}^{\infty} f(k)$ converges.

(ii) If the improper integral $\int_{1}^{\infty} f(x)\, dx$ diverges, then the infinite series $\sum_{k=1}^{\infty} f(k)$ diverges.

PROOF

(i) We have seen that the *n*th partial sum $s_n = f(1) + f(2) + \cdots + f(n)$ of the series $\sum_{k=1}^{\infty} f(k)$ satisfies $\int_1^{n+1} f(x)\, dx \leq s_n \leq f(1) + \int_1^n f(x)\, dx$. If $\int_1^\infty f(x)\, dx$ converges, then $s_n \leq f(1) + \int_1^n f(x)\, dx \leq f(1) + \int_1^\infty f(x)\, dx$, so that $\{s_n\}$ has the upper bound $M = f(1) + \int_1^\infty f(x)\, dx$, and consequently, $\sum_{k=1}^{\infty} f(k)$ converges by Theorem 6 of Section 11.3.

(ii) If $\int_1^\infty f(x)\, dx$ diverges, then $\int_1^{n+1} f(x)\, dx$ grows large without bound as $n \to +\infty$; hence, since $\int_1^{n+1} f(x)\, dx \leq s_n$, it follows that s_n also grows large without bound as $n \to +\infty$. Therefore, in this case, $\{s_n\}$ diverges, and so $\sum_{k=1}^{\infty} f(k)$ diverges. ∎

In the integral test, there is no necessity to start the infinite series at $k = 1$. For instance, to test for convergence or divergence of the series $\sum_{k=2}^{\infty} f(k)$, we would use the improper integral $\int_2^\infty f(x)\, dx$.

In Examples 1 and 2, use the integral test to determine whether the given series converges or diverges.

EXAMPLE 1 $\displaystyle\sum_{k=1}^{\infty} \frac{1}{k^2 + 1}$

SOLUTION   The function f defined by $f(x) = \dfrac{1}{x^2 + 1}$ is continuous, decreasing, and nonnegative on the interval $[1, \infty)$. Also,

$$\int_1^\infty \frac{1}{x^2 + 1}\, dx = \lim_{b \to +\infty} \int_1^b \frac{dx}{x^2 + 1} = \lim_{b \to +\infty} \left[(\tan^{-1} x) \Big|_1^b \right]$$

$$= \lim_{b \to +\infty} (\tan^{-1} b - \tan^{-1} 1) = \frac{\pi}{2} - \frac{\pi}{4} = \frac{\pi}{4}$$

Thus, the improper integral $\displaystyle\int_1^\infty \frac{1}{x^2 + 1}\, dx$ converges, and so the series $\displaystyle\sum_{k=1}^{\infty} \frac{1}{k^2 + 1}$ converges. ∎

EXAMPLE 2 $\displaystyle\sum_{k=2}^{\infty} \frac{1}{k(\ln k)^{1/4}}$

SOLUTION   The function f defined by $f(x) = \dfrac{1}{x(\ln x)^{1/4}}$ is continuous, decreasing, and nonnegative on the interval $[2, \infty)$. Here, by the change of variable $u = \ln x$, we have

$$\int \frac{1}{x(\ln x)^{1/4}}\, dx = \int u^{-1/4}\, du = \frac{4}{3} u^{3/4} + C = \frac{4}{3}(\ln x)^{3/4} + C$$

so

$$\lim_{b \to +\infty} \int_2^b \frac{1}{x(\ln x)^{1/4}}\, dx = \lim_{b \to +\infty} \left[\frac{4}{3}(\ln b)^{3/4} - \frac{4}{3}(\ln 2)^{3/4} \right] = +\infty$$

Thus, the improper integral $\displaystyle\int_2^\infty \frac{1}{x(\ln x)^{1/4}}\, dx$ diverges, and so the series $\displaystyle\sum_{k=2}^\infty \frac{1}{k(\ln k)^{1/4}}$ diverges. ∎

The integral test makes it especially easy to study the convergence or divergence of a **p series,** which is by definition a series of the form

$$\sum_{k=1}^\infty \frac{1}{k^p}$$

where p is a constant. When $p = 1$, the p series becomes

$$\sum_{k=1}^\infty \frac{1}{k} = 1 + \frac{1}{2} + \frac{1}{3} + \frac{1}{4} + \frac{1}{5} + \cdots$$

and is called the **harmonic series.**

THEOREM 2   **Convergence and Divergence of the p Series**

> The p series $\displaystyle\sum_{k=1}^\infty \frac{1}{k^p}$ converges if $p > 1$ and diverges if $p \le 1$. In particular, the harmonic series $\displaystyle\sum_{k=1}^\infty \frac{1}{k}$ diverges.

PROOF   If $p < 0$, then $\displaystyle\lim_{n\to+\infty} \frac{1}{n^p} = +\infty$ (why?), so that $\displaystyle\sum_{k=1}^\infty \frac{1}{k^p}$ diverges, by Theorem 2 of Section 11.3. Thus, we can assume that $p \ge 0$. The function f defined by $f(x) = 1/x^p$ is continuous, decreasing, and nonnegative on $[1, \infty)$, and

$$\int_1^b \frac{1}{x^p}\, dx = \begin{cases} \dfrac{b^{1-p} - 1}{1 - p} & \text{if } p \ne 1 \\ \ln b & \text{if } p = 1 \end{cases}$$

Thus, for $p \le 1$, $\displaystyle\lim_{b\to+\infty} \int_1^b \frac{1}{x^p}\, dx = +\infty$, so that the improper integral $\displaystyle\int_1^\infty \frac{1}{x^p}\, dx$ diverges, and so does the series $\displaystyle\sum_{k=1}^\infty \frac{1}{k^p}$. However, for $p > 1$,

$$\lim_{b\to+\infty} \int_1^b \frac{1}{x^p}\, dx = \lim_{b\to+\infty} \frac{b^{1-p} - 1}{1 - p} = \frac{1}{p - 1}$$

so that $\displaystyle\int_1^\infty \frac{1}{x^p}\, dx$—hence also $\displaystyle\sum_{k=1}^\infty \frac{1}{k^p}$—is convergent. ∎

EXAMPLE 3 Test for convergence or divergence.

(a) $\displaystyle\sum_{k=1}^{\infty} \frac{1}{k^3}$

(b) $\displaystyle\sum_{k=1}^{\infty} \frac{1}{\sqrt[5]{k}}$

SOLUTION

(a) $\displaystyle\sum_{k=1}^{\infty} \frac{1}{k^3}$ is a p series with $p = 3 > 1$; hence, it converges.

(b) $\displaystyle\sum_{k=1}^{\infty} \frac{1}{\sqrt[5]{k}} = \sum_{k=1}^{\infty} \frac{1}{k^{1/5}}$ is a p series with $p = \frac{1}{5} < 1$; hence, it diverges. ∎

The harmonic series $1 + \frac{1}{2} + \frac{1}{3} + \frac{1}{4} + \cdots$ is a particularly intriguing series, since it marks the boundary between the convergent and the divergent p series. Although its partial sums $s_n = 1 + \frac{1}{2} + \frac{1}{3} + \cdots + (1/n)$ become large without bound as $n \to +\infty$, they do so rather slowly. To see this, consider the continuous, decreasing, nonnegative function f defined by $f(x) = 1/x$ for $x \geq 1$. For this function, the inequality

$$\int_1^{n+1} f(x)\, dx \leq f(1) + f(2) + \cdots + f(n) \leq f(1) + \int_1^{n} f(x)\, dx$$

(Figure 1) becomes

$$\ln(n + 1) \leq s_n \leq 1 + \ln n$$

If we put $n = 1{,}000{,}000$, we obtain

$$13.82 \leq s_{1,000,000} \leq 14.82$$

so that the sum of the first million terms of the harmonic series is less than 15.

There is no "nice" formula for the sum of the p series $\displaystyle\sum_{k=1}^{\infty} \frac{1}{k^p}$ with $p > 1$. The function ζ defined on $(1, \infty)$ by $\zeta(p) = \displaystyle\sum_{k=1}^{\infty} \frac{1}{k^p}$ is called the **Riemann zeta function** and plays an important role in analytic number theory.

Comparison Tests

The most practical tests for convergence or divergence of infinite series are based on the idea of comparing a given series with a series that is known to converge or diverge. Geometric series and p series are especially useful in such comparison tests.

We begin with the following definition.

<u>DEFINITION 1</u> **Domination of Series**

Let $\sum_{k=1}^{\infty} a_k$ and $\sum_{k=1}^{\infty} b_k$ be two series whose terms are nonnegative. We say that the series $\sum_{k=1}^{\infty} b_k$ **dominates** the series $\sum_{k=1}^{\infty} a_k$ if $a_k \leq b_k$ holds for all positive integer values of k.

More generally, if there is a positive integer N such that $a_k \leq b_k$ holds for all integers $k \geq N$, we say that the series $\sum_{k=1}^{\infty} b_k$ **eventually dominates** the series $\sum_{k=1}^{\infty} a_k$.

<u>THEOREM 3</u> **Direct Comparison Test**

Let $\sum_{k=1}^{\infty} a_k$ and $\sum_{k=1}^{\infty} b_k$ be series all of whose terms are nonnegative, and suppose that $\sum_{k=1}^{\infty} b_k$ dominates $\sum_{k=1}^{\infty} a_k$ (or that $\sum_{k=1}^{\infty} b_k$ eventually dominates $\sum_{k=1}^{\infty} a_k$).

(i) If $\sum_{k=1}^{\infty} b_k$ converges, then $\sum_{k=1}^{\infty} a_k$ converges.

(ii) If $\sum_{k=1}^{\infty} a_k$ diverges, then $\sum_{k=1}^{\infty} b_k$ diverges.

PROOF We prove the theorem under the hypothesis that $\sum_{k=1}^{\infty} b_k$ dominates $\sum_{k=1}^{\infty} a_k$. Since the convergence or divergence of an infinite series is controlled by its "tail end," conclusions (i) and (ii) must still hold if $\sum_{k=1}^{\infty} b_k$ eventually dominates $\sum_{k=1}^{\infty} a_k$ (Problem 70).

(i) Assume that $\sum_{k=1}^{\infty} b_k$ converges to the sum B. Then for any positive integer n, $\sum_{k=1}^{n} b_k \leq \sum_{k=1}^{\infty} b_k = B$ (see Problem 28 of Problem Set 11.3). Since $a_k \leq b_k$ holds for all positive integer values of k, we have $\sum_{k=1}^{n} a_k \leq \sum_{k=1}^{n} b_k \leq \sum_{k=1}^{\infty} b_k = B$; hence, the sequence of partial sums of $\sum_{k=1}^{\infty} a_k$ is bounded above by B. It follows from Theorem 6 of Section 11.3 that $\sum_{k=1}^{\infty} a_k$ is convergent.

(ii) Assume that $\sum_{k=1}^{\infty} a_k$ is divergent. Then the partial sums $\sum_{k=1}^{n} a_k$ become large without bound as $n \to +\infty$ (see Problem 71). Since $\sum_{k=1}^{n} a_k \leq \sum_{k=1}^{n} b_k$, it follows that the partial sums $\sum_{k=1}^{n} b_k$ become large without bound as $n \to +\infty$; hence, the series $\sum_{k=1}^{\infty} b_k$ cannot be convergent. ■

The choice of a suitable series with which to compare a given series is not always obvious and may require some trial and error; however, *a geometric series or a constant multiple of a p series whose form is similar to the given series often works.* For example, to test the series

$$\sum_{k=1}^{\infty} \frac{1}{5^k + 1}$$

you might try the convergent geometric series

$$\sum_{k=1}^{\infty} \frac{1}{5^k}$$

for comparison, because it closely resembles the given series. If the general term a_k of a given series is a fraction, then an appropriate p series to use for comparison is often obtained by deleting all but the highest powers of k in the numerator and denominator of a_k. For instance,

Series to Be Tested	Suggested Series for Comparison
$\displaystyle\sum_{k=1}^{\infty} \frac{1}{5k^2 + 7k + 1}$	$\displaystyle\sum_{k=1}^{\infty} \frac{1}{5k^2} = \frac{1}{5}\sum_{k=1}^{\infty}\frac{1}{k^2}$
$\displaystyle\sum_{k=1}^{\infty} \frac{1}{(k+4)^3}$	$\displaystyle\sum_{k=1}^{\infty} \frac{1}{k^3}$
$\displaystyle\sum_{k=1}^{\infty} \frac{5k^3 + 2k + 6}{k^4 - 3k + 4}$	$\displaystyle\sum_{k=1}^{\infty} \frac{5k^3}{k^4} = 5\sum_{k=1}^{\infty}\frac{1}{k}$
$\displaystyle\sum_{k=1}^{\infty} \frac{1}{\sqrt{4k^3 + k}}$	$\displaystyle\sum_{k=1}^{\infty} \frac{1}{2k^{3/2}} = \frac{1}{2}\sum_{k=1}^{\infty}\frac{1}{k^{3/2}}$

Sometimes it may be necessary to introduce or delete constant multipliers to obtain a suitable series for comparison.

In Examples 4 to 6, use the direct comparison test to determine whether the given series converges or diverges.

<u>EXAMPLE 4</u> $\displaystyle\sum_{k=1}^{\infty} \frac{1}{7k^2 - 1}$

SOLUTION   Let's try comparing the given series with the series

$$\sum_{k=1}^{\infty} \frac{1}{7k^2} = \frac{1}{7}\sum_{k=1}^{\infty}\frac{1}{k^2}$$

Here our comparison series converges, since it is a constant multiple of a p series for $p = 2 > 1$. Unfortunately, however, our proposed comparison series *does not* dominate the series to be tested; indeed,

$$\frac{1}{7k^2 - 1} \quad \text{is } greater \text{ than} \quad \frac{1}{7k^2}$$

Perhaps, if we *remove* the factor $\frac{1}{7}$, we will obtain a suitable comparison series. This depends on whether

$$\frac{1}{7k^2 - 1} \le \frac{1}{k^2} \qquad \text{for } k \ge 1$$

that is, whether

$$7k^2 - 1 \ge k^2 \qquad \text{for } k \ge 1$$

But the last condition is equivalent to

$$6k^2 - 1 \ge 0 \quad \text{or} \quad 6k^2 \ge 1 \qquad \text{for } k \ge 1$$

which is obviously true. Therefore, the convergent p series $\displaystyle\sum_{k=1}^{\infty} \frac{1}{k^2}$ dominates the given series $\displaystyle\sum_{k=1}^{\infty} \frac{1}{7k^2 - 1}$ and it follows from Theorem 3 that the given series converges.

EXAMPLE 5 $\displaystyle\sum_{k=1}^{\infty} \frac{1}{\sqrt{k+2}}$

SOLUTION   We are going to show that the given series dominates the series

$$\frac{1}{2}\sum_{k=1}^{\infty}\frac{1}{\sqrt{k}} = \sum_{k=1}^{\infty}\frac{1}{2}\cdot\frac{1}{\sqrt{k}}$$

which diverges because the p series $\sum_{k=1}^{\infty}(1/\sqrt{k})$ diverges. Thus, we wish to prove that

$$\frac{1}{2}\cdot\frac{1}{\sqrt{k}} \le \frac{1}{\sqrt{k+2}}$$

that is,

$$\frac{1}{4k} \le \frac{1}{k+2} \qquad \text{or} \qquad k+2 \le 4k \qquad \text{for } k \ge 1$$

Since $k + 2 \le 4k$ is equivalent to $\frac{2}{3} \le k$, it follows that

$$\frac{1}{2}\cdot\frac{1}{\sqrt{k}} \le \frac{1}{\sqrt{k+2}}$$

holds for $k \ge 1$, and the given series diverges. ∎

EXAMPLE 6 $\displaystyle\sum_{k=2}^{\infty} \frac{1}{\ln k}$

SOLUTION   Because $0 < \ln k < k$ for $k \ge 2$, we have $1/k < 1/(\ln k)$, so that the series $\displaystyle\sum_{k=2}^{\infty}\frac{1}{\ln k}$ dominates the series $\displaystyle\sum_{k=2}^{\infty}\frac{1}{k}$. Since the harmonic series $\displaystyle\sum_{k=1}^{\infty}\frac{1}{k}$ diverges, so does the series $\displaystyle\sum_{k=2}^{\infty}\frac{1}{k}$, by Theorem 5 of Section 11.3. It follows that the given series diverges. ∎

The following test is essentially another version of the direct comparison test, but it is sometimes easier to apply.

THEOREM 4   **Limit Comparison Test**

> Let $\sum_{k=1}^{\infty} a_k$ be a series of nonnegative terms and suppose that $\sum_{k=1}^{\infty} b_k$ is a series of positive terms such that $\displaystyle\lim_{n\to+\infty}\frac{a_n}{b_n} = c$, where $c > 0$. Then either both series converge or else both series diverge.

PROOF   Since $\displaystyle\lim_{n\to+\infty}\frac{a_n}{b_n} = c$, it follows that given any positive number ϵ, there exists a positive integer N such that

$$\left|\frac{a_n}{b_n} - c\right| < \epsilon \qquad \text{holds whenever} \qquad n \ge N$$

The condition $\left| \dfrac{a_n}{b_n} - c \right| < \epsilon$ can be rewritten as $-\epsilon < \dfrac{a_n}{b_n} - c < \epsilon$, or as

$c - \epsilon < \dfrac{a_n}{b_n} < c + \epsilon$. Putting $\epsilon = \dfrac{c}{2}$, we see that $\dfrac{c}{2} < \dfrac{a_n}{b_n} < \dfrac{3c}{2}$ holds

for all integers $n \ge N$. Therefore, if $n \ge N$, it follows that $\dfrac{c}{2}b_n < a_n < \dfrac{3c}{2}b_n$;

hence, the series $\displaystyle\sum_{k=1}^{\infty} a_k$ eventually dominates the series $\displaystyle\sum_{k=1}^{\infty} \dfrac{c}{2}b_k$, while the series

$\displaystyle\sum_{k=1}^{\infty} \dfrac{3c}{2}b_k$ eventually dominates the series $\displaystyle\sum_{k=1}^{\infty} a_k$. Consequently, if the series

$\displaystyle\sum_{k=1}^{\infty} b_k$ converges, then the series $\displaystyle\sum_{k=2}^{\infty} \dfrac{3c}{2}b_k$ converges [Theorem 3(ii) in Sec-

tion 11.3], and so the series $\displaystyle\sum_{k=1}^{\infty} a_k$ converges by the direct comparison test. On

the other hand, if the series $\displaystyle\sum_{k=1}^{\infty} b_k$ diverges, then the series $\displaystyle\sum_{k=1}^{\infty} \dfrac{c}{2}b_k$ diverges

[Theorem 3(ii) in Section 11.3 again], and so the series $\displaystyle\sum_{k=1}^{\infty} a_k$ diverges, by the

direct comparison test.   ∎

In Examples 7 and 8, use the limit comparison test to determine whether the given series converges or diverges.

EXAMPLE 7 $\displaystyle\sum_{k=1}^{\infty} \dfrac{1}{\sqrt[4]{k^3 + 1}}$

SOLUTION   We use the *divergent p* series $\displaystyle\sum_{k=1}^{\infty} \dfrac{1}{\sqrt[4]{k^3}}$ for the limit comparison

test. Let a_n be the nth term of the given series and let b_n be the nth term of the series

$\displaystyle\sum_{k=1}^{\infty} \dfrac{1}{\sqrt[4]{k^3}}$. Then

$$\lim_{n \to +\infty} \frac{a_n}{b_n} = \lim_{n \to +\infty} \frac{1/\sqrt[4]{n^3 + 1}}{1/\sqrt[4]{n^3}} = \lim_{n \to +\infty} \frac{\sqrt[4]{n^3}}{\sqrt[4]{n^3 + 1}} = \lim_{n \to +\infty} \sqrt[4]{\frac{n^3}{n^3 + 1}}$$

$$= \lim_{n \to +\infty} \sqrt[4]{\frac{1}{1 + (1/n)^3}} = 1$$

It follows from the limit comparison test that the given series diverges.   ∎

EXAMPLE 8 $\displaystyle\sum_{k=1}^{\infty} \dfrac{7k + 3}{(5k + 1) \cdot 3^k}$

SOLUTION   We use the *convergent* geometric series $\sum_{k=1}^{\infty} \dfrac{1}{3^k}$ for the limit comparison test. Thus, if a_n is the nth term of the given series and b_n is the nth term of the series $\sum_{k=1}^{\infty} \dfrac{1}{3^k}$, then

$$\lim_{n \to +\infty} \frac{a_n}{b_n} = \lim_{n \to +\infty} \frac{\left[\dfrac{7n + 3}{(5n + 1) \cdot 3^n}\right]}{1/3^n} = \lim_{n \to +\infty} \frac{7n + 3}{5n + 1} = \lim_{n \to +\infty} \frac{7 + (3/n)}{5 + (1/n)} = \frac{7}{5}$$

and so the given series converges by Theorem 4.   ∎

The following theorem can be proved by slightly modifying the proof of Theorem 4. Its proof is left as an exercise (Problem 72).

THEOREM 5   **Modified Limit Comparison Test**

> Let $\sum_{k=1}^{\infty} a_k$ be a series of nonnegative terms, and suppose that $\sum_{k=1}^{\infty} b_k$ is a series of positive terms.
>
> **(i)** If $\lim\limits_{n \to +\infty} \dfrac{a_n}{b_n} = 0$ and $\sum_{k=1}^{\infty} b_k$ converges, then $\sum_{k=1}^{\infty} a_k$ converges.
>
> **(ii)** If $\lim\limits_{n \to +\infty} \dfrac{a_n}{b_n} = +\infty$ and $\sum_{k=1}^{\infty} b_k$ diverges, then $\sum_{k=1}^{\infty} a_k$ diverges.

In Examples 9 and 10, determine whether the given series converges or diverges by using the modified limit comparison test.

EXAMPLE 9 $\sum_{k=1}^{\infty} \dfrac{\ln k}{k^4}$

SOLUTION   We use the *convergent* p series $\sum_{k=1}^{\infty} \dfrac{1}{k^3}$ for the modified limit comparison test. If a_n is the nth term of the given series and b_n is the nth term of the series $\sum_{k=1}^{\infty} \dfrac{1}{k^3}$, then

$$\lim_{n \to +\infty} \frac{a_n}{b_n} = \lim_{n \to +\infty} \frac{(\ln n)/n^4}{1/n^3} = \lim_{n \to +\infty} \frac{\ln n}{n} = \lim_{x \to +\infty} \frac{\ln x}{x}$$

$$= \lim_{x \to +\infty} \frac{1/x}{1} = \lim_{x \to +\infty} \frac{1}{x} = 0$$

where we have used Theorem 1 in Section 11.1 and L'Hôpital's rule to evaluate the limit. By part (i) of Theorem 5, the given series converges.

EXAMPLE 10 $\sum_{k=1}^{\infty} \dfrac{1}{\sqrt{2k + 1}}$

SOLUTION   We use the *divergent* harmonic series $\sum_{k=1}^{\infty} \frac{1}{k}$ for the modified limit comparison test. If a_n is the nth term of the given series and b_n is the nth term of the series $\sum_{k=1}^{\infty} \frac{1}{k}$, then

$$\lim_{n \to +\infty} \frac{a_n}{b_n} = \lim_{n \to +\infty} \frac{1/\sqrt{2n+1}}{1/n} = \lim_{n \to +\infty} \frac{n}{\sqrt{2n+1}}$$

$$= \lim_{n \to +\infty} \sqrt{n} \sqrt{\frac{n}{2n+1}} = \lim_{n \to +\infty} \sqrt{n} \sqrt{\frac{1}{2+(1/n)}} = +\infty$$

By part (ii) of Theorem 5, the given series diverges.   ■

Problem Set 11.4

In Problems 1 to 20, use the integral test to determine whether each series converges or diverges.

1 $\sum_{k=1}^{\infty} \frac{1}{k\sqrt[3]{k}}$

2 $\sum_{k=1}^{\infty} \frac{1}{k^2+4}$

3 $\sum_{k=1}^{\infty} \frac{3k^2}{k^3+16}$

4 $\sum_{k=1}^{\infty} \frac{3k}{k^2+8}$

5 $\sum_{k=1}^{\infty} \frac{2k}{(5+3k^2)^{3/2}}$

6 $\sum_{k=2}^{\infty} \frac{1}{k\sqrt{k^2-1}}$

7 $\sum_{n=1}^{\infty} \left(\frac{1000}{n}\right)^2$

8 $\sum_{m=1}^{\infty} e^{-m}$

9 $\sum_{k=2}^{\infty} \frac{\ln k}{k}$

10 $\sum_{k=2}^{\infty} \frac{1}{k \ln k}$

11 $\sum_{j=1}^{\infty} je^{-j}$

12 $\sum_{k=1}^{\infty} ke^{-k^2}$

13 $\sum_{m=1}^{\infty} \frac{\tan^{-1} m}{1+m^2}$

14 $\sum_{r=1}^{\infty} \frac{r}{2^r}$

15 $\sum_{k=1}^{\infty} \frac{1}{(2k+1)(3k+1)}$

16 $\sum_{n=1}^{\infty} \frac{1}{n(n+1)(n+2)}$

17 $\sum_{n=1}^{\infty} \coth n$

18 $\sum_{k=1}^{\infty} \frac{1}{1+\sqrt{k}}$

19 $\sum_{k=2}^{\infty} \frac{1}{k\sqrt{\ln k}}$

20 $\sum_{k=3}^{\infty} \frac{1}{k \ln k \ln (\ln k)}$

In Problems 21 to 32, use the direct comparison test with either a p series or a geometric series to determine whether each series converges or diverges.

21 $\sum_{k=1}^{\infty} \frac{k^2}{k^4+3k+1}$

22 $\sum_{k=1}^{\infty} \frac{k}{k^3+2k+7}$

23 $\sum_{k=1}^{\infty} \frac{1}{k \cdot 5^k}$

24 $\sum_{n=1}^{\infty} \frac{5}{(n+1)3^n}$

25 $\sum_{j=1}^{\infty} \frac{j+1}{(j+2) \cdot 7^j}$

26 $\sum_{r=1}^{\infty} \frac{5r}{\sqrt[3]{r^7+3}}$

27 $\sum_{k=1}^{\infty} \frac{8}{\sqrt[3]{k+1}}$

28 $\sum_{k=1}^{\infty} \frac{1}{4k+6}$

29 $\sum_{j=1}^{\infty} \frac{j^2}{j^3+4j+3}$

30 $\sum_{k=2}^{\infty} \frac{\ln k}{k}$

31 $\sum_{q=1}^{\infty} \frac{\sqrt{q}}{q+2}$

32 $\sum_{j=1}^{\infty} \frac{1+e^{-j}}{e^j}$

In Problems 33 to 38, use a limit comparison test with either a p series or a geometric series to determine whether each series converges or diverges.

33 $\sum_{k=1}^{\infty} \frac{1}{\sqrt[3]{k^2+5}}$

34 $\sum_{k=1}^{\infty} \frac{1}{3 \cdot 2^k+2}$

35 $\sum_{k=1}^{\infty} \frac{5k^2}{(k+1)(k+2)(k+3)(k+4)}$

36 $\displaystyle\sum_{j=1}^{\infty} \frac{1 + e^j}{j + 5^j}$

37 $\displaystyle\sum_{k=1}^{\infty} \frac{k^2}{1 + k^3}$

38 $\displaystyle\sum_{j=1}^{\infty} \frac{1}{j\sqrt{2j^3 + 5}}$

In Problems 39 to 64, use any appropriate test to determine whether the given series converges or diverges.

39 $\displaystyle\sum_{k=2}^{\infty} \frac{k - 1}{k \cdot 2^k}$

40 $\displaystyle\sum_{k=2}^{\infty} \frac{1}{k(\ln k)^2}$

41 $\displaystyle\sum_{k=1}^{\infty} \frac{1}{(3k - 1)3^k}$

42 $\displaystyle\sum_{k=1}^{\infty} \frac{1 + 3^k}{1 + 5^k}$

43 $\displaystyle\sum_{k=1}^{\infty} \frac{1}{1 + 7^k}$

44 $\displaystyle\sum_{k=1}^{\infty} \frac{5\sqrt{k}}{2k + 5}$

45 $\displaystyle\sum_{k=1}^{\infty} \frac{1}{\sqrt[k]{e}}$

46 $\displaystyle\sum_{k=1}^{\infty} k^3 e^{-k^4}$

47 $\displaystyle\sum_{k=2}^{\infty} \frac{k}{(k - 1)(k + 2)}$

48 $\displaystyle\sum_{k=1}^{\infty} \frac{1}{(8k^3 + 7k + 1)^{4/3}}$

49 $\displaystyle\sum_{k=1}^{\infty} \frac{k^2 + 1}{k^2 + 4}$

50 $\displaystyle\sum_{k=5}^{\infty} \frac{2\sqrt{k} + 3}{\sqrt{k^3 - 5k^2 + 1}}$

51 $\displaystyle\sum_{j=1}^{\infty} \frac{1}{7^j - \cos j}$

52 $\displaystyle\sum_{k=2}^{\infty} \frac{\ln k}{k^2 + 4}$

53 $\displaystyle\sum_{k=2}^{\infty} \frac{1}{k\sqrt{k^2 - 1}}$

54 $\displaystyle\sum_{k=1}^{\infty} \ln\left(1 + \frac{1}{k}\right)$

55 $\displaystyle\sum_{k=1}^{\infty} \frac{2k + 1}{(k^3 + 1)^3}$

56 $\displaystyle\sum_{k=1}^{\infty} \frac{k + 3}{k!}$

57 $\displaystyle\sum_{n=1}^{\infty} \frac{\sqrt{n}}{n + 1}$

58 $\displaystyle\sum_{k=1}^{\infty} \frac{\sin^2 k}{3^k}$

59 $\displaystyle\sum_{k=1}^{\infty} \frac{1}{\sqrt[3]{k(k^2 + 1)(2k - 1)}}$

60 $\displaystyle\sum_{k=1}^{\infty} \sin^{-1} \frac{1}{2^k}$

61 $\displaystyle\sum_{k=1}^{\infty} \sin \frac{1}{k}$

62 $\displaystyle\sum_{k=1}^{\infty} k^2 \sin^2\left(\frac{1}{k}\right)$

63 $\displaystyle\sum_{k=1}^{\infty} \tan \frac{1}{k}$

64 $\displaystyle\sum_{j=1}^{\infty} \frac{j!}{(2j)!}$

65 Assume that f is a continuous and decreasing function on the interval $[k - 1, k]$.

(a) Use the mean-value theorem for integrals (Theorem 10 in Section 5.3) to show that there exists a number c with $k - 1 \le c \le k$ such that

$$\int_{k-1}^{k} f(x)\, dx = f(c)$$

(b) Explain why $f(k) \le f(c) \le f(k - 1)$.

(c) Conclude that $f(k) \le \int_{k-1}^{k} f(x)\, dx \le f(k - 1)$.

66 Assume that the function f is a continuous and decreasing function on the interval $[1, n + 1]$, where n is a positive integer.

(a) Use part (c) of Problem 65 to show that

$$\sum_{k=2}^{n} f(k) \le \int_{1}^{n} f(x)\, dx$$

(b) Use part (c) of Problem 65 to show that

$$\int_{1}^{n+1} f(x)\, dx \le \sum_{k=2}^{n+1} f(k - 1)$$

(c) Conclude that

$$\int_{1}^{n+1} f(x)\, dx \le \sum_{k=1}^{n} f(k) \le f(1) + \int_{1}^{n} f(x)\, dx$$

67 Suppose that the function f is continuous, decreasing, and nonnegative on the interval $[1, \infty)$ and that the improper integral $\int_{1}^{\infty} f(x)\, dx$ converges. By the integral test, $\sum_{k=1}^{\infty} f(k)$ converges. Using part (c) of Problem 66, show that

$$\int_{1}^{\infty} f(x)\, dx \le \sum_{k=1}^{\infty} f(k) \le f(1) + \int_{1}^{\infty} f(x)\, dx$$

68 Use the result of Problem 67 to prove that

$$\frac{\pi}{4} \le \sum_{k=1}^{\infty} \frac{1}{k^2 + 1} \le \frac{\pi}{4} + \frac{1}{2}$$

69 Give an example to show that a series $\sum_{k=1}^{\infty} a_k$ with positive terms can be convergent and yet the series $\sum_{k=1}^{\infty} \sqrt{a_k}$ can be divergent.

70 Show that the conclusions of the direct comparison test (Theorem 3) still hold if $\sum_{k=1}^{\infty} b_k$ eventually dominates $\sum_{k=1}^{\infty} a_k$.

71 Suppose that the series $\sum_{k=1}^{\infty} a_k$ diverges and that its terms are nonnegative. Prove that the partial sums $s_n = \sum_{k=1}^{n} a_k$ become large without bound as $n \to +\infty$.

72 Prove the modified limit comparison test (Theorem 5).

73 Suppose that f is a continuous, decreasing, nonnegative function on the interval $[m, M]$, where m and M are positive integers and $m < M$. Prove that

$$f(M) + \int_{m}^{M} f(x)\, dx \le \sum_{k=m}^{M} f(k) \le f(m) + \int_{m}^{M} f(x)\, dx$$

11.5   Series Whose Terms Change Sign

The tests developed in Section 11.4 allow us to handle series whose terms do not change sign. (If all terms are nonpositive, we just multiply by -1 to convert to a series whose terms are nonnegative.) In this section we consider series whose terms change sign. The simplest such series is an **alternating series** whose terms alternate in sign, for instance, the series

$$\sum_{k=1}^{\infty} (-1)^{k+1} \frac{1}{k} = 1 - \frac{1}{2} + \frac{1}{3} - \frac{1}{4} + \cdots + (-1)^{n+1} \frac{1}{n} + \cdots$$

which is called the **alternating harmonic series.** Note that a geometric series with a negative ratio r, such as

$$\sum_{k=1}^{\infty} (-1)\left(-\frac{1}{2}\right)^{k-1} = -1 + \frac{1}{2} - \frac{1}{4} + \frac{1}{8} - \cdots + (-1)\left(\frac{-1}{2}\right)^{n-1} + \cdots$$

is an alternating series.

The following theorem, which exhibits an important feature of an alternating series whose terms decrease in absolute value, will be used to prove a test for convergence of such a series.

THEOREM 1   **Alternating Series Whose Terms Decrease in Absolute Value**

Let $\{a_n\}$ be a decreasing sequence of *positive* terms. Then the partial sums s_n of the alternating series

$$a_1 - a_2 + a_3 - a_4 + \cdots + (-1)^{n+1} a_n + \cdots$$

satisfy the following conditions:

(i) $0 \leq s_2 \leq s_4 \leq s_6 \leq s_8 \leq \cdots$.

(ii) $s_1 \geq s_3 \geq s_5 \geq s_7 \geq s_9 \geq \cdots$.

(iii)   If n is an even positive integer, then $s_{n+1} - s_n = a_{n+1}$.

(iv)   If n is an even positive integer, then $0 \leq s_n \leq s_{n+1} \leq s_1$.

PROOF

(i) If n is an *even* positive integer, then we can form the partial sum $s_n = a_1 - a_2 + a_3 - a_4 + \cdots + a_{n-1} - a_n$ and group the terms in pairs to obtain

$$s_n = (a_1 - a_2) + (a_3 - a_4) + \cdots + (a_{n-1} - a_n)$$

Since $a_1 \geq a_2 \geq a_3 \geq a_4 \geq \cdots$, it follows that each quantity enclosed in parentheses is nonnegative. The next *even* integer after n is $n + 2$, and we have $s_{n+2} = s_n + (a_{n+1} - a_{n+2}) \geq s_n$. It follows that

$$0 \leq s_2 \leq s_4 \leq s_6 \leq s_8 \leq \cdots$$

(ii) Similarly, if m is an *odd* positive integer, we can write

$$s_m = a_1 - a_2 + a_3 - a_4 + \cdots - a_{m-1} + a_m$$
$$= a_1 - (a_2 - a_3) - (a_4 - a_5) - \cdots - (a_{m-1} - a_m)$$

where, again, each quantity enclosed in parentheses is nonnegative. The next *odd* integer after m is $m + 2$, and we have

$$s_{m+2} = s_m - a_{m+1} + a_{m+2} = s_m - (a_{m+1} - a_{m+2}) \leq s_m$$

Hence,
$$s_1 \geq s_3 \geq s_5 \geq s_7 \geq s_9 \geq \cdots$$

(iii) If n is even, then $n + 1$ is odd and $s_{n+1} = s_n + a_{n+1}$. Therefore,

$$s_{n+1} - s_n = a_{n+1}$$

(iv) Again, if n is even, then by (iii), $s_{n+1} - s_n = a_{n+1} \geq 0$; hence, $s_n \leq s_{n+1}$. By (i), $0 \leq s_n$, and by (ii), $s_{n+1} \leq s_1$, so that

$$0 \leq s_n \leq s_{n+1} \leq s_1 \qquad \blacksquare$$

In Theorem 1, if $\{a_n\}$ is a strictly decreasing sequence, so that $a_1 > a_2 > a_3 > \cdots$, then all inequalities appearing in the proof and in the conclusions can be made strict.

The following theorem, discovered by Leibniz, provides a useful test for convergence of alternating series.

THEOREM 2   **Leibniz's Alternating-Series Test**

> If $\{a_n\}$ is a decreasing sequence of *positive* terms with $\lim\limits_{n \to +\infty} a_n = 0$, then the alternating series
> $$a_1 - a_2 + a_3 - a_4 + \cdots + (-1)^{n+1} a_n + \cdots$$
> is convergent. Moreover, if S is its sum and if s_n is its nth partial sum, then
> $$0 \leq (-1)^n (S - s_n) \leq a_{n+1}$$

PROOF   Putting $n = 2j$ in part (iv) of Theorem 1, we see that $0 \leq s_{2j} \leq s_{2j+1} \leq s_1$ holds for all positive integers j. By parts (i), (ii), and (iv) of Theorem 1, $\{s_{2j}\}$ is an increasing sequence bounded above by s_1, and $\{s_{2j+1}\}$ is a decreasing sequence bounded below by 0. By Theorem 2 of Section 11.1, it follows that both of the sequences $\{s_{2j}\}$ and $\{s_{2j+1}\}$ converge. By part (iii) of Theorem 1, $s_{2j+1} - s_{2j} = a_{2j+1}$; hence,

$$0 = \lim_{j \to +\infty} a_{2j+1} = \lim_{j \to +\infty} (s_{2j+1} - s_{2j}) = \lim_{j \to +\infty} s_{2j+1} - \lim_{j \to +\infty} s_{2j}$$

and it follows that $\lim\limits_{j \to +\infty} s_{2j+1} = \lim\limits_{j \to +\infty} s_{2j}$. Since the terms in the sequence $\{s_n\}$ whose indices are even and also the terms in this sequence whose indices are odd converge to the same limit, call it S, it follows (Problem 63) that the entire sequence $\{s_n\}$ must converge to S. Therefore, $\sum_{k=1}^{\infty} (-1)^{k+1} a_k$ converges, and its sum is S.

To prove the second part of the theorem, notice that $s_2, s_4, s_6, s_8, \ldots$ is an increasing sequence and converges to S; hence, $s_n \leq S$ holds for all even positive integers n by Theorem 3 in Section 11.1. Similarly, since $s_1, s_3, s_5, s_7, \ldots$ is a decreasing sequence and converges to S, then $S \leq s_m$ holds for all odd positive integers m. If n is even, then $n + 1$ is odd and so $s_n \leq S \leq s_{n+1}$. Subtracting s_n and noting that $s_{n+1} - s_n = a_{n+1}$, we have $0 \leq S - s_n \leq a_{n+1}$ when n is even. If m is odd, then $m + 1$ is even and so $s_{m+1} \leq S \leq s_m$. Subtracting s_m and noting that $s_{m+1} - s_m = -a_{m+1}$, we have $-a_{m+1} \leq S - s_m \leq 0$ when m is odd. It follows that $0 \leq (-1)^n (S - s_n) \leq a_{n+1}$ holds for every positive integer n, whether even or odd, and the proof is complete. $\blacksquare$

EXAMPLE 1 Test the alternating harmonic series $\displaystyle\sum_{k=1}^{\infty} (-1)^{k+1}\frac{1}{k}$ for convergence.

SOLUTION The sequence $\left\{\dfrac{1}{n}\right\}$ is decreasing, and $\displaystyle\lim_{n\to+\infty} \frac{1}{n} = 0$; hence the alternating harmonic series

$$1 - \frac{1}{2} + \frac{1}{3} - \frac{1}{4} + \cdots + (-1)^{n+1}\frac{1}{n} + \cdots$$

converges by Leibniz's alternating-series test. ∎

In Leibniz's theorem (Theorem 2), notice that $S - s_n$ is the *error* involved in estimating the sum S of the alternating series by the nth partial sum s_n. If n is *even*, then

$$0 \le (-1)^n(S - s_n) = S - s_n \qquad \text{so} \qquad s_n \le S$$

If n is *odd*, then

$$0 \le (-1)^n(S - s_n) = -(S - s_n) \qquad \text{so} \qquad S \le s_n$$

Therefore we have the following important result:

> Suppose that the hypotheses of Leibniz's theorem (Theorem 2) are true, so that the alternating series converges to the sum S. If n is an even positive integer, then s_n *underestimates* S; if n is an odd positive integer, then s_n *overestimates* S. In any case,
>
> $$|S - s_n| \le a_{n+1}$$
>
> that is, *the absolute value of the error of estimation does not exceed the absolute value of the first neglected term.**

For example, if we wish to estimate the sum

$$S = 1 - \frac{1}{2} + \frac{1}{3} - \frac{1}{4} + \cdots + (-1)^{n+1}\frac{1}{n} + \cdots$$

of the alternating harmonic series by a partial sum s_n, then the error cannot exceed the absolute value of the first neglected term a_{n+1}. For instance,

$$S \approx 1 - \tfrac{1}{2} + \tfrac{1}{3} - \tfrac{1}{4} + \tfrac{1}{5} - \tfrac{1}{6} + \tfrac{1}{7} - \tfrac{1}{8} + \tfrac{1}{9} = 0.7456 \ldots$$

with an error that does not exceed $\tfrac{1}{10}$. Moreover, since we have an *odd* number of terms in this estimate, we have *overestimated* S, so that S is *less* than $0.7456\ldots$. Actually, it can be shown† that the sum S of the alternating harmonic series is equal to the natural logarithm of 2:

$$S = \ln 2 = 0.6931 \ldots$$

In Examples 2 and 3, **(a)** *show that the given series is convergent,* **(b)** *find the partial sum s_4 of its first four terms, and* **(c)** *find a bound on the absolute value of the error involved in estimating its sum by s_4.*

---

*If $\{a_n\}$ is strictly decreasing, so that $a_1 > a_2 > a_3 \cdots$, then the conclusion $|S - s_n| \le a_{n+1}$ can be sharpened to $0 < |S - s_n| < a_{n+1}$.

†See Problems 41 to 44 in Problem Set 11.7.

EXAMPLE 2 $\displaystyle\sum_{k=1}^{\infty} (-1)^{k+1} \frac{(k + 3)}{k(k + 2)}$

SOLUTION

(a) Let f be the function defined by $f(x) = \dfrac{x + 3}{x(x + 2)}$. Then

$$f'(x) = -\frac{x^2 + 6x + 6}{x^2(x + 2)^2} < 0 \qquad \text{for } x \geq 1$$

Hence, the function f is decreasing on $[1, \infty)$. It follows that the sequence $\{f(n)\}$—that is, the sequence $\left\{\dfrac{n + 3}{n(n + 2)}\right\}$—is decreasing. Since $\displaystyle\lim_{n \to +\infty} \frac{n + 3}{n(n + 2)} = 0$ (why?) and $\dfrac{n + 3}{n(n + 2)} \geq 0$, the given alternating series converges, by Theorem 2.

(b) $s_4 = \dfrac{4}{3} - \dfrac{5}{8} + \dfrac{6}{15} - \dfrac{7}{24} = \dfrac{49}{60}$

(c) The absolute value of the error of the estimate

$$s_4 = \frac{49}{60} \approx \sum_{k=1}^{\infty} (-1)^{k+1} \frac{k + 3}{k(k + 2)}$$

does not exceed the fifth term, $\dfrac{5 + 3}{5(5 + 2)} = \dfrac{8}{35}$. Here s_4 involves an *even* number of terms, so that $\dfrac{49}{60}$ *underestimates* $\displaystyle\sum_{k=1}^{\infty} (-1)^{k+1} \frac{k + 3}{k(k + 2)}$.   ∎

EXAMPLE 3 $\displaystyle\sum_{k=1}^{\infty} \frac{(-1)^k}{k!}$

SOLUTION

(a) This series begins with a negative term, $-1/1!$, whereas Leibniz's theorem as stated above concerns alternating series which begin with a positive term. However, we can write

$$\sum_{k=1}^{\infty} \frac{(-1)^k}{k!} = -\sum_{k=1}^{\infty} \frac{(-1)^{k+1}}{k!}$$

and apply Leibniz's theorem to the series $\displaystyle\sum_{k=1}^{\infty} \frac{(-1)^{k+1}}{k!}$. Here $\left\{\dfrac{1}{n!}\right\}$ is a decreasing sequence of nonnegative terms, and $\displaystyle\lim_{n \to +\infty} \frac{1}{n!} = 0$, so that $\displaystyle\sum_{k=1}^{\infty} \frac{(-1)^{k+1}}{k!}$ converges; hence $\displaystyle\sum_{k=1}^{\infty} \frac{(-1)^k}{k!}$ also converges.

(b) and **(c)** Here, $1 - \dfrac{1}{2} + \dfrac{1}{6} - \dfrac{1}{24} = \dfrac{5}{8}$ *underestimates* $\displaystyle\sum_{k=1}^{\infty} \dfrac{(-1)^{k+1}}{k!}$

with an error no more than $\dfrac{1}{5!} = \dfrac{1}{120}$, and therefore, $-\dfrac{5}{8}$ *overestimates*

$$-\sum_{k=1}^{\infty} \dfrac{(-1)^{k+1}}{k!} = \sum_{k=1}^{\infty} \dfrac{(-1)^k}{k!}$$

with an error whose absolute value does not exceed $\frac{1}{120}$. ∎

Absolute and Conditional Convergence

Consider the alternating geometric series

$$1 - \tfrac{1}{2} + \tfrac{1}{4} - \tfrac{1}{8} + \tfrac{1}{16} - \tfrac{1}{32} + \cdots$$

with ratio $r = -\frac{1}{2}$. Not only does this series converge, but so does the corresponding series

$$|1| + \left|-\tfrac{1}{2}\right| + \left|\tfrac{1}{4}\right| + \left|-\tfrac{1}{8}\right| + \left|\tfrac{1}{16}\right| + \left|-\tfrac{1}{32}\right| + \cdots$$

of absolute values; that is, the geometric series $1 + \tfrac{1}{2} + \tfrac{1}{4} + \tfrac{1}{8} + \tfrac{1}{16} + \tfrac{1}{32} + \cdots$ is also convergent. Such a series is called *absolutely convergent*.

On the other hand, consider the alternating harmonic series

$$1 - \tfrac{1}{2} + \tfrac{1}{3} - \tfrac{1}{4} + \tfrac{1}{5} - \tfrac{1}{6} + \cdots$$

which converges by Leibniz's theorem. The corresponding series of absolute values is the harmonic series

$$1 + \tfrac{1}{2} + \tfrac{1}{3} + \tfrac{1}{4} + \tfrac{1}{5} + \tfrac{1}{6} + \cdots$$

which diverges. Thus, the convergence of the alternating harmonic series actually depends on the fact that its terms change sign. Such a series is called *conditionally convergent*. More generally, we make the following definition.

DEFINITION 1   **Absolute and Conditional Convergence**

(i) If the series $\sum_{k=1}^{\infty} |a_k|$ converges, we say that the series $\sum_{k=1}^{\infty} a_k$ is **absolutely convergent.**

(ii) If the series $\sum_{k=1}^{\infty} a_k$ is convergent but the series $\sum_{k=1}^{\infty} |a_k|$ is divergent, we say that the series $\sum_{k=1}^{\infty} a_k$ is **conditionally convergent**.

In Examples 4 to 6, determine whether the given series is divergent, conditionally convergent, or absolutely convergent.

EXAMPLE 4 $\displaystyle\sum_{k=1}^{\infty} \dfrac{(-1)^k}{k^2 + 1}$

SOLUTION   Since the series

$$\sum_{k=1}^{\infty} \left| \dfrac{(-1)^k}{k^2 + 1} \right| = \sum_{k=1}^{\infty} \dfrac{1}{k^2 + 1}$$

converges by comparison with the convergent p series $\sum_{k=1}^{\infty} \dfrac{1}{k^2}$, it follows that

$\sum_{k=1}^{\infty} \dfrac{(-1)^k}{k^2 + 1}$ is absolutely convergent. ∎

EXAMPLE 5 $\displaystyle\sum_{k=1}^{\infty} (-1)^{k+1} \dfrac{k + 1}{k + 2}$

SOLUTION Since $\displaystyle\lim_{n \to +\infty} (-1)^{n+1} \dfrac{n + 1}{n + 2}$ does not exist (why?), the given series is divergent (Section 11.3, Theorem 2). ∎

EXAMPLE 6 $\displaystyle\sum_{k=2}^{\infty} \dfrac{(-1)^k}{\ln k}$

SOLUTION The given series converges, by Leibniz's theorem. However, $\displaystyle\sum_{k=2}^{\infty} \left| \dfrac{(-1)^k}{\ln k} \right| = \sum_{k=2}^{\infty} \dfrac{1}{\ln k}$ diverges since $\dfrac{1}{\ln k} \geq \dfrac{1}{k}$ holds for $k \geq 2$ and $\displaystyle\sum_{k=2}^{\infty} \dfrac{1}{k}$

diverges. Hence, $\displaystyle\sum_{k=2}^{\infty} \dfrac{(-1)^k}{\ln k}$ is conditionally convergent. ∎

THEOREM 3

Absolute Convergence Implies Convergence

If a series $\sum_{k=1}^{\infty} a_k$ is absolutely convergent, then it is convergent.

PROOF Assume that $\sum_{k=1}^{\infty} |a_k|$ is convergent. The inequalities $-|a_k| \leq a_k \leq |a_k|$ can be rewritten as

$$0 \leq a_k + |a_k| \leq 2|a_k|$$

Now, $\sum_{k=1}^{\infty} 2|a_k|$ converges since $\sum_{k=1}^{\infty} |a_k|$ converges [part (ii) of Theorem 3 in Section 11.3]; hence, by the comparison test, $\sum_{k=1}^{\infty} (a_k + |a_k|)$ converges. Therefore,

$$\sum_{k=1}^{\infty} [(a_k + |a_k|) - |a_k|] = \sum_{k=1}^{\infty} a_k$$

converges [part (i) of Theorem 3 in Section 11.3]. ∎

EXAMPLE 7 Determine whether the series $\displaystyle\sum_{k=1}^{\infty} \dfrac{\sin k}{k^3 + 4}$ converges or diverges.

SOLUTION Although the given series contains both positive and negative terms, it is *not* an alternating series (why?). However, we have

$$\left| \dfrac{\sin k}{k^3 + 4} \right| = \dfrac{|\sin k|}{k^3 + 4} \leq \dfrac{1}{k^3}$$

for every positive integer k, so that $\displaystyle\sum_{k=1}^{\infty} \left| \dfrac{\sin k}{k^3 + 4} \right|$ is dominated by the convergent

p series $\displaystyle\sum_{k=1}^{\infty} \frac{1}{k^3}$. Therefore, the given series is absolutely convergent and hence convergent by Theorem 3. $\blacksquare$

The following theorem gives one of the most practical tests for absolute convergence.

THEOREM 4   **The Ratio Test**

Let $\displaystyle\sum_{k=1}^{\infty} a_k$ be a given series of nonzero terms.

(i) If $\displaystyle\lim_{n\to+\infty} \left|\frac{a_{n+1}}{a_n}\right| < 1$, then the series converges absolutely.

(ii) If $\displaystyle\lim_{n\to+\infty} \left|\frac{a_{n+1}}{a_n}\right| > 1$ or $\displaystyle\lim_{n\to+\infty} \left|\frac{a_{n+1}}{a_n}\right| = +\infty$, then the series diverges.

(iii) If $\displaystyle\lim_{n\to+\infty} \left|\frac{a_{n+1}}{a_n}\right| = 1$, then the test is inconclusive.

PROOF   **(i)** Suppose that $\displaystyle\lim_{n\to+\infty} \left|\frac{a_{n+1}}{a_n}\right| = L < 1$. Choose and fix a number r with

$L < r < 1$ $\left(\text{for instance, } r = \dfrac{L+1}{2}\right)$. Let $\epsilon = r - L$, noting that $\epsilon > 0$. Since

$\displaystyle\lim_{n\to+\infty} \left|\frac{a_{n+1}}{a_n}\right| = L$, there exists a positive integer N such that

$$\left|\left|\frac{a_{n+1}}{a_n}\right| - L\right| < \epsilon$$

for all integers $n \geq N$; that is, $-\epsilon < \left|\dfrac{a_{n+1}}{a_n}\right| - L < \epsilon$, or

$$L - \epsilon < \left|\frac{a_{n+1}}{a_n}\right| < L + \epsilon$$

for all $n \geq N$. Since $L + \epsilon = L + (r - L) = r$, then $\left|\dfrac{a_{n+1}}{a_n}\right| < r$, or

$|a_{n+1}| < |a_n|r$ holds for $n \geq N$. Therefore,

$$|a_{N+1}| < |a_N|r$$
$$|a_{N+2}| < |a_{N+1}|r < |a_N|r^2$$
$$|a_{N+3}| < |a_{N+2}|r < |a_N|r^3$$

and so forth. In fact,

$$|a_{N+j}| < |a_N|r^j$$

holds for all positive integers j. (For an inductive proof, see Problem 64.) Therefore, the geometric series $\sum_{j=1}^{\infty} |a_N|r^j$ dominates the series $\sum_{j=1}^{\infty} |a_{N+j}|$. Since $0 < r < 1$, the geometric series converges; hence,

$$\sum_{j=1}^{\infty} |a_{N+j}| = \sum_{k=N+1}^{\infty} |a_k|$$

converges by the direct comparison test. Therefore, by Theorem 5 of Section 11.3, $\sum_{k=1}^{\infty} |a_k|$ converges; that is, $\sum_{k=1}^{\infty} a_k$ is absolutely convergent.

(ii) Assume that $\lim_{n \to +\infty} \left| \dfrac{a_{n+1}}{a_n} \right| = L > 1$. Choose and fix a number r with $1 < r < L$ and put $\epsilon = L - r$. Thus, there exists a positive integer N such that $L - \epsilon < \left| \dfrac{a_{n+1}}{a_n} \right| < L + \epsilon$ holds for $n \geq N$. Hence, $1 < r =$

$$L - \epsilon < \left| \frac{a_{n+1}}{a_n} \right| = \frac{|a_{n+1}|}{|a_n|},$$ so that $|a_n| < |a_{n+1}|$ holds for $n \geq N$. Therefore,

$$|a_N| < |a_{N+1}| < |a_{N+2}| < |a_{N+3}| < \cdots$$

so that $0 < |a_N| < |a_n|$ holds for all $n \geq N$. This shows that the condition $\lim_{n \to +\infty} a_n = 0$ cannot hold; hence, $\sum_{k=1}^{\infty} a_k$ is divergent (Theorem 2 in Section 11.3).

Similarly, if $\lim_{n \to +\infty} \left| \dfrac{a_{n+1}}{a_n} \right| = +\infty$, then there exists a positive integer N such that $1 < \left| \dfrac{a_{n+1}}{a_n} \right|$ holds for all $n \geq N$, and we can complete the argument just as above and see that $\sum_{k=1}^{\infty} a_k$ diverges.

(iii) To see that the test really is inconclusive if $\lim_{n \to +\infty} \left| \dfrac{a_{n+1}}{a_n} \right| = 1$, consider

(a) $\displaystyle\sum_{k=1}^{\infty} \frac{1}{k^2}$ and (b) $\displaystyle\sum_{k=1}^{\infty} \frac{1}{k}$. Series (a) is convergent, but

$$\lim_{n \to +\infty} \frac{1/(n+1)^2}{1/n^2} = \lim_{n \to +\infty} \frac{n^2}{n^2 + 2n + 1} = 1$$

Series (b) is divergent, but

$$\lim_{n \to +\infty} \frac{1/(n+1)}{1/n} = \lim_{n \to +\infty} \frac{n}{n+1} = 1$$

■

In Examples 8 to 10, use the ratio test to determine whether each series converges absolutely or diverges.

EXAMPLE 8 $\displaystyle\sum_{k=1}^{\infty} \frac{2^k}{7^k(k+1)}$

SOLUTION   Here

$$a_n = \frac{2^n}{7^n(n+1)} \qquad \text{and} \qquad a_{n+1} = \frac{2^{n+1}}{7^{n+1}(n+2)}$$

Therefore,

$$\lim_{n \to +\infty} \left| \frac{a_{n+1}}{a_n} \right| = \lim_{n \to +\infty} \left(\frac{2^{n+1}}{7^{n+1}(n+2)} \cdot \frac{7^n(n+1)}{2^n} \right) = \lim_{n \to +\infty} \frac{2(n+1)}{7(n+2)}$$

$$= \lim_{n \to +\infty} \frac{2[1 + (1/n)]}{7[1 + (2/n)]} = \frac{2}{7} < 1$$

so the series converges absolutely by the ratio test.   ■

EXAMPLE 9 $\displaystyle\sum_{k=0}^{\infty} \frac{(-1)^{k+1}5^k}{k!}$

SOLUTION   Here

$$a_n = \frac{(-1)^{n+1}5^n}{n!} \qquad \text{and} \qquad a_{n+1} = \frac{(-1)^{n+2}5^{n+1}}{(n+1)!}$$

Therefore,

$$\lim_{n\to+\infty} \left|\frac{a_{n+1}}{a_n}\right| = \lim_{n\to+\infty} \left(\frac{5^{n+1}}{(n+1)!} \cdot \frac{n!}{5^n}\right) = \lim_{n\to+\infty} \frac{5}{n+1} = 0 < 1$$

so the series converges absolutely.   ∎

EXAMPLE 10 $\displaystyle\sum_{k=1}^{\infty} \frac{1\cdot 3\cdot 5\cdots (2k-1)}{k!}$

SOLUTION   Here

$$a_n = \frac{1\cdot 3\cdot 5\cdots (2n-1)}{n!} \qquad \text{and} \qquad a_{n+1} = \frac{1\cdot 3\cdot 5\cdots (2n-1)(2n+1)}{(n+1)!}$$

Therefore,

$$\lim_{n\to+\infty} \left|\frac{a_{n+1}}{a_n}\right| = \lim_{n\to+\infty} \frac{(2n+1)n!}{(n+1)!} = \lim_{n\to+\infty} \frac{2n+1}{n+1} = 2 > 1$$

so the given series diverges.   ∎

Another useful test for absolute convergence is given by the following theorem.

THEOREM 5   **The Root Test**

Let $\displaystyle\sum_{k=1}^{\infty} a_k$ be a given series.

(i) If $\displaystyle\lim_{n\to+\infty} \sqrt[n]{|a_n|} < 1$, then the series converges absolutely.

(ii) If $\displaystyle\lim_{n\to+\infty} \sqrt[n]{|a_n|} > 1$ or if $\displaystyle\lim_{n\to+\infty} \sqrt[n]{|a_n|} = +\infty$, then the series diverges.

(iii) If $\displaystyle\lim_{n\to+\infty} \sqrt[n]{|a_n|} = 1$, then the test is inconclusive.

PROOF   The proof is very similar to the proof of the ratio test, so we only sketch it here and leave the details as an exercise (Problem 66). If $\displaystyle\lim_{n\to+\infty} \sqrt[n]{|a_n|} = L < 1$, choose r with $L < r < 1$. Then for large enough values of n, $\sqrt[n]{|a_n|} < r$, or $|a_n| < r^n$; hence, $\sum_{k=1}^{\infty} |a_k|$ converges by comparison with $\sum_{k=1}^{\infty} r^k$. If the hypothesis of (ii) holds, then $\sqrt[n]{|a_n|} > 1$ holds for large enough values of n so that $\displaystyle\lim_{n\to+\infty} a_n = 0$ cannot hold. For (iii), the same examples used in the proof of the ratio test are still effective.   ∎

EXAMPLE 11   Use the root test to decide whether the series

$$\sum_{k=1}^{\infty} \frac{(-1)^k}{[\ln (k + 1)]^k}$$

converges or diverges.

SOLUTION   Here

$$a_n = \frac{(-1)^n}{[\ln (n + 1)]^n}$$

Therefore,

$$\lim_{n \to +\infty} \sqrt[n]{|a_n|} = \lim_{n \to +\infty} \sqrt[n]{\left| \frac{(-1)^n}{[\ln (n + 1)]^n} \right|} = \lim_{n \to +\infty} \frac{1}{\ln (n + 1)} = 0 < 1$$

so the given series converges absolutely by the root test and thus converges. ■

Problem Set 11.5

In Problems 1 to 14, determine whether the given alternating series converges or diverges. Use Leibniz's alternating-series test to establish convergence whenever it applies.

1 $\sum_{k=1}^{\infty} \frac{(-1)^{k+1}}{k^2}$

2 $\sum_{k=1}^{\infty} \frac{(-1)^{k+1}}{(2k)!}$

3 $\sum_{k=1}^{\infty} \frac{(-1)^{k+1}k}{k^3 + 2}$

4 $\sum_{k=1}^{\infty} \frac{-\cos k\pi}{k^3}$

5 $\sum_{k=1}^{\infty} \frac{(-1)^k k}{\sqrt{k^5 + 7}}$ $\left[Hint: \text{ First consider } \sum_{k=1}^{\infty} \frac{(-1)^{k+1} \; k}{\sqrt{k^5 + 7}}. \right]$

6 $\sum_{k=1}^{\infty} \frac{(-1)^{k+1}}{k^2 - 10k + 26}$ $\left[Hint: \text{First consider } \sum_{k=5}^{\infty} \frac{(-1)^{k+1}}{k^2 - 10k + 26}. \right]$

7 $\sum_{k=1}^{\infty} (-1)^{k+1} \frac{k + 1}{k + 7}$

8 $\sum_{k=1}^{\infty} (-1)^k \frac{3k^2}{4k^2 + 1}$

9 $\sum_{k=0}^{\infty} \frac{(-1)^k}{\ln (k + 2)}$

10 $\sum_{k=1}^{\infty} (-1)^{k+1} \frac{\ln (k + 1)}{k\sqrt{k}}$

11 $\sum_{k=1}^{\infty} (-1)^{k+1} \sin \frac{\pi}{k}$

12 $\sum_{k=2}^{\infty} (-1)^{k+1} \frac{k}{\ln k}$

13 $\sum_{k=1}^{\infty} \frac{(-1)^k \sqrt{k}}{k + 3}$

14 $\sum_{k=1}^{\infty} \ln k \cos k\pi$

In Problems 15 to 26, apply the ratio test to determine whether each series converges absolutely or diverges.

15 $\sum_{k=1}^{\infty} k^2 (\frac{3}{7})^k$

16 $\sum_{k=1}^{\infty} \frac{(-1)^{k+1}3^k}{k!}$

17 $\sum_{k=1}^{\infty} \frac{(-1)^{k+1}5^k}{k \cdot 4^k}$

18 $\sum_{k=1}^{\infty} \frac{(-1)^{k+1}(k^3 + 1)}{k!}$

19 $\sum_{k=1}^{\infty} \frac{(-1)^{k+1}k}{e^k}$

20 $\sum_{k=1}^{\infty} \frac{1}{(5k + 2)3^k}$

21 $\sum_{k=1}^{\infty} \frac{k!}{(2k)!}$

22 $\sum_{k=1}^{\infty} \frac{(k + 1)!}{7^k}$

23 $\sum_{k=1}^{\infty} (-1)^{k+1} \frac{7^k}{(3k)!}$

24 $\sum_{k=1}^{\infty} \frac{(-1)^k(2k - 1)!}{e^k}$

25 $\sum_{k=1}^{\infty} \frac{(-1)^{k+1}k^4}{(1.02)^k}$

26 $\sum_{k=1}^{\infty} (-1)^k \frac{1 + e^k}{2^k}$

In Problems 27 to 34, apply the root test to determine whether each series converges absolutely or diverges.

27 $\sum_{k=1}^{\infty} \frac{8^k}{k^k}$

28 $\sum_{k=1}^{\infty} \left(\frac{1}{2} + \frac{1}{k} \right)^k$

29 $\sum_{k=1}^{\infty} \left(\frac{7k}{5k + 1} \right)^k$

30 $\sum_{k=1}^{\infty} k^k (\frac{2}{3})^k$

31 $\displaystyle\sum_{k=1}^{\infty} (-1)^{k+1}\left(\frac{k}{3k+1}\right)^k$

32 $\displaystyle\sum_{k=2}^{\infty} \frac{(-1)^k k^k}{(\ln k)^k}$

33 $\displaystyle\sum_{k=1}^{\infty} (\sqrt[k]{k} - 1)^k$

34 $\displaystyle\sum_{k=1}^{\infty} \frac{k^k}{[2k + (1/k)]^k}$

In Problems 35 to 54, determine whether each series is divergent, conditionally convergent, or absolutely convergent. Use whatever tests or theorems seem most appropriate to justify your answer.

35 $\displaystyle\sum_{k=1}^{\infty} (-1)^k \frac{3^k}{k!}$

36 $\displaystyle\sum_{k=1}^{\infty} k(\tfrac{3}{5})^k$

37 $\displaystyle\sum_{k=1}^{\infty} \frac{(-1)^{k+1}}{\ln (k+1)}$

38 $\displaystyle\sum_{k=1}^{\infty} \frac{(-1)^{k+1} k^2}{k^3 + 10}$

39 $\displaystyle\sum_{n=1}^{\infty} \frac{(-1)^{n+1} \ln n}{n}$

40 $\displaystyle\sum_{k=1}^{\infty} (-1)^k \frac{k!}{(2k+1)!}$

41 $\displaystyle\sum_{j=1}^{\infty} \frac{(-1)^j}{j^2 + 1}$

42 $\displaystyle\sum_{k=1}^{\infty} \frac{2 \cdot 4 \cdot 6 \cdots (2k)}{1 \cdot 4 \cdot 7 \cdots (3k-2)}$

43 $\displaystyle\sum_{k=1}^{\infty} \frac{(-1)^k k!}{1 \cdot 3 \cdot 5 \cdots (2k-1)}$

44 $\displaystyle\sum_{k=1}^{\infty} \left(\frac{1}{3} + \frac{1}{2k}\right)^k$

45 $\displaystyle\sum_{k=1}^{\infty} \frac{k!}{k^2(k+1)^2}$

46 $\displaystyle\sum_{k=1}^{\infty} \frac{(2k)!}{4^k 3^{k-1}}$

47 $\displaystyle\sum_{k=1}^{\infty} \frac{(k+2)!}{3^k k!}$

48 $\displaystyle\sum_{k=1}^{\infty} \frac{k^2 k!}{(2k)!}$

49 $\displaystyle\sum_{k=1}^{\infty} (-1)^{k+1} \left(\frac{k+2}{3k+1}\right)^k$

50 $\displaystyle\sum_{k=0}^{\infty} (-1)^k \frac{(4k)!}{(k!)^2}$

51 $\displaystyle\sum_{k=1}^{\infty} \frac{k^k}{k!}$

52 $\displaystyle\sum_{k=1}^{\infty} \frac{(k!)^2}{(2k)!}$

53 $\displaystyle\sum_{k=1}^{\infty} (-1)^k \frac{e^k}{k}$

54 $\displaystyle\sum_{k=1}^{\infty} \left(\frac{k^k}{k!}\right)^k$

In Problems 55 to 60, approximate the sum of each series by finding the partial sum of its first n terms for the indicated value of n. Also give a bound on the absolute value of the error involved in this

approximation, and state whether the approximation overestimates or underestimates the true value.

55 $\displaystyle\sum_{k=1}^{\infty} \frac{(-1)^{k+1}}{3k-1}, \; n = 5$

56 $\displaystyle\sum_{k=1}^{\infty} \frac{(-1)^{k+1}}{2^k}, \; n = 100$

57 $\displaystyle\sum_{k=1}^{\infty} \frac{(-1)^{k+1}}{k^2}, \; n = 4$

58 $\displaystyle\sum_{k=1}^{\infty} \frac{(-1)^k}{k^3 + 1}, \; n = 4$

59 $\displaystyle\sum_{k=1}^{\infty} \frac{(-1)^k}{k \cdot 5^k}, \; n = 3$

60 $\displaystyle\sum_{k=1}^{\infty} \frac{\sin (k + \frac{1}{2})\pi}{2k!}, \; n = 3$

[C] In Problems 61 and 62, find the sum of each series with an error not exceeding 5×10^{-4} in absolute value, and write your answer to three decimal places.

61 $\displaystyle\sum_{k=1}^{\infty} \frac{(-1)^{k+1}}{k \cdot 2^k}$

62 $\displaystyle\sum_{k=1}^{\infty} \frac{(-1)^k k}{(2k)!}$

63 Suppose that the sequence $s_2, s_4, s_6, s_8, \ldots$ converges to the limit S and that the sequence $s_1, s_3, s_5, s_7, \ldots$ converges to the same limit S. Prove that the sequence $s_1, s_2, s_3, s_4, s_5, s_6, \ldots$ converges to the limit S.

64 (a) If $|a_{n+1}| < |a_n|r$ holds for all integers $n \geq N$, where r is a positive constant, prove that $|a_{N+j}| < |a_N|r^j$ holds for all positive integers j. (Use mathematical induction.) (b) If $|a_n|r < |a_{n+1}|$ holds for all integers $n \geq N$, where r is a positive constant, prove that $|a_N|r^j < |a_{N+j}|$ holds for all positive integers j.

65 Is it true that if the series $\displaystyle\sum_{k=1}^{\infty} a_k$ converges absolutely, then the series $\displaystyle\sum_{k=1}^{\infty} \frac{a_k^2}{1 + a_k^2}$ also converges? Why?

66 Fill in the details in the proof of the root test (Theorem 5).

67 If the series $\displaystyle\sum_{k=1}^{\infty} a_k$ converges absolutely, show that

$$\left|\sum_{k=1}^{\infty} a_k\right| \leq \sum_{k=1}^{\infty} |a_k|$$

11.6   Power Series

An infinite series of the form

$$\sum_{k=0}^{\infty} c_k(x - a)^k = c_0 + c_1(x - a) + c_2(x - a)^2 + c_3(x - a)^3 + \cdots$$

is called a **power series in $x - a$,** or simply a **power series.** The constants c_0, c_1, c_2, c_3, . . . are called the **coefficients** of the power series, and the constant a is called its **center.** A power series with center $a = 0$ has the form

$$\sum_{k=0}^{\infty} c_k x^k = c_0 + c_1 x + c_2 x^2 + c_3 x^3 + \cdots$$

and thus generalizes the idea of a polynomial in x.

In a power series $\sum_{k=0}^{\infty} c_k(x - a)^k$ we usually think of x as a quantity that can be varied at will. The series may converge for some values of x, but not for others. Naturally, *when $x = a$, we understand that the series converges and that its sum is c_0.* The following three examples show that the ratio test (Theorem 4 in Section 11.5) can be useful in determining the values of x for which a power series converges.

In Examples 1 to 3, find the values of x for which the given power series converges.

EXAMPLE 1 $\displaystyle\sum_{k=0}^{\infty} (-1)^k \frac{k}{3^k} x^k = 0 - \frac{1}{3}x + \frac{2}{9}x^2 - \frac{3}{27}x^3 + \cdots$

SOLUTION   Of course, the series converges for $x = 0$. For $x \neq 0$, we use the ratio test with

$$a_n = \frac{(-1)^n n x^n}{3^n} \qquad \text{and} \qquad a_{n+1} = \frac{(-1)^{n+1}(n + 1)x^{n+1}}{3^{n+1}}$$

Here,

$$\lim_{n \to +\infty} \left| \frac{a_{n+1}}{a_n} \right| = \lim_{n \to +\infty} \left| \frac{(-1)^{n+1}(n + 1)x^{n+1}}{3^{n+1}} \cdot \frac{3^n}{(-1)^n n x^n} \right|$$

$$= \lim_{n \to +\infty} \frac{n + 1}{3n} |x| = |x| \lim_{n \to +\infty} \frac{n + 1}{3n} = \frac{|x|}{3}$$

so the series converges for $|x|/3 < 1$, that is, for $-3 < x < 3$. If $x < -3$ or if $x > 3$, then $|x|/3 > 1$ and the series diverges. When $|x| = 3$, we have

$$|a_n| = \left| (-1)^n \frac{n}{3^n} x^n \right| = \frac{n}{3^n} |x|^n = \frac{n}{3^n} 3^n = n$$

so that $\displaystyle\lim_{n \to +\infty} a_n \neq 0$ and the series must diverge. Therefore, the series converges for values of x in the open interval $(-3, 3)$ and *only* for such values of x.   ∎

EXAMPLE 2 $\displaystyle\sum_{k=0}^{\infty} \frac{(x-5)^{2k}}{k!} = 1 + (x-5)^2 + \frac{1}{2}(x-5)^4 + \frac{1}{6}(x-5)^6 + \cdots$

SOLUTION   The series converges for $x = 5$. For $x \neq 5$, we use the ratio test with

$$a_n = \frac{(x-5)^{2n}}{n!} \quad \text{and} \quad a_{n+1} = \frac{(x-5)^{2(n+1)}}{(n+1)!} = \frac{(x-5)^{2n+2}}{(n+1)!}$$

Here,

$$\lim_{n \to +\infty} \left| \frac{a_{n+1}}{a_n} \right| = \lim_{n \to +\infty} \left| \frac{(x-5)^{2n+2}}{(n+1)!} \cdot \frac{n!}{(x-5)^{2n}} \right|$$

$$= \lim_{n \to +\infty} \frac{(x-5)^2}{n+1} = 0 < 1$$

for *all* values of x; hence, the series converges for *all* values of x.   ■

EXAMPLE 3 $\displaystyle\sum_{k=0}^{\infty} (k!)(x+2)^k = 1 + (x+2) + 2(x+2)^2 + 6(x+2)^3 + \cdots$

SOLUTION   Here we have $a_n = (n!)(x+2)^n$ and $a_{n+1} = [(n+1)!](x+2)^{n+1}$. For $x \neq -2$, the ratio test gives

$$\lim_{n \to +\infty} \left| \frac{a_{n+1}}{a_n} \right| = \lim_{n \to +\infty} \left| \frac{[(n+1)!](x+2)^{n+1}}{(n!)(x+2)^n} \right| = \lim_{n \to +\infty} (n+1)|x+2| = +\infty$$

Hence, the series diverges. Of course, the series converges for $x = -2$, and so it converges *only* when $x = -2$.   ■

The set I of all numbers x for which a power series $\sum_{k=0}^{\infty} c_k(x-a)^k$ converges is called its **interval of convergence**. In Example 1, $I = (-3, 3)$, in Example 2, $I = (-\infty, \infty)$, and in Example 3, I is the "interval" containing the single number -2.

For any power series $\sum_{k=1}^{\infty} c_k(x-a)^k$, the interval of convergence I always has one of the following forms:

Case 1 *I is a bounded interval with center a and with endpoints $a - R$ and $a + R$, where R is a positive real number.*

Case 2 *$I = (-\infty, \infty)$.*

Case 3 *I consists of the single number a.*

In case 1, we call the number R the **radius of convergence** of the power series. In case 2, it is convenient to say that the radius of convergence of the power series is *infinite* and to write $R = +\infty$. Naturally, in case 3, we say that the power series has *radius of convergence zero*, and we write $R = 0$. Examples 1 through 3 above illustrate these three possibilities.

In case 1, *the endpoints $a - R$ and $a + R$ of the interval of convergence I may or may not belong to I*. In Example 1, neither endpoint belongs to I, so that I is an open interval. In general, anything can happen—the series may diverge, converge conditionally, or converge absolutely at an endpoint of I. Thus, in

Figure 1

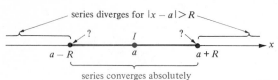

series diverges for $|x - a| > R$

series converges absolutely
for $|x - a| < R$

case 1, the interval of convergence can be any one of the four sets $I = [a - R, \ a + R]$, $I = [a - R, \ a + R)$, $I = (a - R, a + R]$, or $I = (a - R, a + R)$ (Figure 1). *The power series always converges absolutely on the open interval* $(a - R, a + R)$.

The following theorem provides an efficient means for finding the radius of convergence of a power series.

THEOREM 1

Radius of Convergence of a Power Series

Let $\sum_{k=0}^{\infty} c_k(x - a)^k$ be a power series with radius of convergence R. Suppose that $\lim_{n \to +\infty} \left| \dfrac{c_{n+1}}{c_n} \right| = L$, where L is either a nonnegative real number or $L = +\infty$.

(i) If L is a positive real number, then $R = 1/L$.

(ii) If $L = 0$, then $R = +\infty$.

(iii) If $L = +\infty$, then $R = 0$.

PROOF

We attend to part (i) here. Parts (ii) and (iii), which are handled similarly, are left as an exercise (Problem 46). Thus, suppose that

$$\lim_{n \to +\infty} \left| \frac{c_{n+1}}{c_n} \right| = L$$

where L is a positive real number. We apply the ratio test to the infinite series $\sum_{k=0}^{\infty} c_k(x - a)^k$. Here, $a_n = c_n(x - a)^n$ and $a_{n+1} = c_{n+1}(x - a)^{n+1}$, so that

$$\lim_{n \to +\infty} \left| \frac{a_{n+1}}{a_n} \right| = \lim_{n \to +\infty} \left| \frac{c_{n+1}(x - a)^{n+1}}{c_n(x - a)^n} \right| = \lim_{n \to +\infty} \left| \frac{c_{n+1}}{c_n} \right| \cdot |x - a| = L|x - a|$$

By the ratio test, the series converges absolutely for $L|x - a| < 1$, and it diverges for $L|x - a| > 1$; that is, it converges for $|x - a| < 1/L$, and it diverges for $|x - a| > 1/L$. Thus, $1/L = R$, the radius of convergence of the power series. ∎

Several important comments need to be made in connection with Theorem 1:

1 Notice that the ratio c_{n+1}/c_n in Theorem 1 is a ratio of *coefficients*—not terms—of the power series. *Do not confuse Theorem 1 with the original ratio test* (Theorem 4 in Section 11.5), which involves the terms of a series.

2 Theorem 1 is easy to remember if, *for purposes of this theorem only*, we agree that $1/L = +\infty$ when $L = 0$ and that $1/L = 0$ when $L = +\infty$. Then the theorem simply says that $R = 1/L$ in every case.

3 Theorem 1 may not apply in certain cases because the sequence $|c_{n+1}/c_n|$ has no real number as a limit and does not approach $+\infty$ as $n \to +\infty$. In such a case, the power series still has a definite radius of convergence R, but methods beyond the scope of this book may be required to find it.

4 Theorem 1 does not apply to a power series of the form

$$\sum_{k=0}^{\infty} c_k(x - a)^{kp} = c_0 + c_1(x - a)^p + c_2(x - a)^{2p} + c_3(x - a)^{3p} + \cdots$$

where p is a constant integer greater than 1, since c_k is not the coefficient of the

kth power of $x - a$. In such a case, the radius of convergence can often be found by applying the original ratio test (Theorem 4 in Section 11.5) directly to the *terms* of the series, as in Example 2 on page 681.

5 Theorem 1 says nothing, one way or the other, about whether the power series converges at the *endpoints* of its interval of convergence. This has to be checked by substituting $x = a - R$ and $x = a + R$ in the power series and using the standard tests for convergence of a series.

6 Theorem 1 is still valid for a power series such as $\sum_{k=M}^{\infty} c_k(x - a)^k$, where M is a positive integer and the summation starts at $k = M$ rather than at $k = 0$. (Why?)

In Examples 4 to 8, find the center a, the radius of convergence R, and the interval of convergence I of the given power series. Be sure to check the endpoints of I for divergence, absolute convergence, or conditional convergence.

EXAMPLE 4 $\displaystyle\sum_{k=1}^{\infty} \frac{1}{k} x^k$

SOLUTION   Here the center is $a = 0$, and $c_k = 1/k$. In Theorem 1, we put $c_n = 1/n$ and $c_{n+1} = 1/(n + 1)$, so that

$$\lim_{n \to +\infty} \left| \frac{c_{n+1}}{c_n} \right| = \lim_{n \to +\infty} \left| \frac{1/(n + 1)}{1/n} \right| = \lim_{n \to +\infty} \frac{n}{n + 1} = 1 = L$$

Hence, $R = 1/L = 1$. Therefore, the series converges absolutely for values of x in the open interval $(a - R, a + R) = (0 - 1, 0 + 1) = (-1, 1)$, and it diverges for values of x outside the closed interval $[-1, 1]$. Substituting $x = 1$ in the series, we obtain the harmonic series

$$\sum_{k=1}^{\infty} \frac{1}{k}$$

which diverges. For $x = -1$, the series becomes the alternating harmonic series

$$\sum_{k=1}^{\infty} \frac{1}{k}(-1)^k$$

which converges by Leibniz's test. Hence, we have divergence at the endpoint 1 and conditional convergence at the endpoint -1. The interval of convergence is $I = [-1, 1)$. ∎

EXAMPLE 5 $\displaystyle\sum_{k=0}^{\infty} \frac{(x + 3)^k}{3^k}$

SOLUTION   The power series is centered at $a = -3$. We have $c_n = 1/3^n$ and $c_{n+1} = 1/3^{n+1}$, so that

$$\lim_{n \to +\infty} \left| \frac{c_{n+1}}{c_n} \right| = \lim_{n \to +\infty} \left| \frac{1/3^{n+1}}{1/3^n} \right| = \lim_{n \to +\infty} \frac{1}{3} = \frac{1}{3} = L$$

Hence, $R = 1/L = 3$ by Theorem 1. Therefore, the series converges absolutely on the open interval $(a - R, a + R) = (-3 - 3, -3 + 3) = (-6, 0)$. When $x = -6$, the series becomes

$$\sum_{k=0}^{\infty} \frac{(-3)^k}{3^k} = 1 - 1 + 1 - 1 + \cdots$$

which diverges because the general term does not approach zero. When $x = 0$, the series becomes

$$\sum_{k=0}^{\infty} \frac{3^k}{3^k} = 1 + 1 + 1 + \cdots$$

which also diverges. Therefore, $I = (-6, 0)$.   ∎

EXAMPLE 6 $\displaystyle\sum_{k=0}^{\infty} \frac{(-1)^k}{k!}(x - 17)^k$

SOLUTION   The power series is centered at $a = 17$. We have

$$c_n = \frac{(-1)^n}{n!} \qquad \text{and} \qquad c_{n+1} = \frac{(-1)^{n+1}}{(n + 1)!}$$

so that

$$\lim_{n \to +\infty} \left| \frac{c_{n+1}}{c_n} \right| = \lim_{n \to +\infty} \left| \frac{(-1)^{n+1}}{(n + 1)!} \cdot \frac{n!}{(-1)^n} \right| = \lim_{n \to +\infty} \frac{1}{n + 1} = 0 = L$$

Hence, $R = +\infty$ by part (ii) of Theorem 1. Therefore, $I = (-\infty, \infty)$.   ∎

EXAMPLE 7 $\displaystyle\sum_{k=1}^{\infty} k^k x^k$

SOLUTION   The power series is centered at $a = 0$. We have $c_n = n^n$ and $c_{n+1} = (n + 1)^{n+1}$, so that

$$\lim_{n \to +\infty} \left| \frac{c_{n+1}}{c_n} \right| = \lim_{n \to +\infty} \left| \frac{(n + 1)^{n+1}}{n^n} \right| = \lim_{n \to +\infty} \left[\left(\frac{n + 1}{n} \right)^n (n + 1) \right]$$

$$= \left[\lim_{n \to +\infty} \left(1 + \frac{1}{n} \right)^n \right] \left[\lim_{n \to +\infty} (n + 1) \right] = e \left[\lim_{n \to +\infty} (n + 1) \right] = +\infty$$

Thus, $R = 0$ by part (iii) of Theorem 1, and so I consists of the single number 0.   ∎

EXAMPLE 8 $\displaystyle\sum_{k=1}^{\infty} \frac{3^k (x - 4)^{2k}}{k^2}$

SOLUTION   The power series is centered at $a = 4$. Here we cannot use Theorem 1 since $3^k/k^2$ is not the coefficient of the kth power of $x - 4$. Thus, we resort to the original ratio test (Theorem 4 in Section 11.5). The nth *term* (not coefficient!) of the series is

$$a_n = \frac{3^n (x - 4)^{2n}}{n^2} \qquad \text{so} \qquad a_{n+1} = \frac{3^{n+1}(x - 4)^{2(n+1)}}{(n + 1)^2} = \frac{3^{n+1}(x - 4)^{2n+2}}{(n + 1)^2}$$

Therefore,

$$\lim_{n \to +\infty} \left| \frac{a_{n+1}}{a_n} \right| = \lim_{n \to +\infty} \left| \frac{3^{n+1}(x - 4)^{2n+2}}{(n + 1)^2} \cdot \frac{n^2}{3^n(x - 4)^{2n}} \right|$$

$$= \lim_{n \to +\infty} 3 \left(\frac{n}{n + 1} \right)^2 |x - 4|^2 = 3|x - 4|^2$$

It follows that the series converges absolutely when $3|x - 4|^2 < 1$, that is, when $|x - 4| < 1/\sqrt{3}$. It diverges when $3|x - 4|^2 > 1$, that is, when $|x - 4| > 1/\sqrt{3}$. Therefore, $R = 1/\sqrt{3}$. When

$$x = 4 - \frac{1}{\sqrt{3}} \qquad \text{the series becomes} \quad \sum_{k=1}^{\infty} \frac{1}{k^2}$$

which converges absolutely. (Why?) Similarly, when

$$x = 4 + \frac{1}{\sqrt{3}} \qquad \text{the series becomes} \quad \sum_{k=1}^{\infty} \frac{1}{k^2}$$

which converges absolutely. It follows that the series converges absolutely on its entire interval of convergence

$$I = \left[4 - \frac{1}{\sqrt{3}}, 4 + \frac{1}{\sqrt{3}} \right] \qquad \blacksquare$$

Problem Set 11.6

In Problems 1 to 40, find the center a, the radius of convergence R, and the interval of convergence I of the given power series. Be sure to check the endpoints of I for divergence, absolute convergence, or conditional convergence.

1 $\displaystyle\sum_{k=0}^{\infty} 7^k x^k$

2 $\displaystyle\sum_{k=0}^{\infty} \frac{x^{k+1}}{\sqrt{k+1}}$

3 $\displaystyle\sum_{k=0}^{\infty} \frac{x^k}{k!}$

4 $\displaystyle\sum_{k=1}^{\infty} \frac{x^k}{3^k \sqrt{k}}$

5 $\displaystyle\sum_{k=0}^{\infty} k! x^k$

6 $\displaystyle\sum_{k=0}^{\infty} \frac{k^3 x^k}{5^k}$

7 $\displaystyle\sum_{k=0}^{\infty} \frac{3^k x^{2k}}{k+2}$

8 $\displaystyle\sum_{k=0}^{\infty} \frac{x^k}{(k+2)3^k}$

9 $\displaystyle\sum_{k=1}^{\infty} \frac{x^k}{3+k^2}$

10 $\displaystyle\sum_{k=0}^{\infty} \frac{(-1)^k x^{2k}}{(2k)!}$

11 $\displaystyle\sum_{k=1}^{\infty} \frac{(-1)^{k+1} x^{2k-1}}{(2k-1)!}$

12 $\displaystyle\sum_{k=1}^{\infty} \frac{(-1)^k x^{5k}}{k^{5/2}}$

13 $\displaystyle\sum_{k=0}^{\infty} \frac{(x-2)^k}{3^k}$

14 $\displaystyle\sum_{k=0}^{\infty} \frac{(-1)^k (x-1)^k}{k+1}$

15 $\displaystyle\sum_{k=1}^{\infty} \frac{(-1)^{k+1} k (x+3)^{k-1}}{7^{k-1}}$

16 $\displaystyle\sum_{k=1}^{\infty} \frac{(x+2)^{k-1}}{k^2}$

17 $\displaystyle\sum_{k=1}^{\infty} \frac{2^k (x+1)^k}{\ln(k+1)}$

18 $\displaystyle\sum_{k=1}^{\infty} \frac{(2x-3)^k}{4^{2k}}$

19 $\displaystyle\sum_{k=1}^{\infty} \frac{(x+1)^k}{\sqrt{k}}$

20 $\displaystyle\sum_{k=1}^{\infty} \frac{(x+1)^k}{k\sqrt{k+1}}$

21 $\displaystyle\sum_{k=1}^{\infty} \frac{(x+5)^k}{(2k-1)(2k)}$

22 $\displaystyle\sum_{k=2}^{\infty} \frac{(x-1)^{2k-2}}{(2k-4)!}$

23 $\displaystyle\sum_{j=0}^{\infty} \frac{(-1)^j 2^j (x+2)^j}{(j+1)^3}$

24 $\displaystyle\sum_{j=0}^{\infty} \frac{\sqrt{j} (x+1)^j}{1 \cdot 3 \cdot 5 \cdots (2j+1)}$

25 $\displaystyle\sum_{n=0}^{\infty} \frac{(x-1)^n}{(n+2)!}$

26 $\displaystyle\sum_{k=0}^{\infty} \frac{(-1)^k}{2k+1} \left(\frac{x}{2}\right)^{2k}$

27 $\displaystyle\sum_{k=0}^{\infty} \frac{(x+1)^{5k}}{(k+1) \cdot 5^k}$

28 $\displaystyle\sum_{k=0}^{\infty} \frac{(1-x)^k}{(k+1) \cdot 3^k}$

29 $\displaystyle\sum_{k=1}^{\infty} \frac{1}{k} \left(\frac{x}{4} - 1\right)^k$

30 $\displaystyle\sum_{n=0}^{\infty} \frac{1}{3n-1} \left(\frac{x}{3} + \frac{2}{3}\right)^n$

31 $\displaystyle\sum_{k=1}^{\infty} \frac{(-1)^k 2^k (x-5)^{2k}}{k^3}$

32 $\displaystyle\sum_{j=1}^{\infty} \frac{(3-x)^{j-1}}{\sqrt{j}}$

33 $\displaystyle\sum_{k=1}^{\infty} (\tan^{-1} k)(x-1)^k$

34 $\displaystyle\sum_{k=1}^{\infty} \frac{(x-3)^{4k}}{\sqrt[k]{k}}$

35 $\displaystyle\sum_{k=1}^{\infty} k^k (x-2)^k$

36 $\displaystyle\sum_{k=1}^{\infty} (5^k + 5^{-k})(x+1)^{3k-2}$

37 $\displaystyle\sum_{k=1}^{\infty} \frac{(-1)^k (x-3)^{2k}}{9^k}$

38 $\displaystyle\sum_{k=1}^{\infty} \frac{k!}{k^k} x^k$

39 $\displaystyle\sum_{k=1}^{\infty} (-1)^k \frac{1 \cdot 3 \cdot 5 \cdots (2k-1)}{2 \cdot 4 \cdot 6 \cdots (2k)} x^{2k+1}$

40 $(x-8) + (x-8)^2 + 2!(x-8)^3 + 3!(x-8)^4 + \cdots$

41 If R is the radius of convergence of $\sum_{k=0}^{\infty} c_k(x-a)^k$ and p is a positive integer, find the radius of convergence of $\sum_{k=0}^{\infty} c_k(x-a)^{p+k}$.

42 If R is the radius of convergence of $\sum_{k=0}^{\infty} c_k(x-a)^k$, $0 < R < +\infty$, and p is a positive integer, show that the radius of convergence of $\sum_{k=0}^{\infty} c_k(x-a)^{pk}$ is $\sqrt[p]{R}$.

43 Let c be a constant real number.

(a) Show that the series

$$\sum_{n=1}^{\infty} \frac{c(c-1)(c-2)\cdots(c-n)}{n!} x^n$$

converges for $|x| < 1$.

(b) Use the result of part (a) to show that if $|x| < 1$, then

$$\lim_{n \to +\infty} \frac{c(c-1)(c-2)\cdots(c-n)}{n!} x^n = 0$$

44 Let $\sum_{k=0}^{\infty} c_k(x-a)^k$ be a given power series.

(a) If $\lim_{n \to +\infty} \sqrt[n]{|c_n|} = +\infty$, prove that the radius of convergence of the power series is zero.

(b) If $\lim_{n \to +\infty} \sqrt[n]{|c_n|} = 0$, prove that the power series has an infinite radius of convergence.

(c) If $\lim_{n \to +\infty} \sqrt[n]{|c_n|} = L \neq 0$, prove that the radius of convergence of the power series is given by $R = 1/L$.

45 Suppose that b is a constant greater than 1. Find the radius of convergence and the interval of convergence of the power series

$$\sum_{k=0}^{\infty} \frac{x^k}{1 + b^k}$$

46 Complete the proof of Theorem 1 by considering parts (ii) and (iii).

47 If $a > b \geq 0$, find the radius of convergence of the power series

$$\sum_{k=0}^{\infty} \frac{x^k}{a^k + b^k}$$

11.7   Differentiation and Integration of Power Series

In this section we study functions of the form

$$f(x) = \sum_{k=0}^{\infty} c_k(x-a)^k$$

where $\sum_{k=0}^{\infty} c_k(x-a)^k$ is a given power series. Here it is understood that *the domain of f is the interval of convergence of the power series.*

Since a *finite* sum can be differentiated term by term and since

$$f(x) = c_0 + c_1(x-a) + c_2(x-a)^2 + c_3(x-a)^3 + \cdots$$

you might guess that the derivative $D_x f(x)$ can be obtained by differentiating term by term; that is,

$$D_x f(x) = D_x c_0 + D_x c_1(x-a) + D_x c_2(x-a)^2 + D_x c_3(x-a)^3 + \cdots$$
$$= 0 + c_1 + 2c_2(x-a) + 3c_3(x-a)^2 + \cdots$$

Likewise, you might guess that the integral $\int f(x)\, dx$ can be obtained by integrating term by term; that is,

$$\int f(x)\, dx = \int c_0\, dx + \int c_1(x-a)\, dx + \int c_2(x-a)^2\, dx + \int c_3(x-a)^3\, dx + \cdots$$
$$= \left[c_0(x-a) + \frac{c_1}{2}(x-a)^2 + \frac{c_2}{3}(x-a)^3 + \frac{c_3}{4}(x-a)^4 + \cdots \right] + C$$

It turns out that such "calculations" are quite legitimate, provided that $|x-a| < R$, where R is the radius of convergence of the power series.

Notice that term-by-term differentiation or integration of a power series

$$\sum_{k=0}^{\infty} c_k(x - a)^k$$

produces a new power series,

$$\sum_{k=0}^{\infty} D_x[c_k(x - a)^k] = \sum_{k=1}^{\infty} kc_k(x - a)^{k-1}$$

or

$$\sum_{k=0}^{\infty} \int c_k(x - a)^k \, dx = \sum_{k=0}^{\infty} \frac{c_k}{k + 1}(x - a)^{k+1}$$

respectively. (In the second equation, we have suppressed the constants of integration.) Also, the following holds, although we omit the somewhat technical proof.

The three power series

$$\sum_{k=0}^{\infty} c_k(x - a)^k \qquad \sum_{k=0}^{\infty} D_x[c_k(x - a)^k] \qquad \text{and} \qquad \sum_{k=0}^{\infty} \int c_k(x - a)^k \, dx$$

all have the same radius of convergence.

Term-by-term differentiation or integration of a power series is justified by the following properties, whose proofs we also omit. Suppose that

$$f(x) = \sum_{k=0}^{\infty} c_k(x - a)^k$$

where the power series has *radius of convergence R*. Then

*Property I For $|x - a| < R$,

$$f'(x) = D_x\left[\sum_{k=0}^{\infty} c_k(x - a)^k\right] = \sum_{k=1}^{\infty} kc_k(x - a)^{k-1}$$

Property II For $|x - a| < R$,

$$\int f(x) \, dx = \int \left[\sum_{k=0}^{\infty} c_k(x - a)^k\right] dx = \sum_{k=0}^{\infty} \frac{c_k}{k + 1}(x - a)^{k+1} + C$$

Property III For $|b - a| < R$,

$$\int_a^b f(x) \, dx = \int_a^b \left[\sum_{k=0}^{\infty} c_k(x - a)^k\right] dx = \sum_{k=0}^{\infty} \int_a^b c_k(x - a)^k \, dx$$

$$= \sum_{k=0}^{\infty} \frac{c_k}{k + 1}(b - a)^{k+1}$$

*Because a differentiable function is continuous, it follows from Property I that f is continuous on the open interval $(a - R, a + R)$.

EXAMPLE 1 Find $D_x(1 + 2x + 3x^2 + 4x^3 + \cdots)$.

SOLUTION   By Theorem 1 in Section 11.6, the radius of convergence of the power series $\sum_{k=0}^{\infty} (k + 1)x^k$ is $R = 1$; hence, by Property I,

$$D_x(1 + 2x + 3x^2 + 4x^3 + \cdots) = 0 + 2 + 6x + 12x^2 + \cdots$$

that is,

$$D_x\left[\sum_{k=0}^{\infty} (k + 1)x^k\right] = \sum_{k=1}^{\infty} k(k + 1)x^{k-1} \qquad \text{for } |x| < 1 \qquad \blacksquare$$

If

$$f(x) = \sum_{k=0}^{\infty} c_k(x - a)^k$$

for x in the interval I, we say that the power series

$$\sum_{k=0}^{\infty} c_k(x - a)^k$$

represents the function f on the interval I. For instance, by Theorem 1 in Section 11.2, the geometric series $1 + x + x^2 + x^3 + \cdots$ converges for $|x| < 1$, and its sum is $1/(1 - x)$; that is, the function $f(x) = 1/(1 - x)$ is represented by the power series $1 + x + x^2 + x^3 + \cdots$ on the interval $(-1, 1)$. By starting with a known power series representation such as

$$\frac{1}{1 - x} = 1 + x + x^2 + x^3 + \cdots = \sum_{k=0}^{\infty} x^k \qquad \text{for } -1 < x < 1$$

and using Property I or II to differentiate or integrate the power series term by term, it is often possible to obtain interesting new power series representations. In this connection, *substitution* into a known power series representation is also useful. The following examples illustrate the techniques involved.

In Examples 2 to 7, starting with the power series representation given above for $1/(1 - x)$, verify each of the indicated power series representations by using differentiation, integration, or substitution.

EXAMPLE 2 $\dfrac{1}{(1 - x)^2} = 1 + 2x + 3x^2 + 4x^3 + \cdots = \sum_{k=1}^{\infty} kx^{k-1} \qquad \text{for } |x| < 1$

SOLUTION $\dfrac{1}{(1 - x)^2} = D_x\left(\dfrac{1}{1 - x}\right) = D_x\left(\sum_{k=0}^{\infty} x^k\right) = \sum_{k=1}^{\infty} kx^{k-1} \qquad \text{for } |x| < 1 \qquad \blacksquare$

EXAMPLE 3 $\dfrac{1}{1 + t} = 1 - t + t^2 - t^3 + \cdots = \sum_{k=0}^{\infty} (-1)^k t^k \qquad \text{for } |t| < 1$

SOLUTION   Substituting $-t$ for x in the equation

$$\frac{1}{1-x} = \sum_{k=0}^{\infty} x^k$$

we obtain

$$\frac{1}{1+t} = \sum_{k=0}^{\infty} (-t)^k = \sum_{k=0}^{\infty} (-1)^k t^k = 1 - t + t^2 - t^3 + \cdots \qquad \text{for } |-t| < 1$$

that is, for $|t| < 1$. ∎

EXAMPLE 4 $\ln (1 + x) = x - \dfrac{x^2}{2} + \dfrac{x^3}{3} - \dfrac{x^4}{4} + \cdots$

$$= \sum_{k=0}^{\infty} (-1)^k \frac{x^{k+1}}{k+1} \qquad \text{for } |x| < 1$$

SOLUTION   Using the result of Example 3 and Property III, we have

$$\ln (1 + x) = \int_0^x \frac{dt}{1+t} = \int_0^x (1 - t + t^2 - t^3 + \cdots) \, dt$$

$$= \int_0^x dt - \int_0^x t \, dt + \int_0^x t^2 \, dt - \int_0^x t^3 \, dt + \cdots$$

$$= x - \frac{x^2}{2} + \frac{x^3}{3} - \frac{x^4}{4} + \cdots$$

$$= \sum_{k=0}^{\infty} (-1)^k \frac{x^{k+1}}{k+1} \qquad \text{for } |x| < 1$$ ∎

EXAMPLE 5 $\dfrac{1}{1+x^2} = 1 - x^2 + x^4 - x^6 + \cdots = \displaystyle\sum_{k=0}^{\infty} (-1)^k x^{2k} \qquad \text{for } |x| < 1$

SOLUTION   Substituting $-x^2$ for x in the equation

$$\frac{1}{1-x} = \sum_{k=0}^{\infty} x^k$$

we obtain

$$\frac{1}{1+x^2} = \sum_{k=0}^{\infty} (-x^2)^k = \sum_{k=0}^{\infty} (-1)^k x^{2k} = 1 - x^2 + x^4 - x^6 + \cdots \qquad \text{for } |x| < 1$$

(Note that we get convergence of the series when $|-x^2| < 1$; however, $|-x^2| = |x^2| = |x|^2$, and $|x|^2 < 1$ holds exactly when $|x| < 1$.) ∎

EXAMPLE 6 $\dfrac{1}{3-x} = \dfrac{1}{3} + \dfrac{x}{9} + \dfrac{x^2}{27} + \dfrac{x^3}{81} + \cdots = \displaystyle\sum_{k=0}^{\infty} \frac{x^k}{3^{k+1}} \qquad \text{for } |x| < 3$

SOLUTION $\dfrac{1}{3-x} = \dfrac{1}{3} \cdot \dfrac{1}{1-(x/3)} = \dfrac{1}{3} \displaystyle\sum_{k=0}^{\infty} \left(\dfrac{x}{3}\right)^{k} = \displaystyle\sum_{k=0}^{\infty} \dfrac{1}{3} \cdot \dfrac{x^{k}}{3^{k}}$

$$= \sum_{k=0}^{\infty} \dfrac{x^{k}}{3^{k+1}} \qquad \text{for } \left|\dfrac{x}{3}\right| < 1$$

that is, for $|x| < 3$.   ■

EXAMPLE 7 $\dfrac{2}{x^{2}-4x+3} = \dfrac{2}{3} + \dfrac{8}{9}x + \dfrac{26}{27}x^{2} + \dfrac{80}{81}x^{3} + \cdots$

$$= \sum_{k=0}^{\infty} \left(\dfrac{3^{k+1}-1}{3^{k+1}}\right)x^{k} \qquad \text{for } |x| < 1$$

SOLUTION   By partial fractions,

$$\dfrac{2}{x^{2}-4x+3} = \dfrac{2}{(x-1)(x-3)} = \dfrac{-1}{x-1} + \dfrac{1}{x-3} = \dfrac{1}{1-x} - \dfrac{1}{3-x}$$

Therefore, using the result of Example 6, we have

$$\dfrac{2}{x^{2}-4x+3} = \dfrac{1}{1-x} - \dfrac{1}{3-x} = \sum_{k=0}^{\infty} x^{k} - \sum_{k=0}^{\infty} \dfrac{x^{k}}{3^{k+1}}$$

$$= \sum_{k=0}^{\infty} \left(x^{k} - \dfrac{x^{k}}{3^{k+1}}\right) = \sum_{k=0}^{\infty} \left(\dfrac{3^{k+1}-1}{3^{k+1}}\right)x^{k} \qquad \text{for } |x| < 1$$

(Note that if $|x| < 1$, then the condition $|x| < 3$ in Example 6 is automatically satisfied.)   ■

Using the result of Example 5 and Property III, we have

$$\tan^{-1} x = \int_{0}^{x} \dfrac{dt}{1+t^{2}} = \int_{0}^{x} (1 - t^{2} + t^{4} - t^{6} + \cdots) \, dt$$

$$= \int_{0}^{x} dt - \int_{0}^{x} t^{2} \, dt + \int_{0}^{x} t^{4} \, dt - \int_{0}^{x} t^{6} \, dt + \cdots$$

$$= x - \dfrac{x^{3}}{3} + \dfrac{x^{5}}{5} - \dfrac{x^{7}}{7} + \cdots$$

$$= \sum_{k=0}^{\infty} (-1)^{k}\dfrac{x^{2k+1}}{2k+1} \qquad \text{for } |x| < 1$$

The power series in the representation

$$\boxed{\tan^{-1} x = x - \dfrac{x^{3}}{3} + \dfrac{x^{5}}{5} - \dfrac{x^{7}}{7} + \cdots = \sum_{k=0}^{\infty} (-1)^{k}\dfrac{x^{2k+1}}{2k+1} \qquad \text{for } |x| < 1}$$

James Gregory

is called **Gregory's series** in honor of its discoverer James Gregory (1638–1675), a Scottish mathematician and astronomer who was one of the first to make a clear distinction between convergent and divergent series.

It can be shown that for $x = 1$, Gregory's series still converges to

$$\tan^{-1} 1 = \frac{\pi}{4}$$

Hence,

$$\pi = 4(1 - \tfrac{1}{3} + \tfrac{1}{5} - \tfrac{1}{7} + \cdots)$$

Although this equation shows that there is an interesting connection between π and the reciprocals of the odd positive integers, it is of little practical importance for finding the (approximate) numerical value of π because the series converges too slowly.*

*A more efficient method for estimating the numerical value of π is indicated in Problem 40.

Problem Set 11.7

In Problems 1 to 4, differentiate or integrate the power series as indicated, and find the radius of convergence of the resulting power series.

1 $D_x\left(x + \dfrac{x^2}{2} + \dfrac{x^3}{3} + \dfrac{x^4}{4} + \cdots\right)$

2 $D_x\left(x - \dfrac{x^3}{3!} + \dfrac{x^5}{5!} - \dfrac{x^7}{7!} + \dfrac{x^9}{9!} - \dfrac{x^{11}}{11!} + \cdots\right)$

3 $\displaystyle\int (1 + 2x + 3x^2 + 4x^3 + \cdots)\, dx$

4 $\displaystyle\int \left(x - \dfrac{x^3}{3!} + \dfrac{x^5}{5!} - \dfrac{x^7}{7!} + \dfrac{x^9}{9!} - \dfrac{x^{11}}{11!} + \cdots\right) dx$

In Problems 5 to 20, starting with the known power series representation for $1/(1 - x)$, $|x| < 1$, obtain a power series representation for each expression by using differentiation, integration, substitution, or other suitable procedures. In each case, find the radius of convergence of the resulting power series.

5 $\dfrac{1}{1 - x^4}$

6 $\dfrac{x}{1 - x^4}$

7 $\dfrac{1}{1 - 4x}$

8 $\dfrac{x^3}{(1 - x^4)^2}$

9 $\dfrac{x}{1 - x^2}$

10 $\displaystyle\int_0^x \dfrac{t\, dt}{1 - t^2}$

11 $\dfrac{1}{2 + x}$

12 $\dfrac{1 + x^2}{(1 - x^2)^2}$

13 $\ln (1 - x)$

14 $\ln \dfrac{1 + x}{1 - x}$

15 $\displaystyle\int_0^x \ln (1 - t)\, dt$

16 $\tanh^{-1} x$

17 $\displaystyle\int_0^x \tan^{-1} t\, dt$

18 $\dfrac{1}{6 - x - x^2}$

19 $\displaystyle\int_0^x \dfrac{dt}{6 - t - t^2}$

20 $\displaystyle\int_0^x \tanh^{-1} t\, dt$

In Problems 21 to 26, use the known power series representation for $1/(1 - x)$, $|x| < 1$, to find a function that is represented by the given power series. In each case, specify the radius of convergence of the power series.

21 $\displaystyle\sum_{k=0}^{\infty} (-1)^{k+1} x^k$

22 $\displaystyle\sum_{k=0}^{\infty} (-1)^k (x - 1)^k$

23 $\displaystyle\sum_{k=0}^{\infty} x^{2k}$

24 $\displaystyle\sum_{k=0}^{\infty} x^{k+1}$

25 $\displaystyle\sum_{k=1}^{\infty} k x^{2k-1}$

26 $\displaystyle\sum_{k=0}^{\infty} \dfrac{x^{k+1}}{k + 1}$

27 Find the sum of the series $\sum_{k=1}^{\infty} (k/2^k)$. [*Hint:* Use the infinite series representation of $1/(1 - x)^2$ obtained in Example 2.]

28 In probability theory it is necessary to evaluate the quantity $\sum_{k=1}^{\infty} kp(1 - p)^{k-1}$, where $0 \le p \le 1$, in order to find the mean value of a geometric random variable. Evaluate this infinite sum.

In Problems 29 to 34, let the function f be defined by the given power series. Write a power series for $f'(x)$, and find its radius of convergence.

29 $f(x) = \displaystyle\sum_{k=0}^{\infty} k^2 x^k$

30 $f(x) = \displaystyle\sum_{k=1}^{\infty} (-1)^{k+1} k^2 (x - 2)^k$

31 $f(x) = \displaystyle\sum_{k=0}^{\infty} \dfrac{x^k}{k!}$

32 $f(x) = \displaystyle\sum_{k=0}^{\infty} \frac{(-1)^{k+1}x^{2k+1}}{(2k+1)!}$

33 $f(x) = \displaystyle\sum_{k=0}^{\infty} 2^{k/2}(x+1)^{2k}$

34 $f(x) = \displaystyle\sum_{k=1}^{\infty} \frac{(x-1)^{k^3}}{k^3}$

In Problems 35 to 38, let the function f be defined by the given power series. Write a power series for $\displaystyle\int_{0}^{x} f(t)\,dt$, and find its radius of convergence.

35 $f(t) = \displaystyle\sum_{k=0}^{\infty} \frac{(-1)^k t^{2k}}{(2k)!}$

36 $f(t) = \displaystyle\sum_{k=0}^{\infty} \frac{t^k}{2^{k+1}}$

37 $f(t) = \displaystyle\sum_{k=0}^{\infty} \frac{t^{2k+1}}{(2k+1)!}$

38 $f(t) = \displaystyle\sum_{k=1}^{\infty} \frac{t^k}{k^3}$

39 Given $f(x) = \sum_{k=0}^{\infty} [(-1)^k x^{2k}/(2k)!]$, find
(a) $f(0)$ (b) $f'(0)$ (c) $f''(0)$ (d) $f'''(0)$

Ⓒ**40** Use the identity $\pi/4 = \tan^{-1}\frac{1}{7} + 2\tan^{-1}\frac{1}{3}$ and Gregory's series to approximate the value of π to four decimal places.

41 For each fixed value of x, let $s_n(x) = \sum_{k=0}^{n} x^k$ be the nth partial sum of the infinite series $\sum_{k=0}^{\infty} x^k$, and let $R_n(x) =$

$1/(1-x) - s_n(x)$. Show that if $x \neq 1$, then

$$R_n(x) = \frac{x^{n+1}}{1-x}$$

(*Hint:* $\sum_{k=0}^{\infty} x^k$ is a geometric series with initial term 1 and ratio x.)

42 Using the result $1/(1-x) = s_n(x) + R_n(x)$ from Problem 41 and integrating both sides of this equation from $x = -1$ to $x = 0$, conclude that

$$\ln 2 = 1 - \frac{1}{2} + \frac{1}{3} - \frac{1}{4} + \cdots + \frac{(-1)^n}{n+1} + \int_{-1}^{0} \frac{x^{n+1}}{1-x}\,dx$$

43 Criticize the following argument: By Example 4,

$$\ln(1+x) = x - \frac{x^2}{2} + \frac{x^3}{3} - \frac{x^4}{4} + \cdots \qquad \text{for } |x| < 1$$

When $x = 1$, the series on the right is the alternating harmonic series, which converges by Leibniz's test. Therefore, ln 2 is the sum of the alternating harmonic series.

44 Prove that ln 2 really is the sum of the alternating harmonic series (see Problem 43) by using the result of Problem 42 and showing that

$$\lim_{n \to +\infty} \int_{-1}^{0} \frac{x^{n+1}}{1-x}\,dx = 0$$

[*Hint:* For $-1 \leq x \leq 1$, we have $0 \leq 1/(1-x) \leq 1$.]

11.8 Taylor and Maclaurin Series

We have seen in Section 11.7 that a power series $\sum_{k=0}^{\infty} c_k(x-a)^k$ defines a function f by $f(x) = \sum_{k=0}^{\infty} c_k(x-a)^k$. Thus, *starting with a power series,* we obtained a function f. In this section, we study the reverse procedure—*starting with a function f, we try to find a power series that represents it;* that is, we try to **expand** f into a power series. Although a power series expansion cannot *always* be obtained, most of the familiar functions in calculus can be represented as the sum of a convergent power series.

Suppose that f is a function that can be expanded into a power series with a positive radius of convergence R, so that

$$f(x) = \sum_{k=0}^{\infty} c_k(x-a)^k \qquad \text{for } a - R < x < a + R$$

Then, by Property I on page 687, f is differentiable on the open interval $(a - R, a + R)$ and we have

$$f'(x) = \sum_{k=1}^{\infty} kc_k(x-a)^{k-1} \qquad \text{for } a - R < x < a + R$$

Thus, not only is f differentiable, but also the derivative f' of f can itself be expanded into a power series. Hence, we can apply Property I once again to obtain

$$f''(x) = \sum_{k=2}^{\infty} k(k-1)c_k(x-a)^{k-2} \qquad \text{for } a - R < x < a + R$$

Continuing in this way, we have

$$f'''(x) = \sum_{k=3}^{\infty} k(k-1)(k-2)c_k(x-a)^{k-3}$$

$$f^{(4)}(x) = \sum_{k=4}^{\infty} k(k-1)(k-2)(k-3)c_k(x-a)^{k-4}$$

$$f^{(5)}(x) = \sum_{k=5}^{\infty} k(k-1)(k-2)(k-3)(k-4)c_k(x-a)^{k-5}$$

and so on, for $a - R < x < a + R$. The pattern here is obvious—evidently

$$f^{(n)}(x) = \sum_{k=n}^{\infty} k(k-1)(k-2) \cdots (k-n+1)c_k(x-a)^{k-n}$$

$$\text{for } a - R < x < a + R$$

(See Problem 51.)

Putting $x = a$ in the formula just obtained, we find that all terms of the series reduce to zero except for the first term (for which $k = n$), since all the terms after the first contain a factor $(x - a)$. Therefore,

$$f^{(n)}(a) = n(n-1)(n-2) \cdots 3 \cdot 2 \cdot 1 \cdot c_n = (n!)c_n$$

It follows that

$$c_n = \frac{f^{(n)}(a)}{n!}$$

Hence, the coefficients of the power series are given by the same formula as the coefficients of the Taylor polynomial for f (Definition 1, page 618).

These considerations lead us to the following definitions.

DEFINITION 1   **Infinitely Differentiable Function**

A function f defined on an open interval J is said to be **infinitely differentiable** on J if f has derivatives $f^{(n)}$ of all orders $n \geq 1$ on J.

DEFINITION 2   **Taylor Series**

Let the function f be infinitely differentiable on an open interval J, and let a be a number in J. Then, the **Taylor series for f at a** is the power series

$$\sum_{k=0}^{\infty} c_k(x-a)^k \qquad \text{where} \qquad c_k = \frac{f^{(k)}(a)}{k!} \qquad \text{for } k = 0, 1, 2, 3, \ldots$$

Colin Maclaurin

The Taylor series for f at $a = 0$ is called the **Maclaurin series*** for f. Note that there is no implication intended in Definition 2 that the Taylor series for f actually converges to f. Also note the distinction between the nth Taylor polynomial for f at a,

$$P_n(x) = \sum_{k=0}^{n} \frac{f^{(k)}(a)}{k!}(x-a)^k$$

and the Taylor series for f at a,

$$\sum_{k=0}^{\infty} \frac{f^{(k)}(a)}{k!}(x-a)^k$$

The former is a *polynomial* of degree at most n, while the latter is an infinite series. In fact, $P_n(x)$ *is the nth partial sum of the Taylor series.*

<u>EXAMPLE 1</u>   Find the Taylor series for $f(x) = \sin x$ at $a = \pi/4$.

SOLUTION   We arrange our work as follows:

$$f(x) = \sin x \qquad f\left(\frac{\pi}{4}\right) = \sin\frac{\pi}{4} = \frac{\sqrt{2}}{2}$$

$$f'(x) = \cos x \qquad f'\left(\frac{\pi}{4}\right) = \cos\frac{\pi}{4} = \frac{\sqrt{2}}{2}$$

$$f''(x) = -\sin x \qquad f''\left(\frac{\pi}{4}\right) = -\sin\frac{\pi}{4} = -\frac{\sqrt{2}}{2}$$

$$f'''(x) = -\cos x \qquad f'''\left(\frac{\pi}{4}\right) = -\cos\frac{\pi}{4} = -\frac{\sqrt{2}}{2}$$

$$f^{(4)}(x) = \sin x \qquad f^{(4)}\left(\frac{\pi}{4}\right) = \sin\frac{\pi}{4} = \frac{\sqrt{2}}{2}$$

and so forth. Therefore, the coefficients of the Taylor series are

$$c_k = \frac{f^{(k)}(\pi/4)}{k!} = \frac{\pm\sqrt{2}/2}{k!} = \pm\frac{\sqrt{2}}{2 \cdot k!}$$

where the plus and minus signs alternate *in pairs*. The Taylor series

$$\sum_{k=0}^{\infty} c_k\left(x - \frac{\pi}{4}\right)^k \qquad \text{for } \sin x \text{ at } a = \frac{\pi}{4}$$

is accordingly given by

$$\frac{\sqrt{2}}{2} + \frac{\sqrt{2}}{2}\left(x - \frac{\pi}{4}\right) - \frac{\sqrt{2}}{2 \cdot 2!}\left(x - \frac{\pi}{4}\right)^2 - \frac{\sqrt{2}}{2 \cdot 3!}\left(x - \frac{\pi}{4}\right)^3$$
$$+ \frac{\sqrt{2}}{2 \cdot 4!}\left(x - \frac{\pi}{4}\right)^4 + \cdots \ \blacksquare$$

---

*This special case of a Taylor series is named in honor of the Scottish mathematician Colin Maclaurin (1698–1746), who developed and extended Newton's work in calculus.

EXAMPLE 2   Find the Maclaurin series for $f(x) = e^x$.

SOLUTION   The Maclaurin series for e^x is just the Taylor series for e^x at $a = 0$. The work is arranged as follows:

$$f(x) = e^x \qquad f(0) = e^0 = 1$$

$$f'(x) = e^x \qquad f'(0) = e^0 = 1$$

$$f''(x) = e^x \qquad f''(0) = e^0 = 1$$

$$f'''(x) = e^x \qquad f'''(0) = e^0 = 1$$

and so forth. Evidently, $f^{(k)}(0) = e^0 = 1$ holds for all $k = 0, 1, 2, 3, \ldots$. Therefore, the coefficients of the Maclaurin series are given by

$$c_k = \frac{f^{(k)}(0)}{k!} = \frac{1}{k!}$$

Hence, the Maclaurin series $\sum_{k=0}^{\infty} c_k(x - 0)^k = \sum_{k=0}^{\infty} c_k x^k$ for e^x is

$$1 + x + \frac{x^2}{2!} + \frac{x^3}{3!} + \frac{x^4}{4!} + \cdots + \frac{x^n}{n!} + \cdots = \sum_{k=0}^{\infty} \frac{x^k}{k!}$$

∎

As we have seen, if a function can be expanded into a power series, then the function must be infinitely differentiable and the power series must be its Taylor series. However, even if a function is infinitely differentiable, there is no automatic guarantee that it can be expanded into a power series at all! In other words, *although an infinitely differentiable function has a Taylor series, this Taylor series need not converge to the function.* (See Problem 50 for an example.) The following theorem, which is a consequence of Taylor's formula (Theorem 2, page 622), gives a condition under which the Taylor series for a function actually converges to the function.

THEOREM 1   **Expansion of a Function into Its Taylor Series**

Let the function f be infinitely differentiable on some open interval containing the number a. Suppose that there exists a positive number r and a positive constant M such that

$$|f^{(n)}(x)| \leq M$$

holds for all values of x in the interval $(a - r, a + r)$ and all positive integers n. Then f can be expanded into a Taylor series; that is,

$$f(x) = \sum_{k=0}^{\infty} \frac{f^{(k)}(a)}{k!}(x - a)^k$$

holds for all values of x in the interval $(a - r, a + r)$.

PROOF   Let

$$P_n(x) = \sum_{k=0}^{n} \frac{f^{(k)}(a)}{k!}(x - a)^k$$

be the nth Taylor polynomial for f at a, and let $R_n(x) = f(x) - P_n(x)$ be the corresponding Taylor remainder. Notice that $P_n(x)$ is just the nth partial sum of the Taylor series

$$\sum_{k=0}^{\infty} \frac{f^{(k)}(a)}{k!}(x - a)^k$$

Hence, this Taylor series converges to $f(x)$ if and only if $f(x) = \lim_{n \to +\infty} P_n(x)$. The latter condition is equivalent to $\lim_{n \to +\infty} [f(x) - P_n(x)] = 0$; that is, it is equivalent to $\lim_{n \to +\infty} R_n(x) = 0$.

Now assume that $|f^{(n)}(x)| \le M$ holds for $a - r < x < a + r$ and for all positive integers n. By Taylor's formula with Lagrange remainder

$$R_n(x) = \frac{f^{(n+1)}(c)}{(n + 1)!}(x - a)^{n+1}$$

holds for some c between a and x. Hence, for x in $(a - r, a + r)$, we have

$$|R_n(x)| = |f^{(n+1)}(c)|\frac{|x - a|^{n+1}}{(n + 1)!} \le M\frac{|x - a|^{n+1}}{(n + 1)!}$$

For x in $(a - r, a + r)$, the infinite series

$$\sum_{k=0}^{\infty} \frac{|x - a|^{k+1}}{(k + 1)!}$$

converges by the ratio test, and therefore its general term approaches zero; that is,

$$\lim_{n \to +\infty} \frac{|x - a|^{n+1}}{(n + 1)!} = 0$$

Hence, since M is a constant, the inequality

$$|R_n(x)| \le M\frac{|x - a|^{n+1}}{(n + 1)!}$$

shows that $\lim_{n \to +\infty} R_n(x) = 0$ holds for all values of x in the interval $(a - r, a + r)$.

Therefore, the Taylor series converges to $f(x)$, and the proof is complete. ∎

The following examples not only illustrate the use of Theorem 1, but also provide power series expansions that are so important that they should be memorized for future use.

In Examples 3 to 5, justify the following power series expansions.

<u>EXAMPLE 3</u> $e^x = 1 + x + \dfrac{x^2}{2!} + \dfrac{x^3}{3!} + \dfrac{x^4}{4!} + \cdots$ tor all values of x.

SOLUTION   In Theorem 1, we let $f(x) = e^x$ and $a = 0$, noting that f is infinitely differentiable on $(-\infty, \infty)$ and that $f^{(n)}(x) = e^x$ holds for all values of $n \ge 1$. If r is any positive number, then for x in the interval $(-r, r)$, we have

$$|f^{(n)}(x)| = |e^x| = e^x \le e^r$$

since $x < r$. Thus, taking $M = e^r$ in Theorem 1, we conclude that

$$e^x = \sum_{k=0}^{\infty} \frac{f^{(k)}(0)}{k!} x^k = \sum_{k=0}^{\infty} \frac{e^0}{k!} x^k = \sum_{k=0}^{\infty} \frac{x^k}{k!} = 1 + x + \frac{x^2}{2!} + \frac{x^3}{3!} + \cdots$$

holds for all values of x between $-r$ and r. Since we can choose r as large as we please, $e^x = 1 + x + (x^2/2!) + (x^3/3!) + \cdots$ holds for all values of x. ∎

__EXAMPLE 4__ $\sin x = x - \dfrac{x^3}{3!} + \dfrac{x^5}{5!} - \dfrac{x^7}{7!} + \dfrac{x^9}{9!} - \dfrac{x^{11}}{11!} + \cdots$ for all values of x.

SOLUTION   Successive derivatives of $\sin x$ give only $\pm\sin x$ or $\pm\cos x$. In any case, $|D_x^n \sin x| \le 1$, so we can take $M = 1$ in Theorem 1, and r can be chosen as large as we please. Then

$$\sin x = \sin 0 + \frac{\cos 0}{1!} x - \frac{\sin 0}{2!} x^2 - \frac{\cos 0}{3!} x^3 + \frac{\sin 0}{4!} x^4 + \frac{\cos 0}{5!} x^5 - \cdots$$

that is, $\sin x = x - \dfrac{x^3}{3!} + \dfrac{x^5}{5!} - \dfrac{x^7}{7!} + \cdots$ holds for all values of x. ∎

__EXAMPLE 5__ $\cos x = 1 - \dfrac{x^2}{2!} + \dfrac{x^4}{4!} - \dfrac{x^6}{6!} + \cdots$ for all values of x.

SOLUTION   This expansion can be obtained in essentially the same way as the expansion for $\sin x$ was obtained in Example 4. However, it is more interesting to obtain the power series for the cosine by differentiating the power series for the sine as follows:

$$\cos x = D_x \sin x = D_x \left(x - \frac{x^3}{3!} + \frac{x^5}{5!} - \frac{x^7}{7!} + \cdots \right)$$

$$= 1 - \frac{3x^2}{3!} + \frac{5x^4}{5!} - \frac{7x^6}{7!} + \cdots = 1 - \frac{x^2}{2!} + \frac{x^4}{4!} - \frac{x^6}{6!} + \cdots$$ ∎

Some elementary consequences of the expansions obtained above are shown in the following examples.

__EXAMPLE 6__   Find a power series expansion for $\dfrac{1 - \cos x}{x}$, $x \neq 0$.

SOLUTION

$$\cos x = 1 - \frac{x^2}{2!} + \frac{x^4}{4!} - \frac{x^6}{6!} + \cdots$$

so that

$$1 - \cos x = \frac{x^2}{2!} - \frac{x^4}{4!} + \frac{x^6}{6!} - \frac{x^8}{8!} + \cdots \qquad \text{for all values of } x$$

Hence, for $x \neq 0$,

$$\frac{1 - \cos x}{x} = \frac{1}{x}\left(\frac{x^2}{2!} - \frac{x^4}{4!} + \frac{x^6}{6!} - \frac{x^8}{8!} + \cdots \right)$$

$$= \frac{x}{2!} - \frac{x^3}{4!} + \frac{x^5}{6!} - \frac{x^7}{8!} + \cdots$$ ∎

© **EXAMPLE 7**   Estimate e^{-1} with an error of no more than $5/10^4$. Round off your answer to three decimal places.

SOLUTION

$$e^x = \sum_{k=0}^{\infty} \frac{x^k}{k!}$$

so that

$$e^{-1} = \sum_{k=0}^{\infty} \frac{(-1)^k}{k!} = 1 - 1 + \frac{1}{2!} - \frac{1}{3!} + \frac{1}{4!} - \frac{1}{5!} + \frac{1}{6!} - \frac{1}{7!} + \cdots$$

By estimating e^{-1} using only the first seven terms, it follows from Leibniz's theorem that we make an error not exceeding the absolute value of the first omitted term

$$\frac{1}{7!} = \frac{1}{5040} < \frac{1}{5(10^3)} = \frac{10}{5(10^4)} = \frac{2}{10^4} < \frac{5}{10^4}$$

Hence, $e^{-1} \approx 1 - 1 + \frac{1}{2} - \frac{1}{6} + \frac{1}{24} - \frac{1}{120} + \frac{1}{720} \approx 0.368$ ■

Example 7 illustrates the following conventional **rounding-off rule** * for numerical approximations:

> If you intend to round off a numerical approximation to n decimal places, make sure that the error involved in the approximation does not exceed $5 \times 10^{-(n+1)}$.

For instance, in Example 7, we rounded off to $n = 3$ decimal places; our error did not exceed $5 \times 10^{-(3+1)} = 5/10^4$. Use of this rounding-off rule is further illustrated in the following example.

© **EXAMPLE 8**   Estimate $\int_0^1 \sin x^2 \, dx$. Round off your answer to four decimal places.

SOLUTION   Since $\sin x^2$ has no elementary antiderivative, we cannot evaluate the integral by using the fundamental theorem of calculus. However, if we replace x by x^2 in the Maclaurin series

$$\sin x = x - \frac{x^3}{3!} + \frac{x^5}{5!} - \frac{x^7}{7!} + \cdots$$

we obtain the Maclaurin series

$$\sin x^2 = x^2 - \frac{(x^2)^3}{3!} + \frac{(x^2)^5}{5!} - \frac{(x^2)^7}{7!} + \cdots$$

$$= x^2 - \frac{x^6}{3!} + \frac{x^{10}}{5!} - \frac{x^{14}}{7!} + \cdots$$

Integrating term by term, we obtain

$$\int_0^{1^*} \sin x^2 \, dx = \left(\frac{x^3}{3} - \frac{x^7}{7(3!)} + \frac{x^{11}}{11(5!)} - \frac{x^{15}}{15(7!)} + \cdots \right) \Bigg|_0^1$$

$$= \frac{1}{3} - \frac{1}{7(3!)} + \frac{1}{11(5!)} - \frac{1}{15(7!)} + \cdots$$

*For an indication of why the rounding-off rule is reasonable, see Problem 53.

If we use only the first three terms of the series, then by Leibniz's theorem, our error will be less than the absolute value of the fourth term,

$$\frac{1}{15(7!)} \approx 0.000013 = 1.3 \times 10^{-5} < 5 \times 10^{-5}$$

Thus, rounding off to four decimal places, we have

$$\int_0^1 \sin x^2 \, dx \approx \frac{1}{3} - \frac{1}{7(3!)} + \frac{1}{11(5!)} \approx 0.3103 \qquad \blacksquare$$

For convenience, we now assemble some of the more important power series expansions obtained in this and the previous section.

---

1 $e^x = 1 + x + \dfrac{x^2}{2!} + \dfrac{x^3}{3!} + \dfrac{x^4}{4!} + \cdots = \displaystyle\sum_{k=0}^{\infty} \dfrac{x^k}{k!}$ for all x

2 $\sin x = x - \dfrac{x^3}{3!} + \dfrac{x^5}{5!} - \dfrac{x^7}{7!} + \cdots = \displaystyle\sum_{k=0}^{\infty} \dfrac{(-1)^k x^{2k+1}}{(2k+1)!}$ for all x

3 $\cos x = 1 - \dfrac{x^2}{2!} + \dfrac{x^4}{4!} - \dfrac{x^6}{6!} + \cdots = \displaystyle\sum_{k=0}^{\infty} \dfrac{(-1)^k x^{2k}}{(2k)!}$ for all x

4 $\ln(1+x) = x - \dfrac{x^2}{2} + \dfrac{x^3}{3} - \dfrac{x^4}{4} + \cdots$

             $= \displaystyle\sum_{k=0}^{\infty} \dfrac{(-1)^k x^{k+1}}{k+1}$ for $|x| < 1$

5 $\tan^{-1} x = x - \dfrac{x^3}{3} + \dfrac{x^5}{5} - \dfrac{x^7}{7} + \cdots$

             $= \displaystyle\sum_{k=0}^{\infty} \dfrac{(-1)^k x^{2k+1}}{2k+1}$ for $|x| < 1$

Problem Set 11.8

In Problems 1 to 10, find the Taylor series for each function f at the indicated value of a.

1 $f(x) = \sin x$ at $a = \pi/6$ **2** $f(x) = \sqrt{x}$ at $a = 9$

3 $f(x) = 1/x$ at $a = 2$ **4** $f(x) = \sqrt{x^3}$ at $a = 1$

5 $f(x) = e^x$ at $a = 4$ **6** $f(x) = \cos x$ at $a = \pi/6$

7 $f(x) = \sqrt{x-1}$ at $a = 2$

8 $f(x) = \cos x$ at $a = \pi/3$

9 $f(x) = \dfrac{1}{(1+x)^2}$ at $a = -2$

10 $f(x) = \ln(1+x)$ at $a = 1$

In Problems 11 to 14, find the first four terms of the Taylor series for each function f at the indicated value of a.

11 $f(x) = \csc x$ at $a = \pi/6$ **12** $f(x) = \cot x$ at $a = \pi/4$

13 $f(x) = \tan x$ at $a = \pi/4$ **14** $f(x) = \sec x$ at $a = \pi/3$

In Problems 15 to 30, use the known Maclaurin series expansion for a related function to find the Maclaurin series expansion for the given function.

15 $f(x) = \sinh x$ **16** $f(x) = \cosh x$

17 $f(x) = e^{-x^2}$ **18** $f(x) = \ln \dfrac{1+x}{1-x}$

19 $f(x) = \cos x^2$ **20** $f(x) = xe^{-3x}$

21 $f(x) = x^2 \sin 2x$

22 $f(x) = \ln (1 + x^2)$

23 $f(x) = \sin^2 x$. [*Hint:* $\sin^2 x = \frac{1}{2}(1 - \cos 2x)$.]

24 $f(x) = x^2 e^{-x^3}$

25 $f(x) = \begin{cases} \dfrac{\sin x}{x} & \text{if } x \neq 0 \\ 1 & \text{if } x = 0 \end{cases}$

26 $f(x) = \begin{cases} \dfrac{\tan^{-1} x}{x} & \text{if } x \neq 0 \\ 1 & \text{if } x = 0 \end{cases}$

27 $f(x) = \displaystyle\int_0^x \sin t^2 \, dt$

28 $f(x) = \begin{cases} \ln (1 + x)^{1/x} & \text{if } x \neq 0 \\ 1 & \text{if } x = 0 \end{cases}$

29 $f(x) = \displaystyle\int_0^x e^{-t^2} \, dt$

30 $f(x) = \displaystyle\int_0^x g(t) \, dt, \; g(t) = \begin{cases} \dfrac{\sin t}{t} & \text{if } t \neq 0 \\ 1 & \text{if } t = 0 \end{cases}$

C In Problems 31 to 42, use power series to estimate the numerical value of each expression with an error not exceeding 5×10^{-5}. Round off your answer to four decimal places.

31 $e^{-0.02}$

32 $\sin 0.1$

33 $\ln 0.9$

34 $\cos 59°$

35 $\displaystyle\int_0^{0.5} \cos x^2 \, dx$

36 $\displaystyle\int_0^{0.25} x \cos \sqrt{x} \, dx$

37 $\displaystyle\int_0^1 e^{-x^2} \, dx$

38 $\displaystyle\int_{0.1}^1 \dfrac{\sin x}{x} \, dx$

39 $\displaystyle\int_{0.1}^1 \dfrac{1 - e^{-x}}{x} \, dx$

40 $\displaystyle\int_{0.2}^{0.5} \dfrac{\ln (1 + x)}{x} \, dx$

41 $\displaystyle\int_{0.3}^{1.1} \dfrac{1 - \cos x}{x} \, dx$

42 $\displaystyle\int_0^{0.2} \ln (1 + \sin x) \, dx$

43 If $f(x) = \tan^{-1} x$, find a formula for $f^{(n)}(0)$. (*Hint:* Use the known Maclaurin series expansion for $\tan^{-1} x$; that is, use Gregory's series.)

44 If f is a polynomial function of degree n and a is any constant, prove that (a) f can be represented in the form

$$f(x) = c_0 + c_1(x - a) + c_2(x - a)^2 + \cdots + c_n(x - a)^n$$

and (b) the coefficients $c_0, c_1, \ldots, c_n$ in part (a) are uniquely determined by the requirement that

$$c_k = \frac{f^{(k)}(a)}{k!} \qquad \text{for } k = 0, 1, 2, \ldots, n$$

(c) Use parts (a) and (b) to rewrite the polynomial function $f(x) = 5x^4 - 11x^3 + 21x^2 + 29x + 19$ in the form $f(x) = c_0 + c_1(x - 2) + c_2(x - 2)^2 + c_3(x - 2)^3 + c_4(x - 2)^4$.

In Problems 45 to 49, use the Maclaurin series expansion $f(x) = \sum_{k=0}^{\infty} c_k x^k$ and the fact that $c_k = f^{(k)}(0)/k!$ to find the value of the indicated higher-order derivative.

45 $f^{(15)}(0)$, where $f(x) = x \sin x$

46 $f^{(16)}(0)$, where $f(x) = \cos x^2$

47 $f^{(17)}(0)$, where $f(x) = \int_0^x e^{-t^2} \, dt$

48 $f^{(19)}(0)$, where $f(x) = xe^{-x}$

49 $f^{(20)}(0)$, where $f(x) = \ln (1 + x^2)$

50 Let f be the function defined by

$$f(x) = \begin{cases} e^{-1/x^2} & \text{for } x \neq 0 \\ 0 & \text{for } x = 0 \end{cases}$$

(a) Sketch the graph of f.

(b) Find $f'(x)$, $f''(x)$, and $f'''(x)$ for $x \neq 0$.

(c) Prove by induction that for every positive integer n, there exists a polynomial function P (depending on n) such that $f^{(n)}(x) = P(1/x) \cdot f(x)$ holds for $x \neq 0$.

(d) Using (c), prove that $\lim_{n \to +\infty} [f^{(n)}(x)/x] = 0$ for all positive integers n.

(e) Show that f is infinitely differentiable on $(-\infty, \infty)$ and that $f^{(n)}(0) = 0$ for all positive integers n.

(f) Show that f cannot be expanded into a power series about 0.

51 Suppose that the power series $\sum_{k=0}^{\infty} c_k(x - a)^k$ converges at least for x in the interval $(a - r, a + r)$, where $r > 0$. Define the function f by $f(x) = \sum_{k=0}^{\infty} c_k(x - a)^k$ for $a - r < x < a + r$. Prove by induction that for any positive integer n,

$$f^{(n)}(x) = \sum_{k=n}^{\infty} k(k - 1)(k - 2) \cdots (k - n + 1)c_k(x - a)^{k-n}$$

for $a - r < x < a + r$.

52 Suppose that $\sum_{k=0}^{\infty} b_k(x - a)^k$ and $\sum_{k=0}^{\infty} c_k(x - a)^k$ are two power series with positive radii of convergence. If there exists $\epsilon > 0$ such that

$$\sum_{k=0}^{\infty} b_k(x - a)^k = \sum_{k=0}^{\infty} c_k(x - a)^k$$

holds for all values of x with $|x - a| < \epsilon$, prove that $b_k = c_k$ for all nonnegative integers k.

53 In order to better understand the rounding-off rule on page 698, prove the following: If x is a positive real number and r is the result of rounding off x to the nearest 10^{-n}, where n is an integer, then the error $|x - r|$ made in estimating x by r cannot exceed $5 \times 10^{-(n+1)}$.

11.9   Binomial Series

We now set for ourselves the problem of finding the Maclaurin series expansion for the function f defined by $f(x) = (1 + x)^p$, where p is a constant and $1 + x > 0$. Since $f'(x) = p(1 + x)^{p-1}$, we have

$$(1 + x)f'(x) = pf(x)$$

Suppose that f can be expanded into a Maclaurin series

$$f(x) = c_0 + c_1 x + c_2 x^2 + c_3 x^3 + \cdots$$

Then

$$f'(x) = c_1 + 2c_2 x + 3c_3 x^2 + \cdots$$

so that

$$
\begin{aligned}
(1 + x)f'(x) &= f'(x) + xf'(x) \\
&= (c_1 + 2c_2 x + 3c_3 x^2 + \cdots) + (c_1 x + 2c_2 x^2 + 3c_3 x^3 + \cdots) \\
&= c_1 + (2c_2 + c_1)x + (3c_3 + 2c_2)x^2 + (4c_4 + 3c_3)x^3 + \cdots
\end{aligned}
$$

The condition $(1 + x)f'(x) = pf(x)$ therefore becomes

$$c_1 + (2c_2 + c_1)x + (3c_3 + 2c_2)x^2 + (4c_4 + 3c_3)x^3 + \cdots$$
$$= pc_0 + pc_1 x + pc_2 x^2 + pc_3 x^3 + \cdots$$

Equating coefficients of like powers of x in this equation (see Problem 52 in Section 11.8), we have

$$c_1 = pc_0 \qquad 2c_2 + c_1 = pc_1 \qquad 3c_3 + 2c_2 = pc_2 \qquad 4c_4 + 3c_3 = pc_3$$

and so forth. Evidently, $(n + 1)c_{n+1} + nc_n = pc_n$; that is,

$$c_{n+1} = \frac{p - n}{n + 1} c_n$$

holds for all $n \geq 0$. Hence,

$$c_1 = pc_0 \qquad c_2 = \frac{p - 1}{2} c_1 = \frac{(p - 1)p}{2} c_0$$

$$c_3 = \frac{p - 2}{3} c_2 = \frac{(p - 2)(p - 1)p}{3 \cdot 2} c_0$$

$$c_4 = \frac{p - 3}{4} c_3 = \frac{(p - 3)(p - 2)(p - 1)p}{4 \cdot 3 \cdot 2} c_0$$

and so forth. Since $f(0) = (1 + 0)^p = 1^p = 1$, it follows that $c_0 = 1$.

The calculations above suggest the general formula

$$c_n = \frac{\overbrace{p(p - 1)(p - 2) \cdots (p - n + 1)}^{n \text{ factors}}}{n!}$$

for $n \geq 1$, and this is easily confirmed by mathematical induction (Problem 27).

Thus, putting $c_0 = 1$, we obtain the power series

$$\sum_{k=0}^{\infty} c_k x^k = 1 + \sum_{k=1}^{\infty} \frac{p(p-1)(p-2)\cdots(p-k+1)}{k!} x^k$$

which is called a **binomial series.**

Note that we have *not* proved that the equation

$$(1+x)^p = \sum_{k=0}^{\infty} c_k x^k$$

holds. What we have shown is that *if the function f can be expanded into a power series at all,* the above equation holds at least for values of x in an open interval around 0. However, we can now prove the following theorem.

THEOREM 1   **Binomial Series Expansion**

Let p be any constant other than 0 or a positive integer. Define

$$c_0 = 1 \qquad \text{and} \qquad c_n = \frac{p(p-1)(p-2)\cdots(p-n+1)}{n!}$$

for each positive integer n. Then:

(i)   For $n \geq 0$,

$$c_{n+1} = \frac{p-n}{n+1} c_n$$

(ii)   The binomial series $\sum_{k=0}^{\infty} c_k x^k$ has radius of convergence $R = 1$.

(iii)   For $|x| < 1$,

$$(1+x)^p = \sum_{k=0}^{\infty} c_k x^k$$

$$= 1 + px + \frac{p(p-1)}{2!} x^2 + \frac{p(p-1)(p-2)}{3!} x^3 + \cdots$$

PROOF

(i)   None of the coefficients $c_0, c_1, c_2, \ldots$ is equal to zero since if $c_n = 0$, one of the factors $p, p-1, p-2, \ldots, p-n+1$ in the numerator of the fraction defining c_n would have to be 0, so that p would be a nonnegative integer—contrary to the hypothesis. The relation

$$c_{n+1} = \frac{p-n}{n+1} c_n$$

surely holds since it was originally used to generate the numbers $c_0, c_1, c_2, \ldots$. (See Problem 28 for direct verification.)

(ii) $\displaystyle\lim_{n\to+\infty} \left| \frac{c_{n+1}}{c_n} \right| = \lim_{n\to+\infty} \left| \frac{p-n}{n+1} \right| = \lim_{n\to+\infty} \frac{\left| \dfrac{p}{n} - 1 \right|}{1 + (1/n)} = 1$

Hence, the radius of convergence of the power series is given by $R = 1/1 = 1$ (Theorem 1 of Section 11.6).

(iii) Define the function g by $g(x) = \sum_{k=0}^{\infty} c_k x^k$. The power series $\sum_{k=0}^{\infty} c_k x^k$ was set up in the first place so that

$$(1 + x)D_x \sum_{k=0}^{\infty} c_k x^k = p \sum_{k=0}^{\infty} c_k x^k$$

for $|x| < R$; hence, $(1 + x)g'(x) = pg(x)$ holds for $|x| < 1$. (For the details, see Problem 28.) Now we have

$$D_x \frac{g(x)}{(1 + x)^p} = \frac{(1 + x)^p g'(x) - g(x)p(1 + x)^{p-1}}{(1 + x)^{2p}}$$

$$= \frac{(1 + x)^{p-1}}{(1 + x)^{2p}} [(1 + x)g'(x) - pg(x)] = 0$$

for $|x| < 1$. It follows that $\dfrac{g(x)}{(1 + x)^p}$ is a constant, say

$$\frac{g(x)}{(1 + x)^p} = K \qquad \text{for } |x| < 1$$

Putting $x = 0$, we find that

$$K = \frac{g(0)}{(1 + 0)^p} = g(0) = c_0 = 1$$

Therefore,

$$g(x) = K(1 + x)^p = 1 \cdot (1 + x)^p = (1 + x)^p \qquad \text{for } |x| < 1$$

that is,

$$(1 + x)^p = g(x) = \sum_{k=0}^{\infty} c_k x^k \qquad \text{holds for} \qquad |x| < 1 \qquad\blacksquare$$

EXAMPLE 1   Find a power series expansion for $\sqrt[3]{1 + x}$, $|x| < 1$.

SOLUTION   Take $p = \frac{1}{3}$ in Theorem 1. Thus, $c_0 = 1$, $c_1 = \frac{1}{3}$, $c_2 = -\frac{1}{9}$, $c_3 = \frac{5}{81}$, and so forth. In general,

$$c_n = \frac{\frac{1}{3}(\frac{1}{3} - 1)(\frac{1}{3} - 2) \cdots (\frac{1}{3} - n + 1)}{n!} = \frac{1(1 - 3)(1 - 6) \cdots [1 - 3(n - 1)]}{3^n n!}$$

$$= \frac{(-1)^{n+1} 2 \cdot 5 \cdot 8 \cdot 11 \cdots (3n - 4)}{3^n n!} \qquad \text{for } n \geq 2$$

and therefore

$$\sqrt[3]{1 + x} = 1 + \frac{1}{3}x - \frac{1}{9}x^2 + \frac{5}{81}x^3 - \cdots$$

$$= 1 + \frac{1}{3}x + \sum_{k=2}^{\infty} \frac{(-1)^{k+1} 2 \cdot 5 \cdot 8 \cdot 11 \cdots (3k - 4)}{3^k k!} x^k \qquad \text{for } |x| < 1 \quad\blacksquare$$

© **EXAMPLE 2**   Use the first three terms of the expansion obtained in Example 1 to approximate $\sqrt[3]{28}$. Give a bound on the error involved.

SOLUTION   Naturally, we wish to use the fact that $\sqrt[3]{27} = 3$. Thus, we write

$$\sqrt[3]{28} = \sqrt[3]{27 + 1} = \sqrt[3]{27(1 + \tfrac{1}{27})} = \sqrt[3]{27}\,\sqrt[3]{1 + \tfrac{1}{27}} = 3\sqrt[3]{1 + \tfrac{1}{27}}$$

Putting $x = \tfrac{1}{27}$ in Example 1, we obtain

$$\sqrt[3]{1 + \tfrac{1}{27}} \approx 1 + \tfrac{1}{3}(\tfrac{1}{27}) - \tfrac{1}{9}(\tfrac{1}{27})^2 = \tfrac{6641}{6561}$$

with an error no larger than $\tfrac{5}{81}(\tfrac{1}{27})^3$, since the series (apart from the first term) is alternating and Leibniz's theorem applies. Therefore,

$$\sqrt[3]{28} = 3\sqrt[3]{1 + \tfrac{1}{27}} \approx 3(\tfrac{6641}{6561}) = 3.036579 \ldots$$

This approximation *underestimates* $\sqrt[3]{28}$ with an error that does not exceed $3 \cdot \tfrac{5}{81}(\tfrac{1}{27})^3 < 0.0000095$. (The true value of $\sqrt[3]{28}$ is $3.03658897. \ldots$) ∎

ⓒ **EXAMPLE 3**   Estimate $\displaystyle\int_0^{1/2} \sqrt[3]{1 + x^3}\; dx$, and place a bound on the error involved.

SOLUTION   By Example 1, for $|x| < 1$, we have

$$\sqrt[3]{1 + x} = 1 + \tfrac{1}{3}x - \tfrac{1}{9}x^2 + \tfrac{5}{81}x^3 - \cdots$$

Replacing x by x^3, we obtain

$$\sqrt[3]{1 + x^3} = 1 + \tfrac{1}{3}x^3 - \tfrac{1}{9}x^6 + \tfrac{5}{81}x^9 - \cdots$$

Therefore,

$$\int_0^{1/2} \sqrt[3]{1 + x^3}\; dx$$

$$= \int_0^{1/2} 1\; dx + \int_0^{1/2} \tfrac{1}{3}x^3\; dx - \int_0^{1/2} \tfrac{1}{9}x^6\; dx + \int_0^{1/2} \tfrac{5}{81}x^9\; dx - \cdots$$

$$= \tfrac{1}{2} + (\tfrac{1}{3})(\tfrac{1}{4})(\tfrac{1}{2})^4 - (\tfrac{1}{9})(\tfrac{1}{7})(\tfrac{1}{2})^7 + (\tfrac{5}{81})(\tfrac{1}{10})(\tfrac{1}{2})^{10} - \cdots$$

Apart from the first term, the last series is alternating and the terms are decreasing in absolute value (see Problems 31 and 32). Hence, Leibniz's theorem applies, so that using (say) the first three terms of the series, we have

$$\int_0^{1/2} \sqrt[3]{1 + x^3}\; dx \approx \tfrac{1}{2} + (\tfrac{1}{3})(\tfrac{1}{4})(\tfrac{1}{2})^4 - (\tfrac{1}{9})(\tfrac{1}{7})(\tfrac{1}{2})^7 = 0.505084 \ldots$$

with an error whose absolute value does not exceed $(\tfrac{5}{81})(\tfrac{1}{10})(\tfrac{1}{2})^{10} < 0.000007$. Therefore, rounding off to four decimal places, we have $\int_0^{1/2} \sqrt[3]{1 + x^3}\; dx \approx 0.5051$. ∎

Problem Set 11.9

In Problems 1 to 12, use the binomial series expansion (Theorem 1) to find a Maclaurin series expansion for each expression. Specify the range of values of x for which the expansion is correct.

1 $\sqrt[4]{1 + x}$

2 $\sqrt{1 + x^2}$

3 $\dfrac{1}{\sqrt[3]{1 - x^2}}$

4 $\dfrac{2x}{\sqrt{1 - x}}$

5 $\dfrac{1}{\sqrt[3]{1 + x}}$

6 $\dfrac{x}{\sqrt[3]{1 - x^2}}$

7 $\dfrac{x}{(1 + 2x)^2}$

8 $(9 + x)^{3/2}$

9 $\sqrt[3]{27 + x}$

10 $\dfrac{1}{\sqrt{1 + x^3}}$

11 $\sqrt[5]{1 + x^3}$

12 $\dfrac{x}{\sqrt[3]{1 + x^2}}$

© In Problems 13 to 18, use the first three terms of an appropriate binomial series to estimate each number. Give an upper bound for the absolute value of the error.

13 $\sqrt{101}$

14 $\sqrt{99}$

15 $\sqrt{1.03}$

16 $\sqrt[5]{33}$

17 $\sqrt[4]{17}$

18 $\dfrac{1}{\sqrt[3]{100}}$

© In Problems 19 to 26, estimate each quantity, rounded off to three decimal places. (Take sufficiently many terms so that the absolute value of the error does not exceed 5×10^{-4}.)

19 $\displaystyle\int_0^{2/3} \sqrt{1 + x^3}\, dx$

20 $\displaystyle\int_0^{1/2} \dfrac{dx}{\sqrt{1 + x^3}}$

21 $\displaystyle\int_0^{0.5} \sqrt{1 - x^4}\, dx$

22 $\displaystyle\int_0^{0.5} \dfrac{dx}{\sqrt{1 + x^4}}$

23 $\displaystyle\int_0^{0.4} \sqrt[3]{1 + x^4}\, dx$

24 $\displaystyle\int_0^{0.2} \dfrac{dx}{\sqrt[4]{1 + x^2}}$

25 $\displaystyle\int_0^{1} \sqrt[3]{27 + x^3}\, dx$

26 $\displaystyle\int_0^{1} \dfrac{dx}{\sqrt[4]{16 - x^2}}$

27 Given a sequence $c_0, c_1, c_2, c_3, c_4, \ldots$ such that

$$c_0 = 1 \qquad \text{and} \qquad c_{n+1} = \dfrac{p - n}{n + 1}\, c_n \qquad \text{for } n \geq 0$$

where p is a constant, prove by mathematical induction that

$$c_n = \dfrac{1}{n!}\, p(p - 1)(p - 2) \cdots (p - n + 1) \qquad \text{for } n \geq 1$$

28 Let p be a given constant and define $c_0 = 1$ and

$$c_n = (1/n!)p(p - 1)(p - 2) \cdots (p - n + 1) \qquad \text{for } n \geq 1$$

(a) Prove that $c_{n+1} = \dfrac{p - n}{n + 1}\, c_n$ for $n \geq 0$.

(b) Prove that $(1 + x)D_x \displaystyle\sum_{k=0}^{\infty} c_k x^k = p \displaystyle\sum_{k=0}^{\infty} c_k x^k$ for $|x| < 1$.

29 Compare the binomial series expansion of $(1 + x)^{-1}$ with the geometric series expansion of the same expression.

30 If a is a positive constant and p is any constant, show that

$$(a + x)^p = a^p + pa^{p-1}\, x + \dfrac{p(p - 1)}{2!}\, a^{p-2}x^2$$

$$+ \dfrac{p(p - 1)(p - 2)}{3!}\, a^{p-3}x^3 + \cdots$$

$$= a^p + \sum_{k=1}^{\infty} \dfrac{p(p - 1) \cdots (p - k + 1)}{k!}\, a^{p-k}x^k$$

for $|x| < a$.

31 Let $(1 + x)^p = \displaystyle\sum_{k=0}^{\infty} c_k x^k$ for $|x| < 1$ be the binomial series expansion. Show that if $0 \leq x < 1$ and $n > p$, then $\displaystyle\sum_{k=n+1}^{\infty} c_k x^k$ is an alternating series. $\left(\textit{Hint: } c_{n+1} = \dfrac{p - n}{n + 1}\, c_n.\right)$

32 In Problem 31, assume that $p > -1$ and prove that the terms in the series $\sum_{k=n+1}^{\infty} c_k x^k$ are decreasing in absolute value (so that Leibniz's theorem is applicable).

33 Exactly what happens in the binomial series expansion when the exponent p is a positive integer? Is the expansion still correct? For what values of x is it correct? Why?

34 From the binomial series expansion for $\dfrac{1}{\sqrt{1 + x^2}}$ and the fact that $\sinh^{-1} x = \displaystyle\int_0^x \dfrac{dt}{\sqrt{1 + t^2}}$, find a power series expansion for $\sinh^{-1} x$.

35 From the binomial series expansion for $\dfrac{1}{\sqrt{1 - x^2}}$ and the fact that $\sin^{-1} x = \displaystyle\int_0^x \dfrac{dt}{\sqrt{1 - t^2}}$, find a power series expansion for $\sin^{-1} x$.

Review Problem Set, Chapter 11

In Problems 1 to 12, determine whether each sequence converges or diverges. If the sequence converges, find its limit.

1 $\left\{ \dfrac{n(n + 1)}{3n^2 + 7n} \right\}$

2 $\left\{ \dfrac{\sin n}{n} \right\}$

3 $\left\{ \dfrac{\sqrt{n + 1}}{\sqrt{3n + 1}} \right\}$

4 $\left\{ \dfrac{7n^3 + 3n^2 - n^3(\frac{1}{2})^n}{3n^2 + n^2(\frac{3}{4})^n} \right\}$

5 $\left\{ \dfrac{1 + (-1)^n}{n} \right\}$

6 $\left\{ \left(50 + \dfrac{1}{n}\right)^2 \cdot \left(1 + \dfrac{n-1}{n^2}\right)^{50} \right\}$

7 $\left\{ \dfrac{\cos(n\pi/2)}{\sqrt{n}} \right\}$ **8** $\{n[1 + (-1)^n]\}$

9 $\{n^2 + (-1)^n 2n\}$

10 $\left\{ \dfrac{1}{(n+1) + (-1)^n(1-n)} \right\}$

11 $\left\{ 1 - \dfrac{3^n}{n!} \right\}$ **12** $\left\{ \dfrac{2^n n!}{(2n+1)!} \right\}$

In Problems 13 to 16, indicate whether each sequence is increasing, decreasing, or nonmonotone.

13 $\{2^n\}$ **14** $\left\{ \dfrac{1}{2^n} \right\}$

15 $\left\{ \dfrac{(-1)^n}{n} \right\}$ **16** $\{(-1)^n\}$

17 Is the sequence $\{n - (2^n/n)\}$ monotone? Why?

18 For each positive integer n, let

$$a_n = \frac{1}{n} + \frac{1}{n+1} + \frac{1}{n+2} + \cdots + \frac{1}{2n}$$

(a) Show that $\{a_n\}$ is a monotone sequence. Is it increasing or decreasing?

(b) Does $\{a_n\}$ converge or diverge? Why?

19 Indicate whether the sequence

$$\left\{ 1 - \frac{1}{4} + \frac{1}{16} - \cdots + \frac{(-1)^{n-1}}{4^{n-1}} \right\}$$

is bounded or unbounded; increasing, decreasing, or non-monotone; convergent or divergent.

20 If $\{a_n\}$ and $\{b_n\}$ are convergent sequences and $a_n \leq b_n$ holds for all positive integers n, prove that $\lim\limits_{n \to +\infty} a_n \leq \lim\limits_{n \to +\infty} b_n$.

21 Explain carefully the distinction between a *sequence* and a *series*.

In Problems 22 to 25, find the sum of each series by forcing the terms in the partial sums to telescope.

22 $\displaystyle\sum_{k=1}^{\infty} \dfrac{k}{(k+1)(k+2)(k+3)}$

23 $\displaystyle\sum_{k=1}^{\infty} \dfrac{\sqrt{k+1} - \sqrt{k}}{\sqrt{k^2+k}}$

24 $\displaystyle\sum_{k=1}^{\infty} \dfrac{4}{(2k-1)(2k+3)}$

25 $\displaystyle\sum_{k=1}^{\infty} \left(\sin\dfrac{1}{k} - \sin\dfrac{1}{k+1} \right)$

26 If $\{b_n\}$ is a given sequence and p is a fixed positive integer, find a formula for the nth partial sum of the series $\sum_{k=1}^{\infty}(b_k - b_{k+p})$. Assuming that $\lim\limits_{n \to +\infty} b_n = L$, find a formula for the sum of the series $\sum_{k=1}^{\infty}(b_k - b_{k+p})$.

27 Find an infinite series whose nth partial sum is given by

$$s_n = \frac{3n}{2n+5}$$

Determine whether the resulting series converges, and if it does, find its sum.

In Problems 28 to 31, use the facts concerning geometric series to find the sum of each series.

28 $\displaystyle\sum_{k=2}^{\infty} \left[5\left(\dfrac{1}{2}\right)^k + 3\left(\dfrac{1}{3}\right)^k \right]$ **29** $\displaystyle\sum_{k=1}^{\infty} \dfrac{3}{10^k}$

30 $\displaystyle\sum_{k=1}^{\infty} 2\left(-\dfrac{1}{3}\right)^{k+7}$ **31** $\displaystyle\sum_{k=0}^{\infty} \left[2\left(\dfrac{1}{4}\right)^k + 7\left(\dfrac{1}{7}\right)^{k+1} \right]$

32 Assume that A and B are positive constants and that the sequence $\{a_n\}$ satisfies $|a_n| \leq AB^n$ for all positive integers n. Prove that the series $\sum_{k=0}^{\infty} a_k x^k$ converges if $|x| < 1/B$.

33 Criticize the following argument: Since $\lim\limits_{n \to +\infty} \ln\dfrac{2n+1}{2n-1} = 0$, the series $\displaystyle\sum_{k=1}^{\infty} \ln\dfrac{2k+1}{2k-1}$ must be convergent.

34 Make an informal argument to show that if the series $\sum_{k=1}^{\infty} b_k$ converges and if the sequence $\{a_n\}$ converges, then the series $\sum_{k=1}^{\infty} a_k b_k$ must converge.

In Problems 35 to 38, use the integral test to determine whether each series converges.

35 $\displaystyle\sum_{k=2}^{\infty} \dfrac{1}{k(\ln k)^6}$ **36** $\displaystyle\sum_{k=1}^{\infty} \dfrac{k}{10+k^2}$

37 $\displaystyle\sum_{k=1}^{\infty} \dfrac{k^2}{e^k}$ **38** $\displaystyle\sum_{k=2}^{\infty} \dfrac{\ln k}{k^2}$

In Problems 39 to 42, use the direct comparison test to determine the convergence of each series.

39 $\displaystyle\sum_{k=1}^{\infty} \dfrac{k^2}{k^2+2} \left(\dfrac{1}{3}\right)^k$ **40** $\displaystyle\sum_{k=0}^{\infty} \dfrac{1}{3+k!}$

41 $\displaystyle\sum_{k=1}^{\infty} \frac{1}{5k + 1}$

42 $\displaystyle\sum_{k=1}^{\infty} \frac{1}{\sqrt{10k}}$

In Problems 43 to 52, use any appropriate method to decide whether each series converges or diverges. If it converges, determine whether the convergence is absolute or conditional.

43 $\displaystyle\sum_{k=1}^{\infty} \frac{(-1)^k \sqrt{k}}{k + 10}$

44 $\displaystyle\sum_{k=2}^{\infty} \frac{(-1)^k}{k^2 + (-1)^k}$

45 $\displaystyle\sum_{k=1}^{\infty} \frac{k}{k + 1} \left(\frac{1}{9}\right)^k$

46 $\displaystyle\sum_{k=1}^{\infty} \frac{(-1)^k}{\ln (1 + 1/k)}$

47 $\displaystyle\sum_{k=1}^{\infty} \frac{1 + (-1)^k}{k}$

48 $\displaystyle\sum_{k=1}^{\infty} \frac{(-1)^k}{\ln (e^k + e^{-k})}$

49 $\displaystyle\sum_{k=1}^{\infty} \frac{1 \cdot 3 \cdot 5 \cdots (2k - 1)}{3^k k!}$

50 $\displaystyle\sum_{k=1}^{\infty} \left[\frac{2 \cdot 4 \cdot 6 \cdots (2k)}{1 \cdot 3 \cdot 5 \cdots (2k - 1)}\right]^2$

51 $\displaystyle\sum_{k=1}^{\infty} (-1)^{k+1} e^{-k^2}$

52 $\displaystyle\sum_{k=2}^{\infty} \sin \left(\pi k + \frac{1}{\ln k}\right)$

C **53** Estimate the sum of the given series with an error not exceeding 5×10^{-4} in absolute value.

(a) $\displaystyle\sum_{k=1}^{\infty} (-1)^{k+1} \frac{1}{k \cdot 2^k}$

(b) $\displaystyle\sum_{k=1}^{\infty} (-1)^{k+1} \frac{1}{(3k)^3}$

54 Give an example of a convergent series $\sum_{k=1}^{\infty} a_k$ of positive terms for which $\displaystyle\lim_{k \to +\infty} \frac{a_{k+1}}{a_k}$ does not exist.

In Problems 55 to 64, find the center a, the radius R, and the interval I of convergence of the given power series.

55 $\displaystyle\sum_{k=1}^{\infty} \frac{(x - 1)^{2k}}{k \cdot 5^k}$

56 $\displaystyle\sum_{k=0}^{\infty} \left(\sin \frac{\pi k}{2}\right) x^k$

57 $\displaystyle\sum_{k=0}^{\infty} (\cos \pi k)(x + 2)^k$

58 $\displaystyle\sum_{k=1}^{\infty} 1 \cdot 3 \cdot 5 \cdot 7 \cdots (2k - 1) x^k$

59 $\displaystyle\sum_{k=1}^{\infty} \frac{1 \cdot 3 \cdot 5 \cdot 7 \cdots (2k - 1)}{2^{3k+1}} (x - 10)^k$

60 $\displaystyle\sum_{k=0}^{\infty} 2^k (x + 4)^k$

61 $\displaystyle\sum_{k=0}^{\infty} (-1)^k \frac{10^k}{k!} (x + \pi)^k$

62 $\displaystyle\sum_{k=1}^{\infty} \frac{1 \cdot 5 \cdot 9 \cdot 13 \cdots (4k - 3)}{2 \cdot 4 \cdot 6 \cdot 8 \cdots (2k)} (x + 6)^k$

63 $\displaystyle\sum_{k=0}^{\infty} \frac{(-1)^k 2^{2k+1}}{2k + 1} (x - 3)^{2k}$

64 $\displaystyle\sum_{k=1}^{\infty} (1 + 2 + 3 + 4 + \cdots + k) x^{2k-1}$

In Problems 65 and 66, find the Taylor series expansion for each function about the indicated center a. Give the range of values of x for which the expansion is correct.

65 $f(x) = \ln x$, $a = 1$

66 $f(x) = \sqrt{x}$, $a = 4$

In Problems 67 to 72, find the first four terms of the Taylor series for each function at the given value of a.

67 $f(x) = e^x$, $a = -1$

68 $f(x) = \tan x$, $a = \pi/4$

69 $f(x) = \sqrt{x}$, $a = 1$

70 $f(x) = \ln (1/x)$, $a = 2$

71 $g(x) = \sin 2x$, $a = \pi/4$

72 $h(x) = \sec x$, $a = \pi/6$

73 Show that the Taylor series in Problem 67 actually converges to the function for all values of x by direct use of Theorem 1 in Section 11.8.

74 (a) Let f be the function defined by

$$f(x) = \begin{cases} \dfrac{e^x - 1}{x} & \text{for } x \neq 0 \\ 1 & \text{for } x = 0 \end{cases}$$

Find the Maclaurin series expansion for $f(x)$ and indicate the values of x for which it represents the function. Show that f is continuous.

(b) Find the Maclaurin series expansion for f'.

(c) Use the result of part (b) to find the sum of the series

$$\sum_{k=1}^{\infty} \frac{k}{(k + 1)!}$$

75 Use a power series to prove that

$$\int_0^x \tan^{-1} t \, dt = x \tan^{-1} x - \tfrac{1}{2} \ln (1 + x^2) \qquad \text{for } |x| < 1$$

In Problems 76 to 81, use the binomial series to find a power series expansion for each expression. In each case, specify the range of values of x for which the expansion works.

76 $\displaystyle\int_0^x \sqrt{1 + t^2}\, dt$

77 $\displaystyle\frac{1}{\sqrt{1 + x^2}}$

78 $D_x \sqrt[3]{1 + x^2}$

79 $(1 - 2x)^{2/3}$

80 $(16 + x^4)^{1/4}$

81 $\displaystyle\int_0^x \sqrt[3]{1 + t^3}\, dt$

82 Find a function f such that $f''(x) + af(x) = 0$, where a is a positive constant, $f(0) = 0$, and $f'(0) = \sqrt{a}$. (*Hint:* Expand f into a Maclaurin series with unknown coefficients. Then determine these coefficients.)

ⓒ In Problems 83 and 84, use the first three terms of an appropriate binomial series to estimate each number. Give an upper bound for the absolute value of the error.

83 $\sqrt[5]{30}$

84 $\displaystyle\int_0^{1/2} \sqrt[3]{1 + x^2}\, dx$

85 Find Maclaurin series expansions for
  (a) $\sin x + \cos x$ (b) $\cos^2 x - \sin^2 x$
  (c) $\tan^{-1} x^3$ (d) 10^x

86 Suppose that the power series $\sum_{k=0}^{\infty} c_k x^k$ has a positive radius of convergence R and that the function f defined by $f(x) = \sum_{k=0}^{\infty} c_k x^k$ for $|x| < R$ is an even function. Show that $c_k = 0$ for every odd positive integer k.

In Problems 87 to 92, a function f is defined in terms of a power series. Write a formula for f in finite terms.

87 $f(x) = x - \dfrac{x^3}{3!} + \dfrac{x^5}{5!} - \dfrac{x^7}{7!} + \cdots$

88 $f(x) = \displaystyle\sum_{k=0}^{\infty} (-1)^k \frac{x^{2k}}{(2k)!}$

89 $f(x) = x - \dfrac{x^2}{3!} + \dfrac{x^4}{5!} - \dfrac{x^6}{7!} + \dfrac{x^8}{9!} - \cdots$

90 $f(x) = 1 + \sin x + \dfrac{\sin^2 x}{2!} + \dfrac{\sin^3 x}{3!} + \dfrac{\sin^4 x}{4!} + \cdots$

91 $f(x) = 1 + x \ln 2 + \dfrac{(\ln 2)^2}{2!} x^2 + \dfrac{(\ln 2)^3}{3!} x^3 + \dfrac{(\ln 2)^4}{4!} x^4 + \cdots$

92 $f(x) = \displaystyle\sum_{k=1}^{\infty} (-1)^k \frac{3^k + 1}{2^k k!} x^k$

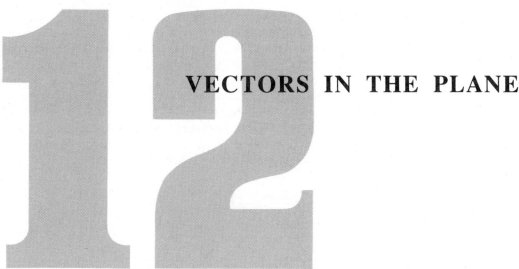

VECTORS IN THE PLANE

Until now, we have dealt exclusively with quantities—such as length, area, volume, angle, mass, density, speed, time, temperature, and probability—that can be measured or represented by real numbers. Since real numbers can be represented by points on a number scale, such quantities are often called **scalars.**

On the other hand, mathematicians and scientists often must deal with quantities that cannot be described or represented by a single real number—quantities that have both magnitude and direction. Such quantities are called **vectors** and include force, velocity, acceleration, displacement, momentum, electric field, magnetic field, gravitational field, angular velocity, and angular momentum.*

Actually, a vector is more than just a quantity with magnitude and direction, since it can interact with other vectors and with scalars in certain well-defined ways. Physicists define vectors to be quantities that transform in a particular manner under groups of symmetries, while mathematicians use an even more abstract definition of a vector as an element of a "linear space." However, for our purposes, a simple geometric definition is quite adequate.

12.1 Vectors

By definition a **vector** is a directed line segment; that is, in everyday language, an arrow. Each vector has a **tail end** (also called an **initial point**) and a **head end** (also called a **terminal point**) and is understood to be directed from its tail end to its head end. Reversing the arrow gives us a vector in the opposite direction.

It is essential to use some special notation for vectors so that they can be distinguished from scalars. In this book we use **A**, **B**, **C**, and so forth to denote vectors.

---

*We have previously treated some of these quantities, for instance, force, velocity, and acceleration, as if they were scalars—but only in connection with straight-line motion where the direction was understood.

Figure 1

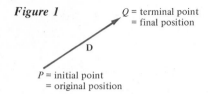

Q = terminal point
= final position

D

P = initial point
= original position

Figure 2

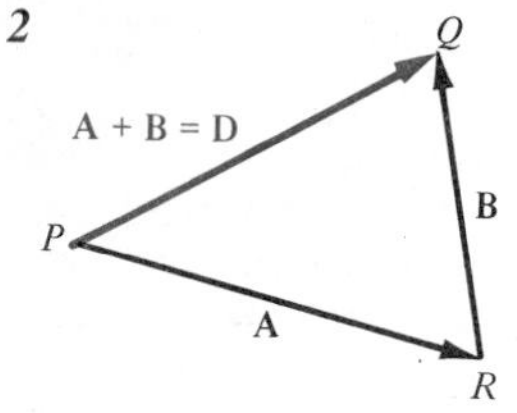

A + B = D

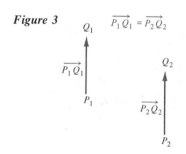

Figure 3

Q_1 $\overrightarrow{P_1Q_1} = \overrightarrow{P_2Q_2}$

$\overrightarrow{P_1Q_1}$

P_1

Q_2

$\overrightarrow{P_2Q_2}$

P_2

Figure 4

Figure 5

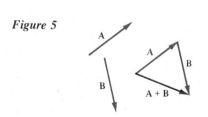

Figure 6

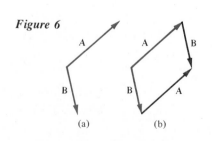

(a)   (b)

If a particle is moved from the point P to the point Q, we say that the particle has undergone a **displacement** from P to Q (Figure 1). Such a displacement can be represented by the vector **D** whose tail end (initial point) is P and whose head end (terminal point) is Q. When we say that the particle has undergone the displacement **D** in Figure 1, we simply mean that it started at P and finished at Q—we do not care how it got from P to Q. For instance, it may have gone first from P to R, then from R to Q as in Figure 2. If the particle is first displaced from P to R—call this displacement **A**—and is then displaced from R to Q—call this displacement **B**—then the resultant displacement is just **D**, and we write **A** + **B** = **D** (Figure 2). This illustrates the **head-to-tail rule** for adding vectors:

> If the head end of **A** coincides with the tail end of **B**, then the vector **D** whose tail end is the tail end of **A** and whose head end is the head end of **B** is called the **sum** of **A** and **B** and written as
>
> $$\mathbf{D} = \mathbf{A} + \mathbf{B}$$

A vector **D** whose tail end is at the point P and whose head end is at the point Q is written as

$$\mathbf{D} = \overrightarrow{PQ}$$

With this notation, the head-to-tail rule can be written

$$\overrightarrow{PQ} = \overrightarrow{PR} + \overrightarrow{RQ}$$

which is quite easy to remember.

If two particles, starting from different points P_1 and P_2, are moved due north through the same distance, it is traditional to say that they have undergone the *same displacement* (Figure 3). More generally, *two vectors that are parallel, point in the same direction, and have the same length are usually regarded as being* **equal.** Following this convention, the two vectors in Figure 3 are regarded as being equal, and we write

$$\overrightarrow{P_1Q_1} = \overrightarrow{P_2Q_2}$$

Actually, the equality convention for vectors is extremely useful, since it implies that a vector may be moved around freely without being changed—provided only that it is always kept parallel to its original position and that its direction and length remain unchanged. Thus, for instance, all the arrows in Figure 4 represent the same vector **A**.

The fact that vectors can be moved around as specified above implies that *any two vectors* **A** *and* **B** *can be brought into a head-to-tail position* (Figure 5); hence, *any two vectors have a sum.*

We can also move any two vectors **A** and **B** into a tail-to-tail position (Figure 6a) so that they form two adjacent sides of a parallelogram (Figure 6b). We refer to this parallelogram as the *parallelogram* **spanned** *by* **A** *and* **B**.

Notice that the diagonal vector whose tail end coincides with the common tail ends of **A** and **B** in the parallelogram spanned by **A** and **B** is the sum of **A** and **B**, as can be seen by looking at the top half of the parallelogram in Figure 7. Glancing at the bottom half of this parallelogram, we see that the same diagonal vector is the sum of **B** and **A**; that is, we see that

$$\mathbf{A} + \mathbf{B} = \mathbf{B} + \mathbf{A}$$

Figure 7

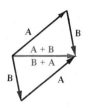

This equation expresses the **commutative law of vector addition.** The technique of finding the sum of two vectors by forming the diagonal of a parallelogram is called the **parallelogram rule** for adding vectors.

Vectors also satisfy the **associative law of vector addition**

$$\mathbf{A} + (\mathbf{B} + \mathbf{C}) = (\mathbf{A} + \mathbf{B}) + \mathbf{C}$$

as can be seen by examining Figure 8.

For many of the same reasons that make it useful to have a zero scalar, it is also convenient to introduce a **zero vector,** written **0**. Intuitively, **0** can be thought of as an arrow that has shrunk to a single point. Alternatively, it can be regarded as a vector $\overrightarrow{QQ}$ that starts and ends at the same point Q. Just as nonzero vectors can be moved around subject to the conditions given above, the zero vector can also be moved around, and we have $\mathbf{0} = \overrightarrow{QQ} = \overrightarrow{RR} = \overrightarrow{SS} = \cdots$. Of course, the zero vector is the only vector whose length is 0, and it is the only vector whose direction is indeterminate. Notice that $\overrightarrow{PQ} + \overrightarrow{QQ} = \overrightarrow{PQ}$; that is, we have the law

Figure 8

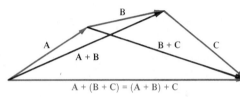

$$\mathbf{A} + \mathbf{0} = \mathbf{A}$$

which, in algebraic parlance, says that **0** is the **additive identity** for vectors.

If we turn a vector **A** around by interchanging its tail and head ends, we obtain a vector in the opposite direction called the **negative** of **A** and written $-\mathbf{A}$ (Figure 9). Evidently, if $\mathbf{A} = \overrightarrow{PQ}$, then $-\mathbf{A} = \overrightarrow{QP}$. Notice that the sum $\overrightarrow{PQ} + \overrightarrow{QP} = \overrightarrow{PP} = \mathbf{0}$; that is, we have the law

Figure 9

$$\mathbf{A} + (-\mathbf{A}) = \mathbf{0}$$

which expresses the fact that $-\mathbf{A}$ is the **additive inverse** of **A**.

We now define the operation of vector subtraction by the equation

$$\mathbf{A} - \mathbf{B} = \mathbf{A} + (-\mathbf{B})$$

Figure 10

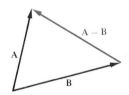

Notice that $\mathbf{A} - \mathbf{B}$ is the solution **X** of the vector equation $\mathbf{X} + \mathbf{B} = \mathbf{A}$ (Problem 15); that is, $\mathbf{A} - \mathbf{B}$ is the vector which, when added to **B**, gives **A** (Figure 10). Thus, if **A** and **B** have the same tail end, as in Figure 10, then $\mathbf{A} - \mathbf{B}$ is the vector running between their head ends and *pointing toward the head end of* **A**. We refer to this method of finding the vector difference $\mathbf{A} - \mathbf{B}$ geometrically as the **triangle rule for vector subtraction.**

If **A** is a vector, it seems reasonable on algebraic grounds to define 2**A** by the equation $2\mathbf{A} = \mathbf{A} + \mathbf{A}$. Using the head-to-tail rule to add **A** to itself, we find that 2**A** has the same direction as **A** but is twice as long (Figure 11). More generally, we make the following **rules for multiplication** of a vector **A** by a scalar s:

Figure 11

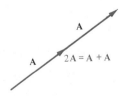

1 If $s > 0$, then $s\mathbf{A}$ is the vector in the *same direction* as **A**, but s times as long.

2 If $s < 0$, then $s\mathbf{A}$ is the vector in the *direction opposite* to **A**, but $|s|$ times as long.

Naturally, we understand that

$$s\mathbf{A} = \mathbf{0} \text{ if either } s = 0 \text{ or } \mathbf{A} = \mathbf{0} \text{ (or both)}$$

Figure 12

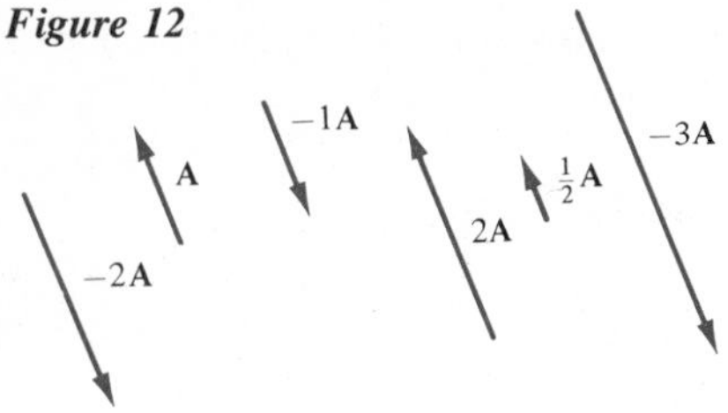

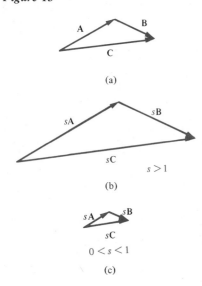

Figure 13

(a)

$s > 1$

(b)

$0 < s < 1$

(c)

Figure 12 shows a vector **A** and some of its scalar multiples. Notice that $(-1)\mathbf{A} = -\mathbf{A}$. More generally,

$$(-s)\mathbf{A} = s(-\mathbf{A})$$

holds for all scalars s and all vectors **A**. (Why?)

Multiplication of a vector by a scalar $s > 1$ "stretches" the vector by a factor of s, while multiplication by a scalar s between 0 and 1 "shrinks" it by a factor of s. If all vectors appearing in a diagram (Figure 13a) are multiplied by the same positive scalar s, $s \neq 1$, then the resulting diagram is either a magnification (Figure 13b) or a compression (Figure 13c) of the original diagram, according to whether $s > 1$ or $s < 1$. In Figure 13 notice that $\mathbf{A} + \mathbf{B} = \mathbf{C}$ by the head-to-tail rule; likewise, $s\mathbf{A} + s\mathbf{B} = s\mathbf{C}$. Substituting from the first equation into the second, we obtain the **first distributive law** $s\mathbf{A} + s\mathbf{B} = s(\mathbf{A} + \mathbf{B})$, $s > 0$. Now suppose that $s > 0$, so that $-s < 0$. Using the first distributive law above, we have

$$(-s)\mathbf{A} + (-s)\mathbf{B} = s(-\mathbf{A}) + s(-\mathbf{B}) = s[(-\mathbf{A}) + (-\mathbf{B})] = s[-(\mathbf{A} + \mathbf{B})]$$

$$= (-s)(\mathbf{A} + \mathbf{B})$$

Hence, the first distributive law works for negative scalars as well. Since it obviously works when $s = 0$, we have

$$s\mathbf{A} + s\mathbf{B} = s(\mathbf{A} + \mathbf{B})$$

for all scalars s and for all vectors **A** and **B**.

There is a **second distributive law** involving two scalars and one vector:

$$s\mathbf{A} + t\mathbf{A} = (s + t)\mathbf{A}$$

This is an obvious consequence of the rules for multiplication by scalars when both s and t are nonnegative. If you check the other possible cases, it is clear that the equation holds for all scalars s and t.

In addition to the two distributive laws, there is an **associative** type of law for scalar multiplication:

$$(st)\mathbf{A} = s(t\mathbf{A})$$

which can be confirmed geometrically simply by considering the various possible cases according to whether s and t are positive, negative, or zero.

If s is a nonzero scalar and **A** is a vector, the expression $(1/s)\mathbf{A}$ is often written as $\mathbf{A}/s$. Thus, when we speak of **dividing** the vector **A** by the nonzero scalar s, we simply mean to multiply **A** by $1/s$.

In order to show that two vectors **A** and **B** are *parallel*, you must show that one of the two vectors (either one) is a scalar multiple of the other. Indeed, we take this as our official definition.

DEFINITION 1   **Parallel Vectors**

We say that the vectors **A** and **B** are **parallel** if there is a scalar s such that $\mathbf{A} = s\mathbf{B}$ or if there is a scalar t such that $\mathbf{B} = t\mathbf{A}$. Two vectors **A** and **B** that are not parallel are said to be **linearly independent**.

If $A = sB$, where $s > 0$, then we say not only that **A** and **B** are parallel, but also that they have the **same direction;** however, if $A = sB$, where $s < 0$, then we say that **A** and **B** are parallel, but have **opposite directions.**

We notice three simple consequences of Definition 1. First, *any vector is parallel to itself.* (Why?) Second, *the zero vector is parallel to every vector* because, if **A** is any vector, then $0 = 0A$. (This is not unreasonable since the zero vector has an indeterminate direction anyway.) Third, *if two vectors are linearly independent, then neither of them can be zero* since the zero vector *is* parallel to every other vector.

The following theorem gives an important property of two linearly independent (that is, nonparallel) vectors.*

THEOREM 1

Linear Independence of Two Vectors

> If **A** and **B** are linearly independent vectors, then the only scalars s and t for which $sA + tB = 0$ are $s = 0$ and $t = 0$.

An expression of the form $sA + tB$ is called a **linear combination** of the vectors **A** and **B** with **coefficients** s and t. To say that **A** and **B** are linearly independent is equivalent to saying that the only way to obtain **0** as a linear combination of **A** and **B** is to make both of the coefficients equal to zero.

Standard Basis Vectors

In dealing with vectors lying in a plane, it is often useful to choose two perpendicular vectors of length 1 and to write all other vectors in the plane as linear combinations of these two vectors. The two chosen vectors, called **basis vectors** are usually denoted by **i** and **j** (Figure 14). In our diagrams, we usually take **i** to be a horizontal vector pointing to the right and we take **j** to be a vertical vector pointing upward. Although **i** and **j** can be moved around in the plane, as all vectors can, we usually draw them with a common tail end O. Naturally, if we have a Cartesian or a polar coordinate system already established in the plane, we ordinarily take O to be the origin or the pole. The basis **i**, **j** is called the **standard basis.** Notice that **i** and **j** are linearly independent.

Figure 14

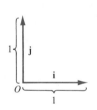

To see that any vector **R** in the plane can be written as a linear combination of **i** and **j**, we argue as follows: Establish a Cartesian coordinate system so that **i** lies along the positive x axis and **j** lies along the positive y axis. Move the given vector **R** so that its tail end is at the origin O, and let its head end be the point $P = (x, y)$. Drop perpendiculars from P to the x and y axes, meeting these axes, respectively, at the points $S = (x, 0)$ and $T = (0, y)$ (Figure 15). If $x > 0$, then $x\mathbf{i}$ is a vector in the same direction as **i**, but x times as long; hence, $x\mathbf{i} = \overrightarrow{OS}$. If $x \leq 0$, we also have $x\mathbf{i} = \overrightarrow{OS}$. (Why?) Similarly, $y\mathbf{j} = \overrightarrow{OT}$. By the parallelogram rule,

$$\mathbf{R} = \overrightarrow{OP} = \overrightarrow{OS} + \overrightarrow{OT}$$

Figure 15

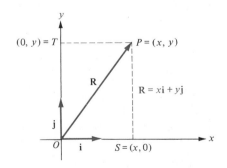

so that

> $$\mathbf{R} = x\mathbf{i} + y\mathbf{j}$$

and **R** is expressed as a linear combination of **i** and **j**. The numbers x and y are called the (scalar) **components** of the vector **R** with respect to the standard **i**, **j** basis.

*For the proof, see Problem 56.

Figure 16

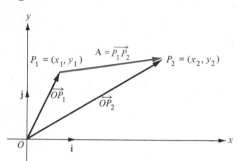

Notice that *these components are just the coordinates of the head end of* **R** *when its tail end is at the origin.*

Now suppose that $P_1 = (x_1, y_1)$ and $P_2 = (x_2, y_2)$ (Figure 16). Then we have $\overrightarrow{OP_1} = x_1\mathbf{i} + y_1\mathbf{j}$ and $\overrightarrow{OP_2} = x_2\mathbf{i} + y_2\mathbf{j}$; hence, by the triangle rule for subtraction,

$$\overrightarrow{P_1P_2} = \overrightarrow{OP_2} - \overrightarrow{OP_1}$$

$$= (x_2\mathbf{i} + y_2\mathbf{j}) - (x_1\mathbf{i} + y_1\mathbf{j})$$

$$= x_2\mathbf{i} + y_2\mathbf{j} - x_1\mathbf{i} - y_1\mathbf{j}$$

$$= x_2\mathbf{i} - x_1\mathbf{i} + y_2\mathbf{j} - y_1\mathbf{j}$$

and

$$\boxed{\overrightarrow{P_1P_2} = (x_2 - x_1)\mathbf{i} + (y_2 - y_1)\mathbf{j}}$$

EXAMPLE 1   If $P = (-2, 5)$ and $Q = (7, 3)$, write the vectors $\overrightarrow{OP}$ and $\overrightarrow{PQ}$ in component form.

SOLUTION

$$\overrightarrow{OP} = -2\mathbf{i} + 5\mathbf{j} \qquad \text{and} \qquad \overrightarrow{PQ} = [7 - (-2)]\mathbf{i} + [3 - 5]\mathbf{j} = 9\mathbf{i} - 2\mathbf{j} \qquad ■$$

EXAMPLE 2   Use vectors to find the coordinates of a point three-fourths of the way from $(-5, -9)$ to $(7, 7)$.

Figure 17

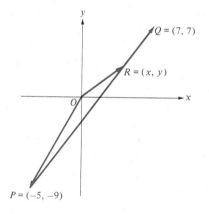

SOLUTION   Put $P = (-5, -9)$ and $Q = (7, 7)$. Let $R = (x, y)$ be the desired point (Figure 17), so that $\overrightarrow{PR} = \frac{3}{4}\overrightarrow{PQ}$. Now, $\overrightarrow{OR} = \overrightarrow{OP} + \overrightarrow{PR} = \overrightarrow{OP} + \frac{3}{4}\overrightarrow{PQ}$. Since

$$\overrightarrow{OR} = x\mathbf{i} + y\mathbf{j} \qquad \overrightarrow{OP} = -5\mathbf{i} - 9\mathbf{j}$$

and

$$\overrightarrow{PQ} = [7 - (-5)]\mathbf{i} + [7 - (-9)]\mathbf{j} = 12\mathbf{i} + 16\mathbf{j}$$

the equation $\overrightarrow{OR} = \overrightarrow{OP} + \frac{3}{4}\overrightarrow{PQ}$ can be rewritten as

$$x\mathbf{i} + y\mathbf{j} = -5\mathbf{i} - 9\mathbf{j} + \tfrac{3}{4}(12\mathbf{i} + 16\mathbf{j})$$

$$= -5\mathbf{i} - 9\mathbf{j} + 9\mathbf{i} + 12\mathbf{j}$$

$$= 4\mathbf{i} + 3\mathbf{j}$$

Therefore,

$$(x - 4)\mathbf{i} + (y - 3)\mathbf{j} = \mathbf{0}$$

Since $\mathbf{i}$ and $\mathbf{j}$ are linearly independent, the last equation implies that $x - 4 = 0$ and $y - 3 = 0$. Therefore, $(x, y) = (4, 3)$. ■

EXAMPLE 3   Show that, *if two vectors are equal, then their components are equal.*

SOLUTION   If $x\mathbf{i} + y\mathbf{j} = a\mathbf{i} + b\mathbf{j}$, then $(x - a)\mathbf{i} + (y - b)\mathbf{j} = \mathbf{0}$. Since $\mathbf{i}$ and $\mathbf{j}$ are linearly independent, the last equation implies that $x - a = 0$ and $y - b = 0$; that is, $x = a$ and $y = b$. ■

Calculation with vectors in component form is really very simple because of the fact that vectors can be added, subtracted, or multiplied by scalars in a "componentwise" manner. In fact, if

$$\mathbf{A} = a\mathbf{i} + b\mathbf{j} \qquad \text{and} \qquad \mathbf{B} = c\mathbf{i} + d\mathbf{j}$$

we have

> 1 $\mathbf{A} + \mathbf{B} = (a + c)\mathbf{i} + (b + d)\mathbf{j}$
>
> 2 $\mathbf{A} - \mathbf{B} = (a - c)\mathbf{i} + (b - d)\mathbf{j}$
>
> 3 $t\mathbf{A} = (ta)\mathbf{i} + (tb)\mathbf{j}$

These facts are easily verified by using the identities for vector addition and multiplication by scalars; for instance, to confirm the second equation, we calculate as follows:

$$\mathbf{A} - \mathbf{B} = \mathbf{A} + (-\mathbf{B}) = a\mathbf{i} + b\mathbf{j} - c\mathbf{i} - d\mathbf{j} = a\mathbf{i} - c\mathbf{i} + b\mathbf{j} - d\mathbf{j} = (a - c)\mathbf{i} + (b - d)\mathbf{j}$$

EXAMPLE 4 If $\mathbf{A} = 2\mathbf{i} + 17\mathbf{j}$ and $\mathbf{B} = 13\mathbf{i} - 3\mathbf{j}$, find **(a)** $\mathbf{A} + \mathbf{B}$, **(b)** $-7\mathbf{A}$, and **(c)** $-7\mathbf{A} - \mathbf{B}$ in component form.

SOLUTION

(a) $\mathbf{A} + \mathbf{B} = (2 + 13)\mathbf{i} + (17 - 3)\mathbf{j} = 15\mathbf{i} + 14\mathbf{j}$

(b) $-7\mathbf{A} = -7(2\mathbf{i} + 17\mathbf{j}) = -14\mathbf{i} - 119\mathbf{j}$

(c) $-7\mathbf{A} - \mathbf{B} = (-14\mathbf{i} - 119\mathbf{j}) - (13\mathbf{i} - 3\mathbf{j}) = (-14 - 13)\mathbf{i} + (-119 + 3)\mathbf{j}$
$= -27\mathbf{i} - 116\mathbf{j}$

If we fix a standard basis **i**, **j**, then a vector $\mathbf{A} = x\mathbf{i} + y\mathbf{j}$ both determines and is determined by its scalar components x and y. For this reason, *the vector $\mathbf{A} = x\mathbf{i} + y\mathbf{j}$ is sometimes denoted by the ordered pair $\langle x, y \rangle$ of its components and is written*

$$\mathbf{A} = \langle x, y \rangle$$

Thus, in the previous example, we have $\mathbf{A} = \langle 2, 17 \rangle$ and $\mathbf{B} = \langle 13, -3 \rangle$; hence,

$$\mathbf{A} + \mathbf{B} = \langle 2 + 13, 17 - 3 \rangle = \langle 15, 14 \rangle \qquad -7\mathbf{A} = \langle -14, -119 \rangle$$

$$-7\mathbf{A} - \mathbf{B} = \langle -14 - 13, -119 + 3 \rangle = \langle -27, -116 \rangle$$

We encourage you to use ordered-pair notation for vectors whenever it seems convenient.

Problem Set 12.1

In Problems 1 to 12, copy the appropriate vectors from Figure 18 onto your paper, and then find each vector by means of a suitable geometric construction.

Figure 18

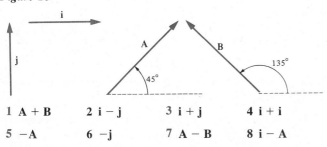

1 $\mathbf{A} + \mathbf{B}$ 2 $\mathbf{i} - \mathbf{j}$ 3 $\mathbf{i} + \mathbf{j}$ 4 $\mathbf{i} + \mathbf{i}$

5 $-\mathbf{A}$ 6 $-\mathbf{j}$ 7 $\mathbf{A} - \mathbf{B}$ 8 $\mathbf{i} - \mathbf{A}$

9 $2\mathbf{i}$ 10 $2\mathbf{i} + 3\mathbf{j}$ 11 $\mathbf{i} - \dfrac{1}{2}\mathbf{j}$ 12 $\dfrac{\mathbf{A} + \mathbf{B}}{2}$

In Problems 13 and 14, solve each equation for the vector **X**.

13 $\mathbf{A} + \mathbf{X} + \mathbf{B} = \mathbf{0}$

14 $(\mathbf{A} - \mathbf{X}) - (\mathbf{B} - \mathbf{X}) = \mathbf{C} - \mathbf{X}$

15 Show by a diagram or otherwise that the *law of transposition* works for vectors; that is, $\mathbf{X} + \mathbf{B} = \mathbf{A}$ holds if and only if $\mathbf{X} = \mathbf{A} - \mathbf{B}$.

16 What is wrong with the following equations?

(a) $\mathbf{A} - \mathbf{B} = 5$ **(b)** $\mathbf{A} + 3 = \mathbf{B}$ **(c)** $\mathbf{A} + \mathbf{B} = 0$

In Problems 17 to 20, write the vectors $\overrightarrow{PQ}$ and $\overrightarrow{QP}$ in component form.

17 $P = (2, 4)$ and $Q = (6, -1)$

18 $P = (5, 0)$ and $Q = (-1, 7)$

19 $P = (-3, 0)$ and $Q = (-1, 0)$

20 $P = (0, 0)$ and $Q = (x, y)$

In Problems 21 to 30, let $\mathbf{A} = 4\mathbf{i} + 2\mathbf{j}$, $\mathbf{B} = -3\mathbf{i} + 4\mathbf{j}$, and $\mathbf{C} = -5\mathbf{i} + 7\mathbf{j}$. Evaluate each expression.

21 $\mathbf{A} + \mathbf{B}$ **22** $\mathbf{A} - \mathbf{B}$

23 $2\mathbf{A} + 3\mathbf{B}$ **24** $7\mathbf{A} - 5\mathbf{C}$

25 $7\mathbf{B} + 3\mathbf{C}$ **26** $-3\mathbf{A} - 8\mathbf{C}$

27 $3(\mathbf{B} - 7\mathbf{C})$ **28** $2\mathbf{A} - (\mathbf{B} + \mathbf{C})$

29 $3\mathbf{B} - \mathbf{C} + 2\mathbf{A}$ **30** $\dfrac{\mathbf{A} + \mathbf{B}}{2} - \dfrac{2}{5}\mathbf{C}$

In Problems 31 to 38, let $\mathbf{A} = \langle 4, 2 \rangle$, $\mathbf{B} = \langle -3, 4 \rangle$, and $\mathbf{C} = \langle -5, 7 \rangle$. Evaluate each expression.

31 $5\mathbf{B} + 2\mathbf{C}$ **32** $4\mathbf{B} - \mathbf{C}$

33 $2\mathbf{A} + 3\mathbf{B}$ **34** $5\mathbf{A} - 2\mathbf{C}$

35 $-3\mathbf{A} + 2\mathbf{B}$ **36** $2\mathbf{A} - \mathbf{B}$

37 $\frac{2}{3}\mathbf{A} - \mathbf{B}$ **38** $\frac{1}{5}\mathbf{A} - \frac{3}{5}\mathbf{C}$

In Problems 39 to 44, let $\mathbf{A} = 2\mathbf{i} + 7\mathbf{j}$, $\mathbf{B} = \mathbf{i} - 6\mathbf{j}$, and $\mathbf{C} = -5\mathbf{i} + 10\mathbf{j}$.

39 Solve the vector equation $(3x + 4y - 12)\mathbf{A} + (3x - 8y)\mathbf{B} = \mathbf{0}$ for x and y.

40 Find a scalar s such that $s\mathbf{A} + \mathbf{B}$ is parallel to $\mathbf{C}$.

41 Solve the vector equation $(x - y)\mathbf{A} = (3x + 2y)\mathbf{B} - \mathbf{A}$ for x and y.

42 Find scalars s and t such that $s\mathbf{A} + t\mathbf{B} = \mathbf{C}$.

43 Find scalars C_1 and C_2 such that $C_1\mathbf{A} + C_2\mathbf{B} = -3\mathbf{j}$.

44 Show that all vectors in the plane can be obtained by forming linear combinations of $\mathbf{A}$ and $\mathbf{B}$.

45 Show geometrically that $-(\mathbf{A} - \mathbf{B}) = \mathbf{B} - \mathbf{A}$.

46 One of the diagonals of the parallelogram spanned by two vectors $\mathbf{A}$ and $\mathbf{B}$ is $\mathbf{A} + \mathbf{B}$. Express the other diagonal in terms of $\mathbf{A}$ and $\mathbf{B}$.

47 Suppose that the four distinct points P, Q, R, and S all lie on the same line. If the vector $\overrightarrow{PQ}$ has the same direction as the vector $\overrightarrow{RS}$, what are the possible orders in which the points lie on the line? (For instance, S, Q, R, P—meaning first S, then Q, then R, and finally P—is one possibility.)

48 A car travels 30 miles due east, then 40 miles due north, then 30 miles due east again. Draw a scale diagram and represent by vectors the successive displacements of the car. Add these vectors, using the head-to-tail rule, and thus find the resultant displacement of the car.

49 In Figure 19, $PQRS$ is a square, and T is the midpoint of $\overline{SR}$. Express $\overrightarrow{PT}$ in terms of $\overrightarrow{PQ}$ and $\overrightarrow{QR}$.

50 Let $\mathbf{A} = \overrightarrow{PQ}$, $\mathbf{B} = \overrightarrow{PR}$, and $\mathbf{C} = \overrightarrow{PS}$, where the point S is on the line segment QR and three-eighths of the way from Q to R (Figure 20). Express $\mathbf{C}$ in terms of $\mathbf{A}$ and $\mathbf{B}$.

Figure 19

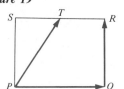

Figure 20

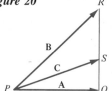

51 Let $O = (0, 0)$, $P = (3, 6)$, $Q = (-1, 3)$, $R = (-7, -1)$, and $S = (3, -6)$ be points in the xy plane. Find the (scalar) components of each of the following vectors with respect to the standard $\mathbf{i}$, $\mathbf{j}$ basis.

(a) $\overrightarrow{OP}$ (b) $\overrightarrow{OQ}$ (c) $\overrightarrow{PQ}$ (d) $\overrightarrow{QP}$

(e) $\overrightarrow{PR}$ (f) $\overrightarrow{RS}$ (g) $\overrightarrow{PQ} + \overrightarrow{RS}$ (h) $3\overrightarrow{QR} - 5\overrightarrow{SP}$

52 In Figure 21, let $\mathbf{A} = \overrightarrow{PQ}$, $\mathbf{B} = \overrightarrow{PR}$, $\overrightarrow{PS} = 5\mathbf{A}$, and $\overrightarrow{PT} = 3\mathbf{B}$. Express each of the following vectors in terms of $\mathbf{A}$ and $\mathbf{B}$:

(a) $\overrightarrow{QR}$ (b) $\overrightarrow{QS}$ (c) $\overrightarrow{ST}$

(d) $\overrightarrow{SR}$ (e) $\overrightarrow{QM}$ (f) $\overrightarrow{MS}$

Figure 21

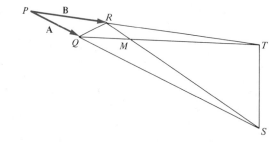

53 Let $ABCD$ be a quadrilateral, and let P, Q, R, and S be the midpoints of $\overline{AB}$, $\overline{BC}$, $\overline{CD}$, and $\overline{DA}$, respectively. Use vectors to show that $PQRS$ is a parallelogram.

54 If $\mathbf{A}$ and $\mathbf{B}$ are two fixed linearly independent vectors in the plane, give a geometric construction to show how to find scalars s and t so that $s\mathbf{A} + t\mathbf{B} = \mathbf{C}$, where $\mathbf{C}$ is an arbitrary given vector in the plane.

55 If $\mathbf{A} = u\mathbf{B}$, where u is a scalar, show that there are scalars s and t, *not both zero*, such that $s\mathbf{A} + t\mathbf{B} = \mathbf{0}$.

56 Show that if $\mathbf{A}$ and $\mathbf{B}$ are linearly independent vectors, then the only scalars s and t for which $s\mathbf{A} + t\mathbf{B} = \mathbf{0}$ are $s = 0$ and $t = 0$.

12.2   The Dot Product

In Section 12.1 we saw that vectors can be added, subtracted, and multiplied by scalars. Vectors can also be multiplied by each other; in fact, there are several different products defined for vectors, each with its own special notation. One of the most useful is the *dot product* of two vectors **A** and **B**, which is so called because of the notation "**A · B**" traditionally used to denote it. In this section we define and study the dot product (also called the *inner product*, or the *scalar product*) and calculate lengths of vectors, angles between vectors, and projections of vectors upon other vectors using dot products.

We begin with a geometric definition of the dot product.

DEFINITION 1

Figure 1

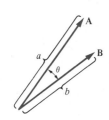

Dot Product of Vectors

> Let **A** and **B** be two vectors with lengths (magnitudes) a and b, respectively, and let θ be the angle between **A** and **B** (Figure 1). Then the **dot product** of **A** and **B** is defined by
> $$\mathbf{A} \cdot \mathbf{B} = ab \cos \theta$$

(If either **A** or **B** is the zero vector, then the angle θ is indeterminate; but, in this case $a = 0$ or $b = 0$, so that $\mathbf{A} \cdot \mathbf{B} = 0$.)

Note that the dot product of two vectors is a *scalar—not a vector*. In order to measure the angle θ between two vectors **A** and **B**, we can always move **A** and **B** so that they have a common tail end (as in Figure 1) and then use a protractor in the usual way. We express the angle either in degrees or in radians.

EXAMPLE 1   If the length of **A** is 5 units and the length of **B** is 4 units, find **A · B** given that the angle θ between **A** and **B** is **(a)** $\theta = 0$, **(b)** $\theta = \pi/2$, and Ⓒ**(c)** $\theta = 40°$.

SOLUTION

(a) $\mathbf{A} \cdot \mathbf{B} = (5)(4) \cos 0 = (5)(4)(1) = 20$

(b) $\mathbf{A} \cdot \mathbf{B} = (5)(4) \cos (\pi/2) = (5)(4)(0) = 0$

(c) $\mathbf{A} \cdot \mathbf{B} = (5)(4) \cos 40° \approx 15.32$ ∎

Now, suppose that **A** is a nonzero vector of length a. Since the angle between **A** and itself is zero, $\mathbf{A} \cdot \mathbf{A} = aa \cos 0 = a^2$; hence, $a = \sqrt{\mathbf{A} \cdot \mathbf{A}}$. On the other hand, if $\mathbf{A} = \mathbf{0}$, then $\mathbf{A} \cdot \mathbf{A} = 0$ and $\sqrt{\mathbf{A} \cdot \mathbf{A}}$ still gives the length of **A**. We use the symbol $|\mathbf{A}|$ for the **length,** or **magnitude,** of a vector **A**; hence, we have the formulas

$$\mathbf{A} \cdot \mathbf{A} = |\mathbf{A}|^2 \qquad \text{and} \qquad |\mathbf{A}| = \sqrt{\mathbf{A} \cdot \mathbf{A}}$$

Using the notation $|\mathbf{A}|$ and $|\mathbf{B}|$ for the lengths of the vectors **A** and **B**, we can now rewrite the formula for the dot product as

$$\mathbf{A} \cdot \mathbf{B} = |\mathbf{A}| \, |\mathbf{B}| \cos \theta$$

where θ is the angle between **A** and **B**.

EXAMPLE 2   Let **i** and **j** be the standard basis vectors in the xy plane. Find **i · i**, **j · j**, and **i · j**.

SOLUTION   Since $|\mathbf{i}| = 1$, it follows that $\mathbf{i} \cdot \mathbf{i} = |\mathbf{i}|^2 = 1^2 = 1$. Similarly, $\mathbf{j} \cdot \mathbf{j} = 1$. The angle between $\mathbf{i}$ and $\mathbf{j}$ is $\pi/2$ radians; hence, $\mathbf{i} \cdot \mathbf{j} = |\mathbf{i}|\,|\mathbf{j}| \cos (\pi/2) = (1)(1)(0) = 0.$ ∎

Properties of the Dot Product

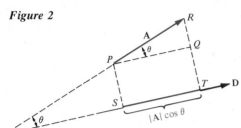

Figure 2

Let $\mathbf{A}$ and $\mathbf{D}$ be nonzero vectors making an acute angle θ as in Figure 2. Then perpendiculars dropped from the endpoints of $\mathbf{A}$ to the line through $\mathbf{D}$ cut off a segment $\overline{ST}$ of length

$$|\overline{ST}| = |\overline{PQ}| = |\overline{PR}| \cos \theta = |\mathbf{A}| \cos \theta$$

The number $|\mathbf{A}| \cos \theta$ is called the **scalar component of A** *in the direction of* $\mathbf{D}$ or the **scalar projection of A** *in the direction of* $\mathbf{D}$. More generally, if θ is the angle between the vectors $\mathbf{A}$ and $\mathbf{D}$, we define

$$\text{comp}_{\mathbf{D}}\mathbf{A} = |\mathbf{A}| \cos \theta$$

even if θ is not an acute angle. Note that, if $\mathbf{A}$ and $\mathbf{D}$ make an *obtuse angle*, so that $\pi/2 < \theta < \pi$, then $\cos \theta < 0$ and $\text{comp}_{\mathbf{D}}\mathbf{A}$ is *negative*. Of course, $\text{comp}_{\mathbf{D}}\mathbf{A} = 0$ when $\theta = \pi/2$.

Because

$$\mathbf{A} \cdot \mathbf{D} = |\mathbf{A}|\,|\mathbf{D}| \cos \theta = (|\mathbf{A}| \cos \theta)|\mathbf{D}| = (\text{comp}_{\mathbf{D}}\mathbf{A})\,|\mathbf{D}|$$

it follows that *the dot product of two vectors is the scalar component of the first vector in the direction of the second vector times the length of the second vector.* If we solve the last equation for $\text{comp}_{\mathbf{D}}\mathbf{A}$, we obtain the useful formula

$$\text{comp}_{\mathbf{D}}\mathbf{A} = \frac{\mathbf{A} \cdot \mathbf{D}}{|\mathbf{D}|}$$

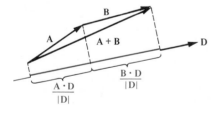

Figure 3

A glance at Figure 3 shows that the sum of the scalar components of $\mathbf{A}$ and $\mathbf{B}$ in the direction of $\mathbf{D}$ is equal to the scalar component of $\mathbf{A} + \mathbf{B}$ in the direction of $\mathbf{D}$, so that

$$\frac{\mathbf{A} \cdot \mathbf{D}}{|\mathbf{D}|} + \frac{\mathbf{B} \cdot \mathbf{D}}{|\mathbf{D}|} = \frac{(\mathbf{A} + \mathbf{B}) \cdot \mathbf{D}}{|\mathbf{D}|}$$

that is, scalar components are **additive.** (You should check appropriate diagrams to see that scalar components are additive in all cases, even when some of the angles involved are obtuse.) Multiplying the last equation on both sides by $|\mathbf{D}|$, we obtain the **distributive law** for the dot product,

$$(\mathbf{A} + \mathbf{B}) \cdot \mathbf{D} = \mathbf{A} \cdot \mathbf{D} + \mathbf{B} \cdot \mathbf{D}$$

From the rule for the product of a vector by a scalar, we have

$$|s\mathbf{A}| = |s|\,|\mathbf{A}|$$

for all scalars s and all vectors $\mathbf{A}$. If θ is the angle between $\mathbf{A}$ and $\mathbf{B}$ and if s is a positive scalar, then θ is still the angle between $s\mathbf{A}$ and $\mathbf{B}$; hence,

$$(s\mathbf{A}) \cdot \mathbf{B} = |s\mathbf{A}|\,|\mathbf{B}| \cos \theta = |s|\,|\mathbf{A}|\,|\mathbf{B}| \cos \theta = |s|(\mathbf{A} \cdot \mathbf{B}) = s(\mathbf{A} \cdot \mathbf{B})$$

Figure 4

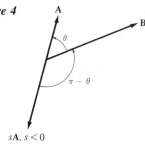

$sA, s < 0$

On the other hand, if s is a negative scalar, then $s\mathbf{A}$ makes an angle of $\pi - \theta$ with $\mathbf{B}$ (Figure 4), and we have

$$(s\mathbf{A}) \cdot \mathbf{B} = |s\mathbf{A}|\,|\mathbf{B}|\cos(\pi - \theta) = |s|\,|\mathbf{A}|\,|\mathbf{B}|(-\cos\theta)$$

$$= (-s)|\mathbf{A}|\,|\mathbf{B}|(-\cos\theta) = s|\mathbf{A}|\,|\mathbf{B}|\cos\theta = s(\mathbf{A}\cdot\mathbf{B})$$

Consequently, $(s\mathbf{A}) \cdot \mathbf{B} = s(\mathbf{A} \cdot \mathbf{B})$ holds for $s \neq 0$. Since this equation also holds when $s = 0$, we have the **homogeneous law** for the dot product:

$$(s\mathbf{A}) \cdot \mathbf{B} = s(\mathbf{A} \cdot \mathbf{B})$$

Because of this law, we may simply write $s\mathbf{A} \cdot \mathbf{B}$, without parentheses.

We can now present the four basic laws governing the behavior of the dot product:

1 $\mathbf{A}\cdot\mathbf{B} = \mathbf{B}\cdot\mathbf{A}$	2 $(\mathbf{A} + \mathbf{B})\cdot\mathbf{D} = \mathbf{A}\cdot\mathbf{D} + \mathbf{B}\cdot\mathbf{D}$
3 $(s\mathbf{A})\cdot\mathbf{B} = s(\mathbf{A}\cdot\mathbf{B})$	4 $\mathbf{A}\cdot\mathbf{A} \geq 0$ and $\mathbf{A}\cdot\mathbf{A} = 0$ only if $\mathbf{A} = \mathbf{0}$

The first law follows from the fact that $|\mathbf{A}|\,|\mathbf{B}|\cos\theta = |\mathbf{B}|\,|\mathbf{A}|\cos\theta$, and the fourth law is a consequence of the fact that $\mathbf{A}\cdot\mathbf{A} = |\mathbf{A}|^2$.

All the properties of the dot product and of lengths of vectors can be derived deductively from the four basic laws together with the identities developed in Section 12.1. Some of these properties are as follows:

5 $\mathbf{D}\cdot(\mathbf{A} + \mathbf{B}) = \mathbf{D}\cdot\mathbf{A} + \mathbf{D}\cdot\mathbf{B}$	6 $(\mathbf{A} - \mathbf{B})\cdot\mathbf{D} = \mathbf{A}\cdot\mathbf{D} - \mathbf{B}\cdot\mathbf{D}$
7 $\mathbf{D}\cdot(\mathbf{A} - \mathbf{B}) = \mathbf{D}\cdot\mathbf{A} - \mathbf{D}\cdot\mathbf{B}$	8 $(s\mathbf{A})\cdot(t\mathbf{B}) = st(\mathbf{A}\cdot\mathbf{B})$
9 $|\mathbf{A}\cdot\mathbf{B}| \leq |\mathbf{A}|\,|\mathbf{B}|$	(Schwarz inequality)
10 $|\mathbf{A} + \mathbf{B}| \leq |\mathbf{A}| + |\mathbf{B}|$	(Triangle inequality)

The following theorem gives a useful formula for calculating the dot product of two vectors given in component form.

THEOREM 1   **Dot Products of Vectors in Component Form**

If $\mathbf{A} = a\mathbf{i} + b\mathbf{j}$ and $\mathbf{B} = c\mathbf{i} + d\mathbf{j}$, then

$$\mathbf{A} \cdot \mathbf{B} = ac + bd$$

PROOF   Using the distributive and homogeneous laws, we have

$$\mathbf{A} \cdot \mathbf{B} = (a\mathbf{i} + b\mathbf{j}) \cdot \mathbf{B} = (a\mathbf{i}) \cdot \mathbf{B} + (b\mathbf{j}) \cdot \mathbf{B} = a(\mathbf{i} \cdot \mathbf{B}) + b(\mathbf{j} \cdot \mathbf{B})$$

In order to calculate $\mathbf{i} \cdot \mathbf{B}$, we observe that $\mathbf{i} \cdot \mathbf{i} = 1$ and $\mathbf{i} \cdot \mathbf{j} = 0$; hence,

$$\mathbf{i} \cdot \mathbf{B} = \mathbf{i} \cdot (c\mathbf{i} + d\mathbf{j}) = \mathbf{i} \cdot (c\mathbf{i}) + \mathbf{i} \cdot (d\mathbf{j}) = c(\mathbf{i} \cdot \mathbf{i}) + d(\mathbf{i} \cdot \mathbf{j}) = c$$

Similarly,

$$\mathbf{j} \cdot \mathbf{B} = \mathbf{j} \cdot (c\mathbf{i} + d\mathbf{j}) = \mathbf{j} \cdot (c\mathbf{i}) + \mathbf{j} \cdot (d\mathbf{j}) = c(\mathbf{j} \cdot \mathbf{i}) + d(\mathbf{j} \cdot \mathbf{j}) = d$$

since $\mathbf{j} \cdot \mathbf{i} = 0$ and $\mathbf{j} \cdot \mathbf{j} = 1$. It follows that

$$\mathbf{A} \cdot \mathbf{B} = a(\mathbf{i} \cdot \mathbf{B}) + b(\mathbf{j} \cdot \mathbf{B}) = ac + bd$$

In words, Theorem 1 says that *the dot product of two vectors is the sum of the products of their corresponding scalar components*. Using the ordered-pair notation, we have

$$\langle a, b \rangle \cdot \langle c, d \rangle = ac + bd$$

EXAMPLE 3   Let $\mathbf{A} = 2\mathbf{i} + 3\mathbf{j}$ and $\mathbf{B} = 4\mathbf{i} - 5\mathbf{j}$. Calculate

   (a) $\mathbf{A} \cdot \mathbf{B}$ **(b)** $\mathbf{A} \cdot (2\mathbf{A} - 3\mathbf{B})$

SOLUTION   Using Theorem 1, we have

   (a) $\mathbf{A} \cdot \mathbf{B} = (2)(4) + (3)(-5) = -7$

   (b) $\mathbf{A} \cdot (2\mathbf{A} - 3\mathbf{B}) = (2\mathbf{i} + 3\mathbf{j}) \cdot (-8\mathbf{i} + 21\mathbf{j}) = (2)(-8) + (3)(21) = 47$ ∎

An important consequence of Theorem 1 is the formula

$$|x\mathbf{i} + y\mathbf{j}| = \sqrt{x^2 + y^2}$$

which says that *the length of a vector is the square root of the sum of the squares of its scalar components*. Indeed, if $\mathbf{A} = x\mathbf{i} + y\mathbf{j}$, then by Theorem 1, $\mathbf{A} \cdot \mathbf{A} = x^2 + y^2$, and so

$$|\mathbf{A}| = \sqrt{\mathbf{A} \cdot \mathbf{A}} = \sqrt{x^2 + y^2}$$

Notice that if $\mathbf{A} \neq \mathbf{0}$, then

$$\left| \frac{\mathbf{A}}{|\mathbf{A}|} \right| = \left| \frac{1}{|\mathbf{A}|} \mathbf{A} \right| = \left| \frac{1}{|\mathbf{A}|} \right| |\mathbf{A}| = \frac{1}{|\mathbf{A}|} |\mathbf{A}| = 1$$

Hence,

$$\frac{\mathbf{A}}{|\mathbf{A}|} \text{ is a vector of unit length in the same direction as } \mathbf{A}$$

A vector of unit length is called a **unit vector,** and the procedure of dividing a nonzero vector by its own length to obtain a unit vector in the same direction is called **normalizing.**

EXAMPLE 4   If $\mathbf{A} = 3\mathbf{i} + 4\mathbf{j}$, find

   (a) $|\mathbf{A}|$ **(b)** A unit vector in the same direction as $\mathbf{A}$

SOLUTION

   (a) $|\mathbf{A}| = \sqrt{3^2 + 4^2} = \sqrt{25} = 5$ units

   (b) Normalizing $\mathbf{A}$, we obtain

$$\frac{\mathbf{A}}{|\mathbf{A}|} = \frac{3\mathbf{i} + 4\mathbf{j}}{5} = \frac{3}{5}\mathbf{i} + \frac{4}{5}\mathbf{j}$$ ∎

EXAMPLE 5   If $\mathbf{A} = 3\mathbf{i} - 5\mathbf{j}$ and $\mathbf{D} = 4\mathbf{i} + 3\mathbf{j}$, find the scalar component of $\mathbf{A}$ in the direction of $\mathbf{D}$.

SOLUTION $\text{comp}_{\mathbf{D}}\mathbf{A} = \dfrac{\mathbf{A} \cdot \mathbf{D}}{|\mathbf{D}|} = \dfrac{(3)(4) + (-5)(3)}{\sqrt{4^2 + 3^2}} = -\dfrac{3}{5}$ ∎

If **A** and **B** are nonzero vectors, then the formula $\mathbf{A} \cdot \mathbf{B} = |\mathbf{A}|\,|\mathbf{B}|\cos\theta$ can be rewritten in the form

$$\cos\theta = \frac{\mathbf{A} \cdot \mathbf{B}}{|\mathbf{A}|\,|\mathbf{B}|}$$

and used to find the cosine of the angle θ between **A** and **B**. Note that $\theta = \pi/2$ if and only if $\mathbf{A} \cdot \mathbf{B} = 0$; that is,

A and **B** are **perpendicular*** if and only if $\mathbf{A} \cdot \mathbf{B} = 0$

Since the zero vector has an indeterminate direction, it is convenient to say, by definition, that **0** *is perpendicular to every vector, even to itself.*

ⓒ **EXAMPLE 6**   Find the angle θ between $\mathbf{A} = \langle 2, 3 \rangle$ and $\mathbf{B} = \langle 3, -1 \rangle$.

SOLUTION

$$\cos\theta = \frac{\mathbf{A} \cdot \mathbf{B}}{|\mathbf{A}|\,|\mathbf{B}|} = \frac{(2)(3) + (3)(-1)}{\sqrt{2^2 + 3^2}\sqrt{3^2 + (-1)^2}} = \frac{3}{\sqrt{13}\sqrt{10}} = \frac{3}{\sqrt{130}}$$

Hence, $\theta = \cos^{-1}(3/\sqrt{130}) \approx 74.74°$. ■

EXAMPLE 7   Find a value of the scalar t so that $\mathbf{A} = -6\mathbf{i} + 3\mathbf{j}$ and $\mathbf{B} = 4\mathbf{i} + t\mathbf{j}$ are perpendicular vectors.

SOLUTION

$$\mathbf{A} \cdot \mathbf{B} = (-6)(4) + 3t = 3t - 24$$

Since **A** and **B** are perpendicular if and only if $\mathbf{A} \cdot \mathbf{B} = 0$, we require that $3t - 24 = 0$; that is, $t = 8$. ■

All the propositions of Euclidean geometry can be proved algebraically by using vectors. The following examples indicate how this is done.

EXAMPLE 8   Use vectors to demonstrate the Pythagorean theorem.

SOLUTION   Consider the right triangle made up of the perpendicular vectors **A** and **B** with hypotenuse $\mathbf{A} + \mathbf{B}$ (Figure 5). We have

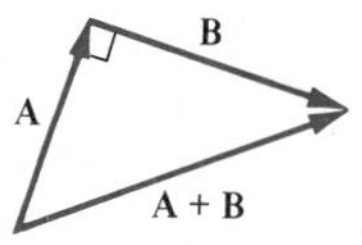

$$|\mathbf{A} + \mathbf{B}|^2 = (\mathbf{A} + \mathbf{B}) \cdot (\mathbf{A} + \mathbf{B}) = \mathbf{A} \cdot (\mathbf{A} + \mathbf{B}) + \mathbf{B} \cdot (\mathbf{A} + \mathbf{B})$$
$$= \mathbf{A} \cdot \mathbf{A} + \mathbf{A} \cdot \mathbf{B} + \mathbf{B} \cdot \mathbf{A} + \mathbf{B} \cdot \mathbf{B}$$
$$= \mathbf{A} \cdot \mathbf{A} + 2\mathbf{A} \cdot \mathbf{B} + \mathbf{B} \cdot \mathbf{B}$$
$$= |\mathbf{A}|^2 + 2\mathbf{A} \cdot \mathbf{B} + |\mathbf{B}|^2$$

Since **A** and **B** are perpendicular, $\mathbf{A} \cdot \mathbf{B} = 0$; hence, $|\mathbf{A} + \mathbf{B}|^2 = |\mathbf{A}|^2 + |\mathbf{B}|^2$, which is the Pythagorean theorem. ■

EXAMPLE 9   Show that the points $P = (4, 3)$, $Q = (1, 2)$, and $R = (4, -7)$ form the vertices of a right triangle.

Figure 5

*Some authors use the word "orthogonal" rather than the word "perpendicular."

Figure 6

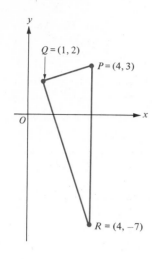

SOLUTION   Figure 6 leads us to suspect that the angle PQR is a right angle. We propose to confirm this by showing that $(\overrightarrow{QP}) \cdot (\overrightarrow{QR}) = 0$. Here,

$$\overrightarrow{QP} = (4 - 1)\mathbf{i} + (3 - 2)\mathbf{j} = 3\mathbf{i} + \mathbf{j}$$

$$\overrightarrow{QR} = (4 - 1)\mathbf{i} + (-7 - 2)\mathbf{j} = 3\mathbf{i} - 9\mathbf{j}$$

Hence,

$$(\overrightarrow{QP}) \cdot (\overrightarrow{QR}) = (3\mathbf{i} + \mathbf{j}) \cdot (3\mathbf{i} - 9\mathbf{j})$$

$$= (3)(3) + (1)(-9) = 0 \qquad \blacksquare$$

Dot products are useful in mechanics for calculating the work done by a force acting on a particle in a direction other than the direction of displacement. Indeed, suppose that a constant force, represented by the vector $\mathbf{F}$, acts on a particle, starting at the point P, and moves it along a straight line to the point Q. Then the work W done by $\mathbf{F}$ in producing the displacement $\mathbf{D} = \overrightarrow{PQ}$ is defined to be the product of the scalar component of $\mathbf{F}$ in the direction of $\mathbf{D}$ and the magnitude $|\mathbf{D}|$ of the displacement (Figure 7). Thus,

$$W = \frac{\mathbf{F} \cdot \mathbf{D}}{|\mathbf{D}|} |\mathbf{D}| = \mathbf{F} \cdot \mathbf{D}$$

Notice that, when $\mathbf{F}$ and $\mathbf{D}$ have the *same direction*, $W = \mathbf{F} \cdot \mathbf{D} = |\mathbf{F}| |\mathbf{D}|$ as usual.

Figure 7

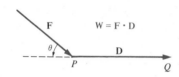

EXAMPLE 10   A person pushes on a lawnmower handle with a force of 35 pounds and moves the mower through a distance of 100 feet. How much work is done if the handle makes an angle of 30° with the ground (Figure 8)?

SOLUTION   For the force vector $\mathbf{F}$ we have $|\mathbf{F}| = 35$ pounds, while for the displacement vector $\mathbf{D}$ we have $|\mathbf{D}| = 100$ feet. Since the angle θ between $\mathbf{F}$ and $\mathbf{D}$ is $\theta = 30°$,

$$W = \mathbf{F} \cdot \mathbf{D} = |\mathbf{F}| |\mathbf{D}| \cos \theta = (35)(100)\left(\frac{\sqrt{3}}{2}\right) = 1750\sqrt{3} \text{ foot-pounds} \qquad \blacksquare$$

Figure 8

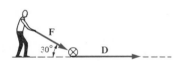

Problem Set 12.2

In Problems 1 to 8, a and b denote the lengths (magnitudes) of vectors $\mathbf{A}$ and $\mathbf{B}$, respectively, while θ denotes the angle between $\mathbf{A}$ and $\mathbf{B}$.

1 If $a = 15$, $b = 2$, and $\theta = 60°$, find $\mathbf{A} \cdot \mathbf{B}$.

2 If $a = 7$, $b = 4$, and $\theta = 5\pi/6$, find $\mathbf{A} \cdot \mathbf{B}$.

C **3** If $a = 1$, $b = 3$, and $\theta = 47°$, find $\mathbf{A} \cdot \mathbf{B}$.

C **4** If $a = 3$, $b = 4$, and $\theta = 10°$, find $\mathbf{A} \cdot \mathbf{B}$.

5 If $a \neq 0$, $b \neq 0$, and $\mathbf{A} \cdot \mathbf{B} = 0$, find θ.

6 If $a = 2$, $b = 3$, and $\mathbf{A} \cdot \mathbf{B} = -6$, find θ.

7 If $a = 10$, $\theta = \pi/3$, and $\mathbf{A} \cdot \mathbf{B} = 30$, find b.

8 If $a = 2$ and $b = 3$, explain why $|\mathbf{A} \cdot \mathbf{B}| \leq 6$.

In Problems 9 to 12, use the information to find the scalar component $\text{comp}_\mathbf{D}\mathbf{A}$ (scalar projection) of $\mathbf{A}$ in the direction of $\mathbf{D}$. Denote the angle between $\mathbf{A}$ and $\mathbf{D}$ by θ.

9 $|\mathbf{A}| = 10$, $\theta = 45°$

10 $\mathbf{A} \cdot \mathbf{D} = 5$, $|\mathbf{D}| = 2$

11 $|\mathbf{A}| = 6$, $\theta = \pi/2$

12 $\mathbf{A} \cdot \mathbf{D} = -7$, $|\mathbf{D}| = \frac{1}{2}$

In Problems 13 to 24, let $\mathbf{A} = \mathbf{i} - 3\mathbf{j}$, $\mathbf{B} = 2\mathbf{i} + \mathbf{j}$, and $\mathbf{C} = 2\mathbf{i} - 4\mathbf{j}$. Evaluate each expression.

13 $|\mathbf{A}|$ and $|\mathbf{B}|$

14 $|3\mathbf{A}|$ and $|-\mathbf{C}|$

15 $|\mathbf{A} + \mathbf{B}|$

16 $|\mathbf{A} - 2\mathbf{B}|$

17 $|\mathbf{A}| + |\mathbf{B}|$

18 $|-2\mathbf{B}| + |3\mathbf{C}|$

19 $\dfrac{\mathbf{A}}{|\mathbf{A}|}$

20 $\left|\dfrac{\mathbf{B}}{|\mathbf{B}|}\right|$

21 $\dfrac{\mathbf{A} + 3\mathbf{B}}{|\mathbf{A} + 3\mathbf{B}|}$

22 $|3\mathbf{A} - 4\mathbf{B} + \mathbf{C}|$

23 A unit vector having the same direction as $\mathbf{A} - \mathbf{B}$

24 A unit vector having the same direction as $\mathbf{A} + \mathbf{B}$

In Problems 25 to 32, find (a) $\mathbf{A} \cdot \mathbf{B}$, (b) $\cos \theta$, where θ is the angle between $\mathbf{A}$ and $\mathbf{B}$, (c) $\text{comp}_{\mathbf{B}}\mathbf{A}$.

25 $\mathbf{A} = \mathbf{i} + \mathbf{j}, \mathbf{B} = \mathbf{i} - \mathbf{j}$

26 $\mathbf{A} = 3\mathbf{i} + 2\mathbf{j}, \mathbf{B} = 4\mathbf{i} - 5\mathbf{j}$

27 $\mathbf{A} = 4\mathbf{i} + \mathbf{j}, \mathbf{B} = \mathbf{i} - 2\mathbf{j}$

28 $\mathbf{A} = -\dfrac{\mathbf{i}}{2} + \dfrac{\mathbf{j}}{3}, \mathbf{B} = \dfrac{4\mathbf{i}}{3} + \dfrac{7\mathbf{j}}{2}$

29 $\mathbf{A} = 2\mathbf{i} + 4\mathbf{j}, \mathbf{B} = -3\mathbf{j}$

30 $\mathbf{A} = \dfrac{\mathbf{i}}{3}, \mathbf{B} = \dfrac{\mathbf{j}}{5}$

31 $\mathbf{A} = \langle -3, 2 \rangle, \mathbf{B} = \langle 2, 1 \rangle$

32 $\mathbf{A} = \langle 4, -1 \rangle, \mathbf{B} = \langle -3, 4 \rangle$

In Problems 33 to 38, let $\mathbf{A} = \langle -1, 3 \rangle$, $\mathbf{B} = \langle 5, 3 \rangle$, and $\mathbf{C} = \langle -2, -3 \rangle$. Find the value of each expression.

33 $\mathbf{A} \cdot (\mathbf{B} + \mathbf{C})$

34 $(2\mathbf{A}) \cdot (3\mathbf{B})$

35 $(-\mathbf{A}) \cdot (4\mathbf{B} - 2\mathbf{C})$

36 $\dfrac{\mathbf{A}}{|\mathbf{A}|} \cdot \dfrac{\mathbf{B}}{|\mathbf{B}|}$

37 $(\mathbf{A} \cdot \mathbf{C})\mathbf{B} - (\mathbf{A} \cdot \mathbf{B})\mathbf{C}$

38 $(\mathbf{i} + \mathbf{j}) \cdot (\mathbf{A} - \mathbf{B} + 2\mathbf{C})$

39 Let $\mathbf{A} = t\mathbf{i} - 3\mathbf{j}$ and $\mathbf{B} = 5\mathbf{i} + 7\mathbf{j}$, where t is a scalar. Find t so that $\mathbf{A}$ and $\mathbf{B}$ are perpendicular.

40 Show that for any real numbers s and t, the vectors $\mathbf{A} = \langle s, t \rangle$ and $\mathbf{B} = \langle t, -s \rangle$ are perpendicular.

41 A triangle has vertices $\mathbf{A} = (-3, 4)$, $\mathbf{B} = (4, -1)$, and $\mathbf{C} = (1, 3)$. Find each of its angles.

42 Let $\mathbf{D}$ be a nonzero vector.

(a) The **vector projection** of $\mathbf{A}$ onto $\mathbf{D}$ is defined to be the vector
$$\mathbf{P} = \dfrac{\mathbf{A} \cdot \mathbf{D}}{\mathbf{D} \cdot \mathbf{D}}\mathbf{D}$$
Find $|\mathbf{P}|$, and describe the direction of $\mathbf{P}$ in terms of the direction of $\mathbf{D}$.

(b) Let $\mathbf{A} = 3\mathbf{i} + 5\mathbf{j}$ and $\mathbf{D} = -4\mathbf{i} + 3\mathbf{j}$. Find the vector projection $\mathbf{P}$ of $\mathbf{A}$ onto $\mathbf{D}$, and find $|\mathbf{P}|$.

43 Explain the geometric meaning of the following conditions:

(a) $\mathbf{A} \cdot \mathbf{B} > 0$ (b) $\mathbf{A} \cdot \mathbf{B} = 0$ (c) $\mathbf{A} \cdot \mathbf{B} < 0$

44 If $\mathbf{A}$ is a vector in the Cartesian plane and if $\mathbf{i}, \mathbf{j}$ is the standard basis, prove that $\mathbf{A} = (\mathbf{A} \cdot \mathbf{i})\mathbf{i} + (\mathbf{A} \cdot \mathbf{j})\mathbf{j}$.

45 Suppose that $\mathbf{A}$ and $\mathbf{B}$ are perpendicular vectors. Write each of the following expressions in terms of $|\mathbf{A}|$ and $|\mathbf{B}|$.

(a) $|\mathbf{A} + \mathbf{B}|$ (b) $|2\mathbf{A} - 3\mathbf{B}|$ (c) $|\mathbf{A} - \mathbf{B}|$

46 Find a scalar t so that the vectors $\mathbf{A} = \mathbf{i} + \mathbf{j}$ and $\mathbf{B} = t\mathbf{i} - \mathbf{j}$ make an angle of $3\pi/4$ radians.

47 Use the dot product to show that the triangle with vertices $A = (2, 1)$, $B = (6, 3)$, and $C = (4, 7)$ is a right triangle.

48 Use the dot product to prove that the diagonals of a rhombus are perpendicular to each other.

49 Use the dot product to prove that the four points $P = (1, 2)$, $Q = (2, 3)$, $R = (1, 4)$, and $S = (0, 3)$ are vertices of a square.

50 Using vectors, show that the medians of a triangle meet at a point two-thirds of the distance from any vertex of the triangle to the midpoint of the opposite side.

In Problems 51 to 55, use the properties of the dot product to prove each identity.

51 $(\mathbf{A} + \mathbf{B}) \cdot (\mathbf{A} - \mathbf{B}) = |\mathbf{A}|^2 - |\mathbf{B}|^2$

52 $|\mathbf{A} + \mathbf{B}|^2 = |\mathbf{A}|^2 + 2\mathbf{A} \cdot \mathbf{B} + |\mathbf{B}|^2$

53 $|s\mathbf{A} + t\mathbf{B}|^2 = s^2|\mathbf{A}|^2 + 2st(\mathbf{A} \cdot \mathbf{B}) + t^2|\mathbf{B}|^2$

54 $|\mathbf{A} + \mathbf{B}|^2 + |\mathbf{A} - \mathbf{B}|^2 = 2|\mathbf{A}|^2 + 2|\mathbf{B}|^2$

55 $\mathbf{A} \cdot \mathbf{B} = \frac{1}{2}(|\mathbf{A} + \mathbf{B}|^2 - |\mathbf{A}|^2 - |\mathbf{B}|^2)$

56 Prove the converse of the Pythagorean theorem; that is, prove that if the equation $|\mathbf{A}|^2 + |\mathbf{B}|^2 = |\mathbf{A} + \mathbf{B}|^2$ holds, then $\mathbf{A}$ is perpendicular to $\mathbf{B}$.

57 A block weighing 15 newtons slides 6 meters down an incline making a 60° angle with the horizontal. Find the work done by the force of gravity.

58 How much work is done by the force of gravity in moving a particle all the way around a vertical triangle with vertices P, Q, and R, starting and ending at vertex P? Assume that the mass of the particle is m so that its weight is mg, with g being the acceleration of gravity.

59 How much work is done by the constant force vector $\mathbf{F}$ in moving a particle first along the line segment from point P to point Q, then along the line segment from point Q to point R? How much work is done by $\mathbf{F}$ in moving the particle directly from P to R?

60 Find the work done by a force $\mathbf{F}$ in the xy plane whose magnitude is 10 units of force and which makes an angle of 60° with the positive x axis if $\mathbf{F}$ moves an object along a straight line from the origin O to the point $Q = (6, 4)$.

61 Let $\mathbf{A}$ and $\mathbf{B}$ be unit vectors that have initial points at the origin and that make angles α and β, respectively, with the positive x axis. (a) Write $\mathbf{A}$ and $\mathbf{B}$ in component form. (b) Compute $\mathbf{A} \cdot \mathbf{B}$ to derive a formula for $\cos(\alpha - \beta)$.

12.3   Vector and Parametric Equations

Figure 1

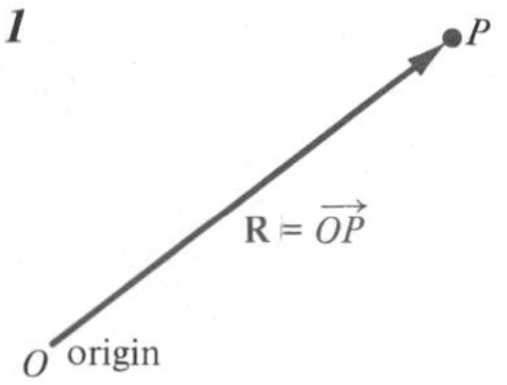

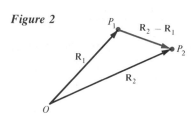

In this section we use the vector algebra developed in Sections 12.1 and 12.2 to obtain the equations of curves in *vector* and *parametric* forms. This is accomplished by using the idea of the "position vector" of a point.

We begin by choosing and fixing a point O in the plane called the **origin.** If we already have a Cartesian or polar coordinate system, we naturally take O to be the origin or the pole of the coordinate system. If P is any point, then the vector $\mathbf{R} = \overrightarrow{OP}$ is called the **position vector** of P (Figure 1). Notice that $|\mathbf{R}| = |\overrightarrow{OP}|$ gives the distance between the point P and the origin O.

Evidently, the point P determines the position vector $\mathbf{R} = \overrightarrow{OP}$ uniquely (since O is fixed beforehand). Conversely, the position vector $\mathbf{R}$ determines the point P uniquely; in fact, if the tail end of $\mathbf{R}$ is placed at the origin, then the vector $\mathbf{R}$ will point to P. For this reason, we ordinarily place all position vectors with their tail ends at the origin.

Suppose that $\mathbf{R}_1 = \overrightarrow{OP}_1$ and $\mathbf{R}_2 = \overrightarrow{OP}_2$ are the position vectors of the points P_1 and P_2, respectively. Then $\mathbf{R}_2 - \mathbf{R}_1$ is the vector $\overrightarrow{P_1P_2}$ from P_1 to P_2 (Figure 2). We usually regard the difference $\mathbf{R}_2 - \mathbf{R}_1$ of two position vectors as representing a displacement from P_1 to P_2. Notice that $|\mathbf{R}_2 - \mathbf{R}_1|$, the magnitude of this displacement, is just the distance between the points P_1 and P_2.

If we have established a Cartesian or polar coordinate system in the plane, then the point P with Cartesian coordinates $P = (x, y)$ has the position vector

Figure 2

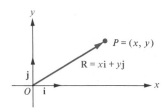

Figure 3

$$\mathbf{R} = \overrightarrow{OP} = x\mathbf{i} + y\mathbf{j}$$

(Figure 3).

© **EXAMPLE 1**   Find the position vectors $\mathbf{R}_1$ and $\mathbf{R}_2$ of the Cartesian points $P_1 = (7, 13)$ and $P_2 = (-3, 1)$, respectively. Also find the displacement vector $\mathbf{R}_2 - \mathbf{R}_1$ and the distance between P_1 and P_2.

SOLUTION   The position vectors $\mathbf{R}_1$ and $\mathbf{R}_2$ are

$$\mathbf{R}_1 = \overrightarrow{OP}_1 = 7\mathbf{i} + 13\mathbf{j} \qquad \text{and} \qquad \mathbf{R}_2 = \overrightarrow{OP}_2 = -3\mathbf{i} + \mathbf{j}$$

Hence, the displacement vector $\mathbf{R}_2 - \mathbf{R}_1$ is given by

$$\mathbf{R}_2 - \mathbf{R}_1 = (-3\mathbf{i} + \mathbf{j}) - (7\mathbf{i} + 13\mathbf{j}) = -10\mathbf{i} - 12\mathbf{j}$$

and the distance between P_1 and P_2 is

$$|\mathbf{R}_2 - \mathbf{R}_1| = \sqrt{(-10)^2 + (-12)^2} = \sqrt{244} \approx 15.62 \text{ units}$$

Figure 4

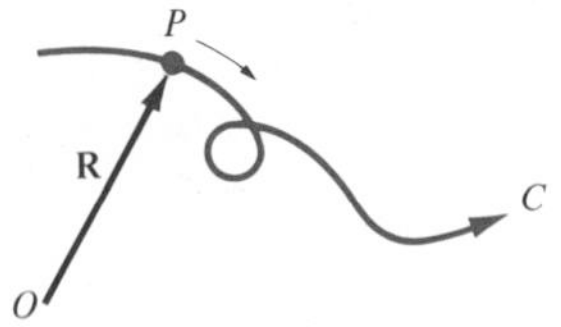

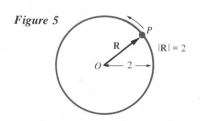

A continuously moving point P traces out a curve C (Figure 4). As P moves, its position vector $\mathbf{R} = \overrightarrow{OP}$ changes, generally in both length and direction. If the head end (terminal point) P of a variable position vector $\mathbf{R}$ traces out a curve C, then, for simplicity, we say that "$\mathbf{R}$ *traces out the curve* C."

The idea of a position vector $\mathbf{R}$ varying and thus tracing out a curve leads naturally to the notion of a **vector equation** of the curve. For instance, if $\mathbf{R}$ is a variable position vector in the plane, then as $\mathbf{R}$ varies subject to the condition $|\mathbf{R}| = 2$, it traces out a circle of radius 2 units with center at the origin (Figure 5). Therefore, we say that $|\mathbf{R}| = 2$ is a *vector equation* of this circle. Notice that the circle consists of all points P and only those points P whose position vector $\mathbf{R}$ satisfies the equation $|\mathbf{R}| = 2$.

Figure 5

More generally, we make the following definition.

DEFINITION 1   **Graph of a Vector Equation**

> The **graph** of an equation involving the position vector $\mathbf{R}$ is the set of all points P, and only those points P, whose position vector $\mathbf{R}$ satisfies the equation.

Figure 6

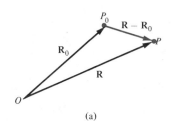

(a)

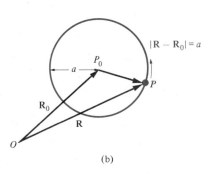

(b)

EXAMPLE 2   Sketch a graph of the vector equation

$$|\mathbf{R} - \mathbf{R}_0| = a$$

where all vectors lie in the same plane, a is a positive constant, and $\mathbf{R}_0$ is a constant position vector.

SOLUTION   Let $\mathbf{R} = \overrightarrow{OP}$ and $\mathbf{R}_0 = \overrightarrow{OP_0}$, so that $\mathbf{R} - \mathbf{R}_0 = \overrightarrow{P_0P}$ (Figure 6a). The condition $|\mathbf{R} - \mathbf{R}_0| = a$ means that the distance between the points P_0 and P is a units. Thus, as $\mathbf{R}$ varies subject to the condition $|\mathbf{R} - \mathbf{R}_0| = a$, the point P traces out a circle of radius a with center at the point P_0 (Figure 6b). ■

If $\mathbf{N}$ is a fixed nonzero vector and $\mathbf{R}_0 = \overrightarrow{OP_0}$, then the condition $\mathbf{N} \cdot (\mathbf{R} - \mathbf{R}_0) = 0$ means that the vectors $\mathbf{N}$ and $\mathbf{R} - \mathbf{R}_0$ are perpendicular (Figure 7). As $\mathbf{R} = \overrightarrow{OP}$ varies, subject to the condition that $\mathbf{R} - \mathbf{R}_0$ is perpendicular to $\mathbf{N}$, it is geometrically clear that P traces out a line through P_0 perpendicular to $\mathbf{N}$. We say that the vector $\mathbf{N}$ is **normal** to this line. Therefore,

Figure 7

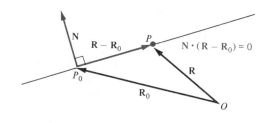

$$\mathbf{N} \cdot (\mathbf{R} - \mathbf{R}_0) = 0$$

is a vector equation of the line containing the point whose position vector is $\mathbf{R}_0$ and having $\mathbf{N}$ as a normal vector.

The equation $\mathbf{N} \cdot (\mathbf{R} - \mathbf{R}_0) = 0$ can be written as $\mathbf{N} \cdot \mathbf{R} - \mathbf{N} \cdot \mathbf{R}_0 = 0$ or as $\mathbf{N} \cdot \mathbf{R} = \mathbf{N} \cdot \mathbf{R}_0$. Since $\mathbf{N} \cdot \mathbf{R}_0$ is a constant K, the equation can also be written as

$$\mathbf{N} \cdot \mathbf{R} = K$$

We can convert the last equation into Cartesian scalar form simply by writing $\mathbf{R} = x\mathbf{i} + y\mathbf{j}$ and $\mathbf{N} = A\mathbf{i} + B\mathbf{j}$, so that $\mathbf{N} \cdot \mathbf{R} = Ax + By$, and the equation takes the form $Ax + By = K$. This equation can also be written as

$$Ax + By + C = 0$$

where we have put $C = -K$. In particular, notice that

$$\mathbf{N} = A\mathbf{i} + B\mathbf{j}$$

gives a normal vector to the line $Ax + By + C = 0$.

EXAMPLE 3   Find a normal vector to the line $y = 2x - 3$, and write an equation of this line in vector form.

SOLUTION   The given equation can be rewritten as $2x + (-1)y - 3 = 0$; hence, $\mathbf{N} = 2\mathbf{i} + (-1)\mathbf{j}$ is a normal vector to the given line. Let the position vector $\mathbf{R}$ be given by $\mathbf{R} = x\mathbf{i} + y\mathbf{j}$. Then $\mathbf{N} \cdot \mathbf{R} = (2\mathbf{i} - \mathbf{j}) \cdot (x\mathbf{i} + y\mathbf{j}) = 2x - y$, and the given equation can be rewritten in vector form as $\mathbf{N} \cdot \mathbf{R} = 3$. ■

EXAMPLE 4   Convert the vector equation $\mathbf{N} \cdot (\mathbf{R} - \mathbf{R}_0) = 0$ into Cartesian scalar form if $\mathbf{N} = -\mathbf{i} + 3\mathbf{j}$ and $\mathbf{R}_0 = 7\mathbf{i} - 2\mathbf{j}$.

SOLUTION   With $\mathbf{R} = x\mathbf{i} + y\mathbf{j}$, we have

$$\mathbf{R} - \mathbf{R}_0 = (x\mathbf{i} + y\mathbf{j}) - (7\mathbf{i} - 2\mathbf{j}) = (x - 7)\mathbf{i} + (y + 2)\mathbf{j}$$

Hence,

$$\mathbf{N} \cdot (\mathbf{R} - \mathbf{R}_0) = (-\mathbf{i} + 3\mathbf{j}) \cdot [(x - 7)\mathbf{i} + (y + 2)\mathbf{j}] = -(x - 7) + 3(y + 2)$$

The condition $\mathbf{N} \cdot (\mathbf{R} - \mathbf{R}_0) = 0$ is therefore equivalent to

$$-(x - 7) + 3(y + 2) = 0 \qquad \text{or} \qquad 3y - x + 13 = 0 \qquad ∎$$

Perpendicular Distance from a Point to a Line

We now use the vector equation of a line to derive a formula for the perpendicular distance from a point to a line.

THEOREM 1   **Distance from a Point to a Line**

> Let P_1 be a point in the plane, let L be a line with vector equation $\mathbf{N} \cdot \mathbf{R} = K$, and let $\mathbf{R}_1$ be the position vector of P_1. Then the perpendicular distance d from P_1 to L is given by
>
> $$d = \frac{|\mathbf{N} \cdot \mathbf{R}_1 - K|}{|\mathbf{N}|}$$

PROOF   Let P_0 be any point on the line L, and let $\mathbf{R}_0 = \overrightarrow{OP_0}$ be the position vector of P_0, so that

$$\mathbf{N} \cdot \mathbf{R}_0 = K$$

The vector

$$\overrightarrow{P_0P_1} = \overrightarrow{OP_1} - \overrightarrow{OP_0} = \mathbf{R}_1 - \mathbf{R}_0$$

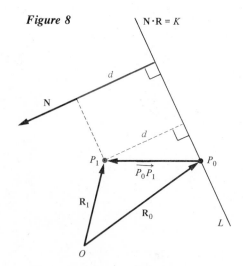

Figure 8

extends from the point P_0 on L to the point P_1. As Figure 8 shows, the perpendicular distance from P_1 to L is the absolute value* of the component of $\overrightarrow{P_0P_1}$ in the direction of $\mathbf{N}$; that is

$$d = |\text{comp}_{\mathbf{N}}(\overrightarrow{P_0P_1})| = \left| \frac{\mathbf{N} \cdot (\overrightarrow{P_0P_1})}{|\mathbf{N}|} \right|$$

$$= \frac{|\mathbf{N} \cdot (\mathbf{R}_1 - \mathbf{R}_0)|}{|\mathbf{N}|} = \frac{|\mathbf{N} \cdot \mathbf{R}_1 - \mathbf{N} \cdot \mathbf{R}_0|}{|\mathbf{N}|}$$

$$= \frac{|\mathbf{N} \cdot \mathbf{R}_1 - K|}{|\mathbf{N}|} \qquad ∎$$

We now recast Theorem 1 in Cartesian form with $P_1 = (x_1, y_1)$, $\mathbf{N} = A\mathbf{i} + B\mathbf{j}$, $\mathbf{R} = x\mathbf{i} + y\mathbf{j}$, $\mathbf{R}_1 = x_1\mathbf{i} + y_1\mathbf{j}$, and $K = -C$. Then

$$\mathbf{N} \cdot \mathbf{R}_1 - K = Ax_1 + By_1 + C$$

*The absolute value is necessary because if $\overrightarrow{P_0P_1}$ makes an obtuse angle with $\mathbf{N}$, then $\text{comp}_{\mathbf{N}}(\overrightarrow{P_0P_1})$ is negative.

and
$$|\mathbf{N}| = \sqrt{A^2 + B^2}$$

so that *the perpendicular distance d from the point* (x_1, y_1) *to the line* $Ax + By + C = 0$ *is*

$$d = \frac{|Ax_1 + By_1 + C|}{\sqrt{A^2 + B^2}}$$

© **EXAMPLE 5**   Find the perpendicular distance from the point $P_1 = (3, 7)$ to the line $y = 2x - 5$.

SOLUTION   We rewrite the equation $y = 2x - 5$ in the form $2x - y - 5 = 0$, that is, in the form $Ax + By + C = 0$ with $A = 2$, $B = -1$, and $C = -5$. Here $P_1 = (x_1, y_1) = (3, 7)$, so the desired distance d is

$$d = \frac{|Ax_1 + By_1 + C|}{\sqrt{A^2 + B^2}} = \frac{|(2)(3) + (-1)(7) + (-5)|}{\sqrt{2^2 + (-1)^2}} = \frac{6}{\sqrt{5}} \approx 2.68 \text{ units}$$ ∎

Parametric Equations

Often the variable position vector that traces out a curve is controlled by a variable scalar called a **parameter.** Equations involving parameters in an explicit way are called **parametric equations.**

As an indication of the way in which parameters are used, we begin by deriving a vector equation for a line L, not in terms of a normal vector $\mathbf{N}$, but rather in terms of a **direction vector M** *parallel* to the line. Figure 9 shows a line L containing the point P_0 and parallel to the nonzero vector $\mathbf{M}$. Here, $\mathbf{R}_0$ is the position vector of P_0.

Evidently, the point P whose position vector is $\mathbf{R}$ belongs to the line L if and only if $\overrightarrow{P_0P}$ is parallel to $\mathbf{M}$, that is, if and only if $\mathbf{R} - \mathbf{R}_0$ is parallel to $\mathbf{M}$. By Definition 1 in Section 12.1, $\mathbf{R} - \mathbf{R}_0$ is parallel to $\mathbf{M}$ if and only if there is a scalar t such that $\mathbf{R} - \mathbf{R}_0 = t\mathbf{M}$, that is, $\mathbf{R} = \mathbf{R}_0 + t\mathbf{M}$. As the scalar t varies over the real numbers, the position vector $\mathbf{R} = \mathbf{R}_0 + t\mathbf{M}$ traces out the straight line L. Here t is the *parameter* whose value determines the position vector $\mathbf{R}$. The equation

$$\mathbf{R} = \mathbf{R}_0 + t\mathbf{M}$$

is therefore called a **vector parametric equation** for L.

Figure 9

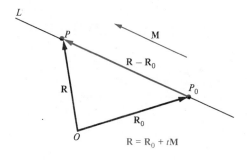

Figure 10

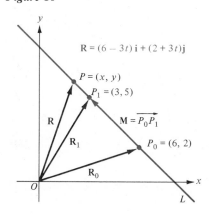

EXAMPLE 6   Let L be the line in the xy plane that contains the two points $P_0 = (6, 2)$ and $P_1 = (3, 5)$ (Figure 10).

   (a) Find a vector $\mathbf{M}$ that is parallel to L.

   (b) Find a vector parametric equation for L.

SOLUTION   Put $\mathbf{R} = \overrightarrow{OP} = x\mathbf{i} + y\mathbf{j}$, $\mathbf{R}_0 = \overrightarrow{OP_0} = 6\mathbf{i} + 2\mathbf{j}$,   and $\mathbf{R}_1 = \overrightarrow{OP_1} = 3\mathbf{i} + 5\mathbf{j}$.

   (a) $\mathbf{M} = \overrightarrow{P_0P_1} = \mathbf{R}_1 - \mathbf{R}_0 = -3\mathbf{i} + 3\mathbf{j}$

   (b) $\mathbf{R} = \mathbf{R}_0 + t\mathbf{M} = (6\mathbf{i} + 2\mathbf{j}) + t(-3\mathbf{i} + 3\mathbf{j})$
$$= (6 - 3t)\mathbf{i} + (2 + 3t)\mathbf{j}$$

Hence, $\mathbf{R} = (6 - 3t)\mathbf{i} + (2 + 3t)\mathbf{j}$ is a vector parametric equation for L.   ∎

To obtain a vector parametric equation for a curve, it is necessary to select an appropriate quantity t to use as a parameter. The selected quantity t should determine the position of a point P on the curve, and as t is varied, P should trace out the entire curve. Sometimes a suitable parameter suggests itself on algebraic grounds, while other times geometric or physical quantities such as distances, angles, or time can be used.*

When angles are used as parameters in the xy plane, it is important to keep the following fact in mind. If $\mathbf{A}$ is a vector in the xy plane, making an angle t with the positive x axis, and if $a = |\mathbf{A}|$, then

Figure 11

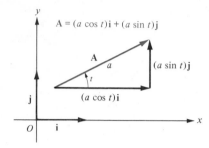

$$\mathbf{A} = (a \cos t)\mathbf{i} + (a \sin t)\mathbf{j}$$

(Figure 11).

EXAMPLE 7   Find a vector parametric equation of the circle in the xy plane with radius a and center P_0. For the parameter, use the angle t in Figure 12 between the radius vector $\mathbf{A} = \overrightarrow{P_0P}$ and the horizontal line through P_0.

SOLUTION   Here $a = |\mathbf{A}|$, so that $\mathbf{A} = (a \cos t)\mathbf{i} + (a \sin t)\mathbf{j}$. We put $\mathbf{R} = \overrightarrow{OP}$ and $\mathbf{R}_0 = \overrightarrow{OP_0}$ as in Figure 12. Then $\mathbf{R} = \mathbf{R}_0 + \mathbf{A}$, so that

$$\mathbf{R} = \mathbf{R}_0 + (a \cos t)\mathbf{i} + (a \sin t)\mathbf{j}$$

Figure 12

$\mathbf{R} = \mathbf{R}_0 + (a \cos t)\,\mathbf{i} + (a \sin t)\,\mathbf{j}$

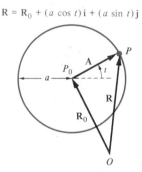

As the parameter t goes from 0 to 2π, the position vector $\mathbf{R}$ traces out the circle once. As a special case of this equation, notice that *a vector parametric equation of the circle with radius a and center O is*

$$\mathbf{R} = (a \cos t)\mathbf{i} + (a \sin t)\mathbf{j}$$

If we have a vector parametric equation giving the variable position vector $\mathbf{R} = x\mathbf{i} + y\mathbf{j} = \overrightarrow{OP}$ in terms of a parameter, then we can always find *two* equations giving the Cartesian coordinates x and y of the point $P = (x, y)$ in terms of the parameter. These two equations are called (*Cartesian*) **scalar parametric equations** of the curve under consideration.

In Examples 8 and 9, find scalar parametric equations for the given curves.

EXAMPLE 8   The circle with radius a and center O.

SOLUTION   We begin by rewriting the vector parametric equation $\mathbf{R} = (a \cos t)\mathbf{i} + (a \sin t)\mathbf{j}$ as

$$x\mathbf{i} + y\mathbf{j} = (a \cos t)\mathbf{i} + (a \sin t)\mathbf{j}$$

Since equal vectors have equal components, it follows that scalar parametric equations for the circle with radius a and center O are

$$\begin{cases} x = a \cos t \\ y = a \sin t \end{cases}$$

EXAMPLE 9   The line containing the distinct points $P_0 = (x_0, y_0)$ and $P_1 = (x_1, y_1)$.

---

*Note that a Cartesian equation $y = f(x)$ can be rewritten in vector parametric form by taking $t = x$ as the parameter. The result is $\mathbf{R} = t\mathbf{i} + f(t)\mathbf{j}$.

SOLUTION   Let $\mathbf{R} = \overrightarrow{OP}$ denote the position vector of a point on the line, and let $\mathbf{R}_0 = \overrightarrow{OP_0}$. The line is parallel to the vector $\mathbf{M} = \overrightarrow{P_0P_1} = (x_1 - x_0)\mathbf{i} + (y_1 - y_0)\mathbf{j}$; hence, the vector parametric equation is $\mathbf{R} = \mathbf{R}_0 + t\mathbf{M}$, or

$$x\mathbf{i} + y\mathbf{j} = x_0\mathbf{i} + y_0\mathbf{j} + t[(x_1 - x_0)\mathbf{i} + (y_1 - y_0)\mathbf{j}]$$
$$= [x_0 + t(x_1 - x_0)]\mathbf{i} + [y_0 + t(y_1 - y_0)]\mathbf{j}$$

The resulting scalar parametric equations are

$$\begin{cases} x = x_0 + t(x_1 - x_0) \\ y = y_0 + t(y_1 - y_0) \end{cases}$$

Parametric equations can provide simple descriptions of curves generated by physical motion in cases where it may be difficult to find a Cartesian equation. One such curve, called a **cycloid,** is traced out by a tack P stuck on the rim of a wheel of radius a as the wheel rolls without slipping along a straight line—say the x axis. The cycloid consists of a sequence of arches, one arch for each revolution of the wheel (Figure 13).

Figure 13

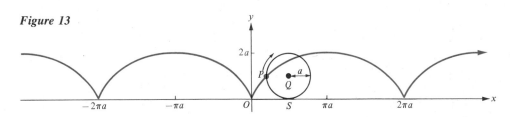

EXAMPLE 10   Find a vector parametric equation and scalar parametric equations for the cycloid in Figure 13.

SOLUTION   We use the angle t in Figure 14 as the parameter and assume that $P = O$ when $t = 0$. We measure t in radians, so that the arc length of the portion of the circle between P and S is ta units. Since this portion of the circle has rolled along the segment from O to S, it follows that $|\overrightarrow{OS}| = ta$; hence, $\overrightarrow{OS} = ta\mathbf{i}$. From Figure 14,

Figure 14

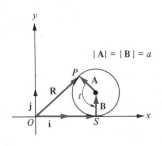

$$\mathbf{R} = \overrightarrow{OS} + \mathbf{B} + \mathbf{A} = ta\mathbf{i} + \mathbf{B} + \mathbf{A}$$

We must write the vectors $\mathbf{B}$ and $\mathbf{A}$ in terms of the parameter t. Since $\mathbf{B}$ has the same direction as $\mathbf{j}$ and $|\mathbf{B}| = a$, it follows that $\mathbf{B} = a\mathbf{j}$. Let θ be the angle between $\mathbf{A}$ and the positive x axis. By Figure 15,

$$\theta + t + \frac{\pi}{2} = 2\pi \qquad \text{so that} \qquad \theta = (2\pi - t) - \frac{\pi}{2}$$

Figure 15

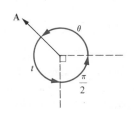

It follows that

$$\cos\theta = \sin(2\pi - t) = -\sin t \qquad \sin\theta = -\cos(2\pi - t) = -\cos t$$

and $\mathbf{A} = (a\cos\theta)\mathbf{i} + (a\sin\theta)\mathbf{j} = (-a\sin t)\mathbf{i} + (-a\cos t)\mathbf{j}$

Since we have

$$\overrightarrow{OS} = ta\mathbf{i} \qquad \mathbf{B} = a\mathbf{j} \qquad \text{and} \qquad \mathbf{A} = (-a\sin t)\mathbf{i} + (-a\cos t)\mathbf{j}$$

we can rewrite $\mathbf{R} = \overrightarrow{OS} + \mathbf{B} + \mathbf{A}$ as

$$\mathbf{R} = ta\mathbf{i} + a\mathbf{j} - (a\sin t)\mathbf{i} - (a\cos t)\mathbf{j}$$

or $$\mathbf{R} = a(t - \sin t)\mathbf{i} + a(1 - \cos t)\mathbf{j}$$

The scalar parametric equations are

$$\begin{cases} x = a(t - \sin t) \\ y = a(1 - \cos t) \end{cases}$$

Sometimes it is desirable to convert a parametric equation to a *nonparametric* equation by **eliminating the parameter.** The technique is illustrated by the following examples.

EXAMPLE 11   Find a scalar nonparametric equation for the curve $\mathbf{R} = 3t\mathbf{i} + t^2\mathbf{j}$, and sketch the graph.

Figure 16

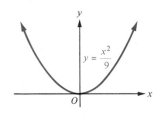

SOLUTION   With $\mathbf{R} = x\mathbf{i} + y\mathbf{j}$, we have $x\mathbf{i} + y\mathbf{j} = 3t\mathbf{i} + t^2\mathbf{j}$; hence,

$$\begin{cases} x = 3t \\ y = t^2 \end{cases}$$

To eliminate the parameter t from the scalar parametric equations, we solve the first equation for t to obtain $t = x/3$ and substitute into the second equation, so that

$$y = \left(\frac{x}{3}\right)^2 \qquad \text{or} \qquad y = \frac{x^2}{9}$$

(Figure 16).

EXAMPLE 12   Find a scalar nonparametric equation for the curve

$$\begin{cases} x = \dfrac{1}{t - 1} \\ y = 3t + 1 \end{cases}$$

Figure 17

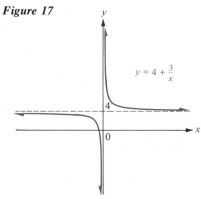

and sketch the graph.

SOLUTION   From the first equation, we have $t - 1 = 1/x$, or $t = 1 + (1/x)$. Substitution into the second equation yields

$$y = 3\left(1 + \frac{1}{x}\right) + 1 \qquad \text{or} \qquad y = 4 + \frac{3}{x}$$

(Figure 17).

EXAMPLE 13   Find a scalar nonparametric equation for $\mathbf{R} = (\cosh t)\mathbf{i} + (\sinh t)\mathbf{j}$.

SOLUTION   In scalar parametric form, we have

$$\begin{cases} x = \cosh t \\ y = \sinh t \end{cases}$$

Here the parameter t can be eliminated by using the identity $\cosh^2 t - \sinh^2 t = 1$ to obtain $x^2 - y^2 = 1$, $x > 0$.

Given the equation of a curve in scalar parametric form, you can find the derivative dy/dx by eliminating the parameter and then proceeding as usual; however, it is possible to find dy/dx directly from the scalar parametric equations. Indeed, assuming the existence of the required derivatives, we have

$$\frac{dy}{dx} \cdot \frac{dx}{dt} = \frac{dy}{dt}$$

by the chain rule. Hence, if $dx/dt \neq 0$, then

$$\frac{dy}{dx} = \frac{dy/dt}{dx/dt}$$

Similarly, if we put $y' = dy/dx$, then we can obtain the second derivative d^2y/dx^2 by using the chain rule again; hence,

$$\frac{d^2y}{dx^2} = \frac{dy'}{dx} = \frac{dy'/dt}{dx/dt}$$

<u>EXAMPLE 14</u>   Given

$$\begin{cases} x = t^2 - 6 \\ y = t^3 + 5 \end{cases}$$

find dy/dx and d^2y/dx^2.

SOLUTION $y' = \dfrac{dy}{dx} = \dfrac{dy/dt}{dx/dt} = \dfrac{3t^2}{2t} = \dfrac{3}{2}t$ for $t \neq 0$

and

$$\frac{d^2y}{dx^2} = \frac{dy'}{dx} = \frac{dy'/dt}{dx/dt} = \frac{\frac{3}{2}}{2t} = \frac{3}{4t}$$ for $t \neq 0$ ■

Problem Set 12.3

In Problems 1 to 4, find the position vector $\mathbf{R}$ of each point P. Express $\mathbf{R}$ in terms of $\mathbf{i}$ and $\mathbf{j}$.

1 $P = (5, -3)$ **2** $P = (0, 7)$

3 $P = O$ **4** $P = (a, a^2)$

In Problems 5 to 8, find the vector $\mathbf{R}_2 - \mathbf{R}_1$ if $\mathbf{R}_1$ and $\mathbf{R}_2$ are the position vectors of P_1 and P_2, respectively. Also find $|\mathbf{R}_2 - \mathbf{R}_1|$.

5 $P_1 = (2, -1), P_2 = (7, 2)$ **6** $P_1 = (5, 1), P_2 = (-3, 4)$

7 $P_1 = (2, -4), P_2 = (-2, -1)$ **8** $P_1 = (a, a^2), P_2 = (b, b^2)$

In Problems 9 to 20, convert each vector equation to Cartesian scalar nonparametric form, and sketch the graph.

9 $(\mathbf{i} - 4\mathbf{j}) \cdot \mathbf{R} = 1$

10 $(2\mathbf{i} + 3\mathbf{j}) \cdot \mathbf{R} = (2\mathbf{i} + 3\mathbf{j}) \cdot \mathbf{j}$

11 $\mathbf{i} \cdot \mathbf{R} = 2 + \mathbf{j} \cdot \mathbf{R}$ **12** $\mathbf{R} \cdot \mathbf{j} = 0$

13 $(\mathbf{i} + \mathbf{j}) \cdot \mathbf{R} = 0$ **14** $(2\mathbf{i} + 3\mathbf{j}) \cdot (\mathbf{R} - \mathbf{i}) = 0$

15 $|\mathbf{R}| = 3$ **16** $\mathbf{R} \cdot \mathbf{R} = 16$

17 $|\mathbf{R} - (2\mathbf{i} + 3\mathbf{j})| = 4$ **18** $|\mathbf{R} + \mathbf{i} - \mathbf{j}| = 3$

19 $|\mathbf{R} - \mathbf{i}| + |\mathbf{R} + \mathbf{i}| = 6$ **20** $||\mathbf{R} - \mathbf{i}| - |\mathbf{R} + \mathbf{i}|| = 1$

In Problems 21 to 25, find the perpendicular distance from the indicated point $P_1 = (x_1, y_1)$ to the line whose equation is given.

21 $P_1 = (1, 2); 3x - y = 4$

22 $P_1 = (-7, 3); y = \frac{1}{2}x + 3$

23 $P_1 = (1, 2); (\mathbf{i} + \mathbf{j}) \cdot (\mathbf{R} - 2\mathbf{i} + \mathbf{j}) = 0$

24 $P_1 = (0, 0); Ax + By + C = 0$

25 $P_1 = (4, 0); y = 2x - 5$

26 Find a formula for the perpendicular distance from the origin to the line $\mathbf{N} \cdot \mathbf{R} = K$.

In Problems 27 to 32, (a) find a vector $\mathbf{M}$ parallel to the line L containing the two given points P_0 and P_1; (b) write a vector parametric equation for L; (c) write scalar parametric equations for L; and (d) sketch a graph of L.

27 $P_0 = (1, 2), P_1 = (3, 4)$

28 $P_0 = (0, 0), P_1 = (-1, 3)$

29 $P_0 = (-\frac{3}{2}, \frac{5}{2}), P_1 = (\frac{2}{3}, -\frac{1}{3})$

30 $P_0 = (0, a)$, $P_1 = (b, 0)$, $a \neq 0$, $b \neq 0$

31 $P_0 = (\pi, e)$, $P_1 = (-\sqrt{2}, \sqrt{3})$

32 $P_0 = (x_0, y_0)$, $P_1 = (x_1, y_1)$

In Problems 33 to 36, find a vector parametric equation for each curve.

33 The line through the point whose position vector is $\mathbf{R}_0 = 7\mathbf{i} - 2\mathbf{j}$ and which is parallel to the vector $\mathbf{M} = -\mathbf{i} + 3\mathbf{j}$

34 The line through the point whose position vector is $\mathbf{R}_0 = -\mathbf{i}/7 + \mathbf{j}/3$ and which has $\mathbf{N} = 3\mathbf{i} - 7\mathbf{j}$ as a normal vector

35 The circle with center at the point whose position vector is $\mathbf{R}_0 = 3\mathbf{i} + 4\mathbf{j}$ and whose radius is 4 units

36 The cycloid generated by a point P on the circumference of a rolling circle of radius 5 units

In Problems 37 to 44, (a) convert the vector parametric equation to scalar parametric form, (b) eliminate the parameter t and thus obtain a scalar nonparametric equation, and (c) sketch the graph.

37 $\mathbf{R} = (2 \cos t)\mathbf{i} + (2 \sin t)\mathbf{j}$ for $0 \leq t \leq 2\pi$

38 $\mathbf{R} = (3 \cos t)\mathbf{i} + (4 \sin t)\mathbf{j}$ for $0 \leq t \leq 2\pi$

39 $\mathbf{R} = (5 \cos 2t)\mathbf{i} - (5 \sin 2t)\mathbf{j}$ for $0 \leq t \leq \pi/2$

40 $\mathbf{R} = t^2\mathbf{i} + t\mathbf{j}$ for $-1 \leq t \leq 1$

41 $\mathbf{R} = 4t\mathbf{i} + (3t + 5)\mathbf{j}$ for $-\infty < t < \infty$

42 $\mathbf{R} = e^t\mathbf{i} + e^{-t}\mathbf{j}$ for $-\infty < t < \infty$

43 $\mathbf{R} = \dfrac{1}{t - 2}\mathbf{i} + (2t + 1)\mathbf{j}$ for $0 \leq t < 2$

44 $\mathbf{R} = (\sin t)\mathbf{i} + (\sin t)\mathbf{j}$ for $0 \leq t \leq \pi/2$

In Problems 45 to 52, (a) eliminate the parameter t, (b) write the equation in scalar nonparametric form, and (c) express dy/dx and d^2y/dx^2 as functions of t.

45 $\begin{cases} x = 3t + 1 \\ y = 2 - t \end{cases}$

46 $\begin{cases} x = \dfrac{1}{t^2 - 1} \\ y = 3t + 1 \end{cases}$

47 $\begin{cases} x = t^2 - 2 \\ y = 5t \end{cases}$

48 $\begin{cases} x = \dfrac{1}{(t - 2)^2} \\ y = -3t + 2 \end{cases}$

49 $\begin{cases} x = 2 - \dfrac{1}{t} \\ y = 2t + \dfrac{1}{t} \end{cases}$

50 $\begin{cases} x = 3 \cos^2 t \\ y = 4 \sin^2 t \end{cases}$

51 $\begin{cases} x = t^2 - 2t \\ y = t^3 - 3t \end{cases}$

52 $\begin{cases} x = 4 \sin^2 t \cos t \\ y = 4 \sin t \cos^2 t \end{cases}$

53 Let $\mathbf{N} \neq \mathbf{0}$, $K \neq 0$, and suppose that L is the line whose vector equation is $\mathbf{N} \cdot \mathbf{R} = K$. Put $\mathbf{D} = (K/|\mathbf{N}|^2)\mathbf{N}$. Show that L also has the vector equation $\mathbf{D} \cdot (\mathbf{R} - \mathbf{D}) = 0$.

54 Find a vector equation of the parabola with focus at the origin whose directrix is the line $\mathbf{D} \cdot (\mathbf{R} - \mathbf{D}) = 0$, where $\mathbf{D}$ is a fixed nonzero vector.

55 Find a formula for the perpendicular distance from the point with position vector $\mathbf{R}_1$ to the line $\mathbf{D} \cdot (\mathbf{R} - \mathbf{D}) = 0$, where $\mathbf{D}$ is a constant nonzero vector.

56 Show that the vector $\mathbf{M} = -B\mathbf{i} + A\mathbf{j}$ is parallel to the line $Ax + By + C = 0$.

57 A point P is located on a spoke of a wheel of radius $a > 0$ at a distance $b > 0$ from the center. Derive vector and scalar parametric equations for the curve traced out by P as the wheel rolls without slipping along the x axis. Assume that when $x = 0$, $P = (0, a - b)$.

58 Find a vector equation for the ellipse whose foci are at the points F_1 and F_2 and whose semimajor axis is $a > 0$ (Figure 18).

Figure 18

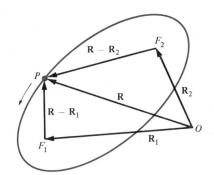

59 Derive a vector parametric equation for the curve traced out by the point P halfway between A and B in Figure 19 as t varies with $0 < t < \pi$. Sketch the graph.

Figure 19

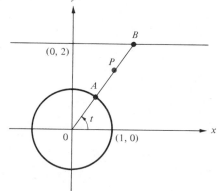

12.4   Vector-Valued Functions of a Scalar

The vector parametric equations developed in Section 12.3 can be better understood in terms of the idea of a **vector-valued function.** Thus, a vector parametric equation has the form

$$\mathbf{R} = \mathbf{F}(t)$$

where the "independent variable" is the parameter t, $\mathbf{F}$ is the vector-valued function, and the "dependent variable" is the position vector $\mathbf{R}$. The following definition makes this idea more precise.

DEFINITION 1   **Vector-Valued Function**

> A **vector-valued function** $\mathbf{F}$ assigns a unique vector $\mathbf{F}(t)$ to each real number t in its **domain.** The **range** of $\mathbf{F}$ is the set of all vectors of the form $\mathbf{F}(t)$ as t runs through the domain of $\mathbf{F}$.

EXAMPLE 1   Let $\mathbf{F}$ be the vector-valued function defined by the equation

$$\mathbf{F}(t) = (2 \cos t)\mathbf{i} + (2 \sin t)\mathbf{j} \qquad 0 \le t \le 2\pi$$

(a) What is the domain of $\mathbf{F}$?

(b) What is the range of $\mathbf{F}$?

(c) Find $\mathbf{F}(\pi)$.

SOLUTION

(a) The domain of $\mathbf{F}$ is the interval $[0, 2\pi]$.

(b) As t runs from 0 to 2π, $\mathbf{F}(t) = (2 \cos t)\mathbf{i} + (2 \sin t)\mathbf{j}$, regarded as a position vector, traces out a circle in the xy plane with radius 2 units and center at the origin. Therefore, the range of $\mathbf{F}$ is the set of all position vectors of points on this circle.

(c) $\mathbf{F}(\pi) = (2 \cos \pi)\mathbf{i} + (2 \sin \pi)\mathbf{j} = -2\mathbf{i}$.   ■

A vector-valued function $\mathbf{F}$ whose range is contained in the xy plane can always be defined by an equation of the form

$$\boxed{\mathbf{F}(t) = g(t)\mathbf{i} + h(t)\mathbf{j}}$$

where g and h are ordinary real-valued functions called the **scalar component functions of F.**

Let $\mathbf{F}$ be a vector-valued function, and let $\mathbf{A}$ be a fixed vector. The scalar $|\mathbf{F}(t) - \mathbf{A}|$ represents the distance between the head ends (terminal points) of $\mathbf{F}(t)$ and $\mathbf{A}$ when their tail ends (initial points) are at the origin (Figure 1). Evidently, $|\mathbf{F}(t) - \mathbf{A}|$ is close to zero when $\mathbf{F}(t)$ is close to $\mathbf{A}$, in both magnitude and direction. Thus, we make the following definition.

Figure 1

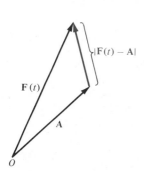

DEFINITION 2   **Limit of a Vector-Valued Function**

> Suppose that the number c belongs to an open interval I and that every number in I, except possibly c, belongs to the domain of the vector-valued function $\mathbf{F}$. If $\mathbf{A}$ is a vector such that $\lim_{t \to c} |\mathbf{F}(t) - \mathbf{A}| = 0$, then we say that $\mathbf{F}(t)$ *approaches* $\mathbf{A}$ *as a* **limit** *when t approaches c*, and we write $\lim_{t \to c} \mathbf{F}(t) = \mathbf{A}$.

Limits of vector-valued functions can be calculated "componentwise," as is shown by the following theorem, whose proof is left as an exercise (Problem 31).

THEOREM 1   **Limits of Vector-Valued Functions**

> Suppose that $\mathbf{F}(t) = g(t)\mathbf{i} + h(t)\mathbf{j}$ and $\mathbf{A} = a\mathbf{i} + b\mathbf{j}$. Then $\lim_{t \to c} \mathbf{F}(t) = \mathbf{A}$ if and only if
> $$\lim_{t \to c} g(t) = a \qquad \text{and} \qquad \lim_{t \to c} h(t) = b$$

EXAMPLE 2   If $\mathbf{F}(t) = (t^2 + 1)\mathbf{i} + (3t - 2)\mathbf{j}$, find $\lim_{t \to 1} \mathbf{F}(t)$.

SOLUTION $\lim_{t \to 1} \mathbf{F}(t) = [\lim_{t \to 1} (t^2 + 1)]\mathbf{i} + [\lim_{t \to 1} (3t - 2)]\mathbf{j} = 2\mathbf{i} + \mathbf{j}$ ∎

By direct analogy with the definition of continuity for real-valued functions, we make the following definition for vector-valued functions.

DEFINITION 3   **Continuity of a Vector-Valued Function**

> A vector-valued function $\mathbf{F}$ is said to be **continuous** at the number c if the following conditions hold:
>
> **(i)** c belongs to the domain of $\mathbf{F}$
>
> **(ii)** $\lim_{t \to c} \mathbf{F}(t)$ exists
>
> **(iii)** $\lim_{t \to c} \mathbf{F}(t) = \mathbf{F}(c)$.

A vector-valued function is said to be **continuous** provided that it is continuous at every number in its domain. One can also define right- and left-sided limits and right- and left-sided continuity for vector-valued functions in essentially the same way as for real-valued functions. There are no surprises, and everything works as expected.

Using Theorem 1, we easily establish the following result (Problem 32).

THEOREM 2   **Continuity of Vector-Valued Functions**

> A vector-valued function $\mathbf{F}$ is continuous at a number c if and only if both of its scalar component functions are continuous at c.

EXAMPLE 3   Discuss the continuity of $\mathbf{F}(t) = \dfrac{1}{t - 2}\mathbf{i} + (5t + 1)\mathbf{j}$.

SOLUTION   The scalar component functions of **F** are

$$g(t) = \frac{1}{t - 2} \quad \text{and} \quad h(t) = 5t + 1$$

where g is continuous for every value of t except for $t = 2$ and h is continuous for every value of t. The domain of **F** consists of all real numbers different from 2, and **F** is continuous at every number t in its domain.   ∎

Given a vector-valued function **F**, we can form the **difference quotient**

$$\frac{\mathbf{F}(t + \Delta t) - \mathbf{F}(t)}{\Delta t} = \frac{1}{\Delta t}[\mathbf{F}(t + \Delta t) - \mathbf{F}(t)]$$

by direct analogy with the difference quotient of a real-valued function. Notice, however, that $(1/\Delta t)[\mathbf{F}(t + \Delta t) - \mathbf{F}(t)]$ is the product of a scalar $1/\Delta t$ and a vector $\mathbf{F}(t + \Delta t) - \mathbf{F}(t)$; hence, it is a *vector*. Pursuing the analogy with real-valued functions, we make the following definition.

DEFINITION 4   **Derivative of a Vector-Valued Function**

> Let t be a number belonging to an open interval contained in the domain of a vector-valued function **F**. We define the **derivative** of **F** at t, in symbols, $\mathbf{F}'(t)$, by
>
> $$\mathbf{F}'(t) = \lim_{\Delta t \to 0} \frac{\mathbf{F}(t + \Delta t) - \mathbf{F}(t)}{\Delta t}$$
>
> provided that this limit exists.

By using Definition 4 and Theorem 1, it is easy to prove the following theorem (Problem 38).

THEOREM 3   **Derivatives of Vector-Valued Functions**

> Let $\mathbf{F}(t) = g(t)\mathbf{i} + h(t)\mathbf{j}$ and suppose that t belongs to an open interval contained in the domain of **F**. Then:
>
> **(i)** **F** has a derivative at t if and only if both of the scalar component functions g and h have derivatives at t.
>
> **(ii)** If $g'(t)$ and $h'(t)$ both exist, then $\mathbf{F}'(t) = g'(t)\mathbf{i} + h'(t)\mathbf{j}$.

<u>EXAMPLE 4</u>   If $\mathbf{F}(t) = e^{2t}\mathbf{i} - (\sinh t)\mathbf{j}$, find $\mathbf{F}'(t)$.

SOLUTION $\mathbf{F}'(t) = (D_t e^{2t})\mathbf{i} + [D_t(-\sinh t)]\mathbf{j} = 2e^{2t}\mathbf{i} - (\cosh t)\mathbf{j}$ ∎

The notation and terminology used in connection with derivatives of vector-valued functions are the direct analogs of those used for real-valued functions. For instance, if **F** is differentiable for all values of t in some open interval, then $\mathbf{F}'$, the derived vector-valued function, is defined on this interval, and we can ask whether $\mathbf{F}'$ has a derivative. If it does, we denote this **second derivative** by $\mathbf{F}''$. Third- and higher-order derivatives of **F** are treated similarly.

<u>EXAMPLE 5</u>   Let $\mathbf{F}(t) = (\cos t)\mathbf{i} + (\sin t)\mathbf{j}$. Find

(a) $\mathbf{F}'(t)$ **(b)** $\mathbf{F}''(t)$ **(c)** $\mathbf{F}'(t) \cdot \mathbf{F}''(t)$

SOLUTION

(a) $\mathbf{F}'(t) = (-\sin t)\mathbf{i} + (\cos t)\mathbf{j}$

(b) $\mathbf{F}''(t) = (-\cos t)\mathbf{i} - (\sin t)\mathbf{j}$

(c) $\mathbf{F}'(t) \cdot \mathbf{F}''(t) = \sin t \cos t - \sin t \cos t = 0$. ∎

The Leibniz differential notation is used in connection with vector-valued functions in much the same way as for real-valued functions. For instance, if $\mathbf{F}$ is a differentiable vector-valued function and the variable vector $\mathbf{R}$ is defined by $\mathbf{R} = \mathbf{F}(t)$, we write

$$\frac{d\mathbf{R}}{dt} = \mathbf{F}'(t) \qquad \text{and} \qquad d\mathbf{R} = \mathbf{F}'(t)\, dt$$

Note that the differential $d\mathbf{R}$ is a *vector* since it is a product of the scalar dt and the vector $\mathbf{F}'(t)$. Thus, we can rewrite Theorem 3 as follows. If $\mathbf{R} = u\mathbf{i} + v\mathbf{j}$, where the scalars u and v are differentiable functions of t, then

$$\frac{d\mathbf{R}}{dt} = \frac{d}{dt}(u\mathbf{i} + v\mathbf{j}) = \frac{du}{dt}\mathbf{i} + \frac{dv}{dt}\mathbf{j} \qquad \text{or} \qquad d\mathbf{R} = du\,\mathbf{i} + dv\,\mathbf{j}$$

EXAMPLE 6   If $\mathbf{R} = t^2\mathbf{i} + (\tan t)\mathbf{j}$, find $\dfrac{d\mathbf{R}}{dt}$ and $d\mathbf{R}$.

SOLUTION $\dfrac{d\mathbf{R}}{dt} = 2t\mathbf{i} + (\sec^2 t)\mathbf{j}$ and $d\mathbf{R} = (2t\,dt)\mathbf{i} + (\sec^2 t\,dt)\mathbf{j}$ ∎

Using the Leibniz notation and Theorem 3, we now establish the basic properties of derivatives of vector-valued functions.

THEOREM 4   **Properties of Derivatives of Vector-Valued Functions**

Let $\mathbf{R}$ and $\mathbf{S}$ be differentiable vector-valued functions of the scalar t, and let w be a differentiable scalar function of t. Then

(i) $\dfrac{d}{dt}(\mathbf{R} + \mathbf{S}) = \dfrac{d\mathbf{R}}{dt} + \dfrac{d\mathbf{S}}{dt}$

(ii) $\dfrac{d}{dt}(\mathbf{R} - \mathbf{S}) = \dfrac{d\mathbf{R}}{dt} - \dfrac{d\mathbf{S}}{dt}$

(iii) $\dfrac{d}{dt}(w\mathbf{R}) = \dfrac{dw}{dt}\mathbf{R} + w\dfrac{d\mathbf{R}}{dt}$

(iv) $\dfrac{d}{dt}(\mathbf{R} \cdot \mathbf{S}) = \dfrac{d\mathbf{R}}{dt} \cdot \mathbf{S} + \mathbf{R} \cdot \dfrac{d\mathbf{S}}{dt}$

(v) $\dfrac{d}{dt}|\mathbf{R}| = \dfrac{\mathbf{R}}{|\mathbf{R}|} \cdot \dfrac{d\mathbf{R}}{dt} \qquad$ if $\mathbf{R} \neq \mathbf{0}$

(vi) $\dfrac{d}{dt}\left(\dfrac{\mathbf{R}}{w}\right) = \dfrac{w(d\mathbf{R}/dt) - (dw/dt)\mathbf{R}}{w^2} \qquad$ if $w \neq 0$

PROOF   Parts (i) through (iv) can be proved by expanding the vectors in terms of their components and using Theorem 3. We leave parts (i), (ii), and (iii) as exercises (Problems 39 and 40) and attend here to parts (iv), (v), and (vi).

(iv) Let $\mathbf{R} = x\mathbf{i} + y\mathbf{j}$, and let $\mathbf{S} = u\mathbf{i} + v\mathbf{j}$. Then, $\mathbf{R} \cdot \mathbf{S} = xu + yv$ and

$$\frac{d}{dt}(\mathbf{R} \cdot \mathbf{S}) = \frac{d}{dt}(xu + yv) = \frac{dx}{dt}u + x\frac{du}{dt} + y\frac{dv}{dt} + \frac{dy}{dt}v$$

$$= \left(\frac{dx}{dt}u + \frac{dy}{dt}v\right) + \left(x\frac{du}{dt} + y\frac{dv}{dt}\right)$$

$$= \left(\frac{dx}{dt}\mathbf{i} + \frac{dy}{dt}\mathbf{j}\right) \cdot (u\mathbf{i} + v\mathbf{j}) + (x\mathbf{i} + y\mathbf{j}) \cdot \left(\frac{du}{dt}\mathbf{i} + \frac{dv}{dt}\mathbf{j}\right)$$

$$= \frac{d\mathbf{R}}{dt} \cdot \mathbf{S} + \mathbf{R} \cdot \frac{d\mathbf{S}}{dt}$$

(v) Assume that $\mathbf{R} \neq \mathbf{0}$. By part (iv), we have

$$\frac{d}{dt}|\mathbf{R}| = \frac{d}{dt}\sqrt{\mathbf{R} \cdot \mathbf{R}} = \frac{1}{2\sqrt{\mathbf{R} \cdot \mathbf{R}}} \cdot \frac{d}{dt}(\mathbf{R} \cdot \mathbf{R}) = \frac{1}{2|\mathbf{R}|}\left(\frac{d\mathbf{R}}{dt} \cdot \mathbf{R} + \mathbf{R} \cdot \frac{d\mathbf{R}}{dt}\right)$$

$$= \frac{1}{2|\mathbf{R}|}\left(2\mathbf{R} \cdot \frac{d\mathbf{R}}{dt}\right) = \frac{\mathbf{R}}{|\mathbf{R}|} \cdot \frac{d\mathbf{R}}{dt}$$

(vi) Assume that $w \neq 0$. By part (iii), we have

$$\frac{d}{dt}\left(\frac{\mathbf{R}}{w}\right) = \frac{d}{dt}(w^{-1}\mathbf{R}) = \left(\frac{d}{dt}w^{-1}\right)\mathbf{R} + w^{-1}\frac{d\mathbf{R}}{dt} = -w^{-2}\frac{dw}{dt}\mathbf{R} + w^{-1}\frac{d\mathbf{R}}{dt}$$

$$= \frac{w(d\mathbf{R}/dt) - (dw/dt)\mathbf{R}}{w^2}$$ ∎

EXAMPLE 7   Let $\mathbf{R} = (5 \sin 2t)\mathbf{i} + (5 \cos 2t)\mathbf{j}$, $\mathbf{S} = e^{2t}\mathbf{i} + e^{-2t}\mathbf{j}$, and $w = e^{-5t}$. Find

(a) $\dfrac{d}{dt}|\mathbf{R}|$ (b) $\dfrac{d}{dt}(\mathbf{R} \cdot \mathbf{S})$ (c) $\dfrac{d}{dt}(w\mathbf{S})$

SOLUTION   By Theorem 4, we have

(a) $\dfrac{d}{dt}|\mathbf{R}| = \dfrac{\mathbf{R}}{|\mathbf{R}|} \cdot \dfrac{d\mathbf{R}}{dt}$

$$= \frac{(5 \sin 2t)\mathbf{i} + (5 \cos 2t)\mathbf{j}}{\sqrt{(5 \sin 2t)^2 + (5 \cos 2t)^2}} \cdot [(10 \cos 2t)\mathbf{i} - (10 \sin 2t)\mathbf{j}]$$

$$= \frac{50 \sin 2t \cos 2t - 50 \cos 2t \sin 2t}{5} = 0$$

(b) $\dfrac{d}{dt}(\mathbf{R} \cdot \mathbf{S}) = \dfrac{d\mathbf{R}}{dt} \cdot \mathbf{S} + \mathbf{R} \cdot \dfrac{d\mathbf{S}}{dt}$

$$= [(10 \cos 2t)\mathbf{i} - (10 \sin 2t)\mathbf{j}] \cdot [e^{2t}\mathbf{i} + e^{-2t}\mathbf{j}]$$
$$\quad + [(5 \sin 2t)\mathbf{i} + (5 \cos 2t)\mathbf{j}] \cdot [2e^{2t}\mathbf{i} - 2e^{-2t}\mathbf{j}]$$

$$= 10 \cos 2t\, e^{2t} - 10 \sin 2t\, e^{-2t} + 10 \sin 2t\, e^{2t} - 10 \cos 2t\, e^{-2t}$$

$$= 10(\cos 2t + \sin 2t)(e^{2t} - e^{-2t})$$

(c) $\dfrac{d}{dt}(w\mathbf{S}) = \dfrac{dw}{dt}\mathbf{S} + w\dfrac{d\mathbf{S}}{dt}$

$= -5e^{-5t}(e^{2t}\mathbf{i} + e^{-2t}\mathbf{j}) + e^{-5t}(2e^{2t}\mathbf{i} - 2e^{-2t}\mathbf{j})$

$= -5e^{-3t}\mathbf{i} - 5e^{-7t}\mathbf{j} + 2e^{-3t}\mathbf{i} - 2e^{-7t}\mathbf{j} = -3e^{-3t}\mathbf{i} - 7e^{-7t}\mathbf{j}$ ∎

The chain rule also works for vector-valued functions, as the following theorem shows.

<u>THEOREM 5</u>   **Chain Rule for Vector-Valued Functions**

> Let $\mathbf{R}$ be a differentiable vector-valued function of the scalar t, and let t be a differentiable function of the scalar s. Then, regarding $\mathbf{R}$ as a function of s, we have
>
> $$\frac{d\mathbf{R}}{ds} = \frac{d\mathbf{R}}{dt}\frac{dt}{ds}$$

PROOF   Let $\mathbf{R} = u\mathbf{i} + v\mathbf{j}$, where the scalar components u and v are functions of t. By the usual chain rule for real-valued functions,

$$\frac{du}{ds} = \frac{du}{dt}\frac{dt}{ds} \qquad \text{and} \qquad \frac{dv}{ds} = \frac{dv}{dt}\frac{dt}{ds}$$

Hence, by Theorem 3,

$$\frac{d\mathbf{R}}{ds} = \frac{du}{ds}\mathbf{i} + \frac{dv}{ds}\mathbf{j} = \frac{du}{dt}\frac{dt}{ds}\mathbf{i} + \frac{dv}{dt}\frac{dt}{ds}\mathbf{j} = \left(\frac{du}{dt}\mathbf{i} + \frac{dv}{dt}\mathbf{j}\right)\frac{dt}{ds} = \frac{d\mathbf{R}}{dt}\frac{dt}{ds}$$ ∎

<u>EXAMPLE 8</u>   Given that $\mathbf{R} = \mathbf{F}(t)$, $\mathbf{F}'(t) = 2t\mathbf{i} - e^{-t}\mathbf{j}$, and $t = \sin\theta$, find

(a) $\dfrac{d\mathbf{R}}{d\theta}$ (b) $\dfrac{d^2\mathbf{R}}{d\theta^2}$

SOLUTION

(a) By Theorem 5,

$$\frac{d\mathbf{R}}{d\theta} = \frac{d\mathbf{R}}{dt}\frac{dt}{d\theta} = \mathbf{F}'(t)\frac{d}{d\theta}(\sin\theta) = (2t\mathbf{i} - e^{-t}\mathbf{j})\cos\theta$$

$$= [(2\sin\theta)\mathbf{i} - e^{-\sin\theta}\mathbf{j}]\cos\theta = (2\sin\theta\cos\theta)\mathbf{i} - (e^{-\sin\theta}\cos\theta)\mathbf{j}$$

$$= (\sin 2\theta)\mathbf{i} - (e^{-\sin\theta}\cos\theta)\mathbf{j}$$

(b) Using the result of part (a), we have

$$\frac{d^2\mathbf{R}}{d\theta^2} = (2\cos 2\theta)\mathbf{i} + e^{-\sin\theta}(\sin\theta + \cos^2\theta)\mathbf{j}$$ ∎

Problem Set 12.4

In Problems 1 to 6, (a) find the domain of $\mathbf{F}$, (b) find $\mathbf{F}(t_0)$ if t_0 belongs to the domain of $\mathbf{F}$, (c) find $\lim\limits_{t\to t_0}\mathbf{F}(t)$ if it exists, and (d) determine where $\mathbf{F}$ is continuous.

1 $\mathbf{F}(t) = (3t + 2)\mathbf{i} + \dfrac{5}{t^2 - 1}\mathbf{j}$; $t_0 = 2$

2 $\mathbf{F}(t) = \sqrt{t - 1}\,\mathbf{i} + \sqrt{5 - t}\,\mathbf{j}$; $t_0 = 1$

3 $F(t) = \dfrac{t^2 - 5t + 6}{t - 3} i + \dfrac{t^2 + 7t - 30}{t - 3} j$; $t_0 = 3$

4 $F(t) = \dfrac{t^2 + 2t + 1}{t - 1} i + (\sin^{-1} t)j$; $t_0 = 1$

5 $F(t) = \dfrac{1}{t} i + (\sin 3t)j$; $t_0 = \dfrac{\pi}{6}$

6 $F(t) = \ln(t + 1)i + e^{-2t}j$; $t_0 = 0$

In Problems 7 to 12, find $F'(t)$ and $F''(t)$.

7 $F(t) = (3t^2 - 1)i + (9t^4 + 5)j$

8 $F(t) = \ln(3 + t)i + (\sin t)j$

9 $F(t) = e^{3t}i + (\ln 2t)j$

10 $F(t) = (3 \sec t)i + (4 \tan t)j$

11 $F(t) = (5 \cos t)i + (3 \sin t)j$

12 $F(t) = (\tan^{-1} 2t)i + e^{7t}j$

In Problems 13 to 16, find dR/dt and d^2R/dt^2.

13 $R = (\cos t^2)i + (\sin t^2)j$

14 $R = te^{-2t}i + te^{2t}j$

15 $R = \dfrac{1}{t} i + \dfrac{1}{t^2} j$

16 $R = \dfrac{t - 2}{t + 2} i + \dfrac{t - 3}{t + 3} j$

In Problems 17 to 20, find (a) $F'(t)$, (b) $F''(t)$, (c) $F'(t) \cdot F''(t)$, and (d) $\dfrac{d}{dt} |F(t)|$.

17 $F(t) = (e^t + 3)i + (e^t + 7)j$

18 $F(t) = (3 \cos 2t)i + (3 \sin 2t)j$

19 $F(t) = (4 \sin 3t)i - (4 \sin 3t)j$

20 $F(t) = e^{-5t}i + e^{5t}j$

In Problems 21 to 24, find (a) $\dfrac{d}{dt}(R \cdot S)$ and (b) $\dfrac{d}{dt}(wR)$.

21 $R = e^{2t}i + e^{-4t}j$, $S = (\cos 2t)i + (\sin 2t)j$, $w = e^{-7t}$

22 $R = 5ti + t^2j$, $S = (\sec t)i + (3 \sin t)j$, $w = \cot t$

23 $R = (\ln t)i + \dfrac{1}{t - 1}j$, $S = \dfrac{i}{t} + \dfrac{3j}{t^2}$, $w = \cos 7t$

24 $R = \dfrac{i}{t^2 + 4} + \dfrac{2j}{1 - t^2}$, $S = (\ln t)i + t^2j$, $w = 5t^2 + 8$

25 Let $F(t) = (\sin t)i + (\cos t)j$ and $t = s^3$. Use Theorems 4 and 5 to find

(a) $\dfrac{d}{ds} F(t)$ (b) $\dfrac{d}{dt} |F(t)|^2$

26 Show that, if dw/dt exists and A is a constant vector, then

$$\frac{d}{dt}(wA) = \frac{dw}{dt} A$$

27 Show that, if dR/dt exists and c is a constant scalar, then

$$\frac{d}{dt}(cR) = c\frac{dR}{dt}$$

28 Suppose that dR/dt exists and that $|R|$ is constant. Show that R and dR/dt are perpendicular.

29 Suppose that R is a variable vector that is always parallel to a fixed nonzero vector A. If dR/dt exists, show that dR/dt is always parallel to the fixed vector A.

30 If $R \neq 0$ and dR/dt exists, find a formula for

$$\frac{d}{dt}\left(\frac{R}{|R|}\right)$$

31 Prove Theorem 1.

32 Prove Theorem 2.

33 Prove that the sum or difference of continuous vector-valued functions is again continuous.

34 Prove that the dot product of continuous vector-valued functions is a continuous real-valued function.

35 If F is a continuous vector-valued function, show that the real-valued function f defined by $f(t) = |F(t)|$ is also continuous.

36 If F is a continuous vector-valued function and if f is a continuous real-valued function, both of which have the same domain, show that the vector-valued function G defined by $G(t) = f(t)F(t)$ is continuous.

37 Give an appropriate definition of the following one-sided limits:

(a) $\lim\limits_{t \to c^+} F(t) = A$ (b) $\lim\limits_{t \to c^-} F(t) = A$

38 Prove Theorem 3.

39 Prove parts (i) and (ii) of Theorem 4.

40 Prove part (iii) of Theorem 4.

41 Prove that a differentiable vector-valued function is continuous.

42 Give an appropriate definition of the following: The vector-valued function F is continuous on the closed interval $[a, b]$.

43 Assuming the existence of the required derivatives, show that

$$\frac{d^2}{dt^2}(R \cdot S) = \frac{d^2R}{dt^2} \cdot S + 2\frac{dR}{dt} \cdot \frac{dS}{dt} + R \cdot \frac{d^2S}{dt^2}$$

12.5 Velocity, Acceleration, and Arc Length

Figure 1

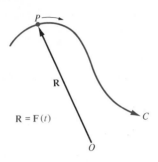

Figure 1

A vector parametric equation $\mathbf{R} = \mathbf{F}(t)$ can be regarded as giving the position vector $\mathbf{R}$ of a moving point P at the time t (Figure 1). As the time t varies, the point P traces out a curve C. Even if the parameter t actually represents some quantity other than time, it may be useful to think of it—at least for the purposes of this section—as corresponding to the elapsed time since some arbitrary (but fixed) initial instant. This point of view enables us to study the vector-valued function $\mathbf{F}$ in terms of the intuitively appealing idea of physical motion.

For the remainder of this section, we assume that the moving point P traces out a curve C in the plane and that the variable position vector $\mathbf{R}$ of P at time t is given by $\mathbf{R} = \mathbf{F}(t)$. Furthermore, we assume that $\mathbf{F}$ has a first derivative $\mathbf{F}'$ and a second derivative $\mathbf{F}''$. We imagine that the moving point P is equipped with an odometer to measure the **distance** s that it has traveled along the path C, starting at P_0 when $t = t_0$, and that it is equipped with a speedometer to measure its **instantaneous speed**

Figure 2

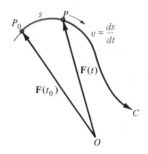

$$v = \frac{ds}{dt}$$

at any time t (Figure 2). Here s is just the arc length of the curve C between P_0 and P, and v is the instantaneous rate of change of s with respect to the parameter t.

The **velocity vector** of the moving point P is defined to be the vector $\mathbf{V}$ whose direction is parallel to the tangent line to the curve C at P and whose length is the speed of P, so that

Figure 3

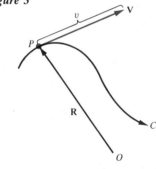

$$v = |\mathbf{V}|$$

(Figure 3). Here we understand that $\mathbf{V}$ points in the "instantaneous direction of motion" of P, so that if the constraints causing P to follow the curve C were suddenly removed, then P would "fly off along the tangent line" in the direction of $\mathbf{V}$.

We are now going to give an informal argument to show that *the velocity vector of a moving point is the derivative with respect to time of the position vector of the point*. To begin the argument, let $\Delta t = t_2 - t_1$ denote a small positive interval of time, and put

$$\mathbf{R}_1 = \overrightarrow{OP_1} = \mathbf{F}(t_1) \qquad \text{and} \qquad \mathbf{R}_2 = \overrightarrow{OP_2} = \mathbf{F}(t_2) = \mathbf{F}(t_1 + \Delta t)$$

(Figure 4). During the time interval Δt from t_1 to t_2, the point P moves along a small portion of the curve C between P_1 and P_2. Denote the arc length of this portion of C by Δs. In moving from P_1 to P_2, the point P undergoes the **displacement**

$$\Delta \mathbf{R} = \mathbf{R}_2 - \mathbf{R}_1 = \mathbf{F}(t_1 + \Delta t) - \mathbf{F}(t_1)$$

Evidently,

$$\Delta s \approx |\Delta \mathbf{R}|$$

with better and better approximation as $\Delta t \to 0^+$. Dividing by Δt, we obtain

$$\frac{\Delta s}{\Delta t} \approx \frac{1}{\Delta t} |\Delta \mathbf{R}| = \left| \frac{\Delta \mathbf{R}}{\Delta t} \right|$$

Figure 4

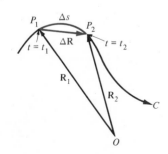

Hence, taking the limit as $\Delta t \to 0^+$, we have

$$|\mathbf{V}| = v = \frac{ds}{dt} = \left|\frac{d\mathbf{R}}{dt}\right|$$

Therefore, $d\mathbf{R}/dt$ has the same *length* as the velocity vector $\mathbf{V}$.

To finish the argument, we only need to show that $d\mathbf{R}/dt$ has the same *direction* as the velocity vector $\mathbf{V}$. If the time interval Δt is small, then $1/\Delta t$ is large; hence, the difference quotient

$$\frac{\Delta\mathbf{R}}{\Delta t} = \frac{\mathbf{F}(t_1 + \Delta t) - \mathbf{F}(t_1)}{\Delta t} = \frac{1}{\Delta t}\Delta\mathbf{R}$$

not only has the same direction as the displacement vector $\Delta\mathbf{R}$ but also is considerably longer than $\Delta\mathbf{R}$ (Figure 5).

As $\Delta t \to 0^+$, the displacement vector $\Delta\mathbf{R}$ becomes shorter and shorter; however, the difference quotient $\Delta\mathbf{R}/\Delta t$ approaches $d\mathbf{R}/dt$, which is not necessarily $\mathbf{0}$. At the same time, $\Delta\mathbf{R}/\Delta t$ pivots about the point P_1 and approaches the direction of the tangent line to the curve C at P_1 (Figure 6). Since $d\mathbf{R}/dt = \lim_{\Delta t \to 0} \Delta\mathbf{R}/\Delta t$, it follows that $d\mathbf{R}/dt$ is parallel to this tangent line; hence, $d\mathbf{R}/dt$ has the same *direction* as the velocity vector $\mathbf{V}$. Since $d\mathbf{R}/dt$ has the same length and direction as $\mathbf{V}$, we conclude that

$$\frac{d\mathbf{R}}{dt} = \mathbf{V}$$

Figure 5

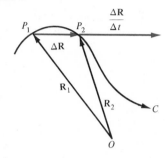

Figure 6

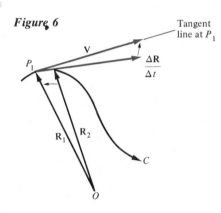

Tangent line at P_1

As the point P moves along the curve C, its velocity vector $\mathbf{V}$ can change in length, in direction, or in both. The instantaneous rate of change of $\mathbf{V}$ with respect to time is a vector $d\mathbf{V}/dt$ called the **acceleration vector** of the moving point P and denoted by $\mathbf{A}$. Thus,

$$\mathbf{V} = \frac{d\mathbf{R}}{dt} = \mathbf{F}'(t) \quad \text{and} \quad \mathbf{A} = \frac{d\mathbf{V}}{dt} = \frac{d^2\mathbf{R}}{dt^2} = \mathbf{F}''(t)$$

__EXAMPLE 1__   A point P is moving on an elliptical path according to the equation $\mathbf{R} = (5\cos 2t)\mathbf{i} + (3\sin 2t)\mathbf{j}$. Distances are measured in meters and time in seconds. Find **(a)** the velocity vector $\mathbf{V}$, **(b)** the acceleration vector $\mathbf{A}$, **(c)** the speed v at the instant t.

SOLUTION

(a) $\mathbf{V} = \dfrac{d\mathbf{R}}{dt} = (-10\sin 2t)\mathbf{i} + (6\cos 2t)\mathbf{j}$

(b) $\mathbf{A} = \dfrac{d\mathbf{V}}{dt} = (-20\cos 2t)\mathbf{i} + (-12\sin 2t)\mathbf{j}$

(c) $v = |\mathbf{V}| = \sqrt{100\sin^2 2t + 36\cos^2 2t}$ meters per second   ∎

__EXAMPLE 2__   A moving point P has position vector $\mathbf{R} = (t^2/2)\mathbf{i} - t^3\mathbf{j}$ at time t. Find **(a)** the velocity vector $\mathbf{V}$, **(b)** the acceleration vector $\mathbf{A}$, and **(c)** the speed v of the moving point at the instant when $t = 10$.

SOLUTION

(a) $\mathbf{V} = t\mathbf{i} - 3t^2\mathbf{j}$; hence, when $t = 10$, $\mathbf{V} = 10\mathbf{i} - 300\mathbf{j}$.

(b) $\mathbf{A} = \mathbf{i} - 6t\mathbf{j}$; hence, when $t = 10$, $\mathbf{A} = \mathbf{i} - 60\mathbf{j}$.

(c) $v = |\mathbf{V}| = \sqrt{t^2 + 9t^4}$; hence, when $t = 10$, $v = \sqrt{100(1 + 900)} = 10\sqrt{901}$.

If the velocity vector $\mathbf{V}$ is not the zero vector, then we can normalize $\mathbf{V}$ by dividing it by its own length $|\mathbf{V}|$. The resulting vector $\mathbf{T}$ has the same direction as the velocity vector, but has *unit length* (Figure 7). Since $\mathbf{V}$ is parallel to the tangent line to C at P, so is $\mathbf{T}$; hence, we call $\mathbf{T}$ the **unit tangent vector** to the curve C at the point P. Evidently,

Figure 7

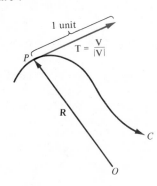

$$\mathbf{T} = \frac{\mathbf{V}}{|\mathbf{V}|} = \frac{\mathbf{V}}{v} = \frac{\dfrac{d\mathbf{R}}{dt}}{\left|\dfrac{d\mathbf{R}}{dt}\right|} = \frac{\dfrac{d\mathbf{R}}{dt}}{\dfrac{ds}{dt}}$$

EXAMPLE 3   Find a unit tangent vector $\mathbf{T}$ to the curve C whose vector parametric equation is $\mathbf{R} = (t^2 - 4t)\mathbf{i} + (\tfrac{3}{4}t^4)\mathbf{j}$.

SOLUTION   Here we have

$$\frac{d\mathbf{R}}{dt} = (2t - 4)\mathbf{i} + (3t^3)\mathbf{j}$$

so that $\left|\dfrac{d\mathbf{R}}{dt}\right| = \sqrt{(2t - 4)^2 + (3t^3)^2} = \sqrt{9t^6 + 4t^2 - 16t + 16}$

and $\mathbf{T} = \dfrac{\dfrac{d\mathbf{R}}{dt}}{\left|\dfrac{d\mathbf{R}}{dt}\right|} = \dfrac{(2t - 4)\mathbf{i} + (3t^3)\mathbf{j}}{\sqrt{9t^6 + 4t^2 - 16t + 16}}$

Since $ds/dt = v = |d\mathbf{R}/dt|$, we have the differential equation

$$ds = \left|\frac{d\mathbf{R}}{dt}\right| dt$$

for the arc length s of the curve whose vector parametric equation is $\mathbf{R} = \mathbf{F}(t)$. Integrating, we find that the arc length s between the point corresponding to $t = a$ and the point corresponding to $t = b$ is

Figure 8

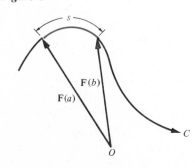

$$s = \int_a^b \left|\frac{d\mathbf{R}}{dt}\right| dt$$

(Figure 8). If $\mathbf{R} = \mathbf{F}(t) = g(t)\mathbf{i} + h(t)\mathbf{j}$, then

$$\left|\frac{d\mathbf{R}}{dt}\right| = |g'(t)\mathbf{i} + h'(t)\mathbf{j}|$$

$$= \sqrt{[g'(t)]^2 + [h'(t)]^2}$$

and

$$s = \int_a^b \sqrt{[g'(t)]^2 + [h'(t)]^2}\ dt$$

In Examples 4 to 6, find the arc length s of the given curve between the indicated values of the parameter.

EXAMPLE 4 $\mathbf{R} = (3t^2)\mathbf{i} + (t^3 - 3t)\mathbf{j}$ between $t = 0$ and $t = 1$

SOLUTION   Here, $d\mathbf{R}/dt = (6t)\mathbf{i} + (3t^2 - 3)\mathbf{j}$ and

$$\left|\frac{d\mathbf{R}}{dt}\right| = \sqrt{(6t)^2 + (3t^2 - 3)^2} = \sqrt{36t^2 + 9t^4 - 18t^2 + 9}$$
$$= \sqrt{9t^4 + 18t^2 + 9} = \sqrt{(3t^2 + 3)^2} = 3t^2 + 3$$

Therefore,

$$s = \int_0^1 \left|\frac{d\mathbf{R}}{dt}\right|\ dt = \int_0^1 (3t^2 + 3)\ dt = (t^3 + 3t)\Big|_0^1 = 4 \text{ units} \qquad \blacksquare$$

© **EXAMPLE 5** $\begin{cases} x = e^{-t}\cos t \\ y = e^{-t}\sin t \end{cases}$ between $t = 0$ and $t = \pi$

SOLUTION   The given scalar parametric equations are equivalent to the vector parametric equation $\mathbf{R} = g(t)\mathbf{i} + h(t)\mathbf{j}$, where $g(t) = e^{-t}\cos t$ and $h(t) = e^{-t}\sin t$. Thus,

$$g'(t) = -e^{-t}(\cos t + \sin t) \qquad h'(t) = -e^{-t}(\sin t - \cos t)$$

and

$$s = \int_0^\pi \sqrt{[g'(t)]^2 + [h'(t)]^2}\ dt$$
$$= \int_0^\pi \sqrt{e^{-2t}(\cos^2 t + 2\cos t \sin t + \sin^2 t) + e^{-2t}(\sin^2 t - 2\sin t \cos t + \cos^2 t)}\ dt$$
$$= \int_0^\pi e^{-t}\sqrt{2(\cos^2 t + \sin^2 t)}\ dt$$
$$= \int_0^\pi \sqrt{2}e^{-t}\ dt$$
$$= (-\sqrt{2}e^{-t})\Big|_0^\pi$$
$$= \sqrt{2}(1 - e^{-\pi}) \approx 1.35 \text{ units} \qquad \blacksquare$$

EXAMPLE 6 $y = f(x)$ between $x = a$ and $x = b$

SOLUTION   With x as the parameter, the scalar equation $y = f(x)$ is equivalent to the vector parametric equation $\mathbf{R} = x\mathbf{i} + f(x)\mathbf{j}$. Here,

$$\frac{d\mathbf{R}}{dx} = \mathbf{i} + f'(x)\mathbf{j} \qquad \left|\frac{d\mathbf{R}}{dx}\right| = \sqrt{1^2 + [f'(x)]^2} \quad \text{and} \quad s = \int_a^b \sqrt{1 + [f'(x)]^2}\ dx$$

which coincides with the formula for the arc length in Section 6.4. $\blacksquare$

Problem Set 12.5

In Problems 1 to 10, the vector equation of motion of a point P in the xy plane is given. Find (a) the velocity vector $\mathbf{V}$, (b) the acceleration vector $\mathbf{A}$, (c) the speed v of the moving point at the time t.

1 $\mathbf{R} = (3t^2 - 1)\mathbf{i} + (2t + 5)\mathbf{j}$

2 $\mathbf{R} = (2 \cos t)\mathbf{i} - (7 \sin t)\mathbf{j}$

3 $\mathbf{R} = t^2\mathbf{i} + (\ln t)\mathbf{j}$

4 $\mathbf{R} = (t^2 + 1)\mathbf{i} + t^3\mathbf{j}$

5 $\mathbf{R} = 2(t - \sin t)\mathbf{i} + 2(1 - \cos t)\mathbf{j}$

6 $\mathbf{R} = (4 \cos^3 t)\mathbf{i} + (4 \sin^3 t)\mathbf{j}$

7 $\mathbf{R} = (\cos t + \sin t)\mathbf{i} + (\cos t - \sin t)\mathbf{j}$

8 $\mathbf{R} = (2 \cot t)\mathbf{i} + (2 \sin^2 t)\mathbf{j}$

9 $\mathbf{R} = (e^t \cos t)\mathbf{i} + (e^t \sin t)\mathbf{j}$

10 $\mathbf{R} = \phi(t)\mathbf{i} + f(\phi(t))\mathbf{j}$

In Problems 11 to 16, the vector equation of motion of a point P in the xy plane is given. Find (a) the velocity vector $\mathbf{V}$, (b) the acceleration vector $\mathbf{A}$, and (c) the speed v of the moving point at the instant when t has the indicated value t_1.

11 $\mathbf{R} = (7t^2 - t)\mathbf{i} + (5t - 7)\mathbf{j}$; $t_1 = 2$

12 $\mathbf{R} = (t \sin t)\mathbf{i} + (\cos t)\mathbf{j}$; $t_1 = \pi/2$

13 $\mathbf{R} = 3(1 + \cos \pi t)\mathbf{i} + 4(1 + \sin \pi t)\mathbf{j}$; $t_1 = \frac{5}{4}$

14 $\mathbf{R} = t\mathbf{i} + e^t\mathbf{j}$; $t_1 = 0$

15 $\mathbf{R} = (\ln \sin t)\mathbf{i} + (\ln \cos t)\mathbf{j}$; $t_1 = \pi/6$

16 $\mathbf{R} = (4 \cos t^2)\mathbf{i} + (4 \sin t^2)\mathbf{j}$; $t_1 = \sqrt{\pi}/2$

In Problems 17 to 24, find (a) the velocity vector $\mathbf{V}$, (b) the speed v at time t, and (c) the unit tangent vector $\mathbf{T}$ at the indicated time t_1.

17 $\mathbf{R} = (12t - 3)\mathbf{i} - (7t + 9)\mathbf{j}$; $t_1 = 3$

18 $\mathbf{R} = (2t - 1)\mathbf{i} + (3t^2 + 7)\mathbf{j}$; $t_1 = 2$

19 $\mathbf{R} = (t - \sin t)\mathbf{i} + (1 - \cos t)\mathbf{j}$; $t_1 = \pi/2$

20 $\mathbf{R} = (\cos^3 t)\mathbf{i} + (\sin^3 t)\mathbf{j}$; $t_1 = 3\pi/4$

21 $\mathbf{R} = (-7 + \cos 2t)\mathbf{i} + (5 + \sin 2t)\mathbf{j}$; $t_1 = \pi/6$

22 $\mathbf{R} = (t \cos t)\mathbf{i} + (t \sin t)\mathbf{j}$; $t_1 = 0$

$\boxed{\text{C}}$ **23** $\mathbf{R} = \dfrac{\cos t}{1 + \cos t}\mathbf{i} + \dfrac{\sin t}{1 + \cos t}\mathbf{j}$; $t_1 = \pi/5$

$\boxed{\text{C}}$ **24** $\mathbf{R} = \ln (t^2 + 1)\mathbf{i} + e^{2t}\mathbf{j}$; $t_1 = \sqrt{3}$

In Problems 25 to 38, find the arc length s of each curve between the indicated values of the parameter.

25 $\mathbf{R} = (7t - 9)\mathbf{i} - (5t + 4)\mathbf{j}$; $t = 0$ to $t = 2$

26 $\mathbf{R} = (at + b)\mathbf{i} + (ct + d)\mathbf{j}$; $t = t_1$ to $t = t_2$

27 $\mathbf{R} = e^{2t}\mathbf{i} + (\frac{1}{4}e^{4t} - t)\mathbf{j}$; $t = 0$ to $t = \frac{1}{2}$

28 $\mathbf{R} = t^2\mathbf{i} + t^3\mathbf{j}$; $t = 0$ to $t = 4$

29 $\mathbf{R} = (3 \cos 2t)\mathbf{i} + (3 \sin 2t)\mathbf{j}$; $t = 0$ to $t = \pi$

30 $\begin{cases} x = 3 \cos t - \cos 3t \\ y = 3 \sin t - \sin 3t \end{cases}$; $t = 0$ to $t = \pi$

31 $\begin{cases} x = e^{-t} \cos t \\ y = e^{-t} \sin t \end{cases}$; $t = 0$ to $t = 2\pi$

32 $\begin{cases} x = t - \sin t \\ y = 1 - \cos t \end{cases}$; $t = 0$ to $t = 2\pi$

33 $\mathbf{R} = (\cos t + t \sin t)\mathbf{i} + (\sin t - t \cos t)\mathbf{j}$; $t = 0$ to $t = \pi/4$

34 $\mathbf{R} = t\mathbf{i} + t^2\mathbf{j}$; $t = 0$ to $t = a$

35 $\mathbf{R} = (\ln \sqrt{1 + t^2})\mathbf{i} + (\tan^{-1} t)\mathbf{j}$; $t = 0$ to $t = 1$

36 $\mathbf{R} = \dfrac{t^2}{2}\mathbf{i} + \left[\dfrac{(6t + 9)^{3/2}}{9} \right]\mathbf{j}$; $t = 0$ to $t = 4$

$\boxed{\text{C}}$ **37** $\mathbf{R} = (\cos^3 t)\mathbf{i} + (\sin^3 t)\mathbf{j}$; $t = 0$ to $t = \pi/9$

$\boxed{\text{C}}$ **38** $\mathbf{R} = t\mathbf{i} + (\cosh t)\mathbf{j}$; $t = 0$ to $t = 4$

39 The curve C whose equation in polar coordinates is $r = f(\theta)$ can be expressed in parametric Cartesian form as

$$\begin{cases} x = f(\theta) \cos \theta \\ y = f(\theta) \sin \theta \end{cases}$$

by using θ as parameter. Using the methods of this section, find a formula for the arc length of the portion of C between the point where $\theta = \theta_1$ and the point where $\theta = \theta_2$.

40 Find a formula for the unit tangent vector $\mathbf{T}$ to the curve C in Problem 39.

41 The curve C whose Cartesian equation is $y = f(x)$ can be expressed in vector parametric form as $\mathbf{R} = x\mathbf{i} + f(x)\mathbf{j}$, by using x as the parameter. Find a formula for the unit tangent vector $\mathbf{T}$ to the curve C.

42 If $\mathbf{R} = \mathbf{F}(s)$, where the parameter s is the arc length, find a formula for the unit tangent vector $\mathbf{T}$.

12.6   Normal Vectors and Curvature

In this section we continue the study, initiated in Section 12.5, of the curve C traced out by a moving point P in the plane according to the vector parametric equation

$$\mathbf{R} = \overrightarrow{OP} = \mathbf{F}(t)$$

Figure 1

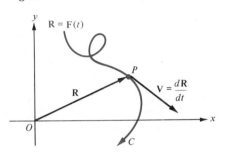

(Figure 1). As in Section 12.5, the parameter t can be thought of as time, so that the vector derivatives

$$\mathbf{V} = \mathbf{F}'(t) \qquad \text{and} \qquad \mathbf{A} = \mathbf{F}''(t)$$

which we assume exist, represent the velocity and acceleration vectors of P, respectively, at the time t. We continue to denote by s the arc length traced out by P along the curve C since some arbitrary (but fixed) initial time t_0, so that

$$v = \frac{ds}{dt} = |\mathbf{V}|$$

is the instantaneous speed of the moving point P.

For the remainder of this section we assume that *the instantaneous speed of the moving point P is never zero*, $v \neq 0$, so that we can form the unit tangent vector

Figure 2

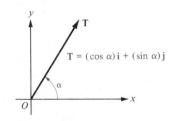

$$\mathbf{T} = \frac{\mathbf{V}}{v} = \frac{\dfrac{d\mathbf{R}}{dt}}{\left|\dfrac{d\mathbf{R}}{dt}\right|}$$

to the curve C at the point P (Figure 2). Using the chain rule, we have

$$\frac{d\mathbf{R}}{ds} = \frac{d\mathbf{R}}{dt}\frac{dt}{ds} = \frac{d\mathbf{R}/dt}{ds/dt} = \frac{\mathbf{V}}{v} = \mathbf{T}$$

that is, *the unit tangent vector $\mathbf{T}$ is the derivative $d\mathbf{R}/ds$ of the position vector with respect to arc length.*

As the moving point P traces out the curve C, the unit tangent vector $\mathbf{T}$ at P can change its direction but not its length (which is always 1 unit). In order to study the change in direction of $\mathbf{T}$, we denote by α the angle in radians from the positive x axis to $\mathbf{T}$ when the initial point of $\mathbf{T}$ is moved to the origin (Figure 3). Thus, since $|\mathbf{T}| = 1$, we have

Figure 3

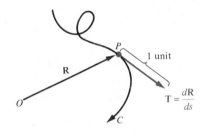

$$\mathbf{T} = (\cos \alpha)\mathbf{i} + (\sin \alpha)\mathbf{j}$$

(see Figure 11, page 728).

Any vector perpendicular to the unit tangent vector $\mathbf{T}$ at the point P on the curve C is called a **normal vector** to C at P. Of course, a normal vector whose length is 1 unit is called a **unit normal vector.** If you were to sit on the moving point P, facing in the direction of the unit tangent vector $\mathbf{T}$, as P traces out the curve C, you would find that one of the unit normal vectors, $\mathbf{N}_l$, is always on your left and the

Figure 4

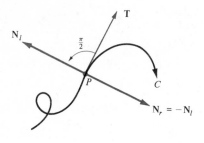

other unit normal vector, $\mathbf{N}_r$, is always on your right (Figure 4). Evidently,

$$\mathbf{N}_r = -\mathbf{N}_l$$

Since $\mathbf{N}_l$ is obtained by rotating $\mathbf{T}$ *counterclockwise* through $\pi/2$ radians, it follows that $\mathbf{N}_l$ makes an angle of $\alpha + (\pi/2)$ radians with the positive x axis. Therefore, since $|\mathbf{N}_l| = 1$, we have

$$\mathbf{N}_l = \cos\left(\alpha + \frac{\pi}{2}\right)\mathbf{i} + \sin\left(\alpha + \frac{\pi}{2}\right)\mathbf{j}$$

$$= (-\sin \alpha)\mathbf{i} + (\cos \alpha)\mathbf{j}$$

Hence,

$$\mathbf{N}_r = -\mathbf{N}_l = (\sin \alpha)\mathbf{i} - (\cos \alpha)\mathbf{j}$$

If g and h are the component functions of $\mathbf{F}$, so that

$$\mathbf{R} = \mathbf{F}(t) = g(t)\mathbf{i} + h(t)\mathbf{j}$$

then, as we have seen in Section 12.5,

$$\mathbf{T} = \frac{d\mathbf{R}/dt}{|d\mathbf{R}/dt|} = \frac{g'(t)}{\sqrt{[g'(t)]^2 + [h'(t)]^2}}\,\mathbf{i} + \frac{h'(t)}{\sqrt{[g'(t)]^2 + [h'(t)]^2}}\,\mathbf{j}$$

Also, $\mathbf{T} = (\cos \alpha)\mathbf{i} + (\sin \alpha)\mathbf{j}$; hence, it follows that

$$\cos \alpha = \frac{g'(t)}{\sqrt{[g'(t)]^2 + [h'(t)]^2}} \qquad \text{and} \qquad \sin \alpha = \frac{h'(t)}{\sqrt{[g'(t)]^2 + [h'(t)]^2}}$$

Therefore, since $\mathbf{N}_l = (-\sin \alpha)\mathbf{i} + (\cos \alpha)\mathbf{j}$, we have

$$\mathbf{N}_l = \frac{-h'(t)}{\sqrt{[g'(t)]^2 + [h'(t)]^2}}\,\mathbf{i} + \frac{g'(t)}{\sqrt{[g'(t)]^2 + [h'(t)]^2}}\,\mathbf{j}$$

EXAMPLE 1   Find the unit normal vector $\mathbf{N}_l$ to the curve $\mathbf{R} = t\mathbf{i} + t^2\mathbf{j}$ at the point P whose position vector is $\mathbf{R}$.

SOLUTION   Here the component functions g and h are given by $g(t) = t$ and $h(t) = t^2$. Thus,

$$g'(t) = 1 \qquad h'(t) = 2t \qquad \sqrt{[g'(t)]^2 + [h'(t)]^2} = \sqrt{1 + 4t^2}$$

and

$$\mathbf{N}_l = \frac{-2t}{\sqrt{1 + 4t^2}}\,\mathbf{i} + \frac{1}{\sqrt{1 + 4t^2}}\,\mathbf{j} \qquad\blacksquare$$

When the curve C bends rather sharply, the unit tangent vector $\mathbf{T}$ changes its direction quite rapidly; that is, the rate of change $d\alpha/ds$ of the angle α (Figure 3) with respect to arc length s is large in absolute value. For this reason, the instantaneous rate $d\alpha/ds$ at which the tangent vector is turning, in radians per unit of arc length, is called the **curvature** of C at the point P and is traditionally denoted by the Greek letter κ (called "kappa"). Thus, by definition,

$$\kappa = \frac{d\alpha}{ds}$$

Figure 5

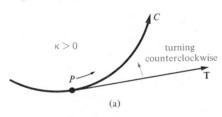

$\kappa > 0$

turning counterclockwise

(a)

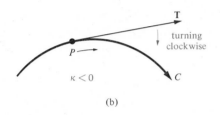

turning clockwise

$\kappa < 0$

(b)

Notice that $\kappa > 0$ if α is increasing as P moves to trace out the curve; that is, $\kappa > 0$ if the unit tangent vector turns counterclockwise as s is increased (Figure 5a). Similarly, $\kappa < 0$ if the unit tangent vector turns clockwise as s is increased (Figure 5b).

If we differentiate the equation $\mathbf{T} = (\cos \alpha)\mathbf{i} + (\sin \alpha)\mathbf{j}$ on both sides with respect to arc length s, we obtain

$$\frac{d\mathbf{T}}{ds} = \left(-\sin \alpha \, \frac{d\alpha}{ds}\right)\mathbf{i} + \left(\cos \alpha \, \frac{d\alpha}{ds}\right)\mathbf{j} = \frac{d\alpha}{ds}\left[(-\sin \alpha)\mathbf{i} + (\cos \alpha)\mathbf{j}\right]$$

Hence,

$$\frac{d\mathbf{T}}{ds} = \kappa \mathbf{N}_l$$

The equation $d\mathbf{T}/ds = \kappa \mathbf{N}_l$ is extremely important in the theory of curves in the plane. Since $\mathbf{N}_l$ is normal to the curve C and κ is a scalar, it implies that *the vector $d\mathbf{T}/ds$ is always a normal vector to the curve C at the point P.* Furthermore, since $|\mathbf{N}_l| = 1$, it follows that

$$\left|\frac{d\mathbf{T}}{ds}\right| = |\kappa \mathbf{N}_l| = |\kappa| \, |\mathbf{N}_l| = |\kappa|$$

Figure 6

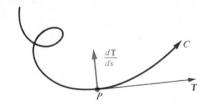

Thus, *the vector $d\mathbf{T}/ds$ is long where C is rather sharply curved and not so long where C bends slowly.* Finally, it can be shown that $d\mathbf{T}/ds$ *always points in the direction of concavity of the curve C* (Figure 6) (see Problem 30).

The vector $d\mathbf{T}/ds$ is called the **curvature vector** of the curve C at the point P. If the curvature vector is nonzero, that is, if $|\kappa| = |d\mathbf{T}/ds| \neq 0$, then the vector $\mathbf{N}$, defined by

$$\mathbf{N} = \frac{\dfrac{d\mathbf{T}}{ds}}{\left|\dfrac{d\mathbf{T}}{ds}\right|} = \frac{1}{|\kappa|}\frac{d\mathbf{T}}{ds}$$

Figure 7

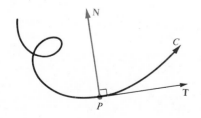

is called the **principal normal vector** to C at P (Figure 7). Notice that $\mathbf{N}$ *is a unit normal vector to the curve C and points in the direction of concavity of C.* (Why?) It follows that

$$\mathbf{N} = \mathbf{N}_l \qquad \text{if } \kappa > 0$$

while

$$\mathbf{N} = \mathbf{N}_r \qquad \text{if } \kappa < 0$$

Evidently, the equation

$$\frac{d\mathbf{T}}{ds} = |\kappa|\mathbf{N}$$

always holds, provided that $\kappa \neq 0$. If $\kappa = 0$, the principal normal vector is undefined.

The following important theorem relates the acceleration vector $\mathbf{A}$, the unit tangent vector $\mathbf{T}$, the speed v, the curvature vector $d\mathbf{T}/ds$, the principal normal vector $\mathbf{N}$, and the curvature κ.

THEOREM 1   **Resolution of the Acceleration Vector into Tangential and Normal Components**

$$\textbf{(i)} \ \ \mathbf{A} = \frac{dv}{dt}\mathbf{T} + v^2\frac{d\mathbf{T}}{ds} \qquad \textbf{(ii)} \ \ \mathbf{A} = \frac{dv}{dt}\mathbf{T} + v^2|\kappa|\mathbf{N} \qquad \text{if } \kappa \neq 0$$

PROOF   Since $\mathbf{T} = \mathbf{V}/v$, we have $\mathbf{V} = v\mathbf{T}$. Differentiating both sides of the last equation with respect to t, we obtain

$$\mathbf{A} = \frac{d\mathbf{V}}{dt} = \frac{dv}{dt}\mathbf{T} + v\frac{d\mathbf{T}}{dt}$$

By the chain rule, $$\frac{d\mathbf{T}}{dt} = \frac{ds}{dt}\frac{d\mathbf{T}}{ds} = v\frac{d\mathbf{T}}{ds}$$

Hence, $$\mathbf{A} = \frac{dv}{dt}\mathbf{T} + v\left(v\frac{d\mathbf{T}}{ds}\right) = \frac{dv}{dt}\mathbf{T} + v^2\frac{d\mathbf{T}}{ds}$$

and part (i) is proved. To obtain part (ii), we substitute $d\mathbf{T}/ds = |\kappa|\mathbf{N}$ into part (i).   ∎

Theorem 1 expresses the acceleration vector $\mathbf{A}$ as a sum of its **tangential component vector** $(dv/dt)\mathbf{T}$ and its **normal component vector** $v^2(d\mathbf{T}/ds)$ (or $v^2|\kappa|\mathbf{N}$ if $\kappa \neq 0$) and has useful applications to mechanics (see Section 12.7). It also has the following important consequence.

THEOREM 2   **Formula for Curvature**

$$\kappa = \frac{\mathbf{A}\cdot\mathbf{N}_l}{v^2}$$

PROOF   We have $d\mathbf{T}/ds = \kappa\mathbf{N}_l$; hence,

$$\frac{d\mathbf{T}}{ds}\cdot\mathbf{N}_l = \kappa\mathbf{N}_l\cdot\mathbf{N}_l = \kappa|\mathbf{N}_l|^2 = \kappa$$

since $|\mathbf{N}_l| = 1$. Therefore, taking the dot product on both sides of the equation in part (i) of Theorem 1 with $\mathbf{N}_l$, we obtain

$$\mathbf{A}\cdot\mathbf{N}_l = \frac{dv}{dt}\mathbf{T}\cdot\mathbf{N}_l + v^2\frac{d\mathbf{T}}{ds}\cdot\mathbf{N}_l = \frac{dv}{dt}(0) + v^2\kappa = v^2\kappa$$

since the vectors $\mathbf{T}$ and $\mathbf{N}_l$ are perpendicular. Solving the last equation for κ, we obtain the formula $\kappa = (\mathbf{A}\cdot\mathbf{N}_l)/v^2$, as desired.   ∎

The formula for curvature in Theorem 2 can be rewritten in terms of the component functions g and h of $\mathbf{R}$ (Problem 22). Thus, if $\mathbf{R} = g(t)\mathbf{i} + h(t)\mathbf{j}$, then

$$\kappa = \frac{g'(t)h''(t) - g''(t)h'(t)}{\{[g'(t)]^2 + [h'(t)]^2\}^{3/2}}$$

EXAMPLE 2   Find the curvature κ of $\mathbf{R} = t\mathbf{i} + t^2\mathbf{j}$.

SOLUTION   Here the component functions g and h are given by $g(t) = t$ and $h(t) = t^2$. Using the formula above and noting that

$$g'(t) = 1 \qquad g''(t) = 0 \qquad h'(t) = 2t \qquad \text{and} \qquad h''(t) = 2$$

we obtain $$\kappa = \frac{(1)(2) - (0)(2t)}{[1^2 + (2t)^2]^{3/2}} = \frac{2}{(1 + 4t^2)^{3/2}}$$ ■

EXAMPLE 3   Find a formula for *the curvature κ of the graph of $y = f(x)$*, using x as the parameter. Then find the curvature κ of the sine curve $y = \sin x$ at the point $(\pi/2,\ 1)$.

SOLUTION   In vector parametric form, we have $\mathbf{R} = x\mathbf{i} + f(x)\mathbf{j}$, so the component functions g and h are given by $g(x) = x$ and $h(x) = f(x)$. Thus, replacing t by x in the formula for κ, we have $g'(x) = 1, g''(x) = 0, h'(x) = f'(x)$, and $h''(x) = f''(x)$; hence,

$$\kappa = \frac{g'(x)h''(x) - g''(x)h'(x)}{\{[g'(x)]^2 + [h'(x)]^2\}^{3/2}}$$

or

$$\kappa = \frac{f''(x)}{\{1 + [f'(x)]^2\}^{3/2}}$$

Using this formula with $f(x) = \sin x$, $f'(x) = \cos x$, and $f''(x) = -\sin x$, we have

$$\kappa = \frac{-\sin x}{(1 + \cos^2 x)^{3/2}}$$

Thus, when $x = \pi/2$, $\kappa = -1/1 = -1$.   ■

Problem Set 12.6

In Problems 1 to 16, find (a) the unit tangent vector $\mathbf{T}$, (b) the unit normal vector $\mathbf{N}_l$, (c) the curvature κ, and (d) the principal unit normal vector $\mathbf{N}$ for each curve.

1 $\mathbf{R} = (7t - 4)\mathbf{i} + (9 - 3t)\mathbf{j}$ **2** $\mathbf{R} = (at + b)\mathbf{i} + (ct + d)\mathbf{j}$

3 $\begin{cases} x = 3\cos t \\ y = 3\sin t \end{cases}$

4 $\mathbf{R} = (a\cos t)\mathbf{i} + (a\sin t)\mathbf{j}$, where a is a positive constant

5 $\mathbf{R} = 3(1 + \cos \pi t)\mathbf{i} + 5(1 + \sin \pi t)\mathbf{j}$

6 $\begin{cases} x = t^2 - 2t \\ y = 1 - 7t \end{cases}$

7 $\mathbf{R} = t\mathbf{i} + e^t\mathbf{j}$

8 $\mathbf{R} = (2\cos \theta)\mathbf{i} + (5\sin \theta)\mathbf{j}$

9 $\begin{cases} x = 3t^2 - 1 \\ y = 2t^2 + 7 \end{cases}$ **10** $\mathbf{R} = t^2\mathbf{i} + (\ln \sec t)\mathbf{j}$

11 $\mathbf{R} = \dfrac{1}{t}\mathbf{i} + (t^2 + 1)\mathbf{j}$ **12** $\begin{cases} x = e^t \\ y = e^{-t} \end{cases}$

13 $R = (\cos 2\theta)i + (\sin \theta)j$ **14** $R = (\ln u)i + e^{2u}j$

15 $y = \ln x$, with x as the parameter

16 $y = 4x^2$, with x as the parameter

In Problems 17 to 21, find (a) the unit tangent vector **T**, (b) the unit normal vector N_l, (c) the curvature κ, and (d) the principal unit normal vector **N** for each curve at the point where the parameter has the indicated value.

17 $R = 3(1 - \cos t)i + 3(1 - \sin t)j$ when $t = \pi/6$

18 $R = (\sin t)i + (\tan t)j$ when $t = 5\pi/6$

19 $\begin{cases} x = \ln (t + 3) \\ y = \dfrac{1}{4} t^4 \end{cases}$ when $t = 1$

20 $\begin{cases} x = ue^u \\ y = ue^{-u} \end{cases}$ when $u = 0$

21 $R = \dfrac{u^2 + 1}{u^2 - 1} i + \dfrac{1}{u^2} j$ when $u = 2$

22 Using Theorem 2, derive the formula for the curvature κ of the curve $R = g(t)i + h(t)j$.

23 Consider the curve C whose equation in polar coordinates is $r = f(\theta)$. With θ as a parameter, the corresponding vector parametric equation is $R = [f(\theta) \cos \theta]i + [f(\theta) \sin \theta]j$, and the component functions g and h are given by $g(\theta) = f(\theta) \cos \theta$ and $h(\theta) = f(\theta) \sin \theta$. Find formulas for (a) **T**, (b) N_l, (c) κ, and (d) **N**.

24 Show that the principal normal vector for a circle always points toward the center of the circle.

In Problems 25 to 28, use the results of Problem 23 to find (a) **T**, (b) N_l, (c) κ, and (d) **N** for the curve C whose equation is given in polar coordinates.

25 $r = 1 - \cos \theta$ **26** $r = 1 + 2 \cos \theta$

27 $r = \dfrac{ed}{1 + e \sin \theta}$, $d > 0$

28 $r = \theta$

29 Show that the absolute value of the curvature of a circle is a constant equal to the reciprocal of its radius.

30 Explain why the curvature vector dT/ds and therefore also the principal unit normal vector **N** always point in the direction of concavity of the curve. (*Hint:* For s small, dT/ds is approximated by $\Delta T/\Delta s$.)

31 If we use x as the parameter, explain the geometric significance of the algebraic sign of the curvature κ of the graph of $y = f(x)$.

32 Prove that $dN/ds = -|\kappa|T$ [*Hint:* Begin by noticing that $N = (\kappa/|\kappa|) N_l$ where $N_l = (-\sin \alpha)i + (\cos \alpha)j$.]

33 Prove that $|\kappa| = A \cdot N/v^2$.

34 Assume that $R = F(t)$ is the vector parametric equation of a curve with constant nonzero curvature κ. Prove that the curve is a circle (or a portion of a circle) by carrying out the following steps:

(a) Show that $R + (1/|\kappa|)N$ is a constant vector by proving that its derivative is zero (use Problem 32).

(b) Put $R_0 = R + (1/|\kappa|)N$.

(c) Show that R satisfies an equation of the form $|R - R_0| = $ constant.

35 Assume that $R = F(t)$ is the vector parametric equation of a curve whose curvature is equal to zero at all points. Prove that the curve is a line (or a portion of a line). [*Hint:* First use the relation $|dT/ds| = |\kappa|$ to prove that **T** is a constant. Then select a fixed vector $N \neq 0$ such that $N \cdot T = 0$. Choose and fix a value of R, say R_0. Show that

$$\frac{d}{ds} [N \cdot (R - R_0)] = 0$$

conclude that $N \cdot (R - R_0)$ is a constant, and then show that $N \cdot (R - R_0) = 0$.]

12.7   Applications to Mechanics

Figure 1

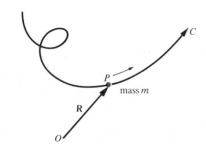

In this section we use the ideas and techniques developed in Sections 12.5 and 12.6 to study the motion of a particle in the *xy* plane. Briefly, the idea here is to consider a curve that is traced out by a physical particle of mass m, rather than by a geometric point.

Thus, suppose that a particle P of mass m moves in the *xy* plane and traces out a curve C (Figure 1). The curve C is called the **path**, or the **trajectory**, or the **orbit** of the particle P. As before, we can locate the particle at any given time t by the position vector $R = \overrightarrow{OP}$, and we have

$$R = F(t)$$

where we assume that $\mathbf{F}$ is a twice-differentiable vector-valued function. The equation $\mathbf{R} = \mathbf{F}(t)$ is called the **equation of motion** of the particle P.

By the results in Section 12.5, the velocity of the particle at any instant is the vector

$$\mathbf{V} = \frac{d\mathbf{R}}{dt} = v\mathbf{T}$$

where v is the instantaneous speed of the particle and $\mathbf{T}$ is the unit tangent vector to the path of the particle.

EXAMPLE 1   A particle moves along the parabola $y^2 = 2x$ from left to right away from the vertex at a speed of 5 units per second. Find the velocity vector $\mathbf{V}$ as the particle moves through the point $(2, 2)$.

SOLUTION   Using y (not t) as a parameter, we can write a vector parametric equation of the parabola as

$$\mathbf{R} = \frac{y^2}{2}\,\mathbf{i} + y\mathbf{j}$$

since $x = y^2/2$. Here we have

$$\frac{d\mathbf{R}}{dy} = y\mathbf{i} + \mathbf{j} \qquad \text{and} \qquad \left|\frac{d\mathbf{R}}{dy}\right| = \sqrt{y^2 + 1}$$

Hence,

$$\mathbf{T} = \frac{\dfrac{d\mathbf{R}}{dy}}{\left|\dfrac{d\mathbf{R}}{dy}\right|} = \frac{y\mathbf{i} + \mathbf{j}}{\sqrt{y^2 + 1}}$$

is the unit tangent vector to the parabola at the point $(y^2/2, y)$. When $y = 2$, we have

$$\mathbf{T} = \frac{2\mathbf{i} + \mathbf{j}}{\sqrt{5}} \qquad \text{and} \qquad \mathbf{V} = v\mathbf{T} = (5)\,\frac{2\mathbf{i} + \mathbf{j}}{\sqrt{5}} = \sqrt{5}(2\mathbf{i} + \mathbf{j}) \qquad \blacksquare$$

In the example above, we were not given the equation of motion of the particle—we were given only the equation of its *path* and the *speed* of the particle at a certain instant. Thus, we were not able to calculate $\mathbf{V}$ directly as $d\mathbf{R}/dt$ and had to rely instead on the formula $\mathbf{V} = v\mathbf{T}$.

Newton's second law of motion can now be expressed in the form

$$\mathbf{f} = m\mathbf{A}$$

where the vector $\mathbf{f}$ represents the force acting on the particle and

$$\mathbf{A} = \frac{d\mathbf{V}}{dt} = \frac{d^2\mathbf{R}}{dt^2}$$

is the acceleration of the particle. (Here we are assuming that the mass m is constant and we are neglecting relativistic effects.)

Tangential and Normal Components of Acceleration

By part (ii) of Theorem 1 in Section 12.6, we have

$$\mathbf{A} = \frac{dv}{dt}\mathbf{T} + |\kappa|v^2\mathbf{N} \qquad \text{if } \kappa \neq 0$$

Geometrically, this equation says that the acceleration vector $\mathbf{A}$ can be resolved into a sum of two perpendicular vectors, $(dv/dt)\mathbf{T}$ and $|\kappa|v^2\mathbf{N}$, the first of which is tangent to the trajectory and the second of which is normal to the trajectory (Figure 2). (If $\kappa = 0$, then the normal component vector of the acceleration is $\mathbf{0}$, and we have $\mathbf{A} = dv/dt\ \mathbf{T}$.) The normal component vector $|\kappa|v^2\mathbf{N}$ of the acceleration is also called the **centripetal acceleration.** It can be regarded as that part of the acceleration $\mathbf{A}$ caused by the change in *direction* of the velocity vector. If P moves along a straight line, then $\kappa = 0$, so that the centripetal acceleration is $\mathbf{0}$. On the other hand, if the speed v of the particle is constant, then $dv/dt = 0$, the tangential component of the acceleration is $\mathbf{0}$, and the acceleration is entirely centripetal.

Figure 2

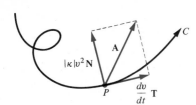

EXAMPLE 2 A particle P is moving with a constant speed of 10 units per second in a counterclockwise direction around the ellipse $(x^2/4) + (y^2/9) = 1$. Find the acceleration vector $\mathbf{A}$ at the moment when the particle passes through the vertex $(0, 3)$.

SOLUTION We find $\mathbf{A}$ by using the formula

$$\mathbf{A} = \frac{dv}{dt}\mathbf{T} + |\kappa|v^2\mathbf{N}$$

Since v is constant, we have $dv/dt = 0$ and $\mathbf{A} = |\kappa|v^2\mathbf{N}$. To find κ, we begin by implicitly differentiating the equation of the ellipse to obtain

$$\frac{2x}{4} + \frac{2y}{9}\frac{dy}{dx} = 0$$

so that

$$\frac{dy}{dx} = -\frac{9x}{4y} \qquad \text{and} \qquad \frac{d^2y}{dx^2} = \frac{-36y + 36x\dfrac{dy}{dx}}{16y^2}$$

Thus, when $x = 0$ and $y = 3$, we have

$$\frac{dy}{dx} = 0 \qquad \text{and} \qquad \frac{d^2y}{dx^2} = \frac{-(36)(3) + 0}{(16)(3)^2} = -\frac{3}{4}$$

Using the formula obtained in Example 3 Section 12.6, we have

$$\kappa = \frac{d^2y/dx^2}{[1 + (dy/dx)^2]^{3/2}} = \frac{-\frac{3}{4}}{(1 + 0)^{3/2}} = -\frac{3}{4}$$

When $(x, y) = (0, 3)$, the tangent vector is horizontal, so that the principal unit normal vector points straight *down* (in the direction of concavity of the ellipse at this point). Therefore, at the point $(0, 3)$, $\mathbf{N} = -\mathbf{j}$, and we have

$$\mathbf{A} = |\kappa|v^2\mathbf{N} = |-\tfrac{3}{4}|(10)^2(-\mathbf{j}) = -75\mathbf{j}$$

Centripetal Force

If we combine Newton's law $\mathbf{f} = m\mathbf{A}$ with the equation $\mathbf{A} = (dv/dt)\mathbf{T} + |\kappa|v^2\mathbf{N}$, we obtain the equation

$$\mathbf{f} = m\,\frac{dv}{dt}\,\mathbf{T} + |\kappa|mv^2\mathbf{N}$$

that is, *the force* $\mathbf{f}$ *acting on the particle can always be resolved into a sum of two perpendicular component vectors, a* **tangential component vector** $m\,(dv/dt)\,\mathbf{T}$ *of magnitude* $m\,|dv/dt|$ *and a* **normal component vector** $|\kappa|mv^2\mathbf{N}$ *of magnitude* $|\kappa|mv^2$. The normal component vector $|\kappa|mv^2\mathbf{N}$ is called the **centripetal force vector,** and its length $|\kappa|mv^2$ is called the **magnitude of the centripetal force,** or sometimes simply the **centripetal force.**

EXAMPLE 3   A particle P is moving on a circle of radius r with center at the point P_0. Find the tangential and the normal component vectors of the force $\mathbf{f}$ acting on the particle in terms of the instantaneous speed v and its time derivative dv/dt. Discuss the case in which the speed v of the particle is a constant (Figure 3).

SOLUTION   By Problem 29 in Problem Set 12.6, the absolute value of the curvature of the circle is given by $|\kappa| = 1/r$. Thus,

$$\mathbf{f} = m\,\frac{dv}{dt}\,\mathbf{T} + \frac{mv^2}{r}\,\mathbf{N}$$

where the principal normal unit vector $\mathbf{N}$ points directly toward the center of the circle and the unit tangent vector $\mathbf{T}$ points in the instantaneous direction of motion of the particle. If the particle moves with constant speed v, then $dv/dt = 0$, and the force vector $\mathbf{f}$ coincides with the centripetal force vector $(mv^2/r)\mathbf{N}$. In this case the centripetal force is

$$|\mathbf{f}| = \left|\frac{mv^2}{r}\,\mathbf{N}\right| = \frac{mv^2}{r}$$

The negative of the centripetal force vector in the example above—that is, the vector $-(mv^2/r)\mathbf{N}$—points directly *away* from the center of the circle and is called the **centrifugal force vector.** If a stone is whirled at the end of a string, the inertia of the stone causes it to pull away from the center of rotation, and this pull is represented by the centrifugal force vector. The magnitude mv^2/r of the centrifugal force vector is often called the **centrifugal force;** it manifests itself as tension in the string. Notice that the centrifugal force and the centripetal force are numerically equal, but the corresponding force vectors point in opposite directions.

The formula mv^2/r for centripetal (or centrifugal) force gives the force in newtons when the mass m is in kilograms, v is in meters per second, and r is in meters.*

Ⓒ EXAMPLE 4   A 0.1-kilogram mass is whirled around in a circle, 3 times per second, at the end of a string 0.7 meter long. Find the tension in the string.

---

*If m is expressed in pounds, v in feet per second, and r in feet, you must divide mv^2/r by $g = 32$ feet per second squared if you want the result in pounds of force.

Figure 3

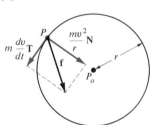

SOLUTION   The tension in the string is equal to the magnitude of the centrifugal force, mv^2/r. Since the mass travels 3 times around the circle each second, its speed v is given by

$$v = (2\pi r)(3) = 6\pi r = (6\pi)(0.7) = 4.2\pi \text{ meters per second}$$

Thus, the tension is given by

$$\frac{mv^2}{r} = \frac{(0.1)(4.2\pi)^2}{0.7} = 2.52\pi^2 \text{ newtons} \approx 24.87 \text{ newtons}$$ ■

Kepler's Laws of Planetary Motion

Tycho Brahe in his observatory

One of the most interesting applications of vectors to mechanics is the derivation of **Kepler's laws of planetary motion.** Johannes Kepler (1571–1630), after learning the Copernican theory of planetary motion at the University of Tübingen, was invited to join the staff of the Danish astronomer Tycho Brahe (1546–1601) at his observatory near Prague. When Brahe died, Kepler was appointed director of the observatory and thus inherited Brahe's highly accurate data on the motions of the planets. From these data, Kepler was able to formulate his three laws:

> **1** Each planet moves in a plane, and its orbit is an ellipse with the sun as one focus.
>
> **2** A radius vector from the sun to a planet sweeps out equal area in equal time intervals.
>
> **3** The square of the period of revolution of a planet is proportional to the cube of the semimajor axis of its elliptical orbit.

The simplicity and elegance of these laws strongly motivated Newton, who, using his newly invented calculus, was able to show that they could be derived from his second law of mechanics and his law of universal gravitation. This derivation,* using modern vector notation, is outlined in Problems 15 to 22.

*For a proof that each planet moves in a plane, see Problem 48 on page 806.

Problem Set 12.7

1 A particle moves along the upper branch of the hyperbola $(y^2/4) - (x^2/9) = 1$ from left to right, with a constant speed of 5 units of distance per second. Find (a) the velocity vector and (b) the acceleration vector at the instant when the particle passes through the vertex $(0, 2)$.

2 A particle moves along the parabola $y = x^2$ from left to right. As it passes through the point $(2, 4)$, v has the value 3 units per second and dv/dt has the value 7 units per second squared. Find (a) the velocity vector and (b) the acceleration vector at this point.

3 A particle moves along an ellipse according to the equations

$$\begin{cases} x = \cos 2\pi t \\ y = 3 \sin 2\pi t \end{cases}$$

Find the speed v and the rate of change of the speed dv/dt of the particle at time t.

4 A particle moves along a curve in the xy plane according to the equations

$$\begin{cases} x = g(t) \\ y = h(t) \end{cases}$$

Derive the following formulas:

(a) $\mathbf{V} = \dfrac{dx}{dt}\,\mathbf{i} + \dfrac{dy}{dt}\,\mathbf{j}$ (b) $\dfrac{ds}{dt} = \sqrt{\left(\dfrac{dx}{dt}\right)^2 + \left(\dfrac{dy}{dt}\right)^2}$

(c) $\mathbf{A} = \dfrac{d^2x}{dt^2}\,\mathbf{i} + \dfrac{d^2y}{dt^2}\,\mathbf{j}$ (d) $\dfrac{dv}{dt} = \dfrac{\dfrac{dx}{dt}\dfrac{d^2x}{dt^2} + \dfrac{dy}{dt}\dfrac{d^2y}{dt^2}}{\sqrt{\left(\dfrac{dx}{dt}\right)^2 + \left(\dfrac{dy}{dt}\right)^2}}$

(e) $\dfrac{dv}{dt}\,\mathbf{T} = \dfrac{\dfrac{dx}{dt}\dfrac{d^2x}{dt^2} + \dfrac{dy}{dt}\dfrac{d^2y}{dt^2}}{(dx/dt)^2 + (dy/dt)^2}\left(\dfrac{dx}{dt}\,\mathbf{i} + \dfrac{dy}{dt}\,\mathbf{j}\right)$

(f) $|\kappa|v^2\mathbf{N} =$
$$\dfrac{\left[\dfrac{dy}{dt}\left(\dfrac{dy}{dt}\dfrac{d^2x}{dt^2} - \dfrac{dx}{dt}\dfrac{d^2y}{dt^2}\right)\mathbf{i} + \dfrac{dx}{dt}\left(\dfrac{dx}{dt}\dfrac{d^2y}{dt^2} - \dfrac{dy}{dt}\dfrac{d^2x}{dt^2}\right)\mathbf{j}\right]}{\left(\dfrac{dx}{dt}\right)^2 + \left(\dfrac{dy}{dt}\right)^2}$$

5 A particle moves along a curve whose parametric equations are

$$\begin{cases} x = \frac{3}{2}t^2 \\ y = \frac{4}{3}t^3 \end{cases}$$

Find the tangential and normal components of its acceleration when $t = 1$.

6 The position vector of a moving particle is given by $\mathbf{R}(t) = (t \cos t)\mathbf{i} + (t \sin t)\mathbf{j}$. Find the tangential and normal components of the acceleration when $t = \pi/3$.

7 Suppose that a projectile is fired at an angle of 60° with the level ground at an initial speed of 300 meters per second. Assuming that the only force acting on the projectile is the force of gravity,

$$\mathbf{f} = -mg\mathbf{j}$$

and that the projectile is fired from the origin O at time $t = 0$, find (a) the velocity vector $\mathbf{V}$ and (b) the position vector $\mathbf{R}$ at time t (Figure 4).

Figure 4

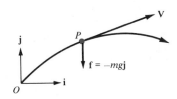

8 A projectile is fired from the point (a, b) in the xy plane in a direction making an angle θ with the positive x axis at an initial speed of v_0 units of distance per second. Neglect air resistance. (a) Derive the vector equation of motion of the projectile. (b) Show that the projectile moves in a parabola.

9 A particle of mass 2 kilograms is whirled in a horizontal circle of radius 3 meters. The particle makes 4 revolutions per second. Find the centripetal force on the particle.

10 A horizontal centrifuge rotates at 4300 revolutions per minute and has a radius of 0.08 meter. The centrifugal force developed by an object in the centrifuge is how many times the force of gravity on the object?

11 A jet plane is flying at a speed of 600 miles per hour in a horizontal circle. Find the radius of the circle in miles if the pilot feels a centrifugal force of "3 g's," that is, 3 times her own weight.

12 A particle of mass m is moving in a circle of radius r at a uniform speed. If the particle makes N revolutions per second, show that the centrifugal force is given by $4\pi^2N^2mr$.

13 A child whirls an open pail of water in a vertical circle. What is the minimum number of revolutions per second that will keep the water in the pail if the radius of the circle is 60 centimeters?

14 A 9-ton truck goes around a curve with a 100-foot radius at a speed of 15 miles per hour. How much centripetal force is required?

15 Suppose that the position of a moving point P in the plane is given by *polar* coordinates (r, θ), where r and θ are functions of the parameter t. Then the position vector $\mathbf{R}$ of P is given by $\mathbf{R} = (r \cos \theta)\mathbf{i} + (r \sin \theta)\mathbf{j}$ (Figure 5). Let $\mathbf{u} = (\cos \theta)\mathbf{i} + (\sin \theta)\mathbf{j}$ be a unit vector in the direction of $\mathbf{R}$, and let

$$\mathbf{w} = \left[\cos\left(\theta + \frac{\pi}{2}\right)\right]\mathbf{i} + \left[\sin\left(\theta + \frac{\pi}{2}\right)\right]\mathbf{j}$$
$$= (-\sin \theta)\mathbf{i} + (\cos \theta)\mathbf{j}$$

be the unit vector obtained by rotating $\mathbf{u}$ counterclockwise through $\pi/2$ radians.

(a) Show that

$$\mathbf{V} = \frac{dr}{dt}\,\mathbf{u} + r\,\frac{d\theta}{dt}\,\mathbf{w}$$

(b) Show that

$$\mathbf{A} = \left[\frac{d^2r}{dt^2} - r\left(\frac{d\theta}{dt}\right)^2\right]\mathbf{u} + \left[2\,\frac{dr}{dt}\,\frac{d\theta}{dt} + r\,\frac{d^2\theta}{dt^2}\right]\mathbf{w}$$

Figure 5

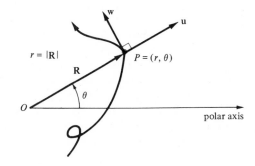

16 In Problem 15, suppose that P is a particle of mass m. We say that P is moving under the influence of a **central force f** if there is a function g such that $\mathbf{f} = g(r)\mathbf{u}$ holds for all values of time t. Using Newton's second law of motion, $\mathbf{f} = m\mathbf{A}$, show that in this case:

(a) $g(r) = m\left[\dfrac{d^2r}{dt^2} - r\left(\dfrac{d\theta}{dt}\right)^2\right]$

(b) $2\dfrac{dr}{dt}\dfrac{d\theta}{dt} + r\dfrac{d^2\theta}{dt^2} = 0$

(c) $\dfrac{d}{dt}\left(r^2\dfrac{d\theta}{dt}\right) = 0$ [*Hint:* Multiply both sides of the equation in part (b) by r.]

(d) $r^2\dfrac{d\theta}{dt} = K$, K a constant

17 In Problem 16, let A denote the area swept out by the radius vector $\mathbf{R}$ during time t, starting from some arbitrary but fixed value of t. Show that for motion under the influence of a central force, *the radius vector sweeps out equal area in equal time intervals*. This result generalizes **Kepler's second law** of planetary motion. [*Hint:* Use the result in part (d) of Problem 16 and the fact that $dA = \frac{1}{2}r^2\,d\theta$ from Section 9.3.]

18 In Problem 16, let $q = 1/r$, and assume that the particle P is moving under the influence of a central force $\mathbf{f} = g(r)\mathbf{u}$. Show that

(a) $\dfrac{d\theta}{dt} = Kq^2$ (b) $\dfrac{dr}{dq} = -\dfrac{1}{q^2}$

(c) $\dfrac{dr}{dt} = -K\dfrac{dq}{d\theta}$ $\left[\textit{Hint: } \dfrac{dr}{dt} = \dfrac{dr}{dq}\dfrac{dq}{d\theta}\dfrac{d\theta}{dt}.\right]$

(d) $\dfrac{d^2r}{dt^2} = -K^2q^2\dfrac{d^2q}{d\theta^2}$ $\left[\textit{Hint: } \dfrac{d}{dt}\left(\dfrac{dr}{dt}\right) = \dfrac{d}{d\theta}\left(\dfrac{dr}{dt}\right)\dfrac{d\theta}{dt}.\right]$

(e) $g(r) = -mq^2K^2\left(\dfrac{d^2q}{d\theta^2} + q\right)$ [*Hint:* Use part (a) of Problem 16.]

19 In Problem 16, the particle P is said to be moving under the influence of an **inverse-square force f** if there is a constant k such that $\mathbf{f} = (k/r^2)\mathbf{u}$. (Note that an inverse-square force is a central force.) In this case, show that $q = 1/r$ satisfies the differential equation

$$\dfrac{d^2q}{d\theta^2} + \left(q + \dfrac{k}{mK^2}\right) = 0$$

[*Hint:* Use part (e) of Problem 18.]

20 In Problem 19, let $y = q + [k/(mK^2)]$.

(a) Show that y satisfies the equation $(d^2y/d\theta^2) + y = 0$. (Note that this is the harmonic oscillator equation with θ playing the role of t and with $\omega = 1$.)

(b) Using Theorem 5 on page 285 to solve the differential equation in part (a), show that there are constants A and ϕ such that $y = A\cos(\theta - \phi)$.

(c) Show that by a suitable rotation of the polar axis (page 577), we may asume that $\phi = 0$ in part (c).

(d) Using the results of parts (b) and (c), show that

$$r = \dfrac{\alpha}{1 - \beta\cos\theta} \quad \text{where } \alpha = -\dfrac{mK^2}{k} \text{ and } \beta = \dfrac{AmK^2}{k}$$

(e) Using part (d) and the results in Section 9.8, show that *a particle moving under the influence of an inverse-square force moves in a conic section.*

21 By Newton's law of universal gravitation, the gravitational force $\mathbf{f}$ exerted on a planet P by the sun is given by

$$\mathbf{f} = -G\dfrac{mM}{r^2}\mathbf{u}$$

where G is the gravitational constant, m is the mass of the planet, M is the mass of the sun, r is the distance between the sun and the planet, and $\mathbf{u}$ is a unit vector in the direction from the sun to the planet. (The minus sign indicates a force of attraction.) Using the result of part (e) of Problem 20 and the observed fact that the orbits of the planets are closed curves, show that **Kepler's first law** follows from Newton's law of universal gravitation.

22 Let a and b be the semimajor and semiminor axes of the elliptical orbit of a planet P, and suppose that T is the period of revolution of P. Let A be the area enclosed by the elliptical orbit of P, so that $A = \pi ab$ (see Problem 55 on page 563).

(a) Using part (d) of Problem 16 and Problem 17, show that $A = \int_0^T \frac{1}{2}K\,dt = \frac{1}{2}KT$.

(b) Using part (a), show that

$$T^2 = \dfrac{4\pi^2a^2b^2}{K^2}$$

(c) In part (d) of Problem 20, show that

$$\alpha = \dfrac{K^2}{GM}$$

(d) Using part (c) and Problems 34 and 36 on page 591, show that

$$\dfrac{K^2}{GM} = \dfrac{b^2}{a}$$

(e) Combine parts (b) and (d) to show that

$$T^2 = \dfrac{4\pi^2a^3}{GM}$$

(f) Explain why **Kepler's third law** follows from part (e).

© **23** The gravitational constant has the value $G = 6.672 \times 10^{-11}$ $N \cdot m^2/kg^2$, the period of revolution of the earth around the sun is $T = 3.156 \times 10^7$ seconds, and the semimajor axis of the earth's elliptical orbit is $a = 1.496 \times 10^{11}$ meters. Using this information and part (e) of Problem 22, find the mass M of the sun.

24 Explain why the speed of a planet in its orbit is maximum when the planet is closest to the sun.

C **25** The eccentricity of the earth's orbit is $e = 1.673 \times 10^{-2}$. Find the semiminor axis b of the earth's orbit. (See Problem 23 for the semimajor axis.)

C **26** Using Problems 23 to 25, find the maximum speed of the earth in its orbit.

C **27** The period of revolution of Venus in its elliptical orbit about the sun is 1.941×10^7 seconds. Using the information in Problem 23, find the semimajor axis of the orbit of Venus.

Review Problem Set, Chapter 12

1 If A, B, C, and D are points in the xy plane such that $\overrightarrow{OB} - \overrightarrow{OA} = \overrightarrow{OC} - \overrightarrow{OD}$, show that $ABCD$ is a parallelogram.

2 Let $\mathbf{A}$ and $\mathbf{B}$ be position vectors of points P and Q, respectively, in the plane. Find the position vector of the point four-fifths of the way from P to Q.

3 If $OABC$ is a parallelogram in the xy plane with A and C as opposite vertices, show that

$$\overrightarrow{OA} + \tfrac{1}{2}(\overrightarrow{OC} - \overrightarrow{OA}) = \tfrac{1}{2}\overrightarrow{OB}$$

4 Explain the geometric significance of the condition that $\mathbf{A} + \mathbf{B} + \mathbf{C} = \mathbf{0}$, where $\mathbf{A}$, $\mathbf{B}$, and $\mathbf{C}$ are vectors in the plane.

5 If u and v are scalars such that $u\mathbf{A} = v\mathbf{A}$, where $\mathbf{A}$ is a nonzero vector, is it true that $u = v$? Why?

6 Show that if $\mathbf{u} = a\mathbf{i} + b\mathbf{j}$, $\mathbf{u} \neq \mathbf{0}$, then $a/\sqrt{a^2 + b^2} = \cos\theta$ and $b/\sqrt{a^2 + b^2} = \sin\theta$, where θ is the angle between $\mathbf{u}$ and the positive x axis (Figure 1.)

Figure 1

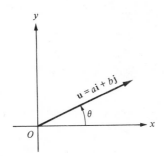

7 If s is a nonzero scalar and if $\mathbf{A}$ and $\mathbf{B}$ are vectors such that $s\mathbf{A} = s\mathbf{B}$, is it true that $\mathbf{A} = \mathbf{B}$? Why?

8 If $\mathbf{i}$, $\mathbf{j}$ is the standard basis in the xy plane, write $\mathbf{A}$ as a linear combination of $\mathbf{i}$ and $\mathbf{j}$ and find $|\mathbf{A}|$, where $\mathbf{A}$ is the vector whose tail end is $(2, -1)$ and whose head end is $(-1, -2)$.

9 Let $\mathbf{A} = 2\mathbf{i} - 3\mathbf{j}$ and $\mathbf{B} = 4\mathbf{i} + \mathbf{j}$ be position vectors in the xy plane. Draw the vectors $\mathbf{A}$, $\mathbf{B}$, and $\mathbf{C} = \mathbf{A} + t\mathbf{B}$ on the same diagram for (a) $t = \tfrac{1}{2}$ and (b) $t = -2$.

10 Let $\mathbf{A} = 2\mathbf{i} + \mathbf{j}$ and $\mathbf{B} = \mathbf{i} + 3\mathbf{j}$ be position vectors in the xy plane. Draw the vectors $\mathbf{A}$, $\mathbf{B}$, and $\mathbf{C} = s\mathbf{A} + t\mathbf{B}$ on the same diagram for (a) $s = t = \tfrac{1}{2}$ and (b) $s = -1$ and $t = 2$.

11 Find the midpoint of the line segment containing the terminal points of the position vectors $\mathbf{A} = -\mathbf{i} + 3\mathbf{j}$ and $\mathbf{B} = 2\mathbf{i} + 7\mathbf{j}$.

12 Find the coordinates of the point that is seven-tenths of the way from $P = (-5, -9)$ to $Q = (7, 7)$.

13 Let $\mathbf{A} = 2\mathbf{i} - \mathbf{j}$ and $\mathbf{B} = 2\mathbf{i} - 3\mathbf{j}$. Find each of the following expressions:
(a) $5\mathbf{A}$　　　　　　　　(b) $-4\mathbf{B}$
(c) $\mathbf{A} + \mathbf{B}$　　　　　　(d) $\mathbf{A} - \mathbf{B}$
(e) $2\mathbf{A} + 3\mathbf{B}$　　　　(f) $\mathbf{A} \cdot \mathbf{B}$
(g) $(2\mathbf{A} - 3\mathbf{B}) \cdot (2\mathbf{A} + 3\mathbf{B})$　(h) $|\mathbf{A}|$
(i) $|\mathbf{A} - \mathbf{B}|$　　　　　(j) $|2\mathbf{A}| + |3\mathbf{B}|$
(k) $\text{comp}_{\mathbf{B}}\,\mathbf{A}$

14 Let $\mathbf{A} = \mathbf{i} + 2\mathbf{j}$, $\mathbf{B} = 2\mathbf{i} - 4\mathbf{j}$, and $\mathbf{C} = 3\mathbf{i} - 5\mathbf{j}$. Find scalars s and t such that $\mathbf{C} = s\mathbf{A} + t\mathbf{B}$.

15 Use dot products to find the three vertex angles of the triangle ABC where $A = (-2, -1)$, $B = (-1, 6)$, and $C = (2, 2)$.

16 Find two vectors $\mathbf{X}$ and $\mathbf{Y}$ in the plane such that both $\mathbf{X}$ and $\mathbf{Y}$ are perpendicular to the vector $\mathbf{A} = 2\mathbf{i} - 3\mathbf{j}$ and $|\mathbf{X}| = |\mathbf{Y}| = |\mathbf{A}|$.

17 Find the scalar components with respect to the standard $\mathbf{i}$, $\mathbf{j}$ basis of the vector obtained by rotating the vector $\mathbf{A} = x\mathbf{i} + y\mathbf{j}$ counterclockwise through $90°$.

18 Find the scalar components with respect to the standard $\mathbf{i}$, $\mathbf{j}$ basis of the vector obtained by rotating the vector $\mathbf{A} = x\mathbf{i} + y\mathbf{j}$ counterclockwise through θ radians.

19 Given that $|\mathbf{A} + \mathbf{B}|^2 - |\mathbf{A} - \mathbf{B}|^2 = 0$, show that $\mathbf{A}$ must be perpendicular to $\mathbf{B}$.

20 Let $\mathbf{A}$ and $\mathbf{B}$ be two given vectors in the plane such that $\mathbf{A} \cdot \mathbf{B} = 0$, $|\mathbf{A}| \neq 0$, and $|\mathbf{B}| \neq 0$. Show that if $\mathbf{C}$ is any vector in the plane, then

$$\mathbf{C} = \frac{\mathbf{A} \cdot \mathbf{C}}{\mathbf{A} \cdot \mathbf{A}}\,\mathbf{A} + \frac{\mathbf{B} \cdot \mathbf{C}}{\mathbf{B} \cdot \mathbf{B}}\,\mathbf{B}$$

21 Find the scalar component $\text{comp}_{\mathbf{B}}\mathbf{A}$ of $\mathbf{A}$ in the direction of $\mathbf{B}$ if $\mathbf{A} = 2\mathbf{i} + 3\mathbf{j}$ and $\mathbf{B} = -3\mathbf{i} + 4\mathbf{j}$.

22 A manufacturer sells x sofas at a dollars per sofa and y chairs at b dollars per chair. The vector $\mathbf{D} = \langle x, y \rangle$ is called the *demand vector*, while the vector $\mathbf{R} = \langle a, b \rangle$ is called the *revenue vector*. The *cost vector* is defined to be $\mathbf{C} = \langle c, d \rangle$, where c is the cost of manufacturing one sofa and d is the cost of manufacturing one chair. Give the economic interpretation of

(a) $\mathbf{D} \cdot \mathbf{R}$ (b) $\mathbf{D} \cdot \mathbf{C}$ (c) $\mathbf{D} \cdot (\mathbf{R} - \mathbf{C})$

23 Draw two vectors $\mathbf{A}$ and $\mathbf{B}$ in the plane such that

(a) $\mathbf{A} \cdot \mathbf{B} > 0$ (b) $\mathbf{A} \cdot \mathbf{B} = 0$ (c) $\mathbf{A} \cdot \mathbf{B} < 0$

24 Sketch the graph of the given vector nonparametric equation, where $\mathbf{A} = \mathbf{i} - 3\mathbf{j}$ and $\mathbf{B} = 2\mathbf{i} + \mathbf{j}$.

(a) $\mathbf{A} \cdot (\mathbf{R} - \mathbf{B}) = 0$ (b) $(\mathbf{R} - \mathbf{B}) \cdot (\mathbf{R} - \mathbf{B}) = 9$

(c) $\mathbf{A} \cdot \mathbf{R} = 8$ (d) $\mathbf{R} \cdot (\mathbf{R} - 2\mathbf{B}) = 0$

25 Find the equation of the line that contains the point $(3, 4)$ and has slope $-\frac{2}{5}$ (a) in scalar nonparametric form, (b) in vector nonparametric form, (c) in vector parametric form, and (d) in scalar parametric form.

26 Rewrite equations (a) through (d) in Problem 24 in vector parametric form.

27 Show that the following two vector parametric equations have the same graph:

(a) $\mathbf{R} = (3 \sin t)\mathbf{i} + (3 \cos t)\mathbf{j}$

(b) $\mathbf{R} = (3 \cos 2t)\mathbf{i} + (3 \sin 2t)\mathbf{j}$

28 Let $\mathbf{F}(t) = \sqrt{t - 1}\,\mathbf{i} + \ln(2 - t)\mathbf{j}$. (a) What is the domain of $\mathbf{F}$? (b) Find $\lim_{t \to 1.5} \mathbf{F}(t)$. (c) Is $\mathbf{F}$ continuous? Why?

29 Find a vector parametric equation for the hyperbola $x^2 - y^2 = 1$. (*Hint:* Use hyperbolic functions.)

In Problems 30 to 34, find (a) $\mathbf{F}'(t)$ and (b) $\mathbf{F}''(t)$.

30 $\mathbf{F}(t) = (t^2 + 7)\mathbf{i} + (t + t^3)\mathbf{j}$

31 $\mathbf{F}(t) = (t - 1)^{-1}\mathbf{i} + 3(t^2 - 2)^{-1}\mathbf{j}$

32 $\mathbf{F}(t) = (2 \cos 5t)\mathbf{i} + (5 \sin 5t)\mathbf{j}$

33 $\mathbf{F}(t) = e^{5t}\mathbf{i} + (e^{-3t} + 7)\mathbf{j}$

34 $\mathbf{F}(t) = (\cos t^2)\mathbf{i} + (\cos^2 t)\mathbf{j}$

In Problems 35 to 38, find (a) $d|\mathbf{F}(t)|/dt$ and (b) $\mathbf{F}'(t) \cdot \mathbf{F}''(t)$.

35 $\mathbf{F}(t) = (t - \sin t)\mathbf{i} + (t + \cos t)\mathbf{j}$

36 $\mathbf{F}(t) = (e^{-t} \cos t)\mathbf{i} + (e^{-t} \sin t)\mathbf{j}$

37 $\mathbf{F}(t) = e^{4t}\mathbf{i} + e^{-4t}\mathbf{j}$ **38** $\mathbf{F}(t) = te^{-3t}\mathbf{i} + te^{3t}\mathbf{j}$

In Problems 39 to 44, suppose that $\mathbf{F}(t) = e^t\mathbf{i} + (t^2 - 1)\mathbf{j}$, $\mathbf{G}(t) = (\cos^2 t)\mathbf{i} + (\sin^2 t)\mathbf{j}$, and $h(t) = \ln(t + 1)$. Evaluate each expression.

39 $\dfrac{d}{dt}[\mathbf{F}(t) + \mathbf{G}(t)]$ **40** $\dfrac{d}{dt}[\mathbf{F}(t) \cdot \mathbf{G}(t)]$

41 $\dfrac{d}{dt}|\mathbf{F}(t)|$ **42** $\dfrac{d}{dt}[h(t)\mathbf{F}(t)]$

43 $\mathbf{F}'(t) \cdot \mathbf{F}''(t)$ **44** $\dfrac{d}{dt}\left[\dfrac{\mathbf{F}(t)}{|\mathbf{F}(t)|}\right]$

45 Find the arc length of the curve whose vector equation is $\mathbf{R} = 3(\sin t - 1)\mathbf{i} + 3(\cos t - 1)\mathbf{j}$ between the point where $t = 0$ and the point where $t = 2$.

46 Find the arc length of the curve

$$\begin{cases} x = 4 \cos^3 t \\ y = 4 \sin^3 t \end{cases}$$

between the point where $t = 0$ and the point where $t = 2\pi$.

47 The position vector of a particle at time t is given by

$$\mathbf{R} = t\mathbf{i} + \left(\frac{t^3}{6} + \frac{1}{2t}\right)\mathbf{j}$$

Find the distance traveled by the particle during the interval from $t = 1$ to $t = 4$.

48 A ball is moving in accordance with the equations

$$\begin{cases} x = 32t \\ y = -16t^2 \end{cases}$$

where t is the time in seconds. How far does the ball move along its path during the first 3 seconds?

In Problems 49 to 52, find (a) the unit tangent vector $\mathbf{T}$, (b) the curvature κ of the given curve at the given value of t, and (c) the principal unit normal vector $\mathbf{N}$.

49 $\mathbf{R}(t) = (3 \cos t)\mathbf{i} + (\sin t)\mathbf{j}$ at $t = 0$

50 $\mathbf{R}(t) = (\tan 2t)\mathbf{i} + (\cot 2t)\mathbf{j}$ at $t = \pi/8$

51 $\begin{cases} x = t^2 \\ y = t^3 \end{cases}$ at $t = 1$

52 $\begin{cases} x = t \\ y = \ln \sec t \end{cases}$ at $t = \pi/4$

In Problems 53 to 57, find (a) the velocity vector $\mathbf{V}$, (b) the acceleration vector $\mathbf{A}$, (c) the speed v, (d) the rate of change dv/dt of the speed, (e) the tangential component vector of the acceleration, and (f) the normal component vector of the acceleration for a particle at time t moving according to the given equation of motion.

53 $\mathbf{R}(t) = (t^3 + 6)\mathbf{i} + (2t^4 - 1)\mathbf{j}$

54 $\mathbf{R}(t) = e^{2t}\mathbf{i} + e^{-3t}\mathbf{j}$

55 $\mathbf{R}(t) = (3 \cos 7t)\mathbf{i} + (3 \sin 7t)\mathbf{j}$

56 $\mathbf{R}(t) = (t \cos t)\mathbf{i} + (t \sin t)\mathbf{j}$

57 $\mathbf{R}(t) = \left(1 + \dfrac{t^2}{2}\right)\mathbf{i} + \dfrac{t^3}{3}\mathbf{j}$

58 A particle is moving along the branch of the hyperbola $xy = 1$ that lies in the first quadrant in such a way that its abscissa is increasing in time. At the instant when the particle passes through the point $(1, 1)$, its speed is given by $v = 2$ units per second, and the time rate of change of the speed is given by $dv/dt = -3$ units per second squared. Find (a) the velocity vector $\mathbf{V}$ and (b) the acceleration vector $\mathbf{A}$ at this instant.

59 A pilot is pulling out of a vertical dive by following an arc of a circle of radius 1 mile. The speed of the plane is a constant 500 miles per hour. With how many times his usual weight is the pilot being pressed into the seat at the lowest point on the arc of the circle?

60 Let $\mathbf{R} = g(t)\mathbf{i} + h(t)\mathbf{j}$ be the equation of motion of a particle P in the xy plane. If $A(t)$ denotes the area swept out by the variable

position vector $\mathbf{R}$ as a function of the time t, show that

$$\frac{dA}{dt} = \left| \frac{g(t)h'(t) - h(t)g'(t)}{2} \right|$$

Ⓒ 61 The period of Neptune is 164.8 years. Show that it is about 30 times as far from the sun as the earth is.

62 Suppose that a particle P, moving under the influence of a central force, has a spiral trajectory of the form $r = b^{-\theta}$, where b is a positive constant. Show that there are constants γ and r_0 such that $r = (\gamma t + r_0^2)^{1/2}$.

Ⓒ 63 The period of Mars is 1.881 years. How much farther is Mars from the sun than the earth is?

64 The **angular momentum** L of a particle P of mass m moving in the plane according to an equation of motion $r = f(\theta)$ is defined by $L = mr^2 \, d\theta/dt$. Prove that L is a constant. (This is the two-dimensional version of the law of *conservation of angular momentum*.)

13

COORDINATE SYSTEMS IN THREE-DIMENSIONAL SPACE

In this chapter we study three-dimensional geometry with the aid of coordinate systems and vectors in three-dimensional space. All the operations defined for vectors in the plane—sums, differences, multiplication by scalars, and dot products—carry over directly to vectors in three-dimensional space. In addition, we define the "cross product" of vectors in space and use vectors to obtain the equations of lines and planes in space. The chapter also includes vector-valued functions, curves in space, surfaces of revolution, and quadric surfaces.

13.1 Cartesian Coordinates in Space

In order to assign Cartesian coordinates to a point P in three-dimensional space, we begin by choosing an origin O as usual. We then set up a Cartesian coordinate system in the horizontal plane passing through O, with the positive x axis pointing toward us and the positive y axis extending to our right (Figure 1). To find the Cartesian coordinates of a point P in space, we drop a perpendicular from P to the xy plane, and we denote the foot of this perpendicular by $Q = (x, y)$. We define the **z coordinate** of P by $z = \pm |\overline{PQ}|$, where we use the plus sign if P is above the xy plane and the minus sign if P is below the xy plane. If P lies on the xy plane, then $P = Q$ and $z = 0$. Thus, z is the **directed distance** from the xy plane to P. The point P is understood to have *three* Cartesian coordinates x, y, and z, and we write

$$P = (x, y, z)$$

In dealing with Cartesian coordinates in three-dimensional space, it is customary to introduce a third coordinate axis, called the z **axis**, perpendicular to the xy plane, passing through the origin O, and directed upward (Figure 2). The z axis is equipped with its own number scale, just as the x and y axes are, and the unit of distance on the z axis is usually chosen to be the same as the unit of distance on the x and y axes.

Figure 1

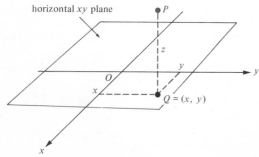

Figure 2

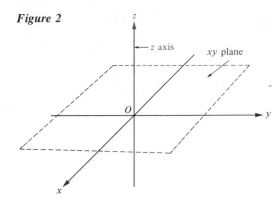

Figure 3

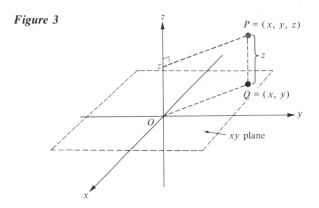

Figure 4

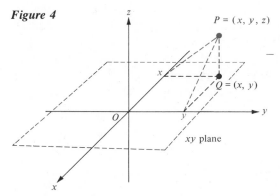

If P is any point in space, then the z coordinate of P can be determined not only by measuring the directed distance from the xy plane to P, but also by dropping a perpendicular from P to the z axis (Figure 3). The scale value at the foot of this perpendicular is evidently the z coordinate of P. Similarly, the x and y coordinates of P can be found by dropping perpendiculars to the x and y axes (Figure 4). Since all three coordinates of P can be found by dropping perpendiculars to the three coordinate axes, it is not necessary to draw the xy plane at all when sketching three-dimensional diagrams—the three coordinate axes are sufficient.

Figure 5

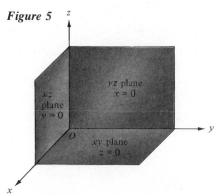

In the same way that the x and y axes determine the xy plane, the remaining pairs of axes determine planes, namely the **xz plane** and the **yz plane.** The xy, xz, and yz planes are called the three **coordinate planes.** You can visualize these planes by thinking of the origin as a corner of a room with the floor as the xy plane, the wall to your left as the xz plane, and the wall facing you as the yz plane (Figure 5). Notice that:

1 The xy plane consists of all points of the form $(x, y, 0)$ and is perpendicular to the z axis.

2 The yz plane consists of all points of the form $(0, y, z)$ and is perpendicular to the x axis.

3 The xz plane consists of all points of the form $(x, 0, z)$ and is perpendicular to the y axis.

The three coordinate planes divide the space into eight parts called **octants.** The octant in which all three coordinates are positive—that is, above the xy plane, to the right of the xz plane, and in front of the yz plane in Figure 5—is called the **first octant.** Figure 6 shows the point $(2, 3, 5)$ in the first octant and all other points obtained by projecting it perpendicularly onto the coordinate planes and the coordinate axes. Notice that the x axis consists of all points of the form $(x, 0, 0)$, the y axis consists of all points of the form $(0, y, 0)$, and the z axis consists of all points of the form $(0, 0, z)$.

The coordinate system described here is called a **right-handed,*** **three-dimensional, Cartesian** (or **rectangular**) coordinate system. If such a coordinate system has been set up, we refer to three-dimensional space as **xyz space.**

Figure 6

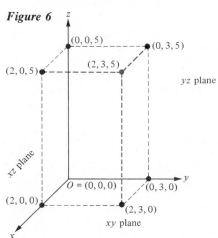

*A *left-handed* Cartesian coordinate system would have the x and y axes interchanged; however, we do not use left-handed coordinate systems in this book.

Figure 7

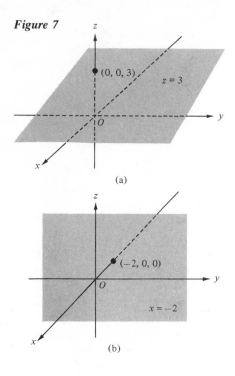

(a)

(b)

Figure 8

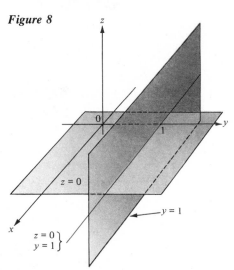

Figure 9

By the **graph** in xyz space of an equation (or of a set of simultaneous equations) involving one or more of the variables x, y, or z, we mean the set of all points $P = (x, y, z)$ whose coordinates x, y, and z satisfy the equation (or the equations). For instance, the graph of the equation $z = 0$ is the set of all points $P = (x, y, 0)$ and therefore consists of all points in the xy plane. Similarly, the graph of $x = 0$ is the yz plane, and the graph of $y = 0$ is the xz plane.

EXAMPLE 1   Sketch the graph in xyz space of

(a) $z = 3$ (b) $x = -2$

SOLUTION

(a) The graph of $z = 3$ is the set of all points whose z coordinate is 3, that is, all points 3 units above the xy plane. Thus, the graph of $z = 3$ is a plane parallel to the xy plane and 3 units above it (Figure 7a).

(b) The graph of $x = -2$ is the set of all points whose x coordinate is -2, that is, all points 2 units behind the yz plane. Thus, the graph of $x = -2$ is a plane parallel to the yz plane and 2 units behind it (Figure 7b). ■

EXAMPLE 2   Sketch the graph of $z = 0$ and the graph of $y = 1$ in xyz space, and show the intersection of these two graphs.

SOLUTION   The graph of $z = 0$ is the xy plane, and the graph of $y = 1$ is a plane parallel to the xz plane and 1 unit to the right of it (Figure 8). These two planes intersect in a *line* parallel to the x axis, in the xy plane, and 1 unit to the right of the origin (Figure 8). ■

Any two nonparallel planes, such as the planes $z = 0$ and $y = 1$ in Figure 8, intersect in a line. For instance, if b and c are constants, then the graph of $z = b$ is a plane perpendicular to the z axis, and the graph of $y = c$ is a plane perpendicular to the y axis. Thus, the intersection of the plane $z = b$ and the plane $y = c$ is a line parallel to the x axis and consisting of all points of the form $P = (x, c, b)$, where x can be any real number. This line is the graph of the simultaneous equations

$$\begin{cases} z = b \\ y = c \end{cases}$$

EXAMPLE 3   Write a pair of simultaneous equations whose graph is the line containing the point $(-3, 2, 1)$ and perpendicular to the xy plane (Figure 9). Using these equations, describe two planes whose intersection is this line.

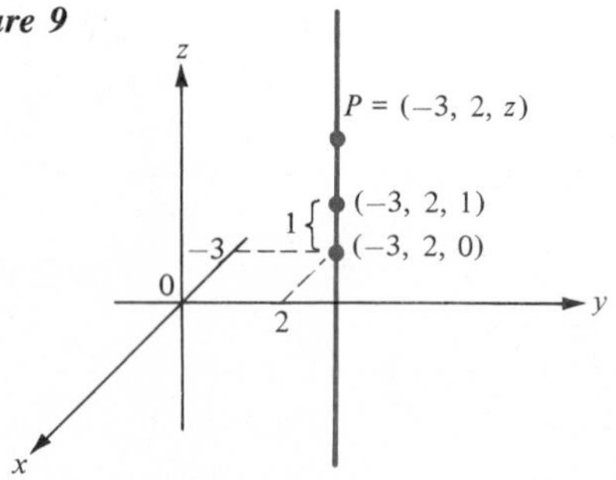

SOLUTION   The line in question contains the point $(-3, 2, 0)$ in the xy plane. A point P is on this line if and only if the x coordinate of P is -3 and the y coordinate of P is 2. There is no restriction on the z coordinate of P. Therefore, the given line is the graph of the simultaneous equations

$$\begin{cases} x = -3 \\ y = 2 \end{cases}$$

This line is the intersection of the plane $x = -3$ and the plane $y = 2$. The plane $x = -3$ is perpendicular to the x axis and contains the point $(-3, 0, 0)$. The plane $y = 2$ is perpendicular to the y axis and contains the point $(0, 2, 0)$. ■

Figure 10

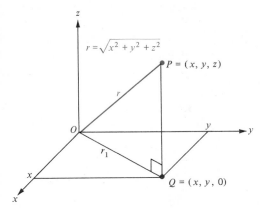

EXAMPLE 4   Find an equation of the plane containing the point (a, b, c) and ·parallel to the *yz* plane.

SOLUTION   On the plane in question, all points have the same *x* coordinate, $x = a$. Conversely, any point *P* of the form $P = (a, y, z)$ belongs to the plane. Therefore, an equation of the plane is $x = a$.   ∎

We now derive a formula for the distance *r* between the point $P = (x, y, z)$ and the origin *O* in *xyz* space. In Figure 10, let r_1 denote the distance in the *xy* plane between the point $Q = (x, y, 0)$ and the origin *O*. By the usual formula for the distance between two points in the *xy* plane, $r_1^2 = x^2 + y^2$. Applying the Pythagorean theorem to the triangle *OQP*, we have $r^2 = r_1^2 + |\overline{PQ}|^2 = x^2 + y^2 + z^2$; hence,

$$r = \sqrt{x^2 + y^2 + z^2}$$

EXAMPLE 5   Find the distance *r* between the origin and the point $(2, 3, -1)$.

SOLUTION   From the formula above, we have

$$r = \sqrt{2^2 + 3^2 + (-1)^2} = \sqrt{14} \text{ units}$$ ∎

EXAMPLE 6   Describe and sketch the graph in *xyz* space of the equation $x^2 + y^2 + z^2 = 4$.

SOLUTION   The point $P = (x, y, z)$ belongs to the graph if and only if $\sqrt{x^2 + y^2 + z^2} = 2$, that is, if and only if the distance from *O* to *P* is 2 units. Therefore, the graph is the surface of a sphere of radius 2 units with center at the origin *O* (Figure 11).   ∎

Figure 11

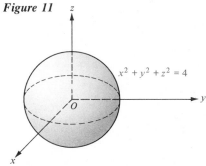

Problem Set 13.1

1 The points $(3, 5, 7)$, $(3, 0, 7)$, $(0, 0, 7)$, $(0, 5, 7)$, $(3, 5, 0)$, $(3, 0, 0)$, $(0, 0, 0)$, and $(0, 5, 0)$, form the vertices of a rectangular box in *xyz* space. Sketch this box and label the vertices.

2 A perpendicular is dropped from the point $P = (-1, 2, 5)$ to the *yz* plane. Find the coordinates of the point *Q* at the foot of this perpendicular.

3 Plot each point in *xyz* space:
(a) $(2, -1, 3)$ (b) $(-2, 5, 3)$ (c) $(2, -1, -3)$
(d) $(\frac{5}{2}, 0, -1)$ (e) $(4, 3, -2)$ (f) $(-1, -2, -1)$

4 Two points *A* and *B* in space are said to be **symmetrically located with respect to the plane** that bisects the line segment $\overline{AB}$ and is perpendicular to it. If $A = (x, y, z)$, find the coordinates of the point *B* that is symmetrically located to *A* with respect to (a) the *xy* plane, (b) the *xz* plane, and (c) the *yz* plane.

5 In each of the following cases, find the coordinates of the point *T* that lies on the line through *P* parallel to the *y* axis but is 2 units farther to the right than *P*.

(a) $P = (1, 2, 4)$ (b) $(3, 5, 0)$ (c) $(2, -1, -3)$

6 Two points *A* and *B* in space are said to be **symmetrically located with respect to a line** that bisects the line segment $\overline{AB}$ and is perpendicular to it. If $A = (x, y, z)$, find the coordinates of the point *B* that is symmetrically located to *A* with respect to (a) the *x* axis, (b) the *y* axis, and (c) the *z* axis.

7 A cube each side of which is 6 units long has its center at the origin, and its faces are perpendicular to the coordinate axes that intersect them. Sketch the cube, and find the coordinates of the eight vertices.

8 Two points *A* and *B* in space are said to be **symmetrically located with respect to the midpoint *M* of the line segment $\overline{AB}$.** If $A = (x, y, z)$, find the coordinates of the point *B* that is symmetrically located to *A* with respect to the origin.

9 A perpendicular is dropped from the point $P = (3, 5, 7)$ to the plane $y = -1$. Find the coordinates of the point *Q* at the foot of this perpendicular.

10 If $P = (3, 4, 5)$, find the coordinates of the indicated points (Figure 12).

(a) The point Q symmetric to P with respect to the xy plane

(b) The point R symmetric to P with respect to the y axis

(c) The point S symmetric to P with respect to the origin

(d) The point T that is 5 units directly below P

(See Problems 4, 6, and 8.)

Figure 12

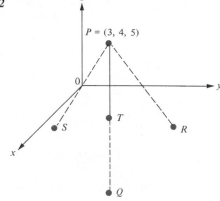

In Problems 11 to 18, describe and sketch the set of points $P = (x, y, z)$ that satisfy the given conditions.

11 $y = 5$ **12** $x = y$ **13** $x = 0$

14 $x > 0$ **15** $x = -3$ **16** $z \leq 2$

17 $y = -4$ **18** $x > 0$, $y < 0$, and $z > 0$

In Problems 19 to 28, give an equation or a pair of simultaneous equations for the plane or line.

19 The horizontal plane 3 units below the origin

20 The line parallel to the x axis and containing the point $(3, 4, 5)$

21 The plane parallel to the xz plane and containing the point $(-1, -2, 3)$

22 The plane perpendicular to the y axis and intersecting it at the point $(0, -5, 0)$

23 The x axis

24 The line parallel to the z axis and containing the point $(-2, -7, -15)$

25 The line parallel to the y axis and containing the point $(5, \frac{16}{3}, -\frac{7}{2})$

26 The line containing the two points $(5, 0, 7)$ and $(5, -1, 7)$

27 The line perpendicular to the yz plane and containing the point $(\frac{2}{5}, -\frac{3}{4}, \frac{4}{7})$

28 The plane containing the z axis and making a 45° angle with the xz plane and the yz plane

In Problems 29 to 36, sketch a graph in xyz space of each set of simultaneous equations.

29 $\begin{cases} y = -2 \\ z = 0 \end{cases}$ **30** $\begin{cases} x = 3 \\ y = 4 \end{cases}$ **31** $\begin{cases} x = -2 \\ y = -5 \end{cases}$

32 $\begin{cases} x = -3 \\ y = 0 \end{cases}$ **33** $\begin{cases} x^2 + y^2 + z^2 = 1 \\ y = 1 \end{cases}$ **34** $\begin{cases} x = y \\ y = z \end{cases}$

35 $\begin{cases} x = 1 \\ y = 2 \\ z = 3 \end{cases}$ **36** $\begin{cases} x^2 + y^2 + z^2 = 1 \\ x = y \end{cases}$

37 Write an equation of the sphere of radius 5 units with center at the origin in xyz space.

38 Write an equation of the sphere with center at the origin and containing the point $(1, 2, -1)$.

In Problems 39 to 42, find the distance r between the origin and the point P.

39 $P = (-2, 1, 2)$ **40** $P = (8, 0, -6)$

41 $P = (-4, -3, 0)$ **42** $P = (1, 4, 5)$

13.2   Vectors in Three-Dimensional Space

The concept of a vector and the algebra of vectors in two-dimensional space carry over almost without change to vectors in three-dimensional space. Thus, a vector in space is defined to be a directed line segment—that is, an arrow—and the directed line segment from P to Q is denoted by $\overrightarrow{PQ}$. Two vectors in space are regarded as being equal if they have the same length, are parallel, and point in the same direction. The definitions of sums and differences of vectors, the negative of a vector, the product of a scalar and a vector, and the dot product of two vectors all carry over verbatim to vectors in space.

The basic algebraic properties of vectors in space are exactly the same as the corresponding properties of vectors in the plane. We encourage you to sketch diagrams illustrating the following.

Figure 1

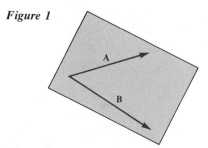

Figure 2

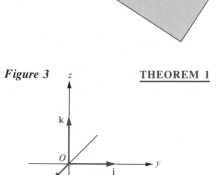

coplanar vectors

Basic Algebraic Properties of Vectors in Space

Let **A**, **B**, and **C** be vectors in three-dimensional space, and let s and t be scalars. Then

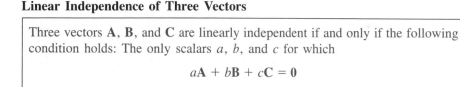

1 $\mathbf{A} + \mathbf{B} = \mathbf{B} + \mathbf{A}$	2 $\mathbf{A} + (\mathbf{B} + \mathbf{C}) = (\mathbf{A} + \mathbf{B}) + \mathbf{C}$
3 $\mathbf{A} + \mathbf{0} = \mathbf{A}$	4 $\mathbf{A} + (-\mathbf{A}) = \mathbf{0}$
5 $(st)\mathbf{A} = s(t\mathbf{A})$	6 $s(\mathbf{A} + \mathbf{B}) = s\mathbf{A} + s\mathbf{B}$
7 $(s + t)\mathbf{A} = s\mathbf{A} + t\mathbf{A}$	8 $1\mathbf{A} = \mathbf{A}$

If just two vectors **A** and **B** in space are involved, they can always be placed so that they have the same tail end (initial point) and lie in the same plane (Figure 1). Notice that *all linear combinations* $s\mathbf{A} + t\mathbf{B}$ *of* **A** *and* **B** *lie in the same plane*. (Why?)

Three vectors **A**, **B**, and **C** in space are called **coplanar** (or **linearly dependent**) if, when they are placed so as to have a common tail end, they all lie in the same plane (Figure 2). Three vectors in space are coplanar if and only if one of them can be expressed as a linear combination of the other two (Problem 70).

Three vectors **A**, **B**, and **C** are said to be **linearly independent** if they are not coplanar. The following theorem provides a handy test for the linear independence of three vectors. (For the proof, see Problem 72.)

Figure 3

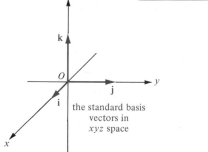

the standard basis
vectors in
xyz space

THEOREM 1

Linear Independence of Three Vectors

Three vectors **A**, **B**, and **C** are linearly independent if and only if the following condition holds: The only scalars a, b, and c for which

$$a\mathbf{A} + b\mathbf{B} + c\mathbf{C} = \mathbf{0}$$

are $a = 0$, $b = 0$, and $c = 0$.

The unit vectors **i**, **j**, and **k** pointing in the positive directions along the x, y, and z axes, respectively, in *xyz* space are three noncoplanar—hence, linearly independent—vectors (Figure 3). We call **i**, **j**, and **k** the **standard basis vectors** in *xyz* space.

Just as every vector in the xy plane can be written as a linear combination of **i** and **j**, every vector in *xyz* space can be written as a linear combination of **i**, **j**, and **k**. In fact, given a vector **R** in *xyz* space, we can move the tail end of **R** to the origin O, so that $\mathbf{R} = \overrightarrow{OP}$ is the *position vector* of the point $P = (x, y, z)$ (Figure 4). Let $Q = (x, y, 0)$ be the point at the foot of the perpendicular from P to the xy plane. Since $\overrightarrow{OQ}$ is the position vector of Q in the xy plane, it follows that $\overrightarrow{OQ} = x\mathbf{i} + y\mathbf{j}$. Because $\overrightarrow{QP}$ is parallel to **k**, it is clear that $\overrightarrow{QP} = z\mathbf{k}$. Therefore, since $\mathbf{R} = \overrightarrow{OQ} + \overrightarrow{QP}$, we have

Figure 4

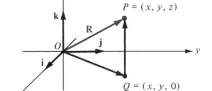

$$\mathbf{R} = x\mathbf{i} + y\mathbf{j} + z\mathbf{k}$$

As before, x, y, and z are called the (*scalar*) **components** of **R** with respect to the standard **i**, **j**, **k** basis.

The sum $\mathbf{A} + \mathbf{B}$, the difference $\mathbf{A} - \mathbf{B}$, and the product $s\mathbf{A}$ are calculated ''component by component'' for vectors $\mathbf{A}$ and $\mathbf{B}$ in xyz space just as they are for vectors in the plane. Thus, if

$$\mathbf{A} = a_1\mathbf{i} + a_2\mathbf{j} + a_3\mathbf{k} \qquad \text{and} \qquad \mathbf{B} = b_1\mathbf{i} + b_2\mathbf{j} + b_3\mathbf{k}$$

then

> **1** $\mathbf{A} + \mathbf{B} = (a_1 + b_1)\mathbf{i} + (a_2 + b_2)\mathbf{j} + (a_3 + b_3)\mathbf{k}$
>
> **2** $\mathbf{A} - \mathbf{B} = (a_1 - b_1)\mathbf{i} + (a_2 - b_2)\mathbf{j} + (a_3 - b_3)\mathbf{k}$
>
> **3** $s\mathbf{A} = (sa_1)\mathbf{i} + (sa_2)\mathbf{j} + (sa_3)\mathbf{k}$

EXAMPLE 1   Let $\mathbf{A} = 2\mathbf{i} - 3\mathbf{j} + \mathbf{k}$ and $\mathbf{B} = \mathbf{i} + 2\mathbf{j} + 5\mathbf{k}$. Find

   (a) $\mathbf{A} + \mathbf{B}$ **(b)** $\mathbf{A} - \mathbf{B}$ **(c)** $7\mathbf{A}$ **(d)** $7\mathbf{A} - \mathbf{B}$

SOLUTION

   (a) $\mathbf{A} + \mathbf{B} = (2 + 1)\mathbf{i} + (-3 + 2)\mathbf{j} + (1 + 5)\mathbf{k} = 3\mathbf{i} - \mathbf{j} + 6\mathbf{k}$

   (b) $\mathbf{A} - \mathbf{B} = (2 - 1)\mathbf{i} + (-3 - 2)\mathbf{j} + (1 - 5)\mathbf{k} = \mathbf{i} - 5\mathbf{j} - 4\mathbf{k}$

   (c) $7\mathbf{A} = (7)(2)\mathbf{i} + (7)(-3)\mathbf{j} + (7)(1)\mathbf{k} = 14\mathbf{i} - 21\mathbf{j} + 7\mathbf{k}$

   (d) $7\mathbf{A} - \mathbf{B} = (14 - 1)\mathbf{i} + (-21 - 2)\mathbf{j} + (7 - 5)\mathbf{k} = 13\mathbf{i} - 23\mathbf{j} + 2\mathbf{k}$ ∎

The three scalar components of a vector in xyz space are uniquely determined by the vector. To see this, assume that

$$x_1\mathbf{i} + y_1\mathbf{j} + z_1\mathbf{k} = x_2\mathbf{i} + y_2\mathbf{j} + z_2\mathbf{k}$$

Then $$(x_1 - x_2)\mathbf{i} + (y_1 - y_2)\mathbf{j} + (z_1 - z_2)\mathbf{k} = \mathbf{0}$$

Hence, since $\mathbf{i}$, $\mathbf{j}$, and $\mathbf{k}$ are linearly independent, it follows that $x_1 - x_2 = 0$, $y_1 - y_2 = 0$, and $z_1 - z_2 = 0$. Therefore, $x_1 = x_2$, $y_1 = y_2$, and $z_1 = z_2$.

Since a vector $\mathbf{R} = x\mathbf{i} + y\mathbf{j} + z\mathbf{k}$ both determines and is determined by its three scalar components x, y, and z, it is customary to represent $\mathbf{R}$ by the *ordered triple* $\langle x, y, z \rangle$ and to write $\mathbf{R} = \langle x, y, z \rangle$. Using this notation, we can write the solution to the last example as

   (a) $\langle 2, -3, 1 \rangle + \langle 1, 2, 5 \rangle = \langle 3, -1, 6 \rangle$

   (b) $\langle 2, -3, 1 \rangle - \langle 1, 2, 5 \rangle = \langle 1, -5, -4 \rangle$

   (c) $7\langle 2, -3, 1 \rangle = \langle 14, -21, 7 \rangle$

   (d) $7\langle 2, -3, 1 \rangle - \langle 1, 2, 5 \rangle = \langle 13, -23, 2 \rangle$

We encourage you to use ordered-triple notation for vectors in xyz space whenever it seems convenient.

The angle θ between two vectors $\mathbf{A}$ and $\mathbf{B}$ in three-dimensional space is measured by bringing the tail ends of the vectors together, so that the two vectors lie in the same plane (Figure 1), and then measuring the angle as usual. The dot product of $\mathbf{A}$ and $\mathbf{B}$ is defined, just as for vectors in the plane, by

> $$\mathbf{A} \cdot \mathbf{B} = |\mathbf{A}|\,|\mathbf{B}|\cos\theta$$

where $|\mathbf{A}|$ and $|\mathbf{B}|$ denote the lengths (magnitudes) of $\mathbf{A}$ and $\mathbf{B}$, respectively.

The basic properties of the dot product in three-dimensional space are established by using virtually the same arguments as for vectors in the plane.

Basic Properties of the Dot Product in Space

If **A**, **B**, and **C** are vectors in three-dimensional space and s is a scalar, then

1 $\mathbf{A} \cdot \mathbf{B} = \mathbf{B} \cdot \mathbf{A}$	**2** $(\mathbf{A} + \mathbf{B}) \cdot \mathbf{C} = \mathbf{A} \cdot \mathbf{C} + \mathbf{B} \cdot \mathbf{C}$
3 $(s\mathbf{A}) \cdot \mathbf{B} = s(\mathbf{A} \cdot \mathbf{B}) = \mathbf{A} \cdot (s\mathbf{B})$	**4** $\mathbf{A} \cdot \mathbf{A} = |\mathbf{A}|^2 \geq 0$
5 If $\mathbf{A} \cdot \mathbf{A} = 0$, then $\mathbf{A} = \mathbf{0}$	**6** $\mathbf{A} \cdot \mathbf{B} = 0$ if and only if **A** and **B** are perpendicular

Furthermore, in xyz space, we have

7 $\mathbf{i} \cdot \mathbf{i} = \mathbf{j} \cdot \mathbf{j} = \mathbf{k} \cdot \mathbf{k} = 1$	**8** $\mathbf{i} \cdot \mathbf{j} = \mathbf{i} \cdot \mathbf{k} = \mathbf{j} \cdot \mathbf{k} = 0$

Properties 7 and 8 follow from the facts that **i**, **j**, and **k** are unit vectors and that they are mutually perpendicular.

The following useful theorem is the analog for xyz space of Theorem 1 in Section 12.2.

THEOREM 2   **Dot Product of Vectors in *xyz* Space**

> Let $\mathbf{A} = a\mathbf{i} + b\mathbf{j} + c\mathbf{k}$ and $\mathbf{B} = x\mathbf{i} + y\mathbf{j} + z\mathbf{k}$. Then
> $$\mathbf{A} \cdot \mathbf{B} = ax + by + cz$$

The proof of Theorem 2 is accomplished by expanding the dot product $(a\mathbf{i} + b\mathbf{j} + c\mathbf{k}) \cdot (x\mathbf{i} + y\mathbf{j} + z\mathbf{k})$ and using Properties 7 and 8 (Problem 74). An immediate consequence of Theorem 2 is that if $\mathbf{B} = x\mathbf{i} + y\mathbf{j} + z\mathbf{k}$, then $\mathbf{B} \cdot \mathbf{B} = x^2 + y^2 + z^2$. Since $|\mathbf{B}| = \sqrt{\mathbf{B} \cdot \mathbf{B}}$ by Property 4, we have the following formula for the length of a vector in xyz space: If $\mathbf{B} = x\mathbf{i} + y\mathbf{j} + z\mathbf{k}$, then

$$|\mathbf{B}| = \sqrt{x^2 + y^2 + z^2}$$

If **A** and **B** are vectors in three-dimensional space, then, by the same argument given in Section 12.2 for vectors in the plane, the **scalar component (scalar projection)**, $\text{comp}_\mathbf{B}\mathbf{A}$, of **A** in the direction of **B** is given by

$$\text{comp}_\mathbf{B}\mathbf{A} = \frac{\mathbf{A} \cdot \mathbf{B}}{|\mathbf{B}|}$$

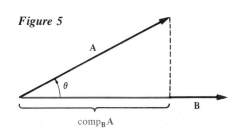

Figure 5

provided that $\mathbf{B} \neq \mathbf{0}$ (Figure 5).*

EXAMPLE 2   Let $\mathbf{A} = 3\mathbf{i} - 2\mathbf{j} + \mathbf{k}$ and $\mathbf{B} = 2\mathbf{i} + 5\mathbf{j} - 2\mathbf{k}$. Find

    (a) $\mathbf{A} \cdot \mathbf{B}$ **(b)** $|\mathbf{A}|$ **(c)** $|\mathbf{A} - \mathbf{B}|$ **(d)** $\text{comp}_\mathbf{B}\mathbf{A}$

SOLUTION

    (a) $\mathbf{A} \cdot \mathbf{B} = (3)(2) + (-2)(5) + (1)(-2) = -6$
    (b) $|\mathbf{A}| = \sqrt{3^2 + (-2)^2 + 1^2} = \sqrt{14}$

---

*Figure 5 shows a case in which the angle θ between **A** and **B** is acute. If the angle is obtuse, then $\text{comp}_\mathbf{B}\mathbf{A}$ is negative.

(c) $|\mathbf{A} - \mathbf{B}| = |\mathbf{i} - 7\mathbf{j} + 3\mathbf{k}| = \sqrt{1^2 + (-7)^2 + 3^2} = \sqrt{59}$

(d) $\text{comp}_\mathbf{B}\mathbf{A} = \dfrac{\mathbf{A} \cdot \mathbf{B}}{|\mathbf{B}|} = \dfrac{-6}{\sqrt{2^2 + 5^2 + (-2)^2}} = \dfrac{-6}{\sqrt{33}}$

Suppose that θ is the angle between the vectors $\mathbf{A}$ and $\mathbf{B}$ in three-dimensional space. Then, since $\mathbf{A} \cdot \mathbf{B} = |\mathbf{A}| \, |\mathbf{B}| \cos \theta$, we have

$$\cos \theta = \frac{\mathbf{A} \cdot \mathbf{B}}{|\mathbf{A}| \, |\mathbf{B}|}$$

This formula enables us to find the cosine of θ, and hence the angle θ, in terms of the dot product.

© **EXAMPLE 3**   Find the angle θ between $\mathbf{A} = 2\mathbf{i} - 3\mathbf{j} + \mathbf{k}$ and $\mathbf{B} = \mathbf{i} + 2\mathbf{j} + 5\mathbf{k}$.

SOLUTION $\cos \theta = \dfrac{\mathbf{A} \cdot \mathbf{B}}{|\mathbf{A}| \, |\mathbf{B}|} = \dfrac{(2)(1) + (-3)(2) + (1)(5)}{\sqrt{2^2 + (-3)^2 + 1^2}\sqrt{1^2 + 2^2 + 5^2}} = \dfrac{1}{2\sqrt{105}}$

Hence, $\theta = \cos^{-1}\dfrac{1}{2\sqrt{105}} \approx \cos^{-1} 0.0488 \approx 87.2°$

Distance Between Points in Space

Figure 6

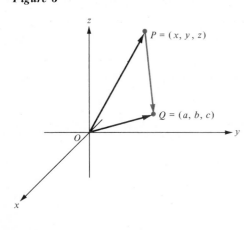

We can now use vectors to derive a formula for the distance between two points $P = (x, y, z)$ and $Q = (a, b, c)$ in xyz space (Figure 6). Here,

$\overrightarrow{OP} = x\mathbf{i} + y\mathbf{j} + z\mathbf{k}$ $\overrightarrow{OQ} = a\mathbf{i} + b\mathbf{j} + c\mathbf{k}$ and $\overrightarrow{PQ} = \overrightarrow{OQ} - \overrightarrow{OP}$

It follows that

$$\overrightarrow{PQ} = (a - x)\mathbf{i} + (b - y)\mathbf{j} + (c - z)\mathbf{k}$$

Thus, *the scalar components of the vector $\overrightarrow{PQ}$ in xyz space are the differences of the coordinates of Q and the corresponding coordinates of P.* Therefore, the distance between P and Q is given by

$$|\overline{PQ}| = |\overrightarrow{PQ}| = \sqrt{(a - x)^2 + (b - y)^2 + (c - z)^2}$$

EXAMPLE 4   If $P = (2, 3, -2)$ and $Q = (-1, 1, 5)$, find **(a)** the scalar components of $\overrightarrow{PQ}$ and **(b)** the distance between P and Q.

SOLUTION

(a) $\overrightarrow{PQ} = [(-1) - 2]\mathbf{i} + [1 - 3]\mathbf{j} + [5 - (-2)]\mathbf{k} = -3\mathbf{i} - 2\mathbf{j} + 7\mathbf{k}$

(b) Using the distance formula above, we have

$|\overline{PQ}| = \sqrt{[(-1) - 2]^2 + [1 - 3]^2 + [5 - (-2)]^2} = \sqrt{62}$ units

Direction Cosines of a Vector

Consider a nonzero vector $\mathbf{A}$ in xyz space. Move $\mathbf{A}$, if necessary, so that its tail end is the origin O, and let α, β, and γ be the angles between $\mathbf{A}$ and the positive x, y,

Figure 7

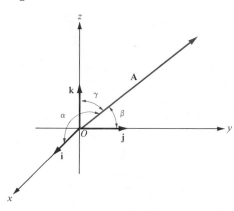

and z axes, respectively (Figure 7). The angles α, β, and γ, which are the same as the angles between **A** and the standard **i**, **j**, and **k** basis vectors, respectively, are called the **direction angles** of the vector **A**. The cosines of the three direction angles are called the **direction cosines** of the vector **A** and are given by the formulas

$$\cos \alpha = \frac{\mathbf{A} \cdot \mathbf{i}}{|\mathbf{A}|} \qquad \cos \beta = \frac{\mathbf{A} \cdot \mathbf{j}}{|\mathbf{A}|} \qquad \text{and} \qquad \cos \gamma = \frac{\mathbf{A} \cdot \mathbf{k}}{|\mathbf{A}|}$$

(Problem 54).

Now, suppose that $\mathbf{A} = a\mathbf{i} + b\mathbf{j} + c\mathbf{k}$, so that

$$\mathbf{A} \cdot \mathbf{i} = a \qquad \mathbf{A} \cdot \mathbf{j} = b \qquad \mathbf{A} \cdot \mathbf{k} = c$$

and
$$|\mathbf{A}| = \sqrt{a^2 + b^2 + c^2}$$

Then,

$$\cos \alpha = \frac{a}{\sqrt{a^2 + b^2 + c^2}} \qquad \cos \beta = \frac{b}{\sqrt{a^2 + b^2 + c^2}}$$

$$\cos \gamma = \frac{c}{\sqrt{a^2 + b^2 + c^2}}$$

Notice that

$$(\cos \alpha)\mathbf{i} + (\cos \beta)\mathbf{j} + (\cos \gamma)\mathbf{k} = \frac{1}{\sqrt{a^2 + b^2 + c^2}}(a\mathbf{i} + b\mathbf{j} + c\mathbf{k}) = \frac{\mathbf{A}}{|\mathbf{A}|}$$

Hence, *the direction cosines of a nonzero vector* **A** *are the scalar components of the unit vector* $\mathbf{A}/|\mathbf{A}|$ *in the same direction as* **A**. Since $(\cos \alpha)\mathbf{i} + (\cos \beta)\mathbf{j} + (\cos \gamma)\mathbf{k}$ is a unit vector, it follows that

$$\cos^2 \alpha + \cos^2 \beta + \cos^2 \gamma = 1$$

that is, *the sum of the squares of the direction cosines of a nonzero vector is always equal to 1* (Problem 59).

EXAMPLE 5   Let $P = (-1, 2, 3)$ and $Q = (-3, 3, 5)$. Find the direction cosines of the vector $\mathbf{A} = \overrightarrow{PQ}$, and verify that the sum of the squares of these direction cosines is equal to 1.

SOLUTION $\mathbf{A} = \overrightarrow{PQ} = (-3 + 1)\mathbf{i} + (3 - 2)\mathbf{j} + (5 - 3)\mathbf{k} = -2\mathbf{i} + \mathbf{j} + 2\mathbf{k}$

Hence,
$$\cos \alpha = \frac{-2}{\sqrt{(-2)^2 + 1^2 + 2^2}} = -\frac{2}{3}$$

$$\cos \beta = \frac{1}{\sqrt{(-2)^2 + 1^2 + 2^2}} = \frac{1}{3}$$

$$\cos \gamma = \frac{2}{\sqrt{(-2)^2 + 1^2 + 2^2}} = \frac{2}{3}$$

and we have

$$\cos^2 \alpha + \cos^2 \beta + \cos^2 \gamma = (-\tfrac{2}{3})^2 + (\tfrac{1}{3})^2 + (\tfrac{2}{3})^2 = \tfrac{4}{9} + \tfrac{1}{9} + \tfrac{4}{9} = 1 \qquad ■$$

Problem Set 13.2

In Problems 1 to 46, suppose $\mathbf{A} = 3\mathbf{i} + \mathbf{j} - 4\mathbf{k}$, $\mathbf{B} = 4\mathbf{i} + \mathbf{j} - \mathbf{k}$, $\mathbf{C} = 2\mathbf{i} - 5\mathbf{j} + 4\mathbf{k}$, $\mathbf{D} = \langle -2, 1, 7 \rangle$, and $\mathbf{E} = \langle 2, 1, 1 \rangle$ are vectors in space. Evaluate each of the following.

1 $\mathbf{A} + \mathbf{B}$

2 $3\mathbf{A} + 2\mathbf{B}$

3 $3\mathbf{C} - 2\mathbf{A}$

4 $\mathbf{D} - \mathbf{E}$

5 $\mathbf{E} - 2\mathbf{D}$

6 $\mathbf{A} - \mathbf{B} + 2\mathbf{C}$

7 $\frac{1}{3}(\mathbf{A} + \mathbf{B} - 6\mathbf{C})$

8 $-3\mathbf{A} - 4(\mathbf{B} + \mathbf{C})$

9 $2\mathbf{A} - (\mathbf{B} - \mathbf{C})$

10 $\frac{1}{2}\mathbf{D} - \frac{2}{3}\mathbf{E}$

11 $\mathbf{A} \cdot \mathbf{B}$

12 $\mathbf{D} \cdot \mathbf{E}$

13 $\mathbf{A} \cdot (\mathbf{B} + \mathbf{C})$

14 $\mathbf{B} \cdot (2\mathbf{A} - \mathbf{B} + \mathbf{C})$

15 $\mathbf{A} \cdot (-\mathbf{B}) + \mathbf{A} \cdot \mathbf{C}$

16 $(\mathbf{A} \cdot \mathbf{B})\mathbf{C} - (\mathbf{A} \cdot \mathbf{C})\mathbf{B}$

17 $(\mathbf{D} + \mathbf{E}) \cdot (\mathbf{D} - \mathbf{E})$

18 $(2\mathbf{A} + \mathbf{B}) \cdot (\mathbf{C} - \mathbf{A})$

19 $(\mathbf{E} \cdot \mathbf{D})\mathbf{E} - (\mathbf{E} \cdot \mathbf{E})\mathbf{D}$

20 $2|\mathbf{A}| + 2|\mathbf{C}|$

21 $|\mathbf{D}|$

22 $|-2\mathbf{B}| + 2|\mathbf{C}|$

23 $|\mathbf{A} + \mathbf{C}|$

24 $|\mathbf{B}|\mathbf{A} - |\mathbf{A}|\mathbf{B}$

25 $|\mathbf{A}| + |\mathbf{C}|$

26 $|\mathbf{D} + \mathbf{E}| - |\mathbf{D}| - |\mathbf{E}|$

27 $\mathbf{B}/|\mathbf{B}|$

28 $\left| \dfrac{\mathbf{B}}{|\mathbf{B}|} \right| + 3 \left| \dfrac{\mathbf{A}}{|\mathbf{A}|} \right|$

29 $\text{comp}_{\mathbf{B}}\mathbf{A}$

30 $\text{comp}_{\mathbf{A}}\mathbf{B}$

31 $\text{comp}_{\mathbf{B}}\mathbf{C}$

32 $\text{comp}_{\mathbf{D}}\mathbf{E}$

33 $\text{comp}_{\mathbf{B}}(\mathbf{A} + \mathbf{C})$

34 $\text{comp}_{\mathbf{B}}(\mathbf{A} - \mathbf{C})$

35 $\text{comp}_{\mathbf{E}}\mathbf{D}$

36 $\text{comp}_{\mathbf{A}}\mathbf{A}$

37 The angle θ_1 between $\mathbf{A}$ and $\mathbf{B}$

38 The angle θ_2 between $\mathbf{A}$ and $\mathbf{B} - \mathbf{A}$

39 The angle θ_3 between $\mathbf{A} - \mathbf{B}$ and $\mathbf{A} + \mathbf{B}$

40 The angle θ_4 between $\mathbf{D}$ and $\mathbf{E}$

41 The direction cosines of $\mathbf{A}$

42 The direction cosines of $\mathbf{E}$

43 The direction cosines of $\mathbf{A} + \mathbf{B}$

©**44** The direction angles of $\mathbf{E}$

45 A scalar a such that $\mathbf{A}$ is perpendicular to $\mathbf{F} = a\mathbf{i} + \mathbf{j} - 5\mathbf{k}$

46 A scalar b such that $\mathbf{D}$ is perpendicular to $\mathbf{G} = \langle 1, b, 4 \rangle$.

47 Use vectors to show that $P = (1, 7, 2)$, $Q = (0, 7, -2)$, and $R = (-1, 6, 1)$ are vertices of a right triangle. At which vertex is the right angle? (*Hint:* Show that two of the three vectors $\overrightarrow{PQ}$, $\overrightarrow{QR}$, and $\overrightarrow{RP}$ are perpendicular.)

48 If $\mathbf{D} \neq \mathbf{0}$, show that $\text{comp}_{\mathbf{D}}(\mathbf{A} + \mathbf{B}) = \text{comp}_{\mathbf{D}}\mathbf{A} + \text{comp}_{\mathbf{D}}\mathbf{B}$ holds for all vectors $\mathbf{A}$, $\mathbf{B}$, and $\mathbf{D}$.

49 Find the distance between the points P_1 and P_2 if
(a) $P_1 = (7, -1, 4)$; $P_2 = (8, 1, 6)$
(b) $P_1 = (1, 1, 5)$; $P_2 = (1, -2, 3)$
(c) $P_1 = (2, 3, -3)$; $P_2 = (2, 1, -2)$

50 Under what conditions is it true that $|\mathbf{A} + \mathbf{B}| = |\mathbf{A}| + |\mathbf{B}|$?

In Problems 51 to 53, find the direction cosines of the vector $\mathbf{A} = \overrightarrow{QP}$. ©Also find the direction angles of $\mathbf{A}$.

51 $P = (6, -1, -2)$ and $Q = (4, 9, 9)$

52 $P = (3, 8, 1)$ and $Q = (1, 2, 10)$

53 $P = (9, 1, -7)$ and $Q = (1, 0, -1)$

54 Verify the formulas given on page 769 for the direction cosines of a vector.

In Problems 55 to 58, determine whether each triple α, β, γ could possibly be direction angles of a vector $\mathbf{A}$.

55 $\alpha = 90°$, $\beta = 135°$, $\gamma = 45°$

56 $\alpha = 2\pi/3$, $\beta = 3\pi/4$, $\gamma = \pi/3$

57 $\alpha = \pi/3$, $\beta = \pi/6$, $\gamma = \pi/4$

58 $\alpha = 120°$, $\beta = 45°$, $\gamma = 60°$

59 Prove that the sum of the squares of the direction cosines of a nonzero vector is 1.

60 If $\alpha = \pi/3$ and $\beta = \pi/3$ are two direction angles of a vector $\mathbf{A}$, find the possible values of the third direction angle γ.

61 Show that a nonzero vector $\mathbf{A}$ in space is completely determined by its length l and its three direction cosines: $\cos \alpha$, $\cos \beta$, and $\cos \gamma$. [*Hint:* Show that $\mathbf{A} = (l \cos \alpha)\mathbf{i} + (l \cos \beta)\mathbf{j} + (l \cos \gamma)\mathbf{k}$.]

62 Show that $\mathbf{A} = 7\mathbf{i} + 2\mathbf{j} - 10\mathbf{k}$, $\mathbf{B} = 4\mathbf{i} - 6\mathbf{j} + 5\mathbf{k}$, and $\mathbf{C} = 3\mathbf{i} - 2\mathbf{j} - 3\mathbf{k}$ are position vectors of the vertices of an isosceles triangle.

63 Let $|\mathbf{A}| = 2$, $|\mathbf{B}| = 6$, and suppose that the angle between $\mathbf{A}$ and $\mathbf{B}$ is 30°. Find (a) $|\mathbf{A} + \mathbf{B}|$ and (b) $|\mathbf{A} - \mathbf{B}|$.

64 Give a geometric interpretation of the "average" $(\mathbf{A} + \mathbf{B})/2$ of two vectors $\mathbf{A}$ and $\mathbf{B}$.

65 Let $\mathbf{A} = \overrightarrow{PQ}$ where $P = (-1, 2, -3)$ and $Q = (11, 11, 3)$. Find a point $S = (x, y, z)$ such that $\overrightarrow{PS} = 2\mathbf{A}$.

66 If $\mathbf{A}$ is a vector and $\mathbf{E}$ is a vector with $|\mathbf{E}| = 1$, show that $\mathbf{E}$ is perpendicular to $\mathbf{A} - (\text{comp}_E\mathbf{A})\mathbf{E}$. Sketch a figure showing $\mathbf{A}$, $\mathbf{E}$, $(\text{comp}_E\mathbf{A})\mathbf{E}$, and $\mathbf{A} - (\text{comp}_E\mathbf{A})\mathbf{E}$.

67 If $\mathbf{A}$ is a vector in space, show that

$$\mathbf{A} = (\mathbf{A} \cdot \mathbf{i})\mathbf{i} + (\mathbf{A} \cdot \mathbf{j})\mathbf{j} + (\mathbf{A} \cdot \mathbf{k})\mathbf{k}$$

(*Hint:* Begin by writing $\mathbf{A}$ as $\mathbf{A} = x\mathbf{i} + y\mathbf{j} + z\mathbf{k}$. Then take the dot product on both sides of the equation with $\mathbf{i}$, with $\mathbf{j}$, and with $\mathbf{k}$.)

68 Show that $\mathbf{D} = (\mathbf{B} \cdot \mathbf{B})\mathbf{A} - (\mathbf{A} \cdot \mathbf{B})\mathbf{B}$ is perpendicular to $\mathbf{B}$.

69 Are the vectors $\mathbf{A} = \langle 1, 1, 0 \rangle$, $\mathbf{B} = \langle 2, -1, 0 \rangle$, and $\mathbf{C} = \langle 1, 1, 1 \rangle$ coplanar? Explain.

70 Show that three vectors are coplanar if and only if one of them can be expressed as a linear combination of the other two.

71 Are the vectors $\mathbf{A} = 2\mathbf{i} - 3\mathbf{j} + 5\mathbf{k}$, $\mathbf{B} = 3\mathbf{i} + \mathbf{j} - 2\mathbf{k}$, and $\mathbf{C} = \mathbf{i} - 7\mathbf{j} + 12\mathbf{k}$ linearly independent? Explain.

72 Prove Theorem 1.

73 Suppose that $\mathbf{A}$, $\mathbf{B}$, and $\mathbf{C}$ are linearly independent vectors and that $x\mathbf{A} + y\mathbf{B} + z\mathbf{C} = a\mathbf{A} + b\mathbf{B} + c\mathbf{C}$. Prove that $x = a$, $y = b$, and $z = c$.

74 Prove Theorem 2.

13.3   Cross and Box Products of Vectors in Space

In this section we define and study a product that is available for vectors in three-dimensional space but not for vectors in the plane. This product is called the *cross product* (*outer product* or *vector product*). By combining the dot and cross products, we obtain the *box product* (or *determinant*) of three vectors in space.

The cross product of two vectors in space is defined to be a *vector* with a certain magnitude and direction. In order to specify the direction of the cross product of two vectors, we need the following concept: Three linearly independent vectors $\mathbf{A}$, $\mathbf{B}$, and $\mathbf{C}$ are said to form a **right-handed triple** in the order $\mathbf{A}$, $\mathbf{B}$, $\mathbf{C}$ if the thumb, index finger, and middle finger of the right hand (held in a "natural" position) can be aligned simultaneously with $\mathbf{A}$, $\mathbf{B}$, and $\mathbf{C}$, respectively (Figure 1). For instance, the standard basis vectors in the order $\mathbf{i}$, $\mathbf{j}$, $\mathbf{k}$ form a right-handed triple, but in the order $\mathbf{j}$, $\mathbf{i}$, $\mathbf{k}$ they do not. (Why?)

We can now give the definition of the cross product.

DEFINITION 1   **Cross Product of Vectors**

> Let $\mathbf{A}$ and $\mathbf{B}$ be two nonparallel vectors in three-dimensional space. Then the **cross product** of $\mathbf{A}$ and $\mathbf{B}$, denoted $\mathbf{A} \times \mathbf{B}$, is the *vector* whose length is numerically equal to the area of the parallelogram spanned by $\mathbf{A}$ and $\mathbf{B}$ and whose direction is perpendicular to both $\mathbf{A}$ and $\mathbf{B}$ in such a way that $\mathbf{A}$, $\mathbf{B}$, $\mathbf{A} \times \mathbf{B}$ is a right-handed triple (Figure 2). If $\mathbf{A}$ and $\mathbf{B}$ are parallel vectors, we define $\mathbf{A} \times \mathbf{B} = \mathbf{0}$. In particular, then, $\mathbf{A} \times \mathbf{A} = \mathbf{0}$.

Figure 1

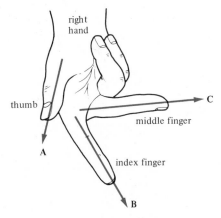

Figure 2

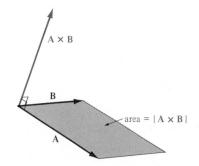

Figure 3

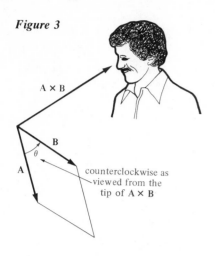

A × B

B

θ

A

counterclockwise as
viewed from the
tip of A × B

The condition that **A**, **B**, **A** × **B** is a right-handed triple can be rephrased in various equivalent ways. For example, a right-handed screw turned through the smaller angle *from* **A** *to* **B** advances in the direction of **A** × **B**. Equivalently, if the parallelogram spanned by **A** and **B** is viewed from the tip of **A** × **B**, then the angle θ from **A** to **B** is generated by a *counterclockwise* rotation (Figure 3). (Also, see Problem 38.)

Notice that the direction of the cross product is reversed when the two factors are interchanged; that is,

$$\mathbf{B} \times \mathbf{A} = -(\mathbf{A} \times \mathbf{B})$$

Thus, *the cross product is not a commutative operation*. Consequently, in calculating with cross products, you must be very careful about the *order* in which the vectors are multiplied.

Figure 4

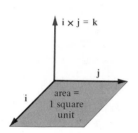

i × j = k

j

i

area =
1 square
unit

EXAMPLE 1   Find

   (a) **i** × **j**   (b) **j** × **i**

SOLUTION

   (a) Because the two perpendicular unit vectors **i** and **j** span a square of unit area, it follows that **i** × **j** is a unit vector perpendicular to both **i** and **j**. Since **i**, **j**, **i** × **j** is a right-handed triple, we conclude that **i** × **j** = **k** (Figure 4).

   (b) Using the result in part (a), we find that **j** × **i** = −(**i** × **j**) = −**k**.   ∎

Proceeding as in the example above, we can find the cross products of all pairs of standard basis vectors:

$$\mathbf{i} \times \mathbf{j} = \mathbf{k} \qquad \mathbf{j} \times \mathbf{k} = \mathbf{i} \qquad \mathbf{k} \times \mathbf{i} = \mathbf{j}$$
$$\mathbf{j} \times \mathbf{i} = -\mathbf{k} \qquad \mathbf{k} \times \mathbf{j} = -\mathbf{i} \qquad \mathbf{i} \times \mathbf{k} = -\mathbf{j}$$

We also have

$$\mathbf{i} \times \mathbf{i} = \mathbf{0} \qquad \mathbf{j} \times \mathbf{j} = \mathbf{0} \qquad \mathbf{k} \times \mathbf{k} = \mathbf{0}$$

Figure 5

k

i

j

In calculating cross products of the standard basis vectors, the following "memory device" may prove useful. Figure 5 shows **i**, **j**, and **k** as symbols on the rim of a wheel. If, in naming two of these symbols successively, you go *clockwise* on the wheel, the cross product of the two named symbols is the remaining symbol; if you go *counterclockwise*, it is the negative of the remaining symbol. For instance, in moving from **k** to **j** we go counterclockwise; hence, **k** × **j** = −**i**.

If *s* is a positive scalar, then *s***A** has the same direction as **A**, but is *s* times as long. It follows that the parallelogram spanned by *s***A** and **B** has an area that is *s* times the area of the parallelogram spanned by **A** and **B**; that is,

$$|(s\mathbf{A}) \times \mathbf{B}| = s|\mathbf{A} \times \mathbf{B}| \qquad \text{for } s > 0$$

Obviously, the direction of (*s***A**) × **B** is the same as the direction of **A** × **B** (why?), and it follows that

$$(s\mathbf{A}) \times \mathbf{B} = s(\mathbf{A} \times \mathbf{B})$$

We leave it to you to make the simple geometric argument showing that the last equation holds even when s is not positive (Problem 43). Similar considerations (Problem 44) show that

$$\mathbf{A} \times (t\mathbf{B}) = t(\mathbf{A} \times \mathbf{B})$$

Putting s or t equal to -1 gives

$$(-\mathbf{A}) \times \mathbf{B} = -(\mathbf{A} \times \mathbf{B}) = \mathbf{A} \times (-\mathbf{B})$$

The Box Product

The cross product $\mathbf{A} \times \mathbf{B}$ of vectors $\mathbf{A}$ and $\mathbf{B}$ is a vector, and it is therefore possible to form its dot product $(\mathbf{A} \times \mathbf{B}) \cdot \mathbf{C}$ with a third vector $\mathbf{C}$. The *scalar*

$$(\mathbf{A} \times \mathbf{B}) \cdot \mathbf{C}$$

Figure 6

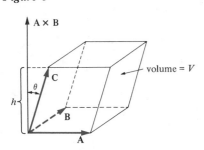

is called the **box product** (**triple scalar product,** or **determinant**) of the vectors $\mathbf{A}$, $\mathbf{B}$, and $\mathbf{C}$. Suppose that $\mathbf{A}$, $\mathbf{B}$, $\mathbf{C}$ is a right-handed triple. Notice that $\mathbf{A}$, $\mathbf{B}$, and $\mathbf{C}$ form three adjacent edges of a "box" whose six faces are parallelograms and whose opposite pairs of faces are parallel and congruent (Figure 6). Such a box is called a **parallelepiped,** and its volume V is given by the area $|\mathbf{A} \times \mathbf{B}|$ of its base times its height h. If θ is the angle between $\mathbf{C}$ and $\mathbf{A} \times \mathbf{B}$, then $h = |\mathbf{C}| \cos \theta$, and by the definition of the dot product,

$$V = |\mathbf{A} \times \mathbf{B}|h = |\mathbf{A} \times \mathbf{B}| \, |\mathbf{C}| \cos \theta = (\mathbf{A} \times \mathbf{B}) \cdot \mathbf{C}$$

Thus, the box product $(\mathbf{A} \times \mathbf{B}) \cdot \mathbf{C}$ gives the volume of the box (Figure 6) spanned by $\mathbf{A}$, $\mathbf{B}$, and $\mathbf{C}$.

If $\mathbf{A}$, $\mathbf{B}$, and $\mathbf{C}$ are linearly independent vectors that do not form a right-handed triple $\mathbf{A}$, $\mathbf{B}$, $\mathbf{C}$, we say that they form a **left-handed triple.** If $\mathbf{A}$, $\mathbf{B}$, $\mathbf{C}$ is a left-handed triple, it is easy to see that $\mathbf{B}$, $\mathbf{A}$, $\mathbf{C}$ is a right-handed triple; hence, if V is the volume of the box spanned by $\mathbf{B}$, $\mathbf{A}$, and $\mathbf{C}$, then

$$(\mathbf{A} \times \mathbf{B}) \cdot \mathbf{C} = [-(\mathbf{B} \times \mathbf{A})] \cdot \mathbf{C} = -[(\mathbf{B} \times \mathbf{A}) \cdot \mathbf{C}] = -V$$

It follows from these considerations that *the volume V of the box (parallelepiped) spanned by the three linearly independent vectors* $\mathbf{A}$, $\mathbf{B}$, *and* $\mathbf{C}$ *is given by*

$$V = \pm(\mathbf{A} \times \mathbf{B}) \cdot \mathbf{C}$$

where the plus or minus sign is used according to whether the triple $\mathbf{A}$, $\mathbf{B}$, $\mathbf{C}$ is right-handed or left-handed, respectively.

If $\mathbf{A}$, $\mathbf{B}$, $\mathbf{C}$ is a right-handed triple (as in Figure 6), then—as you can easily see—$\mathbf{B}$, $\mathbf{C}$, $\mathbf{A}$ is also a right-handed triple. Thus, since the volume V of the box spanned by the three vectors is the same regardless of which face is called the base, we must have

$$V = (\mathbf{B} \times \mathbf{C}) \cdot \mathbf{A} = (\mathbf{A} \times \mathbf{B}) \cdot \mathbf{C}$$

Similarly, if $\mathbf{A}$, $\mathbf{B}$, $\mathbf{C}$ is a left-handed triple, we have

$$V = -(\mathbf{B} \times \mathbf{C}) \cdot \mathbf{A} = -(\mathbf{A} \times \mathbf{B}) \cdot \mathbf{C}$$

In any case, then,

$$(\mathbf{B} \times \mathbf{C}) \cdot \mathbf{A} = (\mathbf{A} \times \mathbf{B}) \cdot \mathbf{C}$$

Since the dot product *is* commutative, we have $(\mathbf{B} \times \mathbf{C}) \cdot \mathbf{A} = \mathbf{A} \cdot (\mathbf{B} \times \mathbf{C})$, so the last equation can be rewritten as

$$\mathbf{A} \cdot (\mathbf{B} \times \mathbf{C}) = (\mathbf{A} \times \mathbf{B}) \cdot \mathbf{C}$$

Thus, *in the box product, the dot and cross can be interchanged,* provided that the three vectors are kept in the same order.

The Distributive Law and the Cross-Product Formula

Although the preceding discussion describes the cross product $\mathbf{A} \times \mathbf{B}$ geometrically and relates it to the volume of a parallelepiped, it does not show how $\mathbf{A} \times \mathbf{B}$ can be found directly from $\mathbf{A}$ and $\mathbf{B}$ when $\mathbf{A}$ and $\mathbf{B}$ are given in component form. A "cross-product formula" that gives the components of $\mathbf{A} \times \mathbf{B}$ in terms of the components of $\mathbf{A}$ and of $\mathbf{B}$ can be derived on the basis of the following theorem.

THEOREM 1   **Distributive Law for the Cross Product**

For any three vectors $\mathbf{A}$, $\mathbf{B}$, and $\mathbf{C}$ in three-dimensional space,

$$\mathbf{A} \times (\mathbf{B} + \mathbf{C}) = (\mathbf{A} \times \mathbf{B}) + (\mathbf{A} \times \mathbf{C})$$

PROOF   Let

$$\mathbf{D} = \mathbf{A} \times (\mathbf{B} + \mathbf{C}) - (\mathbf{A} \times \mathbf{B}) - (\mathbf{A} \times \mathbf{C})$$

and note that it is sufficient to prove $\mathbf{D} = \mathbf{0}$. We do this by showing that $\mathbf{D} \cdot \mathbf{D} = |\mathbf{D}|^2 = 0$. We have

$$\mathbf{D} \cdot \mathbf{D} = \mathbf{D} \cdot [\mathbf{A} \times (\mathbf{B} + \mathbf{C}) - (\mathbf{A} \times \mathbf{B}) - (\mathbf{A} \times \mathbf{C})]$$

$$= \mathbf{D} \cdot [\mathbf{A} \times (\mathbf{B} + \mathbf{C})] - \mathbf{D} \cdot (\mathbf{A} \times \mathbf{B}) - \mathbf{D} \cdot (\mathbf{A} \times \mathbf{C})$$

Interchanging the dot and cross in the three box products, we obtain

$$\mathbf{D} \cdot \mathbf{D} = (\mathbf{D} \times \mathbf{A}) \cdot (\mathbf{B} + \mathbf{C}) - (\mathbf{D} \times \mathbf{A}) \cdot \mathbf{B} - (\mathbf{D} \times \mathbf{A}) \cdot \mathbf{C}$$

$$= (\mathbf{D} \times \mathbf{A}) \cdot \mathbf{B} + (\mathbf{D} \times \mathbf{A}) \cdot \mathbf{C} - (\mathbf{D} \times \mathbf{A}) \cdot \mathbf{B} - (\mathbf{D} \times \mathbf{A}) \cdot \mathbf{C} = 0$$

and the proof is complete.

Naturally, the distributive law for the cross product extends to three or more summands; for instance,

$$\mathbf{A} \times (\mathbf{B} + \mathbf{C} + \mathbf{D}) = [\mathbf{A} \times (\mathbf{B} + \mathbf{C})] + (\mathbf{A} \times \mathbf{D})$$

$$= (\mathbf{A} \times \mathbf{B}) + (\mathbf{A} \times \mathbf{C}) + (\mathbf{A} \times \mathbf{D})$$

Also, the distributive law works "from the right"; that is,

$$(\mathbf{A} + \mathbf{B}) \times \mathbf{C} = -[\mathbf{C} \times (\mathbf{A} + \mathbf{B})] = -[(\mathbf{C} \times \mathbf{A}) + (\mathbf{C} \times \mathbf{B})]$$

$$= -(\mathbf{C} \times \mathbf{A}) - (\mathbf{C} \times \mathbf{B}) = (\mathbf{A} \times \mathbf{C}) + (\mathbf{B} \times \mathbf{C})$$

THEOREM 2   **The Cross-Product Formula**

If $\mathbf{A} = a_1\mathbf{i} + a_2\mathbf{j} + a_3\mathbf{k}$ and $\mathbf{B} = b_1\mathbf{i} + b_2\mathbf{j} + b_3\mathbf{k}$, then

$$\mathbf{A} \times \mathbf{B} = (a_2b_3 - a_3b_2)\mathbf{i} - (a_1b_3 - a_3b_1)\mathbf{j} + (a_1b_2 - a_2b_1)\mathbf{k}$$

PROOF   Using the distributive law for the cross product, we have

$$\mathbf{A} \times \mathbf{B} = \mathbf{A} \times (b_1\mathbf{i} + b_2\mathbf{j} + b_3\mathbf{k})$$

$$= [\mathbf{A} \times (b_1\mathbf{i})] + [\mathbf{A} \times (b_2\mathbf{j})] + [\mathbf{A} \times (b_3\mathbf{k})]$$

$$= b_1(\mathbf{A} \times \mathbf{i}) + b_2(\mathbf{A} \times \mathbf{j}) + b_3(\mathbf{A} \times \mathbf{k})$$

Here,

$$\mathbf{A} \times \mathbf{i} = (a_1\mathbf{i} + a_2\mathbf{j} + a_3\mathbf{k}) \times \mathbf{i}$$

$$= [(a_1\mathbf{i}) \times \mathbf{i}] + [(a_2\mathbf{j}) \times \mathbf{i}] + [(a_3\mathbf{k}) \times \mathbf{i}]$$

$$= a_1(\mathbf{i} \times \mathbf{i}) + a_2(\mathbf{j} \times \mathbf{i}) + a_3(\mathbf{k} \times \mathbf{i})$$

$$= a_1\mathbf{0} + a_2(-\mathbf{k}) + a_3\mathbf{j} = a_3\mathbf{j} - a_2\mathbf{k}$$

Similar calculations yield

$$\mathbf{A} \times \mathbf{j} = a_1\mathbf{k} - a_3\mathbf{i} \qquad \text{and} \qquad \mathbf{A} \times \mathbf{k} = a_2\mathbf{i} - a_1\mathbf{j}$$

Therefore, substituting into the original equation, we have

$$\mathbf{A} \times \mathbf{B} = b_1(a_3\mathbf{j} - a_2\mathbf{k}) + b_2(a_1\mathbf{k} - a_3\mathbf{i}) + b_3(a_2\mathbf{i} - a_1\mathbf{j})$$

$$= b_1a_3\mathbf{j} - b_1a_2\mathbf{k} + b_2a_1\mathbf{k} - b_2a_3\mathbf{i} + b_3a_2\mathbf{i} - b_3a_1\mathbf{j}$$

$$= (a_2b_3 - a_3b_2)\mathbf{i} - (a_1b_3 - a_3b_1)\mathbf{j} + (a_1b_2 - a_2b_1)\mathbf{k} \qquad \blacksquare$$

The formula in Theorem 2 is not easily recalled as it stands; however, by using *determinants*, it can be rewritten in a much simpler form. A **determinant of order 2** is defined by

$$\begin{vmatrix} a & b \\ c & d \end{vmatrix} = ad - bc$$

and a **determinant of order 3** is defined by

$$\begin{vmatrix} a & b & c \\ x & y & z \\ u & v & w \end{vmatrix} = a\begin{vmatrix} y & z \\ v & w \end{vmatrix} - b\begin{vmatrix} x & z \\ u & w \end{vmatrix} + c\begin{vmatrix} x & y \\ u & v \end{vmatrix}$$

In what follows, we often find it convenient to use determinant notation in which the first row contains vectors rather than scalars, so that, for instance,

$$\begin{vmatrix} \mathbf{i} & \mathbf{j} & \mathbf{k} \\ x & y & z \\ u & v & w \end{vmatrix} = \begin{vmatrix} y & z \\ v & w \end{vmatrix}\mathbf{i} - \begin{vmatrix} x & z \\ u & w \end{vmatrix}\mathbf{j} + \begin{vmatrix} x & y \\ u & v \end{vmatrix}\mathbf{k}$$

$$= (yw - zv)\mathbf{i} - (xw - zu)\mathbf{j} + (xv - yu)\mathbf{k}$$

Using determinant notation, we can now rewrite the cross-product formula in Theorem 2 in the following easily remembered form. If $\mathbf{A} = a_1\mathbf{i} + a_2\mathbf{j} + a_3\mathbf{k}$ and $\mathbf{B} = b_1\mathbf{i} + b_2\mathbf{j} + b_3\mathbf{k}$, then

$$\mathbf{A} \times \mathbf{B} = \begin{vmatrix} \mathbf{i} & \mathbf{j} & \mathbf{k} \\ a_1 & a_2 & a_3 \\ b_1 & b_2 & b_3 \end{vmatrix}$$

EXAMPLE 2   Evaluate **(a)** $\mathbf{A} \times \mathbf{B}$ and **(b)** $\mathbf{B} \times \mathbf{A}$ if $\mathbf{A} = 2\mathbf{i} + 3\mathbf{j} - 4\mathbf{k}$ and $\mathbf{B} = 5\mathbf{i} - 2\mathbf{j} - \mathbf{k}$.

SOLUTION

(a) $\mathbf{A} \times \mathbf{B} = \begin{vmatrix} \mathbf{i} & \mathbf{j} & \mathbf{k} \\ 2 & 3 & -4 \\ 5 & -2 & -1 \end{vmatrix} = \begin{vmatrix} 3 & -4 \\ -2 & -1 \end{vmatrix} \mathbf{i} - \begin{vmatrix} 2 & -4 \\ 5 & -1 \end{vmatrix} \mathbf{j} + \begin{vmatrix} 2 & 3 \\ 5 & -2 \end{vmatrix} \mathbf{k}$

$= [(3)(-1) - (-4)(-2)]\mathbf{i} - [(2)(-1) - (-4)(5)]\mathbf{j}$
$\qquad\qquad\qquad\qquad\qquad\qquad\qquad + [(2)(-2) - (3)(5)]\mathbf{k}$

$= -11\mathbf{i} - 18\mathbf{j} - 19\mathbf{k}$

(b) Since $\mathbf{B} \times \mathbf{A} = -(\mathbf{A} \times \mathbf{B})$, we have

$$\mathbf{B} \times \mathbf{A} = -(-11\mathbf{i} - 18\mathbf{j} - 19\mathbf{k}) = 11\mathbf{i} + 18\mathbf{j} + 19\mathbf{k}$$ ∎

By combining the cross-product formula with the formula for calculating dot products, we obtain, in the following theorem, a formula for the box product. The proof is left as an exercise (Problem 42).

THEOREM 3

The Box-Product Formula

If $\mathbf{A} = a_1\mathbf{i} + a_2\mathbf{j} + a_3\mathbf{k}$, $\mathbf{B} = b_1\mathbf{i} + b_2\mathbf{j} + b_3\mathbf{k}$, and $\mathbf{C} = c_1\mathbf{i} + c_2\mathbf{j} + c_3\mathbf{k}$, then

$$(\mathbf{A} \times \mathbf{B}) \cdot \mathbf{C} = \begin{vmatrix} a_1 & a_2 & a_3 \\ b_1 & b_2 & b_3 \\ c_1 & c_2 & c_3 \end{vmatrix}$$

In Examples 3 and 4, evaluate $(\mathbf{A} \times \mathbf{B}) \cdot \mathbf{C}$.

EXAMPLE 3 $\mathbf{A} = \mathbf{i} + \mathbf{j}$, $\mathbf{B} = 2\mathbf{i} - 3\mathbf{j} + \mathbf{k}$, and $\mathbf{C} = \mathbf{j} - 4\mathbf{k}$

SOLUTION

$(\mathbf{A} \times \mathbf{B}) \cdot \mathbf{C} = \begin{vmatrix} 1 & 1 & 0 \\ 2 & -3 & 1 \\ 0 & 1 & -4 \end{vmatrix} = 1\begin{vmatrix} -3 & 1 \\ 1 & -4 \end{vmatrix} - 1\begin{vmatrix} 2 & 1 \\ 0 & -4 \end{vmatrix} + 0\begin{vmatrix} 2 & -3 \\ 0 & 1 \end{vmatrix}$

$= 11 - (-8) + 0 = 19$ ∎

EXAMPLE 4 $\mathbf{A} = \langle 1, 0, -1 \rangle$, $\mathbf{B} = \langle 2, 1, 3 \rangle$, and $\mathbf{C} = \langle 5, -1, 0 \rangle$

SOLUTION

$(\mathbf{A} \times \mathbf{B}) \cdot \mathbf{C} = \begin{vmatrix} 1 & 0 & -1 \\ 2 & 1 & 3 \\ 5 & -1 & 0 \end{vmatrix} = 1\begin{vmatrix} 1 & 3 \\ -1 & 0 \end{vmatrix} - 0\begin{vmatrix} 2 & 3 \\ 5 & 0 \end{vmatrix} + (-1)\begin{vmatrix} 2 & 1 \\ 5 & -1 \end{vmatrix}$

$= 3 - 0 + 7 = 10$ ∎

Identities Involving the Cross Product

Using Theorem 2, we can establish a number of useful identities involving the cross product. For instance, the identities in the following theorem can be verified by expanding both sides in terms of the scalar components of the three vectors (Problem 53).

THEOREM 4 **Triple Vector Product Identities**

> For any three vectors $\mathbf{A}$, $\mathbf{B}$, and $\mathbf{C}$ in three-dimensional space,
>
> (i) $\mathbf{A} \times (\mathbf{B} \times \mathbf{C}) = (\mathbf{A} \cdot \mathbf{C})\mathbf{B} - (\mathbf{A} \cdot \mathbf{B})\mathbf{C}$
>
> (ii) $(\mathbf{A} \times \mathbf{B}) \times \mathbf{C} = (\mathbf{A} \cdot \mathbf{C})\mathbf{B} - (\mathbf{B} \cdot \mathbf{C})\mathbf{A}$

One immediate—and important—consequence of the triple vector product identities is that *the associative law does not hold for cross products;* that is, in general,

$$(\mathbf{A} \times \mathbf{B}) \times \mathbf{C} \neq \mathbf{A} \times (\mathbf{B} \times \mathbf{C})$$

EXAMPLE 5 If $\mathbf{A} = 7\mathbf{i} - \mathbf{j} + 2\mathbf{k}$, $\mathbf{B} = -3\mathbf{i} + 2\mathbf{j} - \mathbf{k}$, and $\mathbf{C} = 5\mathbf{j} - 3\mathbf{k}$, find $(\mathbf{A} \times \mathbf{B}) \times \mathbf{C}$ and $\mathbf{A} \times (\mathbf{B} \times \mathbf{C})$.

SOLUTION

$$(\mathbf{A} \times \mathbf{B}) \times \mathbf{C} = (\mathbf{A} \cdot \mathbf{C})\mathbf{B} - (\mathbf{B} \cdot \mathbf{C})\mathbf{A} = (-11)\mathbf{B} - 13\mathbf{A} = -58\mathbf{i} - 9\mathbf{j} - 15\mathbf{k}$$

$$\mathbf{A} \times (\mathbf{B} \times \mathbf{C}) = (\mathbf{A} \cdot \mathbf{C})\mathbf{B} - (\mathbf{A} \cdot \mathbf{B})\mathbf{C} = (-11)\mathbf{B} + 25\mathbf{C} = 33\mathbf{i} + 103\mathbf{j} - 64\mathbf{k} \quad \blacksquare$$

Using the triple vector product identities and the fact that the cross and dot can be exchanged in the box product, we can prove the following theorem.

THEOREM 5 **Lagrange's Identity**

> For any four vectors $\mathbf{A}$, $\mathbf{B}$, $\mathbf{C}$, and $\mathbf{D}$ in three-dimensional space,
>
> $$(\mathbf{A} \times \mathbf{B}) \cdot (\mathbf{C} \times \mathbf{D}) = \begin{vmatrix} \mathbf{A} \cdot \mathbf{C} & \mathbf{A} \cdot \mathbf{D} \\ \mathbf{B} \cdot \mathbf{C} & \mathbf{B} \cdot \mathbf{D} \end{vmatrix}$$

PROOF

$$(\mathbf{A} \times \mathbf{B}) \cdot (\mathbf{C} \times \mathbf{D}) = \mathbf{A} \cdot [\mathbf{B} \times (\mathbf{C} \times \mathbf{D})] = \mathbf{A} \cdot [(\mathbf{B} \cdot \mathbf{D})\mathbf{C} - (\mathbf{B} \cdot \mathbf{C})\mathbf{D}]$$

$$= (\mathbf{B} \cdot \mathbf{D})(\mathbf{A} \cdot \mathbf{C}) - (\mathbf{B} \cdot \mathbf{C})(\mathbf{A} \cdot \mathbf{D}) = \begin{vmatrix} \mathbf{A} \cdot \mathbf{C} & \mathbf{A} \cdot \mathbf{D} \\ \mathbf{B} \cdot \mathbf{C} & \mathbf{B} \cdot \mathbf{D} \end{vmatrix} \quad \blacksquare$$

EXAMPLE 6 Use Lagrange's identity to find a formula for $|\mathbf{A} \times \mathbf{B}|^2$.

SOLUTION

$$|\mathbf{A} \times \mathbf{B}|^2 = (\mathbf{A} \times \mathbf{B}) \cdot (\mathbf{A} \times \mathbf{B}) = \begin{vmatrix} \mathbf{A} \cdot \mathbf{A} & \mathbf{A} \cdot \mathbf{B} \\ \mathbf{B} \cdot \mathbf{A} & \mathbf{B} \cdot \mathbf{B} \end{vmatrix} = (\mathbf{A} \cdot \mathbf{A})(\mathbf{B} \cdot \mathbf{B}) - (\mathbf{A} \cdot \mathbf{B})^2$$

$$= |\mathbf{A}|^2 |\mathbf{B}|^2 - (\mathbf{A} \cdot \mathbf{B})^2 \quad \blacksquare$$

From the identity obtained in the last example, we have the formula

$$|\mathbf{A} \times \mathbf{B}| = \sqrt{|\mathbf{A}|^2 |\mathbf{B}|^2 - (\mathbf{A} \cdot \mathbf{B})^2}$$

for the length of the cross product of $\mathbf{A}$ and $\mathbf{B}$. If θ is the angle between $\mathbf{A}$ and $\mathbf{B}$, $0 \leq \theta \leq \pi$, then

$$|\mathbf{A} \times \mathbf{B}| = \sqrt{|\mathbf{A}|^2 |\mathbf{B}|^2 - (\mathbf{A} \cdot \mathbf{B})^2} = \sqrt{|\mathbf{A}|^2 |\mathbf{B}|^2 - |\mathbf{A}|^2 |\mathbf{B}|^2 \cos^2 \theta}$$

$$= |\mathbf{A}| \, |\mathbf{B}| \, \sqrt{1 - \cos^2 \theta}$$

and it follows that

$$|\mathbf{A} \times \mathbf{B}| = |\mathbf{A}| \, |\mathbf{B}| \sin \theta$$

Geometric Applications of the Cross and Box Products

Many geometric applications of the cross product follow directly from its definition.

EXAMPLE 7   Find the area of the parallelogram spanned by the vectors $\mathbf{A} = \mathbf{i} - 2\mathbf{j} + 3\mathbf{k}$ and $\mathbf{B} = -\mathbf{i} + \mathbf{j} + 2\mathbf{k}$.

SOLUTION   The desired area is numerically equal to the length $|\mathbf{A} \times \mathbf{B}|$ of the cross product $\mathbf{A} \times \mathbf{B}$. Using the formula obtained above, we have

$$|\mathbf{A} \times \mathbf{B}| = \sqrt{|\mathbf{A}|^2|\mathbf{B}|^2 - (\mathbf{A} \cdot \mathbf{B})^2} = \sqrt{(14)(6) - 3^2} = 5\sqrt{3} \text{ square units} \quad \blacksquare$$

Naturally, the box product is useful for computing the volume V of the parallelepiped spanned by the vectors $\mathbf{A}$, $\mathbf{B}$, and $\mathbf{C}$; in fact,

$$V = |(\mathbf{A} \times \mathbf{B}) \cdot \mathbf{C}|$$

EXAMPLE 8   Find the volume V of the parallelepiped spanned by the vectors $\mathbf{A} = 3\mathbf{i} + \mathbf{j}$, $\mathbf{B} = \mathbf{i} + 2\mathbf{k}$, and $\mathbf{C} = \mathbf{i} + 2\mathbf{j} + 3\mathbf{k}$.

SOLUTION   Here,

$$(\mathbf{A} \times \mathbf{B}) \cdot \mathbf{C} = \begin{vmatrix} 3 & 1 & 0 \\ 1 & 0 & 2 \\ 1 & 2 & 3 \end{vmatrix} = 3\begin{vmatrix} 0 & 2 \\ 2 & 3 \end{vmatrix} - 1\begin{vmatrix} 1 & 2 \\ 1 & 3 \end{vmatrix} + 0\begin{vmatrix} 1 & 0 \\ 1 & 2 \end{vmatrix}$$

$$= 3(-4) - 1(1) + 0 = -13$$

Hence, $V = |-13| = 13$ cubic units. $\quad \blacksquare$

The algebraic sign of the box product can be used to determine whether a triple of vectors is right-handed or left-handed. In fact, $\mathbf{A}$, $\mathbf{B}$, $\mathbf{C}$ is a right-handed or left-handed triple according to whether $(\mathbf{A} \times \mathbf{B}) \cdot \mathbf{C}$ is positive or negative, respectively (Problem 48b). Also, $(\mathbf{A} \times \mathbf{B}) \cdot \mathbf{C} = 0$ if and only if the vectors $\mathbf{A}$, $\mathbf{B}$, and $\mathbf{C}$ are coplanar—that is, linearly dependent (Problem 48a).

EXAMPLE 9   Determine whether $\mathbf{A}$, $\mathbf{B}$, $\mathbf{C}$ is a right-handed or left-handed triple if $\mathbf{A} = -\mathbf{i} + 7\mathbf{j} + 2\mathbf{k}$, $\mathbf{B} = 2\mathbf{i} + 3\mathbf{j} + \mathbf{k}$, and $\mathbf{C} = -\mathbf{i} + 5\mathbf{j} + 2\mathbf{k}$.

SOLUTION

$$(\mathbf{A} \times \mathbf{B}) \cdot \mathbf{C} = \begin{vmatrix} -1 & 7 & 2 \\ 2 & 3 & 1 \\ -1 & 5 & 2 \end{vmatrix} = (-1)\begin{vmatrix} 3 & 1 \\ 5 & 2 \end{vmatrix} - 7\begin{vmatrix} 2 & 1 \\ -1 & 2 \end{vmatrix} + 2\begin{vmatrix} 2 & 3 \\ -1 & 5 \end{vmatrix} = -10 < 0$$

Hence, $\mathbf{A}$, $\mathbf{B}$, $\mathbf{C}$ is a left-handed triple. $\quad \blacksquare$

EXAMPLE 10   Show that the three vectors $\mathbf{A} = \langle 1, -1, 1 \rangle$, $\mathbf{B} = \langle 2, 1, -1 \rangle$, and $\mathbf{C} = \langle 0, -1, 1 \rangle$ are coplanar (linearly dependent).

SOLUTION

$$(\mathbf{A} \times \mathbf{B}) \cdot \mathbf{C} = \begin{vmatrix} 1 & -1 & 1 \\ 2 & 1 & -1 \\ 0 & -1 & 1 \end{vmatrix}$$

$$= 1\begin{vmatrix} 1 & -1 \\ -1 & 1 \end{vmatrix} - (-1)\begin{vmatrix} 2 & -1 \\ 0 & 1 \end{vmatrix} + 1\begin{vmatrix} 2 & 1 \\ 0 & -1 \end{vmatrix} = 0$$

Hence, $\mathbf{A}$, $\mathbf{B}$, and $\mathbf{C}$ are coplanar. $\quad \blacksquare$

Many geometric objects can be built up from nonoverlapping triangles; hence, the following theorem, which gives a formula for the area of a triangle in terms of the coordinates of its vertices, can be quite useful.

THEOREM 6   **Area of a Triangle in Space**

Let PQR be a triangle in xyz space, where $P = (x_1, y_1, z_1)$, $Q = (x_2, y_2, z_2)$, and $R = (x_3, y_3, z_3)$. Then the area a of PQR is given by the formula

$$a = \tfrac{1}{2} |\overrightarrow{PQ} \times \overrightarrow{PR}|$$

PROOF   The area of triangle PQR is half the area of the parallelogram spanned by the vectors $\overrightarrow{PQ}$ and $\overrightarrow{PR}$ (Figure 7).   ∎

Figure 7

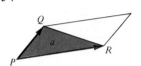

EXAMPLE 11   Find the area a of the triangle whose vertices are $P = (2, 1, 5)$, $Q = (4, 0, 2)$, and $R = (-1, 0, -1)$.

SOLUTION   Here,

$$\overrightarrow{PQ} = (4 - 2)\mathbf{i} + (0 - 1)\mathbf{j} + (2 - 5)\mathbf{k} = 2\mathbf{i} - \mathbf{j} - 3\mathbf{k}$$

and

$$\overrightarrow{PR} = (-1 - 2)\mathbf{i} + (0 - 1)\mathbf{j} + (-1 - 5)\mathbf{k} = -3\mathbf{i} - \mathbf{j} - 6\mathbf{k}$$

Hence,

$$\overrightarrow{PQ} \times \overrightarrow{PR} = \begin{vmatrix} \mathbf{i} & \mathbf{j} & \mathbf{k} \\ 2 & -1 & -3 \\ -3 & -1 & -6 \end{vmatrix} = 3\mathbf{i} + 21\mathbf{j} - 5\mathbf{k}$$

Therefore,

$$a = \frac{1}{2}|\overrightarrow{PQ} \times \overrightarrow{PR}| = \frac{1}{2}\sqrt{3^2 + (21)^2 + (-5)^2} = \frac{\sqrt{475}}{2} = \frac{5}{2}\sqrt{19} \text{ square units}$$ ∎

Problem Set 13.3

1 Determine geometrically whether each triple of vectors is right-handed or left-handed.

(a) $\mathbf{i}, -\mathbf{j}, \mathbf{k}$ (b) $\mathbf{i}, -\mathbf{j}, -\mathbf{k}$ (c) $\mathbf{j}, \mathbf{k}, \mathbf{i}$

(d) $\mathbf{j}, -\mathbf{i}, -\mathbf{k}$ (e) $-\mathbf{i}, -\mathbf{j}, -\mathbf{k}$ (f) $\mathbf{i} + \mathbf{j}, \mathbf{j} - \mathbf{i}, \mathbf{k}$

2 Suppose that $\mathbf{A}, \mathbf{B}, \mathbf{C}$ is a left-handed triple. Determine whether each triple is right- or left-handed.

(a) $\mathbf{B}, \mathbf{A}, \mathbf{C}$ (b) $-\mathbf{A}, \mathbf{B}, -\mathbf{C}$

(c) $\mathbf{C}, \mathbf{A}, \mathbf{B}$ (d) $\mathbf{C}, \mathbf{B}, \mathbf{A}$

3 Find each cross product by sketching an appropriate figure and using only geometric facts. (Do not use the memory device of Figure 5 or the cross-product formula.)

(a) $\mathbf{i} \times \mathbf{k}$ (b) $(-\mathbf{i}) \times \mathbf{j}$

(c) $\mathbf{k} \times \mathbf{i}$ (d) $(\mathbf{i} + \mathbf{j}) \times (\mathbf{j} - \mathbf{i})$

(e) $(-\mathbf{i}) \times (-\mathbf{j})$ (f) $(\mathbf{i} + \mathbf{j}) \times \mathbf{k}$

4 Using only the memory device of Figure 5, calculate (a) $(\mathbf{i} \times \mathbf{j}) \times \mathbf{j}$ and (b) $\mathbf{i} \times (\mathbf{j} \times \mathbf{j})$. Conclude from your result that the associative law does not work for the cross product.

In Problems 5 to 34, let $\mathbf{A} = \mathbf{i} + \mathbf{j} + \mathbf{k}$, $\mathbf{B} = -2\mathbf{j}$, $\mathbf{C} = \mathbf{k} - 3\mathbf{j}$, $\mathbf{D} = 3\mathbf{i} - 2\mathbf{j} + \mathbf{k}$, $\mathbf{E} = -3\mathbf{k}$, $\mathbf{F} = \langle -1, 1, -4 \rangle$, $\mathbf{G} = \langle 0, -1, 1 \rangle$, and $\mathbf{H} = \langle 1, -1, -2 \rangle$. Evaluate each expression using the identities for cross and dot products whenever possible to simplify your calculations.

5 $\mathbf{A} \times \mathbf{B}$ **6** $\mathbf{C} \times \mathbf{B}$

7 $\mathbf{C} \times \mathbf{D}$ **8** $\mathbf{F} \times \mathbf{H}$

9 $\mathbf{E} \times \mathbf{B}$ **10** $\mathbf{D} \times \mathbf{A}$

11 $\mathbf{F} \times \mathbf{G}$ **12** $\mathbf{D} \times \mathbf{C} + \mathbf{C} \times \mathbf{D}$

13 $\mathbf{F} \times \mathbf{G} - \mathbf{G} \times \mathbf{F}$ **14** $\mathbf{G} \times \mathbf{H} - \mathbf{G} \times \mathbf{F}$

15 $\mathbf{B} \times \mathbf{D} + \mathbf{B} \times \mathbf{E}$

16 $\mathbf{C} \times \mathbf{A} - \mathbf{E} \times \mathbf{C} + \mathbf{C} \times \mathbf{F}$

17 $(-2\mathbf{A}) \times (3\mathbf{B})$ **18** $(\mathbf{A} + \mathbf{E}) \times (\mathbf{A} - \mathbf{E})$

19 $(\mathbf{A} \times \mathbf{B}) \cdot \mathbf{C}$ **20** $\mathbf{C} \cdot (\mathbf{D} \times \mathbf{E})$

21 $\mathbf{F} \cdot (\mathbf{G} \times \mathbf{H})$ **22** $\mathbf{F} \cdot (\mathbf{F} \times \mathbf{H})$

23 $(\mathbf{B} \times \mathbf{C}) \cdot \mathbf{E}$

24 $\mathbf{H} \cdot (\mathbf{F} \times \mathbf{H})$

25 $(\mathbf{A} \times \mathbf{A}) \cdot \mathbf{C}$

26 $(\mathbf{A} \times \mathbf{C}) \cdot (\mathbf{A} \times \mathbf{C})$

27 $\mathbf{A} \times (\mathbf{B} \times \mathbf{D})$

28 $(\mathbf{F} \times \mathbf{G}) \times \mathbf{H}$

29 $(\mathbf{A} \times \mathbf{B}) \times \mathbf{D}$

30 $(\mathbf{A} \times \mathbf{B}) \cdot (\mathbf{C} \times \mathbf{D})$

31 $(\mathbf{A} \times \mathbf{D}) \cdot (\mathbf{A} \times \mathbf{C})$

32 $(\mathbf{A} \times \mathbf{B}) \times (\mathbf{B} \times \mathbf{A})$

33 $|\mathbf{A} \times \mathbf{D}|$

34 $|\mathbf{F} \times \mathbf{H}|^2 - |\mathbf{F}|^2 |\mathbf{H}|^2$

35 Find the area of the parallelogram spanned by $\mathbf{A}$ and $\mathbf{B}$:

(a) $\mathbf{A} = \mathbf{i} + \mathbf{j} + \mathbf{k}$, $\mathbf{B} = \mathbf{i} - \mathbf{j} - \mathbf{k}$

(b) $\mathbf{A} = \langle 2, 5, -1 \rangle$, $\mathbf{B} = \langle \frac{1}{2}, -1, \frac{1}{4} \rangle$

36 Suppose that $(\mathbf{A} \times \mathbf{B}) \cdot \mathbf{C} = 4$. Find

(a) $\mathbf{A} \cdot (\mathbf{B} \times \mathbf{C})$

(b) $\mathbf{B} \cdot (\mathbf{C} \times \mathbf{A})$

(c) $\mathbf{B} \cdot (\mathbf{A} \times \mathbf{C})$

(d) $\mathbf{C} \cdot (\mathbf{A} \times \mathbf{B})$

37 Find the volume of the box (parallelepiped) spanned by $\mathbf{A}$, $\mathbf{B}$, and $\mathbf{C}$:

(a) $\mathbf{A} = \mathbf{i} + 2\mathbf{j}$, $\mathbf{B} = 2\mathbf{i} - \mathbf{j}$, $\mathbf{C} = \mathbf{i} + \mathbf{j} + \mathbf{k}$

(b) $\mathbf{A} = \mathbf{i} - 2\mathbf{j} + \mathbf{k}$, $\mathbf{B} = 3\mathbf{i} - 4\mathbf{j} - \mathbf{k}$, $\mathbf{C} = 4\mathbf{i} - \mathbf{j} + 3\mathbf{k}$

(c) $\mathbf{A} = \langle 1, 2, -1 \rangle$, $\mathbf{B} = \langle 0, 1, -1 \rangle$, $\mathbf{C} = \langle 3, 0, 1 \rangle$

38 If you grab the vector $\mathbf{A} \times \mathbf{B}$ with your right hand so that the fingers curl in the direction of the angle from $\mathbf{A}$ to $\mathbf{B}$ and your thumb lies along the shaft of the vector $\mathbf{A} \times \mathbf{B}$ (pointing away from your fist), will your thumb point in the same direction as the vector $\mathbf{A} \times \mathbf{B}$ or will it point in the opposite direction?

39 Determine whether the three vectors $\mathbf{A}$, $\mathbf{B}$, and $\mathbf{C}$ are coplanar or linearly independent. If the vectors are linearly independent, determine whether the triple $\mathbf{A}$, $\mathbf{B}$, $\mathbf{C}$ is right- or left-handed.

(a) $\mathbf{A} = \mathbf{i} + 2\mathbf{j} + 3\mathbf{k}$, $\mathbf{B} = 2\mathbf{i} - \mathbf{j} - \mathbf{k}$, $\mathbf{C} = -2\mathbf{i} + \mathbf{j} - \mathbf{k}$

(b) $\mathbf{A} = 2\mathbf{i} - \mathbf{j} + 7\mathbf{k}$, $\mathbf{B} = \mathbf{j} + 3\mathbf{k}$, $\mathbf{C} = 2\mathbf{i} - 3\mathbf{j} + \mathbf{k}$

(c) $\mathbf{A} = \langle 1, 1, -1 \rangle$, $\mathbf{B} = \langle 1, -1, 1 \rangle$, $\mathbf{C} = \langle -1, 1, 1 \rangle$

40 Suppose that $\mathbf{A}$ is perpendicular to $\mathbf{B}$. (a) Explain geometrically why $|\mathbf{A} \times \mathbf{B}| = |\mathbf{A}| \, |\mathbf{B}|$. (b) Prove algebraically that $|\mathbf{A} \times \mathbf{B}| = |\mathbf{A}| \, |\mathbf{B}|$.

41 Find the area a of triangle PQR in xyz space:

(a) $P = (1, 7, 2)$, $Q = (0, 7, -2)$, $R = (-1, 6, 1)$

(b) $P = (-1, 0, 1)$, $Q = (0, 1, 0)$, $R = (1, 1, 1)$

42 Prove Theorem 3.

43 Show that $(s\mathbf{A}) \times \mathbf{B} = s(\mathbf{A} \times \mathbf{B})$ even if the scalar s is not positive. (Make direct use of the definition of the cross product.)

44 Prove that $\mathbf{A} \times (t\mathbf{B}) = t(\mathbf{A} \times \mathbf{B})$.

45 Prove that $(\mathbf{A} - \mathbf{B}) \times \mathbf{C} = (\mathbf{A} \times \mathbf{C}) - (\mathbf{B} \times \mathbf{C})$. [*Hint:* $\mathbf{A} - \mathbf{B} = \mathbf{A} + (-1)\mathbf{B}$.]

46 Prove the following identity: $(\mathbf{A} - \mathbf{B}) \times (\mathbf{C} - \mathbf{D}) = (\mathbf{A} \times \mathbf{C}) - (\mathbf{A} \times \mathbf{D}) - (\mathbf{B} \times \mathbf{C}) + (\mathbf{B} \times \mathbf{D})$.

47 True or false: $(\mathbf{A} + \mathbf{B}) \times (\mathbf{A} - \mathbf{B}) = (\mathbf{A} \times \mathbf{A}) - (\mathbf{B} \times \mathbf{B})$?

48 (a) Show that $\mathbf{A}$, $\mathbf{B}$, and $\mathbf{C}$ are coplanar (linearly dependent) if and only if $(\mathbf{A} \times \mathbf{B}) \cdot \mathbf{C} = 0$.

(b) If $\mathbf{A}$, $\mathbf{B}$, and $\mathbf{C}$ are linearly independent, show that $(\mathbf{A} \times \mathbf{B}) \cdot \mathbf{C}$ is positive or negative according to whether $\mathbf{A}, \mathbf{B}, \mathbf{C}$ forms a right- or left-handed triple, respectively.

49 Let PQR be a triangle in the xy plane with $P = (x_1, y_1)$, $Q = (x_2, y_2)$, and $R = (x_3, y_3)$. Show that *the area a of PQR is given by*

$$a = \text{absolute value of } \frac{1}{2} \begin{vmatrix} x_1 & y_1 & 1 \\ x_2 & y_2 & 1 \\ x_3 & y_3 & 1 \end{vmatrix}$$

50 Using Problem 49, explain why the equation

$$\begin{vmatrix} x & y & 1 \\ x_1 & y_1 & 1 \\ x_2 & y_2 & 1 \end{vmatrix} = 0$$

represents the line in the xy plane containing the two points (x_1, y_1) and (x_2, y_2).

51 Find the area of triangle PQR in the xy plane:

(a) $P = (1, -5)$, $Q = (-3, -4)$, $R = (6, 2)$

(b) $P = (-5, 7)$, $Q = (3, 6)$, $R = (2, 1)$

52 (a) Show that the triangle in the xy plane two of whose adjacent edges are formed by the vectors $\mathbf{A} = x_1\mathbf{i} + y_1\mathbf{j}$ and $\mathbf{B} = x_2\mathbf{i} + y_2\mathbf{j}$ (Figure 8) has an area given by the absolute value of

$$\frac{1}{2} \begin{vmatrix} x_1 & y_1 \\ x_2 & y_2 \end{vmatrix}$$

(b) Give a geometric interpretation of the determinant

$$\begin{vmatrix} x_1 & y_1 \\ x_2 & y_2 \end{vmatrix}$$

Figure 8

53 Prove Theorem 4. (For simplicity, choose the x axis in the direction of the vector $\mathbf{C}$, and choose the y axis so that the vector $\mathbf{B}$ is contained in the xy plane.)

54 From part (i) of Theorem 4, it follows that the three vectors $\mathbf{B}$, $\mathbf{C}$, and $\mathbf{A} \times (\mathbf{B} \times \mathbf{C})$ are always coplanar. Explain *geometrically* why this should be so.

55 Let $\mathbf{A} = \overrightarrow{OP}$, $\mathbf{B} = \overrightarrow{OQ}$, and $\mathbf{C} = \overrightarrow{OR}$. Prove that the area of triangle PQR is given by $\frac{1}{2}|(\mathbf{A} \times \mathbf{B}) + (\mathbf{B} \times \mathbf{C}) + (\mathbf{C} \times \mathbf{A})|$.

56 Use the cross product to prove the law of sines; that is, for any triangle ABC with $a = |\overline{BC}|$, $b = |\overline{AC}|$, and $c = |\overline{AB}|$

$$\frac{\sin A}{a} = \frac{\sin B}{b} = \frac{\sin C}{c}$$

57 Prove that

$$(A - B) \times (A + B) = 2(A \times B)$$

58 Prove the **Jacobi identity:**

$$A \times (B \times C) + B \times (C \times A) + C \times (A \times B) = 0$$

59 If **u** is a unit vector perpendicular to **A**, show that

$$u \times (A \times u) = A$$

60 If **u** is a unit vector, show that

$$[(A \times u) \times (B \times u)] \cdot u = (A \times B) \cdot u$$

13.4 Equations of Lines and Planes in Space

In this section we make use of the vector algebra developed in Sections 13.2 and 13.3 to derive vector equations for lines and planes in space. By considering the scalar components of the vectors in these equations, we also obtain scalar Cartesian equations for lines and planes in space.

As before, $R = \overrightarrow{OP}$ denotes the variable position vector of the point P, and the graph of an equation involving the vector R is the set of all points P (now in three-dimensional space) whose position vector R satisfies the equation.

Figure 1

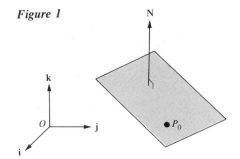

Planes in Space

One of the simplest ways to specify a particular plane in space is to give a point P_0 contained in the plane and to give a nonzero vector **N** perpendicular to the plane. Figure 1 shows a portion of such a plane. The vector **N** is called a **normal** vector to the plane. *Two planes with the same normal vector* **N** *are parallel to each other;* hence, all the planes that have the same normal vector **N** form a "stack" of mutually parallel planes (Figure 2). Specification of the point P_0 chooses from this stack the one and only plane that contains P_0.

It is geometrically clear (Figure 3) that a point P belongs to the plane containing the point P_0 and having the normal vector **N** if and only if the vector $\overrightarrow{P_0P}$ is perpendicular to **N**; that is, $(\overrightarrow{P_0P}) \cdot N = 0$. If R is the position vector of P and R_0 is the position vector of P_0, then $\overrightarrow{P_0P} = R - R_0$; hence, the condition that $\overrightarrow{P_0P}$ is perpendicular to **N** can be written

Figure 2

$$\boxed{N \cdot (R - R_0) = 0}$$

This is a **vector equation** *of the plane with normal vector* **N** *and containing the point whose position vector is* R_0.

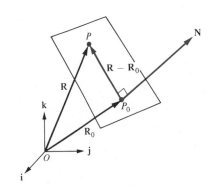

Figure 3

Parallel planes formed by venetian blinds

The equation $\mathbf{N} \cdot (\mathbf{R} - \mathbf{R}_0) = 0$ is easily converted to **scalar Cartesian form** by putting

$$\mathbf{N} = a\mathbf{i} + b\mathbf{j} + c\mathbf{k} \qquad \mathbf{R} = x\mathbf{i} + y\mathbf{j} + z\mathbf{k} \qquad \text{and} \qquad \mathbf{R}_0 = x_0\mathbf{i} + y_0\mathbf{j} + z_0\mathbf{k}$$

Then, $\qquad\qquad \mathbf{R} - \mathbf{R}_0 = (x - x_0)\mathbf{i} + (y - y_0)\mathbf{j} + (z - z_0)\mathbf{k}$

$$\mathbf{N} \cdot (\mathbf{R} - \mathbf{R}_0) = a(x - x_0) + b(y - y_0) + c(z - z_0)$$

and the equation of the plane takes the form

$$\boxed{a(x - x_0) + b(y - y_0) + c(z - z_0) = 0}$$

EXAMPLE 1   Find the scalar Cartesian equation of the plane containing the point $P_0 = (-3, 1, 7)$ and having $\mathbf{N} = 2\mathbf{i} + 3\mathbf{j} - \mathbf{k}$ as a normal vector.

SOLUTION   Here we have $x_0 = -3$, $y_0 = 1$, and $z_0 = 7$. Also, $a = 2$, $b = 3$, and $c = -1$. Therefore, the equation $a(x - x_0) + b(y - y_0) + c(z - z_0) = 0$ becomes

$$2(x + 3) + 3(y - 1) + (-1)(z - 7) = 0 \qquad \text{or} \qquad 2x + 3y - z + 10 = 0 \qquad \blacksquare$$

Given a scalar Cartesian equation of a plane in xyz space,

$$a(x - x_0) + b(y - y_0) + c(z - z_0) = 0$$

we can always carry out the indicated multiplications and thus rewrite the equation in the form

$$ax + by + cz - (ax_0 + by_0 + cz_0) = 0 \qquad \text{or} \qquad ax + by + cz = D$$

where D is a constant equal to $ax_0 + by_0 + cz_0$. In the scalar Cartesian equation

$$\boxed{ax + by + cz = D}$$

for a plane in xyz space, the coefficients a, b, and c—which cannot all be zero—form the scalar components of a normal vector

$$\boxed{\mathbf{N} = a\mathbf{i} + b\mathbf{j} + c\mathbf{k}}$$

As the constant D is assigned different values, we obtain different planes in space, all of which have the same normal vector and therefore are mutually parallel. The value assigned to the constant D determines exactly one plane in this stack of mutually parallel planes.

EXAMPLE 2   Find a scalar Cartesian equation of the plane that contains the point $(1, 2, 3)$ and is parallel to the plane $3x - y - 2z = 14$.

SOLUTION   The desired plane has an equation of the form $3x - y - 2z = D$, where the constant D has to be determined. Since the point $(1, 2, 3)$ belongs to the plane, we have $3(1) - 2 - 2(3) = D$, so $D = -5$. Therefore, an equation of the plane is

$$3x - y - 2z = -5 \qquad \text{or} \qquad 3x - y - 2z + 5 = 0 \qquad \blacksquare$$

A plane $ax + by + cz = D$ that is not parallel to any of the three coordinate axes must intersect each of these axes (Figure 4). The x, y, and z coordinates of the three points of intersection are called the x, y, and z **intercepts** of the plane, respectively.

Figure 4

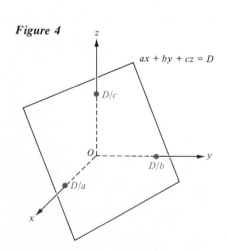

For instance, the point $(x, 0, 0)$ where the plane intersects the x axis satisfies $ax + b(0) + c(0) = D$, or $ax = D$. Hence, the x intercept is given by $x = D/a$. (Note that $a \neq 0$; otherwise, the plane is parallel to the x axis.) Similarly, the y and z intercepts are given by $y = D/b$ and $z = D/c$, respectively. Since there is exactly one plane containing three noncollinear points, the three intercepts determine the plane uniquely.

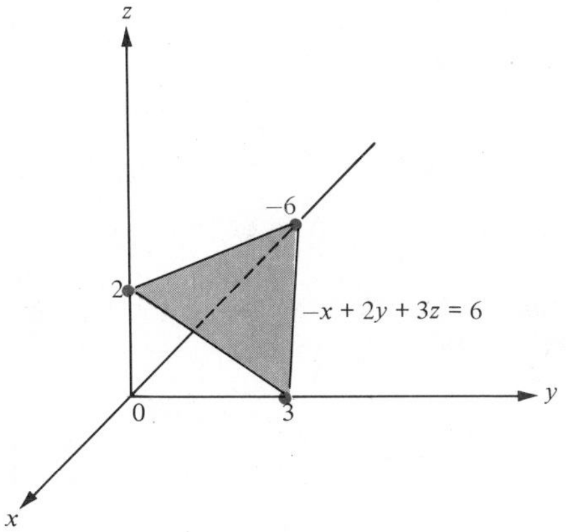

EXAMPLE 3 Consider the plane $-x + 2y + 3z = 6$.

(a) Find a unit normal vector to the plane.

(b) Find the intercepts of the plane.

(c) Sketch a portion of the plane.

SOLUTION

(a) We can take $\mathbf{N} = -\mathbf{i} + 2\mathbf{j} + 3\mathbf{k}$, so that $|\mathbf{N}| = \sqrt{(-1)^2 + 2^2 + 3^2} = \sqrt{14}$. Therefore, a unit normal vector to the plane is given by

$$\frac{\mathbf{N}}{|\mathbf{N}|} = \frac{-\mathbf{i} + 2\mathbf{j} + 3\mathbf{k}}{\sqrt{14}}$$

(b) The x, y, and z intercepts are given by $x = 6/(-1) = -6$, $y = 6/2 = 3$, and $z = 6/3 = 2$, respectively.

(c) The shaded triangle in Figure 5 is a portion of the plane. ■

In general, to find an equation of the plane containing three given noncollinear points $A = (x_0, y_0, z_0)$, $B = (x_1, y_1, z_1)$, and $C = (x_2, y_2, z_2)$, notice that the vectors $\overrightarrow{AB}$ and $\overrightarrow{AC}$ are parallel to the plane; hence,

$$\mathbf{N} = \overrightarrow{AB} \times \overrightarrow{AC}$$

gives a normal vector to the plane (Figure 6). Since (x_0, y_0, z_0) lies on the plane, we can find either a vector equation or a scalar Cartesian equation of the plane just as before.

In Examples 4 and 5, find an equation in scalar Cartesian form of the plane satisfying the given conditions.

EXAMPLE 4 Containing the three points $A = (1, 1, -1)$, $B = (3, 3, 2)$, and $C = (3, -1, -2)$

SOLUTION   Here,

$$\overrightarrow{AB} = (3 - 1)\mathbf{i} + (3 - 1)\mathbf{j} + (2 + 1)\mathbf{k} = 2\mathbf{i} + 2\mathbf{j} + 3\mathbf{k}$$

and $\overrightarrow{AC} = (3 - 1)\mathbf{i} + (-1 - 1)\mathbf{j} + (-2 + 1)\mathbf{k} = 2\mathbf{i} - 2\mathbf{j} - \mathbf{k}$

Hence,

$$\mathbf{N} = \overrightarrow{AB} \times \overrightarrow{AC} = \begin{vmatrix} \mathbf{i} & \mathbf{j} & \mathbf{k} \\ 2 & 2 & 3 \\ 2 & -2 & -1 \end{vmatrix} = 4\mathbf{i} + 8\mathbf{j} - 8\mathbf{k}$$

gives a normal vector to the plane. Since the point $(1, 1, -1)$ belongs to the plane, a scalar Cartesian equation is

$$4(x - 1) + 8(y - 1) - 8(z + 1) = 0$$

This equation can also be written in the form

$$4x + 8y - 8z = 20 \qquad \text{or} \qquad x + 2y - 2z = 5$$ ■

Figure 5

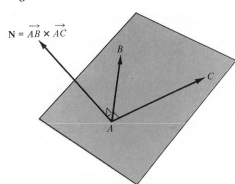

$-x + 2y + 3z = 6$

Figure 6

$\mathbf{N} = \overrightarrow{AB} \times \overrightarrow{AC}$

EXAMPLE 5   Containing the points (3, 2, 1) and (−5, 1, 2), but not intersecting the y axis

SOLUTION   Because the plane is parallel to the y axis, its equation has the form $ax + cz = D$. (Why?) Therefore, since the point (3, 2, 1) satisfies the equation, so does the point (3, 0, 1). Letting $A = (3, 2, 1)$, $B = (-5, 1, 2)$, and $C = (3, 0, 1)$, we have $\overrightarrow{AB} = -8\mathbf{i} - \mathbf{j} + \mathbf{k}$ and $\overrightarrow{AC} = -2\mathbf{j}$, so that a normal vector $\mathbf{N}$ to the plane is given by

$$\mathbf{N} = \overrightarrow{AB} \times \overrightarrow{AC} = \begin{vmatrix} \mathbf{i} & \mathbf{j} & \mathbf{k} \\ -8 & -1 & 1 \\ 0 & -2 & 0 \end{vmatrix} = 2\mathbf{i} + 16\mathbf{k}$$

Therefore, an equation of the plane is

$$2(x - 3) + 0(y - 2) + 16(z - 1) = 0 \qquad \text{or} \qquad x + 8z = 11 \qquad \blacksquare$$

An equation of a plane can also be put into vector or scalar *parametric* form; since a plane is two-dimensional, *two* independent parameters are required. We do not take up this matter here.

Lines in Space

Figure 7

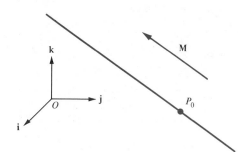

A line in space can be specified by giving a nonzero vector $\mathbf{M}$ parallel to the line and a point $P_0 = (x_0, y_0, z_0)$ on the line (Figure 7). The vector $\mathbf{M}$ is called a **direction vector** for the line. Two lines with the same direction vector are parallel to each other; hence all the lines that have the same direction vector $\mathbf{M}$ form a "bundle" of mutually parallel lines. Specification of the point P_0 chooses from this bundle the one and only line that contains P_0 (Figure 8).

It is geometrically clear (Figure 9) that a point P belongs to the line containing the point P_0 and having the direction vector $\mathbf{M}$ if and only if the vector $\overrightarrow{P_0P}$ is parallel to $\mathbf{M}$, that is,

$$\mathbf{M} \times \overrightarrow{P_0P} = \mathbf{0}$$

(Recall that two vectors are parallel if and only if their cross product is the zero vector.) If $\mathbf{R}$ is the position vector of P and $\mathbf{R}_0$ is the position vector of P_0, then $\overrightarrow{P_0P} = \mathbf{R} - \mathbf{R}_0$; hence, the condition that $\mathbf{M}$ be parallel to $\overrightarrow{P_0P}$ can be written

Figure 8

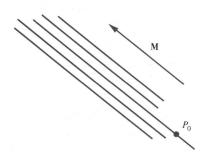

$$\boxed{\mathbf{M} \times (\mathbf{R} - \mathbf{R}_0) = \mathbf{0}}$$

This is a **vector nonparametric equation** *of the line with direction vector* $\mathbf{M}$ *and containing the point whose position vector is* $\mathbf{R}_0$.

To convert the vector equation $\mathbf{M} \times (\mathbf{R} - \mathbf{R}_0) = \mathbf{0}$ to scalar Cartesian form, let

$$\mathbf{M} = a\mathbf{i} + b\mathbf{j} + c\mathbf{k} \qquad \mathbf{R} = x\mathbf{i} + y\mathbf{j} + z\mathbf{k} \qquad \text{and} \qquad \mathbf{R}_0 = x_0\mathbf{i} + y_0\mathbf{j} + z_0\mathbf{k}$$

Then $\mathbf{M} \times (\mathbf{R} - \mathbf{R}_0) = \mathbf{0}$ can be rewritten as

Figure 9

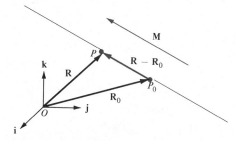

$$\begin{vmatrix} \mathbf{i} & \mathbf{j} & \mathbf{k} \\ a & b & c \\ x - x_0 & y - y_0 & z - z_0 \end{vmatrix} = 0\mathbf{i} + 0\mathbf{j} + 0\mathbf{k}$$

that is,

$$[b(z - z_0) - c(y - y_0)]\mathbf{i} - [a(z - z_0) - c(x - x_0)]\mathbf{j}$$
$$+ [a(y - y_0) - b(x - x_0)]\mathbf{k} = 0\mathbf{i} + 0\mathbf{j} + 0\mathbf{k}$$

Equating the three scalar components, we obtain three simultaneous scalar equations

$$b(z - z_0) - c(y - y_0) = 0 \qquad a(z - z_0) - c(x - x_0) = 0$$

$$a(y - y_0) - b(x - x_0) = 0$$

or

$$b(z - z_0) = c(y - y_0) \qquad a(z - z_0) = c(x - x_0) \qquad a(y - y_0) = b(x - x_0)$$

These three simultaneous equations give a scalar Cartesian nonparametric form for the line.

If the coefficients a, b, and c in the equations above are nonzero, we can rewrite the equations in the form

$$\frac{z - z_0}{c} = \frac{y - y_0}{b} \qquad \frac{z - z_0}{c} = \frac{x - x_0}{a} \qquad \frac{y - y_0}{b} = \frac{x - x_0}{a}$$

or

$$\frac{x - x_0}{a} = \frac{y - y_0}{b} = \frac{z - z_0}{c}$$

These simultaneous scalar nonparametric equations are referred to as **symmetric equations** of the line. Here, (x_0, y_0, z_0) is a point on the line, and the denominators a, b, and c are the scalar components of a vector $\mathbf{M}$ parallel to the line. If any one of the denominators in the symmetric equations of the line is zero, the understanding is that the corresponding numerator is also zero (Problem 50). For instance, if $a = 0$ but $b \neq 0$ and $c \neq 0$, the symmetric equations would be written

$$x - x_0 = 0 \qquad \frac{y - y_0}{b} = \frac{z - z_0}{c}$$

EXAMPLE 6 Find symmetric equations of the line containing the point $P_0 = (x_0, y_0, z_0) = (3, 1, -1)$ and parallel to the vector $\mathbf{M} = 5\mathbf{i} - 7\mathbf{j} + 9\mathbf{k}$.

SOLUTION   Symmetric equations of the line are given by

$$\frac{x - 3}{5} = \frac{y - 1}{-7} = \frac{z + 1}{9}$$

∎

EXAMPLE 7 Write an equation of the line in Example 6 in vector nonparametric form.

SOLUTION   An equation of the line in vector nonparametric form is given by

$$\mathbf{M} \times (\mathbf{R} - \mathbf{R}_0) = \mathbf{0} \qquad \text{where } \mathbf{M} = 5\mathbf{i} - 7\mathbf{j} + 9\mathbf{k} \qquad \text{and} \qquad \mathbf{R}_0 = 3\mathbf{i} + \mathbf{j} - \mathbf{k}$$

Therefore, an equation of the line is

$$(5\mathbf{i} - 7\mathbf{j} + 9\mathbf{k}) \times [(x - 3)\mathbf{i} + (y - 1)\mathbf{j} + (z + 1)\mathbf{k}] = \mathbf{0}$$

∎

If $A = (x_0, y_0, z_0)$ and $B = (x_1, y_1, z_1)$ are two distinct points in three-dimensional space, there is one and only one line containing these points. Obviously,

$$\mathbf{M} = \overrightarrow{AB} = (x_1 - x_0)\mathbf{i} + (y_1 - y_0)\mathbf{j} + (z_1 - z_0)\mathbf{k}$$

is parallel to the line; hence, an equation of the line in vector form is

$$\overrightarrow{AB} \times (\mathbf{R} - \mathbf{R}_0) = \mathbf{0}$$

where $\mathbf{R}_0 = x_0\mathbf{i} + y_0\mathbf{j} + z_0\mathbf{k}$. Thus, *in scalar symmetric form, equations of the line through $A = (x_0, y_0, z_0)$ and $B = (x_1, y_1, z_1)$ are*

$$\frac{x - x_0}{x_1 - x_0} = \frac{y - y_0}{y_1 - y_0} = \frac{z - z_0}{z_1 - z_0}$$

We have seen that the equation $\mathbf{M} \times (\mathbf{R} - \mathbf{R}_0) = \mathbf{0}$ expresses the condition that $\mathbf{R} - \mathbf{R}_0$ is parallel to $\mathbf{M}$. This condition can be expressed equally well by the equation

$$\mathbf{R} - \mathbf{R}_0 = t\mathbf{M}$$

where t is a variable scalar. Solving the last equation for the position vector $\mathbf{R}$, we obtain a **vector parametric equation**

$$\mathbf{R} = \mathbf{R}_0 + t\mathbf{M}$$

of the line parallel to the nonzero direction vector $\mathbf{M}$ and containing the point whose position vector is $\mathbf{R}_0$. As the parameter t varies, the position vector $\mathbf{R}$ traces out the line.

Letting $\mathbf{R} = x\mathbf{i} + y\mathbf{j} + z\mathbf{k}$, $\mathbf{R}_0 = x_0\mathbf{i} + y_0\mathbf{j} + z_0\mathbf{k}$, and $\mathbf{M} = a\mathbf{i} + b\mathbf{j} + c\mathbf{k}$ in the vector parametric equation $\mathbf{R} = \mathbf{R}_0 + t\mathbf{M}$, we have

$$x\mathbf{i} + y\mathbf{j} + z\mathbf{k} = (x_0\mathbf{i} + y_0\mathbf{j} + z_0\mathbf{k}) + t(a\mathbf{i} + b\mathbf{j} + c\mathbf{k})$$

or $\qquad x\mathbf{i} + y\mathbf{j} + z\mathbf{k} = (x_0 + at)\mathbf{i} + (y_0 + bt)\mathbf{j} + (z_0 + ct)\mathbf{k}$

Equating scalar components, we obtain **scalar parametric equations** of the line:

$$\begin{cases} x = x_0 + at \\ y = y_0 + bt \\ z = z_0 + ct \end{cases}$$

As the parameter t varies, the point (x, y, z) given by these simultaneous equations traces out the line. Notice that the point (x_0, y_0, z_0) corresponds to the parameter value $t = 0$.

EXAMPLE 8 Let $A = (-2, -1, -5)$ and $B = (2, 1, 5)$. Find **(a)** an equation in vector parametric form and **(b)** equations in scalar parametric form for the line that contains the point $P_0 = (3, 2, 5)$ and is parallel to the line containing A and B.

SOLUTION   The vector

$$\mathbf{M} = \overrightarrow{AB} = (2 + 2)\mathbf{i} + (1 + 1)\mathbf{j} + (5 + 5)\mathbf{k} = 4\mathbf{i} + 2\mathbf{j} + 10\mathbf{k}$$

is parallel to the desired line, and $\mathbf{R}_0 = 3\mathbf{i} + 2\mathbf{j} + 5\mathbf{k}$ is the position vector of the point P_0 on the line.

(a) $\mathbf{R} = \mathbf{R}_0 + t\mathbf{M} = 3\mathbf{i} + 2\mathbf{j} + 5\mathbf{k} + t(4\mathbf{i} + 2\mathbf{j} + 10\mathbf{k})$

$\qquad = (3 + 4t)\mathbf{i} + (2 + 2t)\mathbf{j} + (5 + 10t)\mathbf{k}$

$$\text{(b)} \begin{cases} x = 3 + 4t \\ y = 2 + 2t \\ z = 5 + 10t \end{cases}$$

EXAMPLE 9   Find equations in scalar parametric form for the line containing the point $P_0 = (3, -2, 5)$ and perpendicular to the lines

$$\frac{x - 1}{3} = \frac{y + 2}{1} = \frac{z - 3}{-2} \quad \text{and} \quad \frac{x + 7}{1} = \frac{y - 5}{2} = \frac{z}{1}$$

SOLUTION   Here, vectors parallel to the indicated lines are given by

$$\mathbf{M}_1 = 3\mathbf{i} + \mathbf{j} - 2\mathbf{k} \quad \text{and} \quad \mathbf{M}_2 = \mathbf{i} + 2\mathbf{j} + \mathbf{k}$$

so that

$$\mathbf{M} = \mathbf{M}_1 \times \mathbf{M}_2 = \begin{vmatrix} \mathbf{i} & \mathbf{j} & \mathbf{k} \\ 3 & 1 & -2 \\ 1 & 2 & 1 \end{vmatrix} = 5\mathbf{i} - 5\mathbf{j} + 5\mathbf{k}$$

is perpendicular to both lines. Therefore, $\mathbf{M}$ is parallel to the desired line, and scalar parametric equations for the line are

$$\begin{cases} x = 3 + 5t \\ y = -2 - 5t \\ z = 5 + 5t \end{cases}$$

Problem Set 13.4

In Problems 1 to 4, find scalar Cartesian equations of the plane containing the point P_0 and having the given normal vector $\mathbf{N}$.

1 $P_0 = (1, -1, 2)$ and $\mathbf{N} = \mathbf{i} + 2\mathbf{j} - 3\mathbf{k}$

2 $P_0 = (1, 3, -1)$ and $\mathbf{N} = 2\mathbf{i} + \mathbf{j} - \mathbf{k}$

3 $P_0 = (0, 0, 0)$ and $\mathbf{N} = 5\mathbf{i} - 2\mathbf{j} + 10\mathbf{k}$

4 $P_0 = (0, 0, 1)$ and $\mathbf{N} = \mathbf{j} + \mathbf{k}$

In Problems 5 and 6, give scalar Cartesian equations of the plane containing the points A, B, and C.

5 $A = (2, -1, 0)$, $B = (-3, -4, -5)$, and $C = (0, 8, 0)$

6 $A = (2, 2, -2)$, $B = (4, 6, 4)$, and $C = (8, -1, 2)$

In Problems 7 to 12, find (a) a unit normal vector to the plane and (b) the intercepts of the plane. (c) Sketch a portion of the plane.

7 $2x + 3y + 6z = 12$ **8** $x - 4y + 8z = 8$

9 $\mathbf{N} \cdot (\mathbf{R} - \mathbf{R}_0) = 0$, where $\mathbf{N} = 12\mathbf{j} - 5\mathbf{k}$ and $\mathbf{R}_0 = 5\mathbf{j}$

10 $\mathbf{N} \cdot (\mathbf{R} - \mathbf{R}_0) = 0$, where $\mathbf{N} = \mathbf{k}$ and $\mathbf{R}_0 = \mathbf{i} + \mathbf{j} + 3\mathbf{k}$

11 $5x = 3y + 4z$

12 $3x = 4z + 12$

In Problems 13 to 16, determine whether the planes are parallel.

13 $3x - y + 2z = 5$ and $6x - 2y + 4z = 10$

14 $-x + y + 3z = 4$ and $5x + 5y + 15z = 21$

15 $2x + y - z = 3$ and $4x + y - 3z = 3$

16 $x + 2y - 2z = 3$ and $-x - 2y + 2z = 3$

In Problems 17 to 26, find a scalar Cartesian equation of the plane satisfying the given conditions.

17 Containing the point $(-1, 3, 5)$ and parallel to the plane $6x - 3y - 2z + 9 = 0$

18 Containing the point $(4, -1, 3)$ and parallel to the plane $2x - y + 5z = 4$

19 Containing the point $(-1, 3, 2)$ and parallel to the yz plane

20 Containing the point $(\frac{2}{5}, \frac{3}{4}, -1)$ and parallel to the xz plane

21 Containing the origin and parallel to the plane $3x - 7y + z = 4$

22 Containing the origin and the points $P = (a, b, c)$ and $Q = (p, q, r)$

23 Containing the points $(1, 2, 3)$ and $(-2, 1, 1)$, but not intersecting the x axis. (*Hint:* Begin by arguing that if the plane $ax + by + cz = D$ is parallel to the x axis, then $a = 0$.)

24 Containing the point $(4, -2, 1)$ and perpendicular to the line containing the two points $A = (2, -1, 2)$ and $B = (3, 2, -1)$

25 Containing the point $(1, 1, 1)$ and perpendicular to the vector whose direction angles are $\pi/3$, $\pi/4$, and $\pi/3$

26 Containing the point $(2, 3, 1)$ and parallel to the plane containing the origin and the points $P = (2, 0, -2)$ and $Q = (1, 1, 1)$

In Problems 27 to 32, give a geometric interpretation of the condition imposed on the plane whose equation is $ax + by + cz = D$ by the given equation or equations.

27 $D = 0$

28 $a = 0$

29 $a = 0$ and $b = 0$

30 $2a + 3b - 4c = 0$

31 $b = 0$

32 $b = 0$ and $c = 0$

In Problems 33 to 40, find an equation or equations of the line satisfying the given conditions in (a) vector nonparametric, (b) scalar symmetric, (c) vector parametric, and (d) scalar parametric form.

33 Containing the point $(-1, 1, 4)$ and parallel to the vector $\mathbf{M} = \mathbf{i} + \mathbf{j} - 2\mathbf{k}$

34 Containing the point $(5, 7, -1)$ and parallel to the vector $\mathbf{M} = 3\mathbf{i} - 2\mathbf{j}$

35 Containing the point $(3, 1, -4)$ and parallel to the vector $\mathbf{M} = -2\mathbf{j} + 5\mathbf{k}$

36 Containing the point $(1, 3, -2)$ and perpendicular to the plane $x - y + 2z = 5$

37 Containing the point $(1, -1, 2)$ and perpendicular to the plane $5x - y + 3z = 7$

38 Containing the point $(0, 0, 1)$ and parallel to the vector with direction angles $2\pi/3$, $\pi/3$, and $\pi/4$

39 Containing the points $(1, 3, -2)$ and $(2, 2, 0)$

40 Containing the points $(-1, 3, 4)$ and $(4, 3, 9)$

In Problems 41 to 48, find an equation or equations of the line L in the indicated form.

41 Symmetric form: L contains the point $(0, 4, 5)$ and is parallel to the line through the points $(1, 5, -2)$ and $(7, 7, 1)$.

42 Scalar parametric form: L contains the origin and is perpendicular to both of the vectors $-3\mathbf{i} + 2\mathbf{j} + \mathbf{k}$ and $6\mathbf{i} - 5\mathbf{j} + 2\mathbf{k}$.

43 Vector nonparametric form: L contains the point $(3, -1, 2)$ and is parallel to the line
$$\frac{x - 1}{3} = \frac{y + 1}{-4} = \frac{z + 3}{7}$$

44 Vector parametric form: L contains the point $(-1, 3, 5)$ and is parallel to the line $x = 3t - 1$, $y = 5t + 7$, $z = -7t + 1$.

45 Symmetric form: L contains the point $(7, 1, -4)$ and is parallel to the x axis.

46 Scalar parametric form: L contains the origin and is perpendicular to the line
$$x - 2 = \frac{y - 3}{2} = \frac{z}{7}$$
at their point of intersection.

47 Symmetric form: L contains the point $(2, 3, -1)$ and is perpendicular to the two lines
$$\frac{x - 3}{4} = \frac{y - 1}{5} = \frac{z + 1}{-3} \quad \text{and} \quad \frac{x}{7} = \frac{y + 1}{3} = \frac{z - 1}{-1}$$

48 Scalar parametric form: L contains the origin and is perpendicular to the two lines
$$\frac{x}{6} = \frac{y - 3}{5} = \frac{z - 4}{4} \quad \text{and} \quad x + 3 = \frac{y - 2}{4} = \frac{z - 5}{-2}$$

49 Show that the points $(2, 0, 0)$, $(-2, -22, -10)$, and $(4, 11, 5)$ lie on a line, and find equations of this line in symmetric form.

50 Assuming that a, b, and c are nonzero, derive equations in the symmetric form by eliminating the parameter t from the scalar parametric equations
$$\begin{cases} x = x_0 + at \\ y = y_0 + bt \\ z = z_0 + ct \end{cases}$$
Exactly what happens if, say, $a = 0$, but b and c are not zero?

In Problems 51 to 56, give a geometric interpretation of the condition imposed on the line
$$\begin{cases} x = x_0 + at \\ y = y_0 + bt \\ z = z_0 + ct \end{cases}$$
by the given equation or equations.

51 $x_0 = y_0 = z_0 = 0$

52 $a = 0$

53 $a = 0$ and $b = 0$

54 $5a - 3b + 7c = 0$

55 $b = 0$

56 $b = 0$ and $c = 0$

57 Find the coordinates of the point on the line
$$\frac{x - 3}{4} = \frac{y - 2}{2} = \frac{z - 5}{10}$$
where this line intersects the plane $x = 5$.

58 Find the coordinates of the point where the line $x = 2 - 3t$, $y = 1 + t$, $z = 4 - t$ intersects the plane $2x - 3y + z = -16$.

13.5   The Geometry of Lines and Planes in Space

Lines in space formed by supporting cables of a bridge

In Section 13.4 we developed equations in both vector and scalar forms for lines and planes in three-dimensional space. In this section we obtain formulas for the distance from a point to a plane, the distance between two lines, the angle between two planes, and the angle between two lines.

The Distance from a Point to a Plane

The argument used in Chapter 12 to derive a formula for the distance from a point to a line (Theorem 1, page 726) can also be used to prove the following theorem.

THEOREM 1   **Distance from a Point to a Plane**

> Let $\mathbf{R}_1 = \overrightarrow{OP_1}$ be the position vector of a point P_1 in space. Then the perpendicular distance d from P_1 to the plane with vector equation $\mathbf{N} \cdot (\mathbf{R} - \mathbf{R}_0) = 0$ is given by
>
> $$d = \frac{|\mathbf{N} \cdot (\mathbf{R}_1 - \mathbf{R}_0)|}{|\mathbf{N}|}$$

PROOF   Let P_0 be the point with position vector $\mathbf{R}_0$, so that $\mathbf{R}_0 = \overrightarrow{OP_0}$, and P_0 lies on the plane $\mathbf{N} \cdot (\mathbf{R} - \mathbf{R}_0) = 0$. The distance d is the absolute value of the component of

$$\overrightarrow{P_0P_1} = \overrightarrow{OP_1} - \overrightarrow{OP_0} = \mathbf{R}_1 - \mathbf{R}_0$$

in the direction of the normal vector $\mathbf{N}$ (Figure 1); thus,

$$d = |\text{comp}_{\mathbf{N}}(\overrightarrow{P_0P_1})| = \left| \frac{\mathbf{N} \cdot (\mathbf{R}_1 - \mathbf{R}_0)}{|\mathbf{N}|} \right| = \frac{|\mathbf{N} \cdot (\mathbf{R}_1 - \mathbf{R}_0)|}{|\mathbf{N}|}$$

Figure 1

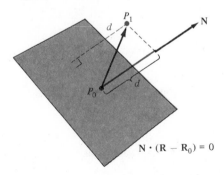

$\mathbf{N} \cdot (\mathbf{R} - \mathbf{R}_0) = 0$

The distance formula in Theorem 1 can be converted to scalar Cartesian form as follows: Suppose that a scalar Cartesian equation of the plane is $ax + by + cz = D$. Then

$$\mathbf{N} = a\mathbf{i} + b\mathbf{j} + c\mathbf{k} \qquad \mathbf{R} = x\mathbf{i} + y\mathbf{j} + z\mathbf{k} \qquad \text{and} \qquad \mathbf{N} \cdot \mathbf{R}_0 = D$$

If $P_1 = (x_1, y_1, z_1)$, so that $\mathbf{R}_1 = x_1\mathbf{i} + y_1\mathbf{j} + z_1\mathbf{k}$, we have

$$\mathbf{N} \cdot (\mathbf{R}_1 - \mathbf{R}_0) = \mathbf{N} \cdot \mathbf{R}_1 - \mathbf{N} \cdot \mathbf{R}_0 = ax_1 + by_1 + cz_1 - D$$

and

$$|\mathbf{N}| = \sqrt{a^2 + b^2 + c^2}$$

Now, substituting into the formula in Theorem 1, we find that *the distance d from the point $P_1 = (x_1, y_1, z_1)$ to the plane $ax + by + cz = D$ is given by*

$$d = \frac{|ax_1 + by_1 + cz_1 - D|}{\sqrt{a^2 + b^2 + c^2}}$$

__EXAMPLE 1__   Find the distance d from the point $P_1 = (1, 3, -4)$ to the plane $2x + 2y - z = 6$.

SOLUTION   Here $a = 2$, $b = 2$, $c = -1$, $D = 6$, $x_1 = 1$, $y_1 = 3$, $z_1 = -4$, and the distance d is given by

$$d = \frac{|ax_1 + by_1 + cz_1 - D|}{\sqrt{a^2 + b^2 + c^2}}$$

$$= \frac{|(2)(1) + (2)(3) + (-1)(-4) - 6|}{\sqrt{2^2 + 2^2 + (-1)^2}}$$

$$= \frac{6}{\sqrt{9}} = 2 \text{ units}$$ ■

__EXAMPLE 2__   Find the perpendicular distance between the two parallel planes $2x - y + 2z = -11$ and $2x - y + 2z + 2 = 0$.

SOLUTION   The required distance is the distance from *any* point on the plane $2x - y + 2z = -11$ to the plane $2x - y + 2z + 2 = 0$. For instance, the point $(-6, -1, 0)$ belongs to the plane $2x - y + 2z = -11$, and its distance from the plane $2x - y + 2z + 2 = 0$ is given by

$$d = \frac{|(2)(-6) + (-1)(-1) + (2)(0) + 2|}{\sqrt{2^2 + (-1)^2 + 2^2}} = \frac{|-9|}{3} = 3 \text{ units}$$ ■

The Angle Between Two Planes

It is geometrically clear that *two planes are* **parallel** *if and only if their normal vectors are parallel* (see Figure 2 in Section 13.4). By definition, we say that *two planes are* **perpendicular** *if and only if their normal vectors are perpendicular* (Figure 2).

Figure 2

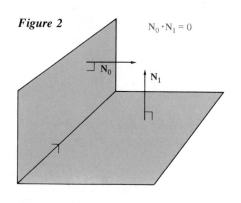

$\mathbf{N}_0 \cdot \mathbf{N}_1 = 0$

perpendicular planes

__EXAMPLE 3__   Find an equation of the plane containing the points $P_0 = (1, 0, 3)$ and $P_1 = (0, 1, -2)$ that is perpendicular to the plane $2x + 3y + z + 4 = 0$.

SOLUTION   The point $P_0 = (1, 0, 3)$ belongs to the plane in question, so the desired equation has the form

$$\mathbf{N} \cdot (\mathbf{R} - \mathbf{R}_0) = 0$$

where $\mathbf{R}_0 = \overrightarrow{OP_0} = \mathbf{i} + 3\mathbf{k}$ and the normal vector $\mathbf{N}$ has yet to be determined. Since P_0 and P_1 both lie on the plane, $\overrightarrow{P_0P_1}$ is perpendicular to $\mathbf{N}$. The plane $2x + 3y + z + 4 = 0$ has normal vector $\mathbf{N}_1 = 2\mathbf{i} + 3\mathbf{j} + \mathbf{k}$; hence, $\mathbf{N}$ is perpendicular to $\mathbf{N}_1$, since the two planes are perpendicular. Thus, $\mathbf{N}$ will be perpendicular to both of the vectors $\overrightarrow{P_0P_1}$ and $\mathbf{N}_1$ provided that we take

$$\mathbf{N} = \overrightarrow{P_0P_1} \times \mathbf{N}_1$$

$$= (-\mathbf{i} + \mathbf{j} - 5\mathbf{k}) \times (2\mathbf{i} + 3\mathbf{j} + \mathbf{k}) = \begin{vmatrix} \mathbf{i} & \mathbf{j} & \mathbf{k} \\ -1 & 1 & -5 \\ 2 & 3 & 1 \end{vmatrix} = 16\mathbf{i} - 9\mathbf{j} - 5\mathbf{k}$$

The desired equation is therefore

$$(16\mathbf{i} - 9\mathbf{j} - 5\mathbf{k}) \cdot [\mathbf{R} - (\mathbf{i} + 3\mathbf{k})] = 0 \qquad \text{or} \qquad 16x - 9y - 5z - 1 = 0$$ ■

Figure 3

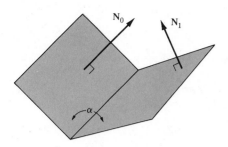

More generally, we can determine the angle between two planes by finding the angle between their normal vectors (Figure 3). However, some care must be taken to ensure that the angle is uniquely defined. Thus, let $\mathbf{N}_0 \cdot (\mathbf{R} - \mathbf{R}_0) = 0$ and $\mathbf{N}_1 \cdot (\mathbf{R} - \mathbf{R}_1) = 0$ be equations of two planes having normal vectors $\mathbf{N}_0$ and $\mathbf{N}_1$. We define **the angle α between the two planes** to be the *smaller* of the two angles θ and $\pi - \theta$, where θ is the angle between $\mathbf{N}_0$ and $\mathbf{N}_1$ (Figure 4).* Since

$$\cos \theta = \frac{\mathbf{N}_0 \cdot \mathbf{N}_1}{|\mathbf{N}_0|\,|\mathbf{N}_1|}$$

it is easy to check that the smaller of the angles θ and $\pi - \theta$ is given by

$$\alpha = \cos^{-1} \frac{|\mathbf{N}_0 \cdot \mathbf{N}_1|}{|\mathbf{N}_0|\,|\mathbf{N}_1|}$$

Figure 4

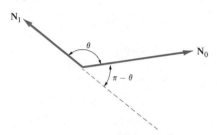

(Problem 52).

ⓒ **EXAMPLE 4**   Find the angle α between the two planes $3x - 2y + z = 5$ and $x + 2y - z = 17$.

SOLUTION   Here, the normal vectors $\mathbf{N}_0$ and $\mathbf{N}_1$ to the two planes are $\mathbf{N}_0 = 3\mathbf{i} - 2\mathbf{j} + \mathbf{k}$ and $\mathbf{N}_1 = \mathbf{i} + 2\mathbf{j} - \mathbf{k}$. Therefore,

$$\alpha = \cos^{-1} \frac{|\mathbf{N}_0 \cdot \mathbf{N}_1|}{|\mathbf{N}_0|\,|\mathbf{N}_1|} = \cos^{-1} \frac{|-2|}{\sqrt{14}\sqrt{6}} = \cos^{-1} \frac{1}{\sqrt{21}} \approx 77.4° \qquad ■$$

The Distance from a Point to a Line

The following theorem provides a handy formula for the distance from a point to a line in space.

THEOREM 2

The Distance from a Point to a Line

The perpendicular distance d from the point P_1 whose position vector is $\mathbf{R}_1 = \overrightarrow{OP_1}$ to the line with vector equation $\mathbf{M} \times (\mathbf{R} - \mathbf{R}_0) = \mathbf{0}$ is given by

$$d = \frac{|\mathbf{M} \times (\mathbf{R}_1 - \mathbf{R}_0)|}{|\mathbf{M}|}$$

PROOF   In Figure 5, we have

$$d = |\mathbf{R}_1 - \mathbf{R}_0|\,\sin \theta$$

$$= \frac{|\mathbf{M}|\,|\mathbf{R}_1 - \mathbf{R}_0|\,\sin \theta}{|\mathbf{M}|}$$

$$= \frac{|\mathbf{M} \times (\mathbf{R}_1 - \mathbf{R}_0)|}{|\mathbf{M}|}$$

Figure 5

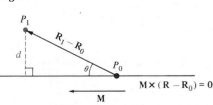

by the formula on page 777 for the length of the cross product of two vectors.   ■

*Notice that the two planes are perpendicular if $\alpha = \pi/2$ and parallel if $\alpha = 0$.

ⓒ **EXAMPLE 5**   Find the distance d from the point $P_1 = (1, 7, -3)$ to the line

$$\frac{x - 5}{2} = \frac{y + 4}{-1} = \frac{z + 6}{2}$$

SOLUTION   A direction vector for the line is $\mathbf{M} = 2\mathbf{i} - \mathbf{j} + 2\mathbf{k}$, and $\mathbf{R}_0 = 5\mathbf{i} - 4\mathbf{j} - 6\mathbf{k}$ is the position vector of a point on the line. Thus, $\mathbf{M} \times (\mathbf{R} - \mathbf{R}_0) = \mathbf{0}$ is a vector equation of the line. Putting $\mathbf{R}_1 = \overrightarrow{OP}_1$, we have $\mathbf{R}_1 = \mathbf{i} + 7\mathbf{j} - 3\mathbf{k}$, and so $\mathbf{R}_1 - \mathbf{R}_0 = -4\mathbf{i} + 11\mathbf{j} + 3\mathbf{k}$. Therefore,

$$\mathbf{M} \times (\mathbf{R}_1 - \mathbf{R}_0) = \begin{vmatrix} \mathbf{i} & \mathbf{j} & \mathbf{k} \\ 2 & -1 & 2 \\ -4 & 11 & 3 \end{vmatrix} = -25\mathbf{i} - 14\mathbf{j} + 18\mathbf{k}$$

$$|\mathbf{M} \times (\mathbf{R}_1 - \mathbf{R}_0)| = \sqrt{(-25)^2 + (-14)^2 + (18)^2} = \sqrt{1145}$$

and

$$d = \frac{|\mathbf{M} \times (\mathbf{R}_1 - \mathbf{R}_0)|}{|\mathbf{M}|} = \frac{\sqrt{1145}}{\sqrt{2^2 + (-1)^2 + 2^2}} = \frac{1}{3}\sqrt{1145} \approx 11.28 \text{ units} \quad \blacksquare$$

The Distance Between Two Lines in Space

The distance between two *parallel* lines in space is easy to find by using the formula in Theorem 2. Indeed, it is only necessary to select a point (any point will do!) on one line and then find the distance from that point to the other line. The distance between two nonparallel lines in space can be found by using the following theorem.

THEOREM 3   **The Distance Between Two Nonparallel Lines in Space**

> Let
>
> $$\mathbf{M}_0 \times (\mathbf{R} - \mathbf{R}_0) = \mathbf{0} \qquad \text{and} \qquad \mathbf{M}_1 \times (\mathbf{R} - \mathbf{R}_1) = \mathbf{0}$$
>
> be the vector equations of two lines in space, and suppose that $\mathbf{M}_0 \times \mathbf{M}_1 \neq \mathbf{0}$, so that the two lines are not parallel. Then, the distance d between the two lines is given by
>
> $$d = \frac{|(\mathbf{M}_0 \times \mathbf{M}_1) \cdot (\mathbf{R}_1 - \mathbf{R}_0)|}{|\mathbf{M}_0 \times \mathbf{M}_1|}$$

PROOF   Notice that the vector $\mathbf{M}_0 \times \mathbf{M}_1$ is perpendicular to *both* of the lines and that the desired distance d is the absolute value of the component of

$$\overrightarrow{P_0 P_1} = \mathbf{R}_1 - \mathbf{R}_0$$

in the direction of $\mathbf{M}_0 \times \mathbf{M}_1$ (Figure 6). Thus,

$$d = |\text{comp}_{\mathbf{M}_0 \times \mathbf{M}_1}(\mathbf{R}_1 - \mathbf{R}_0)| = \frac{|(\mathbf{M}_0 \times \mathbf{M}_1) \cdot (\mathbf{R}_1 - \mathbf{R}_0)|}{|\mathbf{M}_0 \times \mathbf{M}_1|} \quad \blacksquare$$

Figure 6

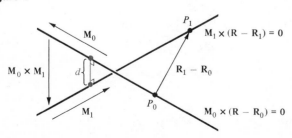

ⓒ **EXAMPLE 6**   Find the distance between the lines

$$\frac{x + 2}{3} = y - 1 = \frac{z + 1}{-2} \qquad \text{and} \qquad \frac{x - 3}{-1} = \frac{y - 1}{4} = z$$

SOLUTION   Here $M_0 = 3i + j - 2k$, $R_0 = -2i + j - k$, $M_1 = -i + 4j + k$, and $R_1 = 3i + j + 0k$; hence,

$$M_0 \times M_1 = \begin{vmatrix} i & j & k \\ 3 & 1 & -2 \\ -1 & 4 & 1 \end{vmatrix} = 9i - j + 13k \qquad R_1 - R_0 = 5i + 0j + k$$

and $(M_0 \times M_1) \cdot (R_1 - R_0) = (9i - j + 13k) \cdot (5i + 0j + k) = 58$

Hence,

$$d = \frac{|(M_0 \times M_1) \cdot (R_1 - R_0)|}{|M_0 \times M_1|} = \frac{|58|}{\sqrt{9^2 + (-1)^2 + 13^2}} = \frac{58}{\sqrt{251}} \approx 3.66 \text{ units} \quad \blacksquare$$

Theorem 3 has the following useful consequence:

> Two nonparallel lines $M_0 \times (R - R_0) = 0$ and $M_1 \times (R - R_1) = 0$ intersect each other if and only if
> $$(M_0 \times M_1) \cdot (R_1 - R_0) = 0$$

EXAMPLE 7

(a) Show that the two lines

$$\frac{x + 9}{5} = \frac{y + 11}{7} = -z \qquad \text{and} \qquad x - 2 = \frac{y - 5}{2} = \frac{z - 1}{3}$$

meet in space.

(b) Find the coordinates of their point of intersection.

SOLUTION

(a) Direction vectors M_0 and M_1 and position vectors R_0 and R_1 of points on the two lines, respectively, are given by

$$M_0 = 5i + 7j - k \qquad R_0 = -9i - 11j$$
$$M_1 = i + 2j + 3k \qquad R_1 = 2i + 5j + k$$

Here we have $M_0 \times M_1 = \begin{vmatrix} i & j & k \\ 5 & 7 & -1 \\ 1 & 2 & 3 \end{vmatrix} = 23i - 16j + 3k$

so that
$$(M_0 \times M_1) \cdot (R_1 - R_0) = (23i - 16j + 3k) \cdot (11i + 16j + k)$$
$$= 253 - 256 + 3 = 0$$

Therefore, the two lines do meet in space.

(b) We have four simultaneous equations

$$\frac{x + 9}{5} = -z \qquad \frac{y + 11}{7} = -z \qquad x - 2 = \frac{y - 5}{2} \qquad x - 2 = \frac{z - 1}{3}$$

in the three unknowns x, y, z. Choosing any three of these equations and solving them, we find that $x = 1$, $y = 3$, and $z = -2$; hence, $(1, 3, -2)$ is the desired point of intersection.* $\blacksquare$

*As a check, we could substitute $x = 1$, $y = 3$, $z = -2$ into the fourth equation.

The Angle Between Two Lines

Consider the two lines $\mathbf{M}_0 \times (\mathbf{R} - \mathbf{R}_0) = \mathbf{0}$ and $\mathbf{M}_1 \times (\mathbf{R} - \mathbf{R}_1) = \mathbf{0}$ having direction vectors $\mathbf{M}_0$ and $\mathbf{M}_1$, respectively. If θ is the angle between $\mathbf{M}_0$ and $\mathbf{M}_1$, we define **the angle α between the two lines** to be the *smaller* of the angles θ and $\pi - \theta$, so that

$$\alpha = \cos^{-1} \frac{|\mathbf{M}_0 \cdot \mathbf{M}_1|}{|\mathbf{M}_0|\,|\mathbf{M}_1|}$$

Note that, by definition, we speak of the angle α between the two lines even if these lines do not meet in space. In particular, the two lines are said to be **perpendicular** if $\alpha = \pi/2$; of course, they are **parallel** if $\alpha = 0$.

ⓒ **EXAMPLE 8**   Find the angle α between the two lines

$$\frac{x - 5}{-7} = y + 3, z = 2 \quad \text{and} \quad \frac{x - 3}{5} = \frac{y - 5}{3} = \frac{z - 7}{-2}$$

SOLUTION   Here $\mathbf{M}_0 = -7\mathbf{i} + \mathbf{j} + 0\mathbf{k}$ and $\mathbf{M}_1 = 5\mathbf{i} + 3\mathbf{j} - 2\mathbf{k}$; hence,

$$\alpha = \cos^{-1} \frac{|\mathbf{M}_0 \cdot \mathbf{M}_1|}{|\mathbf{M}_0|\,|\mathbf{M}_1|} = \cos^{-1} \frac{|-32|}{\sqrt{50}\sqrt{38}} = \cos^{-1} \frac{16}{5\sqrt{19}} \approx 42.77°$$ ∎

Intersecting Planes

Figure 7

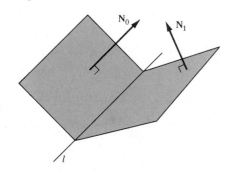

Two nonparallel planes $a_0 x + b_0 y + c_0 z = D_0$ and $a_1 x + b_1 y + c_1 z = D_1$ intersect in a line l (Figure 7). The normal vectors to the two planes are given by

$$\mathbf{N}_0 = a_0\mathbf{i} + b_0\mathbf{j} + c_0\mathbf{k} \quad \text{and} \quad \mathbf{N}_1 = a_1\mathbf{i} + b_1\mathbf{j} + c_1\mathbf{k}$$

respectively. Since l is contained in both planes, it is perpendicular to both of the normal vectors $\mathbf{N}_0$ and $\mathbf{N}_1$; hence, l *is parallel to the vector*

$$\mathbf{M} = \mathbf{N}_0 \times \mathbf{N}_1$$

An equation of l, in vector form, is therefore

$$\mathbf{M} \times (\mathbf{R} - \mathbf{R}_0) = \mathbf{0}$$

where $\mathbf{R}_0$ is the position vector of any particular point P_0 on l. To find such a point P_0, we just find any solution (x_0, y_0, z_0) of the two simultaneous equations

$$\begin{cases} a_0 x + b_0 y + c_0 z = D_0 \\ a_1 x + b_1 y + c_1 z = D_1 \end{cases}$$

This can be accomplished by setting one of the three variables x, y, or z equal to 0 and solving the resulting pair of equations for the remaining two variables.

EXAMPLE 9   Find equations in symmetric form of the line in which the planes $3x + 2y + z = 4$ and $x - 3y + 5z = 7$ intersect.

SOLUTION   The normal vectors to the two planes are given by $\mathbf{N}_0 = 3\mathbf{i} + 2\mathbf{j} + \mathbf{k}$ and $\mathbf{N}_1 = \mathbf{i} - 3\mathbf{j} + 5\mathbf{k}$; hence, a direction vector $\mathbf{M}$ for the line of intersection is given by

$$\mathbf{M} = \mathbf{N}_0 \times \mathbf{N}_1 = \begin{vmatrix} \mathbf{i} & \mathbf{j} & \mathbf{k} \\ 3 & 2 & 1 \\ 1 & -3 & 5 \end{vmatrix} = 13\mathbf{i} - 14\mathbf{j} - 11\mathbf{k}$$

To find a point (x_0, y_0, z_0) common to both planes, we put $y = y_0 = 0$ in the simultaneous equations

$$\begin{cases} 3x + 2y + z = 4 \\ x - 3y + 5z = 7 \end{cases}$$

and then solve the resulting equations

$$\begin{cases} 3x + z = 4 \\ x + 5z = 7 \end{cases}$$

to obtain $x = x_0 = \frac{13}{14}$ and $z = z_0 = \frac{17}{14}$. The point $(x_0, y_0, z_0) = (\frac{13}{14}, 0, \frac{17}{14})$ belongs to both planes; hence, it belongs to the line in which they intersect. Equations in symmetric form for this line are accordingly

$$\frac{x - \frac{13}{14}}{13} = \frac{y - 0}{-14} = \frac{z - \frac{17}{14}}{-11}$$

∎

Problem Set 13.5

In Problems 1 to 6, find the distance from the given point to the indicated plane.

1 $(3, 2, -1)$; $7x - 6y + 6z = 8$

2 $(0, 0, 0)$; $2x + y - 2z + 7 = 0$

3 $(-2, 8, -3)$; $9x - y - 4z = 0$

4 $(-1, 2, -1)$; $x + y = 6$

5 $(0, -3, 0)$; $2x - 7y + z - 1 = 0$

6 $(4, -4, 1)$; $2x + y - z = 3$

In Problems 7 and 8, find the perpendicular distance between the two given parallel planes.

7 $8x + y - 4z + 6 = 0$; $8x + y - 4z = 24$.

8 $2x + y - z = 12$; $8x + 4y - 4z = 7$

©In Problems 9 to 12, find the angle α between the two planes.

9 $x - 2y + 3z = 4$; $-2x + 2y - z = 3$

10 $3x - z = 7$; $2y + 4z = 5$

11 $4x - 4y + 7z = 31$; $3x - 2y - 6z = 11$

12 $2x + 3y + 5z = 17$; $x = 4$

In Problems 13 to 16, determine whether the planes are perpendicular.

13 $2x - y + 7z + 5 = 0$ and $9x + 4y - 2z = 13$

14 $6x + 3y + z - 2 = 0$ and $2x - 5y + 3z = 8$

15 $2x + 3y - z + 8 = 0$ and $5x - 3y + 2z + 1 = 0$

16 $4x - y + 2z = 1$ and $6x - 4y - z = 7$

17 Find a number c such that the plane $(c + 1)x - y + (2 - c)z = 5$ is perpendicular to the plane $2x + 6y - z + 3 = 0$.

18 Find the equation of the plane parallel to the two planes $8x - y + 3z = 12$ and $8x - y + 3z - 17 = 0$ but lying midway between them.

In Problems 19 to 21, find an equation in scalar Cartesian form for the plane satisfying the conditions given.

19 Containing the points $(-1, 3, 2)$ and $(-2, 0, 1)$ and perpendicular to the plane $3x - 2y + z = 15$

20 Containing the origin, perpendicular to the plane whose equation is $14x + 2y + 11 = 0$, and having a normal vector making a 45° angle with the vector $\mathbf{k}$

21 Containing the point $(2, 0, 1)$ and perpendicular to the two planes $2x - 4y - z = 7$ and $x - y + z = 1$

22 Show that the distance d between the two parallel planes $ax + by + cz = D_1$ and $ax + by + cz = D_2$ is given by

$$d = \frac{|D_1 - D_2|}{\sqrt{a^2 + b^2 + c^2}}$$

23 Find a formula for the distance from the origin to the plane $ax + by + cz = D$.

24 Show that if the point P_1 with position vector $\mathbf{R}_1$ is on the same side of the plane $\mathbf{N} \cdot (\mathbf{R} - \mathbf{R}_0) = 0$ as the normal vector $\mathbf{N}$ (when the tail end of $\mathbf{N}$ is on the plane), then $\mathbf{N} \cdot (\mathbf{R}_1 - \mathbf{R}_0)/|\mathbf{N}|$ is positive.

In Problems 25 to 28, find the distance d from the given point to the indicated line.

25 $(1, 2, 3)$; $\dfrac{x + 1}{2} = \dfrac{y - 3}{-1} = \dfrac{z - 4}{3}$

26 $(0, 0, 0)$; $x = \dfrac{y - 1}{2}$ and $z = 4$

27 $(-1, -1, -1)$; $x - 2 = \dfrac{y - 2}{3} = \dfrac{z - 2}{-3}$

28 $(0, 0, 0)$; $\dfrac{x - x_0}{a} = \dfrac{y - y_0}{b} = \dfrac{z - z_0}{c}$

In Problems 29 and 30, find the distance d between the two lines.

29 $\dfrac{x - 1}{2} = \dfrac{y - 3}{-2} = \dfrac{z + 4}{-1}$ and $\dfrac{x}{2} = \dfrac{y + 1}{3} = \dfrac{z + 2}{6}$

30 $\mathbf{R} = (-4\mathbf{i} + 4\mathbf{j} - 7\mathbf{k})t + 4\mathbf{i} - 4\mathbf{j} + 7\mathbf{k}$, and $\mathbf{R} = (2\mathbf{i} + 2\mathbf{j} + \mathbf{k})t + \mathbf{i} - 2\mathbf{j} + 3\mathbf{k}$

In Problems 31 to 34, determine whether the lines are parallel or perpendicular.

31 $\dfrac{x - 3}{4} = \dfrac{y + 1}{-2} = \dfrac{z + 7}{3}$ and $\dfrac{x + 1}{3} = \dfrac{y - 1}{3} = \dfrac{z + 1}{-2}$

32 $\dfrac{x + 1}{3} = \dfrac{y - 1}{-2} = \dfrac{z + 2}{5}$ and $\dfrac{x - 4}{6} = \dfrac{y - 2}{-4} = \dfrac{z + 1}{10}$

33 $\dfrac{x}{2} = \dfrac{y + 3}{4} = \dfrac{z - 1}{-3}$ and $\dfrac{x + 7}{-6} = \dfrac{y + 3}{-12} = \dfrac{z + 4}{9}$

34 $\dfrac{x + 3}{2} = \dfrac{y - 3}{1} = \dfrac{z + 5}{1}$ and $x = 2$, $\dfrac{y - 4}{1} = \dfrac{z + 3}{-1}$

In Problems 35 and 36, determine whether the two lines meet in space. If they do meet, find their point of intersection.

35 $\dfrac{x - 5}{2} = \dfrac{y - 3}{-1} = \dfrac{z - 1}{-3}$ and $\dfrac{x - 11}{7} = \dfrac{y}{8} = \dfrac{z + 8}{-2}$

36 $\dfrac{x - 3}{5} = \dfrac{y - 5}{3} = z$ and $\dfrac{x - 7}{6} = \dfrac{y + 6}{-7} = \dfrac{z - 2}{3}$

[C] In Problems 37 and 38, find the angle α between the two lines.

37 $\dfrac{x}{2} = \dfrac{y}{2} = \dfrac{z}{1}$ and $\dfrac{x - 3}{5} = \dfrac{y - 2}{4} = \dfrac{z + 1}{-3}$

38 $\dfrac{x - 1}{6} = \dfrac{y - 2}{1} = \dfrac{z - 3}{10}$ and $\dfrac{x}{-2} = \dfrac{y + 1}{3} = \dfrac{z}{-6}$

39 Consider the two lines

$$\frac{x}{1} = \frac{y + 3}{2} = \frac{z + 1}{3} \quad \text{and} \quad \frac{x - 3}{2} = \frac{y}{1} = \frac{z - 1}{-1}$$

(a) Find the angle α between the two lines. (b) Show that the two lines intersect in space. (c) Find the point of intersection of the two lines.

40 Find a necessary and sufficient condition for the line

$$\begin{cases} x = x_0 + at \\ y = y_0 + bt \\ z = z_0 + ct \end{cases}$$

to intersect the x axis.

41 Determine whether the line

$$\frac{x + 3}{2} = \frac{y}{-1} = \frac{z + 7}{4}$$

is perpendicular to the plane $3x + 2y - z - 11 = 0$, and find the coordinates of the point where the line intersects the plane.

42 A certain plane contains the two points $(1, 0, 1)$ and $(1, 1, 0)$, but does not contain any of the points on the line

$$\frac{x}{1} = \frac{y}{1} = \frac{z - 1}{1}$$

Find a scalar Cartesian equation of the plane.

In Problems 43 to 46, determine whether the two planes are parallel. If they are not parallel, find equations in symmetric scalar form of their line of intersection.

43 $\begin{cases} 2x - y - z = 3 \\ x + 2y + 3z = 7 \end{cases}$ **44** $\begin{cases} x + 2y + 3z = 1 \\ -3x - 6y - 9z = 2 \end{cases}$

45 $\begin{cases} 2x - 3y + 4z = 5 \\ 6x - 9y + 8z = 1 \end{cases}$ **46** $\begin{cases} 3x + y - 2z = 7 \\ 6x - 5y - 4z = 7 \end{cases}$

47 Find an equation of the plane that contains the points $P_1 = (1, 0, -1)$ and $P_2 = (-1, 2, 1)$ and is parallel to the line of intersection of the plane $3x + y - 2z = 6$ and $4x - y + 3z = 0$.

48 Show that the two lines

$$\begin{cases} x = -1 + 2t \\ y = 1 + t \\ z = -1 - 2t \end{cases} \quad \text{and} \quad \begin{cases} x = -7 + 3t \\ y = -3 + 2t \\ z = 1 - t \end{cases}$$

meet in space, and find an equation in scalar Cartesian form of the plane containing them.

49 Find an equation of the plane that contains the line

$$\frac{x - 2}{3} = \frac{y}{-1} = \frac{z + 3}{2}$$

but does not intersect the line

$$\begin{cases} x = 4t - 1 \\ y = 2t + 2 \\ z = 3t \end{cases}$$

50 Let l_1 be the line of intersection of the two planes $x + y - 3z = 0$ and $2x + 3y - 8z = 1$, and let l_2 be the line of intersection of the two planes $3x - y - z = 3$ and $x + y - 3z = 5$. (a) Show that l_1 and l_2 are parallel lines. (b) Find a scalar Cartesian equation of the plane containing l_1 and l_2.

51 (a) Give a reasonable definition of the angle α between the line $\mathbf{M}_0 \times (\mathbf{R} - \mathbf{R}_0) = \mathbf{0}$ and the plane $\mathbf{N}_1 \cdot (\mathbf{R} - \mathbf{R}_1) = 0$.
(b) Show that the angle α of part (a) is given by

$$\alpha = \sin^{-1} \frac{|\mathbf{M}_0 \cdot \mathbf{N}_1|}{|\mathbf{M}_0| \, |\mathbf{N}_1|}$$

(c) Find the angle between the line

$$\frac{x}{2} = \frac{y - 1}{-2} = \frac{z - 3}{2}$$

and the plane $2x + 2y - z = 6$.

(d) Find the angle between the line

$$\frac{x - 1}{3} = \frac{y}{1}, z = 0$$

and the plane $x + 2y = 7$.

52 If θ is the angle between the nonzero vectors $\mathbf{N}_0$ and $\mathbf{N}_1$, show that

$$\alpha = \cos^{-1} \frac{|\mathbf{N}_0 \cdot \mathbf{N}_1|}{|\mathbf{N}_0| \, |\mathbf{N}_1|}$$

always gives the smaller of the two angles θ and $\pi - \theta$.

53 Find a point on the line $x = y = z$ that is equidistant from the points $(3, 0, 5)$ and $(1, -1, 4)$.

54 A **tetrahedron** is a pyramid with four vertices P_0, P_1, P_2, and P_3 and four triangular faces (Figure 8). Its volume V is one-third of the distance from P_0 to the base $P_1P_2P_3$ times the area of the base. Find a formula for the volume V in terms of the coordinates of the vertices, $P_0 = (x_0, y_0, z_0)$, $P_1 = (x_1, y_1, z_1)$, $P_2 = (x_2, y_2, z_2)$, and $P_3 = (x_3, y_3, z_3)$.

Figure 8

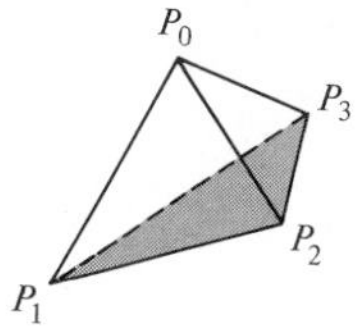

13.6   Curves in Space

The concept of a vector-valued function and the results obtained in Sections 12.4 to 12.7 can be extended to vectors in three-dimensional space. Indeed, all the definitions in those sections were so formulated as to be applicable not only to vectors in the xy plane, but also to vectors in space. Furthermore, all the facts about continuity and differentiability of vector-valued functions carry over immediately to xyz space, provided that the vectors are written in terms of *three* scalar components. For instance, if $\mathbf{F}$ is a vector-valued function of the scalar t, then

$$\mathbf{F}(t) = u(t)\mathbf{i} + v(t)\mathbf{j} + w(t)\mathbf{k}$$

where u, v, and w are the three scalar component functions of $\mathbf{F}$; furthermore, $\mathbf{F}$ *is differentiable if and only if u, v, and w are differentiable. If u, v, and w are differentiable, then*

$$\mathbf{F}'(t) = u'(t)\mathbf{i} + v'(t)\mathbf{j} + w'(t)\mathbf{k}$$

and so forth.

The facts concerning the cross product of two vector-valued functions are exactly as might be expected. Indeed, if **F** and **G** are vector-valued functions, then:

---

1 If $\lim\limits_{t \to c} \mathbf{F}(t)$ and $\lim\limits_{t \to c} \mathbf{G}(t)$ both exist, it follows that $\lim\limits_{t \to c} [\mathbf{F}(t) \times \mathbf{G}(t)]$ exists and

$$\lim_{t \to c} [\mathbf{F}(t) \times \mathbf{G}(t)] = [\lim_{t \to c} \mathbf{F}(t)] \times [\lim_{t \to c} \mathbf{G}(t)]$$

2 If **F** and **G** are continuous, then so is the function **H** defined by

$$\mathbf{H}(t) = \mathbf{F}(t) \times \mathbf{G}(t)$$

3 If **F** and **G** are differentiable and **H** is defined by $\mathbf{H}(t) = \mathbf{F}(t) \times \mathbf{G}(t)$, then **H** is differentiable and

$$\mathbf{H}'(t) = \mathbf{F}'(t) \times \mathbf{G}(t) + \mathbf{F}(t) \times \mathbf{G}'(t)$$

These facts are easy to verify by considering the scalar component functions of **F** and **G** (Problems 38 to 40). Using Leibniz notation, we can rewrite the equation for the derivative of the cross product as

$$\frac{d}{dt}(\mathbf{F} \times \mathbf{G}) = \frac{d\mathbf{F}}{dt} \times \mathbf{G} + \mathbf{F} \times \frac{d\mathbf{G}}{dt}$$

Motion in Space

Figure 1

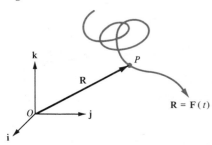

$\mathbf{R} = \mathbf{F}(t)$

If **F** is a three-dimensional vector-valued function, then a vector equation of the form $\mathbf{R} = \mathbf{F}(t)$ can be regarded as specifying a variable position vector **R** in terms of a parameter t. As t varies, **R** traces out a curve in space (Figure 1). For instance, if **M** is a nonzero constant vector, then the position vector **R** given by

$$\mathbf{R} = \mathbf{R}_0 + t\mathbf{M}$$

traces out a line in space, as seen in Section 13.4.

We can always regard $\mathbf{R} = \mathbf{F}(t)$ as giving the position vector of a moving particle P at time t. By the same considerations as those in Section 12.5, a particle P moving in space according to the **equation of motion**

$$\mathbf{R} = \mathbf{F}(t) = x\mathbf{i} + y\mathbf{j} + z\mathbf{k}$$

where x, y, and z are functions of time t, has a **velocity vector V** given by

$$\mathbf{V} = \frac{d\mathbf{R}}{dt} = \frac{dx}{dt}\mathbf{i} + \frac{dy}{dt}\mathbf{j} + \frac{dz}{dt}\mathbf{k} = \mathbf{F}'(t)$$

and an **acceleration vector A** given by

$$\mathbf{A} = \frac{d\mathbf{V}}{dt} = \frac{d^2\mathbf{R}}{dt^2} = \frac{d^2x}{dt^2}\mathbf{i} + \frac{d^2y}{dt^2}\mathbf{j} + \frac{d^2z}{dt^2}\mathbf{k} = \mathbf{F}''(t)$$

Reasoning exactly as in two dimensions, we find that the velocity vector **V** is always tangent to the path traced out by P. Also, the length $|\mathbf{V}|$ of the velocity vector

Figure 2

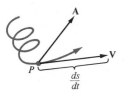

gives the instantaneous **speed** v of P; that is,

$$v = \frac{ds}{dt} = |\mathbf{V}| = \left|\frac{d\mathbf{R}}{dt}\right| = \sqrt{\left(\frac{dx}{dt}\right)^2 + \left(\frac{dy}{dt}\right)^2 + \left(\frac{dz}{dt}\right)^2}$$

where s is the arc length measured along the path of P (Figure 2).

EXAMPLE 1 A particle P is moving in space according to the equation of motion

$$\mathbf{R} = (4 \cos t)\mathbf{i} + (4 \sin t)\mathbf{j} + t^2\mathbf{k}$$

At the instant when $t = \pi/2$, find **(a)** the velocity vector $\mathbf{V}$, **(b)** the speed v, and **(c)** the acceleration vector $\mathbf{A}$.

SOLUTION   At any time t we have $\mathbf{V} = d\mathbf{R}/dt = (-4 \sin t)\mathbf{i} + (4 \cos t)\mathbf{j} + 2t\mathbf{k}$, and $\mathbf{A} = d\mathbf{V}/dt = (-4 \cos t)\mathbf{i} + (-4 \sin t)\mathbf{j} + 2\mathbf{k}$. Therefore, at the instant $t = \pi/2$,

(a) $\mathbf{V} = \left(-4 \sin \dfrac{\pi}{2}\right)\mathbf{i} + \left(4 \cos \dfrac{\pi}{2}\right)\mathbf{j} + 2\dfrac{\pi}{2}\mathbf{k} = -4\mathbf{i} + \pi\mathbf{k}$

(b) $v = |\mathbf{V}| = \sqrt{16 + \pi^2}$

(c) $\mathbf{A} = \left(-4 \cos \dfrac{\pi}{2}\right)\mathbf{i} + \left(-4 \sin \dfrac{\pi}{2}\right)\mathbf{j} + 2\mathbf{k} = -4\mathbf{j} + 2\mathbf{k}$

Let $\mathbf{R} = \mathbf{F}(t)$ be a vector parametric equation of a curve in space. Then, provided that $|d\mathbf{R}/dt| \neq 0$, a unit tangent vector $\mathbf{T}$ to the curve is given by

$$\mathbf{T} = \frac{d\mathbf{R}/dt}{|d\mathbf{R}/dt|} = \frac{d\mathbf{R}/dt}{ds/dt} = \frac{\mathbf{V}}{v}$$

Notice that

$$\mathbf{V} = \frac{ds}{dt}\mathbf{T} = v\mathbf{T}$$

so that the velocity vector is obtained by multiplying the unit tangent vector by the speed. Henceforth, we assume that $|d\mathbf{R}/dt| \neq 0$, so that $\mathbf{T}$ is defined as above.
   Since

$$\frac{ds}{dt} = \left|\frac{d\mathbf{R}}{dt}\right| = \sqrt{\left(\frac{dx}{dt}\right)^2 + \left(\frac{dy}{dt}\right)^2 + \left(\frac{dz}{dt}\right)^2}$$

the arc length of the portion of the curve $\mathbf{R} = \mathbf{F}(t)$ between the points corresponding to $t = a$ and $t = b$ is given by

$$s = \int_a^b \left|\frac{d\mathbf{R}}{dt}\right| dt = \int_a^b \sqrt{\left(\frac{dx}{dt}\right)^2 + \left(\frac{dy}{dt}\right)^2 + \left(\frac{dz}{dt}\right)^2}\, dt$$

EXAMPLE 2 Let C be the curve whose vector parametric equation is $\mathbf{R} = \sqrt{6}t^2\mathbf{i} + 3t\mathbf{j} + \frac{4}{3}t^3\mathbf{k}$. Find **(a)** the unit tangent vector $\mathbf{T}$ to C when $t = 1$ and **(b)** the arc length s of C between the points where $t = 0$ and where $t = 1$.

SOLUTION   Here $d\mathbf{R}/dt = 2\sqrt{6}t\mathbf{i} + 3\mathbf{j} + 4t^2\mathbf{k}$, so that

$$\left|\frac{d\mathbf{R}}{dt}\right| = \sqrt{24t^2 + 9 + 16t^4} = \sqrt{(4t^2 + 3)^2} = 4t^2 + 3$$

and $\qquad \mathbf{T} = \dfrac{d\mathbf{R}/dt}{|d\mathbf{R}/dt|} = \dfrac{2\sqrt{6}t\mathbf{i} + 3\mathbf{j} + 4t^2\mathbf{k}}{4t^2 + 3} \qquad$ for any value of t

(a) For $t = 1$, we have $\mathbf{T} = \dfrac{2\sqrt{6}\mathbf{i} + 3\mathbf{j} + 4\mathbf{k}}{7}$.

(b) $s = \displaystyle\int_0^1 \left|\frac{d\mathbf{R}}{dt}\right| dt = \int_0^1 (4t^2 + 3)\, dt = \frac{13}{3}$ units ∎

Normal Vector and Curvature

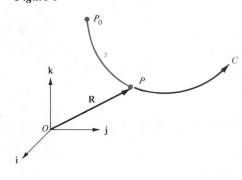

Figure 3

In what follows we denote by s the arc length measured along the curve C between a fixed starting point P_0 and the point P whose position vector is $\mathbf{R} = \mathbf{F}(t)$ (Figure 3). Then, by the chain rule,

$$\frac{d\mathbf{R}}{ds} = \frac{d\mathbf{R}}{dt}\frac{dt}{ds} = \frac{d\mathbf{R}/dt}{ds/dt} = \mathbf{T}$$

The vector $d\mathbf{T}/ds$ is still called the **curvature vector,** just as it was in two dimensions. In three-dimensional space, we do not attempt to give the **curvature** κ an algebraic sign, and we simply define

$$\kappa = \left|\frac{d\mathbf{T}}{ds}\right|$$

so that $\kappa \geq 0$. Because $\mathbf{T}$ is a unit vector, $\mathbf{T} \cdot \mathbf{T} = 1$. Differentiating with respect to s, we obtain

$$\frac{d}{ds}(\mathbf{T} \cdot \mathbf{T}) = \frac{d\mathbf{T}}{ds} \cdot \mathbf{T} + \mathbf{T} \cdot \frac{d\mathbf{T}}{ds} = 2\mathbf{T} \cdot \frac{d\mathbf{T}}{ds} = 0$$

That is,

$$\mathbf{T} \cdot \frac{d\mathbf{T}}{ds} = 0$$

Therefore, *the curvature vector $d\mathbf{T}/ds$ is always perpendicular to the tangent vector* $\mathbf{T}$. If $\kappa \neq 0$, then the vector $\mathbf{N}$ defined by

$$\mathbf{N} = \frac{d\mathbf{T}/ds}{|d\mathbf{T}/ds|} = \frac{d\mathbf{T}/ds}{\kappa}$$

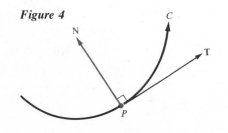

Figure 4

is a unit vector perpendicular to the tangent vector $\mathbf{T}$. Just as in two dimensions, we refer to the vector $\mathbf{N}$ as the **principal unit normal vector** to the curve C (Figure 4). Notice that

$$\frac{d\mathbf{T}}{ds} = \kappa\mathbf{N}$$

If we differentiate both sides of the equation $\mathbf{V} = v\mathbf{T}$ with respect to t, we obtain

$$\mathbf{A} = \frac{d\mathbf{V}}{dt} = \frac{dv}{dt}\mathbf{T} + v\frac{d\mathbf{T}}{dt}$$

$$= \frac{dv}{dt}\mathbf{T} + v\frac{ds}{dt}\frac{d\mathbf{T}}{ds}$$

or

$$\mathbf{A} = \frac{dv}{dt}\mathbf{T} + v^2\kappa\mathbf{N}$$

Hence, even in three-dimensional space, the acceleration vector $\mathbf{A}$ decomposes into the sum of a **tangential component vector** $(dv/dt)\mathbf{T}$ and a **normal component vector** $v^2\kappa\mathbf{N}$.

Taking the cross product with $\mathbf{V}$ on both sides of the last equation $\mathbf{A} = (dv/dt)\mathbf{T} + v^2\kappa\mathbf{N}$, and using the fact that $\mathbf{V} = v\mathbf{T}$, we obtain

$$\mathbf{V} \times \mathbf{A} = \frac{dv}{dt}(\mathbf{V} \times \mathbf{T}) + v^2\kappa(\mathbf{V} \times \mathbf{N}) = \frac{dv}{dt}v(\mathbf{T} \times \mathbf{T}) + v^2\kappa v(\mathbf{T} \times \mathbf{N})$$

$$= \mathbf{0} + v^3\kappa(\mathbf{T} \times \mathbf{N}) = v^3\kappa(\mathbf{T} \times \mathbf{N})$$

In what follows, we make extensive use of the equation

$$\mathbf{V} \times \mathbf{A} = v^3\kappa(\mathbf{T} \times \mathbf{N})$$

Since $\mathbf{T}$ and $\mathbf{N}$ are perpendicular unit vectors, it follows that $|\mathbf{T} \times \mathbf{N}| = |\mathbf{T}|\,|\mathbf{N}| = 1$; hence, by the equation above,

$$|\mathbf{V} \times \mathbf{A}| = |v^3\kappa(\mathbf{T} \times \mathbf{N})| = v^3\kappa|\mathbf{T} \times \mathbf{N}| = v^3\kappa$$

Solving for κ, we obtain

$$\kappa = \frac{|\mathbf{V} \times \mathbf{A}|}{v^3}$$

Because $v\mathbf{T} = \mathbf{V}$, the equation $\mathbf{V} \times \mathbf{A} = v^3\kappa(\mathbf{T} \times \mathbf{N})$ can be rewritten as

$$\mathbf{V} \times \mathbf{A} = v^2\kappa(\mathbf{V} \times \mathbf{N})$$

Taking the cross product with $\mathbf{V}$, we obtain

$$(\mathbf{V} \times \mathbf{A}) \times \mathbf{V} = v^2\kappa(\mathbf{V} \times \mathbf{N}) \times \mathbf{V} = v^2\kappa[(\mathbf{V} \cdot \mathbf{V})\mathbf{N} - (\mathbf{N} \cdot \mathbf{V})\mathbf{V}]$$

where we have used the triple vector-product identity (Theorem 4 in Section 13.3) to expand $(\mathbf{V} \times \mathbf{N}) \times \mathbf{V}$. Since $\mathbf{V} \cdot \mathbf{V} = |\mathbf{V}|^2 = v^2$ and

$$\mathbf{N} \cdot \mathbf{V} = \mathbf{N} \cdot (v\mathbf{T}) = v\mathbf{N} \cdot \mathbf{T} = v(0) = 0$$

the equation above can be rewritten as

$$(\mathbf{V} \times \mathbf{A}) \times \mathbf{V} = v^2\kappa(v^2\mathbf{N} - 0\mathbf{V}) = v^4\kappa\mathbf{N}$$

Thus,

$$\mathbf{N} = \frac{(\mathbf{V} \times \mathbf{A}) \times \mathbf{V}}{v^4\kappa}$$

The equations $\kappa = |\mathbf{V} \times \mathbf{A}|/v^3$ and $\mathbf{N} = (\mathbf{V} \times \mathbf{A}) \times \mathbf{V}/(v^4\kappa)$ allow us to find the curvature and the principal unit normal vector directly in terms of the original parameter t.

EXAMPLE 3   Find $\mathbf{T}$, $\mathbf{N}$, and κ if $\mathbf{R} = (2t^2 + 1)\mathbf{i} + (t^2 - 2t)\mathbf{j} + (t^2 - 1)\mathbf{k}$.

SOLUTION   Here, $\mathbf{V} = d\mathbf{R}/dt = 4t\mathbf{i} + (2t - 2)\mathbf{j} + 2t\mathbf{k}$, so that

$$v = |\mathbf{V}| = \sqrt{16t^2 + (2t - 2)^2 + 4t^2} = 2\sqrt{6t^2 - 2t + 1}$$

and

$$\mathbf{A} = \frac{d\mathbf{V}}{dt} = 4\mathbf{i} + 2\mathbf{j} + 2\mathbf{k}$$

Thus,

$$\mathbf{T} = \frac{\mathbf{V}}{v} = \frac{2t\mathbf{i} + (t - 1)\mathbf{j} + t\mathbf{k}}{\sqrt{6t^2 - 2t + 1}}$$

$$\mathbf{V} \times \mathbf{A} = \begin{vmatrix} \mathbf{i} & \mathbf{j} & \mathbf{k} \\ 4t & 2t - 2 & 2t \\ 4 & 2 & 2 \end{vmatrix} = -4\mathbf{i} + 0\mathbf{j} + 8\mathbf{k} = -4(\mathbf{i} - 2\mathbf{k})$$

$$(\mathbf{V} \times \mathbf{A}) \times \mathbf{V} = \begin{vmatrix} \mathbf{i} & \mathbf{j} & \mathbf{k} \\ -4 & 0 & 8 \\ 4t & 2t - 2 & 2t \end{vmatrix} = (-16t + 16)\mathbf{i} + 40t\mathbf{j} + (-8t + 8)\mathbf{k}$$

$$\kappa = \frac{|\mathbf{V} \times \mathbf{A}|}{v^3} = \frac{4|\mathbf{i} - 2\mathbf{k}|}{8(6t^2 - 2t + 1)^{3/2}} = \frac{\sqrt{5}}{2(6t^2 - 2t + 1)^{3/2}}$$

$$\mathbf{N} = \frac{(\mathbf{V} \times \mathbf{A}) \times \mathbf{V}}{v^4\kappa} = \frac{(2 - 2t)\mathbf{i} + 5t\mathbf{j} + (1 - t)\mathbf{k}}{\sqrt{5}\sqrt{6t^2 - 2t + 1}}$$

Binormal Vector and Torsion

The vector $\mathbf{T} \times \mathbf{N}$ that appears in the equation

$$\mathbf{V} \times \mathbf{A} = v^3\kappa(\mathbf{T} \times \mathbf{N})$$

is called the **unit binormal vector** to the curve and is symbolized by $\mathbf{B}$. Thus, by definition,

$$\boxed{\mathbf{B} = \mathbf{T} \times \mathbf{N}}$$

that is,

$$\boxed{\mathbf{B} = \frac{\mathbf{V} \times \mathbf{A}}{v^3\kappa} = \frac{\mathbf{V} \times \mathbf{A}}{|\mathbf{V} \times \mathbf{A}|}}$$

Figure 5

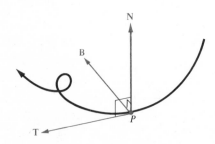

Note that $\mathbf{B}$ *is a unit vector, and it is perpendicular to both* $\mathbf{T}$ *and* $\mathbf{N}$; furthermore, $\mathbf{T}$, $\mathbf{N}$, $\mathbf{B}$ is a right-handed triple, just as $\mathbf{i}$, $\mathbf{j}$, $\mathbf{k}$ is. However, $\mathbf{T}$, $\mathbf{N}$, and $\mathbf{B}$ are variable vectors, moving along with the point P as it traces out the curve (Figure 5).

Just as $\mathbf{i}$, $\mathbf{j}$, $\mathbf{k}$ forms a basis, so does any particular triple $\mathbf{T}$, $\mathbf{N}$, $\mathbf{B}$; hence, any vector $\mathbf{X}$ can be expressed as a linear combination of $\mathbf{T}$, $\mathbf{N}$, and $\mathbf{B}$ as follows:

$$\mathbf{X} = (\mathbf{X} \cdot \mathbf{T})\mathbf{T} + (\mathbf{X} \cdot \mathbf{N})\mathbf{N} + (\mathbf{X} \cdot \mathbf{B})\mathbf{B}$$

(see Problem 67 in Section 13.2). In particular,

$$\frac{d\mathbf{N}}{ds} = \left(\frac{d\mathbf{N}}{ds} \cdot \mathbf{T}\right)\mathbf{T} + \left(\frac{d\mathbf{N}}{ds} \cdot \mathbf{N}\right)\mathbf{N} + \left(\frac{d\mathbf{N}}{ds} \cdot \mathbf{B}\right)\mathbf{B}$$

Differentiation of both sides of the equation $\mathbf{N} \cdot \mathbf{T} = 0$ with respect to s gives $(d\mathbf{N}/ds) \cdot \mathbf{T} + \mathbf{N} \cdot (d\mathbf{T}/ds) = 0$, so that

$$\frac{d\mathbf{N}}{ds} \cdot \mathbf{T} = -\mathbf{N} \cdot \frac{d\mathbf{T}}{ds} = -\mathbf{N} \cdot (\kappa\mathbf{N}) = -\kappa\mathbf{N} \cdot \mathbf{N} = -\kappa(1) = -\kappa$$

Similarly, differentiation of both sides of the equation $\mathbf{N} \cdot \mathbf{N} = 1$ gives $2\,(d\mathbf{N}/ds) \cdot \mathbf{N} = 0$, so that

$$\frac{d\mathbf{N}}{ds} \cdot \mathbf{N} = 0$$

Therefore, substituting into the equation above for $d\mathbf{N}/ds$, we obtain

$$\frac{d\mathbf{N}}{ds} = -\kappa\mathbf{T} + \left(\frac{d\mathbf{N}}{ds} \cdot \mathbf{B}\right)\mathbf{B}$$

The scalar $(d\mathbf{N}/ds) \cdot \mathbf{B}$ in the last equation is called the **torsion** of the curve and is traditionally denoted by the Greek letter τ (called "tau"). Thus, by definition,

$$\tau = \frac{d\mathbf{N}}{ds} \cdot \mathbf{B}$$

and the equation $d\mathbf{N}/ds = -\kappa\mathbf{T} + [(d\mathbf{N}/ds) \cdot \mathbf{B}]\mathbf{B}$ can be written

$$\frac{d\mathbf{N}}{ds} = -\kappa\mathbf{T} + \tau\mathbf{B}$$

The torsion τ is a measure of how rapidly the curve is "twisting" in space. The three equations

$$\frac{d\mathbf{T}}{ds} = \kappa\mathbf{N} \qquad \frac{d\mathbf{N}}{ds} = -\kappa\mathbf{T} + \tau\mathbf{B} \qquad \text{and} \qquad \frac{d\mathbf{B}}{ds} = -\tau\mathbf{N}$$

are called the **Frenet formulas.** (For a proof of the third equation, see Problem 44.) These Frenet formulas—which express the rates of change of $\mathbf{T}$, $\mathbf{N}$, and $\mathbf{B}$ with respect to arc length in terms of $\mathbf{T}$, $\mathbf{N}$, $\mathbf{B}$, the curvature, and the torsion—are the fundamental equations for the study of curves in space. It can be shown that *all* the intrinsic geometric features of the curve are implicit in these three equations.

If we differentiate both sides of the identity $\mathbf{A} = (dv/dt)\mathbf{T} + v^2\kappa\mathbf{N}$ with respect to arc length s and use the identities $d\mathbf{T}/ds = \kappa\mathbf{N}$ and $d\mathbf{N}/ds = -\kappa\mathbf{T} + \tau\mathbf{B}$, we obtain

$$\frac{d\mathbf{A}}{ds} = \frac{d(dv/dt)}{ds}\mathbf{T} + \frac{dv}{dt}\frac{d\mathbf{T}}{ds} + \frac{d(v^2\kappa)}{ds}\mathbf{N} + v^2\kappa\frac{d\mathbf{N}}{ds}$$

$$= \frac{d(dv/dt)}{ds}\mathbf{T} + \frac{dv}{dt}\kappa\mathbf{N} + \frac{d(v^2\kappa)}{ds}\mathbf{N} + v^2\kappa(-\kappa\mathbf{T}) + v^2\kappa(\tau\mathbf{B})$$

Taking the dot product on both sides of the last equation with $\mathbf{B}$ and using the facts that $\mathbf{T} \cdot \mathbf{B} = \mathbf{N} \cdot \mathbf{B} = 0$ and $\mathbf{B} \cdot \mathbf{B} = 1$, we obtain $(d\mathbf{A}/ds) \cdot \mathbf{B} = v^2\kappa\tau$. Since

$v(d\mathbf{A}/ds) = (ds/dt)(d\mathbf{A}/ds) = d\mathbf{A}/dt$, we can multiply the last equation by v to obtain

$$\left(v\,\frac{d\mathbf{A}}{ds}\right)\cdot\mathbf{B} = v^3\kappa\tau \qquad \text{or} \qquad \frac{d\mathbf{A}}{dt}\cdot\mathbf{B} = v^3\kappa\tau$$

Since $\mathbf{B} = (\mathbf{V}\times\mathbf{A})/(v^3\kappa)$, we have

$$\frac{d\mathbf{A}}{dt}\cdot\left(\frac{\mathbf{V}\times\mathbf{A}}{v^3\kappa}\right) = v^3\kappa\tau$$

or

$$\tau = \frac{(d\mathbf{A}/dt)\cdot(\mathbf{V}\times\mathbf{A})}{v^6\kappa^2}$$

This formula gives the torsion τ in terms of quantities that can be found directly in terms of the parameter t.

The following example involves a curve $\mathbf{R} = (a\cos t)\mathbf{i} + (a\sin t)\mathbf{j} + bt\mathbf{k}$, where a and b are constants and $a > 0$. In scalar parametric form, $x = a\cos t$, $y = a\sin t$, and $z = bt$. As the parameter t increases, the point $(x,\ y,\ 0) = (a\cos t,\ a\sin t,\ 0)$ in the xy plane traces out a circle of radius a. Meanwhile, the z coordinate $z = bt$ changes uniformly, so that the point $(x,\ y,\ z)$ traces out a **circular helix** (Figure 6). When t increases by 2π, x and y return to their previous values while z increases by $2\pi b$. We say that the helix has **radius** a and **pitch** $2\pi|b|$.

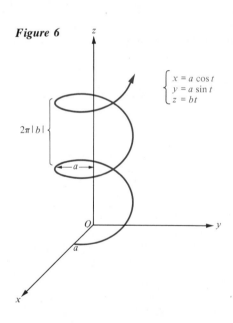

Circular helix: the track of a roller coaster in an amusement park

Figure 6

$$\begin{cases} x = a\cos t \\ y = a\sin t \\ z = bt \end{cases}$$

EXAMPLE 4 If $\mathbf{R} = (2\cos t)\mathbf{i} + (2\sin t)\mathbf{j} + 3t\mathbf{k}$, find $\mathbf{V}$, v, $\mathbf{A}$, $\mathbf{T}$, $\mathbf{V}\times\mathbf{A}$, κ, $\mathbf{N}$, $\mathbf{B}$, $d\mathbf{A}/dt$, and τ.

SOLUTION

$$\mathbf{V} = \frac{d\mathbf{R}}{dt} = (-2\sin t)\mathbf{i} + (2\cos t)\mathbf{j} + 3\mathbf{k}$$

$$v = |\mathbf{V}| = \sqrt{4\sin^2 t + 4\cos^2 t + 9} = \sqrt{13}$$

$$\mathbf{A} = \frac{d\mathbf{V}}{dt} = (-2\cos t)\mathbf{i} - (2\sin t)\mathbf{j}$$

so that

$$\mathbf{T} = \frac{1}{v}\mathbf{V} = \frac{1}{\sqrt{13}}[(-2\sin t)\mathbf{i} + (2\cos t)\mathbf{j} + 3\mathbf{k}]$$

$$\mathbf{V}\times\mathbf{A} = \begin{vmatrix} \mathbf{i} & \mathbf{j} & \mathbf{k} \\ -2\sin t & 2\cos t & 3 \\ -2\cos t & -2\sin t & 0 \end{vmatrix} = (6\sin t)\mathbf{i} - (6\cos t)\mathbf{j} + 4\mathbf{k}$$

Thus,

$$\kappa = \frac{|\mathbf{V}\times\mathbf{A}|}{v^3} = \frac{\sqrt{36\sin^2 t + 36\cos^2 t + 16}}{(13)^{3/2}} = \frac{\sqrt{52}}{13\sqrt{13}} = \frac{2}{13}$$

$$\mathbf{N} = \frac{(\mathbf{V}\times\mathbf{A})\times\mathbf{V}}{v^4\kappa} = \frac{1}{(13)^2\cdot\frac{2}{13}}\begin{vmatrix} \mathbf{i} & \mathbf{j} & \mathbf{k} \\ 6\sin t & -6\cos t & 4 \\ -2\sin t & 2\cos t & 3 \end{vmatrix}$$

$$= (-\cos t)\mathbf{i} - (\sin t)\mathbf{j}$$

and

$$\mathbf{B} = \frac{\mathbf{V}\times\mathbf{A}}{|\mathbf{V}\times\mathbf{A}|} = \frac{(6\sin t)\mathbf{i} - (6\cos t)\mathbf{j} + 4\mathbf{k}}{\sqrt{36\sin^2 t + 36\cos^2 t + 16}} = \frac{(3\sin t)\mathbf{i} - (3\cos t)\mathbf{j} + 2\mathbf{k}}{\sqrt{13}}$$

Also,
$$\frac{d\mathbf{A}}{dt} = (2 \sin t)\mathbf{i} - (2 \cos t)\mathbf{j}$$

so that

$$\tau = \frac{1}{v^6 \kappa^2} \frac{d\mathbf{A}}{dt} \cdot (\mathbf{V} \times \mathbf{A})$$

$$= \frac{1}{52} [(2 \sin t)\mathbf{i} - (2 \cos t)\mathbf{j}] \cdot [(6 \sin t)\mathbf{i} - (6 \cos t)\mathbf{j} + 4\mathbf{k}] = \frac{12}{52} = \frac{3}{13} \quad \blacksquare$$

Problem Set 13.6

In Problems 1 to 6, find (a) $\mathbf{F}'(t_0)$ and (b) $\mathbf{F}''(t_0)$.

1 $\mathbf{F}(t) = 5t^2\mathbf{i} + (3t + 1)\mathbf{j} + (3 - t^2)\mathbf{k}$, $t_0 = 1$

2 $\mathbf{F}(t) = \sqrt{t}\,\mathbf{i} + e^{-3t}\mathbf{j} + t\mathbf{k}$, $t_0 = 4$

3 $\mathbf{F}(t) = (1 + t)^2\mathbf{i} + (\cos t)\mathbf{j} + \ln(1 - t)\mathbf{k}$, $t_0 = 0$

4 $\mathbf{F}(t) = (2 + \sin t)\mathbf{i} + (\cos t)\mathbf{j} + t\mathbf{k}$, $t_0 = \pi/2$

5 $\mathbf{F}(t) = (\cos 2t)\mathbf{i} + (\sin 2t)\mathbf{j} + t^3\mathbf{k}$, $t_0 = \pi/4$

6 $\mathbf{F}(t) = e^{-3t}\mathbf{i} + e^{3t}\mathbf{j} + t\mathbf{k}$, $t_0 = 0$

In Problems 7 to 12, find (a) the velocity vector $\mathbf{V}$, (b) the speed v, and (c) the acceleration vector $\mathbf{A}$ at the time t of the particle whose equation of motion is given.

7 $\mathbf{R} = (2 \cos 2t)\mathbf{i} + (3 \sin 2t)\mathbf{j} + t\mathbf{k}$

8 $\mathbf{R} = (2 + t)\mathbf{i} + 3t\mathbf{j} + (5 - 3t)\mathbf{k}$

9 $\mathbf{R} = e^{-t}\mathbf{i} + 2e^t\mathbf{j} + t^3\mathbf{k}$

10 $\mathbf{R} = (\ln \cos t)\mathbf{i} + (\sin t)\mathbf{j} + (\cos t)\mathbf{k}$

11 $\mathbf{R} = (t \sin t)\mathbf{i} + (t \cos t)\mathbf{j} + t^2\mathbf{k}$

12 $\mathbf{R} = (e^{-t} \cos t)\mathbf{i} + (e^{-t} \sin t)\mathbf{j} + e^{-t}\mathbf{k}$

In Problems 13 to 16, find (a) $\mathbf{V}$, (b) v, and (c) $\mathbf{A}$ for the indicated value of t.

13 $\mathbf{R} = (\cos 2t)\mathbf{i} + (3e^{-t})\mathbf{j} + (\sin 2t)\mathbf{k}$ when $t = 0$

14 $\mathbf{R} = (t \sin t)\mathbf{i} + e^t\mathbf{j} + (\cos t)\mathbf{k}$ when $t = 0$

15 $\mathbf{R} = (4 \cos t)\mathbf{i} + (4 \sin t)\mathbf{j} + e^t\mathbf{k}$ when $t = \pi/2$

16 $\mathbf{R} = (2t - t^2)\mathbf{i} + 3t\mathbf{j} + (t^3 + 1)\mathbf{k}$ when $t = 1$

In Problems 17 to 26, find the arc length of the portion of the given curve between the point where $t = a$ and the point where $t = b$.

17 $\mathbf{R} = (3 \cos t)\mathbf{i} + (4 \cos t)\mathbf{j} + (5 \sin t)\mathbf{k}$; $a = 0$, $b = 2$

18 $\mathbf{R} = (\cos 3t)\mathbf{i} + (\sin 3t)\mathbf{j} + 5t\mathbf{k}$; $a = 0$, $b = 2\pi$

19 $\mathbf{R} = 6t^3\mathbf{i} + 3t^2\mathbf{j} + \mathbf{k}$; $a = 0$, $b = 3$

20 $\mathbf{R} = 2t^{3/2}\mathbf{i} + (t - 3)\mathbf{j} + 3t\mathbf{k}$; $a = 6$, $b = 10$

21 $\mathbf{R} = e^t\mathbf{i} + e^{-t}\mathbf{j} + \sqrt{2}t\mathbf{k}$; $a = 0$, $b = 1$

22 $\mathbf{R} = (t \sin t)\mathbf{i} + (t \cos t)\mathbf{j} + \mathbf{k}$; $a = 0$, $b = 4$

23 $\mathbf{R} = (e^{-t} \cos t)\mathbf{i} + (e^{-t} \sin t)\mathbf{j} + e^{-t}\mathbf{k}$; $a = 0$, $b = \pi$

24 $\mathbf{R} = (\cos t)\mathbf{i} + (\ln \cos t)\mathbf{j} + (\sin t)\mathbf{k}$; $a = 0$, $b = \pi/4$

25 $\mathbf{R} = (t^3/3 - 1/t)\mathbf{i} + (t^3/3 - 7/t)\mathbf{j} + 2t\mathbf{k}$; $a = 1$, $b = 4$

26 $\mathbf{R} = (2t^2 + 3)\mathbf{i} + (3 - 2t^2)\mathbf{j} + 4t\mathbf{k}$; $a = 0$, $b = 2$

In Problems 27 to 32, find (a) $\mathbf{V}$, (b) v, (c) $\mathbf{A}$, (d) $\mathbf{T}$, (e) $\mathbf{V} \times \mathbf{A}$, (f) κ, (g) $\mathbf{N}$, (h) $\mathbf{B}$, (i) $d\mathbf{A}/dt$, and (j) τ for each curve in space.

27 $\mathbf{R} = (2 \cos t)\mathbf{i} + (3 \sin t)\mathbf{j} + t\mathbf{k}$

28 $\mathbf{R} = (t^2 - 3)\mathbf{i} + (t^2 + 7)\mathbf{j} + (t^2 + t)\mathbf{k}$

29 $\mathbf{R} = (e^{2t} + e^{-2t})\mathbf{i} + (e^{2t} - e^{-2t})\mathbf{j} + 5t\mathbf{k}$

30 $\mathbf{R} = e^t\mathbf{i} + 2te^t\mathbf{j} + 3e^t\mathbf{k}$

31 $\mathbf{R} = t\mathbf{i} + \frac{1}{\sqrt{2}}t^2\mathbf{j} + \frac{1}{3}t^3\mathbf{k}$

32 $\mathbf{R} = \frac{\cos t}{4}\mathbf{i} + \frac{\ln \cos t}{4}\mathbf{j} + \frac{\sin t}{4}\mathbf{k}$

In Problems 33 to 36, let $\mathbf{F}(t) = 5t\mathbf{i} + (t^2 - 2)\mathbf{j} + t^3\mathbf{k}$, $\mathbf{G}(t) = t\mathbf{i} + (1/t)\mathbf{j} + t^2\mathbf{k}$, and $h(t) = 3t$.

33 Find $\dfrac{d}{dt}[h(t)\mathbf{F}(t)]$:

(a) By calculating $h(t)\mathbf{F}(t)$ and differentiating

(b) By using the formula

$$\frac{d}{dt}[h(t)\mathbf{F}(t)] = \left[\frac{d}{dt}h(t)\right]\mathbf{F}(t) + h(t)\left[\frac{d}{dt}\mathbf{F}(t)\right]$$

34 Find $\dfrac{d}{dt}[\mathbf{F}(t) \cdot \mathbf{G}(t)]$:

(a) By calculating $\mathbf{F}(t) \cdot \mathbf{G}(t)$ and differentiating

(b) By using the formula

$$\frac{d}{dt}[\mathbf{F}(t) \cdot \mathbf{G}(t)] = \left[\frac{d}{dt}\mathbf{F}(t)\right] \cdot \mathbf{G}(t) + \mathbf{F}(t) \cdot \left[\frac{d}{dt}\mathbf{G}(t)\right]$$

35 Find $\dfrac{d}{dt}[\mathbf{F}(t) \times \mathbf{G}(t)]$:

(a) By calculating $\mathbf{F}(t) \times \mathbf{G}(t)$ and differentiating

(b) By using the formula

$$\frac{d}{dt}[\mathbf{F}(t) \times \mathbf{G}(t)] = \left[\frac{d}{dt}\mathbf{F}(t)\right] \times \mathbf{G}(t) + \mathbf{F}(t) \times \left[\frac{d}{dt}\mathbf{G}(t)\right]$$

36 Find (a) $\dfrac{d}{dt}|\mathbf{F}(t)|$ and (b) $\dfrac{d}{dt}\mathbf{F}(t^3)$.

37 Find $\mathbf{T}$, $\mathbf{N}$, $\mathbf{B}$, κ, and τ for the circular helix

$$\begin{cases} x = a \cos t \\ y = a \sin t \\ z = bt \end{cases}$$

38 Prove that if $\lim\limits_{t \to c} \mathbf{F}(t)$ and $\lim\limits_{t \to c} \mathbf{G}(t)$ both exist, then $\lim\limits_{t \to c} [\mathbf{F}(t) \times \mathbf{G}(t)]$ exists and equals $[\lim\limits_{t \to c} \mathbf{F}(t)] \times [\lim\limits_{t \to c} \mathbf{G}(t)]$.

39 If $\mathbf{F}$ and $\mathbf{G}$ are continuous vector-valued functions, show that $\mathbf{F} \times \mathbf{G}$ is also continuous.

40 Suppose that $\mathbf{F}$ and $\mathbf{G}$ are differentiable vector-valued functions of t. Prove that $\mathbf{F} \times \mathbf{G}$ is differentiable and that

$$\frac{d}{dt}(\mathbf{F} \times \mathbf{G}) = \frac{d\mathbf{F}}{dt} \times \mathbf{G} + \mathbf{F} \times \frac{d\mathbf{G}}{dt}$$

41 Evaluate $\dfrac{d^2(\mathbf{F} \times \mathbf{G})}{dt^2}$.

42 Suppose that the curve $\mathbf{R} = \mathbf{F}(t)$ in space has curvature $\kappa = 0$ at all points. (a) Show that $\mathbf{T}$ is a constant vector. (b) Show that $\mathbf{T} \times \mathbf{R}$ is a constant vector. (c) Show that the curve is a portion of a line.

43 Prove the formula

$$\tau = \frac{1}{\kappa^2}\left(\frac{d\mathbf{R}}{ds} \times \frac{d^2\mathbf{R}}{ds^2}\right) \cdot \frac{d^3\mathbf{R}}{ds^3}$$

$$\left[\textit{Hint: Take } t = s \text{ in the formula } \tau = \frac{1}{v^6\kappa^2}\frac{d\mathbf{A}}{dt} \cdot (\mathbf{V} \times \mathbf{A}).\right]$$

44 Prove the third Frenet formula:

$$\frac{d\mathbf{B}}{ds} = -\tau\mathbf{N}$$

$$\left[\textit{Hint: } \frac{d\mathbf{B}}{ds} = \left(\frac{d\mathbf{B}}{ds} \cdot \mathbf{T}\right)\mathbf{T} + \left(\frac{d\mathbf{B}}{ds} \cdot \mathbf{N}\right)\mathbf{N} + \left(\frac{d\mathbf{B}}{ds} \cdot \mathbf{B}\right)\mathbf{B}.\right]$$

45 Suppose that the curve $\mathbf{R} = \mathbf{F}(t)$ in space lies in the plane $\mathbf{N}_0 \cdot (\mathbf{R} - \mathbf{R}_0) = 0$. Prove that the torsion τ of the curve is identically zero. (*Caution:* Here $\mathbf{N}_0$ is the *constant* normal vector *to the plane*—do not confuse it with $\mathbf{N}$, the variable principal normal vector *to the curve*.)

46 Suppose that the curve $\mathbf{R} = \mathbf{F}(t)$ in space has torsion $\tau = 0$ at all points. Prove that the curve lies in a plane.

47 Prove that $\tau = \pm\left|\dfrac{d\mathbf{B}}{ds}\right|$.

48 Let $\mathbf{R} = \mathbf{F}(t)$ be the equation of motion of a particle P of mass m. We say that P is moving under the influence of a **central force f** if there is a function g such that $\mathbf{f} = g(|\mathbf{R}|)\mathbf{R}$ holds for all values of time t. Assuming that this is so and using Newton's second law of motion, $\mathbf{f} = m\mathbf{A}$, (a) show that $\mathbf{R} \times \mathbf{A} = \mathbf{0}$ for all values of t, (b) show that the orbit of P has torsion $\tau = 0$ at all points, and (c) conclude that P moves in a plane. (See Problem 46.)

13.7 Spheres, Cylinders, and Surfaces of Revolution

Figure 1

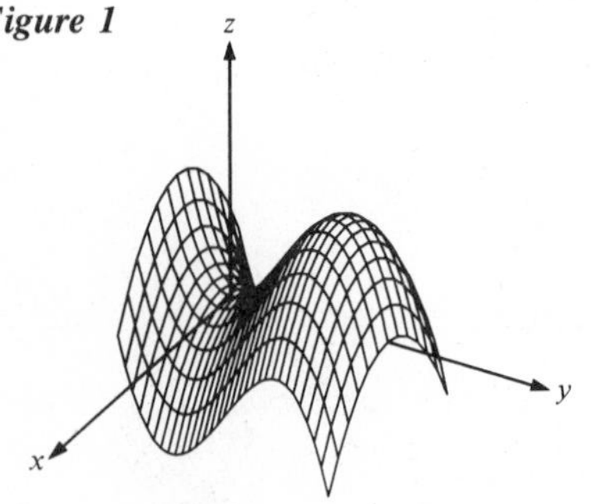

In Section 13.4 we found that a plane in xyz space has a scalar Cartesian equation $ax + by + cz = D$. In general, the graph of an equation involving x, y, and z is a two-dimensional surface in xyz space. For instance, Figure 1 shows a computer-generated graph of the equation

$$z = -\frac{1}{2}(y - 1)^3 + \frac{3}{2}(y - 1) - \frac{1}{2}x^2 + 1$$

In this section and the next, we study some special types of surfaces and their equations.

Figure 2

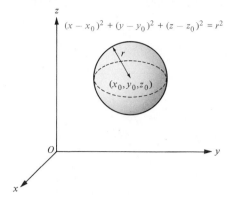

Spheres

If $P = (x, y, z)$ and $P_0 = (x_0, y_0, z_0)$, then, as we saw in Section 13.2, the distance between P and P_0 is given by

$$|\overrightarrow{PP_0}| = \sqrt{(x - x_0)^2 + (y - y_0)^2 + (z - z_0)^2}$$

Therefore, the equation

$$(x - x_0)^2 + (y - y_0)^2 + (z - z_0)^2 = r^2$$

where r is a positive constant, means that the distance between P and P_0 is r units; hence, *this equation represents the surface of a sphere of radius r with center at the point $P_0 = (x_0, y_0, z_0)$* (Figure 2). In particular, if $P_0 = (0, 0, 0)$, then the equation of the sphere becomes

$$x^2 + y^2 + z^2 = r^2$$

EXAMPLE 1   Find an equation of the sphere with radius 5 units and center at $(-1, 2, 4)$.

SOLUTION   An equation of the sphere is $(x + 1)^2 + (y - 2)^2 + (z - 4)^2 = 25$. ■

EXAMPLE 2   Find the radius and the center of the sphere whose equation is $x^2 + y^2 + z^2 - 6x + 2y + 4z - 11 = 0$.

SOLUTION   Regrouping terms and completing the squares, we obtain

$$(x^2 - 6x + 9) + (y^2 + 2y + 1) + (z^2 + 4z + 4) = 9 + 1 + 4 + 11$$

or

$$(x - 3)^2 + (y + 1)^2 + (z + 2)^2 = 25$$

which represents a sphere of radius 5 with center $(3, -1, -2)$. ■

Figure 3

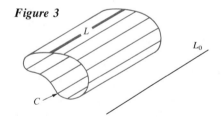

Figure 4

Cylinders

A line L in space which moves so as to remain parallel to a fixed line L_0 while intersecting a fixed curve C is said to sweep out a **cylindrical surface**—or just a **cylinder** for short (Figure 3). Any particular position of L is called a **generator** of the cylinder, and the curve C is called the **base curve.** If the base curve C is a circle and all the generators are perpendicular to the plane of the circle, then the surface is called a **right circular cylinder.** A cylinder whose base curve is an ellipse, parabola, or hyperbola is naturally called an **elliptic, parabolic,** or **hyperbolic** cylinder, respectively.

In what follows we consider only cylinders whose generators are parallel to one of the coordinate axes and whose base curve lies in the coordinate plane perpendicular to that axis. For instance, the cylinder in Figure 4 has generators parallel to the z axis; hence, if the point $Q = (x_0, y_0, 0)$ is on the base curve C, then all points of the form $P = (x_0, y_0, z)$ also belong to the cylinder. It follows that the equation of this cylinder cannot place any restriction on z whatsoever—that is, either z is not present in the equation, or else it can be removed from the equation by algebraic manipulation.

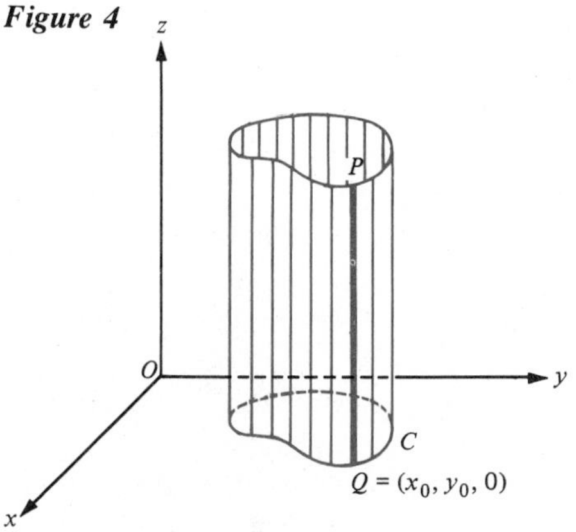

To sketch the graph of a cylindrical surface whose equation is missing one of the variables x, y, or z, begin by sketching the graph of the equation in the coordinate plane involving the variables that are present, thus obtaining the base curve C. The surface is then swept out by a line perpendicular to this coordinate plane and intersecting the base curve.

In Examples 3 and 4, the graph of the given equation is a cylinder. **(a)** *Identify the coordinate axis that is parallel to the generators.* **(b)** *Specify the coordinate plane containing the base curve.* **(c)** *Describe the base curve.* **(d)** *Sketch the cylinder.*

EXAMPLE 3 $y^2 = x$

SOLUTION

(a) The variable z is missing; hence, the generators are parallel to the z axis.

(b) The base curve is contained in the xy plane.

(c) The base curve is a parabola.

(d) The graph appears in Figure 5. ∎

EXAMPLE 4 $z = \sin x$

SOLUTION

(a) The variable y is missing; hence, the generators are parallel to the y axis.

(b) The base curve is contained in the xz plane.

(c) The base curve is a sine curve.

(d) The graph appears in Figure 6. ∎

Figure 5

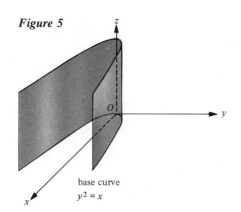

base curve
$y^2 = x$

Figure 6

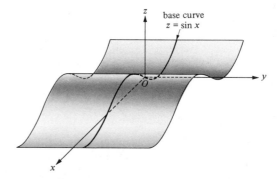

base curve
$z = \sin x$

Figure 7

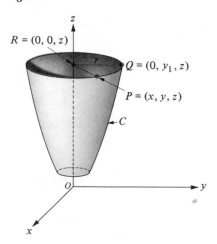

$R = (0, 0, z)$
$Q = (0, y_1, z)$
$P = (x, y, z)$
C

Surfaces of Revolution

A **surface of revolution** is defined to be a surface generated by revolving a plane curve C about a line L lying in the same plane as the curve. The line L is called the **axis of revolution.** For instance, a sphere is the surface of revolution generated by revolving a circle C about an axis L passing through its center. Notice that a surface of revolution intersects a plane perpendicular to the axis of revolution in a circle or a single point, provided that it intersects the plane at all.

Figure 7 shows a portion of a surface of revolution obtained by revolving the curve C in the yz plane about the z axis. In order to find an equation of this surface, consider a general point $P = (x, y, z)$ on the surface, noting that P is obtained by revolving some point Q on the original curve C about the z axis as in Figure 7. Perpendiculars dropped from P and Q to the z axis meet the z axis at the same point

$R = (0, 0, z)$. Notice that P, Q, and R have the same z coordinate. Let $r = |\overline{PR}| = |\overline{QR}|$. Since Q lies in the yz plane, its x coordinate is zero. Let $Q = (0, y_1, z)$, noting that $|y_1| = r$.

Now,

$$y_1^2 = r^2 = |\overline{PR}|^2 = (x - 0)^2 + (y - 0)^2 + (z - z)^2 = x^2 + y^2$$

Hence, $y_1 = \pm\sqrt{x^2 + y^2}$. We conclude that if $P = (x, y, z)$ belongs to the surface of revolution, then either $Q = (0, \sqrt{x^2 + y^2}, z)$ or $Q = (0, -\sqrt{x^2 + y^2}, z)$ belongs to the generating curve. Conversely, if $Q = (0, y_1, z)$ is a point on the generating curve C, then any point $P = (x, y, z)$ such that $\sqrt{x^2 + y^2} = |y_1|$ lies on a horizontal circle of radius $r = |y_1|$ with center at $R = (0, 0, z)$; hence, such a point P belongs to the surface of revolution. We conclude that $P = (x, y, z)$ belongs to the surface of revolution if and only if $Q = (0, \pm\sqrt{x^2 + y^2}, z)$ belongs to the generating curve. In other words, *an equation of the surface generated by revolving the curve C in the yz plane about the z axis is obtained by replacing the variable y in an equation of C by the expression $\pm\sqrt{x^2 + y^2}$.*

EXAMPLE 5   The parabola $z = y^2$ in the yz plane is revolved about the z axis to form a surface of revolution. A computer-generated graph of this surface is shown in Figure 8. Find an equation of this surface.

SOLUTION   Replacing y in the equation $z = y^2$ by $\pm\sqrt{x^2 + y^2}$, we obtain

$$z = (\pm\sqrt{x^2 + y^2})^2 \quad \text{or} \quad z = x^2 + y^2 \qquad \blacksquare$$

The surface in Figure 8 is called a **paraboloid of revolution.** Similarly, a surface of revolution obtained by revolving a hyperbola or an ellipse about one of its own axes of symmetry is called a **hyperboloid of revolution** or an **ellipsoid of revolution,** respectively.

Figure 8

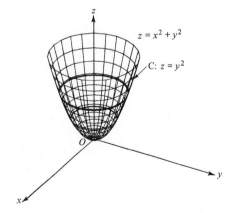

The preceding considerations apply to curves lying in any coordinate plane, and we have the following general rule for finding an equation of a surface of revolution:

> **1** List the three variables in such an order that the first variable represents the axis about which the generating curve C is revolved and the first two variables represent the plane in which C lies.
>
> **2** In an equation for C, replace the second listed variable by plus or minus the square root of the sum of the squares of the second and third listed variables.

This procedure is summarized by the following table.

Axis of revolution	Plane curve C lies in the:	In an equation for C, replace:
x axis	xy plane	y by $\pm\sqrt{y^2 + z^2}$
x axis	xz plane	z by $\pm\sqrt{z^2 + y^2}$
y axis	yx plane	x by $\pm\sqrt{x^2 + z^2}$
y axis	yz plane	z by $\pm\sqrt{z^2 + x^2}$
z axis	zx plane	x by $\pm\sqrt{x^2 + y^2}$
z axis	zy plane	y by $\pm\sqrt{y^2 + x^2}$

EXAMPLE 6 Find an equation of the surface of revolution generated by revolving the curve $y^2 = 3x$ in the xy coordinate plane about the x axis. Sketch a graph of the surface.

Figure 9

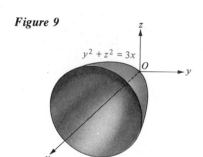

$y^2 + z^2 = 3x$

SOLUTION   Following step 1 of the rule above, we list the three variables in the order x, y, z. Carrying out step 2, we replace y in the equation $y^2 = 3x$ by $\pm\sqrt{y^2 + z^2}$ to obtain $(\pm\sqrt{y^2 + z^2})^2 = 3x$, or $y^2 + z^2 = 3x$ (Figure 9). ∎

Note that *an equation represents a surface of revolution about one of the coordinate axes if it contains the variables corresponding to the other two coordinate axes only in combination as the sum of their squares*. For instance, the equation $y^2 = 4(z^2 + x^2)$ contains the two variables x and z only in the combination $z^2 + x^2$ and so represents a surface of revolution obtained by revolving a curve about the y axis.

If a surface is known to be a surface of revolution about one of the coordinate axes, a generating curve C can be found by intersecting the surface with either one of the coordinate planes containing that coordinate axis. Of course, in order to find an equation of the intersection of a surface with the xy, xz, or yz coordinate plane, you may set z, y, or x, respectively, equal to 0 in the equation of the surface.

EXAMPLE 7 For the surface of revolution $x^2 - y^2 + z^2 = 1$:

(a) Find the axis of revolution.

(b) Find a generating curve in either one of the coordinate planes containing the axis of revolution.

(c) Sketch the graph.

SOLUTION

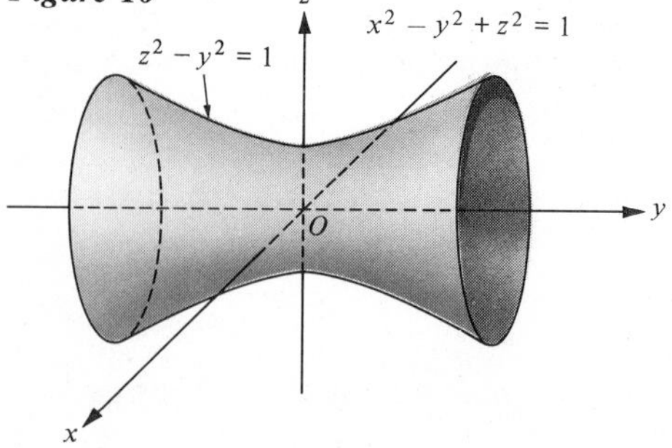

(a) The equation can be rewritten as $(x^2 + z^2) - y^2 = 1$ and thus contains the variables x and z only in the combination $x^2 + z^2$. It is therefore a surface of revolution about the y axis.

(b) We can find a generating curve by intersecting the surface with either the xy plane or the yz plane. The intersection with the yz plane is found by setting $x = 0$ in the equation $x^2 - y^2 + z^2 = 1$ to obtain $z^2 - y^2 = 1$, the equation of a hyperbola.

(c) The surface, sketched in Figure 10, is generated by revolving the hyperbola $z^2 - y^2 = 1$ about the y axis and thus is a hyperboloid of revolution. ∎

Figure 10

$x^2 - y^2 + z^2 = 1$

$z^2 - y^2 = 1$

Problem Set 13.7

In Problems 1 to 4, find an equation of the sphere satisfying the given conditions.

1 Center at $(2, -1, 3)$ and radius 4 units

2 Its diameter is the line segment joining the points $(8, -1, 7)$ and $(-2, -5, 5)$.

3 Center at $(4, 1, 3)$ and containing the point $(2, -1, 2)$

4 Containing the points $(3, 0, 4)$, $(3, 4, 0)$, $(9, 0, \sqrt{7})$, and $(-3, 0, 4)$

In Problems 5 to 9, find the center and radius of the sphere.

5 $x^2 + y^2 + z^2 - 2x + 4z - 4 = 0$

6 $x^2 + y^2 + z^2 + 2x + 14y - 10z + 74 = 0$

7 $2x^2 + 2y^2 + 2z^2 + 4x - 4y - 14 = 0$

8 $x^2 + y^2 + z^2 - 8x + 4y + 10z + 9 = 0$

9 $x^2 + y^2 + z^2 - 6x + 2y - 4z - 19 = 0$

10 Find the set of points that are equidistant from the points $(1, 3, -1)$ and $(2, -1, 2)$.

In Problems 11 to 30, (a) identify the coordinate axis to which the generators of the cylinder are parallel, (b) specify the coordinate plane containing the base curve, (c) describe the base curve, and (d) sketch the cylinder.

11 $y = z^2$

12 $xy = 3$

13 $yz = 1$

14 $y^2 + z^2 = 4$

15 $x^2 + y^2 = 9$

16 $\dfrac{x^2}{4} - \dfrac{y^2}{1} = 1$

17 $\dfrac{x^2}{9} + \dfrac{y^2}{4} = 1$

18 $y^2 = 4 - z$

19 $x^2 - z = 0$

20 $y = e^z$

21 $y + z = 3$

22 $3x + 2y = 6$

23 $9y^2 - 4z^2 = 36$

24 $z = |x|$

25 $z = \sin 2x$

26 $y = e^x$

27 $x^2 + 4y = 0$

28 $x^2 + y^2 = 4x$

29 $x^2 = z^2$

30 $|y| + |z| = 1$

In Problems 31 to 40, find the equation of the surface that results when the graph of the given equation is revolved about the indicated axis. Sketch the surface.

31 $y = x$ in the xy plane about the x axis

32 $y = 3x + 2$ in the xy plane about the y axis

33 $z^2 = 8x$ in the xz plane about the x axis

34 $4x^2 - 9z^2 = 5$ in the xz plane about the z axis

35 $z = x^2$ in the xz plane about the z axis

36 $\dfrac{y^2}{9} + \dfrac{z^2}{4} = 1$ in the yz plane about the z axis

37 $6y^2 + 6z^2 = 7$ in the yz plane about the y axis

38 $x^2 - 4z^2 = 4$ in the xz plane about the x axis

39 $z = 1 - y^2$ in the yz plane about the z axis

40 $y = \ln x$ in the xy plane about the x axis

In Problems 41 to 46, each equation represents a surface of revolution about one of the coordinate axes. Identify this axis, find the equation of a generating curve in the indicated coordinate plane, and sketch the graph.

41 $x^2 + y^2 + z = 3$; xz plane

42 $9z^2 = x^2 + y^2$; xz plane

43 $x^2 - 9y^2 - 9z^2 = 18$; xy plane

44 $4 - y = 3(x^2 + z^2)$; yz plane

45 $x^2 + y^2 = \sin^2 z$; xz plane

46 $9x^2 + 9y^2 + 4z^2 = 36$; yz plane

47 If $A^2 + B^2 + C^2 > 4D$, by completing the squares show that the equation $x^2 + y^2 + z^2 + Ax + By + Cz + D = 0$ can be rewritten in the form $(x - x_0)^2 + (y - y_0)^2 + (z - z_0)^2 = r^2$.

13.8   Quadric Surfaces

In Chapter 9 we showed that, apart from certain degenerate cases, the graph in the xy plane of a second-degree equation

$$Ax^2 + Bxy + Cy^2 + Dx + Ey + F = 0$$

is a circle, an ellipse, a parabola, or a hyperbola; in any case, it is a conic section. A second-degree equation in the *three* variables x, y, and z has the form

$$Ax^2 + By^2 + Cz^2 + Dxy + Exz + Fyz + Gx + Hy + Iz + K = 0$$

and the graph of such an equation in xyz space is called a **quadric surface.** By rotating and translating the coordinate system in three-dimensional space, it is possible to bring the second-degree equation into certain standard forms and thus to classify the quadric surfaces. Figures 1 through 6 show computer-generated graphs of some quadric surfaces.

In this section we do not attempt a detailed study and classification of quadric surfaces. Rather, our intention is to familiarize you with the general shapes and the

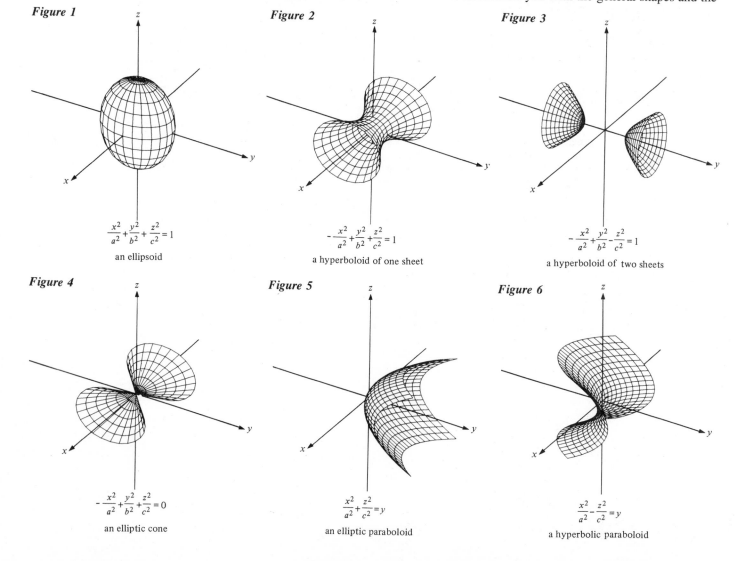

Figure 1

$$\frac{x^2}{a^2} + \frac{y^2}{b^2} + \frac{z^2}{c^2} = 1$$

an ellipsoid

Figure 2

$$-\frac{x^2}{a^2} + \frac{y^2}{b^2} + \frac{z^2}{c^2} = 1$$

a hyperboloid of one sheet

Figure 3

$$-\frac{x^2}{a^2} + \frac{y^2}{b^2} - \frac{z^2}{c^2} = 1$$

a hyperboloid of two sheets

Figure 4

$$-\frac{x^2}{a^2} + \frac{y^2}{b^2} + \frac{z^2}{c^2} = 0$$

an elliptic cone

Figure 5

$$\frac{x^2}{a^2} + \frac{z^2}{c^2} = y$$

an elliptic paraboloid

Figure 6

$$\frac{x^2}{a^2} - \frac{z^2}{c^2} = y$$

a hyperbolic paraboloid

standard equations of the common quadric surfaces. We begin by mentioning, very briefly, some of the basic techniques for visualizing or graphing surfaces in space. These involve (1) **cross sections,** (2) **traces** and **intercepts,** and (3) **symmetries.**

Cross Sections

The simplest (but in many cases the most effective) technique for visualizing, recognizing, or graphing surfaces in space is to find the **cross sections** formed by intersecting the surface with planes—especially by planes perpendicular to the coordinate axes or perpendicular to the axes of symmetry of the surface. For example, Figure 7 shows the cross section cut from a surface by a plane $y = a$ perpendicular to the y axis. An equation of the cross section *in the cutting plane* $y = a$ can be found simply by putting $y = a$ in the equation of the surface. The resulting equation, which involves only x and z, is an equation of the cross section relative to copies of the x and z axes in the cutting plane as shown in Figure 7. Exactly the same idea applies to cross sections cut by planes perpendicular to the other coordinate axes.

Figure 7

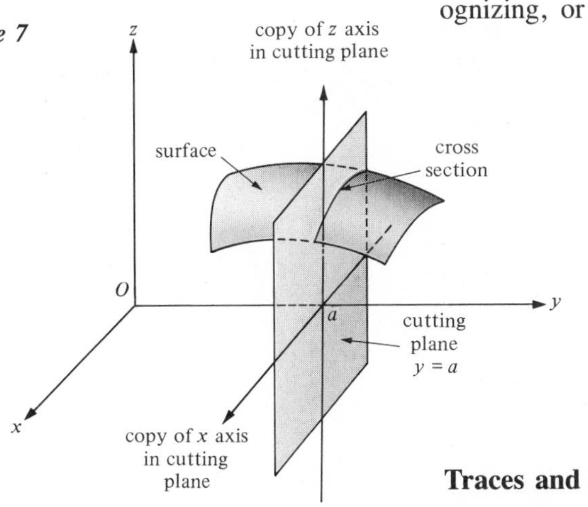

Traces and Intercepts

The cross sections cut from a surface by the coordinate planes themselves are called the **traces** of the surface. For instance, the **yz trace** of the surface (that is, the cross section cut from the surface by the yz plane) is obtained by putting $x = 0$ in an equation of the surface. The **xy trace** and the **xz trace** are defined similarly.

The **x, y,** and **z intercepts** of the surface are defined to be the points (if any) in which the x, y, and z axes, respectively, intersect the surface. For instance, to find the x intercept, we set $y = 0$ and $z = 0$ in an equation of the surface.

Symmetries

Surfaces often display symmetries with respect to points, lines, or planes. Of course, a surface is said to be **symmetric** with respect to a point, line, or plane provided that whenever a point P lies on the surface, so does the point Q, which is symmetrically located with respect to the point, line, or plane. For instance, if an equation equivalent to the original equation of the surface is obtained when y is replaced by $-y$, then the surface is symmetric with respect to the xz plane. Symmetry with respect to the z axis can be tested by replacing *both x and y* by $-x$ and $-y$, respectively, and checking whether the resulting equation is equivalent to the original one. Similar tests are easily discovered for symmetries with respect to the other coordinate planes or coordinate axes as well as for symmetry with respect to the origin (Problems 25 to 27).

Central Quadric Surfaces

The graph in xyz space of an equation of the form

$$\pm\frac{x^2}{a^2} \pm \frac{y^2}{b^2} \pm \frac{z^2}{c^2} = 1$$

where a, b, and c are positive constants and not all three algebraic signs are negative, is called a **central quadric** surface. Since only even powers of the variables appear, a central quadric surface is symmetric with respect to all three coordinate planes, all three coordinate axes, and the origin. The central quadric surfaces are classified as follows:

1 If all three algebraic signs are positive, then the surface is called an **ellipsoid.**

2 If two algebraic signs are positive and one is negative, then the surface is called a **hyperboloid of one sheet.**

3 If one algebraic sign is positive and the other two are negative, then the surface is called a **hyperboloid of two sheets.**

We now discuss each of the central quadrics briefly.

The Ellipsoid

The x, y, and z intercepts of the ellipsoid

$$\frac{x^2}{a^2} + \frac{y^2}{b^2} + \frac{z^2}{c^2} = 1$$

are $(\pm a, 0, 0)$, $(0, \pm b, 0)$, and $(0, 0, \pm c)$, and the traces on the coordinate planes are the ellipses (or circles) $(x^2/a^2) + (y^2/b^2) = 1$, $(x^2/a^2) + (z^2/c^2) = 1$, and $(y^2/b^2) + (z^2/c^2) = 1$. In fact, all cross sections cut by planes perpendicular to the coordinate axes are ellipses (or circles) (Problem 29). The graph is sketched in Figure 8.

Figure 8

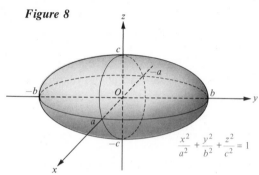

$$\frac{x^2}{a^2} + \frac{y^2}{b^2} + \frac{z^2}{c^2} = 1$$

The Hyperboloid of One Sheet

Suppose, for definiteness, that the term involving z^2 carries the negative sign, so that the equation has the form

$$\frac{x^2}{a^2} + \frac{y^2}{b^2} - \frac{z^2}{c^2} = 1$$

The x and y intercepts are $(\pm a, 0, 0)$ and $(0, \pm b, 0)$, but there is no z intercept since the equation $-z^2/c^2 = 1$ cannot be satisfied by any real number z. The traces on the coordinate planes are as follows:

1 The xy trace is the ellipse (or circle) $(x^2/a^2) + (y^2/b^2) = 1$.

2 The xz trace is the hyperbola $(x^2/a^2) - (z^2/c^2) = 1$.

3 The yz trace is the hyperbola $(y^2/b^2) - (z^2/c^2) = 1$.

Indeed, all cross sections cut by planes perpendicular to the x or y axes are hyperbolas or pairs of lines, and all cross sections cut by horizontal planes are ellipses (Problem 30). A portion of the surface is sketched in Figure 9.

The Hyperboloid of Two Sheets

Suppose, for definiteness, that the terms involving y^2 and z^2 carry the negative signs, so that the equation has the form

$$\frac{x^2}{a^2} - \frac{y^2}{b^2} - \frac{z^2}{c^2} = 1$$

Figure 9

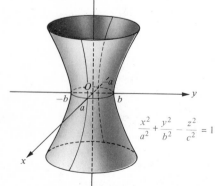

$$\frac{x^2}{a^2} + \frac{y^2}{b^2} - \frac{z^2}{c^2} = 1$$

The x intercepts are $(\pm a, 0, 0)$, but there are no y and no z intercepts. (Why?) The traces on the coordinate planes are as follows:

1 The xy trace is the hyperbola $(x^2/a^2) - (y^2/b^2) = 1$.

2 The xz trace is the hyperbola $(x^2/a^2) - (z^2/c^2) = 1$.

3 There is no yz trace.

Figure 10

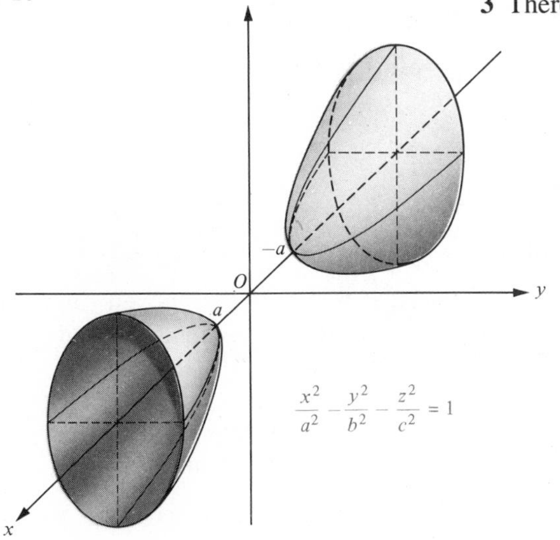

$$\frac{x^2}{a^2} - \frac{y^2}{b^2} - \frac{z^2}{c^2} = 1$$

Although the hyperboloid of two sheets has no trace on the yz plane, it does have cross sections cut by certain planes $x = k$ parallel to the yz plane. Indeed, the cross section cut by the plane $x = k$ has the equation

$$\frac{k^2}{a^2} - \frac{y^2}{b^2} - \frac{z^2}{c^2} = 1 \qquad \text{or} \qquad \frac{y^2}{b^2} + \frac{z^2}{c^2} = \frac{k^2}{a^2} - 1$$

provided that $|k| > a$. The last equation can be rewritten as

$$\frac{y^2}{b^2[(k^2/a^2) - 1]} + \frac{z^2}{c^2[(k^2/a^2) - 1]} = 1$$

Therefore, the cross sections cut by planes perpendicular to the x axis and at least a units from the origin are ellipses. A portion of the surface, which consists of two separated "parts," or "sheets," is shown in Figure 10.

Elliptic Cones

The graph in xyz space of an equation of the form

$$\pm\frac{x^2}{a^2} \pm \frac{y^2}{b^2} \pm \frac{z^2}{c^2} = 0$$

where a, b, and c are positive constants and not all three algebraic signs are the same, is called an **elliptic cone.** By multiplying by -1, if necessary, we can arrange that two of the algebraic signs are positive and the remaining one is negative.

Suppose, for definiteness, that the equation has the form

$$\frac{x^2}{a^2} + \frac{y^2}{b^2} - \frac{z^2}{c^2} = 0$$

Again, since only even powers of the variables occur, the elliptic cone is symmetric with respect to all coordinate planes, all coordinate axes, and the origin. If two of the variables are set equal to zero in the equation, then the third variable must also be zero; hence, the only x, y, or z intercept of the elliptic cone is the origin. The traces on the coordinate planes are as follows:

1 The xy trace is $(x^2/a^2) + (y^2/b^2) = 0$, or $(x, y) = (0, 0)$, just one point at the origin.

2 The xz trace is $(x^2/a^2) - (z^2/c^2) = 0$, or $z = \pm(c/a)x$, a pair of intersecting lines.

3 The yz trace is $(y^2/b^2) - (z^2/c^2) = 0$, or $z = \pm(c/b)x$, a pair of intersecting lines.

Figure 11

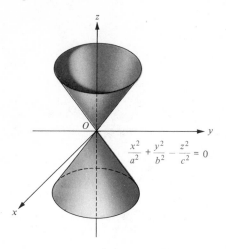

Cross sections cut by planes perpendicular to the x or y axis that do not pass through the origin are hyperbolas, and cross sections cut by horizontal planes that do not pass through the origin are ellipses (or circles) (Problem 31). A portion of the elliptic cone is shown in Figure 11. Notice that if $a = b$, the elliptic cone becomes an ordinary right circular cone.

Elliptic and Hyperbolic Paraboloids

We now consider the graph in xyz space of an equation having one of the forms

$$\pm\frac{x^2}{a^2} \pm \frac{y^2}{b^2} = z \quad \text{or} \quad \pm\frac{y^2}{b^2} \pm \frac{z^2}{c^2} = x \quad \text{or} \quad \pm\frac{z^2}{c^2} \pm \frac{x^2}{a^2} = y$$

where a, b, and c are positive constants. If both terms on the left carry the same algebraic sign, then the graph of any one of these equations is called an **elliptic paraboloid.** On the other hand, if the terms on the left carry opposite algebraic signs, then the graph of any one of the equations is called a **hyperbolic paraboloid.** We discuss these two cases briefly, considering only the first equation, $\pm x^2/a^2 \pm y^2/b^2 = z$. The other equations are handled similarly.

The Elliptic Paraboloid

For definiteness, we assume that the coefficients of x^2 and y^2 are both positive, so that the equation can be written in the form

$$\frac{x^2}{a^2} + \frac{y^2}{b^2} = z$$

where $a, b > 0$. Evidently, this surface intersects the xy plane only at the origin and otherwise lies above the xy plane. It is symmetric with respect to the xz plane, the yz plane, and the z axis, because only even powers of x and y occur. The traces in the xz and yz planes are the parabolas $z = x^2/a^2$ and $z = y^2/b^2$, respectively. The cross sections cut by horizontal planes above the origin are ellipses (or circles) (Problem 32). A sketch of a portion of the elliptic paraboloid appears in Figure 12. Notice that in the special case in which $a = b$, the elliptic paraboloid becomes a paraboloid of revolution (about the z axis).

Figure 12

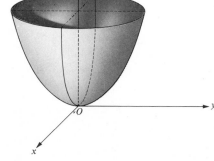

The Hyperbolic Paraboloid

For definiteness, we assume that the coefficient of y^2 is positive while the coefficient of x^2 is negative, so that the equation can be written in the form

$$\frac{y^2}{b^2} - \frac{x^2}{a^2} = z$$

where $a, b > 0$. The coordinate axes intersect this surface only at the origin, and the surface is symmetric with respect to the xz plane, the yz plane, and the z axis. The traces are as follows:

1 The xy trace, $(y^2/b^2) - (x^2/a^2) = 0$, or $y = \pm(b/a)x$, is a pair of intersecting lines.

2 The xz trace, $-x^2/a^2 = z$, is a parabola opening downward.

3 The yz trace, $y^2/b^2 = z$, is a parabola opening upward.

Moreover, cross sections cut by horizontal planes above the origin are hyperbolas with transverse axes parallel to the y axis, and cross sections cut by horizontal planes below the origin are hyperbolas with transverse axes parallel to the x axis (Problem 33). All cross sections cut by planes perpendicular to the x axis or y axis are parabolas opening upward or downward, respectively (Problem 34). A sketch of a portion of the hyperbolic paraboloid appears in Figure 13. Near the origin, the hyperbolic paraboloid has the shape of a saddle.*

Figure 13

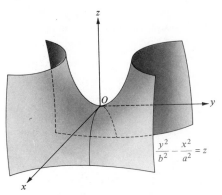

$$\frac{y^2}{b^2} - \frac{x^2}{a^2} = z$$

Examples of Quadric Surfaces

The surfaces discussed above, together with cylinders whose base curves are conic sections, exhaust all possibilities for quadric surfaces with the exception of certain degenerate cases which we do not consider here. In the following examples we illustrate the technique of identifying and graphing quadrics whose equations are in standard form.

In Examples 1 to 4, for the given quadric surface, (a) find the intercepts, (b) discuss the symmetry, (c) find the cross sections perpendicular to the coordinate axes, (d) find the traces, (e) identify the surface, and (f) sketch the graph.

<u>EXAMPLE 1</u> $x^2 - 9y^2 + z^2 = 81$

SOLUTION   After dividing by 81, we see that the equation has the form $(x^2/a^2) - (y^2/b^2) + (z^2/c^2) = 1$, where $a = 9$, $b = 3$, and $c = 9$.

(a) The x intercepts are $(\pm 9, 0, 0)$. There are no y intercepts. The z intercepts are $(0, 0, \pm 9)$.

(b) All variables are squared, so the surface is symmetric with respect to all three coordinate planes, all three coordinate axes, and the origin.

(c) The intersection with the plane $x = k$ is the curve $z^2 - 9y^2 = 81 - k^2$, which is a hyperbola except when $k = \pm 9$. When $k = \pm 9$, the intersection with the plane $x = k$ is the pair of intersecting lines $z = \pm 3y$. The intersection with the plane $y = k$ is the circle $x^2 + z^2 = 81 + 9k^2$ of radius $\sqrt{81 + 9k^2}$. The intersection with the plane $z = k$ is the curve $x^2 - 9y^2 = 81 - k^2$, which is a hyperbola except when $k = \pm 9$, in which case it is a pair of intersecting lines $x = \pm 3y$.

(d) The traces are found by putting $k = 0$ in (c). The yz trace is the hyperbola $z^2 - 9y^2 = 81$. The xz trace is the circle $x^2 + z^2 = 81$. The xy trace is the hyperbola $x^2 - 9y^2 = 81$.

(e) The surface is a hyperboloid of one sheet—in fact, a hyperboloid of revolution about the y axis.

(f) The graph is sketched in Figure 14.   ∎

Figure 14

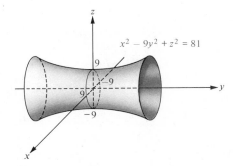

$$x^2 - 9y^2 + z^2 = 81$$

*For this reason, the origin O is called a *saddle point* of this surface.

EXAMPLE 2 $-9x^2 - 16y^2 + z^2 = 144$

SOLUTION   The equation has the form $-(x^2/a^2) - (y^2/b^2) + (z^2/c^2) = 1$ with $a = 4$, $b = 3$, and $c = 12$.

(a) There are no x intercepts. There are no y intercepts. The z intercepts are $(0, 0, \pm12)$.

(b) All variables are squared, so that the surface is symmetric with respect to all three coordinate planes, all three coordinate axes, and the origin.

(c) The intersection with the plane $x = k$ is the hyperbola $z^2 - 16y^2 = 144 + 9k^2$. The intersection with the plane $y = k$ is the hyperbola $z^2 - 9x^2 = 144 + 16k^2$. For $|k| > 12$, the intersection with the plane $z = k$ is the ellipse $9x^2 + 16y^2 = k^2 - 144$.

(d) Putting $k = 0$ in (c), we find that the yz and xz traces are the hyperbolas $z^2 - 16y^2 = 144$ and $z^2 - 9x^2 = 144$, respectively; there is no xy trace.

(e) The surface is a hyperboloid of two sheets.

(f) The graph is sketched in Figure 15.

Figure 15

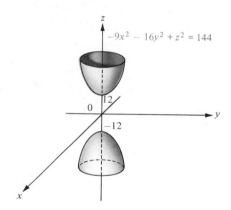

EXAMPLE 3 $\dfrac{x^2}{16} - \dfrac{y^2}{9} + \dfrac{z^2}{4} = 0$

SOLUTION

(a) The only intercept along any coordinate axis is the origin $(0, 0, 0)$.

(b) The surface is symmetric with respect to all three coordinate planes, all three coordinate axes, and the origin.

(c) The hyperbola $(y^2/9) - (z^2/4) = k^2/16$ is the intersection with the plane $x = k$ for $k \neq 0$. For $k \neq 0$, the intersection with the plane $y = k$ is the ellipse $(x^2/16) + (z^2/4) = k^2/9$. For $k \neq 0$, the intersection with the plane $z = k$ is the hyperbola $(y^2/9) - (x^2/16) = k^2/4$.

(d) The yz trace consists of two intersecting lines, $y = \pm\frac{3}{2}z$. The xz trace consists just of the origin $(0, 0, 0)$. The xy trace consists of two intersecting lines, $y = \pm\frac{3}{4}x$.

(e) The surface is an elliptic cone.

(f) The graph is sketched in Figure 16.

Figure 16

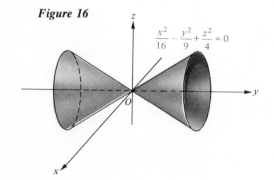

EXAMPLE 4 $9y^2 - 25z^2 = x$

SOLUTION   The equation has the form $(y^2/b^2) - (z^2/c^2) = x$, where $b = \frac{1}{3}$ and $c = \frac{1}{5}$.

(a) The only intercept along any coordinate axis is the origin $(0, 0, 0)$.

(b) The squared variables are y and z, so the surface is symmetric with respect to the xy plane, the xz plane, and the x axis.

(c) For $k < 0$, the intersection with the plane $x = k$ is the hyperbola $25z^2 - 9y^2 = -k$ whose transverse axis is parallel to the z axis. For $k > 0$, the intersection with the plane $x = k$ is the hyperbola $9y^2 - 25z^2 = k$ whose transverse axis is parallel to the y axis. The intersection with the plane $y = k$ is the parabola $x = 9k^2 - 25z^2$, opening in the direction of the negative x axis. The intersection with the plane $z = k$ is the parabola $x = 9y^2 - 25k^2$, opening in the direction of the positive x axis.

(d) The yz trace is the pair of intersecting lines $y = \pm\frac{5}{3}z$. The xz trace is the parabola $x = -25z^2$. The xy trace is the parabola $x = 9y^2$.

(e) The surface is a hyperbolic paraboloid.

(f) The graph appears in Figure 17.

Figure 17

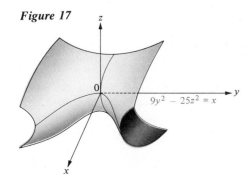

$9y^2 - 25z^2 = x$

Problem Set 13.8

In Problems 1 to 8, identify the intersection of each quadric surface with the indicated plane.

1 $2x^2 + 3y^2 + z^2 = 6$; $x = 1$

2 $\dfrac{x^2}{9} - \dfrac{y^2}{4} + \dfrac{z^2}{25} = 1$; $z = 4$

3 $z^2 - \dfrac{y^2}{9} - \dfrac{x^2}{16} = 1$; $y = 2$

4 $3x^2 + 4y^2 = z$; $x = 2$

5 $25x^2 + 4y^2 - 100z^2 = 0$; $z = -1$

6 $3x^2 - 4z^2 + 5y - z - 2 = 0$; $z = 4$

7 $4x^2 - 16y^2 = z$; $x = \frac{3}{2}$

8 $\dfrac{x^2}{4} - \dfrac{y^2}{9} + \dfrac{z^2}{25} - x + y + xy = 4$; $x = 5$

In Problems 9 to 24, (a) find the intercepts, (b) discuss the symmetry, (c) find the cross sections perpendicular to the coordinate axes, (d) find the traces, (e) identify the quadric surface, and (f) sketch a graph.

9 $x^2 + 3y^2 + 2z^2 = 6$

10 $144x^2 + 9y^2 + 16z^2 = 144$

11 $4x^2 - 9y^2 + 9z^2 = 36$

12 $x^2 + 2xy + y^2 = 1$

13 $x^2 - 4y^2 - 4z^2 = 4$

14 $z^2 = 1 - 2y + y^2$

15 $y^2 - 9x^2 - 9z^2 = 9$

16 $x^2 + y^2 + z^2 - 2y + 2z = 0$

17 $x^2 + 5y^2 - 8z^2 = 0$

18 $-x^2 - 25y^2 - 25z^2 = 25$

19 $3z = x^2 + y^2$

20 $4y = \dfrac{x^2}{9} + y^2$

21 $\dfrac{x^2}{16} - \dfrac{y^2}{9} = 3z$

22 $4y^2 - 9z^2 - 18x = 0$

23 $x^2 = y + z^2$

24 $\dfrac{x^2}{16} - \dfrac{y^2}{9} = 1$

25 Show that the graph of an equation is symmetric with respect to the xy plane provided that when z is replaced by $-z$ in the equation, an equivalent equation is obtained.

26 Find the conditions for symmetry of a graph with respect to (a) the yz plane, (b) the x axis, and (c) the y axis.

27 Find the condition for the symmetry of a graph with respect to the origin.

28 Discuss the symmetry of the graph of the equation $2xy + 3xz - 4yz = 24$.

29 Find all cross sections cut from the ellipsoid $(x^2/a^2) + (y^2/b^2) + (z^2/c^2) = 1$ by (a) the planes $x = k$, (b) the planes $y = k$, and (c) the planes $z = k$.

30 Find all cross sections cut from the hyperboloid of one sheet $(x^2/a^2) + (y^2/b^2) - (z^2/c^2) = 1$ by planes perpendicular to the coordinate axes.

31 Find all cross sections cut from the elliptic cone $(x^2/a^2) + (y^2/b^2) - (z^2/c^2) = 0$ by planes perpendicular to the coordinate axes.

32 Find all cross sections cut from the elliptic paraboloid $(x^2/a^2) + (y^2/b^2) = z$ by the planes perpendicular to the coordinate axes.

33 Find all cross sections cut by planes $z = k$ from the hyperbolic paraboloid $(y^2/b^2) - (x^2/a^2) = z$. Discuss the cases $k < 0$, $k = 0$, and $k > 0$ separately.

34 Find all cross sections cut by planes perpendicular to the x or y axes from the hyperbolic paraboloid $(y^2/b^2) - (x^2/a^2) = z$.

35 Write an equation that describes the surface consisting of all points $P = (x, y, z)$ such that the distance from P to the point $(0, 0, -1)$ is the same as the distance from P to the plane $z = 1$. Identify this surface.

36 Prove that if (x_0, y_0, z_0) is any point on a hyperboloid of one sheet, there are two lines in space, passing through (x_0, y_0, z_0), both of which lie on the hyperboloid.

37 Find an equation that describes the surface consisting of all points $P = (x, y, z)$ whose distance from the y axis is $\frac{2}{3}$ of its distance from P to the xz plane.

38 Prove that if (x_0, y_0, z_0) is any point on a hyperbolic paraboloid, there are two lines in space, passing through (x_0, y_0, z_0), both of which lie on the hyperbolic paraboloid.

39 A central quadric surface $Ax^2 + By^2 + Cz^2 + K = 0$ contains the points $(3, -2, -1)$, $(0, 1, -3)$, and $(3, 0, 2)$. Find the equation of the surface, and identify the surface.

40 A quadric surface $Ax^2 + By^2 + Cz = 0$ contains the points $(1, 0, 1)$ and $(0, 2, 1)$. Find the equation of the surface, and identify the surface.

13.9 Cylindrical and Spherical Coordinates

In Chapter 9 we found that some problems in the plane were easier to formulate and to solve if polar coordinates, rather than Cartesian coordinates, were used. Similarly, there are situations in which problems in three-dimensional space become more tractable if non-Cartesian coordinate systems are introduced. Two of the most important non-Cartesian coordinate systems in space, the **cylindrical** and the **spherical**, are discussed in this section.

Cylindrical Coordinates

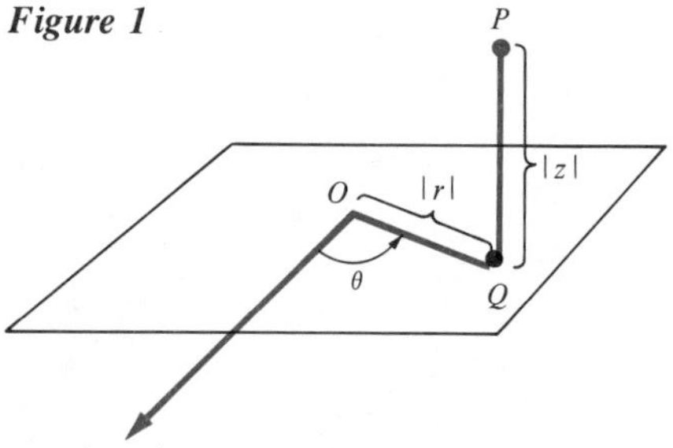

In Section 13.1 we obtained the Cartesian coordinates of a point P in three-dimensional space by dropping a perpendicular $\overline{PQ}$ to the horizontal plane through the origin O and using the Cartesian coordinates of Q in this plane together with the directed distance $z = \pm|\overline{PQ}|$. The **cylindrical coordinate system** is quite similar, except that polar coordinates are used for the point Q in the horizontal plane (Figure 1). Thus, the **cylindrical coordinates** of the point P in Figure 1 are (r, θ, z).

Figure 2 shows the point P with respect to the Cartesian coordinate system and with respect to the cylindrical coordinate system. If the polar coordinates of Q (with the x axis as the polar axis) are (r, θ), then the Cartesian coordinates of Q are $x = r \cos \theta$ and $y = r \sin \theta$; hence, the Cartesian coordinates of P are given by the equations

$$x = r \cos \theta \qquad y = r \sin \theta \qquad z = z$$

Since the point Q at the foot of the perpendicular from point P to the xy plane has an unlimited number of different representations in the polar coordinate system, it follows that P has an unlimited number of different representations in the cylindrical coordinate system. For instance, if $P = (r, \theta, z)$, then also $P = (-r, \theta + \pi, z)$. In any case, if $P = (x, y, z)$ in Cartesian coordinates, then the cylindrical coordinates

Figure 1

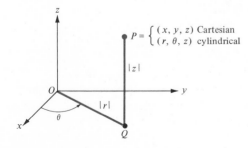

polar axis

Figure 2

Figure 3

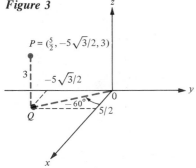

$P = (\frac{5}{2}, -5\sqrt{3}/2, 3)$

3 $-5\sqrt{3}/2$

$60°$

$5/2$

Q

Figure 4

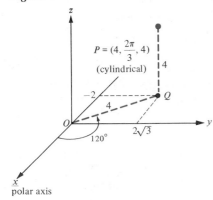

$P = (4, \frac{2\pi}{3}, 4)$
(cylindrical) 4

-2

4 Q

$2\sqrt{3}$

$120°$

x
polar axis

(r, θ, z) of P must satisfy

$$r = \pm\sqrt{x^2 + y^2} \quad \text{and} \quad \tan \theta = \frac{y}{x} \quad \text{for } x \neq 0$$

If $x = 0$, then $\theta = \pi/2$ when $y > 0$ and $\theta = 3\pi/2$ when $y < 0$.

<u>EXAMPLE 1</u> Find Cartesian coordinates of the point P whose cylindrical coordinates are $(5, -\pi/3, 3)$, and plot the point P showing both coordinate systems.

SOLUTION   Here,

$$x = r \cos \theta = 5 \cos\left(-\frac{\pi}{3}\right) = \frac{5}{2}$$

$$y = 5 \sin\left(-\frac{\pi}{3}\right) = -\frac{5\sqrt{3}}{2}$$

and $z = 3$

so that the point P has Cartesian coordinates $(\frac{5}{2}, -5\sqrt{3}/2, 3)$ (Figure 3).   ■

<u>EXAMPLE 2</u> Find cylindrical coordinates of the point P whose Cartesian coordinates are $(-2, 2\sqrt{3}, 4)$, and plot the point P showing both coordinate systems.

SOLUTION   Here, $r = \pm\sqrt{(-2)^2 + (2\sqrt{3})^2} = \pm\sqrt{16} = \pm 4$, and we take $r = 4$. Since $x = -2 < 0$ and $y = 2\sqrt{3} > 0$, it follows that θ is a second quadrant angle with $\tan \theta = y/x = -2\sqrt{3}/2 = -\sqrt{3}$; hence, we take $\theta = 120° = 2\pi/3$ radians. Thus, the cylindrical coordinates of P are $(4, 2\pi/3, 4)$ (Figure 4).   ■

Figure 5

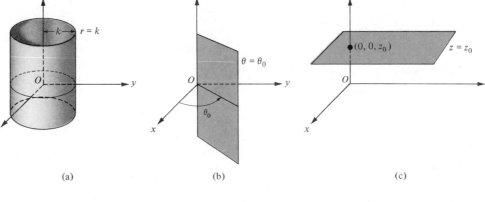

$r = k$

$\theta = \theta_0$

$(0, 0, z_0)$ $z = z_0$

θ_0

(a)   (b)   (c)

Figure 6

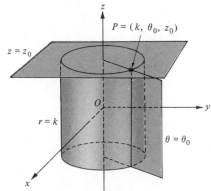

$P = (k, \theta_0, z_0)$

$z = z_0$

O

$r = k$

$\theta = \theta_0$

In cylindrical coordinates, the graph of the equation $r = k$ is a circular cylinder of radius $|k|$ with the z axis as its central axis (Figure 5a). Likewise, the condition $\theta = \theta_0$ describes a plane through the z axis making an angle θ_0 with the positive x axis (Figure 5b), while the equation $z = z_0$ represents a horizontal plane intersecting the z axis at the point $(0, 0, z_0)$ (Figure 5c). The point P with cylindrical coordinates $P = (k, \theta_0, z_0)$ is a point at which the circular cylinder $r = k$, the plane $\theta = \theta_0$, and the plane $z = z_0$ intersect (Figure 6).

Since the equation of a circular cylinder is exceptionally simple in cylindrical coordinates, such coordinates are naturally adapted to the solutions of problems involving such cylinders. More generally, an equation of a surface of revolution about the z axis is usually simpler in cylindrical coordinates than in Cartesian coordinates.

EXAMPLE 3   Write an equation in cylindrical coordinates of the right circular cylinder of radius 17 having the z axis as its central axis.

SOLUTION   An equation of the right circular cylinder is $r = 17$.   ■

EXAMPLE 4   Find the equation in cylindrical coordinates for the paraboloid of revolution whose Cartesian equation is $x^2 + y^2 = z$.

SOLUTION   Since $r^2 = x^2 + y^2$, the desired equation is $r^2 = z$.   ■

EXAMPLE 5   Find an equation in Cartesian coordinates of the surface whose equation in cylindrical coordinates is $z = 3r$, identify the surface, and sketch its graph.

SOLUTION   Squaring both sides of the equation, we obtain $z^2 = 9r^2$, or $z^2 = 9(x^2 + y^2)$. The last equation can be rewritten as $x^2 + y^2 - (z^2/9) = 0$, which represents an elliptic cone. In fact, since the coefficients of x^2 and y^2 are the same, the graph is a circular cone (Figure 7).   ■

Figure 7

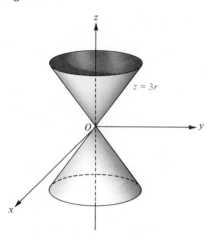

A curve in space can often be expressed parametrically by giving the cylindrical coordinates (r, θ, z) of a point P on the curve in terms of a parameter t. Thus, if

$$\begin{cases} r = f(t) \\ \theta = g(t) \\ z = h(t) \end{cases}$$

where f, g, and h are continuous functions, then the point $P = (r, \theta, z)$ traces out the curve as t varies. If f', g', and h' exist and are continuous, then from the equations

$$x = r \cos \theta \qquad y = r \sin \theta \qquad \text{and} \qquad z = z$$

we obtain

$$\frac{dx}{dt} = \frac{dr}{dt} \cos \theta - r \sin \theta \, \frac{d\theta}{dt} \qquad \text{and} \qquad \frac{dy}{dt} = \frac{dr}{dt} \sin \theta + r \cos \theta \, \frac{d\theta}{dt}$$

Thus,

$$\left(\frac{ds}{dt} \right)^2 = \left(\frac{dx}{dt} \right)^2 + \left(\frac{dy}{dt} \right)^2 + \left(\frac{dz}{dt} \right)^2$$

$$= \left(\frac{dr}{dt} \right)^2 + r^2 \left(\frac{d\theta}{dt} \right)^2 + \left(\frac{dz}{dt} \right)^2 \qquad \text{(Problem 39)}$$

It follows that the arc length of the curve between the point where the parameter has the value $t = a$ and the point where it has the value $t = b$ is given by

$$s = \int_a^b \sqrt{\left(\frac{dr}{dt} \right)^2 + r^2 \left(\frac{d\theta}{dt} \right)^2 + \left(\frac{dz}{dt} \right)^2} \, dt$$

ⓒ **EXAMPLE 6**   Find the arc length of the curve

$$\begin{cases} r = 5 \\ \theta = 2\pi t \\ z = 3t \end{cases}$$

between the point where $t = 0$ and the point where $t = 1$.

SOLUTION   Here $dr/dt = 0$, $d\theta/dt = 2\pi$, $dz/dt = 3$, and

$$s = \int_0^1 \sqrt{0^2 + 5^2(2\pi)^2 + 3^2} \, dt = \int_0^1 \sqrt{100\pi^2 + 9} \, dt$$

$$= \sqrt{100\pi^2 + 9} \int_0^1 dt = \sqrt{100\pi^2 + 9} \approx 31.56 \text{ units}$$ ∎

Spherical Coordinates

Figure 8

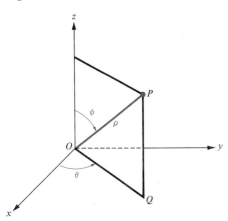

In the **spherical coordinate system,** the angle θ has exactly the same meaning as it does in the cylindrical coordinate system. Thus, θ locates the point P on a plane containing the z axis and making the angle θ with the positive x axis (Figure 8). The distance between P and the origin O is denoted by the Greek letter ρ (called "rho"); thus, $\rho = |\overline{OP}|$. Finally, the angle from the positive z axis to the line segment $\overline{OP}$ is denoted by the Greek letter ϕ (called "phi"). The spherical coordinates of the point P are customarily written in the order (ρ, θ, ϕ), and they are usually chosen so that

$$\rho \geq 0 \qquad 0 \leq \theta < 2\pi \qquad \text{and} \qquad 0 \leq \phi \leq \pi$$

A point P with spherical coordinates $(\rho_0, \theta_0, \phi_0)$ is ρ_0 units from the origin; hence, it is located on a sphere of radius ρ_0 with center at O (Figure 9). The z axis intersects this sphere at the "north and south poles," and the xy plane intersects it in the "equator." Semicircles cut by half planes passing through the north and south poles are called **meridians,** and the meridian that intersects the positive x axis is called the **prime meridian.** The angles θ_0 and ϕ_0 locate P_0 on the surface of the sphere, as shown in Figure 9. The spherical coordinate θ_0, which is called the **longitude** of P_0, measures the angle between the prime meridian and the meridian passing through P_0. (On the surface of the earth, the meridian passing through Greenwich, England, is designated as the prime meridian.)

Circles cut on the surface $\rho = \rho_0$ by planes perpendicular to the z axis (hence, parallel to the equatorial plane) are called **parallels,** and the angle measured from the equator to a parallel is called the **latitude** of that parallel (or of any point on the parallel) (Figure 10). Notice that the point P_0 with spherical coordinates $(\rho_0, \theta_0, \phi_0)$ has latitude $(\pi/2) - \phi_0$; in other words, the angle ϕ_0 is the complement of the latitude of P_0. For this reason, ϕ_0 is called the **colatitude** of P_0. For instance, the latitude of Boston, Massachusetts, is approximately 42.4°, so its colatitude is approximately 47.6°.

Figure 9

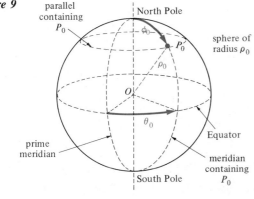

Figure 10

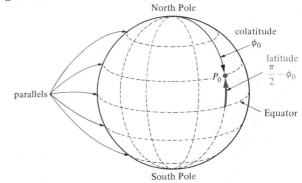

Figure 11

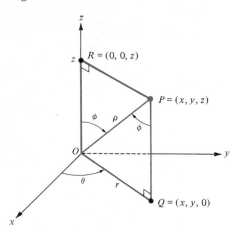

In Figure 11, suppose that the point P has Cartesian coordinates (x, y, z) and spherical coordinates (ρ, θ, ϕ). The line segment $\overline{PQ}$ is parallel to the z axis, so that angle OPQ is equal to ϕ. Since OQP is a right triangle,

$$\sin \phi = \frac{|\overline{OQ}|}{|\overline{OP}|} = \frac{r}{\rho}$$

Hence,
$$r = \rho \sin \phi$$

Therefore, we can rewrite the equations

$$x = r \cos \theta \quad \text{and} \quad y = r \sin \theta$$

as
$$x = \rho \sin \phi \cos \theta \quad \text{and} \quad y = \rho \sin \phi \sin \theta$$

Triangle ORP in Figure 11 is a right triangle, so that

$$\cos \phi = \frac{|\overline{OR}|}{|\overline{OP}|} = \frac{z}{\rho}$$

Hence,
$$z = \rho \cos \phi$$

The equations just derived,

$$x = \rho \sin \phi \cos \theta \qquad y = \rho \sin \phi \sin \theta \qquad z = \rho \cos \phi$$

give the Cartesian coordinates of the point P whose spherical coordinates are (ρ, θ, ϕ). The argument just given applies if P lies above the xy plane, and (Problem 38) the same equations are effective even when P lies on or below the xy plane. In some applications of spherical coordinates, the conditions $\rho \geq 0$, $0 \leq \theta < 2\pi$, and $0 \leq \phi \leq \pi$ are dropped, and (ρ, θ, ϕ) is understood to locate P in spherical coordinates if x, y, and z are given by the equations above.

Since ρ is the distance between $P = (x, y, z)$ and the origin O,

$$\rho^2 = x^2 + y^2 + z^2$$

Also, if $x \neq 0$, then

$$\frac{y}{x} = \frac{\sin \theta}{\cos \theta} = \tan \theta$$

Finally, if $\rho \neq 0$, then

$$\frac{z}{\rho} = \cos \phi$$

Therefore, if we choose the spherical coordinates (ρ, θ, ϕ) so that $\rho \geq 0$, $0 \leq \theta < 2\pi$, and $0 \leq \phi \leq \pi$, we have the formulas

$$\rho = \sqrt{x^2 + y^2 + z^2} \qquad \tan \theta = \frac{y}{x} \quad \text{for } x \neq 0$$

and

$$\phi = \cos^{-1} \frac{z}{\rho} = \cos^{-1} \frac{z}{\sqrt{x^2 + y^2 + z^2}} \quad \text{for } \rho \neq 0$$

EXAMPLE 7 Find Cartesian coordinates of the point P whose spherical coordinates are $(2, \pi/3, 2\pi/3)$, and plot the point P showing both coordinate systems.

Figure 12

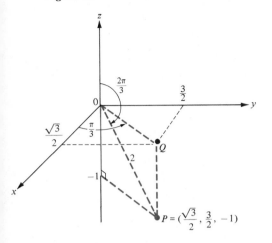

SOLUTION   Here we have $\rho = 2$, $\theta = \pi/3$, and $\phi = 2\pi/3$, so that

$$x = \rho \sin \phi \cos \theta = 2 \sin \frac{2\pi}{3} \cos \frac{\pi}{3}$$

$$= 2\left(\frac{\sqrt{3}}{2}\right)\left(\frac{1}{2}\right) = \frac{\sqrt{3}}{2}$$

$$y = \rho \sin \phi \sin \theta = 2 \sin \frac{2\pi}{3} \sin \frac{\pi}{3}$$

$$= 2\left(\frac{\sqrt{3}}{2}\right)\left(\frac{\sqrt{3}}{2}\right) = \frac{3}{2}$$

and $$z = \rho \cos \phi = 2 \cos \frac{2\pi}{3} = 2\left(-\frac{1}{2}\right) = -1$$

Thus, the Cartesian coordinates of P are $(\sqrt{3}/2, \frac{3}{2}, -1)$ (Figure 12). ∎

__EXAMPLE 8__   Find spherical coordinates of the point P whose Cartesian coordinates are $(\sqrt{3}, -1, 2)$, and plot the point P showing both coordinate systems.

SOLUTION   Since $\tan \theta = y/x$, we have

$$\theta = \tan^{-1} \frac{-1}{\sqrt{3}} = -\frac{\pi}{6} = -30°$$

If we require $0 \le \theta < 2\pi$, we can take

$$\theta = 2\pi + \left(\frac{-\pi}{6}\right) = \frac{11\pi}{6} = 330°$$

Here, $$\rho = \sqrt{(\sqrt{3})^2 + (-1)^2 + 2^2} = \sqrt{8} = 2\sqrt{2}$$

and

$$\phi = \cos^{-1} \frac{z}{\rho} = \cos^{-1} \frac{2}{2\sqrt{2}} = \cos^{-1} \frac{\sqrt{2}}{2} = \frac{\pi}{4} = 45°$$

Therefore, the spherical coordinates of P are $(\sqrt{8}, 11\pi/6, \pi/4)$ (Figure 13). ∎

Figure 13

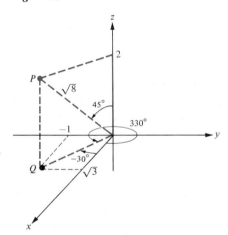

__EXAMPLE 9__   Convert the equation $\phi = \pi/4$ in spherical coordinates into (a) Cartesian coordinates and (b) cylindrical coordinates.

SOLUTION

(a) $$x = \rho \sin \phi \cos \theta = \left(\sin \frac{\pi}{4}\right)\rho \cos \theta = \frac{\sqrt{2}}{2} \rho \cos \theta$$

$$y = \rho \sin \phi \sin \theta = \left(\sin \frac{\pi}{4}\right)\rho \sin \theta = \frac{\sqrt{2}}{2} \rho \sin \theta$$

and $$z = \rho \cos \phi = \rho \cos \frac{\pi}{4} = \frac{\sqrt{2}}{2} \rho$$

Since $z = (\sqrt{2}/2)\rho$, we can rewrite the first two equations as $x = z \cos \theta$ and $y = z \sin \theta$, so that

$$x^2 + y^2 = z^2 \cos^2 \theta + z^2 \sin^2 \theta = z^2(\cos^2 \theta + \sin^2 \theta) = z^2$$

Thus, the equation in Cartesian coordinates is $x^2 + y^2 = z^2$; hence the graph of the equation is a right circular cone.

(b) In cylindrical coordinates $r^2 = x^2 + y^2$, so the equation obtained in part (a) can be written as $r^2 = z^2$. ∎

EXAMPLE 10   Rewrite the equation of the paraboloid $x^2 + y^2 = z$ in spherical coordinates.

SOLUTION   Here we have

$$x^2 + y^2 = (\rho \sin \phi \cos \theta)^2 + (\rho \sin \phi \sin \theta)^2$$
$$= \rho^2 \sin^2 \phi(\cos^2 \theta + \sin^2 \theta) = \rho^2 \sin^2 \phi$$

Thus, since $z = \rho \cos \phi$, the equation $x^2 + y^2 = z$ becomes

$$\rho^2 \sin^2 \phi = \rho \cos \phi$$

We lose no points on the graph by canceling ρ to obtain $\rho \sin^2 \phi = \cos \phi$ since the latter equation is still satisfied when $\rho = 0$ and $\phi = \pi/2$. Therefore, in spherical coordinates, the paraboloid has the equation $\rho \sin^2 \phi = \cos \phi$. ∎

Problem Set 13.9

In Problems 1 to 4, find the Cartesian coordinates of the point whose cylindrical coordinates are given, and plot the point.

1 $(4, \pi/3, 1)$ **2** $(3, \pi/2, 4)$

3 $(5, \pi/6, -2)$ **4** $(2, 2, 2)$

In Problems 5 to 8, find cylindrical coordinates (r, θ, z) with $r \geq 0$ and $0 \leq \theta < 2\pi$ for the point whose Cartesian coordinates are given, and plot the point.

5 $(4, 0, 1)$ **6** $(-2\sqrt{3}, -6, 0)$

7 $(-3\sqrt{3}, 3, 6)$ **8** $(1, 1, -1)$

In Problems 9 to 12, find the Cartesian coordinates of the point whose spherical coordinates are given, and plot each point.

9 $(2, \pi/6, \pi/3)$ **10** $(7, \pi/2, \pi)$

11 $(12, 5\pi/6, 2\pi/3)$ **12** (π, π, π)

In Problems 13 to 16, find spherical coordinates (ρ, θ, ϕ) with $\rho \geq 0$, $0 \leq \theta < 2\pi$, and $0 \leq \phi \leq \pi$ for the points whose Cartesian coordinates are given, and plot each point.

13 $(0, -1, 0)$ **14** $(0, 0, 5)$

15 $(1, 2, -3)$ **16** $(0, 0, 0)$

In Problems 17 to 26, convert each equation to an equivalent equation (a) in cylindrical coordinates and (b) in spherical coordinates. Sketch a graph of the surface.

17 $z = 2(x^2 + y^2)$ **18** $x^2 + y^2 - 4x = 0$

19 $x = 2$ **20** $y = \sqrt{3}x$

21 $x^2 + y^2 = 5z^2$ **22** $x^2 + y^2 + (z - 1)^2 = 1$

23 $x^2 + y^2 = 25$ **24** $x + 2y = 0$

25 $x^2 + y^2 - z^2 = 1$ **26** $x^2 + y^2 + z^2 - 6z = 0$

In Problems 27 to 32, each equation is written in cylindrical coordinates. Convert each equation to an equivalent equation in (a) Cartesian coordinates and (b) spherical coordinates. Sketch a graph of the surface.

27 $z = r^2$ **28** $\theta = \dfrac{3\pi}{4}$

29 $\dfrac{r^2}{9} + \dfrac{z^2}{4} = 1$ **30** $z = \dfrac{1}{2}r^2 \sin 2\theta$

31 $r = 4 \cos \theta$

32 $r \cos \theta + 3r \sin \theta + 2z = 6$

In Problems 33 to 37, each equation is written in spherical coordinates. Express each equation as an equivalent equation (a) in Cartesian coordinates and (b) in cylindrical coordinates. Sketch a graph of the surface.

33 $\rho = 2$ **34** $\theta = \pi/3$

35 $\rho \sin \phi = 3$ **36** $\rho = 2 \cos \phi$

37 $\phi = \pi/3$

38 Show that the equations given on page 824 for converting from spherical to Cartesian coordinates are correct even when the point P lies on or below the xy plane.

39 Complete the proof of

$$\left(\frac{ds}{dt}\right)^2 = \left(\frac{dx}{dt}\right)^2 + \left(\frac{dy}{dt}\right)^2 + \left(\frac{dz}{dt}\right)^2 = \left(\frac{dr}{dt}\right)^2 + r^2\left(\frac{d\theta}{dt}\right)^2 + \left(\frac{dz}{dt}\right)^2$$

on page 822.

40 Consider the curve in space whose equation is given parametrically in spherical coordinates by $\rho = f(t)$, $\theta = g(t)$, and $\phi = h(t)$, where f', g', and h' exist and are continuous.

(a) Prove that

$$\left(\frac{dx}{dt}\right)^2 + \left(\frac{dy}{dt}\right)^2 + \left(\frac{dz}{dt}\right)^2 = \left(\frac{d\rho}{dt}\right)^2 + \rho^2 \sin^2\phi\left(\frac{d\theta}{dt}\right)^2 + \rho^2\left(\frac{d\phi}{dt}\right)^2$$

(b) Show that the arc length of the curve between the point where $t = a$ and the point where $t = b$ is given by

$$s = \int_a^b \sqrt{\left(\frac{d\rho}{dt}\right)^2 + \rho^2 \sin^2\phi\left(\frac{d\theta}{dt}\right)^2 + \rho^2\left(\frac{d\phi}{dt}\right)^2}\, dt$$

41 Find the arc length of the curve whose equation is given parametrically in cylindrical coordinates by

$$r = \frac{\sqrt{2}}{2}\, t \qquad \theta = \sqrt{2}t \qquad \text{and} \qquad z = \frac{\sqrt{2}}{2}\, t$$

between $t = 0$ and $t = \sqrt{2}\pi$

©**42** A Cartesian coordinate system is established with the origin at the center of the earth, the positive x axis passing through the point where the equator intersects the prime meridian, and the positive z axis passing through the North Pole. Taking the radius of the earth to be 3959 miles, find:

(a) The Cartesian coordinates of New York if the latitude of New York is 40.7° north of the equator and the longitude of New York is 74.02° west of the prime meridian

(b) The Cartesian coordinates of San Francisco if the latitude of San Francisco is 37.81° north of the equator and the longitude of San Francisco is 122.4° west of the prime meridian

(c) The angle between the position vector of New York and the position vector of San Francisco

(d) The great-circle distance between New York and San Francisco

Review Problem Set, Chapter 13

In Problems 1 to 6, for the given point P, find another point Q which is symmetric to P with respect to the given plane or axis.

1 $P = (2, -1, -3)$; yz plane

2 $P = (5, 6, 3)$; x axis

3 $P = (-3, 2, -1)$; xy plane

4 $P = (-1, -2, 3)$; z axis

5 $P = (3, -1, -5)$; y axis

6 $P = (1, 2, 5)$; xz plane

7 Find the distance from the point $P = (3, -1, 6)$ to (a) the origin, (b) the x axis, (c) the xy plane, (d) the point $(2, -3, 7)$, and (e) the point Q which is symmetric to P with respect to the origin.

8 Use the distance formula to show that the triangle with vertices $P = (1, 3, 3)$, $Q = (2, 2, 1)$, and $R = (3, 4, 2)$ is equilateral. Also, find the coordinates of the point where the medians of the triangle intersect.

9 Use vectors to determine whether the points $P = (-5, -10, 9)$ $Q = (-1, -5, 5)$, and $R = (11, 10, -9)$ are collinear (that is, lie on the same line).

10 Use vectors to determine whether the quadrilateral with vertices $P = (3, 2, 5)$, $Q = (1, 1, 1)$, $R = (4, 0, 3)$, and $S = (6, 1, 7)$ is a parallelogram.

11 A vector $\mathbf{R}$ makes an acute angle θ with the x axis, an angle $\pi/4$ with the y axis, and an angle $\pi/3$ with the z axis. Find θ.

12 If $P = (1, 2, 3)$ and $Q = (2, 5, 7)$, find (a) the scalar components of the vector $\mathbf{A} = \overrightarrow{PQ}$ and (b) the direction cosines of $\mathbf{A}$.

In Problems 13 to 30, evaluate each of the following, using

$$\mathbf{A} = \mathbf{i} - 2\mathbf{j} + 3\mathbf{k} \qquad \mathbf{B} = 3\mathbf{i} + 2\mathbf{j} + \mathbf{k} \qquad \mathbf{C} = 2\mathbf{i} + 3\mathbf{j}$$
$$\mathbf{D} = -3\mathbf{i} - 5\mathbf{j} + 6\mathbf{k} \qquad \mathbf{E} = \mathbf{i} + 3\mathbf{j} - 2\mathbf{k} \qquad \mathbf{F} = 3\mathbf{i} + 5\mathbf{j} + 6\mathbf{k}$$

13 $\mathbf{A} - 3\mathbf{B} + \mathbf{C}$

14 $|2\mathbf{C} - \mathbf{F}|$

15 $|\mathbf{A} - 3\mathbf{B} + \mathbf{C}|$

16 $(\mathbf{E} \cdot \mathbf{F})\mathbf{C} - (\mathbf{C} \cdot \mathbf{F})\mathbf{A}$

17 $(3\mathbf{A}) \cdot (\mathbf{F} + \mathbf{E})$

18 $\mathbf{A} \times \mathbf{D}$

19 $\mathbf{E} \cdot (\mathbf{D} \times \mathbf{E})$

20 $\mathbf{B} \cdot (\mathbf{C} \times \mathbf{E})$

21 $\mathbf{A} \times \mathbf{B}$

22 $\mathbf{A} \times (\mathbf{B} \times \mathbf{D})$

23 $(\mathbf{A} + \mathbf{B}) \times (\mathbf{A} - \mathbf{B})$

24 $(\mathbf{A} \times \mathbf{B}) \times \mathbf{D}$

25 $\mathbf{A} \cdot (\mathbf{B} \times \mathbf{C})$

26 $(3\mathbf{A} - \mathbf{D}) \times (\mathbf{E} - 2\mathbf{C})$

27 The angle between $\mathbf{C}$ and $\mathbf{D}$

28 The scalar component of $\mathbf{C}$ in the direction of $\mathbf{F}$

29 The volume of the parallelepiped whose edges are the vectors $\mathbf{A}$, $\mathbf{B}$, and $\mathbf{C}$

30 $(\mathbf{A} \times \mathbf{B}) \cdot (\mathbf{C} \times \mathbf{D})$

31 Do the vectors $\mathbf{A} = \mathbf{i} - 2\mathbf{j} + 3\mathbf{k}$, $\mathbf{B} = 3\mathbf{i} + 2\mathbf{j} + \mathbf{k}$, and $\mathbf{C} = 2\mathbf{i} + 3\mathbf{j}$ form a right-handed triple $\mathbf{A}$, $\mathbf{B}$, $\mathbf{C}$? Why?

32 Find the area of the triangle whose vertices are $A = (2, 0, -1)$, $B = (5, 3, 3)$, and $C = (-1, 1, 2)$.

In Problems 33 to 40, find an equation in scalar form of the plane satisfying the conditions given.

33 Containing the point $P = (1, 2, 3)$ and perpendicular to the radius vector $\overrightarrow{OP}$ at the point P

34 Containing the three points $(1, 7, 2)$, $(3, 5, 1)$, and $(6, 3, -1)$

35 Containing the point $P = (2, 0, -1)$ and perpendicular to the line joining the points $Q = (3, 4, 4)$ and $R = (-1, 2, 1)$

36 At a distance of 4 units from the origin and perpendicular to the line joining the points $P = (-2, 3, 1)$ and $Q = (-5, 1, -5)$

37 Containing the point $(4, 3, 1)$ and parallel to the plane $x + 3z = 8$

38 Containing the points $(1, 3, 1)$ and $(4, 6, -2)$ and perpendicular to the plane $x + y - z = 3$

39 Containing the point $(2, -1, 3)$ and perpendicular to the line $x = 3t + 5$, $y = 8t - 4$, $z = -7t + 16$

40 Containing the point $(1, 2, 10)$ and the line

$$\frac{x - 1}{5} = \frac{y - 1}{2} = \frac{z - 6}{-7}$$

In Problems 41 to 48, find the equations of the line (a) in scalar parametric form and (b) in symmetric form for which the given conditions hold.

41 Containing the points $(5, 6, -4)$ and $(2, -1, 1)$

42 Containing the point $(-1, 2, 3)$ and perpendicular to the plane $-2x + 3y - z = -1$

43 Containing the point $(2, -3, -2)$ and parallel to the line

$$\frac{x - 1}{3} = \frac{y + 7}{-2} = \frac{z}{7}$$

44 Containing the origin, perpendicular to the line of intersection of the two planes $x = y - 5$ and $z = 2y - 3$, and meeting the line of intersection of the two planes $y = 2x + 1$ and $z = x + 2$

45 Formed by the intersection of the two planes $2x + y - z = 1$ and $x - y + 3z = 10$

46 Containing the point $(3, 6, 4)$, parallel to the plane $x - 3y + 5z - 6 = 0$, and intersecting the z axis

47 Containing the point $(2, 1, 4)$ and perpendicular to both the x and the y axes

48 Containing the point $(2, -1, 4)$ and perpendicular to the lines $x = 5t + 1$, $y = -3t$, $z = t - 2$ and $x = 2t - 1$, $y = t + 1$, $z = -t$

49 Find the distance from the point $(2, -3, 4)$ to the plane $2x - 2y + z = 5$.

50 Find the distance between the two parallel planes $x - 2y + 2z = 5$ and $x - 2y + 2z = 17$.

51 Show that the lines

$$\frac{x - 100}{99} = \frac{y + 94}{-97} = \frac{z - 51}{50}$$

and

$$\frac{x - 102}{-101} = \frac{y + 96}{99} = \frac{z - 52}{-51}$$

intersect, and find an equation of the plane containing the two lines.

52 Given the equations

$$\frac{x - x_0}{a} = \frac{y - y_0}{b} = \frac{z - z_0}{c}$$

of a line, explain how you would go about finding several points on this line.

53 Find the distance between the two lines

$$x = y + 3 = \frac{z - 2}{2} \quad \text{and} \quad \frac{x - 3}{-1} = \frac{y + 1}{2} = z - 1$$

54 Let P be a variable point on a first line in space, and let Q be a variable point on a second line in space. If $\mathbf{M}_1$ and $\mathbf{M}_2$ are direction vectors for the first and second lines, respectively, show that the quantity $(\mathbf{M}_1 \times \mathbf{M}_2) \cdot \overrightarrow{PQ}$ remains constant as P and Q vary.

55 Show that the lines

$$\frac{x - x_0}{a_0} = \frac{y - y_0}{b_0} = \frac{z - z_0}{c_0}$$

and

$$\frac{x - x_1}{a_1} = \frac{y - y_1}{b_1} = \frac{z - z_1}{c_1}$$

meet in space if and only if

$$\begin{vmatrix} a_0 & b_0 & c_0 \\ a_1 & b_1 & c_1 \\ x_1 - x_0 & y_1 - y_0 & z_1 - z_0 \end{vmatrix} = 0$$

56 Let $\mathbf{R}_1$ be the position vector for a point P_1 in space.

(a) Show that

$$\mathbf{R}_1 - \frac{\mathbf{N} \cdot (\mathbf{R}_1 - \mathbf{R}_0)}{|\mathbf{N}|^2} \mathbf{N}$$

is the position vector of the point in the plane $\mathbf{N} \cdot (\mathbf{R} - \mathbf{R}_0) = 0$ that is closest to P_1.

(b) Show that

$$\mathbf{R}_0 + \frac{\mathbf{M} \cdot (\mathbf{R}_1 - \mathbf{R}_0)}{|\mathbf{M}|^2} \mathbf{M}$$

is the position vector of the point on the line $\mathbf{M} \times (\mathbf{R} - \mathbf{R}_0) = \mathbf{0}$ that is closest to P_1.

57 Suppose that **A** is a known nonzero vector, that **X** is an unknown vector, but that the scalar $a = \mathbf{A} \cdot \mathbf{X}$ and vector $\mathbf{B} = \mathbf{A} \times \mathbf{X}$ are both known. Show that **X** is then determined by the equation

$$\mathbf{X} = \frac{a}{|\mathbf{A}|^2}\,\mathbf{A} - \frac{\mathbf{A} \times \mathbf{B}}{|\mathbf{A}|^2}$$

(*Hint:* Start by expanding $\mathbf{A} \times \mathbf{B}$.)

58 If **A** is a nonzero vector and $\mathbf{A} \times \mathbf{X} = \mathbf{A} \times \mathbf{Y}$, can we "cancel" and conclude that $\mathbf{X} = \mathbf{Y}$? Explain.

In Problems 59 and 60, find $\mathbf{F}'(t)$, $\mathbf{F}''(t)$, and $\dfrac{d}{dt}\,|\mathbf{F}(t)|$.

59 $\mathbf{F}(t) = e^{2t^2}\mathbf{i} - e^{-2t^2}\mathbf{j} + t\mathbf{k}$

60 $\mathbf{F}(t) = \tan\left(t + \dfrac{\pi}{2}\right)\mathbf{i} + (\tan t)\mathbf{j} + \tan\left(t - \dfrac{\pi}{2}\right)\mathbf{k}$

In Problems 61 and 62, assume that a particle P moves according to the given equation of motion. Find (a) the velocity vector **V**, (b) the acceleration vector **A**, (c) the speed v, (d) the unit tangent vector **T**, (e) the unit normal vector **N**, (f) the unit binormal vector **B**, and (g) the distance traveled by the particle along its path from the instant $t = 0$ to the instant $t = 1$.

61 $\mathbf{R} = (3 \cos 2\pi t)\mathbf{i} + (3 \sin 2\pi t)\mathbf{j} + 2t\mathbf{k}$

62 $\mathbf{R} = (e^{4t} \cos t)\mathbf{i} + (e^{4t} \sin t)\mathbf{j} + e^{4t}\mathbf{k}$

In Problems 63 to 66, find (a) **V**, (b) v, (c) **A**, (d) **T**, (e) $\mathbf{V} \times \mathbf{A}$, (f) κ, (g) **N**, (h) **B**, (i) $d\mathbf{A}/dt$, and (j) τ for each curve in xyz space.

63 $\mathbf{R} = t\mathbf{i} + t^2\mathbf{j} + t^3\mathbf{k}$

64 $\mathbf{R} = at\mathbf{i} + bt^2\mathbf{j} + ct^3\mathbf{k}$, where a, b, and c are constants

65 $\mathbf{R} = (\sin t \cos t)\mathbf{i} + (\sin^2 t)\mathbf{j} + (\cos t)\mathbf{k}$

66 $\mathbf{R} = (t \sin t)\mathbf{i} + (t \cos t)\mathbf{j} + t\mathbf{k}$

67 A particle moves according to an equation of motion $\mathbf{R} = \mathbf{F}(t)$. Expand and simplify the following expressions:

(a) $\dfrac{d}{dt}\,(\mathbf{R} \cdot \mathbf{V})$ (b) $\dfrac{d}{dt}\,(\mathbf{R} \times \mathbf{V})$

(c) $\dfrac{d}{dt}\,(\mathbf{V} \cdot \mathbf{A})$ (d) $\dfrac{d}{dt}\,(\mathbf{V} \times \mathbf{A})$

(e) $\dfrac{d\mathbf{T}}{ds} \cdot \dfrac{d\mathbf{B}}{ds}$ (f) $\left(\dfrac{d\mathbf{R}}{ds} \times \dfrac{d^2\mathbf{R}}{ds^2}\right) \cdot \dfrac{d^3\mathbf{R}}{ds^3}$

68 (a) Sketch the curve $z = \cos(\pi y)$ in the yz plane. (b) Sketch the cylinder with generators parallel to the x axis having the curve in

part (a) as its base curve. (c) Write the equation of the cylinder in part (b).

69 Find an equation of the surface of revolution generated by rotating the curve $z = 2(x - 3)^2$ in the xz plane about the x axis.

70 Find an equation of the **torus** (doughnut-shaped surface) that results when the circle $y^2 + (x - a)^2 = r^2$ $(a > r > 0)$ in the xy plane is revolved about the y axis. Sketch the surface.

71 Find the generating curve in the xz plane and the axis of revolution of the surface of revolution $y^2 + z^2 = e^{-2x}$. Sketch the surface.

72 Find an equation of the surface of revolution generated by revolving the cardioid $r = 1 - \cos \theta$ in the xy plane about the x axis.

In Problems 73 to 78, identify and sketch a graph for each quadric surface.

73 $x^2 + 4y^2 + 4z^2 = 16$ **74** $x^2 + 4y^2 + 4z^2 = 16x$

75 $x^2 + z^2 = 4 + y$ **76** $9x^2 + z^2 = y$

77 $y^2 - z^2 + 9x^2 = 1$ **78** $y^2 - z^2 + 9x^2 = 0$

79 A quadric surface has an equation $Ax^2 + By^2 + Cz = 0$. Identify the surface if it contains the points $(3, 5, 8)$ and $(4, -2, -6)$.

80 A quadric surface has an equation $Ax^2 + By^2 + Cz^2 = 1$ and contains the points $(2, -1, 1)$, $(-3, 0, 0)$, and $(1, -1, -2)$. Identify the surface.

81 Describe and sketch the surface whose equation in cylindrical coordinates is (a) $r = 2$; (b) $\theta = \pi/6$; (c) $r = \sin \theta$.

82 Describe and sketch the surface whose equation in spherical coordinates is $\rho = 5 \cos \phi$.

83 Find an equation in cylindrical coordinates of the surface obtained by revolving the curve $z = f(x)$ in the xz plane about the z axis.

84 Find the arc length of the curve whose parametric equations in spherical coordinates are $\rho = t^2$, $\theta = \pi/6$, $\phi = t$ as t varies over the interval $[0, \sqrt{5}]$.

85 Convert the equation $\rho^2 \sin 2\phi = 4$ in spherical coordinates to Cartesian coordinates, and sketch the graph of the surface.

© **86** Find the great-circle distance between Honolulu, with latitude 21.31°N, longitude 157.87°W, and Chicago, with latitude 41.83°N, longitude 87.62°W. Assume that the radius of the earth is 3959 miles.

14

PARTIAL DIFFERENTIATION

In the preceding chapters we dealt exclusively with functions of a single real variable; however, there are many practical situations in which functions depend on *several* variables. For instance, the frequency of a tuned circuit depends on its capacitance, its inductance, and its resistance; the pressure of a gas depends on its temperature and its volume; the demand for a commodity may depend not only on its price but also on the prices of related commodities, on income level, and on time; a person's income tax liability depends not only on income but also on several itemized deductions and the number of dependents; and so forth.

In this chapter we study functions of more than one variable, we see that the concepts of limits and continuity are applicable to such functions, and we investigate their "partial" derivatives. Chain rules are developed for functions of several variables. The chapter also includes a study of directional derivatives, tangent planes and normal lines to surfaces, and maxima and minima of functions of several variables.

14.1 Functions of Several Variables

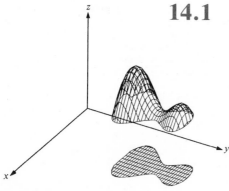

The graph of a function of two variables is a surface.

A right circular cylinder, closed at the top and bottom, with base radius r and height h, has total surface area S given by

$$S = 2\pi rh + 2\pi r^2$$

Here we say that the (dependent) variable S is a function of the two (independent) variables r and h, and we write

$$S = f(r, h)$$

For example, if $r = 11$ cm and $h = 5$ cm, then

$$S = f(11, 5) = 2\pi(11)(5) + 2\pi(11)^2 = 352\pi \text{ cm}^2$$

Proceeding somewhat more formally, we make the following definition.

DEFINITION 1   **Function of Two Variables**

> A **real-valued function** f of **two real variables** is a rule that assigns a unique real number $z = f(x, y)$ to each ordered pair (x, y) of real numbers in a certain set D called the **domain** of f.

In the equation $z = f(x, y)$, we call z the **dependent** variable, and we refer to x and y as the **independent** variables. The set of all possible values of z that can be obtained by applying the rule f to ordered pairs (x, y) in D is called the **range** of the function f. If f is defined by an equation, or a formula, then (unless we make a statement to the contrary) the domain of f is understood to be the set of all ordered pairs of independent variables for which the equation, or formula, makes sense.

Figure 1

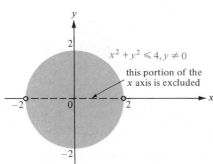

In EXAMPLE 1   Find and sketch the *domain* of $f(x, y) = \dfrac{\sqrt{4 - x^2 - y^2}}{y}$.

SOLUTION   The domain of f consists of all ordered pairs (x, y) for which $x^2 + y^2 \leq 4$ and $y \neq 0$. This is the set of all points not on the x axis that are either on the circle $x^2 + y^2 = 4$ or in the interior region bounded by the circle (Figure 1). ∎

EXAMPLE 2   Find the domain of $f(x, y, z) = \dfrac{\sin^{-1} z}{x + y}$.

SOLUTION   Since $\sin^{-1} z$ is defined only when $|z| \leq 1$, the domain of f consists of all ordered triples (x, y, z) such that $x + y \neq 0$ and $|z| \leq 1$. ∎

We define the **graph** of a function f of two variables to be the graph of the equation $z = f(x, y)$; that is, the set of all points (x, y, z) in Cartesian three-dimensional space such that (x, y) belongs to the domain D of f and $z = f(x, y)$. The domain D can be pictured as a set of points in the xy plane and the graph of f as a surface whose perpendicular projection on the xy plane is D (Figure 2).* Notice that as the point (x, y) varies in D, the corresponding point

$$(x, y, z) = (x, y, f(x, y))$$

varies over the surface. The methods introduced in Sections 13.7 and 13.8 are useful for sketching graphs of functions of two variables.

Figure 2

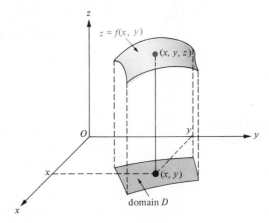

*In Figure 2, the point shown as (x, y) in D is really $(x, y, 0)$; however, the third coordinate has been purposely omitted.

Figure 3

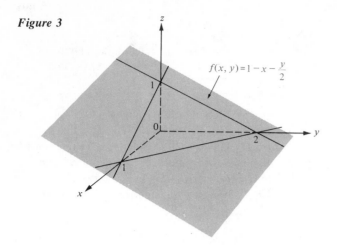

$$f(x, y) = 1 - x - \frac{y}{2}$$

In Examples 3 to 5, sketch the graph of each function of two variables.

<u>EXAMPLE 3</u> $f(x, y) = 1 - x - (y/2)$

SOLUTION   The domain D of f is the entire xy plane. The point (x, y, z) belongs to the graph of f if and only if $z = 1 - x - (y/2)$; that is,

$$2x + y + 2z = 2$$

Therefore, the graph of f is a plane with intercepts $(1, 0, 0)$, $(0, 2, 0)$, and $(0, 0, 1)$. A portion of this plane, showing its traces with the xy, xz, and yz planes, appears in Figure 3. ∎

<u>EXAMPLE 4</u> $f(x, y) = \sqrt{1 - x^2 - y^2}$

SOLUTION   The domain D of f is the circular disk consisting of all points (x, y) with $x^2 + y^2 \leq 1$. A point (x, y, z) belongs to the graph of f if and only if $z = f(x, y)$, that is, $z = \sqrt{1 - x^2 - y^2}$. The condition $z = \sqrt{1 - x^2 - y^2}$ is equivalent to the two conditions $z \geq 0$ and $x^2 + y^2 + z^2 = 1$. Thus, the graph consists of the portion of the sphere $x^2 + y^2 + z^2 = 1$ lying on or above the xy plane (Figure 4). ∎

Figure 4

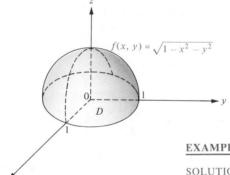

$$f(x, y) = \sqrt{1 - x^2 - y^2}$$

<u>EXAMPLE 5</u> $g(x, y) = x^2 + 3y^2$

SOLUTION   The domain D of g is the entire xy plane, and the graph of g is the graph of the equation

$$z = x^2 + 3y^2$$

By the discussion in Section 13.8 (page 816), this graph is an elliptic paraboloid (Figure 5). ∎

Functions of three or more variables are defined by an obvious extension of Definition 1 as follows.

DEFINITION 2

Function of Several Variables

> A **real-valued function** f of n **real variables** is a rule that assigns a unique real number $w = f(x_1, x_2, x_3, \ldots, x_n)$ to each ordered n-tuple $(x_1, x_2, x_3, \ldots, x_n)$ of real numbers in a certain set D called the **domain** of f.

Figure 5

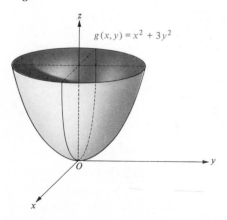

$$g(x, y) = x^2 + 3y^2$$

In the equation $w = f(x_1, x_2, x_3, \ldots, x_n)$, we call w the **dependent** variable, and we refer to $x_1, x_2, x_3, \ldots, x_n$ as the **independent** variables. The set of all possible values of w that can be obtained by applying the rule f to ordered n-tuples in D is called the **range** of the function f. In case $n = 2$, $w = f(x_1, x_2)$ is usually written $z = f(x, y)$ as in Definition 1. In case $n = 3$, $w = f(x_1, x_2, x_3)$ is usually written $w = f(x, y, z)$.

<u>EXAMPLE 6</u>   If $g(x, y, z) = \dfrac{xy}{x^2 + y^2 - z}$ for all values of x, y, and z except those that make the denominator equal to zero, find

(a) $g(2, 3, 7)$ (b) $g(\sin t, \cos t, 0)$

SOLUTION

(a) $g(2, 3, 7) = \dfrac{(2)(3)}{2^2 + 3^2 - 7} = 1$

(b) $g(\sin t, \cos t, 0) = \dfrac{\sin t \cos t}{\sin^2 t + \cos^2 t - 0} = \sin t \cos t$ ■

EXAMPLE 7   If $f(x_1, x_2, x_3, \ldots, x_n) = x_1^2 + x_2^2 + x_3^2 + \cdots + x_n^2$ for all values of $x_1, x_2, x_3, \ldots, x_n$, find $f(1, 2, 3, \ldots, n)$.

SOLUTION   Using the formula for the sum of successive squares (Section 5.1), we have

$$f(1, 2, 3, \ldots, n) = 1^2 + 2^2 + 3^2 + \cdots + n^2 = \sum_{k=1}^{n} k^2 = \frac{n(n + 1)(2n + 1)}{6}$$ ■

Scalar Fields

We have seen that a function f of two independent variables can be thought of in terms of its graph, which is a surface in xyz space. There is a second way of picturing such a function, which, for some purposes, is even more suggestive; namely, the function f is regarded as a **scalar field.** The idea is as follows: The domain D of f is visualized as a set of points (x, y) lying in a certain region in the xy plane, and each point (x, y) in this region is assigned a corresponding scalar $f(x, y)$ by the function f (Figure 6). The scalar value $f(x, y)$ corresponding to the point (x, y) in D is shown in Figure 6 on a "flag" attached to the point. As the point (x, y) is moved around in the region D, the flag moves with it, and the number $f(x, y)$ on the flag changes.

The scalar $f(x, y)$ assigned to the point (x, y) might represent, for instance, the temperature at (x, y), or the atmospheric pressure at (x, y), or the wind speed at (x, y), or the intensity of the magnetic field at (x, y), and so forth.

Figure 6

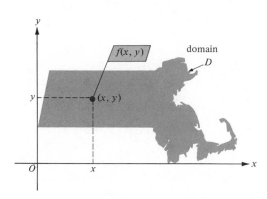

Figure 7

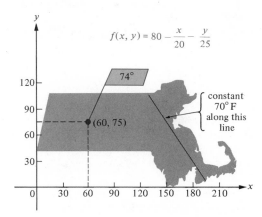

EXAMPLE 8   In Figure 7, suppose that $f(x, y) = 80 - (x/20) - (y/25)$ gives the temperature in degrees Fahrenheit at the point with Cartesian coordinates (x, y), where x and y are measured in miles.

(a) Find the temperature at the point $(60, 75)$.

(b) Find an equation of the curve along which the temperature has a constant value of $70°F$.

(c) Sketch the curve in part (b).

Figure 8

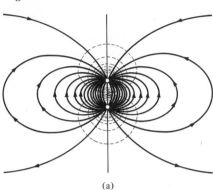

(a)

SOLUTION

(a) $f(60, 75) = 80 - \dfrac{60}{20} - \dfrac{75}{25} = 74°F$

(b) An equation is $f(x, y) = 70$; that is,

$$80 - \frac{x}{20} - \frac{y}{25} = 70 \qquad \text{or} \qquad 5x + 4y = 1000$$

(c) The curve $5x + 4y = 1000$ is a line (Figure 7).

A curve along which a scalar field has constant value (such as the curve along which the temperature field in the previous example maintains the constant value 70°F) is called a **level curve** of the field or of the function f that describes the field. An equation of the level curve along which the function f assumes the constant value k is

$$f(x, y) = k$$

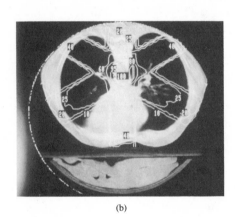

(b)

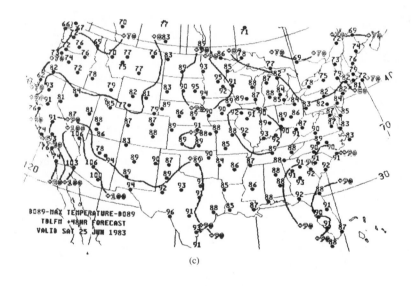

8089-MAX TEMPERATURE-8089
TBLFM +48HR FORECAST
VALID SAT 25 JUN 1983

(c)

Level curves for various scalar fields: (a) equipotential lines for the electric field of a dipole; (b) Computerized Tomography scan of a chest section of a patient undergoing radiation therapy showing isodose curves (level curves for radiation absorbed); (c) weather map showing isotherms (level curves for maximum temperature).

The level curves for various scalar fields are customarily given special names depending on the nature of the field—**isotherms** for the level curves of a temperature field, **equipotential lines** for the level curves of an electric potential field, and so forth. Figure 8 illustrates a number of practical applications of level curves in fields ranging from physics to meteorology.

Suppose that the function f gives the *height* $z = f(x, y)$ of a certain surface S above the xy plane at the point (x, y). (Then S is the graph of the function f.) The intersection of the surface S with a horizontal plane $z = k$ produces a curve C consisting of all points on the surface that are k units above the xy plane (Figure 9a). The perpendicular projection of the curve C on the xy plane gives a level curve for the function f. Such a level curve, whose equation in the xy plane is

$$f(x, y) = k$$

is called a **contour curve** for the surface S. By plotting a number of different contour curves, each labeled with its own value of k, we obtain a **contour map** of

Figure 9

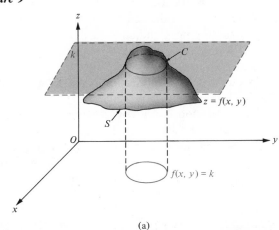

(a)

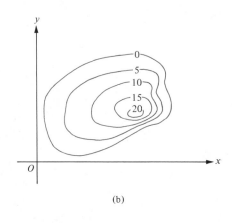

(b)

the surface S (Figure 9b). Such a contour map enables you to visualize the surface as if you were looking down upon it from above and seeing its intersections with horizontal cutting planes at various heights. If these heights are made to differ by equal amounts, then a crowding of successive contour curves indicates a relatively steep part of the surface. Figure 10 shows a contour map of Dutch Flat, N.W., Arizona.

Figure 10

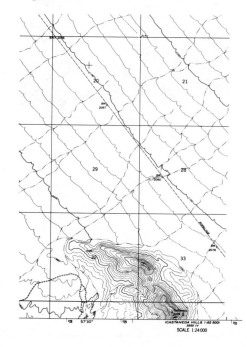

Figure 11

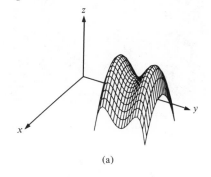

(a)

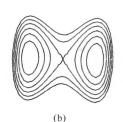

(b)

Not only are contour maps used for representing the surface features of a geographical region, but they are also used to help us visualize complicated graphs of mathematical functions. Figure 11 shows a computer-generated graph of a surface $z = f(x, y)$ and its corresponding contour map.

EXAMPLE 9   Let S be the surface $z = x^2 - y^2 + 20$ for $x \geq 0$. Plot the contour curves for this surface* corresponding to $z = 0$, $z = 10$, $z = 20$, $z = 30$, and $z = 40$.

SOLUTION   For $z = 0$ we obtain $0 = x^2 - y^2 + 20$, or $y^2 - x^2 = 20$, the equation of a hyperbola with a vertical transverse axis. Since $x \geq 0$, we get only the portion of this hyperbola lying in the first and fourth quadrants as the contour

---

*Note that this surface is a portion of a hyperbolic paraboloid.

Figure 12

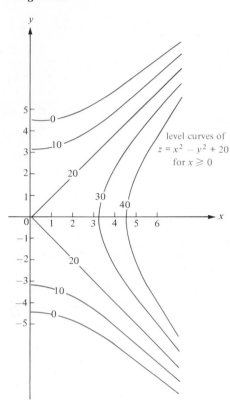

level curves of
$z = x^2 - y^2 + 20$
for $x \geq 0$

curve corresponding to $z = 0$. For $z = 10$, we obtain $10 = x^2 - y^2 + 20$, or $y^2 - x^2 = 10$, another hyperbola. For $z = 20$, the equation is $20 = x^2 - y^2 + 20$, or $x = \pm y$, two lines meeting at the origin. Continuing in this way, we obtain the desired contour map (Figure 12).

The idea of a field as a continuous spatial distribution of a physical quantity was originally conceived by the English experimental physicist Michael Faraday (1791–1867). Faraday's brilliant conception had a powerful impact on the later work of the English mathematical physicist James Clerk Maxwell (1831–1879), who developed the fundamental equations for electromagnetic fields. The concept of a field has become more and more important in physics—indeed, many modern physicists regard physical reality as a network of interacting fields.

Michael Faraday

James Clerk Maxwell

Problem Set 14.1

In Problems 1 to 14, evaluate each expression, using the functions f, g, and h defined by $f(x, y) = 5x^2 + 7xy$, $g(x, y) = \sqrt{xy}$, and $h(x, y, z) = (2xy + z)/(x^2 + y^2 - z^2)$.

1 $f(3, -4)$

2 $f(2, 0)g(2, 0)$

3 $g(k, k)$

4 $g(a^2, b^2)$

5 $h(1, 2, 3)$

6 $h(x, z, y)$

7 $f(\sqrt{a}, b)$

8 $g(\sin \theta, 2 \cos \theta)$

9 $h(\sin t, \cos t, 0)$

10 $\dfrac{f(x, y)}{[g(x, y)]^2}$

11 $f(x, y) + g(x, y)$

12 $f(x + h, y) - f(x, y)$

13 $h(x, y, 0)$

14 $\dfrac{g(x, y + k) - g(x, y)}{k}$

In Problems 15 to 26, find and sketch the domain of each function.

15 $f(x, y) = \dfrac{\sqrt{x^2 + y^2 - 9}}{y}$

16 $g(x, y) = x\sqrt{1 - x^2 - y^2}$

17 $H(x, y) = \dfrac{y}{\sqrt{x^2 + y^2 - 16}}$

18 $F(x, y) = \dfrac{\sqrt{36 - x^2 - y^2}}{x}$

19 $f(x, y) = \ln(x^2 - y)$

20 $G(x, y) = \ln(4 - x^2 - y^2)$

21 $g(x, y) = \sqrt{x + y - 4}$

22 $h(x, y) = \dfrac{y}{\sqrt{36 - 4x^2 - 9y^2}}$

23 $h(x, y) = \ln(xy - 2)$

24 $f(x, y) = \sin^{-1}(2x + y)$

25 $F(x, y) = \dfrac{4x^2 - y^2}{2x - y}$

26 $F(x, y) = \sqrt{\dfrac{x^2 + 9y^2}{x^2 - 9y^2}}$

In Problems 27 to 40, find the domain of each function of two variables, and sketch the graph of the function.

27 $f(x, y) = x + y$

28 $g(x, y) = 2x - 3y + 6$

29 $h(x, y) = \sqrt{4 - x^2 - y^2}$

30 $h(x, y) = 3 + \sqrt{9 - x^2 - y^2}$

31 $F(x, y) = x^2 + y^2$

32 $f(x, y) = \sqrt{1 + x^2 + y^2}$

33 $f(x, y) = 4$

34 $g(x, y) = x^2$

35 $f(x, y) = \sqrt{x^2 + y^2}$

36 $h(x, y) = -\sqrt{x^2 + y^2}$

37 $H(x, y) = 4 - x^2 - y^2$

38 $g(x, y) = y^2$

39 $f(x, y) = -2y + 1$

40 $f(x, y) = \sqrt{x^2 + y^2 - 1}$

41 If $f(x_1, x_2, x_3, \ldots, x_n) = x_1 + x_2 + x_3 + \cdots + x_n$, evaluate and simplify $f(1, 2, 3, \ldots, n)$.

C **42** The present value P of A dollars to be paid t years in the future (assuming an 8 percent continuous interest rate) is $P(A, t) = Ae^{-0.08t}$. Find and interpret $P(100, 13.8)$.

In Problems 43 to 50, plot a contour map of the graph of $z = f(x, y)$ showing the contour curves corresponding to the given values of z.

43 $f(x, y) = 3x + 2y - 1$; $z = -1, z = 0, z = 1, z = 2$

44 $f(x, y) = \sqrt{9 - x^2 - y^2}$; $z = 0, z = 1, z = 2, z = 3$

45 $f(x, y) = x^2 - y^2$; $z = -2, z = -1, z = 0, z = 1, z = 2$

46 $f(x, y) = xy$; $z = 0, z = 1, z = 2, z = 3$

47 $f(x, y) = x + y^2$; $z = -1, z = 0, z = 1, z = 2$

48 $f(x, y) = \dfrac{2x}{x^2 + y^2}$; $z = 0, z = 1, z = 2, z = 3$

49 $f(x, y) = \sqrt{1 - \dfrac{x^2}{4} - \dfrac{y^2}{9}}$; $z = 0, z = 1, z = \dfrac{1}{2}$

50 $f(x, y) = \sqrt{\dfrac{x + y}{x - y}}$; $z = 0, z = 1, z = 2, z = 3$

51 Suppose that the electric potential V in volts at a point (x, y) in the xy plane is given by $V(x, y) = 4/\sqrt{9 - x^2 - y^2}$. The level curves of V are called *equipotential curves*. Plot the equipotential curves corresponding to $V = 16$, $V = 8$, $V = 4$, and $V = \frac{1}{2}$.

52 Suppose that the temperature T in degrees Celsius at a point (x, y) in the xy plane is given by $T = 2x^2 + y^2$. The level curves of T are called *isotherms*. Plot the isotherms corresponding to $T = 6$, $T = 4$, $T = 2$, and $T = 0$.

53 What does it mean if the isotherms (level curves) of a temperature field tend to crowd together near a certain point $P_0 = (x_0, y_0)$?

54 Given a map showing the isotherms of a temperature field, explain how you would plot a path, starting at a point P_0, along which the temperature would increase most rapidly.

55 A person is driving at 50 miles per hour from left to right along the line $y = 75$ in Figure 7. How rapidly does this person feel the temperature dropping, in degrees Fahrenheit per hour, if the temperature at the point (x, y) is given by $f(x, y) = 80 - (x/20) - (y/25)$?

56 Look up the following words in a dictionary, and explain how each refers to the level curves of a certain scalar field: (a) isobar, (b) isocheim, (c) isoclinal line, and (d) isodynamic line.

14.2 Limits and Continuity

The idea of a limit can be extended to functions of two or more variables. For instance, if we say that $f(x, y)$ approaches the limit L as (x, y) approaches (x_0, y_0), we mean that the number $f(x, y)$ can be made to come as close to the number L as we wish by choosing the point (x, y) to be sufficiently close to the point (x_0, y_0), provided that $(x, y) \neq (x_0, y_0)$. The notation is

$$\lim_{(x,y) \to (x_0, y_0)} f(x, y) = L$$

As a specific example, notice that

$$\lim_{(x,y)\to(0,0)} \frac{1}{\sqrt{4 - x^2 - y^2}} = \frac{1}{\sqrt{4}} = \frac{1}{2}$$

since the quantity $1/\sqrt{4 - x^2 - y^2}$ comes closer and closer to $\frac{1}{2}$ as the point (x, y) comes closer and closer to $(0, 0)$. Later in the section we give a formal definition of a limit of a function of two variables; however, for now, we prefer to proceed informally.

It can be shown* that all the properties of limits of functions of single variables extend to functions of several variables; for instance, the limit of a sum, difference, product, or quotient is the sum, difference, product, or quotient of the limits, respectively, provided that these limits exist and that zeros do not appear in denominators. Also, if $f(x, y)$ depends only on x and not on y, so that

$$f(x, y) = h(x)$$

then

$$\lim_{(x,y)\to(x_0,y_0)} f(x, y) = \lim_{x\to x_0} h(x)$$

provided that the last limit exists. We abbreviate this by writing

$$\lim_{(x,y)\to(x_0,y_0)} h(x) = \lim_{x\to x_0} h(x)$$

Similarly, we have

$$\lim_{(x,y)\to(x_0,y_0)} g(y) = \lim_{y\to y_0} g(y)$$

provided that the last limit exists.

In Examples 1 and 2, evaluate the limit.

EXAMPLE 1 $\quad \lim_{(x,y)\to(-1,2)} \left(5x^2y + 2xy - \frac{3y^2}{x + y}\right)$

SOLUTION   Applying the limit properties for sums and products, and abbreviating $\lim_{(x,y)\to(-1,2)}$ as lim, we have

$$\lim \left(5x^2y + 2xy - \frac{3y^2}{x + y}\right)$$

$$= \lim 5x^2y + \lim 2xy - \lim \frac{3y^2}{x + y}$$

$$= 5 \lim x^2 \lim y + 2 \lim x \lim y - \frac{3 \lim y^2}{\lim x + \lim y}$$

$$= 5(-1)^2(2) + 2(-1)(2) - \frac{3(2)^2}{-1 + 2} = -6 \qquad \blacksquare$$

EXAMPLE 2 $\quad \lim_{(x,y)\to(0,0)} (e^{\sin(5x^2+y)} + \cos 3xy)$

SOLUTION $\quad \lim_{(x,y)\to(0,0)} (e^{\sin(5x^2+y)} + \cos 3xy) = e^{\sin[5(0)^2+0]} + \cos [3(0)(0)]$

$$= e^0 + \cos 0 = 2 \qquad \blacksquare$$

*The proofs are similar to the proofs of the properties of limits of functions of a single variable given in Appendix B.

Figure 1

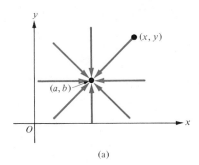

(a)

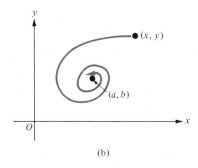

(b)

In Chapter 1 we observed that $\lim\limits_{x \to a} f(x)$ exists if and only if both one-sided limits, $\lim\limits_{x \to a^+} f(x)$ and $\lim\limits_{x \to a^-} f(x)$, exist and are the same. In dealing with a limit of a function f of two variables, say, $\lim\limits_{(x,y) \to (a,b)} f(x, y)$, we can allow the point (x, y) to approach the point (a, b) not only from the right or from the left, but also from above, below, or indeed, any direction (Figure 1a). We can even allow (x, y) to approach (a, b) along a curved path (Figure 1b).

> To say that $\lim\limits_{(x,y) \to (a,b)} f(x, y) = L$ is to require that as (x, y) approaches (a, b) *along any path whatsoever*, $f(x, y)$ approaches the same limit L.

Thus, a convenient way to show that a particular limit $\lim\limits_{(x,y) \to (a,b)} f(x, y)$ *does not exist* is to show that $f(x, y)$ can be made to approach two different limits as (x, y) approaches (a, b) along two different paths.

EXAMPLE 3   Let $f(x, y) = \dfrac{2x^2 y}{3x^2 + 3y^2}$.

(a) Evaluate the limit of $f(x, y)$ as (x, y) approaches $(0, 0)$ along each of the following paths: (i) the x axis, (ii) the y axis, (iii) the line $y = x$, (iv) the parabola $y = x^2$.

(b) Does $\lim\limits_{(x,y) \to (0,0)} f(x, y)$ exist? If so, what is its value?

SOLUTION

(a) (i) On the x axis, $y = 0$ and $f(x, y) = f(x, 0) = \dfrac{2x^2(0)}{3x^2 + 0} = 0$ for $x \neq 0$. Therefore, $\lim\limits_{x \to 0} f(x, 0) = \lim\limits_{x \to 0} 0 = 0$.

(ii) On the y axis, $x = 0$ and $f(x, y) = f(0, y) = \dfrac{2(0)^2 y}{0 + 3y^2} = 0$ for $y \neq 0$. Therefore, $\lim\limits_{y \to 0} f(0, y) = \lim\limits_{y \to 0} 0 = 0$.

(iii) On the line $y = x$, $f(x, y) = f(x, x) = \dfrac{2x^2 x}{3x^2 + 3x^2} = \dfrac{2x^3}{6x^2} = \dfrac{x}{3}$ for $x \neq 0$. Therefore, $\lim\limits_{x \to 0} f(x, x) = \lim\limits_{x \to 0} \dfrac{x}{3} = 0$.

(iv) On the parabola $y = x^2$, $f(x, y) = f(x, x^2) = \dfrac{2x^2 x^2}{3x^2 + 3x^4} = \dfrac{2x^2}{3 + 3x^2}$ for $x \neq 0$. Therefore, $\lim\limits_{x \to 0} f(x, x^2) = \lim\limits_{x \to 0} \dfrac{2x^2}{3 + 3x^2} = 0$.

(b) Along all the paths in part (a), the limit is the same, 0. Although there might be some *other path* along which the limit is not 0, the evidence leads us to *suspect* that the limit is 0. Our suspicion is confirmed as follows: For $x^2 + y^2 \neq 0$, $x^2 \leq x^2 + y^2$, so $x^2|y| \leq (x^2 + y^2)|y|$; hence, $x^2|y|/(x^2 + y^2) \leq |y|$. It follows that

$$|f(x, y)| = \left| \frac{2x^2 y}{3x^2 + 3y^2} \right| = \frac{2}{3} \frac{x^2|y|}{x^2 + y^2} \leq \frac{2}{3}|y|$$

Therefore, as (x, y) approaches $(0, 0)$, $f(x, y)$ must approach 0, since $|f(x, y)|$ is no larger than $\frac{2}{3}|y|$, which approaches 0. It follows that

$$\lim_{(x,y)\to(0,0)} \frac{2x^2y}{3x^2 + 3y^2} = 0$$ ∎

EXAMPLE 4   Follow the directions in Example 3 for the function

$$f(x, y) = \frac{x^2 - y^2}{x^2 + y^2}$$

SOLUTION

(a) (i) On the x axis, $y = 0$ and $f(x, y) = f(x, 0) = \dfrac{x^2 - 0}{x^2 + 0} = 1$ for $x \neq 0$.

Therefore, $\lim\limits_{x\to 0} f(x, 0) = \lim\limits_{x\to 0} 1 = 1$.

(ii) On the y axis, $x = 0$ and $f(x, y) = f(0, y) = \dfrac{0 - y^2}{0 + y^2} = -1$ for $y \neq 0$.

Therefore, $\lim\limits_{y\to 0} f(0, y) = \lim\limits_{y\to 0} (-1) = -1$.

(iii) On the line $y = x$, $f(x, y) = f(x, x) = \dfrac{x^2 - x^2}{x^2 + x^2} = 0$ for $x \neq 0$. Therefore, $\lim\limits_{x\to 0} f(x, x) = \lim\limits_{x\to 0} 0 = 0$.

(iv) On the parabola $y = x^2$, $f(x, y) = f(x, x^2) = \dfrac{x^2 - x^4}{x^2 + x^4} = \dfrac{1 - x^2}{1 + x^2}$ for $x \neq 0$. Therefore, $\lim\limits_{x\to 0} f(x, x^2) = \lim\limits_{x\to 0} \dfrac{1 - x^2}{1 + x^2} = 1$.

(b) Since the limits in (i) and (ii) are different, $\lim\limits_{(x,y)\to(0,0)} f(x, y)$ does not exist. ∎

The following example shows that $f(x, y)$ can approach the same limit as (x, y) approaches (a, b) along *every* line through (a, b), and yet $\lim\limits_{(x,y)\to(a,b)} f(x, y)$ can fail to exist.

EXAMPLE 5   Let $f(x, y) = \begin{cases} \left(x + \dfrac{1}{x}\right)y^2 & \text{if } x \neq 0 \\ 0 & \text{if } x = 0 \end{cases}$

(a) Calculate the limit of $f(x, y)$ as (x, y) approaches $(0, 0)$ along the line $y = mx$.

(b) Calculate the limit of $f(x, y)$ as (x, y) approaches $(0, 0)$ along the parabola $x = y^2$.

(c) Does $\lim\limits_{(x,y)\to(0,0)} f(x, y)$ exist?

SOLUTION

(a) On the line $y = mx$, $f(x, y) = f(x, mx) = [x + (1/x)](mx)^2$ for $x \neq 0$, so that $\lim\limits_{x\to 0} f(x, mx) = \lim\limits_{x\to 0} [x + (1/x)](mx)^2 = \lim\limits_{x\to 0} (m^2x^3 + m^2x) = 0$.

(b) Along the parabola $x = y^2$, $f(x, y) = f(y^2, y) = [y^2 + (1/y^2)]y^2$ for $y \neq 0$, so that $\lim\limits_{y\to 0} f(y^2, y) = \lim\limits_{y\to 0} [y^2 + (1/y^2)]y^2 = \lim\limits_{y\to 0} (y^4 + 1) = 1$.

(c) Since the limits in parts (a) and (b) are different, $\lim\limits_{(x,y)\to(0,0)} f(x, y)$ does not exist. ∎

The formal definition of the limit of a function of two variables is patterned on the definition in Section 1.7 of the limit of a function of a single variable.

DEFINITION 1

Limit of a Function of Two Variables

Let f be a function of two variables, and let (x_0, y_0) be a point in the xy plane. Assume that there exists a circular disk with center at (x_0, y_0) and with positive radius such that every point within the disk, except possibly for the center (x_0, y_0), belongs to the domain of f. Then we say that *the **limit** as (x, y) approaches (x_0, y_0) of $f(x, y)$ is the number L*, and we write

$$\lim_{(x,y)\to(x_0,y_0)} f(x, y) = L$$

provided that for each positive number ϵ, there is a positive number δ such that $|f(x, y) - L| < \epsilon$ holds whenever $(x, y) \neq (x_0, y_0)$ and the distance between (x, y) and (x_0, y_0) is less than δ.

The condition given in this definition can be written as follows: For every $\epsilon > 0$, there exists $\delta > 0$ such that

$$0 < (x - x_0)^2 + (y - y_0)^2 < \delta^2 \quad \text{implies that} \quad |f(x, y) - L| < \epsilon$$

EXAMPLE 6 Show that $\lim\limits_{(x,y)\to(1,2)} (3x + 2y) = 7$ by direct application of Definition 1.

SOLUTION   Let $\epsilon > 0$ be given. We must find $\delta > 0$ such that $|3x + 2y - 7| < \epsilon$ holds whenever $0 < (x - 1)^2 + (y - 2)^2 < \delta^2$. Now,

$$|3x + 2y - 7| = |3x - 3 + 2y - 4| \leq |3x - 3| + |2y - 4|$$

$$= |3(x - 1)| + |2(y - 2)| = 3|x - 1| + 2|y - 2|$$

Hence, if $3|x - 1| < \epsilon/2$ and $2|y - 2| < \epsilon/2$, then

$$|3x + 2y - 7| \leq 3|x - 1| + 2|y - 2| < \frac{\epsilon}{2} + \frac{\epsilon}{2} = \epsilon$$

The condition $3|x - 1| < \epsilon/2$ is equivalent to $9(x - 1)^2 < \epsilon^2/4$, or to $(x - 1)^2 < \epsilon^2/36$. Similarly, the condition $2|y - 2| < \epsilon/2$ is equivalent to $4(y - 2)^2 < \epsilon^2/4$, or to $(y - 2)^2 < \epsilon^2/16$. Therefore, if $(x - 1)^2 < \epsilon^2/36$ and $(y - 2)^2 < \epsilon^2/16$, it will follow that $|3x + 2y - 7| < \epsilon$. Thus, choose $\delta = \epsilon/6$, and note that if

$$0 < (x - 1)^2 + (y - 2)^2 < \delta^2 = \frac{\epsilon^2}{36}$$

then

$$(x - 1)^2 \leq (x - 1)^2 + (y - 2)^2 < \frac{\epsilon^2}{36}$$

and

$$(y - 2)^2 \leq (x - 1)^2 + (y - 2)^2 < \frac{\epsilon^2}{36} < \frac{\epsilon^2}{16}$$

Hence, $(x - 1)^2 < \epsilon^2/36$ and $(y - 2)^2 < \epsilon^2/16$, so that $|3x + 2y - 7| < \epsilon$. ■

The definition and the properties of limits can be extended to functions of three or more variables. For instance, if the number $f(x, y, z)$ can be made to come as close to the number L as we wish by taking the point (x, y, z) sufficiently close to the point (x_0, y_0, z_0) but different from this point, then we write

$$\lim_{(x,y,z)\to(x_0,y_0,z_0)} f(x, y, z) = L$$

Continuity

The definition of continuity for a function of two variables is patterned after the definition in Section 1.9 for continuity of a function of a single variable.

DEFINITION 2

Continuity of a Function of Two Variables

Suppose that f is a function of two variables and that the point (x_0, y_0) is the center of a circular disk of positive radius contained in the domain of f. We say that f is **continuous** at (x_0, y_0) if both of the following conditions hold:

(i) $\lim\limits_{(x,y)\to(x_0,y_0)} f(x, y)$ exists

(ii) $\lim\limits_{(x,y)\to(x_0,y_0)} f(x, y) = f(x_0, y_0)$

In Examples 7 and 8, decide whether the given function is continuous at the indicated point.

EXAMPLE 7 $f(x, y) = 3x^2 + 2xy$ at $(-1, 3)$

SOLUTION   By the properties of limits,

$$\lim_{(x,y)\to(-1,3)} (3x^2 + 2xy) = 3(-1)^2 + 2(-1)(3) = -3 = f(-1, 3)$$

Hence, f is continuous at $(-1, 3)$.   ∎

EXAMPLE 8 $f(x, y) = \begin{cases} \dfrac{4xy}{x^2 + y^2} & \text{if } (x, y) \neq (0, 0) \\ 0 & \text{if } (x, y) = (0, 0) \end{cases}$ at $(0, 0)$

SOLUTION   Along the x axis, we have

$$f(x, y) = f(x, 0) = \begin{cases} 0 & \text{if } x \neq 0 \\ 0 & \text{if } x = 0 \end{cases}$$

so that $\lim\limits_{x\to 0} f(x, 0) = 0$. Along the line $y = x$, we have

$$f(x, y) = f(x, x) = \begin{cases} 2 & \text{if } x \neq 0 \\ 0 & \text{if } x = 0 \end{cases}$$

so that $\lim\limits_{x\to 0} f(x, x) = 2$. Thus, $\lim\limits_{(x,y)\to(0,0)} f(x, y)$ does not exist, and so f is not continuous at $(0, 0)$.   ∎

Figure 2

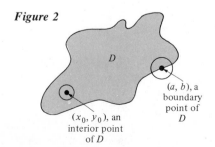

(a, b), a boundary point of D

(x_0, y_0), an interior point of D

Let D be a set of points in the xy plane. A point (x_0, y_0) is called an **interior point** of D if all points in some circular disk of positive radius with (x_0, y_0) as its center are contained in D (Figure 2). On the other hand, a point (a, b) is called a **boundary point** of D if every circular disk of positive radius with (a, b) as its center contains at least one point that belongs to D and at least one point that does not belong to D (Figure 2). There is no requirement that a boundary point (a, b) of D belong to the set D. Note that a point cannot be an interior point of D and also a boundary point of

D. A set *D* is said to be **open** if it contains none of its own boundary points, while a set *D* is said to be **closed** if it contains all its own boundary points. (This terminology is suggested by the ideas of open and closed intervals.)

Notice that Definition 2 applies only to the continuity of a function at an interior point of its domain. A modified definition of continuity, which we do not set forth here, is required for continuity of a function at a boundary point of its domain. Naturally, a function is said to be **continuous** if it is continuous at every point in its domain.

In order to show that a set *D* of points in the plane is open, it is enough to show that every point in *D* is an interior point of *D* (Problem 42).

EXAMPLE 9   Show that the domain of the function $f(x, y) = xy/(x - y)$ is an open set, illustrate this domain with a sketch, and show that f is continuous.

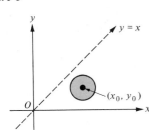

Figure 3

SOLUTION   The domain of f consists of all points in the xy plane except those lying on the line $y = x$ (Figure 3). Suppose that (x_0, y_0) is a point in the domain D of f, so that (x_0, y_0) is not on the line $y = x$. If d is the distance from (x_0, y_0) to the line $y = x$, then any circular disk of radius r with $0 < r < d$ is contained in D; hence, (x_0, y_0) lies in the interior of D. If follows that the domain D is open. Since, for $x_0 \neq y_0$,

$$\lim_{(x,y) \to (x_0,y_0)} \frac{xy}{x - y} = \frac{x_0 y_0}{x_0 - y_0} = f(x_0, y_0)$$

it follows that f is continuous at every point (x_0, y_0) of its domain. Therefore, f is a continuous function.

Functions of two variables have many of the same properties with regard to continuity as do functions of single variables. Among these are the following.

Continuity Properties for Functions of Two Variables

Suppose that (x_0, y_0) is an interior point of the domains of the functions f and g of two variables, and assume that f and g are continuous at (x_0, y_0). Then:

> **1** $h(x, y) = f(x, y) + g(x, y)$ is continuous at (x_0, y_0).
>
> **2** $k(x, y) = f(x, y) - g(x, y)$ is continuous at (x_0, y_0).
>
> **3** $p(x, y) = f(x, y)g(x, y)$ is continuous at (x_0, y_0).
>
> **4** If $g(x_0, y_0) \neq 0$, then $q(x, y) = \dfrac{f(x, y)}{g(x, y)}$ is continuous at (x_0, y_0).
>
> **5** If w is a function of a single variable that is defined and continuous at the number $f(x_0, y_0)$ and if (x_0, y_0) is an interior point of the domain of $v(x, y) = w[f(x, y)]$, then v is continuous at (x_0, y_0).

EXAMPLE 10   What are the interior points of the domain of the function u defined by $u(x, y) = \sqrt{1 - x^2 - y^2} \ln (x + y)$? Is u continuous at such points? Illustrate with a sketch of the domain of u.

SOLUTION   The expression $\sqrt{1 - x^2 - y^2}$ is defined only for $1 - x^2 - y^2 \geq 0$, that is, for $x^2 + y^2 \leq 1$. The condition $x^2 + y^2 \leq 1$ holds exactly when (x, y) is on or inside the circle $x^2 + y^2 = 1$ of radius 1 with center at $(0, 0)$. The expression $\ln (x + y)$ is defined only for $x + y > 0$, that is, only when the point (x, y) lies

Figure 4

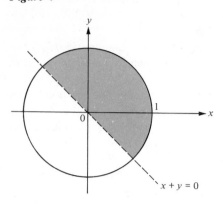

above the line $x + y = 0$. The domain of u is therefore the shaded region shown in Figure 4 inside or on the circle $x^2 + y^2 = 1$ but above the line $x + y = 0$. The interior points of this domain are just those points not lying on the rim of the circle (nor on the line $x + y = 0$). Since the square-root function is continuous, the function $f(x, y) = \sqrt{1 - x^2 - y^2}$ is continuous inside the circle $x^2 + y^2 = 1$, by Property 5 above. Similarly, the function $g(x, y) = \ln(x + y)$ is continuous above the line $x + y = 0$. By Property 3, the function $u(x, y) = f(x, y)g(x, y)$ is continuous at all interior points of its domain. ∎

The definitions and properties of continuous functions are extended in an obvious way to functions of three or more variables. Naturally, *every polynomial function in several variables is continuous*. Also, *a ratio of such polynomial functions—that is, a **rational function** of several variables—has an open domain and is continuous at every point of its domain*. For instance, the function

$$f(x, y, z) = \frac{5x^{17}y^{12}z^7 - 7xy^5z + x^2y^2z - x + 12y + 1984}{25x^3y^3z^2 - 37x + 15y - z + 33}$$

is continuous at every point (x, y, z) for which the denominator is not zero. Roughly speaking, any function built up in a "reasonable way" from continuous functions is continuous at every interior point of its domain.

Problem Set 14.2

In Problems 1 to 12, evaluate each limit.

1 $\lim\limits_{(x,y)\to(1,-2)} (5x^2 + 3xy)$

2 $\lim\limits_{(x,y)\to(\pi/4,\pi)} (\sin 2x + \sin 2y)$

3 $\lim\limits_{(x,y)\to(0,0)} (e^{\cos(3x+y)} + \sec 5xy)$

4 $\lim\limits_{(x,y)\to(0,0)} \dfrac{5}{\ln(x + 3y + e)}$

5 $\lim\limits_{(x,y)\to(0,0)} \dfrac{\cos x + \cos y}{e^{-x} + 3e^y}$

6 $\lim\limits_{(x,y)\to(-1,2)} [\![7x + \frac{1}{3}y^3]\!]$

7 $\lim\limits_{(x,y)\to(-1,2)} e^{3x-2y^2}$

8 $\lim\limits_{(u,v)\to(0,0)} \dfrac{\sin uv}{uv}$

9 $\lim\limits_{(x,y)\to(0,0)} \ln(1 + 4x^3y^2)$

10 $\lim\limits_{(s,t)\to(4,-2)} s\sqrt[3]{t^3 + 2s}$

11 $\lim\limits_{(x,y,z)\to(-1,2,0)} \dfrac{xyz - x + y - z}{x^2 + y^2 + z^2 - 4}$

12 $\lim\limits_{(x,y,z)\to(0,0,0)} z \sin\left(\dfrac{1}{\sqrt{x^2 + y^2 + z^2}}\right)$

In Problems 13 to 18, (a) evaluate the limit of $f(x, y)$ as (x, y) approaches $(0, 0)$ along each of the indicated paths (i), (ii), (iii), and (iv); (b) determine $\lim\limits_{(x,y)\to(0,0)} f(x, y)$, if it exists.

13 $f(x, y) = \dfrac{5xy^2}{x^2 + y^2}$ as (x, y) approaches $(0, 0)$ (i) along the x axis, (ii) along the y axis, (iii) along the line $y = 5x$, (iv) along the parabola $y = x^2$

14 $f(x, y) = \dfrac{\sin(x^3 + y^3)}{x^3 + y^3}$ as (x, y) approaches $(0, 0)$ (i) along the x axis, (ii) along the y axis, (iii) along the line $y = 10x$, (iv) along the curve $y = x^4$

15 $f(x, y) = \dfrac{xy}{x^2 + y^2}$ as (x, y) approaches $(0, 0)$ (i) along the x axis, (ii) along the y axis, (iii) along the line $y = x$, (iv) along the line $y = mx$

16 $f(x, y) = \dfrac{3x^4y^4}{(x^4 + y^2)^3}$ as (x, y) approaches $(0, 0)$ (i) along the x axis, (ii) along the y axis, (iii) along the line $y = x$, (iv) along the curve $y = -x^4$

17 $f(x, y) = \begin{cases} (x + y)\sin\dfrac{1}{y} & \text{if } y \neq 0 \\ 0 & \text{if } y = 0 \end{cases}$ as (x, y) approaches $(0, 0)$ (i) along the x axis (ii) along the y axis, (iii) along the curve $y = x^3$, (iv) along the curve $y = -x^5$

18 $f(x, y) = \begin{cases} \dfrac{1}{y}\sin xy & \text{if } y \neq 0 \\ x & \text{if } y = 0 \end{cases}$ as (x, y) approaches $(0, 0)$ (i) along the x axis, (ii) along the line $y = 2x$, (iii) along the parabola $y = x^2$, (iv) along the curve $y = 5x^3$

In Problems 19 and 20, prove each statement by making direct use of the definition of limit; that is, show that for each given $\epsilon > 0$, there is a $\delta > 0$ so that the conditions of the definition hold.

19 $\lim\limits_{(x,y)\to(2,1)} (5x + 3y) = 13$ **20** $\lim\limits_{(x,y)\to(-4,2)} (-x + 2y) = 8$

In Problems 21 to 28, decide whether each function is continuous at the point indicated. Justify your answer.

21 $f(x, y) = \sqrt{25 - x^2 - y^2}$ at $(-3, 3)$

22 $f(x, y) = e^{-xy} \ln (x^2 - 2y + 7)$ at $(0, 0)$

23 $g(x, y) = \begin{cases} \dfrac{xy}{y - 2x} & \text{if } y \neq 2x \\ 1 & \text{if } y = 2x \end{cases}$ at $(1, 2)$

24 $f(x, y, z) = \dfrac{1}{x^2 + y^2 + z^2}$ at $(0, 0, 1)$

25 $g(x, y) = \dfrac{\sin x}{\sin y}$ at $\left(\pi, \dfrac{\pi}{2} \right)$

26 $h(x, y) = \dfrac{xy}{1 + e^x}$ at $(0, 0)$

27 $f(x, y) = \begin{cases} (x + y) \sin \dfrac{1}{x} & \text{if } x \neq 0 \\ 0 & \text{if } x = 0 \end{cases}$ at $(0, 0)$

28 $f(x, y) = \begin{cases} \dfrac{7x^2 y}{x^2 + y^2} & \text{if } (x, y) \neq (0, 0) \\ 0 & \text{if } (x, y) = (0, 0) \end{cases}$ at $(0, 0)$

In Problems 29 to 39, (a) sketch a diagram showing the domain of each function in the xy plane, (b) specify which points of the domain are interior points, and (c) determine those interior points of the domain—if any—at which the function is discontinuous.

29 $f(x, y) = \sqrt{xy}$ **30** $g(x, y) = \sin^{-1} (2x + y)$

31 $h(x, y) = \dfrac{4x^2 - y^2}{2x - y}$ **32** $F(x, y) = \ln (xy - 2)$

33 $f(x, y) = \dfrac{x}{y^2 - 1}$ **34** $H(x, y) = \ln (x^2 + y^2)$

35 $f(x, y) = \dfrac{xy}{\sqrt{9 - x^2 - y^2}}$ **36** $G(x, y) = \sin^{-1} \left(\dfrac{y}{x} \right)$

37 $g(x, y) = \begin{cases} \dfrac{x - y}{x + y} & \text{if } (x, y) \neq (0, 0) \\ 0 & \text{if } (x, y) = (0, 0) \end{cases}$

38 $h(x, y) = \begin{cases} 4x^2 + 9y^2 & \text{if } 4x^2 + 9y^2 \leq 1 \\ \dfrac{1}{4x^2 + 9y^2} & \text{if } 4x^2 + 9y^2 > 1 \end{cases}$

39 $F(x, y) = \begin{cases} x^2 + y^2 & \text{if } x^2 + y^2 \geq 4 \\ 0 & \text{if } x^2 + y^2 < 4 \end{cases}$

40 Prove that the limit of the sum is the sum of the limits for functions of two variables. Use the precise definition of limits in terms of ϵ and δ, and give a clear statement of whatever assumptions you need to make.

41 Write out a precise definition, in terms of ϵ and δ, for the statement
$$\lim_{(x,y,z)\to(x_0,y_0,z_0)} f(x, y, z) = L$$

42 Prove that a set of points D in the xy plane is an open set if and only if every point in D is an interior point of D.

14.3   Partial Derivatives

The techniques, rules, and formulas developed in Chapter 2 for differentiating functions of a single variable apply to functions of two or more variables, provided that we hold all but one of the independent variables constant and then differentiate with respect to the remaining variable.

For instance, consider the function f of two variables given by
$$f(x, y) = x^2 + 3xy - 4y^2$$

Let's agree, temporarily, to hold the second variable y constant and differentiate with respect to the first variable x. Then since y is constant,
$$\frac{d}{dx} (3xy) = 3y \frac{d}{dx} (x) = 3y \quad \text{and} \quad \frac{d}{dx} (-4y^2) = 0$$

Hence,

$$\frac{d}{dx} f(x, y) = \frac{d}{dx} (x^2) + \frac{d}{dx} (3xy) + \frac{d}{dx} (-4y^2) = 2x + 3y + 0 = 2x + 3y$$

In order to emphasize that only x is being allowed to vary, so that y is held fixed while the derivative is calculated, it is traditional to modify the symbol d/dx and write $\partial/\partial x$ instead. (The symbol ∂ is called the "round d.") Thus, we write the equation above as

$$\frac{\partial}{\partial x} f(x, y) = \frac{\partial}{\partial x} (x^2 + 3xy - 4y^2) = 2x + 3y$$

The derivative calculated with respect to x while temporarily holding y constant is called the **partial derivative with respect to x**, and $\partial/\partial x$ is called the **partial derivative operator** with respect to x. Similarly, if we wish to hold the variable x fixed and differentiate with respect to y, we use the symbol $\partial/\partial y$. Thus, for the function $f(x, y) = x^2 + 3xy - 4y^2$, we have

$$\frac{\partial}{\partial y} f(x, y) = \frac{\partial}{\partial y} (x^2 + 3xy - 4y^2) = \frac{\partial}{\partial y} (x^2) + \frac{\partial}{\partial y} (3xy) + \frac{\partial}{\partial y} (-4y^2)$$

$$= 0 + 3x - 8y = 3x - 8y$$

More formally, we have the following definition.

DEFINITION 1   **Partial Derivatives of Functions of Two Variables**

If f is a function of two variables and (x, y) is a point in the domain of f, then the **partial derivatives** $\dfrac{\partial f(x, y)}{\partial x}$ and $\dfrac{\partial f(x, y)}{\partial y}$ of f at (x, y) with respect to the first and second variables are defined by

$$\frac{\partial f(x, y)}{\partial x} = \lim_{\Delta x \to 0} \frac{f(x + \Delta x, y) - f(x, y)}{\Delta x}$$

and

$$\frac{\partial f(x, y)}{\partial y} = \lim_{\Delta y \to 0} \frac{f(x, y + \Delta y) - f(x, y)}{\Delta y}$$

provided that the limits exist. The procedure of finding partial derivatives is called **partial differentiation.**

EXAMPLE 1   Use Definition 1 directly to find $\partial z/\partial x$ if $z = f(x, y) = 5x^2 - 7xy + 2y^2$.

SOLUTION

$$\frac{\partial z}{\partial x} = \lim_{\Delta x \to 0} \frac{f(x + \Delta x, y) - f(x, y)}{\Delta x}$$

$$= \lim_{\Delta x \to 0} \frac{[5(x + \Delta x)^2 - 7(x + \Delta x)y + 2y^2] - (5x^2 - 7xy + 2y^2)}{\Delta x}$$

$$= \lim_{\Delta x \to 0} \frac{5[x^2 + 2x \Delta x + (\Delta x)^2] - 7xy - 7y \Delta x + 2y^2 - 5x^2 + 7xy - 2y^2}{\Delta x}$$

$$= \lim_{\Delta x \to 0} \frac{10x \Delta x + 5(\Delta x)^2 - 7y \Delta x}{\Delta x} = \lim_{\Delta x \to 0} (10x - 7y + 5 \Delta x)$$

$$= 10x - 7y$$

It is convenient to have a notation for partial derivatives that is analogous to the notation $f'(x)$ for functions of one variable. Thus, if $z = f(x, y)$, we often write $f_1(x, y)$ or $f_x(x, y)$ rather than $\partial z/\partial x$ or $\partial f(x, y)/\partial x$ for the partial derivative of f with respect to x. The subscript 1 (or the subscript x) denotes partial differentiation with respect to the first independent variable (or with respect to x). Thus

$$\frac{\partial z}{\partial x} = \frac{\partial}{\partial x} f(x, y) = f_1(x, y) = f_x(x, y)$$

Similarly, the subscript 2 (or the subscript y) denotes partial differentiation with respect to the second independent variable (or with respect to y), so that

$$\frac{\partial z}{\partial y} = \frac{\partial}{\partial y} f(x, y) = f_2(x, y) = f_y(x, y)$$

Partial derivatives of functions of more than two independent variables are defined by an obvious extension of Definition 1. For instance, if $w = f(x, y, z)$, then

$$\frac{\partial w}{\partial x} = f_1(x, y, z) = f_x(x, y, z) = \lim_{\Delta x \to 0} \frac{f(x + \Delta x, y, z) - f(x, y, z)}{\Delta x}$$

$$\frac{\partial w}{\partial y} = f_2(x, y, z) = f_y(x, y, z) = \lim_{\Delta y \to 0} \frac{f(x, y + \Delta y, z) - f(x, y, z)}{\Delta y}$$

and

$$\frac{\partial w}{\partial z} = f_3(x, y, z) = f_z(x, y, z) = \lim_{\Delta z \to 0} \frac{f(x, y, z + \Delta z) - f(x, y, z)}{\Delta z}$$

More generally, if $w = f(x_1, x_2, x_3, \ldots, x_n)$, then we use the notations

$$\frac{\partial w}{\partial x_k} = \frac{\partial}{\partial x_k} f(x_1, x_2, x_3, \ldots, x_n) = f_k(x_1, x_2, x_3, \ldots, x_n)$$
$$= f_{x_k}(x_1, x_2, x_3, \ldots, x_n)$$

for the partial derivative of f with respect to the kth independent variable x_k. We use the words "partial derivative" to refer both to the function f_k and to the value $f_k(x_1, x_2, \ldots, x_k, \ldots, x_n)$ of this function—you can always tell from the context what is intended.

Partial derivatives can be calculated by using the same techniques that are effective for ordinary derivatives, except that *all independent variables other than the variable with respect to which the differentiation takes place must temporarily be regarded as constants*.

<u>EXAMPLE 2</u>   If $w = xy^2z^3$, find

   (a) $\partial w/\partial y$ **(b)** $\partial w/\partial z$

SOLUTION

   (a) Holding x and z constant, and differentiating with respect to y, we obtain

$$\frac{\partial w}{\partial y} = \frac{\partial}{\partial y} (xy^2z^3) = xz^3 \frac{d}{dy} (y^2) = xz^3(2y) = 2xyz^3$$

(b) Holding x and y constant and differentiating with respect to z, we have

$$\frac{\partial w}{\partial z} = \frac{\partial}{\partial z}(xy^2z^3) = xy^2\frac{d}{dz}(z^3) = xy^2(3z^2) = 3xy^2z^2 \qquad \blacksquare$$

EXAMPLE 3   If $f(x, y) = xe^y + y \ln x$, find

(a) $f_1(x, y)$ **(b)** $f_2(x, y)$

SOLUTION

(a) $f_1(x, y) = \dfrac{\partial}{\partial x}(xe^y + y \ln x) = \dfrac{\partial}{\partial x}(xe^y) + \dfrac{\partial}{\partial x}(y \ln x)$

$$= e^y\frac{dx}{dx} + y\frac{d}{dx}(\ln x) = e^y + \frac{y}{x}$$

(b) $f_2(x, y) = \dfrac{\partial}{\partial y}(xe^y + y \ln x) = \dfrac{\partial}{\partial y}(xe^y) + \dfrac{\partial}{\partial y}(y \ln x)$

$$= x\frac{d}{dy}(e^y) + \ln x\frac{dy}{dy} = xe^y + \ln x \qquad \blacksquare$$

EXAMPLE 4   If $f(x, y, z) = e^{xy^2}z^3$, find $f_z(1, 2, 3)$.

SOLUTION   Holding x and y constant and differentiating with respect to z, we have

$$f_z(x, y, z) = \frac{\partial}{\partial z}(e^{xy^2}z^3) = e^{xy^2}\frac{d}{dz}(z^3) = 3e^{xy^2}z^2$$

Therefore,

$$f_z(1, 2, 3) = 3e^{(1)(2)^2}3^2 = 27e^4$$

Note that to find $f_z(1, 2, 3)$, we *first* differentiate to find $f_z(x, y, z)$ and *then* substitute $x = 1$, $y = 2$, and $z = 3$. $\qquad \blacksquare$

There are many versions of the **chain rule** applicable to partial derivatives, the simplest one being virtually a direct transcription of the ordinary chain rule for functions of a single variable. Thus, let g be a function of more than one variable—say two variables for definiteness. If $w = f(v)$ and $v = g(x, y)$, so that $w = f[g(x, y)]$, then, holding y constant and using the usual chain rule, we have

$$\frac{\partial w}{\partial x} = f'[g(x, y)]g_x(x, y) = f'(v)\frac{\partial v}{\partial x} = \frac{dw}{dv}\frac{\partial v}{\partial x}$$

provided that the derivatives dw/dv and $\partial v/\partial x$ exist. Similarly, holding x constant and using the usual chain rule, we have

$$\frac{\partial w}{\partial y} = f'[g(x, y)]g_y(x, y) = f'(v)\frac{\partial v}{\partial y} = \frac{dw}{dv}\frac{\partial v}{\partial y}$$

provided that the derivatives dw/dv and $\partial v/\partial y$ exist. In summary, if dw/dv, $\partial v/\partial x$, and $\partial v/\partial y$ exist, then

$$\boxed{\frac{\partial w}{\partial x} = \frac{dw}{dv}\frac{\partial v}{\partial x} \qquad \text{and} \qquad \frac{\partial w}{\partial y} = \frac{dw}{dv}\frac{\partial v}{\partial y}}$$

EXAMPLE 5   If $w = \sqrt{1 - x^2 - y^2}$, find $\partial w/\partial x$ and $\partial w/\partial y$.

SOLUTION   Put $v = 1 - x^2 - y^2$, so that $w = \sqrt{v}$. By the chain rule,

$$\frac{\partial w}{\partial x} = \frac{dw}{dv}\frac{\partial v}{\partial x} = \frac{1}{2\sqrt{v}}\frac{\partial}{\partial x}(1 - x^2 - y^2) = \frac{1}{2\sqrt{v}}(-2x) = \frac{-x}{\sqrt{1 - x^2 - y^2}}$$

$$\frac{\partial w}{\partial y} = \frac{dw}{dv}\frac{\partial v}{\partial y} = \frac{1}{2\sqrt{v}}\frac{\partial}{\partial y}(1 - x^2 - y^2) = \frac{1}{2\sqrt{v}}(-2y) = \frac{-y}{\sqrt{1 - x^2 - y^2}} \quad\blacksquare$$

In using the version of the chain rule, given above, you proceed in much the same way as in Section 2.7, except that you multiply the *ordinary* derivative of the "outside function" by the appropriate *partial* derivative of the "inside function."

<u>EXAMPLE 6</u>   If $f(x, y) = \tan(x^2 - y^2)$, find $f_x(x, y)$ and $f_y(x, y)$.

SOLUTION   Here the "outside function" is the tangent function, and the "inside function" is the function $g(x, y) = x^2 - y^2$. Thus,

$$f_x(x, y) = \sec^2(x^2 - y^2)\frac{\partial}{\partial x}(x^2 - y^2) = [\sec^2(x^2 - y^2)](2x)$$

$$= 2x \sec^2(x^2 - y^2)$$

and $f_y(x, y) = \sec^2(x^2 - y^2)\dfrac{\partial}{\partial y}(x^2 - y^2) = [\sec^2(x^2 - y^2)](-2y)$

$$= -2y \sec^2(x^2 - y^2) \quad\blacksquare$$

<u>EXAMPLE 7</u>   If $w = e^{x/y}$, find $\partial w/\partial x$ and $\partial w/\partial y$.

SOLUTION $\dfrac{\partial w}{\partial x} = e^{x/y}\dfrac{\partial}{\partial x}\left(\dfrac{x}{y}\right) = e^{x/y}\dfrac{1}{y} = \dfrac{e^{x/y}}{y}$

$$\frac{\partial w}{\partial y} = e^{x/y}\frac{\partial}{\partial y}\left(\frac{x}{y}\right) = e^{x/y}\left(-\frac{x}{y^2}\right) = \frac{-xe^{x/y}}{y^2} \quad\blacksquare$$

Geometric Interpretation of Partial Derivatives

Suppose that f is a function of two variables and that f has partial derivatives f_1 and f_2. The graph of f is a surface with the equation $z = f(x, y)$ (Figure 1). Let $z_0 = f(x_0, y_0)$, so that $P = (x_0, y_0, z_0)$ is a point on this surface. The plane $y = y_0$ cuts a cross section APB from the surface, while the plane $x = x_0$ cuts a cross section CPD from the surface. As a point moves along the curve APB, its x and z coordinates vary according to the equation $z = f(x, y_0)$, while its y coordinate remains constant with $y = y_0$. The slope of the tangent line to APB at any point is the rate at which the z coordinate changes with respect to x; hence, this slope is given by $\partial z/\partial x = f_1(x, y_0)$. In particular, $f_1(x_0, y_0)$ represents the slope of the tangent line to APB at the point P. Similarly, $f_2(x_0, y_0)$ represents the slope of the tangent line to CPD at the point P. Thus, in Figure 1, we have

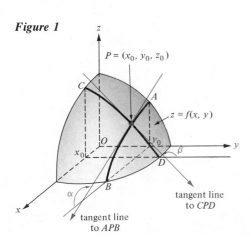

Figure 1

$$\tan \alpha = f_1(x_0, y_0) = f_x(x_0, y_0) = \frac{\partial z}{\partial x} \text{ evaluated at } (x_0, y_0)$$

and $\tan \beta = f_2(x_0, y_0) = f_y(x_0, y_0) = \dfrac{\partial z}{\partial y}$ evaluated at (x_0, y_0)

EXAMPLE 8   Find the slope of the tangent line to the cross section cut from the surface $z = 4x^2y - xy^3$ by the plane $y = 2$ at the point $P = (3, 2, 48)$.

SOLUTION   Here we hold y constant and find $\partial z/\partial x$. We have

$$\frac{\partial z}{\partial x} = \frac{\partial}{\partial x}(4x^2y) - \frac{\partial}{\partial x}(xy^3) = 8xy - y^3$$

so that when $x = 3$ and $y = 2$,

$$\frac{\partial z}{\partial x} = 8(3)(2) - 2^3 = 40$$ ∎

Rates of Change

Suppose that a variable quantity, say w, depends on a number of other quantities, say $x_1, x_2, \ldots, x_n$. If all but one of the independent variables $x_1, x_2, \ldots, x_n$ are held constant, then w depends only on the remaining variable, and the instantaneous rate of change of w with respect to this variable is given as usual by a derivative—in this case, the partial derivative of w with respect to the variable in question.

ⓒ **EXAMPLE 9**   The volume V of a right circular cylinder is given by $V = \pi r^2 h$, where r is the radius of the base and h is the height.

(a) Find the instantaneous rate of change of the volume with respect to the height if the radius remains constant with the value $r = 6$ centimeters while the height h varies. Evaluate this rate of change at the instant when $h = 10$ centimeters.

(b) Suppose that the height remains constant with the value $h = 10$ centimeters. Assuming that the radius is variable, find the rate of change of the volume with respect to the radius at the instant when $r = 6$ centimeters.

SOLUTION

(a) $\dfrac{\partial V}{\partial h} = \dfrac{\partial}{\partial h}(\pi r^2 h) = \pi r^2$

When $r = 6$ and $h = 10$ centimeters, we have $\partial V/\partial h = \pi(36) \approx 113$ cubic centimeters per centimeter.

(b) $\dfrac{\partial V}{\partial r} = \dfrac{\partial}{\partial r}(\pi r^2 h) = 2\pi r h$

When $r = 6$ and $h = 10$ centimeters, we have $\partial V/\partial r = 2\pi(6)(10) \approx 377$ cubic centimeters per centimeter.   ∎

Problem Set 14.3

In Problems 1 to 6, find each partial derivative by direct application of Definition 1.

1 $f_x(x, y)$, where $f(x, y) = 8x - 2y + 13$

2 $\dfrac{\partial}{\partial x} f(x, y)$, where $f(x, y) = 3x^2 + 5xy + 7y^3$

3 $f_1(-1, 2)$, where $f(x, y) = -7x^2 + 8xy^2$

4 $\dfrac{\partial}{\partial z} f(x, y, z)$, where $f(x, y, z) = 2xy^2 - 7xz + 3xyz^2$

5 $f_2(1, -1)$, where $f(x, y) = 5xy^3 + 6x^2 + 11$

6 $\dfrac{\partial z}{\partial y}$ at the point $(3, -5)$, where $z = -7x^3y + xy + 3$

In Problems 7 to 22, find each partial derivative by treating all but one of the independent variables as constants and applying the rules of ordinary differentiation.

7 $\dfrac{\partial}{\partial x} f(x, y)$, where $f(x, y) = 7x^2 + 5x^2y + 2$

8 $\dfrac{\partial}{\partial x} (x^2 \sin y)$

9 $h_x(x, y)$, where $h(x, y) = \sin x \cos 7y$

10 $f_2(x, y)$, where $f(x, y) = e^{-2x} \tan y$

11 $\dfrac{\partial w}{\partial x}$, where $w = \dfrac{x^2 + y^2}{y^2 - x^2}$

12 $\dfrac{\partial}{\partial x} (x \sin y - y \ln x)$

13 $f_1(r, \theta)$, where $f(r, \theta) = r^2 \cos 7\theta$

14 $g_2\left(0, \dfrac{\pi}{2}\right)$, where $g(\theta, \phi) = \sin 2\theta \cos \phi$

15 $f_1(2, -3)$, where $f(x, y) = x^2y^3 - 3xy^2 + 2y$

16 $g_2(x, y)$, where $g(x, y) = \displaystyle\int_y^x e^{-t^2} \, dt$

17 $f_z(x, y, z)$, where $f(x, y, z) = 6xyz + 3x^2y + 7z$

18 $\dfrac{\partial}{\partial y} g(x, y, z)$, where $g(x, y, z) = 3x^2y^3z^4 - 4xyz^3$

19 $\dfrac{\partial w}{\partial x}$, where $w = xy^2 + yz^2 + x^2y$

20 $\dfrac{\partial w}{\partial z}$, where $w = x \cos z + y \sin x + xe^z$

21 $g_1(2, 3, -4)$, where $g(x_1, x_2, x_3) = 2x_1e^{x_3} + 2x_2e^{x_1} + 2x_3e^{x_2}$

22 $p_2(x_1, x_2, x_3, x_4)$, where p is defined by $p(x_1, x_2, x_3, x_4) = x_1x_2x_3 \sin x_4 - x_2x_3x_4 \sin x_1$

In Problems 23 to 44, find each partial derivative by using the chain rule.

23 $\dfrac{\partial w}{\partial x}$, where $w = \sqrt{u}$ and $u = 3x^2 + y^2$

24 $\dfrac{\partial}{\partial x} g(x, y)$, where $g(x, y) = \sin (xy)^2$

25 $\dfrac{\partial w}{\partial x}$, where $w = \ln u$ and $u = 7x^2 + 4y^3$

26 $\dfrac{\partial}{\partial y} \ln \left(\dfrac{x^2}{y}\right)$

27 $h_1(x, y)$, where $h(x, y) = \tan^{-1} (xy)$

28 $f_1(x, y)$, where $f(x, y) = e^{x^2 + y^2}$

29 $\dfrac{\partial w}{\partial y}$, where $w = \sin^{-1} u$ and $u = \sqrt{xy}$

30 $\dfrac{\partial}{\partial y} \displaystyle\int_2^{\sin (x+y)} e^{t^3} \, dt$

31 $f_1(2, -1)$, where $f(x, y) = \sqrt{x^2 + y^2 - 1}$

32 $f_z(x, y, z)$, where $f(x, y, z) = x^2/\sqrt{y^2 + z^2}$

33 $g_3(x, y, z)$, where $g(x, y, z) = xz^2e^{xy} \cos yz$

34 $\dfrac{\partial}{\partial y} [z \sin (xz) \cos (xy)]$

35 $\dfrac{\partial w}{\partial x}$, where $w = (x^2 + y^2 + z^2)^{-3/2}$

36 $\dfrac{\partial w}{\partial s}$, where $w = e^{-t} \sin (s + u)$

37 $f_x(x, y)$, where $f(x, y) = \cos [\ln (x^2 + y^2)]$

38 $\dfrac{\partial w}{\partial y}$, where $w = (y^2 \tan x)^{-3/5}$

39 $g_y(x, y)$, where $g(x, y) = \cos^2 (3x^2y^4)$

40 $\dfrac{\partial w}{\partial s}$, where $w = \sinh \sqrt{s} \cosh^2 (4s^3t^5)$

41 $f_3(1, 1, 5\pi/4)$, where $f(x, y, z) = (xy + 3)^{\cos (z - \pi)}$

42 $\dfrac{\partial s}{\partial u}$, where $s = (\sin u)^{uvw}$

43 $\dfrac{\partial p}{\partial v}$, where $p = (3u - 2v)^{\sin v}$

44 $\dfrac{\partial x}{\partial u}$, where $x = (stu)^{st \ln u}$

45 Given that $w = x^3y^2 - 2xy^4 + 3x^2y^3$, verify that
$$x \frac{\partial w}{\partial x} + y \frac{\partial w}{\partial y} = 5w$$

46 Given that $w = (ax + by + cz)^n$, where a, b, and c are constants, verify that
$$x \frac{\partial w}{\partial x} + y \frac{\partial w}{\partial y} + z \frac{\partial w}{\partial z} = nw$$

47 Given that $w = t^2 + \tan te^{1/s}$, verify that
$$s^2 \frac{\partial w}{\partial s} + t \frac{\partial w}{\partial t} = 2t^2$$

48 Let f be a differentiable function, and let $w = f(\tfrac{5}{2}s^2 - \tfrac{7}{2}t^2)$. Verify that
$$7t \frac{\partial w}{\partial s} + 5s \frac{\partial w}{\partial t} = 0$$

49 Given that $w = x^2 \sin \dfrac{y}{z} + y^2 \ln \dfrac{z}{x} + z^2 e^{x/y}$, verify that

$$x \frac{\partial w}{\partial x} + y \frac{\partial w}{\partial y} + z \frac{\partial w}{\partial z} = 2w$$

50 Given that $w = e^{x/y} + e^{y/z} + e^{z/x}$, verify that

$$x \frac{\partial w}{\partial x} + y \frac{\partial w}{\partial y} + z \frac{\partial w}{\partial z} = 0$$

51 Given that $w = \ln(x^3 + 5x^2y + 6xy^2 + 7y^3)$, verify that

$$x \frac{\partial w}{\partial x} + y \frac{\partial w}{\partial y} = 3$$

52 Given that

$$z = \frac{e^{x^2 + y^2}}{x^2 + y^2}$$

show that

$$y \frac{\partial z}{\partial x} - x \frac{\partial z}{\partial y} = 0$$

In Problems 53 to 60, find the slope of the tangent line to the cross section cut from the given surface by the given plane at the indicated point.

53 The surface $z = 3x - 5y + 7$ and the plane $y = 2$ at the point $(1, 2, 0)$

54 The surface $z = \sqrt{4 - x^2 - y^2}$ and the plane $x = 1$ at the point $(1, 1, \sqrt{2})$

55 The surface $z = \sqrt{31 - 2x^2 - 3y^2}$ and the plane $y = 2$ at the point $(3, 2, 1)$

56 The surface $x^2 + y^2 + z^2 = 14$ and the plane $x = 1$ at the point $(1, 3, 2)$

57 The surface $z = e^{-x^2} \sin 3y$ and the plane $x = 1$ at the point $(1, 0, 0)$

58 The surface $6x^2 + 9y^2 + 4z^2 = 61$ and the plane $y = -1$ at the point $(1, -1, 2)$

59 The surface $z = 2xy/(x^2 + y^2)$ and the plane $y = 4$ at the point $(3, 4, \frac{24}{25})$

60 The surface $z = e^{x/y} - e^{y/x}$ and the plane $y = 2$ at the point $(2, 2, 0)$

©61 The lateral surface area A of a right circular cone of height h and with base radius r is given by $A = \pi r \sqrt{h^2 + r^2}$.

(a) If r is held fixed, with $r = 3$ centimeters while h is varied, find the rate of change of A with respect to h at the instant when $h = 7$ centimeters.

(b) If h is held fixed with $h = 7$ centimeters while r is varied, find the rate of change of A with respect to r at the instant when $r = 3$ centimeters.

62 If two sides x and y of a triangle are held fixed while the angle θ between them is varied, find the rate of change of the third side z with respect to θ. (Use the law of cosines.)

©63 *Clairaut's formula* for the weight W in dynes of a 1-gram mass at latitude L degrees and at height h centimeters above sea level is

$$W = 980.6056 - 2.5028 \cos \frac{\pi L}{90} - \frac{h}{3{,}000{,}000}$$

(a) If L is held constant at 40°N while h is varied, find the rate of change of W with respect to h at the instant when the mass is 6 kilometers above sea level.

(b) If the mass remains at a constant altitude of 6 kilometers above sea level while its latitude L is varied, find the rate of change of W per degree of latitude at the instant when $L = 40°$N.

64 A circular hill has a central vertical section in the form of a curve whose equation is $z = 10 - (x^2/160)$, where the units are in meters. The top is being cut down in horizontal layers at the constant rate of 100 cubic meters per day. How fast is the area of horizontal cross section increasing when the top has been cut down a vertical distance of 4 meters?

65 The resistance R ohms of an electric circuit is given by the formula $R = E/I$, where I is the current in amperes and E is the electromotive force in volts. Calculate $\partial R/\partial I$ and $\partial R/\partial E$ when $I = 15$ amperes and $E = 110$ volts, and give an interpretation of these two partial derivatives in terms of rates of change.

66 A mortgage company computes the maximum size of a 25-year mortgage that it is willing to give on a one-family house by using the formula

$$M = 3x + 0.025y^2$$

where x is the size of the down payment in dollars and y is the applicant's monthly salary in dollars. Find $\partial M/\partial x$ and $\partial M/\partial y$, and interpret their meanings in terms of rate of change.

67 Psychologists define the *intelligence quotient* $Q(M, C)$ of a person to be 100 times the person's mental age M divided by the chronological age C, so that $Q(M, C) = 100M/C$. Find $\partial Q/\partial M$ and $\partial Q/\partial C$ when $M = 12$ years and $C = 10$ years, and interpret the results.

©68 Medical researchers use the body surface area of a patient to calibrate the results of tests of kidney performance. The formula

$$A = 0.007W^{0.425}H^{0.725}$$

is often used to calculate the body surface area A in square centimeters of a person whose weight is W kilograms and whose height is H centimeters. Find $\partial A/\partial W$ and $\partial A/\partial H$ when $W = 58$ kilograms and $H = 168$ centimeters, and interpret the results.

69 The economists Paul Douglas and Charles Cobb proposed the theoretical relationship $P = AL^\alpha K^\beta$ among the total *production P* of a firm or industry, the input L of *labor* in person-hours, and the *investment K* in capital equipment. Here α, β, and A are constants depending on the firm or industry in question and the units in which P is measured. Find formulas for the *marginal* *productivity of labor* $\partial P/\partial L$ and for the *marginal productivity of capital* $\partial P/\partial K$.

70 The formula $H = (2.14 + 1.37 \sqrt{V} - 0.0916V)(91.4 - T)$ gives the *heat loss (wind chill)* H in Btu's per square foot of skin per hour if the air temperature is T degrees Fahrenheit and the wind speed is V miles per hour. Find $\partial H/\partial V$ and $\partial H/\partial T$ when $V = 20$ miles per hour and $T = 25°F$, and interpret the results.

©**71** The *ideal gas law* $PV = nRT$ is used to describe the relationship among the pressure P, the temperature T, and the volume V of a gas, where n is the number of moles of the gas and R is the universal gas constant.

(a) Find $\partial P/\partial T$ when $nR = 10$, $T = 80°F$, and $V = 50$ cubic inches.

(b) Show that

$$\frac{\partial V}{\partial T} \cdot \frac{\partial T}{\partial P} \cdot \frac{\partial P}{\partial V} = -1$$

©**72** According to *Poiseuille's law*, the speed V in centimeters per second of blood flow in a vessel is given by $V = P(R^2 - r^2)/(4L\nu)$, where R is the radius of the vessel, r is the distance from the center of the vessel, L is the length of the vessel, P is the blood pressure, and ν is the viscosity of the blood. Find $\partial V/\partial P$ and $\partial V/\partial R$ when $L = 10$ centimeters, $P = 100$ grams per square centimeter, $R = 0.0075$ centimeter, $r = 0.0025$ centimeter, and $\nu = 0.05$.

73 If $x = r \cos \theta$ and $y = r \sin \theta$, show that

(a) $\begin{vmatrix} \dfrac{\partial x}{\partial r} & \dfrac{\partial x}{\partial \theta} \\ \dfrac{\partial y}{\partial r} & \dfrac{\partial y}{\partial \theta} \end{vmatrix} = r$ (b) $\begin{vmatrix} \dfrac{\partial x}{\partial r} & \dfrac{\partial x}{\partial \theta} & \dfrac{\partial x}{\partial z} \\ \dfrac{\partial y}{\partial r} & \dfrac{\partial y}{\partial \theta} & \dfrac{\partial y}{\partial z} \\ \dfrac{\partial z}{\partial r} & \dfrac{\partial z}{\partial \theta} & \dfrac{\partial z}{\partial z} \end{vmatrix} = r$

74 If $x = \rho \sin \phi \cos \theta$, $y = \rho \sin \phi \sin \theta$, and $z = \rho \cos \phi$, show that

$$\begin{vmatrix} \dfrac{\partial x}{\partial \rho} & \dfrac{\partial x}{\partial \theta} & \dfrac{\partial x}{\partial \phi} \\ \dfrac{\partial y}{\partial \rho} & \dfrac{\partial y}{\partial \theta} & \dfrac{\partial y}{\partial \phi} \\ \dfrac{\partial z}{\partial \rho} & \dfrac{\partial z}{\partial \theta} & \dfrac{\partial z}{\partial \phi} \end{vmatrix} = \rho^2 \sin \phi$$

14.4 Linear Approximation and Differentiable Functions

In Section 14.3 we saw that the partial derivatives $f_1(x_0, y_0)$ and $f_2(x_0, y_0)$ involve only cross sections cut from the surface $z = f(x, y)$ by the two perpendicular planes $y = y_0$ and $x = x_0$; hence, in general, these two partial derivatives tell us very little about the shape of the surface "in between" the two cross sections. For this reason it is not appropriate to call a function of two (or more) variables "differentiable" just because its partial derivatives exist. The key to the proper definition of "differentiability" for a function of more than one variable is the important idea of *linear approximation*.

Linear Approximation

As we saw in Section 4.1, if f is a differentiable function of one variable, then the linear approximation

$$f(x + \Delta x) \approx f(x) + f'(x) \, \Delta x$$

is often quite accurate, provided that $|\Delta x|$ is sufficiently small. The analogous **linear approximation procedure** for a function f of *two* variables involves the partial derivatives f_1 and f_2 and has the form

$$f(x + \Delta x, y + \Delta y) \approx f(x, y) + f_1(x, y) \, \Delta x + f_2(x, y) \, \Delta y$$

Again, if $|\Delta x|$ and $|\Delta y|$ are sufficiently small, this linear approximation can be quite accurate.

© **EXAMPLE 1**   If $f(x, y) = \sqrt{x^2 + y^2 + 2}$, use the linear approximation procedure to estimate $f(3.01, 4.96)$ and check the accuracy of your result by using a calculator.

SOLUTION   In the linear approximation

$$f(x + \Delta x, y + \Delta y) \approx f(x, y) + f_1(x, y)\, \Delta x + f_2(x, y)\, \Delta y$$

we put $x = 3$, $y = 5$, $\Delta x = 0.01$, and $\Delta y = -0.04$, to obtain

$$f(3.01, 4.96) \approx \sqrt{3^2 + 5^2 + 2} + f_1(3, 5)(0.01) + f_2(3, 5)(-0.04)$$

Here,
$$f_1(x, y) = \frac{\partial}{\partial x} \sqrt{x^2 + y^2 + 2} = \frac{x}{\sqrt{x^2 + y^2 + 2}}$$

$$f_2(x, y) = \frac{\partial}{\partial y} \sqrt{x^2 + y^2 + 2} = \frac{y}{\sqrt{x^2 + y^2 + 2}}$$

so that
$$f_1(3, 5) = \frac{3}{\sqrt{36}} = \frac{1}{2} \quad \text{and} \quad f_2(3, 5) = \frac{5}{\sqrt{36}} = \frac{5}{6}$$

Thus,
$$f(3.01, 4.96) \approx \sqrt{36} + \tfrac{1}{2}(0.01) + \tfrac{5}{6}(-0.04) = 5.97166 \cdots$$

The true value of $f(3.01, 4.96)$, correct to five decimal places, is 5.97174, so the linear approximation procedure has given a result that is correct to the first three decimal places.   ∎

Differentiable Functions of Two Variables

The **error** E involved in the linear approximation

$$f(x + \Delta x, y + \Delta y) \approx f(x, y) + f_1(x, y)\, \Delta x + f_2(x, y)\, \Delta y$$

is given by

$$E = f(x + \Delta x, y + \Delta y) - [f(x, y) + f_1(x, y)\, \Delta x + f_2(x, y)\, \Delta y]$$
$$= f(x + \Delta x, y + \Delta y) - f(x, y) - f_1(x, y)\, \Delta x - f_2(x, y)\, \Delta y$$

Roughly speaking, the condition that f is *differentiable* is that the error E approach zero very rapidly as Δx and Δy approach zero, so that the linear approximation is very accurate when $|\Delta x|$ and $|\Delta y|$ are sufficiently small. More formally, suppose that (x_0, y_0) is an interior point of the domain of f and that the two partial derivatives $f_1(x_0, y_0)$ and $f_2(x_0, y_0)$ exist. Then we make the following definition.

DEFINITION 1

Differentiable Function of Two Variables

We say that f is **differentiable** at (x_0, y_0) if the error E involved in the linear approximation

$$f(x_0 + \Delta x, y_0 + \Delta y) \approx f(x_0, y_0) + f_1(x_0, y_0)\, \Delta x + f_2(x_0, y_0)\, \Delta y$$

has the form

$$E = \epsilon_1\, \Delta x + \epsilon_2\, \Delta y$$

where
$$\lim_{(\Delta x, \Delta y) \to (0,0)} \epsilon_1 = 0 \quad \text{and} \quad \lim_{(\Delta x, \Delta y) \to (0,0)} \epsilon_2 = 0$$

If the domain of f is an open set, then f is called a **differentiable function** provided that it is differentiable at every point in its domain. Note the analogy between the condition on the error E in Definition 1 and the condition on the linear-approximation error for functions of a single variable in Theorem 1 on page 249.

Recall that a differentiable function of a single variable is necessarily continuous (Theorem 1, page 96). The following theorem establishes an analogous result for a function of two variables.

THEOREM 1

Continuity of a Differentiable Function

Let f be a function of two variables, and let (x_0, y_0) be an interior point of the domain of f. Then, if f is differentiable at (x_0, y_0), it follows that f is continuous at (x_0, y_0).

PROOF

Let (x, y) be a point in the domain of f, and let

$$\Delta x = x - x_0 \qquad \Delta y = y - y_0$$

Note that the condition $(x, y) \to (x_0, y_0)$ is equivalent to the condition $(\Delta x, \Delta y) \to (0, 0)$. Because $x = x_0 + \Delta x$ and $y = y_0 + \Delta y$, we have

$$f(x, y) = f(x_0 + \Delta x, y_0 + \Delta y)$$
$$= f(x_0, y_0) + f_1(x_0, y_0)\, \Delta x + f_2(x_0, y_0)\, \Delta y + E$$

where E is the error as in Definition 1. Thus, we have

$$\lim_{(x,y) \to (x_0,y_0)} E = \lim_{(\Delta x, \Delta y) \to (0,0)} (\epsilon_1\, \Delta x + \epsilon_2\, \Delta y) = 0$$

since both ϵ_1 and ϵ_2 approach zero as $(\Delta x, \Delta y) \to (0, 0)$. Therefore,

$$\lim_{(x,y) \to (x_0,y_0)} f(x, y) = \lim_{(x,y) \to (x_0,y_0)} [f(x_0, y_0) + f_1(x_0, y_0)\, \Delta x + f_2(x_0, y_0)\, \Delta y + E]$$

$$= f(x_0, y_0) + f_1(x_0, y_0)(0) + f_2(x_0, y_0)(0) + 0 = f(x_0, y_0)$$

and it follows that f is continuous at (x_0, y_0). ■

The mere existence of the partial derivatives $f_1(x_0, y_0)$ and $f_2(x_0, y_0)$ of a function f of two variables does not guarantee that f is differentiable at (x_0, y_0) (see Problem 46). A condition that does ensure the differentiability of f is provided by the following definition.

DEFINITION 2

Continuous Differentiability

Let f be a function of two variables, and let U be an open set of points contained in the domain of f. We say that f is **continuously differentiable on** U if the partial derivatives $f_1(x, y)$ and $f_2(x, y)$ exist at every point (x, y) in U and the functions f_1 and f_2 are continuous on U.

If the domain D of f is an open set and f is continuously differentiable on D, then we say that f is **continuously differentiable.** The following theorem shows that the term "continuously differentiable" is not a misnomer.

THEOREM 2

A Continuously Differentiable Function Is Differentiable

Let U be an open set contained in the domain of the function f of two variables. If f is continuously differentiable on U, then f is differentiable at each point (x_0, y_0) in U.

PROOF*

Let (x_0, y_0) be a point in U. Since U is open, there is a circular disk of positive radius with center (x_0, y_0) such that all points within this disk are contained in U. In what follows we assume that $|\Delta x|$ and $|\Delta y|$ are sufficiently small that $(x_0 + \Delta x, y_0 + \Delta y)$ belongs to this disk and hence to the set U. Define

$$(1) \quad \epsilon_1 = \begin{cases} \dfrac{f(x_0 + \Delta x, y_0) - f(x_0, y_0)}{\Delta x} - f_1(x_0, y_0) & \text{if } \Delta x \neq 0 \\ 0 & \text{if } \Delta x = 0 \end{cases}$$

and

$$(2) \quad \epsilon_2 = \begin{cases} \dfrac{f(x_0 + \Delta x, y_0 + \Delta y) - f(x_0 + \Delta x, y_0)}{\Delta y} - f_2(x_0, y_0) & \text{if } \Delta y \neq 0 \\ 0 & \text{if } \Delta y = 0 \end{cases}$$

Note that

$$(3) \quad \epsilon_1 \, \Delta x = f(x_0 + \Delta x, y_0) - f(x_0, y_0) - f_1(x_0, y_0) \, \Delta x$$

even if $\Delta x = 0$, and

$$(4) \quad \epsilon_2 \, \Delta y = f(x_0 + \Delta x, y_0 + \Delta y) - f(x_0 + \Delta x, y_0) - f_2(x_0, y_0) \, \Delta y$$

even if $\Delta y = 0$. Furthermore,

$$(5) \quad \epsilon_1 \, \Delta x + \epsilon_2 \, \Delta y$$

$$= f(x_0 + \Delta x, y_0 + \Delta y) - f(x_0, y_0) - f_1(x_0, y_0) \, \Delta x - f_2(x_0, y_0) \, \Delta y$$

Also, because

$$\lim_{\Delta x \to 0} \frac{f(x_0 + \Delta x, y_0) - f(x_0, y_0)}{\Delta x} = f_1(x_0, y_0)$$

it follows that

$$(6) \quad \lim_{(\Delta x, \Delta y) \to (0,0)} \epsilon_1 = 0$$

Now choose and temporarily fix Δx and Δy with $\Delta y \neq 0$ and define a function ϕ of one variable by

$$\phi(y) = f(x_0 + \Delta x, y)$$

Note that

$$\phi'(y) = f_2(x_0 + \Delta x, y)$$

By the mean-value theorem (Theorem 3, page 169), there exists a number c between y_0 and $y_0 + \Delta y$ such that

$$\phi'(c) = \frac{\phi(y_0 + \Delta y) - \phi(y_0)}{(y_0 + \Delta y) - y_0} = \frac{f(x_0 + \Delta x, y_0 + \Delta y) - f(x_0 + \Delta x, y_0)}{\Delta y}$$

that is,

$$(7) \quad f_2(x_0 + \Delta x, c) = \frac{f(x_0 + \Delta x, y_0 + \Delta y) - f(x_0 + \Delta x, y_0)}{\Delta y}$$

---

*This proof, which is somewhat technical, may be omitted on a first reading of the section. It is included here because the theorem itself is rather important and is used extensively in what follows.

Note here that the value of c may depend on our choice of Δx and Δy; however, since c is between y_0 and $y_0 + \Delta y$, it follows that $c \to y_0$ as $\Delta y \to 0$. From Equations (2) and (7), we have

(8) $\quad \epsilon_2 = f_2(x_0 + \Delta x, c) - f_2(x_0, y_0)$

Because f_2 is continuous and $c \to y_0$ as $\Delta y \to 0$, Equation (8) implies that

(9) $\quad \lim_{(\Delta x, \Delta y) \to (0,0)} \epsilon_2 = 0$

Now let E be the error involved in the linear approximation

$$f(x_0 + \Delta x, y_0 + \Delta y) \approx f(x_0, y_0) + f_1(x_0, y_0)\, \Delta x + f_2(x_0, y_0)\, \Delta y$$

so that

(10) $\quad E = \epsilon_1\, \Delta x + \epsilon_2\, \Delta y$

by Equation (5). From Definition 2 and Equations (6), (9), and (10), it follows that f is differentiable at (x_0, y_0). ∎

EXAMPLE 2 Is the function $f(x, y) = e^{3x+4y}$ differentiable?

SOLUTION The domain of f is the entire xy plane, so it is an open set. (Why?) Here,

$$f_1(x, y) = \frac{\partial}{\partial x}\, (e^{3x+4y}) = e^{3x+4y}\, \frac{\partial}{\partial x}\, (3x + 4y) = 3e^{3x+4y}$$

$$f_2(x, y) = \frac{\partial}{\partial y}\, (e^{3x+4y}) = e^{3x+4y}\, \frac{\partial}{\partial y}\, (3x + 4y) = 4e^{3x+4y}$$

so that f_1 and f_2 are continuous on the entire xy plane; that is, f is continuously differentiable. It follows from Theorem 2 that f is differentiable at each point in its domain; hence, f is differentiable. ∎

EXAMPLE 3 Is the function $f(x, y) = |xy|$ differentiable **(a)** at $(1, 1)$ and **(b)** at $(0, 1)$?

SOLUTION

(a) Inside a circular disk of radius, say, $\frac{1}{2}$ with center at $(1, 1)$, both x and y are positive, so that $f(x, y) = |xy| = xy$. Thus, inside this disk, $f_1(x, y) = \partial f(x, y)/\partial x = y$ and $f_2(x, y) = \partial f(x, y)/\partial y = x$; hence f is continuously differentiable. It follows that f is differentiable at each point in the disk. In particular, it is differentiable at the center point $(1, 1)$ of the disk.

(b) Since $f(x, 1) = |x|$ and since the absolute-value function (of one variable) is not differentiable at 0, it follows that the partial derivative $f_1(0, 1)$ does not exist. However, by definition, a differentiable function must have both of its partial derivatives. Consequently, f cannot be differentiable at $(0, 1)$. ∎

It can be shown that *the sum, difference, product, and quotient of continuously differentiable functions are continuously differentiable; hence, they are differentiable at each point of their domains.* In particular, every polynomial function in two variables is continuously differentiable, and so is every rational function in two variables.

The Total Differential

Suppose that f is a function of two variables, and let

$$z = f(x, y)$$

If x and y are changed by small amounts Δx and Δy, respectively, then z changes by an amount Δz given by

$$\Delta z = f(x + \Delta x, y + \Delta y) - f(x, y)$$

Assuming that f is differentiable at (x, y), we know that the error involved in the linear approximation

$$f(x + \Delta x, y + \Delta y) \approx f(x, y) + f_1(x, y)\, \Delta x + f_2(x, y)\, \Delta y$$

will be small, and it follows that we can approximate Δz as

$$\Delta z \approx f_1(x, y)\, \Delta x + f_2(x, y)\, \Delta y$$

Using the alternative notation $\partial z/\partial x$ and $\partial z/\partial y$ for the partial derivatives $f_1(x, y)$ and $f_2(x, y)$, we can write the approximation as

$$\Delta z \approx \frac{\partial z}{\partial x}\, \Delta x + \frac{\partial z}{\partial y}\, \Delta y$$

By analogy with the notation used for functions of one variable in Section 4.1, the changes Δx and Δy in the two independent variables x and y are sometimes called the **differentials** of these variables and written as **dx** and **dy,** respectively. Thus, if dx and dy are small, then the change Δz in the value of z caused by changing x to $x + dx$ and changing y to $y + dy$ is approximated as

$$\Delta z \approx \frac{\partial z}{\partial x}\, dx + \frac{\partial z}{\partial y}\, dy$$

Pursuing the analogy with functions of a single variable one step further, we define the **total differential dz** of the dependent variable z by

$$\boxed{dz = \frac{\partial z}{\partial x}\, dx + \frac{\partial z}{\partial y}\, dy}$$

Thus, if dx and dy are small, then

$$\Delta z \approx dz$$

Since $z = f(x, y)$, we also write dz as df, so that

$$\boxed{df = f_1(x, y)\, dx + f_2(x, y)\, dy}$$

<u>EXAMPLE 4</u>   If $f(x, y) = 3x^3y^2 - 2xy^3 + xy - 1$, find the total differential df.

SOLUTION   Here,

$$f_1(x, y) = 9x^2y^2 - 2y^3 + y \qquad \text{and} \qquad f_2(x, y) = 6x^3y - 6xy^2 + x$$

Hence,
$$df = f_1(x, y)\, dx + f_2(x, y)\, dy$$
$$= (9x^2y^2 - 2y^3 + y)\, dx + (6x^3y - 6xy^2 + x)\, dy$$

■

ⓒ **EXAMPLE 5**   The volume V of a right circular cone of height h and base radius r is given by $V = \frac{1}{3}\pi r^2 h$. If the height is increased from 5 to 5.01 centimeters while the base radius is decreased from 4 to 3.98 centimeters, find an approximation to the change ΔV in the volume.

SOLUTION   We use the total differential dV to approximate ΔV. Here,

$$dV = \frac{\partial V}{\partial h}\,dh + \frac{\partial V}{\partial r}\,dr = \frac{\partial}{\partial h}\left(\frac{1}{3}\pi r^2 h\right)dh + \frac{\partial}{\partial r}\left(\frac{1}{3}\pi r^2 h\right)dr$$

$$= \frac{1}{3}\pi r^2\,dh + \frac{2}{3}\pi rh\,dr$$

When $h = 5$, $dh = 0.01$, $r = 4$, and $dr = -0.02$, we have

$$\Delta V \approx dV = \frac{1}{3}\pi(4)^2(0.01) + \frac{2}{3}\pi(4)(5)(-0.02)$$

that is, $$\Delta V \approx -0.6702 \text{ cubic centimeter}$$

(The true value of ΔV, correct to five decimal places, is -0.66978.)   ∎

Functions of Three or More Variables

The notions of differentiability, continuous differentiability, and the total differential extend in an obvious way to functions of three or more variables. For instance, if f is a function of three variables, then a point (x_0, y_0, z_0) is called an **interior point** of the domain D of f if there is a positive number r such that all points (x, y, z) within a sphere of radius r with center at (x_0, y_0, z_0) belong to the domain D; that is, if $(x - x_0)^2 + (y - y_0)^2 + (z - z_0)^2 < r^2$, then (x, y, z) belongs to D. Similarly, a point (a, b, c) is called a **boundary point** of D if every sphere with positive radius r and center (a, b, c) contains a point that belongs to D and a point that does not. We say that D is **open** if it contains none of its own boundary points, that is, if it consists entirely of interior points. Similarly, D is called **closed** if it contains all its boundary points.

Now assume that (x_0, y_0, z_0) is an interior point of the domain D of f and that the three partial derivatives

$$A = f_1(x_0, y_0, z_0) \qquad B = f_2(x_0, y_0, z_0) \qquad C = f_3(x_0, y_0, z_0)$$

exist. Let E denote the error involved in the linear approximation

$$f(x_0 + \Delta x, y_0 + \Delta y, z_0 + \Delta z) \approx f(x_0, y_0, z_0) + A\,\Delta x + B\,\Delta y + C\,\Delta z$$

Then we say that f is *differentiable* at (x_0, y_0, z_0) if there exist three functions ϵ_1, ϵ_2, and ϵ_3 such that

$$E = \epsilon_1\,\Delta x + \epsilon_2\,\Delta y + \epsilon_3\,\Delta z$$

and as $(\Delta x, \Delta y, \Delta z) \to (0, 0, 0)$,

$$\epsilon_1 \to 0 \qquad \epsilon_2 \to 0 \qquad \text{and} \qquad \epsilon_3 \to 0$$

If the domain D of f is open, we say that f is **continuously differentiable** if the partial derivatives f_1, f_2, and f_3 are defined and continuous on D. By arguments similar to those given above for functions of two variables, it can be shown that a continuously differentiable function of three variables is differentiable at each point of its domain and that a differentiable function of three variables is automatically continuous. Again, sums, products, differences, and quotients of continuously differentiable functions of three variables are continuously differentiable. Any polynomial function in three variables is continuously differentiable, as is any rational function in three variables.

If $w = f(x, y, z)$, where f is differentiable at an interior point (x, y, z) of its domain, we define the **total differential** dw, or df, at (x, y, z) by

$$dw = df = \frac{\partial w}{\partial x} \, dx + \frac{\partial w}{\partial y} \, dy + \frac{\partial w}{\partial z} \, dz$$

$$= f_1(x, y, z) \, dx + f_2(x, y, z) \, dy + f_3(x, y, z) \, dz$$

where dx, dy, and dz, the **differentials** of the independent variables, can be assigned arbitrary values. The total differential dw provides an approximation to the change Δw in the dependent variable w caused by changing x, y, and z by small amounts dx, dy, and dz, respectively.

EXAMPLE 6 If $w = x^2 y^3 z^4$, find the total differential dw.

SOLUTION

$$dw = \frac{\partial w}{\partial x} \, dx + \frac{\partial w}{\partial y} \, dy + \frac{\partial w}{\partial z} \, dz = 2xy^3z^4 \, dx + 3x^2y^2z^4 \, dy + 4x^2y^3z^3 \, dz \quad \blacksquare$$

EXAMPLE 7 Three resistors, x, y, and z ohms, are connected in parallel to give a net resistance of w ohms, where

$$w = \frac{xyz}{xy + xz + yz}$$

Each resistor is rated at 300 ohms but is subject to 1 percent error. What is the approximate maximum possible error and the approximate maximum possible percentage of error in the value of w?

SOLUTION   We have

$$\Delta w \approx dw = \frac{\partial w}{\partial x} \, dx + \frac{\partial w}{\partial y} \, dy + \frac{\partial w}{\partial z} \, dz$$

where dx, dy, and dz represent the errors in the values of the three resistors. Here $|dx|$, $|dy|$, and $|dz|$ do not exceed 3 ohms (1 percent of 300 ohms). We have

$$\frac{\partial w}{\partial x} = \frac{(xy + xz + yz) \dfrac{\partial}{\partial x}(xyz) - xyz \dfrac{\partial}{\partial x}(xy + xz + yz)}{(xy + xz + yz)^2}$$

$$= \frac{xy^2z + xyz^2 + y^2z^2 - xy^2z - xyz^2}{(xy + xz + yz)^2} = \frac{y^2z^2}{(xy + xz + yz)^2}$$

so that when $x = y = z = 300$, $\partial w/\partial x = \frac{1}{9}$. Similar calculations yield $\partial w/\partial y = \frac{1}{9}$ and $\partial w/\partial z = \frac{1}{9}$ when $x = y = z = 300$. Thus,

$$dw = \tfrac{1}{9} \, dx + \tfrac{1}{9} \, dy + \tfrac{1}{9} \, dz$$

$$|dw| = \tfrac{1}{9} |dx + dy + dz| \le \tfrac{1}{9} (|dx| + |dy| + |dz|)$$

Since $|dx|$, $|dy|$, and $|dz|$ cannot exceed 3, it follows that

$$|dx| + |dy| + |dz| \le 3 + 3 + 3 = 9$$

Hence,

$$|dw| \leq \tfrac{1}{9}(9) = 1 \text{ ohm}$$

The approximate maximum possible error is 1 ohm. Since $w = 100$ ohms when $x = y = z = 300$ ohms, the approximate maximum possible percent error is $\tfrac{1}{100} \times 100$ percent = 1 percent.   ∎

Problem Set 14.4

In Problems 1 to 6, use the linear-approximation procedure to estimate $f(x_0 + \Delta x, y_0 + \Delta y)$ for each function f and the indicated values of x_0, y_0, Δx, and Δy. In each case, check the accuracy of your result by using a calculator.

1 $f(x, y) = x^3 - 5x^2 + 6xy$; $x_0 = 1$, $y_0 = -2$, $\Delta x = 0.07$, $\Delta y = 0.02$

2 $f(x, y) = x^3 - 2xy + 3y^3$; $x_0 = 2$, $y_0 = 1$, $\Delta x = 0.03$, $\Delta y = -0.01$

3 $f(x, y) = x \ln y - y \ln x$; $x_0 = 1$, $y_0 = 1$, $\Delta x = 0.01$, $\Delta y = 0.02$

4 $f(x, y) = x\sqrt{x - y}$; $x_0 = 6$, $y_0 = 2$, $\Delta x = 0.25$, $\Delta y = 0.25$

5 $f(x, y) = y \tan^{-1} xy$; $x_0 = 2$, $y_0 = \tfrac{1}{2}$, $\Delta x = -\tfrac{1}{15}$, $\Delta y = -\tfrac{1}{10}$

6 $f(x, y) = e^{xy} \sin (x + y)$; $x_0 = 2$, $y_0 = 3$, $\Delta x = 0.008$, $\Delta y = 0.01$

In Problems 7 to 12, decide whether each function is differentiable at the indicated point. Give a reason for your answer.

7 $f(x, y) = xe^{-y}$ at (x, y)

8 $f(x, y) = |xy^2|$ at $(0, 1)$

9 $f(x, y) = \dfrac{3xy}{x^3 + y^3}$ at $(1, 2)$

10 $f(x, y) = \begin{cases} \dfrac{1}{xy} & \text{if } x \neq 0 \text{ and } y \neq 0 \\ 1 & \text{if either } x = 0 \text{ or } y = 0 \end{cases}$ at $(0, 0)$

11 $f(x, y) = \dfrac{xy^2}{x + y} \cos e^{x^2 + y^2}$ at $(1, 1)$

12 $f(x, y, z) = |xyz|$ at $(1, 1, 1)$

In Problems 13 to 30, find each total differential.

13 df if $f(x, y) = 5x^3 + 4x^2y - 2y^3$

14 dz if $z = \sqrt{x^2 + y^2}$

15 dw if $w = e^{xy}$

16 dw if $w = y \ln (2 + x)$

17 dz if $z = x \sin y^2 + y \sin x^2$

18 dz if $z = \ln (3x - 4y^2 + 5)$

19 dw if $w = \tan^{-1} (2y - 3x)$

20 df if $f(u, v) = \cos \dfrac{uv}{u^2 + v^2}$

21 df if $f(x, y) = e^{3 - 2xy}$

22 dw if $w = \tan^{-1} \dfrac{y}{x}$

23 dT if $T = \dfrac{PV}{R}$, R is constant

24 df if $f(u, v) = \sin^{-1} \dfrac{u}{v}$

25 df if $f(x, y, z) = xy^2 - 2zx^2 + 3xyz^2$

26 dw if $w = x \sin yz + y \cos xz$

27 dw if $w = xe^{yz} + ye^{xz}$

28 dw if $w = \ln (2x^2 + 3y^2 + 4z^2)$

29 dw if $w = e^x \ln yz + e^y \ln xz$

30 dw if $w = (xyz)^4 e^{4y}$

In Problems 31 to 40, use total differentials to make each required approximation.

31 The power P consumed in an electric resistor is given by $P = E^2/R$ watts, where E is the electromotive force in volts and R is the resistance in ohms. If, at a given moment, $E = 100$ volts and $R = 5$ ohms, approximately how will the power change if E is decreased by 2 volts and R is decreased by 0.3 ohm?

32 Find the approximate area of a right triangle if the length of the long leg is 14.9 centimeters and the length of the hypotenuse is 17.1 centimeters.

33 The length l and the period T of a simple pendulum are connected by the equation $T = 2\pi\sqrt{l/g}$. If l is calculated for $T = 1$ second and $g = 32$ feet per second squared, approximate the error in l if T is really 1.02 seconds and g is really 32.01 feet per second squared. Also find the approximate percentage error.

34 The altitude and diameter of a right circular cylinder are 10 and 6 centimeters, respectively. If a measurement of the diameter produces a figure 4 percent too large, approximately what percent error in the measurement of the altitude will counteract the resulting error in the computed volume?

35 The dimensions of a rectangular box are 5, 6, and 8 inches. If each dimension is increased by 0.01 inch, what is the approximate resulting change in the volume?

36 The acceleration of gravity as determined by an Atwood machine is given by the formula $g = 82s/t^2$. Suppose that the true values of s and t are $s = 48$ centimeters and $t = 2$ seconds. Approximate the maximum possible error in the calculation of g if the measurements of s and t are subject to no more than 1 percent error.

37 Two electric resistors r_1 and r_2 are connected in parallel, so that the net resistance R satisfies the equation $1/R = (1/r_1) + (1/r_2)$. Suppose that $r_1 = 30$ ohms and $r_2 = 50$ ohms originally, but that r_1 increases by 0.03 ohm while r_2 decreases by 0.05 ohm. Approximate the resulting change in R.

38 A toy manufacturer, which makes two kinds of toys, produces x thousand toys of the first kind and y thousand toys of the second kind. They sell for $80 - 2x$ and $60 - 2y$ dollars per toy, respectively. The total sales revenue, in thousands of dollars, is given by the function $f(x, y) = 80x - 2x^2 + 60y - 2y^2$. Both toys are currently selling for $20 per toy. Approximate the change in sales revenue in dollars if the price of the first kind of toy is increased by $0.50 and the price of the second kind of toy is increased by $0.70.

39 Suppose that the concentration C of a drug in the bloodstream after injection into a vein is given by $C(t, d) = [1 + (t^2 + d^2)^{1/2}]^{-1}$, where t is the elapsed time in minutes since the injection and d is the distance in centimeters from the point of injection. If t increases from 3 to 3.01 minutes and d increases from 4 to 4.1 centimeters, find the approximate change in C.

40 Suppose that ϕ is a function of two variables, that $(0, 0)$ is an interior point of the domain of ϕ, and that

$$\lim_{(\Delta x, \Delta y) \to (0,0)} \phi(\Delta x, \Delta y) = 0$$

Define

$$\epsilon_1 = \begin{cases} \dfrac{|\Delta x|}{\Delta x} \cdot \dfrac{\sqrt{(\Delta x)^2 + (\Delta y)^2}}{|\Delta x| + |\Delta y|} \, \phi(\Delta x, \Delta y) & \text{if } \Delta x \neq 0 \\ 0 & \text{if } \Delta x = 0 \end{cases}$$

and

$$\epsilon_2 = \begin{cases} \dfrac{|\Delta y|}{\Delta y} \cdot \dfrac{\sqrt{(\Delta x)^2 + (\Delta y)^2}}{|\Delta x| + |\Delta y|} \, \phi(\Delta x, \Delta y) & \text{if } \Delta y \neq 0 \\ 0 & \text{if } \Delta y = 0 \end{cases}$$

Prove:

(a) $\sqrt{(\Delta x)^2 + (\Delta y)^2} \, \phi(\Delta x, \Delta y) = \epsilon_1 \, \Delta x + \epsilon_2 \, \Delta y$ for all points $(\Delta x, \Delta y)$ in the domain of ϕ.

(b) $\displaystyle\lim_{(\Delta x, \Delta y) \to (0,0)} \epsilon_1 = 0$ and $\displaystyle\lim_{(\Delta x, \Delta y) \to (0,0)} \epsilon_2 = 0$.

41 Suppose that (x_0, y_0) is an interior point of the domain of f and there are constants A and B such that

$$f(x_0 + \Delta x, y_0 + \Delta y) \approx f(x_0, y_0) + A \, \Delta x + B \, \Delta y$$

with an error

$$E = f(x_0 + \Delta x, y_0 + \Delta y) - f(x_0, y_0) - A \, \Delta x - B \, \Delta y$$

that satisfies the conditions in Definition 1. Prove that $A = f_1(x_0, y_0)$, $B = f_2(x_0, y_0)$, and f is differentiable at (x_0, y_0).

42 Let (x_0, y_0) be an interior point of the domain of f, and suppose that there exist constants A and B and a function ϕ of two variables such that

$$f(x_0 + \Delta x, y_0 + \Delta y)$$
$$= f(x_0, y_0) + A \, \Delta x + B \, \Delta y + \sqrt{(\Delta x)^2 + (\Delta y)^2} \, \phi(\Delta x, \Delta y)$$

holds whenever $(x_0 + \Delta x, y_0 + \Delta y)$ is in the domain of f. If $\displaystyle\lim_{(\Delta x, \Delta y) \to (0,0)} \phi(\Delta x, \Delta y) = 0$, use the results of Problems 40 and 41 to prove that f is differentiable at (x_0, y_0).

43 Let f be defined by

$$f(x, y) = \begin{cases} \dfrac{x^2 y^2}{x^2 + y^2} & \text{if } (x, y) \neq (0, 0) \\ 0 & \text{if } (x, y) = (0, 0) \end{cases}$$

(a) Find $f_1(x, y)$ and $f_2(x, y)$ for $(x, y) \neq (0, 0)$.

(b) Find $f_1(0, 0)$ and $f_2(0, 0)$ by direct calculation using the definitions of f_1 and f_2.

(c) Show that f is continuously differentiable.

(d) Explain why f is differentiable at $(0, 0)$.

44 Define ϵ_1 and ϵ_2 by

$$\epsilon_1 = \begin{cases} \dfrac{\Delta x \, |\Delta y|}{|\Delta x| + |\Delta y|} & \text{if } (\Delta x, \Delta y) \neq (0, 0) \\ 0 & \text{if } (\Delta x, \Delta y) = (0, 0) \end{cases}$$

and

$$\epsilon_2 = \begin{cases} \dfrac{|\Delta x| \, \Delta y}{|\Delta x| + |\Delta y|} & \text{if } (\Delta x, \Delta y) \neq (0, 0) \\ 0 & \text{if } (\Delta x, \Delta y) = (0, 0) \end{cases}$$

(a) Prove that $\displaystyle\lim_{(\Delta x, \Delta y) \to (0,0)} \epsilon_1 = 0$ and $\displaystyle\lim_{(\Delta x, \Delta y) \to (0,0)} \epsilon_2 = 0$.

(b) Prove that $|\Delta x \, \Delta y| = \epsilon_1 \, \Delta x + \epsilon_2 \, \Delta y$.

(c) Use parts (a) and (b) to prove that the function $f(x, y) = |xy|$ is differentiable at $(0, 0)$.

45 Prove that $f(x, y) = \sqrt{x^2 + y^2}$ is not differentiable at $(0, 0)$.

46 Let

$$f(x, y) = \begin{cases} \dfrac{xy}{x^2 + y^2} & \text{if } (x, y) \neq (0, 0) \\ 0 & \text{if } (x, y) = (0, 0) \end{cases}$$

Prove that both of the partial derivatives $f_1(0, 0)$ and $f_2(0, 0)$ exist but that f is not differentiable at $(0, 0)$. [*Hint:* Show that f is not continuous at $(0, 0)$, and use Theorem 1.]

47 Suppose that (x_0, y_0) is an interior point of the domain of f. Prove that f is differentiable at (x_0, y_0) if and only if there exists a function ϕ satisfying the conditions of Problem 42.

14.5   The Chain Rules

In Section 14.3 we gave a version of the chain rule for partial derivatives which was an immediate extension of the chain rule for functions of a single variable. In this section we discuss some additional versions of the chain rule for partial derivatives that are not just restatements of the old chain rule.

The simplest such chain rule is suggested by the notation for total differentials introduced in Section 14.4. Indeed, suppose that z is a function of the two variables x and y, say $z = f(x, y)$, while x and y, in turn, are functions of another variable t, so that $x = g(t)$ and $y = h(t)$. Then z becomes a function of the single variable t; that is, $z = f(g(t), h(t))$. Since

$$dz = \frac{\partial z}{\partial x} \, dx + \frac{\partial z}{\partial y} \, dy$$

we might expect that

$$\boxed{\frac{dz}{dt} = \frac{\partial z}{\partial x} \frac{dx}{dt} + \frac{\partial z}{\partial y} \frac{dy}{dt}}$$

This version of the chain rule is, in fact, correct, provided that f, g, and h are differentiable functions. Indeed, let Δt denote a small change in t; and let Δx, Δy, and Δz be the resulting changes in the variables x, y, and z, respectively. Since f is differentiable, we have

$$\Delta z = \frac{\partial z}{\partial x} \, \Delta x + \frac{\partial z}{\partial y} \, \Delta y + \epsilon_1 \, \Delta x + \epsilon_2 \, \Delta y$$

where $\epsilon_1 \, \Delta x + \epsilon_2 \, \Delta y$ is the error involved in the linear approximation

$$\Delta z \approx \frac{\partial z}{\partial x} \, \Delta x + \frac{\partial z}{\partial y} \, \Delta y$$

and

$$\lim_{(\Delta x, \Delta y) \to (0,0)} \epsilon_1 = \lim_{(\Delta x, \Delta y) \to (0,0)} \epsilon_2 = 0$$

Dividing by Δt, we have

$$\frac{\Delta z}{\Delta t} = \frac{\partial z}{\partial x} \frac{\Delta x}{\Delta t} + \frac{\partial z}{\partial y} \frac{\Delta y}{\Delta t} + \epsilon_1 \frac{\Delta x}{\Delta t} + \epsilon_2 \frac{\Delta y}{\Delta t}$$

Taking the limit on both sides as $\Delta t \to 0$, and noting that $\Delta x \to 0$ and $\Delta y \to 0$, so that $\epsilon_1 \to 0$ and $\epsilon_2 \to 0$ as $\Delta t \to 0$, we obtain

$$\frac{dz}{dt} = \frac{\partial z}{\partial x}\frac{dx}{dt} + \frac{\partial z}{\partial y}\frac{dy}{dt} + (0)\frac{dx}{dt} + (0)\frac{dy}{dt} = \frac{\partial z}{\partial x}\frac{dx}{dt} + \frac{\partial z}{\partial y}\frac{dy}{dt}$$

as claimed.

More formally, we have the following theorem.

THEOREM 1   **First Chain Rule**

> Let f be a function of two variables, and let g and h be functions of a single variable. Assume that (x_0, y_0) is an interior point of the domain of f and that f is differentiable at (x_0, y_0). Suppose that $x_0 = g(t_0)$, that $y_0 = h(t_0)$, and that both g and h are differentiable at t_0. Define the function $F(t) = f(g(t), h(t))$. Then F is differentiable at t_0, and
>
> $$F'(t_0) = f_1(x_0, y_0)g'(t_0) + f_2(x_0, y_0)h'(t_0)$$

© **EXAMPLE 1** If $z = \sqrt{x^2 + y^2}$, $x = 2t + 1$, and $y = t^3$, use the first chain rule to find dz/dt.

SOLUTION $\quad \dfrac{dz}{dt} = \dfrac{\partial z}{\partial x}\dfrac{dx}{dt} + \dfrac{\partial z}{\partial y}\dfrac{dy}{dt} = \dfrac{x}{\sqrt{x^2 + y^2}}(2) + \dfrac{y}{\sqrt{x^2 + y^2}}(3t^2)$

$$= \frac{2x + 3t^2 y}{\sqrt{x^2 + y^2}} = \frac{2(2t + 1) + 3t^2(t^3)}{\sqrt{(2t + 1)^2 + (t^3)^2}} = \frac{3t^5 + 4t + 2}{\sqrt{t^6 + 4t^2 + 4t + 1}}$$ ∎

EXAMPLE 2 If $F(t) = f(g(t), h(t))$, where $f(x, y) = e^{xy}$, $g(t) = \cos t$, and $h(t) = \sin t$, find $F'(t)$.

SOLUTION   By Theorem 1,

$$F'(t) = f_1(g(t), h(t))g'(t) + f_2(g(t), h(t))h'(t)$$

Here, $f_1(x, y) = ye^{xy}$, $f_2(x, y) = xe^{xy}$, $g'(t) = -\sin t$, and $h'(t) = \cos t$. Therefore,

$$F'(t) = \sin t\, e^{\cos t \sin t}(-\sin t) + \cos t\, e^{\cos t \sin t}(\cos t)$$

$$= (\cos^2 t - \sin^2 t)e^{\cos t \sin t} = \cos 2t\, e^{\cos t \sin t}$$ ∎

EXAMPLE 3 The resistance R in ohms of a circuit is given by $R = E/I$, where I is the current in amperes and E is the electromotive force in volts. At a certain instant when $E = 120$ volts and $I = 15$ amperes, E is increasing at a rate of $\frac{1}{10}$ volt per second and I is decreasing at the rate of $\frac{1}{20}$ ampere per second. Find the instantaneous rate of change of R.

SOLUTION   Denoting time in seconds by the variable t, we have

$$\frac{dR}{dt} = \frac{\partial R}{\partial E}\frac{dE}{dt} + \frac{\partial R}{\partial I}\frac{dI}{dt}$$

by the first chain rule. Here,

$$\frac{\partial R}{\partial E} = \frac{\partial}{\partial E}\left(\frac{E}{I}\right) = \frac{1}{I} \qquad \text{and} \qquad \frac{\partial R}{\partial I} = \frac{\partial}{\partial I}\left(\frac{E}{I}\right) = \frac{-E}{I^2}$$

Therefore,
$$\frac{dR}{dt} = \frac{1}{I}\frac{dE}{dt} - \frac{E}{I^2}\frac{dI}{dt}$$

When $E = 120$, $I = 15$, $dE/dt = \frac{1}{10}$, and $dI/dt = -\frac{1}{20}$, we have

$$\frac{dR}{dt} = \left(\frac{1}{15}\right)\left(\frac{1}{10}\right) - \left(\frac{120}{225}\right)\left(-\frac{1}{20}\right) = \frac{1}{30} \text{ ohm per second} \qquad \blacksquare$$

The first chain rule (Theorem 1) extends in an obvious way to functions of more than two variables. In fact, if w is a function of n variables $x_1, x_2, \ldots, x_n$ and each of these n variables is, in turn, a function of a single variable t, then

$$\frac{dw}{dt} = \frac{\partial w}{\partial x_1}\frac{dx_1}{dt} + \frac{\partial w}{\partial x_2}\frac{dx_2}{dt} + \cdots + \frac{\partial w}{\partial x_n}\frac{dx_n}{dt}$$

provided that the function giving w in terms of $x_1, x_2, \ldots, x_n$ is differentiable and that the derivatives

$$\frac{dx_1}{dt}, \frac{dx_2}{dt}, \ldots, \frac{dx_n}{dt}$$

exist.

EXAMPLE 4   Let $w = \ln\dfrac{x^2y^2}{4z^3}$, $x = e^t$, $y = \sec t$, and $z = \cot t$. Use the first chain rule to find dw/dt.

SOLUTION   Here, $w = 2 \ln x + 2 \ln y - 3 \ln z - \ln 4$, so that

$$\frac{\partial w}{\partial x} = \frac{2}{x} = \frac{2}{e^t} \qquad \frac{\partial w}{\partial y} = \frac{2}{y} = \frac{2}{\sec t} \qquad \frac{\partial w}{\partial z} = -\frac{3}{z} = \frac{-3}{\cot t}$$

Therefore,

$$\frac{dw}{dt} = \frac{\partial w}{\partial x}\frac{dx}{dt} + \frac{\partial w}{\partial y}\frac{dy}{dt} + \frac{\partial w}{\partial z}\frac{dz}{dt}$$

$$= \frac{2}{e^t}e^t + \frac{2}{\sec t}\sec t \tan t + \frac{-3}{\cot t}(-\csc^2 t) = 2 + 2\tan t + 3\sec t \csc t \qquad \blacksquare$$

Now consider the case in which a dependent variable z is a function of the two variables x and y, say
$$z = f(x, y)$$

while x and y, in turn, are functions of the two variables u and v, so that
$$x = g(u, v) \qquad y = h(u, v)$$

Then z becomes a function of u and v, namely
$$z = f(g(u, v), h(u, v))$$

Suppose that we temporarily hold the variable v constant and ask for the (partial) derivative of z with respect to u. Using Theorem 1, we get

$$\frac{\partial z}{\partial u} = \frac{\partial z}{\partial x}\frac{\partial x}{\partial u} + \frac{\partial z}{\partial y}\frac{\partial y}{\partial u}$$

provided that f is differentiable and that the partial derivatives $\partial x/\partial u$ and $\partial y/\partial u$ exist. Similarly, if f is differentiable and the partial derivatives $\partial x/\partial v$ and $\partial y/\partial v$ both exist, then

$$\frac{\partial z}{\partial v} = \frac{\partial z}{\partial x}\frac{\partial x}{\partial v} + \frac{\partial z}{\partial y}\frac{\partial y}{\partial v}$$

The following theorem expresses the preceding facts more precisely by using subscript notation for partial derivatives.

THEOREM 2

Second Chain Rule

Let f, g, and h be functions of two variables; let (x_0, y_0) be an interior point of the domain of f; and suppose that f is differentiable at (x_0, y_0). Let $x_0 = g(u_0, v_0)$; let $y_0 = h(u_0, v_0)$; and suppose that the partial derivatives $g_1(u_0, v_0)$, $g_2(u_0, v_0)$, $h_1(u_0, v_0)$, and $h_2(u_0, v_0)$ exist. Define the function F by

$$F(u, v) = f(g(u, v), h(u, v))$$

Then F has partial derivatives $F_1(u_0, v_0)$ and $F_2(u_0, v_0)$, and

$$F_1(u_0, v_0) = f_1(x_0, y_0)g_1(u_0, v_0) + f_2(x_0, y_0)h_1(u_0, v_0)$$

$$F_2(u_0, v_0) = f_1(x_0, y_0)g_2(u_0, v_0) + f_2(x_0, y_0)h_2(u_0, v_0)$$

If (u_0, v_0) is an interior point of the domains of both g and h, and if both g and h are differentiable at (u_0, v_0), then it can be shown that the function F of Theorem 2 is differentiable at (u_0, v_0).

In Examples 5 and 6, use the second chain rule.

EXAMPLE 5   Given that $z = x^2 - y^2$, $x = u \cos v$, and $y = v \sin u$, find $\partial z/\partial u$ and $\partial z/\partial v$.

SOLUTION

$$\frac{\partial z}{\partial x} = 2x \qquad \frac{\partial z}{\partial y} = -2y \qquad \frac{\partial x}{\partial u} = \cos v \qquad \frac{\partial x}{\partial v} = -u \sin v$$

$$\frac{\partial y}{\partial u} = v \cos u \qquad \frac{\partial y}{\partial v} = \sin u$$

Applying the second chain rule, we obtain

$$\frac{\partial z}{\partial u} = \frac{\partial z}{\partial x}\frac{\partial x}{\partial u} + \frac{\partial z}{\partial y}\frac{\partial y}{\partial u} = 2x \cos v - 2yv \cos u$$

$$= 2(u \cos v) \cos v - 2(v \sin u)v \cos u$$

$$= 2u \cos^2 v - v^2 \sin 2u$$

$$\frac{\partial z}{\partial v} = \frac{\partial z}{\partial x}\frac{\partial x}{\partial v} + \frac{\partial z}{\partial y}\frac{\partial y}{\partial v} = 2x(-u \sin v) - 2y \sin u$$

$$= 2(u \cos v)(-u \sin v) - 2(v \sin u) \sin u$$

$$= -u^2 \sin 2v - 2v \sin^2 u$$

EXAMPLE 6 Let f be differentiable at $(3, 1)$ and suppose that $f_1(3, 1) = 2$ and $f_2(3, 1) = -5$. If $V = f(2x + 3y, e^x)$, find $\partial V/\partial x$ and $\partial V/\partial y$ when $x = 0$ and $y = 1$.

SOLUTION   Let $s = 2x + 3y$ and $t = e^x$, so that $V = f(s, t)$ and

$$\frac{\partial V}{\partial x} = \frac{\partial V}{\partial s}\frac{\partial s}{\partial x} + \frac{\partial V}{\partial t}\frac{\partial t}{\partial x} = \frac{\partial V}{\partial s}(2) + \frac{\partial V}{\partial t}e^x$$

$$= 2f_1(s, t) + e^x f_2(s, t)$$

When $x = 0$ and $y = 1$, we have $s = 3$, $t = 1$, and

$$\frac{\partial V}{\partial x} = 2f_1(3, 1) + e^0 f_2(3, 1)$$

$$= (2)(2) + (1)(-5) = -1$$

Similarly,

$$\frac{\partial V}{\partial y} = \frac{\partial V}{\partial s}\frac{\partial s}{\partial y} + \frac{\partial V}{\partial t}\frac{\partial t}{\partial y} = \frac{\partial V}{\partial s}(3) + \frac{\partial V}{\partial t}(0) = 3f_1(s, t)$$

Thus, when $x = 0$ and $y = 1$, we have $s = 3$, $t = 1$, and

$$\frac{\partial V}{\partial y} = 3f_1(3, 1) = (3)(2) = 6$$

The second chain rule admits a natural extension to functions of more than two variables. In fact, if w is a differentiable function of m variables $y_1, y_2, \ldots, y_m$, and if each of these variables, in turn, is a function of n variables $x_1, x_2, \ldots, x_n$, then

$$\frac{\partial w}{\partial x_j} = \frac{\partial w}{\partial y_1}\frac{\partial y_1}{\partial x_j} + \frac{\partial w}{\partial y_2}\frac{\partial y_2}{\partial x_j} + \cdots + \frac{\partial w}{\partial y_m}\frac{\partial y_m}{\partial x_j}$$

holds for each $j = 1, 2, \ldots, n$, provided that the partial derivatives $\partial y_1/\partial x_j, \partial y_2/\partial x_j, \ldots, \partial y_m/\partial x_j$ exist. The equations above can be written more compactly as

$$\frac{\partial w}{\partial x_j} = \sum_{k=1}^{m} \frac{\partial w}{\partial y_k}\frac{\partial y_k}{\partial x_j} \qquad \text{for } j = 1, 2, \ldots, n$$

For example, if $w = f(x, y, z)$, $x = g(s, t, u)$, $y = h(s, t, u)$, and $z = p(s, t, u)$, and if f is differentiable, then

$$\frac{\partial w}{\partial s} = \frac{\partial w}{\partial x}\frac{\partial x}{\partial s} + \frac{\partial w}{\partial y}\frac{\partial y}{\partial s} + \frac{\partial w}{\partial z}\frac{\partial z}{\partial s}$$

$$\frac{\partial w}{\partial t} = \frac{\partial w}{\partial x}\frac{\partial x}{\partial t} + \frac{\partial w}{\partial y}\frac{\partial y}{\partial t} + \frac{\partial w}{\partial z}\frac{\partial z}{\partial t}$$

$$\frac{\partial w}{\partial u} = \frac{\partial w}{\partial x}\frac{\partial x}{\partial u} + \frac{\partial w}{\partial y}\frac{\partial y}{\partial u} + \frac{\partial w}{\partial z}\frac{\partial z}{\partial u}$$

provided that all the partial derivatives of x, y, and z with respect to s, t, and u exist.

EXAMPLE 7 Let $w = xy^2 + yz^2 + zx^2$, $x = r \cos \theta \sin \phi$, $y = r \sin \theta \sin \phi$, and $z = r \cos \phi$. Find $\partial w/\partial r$, $\partial w/\partial \theta$, and $\partial w/\partial \phi$.

SOLUTION   Using the chain rule, we have

$$\frac{\partial w}{\partial r} = \frac{\partial w}{\partial x}\frac{\partial x}{\partial r} + \frac{\partial w}{\partial y}\frac{\partial y}{\partial r} + \frac{\partial w}{\partial z}\frac{\partial z}{\partial r}$$

$$= (y^2 + 2xz)\cos \theta \sin \phi + (2xy + z^2)\sin \theta \sin \phi + (2yz + x^2)\cos \phi$$

$$\frac{\partial w}{\partial \theta} = \frac{\partial w}{\partial x}\frac{\partial x}{\partial \theta} + \frac{\partial w}{\partial y}\frac{\partial y}{\partial \theta} + \frac{\partial w}{\partial z}\frac{\partial z}{\partial \theta}$$

$$= (y^2 + 2xz)(-r \sin \theta \sin \phi) + (2xy + z^2)(r \cos \theta \sin \phi) + (2yz + x^2)(0)$$

$$= -(y^2 + 2xz)r \sin \theta \sin \phi + (2xy + z^2)r \cos \theta \sin \phi$$

and

$$\frac{\partial w}{\partial \phi} = \frac{\partial w}{\partial x}\frac{\partial x}{\partial \phi} + \frac{\partial w}{\partial y}\frac{\partial y}{\partial \phi} + \frac{\partial w}{\partial z}\frac{\partial z}{\partial \phi}$$

$$= (y^2 + 2xz)r \cos \theta \cos \phi + (2xy + z^2)r \sin \theta \cos \phi + (2yz + x^2)(-r \sin \phi) \quad \blacksquare$$

EXAMPLE 8 Suppose that f is a differentiable function at $(0, 0, 0)$ and that $f_1(0, 0, 0) = 3$, $f_2(0, 0, 0) = 7$, and $f_3(0, 0, 0) = -2$. If the function g is defined by the equation $g(x, y) = f(x^2 - y^2, 4x - 4y, 5x - 5)$, find $g_1(1, 1)$ and $g_2(1, 1)$.

SOLUTION   Let $u = x^2 - y^2$, $v = 4x - 4y$, and $w = 5x - 5$, and put $z = f(u, v, w)$. Then $z = f(x^2 - y^2, 4x - 4y, 5x - 5) = g(x, y)$, so

$$g_1(x, y) = \frac{\partial z}{\partial x} = \frac{\partial z}{\partial u}\frac{\partial u}{\partial x} + \frac{\partial z}{\partial v}\frac{\partial v}{\partial x} + \frac{\partial z}{\partial w}\frac{\partial w}{\partial x}$$

$$= f_1(u, v, w)(2x) + f_2(u, v, w)(4) + f_3(u, v, w)(5)$$

When $x = 1$ and $y = 1$, we have $u = 0$, $v = 0$, $w = 0$, and

$$g_1(1, 1) = 2f_1(0, 0, 0) + 4f_2(0, 0, 0) + 5f_3(0, 0, 0)$$

$$= 2(3) + 4(7) + 5(-2) = 24$$

Likewise,

$$g_2(x, y) = \frac{\partial z}{\partial y} = \frac{\partial z}{\partial u}\frac{\partial u}{\partial y} + \frac{\partial z}{\partial v}\frac{\partial v}{\partial y} + \frac{\partial z}{\partial w}\frac{\partial w}{\partial y}$$

$$= f_1(u, v, w)(-2y) + f_2(u, v, w)(-4) + f_3(u, v, w)(0)$$

$$= -2yf_1(u, v, w) - 4f_2(u, v, w)$$

and so $g_2(1, 1) = -2f_1(0, 0, 0) - 4f_2(0, 0, 0) = -2(3) - 4(7) = -34$ $\quad\blacksquare$

Implicit Differentiation

The procedure of implicit differentiation, originally discussed in Section 2.8, can be formulated with more precision and generalized by using partial derivatives.

For instance, given an equation involving the variables x and y, we can transpose all terms to the left of the equality sign, so that the equation takes the form

$$f(x, y) = 0$$

where f is a function of two variables. This equation is said to define y **implicitly** as a function g of x if

$$f(x, \, g(x)) = 0$$

holds for all values of x in the domain of g. Assuming that both f and g are differentiable, then by Theorem 1 we can differentiate both sides of the equation $f(x, \, g(x)) = 0$ with respect to x and obtain

$$f_1(x, \, g(x))\frac{dx}{dx} + f_2(x, \, g(x))\frac{d}{dx}g(x) = 0$$

or

$$f_1(x, \, y) + f_2(x, \, y)\frac{dy}{dx} = 0$$

where $y = g(x)$. If $f_2(x, \, y) \neq 0$, we can solve the last equation for dy/dx and thus obtain

$$\frac{dy}{dx} = -\frac{f_1(x, \, y)}{f_2(x, \, y)}$$

EXAMPLE 9   Suppose that y is a function of x given implicitly by the equation $x^3y^2 + 3xy^2 + 5x^4 = 2y + 7$. Find the value of dy/dx when $x = 1$ and $y = 1$.

SOLUTION   By transposing all terms to the left, we put the equation in the form $f(x, \, y) = 0$, where $f(x, \, y) = x^3y^2 + 3xy^2 + 5x^4 - 2y - 7$. Here,

$$f_1(x, \, y) = 3x^2y^2 + 3y^2 + 20x^3 \qquad \text{and} \qquad f_2(x, \, y) = 2x^3y + 6xy - 2$$

Hence,

$$\frac{dy}{dx} = -\frac{f_1(x, \, y)}{f_2(x, \, y)} = -\frac{3x^2y^2 + 3y^2 + 20x^3}{2x^3y + 6xy - 2}$$

Therefore, when $x = 1$ and $y = 1$,

$$\frac{dy}{dx} = -\frac{3 + 3 + 20}{2 + 6 - 2} = -\frac{13}{3}$$

More generally, given an equation of the form

$$f(x, \, y, \, z) = 0$$

involving three variables, it may be possible to solve for one of the variables, say y, in terms of the other two variables x and z. If this solution has the form

$$y = g(x, \, z)$$

then

$$f(x, \, g(x, \, z), \, z) = 0$$

holds for all points $(x, \, z)$ in the domain of the function g. Again, we say that the equation $f(x, \, y, \, z) = 0$ defines y **implicitly** as a function g of x and z. Assuming that the functions f and g are differentiable, we can take the partial derivatives with respect to x and also with respect to z on both sides of the equation $f(x, y, z) = 0$ to obtain

$$f_1(x, y, z) \frac{\partial x}{\partial x} + f_2(x, y, z) \frac{\partial y}{\partial x} + f_3(x, y, z) \frac{\partial z}{\partial x} = 0$$

and

$$f_1(x, y, z) \frac{\partial x}{\partial z} + f_2(x, y, z) \frac{\partial y}{\partial z} + f_3(x, y, z) \frac{\partial z}{\partial z} = 0$$

Since x and z are independent variables, we have $\partial z/\partial x = 0$, $\partial x/\partial z = 0$, $\partial x/\partial x = 1$, and $\partial z/\partial z = 1$. Therefore, we can rewrite the equations above as

$$f_2(x, y, z) \frac{\partial y}{\partial x} = -f_1(x, y, z) \qquad \text{and} \qquad f_2(x, y, z) \frac{\partial y}{\partial z} = -f_3(x, y, z)$$

Hence, if $f_2(x, y, z) \neq 0$, we can solve for $\partial y/\partial x$ and $\partial y/\partial z$ to obtain

$$\frac{\partial y}{\partial x} = -\frac{f_1(x, y, z)}{f_2(x, y, z)} \qquad \text{and} \qquad \frac{\partial y}{\partial z} = -\frac{f_3(x, y, z)}{f_2(x, y, z)}$$

EXAMPLE 10   Suppose that y is a function of x and z given implicitly by the equation $7x^3y - 4xyz^3 + x^2y^3z^2 - z - 14 = 0$. Find $\partial y/\partial x$ and $\partial y/\partial z$ when $x = 1$, $z = 0$, and $y = 2$.

SOLUTION   The equation has the form $f(x, y, z) = 0$, where

$$f(x, y, z) = 7x^3y - 4xyz^3 + x^2y^3z^2 - z - 14$$

Here,

$$f_1(x, y, z) = 21x^2y - 4yz^3 + 2xy^3z^2$$

$$f_2(x, y, z) = 7x^3 - 4xz^3 + 3x^2y^2z^2$$

$$f_3(x, y, z) = -12xyz^2 + 2x^2y^3z - 1$$

Thus,

$$\frac{\partial y}{\partial x} = -\frac{f_1}{f_2} = -\frac{21x^2y - 4yz^3 + 2xy^3z^2}{7x^3 - 4xz^3 + 3x^2y^2z^2}$$

and

$$\frac{\partial y}{\partial z} = -\frac{f_3}{f_2} = -\frac{-12xyz^2 + 2x^2y^3z - 1}{7x^3 - 4xz^3 + 3x^2y^2z^2}$$

Putting $x = 1$, $z = 0$, and $y = 2$, we obtain

$$\frac{\partial y}{\partial x} = -\frac{42}{7} = -6 \qquad \text{and} \qquad \frac{\partial y}{\partial z} = -\frac{-1}{7} = \frac{1}{7} \qquad \blacksquare$$

The considerations above can be generalized in an obvious way to equations involving more than three variables (Problem 44). They can even be extended to simultaneous systems of such equations (Problem 56).

Problem Set 14.5

In Problems 1 to 16, use the first chain rule to find each derivative.

1 $\dfrac{dz}{dt}$, where $z = x^3y^2 - 3xy + y^2$, $x = 2t$, and $y = 6t^2$

2 $\dfrac{dw}{dt}$, where $w = e^{x^2y}$, $x = \sin t$, and $y = \cos t$

3 $\dfrac{dw}{dx}$, where $w = u \sin v + \cos (u - v)$, $u = x^2$, and $v = x^3$

4 $\dfrac{dw}{d\theta}$, where $w = \sqrt{u^2 - v^2}$, $u = \sin \theta$, and $v = \cos \theta$

5 $\dfrac{dz}{dt}$, where $z = 3u^2 + v^2$, $u = te^t$, and $v = te^{-2t}$

6 $\dfrac{dw}{ds}$, where $w = 3x^2 - y^2$, $x = s^2e^{-s}$, and $y = s^3e^{3s}$

7 $\dfrac{dw}{dr}$, where $w = \ln (xy)$, $x = r^4$, and $y = \sqrt[3]{r + 2}$

8 $\dfrac{dz}{dx}$, where $z = e^{u/v}$, $u = \sqrt[3]{x}$, and $v = (x + 3)^4$

9 $\dfrac{dw}{dt}$, where $w = \ln \dfrac{x^3y^2}{5z}$, $x = 7t$, $y = \sec t$, and $z = \cot t$

10 $\dfrac{dw}{dt}$, where $w = x^2y^3z^4$, $x = \cos t$, $y = \sin t$, and $z = e^{-t}$

11 $\dfrac{dH}{du}$, where $H = x^2y + xz^2 - yz^2 + xyz$, $x = e^u$, $y = e^{-u}$, and $z = \cosh u$

12 $\dfrac{dz}{dt}$, where $z = ue^{vw} + ve^{uw}$, $u = \cos t$, $v = \sinh t$, and $w = e^{-2t}$

13 $F'(t)$, where F is defined by $F(t) = f(g(t), h(t))$, $f(x, y) = \sin (x + y) + \sin (x - y)$, $g(t) = 3t$, and $h(t) = t^3$

14 $G'(t)$, where $G(t) = g(f(t), h(t))$, $g(u, v) = \ln (u^3 + v^3)$, $f(t) = e^{3t}$, and $h(t) = e^{-7t}$

15 $F'(t)$, where $F(t) = f(g(t), h(t))$, $f(u, v) = \frac{1}{2}(e^{u/v} - e^{-u/v})$, $g(t) = \sinh t$, and $h(t) = t$

16 $F'(t)$, where $F(t) = f(g(t), h(t), p(t))$, $f(x, y, z) = \tan^{-1} (xyz)$, $g(t) = t^2$, $h(t) = t^3$, and $p(t) = t^4$

In Problems 17 to 30, use the second chain rule to find each partial derivative.

17 $\dfrac{\partial z}{\partial u}$ and $\dfrac{\partial z}{\partial v}$, where $z = 3x^2 - 4y^2$, $y = \cos u + \sin v$, and $x = uv$

18 $\dfrac{\partial w}{\partial r}$ and $\dfrac{\partial w}{\partial s}$, where $w = 4x^2 + 5xy - 2y^3$, $x = 3r + 5s$, and $y = 7r^2s$

19 $\dfrac{\partial z}{\partial u}$ and $\dfrac{\partial z}{\partial v}$, where $z = 4x^3 - 3x^2y^2$, $x = u \cos v$, and $y = v \sin u$

20 $\dfrac{\partial w}{\partial x}$ and $\dfrac{\partial w}{\partial y}$, where $w = u^2 - uv + 5v^2$, $u = x \cos 2y$, and $v = x \sin 2y$

21 $\dfrac{\partial w}{\partial x}$ and $\dfrac{\partial w}{\partial y}$, where $w = \ln (u^2 + v^2)$, $u = x^2 + y^2$, and $v = 2x^2 + 3xy$

22 $\dfrac{\partial z}{\partial x}$ and $\dfrac{\partial z}{\partial y}$, where $z = e^{s/t}$, $s = xy^2$, and $t = 5x + 2y^3$

23 $\dfrac{\partial u}{\partial r}$ and $\dfrac{\partial u}{\partial s}$, where $u = \cosh (3x + 7y)$, $x = r^2e^{-s}$, and $y = re^{3s}$

24 $F_1(r, s)$ and $F_2(r, s)$, where $F(r, s) = f(g(r, s), h(r, s))$, $f(x, y) = \tan^{-1} (y/x)$, $g(r, s) = 2r + s$, and $h(r, s) = r^2s$

25 $F_1(u, v)$ and $F_2(u, v)$, where $F(u, v) = f(g(u, v), h(u, v))$, $f(x, y) = e^{xy^2}$, $g(u, v) = u^2v$, and $h(u, v) = uv^2$

26 $\dfrac{\partial w}{\partial r}$ and $\dfrac{\partial w}{\partial s}$, where $w = 6xyz^2$, $x = rs$, $y = 2r + s$, and $z = 3r^2 - s$

27 $\dfrac{\partial w}{\partial u}$ and $\dfrac{\partial w}{\partial v}$, where $w = 2x^2 + 3y^2 + z^2$, $x = u \cos v$, $y = u \sin v$, and $z = uv$

28 $F_1(r, s)$, and $F_2(r, s)$, where $F(r, s) = q(f(r, s), g(r, s), h(r, s))$, $q(x, y, z) = xy^2 + yz^2$, $f(r, s) = r \cosh s$, $g(r, s) = r \sinh s$, and $h(r, s) = re^s$

29 $\dfrac{\partial w}{\partial \rho}$, $\dfrac{\partial w}{\partial \theta}$, and $\dfrac{\partial w}{\partial \phi}$, where $w = x^2 + y^2 - z^2$, $x = \rho \sin \phi \cos \theta$, $y = \rho \sin \phi \sin \theta$, and $z = \rho \cos \phi$

30 $\dfrac{\partial w}{\partial r}$ and $\dfrac{\partial w}{\partial s}$, where $w = \displaystyle\int_x^y e^{t^2}\, dt$, $x = rs^4$, and $y = r^4s$

31 Given a differentiable function f of three variables such that $f_1(0, 0, 0) = 4$, $f_2(0, 0, 0) = 3$, and $f_3(0, 0, 0) = 5$, let $g(r, s) = f(r - s, r^2 - 1, 3s - 3)$. Find $g_1(1, 1)$ and $g_2(1, 1)$.

32 Let g and h be differentiable functions of two variables, and define $f(r, s) = [g(r, s)]^{h(r,s)}$. Assume that $g(1, 2) = 2$, $h(1, 2) = -2$, $g_1(1, 2) = -1$, $g_2(1, 2) = 3$, $h_1(1, 2) = 5$, and $h_2(1, 2) = 0$. Find (a) $f(1, 2)$, (b) $f_1(1, 2)$, and (c) $f_2(1, 2)$.

In Problems 33 to 36, assume that y is given implicitly as a differentiable function g of x by the given equation $f(x, y) = 0$.

(a) Use the result $\dfrac{dy}{dx} = -\dfrac{f_1(x, y)}{f_2(x, y)}$ to find $\dfrac{dy}{dx}$.

(b) Find the slope of the tangent line to the graph of $y = g(x)$ at the given point (x, y).

33 $6x^2 - 12xy + 4y^2 + 2 = 0$; $(1, 1)$

34 $(x^2 - y^2)^2 - x^2y^2 - 55 = 0$; $(3, 1)$

35 $\sin(x - y) + \cos(x + y) = 0$; $(\pi/4, \pi/4)$

36 $\tan^{-1}(y/x) + 3e^{2x-2y} - 3 = \pi/4$; $(1, 1)$

In Problems 37 and 38, assume that y is given implicitly as a differentiable function of the remaining variables in each equation. Find the value of the indicated partial derivatives when the variables have the values given.

37 If $xy^2z - 3x^2yz + \dfrac{2xz}{y} - z^2 = 0$, find $\dfrac{\partial y}{\partial x}$ and $\dfrac{\partial y}{\partial z}$ when $x = 1$, $z = -1$, and $y = 2$.

38 If $\dfrac{\sin(y - x)}{z^2} = \dfrac{\cos(x + y)}{w^2 - 3}$, find $\dfrac{\partial y}{\partial x}$, $\dfrac{\partial y}{\partial z}$, and $\dfrac{\partial y}{\partial w}$ when $x = \dfrac{\pi}{4}$, $y = \dfrac{\pi}{4}$, $z = 1$, and $w = -2$.

39 If f is a differentiable function of two variables and $w = f(ay - x, x - ay)$, where a is a constant, prove that

$$a\frac{\partial w}{\partial x} + \frac{\partial w}{\partial y} = 0$$

40 If f is a differentiable function of two variables and if a and b are constants, show that $w = f(x + az, y + bz)$ provides a solution of the partial differential equation

$$\frac{\partial w}{\partial z} = a\frac{\partial w}{\partial x} + b\frac{\partial w}{\partial y}$$

41 Let $w = f(x, y)$, where f is a differentiable function. If $x = r\cos\theta$ and $y = r\sin\theta$, show that

$$\left(\frac{\partial w}{\partial x}\right)^2 + \left(\frac{\partial w}{\partial y}\right)^2 = \left(\frac{\partial w}{\partial r}\right)^2 + \frac{1}{r^2}\left(\frac{\partial w}{\partial \theta}\right)^2$$

42 Let $z = f(u, v)$, where f is a differentiable function. If $u = x\cos t - y\sin t$ and $v = x\sin t + y\cos t$, where t is a constant, show that

$$\left(\frac{\partial z}{\partial x}\right)^2 + \left(\frac{\partial z}{\partial y}\right)^2 = \left(\frac{\partial z}{\partial u}\right)^2 + \left(\frac{\partial z}{\partial v}\right)^2$$

43 Let $w = tf(r^2 - t^2)$, where f is a differentiable function. Show that

$$t\frac{\partial w}{\partial r} + r\frac{\partial w}{\partial t} = \frac{rw}{t}$$

44 Suppose that f is a differentiable function of n variables and that it is possible to solve $f(x_1, x_2, \ldots, x_k, \ldots, x_n) = 0$ for the kth variable x_k as a differentiable function of the remaining variables. Prove that

$$\frac{\partial x_k}{\partial x_i} = -\frac{f_i(x_1, x_2, \ldots, x_k, \ldots, x_n)}{f_k(x_1, x_2, \ldots, x_k, \ldots, x_n)}$$

holds for $1 \le i \le n$, $i \ne k$, if $f_k(x_1, x_2, \ldots, x_k, \ldots, x_n) \ne 0$.

45 Let f be a differentiable function of three variables, and let $w = f(x - y, y - z, z - x)$. Show that w satisfies the partial differential equation

$$\frac{\partial w}{\partial x} + \frac{\partial w}{\partial y} + \frac{\partial w}{\partial z} = 0$$

46 Let $w = g(r, \theta)$, where g is a differentiable function. If $r = \sqrt{x^2 + y^2}$ and $\theta = \tan^{-1}(y/x)$, show that

$$\frac{\partial w}{\partial x} = \frac{x}{\sqrt{x^2 + y^2}}\frac{\partial w}{\partial r} - \frac{y}{x^2 + y^2}\frac{\partial w}{\partial \theta}$$

and

$$\frac{\partial w}{\partial y} = \frac{y}{\sqrt{x^2 + y^2}}\frac{\partial w}{\partial r} + \frac{x}{x^2 + y^2}\frac{\partial w}{\partial \theta}$$

©**47** The height of a right circular cylinder is increasing at the rate of 2 centimeters per minute, and the radius is decreasing at the rate of 0.5 centimeter per minute. Is the volume increasing or decreasing at the instant the height is 10 centimeters and the radius is 6 centimeters?

©**48** The volume V of a right circular cone is given by $V = (\pi r^2/3)\sqrt{s^2 - r^2}$, where r is the radius of the base and s is the slant height. At a certain instant when $r = 4$ and $s = 10$ centimeters, r is decreasing at the rate of 2 centimeters per minute and s is increasing at the rate of 3 centimeters per minute. Find the rate of change of V at this instant.

49 At a certain instant, the legs of a right triangle have lengths 2 and 4 feet, and they are increasing at the rates of 1 foot per minute and 2 feet per minute, respectively. (a) How fast is the area of the triangle changing? (b) How fast is the perimeter of the triangle changing?

©**50** A motorcyclist starts at point A and travels toward point B at 25 miles per hour on a straight-line trail $\overline{AB}$ which is 56 miles long. At the same time, a second motorcyclist leaves B in a direction which makes an angle of 60° with $\overline{AB}$ and travels at 30 miles per hour. How fast is the straight-line distance between the two riders changing at the end of 1 hour?

51 Suppose the temperature T at the point (x, y, z) is given by

$$T(x, y, z) = x^2 + 3xy + 3y^2 + 3z^2$$

A point P moves so that its coordinates at time t are $x = 3t + 2$, $y = 5 - 2t$, and $z = 3 - 4t$. What is the rate of change of temperature at the point P when $t = 2$?

© 52 The surface of a mountainous landscape over the xy plane is given by

$$z = \cos x + \sin y + 2$$

A person is walking on this surface so that at time t the person's x and y coordinates are $x = 3t + (\pi/3)$ and $y = 4t + (\pi/3)$. At the instant $t = 0$, is the person's altitude increasing or decreasing?

© 53 The equation of a perfect gas is $PV = kT$, where T is the temperature, P is the pressure, V is the volume, and k is a constant. (The temperature is given in degrees Kelvin, the abbreviation of which is K.) At a certain instant, a sample of gas is under a pressure of 2×10^5 newtons per square meter, its volume is 5×10^{-3} cubic meter, and its temperature is 300 K. If the pressure is increasing at the rate of 1.5×10^4 newtons per square meter per minute and the volume is decreasing at the rate of 7.5×10^{-4} cubic meter per minute, find the rate at which temperature is changing.

© 54 The magnitude F of the force of gravity exerted on a rocket by the earth is given by

$$F = \frac{GmM}{r^2}$$

where G is the universal gravitational constant, m is the mass of the rocket, M (constant) is the mass of the earth, and r is the distance from the center of the earth to the rocket. When the rocket is 6500 kilometers from the center of the earth, its mass is 3700 kilograms, it is rising at the rate of 100 kilometers per second, and it is using fuel at the rate of 45 kilograms per second. How fast is the force F decreasing?

55 Suppose that

$$\mathbf{F}(x, y) = g(x, y)\mathbf{i} + h(x, y)\mathbf{j}$$

where $x = G(t)$ and $y = H(t)$. Show that

$$\frac{d\mathbf{F}}{dt} = \frac{\partial \mathbf{F}}{\partial x}\frac{dx}{dt} + \frac{\partial \mathbf{F}}{\partial y}\frac{dy}{dt}$$

where

$$\frac{\partial \mathbf{F}}{\partial x} = \frac{\partial g}{\partial x}\mathbf{i} + \frac{\partial h}{\partial x}\mathbf{j} \quad \text{and} \quad \frac{\partial \mathbf{F}}{\partial y} = \frac{\partial g}{\partial y}\mathbf{i} + \frac{\partial h}{\partial y}\mathbf{j}$$

56 Suppose that f and g are differentiable functions of three variables and that the simultaneous equations

$$\begin{cases} f(x, y, z) = 0 \\ g(x, y, z) = 0 \end{cases}$$

can be solved for x and y in terms of z. Assuming that x and y are differentiable functions of z and that the determinant

$$\begin{vmatrix} f_1 & f_2 \\ g_1 & g_2 \end{vmatrix}$$

is not zero, show that

$$\frac{dx}{dz} = -\frac{\begin{vmatrix} f_3 & f_2 \\ g_3 & g_2 \end{vmatrix}}{\begin{vmatrix} f_1 & f_2 \\ g_1 & g_2 \end{vmatrix}} \quad \text{and} \quad \frac{dy}{dz} = -\frac{\begin{vmatrix} f_1 & f_3 \\ g_1 & g_3 \end{vmatrix}}{\begin{vmatrix} f_1 & f_2 \\ g_1 & g_2 \end{vmatrix}}$$

14.6 Directional Derivatives, Gradients, Normal Lines, and Tangent Planes

In this section we study the *directional derivative* and the closely related idea of the *gradient* of a scalar field. The section also includes a discussion of the *normal line* and the *tangent plane* to a surface in space.

Directional Derivatives and Gradients

Consider a scalar field in the xy plane described by a differentiable function f of two variables. Thus, if $z = f(x, y)$, then z is the value of the scalar field at the point $P = (x, y)$. Let L denote a line in the xy plane. As P moves along L, z may change, and it makes sense to ask for the rate of change dz/ds of z with respect to distance s measured along L (Figure 1).

Figure 1

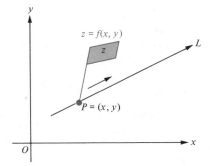

Figure 2

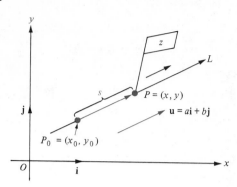

In order to find dz/ds, we introduce a unit vector $\mathbf{u} = a\mathbf{i} + b\mathbf{j}$ parallel to L and in the direction of motion of P along L (Figure 2). If $P = (x, y)$ is s units from a fixed point $P_0 = (x_0, y_0)$ on L, then $\overrightarrow{P_0P} = s\mathbf{u}$; that is,

$$(x - x_0)\mathbf{i} + (y - y_0)\mathbf{j} = as\mathbf{i} + bs\mathbf{j}$$

Equating components, we have $x - x_0 = as$ and $y - y_0 = bs$; that is, $x = x_0 + as$ and $y = y_0 + bs$. Therefore,

$$\frac{dx}{ds} = a \qquad \text{and} \qquad \frac{dy}{ds} = b$$

and it follows from the chain rule that

$$\frac{dz}{ds} = \frac{\partial z}{\partial x}\frac{dx}{ds} + \frac{\partial z}{\partial y}\frac{dy}{ds} = \frac{\partial z}{\partial x}a + \frac{\partial z}{\partial y}b$$

The derivative dz/ds, which is the rate of change of the scalar field z with respect to distance measured in the direction of the unit vector $\mathbf{u}$, is called the **directional derivative** of z (or the **directional derivative** of the function f) *in the direction of* $\mathbf{u}$ and is written as $D_{\mathbf{u}}z$ (or as $D_{\mathbf{u}}f$). Thus, we have

$$D_{\mathbf{u}}z = \frac{\partial z}{\partial x}a + \frac{\partial z}{\partial y}b \qquad \text{or} \qquad D_{\mathbf{u}}f(x, y) = f_1(x, y)a + f_2(x, y)b$$

where

$$\mathbf{u} = a\mathbf{i} + b\mathbf{j}$$

In particular, if $\mathbf{u}$ is the unit vector that makes the angle θ with the positive x axis, then $\mathbf{u} = (\cos \theta)\mathbf{i} + (\sin \theta)\mathbf{j}$ and

$$D_{\mathbf{u}}z = \frac{\partial z}{\partial x}\cos \theta + \frac{\partial z}{\partial y}\sin \theta \qquad \text{or} \qquad D_{\mathbf{u}}f(x, y) = f_1(x, y)\cos \theta + f_2(x, y)\sin \theta$$

In this case we say that $D_{\mathbf{u}}z$ is the directional derivative of z (or f) **in the direction** θ.

EXAMPLE 1 Find the directional derivative of $z = 3x^2y - y^2$ at the point $(-2, 1)$ in the direction $\pi/4$.

SOLUTION A unit vector $\mathbf{u}$ in the direction $\pi/4$ is given by

$$\mathbf{u} = \left(\cos \frac{\pi}{4}\right)\mathbf{i} + \left(\sin \frac{\pi}{4}\right)\mathbf{j} = \frac{\sqrt{2}}{2}\mathbf{i} + \frac{\sqrt{2}}{2}\mathbf{j}$$

Hence,

$$D_{\mathbf{u}}z = \left(\frac{\partial z}{\partial x}\right)\cos \frac{\pi}{4} + \left(\frac{\partial z}{\partial y}\right)\sin \frac{\pi}{4} = (6xy)\frac{\sqrt{2}}{2} + (3x^2 - 2y)\frac{\sqrt{2}}{2}$$

Substituting $x = -2$ and $y = 1$ in this equation, we obtain

$$D_{\mathbf{u}}z = (-12)\frac{\sqrt{2}}{2} + (10)\frac{\sqrt{2}}{2} = -\sqrt{2}$$

The vector $\mathbf{i}$ makes the angle $\theta = 0$ with the positive x axis; hence,

$$D_{\mathbf{i}}z = \frac{\partial z}{\partial x} \cos 0 + \frac{\partial z}{\partial y} \sin 0 = \frac{\partial z}{\partial x}$$

Similarly, the vector $\mathbf{j}$ makes the angle $\theta = \pi/2$; hence,

$$D_{\mathbf{j}}z = \frac{\partial z}{\partial x} \cos \frac{\pi}{2} + \frac{\partial z}{\partial y} \sin \frac{\pi}{2} = \frac{\partial z}{\partial y}$$

Therefore, *the directional derivatives of z in the directions of the positive x and y axes are just the partial derivatives of z with respect to x and y, respectively.*
The directional derivative $D_{\mathbf{u}}z$ can be expressed as a dot product

$$D_{\mathbf{u}}z = \frac{\partial z}{\partial x} a + \frac{\partial z}{\partial y} b = a \frac{\partial z}{\partial x} + b \frac{\partial z}{\partial y}$$

$$= (a\mathbf{i} + b\mathbf{j}) \cdot \left(\frac{\partial z}{\partial x} \mathbf{i} + \frac{\partial z}{\partial y} \mathbf{j} \right)$$

$$= \mathbf{u} \cdot \left(\frac{\partial z}{\partial x} \mathbf{i} + \frac{\partial z}{\partial y} \mathbf{j} \right)$$

The vector $(\partial z/\partial x)\mathbf{i} + (\partial z/\partial y)\mathbf{j}$ whose scalar components are the partial derivatives of z with respect to x and y is called the **gradient** of the scalar field z (or of the function f) and is written as ∇z (or as ∇f). The symbol ∇, an inverted Greek delta, is called "del," or "nabla." Thus, we have

$$\nabla z = \frac{\partial z}{\partial x} \mathbf{i} + \frac{\partial z}{\partial y} \mathbf{j} \qquad \text{or} \qquad \nabla f(x, y) = f_1(x, y)\mathbf{i} + f_2(x, y)\mathbf{j}$$

and we can write the directional derivative as

$$D_{\mathbf{u}}z = \mathbf{u} \cdot \nabla z \qquad \text{or} \qquad D_{\mathbf{u}}f(x, y) = \mathbf{u} \cdot \nabla f(x, y)$$

In words, *the directional derivative of a scalar field in a given direction is the dot product of a unit vector in this direction and the gradient of the scalar field.*

EXAMPLE 2   If $f(x, y) = 4x^2 + xy + 9y^2$, find (a) $\nabla f(1, 2)$ and (b) $D_{\mathbf{u}}f(1, 2)$, where $\mathbf{u}$ is a unit vector in the direction of the vector $\mathbf{v} = 4\mathbf{i} - 3\mathbf{j}$.

SOLUTION   We obtain $\mathbf{u}$ by normalizing $\mathbf{v}$; thus,

$$\mathbf{u} = \frac{\mathbf{v}}{|\mathbf{v}|} = \frac{4\mathbf{i} - 3\mathbf{j}}{\sqrt{4^2 + (-3)^2}} = \frac{4\mathbf{i} - 3\mathbf{j}}{\sqrt{25}} = \frac{4}{5}\mathbf{i} - \frac{3}{5}\mathbf{j}$$

(a) Here, $f_1(x, y) = 8x + y$ and $f_2(x, y) = x + 18y$, so that

$$\nabla f(1, 2) = f_1(1, 2)\mathbf{i} + f_2(1, 2)\mathbf{j} = 10\mathbf{i} + 37\mathbf{j}$$

(b) $D_{\mathbf{u}}f(1, 2) = \mathbf{u} \cdot \nabla f(1, 2) = \left(\frac{4}{5}\mathbf{i} - \frac{3}{5}\mathbf{j} \right) \cdot (10\mathbf{i} + 37\mathbf{j})$

$$= \frac{40}{5} - \frac{111}{5} = -\frac{71}{5}$$

©️ **EXAMPLE 3**   A temperature field in the xy plane is given by

$$z = 60 + \left(\frac{x}{20}\right)^2 + \left(\frac{y}{25}\right)^2$$

where z is the temperature in degrees Fahrenheit at the point (x, y) and distances are measured in miles. How rapidly is the temperature changing in degrees Fahrenheit per mile as we move upward to the right through the point $(60, 75)$ along a line L that is parallel to the vector $\mathbf{u} = (\cos 30°)\mathbf{i} + (\sin 30°)\mathbf{j}$?

SOLUTION   Here $\mathbf{u} = (\cos 30°)\mathbf{i} + (\sin 30°)\mathbf{j} = (\sqrt{3}/2)\mathbf{i} + \frac{1}{2}\mathbf{j}$, and

$$\nabla z = \frac{\partial z}{\partial x}\mathbf{i} + \frac{\partial z}{\partial y}\mathbf{j} = 2\left(\frac{x}{20}\right)\left(\frac{1}{20}\right)\mathbf{i} + 2\left(\frac{y}{25}\right)\left(\frac{1}{25}\right)\mathbf{j}$$

Putting $x = 60$ and $y = 75$, we have

$$\nabla z = \tfrac{3}{10}\mathbf{i} + \tfrac{6}{25}\mathbf{j}$$

Therefore, the rate of change of z as we move through the point $(60, 75)$ in the direction of $\mathbf{u}$ is given by

$$D_{\mathbf{u}}z = \nabla z \cdot \mathbf{u} = \frac{3}{10}\left(\frac{\sqrt{3}}{2}\right) + \frac{6}{25}\left(\frac{1}{2}\right) = \frac{15\sqrt{3} + 12}{100} \approx 0.38 \text{ °F/mi} \quad\blacksquare$$

Figure 3

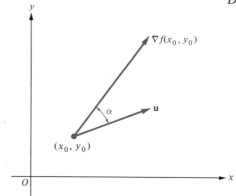

Note that if we fix a point (x_0, y_0) in the xy plane, then the directional derivative

$$D_{\mathbf{u}}f(x_0, y_0) = \mathbf{u} \cdot \nabla f(x_0, y_0)$$

depends only on our choice of the unit vector $\mathbf{u}$, since the gradient vector $\nabla f(x_0, y_0)$ is fixed. If α denotes the angle between $\mathbf{u}$ and $\nabla f(x_0, y_0)$ (Figure 3), then by the definition of the dot product,

$$\mathbf{u} \cdot \nabla f(x_0, y_0) = |\mathbf{u}|\,|\nabla f(x_0, y_0)|\cos \alpha$$

Since $|\mathbf{u}| = 1$, it follows that

$$\boxed{D_{\mathbf{u}}f(x_0, y_0) = |\nabla f(x_0, y_0)|\cos \alpha}$$

As we vary the angle α in the last formula, we obtain the value of the directional derivative in various directions at the point (x_0, y_0). Taking $\alpha = \pi/2$, we have $\cos \alpha = 0$, so that $D_{\mathbf{u}}f(x_0, y_0) = 0$. Therefore:

> **1** The directional derivative is zero when it is taken in a direction perpendicular to the gradient.

Since $\cos \alpha$ assumes its maximum value, namely 1, when $\alpha = 0$, we also obtain the following:

> **2** The directional derivative assumes its maximum value when it is taken in the direction of the gradient, and this maximum value is $|\nabla f(x_0, y_0)|$.

In other words:

The gradient of a scalar field, evaluated at a point P, is a vector that points in the direction in which the scalar field increases most rapidly, and the length of the gradient vector is numerically equal to the instantaneous rate of increase of the field per unit distance in this direction as we move through the point P.

For instance, if you are at a given point in a temperature field and wish to "warm up" most rapidly, then you should move in the direction of the gradient vector at this point. On the other hand, if you move perpendicular to the gradient, the instantaneous rate of change of temperature is zero, and you find yourself moving in the direction of the isotherm through the given point. Moving in the direction opposite to the gradient (that is, in the direction of the negative of the gradient), you "cool off" most rapidly.

EXAMPLE 4   Find **(a)** the maximum value of the directional derivative and **(b)** a unit vector **u** in the direction for which this maximum is attained for $f(x, y) = 2x^2y + xe^{y^2}$ at the point $(1, 0)$.

SOLUTION   Here,

$$\nabla f(x, y) = f_1(x, y)\mathbf{i} + f_2(x, y)\mathbf{j} = (4xy + e^{y^2})\mathbf{i} + (2x^2 + 2xye^{y^2})\mathbf{j}$$

so that
$$\nabla f(1, 0) = \mathbf{i} + 2\mathbf{j}$$

Therefore:

(a) The maximum directional derivative at $(1, 0)$ is

$$|\nabla f(1, 0)| = |\mathbf{i} + 2\mathbf{j}| = \sqrt{1^2 + 2^2} = \sqrt{5}$$

(b) The maximum directional derivative at $(1, 0)$ is attained in the direction of $\nabla f(1, 0) = \mathbf{i} + 2\mathbf{j}$. A unit vector **u** in the same direction is

$$\mathbf{u} = \frac{\nabla f(1, 0)}{|\nabla f(1, 0)|} = \frac{\mathbf{i} + 2\mathbf{j}}{\sqrt{5}}$$ ∎

© **EXAMPLE 5**   The temperature T in degrees Celsius at the point (x, y) on a heated metal plate is given by

$$T = \frac{300}{x^2 + y^2 + 3}$$

where x and y are measured in centimeters.

(a) In what direction should one move from the point $(-4, 3)$ in order that T increase most rapidly?

(b) How rapidly does T increase as one moves through the point $(-4, 3)$ in the direction found in part (a)?

SOLUTION

(a) $\nabla T = \dfrac{\partial T}{\partial x}\mathbf{i} + \dfrac{\partial T}{\partial y}\mathbf{j} = \dfrac{-600x}{(x^2 + y^2 + 3)^2}\mathbf{i} + \dfrac{-600y}{(x^2 + y^2 + 3)^2}\mathbf{j}$

Hence, when $x = -4$ and $y = 3$, we have

$$\nabla T = \tfrac{2400}{784}\,\mathbf{i} - \tfrac{1800}{784}\,\mathbf{j}$$

and

$$|\nabla T| = \sqrt{(\tfrac{2400}{784})^2 + (-\tfrac{1800}{784})^2} = \tfrac{3000}{784} = \tfrac{375}{98}$$

Therefore, in order to maximize $D_{\mathbf{u}}T$ at $(-4, 3)$, we choose $\mathbf{u}$ to be a unit vector in the direction of ∇T; that is, we take

$$\mathbf{u} = \frac{\nabla T}{|\nabla T|} = \frac{4}{5}\,\mathbf{i} - \frac{3}{5}\,\mathbf{j}$$

(b) As we move through the point $(-4, 3)$ in the direction of $\mathbf{u}$, the instantaneous rate of change of T with respect to distance is given by

$$D_{\mathbf{u}}T = |\nabla T| = \frac{375}{98} \approx 3.83 \text{ degrees Celsius per centimeter} \quad \blacksquare$$

Normal Vectors to Level Curves in the Plane

Consider a scalar field in the plane given by $z = f(x, y)$, where f is a differentiable function. Recall from Section 14.1 that a curve in the plane along which z has a constant value, say k, has the equation

$$f(x, y) = k$$

and is called a level curve of the field (Figure 4). Assume that the level curve $f(x, y) = k$ has a unit tangent vector

$$\mathbf{T} = \frac{dx}{ds}\,\mathbf{i} + \frac{dy}{ds}\,\mathbf{j}$$

where s is arc length measured along the curve. Differentiating both sides of the equation $f(x, y) = k$ with respect to s by using the chain rule, we obtain

$$f_1(x, y)\,\frac{dx}{ds} + f_2(x, y)\,\frac{dy}{ds} = 0$$

that is,

$$(\nabla f) \cdot \mathbf{T} = 0$$

Therefore:

> The gradient vector at a point P in a scalar field is normal to the level curve of the field that passes through P.

Here, of course, we are assuming that there *is* a smooth level curve through P.

Since $\nabla f(x_0, y_0) = f_1(x_0, y_0)\mathbf{i} + f_2(x_0, y_0)\mathbf{j}$ is normal to the tangent line to the level curve of the scalar field $z = f(x, y)$ at the point (x_0, y_0), it follows that *an equation of the tangent line is*

> $$f_1(x_0, y_0)(x - x_0) + f_2(x_0, y_0)(y - y_0) = 0$$

EXAMPLE 6   Find a normal vector and an equation of the tangent line to the curve $2x^2 - 4xy^3 + y^5 = 1$ at the point $(2, 1)$.

Figure 4

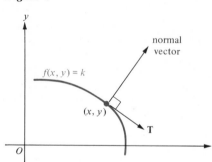

SOLUTION   The curve can be regarded as the level curve $f(x, y) = 1$ of the scalar field $z = f(x, y)$, where $f(x, y) = 2x^2 - 4xy^3 + y^5$. Here

$$f_1(x, y) = 4x - 4y^3 \qquad f_2(x, y) = -12xy^2 + 5y^4$$

and the gradient of f at $(2, 1)$ is given by

$$\nabla f(2, 1) = f_1(2, 1)\mathbf{i} + f_2(2, 1)\mathbf{j} = 4\mathbf{i} - 19\mathbf{j}$$

Thus, $4\mathbf{i} - 19\mathbf{j}$ is normal to the curve at $(2, 1)$. Also, an equation of the tangent line to the curve at $(2, 1)$ is

$$4(x - 2) - 19(y - 1) = 0 \qquad \text{or} \qquad 4x - 19y + 11 = 0 \qquad \blacksquare$$

Directional Derivative and Gradient in Space

Just as a function of two variables can be regarded as specifying a scalar field in the plane, a function f of three variables can be thought of as describing a **scalar field in xyz space;** that is, we can think of f as assigning a scalar w, given by

$$w = f(x, y, z)$$

to each point (x, y, z) in its domain (Figure 5). Examples are temperature fields, pressure fields, density fields, electric potential fields, and so forth.

All the ideas and techniques introduced above for scalar fields in the xy plane extend naturally to scalar fields in xyz space. For example, if $w = f(x, y, z)$, where f is a differentiable function, we define the **gradient** of w (or of f) by

Figure 5

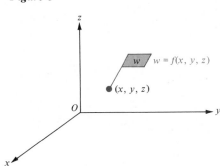

$$\nabla w = \frac{\partial w}{\partial x}\mathbf{i} + \frac{\partial w}{\partial y}\mathbf{j} + \frac{\partial w}{\partial z}\mathbf{k}$$

or $\qquad \nabla f(x, y, z) = f_1(x, y, z)\mathbf{i} + f_2(x, y, z)\mathbf{j} + f_3(x, y, z)\mathbf{k}$

If $\mathbf{u}$ is a unit vector in xyz space, it is easy to show (Problem 64) that the rate of change of the scalar field w with respect to distance measured in the direction of $\mathbf{u}$ is given by the **directional derivative**

$$D_{\mathbf{u}}w = \mathbf{u} \cdot \nabla w \qquad D_{\mathbf{u}}f(x, y, z) = \mathbf{u} \cdot \nabla f(x, y, z)$$

EXAMPLE 7   If $f(x, y, z) = 3x^2 + 8y^2 - 5z^2$, find the directional derivative of f at $(1, -1, 2)$ in the direction of the vector $\mathbf{v} = 2\mathbf{i} - 6\mathbf{j} + 3\mathbf{k}$.

SOLUTION   Here, $\mathbf{v}$ is not a unit vector; however, a unit vector in the direction of $\mathbf{v}$ is given by $\mathbf{u} = \mathbf{v}/|\mathbf{v}| = \frac{2}{7}\mathbf{i} - \frac{6}{7}\mathbf{j} + \frac{3}{7}\mathbf{k}$. We have

$$\nabla f(x, y, z) = f_1(x, y, z)\mathbf{i} + f_2(x, y, z)\mathbf{j} + f_3(x, y, z)\mathbf{k}$$

$$= 6x\mathbf{i} + 16y\mathbf{j} - 10z\mathbf{k}$$

so that

$$\nabla f(1, -1, 2) = 6(1)\mathbf{i} + 16(-1)\mathbf{j} - 10(2)\mathbf{k} = 6\mathbf{i} - 16\mathbf{j} - 20\mathbf{k}$$

and

$$D_{\mathbf{u}}f(1, -1, 2) = \mathbf{u} \cdot \nabla f(1, -1, 2) = \tfrac{2}{7}(6) - \tfrac{6}{7}(-16) + \tfrac{3}{7}(-20) = \tfrac{48}{7} \qquad \blacksquare$$

© **EXAMPLE 8**   The electric potential V in volts at the point $P = (x, y, z)$ in xyz space is given by $V = 100(x^2 + y^2 + z^2)^{-1/2}$, where x, y, and z are measured in centimeters. How rapidly is the potential V changing at the instant when we move through the point $P_0 = (2, 1, -2)$ in the direction of the point $P_1 = (4, 3, 0)$?

SOLUTION   The rate of change of V, in volts per centimeter, is given by $D_{\mathbf{u}}V$, evaluated at $P_0 = (2, 1, -2)$ in the direction of the unit vector

$$\mathbf{u} = \frac{\overrightarrow{P_0P_1}}{|\overrightarrow{P_0P_1}|} = \frac{1}{\sqrt{3}}(\mathbf{i} + \mathbf{j} + \mathbf{k})$$

Here,

$$\frac{\partial V}{\partial x} = -100x(x^2 + y^2 + z^2)^{-3/2} \qquad \frac{\partial V}{\partial y} = -100y(x^2 + y^2 + z^2)^{-3/2}$$

and

$$\frac{\partial V}{\partial z} = -100z(x^2 + y^2 + z^2)^{-3/2}$$

Hence,

$$\nabla V = \frac{\partial V}{\partial x}\mathbf{i} + \frac{\partial V}{\partial y}\mathbf{j} + \frac{\partial V}{\partial z}\mathbf{k} = -100(x^2 + y^2 + z^2)^{-3/2}(x\mathbf{i} + y\mathbf{j} + z\mathbf{k})$$

Putting $x = 2$, $y = 1$, and $z = -2$, we find that the gradient at $P_0 = (2, 1, -2)$ is

$$\nabla V = -\tfrac{100}{27}(2\mathbf{i} + \mathbf{j} - 2\mathbf{k})$$

Hence, the directional derivative at P_0 in the direction of $\mathbf{u}$ is

$$D_{\mathbf{u}}V = \mathbf{u} \cdot \nabla V = \left(\frac{1}{\sqrt{3}}\right)\left(-\frac{100}{27}\right)(\mathbf{i} + \mathbf{j} + \mathbf{k}) \cdot (2\mathbf{i} + \mathbf{j} - 2\mathbf{k})$$

$$= \frac{-100}{27\sqrt{3}}(2 + 1 - 2) = \frac{-100}{27\sqrt{3}} \approx -2.14 \text{ volts per centimeter} \qquad \blacksquare$$

Just as for scalar fields in the xy plane, *the gradient of a scalar field in xyz space points in the direction for which the directional derivative attains its maximum, and its length is numerically equal to this maximum directional derivative* (Problem 66).

EXAMPLE 9   Let $f(x, y, z) = xy + xz + yz + xyz$. Find **(a)** the maximum value of the directional derivative of f at $(8, -1, 4)$ and **(b)** a unit vector in the direction in which this maximum directional derivative is attained.

SOLUTION   Here,

$$\nabla f(x, y, z) = f_1(x, y, z)\mathbf{i} + f_2(x, y, z)\mathbf{j} + f_3(x, y, z)\mathbf{k}$$

$$= (y + z + yz)\mathbf{i} + (x + z + xz)\mathbf{j} + (x + y + xy)\mathbf{k}$$

(a) The maximum value of the directional derivative at $(8, -1, 4)$ is

$$|\nabla f(8, -1, 4)| = |-\mathbf{i} + 44\mathbf{j} - \mathbf{k}| = \sqrt{(-1)^2 + 44^2 + (-1)^2} = \sqrt{1938}$$

(b) The required vector, a unit vector in the direction of $\nabla f(8, -1, 4)$, is given by

$$\frac{\nabla f(8, -1, 4)}{|\nabla f(8, -1, 4)|} = \frac{-\mathbf{i} + 44\mathbf{j} - \mathbf{k}}{\sqrt{1938}} \qquad \blacksquare$$

Normal Lines and Tangent Planes

Let f be a differentiable function of three variables. If k is a constant belonging to the range of f, then the graph in xyz space of the equation

$$f(x, y, z) = k$$

is called a **level surface** for f [or for the scalar field $w = f(x, y, z)$ determined by f].

Earlier in this section we showed that, in the xy plane, the gradient vector at a point P in a scalar field is normal to the level curve of the field at P. This idea can be extended to level surfaces in xyz space as follows: Let (x_0, y_0, z_0) be a point on the surface $f(x, y, z) = k$, and assume that $\nabla f(x_0, y_0, z_0) \neq \mathbf{0}$. Then the line containing the point (x_0, y_0, z_0) and parallel to the gradient vector $\nabla f(x_0, y_0, z_0)$ is called the **normal line** to the surface at (x_0, y_0, z_0) (Figure 6). The plane containing the point (x_0, y_0, z_0) and perpendicular to this normal line is called the **tangent plane** to the surface at (x_0, y_0, z_0) (Figure 7). Because

$$\nabla f(x_0, y_0, z_0) = f_1(x_0, y_0, z_0)\mathbf{i} + f_2(x_0, y_0, z_0)\mathbf{j} + f_3(x_0, y_0, z_0)\mathbf{k}$$

it follows that scalar Cartesian equations of the normal line and the tangent plane are:

Normal line:
$$\frac{x - x_0}{f_1(x_0, y_0, z_0)} = \frac{y - y_0}{f_2(x_0, y_0, z_0)} = \frac{z - z_0}{f_3(x_0, y_0, z_0)}$$

Tangent plane:
$$f_1(x_0, y_0, z_0)(x - x_0) + f_2(x_0, y_0, z_0)(y - y_0) + f_3(x_0, y_0, z_0)(z - z_0) = 0$$

Figure 6

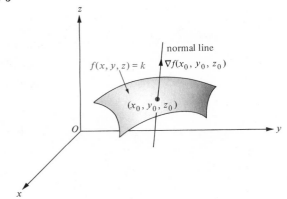

Figure 7

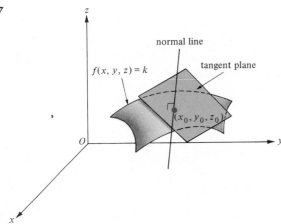

Now, let C be a curve lying on the surface $f(x, y, z) = k$, so that the coordinates x, y, and z of any point $P = (x, y, z)$ on C satisfy the equation

$$f(x, y, z) = k$$

Using the chain rule, we differentiate both sides of the last equation with respect to arc length s measured along the curve C to obtain

$$\frac{\partial f}{\partial x}\frac{dx}{ds} + \frac{\partial f}{\partial y}\frac{dy}{ds} + \frac{\partial f}{\partial z}\frac{dz}{ds} = 0$$

that is, $\nabla f \cdot \mathbf{T} = 0$, where

$$\mathbf{T} = \frac{dx}{ds}\,\mathbf{i} + \frac{dy}{ds}\,\mathbf{j} + \frac{dz}{ds}\,\mathbf{k}$$

It follows that the unit tangent vector $\mathbf{T}$ to a curve C on the surface $f(x, y, z) = k$ is perpendicular to the gradient vector ∇f (Figure 8).

Figure 8

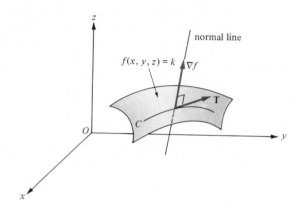

Therefore:

> The tangent plane to the surface $f(x, y, z) = k$ at a point P contains the tangent vectors at P to all (smooth) curves on the surface that pass through P.

In Examples 10 to 12, find equations of (a) the tangent plane and (b) the normal line to the given surface at the indicated point.

EXAMPLE 10   The sphere $x^2 + y^2 + z^2 = 14$ at the point $(-1, 3, 2)$

SOLUTION   Here we put $f(x, y, z) = x^2 + y^2 + z^2$, so that the equation of the sphere is $f(x, y, z) = 14$. We have

$$f_1(x, y, z) = 2x \qquad f_2(x, y, z) = 2y \qquad \text{and} \qquad f_3(x, y, z) = 2z$$

Hence,

$$f_1(-1, 3, 2) = -2 \qquad f_2(-1, 3, 2) = 6 \qquad \text{and} \qquad f_3(-1, 3, 2) = 4$$

(a) An equation of the tangent plane at $(-1, 3, 2)$ is

$$f_1(-1, 3, 2)(x + 1) + f_2(-1, 3, 2)(y - 3) + f_3(-1, 3, 2)(z - 2) = 0$$

that is,

$$-2(x + 1) + 6(y - 3) + 4(z - 2) = 0 \qquad \text{or} \qquad -2x + 6y + 4z = 28$$

(b) Equations of the normal line at $(-1, 3, 2)$ are

$$\frac{x + 1}{f_1(-1, 3, 2)} = \frac{y - 3}{f_2(-1, 3, 2)} = \frac{z - 2}{f_3(-1, 3, 2)}$$

that is,

$$\frac{x + 1}{-2} = \frac{y - 3}{6} = \frac{z - 2}{4} \qquad \text{or} \qquad \frac{x + 1}{-1} = \frac{y - 3}{3} = \frac{z - 2}{2}$$

EXAMPLE 11   The graph of $z = g(x, y)$ at the point $(x_0, y_0, g(x_0, y_0))$, where g is a differentiable function of two variables

SOLUTION   Here we put $f(x, y, z) = g(x, y) - z$, so that the graph of $z = g(x, y)$ is the level surface $f(x, y, z) = 0$ of f. We have

$$f_1(x, y, z) = \frac{\partial}{\partial x}\,[g(x, y) - z] = \frac{\partial}{\partial x}\,g(x, y) = g_1(x, y)$$

$$f_2(x, y, z) = \frac{\partial}{\partial y}\,[g(x, y) - z] = \frac{\partial}{\partial y}\,g(x, y) = g_2(x, y)$$

and

$$f_3(x, y, z) = \frac{\partial}{\partial z}\,[g(x, y) - z] = \frac{\partial}{\partial z}\,(-z) = -1$$

Therefore,

$$f_1(x_0, y_0, g(x_0, y_0)) \;=\; g_1(x_0, y_0)$$
$$f_2(x_0, y_0, g(x_0, y_0)) \;=\; g_2(x_0, y_0)$$

and

$$f_3(x_0, y_0, g(x_0, y_0)) \;=\; -1$$

(a) An equation of the tangent plane at $(x_0, y_0, g(x_0, y_0))$ is

$$g_1(x_0, y_0)(x - x_0) + g_2(x_0, y_0)(y - y_0) + (-1)[z - g(x_0, y_0)] = 0$$

or

$$z = g(x_0, y_0) + g_1(x_0, y_0)(x - x_0) + g_2(x_0, y_0)(y - y_0)$$

(b) Equations of the normal line at $(x_0, y_0, g(x_0, y_0))$ are

$$\frac{x - x_0}{g_1(x_0, y_0)} = \frac{y - y_0}{g_2(x_0, y_0)} = \frac{z - z_0}{-1}$$

EXAMPLE 12   The graph of $g(x, y) = 3x^4y - 7x^3y - x^2 + y + 1$ at the point $(1, 2, -6)$

SOLUTION   We use the results in Example 11. Here we have

$$g_1(x, y) \;=\; 12x^3y - 21x^2y - 2x \quad \text{and} \quad g_2(x, y) \;=\; 3x^4 - 7x^3 + 1$$

Hence, $g_1(1, 2) \;=\; -20$ and $g_2(1, 2) \;=\; -3$

(a) An equation of the tangent plane is

$$z = g(1, 2) + g_1(1, 2)(x - 1) + g_2(1, 2)(y - 2)$$
$$= -6 - 20(x - 1) - 3(y - 2)$$

or $z = 20 - 20x - 3y$

(b) Equations of the normal line are

$$\frac{x - 1}{g_1(1, 2)} = \frac{y - 2}{g_2(1, 2)} = \frac{z - g(1, 2)}{-1} \quad \text{or} \quad \frac{x - 1}{-20} = \frac{y - 2}{-3} = \frac{z + 6}{-1}$$

Problem Set 14.6

In Problems 1 to 22, find (a) the gradient ∇f of each scalar field, (b) the value of ∇f at the given point, and (c) the directional derivative $D_{\mathbf{u}}f$ at the given point in the indicated direction.

1 $f(x, y) = 4x^2y^3$, $(x_0, y_0) = (2, -1)$ in the direction $\theta = \pi/3$

2 $f(x, y) = y^2 \ln x$, $(x_0, y_0) = (1, 4)$ in the direction $\theta = \pi/4$

3 $f(x, y) = xe^y + ye^x$, $(x_0, y_0) = (0, 0)$ in the direction $\theta = 2\pi/3$

4 $f(x, y) = y^2 \tan x$, $(x_0, y_0) = (\pi/4, -3)$ in the direction $\theta = \pi/2$

5 $f(x, y) = x \ln y$, $(x_0, y_0) = (5, 1)$ in the direction $\theta = -\pi/6$

6 $f(x, y) = (1 + xy)^{3/2}$, $(x_0, y_0) = (1, 3)$ in the direction $\theta = -\pi/4$

7 $f(x, y) = 7x^2 - 3y + 4$, $(x_0, y_0) = (1, 1)$ in the direction of $\mathbf{u} = (\sqrt{3}/2)\mathbf{i} + \frac{1}{2}\mathbf{j}$

8 $f(x, y) = xy$, $(x_0, y_0) = (2, -1)$ in the direction of $\mathbf{u} = (\sqrt{2}/2)\mathbf{i} + (\sqrt{2}/2)\mathbf{j}$

9 $f(x, y) = 2x^2 + 3y^2 - 1$, $(x_0, y_0) = (1, 2)$ in the direction of $\mathbf{u} = \frac{1}{2}\mathbf{i} - (\sqrt{3}/2)\mathbf{j}$

10 $f(x, y) = e^{xy}$, $(x_0, y_0) = (1, 1)$ in the direction of $\mathbf{u} = (\sqrt{2}/2)\mathbf{i} - (\sqrt{2}/2)\mathbf{j}$

11 $f(x, y) = x^2y + 2xy^2$, $(x_0, y_0) = (-2, 1)$ in the direction of $\mathbf{u} = (\sqrt{3}/2)\mathbf{i} - \frac{1}{2}\mathbf{j}$

12 $f(x, y) = (x^2 + y^2)^{-1}$, $(x_0, y_0) = (2, 3)$ in the direction of $\mathbf{u} = \frac{5}{13}\mathbf{i} + \frac{12}{13}\mathbf{j}$

13 $f(x, y) = \tan^{-1}(y/x)$, $(x_0, y_0) = (1, -2)$ in the direction of $\mathbf{v} = 4\mathbf{i} - 3\mathbf{j}$

14 $f(x, y) = \ln \sqrt{x^2 + y^2}$, $(x_0, y_0) = (5, -12)$ in the direction of $\mathbf{v} = -5\mathbf{i} - 12\mathbf{j}$

15 $f(x, y) = y \ln (x/y)$, $(x_0, y_0) = (2, 1)$ in the direction from $(2, 1)$ to $(-6, -2)$

16 $f(x, y) = e^{-3y} \cos 4x$, $(x_0, y_0) = (0, 0)$ in the direction from $(0, 0)$ to $(-3, 4)$

17 $f(x, y, z) = x^2y + 3yz^2$, $(x_0, y_0, z_0) = (1, -1, 1)$, in the direction of $\mathbf{u} = \frac{1}{3}\mathbf{i} - \frac{2}{3}\mathbf{j} + \frac{2}{3}\mathbf{k}$

18 $f(x, y, z) = \ln (x^2 + y^2 + z^2)$, $(x_0, y_0, z_0) = (1, 1, 1)$, in the direction of $\mathbf{u} = -\frac{2}{3}\mathbf{i} + \frac{1}{3}\mathbf{j} + \frac{2}{3}\mathbf{k}$

19 $f(x, y, z) = xe^{yz}$, $(x_0, y_0, z_0) = (1, 0, 1)$ in the direction of $\mathbf{v} = 2\mathbf{i} - \mathbf{j} + 2\mathbf{k}$

20 $f(x, y, z) = z \tan^{-1}(y/x)$, $(x_0, y_0, z_0) = (1, 1, 3)$ in the direction of $\mathbf{v} = \mathbf{i} + \mathbf{j} - \mathbf{k}$

21 $f(x, y, z) = z - e^x \sin y$, $(x_0, y_0, z_0) = (\ln 3, 3\pi/2, -3)$ in the direction of $\mathbf{u} = \frac{2}{7}\mathbf{i} + \frac{3}{7}\mathbf{j} + \frac{6}{7}\mathbf{k}$

22 $f(x, y, z) = e^{-y} \sin x + \frac{1}{3}e^{-3y} \sin 3x + z^2$, $(x_0, y_0, z_0) =$

$(\pi/3, 0, 1)$ in the direction of the unit vector $\mathbf{u} = (\cos 2\pi/3)\mathbf{i} + (\cos \pi/4)\mathbf{j} + (\cos \pi/3)\mathbf{k}$

In Problems 23 to 30, find (a) the maximum value of the directional derivative and (b) a unit vector $\mathbf{u}$ in the direction of the maximum directional derivative for each function at the indicated point.

23 $f(x, y) = x^2 - 7xy + 4y^2$ at $(1, -1)$

24 $g(x, y) = (x + y - 2)^2 + (3x - y - 6)^2$ at $(1, 1)$

25 $h(x, y) = x^2 - y^2 - \sin y$ at $(1, \pi/2)$

26 $F(x, y) = e^{-5x} \sin 5y$ at $(0, \pi/20)$

27 $f(x, y, z) = (x^2 + y^2 + z^2)^{-1}$ at $(1, 2, -3)$

28 $h(x, y, z) = (x + y)^2 + (y + z)^2 + (x + z)^2$ at $(2, -1, 2)$

29 $g(x, y, z) = e^x \cos (yz)$ at $(1, 0, \pi)$

30 $f(x, y, z) = \dfrac{x}{x^2 + y^2} + \dfrac{y}{x^2 + z^2}$ at $(3, 1, 1)$

31 Find a unit vector $\mathbf{u}$ that is perpendicular to the level curve of $f(x, y) = 3x^2y - 2xy^2$ at the point $(-2, 3)$.

32 Find a unit vector $\mathbf{u}$ that is perpendicular to the level surface of $f(x, y, z) = 2x^2 + y^2 + z - 8$ at the point $(2, 1, 2)$.

In Problems 33 and 34, find the direction at the point P in which each function increases most rapidly.

33 $z = x^2y^3 + xy^2$, $P = (-2, 1)$

34 $W = 3xy^2 + 3yz^2 + 3zx^2$, $P = (1, 1, -2)$

35 The temperature T at the point (x, y) of a heated circular metal plate with center at the origin is given by

$$T = \frac{400}{2 + x^2 + y^2}$$

where T is measured in degrees Celsius and x and y are measured in centimeters. (a) In what direction should one move through the point $(1, 1)$ in order that T increase most rapidly? (b) How rapidly does T increase as one moves through the point $(1, 1)$ in the direction found in part (a)?

[C] **36** Marine biologists have determined that a shark detecting the presence of blood will respond by moving continually in the direction in which the concentration of blood increases most rapidly. Suppose that the concentration C of blood (in parts per million of water) is given by

$$C(x, y) = e^{(-x^2 - 2y^2)/10^4}$$

where x and y are horizontal coordinates measured in meters from the blood source. In what direction will the shark move through the point $(3, 1)$? Find a unit vector in that direction.

37 Let f be a differentiable function of two variables such that $f_1(-1, 2) = 2$ and $f_2(-1, 2) = -3$. Find the directional derivative $D_{\mathbf{u}}f(-1, 2)$ if $\mathbf{u} = (\sqrt{2}/2)\mathbf{i} - (\sqrt{2}/2)\mathbf{j}$.

38 Suppose that f is a differentiable function of two variables and $D_{\mathbf{u}}f(3, 4) = -4$ if $\mathbf{u} = \frac{5}{13}\mathbf{i} - \frac{12}{13}\mathbf{j}$ and $D_{\mathbf{v}}f(3, 4) = 5$ if $\mathbf{v} = \frac{12}{13}\mathbf{i} + \frac{5}{13}\mathbf{j}$. Find (a) $f_1(3, 4)$ and (b) $f_2(3, 4)$.

39 Suppose that the temperature T (measured in degrees Celsius) at a point (x, y, z) in space is given by

$$T(x, y, z) = 100 - x^2 - y^2 - 2z^2$$

where x, y, and z are measured in centimeters. (a) Find the rate of change of T at the point $P = (2, 1, 1)$ in the direction of the vector $\mathbf{u} = \frac{1}{3}\mathbf{i} + \frac{2}{3}\mathbf{j} - \frac{2}{3}\mathbf{k}$. (b) In what direction should one move from the point P in order to cool off most rapidly? (c) What is the maximum rate of change of T at P?

© **40** Suppose that the electric potential V (in volts) at a point (x, y, z) in space is given by

$$V(x, y, z) = \ln \sqrt{x^2 + y^2 + z^2}$$

(a) Find the rate of change of V at the point $P = (2, -1, 2)$ in the direction from P toward $(0, 0, 0)$. (b) In what direction is the rate of change of V at P a maximum? (c) What is the maximum rate of change of V at P?

In Problems 41 to 62, find equations of (a) the tangent plane and (b) the normal line to each surface at the point indicated.

41 $x^2 + 2y^2 + 3z^2 = 6$ at $(1, 1, 1)$

42 $x^2 - 2y^2 + z^2 = 11$ at $(2, 1, 3)$

43 $xyz = 6$ at $(1, 2, 3)$

44 $8x - y^2 = 0$ at $(2, 4, 7)$

45 $x^3 + y^3 - 6xy + z = 0$ at $(2, 2, 8)$

46 $\sin xz = e^{xy}$ at $(1, 0, \pi/2)$

47 $2x^2 - 3y^2 - z = 0$ at $(2, 1, 5)$

48 $x^2 - 3y^2 - 4z^2 = 2$ at $(3, 1, 1)$

49 $x^2 + 2y^2 - 5z^2 = 4$ at $(1, 2, 1)$

50 $\sqrt{x} + \sqrt{y} - z = 0$ at $(4, 9, 5)$

51 $\cos(xy) + \sin(yz) = 0$ at $(1, \pi/6, -2)$

52 $\ln xy + \sin yz = 2$ at $(e, 1, \pi/2)$

53 $\sqrt{x} + \sqrt{y} + \sqrt{z} = 6$ at $(9, 4, 1)$

54 $\tan^{-1}(y/x) - \ln xyz = \pi/4$ at $(1, 1, 1)$

55 $z = \frac{1}{32}x^7y^{-2}$ at $(2, 1, 4)$

56 $z = 5x^3y^4 - 3y$ at $(1, -1, 8)$

57 $z = \sqrt{9 - x^2 - y^2}$ at $(-1, 2, 2)$

58 $z = x \sin(\pi y/2)$ at $(0, 0, 0)$ **59** $z = x^y$ at $(1, 1, 1)$

60 $z = \ln \cos \sqrt{x^2 + y^2}$ at $(0, 0, 0)$

61 $z = \ln \sqrt{x^2 + y^2}$ at $(0, -1, 0)$

62 $z = xe^{-y}$ at $(1, 0, 1)$

63 Assuming that f and g are differentiable functions of three variables and that a and b are constants, show that $\nabla(af + bg) = a \nabla f + b \nabla g$.

64 Assuming that f is a differentiable function of three variables, prove that the rate of change of the scalar field $w = f(x, y, z)$ with respect to distance measured in the direction of the unit vector $\mathbf{u}$ is given by $\mathbf{u} \cdot \nabla f$.

65 Suppose that f is a differentiable function of a single variable and that $r = \sqrt{x^2 + y^2 + z^2}$. Verify the formula

$$\nabla f(r) = f'(r) \frac{x\mathbf{i} + y\mathbf{j} + z\mathbf{k}}{r}$$

66 Prove that the gradient of a scalar field in xyz space points in the direction for which the directional derivative attains its maximum and that its length is numerically equal to this maximum directional derivative.

67 Assume that f is a function of two variables which is differentiable at (x_0, y_0). Give a geometric interpretation of the differential df in terms of dx, dy, and the tangent plane to the graph of f at $(x_0, y_0, f(x_0, y_0))$.

68 Suppose that a point P is moving along a smooth curve C in a scalar field w. Assuming that the function that gives the scalar field w is differentiable, that $\mathbf{T}$ is the unit tangent vector to C, and that P is moving with the speed ds/dt, prove that the instantaneous rate of change of w with respect to time as measured at the point P is given by

$$\frac{ds}{dt} \mathbf{T} \cdot \nabla w = \frac{ds}{dt} D_{\mathbf{T}}w$$

In Problems 69 to 72, use the result of Problem 68.

69 The temperature T at any point (x, y) of a rectangular plate lying in the xy plane is $T = x \sin 2y$. The point $P = (x, y)$ is moving clockwise around a circle of radius 1 unit with center at the origin at a constant speed of 2 units of arc length per second. How fast is the temperature at the point P changing at the instant when $(x, y) = (\frac{1}{2}, \sqrt{3}/2)$?

70 Suppose that the surface

$$z = \frac{x^3 - 2y^2}{5280}$$

represents an uneven terrain and that a group of tourists is congregated at the origin. Here, x, y, and z are measured in miles. A Moslem tourist sets out for Mecca, going directly east along the positive direction of the x axis. If he travels at a constant speed of 3 miles per hour, how steep is his climb, in feet per minute, at the end of 1 hour?

71 The pressure P, in newtons per square meter, of a certain gas at the point (x, y, z) in space is given by $P = 1000e^{-(x^2 + y^2 + z^2)}$, where x, y, and z are measured in centimeters. A variable point $Q = (x, y, z)$ is moving along a certain curve C in space. At the instant when $Q = (1, -2, 1)$, its speed is given by $ds/dt = 50$ centimeters per second. If the unit tangent vector $\mathbf{T}$ to C at $(1, -2, 1)$ is given by

$$\mathbf{T} = \left(\cos \frac{2\pi}{3}\right)\mathbf{i} + \left(\cos \frac{\pi}{4}\right)\mathbf{j} + \left(\cos \frac{\pi}{3}\right)\mathbf{k}$$

find the instantaneous time rate of change of pressure at Q just as Q passes through $(1, -2, 1)$.

72 The electric potential V in volts at the point (x, y, z) in space caused by placing a charge q at the origin is given by $V = q/r$, where $r = \sqrt{x^2 + y^2 + z^2}$. (Assume that distances are measured in centimeters.) Let C be a curve in space having a unit tangent vector $\mathbf{T}$ at the point P. If P moves along C in the direction of $\mathbf{T}$ at a constant speed of v centimeters per second, write an expression for the instantaneous time rate of change of electric potential at P.

73 Suppose that two surfaces in xyz space, say $f(x, y, z) = k_1$ and $g(x, y, z) = k_2$, intersect in a smooth curve C (Figure 9). Assuming that f and g are differentiable and that (x_0, y_0, z_0) is a point on C, show that the vector

$$\nabla f(x_0, y_0, z_0) \times \nabla g(x_0, y_0, z_0)$$

is tangent to the curve C at the point (x_0, y_0, z_0).

Figure 9

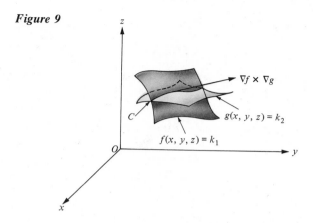

In Problems 74 to 77, use the result of Problem 73 to find a tangent vector at the given point to the curve C in which the two surfaces intersect.

74 $x^2 - y^2 - z^2 + 12 = 0$ and $3x^2 + y^2 + z = 4$ at $(1, 2, -3)$

75 $xz + 2x + 4z = 5$ and $4xy + 3y + 6z = 56$ at $(2, 5, \frac{1}{6})$

76 $x^2 + (y^2/4) - (z^2/9) = 1$ and $x^2 + y^2 + z^2 = 14$ at $(-1, 2, 3)$

77 $x^2 - 2xz + y^2z = 1$ and $3xy + 2yz = -6$ at $(1, -2, 0)$

78 Let $f(x, y) = xy^2/(x^2 + y^4)$ for $(x, y) \neq (0, 0)$ and $f(0, 0) = 0$. Show that $D_{\mathbf{u}}f(0, 0)$ exists for all unit vectors $\mathbf{u}$, but that f is not continuous at $(0, 0)$.

14.7 Higher-Order Partial Derivatives

In studying functions of a single variable, we found it useful to consider not only the first derivative but also higher-order derivatives. Similarly, in studying functions of several variables, it is useful to consider *higher-order partial derivatives*.

Thus, consider a function f of two variables having partial derivatives f_1 and f_2, so that

$$f_1(x, y) = f_x(x, y) = \frac{\partial}{\partial x} f(x, y) \qquad \text{and} \qquad f_2(x, y) = f_y(x, y) = \frac{\partial}{\partial y} f(x, y)$$

The functions f_1 and f_2 are functions of two variables, and they may themselves have partial derivatives. For instance, if $f(x, y) = 3x^2y^3 + 6xy^2$, then

$$f_1(x, y) = f_x(x, y) = \frac{\partial}{\partial x} (3x^2y^3 + 6xy^2) = 6xy^3 + 6y^2$$

$$f_2(x, y) = f_y(x, y) = \frac{\partial}{\partial y} (3x^2y^3 + 6xy^2) = 9x^2y^2 + 12xy$$

Therefore,

$$\frac{\partial}{\partial x} f_1(x, y) = \frac{\partial}{\partial x}\left[\frac{\partial}{\partial x} f(x, y)\right] = \frac{\partial}{\partial x}(6xy^3 + 6y^2) = 6y^3$$

$$\frac{\partial}{\partial y} f_1(x, y) = \frac{\partial}{\partial y}\left[\frac{\partial}{\partial x} f(x, y)\right] = \frac{\partial}{\partial y}(6xy^3 + 6y^2) = 18xy^2 + 12y$$

$$\frac{\partial}{\partial x} f_2(x, y) = \frac{\partial}{\partial x}\left[\frac{\partial}{\partial y} f(x, y)\right] = \frac{\partial}{\partial x}(9x^2y^2 + 12xy) = 18xy^2 + 12y$$

$$\frac{\partial}{\partial y} f_2(x, y) = \frac{\partial}{\partial y}\left[\frac{\partial}{\partial y} f(x, y)\right] = \frac{\partial}{\partial y}(9x^2y^2 + 12xy) = 18x^2y + 12x$$

The four partial derivatives of partial derivatives found above are called the **second-order partial derivatives** of the original function f. Naturally, we can denote the partial derivatives of the function f_1 with respect to the first and second variables as $(f_1)_1$ and $(f_1)_2$, respectively. However, for simplicity, we omit the parentheses and write these second-order partial derivatives as f_{11} and f_{12}, respectively. Likewise, we write f_{xx} and f_{xy} rather than $(f_x)_x$ and $(f_x)_y$, respectively. For instance,

$$f_{12} = f_{xy} = \frac{\partial}{\partial y}\left(\frac{\partial f}{\partial x}\right)$$

The symbolism $\dfrac{\partial}{\partial y}\left(\dfrac{\partial f}{\partial x}\right)$ is also abbreviated $\dfrac{\partial^2 f}{\partial y\, \partial x}$, much in the same way that $\dfrac{d^2 y}{dx^2}$ is used as an abbreviation for the ordinary second derivative. Similarly, we write $\dfrac{\partial^2 f}{\partial x^2}$ for the second-order partial derivative $\dfrac{\partial}{\partial x}\left(\dfrac{\partial f}{\partial x}\right)$, and so forth. In summary, the four second-order partial derivatives of f can be written as follows:

$$f_{11} = f_{xx} = \frac{\partial^2 f}{\partial x^2} = \frac{\partial}{\partial x}\left(\frac{\partial f}{\partial x}\right)$$

$$f_{12} = f_{xy} = \frac{\partial^2 f}{\partial y\, \partial x} = \frac{\partial}{\partial y}\left(\frac{\partial f}{\partial x}\right)$$

$$f_{21} = f_{yx} = \frac{\partial^2 f}{\partial x\, \partial y} = \frac{\partial}{\partial x}\left(\frac{\partial f}{\partial y}\right)$$

$$f_{22} = f_{yy} = \frac{\partial^2 f}{\partial y^2} = \frac{\partial}{\partial y}\left(\frac{\partial f}{\partial y}\right)$$

In the subscript notation, $f_{12} = f_{xy}$ indicates an initial partial differentiation with respect to the first variable x followed by a partial differentiation with respect to the second variable y. However, the symbolism

$$\frac{\partial^2 f}{\partial x\, \partial y} = \frac{\partial}{\partial x}\left(\frac{\partial f}{\partial y}\right)$$

indicates an initial partial differentiation with respect to y followed by a partial differentiation with respect to x. In the subscript notation, the order of the subscripts from *left to right* shows the order of partial differentiation, while in the notation

$$\frac{\partial^2 f}{\partial x \, \partial y}$$

we must read the "denominator" from *right to left*.

EXAMPLE 1   If $f(x, y) = 7x^2 - 13xy + 18y^2$, find f_1, f_2, f_{11}, f_{12}, f_{21}, and f_{22}.

SOLUTION   Here,

$$f_1(x, y) = \frac{\partial}{\partial x} (7x^2 - 13xy + 18y^2) = 14x - 13y$$

$$f_2(x, y) = \frac{\partial}{\partial y} (7x^2 - 13xy + 18y^2) = -13x + 36y$$

Hence, $$f_{11}(x, y) = \frac{\partial}{\partial x} [f_1(x, y)] = \frac{\partial}{\partial x} (14x - 13y) = 14$$

$$f_{12}(x, y) = \frac{\partial}{\partial y} [f_1(x, y)] = \frac{\partial}{\partial y} (14x - 13y) = -13$$

$$f_{21}(x, y) = \frac{\partial}{\partial x} [f_2(x, y)] = \frac{\partial}{\partial x} (-13x + 36y) = -13$$

$$f_{22}(x, y) = \frac{\partial}{\partial y} [f_2(x, y)] = \frac{\partial}{\partial y} (-13x + 36y) = 36$$

EXAMPLE 2   If $f(x, y) = 2e^{2x} \cos y$, find $\dfrac{\partial f}{\partial x}$, $\dfrac{\partial f}{\partial y}$, $\dfrac{\partial^2 f}{\partial x^2}$, $\dfrac{\partial^2 f}{\partial y \, \partial x}$, $\dfrac{\partial^2 f}{\partial x \, \partial y}$, and $\dfrac{\partial^2 f}{\partial y^2}$.

SOLUTION   Here,

$$\frac{\partial f}{\partial x} = \frac{\partial}{\partial x} (2e^{2x} \cos y) = 4e^{2x} \cos y$$

$$\frac{\partial f}{\partial y} = \frac{\partial}{\partial y} (2e^{2x} \cos y) = -2e^{2x} \sin y$$

Hence, $$\frac{\partial^2 f}{\partial x^2} = \frac{\partial}{\partial x} \left(\frac{\partial f}{\partial x} \right) = \frac{\partial}{\partial x} (4e^{2x} \cos y) = 8e^{2x} \cos y$$

$$\frac{\partial^2 f}{\partial y \, \partial x} = \frac{\partial}{\partial y} \left(\frac{\partial f}{\partial x} \right) = \frac{\partial}{\partial y} (4e^{2x} \cos y) = -4e^{2x} \sin y$$

$$\frac{\partial^2 f}{\partial x \, \partial y} = \frac{\partial}{\partial x} \left(\frac{\partial f}{\partial y} \right) = \frac{\partial}{\partial x} (-2e^{2x} \sin y) = -4e^{2x} \sin y$$

$$\frac{\partial^2 f}{\partial y^2} = \frac{\partial}{\partial y} \left(\frac{\partial f}{\partial y} \right) = \frac{\partial}{\partial y} (-2e^{2x} \sin y) = -2e^{2x} \cos y$$

EXAMPLE 3   Let n, m, and k be constants, and suppose that $f(x, y) = kx^n y^m$. Find $\dfrac{\partial f}{\partial x}$, $\dfrac{\partial f}{\partial y}$, $\dfrac{\partial^2 f}{\partial x^2}$, $\dfrac{\partial^2 f}{\partial y \, \partial x}$, $\dfrac{\partial^2 f}{\partial x \, \partial y}$, and $\dfrac{\partial^2 f}{\partial y^2}$.

SOLUTION   Here,

$$\frac{\partial f}{\partial x} = \frac{\partial}{\partial x}(kx^n y^m) = knx^{n-1}y^m \quad \text{and} \quad \frac{\partial f}{\partial y} = \frac{\partial}{\partial y}(kx^n y^m) = kmx^n y^{m-1}$$

Hence,

$$\frac{\partial^2 f}{\partial x^2} = \frac{\partial}{\partial x}\left(\frac{\partial f}{\partial x}\right) = \frac{\partial}{\partial x}(knx^{n-1}y^m) = kn(n-1)x^{n-2}y^m$$

$$\frac{\partial^2 f}{\partial y\,\partial x} = \frac{\partial}{\partial y}\left(\frac{\partial f}{\partial x}\right) = \frac{\partial}{\partial y}(knx^{n-1}y^m) = knmx^{n-1}y^{m-1}$$

$$\frac{\partial^2 f}{\partial x\,\partial y} = \frac{\partial}{\partial x}\left(\frac{\partial f}{\partial y}\right) = \frac{\partial}{\partial x}(kmx^n y^{m-1}) = kmnx^{n-1}y^{m-1}$$

$$\frac{\partial^2 f}{\partial y^2} = \frac{\partial}{\partial y}\left(\frac{\partial f}{\partial y}\right) = \frac{\partial}{\partial y}(kmx^n y^{m-1}) = km(m-1)x^n y^{m-2}$$

The second-order partial derivatives

$$\frac{\partial^2 f}{\partial y\,\partial x} \quad \text{and} \quad \frac{\partial^2 f}{\partial x\,\partial y}$$

are called the **mixed** second-order partial derivatives of f. In all three examples above, notice that the mixed partial derivatives are the same. By Example 3, the mixed partial derivatives of any term in a polynomial function of two variables are the same; hence,

$$\frac{\partial^2 f}{\partial y\,\partial x} = \frac{\partial^2 f}{\partial x\,\partial y}$$

holds for any polynomial function f of two variables. As a matter of fact, the two mixed partial derivatives are equal for a wide class of functions.*

THEOREM 1   **Equality of Mixed Second-Order Partial Derivatives**

Let f be a function of two variables, and suppose that U is an open set of points contained in the domain of f. Then if both of the mixed partial derivatives f_{12} and f_{21} exist and are continuous on U, it follows that $f_{12}(x, y) = f_{21}(x, y)$ for all points (x, y) in U.

The proof of this theorem, which depends on the mean-value theorem for functions of one variable, is usually presented in advanced calculus courses and is not given here.

The notation used for partial derivatives of order higher than 2 is almost self-explanatory. Thus,

$$\frac{\partial^3 f}{\partial x\,\partial y^2} = \frac{\partial}{\partial x}\left(\frac{\partial^2 f}{\partial y^2}\right) = f_{221} = f_{yyx}$$

$$\frac{\partial^5 f}{\partial y^2\,\partial x\,\partial y^2} = \frac{\partial^2}{\partial y^2}\left(\frac{\partial^3 f}{\partial x\,\partial y^2}\right) = f_{22122} = f_{yyxyy}$$

*In Problem 54 we give an example of a function for which the mixed partial derivatives are *not* equal.

and so forth. Theorem 1 on the equality of mixed partial derivatives extends to such higher-order cases. For instance,

$$\frac{\partial^3 f}{\partial x \, \partial y^2} = \frac{\partial^3 f}{\partial y \, \partial x \, \partial y} = \frac{\partial^3 f}{\partial y^2 \, \partial x}$$

holds on a set such as U in the theorem, provided that all three mixed partial derivatives exist and are continuous on U. In summary:

> The order in which successive partial derivatives are taken when you form higher-order partial derivatives is irrelevant, provided that all the partial derivatives in question exist and are continuous.

EXAMPLE 4   Let $f(x, y) = e^{xy} + \sin(x + y)$. Find

(a) f_{xxy} **(b)** f_{yyx} **(c)** f_{xxyxx}

SOLUTION

(a) $f_x(x, y) = ye^{xy} + \cos(x + y)$, and

$$f_{xx}(x, y) = \frac{\partial}{\partial x}[ye^{xy} + \cos(x + y)] = y^2 e^{xy} - \sin(x + y)$$

so that

$$f_{xxy}(x, y) = \frac{\partial}{\partial y}[y^2 e^{xy} - \sin(x + y)]$$
$$= 2ye^{xy} + xy^2 e^{xy} - \cos(x + y)$$

(b) $f_y(x, y) = xe^{xy} + \cos(x + y)$, and

$$f_{yy}(x, y) = \frac{\partial}{\partial y}[xe^{xy} + \cos(x + y)] = x^2 e^{xy} - \sin(x + y)$$

so that

$$f_{yyx}(x, y) = \frac{\partial}{\partial x}[x^2 e^{xy} - \sin(x + y)]$$
$$= 2xe^{xy} + yx^2 e^{xy} - \cos(x + y)$$

(c) Since $f_{xxy}(x, y) = 2ye^{xy} + xy^2 e^{xy} - \cos(x + y)$, we have

$$f_{xxyx}(x, y) = \frac{\partial}{\partial x}[2ye^{xy} + xy^2 e^{xy} - \cos(x + y)]$$
$$= 2y^2 e^{xy} + y^2 e^{xy} + xy^3 e^{xy} + \sin(x + y)$$
$$= 3y^2 e^{xy} + xy^3 e^{xy} + \sin(x + y)$$

so that

$$f_{xxyxx}(x, y) = \frac{\partial}{\partial x}[3y^2 e^{xy} + xy^3 e^{xy} + \sin(x + y)]$$
$$= 3y^3 e^{xy} + y^3 e^{xy} + xy^4 e^{xy} + \cos(x + y)$$
$$= 4y^3 e^{xy} + xy^4 e^{xy} + \cos(x + y)$$

Naturally, the concept of higher-order partial derivatives extends immediately to functions of more than two variables. For instance, if $w = f(x, y, z)$, then

$$\frac{\partial^3 w}{\partial x\, \partial y\, \partial z} = \frac{\partial}{\partial x}\left[\frac{\partial}{\partial y}\left(\frac{\partial w}{\partial z}\right)\right] = f_{zyx} = f_{321}$$

$$\frac{\partial^4 w}{\partial x\, \partial z^2\, \partial y} = \frac{\partial}{\partial x}\left[\frac{\partial^2}{\partial z^2}\left(\frac{\partial w}{\partial y}\right)\right] = f_{yzzx} = f_{2331}$$

and so forth. Furthermore, the conditions for the equality of mixed partial derivatives remain effective.

EXAMPLE 5   Let $w = x^4 y^2 z + \sin xy$. Verify by direct calculation that

$$\frac{\partial^3 w}{\partial x\, \partial y\, \partial z} = \frac{\partial^3 w}{\partial z\, \partial y\, \partial x}$$

SOLUTION   Here, $\dfrac{\partial w}{\partial z} = x^4 y^2$ and $\dfrac{\partial^2 w}{\partial y\, \partial z} = 2x^4 y$

so that $\dfrac{\partial^3 w}{\partial x\, \partial y\, \partial z} = 8x^3 y$. On the other hand,

$$\frac{\partial w}{\partial x} = 4x^3 y^2 z + y \cos xy \qquad \text{and} \qquad \frac{\partial^2 w}{\partial y\, \partial x} = 8x^3 yz + \cos xy - xy \sin xy$$

so that $\dfrac{\partial^3 w}{\partial z\, \partial y\, \partial x} = 8x^3 y$. Therefore,

$$\frac{\partial^3 w}{\partial x\, \partial y\, \partial z} = 8x^3 y = \frac{\partial^3 w}{\partial z\, \partial y\, \partial x} \qquad\blacksquare$$

Higher-order partial derivatives are usually calculated by successive differentiation, as in the last example. At each stage of such a calculation, the standard differentiation rules, including the chain rules, are effective. The use of the chain rules in this connection is illustrated by the following example.

EXAMPLE 6   Let $w = u^2 v^3$, where $u = x^2 + 3y^2$ and $v = 2x^2 - y^2$. Find

$$\frac{\partial^2 w}{\partial x\, \partial y}$$

SOLUTION   By the chain rule,

$$\frac{\partial w}{\partial y} = \frac{\partial w}{\partial u}\frac{\partial u}{\partial y} + \frac{\partial w}{\partial v}\frac{\partial v}{\partial y} = (2uv^3)(6y) + (3u^2 v^2)(-2y) = (12uv^3 - 6u^2 v^2)y$$

Therefore,

$$\frac{\partial^2 w}{\partial x\, \partial y} = \frac{\partial}{\partial x}[(12uv^3 - 6u^2 v^2)]y = \left[12\frac{\partial}{\partial x}(uv^3) - 6\frac{\partial}{\partial x}(u^2 v^2)\right]y$$

$$= \left[12\left(\frac{\partial u}{\partial x}v^3 + 3uv^2\frac{\partial v}{\partial x}\right) - 6\left(2u\frac{\partial u}{\partial x}v^2 + 2u^2 v\frac{\partial v}{\partial x}\right)\right]y$$

$$= [12(2xv^3 + 12uv^2 x) - 6(4uxv^2 + 8u^2 vx)]y$$

$$= 24(v^3 + 5uv^2 - 2u^2 v)xy \qquad\blacksquare$$

Problem Set 14.7

In Problems 1 to 18, (a) find $\dfrac{\partial^2 f}{\partial x^2}$, (b) find $\dfrac{\partial^2 f}{\partial y^2}$, (c) find $\dfrac{\partial^2 f}{\partial y\,\partial x}$, (d) find $\dfrac{\partial^2 f}{\partial x\,\partial y}$, and (e) verify that $\dfrac{\partial^2 f}{\partial y\,\partial x} = \dfrac{\partial^2 f}{\partial x\,\partial y}$.

1 $f(x, y) = 6x^2 + 7xy + 5y^2$

2 $f(x, y) = 4x^2 y^3 - x^4 y + x$

3 $f(x, y) = x \cos y - y^2$

4 $f(x, y) = \tan^{-1} \dfrac{y}{x}$

5 $f(x, y) = (x^2 + y^2)^{3/2}$

6 $f(x, y) = \dfrac{x + y}{x - y}$

7 $f(x, y) = \dfrac{y^2}{x^3}$

8 $f(x, y) = \ln \dfrac{x}{y}$

9 $f(x, y) = y \cos x - xe^{2y}$

10 $f(x, y) = e^{\sqrt{x^2 + y^2}}$

11 $f(x, y) = \sin (x + 2y)$

12 $f(x, y) = e^{3x^2} - 2y^3$

13 $f(x, y) = 5x \cosh 2y$

14 $f(x, y) = \ln \sin \sqrt{x^2 + y^2}$

15 $f(x, y) = 3e^{x^2 y^5}$

16 $f(x, y) = \ln \sqrt{x^3 + y^3}$

17 $f(x, y) = e^{xy} \sin y$

18 $f(x, y) = \sin^3 xy$

In Problems 19 to 26, find each partial derivative.

19 $f(x, y, z) = xy^2 z + 3x^2 e^y$; (a) $f_{xy}(x, y, z)$, (b) $f_{yz}(x, y, z)$

20 $g(x, y, z) = \cos (xyz^2)$; (a) $g_{xz}(x, y, z)$, (b) $g_{yz}(x, y, z)$

21 $h(x, y, z) = xe^y + ze^x + e^{-3z}$; (a) $h_{112}(x, y, z)$, (b) $h_{213}(x, y, z)$

22 $f(x, y, z) = x^2 e^{-2y} \sin z$;
    (a) $f_{113}(1, -1, \pi/4)$,   (b) $f_{2231}(1, -1, \pi/4)$

23 $f(x, y, z) = e^{xyz}$; (a) $f_{xyzx}(1, 2, 3)$, (b) $f_{xxyz}(1, 2, 3)$

24 $f(r, s, t) = \ln (5r + 8s - 6t^2)$; (a) $f_{21}(r, s, t)$, (b) $f_{312}(r, s, t)$

25 $h(x, y, z) = \sin x^2 y - z^2$; (a) $\dfrac{\partial^3 h}{\partial x\,\partial y\,\partial z}$, (b) $\dfrac{\partial^4 h}{\partial x\,\partial y\,\partial z^2}$

26 $f(x, y, z) = y \ln (x - \csc z)$; (a) $\dfrac{\partial^2 f}{\partial y\,\partial z}$, (b) $\dfrac{\partial^2 f}{\partial x\,\partial z}$

27 If $w = (Ax^2 + By^2)^3$, where A and B are constants, verify that
$$\frac{\partial^3 w}{\partial x^2\,\partial y} = \frac{\partial^3 w}{\partial y\,\partial x^2}$$
by direct calculation.

28 If $w = \ln (x - y) + \tan (x + y)$, show that $\dfrac{\partial^2 w}{\partial x^2} = \dfrac{\partial^2 w}{\partial y^2}$.

In Problems 29 to 32, show that each function f satisfies **Laplace's partial differential equation** *in two dimensions*, namely,
$$\frac{\partial^2 f}{\partial x^2} + \frac{\partial^2 f}{\partial y^2} = 0$$

29 $f(x, y) = e^x \sin y + e^y \sin x$

30 $f(x, y) = e^{2x}(\cos^2 y - \sin^2 y)$

31 $f(x, y) = \ln (x^2 + y^2)$

32 $f(x, y) = \tan^{-1} \dfrac{y}{x}$

In Problems 33 and 34, show that each function f satisfies **Laplace's partial differential equation** *in three dimensions*, namely,
$$\frac{\partial^2 f}{\partial x^2} + \frac{\partial^2 f}{\partial y^2} + \frac{\partial^2 f}{\partial z^2} = 0$$

33 $f(x, y, z) = 3x^2 - 2y^2 + 5xy + 8xz - z^2$

34 $f(x, y, z) = (x^2 + y^2 + z^2)^{-1/2}$

35 Let $f(x, y) = \cos (x + y) + \cos (x - y)$. Show that $f_{xx}(x, y) = f_{yy}(x, y)$.

36 Let $f(x, y) = \sin (x - y) + \ln (x + y)$. Show that $f_{11}(x, y) = f_{22}(x, y)$.

37 Show that $w = e^{-y^2} \cos x$ satisfies the partial differential equation
$$\frac{\partial^2 w}{\partial y^2} - 2\frac{\partial^2 w}{\partial x^2} = 4y^2 w$$

38 Assuming that the functions f and g are twice differentiable, show that, if $w = f(x + y) + g(x - y)$, then
$$\frac{\partial^2 w}{\partial x^2} = \frac{\partial^2 w}{\partial y^2}$$

39 If $w = Ax^2 + Bxy + Cy^2 + Dx + Ey + F$, determine the conditions on the constants A, B, C, D, E, and F so that w satisfies Laplace's partial differential equation in two dimensions, namely
$$\frac{\partial^2 w}{\partial x^2} + \frac{\partial^2 w}{\partial y^2} = 0$$

40 A function f of two variables is said to be **homogeneous of degree n** if $f(tx, ty) = t^n f(x, y)$ holds for all values of x, y, and t. (a) Give an example of a function that is homogeneous of degree 1. (b) Give an example of a function that is homogeneous of degree 2. (c) Suppose that f is differentiable and that it is homogeneous of degree n. Prove that $xf_1(x, y) + yf_2(x, y) = nf(x, y)$.

41 Let $w = f(u, v)$, where $u = x + y + z$ and $v = 2x + y - z$. Assuming the existence and continuity of the required partial derivatives of f, find a formula for $\dfrac{\partial^2 w}{\partial x^2} + \dfrac{\partial^2 w}{\partial y^2} + \dfrac{\partial^2 w}{\partial z^2}$.

42 If f is a twice-differentiable function and c is a constant, show that $w = f(x - ct)$ is a solution of the **wave equation,** $\dfrac{\partial^2 w}{\partial t^2} = c^2 \dfrac{\partial^2 w}{\partial x^2}$. Also, show that $w = f(x + ct)$ is a solution of this equation.

43 Show that $w = (a \cos cx + b \sin cx)e^{-kc^2 t}$, where $a, b, c,$ and k are constants, is a solution of the **heat equation** $\dfrac{\partial w}{\partial t} = k \dfrac{\partial^2 w}{\partial x^2}$.

44 Suppose that $w = f(x, y, z)$ and let $x = \rho \sin \phi \cos \theta$, $y = \rho \sin \phi \sin \theta$, and $z = \rho \cos \phi$. Thus, if $(\rho, \theta, \phi) = (4, \pi/3, \pi/6)$, then $(x, y, z) = (1, \sqrt{3}, 2\sqrt{3})$. Suppose that f has continuous partial derivatives and that

$$f_x(1, \sqrt{3}, 2\sqrt{3}) = \sqrt{3} \qquad f_y(1, \sqrt{3}, 2\sqrt{3}) = 2$$
$$f_{xx}(1, \sqrt{3}, 2\sqrt{3}) = 4 \qquad f_{xy}(1, \sqrt{3}, 2\sqrt{3}) = -1$$
$$f_{yy}(1, \sqrt{3}, 2\sqrt{3}) = 4 \qquad f_{yz}(1, \sqrt{3}, 2\sqrt{3}) = -\sqrt{3}$$
$$f_z(1, \sqrt{3}, 2\sqrt{3}) = 1$$
$$f_{xz}(1, \sqrt{3}, 2\sqrt{3}) = 2$$
$$f_{zz}(1, \sqrt{3}, 2\sqrt{3}) = -1$$

Find

(a) $\dfrac{\partial w}{\partial \phi}$ at $\left(4, \dfrac{\pi}{3}, \dfrac{\pi}{6}\right)$ (b) $\dfrac{\partial^2 w}{\partial \rho \, \partial \phi}$ at $\left(4, \dfrac{\pi}{3}, \dfrac{\pi}{6}\right)$

45 If $w = f(x, y)$, $x = r \cos \theta$, and $y = r \sin \theta$, show that

$$\frac{\partial^2 w}{\partial x^2} + \frac{\partial^2 w}{\partial y^2} = \frac{\partial^2 w}{\partial r^2} + \frac{1}{r^2} \frac{\partial^2 w}{\partial \theta^2} + \frac{1}{r} \frac{\partial w}{\partial r}$$

46 If $z = f(x, y)$, $x = g(t)$, and $y = h(t)$, show that

$$\frac{d^2 z}{dt^2} = \frac{\partial^2 z}{\partial x^2}\left(\frac{dx}{dt}\right)^2 + 2 \frac{\partial^2 z}{\partial x \, \partial y} \frac{dx}{dt} \frac{dy}{dt} + \frac{\partial^2 z}{\partial y^2}\left(\frac{dy}{dt}\right)^2$$
$$+ \frac{\partial z}{\partial x} \frac{d^2 x}{dt^2} + \frac{\partial z}{\partial y} \frac{d^2 y}{dt^2}.$$

47 If $z = f(x, y)$, $x = e^r \cos \theta$, and $y = e^r \sin \theta$, show that

$$\frac{\partial^2 z}{\partial x^2} + \frac{\partial^2 z}{\partial y^2} = e^{-2r}\left(\frac{\partial^2 w}{\partial r^2} + \frac{\partial^2 w}{\partial \theta^2}\right)$$

48 If $w = f(u, v)$, $u = g(x, y)$, and $v = h(x, y)$, show that

$$\frac{\partial^2 w}{\partial x^2} = \frac{\partial^2 w}{\partial u^2}\left(\frac{\partial u}{\partial x}\right)^2 + \left(\frac{\partial^2 w}{\partial v \, \partial u} + \frac{\partial^2 w}{\partial u \, \partial v}\right)\frac{\partial u}{\partial x} \frac{\partial v}{\partial x}$$
$$+ \frac{\partial^2 w}{\partial v^2}\left(\frac{\partial v}{\partial x}\right)^2 + \frac{\partial w}{\partial u} \frac{\partial^2 u}{\partial x^2} + \frac{\partial w}{\partial v} \frac{\partial^2 v}{\partial x^2}$$

49 Let $w = f(u)$, where $u = g(x, y)$. Assuming that the required derivatives exist and are continuous, express

$$\frac{\partial^2 w}{\partial x^2} + \frac{\partial^2 w}{\partial y^2}$$

in terms of f', f'', and partial derivatives of g.

50 If $w = f(u, v)$, $u = g(x, y)$, and $v = h(x, y)$, show that

$$\frac{\partial^2 w}{\partial y \, \partial x} = \frac{\partial^2 w}{\partial u^2} \frac{\partial u}{\partial x} \frac{\partial u}{\partial y} + \frac{\partial^2 w}{\partial v \, \partial u} \frac{\partial u}{\partial x} \frac{\partial v}{\partial y} + \frac{\partial^2 w}{\partial u \, \partial v} \frac{\partial u}{\partial y} \frac{\partial v}{\partial x}$$
$$+ \frac{\partial^2 w}{\partial v^2} \frac{\partial v}{\partial x} \frac{\partial v}{\partial y} + \frac{\partial w}{\partial u} \frac{\partial^2 u}{\partial y \, \partial x} + \frac{\partial w}{\partial v} \frac{\partial^2 v}{\partial y \, \partial x}$$

51 Let $w = f(r)$, where $r = \sqrt{x^2 + y^2}$. Show that

$$\frac{\partial^2 w}{\partial x^2} + \frac{\partial^2 w}{\partial y^2} = \frac{d^2 w}{dr^2} + \frac{1}{r} \frac{dw}{dr}$$

52 Suppose that $f(x, y) = 0$ defines y implicitly as a function of f. Assuming the existence and continuity of the required derivatives and that $\partial f/\partial y \neq 0$, show that

$$\frac{d^2 y}{dx^2} = -\frac{\dfrac{\partial^2 f}{\partial x^2}\left(\dfrac{\partial f}{\partial y}\right)^2 - 2 \dfrac{\partial f}{\partial x} \cdot \dfrac{\partial f}{\partial y} \cdot \dfrac{\partial^2 f}{\partial x \, \partial y} + \dfrac{\partial^2 f}{\partial y^2}\left(\dfrac{\partial f}{\partial x}\right)^2}{\left(\dfrac{\partial f}{\partial y}\right)^3}$$

53 If u and v are functions of x and y, and if u and v satisfy the **Cauchy-Riemann equations,** namely,

$$\frac{\partial u}{\partial x} = \frac{\partial v}{\partial y} \qquad \text{and} \qquad \frac{\partial u}{\partial y} = -\frac{\partial v}{\partial x}$$

show that

$$\frac{\partial^2 u}{\partial x^2} + \frac{\partial^2 u}{\partial y^2} = 0 \qquad \text{and} \qquad \frac{\partial^2 v}{\partial x^2} + \frac{\partial^2 v}{\partial y^2} = 0$$

Assume the existence and continuity of the required partial derivatives.

54 Let

$$f(x, y) = \begin{cases} xy\left(\dfrac{x^2 - y^2}{x^2 + y^2}\right) & \text{if } (x, y) \neq (0, 0) \\ 0 & \text{if } (x, y) = (0, 0) \end{cases}$$

(a) Compute $f_1(0, y)$ and $f_2(x, 0)$ by direct use of the definition of partial derivatives. (b) Show that $f_{12}(0, y) = -1$ and $f_{21}(x, 0) = 1$. (c) Show that $f_{12}(0, 0) \neq f_{21}(0, 0)$. (d) Some hypothesis of Theorem 1 must be violated here because of part (c). What is it?

55 If $w = e^{a_1 x_1 + a_2 x_2 + \cdots + a_n x_n}$ and $a_1^2 + a_2^2 + \cdots + a_n^2 = 1$, show that

$$\frac{\partial^2 w}{\partial x_1^2} + \frac{\partial^2 w}{\partial x_2^2} + \cdots + \frac{\partial^2 w}{\partial x_n^2} = w$$

14.8   Extrema for Functions of More Than One Variable

In Chapter 3 we saw that the derivative is an indispensable tool for solving problems involving extrema of functions of single variables. In this section we study the problem of finding maximum and minimum values of functions of more than one variable, and we see that partial derivatives are especially useful in this connection.

The basic concepts, such as relative extrema, absolute extrema, critical points, and so forth for functions of several variables are the natural analogs of the corresponding concepts for functions of one variable. For simplicity, we formulate these concepts for functions of two variables; it should be obvious how they can be extended to functions of more than two variables.

DEFINITION 1   **Relative Extrema**

> **(i)** A function f of two variables has a **relative maximum value** $f(a, b)$ at the point (a, b) if there is a circular disk with positive radius and with center at (a, b) such that if (x, y) is a point in this disk, then (x, y) is in the domain of f and $f(x, y) \leq f(a, b)$.
>
> **(ii)** Similarly, a function f of two variables has a **relative minimum value** $f(a, b)$ at the point (a, b) if there is a circular disk with positive radius and with center at (a, b) such that if (x, y) is a point in this disk, then (x, y) is in the domain of f and $f(x, y) \geq f(a, b)$.
>
> **(iii)** A relative maximum or minimum value of a function is called a **relative extreme value,** or a **relative extremum,** of the function.

Figure 1

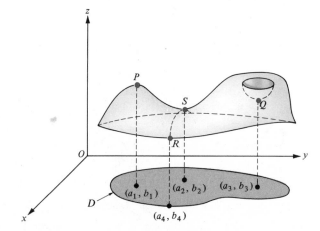

Consider, for instance, the surface in Figure 1 as the graph of a function f of two variables. The point P at the apex of the "hill" represents a relative maximum of f, since $f(a_1, b_1)$—the height of P above the xy plane—is larger than all *nearby* values of $f(x, y)$. Similarly, the point Q at the bottom of the "hollow" corresponds to a relative minimum, $f(a_3, b_3)$.

The point S in the "pass" is neither a relative maximum nor a relative minimum; indeed, we *increase* our height above the xy plane as we move from S up the hill toward P, whereas we *decrease* our height above the xy plane as we move from S down toward the point R on the "rim" of the surface. Incidentally, R does *not* represent a relative extremum of f because there is no circular disk with positive radius and with center at (a_4, b_4) that is contained in the domain D of f. Notice that, by definition, a relative extremum of a function f can occur only at an interior point of the domain D of f—never at a boundary point of D.

Sometimes it is possible to locate the relative extrema of a function f of more than one variable by algebraic considerations or by sketching the graph of f. This is illustrated by the following example.

<u>EXAMPLE 1</u>   If $f(x, y) = 1 - x^2 - y^2$, find all relative extrema of f.

SOLUTION   Here, $f(x, y) = 1 - (x^2 + y^2)$ and $x^2 + y^2 \geq 0$, so that $f(x, y) \leq 1$ for all values of x and y. Since $f(0, 0) = 1$, f attains a relative maximum value of 1 at $(0, 0)$. The graph of f, a paraboloid of revolution about the z axis, clearly shows that f has no other relative extrema (Figure 2).

Figure 2

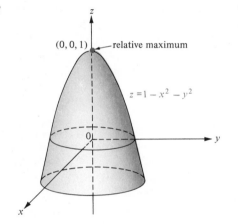

Figure 3

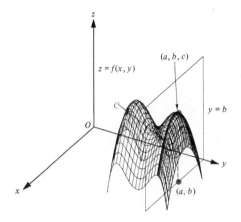

Suppose that f is a function of two variables which has a relative maximum $c = f(a, b)$ at the point (a, b) (Figure 3). Let C be the cross section cut from the graph of f by the plane $y = b$, and notice that the curve C attains a relative maximum at the point (a, b). Therefore, if C is a smooth curve, it must have a horizontal tangent line at (a, b, c), by Theorem 1 on page 112. Since the slope of the tangent line to C at (a, b, c) is given by the partial derivative $f_1(a, b)$, then $f_1(a, b) = 0$.

By cutting the surface in Figure 3 with the plane $x = a$ and reasoning similarly, we can also conclude that if $f_2(a, b)$ exists, then $f_2(a, b) = 0$. Likewise, if f has a relative minimum at (a, b) and if $f_1(a, b)$ and $f_2(a, b)$ exist, then $f_1(a, b) = 0$ and $f_2(a, b) = 0$. Thus, we have the following theorem.

THEOREM 1   **Necessary Condition for Relative Extrema**

Let (a, b) be an interior point of the domain of a function f, and suppose that the two partial derivatives $f_1(a, b)$ and $f_2(a, b)$ exist. Then if f has a relative extremum at (a, b), it is necessary that both $f_1(a, b) = 0$ and $f_2(a, b) = 0$.

Note that the condition $f_1(a, b) = f_2(a, b) = 0$ in Theorem 1 can also be expressed as $\nabla f(a, b) = \mathbf{0}$, since $\nabla f(a, b) = f_1(a, b)\mathbf{i} + f_2(a, b)\mathbf{j}$. Therefore, *at a point where a function attains a relative extremum, its gradient must either fail to exist or be the zero vector.* Thus, we make the following definition.

DEFINITION 2   **Critical Point**

> Let f be a function of two variables. A point (a, b) in the domain of f such that either $\nabla f(a, b)$ fails to exist or $\nabla f(a, b) = \mathbf{0}$, is called a **critical point** of f. A critical point of f which is in the interior of the domain of f is called an **interior critical point** of the domain of f.

Now, Theorem 1 can be restated thus: *If a function f has a relative extremum at a point (a, b), then (a, b) must be an interior critical point of the domain of f.*

Therefore, to locate all the relative extrema of f, begin by finding all the interior critical points of its domain. Some of these critical points may correspond to relative extrema; however, some may occur at "saddle points," that is, critical points where the function has neither a relative maximum nor a relative minimum. The point S on the surface in Figure 1 corresponds to a saddle point; thus, the interior critical point (a_2, b_2) in Figure 1 would not give a relative extremum for the function. (Notice that the graph of the function has the shape of a saddle near the point S in Figure 1.)

In order to sort out the interior critical points of a function and determine which correspond to relative maxima, relative minima, or saddle points, the following theorem (whose proof is best left to a course in advanced calculus) may be helpful.

THEOREM 2   **Second-Derivative Test**

> Let (a, b) be an interior point of the domain of a function f, and suppose that the first and second partial derivatives of f exist and are continuous on some circular disk with (a, b) as its center and contained in the domain of f. Assume that (a, b) is a critical point of f, so that $f_1(a, b) = f_2(a, b) = 0$. Let
>
> $$\Delta = \begin{vmatrix} f_{11}(a, b) & f_{12}(a, b) \\ f_{12}(a, b) & f_{22}(a, b) \end{vmatrix} = f_{11}(a, b)f_{22}(a, b) - [f_{12}(a, b)]^2$$
>
> Then
>
> (i)* If $\Delta > 0$ and $f_{11}(a, b) < 0$, then f has a relative maximum at (a, b).
>
> (ii)* If $\Delta > 0$ and $f_{11}(a, b) > 0$, then f has a relative minimum at (a, b).
>
> (iii) If $\Delta < 0$, then f has a saddle point at (a, b).
>
> (iv) If $\Delta = 0$, then the test is inconclusive, and other methods must be used.

In Examples 2 to 4, find all the interior critical points of the domain of the given function, and then apply the second-derivative test to decide (if possible) at which of these points relative maxima, relative minima, or saddle points occur.

EXAMPLE 2 $f(x, y) = x^2 + y^2 - 2x + 4y + 2$

---

*If $\Delta = f_{11}(a, b)f_{22}(a, b) - [f_{12}(a, b)]^2 > 0$, then $f_{11}(a, b)f_{22}(a, b)$ must be positive, so that $f_{11}(a, b)$ and $f_{22}(a, b)$ have the *same algebraic sign*. Therefore, in parts (i) and (ii) of Theorem 2, we can replace $f_{11}(a, b)$ by $f_{22}(a, b)$.

SOLUTION   Here,

$$f_1(x, y) = 2x - 2 \qquad f_2(x, y) = 2y + 4$$

$$f_{11}(x, y) = 2 \qquad f_{12}(x, y) = f_{21}(x, y) = 0 \qquad f_{22}(x, y) = 2$$

To find all critical points, we must solve the simultaneous equations

$$\begin{cases} f_1(x, y) = 0 \\ f_2(x, y) = 0 \end{cases} \quad \text{that is,} \quad \begin{cases} 2x - 2 = 0 \\ 2y + 4 = 0 \end{cases}$$

In this case, the only solution is $x = 1$, $y = -2$; hence, $(1, -2)$ is the only critical point of f. Calculating the quantity Δ of Theorem 2 at the critical point $(1, -2)$, we obtain

$$\Delta = \begin{vmatrix} f_{11}(1, -2) & f_{12}(1, -2) \\ f_{12}(1, -2) & f_{22}(1, -2) \end{vmatrix} = \begin{vmatrix} 2 & 0 \\ 0 & 2 \end{vmatrix} = 4 - 0 = 4 > 0$$

Also, $f_{11}(1, -2) = 2 > 0$. Therefore, by Theorem 2, f has a relative minimum at $(1, -2)$. ∎

EXAMPLE 3 $f(x, y) = 12xy - 4x^2y - 3xy^2$

SOLUTION   Here,

$$f_1(x, y) = 12y - 8xy - 3y^2 = (12 - 8x - 3y)y$$

$$f_2(x, y) = 12x - 4x^2 - 6xy = (12 - 4x - 6y)x$$

$$f_{11}(x, y) = -8y$$

$$f_{12}(x, y) = f_{21}(x, y) = 12 - 8x - 6y$$

and $$f_{22}(x, y) = -6x$$

To find all critical points (x, y), we must find all solutions of the simultaneous equations

$$\begin{cases} (12 - 8x - 3y)y = 0 \\ (12 - 4x - 6y)x = 0 \end{cases}$$

Obviously, $x = 0$, $y = 0$ gives a solution, so that $(0, 0)$ is a critical point. If $x \neq 0$ and $y = 0$, the equations become

$$\begin{cases} 0 = 0 \\ 12 - 4x = 0 \end{cases}$$

so that $x = 3$, $y = 0$ gives a solution, and $(3, 0)$ is a critical point. If $x = 0$ and $y \neq 0$, the equations become

$$\begin{cases} 12 - 3y = 0 \\ 0 = 0 \end{cases}$$

giving the critical point $(0, 4)$. Finally, if $x \neq 0$ and $y \neq 0$, the simultaneous equations become

$$\begin{cases} 12 - 8x - 3y = 0 \\ 12 - 4x - 6y = 0 \end{cases}$$

Solving the latter equations as usual (say, by elimination of variables), we obtain $x = 1$, $y = \frac{4}{3}$; hence, $(1, \frac{4}{3})$ is also a critical point.

The critical points of f are therefore $(0, 0)$, $(3, 0)$, $(0, 4)$, and $(1, \frac{4}{3})$. Evaluating the quantity Δ at each of these critical points, we have:

1 At $(0, 0)$, $\Delta = f_{11}(0, 0)f_{22}(0, 0) - [f_{12}(0, 0)]^2 = 0 - (12)^2 = -144$.

2 At $(3, 0)$, $\Delta = f_{11}(3, 0)f_{22}(3, 0) - [f_{12}(3, 0)]^2 = 0 - (-12)^2 = -144$.

3 At $(0, 4)$, $\Delta = f_{11}(0, 4)f_{22}(0, 4) - [f_{12}(0, 4)]^2 = 0 - (-12)^2 = -144$.

4 At $(1, \frac{4}{3})$, $\Delta = f_{11}(1, \frac{4}{3})f_{22}(1, \frac{4}{3}) - [f_{12}(1, \frac{4}{3})]^2 = 64 - (-4)^2 = 48$.

Since $\Delta < 0$ at the first three critical points $(0, 0)$, $(3, 0)$, and $(0, 4)$, these are saddle points. At $(1, \frac{4}{3})$, we have $\Delta > 0$ and $f_{11}(1, \frac{4}{3}) = -\frac{32}{3} < 0$; hence, the function has a relative maximum at $(1, \frac{4}{3})$. ■

EXAMPLE 4 $f(x, y) = x^4 + y^4$

SOLUTION   Here,

$$f_1(x, y) = 4x^3 \qquad f_2(x, y) = 4y^3$$

$$f_{11}(x, y) = 12x^2 \qquad f_{12}(x, y) = f_{21}(x, y) = 0 \qquad f_{22}(x, y) = 12y^2$$

The only solution of the simultaneous equations

$$\begin{cases} 4x^3 = 0 \\ 4y^3 = 0 \end{cases}$$

is $x = 0$, $y = 0$; hence, $(0, 0)$ is the only critical point of f. Thus,

$$\Delta = f_{11}(0, 0)f_{22}(0, 0) - [f_{12}(0, 0)]^2 = 0$$

and therefore the second-derivative test is inconclusive at $(0, 0)$. However, since $f(0, 0) = 0$ and since $f(x, y) = x^4 + y^4 > 0$ when $(x, y) \neq (0, 0)$, it is clear on purely algebraic grounds that f attains a relative minimum at $(0, 0)$ and that it has no relative maximum. ■

Absolute Extrema

In contrast to the notion of a relative extremum given in Definition 1, we now introduce the idea of an absolute extremum.

DEFINITION 3   **Absolute Extrema**

A function f of two variables has an **absolute maximum value** $f(a, b)$ at the point (a, b) of its domain D if $f(x, y) \leq f(a, b)$ holds for every point (x, y) in D. Similarly, f has an **absolute minimum value** $f(c, d)$ at the point (c, d) of its domain D if $f(x, y) \geq f(c, d)$ holds for every point (x, y) in D. An absolute maximum or minimum value of f is called an **absolute extremum** of f.

In Figure 2, the function $f(x, y) = 1 - x^2 - y^2$ actually attains not only a relative maximum but also an absolute maximum at $(0, 0)$. However, in Figure 1, the point P does not represent an absolute maximum, since the surface climbs even higher than P as we move to the right of the saddle point S.

It is intuitively clear that a continuous and bounded surface must have a highest point and a lowest point (provided that the rim of the surface is considered part of the surface). For instance, the point R on the surface in Figure 1 appears to be a lowest point, while a highest point would be found somewhere along the upper edge of the hollow surrounding the point Q. The following theorem confirms our intuition about these matters; however, its proof is best left to courses in advanced calculus or analysis.

THEOREM 3   **Existence of Absolute Extrema**

> Let f be a continuous function of two variables whose domain D not only is bounded but also contains all its own boundary points.* Then f attains an absolute maximum value and an absolute minimum value.

Note that an absolute extremum which occurs at an interior point of the domain D of a function f is automatically a relative extremum of f. (Why?) Consequently, an absolute extremum of f which is not a relative extremum must occur at a boundary point of D (that is, at a point of D which is not an interior point). Hence, *in order to locate the absolute extrema of f, you must first find all the relative extrema and then compare the largest and the smallest of these with the values of f around the boundary of D*. The technique is quite similar to that used in Chapter 3 to find absolute extrema of functions of one variable and is illustrated by the following example.

© **EXAMPLE 5**   A circular metal plate of radius 1 meter is placed with its center at the origin in the xy plane and heated so that its temperature at the point (x, y) is given by $T = 64(3x^2 - 2xy + 3y^2 + 2y + 5)$ degrees Celsius, where x and y are measured in meters. Find the highest and the lowest temperatures on the plate.

SOLUTION   Since

$$\frac{\partial T}{\partial x} = 64(6x - 2y) \qquad \text{and} \qquad \frac{\partial T}{\partial y} = 64(-2x + 6y + 2)$$

the critical points inside the circular disk are found by solving the simultaneous equations

$$\begin{cases} 64(6x - 2y) = 0 \\ 64(-2x + 6y + 2) = 0 \end{cases}$$

The only solution, $x = -\frac{1}{8}$, $y = -\frac{3}{8}$, gives exactly one internal critical point on the plate, namely $(-\frac{1}{8}, -\frac{3}{8})$. For the second-derivative test at $(-\frac{1}{8}, -\frac{3}{8})$, we need the quantities

$$\frac{\partial^2 T}{\partial x^2} = (64)(6) \qquad \frac{\partial^2 T}{\partial x\, \partial y} = (64)(-2) \qquad \text{and} \qquad \frac{\partial^2 T}{\partial y^2} = (64)(6)$$

Thus, at the critical point,

$$\Delta = \frac{\partial^2 T}{\partial x^2} \frac{\partial^2 T}{\partial y^2} - \left(\frac{\partial^2 T}{\partial x\, \partial y}\right)^2 = (384)(384) - (-128)^2 = 131{,}072 > 0$$

and $\partial^2 T/\partial x^2 = 384 > 0$; hence, there is a relative minimum temperature of

$$64[3(-\tfrac{1}{8})^2 - 2(-\tfrac{1}{8})(-\tfrac{3}{8}) + 3(-\tfrac{3}{8})^2 + 2(-\tfrac{3}{8}) + 5] = 296°C$$

at the point $(-\frac{1}{8}, -\frac{3}{8})$.

*That is, the domain D is a closed and bounded set.

Now, we must examine the value of T on the boundary $x^2 + y^2 = 1$ of the circular plate. If we put

$$\begin{cases} x = \cos \theta \\ y = \sin \theta \end{cases}$$

then as θ varies from 0 to 2π, the point (x, y) traverses the boundary of the plate. The temperature at the point corresponding to θ is given by

$$\begin{aligned} T &= 64(3 \cos^2 \theta - 2 \cos \theta \sin \theta + 3 \sin^2 \theta + 2 \sin \theta + 5) \\ &= 64(3 - 2 \cos \theta \sin \theta + 2 \sin \theta + 5) \\ &= 64(8 - 2 \cos \theta \sin \theta + 2 \sin \theta) \\ &= 128(4 - \cos \theta \sin \theta + \sin \theta) \end{aligned}$$

Thus, on the boundary of the plate,

$$\frac{dT}{d\theta} = 128(\sin^2 \theta - \cos^2 \theta + \cos \theta) = 128(1 - 2 \cos^2 \theta + \cos \theta)$$

The critical values of θ on the boundary are the solutions of

$$1 - 2 \cos^2 \theta + \cos \theta = 0 \qquad \text{or} \qquad 2 \cos^2 \theta - \cos \theta - 1 = 0$$

that is, $(2 \cos \theta + 1)(\cos \theta - 1) = 0$

Thus, $\cos \theta = -\frac{1}{2}$ or $\cos \theta = 1$, so the critical values of θ are

$$\theta = \frac{2\pi}{3} \qquad \theta = \frac{4\pi}{3} \qquad \theta = 0$$

When $\theta = 2\pi/3$, we have

$$T = 128\left(4 - \cos \frac{2\pi}{3} \sin \frac{2\pi}{3} + \sin \frac{2\pi}{3}\right) = 32(16 + 3\sqrt{3}) \approx 678.28°C$$

When $\theta = 4\pi/3$, then

$$T = 128\left(4 - \cos \frac{4\pi}{3} \sin \frac{4\pi}{3} + \sin \frac{4\pi}{3}\right) = 32(16 - 3\sqrt{3}) \approx 345.72°C$$

When $\theta = 0$, we have

$$T = 128(4 - \cos 0 \sin 0 + \sin 0) = 512°C$$

On the boundary of the plate, the maximum temperature is therefore $32(16 + 3\sqrt{3}) \approx 678.28$ degrees Celsius and the minimum temperature is $32(16 - 3\sqrt{3}) \approx 345.72$ degrees Celsius. The relative minimum of 296 degrees Celsius on the interior of the plate is smaller then the minimum on the boundary; hence, it is the absolute minimum temperature on the plate. Therefore, the absolute minimum temperature on the plate is 296 degrees Celsius, and the absolute maximum temperature on the plate is $32(16 + 3\sqrt{3}) \approx 678.28$ degrees Celsius. ∎

In applied problems, it is usually the *absolute* extrema that are desired, and a rigorous treatment ordinarily requires an exhaustive analysis, as illustrated in the last example. However, in many cases, people rely on their physical or geometric intuition and carry out an informal analysis involving only an examination of interior critical points. Such an informal procedure is illustrated in the following example.

EXAMPLE 6   A long piece of tin 12 inches wide is to be made into a trough by bending up strips of equal width along the edges so that the strips make equal angles with the horizontal. How wide should these strips be, and what angle should they make with the horizontal so that the trough has a maximum carrying capacity?

Figure 4

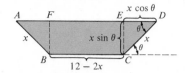

SOLUTION   Figure 4 shows a cross section of the trough. Here x denotes the width of the strips, and θ denotes their angle with the horizontal. We must maximize the area z of the cross section. Since the triangles ABF and ECD each have areas of $\frac{1}{2}(x \cos \theta)(x \sin \theta)$ square inches and the rectangle $BCEF$ has an area of $x \sin \theta(12 - 2x)$ square inches, it follows that

$$z = x^2 \cos \theta \sin \theta + x \sin \theta (12 - 2x) \qquad \text{square inches}$$

Here, $$\frac{\partial z}{\partial x} = \sin \theta \, [x(2 \cos \theta - 4) + 12]$$

$$\frac{\partial z}{\partial \theta} = x[x(2 \cos^2 \theta - 2 \cos \theta - 1) + 12 \cos \theta]$$

Hence, critical points will correspond to solutions of the simultaneous equations

$$\begin{cases} x(2 \cos \theta - 4) + 12 = 0 \\ x(2 \cos^2 \theta - 2 \cos \theta - 1) + 12 \cos \theta = 0 \end{cases}$$

Solving the first equation for $\cos \theta$, we obtain

$$\cos \theta = 2 - \frac{6}{x}$$

Substituting this value of $\cos \theta$ into the second equation, we have

$$x\left[2\left(4 - \frac{24}{x} + \frac{36}{x^2}\right) - \left(4 - \frac{12}{x}\right) - 1\right] + 24 - \frac{72}{x} = 0 \qquad \text{or} \qquad 3x - 12 = 0$$

Hence, $x = 4$ and $\cos \theta = 2 - \frac{6}{4} = \frac{1}{2}$. The critical point, hence (we assume) the desired solution, is therefore given by

$$x = 4 \text{ inches} \qquad \text{and} \qquad \theta = 60° \qquad\qquad \blacksquare$$

The solution $x = 4$ inches and $\theta = 60°$ in the last example seems so reasonable on geometric grounds that we have not even invoked the second-derivative test for a *relative* maximum. (We invite you to do so.) Of course, if great importance were somehow associated with this problem—for instance, if we were going into the trough business—we not only would test for a relative maximum, but also we would go on to check that our solution represents the *absolute* maximum.

The facts concerning maxima and minima for functions of three or more variables are analogous to the facts set forth above for functions of two variables, except that the second-derivative test becomes more complicated as the number of independent variables increases. Again, these matters are best left to advanced calculus.

The Method of Least Squares

In the experimental sciences it is often necessary to determine the extent to which two quantities x and y are **correlated.** This is done by collecting data in the form of paired measurements, or observations, of values of x and y. If n such measurements are made and the corresponding points

$$(x_1, y_1), (x_2, y_2), (x_3, y_3), \ldots, (x_n, y_n)$$

are plotted on an xy coordinate system, then the resulting graph is called a **scattergram** (Figure 5). A **perfect linear correlation** between x and y is indicated when all points in the scattergram fall on a line. Even if no such perfect correlation exists, it is possible to draw a line $y = mx + b$ that "best fits" the scattergram in a suitable sense.

The usual criterion for determining how well the line $y = mx + b$ fits the scattergram is based on the sequence

$$E_1, E_2, E_3, \ldots, E_n$$

of vertically measured distances, or deviations, between the points in the scattergram and the line $y = mx + b$. The ith deviation is given by

$$E_i = |y_i - (mx_i + b)| = |y_i - mx_i - b|$$

for $i = 1, 2, 3, \ldots, n$ (Figure 5). The **method of least squares** uses the sum of the squares of these deviations

$$f(m, b) = E_1^2 + E_2^2 + E_3^2 + \cdots + E_n^2$$

as a measure of how well the line $y = mx + b$ fits the scattergram.* The line that

Figure 5

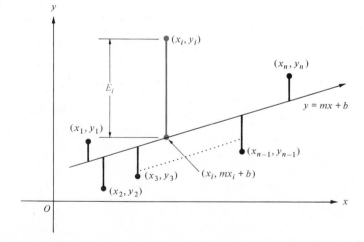

best fits, that is, the line $y = mx + b$ for which $f(m, b)$ is a minimum, is called the **regression line** for the scattergram. Finding the regression line for a scattergram is one of the practical applications of the methods introduced in this section. The technique and the results are outlined in Problems 33 to 38.

*The reasons for this are discussed in textbooks on statistics.

Problem Set 14.8

In Problems 1 to 18, find all critical points of each function, and then test each critical point to see whether it corresponds to a relative maximum, a relative minimum, or a saddle point.

1 $f(x, y) = x^2 + (y - 1)^2$

2 $f(x, y) = x^2 + 4xy - y^2 - 8x - 6y$

3 $f(x, y) = (x - y + 1)^2$ **4** $g(x, y) = x^3 - 3xy^2 + y^3$

5 $f(x, y) = x^2 + y^2 + [2/(xy)]$ **6** $f(x, y) = x \sin y$

7 $g(x, y) = xy - x - y$ **8** $h(x, y) = e^x + e^y - e^{x+y}$

9 $f(x, y) = x^4 + y^4 + 4x + 4y$ **10** $f(x, y) = xy^2 + x^2 + y$

11 $F(x, y) = x^3y + 3x + y$

12 $G(x, y) = \sin x + \sin y + \cos (x + y)$

13 $H(x, y) = x^3 + y^3 - 3x$

14 $f(x, y) = xy(12 - 4x - 3y)$

15 $F(x, y) = x^2 + y^2 + (3x + 4y - 26)^2$

16 $f(x, y, z) = e^{-(x^2+y^2+z^2)}$

17 $f(x, y) = x^4 + y^4 + 32x - 4y + 52$

18 $f(x, y, z) = xy + xz$

In Problems 19 to 22, find the absolute maximum and the absolute minimum of each function (if they exist).

19 $f(x, y) = \sqrt{1 - x^2 - y^2}$ for $x^2 + y^2 \leq 1$

20 $f(x, y) = 3x^2 + 2xy + 4y^2 + 2x - 3y + 1$ for $0 \leq x \leq 1$ and $0 \leq y \leq 1$

21 $f(x, y) = xy + 12(x + y) - (x + y)^2$ for $0 \leq x \leq 12$ and $0 \leq y \leq 12$

22 $f(x, y) = 5x - 2y + 7$ for (x, y) on or inside the ellipse

$$\begin{cases} x = 3 \cos \theta \\ y = 4 \sin \theta \end{cases} \quad 0 \leq \theta \leq 2\pi$$

23 A rectangular box is to have a volume of 20 cubic meters. The material used for the sides costs 1 dollar per square meter, the material used for the bottom costs 2 dollars per square meter, and the material used for the top costs 3 dollars per square meter. What are the dimensions of the cheapest box?

24 A manufacturer produces two grades of alloy in quantities of x and y tons. If the total cost of production is expressed by the function $C(x, y) = x^2 + 100x + y^2 - xy$ and the total revenue is given by the function $R(x, y) = 100x - x^2 + 2000y + xy$, find the production level that maximizes the profit.

25 The cost of paying for the inspection of an assembly-line operation depends on the number of inspections x and y at

each of two sites according to the following function: $C(x, y) = x^2 + y^2 + xy - 20x - 25y + 1500$. How many inspections should be made at each site in order to minimize the cost?

26 One end of a house is to consist of solar panels in the shape of a rectangle surmounted by an isosceles triangle and is to have a perimeter p feet, where p is a given constant. If the house is to be constructed so as to collect a maximum amount of solar energy, show that the slope of the roof must be $\sqrt{3}/3$.

27 The temperature T in degrees Celsius at each point in the region $x^2 + y^2 \leq 1$ is given by $T = 16x^2 + 24x + 40y^2$. Find the temperatures at the hottest and coldest points in the region.

28 Suppose that the function f is continuous and that its domain D is bounded and contains all its own boundary points. Suppose also that f never takes on any negative values and that $f(x, y) = 0$ whenever (x, y) is on the boundary of D. If f has exactly one critical point in the interior of D, show that it takes on its absolute maximum at this point.

29 Find three nonnegative numbers x, y, and z such that $x + y + z = 2001$ and xyz is as large as possible. Show carefully that your answer actually represents an absolute maximum.

30 The electric potential V at each point (x, y) in the region $0 \leq x \leq 1$ and $0 \leq y \leq 1$ is given by $V = 48xy - 32x^3 - 24y^2$. Find the maximum and minimum potential in this region.

31 Explain why an absolute extremum that occurs at an interior point of the domain D of a function f must be a relative extremum of f.

32 A certain state plans to supplement its revenue by selling weekly lottery tickets. Opinion polls show that a potential 1 million tickets will be purchased per week at \$1 per ticket, but that 130,000 fewer tickets will be purchased per week for every \$0.25 increase in the price per ticket. Fixed costs such as printing and distributing tickets, paying salaries of lottery officials, and advertising are expected to run \$140,000 per week. Regardless of the price per ticket, it is estimated that each additional dollar (over the basic allotment provided for in fixed costs) spent in advertising per week will result in the sale of one additional ticket per week. The state, by law, must return one-third of its weekly revenue from ticket sales as prizes to the purchasers. How much should the state charge for a lottery ticket in order to maximize its profits, and what maximum weekly profit from the lottery can it expect?

33 Consider a scattergram with the points (x_1, y_1), (x_2, y_2), (x_3, y_3), . . . , (x_n, y_n) (Figure 5). If $E_i = |y_i - mx_i - b|$ for $i = 1, 2, 3$, . . . , n, show that

(a) $\dfrac{\partial}{\partial m} E_i^2 = -2(x_iy_i - mx_i^2 - bx_i)$

(b) $\dfrac{\partial}{\partial b} E_i^2 = -2(y_i - mx_i - b)$

34 In Problem 33, let $f(m, b) = \sum_{i=1}^{n} E_i^2$. Assuming that f has a relative minimum at (m, b), show that

(a) $\sum_{i=1}^{n} x_i y_i = m \sum_{i=1}^{n} x_i^2 + b \sum_{i=1}^{n} x_i$ (b) $\sum_{i=1}^{n} y_i = m \sum_{i=1}^{n} x_i + nb$

Equations (a) and (b) are called the **normal equations** for m and b.

35 If we abbreviate $\sum_{i=1}^{n}$ by Σ, show that the solution of the normal equations in Problem 34 is

(a) $m = \dfrac{\Sigma x_i y_i - (1/n)(\Sigma x_i)(\Sigma y_i)}{\Sigma x_i^2 - (1/n)(\Sigma x_i)^2}$ (b) $b = \dfrac{\Sigma y_i - m\Sigma x_i}{n}$

36 In Problem 35, let $\bar{x} = (\Sigma x_i)/n$, and let $\bar{y} = (\Sigma y_i)/n$. Show that

(a) $m = \dfrac{\Sigma(x_i - \bar{x})(y_i - \bar{y})}{\Sigma(x_i - \bar{x})^2}$ (b) $b = \bar{y} - m\bar{x}$

37 Using the results of either Problem 35 or 36, find an equation $y = mx + b$ of the regression line for the scattergram consisting of the points $(1, 1)$, $(2, 3)$, $(3, 2)$, and $(4, 3)$. Sketch a graph showing the scattergram and the regression line.

© **38** The following table shows the cumulative grade point average (GPA) for eight randomly selected graduates of a school of business management and their corresponding salaries in thousands of dollars during their first year of employment after graduation.

GPA	1.2	2.3	2.7	3.0	3.0	3.4	3.5	4.0
Salary	17	16	20	19	21	20	25	29

With x as the GPA and y as the annual salary, plot a scattergram for these data, and find the regression line. Use the regression line to estimate the value of y when $x = 2.0$.

14.9 Lagrange Multipliers

In Section 14.8 we presented a routine for locating the extrema of functions of more than one variable. In this section we study the method of *Lagrange multipliers* for solving extremum problems that involve *constraints*.

A typical **constrained extremum problem** requires us to find the extrema of a function f of several variables when these variables are not independent, but satisfy one or more given conditions called **constraints.** The constraints are ordinarily specified by equations, called **constraint equations,** involving the variables in question. For instance, consider the following example.

EXAMPLE 1   Maximize the value of the function f given by

$$f(x, y, z) = xyz$$

subject to the constraint

$$g(x, y, z) = 42$$

where g is the function defined by $g(x, y, z) = x + y + z$.

SOLUTION   Here the constraint equation $g(x, y, z) = 42$, or $x + y + z = 42$, can be solved for z in terms of x and y to obtain

$$z = 42 - x - y$$

Therefore, the quantity to be maximized becomes

$$f(x, y, z) = f(x, y, 42 - x - y) = xy(42 - x - y)$$

If we let F be the function of x and y defined by

$$F(x, y) = xy(42 - x - y)$$

then our problem is simply to maximize $F(x, y)$, and we proceed as in Section 14.8.

Thus,

$$\frac{\partial F}{\partial x} = y(42 - x - y) + xy(-1) = y(42 - 2x - y)$$

and

$$\frac{\partial F}{\partial y} = x(42 - x - y) + xy(-1) = x(42 - x - 2y)$$

so that the critical points of F are given by the solutions of the simultaneous equations

$$\begin{cases} y(42 - 2x - y) = 0 \\ x(42 - x - 2y) = 0 \end{cases}$$

Solving these equations, we obtain the critical points $(0, 0)$, $(0, 42)$, $(42, 0)$, and $(14, 14)$. The second-derivative test shows a relative maximum only at $(14, 14)$. Actually, $x = 14$, $y = 14$ gives an absolute maximum value of

$$F(14, 14) = (14)(14)(42 - 14 - 14) = 14^3 \qquad \blacksquare$$

The technique used to solve the constrained extremum problems above can be generalized as follows: Suppose that we require the extrema of

$$f(x, y, z)$$

subject to the constraint

$$g(x, y, z) = k$$

where k is a constant and the functions f and g are differentiable. Assume that the constraint equation $g(x, y, z) = k$ can be solved for, say, z as a function of x and y, so that

$$z = h(x, y) \qquad \text{where } g(x, y, h(x, y)) = k$$

The quantity $f(x, y, z)$ whose extrema are sought can now be written as

$$f(x, y, z) = f(x, y, h(x, y))$$

If we define the function F of two variables by

$$F(x, y) = f(x, y, h(x, y))$$

then our problem is to find the extrema of F, and we proceed as follows.

Assuming that the function h has partial derivatives, we can apply the chain rule to the equation $F(x, y) = f(x, y, h(x, y))$ to obtain

$$\frac{\partial F}{\partial x} = \frac{\partial f}{\partial x}\frac{\partial x}{\partial x} + \frac{\partial f}{\partial y}\frac{\partial y}{\partial x} + \frac{\partial f}{\partial z}\frac{\partial h}{\partial x} = \frac{\partial f}{\partial x} + \frac{\partial f}{\partial z}\frac{\partial h}{\partial x}$$

and

$$\frac{\partial F}{\partial y} = \frac{\partial f}{\partial x}\frac{\partial x}{\partial y} + \frac{\partial f}{\partial y}\frac{\partial y}{\partial y} + \frac{\partial f}{\partial z}\frac{\partial h}{\partial y} = \frac{\partial f}{\partial y} + \frac{\partial f}{\partial z}\frac{\partial h}{\partial y}$$

Thus, the critical points of F are the solutions of the simultaneous equations

$$\begin{cases} \dfrac{\partial f}{\partial x} + \dfrac{\partial f}{\partial z}\dfrac{\partial h}{\partial x} = 0 \\[2mm] \dfrac{\partial f}{\partial y} + \dfrac{\partial f}{\partial z}\dfrac{\partial h}{\partial y} = 0 \end{cases}$$

Differentiating both sides of the equation $g(x, y, h(x, y)) = k$ with respect to x and with respect to y using the chain rule, we obtain a second pair of simultaneous equations:

$$\begin{cases} \dfrac{\partial g}{\partial x}\dfrac{\partial x}{\partial x} + \dfrac{\partial g}{\partial y}\dfrac{\partial y}{\partial x} + \dfrac{\partial g}{\partial z}\dfrac{\partial h}{\partial x} = 0 \\[2mm] \dfrac{\partial g}{\partial x}\dfrac{\partial x}{\partial y} + \dfrac{\partial g}{\partial y}\dfrac{\partial y}{\partial y} + \dfrac{\partial g}{\partial z}\dfrac{\partial h}{\partial y} = 0 \end{cases} \quad \text{or} \quad \begin{cases} \dfrac{\partial g}{\partial x} + \dfrac{\partial g}{\partial z}\dfrac{\partial h}{\partial x} = 0 \\[2mm] \dfrac{\partial g}{\partial y} + \dfrac{\partial g}{\partial z}\dfrac{\partial h}{\partial y} = 0 \end{cases}$$

Therefore the desired critical points of F are to be found among the solutions of the four simultaneous equations:

$$\begin{cases} \dfrac{\partial f}{\partial x} + \dfrac{\partial f}{\partial z}\dfrac{\partial h}{\partial x} = 0 \\[3mm] \dfrac{\partial f}{\partial y} + \dfrac{\partial f}{\partial z}\dfrac{\partial h}{\partial y} = 0 \\[3mm] \dfrac{\partial g}{\partial x} + \dfrac{\partial g}{\partial z}\dfrac{\partial h}{\partial x} = 0 \\[3mm] \dfrac{\partial g}{\partial y} + \dfrac{\partial g}{\partial z}\dfrac{\partial h}{\partial y} = 0 \end{cases}$$

Now, assume that $\partial g/\partial z \neq 0$, and define λ to be the ratio

$$\lambda = -\frac{\partial f/\partial z}{\partial g/\partial z}$$

(λ is the Greek letter "lambda"). Solving the third equation in the last set for $\partial h/\partial x$, we obtain

$$\frac{\partial h}{\partial x} = -\frac{\partial g/\partial x}{\partial g/\partial z}$$

Hence, substituting this value of $\partial h/\partial x$ into the first of the simultaneous equations, we have

$$\frac{\partial f}{\partial x} + \frac{\partial f}{\partial z}\left(-\frac{\partial g/\partial x}{\partial g/\partial z}\right) = 0 \quad \text{or} \quad \text{since } \lambda = -\frac{\partial f/\partial z}{\partial g/\partial z}$$

$$\frac{\partial f}{\partial x} + \lambda\frac{\partial g}{\partial x} = 0$$

Likewise, solving the fourth equation in the set above for $\partial h/\partial y$ and substituting into the second equation in the set, we obtain

$$\frac{\partial f}{\partial y} + \lambda\frac{\partial g}{\partial y} = 0$$

From these calculations, we conclude that the *critical points of F are among the solutions of the simultaneous equations*

$$\begin{cases} \dfrac{\partial f}{\partial x} + \lambda\dfrac{\partial g}{\partial x} = 0 \\[3mm] \dfrac{\partial f}{\partial y} + \lambda\dfrac{\partial g}{\partial y} = 0 \\[3mm] \dfrac{\partial f}{\partial z} + \lambda\dfrac{\partial g}{\partial z} = 0 \end{cases}$$

(The third equation comes directly from the definition of λ.) Notice that these three simultaneous equations can be written as the single vector equation

$$\nabla f + \lambda \nabla g = \mathbf{0} \qquad \text{or} \qquad \nabla f = -\lambda \nabla g$$

The argument above shows that at a point where $f(x, y, z)$ attains a relative extremum, subject to the constraint $g(x, y, z) = k$, *the gradient of f must be parallel to the gradient of g.* A geometric interpretation of this condition is discussed in Problem 19.

We can now formulate the main theorem of this section.

THEOREM 1　**Lagrange Method of Multipliers**

Assume that the functions f and g are defined and have continuous partial derivatives on the subset D of xyz space, where D consists entirely of interior points. Suppose that at each point (x, y, z) in D at least one of the three partial derivatives $g_1(x, y, z)$, $g_2(x, y, z)$, and $g_3(x, y, z)$ is different from zero. Then the points (x, y, z) in D at which f attains relative extrema subject to the constraint

$$g(x, y, z) = k$$

where k is a constant, can be found as follows:

Form the function u defined by

$$u(x, y, z) = f(x, y, z) + \lambda g(x, y, z)$$

for (x, y, z) in D, where λ (which is called the **Lagrange multiplier**) represents a constant yet to be determined. Then solve the simultaneous equations

$$\begin{cases} \dfrac{\partial u}{\partial x} = 0 \\[2mm] \dfrac{\partial u}{\partial y} = 0 \\[2mm] \dfrac{\partial u}{\partial z} = 0 \\[2mm] g(x, y, z) = k \end{cases}$$

for x, y, z, and λ. Several solutions may be obtained, but the desired points (x, y, z), where f attains its extrema subject to the constraint, are among these solutions.

INDICATION OF
THE PROOF

A rigorous proof requires techniques beyond the scope of this book; however, we give an indication of why the theorem is true. If $u(x, y, z) = f(x, y, z) + \lambda g(x, y, z)$, then

$$\nabla u = \nabla f + \lambda \nabla g$$

Hence, the simultaneous equations $\partial u/\partial x = 0$, $\partial u/\partial y = 0$, and $\partial u/\partial z = 0$ are equivalent to $\nabla u = \mathbf{0}$, that is, to

$$\nabla f = -\lambda \nabla g$$

However, the last equation means that ∇f is parallel to ∇g, a condition that we previously found must hold at a point (x, y, z) where $f(x, y, z)$ attains a relative extremum subject to the constraint $g(x, y, z) = k$.

In Examples 2 to 4, use the Lagrange method of multipliers to work the given constrained extremum problem.

<u>EXAMPLE 2</u>   Find the extrema of the function f given by $f(x, y, z) = x^2 + y^2 + z^2$ subject to the constraint $x^2 + 2y^2 - z^2 = 1$.

SOLUTION   We let g be the function defined by $g(x, y, z) = x^2 + 2y^2 - z^2$ and form the quantity

$$u = f(x, y, z) + \lambda g(x, y, z)$$

that is,

$$u = x^2 + y^2 + z^2 + \lambda(x^2 + 2y^2 - z^2)$$

as in Theorem 1. Here,

$$\frac{\partial u}{\partial x} = 2x + 2\lambda x = 2(1 + \lambda)x$$

$$\frac{\partial u}{\partial y} = 2y + 4\lambda y = 2(1 + 2\lambda)y$$

and

$$\frac{\partial u}{\partial z} = 2z - 2\lambda z = 2(1 - \lambda)z$$

After dividing by 2, we can write the simultaneous equations $\partial u/\partial x = 0$, $\partial u/\partial y = 0$, and $\partial u/\partial z = 0$ as $(1 + \lambda)x = 0$, $(1 + 2\lambda)y = 0$, and $(1 - \lambda)z = 0$. Appending the equation of constraint, we have

$$\begin{cases} (1 + \lambda)x = 0 \\ (1 + 2\lambda)y = 0 \\ (1 - \lambda)z = 0 \\ x^2 + 2y^2 - z^2 = 1 \end{cases}$$

Notice that $x = y = z = 0$ is not a solution of these simultaneous equations; hence, at least one of x, y, or z must be nonzero, and it follows that $1 + \lambda = 0$, or else $1 + 2\lambda = 0$, or else $1 - \lambda = 0$; that is, $\lambda = -1$, or else $\lambda = -\frac{1}{2}$, or else $\lambda = 1$. For $\lambda = -1$, the equations become

$$\begin{cases} 0 = 0 \\ -y = 0 \\ 2z = 0 \\ x^2 + 2y^2 - z^2 = 1 \end{cases}$$

so that $y = 0$, $z = 0$, and $x^2 = 1$. Thus, when $\lambda = -1$, we obtain the solutions $(1, 0, 0)$ and $(-1, 0, 0)$.

Similarly, when $\lambda = -\frac{1}{2}$, we get $x = 0$, $z = 0$, and $2y^2 = 1$; that is, we obtain the solutions $(0, \sqrt{2}/2, 0)$ and $(0, -\sqrt{2}/2, 0)$. Finally, when $\lambda = 1$, we get $x = 0$, $y = 0$, and $-z^2 = 1$. Since there are no real numbers z for which $-z^2 = 1$, there are no solutions corresponding to $\lambda = 1$. Thus, the critical points are $(\pm 1, 0, 0)$ and $(0, \pm\sqrt{2}/2, 0)$. At $(\pm 1, 0, 0)$, we have

$$f(\pm 1, 0, 0) = (\pm 1)^2 + 0^2 + 0^2 = 1$$

and at $(0, \pm\sqrt{2}/2, 0)$, we have

$$f(0, \pm\sqrt{2}/2, 0) = 0^2 + (\pm\sqrt{2}/2)^2 + 0^2 = \frac{1}{2}$$

Figure 1

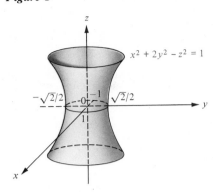

$x^2 + 2y^2 - z^2 = 1$

Since we have given no second-derivative test for constrained extrema problems, we must use algebraic or geometric means to decide whether the critical points obtained above correspond to relative maxima, relative minima, or saddle points.

Actually, $x^2 + 2y^2 - z^2 = 1$ is an equation of a hyperboloid of one sheet (Figure 1), and $f(x, y, z) = x^2 + y^2 + z^2$ gives the square of the distance from the origin to the point (x, y, z). Hence, it is geometrically evident that $(0, \pm\sqrt{2}/2, 0)$ are points where $f(x, y, z)$ takes on its absolute minimum value, namely $f(0, \pm\sqrt{2}/2, 0) = \frac{1}{2}$. Evidently, f has no absolute maximum, and $(\pm 1, 0, 0)$ are saddle points.　■

EXAMPLE 3　Find the dimensions of the rectangular box of largest volume which can be inscribed in the ellipsoid $(x^2/9) + (y^2/4) + z^2 = 1$, assuming that the edges of the box are parallel to the coordinate axes.

SOLUTION　Let (x, y, z) be the corner of the box in the first octant, so that the dimensions of the box are $2x$, $2y$, and $2z$ and its volume is given by

$$V = (2x)(2y)(2z) = 8xyz$$

Thus, we must maximize V subject to the constraint

$$\frac{x^2}{9} + \frac{y^2}{4} + z^2 = 1$$

We form the quantity

$$u = 8xyz + \lambda\left(\frac{x^2}{9} + \frac{y^2}{4} + z^2\right)$$

noting that

$$\frac{\partial u}{\partial x} = 8yz + \frac{2\lambda x}{9}$$

$$\frac{\partial u}{\partial y} = 8xz + \frac{2\lambda y}{4}$$

$$\frac{\partial u}{\partial z} = 8xy + 2\lambda z$$

Setting these partial derivatives equal to zero and appending the equation of constraint, we have

$$\begin{cases} 8yz + \dfrac{2}{9}\lambda x = 0 \\[2mm] 8xz + \dfrac{2}{4}\lambda y = 0 \\[2mm] 8xy + 2\lambda z = 0 \\[2mm] \dfrac{x^2}{9} + \dfrac{y^2}{4} + z^2 = 1 \end{cases}$$

Multiplying the first three equations by $x/2$, $y/2$, and $z/2$, respectively, we have

$$4xyz + \lambda\frac{x^2}{9} = 0$$

$$4xyz + \lambda\frac{y^2}{4} = 0$$

$$4xyz + \lambda z^2 = 0$$

Adding these three equations and using the fact that $(x^2/9) + (y^2/4) + z^2 = 1$, we obtain

$$12xyz + \lambda = 0$$

that is,

$$\lambda = -12xyz$$

Substituting this value of λ into the first three of the original simultaneous equations and simplifying yields

$$\begin{cases} yz(3 - x^2) = 0 \\ xz(4 - 3y^2) = 0 \\ xy(1 - 3z^2) = 0 \end{cases}$$

For a maximum volume, it is clear that x, y, and z must be positive; hence, we can cancel the factors yz, xz, and xy from the equations above to obtain

$$3 - x^2 = 0 \qquad 4 - 3y^2 = 0 \qquad \text{and} \qquad 1 - 3z^2 = 0$$

Therefore,

$$x = \sqrt{3} \qquad y = \frac{2}{\sqrt{3}} \qquad \text{and} \qquad z = \frac{1}{\sqrt{3}}$$

■

EXAMPLE 4 Find all critical points of the function f given by $f(x, y) = 3x^2 - 2xy + 5y^2$ subject to the constraint that $x^2 + 2y^2 = 6$.

SOLUTION   Here, f is a function of only two variables, but the method is still the same (Problem 18). We form the quantity

$$u = 3x^2 - 2xy + 5y^2 + \lambda(x^2 + 2y^2)$$

so that

$$\frac{\partial u}{\partial x} = 6x - 2y + 2\lambda x = 2[(3 + \lambda)x - y]$$

and

$$\frac{\partial u}{\partial y} = -2x + 10y + 4\lambda y = 2[-x + (5 + 2\lambda)y]$$

Thus, the desired critical points are found by solving the simultaneous equations

$$\begin{cases} (3 + \lambda)x - y = 0 \\ -x + (5 + 2\lambda)y = 0 \\ x^2 + 2y^2 = 6 \end{cases}$$

Multiplying the second equation by $3 + \lambda$ and adding it to the first, we cancel the x term and obtain

$$(2\lambda^2 + 11\lambda + 14)y = 0$$

Similarly, multiplying the first equation by $5 + 2\lambda$ and adding it to the second, we cancel the y term and obtain

$$(2\lambda^2 + 11\lambda + 14)x = 0$$

Since the equation of constraint, $x^2 + 2y^2 = 6$, must hold, we cannot have both x and y equal to zero; therefore, the coefficient $(2\lambda^2 + 11\lambda + 14)$ in the last two equations must vanish, and we have

$$2\lambda^2 + 11\lambda + 14 = 0 \qquad \text{that is} \qquad (2\lambda + 7)(\lambda + 2) = 0$$

It follows that

$$\lambda = -\tfrac{7}{2} \quad \text{or} \quad \lambda = -2$$

Putting $\lambda = -\tfrac{7}{2}$ in the first of the original simultaneous equations, $(3 + \lambda)x - y = 0$, we find that

$$y = -\frac{x}{2} \quad \text{when} \quad \lambda = -\frac{7}{2}$$

Substituting $y = -x/2$ into the constraint equation $x^2 + 2y^2 = 6$, we obtain

$$x^2 + \frac{x^2}{2} = 6 \quad \text{so} \quad x = \pm 2$$

When $x = 2$, $y = -2/2 = -1$; when $x = -2$, $y = -(-2/2) = 1$.
Similarly, putting $\lambda = -2$ in the equation $(3 + \lambda)x - y = 0$, we find that

$$y = x \quad \text{when} \quad \lambda = -2$$

Substituting $y = x$ into the constraint equation $x^2 + 2y^2 = 6$, we obtain

$$x^2 + 2x^2 = 6 \quad \text{so} \quad x = \pm\sqrt{2}$$

When $x = \sqrt{2}$, $y = \sqrt{2}$; when $x = -\sqrt{2}$, $y = -\sqrt{2}$. Therefore, the desired critical points are $(2, -1)$, $(-2, 1)$, $(\sqrt{2}, \sqrt{2})$, and $(-\sqrt{2}, -\sqrt{2})$. ■

The method of Lagrange multipliers is effective when there is more than one constraint; however, more than one multiplier must then be used. Although we give an indication of this technique in Problem 20, the details are best left to an advanced course.

Problem Set 14.9

In Problems 1 to 9, use the Lagrange multiplier method to find all critical points of each function f subject to the indicated constraint.

1 $f(x, y) = x^2 - y^2 - y$ with the constraint $x^2 + y^2 = 1$

2 $f(x, y) = 3x^2 + 2\sqrt{2}\,xy + 4y^2$ with the constraint $x^2 + y^2 = 9$

3 $f(x, y) = x^2 + y^2$ with the constraint $5x^2 + 6xy + 5y^2 = 8$

4 $f(x, y) = x^2 + y^2$ with the constraint $2x^2 + y^2 = 1$

5 $f(x, y) = x + y^2$ with the constraint $2x^2 + y^2 = 1$

6 $f(x, y, z) = x^2 + y^2 + z^2$ with the constraint

$$x^2 + \frac{y^2}{4} + \frac{z^2}{9} = 1$$

7 $f(x, y, z) = xyz$ with the constraint

$$x^2 + \frac{y^2}{12} + \frac{z^2}{3} = 1$$

8 $f(x, y, z) = x^2 + y^2 + z^2$ with the constraint $x + 3y - 2z = 4$

9 $f(x, y, z) = 3x^2 + 2y^2 + 4z^2$ with the following constraint: $2x + 4y - 6z = -5$

10 Use the Lagrange multiplier method to find a point on the plane $3x - 4y + z = 2$ which is closest to the origin.

11 The cost of audit for tax purposes of a certain organization depends on the number of audits x and y at each of two headquarters according to the formula $C(x, y) = 2x^2 + xy + y^2 + 100$. How many audits should be made at each of the headquarters in order to minimize cost if the total number of audits must be 16?

12 A sheet-metal container is to be made of a right circular cylinder with equal right circular conical caps on the ends. Show that for a fixed volume, the total surface area is the smallest when the length of the cylinder is the same as the altitude of each cone and the diameter of the cylinder is $\sqrt{5}$ times its length.

13 The Postal Service specifies that the sum of the length and the perimeter of a cross section of a rectangular box accepted for parcel post must not exceed 100 inches. What are the dimensions of such a box that enable you to send the most by parcel post?

14 A manufacturer wants to make a rectangular box to hold a fixed volume V cubic units of a product. The material used for the sides costs a dollars per square unit, the material used for the bottom costs b dollars per square unit, and the material used for the top costs c dollars per square unit. What are the dimensions of the cheapest box?

15 The base of a rectangular box, open at the top, costs half as much per square foot as the sides. Find the dimensions of the box of largest volume that can be made for a fixed cost if the height of the box is to be 2 feet.

16 Show that the extreme values of $Lx^2 + 2Mxy + Ny^2$, subject to the constraint $Ex^2 + 2Fxy + Gy^2 = 1$, where L, M, N, E, F, and G are constants and $EG - F^2 > 0$, are the two roots of the quadratic equation

$$(EG - F^2)t^2 - (LG - 2MF + NE)t + (LN - M^2) = 0$$

17 Suppose that we wish to find the critical values of the function f of two variables subject to the constraint $g(x, y) = k$. Assuming that f and g have continuous partial derivatives, that $g_2(x, y) \neq 0$, and that the equation $g(x, y) = k$ can be solved for y in terms of x, show that ∇f and ∇g are parallel at a critical point.

18 Using the result of Problem 17 as a guide, give a careful statement of a "Lagrange method of multipliers theorem" for functions of two variables.

19 In Figure 2, assume that the curve $g(x, y) = k$ lies in a scalar field given by $w = f(x, y)$. At the point (a, b) on the curve, the normal vector is $\nabla g(a, b)$. Suppose that $\nabla f(a, b)$ is not parallel to $\nabla g(a, b)$. In what direction should one move through (a, b)

along the curve in order to (a) increase the value of w and (b) decrease the value of w? Thus explain geometrically why at an extremum of f subject to the constraint $g(x, y) = k$, ∇f must be parallel to ∇g.

Figure 2

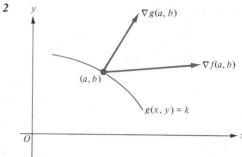

20 Suppose that we wish to find extrema of $f(x, y, z)$ subject to *two* constraints, namely $g(x, y, z) = k$ and $G(x, y, z) = K$. Show that at such an extremum, there must exist two constants λ and μ (the Lagrange multipliers!) such that

$$\nabla f + \lambda \nabla g + \mu \nabla G = 0$$

Assume that f and g have continuous partial derivatives and that the two simultaneous constraint equations can be solved for two of the variables x, y, and z in terms of the third.

Review Problem Set, Chapter 14

In Problems 1 to 4, find the domain of each function, sketch the domain, and identify the interior points of the domain.

1 $f(x, y) = \sqrt{64 - x^2 - y^2}$

2 $g(x, y) = \ln(x^2 + y^2 - 4)$

3 $g(x, y) = \dfrac{x^2y - x + 1}{1 - 2x - y}$

4 $f(x, y) = \dfrac{\sqrt{9 - x^2 - y^2}}{3y}$

5 Let f be a function of four variables defined by $f(x, y, z, w) = 3xyzw^2$. Find

(a) $f(1, -1, 2, 1)$ (b) $f(3, 2, -1, 2)$

(c) $f(a, b^2, c, d^3)$ (d) the domain of f

In Problems 6 to 8, evaluate the limit, if it exists.

6 $\displaystyle\lim_{(x,y)\to(0,0)} \dfrac{x^2 - y^2}{1 + x^2 + y^2}$

7 $\displaystyle\lim_{(x,y)\to(0,0)} \dfrac{6xy}{x^2 + y^2}$

8 $\displaystyle\lim_{(x,y)\to(0,0)} \dfrac{(1 + y^2) \sin x}{x}$

In Problems 9 and 10, show that each function is discontinuous at the point $(0, 0)$.

9 $f(x, y) = \begin{cases} \dfrac{x}{x - y} & \text{for } x \neq y \\ 0 & \text{for } x = y \end{cases}$

10 $f(x, y) = \begin{cases} \ln(x^2 + y^2) & \text{for } (x, y) \neq (0, 0) \\ 0 & \text{for } (x, y) = (0, 0) \end{cases}$

In Problems 11 to 24, find the first partial derivatives of each function.

11 $f(x, y) = x^3 + 7xy^2 - 8y^3$

12 $g(x, y) = y^2/(y - x)$

13 $f(x, y) = y^3e^x + x^3e^y$

14 $f(x, y) = x^4y - \sin(xy^4)$

15 $g(x, y) = \ln(x^3 - y^3)$

16 $h(x, y) = \displaystyle\int_x^y e^{-7t^2} \, dt$

17 $g(x, y) = \ln(x + \sqrt[3]{x^3 + y^3})$

18 $f(x, y) = \sin^{-1} \sqrt{1 - x^2 y^2}$

19 $g(x, y) = e^{-7x} \tan (x + y)$

20 $f(r, \theta, z) = zr^2 \sin \theta$

21 $g(x, y, z) = x^3 + y^3 + z^3 - 13x^2 y^2 z^2$

22 $h(x, y, z) = \tan \left(\dfrac{x}{y} + \dfrac{y}{z} + \dfrac{z}{x} \right)$

23 $f(x, y, z) = z \coth (xy)$

24 $g(x, y, z) = (\ln y)^x + \cos z$

25 Let $w = \dfrac{e^{x+y}}{e^x + e^y}$. Show that $\dfrac{\partial w}{\partial x} + \dfrac{\partial w}{\partial y} = w$.

26 If $w = x^2 y + y^2 z + xz^2$, show that

$$\frac{\partial w}{\partial x} + \frac{\partial w}{\partial y} + \frac{\partial w}{\partial z} = (x + y + z)^2$$

27 Let $w = \cos (5x - 8y + 7z)$ and $\mathbf{A} = \dfrac{\partial w}{\partial x}\mathbf{i} + \dfrac{\partial w}{\partial y}\mathbf{j} + \dfrac{\partial w}{\partial z}\mathbf{k}$. Show that $\mathbf{A}$ is perpendicular to $\mathbf{B} = 10\mathbf{i} + \mathbf{j} - 6\mathbf{k}$.

28 Find $(2x\mathbf{i} + y\mathbf{j} + 2z\mathbf{k}) \cdot \nabla f$ if $f(x, y, z) = x \sin \dfrac{y^2}{z} - y^2 \tan \dfrac{z}{y^2}$.

29 Let $f(x, y, z) = \tan y^2 e^{1/x}$. Verify that $(x^2\mathbf{i} + \frac{1}{2}y\mathbf{j}) \cdot \nabla f = 0$.

30 At the point $(2, \sqrt{5}, 4)$ on the sphere $x^2 + y^2 + z^2 = 25$, a tangent line is drawn parallel to the xz plane. Find the angle between this tangent line and the xy plane.

31 Find the equation of (a) the normal line and (b) the tangent plane to the surface $xyz = 8$ at the point $(2, 2, 2)$.

32 Two sides and the included angle of a triangle are given by x, y, and θ, respectively. Find the rate of change of its area with respect to each of these quantities when the other two are held constant.

© **33** Let $f(x, y) = x^2 y$. (a) Find df. (b) Use the result of part (a) to find an approximation for $(5.04)^2(2.98)$. (c) Use a calculator to check the accuracy of the approximation.

34 Find the linear approximation for the function f defined by

$$f(x, y) = \frac{xy - 1}{xy + 1}$$

near the point $(x_0, y_0) = (\frac{1}{2}, \frac{1}{4})$.

© **35** A solid metal cylinder with radius 4.02 and height 8.03 centimeters is cut down to a cylinder of radius 4 and height 8 centimeters. Approximately how much metal is removed?

36 A fence 4 feet high runs parallel to the wall of a building and 3 feet from it. A person standing at a window in the building looks directly over the fence at a point P on the ground. The person's eyes are 8 feet above the ground. If the fence were

3 inches higher and 4 inches farther from the building, approximately how much farther would the new point P be (a) from the building and (b) from the person's eyes?

37 If $r = \sqrt{x^2 + y^2 + z^2}$ and $\mathbf{R} = x\mathbf{i} + y\mathbf{j} + z\mathbf{k}$, show that (a) $\mathbf{R} \cdot \nabla r = r$ and (b) $\mathbf{R} \cdot \nabla(r^2) = 2r^2$.

38 Suppose that $y = r \sin \theta$. If erroneous values are used for r and θ, explain why the approximate error in the resulting value of y should not exceed $|r \cos \theta| |\Delta\theta| + |\sin \theta| |\Delta r|$, where $\Delta\theta$ and Δr are the errors in θ and r, respectively, provided that the errors are small.

39 Compute $d\rho$ if $\rho = e^{\theta/2} \sin (\theta - \phi)$, $\theta = 0$, $\phi = \pi/2$, $d\theta = 0.2$, and $d\phi = -0.2$.

In Problems 40 to 46, use the chain rule to find the indicated derivative(s).

40 $w = x^3 + 5xy - y^3$, $x = r^2 + s$, $y = r - s^2$; $\partial w/\partial r$ and $\partial w/\partial s$

41 $w = 2x^4 - 3x^2 y^2 + y^4$, $x = 3u + v$, $y = u - 2v$; $\partial w/\partial u$ and $\partial w/\partial v$

42 $f(u, v) = \cos (u + v)$, $g(x, y) = x^2 + y^2$, $h(x, y) = x^2 - y^2$, $F(x, y) = f(g(x, y), h(x, y))$; F_1 and F_2

43 $w = \displaystyle\int_x^y e^{t^4} \, dt$, $x = 2r + s$, $y = r - 3s$; $\partial w/\partial r$ and $\partial w/\partial s$

44 $u = f(x - y, x + y)$; $\partial u/\partial x$ and $\partial u/\partial y$

45 $f(u, v) = \cos uv$, $g(x) = \sqrt[3]{x}$, $h(x) = \sqrt[4]{x}$, $F(x) = f(g(x), h(x))$; F'

46 $w = \tan^{-1} uv$, $u = e^x$, $v = e^{-5x}$; dw/dx

47 Let f be the function defined by $f(x, y) = e^{-xy}$, and let g and h be functions such that $g(3) = 5$, $g'(3) = -2$, $h(3) = 4$, and $h'(3) = 7$. If $K(t) = f(g(t), h(t))$, find $K'(3)$.

48 If f is a differentiable function and

$$w = f\left(\frac{y - x}{xy}, \frac{z - x}{xy} \right)$$

find the total differential dw.

49 Suppose that $w = f(x, y)$, $x = r \cos \theta$, and $y = r \sin \theta$. If $f_x(0, 1) = 2$ and $f_y(0, 1) = -3$, find

(a) $\dfrac{\partial w}{\partial r}$ when $(r, \theta) = \left(1, \dfrac{\pi}{2} \right)$

(b) $\dfrac{\partial w}{\partial \theta}$ when $(r, \theta) = \left(1, \dfrac{\pi}{2} \right)$

50 If f is a differentiable function of two variables such that $f(1, 1) = 1$, $f_1(1, 1) = a$, and $f_2(1, 1) = b$, and if $g(x) = f(x, f(x, f(x, x)))$, find (a) $g(1)$ and (b) $g'(1)$.

51 If f is the function defined by $f(x, y) = x \sin y$, find (a) $f_1(y, x)$ and (b) $f_2(y, x)$.

52 If $w = 4x^3 + 3x^2y - 3zy^2$, find the directional derivative of w in the direction of the vector

$$\mathbf{u} = \frac{\sqrt{3}}{2\sqrt{2}}\mathbf{i} + \frac{1}{2\sqrt{2}}\mathbf{j} - \frac{1}{\sqrt{2}}\mathbf{k}$$

at the point $(1, -1, 3)$.

53 (a) Find the directional derivative of $w = x^2y$ at the point $(1, -3)$ in the direction of $\mathbf{u} = [\cos(5\pi/6)]\mathbf{i} + [\sin(5\pi/6)]\mathbf{j}$. (b) What is the maximum value of the directional derivative of $w = x^2y$ at $(1, -3)$, and in what direction is it attained?

54 Find the angle θ such that the directional derivative of $w = x^2 + \frac{1}{4}y^2$ at the point $(1, 2)$ is a maximum in the direction θ, and find this maximum value.

55 Find dy/dx if $y \sin x - x \cos y = 0$.

56 If f is a differentiable function and $f(x, y) = 0$, find the value of dy/dx when $x = 3$ and $y = -7$ if $f_1(3, -7) = 2$ and $f_2(3, -7) = 5$.

57 Sand is being poured on a conical pile at the rate of 4 cubic feet per minute. At a given instant, the pile is 6 feet in diameter and 5 feet high. At what rate is the height increasing at this instant if the diameter is increasing at the rate of 2 inches per minute?

58 A piece of copper in the form of a rectangular parallelepiped has edges of lengths 2, 4, and 8 centimeters. Because of heating, the edges are increasing at 0.001 centimeter per minute. At what rate is the volume changing?

In Problems 59 to 62, find the equation of the tangent plane and the normal line to each surface at the point indicated.

59 $z^3 + y^3 + x^3 - 3xyz = 8$; $(3, 3, 2)$

60 $\dfrac{x^2}{16} + z = \dfrac{y^2}{9}$; $(15, \frac{75}{4}, 25)$

61 $z^3 + 3xz - 2y = 0$; $(1, 7, 2)$

62 $5x^2 + 4y^2 + 2z^2 = 17$; $(-1, 1, 2)$

63 Let $w = e^{xy}$, $x = s^2 + 2st$, and $y = 2st + t^2$. Find

(a) $\dfrac{\partial^2 w}{\partial s^2}$ (b) $\dfrac{\partial^2 w}{\partial t^2}$

64 Let $w = f(u, v)$, $u = g(x, y)$ and $v = h(x, y)$. Assume that f, g, and h have continuous second partial derivatives and that the conditions $\dfrac{\partial u}{\partial x} = \dfrac{\partial v}{\partial y}$ and $\dfrac{\partial u}{\partial y} = -\dfrac{\partial v}{\partial x}$ hold. Show that

$$\frac{\partial^2 w}{\partial x^2} + \frac{\partial^2 w}{\partial y^2} = \left[\left(\frac{\partial u}{\partial x}\right)^2 + \left(\frac{\partial v}{\partial x}\right)^2\right] \cdot \left(\frac{\partial^2 w}{\partial u^2} + \frac{\partial^2 w}{\partial v^2}\right)$$

In Problems 65 to 68, find the relative maxima and minima for each function.

65 $f(x, y) = x^2 - y^2 + 2x - 4y + 3$

66 $g(x, y) = x^3 - 4y^2$

67 $h(x, y) = xy(3 - x - y)$

68 $F(x, y) = 3x^2 + 2y^2 + 3xy - 66x - 58y + 1600$

In Problems 69 to 71, use the method of Lagrange multipliers to find the critical points of each function subject to the indicated constraint. In each case, determine whether the critical point corresponds to a relative (or absolute) maximum or minimum or whether it is a saddle point.

69 $f(x, y) = x^2 + y^2$ with the constraint $x + y - 1 = 0$

70 $f(x, y, z) = x^2 + y^2 + z^2$ with the following constraint: $ax + by + cz + d = 0$

71 $f(x, y) = x + y$ with the constraint $x^2 + y^2 = 1$

72 A rectangular sheet-metal tank is open at the top and is filled with 9 cubic meters of liquid. Find the dimensions of the tank so that the metal surface in contact with the liquid is minimal.

73 Show that a rectangular box (with top) made out of S square units of material has maximum volume when it is a cube.

74 There will be N lottery tickets sold at a price of p dollars per ticket provided that A dollars is spent for publicity. Every x-dollar increase in the price per ticket will result in Bx fewer tickets being sold. Apart from publicity costs, a fixed overhead of K dollars is required to operate the lottery. Total prize money distributed to winners of the lottery will be $100k$ percent of the revenue from ticket sales. It is estimated that each additional dollar spent for publicity will result in C additional tickets being sold, regardless of the price per ticket. Find (a) the price per ticket and (b) the expenditure for publicity that will maximize the profit from the lottery. Also find (c) the number of tickets sold when the maximum profit is realized and (d) the maximum profit. (Your answers will be in terms of the constants p, A, B, K, k, and C.)

C **75** A manufacturer produces razors and razor blades at a cost of \$0.60 per razor and \$0.30 per dozen for the blades. If she charges x cents per razor and y cents per dozen for the blades, she will sell $6 \times 10^8/(x^2y)$ razors and $48 \times 10^7/(xy^2)$ dozen blades per day. What should she charge for razors and for blades so as to maximize her profit?

C In Problems 76 to 80, find the regression line for each set of points.

76 $(1, 6), (3, 0), (6, 10)$ **77** $(1, 1), (3, 2), (5, 4)$

78 $(0, 7), (-1, 5), (1, 9), (3, 10)$

79 $(-1, 1), (4, -2), (3, 5), (6, 7)$

80 $(1, 38), (2, 40), (3, 39), (5, 45)$

15
MULTIPLE INTEGRATION AND VECTOR CALCULUS

In Chapter 5 we introduced and studied the definite (Riemann) integral for a function of a single variable. In this chapter we extend the notion of a definite integral to functions of two or more variables in a natural and useful way to obtain *multiple integrals*. We find that most of the multiple integrals encountered in elementary applications to geometry and the physical sciences can be evaluated in terms of *iterated integrals*—that is, repeated definite integrals—in the Cartesian, polar, cylindrical, or spherical coordinate system. The chapter also includes *line integrals, surface integrals, Green's theorem, Stokes' theorem,* and *the divergence theorem of Gauss.*

15.1   Iterated Integrals

In Section 14.3 we calculated partial derivatives of functions of several variables by regarding all but one of the independent variables as being constant and differentiating with respect to the remaining variable. Likewise, it is possible to take an indefinite *integral* of such a function with respect to one of its variables, while temporarily regarding the remaining variables as being held constant. For instance, holding y constant, we have

$$\int x^2y^3 \, dx = y^3 \int x^2 \, dx = y^3\frac{x^3}{3} + C$$

and holding x constant, we have

$$\int x^2y^3 \, dy = x^2 \int y^3 \, dy = x^2\frac{y^4}{4} + K$$

Notice how the variable of integration is clearly indicated by the differential dx or dy under the integral sign.

In the calculation of $\int x^2y^3 \, dx$, we temporarily held y constant; however, for different fixed values of y we could use different values of the constant of integration C. The possible dependence of C on y can be indicated by writing $C(y)$ rather than C; that is, we can regard the "constant" of integration as a function of y and write

$$\int x^2 y^3 \, dx = \frac{x^3 y^3}{3} + C(y)$$

Similarly, when integrating with respect to y, we should write

$$\int x^2 y^3 \, dy = \frac{x^2 y^4}{4} + K(x)$$

The indefinite integrals above are just the analogs for antidifferentiation of partial derivatives for differentiation, and they could be called "partial antiderivatives." Thus, the equation

$$\int x^2 y^3 \, dx = \frac{x^3 y^3}{3} + C(y)$$

means that

$$\frac{\partial}{\partial x}\left[\frac{x^3 y^3}{3} + C(y)\right] = x^2 y^3$$

and the equation

$$\int x^2 y^3 \, dy = \frac{x^2 y^4}{4} + K(x)$$

means that

$$\frac{\partial}{\partial y}\left[\frac{x^2 y^4}{4} + K(x)\right] = x^2 y^3$$

EXAMPLE 1   If $f(x, y) = x \cos y$, find $\int f(x, y) \, dx$ and $\int f(x, y) \, dy$.

SOLUTION $\displaystyle \int f(x, y) \, dx = \int x \cos y \, dx = \cos y \int x \, dx = \frac{x^2}{2} \cos y + C(y)$

$\displaystyle \int f(x, y) \, dy = \int x \cos y \, dy = x \int \cos y \, dy = x \sin y + K(x)$ ∎

Now suppose that f is a function of two variables such that for each *fixed* value of $y, f(x, y)$ is an integrable function of x. Then, *for each fixed value of y,* we can form the definite integral

$$\int_{x=a}^{x=b} f(x, y) \, dx$$

Furthermore, for different fixed values of y, we can use different limits of integration a and b; that is, *a and b can depend on y.* Such dependence can be indicated by the usual function notation, and the integral becomes

$$\int_{x=a(y)}^{x=b(y)} f(x, y) \, dx$$

EXAMPLE 2   Evaluate $\displaystyle \int_{x=\ln y}^{x=y^2} y e^{xy} \, dx$ for $y > 0$.

SOLUTION   Holding y temporarily constant and integrating with respect to x, we obtain

$$\int ye^{xy}\, dx = y\int e^{xy}\, dx = y\left(\frac{e^{xy}}{y}\right) + C(y) = e^{xy} + C(y)$$

Therefore,

$$\int_{x=\ln y}^{x=y^2} ye^{xy}\, dx = [e^{xy} + C(y)]\Big|_{x=\ln\, y}^{x=y^2}$$

$$= [e^{y^2 y} + C(y)] - [e^{(\ln y)y} + C(y)]$$

$$= e^{y^3} - e^{y\ln y} + C(y) - C(y) = e^{y^3} - y^y \qquad \blacksquare$$

In the preceding example, notice that the constant of integration, $C(y)$, cancels out as usual during the definite integration. Therefore, when dealing with *definite* integrals, there is no necessity to write the constant of integration at all. Also observe that the integration takes place *with respect to* x; hence, the limits of integration must be *substituted for* x after the indefinite integration is performed. Thus, there is no necessity for writing the limits of integration in the form

$$\int_{x=\ln y}^{x=y^2} ye^{xy}\, dx$$

and the abbreviated form

$$\int_{\ln y}^{y^2} ye^{xy}\, dx$$

is quite acceptable. Similar remarks apply to integration with respect to y.

<u>**EXAMPLE 3**</u>   Evaluate $\displaystyle\int_x^{\pi x} \cos xy\, dy$.

SOLUTION   The differential dy indicates that the integration is to be performed with respect to y while holding x constant and that the limits of integration are to be substituted for y. Thus, for $x \neq 0$,

$$\int_x^{\pi x} \cos xy\, dy = \left(\frac{\sin xy}{x}\right)\Big|_{y=x}^{y=\pi x} = \frac{\sin \pi x^2}{x} - \frac{\sin x^2}{x}$$

For $x = 0$, we have

$$\int_x^{\pi x} \cos xy\, dy = \int_0^0 1\, dy = 0 \qquad \blacksquare$$

Notice that the quantity

$$\int_{g(x)}^{h(x)} f(x,\, y)\, dy = \int_{y=g(x)}^{y=h(x)} f(x,\, y)\, dy$$

depends only on x. Similarly, the quantity

$$\int_{a(y)}^{b(y)} f(x,\, y)\, dx = \int_{x=a(y)}^{x=b(y)} f(x,\, y)\, dx$$

depends only on y. Consequently, we can define functions F and G of the single variables x and y, respectively, by

$$F(x) = \int_{y=g(x)}^{y=h(x)} f(x,\, y)\, dy \qquad \text{and} \qquad G(y) = \int_{x=a(y)}^{x=b(y)} f(x,\, y)\, dx$$

In many cases, the functions F and G are themselves integrable, and we can write

$$\int_{x=c}^{x=d} F(x)\,dx = \int_{x=c}^{x=d} \left[\int_{y=g(x)}^{y=h(x)} f(x,\,y)\,dy \right] dx$$

and

$$\int_{y=c}^{y=d} G(y)\,dy = \int_{y=c}^{y=d} \left[\int_{x=a(y)}^{x=b(y)} f(x,\,y)\,dx \right] dy$$

These are called **iterated** (or **repeated**) integrals and are customarily written without the brackets and with the abbreviated notation for the limits of integration:

$$\int_c^d \int_{g(x)}^{h(x)} f(x,\,y)\,dy\,dx = \int_{x=c}^{x=d} \left[\int_{y=g(x)}^{y=h(x)} f(x,\,y)\,dy \right] dx$$

and

$$\int_c^d \int_{a(y)}^{b(y)} f(x,\,y)\,dx\,dy = \int_{y=c}^{y=d} \left[\int_{x=a(y)}^{x=b(y)} f(x,\,y)\,dx \right] dy$$

In order to evaluate an iterated integral, you must first perform an "inside integration" and then an "outside integration" as indicated by the following diagrams:

$$\int_c^d \int_{g(x)}^{h(x)} f(x,\,y)\,dy\,dx \qquad \int_c^d \int_{a(y)}^{b(y)} f(x,\,y)\,dx\,dy$$

The following examples illustrate the technique.

In Examples 4 to 6, evaluate the iterated integrals.

__EXAMPLE 4__ $\displaystyle \int_{-1}^{2} \int_0^2 x^2 y^3\,dy\,dx$

SOLUTION

$$\int_{-1}^{2} \int_0^2 x^2 y^3\,dy\,dx = \int_{x=-1}^{x=2} \left(\int_{y=0}^{y=2} x^2 y^3\,dy \right) dx = \int_{x=-1}^{x=2} \left(\frac{x^2 y^4}{4} \Big|_{y=0}^{y=2} \right) dx$$

$$= \int_{x=-1}^{x=2} \left(\frac{16x^2}{4} - 0 \right) dx = \frac{4}{3} x^3 \Big|_{-1}^{2} = \frac{32}{3} - \left(-\frac{4}{3} \right) = 12 \quad \blacksquare$$

__EXAMPLE 5__ $\displaystyle \int_0^4 \int_0^{3x/2} \sqrt{16 - x^2}\,dy\,dx$

SOLUTION $\displaystyle \int_0^4 \int_0^{3x/2} \sqrt{16 - x^2}\,dy\,dx = \int_{x=0}^{x=4} \left(\int_{y=0}^{y=3x/2} \sqrt{16 - x^2}\,dy \right) dx$

$$= \int_{x=0}^{x=4} \left(\sqrt{16 - x^2} \int_{y=0}^{y=3x/2} dy \right) dx$$

$$= \int_{x=0}^{x=4} \sqrt{16 - x^2} \left(y \Big|_{y=0}^{y=3x/2} \right) dx$$

$$= \int_0^4 \sqrt{16 - x^2} \left(\frac{3x}{2} \right) dx$$

The last integral can be evaluated by using the substitution $u = 16 - x^2$, so that $du = -2x\,dx$, $x\,dx = -\frac{1}{2}\,du$, and $(3x/2)\,dx = -\frac{3}{4}\,du$. Since $u = 16$ when $x = 0$ and $u = 0$ when $x = 4$, we have

$$\int_0^4 \sqrt{16 - x^2}\left(\frac{3x}{2}\right) dx = \int_{16}^0 \sqrt{u}\left(-\frac{3}{4}\right) du = \frac{3}{4}\int_0^{16} \sqrt{u}\,du$$

$$= \frac{3}{4}\left(\frac{2}{3}\,u^{3/2}\Big|_0^{16}\right) = 32 \qquad \blacksquare$$

<u>EXAMPLE 6</u> $\displaystyle\int_0^\pi \int_0^{y^2} \sin\frac{x}{y}\,dx\,dy$

SOLUTION

$$\int_0^\pi \int_0^{y^2} \sin\frac{x}{y}\,dx\,dy = \int_0^\pi \left(\int_0^{y^2} \sin\frac{x}{y}\,dx\right) dy = \int_0^\pi \left(-y\cos\frac{x}{y}\Big|_0^{y^2}\right) dy$$

$$= \int_0^\pi \left(-y\cos\frac{y^2}{y} + y\cos\frac{0}{y}\right) dy = \int_0^\pi (y - y\cos y)\,dy$$

$$= \int_0^\pi y\,dy - \int_0^\pi y\cos y\,dy = \frac{y^2}{2}\Big|_0^\pi - (y\sin y + \cos y)\Big|_0^\pi$$

$$= \frac{\pi^2}{2} - (\pi\sin\pi + \cos\pi) + (0\sin 0 + \cos 0) = \frac{\pi^2}{2} + 2$$

(The integral $\int_0^\pi y\cos y\,dy$ was evaluated by using integration by parts.) $\blacksquare$

Problem Set 15.1

In Problems 1 to 6, evaluate each integral.

1 $\displaystyle\int \sin(xy)\,dx$

2 $\displaystyle\int \frac{dy}{x^2 + y^2}$

3 $\displaystyle\int \left(x\sqrt{y} - \frac{y}{x} + \frac{1}{x\sqrt{y}}\right) dy$

4 $\displaystyle\int \sqrt{y^2 - x^2}\,dx$

5 $\displaystyle\int_{x=\pi/2}^{x=y^3} y^2 \sin x\,dx$

6 $\displaystyle\int_{y=2}^{y=\ln x} y e^{xy}\,dy$

In Problems 7 to 42, evaluate each iterated integral.

7 $\displaystyle\int_0^1 \int_0^2 x^4 y^2\,dy\,dx$

8 $\displaystyle\int_0^2 \int_0^3 y^3\,dy\,dx$

9 $\displaystyle\int_0^1 \int_0^2 (x + 3y^2)\,dx\,dy$

10 $\displaystyle\int_0^1 \int_1^y \left[2\left(\frac{y}{x}\right)^2 + 2x\right] dx\,dy$

11 $\displaystyle\int_0^3 \int_0^1 x e^{xy}\,dy\,dx$

12 $\displaystyle\int_0^4 \int_0^1 y e^{xy}\,dx\,dy$

13 $\displaystyle\int_0^{\pi/2} \int_0^1 xy\cos(xy^2)\,dy\,dx$

14 $\displaystyle\int_0^\pi \int_0^{\pi/2} \sin x\cos y\,dx\,dy$

15 $\displaystyle\int_{-3}^5 \int_0^1 \frac{y^2}{1 + x^2}\,dx\,dy$

16 $\displaystyle\int_0^1 \int_1^2 u e^v\,dv\,du$

17 $\displaystyle\int_0^2 \int_0^{\sqrt{y}} x^3\,dx\,dy$

18 $\displaystyle\int_0^1 \int_0^{x^4} \sin(\pi x^5)\,dy\,dx$

19 $\displaystyle\int_0^1 \int_0^{1-x^2} x\sqrt{y}\,dy\,dx$

20 $\displaystyle\int_0^2 \int_0^y (y - x)^4\,dx\,dy$

21 $\displaystyle\int_0^2 \int_{x^3}^{4x} (1 + 2y)\,dy\,dx$

22 $\displaystyle\int_0^4 \int_{y^2/4}^{2\sqrt{y}} (1 + 2xy)\,dx\,dy$

23 $\displaystyle\int_1^5 \int_{\sqrt[3]{x}}^x \frac{1}{x}\,dy\,dx$

24 $\displaystyle\int_1^4 \int_{x/3}^{x/27} \sqrt[3]{\frac{x}{3y}}\,dy\,dx$

25 $\displaystyle\int_0^{\pi/3} \int_0^{\sin x} \frac{dy\,dx}{\sqrt{1 - y^2}}$

26 $\displaystyle\int_1^4 \int_1^t s^2 \ln t^4\,ds\,dt$

27 $\displaystyle\int_0^1 \int_0^r \sqrt{1 - r^2}\,d\theta\,dr$

28 $\displaystyle\int_0^{\pi/2} \int_0^{\sin\theta} r\theta\,dr\,d\theta$

29 $\displaystyle\int_0^1 \int_{4u}^{6u} e^{u+v}\,dv\,du$

30 $\displaystyle\int_0^{\sqrt{\ln 2}} \int_0^y xy^5 e^{x^2 y^2}\,dx\,dy$

31 $\displaystyle\int_0^{\pi/2} \int_0^{2\cos\theta} \phi\cos\theta\,d\phi\,d\theta$

32 $\displaystyle\int_0^{\pi/6} \int_0^{\sec y\tan y} x^3\cos^4 y\,dx\,dy$ **33** $\displaystyle\int_1^2 \int_0^{x^2} e^{y/x}\,dy\,dx$

34 $\displaystyle\int_0^{\pi/4}\int_0^{\sin x} 4e^{-y}\cos x\, dy\, dx$

35 $\displaystyle\int_0^{\pi/2}\int_4^{4+3\cos t} s\, ds\, dt$

36 $\displaystyle\int_0^{\pi}\int_0^{3\cos\phi}\theta\sin\phi\, d\theta\, d\phi$

37 $\displaystyle\int_0^{\sqrt{\pi}}\int_0^{r^2} r\sin\theta\, d\theta\, dr$

38 $\displaystyle\int_0^1\int_v^{\sqrt{v}}(u^2+v^2)\, du\, dv$

39 $\displaystyle\int_0^1\int_0^{2x} ye^{x^3}\, dy\, dx$

40 $\displaystyle\int_0^{\pi}\int_0^{\sin x} y\, dy\, dx$

41 $\displaystyle\int_1^{e^2}\int_0^{1/y} e^{xy}\, dx\, dy$

42 $\displaystyle\int_{\ln(\pi/6)}^{\ln(\pi/2)}\int_0^{e^{\theta}}\cos e^{\theta}\, dr\, d\theta$

43 Let a, b, c, and d be constants, and let f, F, and ϕ be functions of two variables such that $\partial F(x,y)/\partial x = f(x,y)$ and $\partial\phi(x,y)/\partial y = F(x,y)$. Show that

$$\int_a^b\int_c^d f(x,y)\, dx\, dy = \phi(d,b) - \phi(d,a) - \phi(c,b) + \phi(c,a)$$

44 With the notation of Problem 43 and by making the necessary assumptions about the existence and continuity of the required partial derivatives, show that

$$\int_a^b\int_c^d f(x,y)\, dx\, dy = \int_c^d\int_a^b f(x,y)\, dy\, dx$$

15.2   The Double Integral

Figure 1

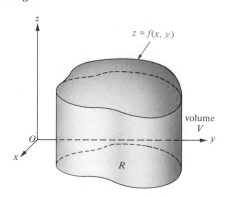

In Section 5.2 we defined the definite (Riemann) integral of a function f over a closed interval $[a, b]$ as a limit of Riemann sums. This definition was suggested by the problem of calculating the area under the graph of f between $x = a$ and $x = b$. Now, if f is a function of *two* variables and R is a plane region contained in the domain of f, we can formulate an analogous problem in three-dimensional space by asking for the volume V shown in Figure 1. Thus, if $f(x, y) \geq 0$ for (x, y) in R, we are asking for the volume of the solid that is bounded above by the graph of f, below by the region R, and laterally by the cylinder over the boundary of R whose generators are parallel to the z axis. We speak of this solid as "the solid below the graph of f and above the region R."

In this section we attack the problem of finding the volume V in the same way that we attacked the analogous two-dimensional problem in Chapter 5, namely, by setting up better and better approximations to V (called *Riemann sums*) and obtaining V as a limit of such approximations. This limit is called the *double integral of f over the region R* and is written as

$$\iint\limits_R f(x, y)\, dA$$

Figure 2

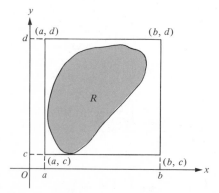

The problem of defining the double integral in all generality is best left to more advanced courses. For our purposes we assume that R is an admissible two-dimensional region (Section 5.6), that R contains all its own boundary points, and that f is a continuous function on R. We partition R into small rectangles in much the same way that we partitioned the interval of integration into small subintervals in Chapter 5; however, we consider only "regular" partitions in which all the small rectangles are congruent.

Since R is an admissible region, it is bounded and can therefore be enclosed in a rectangle $a \leq x \leq b$, $c \leq y \leq d$ in the xy plane (Figure 2). Given a positive integer n, we partition this rectangle into n^2 congruent subrectangles as follows:

1 Divide the interval $[a, b]$ into n subintervals of equal length $\Delta x = (b - a)/n$ by means of the points $x_0 = a$, $x_1 = x_0 + \Delta x$, $x_2 = x_1 + \Delta x$, ..., $x_n = x_{n-1} + \Delta x = b$.

Figure 3

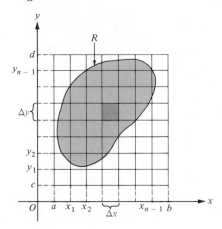

2 Divide the interval $[c, d]$ into n subintervals of equal length $\Delta y = (d - c)/n$ by means of the points $y_0 = c$, $y_1 = y_0 + \Delta y$, $y_2 = y_1 + \Delta y$, . . . , $y_n = y_{n-1} + \Delta y = d$.

3 Within the rectangle $a \le x \le b$, $c \le y \le d$, form a grid consisting of vertical line segments $x = x_0$, $x = x_1$, $x = x_2$, . . . , $x = x_n$ and horizontal line segments $y = y_0$, $y = y_1$, $y = y_2$, . . . , $y = y_n$ (Figure 3). This grid divides the rectangle into n^2 congruent subrectangles, each of which has area $\Delta A = \Delta x \, \Delta y$. One of these subrectangles is shown shaded in Figure 3.

We call this decomposition of the rectangle $a \le x \le b$, $c \le y \le d$ into n^2 congruent subrectangles a **regular partition,** and we refer to each of the n^2 subrectangles as a **cell** of the partition. Some of these cells may be contained in the region R, others may lie outside, and still others may extend across part of the boundary of R. We now discard all cells that do not touch the region R and number the remaining cells (that do touch R) in some convenient manner, calling them (say) ΔR_1, ΔR_2, ΔR_3, . . . , ΔR_m. Of course, each of these cells has an area $\Delta A = \Delta x \, \Delta y$, and when taken together, they contain the region R and approximate its shape and its area (Figure 4). As n increases, the grid becomes finer and the approximation improves.

By analogy with the notion of an augmented partition considered in Chapter 5, we now choose a point inside each of the cells ΔR_1, ΔR_2, . . . , ΔR_m, making certain that each chosen point belongs to the region R. For definiteness, denote the point chosen from the kth cell ΔR_k by (x_k^*, y_k^*) for $k = 1, 2, . . . , m$.

Now, consider the solid below the graph of f and above the cell ΔR_k (Figure 5), noting that this solid is approximately a rectangular parallelepiped with base ΔR_k of area ΔA and with height $f(x_k^*, y_k^*)$. Its volume is approximately

$$f(x_k^*, y_k^*) \, \Delta A$$

Adding the approximate volumes corresponding to each cell ΔR_1, ΔR_2, . . . , ΔR_m, we obtain the approximation

$$V \approx \sum_{k=1}^{m} f(x_k^*, y_k^*) \, \Delta A$$

for the total volume below the graph of f and above the region R.

By analogy with the terminology introduced in Chapter 5, the sum

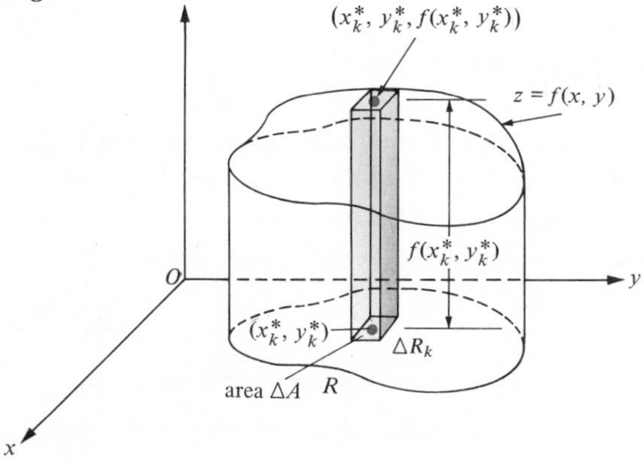

$$\sum_{k=1}^{m} f(x_k^*, y_k^*) \, \Delta A$$

is called a **Riemann sum** corresponding to the given regular augmented partition. The limit of such Riemann sums, as the partition becomes finer and finer (in the present case, as $n \to +\infty$), is called the **double integral of f over R** and is written as $\displaystyle\iint_R f(x, y) \, dA$. Thus, by definition,

Figure 4

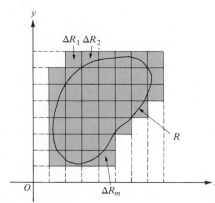

Figure 5

$$V = \iint_R f(x, y) \, dA = \lim_{n \to +\infty} \sum_{k=1}^{m} f(x_k^*, y_k^*) \, \Delta A$$

provided that the limit exists. If we write

$$dA = dx\, dy$$

as suggested by the fact that $\Delta A = \Delta x\, \Delta y$, then we can write the double integral of f over R in the alternative form:

$$\iint\limits_{R} f(x,\, y)\, dA = \iint\limits_{R} f(x,\, y)\, dx\, dy$$

However, we usually prefer to avoid this notation since it is so easily confused with the notation for iterated integrals introduced in Section 15.1.

Figure 6

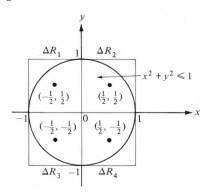

EXAMPLE 1   Approximate the double integral $\displaystyle\iint\limits_{R} x^2 y^4\, dA$, where R is the region inside the circle $x^2 + y^2 = 1$. Use the regular partition of the rectangle $-1 \le x \le 1,\ -1 \le y \le 1$ into four congruent cells, and use the midpoints of the cells to augment the partition (Figure 6).

SOLUTION   Each of the four cells has dimensions $\Delta x = 1$ and $\Delta y = 1$. Here, $f(x,\, y) = x^2 y^4$, $\Delta A = \Delta x\, \Delta y = 1$,

$$(x_1^*,\, y_1^*) = (-\tfrac{1}{2},\, \tfrac{1}{2}) \qquad (x_2^*,\, y_2^*) = (\tfrac{1}{2},\, \tfrac{1}{2})$$

$$(x_3^*,\, y_3^*) = (-\tfrac{1}{2},\, -\tfrac{1}{2}) \qquad (x_4^*,\, y_4^*) = (\tfrac{1}{2},\, -\tfrac{1}{2})$$

and the Riemann sum

$$\sum_{k=1}^{4} f(x_k^*,\, y_k^*)\, \Delta A = \sum_{k=1}^{4} (x_k^*)^2 (y_k^*)^4 (1)$$

is given by

$$(-\tfrac{1}{2})^2(\tfrac{1}{2})^4 + (\tfrac{1}{2})^2(\tfrac{1}{2})^4 + (-\tfrac{1}{2})^2(-\tfrac{1}{2})^4 + (\tfrac{1}{2})^2(-\tfrac{1}{2})^4 = \tfrac{1}{16}$$

Therefore,

$$\iint\limits_{R} x^2 y^4\, dA \approx \sum_{k=1}^{4} (x_k^*)^2 (y_k^*)^4 = \tfrac{1}{16}$$

Now consider a region R in the xy plane, and let f be the constant function defined by $f(x,\, y) = 1$ for all values of x and y. The graph of f is the horizontal plane $z = 1$. In this case, the double integral

$$\iint\limits_{R} f(x,\, y)\, dA = \iint\limits_{R} 1\, dA = \iint\limits_{R} dA$$

represents the volume of the cylinder of height $h = 1$ with base R (Figure 7). If A is the area of the region R, then the cylinder has volume $hA = 1 \cdot A = A$, so

$$A = \iint\limits_{R} dA$$

Figure 7

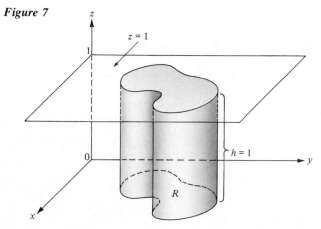

Thus, $\displaystyle\iint\limits_{R} dA$ *is numerically equal to the area of the region R.*

EXAMPLE 2   Let R be the interior of the triangle BCD in the plane, where $B = (0, 0)$, $C = (4, 0)$, and $D = (4, 15)$. Interpret the double integral $\iint\limits_R dA$ geometrically and thus evaluate the integral.

SOLUTION   The double integral $\iint\limits_R dA$ represents the area of the triangle BCD. Because BCD is a right triangle with base 4 units and height 15 units, its area is given by

$$A = \tfrac{1}{2}(4)(15) = 30 \text{ square units}$$

Therefore, $\iint\limits_R dA = 30$. ∎

If a function f has only nonnegative values over the region R, then the double integral

$$\iint\limits_R f(x, y)\, dA$$

can be interpreted as the volume V of the solid below the graph of f and above the region R. Often V can be found by the methods presented in Chapter 6 (slicing, circular disks, cylindrical shells, and so forth), and thus the double integral can be evaluated.

In Examples 3 and 4, interpret the double integral $\iint\limits_R f(x, y)\, dA$ geometrically and thus evaluate the integral.

EXAMPLE 3   Let R be the region inside the circle $x^2 + y^2 = 4$, and let $f(x, y) = \sqrt{4 - x^2 - y^2}$.

SOLUTION   The graph of f is a hemisphere of radius $r = 2$ units, and the region R forms the base of this hemisphere. The solid above R and below the graph of f is a hemispherical solid of radius $r = 2$ units (Figure 8), and $\iint\limits_R f(x, y)\, dA$ is the volume of this solid. Because the volume V of the hemisphere is given by

$$V = \frac{1}{2}\left(\frac{4}{3}\pi r^3\right) = \frac{16\pi}{3} \text{ cubic units}$$

we have

$$\iint\limits_R \sqrt{4 - x^2 - y^2}\, dA = \frac{16\pi}{3}$$ ∎

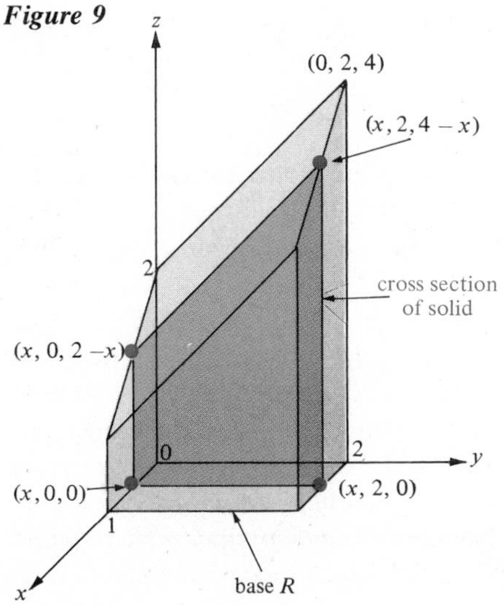

EXAMPLE 4   Let R be the rectangular region consisting of all points (x, y) such that $0 \le x \le 1$ and $0 \le y \le 2$, and let $f(x, y) = 2 + y - x$.

SOLUTION   The graph of f is the plane $z = 2 + y - x$ and $\iint\limits_R f(x, y)\, dA$ is the volume of the solid below this plane and above the rectangle R (Figure 9). We determine this volume by the method of slicing, using the x axis as our reference axis and taking cross sections perpendicular to the reference axis as in Figure 9. The cross section x units from the origin is a trapezoid with vertices $(x, 0, 0)$, $(x, 2, 0)$,

Figure 8

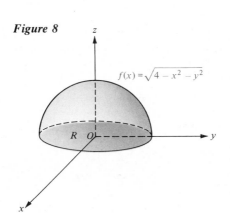

Figure 9

$(x, 2, 4 - x)$, and $(x, 0, 2 - x)$. The two parallel bases of this trapezoid have lengths $2 - x$ and $4 - x$ units, and the distance between these bases is 2 units; hence, the area of the trapezoid is

$$A(x) = 2 \frac{(2 - x) + (4 - x)}{2} = 6 - 2x \text{ square units}$$

The volume V of the solid is given by

$$V = \int_0^1 A(x)\, dx = \int_0^1 (6 - 2x)\, dx$$

$$= (6x - x^2) \Big|_0^1 = 5 \text{ cubic units}$$

Consequently,

$$\iint_R (2 + y - x)\, dA = 5 \qquad \blacksquare$$

Basic Properties of the Double Integral

In the discussion above, we considered only double integrals $\iint_R f(x, y)\, dA$ for which $f(x, y) \geq 0$ holds for all points (x, y) in R. If this condition does not hold, we can still form Riemann sums corresponding to regular augmented partitions, and we can still ask for the limit of such Riemann sums as the partitions become finer and finer. Thus, we define the **double integral,**

$$\iint_R f(x, y)\, dA = \lim_{n \to +\infty} \sum_{k=1}^m f(x_k^*, y_k^*)\, \Delta A$$

just as before, for *any* continuous function f defined over an admissible region R, provided that the limit exists.

If the condition $f(x, y) \geq 0$ fails to hold for some points (x, y) in the region R, then part of the graph of f falls below the xy plane. In this case, the region R can be decomposed into two subregions R_1 and R_2 so that $f(x, y) \geq 0$ for (x, y) in R_1 and $f(x, y) \leq 0$ for (x, y) in R_2. The double integral

$$\iint_R f(x, y)\, dA$$

can then be interpreted as the difference $V_1 - V_2$ between the volume V_1 of the solid below the graph of f and above R_1 and the volume V_2 of the solid above the graph of f and below R_2. (This is perfectly analogous to the interpretation of a definite integral as a difference of two areas.)

In advanced calculus, a general definition is given of a double integral, similar to our definition above, but allowing partitions in which the cells are not all congruent (that is, nonregular partitions). Although the region R is usually required to have a "reasonably nice" shape (for instance, to be admissible in our sense), the function f is not required to be continuous. The question of whether the Riemann sums have a limit as the partitions become finer and finer has to be handled with considerable finesse. If such a limit exists, then one says that f is a *(Riemann) integrable function of two variables over the region R.* The following basic properties of the double integral are established in advanced calculus.

Basic Properties of the Double Integral

1 *Existence* If f is continuous on the admissible region R, then f is integrable over R; that is,

$$\iint\limits_{R} f(x,\,y)\,dA \text{ exists}$$

2 *Interpretation as an Area* If R is an admissible region of area A, then

$$\iint\limits_{R} dA = A$$

3 *Homogeneous Property* If f is an integrable function on the admissible region R and K is a constant, then Kf is also integrable on R and

$$\iint\limits_{R} Kf(x,\,y)\,dA = K \iint\limits_{R} f(x,\,y)\,dA$$

4 *Additive Property* If f and g are integrable functions on the admissible region R, then $f + g$ is also integrable on R and

$$\iint\limits_{R} [f(x,\,y) + g(x,\,y)]\,dA = \iint\limits_{R} f(x,\,y)\,dA + \iint\limits_{R} g(x,\,y)\,dA$$

5 *Linear Property* If f and g are integrable functions on the admissible region R and if A and B are constants, then $Af \pm Bg$ is also integrable on R and

$$\iint\limits_{R} [Af(x,\,y) \pm Bg(x,\,y)]\,dA = A \iint\limits_{R} f(x,\,y)\,dA \pm B \iint\limits_{R} g(x,\,y)\,dA$$

6 *Positivity* If f is integrable on the admissible region R and if $f(x,\,y) \geq 0$ for all points $(x,\,y)$ in R, then

$$\iint\limits_{R} f(x,\,y)\,dA \geq 0$$

7 *Comparison* If f and g are integrable functions on the admissible region R, and if $f(x,\,y) \leq g(x,\,y)$ holds for all points $(x,\,y)$ in R, then

$$\iint\limits_{R} f(x,\,y)\,dA \leq \iint\limits_{R} g(x,\,y)\,dA$$

8 *Additivity with Respect to the Region of Integration* Let R be an admissible region, and suppose that R can be decomposed into two nonoverlapping admissible regions R_1 and R_2. (*Note:* The regions can share common boundary points.) If f is integrable on both R_1 and R_2, then f is integrable on R and

$$\iint\limits_{R} f(x,\,y)\,dA = \iint\limits_{R_1} f(x,\,y)\,dA + \iint\limits_{R_2} f(x,\,y)\,dA$$

EXAMPLE 5   Let R be the rectangle $0 \leq x \leq 1$, $0 \leq y \leq 1$; let R_1 be the portion of R above or on the diagonal $y = x$; and let R_2 be the portion of R below or on the diagonal $y = x$. Suppose that f and g are functions such that

$$\iint_{R_1} f(x, y)\, dA = 3 \qquad \iint_{R_1} g(x, y)\, dA = -2$$

$$\iint_{R_2} f(x, y)\, dA = 5 \qquad \iint_{R_2} g(x, y)\, dA = 1$$

Find

(a) $\displaystyle\iint_R f(x, y)\, dA$ **(b)** $\displaystyle\iint_R g(x, y)\, dA$ **(c)** $\displaystyle\iint_R [4f(x, y) - 3g(x, y)]\, dA$

SOLUTION

(a) By Property 8,

$$\iint_R f(x, y)\, dA = \iint_{R_1} f(x, y)\, dA + \iint_{R_2} f(x, y)\, dA = 3 + 5 = 8$$

(b) By Property 8,

$$\iint_R g(x, y)\, dA = \iint_{R_1} g(x, y)\, dA + \iint_{R_2} g(x, y)\, dA = -2 + 1 = -1$$

(c) Using parts (a) and (b) and Property 5, we have

$$\iint_R [4f(x, y) - 3g(x, y)]\, dA = 4 \iint_R f(x, y)\, dA - 3 \iint_R g(x, y)\, dA$$
$$= (4)(8) - (3)(-1) = 35 \qquad \blacksquare$$

Problem Set 15.2

In Problems 1 to 4, approximate the given double integral over the indicated rectangular region R. Use the regular partition of R into n^2 cells for the given value of n, and use the midpoints of the cells to augment the partition.

1 $\displaystyle\iint_R xy\, dA$; R: $0 \leq x \leq 2$ and $0 \leq y \leq 4$; $n = 2$

2 $\displaystyle\iint_R (|x| + |y|)\, dA$; R: $-6 \leq x \leq 0$ and $-1 \leq y \leq 2$; $n = 3$

3 $\displaystyle\iint_R (3x + 7y)\, dA$; R: $0 \leq x \leq 3$ and $2 \leq y \leq 5$; $n = 3$

4 $\displaystyle\iint_R (x - y^2)\, dA$; R: $-2 \leq x \leq 1$ and $0 \leq y \leq 2$; $n = 2$

5 Approximate the double integral $\displaystyle\iint_R 4xy\, dA$, where R is the region inside the circle $x^2 + y^2 = 1$. Use the regular partition of the rectangle $-1 \leq x \leq 1$, $-1 \leq y \leq 1$ into four congruent cells, and use the midpoints of the cells to augment the partition.

6 Approximate the double integral $\displaystyle\iint_R 5x^2y\, dA$, where R is the region in the first quadrant inside the circle $x^2 + y^2 = 1$. Use the regular partition of the rectangle $0 \leq x \leq 1$, $0 \leq y \leq 1$ into nine congruent cells, and use the point in each cell that is nearest to the origin to augment the partition.

7 Approximate the double integral $\displaystyle\iint_R x^3y^2\, dA$, where R is the region $0 \leq x \leq 3$, $0 \leq y \leq 9 - x^2$. Use the regular partition of the rectangle $0 \le x \le 3$, $0 \le y \le 9$ into nine congruent cells. Augment the partition by choosing *for each cell not discarded* the lower left corner.

8 Approximate the double integral $\iint\limits_{R} (3x^2 - 2y)\, dA$, where R is the region $-1 \le x \le 1$, $0 \le y \le 1 - x^2$. Use the regular partition of the rectangle $-1 \le x \le 1$, $0 \le y \le 1$ into nine congruent cells. Augment the partition by choosing for each cell not discarded the point nearest the origin.

In Problems 9 to 20, interpret each double integral as a volume or as an area. Evaluate this volume or area, and thus the integral, by whatever method seems appropriate.

9 $\iint\limits_{R} \sqrt{25 - x^2 - y^2}\, dA$; R: $x^2 + y^2 \le 25$

10 $\iint\limits_{R} \sqrt{9 - x^2 - y^2}\, dA$; R: $x^2 + y^2 \le 9$, $x \ge 0$, $y \ge 0$

11 $\iint\limits_{R} (4 + \sqrt{2 - x^2 - y^2})\, dA$; R: $x^2 + y^2 \le 2$

12 $\iint\limits_{R} (x - y + 1)\, dA$; R: $0 \le x \le 1$, $0 \le y \le 1$

13 $\iint\limits_{R} (5 + 2x + 3y)\, dA$; R: $0 \le x \le 1$, $0 \le y \le 2$

14 $\iint\limits_{R} (1 - x - y)\, dA$; R: $0 \le y \le 1 - x$, $x \ge 0$

15 $\iint\limits_{R} dA$; R: $x^2 + y^2 \le r^2$

16 $\iint\limits_{R} dA$; R: $0 \le y \le 1 - x$, $x \ge 0$

17 $\iint\limits_{R} dA$; R: the interior of the triangle with vertices (x_1, y_1), (x_2, y_2), (x_3, y_3)

18 $\iint\limits_{R} (1 - x^2 - y^2)\, dA$; R: $x^2 + y^2 \le 1$

19 $\iint\limits_{R} (1 - \sqrt{x^2 + y^2})\, dA$; R: $x^2 + y^2 \le 1$

20 $\iint\limits_{R} (5 - \sqrt{x^2 + y^2})\, dA$; R: $x^2 + y^2 \le 1$

In Problems 21 to 28, let R be the circular disk $x^2 + y^2 \le 1$, let R_1 be the upper half of R, and let R_2 be the lower half of R. Assume that

$$\iint\limits_{R_1} f(x, y)\, dA = 7, \quad \iint\limits_{R_2} f(x, y)\, dA = -5, \quad \iint\limits_{R_1} g(x, y)\, dA = -2,$$

and $\iint\limits_{R_2} g(x, y)\, dA = 4$. Evaluate the given double integral using Properties 1 to 8.

21 $\iint\limits_{R} f(x, y)\, dA$

22 $\iint\limits_{R_1} [g(x, y) - f(x, y)]\, dA$

23 $\iint\limits_{R} g(x, y)\, dA$

24 $\iint\limits_{R_2} [1 - f(x, y) + g(x, y)]\, dA$

25 $\iint\limits_{R} [4g(x, y) - 6f(x, y)]\, dA$

26 $\iint\limits_{R} [3f(x, y) - 5g(x, y) + 9]\, dA$

27 $\iint\limits_{R} 8\, dA$

28 $\iint\limits_{R} [f(x, y) - \sqrt{1 - x^2 - y^2}]\, dA$

29 Give a geometric explanation of Property 6.

30 Prove Property 7 by using Properties 5 and 6.

31 Give a geometric explanation of Property 8, assuming that the integrand is nonnegative.

32 Suppose that M is the maximum value of the continuous function f and that m is the minimum value of f on the admissible region R. Using the properties of double integrals, prove

(a) $\left| \iint\limits_{R} f(x, y)\, dA \right| \le \iint\limits_{R} |f(x, y)|\, dA$

(b) $m \le \dfrac{\iint\limits_{R} f(x, y)\, dA}{\iint\limits_{R} dA} \le M$, if $\iint\limits_{R} dA \ne 0$

33 Suppose that f is a continuous function on the admissible region R and that (a, b) is a point in R. Let R_1 be an admissible subregion of R such that the point (a, b) belongs to the interior of R_1. Denote by A_1 the area of R_1, and denote by δ_1 the diameter of R_1 (that is, the maximum distance between any two points in R_1). Make informal arguments to show that

(a) $\iint\limits_{R_1} f(x, y)\, dA \approx f(a, b)A_1$ if δ_1 is small

(b) $\lim\limits_{\delta_1 \to 0} \dfrac{\iint\limits_{R_1} f(x, y)\, dA}{A_1} = f(a, b)$

15.3   Evaluation of Double Integrals by Iteration

In Section 15.1 we considered the *iterated* integral

$$\int_a^b \int_{g(y)}^{h(y)} f(x, y) \, dx \, dy$$

of a function f of two variables; in Section 15.2 we introduced the *double* integral

$$\iint_R f(x, y) \, dA$$

of f over a region R in the xy plane. These two kinds of integrals are defined in completely different ways, and *it is important not to confuse them*. However, as we see in this section, it is sometimes possible to convert a double integral to an equivalent iterated integral, and vice versa.

Consider, for instance, the double integral

$$\iint_R f(x, y) \, dA$$

where R is the specially shaped region shown in Figure 1, f is continuous on R, and $f(x, y) \geq 0$ for (x, y) in R. Notice that R is bounded below by the line $y = a$, above by the line $y = b$, on the left by the graph of $x = g(y)$, and on the right by the graph of $x = h(y)$. Thus R is described by the inequalities

$$R: a \leq y \leq b, \ g(y) \leq x \leq h(y)$$

Figure 1

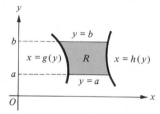

We assume that g and h are continuous functions defined on $[a, b]$ and that $g(y) \leq h(y)$ for $a \leq y \leq b$.

Since $f(x, y) \geq 0$ for (x, y) in R, the double integral

$$\iint_R f(x, y) \, dA$$

can be interpreted as the volume V of the solid under the graph of f and above the region R (Figure 2). We propose to find the volume V by the method of slicing, using the y axis as our reference axis. Figure 2 shows the cross section $ABCD$ cut from the solid by the plane perpendicular to the y axis and y units from the origin. If we denote the area of $ABCD$ by $F(y)$, then by the method of slicing,

$$\iint_R f(x, y) \, dA = V = \int_a^b F(y) \, dy$$

Figure 2

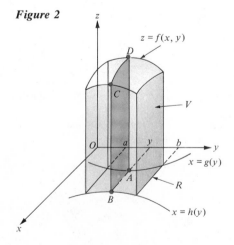

Figure 3

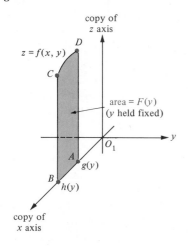

We now find a formula for the cross-sectional area $F(y)$. Thus, we temporarily fix a value of y between a and b and set up parallel copies of the x and z axes in the same plane as the cross section $ABCD$ (Figure 3). The equation of the curve DC is $z = f(x, y)$. In Figure 2, the points A and B lie on the curves $x = g(y)$ and $x = h(y)$, respectively; hence, in Figure 3, the points A and B have x coordinates $g(y)$ and $h(y)$, respectively. Figure 4 is obtained from Figure 3 by rotating the xz plane about the z axis so that the x axis extends to our right as usual. From Figure 4, it is now clear that the desired area $F(y)$ is just the area under the curve $z = f(x, y)$ between $x = g(y)$ and $x = h(y)$ (y being held fixed). Hence,

$$F(y) = \int_{g(y)}^{h(y)} f(x, y)\, dx$$

It follows that

$$\iint_R f(x, y)\, dA = \int_a^b F(y)\, dy$$

Figure 4

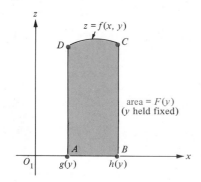

$$= \int_a^b \left[\int_{g(y)}^{h(y)} f(x, y)\, dx \right] dy$$

that is,

$$\boxed{\iint_R f(x, y)\, dA = \int_a^b \int_{g(y)}^{h(y)} f(x, y)\, dx\, dy}$$

The following example illustrates the use of the last equation for evaluating double integrals.

EXAMPLE 1 Let R be the region inside the trapezoid whose vertices are $(2, 2)$, $(4, 2)$, $(5, 4)$, and $(1, 4)$ (Figure 5). Evaluate $\iint_R 8xy\, dA$ by converting it to an iterated integral.

SOLUTION   The equation of the line through the points $(2, 2)$ and $(1, 4)$ is $y = 6 - 2x$, or $x = (6 - y)/2$. Similarly, the equation of the line through the points $(4, 2)$ and $(5, 4)$ is $y = 2x - 6$, or $x = (6 + y)/2$. Using the result obtained above, we have

Figure 5

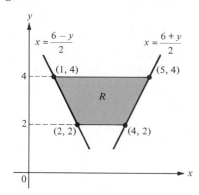

$$\iint_R 8xy\, dA = \int_2^4 \int_{(6-y)/2}^{(6+y)/2} 8xy\, dx\, dy$$

$$= \int_2^4 \left(8y\, \frac{x^2}{2} \bigg|_{(6-y)/2}^{(6+y)/2} \right) dy$$

$$= \int_2^4 4y \left[\left(\frac{6+y}{2} \right)^2 - \left(\frac{6-y}{2} \right)^2 \right] dy$$

$$= \int_2^4 24y^2\, dy = \frac{24y^3}{3} \bigg|_2^4 = 448 \qquad \blacksquare$$

In our derivation of the equation for converting a double integral to an iterated integral, we used the method of slicing and took cross sections perpendicular to the y axis. We can also use the method of slicing with cross sections perpendicular to the x axis, provided that the region R has the following shape: R is bounded on the

Figure 6

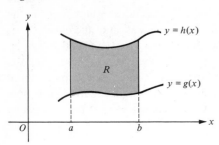

left by the vertical line $x = a$, on the right by the vertical line $x = b$, above by the graph of an equation $y = h(x)$, and below by the graph of an equation $y = g(x)$, where g and h are continuous on $[a, b]$ and $g(x) \leq h(x)$ for $a \leq x \leq b$ (Figure 6). Such a region, which is described by the inequalities

$$R: a \leq x \leq b, \ g(x) \leq y \leq h(x)$$

is said to be of *type I*. On the other hand, a region described by the inequalities

$$R: a \leq y \leq b, \ g(y) \leq x \leq h(y)$$

as in Figure 1, is said to be of *type II*.

As is indicated by the discussion above, *a double integral of a continuous function f over a region R of type I or of type II* can be converted to an iterated integral. In more advanced courses, this is proved rigorously, even for the case in which the function f takes on negative values. Thus, we have the following method, called the **method of iteration,** for evaluating double integrals over special regions.

The Method of Iteration

Suppose that R is either a type I or a type II region in the plane and that the function f is continuous on R. In order to evaluate the double integral $\iint\limits_{R} f(x, y) \, dA$ by the method of iteration, proceed as follows:

Figure 7

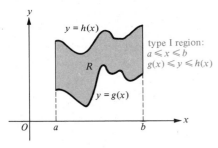

Case 1 If the region R is of type I, find the equations of the continuous curves $y = g(x)$ and $y = h(x)$ bounding R below and above, respectively. Also, find the constants a and b for which the vertical lines $x = a$ and $x = b$ bound R on the left and on the right, respectively (Figure 7). Then

$$\iint\limits_{R} f(x, y) \, dA = \int_{x=a}^{x=b} \left[\int_{y=g(x)}^{y=h(x)} f(x, y) \, dy \right] dx = \int_{a}^{b} \int_{g(x)}^{h(x)} f(x, y) \, dy \, dx$$

Case 2 If the region R is of type II, find the equations of the continuous curves $x = g(y)$ and $x = h(y)$ bounding R on the left and on the right, respectively. Also, find the constants a and b for which the horizontal lines $y = a$ and $y = b$ bound R below and above, respectively (Figure 8). Then

$$\iint\limits_{R} f(x, y) \, dA = \int_{y=a}^{y=b} \left[\int_{x=g(y)}^{x=h(y)} f(x, y) \, dx \right] dy = \int_{a}^{b} \int_{g(y)}^{h(y)} f(x, y) \, dx \, dy$$

Figure 8

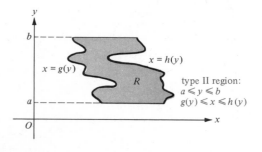

In Examples 2 and 3, evaluate the given double integral by the method of iteration.

EXAMPLE 2 $\iint\limits_R x \cos xy \, dA;\ R:\ 1 \le x \le 2$ and $\pi/2 \le y \le 2\pi/x$

Figure 9

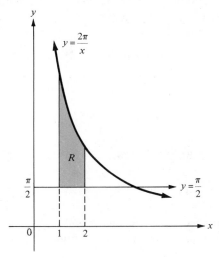

SOLUTION   The region is evidently of type I (Figure 9); hence,

$$\iint\limits_R x \cos xy \, dA = \int_{x=1}^{x=2} \left(\int_{y=\pi/2}^{y=2\pi/x} x \cos xy \, dy \right) dx$$

$$= \int_{x=1}^{x=2} \left(\sin xy \bigg|_{y=\pi/2}^{y=2\pi/x} \right) dx$$

$$= \int_{1}^{2} \left(\sin 2\pi - \sin \frac{\pi}{2} x \right) dx$$

$$= \int_{1}^{2} \left(-\sin \frac{\pi x}{2} \right) dx = \frac{2}{\pi} \cos \frac{\pi x}{2} \bigg|_{1}^{2}$$

$$= \frac{2}{\pi} \cos \pi - \frac{2}{\pi} \cos \frac{\pi}{2} = -\frac{2}{\pi}$$ ∎

EXAMPLE 3 $\iint\limits_R (x + y) \, dA$, where R is the region in the first quadrant above the curve $y = x^2$ and below the curve $y = \sqrt{x}$

SOLUTION   In this case the region R is both of type I and of type II (Figure 10). As a type II region, R is bounded on the left by the curve $x = y^2$, on the right by the curve $x = \sqrt{y}$, below by the line $y = 0$, and above by the line $y = 1$. Therefore,

Figure 10

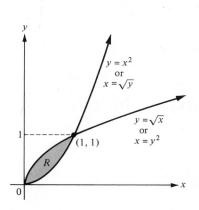

$$\iint\limits_R (x + y) \, dA = \int_{y=0}^{y=1} \left[\int_{x=y^2}^{x=\sqrt{y}} (x + y) \, dx \right] dy$$

$$= \int_{y=0}^{y=1} \left[\left(\frac{x^2}{2} + yx \right) \bigg|_{x=y^2}^{x=\sqrt{y}} \right] dy$$

$$= \int_{0}^{1} \left(\frac{y}{2} + y\sqrt{y} - \frac{y^4}{2} - y^3 \right) dy$$

$$= \left(\frac{y^2}{4} + \frac{2}{5} y^{5/2} - \frac{y^5}{10} - \frac{y^4}{4} \right) \bigg|_{0}^{1} = \frac{3}{10}$$ ∎

Example 3 above illustrates two important facts about the method of iteration. First, the horizontal lines bounding a type II region below and above (and, likewise, the vertical lines bounding a type I region on the left and on the right) are permitted to touch the region R in *single points*, rather than along line segments. Second, *there are regions that are both of type I and of type II.*

Since the region R in Figure 10 is of type I, we can also iterate the integral of Example 3 as follows:

$$\iint\limits_R (x + y) \, dA = \int_{x=0}^{x=1} \left[\int_{y=x^2}^{y=\sqrt{x}} (x + y) \, dy \right] dx = \frac{3}{10}$$

Figure 11

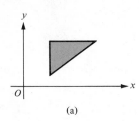

(a)

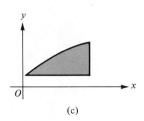

(b)

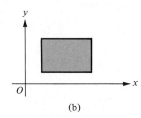

(c)

Additional examples of regions that are both of type I and of type II appear in Figure 11. A double integral over any such region can be iterated in two different ways, resulting in two iterated integrals with opposite orders of integration, but with the same value. These two iterated integrals are said to be obtained from each other by **reversing the order of integration**. A reversal of the order of integration often converts a complicated iterated integral to a simpler one.

In Examples 4 and 5, reverse the order of integration, and then evaluate the resulting integral.

© **EXAMPLE 4** $\displaystyle\int_0^1 \int_x^1 x \sin y^3 \, dy \, dx$ (Round off your answer to three decimal places.)

SOLUTION The given iterated integral is equivalent to the double integral

$$\iint_R x \sin y^3 \, dA$$

over the type I region R determined by the inequalities $0 \le x \le 1$ and $x \le y \le 1$ (Figure 12). Since R is also of type II, we have from Figure 12,

$$\int_0^1 \int_x^1 x \sin y^3 \, dy \, dx = \iint_R x \sin y^3 \, dA$$

$$= \int_{y=0}^{y=1} \left(\int_{x=0}^{x=y} x \sin y^3 \, dx \right) dy$$

$$= \int_{y=0}^{y=1} \left(\frac{x^2 \sin y^3}{2} \Big|_{x=0}^{x=y} \right) dy = \frac{1}{2} \int_0^1 y^2 \sin y^3 \, dy$$

$$= \left(-\frac{\cos y^3}{6} \right) \Big|_0^1 = -\frac{1}{6}(\cos 1 - \cos 0)$$

$$\approx -\tfrac{1}{6}(0.5403 - 1) \approx 0.077$$

Figure 12

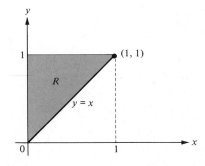

EXAMPLE 5 $\displaystyle\int_0^3 \int_{y^2}^9 y e^{-x^2} \, dx \, dy$

SOLUTION The given iterated integral is equivalent to the double integral

$$\iint_R y e^{-x^2} \, dA$$

over the type II region R determined by the inequalities $0 \le y \le 3$ and $y^2 \le x \le 9$ (Figure 13). Since R is also of type I, we have from Figure 13,

$$\int_0^3 \int_{y^2}^9 y e^{-x^2} \, dx \, dy = \iint_R y e^{-x^2} \, dA$$

$$= \int_{x=0}^{x=9} \left(\int_{y=0}^{y=\sqrt{x}} y e^{-x^2} \, dy \right) dx$$

$$= \int_{x=0}^{x=9} \left(\frac{y^2 e^{-x^2}}{2} \Big|_{y=0}^{y=\sqrt{x}} \right) dx = \int_0^9 \frac{x e^{-x^2}}{2} \, dx$$

$$= \left(-\frac{e^{-x^2}}{4} \right) \Big|_0^9 = -\frac{e^{-81}}{4} + \frac{1}{4} = \frac{1}{4}(1 - e^{-81})$$

Figure 13

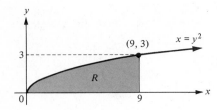

Although there are regions that are neither of type I nor of type II, it is usually possible to cut such a region into nonoverlapping subregions, each of which is of type I or of type II. The double integral of a function over the large region can then be evaluated by integrating the function over each subregion and adding the resulting values. (See Problems 35 and 36.)

Areas by Double Integration

As we noticed in Section 15.2, the area A of a region R in the xy plane is given by

$$A = \iint_R dA$$

Using the method of iteration, we can evaluate the double integral and thus determine the area A.

Figure 14

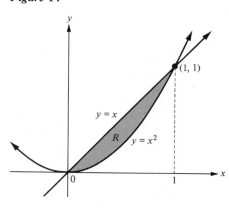

EXAMPLE 6   Find the area A of the region R bounded by the curves $y = x^2$ and $y = x$.

SOLUTION   The region R (Figure 14) is of type I; hence, its area A is given by

$$A = \iint_R dA = \int_{x=0}^{x=1} \left(\int_{y=x^2}^{y=x} dy \right) dx = \int_{x=0}^{x=1} \left(y \Big|_{x^2}^{x} \right) dx = \int_0^1 (x - x^2)\, dx$$

$$= \left(\frac{x^2}{2} - \frac{x^3}{3} \right) \Big|_0^1 = \frac{1}{6} \text{ square unit} \qquad \blacksquare$$

Volumes by Double Integration

Using the double integral, we can express the volume V under the graph of a continuous nonnegative function f over a region R (as we did in Section 15.2) by

$$V = \iint_R f(x, y)\, dA$$

Using the method of iteration, we can evaluate the double integral and thus determine the volume V.

Figure 15

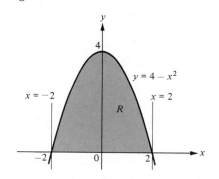

EXAMPLE 7   Let R be the region in the xy plane bounded above by the parabola $y = 4 - x^2$ and bounded below by the x axis (Figure 15). Find the volume V under the graph of $f(x, y) = x + 2y + 3$ and above the region R.

SOLUTION   The region R is of type I, since it is bounded on the left by the vertical line $x = -2$, on the right by the vertical line $x = 2$, above by the parabola $y = 4 - x^2$, and below by the line $y = 0$. Therefore, by the method of iteration

$$V = \iint_R (x + 2y + 3)\, dA = \int_{-2}^{2} \left[\int_0^{4-x^2} (x + 2y + 3)\, dy \right] dx$$

$$= \int_{-2}^{2} \left[(xy + y^2 + 3y) \Big|_0^{4-x^2} \right] dx = \int_{-2}^{2} [x(4 - x^2) + (4 - x^2)^2 + 3(4 - x^2)]\, dx$$

$$= \int_{-2}^{2} (x^4 - x^3 - 11x^2 + 4x + 28)\, dx = \left(\frac{x^5}{5} - \frac{x^4}{4} - \frac{11x^3}{3} + 2x^2 + 28x \right) \Big|_{-2}^{2}$$

$$= \frac{992}{15} \text{ cubic units} \qquad \blacksquare$$

Figure 16

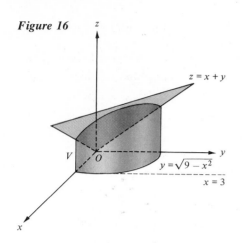

EXAMPLE 8   Find the volume V of the solid in the first octant bounded laterally by the cylinder $x^2 + y^2 = 9$ and bounded above by the plane $z = x + y$.

SOLUTION   As shown in Figure 16, the base of the solid is the type I region R in the xy plane bounded by $x = 0$, $x = 3$, $y = 0$, and $y = \sqrt{9 - x^2}$. Therefore, the volume V is given by

$$V = \iint_R (x + y)\, dA = \int_0^3 \left[\int_0^{\sqrt{9 - x^2}} (x + y)\, dy \right] dx$$

$$= \int_0^3 \left[\left(xy + \frac{y^2}{2} \right) \Big|_0^{\sqrt{9 - x^2}} \right] dx = \int_0^3 \left(x\sqrt{9 - x^2} + \frac{9 - x^2}{2} \right) dx$$

$$= \left[-\frac{1}{3}(9 - x^2)^{3/2} + \frac{9}{2}x - \frac{x^3}{6} \right] \Big|_0^3 = 9 + 9 = 18 \text{ cubic units} \qquad \blacksquare$$

Problem Set 15.3

In Problems 1 to 22, (a) sketch the region R, (b) decide whether R is of type I or type II (or both), and (c) evaluate each double integral by using the method of iteration.

1 $\displaystyle\iint_R x \sin (xy)\, dA$; R: $0 \leq x \leq \pi$, $0 \leq y \leq 1$

2 $\displaystyle\iint_R \frac{xe^{x/y}}{y^2}\, dA$; R: $0 \leq x \leq 1$, $1 \leq y \leq 2$

3 $\displaystyle\iint_R x \sin y\, dA$; R: $0 \leq x \leq \pi$, $0 \leq y \leq x$

4 $\displaystyle\iint_R (x + y + 2)\, dA$; R: $0 \leq x \leq 2$, $-\sqrt{x} \leq y \leq \sqrt{x}$

5 $\displaystyle\iint_R x^2 y\, dA$; R: $0 \leq y \leq \sqrt{4 - x^2}$, $0 \leq x \leq 2$

6 $\displaystyle\iint_R xe^y\, dA$; R: $0 \leq y \leq x^2$, $0 \leq x \leq 1$

7 $\displaystyle\iint_R y\, dA$; R: $0 \leq x \leq \sqrt{1 - y^2}$, $0 \leq y \leq 1$

8 $\displaystyle\iint_R e^x\, dA$; R: $1 \leq x \leq \ln y$, $1 \leq y \leq e$

9 $\displaystyle\iint_R y\, dA$; R: $0 \leq x \leq \pi$, $0 \leq y \leq \sin x$

10 $\displaystyle\iint_R y \sin^{-1} x\, dA$; R: $0 \leq x \leq \frac{1}{2}$, $-1 \leq y \leq 1$

11 $\displaystyle\iint_R 3y\, dA$; R: $x^2 + y^2 \leq 1$

12 $\displaystyle\iint_R e^{x+y}\, dA$; R: $|x| + |y| \leq 1$

13 $\displaystyle\iint_R x\, dA$; R: the region in the first quadrant bounded by $x = y - 2$, $y = x^2$, and $x = 0$

14 $\displaystyle\iint_R (6x + 5y)\, dA$; R: the region between the curves $y = 2\sqrt{x}$ and $y = 2x^3$

15 $\displaystyle\iint_R xy\, dA$; R: the region bounded by $y = x$ and $y = x^2$

16 $\displaystyle\iint_R 3x\, dA$; R: the region bounded by $x = y^2$ and $x = y + 6$

17 $\displaystyle\iint_R y^2\, dA$; R: the region bounded by $y = x^2$ and $y = 2 - x$

18 $\displaystyle\iint_R (2 - x^2)\, dA$; R: the region bounded by $y = 1 + x^2$ and $y = 9 - x^2$

19 $\displaystyle\iint_R (7xy - 2y^2)\, dA$; R: the region in the first quadrant bounded by $y = x$, $y = x/3$, $x = 3$

20 $\displaystyle\iint_R \sqrt{1 + x^2}\, dA$; R: the inside of the triangle whose vertices are $(0, 0)$, $(0, 1)$, and $(1, 1)$

21 $\displaystyle\iint_R \frac{1}{1 + y^2}\, dA$; R: the region in the first quadrant bounded by $y = x$, $y = 0$, and $x = 1$

22 $\displaystyle\iint_R x\, dA$; R: the smaller region cut from the disk $x^2 + y^2 \leq 9$ by the line $x + y = 3$

In Problems 23 to 34, evaluate each iterated integral by reversing the order of integration. In each case, sketch an appropriate region in the xy plane over which the double integral corresponding to the given iterated integral is evaluated.

23 $\int_0^1 \int_y^1 e^{-3x^2} \, dx \, dy$

24 $\int_1^e \int_0^{\ln x} y \, dy \, dx$

25 $\int_0^2 \int_{3y}^6 \sin \frac{\pi x^2}{6} \, dx \, dy$

26 $\int_0^{\sqrt{2}/2} \int_0^{\sin^{-1} x} x \, dy \, dx$

27 $\int_0^1 \int_y^{\sqrt{y}} \frac{\sin x}{x} \, dx \, dy$

28 $\int_0^1 \int_{\sqrt{y}}^1 \sqrt{1 - x^3} \, dx \, dy$

29 $\int_0^2 \int_{x^3}^8 e^{x/\sqrt[3]{y}} \, dy \, dx$

30 $\int_0^1 \int_{-1}^{-\sqrt{y}} \sqrt[5]{x^3 + 1} \, dx \, dy$

31 $\int_0^4 \int_{\sqrt{y}}^2 \frac{y \, dx \, dy}{\sqrt{1 + x^5}}$

32 $\int_0^4 \int_{\sqrt{y}}^2 \cos\left(\frac{x^3}{3} + \pi\right) dx \, dy$

33 $\int_0^{1/2} \int_{\sin^{-1} y}^{\pi/6} \csc^2 (\cos x) \, dx \, dy$ **34** $\int_{-1}^1 \int_{-\sqrt{1-x^2}}^{\sqrt{1-x^2}} \frac{x \, dy \, dx}{\sqrt{y^4 + 1}}$

35 Evaluate $\iint_R (1 - x + y) \, dx \, dy$ over the region R shown in Figure 17 by dividing R into rectangles.

Figure 17

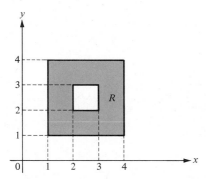

36 Suppose that R is the region bounded by the lines $y = 1$, $y = x + 6$, and the parabola $y = x^2$ (Figure 18). Evaluate $\iint_R xy \, dA$ over the region R by (a) dividing R into type II regions and (b) dividing R into type I regions.

Figure 18

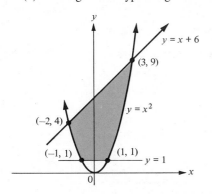

In Problems 37 to 50, use double integration to find the area of the region R in the xy plane bounded by the given curves.

37 $y = x^2$ and $y = 5x$

38 $xy = 3$ and $x + y = 4$

39 $y = x$ and $y = 3x - x^2$

40 $y = \sqrt[3]{x}$, $y = 0$, and $x = -8$

41 $y = \sqrt{x}$, $y = 6 - x$, and $y = 0$

42 $y^2 = -x$ and $3y - x = 4$

43 $x = y^2$ and $x = 32 - y^2$

44 $y^2 = 9 - x$ and $y^2 = 9(1 - x)$

45 $x^2 = 16 - 2y$ and $x = 4 - 2y$

46 $y = xe^{-x}$, $y = x$, and $x = 2$

47 $y = \cos x$, $y = 0$, $x = -\pi/2$, and $x = \pi/2$

48 $y = \sin x$, $y = \cos x$, $x = 0$, and $x = \pi/4$

49 $y = \ln |x|$, $y = 1$, and $y = 0$

50 $y = \cosh x$, $y = \sinh x$, $x = -1$, and $x = 1$

In Problems 51 to 56, use double integration to find the volume under the graph of each function f and above the indicated region R.

51 $f(x, y) = 12 + y + x^2$; R: $0 \le x \le 1$ and $x^2 \le y \le \sqrt{x}$

52 $f(x, y) = \sqrt{9 - y^2}$; R: $0 \le y \le 3$ and $0 \le x \le \sqrt{9 - y^2}$

53 $f(x, y) = 1 - x$; R: $0 \le x \le 1$ and $x^3 \le y \le x$

54 $f(x, y) = Ax + By + C$; R: $a \le x \le b$ and $c \le y \le d$; $Ax + By + C \ge 0$ on R

55 $f(x, y) = xy$; R: $x^2 + y^2 \le 4$, $x \ge 0$, $y \ge 0$

56 $f(x, y) = x^2 + 9y^2$; R: the region inside the triangle whose vertices are $(0, 0, 0)$, $(0, 1, 0)$, and $(1, 1, 0)$

In Problems 57 to 66, use double integration to find the volume of each solid.

57 The solid below the paraboloid $z = 4 - x^2 - y^2$, above the plane $z = 0$, and bounded laterally by the planes $x = 0$, $y = 0$, $x = 1$, and $y = 1$

58 The solid in the first octant bounded by the paraboloid $z = 3(x^2 + y^2)$, the plane $x + y = 2$, and the coordinate planes

59 The solid in the first octant under the plane $x + y + z = 6$ and inside the parabolic cylinder $y = 4 - x^2$

60 The solid in the first octant bounded by the graphs of $z = e^{x+y}$, $y = \ln x$, $x = 2$, $y = 0$, and $z = 0$

61 The solid in the first octant bounded by the cylinder $x^2 + z = 1$, the plane $x + y = 1$, and the coordinate planes

62 The solid in the first octant bounded by the cylinder $x^2 + y^2 = 4$ and the planes $z - y = 0$, $z = 0$, and $x = 0$

63 The solid in the first octant bounded above by the paraboloid $z = x^2 + y^2$, below by the plane $z = 0$, and laterally by the graphs $y = x^2$ and $y = x$

64 The solid in the first octant common to the two cylinders $x^2 + y^2 = 1$ and $x^2 + z^2 = 1$

65 The solid bounded by the cylinder $y^2 + z = 9$ and the planes $x = y$, $z = 0$, and $x = 0$

66 The solid bounded above by the plane $z + y = 2$, below by the plane $z = 0$, and laterally by the right circular cylinder $x^2 + y^2 = 4$

15.4   Double Integrals in Polar Coordinates

Figure 1

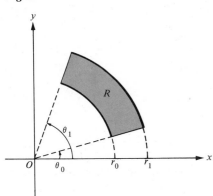

Often the region R over which a double integral is to be evaluated is more easily described by polar coordinates than by Cartesian coordinates. For instance, the region R in Figure 1 is described in polar coordinates by the conditions $r_0 \leq r \leq r_1$ and $\theta_0 \leq \theta \leq \theta_1$; however, its description in Cartesian coordinates is considerably more complicated. In this section we set forth a method for converting a double integral in Cartesian coordinates to an equivalent iterated integral in polar coordinates. The technique is analogous to a change of variable for the ordinary definite integral.

The clue to the appropriate method for changing from Cartesian to polar coordinates can be found in Figure 2, which shows an "infinitesimal" portion dA of the area of the region R in Figure 1 corresponding to infinitesimal changes dr in r and $d\theta$ in θ. Evidently, dA is virtually the area of a rectangle of dimensions $r\, d\theta$ and dr, so that

$$dA = (r\, d\theta)\, dr = r\, dr\, d\theta$$

Figure 2

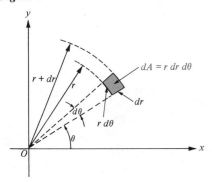

Thus, whereas in Cartesian coordinates the area of an infinitesimal rectangle of dimensions dx and dy is given by $dA = dx\, dy$, the analogous infinitesimal area in polar coordinates is given by $dA = r\, dr\, d\theta$.

In view of the argument above, it seems plausible that the integral

$$\iint_R f(x, y)\, dA$$

can be converted to an iterated integral in polar coordinates by putting $x = r\cos\theta$, $y = r\sin\theta$, and $dA = r\, dr\, d\theta$. The following theorem shows exactly how such a conversion to polar coordinates is accomplished.

THEOREM 1   **Change to Polar Coordinates in a Double Integral**

Suppose that the function f is continuous on the region R consisting of all points in the xy plane of the form $(x, y) = (r\cos\theta, r\sin\theta)$, where $0 \leq r_0 \leq r \leq r_1$ and $\theta_0 \leq \theta \leq \theta_1$ with $0 < \theta_1 - \theta_0 \leq 2\pi$ (Figure 1). Then

$$\iint_R f(x, y)\, dA = \int_{\theta = \theta_0}^{\theta = \theta_1} \left[\int_{r = r_0}^{r = r_1} f(r\cos\theta, r\sin\theta) r\, dr \right] d\theta$$

$$= \int_{\theta_0}^{\theta_1} \int_{r_0}^{r_1} f(r\cos\theta, r\sin\theta) r\, dr\, d\theta$$

Figure 3

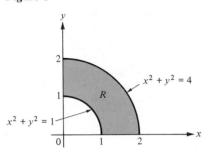

EXAMPLE 1 By changing to polar coordinates as in Theorem 1, evaluate $\iint_R e^{x^2+y^2} \, dA$, where R is the region in the first quadrant inside the circle $x^2 + y^2 = 4$ and outside the circle $x^2 + y^2 = 1$ (Figure 3).

SOLUTION The region R is described in polar coordinates by $1 \le r \le 2$ and $0 \le \theta \le \pi/2$; hence, by Theorem 1,

$$\iint_R e^{x^2+y^2} \, dA = \int_{\theta=0}^{\theta=\pi/2} \left(\int_{r=1}^{r=2} e^{r^2} r \, dr \right) d\theta = \int_{\theta=0}^{\theta=\pi/2} \left(\frac{1}{2} e^{r^2} \Big|_{r=1}^{r=2} \right) d\theta$$

$$= \int_0^{\pi/2} \left(\frac{1}{2} e^4 - \frac{1}{2} e \right) d\theta$$

$$= \left(\frac{1}{2} e^4 - \frac{1}{2} e \right) \left(\theta \Big|_0^{\pi/2} \right) = \frac{\pi e}{4} (e^3 - 1) \qquad \blacksquare$$

Sometimes it is useful to rewrite a given iterated integral as an equivalent double integral and then to evaluate the double integral by changing to polar coordinates. The following example illustrates the technique.

EXAMPLE 2 Evaluate the iterated integral $\int_{-3}^{3} \int_0^{\sqrt{9-x^2}} (2x + y) \, dy \, dx$ by switching to polar coordinates.

Figure 4

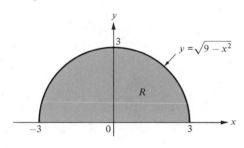

SOLUTION The iterated integral $\int_{-3}^{3} \int_0^{\sqrt{9-x^2}} (2x + y) \, dy \, dx$ is equivalent to the double integral $\iint_R (2x + y) \, dA$ over the region R: $-3 \le x \le 3$, $0 \le y \le \sqrt{9 - x^2}$ (Figure 4). This region can also be described in polar coordinates by $0 \le r \le 3$ and $0 \le \theta \le \pi$. Using Theorem 1, we have

$$\int_{-3}^{3} \int_0^{\sqrt{9-x^2}} (2x + y) \, dy \, dx = \iint_R (2x + y) \, dA$$

$$= \int_0^{\pi} \int_0^3 (2r \cos \theta + r \sin \theta) r \, dr \, d\theta$$

$$= \int_0^{\pi} \left[\int_0^3 (2 \cos \theta + \sin \theta) r^2 \, dr \right] d\theta$$

$$= \int_0^{\pi} \left[(2 \cos \theta + \sin \theta) \frac{r^3}{3} \Big|_0^3 \right] d\theta$$

$$= 9 \int_0^{\pi} (2 \cos \theta + \sin \theta) \, d\theta$$

$$= 9(2 \sin \theta - \cos \theta) \Big|_0^{\pi} = 18 \qquad \blacksquare$$

Theorem 1 can be generalized in a number of useful ways. For instance, consider the region R in the xy plane consisting of all points whose polar coordinates satisfy the conditions

$$\theta_0 \le \theta \le \theta_1 \qquad \text{and} \qquad g(\theta) \le r \le h(\theta)$$

where $0 < \theta_1 - \theta_0 \le 2\pi$ and g and h are continuous functions defined on the closed

Figure 5

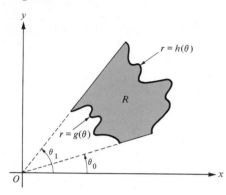

interval $[\theta_0, \theta_1]$ such that $0 \le g(\theta) \le h(\theta)$ holds for all values of θ in $[\theta_0, \theta_1]$ (Figure 5). Then, if f is a continuous function of two variables defined on the region R, we have

$$
\iint\limits_{R} f(x, y)\, dA = \int_{\theta=\theta_0}^{\theta=\theta_1} \left[\int_{r=g(\theta)}^{r=h(\theta)} f(r\cos\theta, r\sin\theta) r\, dr \right] d\theta
$$
$$
= \int_{\theta_0}^{\theta_1} \int_{g(\theta)}^{h(\theta)} f(r\cos\theta, r\sin\theta) r\, dr\, d\theta
$$

In Examples 3 to 5, use the preceding formula.

<u>EXAMPLE 3</u>   Evaluate $\iint\limits_{R} x\, dA$ over the region R consisting of all points whose polar coordinates satisfy the conditions $0 \le \theta \le \pi/4$ and $2\cos\theta \le r \le 2$ (Figure 6).

Figure 6

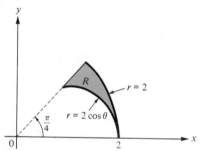

SOLUTION

$$
\iint\limits_{R} x\, dA = \int_{\theta=0}^{\theta=\pi/4} \left(\int_{r=2\cos\theta}^{r=2} r\cos\theta\, r\, dr \right) d\theta = \int_{0}^{\pi/4} \cos\theta \left(\frac{r^3}{3} \bigg|_{2\cos\theta}^{2} \right) d\theta
$$
$$
= \int_{0}^{\pi/4} \left(\frac{8}{3}\cos\theta - \frac{8\cos^4\theta}{3} \right) d\theta
$$
$$
= \frac{8}{3} \int_{0}^{\pi/4} \cos\theta\, d\theta - \frac{8}{3} \int_{0}^{\pi/4} \cos^4\theta\, d\theta
$$
$$
= \frac{8}{3} \int_{0}^{\pi/4} \cos\theta\, d\theta - \frac{1}{3} \int_{0}^{\pi/4} (3 + 4\cos 2\theta + \cos 4\theta)\, d\theta
$$
$$
= \frac{8}{3}\sin\theta \bigg|_{0}^{\pi/4} - \frac{1}{3}\left(3\theta + 2\sin 2\theta + \frac{1}{4}\sin 4\theta \right) \bigg|_{0}^{\pi/4}
$$
$$
= \frac{16\sqrt{2} - 3\pi - 8}{12}
$$

Figure 7

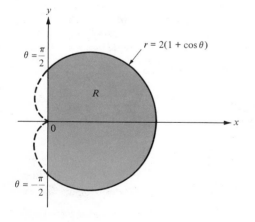

<u>EXAMPLE 4</u>   Find the area enclosed by the y axis and the portion of the cardioid $r = 2(1 + \cos\theta)$ lying in the fourth and first quadrants (Figure 7).

SOLUTION   The region R whose area is desired can be described in polar coordinates by the conditions $-\pi/2 \le \theta \le \pi/2$ and $0 \le r \le 2(1 + \cos\theta)$. Therefore, the area A of R is given by

$$
A = \iint\limits_{R} dA = \int_{\theta=-\pi/2}^{\theta=\pi/2} \left(\int_{r=0}^{r=2(1+\cos\theta)} r\, dr \right) d\theta = \int_{-\pi/2}^{\pi/2} \left(\frac{r^2}{2} \bigg|_{0}^{2(1+\cos\theta)} \right) d\theta
$$
$$
= \int_{-\pi/2}^{\pi/2} 2(1 + \cos\theta)^2\, d\theta = \int_{-\pi/2}^{\pi/2} 2(1 + 2\cos\theta + \cos^2\theta)\, d\theta
$$
$$
= 2\left(\theta + 2\sin\theta + \frac{\theta}{2} + \frac{\sin 2\theta}{4} \right) \bigg|_{-\pi/2}^{\pi/2}
$$
$$
= 3\pi + 8 \text{ square units}
$$

<u>EXAMPLE 5</u>   Find the volume of the solid in the first octant bounded by the cone $z = r$ and the cylinder $r = 4\sin\theta$.

Figure 8

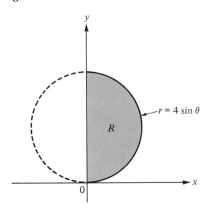

SOLUTION   The cone $z = r$ has the Cartesian equation $z = \sqrt{x^2 + y^2}$; hence, if R is the region consisting of all points whose polar coordinates satisfy $0 \leq \theta \leq \pi/2$ and $0 \leq r \leq 4 \sin \theta$ (Figure 8), then the required volume V is given by

$$V = \iint_R \sqrt{x^2 + y^2}\, dA = \int_0^{\pi/2} \int_0^{4 \sin \theta} r \cdot r\, dr\, d\theta = \int_0^{\pi/2} \left(\frac{r^3}{3} \Big|_0^{4 \sin \theta} \right) d\theta$$

$$= \int_0^{\pi/2} \tfrac{64}{3} \sin^3 \theta\, d\theta = \tfrac{64}{3} \int_0^{\pi/2} \sin \theta (1 - \cos^2 \theta)\, d\theta$$

Thus, making the change of variable $u = \cos \theta$, we obtain

$$V = \tfrac{64}{3} \int_1^0 (1 - u^2)(-du) = \tfrac{64}{3} \int_0^1 (1 - u^2)\, du$$

$$= \frac{64}{3} \left(u - \frac{u^3}{3} \right) \Big|_0^1 = \frac{128}{9} \text{ cubic units} \qquad ■$$

Change of Variables in a Double Integral

The method used above for changing from Cartesian to polar coordinates is just a special case of a more general procedure for changing variables in a double integral. A rigorous treatment of this procedure may be found in Kreyszig, *Advanced Engineering Mathematics*, 5th ed., John Wiley & Sons, New York, page 416. A brief informal discussion is sufficient for our present purposes.

Let f be a continuous function defined on an admissible region R in the xy plane, and suppose that we wish to evaluate the double integral of f over R. By introducing new variables, say, u and v, we may be able to simplify our problem. We begin by expressing x and y in terms of the new variables, so that

$$x = g(u, v) \qquad \text{and} \qquad y = h(u, v)$$

where g and h are suitable functions. These equations may be thought of as describing a **mapping,** or **transformation,** T from a region R_{uv} in the uv plane onto the region R in the xy plane (Figure 9). Thus, each point (u, v) in R_{uv} is **mapped,** or **transformed,** by T into a corresponding point (x, y) in R; in symbols,

$$(u, v) \overset{T}{\longmapsto} (x, y) = (g(u, v), h(u, v))$$

Figure 9

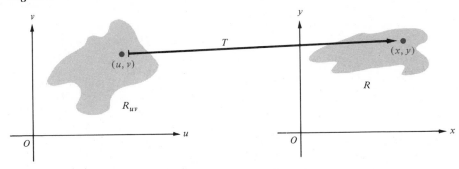

We assume that R_{uv} is an admissible region in the uv plane and that the transformation T establishes a one-to-one correspondence between the points (u, v) in R_{uv} and the points (x, y) in R. Furthermore, we assume that the functions g and h are continuously differentiable.

Figure 10

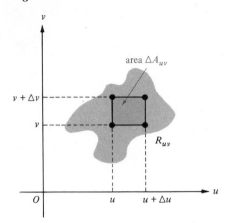

Consider a small rectangle of dimensions Δu and Δv (Figure 10) with vertices (u, v), $(u + \Delta u, v)$, $(u + \Delta u, v + \Delta v)$, $(u, v + \Delta v)$ within the region R_{uv}. The area ΔA_{uv} of this rectangle is given by

$$\Delta A_{uv} = \Delta u \Delta v$$

Under the transformation T, this rectangle is transformed into a subregion $PQVW$ of R, where

$$P = (g(u, v), h(u, v))$$
$$Q = (g(u + \Delta u, v), h(u + \Delta u, v))$$
$$V = (g(u + \Delta u, v + \Delta v), h(u + \Delta u, v + \Delta v))$$
$$W = (g(u, v + \Delta v), h(u, v + \Delta v))$$

(Figure 11). Notice that the area ΔA of the subregion $PQVW$ is approximately the area of the parallelogram spanned by the vectors $\overrightarrow{PQ}$ and $\overrightarrow{PW}$, so that

$$\Delta A \approx |\overrightarrow{PQ} \times \overrightarrow{PW}|$$

Figure 11

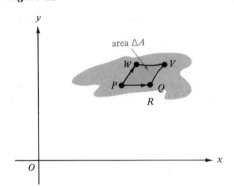

By the definition of a partial derivative,

$$\frac{g(u + \Delta u, v) - g(u, v)}{\Delta u} \approx g_1(u, v)$$

with better and better approximation as $\Delta u \to 0$; hence,

$$g(u + \Delta u, v) - g(u, v) \approx g_1(u, v) \, \Delta u$$

Similarly,

$$h(u + \Delta u, v) - h(u, v) \approx h_1(u, v) \, \Delta u$$

Therefore,

$$\overrightarrow{PQ} = [g(u + \Delta u, v) - g(u, v)]\mathbf{i} + [h(u + \Delta u, v) - h(u, v)]\mathbf{j}$$
$$\approx g_1(u, v) \, \Delta u \, \mathbf{i} + h_1(u, v) \, \Delta u \, \mathbf{j} = \frac{\partial x}{\partial u} \Delta u \, \mathbf{i} + \frac{\partial y}{\partial u} \Delta u \, \mathbf{j}$$

Likewise,

$$\overrightarrow{PW} = [g(u, v + \Delta v) - g(u, v)]\mathbf{i} + [h(u, v + \Delta v) - h(u, v)]\mathbf{j}$$
$$\approx g_2(u, v) \, \Delta v \, \mathbf{i} + h_2(u, v) \, \Delta v \, \mathbf{j} = \frac{\partial x}{\partial v} \Delta v \, \mathbf{i} + \frac{\partial y}{\partial v} \Delta v \, \mathbf{j}$$

It follows that

$$\overrightarrow{PQ} \times \overrightarrow{PW} \approx \begin{vmatrix} \mathbf{i} & \mathbf{j} & \mathbf{k} \\ \dfrac{\partial x}{\partial u} \Delta u & \dfrac{\partial y}{\partial u} \Delta u & 0 \\ \dfrac{\partial x}{\partial v} \Delta v & \dfrac{\partial y}{\partial v} \Delta v & 0 \end{vmatrix} = \begin{vmatrix} \dfrac{\partial x}{\partial u} \Delta u & \dfrac{\partial y}{\partial u} \Delta u \\ \dfrac{\partial x}{\partial v} \Delta v & \dfrac{\partial y}{\partial v} \Delta v \end{vmatrix} \mathbf{k}$$

Hence,

$$\overrightarrow{PQ} \times \overrightarrow{PW} \approx \begin{vmatrix} \dfrac{\partial x}{\partial u} & \dfrac{\partial y}{\partial u} \\ \dfrac{\partial x}{\partial v} & \dfrac{\partial y}{\partial v} \end{vmatrix} \Delta u \, \Delta v \, \mathbf{k} = \begin{vmatrix} \dfrac{\partial x}{\partial u} & \dfrac{\partial y}{\partial u} \\ \dfrac{\partial x}{\partial v} & \dfrac{\partial y}{\partial v} \end{vmatrix} \Delta A_{uv} \, \mathbf{k}$$

Karl Gustav Jacob Jacobi

The determinant of partial derivatives that appears in the last equation is called the **Jacobian** of the transformation T in honor of the German mathematician Karl Gustav Jacob Jacobi (1804–1851). The special symbol $\partial(x, y)/\partial(u, v)$ is used for the Jacobian, so that by definition

$$\frac{\partial(x, y)}{\partial(u, v)} = \begin{vmatrix} \dfrac{\partial x}{\partial u} & \dfrac{\partial y}{\partial u} \\[2mm] \dfrac{\partial x}{\partial v} & \dfrac{\partial y}{\partial v} \end{vmatrix}$$

Using this notation, we can write the result of our previous calculation as

$$\overrightarrow{PQ} \times \overrightarrow{PW} \approx \frac{\partial(x, y)}{\partial(u, v)} \Delta A_{uv}\, \mathbf{k}$$

Therefore, in Figure 11, we have

$$\Delta A \approx |\overrightarrow{PQ} \times \overrightarrow{PW}| \approx \left| \frac{\partial(x, y)}{\partial(u, v)} \right| \Delta A_{uv}\, |\mathbf{k}| = \left| \frac{\partial(x, y)}{\partial(u, v)} \right| \Delta A_{uv}$$

so that

$$\frac{\Delta A}{\Delta A_{uv}} \approx \left| \frac{\partial(x, y)}{\partial(u, v)} \right|$$

Presumably, the approximation derived above should become more and more accurate as $\Delta A_{uv} \to 0$, and in the limit we should have

$$\frac{dA}{dA_{uv}} = \left| \frac{\partial(x, y)}{\partial(u, v)} \right| \qquad \text{or} \qquad dA = \left| \frac{\partial(x, y)}{\partial(u, v)} \right| dA_{uv}$$

In other words, the absolute value of the Jacobian may be interpreted as the ratio by which infinitesimal areas are transformed by T. In particular, then, it seems reasonable to expect that

$$\iint\limits_{R} f(x, y)\, dA = \iint\limits_{R_{uv}} f(g(u, v), h(u, v)) \left| \frac{\partial(x, y)}{\partial(u, v)} \right| dA_{uv}$$

Indeed, this is the correct *formula for changing variables in a double integral*. This formula is often written in the alternative form

$$\iint\limits_{R} f(x, y)\, dx\, dy = \iint\limits_{R_{uv}} f(g(u, v), h(u, v)) \left| \frac{\partial(x, y)}{\partial(u, v)} \right| du\, dv$$

EXAMPLE 6 Use the formula for changing variables in a double integral to confirm the rule for changing a double integral to polar coordinates.

SOLUTION   Here the polar coordinates r and θ play the roles of the variables u and v. Because

$$x = r \cos\theta \qquad \text{and} \qquad y = r \sin\theta$$

we have

$$\frac{\partial x}{\partial r} = \cos\theta \qquad\qquad \frac{\partial y}{\partial r} = \sin\theta$$

$$\frac{\partial x}{\partial \theta} = -r \sin\theta \qquad\qquad \frac{\partial y}{\partial \theta} = r \cos\theta$$

and it follows that

$$\frac{\partial(x, y)}{\partial(r, \theta)} = \begin{vmatrix} \cos \theta & \sin \theta \\ -r \sin \theta & r \cos \theta \end{vmatrix} = r(\cos^2 \theta + \sin^2 \theta) = r$$

Therefore, assuming that r is nonnegative, we have

$$\iint\limits_R f(x, y) \, dx \, dy = \iint\limits_{R_{r\theta}} f(r \cos \theta, r \sin \theta) \, r \, dr \, d\theta$$

in conformity with our earlier results.

Figure 12

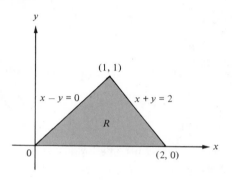

EXAMPLE 7   Evaluate $\iint\limits_R \sin [(x + y)/2] \cos [(x - y)/2] \, dA$ over the triangular region R shown in Figure 12 by using the change of variables $u = (x + y)/2$, $v = (x - y)/2$.

SOLUTION   We begin by solving for x and y in terms of the new variables u and v to obtain

$$x = u + v \qquad \text{and} \qquad y = u - v$$

Thus,

$$\frac{\partial(x, y)}{\partial(u, v)} = \begin{vmatrix} \dfrac{\partial x}{\partial u} & \dfrac{\partial y}{\partial u} \\ \dfrac{\partial x}{\partial v} & \dfrac{\partial y}{\partial v} \end{vmatrix} = \begin{vmatrix} 1 & 1 \\ 1 & -1 \end{vmatrix} = -2$$

Figure 13

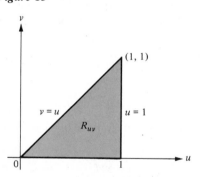

Because $v = (x - y)/2$, the condition $x - y = 0$, which holds on the upper left side of the triangular region R, is equivalent to the condition $v = 0$. Likewise, since $u = (x + y)/2$, the condition $x + y = 2$, which holds on the upper right side of R, is equivalent to $u = 1$. Finally, since $y = u - v$, the condition $y = 0$, which holds on the bottom side of R, is equivalent to $v = u$. Therefore, under the given change of variables, the triangular boundary of R in the xy plane corresponds to the right triangle with vertices $(0, 0)$, $(1, 0)$, and $(1, 1)$ in the uv plane (Figure 13). We leave it as an exercise for you to verify that the interior of R corresponds to the interior of the triangle R_{uv} in Figure 13. Notice that R_{uv} is a type I region in the uv plane. Therefore, we have

$$\iint\limits_R \sin \frac{x + y}{2} \cos \frac{x - y}{2} \, dA = \iint\limits_{R_{uv}} \sin u \cos v \, |-2| \, dA_{uv}$$

$$= \int_{u=0}^{u=1} \left(\int_{v=0}^{v=u} 2 \sin u \cos v \, dv \right) du$$

$$= \int_{u=0}^{u=1} \left(2 \sin u \sin v \, \Big|_{v=0}^{v=u} \right) du$$

$$= \int_0^1 2 \sin^2 u \, du$$

$$= \int_0^1 (1 - \cos 2u) \, du$$

$$= \left(u - \frac{\sin 2u}{2} \right) \Big|_0^1 = 1 - \frac{\sin 2}{2}$$

Problem Set 15.4

In Problems 1 to 8, use polar coordinates to evaluate each double integral over the region indicated.

1 $\displaystyle\iint_R \sqrt{4 - x^2 - y^2}\, dA$; R: $x^2 + y^2 \leq 4$, $x \geq 0$, $y \geq 0$

2 $\displaystyle\iint_R \frac{1}{(x^2 + y^2)^3}\, dA$; R: $4 \leq x^2 + y^2 \leq 9$

3 $\displaystyle\iint_R \sqrt{x^2 + y^2}\, dA$; R: $1 \leq x^2 + y^2 \leq 2$, $0 \leq y \leq \sqrt{3}x$

4 $\displaystyle\iint_R \frac{dA}{1 + x^2 + y^2}$; R: $x^2 + y^2 \leq 1$

5 $\displaystyle\iint_R (x - y)\, dA$; R: $x^2 + y^2 \leq 9$, $x \geq 0$, $y \geq 0$

6 $\displaystyle\iint_R 2x\, dA$; R: $0 \leq \theta \leq \pi/4$, $0 \leq r \leq 2 \sin \theta$

7 $\displaystyle\iint_R 3xy\, dA$; R: $0 \leq \theta \leq \pi/2$, $0 \leq r \leq 2$

8 $\displaystyle\iint_R dA$; R: $-\pi/2 \leq \theta \leq \pi/2$, $3 \leq r \leq 3(1 + \cos \theta)$

In Problems 9 to 20, evaluate each iterated integral by switching to polar coordinates.

9 $\displaystyle\int_{-3}^{3} \int_{-\sqrt{9-x^2}}^{\sqrt{9-x^2}} e^{-x^2-y^2}\, dy\, dx$

10 $\displaystyle\int_{0}^{2} \int_{0}^{\sqrt{4-x^2}} \sqrt{x^2 + y^2}\, dy\, dx$

11 $\displaystyle\int_{0}^{a} \int_{0}^{\sqrt{a^2-y^2}} (x^2 + y^2)^{3/2}\, dx\, dy$

12 $\displaystyle\int_{0}^{2} \int_{0}^{\sqrt{4-x^2}} \frac{dy\, dx}{4 + \sqrt{x^2 + y^2}}$

13 $\displaystyle\int_{0}^{3/\sqrt{2}} \int_{y}^{\sqrt{9-y^2}} x\, dx\, dy$

14 $\displaystyle\int_{0}^{a/\sqrt{2}} \int_{-\sqrt{a^2-x^2}}^{-x} y\, dy\, dx$

15 $\displaystyle\int_{-2}^{0} \int_{-\sqrt{4-y^2}}^{\sqrt{4-y^2}} x^2\, dx\, dy$

16 $\displaystyle\int_{0}^{1} \int_{y}^{\sqrt{y}} \sqrt{x^2 + y^2}\, dx\, dy$

17 $\displaystyle\int_{0}^{1} \int_{0}^{\sqrt{1-x^2}} \sin (x^2 + y^2)\, dy\, dx$

18 $\displaystyle\int_{0}^{1} \int_{0}^{\sqrt{2y-y^2}} (x^2 + y^2)^{3/2}\, dx\, dy$

19 $\displaystyle\int_{0}^{1} \int_{0}^{\sqrt{1-x^2}} e^{\sqrt{x^2+y^2}}\, dy\, dx$

20 $\displaystyle\int_{0}^{3} \int_{0}^{\sqrt{9-y^2}} \frac{y\, dx\, dy}{\sqrt{x^2 + y^2}}$

In Problems 21 to 28, find the area of each region R by setting up and evaluating a suitable double integral.

21 R: $0 \leq \theta \leq \pi$, $0 \leq r \leq 4(1 - \cos \theta)$

22 R: $0 \leq \theta \leq \pi/4$, $0 \leq r \leq 3 \cos 2\theta$

23 R: $0 \leq \theta \leq \pi$, $1 \leq r \leq 1 + \sin \theta$

24 R: $0 \leq \theta \leq \pi$, $r = 2 + \cos \theta$

25 R: the region enclosed by the graph of $r = 2 \sin^2 (\theta/2)$

26 R: the region enclosed by the lemniscate $r^2 = 2a^2 \cos 2\theta$

27 R: the region inside the circle $r = 2\sqrt{3} \sin \theta$ and outside the circle $r = 3$

28 R: the region inside the circle $r = 1$ and outside the cardioid $r = 1 - \cos \theta$

29 Find the volume of the solid in the first octant bounded by the paraboloid $z = 1 - r^2$ and the three coordinate planes.

30 Find the volume of the solid bounded above and below by the sphere $r^2 + z^2 = 4$ and bounded laterally by the cylinder $r = 1$.

31 Find the volume of the solid in the first octant bounded above by the plane $z = r \sin \theta$ and bounded laterally by the coordinate planes and by the cylinder $r = 2 \sin \theta$.

32 Find the volume of the solid in the first octant bounded by the paraboloid $z = \frac{1}{4}r^2$ and the planes $r = 2 \sec \theta$, $\theta = 0$, $\theta = \pi/4$, and $z = 0$.

33 Find the volume of the solid bounded above by the paraboloid $z = x^2 + y^2$, below by the xy plane, and inside the cylinder $x^2 + y^2 - 2y = 0$.

34 Find the volume of the solid bounded above by the ellipsoid $x^2 + y^2 + 16z^2 = 16$, below by the xy plane, and laterally by the cylinder $x^2 + y^2 = 4$.

35 Find the volume of the solid common to the sphere $x^2 + y^2 + z^2 = 16$ and the cylinder $x^2 + y^2 = 4$.

36 Find the volume of the solid bounded above by the cone $z = \sqrt{x^2 + y^2}$, below by the xy plane, and laterally by the cylinder $x^2 + y^2 = 4y$.

37 Let R be the region consisting of all points whose polar coordinates satisfy $\theta_0 \leq \theta \leq \theta_1$ and $r_0 \leq r \leq r_1$. Assume that F and G are continuous functions defined on the intervals $[\theta_0, \theta_1]$ and $[r_0, r_1]$, respectively. If (x, y) is a point in R whose polar coordinates are r and θ, define $f(x, y) = F(\theta)G(r)$. Prove that

$$\iint_R f(x, y)\, dA = \left[\int_{\theta_0}^{\theta_1} F(\theta)\, d\theta\right] \cdot \left[\int_{r_0}^{r_1} G(r)r\, dr\right]$$

38 The improper integral $\displaystyle\int_{-\infty}^{\infty} e^{-x^2}\, dx$ is important in the theory of probability. Its exact value can be found by using polar coordinates and a clever trick. Write $A = \int_{-\infty}^{\infty} e^{-x^2} \, dx$, so that

$$A^2 = \left(\int_{-\infty}^{\infty} e^{-x^2} \, dx \right) \cdot \left(\int_{-\infty}^{\infty} e^{-y^2} \, dy \right)$$

(a) Show that $A^2 = \iint\limits_{R} e^{-(x^2+y^2)} \, dx \, dy$, where R is the entire xy plane.

(b) Convert the (improper) double integral in part (a) to an equivalent integral in polar coordinates.

(c) Conclude that $\int_{-\infty}^{\infty} e^{-x^2} \, dx = \sqrt{\pi}$.

In Problems 39 to 42, find the Jacobian $\partial(x, y)/\partial(u, v)$.

39 $x = \frac{1}{2}(u - v)$, $y = \frac{1}{2}(u + v)$ **40** $x = uv$, $y = u^2 - v^2$

41 $x = ve^u$, $y = ue^v$

42 $x = au + bv + c$, $y = Au + Bv + C$

In Problems 43 and 44, evaluate the double integral by making the indicated change of variables.

43 $\iint\limits_{R} e^{x+y} \sin(x - y) \, dA$, where R is the triangular region with vertices $(0, -\pi)$, $(\pi, 0)$, $(0, \pi)$; $x = \frac{1}{2}(u + v)$, $y = \frac{1}{2}(u - v)$

44 $\iint\limits_{R} e^{x^2-y^2} \, dA$, where R is the region bounded by the curves $x = \sqrt{1 + y^2}$, $x = \sqrt{4 + y^2}$, $y = 0$, $y = (\tanh 3)x$; $x = u \cosh v$, $y = u \sinh v$

15.5   Applications of Double Integrals

We have already seen applications of double integrals in connection with areas and volumes. Double integrals also find numerous applications in engineering and physics in problems involving *density, center of mass, centroids,* and *moments*. Because many physical quantities can be approximated by Riemann sums, it is possible to define these quantities rigorously and to develop their mathematical theory on the basis of the definition of the double integral as a limit of such sums. However, practicing engineers and physicists usually deal with such quantities more informally by using the physically appealing idea of infinitesimals.

Density and Double Integrals

Consider a quantity q distributed in a continuous but perhaps nonuniform manner over a region R in the xy plane. For instance, q could be mass, electric charge, energy, or even a measure of the probability of finding a particle in various subregions of R. Let (x, y) be a point in R, and consider a small subregion ΔR of R with (x, y) in ΔR. If Δq represents the amount of the quantity q contained in ΔR and if ΔA is the area of ΔR, we define $\Delta q / \Delta A$ to be the **average density** of q over the region ΔR (Figure 1). Now imagine that ΔR "shrinks down toward the point (x, y)" so that $\Delta A \to 0$. The **density** of q at the point (x, y) is defined to be the limit (if it exists) of the average density $\Delta q / \Delta A$ as $\Delta A \to 0$. In what follows, we use the symbol $\sigma(x, y)$ to denote this density (σ is the small Greek letter "sigma"), so that

$$\sigma(x, y) = \lim_{\Delta A \to 0} \frac{\Delta q}{\Delta A}$$

Using differential notation, we can write the last equation as $\sigma(x, y) = dq/dA$, or as

$$dq = \sigma(x, y) \, dA$$

Figure 1

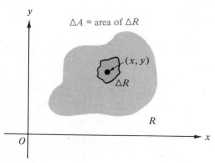

We regard this formula as giving *the infinitesimal amount dq of the quantity q contained in an infinitesimal region of area dA around the point (x, y).* We also refer to dq as an **element** of the quantity q. By "summing," or, more accurately, integrating all these elements of q, we obtain *the total amount of the quantity q contained in the region R:*

$$q = \iint\limits_{R} \sigma(x, y) \, dA$$

Figure 2

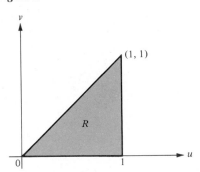

EXAMPLE 1 An electric charge of q coulombs is distributed over the triangular region R of Figure 2 so that the charge density at any point (x, y) in R is given by

$$\sigma(x, y) = (x - x^2)(y - y^2) \qquad \text{coulombs per square meter}$$

Find the total electric charge q on the region R.

SOLUTION   The total charge on R is given by

$$q = \iint\limits_{R} \sigma(x, y) \, dA = \int_{0}^{1} \int_{0}^{x} (x - x^2)(y - y^2) \, dy \, dx$$

$$= \int_{0}^{1} (x - x^2)\left(\frac{y^2}{2} - \frac{y^3}{3}\right) \Big|_{0}^{x} dx$$

$$= \int_{0}^{1} \left(\frac{x^3}{2} - \frac{5x^4}{6} + \frac{x^5}{3}\right) dx$$

$$= \left(\frac{x^4}{8} - \frac{x^5}{6} + \frac{x^6}{18}\right) \Big|_{0}^{1} = \frac{1}{72} \text{ coulomb} \qquad \blacksquare$$

Moments and Center of Mass

Figure 3

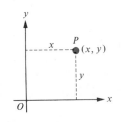

Suppose that a particle P of mass m is situated at the point (x, y) in the xy plane (Figure 3). Then the product mx, the mass m of the particle times its signed distance x from the y axis, is called the **moment of P about the y axis.** Similarly, the product my is called the **moment of P about the x axis.**

Now, suppose that a total mass m is continuously distributed over an admissible plane region R, say in the form of a thin sheet of material. Such a thin sheet is called a **lamina** (Figure 4). Let σ be the density function for this mass distribution.

If (x, y) is a point in R, consider an infinitesimal region of area dA containing (x, y). The element of mass contained in this infinitesimal region is given by

$$dm = \sigma(x, y) \, dA$$

Figure 4

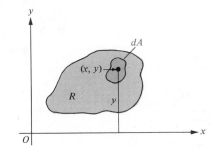

and the signed distance of dm from the x axis is y units; hence, the moment of this element of mass about the x axis is given by

$$dM_x = (dm)y = \sigma(x, y)y \, dA$$

The total **moment** M_x *of the lamina about the x axis* is obtained by "summing," or, more accurately, intregating all such infinitesimal moments, so that

$$M_x = \iint\limits_{R} \sigma(x, y)y \, dA$$

Similarly, the **moment** M_y *of the lamina about the y axis* is given by

$$M_y = \iint\limits_R \sigma(x, y)x \, dA$$

Also, the *total mass m of the lamina* is given by

$$m = \iint\limits_R \sigma(x, y) \, dA$$

By definition, the coordinates $\bar{x}$ and $\bar{y}$ of the **center of mass** of the lamina are given by the equations

$$\bar{x} = \frac{M_y}{m} = \frac{\iint\limits_R \sigma(x, y)x \, dA}{\iint\limits_R \sigma(x, y) \, dA} \quad \text{and} \quad \bar{y} = \frac{M_x}{m} = \frac{\iint\limits_R \sigma(x, y)y \, dA}{\iint\limits_R \sigma(x, y) \, dA}$$

Notice that

$$m\bar{x} = M_y \quad \text{and} \quad m\bar{y} = M_x$$

In other words:

> If all the mass m of the lamina were concentrated in a particle P at the center of mass, then the moments of P about the x and y axes would be the same as the moments of the whole lamina about the x and y axes, respectively.

In physics it is shown that *a horizontal lamina balances perfectly on a sharp point placed at its center of mass* (Figure 5).

Figure 5

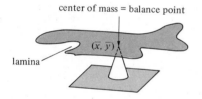

center of mass = balance point

$(\bar{x}, \bar{y})$

lamina

EXAMPLE 2   A lamina R is bounded above by the graph of $y = \sqrt[3]{x}$, below by the x axis, and on the right by the vertical line $x = 8$. The mass density of the lamina at the point (x, y) is given by $\sigma(x, y) = kx$, where k is a positive constant. Find **(a)** the total mass m of the lamina, **(b)** the moments M_x and M_y, and **(c)** the center of mass $(\bar{x}, \bar{y})$.

Figure 6

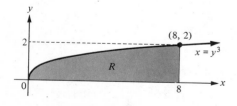

SOLUTION   The region R occupied by the lamina is of both type I and type II (Figure 6). We regard it as type II in iterating our double integrals.

(a) $m = \iint\limits_R \sigma(x, y) \, dA = \int_0^2 \int_{y^3}^8 kx \, dx \, dy = \int_0^2 \left(\frac{kx^2}{2} \Big|_{y^3}^8 \right) dy$

$= k \int_0^2 \left(32 - \frac{y^6}{2} \right) dy = k \left(32y - \frac{y^7}{14} \right) \Big|_0^2 = \frac{384k}{7}$ mass units

(b) $M_x = \iint\limits_R \sigma(x, y)y \, dA = \int_0^2 \int_{y^3}^8 kxy \, dx \, dy = k \int_0^2 \left(\frac{yx^2}{2} \Big|_{y^3}^8 \right) dy$

$= k \int_0^2 \left(32y - \frac{y^7}{2} \right) dy = k \left(16y^2 - \frac{y^8}{16} \right) \Big|_0^2 = 48k$

and $M_y = \iint\limits_R \sigma(x, y)x\, dA = \int_0^2 \int_{y^3}^8 kx^2\, dx\, dy = k\int_0^2 \left(\frac{x^3}{3}\,\Big|_{y^3}^8\right) dy$

$= k\int_0^2 \left(\frac{512}{3} - \frac{y^9}{3}\right) dy = k\left(\frac{512}{3}\,y - \frac{y^{10}}{30}\right)\Big|_0^2 = \frac{1536k}{5}$

(c) $\bar{x} = \dfrac{M_y}{m} = \dfrac{1536k}{5}\cdot\dfrac{7}{384k} = \dfrac{28}{5}$ and $\bar{y} = \dfrac{M_x}{m} = 48k\cdot\dfrac{7}{384k} = \dfrac{7}{8}$

Hence, the center of mass is $(\bar{x}, \bar{y}) = (\frac{28}{5}, \frac{7}{8})$. ■

Centroids

A distribution of mass (or any other quantity) whose density function σ is constant on a region R is said to be **uniform,** or **homogeneous,** on R. If a quantity is distributed uniformly on R, then the amount of this quantity in any subregion R_1 of R is proportional to the area of R_1 (Problem 18).

The **centroid** of a planar region R is defined to be the center of mass of a *uniform* mass distribution on R. If the density of such a uniform distribution is given by $\sigma(x, y) = k$ for all (x, y) in R, where k is a constant, then the coordinates $(\bar{x}, \bar{y})$ of the centroid of R are

$$\bar{x} = \frac{\iint\limits_R \sigma(x, y)x\, dA}{\iint\limits_R \sigma(x, y)\, dA} = \frac{k\iint\limits_R x\, dA}{k\iint\limits_R dA} = \frac{\iint\limits_R x\, dA}{\iint\limits_R dA}$$

$$\bar{y} = \frac{\iint\limits_R \sigma(x, y)y\, dA}{\iint\limits_R \sigma(x, y)\, dA} = \frac{k\iint\limits_R y\, dA}{k\iint\limits_R dA} = \frac{\iint\limits_R y\, dA}{\iint\limits_R dA}$$

Since $A = \iint\limits_R dA$ is the area of the region R, we can also write

$$\bar{x} = \frac{1}{A}\iint\limits_R x\, dA \qquad\text{and}\qquad \bar{y} = \frac{1}{A}\iint\limits_R y\, dA$$

Figure 7

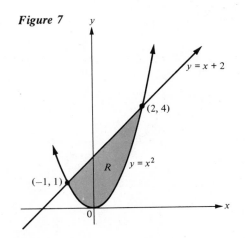

EXAMPLE 3 Find the centroid of the region R bounded by $y = x + 2$ and $y = x^2$.

SOLUTION   The region R is of type I (Figure 7), and we have

$$A = \int_{-1}^2 \int_{x^2}^{x+2} dy\, dx = \int_{-1}^2 (x + 2 - x^2)\, dx$$

$$= \left(\frac{x^2}{2} + 2x - \frac{x^3}{3}\right)\Big|_{-1}^2$$

$$= \tfrac{9}{2} \text{ square units}$$

Therefore,

$$\bar{x} = \frac{1}{A} \iint_R x \, dA = \frac{2}{9} \int_{-1}^{2} \int_{x^2}^{x+2} x \, dy \, dx$$

$$= \frac{2}{9} \int_{-1}^{2} [x(x+2) - x^3] \, dx = \frac{2}{9} \left(\frac{x^3}{3} + x^2 - \frac{x^4}{4} \right) \Big|_{-1}^{2}$$

$$= \frac{2}{9} \left(\frac{8}{3} - \frac{5}{12} \right) = \frac{1}{2}$$

and

$$\bar{y} = \frac{1}{A} \iint_R y \, dA = \frac{2}{9} \int_{-1}^{2} \int_{x^2}^{x+2} y \, dy \, dx = \frac{2}{9} \int_{-1}^{2} \frac{1}{2} [(x+2)^2 - x^4] \, dx$$

$$= \frac{1}{9} \left(\frac{x^3}{3} + 2x^2 + 4x - \frac{x^5}{5} \right) \Big|_{-1}^{2} = \frac{1}{9} \left[\frac{184}{15} - \left(-\frac{32}{15} \right) \right] = \frac{8}{5}$$

Consequently, the centroid of R is $(\bar{x}, \bar{y}) = (\frac{1}{2}, \frac{8}{5})$. ■

The centroid of a plane region is a purely geometric notion and is independent of the physical concept of mass. Indeed, if $(\bar{x}, \bar{y})$ is the centroid of a region R, then $\bar{x}$ should be thought of as the "average" x coordinate of points in R, and $\bar{y}$ should be regarded as the "average" y coordinate of points in R. However, as can be confirmed by experiment, if a thin sheet of metal of uniform density is cut in the shape of R, it will balance perfectly on a sharp point placed at the centroid of R (Figure 8). Thus, if R has an axis of symmetry, then the centroid of R must lie on this axis. Also, the position of the centroid of a region is independent of the choice of the coordinate system; hence, in finding the centroid, the coordinate axes can be chosen for convenience of calculation.

Figure 8

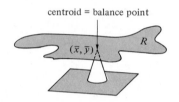
centroid = balance point

EXAMPLE 4   Find the centroid of the region R inside the circle $r = 4 \sin \theta$ and outside the circle $r = 2$ (Figure 9).

SOLUTION   Solving for the points of intersection of the circles, we obtain $(2, \pi/6)$ and $(2, 5\pi/6)$ (in polar coordinates). Thus, the region R can be described in polar coordinates by the conditions $\pi/6 \leq \theta \leq 5\pi/6$ and $2 \leq r \leq 4 \sin \theta$. By symmetry, the centroid of R lies on the y axis; hence, $\bar{x} = 0$ and it is only necessary to find $\bar{y}$. The area A of R is given by

Figure 9

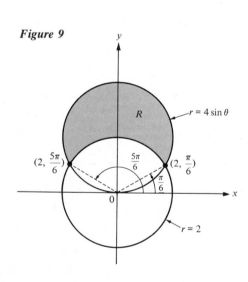

$$A = \iint_R dA = \int_{\pi/6}^{5\pi/6} \int_{2}^{4 \sin \theta} r \, dr \, d\theta$$

$$= \frac{1}{2} \int_{\pi/6}^{5\pi/6} \left(r^2 \Big|_{2}^{4 \sin \theta} \right) d\theta = \int_{\pi/6}^{5\pi/6} (8 \sin^2 \theta - 2) \, d\theta$$

$$= \int_{\pi/6}^{5\pi/6} \left[8 \left(\frac{1 - \cos 2\theta}{2} \right) - 2 \right] d\theta$$

$$= \int_{\pi/6}^{5\pi/6} (2 - 4 \cos 2\theta) \, d\theta$$

$$= (2\theta - 2 \sin 2\theta) \Big|_{\pi/6}^{5\pi/6}$$

$$= \frac{4\pi}{3} + 2\sqrt{3} \text{ square units}$$

Therefore,

$$\bar{y} = \frac{1}{A} \iint_R y \, dA = \frac{1}{A} \int_{\pi/6}^{5\pi/6} \int_{2}^{4\sin\theta} r \sin\theta \, r \, dr \, d\theta$$

$$= \frac{1}{A} \int_{\pi/6}^{5\pi/6} \left[\sin\theta \left(\frac{r^3}{3} \right) \Big|_{2}^{4\sin\theta} \right] d\theta$$

$$= \frac{1}{A} \int_{\pi/6}^{5\pi/6} \frac{1}{3} (64 \sin^4\theta - 8 \sin\theta) \, d\theta$$

$$= \frac{1}{3A} \int_{\pi/6}^{5\pi/6} \left[64 \left(\frac{3 - 4\cos 2\theta + \cos 4\theta}{8} \right) - 8 \sin\theta \right] d\theta$$

$$= \frac{8}{3A} \int_{\pi/6}^{5\pi/6} (3 - 4\cos 2\theta + \cos 4\theta - \sin\theta) \, d\theta$$

$$= \frac{8}{3A} \left(3\theta - 2\sin 2\theta + \frac{\sin 4\theta}{4} + \cos\theta \right) \Big|_{\pi/6}^{5\pi/6}$$

$$= \frac{8}{3A} \left[\left(\frac{5\pi}{2} + \frac{3\sqrt{3}}{8} \right) - \left(\frac{\pi}{2} - \frac{3\sqrt{3}}{8} \right) \right] = \frac{1}{3A} (16\pi + 6\sqrt{3})$$

$$= \frac{16\pi + 6\sqrt{3}}{4\pi + 6\sqrt{3}} = \frac{8\pi + 3\sqrt{3}}{2\pi + 3\sqrt{3}}$$

Thus, the centroid of R is

$$(\bar{x}, \bar{y}) = \left(0, \frac{8\pi + 3\sqrt{3}}{2\pi + 3\sqrt{3}} \right)$$

The following theorem provides an interesting connection between centroids and volumes of solids of revolution.

THEOREM 1

Theorem of Pappus for Volumes of Solids of Revolution

Let R be an admissible planar region lying in the same plane as a line L and entirely on one side of L. Let r be the distance from the centroid of R to the line L, and denote the area of R by A. Then, the volume V of the solid of revolution generated by revolving R about the line L is given by $V = 2\pi rA$.

Figure 10

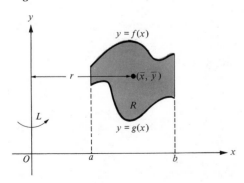

For an indication of a proof of Pappus' theorem, consider the case in which R is a type I region in the first quadrant (Figure 10) and L is the y axis. Notice that $r = \bar{x}$, so we must prove that $V = 2\pi\bar{x}A$. By definition of $\bar{x}$, we have

$$\bar{x}A = \iint_R x \, dA = \int_a^b \int_{g(x)}^{f(x)} x \, dy \, dx$$

$$= \int_a^b x[f(x) - g(x)] \, dx$$

Therefore, using the method of cylindrical shells (Section 6.2), we obtain

$$V = 2\pi \int_a^b x[f(x) - g(x)] \, dx = 2\pi\bar{x}A$$

as desired.

Figure 11

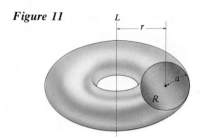

EXAMPLE 5 A circular disk R of radius a units is revolved about an axis L which is in the same plane as R and r units from the center of R, $r > a$. Find the volume V of the doughnut-shaped solid (*torus*) thus generated (Figure 11).

SOLUTION   By symmetry, the centroid of R is its center. Since the area of R is given by $A = \pi a^2$, Pappus' theorem gives $V = 2\pi r A = 2\pi^2 r a^2$. ∎

Moment of Inertia

Consider a force $\mathbf{F}$ acting at a point P in a rigid body, and let $\overline{AB}$ be an axis not passing through P. Let O be the point at the foot of a perpendicular dropped from P to $\overline{AB}$ (Figure 12). Denote the absolute value of the scalar component of $\mathbf{F}$ in the direction perpendicular to the plane containing $\overline{AB}$ and $\overline{OP}$ by F_p. If $F_p \neq 0$, then the force $\mathbf{F}$ tends to cause the body to turn about the axis $\overline{AB}$; in fact, it produces a definite angular acceleration α radians per second squared about this axis. The quantity L defined by

Figure 12

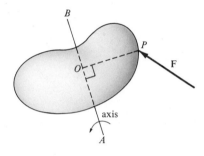

$$L = F_p|\overline{OP}|$$

is called the magnitude of the **torque** caused by the application of the force $\mathbf{F}$. It is shown in elementary mechanics that L is proportional to the angular acceleration,

$$L = I_{\overline{AB}}\,\alpha$$

where the constant of proportionality $I_{\overline{AB}}$, which is called the **moment of inertia** of the body about the axis $\overline{AB}$, depends only on the axis $\overline{AB}$ and the distribution of mass in the body.

In Figure 13, a particle P of mass m is attached to the origin O by a rigid massless rod of length $r = |\overline{OP}|$ and made to move in a circle in the yz plane by a force $\mathbf{F}$ lying in the yz plane and perpendicular to $\overline{OP}$. If θ denotes the angle in radians between the y axis and $\overline{OP}$, then, by definition, $\alpha = d^2\theta/dt^2$ gives the angular acceleration of P about the x axis. Here it is easy to show that

Figure 13

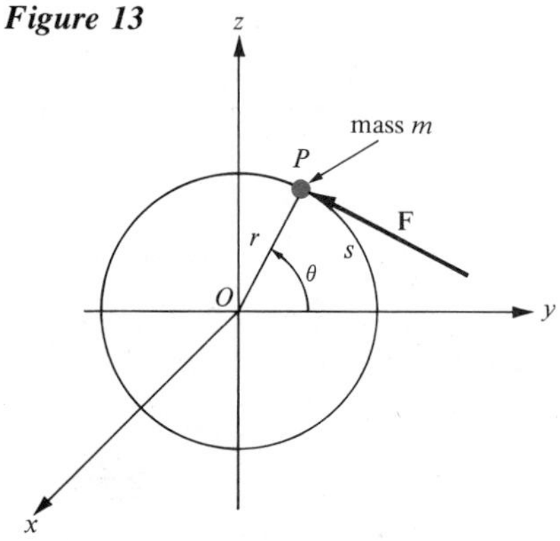

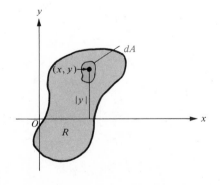

$$|\mathbf{F}| = mr\,\frac{d^2\theta}{dt^2}$$

(Problem 56); hence, multiplying by r and noting that $L = F_p|\overline{OP}| = |\mathbf{F}|r$ and $\alpha = d^2\theta/dt^2$, we obtain

$$L = mr^2\alpha$$

Thus, *the moment of inertia of a particle P about an axis* (in this case, the x axis) is given by

$$\boxed{I = mr^2}$$

where m is the mass of the particle and r is its distance from the axis.

Using the result above, we can now tackle the problem of finding the moment of inertia of a lamina, occupying a region R in the plane, about an axis lying in this plane—say the x axis for definiteness. Suppose that the lamina has density $\sigma(x, y)$ at the point (x, y), and consider an infinitesimal region of area dA containing (x, y) (Figure 14). The element of mass contained in this infinitesimal region is given by

Figure 14

$$dm = \sigma(x, y)\,dA$$

and the distance from dm to the x axis is $|y|$ units; hence its moment of inertia about the x axis is

$$dI_x = |y|^2\,dm = \sigma(x, y)y^2\,dA$$

The total *moment of inertia I_x of the lamina about the x axis* is obtained by "summing," or, more accurately, integrating all such infinitesimal quantities dI_x, so that

$$I_x = \iint\limits_R \sigma(x, y)y^2 \, dA$$

Reasoning in a similar way, we find that the *moment of inertia I_y of the lamina about the y axis* is given by

$$I_y = \iint\limits_R \sigma(x, y)x^2 \, dA$$

More generally, the moment of inertia I_l about the line l: $ax + by + c = 0$ is given by

$$I_l = \iint\limits_R \sigma(x, y) \frac{(ax + by + c)^2}{a^2 + b^2} \, dA$$

(Problem 50).

We can also ask for the moment of inertia of the lamina about an axis perpendicular to the plane of the lamina. If this axis passes through the origin, the result, called the **polar moment of inertia I_o**, is given by

$$I_o = \iint\limits_R \sigma(x, y)(x^2 + y^2) \, dA = I_x + I_y$$

(Problem 57).

<u>EXAMPLE 6</u> Find the moments of inertia I_x, I_y, and I_o of a square lamina whose sides are 2 centimeters long and parallel to the x and y axes and whose center is at the origin. Assume that the lamina is homogeneous (that is, its mass is distributed uniformly) and that its total mass is 8 grams.

SOLUTION   The area of the lamina is 4 square centimeters. Since it is homogeneous, its mass density is a constant, $\frac{8}{4}$ grams per square centimeter. Thus, $\sigma(x, y) = 2$ for all points (x, y) within the lamina. Here

$$I_x = \iint\limits_R \sigma(x, y)y^2 \, dA = 2 \iint\limits_R y^2 \, dA = 2 \int_{-1}^{1} \int_{-1}^{1} y^2 \, dx \, dy$$

$$= 2 \int_{-1}^{1} 2y^2 \, dy = \left(\frac{4y^3}{3} \right) \bigg|_{-1}^{1} = \frac{8}{3} \text{ g} \cdot \text{cm}^2$$

$$I_y = \iint\limits_R \sigma(x, y)x^2 \, dA = 2 \iint\limits_R x^2 \, dA = 2 \int_{-1}^{1} \int_{-1}^{1} x^2 \, dx \, dy$$

$$= \frac{2}{3} \int_{-1}^{1} \left(x^3 \bigg|_{-1}^{1} \right) dy = \frac{2}{3} \int_{-1}^{1} 2 \, dy = \frac{8}{3} \text{ g} \cdot \text{cm}^2$$

and

$$I_o = \iint\limits_R \sigma(x, y)(x^2 + y^2) \, dA = I_x + I_y = \frac{16}{3} \text{ g} \cdot \text{cm}^2$$

EXAMPLE 7 Find the moment of inertia I_x of a lamina occupying the region R: $0 \leq x \leq 1$, $0 \leq y \leq \sqrt{1 - x^2}$ if the mass density at the point (x, y) is given by $\sigma(x, y) = 3y^3$ kilograms per square meter.

SOLUTION

$$I_x = \iint_R \sigma(x, y) y^2 \, dA$$

$$= \int_0^1 \int_0^{\sqrt{1-x^2}} (3y^3) y^2 \, dy \, dx = \int_0^1 \left(\frac{y^6}{2} \bigg|_0^{\sqrt{1-x^2}} \right) dx$$

$$= \frac{1}{2} \int_0^1 (1 - x^2)^3 \, dx = \frac{1}{2} \int_0^1 (1 - 3x^2 + 3x^4 - x^6) \, dx$$

$$= \frac{1}{2} \left(x - x^3 + \frac{3x^5}{5} - \frac{x^7}{7} \right) \bigg|_0^1$$

$$= \tfrac{8}{35} \text{ kg} \cdot \text{m}^2 \qquad \blacksquare$$

Problem Set 15.5

1 Electric charge is distributed over the region R bounded by the parabolas $y^2 = x$ and $x^2 = y$ with a charge density σ given by $\sigma(x, y) = x^2 + 4y^2$ coulombs per square centimeter. Find the total electric charge on the region R.

2 One coulomb of electric charge is distributed uniformly over the region R bounded by the parabola $y = x^2$ and the straight line $y = x + 2$. Find the (constant) value of the charge density if distances are measured in centimeters.

In Problems 3 to 17, assume that a lamina with mass density function σ occupies the region R in the xy plane. Find the total mass m and the coordinates $(\bar{x}, \bar{y})$ of the center of mass of the distribution. Assume that mass is measured in kilograms and the distance in meters.

3 R: $0 \leq x \leq 3$, $0 \leq y \leq 3 - x$; $\sigma(x, y) = x^2 + y^2$

4 R: $2 \leq x \leq 3$, $2 \leq y \leq 5$; $\sigma(x, y) = xy$

5 R: $0 \leq y \leq 4$, $y^2 - 2y \leq x \leq 2y$; $\sigma(x, y) = 6$

6 R: $0 \leq y \leq 2$, $\sqrt{3 - y} \leq x \leq 3 - y$; $\sigma(x, y) = 2x$

7 R: $0 \leq y \leq 1$, $y^2 \leq x \leq 1$; $\sigma(x, y) = 3x + 5y$

8 R: $-3 \leq x \leq 3$, $-3 \leq y \leq \sqrt{9 - x^2}$; $\sigma(x, y) = y + 3$

9 R: the triangle with vertices $(0, 0)$, $(3, 0)$, and $(3, 5)$; $\sigma(x, y) = x$

10 R: the triangle with vertices $(0, 0)$, $(4, 0)$, and $(4, 4)$; $\sigma(x, y) = y^2$

11 R: the region bounded by $y^2 = x$ and $y = x$; $\sigma(x, y) = 2x + 1$

12 R: the region bounded by the curves $y = x - 1$ and $y = 1 - x^2$; $\sigma(x, y) = x^2 + y^2$

13 R: the region in the first quadrant bounded by the curves $y = x + x^2$, $y = 0$, $x = 0$, and $x = 2$; $\sigma(x, y) = y/(1 + x)$

14 R: the region in the first quadrant bounded by the curves $y = e^x$, $x = 0$, $y = 0$, and $x = 2$; $\sigma(x, y)$ is proportional to the distance of the point (x, y) from the x axis and $\sigma(0, 1) = 1$.

15 R: $x^2 + y^2 \leq 1$, $x \geq 0$, $y \geq 0$; $\sigma(x, y) = \sqrt{x^2 + y^2}$

16 R: the region enclosed by the cardioid $r = 1 + \cos \theta$; $\sigma(x, y) = \sqrt{x^2 + y^2}$

17 R: the region enclosed by one leaf of the four-leaved rose $r = \sin 2\theta$; $\sigma(x, y) = \sqrt{x^2 + y^2}$. Use the leaf in the first quadrant.

18 Suppose that mass is distributed uniformly on a plane region R. Show that the mass on an admissible subregion R_1 of R is proportional to the area of R_1.

In Problems 19 to 32, find the centroid $(\bar{x}, \bar{y})$ of the region R bounded by the given curves.

19 $y = x$, $y = 0$, and $x = 1$

20 $y = 6x - x^2$ and $y = x$

21 $y^2 = 2x$ and $y = x$

22 $y = 2\sqrt{x}$ and $y = x^2/4$

23 $y = x^2$ and $y = 4x - x^2$

24 $y = \sqrt{25 - x^2}$ and $xy = 12$

25 $y = \dfrac{x^2}{4}$, $xy = 2$, and $3y = 2x + 4$, $x \geq 1$, $y \geq 1$

26 $y^2 = x^3$, $x + y = 2$, and $x = 4$

27 $y = x^{3/2}$ and $y = x$

28 $y = \sin x$, $y = x$, and $x = \pi/2$

29 Inside the circle $r = 4 \cos \theta$ and outside the circle $r = 2$

30 The first-quadrant region that lies inside the curve $r = 2 \sin 2\theta$ and outside the circle $r = \sqrt{3}$

31 Inside the cardioid $r = 2(1 + \cos \theta)$ and outside the circle $r = 2$

32 The leaf of $r = 4 \cos 2\theta$ that cuts the positive x axis

In Problems 33 to 36, use the theorem of Pappus to find the volume of the solid generated by revolving the region bounded by the given curves about the indicated axis.

33 $y = x^2$ and $y = 2x + 3$ about the x axis

34 $y = x^3$, $x + y = 2$, and $y = 0$ about the line $y = -2$

35 $2x + y = 2$, $x = 0$, and $y = 0$ about the y axis

36 $x = \sqrt{4 - y}$, $x = 0$, and $y = 0$ about the line $x = -2$

In Problems 37 to 48, find the moment of inertia I_x and the moment of inertia I_y about the x and y axes, respectively, of a homogeneous lamina of total mass 1 kilogram occupying each region R. Assume that all distances are measured in meters.

37 R: $0 \leq x \leq 1$, $x^2 \leq y \leq x$

38 R: $0 \leq x \leq 4$, $x^2 + 1 \leq y \leq 4x + 1$

39 R: $0 \leq x \leq 1$, $x^3 \leq y \leq \sqrt{x}$

40 R: $-1 \leq x \leq 8$, $3x - 8 \leq y \leq 10x - x^2$

41 R: $0 \leq x \leq 1$, $x^4 \leq y \leq x^2$

42 R: the region bounded by $y = x^3/4$ and $y = |x|$

43 R: the region bounded by $xy = 4$ and $2x + y = 6$

44 R: the region bounded by $y = e^x$, $y = e$, and $x = 0$

45 R: $0 \leq \theta \leq \pi$, $0 \leq r \leq 1 - \cos \theta$

46 R: $-\dfrac{\pi}{4} \leq \theta \leq \dfrac{\pi}{4}$, $0 \leq r \leq 2 \cos 2\theta$

47 R: $0 \leq \theta \leq \dfrac{\pi}{3}$, $2 \cos \theta \leq r \leq 2$

48 R: $-\dfrac{\pi}{4} \leq \theta \leq \dfrac{\pi}{4}$, $0 \leq r^2 \leq 8 \cos 2\theta$

49 Find the moments of inertia I_x, I_y, and I_o for a lamina occupying the region R: $0 \leq x \leq 2$, $0 \leq y \leq 2 - x$ if the mass density function σ is given by $\sigma(x, y) = x + 2y$ grams per square centimeter.

50 A lamina occupies the admissible region R and has mass density function σ. Show that the moment of inertia of the lamina about the line l: $ax + by + c = 0$ is given by

$$I_l = \iint\limits_{R} \sigma(x, y) \frac{(ax + by + c)^2}{a^2 + b^2} \, dA$$

51 A lamina occupies the region R: $-\pi/2 \leq x \leq \pi/2$, $0 \leq y \leq \cos x$ and has mass density function σ given by $\sigma(x, y) = y$ grams per square centimeter. Find its moment of inertia I_x about the x axis.

52 Find the moment of inertia about the x axis, I_x, about the y axis, I_y, and about the origin, I_o, of a homogeneous lamina in the shape of one leaf of the four-leaved rose $r = a \cos 2\theta$. Use the leaf that cuts the positive x axis, and assume that the total mass of the lamina is m grams.

53 Find the moment of inertia about the y axis, I_y, of a homogeneous lamina of total mass m grams in the shape of a lemniscate $r^2 = 2a^2 \cos 2\theta$. Assume that distances are measured in centimeters.

54 A circular lamina of radius r_1 is centered at the origin. Its total mass is m grams, and its density at a point r units from the origin is proportional to r^n, where n is a constant. Find its moment of inertia I_x about the x axis. Assume that distances are measured in centimeters.

55 Find the polar moment of inertia I_o of a ring-shaped lamina occupying the region $\frac{1}{4} \leq x^2 + y^2 \leq 1$ if the mass density function σ is given by $\sigma(x, y) = (x^2 + y^2)^{-1}$ grams per square centimeter.

56 In Figure 13, let $\mathbf{R}$ be the variable position vector of P, so that $\mathbf{R} = (r \cos \theta)\mathbf{j} + (r \sin \theta)\mathbf{k}$. Here, $s = r\theta$. (a) Show that the tangential component vector of the acceleration is given by $r\alpha\mathbf{T}$, where $\alpha = d^2\theta/dt^2$. (b) Show that the normal component vector of the acceleration is given by $-(d\theta/dt)^2\mathbf{R}$. (c) Since $\mathbf{F}$ is the tangential force acting on P, use part (a) to obtain the result $|\mathbf{F}| = mr\alpha$. (d) Calculate the normal force acting on P, and explain what provides this force.

57 Derive the formula for the moment of inertia of a lamina occupying an admissible region R in the xy plane about an axis perpendicular to the plane of the lamina and passing through the point with xy coordinates (a, b). Assume that σ is the mass density function for the lamina. In particular, show that $I_o = I_x + I_y$.

15.6   Triple Integrals

Triple integrals over solids in xyz space are defined by an obvious analogy with the definition of double integrals over regions in the xy plane. The details of the definition are best left to advanced calculus, so here we simply outline the main ideas.

Given a solid region S in three-dimensional space, such as a rectangular box, a cube, a pyramid, a ball, an ellipsoid, and so forth, and given a function f of three variables defined at each point (x, y, z) in S, we define the **triple integral**

$$\iiint\limits_S f(x, y, z)\, dV$$

(if it exists) as follows.

First we enclose the solid S in a rectangular box B with edges parallel to the coordinate axes (Figure 1). The box B is now partitioned into a large number of smaller boxes by intersecting it with planes parallel to the coordinate planes (Figure 2). These smaller boxes are called the **cells** of the partition. All cells of the partition that do not touch the solid region S are now discarded. The remaining cells, which, taken together, contain the solid S and roughly approximate its shape, are now numbered in some convenient way and called, say, $\Delta S_1, \Delta S_2, \ldots, \Delta S_m$. The maximum diagonal measurement of all these cells is called the **norm** of the partition and denoted by η (the Greek letter "eta").

Points are now chosen, one from each cell ΔS_1, $\Delta S_2, \ldots, \Delta S_m$, in such a way that each chosen point belongs to S, and the point chosen from the kth cell is denoted by (x_k^*, y_k^*, z_k^*) for $k = 1, 2, \ldots, m$. The partition, together with the chosen points, is called an **augmented partition.**

Corresponding to each augmented partition we can form a **Riemann sum**

$$\sum_{k=1}^{m} f(x_k^*, y_k^*, z_k^*)\, \Delta V_k$$

where ΔV_k is the volume of the kth cell ΔS_k. We can now define the triple integral to be the limit (if it exists) of such Riemann sums as the number of cells becomes larger and larger in such a way that the norm η approaches zero; in symbols,

$$\iiint\limits_S f(x, y, z)\, dV = \lim_{\eta \to 0} \sum_{k=1}^{m} f(x_k^*, y_k^*, z_k^*)\, \Delta V_k$$

If we write

$$dV = dx\, dy\, dz$$

Figure 1

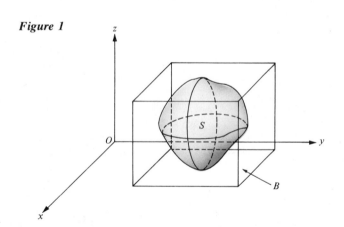

Figure 2

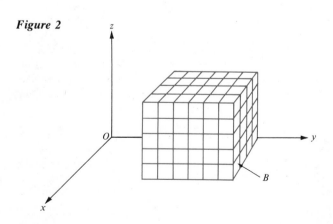

then we can write the triple integral of f over S in the alternative form:

$$\iiint\limits_{S} f(x, y, z)\, dV = \iiint\limits_{S} f(x, y, z)\, dx\, dy\, dz$$

If the limit of Riemann sums that defines the triple integral of the function f exists, then f is said to be *(Riemann) integrable* on the solid S. In advanced calculus, it is shown that if S is an admissible three-dimensional region (Section 6.1) which contains all its own boundary, and if f is continuous on S, then f is integrable on S. Thus, triple integrals satisfy the analog of Property 1 (the existence property) for double integrals set forth in Section 15.2. As a matter of fact, the obvious analogs of Properties 1 through 8 of double integrals given in Section 15.2 hold for triple integrals (Problem 38). For instance, the analog of Property 2 is the following: If S is an admissible three-dimensional region of volume V, then

$$V = \iiint\limits_{S} dV$$

It can be shown that a triple integral of a continuous function over an appropriately shaped solid can be reduced to an equivalent iterated integral—here, however, the iterated integral involves a *double* integral. In fact, we have the following.

Procedure for Evaluation of Triple Integrals by Iteration

Let R be an admissible region in the xy plane which contains all its own boundary, and suppose that g and h are continuous functions defined on R and satisfying $g(x, y) \le h(x, y)$ for all points (x, y) in R. Let S be the solid consisting of all points (x, y, z) satisfying the conditions that (x, y) belongs to R and

$$g(x, y) \le z \le h(x, y)$$

(Figure 3). Then, if f is a continuous function defined on S,

$$\iiint\limits_{S} f(x, y, z)\, dV = \iint\limits_{R} \left[\int_{z=g(x,y)}^{z=h(x,y)} f(x, y, z)\, dz \right] dA$$

Figure 3

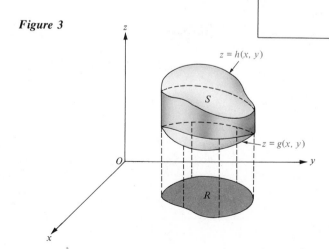

In this iteration procedure, notice that the solid S is bounded *above* by the surface $z = h(x, y)$, *below* by the surface $z = g(x, y)$, and *laterally* by the cylinder over the boundary of R with generators parallel to the z axis. The inside integral

$$\int_{z=g(x,y)}^{z=h(x,y)} f(x, y, z)\, dz$$

is, of course, calculated while x and y are temporarily held constant. After it is calculated, the outside double integral can be evaluated by using the methods given in Sections 15.2, 15.3, and 15.4.

EXAMPLE 1 Evaluate $\iiint_S (x + y + z)\, dV$, where S is the solid bounded above by the plane $z = 2 - x - y$, below by the plane $z = 0$, and laterally by the cylinder over the boundary of the triangular region R: $0 \le x \le 1, 0 \le y \le 1 - x$ (Figure 4).

SOLUTION We use the iteration procedure with $g(x, y) = 0$, $h(x, y) = 2 - x - y$, and R as described. Thus,

$$\begin{aligned}
\iiint_S (x + y + z)\, dV &= \iint_R \left[\int_{z=0}^{z=2-x-y} (x + y + z)\, dz\right] dA \\
&= \iint_R \left[\left(xz + yz + \frac{z^2}{2}\right)\Big|_0^{2-x-y}\right] dA \\
&= \iint_R \left[x(2 - x - y) + y(2 - x - y) + \frac{(2 - x - y)^2}{2}\right] dA \\
&= \iint_R \left(2 - \frac{x^2}{2} - xy - \frac{y^2}{2}\right) dA \\
&= \int_{x=0}^{x=1} \left[\int_{y=0}^{y=1-x} \left(2 - \frac{x^2}{2} - xy - \frac{y^2}{2}\right) dy\right] dx \\
&= \int_{x=0}^{x=1} \left[\left(2y - \frac{x^2 y}{2} - \frac{xy^2}{2} - \frac{y^3}{6}\right)\Big|_{y=0}^{y=1-x}\right] dx \\
&= \int_0^1 \left(\frac{11}{6} - 2x + \frac{x^3}{6}\right) dx \\
&= \left(\frac{11}{6}x - x^2 + \frac{x^4}{24}\right)\Big|_0^1 = \frac{7}{8}
\end{aligned}$$

■

Figure 4

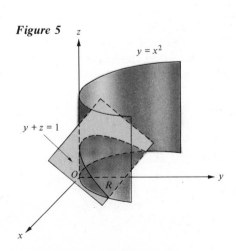

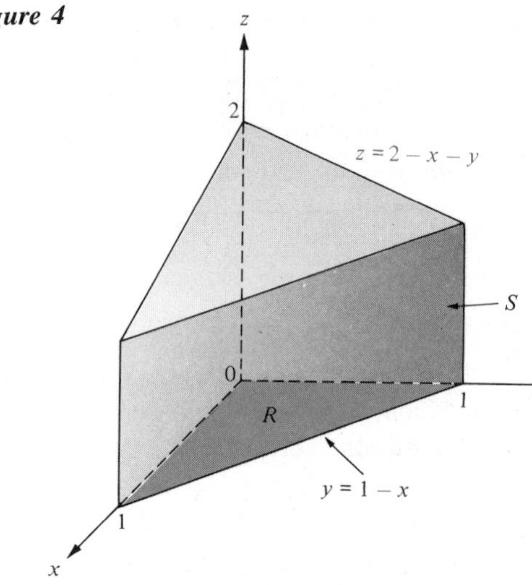

EXAMPLE 2 Find the volume V of the solid bounded by the parabolic cylinder $y = x^2$, the xy plane, and the plane $y + z = 1$.

SOLUTION The volume V is given by

$$V = \iiint_S dV$$

where S is the solid given by

$$S: 0 \le z \le 1 - y \qquad (x, y) \text{ in } R$$

and R is the region in the xy plane given by

$$R: -1 \le x \le 1 \qquad x^2 \le y \le 1$$

(Figure 5). Thus,

$$\begin{aligned}
V &= \iint_R \left(\int_{z=0}^{z=1-y} dz\right) dA = \iint_R (1 - y)\, dA \\
&= \int_{-1}^{1} \int_{x^2}^{1} (1 - y)\, dy\, dx = \int_{-1}^{1} \left(y - \frac{y^2}{2}\right)\Big|_{x^2}^{1} dx \\
&= \int_{-1}^{1} \left(\frac{1}{2} - x^2 + \frac{x^4}{2}\right) dx \\
&= \left(\frac{1}{2}x - \frac{x^3}{3} + \frac{x^5}{10}\right)\Big|_{-1}^{1} = \frac{8}{15} \text{ cubic unit}
\end{aligned}$$

■

Figure 5

Figure 6

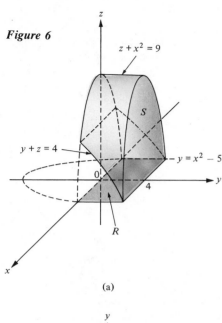

(a)

(b)

© **EXAMPLE 3**   Find the volume V of the solid S bounded by $z + x^2 = 9$, $y + z = 4$, $y = 0$, and $y = 4$. (Round off your answer to two decimal places.)

SOLUTION   The surface $z + x^2 = 9$ is a parabolic cylinder, opening downward, with generators parallel to the y axis; hence, it must form the upper boundary surface of S. The surface $y + z = 4$ is a plane which cuts the plane $z = 0$ in the line $y = 4$ and forms the lower boundary surface of S (Figure 6a). The desired volume V is given by

$$V = \iiint\limits_{S} dV$$

To evaluate the triple integral by iteration, we must determine the region R obtained by projecting the solid S perpendicularly onto the xy plane. Obviously, the lines $y = 0$ and $y = 4$ provide two of the boundaries of R. In order to determine the remainder of the boundary of R, we notice that the upper and lower surfaces bounding the solid S meet in space in a curve consisting of all points (x, y, z) satisfying the simultaneous equations

$$\begin{cases} z + x^2 = 9 \\ y + z = 4 \end{cases}$$

The remaining boundary of R is obtained by projecting this curve perpendicularly on the xy plane, and this can be accomplished algebraically by eliminating the variable z from the two simultaneous equations. Thus, $y + 9 - x^2 = 4$ or $y = x^2 - 5$, so that $x = \pm\sqrt{y + 5}$.

The region R is evidently of type II, and is described by $0 \leq y \leq 4$ and $-\sqrt{y + 5} \leq x \leq \sqrt{y + 5}$ (Figure 6b). Thus,

$$V = \iiint\limits_{S} dV = \iint\limits_{R} \left(\int_{z=4-y}^{z=9-x^2} dz \right) dA = \iint\limits_{R} (5 - x^2 + y)\, dA$$

$$= \int_{y=0}^{y=4} \left[\int_{x=-\sqrt{y+5}}^{x=\sqrt{y+5}} (5 - x^2 + y)\, dx \right] dy$$

$$= \int_{y=0}^{y=4} \left[\left(5x - \frac{x^3}{3} + xy \right) \Big|_{x=-\sqrt{y+5}}^{x=\sqrt{y+5}} \right] dy$$

$$= \int_{0}^{4} \left[10\sqrt{y + 5} - \frac{2}{3}(y + 5)^{3/2} + 2y\sqrt{y + 5} \right] dy$$

$$= \left[\frac{20}{3}(y + 5)^{3/2} - \frac{4}{15}(y + 5)^{5/2} + \frac{4}{5}(y + 5)^{5/2} - \frac{20}{3}(y + 5)^{3/2} \right] \Big|_{0}^{4}$$

$$= \left[\frac{8}{15}(y + 5)^{5/2} \right] \Big|_{0}^{4} = \frac{8}{15}(243 - 25\sqrt{5}) \approx 99.79 \text{ cubic units} \qquad ■$$

In the iteration procedure for triple integrals, we can change the roles of the variables x, y, and z. For instance, in Figure 7 we have a solid S bounded on the left and right by surfaces $y = g(x, z)$ and $y = h(x, z)$, respectively, and bounded laterally by the cylinder over the boundary of the region R with generators parallel to the y axis. Here, the admissible region R is contained in the xz plane, and we have

Figure 7

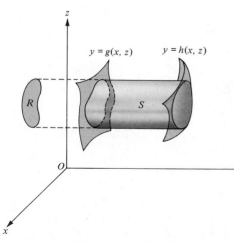

$$\iiint\limits_{S} f(x, y, z)\, dV = \iint\limits_{R} \left[\int_{y=g(x,z)}^{y=h(x,z)} f(x, y, z)\, dy \right] dA$$

provided that f is a continuous function on S. A similar result obtains if S is bounded behind by $x = g(y, z)$, in front by $x = h(y, z)$, and laterally by a cylinder with generators parallel to the x axis.

In summary, then, there are really three cases of the iteration theorem for triple integrals of continuous functions over solid regions, depending on the shape of the solid:

Case 1 R is an admissible region in the xy plane, and the solid S consists of all points (x, y, z) such that (x, y) is in R and $g(x, y) \le z \le h(x, y)$. Then

$$\iiint\limits_{S} f(x,\ y,\ z)\ dV = \iint\limits_{R} \left[\int_{z=g(x,y)}^{z=h(x,y)} f(x,\ y,\ z)\ dz \right] dA$$

Case 2 R is an admissible region in the xz plane, and the solid S consists of all points (x, y, z) such that (x, z) is in R and $g(x, z) \le y \le h(x, z)$. Then

$$\iiint\limits_{S} f(x,\ y,\ z)\ dV = \iint\limits_{R} \left[\int_{y=g(x,z)}^{y=h(x,z)} f(x,\ y,\ z)\ dy \right] dA$$

Case 3 R is an admissible region in the yz plane, and the solid S consists of all points (x, y, z) such that (y, z) is in R and $g(y, z) \le x \le h(y, z)$. Then

$$\iiint\limits_{S} f(x,\ y,\ z)\ dV = \iint\limits_{R} \left[\int_{x=g(y,z)}^{x=h(y,z)} f(x,\ y,\ z)\ dx \right] dA$$

EXAMPLE 4   Evaluate $\displaystyle\iiint\limits_{S} 3z\ dV$ if S is the solid bounded by $x = 0$, $y = 0$, $z = 1$, $z = 0$, and $x + y + z = 2$.

SOLUTION   The surface $x + y + z = 2$ is a plane with x, y, and z intercepts all equal to 2, and it forms the front boundary surface of S. The remaining boundary surfaces are the coordinate planes and the plane $z = 1$ (Figure 8). We treat S as in case 3, with the region R in the yz plane described by

$$R: 0 \le z \le 1 \qquad 0 \le y \le 2 - z$$

and the solid S described by

$$S: 0 \le x \le 2 - y - z \qquad (y, z)\ \text{in}\ R$$

Thus,

$$\iiint\limits_{S} 3z\ dV = \iint\limits_{R} \left(\int_{x=0}^{x=2-y-z} 3z\ dx \right) dA = \iint\limits_{R} 3z(2 - y - z)\ dA$$

$$= \int_{z=0}^{z=1} \left[\int_{y=0}^{y=2-z} 3z(2 - y - z)\ dy \right] dz$$

$$= \int_{z=0}^{z=1} \left[\left(6yz - \frac{3}{2} y^2 z - 3yz^2 \right) \Big|_{y=0}^{y=2-z} \right] dz$$

$$= \int_{0}^{1} \left(\frac{3}{2} z^3 - 6z^2 + 6z \right) dz = \left(\frac{3z^4}{8} - 2z^3 + 3z^2 \right) \Big|_{0}^{1} = \frac{11}{8} \quad \blacksquare$$

The integral in the previous example can be written as

$$\iint\limits_{R} \left(\int_{x=0}^{x=2-y-z} 3z\ dx \right) dy\ dz = \int_{z=0}^{z=1} \left[\int_{y=0}^{y=2-z} \left(\int_{x=0}^{x=2-y-z} 3z\ dx \right) dy \right] dz$$

Figure 8

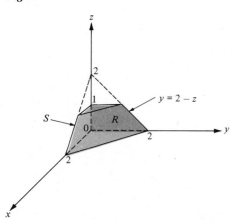

and, in this form, it is called a **threefold iterated integral.** Unless confusion threatens, the brackets and the detailed information about the limits of integration are usually omitted, and the threefold iterated integral is simply written as

$$\int_0^1 \int_0^{2-z} \int_0^{2-y-z} 3z \, dx \, dy \, dz$$

The order of integration is determined by the order of the differentials, reading from left to right, just as for the twice-iterated integrals considered in Section 15.1.

<u>EXAMPLE 5</u> Evaluate the threefold iterated integral $\int_0^{\pi/2} \int_0^1 \int_0^{x^2} x \cos y \, dz \, dx \, dy$.

SOLUTION

$$\int_0^{\pi/2} \int_0^1 \int_0^{x^2} x \cos y \, dz \, dx \, dy = \int_0^{\pi/2} \left[\int_0^1 \left(\int_0^{x^2} x \cos y \, dz \right) dx \right] dy$$

$$= \int_0^{\pi/2} \left\{ \int_0^1 \left[(x \cos y) z \Big|_0^{x^2} \right] dx \right\} dy$$

$$= \int_0^{\pi/2} \left(\int_0^1 x^3 \cos y \, dx \right) dy$$

$$= \int_0^{\pi/2} \left(\frac{x^4}{4} \cos y \Big|_0^1 \right) dy$$

$$= \int_0^{\pi/2} \frac{\cos y}{4} \, dy = \frac{\sin y}{4} \Big|_0^{\pi/2} = \frac{1}{4} \quad \blacksquare$$

The evaluation of a triple integral by iteration always leads to an iterated integral of the form

$$\iint\limits_{R} \left[\int_{w=g(u,v)}^{w=h(u,v)} f(u, v, w) \, dw \right] dA$$

where u, v, and w represent the variables x, y, and z in some order and where R is an admissible region in the uv plane. If the plane region R is of type I or of type II, the last integral can be rewritten as a threefold iterated integral having either the form

$$\int_{u=a}^{u=b} \int_{v=G(u)}^{v=H(u)} \int_{w=g(u,v)}^{w=h(u,v)} f(u, v, w) \, dw \, dv \, du$$

or the form

$$\int_{v=a}^{v=b} \int_{u=G(v)}^{u=H(v)} \int_{w=g(u,v)}^{w=h(u,v)} f(u, v, w) \, dw \, du \, dv$$

Problem Set 15.6

In Problems 1 to 6, evaluate each threefold iterated integral.

1 $\int_0^1 \int_0^{3x} \int_0^{2x+y} y \, dz \, dy \, dx$ **2** $\int_0^2 \int_0^{2y} \int_1^{y+3z} 5x \, dx \, dz \, dy$ **4** $\int_0^2 \int_0^{\sqrt{4-y^2}} \int_0^y xz \, dz \, dx \, dy$ **5** $\int_0^{2\pi} \int_0^\pi \int_1^2 z^4 \sin y \, dz \, dy \, dx$

3 $\int_0^1 \int_0^x \int_0^z (x + z) \, dy \, dz \, dx$ **6** $\int_0^{\pi/2} \int_0^{2\sin y} \int_0^{2-(z^2/2)} z \, dx \, dz \, dy$

In Problems 7 to 10, (a) evaluate each threefold iterated integral; (b) rewrite the integral as an iterated integral of the form

$$\iint_R \left[\int_{g(u,v)}^{h(u,v)} f(u, v, w) \, dw \right] dA,$$ where u, v, and w are the variables

x, y, and z in some order; and (c) rewrite the integral as a triple integral.

7 $\int_0^2 \int_0^{2-y} \int_0^{\sqrt{x}} y \, dz \, dx \, dy$ **8** $\int_0^1 \int_0^y \int_0^{\sqrt{1-y^2}} z \, dz \, dx \, dy$

9 $\int_0^1 \int_0^{\sqrt{1-z^2}} \int_0^{\sqrt{1-z^2}} xyz \, dy \, dx \, dz$

10 $\int_0^{\pi} \int_0^y \int_z^{z+y} \sin(x+y) \, dx \, dz \, dy$

In Problems 11 to 20, sketch the solid S, and evaluate each triple integral.

11 $\iiint_S (3x + 2y) \, dV$, where S is the solid bounded above by the plane $z = 4$, below by the plane $z = 0$, and laterally by the cylinder with generators parallel to the z axis over the boundary of the square region R: $-1 \le x \le 1$, $-1 \le y \le 3$

12 $\iiint_S 3xy \, dV$, where S is the solid in the first octant bounded by the coordinate planes and the plane $x + y + z = 6$

13 $\iiint_S y \, dV$, where S is the solid bounded by the cylinder $x = y^2$ and the planes $x + z = 1$ and $z = 0$

14 $\iiint_S xy^2z^3 \, dV$, where S is the solid determined by the conditions $0 \le x \le 1$, $0 \le y \le x$, and $0 \le z \le xy$

15 $\iiint_S yz \, dV$, where S is the solid bounded by the cylinder $y = 9 - x^2$ and the planes $y + z = 9$, $y = 0$, and $z = 0$

16 $\iiint_S (xy + 2y) \, dV$, where S is the solid bounded by the cylinder $x^2 + y^2 = 4$ and the planes $x + z = 2$, $y = 0$, and $z = 0$

17 $\iiint_S \sqrt{x^2 + y^2} \, dV$, where S is the solid determined by the conditions $x^2 + y^2 \le 1$ and $0 \le z \le \sqrt{x^2 + y^2}$

18 $\iiint_S x^2y^2 \, dV$, where S is the solid bounded by the planes $z = 0$, $z = 1$, $x + y = 0$, $x + y = 1$, $x - y = 0$, and $x - y = 1$

19 $\iiint_S x \, dV$, where S is the solid bounded by the cylinder $z = 1 - x^2$ and the planes $y = -1$, $y = 1$, and $z = 0$

© **20** $\iiint_S z \sin(x + y) \, dV$, where S is the solid bounded by the planes $z = x - y$, $z = x + y$, $y = x$, and $x = 1$ (Round off your answer to two decimal places.)

In Problems 21 to 28, use triple integration to find the volume of each solid S.

21 S is the solid bounded above by $z = y$, below by $z = 0$, and laterally by the cylinder $y = 1 - x^2$.

22 S is the solid in the first octant bounded by the cylinder $y = 4 - x^2$ and the planes $z = x$, $y = 0$, and $z = 0$.

23 S is the solid bounded by the planes $x + z = 1$, $y = x$, $y = 0$, and $z = 0$.

24 S is the solid bounded by the cylinder $z = 4 - y^2$ and the planes $x + z = 9$, $x = 0$, and $z = 0$.

25 S is the solid bounded by the parabolic cylinder $z - y^2 = 0$ and the planes $z = y - x$ and $x = 0$.

26 S is the solid bounded by the elliptic paraboloids $z = 18 - x^2 - y^2$ and $z = x^2 + 5y^2$ (Figure 9).

Figure 9

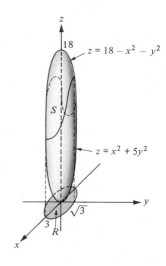

27 S is the solid bounded by the parabolic cylinders $z = 5 - x^2$ and $z = 4x^2$ and by the planes $x + y = 1$ and $y = 0$.

28 S is the solid bounded by the paraboloid of revolution $z = x^2 + y^2$ and the plane $z = 2 - x$.

In Problems 29 to 32, express the triple integral $\iiint_S f(x, y, z) \, dV$ as a threefold iterated integral in six different ways (that is, with six different orders of integration) for the given solid.

29 S: $0 \le x \le 1$, $0 \le y \le 2$, $0 \le z \le 3$

30 S: $0 \le x \le 1$, $0 \le y \le 2$, $-3 \le z \le 3$

31 S: $0 \le x \le 1$, $0 \le y \le 1 - x$, $0 \le z \le 1 - x - y$

32 S: $-1 \le x \le 1$, $-\sqrt{1 - x^2} \le y \le \sqrt{1 - x^2}$,
$0 \le z \le 1 - x^2 - y^2$

In Problems 33 to 36, describe a solid S whose volume is given by the threefold iterated integral, and evaluate the integral to find this volume.

33 $\displaystyle\int_0^1 \int_0^{1-z} \int_0^y dx\, dy\, dz$

34 $\displaystyle\int_0^2 \int_{x^2}^4 \int_3^6 dz\, dy\, dx$

35 $\displaystyle\int_0^2 \int_0^{\sqrt{4-x^2}} \int_0^4 dy\, dz\, dx$

36 $\displaystyle\int_{-1}^1 \int_0^{1-x^2} \int_0^6 dy\, dz\, dx$

37 Express the threefold iterated integral

$$\int_0^1 \int_0^{\sqrt{1-x^2}} \int_0^{\sqrt{1-x^2-y^2}} f(x,\, y,\, z)\, dz\, dy\, dx$$

as an equivalent threefold iterated integral in which the order of integration is $dx\, dy\, dz$.

38 State the analogs for triple integrals of Properties 3 to 8 of double integrals given in Section 15.2.

39 Express the threefold iterated integral

$$\int_0^3 \int_0^{\sqrt{9-z^2}} \int_0^x f(x,\, y,\, z)\, dy\, dx\, dz$$

as an equivalent threefold iterated integral in which the order of integration is $dz\, dy\, dx$.

40 Suppose that f is a continuous function defined on all xyz space and that $\displaystyle\iiint_S f(x,\, y,\, z)\, dV = 0$ for every solid S. Prove that f is the constant zero function.

15.7 Triple Integrals in Cylindrical and Spherical Coordinates

In Section 15.4 we found that conversion to polar coordinates can render certain double integrals easier to evaluate. Similarly, as we show in this section, conversion to cylindrical or to spherical coordinates can be advantageous in evaluating triple integrals.

Conversion to Cylindrical Coordinates

A triple integral can be converted to a threefold iterated integral in cylindrical coordinates according to the following procedure.

Procedure for Conversion of a Triple Integral to Cylindrical Coordinates

Let θ_0 and θ_1 be constants such that $0 < \theta_1 - \theta_0 \le 2\pi$, and suppose that G and H are continuous functions such that $0 \le G(\theta) \le H(\theta)$ holds for all values of θ in $[\theta_0, \theta_1]$. Let g and h be continuous functions such that $g(r, \theta) \le h(r, \theta)$ holds for all values of r and θ with $\theta_0 \le \theta \le \theta_1$ and $G(\theta) \le r \le H(\theta)$. Denote by S the solid consisting of all points whose cylindrical coordinates (r, θ, z) satisfy the conditions

$$\theta_0 \le \theta \le \theta_1 \qquad G(\theta) \le r \le H(\theta) \qquad g(r, \theta) \le z \le h(r, \theta)$$

(Figure 1). Then, if f is a continuous function defined for all points (x, y, z) in the solid S,

$$\iiint_S f(x,\, y,\, z)\, dV = \int_{\theta_0}^{\theta_1} \int_{G(\theta)}^{H(\theta)} \int_{g(r,\theta)}^{h(r,\theta)} f(r \cos\theta,\, r \sin\theta,\, z) r\, dz\, dr\, d\theta$$

To see why this procedure works, denote by R the region in the xy plane consisting of all points whose polar coordinates (r, θ) satisfy $\theta_0 \le \theta \le \theta_1$ and $G(\theta) \le r \le H(\theta)$. Also, let $z = a(x, y)$ and $z = b(x, y)$ be equations of the lower

Figure 1

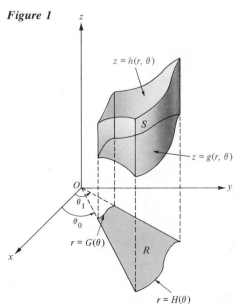

and upper boundary surfaces of S, respectively, written in Cartesian rather than cylindrical coordinates. Then, by the iteration procedure of Section 15.6,

$$\iiint_S f(x,\ y,\ z)\ dV = \iint_R \left[\int_{z=a(x,y)}^{z=b(x,y)} f(x,\ y,\ z)\ dz \right] dA$$

Now, if the double integral over the region R in the last equation is converted to polar coordinates as in Section 15.4, the result is the formula shown in the procedure above (Problem 34).

__EXAMPLE 1__   Express the integral $\displaystyle\int_0^2 \int_0^{\sqrt{4-x^2}} \int_0^6 \sqrt{x^2+y^2}\ dz\ dy\ dx$ as an equivalent threefold iterated integral in cylindrical coordinates, and then evaluate the integral obtained.

SOLUTION   The given threefold iterated integral is equivalent to the triple integral $\displaystyle\iiint_S \sqrt{x^2+y^2}\ dV$, where

$$S: 0 \le x \le 2 \qquad 0 \le y \le \sqrt{4-x^2} \qquad 0 \le z \le 6$$

(Figure 2). The region

$$R: 0 \le x \le 2 \qquad 0 \le y \le \sqrt{4-x^2}$$

is the portion of the circular disk $x^2 + y^2 \le 4$ lying in the first quadrant; hence, its description in polar coordinates is

$$R: 0 \le \theta \le \frac{\pi}{2} \qquad 0 \le r \le 2$$

Using the procedure for converting a triple integral to cylindrical coordinates, we have

$$\iiint_S \sqrt{x^2+y^2}\ dV = \int_0^{\pi/2} \int_0^2 \int_0^6 \sqrt{(r\cos\theta)^2 + (r\sin\theta)^2}\ r\ dz\ dr\ d\theta$$

$$= \int_0^{\pi/2} \int_0^2 \int_0^6 \sqrt{r^2}\ r\ dz\ dr\ d\theta = \int_0^{\pi/2} \int_0^2 \left(r^2 z \Big|_0^6 \right) dr\ d\theta$$

$$= \int_0^{\pi/2} \int_0^2 6r^2\ dr\ d\theta = \int_0^{\pi/2} \left(2r^3 \Big|_0^2 \right) d\theta$$

$$= \int_0^{\pi/2} 16\ d\theta = 16\theta \Big|_0^{\pi/2} = 8\pi$$

__EXAMPLE 2__   Use cylindrical coordinates to find the volume of a right circular cone whose base radius is a units and whose altitude is h units (Figure 3).

SOLUTION   If we denote the solid cone by S and place its base R on the xy plane and its vertex on the z axis at the point $(0,\ 0,\ h)$, then its volume is given by

$$V = \iiint_S dV$$

The upper boundary surface of S is formed by revolving the line in the yz plane whose equation is $(y/a) + (z/h) = 1$ about the z axis; hence, the equation in Cartesian coordinates of the upper boundary surface is

$$\frac{\pm\sqrt{x^2+y^2}}{a} + \frac{z}{h} = 1 \qquad \text{or} \qquad z = h \mp \frac{h}{a}\sqrt{x^2+y^2}$$

Figure 2

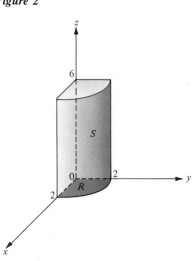

Figure 3

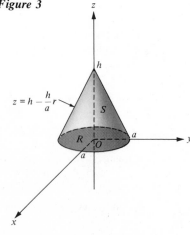

(Section 13.8). Since $z \leq h$ holds for the portion of the surface that actually bounds S, we must use the minus sign in the above equation, so that

$$z = h - \frac{h}{a}\sqrt{x^2 + y^2}$$

Converting the last equation to cylindrical coordinates, using $r = \sqrt{x^2 + y^2}$, we obtain

$$z = h - \frac{h}{a}r$$

for the equation of the upper boundary surface of S.

Since the base R of the solid cone S can be described in polar coordinates by $R: 0 \leq \theta \leq 2\pi$, $0 \leq r \leq a$, it follows that

$$V = \iiint_S dV = \int_0^{2\pi} \int_0^a \int_0^{h-(h/a)r} r \, dz \, dr \, d\theta = \int_0^{2\pi} \int_0^a \left(h - \frac{h}{a}r\right)r \, dr \, d\theta$$

$$= \int_0^{2\pi} \left(\frac{ha^2}{2} - \frac{ha^3}{3a}\right) d\theta = 2\pi\left(\frac{ha^2}{2} - \frac{ha^2}{3}\right) = \frac{h}{3}\pi a^2 \quad \text{cubic units}$$

This reconfirms the familiar result that the volume of a cone is one-third its height times the area of its base. ∎

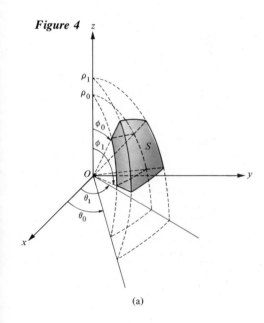

Figure 4

(a)

Conversion to Spherical Coordinates

A triple integral can be converted to spherical coordinates according to the following procedure.

Procedure for Conversion of a Triple Integral to Spherical Coordinates

Let θ_0, θ_1, ϕ_0, ϕ_1, ρ_0, and ρ_1 be constants such that $0 < \theta_1 - \theta_0 \leq 2\pi$ and $0 \leq \rho_0 < \rho_1$. Suppose that the solid S consists of all points whose spherical coordinates (ρ, θ, ϕ) satisfy the conditions

$$\rho_0 \leq \rho \leq \rho_1 \qquad \theta_0 \leq \theta \leq \theta_1 \qquad \phi_0 \leq \phi \leq \phi_1$$

(Figure 4a). Then, if f is a continuous function defined for all points (x, y, z) in the solid S,

$$\iiint_S f(x, y, z) \, dV$$

$$= \int_{\phi_0}^{\phi_1} \int_{\theta_0}^{\theta_1} \int_{\rho_0}^{\rho_1} f(\rho \sin \phi \cos \theta, \rho \sin \phi \sin \theta, \rho \cos \phi) \, \rho^2 \sin \phi \, d\rho \, d\theta \, d\phi$$

To see why this procedure works, consider Figure 4b, which shows a portion ΔV of the volume V of the solid S in Figure 4a corresponding to small changes in ρ, θ, and ϕ of amounts $\Delta\rho$, $\Delta\theta$, and $\Delta\phi$, respectively. Thus, in spherical coordinates,

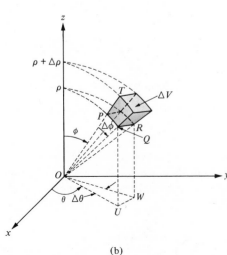

(b)

$$P = (\rho, \theta, \phi) \qquad\qquad R = (\rho, \theta + \Delta\theta, \phi + \Delta\phi)$$

$$Q = (\rho, \theta, \phi + \Delta\phi) \qquad T = (\rho + \Delta\rho, \theta, \phi)$$

Notice that ΔV is approximately the volume of a box with dimensions $|\overline{PQ}|$, $|\overline{QR}|$, and $|\overline{PT}|$. Clearly,

$$|\overline{PT}| = \Delta\rho$$

Since P and Q lie on a circle of radius $|\overline{OP}| = |\overline{OQ}| = \rho$, and since the arc $\widehat{PQ}$ subtends the angle $\Delta\phi$, we have

$$|\overline{PQ}| \approx |\widehat{PQ}| = \rho \, \Delta\phi$$

In Figure 4b, we see that

$$\phi + \Delta\phi + \text{angle } QOU = \frac{\pi}{2} \qquad \text{so} \qquad \text{angle } QOU = \frac{\pi}{2} - (\phi + \Delta\phi)$$

Thus, considering right triangle OUQ, we have

$$|\overline{OU}| = |\overline{OQ}| \cos\left[\frac{\pi}{2} - (\phi + \Delta\phi)\right] = \rho \sin(\phi + \Delta\phi) \approx \rho \sin\phi$$

Notice that $|\overline{QR}| = |\overline{UW}|$ and that U and W lie on a circle of radius $|\overline{OU}| \approx \rho \sin\phi$. Thus, since the arc $\widehat{UW}$ subtends the angle $\Delta\theta$, we have

$$|\overline{QR}| = |\overline{UW}| \approx |\overline{OU}| \, \Delta\theta \approx \rho \sin\phi \, \Delta\theta$$

Hence,

$$\Delta V \approx |\overline{PQ}| \, |\overline{QR}| \, |\overline{PT}| \approx (\rho \, \Delta\phi)(\rho \sin\phi \, \Delta\theta)(\Delta\rho) = \rho^2 \sin\phi \, \Delta\rho \, \Delta\theta \, \Delta\phi$$

As $\Delta\rho$, $\Delta\theta$, and $\Delta\phi$ become smaller and smaller, the approximation $\Delta V \approx \rho^2 \sin\phi \, \Delta\rho \, \Delta\theta \, \Delta\phi$ becomes more and more accurate, which suggests that the element of volume dV in spherical coordinates should be given by

$$\boxed{dV = \rho^2 \sin\phi \, d\rho \, d\theta \, d\phi}$$

These considerations should make the formula for converting a triple integral to spherical coordinates appear plausible.

EXAMPLE 3   Express the threefold iterated integral

$$\int_0^3 \int_0^{\sqrt{9-x^2}} \int_0^{\sqrt{9-x^2-y^2}} (x^2 + y^2 + z^2)^3 \, dz \, dy \, dx$$

as an equivalent threefold iterated integral in spherical coordinates, and then evaluate the integral obtained.

SOLUTION   The given threefold iterated integral is equivalent to the triple integral

$$\cdot \iiint_S (x^2 + y^2 + z^2)^3 \, dV$$

where S is bounded by the first octant portion of the sphere $x^2 + y^2 + z^2 = 9$ and the three coordinate planes (Figure 5). Switching to spherical coordinates, and noting that S is described by the conditions

$$S: 0 \leq \phi \leq \frac{\pi}{2} \qquad 0 \leq \theta \leq \frac{\pi}{2} \qquad 0 \leq \rho \leq 3$$

Figure 5

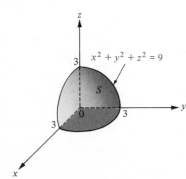

we have

$$\iiint_S (x^2 + y^2 + z^2)^3 \, dV = \int_0^{\pi/2} \int_0^{\pi/2} \int_0^3 (\rho^2)^3 \, \rho^2 \sin \phi \, d\rho \, d\theta \, d\phi$$

$$= \int_0^{\pi/2} \int_0^{\pi/2} \left(\frac{\rho^9}{9} \sin \phi \, \bigg|_0^3 \right) d\theta \, d\phi$$

$$= 2187 \int_0^{\pi/2} \left(\theta \sin \phi \, \bigg|_0^{\pi/2} \right) d\phi$$

$$= \frac{2187\pi}{2} \int_0^{\pi/2} \sin \phi \, d\phi$$

$$= \frac{2187\pi}{2} \left(-\cos \phi \, \bigg|_0^{\pi/2} \right) = \frac{2187\pi}{2} \qquad \blacksquare$$

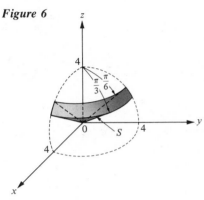

Figure 6

© **EXAMPLE 4** Find the volume V of the solid S in the first octant bounded by the sphere $\rho = 4$, the coordinate planes, the cone $\phi = \pi/6$, and the cone $\phi = \pi/3$ (Figure 6). (Round off your answer to two decimal places.)

SOLUTION

$$V = \iiint_S dV = \int_{\pi/6}^{\pi/3} \int_0^{\pi/2} \int_0^4 \rho^2 \sin \phi \, d\rho \, d\theta \, d\phi$$

$$= \int_{\pi/6}^{\pi/3} \int_0^{\pi/2} \frac{4^3}{3} \sin \phi \, d\theta \, d\phi = \int_{\pi/6}^{\pi/3} \frac{64}{3} (\sin \phi) \left(\frac{\pi}{2} \right) d\phi$$

$$= \frac{32\pi}{3} (-\cos \phi) \, \bigg|_{\pi/6}^{\pi/3} = \frac{-32\pi}{3} \left(\cos \frac{\pi}{3} - \cos \frac{\pi}{6} \right)$$

$$= \frac{16\pi}{3} (\sqrt{3} - 1) \approx 12.27 \text{ cubic units} \qquad \blacksquare$$

Change of Variables in a Triple Integral

In Section 15.4, we introduced a general formula for changing variables in a *double* integral. The corresponding formula for changing variables in a *triple* integral is as follows: To change from the variables x, y, and z to new variables u, v, and w, begin by expressing x, y, and z in terms of the new variables, so that

$$x = g(u, v, w) \qquad y = h(u, v, w) \qquad \text{and} \qquad z = k(u, v, w)$$

The corresponding **Jacobian** is defined by

$$\frac{\partial(x, y, z)}{\partial(u, v, w)} = \begin{vmatrix} \dfrac{\partial x}{\partial u} & \dfrac{\partial y}{\partial u} & \dfrac{\partial z}{\partial u} \\[2mm] \dfrac{\partial x}{\partial v} & \dfrac{\partial y}{\partial v} & \dfrac{\partial z}{\partial v} \\[2mm] \dfrac{\partial x}{\partial w} & \dfrac{\partial y}{\partial w} & \dfrac{\partial z}{\partial w} \end{vmatrix}$$

Let S be a solid in xyz space, and let S_{uvw} be the corresponding solid in uvw space. Then, if f is a continuous function on S,

$$\iiint_S f(x, y, z) \, dV =$$
$$\iiint_{S_{uvw}} f(g(u, v, w), h(u, v, w), k(u, v, w)) \left| \frac{\partial(x, y, z)}{\partial(u, v, w)} \right| dV_{uvw}$$

For instance, the change to cylindrical coordinates

$$x = r \cos \theta \qquad y = r \sin \theta \qquad \text{and} \qquad z = z$$

yields the Jacobian

$$\frac{\partial(x, y, z)}{\partial(r, \theta, z)} = r$$

(Problem 35), and the change to spherical coordinates

$$x = \rho \sin \phi \cos \theta \qquad y = \rho \sin \phi \sin \theta \qquad \text{and} \qquad z = \rho \cos \phi$$

yields the Jacobian

$$\frac{\partial(x, y, z)}{\partial(\rho, \phi, \theta)} = \rho^2 \sin \phi$$

(Problem 37), in conformity with the formulas given earlier in this section.

Problem Set 15.7

In Problems 1 to 4, evaluate each threefold iterated integral.

1 $\int_0^\pi \int_0^{3\cos\theta} \int_0^3 r \sin \theta \, dz \, dr \, d\theta$

2 $\int_0^{\pi/2} \int_0^{\sin\theta} \int_0^{r^2} r \cos \theta \, dz \, dr \, d\theta$

3 $\int_0^{\pi/6} \int_0^{2\pi} \int_0^4 \rho^2 \sin \phi \, d\rho \, d\theta \, d\phi$

4 $\int_0^{\pi/2} \int_{\pi/4}^{\phi} \int_0^{2\csc\theta} \rho^3 \sin^2 \theta \sin \phi \, d\rho \, d\theta \, d\phi$

In Problems 5 to 8, (a) rewrite each threefold iterated integral in Cartesian coordinates as an equivalent triple integral over an appropriate solid S and sketch the solid; (b) rewrite the triple integral obtained in (a) as an equivalent threefold iterated integral in cylindrical coordinates in Problems 5 and 6 and in spherical coordinates in Problems 7 and 8; and (c) evaluate the integral obtained in part (b).

5 $\int_0^5 \int_0^{\sqrt{25-x^2}} \int_0^6 \frac{dz \, dy \, dx}{\sqrt{x^2 + y^2}}$

6 $\int_0^2 \int_0^{\sqrt{4-x^2}} \int_0^{(x^2+y^2)/2} \frac{z \, dz \, dy \, dx}{\sqrt{x^2 + y^2}}$

7 $\int_0^1 \int_0^{\sqrt{1-x^2}} \int_0^{\sqrt{1-x^2-y^2}} \frac{dz \, dy \, dx}{1 + x^2 + y^2 + z^2}$

8 $\int_0^3 \int_0^{\sqrt{9-x^2}} \int_0^{\sqrt{9-x^2-y^2}} xz \, dz \, dy \, dx$

In Problems 9 to 12, convert to cylindrical coordinates, and evaluate the integral.

9 $\iiint_S \sqrt{x^2 + y^2} \, dV$, where S is the solid in the first octant bounded by the coordinate planes, the plane $z = 4$, and the cylinder $x^2 + y^2 = 25$

10 $\iiint_S (x^2 + y^2)^{3/2} \, dV$, where S is the solid bounded above by the paraboloid of revolution $z = \frac{1}{2}(x^2 + y^2)$, below by the xy plane, and laterally by the cylinder $x^2 + y^2 = 4$

11 $\iiint_S \frac{dV}{\sqrt{x^2 + y^2}}$, where S is the solid bounded above by the plane $z = 4$, below by the plane $z = 1$, and laterally by the cylinder $x^2 + y^2 = 16$

12 $\iiint_S dV$, where S is the cylindrical shell of height $h > 0$ with base on the xy plane between the cylinders $x^2 + y^2 = a^2$ and $x^2 + y^2 = b^2$, where $b > a > 0$

In Problems 13 to 16, convert to spherical coordinates, and evaluate the integral.

13 $\iiint_S \sqrt{x^2 + y^2 + z^2} \, dV$, where S is the region bounded by the sphere of radius 3 with center at the origin

14 $\iiint_S dV$, where S is the spherical shell determined by the condition $0 < a \le x^2 + y^2 + z^2 \le b$

15 $\iiint_S (x^2 + y^2 + z^2)^{3/2} \, dV$, where S is the solid in the first octant bounded by the sphere $x^2 + y^2 + z^2 = 25$, the cone $z = \sqrt{x^2 + y^2}$, and the cone $z = 2\sqrt{x^2 + y^2}$

16 $\displaystyle\iiint\limits_{S} \sin \sqrt{x^2 + y^2 + z^2}\, dV$, where S is the solid bounded above by the sphere $x^2 + y^2 + z^2 = 49$ and below by the cone $z = \sqrt{x^2 + y^2}$

In Problems 17 to 24, use an appropriate integration in cylindrical coordinates to find the volume V of each solid S.

17 S is the solid bounded above by the paraboloid of revolution $z = 1 - (x^2 + y^2)$ and below by the plane $z = 0$ (Figure 7).

Figure 7

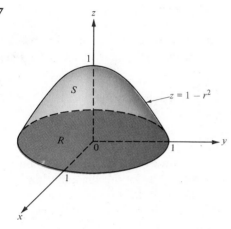

18 S is the solid bounded above by the paraboloid of revolution $z = 8 - (x^2 + y^2)$ and below by the paraboloid of revolution $z = x^2 + y^2$.

19 S is the solid in the first octant bounded by the cylinder $x^2 + y^2 = 9$, the plane $z = y$, and the coordinate planes.

20 S is the solid bounded by the plane $z = x$ and the paraboloid of revolution $z = x^2 + y^2$.

21 S is the solid in the first octant bounded by the coordinate planes and the surfaces $x^2 + y^2 = z$ and $x^2 + y^2 = 2y$.

22 S is the solid bounded above by the cone $z^2 = x^2 + y^2$, below by the plane $z = 0$, and laterally by the cylinder $x^2 + y^2 = 4x$.

23 S is the solid bounded above by the sphere $x^2 + y^2 + z^2 = 8$ and below by the paraboloid $x^2 + y^2 = 2z$.

24 S is the solid bounded below by the cone $z = \sqrt{x^2 + y^2}$ and above by the sphere $x^2 + y^2 + (z - \frac{1}{2})^2 = \frac{1}{4}$.

In Problems 25 to 32, use an appropriate integration in spherical coordinates to find the volume V of each solid S.

25 S is a solid sphere of radius a.

26 S is the solid bounded above by the sphere $\rho = 3$ and below by the cone $\phi = \pi/3$.

27 S: $0 \le \phi \le \pi/4$, $\pi/6 \le \theta \le \pi/3$, $2 \le \rho \le 4$

28 S is the solid inside the cone $\phi = \pi/6$ and between the spheres $\rho = 2$ and $\rho = 4$.

29 S is the solid inside both the sphere $x^2 + y^2 + z^2 = 1$ and the cone $z = \sqrt{x^2 + y^2}$.

30 S is bounded laterally by the sphere $x^2 + y^2 + z^2 = a^2$ and lies between the lower and upper nappes of the right circular cone $z^2 = b^2(x^2 + y^2)$, where a and b are positive constants.

©31 S is bounded above by the sphere $x^2 + y^2 + z^2 = 9$ and below by the cone $z = 2\sqrt{x^2 + y^2}$. (Round off your answer to two decimal places.)

32 S is the solid consisting of all points whose spherical coordinates satisfy $0 \le \theta \le 2\pi$, $0 \le \phi \le \pi$, and $0 \le \rho \le 2(1 - \cos \phi)$.

33 Find the volume of each of the two parts into which a solid sphere of radius $a > 0$ is cut by a plane passing k units from its center, where $0 < k < a$.

34 Complete the argument making the procedure for conversion of a triple integral to cylindrical coordinates plausible by actually iterating the double integral

$$\iint\limits_{R} \left[\int_{z=a(x,y)}^{z=b(x,y)} f(x, y, z)\, dz \right] dA$$

in polar coordinates as suggested in the text. What assumptions do you need to make?

35 Let $x = r \cos \theta$, $y = r \sin \theta$, and $z = z$. Show that

$$\left| \frac{\partial(x, y, z)}{\partial(r, \theta, z)} \right| = r$$

Then express $\displaystyle\iiint\limits_{S} f(x, y, z)\, dV$ in cylindrical coordinates.

36 Suppose that S is a solid consisting of all points whose spherical coordinates satisfy the conditions $\theta_0 \le \theta \le \theta_1$, $G(\theta) \le \phi \le H(\theta)$, and $g(\phi, \theta) \le \rho \le h(\phi, \theta)$, where G, H, g, and h are continuous functions. If f is a continuous function defined on S, find a threefold iterated integral in spherical coordinates equivalent to $\displaystyle\iiint\limits_{S} f(x, y, z)\, dV$.

37 Let $x = \rho \sin \phi \cos \theta$, $y = \rho \sin \phi \sin \theta$, and $z = \rho \cos \phi$. Show that

$$\left| \frac{\partial(x, y, z)}{\partial(\rho, \phi, \theta)} \right| = \rho^2 \sin \phi$$

Then express $\displaystyle\iiint\limits_{S} f(x, y, z)\, dV$ in spherical coordinates.

38 The "hypervolume" of a "four-dimensional hypersphere" of radius a is given by $2\displaystyle\iiint\limits_{S} \sqrt{a^2 - x^2 - y^2 - z^2}\, dV$, where S is the ordinary solid sphere of radius a in xyz space. Evaluate this hypervolume.

15.8   Applications of Triple Integrals

In Section 15.5 we applied the double integral to the solution of problems involving density, moments, centers of mass, centroids, and moments of inertia. In this section we present similar applications of triple integrals.

Density and Triple Integrals

Consider a quantity, such as mass or electric charge, which is distributed in a continuous, but perhaps nonuniform, manner over a portion of three-dimensional *xyz* space. By analogy with the two-dimensional case considered in Section 15.5, a function σ of three variables is called a **density function** for this three-dimensional distribution if, for every solid three-dimensional region S, the triple integral

$$q = \iiint\limits_{S} \sigma(x, y, z)\, dV$$

exists and gives the amount q of the quantity contained in the region S.

The value of the density $\sigma(a, b, c)$ of a three-dimensional distribution at the point (a, b, c) can be interpreted as the limiting value of the amount of the quantity per unit volume in a small three-dimensional region ΔS around the point (a, b, c) as ΔS "shrinks to zero" in the sense that the maximum distance between any two points in ΔS approaches zero (Problem 41). It follows that an element dq of the quantity q is given by

$$dq = \sigma(x, y, z)\, dV$$

Moments and Center of Mass

Consider a continuous distribution of mass over a solid S with mass density function σ. The total mass of S is given by

$$m = \iiint\limits_{S} \sigma(x, y, z)\, dV$$

By analogy with the definitions given for two-dimensional distributions in Section 15.5, we define the **moments** of the distribution *about the xy plane, the xz plane,* and *the yz plane* by

$$M_{xy} = \iiint\limits_{S} z\sigma(x, y, z)\, dV$$

$$M_{xz} = \iiint\limits_{S} y\sigma(x, y, z)\, dV$$

and $\quad M_{yz} = \iiint\limits_{S} x\sigma(x, y, z)\, dV$

respectively. Similarly, the point

$$(\bar{x}, \bar{y}, \bar{z}) = \left(\frac{M_{yz}}{m}, \frac{M_{xz}}{m}, \frac{M_{xy}}{m} \right)$$

is called the **center of mass** of the distribution.

<u>EXAMPLE 1</u>   Find the mass m and the center of mass $(\bar{x}, \bar{y}, \bar{z})$ of a solid S in the first octant bounded by the coordinate planes, the plane $z = 2$, and the cylinder $x^2 + y^2 = 9$ if its density at a point P is proportional to the distance of P from the z axis.

SOLUTION   The density at the point (x, y, z) is given by $\sigma(x, y, z) = k\sqrt{x^2 + y^2}$, where k is the constant of proportionality. We have

$$m = \iiint_S \sigma(x, y, z)\, dV = \iiint_S k\sqrt{x^2 + y^2}\, dV$$

Using cylindrical coordinates, we have

$$m = \int_0^{\pi/2} \int_0^3 \int_0^2 (kr)r\, dz\, dr\, d\theta = \int_0^{\pi/2} \int_0^3 2kr^2\, dr\, d\theta$$

$$= \int_0^{\pi/2} \left(\frac{2kr^3}{3} \Big|_0^3 \right) d\theta = 18k \int_0^{\pi/2} d\theta = 9\pi k \qquad \text{mass units}$$

Here,

$$M_{xy} = \iiint_S z\sigma(x, y, z)\, dV = \iiint_S zk\sqrt{x^2 + y^2}\, dV$$

$$= \int_0^{\pi/2} \int_0^3 \int_0^2 (zkr)r\, dz\, dr\, d\theta = \int_0^{\pi/2} \int_0^3 \left(\frac{z^2 kr^2}{2} \Big|_0^2 \right) dr\, d\theta$$

$$= \int_0^{\pi/2} \int_0^3 2kr^2\, dr\, d\theta = 9\pi k$$

just as above. Also,

$$M_{xz} = \iiint_S y\sigma(x, y, z)\, dV = \iiint_S yk\sqrt{x^2 + y^2}\, dV$$

$$= \int_0^{\pi/2} \int_0^3 \int_0^2 [(r \sin \theta)kr]r\, dz\, dr\, d\theta = \int_0^{\pi/2} \int_0^3 2kr^3 \sin \theta\, dr\, d\theta$$

$$= \int_0^{\pi/2} \left(\frac{kr^4 \sin \theta}{2} \Big|_0^3 \right) d\theta = \frac{81k}{2} \int_0^{\pi/2} \sin \theta\, d\theta$$

$$= \frac{81k}{2} (-\cos \theta) \Big|_0^{\pi/2} = \frac{81k}{2}$$

Similarly,

$$M_{yz} = \iiint_S x\sigma(x, y, z)\, dV = \iiint_S xk\sqrt{x^2 + y^2}\, dV$$

$$= \int_0^{\pi/2} \int_0^3 \int_0^2 [(r \cos \theta)kr]r\, dz\, dr\, d\theta = \frac{81k}{2}$$

Therefore,

$$\bar{x} = \frac{M_{yz}}{m} = \frac{81k/2}{9\pi k} = \frac{9}{2\pi}$$

$$\bar{y} = \frac{M_{xz}}{m} = \frac{81k/2}{9\pi k} = \frac{9}{2\pi}$$

and

$$\bar{z} = \frac{M_{xy}}{m} = \frac{9\pi k}{9\pi k} = 1$$

Hence, the center of mass of the distribution is

$$(\bar{x}, \bar{y}, \bar{z}) = \left(\frac{9}{2\pi}, \frac{9}{2\pi}, 1\right)$$

Centroids

The center of mass of a uniform distribution (that is, a distribution with constant mass density) over a solid S is called the **centroid** of S. If we denote the volume of S by V, then we have the following formulas for the centroid $(\bar{x}, \bar{y}, \bar{z})$ of S (Problem 43):

$$\bar{x} = \frac{1}{V} \iiint_S x \, dV \qquad \bar{y} = \frac{1}{V} \iiint_S y \, dV \qquad \bar{z} = \frac{1}{V} \iiint_S z \, dV$$

As before, if S has an axis of symmetry, then its centroid must lie on this axis.

Figure 1

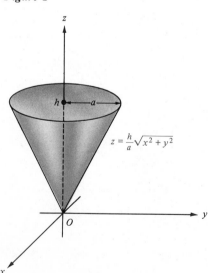

EXAMPLE 2   Find the centroid of the solid right circular cone bounded above by $z = h$ and bounded below by $z = (h/a)\sqrt{x^2 + y^2}$ (Figure 1).

SOLUTION   The volume V of the cone is given by

$$V = \frac{h}{3}\pi a^2$$

By symmetry, the centroid must lie on the z axis; hence, $\bar{x} = 0, \bar{y} = 0$, and it only remains to find $\bar{z}$. Using cylindrical coordinates, we have

$$\bar{z} = \frac{1}{V} \iiint_S z \, dV = \frac{1}{V} \int_{\theta=0}^{\theta=2\pi} \int_{r=0}^{r=a} \int_{z=(h/a)r}^{z=h} zr \, dz \, dr \, d\theta$$

$$= \frac{1}{V} \int_0^{2\pi} \int_0^a \left[\frac{h^2}{2} - \frac{(h/a)^2 r^2}{2}\right] r \, dr \, d\theta = \frac{h^2}{2V} \int_0^{2\pi} \int_0^a \left(r - \frac{r^3}{a^2}\right) dr \, d\theta$$

$$= \frac{h^2}{2V} \int_0^{2\pi} \left(\frac{a^2}{2} - \frac{a^4}{4a^2}\right) d\theta = \frac{h^2}{2V} \int_0^{2\pi} \frac{a^2}{4} \, d\theta$$

$$= \frac{2\pi h^2 a^2}{8V} = \frac{2\pi h^2 a^2}{8(h/3)\pi a^2}$$

$$= \frac{3h}{4}$$

Therefore, the centroid of the cone is $(\bar{x}, \bar{y}, \bar{z}) = (0, 0, 3h/4)$.

Figure 2

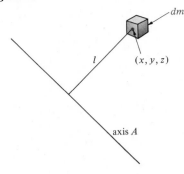

Moment of Inertia

Again consider a continuous distribution of mass over a solid S with mass density function σ. Let A be a fixed line in space, and consider an element dm of the mass m at the point (x, y, z) (Figure 2). Here,

$$dm = \sigma(x, y, z)\, dV$$

and the moment of inertia dI_A of dm about the axis A is given by

$$dI_A = l^2\, dm = l^2\sigma(x, y, z)\, dV$$

where l is the distance from (x, y, z) to the line A. Integrating over the solid S, we obtain the formula

$$I_A = \iiint_S l^2\sigma(x, y, z)\, dV$$

for the moment of inertia S about the axis A.

In particular, the moments of inertia I_x, I_y, and I_z of S about the x, y, and z axes are given by

$$I_x = \iiint_S (y^2 + z^2)\sigma(x, y, z)\, dV$$

$$I_y = \iiint_S (x^2 + z^2)\sigma(x, y, z)\, dV$$

$$\text{and } \quad I_z = \iiint_S (x^2 + y^2)\sigma(x, y, z)\, dV$$

EXAMPLE 3   Consider the spherical shell S: $10 \le \sqrt{x^2 + y^2 + z^2} \le 11$ with mass density at a point P inversely proportional to the square of the distance from P to the origin. Find its moment of inertia I_z.

SOLUTION   Here, $\sigma(x, y, z) = k/(x^2 + y^2 + z^2)$, where k is the constant of proportionality; hence,

$$I_z = \iiint_S (x^2 + y^2)\frac{k}{x^2 + y^2 + z^2}\, dV$$

Switching to spherical coordinates, we have

$$I_z = \int_0^\pi \int_0^{2\pi} \int_{10}^{11} \left(\rho^2 \sin^2 \phi\, \frac{k}{\rho^2}\right)\rho^2 \sin \phi\, d\rho\, d\theta\, d\phi$$

$$= \int_0^\pi \int_0^{2\pi} \left(k\, \frac{\rho^3 \sin^3 \phi}{3}\, \bigg|_{10}^{11}\right) d\theta\, d\phi = \frac{331k}{3} \int_0^\pi \int_0^{2\pi} \sin^3 \phi\, d\theta\, d\phi$$

$$= \frac{662\pi k}{3} \int_0^\pi \sin^3 \phi\, d\phi = \frac{662\pi k}{3} \int_0^\pi \sin \phi(1 - \cos^2 \phi)\, d\phi$$

$$= \frac{662\pi k}{3} \left(-\cos \phi + \frac{1}{3} \cos^3 \phi\right) \bigg|_0^\pi$$

$$= \frac{2648\pi k}{9} \qquad\qquad\blacksquare$$

Problem Set 15.8

In Problems 1 to 8, find the mass m and the coordinates $(\bar{x}, \bar{y}, \bar{z})$ of the center of mass of the given solid S with the indicated mass density function σ.

1 S: $0 \le x \le 1$, $x^2 \le y \le x$, $x - y \le z \le x + y$; $\sigma(x, y, z) = x + 3y$

2 S: $0 \le x \le 1$, $0 \le y \le 1 - x^2$, $0 \le z \le x^2 + y^2$; $\sigma(x, y, z) = x + 1$

3 S: $0 \le x \le 1$, $0 \le y \le 1$, $0 \le z \le 1$; $\sigma(x, y, z) = 3 - x - y - z$

4 S: a tetrahedron cut from the first octant by the plane $x + y + z = 1$ if $\sigma(x, y, z) = x$

5 S: $x^2 + y^2 \le 16$, $0 \le z \le 10$; $\sigma(x, y, z) = \sqrt{x^2 + y^2}/40$

6 S: $x^2 + y^2 \le a^2$, $(h/a)\sqrt{x^2 + y^2} \le z \le h$; $h > 0$, $a > 0$; $\sigma(x, y, z) = x^2 + y^2 + z^2$

7 S: $x^2 + y^2 \le 9$, $0 \le z \le 9 - x^2 - y^2$; $\sigma(x, y, z) = z$

8 S: $x^2 + y^2 \le a^2$, $0 \le z \le h - (h/a)\sqrt{x^2 + y^2}$; $h > 0$, $a > 0$; $\sigma(x, y, z) = \sqrt{x^2 + y^2}$

9 Mass is distributed throughout the interior of a sphere of radius 1 meter in such a way that the density at a point P is inversely proportional to the distance of P from the center and at the surface of the sphere the density is 1 gram per cubic centimeter. Find the total mass in kilograms inside the sphere.

10 One kilogram of mass is distributed throughout a spherical ball in such a way that the density at a point is proportional to the distance of the point from the surface of the sphere. If the sphere is 10 centimeters in diameter, find the density in grams per cubic centimeter at the center of the sphere.

11 Mass is distributed throughout the interior of a right circular cylinder of radius 4 meters and height 6 meters in such a way that the density at a point P is proportional to the square of the distance from the axis of the cylinder, and at the surface of the cylinder the density is 1 kilogram per cubic meter. Find the total mass inside the cylinder.

12 One coulomb of electric charge is distributed throughout a solid right circular cone of height 10 centimeters and base diameter 4 centimeters in such a way that the charge density is proportional to the square of the distance from the vertex of the cone. Find the charge density at the center of the base of the cone.

In Problems 13 to 22, find the coordinates of the centroid of each solid S.

13 S: $0 \le x \le 2$, $0 \le y \le 3$, $0 \le z \le 4$

14 S: $-1 \le x \le 1$, $y \ge 0$, $0 \le z \le 1 - y^2$

15 S: $0 \le x \le 1$, $y^2 \le z \le 1$

16 S: $0 \le x \le 3$, $0 \le y \le \sqrt{9 - x^2}$, $0 \le z \le 3 - x$

17 S: $x^2 + y^2 \le 1$, $x \ge 0$, $y \ge 0$, $0 \le z \le xy$

18 S: $0 \le \theta \le \dfrac{\pi}{3}$, $0 \le r \le \sin \theta$, $0 \le z \le r$

19 S: $x^2 + y^2 + z^2 \le 2$, $z^2 \le x^2 + y^2$

20 S: $x^2 + y^2 \le a^2$, $(h/a)\sqrt{x^2 + y^2} \le z \le h$, where h and a are positive constants

21 S: A tetrahedron in the first octant bounded by the coordinate planes and the plane $(x/a) + (y/b) + (z/c) = 1$, where a, b, and c are positive constants

22 S: A solid bounded by the xy plane and the hemisphere $z = \sqrt{4 - x^2 - y^2}$

In Problems 23 to 32, find the designated moment or moment of inertia for the given solid S with the indicated density function σ.

23 M_{xy}; S: $0 \le x \le 1$, $0 \le y \le 1 - x$, $0 \le z \le 1$; $\sigma(x, y, z) = 3xy$

24 M_{xz}; S: $0 \le x \le 1$, $0 \le y \le 1$, $0 \le z \le xy$; $\sigma(x, y, z) = 2(x + y)$

25 M_{yz}; S: $0 \le y \le 4$, $0 \le x \le \dfrac{y}{2}$, $0 \le z \le \sqrt{8x}$; $\sigma(x, y, z) = 2$

26 M_{xz}; S is the solid in the first octant bounded by $x + z = 1$, $x = y$, $x = 0$, $y = 0$, and $z = 0$; $\sigma(x, y, z) = 5y$

27 I_z; S: $0 \le x \le 1$, $0 \le y \le 1 - x$, $0 \le z \le 5$; $\sigma(x, y, z) = 3$

28 I_z; S is the first octant solid bounded by $x + z = 1$, $y + z = 1$, and the coordinate planes; $\sigma(x, y, z) = z$

29 I_x; S: $0 \le x \le 1$, $0 \le y \le 1$, $0 \le z \le 1$; $\sigma(x, y, z) = 8z$

30 I_y; S is the first octant solid bounded by $x^2 + z^2 = 1$, $y = x$, $y = 0$, $z = 0$; $\sigma(x, y, z) = 2z$

31 M_{xy}; S is the solid bounded by the cylinder $x^2 + y^2 = 4$ and the planes $z = 0$ and $z = 3$; $\sigma(x, y, z) = 4z$

32 I_z; S is the solid bounded above by the paraboloid $z = 9 - x^2 - y^2$ and below by the plane $z = 0$; $\sigma(x, y, z) = 3/(x^2 + y^2)$

33 Find the moment of inertia with respect to the central axis of a homogeneous solid right circular cylinder S of radius a, height h, and total mass m.

34 The spherical shell S in Example 3 has a total mass of 18 grams. Assuming that distances are measured in centimeters, find its moment of inertia I_z.

35 Find the moment of inertia about the central axis of a homogeneous right circular cylindrical shell with inner radius a, outer radius b, and height h.

36 Find the moment of inertia of a homogeneous solid right circular cylinder about an axis perpendicular to the central axis and passing through the center of gravity of the solid cylinder. Assume that the height of the cylinder is h, that its radius is a, and that its mass is m.

37 Find the moment of inertia of a homogeneous solid spherical ball of radius a about an axis through the center of the ball. Let m denote the mass of the ball.

38 Find the moment of inertia of a homogeneous solid rectangular parallelepiped with sides a, b, and c about an axis passing through the center of gravity and parallel to the side of length c. Assume that the total mass is m.

39 Find the moment of inertia of a homogeneous spherical shell of inner radius a and outer radius b about an axis passing through its center. Assume that the total mass of the shell is m.

40 Let A be an axis passing through the center of gravity of a homogeneous solid S of total mass m, and let B be an axis parallel to A. If h denotes the distance between the axes A and B, prove that

$$I_B = I_A + mh^2$$

This result is known as the **parallel-axis theorem.**

41 Give an informal argument to show that the density of a three-dimensional distribution at a point is the limit of the amount of the quantity per unit volume in a small region about the point as the region "shrinks to zero."

42 Combine the results of Problems 36 and 40 to determine the moment of inertia of a homogeneous solid right circular cylinder about an axis passing through the center of one of the bases and perpendicular to the central axis.

43 Derive the formulas for the centroid of a solid S.

44 The temperature of a hollow spherical shell with inner radius a and outer radius b is inversely proportional to the distance from the center and has the value $T_0 > 0$ on the inner surface. The quantity of heat required to raise any portion of the shell from one uniform temperature to another is proportional jointly to the volume of this portion and to the rise in temperature. Assume that C units of heat are required to raise the temperature of 1 cubic unit of the shell by $1°$. How much heat will the shell give out if it is cooled to a uniform temperature of $0°$?

15.9   Line Integrals and Green's Theorem

In Section 15.2 we introduced double integrals over two-dimensional regions R, while in Section 15.6 we introduced triple integrals over three-dimensional solids S. In this section we round out the picture by considering integrals over one-dimensional curves. It is customary to refer to a curve over which an integral is to be taken as a *path*, or *line* (not necessarily a straight line!), and to call an integral over such a curve a *line integral*. Later in this section we present an important theorem, called *Green's theorem*, which relates line integrals in the plane and double integrals.

Line Integrals

In more advanced courses, line integrals are defined formally in terms of limits of Riemann sums in a manner similar to the definition of the definite integral. For our purposes, it is simpler to adopt the following definition, which is equivalent to the formal one for the kinds of curves and functions that we consider.

Line Integral in the Plane

Let C be a curve in the xy plane with the parametric equations

$$C: \begin{cases} x = f(t) \\ y = g(t) \end{cases} \qquad a \le t \le b$$

where f and g have continuous first derivatives. Suppose that P and Q are continuous functions of two variables whose domains contain the curve C. Then the **line integral** $\int_C P(x, y) \, dx + Q(x, y) \, dy$ is defined by

$$\int_C P(x, y) \, dx + Q(x, y) \, dy = \int_a^b P(f(t), g(t))f'(t) \, dt + Q(f(t), g(t))g'(t) \, dt$$

Thus, to evaluate the line integral

$$\int_C P(x, y) \, dx + Q(x, y) \, dy$$

you simply make the substitutions $x = f(t)$, $dx = f'(t) \, dt$, $y = g(t)$, and $dy = g'(t) \, dt$, and then integrate from $t = a$ to $t = b$.

EXAMPLE 1   Evaluate the line integral $\int_C (x^2 + 3y) \, dx + (y^2 + 2x) \, dy$ if

$$C: \begin{cases} x = t \\ y = t^2 + 1 \end{cases} \qquad 0 \le t \le 1$$

SOLUTION   Making the substitutions $x = t$, $dx = dt$, $y = t^2 + 1$, and $dy = 2t \, dt$, we have

$$\int_C (x^2 + 3y) \, dx + (y^2 + 2x) \, dy = \int_0^1 [t^2 + 3(t^2 + 1)] \, dt + [(t^2 + 1)^2 + 2t]2t \, dt$$

$$= \int_0^1 [(4t^2 + 3) + (t^4 + 2t^2 + 2t + 1)(2t)] \, dt$$

$$= \int_0^1 (2t^5 + 4t^3 + 8t^2 + 2t + 3) \, dt$$

$$= \left(\frac{t^6}{3} + t^4 + \frac{8}{3}t^3 + t^2 + 3t \right) \Big|_0^1 = 8 \quad \blacksquare$$

Although Definition 1 makes it appear as if the line integral depends on the choice of the parameter t, it can be shown that:

> The value of the line integral is unaffected by the choice of parameter as long as the direction along the geometric curve C corresponding to increasing values of the parameter is kept the same.

Thus, line integrals are actually taken over **directed,** or **oriented,** curves; that is, over curves for which one of the two endpoints is understood to be the *starting,* or *initial, point,* while the other endpoint is the *ending,* or *terminal, point.* In dealing with line integrals, *the curve obtained from a given curve C by reversing its direction is usually denoted by* $-C$ (Figure 1).

Figure 1

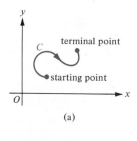

(a)

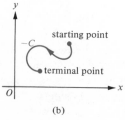

(b)

In Examples 2 and 3, evaluate the given line integral.

EXAMPLE 2 $\int_C (x + y)\, dx + (y - x)\, dy$ if C is the line segment from $(1, 1)$ to $(4, 2)$

SOLUTION   Using the methods of Section 12.3, we obtain the scalar parametric equations

$$C: \begin{cases} x = 1 + 3t \\ y = 1 + t \end{cases} \qquad 0 \le t \le 1$$

for the line segment from $(1, 1)$ to $(4, 2)$. Notice that $(x, y) = (1, 1)$ when $t = 0$, and $(x, y) = (4, 2)$ when $t = 1$, so that $(1, 1)$ is the initial point and $(4, 2)$ is the terminal point of C. Making the substitutions $x = 1 + 3t$, $dx = 3\, dt$, $y = 1 + t$, and $dy = dt$, we have

$$\int_C (x + y)\, dx + (y - x)\, dy = \int_0^1 [(1 + 3t) + (1 + t)]3\, dt + [(1 + t) - (1 + 3t)]\, dt$$

$$= \int_0^1 (10t + 6)\, dt = (5t^2 + 6t)\Big|_0^1 = 11 \qquad \blacksquare$$

EXAMPLE 3 $\int_C (x + 2y)\, dy$ if C is the arc of the parabola $x = y^2$ from $(1, -1)$ to $(9, -3)$

SOLUTION   Here the line integral has the form $\int_C P(x, y)\, dx + Q(x, y)\, dy$ with $P(x, y) = 0$ and $Q(x, y) = x + 2y$. The arc of the parabola is described parametrically by

$$C: \begin{cases} x = t^2 \\ y = -t \end{cases} \qquad 1 \le t \le 3$$

in such a way that as t goes from 1 to 3, (x, y) goes from $(1, -1)$ to $(9, -3)$. Making the substitutions $x = t^2$, $y = -t$, and $dy = -dt$, we have

$$\int_C (x + 2y)\, dy = \int_1^3 (t^2 - 2t)(-dt) = \int_1^3 (2t - t^2)\, dt = \left(t^2 - \frac{t^3}{3}\right)\Big|_1^3 = -\frac{2}{3} \qquad \blacksquare$$

Definition 1 extends in an obvious way to line integrals over curves in three-dimensional space. Indeed, suppose that C is such a curve, say with scalar parametric equations

$$C: \begin{cases} x = f(t) \\ y = g(t) \\ z = h(t) \end{cases} \qquad a \le t \le b$$

where f, g, and h have continuous first derivatives. Suppose also that P, Q, and T are continuous functions of three variables whose domains contain the curve C. Then the line integral

$$\int_C P(x, y, z)\, dx + Q(x, y, z)\, dy + T(x, y, z)\, dz$$

is evaluated by making the substitutions $x = f(t)$, $dx = f'(t)\, dt$, $y = g(t)$, $dy = g'(t)\, dt$, $z = h(t)$, and $dz = h'(t)\, dt$ and then by integrating from a to b.

<u>**EXAMPLE 4**</u>   Evaluate $\int_C yz\ dx + xz\ dy + xy\ dz$ if

$$C: \begin{cases} x = t \\ y = t^2 \\ z = t^3 \end{cases} \quad -1 \le t \le 1$$

SOLUTION   Making the substitutions $x = t$, $dx = dt$, $y = t^2$, $dy = 2t\ dt$, $z = t^3$, and $dz = 3t^2\ dt$, we have

$$\int_C yz\ dx + xz\ dy + xy\ dz = \int_{-1}^{1} t^2 t^3\ dt + tt^3(2t\ dt) + tt^2(3t^2\ dt)$$

$$= \int_{-1}^{1} (t^5 + 2t^5 + 3t^5)\ dt = \int_{-1}^{1} 6t^5\ dt = t^6 \Big|_{-1}^{1} = 0 \quad \blacksquare$$

Vector Notation and Work

Consider the line integral $\int_C P(x, y)\ dx + Q(x, y)\ dy$, where the curve C is given by the parametric equations

$$C: \begin{cases} x = f(t) \\ y = g(t) \end{cases} \quad a \le t \le b$$

If we let the vector

$$\mathbf{R} = x\mathbf{i} + y\mathbf{j} = f(t)\mathbf{i} + g(t)\mathbf{j} \quad a \le t \le b$$

denote the variable position vector of a point (x, y) on C, then

$$\frac{d\mathbf{R}}{dt} = \frac{dx}{dt}\mathbf{i} + \frac{dy}{dt}\mathbf{j}$$

or in differential form,

$$d\mathbf{R} = dx\ \mathbf{i} + dy\ \mathbf{j}$$

Now, putting

$$\mathbf{F} = P(x, y)\mathbf{i} + Q(x, y)\mathbf{j}$$

we have

$$\mathbf{F} \cdot d\mathbf{R} = P(x, y)\ dx + Q(x, y)\ dy$$

so that

$$\boxed{\int_C P(x, y)\ dx + Q(x, y)\ dy = \int_C \mathbf{F} \cdot d\mathbf{R}}$$

Similarly, if C is a curve in three-dimensional space swept out by a variable position vector $\mathbf{R}$, then

$$\int_C P(x, y, z)\ dx + Q(x, y, z)\ dy + T(x, y, z)\ dz = \int_C \mathbf{F} \cdot d\mathbf{R}$$

where

$$\mathbf{F} = P(x, y, z)\mathbf{i} + Q(x, y, z)\mathbf{j} + T(x, y, z)\mathbf{k} \quad \text{and} \quad d\mathbf{R} = dx\ \mathbf{i} + dy\ \mathbf{j} + dz\ \mathbf{k}$$

Figure 2

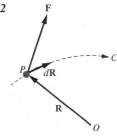

The vector notation for the line integral not only has the advantage of brevity but also suggests an important physical interpretation. Suppose that **F** represents a variable force acting on a particle P which is moving along the curve C. If **R** is the variable position vector of P, then we can regard $d\mathbf{R}$ as representing an infinitesimal displacement of the particle, so that $\mathbf{F} \cdot d\mathbf{R}$ represents the element of work done on the particle by the force **F** during this displacement (Figure 2). Summing up—that is, integrating—all these elements of work, we obtain $\displaystyle\int_C \mathbf{F} \cdot d\mathbf{R}$. Therefore:

> The line integral $\displaystyle\int_C \mathbf{F} \cdot d\mathbf{R}$ represents the net work done by the force **F** in moving a particle along the curve C from its initial point to its terminal point.

EXAMPLE 5 The variable force $\mathbf{F} = (3x - 4y)\mathbf{i} + (4x + 2y)\mathbf{j}$ moves a particle along the curve

$$C: \begin{cases} x = 4t + 1 \\ y = 3t^2 \end{cases} \qquad 0 \le t \le 2$$

from $(1, 0)$ to $(9, 12)$. Find the work done if distances are measured in meters and force is measured in newtons.

SOLUTION   The work is given by the line integral $\displaystyle\int_C \mathbf{F} \cdot d\mathbf{R}$, where $\mathbf{R} = x\mathbf{i} + y\mathbf{j}$, so that $d\mathbf{R} = dx\,\mathbf{i} + dy\,\mathbf{j}$. Thus,

$$\int_C \mathbf{F} \cdot d\mathbf{R} = \int_C (3x - 4y)\,dx + (4x + 2y)\,dy$$

Making the substitutions $x = 4t + 1$, $dx = 4\,dt$, $y = 3t^2$, and $dy = 6t\,dt$, we have

$$\int_C \mathbf{F} \cdot d\mathbf{R} = \int_0^2 [3(4t + 1) - 4(3t^2)](4\,dt) + [4(4t + 1) + 2(3t^2)](6t\,dt)$$

$$= \int_0^2 (36t^3 + 48t^2 + 72t + 12)\,dt = (9t^4 + 16t^3 + 36t^2 + 12t)\Big|_0^2$$

$$= 440 \text{ joules}$$

Figure 3

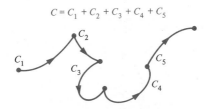

$$C = C_1 + C_2 + C_3 + C_4 + C_5$$

Suppose that $C_1, C_2, \ldots, C_n$ are curves such that the terminal point of each curve (except for C_n) is the initial point of the next curve. Figure 3 illustrates the case $n = 5$. If C is the curve formed by joining the successive curves $C_1, C_2, \ldots, C_n$, then we write

$$C = C_1 + C_2 + \cdots + C_n$$

A curve that can be described parametrically by functions with continuous first derivatives is said to be **smooth,** and a curve that is obtained by joining successive smooth curves is said to be **piecewise smooth.**

If $C = C_1 + C_2 + \cdots + C_n$, where $C_1, C_2, \ldots, C_n$ are piecewise smooth curves, then C is piecewise smooth. Furthermore, if the variable vector **F** is continuous on each of the curves $C_1, C_2, \ldots, C_n$, then it can be shown that

$$\int_C \mathbf{F} \cdot d\mathbf{R} = \int_{C_1} \mathbf{F} \cdot d\mathbf{R} + \int_{C_2} \mathbf{F} \cdot d\mathbf{R} + \cdots + \int_{C_n} \mathbf{F} \cdot d\mathbf{R}$$

Piecewise smooth curves used in sculpture

Figure 4

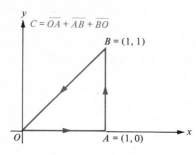

EXAMPLE 6   Let $A = (1, 0)$ and $B = (1, 1)$. Calculate

$$\int_C (x^2 - y)\, dx + (x + y^2)\, dy$$

if $C = \overline{OA} + \overline{AB} + \overline{BO}$; that is, C is the perimeter of the triangle OAB taken in a counterclockwise direction (Figure 4).

SOLUTION   The line segments $\overline{OA}$, $\overline{AB}$, and $\overline{BO}$ are given parametrically by

$$\overline{OA}: \begin{cases} x = t \\ y = 0 \end{cases} \qquad \overline{AB}: \begin{cases} x = 1 \\ y = t \end{cases} \qquad \overline{BO}: \begin{cases} x = 1 - t \\ y = 1 - t \end{cases}$$

where $0 \le t \le 1$ in each case. Thus,

$$\int_{\overline{OA}} (x^2 - y)\, dx + (x + y^2)\, dy = \int_0^1 (t^2 - 0)\, dt + (t + 0)(0)\, dt$$

$$= \int_0^1 t^2\, dt = \frac{t^3}{3}\bigg|_0^1 = \frac{1}{3}$$

$$\int_{\overline{AB}} (x^2 - y)\, dx + (x + y^2)\, dy = \int_0^1 (1 - t)(0)\, dt + (1 + t^2)\, dt$$

$$= \int_0^1 (1 + t^2)\, dt = \left(t + \frac{t^3}{3}\right)\bigg|_0^1 = \frac{4}{3}$$

and

$$\int_{\overline{BO}} (x^2 - y)\, dx + (x + y^2)\, dy$$

$$= \int_0^1 [(1 - t)^2 - (1 - t)](-dt) + [(1 - t) + (1 - t)^2](-dt)$$

$$= \int_0^1 (-2t^2 + 4t - 2)\, dt = \left(-2\frac{t^3}{3} + 2t^2 - 2t\right)\bigg|_0^1 = -\frac{2}{3}$$

Therefore,

$$\int_C (x^2 - y)\, dx + (x + y^2)\, dy = \tfrac{1}{3} + \tfrac{4}{3} - \tfrac{2}{3} = 1 \qquad \blacksquare$$

A change in the direction (orientation) of the curve over which a line integral is taken results in a change in the algebraic sign of the integral; that is,

$$\boxed{\int_{-C} \mathbf{F} \cdot d\mathbf{R} = -\int_C \mathbf{F} \cdot d\mathbf{R}}$$

(For an indication of the proof, see Problem 34.) For instance, in the preceding example, we found that

$$\int_{\overline{OA}} (x^2 - y)\, dx + (x + y^2)\, dy = \tfrac{1}{3}$$

and since $-(\overline{OA}) = \overline{AO}$, it follows that

$$\int_{\overline{AO}} (x^2 - y)\, dx + (x + y^2)\, dy = -\tfrac{1}{3}$$

Of course, line integrals, like ordinary integrals, are additive and homogeneous with respect to the integrand, so that

$$\int_C (\mathbf{F} + \mathbf{G}) \cdot d\mathbf{R} = \int_C \mathbf{F} \cdot d\mathbf{R} + \int_C \mathbf{G} \cdot d\mathbf{R}$$

$$\int_C (\mathbf{F} - \mathbf{G}) \cdot d\mathbf{R} = \int_C \mathbf{F} \cdot d\mathbf{R} - \int_C \mathbf{G} \cdot d\mathbf{R}$$

and

$$\int_C (a\mathbf{F}) \cdot d\mathbf{R} = a \int_C \mathbf{F} \cdot d\mathbf{R}$$

for any two continuous vector functions $\mathbf{F}$ and $\mathbf{G}$ and any constant scalar a. In particular, if M, N, and P are continuous functions of x, y, and z, then

$$\int_C M\, dx + N\, dy + P\, dz = \int_C M\, dx + \int_C N\, dy + \int_C P\, dz$$

Green's Theorem

A remarkable theorem connecting a line integral around a closed curve in the plane with a double integral over the region enclosed by this curve is traditionally associated with the English mathematician George Green (1793–1841), although the result was actually obtained earlier by the German mathematician Karl Friedrich Gauss (1777–1855). Here we give a somewhat informal statement of the theorem. Later in this section we give an argument to make the result plausible.

By a **closed curve,** we mean a curve whose initial point is the same as its terminal point. A **simple** closed curve is one that does not intersect itself, except at its common initial and terminal point.

THEOREM 1

Green's Theorem

Let C be a piecewise smooth, simple, closed curve in the xy plane, and suppose that C forms the boundary of a two-dimensional region R. Assume that C is directed so that it turns counterclockwise around R (Figure 5). Suppose that P and Q are continuous functions of two variables having continuous partial derivatives $\partial Q/\partial x$ and $\partial P/\partial y$ on R and its boundary C. Then

$$\int_C P(x, y)\, dx + Q(x, y)\, dy = \iint_R \left(\frac{\partial Q}{\partial x} - \frac{\partial P}{\partial y} \right) dA$$

Figure 5

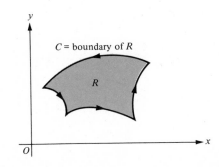

C = boundary of R

R

In Examples 7 to 10, use Green's theorem to evaluate the given line integral around the indicated simple closed curve C.

__EXAMPLE 7__ $\int_C (x^2 - y)\, dx + (x + y^2)\, dy$, where C is the perimeter of the triangle OAB taken in a counterclockwise sense, with $A = (1, 0)$ and $B = (1, 1)$ (Figure 4).

SOLUTION   Here $P(x, y) = x^2 - y$, $Q(x, y) = x + y^2$, and R is the triangular region enclosed by OAB. Thus,

$$\frac{\partial Q}{\partial x} = \frac{\partial}{\partial x}(x + y^2) = 1 \quad \text{and} \quad \frac{\partial P}{\partial y} = \frac{\partial}{\partial y}(x^2 - y) = -1$$

Hence, by Green's theorem,

$$\int_C (x^2 - y)\, dx + (x + y^2)\, dy = \iint_R \left(\frac{\partial Q}{\partial x} - \frac{\partial P}{\partial y}\right) dA = \iint_R [1 - (-1)]\, dA$$

$$= \iint_R 2\, dA = 2 \iint_R dA = 2(\tfrac{1}{2}) = 1$$

since the area A of the triangular region R is $\frac{1}{2}$ square unit. (This result is consistent with our previous solution of Example 6.) ∎

__EXAMPLE 8__ $\int_C (y + 3x)\, dx + (2y - x)\, dy$, where C is the circle $x^2 + y^2 = 9$ with the counterclockwise orientation

SOLUTION   Here the region R enclosed by C is a circular disk of radius 3 and area $A = \pi 3^2 = 9\pi$ square units. Thus,

$$\int_C (y + 3x)\, dx + (2y - x)\, dy = \iint_R \left[\frac{\partial}{\partial x}(2y - x) - \frac{\partial}{\partial y}(y + 3x)\right] dA$$

$$= \iint_R [(-1) - 1]\, dx\, dy = -2 \iint_R dA$$

$$= -2(9\pi) = -18\pi$$ ∎

__EXAMPLE 9__   Let R be the region bounded by the three curves whose polar equations are $\theta = \pi/4$, $r = 2$, and $\theta = 3\pi/4$ (Figure 6), and let C be the boundary of R taken in a counterclockwise sense. Evaluate $\int_C xy\, dx + x^2\, dy$.

SOLUTION   Using Green's theorem and then using polar coordinates to evaluate the resulting double integral, we have

$$\int_C xy\, dx + x^2\, dy = \iint_R \left[\frac{\partial}{\partial x}(x^2) - \frac{\partial}{\partial y}(xy)\right] dA = \iint_R (2x - x)\, dA$$

$$= \iint_R x\, dA = \int_{\pi/4}^{3\pi/4} \int_0^2 (r\cos\theta) r\, dr\, d\theta$$

$$= \int_{\pi/4}^{3\pi/4} \left(\frac{r^3}{3}\cos\theta \Big|_0^2\right) d\theta = \int_{\pi/4}^{3\pi/4} \frac{8}{3}\cos\theta\, d\theta$$

$$= \frac{8}{3}\sin\theta \Big|_{\pi/4}^{3\pi/4} = \frac{8}{3}\left(\sin\frac{3\pi}{4} - \sin\frac{\pi}{4}\right) = 0$$ ∎

__Figure 6__

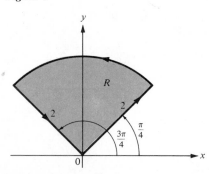

EXAMPLE 10 $\int_C \mathbf{F} \cdot d\mathbf{R}$, where $\mathbf{F} = (2xy - x^2)\mathbf{i} + (x + y^2)\mathbf{j}$ and C is the boundary, taken counterclockwise, of the region R bounded by the parabola $y = x^2$ and the line $y = x$.

SOLUTION

$$\int_C \mathbf{F} \cdot d\mathbf{R} = \iint_R \left[\frac{\partial}{\partial x}(x + y^2) - \frac{\partial}{\partial y}(2xy - x^2) \right] dA = \iint_R (1 - 2x)\, dA$$

$$= \int_0^1 \int_{x^2}^{x} (1 - 2x)\, dy\, dx = \int_0^1 \left[(1 - 2x)y \Big|_{x^2}^{x} \right] dx$$

$$= \int_0^1 (1 - 2x)(x - x^2)\, dx$$

$$= \int_0^1 (2x^3 - 3x^2 + x)\, dx$$

$$= \left(\frac{x^4}{2} - x^3 + \frac{x^2}{2} \right) \Bigg|_0^1 = 0$$

By considering a particular line integral and evaluating it by Green's theorem, we can obtain an interesting formula for the area A of the region R enclosed by a piecewise smooth, simple, closed curve C (Figure 7). Indeed, we have

$$\int_C -y\, dx + x\, dy = \iint_R \left(\frac{\partial x}{\partial x} - \frac{\partial(-y)}{\partial y} \right) dA = \iint_R 2\, dA = 2 \iint_R dA = 2A$$

Hence,

$$\boxed{A = \tfrac{1}{2} \int_C -y\, dx + x\, dy}$$

Figure 7

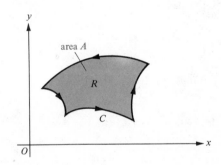

EXAMPLE 11 Use the formula above to find the area A of the region R enclosed by the ellipse $(x^2/a^2) + (y^2/b^2) = 1$, where a and b are positive constants.*

SOLUTION   Parametric equations of the ellipse are

$$C: \begin{cases} x = a\cos t \\ y = b\sin t \end{cases} \qquad 0 \le t \le 2\pi$$

Thus,

$$A = \frac{1}{2} \int_C -y\, dx + x\, dy$$

$$= \frac{1}{2} \int_0^{2\pi} [-b\sin t(-a\sin t)\, dt + a\cos t\, (b\cos t)\, dt]$$

$$= \frac{1}{2} \int_0^{2\pi} ab(\sin^2 t + \cos^2 t)\, dt = \frac{ab}{2} \int_0^{2\pi} dt = \pi ab$$

*See Problem 55 on page 563 for an alternative derivation of the formula for the area enclosed by an ellipse.

We now give an informal proof of Green's theorem for the special case in which the region R enclosed by the simple closed curve C is both of type I and of type II. The result for more general regions that can be decomposed into nonoverlapping subregions of both types then follows from this special case.

In Figure 8, the simple closed curve C bounds the region R and is directed in the counterclockwise sense. Here, C is the sum of the directed arcs ATB and BSA, and it is also the sum of the directed arcs SAT and TBS; that is,

$$C = ATB + BSA \qquad \text{and} \qquad C = SAT + TBS$$

Figure 8

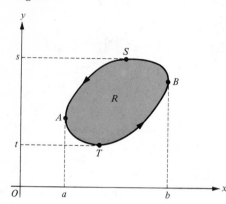

Suppose that the directed arcs ATB, $-(BSA)$, $-(SAT)$, and TBS are described as follows:

$$\begin{aligned}
ATB: & \quad y = g(x) & a \leq x \leq b \\
-(BSA): & \quad y = h(x) & a \leq x \leq b \\
-(SAT): & \quad x = G(y) & t \leq y \leq s \\
TBS: & \quad x = H(y) & t \leq y \leq s
\end{aligned}$$

Now, assume that P and Q are continuous functions and that $\partial Q/\partial x$ and $\partial P/\partial y$ are defined and continuous on the region R and on its boundary curve C. We have

$$\iint_R \frac{\partial Q}{\partial x}\, dA = \int_t^s \left(\int_{G(y)}^{H(y)} \frac{\partial Q}{\partial x}\, dx \right) dy = \int_t^s \left[Q(x, y) \Big|_{G(y)}^{H(y)} \right] dy$$

$$= \int_t^s [Q(H(y), y) - Q(G(y), y)]\, dy$$

$$= \int_t^s Q(H(y), y)\, dy - \int_t^s Q(G(y), y)\, dy$$

$$= \int_{TBS} Q(x, y)\, dy - \int_{-(SAT)} Q(x, y)\, dy$$

$$= \int_{TBS} Q(x, y)\, dy + \int_{SAT} Q(x, y)\, dy = \int_{TBS+SAT} Q(x, y)\, dy$$

$$= \int_C Q(x, y)\, dy$$

Similarly,

$$-\iint_R \frac{\partial P}{\partial y}\, dA = -\int_a^b \left(\int_{g(x)}^{h(x)} \frac{\partial P}{\partial y}\, dy \right) dx = -\int_a^b \left[P(x, y) \Big|_{g(x)}^{h(x)} \right] dx$$

$$= -\int_a^b [P(x, h(x)) - P(x, g(x))]\, dx$$

$$= \int_a^b P(x, g(x))\, dx - \int_a^b P(x, h(x))\, dx$$

$$= \int_{ATB} P(x, y)\, dx - \int_{-(BSA)} P(x, y)\, dx$$

$$= \int_{ATB} P(x, y)\, dx + \int_{BSA} P(x, y)\, dx = \int_{ATB+BSA} P(x, y)\, dx$$

$$= \int_C P(x, y)\, dx$$

Combining the two results, we obtain

$$\int_C P(x, y)\, dx + Q(x, y)\, dy = \int_C P(x, y)\, dx + \int_C Q(x, y)\, dy$$

$$= -\iint_R \frac{\partial P}{\partial y}\, dA + \iint_R \frac{\partial Q}{\partial x}\, dA$$

$$= \iint_R \left(\frac{\partial Q}{\partial x} - \frac{\partial P}{\partial y} \right) dA$$

and our argument is complete.

Problem Set 15.9

In Problems 1 to 10, evaluate each line integral by using Definition 1.

1 $\displaystyle\int_C (3x^2 - 6y)\, dx + (3x + 2y)\, dy$, where

$$C: \begin{cases} x = t \\ y = t^2 \end{cases} \quad 0 \le t \le 1$$

2 $\displaystyle\int_C y^2\, dx + x^2\, dy$, where $C: \begin{cases} x = t^2 \\ y = t + 1 \end{cases} \quad 0 \le t \le 1$

3 $\displaystyle\int_C x\, dy - y\, dx$, where $C: \begin{cases} x = 2 + \cos t \\ y = \sin t \end{cases} \quad 0 \le t \le 2\pi$

4 $\displaystyle\int_C 2xy^3\, dx + 3x^2y^2\, dy$, where C is the portion of the parabola $y = x^2$ from $(0, 0)$ to $(1, 1)$

5 $\displaystyle\int_C \mathbf{F} \cdot d\mathbf{R}$, where $\mathbf{F} = (x^2 + y^2)\mathbf{i} + 2xy\mathbf{j}$ and C is the portion of the parabola $y = x^2$ from $(0, 0)$ to $(1, 1)$

6 $\displaystyle\int_C \mathbf{F} \cdot d\mathbf{R}$, where $\mathbf{F} = (\cos y)\mathbf{i} + (\sin x)\mathbf{j}$ and C is the triangle with vertices $(1, 0)$, $(1, 1)$, and $(0, 0)$ taken in the counterclockwise sense

7 $\displaystyle\int_C \mathbf{F} \cdot d\mathbf{R}$, where $\mathbf{F} = xy\mathbf{i} + x^2\mathbf{j}$ and C is the circle of radius 2 with center at the origin taken in the counterclockwise sense

8 $\displaystyle\int_C (\nabla f) \cdot d\mathbf{R}$, where $f(x, y) = x^2 - 2xy$ and C is the path from $(0, 0)$ to $(2, 1)$, then from $(2, 1)$ to $(2, 3)$, then from $(2, 3)$ to $(1, 3)$

9 $\displaystyle\int_C (y + z)\, dx + (2xz)\, dy + (x + z)\, dz$, where

$$C: \begin{cases} x = 2t + 1 \\ y = 3t - 1 \\ z = -t + 2 \end{cases} \quad 0 \le t \le 2$$

10 $\displaystyle\int_C \mathbf{F} \cdot d\mathbf{R}$, where $\mathbf{F}$ is given by $\mathbf{F} = 2zy\mathbf{i} + x^2\mathbf{j} + (x + y)\mathbf{k}$, $C = \overline{OA} + \overline{AB} + \overline{BO}$, $A = (1, 2, 3)$, and $B = (2, 0, -1)$

In Problems 11 to 24, use Green's theorem to evaluate each line integral. In each case, assume that the closed curve C is taken in the counterclockwise sense.

11 $\displaystyle\int_C (x^2 - xy^3)\, dx + (y^2 - 2xy)\, dy$, where C is the square with vertices $(0, 0)$, $(3, 0)$, $(3, 3)$, and $(0, 3)$

12 $\displaystyle\int_C (x^2 + 3y^2)\, dx$, where C is the square with vertices $(-2, -2)$, $(2, -2)$, $(2, 2)$, and $(-2, 2)$

13 $\displaystyle\int_C -x^2y\, dx + xy^2\, dy$, where C is the circle $x^2 + y^2 = 1$

14 $\displaystyle\int_C y\, dx + x^3\sqrt{4 - y^2}\, dy$, where C is the circle $x^2 + y^2 = 4$

15 $\displaystyle\int_C (x^2 + y^2)\, dx - 2xy\, dy$, where C is the circle $x^2 + y^2 = 25$

16 $\displaystyle\int_C (xy^3 + \sin x)\, dx + (x^2y^2 + 3x)\, dy$, where C is the circle $x^2 + y^2 = 16$

17 $\displaystyle\int_C (x^2 + y)\, dx + (x - y^2)\, dy$, where C encloses the region R bounded by $y = x^3$, $y = 0$, and $x = 1$

18 $\displaystyle\int_C x \cos y\, dx - y \sin x\, dy$, where C is the square with vertices $(0, 0)$, $(1, 0)$, $(1, 1)$, and $(0, 1)$

19 $\displaystyle\int_C \mathbf{F} \cdot d\mathbf{R}$, where $\mathbf{F} = (x^2 + y^2)\mathbf{i} + 3xy^2\mathbf{j}$ and C is the circle $x^2 + y^2 = 9$

20 $\int_C \mathbf{F} \cdot d\mathbf{R}$, where $\mathbf{F} = (3x^2 - 8y^2)\mathbf{i} + (4y - 6xy)\mathbf{j}$ and C encloses the region R: $x \geq 0$, $y \geq 0$, $x + y \leq 2$

21 $\int_C (\sin x^2 + y^2)\, dx + [\ln (2y^2 + 5) - x^2]\, dy$, where C encloses the region R bounded by $y = 2x^2$ and $y = 4x$

22 $\int_C (x^3 - x^2y)\, dx + xy^2\, dy$, where C encloses the region R bounded by $y = x^2$ and $x = y^2$

23 $\int_C (e^{x^2} + y^2)\, dx + (e^{y^2} + x^2)\, dy$, where C encloses the region R bounded by $y = x^2$ and $y = x + 12$

24 $\int_C e^y\, dx + xe^y\, dy$, where C encloses the region R: $0 \leq x \leq 1$, $\sin^{-1} x \leq y \leq \pi/2$

25 Find the work done in moving a particle under the action of a force $\mathbf{F} = 3x^2\mathbf{i} + (2xy - y^2)\mathbf{j}$ newtons along the line segment from $(0, 0)$ to $(2, 5)$. Assume that distances are measured in meters.

26 Find the work done in moving a particle under the action of a force $\mathbf{F} = xy\mathbf{i} + [(x^2/2) + xy]\mathbf{j}$ newtons along the upper half of the circle $x^2 + y^2 = 16$. Start at the point $(4, 0)$, and return to the starting point along the x axis. Measure the distance in meters.

27 Find the work done in moving a particle under the action of a force $\mathbf{F} = (2x + 3y)\mathbf{i} + xy\mathbf{j}$ newtons from $(0, 0)$ to $(1, 1)$ (a) along the line segment from $(0, 0)$ to $(1, 1)$, (b) along the parabola $y = x^2$, and (c) along the arc of the circle $x^2 + (y - 1)^2 = 1$ from $(0, 0)$ to $(1, 1)$. Assume that distances are measured in meters.

28 Find the work done in moving a particle under the action of a force $\mathbf{F} = y\mathbf{i} - x\mathbf{j}$ newtons around the cardioid $r = 2(1 + \cos \theta)$ in a counterclockwise direction, where the distance is measured in meters.

In Problems 29 to 32, use the formula on page 981 to find the area of the enclosed region.

29 $x^2 + y^2 = a^2$

30 $y = x^2$, $y = x + 2$

31 $x = \cos^3 t$, $y = \sin^3 t$, $0 \leq t \leq \pi/2$

32 $y = \sqrt{x}$, $y = 0$, $x = 4$

33 Criticize the following argument: Let $P(x, y) = -y/(x^2 + y^2)$ and let $Q(x, y) = x/(x^2 + y^2)$. Then $(\partial Q/\partial x) - (\partial P/\partial y) = 0$, so that

$$\int_C \frac{-y}{x^2 + y^2}\, dx + \frac{x}{x^2 + y^2}\, dy = \iint_R \left(\frac{\partial Q}{\partial x} - \frac{\partial P}{\partial y} \right) dA = 0$$

where C is the circle

$$C: \begin{cases} x = \cos t \\ y = \sin t \end{cases} \quad 0 \leq t \leq 2\pi$$

and R is the disk bounded by C. (*Note:* Something must be wrong, because direct calculation of the line integral gives 2π, not 0.)

34 Let C be a curve with the parametric equations

$$C: \begin{cases} x = f(t) \\ y = g(t) \end{cases} \quad a \leq t \leq b$$

where f and g have continuous first derivatives. Suppose that P and Q are continuous functions of two variables whose domains contain the curve C. Define functions F and G on $[a, b]$ by $F(t) = f(a + b - t)$ and $G(t) = g(a + b - t)$.

(a) Show that

$$\begin{cases} x = F(t) \\ y = G(t) \end{cases} \quad a \leq t \leq b$$

are parametric equations for $-C$.

(b) Use the result of part (a) to prove that

$$\int_{-C} P(x, y)\, dx + Q(x, y)\, dy = -\int_C P(x, y)\, dx + Q(x, y)\, dy$$

35 The line integral $\int_C \mathbf{F} \cdot d\mathbf{R}$ is sometimes written as $\int_C (\mathbf{F} \cdot \mathbf{T})\, ds$, where $\mathbf{T}$ denotes the unit tangent vector to C and s is the arc length measured along C. Explain why this notation is reasonable.

36 Let $\mathbf{T}$ denote the unit tangent vector to the oriented curve C and let $\mathbf{N}_r$ denote the normal vector to C obtained by rotating $\mathbf{T}$ to the right through 90° (Section 12.6). If $\mathbf{V} = f(x, y)\mathbf{i} + g(x, y)\mathbf{j}$, then the **flux** of $\mathbf{V}$ across C in the direction of $\mathbf{N}_r$ is defined to be $\int_C (\mathbf{V} \cdot \mathbf{N}_r)\, ds$. The scalar field $(\partial f/\partial x) + (\partial g/\partial y)$ is called the **divergence** of $\mathbf{V}$. Prove the following theorem, called the **divergence theorem** (in two dimensions): The outward flux of $\mathbf{V}$ across a simple closed curve C is the double integral of the divergence of $\mathbf{V}$ over the region R enclosed by C. (State carefully whatever assumptions about continuity and so forth that you need.)

37 In our derivation of the formula $A = \pi ab$ for the area of an ellipse (Example 11), exactly where did we use the hypothesis that a and b are *positive*? (Note that the given parametric equations describe an ellipse even if a or b or both are negative.)

38 Suppose that a fluid is flowing in a thin sheet over the xy plane and that the velocity of a particle of this fluid at the point (x, y) is given by $\mathbf{V} = f(x, y)\mathbf{i} + g(x, y)\mathbf{j}$. Assume that the density of the fluid at the point (x, y) is given by $\sigma(x, y)$ mass units per unit of area. Show that the flux of $\sigma(x, y)\mathbf{V}$ across a curve C in the direction of the normal $\mathbf{N}_r$ (Problem 36) gives the mass of fluid, per unit time, flowing across C in the direction of $\mathbf{N}_r$.

15.10   Conservative Vector Fields

In this section we introduce the idea of a *vector field*, and we use line integrals to study the important concept of a *conservative* field. The section includes a discussion of line integrals that are *independent of the path* and the idea of a *potential*.

A continuous mapping or function

$$A \longmapsto \mathbf{F}$$

Figure 1

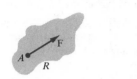

that assigns a vector $\mathbf{F}$ to each point A in a region R is called a **vector field** on R. We usually picture A as the tail end (initial point) of the vector $\mathbf{F}$ (Figure 1). Vectors attached to different points in R may have different lengths and directions. A vector field that represents the velocity at each point of a flowing fluid is called a **velocity field,** and a vector field that represents the force that would act on a test particle placed at each point is called a **force field.** Figure 2 shows some of the vectors in a velocity field for the flow of water from a tributary into a river, and Figure 3 shows some of the vectors in a force field for a positive electric charge placed at a central point.

Figure 2

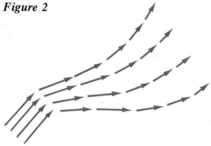

As we saw in Section 15.9, if $\mathbf{F}$ is a force field on the region R and if C is a curve contained in R, then

$$W = \int_C \mathbf{F} \cdot d\mathbf{R}$$

gives the work W done by the field in carrying a test particle from the initial point A to the terminal point B of the curve C. If $A = B$, so that C is a closed curve, then W is called the **circulation** of the vector field $\mathbf{F}$ around the curve C.

Ordinarily, the work W done by the field $\mathbf{F}$ in carrying a test particle from the initial point A to the terminal point B will depend on the choice of the curve, or path, C connecting A and B. However, for certain special vector fields, W depends only on A and B and not on the choice of the path C.

DEFINITION 1

Figure 3

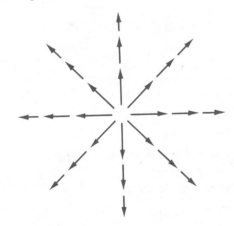

Conservative Field

A vector field $\mathbf{F}$ on a region R is said to be **conservative*** if for any two points A and B in R and any two piecewise smooth curves C_1 and C_2 contained in R with A as initial point and B as terminal point,

$$\int_{C_1} \mathbf{F} \cdot d\mathbf{R} = \int_{C_2} \mathbf{F} \cdot d\mathbf{R}$$

If $\mathbf{F}$ is a conservative vector field on the region R and if A and B are two points in R, we define

$$\int_A^B \mathbf{F} \cdot d\mathbf{R} = \int_C \mathbf{F} \cdot d\mathbf{R}$$

where C is any piecewise smooth curve in R with A as initial point and B as terminal point. Here we say that the integral on the right is **independent of the path** C from A to B.

*The word "conservative" comes from mechanics, where it is shown that for motion under the influence of a conservative force field, a law of conservation of energy prevails. (See Problem 32.)

Figure 4

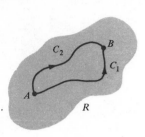

Figure 4 shows two piecewise smooth curves C_1 and C_2 from A to B as in Definition 1. If we let

$$C_0 = C_1 - C_2$$

then C_0 is a closed curve with A as its common initial and terminal point, and we have

$$\int_{C_0} \mathbf{F} \cdot d\mathbf{R} = \int_{C_1} \mathbf{F} \cdot d\mathbf{R} - \int_{C_2} \mathbf{F} \cdot d\mathbf{R}$$

From this observation, it follows that *the vector field* $\mathbf{F}$ *on the region* R *is conservative if and only if the circulation of* $\mathbf{F}$ *is zero around every closed curve* C_0 *in* R.

Suppose that the region R is open and that f is a continuous function defined on R. As we mentioned in Section 14.1, f can be regarded as a *scalar field* on R. If f is continuously differentiable on R, then the gradient

$$\mathbf{F} = \nabla f$$

is defined at each point A in R. In other words, *the gradient of a (continuously differentiable) scalar field is a vector field.* If $\mathbf{F} = \nabla f$, we call f a **potential function** for the vector field $\mathbf{F}$. The term "potential" was introduced in a memoir on electricity and magnetism written by George Green in 1828. (Green's theorem was announced in the same memoir.)

DEFINITION 2

Fields Derivable from a Potential

> If $\mathbf{F}$ is a vector field on an open region R, we say that $\mathbf{F}$ is *derivable from a potential* if there exists a continuously differentiable function f on R such that
>
> $$\mathbf{F} = \nabla f$$

An open region R is said to be **connected** if for each pair of points A and B in R, there exists a piecewise smooth curve C contained in R with A as its initial point and B as its terminal point.

THEOREM 1

Conservative Fields and Potentials

> Let $\mathbf{F}$ be a vector field on an open connected region R. Then $\mathbf{F}$ is conservative if and only if it is derivable from a potential.

PROOF

We give the proof for the case in which R is a region in the xy plane; the proof for the case in which R is a three-dimensional region is similar. Denote by $P(x, y)$ and $Q(x, y)$ the scalar components of $\mathbf{F}$ at the point $A = (x, y)$, so that

$$\mathbf{F} = P(x, y)\mathbf{i} + Q(x, y)\mathbf{j}$$

Since $\mathbf{F}$ is a continuous vector-valued function of x and y, it follows that P and Q are continuous functions of two variables on the region R.

Suppose first that $\mathbf{F}$ is conservative. Choose and fix a point (a, b) in R, and define the function f on R by

$$f(x, y) = \int_{(a,b)}^{(x,y)} \mathbf{F} \cdot d\mathbf{R}$$

Recall that the integral in this equation is the line integral of $\mathbf{F} \cdot d\mathbf{R}$ along *any* path in R from (a, b) to (x, y).

Figure 5

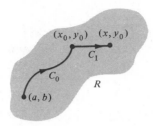

Now, let (x, y_0) be a point in R, and choose a point (x_0, y_0) in R and sufficiently close to (x, y_0) so that the line segment C_1 from (x_0, y_0) to (x, y_0) is contained in R (Figure 5). Denoting by C_0 any curve in R from (a, b) to (x_0, y_0), we have

$$f(x, y_0) = \int_{C_0 + C_1} \mathbf{F} \cdot d\mathbf{R} = \int_{C_0} \mathbf{F} \cdot d\mathbf{R} + \int_{C_1} \mathbf{F} \cdot d\mathbf{R}$$

$$= \int_{(a,b)}^{(x_0,y_0)} \mathbf{F} \cdot d\mathbf{R} + \int_{(x_0,y_0)}^{(x,y_0)} \mathbf{F} \cdot d\mathbf{R}$$

$$= f(x_0, y_0) + \int_{(x_0,y_0)}^{(x,y_0)} P \, dx + Q \, dy$$

Because $dy = 0$ on the horizontal line segment C_1 from (x_0, y_0) to (x, y_0), we have

$$\int_{(x_0,y_0)}^{(x,y_0)} P \, dx + Q \, dy = \int_{x_0}^{x} P(t, y_0) \, dt$$

Therefore, from the preceding equations,

$$f(x, y_0) = f(x_0, y_0) + \int_{x_0}^{x} P(t, y_0) \, dt$$

Now, by the fundamental theorem of calculus (in the form of Theorem 1 on page 334),

$$\frac{\partial}{\partial x} f(x, y_0) = \frac{\partial}{\partial x} f(x_0, y_0) + \frac{\partial}{\partial x} \int_{x_0}^{x} P(t, y_0) \, dt = 0 + P(x, y_0) = P(x, y_0)$$

Because the last result holds for all points (x, y_0) in R, we have

$$\frac{\partial}{\partial x} f(x, y) = P(x, y) \qquad (x, y) \text{ in } R$$

By an analogous argument (Problem 26), we can show that

$$\frac{\partial}{\partial y} f(x, y) = Q(x, y) \qquad (x, y) \text{ in } R$$

Since P and Q are continuous on R, it follows that f is continuously differentiable on R. Furthermore,

$$\mathbf{F} = P\mathbf{i} + Q\mathbf{j} = \frac{\partial f}{\partial x} \mathbf{i} + \frac{\partial f}{\partial y} \mathbf{j} = \nabla f$$

Thus, if $\mathbf{F}$ is conservative, it is derivable from a potential.

Conversely, suppose that $\mathbf{F}$ is derivable from a potential, so that

$$\mathbf{F} = P\mathbf{i} + Q\mathbf{j} = \nabla f = \frac{\partial f}{\partial x} \mathbf{i} + \frac{\partial f}{\partial y} \mathbf{j} = f_1\mathbf{i} + f_2\mathbf{j}$$

where f is a continuously differentiable function on R. Let C be any piecewise smooth curve in R from the initial point $A = (x_1, y_1)$ to the terminal point $B = (x_2, y_2)$. Let C be described by the parametric equations

$$x = g(t) \qquad y = h(t) \qquad a \le t \le b$$

Then

$$\int_{C} \mathbf{F} \cdot d\mathbf{R} = \int_{a}^{b} [f_1(g(t), h(t))g'(t) + f_2(g(t), h(t))h'(t)] \, dt$$

By the first chain rule for functions of two variables, the integrand on the right is $df(g(t), h(t))/dt$, so that

$$\int_C \mathbf{F} \cdot d\mathbf{R} = \int_a^b \left[\frac{d}{dt} f(g(t), h(t)) \right] dt$$

$$= f(g(t), h(t)) \Big|_a^b$$

$$= f(g(b), h(b)) - f(g(a), h(a))$$

$$= f(x_2, y_2) - f(x_1, y_1)$$

This shows that the line integral depends only on the coordinates of A and B and is independent of the path C; hence, $\mathbf{F}$ is a conservative vector field. ∎

The last result obtained in the proof of Theorem 1 is of importance in its own right. It can be stated in words as follows:

> If a vector field on an open connected region is the gradient of a potential, then the line integral of the field along any path from point A to point B in the region is the difference in the value of the potential at B and the potential at A.

The difference in the value of the potential referred to in the statement above is usually called the **potential difference** between the two points. In particular, we may conclude that *a conservative force field is derivable from a potential* and that *the work done by such a field in moving a test particle along a path is the potential difference between the terminal and initial points of the path.*

Suppose that the vector field

$$\mathbf{F} = P\mathbf{i} + Q\mathbf{j}$$

is derivable from a potential on an open region R, so that

$$\mathbf{F} = P\mathbf{i} + Q\mathbf{j} = \nabla f = \frac{\partial f}{\partial x}\mathbf{i} + \frac{\partial f}{\partial y}\mathbf{j} \qquad P = \frac{\partial f}{\partial x} \qquad Q = \frac{\partial f}{\partial y}$$

for some continuously differentiable function f on R. If the component functions P and Q of $\mathbf{F}$ are continuously differentiable, we say that the vector field $\mathbf{F}$ is **continuously differentiable.** In this case, f has continuous second partial derivatives, and it follows that

$$\frac{\partial P}{\partial y} = \frac{\partial}{\partial y}\left(\frac{\partial f}{\partial x}\right) = \frac{\partial^2 f}{\partial y\, \partial x} = \frac{\partial^2 f}{\partial x\, \partial y} = \frac{\partial}{\partial x}\left(\frac{\partial f}{\partial y}\right) = \frac{\partial Q}{\partial x}$$

Thus:

> If a continuously differentiable vector field
>
> $$\mathbf{F} = P\mathbf{i} + Q\mathbf{j}$$
>
> is derivable from a potential on an open region R, then
>
> $$\frac{\partial P}{\partial y} = \frac{\partial Q}{\partial x}$$
>
> holds at every point (x, y) in R.

A similar result can be obtained for a vector field on a three-dimensional region R (Problem 33):

> If a continuously differentiable vector field
>
> $$\mathbf{F} = P\mathbf{i} + Q\mathbf{j} + T\mathbf{k}$$
>
> is derivable from a potential on an open region R, then
>
> $$\frac{\partial P}{\partial y} = \frac{\partial Q}{\partial x} \qquad \frac{\partial Q}{\partial z} = \frac{\partial T}{\partial y} \qquad \text{and} \qquad \frac{\partial T}{\partial x} = \frac{\partial P}{\partial z}$$
>
> hold at every point (x, y, z) in R.

Figure 6

R is simply connected

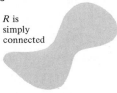

Figure 7

R is not simply connected

It can be shown that the converses of the statements given above are also true provided that the region R has certain properties. A two-dimensional region R is said to be **simply connected** if it is connected and has the property that every simple closed curve in R forms the boundary of a region contained in R. Loosely speaking, this means that there are no "holes" in R. For instance, the region in Figure 6 is simply connected, but the region in Figure 7 is not. The formal definition of a simply connected three-dimensional region is too technical to be stated here. *Informally*, a three-dimensional region R is said to be **simply connected** if it is connected and every simple closed curve in R forms the boundary of a piecewise smooth surface contained in R. Roughly speaking, this means that although there may be "hollows" inside of R, there are no "tunnels" running all the way through R.

THEOREM 2

Condition for Derivability from a Potential

> Suppose that $\mathbf{F} = P\mathbf{i} + Q\mathbf{j}$ is a continuously differentiable vector field on an open, simply connected region R. Then $\mathbf{F}$ is derivable from a potential on R if and only if
>
> $$\frac{\partial P}{\partial y} = \frac{\partial Q}{\partial x}$$
>
> holds at every point (x, y) in R.

A formal proof of Theorem 2 is too technical to be given here, but we can give the following *informal* proof. We have already shown that if $\mathbf{F}$ is derivable from a potential, then $\partial P/\partial y = \partial Q/\partial x$ holds at every point (x, y) in R. Conversely, suppose that $\partial P/\partial y = \partial Q/\partial x$ holds at every point (x, y) in R. To show that $\mathbf{F}$ is derivable from a potential, it is sufficient to show that $\mathbf{F}$ is conservative (Theorem 1). As we mentioned earlier, $\mathbf{F}$ will be conservative if and only if the circulation of $\mathbf{F}$ is zero around every closed curve C in R. Here we verify this for every *simple* closed curve C in R. (The extension to closed curves that are not simple is the technical part of the formal proof.) Thus, suppose that C is a simple closed curve in R enclosing, say, the subregion R_C of R. Then by Green's theorem (Theorem 1 in Section 15.9)

$$\int_C \mathbf{F} \cdot d\mathbf{R} = \int_C P\, dx + Q\, dy = \iint_{R_C} \left(\frac{\partial Q}{\partial x} - \frac{\partial P}{\partial y} \right) dA = 0$$

since $\partial P/\partial y = \partial Q/\partial x$ holds at every point (x, y) in R_C.

<u>**EXAMPLE 1**</u>   Determine whether $\int_C e^x \sin y \, dx + e^x \cos y \, dy$ is independent of the path C.

SOLUTION   Let $P = e^x \sin y$ and $Q = e^x \cos y$. The given line integral $\int_C P \, dx + Q \, dy$ will be independent of the path if and only if the vector field $\mathbf{F} = P\mathbf{i} + Q\mathbf{j}$ is conservative, that is, if and only if $\mathbf{F}$ is derivable from a potential. Here,

$$\frac{\partial P}{\partial y} = e^x \cos y \qquad \text{and} \qquad \frac{\partial Q}{\partial x} = e^x \cos y$$

Hence, $\partial P/\partial y = \partial Q/\partial x$, so $\mathbf{F}$ is derivable from a potential by Theorem 2. Therefore, the line integral is independent of the path. ∎

<u>**EXAMPLE 2**</u>   Let $\mathbf{F} = (x^3 + 5xy^2)\mathbf{i} + (5x^2y + 2y^3)\mathbf{j}$

(a) Show that $\mathbf{F}$ is a conservative vector field.

(b) Find a potential function f for $\mathbf{F}$.

(c) Evaluate $\displaystyle\int_{(0,2)}^{(2,4)} \mathbf{F} \cdot d\mathbf{R}$.

SOLUTION

(a) Let $P = x^3 + 5xy^2$ and $Q = 5x^2y + 2y^3$. Then

$$\frac{\partial P}{\partial y} = 10xy \qquad \text{and} \qquad \frac{\partial Q}{\partial x} = 10xy$$

Hence, $\mathbf{F}$ is derivable from a potential, and therefore it is conservative.

(b) If f is a potential function for $\mathbf{F}$, then

$$\frac{\partial f}{\partial x} = P(x, y) \qquad \text{and} \qquad \frac{\partial f}{\partial y} = Q(x, y)$$

Antidifferentiating both sides of the first equation with respect to x, we find that

$$f = \int P(x, y) \, dx = \int (x^3 + 5xy^2) \, dx = \frac{x^4}{4} + \frac{5x^2y^2}{2} + K(y)$$

where $K(y)$ must still be determined. Differentiation of both sides of the last equation with respect to y gives us

$$\frac{\partial f}{\partial y} = \frac{\partial}{\partial y}\left[\frac{x^4}{4} + \frac{5x^2y^2}{2} + K(y)\right] = 5x^2y + K'(y)$$

Now, using the fact that $\partial f/\partial y = Q$, we find that

$$5x^2y + K'(y) = \frac{\partial f}{\partial y} = Q = 5x^2y + 2y^3$$

and it follows that

$$K'(y) = 2y^3$$

Therefore,

$$K(y) = \int K'(y) \, dy = \int 2y^3 \, dy = \frac{y^4}{2} + C$$

and so $\qquad f = \dfrac{x^4}{4} + \dfrac{5x^2y^2}{2} + K(y) = \dfrac{x^4}{4} + \dfrac{5x^2y^2}{2} + \dfrac{y^4}{2} + C$

The presence of the constant C indicates that the potential f is not unique.

(c) Because $\mathbf{F}$ is the gradient of the potential f, it follows that

$$\int_{(0,2)}^{(2,4)} \mathbf{F} \cdot d\mathbf{R} = f(2, 4) - f(0, 2) = 292 - 8 = 284 \qquad \blacksquare$$

Theorem 2 has the following extension to vector fields in three-dimensional space.

THEOREM 3

Condition for Derivability from a Potential*

Suppose that $\mathbf{F} = P\mathbf{i} + Q\mathbf{j} + T\mathbf{k}$ is a continuously differentiable vector field on an open, simply connected region R. Then $\mathbf{F}$ is derivable from a potential on R if and only if

$$\frac{\partial P}{\partial y} = \frac{\partial Q}{\partial x} \qquad \frac{\partial Q}{\partial z} = \frac{\partial T}{\partial y} \qquad \text{and} \qquad \frac{\partial T}{\partial x} = \frac{\partial P}{\partial z}$$

hold at every point (x, y, z) in R.

EXAMPLE 3   Let $P = z^2 - y$, $Q = 2y - x$, and $T = 2xz$. Determine whether $\int_C P\,dx + Q\,dy + T\,dz$ is independent of the path. If it is, find a potential f for the vector field $\mathbf{F} = P\mathbf{i} + Q\mathbf{j} + T\mathbf{k}$.

SOLUTION   We have

$$\frac{\partial P}{\partial y} = -1 = \frac{\partial Q}{\partial x} \qquad \frac{\partial Q}{\partial z} = 0 = \frac{\partial T}{\partial y} \qquad \text{and} \qquad \frac{\partial T}{\partial x} = 2z = \frac{\partial P}{\partial z}$$

Hence, the vector field $\mathbf{F}$ is derivable from a potential by Theorem 3. Therefore, by Theorem 1,

$$\int_C P\,dx + Q\,dy + T\,dz = \int_C \mathbf{F} \cdot d\mathbf{R}$$

is independent of the path. To find the potential f, we require that

$$\frac{\partial f}{\partial x} = P = z^2 - y \qquad \frac{\partial f}{\partial y} = Q = 2y - x \qquad \text{and} \qquad \frac{\partial f}{\partial z} = T = 2xz$$

From the first of these three equations, we have

$$f = \int (z^2 - y)\,dx = xz^2 - xy + K(y, z)$$

[Note that our "constant" of integration $K(y, z)$ could depend on *both* y and z.] Differentiating both sides of the last equation with respect to y and setting the result equal to Q, we find that

$$\frac{\partial f}{\partial y} = \frac{\partial}{\partial y}[xz^2 - xy + K(y, z)] = -x + \frac{\partial}{\partial y}K(y, z) = 2y - x$$

from which it follows that $\qquad \dfrac{\partial}{\partial y}K(y, z) = 2y$

---

*See Problem 24 in Problem Set 15.12.

Antidifferentiating the last equation with respect to y, we obtain

$$K(y, z) = \int 2y \, dy = y^2 + k(z)$$

Therefore,

$$f = xz^2 - xy + y^2 + k(z)$$

Now we differentiate with respect to z and set the result equal to T to obtain

$$\frac{\partial f}{\partial z} = 2xz + k'(z) = 2xz$$

from which it follows that

$$k'(z) = 0 \quad \text{and so} \quad k(z) = C$$

where C is a constant. Therefore

$$f = xz^2 - xy + y^2 + C$$

is a potential function for $\mathbf{F}$.

Problem Set 15.10

In Problems 1 to 10, determine whether each line integral is independent of the path.

1 $\displaystyle\int_C (7x^2 + 4y) \, dx + (4x - 7y^3) \, dy$

2 $\displaystyle\int_C (\cos x \sin x - xy^2) \, dx + y(1 - x^2) \, dy$

3 $\displaystyle\int_C 2y(y - 1) \, dx + x(2y - 1) \, dy$

4 $\displaystyle\int_C (x + y)(x - y) \, dx + x(x - 2y) \, dy$

5 $\displaystyle\int_C (1 + \ln x + y/x) \, dx - (1 - \ln x) \, dy$ on the region R: $x > 0$

6 $\displaystyle\int_C (1 + \ln x + \ln y) \, dx + (\sin x - x/y) \, dy$ on the region R: $x > 0, \ y > 0$

7 $\displaystyle\int_C (1 + 2xy^3z^4) \, dx + (3x^2y^2z^4 - 2) \, dy + (3 + 4x^2y^3z^3) \, dz$

8 $\displaystyle\int_C y \sin z \, dx + x \sin z \, dy + x \sin y \, dz$

9 $\displaystyle\int_C (x - 2y^2) \, dx + (5y + x^2) \, dy + x^3 \, dz$

10 $\displaystyle\int_C e^x \, dx + \tan y \, dy + \ln z \, dz$

In Problems 11 to 24, (a) show that $\mathbf{F}$ is a conservative vector field, (b) find a potential function f for $\mathbf{F}$, and (c) evaluate the given line integral.

11 $\mathbf{F} = 2xy\mathbf{i} + (x^2 - 1)\mathbf{j}; \displaystyle\int_{(0,1)}^{(3,2)} \mathbf{F} \cdot d\mathbf{R}$

12 $\mathbf{F} = (2x - 5y)\mathbf{i} + (-5x + 3y^2)\mathbf{j}; \displaystyle\int_{(1,1)}^{(3,2)} \mathbf{F} \cdot d\mathbf{R}$

13 $\mathbf{F} = 3x^2y^3\mathbf{i} + 3x^3y^2\mathbf{j}; \displaystyle\int_{(-1,-1)}^{(3,4)} \mathbf{F} \cdot d\mathbf{R}$

14 $\mathbf{F} = 2xy(\sin x^2y)\mathbf{i} + (1 + x^2 \sin x^2y)\mathbf{j}; \displaystyle\int_{(0,0)}^{(1,\pi/2)} \mathbf{F} \cdot d\mathbf{R}$

15 $\mathbf{F} = (e^{2y} - y \cos xy)\mathbf{i} + (2xe^{2y} - x \cos xy + 2y)\mathbf{j}; \displaystyle\int_{(-1,0)}^{(4,0)} \mathbf{F} \cdot d\mathbf{R}$

16 $\mathbf{F} = (x - \sin x \sin y)\mathbf{i} + (\cos x \cos y)\mathbf{j}; \displaystyle\int_{(0,1)}^{(2\pi,\pi/2)} \mathbf{F} \cdot d\mathbf{R}$

17 $\mathbf{F} = (e^x + y)\mathbf{i} + (2 + x + ye^y)\mathbf{j}; \displaystyle\int_{(0,1)}^{(1,5)} \mathbf{F} \cdot d\mathbf{R}$

18 $\mathbf{F} = (y^2 \cos x - 3x^2y - 2x)\mathbf{i} + (2y \sin x - x^3 + \ln y)\mathbf{j};$
$\displaystyle\int_{(\pi/2,1)}^{(0,e)} \mathbf{F} \cdot d\mathbf{R}$

19 $\mathbf{F} = x\mathbf{i} + y\mathbf{j} - 2\mathbf{k}; \displaystyle\int_{(1,2,3)}^{(2,4,6)} \mathbf{F} \cdot d\mathbf{R}$

20 $\mathbf{F} = (y^2 + 2xz^2 - 1)\mathbf{i} + 2xy\mathbf{j} + (2x^2z + z^3)\mathbf{k}; \displaystyle\int_{(0,-1,2)}^{(1,2,4)} \mathbf{F} \cdot d\mathbf{R}$

21 $\mathbf{F} = (\cos x)\mathbf{i} + e^{-y}\mathbf{j} + z^2\mathbf{k}; \displaystyle\int_{(-1,2,4)}^{(6,0,2)} \mathbf{F} \cdot d\mathbf{R}$

22 $\mathbf{F} = (y + z^2)\mathbf{i} + (x + 2yz^3)\mathbf{j} + (3y^2z^2 + 2xz)\mathbf{k}; \displaystyle\int_{(0,0,0)}^{(1,-2,3)} \mathbf{F} \cdot d\mathbf{R}$

23 $\mathbf{F} = yz\mathbf{i} + xz\mathbf{j} + xy\mathbf{k}; \displaystyle\int_{(1,-1,1)}^{(2,3,4)} \mathbf{F} \cdot d\mathbf{R}$

24 $\mathbf{F} = y^2z^3\mathbf{i} + 2xyz^3\mathbf{j} + 3xy^2z^2\mathbf{k}; \displaystyle\int_{(-1,4,1)}^{(2,1,3)} \mathbf{F} \cdot d\mathbf{R}$

25 If P and Q are continuous functions of two variables defined on an open region R in the xy plane, the expression $P\,dx + Q\,dy$ is said to be an **exact differential** if there exists a continuously differentiable function f on R such that $df = P\,dx + Q\,dy$. (a) Show that $P\,dx + Q\,dy$ is an exact differential if and only if the vector field $\mathbf{F} = P\mathbf{i} + Q\mathbf{j}$ is conservative. (b) If $P\,dx + Q\,dy$ is an exact differential and $\mathbf{F} = P\mathbf{i} + Q\mathbf{j}$, show that f is a potential for $\mathbf{F}$ if and only if $df = P\,dx + Q\,dy$.

26 Complete the proof of Theorem 1 by showing that if the vector field $\mathbf{F} = P\mathbf{i} + Q\mathbf{j}$ is conservative on an open connected region R in the plane and $f(x, y) = \displaystyle\int_{(a,b)}^{(x,y)} \mathbf{F} \cdot d\mathbf{R}$, then $\partial f / \partial y = Q$.

27 A differential equation of the form $P(x, y)\,dx + Q(x, y)\,dy = 0$ is said to be **exact** if $P\,dx + Q\,dy$ is an exact differential (see Problem 25). If $P\,dx + Q\,dy = 0$ is an exact differential equation, so that $P\,dx + Q\,dy = df$ for some continuously differentiable function f, show that for any constant C, $f = C$ is a *solution* of the differential equation.

28 Consider the vector field $\mathbf{r} = x\mathbf{i} + y\mathbf{j}$, and let r be the scalar field defined by $r = |\mathbf{r}|$. (a) Show that for any constant real number n, $\nabla r^n = nr^{n-2}\mathbf{r}$. (b) If R is the region in the xy plane consisting of all points other than the origin, c is a constant, and $\mathbf{F} = cr^n\mathbf{r}$, for what values of n (if any) is $\mathbf{F}$ derivable from a potential on R?

29 (a) Show that $(3x^2y - x)\,dx + (x^3 + y)\,dy = 0$ is an exact dif-

ferential equation. (See Problem 25.) (b) Solve the differential equation in part (a). (See Problem 27.)

30 Let $\mathbf{F}$ be a vector field defined everywhere except possibly at the origin O. Let $\mathbf{u}$ be the vector field defined everywhere except at the origin by the condition that $|\mathbf{u}| = 1$ and $\mathbf{u}$ always points directly away from the origin. The vector field $\mathbf{F}$ is called an **inverse-square field** if at each point $A \neq O$, $\mathbf{F}$ is parallel to $\mathbf{u}$ and $|\mathbf{F}|$ is proportional to $|\overline{OA}|^{-2}$. Prove that an inverse square field $\mathbf{F}$ is derivable from a potential, and find such a potential f.

31 Let $\mathbf{F} = P\mathbf{i} + Q\mathbf{j}$, where $P = -y/(x^2 + y^2)$ and $Q = x/(x^2 + y^2)$. (a) Show that $\partial P/\partial y = \partial Q/\partial x$. (b) Show that $\mathbf{F}$ is *not* conservative by calculating $\int_C \mathbf{F} \cdot d\mathbf{R}$, where C is the circle $x^2 + y^2 = 1$ and the integral is taken around the circle in the counterclockwise sense. (c) Explain why the results of parts (a) and (b) do not contradict Theorems 1 and 2. (d) Show that if $\mathbf{F}$ is restricted to the half plane $R: x > 0$, then it is derivable from a potential f; find such a potential.

32 Let $\mathbf{F}$ be a force field defined on an open connected region R, and suppose that $\mathbf{F}$ is derivable from a potential f, so that $\mathbf{F} = \nabla f$ on R. Assume that a particle of mass m moves from the point A in R to the point B in R under the influence of the force $\mathbf{F}$ and traces out a path C from A to B in R. Let $\mathbf{R}$ denote the position vector of the particle at time t, and assume that $\mathbf{R} = \overrightarrow{OA}$ when $t = a$ and $\mathbf{R} = \overrightarrow{OB}$ when $t = b$. The **potential energy** $p(t)$ of the particle at time t is defined to be the *negative* of the value of the potential function f at the point where the particle is located at time t. If $\mathbf{V} = d\mathbf{R}/dt$ is the velocity vector of the particle at time t, then the **kinetic energy** $K(t)$ of the particle at time t is defined by $K(t) = \frac{1}{2}m\mathbf{V} \cdot \mathbf{V}$. The **total energy** $E(t)$ of the particle at time t is defined to be the sum of the potential and kinetic energies: $E(t) = p(t) + K(t)$. Assuming Newton's (second) law of motion, $\mathbf{F} = m(d\mathbf{V}/dt)$, prove that $dE(t)/dt = 0$ for $a \leq t \leq b$; that is, the total energy of the particle is a constant as it moves along its path. This is the **law of conservation of energy** for a conservative force field.

33 Prove that if a continuously differentiable vector field $\mathbf{F} = P\mathbf{i} + Q\mathbf{j} + T\mathbf{k}$ is derivable from a potential on an open region R, then $\partial P/\partial y = \partial Q/\partial x$, $\partial Q/\partial z = \partial T/\partial y$, and $\partial T/\partial x = \partial P/\partial z$.

15.11   Surface Area and Surface Integrals

In Section 15.9 we considered line integrals, that is, integrals over curves. In this section, we study the analogous* notion of *surface integrals,* that is, integrals over surfaces. A special case of the surface integral provides a technique for finding the area of a surface. The section also includes a discussion of the *flux* of a vector field through a surface.

*To obtain a strict analogy, we would have to define surface integrals over *oriented* surfaces; however, we do not consider the question of surface orientation in this section.

Figure 1

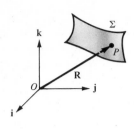

Area of a Surface Described by Vector Parametric Equations

Since we used the letter S to denote solids in Section 15.6, we use the Greek letter Σ to denote surfaces in this section. A surface Σ in xyz space can be described by a variable position vector $\mathbf{R}$ whose endpoint P traces out the surface (Figure 1). If we wish to express $\mathbf{R}$ parametrically, it is necessary to use *two* independent parameters, since Σ is "two-dimensional." Thus, if the two parameters are denoted by u and v, the vector parametric equation of Σ can be expressed as

$$\mathbf{R} = f(u, v)\mathbf{i} + g(u, v)\mathbf{j} + h(u, v)\mathbf{k}$$

Figure 2

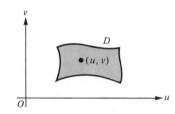

Here we assume that the functions f, g, and h are continuously differentiable and defined on an admissible region D in the uv plane (Figure 2). As the point (u, v) moves through the region D, the position vector $\mathbf{R}$ traces out the surface Σ.

It is not necessary to denote the surface parameters by u and v; they could just as easily be denoted by θ and ϕ, or by x and y, or by s and t, and so forth. For instance, a vector parametric equation of a sphere of radius a can be written in terms of the longitude θ and colatitude ϕ as

$$\mathbf{R} = (a \sin \phi \cos \theta)\mathbf{i} + (a \sin \phi \sin \theta)\mathbf{j} + (a \cos \phi)\mathbf{k}$$

where (θ, ϕ) ranges over the rectangular region

$$D: 0 \leq \theta \leq 2\pi \qquad 0 \leq \phi \leq \pi$$

in the $\theta\phi$ plane.

Now, let Σ be a surface described vector parametrically by

Figure 3

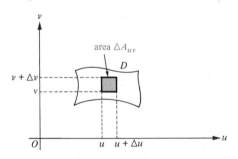

$$\mathbf{R} = f(u, v)\mathbf{i} + g(u, v)\mathbf{j} + h(u, v)\mathbf{k}$$

where (u, v) ranges over the region D in the uv plane. Consider a small rectangle of dimensions Δu and Δv with vertices (u, v), $(u + \Delta u, v)$, $(u + \Delta u, v + \Delta v)$, $(u, v + \Delta v)$ within the region D (Figure 3). The area ΔA_{uv} of this rectangle is given by

$$\Delta A_{uv} = \Delta u \, \Delta v$$

Let the four vertices of the rectangle, in the order given above, correspond to the four points P, Q, V, and W, respectively, on the surface Σ (Figure 4), so that

$$P = (f(u, v), g(u, v), h(u, v))$$

$$Q = (f(u + \Delta u, v), g(u + \Delta u, v), h(u + \Delta u, v))$$

$$V = (f(u + \Delta u, v + \Delta v), g(u + \Delta u, v + \Delta v), h(u + \Delta u, v + \Delta v))$$

Figure 4

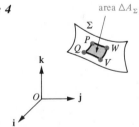

and

$$W = (f(u, v + \Delta v), g(u, v + \Delta v), h(u, v + \Delta v))$$

Notice that the area ΔA_{Σ} of the region $PQVW$ on the surface Σ is approximately the area of the parallelogram spanned by the vectors $\overrightarrow{PQ}$ and $\overrightarrow{PW}$, so that

$$\Delta A_{\Sigma} \approx |\overrightarrow{PQ} \times \overrightarrow{PW}|$$

By the definition of the partial derivatives f_1, g_1, and h_1,

$$f(u + \Delta u, v) - f(u, v) \approx f_1(u, v) \, \Delta u$$

$$g(u + \Delta u, v) - g(u, v) \approx g_1(u, v) \, \Delta u$$

and

$$h(u + \Delta u, v) - h(u, v) \approx h_1(u, v) \, \Delta u$$

with better and better approximations as $\Delta u \to 0$. Therefore,

$$\overrightarrow{PQ} = [f(u + \Delta u, v) - f(u, v)]\mathbf{i} + [g(u + \Delta u, v) - g(u, v)]\mathbf{j}$$
$$+ [h(u + \Delta u, v) - h(u, v)]\mathbf{k}$$

$$\approx [f_1(u, v) \, \Delta u]\mathbf{i} + [g_1(u, v) \, \Delta u]\mathbf{j} + [h_1(u, v) \, \Delta u]\mathbf{k}$$

$$= \left(\frac{\partial f}{\partial u} \, \Delta u\right)\mathbf{i} + \left(\frac{\partial g}{\partial u} \, \Delta u\right)\mathbf{j} + \left(\frac{\partial h}{\partial u} \, \Delta u\right)\mathbf{k} = \left(\frac{\partial f}{\partial u} \, \mathbf{i} + \frac{\partial g}{\partial u} \, \mathbf{j} + \frac{\partial h}{\partial u} \, \mathbf{k}\right) \Delta u$$

$$= \frac{\partial \mathbf{R}}{\partial u} \, \Delta u$$

where we have used the obvious symbolism $\partial \mathbf{R}/\partial u$ for the vector obtained by partially differentiating $\mathbf{R}$, component by component, with respect to u. Similarly, we have

$$\overrightarrow{PW} = [f(u, v + \Delta v) - f(u, v)]\mathbf{i} + [g(u, v + \Delta v) - g(u, v)]\mathbf{j}$$
$$+ [h(u, v + \Delta v) - h(u, v)]\mathbf{k}$$

$$\approx [f_2(u, v) \, \Delta v]\mathbf{i} + [g_2(u, v) \, \Delta v]\mathbf{j} + [h_2(u, v) \, \Delta v]\mathbf{k}$$

$$= \left(\frac{\partial f}{\partial v} \, \Delta v\right)\mathbf{i} + \left(\frac{\partial g}{\partial v} \, \Delta v\right)\mathbf{j} + \left(\frac{\partial h}{\partial v} \, \Delta v\right)\mathbf{k} = \frac{\partial \mathbf{R}}{\partial v} \, \Delta v$$

It follows that

$$\Delta A_\Sigma \approx \left|\frac{\partial \mathbf{R}}{\partial u} \, \Delta u \times \frac{\partial \mathbf{R}}{\partial v} \, \Delta v\right| = \left|\frac{\partial \mathbf{R}}{\partial u} \times \frac{\partial \mathbf{R}}{\partial v}\right| \Delta u \, \Delta v = \left|\frac{\partial \mathbf{R}}{\partial u} \times \frac{\partial \mathbf{R}}{\partial v}\right| \Delta A_{uv}$$

with better and better approximation as $(\Delta u, \Delta v) \to (0, 0)$.

From the result above, we may conclude that the element dA_Σ of surface area is given by

$$dA_\Sigma = \left|\frac{\partial \mathbf{R}}{\partial u} \times \frac{\partial \mathbf{R}}{\partial v}\right| dA_{uv}$$

where dA_{uv} is the element of area in the uv plane. Using the identity $|\mathbf{A} \times \mathbf{B}| = \sqrt{|\mathbf{A}|^2|\mathbf{B}|^2 - (\mathbf{A} \cdot \mathbf{B})^2}$ (page 777), we can rewrite the formula for dA_Σ as

$$dA_\Sigma = \sqrt{\left|\frac{\partial \mathbf{R}}{\partial u}\right|^2 \left|\frac{\partial \mathbf{R}}{\partial v}\right|^2 - \left(\frac{\partial \mathbf{R}}{\partial u} \cdot \frac{\partial \mathbf{R}}{\partial v}\right)^2} \, dA_{uv}$$

Thus, the total area A_Σ of the surface Σ can be obtained by integration:

$$A_\Sigma = \iint_D dA_\Sigma = \iint_D \left|\frac{\partial \mathbf{R}}{\partial u} \times \frac{\partial \mathbf{R}}{\partial v}\right| dA_{uv}$$
$$= \iint_D \sqrt{\left|\frac{\partial \mathbf{R}}{\partial u}\right|^2 \left|\frac{\partial \mathbf{R}}{\partial v}\right|^2 - \left(\frac{\partial \mathbf{R}}{\partial u} \cdot \frac{\partial \mathbf{R}}{\partial v}\right)^2} \, du \, dv$$

where in the last double integral we have written the element of area dA_{uv} in the alternative form $du \, dv$.

EXAMPLE 1   Use the formula above to find the area of a sphere of radius a.

SOLUTION   The sphere is represented by the vector parametric equation

$$\mathbf{R} = (a \sin \phi \cos \theta)\mathbf{i} + (a \sin \phi \sin \theta)\mathbf{j} + (a \cos \phi)\mathbf{k}$$

where (θ, ϕ) ranges over the rectangular region

$$D: 0 \le \theta \le 2\pi \qquad 0 \le \phi \le \pi$$

in the $\theta\phi$ plane. Here,

$$\frac{\partial \mathbf{R}}{\partial \theta} = (-a \sin \phi \sin \theta)\mathbf{i} + (a \sin \phi \cos \theta)\mathbf{j} + 0\mathbf{k}$$

$$\frac{\partial \mathbf{R}}{\partial \phi} = (a \cos \phi \cos \theta)\mathbf{i} + (a \cos \phi \sin \theta)\mathbf{j} - (a \sin \phi)\mathbf{k}$$

Hence,

$$\left| \frac{\partial \mathbf{R}}{\partial \theta} \right|^2 = a^2 \sin^2 \phi \sin^2 \theta + a^2 \sin^2 \phi \cos^2 \theta = a^2 \sin^2 \phi(\sin^2 \theta + \cos^2 \theta)$$

$$= a^2 \sin^2 \phi$$

$$\left| \frac{\partial \mathbf{R}}{\partial \phi} \right|^2 = a^2 \cos^2 \phi \cos^2 \theta + a^2 \cos^2 \phi \sin^2 \theta + a^2 \sin^2 \phi$$

$$= a^2 \cos^2 \phi(\cos^2 \theta + \sin^2 \theta) + a^2 \sin^2 \phi = a^2 \cos^2 \phi + a^2 \sin^2 \phi$$

$$= a^2(\cos^2 \phi + \sin^2 \phi) = a^2$$

and

$$\frac{\partial \mathbf{R}}{\partial \theta} \cdot \frac{\partial \mathbf{R}}{\partial \phi} = -a^2 \sin \phi \cos \phi \sin \theta \cos \theta + a^2 \sin \phi \cos \phi \sin \theta \cos \theta + 0 = 0$$

Therefore,

$$\sqrt{\left| \frac{\partial \mathbf{R}}{\partial \theta} \right|^2 \left| \frac{\partial \mathbf{R}}{\partial \phi} \right|^2 - \left(\frac{\partial \mathbf{R}}{\partial \theta} \cdot \frac{\partial \mathbf{R}}{\partial \phi} \right)^2} = \sqrt{(a^2 \sin^2 \phi)a^2 - 0} = a^2 \sin \phi$$

and it follows that for the sphere of radius a with longitude and colatitude as parameters,

$$\boxed{dA_\Sigma = a^2 \sin \phi \, d\theta \, d\phi}$$

Consequently, the surface area of the sphere is given by

$$A_\Sigma = \iint_D dA_\Sigma = \iint_D a^2 \sin \phi \, d\theta \, d\phi = a^2 \iint_D \sin \phi \, d\theta \, d\phi$$

$$= a^2 \int_0^\pi \int_0^{2\pi} \sin \phi \, d\theta \, d\phi = a^2 \int_0^\pi \left(\sin \phi \int_0^{2\pi} d\theta \right) d\phi$$

$$= a^2 \int_0^\pi \left[\sin \phi \left(\theta \, \Big|_0^{2\pi} \right) \right] d\phi = a^2 \int_0^\pi 2\pi \sin \phi \, d\phi$$

$$= 2\pi a^2(-\cos \phi) \, \Big|_0^\pi = -2\pi a^2(-1 - 1) = 4\pi a^2 \qquad \blacksquare$$

Figure 5

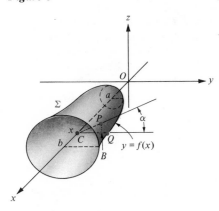

EXAMPLE 2   Use the integral for surface area to rederive the formula for the area of a surface of revolution originally derived in Section 6.4.

SOLUTION   Let the surface Σ be generated by revolving the graph of $y = f(x)$, $a \le x \le b$, about the x axis, where f has a continuous first derivative and $f(x) \ge 0$ for $a \le x \le b$ (Figure 5). A point P on the surface Σ can be located by specifying its x coordinate and the angle α through which P has turned from its original position Q on the curve $y = f(x)$. Thus, we take x and α as our parameters. In Figure 5, notice that

$$|\overline{CP}| = |\overline{CQ}| = f(x)$$

Consideration of the right triangle CBP shows that the y and z coordinates of P are given by

$$y = |\overline{CP}| \cos \alpha = f(x) \cos \alpha \qquad \text{and} \qquad z = |\overline{CP}| \sin \alpha = f(x) \sin \alpha$$

Therefore, the position vector $\mathbf{R}$ of P is given by

$$\mathbf{R} = x\mathbf{i} + y\mathbf{j} + z\mathbf{k} = x\mathbf{i} + f(x)(\cos \alpha)\mathbf{j} + f(x)(\sin \alpha)\mathbf{k}$$

and as (x, α) ranges over the rectangular region

$$D: a \le x \le b \qquad 0 \le \alpha \le 2\pi$$

in the $x\alpha$ plane, the vector $\mathbf{R}$ traces out the surface Σ. Here,

$$\frac{\partial \mathbf{R}}{\partial x} = \mathbf{i} + f'(x)(\cos \alpha)\mathbf{j} + f'(x)(\sin \alpha)\mathbf{k}$$

$$\frac{\partial \mathbf{R}}{\partial \alpha} = 0\mathbf{i} - f(x)(\sin \alpha)\mathbf{j} + f(x)(\cos \alpha)\mathbf{k}$$

Therefore,

$$\left|\frac{\partial \mathbf{R}}{\partial x}\right|^2 = 1^2 + [f'(x)]^2 \cos^2 \alpha + [f'(x)]^2 \sin^2 \alpha = 1 + [f'(x)]^2$$

$$\left|\frac{\partial \mathbf{R}}{\partial \alpha}\right|^2 = 0^2 + [f(x)]^2 \sin^2 \alpha + [f(x)]^2 \cos^2 \alpha = [f(x)]^2$$

and

$$\frac{\partial \mathbf{R}}{\partial x} \cdot \frac{\partial \mathbf{R}}{\partial \alpha} = 0 - f'(x)f(x) \cos \alpha \sin \alpha + f'(x)f(x) \cos \alpha \sin \alpha = 0$$

Hence,

$$dA_\Sigma = \sqrt{\left|\frac{\partial \mathbf{R}}{\partial x}\right|^2 \left|\frac{\partial \mathbf{R}}{\partial \alpha}\right|^2 - \left(\frac{\partial \mathbf{R}}{\partial x} \cdot \frac{\partial \mathbf{R}}{\partial \alpha}\right)^2} \, dx \, d\alpha = \sqrt{\{1 + [f'(x)]^2\}[f(x)]^2 - 0} \, dx \, d\alpha$$

$$= f(x)\sqrt{1 + [f'(x)]^2} \, dx \, d\alpha$$

It follows that

$$A_\Sigma = \iint_D dA_\Sigma = \iint_D f(x)\sqrt{1 + [f'(x)]^2} \, dx \, d\alpha = \int_a^b \int_0^{2\pi} f(x)\sqrt{1 + [f'(x)]^2} \, d\alpha \, dx$$

$$= \int_a^b \left[f(x)\sqrt{1 + [f'(x)]^2} \; \alpha \; \Big|_0^{2\pi} \right] dx = 2\pi \int_a^b f(x)\sqrt{1 + [f'(x)]^2} \, dx \qquad \blacksquare$$

Figure 6

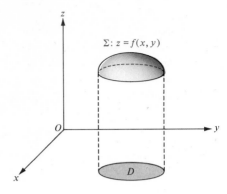

$\Sigma: z = f(x, y)$

Area of the Graph of a Function of Two Variables

Let f be a continuously differentiable function of two variables defined over an admissible region D in the xy plane, and denote the graph of f by Σ (Figure 6). Using x and y as parameters, we find that the vector parametric equation of Σ is

$$\mathbf{R} = x\mathbf{i} + y\mathbf{j} + f(x, y)\mathbf{k}$$

where (x, y) ranges over the region D. Here,

$$\frac{\partial \mathbf{R}}{\partial x} = \mathbf{i} + 0\mathbf{j} + f_1(x, y)\mathbf{k} \qquad \text{and} \qquad \frac{\partial \mathbf{R}}{\partial y} = 0\mathbf{i} + \mathbf{j} + f_2(x, y)\mathbf{k}$$

Hence,

$$\left| \frac{\partial \mathbf{R}}{\partial x} \right|^2 = 1^2 + 0^2 + [f_1(x, y)]^2$$

$$\left| \frac{\partial \mathbf{R}}{\partial y} \right|^2 = 0^2 + 1^2 + [f_2(x, y)]^2$$

and

$$\frac{\partial \mathbf{R}}{\partial x} \cdot \frac{\partial \mathbf{R}}{\partial y} = 0 + 0 + f_1(x, y)f_2(x, y)$$

It follows that

$$dA_\Sigma = \sqrt{\left| \frac{\partial \mathbf{R}}{\partial x} \right|^2 \left| \frac{\partial \mathbf{R}}{\partial y} \right|^2 - \left(\frac{\partial \mathbf{R}}{\partial x} \cdot \frac{\partial \mathbf{R}}{\partial y} \right)^2} \, dx \, dy$$

$$= \sqrt{\{1 + [f_1(x, y)]^2\}\{1 + [f_2(x, y)]^2\} - [f_1(x, y)f_2(x, y)]^2} \, dx \, dy$$

$$= \sqrt{1 + [f_1(x, y)]^2 + [f_2(x, y)]^2} \, dx \, dy$$

Therefore, the area A_Σ of the portion of the graph of $z = f(x, y)$ lying above the region D in the xy plane is given by

$$A_\Sigma = \iint\limits_D \sqrt{1 + [f_1(x, y)]^2 + [f_2(x, y)]^2} \, dx \, dy$$

$$= \iint\limits_D \sqrt{1 + \left(\frac{\partial z}{\partial x} \right)^2 + \left(\frac{\partial z}{\partial y} \right)^2} \, dx \, dy$$

©**EXAMPLE 3**   Find the area of the portion of the surface $z = x^2 + y^2$ that lies over the region D: $x^2 + y^2 \leq 1$. (Round off your answer to two decimal places.)

SOLUTION

$$A_\Sigma = \iint\limits_D \sqrt{1 + \left(\frac{\partial z}{\partial x} \right)^2 + \left(\frac{\partial z}{\partial y} \right)^2} \, dx \, dy$$

$$= \iint\limits_D \sqrt{1 + (2x)^2 + (2y)^2} \, dx \, dy$$

$$= \iint\limits_D \sqrt{4x^2 + 4y^2 + 1} \, dx \, dy$$

Converting the last double integral to polar coordinates, we obtain

$$A_\Sigma = \int_0^{2\pi} \int_0^1 \sqrt{4r^2 \cos^2 \theta + 4r^2 \sin^2 \theta + 1} \, r \, dr \, d\theta = \int_0^{2\pi} \int_0^1 \sqrt{4r^2 + 1} \, r \, dr \, d\theta$$

Making the change of variable $u = 4r^2 + 1$ in the inside integral, we have

$$A = \int_0^{2\pi} \left(\int_1^5 \sqrt{u} \, \frac{du}{8} \right) d\theta = \int_0^{2\pi} \left(\frac{u^{3/2}}{12} \Big|_1^5 \right) d\theta = \int_0^{2\pi} \frac{\sqrt{125} - 1}{12} \, d\theta$$

$$= 2\pi \frac{5\sqrt{5} - 1}{12} \approx 5.33 \text{ square units}$$

■

EXAMPLE 4   Find the area of the portion of the plane $2x + 3y + z = 6$ that is cut off by the three coordinate planes.

SOLUTION   The equation of the plane can be written as $z = 6 - 2x - 3y$, so that

$$\frac{\partial z}{\partial x} = -2 \quad \text{and} \quad \frac{\partial z}{\partial y} = -3$$

The plane $z = 6 - 2x - 3y$ intersects the plane $z = 0$ in the line $2x + 3y = 6$; hence, the portion of the plane cut off by the three coordinate planes lies above the triangular region D in the xy plane bounded by the x axis, the y axis, and the line $2x + 3y = 6$ (Figure 7). The area of D is given by

$$\iint_D dx \, dy = \tfrac{1}{2}(2)(3) = 3 \text{ square units}$$

Thus, the area A_Σ of the portion of the plane $z = 6 - 2x - 3y$ lying above D is given by

$$A_\Sigma = \iint_D \sqrt{1 + \left(\frac{\partial z}{\partial x}\right)^2 + \left(\frac{\partial z}{\partial y}\right)^2} \, dx \, dy = \iint_D \sqrt{1 + (-2)^2 + (-3)^2} \, dx \, dy$$

$$= \sqrt{14} \iint_D dx \, dy = \sqrt{14}(3) = 3\sqrt{14} \text{ square units}$$

■

Figure 7

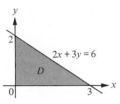

The Surface Integral

Suppose that Σ is a surface described vector parametrically by

$$\Sigma: \mathbf{R} = f(u, v)\mathbf{i} + g(u, v)\mathbf{j} + h(u, v)\mathbf{k}$$

for (u, v) in an admissible region D in the uv plane. We assume that the scalar component functions f, g, and h are continuously differentiable, so that the differential dA_Σ of surface area is given by

$$dA_\Sigma = \left| \frac{\partial \mathbf{R}}{\partial u} \times \frac{\partial \mathbf{R}}{\partial v} \right| dA_{uv} = \sqrt{\left| \frac{\partial \mathbf{R}}{\partial u} \right|^2 \left| \frac{\partial \mathbf{R}}{\partial v} \right|^2 - \left(\frac{\partial \mathbf{R}}{\partial u} \cdot \frac{\partial \mathbf{R}}{\partial v} \right)^2} \, du \, dv$$

If F is a function of three variables, defined and continuous on a subset of xyz space that contains the surface Σ, then the **surface integral** of F over Σ, denoted symbolically by $\iint_\Sigma F \, dA_\Sigma$, is defined by

$$\iint_\Sigma F \, dA_\Sigma$$

$$= \iint_D F(f(u, v), g(u, v), h(u, v)) \sqrt{\left| \frac{\partial \mathbf{R}}{\partial u} \right|^2 \left| \frac{\partial \mathbf{R}}{\partial v} \right|^2 - \left(\frac{\partial \mathbf{R}}{\partial u} \cdot \frac{\partial \mathbf{R}}{\partial v} \right)^2} \, du \, dv$$

EXAMPLE 5   Let Σ denote the surface of a sphere of radius a in xyz space with center at the origin. Evaluate the surface integral

$$\iint_{\Sigma} (x^2 + z)\, dA_{\Sigma}$$

SOLUTION   Using longitude θ and colatitude ϕ as parameters, we can write the vector parametric equation of the sphere as

$$\mathbf{R} = (a \sin \phi \cos \theta)\mathbf{i} + (a \sin \phi \sin \theta)\mathbf{j} + (a \cos \phi)\mathbf{k}$$

where (θ, ϕ) ranges over the rectangular region

$$D: 0 \le \theta \le 2\pi \qquad 0 \le \phi \le \pi$$

in the $\theta\phi$ plane. In Example 1 we found that

$$dA_{\Sigma} = a^2 \sin \phi\, d\theta\, d\phi$$

Therefore,

$$
\begin{aligned}
\iint_{\Sigma} (x^2 + z)\, dA_{\Sigma} &= \iint_{D} [(a \sin \phi \cos \theta)^2 + (a \cos \phi)]a^2 \sin \phi\, d\theta\, d\phi \\
&= \iint_{D} (a^4 \sin^3 \phi \cos^2 \theta + a^3 \sin \phi \cos \phi)\, d\theta\, d\phi \\
&= \int_0^{\pi} \left[\int_0^{2\pi} \left(a^4 \sin^3 \phi \frac{1 + \cos 2\theta}{2} + a^3 \sin \phi \cos \phi \right) d\theta \right] d\phi \\
&= \int_0^{\pi} \left[a^4 \sin^3 \phi \left(\frac{\theta}{2} + \frac{\sin 2\theta}{4} \right) + (a^3 \sin \phi \cos \phi)\theta \Big|_0^{2\pi} \right] d\phi \\
&= \int_0^{\pi} (a^4 \pi \sin^3 \phi + 2a^3 \pi \sin \phi \cos \phi)\, d\phi \\
&= a^4 \pi \int_0^{\pi} \sin^3 \phi\, d\phi + 2a^3 \pi \int_0^{\pi} \sin \phi \cos \phi\, d\phi \\
&= a^4 \pi \int_0^{\pi} (1 - \cos^2 \phi) \sin \phi\, d\phi + 2a^3 \pi \left(\frac{\sin^2 \phi}{2} \Big|_0^{\pi} \right) \\
&= a^4 \pi \int_0^{\pi} \sin \phi\, d\phi + a^4 \pi \int_0^{\pi} \cos^2 \phi(-\sin \phi)\, d\phi + 0 \\
&= a^4 \pi \left(-\cos \phi \Big|_0^{\pi} \right) + a^4 \pi \left(\frac{\cos^3 \phi}{3} \Big|_0^{\pi} \right) \\
&= 2a^4 \pi - \tfrac{2}{3} a^4 \pi = \tfrac{4}{3} a^4 \pi
\end{aligned}
$$

The Flux of a Vector Field Through a Surface

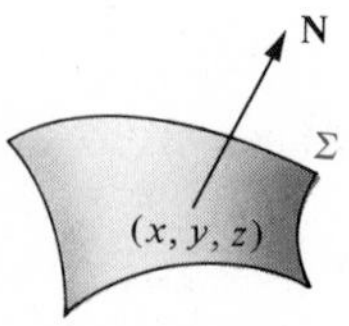

Suppose that Σ is a surface in xyz space and that $\mathbf{N}$ denotes a unit normal vector to Σ at the point (x, y, z) (Figure 8). We assume that as the point (x, y, z) moves around on the surface Σ, the unit normal vector $\mathbf{N}$ varies in a continuous manner. Suppose that Σ is contained in the three-dimensional region S on which the vector field

$$\mathbf{F} = P(x, y, z)\mathbf{i} + Q(x, y, z)\mathbf{j} + T(x, y, z)\mathbf{k}$$

is defined. For definiteness, we visualize $\mathbf{F}$ as the velocity field of a flowing fluid.

Figure 8

Figure 9

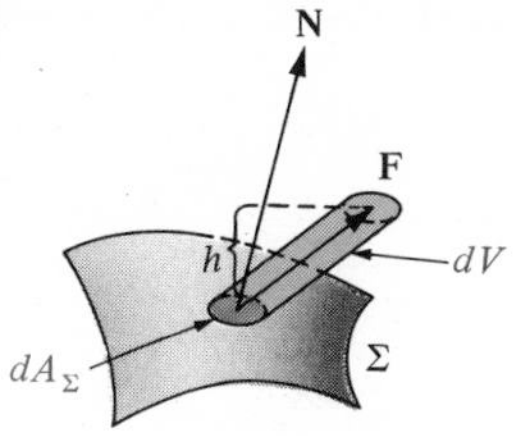

Consider an infinitesimal region of area dA_Σ at a point on the surface Σ, and let $\mathbf{N}$ be the unit normal to the surface at this point (Figure 9). If we let 1 unit of time go by, then the fluid flowing through dA_Σ will fill an infinitesimal solid cylinder of height $h = \mathbf{F} \cdot \mathbf{N}$ and with volume

$$dV = h \, dA_\Sigma = \mathbf{F} \cdot \mathbf{N} \, dA_\Sigma$$

The infinitesimal volume dV of fluid flowing through the infinitesimal region dA_Σ in unit time is called the **flux** through dA_Σ. Integrating dV over the whole surface Σ, we obtain the total volume of fluid flowing through Σ in unit time; hence, we define the **flux of the vector field F through the surface** Σ *in the direction of the unit normal vector* $\mathbf{N}$ to be the surface integral

$$\boxed{\iint_\Sigma dV = \iint_\Sigma \mathbf{F} \cdot \mathbf{N} \, dA_\Sigma}$$

<u>EXAMPLE 6</u>   Let Σ denote the portion of the plane $z = x + 2y + 1$ that lies above the region $D\colon 0 \le x \le 1,\ 0 \le y \le 2$ in the xy plane. Find the flux of the vector field $\mathbf{F} = x^3\mathbf{i} + xy\mathbf{j} + z\mathbf{k}$ through the surface Σ in the direction of the normal $\mathbf{N}$ that makes an acute angle with the z axis.

SOLUTION   Writing the equation of the plane in the form

$$x + 2y - z + 1 = 0$$

we find that the vector $\mathbf{N}_1 = \mathbf{i} + 2\mathbf{j} - \mathbf{k}$ is normal to the plane, as is the vector $\mathbf{N}_2 = -\mathbf{N}_1 = -\mathbf{i} - 2\mathbf{j} + \mathbf{k}$. Since $\mathbf{N}_2$ has a positive $\mathbf{k}$ component, it follows that $\mathbf{N}_2$ makes an acute angle with the positive z axis. (Why?) Normalizing $\mathbf{N}_2$, we obtain the desired unit normal vector

$$\mathbf{N} = \frac{\mathbf{N}_2}{|\mathbf{N}_2|} = \frac{-\mathbf{i} - 2\mathbf{j} + \mathbf{k}}{\sqrt{(-1)^2 + (-2)^2 + 1^2}} = \frac{-\mathbf{i} - 2\mathbf{j} + \mathbf{k}}{\sqrt{6}}$$

Therefore, $\mathbf{F} \cdot \mathbf{N} = (-x^3 - 2xy + z)/\sqrt{6}$; hence the flux through Σ is given by

$$\begin{aligned}
\iint_\Sigma \mathbf{F} \cdot \mathbf{N} \, dA_\Sigma &= \iint_\Sigma \frac{-x^3 - 2xy + z}{\sqrt{6}} \, dA_\Sigma \\
&= \iint_D \frac{-x^3 - 2xy + z}{\sqrt{6}} \sqrt{1 + \left(\frac{\partial z}{\partial x}\right)^2 + \left(\frac{\partial z}{\partial y}\right)^2} \, dx \, dy \\
&= \iint_D \frac{-x^3 - 2xy + x + 2y + 1}{\sqrt{6}} \sqrt{1 + (1)^2 + (2)^2} \, dx \, dy \\
&= \int_0^2 \int_0^1 (-x^3 - 2xy + x + 2y + 1) \, dx \, dy \\
&= \int_0^2 \left[\left(\frac{-x^4}{4} - x^2 y + \frac{x^2}{2} + 2xy + x \right) \Big|_0^1 \right] dy \\
&= \int_0^2 \left(y + \frac{5}{4} \right) dy \\
&= \left(\frac{y^2}{2} + \frac{5}{4} y \right) \Big|_0^2 = \frac{9}{2}
\end{aligned}$$

■

Problem Set 15.11

In Problems 1 to 10, find the area A_Σ of each surface Σ.

1 Σ is the portion of the plane $x + y + z = 5$ that lies above the circular region D: $x^2 + y^2 \le 9$.

2 Σ is the portion of the cylinder $y^2 + z^2 = 16$ that lies above the triangular region D: $0 \le x \le 2$, $0 \le y \le 2 - x$.

3 Σ is the portion of the plane $3x + 2y + z = 7$ that is cut off by the three coordinate planes.

4 Σ is the portion of the parabolic cylinder $z^2 = 8x$ that lies above the region D: $0 \le x \le 1$, $0 \le y \le \frac{1}{4}x^2$.

5 Σ is the portion of the cylinder $x^2 + z^2 = 9$ that lies above the rectangular region D with vertices $(0, 0)$, $(1, 0)$, $(1, 2)$, and $(0, 2)$.

6 Σ is the portion of the sphere $x^2 + y^2 + z^2 = 36$ that lies above the circular region D: $x^2 + y^2 \le 9$.

7 Σ is the portion of the cylinder $y^2 + z^2 = 4$ inside the cylinder $x^2 = 2y + 4$ and above the plane $z = 0$.

8 Σ is the portion of the cone $z = \sqrt{x^2 + y^2}$ inside the cylinder $x^2 + y^2 = 6y$.

9 Σ is generated by revolving the arc of $y^2 = 4x$ from $(0, 0)$ to $(3, 2\sqrt{3})$ about the x axis.

10 Σ is generated by revolving the arc of $x = e^y$ from $(1, 0)$ to $(e, 1)$ about the y axis.

11 Let D be an admissible region of area A in the xy plane, and let Σ be the portion of the plane $ax + by + cz + d = 0$ consisting of all points (x, y, z) on the plane for which (x, y) is in the region D. Assuming that $c > 0$, show that the area A_Σ of Σ is given by

$$A_\Sigma = \frac{A}{c} \sqrt{a^2 + b^2 + c^2}$$

12 Let a and b be positive constants with $a > b$, and suppose that the surface Σ is described vector parametrically by

$$\mathbf{R} = (a + b \sin u)(\cos v)\mathbf{i} + (a + b \sin u)(\sin v)\mathbf{j} + (b \cos u)\mathbf{k}$$

for (u, v) in the rectangular region D: $0 \le u \le 2\pi$, $0 \le v \le 2\pi$ in the uv plane. Find the area A_Σ of the surface Σ.

13 In Problem 11, show that $A = A_\Sigma \cos \alpha$, where α is the angle between the plane containing D and the plane containing Σ.

14 Let $\mathbf{w}$ be a fixed vector of unit length in xyz space such that $\mathbf{w} \cdot \mathbf{k} \ne 0$, and let C be a curve with total arc length L in the xy plane having the scalar parametric equations

$$C: \begin{cases} x = f(s) \\ y = g(s) \end{cases} \quad 0 \le s \le L$$

where the parameter s is arc length. Let Σ be the surface described by the vector parametric equation

$$\mathbf{R} = f(s)\mathbf{i} + g(s)\mathbf{j} + t\mathbf{w}$$

where the parameters s and t satisfy $0 \le s \le L$ and $a \le t \le b$. Show that the area of Σ is given by $A_\Sigma = L(b - a)$.

15 If c is a constant, what is the value of the surface integral

$$\iint_\Sigma c \, dA_\Sigma?$$

16 Let Σ be the portion of the right circular cylinder $x^2 + y^2 = 1$ lying between the plane $z = 0$ and the plane $z = 1$. Evaluate the surface integral $\iint_\Sigma (x + y + z) \, dA_\Sigma$. (*Hint:* Use cylindrical coordinates.)

17 Evaluate $\iint_\Sigma (y + z^2) \, dA_\Sigma$, where Σ is the sphere of radius 1 with center at the origin.

18 Let Σ be the right circular cone (with open base) whose vertex is at the origin and whose base is a circle of radius b which is parallel to the xy plane and h units above it. Evaluate

$$\iint_\Sigma xyz \, dA_\Sigma.$$

19 Let Σ be the portion of the plane $z = x + 4y + 5$ that lies above the region D: $0 \le x \le 1$, $0 \le y \le 1$ in the xy plane. Evaluate

$$\iint_\Sigma (2xy - z) \, dA_\Sigma.$$

20 Suppose that the surface Σ is an admissible region in the xy plane and that F is a function of three variables, defined and continuous on a subset of xyz space that contains Σ. Show that in this case the surface integral $\iint_\Sigma F \, dA_\Sigma$ is the same as the double integral $\iint_\Sigma F(x, y, 0) \, dA$.

In Problems 21 to 25, find the flux of each vector field $\mathbf{F}$ through the indicated surface Σ in the direction of the given unit normal vector $\mathbf{N}$.

21 $\mathbf{F} = yz\mathbf{i} + xz\mathbf{j} + xy\mathbf{k}$; Σ is the portion of the plane $z = x + y + 1$ that lies above the region D: $0 \le x \le 1$, $0 \le y \le 1$ in the xy plane; $\mathbf{N}$ is the unit normal to Σ that makes an acute angle with the positive z axis.

22 $\mathbf{F} = x\mathbf{i} + y\mathbf{j} + z\mathbf{k}$; Σ is the sphere of radius 1 with center at the origin; $\mathbf{N}$ is the unit normal to Σ that points away from the center.

23 $\mathbf{F} = x^2\mathbf{i} + xy\mathbf{j} + z\mathbf{k}$; Σ is the triangle with vertices $(1, 0, 0)$, $(0, 2, 0)$, and $(0, 0, 3)$; $\mathbf{N}$ is the unit normal to Σ that makes an acute angle with the positive x axis.

24 $\mathbf{F} = x^2\mathbf{i} + y^2\mathbf{j} + z^2\mathbf{k}$; Σ is the entire surface of the tetrahedron with vertices $(1, 0, 0)$, $(0, 1, 0)$, $(0, 0, 1)$, and $(0, 0, 0)$; $\mathbf{N}$ is the unit normal that points away from the interior of the tetrahedron. (*Hint:* Integrate over each of the four triangular faces and add the resulting numbers.)

25 $\mathbf{F} = (1/x)\mathbf{i} + (1/y)\mathbf{j} + (1/z)\mathbf{k}$; Σ is the portion of the right circular cylinder $x^2 + y^2 = 3$ that lies between the plane $z = 1$ and

the plane $z = 2$; $\mathbf{N}$ is the unit normal to Σ that points away from the z axis.

26 If $\mathbf{R}$ is a variable position vector sweeping out a surface Σ and if u and v are the parameters, show that $(\partial\mathbf{R}/\partial u) \times (\partial\mathbf{R}/\partial v)$ is normal to Σ, and show that the flux of a vector field $\mathbf{F}$ through Σ in the direction of this normal is given by

$$\iint\limits_{D} \mathbf{F} \cdot \left(\frac{\partial\mathbf{R}}{\partial u} \times \frac{\partial\mathbf{R}}{\partial v} \right) du \, dv$$

15.12   The Divergence Theorem and Stokes' Theorem

The *divergence theorem* and *Stokes' theorem* are generalizations to three-dimensional space of Green's theorem in the plane (Section 15.9). Before we present these theorems, we introduce the ideas of the *divergence* and the *curl* of a vector field.

Divergence and Curl

In dealing with scalar and vector fields it is useful to introduce the "symbolic vector," or **operator**, ∇ defined by

$$\nabla = \mathbf{i} \frac{\partial}{\partial x} + \mathbf{j} \frac{\partial}{\partial y} + \mathbf{k} \frac{\partial}{\partial z}$$

Of course, ∇ is not really a vector at all; however, in a formal sense, it can be "multiplied" by a scalar field or a vector field. For instance, if $w = f(x, y, z)$ is a scalar field, then

$$\nabla w = \mathbf{i} \frac{\partial w}{\partial x} + \mathbf{j} \frac{\partial w}{\partial y} + \mathbf{k} \frac{\partial w}{\partial z} = \frac{\partial w}{\partial x}\mathbf{i} + \frac{\partial w}{\partial y}\mathbf{j} + \frac{\partial w}{\partial z}\mathbf{k} = \text{ the gradient of } w$$

which explains why the notation ∇w is used for the gradient of w.

If $\mathbf{F} = P(x, y, z)\mathbf{i} + Q(x, y, z)\mathbf{j} + T(x, y, z)\mathbf{k}$ is a vector field, then we can form either the "dot product"

$$\nabla \cdot \mathbf{F}$$

or the "cross product"

$$\nabla \times \mathbf{F}$$

of the symbolic vector ∇ with $\mathbf{F}$. The dot product $\nabla \cdot \mathbf{F}$ is called the **divergence** of $\mathbf{F}$, abbreviated **div F**, while the cross product $\nabla \times \mathbf{F}$ is called the **curl** of $\mathbf{F}$, abbreviated **curl F**. Thus, we have

$$\text{div } \mathbf{F} = \nabla \cdot \mathbf{F} = \left(\mathbf{i} \frac{\partial}{\partial x} + \mathbf{j} \frac{\partial}{\partial y} + \mathbf{k} \frac{\partial}{\partial z} \right) \cdot (P\mathbf{i} + Q\mathbf{j} + T\mathbf{k})$$

$$= \frac{\partial P}{\partial x} + \frac{\partial Q}{\partial y} + \frac{\partial T}{\partial z}$$

and

$$\text{curl } \mathbf{F} = \nabla \times \mathbf{F} = \begin{vmatrix} \mathbf{i} & \mathbf{j} & \mathbf{k} \\ \dfrac{\partial}{\partial x} & \dfrac{\partial}{\partial y} & \dfrac{\partial}{\partial z} \\ P & Q & T \end{vmatrix}$$

$$= \left(\frac{\partial T}{\partial y} - \frac{\partial Q}{\partial z} \right) \mathbf{i} + \left(\frac{\partial P}{\partial z} - \frac{\partial T}{\partial x} \right) \mathbf{j} + \left(\frac{\partial Q}{\partial x} - \frac{\partial P}{\partial y} \right) \mathbf{k}$$

If the vector field $\mathbf{F}$ is thought of as the velocity field of a flowing fluid, then $\nabla \cdot \mathbf{F}$ and $\nabla \times \mathbf{F}$ have interesting physical interpretations. The scalar $\nabla \cdot \mathbf{F}$ is a measure of the tendency of the velocity vectors to ''diverge'' from one another. For instance, if the fluid is flowing with a constant velocity (Figure 1a), then the velocity vectors are parallel to one another, and the divergence is zero; however, near a ''source'' of fluid (Figure 1b) the divergence would be rather large. On the other hand, if the tail end of the vector $\nabla \times \mathbf{F}$ were equipped with small vanes (Figure 2), then the flowing fluid would cause the vanes to rotate with an angular speed proportional to $|\nabla \times \mathbf{F}|$.

Figure 1
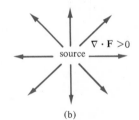
(a)
(b)

Figure 2
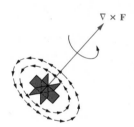

EXAMPLE 1   Let $\mathbf{F} = 2xy\mathbf{i} + 4yz\mathbf{j} - xz\mathbf{k}$. Find

   (a) $\nabla \cdot \mathbf{F}$ **(b)** $\nabla \times \mathbf{F}$

SOLUTION

(a) $\nabla \cdot \mathbf{F} = \dfrac{\partial}{\partial x}(2xy) + \dfrac{\partial}{\partial y}(4yz) + \dfrac{\partial}{\partial z}(-xz) = 2y + 4z - x$

(b) $\nabla \times \mathbf{F} = \begin{vmatrix} \mathbf{i} & \mathbf{j} & \mathbf{k} \\ \dfrac{\partial}{\partial x} & \dfrac{\partial}{\partial y} & \dfrac{\partial}{\partial z} \\ 2xy & 4yz & -xz \end{vmatrix}$

$$= \left[\frac{\partial}{\partial y}(-xz) - \frac{\partial}{\partial z}(4yz) \right] \mathbf{i} + \left[\frac{\partial}{\partial z}(2xy) - \frac{\partial}{\partial x}(-xz) \right] \mathbf{j}$$

$$+ \left[\frac{\partial}{\partial x}(4yz) - \frac{\partial}{\partial y}(2xy) \right] \mathbf{k}$$

$$= (-4y)\mathbf{i} + [-(-z)]\mathbf{j} + (-2x)\mathbf{k} = -4y\mathbf{i} + z\mathbf{j} - 2x\mathbf{k} \quad \blacksquare$$

<u>EXAMPLE 2</u> If **A** is a constant vector and $\mathbf{r} = x\mathbf{i} + y\mathbf{j} + z\mathbf{k}$, find

(a) div **r**   (b) curl (**A** × **r**)

SOLUTION

(a) div $\mathbf{r} = \dfrac{\partial}{\partial x} x + \dfrac{\partial}{\partial y} y + \dfrac{\partial}{\partial z} z = 1 + 1 + 1 = 3$

(b) Let $\mathbf{A} = a\mathbf{i} + b\mathbf{j} + c\mathbf{k}$. Then

$$\mathbf{A} \times \mathbf{r} = \begin{vmatrix} \mathbf{i} & \mathbf{j} & \mathbf{k} \\ a & b & c \\ x & y & z \end{vmatrix} = (bz - cy)\mathbf{i} + (cx - az)\mathbf{j} + (ay - bx)\mathbf{k}$$

Hence,

$$\text{curl } (\mathbf{A} \times \mathbf{r}) = \begin{vmatrix} \mathbf{i} & \mathbf{j} & \mathbf{k} \\ \dfrac{\partial}{\partial x} & \dfrac{\partial}{\partial y} & \dfrac{\partial}{\partial z} \\ bz - cy & cx - az & ay - bx \end{vmatrix} = 2a\mathbf{i} + 2b\mathbf{j} + 2c\mathbf{k} = 2\mathbf{A} \quad \blacksquare$$

The Divergence Theorem

The **divergence theorem,** also called **Gauss's theorem** in honor of the renowned German mathematician Karl Friedrich Gauss (1777–1855), effects a profound connection between the divergence and the flux of a vector field and represents a generalization to three-dimensional space of Green's theorem in the plane (see Problem 36 in Problem Set 15.9). The following is a somewhat informal statement of the theorem.

Karl Friedrich Gauss

THEOREM 1

The Divergence Theorem of Gauss

Let S be a closed and bounded region in xyz space whose boundary is a piecewise smooth surface Σ. Suppose that **F** is a vector field defined on an open set U containing S, and assume that the scalar component functions of **F** are continuously differentiable on U. Let **N** denote the outward-pointing unit normal vector to the surface Σ. Then

$$\iiint_S \nabla \cdot \mathbf{F} \, dV = \iint_\Sigma \mathbf{F} \cdot \mathbf{N} \, dA_\Sigma$$

In words, *the integral over a solid S of the divergence of a vector field is the flux of the field through the boundary of the solid.* In particular, if the integrand of a triple integral can be expressed as the divergence of a vector field, then the value of the integral depends only on the vectors on the surface that encloses the volume! The proof of the divergence theorem is beyond the scope of this book.

Figure 3

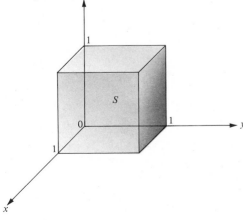

<u>EXAMPLE 3</u> Let $\mathbf{F} = (2x - z)\mathbf{i} + x^2 y\mathbf{j} + xz^2\mathbf{k}$, and suppose that S is the solid cube bounded by the planes $x = 0$, $x = 1$, $y = 0$, $y = 1$, $z = 0$, and $z = 1$ (Figure 3). If Σ denotes the surface of S, use the divergence theorem to evaluate $\iint_\Sigma \mathbf{F} \cdot \mathbf{N} \, dA_\Sigma$, where **N** is the outward-pointing unit normal vector to Σ.

Archimedes

Figure 4

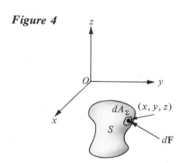

SOLUTION　By the divergence theorem,

$$\iint_\Sigma \mathbf{F} \cdot \mathbf{N} \, dA_\Sigma = \iiint_S \nabla \cdot \mathbf{F} \, dV$$

$$= \iiint_S \left[\frac{\partial}{\partial x} (2x - z) + \frac{\partial}{\partial y} (x^2 y) + \frac{\partial}{\partial z} (xz^2) \right] dV$$

$$= \iiint_S (2 + x^2 + 2xz) \, dV = \int_0^1 \int_0^1 \int_0^1 (2 + x^2 + 2xz) \, dx \, dy \, dz$$

$$= \int_0^1 \int_0^1 \left[\left(2x + \frac{x^3}{3} + x^2 z \right) \Big|_0^1 \right] dy \, dz = \int_0^1 \int_0^1 \left(\frac{7}{3} + z \right) dy \, dz$$

$$= \int_0^1 \left[\left(\frac{7}{3} + z \right) y \Big|_0^1 \right] dz = \int_0^1 \left(\frac{7}{3} + z \right) dz$$

$$= \left(\frac{7}{3} z + \frac{z^2}{2} \right) \Big|_0^1 = \frac{17}{6}$$

■

EXAMPLE 4　Use the divergence theorem to prove **Archimedes' principle:** *The buoyant force on a solid S immersed in a fluid of constant density is equal to the weight of the displaced fluid.*

SOLUTION　Let the *xyz* coordinate system be placed so that the solid S lies below the *xy* plane and the *z* axis points straight upward as usual (Figure 4). Since the pressure P of the fluid varies linearly with the depth, the pressure at the point (x, y, z) on the surface Σ of S is given by

$$P = a - \delta z$$

where a is a constant and δ is the weight of a unit volume of the fluid. The force $d\mathbf{F}$ caused by this pressure on an infinitesimal area dA_Σ has magnitude

$$|d\mathbf{F}| = \text{pressure times area} = (a - \delta z) \, dA_\Sigma$$

and is directed *inward* along the normal to the surface. Thus, if $\mathbf{N}$ is the *outward*-pointing unit normal to Σ, then

$$d\mathbf{F} = |d\mathbf{F}|(-\mathbf{N}) = -(a - \delta z)\mathbf{N} \, dA_\Sigma = (\delta z - a)\mathbf{N} \, dA_\Sigma$$

The vertical component of $d\mathbf{F}$ represents the magnitude of the buoyant force dF_b acting on dA_Σ; hence,

$$dF_b = \mathbf{k} \cdot d\mathbf{F} = \mathbf{k} \cdot [(\delta z - a)\mathbf{N} \, dA_\Sigma] = (\delta z - a)\mathbf{k} \cdot \mathbf{N} \, dA_\Sigma$$

The magnitude of the net buoyant force on the entire surface Σ is obtained by summing—that is, integrating—dF_b over Σ; hence,

$$F_b = \iint_\Sigma (\delta z - a)\mathbf{k} \cdot \mathbf{N} \, dA_\Sigma$$

Here we have　$\nabla \cdot (\delta z - a)\mathbf{k} = \dfrac{\partial}{\partial z} (\delta z - a) = \delta$

Therefore, by the divergence theorem,

$$F_b = \iint_\Sigma (\delta z - a)\mathbf{k} \cdot \mathbf{N} \, dA_\Sigma = \iiint_S \nabla \cdot (\delta z - a)\mathbf{k} \, dV = \iiint_S \delta \, dV$$

$$= \delta \iiint_S dV = \delta V$$

where V is the volume of S. Since δ is the weight of a unit volume of the fluid, it follows that δV is the weight of the displaced fluid.   ■

<u>EXAMPLE 5</u>   Let D be a closed admissible region in the xy plane, and suppose that g and h are continuous functions defined on D and satisfying $g(x, y) \leq h(x, y)$ for all points (x, y) in D. Let S be the solid consisting of all points (x, y, z) satisfying the conditions that (x, y) belongs to D and $g(x, y) \leq z \leq h(x, y)$ (Figure 5). Suppose that f is a function of three variables that is continuously differentiable on some open set U containing S, and let $\mathbf{F}$ be the vector field defined by $\mathbf{F} = f(x, y, z)\mathbf{k}$. Verify the divergence theorem for the vector field $\mathbf{F}$ and the solid S.

SOLUTION   The surface Σ of S consists of the three parts

Σ_1: the graph of $z = g(x, y)$ for (x, y) in D

Σ_2: the graph of $z = h(x, y)$ for (x, y) in D

Σ_3: the portion of the cylinder over the boundary of D, with vertical generators, cut off by Σ_1 and Σ_2

Let $\mathbf{N}$ denote the outward-pointing unit normal vector to the surface Σ. The vector

$$g_1(x, y)\mathbf{i} + g_2(x, y)\mathbf{j} - \mathbf{k}$$

is normal to Σ_1 (Section 14.6, page 883) and points outward from the solid S, since its $\mathbf{k}$ component is negative; hence, on Σ_1, we have

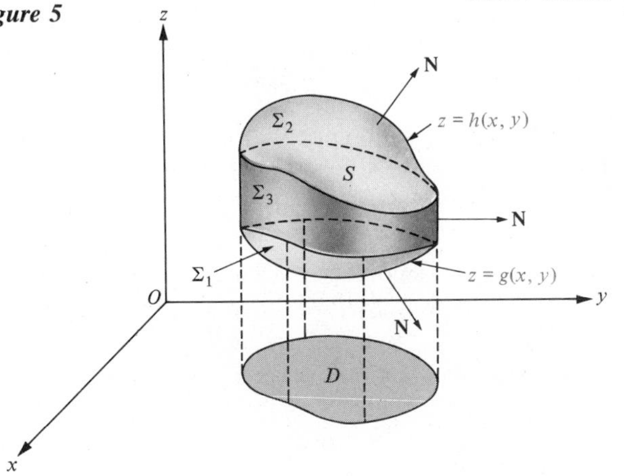

$$\mathbf{N} = \frac{g_1(x, y)\mathbf{i} + g_2(x, y)\mathbf{j} - \mathbf{k}}{\sqrt{[g_1(x, y)]^2 + [g_2(x, y)]^2 + 1}}$$

Also, on Σ_1, the differential of surface area is given by

$$dA_\Sigma = \sqrt{[g_1(x, y)]^2 + [g_2(x, y)]^2 + 1}\ dx\ dy$$

Therefore, on Σ_1,

$$\mathbf{N}\ dA_\Sigma = [g_1(x, y)\mathbf{i} + g_2(x, y)\mathbf{j} - \mathbf{k}]\ dx\ dy$$

and

$$\mathbf{F} \cdot \mathbf{N}\ dA_\Sigma = f(x, y, z)\mathbf{k} \cdot [g_1(x, y)\mathbf{i} + g_2(x, y)\mathbf{j} - \mathbf{k}]\ dx\ dy$$

$$= -f(x, y, z)\ dx\ dy = -f(x, y, g(x, y))\ dA$$

where dA is the element of area in the domain D. Hence,

$$\iint_{\Sigma_1} \mathbf{F} \cdot \mathbf{N}\ dA_\Sigma = -\iint_D f(x, y, g(x, y))\ dA$$

Reasoning in a similar way, but noting that the $\mathbf{k}$ component of $\mathbf{N}$ on Σ_2 is positive, we find that

$$\iint_{\Sigma_2} \mathbf{F} \cdot \mathbf{N}\ dA_\Sigma = \iint_D f(x, y, h(x, y))\ dA$$

Since $\mathbf{N}$ is parallel to the xy plane on the cylinder Σ_3, it follows that

$$\mathbf{F} \cdot \mathbf{N} = f(x, y, z)\mathbf{k} \cdot \mathbf{N} = f(x, y, z)(0) = 0 \qquad \text{on } \Sigma_3$$

Hence, $$\iint_{\Sigma_3} \mathbf{F} \cdot \mathbf{N}\ dA_\Sigma = 0$$

Figure 5

Therefore,

$$\iint_{\Sigma} \mathbf{F} \cdot \mathbf{N} \, dA_{\Sigma} = \iint_{\Sigma_1} \mathbf{F} \cdot \mathbf{N} \, dA_{\Sigma} + \iint_{\Sigma_2} \mathbf{F} \cdot \mathbf{N} \, dA_{\Sigma} + \iint_{\Sigma_3} \mathbf{F} \cdot \mathbf{N} \, dA_{\Sigma}$$

$$= -\iint_{D} f(x, y, g(x, y)) \, dA + \iint_{D} f(x, y, h(x, y)) \, dA + 0$$

$$= \iint_{D} [f(x, y, h(x, y)) - f(x, y, g(x, y))] \, dA$$

$$= \iint_{D} \left[f(x, y, z) \Big|_{z = g(x,y)}^{z = h(x,y)} \right] dA$$

$$= \iint_{D} \left(\int_{g(x,y)}^{h(x,y)} \frac{\partial f}{\partial z} \, dz \right) dA$$

$$= \iiint_{S} \frac{\partial f}{\partial z} \, dV = \iiint_{S} \nabla \cdot \mathbf{F} \, dV$$

in conformity with the divergence theorem. ∎

Stokes' Theorem

We now turn to another generalization of Green's theorem, which is attributed to the Irish mathematical physicist Sir George G. Stokes (1819–1903). Roughly speaking, Stokes' theorem says that the flux of the curl of a vector field **F** through a surface Σ is equal to the line integral of the tangential component of **F** around the boundary of Σ.

To be more precise, suppose that Σ is a smooth surface and that **N** denotes a unit normal vector to Σ which varies continuously as we move around on the surface (Figure 6). Furthermore, we assume that the boundary of Σ consists of a single closed curve C in xyz space. Imagine standing on the boundary curve C with your head pointing in the direction of the normal vector **N** and with the surface Σ on your left. Now, if you walk forward along C, you will, by definition, move in the *positive* direction around the boundary. If we wish to describe C, or a part of C, parametrically, we always choose the parameter t so that when t increases, we move along C in the positive direction. With this understanding, we can now give an informal statement of Stokes' theorem.

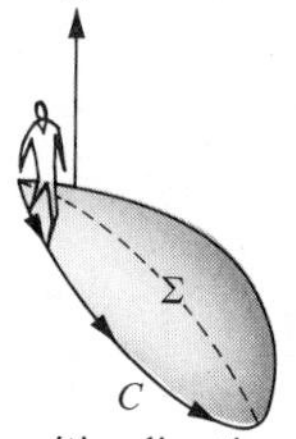

Figure 6

positive direction
around the boundary

THEOREM 2   **Stokes' Theorem**

Let **F** be a vector field whose component functions are continuously differentiable on an open set U containing the surface Σ and its boundary curve C. Then

$$\int_{C} \mathbf{F} \cdot d\mathbf{R} = \iint_{\Sigma} (\nabla \times \mathbf{F}) \cdot \mathbf{N} \, dA_{\Sigma}$$

Recall that the line integral $\int_{C} \mathbf{F} \cdot d\mathbf{R}$ around a *closed* curve C is called the *circulation* of the vector field **F** around C. In particular, if **F** represents a force field, then the circulation of **F** around C is the total work done by the force **F** in carrying a particle once around the closed curve C. Thus, Stokes' theorem says that *the circulation of a vector field around the boundary of a surface in xyz space is equal to the flux of the curl of the field through the surface*. The proof of Stokes' theorem is beyond the scope of this book.

Figure 7

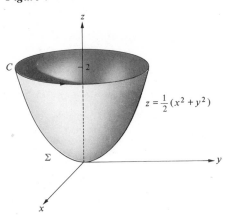

C

$z = \frac{1}{2}(x^2 + y^2)$

Σ

EXAMPLE 6   Let Σ be the portion of the paraboloid of revolution $z = \frac{1}{2}(x^2 + y^2)$ cut off by the plane $z = 2$ and lying below this plane (Figure 7), and let **N** be the unit normal vector to Σ that makes an acute angle with the positive z axis. If C is the boundary curve of Σ and **F** is the vector field $\mathbf{F} = 3y\mathbf{i} - xz\mathbf{j} + yz^2\mathbf{k}$, verify Stokes' theorem for **F** and Σ.

SOLUTION   Proceeding as in Example 5, we find that

$$\mathbf{N}\,dA_\Sigma = -\left(\frac{\partial z}{\partial x}\mathbf{i} + \frac{\partial z}{\partial y}\mathbf{j} - \mathbf{k}\right)dx\,dy = (-x\mathbf{i} - y\mathbf{j} + \mathbf{k})\,dA$$

Here,

$$\nabla \times \mathbf{F} = \begin{vmatrix} \mathbf{i} & \mathbf{j} & \mathbf{k} \\ \dfrac{\partial}{\partial x} & \dfrac{\partial}{\partial y} & \dfrac{\partial}{\partial z} \\ 3y & -xz & yz^2 \end{vmatrix} = (z^2 + x)\mathbf{i} + 0\mathbf{j} + (-z - 3)\mathbf{k}$$

Hence,

$$(\nabla \times \mathbf{F}) \cdot \mathbf{N}\,dA_\Sigma = [(z^2 + x)(-x) + 0(-y) + (-z - 3)(1)]\,dA$$

$$= (-z^2 x - x^2 - z - 3)\,dA$$

Notice that the surface Σ lies over the circular region

$$D: x^2 + y^2 \le 4$$

in the xy plane. Therefore,

$$\iint\limits_\Sigma (\nabla \times \mathbf{F}) \cdot \mathbf{N}\,dA_\Sigma = \iint\limits_D (-z^2 x - x^2 - z - 3)\,dA$$

$$= \iint\limits_D [-\tfrac{1}{4}(x^2 + y^2)^2 x - x^2 - \tfrac{1}{2}(x^2 + y^2) - 3]\,dA$$

Switching to polar coordinates, we find that the flux through Σ of the curl of **F** is given by

$$\iint\limits_\Sigma (\nabla \times \mathbf{F}) \cdot \mathbf{N}\,dA_\Sigma = \int_0^{2\pi} \int_0^2 (-\tfrac{1}{4}r^4 r \cos\theta - r^2 \cos^2\theta - \tfrac{1}{2}r^2 - 3)r\,dr\,d\theta$$

$$= \int_0^{2\pi} \int_0^2 \left(-\frac{r^6}{4}\cos\theta - r^3 \cos^2\theta - \frac{r^3}{2} - 3r\right)dr\,d\theta$$

$$= \int_0^{2\pi} \left[\left(-\frac{r^7}{28}\cos\theta - \frac{r^4}{4}\cos^2\theta - \frac{r^4}{8} - \frac{3r^2}{2}\right)\Big|_0^2\right]d\theta$$

$$= \int_0^{2\pi} (-\tfrac{32}{7}\cos\theta - 4\cos^2\theta - 8)\,d\theta$$

$$= \int_0^{2\pi} [-\tfrac{32}{7}\cos\theta - 2(1 + \cos 2\theta) - 8]\,d\theta$$

$$= \int_0^{2\pi} (-\tfrac{32}{7}\cos\theta - 2\cos 2\theta - 10)\,d\theta$$

$$= \left(-\tfrac{32}{7}\sin\theta - \sin 2\theta - 10\theta\right)\Big|_0^{2\pi} = -20\pi$$

Now we calculate the circulation $\int_C \mathbf{F} \cdot d\mathbf{R}$. Notice that C is described vector parametrically by

$$\mathbf{R} = (2 \cos t)\mathbf{i} + (2 \sin t)\mathbf{j} + 2\mathbf{k} \qquad 0 \le t \le 2\pi$$

so that

$$d\mathbf{R} = [(-2 \sin t)\mathbf{i} + (2 \cos t)\mathbf{j}]\, dt$$

and

$$
\begin{aligned}
\mathbf{F} \cdot d\mathbf{R} &= [(3y)(-2 \sin t) + (-xz)(2 \cos t)]\, dt \\
&= [3(2 \sin t)(-2 \sin t) + (-2 \cos t)(2)(2 \cos t)]\, dt \\
&= (-12 \sin^2 t - 8 \cos^2 t)\, dt
\end{aligned}
$$

Therefore,

$$
\begin{aligned}
\int_C \mathbf{F} \cdot d\mathbf{R} &= \int_0^{2\pi} (-12 \sin^2 t - 8 \cos^2 t)\, dt \\
&= -\int_0^{2\pi} (4 \sin^2 t + 8 \sin^2 t + 8 \cos^2 t)\, dt \\
&= -\int_0^{2\pi} (4 \sin^2 t + 8)\, dt = -\int_0^{2\pi} [2(1 - \cos 2t) + 8]\, dt \\
&= -\int_0^{2\pi} (10 - 2 \cos 2t)\, dt = -(10t - \sin 2t)\Big|_0^{2\pi} = -20\pi
\end{aligned}
$$

in conformity with Stokes' theorem.

Problem Set 15.12

In Problems 1 to 4, find (a) $\nabla \cdot \mathbf{F}$ and (b) $\nabla \times \mathbf{F}$.

1 $\mathbf{F} = xy^2\mathbf{i} - x^2\mathbf{j} + (x + y)\mathbf{k}$

2 $\mathbf{F} = (z^2 - x)\mathbf{i} - xy\mathbf{j} + 3z\mathbf{k}$

3 $\mathbf{F} = 3xyz^2\mathbf{i} + (5x^2y + z)\mathbf{j} + 2y^2z^3\mathbf{k}$

4 $\mathbf{F} = x(\cos y)\mathbf{i} + 3x^2(\sin y)\mathbf{j} + xz^2\mathbf{k}$

In Problems 5 to 8, use the divergence theorem to evaluate the flux $\iint_{\Sigma} \mathbf{F} \cdot \mathbf{N}\, dA_{\Sigma}$ through the boundary Σ of the indicated solid S of the given vector field $\mathbf{F}$. Here, $\mathbf{N}$ denotes the outward-pointing unit normal vector to Σ.

5 $\mathbf{F} = (2xy + z)\mathbf{i} + y^2\mathbf{j} - (x + 3y)\mathbf{k}$; S is the solid bounded by the planes $2x + 2y + z = 6$, $x = 0$, $y = 0$, and $z = 0$.

6 $\mathbf{F} = x^2\mathbf{i} + y^2\mathbf{j} + z^2\mathbf{k}$; S is the solid cube bounded by the planes $x = 0$, $x = 1$, $y = 0$, $y = 1$, $z = 0$, and $z = 1$.

7 $\mathbf{F} = yz\mathbf{i} + xz\mathbf{j} + xy\mathbf{k}$; S is the solid sphere $x^2 + y^2 + z^2 \le 1$.

8 $\mathbf{F} = (z^2 - x)\mathbf{i} - xy\mathbf{j} + 3z\mathbf{k}$; S is the solid bounded above by the parabolic cylinder $z = 9 - y^2$, in back by the plane $x = 0$, in front by the plane $x = 4$, and below by the xy plane.

In Problems 9 to 12, use Stokes' theorem to evaluate the flux $\iint_{\Sigma} (\nabla \times \mathbf{F}) \cdot \mathbf{N}\, dA_{\Sigma}$ of the curl of each vector field $\mathbf{F}$ through the indicated surface Σ in the direction of the given unit normal vector $\mathbf{N}$.

9 $\mathbf{F} = y^2\mathbf{i} + xy\mathbf{j} + xz\mathbf{k}$; Σ is the hemisphere $x^2 + y^2 + z^2 = 9$, $z \ge 0$; $\mathbf{N}$ has a nonnegative $\mathbf{k}$ component.

10 $\mathbf{F} = y\mathbf{i} + z\mathbf{j} + x\mathbf{k}$; Σ is the portion of the paraboloid of revolution $z = 1 - x^2 - y^2$ for which $z \ge 0$; $\mathbf{N}$ has a nonnegative $\mathbf{k}$ component.

11 $\mathbf{F} = (z + y)\mathbf{i} + (z + x)\mathbf{j} + (x + y)\mathbf{k}$; Σ is the triangle with vertices $(1, 0, 0)$, $(0, 1, 0)$, and $(0, 0, 1)$; $\mathbf{N}$ is the unit normal vector whose components are all positive.

12 $\mathbf{F} = \mathbf{r}/r^3$, $\mathbf{r} = x\mathbf{i} + y\mathbf{j} + z\mathbf{k}$, $r = |\mathbf{r}|$; Σ is the portion of the ellipsoid $(x^2/4) + (y^2/4) + (z^2/9) = 1$ for which $z \le 0$; $\mathbf{N}$ is the unit normal vector whose $\mathbf{k}$ component is nonnegative.

In Problems 13 and 14, verify the divergence theorem by direct calculation for the given vector field $\mathbf{F}$ and the indicated solid S.

13 $\mathbf{F} = x\mathbf{i} + y\mathbf{j} + z\mathbf{k}$; S is the solid cube bounded by the planes $x = 0$, $x = 1$, $y = 0$, $y = 1$, $z = 0$, and $z = 1$.

14 $\mathbf{F} = g(x, y, z)\mathbf{j}$, where g is a continuously differentiable function defined on an open set U containing the solid sphere $S: x^2 + y^2 + z^2 \le 1$.

In Problems 15 and 16, verify Stokes' theorem for the given vector field $\mathbf{F}$, the indicated surface Σ, and the given unit normal vector $\mathbf{N}$ to Σ.

15 $\mathbf{F} = (2y + z)\mathbf{i} + (x - z)\mathbf{j} + (y - x)\mathbf{k}$; Σ is the triangle cut from the plane $x + y + z = 1$ by the coordinate planes; $\mathbf{N}$ is the unit normal vector whose components are all positive.

16 $\mathbf{F} = 2y\mathbf{i} - x\mathbf{j} + z\mathbf{k}$; Σ is the hemisphere $x^2 + y^2 + z^2 = 4$, $z \ge 0$; $\mathbf{N}$ is the unit normal vector whose $\mathbf{k}$ component is nonnegative.

17 Prove the divergence theorem for the special case in which S is the solid cube bounded by the planes $x = 0$, $x = 1$, $y = 0$, $y = 1$, $z = 0$, and $z = 1$.

18 Let $\mathbf{F} = f(x, y, z)\mathbf{i} + g(x, y, z)\mathbf{j} + h(x, y, z)\mathbf{k}$. Explain why the surface integral $\iint\limits_{\Sigma} \mathbf{F} \cdot \mathbf{N} \, dA_{\Sigma}$ is sometimes written as

$$\iint\limits_{\Sigma} f(x, y, z) \, dy \, dz + g(x, y, z) \, dz \, dx + h(x, y, z) \, dx \, dy$$

19 Assuming the existence and continuity of the required partial derivatives, prove that (a) *the curl of the gradient of a scalar field is zero* and (b) *the divergence of the curl of a vector field is zero*.

20 Show that the *flux of the curl of a vector field through the surface Σ of a solid S is zero*. State carefully what assumptions you need to make.

21 Suppose that Σ is an admissible region in the xy plane whose boundary C forms a piecewise smooth, simple closed curve. Show that Stokes' theorem for Σ is essentially just Green's theorem in the plane.

22 Show that a solid S immersed in a fluid of constant density experiences no net horizontal force because of the fluid pressure.

23 Show that the condition given in Theorem 3 on page 991—that the vector field $\mathbf{F}$ be derivable from a potential on an open, simply connected region R—is equivalent to the statement that $\nabla \times \mathbf{F} = \mathbf{0}$ on R.

24 Give an informal proof of Theorem 3 on page 991 patterned after the informal argument given for Theorem 2 on page 989, but using Stokes' theorem rather than Green's theorem.

Review Problem Set, Chapter 15

In Problems 1 to 8, evaluate each iterated integral.

1 $\displaystyle\int_0^1 \int_0^y xy^2 \, dx \, dy$

2 $\displaystyle\int_0^4 \int_0^{\sqrt{x}} y\sqrt{x + y^2} \, dy \, dx$

3 $\displaystyle\int_1^2 \int_y^{y^2} (x + 2y) \, dx \, dy$

4 $\displaystyle\int_0^2 \int_0^{\sqrt{4-x^2}} (x + y) \, dy \, dx$

5 $\displaystyle\int_1^2 \int_1^{x^2} x^3 y e^{y^2} \, dy \, dx$

6 $\displaystyle\int_3^5 \int_1^y \frac{y}{x \ln y} \, dx \, dy$

7 $\displaystyle\int_3^5 \int_{\pi/6}^{\pi/3} r^2 \sin \theta \, d\theta \, dr$

8 $\displaystyle\int_0^\pi \int_0^{2 \cos \theta} r \sin \theta \, dr \, d\theta$

In Problems 9 to 12, replace each iterated integral by an equivalent integral with the order of integration reversed, and then evaluate the integral obtained. Sketch the appropriate region.

9 $\displaystyle\int_0^4 \int_{(4-x)/2}^{\sqrt{4-x}} y \, dy \, dx$

10 $\displaystyle\int_0^3 \int_x^{4x-x^2} dy \, dx$

11 $\displaystyle\int_0^6 \int_{(2/3)(6-y)}^{(1/9)(36-y^2)} y^2 x \, dx \, dy$

12 $\displaystyle\int_0^6 \int_{y^3/18}^{2y} xy^2 \, dx \, dy$

In Problems 13 to 16, evaluate each double integral over the region indicated.

13 $\displaystyle\iint\limits_{R} (\sqrt{y} + x - 3xy^2) \, dA$; $R: 0 \le x \le 1, 1 \le y \le 3$

14 $\displaystyle\iint\limits_{R} \sin (x + y) \, dA$; $R: 0 \le x \le \pi/2, 0 \le y \le \pi/2$

15 $\displaystyle\iint\limits_{R} e^{x^2} \, dA$; $R: 0 \le x \le 2, 0 \le y \le x$

16 $\displaystyle\iint\limits_{R} \frac{ds \, dt}{4 - s}$; $R: 2 \le s \le 3, 0 \le t \le s$

In Problems 17 to 20, use double integration to find the area of the region R bounded by the given pair of curves.

17 $y = 4x - x^2$ and $y = x$

18 $y^2 = 4x$ and $2x - y = 4$

19 $4y^2 = x^3$ and $x = y$

20 $y^2 = 4x$ and $x = 12 + 2y - y^2$

In Problems 21 to 26, express each iterated integral as an equivalent iterated integral in polar coordinates, and then evaluate the integral.

21 $\int_0^{10} \int_0^{\sqrt{100-x^2}} \sqrt{x^2 + y^2} \, dy \, dx$ **22** $\int_0^4 \int_0^x (x^2 + y^2)^{3/2} \, dy \, dx$

23 $\int_0^2 \int_0^{\sqrt{4-y^2}} (1 - x^2 - y^2)^2 \, dx \, dy$

24 $\int_0^2 \int_0^{\sqrt{24-y^2}} \sqrt{x^2 + y^2} \, dx \, dy$

25 $\int_{-1}^1 \int_{-\sqrt{1-x^2}}^{\sqrt{1-x^2}} xy \, dy \, dx$

26 $\int_{-3}^3 \int_0^{\sqrt{9-y^2}} \frac{y}{\sqrt{x^2 + y^2}} \, dx \, dy$

In Problems 27 to 30, find the volume V of the solid under the graph of the given function f and above the indicated region R in the xy plane.

27 $f(x, y) = 4 - x^2$; R: $0 \le y \le 2$, $y^2/2 \le x \le 2$

28 $f(x, y) = 2 - x$; R: $x^2 + y^2 \le 4$

29 $f(x, y) = 8 - x - y$; R: the triangular region bounded by $x + y = 8$, $x + 2y = 8$, and $x = 0$

30 $f(x, y) = x^2 + y^2$; R: the region bounded by the cardioid whose polar equation is $r = 1 - \sin \theta$

31 The volume V under the hyperbolic paraboloid $z = xy$ and above a region R in the xy plane is given by

$$V = \int_0^1 \int_0^y xy \, dx \, dy + \int_1^2 \int_0^{2-y} xy \, dx \, dy$$

Sketch the region R in the xy plane, express V as an iterated integral in which the order of integration is reversed, and evaluate V.

32 If $b > a > 0$, show that

$$\int_0^\infty \frac{e^{-ax} - e^{-bx}}{x} \, dx = \ln \frac{b}{a}$$

by using the fact that

$$\int_a^b e^{-xy} \, dy = \frac{e^{-ax} - e^{-bx}}{x}$$

forming a suitable iterated integral, and reversing the order of integration.

In Problems 33 to 36, use double integration to find the centroid of the region R.

33 R: $0 \le x \le \pi/4$, $0 \le y \le \sec^2 x$

34 R: $0 \le y \le \ln 4$, $e^y \le x \le 4$

35 R: $0 \le \theta \le \pi/3$, $0 \le r \le \sin 2\theta$

36 R: the triangular region with vertices $(0, 0)$, $(b, 0)$, and (c, h), where b, c, and h are positive constants

37 Find the moments of inertia I_x, I_y, and I_o of a homogeneous lamina of mass m occupying the triangular region R of Problem 36.

38 A lamina has the shape of a circular disk $x^2 + y^2 \le a^2$, and its density at the point (x, y) is given by $\sigma(x, y) = k(x^2 + y^2)^{3/2}$. Here, a and k are positive constants. Find I_x, I_y, and I_o.

In Problems 39 to 43, evaluate each threefold iterated integral.

39 $\int_0^1 \int_0^{x^2} \int_0^{xy^2} xy^2z^3 \, dz \, dy \, dx$

40 $\int_0^1 \int_{-x}^x \int_0^{x+z} e^{x+y+z} \, dy \, dz \, dx$

41 $\int_0^1 \int_0^{1+x^2} \int_{3x^2+y}^{4-x^2} dz \, dy \, dx$

42 $\int_0^1 \int_0^{\sqrt{3z}} \int_0^{\sqrt{3(z^2+y^2)}} xyz\sqrt{x^2 + y^2 + z^2} \, dx \, dy \, dz$

43 $\int_0^\pi \int_0^{\pi/2} \int_0^{2a \cos \phi} \rho^2 \sin \phi \, d\rho \, d\phi \, d\theta$

44 Rewrite $\int_0^1 \int_0^{2\sqrt{1-z}} \int_0^{\sqrt{1-z}} f(x, y, z) \, dy \, dx \, dz$ as an equivalent triple integral over a solid S, and sketch S. Then rewrite the integral as an equivalent threefold iterated integral with as many different orders of integration as possible.

In Problems 45 and 46, sketch the solid S, and evaluate the triple integral by iteration.

45 $\iiint\limits_S y \, dV$, where S is the solid bounded above by the plane $3x + 2y + z = 6$, below by the plane $z = 0$, and laterally by the cylinder with generators parallel to the z axis over the boundary of the region R: $0 \le x \le 1$, $0 \le y \le 2 - 2x$

46 $\iiint\limits_S xz \, dV$, where S is the solid bounded by the planes $x = 0$, $y = 0$, $z = 1$, $x = 2 - 2z$, and $y = 3 - 3z$

In Problems 47 and 48, use triple integration to find the volume of each solid S.

47 S is the solid bounded above by the parabolic cylinder $x^2 + z = 4$, below by the plane $x + z = 2$, on the left by the plane $y = 0$, and on the right by the plane $y = 3$.

48 S is the solid in the first octant bounded by the circular cylinder $x^2 + z^2 = 9$ and the planes $y = 2x$, $y = 0$, and $z = 0$.

In Problems 49 and 50, express each threefold iterated integral as an equivalent threefold iterated integral in cylindrical coordinates, and then evaluate it.

49 $\int_0^1 \int_0^{\sqrt{1-x^2}} \int_0^{\sqrt{1-x^2-y^2}} z^2 \, dz \, dy \, dx$

50 $\int_0^2 \int_0^{\sqrt{2x-x^2}} \int_0^{9-x^2-y^2} \sqrt{x^2+y^2} \, dz \, dy \, dx$

In Problems 51 and 52, express each threefold iterated integral as an equivalent threefold iterated integral in spherical coordinates, and then evaluate it.

51 $\int_0^1 \int_0^{\sqrt{1-y^2}} \int_{\sqrt{3(x^2+y^2)}}^{\sqrt{4-x^2-y^2}} \sqrt{x^2+y^2+z^2} \, dz \, dx \, dy$

52 $\int_0^{\sqrt{2}/2} \int_y^{\sqrt{1-y^2}} \int_{-\sqrt{8(x^2+y^2)}}^{\sqrt{9-x^2-y^2}} f'[(x^2+y^2+z^2)^{3/2}] \, dz \, dx \, dy$

53 Use cylindrical coordinates to evaluate $\iiint\limits_S z\sqrt{x^2+y^2} \, dV$, where S is the half of the solid right circular cone with vertex at $(0, 0, h)$ and base $x^2 + y^2 \le a^2$ lying to the right of the plane $y = 0$.

54 Use cylindrical coordinates to find the volume of the solid bounded above by the plane $z = ax + by + c$, below by the xy plane, and laterally by the cylinder $r = k \cos \theta$. [Assume that $k > 0$ and that $ax + by + c \ge 0$ for (x, y) inside the circle whose polar equation is $r = k \cos \theta$.]

In Problems 55 to 58, find the coordinates $(\bar{x}, \bar{y}, \bar{z})$ of the centroid of each solid S.

55 S is bounded above by the plane $z = 4$ and below by the paraboloid of revolution $3z = r^2$.

56 S is bounded above by the sphere $x^2 + y^2 + z^2 = a^2$, below by the xy plane, and laterally by the cylinder $[x - (a/2)]^2 + y^2 = (a/2)^2$.

57 S: $\quad 0 \le y \le 1, \quad 0 \le x \le \sqrt{1 - y^2}, \quad \sqrt{3(x^2 + y^2)} \le z \le \sqrt{4 - x^2 - y^2}$.

58 S is bounded above by the paraboloid of revolution $x = 5 - y^2 - z^2$, below by the xy plane, behind by the yz plane, on the left by the xz plane, and on the right by the plane $y = 1$.

In Problems 59 to 62, find the moment of inertia I_z about the z axis of a homogeneous solid S of total mass m as described.

59 S is a solid right circular cone of height h with vertex at the origin, central axis along the positive z axis, and base radius a.

60 S is bounded by the coordinate planes, the planes $x = 3$ and $z = 1$, and the paraboloid $y = 10 - x^2 - z^2$.

61 S: $(x - a)^2 + y^2 + z^2 \le a^2$

62 S: $r \le 2 \cos \theta, \ 0 \le \theta \le \pi, \ 0 \le z \le \sqrt{4 - r^2}$

In Problems 63 to 66, evaluate the line integral directly.

63 $\int_C (x + y) \, dx + (x + y^2) \, dy$; C: $\begin{cases} x = t + 1 \\ y = t^2 \end{cases} \quad 0 \le t \le 1$

64 $\int_C \mathbf{F} \cdot d\mathbf{R}$; $\mathbf{F} = ye^{xy}\mathbf{i} + xe^{xy}\mathbf{j}$; C: $\begin{cases} x = t^3 \\ y = 1 - t^6 \end{cases} \quad -1 \le t \le 1$

65 $\int_C x^2 \, dx$; C: $\begin{cases} x = t \\ y = t^2 \end{cases} \quad 0 \le t \le 1$

66 $\int_C (ax + by) \, dx + (cx + ky) \, dy$; C is the line segment from (x_0, y_0) to (x_1, y_1).

In Problems 67 to 70, use Green's theorem to evaluate each line integral.

67 $\int_C x^2y \, dx + y^3 \, dy$; C is the boundary, taken in the counterclockwise sense, of the region R bounded by the curves $y^3 = x^2$ and $y = x$.

68 $\int_C (x^2 + y) \, dx + (x - y^2) \, dy$; C is the counterclockwise boundary of the region R bounded by the curves $y = 2x^2$ and $y = 4x$.

69 $\int_C y \, dx - x \, dy$; C is the counterclockwise boundary of the triangle with vertices $(0, 0)$, $(b, 0)$, and (c, h), where b, c, and h are positive constants.

70 $\int_C \mathbf{F} \cdot d\mathbf{R}$; $\mathbf{F}(x, y) = f(x)\mathbf{i} + g(y)\mathbf{j}$; C is any simple closed curve; f and g are continuous functions.

In Problems 71 to 76, determine whether each line integral is independent of the path.

71 $\int_C (1 + 2x^3 + 3x^2y) \, dx + (x^3 + y^3 - 1) \, dy$

72 $\int_C [y + (y/x)] \, dx + (x + \ln xy) \, dy$; R: $x > 0, \ y > 0$

73 $\int_C x^2 \sin y \, dx + x^3 \cos y \, dy$

74 $\int_C (3x^2 + 2ye^{2x}) \, dx + (e^{2x} + 3y^2 + 1) \, dy$

75 $\int_C (3x - y + 2z) \, dx + (2y + 3z - x) \, dy + (2x + 3y - z) \, dz$

76 $\int_C (3x^2 + 2y \sin 2x + z^2) \, dx + (2 \sin^2 x + y - yz) \, dy + (xy^2z) \, dz$

In Problems 77 to 80, (a) show that $\mathbf{F}$ is a conservative vector field; (b) find a potential function f for $\mathbf{F}$; (c) evaluate the given line integral.

77 $\mathbf{F} = (3x^2 + 6y)\mathbf{i} + (3y^2 + 6x)\mathbf{j}; \int_{(1,0)}^{(2,1)} \mathbf{F} \cdot d\mathbf{R}$

78 $\mathbf{F} = (2 \sin^2 y + 3x^2 + 1)\mathbf{i} + (2x \sin 2y + 3y^2 - 1)\mathbf{j};$

$\int_{(0,0)}^{(\pi/2, \pi)} \mathbf{F} \cdot d\mathbf{R}$

79 $\mathbf{F} = (2xy + z^2)\mathbf{i} + (2zy + x^2)\mathbf{j} + (2xz + y^2)\mathbf{k}; \int_{(-1,2,0)}^{(1,2,3)} \mathbf{F} \cdot d\mathbf{R}$

80 $\mathbf{F} = 2xyz\mathbf{i} + x^2z\mathbf{j} + x^2y\mathbf{k}; \int_{(2,1,-1)}^{(0,0,2)} \mathbf{F} \cdot d\mathbf{R}$

81 Find the area of the portion of the surface $x^2 + y^2 + z^2 = 25$ that lies outside the paraboloid $z^2 + y^2 = 2x + 10$.

82 Set up an integral which gives the area of the portion of the surface $f(x, y, z) = k$ that lies above the region R in the xy plane. Make whatever assumptions you need concerning continuity, differentiability, and so forth.

83 Two congruent cylinders are tangent to each other externally along a diameter of a sphere whose radius is twice that of the cylinders. Find the area of the portion of the surface of the sphere interior to the cylinders.

84 Suppose that the region R in the xy plane is bounded by the simple closed curve C. Write formulas for the coordinates of the centroid $(\bar{x}, \bar{y})$ of R that involve only line integrals around the curve C.

85 Evaluate $\iint_\Sigma (x^2 + y^2) \, dA_\Sigma$, where Σ is the portion of the surface of the cone $z^2 = x^2 + y^2$ between $z = 0$ and $z = 1$.

86 A thin homogeneous lamina of mass m has the shape of the hemisphere $x^2 + y^2 + z^2 = a^2$. Set up a surface integral that gives the moment of inertia I_y of this lamina about the y axis, and then evaluate this integral.

In Problems 87 and 88, find (a) $\nabla \cdot \mathbf{F}$ and (b) $\nabla \times \mathbf{F}$ for each vector field $\mathbf{F}$.

87 $\mathbf{F} = (x^2 + yz)\mathbf{i} + (y^2 + xz)\mathbf{j} + (z^2 + xy)\mathbf{k}$

88 $\mathbf{F} = (y \cos z)\mathbf{i} + (z \cos x)\mathbf{j} + (x \cos y)\mathbf{k}$

In Problems 89 and 90, use the divergence theorem to calculate the flux $\iint_\Sigma \mathbf{F} \cdot \mathbf{N} \, dA_\Sigma$ of the given vector field $\mathbf{F}$ through the surface Σ of the indicated solid S in the direction of the outward-pointing unit normal vector $\mathbf{N}$.

89 $\mathbf{F} = x\mathbf{i} + y\mathbf{j} - z\mathbf{k}$; S is the solid right circular cylinder bounded above by $z = 2$, below by $z = 1$, and laterally by $x^2 + y^2 = 1$.

90 $\mathbf{F} = x^2\mathbf{i} + 2y^2\mathbf{j} + 3z^2\mathbf{k}$; S is the solid sphere $x^2 + y^2 + z^2 \le 1$.

In Problems 91 and 92, use Stokes' theorem to calculate the flux $\iint_\Sigma (\nabla \times \mathbf{F}) \cdot \mathbf{N} \, dA_\Sigma$ of the curl of the given vector field $\mathbf{F}$ through the indicated surface Σ in the direction of the given unit normal vector $\mathbf{N}$.

91 $\mathbf{F} = y\mathbf{i} + z\mathbf{j} + x\mathbf{k}$; Σ is the hemisphere $x^2 + y^2 + z^2 = 1$, $z \ge 0$; $\mathbf{N}$ is the unit normal vector whose $\mathbf{k}$ component is nonnegative.

92 $\mathbf{F} = (y + 3x)\mathbf{i} + (2y - x)\mathbf{j} + (xy^2 + z^3)\mathbf{k}$; Σ is the portion of the paraboloid of revolution $z = 1 - x^2 - y^2$ for which $z \ge 0$; $\mathbf{N}$ is the unit normal vector whose $\mathbf{k}$ component is nonnegative.

93 Let $\mathbf{r} = x\mathbf{i} + y\mathbf{j} + z\mathbf{k}$. Show that the flux of $\mathbf{r}$ through the surface Σ of any solid S is 3 times the volume of S.

94 Let $\mathbf{r} = x\mathbf{i} + y\mathbf{j} + z\mathbf{k}$, and let $r = |\mathbf{r}|$. If $\mathbf{F} = r^2\mathbf{r}$, verify the divergence theorem for the solid sphere S of radius 1 with center at the origin.

Appendix A FORMULAS FROM GEOMETRY

Figure 1

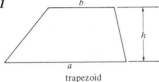

trapezoid

Figure 2

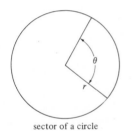

sector of a circle

Figure 3

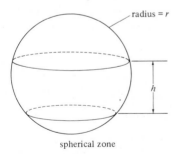

spherical zone

Figure 4

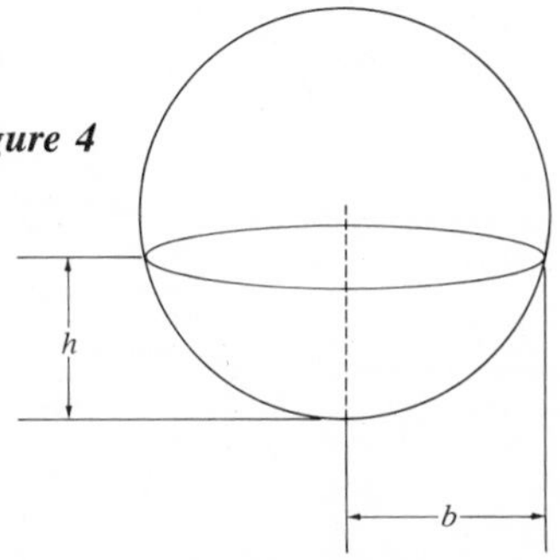

spherical segment

1 Plane Area

  (a) *Square* $A = l^2$; l = length of a side

  (b) *Rectangle* $A = lw$; l = length, w = width

  (c) *Triangle* $A = \frac{1}{2}bh$; b = length of base, h = height

  (d) *Trapezoid* $A = h\left(\dfrac{a + b}{2}\right)$; h = height, a = length of one base, b = length of other base (Figure 1)

  (e) *Circle* $A = \pi r^2$; r = radius

  (f) *Sector of a circle* $A = \frac{1}{2}\theta r^2$; θ = central angle, r = radius (Figure 2)

2 Perimeter

  (a) *Square* $p = 4l$; l = length of a side

  (b) *Rectangle* $p = 2l + 2w$; l = length, w = width

  (c) *Circle* $p = 2\pi r$; r = radius

3 Surface Area

  (a) *Closed rectangular box* $A = 2lw + 2lh + 2wh$; l = length, w = width, h = height

  (b) *Right circular cylinder* (open at top and bottom) $A = 2\pi rh$; r = radius, h = height

  (c) *Sphere* $A = 4\pi r^2$; r = radius

  (d) *Spherical zone* $A = 2\pi rh$; r = radius of sphere, h = height of zone (Figure 3)

  (e) *Right circular cone with open base* $A = \pi rl$; h = height, r = radius of base, l = slant height = $\sqrt{r^2 + h^2}$

4 Volume

  (a) *Rectangular box* $V = lwh$

  (b) *Right circular cylinder* $V = \pi r^2 h$

  (c) *Cylinder or prism* $V = Ah$; A = area of base, h = height

  (d) *Sphere* $V = \frac{4}{3}\pi r^3$

  (e) *Spherical segment of one base* $V = \frac{1}{6}\pi h(3b^2 + h^2)$; h = height of segment, b = radius of base (Figure 4)

  (f) *Right circular cone* $V = \frac{1}{3}\pi r^2 h$

  (g) *Cone or pyramid* $V = \frac{1}{3}Ah$; A = area of base, h = height

Appendix B   PROOFS OF BASIC PROPERTIES OF LIMITS AND OF CONTINUOUS FUNCTIONS

Although rigorous proofs of many of the theorems of calculus are best left to more advanced courses in analysis, it is perhaps not amiss to present a few typical proofs here so that the interested reader can gain some appreciation of the techniques involved. We begin by giving a proof of part of Property 4 in Section 1.8.

THEOREM 1   **Additivity of Limits**

If $\lim_{x \to a} f(x) = L$ and $\lim_{x \to a} g(x) = M$, then

$$\lim_{x \to a} [f(x) + g(x)] = \lim_{x \to a} f(x) + \lim_{x \to a} g(x) = L + M$$

PROOF   According to the definition of limit (Definition 1, page 57), given a positive number ϵ, we must show that there exists a positive number δ such that

$$|[f(x) + g(x)] - (L + M)| < \epsilon \qquad \text{holds whenever} \qquad 0 < |x - a| < \delta$$

By the triangle inequality, we have

$$|[f(x) + g(x)] - (L + M)| = |[f(x) - L] + [g(x) - M]|$$
$$\leq |f(x) - L| + |g(x) - M|$$

Hence, it is enough to find a positive number δ such that

$$|f(x) - L| + |g(x) - M| < \epsilon \qquad \text{holds whenever} \qquad 0 < |x - a| < \delta$$

Now, in order to guarantee that

$$|f(x) - L| + |g(x) - M| < \epsilon$$

it certainly is enough to have

$$|f(x) - L| < \tfrac{1}{2}\epsilon \qquad \text{and} \qquad |g(x) - M| < \tfrac{1}{2}\epsilon$$

Notice that $\tfrac{1}{2}\epsilon$ is a positive number; hence, since $\lim_{x \to a} f(x) = L$, there exists (by definition of limit) a positive number δ_1 such that

$$|f(x) - L| < \tfrac{1}{2}\epsilon \qquad \text{holds whenever} \qquad 0 < |x - a| < \delta_1$$

Similarly, since $\lim\limits_{x \to a} g(x) = M$, there exists a positive number δ_2 such that

$$|g(x) - M| < \tfrac{1}{2}\epsilon \qquad \text{holds whenever} \qquad 0 < |x - a| < \delta_2$$

Let δ be the smaller of the two numbers δ_1 and δ_2 (or their common value if they are equal). Then

$$\delta \leq \delta_1 \qquad \text{and} \qquad \delta \leq \delta_2$$

so that if $0 < |x - a| < \delta$, both of the conditions

$$0 < |x - a| < \delta_1 \qquad \text{and} \qquad 0 < |x - a| < \delta_2$$

hold, and it follows that

$$|[f(x) + g(x)] - (L + M)| < \epsilon$$

as desired. ∎

The following technical theorem can be quite helpful for establishing the remaining properties of limits.

THEOREM 2 **Boundedness Theorem**

> If $\lim\limits_{x \to a} f(x)$ exists, then there exist positive numbers N and δ_1 such that $|f(x)| < N$ holds whenever $0 < |x - a| < \delta_1$.

PROOF Suppose that $\lim\limits_{x \to a} f(x) = L$. Then, since 1 is a positive number, there exists a positive number δ_1 such that

$$|f(x) - L| < 1 \qquad \text{holds whenever} \qquad 0 < |x - a| < \delta_1$$

(There is nothing special about our choice of the number 1 here—any positive number will work.) Now, let $N = 1 + |L|$. If $|f(x) - L| < 1$, then, adding $|L|$ to both sides, we have $|f(x) - L| + |L| < 1 + |L| = N$; therefore,

$$|f(x) - L| + |L| < N \qquad \text{holds whenever} \qquad 0 < |x - a| < \delta_1$$

But, by the triangle inequality,

$$|f(x)| = |f(x) - L + L| \leq |f(x) - L| + |L|$$

and it follows that

$$|f(x)| < N \qquad \text{holds whenever} \qquad 0 < |x - a| < \delta_1 \qquad ∎$$

Theorem 2 is used in the course of the proof of the following theorem, which establishes Property 6 in Section 1.8.

THEOREM 3 **Multiplicativity of Limits**

> If $\lim\limits_{x \to a} f(x) = L$ and $\lim\limits_{x \to a} g(x) = M$, then
>
> $$\lim_{x \to a} [f(x) \cdot g(x)] = [\lim_{x \to a} f(x)] \cdot [\lim_{x \to a} g(x)] = LM$$

PROOF   Let $\epsilon > 0$ be given. We must show that there exists a positive number δ such that

$$|f(x)g(x) - LM| < \epsilon \qquad \text{holds whenever} \qquad 0 < |x - a| < \delta$$

Using some elementary algebra and the triangle inequality, we have

$$
\begin{aligned}
|f(x)g(x) - LM| &= |f(x)g(x) + 0 - LM| \\
&= |f(x)g(x) + [-f(x)M + f(x)M] - LM| \\
&= |f(x)g(x) - f(x)M + f(x)M - LM| \\
&= |f(x)[g(x) - M] + [f(x) - L]M| \\
&\leq |f(x)[g(x) - M]| + |[f(x) - L]M| \\
&= |f(x)|\,|g(x) - M| + |f(x) - L|\,|M|
\end{aligned}
$$

Therefore, to guarantee that $|f(x)g(x) - LM| < \epsilon$, it certainly is enough to have

(i) $|f(x)|\,|g(x) - M| < \tfrac{1}{2}\epsilon$ and   (ii) $|f(x) - L|\,|M| < \tfrac{1}{2}\epsilon$

In order to obtain the first condition, we plan to use Theorem 2 to make $|f(x)|$ smaller than a fixed positive number N and then to make $|g(x) - M|$ less than $1/N$ times $\tfrac{1}{2}\epsilon$. Thus, by Theorem 2, there exist positive numbers N and δ_1 such that

$$|f(x)| < N \qquad \text{holds whenever} \qquad 0 < |x - a| < \delta_1$$

Also, since $\epsilon/(2N)$ is a positive number and $\lim_{x \to a} g(x) = M$, there exists a positive number δ_2 such that

$$|g(x) - M| < \frac{\epsilon}{2N} \qquad \text{holds whenever} \qquad 0 < |x - a| < \delta_2$$

If both inequalities $|f(x)| < N$ and $|g(x) - M| < \epsilon/(2N)$ hold, then

$$|f(x)|\,|g(x) - M| < N\left(\frac{\epsilon}{2N}\right) = \frac{1}{2}\,\epsilon$$

also holds; hence, condition (i) holds if

$$0 < |x - a| < \delta_1$$

and $$0 < |x - a| < \delta_2$$

In order to obtain condition (ii), we plan to use the fact that $\lim_{x \to a} f(x) = L$ to show that there exists a positive number δ_3 such that

$$|f(x) - L|\,|M| < \tfrac{1}{2}\epsilon \qquad \text{holds whenever} \qquad 0 < |x - a| < \delta_3$$

If $M = 0$, the last condition holds for any choice of δ_3, so we need concern ourselves only with the case in which $M \neq 0$. Then, since $\epsilon/(2|M|)$ is a positive number and $\lim_{x \to a} f(x) = L$, there exists a positive number δ_3 such that

$$|f(x) - L| < \frac{\epsilon}{2|M|} \qquad \text{holds whenever} \qquad 0 < |x - a| < \delta_3$$

Therefore, condition (ii) holds if

$$0 < |x - a| < \delta_3$$

Now, to finish the proof, just let δ be the smallest of the three numbers δ_1, δ_2, and δ_3, so that if $0 < |x - a| < \delta$, then

$$0 < |x - a| < \delta_1 \qquad 0 < |x - a| < \delta_2 \qquad \text{and } 0 < |x - a| < \delta_3$$

Consequently, conditions (i) and (ii) hold whenever

$$0 < |x - a| < \delta$$

It follows that

$$|f(x)g(x) - LM| < \epsilon \qquad \text{holds whenever} \qquad 0 < |x - a| < \delta$$

as desired. ∎

By using the properties of limits and the definition of continuity (Definition 1, page 69), it is not difficult to establish the basic properties of continuous functions. The proof of the following theorem illustrates the general technique.

THEOREM 4   **Continuity of a Product**

> If the functions f and g are both continuous at the number a, then the product function $f \cdot g$ is also continuous at a.

PROOF   Since f and g are both continuous at a, they are both defined at a; hence, $f \cdot g$ is defined at a and

$$(f \cdot g)(a) = f(a)g(a)$$

Because f and g are both continuous at a, $\lim_{x \to a} f(x)$ and $\lim_{x \to a} g(x)$ both exist and

$$\lim_{x \to a} f(x) = f(a) \qquad \text{while} \qquad \lim_{x \to a} g(x) = g(a)$$

Therefore, by Theorem 3, $\lim_{x \to a} [f(x) \cdot g(x)]$ exists and we have

$$\lim_{x \to a} (f \cdot g)(x) = \lim_{x \to a} [f(x)g(x)] = [\lim_{x \to a} f(x)]\,[\lim_{x \to a} g(x)]$$

$$= f(a)g(a) = (f \cdot g)(a)$$

It follows that $f \cdot g$ is continuous at a. ∎

Problem Set, Appendix B

1 If c is a constant number and f is the constant function defined by $f(x) = c$, prove that $\lim_{x \to a} f(x) = c$ holds for each number a (Property 1, Section 1.8).

2 If f is the identity function, that is, if f is defined by $f(x) = x$, prove that $\lim_{x \to a} f(x) = a$ holds for each number a (Property 2, Section 1.8).

3 Combine Problem 1 with Theorem 3 to prove that if c is a constant number and if the limit $\lim_{x \to a} g(x)$ exists, then $\lim_{x \to a} [cg(x)] = c \lim_{x \to a} g(x)$ (Property 3, Section 1.8).

4 Prove that if $\lim_{x \to a} f(x)$ exists, then $\lim_{x \to a} |f(x)| = |\lim_{x \to a} f(x)|$ (Property 10, Section 1.8).

5 Prove Property 5, Section 1.8: If $\lim\limits_{x \to a} f(x) = L$ and $\lim\limits_{x \to a} g(x) = M$, then $\lim\limits_{x \to a} [f(x) - g(x)] = L - M$.

6 Assume that $\lim\limits_{x \to a} g(x) = M \neq 0$. Prove that there exist positive numbers N and δ_0 such that (i) $g(x) \neq 0$ holds for $0 < |x - a| < \delta_0$, and (ii) $|1/g(x)| < N$ holds for $0 < |x - a| < \delta_0$. (*Hint:* Select δ_0 such that $|g(x) - M| < |M|/2$ holds whenever $0 < |x - a| < \delta_0$. Put $N = 2/|M|$.)

7 Suppose that $\lim\limits_{x \to a} f(x)$, $\lim\limits_{x \to a} g(x)$, and $\lim\limits_{x \to a} h(x)$ all exist. Prove that $\lim\limits_{x \to a} [f(x) + g(x) + h(x)] = \lim\limits_{x \to a} f(x) + \lim\limits_{x \to a} g(x) + \lim\limits_{x \to a} h(x)$.

8 Assume that $\lim\limits_{x \to a} g(x) = M \neq 0$. Prove that $\lim\limits_{x \to a} 1/g(x) = 1/M$.

[*Hints:* Select N and δ_0 as in Problem 6. Notice that $|1/g(x) - 1/M|$ can be written as

$$\frac{1}{|g(x)|} \cdot \frac{1}{|M|} \cdot |g(x) - M|$$

Given $\epsilon > 0$, select δ_1 such that $|g(x) - M| < (|M|\epsilon)/N$ holds whenever $0 < |x - a| < \delta_1$. Choose δ to be the smaller of the two numbers δ_0 and δ_1.]

9 Combine Theorem 3 and Problem 8 and thus establish Property 7 of limits of functions in Section 1.8.

10 Assume that $\lim\limits_{x \to b} f(x) = f(b) = L$ and that $\lim\limits_{x \to a} g(x) = b$. Prove that $\lim\limits_{x \to a} f(g(x)) = L$.

11 Using the result of Problem 10, prove Property 5 of continuous functions in Section 1.9.

12 By slightly modifying the proof in Problem 10, give a proof of Property 15, Section 1.8.

13 Prove Properties 13 and 14 in Section 1.8.

14 Prove the squeezing property (Property 12, Section 1.8).

Appendix C   MATHEMATICAL INDUCTION

The principle of *mathematical induction* is often illustrated by a row of oblong wooden blocks, called dominoes, set on end as in Figure 1.

Figure 1

If we are guaranteed two conditions,

    (i) that the first domino is toppled over

    (ii) that if any domino topples over, it will hit the next one and topple it over

then we can be certain that *all* the dominoes will topple over.

The principle of mathematical induction concerns a sequence of statements

$$S_1, S_2, S_3, S_4, S_5, S_6, \ldots .$$

Let's think of this sequence as corresponding to a row of dominoes; the first statement S_1 corresponds to the first domino, the second statement S_2 corresponds to the second domino, and so on. Each statement can be either true or false. If a statement proves to be true, let's think of the corresponding domino toppling over. The two conditions given above for the row of dominoes can then be interpreted as follows:

    (i) The first statement S_1 is true.

    (ii) If any statement S_k is true, then the next statement S_{k+1} is also true.

If these two conditions hold, then, by analogy with the toppling over of *all* the dominoes, we conclude that *all statements S_1, S_2, S_3, . . . are true*. That this argument is valid is the **principle of mathematical induction.**

Although the principle of mathematical induction can be established formally, the required argument is beyond the scope of this book. We ask you to accept this principle on the intuitive basis of the domino analogy. Thus, we have the following.

1021

Procedure for Making a Proof by Mathematical Induction

Begin by clearly identifying the sequence S_1, S_2, S_3, . . . of statements to be proved. This is usually done by specifying the meaning of S_n, where n denotes an arbitrary positive integer, $n = 1, 2, 3, \ldots$. Then carry out the following two steps:

> **Step 1** Show that S_1 is true.
>
> **Step 2** Let k denote an arbitrary positive integer. Assume that S_k is true, and show on the basis of this assumption that S_{k+1} is also true.

If both steps can be carried out, conclude that S_n is true for all positive integer values of n.

The assumption in step 2 that S_k is true is called the **induction hypothesis.** When you make the induction hypothesis, you're not saying that S_k is *in fact* true—you're just *supposing* that it is true to see whether the truth of S_{k+1} follows from this supposition. It's as if you were checking to make sure that if the kth domino *were* to topple over, it *would* knock over the next domino.

In Examples 1 and 2, use mathematical induction to prove each result.

EXAMPLE 1   For every positive integer n, $1 + 2 + 3 + \cdots + n = n(n + 1)/2$.

SOLUTION   Let S_n be the statement

$$S_n: 1 + 2 + 3 + \cdots + n = \frac{n(n + 1)}{2}$$

In other words, S_n asserts that the sum of the first n positive integers is $n(n + 1)/2$. For instance,

$$S_2 \qquad \text{says that} \qquad 1 + 2 = \frac{2(2 + 1)}{2} = 3$$

and

$$S_{20} \qquad \text{says that} \qquad 1 + 2 + 3 + \cdots + 20 = \frac{20(20 + 1)}{2} = 210$$

Step 1 The statement

$$S_1: 1 = \frac{1(1 + 1)}{2}$$

is clearly true.

Step 2 Here we must deal with the statements

$$S_k: 1 + 2 + 3 + \cdots + k = \frac{k(k + 1)}{2}$$

and

$$S_{k+1}: 1 + 2 + 3 + \cdots + (k + 1) = \frac{(k + 1)[(k + 1) + 1]}{2}$$

obtained by replacing n in S_n by k and then by $k + 1$. The statement S_k is our induction hypothesis, and we assume, for the sake of argument, that it is true. Our goal is to prove that, on the basis of this assumption, S_{k+1} is true. To this end, we add $k + 1$ to both sides of the equation expressing S_k to obtain

$$1 + 2 + 3 + \cdots + k + (k + 1) = \frac{k(k + 1)}{2} + (k + 1)$$

$$= \frac{k(k + 1) + 2(k + 1)}{2}$$

$$= \frac{(k + 1)(k + 2)}{2}$$

$$= \frac{(k + 1)[(k + 1) + 1]}{2}$$

which is the equation expressing S_{k+1}. This completes the proof by mathematical induction. ■

EXAMPLE 2   If a principal of P dollars is invested at a compound interest rate R per conversion period, the final value F dollars of the investment at the end of n conversion periods is given by $F = P(1 + R)^n$. [For *compound interest*, the interest is periodically calculated and added to the principal. The time interval between successive conversions of interest into principal is called the *conversion period*.]

SOLUTION   Let S_n be the statement

S_n: $F = P(1 + R)^n$ at the end of n conversion periods

Step 1 The statement

S_1: $F = P(1 + R)$ at the end of the first conversion period

is true, because the interest on P dollars for one conversion period is PR dollars, so the final value of the investment at the end of the first conversion period is

$$P + PR = P(1 + R) \qquad \text{dollars}$$

Step 2 Assume that the statement

S_k: $F = P(1 + R)^k$ at the end of k conversion periods

is true. It follows that, over the $(k + 1)$st conversion period, interest at rate R is paid on $P(1 + R)^k$ dollars. Adding this interest, $P(1 + R)^k R$ dollars, to the value of the investment at the beginning of the $(k + 1)$st conversion period, we find that the value of the investment at the end of the $(k + 1)$st conversion period is given by

$$F = P(1 + R)^k + P(1 + R)^k R$$

$$= P(1 + R)^k(1 + R)$$

$$= P(1 + R)^{k+1} \qquad \text{dollars}$$

Hence, S_{k+1} is true, and the proof by mathematical induction is complete. ■

Problem Set, Appendix C

In Problems 1 to 14, use mathematical induction to prove that the assertion is true for all positive integers n.

1 $1 + 3 + 5 + \cdots + (2n - 1) = n^2$; in other words, the sum of the first n odd positive integers is n^2.

2 $2 + 4 + 6 + \cdots + 2n = n(n + 1)$; in other words, the sum of the first n even positive integers is $n(n + 1)$.

3 $1^2 + 2^2 + 3^2 + \cdots + n^2 = \frac{1}{6}n(n + 1)(2n + 1)$; that is, the sum of the first n perfect squares is $\frac{1}{6}n(n + 1)(2n + 1)$.

4 $1 + 5 + 9 + \cdots + (4n - 3) = n(2n - 1)$

5 $1^2 + 3^2 + 5^2 + \cdots + (2n - 1)^2 = \frac{1}{3}n(2n - 1)(2n + 1)$

6 $1^3 + 2^3 + 3^3 + \cdots + n^3 = [n(n + 1)/2]^2$

7 $1 \cdot 2 + 2 \cdot 3 + 3 \cdot 4 + \cdots + n(n + 1) = \frac{1}{3}n(n + 1)(n + 2)$

8 $\dfrac{1}{1 \cdot 2} + \dfrac{1}{2 \cdot 3} + \dfrac{1}{3 \cdot 4} + \cdots + \dfrac{1}{n(n + 1)} = \dfrac{n}{n + 1}$

9 $(ab)^n = a^n b^n$

10 If $h \geq 0$, then $1 + nh \leq (1 + h)^n$.

11 $2 + 2^2 + 2^3 + \cdots + 2^n = 2(2^n - 1)$

12 $r + r^2 + r^3 + \cdots + r^n = \dfrac{r(r^n - 1)}{r - 1}$

13 $\cos n\pi = (-1)^n$

14 $[r(\cos \theta + i \sin \theta)]^n = r^n(\cos n\theta + i \sin n\theta)$ (De Moivre's theorem)

C 15 Using a calculator, verify the identities in odd-numbered Problems 1 to 7 for $n = 15$ by directly calculating both sides.

16 The domino theory was a tenet of U.S. foreign policy subscribed to by the administrations of Presidents Eisenhower, Kennedy, Johnson, and Nixon. Explain the connection, if any, between the domino theory and mathematical induction.

Appendix D   TABLES

Table 1   Trigonometric Functions

Degrees	Radians	Sin	Tan	Cot	Cos		
0	0	0	0	—	1.000	1.5708	90
1	0.0175	0.0175	0.0175	57.290	0.9998	1.5533	89
2	0.0349	0.0349	0.0349	28.636	0.9994	1.5359	88
3	0.0524	0.0523	0.0523	19.081	0.9986	1.5184	87
4	0.0698	0.0698	0.0699	14.301	0.9976	1.5010	86
5	0.0873	0.0872	0.0875	11.430	0.9962	1.4835	85
6	0.1047	0.1045	0.1051	9.5144	0.9945	1.4661	84
7	0.1222	0.1219	0.1228	8.1443	0.9925	1.4486	83
8	0.1396	0.1392	0.1405	7.1154	0.9903	1.4312	82
9	0.1571	0.1564	0.1584	6.3138	0.9877	1.4137	81
10	0.1745	0.1736	0.1763	5.6713	0.9848	1.3963	80
11	0.1920	0.1908	0.1944	5.1446	0.9816	1.3788	79
12	0.2094	0.2079	0.2126	4.7046	0.9781	1.3614	78
13	0.2269	0.2250	0.2309	4.3315	0.9744	1.3439	77
14	0.2443	0.2419	0.2493	4.0108	0.9703	1.3265	76
15	0.2618	0.2588	0.2679	3.7321	0.9659	1.3090	75
16	0.2793	0.2756	0.2867	3.4874	0.9613	1.2915	74
17	0.2967	0.2924	0.3057	3.2709	0.9563	1.2741	73
18	0.3142	0.3090	0.3249	3.0777	0.9511	1.2566	72
19	0.3316	0.3256	0.3443	2.9042	0.9455	1.2392	71
20	0.3491	0.3420	0.3640	2.7475	0.9397	1.2217	70
21	0.3665	0.3584	0.3839	2.6051	0.9336	1.2043	69
22	0.3840	0.3746	0.4040	2.4751	0.9272	1.1868	68
23	0.4014	0.3907	0.4245	2.3559	0.9205	1.1694	67
24	0.4189	0.4067	0.4452	2.2460	0.9135	1.1519	66
25	0.4363	0.4226	0.4663	2.1445	0.9063	1.1345	65
26	0.4538	0.4384	0.4877	2.0503	0.8988	1.1170	64
27	0.4712	0.4540	0.5095	1.9626	0.8910	1.0996	63
28	0.4887	0.4695	0.5317	1.8807	0.8829	1.0821	62
29	0.5061	0.4848	0.5543	1.8040	0.8746	1.0647	61
30	0.5236	0.5000	0.5774	1.7321	0.8660	1.0472	60
31	0.5411	0.5150	0.6009	1.6643	0.8572	1.0297	59
32	0.5585	0.5299	0.6249	1.6003	0.8480	1.0123	58
33	0.5760	0.5446	0.6494	1.5399	0.8387	0.9948	57
34	0.5934	0.5592	0.6745	1.4826	0.8290	0.9774	56
35	0.6109	0.5736	0.7002	1.4281	0.8192	0.9599	55
36	0.6283	0.5878	0.7265	1.3764	0.8090	0.9425	54
37	0.6458	0.6018	0.7536	1.3270	0.7986	0.9250	53
38	0.6632	0.6157	0.7813	1.2799	0.7880	0.9076	52
39	0.6807	0.6293	0.8098	1.2349	0.7771	0.8901	51
40	0.6981	0.6428	0.8391	1.1918	0.7660	0.8727	50
41	0.7156	0.6561	0.8693	1.1504	0.7547	0.8552	49
42	0.7330	0.6691	0.9004	1.1106	0.7431	0.8378	48
43	0.7505	0.6820	0.9325	1.0724	0.7314	0.8203	47
44	0.7679	0.6947	0.9657	1.0355	0.7193	0.8029	46
45	0.7854	0.7071	1.0000	1.0000	0.7071	0.7854	45
		Cos	Cot	Tan	Sin	Radians	Degrees

Table 2   Natural Logarithms, ln t

t	0.00	0.01	0.02	0.03	0.04	0.05	0.06	0.07	0.08	0.09
1.0	0.0000	0.0100	0.0198	0.0296	0.0392	0.0488	0.0583	0.0677	0.0770	0.0862
1.1	0.0953	0.1044	0.1133	0.1222	0.1310	0.1398	0.1484	0.1570	0.1655	0.1740
1.2	0.1823	0.1906	0.1989	0.2070	0.2151	0.2231	0.2311	0.2390	0.2469	0.2546
1.3	0.2624	0.2700	0.2776	0.2852	0.2927	0.3001	0.3075	0.3148	0.3221	0.3293
1.4	0.3365	0.3436	0.3507	0.3577	0.3646	0.3716	0.3784	0.3853	0.3920	0.3988
1.5	0.4055	0.4121	0.4187	0.4253	0.4318	0.4383	0.4447	0.4511	0.4574	0.4637
1.6	0.4700	0.4762	0.4824	0.4886	0.4947	0.5008	0.5068	0.5128	0.5188	0.5247
1.7	0.5306	0.5365	0.5423	0.5481	0.5539	0.5596	0.5653	0.5710	0.5766	0.5822
1.8	0.5878	0.5933	0.5988	0.6043	0.6098	0.6152	0.6206	0.6259	0.6313	0.6366
1.9	0.6419	0.6471	0.6523	0.6575	0.6627	0.6678	0.6729	0.6780	0.6831	0.6881
2.0	0.6931	0.6981	0.7031	0.7080	0.7130	0.7178	0.7227	0.7275	0.7324	0.7372
2.1	0.7419	0.7467	0.7514	0.7561	0.7608	0.7655	0.7701	0.7747	0.7793	0.7839
2.2	0.7885	0.7930	0.7975	0.8020	0.8065	0.8109	0.8154	0.8198	0.8242	0.8286
2.3	0.8329	0.8372	0.8416	0.8459	0.8502	0.8544	0.8587	0.8629	0.8671	0.8713
2.4	0.8755	0.8796	0.8838	0.8879	0.8920	0.8961	0.9002	0.9042	0.9083	0.9123
2.5	0.9163	0.9203	0.9243	0.9282	0.9322	0.9361	0.9400	0.9439	0.9478	0.9517
2.6	0.9555	0.9594	0.9632	0.9670	0.9708	0.9746	0.9783	0.9821	0.9858	0.9895
2.7	0.9933	0.9969	1.0006	1.0043	1.0080	1.0116	1.0152	1.0188	1.0225	1.0260
2.8	1.0296	1.0332	1.0367	1.0403	1.0438	1.0473	1.0508	1.0543	1.0578	1.0613
2.9	1.0647	1.0682	1.0716	1.0750	1.0784	1.0818	1.0852	1.0886	1.0919	1.0953
3.0	1.0986	1.1019	1.1053	1.1086	1.1119	1.1151	1.1184	1.1217	1.1249	1.1282
3.1	1.1314	1.1346	1.1378	1.1410	1.1442	1.1474	1.1506	1.1537	1.1569	1.1600
3.2	1.1632	1.1663	1.1694	1.1725	1.1756	1.1787	1.1817	1.1848	1.1878	1.1909
3.3	1.1939	1.1970	1.2000	1.2030	1.2060	1.2090	1.2119	1.2149	1.2179	1.2208
3.4	1.2238	1.2267	1.2296	1.2326	1.2355	1.2384	1.2413	1.2442	1.2470	1.2499
3.5	1.2528	1.2556	1.2585	1.2613	1.2641	1.2669	1.2698	1.2726	1.2754	1.2782
3.6	1.2809	1.2837	1.2865	1.2892	1.2920	1.2947	1.2975	1.3002	1.3029	1.3056
3.7	1.3083	1.3110	1.3137	1.3164	1.3191	1.3218	1.3244	1.3271	1.3297	1.3324
3.8	1.3350	1.3376	1.3403	1.3429	1.3455	1.3481	1.3507	1.3533	1.3558	1.3584
3.9	1.3610	1.3635	1.3661	1.3686	1.3712	1.3737	1.3762	1.3788	1.3813	1.3838
4.0	1.3863	1.3888	1.3913	1.3938	1.3962	1.3987	1.4012	1.4036	1.4061	1.4085
4.1	1.4110	1.4134	1.4159	1.4183	1.4207	1.4231	1.4255	1.4279	1.4303	1.4327
4.2	1.4351	1.4375	1.4398	1.4422	1.4446	1.4469	1.4493	1.4516	1.4540	1.4563
4.3	1.4586	1.4609	1.4633	1.4656	1.4679	1.4702	1.4725	1.4748	1.4770	1.4793
4.4	1.4816	1.4839	1.4861	1.4884	1.4907	1.4929	1.4952	1.4974	1.4996	1.5019
4.5	1.5041	1.5063	1.5085	1.5107	1.5129	1.5151	1.5173	1.5195	1.5217	1.5239
4.6	1.5261	1.5282	1.5304	1.5326	1.5347	1.5369	1.5390	1.5412	1.5433	1.5454
4.7	1.5476	1.5497	1.5518	1.5539	1.5560	1.5581	1.5602	1.5623	1.5644	1.5665
4.8	1.5686	1.5707	1.5728	1.5748	1.5769	1.5790	1.5810	1.5831	1.5851	1.5872
4.9	1.5892	1.5913	1.5933	1.5953	1.5974	1.5994	1.6014	1.6034	1.6054	1.6074
5.0	1.6094	1.6114	1.6134	1.6154	1.6174	1.6194	1.6214	1.6233	1.6253	1.6273
5.1	1.6292	1.6312	1.6332	1.6351	1.6371	1.6390	1.6409	1.6429	1.6448	1.6467
5.2	1.6487	1.6506	1.6525	1.6544	1.6563	1.6582	1.6601	1.6620	1.6639	1.6658
5.3	1.6677	1.6696	1.6715	1.6734	1.6752	1.6771	1.6790	1.6808	1.6827	1.6845
5.4	1.6864	1.6882	1.6901	1.6919	1.6938	1.6956	1.6974	1.6993	1.7011	1.7029
5.5	1.7047	1.7066	1.7084	1.7102	1.7120	1.7138	1.7156	1.7174	1.7192	1.7210
5.6	1.7228	1.7246	1.7263	1.7281	1.7299	1.7317	1.7334	1.7352	1.7370	1.7387
5.7	1.7405	1.7422	1.7440	1.7457	1.7475	1.7492	1.7509	1.7527	1.7544	1.7561
5.8	1.7579	1.7596	1.7613	1.7630	1.7647	1.7664	1.7682	1.7699	1.7716	1.7733
5.9	1.7750	1.7766	1.7783	1.7800	1.7817	1.7834	1.7851	1.7867	1.7884	1.7901
6.0	1.7918	1.7934	1.7951	1.7967	1.7984	1.8001	1.8017	1.8034	1.8050	1.8066
6.1	1.8083	1.8099	1.8116	1.8132	1.8148	1.8165	1.8181	1.8197	1.8213	1.8229
6.2	1.8245	1.8262	1.8278	1.8294	1.8310	1.8326	1.8342	1.8358	1.8374	1.8390
6.3	1.8406	1.8421	1.8437	1.8453	1.8469	1.8485	1.8500	1.8516	1.8532	1.8547
6.4	1.8563	1.8579	1.8594	1.8610	1.8625	1.8641	1.8656	1.8672	1.8687	1.8703

Table 2   Natural Logarithms, ln t (Continued)

t	0.00	0.01	0.02	0.03	0.04	0.05	0.06	0.07	0.08	0.09
6.5	1.8718	1.8733	1.8749	1.8764	1.8779	1.8795	1.8810	1.8825	1.8840	1.8856
6.6	1.8871	1.8886	1.8901	1.8916	1.8931	1.8946	1.8961	1.8976	1.8991	1.9006
6.7	1.9021	1.9036	1.9051	1.9066	1.9081	1.9095	1.9110	1.9125	1.9140	1.9155
6.8	1.9169	1.9184	1.9199	1.9213	1.9228	1.9242	1.9257	1.9272	1.9286	1.9301
6.9	1.9315	1.9330	1.9344	1.9359	1.9373	1.9387	1.9402	1.9416	1.9430	1.9445
7.0	1.9459	1.9473	1.9488	1.9502	1.9516	1.9530	1.9544	1.9559	1.9573	1.9587
7.1	1.9601	1.9615	1.9629	1.9643	1.9657	1.9671	1.9685	1.9699	1.9713	1.9727
7.2	1.9741	1.9755	1.9769	1.9782	1.9796	1.9810	1.9824	1.9838	1.9851	1.9865
7.3	1.9879	1.9892	1.9906	1.9920	1.9933	1.9947	1.9961	1.9974	1.9988	2.0001
7.4	2.0015	2.0028	2.0042	2.0055	2.0069	2.0082	2.0096	2.0109	2.0122	2.0136
7.5	2.0149	2.0162	2.0176	2.0189	2.0202	2.0215	2.0229	2.0242	2.0255	2.0268
7.6	2.0282	2.0295	2.0308	2.0321	2.0334	2.0347	2.0360	2.0373	2.0386	2.0399
7.7	2.0412	2.0425	2.0438	2.0451	2.0464	2.0477	2.0490	2.0503	2.0516	2.0528
7.8	2.0541	2.0554	2.0567	2.0580	2.0592	2.0605	2.0618	2.0631	2.0643	2.0665
7.9	2.0669	2.0681	2.0694	2.0707	2.0719	2.0732	2.0744	2.0757	2.0769	2.0782
8.0	2.0794	2.0807	2.0819	2.0832	2.0844	2.0857	2.0869	2.0882	2.0894	2.0906
8.1	2.0919	2.0931	2.0943	2.0956	2.0968	2.0980	2.0992	2.1005	2.1017	2.1029
8.2	2.1041	2.1054	2.1066	2.1078	2.1090	2.1102	2.1114	2.1126	2.1138	2.1150
8.3	2.1163	2.1175	2.1187	2.1199	2.1211	2.1223	2.1235	2.1247	2.1258	2.1270
8.4	2.1282	2.1294	2.1306	2.1318	2.1330	2.1342	2.1353	2.1365	2.1377	2.1389
8.5	2.1401	2.1412	2.1424	2.1436	2.1448	2.1459	2.1471	2.1483	2.1494	2.1506
8.6	2.1518	2.1529	2.1541	2.1552	2.1564	2.1576	2.1587	2.1599	2.1610	2.1622
8.7	2.1633	2.1645	2.1656	2.1668	2.1679	2.1691	2.1702	2.1713	2.1725	2.1736
8.8	2.1748	2.1759	2.1770	2.1782	2.1793	2.1804	2.1815	2.1827	2.1838	2.1849
8.9	2.1861	2.1872	2.1883	2.1894	2.1905	2.1917	2.1928	2.1939	2.1950	2.1961
9.0	2.1972	2.1983	2.1994	2.2006	2.2017	2.2028	2.2039	2.2050	2.2061	2.2072
9.1	2.2083	2.2094	2.2105	2.2116	2.2127	2.2138	2.2148	2.2159	2.2170	2.2181
9.2	2.2192	2.2203	2.2214	2.2225	2.2235	2.2246	2.2257	2.2268	2.2279	2.2289
9.3	2.2300	2.2311	2.2322	2.2332	2.2343	2.2354	2.2364	2.2375	2.2386	2.2396
9.4	2.2407	2.2418	2.2428	2.2439	2.2450	2.2460	2.2471	2.2481	2.2492	2.2502
9.5	2.2513	2.2523	2.2534	2.2544	2.2555	2.2565	2.2576	2.2586	2.2597	2.2607
9.6	2.2618	2.2628	2.2638	2.2649	2.2659	2.2670	2.2680	2.2690	2.2701	2.2711
9.7	2.2721	2.2732	2.2742	2.2752	2.2762	2.2773	2.2783	2.2793	2.2803	2.2814
9.8	2.2824	2.2834	2.2844	2.2854	2.2865	2.2875	2.2885	2.2895	2.2905	2.2915
9.9	2.2925	2.2935	2.2946	2.2956	2.2966	2.2976	2.2986	2.2996	2.3006	2.3016

Table 3   Exponential Functions

x	e^x	e^{-x}	x	e^x	e^{-x}
0.00	1.0000	1.0000	3.0	20.086	0.0498
0.05	1.0513	0.9512	3.1	22.198	0.0450
0.10	1.1052	0.9048	3.2	24.533	0.0408
0.15	1.1618	0.8607	3.3	27.113	0.0369
0.20	1.2214	0.8187	3.4	29.964	0.0334
0.25	1.2840	0.7788	3.5	33.115	0.0302
0.30	1.3499	0.7408	3.6	36.598	0.0273
0.35	1.4191	0.7047	3.7	40.447	0.0247
0.40	1.4918	0.6703	3.8	44.701	0.0224
0.45	1.5683	0.6376	3.9	49.402	0.0202
0.50	1.6487	0.6065	4.0	54.598	0.0183
0.55	1.7333	0.5769	4.1	60.340	0.0166
0.60	1.8221	0.5488	4.2	66.686	0.0150
0.65	1.9155	0.5220	4.3	73.700	0.0136
0.70	2.0138	0.4966	4.4	81.451	0.0123
0.75	2.1170	0.4724	4.5	90.017	0.0111
0.80	2.2255	0.4493	4.6	99.484	0.0101
0.85	2.3396	0.4274	4.7	109.95	0.0091
0.90	2.4596	0.4066	4.8	121.51	0.0082
0.95	2.5857	0.3867	4.9	134.29	0.0074
1.0	2.7183	0.3679	5.0	148.41	0.0067
1.1	3.0042	0.3329	5.1	164.02	0.0061
1.2	3.3201	0.3012	5.2	181.27	0.0055
1.3	3.6693	0.2725	5.3	200.34	0.0050
1.4	4.0552	0.2466	5.4	221.41	0.0045
1.5	4.4817	0.2231	5.5	244.69	0.0041
1.6	4.9530	0.2019	5.6	270.43	0.0037
1.7	5.4739	0.1827	5.7	298.87	0.0033
1.8	6.0496	0.1653	5.8	330.30	0.0030
1.9	6.6859	0.1496	5.9	365.04	0.0027
2.0	7.3891	0.1353	6.0	403.43	0.0025
2.1	8.1662	0.1225	6.5	665.14	0.0015
2.2	9.0250	0.1108	7.0	1,096.6	0.0009
2.3	9.9742	0.1003	7.5	1,808.0	0.0006
2.4	11.023	0.0907	8.0	2,981.0	0.0003
2.5	12.182	0.0821	8.5	4,914.8	0.0002
2.6	13.464	0.0743	9.0	8,103.1	0.0001
2.7	14.880	0.0672	9.5	13,360	0.00007
2.8	16.445	0.0608	10.0	22,026	0.00004
2.9	18.174	0.0550			

Table 4   Hyperbolic Functions

x	sinh x	cosh x	tanh x	x	sinh x	cosh x	tanh x
0.0	0.00000	1.0000	0.00000	3.0	10.018	10.068	0.99505
0.1	0.10017	1.0050	0.09967	3.1	11.076	11.122	0.99595
0.2	0.20134	1.0201	0.19738	3.2	12.246	12.287	0.99668
0.3	0.30452	1.0453	0.29131	3.3	13.538	13.575	0.99728
0.4	0.41075	1.0811	0.37995	3.4	14.965	14.999	0.99777
0.5	0.52110	1.1276	0.46212	3.5	16.543	16.573	0.99818
0.6	0.63665	1.1855	0.53705	3.6	18.285	18.313	0.99851
0.7	0.75858	1.2552	0.60437	3.7	20.211	20.236	0.99878
0.8	0.88811	1.3374	0.66404	3.8	22.339	22.362	0.99900
0.9	1.0265	1.4331	0.71630	3.9	24.691	24.711	0.99918
1.0	1.1752	1.5431	0.76159	4.0	27.290	27.308	0.99933
1.1	1.3356	1.6685	0.80050	4.1	30.162	30.178	0.99945
1.2	1.5095	1.8107	0.83365	4.2	33.336	33.351	0.99955
1.3	1.6984	1.9709	0.86172	4.3	36.843	36.857	0.99963
1.4	1.9043	2.1509	0.88535	4.4	40.719	40.732	0.99970
1.5	2.1293	2.3524	0.90515	4.5	45.003	45.014	0.99975
1.6	2.3756	2.5775	0.92167	4.6	49.737	49.747	0.99980
1.7	2.6456	2.8283	0.93541	4.7	54.969	54.978	0.99983
1.8	2.9422	3.1075	0.94681	4.8	60.751	60.759	0.99986
1.9	3.2682	3.4177	0.95624	4.9	67.141	67.149	0.99989
2.0	3.6269	3.7622	0.96403	5.0	74.203	74.210	0.99991
2.1	4.0219	4.1443	0.97045	5.1	82.008	82.014	0.99993
2.2	4.4571	4.5679	0.97574	5.2	90.633	90.639	0.99994
2.3	4.9370	5.0372	0.98010	5.3	100.17	100.17	0.99995
2.4	5.4662	5.5569	0.98367	5.4	110.70	110.71	0.99996
2.5	6.0502	6.1323	0.98661	5.5	122.34	122.35	0.99997
2.6	6.6947	6.7690	0.98903	5.6	135.21	135.22	0.99997
2.7	7.4063	7.4735	0.99101	5.7	149.43	149.44	0.99998
2.8	8.1919	8.2527	0.99263	5.8	165.15	165.15	0.99998
2.9	9.0596	9.1146	0.99396	5.9	182.52	182.52	0.99998

Table 5 *Common Logarithms,* $\log_{10} x$

x	0.00	0.01	0.02	0.03	0.04	0.05	0.06	0.07	0.08	0.09
1.0	.0000	.0043	.0086	.0128	.0170	.0212	.0253	.0294	.0334	.0374
1.1	.0414	.0453	.0492	.0531	.0569	.0607	.0645	.0682	.0719	.0755
1.2	.0792	.0828	.0864	.0899	.0934	.0969	.1004	.1038	.1072	.1106
1.3	.1139	.1173	.1206	.1239	.1271	.1303	.1335	.1367	.1399	.1430
1.4	.1461	.1492	.1523	.1553	.1584	.1614	.1644	.1673	.1703	.1732
1.5	.1761	.1790	.1818	.1847	.1875	.1903	.1931	.1959	.1987	.2014
1.6	.2041	.2068	.2095	.2122	.2148	.2175	.2201	.2227	.2253	.2279
1.7	.2304	.2330	.2355	.2380	.2405	.2430	.2455	.2480	.2504	.2529
1.8	.2553	.2577	.2601	.2625	.2648	.2672	.2695	.2718	.2742	.2765
1.9	.2788	.2810	.2833	.2856	.2878	.2900	.2923	.2945	.2967	.2989
2.0	.3010	.3032	.3054	.3075	.3096	.3118	.3139	.3160	.3181	.3201
2.1	.3222	.3243	.3263	.3284	.3304	.3324	.3345	.3365	.3385	.3404
2.2	.3424	.3444	.3464	.3483	.3502	.3522	.3541	.3560	.3579	.3598
2.3	.3617	.3636	.3655	.3674	.3692	.3711	.3729	.3747	.3766	.3784
2.4	.3802	.3820	.3838	.3856	.3874	.3892	.3909	.3927	.3945	.3962
2.5	.3979	.3997	.4014	.4031	.4048	.4065	.4082	.4099	.4116	.4133
2.6	.4150	.4166	.4183	.4200	.4216	.4232	.4249	.4265	.4281	.4298
2.7	.4314	.4330	.4346	.4362	.4378	.4393	.4409	.4425	.4440	.4456
2.8	.4472	.4487	.4502	.4518	.4533	.4548	.4564	.4579	.4594	.4609
2.9	.4624	.4639	.4654	.4669	.4683	.4698	.4713	.4728	.4742	.4757
3.0	.4771	.4786	.4800	.4814	.4829	.4843	.4857	.4871	.4886	.4900
3.1	.4914	.4928	.4942	.4955	.4969	.4983	.4997	.5011	.5024	.5038
3.2	.5051	.5065	.5079	.5092	.5105	.5119	.5132	.5145	.5159	.5172
3.3	.5185	.5198	.5211	.5224	.5237	.5250	.5263	.5276	.5289	.5302
3.4	.5315	.5328	.5340	.5353	.5366	.5378	.5391	.5403	.5416	.5428
3.5	.5441	.5453	.5465	.5478	.5490	.5502	.5514	.5527	.5539	.5551
3.6	.5563	.5575	.5587	.5599	.5611	.5623	.5635	.5647	.5658	.5670
3.7	.5682	.5694	.5705	.5717	.5729	.5740	.5752	.5763	.5775	.5786
3.8	.5798	.5809	.5821	.5832	.5843	.5855	.5866	.5877	.5888	.5899
3.9	.5911	.5922	.5933	.5944	.5955	.5966	.5977	.5988	.5999	.6010
4.0	.6021	.6031	.6042	.6053	.6064	.6075	.6085	.6096	.6107	.6117
4.1	.6128	.6138	.6149	.6160	.6170	.6180	.6191	.6201	.6212	.6222
4.2	.6232	.6243	.6253	.6263	.6274	.6284	.6294	.6304	.6314	.6325
4.3	.6335	.6345	.6355	.6365	.6375	.6385	.6395	.6405	.6415	.6425
4.4	.6435	.6444	.6454	.6464	.6474	.6484	.6493	.6503	.6513	.6522
4.5	.6532	.6542	.6551	.6561	.6571	.6580	.6590	.6599	.6609	.6618
4.6	.6628	.6637	.6646	.6656	.6665	.6675	.6684	.6693	.6702	.6712
4.7	.6721	.6730	.6739	.6749	.6758	.6767	.6776	.6785	.6794	.6803
4.8	.6812	.6821	.6830	.6839	.6848	.6857	.6866	.6875	.6884	.6893
4.9	.6902	.6911	.6920	.6928	.6937	.6946	.6955	.6964	.6972	.6981
5.0	.6990	.6998	.7007	.7016	.7024	.7033	.7042	.7050	.7059	.7067
5.1	.7076	.7084	.7093	.7101	.7110	.7118	.7126	.7135	.7143	.7152
5.2	.7160	.7168	.7177	.7185	.7193	.7202	.7210	.7218	.7226	.7235
5.3	.7243	.7251	.7259	.7267	.7275	.7284	.7292	.7300	.7308	.7316
5.4	.7324	.7332	.7340	.7348	.7356	.7364	.7372	.7380	.7388	.7396
5.5	.7404	.7412	.7419	.7427	.7435	.7443	.7451	.7459	.7466	.7474
5.6	.7482	.7490	.7497	.7505	.7513	.7520	.7528	.7536	.7543	.7551
5.7	.7559	.7566	.7574	.7582	.7589	.7597	.7604	.7612	.7619	.7627
5.8	.7634	.7642	.7649	.7657	.7664	.7672	.7679	.7686	.7694	.7701
5.9	.7709	.7716	.7723	.7731	.7738	.7745	.7752	.7760	.7767	.7774
6.0	.7782	.7789	.7796	.7803	.7810	.7818	.7825	.7832	.7839	.7846
6.1	.7853	.7860	.7868	.7875	.7882	.7889	.7896	.7903	.7910	.7917
6.2	.7924	.7931	.7938	.7945	.7952	.7959	.7966	.7973	.7980	.7987
6.3	.7993	.8000	.8007	.8014	.8021	.8028	.8035	.8041	.8048	.8055
6.4	.8062	.8069	.8075	.8082	.8089	.8096	.8102	.8109	.8116	.8122

Table 5   Common Logarithms, $log_{10} x$ (Continued)

x	0.00	0.01	0.02	0.03	0.04	0.05	0.06	0.07	0.08	0.09
6.5	.8129	.8136	.8142	.8149	.8156	.8162	.8169	.8176	.8182	.8189
6.6	.8195	.8202	.8209	.8215	.8222	.8228	.8235	.8241	.8248	.8254
6.7	.8261	.8267	.8274	.8280	.8287	.8293	.8299	.8306	.8312	.8319
6.8	.8325	.8331	.8338	.8344	.8351	.8357	.8363	.8370	.8376	.8382
6.9	.8388	.8395	.8401	.8407	.8414	.8420	.8426	.8432	.8439	.8445
7.0	.8451	.8457	.8463	.8470	.8476	.8482	.8488	.8494	.8500	.8506
7.1	.8513	.8519	.8525	.8531	.8537	.8543	.8549	.8555	.8561	.8567
7.2	.8573	.8579	.8585	.8591	.8597	.8603	.8609	.8615	.8621	.8627
7.3	.8633	.8639	.8645	.8651	.8657	.8663	.8669	.8675	.8681	.8686
7.4	.8692	.8698	.8704	.8710	.8716	.8722	.8727	.8733	.8739	.8745
7.5	.8751	.8756	.8762	.8768	.8774	.8779	.8785	.8791	.8797	.8802
7.6	.8808	.8814	.8820	.8825	.8831	.8837	.8842	.8848	.8854	.8859
7.7	.8865	.8871	.8876	.8882	.8887	.8893	.8899	.8904	.8910	.8915
7.8	.8921	.8927	.8932	.8938	.8943	.8949	.8954	.8960	.8965	.8971
7.9	.8976	.8982	.8987	.8993	.8998	.9004	.9009	.9015	.9020	.9025
8.0	.9031	.9036	.9042	.9047	.9053	.9058	.9063	.9069	.9074	.9079
8.1	.9085	.9090	.9096	.9101	.9106	.9112	.9117	.9122	.9128	.9133
8.2	.9138	.9143	.9149	.9154	.9159	.9165	.9170	.9175	.9180	.9186
8.3	.9191	.9196	.9201	.9206	.9212	.9217	.9222	.9227	.9232	.9238
8.4	.9243	.9248	.9253	.9258	.9263	.9269	.9274	.9279	.9284	.9289
8.5	.9294	.9299	.9304	.9309	.9315	.9320	.9325	.9330	.9335	.9340
8.6	.9345	.9350	.9355	.9360	.9365	.9370	.9375	.9380	.9385	.9390
8.7	.9395	.9400	.9405	.9410	.9415	.9420	.9425	.9430	.9435	.9440
8.8	.9445	.9450	.9455	.9460	.9465	.9469	.9474	.9479	.9484	.9489
8.9	.9494	.9499	.9504	.9509	.9513	.9518	.9523	.9528	.9533	.9538
9.0	.9542	.9547	.9552	.9557	.9562	.9566	.9571	.9567	.9581	.9586
9.1	.9590	.9595	.9600	.9605	.9609	.9614	.9619	.9624	.9628	.9633
9.2	.9638	.9643	.9647	.9652	.9657	.9661	.9666	.9671	.9675	.9680
9.3	.9685	.9689	.9694	.9699	.9703	.9708	.9713	.9717	.9722	.9727
9.4	.9731	.9736	.9741	.9745	.9750	.9754	.9759	.9763	.9768	.9773
9.5	.9777	.9782	.9786	.9791	.9795	.9800	.9805	.9809	.9814	.9818
9.6	.9823	.9827	.9832	.9836	.9841	.9845	.9850	.9854	.9859	.9863
9.7	.9868	.9872	.9877	.9881	.9886	.9890	.9894	.9899	.9903	.9908
9.8	.9912	.9917	.9921	.9926	.9930	.9934	.9939	.9943	.9948	.9952
9.9	.9956	.9961	.9965	.9969	.9974	.9978	.9983	.9987	.9991	.9996

Table 6 Powers and Roots

Number	Square	Square root	Cube	Cube root	Number	Square	Square root	Cube	Cube root
1	1	1.000	1	1.000	51	2,601	7.141	132,651	3.708
2	4	1.414	8	1.260	52	2,704	7.211	140,608	3.733
3	9	1.732	27	1.442	53	2,809	7.280	148,877	3.756
4	16	2.000	64	1.587	54	2,916	7.348	157,464	3.780
5	25	2.236	125	1.710	55	3,025	7.416	166,375	3.803
6	36	2.449	216	1.817	56	3,136	7.483	175,616	3.826
7	49	2.646	343	1.913	57	3,249	7.550	185,193	3.849
8	64	2.828	512	2.000	58	3,364	7.616	195,112	3.871
9	81	3.000	729	2.080	59	3,481	7.681	205,379	3.893
10	100	3.162	1,000	2.154	60	3,600	7.746	216,000	3.915
11	121	3.317	1,331	2.224	61	3,721	7.810	226,981	3.936
12	144	3.464	1,728	2.289	62	3,844	7.874	238,328	3.958
13	169	3.606	2,197	2.351	63	3,969	7.937	250,047	3.979
14	196	3.742	2,744	2.410	64	4,096	8.000	262,144	4.000
15	225	3.873	3,375	2.466	65	4,225	8.062	274,625	4.021
16	256	4.000	4,096	2.520	66	4,356	8.124	287,496	4.041
17	289	4.123	4,913	2.571	67	4,489	8.185	300,763	4.062
18	324	4.243	5,832	2.621	68	4,624	8.246	314,432	4.082
19	361	4.359	6,859	2.668	69	4,761	8.307	328,509	4.102
20	400	4.472	8,000	2.714	70	4,900	8.367	343,000	4.121
21	441	4.583	9,261	2.759	71	5,041	8.426	357,911	4.141
22	484	4.690	10,648	2.802	72	5,184	8.485	373,248	4.160
23	529	4.796	12,167	2.844	73	5,329	8.544	389,017	4.179
24	576	4.899	13,824	2.884	74	5,476	8.602	405,224	4.198
25	625	5.000	15,625	2.924	75	5,625	8.660	421,875	4.217
26	676	5.099	17,576	2.962	76	5,776	8.718	438,976	4.236
27	729	5.196	19,683	3.000	77	5,929	8.775	456,533	4.254
28	784	5.292	21,952	3.037	78	6,084	8.832	474,552	4.273
29	841	5.385	24,389	3.072	79	6,241	8.888	493,039	4.291
30	900	5.477	27,000	3.107	80	6,400	8.944	512,000	4.309
31	961	5.568	29,791	3.141	81	6,561	9.000	531,441	4.327
32	1,024	5.657	32,768	3.175	82	6.724	9.055	551,368	4.344
33	1,089	5.745	35,937	3.208	83	6,889	9.110	571,787	4.362
34	1,156	5.831	39,304	3.240	84	7,056	9.165	592,704	4.380
35	1,225	5.916	42,875	3.271	85	7,225	9.220	614,125	4.397
36	1,296	6.000	46,656	3.302	86	7,396	9.274	636,056	4.414
37	1,369	6.083	50,653	3.332	87	7,569	9.327	658,503	4.431
38	1,444	6.164	54,872	3.362	88	7,744	9.381	681,472	4.448
39	1,521	6.245	59,319	3.391	89	7,921	9.434	704,969	4.465
40	1,600	6.325	64,000	3.420	90	8,100	9.487	729,000	4.481
41	1,681	6.403	68,921	3.448	91	8,281	9.539	753,571	4.498
42	1,764	6.481	74,088	3.476	92	8,464	9.592	778,688	4.514
43	1,849	6.557	79,507	3.503	93	8,649	9.644	804,357	4.531
44	1,936	6.633	85,184	3.530	94	8,836	9.695	830,584	4.547
45	2,025	6.708	91,125	3.557	95	9,025	9.747	857,375	4.563
46	2,116	6.782	97,336	3.583	96	9,216	9.798	884,736	4.579
47	2,209	6.856	103,823	3.609	97	9,409	9.849	912,673	4.595
48	2,304	6.928	110,592	3.634	98	9,604	9.899	941,192	4.610
49	2,401	7.000	117,649	3.659	99	9,801	9.950	970,299	4.626
50	2,500	7.071	125,000	3.684	100	10,000	10.000	1,000,000	4.642

ANSWERS TO SELECTED PROBLEMS

Chapter 1

Problem Set 1.1, page 8

1 (a) true; (b) true; (c) false, let $x = 1$, $y = z$; (d) true; (e) true
5 (a) when $x < 0$; (b) when $x > 0$; (c) when $x = 0$
7 No, for instance, let $x = -4$.

9

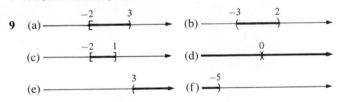

11 $(-\infty, 3)$ **13** $(-2, 1]$ **15** $(-\frac{7}{4}, 1]$
17 $(-\infty, -3)$ together with $(3, \infty)$ **19** $(-1, 2)$
21 $[-2, 2]$ **23** $(-4, \frac{3}{2})$ **25** $1, 5$ **27** $-\frac{7}{3}, 1$
29 $-\frac{3}{4}, \frac{1}{2}$ **31** $(2, 3)$ **33** $(-\infty, 1)$ together with $(\frac{7}{3}, \infty)$
35 $(1.9, 2.1)$ **37** $(-\infty, -3.5)$ together with $(-2.5, \infty)$
39 $(-2.2, -1.8)$ **41** between $\frac{10}{3}$ and $\frac{35}{6}$ m, inclusive
43 between 13 and 20 min, inclusive **45** 6
47 $600 \le R \le 3600$

Problem Set 1.2, page 14

1 (a) Q_I; (b) Q_{II}; (c) Q_I; (d) Q_{IV}; (e) Q_{III}; (f) y axis; (g) x axis; (h) y axis
3 (a) $(3, -2)$; (b) $(-3, 2)$; (c) $(-3, -2)$
5 (a) $(-1, -3)$; (b) $(1, 3)$; (c) $(1, -3)$ **7** 10 **9** 4
11 $\sqrt{145}$ **13** 10 **15** 5 **17** $\sqrt{9 + 4t^2}$ **19** 6.224
21 (b) 12 **23** (b) 6 **27** yes **29** $3, -2$

31 yes **33** no
35 (a) $x^2 + (y - 2)^2 = 9$; (b) $(x + 1)^2 + (y - 4)^2 = 4$;
(c) $(x - 3)^2 + (y - 4)^2 = 25$
37 $(h, k) = (-1, 2)$, $r = 3$ **39** $(h, k) = (-1, -2)$, $r = 1$
41 $(h, k) = (-1, \frac{1}{2})$, $r = 1$ **43** $(h, k) = (-\frac{1}{2}, \frac{1}{2})$, $r = \frac{1}{2}$
45 $(x - 2)^2 + (y - 10)^2 = 106$
47 $(x + 4)^2 + y^2 = 17$ and $(x - 4)^2 + y^2 = 17$
49 Yes, the distance between the centers is less than the sum of the radii.

Problem Set 1.3, page 21

1 $-\frac{5}{3}$ **3** $\frac{1}{2}$ **5** $\frac{5}{11}$ **7** $y - 4 = 2(x - 5)$
9 $y - 2 = \frac{1}{4}(x - 3)$ **11** $y + 2 = -3(x - 7)$
13 $y - \frac{2}{3} = 0$ **15** $y - 2 = 6(x - 3)$
17 $y - 4 = 0$ **19** (a) $y = 3$; (b) $x = -2$
21 $y = \frac{3}{2}x - 3$, $m = \frac{3}{2}$, $b = -3$
23 $y = 0(x) - 1$, $m = 0$, $b = -1$
25 (a) $y - 5 = -3(x - 0)$; (b) $y = -3x + 5$; (c) $3x + y - 5 = 0$
27 (a) $y - 5 = -\frac{5}{3}(x - 0)$; (b) $y = -\frac{5}{3}x + 5$;
(c) $5x + 3y - 15 = 0$
29 (a) $y + 4 = \frac{2}{5}(x - 4)$; (b) $y = \frac{2}{5}x - \frac{28}{5}$; (c) $2x - 5y - 28 = 0$
31 (a) $y - \frac{2}{3} = \frac{3}{5}(x + 3)$; (b) $y = \frac{3}{5}x + \frac{37}{15}$; (c) $9x - 15y + 37 = 0$
33 $B = -\frac{5}{4}$ **35** parallel **37** perpendicular
39 slope $\overline{AB} = \frac{1}{6} =$ slope $\overline{DC}$, slope $\overline{AD} = \frac{5}{3} =$ slope $\overline{BC}$
41 (a) $d = 1$; (b) $k = -\frac{10}{3}$
43 (a) $(\frac{15}{2}, 2)$; (b) $(2, 5)$; (c) $(2, 2)$; (d) $(3, \frac{5}{2})$
45 $y = 22N + 0.2x$ **47** \$200,000
49 $y = -0.75x + 7$, about 1989

Problem Set 1.4, page 29

1 -5 **3** 2 **5** 3 **7** -4 **9** $|5c + 8|$
11 $\dfrac{a - 6}{3a + 21}$ **13** 4.105524

15 (a) $\mathbb{R}$; (b) all real numbers except -2; (c) $[0, \infty)$; (d) $(-\infty, \frac{5}{3}]$; (e) all real numbers except $\frac{5}{6}$; (f) $(-\infty, \frac{4}{5})$

17 (a) and (c)

19 domain: $\mathbb{R}$; range: $\mathbb{R}$

21 domain: $\mathbb{R}$; range: $[0, \infty)$

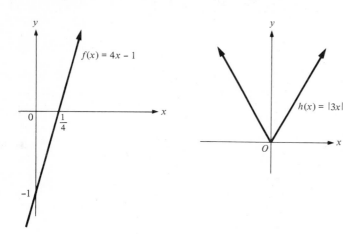

$f(x) = 4x - 1$

$h(x) = |3x|$

23 domain: all real numbers except $\frac{2}{3}$; range: all real numbers except 4

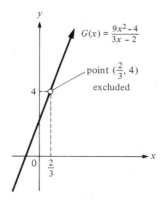

$G(x) = \dfrac{9x^2 - 4}{3x - 2}$

point $(\frac{2}{3}, 4)$ excluded

25 domain: $[1, \infty)$; range: $[0, \infty)$ **27** domain: $\mathbb{R}$; range: $(-\infty, 8]$

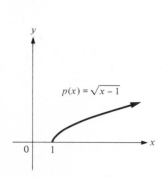

$p(x) = \sqrt{x - 1}$

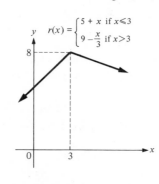

$r(x) = \begin{cases} 5 + x & \text{if } x \leqslant 3 \\ 9 - \dfrac{x}{3} & \text{if } x > 3 \end{cases}$

29 domain: $\mathbb{R}$; range: $[0, \infty)$

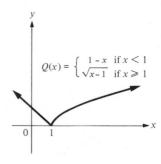

$Q(x) = \begin{cases} 1 - x & \text{if } x < 1 \\ \sqrt{x - 1} & \text{if } x \geqslant 1 \end{cases}$

31 domain: all real numbers except -2; range: all real numbers except 0

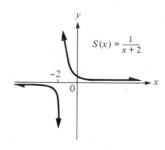

$S(x) = \dfrac{1}{x + 2}$

33 (a) 4; (b) 0; (c) $2x + h$

35 $f(0) = 32$ or $0 \mapsto 32$, $f(15) = 59$ or $15 \mapsto 59$, $f(-10) = 14$ or $-10 \mapsto 14$, $f(55) = 131$ or $55 \mapsto 131$

37 $p(50) = 29$, $p(100) = 26$, $p(200) = 20$, $p(400) = 8$, $p(500) = 2$

39 $T = 65 - \dfrac{h}{250}$, $T = -55°$ when $h = 30{,}000$ ft

41 $P = (9.9 \times 10^4)h + (1.013 \times 10^5)$, $P \approx 10^7$ N when $h = 100$ m

43 $s = 30\sqrt{18 - 6t + t^2}$

45 The graph actually falls *below* the x axis between 0 and approximately 0.312107046.

Problem Set 1.5, page 39

1 (a) domain: $\mathbb{R}$, range: $(-\infty, 2]$, even; (b) domain: $[-5, 5]$, range: $[-3, 3]$, neither even nor odd; (c) domain: $\left[-\dfrac{3\pi}{2}, \dfrac{3\pi}{2}\right]$, range: $[-1, 1]$, odd; (d) domain: $\mathbb{R}$, range: $[-2, 1]$, neither even nor odd; (e) domain: $\mathbb{R}$, range: $\mathbb{R}$, neither even nor odd; (f) domain: $\mathbb{R}$, range: $(-\infty, 2]$, even; (g) domain: $\mathbb{R}$; range: $\mathbb{R}$; neither even nor odd; (h) domain: all real numbers except 0; range: all real numbers except 0; odd

3 even   **5** neither   **7** odd

9 neither   **11** neither

13 polynomial of degree 2; coefficients: 6, -3, -8

15 polynomial of degree 3; coefficients: -1, 1, -5, 6

17 polynomial of degree 4; coefficients: $\sqrt{2}$, -5^{-1}, 0, 0, 20

19 polynomial (degree undefined); coefficient: 0

23 $f(x) = \frac{29}{5} - \frac{2}{5}x$ **25** $f(x) = ax$, where a is any constant

27 If $f(x) = ax + b$ and $a \neq 0$, then $-\dfrac{b}{a}$ is a zero of f.

31 $(x + 1)^2 - 5$ **33** $3\left(x - \frac{5}{3}\right)^2 - \frac{31}{3}$

35 $-2\left(x - \frac{3}{2}\right)^2 + \frac{15}{2}$

37

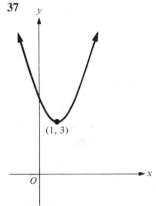

39

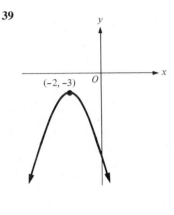

41 (a) $f(x) = -3(x + 2)^2 + 11$; (b) graph is a parabola, opening downward, with vertex $(-2, 11)$

43 (a) $(f + g)(x) = x^2 + 2x - 4$, $(f - g)(x) = -x^2 + 2x - 6$,

$(f \cdot g)(x) = 2x^3 - 5x^2 + 2x - 5$, $\left(\dfrac{f}{g}\right)(x) = \dfrac{2x - 5}{x^2 + 1}$;

(b) $(f + g)(x) = \sqrt{x} + x^2 + 4$, $(f - g)(x) = \sqrt{x} - x^2 - 4$,

$(f \cdot g)(x) = x^2\sqrt{x} + 4\sqrt{x}$, $\left(\dfrac{f}{g}\right)(x) = \dfrac{\sqrt{x}}{x^2 + 4}$;

(c) $(f + g)(x) = 12 - x$, $(f - g)(x) = 7x - 2$,

$(f \cdot g)(x) = -12x^2 + x + 35$, $\left(\dfrac{f}{g}\right)(x) = \dfrac{3x + 5}{7 - 4x}$;

(d) $(f + g)(x) = \sqrt{x + 3} + \dfrac{1}{x}$, $(f - g)(x) = \sqrt{x + 3} - \dfrac{1}{x}$,

$(f \cdot g)(x) = \dfrac{\sqrt{x + 3}}{x}$, $\left(\dfrac{f}{g}\right)(x) = x\sqrt{x + 3}$;

(e) $(f + g)(x) = |x| + |x - 2|$, $(f - g)(x) = |x| - |x - 2|$,

$(f \cdot g)(x) = |x^2 - 2x|$, $\left(\dfrac{f}{g}\right)(x) = \left|\dfrac{x}{x - 2}\right|$;

(f) $(f + g)(x) = (a + c)x + b + d$,

$(f - g)(x) = (a - c)x + b - d$,

$(f \cdot g)(x) = acx^2 + (ad + bc)x + bd$, $\left(\dfrac{f}{g}\right)(x) = \dfrac{ax + b}{cx + d}$

45 (a) rational; (b) not rational; (c) rational; (d) rational; (e) not rational

47 domain: $\mathbb{R}$; range: $[-1, 1]$

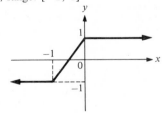

49 domain: $\mathbb{R}$; range: all integers

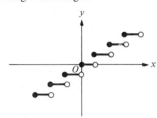

51 domain: $\mathbb{R}$; range: all integers

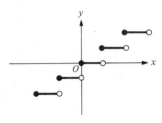

53 domain: $\mathbb{R}$; range: all nonnegative integers

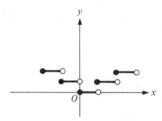

55 (a) sgn $(-2) = -1$, sgn $(-3) = -1$, sgn $(0) = 0$, sgn $(2) = 1$, sgn $(3) = 1$, sgn $(151) = 1$; (b) consider the three cases: $x < 0$, $x = 0$, $x > 0$; (c) consider the three cases: $ab < 0$, $ab = 0$, $ab > 0$; (e) domain: $\mathbb{R}$, range: the three numbers -1, 0, 1; (g) The graph is not one connected piece.

(d)   (f)

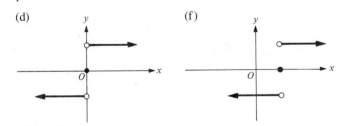

57 (a) $C = F + V$ (b) $C(x) = x^2 + 4x + 500$

Problem Set 1.6, page 50

1 3.3 m   **3** $\dfrac{4}{3}$ rad   **5** $\dfrac{10\pi}{3}$ in

7 (a) $\dfrac{\pi}{6}$; (b) $\dfrac{\pi}{4}$; (c) $\dfrac{\pi}{2}$; (d) $\dfrac{2\pi}{3}$; (e) $-\dfrac{5\pi}{6}$; (f) $\dfrac{26\pi}{9}$; (g) $\dfrac{2\pi}{5}$;

(h) $\dfrac{3\pi}{8}$; (i) $-\dfrac{11\pi}{6}$; (j) $\dfrac{5\pi}{2}$; (k) $\dfrac{7\pi}{60}$; (l) -2π

9 (a) 90°; (b) 60°; (c) 45°; (d) 30°; (e) 120°; (f) −180°;
(g) 108°; (h) −450°; (i) 405°; (j) −67.5°; (k) 1260°;
(l) $-\dfrac{90°}{7}$

11 (a) −135° or $-\dfrac{3\pi}{4}$; (b) 1500° or $\dfrac{25\pi}{3}$; (c) 120° or $\dfrac{2\pi}{3}$

13 (a) $\dfrac{21\pi}{4}$ cm²; (b) $\dfrac{117\pi}{2}$ in²

15 $\sin\dfrac{2\pi}{7} = 0.781831483$, $\csc\dfrac{2\pi}{7} = 1.279048008$,

$\cos\dfrac{2\pi}{7} = 0.623489802$, $\sec\dfrac{2\pi}{7} = 1.603875472$,

$\tan\dfrac{2\pi}{7} = 1.253960338$, $\cot\dfrac{2\pi}{7} = 0.797473389$

17 $\sin\left(-\dfrac{17\pi}{3}\right) = 0.866025404$, $\csc\left(-\dfrac{17\pi}{3}\right) =$

1.154700538, $\cos\left(-\dfrac{17\pi}{3}\right) = 0.500000000$,

$\sec\left(-\dfrac{17\pi}{3}\right) = 2$, $\tan\left(-\dfrac{17\pi}{3}\right) = 1.732050809$,

$\cot\left(-\dfrac{17\pi}{3}\right) = 0.577350269$

19 $\sin 1.7764 = 0.978937918$, $\csc 1.7764 = 1.021515238$,
$\cos 1.7764 = -0.204158156$, $\sec 1.7764 = -4.898163370$,
$\tan 1.7764 = -4.794997853$, $\cot 1.7764 = -0.208550667$

21 $\sin 48° = 0.743144826$, $\csc 48° = 1.345632730$,
$\cos 48° = 0.669130606$, $\sec 48° = 1.494476550$,
$\tan 48° = 1.110612515$, $\cot 48° = 0.900404044$

23 Enter 30 and press SIN. If the display shows 0.5, the
calculator is in degree mode.

25 (a) $\sin^2 t$; (b) $2\cos t$; (c) $\csc^2 t$; (d) $\cot^2 t$; (e) $-\cos t$

27 (a) $\sin 75° = \dfrac{\sqrt{2}(\sqrt{3}+1)}{4}$, $\cos 75° = \dfrac{\sqrt{2}(\sqrt{3}-1)}{4}$
(b) $\tan 75° = 2 + \sqrt{3}$, $\cot 75° = 2 - \sqrt{3}$, $\sec 75° =$
$\sqrt{2}(\sqrt{3}+1)$, $\csc 75° = \sqrt{2}(\sqrt{3}-1)$

29 (a) $\sec^2 t$; (b) $\cot 2t$; (c) $\cos^2 t\,(4\cos^2 t - 3)$
(d) $\cot t$; (e) $\cos s$

31 $\sin\theta = \dfrac{3}{5}$, $\cos\theta = \dfrac{4}{5}$, $\tan\theta = \dfrac{3}{4}$, $\sec\theta = \dfrac{5}{4}$,

$\csc\theta = \dfrac{5}{3}$, $\cot\theta = \dfrac{4}{3}$

33 $\sin\theta = \dfrac{3}{4}$, $\cos\theta = \dfrac{\sqrt{7}}{4}$, $\tan\theta = \dfrac{3\sqrt{7}}{7}$, $\sec\theta = \dfrac{4\sqrt{7}}{7}$,

$\csc\theta = \dfrac{4}{3}$, $\cot\theta = \dfrac{\sqrt{7}}{3}$

35 $\sin\theta = \dfrac{1}{2}$, $\cos\theta = \dfrac{\sqrt{3}}{2}$, $\tan\theta = \dfrac{\sqrt{3}}{3}$, $\sec\theta = \dfrac{2\sqrt{3}}{3}$,

$\csc\theta = 2$, $\cot\theta = \sqrt{3}$

37 $\sin\theta = \dfrac{\sqrt{2}}{2}$, $\cos\theta = \dfrac{\sqrt{2}}{2}$, $\tan\theta = 1$, $\sec\theta = \sqrt{2}$,

$\csc\theta = \sqrt{2}$, $\cot\theta = 1$

39 $\dfrac{2}{\sqrt{4-x^2}}$ **41** $\dfrac{5}{z}$ **45** 187.23 m   **47** 7.597 m²

Problem Set 1.7, page 58

1 12   **3** 8   **5** 6   **7** $\lim\limits_{x\to 2}\dfrac{x^2-5x+6}{x-2} = -1$;
top table: −2, −1.1, −1.01, −1.001, −1.0001;
bottom table: 0, −0.9, −0.99, −0.999, −0.9999

9 $\lim\limits_{x\to -1}\dfrac{x^2-2x-3}{x+1} = -4$; top table: −5, −4.1,
− 4.01, −4.001, −4.0001; bottom table: −3, −3.9, −3.99,
− 3.999, −3.9999

11 $\lim\limits_{t\to 4}\dfrac{\sqrt{t}-2}{t-4} = \dfrac{1}{4} = 0.25$; top table: 0.267949192,
0.251582340, 0.250156400, 0.250016000, 0.25000000;
bottom table: 0.236067977, 0.248456730, 0.249843900,
0.249984000, 0.250000000 (Note: Apparently exact values
of 0.25 are only the result of the calculator's not being able
to resolve the difference between 0.25 and the true value.)

13 $\lim\limits_{h\to 1}\dfrac{1-h}{1-(1/h)} = -1$; top table: undefined, −0.9, −0.99,
− 0.999000001, −0.999900010 (Note calculator errors in
last two numbers.); bottom table: −2, −1.1, −1.01,
−1.000999999, − 1.000100010 (Again, note calculator
errors in last two numbers.)

17 0   **19** 1

21 (a) $\lim\limits_{x\to 2} f(x) = 4 = f(2)$ (b) $\lim\limits_{x\to 2} f(x) = 4$, but $f(2) = 6$;

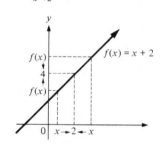

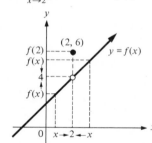

(c) $\lim\limits_{x\to 2} f(x) = 4$, but $f(2)$ is undefined

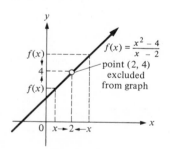

23 $\delta = 0.0025$ **25** $\delta = 0.01$ **27** $\delta = 0.2$

29 Take $\delta = \dfrac{\epsilon}{2}$.   **31** Take $\delta = \dfrac{\epsilon}{4}$.   **33** Take any $\delta > 0$.

Problem Set 1.8, page 68

1 5   **3** π **5** π **7** 14   **9** 1   **11** -5
13 15   **15** -18 **17** 1   **19** 16   **21** 7
23 $\frac{7}{8}$ **25** 6   **27** $\frac{3}{2}$ **29** $-\frac{1}{2}$ **31** $\frac{5}{16}$
33 7   **35** 10   **37** 6   **39** $\frac{1}{4}$ **41** $\frac{1}{2}$
43 $-\frac{1}{16}$ **45** 0   **47** -1 **49** $\sqrt{3}$ **51** 0
53 $\sqrt{3}/2$ **55** 6   **57** $\frac{2}{5}$ **59** 1   **61** 1
63 -1 **65** 2   **67** $2x$ **69** $1/2\sqrt{x}$
77 $20 - (x/500)$

Problem Set 1.9, page 76

1 (b) $\lim_{x \to 3^-} f(x) = 8$, $\lim_{x \to 3^+} f(x) = 6$; (c) does not exist;
(d) discontinuous

3 (b) $\lim_{x \to 1^-} g(x) = 4$, $\lim_{x \to 1^+} g(x) = 2$; (c) does not exist;
(d) discontinuous

5 (b) $\lim_{x \to 5^-} H(x) = 0$, $\lim_{x \to 5^+} H(x) = 0$; (c) $\lim_{x \to 5} H(x) = 0$;
(d) discontinuous

7 (b) $\lim_{x \to 1^-} f(x) = 1$, $\lim_{x \to 1^+} f(x) = 1$; (c) $\lim_{x \to 1} f(x) = 1$;
(d) continuous

9 (b) $\lim_{x \to 3^-} F(x) = 6$, $\lim_{x \to 3^+} F(x) = 6$; (c) $\lim_{x \to 3} F(x) = 6$;
(d) discontinuous

11 (b) $\lim_{x \to 1/2^-} S(x) = 5$, $\lim_{x \to 1/2^+} S(x) = 5$; (c) $\lim_{x \to 1/2} S(x) = 5$;
(d) continuous

13 (b) $\lim_{x \to -1^-} f(x) = -4$, $\lim_{x \to -1^+} f(x) = -4$;
(c) $\lim_{x \to -1} f(x) = -4$; (d) discontinuous

15 continuous everywhere   **17** continuous everywhere
19 continuous everywhere except at $x = 0$
21 continuous everywhere except at $x = 1$
23 continuous everywhere except at $x = 1$ and at $x = -1$
25 continuous everywhere except at $x = 2$
27 continuous everywhere except at integer multiples of π
29 continuous everywhere except at integer multiples of π
31 continuous everywhere except at odd integer multiples of $\pi/2$
33 continuous everywhere except at negative odd integer
multiples of $\pi/2$
35 continuous on $[-2, 2]$ and $(-2, 2)$
37 discontinuous on all indicated intervals
39 continuous on $[-\frac{1}{2}, 0]$ and $(-1, -\frac{3}{4})$
41 $b = aR^3$ **43** $a = 3$
45 $\lim_{x \to 2^+} \sqrt{x - 2} = 0$; $\sqrt{x - 2}$ is undefined for values of x
that are less than 2

Review Problem Set, Chapter 1, page 77

1 necessarily true: (c) and (d) only   **3** $(2, \infty)$ **5** $[-\frac{3}{8}, 1]$
7 $(-5, 4)$ **9** no solution   **11** $(\frac{1}{2}, 6)$ **13** $[-5, -2)$

15 $a = -4$, $b = -2$, $c = -3$, $d = 1$ or $a = 0$, $b = 2$,
$c = -5$, $d = -3$
17 If s is the score on the final exam, the condition is
$73 \le s < 98$.
19 (a) 7; (b) 5; (c) 0.406592654; (d) 7.3; (e) $\frac{19}{6}$;
(f) 0.005786438
21 $-4, 2$ **23** $-3, 2$ **25** $\dfrac{1 \pm \sqrt{17}}{4}$
27 $[-\frac{11}{2}, \frac{1}{2}]$ **29** $(0.99, 1.01)$
31 $(-\infty, \frac{3}{4})$ together with $(1, \infty)$
33 $(-\infty, -\frac{7}{2}]$ together with $[\frac{1}{2}, \infty)$
35 true: (a); (c); (d); (e); (f)   **37** 5 units   **39** 13 units
41 8 units   **43** 30.35 units   **45** $r = 5$, $(h, k) = (2, -3)$
47 $r = \frac{3}{2}$, $(h, k) = (\frac{3}{2}, -2)$
49 slope $= -7$, $y + 5 = -7(x - 3)$ or $y - 2 = -7(x - 2)$
51 slope $= \frac{3}{2}$, $y - 2 = \frac{3}{2}(x - 1)$ or $y + 4 = \frac{3}{2}(x + 3)$
53 $y - 2 = -\frac{3}{5}(x - 5)$
55 (a) $y + 5 = \frac{3}{2}(x - 7)$ or $3x - 2y - 31 = 0$;
(b) $y + 5 = -\frac{2}{3}(x - 7)$ or $2x + 3y + 1 = 0$
57 $y = \frac{4}{3}x + \frac{2}{3}$, $m = \frac{4}{3}$, $b = \frac{2}{3}$
59 $y - 1 = 3(x + 7)$; (b) $y = 3x + 22$; (c) $3x - y + 22 = 0$
61 (a) $y + 2 = \frac{7}{3}(x - 1)$; (b) $y = \frac{7}{3}x - \frac{13}{3}$;
(c) $7x - 3y - 13 = 0$
63 $y - b = -\dfrac{a - h}{b - k}(x - a)$
65 23   **67** 0   **69** $3x^2 - 12$
71 $75x^2 - 180x + 104$ **73** 12   **75** $\dfrac{-1}{x(x + k)}$
77 all real numbers except 1
79 $[-1, \infty)$ **81** $(-\infty, -1]$ together with $[1, \infty)$
83 (a) function; (b) not a function
85 domain: $\mathbb{R}$; range: $\mathbb{R}$ **87** domain: $[2, \infty)$; range: $[0, \infty)$
89 domain: $\mathbb{R}$; range: $\mathbb{R}$ **91** 14.49 horsepower
93 (a) increase, graph rises to the right; (b) $G(p) = pg(p)$; (c)
If the price is p_1 dollars per unit or less, producers are
unwilling to supply any of the commodity.
95 odd, symmetric about origin   **97** neither even nor odd
99 even, symmetric about y axis
101 (a) $4x - 2$; (b) $6 - 2x$; (c) $3x^2 + 2x - 8$; (d) $\dfrac{x + 2}{3x - 4}$
103 (a) $\dfrac{2x}{x^2 - 1}$; (b) $\dfrac{2}{x^2 - 1}$; (c) $\dfrac{1}{x^2 - 1}$; (d) $\dfrac{x + 1}{x - 1}$
105 (a) $x^4 + \sqrt{x + 1}$; (b) $x^4 - \sqrt{x + 1}$; (c) $x^4\sqrt{x + 1}$;
(d) $\dfrac{x^4}{\sqrt{x + 1}}$
107 (a) $|x| - x$; (b) $|x| + x$; (c) $-x|x|$; (d) $-\dfrac{|x|}{x}$
109 (a) $x^{2/3} + 1 + \sqrt{x}$; (b) $x^{2/3} + 1 - \sqrt{x}$; (c) $\sqrt{x}(x^{2/3} + 1)$;
(d) $\dfrac{x^{2/3} + 1}{\sqrt{x}}$
111 vertex: $(0, 0)$; opens upward; domain: $\mathbb{R}$; range: $[0, \infty)$
113 vertex: $(0, 0)$; opens downward; domain: $\mathbb{R}$; range: $(-\infty, 0]$
115 vertex: $(2, 1)$; opens upward; domain: $\mathbb{R}$; range: $[1, \infty)$
117 vertex: $(\frac{3}{2}, -\frac{1}{4})$; opens upward; domain: $\mathbb{R}$; range: $[-\frac{1}{4}, \infty)$

119 vertex: $(-\frac{7}{12}, \frac{529}{24})$; opens downward; domain: $\mathbb{R}$; range: $(-\infty, \frac{529}{24}]$

121 vertex: (5, 0); opens downward; domain: $\mathbb{R}$; range: $(-\infty, 0]$

123 (a) $P(x) = -4x^2 + 240x - 600$; (b) $P(10) = 1{,}400$, $P(35) = 2{,}900$; (c) $30 + 5\sqrt{30} \approx 57.386$, $30 - 5\sqrt{30} \approx 2.614$

125 2.85 m   **127** 3 ft   **129** $\frac{\pi}{2}$ rad

131 (a) $\frac{4\pi}{9}$; (b) $\frac{19\pi}{6}$; (c) $-\frac{71\pi}{36}$; (d) $-\frac{9\pi}{2}$;

(e) $-\frac{31\pi}{18}$; (f) $\frac{17\pi}{4}$

133 (a) 72°; (b) −585°; (c) −157.5°; (d) 2100°; (e) 2295°; (f) 648°

135 $\frac{625\pi}{12}$ cm²   **137** $\frac{2\pi}{25}$ m   **139** 2094 mi

141 0.459114771   **143** 0.595243604

145 −1.625839380   **147** 1.042572391

149 −1.701301619   **151** −26.02388181

153 0.934678015   **155** −0.954561625

157 0.444015840   **159** $-\tan\theta$ **161** $\sin x$

163 $\tan^2 t$ **165** $1 + \tan u$ **167** $\tan\theta$

169 $\cos\theta$ **171** $-\cos\alpha$ **173** $\sin t$

175 $\frac{\sqrt{3}}{2}$ **177** $\sin(x + y)$

179 (a) $\frac{\sqrt{2}}{4}(1 + \sqrt{3})$; (b) $\frac{\sqrt{2}}{4}(1 - \sqrt{3})$; (c) $\frac{1 + \sqrt{3}}{1 - \sqrt{3}}$ or $-(2 + \sqrt{3})$

181 $\cos 4x$ **183** $\sin t$ **185** 1   **187** $\frac{1}{2}\tan 2\omega t$

189 $\frac{\sqrt{7 - x^2}}{x}$ **191** 658.18 m

193 $\lim_{x \to 2} \frac{x^3 - 8}{1 - (2/x)} = 24$; top table: 21.67899998, 23.76079724, 23.97601399, 23.99736016; bottom table: 26.48100001, 24.24080089, 24.02400439, 24.00235211

195 $\delta = 0.005$ **197** $\delta = 0.0004$ **199** $\delta = 0.002$

201 151   **203** $\frac{3}{2}$ **205** −1   **207** $\frac{\sqrt{3}}{3}$

209 2   **211** $\frac{1}{6}$ **213** 0   **215** $\frac{3}{32}$

217 13   **219** $\frac{19}{7}$ **221** 1

223 (a) $|x - a| < \frac{\epsilon}{3}$; (b) $|x - a| < \frac{0.01}{3}$. No, interval is too long.

225 $\frac{1}{6}$ **227** $\frac{3}{2}$

229 $\lim_{x \to (3/2)^-} f(x) = 0$, $\lim_{x \to (3/2)^+} f(x) = 0$, $\lim_{x \to 3/2} f(x) = 0$

231 $\lim_{x \to 2^-} g(x) = 4$, $\lim_{x \to 2^+} g(x) = 4$, $\lim_{x \to 2} g(x) = 4$

233 continuous at 3   **235** continuous at 1

237 discontinuous at $x = \frac{k}{4}$ for $k = 0, 1, 2, 3, \ldots$

239 discontinuous at 10 and 30   **241** $A = -12$, $B = 30$

243 (a) only on $(-1, \frac{1}{2})$; (b) on all intervals

Chapter 2

Problem Set 2.1, page 91

1 $r \approx 54.55$ mi/h   **3** (a) 7.5; (b) 7

5 (a) 30 m/s; (b) 24 m/s   **7** (a) 8 m/s; (b) 7 m/s

9 0   **11** 2   **13** 2   **15** $-\frac{1}{3}$

17 $\frac{1}{4}$ **19** (a) 80 ft/s; (b) 160 ft/s

21 $5\sqrt{3}$ cm²/cm of edge length

23 (a) −0.16 (lb/in²)/in³; (b) −0.2 (lb/in²)/in³

25 0.0044 (breaths/min)/(N/m²)   **27** −1 amp/ohm

Problem Set 2.2, page 97

1 4   **3** 0   **5** $2x + 4$ **7** $6x^2 - 4$ **9** $\frac{-2}{x^2}$

11 $\frac{-3}{(t - 1)^2}$ **13** $\frac{1}{2\sqrt{v - 1}}$ **15** $\frac{-2}{(x + 1)^2}$ **17** 4

19 $-\frac{14}{25}$ **21** $-\frac{1}{9}$ **23** $-\frac{4}{25}$

27 $32t + 30$ **29** (b) continuous at 3; (c) differentiable at 3

31 (b) continuous at 3; (c) not differentiable at 3

33 (b) continuous at 2; (c) not differentiable at 2

35 (b) continuous at 0; (c) differentiable at 0

39 not differentiable at $x = \frac{1}{3}$

41 −0.8661   **43** 0.1667

Problem Set 2.3, page 108

1 $6x$ **3** $-20x^3$ **5** $-4y^2$ **7** 5

9 $5x^4 - 9x^2$ **11** $5x^9 + x^4$ **13** $8t^7 - 14t^6 + 3$

15 $\frac{-6}{x^3} - \frac{4}{3x^2}$ **17** $\frac{25}{y^2} - \frac{25}{y^6}$

19 $-6x^{-3} + 7x^{-2} + \frac{3}{\sqrt{x}}$ **21** $\frac{-2}{5x^2} + \frac{2\sqrt{2}}{3x^3} - \frac{1}{2x\sqrt{x}}$

23 $15x^4 - 2x$ **25** $5x^4 + 12x^3 - 27x^2 - 54x$

27 $10y\sqrt{y} + \frac{7}{2\sqrt{y}}$ **29** $\frac{16}{x^2} + 4x - 3x^2$

31 $-\frac{10}{x^6} - \frac{18}{x^4} - \frac{1}{x^2} + 3$ **33** $\frac{-23}{(3x - 1)^2}$

35 $\frac{-7x^2 + 6x + 5}{(x^2 - 3x + 2)^2}$ **37** $\frac{-20t}{(t^2 - 1)^2}$ **39** $\frac{3x^2 + 12x + 37}{(x + 2)^2}$

41 (a) 4; (b) $-\frac{3}{16}$; (c) −9; (d) −2; (e) $-\frac{1}{18}$; (f) $\frac{64}{81}$

43 (a) $8x^3 - 3x^2 - 24x + 1$; (b) $54x^2 + 66x - 28$; (c) $9x^2 - 20x + 6 - 3x^{-2}$; (d) $12x(2x^2 + 7)^2$

45 (a) −1; (b) 5; (c) −5; (d) −2; (e) 16; (f) −4

47 (a) 16; (b) $\frac{-3}{49}$ **49** −4

51 We must *first* calculate $f'(x) = 4x + 3$, *then* substitute $x = 2$ into this equation to get $f'(2) = 11$.

53 7.5 ft/s   **55** −120 per year

57 −0.0979 (mole/m³)/day

Problem Set 2.4, page 113

1 tangent line: $y = 8x - 15$; normal line: $x + 8y - 10 = 0$

3 tangent line: $y = 3x$; normal line: $x + 3y - 10 = 0$

5 tangent line: $x + 2y - 5 = 0$; normal line: $y = 2x - 5$

7 tangent line: $y = 20 - 7x$; normal line: $x - 7y - 60 = 0$

9 tangent line: $y = 175x - 525$; normal line:
$x + 175y - 3 = 0$

11 tangent line: $y = 7 - 2x$; normal line: $x - 2y + 4 = 0$

13 tangent line: $y = 4$; normal line: $x = 4$

15 (a) $(-1, 0)$; (b) $(0, 1)$ **17** $(8, 72)$; $y = 16x - 56$

19 $(1, 0)$; $y = 1 - x$ **21** $(4, 2)$; $y = 18 - 4x$

23 $(5, 30)$; $y = 10x - 20$ and also $(-1, 6)$; $y = 4 - 2x$

25 $(\frac{5}{2}, -\frac{1}{4})$ **27** $(-\frac{5}{6}, \frac{47}{12})$ **29** $(1, -\frac{11}{3})$ and $(-5, \frac{97}{3})$

31 $(1, 2)$ **33** $(0, 0)$ and $(-2, -4)$ **35** $b = -4$

37 (a) $f'(x) = 3x^2$, so $f'(0) = 0$; (b) $f(x) > 0$ for $x > 0$ and $f(x) < 0$ for $x < 0$, so f has neither a relative maximum nor minimum at $x = 0$; (c) Theorem 1 does *not* say that $f'(c) = 0$ implies that f has a relative extremum at c.

39 $x = c$

Problem Set 2.5, page 119

1 $7 \cos x$ **3** $3 \cos t + t \sin t$

5 $3 \sec^2 x + \sec x \tan x$ **7** $8 \sec y \tan y - 2y^5$

9 $4r^3 \sin r + r^4 \cos r - 4 \csc r \cot r$

11 $-\csc^2 z + \dfrac{\tan z}{2\sqrt{z}} + \sqrt{z} \sec^2 z$

13 $\cos^2 x - \sin^2 x$ or $\cos 2x$

15 $7 \csc y (\csc^2 y + \cot^2 y)$

17 $3 \sec x (\tan x - \tan^2 x - \sec^2 x)$

19 $\dfrac{2\theta \cos \theta + 10 \cos \theta - 2 \sin \theta}{(\theta + 5)^2}$

21 $\dfrac{2x \cos x - 2x + x^2 \sin x + 5 \sin x}{(\cos x - 1)^2}$

23 $\dfrac{3 \sec t (1 + 4 \sec t)}{(\sec t + 4)^2}$ **25** $\dfrac{-6 \sin y}{(3 - \cos y)^2}$

27 $\dfrac{-\csc z (\cot z + \sec^2 z + 1)}{(1 + \tan z)^2}$

29 $\dfrac{-3 \tan^2 \theta + 12 \tan \theta + 3}{(2 \cos \theta - \sin \theta)^2}$

31 tangent line: $y = \sqrt{3}x + \dfrac{6 - \pi\sqrt{3}}{6}$; normal line:
$y = -\dfrac{\sqrt{3}}{3}x + \dfrac{18 + \pi\sqrt{3}}{18}$

33 tangent line: $y = 6x + 3 - \dfrac{3\pi}{2}$; normal line:
$x + 6y - 18 - \dfrac{\pi}{4} = 0$

35 tangent line: $y = 7x - 5\pi$; normal line:
$x + 7y - 15\pi = 0$

37 $\dfrac{\pi}{4}$ and $-\dfrac{3\pi}{4}$ **39** $\dfrac{v^2 \sin \theta \cos \theta}{g}$ **41** 2 units/s

43 $\dfrac{\mu W(\sin \theta - \mu \cos \theta)}{(\cos \theta + \mu \sin \theta)^2}$

Problem Set 2.6, page 123

1 17   **3** 6.9929   **5** -3 **7** 2   **9** $x^2 + 1$

11 (a) $\sin x^2$; (b) $\sin^2 x$; (c) x^4; (d) $(\sin x + \cos x)^2$; (e) $\tan^2 x$; (f) $\tan (\cot x)$; (g) $\sin (\cos^2 x)$; (h) $\sin (\cos^2 x)$

13 (a) x for $x \geq 0$; (b) $|x|$; (c) x^4

15 (a) $\tan \sqrt{x}$; (b) $\sqrt{\tan x}$; (c) $\tan (\tan x)$

17 (a) $|\csc x|$; (b) $\csc |x|$; (c) $|x|$

19 (a) $\cos^2 x + 2 \cos x$; (b) $1 + \cos (x^2 - 1)$; (c) $x^4 - 2x^2$

21 (a) $\dfrac{1}{4x - 9}$; (b) $\dfrac{11 - 6x}{2x - 3}$; (c) $\dfrac{2x - 3}{11 - 6x}$

23 (a) x for $x \neq 3$; (b) x for $x \neq 0$; (c) $\dfrac{8x - 3}{3x - 1}$

25 $F = h \circ g$ **27** $H = h \circ f$ **29** $Q = f \circ h$

31 $r = h \circ h$ **33** $f(x) = x^7$, $g(x) = \cos x$

35 $f(x) = 1 - x^2$; $g(x) = \tan x$

37 $f(x) = \sqrt{x}$; $g(x) = \dfrac{x + 1}{x - 1}$

39 $f(x) = \dfrac{|x|}{x}$; $g(x) = x + 1$

43 $f = g \circ h$, where $g(x) = \sin x$ and $h(x) = |x|$. Because h is continuous at 0 and g is continuous at $0 = h(0)$, it follows from the substitution property for continuous functions that f is continuous at 0.

45 The continuity of g follows from the substitution property for continuous functions and the fact that the sine function is continuous.

49 9, 17, 21, 19, 20, 20 (rounded off to the nearest whole number)

Problem Set 2.7, page 130

1 $\dfrac{2x + 1}{2\sqrt{x^2 + x + 1}}$ **3** $\dfrac{-20x^3}{(x^4 + 1)^6}$ **5** $-20(5 - 2x)^9$

7 $\dfrac{-20}{(4y + 1)^6}$ **9** $(3x^2 + 7)(5 - 3x)^2(-63x^2 + 60x - 63)$

11 $\left(3x + \dfrac{1}{x}\right)(6x - 1)^4\left(126x - 6 + \dfrac{18}{x} + \dfrac{2}{x^2}\right)$

13 $\dfrac{(28y + 38)(2y - 1)^3}{(7y + 3)^3}$ **15** $\dfrac{4(x^2 + x)^3}{(1 - 2x)^5}(1 + 2x - 2x^2)$

17 $\dfrac{-3(3x + 1)^2(3x + 2)}{x^7}$ **19** $\dfrac{-1}{2x\sqrt{x}}$

21 $\dfrac{x+1}{\sqrt{x^2+2x-1}}$　　**23** $\dfrac{2t^3-t}{\sqrt{t^4-t^2+\sqrt{3}}}$

25 $4(x-\sqrt{x})^3\left(1-\dfrac{1}{2\sqrt{x}}\right)$　　**27** $35\cos 7x$

29 $48x\cos 6x^2$　　**31** $\dfrac{\cos\sqrt{x}}{2\sqrt{x}}$　　**33** $12\sin^3 3t\cos 3t$

35 $-\cos x\sin(\sin x)$　　**37** $\dfrac{-5\sin 5x}{2\sqrt{\cos 5x}}$

39 $\dfrac{\cos x+\cos 5x\cos x+5\sin x\sin 5x}{(1+\cos 5x)^2}$

41 $70\sec 2t\tan 2t-54\csc 2t\cot 2t$

43 $-15t^4\csc^2(3t^5)$　　**45** $\dfrac{-u\csc\sqrt{u^2+1}\cot\sqrt{u^2+1}}{\sqrt{u^2+1}}$

47 $\dfrac{5\sec 5x\tan 5x}{2\sqrt{1+\sec 5x}}$　　**49** 0　　**51** $52\sec^2 13s\tan 13s$

53 $x^2\tan^4 2x\,(10x\sec^2 2x+3\tan 2x)$

55 $\dfrac{2\sec 5x-10x\sec 5x\tan 5x+2}{(1+\sec 5x)^2}$

57 $\dfrac{2}{3}x+6\cot^2 2x\csc^2 2x$　　**59** $\dfrac{3\sec^2 3t\,(2t\tan 3t-1)}{t^4}$

61 $10x\sec^2 5x^2\cos(\tan 5x^2)$　　**63** 12　　**65** -18

67 -30　　**69** -1500　　**73** 5　　**75** 3495 amp/s

77 -200 moose/year

79 (a) $u(t)=40{,}000+10{,}000t$; (b) $(F\circ u)(t)=600\sqrt{4+t}$

gives the labor force that will be required t years from now;

(c) $(F\circ u)'(t)=\dfrac{300}{\sqrt{4+t}}$ gives the instantaneous rate of

change of the labor force t years from now;

(d) $(F\circ u)'(5)=100$ persons/year

Problem Set 2.8, page 137

1 $-\dfrac{9x}{4y}$　　**3** $\dfrac{y^2-2xy-2x}{x^2-2xy}$　　**5** $\dfrac{3+3y-2x}{2y-3x}$

7 $\dfrac{-y^2}{8x^2y^2+x^2}$　　**9** $-\dfrac{8x^3y\sqrt{xy}+y}{2x^4\sqrt{xy}+x}$

11 $\dfrac{1-6xy(x^2-y^2)^2}{(x^2-7y^2)(x^2-y^2)^2}$　　**13** $\dfrac{2xy^2-\frac{1}{2}y^3}{x^2y+y^3+8}$

15 $\dfrac{\sqrt{x+y}+\sqrt{x-y}}{\sqrt{x+y}-\sqrt{x-y}}$　　**17** $\dfrac{2\cos(2x+y)}{1-\cos(2x+y)}$　　**19** $-\dfrac{y}{x}$

21 $\dfrac{\sin x\cos x}{\sin y\cos y}$ or $\dfrac{\sin 2x}{\sin 2y}$　　**23** $\dfrac{y\sin xy}{2y-x\sin xy}$

25 $\dfrac{-3\csc^2(3x+y)-5y}{5x+\csc^2(3x+y)}$

27 tangent line: $y=-\frac{1}{2}x+4$; normal line: $y=2x-1$

29 tangent line: $y=-x+5$; normal line: $y=x+1$

31 tangent line: $y=1$; normal line: $x=\dfrac{\pi}{2}$

33 $\dfrac{-5x}{6x+5y}$　　**35** $\dfrac{2y-1}{2x}$　　**37** $y=\dfrac{5x-6}{4}$

39 $y=\dfrac{-x\pm\sqrt{x^2+4x}}{2}$　　**41** $y=\dfrac{x+1}{2-3x}$

43 $y=\pm\sqrt[4]{x}$　　**45** $y=1+\sqrt[3]{3x+2}$

47 Every point on the graph of f is on the graph of

$\dfrac{x^2}{16}+\dfrac{y^2}{9}=1$.

49 (a) $\dfrac{-3x}{4\sqrt{16-x^2}}$; (b) $\dfrac{3x}{4\sqrt{16-x^2}}$; (c) $-\dfrac{9x}{16y}$

51 slope $=-2$　　**53** $I=125$

Problem Set 2.9, page 142

1 $\dfrac{\sqrt[5]{x}}{5x}$　　**3** $-16x^{-13/9}$　　**5** $\dfrac{2}{3}(1-t)^{-5/3}$

7 $\dfrac{-3}{2u^2}\left(1+\dfrac{2}{u}\right)^{-1/4}$　　**9** $-\dfrac{1}{2}x^{-3/2}-\dfrac{1}{3}x^{-4/3}-\dfrac{1}{4}x^{-5/4}$

11 $\dfrac{1}{5}(t^3-t^{1/4})^{-4/5}\left(3t^2-\dfrac{1}{4}t^{-3/4}\right)$　　**13** $\dfrac{\sqrt[10]{x/(x+1)}}{10x(x+1)}$

15 $\dfrac{1-2x}{4}(1+x)^{-7/4}(2x+1)^{-1/2}$

17 $\dfrac{1}{20}(9t+33)(t+2)^{-3/4}(t+5)^{-4/5}$

19 $\dfrac{1}{5}(\sin t)^{-4/5}\cos t$ or $\dfrac{1}{5}\sqrt[5]{\sin t}\cot t$

21 $-\dfrac{3}{4}(\cos x)^{-1/4}\sin x$ or $-\dfrac{3}{4}\cos^{3/4}x\tan x$

23 $12t\sec^{3/2}(4t^2+1)\tan(4t^2+1)$

25 $\dfrac{\sqrt[3]{y}}{3y}\sec^2\sqrt[3]{y}\csc\sqrt[4]{y}-\dfrac{\sqrt[4]{y}}{4y}\tan\sqrt[3]{y}\csc\sqrt[4]{y}\cot\sqrt[4]{y}$

27 $\dfrac{2}{3}z^{-1}(z^{2/3}\cos z^{-2/3}+\sin z^{-2/3})$

29 $-4\sin 3\theta\cos^3 3\theta\,(3+\cos^4 3\theta)^{-2/3}$

31 tangent line: $8x-27y+73=0$; normal line:

$27x+8y-51=0$

33 tangent line: $y=-1$; normal line: $x=0$

35 tangent line: $x+2y-3=0$; normal line: $y=2x-1$

39 9000 people/day

Problem Set 2.10, page 148

1 $v=3t^2+4t+3$ m/s, $a=6t+4$ m/s^2

3 $v=3\pi\cos\pi t+4\pi\sin 2\pi t$ m/s,

$a=-3\pi^2\sin\pi t+8\pi^2\cos 2\pi t$ m/s^2

5 $v=\dfrac{25}{4}t^{3/2}+t^{1/2}$ km/h, $a=\dfrac{75}{8}t^{1/2}+\dfrac{1}{2}t^{-1/2}$ km/h^2

7 problem 1: $v=10$ m/s, $a=10$ m/s^2;

problem 3: $v=-3\pi$ m/s, $a=8\pi^2$ m/s^2;

problem 7: $v=\dfrac{29}{4}$ km/h, $a=\dfrac{79}{8}$ km/h^2

9 $f'(x) = 15x^2 + 4$, $f''(x) = 30x$

11 $f'(t) = 35t^4 - 46t + 1$, $f''(t) = 140t^3 - 46$

13 $G'(x) = 6x^5$, $G''(x) = 30x^4$

15 $g'(t) = \frac{7}{2}t^{5/2} - 5$, $g''(t) = \frac{35}{4}t^{3/2}$

17 $f'(x) = 2x + 3x^{-4}$, $f''(x) = 2 - 12x^{-5}$

19 $f'(u) = 4(2 - u)^{-2}$, $f''(u) = 8(2 - u)^{-3}$

21 $f'(t) = t(t^2 + 1)^{-1/2}$, $f''(t) = (t^2 + 1)^{-3/2}$

23 $F'(r) = 1 - r^{-1/2}$, $F''(r) = \frac{1}{2}r^{-3/2}$

25 $f'(x) = -77 \sin 11x$, $f''(x) = -847 \cos 11x$

27 $F'(\theta) = 2 \cos 2\theta - 3 \sin 3\theta$, $F''(\theta) = -4 \sin 2\theta - 9 \cos 3\theta$

29 $H'(t) = -14 \csc 7t \cot 7t$,
$H''(t) = 98 \csc 7t (\cot^2 7t + \csc^2 7t)$

31 $Q'(\theta) = \cot 3\theta - 3\theta \csc^2 3\theta$,
$Q''(\theta) = 6 \csc^2 3\theta (3\theta \cot 3\theta - 1)$

33 $G'(x) = \dfrac{1}{(x + 1)^2} \cos \dfrac{x}{x + 1}$,

$G''(x) = \dfrac{-2}{(x + 1)^3} \cos \dfrac{x}{x + 1} - \dfrac{1}{(x + 1)^4} \sin \dfrac{x}{x + 1}$

35 $\dfrac{dy}{dx} = -20 \sec 5x \tan 5x$,

$\dfrac{d^2y}{dx^2} = -100 \sec 5x (\tan^2 5x + \sec^2 5x)$

37 $f'(-1) = 1$, $f''(-1) = 14$

39 $F'\left(\dfrac{\pi}{6}\right) = -2$, $F''\left(\dfrac{\pi}{6}\right) = -2\sqrt{3}$

41 $f'(x) = 28x^3 - 15x^2 + 16x - 3$, $f''(x) = 84x^2 - 30x + 16$,
$f'''(x) = 168x - 30$, $f^{(4)}(x) = 168$, and $f^{(n)}(x) = 0$ for $n \geq 5$

43 $\dfrac{dy}{dx} = x(x^2 - 1)^{-1/2}$, $\dfrac{d^2y}{dx^2} = -(x^2 - 1)^{-3/2}$,

$\dfrac{d^3y}{dx^3} = 3x(x^2 - 1)^{-5/2}$

45 $D_x^n\left(\dfrac{1}{x}\right) = (-1)^n n! \, x^{-(n+1)}$

47 0   **49** $\frac{736}{3}$

51 (a) when $t = 2$; (b) a is never zero; (c) a is never zero

55 $-\dfrac{192x^2}{y^7}$ **57** $\dfrac{2 \tan 3y}{3x^2} (1 + 2 \sec^2 3y)$

59 500 m   **61** 6 days

Problem Set 2.11, page 157

1 Let $f(x) = 4x^3 - x^2$. Then $f(0) = 0 < 2 < 3 = f(1)$.

3 Let $f(x) = x^3 + 3x^2 - 9x$. Then $f(2) = 2 < 10 < 27 = f(3)$.

5 Let $f(x) = \dfrac{x^3 + 5}{\sqrt{x + 1}}$. Then $f(0) = 5 > 4 > 3 = f(1)$.

7 Let $f(x) = 2x^3 - 3x^2 - 12x$. Then
$f(-2) = -4 < 1 < 7 = f(-1)$.

9 Let $f(x) = \sin x + 2 \cos 2x$. Then
$f\left(\dfrac{3\pi}{4}\right) = \dfrac{\sqrt{2}}{2} < 1 < 2 = f(\pi)$.

11 Let $f(x) = x + \sin x$. Then $f(0) = 0 < 1 < \dfrac{\pi}{6} + \dfrac{1}{2} = f\left(\dfrac{\pi}{6}\right)$.

13 $f(1) = -2$ and $f(2) = 15$

15 $f(1.5) = -0.15625$ and $f(1.6) = 1.29376$

17 $h\left(\dfrac{\pi}{4}\right) = \dfrac{\sqrt{2}}{2}$ and $h\left(\dfrac{\pi}{3}\right) = \dfrac{\sqrt{3}}{2} - 1 < 0$

19 positive on $(-\infty, 2)$ and $(4, \infty)$; negative on $(2, 4)$

21 positive on $(-5, 0)$ and $(3, \infty)$; negative on $(-\infty, -5)$ and $(0, 3)$

23 positive on $\left(-\frac{1}{2}, \frac{1}{3}\right)$ and $(2, \infty)$; negative on $\left(-\infty, -\frac{1}{2}\right)$ and $\left(\frac{1}{3}, 2\right)$

25 positive on $\left(-\infty, \frac{1}{2}\right)$ and $\left(\frac{2}{3}, \infty\right)$; negative on $\left(\frac{1}{2}, \frac{2}{3}\right)$

27 positive on $(-\infty, -3)$, $(-1, 2)$, and $(4, \infty)$; negative on $(-3, -1)$ and $(2, 4)$

29 positive on $(-\infty, -4)$, $(-3, -1)$, $\left(\frac{1}{2}, 1\right)$, $(1, 2)$, and $(2, \infty)$; negative on $(-4, -3)$ and $\left(-1, \frac{1}{2}\right)$

31 positive on $(-\infty, 0)$ and $(0, 3)$; negative on $(3, \infty)$

33 positive on $(-1, 1)$ and $(1, \infty)$; negative on $(-\infty, -1)$

35 zero in $(1.5, 1.525)$ **37** zero in $\left(\dfrac{5\pi}{16}, \dfrac{\pi}{3}\right)$

39 1.512876397   **41** 1.002966954   **43** -3.048917339

45 1.726279398   **47** 0.876726215   **49** (a) 1.414213563

51 If a is an approximation to $\sqrt[n]{k}$, then $b = \dfrac{(n - 1)a^n + k}{na^{n-1}}$ is
often a better approximation to $\sqrt[n]{k}$.

55 Successive values do not settle down—rather they oscillate between -1 and 1.

57 (b) -0.766044443, -0.173648178, 0.939692621;
(c) $\cos 20° \approx 0.939692621$

61 0.0226 m³

Review Problem Set, Chapter 2, page 159

1 (a) 120 ft/s; (b) 136 ft/s when $t = 2$, 104 ft/s when $t = 3$;
(c) 625 ft

3 (a) 40.2 in²/in; (b) 0.3216 in²/deg

5 tangent line: $y = 4x - 14$; normal line: $4y + x = 12$

7 tangent line: $72x - 16y = 81$; normal line: $144y + 32x = 291$

9 (i) All are continuous except (d). (ii) None are differentiable.
(iii) None are continuous and differentiable.

11 (a) 3; (b) 3   **13** (a) $2x + 1 + \Delta x$; (b) $2x + 1$

15 (a) $x - 2 + \frac{1}{2}\Delta x$; (b) $x - 2$ **17** (a) $\dfrac{-2}{x(x + \Delta x)}$; (b) $\dfrac{-2}{x^2}$

21 (b) at 0 and at -1

23 (a) $\frac{89}{30}$; (b) $\frac{91}{30}$; (c) $\frac{44}{3}$; (d) $\frac{46}{75}$; (e) $-\frac{23}{50}$; (f) $\frac{44}{15}$; (g) $\frac{46}{675}$

25 (a) continuous at 2; (b) $f'_-(2) = -3$, $f'_+(2) = 4$; (c) not differentiable at 2

27 (a) continuous at -3; (b) $f'_-(-3) = -1$, $f'_+(-3) = 0$; (c) not differentiable at -3

29 0.94869   **31** $35x^4$ **33** $-\frac{3}{2}x^{-4}$ **35** $\dfrac{2\sqrt{2}}{3}x^{-1/3}$

37 $10x - 7$ **39** $-27x^2 + 6x - 1$

41 $105t^{20} + 60t^3 + 10t^{-3}$

43 $(2x + 2)(2x^3 + 5) + (x^2 + 2x + 1)(6x^2)$

45 $27(6s^4 + 5s^2 - s^{-1})^{26}(24s^3 + 10s + s^{-2})$

47 $\dfrac{x^2 + 4x + 7}{(x + 2)^2}$ **49** $\dfrac{1}{2\sqrt{x}}(x^2 + 2x + 1) + \sqrt{x}(2x + 2)$

51 $\left(\dfrac{1}{x} - \dfrac{3}{x^2}\right)^2$ **53** $\dfrac{x}{\sqrt{x^2 + 12}}$

55 $\dfrac{8\sqrt[3]{x}}{3x} - \dfrac{\frac{1}{2}\sqrt{x} - 1}{(\sqrt{x} - 1)^2}$ **57** $\dfrac{(x^2 + 7)^2}{2\sqrt{x - 7}}(13x^2 - 84x + 7)$

59 $\dfrac{3}{2}\left(\dfrac{t^2 - 3t + 2}{t^2 + 2t + 5}\right)^{1/2}\dfrac{5t^2 + 6t - 19}{(t^2 + 2t + 5)^2}$ **61** $\dfrac{ad - bc}{(cx + d)^2}$

63 $\cos 2x$ **65** $6 \sin 4t$ **67** $5 \cos (3x - 1)$

69 $\dfrac{\cot^2 t - \sin t - 1}{(1 + \csc t)^2}$ **71** $3 \cos \theta \cos 2\theta - 6 \sin \theta \sin 2\theta$

73 $\sec^2 \frac{3}{2}s + \csc^2 \frac{4}{3}s$

75 $\dfrac{(x + 1)(3 \csc^2 x - 2 \csc x \cot x) + 3 \cot x - 2 \csc x}{(x + 1)^2}$

77 $\dfrac{1 - \theta \sec \theta}{\sec \theta + \tan \theta}$ **79** $-a\omega \sin (\omega t - \phi)$

81 1.307 units/day **85** $\frac{2}{5}$ **87** $(1, 2)$, $y = 4x - 2$

89 $\left(\dfrac{\pi}{6}, \dfrac{\sqrt{3}}{2}\right)$, $y = 2x - \dfrac{\pi}{3} + \dfrac{\sqrt{3}}{2}$ **91** $(1, -4)$, $(-1, 0)$

93 $\left(2, \dfrac{1}{4}\right)$, $\left(-2, -\dfrac{1}{4}\right)$

95 $(-\pi, -1)$, $(0, 1)$, $(\pi, -1)$, $\left(-\dfrac{\pi}{3}, \dfrac{5}{4}\right)$, $\left(\dfrac{\pi}{3}, \dfrac{5}{4}\right)$

97 $f'(2) = 0$ follows from Theorem 1 in Section 2.4.

99 -12 **101** 2

103 $f = g \circ h$, where $h(x) = x^2$ and $g(x) = \frac{1}{2}x$

105 (a) $x_0 = \dfrac{\alpha}{1 - \beta}$; (b) $|f(x) - x_0| = |f(x) - f(x_0)| = |(\alpha + \beta x) - (\alpha + \beta x_0)| = |\beta(x - x_0)| = |\beta| \, |x - x_0|$; (d) If $|\beta| < 1$, then $|\beta|^n$ approaches zero as n gets larger and larger.

107 (a) $7.2, 9.18, 10.962, 12.5658, 14.00922, 15.308298, 16.4774682, 17.52972138, 18.47674924$; (b) $8.2, 11.72, 15.592, 19.8512, 24.53632, 29.689952, 35.3589472, 41.59484192, 48.45432611$

109 $\dfrac{-3x^2}{2y^2}$ **111** $\dfrac{1}{3y^2\sqrt{x + 1}} - \dfrac{5x}{12y^2}$ **113** $\sec y$

115 $\dfrac{\cos y}{1 + x \sin y}$ **117** $\dfrac{-9x^4 - 6xy^3}{2y^5} = \dfrac{-3x}{2y^5}$

119 tangent line: $2x + 5y - 16 = 0$; normal line: $5x - 2y - 11 = 0$

121 tangent line: $y = 6 - 2x$; normal line: $x - 2y + 2 = 0$

123 tangent line: $6x + 3\sqrt{3}y - \sqrt{3}\pi - 3 = 0$; normal line: $6\sqrt{3}x - 12y + 4\pi - 3\sqrt{3} = 0$

125 slope of tangent line at $(a, b) = -\dfrac{a}{b}$; slope of line segment from $(0, 0)$ to $(a, b) = \dfrac{b}{a}$

127 $(\sqrt[3]{2}, \sqrt[3]{4})$ **129** $80x^3 + 6$ **131** $\dfrac{-2}{(t + 1)^3}$

133 $\frac{21}{16}(u^{1/4} + 3)^5 u^{-7/4}(u^{1/4} - 3)$

135 $2 \cos 3\theta - 12 \theta \sin 3\theta - 9\theta^2 \cos 3\theta$

137 $\frac{8}{3}x^{-5/3} - 3x^{-3/2}$ **139** $\dfrac{12}{(x + 1)^4}$ **141** $-a^6 \cos ax$

143 $f'(x) = 30x^4 + 28x^3 - 24x^2 - 18x + 10$, $f''(x) = 120x^3 + 84x^2 - 48x - 18$, $f'''(x) = 360x^2 + 168x - 48$, $f^{(4)}(x) = 720x + 168$, $f^{(5)}(x) = 720$, and $f^{(n)}(x) = 0$ for $n \geq 6$

145 (a) -16; (b) 31; (c) -7; (d) -10; (e) 132; (f) $-\frac{127}{128}$

149 $v = 12t - 6t^2$, $a = 12 - 12t$

151 $v = 6\pi \cos 2\pi t$, $a = -12\pi^2 \sin 2\pi t$

153 $f''(x) = \begin{cases} 6x & \text{if } x < 1 \\ 0 & \text{if } x > 1 \end{cases}$, $f''(1)$ does not exist

155 Let $f(x) = 4x^3 - 7x^2 + 2x$. Then $f(1) = -1 < \sqrt{5} < 8 = f(2)$.

157 positive on $(0, \frac{1}{2})$ and $(\frac{2}{3}, \infty)$; negative on $(-\infty, 0)$ and $(\frac{1}{2}, \frac{2}{3})$

159 positive on $(-\infty, -\frac{5}{2})$, $(-\frac{3}{2}, -\frac{5}{4})$, and $(\frac{1}{3}, \infty)$; negative on $(-\frac{5}{2}, -\frac{3}{2})$ and $(-\frac{5}{4}, \frac{1}{3})$

161 positive on $\left[-\pi, -\dfrac{\pi}{2}\right)$, $\left(-\dfrac{\pi}{4}, \dfrac{\pi}{2}\right)$, and $\left(\dfrac{3\pi}{4}, \pi\right]$; negative on $\left(-\dfrac{\pi}{2}, -\dfrac{\pi}{4}\right)$ and $\left(\dfrac{\pi}{2}, \dfrac{3\pi}{4}\right)$

163 $f(-2) = -1$, $f(-1) = 4$ **165** $h(0) = 1$, $h(1) \approx -0.46$

167 zero on $(-2, -1.75)$ **169** -1.881239402

171 0.641714371 **173** 1.725027751

175 0.682327804 **177** f is discontinuous at $x = 2$

Chapter 3

Problem Set 3.1, page 170

1 no **3** $c = 3$ **5** $c = \sqrt{\dfrac{13}{3}}$ **7** $c = \dfrac{\pi}{2}$

9 $c = \dfrac{3}{2}$ **11** $c = 1 \pm \dfrac{2}{3}\sqrt{3}$ **13** $c = \dfrac{4}{9}$

15 $\dfrac{\pi}{2}, \dfrac{3\pi}{2}, \dfrac{5\pi}{2}, \dfrac{7\pi}{2}$

17 $c = 4$ **19** $c = \dfrac{2}{3}\sqrt{3}$ **21** $c = 1$

23 $c = \pm \dfrac{\sqrt{2}}{2}$ **25** $\pi, 2\pi$ **27** f is not differentiable at 0

29 f is not differentiable at 1 **31** f is not defined at $\dfrac{\pi}{2}$

33 (a) not differentiable on (a, b); (b) not differentiable on (a, b); (c) not defined on $[a, b]$; (d) discontinuous at a; (e) not continuous on $[a, b]$

39 $c \approx 0.796718402$

41 (a) $f(1) = -12$, $f(2) = 28$; (b) $f'(x) = 5x^4 + 6x^2 - 5 = 5(x^2)^2 + 6x^2 - 5$. By the quadratic formula, $f'(x) = 0$ only for $x^2 = \dfrac{\sqrt{34} - 3}{5}$;

that is, $x = \pm \sqrt{\dfrac{\sqrt{34} - 3}{5}}$. But neither of these values

of x is between 1 and 2.

43 $c = \dfrac{a + b}{2}$

Problem Set 3.2, page 179

1 (a) increasing on [0, 2], decreasing on [−2, 0], constant on $(-\infty, -2]$ and $[2, \infty)$; (b) increasing on $[-3, -\frac{2}{3}]$ and $[4, \infty)$, decreasing on $(-\infty, -3]$ and $[-\frac{2}{3}, 2]$, constant on [2, 4]; (c) increasing on $(-\infty, -5]$ and [2, 4], decreasing on [−5, −2] and $[4, \infty)$, constant on [−2, 2]

3 increasing on $[3, \infty)$, decreasing on $(-\infty, 3]$

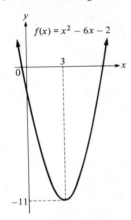

5 increasing on $(-\infty, 1]$, decreasing on $[1, \infty)$

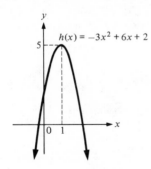

7 increasing on [−2, 2], decreasing on $(-\infty, -2]$ and $[2, \infty)$

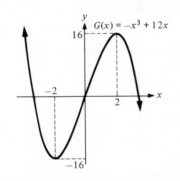

9 increasing on $(-\infty, -2]$ and $[2, \infty)$, decreasing on [−2, 2]
11 decreasing on $\mathbb{R}$
13 increasing on $(-\infty, 1]$ and $[3, \infty)$, decreasing on [1, 3]
15 increasing on [−2, 0] and $[2, \infty)$, decreasing on $(-\infty, -2]$ and [0, 2]
17 increasing on $(-\infty, -1]$ and $[1, \infty)$, decreasing on [−1, 1]
19 decreasing on $\mathbb{R}$
21 increasing on $(-\infty, -2]$ and $(0, \infty)$, decreasing on [−2, 0)
23 increasing on $[4, \infty)$, decreasing on (0, 4]
25 increasing on $[-1, \infty)$, decreasing on $(-\infty, -1]$
27 decreasing on $(-\infty, -1)$, (−1, 1), and $(1, \infty)$
29 increasing on $\left[2\pi k - \dfrac{\pi}{2}, 2\pi k + \dfrac{\pi}{2} \right]$ for k an integer;

decreasing on $\left[2\pi k + \dfrac{\pi}{2}, 2\pi k + \dfrac{3\pi}{2} \right]$ for k an integer

31 increasing on $\left(\pi k - \dfrac{\pi}{2}, \pi k + \dfrac{\pi}{2} \right)$ for k an integer

33 critical number: 3; relative max at $x = 3$

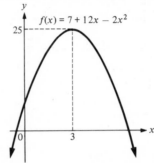

35 critical numbers: −3 and 3; relative max at $x = -3$; relative min at $x = 3$

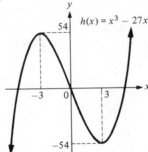

37 critical numbers: $-\frac{1}{3}$ and 1; relative max at $x = -\frac{1}{3}$; relative min at $x = 1$

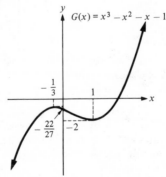

39 critical numbers: -2 and 1; relative max at $x = -2$; relative min at $x = 1$

41 critical number: 1; relative min at $x = 1$

43 critical numbers: 0, 1, and 2; relative max at $x = 1$; relative min at $x = 0$ and $x = 2$

45 critical numbers: 0 and 1; relative max at $x = 1$

47 no critical number; no relative max or min

49 critical numbers: 0 and 4; relative min at $x = 0$; neither relative max nor min at $x = 4$ (f is discontinuous).

51 critical number: 1; relative min at $x = 1$

59 (a) critical numbers: $-\dfrac{11\pi}{6}$, $-\dfrac{3\pi}{2}$, $-\dfrac{7\pi}{6}$, $-\dfrac{\pi}{2}$, $\dfrac{\pi}{6}$, $\dfrac{\pi}{2}$, $\dfrac{5\pi}{6}$, and $\dfrac{3\pi}{2}$; (b) increasing on $\left[-2\pi, -\dfrac{11\pi}{6}\right]$, $\left[-\dfrac{3\pi}{2}, -\dfrac{7\pi}{6}\right]$, $\left[-\dfrac{\pi}{2}, \dfrac{\pi}{6}\right]$, $\left[\dfrac{\pi}{2}, \dfrac{5\pi}{6}\right]$, and $\left[\dfrac{3\pi}{2}, 2\pi\right]$; decreasing on $\left[-\dfrac{11\pi}{6}, -\dfrac{3\pi}{2}\right]$, $\left[-\dfrac{7\pi}{6}, -\dfrac{\pi}{2}\right]$, $\left[\dfrac{\pi}{6}, \dfrac{\pi}{2}\right]$, and $\left[\dfrac{5\pi}{6}, \dfrac{3\pi}{2}\right]$; (c) relative max at $x = -\dfrac{11\pi}{6}$, $x = -\dfrac{7\pi}{6}$, $x = \dfrac{\pi}{6}$, and $x = \dfrac{5\pi}{6}$; relative min at $x = -\dfrac{3\pi}{2}$, $x = -\dfrac{\pi}{2}$, $x = \dfrac{\pi}{2}$, and $x = \dfrac{3\pi}{2}$

61 (a) increasing on $[0, 2]$, decreasing on $[2, \infty)$; (b) when $t = 2$ h; (c) 1 mg/liter

Problem Set 3.3, page 186

1 concave upward on $\mathbb{R}$; no point of inflection

3 concave downward on $\mathbb{R}$; no point of inflection

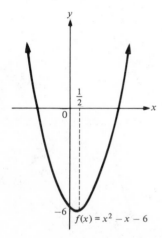

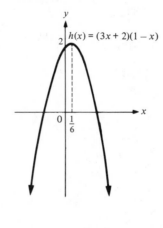

5 concave upward on $(-1, \infty)$; concave downward on $(-\infty, -1)$; point of inflection: $(-1, 0)$

7 concave upward on $(-2, \infty)$; concave downward on $(-\infty, -2)$; point of inflection: $(-2, 2)$

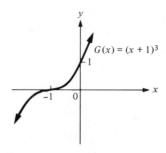

9 concave upward on $(\frac{1}{12}, \infty)$; concave downward on $(-\infty, \frac{1}{12})$; point of inflection: $(\frac{1}{12}, \frac{611}{432})$

11 concave upward on $(2, \infty)$; concave downward on $(-\infty, 2)$; point of inflection: $(2, 3)$

13 concave upward on $\mathbb{R}$, no point of inflection

15 concave downward on $\mathbb{R}$, no point of inflection

17 concave upward on $(-\infty, -1)$ and $(3, \infty)$; concave downward on $(-1, 3)$; points of inflection: $(-1, -13)$ and $(3, -189)$

19 concave upward on $(-\sqrt{3}, 0)$ and $(\sqrt{3}, \infty)$; concave downward on $(-\infty, -\sqrt{3})$ and $(0, \sqrt{3})$; points of inflection: $(-\sqrt{3}, 42\sqrt{3})$, $(0, 0)$, and $(\sqrt{3}, -42\sqrt{3})$

21 concave upward on $(0, \infty)$; concave downward on $(-\infty, 0)$; no point of inflection

23 concave upward on $(-\infty, 0)$ and $(0, \infty)$; no point of inflection

25 concave upward on $(0, \infty)$; concave downward on $(-\infty, 0)$; point of inflection: $(0, 0)$

27 concave upward on $(-\infty, -1)$ and $(0, \infty)$; concave downward on $(-1, 0)$; no point of inflection

29 concave upward on $(-2\sqrt{3}, 0)$ and $(2\sqrt{3}, \infty)$; concave downward on $(-\infty, -2\sqrt{3})$ and $(0, 2\sqrt{3})$; points of inflection: $(-2\sqrt{3}, -\frac{5}{8}\sqrt{3})$, $(0, 0)$, and $(2\sqrt{3}, \frac{5}{8}\sqrt{3})$

31 concave upward on $(0, \infty)$; no point of inflection

33 concave upward on $\left(\dfrac{\pi}{2} + 2\pi k, \dfrac{3\pi}{2} + 2\pi k\right)$ for k an integer; concave downward on $\left(-\dfrac{\pi}{2} + 2\pi k, \dfrac{\pi}{2} + 2\pi k\right)$ for k an integer; points of inflection: $\left((2k + 1)\dfrac{\pi}{2}, 0\right)$ for k an integer

35 relative min at $x = \frac{5}{2}$

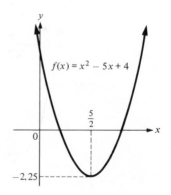

37 relative max at $x = \frac{2}{3}$; relative min at $x = 2$

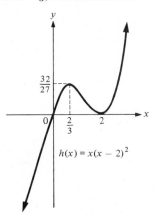

$$h(x) = x(x - 2)^2$$

39 relative max at $x = \frac{2}{3}$; relative min at $x = 0$

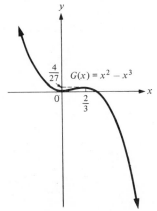

$$G(x) = x^2 - x^3$$

41 relative max at $x = 0$; relative min at $x = -\sqrt{3}$ and $x = \sqrt{3}$

43 relative max at $x = 0$; relative min at $x = 2$

45 relative max at $x = \frac{3\pi}{4} + 2\pi k$ for k an integer; relative min at $x = -\frac{\pi}{4} + 2\pi k$ for k an integer

47 (a) (i) increasing on $[a, b]$ and $[c, d]$, (ii) decreasing on $[b, c]$ and $[d, e]$, (iii) never concave upward, (iv) never concave downward, (v) no inflection points; (b) (i) increasing on $[b, c]$, $[c, d]$, and $[d, e]$, (ii) not decreasing on any displayed subinterval, (iii) concave upward on (c, d), (iv) concave downward on (a, b), (b, c), and (d, e), (v) points of inflection: $(c, f(c))$ and $(d, f(d))$; (c) (i) increasing on $[a, b]$ and $[d, e]$, (ii) decreasing on $[b, c]$, (iii) concave upward on (c, d), (iv) concave downward on (a, b) and (b, c), (v) points of inflection: $(c, f(c))$ and $(d, f(d))$; (d) (i) increasing on $[c, e]$, (ii) decreasing on $[a, b]$ and $[b, c]$, (iii) concave upward on (b, c) and (c, d), (iv) concave downward on (a, b) and (d, e), (v) points of inflection: $(b, f(b))$ and $(d, f(d))$

49 relative max at $x = 0$, relative min at $x = 1$

51 relative min at $x = 1$

53 relative max at $x = 0$; relative min at $x = -1$ and $x = 2$

55 no relative max or min   **57** relative min at $x = 0$

59 relative max at $x = 2$, relative min at $x = 0$

61 relative max at $x = 1$, relative min at $x = 2$

63 relative max at $x = 2\pi k - \frac{\pi}{2}$, for k an integer; relative min at $x = 2\pi k + \frac{\pi}{2}$, for k an integer

65 The equation of the tangent line at $(a, f(a))$ is $y = f(a) + f'(a)(x - a)$. The point $(b, f(b))$ is above this line if $f(b) > f(a) + f'(a)(b - a)$.

67

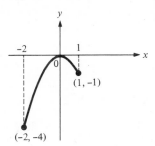

69 Let $P = f(t)$, where P is performance and t is study time. Then $f'(t)$ is the rate of increase of performance per unit of study time. At the point of diminishing returns, $f''(t) = 0$ and $f'(t)$ is a maximum.

Problem Set 3.4, page 193

1 max: 0 at $x = 0$; min: -4 at $x = -2$

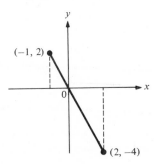

3 max: 2 at $x = -1$; min: -4 at $x = 2$

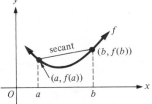

5   max: 2 at $x = 0$; min: 0 at $x = -2$ and $x = 2$

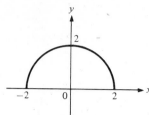

7   max: 4 at $x = 1$; min: 0 at $x = -1$

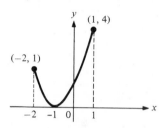

9   max: 15 at $x = 5$; min: -4 at $x = -6$

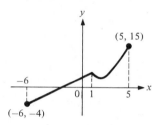

11   no max and no min   **13**   max: 2 at $x = 3$; min: 0 at $x = 1$

15   max: 3 at $x = 8$; min: 0 at $x = -1$

17   max: 76 at $x = 4$; min: -5 at $x = -5$ and $x = 1$

19   max: $\frac{1}{5}$ at $x = -1$; min: -3 at $x = 1$

21   max: 1 at $x = 0$; min: $\frac{1}{5}$ at $x = -2$

23   max: $\dfrac{\sqrt{2}}{4}$ at $x = \sqrt{2}$; min: $-\dfrac{\sqrt{2}}{4}$ at $x = -\sqrt{2}$

25   max: $5^{2/3}$ at $x = 3$; min: 0 at $x = -2$

27   max: 0 at $x = 0$; min: -2 at $x = \dfrac{\pi}{2}$

29   max: $1 - \dfrac{\pi}{4}$ at $x = -\dfrac{\pi}{4}$; min: $\dfrac{\pi}{4} - 1$ at $x = \dfrac{\pi}{4}$

31   max: 4 at $x = 0$; no min   **33**   no max; min: -1 at $x = 1$

35   no max; no min

37   no max; min: -8 at $x = -2$ and $x = 2$

39   no max; no min

41   max: $\frac{1}{4}$ at $x = 2$; min: $-\frac{1}{16}$ at $x = -8$

43   no max; min: 0 at $x = 0$

45   max: 0 at $x = -1$; no min   **47**   no max; min: 0 at $x = 0$

49   max: $\sqrt{2}$ at $x = (2k + 1)\dfrac{\pi}{2}$ for k an integer; min: 1 at

$x = k\pi$ for k an integer

51   5 years   **53** $\dfrac{\pi}{4}$

55   (a) $x = \dfrac{2B}{3}$; (b) $R_{\max} = \dfrac{4AB^3}{27}$; (c) $x = \dfrac{B}{3}$

Problem Set 3.5, page 202

1 6   **3** $\frac{5}{8}$ **5** 0   **7** $\dfrac{8}{\sqrt[4]{3}}$ **9** $+\infty$ **11** $+\infty$

13 $+\infty$ **15** $-\infty$ **17** -1 **19** $-\infty$

21 $+\infty$ **23** $-\infty$

27   horizontal asymptote: $y = \frac{7}{2}$; vertical asymptote: $x = \frac{5}{2}$

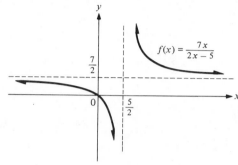

29   horizontal asymptote: $y = -\frac{2}{5}$; vertical asymptote: $x = -\frac{3}{5}$

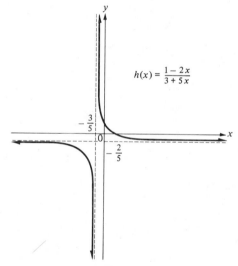

31   horizontal asymptote: $y = 0$; vertical asymptote: $x = 1$

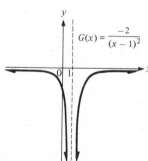

33   horizontal asymptote: $y = 1$; no vertical asymptote

35   horizontal asymptotes: $y = -\dfrac{3}{\sqrt{2}}$ and $y = \dfrac{3}{\sqrt{2}}$; no

vertical asymptote

37 horizontal asymptote: $y = 1$; vertical asymptote: $x = 2$

39 no horizontal asymptote; vertical asymptote: $x = 0$

41 horizontal asymptote: $y = \frac{3}{2}$; vertical asymptotes: $x = 0$ and $x = \frac{7}{2}$

43 no horizontal asymptote; vertical asymptotes: $x = k\pi$ for k an integer

45 no horizontal asymptote; vertical asymptotes: $x = k\pi$ for k an integer

47 oblique asymptote: $y = x$

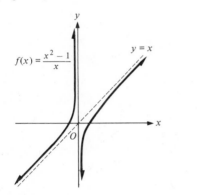

49 oblique asymptote: $y = 3x - 2$

51 oblique asymptote: $y = x + 1$

53 f is continuous at $x = 0$ and $\lim\limits_{\Delta x \to 0} \left| \dfrac{f(0 + \Delta x) - f(0)}{\Delta x} \right| =$

$$\lim\limits_{\Delta x \to 0} \left| \dfrac{1 + (\Delta x)^{1/3} - 1}{\Delta x} \right| = \lim\limits_{\Delta x \to 0} \left| \Delta x^{-2/3} \right| = +\infty$$

55 h is continuous at $x = 1$ and $\lim\limits_{\Delta x \to 0} \left| \dfrac{h(1 + \Delta x) - h(1)}{\Delta x} \right| =$

$$\lim\limits_{\Delta x \to 0} \left| \dfrac{1 + (\Delta x)^{2/3} - 1}{\Delta x} \right| = \lim\limits_{\Delta x \to 0} \left| \Delta x^{-1/3} \right| = +\infty$$

65 (a) horizontal asymptote: $I = b$; (b) For values of I less than b, there is no excitation regardless of the duration of the stimulus.

Problem Set 3.6, page 210

1

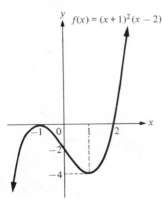

3

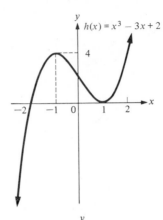

5

7

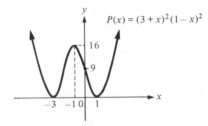

9

11

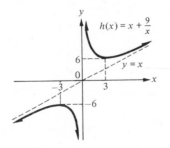

13

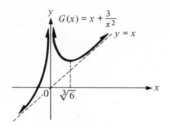

$G(x) = x + \dfrac{3}{x^2}$

$y = x$

$0 \quad \sqrt[3]{6}$

15

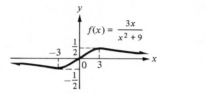

$f(x) = \dfrac{3x}{x^2 + 9}$

$-3 \quad \dfrac{1}{2} \quad 3 \quad -\dfrac{1}{2}$

17

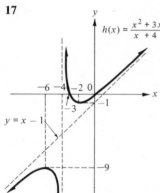

$h(x) = \dfrac{x^2 + 3x}{x + 4}$

$-6 \ -4 \ -2 \ 0 \ -3 \ -1$

$y = x - 1$

-9

19

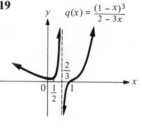

$q(x) = \dfrac{(1 - x)^3}{2 - 3x}$

$0 \ \dfrac{1}{2} \ 1 \quad \dfrac{2}{3}$

21

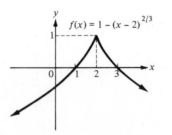

$f(x) = 1 - (x - 2)^{2/3}$

$0 \quad 1 \quad 2 \quad 3$

23

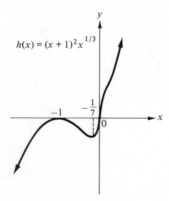

$h(x) = (x + 1)^2 x^{1/3}$

$-1 \quad -\dfrac{1}{7} \quad 0$

25

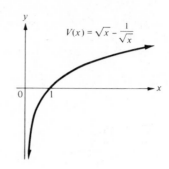

$V(x) = \sqrt{x} - \dfrac{1}{\sqrt{x}}$

$0 \quad 1$

27

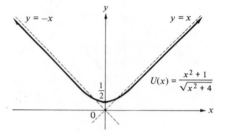

$y = -x \qquad y = x$

$\dfrac{1}{2}$

$U(x) = \dfrac{x^2 + 1}{\sqrt{x^2 + 4}}$

0

29

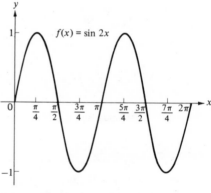

$f(x) = \sin 2x$

1

$0 \ \dfrac{\pi}{4} \ \dfrac{\pi}{2} \ \dfrac{3\pi}{4} \ \pi \ \dfrac{5\pi}{4} \ \dfrac{3\pi}{2} \ \dfrac{7\pi}{4} \ 2\pi$

-1

31

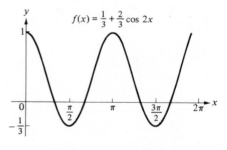

$f(x) = \dfrac{1}{3} + \dfrac{2}{3}\cos 2x$

1

$0 \quad \dfrac{\pi}{2} \quad \pi \quad \dfrac{3\pi}{2} \quad 2\pi$

$-\dfrac{1}{3}$

33

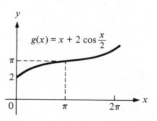

$g(x) = x + 2\cos\dfrac{x}{2}$

π

2

$0 \quad \pi \quad 2\pi$

35

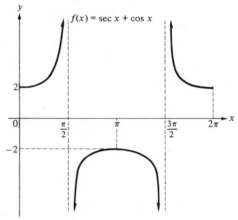

$f(x) = \sec x + \cos x$

37

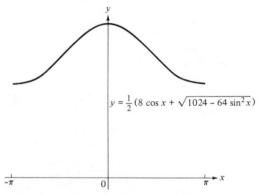

$y = \frac{1}{2}(8\cos x + \sqrt{1024 - 64\sin^2 x})$

39 (a) -0.6678; (b) -0.6668; (c) -0.6680; (d) The calculator has difficulty dealing with the relatively small difference between x and $x - 1$ for $x = 10^{10}$.

Problem Set 3.7, page 217

1 10 m by 10 m **3** (a) 10 and 10; (b) 20 and 0

5 $4(1 + \sqrt{10})$ cm by $10(1 + \sqrt{10})$ cm **7** $\dfrac{125,000}{\pi}$ m²

9 $\dfrac{1}{3}$ m by $\dfrac{1}{3}$ m

11 The minimum amount of metal occurs for:
radius $= \sqrt[3]{2057} \approx 12.72$ inches and
height $= 4114(2057)^{-2/3} \approx 25.44$ inches.

13 (a) $h = 2^{-2/3}$ m, $x = 2^{1/3}$ m; (b) $h = 1$ m, $x = 1$ m

15 $l = 0$ so that the bacterium is a sphere (endpoint extremum).

17 $|\overline{PT}| = \dfrac{9}{4}$ km **19** 0 km (endpoint extremum)

21 The power is maximum for $R = r$. **23** 3 **25** $\dfrac{1}{2}$

29 (a) $x = a\csc\theta$, $y = b\sec\theta$;

(b) $\dfrac{d}{d\theta}(x + y) = -a\csc\theta\cot\theta + b\sec\theta\tan\theta = 0$ when

$\dfrac{a}{b} = \dfrac{\sec\theta\tan\theta}{\csc\theta\cot\theta} = \tan^3\theta$

31 $(3^{2/3} + 8^{2/3})^{3/2} \approx 14.99$ m **33** $\theta = \dfrac{\pi}{3}$

35 The triangle is isosceles with legs of length $\dfrac{c}{\sqrt{2}}$.

35 $\theta = \dfrac{\pi}{3}$ **37** $x = 15\sqrt{2}$ m

Problem Set 3.8, page 225

1 $8.5\pi \approx 26.7$ m²/s **3** $0.16\pi \approx 0.503$ cm²/s

5 (a) decreasing by 12,000 cm³/min; (b) decreasing by 2,400 cm²/min

7 38 m²/min **9** (a) shortening at $\frac{18}{11}$ m/s; (b) $\frac{40}{11}$ m/s

11 $\dfrac{1350}{\sqrt{2089}} \approx 29.54$ knots **13** $\dfrac{21}{\sqrt{113}} \approx 1.976$ km/s

15 $\dfrac{24}{5}\sqrt{10} \approx 15.2$ km/h **17** 78 ft/min

19 $\dfrac{7\sqrt{3}}{10} \approx 1.21$ m/s

21 level dropping at $\dfrac{268}{1750\pi} \approx 0.0487$ ft/min

23 480 in³/min **25** $\dfrac{1}{18\pi} \approx 0.02$ m/min

27 $\dfrac{8}{5\pi} \approx 0.51$ m/min **29** 1.2×10^{-3} cm

31 $\frac{100}{3}$ in³/s **33** $-\frac{8}{5}$ cm/s **35** $\frac{2}{3}$ cm/s

37 $\dfrac{\sqrt{3}}{9} \approx 0.19$ meter/s **39** $\dfrac{\sqrt{3}}{3} \approx 0.58$ rad/s

41 $\dfrac{\pi}{40} \approx 0.079$ cm²/s **43** 105.6 rad/h

45 $\dfrac{56\pi}{45} \approx 3.91$ m/h **47** 10 m/s

Problem Set 3.9, page 233

1 (a) $10,000; (b) $2500x - x^2$; (c) $2500 - 2x$; (d) $699;
(e) $700; (f) $x = 1000$

3 (a) $5000; (b) $(0.00003)x^3 - (0.18)x^2 + 500x$;
(c) $(0.00009)x^2 - (0.36)x + 500$; (d) $248.801;
(e) $248.90; (f) $x = 2000$

5 problem 1: $\overline{C}(900) = 1611.11$ dollars/unit; problem 3:
$\overline{C}(900) = 367.86$ dollars/unit

7 Often, the point of inflection has abscissa x with $C''(x) = 0$,
so x is a critical number for C'. If this critical number
corresponds to a minimum value of C', then it is the most
efficient production level.

9 $\dfrac{x^3}{900} - 3x^2 + 4000x + 100,000$ for $0 \le x \le 1000$;

(b) $\dfrac{x^2}{300} - 6x + 4000$, for $0 < x < 1000$; (c) $1,834.668;

(d) $1,836.00; (e) $x = 900$ tons

11 (a) $(5 \times 10^{-6})x^3 - (1.5 \times 10^{-2})x^2 + 40x + 4000$ for $0 \le x \le 3000$; (b) $(1.5 \times 10^{-5})x^2 - (3 \times 10^{-2})x + 40$ for $0 < x < 3000$; (c) \$34.588005; (d) \$34.60; (e) $x = 1000$

13 (a) $p = 7000 - 5x$; (b) $7000x - 5x^2$; (c) $7000 - 10x$; (d) \$2000; (e) $x = 700$ tons

15 (a) $p = 100 - \dfrac{x}{50}$; (b) $100x - \dfrac{x^2}{50}$; (c) $100 - \dfrac{x}{25}$; (d) \$92; (e) $x = 2500$

17 (a) $-\dfrac{x^3}{900} - 2x^2 + 3000x - 100{,}000$ for $0 \le x \le 1000$; (b) $-\dfrac{x^2}{300} - 4x + 3000$ for $0 < x < 1000$; (c) $x = 522.5$ tons; (d) 4,387.51 dollars/ton; (e) $x = 34.12$ tons and $x = 947.99$ tons

19 (a) $-(5 \times 10^{-6})x^3 - (5 \times 10^{-3})x^2 + 60x - 4000$; (b) $-(1.5 \times 10^{-5})x^2 - 10^{-2}x + 60$; (c) $x = 1694$; (d) \$66.12; (e) $x = 67$ and $x = 2961$

21 $C(x) = \dfrac{23x^3}{10{,}000} - \dfrac{69x^2}{100} + 159x + 1700$ for $0 \le x \le 125$; $C'(x) = \dfrac{69x^2}{10{,}000} - \dfrac{69x}{50} + 159$ for $0 < x < 125$; most efficient production level: $x = 100$ bikes/day; $p = 200 - \frac{2}{3}x$; $R(x) = 200x - \frac{2}{3}x^2$; $R'(x) = 200 - \frac{4}{3}x$; maximum revenue when $x = 125$ bikes/day; $P(x) = -\dfrac{23x^3}{10{,}000} + \dfrac{7x^2}{300} + 41x - 1700$; $P'(x) = -\dfrac{69x^2}{10{,}000} + \dfrac{7x}{150} + 41$; maximum profit when $x = 81$ bikes/day; break even: $x = 46$ and $x = 111$

23 (a) \$12.50; (b) \$14 **25** (d) $R(x) = \beta x - \alpha x^2$; (e) $x = \dfrac{\beta}{2\alpha}$

Review Problem Set, Chapter 3, page 236

1 -3 **3** $0, \frac{7}{2}$ **5** $0, \pi$

7 (b) is not differentiable at 0; (c) is not differentiable at 2

9 (b), (c), (f), (g), and (h) fail to be differentiable on (a, b)

11 $\dfrac{9}{4}$ **13** $\dfrac{4}{3}$ **15** $\dfrac{\pi}{2}, \pi, \dfrac{3\pi}{2}$ **17** 0.497767144

19 Suppose f has two critical numbers c and d with $a < c < d < b$. Then $f'(c) = f'(d) = 0$; so, by Rolle's theorem, there is a number x with $c < x < d$ and $f''(x) = 0$. This contradicts the assumption that f'' has constant algebraic sign on (a, b).

21 (a) increasing on $(-\infty, a]$ and $[b, \infty)$, decreasing on $[a, b]$; (b) increasing on $[0, b]$, decreasing on $(-\infty, 0]$ and $[b, \infty)$; (c) increasing on $(-\infty, a]$, $[b, 0)$, and $[0, \infty)$, decreasing on $[a, b]$; (d) increasing on $(-\infty, a)$ and $[0, b]$, decreasing on $[a, 0]$ and (b, ∞)

23 increasing on $(-\infty, -2]$ and $[0, \infty)$; decreasing on $[-2, 0]$

25 increasing on $(-\infty, \frac{4}{7}]$ and $[4, \infty)$; decreasing on $[\frac{4}{7}, 4]$

27 increasing on $(-\infty, 0]$ and $(0, \infty)$

29 increasing on $\left[-\dfrac{3\pi}{2}, -\pi\right]$, $\left[-\dfrac{\pi}{2}, 0\right]$, $\left[\dfrac{\pi}{2}, \pi\right]$, and $\left[\dfrac{3\pi}{2}, 2\pi\right]$; decreasing on $\left[-2\pi, -\dfrac{3\pi}{2}\right]$, $\left[-\pi, -\dfrac{\pi}{2}\right]$, $\left[0, \dfrac{\pi}{2}\right]$, and $\left[\pi, \dfrac{3\pi}{2}\right]$

31 relative max at $x = 0$; relative min at $x = \frac{2}{3}$

33 relative max at $x = 1$; relative min at $x = 2$

35 relative max at $x = -4$; relative min at $x = 2$

37 relative max at $x = 0$; relative min at $x = -3$ and $x = 3$

39 no critical number; no relative max or min

41 relative min at $x = 1$ **43** relative max at $x = 0$

45 relative min at $x = 0$

47 critical numbers: $-\pi$ and π; no relative max or min

49 relative max at $x = 0$; relative min at $x = -\pi$ and $x = \pi$

51 relative max at $x = 0$; relative min at $x = 1$

53 concave upward on $(0, \infty)$; concave downward on $(-\infty, 0)$; point of inflection: $(0, 0)$

55 concave upward on $(-\infty, \frac{2}{3})$; concave downward on $(\frac{2}{3}, \infty)$; point of inflection: $(\frac{2}{3}, \frac{167}{27})$

57 concave upward on $(-\frac{2}{3}, \infty)$; concave downward on $(-\infty, -\frac{2}{3})$; point of inflection: $(-\frac{2}{3}, \frac{23}{27})$

59 concave upward on $(-\infty, -2)$ and $(1, \infty)$; concave downward on $(-2, 1)$; points of inflection: $(-2, -10)$ and $(1, -20)$

61 concave upward on $(-3, 0)$ and $(3, \infty)$; concave downward on $(-\infty, -3)$ and $(0, 3)$; point of inflection: $(0, 0)$

63 concave upward on $\left(\dfrac{\pi}{4}, \dfrac{3\pi}{4}\right)$ and $\left(\dfrac{5\pi}{4}, \dfrac{7\pi}{4}\right)$; concave downward on $\left(0, \dfrac{\pi}{4}\right)$, $\left(\dfrac{3\pi}{4}, \dfrac{5\pi}{4}\right)$, and $\left(\dfrac{7\pi}{4}, 2\pi\right)$; points of inflection: $\left(\dfrac{\pi}{4}, 0\right)$, $\left(\dfrac{3\pi}{4}, 0\right)$, $\left(\dfrac{5\pi}{4}, 0\right)$, and $\left(\dfrac{7\pi}{4}, 0\right)$

65 concave upward on $\left(k\pi, \dfrac{\pi}{2} + k\pi\right)$ for k an integer; concave downward on $\left(\dfrac{\pi}{2} + k\pi, \pi + k\pi\right)$ for k an integer; points of inflection: $(k\pi, k\pi)$ for k an integer

67 (a) $x = 6$; (b) $x = 12$; (c)

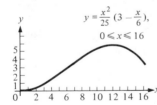

$y = \dfrac{x^2}{25}\left(3 - \dfrac{x}{6}\right),$ $0 \le x \le 16$

71 absolute max: 5 at $x = 1$ and $x = 4$; absolute min: 1 at $x = 0$ and $x = 3$

73 absolute max: 1 at $x = 2$; absolute min: -8 at $x = -1$

75 no absolute max or min

77 absolute max: $\dfrac{5}{4}$ at $x = -\dfrac{\pi}{3}$ and $x = \dfrac{\pi}{3}$; absolute min -1 at $x = -\pi$ and $x = \pi$

79 maximum value: 1.25 when $x = \frac{1}{2}$; no minimum

81 no maximum; minimum value: $-\frac{11}{16}$ when $x = \frac{3}{2}$

83 maximum value: $\dfrac{1 + \sqrt{2}}{2}$ when $x = \sqrt{2} - 1$; minimum

value: $\dfrac{1 - \sqrt{2}}{2}$ when $x = -\sqrt{2} - 1$

85 no maximum; minimum value: $\dfrac{1}{\sqrt[3]{4}} + \sqrt[3]{2}$ when $x = \dfrac{1}{\sqrt[3]{4}}$

87 maxmimum value: 1 when $x = k\pi$ for k an integer;

minimum value: -1 when $x = k\pi + \dfrac{\pi}{2}$ for k an integer

89 2   **91** 0   **93** $\dfrac{\sqrt{5}}{5}$ **95** $+\infty$ **97** $-\infty$

99 $+\infty$ **101** $-\infty$

103 horizontal asymptote: $y = \frac{1}{2}$; vertical asymptote: $x = -\frac{1}{2}$

105 horizontal asymptote: $y = 0$; no vertical asymptote

107 no horizontal asymptote; vertical asymptote: $x = -1$

109 no horizontal asymptote; vertical asymptotes:

$x = (2k - 1)\dfrac{\pi}{4}$ for k an integer

111 $y = 2x + 3$ **113** $y = 1 - x$

115 $\displaystyle\lim_{\Delta x \to 0} \left| \dfrac{f(-1 + \Delta x) - f(-1)}{\Delta x} \right| = \lim_{\Delta x \to 0} \left| \dfrac{\sqrt[3]{\Delta x}}{\Delta x} \right| = +\infty$

117

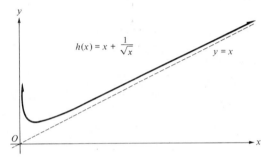

119

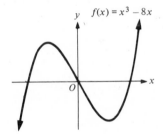

121

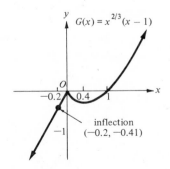

123

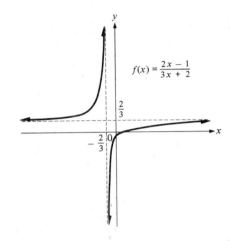

125

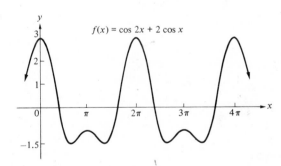

127

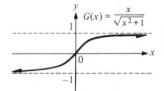

129

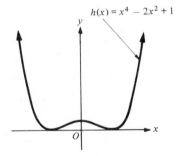

131 (a)

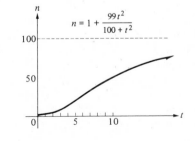

(b) $\dfrac{10\sqrt{3}}{3} \approx 5.77$ h; (c) No matter how well trained a worker is, he or she can't assemble more than 100 calculators per hour.

133 (a) 2; (b) $\frac{5}{2}$; (c) 0; (d) 5

135 $2\sqrt{2}$ cm   **137** $\dfrac{\pi}{4}$ **139** 2 meters by 1 meter

141 $x = \sqrt{\dfrac{nR}{r}}$ **143** $|\overline{BC}| = \frac{5}{4}$ km

145 $\dfrac{2\pi a^3}{9\sqrt{3}}$ **147** $\dfrac{\pi}{3}$ **149** $\dfrac{\pi}{6}$

151 approximately 0.38 rad   **153** \$10 (endpoint extremum)
155 1118   **157** 12.5 persons/year
159 $\frac{3}{400} \approx 0.01$ (dollars/bushel)/day

161 (a) $-\dfrac{6}{\sqrt{91}}$ m/s; (b) $-\frac{3}{2}$ m/s

163 53.6 km/h   **165** $\dfrac{1}{\sqrt{26}}$ m/s

167 $\frac{1}{250}$ rad/s
169 $C(x) = (3.5 \times 10^{-6})x^3 - (1.05 \times 10^{-2})x^2 + 40.5x + 1500$; $C'(x) = (1.05 \times 10^{-5})x^2 - (2.1 \times 10^{-2})x + 40.5$; most efficient production level: $x = 1000$ barrels/day;

$p = 47 - \dfrac{x}{100}$; $R(x) = 47x - \dfrac{x^2}{100}$; $R'(x) = 47 - \dfrac{x}{50}$;

$R(x)$ is maximum when $x = 1400$ barrels/day; $P(x) = -(3.5 \times 10^{-6})x^3 + (5 \times 10^{-4})x^2 + 6.5x - 1500$; $P'(x) = -(1.05 \times 10^{-5})x^2 + 10^{-3}x + 6.5$; $P(x)$ is maximum when $x = 835.85$ barrels/day; break even: $x = 233.42$ and $x = 1310.46$ barrels/day

Chapter 4

Problem Set 4.1, page 250

1 $dy = 6x\,dx$ **3** $dy = (3x^2 - 4)\,dx$
5 $dy = 13(3x + 1)^{-2}\,dx$ **7** $dy = 18x(3x^2 + 2)^2\,dx$
9 $dy = \dfrac{4(1 + x^2)^3(x^2 + 2x - 1)\,dx}{(1 + x)^5}$ **11** $dy = \dfrac{-3x\,dx}{\sqrt{9 - 3x^2}}$
13 $dy = \dfrac{3\,dx}{\sqrt{(x - 3)(x + 3)^3}}$ **15** $dy = \dfrac{-9x^2 - 4x + 21}{2(x^2 + 7)^2\sqrt{3x + 1}}\,dx$
17 $dy = 6 \cos 2x\,dx$ **19** $dy = 4 \tan x \sec^2 x\,dx$
21 $d(\cot u) = -\csc^2 u\,du$
23 $d(\csc u) = -\csc u \cot u\,du$ **25** $x\,dx + y\,dy = 0$
27 $9x\,dx - 16y\,dy = 0$ **29** $x^{-2/3}\,dx = -y^{-2/3}\,dy$
31 $(6x^2 - 4y^2 + y)\,dx + (15y^2 - 8xy + x)\,dy = 0$
33 $[9x^2(x + y)^{2/3} - 1]\,dx + [9y^2(x + y)^{2/3} - 1]\,dy = 0$
35 $\tan x \sec^2 x\,dx + \tan y \sec^2 y\,dy = 0$
37 (a) 0.63; (b) 0.6; (c) 0.03
39 (a) -1.92; (b) -1.6; (c) -0.32

41 (a) 0.181980515; (b) 0.16667; (c) 0.0153
43 (a) -0.034740437; (b) -0.0346; (c) -9.9×10^{-5}
45 3.01   **47** 28.89   **49** 0.28   **51** 1.969
53 0.315   **55** 0.4849   **57** 10.6 in^3
59 6 cm^3 **61** 0.21 m^3 **63** 4.41 cm^3
65 4%   **69** $\pm 7.45 \times 10^{-4}$

Problem Set 4.2, page 257

1 $D_x(4x^3 - 3x^2 + x - 1) = 12x^2 - 6x + 1$
3 $D_u(-\frac{1}{2}\cos u^2) = u \sin u^2$ **5** $x^3 - 2x^2 - 5x + C$
7 $\frac{1}{2}t^4 - \frac{4}{3}t^3 - \frac{5}{2}t^2 + 6t + C$
9 $\frac{3}{5}u^5 - \frac{1}{2}u^4 - \frac{1}{2}u^2 - u + C$
11 $-4x^{-1} + x^{-3} + x + C$ **13** $\dfrac{2}{5}t^5 + 5t - 7t^{-1} + C$
15 $\frac{1}{3}t^3 + \frac{3}{2}t^2 - \frac{1}{t} + C$ **17** $\frac{5}{2}x^4 + x^2 - 5x + \frac{1}{x} + C$
19 $\dfrac{16}{5}t^5 + 8t^3 + 9t + C$
21 $2w\sqrt{w}\,(\frac{1}{5}w - \frac{2}{3}) + C$ **23** $2\sqrt{x}\,(\frac{25}{7}x^3 - 1) + C$
25 $\frac{2}{3}x\sqrt{x} - 2x + 2\sqrt{x} + C$
27 $2 \sin u - 4 \cos u + C$
29 $3 \sec x - 2 \tan x + C$
31 $-2 \csc y + 7 \cot y + C$
33 $-5 \cot t + 7 \tan t + 4t + C$
35 $4 \tan x + \cot x + C$
37 $\frac{3}{5}x^{5/3} + 4 \cos x + 5 \sin x + C$
39 $\frac{4}{3}x\sqrt{x} + 3 \tan x + 4 \csc x + C$
45 Let $f(x) = g(x) = x$ in (a), (b), and (c).

Problem Set 4.3, page 263

1 $\frac{1}{20}(4x + 3)^5 + C$ **3** $\frac{1}{12}(4x^2 + 15)^{3/2} + C$
5 $\frac{3}{20}(5s^2 + 16)^{2/3} + C$ **7** $-\frac{1}{4}(1 - x^{3/2})^{8/3} + C$
9 $-\frac{1}{72}(4x^3 + 1)^{-6} + C$ **11** $\frac{4}{15}(5t^3 + 3t - 2)^{5/4} + C$
13 $-\dfrac{2}{9}(6x^3 - 9x + 1)^{-1/2} + C$ **15** $\dfrac{1}{22}\left(x + \dfrac{5}{x}\right)^{22} + C$
17 $\frac{2}{5}(5 - x)^{5/2} - \frac{10}{3}(5 - x)^{3/2} + C$
19 $\frac{2}{3}(t + 1)^{3/2} - 2(t + 1)^{1/2} + C$
21 $\frac{3}{2}(2 - x)^{4/3} - 12(2 - x)^{1/3} + C$
23 $\frac{1}{42}(3x^2 + 5)^{7/3} - \frac{5}{24}(3x^2 + 5)^{4/3} + C$
25 $\frac{2}{5}(t + 4)^{5/2} - \frac{16}{3}(t + 4)^{3/2} + 32(t + 4)^{1/2} + C$
27 $-\frac{2}{35}\cos 35x + C$ **29** $\frac{1}{2}\sin(16x - 1) + C$
31 $\frac{1}{11}\tan 11x + C$ **33** $-\frac{1}{3}\cot 3t + C$
35 $\frac{1}{2}\sec(2y + 1) + C$ **37** $-5 \sec \dfrac{t}{5} + C$
39 $\sin(\sin x) + C$ **41** $-\frac{1}{28}\cot 7x^4 + C$
43 $\dfrac{1}{2 + \cos x} + C$ **45** $\frac{1}{3}(5 + \sin 2y)^{3/2} + C$

47 $-2 \csc \sqrt{x} + C$ **49** $-\frac{2}{15}\sqrt{1 - 5 \sec 3\theta} + C$

51 (a) $\frac{1}{2} \sin^2 x + C$; (b) $-\frac{1}{4} \cos 2x + C$; (c) $-\frac{1}{4}(\cos 2x) + C = -\frac{1}{4}(1 - 2 \sin^2 x) + C = \frac{1}{2} \sin^2 x + (C - \frac{1}{4})$ and $C - \frac{1}{4}$ is a constant

53 (a) $\frac{2}{3}\sqrt{x + 1}(x - 2) + C$; (b) same as (a)

55 (a) $\dfrac{2}{13,125}\sqrt{5x - 1}(5x - 1)(375x^2 + 60x + 8) + C$;
(b) same as (a)

Problem Set 4.4, page 272

1 $y = x^5 + x^3 + x + C$ **3** $y = -\dfrac{6}{x} + 5x^3 + 10x + C$

5 $y = \dfrac{2\sqrt{7}}{5}x^2\sqrt{x} + C$ **7** $s = -\dfrac{1}{t} - \cos t + C$

9 $y = 5x - \dfrac{3}{2}x^2 + 4$ **11** $y = \dfrac{1}{4}t^4 - \dfrac{1}{t} - \dfrac{7}{2}$

13 $y = 1 + 3\sqrt{3} - 6 \cos \dfrac{x}{2}$ **15** $y = \dfrac{1}{10}(x^2 - 3)^5 + C$

17 $y = \dfrac{2}{3}\sqrt{x^3 + 7} + C$ **19** $y = \dfrac{-1}{\sqrt{2x + 1} + C}$

21 $15(y^4 + 7)^{2/3} = 4x^2 + C$ **23** $2 \sin 3y + 3 \cos 2x = C$

25 $y = \dfrac{1}{\cos x + C}$ **27** $20x - 4x^{5/2} = 5y^2 - 812$

29 $W = -\dfrac{3}{4}(1 - t^{1/3})^4 - \dfrac{1}{4}$ **31** $6 \cos 2s = 4 \sin 3t + 1$

33 $y = \frac{1}{4}x^4 + \frac{1}{3}x^3 + \frac{1}{2}x^2 + C_1x + C_2$

35 $y = \frac{9}{448}(4x + 5)^{7/3} + C_1x + C_2$

37 $s = \frac{1}{15}t^6 + \frac{3}{2}t^2 + C_1t + C_2$ **39** $y = C_1x + C_2$

41 $y = -\frac{1}{4} \cos 2x + C_1x + C_2$ **43** $y = x^3 + \frac{1}{2}x^2 + 3x + 2$

45 $s = t^2 + 2t - 3$

47 $y = \frac{25}{4}x^4 + 10x^3 + 6x^2 - 68x + \frac{191}{4}$

49 $y = -4 \sin \dfrac{x}{2} + 2$ **51** 19 m

53 (a) 4.96 s; (b) 158.68 ft   **55** (a) 2.83 s; (b) 80.62 ft/s

57 $1 + \sqrt{12g} \approx 11.84$ m/s

Problem Set 4.5, page 279

1 $f(x) = x - \frac{3}{2}x^2 + \frac{13}{2}$ **3** $f(x) = \dfrac{2x}{2 + x}$

5 $f(x) = 2 - \sin 3x$ **7** $y^2 - x^2 = 1$ **9** 24 joules

11 18 joules   **13** 2 joules   **15** 10 joules

25 (a) $C(x) = 5x + 16\sqrt{x} + 540$; (b) $P(x) = 16x - 16\sqrt{x} - 540$

27 (a) \$2420; (b) \$2870   **29** 42.5 thousand

31 (a) $R(x) = 13x - \dfrac{x^2}{80}$; (b) $P(x) = -\dfrac{x^2}{80} + 9.5x - 100$; (c) $x = 380$; (d) 8.25 dollars/month

33 (a) 1030 thousand; (b) 330 thousand; (c) 0.865 dollar/dinner

Problem Set 4.6, page 288

1 $y = \cos\left(3t - \dfrac{\pi}{2}\right)$ **3** $f(t) = \cos(\sqrt{3}t - \pi)$

5 $y = \cos\left(t - \dfrac{\pi}{2}\right)$

7 (a) problem 1: $\nu = \dfrac{3}{2\pi}$ Hz, problem 3: $\nu = \dfrac{\sqrt{3}}{2\pi}$ Hz; (b)

problem 1: $T = \dfrac{2\pi}{3}$ s, problem 3: $T = \dfrac{2\pi}{\sqrt{3}}$ s

9

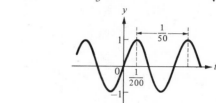

11

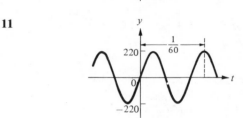

13 (a) $y = A_0 \cos 14t$; (b) $\nu = \dfrac{7}{\pi}$ Hz

15 (a) $y = 0.8 \cos 2\pi t$; (b) $\nu = 1$ Hz   **17** $K = \dfrac{m}{2}$

19 (a) differentiate both sides of $L\dfrac{dI}{dt} + \dfrac{Q}{C} = 0$ with respect to t and use the fact that $I = \dfrac{dQ}{dt}$; (b) substitute $\dfrac{dI}{dt} = \dfrac{d^2Q}{dt^2}$ into $L\dfrac{dI}{dt} + \dfrac{Q}{C} = 0$; (c) $Q = Q_0 \cos \dfrac{t}{\sqrt{LC}}$; (d) $I = \dfrac{-Q_0}{\sqrt{LC}} \sin \dfrac{t}{\sqrt{LC}}$; (e) $\nu = \dfrac{1}{2\pi\sqrt{LC}}$

21 $K = \dfrac{L}{2}$ **25** $y = A \cos(\omega t - \phi) + \dfrac{B}{\omega^2}$

27 $y = A \sin \theta = A \cos\left(\theta - \dfrac{\pi}{2}\right)$; $\theta = \displaystyle\int \omega \, dt = \omega t + C.$

Therefore, $y = A \cos(\omega t - \phi)$, where $\phi = \dfrac{\pi}{2} - C.$

29 $\phi = \dfrac{\pi}{2} - \theta_0$

Problem Set 4.7, page 294

1 $\frac{8}{3}$ square units   **3** $\frac{64}{3}$ square units   **5** 8 square units

7 $\pi - 2$ square units   **9** 8 square units

11 4 square units   **13** 4 square units

15 $\frac{16}{3}$ square units   **17** $\frac{9}{2}$ square units

19 $\frac{27}{4}$ square units   **21** $\frac{125}{6}$ square units

23 $8\sqrt{3}$ square units   **25** $\frac{1}{3}$ square unit

27 $8\sqrt{6}$ square units

Review Problem Set, Chapter 4, page 295

1 $dy = (3x^2 - 1)\, dx$ **3** $dy = \dfrac{2(x^2 + x - 5)}{(2x + 1)^2}\, dx$

5 $dy = -6 \cot x \csc^2 x\, dx$

7 $(x^2 - 2y)\, dx + (y^2 - 2x)\, dy = 0$

9 $\sec^2 x \tan x\, dx - \csc^2 y \cot y\, dy = 0$

11 $(\pi x^3 \cos \pi xy - 12x)\, dy + (\pi x^2 y \cos \pi xy + 12y)\, dx = 0$

13 (a) 0.0401; (b) 0.04; (c) 0.25%

15 (a) 0.2424; (b) 0.24; (c) 0.99%

17 (a) 0.01727 (b) 0.01732 (c) 0.29%

19 $\sqrt{36.1} \approx 6 + \dfrac{0.1}{2\sqrt{36}} \approx 6.00833$

21 $\cos\left(\dfrac{\pi}{3} + 0.1\right) \approx 0.5 - \left(\sin\dfrac{\pi}{3}\right)(0.1) \approx 0.413$

23 $\dfrac{1}{1.02} \approx 1 - \dfrac{1}{1^2}(0.02) = 0.98$ **25** 400 m^2

27 $3x^2\, \Delta x$ cubic units   **29** ± 0.66 cm^2

31 $\frac{3}{5}x^5 + \frac{4}{3}x^3 + 11x + C$ **33** $\frac{9}{7}t^2\sqrt[3]{t} + C$

35 $\frac{7}{24}(3t + 9)^{8/7} + C$ **37** $\frac{1}{54}(x^3 + 8)^{18} + C$

39 $\frac{5}{32}(x^8 + 13)^{4/5} + C$

41 $\frac{2}{5}(7 + x)^{5/2} - \frac{14}{3}(7 + x)^{3/2} + C$

43 $\frac{2}{45}(x^3 + 1)\sqrt{x^3 + 1}(3x^3 - 2) + C$

45 $3 \sin x + 2 \cos x + C$ **47** $-\frac{2}{3} \cos 3x + C$

49 $2t - \frac{3}{4} \sin 4t + C$ **51** $\frac{1}{2} \sec x^2 + C$

53 $-\frac{2}{15}(2 + 3 \cot \beta)^{5/2} + C$ **55** $\frac{1}{3}(1 - \sin v)^{-3} + C$

57 $-\frac{1}{27}(\sec 3x + 8)^{-9} + C$ **59** $-\frac{1}{3}(ax + b \sin x)^{-3} + C$

61 $y = x^2 + x + C$ **63** $s = \dfrac{1}{3 - t} + C$

65 $y = -\frac{1}{24}(1 - x^{3/2})^{16} + C$

67 $y = \frac{3}{2}x^2 - \frac{1}{3}x^3 + \frac{1}{2}x^4 + C_1 x + C_2$

69 $\sqrt{x} + \sqrt{y} = C$ **71** $y = \dfrac{x^2}{2} - 3 \sin x + C$

73 $y = \frac{2}{3}x^{3/2} - \tan x + C$ **75** $\sec x + \tan y = C$

77 $y = \frac{1}{2}x^4 + x^2 + x$ **79** $y = -\sqrt{1 - x^2}$

81 $y = \frac{1}{20}x^5 + \frac{1}{2}x^2 + x$ **83** $s = \frac{1}{2}t^2 + 3 \cos t - 3$

85 $f(x) = \frac{1}{3}(x^2 + 5)^{3/2} - 3$ **89** $\frac{289}{6}$ joules

91 $2\sqrt{3}$ joules   **93** 1000 ft-lb   **95** 4.4083 m/s^2

97 (a) 2s; (b) 160 ft/s   **99** 112,500 joules

101 $\dfrac{GMm}{ab}$ newtons

103 (a) $10x - \dfrac{x^2}{25,000}$ dollars; (b) $400 + 6x$ dollars;

(c) 125,000; (d) 50,000; (e) $8

105 $y = \cos\left(5t - \dfrac{\pi}{2}\right)$ **107** $y = 4 \cos\left(6x - \dfrac{\pi}{6}\right)$

109 (a) $y = 0.03 \cos(10t - 0.88)$; (b) $\dfrac{10}{2\pi} \approx 1.6$ Hz;

(c) $\dfrac{2\pi}{10} \approx 0.63$ s

111 12 square units   **113** $\frac{7}{6}$ square units

115 $\frac{32}{3}$ square units   **117** 4 square units

119 $\frac{2}{3}$ square unit

Chapter 5

Problem Set 5.1, page 306

1 $3 + 5 + 7 + 9 + 11 + 13 = 48$

3 $\sin \pi + \sin \dfrac{\pi}{2} + \sin \dfrac{\pi}{3} + \sin \dfrac{\pi}{4} = \dfrac{2 + \sqrt{3} + \sqrt{2}}{2}$

5 $\frac{1}{2} + \frac{1}{6} + \frac{1}{12} + \frac{1}{20} + \frac{1}{30} = \frac{5}{6}$

7 $\frac{1}{3} + \frac{1}{4} + \frac{1}{7} + \frac{1}{12} = \frac{17}{21}$ **9** 412

11 2,700   **13** $\dfrac{5^{101} - 5}{4}$ **15** $\dfrac{n(n + 1)(n + 2)}{3}$

17 $\dfrac{(n - 1)n(2n - 1)}{6}$ **19** $\dfrac{(n - 1)(2n - 1)n}{6}$ **21** $\frac{7}{3}$

27 $k^2 - (k - 1)^2 = 2k - 1$, and so

$$\sum_{k=1}^{n} (2k - 1) = \sum_{k=1}^{n} [k^2 - (k - 1)^2] = n^2.$$

29 $n^2 = \displaystyle\sum_{k=1}^{n} (2k - 1) = 2\sum_{k=1}^{n} k - \sum_{k=1}^{n} 1 = 2\sum_{k=1}^{n} k - n$; hence,

$$\sum_{k=1}^{n} k = \frac{n^2 - n}{2}.$$

Problem Set 5.2, page 315

1 $\frac{75}{4}$ **3** $\frac{3776}{3465}$ **5** 9   **7** 15

9 $\frac{11}{6}$ **11** 8   **13** -15 **15** exists

17 exists   **19** exists   **21** exists

23 does not exist   **25** exists   **27** does not exist

29 -9 **31** $-\frac{11}{6}$ **33** 0   **35** 7

37 *Every* Riemann sum equals $b - a$.

Problem Set 5.3, page 326

1 2   **3** -9 **5** -1 **7** 4π

9 20   **11** 24   **13** 10   **15** $\frac{23}{6}$

17 35   **19** $-\frac{145}{2}$ **21** $-\frac{70}{3}$ **23** 0

25 12   **27** $\frac{5}{2}$ **29** $\frac{17}{4}$ **31** $3 - \dfrac{\pi}{2}$

33 -3 **35** $\frac{88}{3}$ **37** $\frac{23}{2}$

39 (a) true; (b) false; (c) true; (d) true; (e) false; (f) true

41 7; $c = 2$ **43** 6; $c = -1$ or 3   **45** $\frac{29}{14}$; $c = \frac{29}{14}$

47 $\dfrac{A(b + a)}{2} + B$; $c = \dfrac{b + a}{2}$ **49** $1 + \cos x \geq 0$ for all x

51 same function, same interval (dummy variables)

53 additivity with respect to the interval of integration

55 additivity with respect to the interval of integration

57 additivity with respect to the interval of integration

61 $\dfrac{k}{2}(b + a)$; $c = \dfrac{b + a}{2}$ **63** yes

67 (a) $x - a$; (b) $\frac{1}{2}(x^2 - a^2)$; (c) $\frac{1}{3}(x^3 - a^3)$

Problem Set 5.4, page 336

1 6   **3** $-\frac{15}{2}$ **5** -244 **7** $\frac{23}{2}$

9 36   **11** $\frac{83}{15}$ **13** $\frac{16}{5}$ **15** $\frac{1}{2}$

17 $\sqrt{2} - 1$ **19** $\sqrt{3}$ **21** $\frac{5}{64}$ **23** $\frac{4}{3}\sqrt{2}$

25 $4 - \sqrt{5}$ **27** $\frac{1}{4}(9^{4/3} - 1)$ **29** $\frac{50}{3}$

31 $\frac{4}{3}$ **33** $-\dfrac{4}{\pi}$ **35** $\dfrac{2}{\pi}(3 - \sqrt{3})$

37 $\frac{1}{3}$ **39** $-\frac{3}{16}$ **41** $4\sqrt{3}$ **43** $\frac{8}{3}$

45 $\frac{376}{15}$ **47** $x^2 + 1$ **49** $\dfrac{1}{1 + x^2}$

51 $\sin (x^4)$

53 $\sqrt{x^2 + 4}$ **55** $-(x^{10} + 3)^{25}$ **57** 0

59 $3(135x^3 + 1)^7$ **61** $8(8x - 1)^{15}$ **63** $\sqrt{2 - x}$

65 $6x[(3x^2 + 2)^4 + 17]^{1/4} - (x^4 + 17)^{1/4}$ **71** \$1,860

75 (a) 8; (b) 9; (c) $\frac{13}{6}$; (d) $\dfrac{b|b| - a|a|}{2(b - a)} + 1$

77

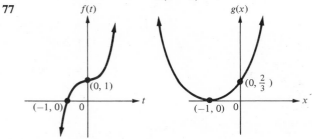

79

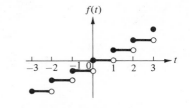

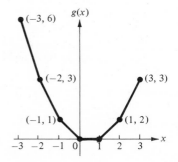

Problem Set 5.5, page 345

1 .7827941   **3** 1.106746   **5** .9246708

7 2.0503025   **9** 0.448082   **11** .3497583

13 $\frac{1}{96}$ **15** Each trapezoid lies under the curve.

17 1.5666667   **19** .9086625   **21** 4.6713533

23 3.6534704   **25** 1.1872812   **27** $n = 4$

29 $\pi \approx 3.08$

Problem Set 5.6, page 351

1

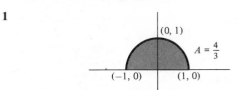

3 $\frac{1}{2}$

5 8   **7** $\frac{3}{2}$ **9** $\frac{11}{12}$

11 1   **13** 2   **15** $\frac{4}{3}$

17 (b) $(-\frac{1}{2}, \frac{1}{4})$ and $(\frac{5}{2}, \frac{25}{4})$; (c) $\frac{9}{2}$

19 (b) $(-2, -8)$ and $(2, -8)$; (c) $\frac{32}{3}$

21 (b) $(0, 0)$ and $(2, 2)$; (c) $\frac{4}{3}$

23 (b) $(0, 0)$ and $(2, 4)$; (c) $\frac{8}{3}$

25 (b) $(-1, -11)$ and $(8, 16)$; (c) $\frac{243}{2}$

27 (b) $(1, 1)$ and $(4, 4)$; (c) $\frac{9}{2}$

29 (b) $(-\frac{3}{2}, \frac{1}{2})$ and $(3, -1)$; (c) $\frac{27}{8}$

31 (b) $(-1, -1)$, $(0, 0)$, and $(1, 1)$; (c) 1

33 (b) $(-1, 1)$, $(0, 0)$, and $(1, 1)$; (c) $\frac{4}{15}$

35 (b) $(2, 2)$ and $(2, -2)$; (c) $\frac{64}{3}$

37 (b) $(-1, -1)$, $(0, 0)$, and $(1, 1)$; (c) $\frac{1}{6}$

39 (b) $\left(-\dfrac{\pi}{3}, \dfrac{1}{2}\right)$ and $\left(\dfrac{\pi}{3}, \dfrac{1}{2}\right)$; (c) $2\left(\sqrt{3} - \dfrac{\pi}{3}\right)$

41 15   **43** $\frac{49}{12}$ **45** $\frac{323}{12}$, $\frac{35}{12}$

47 $\frac{344}{6}$, $\frac{130}{3}$ **49** $\frac{77}{4}$, $\frac{73}{12}$ **51** $\frac{7}{2}$

Review Problem Set, Chapter 5, page 353

1 90   **3** $\frac{163}{60}$ **5** $\frac{1}{2}(\sqrt{2} - 1)$

7 $\dfrac{n(n + 1)(4n - 1)}{6}$ **9** $2(3^{n+1} - 1)$

11 1,001,000   **13** (a) $\displaystyle\sum_{k=0}^{13} 3^k$ cents; (b) \$23,914.84

15 $\frac{21}{64}$ **17** $\frac{21}{2}$ **19** $\frac{45}{2}$ **21** 24

23 24   **25** exists   **27** does not exist

29 does not exist   **31** does not exist   **33** -24

35 0   **37** no; yes   **39** -30

41 31   **43** 13   **45** 8   **47** $\dfrac{5\pi}{4}$

49 (a) true; (b) true   **51** $x \leq x^3$ for $1 \leq x \leq 3$

53 $x^5 \le x$ for $0 \le x \le 1$ **55** $\sin x \le \cos x$ for $0 \le x \le \dfrac{\pi}{4}$

57 $4 \le 20$ **59** $7;\ c = 2$ **61** $\dfrac{8}{3};\ c = \dfrac{2\sqrt{3}}{3}$

63 $\dfrac{5}{4};\ c = \dfrac{5}{4}$ **65** 27   **67** 32

69 2   **71** $\dfrac{166}{15}$ **73** 9   **75** $\dfrac{4\sqrt{2}}{3}$

77 $\dfrac{2}{5}$ **79** 9   **81** $\dfrac{1}{4}$ **83** $-\dfrac{1}{8}$

85 4   **87** $\dfrac{16}{3}$ **89** $(4x + 1)^{300}$

91 $(8x^{17} + 5x^2 - 13)^{40}$ **93** $-\dfrac{x^2}{\sqrt{x^4 + 8}}$

95 0   **97** $\dfrac{2x^5}{1 + x^4}$

99 $(10t + 1) \cos [(5t^2 + t)^5 + 1] - 4 \cos [(4t + 3)^5 + 1]$

101 (a) $\dfrac{16}{3}$; (b) $\dfrac{1}{6}$ **103** false   **105** false

107 true   **109** false   **111** 7.011667

113 68.268172   **115** 1.3256537

117 3.3071124   **119** 5.8922432

121 0.6046034   **123** (b) (0, 0) and (2, 2); (c) 1

125 (b) $\left(-\dfrac{3}{\sqrt{2}}, \dfrac{9}{2}\right)$ and $\left(\dfrac{3}{\sqrt{2}}, \dfrac{9}{2}\right)$; (c) $18\sqrt{2}$

127 (b) (−1, 2) and (3, 18); (c) $\dfrac{32}{3}$

129 (b) (1, −1) and $(-\dfrac{11}{4}, \dfrac{3}{2})$; (c) $\dfrac{125}{48}$

131 (b) (0, 4) and (4, −12); (c) $\dfrac{32}{3}$

133 (b) (3, 3), (−3, −3), and (3, −3); (c) 30   **135** $\dfrac{81}{2}$

Chapter 6

Problem Set 6.1, page 364

1 $\dfrac{2196\pi}{5}$ **3** $\dfrac{80\pi}{3}$ **5** $\dfrac{20\pi}{3}$ **7** π

9 $\dfrac{96\pi}{5}$ **11** $\dfrac{64\pi}{5}$ **13** $\dfrac{384\pi}{7}$ **15** $2\sqrt{2}$

17 $\dfrac{64\pi}{15}$ **19** $\dfrac{\pi}{6}$ **21** $\dfrac{64\pi}{5}$ **23** $\dfrac{1024\pi}{15}$

25 $\dfrac{29\pi}{30}$ **27** $\dfrac{27\pi}{2}$ **29** $\dfrac{\pi}{2}$ **31** $\dfrac{288\pi}{5}$

33 24π **35** 8π **37** $\dfrac{768\pi}{5}$ **39** $8\pi(2a - 3)$

Problem Set 6.2, page 368

1 $\dfrac{3\pi}{2}$ **3** $\dfrac{2\pi}{3}(2\sqrt{2} - 1)$ **5** $\dfrac{11\pi}{10}$

7 $\dfrac{38\pi}{3}$ **9** $\dfrac{\pi}{2}(2 + \sqrt{2})$ **11** $\dfrac{13,122\pi}{7}$

13 $\dfrac{512\pi}{5}$ **17** $\dfrac{20\pi}{3}$ **19** $\dfrac{29\pi}{30}$

21 $\dfrac{8192\pi}{5}$ **23** $\dfrac{10,240\pi}{3}$ **25** $\dfrac{56\pi}{3}$

29 $\dfrac{\pi h}{3}(b - a)(b + 2a)$ **31** $\dfrac{4}{3}\pi ab^2$

Problem Set 6.3, page 373

1 144 cubic units   **3** $10\sqrt{3}$ m³

5 $\dfrac{\pi}{6}(y_2^3 - y_1^3)$ cubic units

7 (a) 135 cubic units; (b) $\dfrac{325}{6}$ cubic units

9 $\dfrac{16}{3}$ ft³   **11** $\dfrac{4347\pi}{5120}$ m³

15 $\pi h^2\left(r - \dfrac{h}{3}\right)$ cubic units   **17** $\dfrac{4a^3}{3}$ cubic units

19 $\dfrac{\sqrt{2}}{3} l^3$ cubic units   **21** 1120 m³

Problem Set 6.4, page 381

1 $2\sqrt{17}$ **3** $a\sqrt{1 + m^2}$ **5** $\dfrac{8}{27}(10\sqrt{10} - 1) \approx 9.0734$

7 $\dfrac{8}{27}\left(10\sqrt{10} - \dfrac{13\sqrt{13}}{8}\right) \approx 7.6337$

9 $\dfrac{2}{3}(2\sqrt{2} - 1) \approx 1.21895$ **11** $\dfrac{53}{6}$

13 $\dfrac{59}{24}$ **15** 1.13   **17** 1.22

19 $\displaystyle\int_a^b \sqrt{[f'(t)]^2 + [g'(t)]^2}\, dt$ **21** $39\pi\sqrt{10}$

23 $\dfrac{\pi}{27}(145\sqrt{145} - 1) \approx 203.0436$

25 $\dfrac{1179\pi}{256} \approx 14.4685$ **27** $\dfrac{\pi}{96}(577^{3/2} - 1) \approx 453.5352$

29 $\dfrac{61\pi}{192}$ **31** 14.69   **33** $\pi r(r + \sqrt{h^2 + r^2})$

35 (a) 6; (b) $\dfrac{12\pi}{5}$ **37** (a) $\sqrt[3]{36n\pi}$; (b) infinity

Problem Set 6.5, page 391

1 $\dfrac{275}{48}$ ft-lb   **3** $\dfrac{409,455\pi}{2}$ joules

5 $88,200\pi$ joules   **7** 81,868.8 ft-lb

9 6250 joules   **11** 1.98559 ft-lb

13 4284.7 joules   **15** 5.292×10^6 N

17 145.824 N   **19** 148,262.4 lb   **21** 156.8 N

23 (a) 5.096×10^{-2} N; (b) 5.096×10^{-2} joules; (c) 35,321,821

25 (a) no; (b) yes   **27** $\dfrac{2\sqrt{5}}{5} \approx 0.8944$ m

29 (a) $v = \pm\sqrt{\dfrac{2}{ms}}$ (m = mass of P);

(b) $s = \left(125 - \dfrac{3}{\sqrt{2m}}t\right)^{2/3}$; (c) $s = \left(125 + \dfrac{3}{\sqrt{2m}}t\right)^{2/3}$; (d) 0

31 $V = \dfrac{8.99 \times 10^9}{s}$ **33** $\frac{1}{2}m(v_0^2 - 2gs)$

Problem Set 6.6, page 397

1 (a) $p = \left(2 - \dfrac{q}{200}\right)^2$; (b) \$266.67; (c) \$266.67

3 $p = \dfrac{3}{2} - \dfrac{q}{10,000}$; \$312.50   **5** 4320 autos

7 1560.77 tons   **9** 0.0015π cm^3/s

Review Problem Set, Chapter 6, page 397

1 $\dfrac{96\pi}{5}$ **3** $\dfrac{16\pi}{15}$ **5** $\dfrac{8\pi}{3}$

7 $\dfrac{\pi}{3}$ **9** $\dfrac{255\pi}{4}$ **11** $\dfrac{160\pi}{3}$

13 $126\pi^2$ cm^2 **15** (b) $\dfrac{\pi b^2 a}{2}$; (c) $\dfrac{\pi b^2 a}{2}$

17 $\dfrac{\pi B^2 A}{2}$; cylinder has twice the volume

19 $\frac{8}{3}$ **21** $\frac{338}{15}$ ft^3 **23** $2\sqrt{3}a^3$

25 $\dfrac{686}{3}$ **27** $\dfrac{13\sqrt{13} - 8}{27}$ **29** $\frac{2}{3}(5\sqrt{5} - 1)$

31 $\frac{2}{3}(2\sqrt{2} - 1)$ **33** $\frac{5}{27}(17\sqrt{85} - 16\sqrt{10})$

35 $\frac{2}{27}(10\sqrt{10} - 1)$ **37** $\frac{2}{9}(10\sqrt{10} - 3\sqrt{3})$

39 $\displaystyle\int_1^4 \sqrt{1 + x^{-1}}\, dx \approx 3.62$ **41** $\displaystyle\int_0^1 \sqrt{1 + 4x^2}\, dx \approx 1.48$

45 $\dfrac{\pi}{2}(125 - 13\sqrt{13})$ **47** $\dfrac{10\pi}{27}(73\sqrt{730} - \sqrt{10})$

49 $\dfrac{\pi}{9}(82\sqrt{82} - 1)$ **51** $\dfrac{8\pi}{3}(5\sqrt{5} - 1)$

53 $\dfrac{\pi}{6}(37\sqrt{37} - 1)$

55 $62.4\pi(\frac{1}{2}r^2h^2 + \frac{2}{3}r^2h + \frac{1}{4}r^4)$ ft-lb

57 13,500 joules   **59** 3,927,363.64 ft-lb

61 117,600 N   **63** 20,000 joules

65 \$2.14   **67** \$13,066.67

69 21.55%

Chapter 7

Problem Set 7.1, page 407

7 (a) invertible; (b) not invertible; (c) not invertible

11 (a) $f^{-1}(x) = \dfrac{x + 5}{2}$ **13** (a) $f^{-1}(x) = \sqrt[3]{x}$

15 (a) $f^{-1}(x) = \sqrt[3]{\dfrac{1 - x}{2}}$ **17** (a) $f^{-1}(x) = \dfrac{-1}{1 + x}$

19 (a) $f^{-1}(x) = \dfrac{7 + x}{3 - x}$ **21** $f^{-1}(x) = \dfrac{-dx + b}{cx - a}$

23 $\frac{1}{5}$ **25** $\frac{1}{4}$ **27** $-\frac{5}{9}$ **29** $-(3^{3/2})$

31 $\frac{1}{2}$ **33** 7   **35** $\sqrt{2}$ **37** (b) -5; (c) -5

39 (a) $f^{-1}(x) = \dfrac{1 + \sqrt{8x - 7}}{4}$ for $x \geq \dfrac{7}{8}$; (b) $\dfrac{1}{3}$; (c) $\dfrac{1}{3}$

41 $\frac{2}{3}$ **45** $C = f^{-1}(t)$

Problem Set 7.2, page 416

1 $\dfrac{\pi}{2}$ **3** $-\dfrac{\pi}{4}$ **5** 0   **7** $\dfrac{\pi}{4}$

9 $\dfrac{\pi}{3}$ **11** $-\dfrac{\pi}{4}$ **13** $-\dfrac{\pi}{6}$ **15** $\dfrac{3\pi}{4}$

17 $\dfrac{2\pi}{3}$ **19** $\dfrac{2\pi}{3}$ **21** $\dfrac{\pi}{6}$

23 0.6999768749   **25** 0.4391814802

27 1.107148718   **29** -1.270032196

31 -0.5829630403 **33** 0.9775965505

35 0.3013796990   **37** 1.194027796

39 -0.6094418025 **43** $\frac{3}{5}$

45 $\dfrac{\pi}{6}$ **47** $-\dfrac{\pi}{4}$ **49** $\dfrac{4}{5}$ **51** $\dfrac{3\sqrt{10}}{10}$

53 $\frac{4}{3}$ **55** $\frac{10}{7}$ **57** $\sqrt{2}$ **59** $4\sqrt{3}$

71 $2x\sqrt{1 - x^2}$ **73** $2x\sqrt{1 - x^2}$ **75** $\dfrac{\sqrt{2}}{2}$

Problem Set 7.3, page 424

1 $\dfrac{3}{\sqrt{1 - 9x^2}}$ **3** $\dfrac{5}{25 + x^2}$ **5** $\dfrac{3}{|t|\sqrt{t^6 - 1}}$

7 $\dfrac{-2t}{t^4 + 6t^2 + 10}$ **9** $\dfrac{2}{\sqrt{9 - 4x^2}}$ **11** $\dfrac{u}{|u|\sqrt{1 - u^2}}$

13 $\dfrac{-2}{|s|\sqrt{s^2 - 4}} - \dfrac{2}{s^2 + 4}$ **15** 0

17 $\dfrac{-3x^2}{\sqrt{1 - 9x^2}} + 2x \cos^{-1} 3x$

19 $\dfrac{-5}{x^2(x^2 + 25)} - \dfrac{2}{x^3} \tan^{-1} \dfrac{5}{x}$

21 $\dfrac{-2x - x\sqrt{x^4 + 2x^2}\, \csc^{-1}(x^2 + 1)}{(x^2 + 1)^{3/2}\sqrt{x^4 + 2x^2}}$

23 $\dfrac{\sqrt{1 - y^2}(1 - \sin^{-1} y)}{x - \sqrt{1 - y^2}}$

25 $\dfrac{1 + y^2}{1 + x^2}$ **29** $\sin^{-1} \dfrac{x}{2} + C$ **31** $\dfrac{1}{2}\sin^{-1} \dfrac{2x}{3} + C$

33 $\dfrac{1}{3}\sin^{-1} 3t + C$ **35** $\dfrac{1}{6}\tan^{-1} \dfrac{3y}{2} + C$

37 $\dfrac{1}{6}\tan^{-1} \dfrac{2x}{3} + C$ **39** $\dfrac{1}{5}\sec^{-1}\left|\dfrac{4t}{5}\right| + C$

41 $\sec^{-1}\left|\dfrac{x}{4}\right| + C$ **43** $\dfrac{2\pi}{3}$ **45** $\dfrac{\sqrt{3}\pi}{9}$

47 $-\dfrac{\pi}{12}$ **49** $\sin^{-1}\left(\dfrac{\sin x}{6}\right) + C$

51 $\dfrac{1}{4}\tan^{-1}\left(\dfrac{x^2}{2}\right) + C$ **53** $2\tan^{-1}\left(\sin\dfrac{x}{2}\right) + C$

55 $\dfrac{1}{3}(\tan^{-1}6 - \tan^{-1}2\sqrt{3})$ **57** $3(\tan^{-1}2) - \dfrac{3\pi}{4}$

59 $\dfrac{1}{b}\sin^{-1}\dfrac{bx}{a} + C$

61 $y = \pi - 2x$ for $0 \le x \le \dfrac{\pi}{2}$; $y = 0$ for $\dfrac{\pi}{2} \le x \le \pi$;

$y = 2x - 2\pi$ for $\pi \le x \le \dfrac{3\pi}{2}$; $y = \pi$ for $\dfrac{3\pi}{2} \le x \le 2\pi$

63 3.65 m **65** $-\frac{1}{3}$ rad/min **67** $\frac{126}{169}$ rad/s

69 $2x - \dfrac{\pi}{2}$ **71** $\dfrac{\pi^2}{4}$ cubic units

Problem Set 7.4, page 430

1 $\dfrac{8x}{4x^2 + 1}$ **3** $\dfrac{\cos(\ln x)}{x}$ **5** $1 - 6\cot 6x$

7 $\dfrac{2t\sin t}{t^2 + 7} + \cos t \ln(t^2 + 7)$ **9** $\dfrac{1}{u\ln u}$

11 $x\tan x + \ln\left(\dfrac{\sec x}{5}\right)$ **13** $\dfrac{5v + 4}{2v(v + 1)}$

15 $-2\tan x$ **17** $\dfrac{1 - 2x^3}{3x(4x^3 + 1)}$

19 $\dfrac{t^3 + 5 - 3t^3\ln t}{t(t^3 + 5)^2}$ **21** $\dfrac{2x\sec^2 x - 2\tan x \ln(\tan^2 x)}{x^3\tan x}$

23 $\dfrac{y}{x}$ **25** $\dfrac{y^2 - y\cot x}{\ln(\sin x) - 2xy}$

27 (a) $\dfrac{\cos[(\ln x)^2]}{x}$; (b) $-\sin x \ln[\tan(\cos^4 x)]$

29 $\frac{1}{5}\ln|7 + 5x| + C$ **31** $\ln|\sin x| + C$

33 $\tan(\ln 4x) + C$ **35** $2\ln(x^2 + 7) + C$

37 $\sin(\ln x) + C$ **39** $\sin^{-1}(\ln x) + C$

41 $\ln\frac{1}{5} - \ln\frac{1}{8}$ **43** $\frac{1}{2}(\ln 9 - \ln 4)$

45 $\sin(\ln 4)$ **47** $\ln 4$ m **49** $\frac{1}{3}(\ln 4)^2$

Problem Set 7.5, page 437

1 2.3025 **3** 4.605 **5** 0.5493 **7** 4 **9** 4

11 domain: $(-\infty, 2)$; range: $\mathbb{R}$; vertical asymptote: $x = 2$

13 domain: all real numbers except -1; range: $\mathbb{R}$; vertical asymptote: $x = -1$

15 domain: $(0, \infty)$; range: $[a, \infty)$, where $\ln a = -1$; minimum at $(a, -a)$

17 domain: $(0, \infty)$; range: $[1, \infty)$; minimum at $(1, 1)$; vertical asymptote: $x = 0$

19 tangent line: $y - 4\ln 2 = (2 + 4\ln 2)(x - 2)$;
normal line: $y - 4\ln 2 = -(2 + 4\ln 2)^{-1}(x - 2)$

21 $\ln\dfrac{7}{5}$ **23** $3\ln 2$ **25** $\dfrac{\pi}{2}\ln\dfrac{17}{2}$ **27** $\dfrac{3\pi}{2}\ln 2$

29 8.325063694 **31** -3.208826489 **33** -15.56881884

35 (a) 0.3365; (b) 3.3616 **37** absolute maximum at $(1, 1)$

39 $x = a$ where $\ln a = -\frac{1}{2}$ **41** 10 days

Problem Set 7.6, page 444

1 (a) 5; (b) $\dfrac{1}{8}$; (c) $16e^3$; (d) $\dfrac{1}{x}$; (e) $x - x^2$; (f) x; (g) $x^2 - 4$;

(h) $\dfrac{x^2}{e^4}$; (i) $x - 2$; (j) $\dfrac{x}{y^3}$

3 (a) 0.3678794412; (b) 0.1353352832; (c) 20.08553692;
(d) 1.648721271; (e) 9.356469012; (f) 15.15426223;
(g) 0.04460095534; (h) 0.06598803588; (i) 23.14069264;
(j) 0.6608598017

5 $7e^{7x}$ **7** $3x^2$ **9** $-e^x\sin(e^x)$

11 $e^{-2t}(\cos t - 2\sin t)$ **13** $e^{x^2 + 5\ln x}\left(2x + \dfrac{5}{x}\right)$

15 $e^{t\ln t}(1 + \ln t)$ **17** $\dfrac{3}{\sqrt{e^{6x} - 1}}$ **19** $-6e^{3s}(1 - e^{3s})$

21 $\dfrac{-x}{\sqrt{2\pi}}\exp\left(-\dfrac{x^2}{2}\right)$ **23** $e^{-x}(-8x^2 + 19x - 4)$

25 $\dfrac{y(y - e^x)}{1 + e^x - 2xy}$ **27** $\dfrac{e^{x+y} - \sin y}{x\cos y - e^{x+y}}$ **31** $\dfrac{1}{3}e^{3x} + C$

33 $\frac{1}{5}e^{5x+3} + C$ **35** $\frac{1}{10}e^{5x^2} + C$ **37** $\tan^{-1}e^x + C$

39 $6\sqrt{e^x + 4} + C$ **41** $-e^{\cot x} + C$

43 $\frac{1}{2}(e^2 - 1)$ **45** $\dfrac{2e}{3}$ **47** 0

49 decreasing on $\mathbb{R}$; concave up on $\mathbb{R}$; horizontal asymptote: $y = 0$

51 increasing on $(-\infty, 1]$; decreasing on $[1, \infty)$; concave up on $(2, \infty)$; concave down on $(-\infty, 2)$; inflection point: $(2, 2e^{-2})$; absolute max at $x = 1$; horizontal asymptote: $y = 0$

53 $\dfrac{e^3}{3} - \dfrac{e^2}{2} + \dfrac{1}{6}$ **59** 606.53 dollars/year

61 -0.1098938 (lb/in^2)/min **63** $4\ln 25 \approx 13$ days

Problem Set 7.7, page 450

1 (a) 2.665144143; (b) 0.37521442272; (c) 8.824977827;
(d) 0.1133147323; (e) 1.632526919; (f) 36.46215961;
(g) 0.2927940321; (h) 20.96434682

3 $-3\pi x^{-3\pi - 1}$ **5** $-5(\ln 6)(6^{-5x})$ **7** $2(\ln 3)(3^{2x+1})$

9 $5^{\sin t}(\ln 5)\cos t$ **11** $2^{-7x^2}[2x - 14x(\ln 2)(x^2 + 5)]$

13 $\dfrac{2^{x+1}[(\ln 2)(x^2 + 5) - 2x]}{(x^2 + 5)^2}$ **15** $(\ln 5)\dfrac{5^x - 5^{-x}}{5^x + 5^{-x}}$

17 $\dfrac{2(\ln 3)3^x}{(3^x + 1)^2}$ **19** $\dfrac{1}{t(\ln 10)(1 + t)}$

21 $3^{\tan u}\left[(\ln 3)(\sec^2 u)\log_8 u + \dfrac{1}{u\ln 8}\right]$

23 $\dfrac{2x(x+2) - (\ln 3)(x^2+5)\log_3(x^2+5)}{(\ln 3)(x^2+5)(x+2)^2}$

25 $\dfrac{-2^y}{x(\ln 2)2^y + 2y}$ **27** $x^{\sqrt{x}}\left(\dfrac{2+\ln x}{2\sqrt{x}}\right)$

29 $(\sin x^2)^{3x}[6x^2 \cot x^2 + 3 \ln(\sin x^2)]$

31 $(x^2+4)^{\ln x}\left[\dfrac{2x\ln x}{x^2+4} + \dfrac{\ln(x^2+4)}{x}\right]$

33 $(x^2+7)(6x^3+1)^3(4x)(24x^3 + 126x^2 + 1)$

35 $\dfrac{(\sin x)\sqrt[3]{1+\cos x}}{\sqrt{\cos x}}\left[\cot x - \dfrac{\sin x}{3(1+\cos x)} + \dfrac{1}{2}\tan x\right]$

37 $\dfrac{x^2\sqrt[5]{x^2+7}}{\sqrt[4]{11x+8}}\left[\dfrac{2}{x} + \dfrac{2x}{5(x^2+7)} - \dfrac{11}{4(11x+8)}\right]$

39 $\dfrac{3^{5x}}{5\ln 3} + C$ **41** $\dfrac{7^{x^4+4x^3}}{4\ln 7} + C$ **43** $\dfrac{-2^{-\ln x}}{\ln 2} + C$

45 $\dfrac{-4^{\cot x}}{\ln 4} + C$ **47** $\dfrac{12}{25\ln 5}$

49 (a) 4.64385619; (b) 0.6309297536; (c) 0.480898347; (d) 1.405954306; (e) −7.55152422

51 $e^\pi > \pi^e$ **61** $\dfrac{40\pi}{\ln 3}$

Problem Set 7.8, page 456

9 (a) 1.509461355; (b) 2.150898465; (c) 0.6043677771; (d) 1.160465504; (e) 0.8435506876; (f) −0.9741682480

11 $6x \cosh(3x^2+5)$ **13** $3t^2 \coth t^3$

15 $3e^{3t}\operatorname{sech}^2 e^{3t}$ **17** $\operatorname{sech} s$ **19** $2x\operatorname{sech} y$

21 $\cos x \operatorname{sech} y$ **23** $\frac{1}{7}\sinh 7x + C$

25 $\frac{1}{3}\tanh 3x + C$ **27** $-\frac{1}{10}\operatorname{sech}^2 5x + C$

29 $\dfrac{1}{4}(\cosh^4 1 - 1)$ **31** $\dfrac{x^2}{4} + \dfrac{1}{2}\ln|x| + C$

35 $\dfrac{3x^2}{\sqrt{1+x^6}}$ **37** $\dfrac{1}{\sqrt{x^2-9}}$ **39** $\dfrac{5}{1-25t^2}$

41 $\dfrac{xe^x}{\sqrt{e^{2x}-1}} + \cosh^{-1} e^x$

43 $\dfrac{1}{1-(\ln u)^2} + \tanh^{-1}(\ln u)$

51 $\sinh^{-1}\dfrac{x}{3} + C$ **53** $2a \sinh\dfrac{b}{a}$

Problem Set 7.9, page 465

1 e^6 **3** e^{-5} **5** e^4 **7** e^{-4}

9 $h = 10$: 109.95116; $h = 100$: 339.30208; $h = 1000$: 396.26044; $h = 10{,}000$: 402.70356; $e^6 \approx 403.4287935$

11 (a) $R = 7\%$; (b) $S = \$2409.85$

13 (a) $R = 7.2458\%$; (b) $S = \$2482.80$

15 (a) $R = 14.3674\%$; (b) $S = \$53{,}471.36$

17 (a) \$1166.40; (b) \$1169.86; (c) \$1171.66; (d) \$1172.89; (e) \$1173.37; (f) \$1173.49; (g) \$1173.51; (h) \$1173.51

19 (a) \$96.12; (b) \$96.09; (c) \$96.08; (d) \$96.08

21 (a) 5.65406%; (b) \$11,082.77

23 (a) 7%; (b) 7.25%

25 \$12,047.77   **27** $q = 2e^{5t}$

29 $N = 40e^{-4t}$ **31** $q = 10 - 7e^{-t}$

33 164.49 min   **35** approximately 292 bears

37 5.12 h   **39** (a) $q = 10(2^{t/20})$; (b) $q = 80$

41 (a) 4 g; (b) 1.7227 years   **43** 605.07 days

45 0.9752 g   **47** \$5206.16   **51** 2209 years

Problem Set 7.10, page 473

1 (a) $N = 10e^{0.03t}$ (million); (b) 18.22 (million); (c) $T \approx 23$ years

3 approximately 2%/yr   **5** approximately 35%/day

7 (a) $C = 700$; (b) 421; (c) $t_I = 2.88$ years

11 (a) $C = 5.92$ (million); (b) 5.74 (million)

17 $\dfrac{d^2N}{dt^2} = -k$; hence, there is no inflection point.

19 (a) 57.5 words/min; (b) 1.06 (words/min)/h

Problem Set 7.11, page 478

1 $y = Ce^{-x^2/2}$ **3** $N = 1000 + Ce^{t/10}$

5 7.87 min   **7** 159.46°C   **9** 61.8 min   **11** 0.78%

15 $y = e^{4x}(x + C)$ **17** $y = e^{-t^2}(t^2 + C)$ **19** $q = e^{-\sin t}(t + C)$

21 (a) $I = \dfrac{E}{R} + Ce^{-Rt/L}$, $\displaystyle\lim_{t\to\infty} I = \dfrac{E}{R}$; (b) $I = \dfrac{1}{Le^{Rt/L}}\displaystyle\int Ee^{Rt/L}\, dt$

23 5.61%

Review Problem Set, Chapter 7, page 480

7 invertible   **9** not invertible

13 (a) $f^{-1}(x) = \dfrac{x+9}{7}$ **15** (a) $f^{-1}(x) = \dfrac{4}{x} - 1$

21 $\dfrac{7}{15}\left(\dfrac{3}{7}\right)^{4/5}$ **23** $\dfrac{5}{18}$ **25** $\dfrac{1}{2}$ **27** $\dfrac{1}{7}$ **29** $\dfrac{1}{\pi}$

31 (a) 0.3843967745; (b) 1.971536521; (c) 0.3836622700; (d) 1.553343046; (e) 0.9588938924; (f) 1.445467496; (g) −0.8480620790; (h) 1.446441332; (i) 2.711892987; (j) 0.3510036017; (k) 1.107148718; (l) −0.2526802551

33 $\dfrac{3}{5}$ **35** $-\dfrac{12}{13}$ **37** $-\dfrac{5\pi}{14}$

41 $t = -\sin^{-1}\dfrac{2}{3}$ or $t = \dfrac{\pi}{6}$

43 $\dfrac{2}{\sqrt{9-x^2}}$ **45** $\dfrac{1}{2(x-1)\sqrt{x-2}}$

47 $\dfrac{-\sqrt{3}}{2\sqrt{t}\sqrt{1-3t}}$ **49** $\dfrac{4u^5}{1+u^8} + 2u\tan^{-1} u^4$

51 $\dfrac{-2\csc^{-1} u}{|u|\sqrt{u^2-1}}$ **53** $\dfrac{-5x^3}{1+25x^2} + 3x^2\cot^{-1} 5x$

55 $\dfrac{[17 + (\sin^{-1} u)^2]^{34}}{\sqrt{1 - u^2}}$ **57** $\dfrac{\sqrt{1 - (x + y)^2}\,\cos^{-1}(x + y) - x}{2y\sqrt{1 - (x + y)^2} + x}$

59 $\dfrac{3x \cos(x - y) + 3 \sin(x - y) - 2xy}{x^2 + 3x \cos(x - y)}$ **61** $\sin^{-1} \dfrac{x}{3} + C$

63 $\dfrac{1}{6} \tan^{-1} \dfrac{x}{6} + C$ **65** $\dfrac{1}{2\sqrt{2}} \sin^{-1}\left(\dfrac{\sqrt{2}x^2}{3}\right) + C$

67 $\dfrac{1}{7} \sec^{-1} \dfrac{|x|}{7} + C$ **69** $-\dfrac{1}{2} \tan^{-1}\left(\dfrac{\cos x}{2}\right) + C$

71 $-\dfrac{1}{2}(\cot^{-1} v)^2 + C$ **73** $\dfrac{\pi}{20}$

75 $t = 1 + n\pi$ and $t = 1 + \dfrac{2n + 1}{2}\pi$ for $n = 0, \pm 1, \pm 2, \ldots$

77 $\dfrac{\pi}{10} \cos \theta$ cm^2/min **79** $\dfrac{2x}{x^2 + 7}$

81 $\dfrac{3r + 4}{2r(r + 2)}$ **83** $\dfrac{x}{1 + (\ln x)^2} + 2x \tan^{-1}(\ln x)$

85 $\dfrac{\ln u\,(2 - \ln u)}{u^2}$ **87** $\cot x - \dfrac{1}{x}$

89 $-12x^2 e^{-4x^3}$ **91** $\dfrac{-2e^{-2x}}{\sqrt{1 - e^{-4x}}}$

93 $\dfrac{-4e^t}{e^{2t} - 4}$ **95** $\dfrac{-2x}{\sqrt{1 - x^4}} - e^{x^3}(3x^3 + 1)$

97 $e^x \sec^2 e^x$ **99** $-16e^{4x}(3 - e^{4x})^3$

101 $\dfrac{2(1 + e^x + e^{-x})}{(e^x + 2)(e^{-x} + 2)}$ **103** $4ex^{4e-1}$

105 $-(\ln 5)(\sin x)5^{\cos x}$ **107** $2x(\ln 7)(\cos x^2)7^{\sin x^2}$

109 $3^{5x}2^{4x^2}(5 \ln 3 + 8x \ln 2)$ **111** $\dfrac{1 - \ln t}{t^2 \ln 7}$

113 $\dfrac{1}{4x(\ln 10)(\log_{10} x)^{3/4}}$ **115** $4e^{4x} \sinh e^{4x}$

117 $e^{-t} \operatorname{csch}(e^{-t}) \coth(e^{-t})$ **119** $-2xe^{\operatorname{sech} x^2} \operatorname{sech} x^2 \tanh x^2$

121 $\dfrac{(\operatorname{sech} t)(\operatorname{sech} t - \tanh t)}{\operatorname{sech} t + \tanh t}$ **123** $\dfrac{3}{\sqrt{9x^2 + 6x + 2}}$

125 $\dfrac{e^x}{1 - e^{2x}}$ **127** No, x could also be 4.

129 $(3x)^x(1 + \ln 3x)$ **131** $(\sin x)^{x^2}[x^2 \cot x + 2x \ln(\sin x)]$

133 $(\tanh^{-1} x)^{x^3}\left[3x^2 \ln(\tanh^{-1} x) + \dfrac{x^3}{(\tanh^{-1} x)(1 - x^2)}\right]$

135 $\dfrac{\cos x \sqrt[3]{1 + \sin^2 x}}{\sin^5 x}\left[-\tan x + \dfrac{\sin 2x}{3(1 + \sin^2 x)} - 5 \cot x\right]$

137 $\dfrac{x^2(x + 5)^3 \sin 2x}{\sec 3x}\left[\dfrac{2}{x} + \dfrac{3}{x + 5} + 2 \cot 2x - 3 \tan 3x\right]$

141 $\dfrac{2}{x^2(\sin y - 4e^{4y})}$ **143** $\dfrac{\sinh(x - y) + \cosh(x + y)}{\sinh(x - y) - \cosh(x + y)}$

145 $\dfrac{1}{x[5 + (\ln x)^3]}$ **147** $\dfrac{1}{2} \ln|8 + 3x| + C$

149 $\dfrac{1}{2}(\ln x)^2 + C$ **151** $2e^{\sqrt{x}} + C$ **153** $\tan(e^x) + C$

155 $\dfrac{10^x}{\ln 10} + C$ **157** $\ln \dfrac{8}{5}$ **159** $-\dfrac{1}{2}(\coth x)^2 + C$

161 $e^{\cosh^{-1} x} + C$ or $x + \sqrt{x^2 - 1} + C$ **163** $\cosh^{-1} x + C$

165 $-\dfrac{1}{4} \operatorname{sech}^{-1} \dfrac{|x|}{2} + C$ **177** 3π

179 max of $\dfrac{1}{\sigma\sqrt{2\pi}}$ at $x = \mu$

181 after 12 years: (a) \$99,603.51; (b) \$167,772.16
185 6.64 h **187** an additional 2.42 min
191 (a) 75,064; (b) 57,624; (c) 8.07 years

Chapter 8

Problem Set 8.1, page 489

1 $\sin x - \dfrac{1}{3} \sin^3 x + C$

3 $-\dfrac{1}{2}\left(\cos 2t - \dfrac{2}{3} \cos^3 2t + \dfrac{1}{5} \cos^5 2t\right) + C$

5 $\dfrac{1}{16} \sin^8 2x - \dfrac{1}{20} \sin^{10} 2x + C$ **7** $\dfrac{1}{3} \sin^3 x - \dfrac{1}{5} \sin^5 x + C$

9 $\dfrac{1}{12}(6x - \sin 6x) + C$ **11** $\dfrac{1}{2}(t - \sin t) + C$

13 $\dfrac{5}{16}u - \dfrac{1}{4} \sin 2u + \dfrac{3}{64} \sin 4u + \dfrac{1}{48} \sin^3 2u + C$

15 $-\dfrac{1}{14} \cos 7x - \dfrac{1}{6} \cos 3x + C$

17 $\dfrac{1}{2} \sin x + \dfrac{1}{14} \sin 7x + C$ **19** $\dfrac{1}{8} \sin 4u - \dfrac{1}{20} \sin 10u + C$

21 $\dfrac{2}{5} \cos^5 x - \dfrac{1}{3} \cos^3 x - \dfrac{1}{7} \cos^7 x + C$

23 $\dfrac{1}{3 \cos^3 x} - \dfrac{1}{\cos x} + C$

25 $\dfrac{5}{16}x + \dfrac{1}{16} \sin 8x + \dfrac{3}{256} \sin 16x - \dfrac{1}{192} \sin^3 8x + C$

27 $\dfrac{8}{5} + \dfrac{2}{5}\left(\dfrac{\sqrt{2}}{2}\right)^{5/2} - 2\left(\dfrac{\sqrt{2}}{2}\right)^{1/2}$ **29** $\dfrac{62}{65\pi}$

31 $\dfrac{3\pi - 4}{384}$ **33** $-\dfrac{4}{5\pi}$ **39** $\dfrac{\pi^2}{2}$ **41** π

43 $f(t) = \dfrac{t}{2} - \dfrac{1}{4\pi} \sin 2\pi t$; $s = 4$ at $t = 8$

Problem Set 8.2, page 494

1 $\dfrac{1}{4} \ln|\sin 4x| + C$ **3** $\dfrac{1}{3} \ln|\sec 3x + \tan 3x| + C$

5 $\dfrac{3}{2} \tan(2x/3) - x + C$

7 $-\dfrac{1}{12} \cot^3 4x + \dfrac{1}{4} \cot 4x + x + C$

9 $-\dfrac{1}{3}\left(\cot 3t + \dfrac{1}{3} \cot^3 3t\right) + C$

11 $\dfrac{1}{10} \tan^5 2t + C$ **13** $\dfrac{1}{35} \sec^7 5x - \dfrac{1}{25} \sec^5 5x + C$

15 $\dfrac{1}{2} \tan 2x - \dfrac{1}{2} \cot 2x + C$ **17** $\dfrac{2}{5} \tan^{5/2} t + 2 \tan^{1/2} t + C$

19 $\dfrac{1}{28} \tan^4 7x + \dfrac{1}{42} \tan^6 7x + C$ **21** $-\dfrac{1}{9} \csc^3 3x + C$

23 $\dfrac{1}{15} \sec^3 5x - \dfrac{1}{5} \sec 5x + C$

25 $\dfrac{1}{5} \sec^{5/2} 2x - \sec^{1/2} 2x + C$

27 $\dfrac{1}{11} \sec^{11} x - \dfrac{2}{9} \sec^9 x + \dfrac{1}{7} \sec^7 x + C$

29 $\dfrac{1}{24 \cot^3 8x} + C$ **31** $\dfrac{1}{6} \csc^3 (x^2) - \dfrac{1}{10} \csc^5 (x^2) + C$

33 $\frac{1}{3} \ln (\sqrt{3}/2)$ **35** $\frac{12}{35}$ **37** $5[(4 - \pi)/2]$

39 $2\sqrt{3}\pi$ **41** $-\ln (\sqrt{2} - 1)$

Problem Set 8.3, page 498

1 $\dfrac{-\sqrt{16 - x^2}}{16x} + C$

3 $-\dfrac{1}{16}\left[\dfrac{1}{3}\left(\dfrac{\sqrt{4 - t^2}}{4}\right)^3 + \dfrac{\sqrt{4 - t^2}}{t}\right] + C$

5 $\dfrac{2}{27}\left[\sin^{-1}\dfrac{3x}{2} - \dfrac{3x\sqrt{4 - 9x^2}}{4}\right] + C$

7 $\dfrac{-(\sqrt{7 - 4t^2})^3}{21t^3} + C$

9 $\sqrt{t^2 - a^2} + C$ **11** $\dfrac{-\sqrt{1 + x^2}}{x} + C$

13 $\dfrac{1}{\sqrt{5}} \ln \left|\dfrac{\sqrt{t^2 + 5} - \sqrt{5}}{t}\right| + C$ **15** $\dfrac{7}{8}\left(\dfrac{2x^2 + 9}{\sqrt{4x^2 + 9}}\right) + C$

17 $\dfrac{\sqrt{x^2 - 4}}{4x} + C$ **19** $\dfrac{1}{3} \ln \left|\dfrac{3t + \sqrt{9t^2 - 4}}{2}\right| + C$

21 $\dfrac{1}{2} \ln \dfrac{3}{2}$ **25** $\dfrac{2 + t}{9\sqrt{5 - 4t - t^2}} + C$

27 $\dfrac{4t^2 + 6t}{7(t^2 + 3t + 4)} - \dfrac{12\sqrt{7}}{49} \tan^{-1}\left(\dfrac{2t + 3}{\sqrt{7}}\right) + C$

29 $\dfrac{1}{3}\sqrt{3x^2 - x + 1} +$

$\qquad\qquad \dfrac{\sqrt{3}}{18} \ln \left|\sqrt{3x^2 - x + 1} + \sqrt{3}x - \dfrac{\sqrt{3}}{6}\right| + C$

31 $\dfrac{-x}{9\sqrt{4x^2 - 9}} + C$

33 $\dfrac{1}{2} \ln (t^2 - 4t + 8) + \tan^{-1}\dfrac{t - 2}{2} + C$

35 $\dfrac{1}{2} \sin^{-1} [2(x - 1)] + C$ **37** $-\sqrt{4 - x^2} + C$

39 $\left(x - \dfrac{3}{2}\right)\sqrt{x^2 - 3x + 2} -$

$\qquad\qquad \dfrac{1}{4} \ln |2x - 3 + 2\sqrt{x^2 - 3x + 2}| + C$

41 $2\sqrt{x^2 + 2x + 2} + C$ **43** $\dfrac{-\cos v}{25\sqrt{25 - \cos^2 v}} + C$

45 $\dfrac{1}{27}\left(\dfrac{3}{4} - \sin^{-1}\dfrac{3}{5}\right)$ **47** $3\sqrt{3} - \pi$

49 $\dfrac{45}{4} \ln \dfrac{7}{5}$ **51** $\dfrac{\pi^2}{2048}$

57 $y = \ln \left|\dfrac{x + \sqrt{x^2 - 9}}{3}\right| - \dfrac{\sqrt{x^2 - 9}}{x} + C$

Problem Set 8.4, page 506

1 $\frac{1}{2}x \sin 2x + \frac{1}{4} \cos 2x + C$ **3** $\frac{1}{3}xe^{3x} - \frac{1}{9}e^{3x} + C$

5 $x \ln 5x - x + C$ **7** $x \cos^{-1} x - \sqrt{1 - x^2} + C$

9 $x \sec^{-1} x - \ln |x + \sqrt{x^2 - 1}| + C$

11 $t \sec t - \ln |\sec t + \tan t| + C$

13 $-\dfrac{x^2}{3} \cos 3x + \dfrac{2}{9} x \sin 3x + \dfrac{2}{27} \cos 3x + C$

15 $(3x^2 - 2x - 5) \sin x + (6x - 2) \cos x + C$

17 $(\frac{1}{4}x^2 + \frac{1}{4}x - \frac{1}{8})e^{2x} + C$

19 $\frac{1}{5}e^{-x}(2 \sin 2x - \cos 2x) + C$

21 $-\frac{1}{2} \csc x \cot x + \frac{1}{2} \ln |\csc x - \cot x| + C$

23 $\frac{1}{2}e^{x^2}(x^2 - 1) + C$

25 $\frac{1}{2}x\sqrt{1 + x^2} + \frac{1}{2} \ln |x + \sqrt{1 + x^2}| + C$

27 $-(2x + 1)e^{-x} + C$

29 $\frac{1}{2}e^x(x \sin x - x \cos x + \cos x) + C$

31 $x(\ln x)^2 - 2x \ln x + 2x$

33 $-x \cot x + \ln |\sin x| + C$

35 $-\frac{1}{3}(x^2 + 2)\sqrt{1 - x^2} + C$

37 $\frac{1}{4}[x^8 \sin (x^4) + 2x^4 \cos (x^4) - 2 \sin (x^4)] + C$

39 $\frac{1}{243}(12\sqrt{3}\pi - 2\pi^2 - 36)$

41 $3 \sec^{-1} 3 - \dfrac{2\pi}{3} + \ln \dfrac{2 + \sqrt{3}}{3 + \sqrt{8}}$

43 $\sqrt{2}\left(3 + \dfrac{13\pi}{8} - \dfrac{5\pi^2}{32}\right) - 9$ **45** $\dfrac{\pi^2 + 4}{16}$

47 $\frac{1}{2}x^4 \sin 2x + x^3 \cos 2x - \frac{3}{2}x^2 \sin 2x - \frac{3}{2}x \cos 2x + \frac{3}{4} \sin 2x + C$

49 $-e^{-t}(t^4 + 4t^3 + 12t^2 + 24t + 24) + C$

55 $\frac{1}{4} \sec^3 x \tan x + \frac{3}{8} \sec x \tan x + \frac{3}{8} \ln |\sec x + \tan x| + C$

57 $x^2 g'(x) - 2xg(x) + 2\int g(x)\, dx$

59 $1 - \dfrac{2}{e}$ **61** $\dfrac{x^{a+1} \ln x}{a + 1} - \dfrac{x^{a+1}}{(a + 1)^2} + C$

Problem Set 8.5, page 515

1 $\ln \dfrac{|x - 2|^{3/2}}{|x|^{1/2}} + C$

3 $\frac{5}{3} \ln |3y - 2| + \frac{7}{2} \ln |2y + 1| + C$

5 $2 \ln |t| + \ln |t - 2| + \ln |t + 1| + C$

7 $\ln \dfrac{|x - 1|}{\sqrt{|x(x + 2)|}} + C$ **9** $\ln \dfrac{\sqrt{|x^2 - 1|}}{|x|} + C$

11 $x + \ln \left|\dfrac{(x - 3)^{9/5}}{(x + 2)^{4/5}}\right| + C$

13 $\frac{1}{2}x^2 - \frac{15}{7} \ln |x + 4| + \frac{6}{7} \ln |x - 3|$

15 $\dfrac{x^2}{2} + 2x + C$ **17** $\ln \left|\dfrac{(x + 4)^{29/5}(x - 1)^{6/5}}{x^2}\right| + C$

19 $x + \ln \left|\dfrac{(x - 2)^{4/5}}{(x + 3)^{9/5}}\right|$ **21** $\ln |y| - \dfrac{3}{y - 1} + \dfrac{2}{(y - 1)^2} + C$

23 $-\dfrac{3}{z} + \ln \left[\left(\dfrac{4z + 1}{z}\right)^{10}\right] + C$

25 $\ln \left(\left| \dfrac{x+1}{x+7} \right|^{1/9} \right) - \dfrac{1}{3(x+1)} + C$

27 $\ln \dfrac{|x+2|^{5/8}}{|3x-2|^{13/72}} - \dfrac{8}{9(3x-2)} + C$

29 $\ln \left(\left| \dfrac{z-3}{z-1} \right|^{9/2} \right) + \dfrac{1}{(z-1)^2} + \dfrac{5}{z-1} + C$

31 $\ln \frac{27}{20}$ **33** $\frac{167}{6} + \ln \frac{3}{2}$ **35** $\frac{10}{3} + 4 \ln 3$

37 $\dfrac{ax}{c} + \dfrac{bc - ad}{c^2} \ln |cx + d| + K$

39 $\ln |(x-3)^{4/5}(x+2)^{1/5}| + C$ **41** $2 - \ln 3$

43 $\dfrac{1}{2\sqrt{aq}} \ln \left| \dfrac{\sqrt{q} + \sqrt{ax})^2}{q - ax^2} \right| + C$

45 $x = a - \dfrac{1}{\sqrt[3]{3kt + C}}$ **47** $C(x) = 100 \ln \dfrac{5x^3(x-1)^2}{8(x+3)}$

49 $y = \dfrac{ACe^{(A+B)kt} - B}{1 + Ce^{(A+B)kt}}$ **51** $p = \dfrac{1}{1 + Ce^{-kt}}$

Problem Set 8.6, page 523

1 $\ln |x-1| - \dfrac{1}{2} \ln (x^2 + 4) - \dfrac{1}{2} \tan^{-1} \dfrac{x}{2} + C$

3 $\ln \dfrac{|x|^3}{(x^2+1)^{3/2}} + \tan^{-1} x + C$

5 $\ln [|x-1|(x^2+1)] + 3 \tan^{-1} x + C$

7 $\ln \dfrac{|x^2 - 1|}{x^2 + 1} + C$

9 $\ln |t^{1/25}(t^2 + 25)^{49/50}| - \dfrac{1}{5} \tan^{-1} \left(\dfrac{t}{5} \right) + C$

11 $\ln \left| \dfrac{x}{\sqrt{x^2 + 4}} \right| + \dfrac{2}{x^2 + 4} + C$

13 $\ln [(y^2 + 4)^2 |3y - 1|] + C$

15 $\ln [|x-2|(2x^2 - 3x + 5)] + \dfrac{12}{\sqrt{31}} \tan^{-1} \dfrac{4x - 3}{\sqrt{31}}$

17 $\ln [(x-2)^8(2x^2 - 3x + 5)] + \frac{1}{2}x^2 + C$

19 $\ln [t^{2/9}(t^2 + 9)^{43/18}] + \dfrac{1}{9t} - \dfrac{26}{27} \tan^{-1} \dfrac{t}{3} + C$

21 $-\dfrac{4}{x} - 6 \tan^{-1} x - \dfrac{1}{2(x^2 + 1)} - \dfrac{2x}{x^2 + 1} + C$

23 $x + \ln \dfrac{|x|^{1/2}}{(x^2 + 2)^{1/4}} - \dfrac{1}{x^2 + 2} - \dfrac{5x}{4(x^2 + 2)} - \dfrac{5\sqrt{2}}{8} \tan^{-1} \dfrac{x}{\sqrt{2}}$

25 $\ln \dfrac{|t|}{(t^2 + 2t + 2)^{1/2}} + \dfrac{1}{t^2 + 2t + 2} + \tan^{-1} (t+1) + C$

27 $\frac{9}{2} \ln 4 - \frac{9}{4} \ln 10 + \frac{11}{2} \tan^{-1} 3$ **29** $\dfrac{\pi - 2 \ln 2}{4}$

31 $\ln \frac{4}{5}$ **33** $u = \sin x$ **35** $u = e^x$

37 $\dfrac{ax + b}{x^2 + 1} + \dfrac{(c - a)x + (d - b)}{(x^2 + 1)^2}$ **39** $\ln |x^5 + 2x^3 + x| + C$

45 (a) $\dfrac{1}{2} \dfrac{w}{w^2 + 1} + \dfrac{1}{2} \tan^{-1} w + C;$

(b) $\dfrac{1}{4} \dfrac{w}{(w^2 + 1)^2} + \dfrac{3}{4} \left[\dfrac{w}{2(w^2 + 1)} + \dfrac{1}{2} \tan^{-1} w \right] + C$

Problem Set 8.7, page 527

1 $-2\sqrt{x} - 2 \ln (1 - \sqrt{x}) + C$

3 $\frac{3}{2}x^{2/3} - 3x^{1/3} + 3 \ln |1 + \sqrt[3]{x}| + C$

5 $\frac{2}{3}x^{3/2} - 2x + 8\sqrt{x} - 16 \ln (2 + \sqrt{x}) + C$

7 $\frac{1}{20}(2x^2 - 1)^{5/2} + \frac{1}{12}(2x^2 - 1)^{3/2} + C$

9 $\frac{1}{21}(3x + 1)^{7/3} - \frac{1}{12}(3x + 1)^{4/3} + C$

11 $\frac{1}{288}(4x + 1)^{9/2} - \frac{1}{112}(4x + 1)^{7/2} + \frac{1}{160}(4x + 1)^{5/2} + C$

13 $\dfrac{1}{4} \ln \left| \dfrac{1 + 3 \tan \frac{x}{2}}{3 + \tan \frac{x}{2}} \right| + C$

15 $\dfrac{1}{2} \ln \left| \tan \dfrac{x}{2} \right| - \dfrac{1}{4} \tan^2 \dfrac{x}{2} + C$

17 $-\dfrac{1}{4 \tan^2 \frac{\theta}{2}} - \dfrac{1}{2} \ln \left| \tan \dfrac{\theta}{2} \right| + C$

19 $-\operatorname{csch}^{-1} x + C$ **21** $t - \frac{4}{5}t^{5/4} + C$

23 $2\sqrt{x} - 4\sqrt[4]{x} + 4 \ln (1 + \sqrt[4]{x}) + C$

25 $\frac{4}{7}(1 - x)^{7/4} - \frac{4}{3}(1 - x)^{3/4} + C$

27 $-4\sqrt[4]{x} - 4 \ln |\sqrt[4]{x} - 1| + C$

29 $\frac{2}{5}(1 + e^x)^{5/2} - \frac{2}{3}(1 + e^x)^{3/2} + C$

31 $x\sqrt{\dfrac{1 - x}{x}} - \tan^{-1} \sqrt{\dfrac{1 - x}{x}} + C$

33 $\frac{5}{9}(w + 32)^{9/5} - 40(w + 32)^{4/5} + C$

35 $\ln \left| 1 + \tan \dfrac{x}{2} \right| + C$

37 $\dfrac{1}{2} \ln \left| \tan \left(\dfrac{t}{2} \right) - 1 \right| - \dfrac{1}{2} \ln \left| \tan \left(\dfrac{t}{2} \right) + 1 \right| -$
$\dfrac{1}{\tan \left(\frac{t}{2} \right) + 1} + \dfrac{1}{\left[\tan \left(\frac{t}{2} \right) + 1 \right]^2} + C$

39 $\dfrac{28\sqrt{3}}{5}$ **41** $2\left(-1 - \dfrac{\pi}{4} + 2 \ln \dfrac{5}{2} + \tan^{-1} 2 \right)$

43 $\dfrac{92 - 14\sqrt{2}}{15}$ **45** $\ln \dfrac{3}{2}$ **47** $\ln 2$

53 $75 - 10 \ln 4$ **51** $\dfrac{2}{1 - \tanh \frac{x}{2}}$

Problem Set 8.8, page 530

1 $\dfrac{1}{125} \left[\dfrac{3}{3 + 5u} + \ln |3 + 5u| \right] + C$ **3** $\dfrac{-\sqrt{5 + x^2}}{5x} + C$

5 $\dfrac{3}{\sqrt{5}} \ln |\sqrt{5y} + \sqrt{11 + 5y^2}| + C$

7 $\dfrac{t}{64}(13 + 16t^2)\sqrt{13 + 8t^2} - \dfrac{169\sqrt{8}}{512} \ln |\sqrt{8}t + \sqrt{13 + 8t^2}| + C$

9 $\dfrac{-\sqrt{3y^2 - 5}}{y} + \sqrt{3}\,\ln\left|\sqrt{3}y + \sqrt{3y^2 - 5}\right| + C$

11 $\dfrac{t}{8}(2t^2 - 5)\sqrt{t^2 - 5} - \dfrac{25}{8}\ln\left|t + \sqrt{t^2 - 5}\right| + C$

13 $\dfrac{1}{7\sqrt{7}}\left[\dfrac{\sqrt{7}x(14x^2 - 5)\sqrt{5 - 7x^2}}{8} + \dfrac{25}{8}\sin^{-1}\sqrt{\dfrac{7}{5}}x\right] + C$

15 $\dfrac{2}{\sqrt{23}}\tan^{-1}\dfrac{8t + 3}{\sqrt{23}} + C$

17 $-\dfrac{1}{\sqrt{5}}\ln\left|\dfrac{\sqrt{5 - 4x + 2x^2}}{x} - \dfrac{2}{\sqrt{5}}\right| + C$

19 $\frac{1}{3}[(3y + 2)\sin^{-1}(3y + 2) + \sqrt{1 - (3y + 2)^2}] + C$

21 $\dfrac{1}{25}\left[\dfrac{50x^2 - 1}{4}\cos^{-1}5x - \dfrac{5x\sqrt{1 - 25x^2}}{4}\right] + C$

23 $\dfrac{x}{2(x^2 + 1)} + \dfrac{1}{2}\tan^{-1}x + C$

25 $4\sqrt{3 + 2v}\left(\dfrac{v^4}{9} - \dfrac{4v^3}{21} + \dfrac{12v^2}{35} - \dfrac{24v}{35} + \dfrac{72}{35}\right) + C$

27 $-\frac{1}{12}\csc^3 3x\cot 3x - \frac{1}{8}\csc 3x\cot 3x + \frac{1}{8}\ln\left|\csc 3x - \cot 3x\right| + C$

29 $\frac{1}{7}\left(\frac{1}{4}\tan^4 7x - \frac{1}{2}\tan^2 7x - \ln\left|\cos 7x\right|\right) + C$

33 (a) $-\dfrac{\sin ax\cos ax}{2a} + \dfrac{1}{2}x + C;$

(b) $-\dfrac{\sin^2 ax\cos ax}{3a} - \dfrac{2}{3a}\cos ax + C;$

(c) $-\dfrac{\sin^3 ax\cos ax}{4a} - \dfrac{3\sin ax\cos ax}{8a} + \dfrac{3}{8}x + C$

Review Problem Set, Chapter 8, page 531

1 $\frac{1}{2}\sin 2x - \frac{1}{6}\sin^3 2x + C$

3 $\frac{1}{3}(\frac{1}{4}\sin^4 3x - \frac{1}{6}\sin^6 3x) + C$

5 $\frac{1}{2}[\cos(1 - 2x) - \frac{1}{3}\cos^3(1 - 2x)] + C$

7 $\frac{3}{5}\sin^{1/3}5x - \frac{3}{35}\sin^{7/3}5x + C$

9 $\frac{1}{2}\sin(4 - 6x) + \frac{1}{2}x + C$ **11** $x - \sin^2 x + C$

13 $\dfrac{1}{8}\left(x - \dfrac{\sin 24x}{24}\right)$ **15** $\sin^{2/3}\dfrac{3t}{2} - \dfrac{1}{4}\sin^{8/3}\dfrac{3t}{2} + C$

17 $\frac{1}{10}\sin 5x - \frac{1}{22}\sin 11x + C$

19 $-\frac{1}{8}\cos 2x - \frac{1}{16}\cos 4x + \frac{1}{24}\cos 6x + C$

21 $\frac{1}{6}\tan^3(2x - 1) - \frac{1}{2}\tan(2x - 1) + x + C$

23 $\frac{1}{20}\tan^2(5x^2) - \frac{1}{10}\ln\left|\sec(5x^2)\right| + C$

25 $2\tan t - 2\sec t - t + C$

27 $\frac{2}{3}\tan^3 x + \tan x + \frac{2}{3}\sec^3 x + C$

29 $\frac{1}{2}[\frac{1}{3}\tan^3(1 + 2x) + \tan(1 + 2x)] + C$

31 $\dfrac{1}{3}\left[\dfrac{\tan^6(2 + 3x)}{6} + \dfrac{\tan^4(2 + 3x)}{4}\right] + C$

33 $\sinh^{-1}\dfrac{x}{8} + C$ **35** $\dfrac{1}{3}\left(\dfrac{x}{\sqrt{1 - x^2}}\right)^3 + \dfrac{x}{\sqrt{1 - x^2}} + C$

37 $\frac{1}{3}(x^2 - 4)^{3/2} + C$ **39** $\sin^{-1}(x - 1) + C$

41 $\sinh^{-1}\dfrac{x + 3}{2}$

43 $-\frac{1}{7}x^2 e^{-7x} - \frac{2}{49}xe^{-7x} - \frac{2}{343}e^{-7x} + C$

45 $\dfrac{t^3}{3}\sin^{-1}2t - \dfrac{1}{24}\left[\dfrac{1}{3}(1 - 4t^2)^{3/2} - (1 - 4t^2)^{1/2}\right] + C$

47 $\frac{1}{3}(x + 2)e^{3x} - \frac{1}{9}e^{3x} + C$

49 $\frac{1}{3}t^3\sin 3t + \frac{1}{3}t^2\cos 3t - \frac{2}{9}t\sin 3t - \frac{2}{27}\cos 3t + C$

51 $x\tan^{-1}\sqrt{x} - \sqrt{x} + \tan^{-1}\sqrt{x} + C$

53 $-\dfrac{1}{x}\tan^{-1}x + \ln\dfrac{|x|}{\sqrt{1 + x^2}} + C$

55 $e^{3x}[\frac{1}{6} + \frac{3}{26}\cos 2x + \frac{1}{13}\sin 2x] + C$

57 $-\frac{1}{4}e^{-x^4}(x^8 + 2x^4 + 2) + C$

59 $\frac{1}{6}x^2\sin(3x^2) + \frac{1}{18}\cos(3x^2) + C$

61 $\ln\dfrac{|y - 1|^{3/2}|y + 1|^{5/2}}{|y|} + C$

63 $\dfrac{1}{x} + 3\ln|x - 1| + C$

65 $\ln\dfrac{t^2 + 1}{t^2 + 4} + \tan^{-1}t + C$

67 $t + 3\ln|t - 1| + \ln|t + 1| + 4\tan^{-1}t + C$

69 $\frac{1}{7}(1 + 2x)^{7/4} - \frac{1}{3}(1 + 2x)^{3/4} + C$

71 $\frac{10}{11}x^{11/10} + \frac{3}{2}x^{2/3} + C$ **73** $\ln\left|\dfrac{\sqrt{e^t + 1} - 1}{\sqrt{e^t + 1} + 1}\right| + C$

75 $\frac{4}{3}(4 + \sqrt{x + 1})^{3/2} - 16\sqrt{4 + \sqrt{x + 1}} + C$

77 $x\ln\sqrt{x^2 + 3} - x + \sqrt{3}\tan^{-1}\dfrac{x}{\sqrt{3}} + C$

79 $-\frac{3}{7}(1 - x^{2/3})^{7/2} + \frac{6}{5}(1 - x^{2/3})^{5/2} - (1 - x^{2/3})^{3/2} + C$

81 $3x^{2/3}\sin\sqrt[3]{x} + 6\sqrt[3]{x}\cos\sqrt[3]{x} - 6\sin\sqrt[3]{x} + C$

83 $\dfrac{1}{\sqrt{21}}\ln\left|\dfrac{\sqrt{21} + \tan\frac{x}{2}}{\sqrt{21} - \tan\frac{x}{2}}\right| + C$

85 $\tan^{-1}\left(1 + \tan\dfrac{y}{2}\right) + C$

87 $\dfrac{1}{2}\ln\left|\dfrac{1 + \tan\frac{x}{2}}{1 - \tan\frac{x}{2}}\right| + \dfrac{\tan\frac{x}{2}}{\left[1 + \tan\frac{x}{2}\right]^2} + C$

89 $\frac{2}{7}(e^{2x} + 1)^{7/4} - \frac{2}{3}(e^{2x} + 1)^{3/4} + C$ **91** $-\frac{1}{12}$

93 $\frac{1}{6} + \frac{1}{4}\ln\frac{2}{3}$ **95** $2(\ln 2)^2 - 4\ln 2 + 2$

97 $4\ln 2 - \frac{15}{16}$ **99** $\frac{1}{2} - \ln\sqrt{2}$

101 $\dfrac{9\pi}{2} - \dfrac{81\sqrt{3}}{8}$ **103** $5\left(\sqrt{3} - \dfrac{\pi}{3}\right)$

105 $3\sqrt{2}$ **107** $4 + 4\ln 3$ **109** $\dfrac{\pi + 3}{4}$

111 $\dfrac{\pi}{6}$ **113** $\frac{3}{2}\ln\frac{5}{2}$ **115** 3 **117** $\frac{13}{6}$

119 $3 - \sqrt{3}$ **121** $\dfrac{1}{2} \ln\left(\dfrac{\tan\dfrac{\pi}{6}}{\tan\dfrac{\pi}{8}}\right) - \dfrac{1}{4}\left(\tan^2\dfrac{\pi}{16} - \tan^2\dfrac{\pi}{8}\right)$

123 $A = \dfrac{ce + df}{e^2 + f^2}$, $B = \dfrac{de - cf}{e^2 + f^2}$;

$\dfrac{ce + df}{e^2 + f^2}\theta + \dfrac{de - cf}{e^2 + f^2} \ln|e \sin\theta + f \cos\theta| + C$

125 $\dfrac{1}{\sqrt{5}} \ln\left|\dfrac{x + \sqrt{x^2 - 2x + 5} - \sqrt{5}}{x + \sqrt{x^2 - 2x + 5} + \sqrt{5}}\right| + C$

131 $\dfrac{27}{4} - 2\ln 4$ **133** $\dfrac{4}{3}$ **135** $8\pi + \dfrac{9\pi^2}{2}$

137 $\ln(2 + \sqrt{3})$ **139** $-\ln(2 - \sqrt{3})$

Chapter 9

Problem Set 9.1, page 540

1 (a) $\left(-3, \dfrac{5\pi}{4}\right)$; (b) $\left(3, -\dfrac{7\pi}{4}\right)$; (c) $\left(-3, -\dfrac{3\pi}{4}\right)$

3 (a) $\left(-2, \dfrac{\pi}{6}\right)$; (b) $\left(2, -\dfrac{5\pi}{6}\right)$; (c) $\left(-2, -\dfrac{11\pi}{6}\right)$

5 (a) $(-4, 0°)$; (b) $(4, -180°)$; (c) $(-4, 0°)$

7 $\left(\dfrac{7}{2}, \dfrac{7\sqrt{3}}{2}\right)$ **9** $(-\sqrt{2}, -\sqrt{2})$ **11** $\left(\dfrac{1}{2}, -\dfrac{\sqrt{3}}{2}\right)$

13 $\left(7\sqrt{2}, \dfrac{\pi}{4}\right)$ **15** $\left(6, -\dfrac{2\pi}{3}\right)$ **17** $\left(7, \dfrac{\pi}{2}\right)$

19 problem 13: $\left(7\sqrt{2}, \dfrac{\pi}{4}\right)$; problem 14: $\left(2, \dfrac{5\pi}{3}\right)$;

problem 15: $\left(6, \dfrac{4\pi}{3}\right)$; problem 16: $\left(5\sqrt{2}, \dfrac{3\pi}{4}\right)$;

problem 17: $\left(7, \dfrac{\pi}{2}\right)$; problem 18: $(2, \pi)$

21 $\left(-3, \dfrac{7\pi}{6}\right)$, $\left(3, \dfrac{\pi}{6}\right)$, $\left(3, -\dfrac{11\pi}{6}\right)$, $\left(3, \dfrac{13\pi}{6}\right)$, $\left(-3, -\dfrac{5\pi}{6}\right)$

23 circle of radius 1, center at pole

25 line through pole, perpendicular to polar axis

27 same as problem 25   **29** polar axis extended

31 $x^2 + 3x + y^2 = 0$; circle with center at $r = -\dfrac{3}{2}$, $\theta = 0$, and radius $\dfrac{3}{2}$

33 $x^2 = y$: parabola, vertex at origin, opening upward

35 $x = -2$: vertical line two units to left of y axis

37 $y = -\sqrt{3}x$: line of slope $-\sqrt{3}$ through origin

39 $(x^2 + y^2)^2 = x^2 - y^2$ **41** $r = 5$

43 $r(2\cos\theta + 3\sin\theta) = 6$ **45** $6r\cos^2\theta = -\sin\theta$

47 $r^2(\cos^2\theta + 4\sin^2\theta) = 4$

49 multiply both sides by r; origin already there

51 divide by r; origin *still* there

53 (a) Cartesian: $x^2 + y^2 = \pm 2ax$; (b) Cartesian: $x^2 + y^2 = \pm 2ay$

Problem Set 9.2, page 548

1 c   **2** f   **3** d   **4** h
5 g   **6** a   **7** e   **8** b

9 symmetric with respect to polar axis; circle

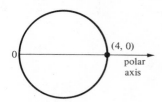

11 symmetric with respect to polar axis; vertical line

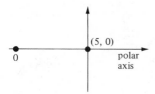

13 symmetric with respect to $\theta = \pm\dfrac{\pi}{2}$; circle

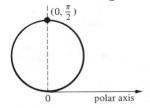

15 symmetric with respect to $\theta = \pm\dfrac{\pi}{2}$

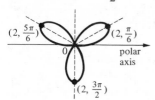

17 symmetric with respect to polar axis, $\theta = \pm\dfrac{\pi}{2}$, and pole

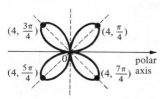

19 symmetric with respect to polar axis

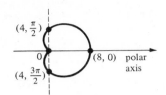

21 symmetric with respect to polar axis

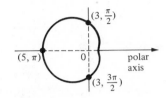

23 symmetric with respect to $\theta = \pm\dfrac{\pi}{2}$

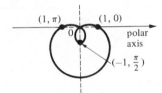

25 symmetric with respect to polar axis, $\theta = \pm\dfrac{\pi}{2}$, and pole

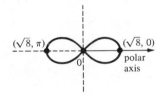

27 1   **29** 0

31 (b) $\dfrac{dr}{d\theta}\sin\theta + r\cos\theta \neq 0$ and $\dfrac{dr}{d\theta}\cos\theta - r\sin\theta = 0$

33 $\dfrac{\sqrt{3}}{3}$ **35** 1   **37** undefined

39 horizontal at $\left(2, -\sin^{-1}\dfrac{2}{3}\right)$, $\left(2, \pi - \sin^{-1}\dfrac{2}{3}\right)$, $\left(7, \dfrac{\pi}{2}\right)$,

and $\left(1, \dfrac{3\pi}{2}\right)$; vertical at $\left(\dfrac{18 + 3\sqrt{22}}{6}, \sin^{-1}\dfrac{\sqrt{22} - 2}{6}\right)$

and $\left(\dfrac{18 + 3\sqrt{22}}{6}, \pi - \sin^{-1}\dfrac{\sqrt{22} - 2}{6}\right)$

41 $a - b$ **45** $r = ce^{\theta/K}$

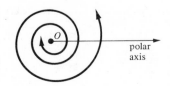

Problem Set 9.3, page 554

1 $\dfrac{31}{6}\pi^3$

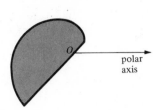

3 $\dfrac{128}{5\pi}$

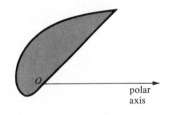

5 $\dfrac{9\sqrt{3}}{2}$

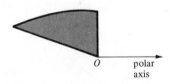

7 4π **9** π

11 $\dfrac{9\pi}{2}$

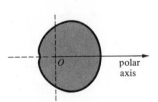

13 6π

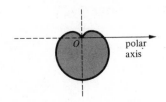

15 $\left(\frac{3}{2}, -\frac{\pi}{6}\right)$ and $\left(\frac{3}{2}, -\frac{5\pi}{6}\right)$

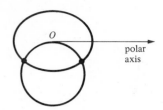

17 $r = 1$; $\theta = \pm\frac{\pi}{9}, \pm\frac{7\pi}{9}, \pm\frac{13\pi}{9}$

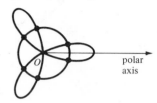

19 $(0, 0)$, $\left(\frac{\sqrt{3}}{2}, \frac{\pi}{3}\right)$, and $\left(\frac{\sqrt{3}}{2}, \frac{2\pi}{3}\right)$

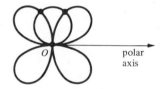

21 $\left(2, \frac{3\pi}{2}\right)$

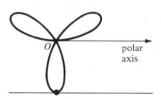

23 $\frac{1}{2}(6\pi + 9\sqrt{3})$

25 $\frac{5\pi}{4}$ **27** $12(3\sqrt{3} - \pi)$ **29** $\frac{25\pi^3}{24}$ **31** 6π

33 $\frac{61}{6}$ **35** $\sqrt{2}(e^{4\pi} - 1)$ **37** $\frac{5\pi}{2}$

39 limits of integration should be 0 to π

43 16π **45** 25π

Problem Set 9.4, page 562

	a	b	Foci	Vertices
1	4	2	$(-2\sqrt{3}, 0), (2\sqrt{3}, 0)$	$(-4, 0), (4, 0), (0, 2), (0, -2)$
3	4	2	$(0, -2\sqrt{3}), (0, 2\sqrt{3})$	$(-2, 0), (2, 0), (0, 4), (0, -4)$
5	4	1	$(-\sqrt{15}, 0), (\sqrt{15}, 0)$	$(-4, 0), (4, 0), (0, -1), (0, 1)$
7	$\frac{2}{3}$	$\frac{1}{3}$	$(-1/\sqrt{3}, 0), (1/\sqrt{3}, 0)$	$(-\frac{2}{3}, 0), (\frac{2}{3}, 0), (0, \frac{1}{3}), (0, -\frac{1}{3})$

9 $\frac{x^2}{25} + \frac{y^2}{9} = 1$ **11** $\frac{x^2}{25} + \frac{y^2}{169} = 1$

13 (a) $(1, -2)$; (b) $(-1, -1)$; (c) $(4, -5)$; (d) $(-2, -4)$; (e) $(6, 3)$; (f) $(7, -2)$

15 (a) $(3, -2)$; (b) $(6, 0)$; (c) $(0, 2)$; (d) $(\sqrt{2} + 3, -4)$; (e) $(3, -\pi - 2)$; (f) $(0, 0)$

17 $\frac{(x + 1)^2}{4} + (y - 1)^2 = 1$: center $(-1, 1)$; foci $(-\sqrt{3} - 1, 0), (\sqrt{3} - 1, 0)$; vertices $(-3, 1), (1, 1), (-1, 2), (-1, 0)$

19 $\frac{(x - 2)^2}{9} + \frac{(y - 3)^2}{6} = 1$: center $(2, 3)$; foci $(-\sqrt{3} + 2, 3), (\sqrt{3} + 2, 3)$; vertices $(-1, 3), (5, 3), (2, \sqrt{6} + 3), (2, -\sqrt{6} + 3)$

21 $\frac{(x - 6)^2}{9} + \frac{(y + 2)^2}{16} = 1$: center $(6, -2)$; foci $(6, \sqrt{7} - 2), (6, -\sqrt{7} - 2)$; vertices $(3, -2), (9, -2), (6, 2), (6, -6)$

	Center	Vertices	Foci
23	$(1, -2)$	$(-2, -2), (4, -2), (1, 0),$ $(1, -4)$	$(1 - \sqrt{5}, -2),$ $(1 + \sqrt{5}, -2)$
25	$(-3, 0)$	$(-6, 0), (0, 0), (-3, 6),$ $(-3, -6)$	$(-3, 3\sqrt{3}),$ $(-3, -3\sqrt{3})$
27	$(-3, 0)$	$(-3 - \sqrt{2}, 0), (-3 + \sqrt{2}, 0),$ $(-3, 1), (-3, -1)$	$(-4, 0), (-2, 0)$
29	$(-5, 3)$	$(-5 - \sqrt{10}, 3), (-5 + \sqrt{10}, 3),$ $(-5, 5), (-5, 1)$	$(-5 - \sqrt{6}, 3),$ $(-5 + \sqrt{6}, 3)$

31 $\frac{x^2}{25} + \frac{(y - 1)^2}{9} = 1$ **33** $\frac{x^2}{36} + \frac{y^2}{64} = 1$

35 $\frac{x^2}{16} + \frac{7y^2}{64} = 1$ **37** $\frac{(x + 2)^2}{25} + \frac{(y - 1)^2}{16} = 1$

39 $\frac{(x - 1)^2}{4} + \frac{(y + 2)^2}{9}$ **41** $\frac{(x - 2)^2}{7} + \frac{(y - 1)^2}{16} = 1$

43 tangent line: $4y + x = 25$; normal line: $4x - y = 32$
45 tangent line: $x + 6y = 15$; normal line: $6x - y = 16$
47 (b) $\frac{9}{2}$ units   **49** $2ab$ square units
51 $l = 2a + 2c = 60 + 40\sqrt{2}$; $2c = 40\sqrt{2}$
53 $\frac{x^2}{50^2} + \frac{y^2}{25^2} = 1$

Problem Set 9.5, page 568

1 vertex $(0, 0)$; focus $(1, 0)$; directrix $x = -1$; focal chord 4
3 vertex $(0, 0)$; focus $(0, \frac{1}{4})$; directrix $y = -\frac{1}{4}$; focal chord 1
5 vertex $(0, 0)$; focus $(0, -\frac{9}{4})$; directrix $y = \frac{9}{4}$; focal chord 9
7 $y = \frac{1}{12}x^2$

	Vertex	Focus	Directrix	Focal chord
9	$(-3, 2)$	$(-1, 2)$	$x = -5$	8
11	$(4, -7)$	$(4, -4)$	$y = -10$	12
13	$(-3, 4)$	$(-\frac{3}{2}, 4)$	$x = -\frac{9}{2}$	6
15	$(3, -1)$	$(3, 1)$	$y = -3$	8

17 $(y - 2)^2 = -4(x - 5)$ **19** $(y + 5)^2 = 32(x + 6)$
21 $(y + 1)^2 = 8(x + \frac{1}{2})$ **23** $\bar{x} = x + \frac{1}{2}, \bar{y} = y + 1; \bar{y}^2 = 8\bar{x}$
25 $\bar{x} = x + 2, \bar{y} = y - 3; \bar{x}^2 = 5\bar{y}$
27 (a) tangent line: $x + y = -2$, normal line: $x - y = 6$;
    (b) tangent line: $3x + 4y = -6$, normal line: $4x - 3y = 17$;
    (c) tangent line: $y - x = 3$, normal line: $x + y = -9$;
    (d) tangent line: $6x + 8y = 21$, normal line: $32x - 24y = -13$;
    (e) tangent line: $x = \frac{9}{2}$, normal line: $y = 1$
29 (a) $(1, 5)$; (b) $(-3, 1)$; (c) $(-1, 5)$; (d) $(2, 7)$
31 base: 4 units; height: 8 units
33 Lengths are 4, 10, 28, and 58 meters.
35 (a) $f(a) = (a^2/4p) + 2p$; (b) $2p$
41 (a) Any two parabolas are *similar* in the sense that one is a magnification of the other.

Problem Set 9.6, page 575

	Vertices	Foci	Asymptotes
1	$(3, 0), (-3, 0)$	$(\sqrt{13}, 0), (-\sqrt{13}, 0)$	$y = \pm\frac{2}{3}x$
3	$(0, 4), (0, -4)$	$(0, -2\sqrt{5}), (0, 2\sqrt{5})$	$y = \pm 2x$
5	$(-4, 0), (4, 0)$	$(-2\sqrt{5}, 0), (2\sqrt{5}, 0)$	$y = \pm\frac{1}{2}x$
7	$(0, -\sqrt{10}), (0, \sqrt{10})$	$(0, -\sqrt{46}), (0, \sqrt{46})$	$y = \pm\frac{\sqrt{10}}{6}x$

9 $\dfrac{x^2}{16} - \dfrac{y^2}{20} = 1$ **11** $\dfrac{x^2}{16} - \dfrac{y^2}{25} = 1$

	Center	Vertices	Foci	Asymptotes
13	$(1, -2)$	$(-2, -2),$ $(4, -2)$	$(1 + \sqrt{13}, -2),$ $(1 - \sqrt{13}, -2)$	$y + 2 = \pm\frac{2}{3}(x - 1)$
15	$(-2, -1)$	$(-2, 3),$ $(-2, -5)$	$(-2, \sqrt{41} - 1),$ $(-2, -\sqrt{41} - 1)$	$y + 1 = \pm\frac{4}{5}(x + 2)$
17	$(2, -1)$	$(0, -1),$ $(4, -1)$	$(2 - \sqrt{5}, -1),$ $(2 + \sqrt{5}, -1)$	$y + 1 = \pm\frac{1}{2}(x - 2)$
19	$(-4, -2)$	$(-4, 1),$ $(-4, -5)$	$(-4, \sqrt{34} - 2),$ $(-4, -\sqrt{34} - 2)$	$y + 2 = \pm\frac{3}{5}(x + 4)$

21 (a) $\dfrac{(x - 4)^2}{1} - \dfrac{(y + 1)^2}{8} = 1$; (b) $\dfrac{(x + 2)^2}{4} - \dfrac{4(y - 3)^2}{9} = 1$;
    (c) $\dfrac{(y - 3)^2}{25} - \dfrac{(x - 2)^2}{11} = 1$
23 1
25 (a) $\dfrac{4x^2}{3} - \dfrac{y^2}{3} = 1$; (b) $\dfrac{(y - 2)^2}{3} - \dfrac{4(x - \frac{1}{2})^2}{3} = 1$;
    (c) $\dfrac{(x + 1)^2}{9} - \dfrac{(y - 4)^2}{9} = 1$
27 $\bar{x} = x + 2, \bar{y} = y - 4$ **29** $\bar{x} = x - 3, \bar{y} = y + 1$
31 $\bar{x} = x + 2, \bar{y} = y - 3$ **33** 8 units/s   **35** $\dfrac{3\sqrt{10}}{2}$ units
37 $\lim\limits_{x_0 \to \infty} m = \dfrac{b}{a} =$ slope of asymptote   **39** $\dfrac{x^2}{a^2} - \dfrac{y^2}{a^2} = 1$

41 It approaches the intersecting asymptotes.
43 any point on the right-hand branch of the hyperbola
$\dfrac{b^2 x^2}{h^2 s^2} - \dfrac{b^2 y^2}{h^2(b^2 - s^2)} = 1$ (*except* at the vertex, where the bullet will pass)

Problem Set 9.7, page 583

1 $\bar{r} = 5$ **3** $\bar{r} = 4 \sin \bar{\theta}$ **5** $\bar{r} = 3 - 5 \sin \bar{\theta}$
7 $\bar{r} = 3 + 5 \cos \bar{\theta}$ **9** $\bar{r}^2 = -25 \cos 2\bar{\theta}$
11 $(-7, -4)$ **13** $\left(\dfrac{3\sqrt{3} - 3}{2}, -\dfrac{3\sqrt{3} + 3}{2}\right)$
15 $(1 - 2\sqrt{3}, -2 - \sqrt{3})$ **17** $(4, 0)$ **19** $(0, 8)$
21 $\bar{x}^2 + 2\sqrt{3}\bar{x}\bar{y} + 3\bar{y}^2 - 6\sqrt{3}\bar{x} + 6\bar{y} = 0$
23 $x^2 + y^2 = 1$ **25** $\bar{x}^2 + \bar{y}^2 = 1$
29 (a) $\phi = \sin^{-1} \dfrac{\sqrt{5}}{5} \approx 26.57°$; (b) $x = \dfrac{\sqrt{5}}{5}(2\bar{x} - \bar{y})$,
    $y = \dfrac{\sqrt{5}}{5}(\bar{x} + 2\bar{y})$; (c) $2\bar{x}^2 - 3\bar{y}^2 = 12$
31 (a) $\phi = \dfrac{\pi}{4} = 45°$; (b) $x = \dfrac{\sqrt{2}}{2}(\bar{x} - \bar{y})$, $y = \dfrac{\sqrt{2}}{2}(\bar{x} + \bar{y})$;
    (c) $2\bar{x}^2 = 1$
33 (a) $\phi = \sin^{-1} \frac{3}{5} \approx 36.87°$; (b) $x = \frac{1}{5}(4\bar{x} - 3\bar{y})$,
    $y = \frac{1}{5}(3\bar{x} + 4\bar{y})$; (c) $25\bar{y}^2 = 144$
35 (a) $\phi = \sin^{-1} \dfrac{\sqrt{10}}{10} \approx 18.43°$; (b) $x = \dfrac{\sqrt{10}}{10}(3\bar{x} - \bar{y})$,
    $y = \dfrac{\sqrt{10}}{10}(\bar{x} + 3\bar{y})$; (c) $\bar{x}^2 + 3\bar{y}^2 = 9$
37 (a) $\phi = \sin^{-1} \dfrac{\sqrt{10}}{10} \approx 18.43°$; (b) $x = \dfrac{\sqrt{10}}{10}(3\bar{x} - \bar{y})$,
    $y = \dfrac{\sqrt{10}}{10}(\bar{x} + 3\bar{y})$; (c) $3\bar{x}^2 - 7\bar{y}^2 + 9\bar{x} + 7\bar{y} - 16 = 0$
41 (a) ellipse; (b) parabola; (c) circle; (d) hyperbola
43 $\phi = \dfrac{\pi}{4} = 45°$; $\dfrac{7}{2}\bar{x}^2 + \dfrac{1}{2}\bar{y}^2 = 4$ (ellipse)
47 (a) $\phi = \sin^{-1} \dfrac{\sqrt{10}}{10}$, $3\bar{x}^2 + 8\bar{y}^2 = 48$; (b) $\phi = \sin^{-1} \dfrac{3}{5}$,
    $16\bar{x}^2 - 9\bar{y}^2 = 144$
49 $y^2 - 9x^2 = 0$
51 $\phi = \sin^{-1} \dfrac{2\sqrt{5}}{5} \approx 63.43°$; $5\bar{x}^2 = 8$ (two parallel lines)
53 $\phi = \dfrac{\pi}{4} = 45°$, $11\bar{y}^2 - 7\bar{x}^2 = 24$ (hyperbola)
55 $\bar{A} = \bar{C}$ only if $A = C$ and $B = 0$

Problem Set 9.8, page 591

1 $\dfrac{(x - 2)^2}{9} + \dfrac{y^2}{5} = 1$ **3** $8(x + 2) = y^2$
5 $\dfrac{(x + 4)^2}{4} - \dfrac{y^2}{12} = 1$

*7 $e = \dfrac{3}{4}$; directrices: $x = -\dfrac{20}{3}$, $x = \dfrac{500}{21}$; center: $\left(\dfrac{60}{7}, 0\right)$;

foci: $(0, 0)$, $\left(\dfrac{120}{7}, 0\right)$; vertices: $\left(-\dfrac{20}{7}, 0\right)$, $(20, 0)$,

$\left(\dfrac{60}{7}, \dfrac{20\sqrt{7}}{7}\right)$, $\left(\dfrac{60}{7}, -\dfrac{20\sqrt{7}}{7}\right)$

9 $e = \dfrac{1}{4}$; directrices: $x = -20$, $x = \dfrac{68}{3}$; center: $\left(\dfrac{4}{3}, 0\right)$; foci

at $(0, 0)$, $\left(\dfrac{8}{3}, 0\right)$; vertices: $(-4, 0)$, $\left(\dfrac{20}{3}, 0\right)$, $\left(\dfrac{4}{3}, \dfrac{4\sqrt{15}}{3}\right)$,

$\left(\dfrac{4}{3}, \dfrac{-4\sqrt{15}}{3}\right)$

11 focus: $(0, 0)$; directrix: $x = -4$; vertex: $(-2, 0)$

13 $e = 2$; directrices: $x = -\dfrac{1}{2}$, $x = -\dfrac{5}{6}$; center: $(-\dfrac{2}{3}, 0)$; foci: $(0, 0)$, $(-\dfrac{4}{3}, 0)$; vertices: $(-\dfrac{1}{3}, 0)$, $(-1, 0)$

15 $e = \dfrac{3}{5}$ (ellipse); directrices: $x = -\dfrac{16}{3}$, $x = \dfrac{34}{3}$; center: $(3, 0)$; foci: $(0, 0)$, $(6, 0)$; vertices: $(-2, 0)$, $(8, 0)$, $(3, 4)$, $(3, -4)$

17 $e = 1$ (parabola); directrix: $x = -6$; focus: $(0, 0)$; vertex: $(-3, 0)$

19 $e = 1$ (parabola); directrix: $y = -10$; focus: $(0, 0)$; vertex: $(0, -5)$

21 $e = z$ (hyperbola); directrices: $y = \dfrac{1}{2}$, $y = \dfrac{5}{6}$; center: $(0, \dfrac{2}{3})$ foci: $(0, 0)$, $(0, \dfrac{4}{3})$; vertices: $(0, \dfrac{1}{3})$, $(0, 1)$

23 $e = 1$ (parabola); directrix: $x = 4$; focus: $(0, 0)$; vertex: $(2, 0)$

25 $r = \dfrac{3}{3 - \cos\theta}$ **27** $r = \dfrac{3}{1 + \cos\theta}$

29 $r = \dfrac{6}{1 - 3\cos\theta}$ **31** $r = \dfrac{5}{3 + \sin\theta}$

47 $e \approx \dfrac{1}{59}$ **49** $r = \dfrac{ed}{1 + e\cos\theta}$

51 shrinks to a point, becoming more circular

53 $\left(2, \dfrac{\pi}{2}\right)$, $\left(2, -\dfrac{\pi}{2}\right)$, $(2, \pi)$, $\left(-2 - 2\sqrt{2}, \dfrac{\pi}{4}\right)$,

$\left(-2 - 2\sqrt{2}, \dfrac{5\pi}{4}\right)$, $\left(-2 + 2\sqrt{2}, \dfrac{3\pi}{4}\right)$, $\left(-2 + 2\sqrt{2}, \dfrac{5\pi}{4}\right)$

55 $x = \pm\dfrac{a^2}{\sqrt{a^2 - b^2}}$ **57** $x = \pm\dfrac{a^2}{\sqrt{a^2 + b^2}}$

*In 7–23, all answers are given in Cartesian coordinates.

Review Problem Set, Chapter 9, page 592

1 (a) $\left(-1, \dfrac{4\pi}{3}\right)$; (b) $\left(1, -\dfrac{5\pi}{3}\right)$; (c) $\left(-1, -\dfrac{2\pi}{3}\right)$

3 (a) $\left(-2, \dfrac{3\pi}{4}\right)$; (b) $\left(2, -\dfrac{\pi}{4}\right)$; (c) $\left(-2, -\dfrac{5\pi}{4}\right)$

5 (a) $(-3, 0)$; (b) $(3, -\pi)$; (c) $(-3, -2\pi)$

7 $(0, 0)$ **9** $(-17, 0)$ **11** $\left(\dfrac{-11\sqrt{2}}{2}, \dfrac{11\sqrt{2}}{2}\right)$

13 $(17, 0)$ **15** $\left(4, -\dfrac{\pi}{3}\right)$ **17** $\left(17\sqrt{2}, \dfrac{3\pi}{4}\right)$

19 $x^2 - y^2 = 1$ **21** $y^2 = \dfrac{1 - x^4}{x^2}$

23 $r = \dfrac{1}{2\cos\theta - \sin\theta}$ **25** $r = 4\cot\theta\csc\theta$

27 3-leaved rose, symmetrical about polar axis

29 straight line, no symmetry

31 parabola, symmetric about $\theta = \dfrac{\pi}{2}$

33 $\dfrac{5}{3}$ **35** $-\dfrac{3}{2}$

37 horizontal at $\left(3 - \dfrac{3 + \sqrt{137}}{4}, \cos^{-1}\dfrac{3 + \sqrt{137}}{16}\right)$,

$\left(3 - \dfrac{3 - \sqrt{137}}{4}, \cos^{-1}\dfrac{3 - \sqrt{137}}{16}\right)$,

$\left(3 - \dfrac{3 + \sqrt{137}}{4}, 2\pi - \cos^{-1}\dfrac{3 + \sqrt{137}}{16}\right)$, and

$\left(3 - \dfrac{3 - \sqrt{137}}{4}, 2\pi - \cos^{-1}\dfrac{3 - \sqrt{137}}{16}\right)$; vertical at

$(-1, 0)$, $(7, \pi)$, $\left(\dfrac{3}{2}, \cos^{-1}\dfrac{3}{8}\right)$, and $\left(\dfrac{3}{2}, 2\pi - \cos^{-1}\dfrac{3}{8}\right)$

39 $\dfrac{1}{3}\tan\theta$

41 $(1, \pi)$, $\left(\dfrac{-1 + \sqrt{17}}{2}, \cos^{-1}\dfrac{3 - \sqrt{17}}{4}\right)$,

$\left(\dfrac{-1 + \sqrt{17}}{2}, 2\pi - \cos^{-1}\dfrac{3 - \sqrt{17}}{4}\right)$

43 $\dfrac{3\pi + 8\sqrt{2} + 2}{16}$ **45** 4π **47** $\dfrac{3\pi}{4}$

49 $\left(2, \dfrac{\pi}{6}\right)$, $\left(2, \dfrac{5\pi}{6}\right)$; area $= \dfrac{4\pi}{3} - \sqrt{3}$ square units

51 $(2a^2 + b^2)\sin^{-1}\dfrac{a}{b} - \dfrac{9}{2}a\sqrt{b^2 - a^2}$

53 $\dfrac{\sqrt{10}}{3}(1 - e^{-6\pi})$ **55** 2

57 $e = 1$ (parabola); directrix: $x = -17$

59 $e = \dfrac{2}{5}$ (ellipse); $y = -5$

61 $e = 2$ (hyperbola); $y = 1$

63 $\dfrac{x^2}{25} + \dfrac{y^2}{16} = 1$ **65** $\dfrac{x^2}{16} + \dfrac{y^2}{64} = 1$

67 $\dfrac{(x - 5)^2}{25} + \dfrac{y^2}{9} = 1$

	Center	Vertices	Foci	e	Directrices
69	$(0, 0)$	$(0, 2\sqrt{3})$, $(0, -2\sqrt{3})$, $(2\sqrt{2}, 0)$, $(-2\sqrt{2}, 0)$	$(0, 2)$, $(0, -2)$	$\dfrac{\sqrt{3}}{3}$	$y = \pm 6$
71	$(-1, 1)$	$(4, 1)$, $(-6, 1)$, $(-1, 4)$, $(-1, -2)$	$(-5, 1)$, $(3, 1)$	$\dfrac{4}{5}$	$x = -1 \pm \dfrac{25}{4}$
73	$(-4, 6)$	$(-4, 12)$, $(-4, 0)$, $(-8, 6)$, $(0, 6)$	$(-4, 6 \pm 2\sqrt{5})$	$\dfrac{\sqrt{5}}{3}$	$y = 30 \pm \dfrac{18\sqrt{5}}{5}$

75 $(3, \dfrac{16}{3})$ and $(-3, -\dfrac{16}{3})$

77 (a) $\bar{x} = x - 1$, $\bar{y} = y + z$, $\dfrac{\bar{x}^2}{4} + \dfrac{\bar{y}^2}{64} = 1$; (b) $\bar{x} = x + 2$,

$\bar{y} = y - 3$, $\dfrac{\bar{x}^2}{36} + \dfrac{\bar{y}^2}{81} = 1$

79 $y^2 = -18x$ **81** $y - 3 = \frac{1}{2}(x - 2)^2$ **83** $x^2 + y^2 = 10x$

85 $8\sqrt{3}$ by 4   **87** $(2, 4)$ **89** $\dfrac{x^2}{4} - \dfrac{y^2}{21} = 1$

91 $\dfrac{(x - 2)^2}{16} - \dfrac{(y - 3)^2}{9} = 1$ **93** $\dfrac{y^2}{36} - \dfrac{x^2}{108} = 1$

	Center	Vertices	Foci	e	Asymptotes
95	$(0, 0)$	$(6\sqrt{2}, 0)$, $(-6\sqrt{2}, 0)$	$(4\sqrt{5}, 0)$, $(-4\sqrt{5}, 0)$	$\dfrac{\sqrt{10}}{3}$	$y = \pm\dfrac{1}{3}x$
97	$(-2, 3)$	$(-6, 3)$, $(2, 3)$	$(-2 + 2\sqrt{5}, 3)$, $(-2 - 2\sqrt{5}, 3)$	$\dfrac{\sqrt{5}}{2}$	$y - 3 = \pm\dfrac{1}{2}(x + 2)$
99	$(1, 3)$	$\left(1, \dfrac{7}{2}\right)$, $\left(1, \dfrac{5}{2}\right)$	$\left(1, -3 + \dfrac{\sqrt{5}}{2}\right)$, $\left(1, -3 - \dfrac{\sqrt{5}}{2}\right)$	$\dfrac{\sqrt{5}}{2}$	$y - 3 = \pm\dfrac{1}{2}(x - 1)$

101 tangent line: $8y - 3x + 1 = 0$; normal line: $8x + 3y - 27 = 0$

103 $-\frac{9}{4}$ units/s   **105** $e \approx 0.52$ **107** 3108π m^2

109 $\bar{r}^2 \cos 2\bar{\theta} = 2$ **111** $5\bar{x}^2 - 3\bar{y}^2 = 15$ (hyperbola)

113 $(\sqrt{3} + 2)\bar{x}^2 - (2 - \sqrt{3})\bar{y}^2 = 22$ (hyperbola)

115 no rotation needed (ellipse)

117 ellipse   **119** ellipse   **121** parabola

Chapter 10

Problem Set 10.1, page 602

1 -3 **3** $\frac{5}{13}$ **5** 1   **7** 0   **9** $+\infty$

11 $-\frac{1}{7}$ **13** $-\frac{1}{2}$ **15** $\frac{1}{6}$ **17** $-\frac{1}{8}$ **19** $\frac{7}{5}$

21 0   **23** 3   **25** $\frac{4}{3}$ **27** $\dfrac{e}{e - 1}$ **29** $\sec^{-1}\dfrac{2}{\sqrt{\pi}}$

31 $a = 4$, $b = 1$

35 $\lim\limits_{x \to 0} \dfrac{\sin x}{x}$ was used to derive $D_x \sin x = \cos x$

37 0   **39** $\dfrac{Et}{L}$ **41** $\dfrac{-At \cos wt}{2w}$

Problem Set 10.2, page 607

1 1   **3** 0   **5** $+\infty$ **7** 0   **9** 0   **11** $-\infty$

13 $+\infty$ **15** 0   **17** -2 **19** $\frac{1}{2}$ **21** $+\infty$

23 $-\frac{1}{2}$ **25** $-\frac{1}{7}$ **27** 1   **29** 1   **31** 1   **33** 1

35 e^2 **37** 1   **39** e **41** e^6 **43** e^{-2}

45 e **47** 1   **49** 1   **53** 3   **55** $+\infty$

Problem Set 10.3, page 612

1 2   **3** $\dfrac{\pi}{12}$ **5** divergent   **7** $\ln 2$ **9** divergent

11 $\dfrac{\pi}{8}$ **13** divergent   **15** $\frac{1}{81}$ **17** -1 **19** $\dfrac{\pi}{2}$

21 0   **23** $\dfrac{\pi}{a}$ **25** $\dfrac{1}{\ln 2}$ **27** $n = 2$; $2 \ln 3$ **31** $\dfrac{\pi}{2}$

33 π **35** $1 - \dfrac{1}{2e^2}$ **37** $\dfrac{A}{r} + \dfrac{B}{r^2}$ **45** $\dfrac{1}{r^2 + 1}$

Problem Set 10.4, page 616

1 4   **3** $\dfrac{27}{2}$ **5** $3 \sin 1$ **7** diverges   **9** $\dfrac{\pi}{2}$

11 $2 - \dfrac{2}{e^2}$ **13** $\dfrac{7}{5}(\ln 2)^{5/7}$ **15** diverges   **17** 0

19 diverges   **21** $2 - 2\sqrt{1 - \sqrt{3}}$

23 $\dfrac{5}{4}\left[\sqrt[5]{(e - 1)^4} - \sqrt[5]{\left(\dfrac{1}{e} - 1\right)^4}\right]$

27 The fundamental theorem of calculus does not apply to improper integrals.

29 π square units   **31** infinite

33 $\dfrac{\pi}{1 - 2p}$ for $0 < p < \dfrac{1}{2}$; otherwise, infinite

35 $\dfrac{2\pi}{2 - p}$ for $0 < p < 2$; infinite for $p \geq 2$

Problem Set 10.5, page 626

1 $\dfrac{1}{2} - \dfrac{(x - 2)}{4} + \dfrac{(x - 2)^2}{8} - \dfrac{(x - 2)^3}{16} + \dfrac{(x - 2)^4}{32} - \dfrac{(x - 2)^5}{64} + \dfrac{(x - 2)^6}{128}$; $-c^{-8}(x - 2)^7$

3 $\dfrac{1}{10} - \dfrac{(x - 100)}{2(10^3)} + \dfrac{3(x - 100)^2}{8(10^5)} - \dfrac{5(x - 100)^3}{16(10^7)} + \dfrac{35(x - 100)^4}{128(10^9)}$; $-\dfrac{63}{256}c^{-11/2}(x - 100)^5$

5 $1 - 2(x - 3) + 3(x - 3)^2 - 4(x - 3)^3 + 5(x - 3)^4 - 6(x - 3)^5$; $\dfrac{7(x - 3)^6}{(c - 2)^8}$

7 $x - \dfrac{x^3}{3!} + \dfrac{x^5}{5!}$; $\dfrac{-\cos c}{7!}x^7$

9 $1 + 2\left(x - \dfrac{\pi}{4}\right) + 2\left(x - \dfrac{\pi}{4}\right)^2 + \dfrac{8}{3}\left(x - \dfrac{\pi}{4}\right)^3 + \dfrac{10}{3}\left(x - \dfrac{\pi}{4}\right)^4$; $\dfrac{16 \sec^2 c \tan^4 c + 88 \sec^4 c \tan^2 c + 16 \sec^6 c}{5!}\left(x - \dfrac{\pi}{4}\right)^5$

11 $e + 2e(x - 1) + \dfrac{3e}{2}(x - 1)^2 + \dfrac{2e}{3}(x - 1)^3; \dfrac{(c + 4)e^c}{4!}(x - 1)^4$

13 $2 + 2(\ln 2)(x - 1) + (\ln 2)^2(x - 1)^2 + \dfrac{(\ln 2)^3}{3}(x - 1)^3;$

$\dfrac{(\ln 2)^4 2^c (x - 1)^4}{4!}$

15 $x + \dfrac{x^3}{3!}; \dfrac{\cosh c}{5!}x^5$ **17** 0.8415

19 2.7183 **21** -0.0202 **23** 3.0067

25 $\dfrac{1}{10^{n+1}(n + 1)!}$ **29** $\dfrac{s^3}{12r}; \dfrac{s^5}{240r^3}$

31 (a) $1 + px$; (b) $1 + px + \dfrac{p(p - 1)}{2}x^2$;

(c) $1 + px + \dfrac{p(p - 1)}{2}x^2 + \dfrac{p(p - 1)(p - 2)x^3}{6}$;

(d) $1 + px + \dfrac{p(p - 1)}{2}x^2 + \dfrac{p(p - 1)(p - 2)}{6}x^3 +$

$\dfrac{p(p - 1)(p - 2)(p - 3)}{24}x^4$

33 $T_s \approx T + \dfrac{w^2 L^2}{8T}$ **35** (d) 0.630

5 $\dfrac{n + 1}{2}$ **7** $\dfrac{1}{n + 1}$ **9** 0 **11** $\dfrac{1}{7}$ **13** diverges

15 π **17** 0 **19** -1 **21** diverges   **23** 1

25 e **27** increasing, bounded, convergent

29 increasing, bounded below but not above, divergent

31 nonmonotonic, bounded, divergent

33 decreasing, bounded, convergent

35 decreasing, bounded, convergent

37 decreasing, bounded, convergent

39 nonmonotonic, bounded, divergent

41 decreasing, bounded above but not below, divergent

43 nonmonotonic, bounded, convergent

45 $a_n = n, b_n = -n$

47 (a) 1, 2, 3, 4, 5, 6; (b) 727; (c) It is difficult to guarantee that the general term of the intended sequence has been determined by an examination of its first few terms.

49 choose $N = \left(\dfrac{|c|}{\epsilon}\right)^{1/k}$ in the definition of $\lim\limits_{n \to \infty} a_n$

51 (a) diverges; (b) diverges; (c) converges; (d) converges; (e) diverges

53 (a) $1, 3, 2, \frac{5}{2}, \frac{9}{4}, \frac{19}{8}, \frac{37}{16}, \frac{75}{32}$; (c) $\frac{7}{3}$

55 $a_n = (-1)^n$ **57** $\dfrac{A}{1 - B}$

Review Problem Set, Chapter 10, page 628

1 -1 **3** 4 **5** $+\infty$ **7** $-\infty$ **9** -1 **11** 0

13 0 **15** $\sin a$ **17** $-\frac{1}{2}$ **19** $+\infty$ **21** $e^{-1/2}$

23 $+\infty$ **25** 1 **27** $e^{-1/2}$ **29** 1 **31** 1

33 3 **35** $\dfrac{1}{3}$ **39** $\dfrac{\pi}{4}$ **41** $\dfrac{e^2 - 1}{4}$ **43** $\dfrac{2}{3}$

45 $\dfrac{2}{5}$ **47** $\dfrac{2e^3}{9}$ **49** diverges   **51** $\dfrac{3}{2}(e - 1)^{2/3}$

53 $3\sqrt[3]{a^2}$ **55** $\dfrac{\pi}{6}$ **57** infinite   **59** (a) $\dfrac{3\pi}{4a^5}$; (b) $\dfrac{12\pi}{a^4}$

61 $2x - \dfrac{8x^3}{3!}; \dfrac{2 \sin 2c}{3}x^4$

63 $1 - x + \dfrac{x^2}{2!} - \dfrac{x^3}{3!} + \dfrac{x^4}{4!} - \dfrac{x^5}{5!} + \dfrac{x^6}{6!} - \dfrac{x^7}{7!}; \dfrac{e^{-c}}{8!}x^8$

65 0.99939 **67** 0.40547 **69** 1.01489

Chapter 11

Problem Set 11.1, page 640

1 2, 5, 10, 17, 26, 37; 10,001

3 $\frac{1}{6}, \frac{2}{9}, \frac{3}{14}, \frac{4}{21}, \frac{5}{30}, \frac{6}{41}; \frac{100}{10,005}$

Problem Set 11.2, page 649

1 $\dfrac{1}{6} + \dfrac{1}{12} + \dfrac{1}{20} + \dfrac{1}{30} + \dfrac{1}{42} + \cdots; \dfrac{1}{6}, \dfrac{1}{4}, \dfrac{3}{10}, \dfrac{1}{3}, \dfrac{5}{14};$

$s_n = \dfrac{1}{2} - \dfrac{1}{n + 2} = \dfrac{n}{2(n + 2)}; S = \dfrac{1}{2}$

3 $2 + 6 + 12 + 20 + 30 + \cdots; 2, 8, 20, 40, 70;$

$s_n = \dfrac{n(n + 1)(n + 2)}{3}$, diverges

5 $-1 + \dfrac{1}{3} + \dfrac{1}{15} + \dfrac{1}{35} + \dfrac{1}{63} + \cdots; -1, -\dfrac{2}{3}, -\dfrac{3}{5},$

$-\dfrac{4}{7}, -\dfrac{5}{9}; s_n = -\dfrac{1}{2} - \dfrac{1}{4n + 2} = -\dfrac{n + 1}{2n + 1}; S = -\dfrac{1}{2}$

7 $\displaystyle\sum_{k=1}^{\infty} \dfrac{1}{k(k + 1)} = 1$ **9** $\displaystyle\sum_{k=1}^{\infty} \dfrac{2(3k^2 + 7k - 5)}{(3k + 2)(3k + 5)}$, diverges

11 $\displaystyle\sum_{k=1}^{\infty} 2(-1)^{k+1}$, diverges

13 $a = \frac{5}{16}, r = \frac{1}{4}$, converges to $\frac{5}{12}$

15 $a = 1, r = -\frac{2}{3}$, converges to $\frac{3}{5}$

17 $a = 1, r = \frac{2}{7}$, converges to $\frac{7}{5}$

19 $a = \frac{7}{6}, r = \frac{7}{6}$, diverges

21 $a = \frac{81}{100}, r = \frac{9}{10}$, converges to $\frac{81}{10}$

23 $a = 1, r = -1$, diverges

25 $a = \frac{1}{5}, r = \frac{1}{5}$, converges to $\frac{1}{4}$

27 $a = \frac{1}{5}, r = \frac{3}{5}$, converges to $\frac{1}{2}$

29 $a = 1, r = -\frac{1}{10}$, converges to $\frac{10}{11}$ **31** $\frac{1}{3}$ **33** $\frac{467}{99}$

35 yes, if it converges   **37** $\frac{244}{495}$ **39** 8 m

Problem Set 11.3, page 656

9 $\frac{5}{6}$ **11** -3 **13** $\frac{31}{21}$ **15** no   **17** $-2\ln 2$

21 $\displaystyle\sum_{j=1}^{\infty} \frac{1}{j(j+1)}$ **23** $\displaystyle\sum_{j=M}^{\infty} a_{j-M+1}$ **25** $\dfrac{1}{M+1}$ **27** $e-1$

Problem Set 11.4, page 667

1 converges   **3** diverges   **5** converges
7 converges   **9** diverges   **11** converges
13 converges   **15** converges   **17** diverges

19 diverges   **21** converges by comparison with $\displaystyle\sum_{k=1}^{\infty}\frac{1}{k^2}$

23 converges by comparison with $\displaystyle\sum_{k=1}^{\infty}\frac{1}{5^k}$

25 converges by comparison with $\displaystyle\sum_{j=1}^{\infty}\frac{1}{7^j}$

27 diverges by comparison with $\displaystyle\sum_{k=1}^{\infty}\frac{1}{k^{1/3}}$

29 diverges by comparison with $\displaystyle\sum_{j=1}^{\infty}\frac{1}{5j}$

31 diverges by comparison with $\displaystyle\sum_{q=1}^{\infty}\frac{1}{3q^{1/2}}$

33 diverges $\left(\text{use }\displaystyle\sum_{k=1}^{\infty}\frac{1}{k^{2/3}}\right)$ **35** converges $\left(\text{use }\displaystyle\sum_{k=1}^{\infty}\frac{1}{k^2}\right)$

37 diverges $\left(\text{use }\displaystyle\sum_{k=1}^{\infty}\frac{1}{k}\right)$ **39** converges   **41** converges

43 converges   **45** diverges   **47** diverges
49 diverges   **51** converges   **53** converges
55 converges   **57** diverges   **59** converges
61 diverges   **63** diverges   **69** $a_k = 1/k^2$

Problem Set 11.5, page 678

1 converges   **3** converges   **5** converges
7 diverges   **9** converges   **11** converges
13 converges   **15** converges absolutely
17 diverges   **19** converges absolutely
21 converges absolutely   **23** converges absolutely
25 converges absolutely   **27** converges absolutely
29 diverges   **31** converges absolutely
33 converges absolutely   **35** converges absolutely
37 converges conditionally   **39** converges conditionally
41 converges absolutely   **43** converges absolutely
45 diverges   **47** converges absolutely
49 converges absolutely   **51** diverges
53 diverges   **55** $\frac{1249}{3080}$ overestimates with error $<\frac{1}{17}$

57 $\frac{115}{144}$ underestimates with error $<\frac{1}{25}$
59 $-\frac{137}{750}$ underestimates with error $<\frac{1}{2500}$
61 0.406   **65** true

Problem Set 11.6, page 685

1 $a = 0, R = \frac{1}{7}, I = \left(-\frac{1}{7}, \frac{1}{7}\right)$
3 $a = 0, R = +\infty, I = (-\infty, +\infty)$
5 $a = 0, R = 0, I$ consists of the single number 0
7 $a = 0, R = \dfrac{1}{\sqrt{3}}, I = \left(-\dfrac{1}{\sqrt{3}}, \dfrac{1}{\sqrt{3}}\right)$
9 $a = 0, R = 1, I = [-1, 1]$
11 $a = 0, R = +\infty, I = (-\infty, +\infty)$
13 $a = 2, R = 3, I = (-1, 5)$
15 $a = -3, R = 7, I = (-10, 4)$
17 $a = -1, R = \frac{1}{2}, I = [-\frac{3}{2}, -\frac{1}{2})$
19 $a = -1, R = 1, I = [-2, 0)$
21 $a = -5, R = 1, I = [-6, -4]$
23 $a = -2, R = \frac{1}{2}, I = [-\frac{5}{2}, -\frac{3}{2}]$
25 $a = 1, R = +\infty, I = (-\infty, +\infty)$
27 $a = -1, R = \sqrt[5]{5}, I = [-1 - \sqrt[5]{5}, -1 + \sqrt[5]{5}]$
29 $a = 4, R = 4, I = [0, 8)$
31 $a = 5, R = \dfrac{1}{\sqrt{2}}, I = \left[5 - \dfrac{1}{\sqrt{2}}, 5 + \dfrac{1}{\sqrt{2}}\right]$
33 $a = 1, R = 1, I = (0, 2)$
35 $a = 2, R = 0, I$ consists of the single number 2
37 $a = 3, R = 3, I = (0, 6)$
39 $a = 0, R = 1, I = [-1, 1)$ **41** R
45 $R = b, I = (-b, b)$ **47** $R = a, I = (-a, a)$

Problem Set 11.7, page 691

1 $\displaystyle\sum_{k=0}^{\infty} x^k, R = 1$ **3** $\displaystyle\sum_{k=0}^{\infty} x^{k+1}, R = 1$

5 $\displaystyle\sum_{k=0}^{\infty} x^{4k}, R = 1$ **7** $\displaystyle\sum_{k=0}^{\infty} (4x)^k, R = \dfrac{1}{4}$

9 $\displaystyle\sum_{k=0}^{\infty} x^{2k+1}, R = 1$ **11** $\displaystyle\sum_{k=0}^{\infty} \frac{(-1)^k x^k}{2^{k+1}}, R = 2$

13 $-\displaystyle\sum_{k=0}^{\infty} \frac{x^{k+1}}{k+1}, R = 1$ **15** $-\displaystyle\sum_{k=0}^{\infty} \frac{x^{k+2}}{(k+1)(k+2)}, R = 1$

17 $\displaystyle\sum_{k=1}^{\infty} \frac{(-1)^{k+1} x^{2k}}{2k(2k-1)}, R = 1$

19 $\displaystyle\sum_{k=0}^{\infty} \frac{(-1)^k 2^{k+1} + 3^{k+1}}{5(k+1)6^{k+1}} x^{k+1}, R = 2$

21 $\dfrac{-1}{1+x}, R = 1$ **23** $\dfrac{1}{1-x^2}, R = 1$

25 $\dfrac{x}{(1-x^2)^2}, R = 1$ **27** 2

29 $\displaystyle\sum_{k=1}^{\infty} k^3 x^{k-1}, R = 1$ **31** $\displaystyle\sum_{k=0}^{\infty} \frac{x^k}{k!}, R = +\infty$

33 $\displaystyle\sum_{k=1}^{\infty} k \cdot 2^{(k+2)/2}(x + 1)^{2k-1}, R = 2^{-1/4}$

35 $\displaystyle\sum_{k=0}^{\infty} \frac{(-1)^k x^{2k+1}}{(2k + 1)!}, R = +\infty$ **37** $\displaystyle\sum_{k=0}^{\infty} \frac{x^{2k+2}}{(2k + 2)!}, R = +\infty$

39 (a) 1; (b) 0; (c) -1; (d) 0

Problem Set 11.8, page 699

1 $\dfrac{1}{2} + \dfrac{\sqrt{3}}{2}\left(x - \dfrac{\pi}{6}\right) - \dfrac{1}{2(2!)}\left(x - \dfrac{\pi}{6}\right)^2 - \dfrac{\sqrt{3}}{2(3!)}\left(x - \dfrac{\pi}{6}\right)^3 +$

$\qquad\qquad\qquad\qquad \dfrac{1}{2(4!)}\left(x - \dfrac{\pi}{6}\right)^4 + \cdots$

3 $\displaystyle\sum_{k=0}^{\infty} (-1)^k \frac{(x - 2)^k}{2^{k+1}}$ **5** $\displaystyle\sum_{k=0}^{\infty} \frac{e^4(x - 4)^k}{k!}$

7 $1 + \dfrac{1}{2}(x - 2) - \dfrac{1}{8}(x - 2)^2 +$

$\qquad\qquad \displaystyle\sum_{k=3}^{\infty} \frac{(-1)^{k+1} 1 \cdot 3 \cdots (2k - 3)}{k! \, 2^k}(x - 2)^k$

9 $\displaystyle\sum_{k=0}^{\infty} (k + 1)(x + 2)^k$

11 $2 - 2\sqrt{3}\left(x - \dfrac{\pi}{6}\right) + 7\left(x - \dfrac{\pi}{6}\right)^2 - \dfrac{23\sqrt{3}}{3}\left(x - \dfrac{\pi}{6}\right)^3 + \cdots$

13 $1 + 2\left(x - \dfrac{\pi}{4}\right) + 2\left(x - \dfrac{\pi}{4}\right)^2 + \dfrac{8}{3}\left(x - \dfrac{\pi}{4}\right)^3 + \cdots$

15 $\displaystyle\sum_{k=0}^{\infty} \frac{x^{2k+1}}{(2k + 1)!}$ **17** $\displaystyle\sum_{k=0}^{\infty} (-1)^k \frac{x^{2k}}{k!}$

19 $\displaystyle\sum_{k=0}^{\infty} \frac{(-1)^k x^{4k}}{(2k)!}$ **21** $\displaystyle\sum_{k=0}^{\infty} \frac{(-1)^k 2^{2k+1} x^{2k+3}}{(2k + 1)!}$

23 $\displaystyle\sum_{k=1}^{\infty} \frac{(-1)^{k+1} 2^{2k-1} x^{2k}}{(2k)!}$ **25** $\displaystyle\sum_{k=0}^{\infty} \frac{(-1)^k x^{2k}}{(2k + 1)!}$

27 $\displaystyle\sum_{k=0}^{\infty} \frac{(-1)^k x^{4k+3}}{(4k + 3)(2k + 1)!}$ **29** $\displaystyle\sum_{k=0}^{\infty} \frac{(-1)^k x^{2k+1}}{(2k + 1)k!}$

31 0.9802   **33** -0.1054 **35** 0.4969

37 0.7468   **39** 0.6990   **41** 0.2652

43 $f^{(n)}(0) = 0$ if n is even; $f^{(n)}(0) = (-1)^{(n+3)/2}(n - 1)!$ if n is odd

45 0   **47** $\dfrac{16!}{8!}$ **49** $(-2)19!$

Problem Set 11.9, page 704

1 $1 + \dfrac{1}{4}x + \displaystyle\sum_{k=2}^{\infty} \frac{(-1)^{k+1} 3 \cdot 7 \cdot 11 \cdots (4k - 5)}{4^k k!} x^k, |x| < 1$

3 $1 + \displaystyle\sum_{k=1}^{\infty} \frac{1 \cdot 4 \cdot 7 \cdots (3k - 2)}{3^k k!} x^{2k}, |x| < 1$

5 $1 + \displaystyle\sum_{k=1}^{\infty} \frac{(-1)^k [1 \cdot 4 \cdot 7 \cdots (3k - 2)]}{3^k k!} x^k, |x| < 1$

7 $\displaystyle\sum_{k=1}^{\infty} (-1)^{k+1} 2^{k-1} k x^k, |x| < \dfrac{1}{2}$

9 $3 + \dfrac{x}{27} + \displaystyle\sum_{k=2}^{\infty} \frac{(-1)^{k+1} 2 \cdot 5 \cdot 8 \cdots (3k - 4)}{3^{4k-1} k!} x^k, |x| < 27$

11 $1 + \dfrac{1}{5}x^3 + \displaystyle\sum_{k=2}^{\infty} \frac{(-1)^k 4 \cdot 9 \cdot 14 \cdots (5k - 6)}{5^k k!} x^{3k}, |x| < 1$

13 10.049875 with $|\text{error}| \le 6.25 \times 10^{-7}$

15 1.0148875 with $|\text{error}| \le 1.7 \times 10^{-6}$

17 2.030518 with $|\text{error}| \le 2.7 \times 10^{-5}$ **19** 0.690

21 0.497   **23** 0.401   **25** 3.009

29 They are the same.

33 The series becomes a finite sum, correct for all values of x.

35 $x + \displaystyle\sum_{k=1}^{\infty} \frac{1 \cdot 3 \cdot 5 \cdots (2k - 1)}{(2k + 1)2^k k!} x^{2k+1}, |x| < 1$

Review Problem Set, Chapter 11, page 705

1 converges; limit is $\frac{1}{3}$ **3** converges; limit is $\sqrt{\frac{1}{3}}$

5 converges; limit is 0   **7** converges; limit is 0

9 diverges   **11** converges; limit is 1   **13** increasing

15 nonmonotonic   **17** no; it first increases and then decreases

19 bounded; nonmonotonic; convergent; limit is $\frac{4}{5}$

23 1   **25** $\sin 1$

27 $\displaystyle\sum_{k=1}^{\infty} \frac{15}{(2k + 3)(2k + 5)}$; converges; sum is $\dfrac{3}{2}$

29 $\frac{1}{3}$ **31** $\frac{23}{6}$

33 $\displaystyle\lim_{n \to \infty} a_n = 0$ does not guarantee that $\displaystyle\sum_{k=1}^{\infty} a_k$ is convergent.

35 converges   **37** converges   **39** converges

41 diverges   **43** conditionally convergent

45 absolutely convergent   **47** diverges

49 absolutely convergent   **51** absolutely convergent

53 (a) 0.4058; (b) 0.0332

55 $a = 1, R = \sqrt{5}, I = (1 - \sqrt{5}, 1 + \sqrt{5})$

57 $a = -2, R = 1, I = (-3, -1)$

59 $a = 10, R = 0, I$ consists of the single number 10

61 $a = -\pi, R = +\infty, I = (-\infty, +\infty)$

63 $a = 3, R = \frac{1}{2}, I = [\frac{5}{2}, \frac{7}{2}]$

65 $\displaystyle\sum_{k=1}^{\infty} (-1)^{k+1} \frac{(x - 1)^k}{k}$ for $0 < x < 2$

67 $\dfrac{1}{e} + \dfrac{1}{e}(x + 1) + \dfrac{1}{2!e}(x + 1)^2 + \dfrac{1}{3!e}(x + 1)^3$

69 $1 + \dfrac{1}{2}(x - 1) - \dfrac{1}{2!2^2}(x - 1)^2 + \dfrac{3}{3!2^3}(x - 1)^3$

71 $1 + 0 - \dfrac{4}{2!}\left(x - \dfrac{\pi}{4}\right)^2 + 0$

77 $1 + \displaystyle\sum_{k=1}^{\infty} \dfrac{(-1)^k 1 \cdot 3 \cdot 5 \cdot 7 \cdots (2k-1)}{2^k k!} x^{2k}$ for $|x| < 1$

79 $1 - \dfrac{4x}{3} - \displaystyle\sum_{k=1}^{\infty} \dfrac{2^{k+2} 1 \cdot 4 \cdot 7 \cdots (3k-2)}{3^{k+1}(k+1)!} x^{k+1}$ for $|x| < \dfrac{1}{2}$

81 $x + \dfrac{x^4}{12} + \displaystyle\sum_{k=2}^{\infty} \dfrac{(-1)^{k+1} 2 \cdot 5 \cdot 8 \cdots (3k-4)}{(3k+1)3^k k!} x^{3k+1}$ for $|x| < 1$

83 1.974375, $|\text{error}| < 2.34 \times 10^{-5}$

85 (a) $1 + x - \dfrac{x^2}{2!} - \dfrac{x^3}{3!} + \dfrac{x^4}{4!} + \dfrac{x^5}{5!} - \dfrac{x^6}{6!} - \dfrac{x^7}{7!} + \cdots$ for

all x; (b) $1 - 2x^2 + \dfrac{4^2 x^4}{4!} - \dfrac{4^3 x^6}{6!} + \dfrac{4^4 x^8}{8!} - \dfrac{4^5 x^{10}}{10!} + \cdots$

for all x; (c) $x^3 - \dfrac{x^9}{3} + \dfrac{x^{15}}{5} - \dfrac{x^{21}}{7} + \dfrac{x^{27}}{9} - \dfrac{x^{33}}{11} + \cdots$ for

$|x| < 1$; (d) $1 + (\ln 10)x + \dfrac{(\ln 10)^2}{2!} x^2 + \dfrac{(\ln 10)^3}{3!} x^3 + \cdots$

for all x

87 $\sin x$ **89** $x - 1 + \dfrac{\sin x}{x}$ **91** 2^x

Chapter 12

Problem Set 12.1, page 715

13 $\mathbf{X} = -(\mathbf{A} + \mathbf{B}) = -\mathbf{A} - \mathbf{B}$

17 $\overrightarrow{PQ} = 4\mathbf{i} - 5\mathbf{j}$; $\overrightarrow{QP} = -4\mathbf{i} + 5\mathbf{j}$

19 $\overrightarrow{PQ} = 2\mathbf{i}$; $\overrightarrow{QP} = -2\mathbf{i}$ **21** $\mathbf{i} + 6\mathbf{j}$ **23** $-\mathbf{i} + 16\mathbf{j}$

25 $-36\mathbf{i} + 49\mathbf{j}$ **27** $96\mathbf{i} - 135\mathbf{j}$ **29** $4\mathbf{i} + 9\mathbf{j}$

31 $\langle -25, 34 \rangle$ **33** $\langle -1, 16 \rangle$ **35** $\langle -18, 2 \rangle$

37 $\langle \frac{17}{3}, -\frac{8}{3} \rangle$ **39** $x = \frac{8}{3}$, $y = 1$

41 $x = -\frac{2}{5}$, $y = \frac{3}{5}$ **43** $C_1 = -\frac{3}{19}$, $C_2 = \frac{6}{19}$

47 $RSPQ$, $RPSQ$, $RPQS$, $PRSQ$, $PRQS$, $PQRS$ and the reverse of each of these

49 $\overrightarrow{QR} + \frac{1}{2}\overrightarrow{PQ}$

51 (a) $3\mathbf{i} + 6\mathbf{j}$; (b) $-\mathbf{i} + 3\mathbf{j}$; (c) $-4\mathbf{i} - 3\mathbf{j}$; (d) $4\mathbf{i} + 3\mathbf{j}$;
  (e) $-10\mathbf{i} - 7\mathbf{j}$; (f) $10\mathbf{i} - 5\mathbf{j}$; (g) $6\mathbf{i} - 8\mathbf{j}$; (h) $-18\mathbf{i} - 72\mathbf{j}$

Problem Set 12.2, page 722

1 15 **3** 2.0459951 **5** $\dfrac{\pi}{2}$ **7** 6

9 $5\sqrt{2}$ **11** 0 **13** $|\mathbf{A}| = \sqrt{10}$; $|\mathbf{B}| = \sqrt{5}$

15 $\sqrt{13}$ **17** $\sqrt{10} + \sqrt{5}$ **19** $\dfrac{1}{\sqrt{10}}\mathbf{i} - \dfrac{3}{\sqrt{10}}\mathbf{j}$

21 $\mathbf{i}$ **23** $\dfrac{-\sqrt{17}}{17}\mathbf{i} - \dfrac{4\sqrt{17}}{17}\mathbf{j}$

25 (a) 0; (b) 0; (c) 0

27 (a) 2; (b) $\dfrac{2\sqrt{85}}{85}$; (c) $\dfrac{2\sqrt{5}}{5}$

29 (a) -12; (b) $\dfrac{-2\sqrt{5}}{5}$; (c) -4

31 (a) -4; (b) $\dfrac{-4\sqrt{65}}{65}$; (c) $\dfrac{-4\sqrt{5}}{5}$

33 -3 **35** -30 **37** $\langle -27, -9 \rangle$

39 $\frac{21}{5}$ **41** $\angle A = 21.501434°$; $\angle B = 17.592425°$; $\angle C = 140.90614°$

43 (a) θ is acute; (b) $\theta = \dfrac{\pi}{2}$; (c) θ is obtuse

45 (a) $\sqrt{|\mathbf{A}|^2 + |\mathbf{B}|^2}$; (b) $\sqrt{4|\mathbf{A}|^2 + 9|\mathbf{B}|^2}$; (c) $\sqrt{|\mathbf{A}|^2 + |\mathbf{B}|^2}$

47 $\overrightarrow{AB} \cdot \overrightarrow{BC} = 0$ so $\angle \mathbf{B} = 90°$

49 $\overrightarrow{PQ} \cdot \overrightarrow{QR} = \overrightarrow{QR} \cdot \overrightarrow{RS} = \overrightarrow{RS} \cdot \overrightarrow{SP} = \overrightarrow{SP} \cdot \overrightarrow{PQ} = 0$

57 $45\sqrt{3}$ joules   **59** $\mathbf{F} \cdot \overrightarrow{PR}$; same

61 (a) $\mathbf{A} = \cos \alpha\, \mathbf{i} + \sin \alpha\, \mathbf{j}$, $\mathbf{B} = \cos \beta\, \mathbf{i} + \sin \beta\, \mathbf{j}$;
  (b) $\mathbf{A} \cdot \mathbf{B} = \cos \alpha \cos \beta + \sin \alpha \sin \beta$ and $|\mathbf{A}| = |\mathbf{B}| = 1$ for $\theta = \alpha - \beta$; so $\cos(\alpha - \beta) = \cos \alpha \cos \beta + \sin \alpha \sin \beta$.

Problem Set 12.3, page 731

1 $5\mathbf{i} - 3\mathbf{j}$ **3** $0\mathbf{i} + 0\mathbf{j}$ **5** $5\mathbf{i} + 3\mathbf{j}$; $\sqrt{34}$

7 $-4\mathbf{i} + 3\mathbf{j}$; 5   **9** $x - 4y = 1$ (line)   **11** $x = 2 + y$ (line)

13 $x + y = 0$ (line)   **15** $x^2 + y^2 = 9$ (circle)

17 $(x - 2)^2 + (y - 3)^2 = 16$ (circle)

19 $\dfrac{x^2}{9} + \dfrac{y^2}{8} = 1$ (ellipse); foci: $(1, 0)$, $(-1, 0)$; vertices: $(3, 0)$, $(-3, 0)$, $(0, 2\sqrt{2})$, $(0, -2\sqrt{2})$

21 $\dfrac{3}{\sqrt{10}}$ **23** $\sqrt{2}$ **25** $\dfrac{3\sqrt{5}}{5}$

27 (a) $\mathbf{M} = 2\mathbf{i} + 2\mathbf{j}$; (b) $\mathbf{R} = (1 + 2t)\mathbf{i} + (2 + 2t)\mathbf{j}$; (c) $\begin{cases} x = 1 + 2t \\ y = 2 + 2t \end{cases}$

29 (a) $\mathbf{M} = 13\mathbf{i} - 17\mathbf{j}$; (b) $\mathbf{R} = (-\frac{3}{2} + 13t)\mathbf{i} + (\frac{5}{2} - 17t)\mathbf{j}$;
  (c) $\begin{cases} x = -\frac{3}{2} + 13t \\ y = \frac{5}{2} - 17t \end{cases}$

31 (a) $\mathbf{M} = (-\sqrt{2} - \pi)\mathbf{i} + (\sqrt{3} - e)\mathbf{j}$;
  (b) $\mathbf{R} = [\pi + t(-\sqrt{2} - \pi)]\mathbf{i} + [e + (\sqrt{3} - e)t]\mathbf{j}$;
  (c) $\begin{cases} x = \pi + (-2 - \pi)t \\ y = e + (\sqrt{3} - e)t \end{cases}$

33 $\mathbf{R} = (7 - t)\mathbf{i} + (-2 + 3t)\mathbf{j}$

35 $\mathbf{R} = (3 + 4\cos t)\mathbf{i} + (4 + 4\sin t)\mathbf{j}$

37 (a) $x = 2\cos t$, $y = 2\sin t$ for $0 \le t \le 2\pi$;
  (b) $x^2 + y^2 = 4$ (circle)

39 (a) $x = 5\cos 2t$, $y = -5\sin 2t$ for $0 \le t \le \pi/2$;
  (b) $x^2 + y^2 = 25$ (circle)

41 (a) $x = 4t$, $y = 5 + 3t$; (b) $y = 5 + \frac{3}{4}x$ (line)

43 (a) $x = \dfrac{1}{t - 2}$, $y = 2t + 1$ for $0 \le t < 2$;
  (b) $y = \dfrac{2}{x} + 5$, $x \le -\dfrac{1}{2}$

45 (b) $x + 3y = 7$; (c) $\dfrac{dy}{dx} = -\dfrac{1}{3}$, $\dfrac{d^2y}{dx^2} = 0$

47 (b) $y^2 = 25(x + 2)$; (c) $\dfrac{dy}{dx} = \dfrac{5}{2t}$, $\dfrac{d^2y}{dt^2} = \dfrac{-5}{4t^3}$

49 (b) $y = \dfrac{6 - 4x + x^2}{2 - x}$; (c) $\dfrac{dy}{dx} = 2t^2 - 1$, $\dfrac{d^2y}{dx^2} = 4t^3$

51 (b) $(y - 3x - 1)^2 = (1 + x)^3$;

(c) $\dfrac{dy}{dx} = \dfrac{3}{2}(t + 1)$, $\dfrac{d^2y}{dt^2} = \dfrac{3}{4(t - 1)}$

53 $\dfrac{|\mathbf{D} \cdot (\mathbf{R}_1 - \mathbf{D})|}{|\mathbf{D}|}$

57 $\mathbf{R} = (at - b \sin t)\mathbf{i} + (a - b \cos t)\mathbf{j}$;
$x = at - b \sin t$, $y = a - b \cos t$

59 $\mathbf{R} = \frac{1}{2}(\cos t + 2 \cot t)\mathbf{i} + \frac{1}{2}(\sin t + 2)\mathbf{j}$

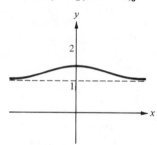

Problem Set 12.4, page 738

1 (a) all real numbers except 1 and -1; (b) $8\mathbf{i} + \frac{5}{3}\mathbf{j}$;
(c) $8\mathbf{i} + \frac{5}{3}\mathbf{j}$; (d) all real numbers except 1 and -1

3 (a) all real numbers except 3; (b) undefined; (c) $\mathbf{i} + 13\mathbf{j}$;
(d) all real numbers except 3

5 (a) all real numbers except 0; (b) $\dfrac{6}{\pi}\mathbf{i} + \mathbf{j}$; (c) $\dfrac{6}{\pi}\mathbf{i} + \mathbf{j}$;

(d) all real numbers except 0

7 $\mathbf{F}'(t) = 6t\mathbf{i} + 36t^3\mathbf{j}$; $\mathbf{F}''(t) = 6\mathbf{i} + 108t^2\mathbf{j}$

9 $\mathbf{F}'(t) = 3e^{3t}\mathbf{i} + \dfrac{1}{t}\mathbf{j}$; $\mathbf{F}''(t) = 9e^{3t}\mathbf{i} - \dfrac{1}{t^2}\mathbf{j}$

11 $\mathbf{F}'(t) = -5 \sin t\, \mathbf{i} + 3 \cos t\, \mathbf{j}$; $\mathbf{F}''(t) = -5 \cos t\, \mathbf{i} - 3 \sin t\, \mathbf{j}$

13 $\dfrac{d\mathbf{R}}{dt} = (-2t \sin t^2)\mathbf{i} + (2t \cos t^2)\mathbf{j}$;

$\dfrac{d^2\mathbf{R}}{dt^2} = (-4t^2 \cos t^2 - 2 \sin t^2)\mathbf{i} + (-4t^2 \sin t^2 + 2 \cos t^2)\mathbf{j}$

15 $\dfrac{d\mathbf{R}}{dt} = -\dfrac{1}{t^2}\mathbf{i} - \dfrac{2}{t^3}\mathbf{j}$; $\dfrac{d^2\mathbf{R}}{dt^2} = \dfrac{2}{t^3}\mathbf{i} + \dfrac{6}{t^4}\mathbf{j}$

17 (a) $e^t\mathbf{i} + e^t\mathbf{j}$; (b) $e^t\mathbf{i} + e^t\mathbf{j}$; (c) $2e^{2t}$; (d) $\dfrac{2e^{2t} + 10e^t}{\sqrt{2e^{2t} + 20e^t + 58}}$

19 (a) $(12 \cos 3t)\mathbf{i} - (12 \cos 3t)\mathbf{j}$;
(b) $(-36 \sin 3t)\mathbf{i} + (36 \sin 3t)\mathbf{j}$;

(c) $-432 \sin 6t$; (d) $12\sqrt{2}\, \dfrac{\sin (3t) \cos (3t)}{|\sin 3t|}$

21 (a) $2e^{2t}(\cos 2t - \sin 2t) + 2e^{-4t}(\cos 2t - 2 \sin 2t)$;
(b) $-5e^{-5t}\mathbf{i} - 11e^{-11t}\mathbf{j}$

23 (a) $\dfrac{1}{t^2} - \dfrac{\ln t}{t^2} - \dfrac{3(3t - 2)}{t^3(t - 1)^2}$;

(b) $\left(\dfrac{\cos 7t}{t} - 7 \sin t \ln t\right)\mathbf{i} + \left(\dfrac{(7 - t) \sin 7t - \cos 7t}{(t - 1)^2}\right)\mathbf{j}$

25 (a) $(3s^2 \cos t)\mathbf{i} - (3s^2 \sin t)\mathbf{j}$; (b) 0

Problem Set 12.5, page 738

1 (a) $6t\mathbf{i} + 2\mathbf{j}$; (b) $6\mathbf{i}$; (c) $2\sqrt{9t^2 + 1}$

3 (a) $2t\mathbf{i} + \dfrac{1}{t}\mathbf{j}$; (b) $2\mathbf{i} - \dfrac{1}{t^2}\mathbf{j}$; (c) $\sqrt{\dfrac{4t^4 + 1}{t^2}}$

5 (a) $2(1 - \cos t)\mathbf{i} + 2 \sin t\, \mathbf{j}$; (b) $2 \sin t\, \mathbf{i} + 2 \cos t\, \mathbf{j}$;
(c) $2\sqrt{2(1 - \cos t)}$

7 (a) $(-\sin t + \cos t)\mathbf{i} + (-\sin t - \cos t)\mathbf{j}$;
(b) $(-\cos t - \sin t)\mathbf{i} + (-\cos t + \sin t)\mathbf{j}$; (c) $\sqrt{2}$

9 (a) $e^t(\cos t - \sin t)\mathbf{i} + e^t(\cos t + \sin t)\mathbf{j}$;
(b) $-2e^t \sin t\, \mathbf{i} + 2e^t \cos t\, \mathbf{j}$; (c) $\sqrt{2}e^t$

11 (a) $27\mathbf{i} + 5\mathbf{j}$; (b) $14\mathbf{i}$; (c) $\sqrt{754}$

13 (a) $\dfrac{3\pi\sqrt{2}}{2}\mathbf{i} - 2\pi\sqrt{2}\mathbf{j}$; (b) $\dfrac{3\pi^2\sqrt{2}}{2}\mathbf{i} + 2\pi^2\sqrt{2}\mathbf{j}$; (c) $\dfrac{5\pi\sqrt{2}}{2}$

15 $\sqrt{3}\mathbf{i} - \dfrac{\sqrt{3}}{3}\mathbf{j}$; (b) $-4\mathbf{i} - \dfrac{4}{3}\mathbf{j}$; (c) $\sqrt{\dfrac{10}{3}}$

17 (a) $12\mathbf{i} - 7\mathbf{j}$; (b) $\sqrt{193}$; (c) $\dfrac{12\sqrt{193}}{193}\mathbf{i} - \dfrac{7\sqrt{193}}{193}\mathbf{j}$

19 (a) $(1 - \cos t)\mathbf{i} + (\sin t)\mathbf{j}$; (b) $\sqrt{2(1 - \cos t)}$;
(c) $\dfrac{\sqrt{2}}{2}\mathbf{i} + \dfrac{\sqrt{2}}{2}\mathbf{j}$

21 (a) $(-2 \sin 2t)\mathbf{i} + (2 \cos 2t)\mathbf{j}$; (b) 2; (c) $-\dfrac{\sqrt{3}}{2}\mathbf{i} + \dfrac{1}{2}\mathbf{j}$

23 (a) $\dfrac{-\sin t}{(1 + \cos t)^2}\mathbf{i} + \dfrac{1}{1 + \cos t}\mathbf{j}$; (b) $\sqrt{\dfrac{2}{(1 + \cos t)^3}}$;
(c) $-0.3090\mathbf{i} + 0.9511\mathbf{j}$

25 $2\sqrt{74}$ **27** $\frac{1}{4}(e^2 + 1)$ **29** 6π **31** $\sqrt{2}(1 - e^{-2\pi})$

33 $\dfrac{\pi^2}{32}$ **35** $\ln (1 + \sqrt{2})$ **37** $\dfrac{3}{2} \sin^2 \dfrac{\pi}{9} \approx 0.175466668$

39 $\displaystyle\int_{\theta_1}^{\theta_2} \sqrt{\left(\dfrac{dr}{d\theta}\right)^2 + r^2}\, d\theta$ **41** $\dfrac{\mathbf{i} + f'(x)\mathbf{j}}{\sqrt{1 + [f'(x)]^2}}$

Problem Set 12.6, page 749

1 (a) $\dfrac{7\mathbf{i} - 3\mathbf{j}}{\sqrt{58}}$; (b) $\dfrac{3\mathbf{i} + 7\mathbf{j}}{\sqrt{58}}$; (c) 0; (d) undefined

3 (a) $(-\sin t)\mathbf{i} + (\cos t)\mathbf{j}$; (b) $(-\cos t)\mathbf{i} - (\sin t)\mathbf{j}$; (c) $\frac{1}{3}$;
(d) $(-\cos t)\mathbf{i} - (\sin t)\mathbf{j}$

5 (a) $\dfrac{(-3 \sin \pi t)\mathbf{i} + (5 \cos \pi t)\mathbf{j}}{\sqrt{9 \sin^2 \pi t + 25 \cos^2 \pi t}}$; (b) $\dfrac{(-5 \cos \pi t)\mathbf{i} - (3 \sin \pi t)\mathbf{j}}{\sqrt{9 \sin^2 \pi t + 25 \cos^2 \pi t}}$;

(c) $\dfrac{15}{(9 \sin^2 \pi t + 25 \cos^2 \pi t)^{3/2}}$; (d) $\mathbf{N} = \mathbf{N}_l$

7 (a) $\dfrac{\mathbf{i} + e^t\mathbf{j}}{\sqrt{1 + e^{2t}}}$; (b) $\dfrac{-e^t\mathbf{i} + \mathbf{j}}{\sqrt{1 + e^{2t}}}$; (c) $\dfrac{e^t}{(1 + e^{2t})^{3/2}}$; (d) $\mathbf{N} = \mathbf{N}_l$

9 (a) $\dfrac{t}{\sqrt{13}|t|}$ $(3\mathbf{i} + 2\mathbf{j})$; (b) $\dfrac{t}{\sqrt{13}|t|}$ $(-2\mathbf{i} + 3\mathbf{j})$; (c) 0;

(d) undefined

11 (a) $\dfrac{-\mathbf{i} + 2t^3\mathbf{j}}{\sqrt{1 + 4t^6}}$; (b) $\dfrac{-2t^3\mathbf{i} - \mathbf{j}}{\sqrt{1 + 4t^6}}$; (c) $\dfrac{-6t^4}{(1 + 4t^6)^{3/2}}$;

(d) $\mathbf{N} = -\mathbf{N}_l$

13 (a) $\dfrac{(-2 \sin 2\theta)\mathbf{i} + (\cos \theta)\mathbf{j}}{\sqrt{4 \sin^2 2\theta + \cos^2 \theta}}$; (b) $\dfrac{(-\cos \theta)\mathbf{i} - (2 \sin 2\theta)\mathbf{j}}{\sqrt{4 \sin^2 2\theta + \cos^2 \theta}}$;

(c) $\dfrac{4}{(16 \sin^2 \theta + 1)^{3/2}}$; (d) $\mathbf{N} = \mathbf{N}_l$

15 (a) $\dfrac{x\mathbf{i} + \mathbf{j}}{\sqrt{1 + x^2}}$; (b) $\dfrac{-\mathbf{i} + x\mathbf{j}}{\sqrt{1 + x^2}}$; (c) $\dfrac{-x}{(1 + x^2)^{3/2}}$; (d) $\mathbf{N} = -\mathbf{N}_l$

17 (a) $\dfrac{1}{2}\mathbf{i} - \dfrac{\sqrt{3}}{2}\mathbf{j}$; (b) $\dfrac{\sqrt{3}}{2}\mathbf{i} + \dfrac{1}{2}\mathbf{j}$; (c) $\dfrac{1}{3}$; (d) $\mathbf{N} = \mathbf{N}_l$

19 (a) $\dfrac{\mathbf{i} + 4\mathbf{j}}{\sqrt{17}}$; (b) $\dfrac{-4\mathbf{i} + \mathbf{j}}{\sqrt{17}}$; (c) $\dfrac{52}{(17)^{3/2}}$; (d) $\mathbf{N} = \mathbf{N}_l$

21 (a) $\dfrac{-32\mathbf{i} - 9\mathbf{j}}{\sqrt{1105}}$; (b) $\dfrac{9\mathbf{i} - 32\mathbf{j}}{\sqrt{1105}}$; (c) $\dfrac{6912}{(1105)^{3/2}}$; (d) $\mathbf{N} = \mathbf{N}_l$

23 (a) $\mathbf{T} = \dfrac{\left(\dfrac{dr}{d\theta} \cos \theta - r \sin \theta\right)\mathbf{i} + \left(\dfrac{dr}{d\theta} \sin \theta + r \cos \theta\right)\mathbf{j}}{\sqrt{\left(\dfrac{dr}{d\theta}\right)^2 + r^2}}$;

(b) $\mathbf{N}_l = \dfrac{-\left(\dfrac{dr}{d\theta} \sin \theta + r \cos \theta\right)\mathbf{i} + \left(\dfrac{dr}{d\theta} \cos \theta - r \sin \theta\right)\mathbf{j}}{\sqrt{\left(\dfrac{dr}{d\theta}\right)^2 + r^2}}$;

(c) $\kappa = \dfrac{r^2 + 2\left(\dfrac{dr}{d\theta}\right)^2 - r \dfrac{d^2r}{d\theta^2}}{\left(r^2 + \left(\dfrac{dr}{d\theta}\right)^2\right)^{3/2}}$;

(d) $\mathbf{N} = \begin{cases} \mathbf{N}_l & \text{if } \kappa > 0 \\ -\mathbf{N}_l & \text{if } \kappa < 0 \end{cases}$

25 (a) $\mathbf{T} = \dfrac{(\sin 2\theta - 1)\mathbf{i} + (1 - \cos 2\theta)\mathbf{j}}{\sqrt{2(1 - \cos \theta)}}$;

(b) $\mathbf{N}_l = \dfrac{(\cos 2\theta - 1)\mathbf{i} + (\sin 2\theta - 1)\mathbf{j}}{\sqrt{2(1 - \cos \theta)}}$;

(c) $\kappa = \dfrac{3}{2\sqrt{2}\sqrt{1 - \cos \theta}}$; (d) $\mathbf{N} = \mathbf{N}_l$

27 (a) $\mathbf{T} = \dfrac{(1 + e \sin \theta)\mathbf{i} + (\cos \theta)\mathbf{j}}{\sqrt{e^2 + 2e \sin \theta + 1}}$;

(b) $\mathbf{N}_l = \dfrac{(-\cos \theta)\mathbf{i} + (1 - e \sin \theta)\mathbf{j}}{\sqrt{e^2 + 2e \sin \theta + 1}}$;

(c) $\kappa = \dfrac{(1 + e \sin \theta)^4(e \sin \theta + e^2 - e^2 d + 1)}{ed(e^2 + 2e \sin \theta + 1)^{3/2}}$;

(d) $\mathbf{N} = \begin{cases} \mathbf{N}_l & \text{if } \kappa > 0 \\ -\mathbf{N}_l & \text{if } \kappa < 0 \end{cases}$

31 If $\kappa > 0$, curve is concave up; if $\kappa < 0$, curve is concave down.

Problem Set 12.7, page 754

1 (a) $\mathbf{V} = 5\mathbf{i}$; (b) $\mathbf{A} = \frac{50}{9}\mathbf{j}$

3 $v = 2\pi\sqrt{\sin^2 2\pi t + 9 \cos^2 2\pi t}$;

$\dfrac{dv}{dt} = \dfrac{-32\pi^2 \sin 2\pi t \cos 2\pi t}{\sqrt{\sin^2 2\pi t + 9 \cos^2 2\pi t}}$

5 $A_T = \frac{41}{5}$, $A_N = \frac{12}{5}$

7 (a) $\mathbf{V} = 150\mathbf{i} + (150\sqrt{3} - gt)\mathbf{j}$;

(b) $\mathbf{R} = (150t)\mathbf{i} + \left(-\dfrac{g}{2}t^2 + 150\sqrt{3}t\right)\mathbf{j}$

9 $384\pi^2$ N   **11** $\frac{55}{36}$ mi   **13** $\dfrac{7}{\sqrt{12\pi}}$ rev/s

23 1.989×10^{30} kg   **25** 1.495×10^{11} m

27 1.082×10^{11} m

Review Problem Set, Chapter 12, page 757

5 yes   **7** yes   **11** $(\frac{1}{2}, 5)$

13 (a) $10\mathbf{i} - 5\mathbf{j}$; (b) $-8\mathbf{i} + 12\mathbf{j}$; (c) $4\mathbf{i} - 4\mathbf{j}$; (d) $2\mathbf{j}$;

(e) $10\mathbf{i} - 11\mathbf{j}$; (f) 7; (g) -97; (h) $\sqrt{5}$; (i) 2;

(j) $2\sqrt{5} + 3\sqrt{13}$; (k) $\dfrac{7\sqrt{13}}{13}$

15 $\dfrac{\pi}{4}, \dfrac{\pi}{4}, \dfrac{\pi}{2}$ **17** $-y\mathbf{i} + x\mathbf{j}$ **21** $\dfrac{6}{5}$

23 (a)   (b)   (c)

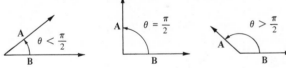

25 (a) $2x + 5y = 26$; (b) $(2\mathbf{i} + 5\mathbf{j}) \cdot [\mathbf{R} - (3\mathbf{i} + 4\mathbf{j})] = 0$;

(c) $\mathbf{R} = (3 + 5t)\mathbf{i} + (4 - 2t)\mathbf{j}$; (d) $\begin{cases} x = 3 + 5t \\ y = 4 - 2t \end{cases}$

29 $\mathbf{R} = (\cosh t)\mathbf{i} + (\sinh t)\mathbf{j}$

31 (a) $-(t - 1)^{-2}\mathbf{i} - 6t(t^2 - 2)^{-2}\mathbf{j}$;

(b) $2(t - 1)^{-3}\mathbf{i} + [-6(t^2 - 2)^{-2} + 24t^2(t^2 - 2)^{-3}]\mathbf{j}$

33 (a) $5e^{5t}\mathbf{i} - 3e^{-3t}\mathbf{j}$; (b) $25e^{5t}\mathbf{i} + 9e^{-3t}\mathbf{j}$

35 (a) $\dfrac{2t - \sin t - t \cos t + \cos t - t \sin t}{\sqrt{1 + 2t^2 - 2t \sin t + 2t \cos t}}$; (b) $\sin t - \cos t$

37 (a) $\dfrac{4(e^{8t} - e^{-8t})}{\sqrt{e^{8t} + e^{-8t}}}$; (b) $64(e^{8t} - e^{-8t})$

39 $(e^t - \sin 2t)\mathbf{i} + (2t + \sin 2t)\mathbf{j}$ **41** $\dfrac{e^{2t} + 2t^3 - 2t}{\sqrt{e^{2t} + (t^2 - 1)^2}}$

43 $e^{2t} + 4t$ **45** 6   **47** $\frac{87}{8}$

49 (a) $\mathbf{j}$; (b) 3; (c) $-\mathbf{i}$

51 (a) $\dfrac{2\mathbf{i} + 3\mathbf{j}}{\sqrt{13}}$; (b) $\dfrac{6}{(13)^{3/2}}$; (c) $\dfrac{-3\mathbf{i} + 2\mathbf{j}}{\sqrt{13}}$

53 (a) $3t^2\mathbf{i} + 8t^3\mathbf{j}$; (b) $6t\mathbf{i} + 24t^2\mathbf{j}$; (c) $t^2\sqrt{9 + 64t^2}$;

(d) $\dfrac{18t + 192t^3}{\sqrt{9 + 64t^2}}$; (e) $\dfrac{(54t + 576t^3)\mathbf{i} + (144t^2 + 1536t^4)\mathbf{j}}{9 + 64t^2}$;

(f) $\dfrac{-192t^3\mathbf{i} + 72t^2\mathbf{j}}{9 + 64t^2}$

55 (a) $(-21 \sin 7t)\mathbf{i} + (21 \cos 7t)\mathbf{j}$;

(b) $(-147 \cos 7t)\mathbf{i} - (147 \sin 7t)\mathbf{j}$; (c) 21; (d) 0; (e) $\mathbf{0}$;

(f) $(-147 \cos 7t)\mathbf{i} - (147 \sin 7t)\mathbf{j}$

57 (a) $t\mathbf{i} + t^2\mathbf{j}$; (b) $\mathbf{i} + 2t\mathbf{j}$; (c) $\sqrt{t^2 + t^4}$; (d) $\dfrac{t + 2t^3}{\sqrt{t^2 + t^4}}$;

(e) $\dfrac{(1 + 2t^2)\mathbf{i} + (t + 2t^3)\mathbf{j}}{1 + t^2}$; (f) $\dfrac{-t^2\mathbf{i} + t\mathbf{j}}{1 + t^2}$

59 $\frac{1807}{432}$

61 semimajor axis of Neptune's orbit is 4.496×10^{12} m

63 about 1.524 times as far

Chapter 13

Problem Set 13.1, page 763

5 (a) $(1, 4, 4)$; (b) $(3, 7, 0)$; (c) $(2, 1, -3)$

7 The coordinates of the vertices are $(3, 3, 3)$, $(3, 3, -3)$, $(3, -3, 3)$, $(3, -3, -3)$, $(-3, 3, 3)$, $(-3, 3, -3)$, $(-3, -3, 3)$, and $(-3, -3, -3)$.

9 $(3, -1, 7)$

11 a plane parallel to the xz plane and containing $(0, 5, 0)$

13 the yz plane

15 a plane parallel to the yz plane and containing $(-3, 0, 0)$

17 a plane parallel to the xz plane and containing $(0, -4, 0)$

19 $z = -3$ **21** $y = -2$ **23** $y = 0$, $z = 0$

25 $x = 5$, $z = -\frac{7}{2}$ **27** $y = -\frac{3}{4}$, $z = \frac{4}{7}$

29 a line parallel to the x axis containing $(0, -2, 0)$

31 a line parallel to the z axis containing $(-2, -5, 0)$

33 the point $(0, 1, 0)$ **35** the point $(1, 2, 3)$

37 $x^2 + y^2 + z^2 = 25$ **39** 3   **41** 5

Problem Set 13.2, page 770

1 $7\mathbf{i} + 2\mathbf{j} - 5\mathbf{k}$ **3** $-17\mathbf{j} + 20\mathbf{k}$ **5** $6\mathbf{i} - \mathbf{j} - 13\mathbf{k}$

7 $-\frac{5}{3}\mathbf{i} + \frac{32}{3}\mathbf{j} - \frac{29}{3}\mathbf{k}$ **9** $4\mathbf{i} - 4\mathbf{j} - 3\mathbf{k}$ **11** 17

13 2   **15** -32 **17** 48   **19** $\langle 20, -2, -38 \rangle$

21 $3\sqrt{6}$ **23** $\sqrt{41}$ **25** $\sqrt{26} + 3\sqrt{5}$

27 $\dfrac{2\sqrt{2}}{3}\mathbf{i} + \dfrac{\sqrt{2}}{6}\mathbf{j} - \dfrac{\sqrt{2}}{6}\mathbf{k}$ **29** $\dfrac{17\sqrt{2}}{6}$ **31** $-\dfrac{\sqrt{2}}{6}$

33 $\dfrac{8\sqrt{2}}{3}$ **35** $\dfrac{2\sqrt{6}}{3}$ **37** $\cos^{-1}\dfrac{17}{6\sqrt{13}}$

39 $\cos^{-1}\dfrac{4}{\sqrt{195}}$ **41** $\dfrac{3}{\sqrt{26}}, \dfrac{1}{\sqrt{26}}, -\dfrac{4}{\sqrt{26}}$

43 $\dfrac{7}{\sqrt{78}}, \dfrac{2}{\sqrt{78}}, -\dfrac{5}{\sqrt{78}}$ **45** $a = -7$

47 $\overrightarrow{QR}$ is perpendicular to $\overrightarrow{RP}$ **49** (a) 3; (b) $\sqrt{13}$; (c) $\sqrt{5}$

51 $\dfrac{2}{15}, -\dfrac{2}{3}, -\dfrac{11}{15}$; $\cos^{-1}\dfrac{2}{15} \approx 82.338°$,

$\cos^{-1}\left(-\dfrac{2}{3}\right) \approx 131.810°$, $\cos^{-1}\left(-\dfrac{11}{15}\right) \approx 137.167°$

53 $\dfrac{8}{\sqrt{101}}, \dfrac{1}{\sqrt{101}}, -\dfrac{6}{\sqrt{101}}$; $\cos^{-1}\dfrac{8}{\sqrt{101}} \approx 37.247°$,

$\cos^{-1}\dfrac{1}{\sqrt{101}} \approx 84.289°$, $\cos^{-1}-\dfrac{6}{\sqrt{101}} \approx 126.657°$

55 yes   **57** no

63 (a) $2\sqrt{10 + 3\sqrt{3}}$; (b) $2\sqrt{10 - 3\sqrt{3}}$

65 $(23, 20, 9)$ **69** no   **71** no

Problem Set 13.3, page 779

1 (a) left; (b) right; (c) right; (d) left; (e) left; (f) right

3 (a) $-\mathbf{j}$; (b) $-\mathbf{k}$; (c) $\mathbf{j}$; (d) $2\mathbf{k}$; (e) $\mathbf{k}$; (f) $\mathbf{i} - \mathbf{j}$;   **5** $2\mathbf{i} - 2\mathbf{k}$

7 $-\mathbf{i} + 3\mathbf{j} + 9\mathbf{k}$ **9** $-6\mathbf{i}$ **11** $-3\mathbf{i} + \mathbf{j} + \mathbf{k}$

13 $-6\mathbf{i} + 2\mathbf{j} + 2\mathbf{k}$ **15** $4\mathbf{i} + 6\mathbf{k}$ **17** $-12\mathbf{i} + 12\mathbf{k}$

19 -2 **21** -6 **23** 0   **25** 0   **27** $6\mathbf{i} - 8\mathbf{j} + 2\mathbf{k}$

29 $-4\mathbf{i} - 8\mathbf{j} - 4\mathbf{k}$ **31** 25   **33** $\sqrt{38}$

35 (a) $2\sqrt{2}$; (b) $\dfrac{\sqrt{341}}{4}$ **37** (a) 5; (b) 26; (c) 2;

39 (a) linearly independent, right-handed; (b) coplanar; (c) linearly independent, left-handed

41 (a) $\dfrac{\sqrt{66}}{2}$; (b) $\dfrac{\sqrt{6}}{2}$ **47** false   **51** (a) $\dfrac{33}{2}$; (b) $\dfrac{41}{2}$

Problem Set 13.4, page 787

1 $x + 2y - 3z + 7 = 0$ **3** $5x - 2y + 10z = 0$

5 $45x + 10y - 51z = 80$ **7** (a) $\frac{2}{7}\mathbf{i} + \frac{3}{7}\mathbf{j} + \frac{6}{7}\mathbf{k}$; (b) 6, 4, 2

9 (a) $\frac{12}{13}\mathbf{j} - \frac{5}{13}\mathbf{k}$; (b) none, 5, -12

11 (a) $\dfrac{1}{\sqrt{2}}\mathbf{i} - \dfrac{3}{5\sqrt{2}}\mathbf{j} - \dfrac{4}{5\sqrt{2}}\mathbf{k}$; (b) 0, 0, 0

13 parallel (in fact, identical)   **15** not parallel

17 $6x - 3y - 2z + 25 = 0$ **19** $x = -1$

21 $3x - 7y + z = 0$ **23** $2y - z = 1$

25 $x + \sqrt{2}y + z = 2 + \sqrt{2}$ **27** $(0, 0, 0)$ is on the plane

29 parallel to the xy plane   **31** parallel to the y axis

33 (a) $(\mathbf{i} + \mathbf{j} - 2\mathbf{k}) \times [(x + 1)\mathbf{i} + (y - 1)\mathbf{j} + (z - 4)\mathbf{k}] = \mathbf{0}$;

(b) $\dfrac{x + 1}{1} = \dfrac{y - 1}{1} = \dfrac{z - 4}{-2}$;

(c) $x\mathbf{i} + y\mathbf{j} + z\mathbf{k} = (-\mathbf{i} + \mathbf{j} + 4\mathbf{k}) + t(\mathbf{i} + \mathbf{j} - 2\mathbf{k})$;

(d) $x = -1 + t$, $y = 1 + t$, $z = 4 - 2t$

35 (a) $(-2\mathbf{j} + 5\mathbf{k}) \times [(x - 3)\mathbf{i} + (y - 1)\mathbf{j} + (z + 4)\mathbf{k}] = \mathbf{0}$;

(b) $x = 3$, $\dfrac{y - 1}{-2} = \dfrac{z + 4}{5}$;

(c) $x\mathbf{i} + y\mathbf{j} + z\mathbf{k} = (3\mathbf{i} + \mathbf{j} - 4\mathbf{k}) + t(-2\mathbf{j} + 5\mathbf{k})$;

(d) $x = 3$, $y = 1 - 2t$, $z = -4 + 5t$

37 (a) $(5\mathbf{i} - \mathbf{j} + 3\mathbf{k}) \times [(x - 1)\mathbf{i} + (y + 1)\mathbf{j} + (z - 2)\mathbf{k}] = \mathbf{0}$;

(b) $\dfrac{x - 1}{5} = \dfrac{y + 1}{-1} = \dfrac{z - 2}{3}$;

(c) $x\mathbf{i} + y\mathbf{j} + z\mathbf{k} = (\mathbf{i} - \mathbf{j} + 2\mathbf{k}) + t(5\mathbf{i} - \mathbf{j} + 3\mathbf{k})$;

(d) $x = 1 + 5t,\ y = -1 - t,\ z = 2 + 3t$

39 (a) $(-\mathbf{i} + \mathbf{j} - 2\mathbf{k}) \times [(x - 1)\mathbf{i} + (y - 3)\mathbf{j} + (z + 2)\mathbf{k}] = \mathbf{0}$;

(b) $\dfrac{x - 1}{-1} = \dfrac{y - 3}{1} = \dfrac{z + 2}{-2}$;

(c) $x\mathbf{i} + y\mathbf{j} + z\mathbf{k} = (\mathbf{i} + 3\mathbf{j} - 2\mathbf{k}) + t(-\mathbf{i} + \mathbf{j} - 2\mathbf{k})$;

(d) $x = 1 - t,\ y = 3 + t,\ z = -2 - 2t$

41 $\dfrac{x}{6} = \dfrac{y - 4}{2} = \dfrac{z - 5}{3}$

43 $(3\mathbf{i} - 4\mathbf{j} + 7\mathbf{k}) \times [(x - 3)\mathbf{i} + (y + 1)\mathbf{j} + (z - 2)\mathbf{k}] = \mathbf{0}$

45 $y - 1 = 0,\ z + 4 = 0$, no restriction on x

47 $\dfrac{x - 2}{4} = \dfrac{y - 3}{-17} = \dfrac{z + 1}{-23}$ **49** $\dfrac{x - 2}{2} = \dfrac{y}{11} = \dfrac{z}{5}$

51 contains $(0, 0, 0)$ **53** parallel to the z axis

55 parallel to the xz plane   **57** $(5, 3, 10)$

Problem Set 13.5, page 795

1 $\frac{5}{11}$ **3** $\sqrt{2}$ **5** $\dfrac{10\sqrt{6}}{9}$ **7** $\dfrac{10}{3}$

9 $\cos^{-1} \dfrac{3}{\sqrt{14}} \approx 36.699°$ **11** $\cos^{-1} \dfrac{22}{63} \approx 69.561°$

13 perpendicular   **15** not perpendicular   **17** 2

19 $5x + 2y - 11z + 21 = 0$ **21** $5x + 3y - 2z = 8$

23 $\dfrac{|D|}{\sqrt{a^2 + b^2 + c^2}}$ **25** $\dfrac{2\sqrt{70}}{7}$

27 $\dfrac{6\sqrt{14}}{\sqrt{19}}$ **29** $\dfrac{85}{\sqrt{377}}$

31 perpendicular   **33** parallel   **35** meet at $(11, 0, -8)$

37 $45°$ **39** (a) $\cos^{-1} \dfrac{1}{2\sqrt{21}}$; (c) $(1, -1, 2)$

41 The line is not perpendicular to the plane and does not intersect the plane.

43 line of intersection: $\dfrac{x}{-1} = \dfrac{y + 16}{-7} = \dfrac{z - 13}{5}$

45 line of intersection: $\dfrac{x + \frac{9}{2}}{12} = \dfrac{y}{8},\ z = \dfrac{7}{2}$

47 $5x - 3y + 8z + 3 = 0$ **49** $7x + y - 10z = 44$

51 (a) $\dfrac{\pi}{2} - \beta$, where β is the angle between the line and the normal to the plane; (c) $\sin^{-1} \dfrac{1}{3\sqrt{3}}$; (d) $45°$

53 $(2, 2, 2)$

Problem Set 13.6, page 805

1 (a) $10\mathbf{i} + 3\mathbf{j} - 2\mathbf{k}$; (b) $10\mathbf{i} - 2\mathbf{k}$

3 (a) $2\mathbf{i} - \mathbf{k}$; (b) $2\mathbf{i} - \mathbf{j} - \mathbf{k}$

5 (a) $-2\mathbf{i} + \dfrac{3\pi^2}{16}\mathbf{k}$; (b) $-4\mathbf{j} + \dfrac{3\pi}{2}\mathbf{k}$

7 (a) $(-4 \sin 2t)\mathbf{i} + (6 \cos 2t)\mathbf{j} + \mathbf{k}$;

(b) $\sqrt{16 \sin^2 2t + 36 \cos^2 2t + 1}$;

(c) $(-8 \cos 2t)\mathbf{i} - (12 \sin 2t)\mathbf{j}$

9 (a) $-e^{-t}\mathbf{i} + 2e^t\mathbf{j} + 3t^2\mathbf{k}$; (b) $\sqrt{e^{-2t} + 4e^{2t} + 9t^4}$;

(c) $e^{-t}\mathbf{i} + 2e^t\mathbf{j} + 6t\mathbf{k}$

11 (a) $(\sin t + t \cos t)\mathbf{i} + (\cos t - t \sin t)\mathbf{j} + 2t\mathbf{k}$;

(b) $\sqrt{5t^2 + 1}$;

(c) $(2 \cos t - t \sin t)\mathbf{i} + (-2 \sin t - t \cos t)\mathbf{j} + 2\mathbf{k}$

13 (a) $-3\mathbf{j} + 2\mathbf{k}$; (b) $\sqrt{13}$; (c) $-4\mathbf{i} + 3\mathbf{j}$

15 (a) $-4\mathbf{i} + e^{\pi/2}\mathbf{k}$; (b) $\sqrt{16 + e^\pi}$; (c) $-4\mathbf{j} + e^{\pi/2}\mathbf{k}$

17 10   **19** $\dfrac{2}{9}(82^{3/2} - 1)$ **21** $\dfrac{e^2 - 1}{e}$

23 $\sqrt{3}(1 - e^{-\pi})$ **25** $\dfrac{99\sqrt{2}}{4}$

27 (a) $(-2 \sin t)\mathbf{i} + (3 \cos t)\mathbf{j} + \mathbf{k}$; (b) $\sqrt{5(1 + \cos^2 t)}$;

(c) $(-2 \cos t)\mathbf{i} - (3 \sin t)\mathbf{j}$; (d) $\dfrac{-2 \sin t\, \mathbf{i} + 3 \cos t\, \mathbf{j} + \mathbf{k}}{\sqrt{5(1 + \cos^2 t)}}$;

(e) $(3 \sin t)\mathbf{i} - (2 \cos t)\mathbf{j} + 6\mathbf{k}$; (f) $\dfrac{\sqrt{5(8 + \sin^2 t)}}{(5 + 5 \cos^2 t)^{3/2}}$;

(g) $\dfrac{(-4 \cos t)\mathbf{i} - (3 \sin t)\mathbf{j} + (\sin t \cos t)\mathbf{k}}{\sqrt{(1 + \cos^2 t)(8 + \sin^2 t)}}$;

(h) $\dfrac{(3 \sin t)\mathbf{i} - (2 \cos t)\mathbf{j} + 6\mathbf{k}}{\sqrt{5(8 + \sin^2 t)}}$; (i) $(2 \sin t)\mathbf{i} - (3 \cos t)\mathbf{j}$;

(j) $\dfrac{6}{5(8 + \sin^2 t)}$

29 (a) $(2e^{2t} - 2e^{-2t})\mathbf{i} + (2e^{2t} + 2e^{-2t})\mathbf{j} + 5\mathbf{k}$;

(b) $\sqrt{8e^{4t} + 8e^{-4t} + 25}$;

(c) $(4e^{2t} + 4e^{-2t})\mathbf{i} + (4e^{2t} - 4e^{-2t})\mathbf{j}$;

(d) $\dfrac{(2e^{2t} - 2e^{-2t})\mathbf{i} + (2e^{2t} + 2e^{-2t})\mathbf{j} + 5\mathbf{k}}{\sqrt{8e^{4t} + 8e^{-4t} + 25}}$;

(e) $20(e^{-2t} - e^{2t})\mathbf{i} + 20(e^{2t} + e^{-2t})\mathbf{j} - 32\mathbf{k}$;

(f) $\dfrac{4\sqrt{50(e^{4t} + e^{-4t}) + 64}}{[8(e^{4t} + e^{-4t}) + 25]^{3/2}}$;

(g) $\dfrac{41(e^{2t} + e^{-2t})\mathbf{i} + 9(e^{2t} - e^{-2t})\mathbf{j} - 20(e^{4t} - e^{-4t})\mathbf{k}}{\sqrt{[50(e^{4t} + e^{-4t}) + 64](8e^{4t} + 8e^{-4t} + 25)}}$;

(h) $\dfrac{5(e^{-2t} - e^{2t})\mathbf{i} + 5(e^{2t} + e^{-2t})\mathbf{j} - 8\mathbf{k}}{\sqrt{50(e^{4t} + e^{-4t}) + 64}}$;

(i) $8(e^{2t} - e^{-2t})\mathbf{i} + 8(e^{2t} + e^{-2t})\mathbf{j}$; (j) $\dfrac{20}{25(e^{4t} + e^{-4t}) + 32}$

31 (a) $\mathbf{i} + \sqrt{2}t\mathbf{j} + t^2\mathbf{k}$; (b) $t^2 + 1$; (c) $\sqrt{2}\mathbf{j} + 2t\mathbf{k}$;

(d) $\dfrac{\mathbf{i} + \sqrt{2}t\mathbf{j} + t^2\mathbf{k}}{t^2 + 1}$; (e) $\sqrt{2}t^2\mathbf{i} - 2t\mathbf{j} + \sqrt{2}\mathbf{k}$; (f) $\dfrac{\sqrt{2}}{(t^2 + 1)^2}$;

(g) $\dfrac{-\sqrt{2}t\mathbf{i} - (t^2 - 1)\mathbf{j} + \sqrt{2}t\mathbf{k}}{t^2 + 1}$; (h) $\dfrac{t^2\mathbf{i} - \sqrt{2}t\mathbf{j} + \mathbf{k}}{t^2 + 1}$;

(i) $2\mathbf{k}$; (j) $\dfrac{\sqrt{2}}{t^2 + 1}$

33 $30t\mathbf{i} + (9t^2 - 6)\mathbf{j} + 12t^3\mathbf{k}$

35 $(4t^3 - 6t)\mathbf{i} + (4t^3 - 15t^2)\mathbf{j} + (2 - 3t^2)\mathbf{k}$

37 $\mathbf{T} = \dfrac{(-a \sin t)\mathbf{i} + (a \cos t)\mathbf{j} + b\mathbf{k}}{\sqrt{a^2 + b^2}};$

$\mathbf{N} = (-\cos t)\mathbf{i} - (\sin t)\mathbf{j};\;\; \mathbf{B} = \dfrac{(b \sin t)\mathbf{i} - (b \cos t)\mathbf{j} + a\mathbf{k}}{\sqrt{a^2 + b^2}};$

$\kappa = \dfrac{a}{a^2 + b^2};\; \tau = \dfrac{b}{a^2 + b^2}$

41 $\dfrac{d^2\mathbf{F}}{dt^2} \times \mathbf{G} + 2\dfrac{d\mathbf{F}}{dt} \times \dfrac{d\mathbf{G}}{dt} + \mathbf{F} \times \dfrac{d^2\mathbf{G}}{dt^2}$

Problem Set 13.7, page 811

1 $(x - 2)^2 + (y + 1)^2 + (z - 3)^2 = 16$
3 $(x - 4)^2 + (y - 1)^2 + (z - 3)^2 = 9$
5 $(1, 0, -2), r = 3$ **7** $(-1, 1, 0), r = 3$
9 $(3, -1, 2), r = \sqrt{33}$
11 (a) x axis; (b) yz plane; (c) parabola

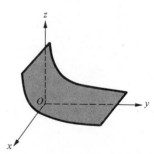

13 (a) x axis; (b) yz plane; (c) hyperbola

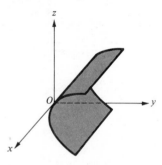

shows only one sheet of the graph

15 (a) z axis; (b) xy plane; (c) circle

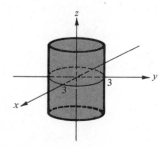

17 (a) z axis; (b) xy plane; (c) ellipse

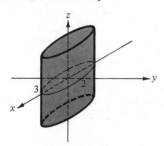

19 (a) y axis; (b) xz plane; (c) parabola

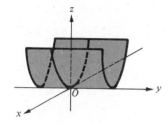

21 (a) x axis: (b) yz plane; (c) line

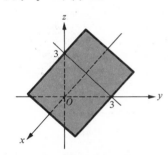

23 (a) x axis; (b) yz plane; (c) ellipse
25 (a) y axis; (b) xz plane; (c) sine curve

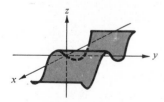

27 (a) z axis; (b) xy plane; (c) parabola

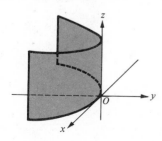

29 (a) y axis; (b) xz plane; (c) two intersecting lines

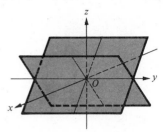

31 $y^2 + z^2 = x^2$ **33** $y^2 + z^2 = 8x$
35 $z = x^2 + y^2$ **37** $6x^2 + 6y^2 + 6z^2 = 7$
39 $z = 1 - y^2 - x^2$
41 z axis, $x^2 + z = 3$

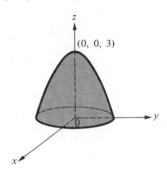

43 x axis, $x^2 - 9y^2 = 18$

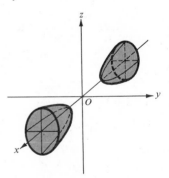

45 z axis, $x^2 = \sin^2 z$

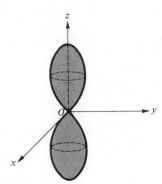

Problem Set 13.8, page 819

1 $3y^2 + z^2 = 4$, ellipse   **3** $z^2 - \dfrac{x^2}{16} = \dfrac{13}{9}$, hyperbola

5 $25x^2 + 4y^2 = 100$, ellipse   **7** $9 - 16y^2 = z$, parabola

9 (a) $(\pm\sqrt{6}, 0, 0)$, $(0, \pm\sqrt{2}, 0)$, $(0, 0, \pm\sqrt{3})$; (b) all symmetries; (c) the ellipses $3y^2 + 2z^2 = 6 - k^2$, $x^2 + 2z^2 = 6 - 3k^2$, $x^2 + 3y^2 = 6 - 2k^2$; (d) the ellipses $3y^2 + 2z^2 = 6$, $x^2 + 2z^2 = 6$, $x^2 + 3y^2 = 6$; (e) ellipsoid

11 (a) $(\pm3, 0, 0)$, no y intercept, $(0, 0, \pm2)$; (b) all symmetries; (c) the hyperbola $9y^2 - 9z^2 = 4k^2 - 36$ (pair of straight lines for $k = \pm3$), the ellipse $4x^2 + 9z^2 = 36 + 9k^2$, the hyperbola $4x^2 - 9y^2 = 36 - 9k^2$ (pair of straight lines for $k = \pm2$); (d) hyperbola $9z^2 - 9y^2 = 36$, ellipse $4x^2 + 9z^2 = 36$, hyperbola $4x^2 - 9y^2 = 36$; (e) hyperboloid of one sheet

13 (a) $(\pm2, 0, 0)$, no y or z intercepts; (b) all symmetries; (c) circle $4y^2 + 4z^2 = k^2 - 4$, hyperbola $x^2 - 4z^2 = 4 + 4k^2$, hyperbola $x^2 - 4y^2 = 4 + 4k^2$; (d) hyperbola $x^2 - 4z^2 = 4$, hyperbola $x^2 - 4y^2 = 4$, no yz trace; (e) hyperboloid of two sheets

15 (a) no x intercept, $(0, \pm3, 0)$, no z intercept; (b) all symmetries; (c) hyperbola $y^2 - 9z^2 = 9 + 9k^2$, circle $9x^2 + 9z^2 = k^2 - 9$, hyperbola $y^2 - 9x^2 = 9 + 9k^2$; (d) hyperbola $y^2 - 9z^2 = 9$, hyperbola $y^2 - 9x^2 = 9$, no xz trace; (e) hyperboloid of two sheets

17 (a) $(0, 0, 0)$; (b) all symmetries; (c) hyperbola $8z^2 - 5y^2 = k^2$, hyperbola $8z^2 - x^2 = 5k^2$, ellipse $x^2 + 5y^2 = 8k^2$; (d) two lines $z = \pm\sqrt{\tfrac{5}{8}}y$, two lines $z = \pm\sqrt{\tfrac{1}{8}}x$, point $(0, 0)$; (e) elliptic cone

19 (a) $(0, 0, 0)$; (b) symmetry with respect to xz plane, yz plane, and z axis; (c) parabola $3z - k^2 = y^2$, parabola $3z - k^2 = x^2$, circle $3k = x^2 + y^2$; (d) parabola $3z = y^2$, parabola $3z = x^2$, point $(0, 0)$; (e) elliptic paraboloid (paraboloid of revolution)

21 (a) $(0, 0, 0)$; (b) symmetry with respect to xz plane, yz plane, and z axis; (c) parabola $\dfrac{k^2}{16} - 3z = \dfrac{y^2}{9}$, parabola $\dfrac{x^2}{16} = 3z + \dfrac{k^2}{9}$, hyperbola $\dfrac{x^2}{16} - \dfrac{y^2}{9} = 3k$; (d) parabola $-3z = \dfrac{y^2}{9}$, parabola $\dfrac{x^2}{16} = 3z$, two lines $y = \pm\dfrac{3}{4}x$; (e) hyperbolic paraboloid

23 (a) $(0, 0, 0)$; (b) symmetry with respect to xy plane, yz plane, and y axis; (c) parabola $k^2 = y + z^2$, hyperbola $x^2 - z^2 = k$, parabola $x^2 = y + k^2$; (d) parabola $y = -z^2$, two lines $x = \pm z$, parabola $x^2 = y$; (e) hyperbolic paraboloid

27 An equivalent equation is obtained when x, y, and z are replaced by $-x$, $-y$, and $-z$, respectively.

29 (a) ellipse or circle $\dfrac{y^2}{b^2} + \dfrac{z^2}{c^2} = 1 - \dfrac{k^2}{a^2}$ for $|k| < a$, points $(\pm a, 0, 0)$ for $k = \pm a$; (b) ellipse or circle $\dfrac{x^2}{a^2} + \dfrac{z^2}{c^2} = 1 - \dfrac{k^2}{b^2}$ for $|k| < b$, points $(0, \pm b, 0)$ for

$k = \pm b$; (c) ellipse or circle $\dfrac{x^2}{a^2} + \dfrac{y^2}{b^2} = 1 - \dfrac{k^2}{c^2}$ for

$|k| < c$, points $(0, 0, \pm c)$ for $k = \pm c$

31 with $x = k \neq 0$, the hyperbola $\dfrac{z^2}{c^2} - \dfrac{y^2}{b^2} = \dfrac{k^2}{a^2}$; with $x = 0$,

the pair of intersecting lines $z = \pm \dfrac{c}{b} y$; with $y = k \neq 0$,

the hyperbola $\dfrac{z^2}{c^2} - \dfrac{x^2}{a^2} = \dfrac{k^2}{b^2}$; with $y = 0$, the pair of

intersecting lines $z = \pm \dfrac{c}{a} x$; with $z = k \neq 0$, the ellipse

$\dfrac{x^2}{a^2} + \dfrac{y^2}{b^2} = \dfrac{k^2}{c^2}$; with $z = 0$, the point $(0, 0, 0)$

33 with $z = k$, $k < 0$, the hyperbola $\dfrac{x^2}{a^2} - \dfrac{y^2}{b^2} = -k$; with

$z = 0$, the pair of intersecting lines $y = \pm \dfrac{b}{a} x$; with $z = k$,

$k > 0$, the hyperbola $\dfrac{y^2}{b^2} - \dfrac{x^2}{a^2} = k$

35 $x^2 + y^2 = -4z$, a paraboloid of revolution about the negative z axis

37 $9x^2 + 9z^2 = 4y^2$; a circular cone

39 $23x^2 + 27y^2 + 36z^2 = 351$, an ellipsoid

Problem Set 13.9, page 826

1 $(2, 2\sqrt{3}, 1)$ **3** $\left(\dfrac{5\sqrt{3}}{2}, \dfrac{5}{2}, -2\right)$ **5** $(4, 0, 1)$

7 $\left(6, \dfrac{5\pi}{6}, 6\right)$ **9** $\left(\dfrac{3}{2}, \dfrac{\sqrt{3}}{2}, 1\right)$ **11** $(-9, 3\sqrt{3}, -6)$

13 $\left(1, \dfrac{3\pi}{2}, \dfrac{\pi}{2}\right)$ **15** $\left(\sqrt{14}, \tan^{-1} 2, \cos^{-1} \dfrac{-3}{\sqrt{14}}\right)$

17 (a) $z = 2r^2$; (b) $2\rho \sin^2 \phi = \cos \phi$

19 (a) $r \cos \theta = 2$; (b) $\rho \sin \phi \cos \theta = 2$

21 (a) $r^2 = 5z^2$; (b) $\tan \phi = \sqrt{5}$

23 (a) $r = 5$; (b) $\rho \sin \phi = 5$

25 (a) $r^2 - z^2 = 1$; (b) $\rho^2 \cos 2\phi = -1$

27 (a) $z = x^2 + y^2$; (b) $\cos \phi = \rho \sin^2 \phi$

29 (a) $\dfrac{x^2}{9} + \dfrac{y^2}{9} + \dfrac{z^2}{4} = 1$; (b) $\rho^2(4 \sin^2 \phi + 9 \cos^2 \phi) = 36$

31 (a) $x^2 + y^2 = 4x$; (b) $\rho \sin \phi = 4 \cos \theta$

33 (a) $x^2 + y^2 + z^2 = 4$; (b) $r^2 + z^2 = 4$

35 (a) $x^2 + y^2 = 9$; (b) $r = 3$

37 (a) $3z^2 = x^2 + y^2$; (b) $3z^2 = r^2$

41 $\frac{1}{2}\sqrt{2}\pi\sqrt{1 + 2\pi^2} + \frac{1}{2} \ln (\sqrt{1 + 2\pi^2} + \sqrt{2}\pi)$

Review Problem Set, Chapter 13, page 827

1 $(-2, -1, -3)$ **3** $(-3, 2, 1)$ **5** $(-3, -1, 5)$

7 (a) $\sqrt{46}$; (b) $\sqrt{37}$; (c) 6; (d) $\sqrt{6}$; (e) $2\sqrt{46}$

9 not colinear   **11** $\dfrac{\pi}{3}$ **13** $-6\mathbf{i} - 5\mathbf{j}$ **15** $\sqrt{61}$

17 0   **19** 0   **21** $-8\mathbf{i} + 8\mathbf{j} + 8\mathbf{k}$ **23** $16\mathbf{i} - 16\mathbf{j} + 16\mathbf{k}$

25 8   **27** $\cos^{-1} \dfrac{-21}{\sqrt{910}}$ **29** 8   **31** yes

33 $x + 2y + 3z = 14$ **35** $4x + 2y + 3z = 5$

37 $x + 3z = 7$ **39** $3x + 8y - 7z + 23 = 0$

41 (a) $x = 5 + 3t$, $y = 6 + 7t$, $z = -4 - 5t$;

(b) $\dfrac{x - 5}{3} = \dfrac{y - 6}{7} = \dfrac{z + 4}{-5}$

43 (a) $x = 2 + 3t$, $y = -3 - 2t$, $z = -2 + 7t$;

(b) $\dfrac{x - 2}{3} = \dfrac{y + 3}{-2} = \dfrac{z + 2}{7}$

45 (a) $x = 3 + 2t$, $y = -4 - 7t$, $z = 1 - 3t$;

(b) $\dfrac{x - 3}{2} = \dfrac{y + 4}{-7} = \dfrac{z - 1}{-3}$

47 (a) $x = 2$, $y = 1$, $z = 4 + t$; (b) $x = 2$, $y = 1$, z arbitrary

49 3   **51** $3x + y - 4z = 2$ **53** $2\sqrt{3}$

59 $4t(e^{2t^2}\mathbf{i} + e^{-2t^2}\mathbf{j}) + \mathbf{k}$, $4e^{2t^2}(1 + 4t^2)\mathbf{i} + 4e^{-2t^2}(1 - 4t^2)\mathbf{j}$,

$\dfrac{4te^{4t^2} - 4te^{-4t^2} + t}{\sqrt{e^{4t^2} + e^{-4t^2} + t^2}}$

61 (a) $(-6\pi \sin 2\pi t)\mathbf{i} + (6\pi \cos 2\pi t)\mathbf{j} + 2\mathbf{k}$;

(b) $(-12\pi^2 \cos 2\pi t)\mathbf{i} - (12\pi^2 \sin 2\pi t)\mathbf{j}$; (c) $2\sqrt{9\pi^2 + 1}$;

(d) $\mathbf{T} = \dfrac{(-3\pi \sin 2\pi t)\mathbf{i} + (3\pi \cos 2\pi t)\mathbf{j} + \mathbf{k}}{\sqrt{9\pi^2 + 1}}$;

(e) $(-\cos 2\pi t)\mathbf{i} - (\sin 2\pi t)\mathbf{j}$;

(f) $\dfrac{(\sin 2\pi t)\mathbf{i} - (\cos 2\pi t)\mathbf{j} + 3\pi\mathbf{k}}{\sqrt{9\pi^2 + 1}}$; (g) $2\sqrt{9\pi^2 + 1}$

63 (a) $\mathbf{i} + 2t\mathbf{j} + 3t^2\mathbf{k}$; (b) $\sqrt{1 + 4t^2 + 9t^4}$; (c) $2\mathbf{j} + 6t\mathbf{k}$;

(d) $\dfrac{\mathbf{i} + 2t\mathbf{j} + 3t^2\mathbf{k}}{\sqrt{1 + 4t^2 + 9t^4}}$; (e) $6t^2\mathbf{i} - 6t\mathbf{j} + 2\mathbf{k}$;

(f) $\dfrac{2\sqrt{9t^4 + 9t^2 + 1}}{(1 + 4t^2 + 9t^4)^{3/2}}$;

(g) $\dfrac{(-9t^3 - 2t)\mathbf{i} + (1 - 9t^4)\mathbf{j} + (6t^3 + 3t)\mathbf{k}}{\sqrt{1 + 4t^2 + 9t^4}\sqrt{9t^4 + 9t^2 + 1}}$;

(h) $\dfrac{3t^2\mathbf{i} - 3t\mathbf{j} + \mathbf{k}}{\sqrt{9t^4 + 9t^2 + 1}}$; (i) $6\mathbf{k}$; (j) $\dfrac{3}{9t^4 + 9t^2 + 1}$

65 (a) $(\cos^2 t - \sin^2 t)\mathbf{i} + (2 \sin t \cos t)\mathbf{j} - (\sin t)\mathbf{k}$;

(b) $\sqrt{1 + \sin^2 t}$;

(c) $(-4 \sin t \cos t)\mathbf{i} + 2(\cos^2 t - \sin^2 t)\mathbf{j} - (\cos t)\mathbf{k}$;

(d) $\dfrac{(\cos^2 t - \sin^2 t)\mathbf{i} + (2 \sin t \cos t)\mathbf{j} - (\sin t)\mathbf{k}}{\sqrt{1 + \sin^2 t}}$;

(e) $(-2 \sin^3 t)\mathbf{i} + (\cos^3 t + 3 \sin^2 t \cos t)\mathbf{j} + 2\mathbf{k}$;

(f) $\dfrac{\sqrt{5 + 3 \sin^2 t}}{(1 + \sin^2 t)^{3/2}}$;

(g) $\dfrac{-\sin t \cos t(5 + 2 \sin^2 t)\mathbf{i} + [4 - 2(1 + \sin^2 t)^2]\mathbf{j} - (\cos t)\mathbf{k}}{\sqrt{1 + \sin^2 t}\sqrt{5 + 3 \sin^2 t}}$;

(h) $\dfrac{(-2 \sin^3 t)\mathbf{i} + (\cos^3 t + 3 \sin^2 t \cos t)\mathbf{j} + 2\mathbf{k}}{\sqrt{5 + 3 \sin^2 t}}$;

(i) $4(\sin^2 t - \cos^2 t)\mathbf{i} - (8 \sin t \cos t)\mathbf{j} + (\sin t)\mathbf{k}$;

(j) $\dfrac{-6 \sin t}{5 + 3 \sin^2 t}$

67 (a) $\mathbf{R} \cdot \mathbf{A} + v^2$; (b) $\mathbf{R} \times \mathbf{A}$; (c) $|\mathbf{A}|^2 + \mathbf{V} \cdot \dfrac{d\mathbf{A}}{dt}$; (d) $\mathbf{V} \times \dfrac{d\mathbf{A}}{dt}$;
(e) $-\kappa\tau$; (f) $\kappa^2\tau$

69 $z^2 + y^2 = 4(x - 3)^4$ **71** x axis, $z^2 = e^{-2x}$

73 ellipsoid of revolution   **75** paraboloid of revolution

77 hyperboloid of one sheet   **79** hyperbolic paraboloid

81 (a) right circular cylinder of radius 2; (b) plane making an angle of $\pi/6$ with the xz plane; (c) right circular cylinder of radius $\frac{1}{2}$ whose central axis is parallel to the z axis and contains $(0, \frac{1}{2}, 0)$

83 $z = f(\pm r)$ **85** $z = \dfrac{\pm 2}{\sqrt{x^2 + y^2}}$

Chapter 14

Problem Set 14.1, page 836

1 -39 **3** $|k|$ **5** $-\frac{7}{4}$ **7** $5a + 7\sqrt{ab}$

9 $\sin 2t$ **11** $5x^2 + 7xy + \sqrt{xy}$ **13** $\dfrac{2xy}{x^2 + y^2}$

15 all points on or outside of the circle $x^2 + y^2 = 9$, *except* points on the x axis

17 all points outside the circle $x^2 + y^2 = 16$

19 points below the parabola $y = x^2$

21 points on or above the line $y = 4 - x$

23 points in the first quadrant *above* the hyperbola $xy = 2$ or in the third quadrant below $xy = 2$

25 all points except those on the line $y = 2x$

27 domain: entire xy plane

29 all points on or inside of the circle $x^2 + y^2 = 4$

31 all points on or inside of the circle $x^2 + y^2 = 9$

33 entire xy plane   **35** entire xy plane

37 entire xy plane   **39** entire xy plane

41 $\dfrac{n(n + 1)}{2}$ **43** contour curves are lines

45 contour curves are hyperbolas, except when $z = 0$: $x = \pm y$

47 contour curves are parabolas

49 $z = 0, \frac{1}{2}$: ellipses; $z = 1$: single point $(0, 0)$

51 equipotential curves are circles centered at the origin; radius decreases as V decreases; no curve for $|v| < \frac{4}{3}$

53 temperature changes rapidly near P

55 2.5 degrees/h (dropping)

Problem Set 14.2, page 844

1 -1 **3** $e + 1$ **5** $\frac{1}{2}$ **7** e^{-11} **9** 0

11 3   **13** (a) (i) 0 (ii) 0 (iii) 0 (iv) 0; (b) 0

15 (a) (i) 0 (ii) 0 (iii) $\dfrac{1}{2}$ (iv) $\dfrac{m}{1 + m^2}$; (b) does not exist

17 (a) (i) 0 (ii) 0 (iii) 0 (iv) 0; (b) 0

19 $\delta \le \dfrac{\epsilon}{8}$ will work   **21** continuous

23 discontinuous; e.g., $\lim g(x, y)$ does not exist as (x, y) approaches $(1, 2)$ along the line $x = 1$.

25 continuous   **27** continuous

29 (a) all points in first and third quadrants, including axes; (b) all points in first and third quadrants, excluding axes; (c) none

31 (a) all points except those on the line $y = 2x$; (b) all points in the domain are interior points; (c) none

33 (a) all points (x, y) such that $y \ne \pm 1$; (b) all points in the domain are interior points; (c) none

35 (a) all points inside the circle $x^2 + y^2 = 9$ (but not on it); (b) all points in the domain are interior points; (c) none

37 (a) all points except those for which $x \ne 0$ and $x = -y$; (b) all points of the domain except $(0, 0)$; (c) none

39 (a) entire plane; (b) all points are interior points; (c) all points on the circle $x^2 + y^2 = 4$

Problem Set 14.3, page 850

1 8   **3** 46   **5** 15   **7** $14x + 10xy$ **9** $\cos x \cos 7y$

11 $\dfrac{4xy^2}{(y^2 - x^2)^2}$ **13** $2r \cos 7\theta$ **15** -135 **17** $6xy + 7$

19 $y^2 + 2xy$ **21** $2e^{-4} + 6e^2$ **23** $\dfrac{3x}{\sqrt{3x^2 + y^2}}$

25 $\dfrac{14x}{7x^2 + 4y^3}$ **27** $\dfrac{y}{1 + x^2y^2}$ **29** $\dfrac{x}{2\sqrt{xy}\sqrt{1 - xy}}$

31 1   **33** $xze^{xy}(2\cos yz - yz \sin yz)$

35 $-3x(x^2 + y^2 + z^2)^{-5/2}$ **37** $\dfrac{-2x \sin[\ln(x^2 + y^2)]}{x^2 + y^2}$

39 $-12x^2y^3 \sin(6x^2y^4)$ **41** $-4^{\sqrt{2}/2}\sqrt{2} \ln 2$

43 $(3u - 2v)^{\sin v}\left[\ln(3u - 2v) \cos v - \dfrac{2 \sin v}{3u - 2v}\right]$

53 3   **55** -6 **57** $\dfrac{3}{e}$ **59** $\dfrac{56}{625}$

61 (a) $\dfrac{21\pi}{\sqrt{58}} \approx 8.663$ cm^2/cm;

(b) $\pi\sqrt{58} + \dfrac{9\pi}{\sqrt{58}} \approx 27.638$ cm^2/cm

63 (a) $\dfrac{-1}{3,000,000}$; (b) $\dfrac{2.5028\pi}{90} \sin \dfrac{4\pi}{9} \approx .086$ dynes/degree

65 $-\dfrac{22}{45}, \dfrac{1}{15}$

67 $\dfrac{\partial Q}{\partial M}\Big|_{\substack{M=12 \\ C=10}} = 10$: at $C = 10$, IQ increases by 10 for each year of increase in M (regardless of the value of M)

$\dfrac{\partial Q}{\partial C}\Big|_{\substack{M=12 \\ C=10}} = -12$: with M fixed, IQ decreases as C increases

69 $\dfrac{\partial P}{\partial L} = A\alpha L^{\alpha-1}K^\beta$; $\dfrac{\partial P}{\partial K} = A\beta L^\alpha K^{\beta-1}$ **71** (a) $\dfrac{1}{5}$

Problem Set 14.4, page 861

1 -17.21 **3** 0.01 **5** $\dfrac{12\pi - 7}{120} \approx 0.2558$

7 differentiable   **9** differentiable   **11** differentiable

13 $(15x^2 + 8xy)dx + (4x^2 - 6y^2)dy$ **15** $ye^{xy}\,dx + xe^{xy}\,dy$

17 $(\sin y^2 + 2xy \cos x^2)\,dx + (2xy \cos y^2 + \sin x^2)\,dy$

19 $\dfrac{-3}{1 + (2y - 3x)^2}\,dx + \dfrac{2}{1 + (2y - 3x)^2}\,dy$

21 $-2ye^{3-2xy}\,dx - 2xe^{3-2xy}\,dy$ **23** $\dfrac{P}{R}\,dV + \dfrac{V}{R}\,dP$

25 $(y^2 - 4xz + 3yz^2)\,dx + (2xy + 3xz^2)\,dy + (-2x^2 + 6xyz)\,dz$

27 $(e^{yz} + yze^{xz})\,dx + (xze^{yz} + e^{xz})\,dy + (xye^{yz} + xye^{xz})\,dz$

29 $\left(e^x \ln yz + \dfrac{e^y}{x}\right)dx + \left(\dfrac{e^x}{y} + e^y \ln xz\right)dy + \left(\dfrac{e^x}{z} + \dfrac{e^y}{z}\right)dz$

31 40 watts increase   **33** $\dfrac{1.29}{4\pi^2} \approx .03268$; about 4%

35 1.18 in^3 **37** $\dfrac{3}{640} \approx 0.0047$ ohms increase

39 $-\dfrac{0.43}{180} \approx -0.0024$

43 (a) $f_1(x, y) = \dfrac{2xy^4}{(x^2 + y^2)^2}$, $f_2(x, y) = \dfrac{2x^4y}{(x^2 + y^2)^2}$;
(b) $f_1(0, 0) = f_2(0, 0) = 0$

Problem Set 14.5, page 871

1 $2(3x^2y^2 - 3y) + 12t(2x^3y - 3x + 2y)$

3 $2x[\sin v - \sin (u - v)] + 3x^2[u \cos v + \sin (u - v)]$

5 $6e^t(t + 1)u + 2e^{-2t}(1 - 2t)v$ **7** $\dfrac{4r^3}{x} + \dfrac{1}{3y(r + 2)^{2/3}}$

9 $\dfrac{21}{x} + \dfrac{2 \sec t \tan t}{y} + \dfrac{5 \csc^2 t}{z}$

11 $e^u(2xy + z^2 + yz) - e^{-u}(x^2 - z^2 + xz) + \sinh u\,(2xz - 2yz + xy)$

13 $3[\cos (x + y) + \cos (x - y)] + 3t^2[\cos (x + y) - \cos (x - y)]$

15 $\dfrac{1}{2v}[e^{u/v} + e^{-u/v}] \cosh t - \dfrac{u}{2v^2}[e^{u/v} + e^{-u/v}]$

17 $6xv + 8y \sin u;\ 6xu - 8y \cos v$

19 $(12x^2 - 6xy^2) \cos v - 6x^2yv \cos u;$
$-(12x^2 - 6xy^2)u \sin v - 6x^2y \sin u$

21 $\dfrac{4x(u + 2v) + 6vy}{u^2 + v^2};\ \dfrac{4uy + 6vx}{u^2 + v^2}$

23 $6re^{-s} \sinh (3x + 7y) + 7e^{3s} \sinh (3x + 7y);$
$-3r^2e^{-s} \sinh (3x + 7y) + 21re^{3s} \sinh (3x + 7y)$

25 $2uvy^2e^{xy^2} + 2v^2xye^{xy^2};\ u^2y^2e^{xy^2} + 4uvxye^{xy^2}$

27 $4x \cos v + 6y \sin v + 2zv;\ -4xu \sin v + 6yu \cos v + 2zu$

29 $2x \sin \phi \cos \theta + 2y \sin \phi \sin \theta - 2z \cos \phi;$
$-2x\rho \sin \phi \sin \theta + 2y\rho \sin \phi \cos \theta = 0;$
$2x\rho \cos \phi \cos \theta + 2y\rho \cos \phi \sin \theta + 2z\rho \sin \phi$

31 $10; 11$ **33** (a) $\dfrac{3x - 3y}{3x - 2y}$; (b) 0

35 (a) $\dfrac{\cos (x - y) - \sin (x + y)}{\cos (x - y) + \sin (x + y)}$; (b) 0 **37** $14; 2$

47 increasing at the rate of $12\pi \approx 37.699$ cm^3/min

49 (a) 4 ft^2/min (b) $3 + \sqrt{5}$ ft/min

51 117 degrees/unit of time   **53** -22.5 degrees/min

Problem Set 14.6, page 884

1 (a) $8xy^3\mathbf{i} + 12x^2y^2\mathbf{j}$; (b) $-16\mathbf{i} + 48\mathbf{j}$; (c) $24\sqrt{3} - 8$

3 (a) $(e^y + ye^x)\mathbf{i} + (xe^y + e^x)\mathbf{j}$; (b) $\mathbf{i} + \mathbf{j}$; (c) $\frac{1}{2}(\sqrt{3} - 1)$

5 (a) $\ln y\,\mathbf{i} + (x/y)\mathbf{j}$; (b) $5\mathbf{j}$; (c) $-\frac{5}{2}$

7 (a) $14x\mathbf{i} + (-3)\mathbf{j}$; (b) $14\mathbf{i} - 3\mathbf{j}$; (c) $7\sqrt{3} - \frac{3}{2}$

9 (a) $4x\mathbf{i} + 6y\mathbf{j}$; (b) $4\mathbf{i} + 12\mathbf{j}$; (c) $2 - 6\sqrt{3}$

11 (a) $(2xy + 2y^2)\mathbf{i} + (x^2 + 4xy)\mathbf{i}$; (b) $-2\mathbf{i} - 4\mathbf{j}$; (c) $\sqrt{3} + 2$

13 (a) $\dfrac{-y}{x^2 + y^2}\mathbf{i} + \dfrac{x}{x^2 + y^2}\mathbf{j}$; (b) $\dfrac{2}{5}\mathbf{i} + \dfrac{1}{5}\mathbf{j}$; (c) $\dfrac{1}{5}$

15 (a) $\dfrac{y}{x}\mathbf{i} + \left(\ln \dfrac{x}{y} - 1\right)\mathbf{j}$; (b) $\dfrac{1}{2}\mathbf{i} + (\ln 2 - 1)\mathbf{j}$; (c) $-\dfrac{1 + 3 \ln 2}{\sqrt{73}}$

17 (a) $2xy\mathbf{i} + (x^2 + 3z^2)\mathbf{j} + 6yz\mathbf{k}$; (b) $-2\mathbf{i} + 4\mathbf{j} - 6\mathbf{k}$; (c) $-\dfrac{22}{3}$

19 (a) $e^{yz}\mathbf{i} + xze^{yz}\mathbf{j} + xye^{yz}\mathbf{k}$; (b) $\mathbf{i} + \mathbf{j}$; (c) $\frac{1}{3}$

21 (a) $-e^x \sin y\,\mathbf{i} - e^x \cos y\,\mathbf{j} + \mathbf{k}$; (b) $3\mathbf{i} + \mathbf{k}$; (c) $\frac{12}{7}$

23 (a) $3\sqrt{34}$; (b) $\dfrac{3\sqrt{34}}{34}\mathbf{i} - \dfrac{5\sqrt{34}}{34}\mathbf{j}$

25 (a) $\sqrt{4 + \pi^2}$; (b) $\dfrac{2\sqrt{4 + \pi^2}}{4 + \pi^2}\mathbf{i} - \dfrac{\pi\sqrt{4 + \pi^2}}{4 + \pi^2}\mathbf{j}$

27 (a) $\dfrac{\sqrt{14}}{98}$; (b) $-\dfrac{\sqrt{14}}{14}\mathbf{i} - \dfrac{\sqrt{14}}{7}\mathbf{j} + \dfrac{3\sqrt{14}}{14}\mathbf{k}$

29 (a) e; (b) $\mathbf{i}$ **31** $-\dfrac{3\sqrt{13}}{13}\mathbf{i} + \dfrac{2\sqrt{13}}{13}\mathbf{j}$

33 $\mathbf{u} = \dfrac{-3}{\sqrt{73}}\mathbf{i} + \dfrac{8}{\sqrt{73}}\mathbf{j}$

35 (a) $-\dfrac{\sqrt{2}}{2}\mathbf{i} - \dfrac{\sqrt{2}}{2}\mathbf{j}$; (b) $50\sqrt{2}$ °C/cm

37 $\dfrac{5\sqrt{2}}{2}$ **39** (a) 0; (b) $\mathbf{u} = \frac{2}{3}\mathbf{i} + \frac{1}{3}\mathbf{j} + \frac{2}{3}\mathbf{k}$ (c) 6 °C/cm

41 (a) $x + 2y + 3z = 6$; (b) $\dfrac{x - 1}{1} = \dfrac{y - 1}{2} = \dfrac{z - 1}{3}$

43 (a) $6x + 3y + 2z = 18$; (b) $\dfrac{x - 1}{6} = \dfrac{y - 2}{3} = \dfrac{z - 3}{2}$

45 (a) $z = 8$; (b) $x = 2,\ y = 2,\ z = t$

47 (a) $8x - 6y - z = 5$; (b) $\dfrac{x - 2}{8} = \dfrac{y - 1}{-6} = \dfrac{z - 5}{-5}$

49 (a) $x + 4y - 5z = 4$; (b) $\dfrac{x - 1}{1} = \dfrac{y - 2}{4} = \dfrac{z - 1}{-5}$

51 (a) $\pi x + 18y - \pi z = 6\pi$; (b) $\dfrac{x - 1}{\pi} = \dfrac{y - (\pi/6)}{18} = \dfrac{z + 2}{-\pi}$

53 (a) $2x + 3y + 6z = 36$; (b) $\dfrac{x - 9}{2} = \dfrac{y - 4}{3} = \dfrac{z - 1}{6}$

55 (a) $14x - 8y - z = 16$; (b) $\dfrac{x - 2}{14} = \dfrac{y - 1}{-8} = \dfrac{z - 4}{-1}$

57 (a) $x - 2y - 2z = -9$; (b) $\dfrac{x+1}{1} = \dfrac{y-2}{-2} = \dfrac{z-2}{-2}$

59 (a) $x - z = 0$; (b) $x - 1 = 1 - z$, $y = 1$

61 (a) $y + z = -1$; (b) $y + 1 = z$, $x = 0$

69 $-\cos \sqrt{3} + \sqrt{3} \sin \sqrt{3}$ **71** $100,000\sqrt{2}e^{-6}$ (m/m²)/s

75 $-396\mathbf{i} + 642\mathbf{j} + 143\mathbf{k}$ **77** $-3\mathbf{i} - 2\mathbf{j} + 3\mathbf{k}$

Problem Set 14.7, page 892

1 (a) 12; (b) 10; (c) 7; (d) 7

3 (a) 0; (b) $-x \cos y - 2$; (c) $-\sin y$ (d) $-\sin y$

5 (a) $\dfrac{6x^2 + 3y^2}{\sqrt{x^2 + y^2}}$; (b) $\dfrac{3x^2 + 6y^2}{\sqrt{x^2 + y^2}}$; (c) and (d) $\dfrac{3xy}{\sqrt{x^2 + y^2}}$

7 (a) $\dfrac{12y^2}{x^5}$; (b) $\dfrac{2}{x^3}$; (c) and (d) $\dfrac{-6y}{x^4}$

9 (a) $-y \cos x$; (b) $-4xe^{2y}$; (c) and (d) $-\sin x - 2e^{2y}$

11 (a) $-\sin (x + 2y)$; (b) $-4 \sin (x + 2y)$;
    (c) and (d) $-2 \sin (x + 2y)$

13 (a) 0; (b) $20x \cosh 2y$; (c) and (d) $10 \sinh 2y$

15 (a) $(6y^5 + 12x^2y^{10})e^{x^2y^5}$; (b) $(60x^2y^3 + 75x^4y^8)e^{x^2y^5}$;
    (c) and (d) $(30xy^4 + 30x^3y^9)e^{x^2y^5}$

17 (a) $y^2e^{xy} \sin y$; (b) $e^{xy}[(x^2 - 1) \sin y + 2x \cos y]$;
    (c) and (d) $e^{xy}[(xy + 1) \sin y + y \cos y]$

19 (a) $2yz + 6xe^y$; (b) $2xy$ **21** (a) 0; (b) 0

23 (a) $420e^6$; (b) $420e^6$

39 $A = -C$; B, D, E, and F arbitrary

41 $3\dfrac{\partial^2 w}{\partial u^2} + 4\dfrac{\partial^2 w}{\partial u \partial v} + 6\dfrac{\partial^2 w}{\partial v^2}$

49 $f''(u)\left[\left(\dfrac{\partial u}{\partial x}\right)^2 + \left(\dfrac{\partial u}{\partial y}\right)^2\right] + f'(u)\left[\dfrac{\partial^2 u}{\partial x^2} + \dfrac{\partial^2 u}{\partial y^2}\right]$

Problem Set 14.8, page 903

1 relative min at $(0, 1)$

3 relative min at all points (x, y) on the line $y = x + 1$

5 relative min at $(1, 1)$ and $(-1, -1)$

7 saddle point at $(1, 1)$

9 relative min at $(-1, -1)$ **11** saddle point at $(-1, -1)$

13 critical points $(\pm 1, 0)$, but second-derivative test inconclusive

15 relative min at $(3, 4)$ **17** relative min at $(-2, 1)$

19 max: $f(0, 0) = 1$; min: $f(x, y) = 0$ for all (x, y) on $x^2 + y^2 = 1$

21 max: $f(4, 4) = 48$; min: $f(12, 12) = -144$

23 2×2 (bottom and top) $\times 5$ (height)   **25** $x = 5$, $y = 10$

27 min: $T\left(-\dfrac{3}{4}, 0\right) = -9°$; max: $T\left(\dfrac{1}{2}, \pm\dfrac{\sqrt{3}}{2}\right) = 46°$

29 $x = y = z = 667$ **37** $y = \dfrac{1}{2}x + 1$

Problem Set 14.9, page 911

1 $(0, -1)$, $(0, 1)$, $\left(\dfrac{\sqrt{15}}{4}, -\dfrac{1}{4}\right)$, $\left(-\dfrac{\sqrt{15}}{4}, -\dfrac{1}{4}\right)$

3 $(\sqrt{2}, -\sqrt{2})$, $(-\sqrt{2}, \sqrt{2})$, $\left(\dfrac{\sqrt{2}}{2}, \dfrac{\sqrt{2}}{2}\right)$, $\left(\dfrac{-\sqrt{2}}{2}, \dfrac{-\sqrt{2}}{2}\right)$

5 $\left(\dfrac{1}{4}, \dfrac{\sqrt{14}}{4}\right)$, $\left(\dfrac{1}{4}, \dfrac{-\sqrt{14}}{4}\right)$, $\left(\dfrac{\sqrt{2}}{2}, 0\right)$, $\left(\dfrac{-\sqrt{2}}{2}, 0\right)$

7 $(\pm 1, 0, 0)$, $(0, \pm 2\sqrt{3}, 0)$, $(0, 0, \pm\sqrt{3})$, $\left(\pm\dfrac{\sqrt{3}}{3}, \pm 2, \pm 1\right)$

9 $\left(-\dfrac{2}{11}, -\dfrac{6}{11}, \dfrac{9}{22}\right)$ **11** $x = 4$, $y = 12$

13 $\dfrac{50}{3}$ by $\dfrac{50}{3}$ by $\dfrac{100}{3}$ inches   **15** length = width

19 (a) in the direction so that the tangent to the curve makes an acute angle with $\nabla f(a, b)$; (b) opposite of (a)

Review Problem Set, Chapter 14, page 912

1 domain: points on or inside the circle $x^2 + y^2 = 64$; interior points are *inside*

3 domain: all points not on the line $y = 1 - 2x$; interior points: all points in the domain

5 (a) -6; (b) -72; (c) $3ab^2cd^6$; (d) entire $xyzw$ space

7 does not exist

9 The limit of $f(x, y)$ as $(x, y) \to (0, 0)$ along the y axis is 1, $\neq f(0, 0)$.

11 $f_x(x, y) = 3x^2 + 7y^2$; $f_y(x, y) = 14xy - 24y^2$

13 $f_x(x, y) = y^3e^x + 3x^2e^y$; $f_y(x, y) = 3y^2e^x + x^3e^y$

15 $g_x(x, y) = \dfrac{3x^2}{x^3 - y^3}$; $g_y(x, y) = \dfrac{-3y^2}{x^3 - y^2}$

17 $g_x = \dfrac{(x^3 + y^3)^{2/3} + x^2}{x(x^3 + y^3)^{2/3} + x^3 + y^3}$; $g_y = \dfrac{y^2}{x(x^3 + y^3)^{2/3} + x^3 + y^3}$

19 $g_x = e^{-7x}[\sec^2 (x + y) - 7 \tan (x + y)]$;
    $g_y = e^{-7x} \sec^2 (x + y)$

21 $g_x = 3x^2 - 26xy^2z^2$; $g_y = 3y^2 - 26x^2yz^2$; $g_z = 3z^2 - 26x^2y^2z$

23 $f_x = -zy \operatorname{csch}^2 (xy)$; $f_y = -zx \operatorname{csch}^2 (xy)$; $f_z = \coth (xy)$

31 (a) $x - 2 = y - 2 = z - 2$; (b) $x + y + z = 6$

33 (a) $2xy \, dx + x^2 \, dy$; (b) 75.7; (c) exact: 75.696768

35 $\approx 1.76\pi \approx 5.53$ cm³   **39** -0.1

41 $\dfrac{\partial w}{\partial u} = 24x^3 - 18xy^2 - 6x^2y + 4y^3$;

    $\dfrac{\partial w}{\partial v} = 8x^3 - 6xy^2 + 12x^2y - 8y^3$

43 $\dfrac{\partial w}{\partial r} = -2e^{x^4} + e^{y^4}$; $\dfrac{\partial w}{\partial s} = -e^{x^4} - 3e^{y^4}$

45 $\dfrac{-v \sin uv}{3x^{2/3}} - \dfrac{u \sin uv}{4x^{3/4}}$ **47** $-27e^{-20}$

49 (a) -3; (b) -2 **51** (a) $\sin x$; (b) $y \cos x$

53 (a) $\dfrac{6\sqrt{3} + 1}{2}$; (b) $\sqrt{37}$, in the direction of

    $\mathbf{u} = -\dfrac{6\sqrt{37}}{37}\mathbf{i} + \dfrac{\sqrt{37}}{37}\mathbf{j}$

55 $\dfrac{\cos y - y \cos x}{\sin x + x \cos y}$

57 $\dfrac{24 - 5\pi}{18\pi}$ ft/min or $\dfrac{48 - 10\pi}{3\pi}$ in/min

59 tangent plane: $3x + 3y - 5z = 8$;

normal line: $\dfrac{x - 3}{3} = \dfrac{y - 3}{3} = \dfrac{z - 2}{-5}$

61 tangent plane: $6x - 2y + 15z = 22$;

normal line: $\dfrac{x - 1}{6} = \dfrac{y - 7}{-2} = \dfrac{z - 2}{15}$

63 (a) $\dfrac{\partial^2 w}{\partial s^2} = e^{xy}[12st + 10t^2 + (6s^2t + 10st^2 + 2t^3)^2]$;

(b) $\dfrac{\partial^2 w}{\partial t^2} = e^{xy}[12st + 10s^2 + (6st^2 + 10s^2t + 2s^3)^2]$

65 no max or min; $(-1, -2)$ is a saddle point

67 relative max at $(1, 1)$; saddle points $(0, 0)$, $(0, 3)$, $(3, 0)$

69 absolute min at $\left(\dfrac{1}{2}, \dfrac{1}{2}\right)$

71 absolute max at $\left(\dfrac{\sqrt{2}}{2}, \dfrac{\sqrt{2}}{2}\right)$, absolute min at

$\left(-\dfrac{\sqrt{2}}{2}, -\dfrac{\sqrt{2}}{2}\right)$

75 1.00 dollar/razor; 0.40 dollar/dozen blades

77 $y = \dfrac{3}{4}x + \dfrac{1}{12}$ **79** $y = \dfrac{15}{26}x + \dfrac{53}{52}$

Chapter 15

Problem Set 15.1, page 919

1 $-\dfrac{1}{y}\cos xy + C(y)$ **3** $\dfrac{2}{3}xy^{3/2} - \dfrac{y^2}{2x} + \dfrac{2\sqrt{y}}{x} + K(x)$

5 $-y^2 \cos y^3$ **7** $\dfrac{8}{15}$ **9** 4 **11** $e^3 - 4$

13 $\dfrac{1}{2}$ **15** $\dfrac{38\pi}{3}$ **17** $\dfrac{2}{3}$ **19** $\dfrac{2}{15}$ **21** $\dfrac{596}{21}$

23 $7 - 3\sqrt[3]{5}$ **25** $\dfrac{\pi^2}{18}$ **27** $\dfrac{1}{3}$ **29** $\dfrac{1}{7}e^7 - \dfrac{1}{5}e^5 + \dfrac{2}{35}$

31 $\dfrac{4}{3}$ **33** $\dfrac{2e^2 - 3}{2}$ **35** $\dfrac{96 + 9\pi}{8}$

37 $\dfrac{\pi}{2}$ **39** $\dfrac{2}{3}(e - 1)$ **41** $2e - 2$

Problem Set 15.2, page 926

1 16 **3** 261 **5** 0 **7** 351 **9** $\dfrac{250\pi}{3}$

11 $\dfrac{24\pi + 4\sqrt{2}\pi}{3}$ **13** 18 **15** πr^2

17 $\dfrac{1}{2}|x_1y_2 + x_2y_3 + x_3y_1 - x_1y_3 - x_2y_1 - x_3y_2|$

19 $\dfrac{\pi}{3}$ **21** 2 **23** 2 **25** -4 **27** 8π

Problem Set 15.3, page 934

1 (b) both; (c) π **3** (b) both; (c) $\dfrac{4 + \pi^2}{2}$

5 (b) both; (c) $\dfrac{32}{15}$ **7** (b) both; (c) $\dfrac{1}{3}$

9 (b) both; (c) $\dfrac{\pi}{4}$ **11** (b) both; (c) 0

13 (b) I; (c) $\dfrac{8}{3}$ **15** (b) both; (c) $\dfrac{1}{24}$

17 (b) I; (c) $\dfrac{423}{28}$ **19** (b) I; (c) 50

21 (b) both; (c) $\dfrac{\pi}{4} - \ln \sqrt{2}$ **23** $\dfrac{1}{6}(1 - e^{-3})$

25 0 **27** $1 - \sin 1$ **29** $12(e - 1)$ **31** $\dfrac{1}{5}(\sqrt{33} - 1)$

33 $\cot \dfrac{\sqrt{3}}{2} - \cot 1$ **35** 8 **37** $\dfrac{125}{6}$ square units

39 $\dfrac{4}{3}$ square units   **41** $\dfrac{22}{3}$ square units

43 $\dfrac{512}{3}$ square units   **45** $\dfrac{343}{12}$ square units

47 2 square units   **49** $2(e - 1)$ square units

51 $\dfrac{593}{140}$ cubic units   **53** $\dfrac{7}{60}$ cubic units

55 2 cubic units   **57** $\dfrac{10}{3}$ cubic units

59 $\dfrac{292}{15}$ cubic units   **61** $\dfrac{5}{12}$ cubic units

63 $\dfrac{3}{35}$ cubic units   **65** $\dfrac{81}{4}$ cubic units

Problem Set 15.4, page 943

1 $\dfrac{4\pi}{3}$ **3** $\dfrac{\pi}{9}(2\sqrt{2} - 1)$ **5** 0 **7** 6 **9** $\dfrac{\pi(e^9 - 1)}{e^9}$

11 $\dfrac{a^5\pi}{10}$ **13** $\dfrac{9\sqrt{2}}{2}$ **15** 2π **17** $\dfrac{(1 - \cos 1)\pi}{4}$

19 $\dfrac{\pi}{2}$ **21** 12π square units   **23** $\dfrac{\pi + 8}{4}$ square units

25 $\dfrac{3\pi}{2}$ square units   **27** $\dfrac{3\sqrt{3} - \pi}{2}$ square units

29 $\dfrac{\pi}{8}$ cubic unit   **31** $\dfrac{\pi}{2}$ cubic units   **33** $\dfrac{3\pi}{2}$ cubic units

35 $\dfrac{32\pi}{3}(8 - 3\sqrt{3})$ cubic units   **39** $\dfrac{1}{2}$

41 $e^{u+v}(uv - 1)$ **43** $\dfrac{e^\pi - e^{-\pi}}{4}$

Problem Set 15.5, page 952

1 $\dfrac{3}{7}$ coulomb   **3** $\dfrac{27}{2}$ kg, $\left(\dfrac{6}{5}, \dfrac{6}{5}\right)$ **5** 64 kg, $\left(\dfrac{12}{5}, 2\right)$

7 $\dfrac{49}{20}$ kg, $\left(\dfrac{710}{1029}, \dfrac{10}{21}\right)$ **9** 15 kg, $\left(\dfrac{9}{4}, \dfrac{15}{8}\right)$

11 $\dfrac{3}{10}$ kg, $\left(\dfrac{29}{63}, \dfrac{5}{9}\right)$ **13** $\dfrac{10}{3}$ kg, $\left(\dfrac{39}{25}, \dfrac{206}{75}\right)$

15 $\dfrac{\pi}{6}$ kg, $\left(\dfrac{3}{2\pi}, \dfrac{3}{2\pi}\right)$ **17** $\dfrac{2}{9}$ kg, $\left(\dfrac{16}{35}, \dfrac{16}{35}\right)$

19 $(\frac{2}{3}, \frac{1}{3})$ **21** $(\frac{4}{5}, 1)$ **23** $(1, 2)$

25 $\left(\dfrac{21}{13 - 6\ln 2}, \dfrac{102}{65 - 30\ln 2}\right)$ **27** $\left(\dfrac{10}{21}, \dfrac{5}{12}\right)$

29 $\left(\dfrac{8\pi + 3\sqrt{3}}{2\pi + 3\sqrt{3}}, 0\right)$ **31** $\left(\dfrac{4}{9}\left(\dfrac{44 + 9\pi}{8 + \pi}\right), 0\right)$

33 $\dfrac{1088\pi}{15}$ **35** $\dfrac{2\pi}{3}$

37 $I_x = \dfrac{3}{14}$ kg · m², $I_y = \dfrac{3}{10}$ kg · m²

39 $I_x = \frac{6}{25}$ kg · m², $I_y = \frac{2}{7}$ kg · m²

41 $I_x = \frac{15}{91}$ kg · m², $I_y = \frac{3}{7}$ kg · m²

43 $I_x = \dfrac{2}{3 - 4\ln 2}$ kg · m², $I_y = \dfrac{1}{6 - 8\ln 2}$

45 $I_x = \frac{7}{16}$ kg · m², $I_y = \frac{49}{48}$ kg · m²

47 $\left(\dfrac{28\pi - 27\sqrt{3}}{16\pi - 12\sqrt{3}}, \dfrac{6\pi + 3\sqrt{3}}{8\pi - 6\sqrt{3}}\right)$

49 $I_x = \frac{56}{15}$ g · cm², $I_y = \frac{8}{3}$ g · cm², $I_O = \frac{32}{5}$ g · cm²

51 $\dfrac{3\pi}{32}$ g · cm²   **53** $\dfrac{a^2m(3\pi + 8)}{24}$ **55** $\dfrac{3\pi}{4}$ g · cm²

57 $\displaystyle\iint\limits_R \sigma(x, y)[(x - a)^2 + (y - b)^2]\, dA$

Problem Set 15.6, page 959

1 $\dfrac{9}{2}$ **3** $\dfrac{5}{24}$ **5** $\dfrac{124\pi}{5}$ **7** (a) $\dfrac{64\sqrt{2}}{105}$

9 (a) $\dfrac{1}{24}$ **11** 64   **13** 0   **15** $\dfrac{8748}{7}$ **17** $\dfrac{\pi}{2}$

19 0   **21** $\frac{8}{15}$ cubic unit   **23** $\frac{1}{6}$ cubic unit

25 $\frac{1}{60}$ cubic unit   **27** $\frac{20}{3}$ cubic units

31 $\displaystyle\int_0^1\int_0^{1-x}\int_0^{1-x-y} f(x, y, z)\, dz\, dy\, dx,$

$\displaystyle\int_0^1\int_0^{1-y}\int_0^{1-x-y} f(x, y, z)\, dz\, dx\, dy,$

$\displaystyle\int_0^1\int_0^{1-x}\int_0^{1-x-z} f(x, y, z)\, dy\, dz\, dx,$

$\displaystyle\int_0^1\int_0^{1-z}\int_0^{1-z-x} f(x, y, z)\, dy\, dx\, dz,$

$\displaystyle\int_0^1\int_0^{1-z}\int_0^{1-y-z} f(x, y, z)\, dx\, dy\, dz,$

$\displaystyle\int_0^1\int_0^{1-y}\int_0^{1-y-z} f(x, y, z)\, dx\, dz\, dy$

33 $\frac{1}{6}$ cubic unit

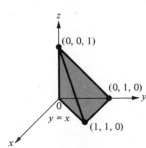

35 4π cubic units

37 $\displaystyle\int_0^1\int_0^{\sqrt{1-z^2}}\int_0^{\sqrt{1-y^2-z^2}} f(x, y, z)\, dx\, dy\, dz$

39 $\displaystyle\int_0^3\int_0^x\int_0^{\sqrt{9-x^2}} f(x, y, z)\, dz\, dy\, dx$

Problem Set 15.7, page 966

1 9   **3** $\dfrac{64\pi(2 - \sqrt{3})}{3}$ **5** 15π **7** $\dfrac{4\pi - \pi^2}{8}$

9 $\dfrac{250\pi}{3}$ **11** 24π **13** 81π **15** $\dfrac{3125\pi(4\sqrt{5} - 5\sqrt{2})}{24}$

17 $\dfrac{\pi}{2}$ cubic units   **19** 9 cubic units

21 $\dfrac{3\pi}{4}$ cubic units   **23** $\dfrac{4\pi}{3}(8\sqrt{2} - 7)$

25 $\dfrac{4}{3}\pi a^3$ **27** $\dfrac{14\pi}{9}(2 - \sqrt{2})$

29 $\dfrac{\pi}{3}(2 - \sqrt{2})$ **31** 5.97 cubic units

33 $\dfrac{\pi}{3}(2a^3 - 3a^2k + k^3)$ and $\dfrac{\pi}{3}(2a^3 + 3a^2k - k^3)$

Problem Set 15.8, page 972

1 $m = \dfrac{25}{84}$; $(\bar{x}, \bar{y}, \bar{z}) = \left(\dfrac{87}{125}, \dfrac{77}{125}, \dfrac{87}{125}\right)$

3 $m = \dfrac{3}{2}$; $(\bar{x}, \bar{y}, \bar{z}) = \left(\dfrac{4}{9}, \dfrac{4}{9}, \dfrac{4}{9}\right)$

5 $m = \dfrac{32\pi}{3}$; $(\bar{x}, \bar{y}, \bar{z}) = (0, 0, 5)$

7 $m = \dfrac{243\pi}{2}$; $(\bar{x}, \bar{y}, \bar{z}) = \left(0, 0, \dfrac{2}{9}\right)$

9 2000π kg   **11** 48π kg   **13** $\left(1, \dfrac{3}{2}, 2\right)$

15 $\left(\dfrac{1}{2}, 0, \dfrac{3}{5}\right)$ **17** $\left(\dfrac{8}{15}, \dfrac{8}{15}, \dfrac{\pi}{24}\right)$ **19** $(0, 0, 0)$

21 $\left(\dfrac{a^2bc}{24}, \dfrac{ab^2c}{24}, \dfrac{abc^2}{24}\right)$ **23** $\dfrac{1}{16}$ **25** $\dfrac{512}{35}$

27 $\dfrac{5}{2}$ **29** $\dfrac{10}{3}$ **31** 144π **33** $\dfrac{ma^2}{2}$

35 $\dfrac{m}{2}(b^2 + a^2)$ **37** $\dfrac{2ma^2}{5}$ **39** $\dfrac{2m}{5}\dfrac{(b^5 - a^5)}{(b^3 - a^3)}$

Problem Set 15.9, page 983

1 2   **3** 2π **5** $\dfrac{4}{3}$ **7** 0   **9** 32   **11** $\dfrac{189}{2}$

13 $\dfrac{\pi}{2}$ **15** 0   **17** 0   **19** $\dfrac{243\pi}{4}$ **21** $-\dfrac{112}{5}$

23 $-\dfrac{24{,}353}{30}$ **25** $-\dfrac{1}{3}$ joule

27 (a) $\dfrac{17}{6}$ joules; (b) $\dfrac{12}{5}$ joules; (c) $\dfrac{22-3\pi}{6}$ joules

29 πa^2 square units   **31** $\dfrac{3\pi}{32}$ square unit

33 P and Q are discontinuous at $(0,\,0)$

Problem Set 15.10, page 992

1 independent   **3** not independent   **5** independent
7 independent   **9** not independent
11 (b) $f = x^2y - y + C$; (c) 17
13 (b) $f = x^3y^3 + C$; (c) 1727
15 (b) $f = xe^{2y} - \sin xy + y^2 + C$; (c) 5
17 (b) $f = e^x + xy + 2y + ye^y - e^y + C$; (c) $4e^5 + e + 12$
19 (b) $f = \dfrac{x^2}{2} + \dfrac{y^2}{2} - 2z + C$; (c) $\dfrac{3}{2}$

21 (b) $f = \sin x - e^{-y} + \dfrac{z^3}{3} + C$;

   (c) $\sin 6 + \sin 1 + e^{-2} - \dfrac{59}{3}$

23 (b) $f = xyz + C$; (c) 25   **29** (b) $x^3y - \dfrac{x^2}{2} + \dfrac{y^2}{2} = C$

31 (b) $\displaystyle\int_C \mathbf{F}\cdot d\mathbf{R} = 20$; (c) $\mathbf{R}$ has discontinuity at $(0,\,0)$;

   (d) $f = -\tan^{-1}\dfrac{x}{y} + C,\ x > 0$

Problem Set 15.11, page 1002

1 $9\sqrt{3}\pi$ square units   **3** $\dfrac{49\sqrt{14}}{2}$ square units

5 $6\sin^{-1}\dfrac{1}{3}$ square units   **7** $16\sqrt{2}$ square units

9 $\dfrac{56\pi}{3}$ square units   **15** $C \cdot$ (area of Σ)

17 $\dfrac{4\pi}{3}$ **19** $-21\sqrt{2}$ **21** $-\dfrac{23}{12}$

23 $\dfrac{15}{4}$ **25** 4π

Problem Set 15.12, page 1010

1 (a) y^2; (b) $\mathbf{i} - \mathbf{j} - (2x + 2xy)\mathbf{k}$
3 (a) $3yz^2 + 5x^2 + 6y^2z^2$;
   (b) $(4yz^3 - 1)\mathbf{i} + (6xyz)\mathbf{j} + (10xy - 3xz^2)\mathbf{k}$
5 27   **7** 0   **9** 0   **11** 0   **13** 3 in both cases
15 $\dfrac{3}{2}$ in both cases

Review Problem Set, Chapter 15, page 1011

1 $\dfrac{1}{10}$ **3** $\dfrac{143}{30}$ **5** $\dfrac{e^{16}}{8} - 2e$ **7** $\dfrac{49}{3}(\sqrt{3} - 1)$

9 $\dfrac{4}{3}$ **11** $\dfrac{2592}{35}$ **13** $\dfrac{6\sqrt{3} - 38}{3}$ **15** $\dfrac{1}{2}(e^4 - 1)$

17 $\dfrac{9}{2}$ square units   **19** $\dfrac{8}{5}$ square units   **21** $\dfrac{500\pi}{3}$

23 $\dfrac{7\pi}{3}$ **25** 0   **27** $\dfrac{128}{21}$ cubic units

29 $\dfrac{64}{3}$ cubic units   **31** $\dfrac{1}{3}$ cubic unit

33 $\left(\dfrac{\pi}{4} + \ln\dfrac{\sqrt{2}}{2},\ \dfrac{2}{3}\right)$ **35** $\left(\dfrac{466}{35(8\pi + 3\sqrt{3})},\ \dfrac{234\sqrt{3}}{35(8\pi + 3\sqrt{3})}\right)$

37 $I_x = \dfrac{h^2m}{6}$; $I_y = \dfrac{m}{6}(b^2 + bc + c^2)$;

   $I_O = \dfrac{m}{6}(h^2 + b^2 + bc + c^2)$

39 $\dfrac{1}{1232}$ **41** $\dfrac{34}{15}$ **43** $\dfrac{2a^3\pi}{3}$ **45** $\dfrac{13}{6}$

47 $\dfrac{27}{2}$ cubic units   **49** $\dfrac{\pi}{30}$ **51** $\pi(2 - \sqrt{3})$

53 $\dfrac{\pi a^3 h^2}{60}$ **55** $\left(0,\ 0,\ \dfrac{8}{3}\right)$

57 $\left(\dfrac{2\pi - 3\sqrt{3}}{4\pi(2 - \sqrt{3})},\ \dfrac{2\pi - 3\sqrt{3}}{4\pi(2 - \sqrt{3})},\ \dfrac{3}{8(2 - \sqrt{3})}\right)$ **59** $\dfrac{3ma^2}{10}$

61 $\dfrac{7a^2m}{5}$ **63** $\dfrac{23}{6}$ **65** $\dfrac{1}{3}$ **67** $-\dfrac{1}{44}$ **69** $-bh$
71 independent   **73** not independent   **75** independent
77 (b) $f(x,\,y) = x^3 + 6xy + y^3 + C$; (c) 20
79 (b) $f(x,\,y,\,z) = x^2y + xz^2 + y^2z + C$; (c) 21
81 80π **83** $4a^2(\pi - 2)$ (a = radius of sphere)
85 $\dfrac{\sqrt{2}\pi}{2}$ **87** (a) $2x + 2y + 2z$; (b) $\mathbf{0}$
89 π **91** $-\pi$

ILLUSTRATION CREDITS

INDEX

Exponential and Logarithmic Forms

43 $\displaystyle\int e^{au} \sin bu \, du = \frac{e^{au}}{a^2 + b^2} [a \sin bu - b \cos bu]$

44 $\displaystyle\int e^{au} \cos bu \, du = \frac{e^{au}}{a^2 + b^2} [a \cos bu + b \sin bu]$

45 $\displaystyle\int u^n e^u \, du = u^n e^u - n \int u^{n-1} e^u \, du$

46 $\displaystyle\int u^n \ln u \, du = \frac{u^{n+1}}{n + 1} \ln u - \frac{u^{n+1}}{(n + 1)^2}$

Forms Involving $\sqrt{a^2 - u^2}$

47 $\displaystyle\int \sqrt{a^2 - u^2} \, du = \frac{u}{2}\sqrt{a^2 - u^2} + \frac{a^2}{2} \sin^{-1} \frac{u}{a}$

48 $\displaystyle\int \frac{\sqrt{a^2 - u^2}}{u} \, du = \sqrt{a^2 - u^2} - a \ln \left| \frac{a + \sqrt{a^2 - u^2}}{u} \right|$

49 $\displaystyle\int \frac{du}{u\sqrt{a^2 - u^2}} = -\frac{1}{a} \ln \left| \frac{a + \sqrt{a^2 - u^2}}{u} \right|$

50 $\displaystyle\int \frac{du}{u^2\sqrt{a^2 - u^2}} = -\frac{\sqrt{a^2 - u^2}}{a^2 u^2}$

51 $\displaystyle\int \frac{\sqrt{a^2 - u^2}}{u^2} \, du = -\frac{\sqrt{a^2 - u^2}}{u} - \sin^{-1} \frac{u}{a}$

52 $\displaystyle\int u^2\sqrt{a^2 - u^2} \, du = \frac{u(2u^2 - a^2)\sqrt{a^2 - u^2}}{8} + \frac{a^4}{8} \sin^{-1} \frac{u}{a}$

53 $\displaystyle\int \frac{du}{(a^2 - u^2)^{3/2}} = \frac{u}{a^2\sqrt{a^2 - u^2}}$

Forms Involving $\sqrt{u^2 - a^2}$

54 $\displaystyle\int \sqrt{u^2 - a^2} \, du = \frac{u}{2}\sqrt{u^2 - a^2} - \frac{a^2}{2} \ln |u + \sqrt{u^2 - a^2}|$

55 $\displaystyle\int u^2\sqrt{u^2 - a^2} \, du = \frac{u}{8}(2u^2 - a^2)\sqrt{u^2 - a^2}$
$$- \frac{a^4}{8} \ln |u + \sqrt{u^2 - a^2}|$$

56 $\displaystyle\int \frac{du}{\sqrt{u^2 - a^2}} = \ln |u + \sqrt{u^2 - a^2}|$

57 $\displaystyle\int \frac{du}{u^2\sqrt{u^2 - a^2}} = \frac{\sqrt{u^2 - a^2}}{a^2 u}$

58 $\displaystyle\int \frac{\sqrt{u^2 - a^2}}{u} \, du = \sqrt{u^2 - a^2} - a \cos^{-1} \left| \frac{a}{u} \right|$

59 $\displaystyle\int \frac{\sqrt{u^2 - a^2}}{u^2} \, du = -\frac{\sqrt{u^2 - a^2}}{u} + \ln |u + \sqrt{u^2 - a^2}|$

60 $\displaystyle\int \frac{du}{(u^2 - a^2)^{3/2}} = \frac{-u}{a^2\sqrt{u^2 - a^2}}$

Forms Involving $\sqrt{a^2 + u^2}$

61 $\displaystyle\int \sqrt{a^2 + u^2} \, du = \frac{u}{2}\sqrt{a^2 + u^2} + \frac{a^2}{2} \ln |u + \sqrt{a^2 + u^2}|$

62 $\displaystyle\int u^2\sqrt{a^2 + u^2} \, du = \frac{u}{8}(a^2 + 2u^2)\sqrt{a^2 + u^2}$
$$- \frac{a^4}{8} \ln |u + \sqrt{a^2 + u^2}|$$

63 $\displaystyle\int \frac{\sqrt{a^2 + u^2}}{u} \, du = \sqrt{a^2 + u^2} - a \ln \left| \frac{a + \sqrt{a^2 + u^2}}{u} \right|$

64 $\displaystyle\int \frac{du}{\sqrt{a^2 + u^2}} = \ln |u + \sqrt{a^2 + u^2}|$

65 $\displaystyle\int \frac{du}{(a^2 + u^2)^{3/2}} = \frac{u}{a^2\sqrt{a^2 + u^2}}$

66 $\displaystyle\int \frac{\sqrt{a^2 + u^2}}{u^2} \, du = -\frac{\sqrt{a^2 + u^2}}{u} + \ln |u + \sqrt{a^2 + u^2}|$

67 $\displaystyle\int \frac{du}{u^2\sqrt{a^2 + u^2}} = -\frac{\sqrt{a^2 + u^2}}{a^2 u}$